Source Book of Food Enzymology

SIGMUND SCHWIMMER, Ph.D.

Consultant in Food Enzymology
Formerly Head, Enzyme Technology Investigations
Western Regional Research Center
United States Department of Agriculture
Berkeley, California

THE AVI PUBLISHING COMPANY, INC.

Westport, Connecticut, U.S.A.

Library of Congress Cataloging in Publication Data

Schwimmer, Sigmund.
Source book of food enzymology.

Bibliography: p.
Includes index.
1. Enzymes. 2. Food—Composition. I. Title.
TX553.E6S38 664'.001'547758 80-25116
ISBN 0-87055-369-0

Manufactured by The Saybrook Press, Inc.
Printed in the United States of America

Source Book of Food Enzymology

Preface

Until quite recently, three major factors, upon which successful metamorphoses of living organisms into foods have hinged, have been economic feasibility, wholesomeness and safety as dictated and defined by governmental monitoring activities and, of course, consumer acceptance. The latter factor has usually been translated into optimization, maximization and retention of product and produce quality with heavy emphasis on sensory attributes and appearance, including packaging. Allied but hitherto subsidiary considerations and goals which now possess equivalent interest in the food research community and are rapidly attaining comparable status in the food industry as a whole are: nutritional quality and nutrient retention and availability; environmental compatibility of the previously mentioned metamorphosis; and energy conservation.

To achieve these past and future aims has required and will in the future demand a steady contribution from applied and basic research in the appropriate associated, relevant scientific and engineering disciplines. Of special relevance are those areas of research and endeavor which seek to influence, rationalize and provide an underlying understanding of events occurring in the organism-*qua*-food transition on the basis of the molecular transformations involved. Among the sciences dealing with these are, principally, microbiology, organic and biochemistry, and the latter's subdiscipline or offshoot, enzymology.

Although the importance of enzyme action as an agent of chemical and physical change in foods was recognized over 150 years ago, almost contemporaneously with the concept of catalysis itself, nonenzymatic aspects of food quality changes have received the lion's share of attention until the last 30 years or so. The post World War II period has witnessed accelerating interest and research in food enzymology. This interest has resulted in a certain degree of maturation of this segment of food science and an ever-widening awareness of enzymes as frequently crucial-factors for the manufacture of foods as well as the creation and retention of food quality and identity. This awareness extends to the necessity of managing the foods' own (endogenous) enzymes by appropriate manipulation and treatment which can either enhance or prevent their action along every segment of the food chain, from the farmer to the consumer. Paralleling this awareness is the increasing recognition that enzyme preparations constitute an important class of biologically-derived food additives and ingredients as well as valuable reagents for food analysis.

In view of these developments, I felt that the time was propitious for comprehensively examining, critically evaluating and providing a syncretic Ariadne's Thread for food enzymology in its labyrinthine ramifications. This *Source Book of Food Enzymology* is intended as such a compendium. Further, it is the hope of this author that this book will serve as a catalyst for utilizing all the earth's resources in producing safe, nutritious, sensorily attractive and universally available food made ready for consumption through a minimum of environmental disturbances and expenditure of dwindling energy reserves.

This source book not only constitutes an assessment of the present status of the advance understanding of enzymology in its diverse food-oriented ramifications, but, believing that the Past is Prologue, also develops for the reader a sound appreciation of the historical foundation of what I consider to be the more promising leads to the food enzymology of tomorrow. This is especially true with regard to topics hardly touched upon by traditional food science, such as the future utility of transition-state analog inhibitors and synzymes, among other topics. Of course, I also show how insight into the nature and behavior of enzymes, attained for its own sake or as part of nonfood mission-oriented research, has entrained and redounded to the benefit of food processing and food quality.

While I trust that I have adequately covered what I perceive to be the proper "provenance" of food enzymology, I should like to point out that inevitably and unavoidably my discourse will lead the reader into other areas of food science and engineering and even into nonfood areas of pure and applied enzymology, since food enzymology does not, so to speak, exist in a vacuum.

This author has also taken the opportunity to allow this book to serve as an outlet for unpublished ideas which have accumulated during his career in pursuit of the elusive grail of food research—ideas which I trust may prove to be of benefit to the reader, whether he/she be a graduate student, advisor, research scientist, food processing technologist or plant manager.

A compelling motivation in writing this book is epitomized in the statement of Nobelist Arthur Kornberg (1976) in his encomium "For the Love of Enzymes," that he had never known a dull enzyme. This ageless attitude of scientific curiosity was expressed by William Blake in an exquisite quatrain, a paraphrasing of which, to render it more relevant (if less exquisite) to the subject matter of this book, follows.

To see a World in a Grain or Gram
And Heaven in a Mild Flour
Behold Infinity in a Palm of Ham
And Eternity in the Happy Hour.

My own abiding interest in enzymes started about 44 years ago upon acceptance of the position of Minor Scientific Helper in the Enzyme Research Laboratory of the Food Research Division of the U.S. Department of Agriculture, some five years after its inception by my mentor, the late Dr. Arnold Kent Balls, who remained its director during its 20 year existence. The creation of this laboratory marked a milestone in food science in that it was probably the first sizeable food research group whose formal designation included the word "enzyme." It thus signaled the preliminary recognition of the central role played by enzymes in foods and in all related biological disciplines and industries, as we shall have ample opportunity to demonstrate in the ensuing exposition. Later, the term "enzyme" was used to designate somewhat smaller units including my own Enzyme Technology Investigations group. The term "enzyme" in the formal designation of a research team has survived in the United States as the Institute for Enzyme Research at the University of Wisconsin.

It is a pleasure to acknowledge here the special association of Maynard Joslyn, Professor Emeritus of Food Science, Department of Nutritional Sciences, University of California, Berkeley, with food enzymology in general and with this book in particular. The launching of the effort which resulted in this book is in no small measure owing to Professor Joslyn's stimulating suggestions vis-à-vis his in-depth, retentive knowledge of the broad sweep of all aspects of food science, as attested to, in part, by citations of his papers profusely distributed throughout the text. These citations provide a uniquely historical backdrop to the application of the discipline of enzymology to fruit and vegetable and allied processing technology. It is owing to the outstanding experimental and didactic output of a few outstanding oracular pioneers such as Professor Joslyn, Dr. Balls and, earlier, Professors Samuel C. Prescott of MIT and Emil Mrak of the University of California that it is now universally accepted that enzymes are of seminal importance in food science and processing. The author fondly trusts that the ensuing account of the fruits which these pioneering efforts have borne constitutes a proper and fitting tribute to Professors Joslyn and Balls' contributions to food enzymology.

This author would also like to thank the late Dr. Donald K. Tressler, Ms. Lisa E. Melilli, and especially Dr. N.W. Desrosier and Barbara J. Flouton of the AVI Publishing Company for their encouragement and assistance in bringing this book into being, and also the individuals who provided photographic prints acknowledged in the appropriate figure legends by the designation *"Courtesy of . . . ."*

Finally, I wish to thank my wife Sylvia in her role as "catalyst," not only for her help in accelerating the manuscript preparation during its bookkeeping and secretarial phases, but also for other roles which define enzyme action: her patience and understanding imparted a sense of direction and specificity, and her occasional reminders (especially during the period of Sisyphean labor in which the chapters and references were multiplying in a sorcerer's apprentice-like fashion) that a world exists outside the realm of enzymology served a regulatory control function.

SIGMUND SCHWIMMER, PH.D.

August 1980
Berkeley, California

Lassen Sie doch die Enzyme für sich arbeiten.

Contents

PREFACE

PART I PERSPECTIVES AND PROSPECTUS 1

1 **Perspectives and Prospectus** 3

Historical Association of Foods and Enzymes
Applications of Enzymology
How Enzyme Action Is Involved in Food Quality—The Operational Approach
Promotion/Potentiation of Desirable Endogenous Enzyme Action
Enzyme *vs* Nonenzyme Reactions
Enzyme Action and Quality Control—Testing
Enzyme Action and Food Quality Attributes

PART II BASICS OF ENZYMOLOGY 25

2 **The Nature of Enzyme Action—Definitions and Specificity** 27

Specificity, Classification and Nomenclature
Latitude of Specificity
Specificity, the Enzyme-Substrate Complex and the Active Site
Elements of Enzyme Kinetics, the Grammar of Enzymology
Other Molecular Determinants of Specificity

3 **The Nature of Enzyme Action—Catalytic Efficiency** 49

Enzymes Are the Most Efficient Catalysts
The Transition State
Tertiary Structure and the Active Site
Model Enzymes

4 **The Nature of Enzyme Action—Regulation, Biosynthesis and Biology** 63

Levels of Biological Regulation
Quaternary Protein Structure, Allostery, Cooperativity
In Vivo Devices for Nongenetic Enzyme Regulation

Genetic Regulation and Biosynthesis of Enzymes
Posttranslational and Other Aspects of Regulation
Relation of Enzymes to Other Biologically Active Proteins

PART III ENZYME PRODUCTION AND RELATED TOPICS 87

5 Industrial Food Enzymes—Production, Stabilization, Costs 89

Production from Nonmicrobial Sources
Production of Microbial Enzymes
Large-scale Operations
Recovery, Separation and Further Handling
Stabilization and Standardization
Economic Considerations

6 Enzyme Purification, Enrichment, Isolation 105

Present Practices for Large-scale Purification
Preliminary Steps
Nonchromatographic Isolation Procedures
Chromatographic and Related Procedures

7 Affinity Chromatography of Enzymes 123

Historical Background
Adsorption
Elution-Desorption
Advantages and Adaptations
Prospects for Scale-up and Food Applications

8 Immmobilization of Enzymes 137

Immobilization Procedures
Properties of Immobilized Enzyme Systems
Applications

9 Enzymology of Sweetener Production 152

Starch-derived Sugar Syrups
Invert Sugar
Fructose Syrups and Glucose Isomerase
Immobilization and Production
Proposed and Potential Enzymatic Sugar Sweetener Production
Cellulase, Xylanase and Xylitol, and Other Enzymes

PART IV NONBIOLOGICAL CONTROL AND MANAGEMENT OF ENZYME ACTION 181

10 Control of Enzymes by Energy Input 183

Control by Thermal Energy: Temperature Optimum and Thermal Stability
Blanching
Control by Nonthermal Energy

11 **Peroxidases, Other Enzymes and Adequacy of Heat Treatment and Regeneration** 202

Biochemistry of the Peroxidases
Thermal Properties of Peroxidase
Peroxidase Assay and Adequacy of Blanching
Other Enzymes as Indicators of Adequacy of Heat Treatment
Regeneration of Enzyme Activity

12 **Action and Control of Enzymes at Low Temperatures** 218

Factors Tending to Minimize Enzyme Action
Persistence of Enzyme Action at Low Temperatures

13 **Action and Control of Enzymes at Low Water Activities** 231

Physics and Engineering of Dehydrated Foods
Assessment of Enzyme and Nonenzyme Action in Dried Foods
Enzyme Action in Dried Foods
Enzyme Reaction Pathways Altered in Water-restricted Food Systems
Approaches to Enzyme Management in Dried Foods

14 **Control of Enzymes by Perturbation of Molecular Environment** 247

Food Additives as Enzyme Inhibitors
Varieties of Inhibition
Active-site Directed Inhibitors
Non-active-site Directed Inhibitors
Hydrogen Ion as Enzyme Inhibitor

PART V ENZYME ACTION AND FOOD COLOR QUALITY 265

15 **Food Color, the Phenolases and Undesirable Enzymatic Browning** 267

Color as a Food Quality Attribute
Phenolase Biochemistry
Phenolase Action in Foods
Prevention and Retardation of Undesirable Phenolase-induced Discoloration

16 **Other Roles of Enzyme Action on Phenols as Determinants of Food Color Quality** 284

Suggested Roles of Nonphenolase Oxidoreductases
Desirable Enzymatic Browning
Decolorization of Anthocyanins

17 **Color Changes Induced by Enzyme Action on Nonphenolic Substrates in Plant Foods** 298

Enzymatic Destruction of Carotenoid Pigments
Color Changes Associated with Chlorophyll Transformations
Other Enzyme-related Color Changes

18 **Enzymatic Aspects of Color Change and Retention in Meat and Fish** 315

Meat Pigments
Enzyme Action and Red Meat Discoloration

Enzyme Action and Fresh Meat Color Retention and Creation
Enzymology of Ham Curing
Nonbrown Discolorations

19 **Enzymatic Aspects of Nonenzymatic Browning** 326

Nonenzymatic Browning
Enzyme Action and Color of Processed Potato Products
Prevention of Browning of Processed Potato Products
Glucose Oxidase and Egg Desugaring

PART VI ENZYME ACTION AND FOOD FLAVOR QUALITY 343

20 **Enzyme Action and Accumulation and Retention of Nonvolatile Flavorants** 345

The Nature of Food Flavor
Enzymes of Food Acid Accumulation
Enzymes of Sugar Sweetener Accumulation
Enzymes of Flavor Potentiator Accumulation
Enzymes of Glutamate Accumulation
Enzymatic Aspects of 5′-Nucleotide Accumulation and Retention

21 **Enzyme Action and Aroma Genesis in Vegetables** 368

Origin of Volatile Flavorants in Foods
Aldehydes and Ketones
Sulfur Volatiles—The Disulfides and Related Volatiles
Isothiocyanates and Glucosinolase
Other Vegetable Aromas

22 **Enzymatic Aroma Genesis in Fruits, Flavorants and Beverages** 388

Fruit Aromas and Ester Biosynthesis
Enzymes of Food Flavorings—Cyanogenic Glycosides and Emulsin
Enzymatic Aspects of Hot Beverage Aroma Genesis
Genesis of the Terpenoids

23 **Enzymatic Aroma Genesis in Meats, Fermented Foods and via Flavorese** 406

Enzymes of Cheese Aroma Genesis
Enzymes Involved in Bread Aroma Genesis
Meat Flavor
Flavorese

24 **Enzyme Involvement in Off-flavor from the Oxidation of Unsaturated Lipids in Nondairy Foods** 421

Contribution of Enzyme Action to Food Off-flavors
Lipoxygenase and Other Modes of Fatty Acid Transformation
Enzymatic Origins of Off-flavor in Frozen Foods
Nonlipoxygenase, Nondairy-mediated Lipid Transformations Which May Lead to Off-flavor in Meats, Vegetables and Fish
Prevention of Nondairy Lipohydroperoxide Associated Off-flavors

25 **Lipase and Other Enzyme Involvement in Dairy Off-flavors, Citrus Bitterness and Other Flavor Defects** 440

The Lipases
Milk Lipase and Hydrolytic Rancidity
Enzyme Involvement in Other Flavor Defects in Dairy Products
Enzyme Involvement in Citrus Bitterness
Other Enzymatic Aspects of Off-flavors

PART VII ENZYME ACTION AND THE TEXTURAL QUALITY OF FOODS 459

26 **Enzyme Action and Food Texture: Muscle to Meat** 461

The Texture of Texture
Muscle—Composition, Structure and Function
Nonproteolytic Postmortem Events
Roles of Nonglycolytic Enzymes in Aging

27 **Applied Enzymology of Meat Texture Optimization** 481

Tenderness Through Endogenous Enzyme Action
Action of the Meat Tenderizing Enzymes
Meat Tenderizer Technology—State-of-the-Art
Future Possibilities

28 **Roles of Enzymes in Fish Texture and Transformation** 497

Enzyme Involvement in Fresh Fish Texture
Frozen Fish Texture
Roles of Proteinases in Fish Texture and Transformation

29 **Enzyme Action and Plant Food Texture** 511

Determinants of Plant Food Texture
Pectin Transforming Enzymes
Enzyme Involvement in Plant Food Texture: Specific Fruits
Vegetable Texture

PART VIII ENZYME ACTION AND FOOD TRANSFORMATIONS: CHANGE-OF-STATE, APPEARANCE AND IDENTITY 533

30 **Roles of Enzyme Action in Wine and Fruit and Vegetable Juice Manufacture and Appearance** 535

Enzymes as Clarification and Cloud Retention Aids
Comminuted Tomato Products
Cellulase and Cell-separating Enzymes

31 **Enzyme Action in Malting and Brewing** 552

α-Amylase
Other Brewing Associated Starch and Other Glycan Hydrolases
Proteinase Action in Brewing

32 Enzyme Action in Bread-making and Other Texture-related Modification of Cereal Foods 572

Starch-transforming Enzymes
Protease Action and Baked Goods Quality
Lipid-transforming Enzymes

33 Enzyme Action in Cheesemaking and Cheese Texture 593

Milk Clotting with Chymosin (Rennin)
The Curdling Step in Cheesemaking
Rennet Alternatives to Chymosin
Postclotting Contribution of Proteinase Action to Cheese Texture

PART IX ENZYME ACTION IN PROTEIN MODIFICATION AND HEALTH-ASSOCIATED ASPECTS OF FOOD QUALITY 611

34 Protein Functionality and Nutritiousness as Modified by Enzyme Action 613

Addition of Proteolytic Enzymes
Protein Modification by Endogenous Proteases—Autolysis
Improvement of Protein Nutrition by Action of Nonproteolytic Enzymes
Protein Impairment by Enzyme Action and Its Prevention

35 Enzymes in Foods as Health and Safety Hazards 634

Generation of Toxic and Unwanted Pharmacologically Active Substances
Loss of Nutrients
Safety of Enzymes and Enzyme Preparations

36 Enzymes and Their Action in Foods as Health and Safety Benefits 650

Enzymes as Detoxicants
Prevention of Microbial Contamination
Enzymes as Digestive and Weight Reducing Aids

PART X ENZYME ACTION IN THE QUALITY CONTROL LABORATORY—ASSAY, TESTING, ANALYSIS 667

37 Enzyme Assays—Principles and Applications in Food Quality Control 669

Use of Enzymes in Official Analyses
Principles of Enzyme Assay
Enzyme Assays as Indicators, Markers, Monitors, Indices and Predictors
Isozyme Patterns
Standardizing and Monitoring Commercial Enzymes

38 Food Anylates as Enzyme Substrates 685

History and Principles
Instrumentation and Methodology
Analysis of Food Constituents

39 Further Aspects of the Utility of Enzymes in Food Analysis 703

Anylate as Enzyme Affector
Adding Enzymes to Foods as Shortcuts to Quality Assessment
Enzymes as Analytical Aids
Enzyme Immunoassay (EIA)

BIBLIOGRAPHY AND SELECTED REFERENCES 717

ENZYME INDEX

INDEX

Part 1

PERSPECTIVES AND PROSPECTUS

1

PERSPECTIVES AND PROSPECTUS

HISTORICAL ASSOCIATION OF FOODS AND ENZYMES

The Alchemist (Archaeus) separates the good from the bad, and changes the good into a tincture which tinges the body with life. This Alchemist dwells in the stomach, and as soon as the food comes into the stomach, the Alchemist is at once there and proceeds to digest it, rejecting that which is not helpful to the body into a special place, in order that the good may go where it belongs . . . .

—Paracelsus (1589)

Paracelsus appears to be the first to have invoked action of a localized and specifically adapted entity, independent and separable from living tissue, to allude to what we know today as enzyme action.

The historical origins of enzymology as a distinct scientific discipline are inextricably associated with foods. Contrary to the speculations of Paracelsus, the prevailing theory in the mid 18th century (consonant with the developments in physics of that time) was a purely mechanistic one in which digestion was due to the friction of the food particles acting against the pressure of the stomach wall. According to Effront and Prescott (1917), this theory first was disproven experimentally by Reamur (1683–1757) who observed that food (meat) contained within pierced metal tubes swallowed by falcons underwent digestion. It remained for Spallanzani (1784) to establish that juice removed from the stomach retained this capacity for digesting meat. He obtained the juice by squeezing out a sponge, retrieved from the stomach of a tame eagle, which had been inserted with a string attached.

Of course, enzyme action was utilized by man for transformation of plants and animals into foods long before there was any scientific knowledge of what was happening. Usually these enzymes were acting as constituents of microorganisms which, together with the enzymes of plants or animals, transformed the identity of grains to alcoholic beverages and baked goods, and the identity of milk to cheese and a variety of other dairy products.

One of the prominent phenomena which led to the concept of catalysis, first enunciated and promulgated by Berzelius in the 1830s, was the by then well-known degradation of starch by extracts of germinated barley, malt. The first attempt to isolate such a catalyst, the "diastase" of malt, was that of Payen and Persoz (1833). In the same paper, they also discussed the industrial application of their findings, thus making it one of the first papers of what now has become a veritable deluge of predictions on the promises of applied enzymology.

Further development of the concepts of "diastases" and "ferments" by Liebig and others during the next 25 years were based largely upon enzyme action in foods such as fruits, almonds, sprouted and unsprouted cereal and cheese, as well as on the action of what we now know as pepsin. Indeed, Liebig (1859) in several of his writings hints at the close association of the ferments with "caseins" and "glutens," thus coming close to identification of enzymes as proteins, and goes on to enunciate his theory of "ferments":

> The phenomena we have described, if considered in their true signification, prove that the decompositions and transformations which occur in the processes of fermentation, are effected by matter, the smallest particles or atoms of which are in a state of motion and transposition, —a state susceptible of being communicated to other atoms in contact with the former, so as to cause the atoms and elements of these latter also, in consequence of the resulting disturbance of the equilibrium of their chemical attraction, to change their position, and to arrange themselves into one or more new groups.

Liebig's concept of how enzymes act have their modern counterpart in the theory that molecules, in general, can be resonantly destabilized by long wave vibrations in the energy range of ultrasonic or Raman spectra (Menefee 1976), or, less speculatively, that an enzyme molecule is a "floppy body" whose rapid conformational fluctuations lead to the lowering of the free energy of the transition state (Chapter 3) and hence facilitate catalysis (Careri *et al.* 1979).

Although Paracelsus' concept may be seen as the forerunning concept for enzymology, his disciples interpreted it as supporting the concept that "vital force" explains many biological phenomena—a misinterpretation which Liebig struggled against for many years.

However, the very term "enzyme" adopted 100 years ago (Kuhne 1876)—ϵν ζυμ, meaning "in leaven" or yeast—arose from the product of food processors, bakers and brewers. Yeast was also the vehicle which provided one of the crucial turning points in the development of enzymology with the demonstration by Buchner (1897) that organized ferments (i.e., alcoholic fermentation by yeast) as well as unorganized ferments (the milk clotting enzyme rennin, pepsin, catalase, the emulsin-amygdalin system in almonds) could proceed in the absence of living cells. Buchner suggested that his "zymase" was protein. It was not long after this that Pekelharing (1902) presented very convincing evidence that at least one enzyme, pepsin, was most probably a protein. However, at least another 25 years elapsed before the scientific community recognized that enzymes were indeed proteins, thanks to the crystallization of jack bean urease by Sumner (1926). Edsall (1976) provides a brief, interesting account of this phase of the development of enzymology.

Since these early years, enzymology has gone far beyond its historical association with foods and is intertwined with all of the myriad facets of modern biology—the "enzyme theory of life" (Fruton 1976).

APPLICATIONS OF ENZYMOLOGY

Medical Associated Applications

While knowledge of enzymes and their role in living processes are still being studied by those investigators in many different disciplines who pursue basic information and understanding of life, this knowledge has been put to practical use in several areas, other than those which are agriculturally oriented. Probably the first and most useful application is within the medical field, and the most beneficial application has been in analyzing patients' tissues and fluids for enzyme activity as diagnostic aids. In addition to being used routinely in thousands of medical examinations, perhaps the most interesting and recent use of enzyme activity is in the determination of specific enzymes in the amniotic fluid surrounding the embryo (via amniocentesis) to detect severe birth defects which have occurred because a certain enzyme is missing or is too low in activity to effectively carry out its function. Enzymes are also helpful as specific analytical reagents for the determination of medically important blood and urine components (see Chapters 38 and 39). To a somewhat more limited extent enzyme preparations have been alleged to have therapeutic properties and thus have been used as pharmaceuticals. In the early days, protein hydrolyzing enzymes were considered to be efficient digestive aids and pepsin is still added to chewing gum. More recently, very careful tests have shown they can be used for effective wound debridement, amelioration of adhesions associated with an operation, and in other skin associated ailments. One highly specific protein-hydrolyzing enzyme, collagenase, is now used for treatment of third degree burns. Chymopapain, one of the proteases of commercial meat tenderizers (Chapter 27), is being widely used outside of the United States for treatment of spine-related pains and injuries. Unfortunately, the treatment of cancer with asparaginase has as of this writing not yet fulfilled its early promise. Holcenberg and Roberts (1977) provide a concise, informative discussion on enzymes as drugs. In the pharmaceutical industry, they have been used, or their use has been proposed, for the preparation of organic chemical medicinals (see Chapter 8, Immobilized Enzymes for Steroid Transformations) and, of course, here too they have been helpful as in analysis of pharmaceuticals (Chapter 37). A useful book which surveys the entire field of enzymes and medicine is that of Moss and Butterworth (1974).

Other Nonfood Applications

Most of the other nonfood applications of enzymology involve the use of commercial enzyme preparations, usually containing protein and/or starch degrading enzymes. One of the most extensive uses of amylases is in the desizing of textiles. In order to impart smoothness and strength, textile fibers are usually sized with enzyme-pretreated starch applied to warp threads before they are woven into fabrics. Starch also serves as a vehicle to keep other loadings (salts, etc.) adhered to the fiber. After the fabric is woven, it is necessary to remove the starch. Although this has sometimes been accomplished with acid or alkali, the mild action of bacterial amylases is much preferred. It is estimated that $1,000,000 worth of desizing enzyme is sold in the United States each year (Wolnak 1972), and 600,000,000 yen worth in Japan (Samejima 1974).

Enzymes have played an important role in fabric cleaning since 1915, according to Tauber (1949). American dry cleaners have regularly used commercial enzyme preparations in their collection of spot cleaners. Several years ago questions on the safety of their use for home laundering arose. These preparations are still used extensively in Europe and undoubtedly will reappear in American home laundering once misunderstandings concerning their safety are cleared up and the stigma attached to their use fades.

Of historical interest is the use of proteolytic enzymes to "degum" raw silk. They are still used in the arts and crafts supply industry to recover silver from photographic films and silk, or rather its modern replacements in serigraphy (silk screen painting), by removing gelatin. This minor use has enjoyed a boomlet as a result of the current fad of silk screening of T-shirts.

Commercial starch-digesting enzymes are also used in the papermaking industries as sizings and to prepare adhesives. Invertase has also been used in the paper industry to prepare invert sugar to be used as a plasticizer. It will be interesting to ascertain whether fructose-containing corn syrups (Chapter 9) are also being used for this purpose. A more modern suggestion is the use of immobilized enzymes (Chapter 8) for clarification of colloidal starch-clay suspensions, "white-water," a waste effluent which creates pollution problems.

Another consumer of commercial enzyme preparations is the leather industry. These preparations are used as aids in removing the hair from hides and for bating, the preparation step in the tannery prior to the actual tanning step. Until about 50 to 60 years ago, the source of the proteolytic enzyme was bird or dog dung usually applied as a warm suspension. This environmental and esthetic problem has been solved by substitutions of pancreatin, pepsin, trypsin and microbial proteolytic enzyme preparations. Again, as was the case for using enzymes for dry cleaning, the basic patents were issued to Röhm (Tauber 1949). In the United States, pancreatin, with an annual value over $200,000, has been the protease of choice in the leather tanning industry (Wolnak 1972). The annual value of bacterial proteinase used in tanneries in Japan amounts to 90,000,000 yen (Samejima 1974).

Proteolytic enzyme preparations have been used as extraction agents in the perfume, plant and animal oil industries, in the rubber industry for treating latex, and along with lipase, in sewage disposal and cleaning of septic tanks. Enzymology undoubtedly will be increasingly useful in solving environmental problems.

Oxidases, especially commercially available glucose oxidase (Chapters 16 and 19) and catalase, have been recommended for a wide variety of nonfood uses where it is desired to maintain an oxygen-free atmosphere (e.g., to keep paper from yellowing) or to decompose hydrogen peroxide. The latter is used for bleaching of furs, textiles, feathers and wood, and in the manufacture of floating soaps and foam rubber. In the future, we predict that superoxide dismutase (Chapter 16) will be used for similar purposes. Oxidoreductases may turn out to be yet another source of useful energy and have already been applied to electric current generation as integral components of fuel cells (Yachiro *et al.* 1964; Varfolomeev and Berezin 1978; Anon. 1980A). Contributions of enzymology to other aspects of energy economy are included in Chapters 9, 10 and 31 where the subject matter is closely related to the enzymology of biomass conversion to fuels.

At the time of this writing it would appear that the first large-scale use of enzymes in the chemical industry will be for the production of ethylene oxide from ethylene using immobilized enzyme(s) (Chapter 8) synthesized via DNA recombinant techniques (Chapter 5). Ethylene oxide is an intermediate in manufacture of the antifreeze ethylene glycol.

HOW ENZYME ACTION IS INVOLVED IN FOOD QUALITY: THE OPERATIONAL APPROACH

Peripheral Topics

While this book deals principally with reactions of enzymes and the consequences thereof on food quality, there are related areas which will also be discussed briefly as separate topics or as part of the main theme where their effects are interwoven. Others will be hardly mentioned at all. This does not mean that these topics are not important determinants of food quality, but limitations of space as well as the main aim of this book, and the traditional assignment of some of these topics to other disciplines dictate these delimitations. A brief outline of these related topics may be helpful in delineating the main objectives of this book.

Enzyme action and its consequences arise as the result of deliberate, natural or adventitious microbial growth and the possible resulting accompanying fermentation in or on the food. There are instances where parallel action (both desirable and undesirable) of the same type of enzyme is derived from both microbial and food sources. These have to be distinguished. One of the goals we have set is to assess the relative contribution of these factors to food quality. Examples are desirable flavor in cheese and the removal of skins in coffee berries. In such cases, the distinction may not be well defined. In other cases, as in the removal of glucose from eggs, yeast fermentation or enzyme addition offers distinct alternatives for quality improvement. As we have mentioned, the creation of many foods, especially those foods of long tradition, such as beer, bread, pickles, sausages, tofu, nuoc-mam and cheese, involves necessary and cooperative enzyme action during microbial fermentation as well as that arising in the food. Three books covering the principles and problems of food microbiology are those of Weiser *et al.* (1971), Nickerson and Sinskey (1972) and Ayres *et al.* (1980).

On the other hand, when food undergoes deterioration in quality, it is not always readily discernible whether the cause is due to the action of the foods' own enzymes or to uncontrolled microbial growth. Thus, while development of off-odors in frozen vegetables is commonly and correctly attributed to endogenous attrite enzyme action (Chapter 24), off-odor in fresh salad vegetables, such as sliced cabbage held at refrigerator temperatures, has usually been attributed to microbial degradation of the tissue. In response to requests for information concerning keeping qualities of delicatessen coleslaw in a sour cream dressing, microbiologically-oriented investigators at the Western Regional Research Center conducted an inquiry as to the cause of the development of an off-odor (King *et al.* 1976). Instead of pinpointing a particular microorganism as the cause of the spoilage, they found that the total microbial count actually decreased with time at the typical display temperature of 7°C, as shown in Fig. 1.1. Furthermore, they observed that the off-odor was no worse at 14°C, where there was evidence for microbial infestation, than it was at 7°C. The authors suggested that the cabbage portion of the coleslaw at first continued to respire and thus used up the oxygen whose access to the tissue was impeded by the plastic bags used and by the oily emulsion (mayonnaise and sour cream) surrounding the cabbage strips. This lack of oxygen resulted in tissue death, disruption of the cells and enzyme degradation. They considered the latter to be the probable main cause of deterioration. On the other hand, the softening of apricots and cucumbers, at first suspected to be due to apricot pectin-degrading enzymes, was traced to fungal enzyme action (Chapter 29).

A detailed critical discussion of the preparation of enzymes, both in their pure state and as commercial enzyme preparations available as food adjuncts to the food processor and to the food ingredient manufacturer, will be taken up in Chapters 5 through 9.

Space does not permit details of the determination of the catalytic activity assay of each enzyme of importance in food quality. We shall, however, outline some of the principles and promises of enzyme assay and other analytically oriented applications of enzymology in Chapters 37 through 39. Furthermore, we shall refer to literature sources of enzyme assay at the appropriate places in the text.

Enzyme action discussed in this book cannot possibly cover even a small fraction of the myriad reactions occurring in live food tissues as part of their intermediary metabolism. Indeed, we cannot even cover all of the final reactions leading to production of the major stable end-products of intermediary metabolism of the food constituents. After all, almost every food constituent arises as the result of differential rates of their formation and transformation by enzymes.

For the most part, we shall confine our attention to those enzymes whose action takes place in the food itself, or is deliberately added, after the plant is harvested or the animal is slaughtered. However, in some instances the final or penultimate *in vivo* enzymes involved in the formation and disappearance of a quality-associated food constituents will be considered in some detail (i.e., acid-producing enzymes in fruits, Chapter 20). In a few instances we shall also refer to the metabolic pathways (intermediary metabolism) associated with the appearance of such constituents. The extent to which we delve into these latter two instances is partially dictated by what circumstances led the food scientist to adopt these areas of investigation as his/her purview, whether or not they would be directly involved in some economical and practical problem involving food quality and whether or not fundamental understanding of what goes on at the enzyme level would hopefully lead to practicable solutions.

With some rare exceptions we shall not consider the effect of food constituents on the food consumers' own enzymes. There are two aspects or ways in which this can occur: One, that a food constituent may inhibit the consumers' own enzymes, i.e., they are acute toxins or otherwise harmful to health, for instance, trypsin inhibitors. The other is annotated by a vast literature showing that pattern and level of activities of the consumers' enzymes are sensitive to dietary constituents and a vast pharmacological literature showing that special "xenobiotic" oxidases are synthesized *de novo* in the microsomes of liver in response to the ingestion of poisonous materials, drugs and other "alien' (=*xeno*) substances. We shall allude to these only where there is some evidence that the food's own enzymes (or added enzyme preparations) can effectively lower the level of these undesirable food constituents or, conversely, result in their formation (Chapters 35 and 36).

Related to this is the area of digestive enzymes. We shall consider them only in conjunction with their action as food adjuncts and not their action after the food enters the gastrointestinal tract.

Approaches to the Subject Matter

The above circumscriptions leave us with those enzyme actions which occur in the chain leading from the farm and slaughterhouse to utilization by the consumer (Fig. 1.2), or are added to the food at one or more of these points. The enzyme action occurring along this chain could be recounted chronologically. Other alternatives include an enzyme-by-enzyme approach, or food industry-by-

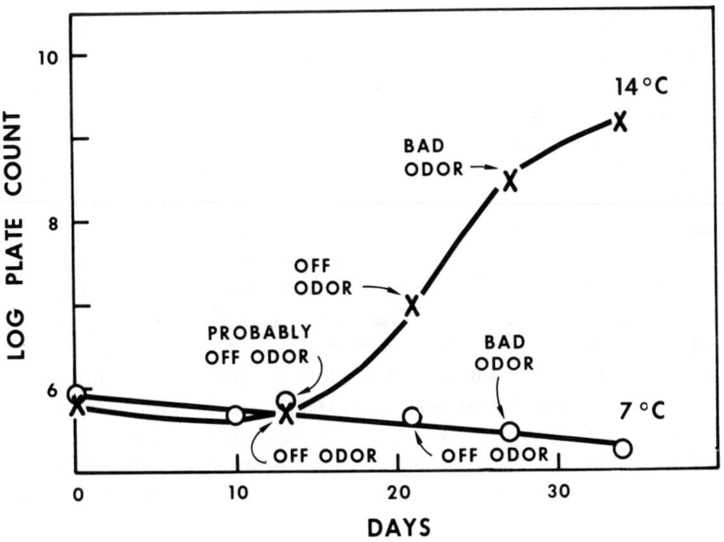

FIG. 1.1. COLESLAW DETERIORATION—MICROBIAL OR ENZYMATIC

At 7°C cold-susceptible organisms gradually die off; off-odor due to cabbage enzyme action.

Courtesy of King et al. (1976)

food industry. Reed (1966,1975) deals almost exclusively with the use of commercially applied enzymes in food processing on an industry-by-industry basis and not on the effect of the food's own enzymes on food quality. We trust that the Index may serve as an adequately useful integrating and cross-referencing device to obtain information on these bases.

Another useful device in presenting the ensuing material is to approach the subject matter from the operational viewpoint, that is, as we shall have occasion to do as a secondary subclassification. It may help to clarify our presentation to briefly dwell on this approach. We ask how the food scientist/technologist actually operates with enzymes. Such an individual is concerned with enzyme action because he/she either wants to promote it, that is, to let it proceed, or wants to prevent enzyme action at the wrong time in the wrong place in the food chain. We may epitomize these operations as:

(1) Enzyme management (Balls 1948; Underkofler 1972; Joslyn 1963).
(2) Addition of a commercial outside source of enzyme. A survey of the production, purification, use and immobilization of external enzyme sources (Chapters 5 through 9).
(3) Enzymes as testing devices (Chapters 37 through 39).

Control of the Food's Own Enzymes—Enzyme Management

Enzyme management of the food's own enzymes comprises either: (1) control inactivation/inhibition or removal to prevent enzyme action, or (2) the promotion of potentiation of the action of the food's own (endogenous) enzymes.

With regard to the management of enzymes, this author wishes to point out that it is not possible to rigidly categorize them as either undesirable and therefore always to be prevented from acting, or beneficial and therefore always to be potentiated, added or otherwise be encouraged to act even if properly controlled. A majority of food-related enzymes can either improve or impair food quality, depending on the extent of enzyme action on the raw food, on the processing, or on the food product, as well as on such variables as time, temperature, pH and past history of the food, etc. Examples will be found throughout the following discourse. Briefly, as examples, phenolase action is needed to impart color and aroma to tea and cocoa, but needs to be stopped in order that off-color of processed fruits and potatoes can be prevented. Lipoxygenase, notorious for the production of off-flavor in legumes and off-color in some cereal products is used to bleach wheat flour and, endogenously at very low levels, contributes to the desirable aroma of fresh-cut vegetables. Endogenous proteinase action can be very helpful in improving the nutritional availability of several high protein sources, especially FPC (fish protein concentrate), but can also in certain FPC processes result in a lowered protein recovery (Chapter 34).

Commercial Food Grade Enzymes

While we cannot provide a critical survey and exposition of the vast areas of the addition of enzymes to foods, we shall discuss them at appropriate times where they are used as food additives and as analytical aids. In Chapters 5 through 7 we shall survey general principles of enzymes as food adjuncts, including a scan of the burgeoning literature on immobilized enzymes (Chapter 8) as well as an assessment of their application to improvement of food quality. It may be well, however, to bring up one general point about adding enzymes at this time. One should distinguish between addition of commercial enzymes directly to a food destined for direct consumer use (or the use of the added enzyme, as in meat tenderizing and in the preparation of milk custards) on one hand, and the use of enzymes in the manufacture of food ingredients such as sugar syrups (Chapter 9), flavor potentiators and functionally modified proteins on the other. Another point concerning commercial enzyme preparations for addition to foods which may be apropos here is that these preparations almost always contain enzymes other than those designated as the major one. Many are sold as secondary mixtures or blends of primary preparations. An exception is glucose isomerase (Chapter 9).

A summary of enzymes now available for addition to foods in the United States is shown below. The applications are, for the most part, those suggested by the enzyme producers. Many other uses have been proposed. It is probably not an infrequent occurrence for an enzyme supplier to provide a special enzyme to a user on an *ad hoc* experimental basis, and continued production would depend on successful adaptation and judgment as to continued and widespread demand. They are designated by their principal and pertinent enzyme components.

Oxidoreductases.—*Catalase (Liver)*.—Removes residual H_2O_2 from cold sterilization of milk, cheese and egg white; added to glucose oxidase.

Glucose Oxidase-catalase.—Desugars eggs to prevent browning and off-flavor during and after drying. Removes O_2 from beverages and salad dressings to prevent off-flavor and improved storage stability.

Lipoxygenase (Soybean Flour).—Improves color and texture of baked goods and machinability of doughs.

Glucan Hydrolases (Polysaccharidases).—*α-Amylase, Malt*.—Liquefies and saccharifies starch in brewing and distilling industry; provides sugar for yeast; reduces drying time of baby foods; improves wheat flavor.

α-Amylase, Fungal.—Provides sugar for yeast in bread making; improves gassing power and loaf texture; retards staling. Converts acid thinned starch into highly fermentable, noncrystalline, nondextrose syrups. Has replaced malt in some breweries in Europe (barley brewing). Removes haze-producing residual starch left in beer. Facilitates filtration in production of starch-containing fruit juices by lowering viscosity. Reduces and controls viscosity and stabilizes chocolate syrup.

α-Amylase, Bacterial.—Liquefies and dextrinizes starch prior to addition of amyloglucosidase for syrup production. Speeds up the liquefaction of mash in brewing; used in recovery of candy scraps in soy sauce manufacture and in the preparation of moist baked goods.

Amyloglucosidase.—Converts properly thinned starches directly to glucose which then can be used as the substrate of glucose isomerase.

Polygalacturonase (Pectinase).—Aids in the clarification of fruit juices and wine; facilitates and speeds filtration; increases yield and processing capacity. Stabilizes and prevents gel formation in fruit juice concentrates and purees. Controls cloud retention in juices. Prevents burning on evaporation coils. Removes pectin in jelly, thus permitting addition of controlled amounts of pectin. With other enzymes, speeds up the manufacture of glazed fruit and separation of mandarin orange segments.

Cellulase.—Supplements action of amylases and pectinases in commercial preparations. Aids in juice clarification. Increases yield in extraction of essential oils and spices. Increases beer "body." Improves cookability and rehydratability of dehydrated vegetables. Aids in the increasing of available protein in seeds. Forms fermentable sugar in waste grape and apple pomace. These uses will be minor if and when a breakthrough in commerically feasible cellulase-sparked production of glucose from cellulosic waste and biomass occurs.

Hemicellulase.—Accompanies most microbial preparations. Aids in dehusking seed coats, especially coffee beans. Used in controlled degradation of food gums such as locust bean, guar, etc. Removes pentosans from bread. Aids in the degermination of corn and probably complements action of other carbohydrases in brewing. Improves nutritional availability of plant proteins. Other carbohydrases for food use in Japan (but not in the United States) include lysozyme for "humanizing" cows' milk and hesperidinase and naringinase for debittering citrus products.

Glycoside Hydrolases.—*Invertase (Sucrose Hydrolase)*.—Preparation of invert sugar. Prevents crystallization and grittiness. Production of creamy centers, prevents fermentation, prolongs shelf life of creams, cordials in confections. Improves texture and appearance of "sugar wall" dates.

Lactase (β-Galactosidase).—Increases sweetness and digestibility of milk-containing products. Hydrolyzes lactose in acid whey. Improves baking quality of milk-containing bread. Prevents crystallization in ice cream, concentrated milk, evaporated milk and concentrated whey products. Used in the preparation of lactose-free milk for consumption by lactose intolerant individuals.

Ester Hydrolases.—*Lipases, Pregastric (Lamb, Kid)*.—Hydrolyzes fat for improvement of cheese flavor, especially Romano and other Italian-type cheeses, and in butter and other dairy product flavors.

Lipases (Other).—Improves flavor of some confections. Improves whipping properties of egg albumin. Removes fats in gelatin manufacture.

Protein Hydrolases (Proteinases, Proteases, Peptidases).—*Papain, Other Plant Proteases*.—Tenderizes meat. Chill-proofs beer. Improves oil, protein extraction from animals and plants. Controls and modifies protein functional properties. Improves fish protein processing, especially FPC. Used in the preparation of protein hydrolysates. Improves cookies and waffles. Enhances diastase activity of malt. Acts as a stockinette release agent in sausage manufacture. Improves hot cereals, marinades.

Fungal Proteases.—Used in bread making: Improves grain texture, flavor, crust color, compressibility, increases loaf volume, controls dough mix time and machineability of doughs; inhibits blistering, curling and burning. Used in meat tenderizing formulations, in liquid meat products and in recovery of meat from bones. In marine products tenderizes clams, used in FPC production, improves evaporation efficiency, reduces scale formation in tubes and decreases sludge in centrifuge. Improves dispersibility of dried milk, stability of evaporated milk and spreadability of cheese spreads.

Bacterial Proteases (Heat Stable).—Improves flavor, texture and keeping qualities of crackers as well as waffles, pancakes and fruit cakes. Aids evaporation of fish water; used in filtration during cane sugar manufacture.

Rennets.—Includes calf rennin (chymosin), bovine pepsin, chicken pepsin, microbial rennets *(Mucor mihei, Endothia parasitica, Mucor pusillus, Bacillus cereus).* Clots milk in cheese manufacture, contributes to flavor of cheese during ripening. Bovine pepsin used as rennet extender.

Other Hydrolases.—In addition to these hydrolases one might add specialized preparations such as dextranase and α-galactosidase (immobilized within fungi), both used in sugar refining to remove dextran and raffinose, respectively, and aminoacylases (which hydrolyze *N*-acyl derivatives of L-but not D-amino acids) used for the manufacture of L-amino acid from mixtures thereof.

Other Commercial Enzymes.—Enzyme classes other than oxidoreductases and hydrolases are sparsely represented among the well established, widely used food grade enzyme preparations. They include:

Glucose Isomerase.—Converts glucose to fructose in production of fructose-containing starch derived sweet syrups.

Pectin Lyase.—Degrades pectin; is probably present along with polygalacturonases in commercial pectinases and is effective for same uses as polygalacturonase in these preparations.

Prevention of Enzyme Action

Before outlining the approaches to prevention of enzyme action in foods, we wish to answer in a general way the questions as to why prevention is at all necessary.

When the food is still a living organism it contains thousands of different kinds of enzymes which are not scattered at random but are organized as members of highly integrated systems, localized and compartmentalized within highly specialized subcellular packets (organelles). Here these enzymes perform their assigned catalytic functions in well-ordered temporal and spacial sequences along extremely specifically ordered metabolic pathways under a hierarchy of fantastically elaborate controls. A glimpse of some of these is afforded by Chapter 4 and elsewhere.

In food handling and in processing the foods are subjected as living organisms to stressful situations (Romani 1972; Schwimmer 1972, 1978). During subsequent processing some of the cells are disrupted. Even if the cells are not actually physically damaged, as viewed under a light microscope, the extremely sensitive membranes surrounding the cells and the above-mentioned organelles upon whose functioning the organism is dependent for continued life, may be ruptured or otherwise damaged. When this happens at the wrong place and time in the food processing chain (Fig. 1.2), the orderly sequence of metabolic events and their component enzyme-catalyzed chemical reactions are disrupted. Some, such as the ligases which are dependent on ATP (the common currency of the living organisms, Chapters 2 and 4), cease to function, while some of the others function completely out of control. This can result in the buildup of nonphysiological substances, some of which may lower food quality, and the disappearance of other constituents whose presence is associated with desirable food properties. This was succinctly and humorously depicted by Balls (1948) as shown in Fig. 1.3.

One of the aims of the following text is to fill out a skeletal sketch of what the enzymes involved are, what the products are, as well as to indicate methods for the prevention of this unwanted enzyme action.

Harvesting - Mechanical, Field Processing

↓

Transportation - Handling, Transfer, Biological

↓

PreProcessing - Conditioning, Controlled Atmosphere

↓

Processing - Freezing, Shear, Temperature
- Drying, Enzymes

↓

Packaging, Storage - Temperature, Container,
- N_2, Enzymes

↓

Distribution - Temperature Handling

↓

Consumer - Mastication, Blends

From Schwimmer (1972)

FIG. 1.2. CELL DISRUPTION AND ITS PREVENTION IN THE FOOD PROCESSING CHAIN

How to Curb and Control Unwanted Enzyme Action.—From the operational viewpoint we may distinguish among the following treatments-approaches which lead to, are being used or have been proposed to prevent, or at least control, unwanted enzyme action:

(1) Low temperature to slow down enzyme action.
(2) High temperature-blanching to inactivate enzymes.
(3) Removal of water.
(4) Extremes of pH—high acidity or alkalinity to inhibit or inactivate enzymes.
(5) Chemicals to inactivate or inhibit enzyme action.
(6) Modification of substrate or prevention of access of substrate to enzyme.
(7) Modification or removal of product of enzyme action.
(8) Preprocessing control.
(9) Ionizing radiation.

From Balls (1948)

FIG. 1.3. FOOD PROCESSING DISRUPTS ORDERLY METABOLIC REACTION SEQUENCES

Figure 1.2 shows in a general manner some stratagems which may counteract the unwanted enzyme action at various stages of the food processing chain.

Although there appears to be a parallelism between the prevention of enzyme activity on the one hand and microbial growth on the other, we shall see that this parallelism breaks down in many instances. The introduction of a step to prevent enzyme action in a processing line will cause a readjustment of the process to take into account a sort of "radiation" of effects, so that the decision to adopt such a step will depend upon trade-off of the benefits (some of which may be serendipitous) against disadvantages. In addition, economic, environmental and health associated considerations have to be taken into account. Sometimes it is accrual of additional benefits, above and beyond those associated with prevention of enzyme action, which tip the scale in favor of adoption of such a step. While the more extensively used approaches to preventing enzyme action, methods 1 through 4, will each receive special attention as general methods in ensuing Chapters 10–14, we should like to briefly dwell in this chapter upon methods 5 through 8, since we shall refer to these hereafter only in specific cases.

Modification or Removal of Substrate.—Conceptually, at least, it should be possible to remove the substrate so that it is no longer accessible to the enzyme. Leaching, used frequently to remove reactants in unwanted nonenzymatic reactions (especially browning), does at times remove unwanted enzyme or substrate. It is also possible to remove oxygen, frequently a cosubstrate in unwanted enzyme action, either by physical means (i.e., flushing with nitrogen) or by scavenging the oxygen with the proper oxidoreductase. Potentiation of endogenous enzymes to modify substrate may take place in at least one procedure suggested for delaying enzymatic browning (Chapter 16).

Modification or Removal of Product.—One way to modify the product of undesirable enzyme action is to add a food grade substance which will cause reversion to substrate. Use of ascorbic acid to prevent or delay enzymatic browning of fruit and potatoes is, as we shall see (Chapter 16), an example of the prevention of unwanted enzyme action by "reprocessing" the primary (but not the ultimate brown) product of enzyme action—quinones—back to the phenolic substrate. However, ascorbic acid, like SO_2, delays enzymatic browning *via* several distinct mechanisms. Such measures may be only "holding actions" but may accomplish enough of a delay until the food is further processed. As many authors have pointed out (Desrosier and Desrosier 1977; Woodroof and Luh 1975; Winter 1976), time is of the essence in both heating and cooling during the processing of fruits and vegetables, and such delayed action can result in considerable improvement in quality. Related to this area is the advantage of keeping microbial populations to a minimum in plant operations, not only for reasons of safety and health, but also to minimize enzymically-induced quality deterioration. Other examples of product modification will be brought out in the ensuing discussion.

Preprocessing Control

Control at the Genetic Level.—Not all enzymes and enzyme systems in intact living tissue are continuously active. As mentioned previously, enzymes may not be active because they are physically separated from their substrate. In other cases the enzymes are under such stringent metabolic regulation that although their substrates are ostensibly available, their action is completely suppressed by various intermediate metabolites (Chapter 4). Furthermore, no living organism possesses at any one time all of the enzymes it has the potential of synthesizing. That is, the DNA (the genes coding for biosynthesis of most of the enzymes) are shut off or repressed. For plants, the highest gene repression, and therefore the lowest and least diverse enzyme activity, occurs in the seed. During germination or sprouting many enzymes, especially those required for the utilization of the reserve substance—be it carbohydrate (potato), protein (some grains) or lipid (oil seeds such as soybean)—increase dramatically or appear *de novo* upon sprouting. In growing plants, enzyme activity is generally greatest in the rapidly growing tissue. Upon senescence, switching on and off of different sets of genes occurs and new enzymes, mostly degradative, are induced.

Breeding.—In addition to the natural genetic controls, man can occasionally intervene at a genetic level of control. Breeders may use the products of gene expression—the enzymes—or products of unwanted enzyme action as a guide to breed out the undesirable action. We shall have occasion to refer to this approach. On the other hand, enzymes can be used in breeding in desirable traits such as high yield and protein (Chapter 34).

Natural Control.—Related to genetic control, yet distinct enough to warrant a special category, is natural control of unwanted enzyme action. It is important for the grower and processor to take advantage of natural control when it presents itself.

Physical Separation of Enzyme and Substrate.—While we have been discussing compartmentation at the microscopic cellular level, in many cases control of enzyme action is achieved by actual separation of the enzyme and its substrate at the macroscopic level, so that ordinary mechanical devices can be employed to ensure that enzyme and substrate do not get together. In wheat seed, some unwanted enzymes, espcially those involved in poor flavor, are confined to the germ and to a lesser extent in the aleurone layer, while the endosperm may contain much of the substrate. In oats, lipase, involved in off-flavor development, is confined entirely to pericarp while other tissues contain most of the lipid. The tassel of sugarcane contains the sucrose-hydrolyzing enzyme invertase, whereas its substrate, sucrose, is confined to cane tissue. Leaves of the cucumber plant may contain pectin degrading enzymes due mostly to the untimely presence of certain fungi, whereas the cucumber fruit itself contains a high level of pectin, an important contributor to the characteristic textural quality of this vegetable.

Absence of Substrate.—Quite aside from the teleological questions which arise, the absence of substrate and the presence of a very active enzyme in a food may be considered to constitute a special kind of natural control. For instance, the jack bean is so rich in urease that it was a comparatively simple task for Sumner (1926) to isolate crystals therefrom. Yet, there is not a trace of urea present. Soybean, too, contains active urease (a good test to determine if it has been heated) but no urea. When soybean is used in mixed feeds containing added urea, the latter hydrolyzes to ammonia and water.

Certain fruits such as figs, pineapple, cantaloupe and other melons and papaya contain high protein-digesting enzyme activity but do not contain large amounts of proteins. The Sunbeam peach and the kiwi (the fruit, not the animal) can be frozen and thawed without darkening because, although they possess very active phenol oxidizing enzymes, they do not contain much, if any, phenolic substrate (Balls 1948; Winter 1976). Contrariwise, enzymatic browning is no problem in the processing of canteloupes, oranges and papayas because while these fruits contain large amounts of phenols and may also contain phenol-oxidizing enzymes the latter are largely inactivated or inactive at low pH of the juice (Bruemmer 1977).

The merliton, or vegetable pear, contains neither phenolase nor lipoxygenase (Flick *et al.* 1977) but, being a cucurbit, probably possesses considerable ascorbate oxidase (Chapter 16). When enzyme action occurs it is not always clear whether the substrate or the enzyme limits the rate and the extent of the enzyme action, but this may be important to know in order to cope with it. Thus, in the case of the avocado, it appears to be the phenolase activity and not the level of phenols, presence of inhibitor, isozyme pattern or change in enzyme properties in different fruit varieties which determines the rate and extent of browning (Kahn 1977). By contrast, the extent of onion flavor development is almost entirely dependent upon the level of flavor precursor or substrate and not on the flavor-forming enzyme (Schwimmer *et al.* 1964A).

Agronomical considerations (soil, variety, climate, prevention of insect and microbial damage, by keeping a clean orchard (Winter 1976); selection of produce during harvesting; harvesting practice, storage conditions (Ryall and Lipton 1974; Haard and Salunkhe 1975); handling, transportation practice) can all contribute to the prevention of unwanted enzyme action. Examples of these and other approaches will be found throughout this book.

Ionizing Radiation.—While ionizing radiation can at appropriately high levels inactivate enzymes, the main thrust in the initial experimentation on its application to food quality improvement was towards achievement of asepsis in packaged foods as an alternative to refrigeration (cold sterilization, radurization). The dosage of irradiation needed to destroy microbes is in the range of about 10^5-10^6 rads, about 1 to 2 orders of magnitude less than that required to inactivate most individual enzymes. At these high levels of ionizing radiation, reactions of the abundant free radicals formed especially in the presence of water and oxygen usually results in gross deterioration of the quality attributes of most foods. Even for sterilization purposes, the food products which have shown the most promise are those with high levels of penetrating flavors, such as bacon, which can be satisfactorily dosed with 6×10^5 rads.

While ionizing radiation may be considered as an alternative to blanching in preventing enzyme action, at a dosage of about 1 or less order of magnitude lower than the microbial sterilization range (*ca* 10^4-10^5 rads), these rays exert some beneficial effects on foods, especially on the storage stability of several fruits and vegetables. The prolonged benefits derived from irradiation at these levels may arise from three independent causes:

(1) A pasteurization effect which prevents the development of fungal and viral plant diseases.
(2) An insecticidal effect which, by killing off larvae and adults, results in deinfestation.
(3) A stress effect in which the plant responds to ionizing radiation by shifting the pattern of enzyme synthesis and degradation so that the level of some enzyme activities rises and others decrease (Schwimmer *et al.* 1958).

At somewhat higher doses some of the delicately balanced regulatory mechanisms of the plant are disturbed and rigid enzyme regulatory control is loosened. Practically, the effects observed range from delay of ripening of tomatoes to prevention or delay of sprouting of potatoes. Where the effects of ionizing radiation bear a clear and close relation to some enzymatic aspect of food quality, it will be mentioned in some detail. An expanded treatment of enzyme action and ionizing radiation will be found in Chapter 10.

PROMOTION/POTENTIATION OF DESIRABLE ENDOGENOUS ENZYME ACTION

Not all endogenous enzyme action in food results in quality deterioration, and not all desirable enzyme action has to be introduced in a food through the agency of added exogenous commercial enzyme preparations. Although the present number of classes of enzymes as food adjuncts is quite limited, using the food's own enzymes to catalyze specific desirable chemical changes in food is by no means so restricted. Not only can such enzyme action in many foods be beneficial, but it is necessary in other foods to impart identity, that is, the distinguishing properties of a given food product which give it its own identity. Without such actions, traditional foods such as cocoa, tea, dates and meats would not possess their typical identifying color, aroma and texture and general appearance. These enzymes were utilized millenia before enzymes were known. Indeed, the enzymatic nature of some of these changes is still far from being understood. One of the aims of this book is to assess the depth of that understanding. For other foods, potentiation of enzymes has also arisen in modern times as a mix of parallel empirical and enzyme oriented investigations. A prime example of this is the firming of many canned fruits and vegetables (Chapter 25) by preblanch potentiation of pectin methylesterase activity within these commodities.

The basic stratagem in effectively using the foods' own enzymes for maximum benefit is to keep them from acting until the opportune temporal and spatial situation. This can be accomplished by careful empirical manipulation, by experience and by judicious knowledge of the biology and chemistry of the enzymes, such as where they are located in the cell. Of course some of the conditions which evoke undesirable enzyme action can also be used at the right place and time to evoke *desirable* enzyme action.

Methods or Approaches to Enzyme Potentiation.—At the biological level these include: (1) derepression of genes coding for the enzyme by means of hormones, (2) horticultural (or animal husbandry) practice, (3) ripening, (4) curing, (5) germination, and (6) sprouting.

Stress and Cellular Modification.—While climatic conditions and agronomic parameters which create stress in a plant food may promote enzyme actions which end up in the production of undesirable quality attributes—i.e., low temperature storage of potatoes results eventually in production of overly brown processed potato foods because of the accumulation of reducing sugars (Chapter 19)—it is also frequently true that stresses to which other plant foods may be subject, in both the pre- and postharvest period, can lead to improved food quality. Sugar buildup is but one specific example of accumulation of low molecular weight metabolites in such stressful situations as extremes of temperature, nutrient deficiency and water deficit. In each case, membrane modification and its entrainment of altered enzyme action are implicated.

Quality improvement *via* water stress can occasionally occur. Thus, Freeman and Mossadeghi (1973) demonstrated that onions grown under water-stress conditions accumulate S-propenyl-L-cysteine sulfoxide, the precursor-substrate of the enzyme responsible for the development of onion odor and that watercress accumulates glucosinolates (mustard oil glucosides) under similar conditions. Prolonged overwinter postharvest storage and sprouting result in further increases in onion strength (Freeman and Whenham 1976A) due to the release of the flavor precursor from γ-glutamyl peptide bondage by the action of a γ-glutamyl transpeptidase (Schwimmer 1973; Schwimmer and Austin 1971A).

Enzyme action during the growth of the tea leaf and the manufacture of the beverage illustrates the beneficial effects, both pre- and postharvest, of water deprivation in relation to cellular organelle damage. According to Wickremasinghe (1974), "flavory" weather with respect to the growth of tea leaves in Sri Lanka corresponds to hot, dry, cloudless days followed by cold nights. The resulting improved flavor of the tea, according to the author, may be traced back to a partial disorganization and injury of the chloroplasts imposed by the lack of adequate water. This in turn results in a shift of some metabolic pathways from the chloroplast to the cytoplasm, so that the aroma-bearing volatile terpenoids, which contribute to the tea flavor, are synthesized at an enhanced rate outside the chloroplasts, apparently from the amino acid leucine (instead of acetate) as the ultimate precursor. Further along in the manufacture of tea, the leaf is subject to a withering process which further promotes or selectively potentiates those enzymes which improve the ultimate quality of the tea (Chapter 16).

In a reversal of their usual role as undesirable agents of cell disruption, certain insects such as the Ceylon mite or the greenfly improve tea flavor also by damaging the chloroplasts. Conditions for operation of home dehydrators afford an example of the beneficial effects of the dehydration regime at relatively low temperatures as advocated by Gee *et al.* (1980).

Biochemical.—At the biochemical level of enzyme control, enzyme action can be rationally evoked by:

(1) conversion of inactive zymogen (proenzyme) to active enzyme
(2) activation by specific activators and coenzymes
(3) decompartmentation of enzyme and substrate via cell and organelle breakage, and
(4) altering the physical character of the substrate such as native starch so that it now becomes susceptible to enzyme attack.

These processes do not result in cessation of viability of the organisms.

A prominent aspect of promotion of endogenous enzyme action is, as we have mentioned, related to the state of the cells of which the food is comprised. More importantly, it is possible to injure the cell membranes without physically disrupting the cell or inactivating any enzyme. The biological functions of these membranes are much more sensitive to heat than are enzymes in general. In particular, in our subsequent discussion we shall recurrently cite examples of potentiation by exposing the food to temperatures in the range of 40°–80°C. In connection with the breakdown of subcellular organelles, the lysosomes, microscopic bags of mostly hydrolytic enzyme acting best in a slightly acid medium, may be of particular importance in the potentiation of desirable enzyme activity. More extreme intermingling of enzyme and substrate which results in cell wall breakage and drastic changes to produce "unicellular" food by the application of cell-separating enzymes may be termed "exolysis" (Schwimmer 1972).

Potentiation and Membrane Modification.—With the possible exception of the action of α-amylase in sweet potatoes (where the potentiation may be due, at least in part, to the accessibility of the enzyme to the substrate, starch, occasioned by the hydration of the latter by heating, thus making it amenable to enzyme attack) the invocation of enzyme action in the examples of Table 1.2 may be interpreted in terms of some modification of cellular envelopes including walls but especially membranes. A clue to the initial events is afforded by the observations of investigators on the altered physical state of the membrane phospholipids as detected, for instance, by spin-labeling of cold-hardy plants as compared to that of cold-sensitive plants (Lyons *et al.* 1979) and the association of lowered phospholipid level to increased tendency of potatoes to undergo enzymatic browning (Mondy and Mueller 1977). In the latter case the deficiency of phospholipid may be indicative of thinner membranes which would be more susceptible to disruptive forces. It is not too farfetched to assume that in the temperature range under consideration some of the membrane lipids "melt" as indicated by the nonaligned phospholipid symbols in Fig. 1.4. This disorientation of the aligned phospholipids would undoubtedly injure the membrane with respect to its normal physiological functions in active transport and hormone transmission but may not, per-

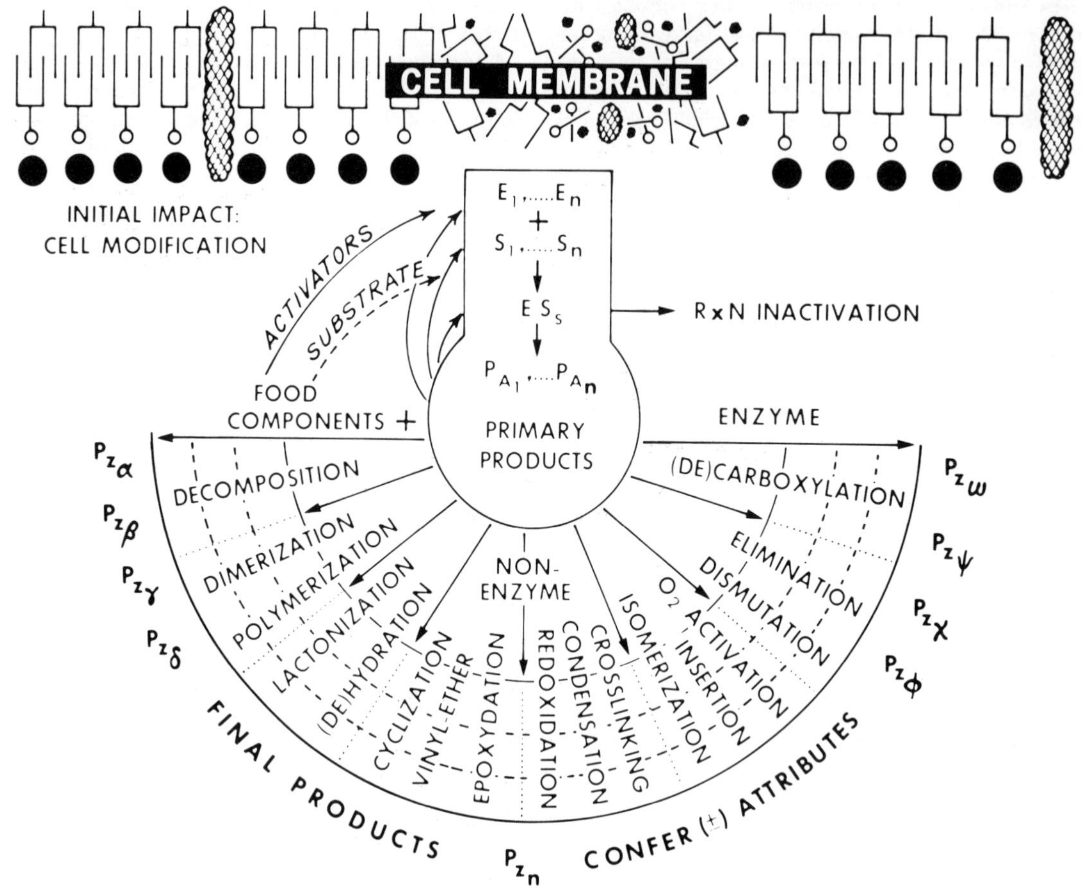

From Schwimmer (1978)

FIG. 1.4. CONSEQUENCES OF ENZYME ACTION IN FOODS

Depiction of endogenous enzyme potentiation following modification of cellular membrane integrity and the formation of unstable primary products which undergo subsequent reactions leading to secondary stable final products which confer attributes upon foods.

haps, render it particularly adapted to enzyme and substrate interaction. It may, however, produce conditions under which lysosomal acylhydrolases (lipases) now can have access to the disoriented polyunsaturated phospholipids and thus produce unsaturated free fatty acids as suggested by Galliard (1980) and others. The cell is thus now susceptible to further degradative modification via several mechanisms. In the presence of oxygen in vegetables the ubiquitous lipoxygenases (Wardale and Galliard 1975) apparently localized in distinct, exceptionally fragile organelles may be considered as mediators for the production of active oxygen in the form of lipohydroperoxides, as discussed recurrently in ensuing chapters. Other means of producing activated oxygen are also available at the site of the membrane disintegration.

As indicated in Fig. 1.4, further membrane disintegration may be promoted by the action of proteolytic enzymes on membrane proteins especially in senescent tissue. Endopeptidase activity (in contrast to exopeptidase which remains at a low but constant level) is induced at the onset of senescence (Feller *et al.* 1977). Such proteinase activity also elicits mitochondrial ATPase whose action would further interfere with the membrane associated action of synthetases (EC 6.) and with other membrane-associated functions such as active transport (Jung and Laties 1975).

The preceding sequence of events thus sets the stage for interaction of the enzyme of interest, by allowing for passage of substrate and activators, if the enzyme is immobilized, or for passage of the enzyme to the site of the substrate. Of course, in the intact cell the membrane is not an inert gossamer web but actively participates in the selective access and egress of activators, inhibitors and enzymes. It is also the site of attachment of hormones which exert their effect *via* the "second messenger," cyclic AMP, which initiates a cascade of sequential activation of a series of enzymes (Chapter 4), winding up finally in the "potentiation" of a specific metabolic enzyme such as phosphorylase or lipase and the accomplishment of a particular physiological function. Here we have a cascade of physical, chemical and enzymatic events leading to an ever-increasing disarray of the structural and functional elements of cellular envelopes, especially the temperature a sensitive organelle membrane, *via* disorientation, desensitization, deregulation, disfunction, malfunction and, finally, disruption. In some ways the events taking place during the thermal potentiation of enzymes in foods may be considered as a quasicascade of disorganization, with death mimicking life. Frequently, this charade of a cascade results in the improvement of food quality.

The Flavorese Approach.—One general approach for using the food's own enzymes to improve quality, which has been suggested in the area of flavor quality but may well find broader applicability, is that of flavorese (Chapter 23). In this approach crude enzyme preparations derived from the food are added to the same food after processing to restore quality. The rationale behind this is that processing destroys enzyme and removes quality (flavor) but not substrate, which is then available for transformation to fresh-quality associated components. As of this writing this approach has not been widely, if at all, adopted by industry. Recent investigators have suggested that it will be necessary to prepare "flavorese" enzymes from suitable microorganisms before flavoreses will be widely used.

Enzyme Regeneration as a Method of Potentiation.—Another possibility of utilizing the food's own enzyme, which we may have stumbled into without realizing it, may be to allow at least some enzymes to regenerate themselves after being heat inactivated. As will be discussed in Chapter 11, regeneration has traditionally been considered undesirable. Of special attention in this regard is the test enzyme for blanching, peroxidase. However, there is no evidence that regeneration results in deterioration or lowering of food quality in frozen foods. Some food processors have empirically noted improvement of quality and storage stability (Winter 1976). There may be a rationale for this. As we shall discuss later on, it is quite possible that such deteriorative enzyme action as that of lipoxygenase and especially superoxide anion-producing enzymes are destroyed irreversibly. Residual enzymes such as peroxidase and superoxide dismutase in the food may actually be beneficial in that they prevent any superoxide anion (which can be enzymatic and nonenzymatic in origin) from reacting with hydrogen peroxide to form the hydroxy free radical, ·OH, and singlet oxygen, O_2^*, which may be highly deteriorative of quality (Chapter 16). Furthermore, it is possible that the regenerated peroxidase may possess, as does the peroxidase of the white blood cell, germicidal properties. Some examples of enzyme potentiation which will be discussed in detail in the following chapters are shown in Table 1.1.

Alternative Sources of Enzyme: Addition *vs* Potentiation.—Occasionally a choice has to or can be made between using the food's own enzyme and adding commercial food grade enzymes to affect a specific quality attribute. Thus, some grape and other fruit juices may possess the right conditions for effective action of pectin degrading enzymes so that the juices clear spontaneously. Hence, it is not necessary to add commercial pectinases. In cheesemaking, flavor-bearing fatty acids may arise from the action of the lipases in the milk, or in the microbial population; or may be added deliberately in the form of oral or pregastric lipase (Chapter 23). More frequently all three sources may act together to yield the desired flavor note contributed by the fatty acids.

Agronomical and horticultural practice may dictate the relative levels of added and endogenous enzymes needed to obtain maximum benefit of enzyme action. Grain harvested in the past always included some germinated grains so that the resulting flour contained needed proteolytic and amylolytic enzymes, if not in optimum amounts, at least in amounts adequate to result in improved baked goods quality. Undoubtedly, some of these enzymes also came from adventitious microbial sources (such as fungi when the grain was stored moist). In modern agriculture these sources of enzymes are usually not present and it is almost universal practice in the baking industry to add commercial enzymes (Chapter 32). When confronted with these alternate or concomitant sources of enzyme, the processor should have available enzyme assay procedures which will serve as guides to the amounts of enzyme needed. A summary of

TABLE 1.1

IMPROVEMENT OF FOOD QUALITY BY ENZYME POTENTIATION, 60° ± 20

Food/Process	Improvement	Enzymes	Chapter
Fruits and vegetables	Firmness	Pectin esterase (+ calcium salt)	29
Fruits and vegetables	Flavor, appearance	Regenerated enzymes	11
Beans, dry wheat	Nutrition, acceptance	Phytase, α-galactosidases	34
Salad vegetables	Flavor	Lipoxygenase	21
Vegetables, green	Color, health (?)	Chlorophyllase (?)	17
Sweet potato, dehydrated	Color, appearance	α-Amylase	17
Mushrooms, potatoes	Flavor (5′-nucleotides)	Ribonucleases	20
Single cell protein (SCP)	Removal of gout-causing purines via nucleotides	Ribonucleases	34
Tea	Flavor, color	Phenolase	16
Dates	Color, texture, flavor	Phenolase, Invertase, Cellulase, PG	
Oranges	Flavor, removes bitterness	Limonin-catabolizing enzymes	25
Unconventional protein sources: seeds, SCP, FPC, forages	Nutrition, texture, flavor, recovery	Proteinases	28, 34
Meat	Texture	Collagenase	27
Meat, aging at low temperature	Texture	Calcium-activated proteinase, cathepsins (?)	26

approaches to potentiation of enzyme action is shown in Table 1.2.

At the processing level, one may achieve optimal action of the foods' own enzymes by control of processing variables; time, temperature, water content and water activity; unit operations such as grinding, shearing, filtration and centrifugation.

ENZYME *VS* NONENZYME REACTIONS

Changes in food quality attributes may arise as the result of chemical reactions carried out by enzymes of microorganisms (fermentative), by the foods' enzymes, by added enzyme or may also occur nonenzymatically. Fermentative changes are, as we have noted, the result of the action of the organized enzyme systems of the microbes in the food. Enzymatic and nonenzymatic modes of the same or similar chemical reactions may take place in a food concurrently and it is helpful to disentangle these modes in order to gain a better understanding of the factors contributing to quality food. In the discussion of the color changes in green vegetables, anthocyanin-containing foods, in the loss of ascorbic acid, and in meat color, enzymatic and nonenzymatic reactions are not always easily distinguishable even when the change is due to only one of these modes. On the other hand, the after-cooking darkening of potatoes appears to be strictly a nonenzymatic phenomenon. One of the aims of this book is to attempt to pinpoint and assess the contribution of enzyme action to food quality.

TABLE 1.2

APPROACHES TO ENZYME POTENTIATION

Chemicals	Preprocessing	Process Variables
Hormones	Ripening	Temperature
Salts	Storage	Time
Coenzymes	Curing	pH
Vitamins	Germination	Water
Activators	Genetic engineering (?)	Mechanical

Enzymatic or Nonenzymatic?—The simplest test to ascertain if a defect in quality is due to enzyme action is to heat the food at 100°C for at least 10 min. The problem is most likely due to enzyme action if this treatment eliminates or lessens the severity of the defect. Of course, the heating does not have to be incorporated into the processing line. Nonenzymatic reactions, as a rule, will be enhanced and the problem will be more severe with increase in temperature. However, heating may be an indecisive test because of the presence in the food of interfering or obscuring substances such as starch which would gel when the food is heated. Another obscuring effect of heating is the nonenzymatic catalytic effect of metals liberated from proteins (including enzymes) during the heating. In most processed foods such reactions probably may be of considerable importance during prolonged storage, especially with regard to lipid oxidation.

Desirable Changes.—Quite aside from the question of the enzymatic nature of undesirable reactions of foods, the same general desirable changes have been and can be carried out by both enzymatic and nonenzymatic modes of action. Perhaps the most familiar example is the use of acid to clot milk in the manufacture of cottage cheese, in contrast to the use of enzymes (rennets or chymosins) in the manufacture of most other cheeses. In the manufacture of food ingredients both enzyme addition and nonenzymatic reactions (as well as fermentation) have been employed. Historically, nonenzymatic methods, especially the use of acids at high temperatures, preceded enzymes used under mild conditions to degrade important food polymers such as starch and protein for the manufacture of sugar syrups and flavor ingredients, respectively.

As discussed in Chapter 9, acid is still the preferred method for the production of invert sugar from sucrose in large quantities in the United States because of the susceptibility to low pH of the glycosidic linkage in the sugar. Until a breakthrough occurs, acid and not cellulase will continue to be used for glucose production by the hydrolysis of cellulose (Chapter 9).

On the other hand, certain chemical changes in food can only be carried out with enzymes alone without unalterably changing the food. This obligatory use of enzymes holds especially where the change is due to a limited transformation, i.e., meat tenderization and chill proofing of beer by proteases, cheese flavor production by lipases, removal of H_2O_2 from cheese by catalase and production of creamy centers in candy by invertase.

On the other hand, alternative choices between chemical/physical approaches are still being made, for example, clarification of various liquid foods by pectic enzymes *vs* "fining" agents (Chapter 30), proteinases *vs* acid (HVP) for protein hydrolysate production and rennet *vs* "non-rennet" cheeses displayed in health food stores (Chapter 33).

Enzyme Action Provides Reactants for Nonenzymatic Changes Which Directly Affect Food Quality

In assessing the relative roles of enzymatic and nonenzymatic reactions it appears that frequently enzyme activity and the subsequent train of events arising as a consequence of cellular disruption is initiated by the formation of relatively unstable primary products as depicted in Fig. 1.4. Some of these primary products formed from the action of enzymes associated with fruit and vegetables are shown in Table 1.3. These unstable products can then usually undergo a plethora of reactions leading to a diversity of products. The primary source of this diversity for a given enzyme reaction type is, as indicated, the presence of numerous substrates. Of secondary importance is the presence of distinct isozymes in a given food. Furthermore, the primary reaction product can then undergo rearrangements and relatively minor reactions, revert back to the original substrate, react with the enzyme usually to inactivate it (reaction inactivation), undergo gross decomposition, or react with other stable food constituents in further complex interrelated reactions ending in stable products which confer quality (desirable or undesirable) on the food. These secondary reactions may be enzymatic or nonenzymatic. Another source of food constituent diversity arising from both enzymatic and nonenzymatic reactions is that due to temperature fluctuation which foods are exposed to, as discussed in Chapter 12.

The same set of primary enzymes initiates alternative pathways leading to alternative quality attributes. It is difficult to know where to draw the line when considering consequences of enzyme action in contrast to strictly nonenzyme action, because initially all substances in fresh produce in which the cells are intact may be considered as primary products of enzyme reaction. Thus, one might consider the Maillard type of nonenzymatic browning as the result of secondary reactions arising from the action of invertase to produce the primary product, glucose. However, it is only when they are heated that such "primary" products become unstable and yield such an abundance of quality-affecting constituents which confer color, flavor and antioxidant properties on the food. Hence we usually include only those primary enzyme products which undergo transformations readily at temperatures at which enzymes usually act. We shall examine the consequence of these transformations both in terms of food quality and in terms of the chemistry involved.

TABLE 1.3

ENZYME-INDUCED QUALITY CHANGES IN FOODS DUE TO SECONDARY TRANSFORMATIONS

Enzyme or Substrate	Enzyme-produced Product (Primary)	Transformation of Enzyme-produced Product (Secondary)	Product or Quality Change
Lipoxygenase	Lipohydroperoxides	Isomerization to ethers and hydroxy acids, polymerization, decomposition and other interaction with protein	Flavor, color, texture, nutritive value
Phenolase	Quinones	Rearrangement polymerization, oxidation of alcohols and aldehydes, oxidation of anthocyanins	Color (melanin), flavor (tea, cocoa), decolorization
Nitrate reductase	Nitrite	Reaction with secondary amines	Carcinogens
Flavoprotein oxidases, photosynthesis	Superoxide anion, O_2^-, hydrogen peroxide, singlet oxygen, (O_2^*)	Formation of ·OH,O_2^*, reduction or oxidation of food essentials	Color, flavor, texture, appearance, nutritive value, labilizes membranes
Rennets	Para-kappa-casein	Micelle aggregation due to hydrophobic interaction, Ca^{2+}	Phase transition
Anthocyanase	Anthocyanidins	Rearrangement	Decolorization
Glucosidases	Glucose	Reacts with primary amines (+ heat)	Browning, flavor

ENZYME ACTION AND QUALITY CONTROL—TESTING

Another general area in which enzymology contributes to improvement of food quality is what we prefer to term "enzyme testing" in the food quality control laboratory. While we can not present details of all the myriad procedures involved in enzyme testing, Chapters 37 through 39 do annotate its versatility and wide-ranging applicability to diverse food quality control problems. We shall present a systematic survey of the application of knowledge of enzymes to the improvement of food quality assurance, in general, and in the food-processing quality control laboratory, in particular. Highlighted are critiques of presently-accepted procedures and the potential application of modern biochemical approaches which, at present, may not be widely known. For that reason, we outline the general ways in which enzyme testing may prove to be of use in quality control and food research laboratories.

Enzymes may be part of the food or may be added for analytically-related purposes. When they are part of the food, either naturally present or added as part of a step in food processing, measurement of their activities may serve the following useful purposes:

(1) As indices, markers, empirical correlates or predictors of parameters such as maturity, suitability for processing, quality attribute, past history, processing analogs (i.e., adequacy of heat treatment for blanching or pasteurization), maturity, protein productivity, crop yield, etc.
(2) As isoenzyme "fingerprints" for similar purposes (i.e., to determine if a meat or seafood has been frozen).
(3) For standardization and monitoring of commercial enzyme preparations.

Standardization and monitoring of commercial enzyme adjuncts is essential in order to use the minimum level in consonance with enzymes already present in the food and with the optimum quality change expected from the action of these enzymes. This is not only to save money, but to avoid poor quality and product rejections which result from enzyme overdosage. In this connection, it behooves the food technologist to disabuse nontechnical assistants of the idea that "if little is good, then more is better."

In this area the future will see the replacement of conventional time-dependent enzyme assay by applications of immunoenzymology (Chapter 39) and active site titration (Chapter 37). The latter, expressed as operational normality of the enzyme, is especially applicable to commercial enzyme preparations both before and after they have been added to foods.

Purified specific enzymes may be added to foods or food extracts for the following analytically associated purposes:

(1) As specific analytical reagents for quantitative estimation of gross food constituents which are substrates for the enzymes, such as sugar, protein, alcohol, etc.
(2) For analysis and detection of food constituents which serve as affectors (inhibitors, activators) of the added enzyme. This analysis is particularly important in health-associated food quality, such as pesticides, harmful adventitious substances and trace metals.
(3) As "short-cuts" for otherwise tedious, time consuming and expensive biological tests (nutritional digestibility, availability of protein/carbohydrates; estimation of crude fiber).
(4) As analytical aids in clean-up procedures and preliminary treatment (preparation of protein hydrolysates for automated amino acid analysis, preliminary treatment in examination for filth).
(5) Finally, as auxiliary enzymes in both enzyme activity assays and analysis of food constituents, as well as in enzyme immunoassay.

ENZYME ACTION AND FOOD QUALITY ATTRIBUTES

A good portion of this book is devoted to what we hope will be a critical assessment of what is known as of this writing about the nature and contribution of enzyme action to the traditional quality attributes of foods (appearance, color, flavor and texture) and health-associated qualities (wholesomeness and nutritional quality). In addition, enzyme action plays a major role in the transformation of raw foodstuffs to completely new foods. In other words, the enzyme action has resulted in a change in appearance of the raw natural products (cheese, bread, beer, juices). These foods have in a sense assumed a new identity largely as the result of enzyme action, or the action of enzyme is essential to yield products of desirable optimal quality in the course of this transformation. In each case there is at some point a drastic change in the appearance of the food. We shall for the most part avoid a discussion of "appearance," although this is undoubtedly an important parameter of food quality and probably includes elements of color and texture as well as overall visual perception.

Further subdivision is dictated by logic of the subject matter and depends also on the voluminousness of the literature. In the case of color, we distinguish in the first instance between enzyme action on phenols and nonphenolic compounds. The major subdivision on the discussion of flavor is based on whether the enzyme results in desirable or undesirable flavor. Wherever enzyme action resulting in undesirable quality is discussed, we also try to assess and suggest various means of preventing this enzyme action. This may lead us to areas peripheral to enzymology such as economics and engineering, but we feel that some such peregrination is unavoidable in order to present an integrated picture of the role of enzyme action in foods.

This leads us to the realization that a proper perspective should be kept of the role of food enzyme action in relation to other food-oriented disciplines. The food technologist/scientist must exercise eclectic judgment in that he/she must select those aspects of the 30-odd disciplines which constitute his/her purview in order to properly handle and understand the agricultural commodities which are processed by a technologist or studied by a food scientist. As we previously pointed out, it is not always easy to distinguish which of the many disciplines is applicable and how to apportion the relative importance of each in a given case. Furthermore, a step adopted in the food chain to stop the deteriorative consequences of enzyme action may also affect other qualities *via* nonenzymatic mechanisms. Hence, a discussion of color, for example, may in part also concern texture and flavor changes in order to give a true picture of a given change in quality.

We would also like to point out that rather than present a separate formal introduction to and discussion of the properties of each of the more important enzymes, we shall usually introduce and integrate the detailed description of each enzyme at the point where it is first discussed with reference to an important quality change.

We assume that the reader has had a background of organic and biological chemistry, including a knowledge of the elements of enzymology. In the ensuing chapters, we shall, nevertheless, review some of these basics, especially with regard to how they fit into current concepts of what an enzyme really is and does and what relation it has to other biologically active proteins, as well as to provide a literature base for those who wish to further pursue a particular aspect or phase of enzymology.

INFORMATION SOURCES

General Enzymology

Enzyme studies have been and are still so intimately entwined with and form vast portions of all facets of biologically-oriented academic disciplines and technologies, that an attempt to annotate even part of the vast literature or even sources of literature, i.e., journals, etc., would be inappropriate and meaningless. However, the tradition of enzymology as a special, separate subdiscipline of biochemistry lives on in the continuing encyclopedia references ("The Enzymes," "Methods in Enzymology" and "Advances in Enzymology") as well as in the now generally accepted scheme of the Commission on Biochemical Nomenclature on the Nomenclature and Classification of Enzymes of the International Union of Biochemistry, the "Enzyme Commission," EC (Florkin and Stotz 1973; Anon. 1979). Barman (1969,1974) has catalogued some 800 distinct enzymes along with data on their characteristics, such as enzyme kinetic constants, activity modifiers, sources, pH optima, as well as data on their protein-associated properties, such as molecular weight and subunit structure.

The three encyclopedic references just mentioned can be traced back to historical antecedents. Each is apparently derived from analogous German texts which appeared in the mid 1920s when the literature began to expand. Of historical interest are the compendia of Oppenheimer (1925–1936) containing some 11,000 references to original investigations. "The Enzymes" can be traced from von Euler and Myrbäck (1927–1934) through Sumner and Somers (1947); Sumner and Myrbäck (1950–1952), four volumes; Boyer, Lardy and Myrbäck (1959–1963), eight volumes; Boyer (1970–1976), 13 volumes altogether. Citations in bibliographies to "Enzymes" refer to this third edition of "The Enzymes."

Methods in Enzymology, edited by Colowick and Kaplan (1955–1980), 69 volumes and continuing, has an antecedent, the four volume "Methoden der Enzymforschung" of Bamann and Myrbäck (1941). Similarly, "Advances in Enzymology" (Nord and Werkman 1941–1943; Nord 1944–1970; Meister 1971–1979 and continuing) is the direct successor to "Ergebnisse der Enzymeforschung" (Nord and Weidenhagen 1932–1939). It is symptomatic of the relation of modern enzymology to other subdisciplines that the subtitle of this series was changed from ". . . Related Subjects Area of Biochemistry" to ". . . Molecular Biology." Similarly, in 1974 the journal *Enzymologia* was renamed *Molecular and Cellular Biochemistry.* On the other hand, the Swiss journal *Enzymologia Biologia et Clinica* was superceded in 1975 by *Enzyme.*

A steady stream of shorter, nonserial books dealing with one or more aspects of enzymology continues to appear. We shall refer to the more specialized ones at the appropriate times. The classical monographs of Haldane (1930), Haldane and Stern (1932) and Northrop *et al.* (1948) can still be read for enlightenment and understanding, as well as for historical interest. In addition, introductory textbooks dealing in general enzymology include those of Neilands and Stumpf (1955), Mehler (1957), Gutfreund (1965), Wynn (1979) and Foster (1979). Gutfreund (1976A) has edited a most interesting series of short, but informative and authoritative essays marking the centenary of the first use of the word "enzyme" by Kühne (1876). Among the non-English enzyme textbooks is one with the intriguing translation (from Danish) "Enzymes—Why Them" (Lundberg 1972). Wynn (1979) has written a precise monograph which emphasizes the relationship between the enzyme function and protein structure. A more comprehensive general treatment of the principles of enzymology is provided by Dixon and Webb (1968). Members of the American Chemical Society may find it of interest to learn that the first of the ACS Monograph series concerns the chemistry of enzyme action (Falk 1924).

Food Related Enzymology

Books.—Samuel C. Prescott, for whom the annual Institute of Food Technologists Award (IFT) is named, translated, from French, the first book on enzyme technology by Effront (1899). He also translated a further monograph on proteolytic enzymes in life and industry (Effront and Prescott 1917). Probably the first complete book in English purporting to deal with enzyme technology, including food, is that of Tauber (1949). However, a great proportion of this monograph is devoted to food microbiology, antibiotics and vitamin analysis. Reed (1966) wrote the first comprehensive and systematic survey and description of the use of commercially available food grade enzymes in various food processing industries. A revised edition has been edited by Reed (1975). Other books on enzyme production, availability and economics are those of Wiseman (1975), Wang *et al.* (1979) and Daheny and Wolnak (1980). Other literature related to production and other aspects of added enzymes will be cited in Chapter 5.

The unique monograph of Whitaker (1972A) on principles of enzymology for the food sciences provides an excellent academic background to the applications of enzymology in food-related fields. Presentation of principles is followed by fairly detailed description and analysis of the mechanism as then conceived of purified enzymes whose action is considered to be important in foods. Actual relations to food quality and processing are dealt with briefly in the introductory chapter. Several books devoted to food biochemistry and technology have good summaries of the role of enzyme action (Potter 1978; Charley 1970; Borgstrom 1968; Lee 1975). A German programmed monograph on enzymes for a food technology course is available (Jürgen 1975). Useful information on the management of enzyme action, especially endogenous action, is discussed in the food biochemistry textbooks of Braverman (1963), a revision of Braverman's textbook by Berk (1976) and Eskin *et al.* (1971), and publication of a symposium on biochemical control systems in food (Hultin and Haard 1972; Hultin and Milner 1978). Synoptic discussions, some of direct relevance to food enzymology, can be found in the journal *Trends in Biochemical Sciences* (TIBS).

Food enzyme-oriented books, which are compendia of papers, usually reviews of selected areas presented at symposia, are appearing at a progressively increasing rate: Schultz (1960), Society of Chemical Industry, London (1961), Whitaker (1974A), Ory and St. Angelo (1977) and Cooler (1976). Several food-related enzymes are covered in the treatise on plant biochemistry edited by Stumpf and Conn (1980).

Enzyme Technology Digest, an information service on enzyme technology including some articles on food application, has been published for the National Science Foundation (RANN, Research applied to national needs) (Roberts 1972–1976 and continuing). *The Catalyst*, a handy, timely, information bulletin is directed specifically to both the food technologist and the academically-oriented food scientist (Whitaker 1975; Swaisgood 1977). Some commercial enzyme producers publish bulletins periodically such as Novo's *News About Enzymes* which include information about the action of enzymes in foods.

Patent Literature.—The Noyes series of digests of patents includes books by Wieland (1972) and Pintauro (1979) on patents dealing with enzymes in food processing. They cover, without any attempt to evaluate, procedures for addition of enzymes and the promotion (potentiation) of the foods' own enzymes. Other Noyes food processing reviews describe patents in which enzyme action is involved. Patent literature is also available in an encyclopedia on enzyme technology (Meltzer 1973). According to Beck and Scott (1974), the International Research Patent Office, P.O. Box 1260, The Hague, the Netherlands, issues periodic lists of patents pertaining to both nonfood and food use of enzymes. Other literature pertaining to "enzyme engineering" will be found in Chapters 5, 8, 9 and elsewhere.

Other General Sources of Information.—A selection of reviews and discussions of enzyme action in foods covering 1932 to 1980 is shown in Table 1.4. They range from scholarly, comprehensive critical reviews in books to editorials and semipopular expositions in trade journals. What they do have in common is an attempt to present the action of enzymes in foods in a fairly broad perspective and many contain elements of prognostication if not prophesy. As an aid in assessing the general level and tenor of each reference, we have included the number of pages and references cited. We also indicate relative emphasis as a critical review or as an informational editorial. While most of the papers deal with more than one aspect of food enzymes, we have codified these aspects to indicate which were stressed. As of this writing the enthusiasm for, and funding of, "enzyme engineering" has tended to overshadow and obscure the problems as well as the achievements and progress in the management of endogenous enzymes. Because of this and because we shall cite further references in the enzyme engineering area in Chapter 8, we have tended to include more of the papers dealing with endogenous enzyme action insofar as they are available. As we previously mentioned, a more thorough understanding of enzyme potentiation will lead to more successful, efficient and deliberate application of the same type of enzyme action as a food adjunct. Again we wish to emphasize that other more specific references will be presented at the appropriate time. Furthermore, we shall have occasion to bring up some of the ideas and viewpoints expressed in the references of Table 1.4.

TABLE 1.4

49 YEARS OF GENERAL REVIEWS, DISCUSSIONS, SUMMARIES, EDITORIALS, TRENDS, PREDICTIONS, PROMISES OF ENZYME ACTION IN/ON FOODS

Author	Publication Type[1]	Pages	No. Ref.	Area[2]
Balls (1932)	J,I	6	0	I,A,F
Hesse (1935)	B,R	123	223	E+,C
Balls (1939)	B,I	6	6	E±
Hesse (1940)	B,R	213	1100	A,P,C
Balls (1942)	J,I	4	11	E−
Balls (1947)	J,I	7	5	E±,F
Balls (1948)	J,I	4	0	E−,F
Balls (1950)	B,R,I	27	76	A,P,C,E±,T
Jansen and Balls (1951)	B,I	6	0	E±
Schwimmer and Curl (1951)	B,I	6	0	A,P
Langlykke *et al.* (1952)	B,R	99	31	P,C
Smyth (1955)	B,R	24	71	E−,(C)
Makower (1956)	J,I	4	14	E−
Joslyn (1957A)	J,I	8	21	A,C,E±
Balls (1962)	J,I	4	11	A−,(C)
Joslyn (1963)	B,I	17	48	A,C,EI,T
Schwimmer (1964A)	J,I	4	0	E±,A,F
DeBecze (1970)	B,I,R	57	163	C
Wiseman and Gould (1968)	J,I	3	0	A,F
Schwimmer (1969A)	J,R	8	203	P,C,E±,T,A
DeBecze (1970)	J,R	39	153	P,C,T
Weiser *et al.* (1971)	B,I	30	20	A,E±,C
Underkofler (1972)	B,R	57	201	P,C,E±
Hultin and Haard (1972)	J,R	32	250	E±,A
Schwimmer (1972)	J,R	6	50	E±
Fox (1974)	B,R	26	87	E±,C
Beck and Scott (1974)	B,R,I	30	43	E±,C,F
Skinner (1975)	J,I	20	0	P,C,F
Urquidi (1975)	J,I	5	0	C
Winter (1976)	B,I	5	0	E−
Reed (1976)	I	4	11	A,E±,F
Wiley (1977)	B,R	16	30	E+
Schwimmer (1978)	B,R	31	108	E±,F
Drapron (1980)	J,I	14	34	
Schwimmer (1980A)	R,I	9	44	A,E±

[1] B, book chapter; J, journal; N, neither; R, critical, fairly comprehensive review; I, interpretive, informational, didactic, editorial, semipopular.

[2] A, principles of enzymology; C, commercial enzyme; E+, favorable endogenous enzyme action (potentiation); E−, undesirable enzyme action and its prevention; F, future, oracular, trends, predictions; P, enzyme production; T, enzyme testing.

Part II

BASICS OF ENZYMOLOGY

2

THE NATURE OF ENZYME ACTION—DEFINITIONS AND SPECIFICITY

DEFINITIONS

In the first chapter we outlined the historical development of the "enzyme" concept. Like other dynamic, pivotal and historically-developed concepts, the idea of what an enzyme really is and does is not static but changes in consonance with ever-deepening insights into the nature of its action. It is not surprising, therefore, that different textbook authors and enzyme authorities emphasize different aspects of enzyme action in their definitions. What they all have in common (with one exception), however, is the firmly established credo that all enzymes have to be proteins. Further delineation involves just how they differ from nonenzyme protein. Each authority puts more emphasis on one aspect and nuance of enzyme action than on another. We therefore find it worthwhile to quote several of these definitions:

> . . . specific protein with catalytic properties due to its specific power of activation.
>
> *Dixon and Webb (1964); Whitaker (1972A)*

> . . . a protein that is synthesized in a living cell and catalyzes or speeds up a thermodynamically possible reaction so that the rate is compatible with the biochemical processes essential for the maintenance of a cell.
>
> *Conn and Stumpf (1972)*

> . . . macromolecular protein catalyst . . . sequence into three dimensions in a precise orientation . . . to . . . effect controlled chemical changes in biological systems.
>
> *Walsh (1979)*

> The essence of an enzyme is its ability to speed up (catalyze) a reaction involving the making and breaking of a specific covalent bond.
>
> *Watson (1976)*

> . . . proteins whose striking characteristics are their catalytic power and specificity . . . regulated . . . intimately involved in transformations of different forms of energy.
>
> *Stryer (1975)*

> . . . proteins . . . specialized to catalyze biological reactions . . . extraordinary catalytic power . . . highly specific.
>
> *Lehninger (1975)*

> . . . molecules of biological origin which increase the rate of specific reactions, although not affecting the final position of the equilibrium established, and which may be recovered from the reaction mixture at the end of the reaction.
>
> *Wynn (1979)*

In most of these definitions there is an implicit propensity to sharply delineate enzymes from other biologically important proteins such as some hormones, hemoglobins, cytochromes and even protein catalysts such as the initiation and termination factors in protein synthesis (Chapter 4).

As we shall see in Chapter 3, there is a trend to find unified principles underlying the action of all biologically active proteins. However, for the time being we believe that the definitions of Conn and Stumpf and that of Stryer come closest to the heart of what an enzyme is and does. Enzymes thus possess three characteristics which distinguish them from other biologically active proteins. These arise directly from their structures and can be defined as follows:

(1) *Specificity.*—The capacity for a given enzyme to sharply discriminate among many possible chemical reactions by limiting its catalytic capacity to one or a group of closely related reactions.
(2) *Catalytic Efficiency.*—As catalysts they should not be used up in the reaction and do not influence the final equilibrium. However, the rate at which the enzyme-catalyzed reaction reaches equilibrium is unmatched by any other known catalyst and in a sense limitation on catalytic efficiency is imposed by the property of specificity since one molecule can be transformed in many ways.
(3) *Regulation.*—We believe that the control and regulation are at least as important and universal properties or characteristic of all enzyme action as are specificity and catalytic efficiency. Without a built-in control mechanism, the extremely complicated life processes which are guided by the consecutive action of teams of enzymes would be impossible.

It is on the basis of these three inextricably-associated concepts of enzymology that we shall proceed with our discussion of the nature of enzyme action. We shall first discuss specificity, partly because it is the first of three properties of enzymes of which we now have, in principle, a fundamental understanding on a molecular basis. We shall also use "specificity" as the vehicle for introduction of some of the traditional topics associated with an elementary discourse of enzymes, although the other two categories might have served equally well.

SPECIFICITY, CLASSIFICATION AND NOMENCLATURE

Specificity, the unique characteristic of an enzyme, is the restriction of the enzyme's capacity to catalyze a limited number of chemical reactions in which the reactants and reaction types are usually closely related in structure. Specificity is also the primary foundation upon which is built the almost universally and generally accepted scheme for enzyme classification and nomenclature. Although it was recognized quite some time ago that substrate specificity was the only basis for a rational classification scheme, some problems which needed to be solved were outlined by Hoffmann-Ostenhoff (1953) in a progress report on a concerted effort to formalize enzyme nomenclature. Before this important progress report, about the only rule, proposed by Duclaux (1883), which was fairly uniformly adopted was that an enzyme should be named by adding the suffix "-ase" to the substrate which undergoes catalytic transformation. Previously many of the "unformed ferments" ended with "-sin" (trypsin, pepsin, emulsin, myrosin) or "-ain" (papain, bromelain). Departure from Duclaux's dictum became widespread with a growing tendency to add "-ase" to the type of reaction carried out (dehydrogenase, oxidase, carboxylase). "-Ase" was even extended to more complicated or obscure biological phenomena (replicase, joinase, permease, sealase, convertase, gyrase, twistase).

In 1955 an International Commission on Enzymes was established to provide a uniform, comprehensive and detailed system of enzyme nomenclature. Published in 1964, it contained some 650 entries. Nine years later a revised edition listed about 1700 enzymes (Florkin and Stotz 1973); in the latest revision (Anon. 1979) some 2300 well-characterized, distinct enzymes have been officially recognized. It should be emphasized that a particular entry encompasses a family of closely related enzymes (see isoenzymes). To each enzyme, or this family of multiple forms of the "same" enzyme, is assigned a four-number code, each number separated by a period and prefaced by EC.

Table 2.1 is an attempt to epitomize the nature of the classification scheme and to show how the scheme is used as a basis for the systematic name. The scheme recognizes six classes of reactions. The subclass (second) numbers are fairly consistent and analogous in going from class to class in that they refer to groups or bonds undergoing transformation (Table 2.2). The third number, however, does not show such inner logic but skips around from

TABLE 2.1

BASIS FOR ENZYME CLASSIFICATION AND SYSTEMATIC NOMENCLATURE

First #, Class, α (Second #, Subclass, β)	Third #, Sub-Subclass, γ (Fourth # = δ)	Basis for Systematics
(1) Oxidoreductases (Group in donor, *D* oxidized)	Acceptor, *A*, reduced	*D:A* Oxidoreductase
(2) Transferases (synthases) (Group transferred from *D* to *A*)	Group transferred (Further delineated)	*D:A* δ [or γ] transferase (P, NH_2, Me, Glucosyl, etc.)
(3) Hydrolases (Bond hydrolyzed—ester, peptide, etc.)	Substrate class (glycoside, peptide, etc.)	Substrate [γ] hydrolase[2]
(4) Lyases Bond cleaved (C-S, C-N, etc.)	Group eliminated	Substrate Product [γ] lyase
(5) Isomerases Type of reaction	A mix of S[1], reaction type, chiral position involved in isomerization[3]	S[P] (γ-δ) [(α-γ)] isomerase
(6) Ligases (Synthetases) (Bond synthesized—C-C, C-O, C-N, etc.)	Substrate S_1, cosubstrate S_2 (third cosubstrate is almost always a nucleoside triphosphate)	S_1:S_2 ligase (NP[4] forming)

[1] S, substrate; P, product.
[2] The proteases (endopeptidases) classified as serine (3.4.21), SH-(3.4.22), acid (3.4.23) and metallo (3.4.24), have not been assigned systematic names.
[3] Isomerases, racemases, epimerases, mutases, ligases (decyclizing, isomerizing).
[4] Nucleoside mono- or diphosphate.

TABLE 2.2

ENZYME SUBCLASSES

Sub-class	1. *Oxidoreductases* Donor	2. *Transferases* Group	3. *Hydrolases* Bond	4. *Lyases* Bond	5. *Isomerases* Type	6. *Ligases* Bond Formed
(1)	CH-OH	1-Carbon	Ester	C-C	Racemases, epimerases	C-O
(2)	CHO, CO	CHO, CO	Glycosyl	C-O	*cis-trans* Isomerases	C-S
(3)	CH-CH	Acyl	Ether	C-N	Intramolecular Oxidoreductases	C-N
(4)	CH-NH_2	Glycosyl	Peptide	C-S	Intramolecular Transferases	C-C
(5)	CH-NH	Alkyl, aryl[2]	C-N	C-X	Intramolecular Lyases	P-O-P[3]
(6)	NAD(P)H	N-contain.	Acid anhydrides	P-O	—	—
(7)	N-cmpds	P-contain.	C-C	—	—	—
(8)	S-cmpds	S-contain.	C[P]-X	—	—	—
(9)	Haem	—	P-N	—	—	—
(10)	Diphenols	—	S-N	—	—	—
(11)[1]	H_2O	—				

[1] 1.12—donor, H; 1.13—single donor, O_2 acceptor; 1.14—paired donors, oxygen acceptor incorporated; 1.15—superoxide, acceptor; 1.16—donor, -CH_2-.
[2] Except methyl.
[3] X = halogen, P-O-P = phosphate bond.

reaction type to substrate used to donor. Within these sub-subclasses is a number, 99, which lumps together singular and ill-defined enzymes. The fourth number donates an arbitrary position within the group defined by the other three numbers.

As shown in Table 2.1, the systematic name of the enzyme, relegated from the first to the last column in the revised edition, is based on the classification. The Enzyme Commission also recommends a shorter or trivial name but includes other names. The EC numbers of enzymes mentioned in this book are listed in the Enzyme Index.

Some Classification Problems

The systematic name is not as internally consistent as one might expect. Attempts to make a completely logical, consistent scheme have encountered difficulties, as illustrated with the problems attendant on classifying what is referred to in Chapter 15 as the "phenolases." In 1964 the classification distinguished between a "catechol" (*o*-diphenol, 1,2-benzenediol) oxidase, EC 1.10.3.2, and a *p*-diphenol oxidase (laccase), EC 1.10.3.2. In 1973, the subclass "10" was abolished and all of the phenolases were categorized as "monophenol monooxygenases, EC 1.14.18.1. In 1979 all three numbers were used, with slight departures from the original specificity assignments. Thus, "laccase," EC 1.10.3, now acts on both *o*- and *p*-diphenols, and EC 1.14.18.1 (the only member of its subclass), still known as "monophenol monooxygenase," will, if substrates are available, catalyze the same reactions as do the 1.10.3.1 enzymes.

Another problem is exemplified by the enzyme activity usually referred to as γ-glutamyl transpeptidase (γ-glutamyl transferase), of potential importance in food quality (Chapter 21) with efficient hydrolytic as well as transferring competence. Conversely, the great majority of the hydrolases discussed in this book also possess considerable transferase activity. Furthermore, they cross over subclass lines. Thus, proteinase not only hydrolyzes peptide bonds, C-N, in proteins but also catalyzes the transfer of an amino acid from one polypeptide and can hydrolyze C-O and C-C bonds. Another dilemma is that in many lyase reactions the reverse reaction, addition, predominates. In such cases the term "synthase" is used; in the case of CO_2, the term carboxylase is used. Synthases are not to be confused with true synthetases which are found in Class 6 (ligases) only. This rule is honored more in the breaking, i.e., "starch synthetase" instead of starch synthase.

An interesting lyase in this regard is the enzyme commonly referred to as carboxydismutase, a key enzyme in the Calvin photosynthetic carbon pathway and as important as the major protein of the new green vegetable protein foods (Chapter 34), as well as an important taste determinant in grapes and other fruit (Chapter 20). Its systematic name is 3-phospho-D-glycerate carboxylase, whereas its recommended name is ribulose bisphosphate carboxylase (dimerizing):

$$\text{D-Ribulose-1,5-diphosphate (RDP)} + CO_2 \rightarrow 2\,(\text{3-phospho-D-glycerate})$$

This enzyme also defies group lines in that it can also employ oxygen as substrate instead of carbon dioxide:

$$\text{RDP} + O_2 \rightarrow \text{3-phospho-D-glycerate} + \text{2-phosphoglycollate}$$

Group 5, the isomerases, provides the main exception to the rule that the last term in the name refers to the reaction catalyzed by the group. This is because the special geometric transformations which a substance can undergo render the term "isomerase" quite uninformative; as for instance, one of the enzymes involved in the biosynthesis of phytic acid, *myo*-inositol synthase, *via* the isomerization of glucose-6-phosphate, which involves an internal oxidation-reduction by NAD. Food-related isomerases retain the term "isomerase" (glucose isomerase, chalcone isomerase and the putative thiocyanate isomerase). The latter, if it exists, like phenolase, is in a subclass of its own.

The sixth group, the ligases, uses the common currency of life, ATP (and occasionally NAD), for synthesis of the peptide bond in all enzymes and in other proteins as well as in the synthesis of nucleic acids and of many regulatory molecules. We shall have but little, if any, occasion to invoke action of the ligases as being directly involved in food quality.

Not all biochemists adhere to the EC classification. Metzler (1977) distinguishes among enzyme catalyzing: nucleophilic displacement; addition to and elimination to create double bonds; isomerizations; C-C cleavage; other rearrangements. Walsh (1979), criticizing EC's 6 divisions as having little didactic value and as not systematically addressing the large numbers of enzymes involved in making and breaking C-C bonds, classifies enzymes according to their catalysis of (1) group transfer via nucleophilic attack (including hydrolysis via the nucleophile water); (2) redox reactions; (3) eliminations, isomerizations, and rearrangements of carbon skeletons; and (4) the making and breaking of C-C bonds, the building blocks of biosynthesis.

LATITUDE OF SPECIFICITY

Most introductions to enzyme specificity distinguish among absolute, group and low categories. Absolute specificity means the restriction of the action of the enzyme to the catalysis of one chemical reaction. This, of course, may involve more than one substrate if the reaction is multimolecular. While the action of a given enzyme will range over comparatively few substrates, we have already by way of illustration encountered some of the problems involved even in assignment of class specificity—that enzymes are not so specific as we might suspect. Some of the now traditional enzymes which were thought to possess absolute specificity have been found to catalyze a rather wide spectrum of reactions. Thus, catalase, which was once thought to catalyze the decomposition of hydrogen peroxide to water and oxygen exclusively, is now considered to be a peroxidase in which hydrogen peroxide is both the preferred but not the only hydrogen donor as well as acceptor. Peroxidase exhibits a most interesting range of specificity. Not only will conventional hydrogen-containing substrates act as hydrogen donors, but the halogen anions, for instance, can also serve in an analogous role. The peroxidases can also catalyze the direct oxidation of phenylacetaldehyde, phenyl pyruvate, dicarboxylic acid, indole acetic acid, NAD(P)H, ferrocytochromes, dihydroxyfumaric acid and glutathione. They even catalyze the decarboxylation of methionine and use chlorine dioxide as both donor and acceptor (Chapter 11).

Another enzyme which was once considered to possess absolute specificity is carbonic anhydrase which also catalyzes reversible hydration of aldehydes and the hydrolysis of a variety of nitrophenyl esters (Pocker and Stone 1967; Schwimmer 1969B). Another enzyme which has so far survived the intense scrutiny of modern enzymology as having absolute specificity is the nickel-containing urease. It would be hazardous to venture a guess, but of the 2000-odd enzyme types so far discovered, only those in Class 6, the ligases, will be found to contain a large number of enzymes which possess absolute specificity.

Group and Low Specificity

In contrast to urease, most of the other hydrolyses exhibit group specificity with varying degrees of latitude: There are fairly rigid specific requirements for the type of bond hydrolyzed and for one-half of the molecule, i.e., for one of the moieties which participate in the formation of the specified bond. Thus, the aminopeptidases require the presence of a free amino group and hydrolyze a polypeptide starting from the amino end, whereas the carboxypeptidases start from the opposite end of the peptide, the end with a free carboxyl group. Lipoxygenases can tolerate changes in the length of the fatty acid or other lipid-like molecule but are quite stringent with respect to requirement of the *cis,cis* penta-1,4-diene grouping.

Lipase action is frequently cited as an example of low specificity. In the equation

$$R_1\text{–}R_2 + H_2O \rightarrow R_1OH + R_2H$$

the lipases will, in contrast with other group-specific enzymes, show no or little discrimination for either R_1 or R_2, only that the bond be a carboxylic ester bond. Furthermore, most lipases will hydrolyze only the fatty acid ester bonds at the 1- and 3-position of a triglyceride molecule.

An example of the practical importance of specificity with regard to food quality is afforded in Fig. 2.1. From an economic point of view, it would have been preferable if the easily available pancreatic ribonuclease had released 5′- instead of 3′-ribonucleotides from nucleic acids, thus making the production of these flavor potentiators much less expensive (Chapter 20).

Stereospecificity

One of the most singularly remarkable specificity-related features of enzymes is their capacity for distinguishing among closely related isomers, especially those which exhibit a chiral relation to each other. Examples abound in which chirality is the decisive factor in determining whether an enzyme will or will not act on a substrate. Thus, almost all enzymes catalyzing the transformation of amino acids, be that transformation hydrolysis, transfer, isomerization, lyase action or synthesis, exhibit exclusive preference for the L-amino acid, whereas the D-isomer may inhibit the reaction.

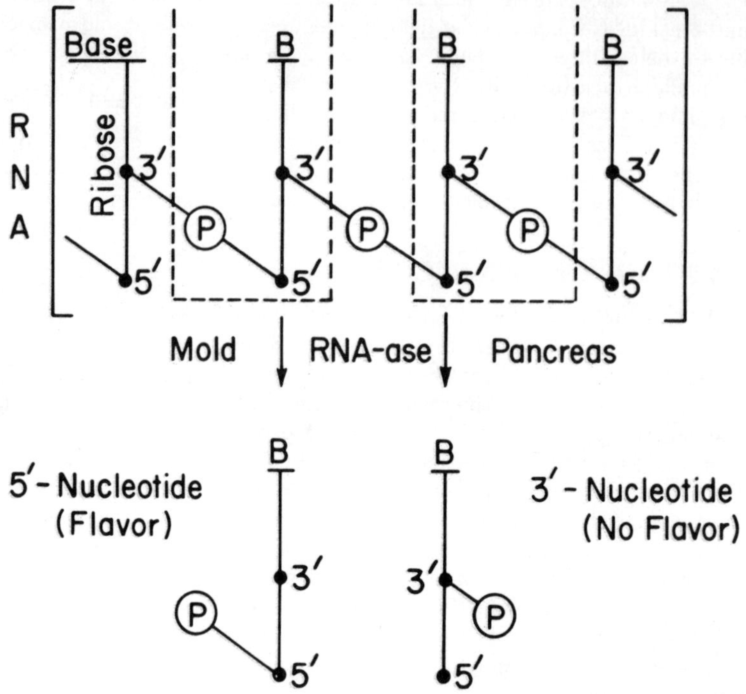

FIG. 2.1. ENZYME SPECIFICITY DETERMINES FLAVOR

From Schwimmer (1969A)

Enzymes attacking glucosides will hydrolyze either the α- or β-glycosidic linkage but seldom, if ever, both.

Carbon-sulfur lyases, involved in flavor production in vegetables (Chapter 21), exhibit typical stereoisomeric specificity. The cysteine sulfoxide lyase from seeds of the acacia shrub *Albizzia lophanta* converts derivatives of S-substituted derivatives of L-cysteine (I) and L-cysteine sulfoxide (II) into pyruvate, ammonia and mercaptan or sulfenic acid, respectively.

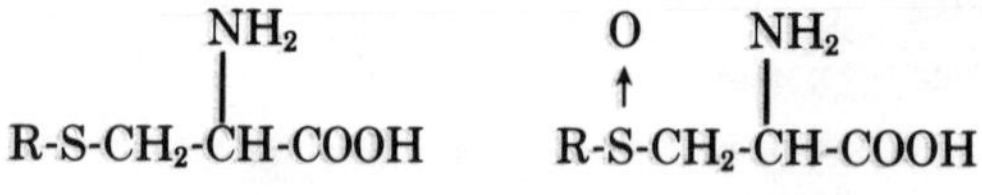

$$\underset{\text{I}}{\text{R-S-CH}_2\text{-}\overset{\overset{\text{NH}_2}{|}}{\text{CH}}\text{-COOH}} \qquad \underset{\text{II}}{\text{R-S-CH}_2\text{-}\overset{\overset{\text{NH}_2}{|}}{\text{CH}}\text{-COOH}} \quad \overset{\text{O}}{\uparrow}$$

$$\underset{\text{III}}{\begin{array}{c} \text{HS-CH}_2\text{CHNH}_2\text{COOH} \\ | \\ \text{CH}_2 \\ | \\ \text{HS-CH}_2\text{CHNH}_2\text{COOH} \end{array}}$$

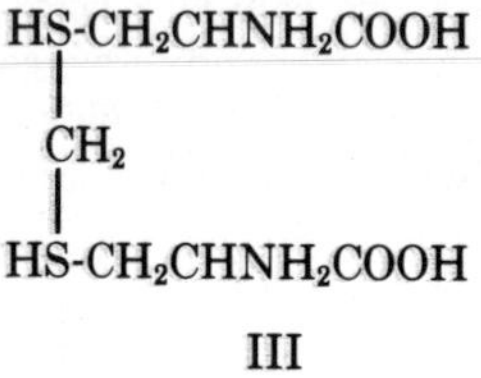

The activities on various substrates related to cysteine relative to its natural substrate, djenkolic acid (III), are shown in Table 2.3. As expected, the L-cysteine moiety is an absolute requirement with no substitutions in any of the atoms other than the sulfur allowed. Neither D-cysteine, nor the higher homologs of L-cysteine can serve as substrates. However, all S-substituted derivatives of both L-cysteine and L-cysteine sulfoxide are substrates, although there is considerable variation in susceptibility to the action of the enzyme. The specificity of the alliinases from alliaceous foods (onions, garlic, etc., see Chapter 21) is similar except that they will act only on the derivatives of L-cysteine sulfoxide and not on derivatives of L-cysteine, which are inhibitors of these enzymes (Schwimmer *et al.* 1964A). The presence of oxygen on the sulfur introduces another center of asymmetry. The onion enzyme can act on both the (+) and (−) diastereomers of derivatives of L-cysteine sulfoxide, but acts more efficaciously on the naturally-occurring (+) diastereomers.

On the other hand, enzymes which convert non-optically active substrates to optically active ones almost invariably convert them to only one of the possible optically active stereoisomers. Thus, when a new amino acid is formed *via* transfer of an α-amino group from an amino acid to a 2-ketodicarboxylic acid (e.g., α-ketoglutaric acid) by the ac-

TABLE 2.3

LATITUDE OF SPECIFICITY: ALBIZZIA C-S-LYASE[1]

Substrate	Rate[2]	Substrate	Rate
D-cy	0	L-Djenkolic acid	100
L-cy	18	Methyl djenkolic acid	51
S-Methyl-L-cy	71	Isobutyldjenkolic acid	3
S-Methyl-L-cy sulfoxide	39	S-(2,4-dinitrophenyl)-L-cy[5]	3
S-Ethyl-L-cy	87	Porphyrin c (synth.)[4]	0.1
S-Ethyl-L-cy sulfone	15	L-Homocysteine	0
S-Propyl-L-cy	96	L-Methionine	0
Felinine[3]	10	L-Penicillamine	0
S-β-Chloroethyl-L-cy	63	α-Methyl-DL-cy	0
S-β-Carboxyethyl-L-cy	4	Glutathione	0
L-Lanthionine	12	S-Methylglutathione	0
β-Methyl lanthionine (ex subtilin)	4		

[1] Source: Adapted from Schwimmer and Kjaer (1960); cy, abbreviation for L-cysteine.
[2] Rate of pyruvate liberation, expressed as % of that of the endogenous substrate, djenkolic acid, CyS-CH_2-SCy.
[3] S-1,1-dimethyl-3-hydroxypropyl-L-cysteine, present in cat urine.
[4] Identical with that obtained from cytochrome *c*.
[5] See Fig. 2.4 and 2.5.

tion of aminotransferase (transaminase), the new amino acid is of the L-configuration. Thus, for aspartate aminotransferase:

L-Aspartate + 2-Oxoglutarate →
Oxalacetate + L-Glutamate

Other well-known enzymes which show similar stereospecific synthase specificity are fumarase and succinate dehydrogenase. In contrast, another member of the tricarboxylic acid cycle, aconitase, forms both citrate and isocitrate *via* the addition of water to aconitate. However, its normal physiological function is considered to be the extraction of a molecule of water from citric acid in the TCA cycle.

Prochirality.—The action of aconitase on citric acid illustrates a specificity not thought of before the advent of isotope tracers in biochemistry. The action implies that citric acid can be dehydrated in two different ways. From the pioneering work of Ogston discussed by Cornforth (1976), enzymes not only discriminate between true diastereomers but also selectively among certain prochiral molecules such as citric acid. In contrast to molecules which possess true chirality and which may be symbolized as Cabcd, citrate is a Caabc compound. The two acetic acid groups (aa) attached to the central carbon atom are identical; it is ostensibly nonasymmetric. Such prochiral molecules are nevertheless asymmetric in the sense that they cannot be divided conceptually into two equal superimposible halves, although they are not true mirror images (Alworth 1973). When viewed from the top, the upper half of the molecule

CH_2-COOH
|
OH-C-COOH
|
- - - - - - - - - - - - - - - -

cannot, without taking it out of the plane of the page, be superimposed on the bottom half, as viewed from below:

- - - - - - - - - - - - - - - -
|
OH-C-COOH
|
CH_2-COOH

From the viewpoint of food quality, we want to know how the action of the enzymes involved in this sort of stereospecificity is regulated so that, for instance, citric acid accumulates to contribute to the taste of fruits (Chapter 20).

In addition to the book by Alworth, other key references, reviews and expositions of chirality and stereospecificity in enzyme reactions include those of Rose (1972) and the 1975 Nobel Prize lecture by Prelog (1976), dealing with chirality in chemistry in general, and Cornforth (1976), dealing more specifically with asymmetry and enzyme action.

SPECIFICITY, THE ENZYME-SUBSTRATE COMPLEX AND THE ACTIVE SITE

The Enzyme-Substrate Complex.—It was quite clear to enzymologists of 100 years ago that the specificity of enzyme action implied some sort of interaction between the enzyme and the substrates. One of the facts supporting this idea was the increased stability of the enzyme in the presence of its substrate. But in the absence of any knowledge of the structure of enzymes, the most convincing evidence was the establishment of the hyperbolic relation between the rate of all the then-known enzyme-catalyzed chemical reactions and the substrate concentration. Figure 2.2 illustrates this relation with the action of barley malt α-amylase acting on a series of starch-related substrates. Henri (1903) and Brown (1902) were probably the first to recognize this relation and to assume that it was due to the formation of an enzyme substrate complex. This idea was quantitated and further developed by Michaelis and Menten (1913) to yield the well-known Michaelis-Menten equation. As pointed out by Fruton (1976), as late as the late 1940s, direct evidence for this complex was not available and investigators considered adsorption based on the "colloid" state of the enzyme, action at a distance, and going back to Liebig, of "field forces" resulting from "extended resonators" (Tauber 1949). Of course it should be pointed out that many adsorptive processes also obey hyperbolic laws. In 1947, Sumner and Somers preferred the ". . . hypothesis that the enzyme combines with the substrate to form a chemical intermediate rather than forming an adsorption complex" since ". . . the enzyme is extraordinarily specific" in its action and that this specificity cannot be readily explained on the assumption that an adsorption complex is formed. As late as 1958, Fruton and Simmonds felt it necessary to point out that "all . . . theories begin with the basic assumption that the enzyme combines with the substrate . . . ."

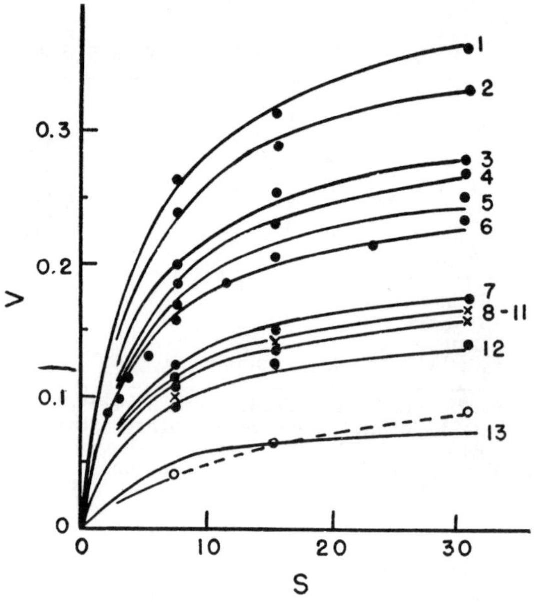

From Schwimmer (1950)

FIG. 2.2. α-AMYLASE ACTION ON AMYLOSES

Typical Michaelis rate (v) *vs* substrate concentration (S) profiles.

Elements of Enzyme Kinetics, the Grammar of Enzymology (Gutfreund 1976B).—Enzyme kinetics is that branch of enzymology in which the primary experimental data are the values obtained from the study of the parameters which influence the rate of an enzyme-catalyzed chemical reaction and the interpretation of this data in terms of the possible molecular events consonant with the data (mechanisms). Gutfreund (1976B) points out that not only are enzyme kinetics the bases of most tools for the elucidation of mechanism, but that in biology a rate measurement is often an end in itself, answering a specific question. Experimentally there are three distinct approaches which differ sharply with respect to the time scale of rate measurement, and the smaller the time scale the greater the potential insight into the underlying mechanism. These are:

(1) *Steady-state kinetics*, extending from about 1 sec and beyond. About all food-related enzymes and perhaps as much as 99% of all other enzyme research utilizes steady-state kinetics.
(2) *Transient state kinetics* utilizing rapid-flow and stopped-flow techniques covers the millisecond range. It was with this approach that the first experimentally direct enzyme-substrate complex was first observed.
(3) *Relaxation spectrometry*, utilizing temperature-jump, reveals events which occur in microseconds. The general approach is to momentarily disturb the reaction and to observe what happens when it returns to its original state. Picosecond events have been recorded.

Formulation of the Simplest Steady-state Kinetic Mechanism.—The enzyme combines with the substrate S to form the enzyme-substrate complex ES. ES then decomposes to yield product P and free E, which is now ready for another round of catalysis.

$$E + S \underset{k_{-1}}{\overset{k_{+1}}{\rightleftharpoons}} ES \xrightarrow{k_{+2}} E + P \qquad \text{(Equ. 2.1)}$$

[S] = initial substrate concentration; it is not measurably changed by combination with E; [E] = concentration of free enzyme; [ES] = concentration of ES; e = total enzyme concentration; k_{+1} = rate constant for the formation of ES; k_{-1} = rate constant for the reversal of the formation of ES; k_{+2} = rate constant for the conversion of ES into E and P.

The observed rate, v, of the enzyme-catalyzed reaction is proportional to the concentration of the enzyme-substrate complex:

$$v = k_{+2}\,[ES];\; [ES] = v/k_{+2} \qquad \text{(Equ. 2.2)}$$

When all the enzyme molecules are tied up as ES (e = [ES]), further addition of substrate will not accelerate the reaction and the rate will be at a maximum, V (frequently denoted as V_{max}) and V = k_{+2}e.

Now the fundamental postulate of steady-state theory (as distinguished from equilibrium) is that [ES] does not change during the time of the measurement, so that the rate at which ES is formed is equal to the rate at which it disappears:

$$k_{+1}\,[E]\,[S] = k_{-1}\,[ES] + k_{+2}\,[ES] = (k_{-1} + k_{+2})\,[ES] \qquad \text{(Equ. 2.3)}$$

Dividing both sides by k_{+1} S and solving for E we obtain

$$[E] = \frac{k_{-1} + k_{+2}}{k_{+1}[S]}\,[ES] = \frac{K_m}{[S]}\,[ES];$$

$$K_m = \frac{k_{-1} + k_{+2}}{k_{+1}} \qquad \text{(Equ. 2.4)}$$

The total enzyme concentration can now be expressed in terms of ES

$$e = [E] + [ES] =$$

$$\frac{K_m}{[S]}\,[ES] + [ES] = \left(\frac{K_m}{[S]} + 1\right)\,[ES] =$$

$$\left(\frac{K_m}{[S]} + 1\right)\,\frac{v}{k_{+2}} \qquad \text{(Equ. 2.5)}$$

Solving for v, we obtain

$$v = \frac{k_{+2}\,e\,[S]}{K_m + S} = \frac{V[S]}{K_m + S} \qquad \text{(Equ. 2.6)}$$

which is the classic Michaelis-Menten equation.

The Michaelis-Menten equation, as expressed in this form, has certain advantages. It will be noted, for instance, that the number of separate terms in the denominator corresponds to the number of forms of the enzyme in the model, no matter how complex the mechanism. This form of the equation also tells us at a glance how the important parameters V and K_m will be affected, if at all, assuming various models or mechanisms. Also note that K_m = S when v = ½V.

It should be stressed that K_m data by themselves are treated more frequently like melting points than like data which give insight into mechanism. Thus, except under very special circumstances, it is not possible to derive the values for the individual rate constants, and K_m is not a true measure of "affinity" in the sense that it invariably measures the equilibrium constant for combination of the enzyme with substrate. However, one does obtain an intuitive feeling that a substrate with a K_m of, for example, $10^{-6}M$ has a greater avidity for the enzyme than one with a K_m of, for example, $10^{-2}M$, although they may be acted on at the same rate at enzyme-saturation concentrations.

Finally, it should be mentioned that the *in vivo* concentration of enzyme may indeed be comparable to that of the substrate. In this case, according to Laidler (1958):

$$v = \tfrac{1}{2}k_{+2} \cdot (K_m + [S] + e) \cdot \{1 - [4e[S]/(K_m + [S])^2]^{1/2}\} \qquad \text{(Equ. 2.7)}$$

and the point of half-saturation is equal to (K_m + ½e). This suggests that it should be possible to determine the molecular weight of the enzyme, if used pure, since enzyme concentration enters into the equation.

Transformations of the Michaelis-Menten Equation.—Any attempt to present even the barest outline of enzyme kinetics would not be complete without a brief discussion of some of the methods traditionally employed to obtain V and K_m. Since, mathematically, saturation of the enzyme occurs at infinite concentration, one must transform the Michaelis-Menten equation to a form that lends itself readily to the determination of these kinetic parameters. Of the many approaches which have been put forward, one that still survives is the double reciprocal plot, first elaborated upon by Lineweaver and Burk (1934). Two years earlier this equation was included in Haldane and Stern (1932) which is essentially a German translation of Haldane (1930). Their chapter on the "Course of Enzymatic Reactions" includes a brief section on graphic methods as an addendum. They attribute these graphic methods to "Woolf" without further citation (presumably B. Woolf, who published with Haldane in papers in which these graphic methods are not in evidence). Other algebraic manipulations of this equation attributed to later investigators were also included.

The reciprocal of the Michaelis-Menten equation

$$\frac{1}{v} = \frac{K_m}{V} \left(\frac{1}{S}\right) + \frac{1}{V} \qquad \text{(Equ. 2.8)}$$

describes a straight line of the form: y = ax + b. V has the value of the reciprocal of b, the y-intercept, i.e., the value of the ordinate y when x = 0. Similarly, the x-intercept has the numerical value of $(-K_m)$. The value of K_m can also be calculated from the slope of this straight line, K_m/V.

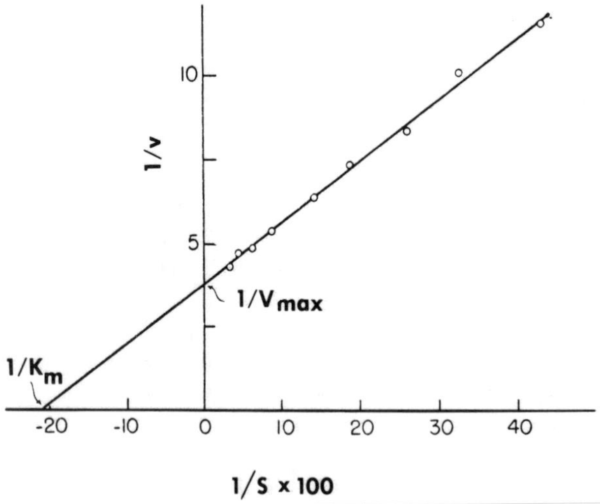

From Schwimmer (1950)

FIG. 2.3. LINEWEAVER–BURK DOUBLE RECIPROCAL PLOT

Shown is 1/v *vs* 1/S for barley malt α-amylase acting on wheat amylose, curve 6, Fig. 2.2.

A typical double reciprocal plot, obtained from the primary data of curve 1 in Fig. 2.2, in this case for the effect of substrate (wheat amylose) concentration on the rate of its hydrolysis by barley malt α-amylase, is shown in Fig. 2.3. Incidentally, except for glycogen, all the other substrates used to obtain the data in Fig. 2.2 have the same K_m as wheat amylose, viz., 4.8 m*M*, expressed as glucoside bonds. For wheat amylose, 68,000 bonds were hydrolyzed per minute per molecule of enzyme. This value, usually referred to as the *turnover number*, is being replaced, if the Enzyme Commission's guidelines are followed, by the *molar activity*, which is defined as katals per mole of enzyme. The *katal*, the unit of enzyme activity, is the amount that converts, or more precisely that catalytic activity that will raise the rate of, the reaction by one mole of substrate per second in a specified system.

Considerable thought has gone into how to get reliable constants from the double-reciprocal plot. Cohen (1968) provided a valuable discussion of "best values" which we have found quite helpful (Schwimmer and Austin 1971A). The reliability of the data are related to the spacings of substrate concentrations and to the statistical weights assigned to each of them. Haldane and Stern (1932) pointed out that even spacing of the substrate concentration results in the crowding of points around low values of 1/[S]. Measurements of v at [S] much greater than K_m are used primarily for accurate estimation of V. Computer programs for the estimation of these kinetic constants have been published (Cleland 1967). Assessments of different methods of treatment of experimental error and curve fitting of the Michaelis-Menten equation are available (Atkins and Nimmo 1975; Nimmo and Mabood 1979). Of course accuracy and reliability and significance of the data depend ultimately on methods for determination of the initial rate of the enzyme reaction. This will be dealt with in Chapter 37.

Another algebraic transformation, also attributed to Woolf by Haldane and Stern, first appearing in the literature in a paper by Augustinsson (1948), but usually referred to as the Hofstee-Edie or Edie plot, is that obtained by multiplying each side of the Michaelis-Menten equation by $(K_m + [S])$, solving for v, and plotting v against v/[S]:

$$v = (-K_m) \frac{v}{[S]} + V \qquad \text{(Equ. 2.9)}$$

Here, the y-intercept is V, the slope is $-K_m$ and the x-intercept is V/K_m. An example of the plot is shown in Fig. 2.4. Curve 1 is the effect of con-

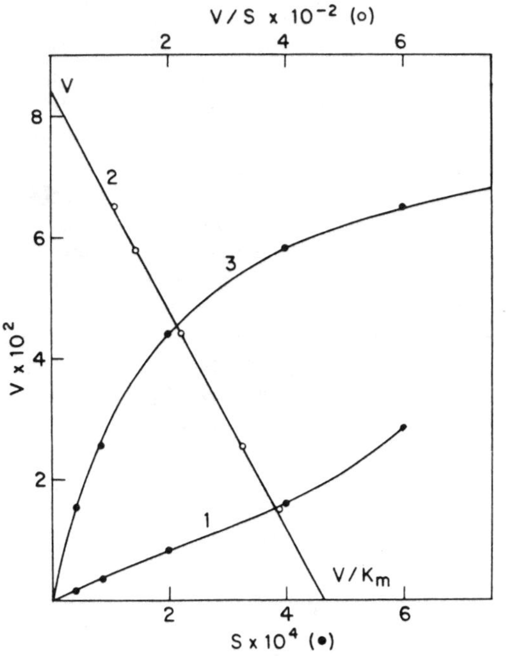

From Hansen et al. (1959)

FIG. 2.4. THE AUGUSTINSSON-EDIE PLOT
For the action of cysteine C-S lyases on S-2,4-dinitrophenyl-L-cysteine: 1, nonenzymatic; 2, v *vs* v/S (the Augustinsson-Edie plot); 3, v *vs* S.

centration of the substrate S-2,4-dinitrophenyl-L-cysteine on the nonenzymatic rate of hydrolysis, curve 3 is for its enzymatic hydrolysis by *Albizzia* cysteine-C-S lyase and curve 2 a plot of v *vs* v/[S]. Haldane and Stern pointed out that even spacing of [S] does not result, as does the double reciprocal plot, in warping of the weights given to the function of v. Furthermore, they found that the plot is more sensitive to real deviation from the straight-line relationship as pointed out by Horvath and Engasser (1974) in connection with the kinetics of immobilized enzymes (Chapter 8). Also, [S] can be much more precisely controlled than can v.

Perhaps the graphic method employing the maximum of "eyeball" judgment most accurately and efficiently in which the raw v and [S] data are plotted directly without any algebraic manipulation except for changing the sign of [S] is that of Eisenthal and Cornish-Bowden (1974). One simply places a point on the negative side of the x-axis corresponding to –[S], another point on the y-axis corresponding to the experimental value of v for that [S] and then draws a straight line. This step is repeated for all experimental values. The point in the "VK_m"-space at which all the straight lines (should) intersect defines the value of K_m ("x"-value) and V ($=V_{max}$) ("y"-value) as shown in Fig. 2.5 using the raw v and [S] data of Fig. 2.4.

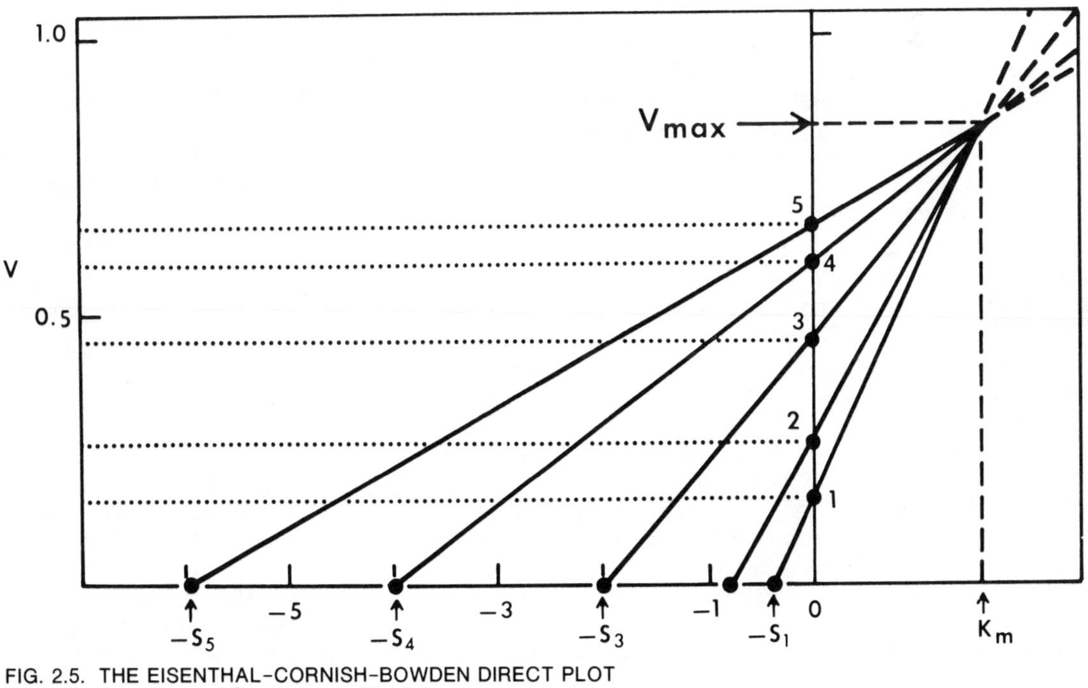

FIG. 2.5. THE EISENTHAL-CORNISH-BOWDEN DIRECT PLOT
Using the "raw" v and S data from Fig. 2.4.

Steady-state kinetics, although an approximation, has yielded considerable insight into reaction mechanism and enzyme specificity. For instance, by a rather simple mathematical analysis, it has been possible to distinguish in two-substrate enzyme reactions among the formation of a ternary complex, such as ES_1S_2 or consecutive binary complexes. If only binary complexes are formed, the order of binding with the enzyme can be ascertained (Dalziel 1957; Cleland 1970; Whitaker 1972A).

Steady-state kinetics of the Michaelis-Menten type is important in analytical applications of enzymes to quality control testing (Chapter 37), and is of economic importance in determining the minimum amount of enzyme needed to perform a given task in a unit food processing operation. Knowledge of the kinetic constants can also be of help in managing the endogenous food enzymes. We shall have occasion to specify some of these food-oriented applications. However, it may be well to point out at this time that from a strictly biological point of view enzymes that obey the hyperbolic v *vs* [S] relationship may actually be in the minority, those elusive ones which escaped early recognition and are involved in intermediary metabolism and control (Chapter 4). But most of the food-related enzymes of traditional interest are still of the conventional type.

Literature.—The basic outline of what is known about steady-state hyperbolic enzyme kinetics can be found in Haldane (1930) and Haldane and Stern (1932). All of the books on enzymes cited in Chapter 1 and all biochemistry and almost all textbooks in other biologically-oriented disciplines have something to say about enzyme kinetics. Of special utility among the enzyme texts with respect to kinetics is that by Dixon and Webb (1964). In the food field, Whitaker (1972A) goes into a fair amount of clear expository detail. Learning programs for relatively simple enzymes are available in English (Christensen and Palmer 1976) and in German (Ahlers 1974). The latter book also deals with practical applications, especially in enzyme analysis.

Books solely devoted to enzyme kinetics have been appearing with decreasing periodicity ever since the classic by Laidler (1958). Comprehensive treatises which deal not only with steady-state hyperbolic kinetics but also with other aspects of enzyme kinetics include those of: Reiner (1969), Westley (1969), Gutfreund (1972), Wong (1975), Plowman (1976), Fromm (1976) and Cornish-Bowden (1979) and Engel (1977). A series of five volumes of "Methods of Enzymology," two of which have been published as of this writing (Purich 1979, 1980), will deal with the methodology of enzyme kinetics and mechanisms. A more generalized treatment of kinetics in biological systems is provided by Johnson-Eyring of the absolute reaction rate theory school (Johnson *et al.* 1974).

Of the over 10,000 papers dealing with enzyme kinetics which have appeared in the scientific literature we should like to mention the following. Much of the increased utility of enzyme kinetics has come about by the application of mathematical concepts from other fields. For instance, Volkenstein and Goldstein (1969) applied the topological theory of graphs, useful in information and electrical network theory. Cleland (1975), who had already provided two-substrate kinetics with new concepts and terminology—bi-bi, ping-pong, uni-uni, etc.; see Cleland (1970)—has more recently advanced, as a powerful time-saving kinetic analysis tool, the concepts of net rate constants and mathematical partition analysis. Of earlier importance are the two-substrate kinetic treatments which include Dalziel (1957), Ingraham and Makower (1954) and Wong and Hanes (1962). Among the many excellent papers providing background for further reading are the older reviews of Dawes (1964) and Segal (1959) and more recently the stimulating discussion of the overall importance of kinetics by Gutfreund (1976). Further application of computer programming in enzyme kinetics is exemplified by the contribution of Osmundsen (1975). Atkins and Nimmo (1980) provide an overview of the current trends in the estimation of Michaelis kinetic parameters.

The Enzyme-Substrate Complex and the Active Site

Direct Evidence for the Complex and Transient-state Kinetics.—It is now some 30 years since Chance (1949) demonstrated spectroscopically the existence of a fleeting enzyme substrate complex for the first time. The catalase-H_2O_2 complex he observed with stopped-flow apparatus completely disappeared within a few seconds after its formation. As previously mentioned, this type of measurement of transient-state kinetics permits direct estimation of the individual rate constants postulated in the formulation of the equations of steady-state kinetics (Greenstein 1956). It has also been used to confirm another fundamental postulate of steady-state kinetics: That after a fraction of a second, the level of the enzyme-substrate remains constant until almost all of the substrate has disappeared. Since this early work, diverse experimental approaches have been used to demonstrate the validity, generality, universality, and ubiquity of the ES complex provenance. Spectroscopy including not only visible but also UV, electron spin resonance, NMR, ORD, circular dichroism and Raman have all been used. In a few instances it has been possible to actually isolate such complexes, especially when k_{+1} is much greater than the other two rate constants (Chapter 6). Also, the complexes can be obtained when only one of two reaction partners are added to the enzyme in two-substrate reactions. Examples of fairly stable isolatable ES complexes in the absence of substrate include:

(1) *Alcohol dehydrogenase*: NADH in the absence of alcohol.
(2) *D-Amino acid oxidase*: Amino acid in the absence of oxygen.
(3) *Tryptophan synthase*: Indole in the absence of serine.
(4) *Aldolase*: Dihydroxyacetone phosphate in the absence of glyceraldehyde phosphate, when the complex is stabilized by reduction.
(5) *Hexokinase:* Glucose (Wilkinson and Rose 1980).

All reactive site titrants (Chapter 37) also form complexes, but these no longer possess catalytic competency. In contrast to the preceding, some substrates split in half and only part of the substrate, or one of the products, will form an isolatable complex with the enzyme, e.g., acyl esters of chymotrypsin.

Specificity and the Active Site.—Parallel to the development and proof of existence of the ES complex, enzymologists engaged in a running debate of just how the enzyme combined with the substrate. Among the ideas, then not directly demonstrable, was that there was one site on the enzyme which could attract and combine with the substrate and carry out the catalytic function—the active site. Even before enzymes were even thought of as proteins, the phenomenon of specificity compelled many investigators to conclude that there had to be a specific three-dimensional locus in or on the enzyme to correspond to the demands imposed by stereospecificity. These thoughts were most forcefully epitomized in the now classical template-mold or lock-and-key metaphor of Fischer (1894). In this analogy (concept, theory, postulate) the key is analogous to the enzyme, the lock to the substrate, and the opening and closing operation to the chemical reaction catalyzed. Not only does this idea utilize the concept of an active site, but it foreshadows modern ideas concerning the mutual interaction of enzyme and substrate.

The Active Site and Enzyme Inhibition.—Another set of experimental data which bespoke strongly of the existence of an active site was the phenomenon of competitive inhibition. Chapter 14 deals with the food-related aspects of enzyme inhibition. In order to explain competitive inhibition it is assumed that the competitive inhibitor competes with the substrate for the same sole active site on the enzyme.

More frequently than not, the competitive inhibitor's structure is similar to that of substrate. Derivation of the steady-state equation for competitive inhibitors follows. One assumes that in addition to equation (2.1) the enzyme can also combine with inhibitor, I

$$E + I \rightleftharpoons EI; \text{ and hence, } e = [E] + [EI] + [ES] \qquad \text{(Equ. 2.10)}$$

At the steady state, the rate of formation of EI = rate of its disappearance, and the rate of formation of ES = rate of its disappearance, as before. Expressing these three forms of the enzyme in terms of ES and assuming K_i to be the steady-state equilibrium constant for the formation of EI, we arrive at the rate equation

$$v = \frac{VS}{K_m + K_m([I]/K_i) + S} = \frac{VS}{K_m(1 + [I]/K_i) + S} \qquad \text{(Equ. 2.11)}$$

Note that the denominator consists of three terms, the same as the number of forms of the enzyme postulated. This is only evident when the equation is cast in the hyperbolic form. Note that there is no change in the position occupied by V in the original M-M equation but that there is in that occupied by what was K_m, now K_m $(1 + [I]/K_i)$. That is to say, V is not changed but the concentration at which the rate = ½ V is increased when [I] is varied. Derivation of noncompetitive inhibition and a transform for estimating K_i will be found later on in this chapter.

Active Site Modification.—Once enzymes were found to be proteins, the search for the active site was undertaken in earnest. The first attempts involved general group reagents such as formaldehyde and substances which combined more or less specifically with amino, SH and other amino acid side groups. The breakthrough arose out of the post World War II investigation by Balls of the poison gases available to the German military, but fortunately never used. One of these was the nerve gas diisopropyl fluorophosphate (DFP). Balls, with the collaboration of Jansen and others (Balls and Jansen 1952), showed that the DFP reacted stoichiometrically with one serine residue in the chymotrypsin molecule. This resulted in the simultaneous release of one molecule of hydrogen fluoride and concomitant complete suppression of enzyme activity. This work demonstrated the general principal that amino acid residues in the active site have special properties, since the nonactive site serine residues did not react with DFP. It is also one of the first examples of active site titration (Chapters 3 and 37).

The next step in the quest for the active site hinged upon two important experimental developments: the application of radioactive tracer methods and elaboration of methods for determining the primary sequence of amino acids along a single polypeptide chain. Thus, in the 1950s and early 1960s ^{32}P-radioactive serine phosphate was found in the active site of chymotrypsinogen, choline aliesterase, phosphoglucomutase and papain (Smith *et al.* 1962). This early work also revealed a common pattern of amino acid sequence around this active site serine. Thus, the sequence Ala-Ser-Gly-Glu-Ala-Val was found to comprise part of the active site of enzymes as disparate in specificity as the proteinase chymotrypsin and isomerase phosphoglucomutase, a glycolytic enzyme which transfers phosphate from the 1- to the 6-position of glucose through glucose-1,6-diphosphate as an intermediate (Chapter 4). Parenthetically it is now possible, from a knowledge of the genetic code, to obtain the amino sequence of an enzyme before it has been isolated, let alone analyzed. This was actually accomplished with an RNA replicase.

This approach to understanding specificity continues to attract investigators, but probably it is more important in the elucidation of mechanism and catalytic efficiency (Chapter 3). Thus, from the viewpoint of explaining the specificity of enzymes from first principles, the importance of modification by active site-directed agents (Chapter 14) and techniques such as affinity labeling lies in their contributions to an understanding of the shape of the active site, which, in Fischer's metaphor, is all-important in noting the specificity of enzymes. Parenthetically, at least one medicine, aspirin, exerts its beneficial effect because it is an inhibitor, more precisely an active site modifier of the lipoxygenase-like enzyme cyclo-oxygenase or prostaglandin synthase (Roth *et al.* 1975). Prostaglandins, biosynthesized from arachidonic acid in mammals, are intimately involved in hormone message transmission.

Tertiary Structure and the Active Site.—While elucidation of primary structure, the sequence of amino acids, emphasized their role in the formation of the active site, it was recognized in the early 1960s that even detailed acquisition of such knowledge would be insufficient to account for the active site. Thus, it was found that amino acids could be removed from enzymes without affecting their activity. More than one-third of the 180-amino acid peptide chain of papain can be removed with retention of full activity and specificity. By contrast, three amino acids at the carboxyl end (C-terminal) of pancreatic ribonuclease can be removed without loss of activity, but removal of the fifth amino acid residue, aspartate, does result in complete loss.

The secondary structures of proteins, this includes most enzymes, come about as a result of the formation of α-helices and β-pleated sheets, usually along one dimension. These structures are, with the exception of lysozyme, the common secondary structural elements in enzymes (Blake 1976). On the other hand, being globular proteins they do not possess the helicity associated with fibrous proteins. Thus, chymotrypsin has very few α-helices. Some of the bonds which participate in the formation of elements of the secondary structure of proteins are shown in Fig. 2.6.

The role of the tertiary structure, further folding of the polypeptide chain to give a protein its three-dimensional space-filling characteristics, was anticipated before the success of X-ray enzymology and it was specificity which held the clues. Thus Eley (1955), from a deductive analysis of the specificity characteristics of certain transferases, calculated that the active site extended over a circular area on the surface of the enzyme some 15–20 Å in diameter. He considered that this area might be associated with a particular folding of the peptide chains or that, as we now know, the

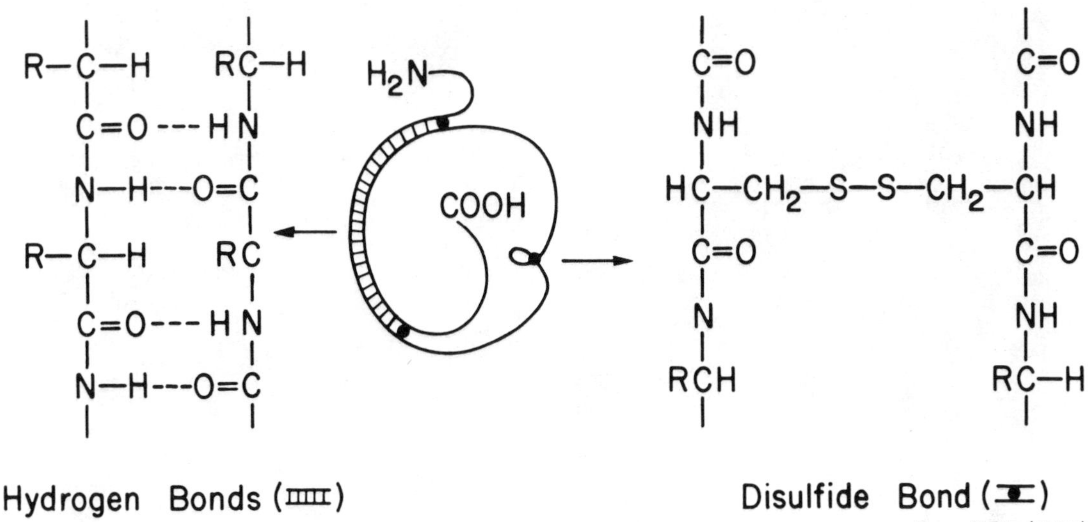

From Balls (1960)

FIG. 2.6. COVALENT (DISULFIDE) AND NONCOVALENT (HYDROGEN) BONDS IN PROTEINS

catalytic site comprised a small number of amino acid residues arranged in a very specific conformation. He concluded that these amino acids not only provide for the catalytic action but also for specificity of the enzyme.

X-Ray Enzymology.—The advent of advanced computers led to the complete elucidation of the tertiary structure of crystalline proteins, gave rise to the now well-established subdiscipline of X-ray enzymology, and has provided incontrovertible proof and promise of detailed understanding of the morphology of the active site of enzymes in relation to specificity (Eisenberg 1970; Blake 1976). The first enzymes whose structures were thus elucidated were mostly low molecular weight proteins and by chance all traditionally food-related (papain, lysozyme and carboxypeptidase). The three-dimensional structure of the latter enzyme in the vicinity of the active site is depicted in Fig. 2.7. This packing model of about 25% of the entire enzyme molecule clearly and strikingly, especially when viewed with a stereoviewer, shows the active site cavity near the center of the photograph. In this enzyme and all others examined, the active sites are manifested as definite loci, usually one for each enzyme molecule or subunit (Chapter 4), which have been described as dimples, grooves, depressions, wedges, slits, slots, pockets, crevices and crevasses, depending upon the judgement and fancy of the observer. The active site is made up of amino acid residues which are not necessarily in juxtapositional sequence in the peptide. Thus, the active site of lysozyme consists of amino acid residues (from the C-terminal) numbers 35, 52, 62, 63, 101 and 108. Most enzyme molecules fold up into two globular units corresponding to the folding of each half of the peptide chain. The active site is situated between these clusters, each of which contributes to the makeup of the active site (Blake 1976).

One possible objection to the use of X-ray analysis for the determination of the active site is that the structure of proteins in a crystal may not be the same as in solution. However, the phenomenal success with which specificity can be analyzed in detail in terms of interaction of enzyme and substrate in enzymes so far studied would suggest otherwise (Lipscomb 1973). Figure 2.7(right) depicts an enzyme-substrate complex of carboxypeptidase in which the H-atoms of the substrate are marked with the label S. We shall examine some of the forces involved in the formation of this complex in the next chapter.

Precise knowledge of the amino acids making up the active site of egg white lysozyme has permitted comparison with other food lysozymes. Thus, Friend *et al.* (1975) found that tryptophan, three

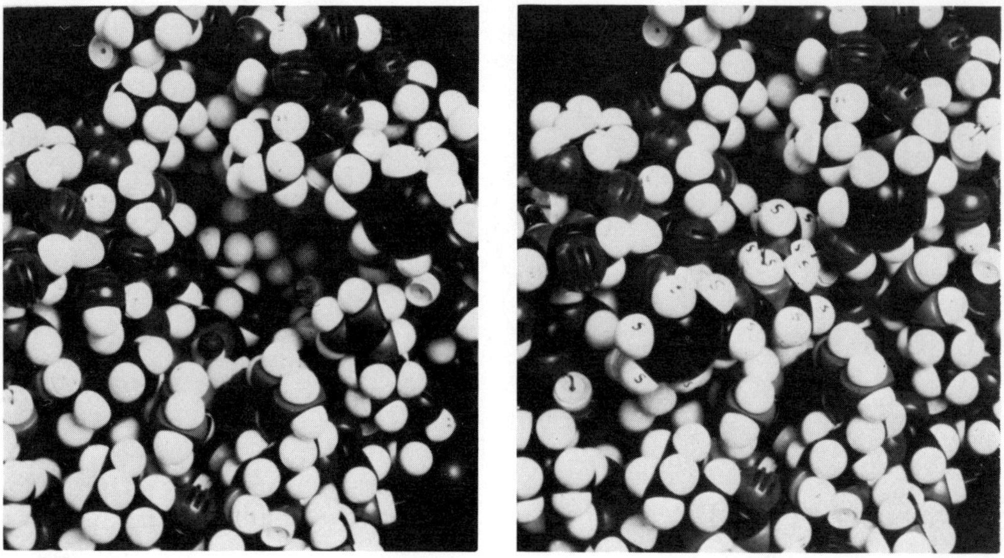

Courtesy of Lipscomb (1972)

FIG. 2.7. ACTIVE SITE (CENTER) OF CARBOXYPEPTIDASE

The figure on the left affords a view of the active site clearly seen as the cavity near the center of the photograph of a packing model representing about half of the enzyme molecule; on the right, with substrate, whose atoms are marked "S."

residues of which are part of the active site of egg white and other animal lysozymes, does not appear to be part of the active site of that of bovine milk, thus resembling plant lysozymes in this regard.

This again illustrates that it is not the same amino acid sequence as such but the final shape and forces associated with the molding of the active site which determine specificity. Just why a cow and a human should differ so with respect to amino acids making up the active site of a minor milk enzyme which would appear to have the same specificity is a mystery which should engage the efforts of students of evolution. From the viewpoint of food quality, such information should be of use in distinguishing between cow and human milk.

OTHER MOLECULAR DETERMINANTS OF SPECIFICITY

Induced Fit: Ligand-induced Conformational Changes

In his classical paper, Fischer (1894) not only conceived of the idea that the enzyme and substrate should fit together like a lock and key, but also that they act reciprocally on each other prior to the actual bond-breaking event which signals the actual chemical reaction. This aspect of Fischer's concept was lost in the intervening decades and the concept of a static, rigid template-mold paradigm was attributed to him. Some 20 years ago, considerations of some problems of specificity led Koshland (1960) to question the mold-template hypothesis and to suggest that not only the substrate underwent a change during the catalysis but that the enzyme itself, or rather the active site, underwent a change in conformation to better accommodate the substrate somewhat as a glove changes shape to accommodate the hand. One of these problems was how to explain noncompetitive inhibition.

Noncompetitive Inhibition.—In the simplest formulation of the equation for noncompetitive inhibition, it is assumed that the inhibitor, in contrast to a competitive inhibitor, combines at a site other than that at which the substrate combines. It does not compete with the substrate for the active site. This means that in addition to ES and EI complexes, a third one, ESI, is also formed, making a total of four forms of the enzyme in a reaction system containing enzyme, substrate and noncompetitive inhibitor: E, EI, ES and ESI

$$e = [E] + [ES] + [EI] + [ESI] \quad \text{(Equ. 2.12)}$$

As in the case of competitive inhibition, we assume that only ES is productive. Each of the concentrations of the four forms of the enzyme is expressed as a function of [ES] and the resulting equation is solved for v:

$$v = \frac{V[S]}{K_m + K_m[I]/K_i + [I]/K_i[S] + [S]} =$$

$$\frac{V/(1 + [I]/K_i)[S]}{K_m + [S]} \quad \text{(Equ. 2.13)}$$

The equation shows at a glance that there are four separable terms in the denominator corresponding to the four forms of the enzyme. Furthermore, it shows that increasing the substrate concentration will not relieve the inhibition. Note

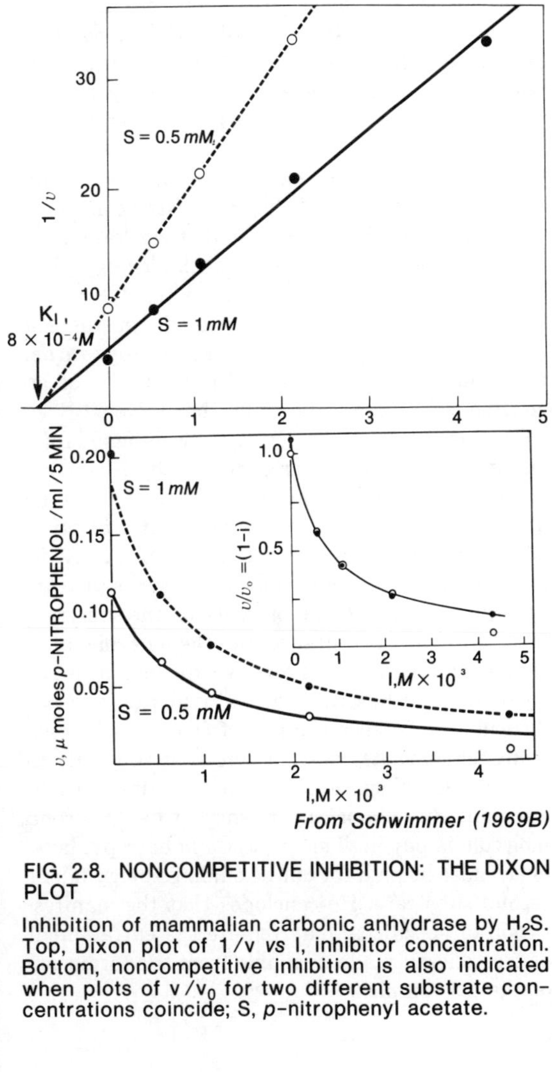

FIG. 2.8. NONCOMPETITIVE INHIBITION: THE DIXON PLOT

Inhibition of mammalian carbonic anhydrase by H_2S. Top, Dixon plot of $1/v$ *vs* I, inhibitor concentration. Bottom, noncompetitive inhibition is also indicated when plots of v/v_0 for two different substrate concentrations coincide; S, *p*-nitrophenyl acetate.

that the apparent K_m is not affected by [I] but that the apparent V is so affected. Further discussion of inhibitor transforms will be found in Chapters 3, 4 and 14. Two of the several useful transforms of the Michaelis-Menten equation for noncompetitive inhibition are shown in Fig. 2.8A and B. Figure 2.8A is the familiar Dixon plot (Dixon and Webb 1964) in which 1/v is plotted against [I] at two set substrate concentrations. In this case the S is nitrophenyl acetate, I is propyl mercaptan, and the enzyme is carbonic anhydrase. As shown, intersection of the two straight lines on the x-axis is indicative of noncompetitive inhibition whereas intersection on the y-axis indicates competitive inhibition. Noncompetitive inhibition is also indicated if a plot of v/v_0 against [I] of the same data yields the same curve for different concentrations oI substrate, as shown in Fig. 2.8 (bottom).

Application of Occam's razor to explain the demonstration of countless examples of noncompetitive inhibition leads compellingly to induced fit. There is no simpler explanation than to assume that when the noncompetitive inhibitor binds, it induces some change in the shape of the active site. As expressed by Citri (1973) in his comprehensive 251-page, 916-reference review, noncompetitive inhibition is the predicted consequence of conformative response to the inhibitor, or interference with the conformative response to the substrate, or both.

Other Relations Between Specificity and Induced Fit.—Although competitive inhibition can be explained, in general, without recourse to induced fit, Citri points out that in many cases an inhibitor can combine at a site other than that of the substrate and still inhibit competitively. This bespeaks a conformational response of the enzyme similar to but counterbalanced by response to the substrate. Competitive inhibition of an important food-related enzyme, α-amylase, by cycloamyloses (Schwimmer and Garibaldi 1952), in which the inhibitor acting as a substrate analog is larger than the substrate, provides further support for the induced fit theory (Thoma and Koshland 1960). Similar considerations would explain why α-amylase is inhibited competitively by long chains of amylase substrate and noncompetitively by one of the reactions' final products—maltose (Schwimmer 1950).

The role of specificity in induced-fit theory is highlighted by the estimation of Pauling, as cited by Steitz (1968), that if an enzyme discriminated between valine (whose side chain may be represented as $CH_3(CH_3)$-CH-) and isoleucine, (CH_3-

$CH_2(CH_3)$-CH-) only on the basis of their ability to bind, they could be distinguished 95% of the time. Experimentally it was found that the enzyme (not specified) could distinguish between the two amino acids 99.67% of the time. According to Koshland (1976) the greatest importance of active site flexibility is not that it can explain the exclusion of substrate-like molecules from the active site, but in its ability to exclude water in the active site designed for a hydroxylic molecule. The point is that the failure to react is not based on steric hindrance or exclusion but is based on kinetics since it is failure of such inhibitors to react after binding which controls the reaction.

Direct proof of the induced fit, like evidence for the active site, comes from X-ray enzymology and somewhat less convincingly but to a greater extent from polarimetric measurements (ORD,CD). Citri lists enzymes in which a significant shift in the position of one or more of the amino acids comprising the active site is observed after combination with inhibitor or with a substrate partner. For instance, combination of glycyl-L-tyrosine with carboxypeptidase results in a 14 Å shift of a tyrosine residue in the active site of the enzyme towards the combined peptides (Lipscomb 1973). This shift with benzoxycarbonyl-Ala-Ala-Phe as substrate is vividly shown in Fig. 2.7. The benzene ring carbon atoms of the tyrosine comprising the large round black object at the extreme right in the upper middle of Fig. 2.7 (left) has shifted to the left about 25% across the figure upon formation of an ES complex shown in Fig. 2.7 (right). Chymotrypsin completely engulfs N-formyl-L-tryptophan. Refinement in X-ray enzymology with respect to the time, details and niceties of induced fit is exemplified by the observation that the side chain of Trp 62 of both the triclinic and the tetragonal forms of lysozyme move towards the inhibitor, N-acetyl glucosamine, when the latter is bound to the enzyme but that there is movement in the lobes only of the tetragonal crystals (Kurachi *et al.* 1976).

Koshland (1976) also points out that flexibility most facilely explains not only inhibition but also such diverse enzymatic behavior, based mostly upon kinetic measurements, as the ordered binding of substrates in two-substrate reactions, control by molecules which themselves are not consumed in the reaction, cooperativity, covalent modification, catalytic efficiency and evolution in complexity of the tasks which enzymes are now required to perform, as well as the behavior of other biologically active proteins. We shall have the opportunity to examine some of these aspects of induced fit in the ensuing chapters.

Specificity and Entropy

While induced fit goes a long way in explaining most of the intricacies of specificity, entropy considerations may also be applied as an alternate, but not necessarily conflicting, approach. The case for entropy as a determinant of specificity is examined by Jencks (1975) whose ideas are interpreted and adumbrated as follows: A small amount of entropy change is associated with a highly specific substrate in the sense that the substrate is the structure nearer to that of the transition state (Chapter 3). Loss of entropy upon binding of substrate to enzyme provides a mechanism in which enzyme specificity may be exhibited in the decomposition of the enzyme-substrate complex (V_{max}, maximal velocity) rather than in the binding of enzyme to the substrate (K_m). In a discussion of the formation of acyl enzymes, Jencks concludes that correct positioning of the substrate as a necessary step for interaction with enzyme serine hydroxyl within the enzyme-ester complex involves a decrease in entropy. This decrease in entropy can be looked upon as "freezing" of a bound, "poor" substrate in which the latter is held loosely so that it can take up a large number of positions of similar energy. The positions left open to a "good" or highly specific substrate are presumably restricted, i.e., there is less entropy change. Jencks invokes entropy rather than induced fit to account for the ability of an enzyme to exclude water from the active site.

Enzymes in which the active site is designed for a "hydroxylic" molecule, as expressed by Koshland, according to Jencks, can discriminate against an unpositioned water molecule by a factor of up to 3000 from the entropy effect alone. Similarly, he suggests that the phenomenon of two-substrate synergism, in which the presence of one substrate facilitates the acceptance by the enzyme of a second substrate can be readily explained by a loss of entropy rather than by induced fit. It is not difficult to visualize the loss of freedom of movement (entropy decrease) of the first substrate in the restricted space of the active site without assuming any change in its shape when a second molecule is put in. This decrease in entropy, however, must be paid for out of binding energy of the second substrate. He concludes that the manifestation of specificity may require the entrainment of free energy made available from the formation of the ES complex.

Specificity and Nonamino Acid Components of Enzymes

As with other proteins, purified enzymes may contain nonprotein components. When these components are involved in the action of the enzymes they are generally known as cofactors or more specifically as coenzymes, prosthetic groups or activators. A prosthetic group is a closely bound, covalently-linked, relatively small organic molecule or metal ion. Coenzymes may be found in the purified enzyme, but they can also serve as cosubstrates in *in vitro* reaction systems. The prime example is the nicotinamide adenine dinucleotides, the so-called coenzymes of the dehydrogenases. Activators are substances which are not present in the isolated enzyme and which are not cosubstrates but, when added to the enzyme reaction system, elicit full activity. From the viewpoint of their role in conferring specificity, these small molecules can usually catalyze the same sort of reaction in the absence of the protein but at a much slower rate and with a greater latitude of specificity. Once they are in the active site their ability to catalyze a wide variety of reactions is much more highly restricted. Thus, pyridoxal phosphate and its congeners can catalyze dehydration (lyase-like action), amino transfer, decarboxylations and a host of other reactions, each of which can also be specifically catalyzed by each of a group of pyridoxal-phosphate-containing enzymes. In this sense the presence of the protein confers specificity as well as catalytic efficiency. Frequently, naturally-occurring coenzymes can be substituted by a similar molecule or ion, i.e., cobalt can replace zinc in some zinc-containing enzymes without changing the specificity. In general, the catalytic and specificity-conferring roles of these cofactors are so closely intertwined and intimately concerned with reaction mechanism that it may not be profitable to discuss them separately from these aspects in Chapter 3. Thus, Dixon *et al.* (1976) support the view that the action of metalloenzymes is primarily related to the chemistry of the metal itself as a Lewis acid, and that special characteristics of the enzyme-metal complex have received undue misplaced attention. This emphasis may hold for cofactors in general.

We categorize these cofactors as follows: (1) coenzymes which are derivatives of vitamins, (2) small organic molecules which are integral parts of the enzyme's active site for which no vitamin function has been discovered, (3) others which are more or less loosely held by the enzyme or actually serve as cosubstrate shuttles in the alternation of one of two enzyme states in intermediary metabolism, (4) metal cations which may be firmly bound or may serve as activators in *in vitro* systems, and (5) anions serving usually as activators.

Vitamins.—The following vitamins or their derivatives have been demonstrated to be parts of enzyme molecules, cosubstrates in the above-mentioned shuttle, or involved in a less direct way in enzyme action. Listed is the vitamin, its form as cofactor, and either examples of enzymes of which it is a part or its general metabolic role.

Vitamin B_1, Thiamin.—Present as thiamin pyrophosphate as part of the pyruvate decarboxylase complex (Chapter 4); in general it is present in the active site of enzymes involved in the decarboxylation, both oxidative and nonoxidative, of keto acids, in "C_2" aldehyde transfer reaction and in the generation of α-ketols.

Vitamin B_2, Riboflavin.—In oxidoreductase enzymes as flavin mononucleotide (FMN) and flavin adenine dinucleotide (FAD), as exemplified by two food-related enzymes, xanthine and glucose oxidases (Singer and Kenney 1974; Yagi and Yamano 1980).

Vitamin B_6, Pyridoxine.—Present as pyridoxal phosphate (PLP) in various amino acid-transforming enzymes and in α-1,4-glucan phosphorylases where as a dianion it serves a completely different role than in amino acid metabolizing enzymes (Helmrich and Klein 1980).

Vitamin B_{12}.—Part of the prosthetic group of cyano-cobalamine, found in enzymes such as diol dehydrase which catalyze interchange of groups attached to adjacent carbon atoms (Abeles and Dolphin 1976). A cobalamine-dependent enzyme, leucine 2,3-aminomutase, whose action yields ammonia, acetic acid and *iso*butyric acid, is present in a food, bean seedlings (Poston 1977).

Vitamin C, Ascorbic Acid.—Although not shown as yet to be part of an enzyme, the vitamin has been implicated in hydroxylation reactions (Chapter 36) and is a specific activator of the food-related enzyme, glucosinolase (Chapter 21).

Vitamin H, Biotin.—Present as lightly bound biocytin in enzymes which catalyze carboxyl transfers in fatty acid metabolism, i.e., acetyl CoA carboxylase.

Vitamin K, Phylloquinone.—Present as such in the relatively recently discovered γ-glutamyl carboxylase for which only glutamate in certain blood-clotting proteins is a substrate. The resulting γ-carboxyglutamic acid possesses properties which account for the role of this vitamin in blood clotting via regulation of prothrombin formation (Suttie 1977, 1980; Stenflo 1978; Sperling *et al.* 1978).

Niacin.—Part of the coenzymes nicotinamide adenine dinucleotide (NAD) and the corresponding phosphate (NADP). This moiety is a cosubstrate *in vitro* rather than a tightly-bound prosthetic group, in at least 200 distinct NAD- or NADP-dependent dehydrogenases and reductases (EC 1.1.1.___).

Folic Acid.—As tetrahydrofolic acid this "shuttle" type coenzyme is a cosubstrate in enzymes responsible for the transformation of methyl groups and C_1 metabolism in general.

Pantothenic Acid.—As part of coenzyme A, CoA, this vitamin is associated with enzymes which catalyze the transfer of acyl groups in a wide variety of enzymes which utilize CoA and its transforms as cosubstrates *in vitro.*

Other Organic Cofactors.—Other small organic molecules which are either lightly bound *in vitro* to the enzyme or serve as cosubstrates (and may be closely associated with the enzyme *in vivo*) include:

Haems, the Fe-porphyrins found in peroxidase and catalase as well as in the globins (Chapter 18).

Coenzyme Q, ubiquinone, is part of the electron transport system in mitochondria and chloroplasts.

Coenzyme M, 2-mercaptoethane sulfonic acid is, along with vitamin B_{12}, associated with enzymes responsible for the biosynthesis of methane via methylcobalamine-coenzyme M methyl transferase and a "coenzyme F420" and thus of potential importance in solving energy problems (Ferry and Wolfe 1977).

Purine/Pyrimidine nucleoside diphosphate sugars, cosubstrates for enzyme reactions resulting in the synthesis of glycosidic foods to form—among other carbohydrates—sucrose, starch, cellulose pectin, hemicelluloses, etc. In conjunction with the nicotinamide coenzymes, it is involved in enzymes which transform monoses into each other via epimerization.

Lipoic acid, a dithiooctanoic acid combined with lysine, is intimately associated with enzymes involved in acyl group transfer, the classical examples being the pyruvate and succinic dehydrogenase complexes (Chapter 4).

Other small molecules which have been considered to be coenzymes in the rather loose sense of the word include 3′-phosphoadenosine-5′-phosphosulfate involved in the biosynthesis of sulfate esters such as heparin and some food colloids, S-adenosyl methionine, involved in methyl group transfer, ferredoxin involved in nitrogen fixation, nitrate reductase and photosynthesis, and perhaps phospholipids. Glutathione is a cofactor for an NADP-dependent alcohol dependent intestinal alcohol dehydrogenase.

Inorganic Ions.—A list of elements whose cations function as loosely bound cofactors of enzymes follows in order of ascending atomic number; metals in parentheses indicate that this metal is also required for full activity. Examples of food-related enzymes are given wherever possible.

1.—H: All enzymes, as H^+ (Chapter 14).
11.—Na: ATPase (membrane associated); ethanolamine kinase (Mg).
12.—Mg: Most phosphatases; PEP carboxylase.
19.—K: Pyruvate kinase (Mg); pectin esterase.
17.—Cl: α-Amylase, mammalian.
20.—Ca: α-Amylase (plant, microbial), calcium-activated factor in muscles (Chapter 26).
25.—Mn: Glucose isomerase; arginase; pyruvate carboxylase; some SODs; amino acid ammonia lyases; some dehydrogenases (Mg).
26.—Fe: Lipoxygenase; some SODs, heme enzymes.
27.—Co: Some dipeptidases; part of cobalamine (see above).
28.—Ni: Urease; methanogenase; CO dehydrogenase.
29.—Cu: Phenolases; laccase; ascorbate oxidase; cytochrome oxidase; a liver hexose oxidase; nitrite reductase; some superoxide dismutases.
30.—Zn: Milk alkaline, phosphatase, carboxypeptidase, carbonic anhydrase; leucine aminopeptidase.
34.—Se: Glutathione peroxidase; formate dehydrogenase, glycine reductase.
42.—Mo: Xanthine oxidase (Fe); nitrate reductase; nitric oxide reductase; nitrogenase (Fe); for others, see Newton and Otsuka (1980).

Carbohydrate.—Aside from the sugars comprising some of the cofactors, many enzymes contain considerable amounts of carbohydrate, usually in the form of oligo- and polysaccharides. By chance it turns out that many of the traditional enzymes related to food quality are quite high in carbohydrate content: invertase, glucose oxidase, etc.

We shall refer to this aspect at the appropriate place. Quite possibly the highest proportion of carbohydrate is possessed by a potential food enzyme additive, tannase. According to Aoki *et al.* (1976) it contains 64% carbohydrate, present mainly as galactosyl and mannosyl glycans.

The role of these polysaccharides in the function of the enzyme remains obscure. Speculation has assigned to them roles in protein conformation and conferring of permeability and delivery across membranes (Chapter 4). It is known that they do confer heat (and "xero," Chapter 13) stability. This property has some practical consequences in commercial enzyme production and will be examined in Chapter 5.

Specificity and Multiple Enzyme Forms: Isozymes

For at least as long as the term "enzyme" came into general use it was generally recognized that different enzymes catalyzing the same chemical reaction could be obtained from different organisms and even from different tissues in the same organisms.

Early hints that multiple forms of the same enzyme were present in a single tissue were based on observations of rather subtle modulations of the specificity of a particular enzyme activity. Thus, variation in the ratio of the values by two slightly different assays (i.e., changing the substrate, the substrate concentration or the pH) upon fractionation during purification or after heating suggested existence of such multiple forms (Schwimmer 1944A). However, it is only in the last 20 years that it has been indisputably established that multiple molecular forms of the enzyme catalyzing the same general reaction did in fact exist within the living cell. About 25 years ago, in a broad review of enzymes, we concluded "that if one were to select a theme that threads through and integrates the ensuing discussion, it would involve the experimental evidence for the multienzyme nature of particular enzyme activities, apparent not only in different organisms, organs, and tissues but also at the intracellular level. Whether this enzymatic microheterogeneity represents a prerequisite function for proper cell metabolism or whether it is a consequence of artifacts arising from maltreatment of the cell by the enzymologist cannot be ascertained from present knowledge of subcellular enzymatic events" (Schwimmer 1957).

It is now abundantly clear that these isoenzymes or isozymes, as they were coined a few years later by Markert (1975), were indeed biological fact and not artifact. This was made possible by the advent of precise electrophoretic separation techniques, especially gel electrophoresis and to some extent also by advances in immunochemical technique (Brewer 1970). At least 15% of all well-characterized enzymes were shown to exist in multiple forms in 1974 (Kenney 1974). This proportion is probably much higher now. Of course, this does not mean that artifacts do not occur. Furthermore, while these artifacts may not be of biological significance, they may form in foods after death of the tissue and cause changes in the properties and quality of the food. However, they are specifically excluded in the classification of the origin of multiple forms of enzymes suggested by the Enzyme Commission (Florkin and Stotz 1973).

(1) Genetically independent proteins such as the mitochondrial and cytosol malate dehydrogenases. Kenney considers that since they arise as the result of convergent evolution, they would most likely have dissimilar amino acid sequences and be immunochemically readily distinguishable. The nuances of specificity modulation would be quite large.

(2) Heteropolymer hybrids consisting of two or more genetically independent polypeptide chains noncovalently bound as exemplified by the well-known case of the muscle and heart lactic dehydrogenases (Chapter 4). Markert (1968) has pointed out that the total number of such isozymes, n, derived from such heteropolymerism is

$$n = \frac{(s + p - 1)!}{p!\,(s - 1)!}$$

where p = the total number of subunits per enzyme molecule and s = the number of different types of subunits. Specificity difference of hybrids would be intermediate. However, it has been found empirically that single subunit enzymes (consisting of a single chain of amino acids) are more polymorphic (more isozyme forms) than multi-subunit enzymes.

(3) Genetic variants such as the glucose-6-phosphate dehydrogenases in man. In this case the amino acid sequences of the isozymes would be quite similar and the specificity differences would not be readily detectable. Only techniques such as electrophoretic separations would readily distinguish such types of isozymes. Genetic variants play a significant role in discussions of evolution. For

instance, changes in a few amino acids due to mutation, which would not affect metabolically significant properties, would be inherited, thus giving rise to evolution *via* genetic "drift" rather than *via* the pressure of natural selection.

The suggestion by the Enzyme Commission that these three classes, which arise from genetically determined differences in primary structure (amino acid sequence), be considered as true isozymes and that others which arise epigenetically be called "multiple forms" of enzymes like the term "synthase" is more honored in the breach than in the observance.

(4) Complex formation with nonprotein substances, as exemplified by the phosphorylases *a* and *b*. It is sometimes difficult to tell whether such complexes are present in the living cell or are artifacts, but they may include coenzymes, lipids and a wide variety of other substances. The question of whether to call them cofactors is a gray area since they are not essential to activity but may change some properties including the latitude of specificity of the enzymes.

(5) Proteins derived from one polypeptide chain as exemplified by the well-studied family of the chymotrypsins, the trypsins and their proenzymes. In this case the posttranslational modifications (Chapter 4) result from limited selective action of the chymotrypsin or trypsin on themselves or that of other proteinases. Modification of the polypeptide chain may also occur by chemical changes in the side-chains of individual amino acids such as the deamidation of an asparagine residue in aldolase. Another possibility is the posttranslational formation of hydroxyprolyl residues, although this usually occurs in nonenzyme proteins such as plant cell wall as well as constituents and collagen. Undoubtedly this type of modification is also a source of artifacts and probably occurs in food-related situations.

(6) Polymers of a single subunit such as glutamate dehydrogenases.

(7) Forms differing in conformation, including all allosteric modifications discussed in Chapter 4 on enzyme regulation and biology. Isoenzyme multiplicity may originate from more than one of the mentioned classes. Thus, according to Markert (1968), just about all of the enumerated classes, in addition to artifacts, are needed to account for the multiplicity of the lactate dehydrogenases.

Just as cofactors are more readily discussed in the context of reaction mechanism and catalytic efficiency, so isozymes are more closely related to those characteristics which relate their regulation and control of their activity within the living cell. With respect to their role in food quality, we shall have occasion to show their importance in: (1) obtaining fundamental information and characterizing their presence in individual foods, (2) assessing the contribution of individual isozymes to certain changes in quality—for instance, not taking isozymes of lipoxygenase into account led to a misapprehension of their role in bleaching of bread flour (Chapter 17); also the fine-tuned isomerism of the primary products and thus the character of the final products of lipoxygenase action and quality attributes conferred by the end products may in part depend on the isozyme pattern, (3) application of isozyme patterns in analytically-oriented and other quality control-related problems in food science, technology and industry (Chapter 37).

Other selected key isozyme references include books edited by Markert (1975) and Rattazzi *et al.* (1979), and reviews by Daugherty (1979) and Miranda *et al.* (1979).

3

THE NATURE OF ENZYME ACTION—CATALYTIC EFFICIENCY

ENZYMES ARE THE MOST EFFICIENT CATALYSTS

While the phenomenon of enzyme specificity can, in principle, be completely understood in terms of the molecular architecture of the active site of the protein, according to most investigators, a complete account of the phenomenon of catalytic efficiency derived from the detailed knowledge of the arrangement and properties of the atoms and domains within the protein molecule has not as yet been attained, and still eludes us. As pointed out by Lumry (1974), just what constitutes an understanding depends in large part upon the orientational biases provided by previous training and experience. As far back as about 17 years ago Bender *et al.* (1964) were able to ostensibly account for the rate constant associated with hydrolysis of N-acetyl-L-tryptophanamide by chymotrypsin on the basis of assigning "reasonable" values to kinetic factors such as orientation, proximity (see below) and assorted histidine-mediated acid and base catalyses. Some investigators feel that, at least from a broad thermodynamic point of view, the key factors which lend singularity to enzyme catalysis are, in principle, understood. Thus, Lienhard (1973) concludes that enzyme catalysis can be perceived as a resultant of the operation of the following factors: changes in substrate transition-state structure, changes in enzyme conformation, interactions with solvent water and enzyme, and entropy changes.

Jencks (1975), though, tells us that efficiency of enzyme catalysis can be accounted for by a combination of the factors that contribute to ordinary chemical catalysis (i.e., decrease of free energy of activation) plus the circumstance that enzymes can exploit the energy of the noncovalent bonding interaction with substrate for further acceleration. Jencks does let us know in the same discussion that there is yet something to be learned since he refers to the relation between enzyme catalysis and binding energy as the "Circe" effect, "the utilization of strong attractive forces to lure a substrate into a site which undergoes an extraordinary transformation of form and structure." By the same token we should refer to an effective inhibitor, especially a transition state analog, as a "moly," a magical herb which prevents such transformations and perhaps induced fit might be called the Procrustean Bed effect.

Substantial progress precedented by the contribution of diversity of factors whose reality is being substantiated experimentally by application of modern theories of theoretical physical chemistry, kinetics, X-ray enzymology and spectroscopic instrumentation should in the not-too-distant future lead to a significant breakthrough. The practical aspects and the significance for the future of food science, technology and processing lie in the potential capability of making inexpensive enzyme-model catalysts, synzymes, (see below) and the presently available capability for standardization of commercial enzyme potency (Chapter 37).

The goal of our theorist brethren is to account for the fact that many enzyme-catalyzed reactions take place up to 10^{20} times faster than the corresponding noncatalyzed reactions. It is pedagogic convention to usually make comparisons of uncatalyzed reactions which proceed measurably fast in the temperature range of enzyme action, with nonenzyme catalyzed reactions, or to present molar activity values. Another useful measure of enzyme "perfection" is to compare k_{cat}/K_m values where k_{cat} is usually the molar activity. For a "perfect" enzyme, one which is evolutionarily stable such as triosephosphate isomerase, this ratio has a value of 10^7-10^8. In general, enzymes with perfunctory regulatory function attain such values. An assortment of multiple restraints prevent most regulatory enzymes (Chapter 4) from approaching such values (Cornish-Bowden 1976; Knowles and Albery 1977).

Comparison with uncatalyzed reactions is usually calculated from energy of activation values (see below). Comparison with nonenzyme catalysts may also be so based but is more frequently expressed as the ratio of the molar activities of the two kinds of catalysts. Examples of traditional food-related enzymes from each of these categories are presented in Table 3.1. We also include carbonic anhydrase, reported to be the most efficient catalyst. However, this is based on *in vitro* experiments. Some of the slower *in vitro* enzyme reactions like those involving coenzymes as substrate (i.e., NAD) and enzymes which are immobilized in the living cell as members of metabolic pathways may proceed much faster *in vivo*. While we cannot present a critical assessment of the status of the depth of understanding, we shall present a brief account of what are considered to be the salient factors contributing to catalytic efficiency of enzymes.

THE TRANSITION STATE

Any explanation of enzyme catalysis has to take into account the fact that before a molecule can undergo a permanent change, that is, to break a covalent chemical bond, its free energy must be raised to a certain critical value, the energy of activation (Fig. 3.1). The free energy of the reaction's products is at a lower level than this critical value, so that the latter constitutes a barrier which has to be surmounted before the reaction can occur. At the top of this barrier the reactant, whether a catalyst is present or not, is said to be in a transition state. In the configuration represented by the transition state, the reaction system possesses the maximum probability of passing over into products.

From transition-state theory the rate of any chemical process, catalyzed or uncatalyzed, is proportional to the number of molecules in the transition state. The amount of free energy required to bring one mole of molecules in a reaction mixture to the transition state is termed the energy of activa-

TABLE 3.1

EFFICIENCY OF ENZYME CATALYSIS

Comparison with Uncatalyzed Reaction; from Energy of Activation			
Catalase:	$1 \cdot 10^8$-fold		
Carbonic anhydrase:	$1 \cdot 10^7$-fold		
Comparison with Nonenzyme Catalysts; from Energy of Activation			
Catalase:	$2 \cdot 10^6$-fold; catalyst, Fe		
Catalase:	$1 \cdot 10^5$-fold; catalyst, I^-		
Catalase:	$5 \cdot 10^1$-fold; catalyst, ferric trimethyl tetraethylamine		
Invertase:	$6 \cdot 10^{10}$-fold; catalyst, H^+		
Urease	$4 \cdot 10^{11}$-fold; catalyst, H^+		
Molar Activity, Molecules Transformed/Sec/Molecule of Enzyme			
Carbonic anhydrase	600,000	Lactase	12,500
Catalase	100,000	Lipoxygenase	3,000
β-amylase	20,000	Phenolase	2,000
Urease	8,000	Invertase	700

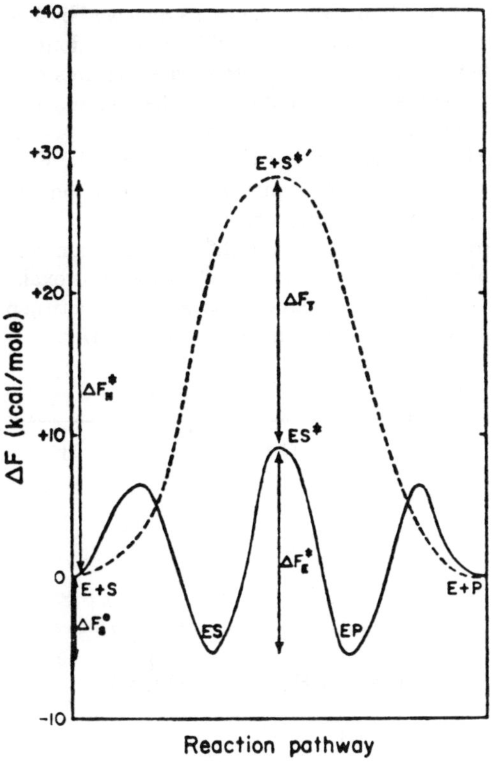

From Lienhard (1973). Copyright 1973 by the Am. Assoc. Adv. Sci.

FIG. 3.1. FREE ENERGY REACTION PATHWAY PROFILES

For an enzyme catalyzed (solid curve) and uncatalyzed (broken curve) reactions; see text for details.

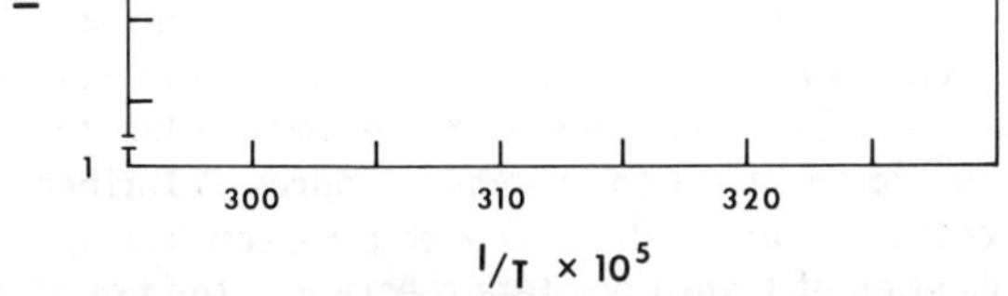

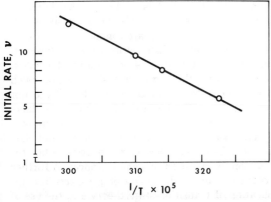

From Chang and Schwimmer (1977)

FIG. 3.2. ARRHENIUS PLOT FOR THE DETERMINATION OF ENERGY OF ACTIVATION

Shown is the enzyme catalyzed hydrolysis of phytic acid by bean enzyme(s).

tion, which is expressed as calories or kilocalories per mole. Experimentally the energy of activation, variously symbolized as A, μ, E_a, $\Delta G^{\ddagger}$, is determined by application of the empirical Arrhenius equation which relates the rate of a chemical reaction, v, to temperature, T:

$$v = A \exp(-E_a/RT)$$

$$\log v = \frac{-E_a}{2.3R}\left(\frac{1}{T}\right) + \log A \qquad \text{(Equ. 3.1)}$$

where A is a constant related to orientation and collision frequency of the reacting molecules. A plot of log v *vs* 1/T yields a straight line from whose slope ($-E_a/2.3R$) the activation energy can be calculated. With respect to foods, Watson (1976) points out that the "best" food molecules are those thermodynamically suitable ones that contain weak covalent bonds and thus large amounts of free energy. The latter can be readily utilized when transformed, i.e., glucose. An example of the validity of the Arrhenius equation for an enzyme reaction is shown in Fig. 3.2. The application of such temperature-related parameters in food processing will be expanded upon in Chapter 10.

The Arrhenius equation and transition-state theory are usually interpreted in terms of the Theory of Absolute Reaction Rates which accounts reasonably well for the empirical constants of the Arrhenius equation (Johnson *et al.* 1974). Thus, from relatively simple correlations one can calculate thermodynamic parameters other than E_a, such as changes in entropy, $\Delta S^{\ddagger}$, and enthalpy, $\Delta H^{\ddagger}$

$$2.3 \log \frac{v}{T} = 2.3 \log \frac{k_B}{h} + \frac{\Delta S^{\ddagger}}{R} - \frac{\Delta H^{\ddagger}}{RT} \qquad \text{(Equ. 3.2)}$$

where the symbol (‡) indicates the transition state; k and h are the Boltzmann and Planck constants:

$$\Delta H^{\ddagger} \text{ and } E_a^{\ddagger} \text{ are related; } E_a = \Delta H^{\ddagger} + RT$$

The significance of these thermodynamic parameters for catalytic efficiency of enzymes is that relatively small changes in their values caused by the presence of the enzyme is reflected in a highly amplified change in reaction rates. Thus, the value for E_a for the catalase-promoted decomposition of H_2O_2 is one-third that of the noncatalyzed reaction. This means that the catalyzed reaction proceeds 10 million times faster than the noncatalyzed.

What effect an alternative to Transition-State Theory, Quantum Statistical Mechanical Theory of Molecular Relaxation featuring the role of "tunneling" (Fong 1976) will have on an understanding of the nature of enzyme action remains for future investigation. Presumably an enzyme would increase this tunneling effect.

Transition State Analogs.—Figure 3.1 depicts a free energy pathway profile for a nonenzymatic reaction as compared with that of an enzyme-catalyzed one. The former may be represented as

$$E + S \underset{}{\overset{K_n^{\ddagger}}{\rightleftharpoons}} E + S^{\ddagger} \longrightarrow E + P$$

where $K_n^{\ddagger}$ is the equilibrium constant. The corresponding free energy change is ΔF_n. For the simplest enzyme reaction

$$E + S \overset{K_s}{\rightleftharpoons} ES \overset{K_E^{\ddagger}}{\rightleftharpoons} ES^{\ddagger} \longrightarrow$$

$$EP \longrightarrow E + P; \qquad E + S^{\ddagger} \overset{K_T}{\rightleftharpoons} ES^{\ddagger}$$

Now transition state theory concludes that the transition state conformation is bound much tighter to the enzyme than is the substrate itself as expressed by the ratio K_T/K_s. Values for this ratio fall in the range of 10^8 to 10^{14}, much higher than values of K_s (similar but not identical to the reciprocal of K_m) in the order of 10^3 to 10^5. About 30 years ago, Pauling (1948) suggested that the efficacy of enzyme action might be explained by the effect of distortion of the substrate which would lead to this enhanced affinity. Parenthetically, a theoretical stereoexplicit detailed analysis of torsional strain accompanying enzymatic peptide bond hydrolysis has revealed that the proteases may induce two distinct modes of torsional activation of their substrate (Mock 1976). The serine endopeptidases (trypsin, chymotrypsin) induce an *anti* distortion (Chapter 21) of *cis* substituents whereas an exopeptidase such as carboxypeptidase induces a *syn* distortion of *trans* substituents. These distortions yield *trans* and *cis* addition forms of HOR, respectively.

Pauling further predicted that substances whose structure resembles more closely that of the transition-state conformation of the molecules will be much better inhibitors than are the traditional substrate analogs. That this idea has validity is vividly exemplified in the increasing discovery of such transition-state analogs, as reviewed by Lienhard (1973), Wolfenden (1976) and Gandour and Schowen (1978). Just what such structures are is not completely revealable, but they should be in between that of the substrate and the product. Again, their existence was foreshadowed by Fischer's concept of the mutual interaction on each other of enzyme and substrate.

An early example of putative transition state analog of an enzyme in a food organism is oxalate (IV) which is, of course, also a food constituent. The K_m with respect to the substrate (I)

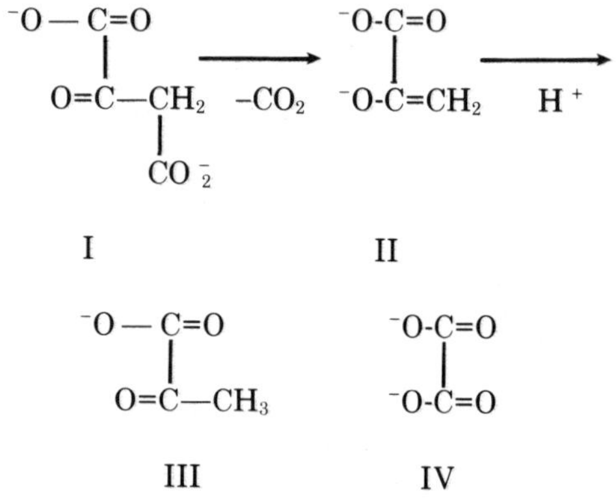

of the enzyme codfish oxalacetate decarboxylase is $1 \times 10^{-3}M$. It is competitively inhibited by oxalate, $K_i = 3.5 \times 10^{-6}$. While this is far from the factor of 10^8 demanded by a "true" transition-state analog—50% inhibition occurs in hundredths of a percentage of solution—the inhibition is considered to be due to the structural similarity to the transition state, which is supposed to be similar to that of pyruvate enolate (II). Note that its structure has features which are common to both substrate and final product, pyruvate (III). However, one should not confuse such a detectable intermediate with the "true" transition state. Such transition state-activated species should also not be confused with the unstable primary products which become intermediates in subsequent secondary reactions discussed in Chapter 1 and elsewhere. They may, however, be responsible for reaction inactivation in some instances. A further treatment and a discussion of prospects for application of transition-state theory and the use of analogs to control food related enzymes will be found in Chapter 14. Discussion of enzyme mechanisms (*c.f.*, Henderson and Wong 1972) and even the biological evolution of enzymes (Cornish-Bowden 1976) frequently begin with transition-state considerations.

Induced Fit

In addition to explaining some of the intricacies of specificity, some but not all investigators believe that it is the conformational adaptability of enzyme molecules which allows for the facilely-attained transition state of the substrate and thus accelerated reaction rates. The manner by which this is accomplished may lead to utilization of energy generated in the act of conformational change for producing strain in the substrate, energy utilization within the enzyme molecule or between neighboring protomers (subunits) in an oligomeric enzyme molecule (Chapter 4). Conformational changes can lead to repulsion of water (desolvation) from the active site—a powerful aid to acceleration of enzyme action. Qualitatively the role of induced fit in catalytic power finds its expression in the idea of its ability to effectively position the "cutting edge" of the enzyme (Fig. 3.3) so as to more readily render the union productive. In an adaptation of the shoelace tying-chair metaphor of Asimov (1959) this particular aspect of induced fit may be likened to the "conformational" changes induced in a recliner which has been modified so that the back tilts forward instead of backward, the footrest motion causes the knee to bend and that of the armrest to move inward when you (the substrate or ligand) sit on (form a complex with) the recliner (enzyme). This "induced fit" thus facilitates and accelerates the shoelace-tying reactions. An ordinary kitchen chair may be likened to a nonenzyme catalyst which also facilitates shoelace tying but much less efficaciously.

Another aspect of induced fit in relationship to catalytic power is its capability of enhancing other factors which can contribute to catalysis in the absence of induced fit. Furthermore, induced fit may contribute to catalytic power by preventing wasteful side reactions.

Some investigators believe that "induced fit" may be overemphasized. As Koshland (1976) himself points out, while these conformational changes are the usual concomitant of ligand binding, they are not required to explain catalytic power in *all* cases. Thus, chymotrypsin's and ribonuclease's catalytic capabilities can, so far as is known, be explained without evoking induced fit, and indeed only very small conformation changes in these enzymes occur when they bind to their respective substrates. It would thus appear that induced fit primarily provides mechanisms for specificity and control (regulation, see Chapter 4) but does not ordinarily or only to a subsidiary degree contribute directly to the catalytic process itself (Jencks 1975). The reciprocal situation, enzyme-induced distortion of the substrate, has been experimentally established using Fourier transform infrared spectroscopy (Belasco and Knowles 1980).

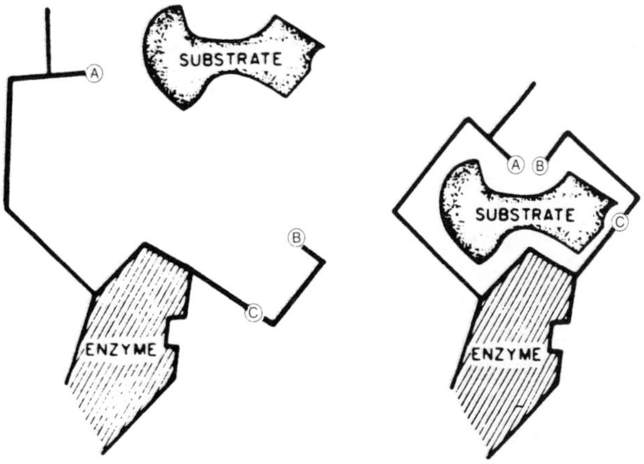

From Koshland (1964). Reprinted from Fed. Proc. 23, 719–726.

FIG. 3.3. INDUCED FIT BY SUBSTRATE BINDING

Aligns the "cutting edge" of the active site of the enzyme to the substrate.

Intermediate Enzyme–Substrate Complexes

As mentioned in the last chapter, the shorter the time interval over which an enzyme reaction is measured, the greater the information retrieval but also the more difficult the interpretation of this data. The shortest time interval over which enzyme reaction has been measured is the nanosecond range (Eigen and Hammes 1963; Hammes 1968, 1972), but relaxation processes down to the picosecond range have been observed for other biological dynamic systems such as the primary events in photosynthesis, ascorbic acid oxidation, immunological events, shape changes of sickling erythrocyte and pulse radiolysis of enzymes (Parsegian 1978). From the viewpoint of what makes enzyme reaction go so fast, the most significant observations are:

(1) Measuring precisely how fast each individual step in the enzyme reaction takes place. Thus, initial encounter of enzyme and substrate is often diffusion-controlled, yield values of the typical second order rate constants of 10^7–10^8 liters mole^{-1} sec^{-1}, some-

what slower than subsidiary reactions such as those involving H^+ and H-bonds.

(2) From the initial successful union between enzyme and substrate until the bond-breaking act, a series of ES intermediates are observed. Thus, eight states of the enzyme are associated with ribonuclease action and 15 with that of aspartate aminotransferase which catalyzes a two-substrate reaction.

These take place in the range of 10^{-4}—10^{-3} sec. The significance of these changes for catalytic efficiency has not been given much attention but they must be important clues. It has been suggested that transformation may involve conformational changes in the substrate and/or the enzyme. Thus, Hammes (1968) and Hammes and Shimmel (1970) suggest that these changes in ES confer stability upon it and that proton-transfer rates are increased. Knowles and Gutfreund (1974) allude to avoiding the necessity for freezing the translational and rotational freedom of many reactants at a single transition state and to lowered increments in energy of transition states between substrate and intermediate and between intermediate and product as compared with the increment between substrate and product energies. That is, a number of such changes are "easier" to make than are large ones. They relate these fleeting intermediates to the more well-defined acyl- and phosphoryl-enzyme intermediates formed in the hydrolysis of proteins and phosphate esters. The role of these changes in ES, played out on a submolecular level, may be a paradigm of the series of enzyme catalyzed-reactions that occur stepwise in intermediary metabolism at the cellular level. Perhaps the existence of these intermediate complexes is more compatible with the recently developed alternative "tunneling" theory to transition state, as briefly alluded to earlier in this chapter.

Although the algebra is not as straightforward as that of simple steady-state kinetics, the calculation of the individual rate constants is based on the following considerations. If the system, $E + S \underset{k_{-1}}{\overset{k_1}{\rightleftharpoons}} ES$ is perturbed but still near equilibrium

$$\frac{d\,\Delta(\overline{ES})}{dt} = \frac{\Delta}{\tau}\ (\overline{E}) + \frac{\Delta\,(\overline{ES})}{\tau} \qquad \text{(Equ. 3.3)}$$

where τ, the time for S to decay to 1/exp of its original value, is $-k_1\ (\overline{E}) + (\overline{S}) + k_{-1}$ and Δ denotes deviation from the unperturbed system. $(\overline{E}) = E_o + (\Delta\overline{E})$; $(\overline{S}) = S_o + (\Delta\overline{S})$; $(\overline{ES}) = (ES_o) + (\Delta\overline{ES})$; the subscript $_o$ is the original stoichiometric concentration and the overbar denotes equilibrium values of the variables. Solutions of this equation involve matrix algebra and setting

$$\frac{\Delta\,(\overline{ES})}{\tau} = 0.$$

Information obtained on this micro- and nanosecond scale is also important in understanding the regulatory properties of enzymes (Chapter 4). The general experimental approach which is used is referred to as relaxation spectroscopy. The principle is to momentarily disturb a steady-state reacting system and to observe the course of reversion ("relaxation") to the original steady state or equilibrium. The most useful perturbation method is the "temperature-jump" method in which the temperature is raised 5°–10°C within microseconds by an electric discharge. Sound absorption/dispersion has also been used. The course back to equilibrium is usually measured spectroscopically but conductometric and resonance methods have also been used. Some of the phenomena revealed by relaxation spectroscopy can now be detected by high-precision calorimetry (Chapter 38).

TERTIARY STRUCTURE AND THE ACTIVE SITE

Active Site Brings Substrate Very Close

Just as the active site of an enzyme is, as we have seen, exquisitely contrived to impart specificity, so is it also structured to accomplish catalytic efficiency. However, the precise nature of the factors and forces which contribute to this catalytic efficiency is more diverse and elusive than those which account for specificity.

In many instances, as we have seen, specificity and catalytic efficiency stem from the same basic factors. While it is not within the scope of this book to account in detail for what is known in this area, a few qualitative silhouettes may give the reader at least an intuitive feeling for factors that facilitate enzyme catalysis.

Proximity.—The demands of specificity require that the substrate lie in close proximity ("propinquity," "togetherness") to the active site so that the substrate concentration is, in effect, increased tremendously—from, for example, 0.01*M* to as much as 100*M*—so that mass-action considerations alone account for an acceleration on the order of 10^5.

Orientation.—The imposition of complementarity implicit in specificity leads to the idea of "orientation" or more picturesquely "orbital steering" (Dafforn and Koshland 1973). Based on a comparison of intramolecular catalysis involving lactones (anchimeric assistance) they suggested that the proper juxtaposition and orientation resulting in alignment of the individual groupings within the substrate with respect to those in the active site ease the entrance of ES into the transition state.

Anchoring and FARCE (Freezing at Reactive Center of Enzyme).—Closely related to proximity and orientation is the concept that optimal disposition, enforced by anchoring to the active site, results in loss of unproductive molecular motions by ostensibly immobilizing the parts of the substrate directly involved in the covalent alteration. This "freezing" results in a favorable entropy, i.e., increases probability for achieving the transition state. This anchoring effect explains why enzymes and coenzymes are relatively large; the highly improbable binding/orientation of active-site reactive groups *vis-a-vis* those of the substrate requires a large interaction area and a high degree of three-dimensional structure (Novak and Mildvan 1972; Jencks 1975).

Water, Hydrophobicity and van der Waals Forces

The influence of and changes in the orientation of the molecules of the solvent also contribute considerably to catalytic efficiency. The cohesiveness or "stickiness" of water brought about by hydrogen bond formation tends to force nonpolar groups to associate, somewhat akin to oil droplets in water tending to coalesce into larger drops. Hydrophobic interactions among nonpolar groups (Tanford, 1980; Ben-Naim 1980), originating from van der Waals forces, result in a gain in free energy which is usable in lowering the activation energy barrier as well as in maintaining tertiary structure of the protein (*c.f.*, Oakenfull and Fenwick 1977). These interactions may also provide the noncovalent binding energy required for active site acylation by substrates, an early step in protease action mechanisms (Smolarsky 1980). Observations from X-ray enzymology that the active site may be devoid of water would indicate that changes in water are important in maintaining its structure and promoting enzyme catalysis.

A somewhat different role assigned to water molecules is that of Low and Somero (1975). They suggest that enhancement of enzyme catalysis may come about as the result of change in organization of water around amino acid side chains and peptide linkages which are transferred during conformational events in enzyme catalysis.

Hydrophobic interaction between, and hydrophobicity of, constituent amino acids of proteins, peptide and lipids have accounted for a number of food properties and changes during manufacture, production, processing and storage of foods and ingredients, as will be noted in the following chapters. These include the following:

A. Protein-protein interactions accounting for:
   (1) Meat texture and textural aberrations involving troponin-T (Chapter 26)
   (2) Oligomericity of regulatory enzymes (Chapter 4)
   (3) Baking properties of dough involving gluten (Chapter 31)
   (4) Behavior of enzymes at low temperatures (Chapter 12)
   (5) Behavior of enzymes at low water activities (Chapter 13)
B. Protein-protein interaction after modification of protein by limited proteolysis accounting for:
   (1) Elicitation of functional properties such as foamability (Chapter 34)
   (2) Clotting of milk by rennet enzymes involving casein submicelles (Chapter 33)
   (3) Improved properties of doughs and baked goods involving gluten (Chapter 31)
C. Protein-nonprotein interactions accounting for:
   (1) Action of lipid transforming enzymes such as lipoxygenase (Chapter 24) and lipase (Chapter 25)
   (2) Fish texture involving free fatty acids (Chapter 28)
   (3) Homologous inhibition regularities (Chapters 14, 21)
   (4) Formation of complexes with tannins (Chapter 34)
   (5) Purification of some enzymes (Chapters 6 and 7)
D. Di- and oligopeptide properties accounting for:
   (1) Bitter taste of hydrophobic peptides (Chapter 34)
   (2) Plastein formation involving protease-produced peptides (Chapter 34)

*Physical-Organic Chemical Theories of Catalysis.—*Quite aside from the special considerations of the protein nature of the enzyme, many of the mechanisms elaborated in theoretical organic chemistry have been applied to the active site constituents to help explain enzyme catalysis. Indeed, enzyme reactions are in principle simpler than other organic reactions because the latter may follow multiple pathways whereas enzyme reactions are monomechanistic (Mildvan 1974). Especially important are the roles of serine and histidine of the active site. The former is intimately involved in the formation of a covalent bond with the substrate as a forerunner to the bond-breaking event which signals the catalysis. Again, as in the formation of successive ES complexes, the net effect here is to lower the transitional barrier. Several proteases and phosphotransferases fall into this class. Covalent intermediates can also be found with nucleophiles such as the SH of cysteine (papain), the ϵ-amino of lysine (aldolase) and the imidazole of histidine (many enzymes).

The latter is especially involved in that class of organic chemical mechanisms referred to as general acid-base catalysis. The net effect is to more fully utilize the catalytic capabilities at room temperatures and neutral pH of the hydrogen ion *via* such mechanisms as "concerted" proton transfer and "push-pull" mechanisms. Dunn and Bruice (1974) provide a particularly excellent tour-de-force in applying such physical organic "modeling" to lysozyme action.

Active Site Modification.—In the last chapter, we pointed out that examination of the consequences of chemical modification of enzymes is probably more important in elucidation of catalytic efficiency than in probing the nature of specificity. Although this approach preceded and has to a certain extent been replaced by X-ray enzymology and magnetic resonance, it still is essential in identifying catalytically important amino acids and in final confirmation of details of mechanism. As described in Chapter 14, it may have some use as an application to food quality control.

Information concerning catalytic efficiency has been obtained with the following classes of chemical modification of the active site (Sigman and Mooser 1975):

(1) Reaction with group specific reagents, the classical approach, whose historical origins were outlined in Chapter 2.
(2) Reaction with pseudosubstrates, reagents whose detailed mechanism of interaction with enzyme closely mimics that of a true substrate.
(3) Formation of affinity labels (active site-directed inhibitors) in which the initial reagent is a pseudosubstrate which contains a group specific reagent. The latter can interact with a specific amino acid residue by virtue of the resemblance of the initial reagent to the substrate.

    Into such affinity labels can be incorporated "reporter" groups such as "spin" labels in order to amplify the information extractable from such experiments. Koshland (1976) has likened affinity labeling experiments to the stopping of a cinema film at a single frame, thus freezing a stage in conformation changes induced by ligand.
(4) Interaction with suicide substrates, pseudosubstrates (Waley 1980) which are converted in the active site into unstable intermediates which interact further with amino acids of the active site. They are thus more specific than other types of active site-directed inhibitors since the amino acid side group reagent incorporated into the latter can also react with amino acids not in the active site before it ever gets there, even with amino acids of foreign proteins (Abeles and Maycock 1976).
(5) Trapping of active intermediates is especially useful for elucidation of catalytic mechanism.
(6) Removal of terminal amino acids from the amino- or carboxyl-end of the polypeptide chain of the enzyme. In this case the influence of changes which may be far removed from the active site is examined.
(7) Synthesis *in vitro* of fragments of the polypeptide chain or modification thereof. This latest approach has been made possible by the development of methods for the stepwise organic chemical synthesis of such fragments as well as entire polypeptide chains of enzymes (Chapter 4).

Figure 3.4 illustrates how a particular mechanism of enzyme catalysis was postulated, as afforded by a detailed but slightly flawed knowledge of the active site. The amino acid sequence common to many "serine" enzymes, discovered some 25 years ago and briefly alluded to in Chapter 2—serine → histidine → carboxylic group (aspartic or glutamic acid)—constitutes a hydrogen-bonded charge-activated relay system. This system facilitates both deprotonization of the serine and stabilization of

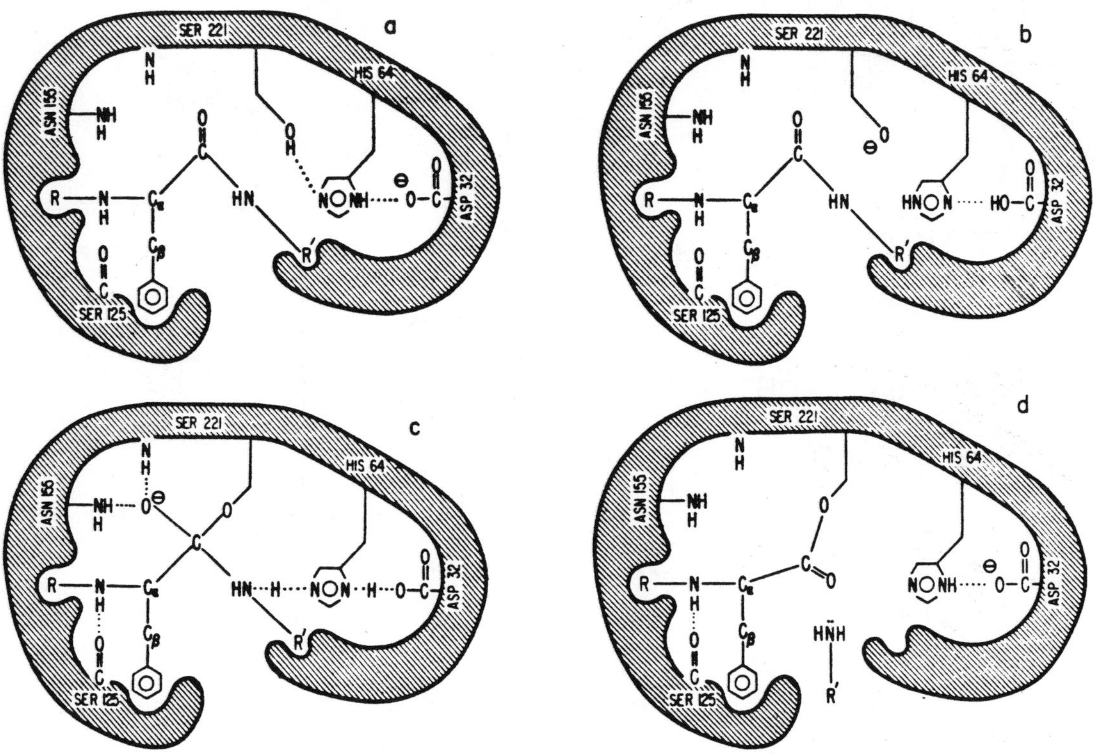

From Robertus et al. (1972). Reprinted with permission from Biochemistry 11 (23) Fig. 8. Copyright by the Am. Chem. Soc.

FIG. 3.4. ROLE OF CHARGE RELAY IN CATALYTIC EFFICIENCY OF ENZYMES

Depicted are substrate interactions with amino acids of active site (shaded) of a protease, subtilisin BPN′: (a) Michaelis complex, (b) activated charge relay system, (c) tetrahedral addition product, (d) acyl intermediate.

the tetrahedral intermediate. This tetrahedral configuration is one that was consonant with that of the transition state.

This "textbook" mechanism has been challenged because it is based on certain assumptions (concerning distances between certain amino acids in the active site) which NMR measurements have shown to be incorrect; again illustrating the kaleidoscopic, rapidly moving horizons of our understanding of how enzymes can act (Fox 1979).

Role of Metals.—Stabilization of special geometrical conformations such as the tetrahedral or trigonal bipyramidal transition state *via* metal ions appears to be a principal mechanism of catalysis for at least half of the known enzymes (Mildvan 1974). Thus, the role of metal ions in enzymes is more intimately related to the catalytic efficiency and less to specificity and regulation than are the other nonamino acid constituents of enzymes such as organic prosthetic groups, some anions or carbohydrates. On the other hand, metal ions can induce reaction specificity in model enzyme systems, as mentioned in Chapter 2. Their contribution to catalysis stems largely from their capacity to form chelates, bridges and coordination complexes among the amino acid residues of the active site, of which they are part, and substrates. In addition to promotion of appropriate geometry of the transition state, in general, each can also promote catalysis by acting as an electron sink, thus facilitating nucleophilic attack by Lewis acids and perhaps by initiating charge-transfer.

More specifically, Mildvan (1974) has assigned various catalytic roles to individual metal ions. Thus zinc, in addition to playing such general roles in carboxypeptidase and carbonic anhydrase (Fig. 3.5), may activate the 3′-OH of growing polynucleotide chains formed by the action of nucleotidyl transfer enzymes such as those catalyzing the synthesis of RNA and DNA. In some iron-containing enzymes, such as milk xanthine oxidase,

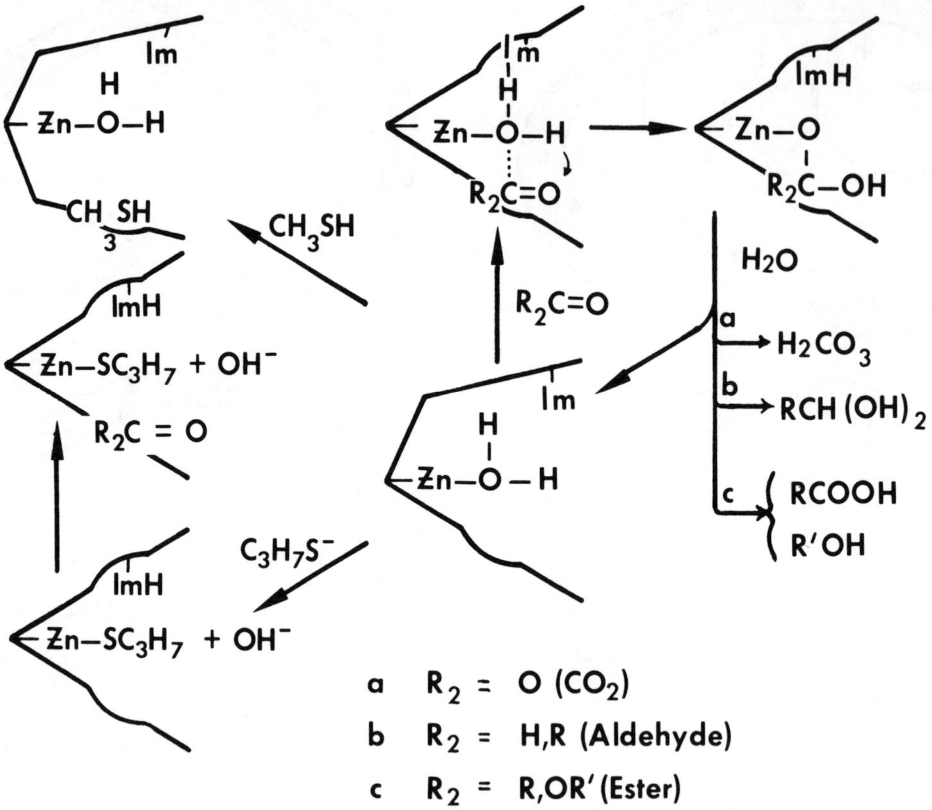

From Schwimmer (1969B)

FIG. 3.5. ROLE OF ACTIVE SITE METAL IN EFFICIENCY OF ENZYME ACTION

As exemplified by zinc in the active site of carbonic anydrase. Also depicted are its roles in determination of specificity and anion inhibition.

the presence of the metal results in the accompanying prosthetic group (i.e., FAD, heme) having an atom(s) with unpaired spin density accessible to solvent. This results in outer sphere electron transfer reactions. Inner and/or second coordination spheres of metal ions are used for catalysis by some enzymes such as that of K^+ by pyruvate kinase (like all kinases, the enzyme also requires Mg^{2+}). In metal-requiring lyases and in certain isomerases, the metal coordinates a basic group (B) in the substrate which is one carbon removed from the atom to be deprotonated (H-C-)

$$\text{H-}\overset{|}{\underset{|}{\text{C}}}\text{-}\overset{|}{\underset{|}{\text{C}}}\text{-B-Metal-Enzyme}$$

Since this has been documented for D-xylose isomerase, it probably also holds for magnesium-requiring glucose isomerase, the key enzyme in the manufacture of fructose-containing corn syrups (Chapter 9).

Dixon *et al.* (1976) postulate that each enzyme which catalyzes a reaction involving ammonia contains a fairly tightly bound transition metal ion which complexes to the ammonia (i.e., nickel in urease). In contrast to other investigators, they believe that too much emphasis has been placed on supposedly special characteristics (strain, "entasis") conferred on the metal ion by association with the active site. They support the view that catalyses by metalloenzymes may be a reflection of the chemistry of the metal ion as a Lewis acid (Chapter 2).

Another way in which metals have contributed to our understanding of the catalytic power of enzymes is *via* their deployment either alone or as constituents of model enzyme systems, whose contribution and significance will be forthwith examined. We subject them to a degree of scrutiny perhaps seemingly out of place in a book on food enzymology because of the promise that research on them holds for their use in place of commercial enzymes in food processing.

MODEL ENZYMES

An alternative attack to the understanding of the nature of enzyme catalysis is to study model enzyme systems (Westheimer 1959; Ingraham 1967), also referred to as artificial enzymes (Martell 1973) or synzymes (Klotz *et al.* 1971). These are usually low-molecular weight molecules of known constitution and chirality, having relatively simple structures, although they may also be incorporated into synthetic polymers. They not only act as catalysts of the same reactions that enzymes catalyze, but are frequently designed to selectively operate *via* factors known or suspected to be involved in enzyme catalysis as well as to exhibit specificity. We previously referred to the observations of Mock (1976) on the role of metal ions on the specificity of such a model system and on the use of lactones in the development of the concept of "orbital steering." A now classical example of a proximity effect is the 53,000-fold stimulation of the general-base intramolecular catalysis by the remaining free COO^-, of the hydrolysis of monosubstituted esters of succinic acid by restricting movement of the COO^- through appropriate substitution (Bruice and Bencovic 1966).

Metal Cations.—Although this approach has been sustained for the most part by organic chemists, some of the most effective insights have been gained by study of inorganic ions and their complexes as model enzyme systems (Martell 1973). Indeed, the deliberate consideration of nonenzymes as clues to enzyme action goes back some 20 years ago to a review by Bamann and Trapmann (1959) who cite work going on in 1930 on the catalysis of the hydrolysis of phosphomono- and especially -diesters by lanthanum salts in slightly alkaline solution—model phosphodiesterases. They are not efficient catalysts. Thus, although Whithey (1969) observed a 36,000-fold increase in the rate of hydrolysis of *p*-nitrophenyl methyl phosphonate, molar activity is much less than unity. Whithey suggested that La^{++} may act as a general acid catalyst or may carry a special water molecule in its hydration sheath which can act as a carrier of the cationic charge to the substrate. As pointed out by Bruice and Bencovic (1966), available data on heterogeneous metal ion catalysis did not yield, at that time, insight into reaction mechanisms.

More fruitful results have been obtained with metal chelates or complexes with small organic molecules. These were thoroughly and systematically reviewed by Martell (1973). He points out that such metal complexes may also serve as models for metal-free enzymes, and set the general tone for reasons why such enzyme model studies are advantageous. He distinguishes among: (1) activation of substrate towards nucleophilic attack by metal ion coordination (chelate ring formation is the driving force but hydrophobic bonding effects may also contribute), (2) electron transfer reactions, and (3) oxidations involving oxygen insertion. Satisfactory models of the latter, which mimic lipoxygenase for instance by inserting both atoms of oxygen into a substrate, were not available in 1973.

As previously mentioned, metal ions can induce reaction specificity in model enzyme systems. Thus, the pathway of certain phosphopyridoxal model enzyme reaction systems in the presence of Zn is different from that in the presence of Cu or in the absence of added metal ion (Blum and Thanassi 1977). The susceptibility to metal catalysis of two similar substrates may vary widely. Thus, DNA is much more susceptible to Zn-catalyzed depolymerization than is RNA (Butzow and Eichhorn 1975).

They and others have discussed the evolutionary implications of selective metal ion catalysis. Thus, from an evolutionary point of view their observations may be related to why DNA rather than RNA is the bearer of primary genetic information (Chapter 4). In a discussion of the early evolution of transition element enzymes, Egami (1975) regards iron complexes as the evolutionary forebearers of electron transfer enzymes (cytochrome), molybdenum complexes as those of enzymes involved in the metabolism of small molecules and zinc complexes as the precursors of hydrolytic and transferring enzymes. Incorporation of copper into enzymes evolved in response to the consequences of photosynthesis, especially the accumulation of atmospheric oxygen, and also presumably to remove superoxide anion (Chapters 16 and 21).

Inorganic Anions and Glucose Isomerase: Models and Application

While inorganic anions have not been extensively studied in the context of model enzyme systems, their utilization as effective catalysts on a commercial scale at least for one purpose appears to be more promising than that of other more thoroughly studied synzymes. This one purpose is the reversible isomerization of glucose to fructose. Starting with the well-known complexing of borate with *cis-cis* adjacent hydroxyl groups in sugars, a phenomenon which has been quite useful for their separation, identification and analysis (*c.f.* Schwimmer *et al.* 1956), Barker and coworkers developed poly(4-vinyl-benzeneboronic acid) columns of resins which they used as model reactions for the manufacture of fructose from glucose (Barker *et al.* 1973). In this case the mechanism appears to be not so much one of selective catalysis but rather preferential selective complexation of an arene boronic acid with fructose as compared with mannose and glucose during the alkaline interconversion of these three sugars.

More recently, Rendleman and Hodge (1975, 1977) have investigated the aluminate ion on ion exchange resin as an effective model of immobilized glucose isomerase. In this case the aluminate itself also appears to catalyze the isomerization thus constituting a more valid model enzyme system than that of the boronic acid derivatives. As a practical means for fructose production, the process has certain advantages over that of immobilized glucose isomerase now being used in the corn syrup industry (Chapter 9). The catalyst is more abundant and less expensive. Up to 70% fructose can be readily prepared on the column; if desirable, glucose can be readily separated from fructose by simply eluting the column with water. Finally, the process may solve one problem that the practical adoption of all model enzyme systems faces, that of health-related safety of the catalyst. In this case, both the innocuousness of aluminate and its immobilization on a resin column and thus separation from the food products should ensure the safety of the products. Like immobilized enzymes in general, column stability still poses a problem.

Organic Molecules as Model Enzymes

Cycloamyloses.—Perhaps the most thoroughly examined single class of organic compounds as enzyme models are the tori of 6–8 α-1,4-D-glucosyl units, cycloamyloses, formerly known as the Schardinger dextrins. They are found as the result of the action of a special enzyme, Schardinger amylase or dextrinogenase (Schwimmer 1953D) or cyclodextrin glucanotransferase. The cycloamyloses exhibit feeble but definite and diverse catalytic properties and have been used as models of ester synthases, hydrolases of various kinds (including, for instance, penicillinase), decarboxylases and enzymes of phosphate transfer. Stereospecificity occurs with substrates containing one or more asymetric carbon atoms. Although not particularly spectacular as catalysts—they are about 0.005% as effective as chymotrypsin in catalyzing the hydrolysis of *p*-nitrophenyl acetate—they do exhibit respectable enzyme-like K_ms, in the order of 10^{-3}–$10^{-4}M$ for many substrates. Like enzymes, a decrease in activation energy (Chapter 2) is observed. It is not surprising, therefore, that attention has been primarily directed to the forces and factors involved in the entry of substrate into the void space of the torus. Table 3.2 illustrates how studies of the cycloamyloses are directed towards an understanding *via* analogization of the factors contributing to the specificity and catalytic efficiency of enzymes. For more examples and discussion of cyclodextrin model enzymes see Cramer and Hettler (1967), Saenger (1976) and Bender and Komiyama (1978).

Imidazole.—Two examples shown in Table 3.2 demonstrate the catalytic effect of imidazole derivatives. Other examples are shown in Table 3.3. Imidazole catalysis (corresponding to the role of histidine in the active site of many hydrolases) is involved in perhaps one of the most impressive accelerations observed for a synzyme system; the over 10^{12}-fold (as compared to unbound imidazole) increase in the rate of hydrolysis of phenolic sulfate esters by a synthetic branched polymer, polyethyleneimine containing long hydrocarbon chains ($C_{12}H_{25}$) for binding sites and covalently bound imidazole groups for catalysis (Kiefer *et al.* 1972). Interestingly, hydrolysis of nitrophenyl esters of carboxylic acids is accelerated only 100-fold. This difference may reflect the real-enzyme circumstance that the carboxylic esters are hydrolyzed by serine proteases (enzymes) in which imidazole plays a peripheral role, whereas imidazole is the essential nucleophile in aryl esterase mechanism.

Other themes of model enzyme research (Table 3.3) include the roles of high molecular weight and/or size, hydrophobic interactions, transition-state conformational restraints, intramolecular general acid/base catalysis, distribution and relay of

TABLE 3.2

CYCLOAMYLOSES AS MODEL ENZYMES

Enzyme or Reaction	Aspect of Enzyme Behavior Being Modeled[12]
Phenyl esterase[1]	Stereoselectivity, a striking model of Fischer's lock-and-key metaphor. (Im)
Phenyl esterase[2]	Attachment of imidazole groups to cycloamylose accelerates reaction 100-fold. (Im)
Benzoyl decarboxylase[3]	Restraint on substrate, conformational catalysis.
Penicillinase[4]	β-Lactam cleavage accelerated 10-fold. (Im)
Enzyme-substrate complexes[5]	Complexing with small molecules creates "induced fit" as determined from X-ray analysis.
Enzyme-*p*-nitrophenol complex[6]	Expulsion of high-energy cavity water as source of binding energy
Acyl hydrolase[7]	Deacylation of benzimidazole derivatives of acyl-cyloamyloses exhibits enzyme-like behavior.
A copper oxidase[8]	Two molecules of dextrin per Cu, may contribute to remarkable rate acceleration.
Phenol formylase[9]	Cyclodextrin selectively formylates phenol in para-position.
Optical resolution[10]	Resolution of chiral sulfoxides, sulfinates and thiosulfinate esters, important in *Allium* flavor (Chapter 21). Other studies show orientation of "substrate" in the doughnut-like hole.
Carbonic anhydrase[11]	Capped with imidazole to yield regiospecifically bifunctionalized model enzyme.

[1] Source: Van Etten *et al.* (1967).
[2] Source: Cramer and Hettler (1967).
[3] Source: Straub and Bender (1972).
[4] Source: Tutt and Schwartz (1971).
[5] Source: Saenger *et al.* (1976).
[6] Source: Bergeron and Channing (1976).
[7] Source: Kurono *et al.* (1976).
[8] Source: Matsui *et al.* (1976).
[9] Source: Ohara and Fukuda (1978).
[10] Source: Mikolajczyk and Drabowicz (1978).
[11] Source: Tabushi *et al.* (1980).
[12] (Im) = imidazole groups present.

TABLE 3.3

SELECTED STUDIES ON IMIDAZOLE AND OTHER ENZYME MODELS (SYNZYMES)

Synzyme	Substrate/Reaction	Take-home Lesson
N-Acyl histidines[1]	Phenyl esterase	Hydrophobic Interaction (Hy-I)
Naphthyl imidazole[2]	Phenyl esterase	Hy-I; ES formation; nucleophilicity
Steroid imidazole[3]	Ketone enolase (β-elimination)	Hy-I; lowers free energy level of transition state.
Polyethyleneimine imidazole[4]	Sulfatase	Catalytic acceleration, 10^{12}-fold; obeys Michaelis kinetics.
Imidazole-benzoate-anion in acetonitrile, $1M$ H_2O[5]	Nitrophenyl acetate	This synzyme system not a model for charge relay effect in serine proteases; catalysis proceeds via nucleophilic attack by H-bonded dimer.
Long chain alkylamines[6]	Long chain carboxy esters of 4-nitrophenol	Billion-fold acceleration; decrease in free energy in accord with Hy-I transition state calculations.
Micelles of N,N-dimethyl dodecyl glycine[7]	Decarboxylation	Acceleration, *ca.* 700-fold; arrangement of zwitterion charges important.
Methyl sulfonium with nucleophiles[8]	Transmethylation	Stringent linear transition state required for catalysis.
Crown ethers[9]	Carbohydrates	Lock and key chemistry yields enzyme analog models.
Fe (III) porphyrin Cu (II)[10]	Cytochrome oxidase	Mimics state of metals found in this enzyme.

[1] Source: Shorenstein *et al.* (1968).
[2] Source: Kunitake *et al.* (1970).
[3] Source: Guthrie and O'Leary (1975).
[4] Source: Kiefer *et al.* (1972).
[5] Source: Hogg *et al.* (1978).
[6] Source: Oakenfull (1973).
[7] Source: Bunton *et al.* (1975).
[8] Source: Lok and Coward (1976).
[9] Source: Coxon *et al.* (1979).
[10] Source: Gunter *et al.* (1980).

charge and anchimeric assistance. This activity constitutes a two-way feedback system in which information obtained from model systems will lead to elucidation of enzyme action and understanding of how and why enzymes act as they do will lead to the production of even more enzyme-like synzymes. In each case the fruits of this activity can be potentially harvested by the food industry to better understand and improve food quality.

Westheimer (1959) was probably the first to direct the attention of physical organic chemists to the study of small organic molecules as "model" enzymes to be replaced in less than two decades by enzyme analogs. The contributions of Westheimer and related investigation were ably summarized in a book by one of his early collaborators (Ingraham 1967; Lowe and Ingraham 1974). A more recent review of model enzymes (Bruice 1976) points out the continued usefulness of the model approach and that "true" enzyme analogs have yet to be synthesized. Perhaps the design of future practicable synzymes will incorporate such model enzymes with *de novo in vitro* enzyme synthesis, alluded to in Chapter 4.

In addition to those already cited in this chapter, in Chapter 1 and in biochemistry textbooks, there are many other broad discussions of the origin of catalytic efficiency in enzymes, including books by Bernhard (1976), Westley (1969), Bender (1971), Bender and Brubacher (1973), Zeffren and Hall (1973), Blackburn (1976), Firsht (1977) and Walsh (1979), many reviews and several symposia.

Finally we should like to point out, as alluded to in the next chapter, the quaternary, as well as the tertiary structure, although principally concerned with regulatory properties of enzymes, may also contribute to catalytic efficiency of enzymes which consist of subunits but which obey Michaelis kinetics.

4

THE NATURE OF ENZYME ACTION—REGULATION, BIOSYNTHESIS AND BIOLOGY

LEVELS OF BIOLOGICAL REGULATION

In Chapter 1 we highlighted the various approaches available to food scientists for controlling and regulating enzyme action for improvement of food quality. These interventions into natural processes constitute a pale, simplified expression of the exquisitely elaborate labyrinthine regulatory mechanisms for control of enzyme action in the living cell. Even in a simple bacterium such as *E. coli*, the action of as many as 3000 enzymes has to be orchestrated and modulated in consonance with the requirements for its existence, development and reproduction in response to changing environment, and integration as members of biochemical pathways. It is not surprising that the capacity for the regulation of the activity of most, if not all, enzymes is as integral a facet of enzyme action as are specificity and catalytic efficiency. The analysis of biological phenomena in terms of regulation of enzyme activity has increasingly occupied the attention of both biologists and chemists. A chronicling of the vast accumulating literature is afforded by the continuing series of published symposia, *Advances in Enzyme Regulation* (Weber 1963–1975). We can only present here highlights of some of these regulatory mechanisms operating in all, including food organisms.

In general the cell has three options for regulating enzymes: (1) control of the activity of the enzymes already in the cell, (2) changing the level of the number of enzymes in the cell and (3) changing the kind of enzyme molecules being made. The first option operates at the somatic level and the latter two at the genetic level of cellular performance. In the somewhat more elaborate scheme shown in Fig. 4.1, the three options comprise regulatory points 1–2, 3–6 and 7–13, respectively. If there is one theme which is constantly repeated in the operation of controls at all of these points of regulation, it is the ligand-induced conformational changes in the enzyme itself or in the proteins (repressors, hormone, receptor sites) which are indirectly associated with this regulation.

In the cells of the living organism as in the test tube, an enzyme is subject to control by the classical factors affecting enzyme activity such as substrate and enzyme concentration and the presence of cofactors and effectors. Although pH and temperature determine the level of enzyme action, the dictum of Claude Bernard that the constancy of the internal milieu is the condition for continued existence suggests that changes in these two variables *in vivo* are not directly important in regulating enzyme action. This is especially true of higher homoiothermic animals.

Regulation of Michaelis Enzymes

Regulation by Substrate.—The fact that enzyme activity varies with substrate concentration constitutes a primitive and gross regulatory mode (Hammes 1972). Thus, with Michaelis enzymes the reaction rate increases with increasing substrate concentration until a limiting value is revealed. As the reaction proceeds, the rate is decreased. This decrease may be reinforced by substrate inhibition (see Chapter 14). The substrate is thus acting as an on-off switch and the product as a rough-tuning device. Further control is afforded by product inhibition and activation. Some stimulation of activity (Massey 1953) due to general anion activation might be mistaken for true allostery (see below). Product inhibition will be further pursued in this and other chapters.

Inhibition by excess substrate, which occurs quite frequently with Michaelis enzymes, is accounted for and obeys equations derived from the assumption that an additional substrate molecule can combine with the enzyme at a site other than the active site.

$$E + S \overset{K_m}{\rightarrow} ES; \quad ES + S \overset{K_n}{\rightarrow} SES$$
$$ES \overset{k_3}{\rightarrow} E + P \qquad \text{(Equ. 4.1)}$$

If SES is an unproductive complex, the equation relating rate v to [S] is

$$v = \frac{k_3 [S]}{K_m + [S] + \frac{[S]^2}{K_n}} \qquad \text{(Equ. 4.2)}$$

so that eventually v approaches zero. Esterases frequently show such behavior (Haldane 1930). With enzymes such as fungal α-galactosidase (Kobayashi and Suzuki 1975) and bean phytase (Chang and Schwimmer 1977) the rate, after reaching a maximum, decreases to a constant but finite value. This is interpreted as being due to the capacity of SES to form a product but at a reduced rate

$$E + S \underset{}{\overset{K_m}{\rightleftharpoons}} SE; \quad SE + S \overset{K_n}{\rightleftharpoons} SES$$
$$SES \overset{k_3'}{\rightarrow} SE + P \qquad \text{(Equ. 4.3)}$$

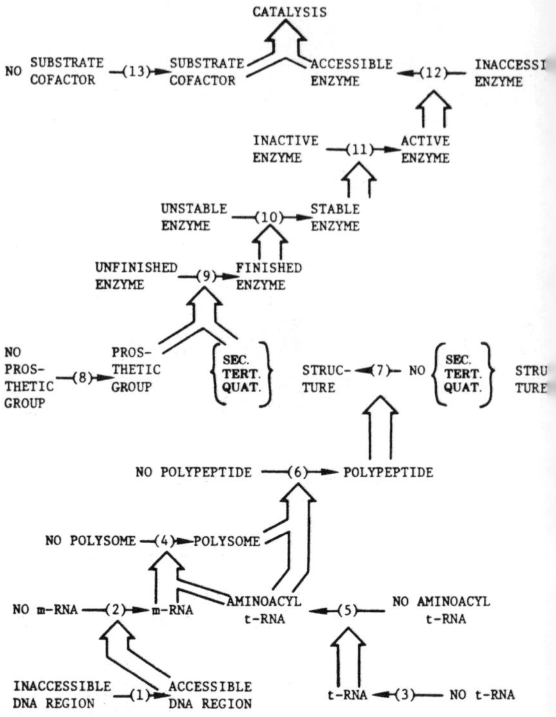

From Filner (196

FIG. 4.1. THE HIERARCHY OF BIOLOGICAL ENZYME REG ULATION

The generalized rate equation is

$$v = \frac{\left(k_3 + \frac{k_3'[S]}{k_n}\right)[S]}{K_m + \left(1 + \frac{K_m}{K_n}\right)[S] + \frac{[S]^2}{K_n}} \qquad \text{(Equ. 4.4)}$$

Both product and excess substrate inhibitions may be functioning as a result of the quaternary structure of many Michaelis enzymes, as discussed later in this chapter.

It will be noted in this formulation of the mechanism that the substrate is assumed to combine with the enzyme to form an unproductive complex, SE. If this prevents S from combining at the active site, formally this would not change the hyperbolic nature of the v *vs* [S] profile but simply shift the apparent K_m and V_{max} (V)

$$v = \frac{k_3e[S]}{K_m + (1 + \frac{K_m}{K_n})[S]} = \frac{\frac{V}{K_c}[S]}{\frac{K_m}{K_c} + [S]} \quad \text{(Equ. 4.5)}$$

where $K_c = (1 + K_m/K_n)$. Formally this resembles "uncompetitive" inhibition in which the inhibitor can combine with ES but not with E. According to Christiansen (1957), a theorem of absolute reaction rate theory demands that such an unproductive complex, SE, inevitably accompany and be in equilibrium with ES. Perhaps the existence of multiple enzyme-substrate intermediates (Chapter 3) is somehow associated with such considerations. However, for the present discussion, it does illustrate how one regulatory mechanism is an inevitable consequence of the action of even the simplest Michaelis enzymes.

Sensitivity to Regulation.—A convenient measure of the regulatory capacity of an enzyme is to determine the ratio of reaction rates at two different S concentrations. Compared with other enzymes, Michaelis enzymes—those which obey the hyperbolic v *vs* [S] profile—are quite insensitive to such changes and do not depend on the values of K_m. Thus, consider the Michaelis equation for two different substrate concentrations S_1 and S_2: $v_1 = V\ S_1/(K_m + S_1)$ and $v_2 = V\ S_2/(K_m + S_2)$. If the ratio $v_1/V = A$, and $v_2/V = B$, then $K_m = S_1(1 - A)/A = S_2\ (1 - B)/B$ so that the ratio $S_2/S_1 = B/A\ [(1 - A)/(1 - B)]$.

Thus, in order to achieve a nine-fold increase in *rate* $(\frac{B}{A} = \frac{0.9}{0.1})$ the *substrate concentration* has to be increased 81-fold. Non-Michaelis enzymes are much more finely tuned.

Other Factors.—Most of the cofactors, especially coenzymes, are probably not present in the living cell in saturating quantities and hence their levels will provide a degree of regulation (Hammes 1972). Naturally-occurring protein inhibitors can also be involved in regulation of enzyme activity. Thus, the appearance of an invertase inhibitor in potatoes can shut off the accumulation of reducing sugars although abundant latent invertase is present (Chapter 19).

The role of the surface in membranes, other cellular structure and nonliving systems as a determinant of enzyme regulation has been pioneered by McLaren and coworkers and will further be explored in conjunction with immobilized enzymes (Chapter 8). Regulation of enzyme action can also be imposed by the interaction of a soluble enzyme with an insoluble substrate as exemplified by, but by no means restricted to, the lipases thoroughly reviewed by Verger and DeHaas (1976) and elaborated upon in Chapter 25.

Regulation of enzymes is also achieved, as previously mentioned, by virtue of their not being scattered at random in the cell but being bound and compartmentalized with respect to substrate and to other enzymes (Schwimmer 1978). Addition of an inhibitor to a bound enzyme system attenuates the diffusional regulation occasioned when the substrate in the microenvironment of free enzymes is transposed to the macroenvironment of bound enzymes (Engasser and Horvath 1974).

Limited Proteolysis.—Another mechanism for regulation is that occasioned by conversion of a proenzyme (zymogen) to active enzyme. Classically this occurs in animal tissue by limited proteolysis of the zymogen. Thus, the conversion of prochymotrypsin (chymotrypsinogen) to active chymotrypsin involves the hydrolysis by trypsin of four specific peptide bonds. Although this results in the formation of five polypeptide chains, only two are released from the zymogen molecule. The remaining three chains are held together by disulfide bonds. According to Shain and Mayer (1968) at least some enzymes which manifest activity upon seed germination are thus activated, but most such enzymes are previously elaborated as a result of gene derepression as discussed below. Newer mechanisms for the conversion of proenzyme to active enzymes is that associated with hormone action and cascade regulation. Limited proteolysis-post-translational cleavage is also the agency for achieving *in vivo* regulation of levels of nonenzyme proteins such as insulin, collagen, blood albumin, immunoglobulin, and intracellular digestive (lysosomal) enzymes in addition to virus proteins and blood- and milk-clotting proteins (Chapter 33) (Hershko and Fry 1975; Reich *et al.* 1975; Neurath 1976). Regulation of an enzyme of gluconeogenesis, fructose diphosphatase, is achieved, in part, by its limited proteolysis (Horecker *et al.* 1975).

QUATERNARY PROTEIN STRUCTURE, ALLOSTERY, COOPERATIVITY

Quaternary Protein Structure

Combinations of single folded polypeptide chains (subunits, protomers), usually held together by hydrophobic interactions and ionic bonding, are collectively referred to as the quaternary structure of proteins. Excluded are enzyme proteins like chymotrypsin in which the polypeptide chains are held together covalently by disulfide bonds.

The combination of the subunits to make a functional protein unit constitutes an example of the principle of self-assembly, albeit at a rather rudimentary level. Self-assembly extends to formation of multienzyme complexes, enzyme complexes comprising metabolic pathways and nonenzyme entities such as collagen and viruses (Perham 1975). Thus, some viruses contain as many as 140 identical subunits, each of which consists of two polypeptide chains.

Quaternary Structure and Catalytic Efficiency.—The completed protein molecules, oligomers, comprise some two-thirds of all enzymes. Just as the tertiary structure determines specificity, catalytic efficiency and crude on-off regulation, so the quaternary structure of enzymes is related to their fine-tuned *in vivo* regulation. Reviews, discussions and symposia in this area include those of Matthews and Bernhard (1973), Ebner (1975) and Klotz *et al.* (1975).

The subunits in a given oligomer may or may not be identical. Of the 300 distinct oligomers (90% of which are enzymes) listed by Klotz *et al.*, most consist of from two to twelve subunits; over half have either two or four subunits. Seven proteins contain three subunits and all the rest contain even numbers of subunits.

Table 4.1 lists multiunit enzymes conventionally associated with food quality. Most of them are dimers and tetramers of identical subunits and, in contrast to what are considered "authentic" regulatory allosteric enzymes, they all normally obey Michaelis kinetics. Most of the known enzyme oligomers do so and one of their salient characteristics is their symmetry. However, the subunit interaction and regulatory properties of such enzymes are expressed as negative cooperativity (see below) explained by "half of the sites"-reactivity (Lazdunski 1974) in which induced fit plays a major role. In these enzymes, exemplified by *E. coli* alkaline phosphatase and γ-glutamyl cysteine synthetase (Sekura and Meister 1977), the allostery (see below) of each half-site is cancelled out by the alternating operation of the two subunits within a functional dimer. This "flip-flop," an integral part of the catalytic mechanism, ensures thermodynamic coupling between successive steps (e.g., substrate binding with phosphorylation) and thus contributes to catalytic efficiency. With respect to *in vivo* regulation, such mechanisms in oligomeric Michaelis enzymes facilitate the previously discussed inhibition/activation by excess substrate and by reaction products. The action of such Michaelis enzymes, as well as that of other oligomers, is also

TABLE 4.1

FOOD SCIENCE ASSOCIATED ENZYMES AS OLIGOMERS[1]

EC No.	Enzyme[2]	Subunits[2]	Molecular Weight[2]
1.1.1.1	Alcohol dehydrogenase	4	100×10^3
1.1.1.5	Diacetyl reductase	2	100
1.1.1.28	Lactate dehydrogenase	4	128
	Prophenolase	2	80
1.10.3.3	Ascorbate oxidase	6	146
1.11.1.6	Catalase	4	232
1.13.11.12	Lipoxygenase	2	108
1.15.1.1	Superoxide dismutase	2	25–40
2.4.1.1	Phosphorylase	4	370
2.8.1.1	Rhodanese	2	37
3.1.1.1	Carboxylesterase	2	160
3.1.3.3	Phosphatase, alkaline, acid	2	85,86
3.2.1.1	α-Amylase	2	50
3.2.12	β-Amylase	4	201
3.2.1.15	Polygalacturonase	2	38
3.2.1.23	Lactase	4	540
3.2.1.26	Invertase	4	210
4.1.1.39	Ribulose bisphosphate carboxylase-oxygenase	8	515

[1] With few exceptions, references can be found in Klotz *et al.* (1975). Others can be found elsewhere in this book.
[2] These are minimum figures. Frequently, further investigation reveals dissociation into more fundamental subunits; also, the number of subunits may depend on the source.

regulated by the circumstance that their functionality is more sensitive to enzyme concentration than are monomeric enzymes. At low concentrations, oligomers will dissociate into inactive afunctional protomers and thus become completely ineffective whereas the activity of monomers will be only decreased in proportion to the dilution.

Non-Michaelis Kinetics: Allostery

One of the characteristics of oligomeric enzymes is that, as expected of self-assembly units, most can be readily dissociated and reassembled into active enzyme. In addition to the mild reagents which can affect this dissociation, the state of the aggregation of such enzymes is frequently affected by the presence of substrate, coenzyme and other effectors (ligands). The state of aggregation is thus involved in their regulation. However, dissociation/association is not the only important feature of regulatory aspects of the enzymes; it is the ability of each protomer (subunit) to undergo a ligand-induced conformational change without changing the degree of aggregation. Such allosteric enzymes obey the following equation:

$$v = \frac{V[S]^n}{K_m + [S]^n} \qquad \text{(Equ. 4.5)}$$

The v *vs* S profile for this equation follows an S-shaped curve as shown in Fig. 4.2.

By multiplying each side of equation (4.5) by (K_m + $[S]^n$) and solving for $[v/(V - v)]$ we obtain the Hill-plot (Hill 1910) first applied to the relation between oxygen pressure and uptake of oxygen by hemoglobin.

$$\log [v/(V - v)] = n \log S - \log K_m$$

Note that the slope of the resulting straight line allows calculation of "n," the number of subunits, which can be compared with independent dissociation measurements. They should agree if perfect cooperativity is obtained. Techniques for improving precision accuracy and speeding computation of data for such allosteric phenomena are available (Wieker *et al.* 1970; Watari and Isogai 1976).

The closer control of enzyme activity afforded by a sigmoidal v *vs* S curve can be seen by comparing the change in substrate concentration [S] required to increase v from a value of 0.1 V to 0.9 V. As noted before, an 81-fold increase in S is needed for a Michaelis enzyme whereas for a typical allosteric enzyme, CTP synthetase, only a 1.5-fold increase in S is needed. Thus, the sensitivity to change in substrate is amplified 55-fold (Koshland 1973).

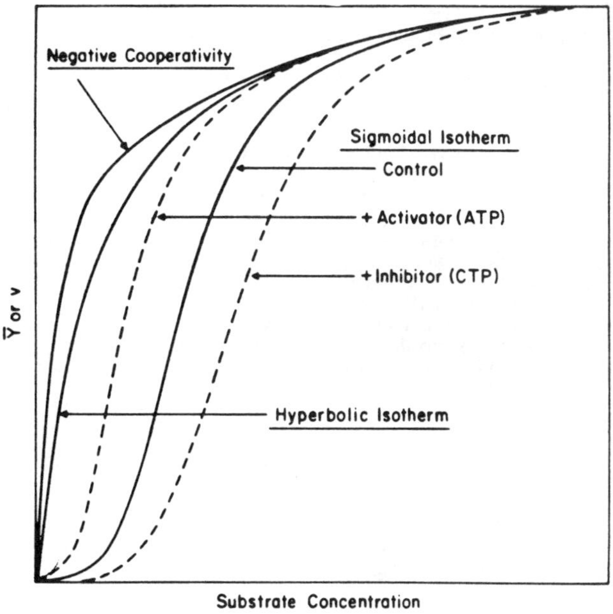

From Hammes (1972)

FIG. 4.2. RATE *VS* SUBSTRATE CONCENTRATION PROFILES
For Michaelis (hyperbolic) and non-Michaelis enzyme kinetics.

As of this writing, two general mechanisms have been advanced to account for the sigmoidal shape, that of Monod *et al.* (1965) and that of Koshland. Both include the concept of positive cooperativity, first applied almost 70 years ago, to explain, as mentioned, the sigmoidal nature of the binding of oxygen by hemoglobin. In general, the phenomenon involves the facilitation, by binding of a ligand to a protein, of the binding of a second and additional ligand(s). In the Monod-Changeux-Wyman model the protomers of an oligomer can exist in one of two configuration states, R and T. Consider an enzyme with four subunits almost all in the T state in absence of ligand/substrate (S). When ligand or substrate S is added, it binds preferentially with the R state:

$$T_4 \underset{- - \rightarrow}{\overset{\text{No S}}{\longleftarrow}} R_4 \overset{nS}{\rightarrow} R_{(4-n)} \cdot S_n \rightarrow \rightarrow \rightarrow R_4 \cdot S_4$$

so that the equilibrium shifts to R by addition of ligands. The R state gives rise to productive enzyme-substrate complexes so that if a nonsub-

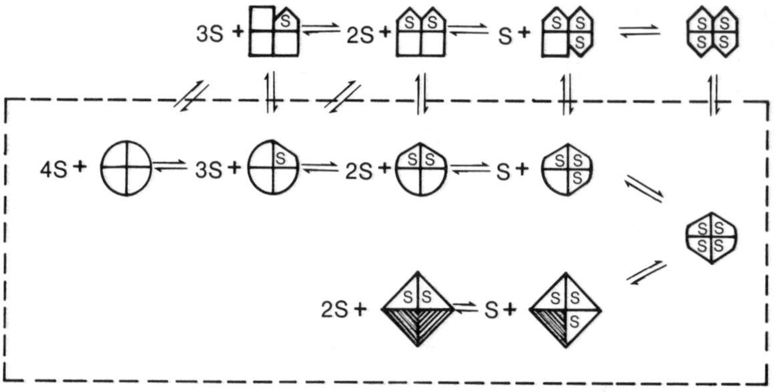

FIG. 4.3. ENZYME-SUBSTRATE BINDING MODES
As revealed and interpreted from ultrafast and fast reaction kinetic measurements. Depicted is the enzyme D-glyceraldehyde-3-phosphate dehydrogenase. Different geometric shapes represent changes of enzyme conformation in response to substrates.

From Hammes (1972)

strate ligand binds preferentially to a T-state subunit, inhibition is obtained. If it binds to an R-state subunit, activation is observed.

By contrast in Koshland's model, S induces a change in the shape of the subunit so that:

$$T_4 \xrightarrow{S} T_3R\cdot S \xrightarrow{S} T_2R_2\cdot S_2 \xrightarrow{S} TR_3\cdot S_3 \xrightarrow{S} R_4\cdot S_4$$

The binding of one ligand facilitates subsequent ligand binding. This is known as positive cooperativity. Only the latter theory explains negative cooperativity; as more S is added it becomes harder rather than easier to add an additional ligand for some enzymes.

In addition to the substrate, naturally occurring metabolites can both activate and inhibit such molecules, giving rise to the different kinds of cooperativity and kinetics. The apparent K_m's of allosteric enzymes are usually smaller than those of other enzymes. In general, the binding (regulatory) site for such small molecule affectors is on a different subunit than is the catalytic site. The regulatory site is thus considered to be the "allosteric" site (the relation of the sites to those associated with noncompetitive inhibition has not been explored). Thus, the first thoroughly dissected allosteric enzyme, aspartate carbamoyltransferase (transcarbamylase) consists of six catalytic subunits and six distinctly different and smaller regulatory subunits. A complex but minimum mechanism for a "simple" enzyme, D-glyceraldehyde-3-P dehydrogenase, deduced from rapid reaction techniques, is shown in Fig. 4.3. It invokes six distinct ligand-induced changes in conformation of the enzyme molecule. Another view of the LDH substrate:NAD^+ complex will be looked at later in this chapter.

Within the cell, the catalytic site of a regulatory enzyme such as phosphorylase a faces organelle-associated substrates whereas the regulatory domain of the molecule is oriented to the cytosol (Fletterick and Madsen 1980).

Evolutionary pressures favoring the elaboration of symmetrical oligomeric enzyme molecules have led to the following advantages as compared with monomers of equivalent weight:

(1) Less DNA is needed to code for identical subunits than for one giant molecule.
(2) They are much more sensitive targets for molecular evolution when there is structural and functional interdependence between the subunits.
(3) There is less likelihood for error to occur in the biosynthesis of the enzyme from constituent amino acids.
(4) The surface-volume area is decreased.
(5) Hydrophobic areas of the monomers are covered up.
(6) Oligomers can diffuse away from the site of synthesis faster (Monod *et al.* 1965; Klotz *et al.* 1975).

IN VIVO DEVICES FOR NONGENETIC ENZYME REGULATION

In this section we survey some of the more well-understood and characterized mechanisms by which enzyme activity, and thereby its functioning, at higher levels of biological organization is regulated. Undoubtedly the actual mechanisms in operation *in vivo* are incomparably more intricate than the labyrinthine network of interlocking devices so far uncovered by researchers.

We do not include at this time the elaborate schemes evolved for regulating the amount and kinds of enzyme molecules synthesized (the genetic level). Most of the "devices" operate through the quaternary and higher self-assembly levels of protein and enzyme organization. One of these devices is the multienzyme complex.

Multienzyme Complexes

As far back as 1948, crystalline proteins were isolated which consisted of complexes such as that of aldolase and glyceraldehyde phosphate dehydrogenase, two enzymes which catalyze consecutive reactions in glycolysis. They were dismissed at that time as examples of the pitfall of using crystallizability as a criterion of enzyme purity (see Chapter 6). Since then, numerous enzyme complexes have been isolated. For example, from yeast two consecutive enzymes of pyrimidine biosynthetic pathway, carbamoyl phosphate (CP) synthetase and aspartate transcarbamylase, can be isolated together along with a "regulatory" protein. In this case the function of the complexing is to channel CP into the pyrimidine pathway since CP constitutes a branch point for the biosynthesis of both the pyrimidines and the amino acid arginine (Lue and Kaplan 1970).

Also, the entire complex of five enzymes and an acyl carrier protein which are responsible for the biosynthesis of fatty acid have been isolated intact. As elaborated upon later in our discussion of immobilized enzymes, when two enzymes catalyzing consecutive reactions are in close association, as parts of metabolic pathways or even when brought together on artificial nonbiological carriers (Chapter 8), the rate of the overall reaction is greater than that observed when the reactions take place in solution.

Pyruvate Dehydrogenase.—Some multienzyme complexes act as a single enzyme, each catalyzing a relatively simple reaction. Perhaps the most spectacular and well worked-out illustration of the self-assembly of a key regulatory multienzyme complex is that of pyruvate dehydrogenase (Reed *et al.* 1976). This enzyme links glycolysis to the tricarboxylic acid (TCA) cycle by catalyzing the overall reaction

$$CH_3COCOO^- + CoA \cdot SH + NAD^+ \rightleftharpoons CH_3CO\text{-}S\text{-}CoA + NADH + CO_2 + H^+$$

This 60-molecule complex comprises three distinct enzymes: E_1, a thiamin-containing enzyme, pyruvate dehydrogenase; E_2, dehydrolipoyl transacetylase; and E_3, dehydrolipoyl dehydrogenase, a flavoprotein containing FAD as a prosthetic group.

$$CH_3 \cdot CO \cdot COO^- \xrightarrow[(NAD^+)]{E_1(TPP)}$$

$$CO_2 + NADH + E_1\text{-}CHOH \cdot CH_3 \xrightarrow{E_2\text{—}LSS}$$

$$E_1 + E_2\text{—}LS \cdot CO \cdot CH_3 \xrightarrow{CoA \cdot SH}$$

$$CoAS \cdot CO \cdot CH_3 + E_2\text{—}LSH \xrightarrow{E_3\text{—}FAD} \ldots$$

where LSS and LSH are the oxidized and reduced forms of lipoic acid, $CH_2SH \cdot CH_2 \cdot CH(SH) \cdot (CH_2)_4 \cdot COOH$.

The initial act is the release of CO_2 and the passing on of the remaining two-carbon fragment to E_2. E_3 regenerates the oxidized lipoate. Of the 60 molecules comprising this complex, 24 are E_2 molecules (MW 65,000) arranged in the form of a cube. Each E_2 molecule is a dimer containing one molecule of lipoic acid per protomer and has sites for binding both E_1 and E_3, each of whose functional molecules can dimerize. The native complex which works most efficiently carries 12 dimers of E_1 and six of E_3. These 18 molecules (made up of 36 enzymatically functional molecules) are distributed so that there is one molecule for each of the 12 edges and six faces of the 24-E_2 cube core. This mode of regulation is frequently encountered with other key enzymes such as the α-ketoglutarate dehydrogenase of the TCA cycle and is subject to several hierarchies of regulatory controls including that associated with metabolic pathways (see below).

Lactose Synthase.—The *in vivo* synthesis of an important food constituent, lactose, illustrates a simpler alternative of a somewhat unique and specialized multienzyme regulation scheme. Two proteins are necessary to catalyze the overall lactose synthase reaction

$$\text{UDPG-Galactose} + \text{D-Glucose} \rightarrow \text{UDP} + \text{Lactose}$$

One of the proteins is α-lactalbumin, long known to be present in milk as just one of several seemingly biologically inactive proteins. Its biological activity is manifested only when it is combined in the mammary gland with a second protein. This second protein by itself is an enzyme which catalyzes the reaction

$$\text{UDP-Galactose} + \text{N-Acetyl-D-glucosamine} \rightarrow \text{UDP} + \text{N-Acetyl-D-lactosamine}$$

but when complexed with the purely "regulatory" α-lactalbumin through one of its histidines (Schindler and Sharon 1976) it catalyzes the synthesis of lactose. Lysozyme, whose primary structure resembles that of α-lactalbumin, can replace the latter as "specifier" protein in the lactose synthase complex. Of course the action of this enzyme can result in the net synthesis and accumulation of lactose only if it is assembled at a site where there is no lactase, an enzyme which in bacteria has played a major role in understanding enzyme regulation at the genetic level. The employment of such "modifier" or "specifier" regulatory proteins which have biological properties only in conjunction with another protein is a device used frequently in more complex schemes of enzyme regulation, i.e., glutamine synthetase. Again it is *via* conformational changes that these proteins exert their effects. This mode of regulation has led to a lively segment of food technology—relating to how to use or get rid of the lactose which accumulates due to the operation of this mode of enzyme regulation—which we shall have occasion to examine in the ensuing chapters.

Regulation of Metabolic Pathways by Enzyme Feedback Control and Elaborations

Feedback Inhibition.—As we have pointed out, the action of a given enzyme in a living cell or microorganism has to be orchestrated in consonance with that of some 3000–30,000 other enzymes. These enzymes are not scattered at random in either real space or "chemical" or "metabolic" space but most of them exert their action as members of teams guiding a series of well-defined chemical changes—metabolic pathways. Most of the enzymes catalyzing the intermediate reactions of metabolic pathways are under allosteric control. Especially sensitive to such control are enzymes whose substrates can be transformed along alternate metabolic pathways at branch points. After the branch point the enzyme which catalyzes the first step unique to a particular metabolic pathway is inhibited by final products of the pathway. This type of regulation *via* feedback inhibition (Fig. 4.4) shuts off the entire pathway until the level of the product drops below a point where the enzyme can

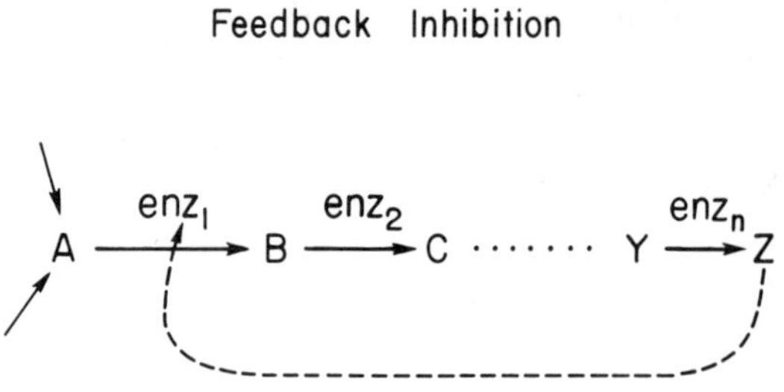

FIG. 4.4. ENZYME REGULATION BY FEEDBACK INHIBITION

again function. Since this is always an allosteric enzyme, it is quite sensitive to changes in substrate concentration so that a very small change in a certain range of concentration of the product can either shut off or completely open up the pathway. Unlike competitive inhibitors, the structure of the final product usually has no resemblance to that of the substrate. One of the tasks of the food sciences is to find the regulatory sequences which result in the accumulation of substances in foods which affect or contribute to their quality attributes. We shall have occasion to scrutinize the regulatory aspects of some of these in relation to sugar and acid accumulation in fruits and vegetables (Chapter 20).

Stimulation of allosteric enzyme activity is frequently referred to as "feed forward" regulation. In addition to strict feedback and feed forward effects, enzymes are affected by other substances involved in the pathways. These are referred to as allosteric affectors. Sometimes these affectors switch an allosteric enzyme to a Michaelis enzyme, i.e., from sigmoidal to hyperbolic kinetics.

Glycolysis/Gluconeogenesis.—The functioning of these inversely related pathways in the progenitor to an important food, mammalian muscle tissue, provides a textbook classical illustration of a rich source of enzymes whose activity is influenced by allosteric affectors. Glycolysis, along with the TCA cycle, is the source of useful biological energy in the form of ATP. Briefly, the reactions in glycolysis and its reverse, gluconeogenesis, starting and ending with the storage glucan, glycogen (starch in plants), may be represented in abbreviated form as follows:

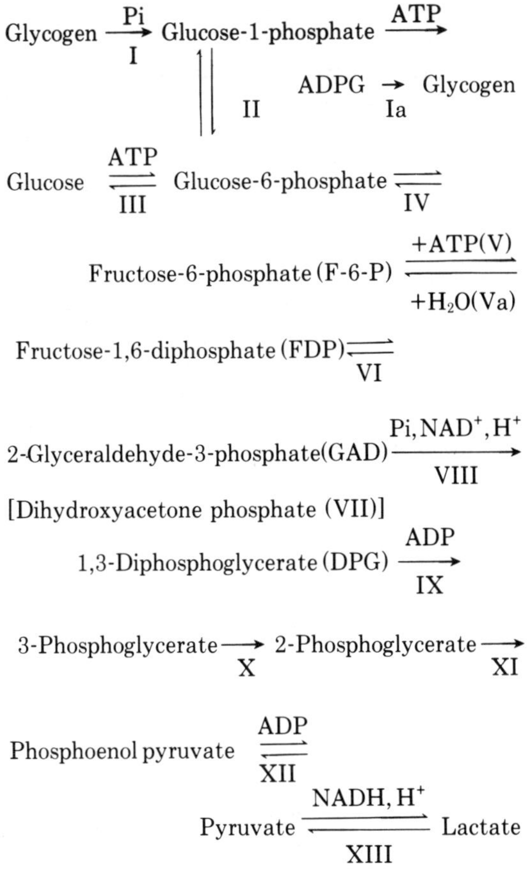

Their names, usual oligometricity (in parentheses) and nonidentity of subunits (*) are as follows: (Klotz *et al.* 1975): I, glycogen (starch) phosphorylase (4); Ia, glycogen synthase (4); II, phosphoglucomutase (2); III, hexokinase (2); IV, phosphohexose isomerase (2); V, phosphofructokinase (2,4,6); Va, hexosediphosphatase (4*); VI, (aldolase, 2,4); VII, triosephosphate isomerase (2); VIII, glyceraldehyde-3-phosphate dehydrogenase (4); IX, phosphoglycerate kinase (monomer); X, phosphoglyceromutase (2,4); XI, enolase (2); XII, pyruvate kinase (4,8); XIII, lactate dehydrogenase (4*).

The glycolytic and gluconeogenetic pathways share the same enzymes which catalyze their reversible reactions, with two exceptions: the formation and breakdown of glycogen and of fructose diphosphate, Ia and Va. Table 4.2 shows the allosteric affectors of these key regulatory enzymes. These allosteric affectors, also intimately associated with the TCA pathways, act in such a way that when the muscle cell is resting, products of the action of the relevant enzymes accumulate. These products are allosteric affectors which encourage storage of energy in the form of glycogen by directing the metabolites through the gluconeogenesis pathway. Inversely, when exercise is performed, the needed useful energy in the form of ATP (ATP is a pivotal feedback end-product inhibitor of key glycolytic enzymes) becomes available by allosteric inhibition of the key gluconeogentic enzymes and stimulation of the corresponding glycolytic enzymes. It should be emphasized that this type of allosteric regulation is but one of the more primitive coarsely-tuned controls to which each enzyme of these and other pathways are subject.

To mention a few types, these include interlocking and entrainment with other pathways (Jensen 1969), concerted cumulative feedback regulation *via* derepression of *de novo* enzyme synthesis by hormones, all acting to influence the pathway by modulating the activity of individual enzymes—not only the key enzymes but almost all of the

TABLE 4.2

RECIPROCITY OF REGULATORY FACTORS ON KEY GLYCOLYTIC ENZYMES

Enzyme	Stimulated by:	Inhibited by:
Glycolytic		
α-1,4-Glucan phosphorylase	Protein kinase, AMP + Pi	Phosphoprotein phosphatase, ATP, AMP, G-6P
Phosphofructokinase	AMP, cAMP, ADP, F-6-P (excess)	ATP, citrate, free fatty acids (fuel)
Gluconeogenic		
α-1,4-Glucan synthase	Phosphoprotein phosphatase, G-6-P	Protein kinase, UDP (a product)
Phosphofructohydrolase = fructose-bisphosphatase	Fuels, histidine, diabetes, fasting[1]	AMP, other nucleotides, insulin (suppresses biosynthesis)

[1] Stimulation due to transformation of enzyme affected by potentiation of lysosomal protease; changes pH optimum, greater stimulation by histidine (Horecker *et al.* 1975).

glycolytic enzymes and of course all metabolic pathways.

Glutamine Synthetase and Cascade Control.— The amino acid glutamine constitutes one of the most highly ramified metabolic pathway branch points in bacteria and probably in higher living forms. It stands at the crossroads of the metabolisms of most if not all nitrogen-containing biological molecules. In turn, the level of glutamine hinges on the regulation of the enzyme catalyzing its synthesis, glutamine synthetase:

$$NH_3 + ATP + Glutamate \rightleftharpoons Glutamine + ADP + Pi$$

This enzyme is an illustration *par excellence* of how regulatory complexity of an enzyme is reflected in complexity of protein structure and allosteric control. Key reviews of this area, developed principally by Stadtman and his colleagues, include those of Stadtman and Ginsburg (1974) and Adler and Stadtman (1974).

The enzyme from *E. coli* comprises 12 identical subunits arranged as a double hexagon discernible in the electron microscope. At the nongenetic level alone, its activity is regulated by five different elaborate mechanisms.

(1) Cumulative feedback regulation; allosteric inhibition by ten different nitrogen-metabolites; each is involved in each of ten pathways; each acts independently and causes only partial inhibition—but cumulatively they can shut down all activity; there is a separate allosteric site for each metabolite.
(2) Metal ion regulation; distortion of the quaternary structure occurs, the giant molecule going from a "taut" to a "relaxed" form with accompanying variations in activity.
(3) Interaction of ATP and metal ion; here divalent cations serve in their traditional role as part of the ATP substrate.
(4) Adenylylation and deadenylylation; in addition to being an allosteric inhibitor, AMP can also inhibit by a covalent link to the enzyme through a tyrosyl residue in each subunit. This "adenylylation," as well as the disruption of the linkage, is catalyzed by two separate enzymes, each under allosteric regulation by substrates, UTP, Pi, α-ketoglutarate.
(5) Cascade control; the adenylylation system constitutes another ubiquitous, sophisticated mechanism for elaborate fine-tuned regulation, that of cascade control. This type of control involves the sequential activation of a series of enzymes in which each enzyme in the series catalyzes the activation of the next member in a cascade-like fashion until the actual metabolite involved is transformed by the final modified (target) enzyme of the series. Thus the adenylylation and thus partial inhibition of glutamine synthetase may be represented by the following simplified presentation:

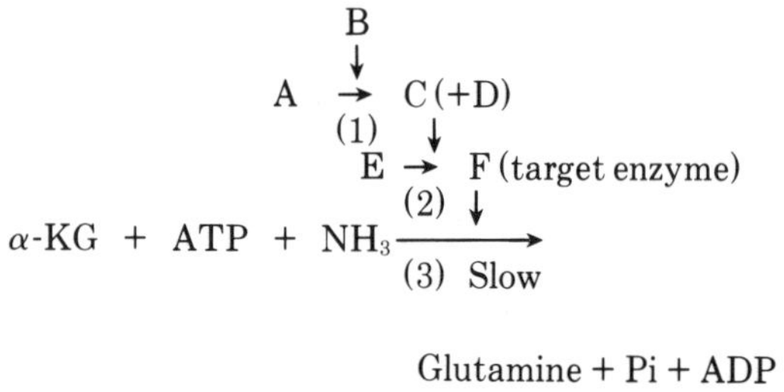

and deadenylylation and thus stimulation of the enzymes as follows:

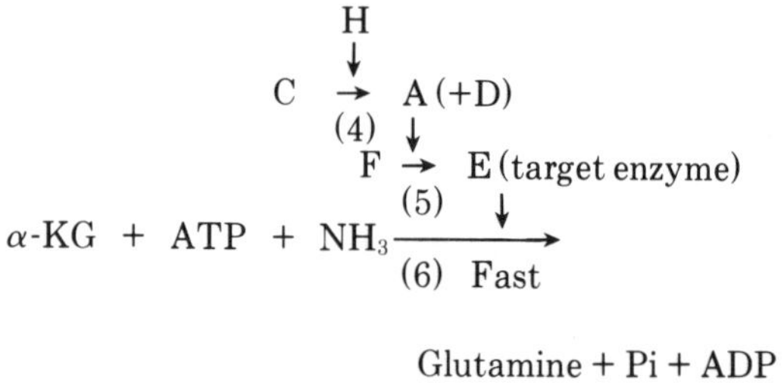

where the uppercase letters represent catalytic protein or enzymes and the vertical arrows indicate catalysis of the reaction to which they point. Reactions (2) and (5) are the adenylylation and deadenylylation reactions, and F and G are the modified forms of enzymes catalyzing transformation of the indicated metabolite, i.e., the synthesis of glutamine, and α-KG is α-ketoglutarate, assigned the role of substrate as well as allosteric effector by Adler and Stadtman (1974). Each of these six reactions is regulated *via* the following allosteric modifiers as indicated.

Effector	*Reaction Numbers* 1	2	3	4	5	6
α-KG	0	–	0	+	+	0
ATP	0	0	0	+	+	0
UTP	0	–	0	0	0	0
Pi	0	0	0	–	0	0
Glutamine	0	+	0	–	–	0
Mn^{2+}	+	+	+	+	+	+
Mg^{2+}	0	+	0	+	+	+
EP	0	0	–	0	0	–

where (+) indicates stimulation, (–) inhibition, (0) either no effect or unknown, and EP are end products of glutamine metabolism, those mentioned previously as contributing to cumulative feedback inhibition. UTP is included because some of the preceding (converter) enzymes are involved in (de)uridylylation in analogy to (de)adenylylation.

Cascade Control and Hormone Regulation

It is becoming increasingly evident that cascade control is a frequently used device in organisms for regulation of enzyme activity and cellular functions. It provides for flexibility and more effective allosteric regulation since points for such control and the primary stimulus, the target enzyme (the last enzyme in the cascade), are amplified. Increased sensitivity is also obtained when more than one step in the cascade is subject to regulation by the same allosteric regulators. As expressed by Stadtman and associates (Adler and Stadtman 1974; Chock and Stadtman 1979), in a discussion of the preeminence of coupled "interconvertible enzyme systems" in the regulation of metabolism, cascade systems are, in essence, physiological computers. The input terminals are the converter enzymes. By means of allosteric and active site interactions, these converter enzymes are programmed to sense fluctuations in the concentrations of a host of metabolites. This leads to automatic adjustments in the activities of the converter enzymes. A massive amount of metabolic information is thus integrated as a single output, i.e., the specific activity of the target enzyme. Cascades are collectively the instrumentality through which many, but not all, animal hormones exert their effects. Historically, their role in this regard precedes that of glutamine synthetase and stems from Nobel Prize-winning investigations of Sutherland in the 1950s on factors affecting mammalian phosphorylase activity. One may represent the hormone-induced cascade as follows.

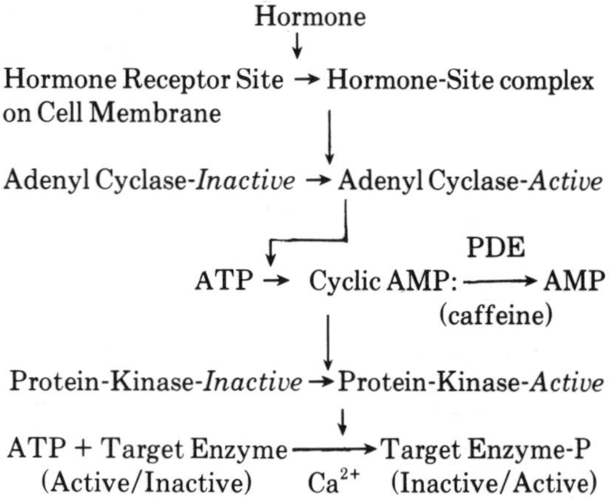

As mentioned, the enzymes of glycolysis and gluconeogenesis are also regulated by hormones. The level of the two corresponding target enzymes, glycogen phosphorylase and glycogen synthetase, are affected in opposite ways by the operation of the converter enzymes associated with the hormone cascade (Hers 1976). Thus the former is activated and the latter is inhibited by phosphorylation *via* the converter enzyme, protein kinase, which in turn is "turned on" initially by the binding of the hormone epinephrine secreted by the adrenal medulla.

Hormones, the "first" messengers; cyclic AMP, the "second" messenger (Finean *et al.* 1974); the prostaglandins, the "third" messenger (Samuelsson *et al.* 1978); and the recently discovered cascade- and hormone-modulating Ca^{2+}-binding protean protein, calmodulin (Klee *et al.* 1980; Marx 1980A) all exert their effects on receptor sites and converter enzymes by inducing allosteric conformational changes and rearrangement of their subunits. An "Advances" series and at least one journal devoted to cyclic nucleotide research have been initiated. By the time this appears, investigation of their role as determinants of quality of meats and other foods will be well underway.

Parenthetically, the rise in free fatty acids in the blood and presumably the stimulating action of coffee is due to the inhibition by caffeine of the indicated phosphodiesterase, PDE. This results in a higher level of cyclic AMP, thus potentiating the stimulatory action of epinephrine which indirectly stimulates blood lipase activity *via* a similar cascade mechanism. The important point we wish to stress is that this regulation is built into the structure of the target enzyme. Blood clotting also proceeds *via* a cascade mechanism involving a series of limited proteolyses by at least 12 highly specific peptidases (Davie and Fujikawa 1975). Cascade mechanisms also operate in immunological mechanisms involving complement.

Regulation via Oscillation.—Another property of enzymes which may underlie diurnal and other long-term biological periodicities is their capacity for periodically varying their action under certain conditions. The period of oscillation covers minutes to hours. Peroxidase and the glycolytic enzyme phosphofructokinase, responsible for the oscillation of glycolysis, exhibit oscillation. Oscillatory enzymes and the reasons why their action oscillates have been reviewed by Goldberger and Caplan (1976).

GENETIC REGULATION AND BIOSYNTHESIS OF ENZYMES

Regulation of Transcription

Heretofore we have explored some of the principal tortuosities so far uncovered by which the activity of a given enzyme molecule in a cell is regulated. But total activity in the cell is also governed by the *total* number of molecules of the enzyme. This means we have to know how and what regulates the *de novo* synthesis and breakdown from and to constituent amino acids. Space and orientation permit only a superficial overview of this wedding of genetics and biochemistry, one of the greatest triumphs in the history of science. The reader is referred to any good recent biochemistry textbook for a more comprehensive overview and also to some of the key discussions cited in the following and in Chapter 6 where we shall discuss how this new knowledge may be used for the production of commercial enzymes for the food industry.

Figure 4.5 represents in a most synoptically abbreviated manner the obligatory events preceding the biosynthesis of almost all polypeptides including all proteins and, of course, enzymes in higher (eukaryotic) organisms.

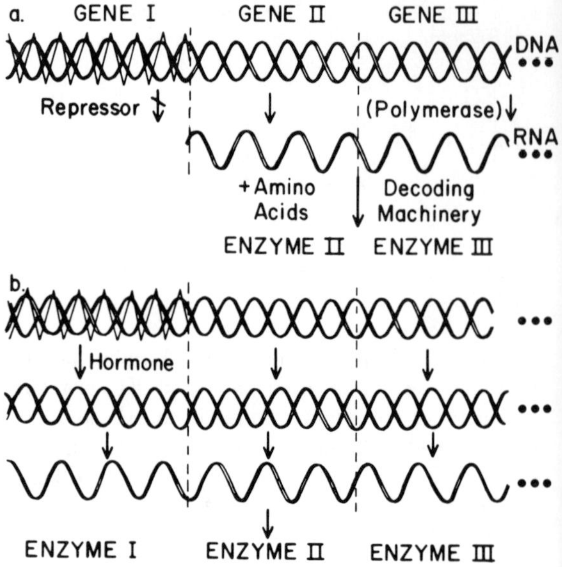

From Schwimmer (1969A)

FIG. 4.5. ENZYME REGULATION VIA CONTROL OF PROTEIN BIOSYNTHESIS
A simplified representation of how the expression of genetic information results in enzyme synthesis and how this synthesis might be regulated at transcriptional and translational levels by hormones.

The Genetic Code.—Information for synthesizing an enzyme from its constituent amino acids is encoded in genes, i.e., in a string of deoxynucleotides, DNA. Each of a limited number of sequences of three pyrimidine or purine bases in three consecutive deoxynucleotides code for a particular amino acid. For instance, only the base sequences (codons) GGU, GGC, GGA and GGG and no others code for the amino acid glycine, where U, C, A and G represent the bases uracil, cytosine, adenine and guanine, respectively. Any given cell has a finite capacity (an almost possible exception are cells responsible for antibody formation) with regard to the total number of different kinds of enzymes and other proteins which can be synthesized. Estimates of the number of enzymes (and other proteins) potentially capable of being synthesized by a single cell range from 3000 in prokaryotes, 14,000 in the sea urchin to as high as 40,000 (or about 15,000 if we disregard multiple forms, isozymes), for man (Lewin 1975A). However, not all these enzymes appear at a given time or even during the lifetime of the cell. A gene or stretch of DNA potentially capable of coding for a given enzyme but which does not do so is said to be "repressed" or "turned off."

Transcription.—Those genes which are not repressed serve as a template for the *transcription* of the *message* in the DNA code to *messenger* RNA. The vehicle by which this transcription is accomplished is the enzyme RNA polymerase, or more formally RNA nucleotidyl transferase.

$$n\left\{\begin{matrix}\text{ATP}\\ \text{UTP}\\ \text{GTP}\\ \text{CTP}\end{matrix}\right\} + \text{DNA}_{\text{(Template)}} \rightarrow \text{RNA} + \text{DNA}_{\text{(Template)}} + n \text{ pyrophosphate}$$

This is not the place to go into base-pairing details but the point is that wherever there is an A in the DNA, there will be a U on the corresponding place on the newly synthesized RNA chain; wherever there is a thymine (T) on DNA, there will be A on the RNA. A similar reciprocal relation holds for the base pair C and G. The messenger RNA goes to the site of protein (enzyme) synthesis and the message is *translated* into a string of amino acids. In prokaryotes (microorganisms), this occurs almost simultaneously with transcription; one end of a message is being transcribed while the other end is being translated. In eukaroytes (higher forms of life whose cells contain true nuclei) the messenger RNA after "maturation" is carried from the nucleus to the rough endoplasmic reticulum. These cells also contain the basic proteins, histones, whose function seems to be primarily to package the long string of DNA into highly condensed hypercoiled modes (supercoiling factor greater than 3000). More recent results (Kornberg 1977; Lilley and Pardon 1979) tend to confirm the author's earlier conclusion (Schwimmer 1967) that DNA may be replicated *in vivo* while still complexed to histones. The latter are probably not involved directly in gene expression, as questioned in Fig. 4.5.

Regulation of Enzyme Synthesis in Response to Food Intake at the Level of Transcription.—It has been well established in both microbes and in man that the level and kinds of some of their enzymes are highly dependent on available food. Certain enzymes—constitutive—are needed at all times for the continued functioning and existence, and are independent of a food source. Others, induced enzymes, are synthesized in response to the type of food supplied. We now know, in principle, for all living things, and in highly perfected detail at the molecular level for one organism and one food, exactly how the biosynthesis of one particular enzyme transforming that food is regulated at the level of transcription. Other examples are and will be increasingly available, but the first breakthrough is related to the utilization of lactose (the inducer) by *E. coli*.

Figure 4.5 is a multifaceted oversimplification in that in reality each gene consists of more DNA than is necessary to code exactly for the given enzyme, in this case that important food-related enzyme, β-galactosidase or lactase. The part of the gene that does code for the amino acid sequence of the enzyme is referred to as the structural gene or in this case designated as the Z-gene. Adjacent to the Z-gene, conventionally shown to the left, is a series of short runs of deoxyribonucleotides, the loci of the regulatory genes; the operator or O-gene and the promoter or P-gene, i.e., P-O-Z. Now the signal for RNA polymerase to proceed lies in this P-gene. In the absence of any impediment the RNA polymerase passes smoothly over the O-gene and starts transcribing the Z-gene according to the equation just depicted.

However, in the absence of a designated food—in this case, lactose—there is indeed an impediment to this passage sitting on the O-gene. This impediment is a specific protein called the *lac* repressor. Lactose, when present in the bacterium's diet, transverses the bacterial wall. Once inside the bacterium it combines specifically with the *lac* repressor. Like so many other biological phenomena, this combination induces a conformational change in the *lac* repressor which prevents the latter from attaching itself to the O-gene DNA. The exact nucleotides in DNA and amino acids in the repressor are now known. The RNA polymerase can now glide over the O-gene and start to transcribe.

To complete the picture, it has been established that the structural gene which codes for the biosynthesis of the repressor protein (I-gene) is located immediately to the left of the P-gene. In addition most enzymes are synthesized together with other closely related enzymes. For this case, these include the "permease" which permits the lactose to transverse the cell wall and membrane and galactoside transacetylase of unknown function. The structural genes for each of these, designated as Y and A, lie to the right of the Z-gene. The entire string of DNA is known as the *operon* and in this particular case, the *lac* operon; there are about five per bacterium.

I	P(85)	O(30)	Z(2000)	Y	A

The numbers next to the P, O and Z genes denote the number of deoxyribonucleotides comprising the stretch.

For eukaryotes the situation is somewhat more complicated as far as the structural gene is concerned. As a result of recent investigations, it is now known that the sequence of base pairs containing the DNA of the structural gene consists largely of nongene or "mini-gene" stretches of DNA interspersed by segments of the structural gene DNA (Kolata 1980). In the expression of eukaryotic genes, this entire order of succession of DNA is transcribed by RNA polymerase. The true gene segments are clipped off the resulting stretch of RNA by appropriate RNases and spliced together, presumably by ligases, to yield the resulting messenger RNA now ready for translation. This dispersal of the gene, together with the actions of the newly discovered topoisomerases, DNA gyrases (Cozzarelli 1980) and synaptases, may be the basis for recombination of classical, pre-DNA genetics, as well as facilitate the hyper-packaging function of the histones.

Catabolite Repression and Cyclic AMP.—Other substances, especially glucose, can act on one operon in such a way as to repress the synthesis of enzymes. The action of glucose in this regard is designated a *catabolite repression*. When the bacterium is "starving" it has to make as many of its own enzymes as possible. To do this, it makes an abundance of cyclic AMP. The latter, by combining with yet another protein (CRP or CAP, cyclic AMP receptor site), induces a change in its shape. This gene-activation protein now combines with a portion of the operon in such a manner as to facilitate the hitching of RNA polymerase to the P-gene, the site of the signal for transcription. However, once food in the form of glucose is introduced, these enzymes are no longer needed. Their synthesis is prevented by suppressing the synthesis of cyclic AMP. The food itself, glucose, accomplishes this suppression (Pastan 1972).

Feedback Gene Repression.—Analogous to product inhibition at the phenotype or substrate level of enzyme regulation, the end product of a metabolic pathway represses expression of the gene coding for enzymes of the pathway at the genetic/transcriptional level. Just as glutamine synthetase is under elaborate regulation at the substrate level, so at the transcription level the enzyme operates under a most versatile regulatory system in which the enzyme itself can act as either activator or repressor of its own biosynthesis or can regulate other genes coding for enzymes involved in the metabolism of other amino acids. It can even take over the activator function of cyclic AMP under certain circumstances. A good discussion providing an interesting historical background of the gene-enzyme relationship with an emphasis on transcriptional regulation and its evolutionary aspects is that of Clarke (1976).

Regulation of Enzyme Synthesis in Mammals.—While detailed knowledge of what occurs at the molecular genetic level is not as advanced as for *E. coli*, the induction and repression of enzyme biosynthesis in mammals and other eukaryotes in response to diet is well-documented as evidenced, for instance, by Schimke (1973). The observed changes in the level of enzymes are less dramatic than those occurring in prokaryotes. One may distinguish between "true" food constituents and "alien" components such as drugs, toxins and adventitious components. The following are a few samplings from this documentation of the effects of nutrition on the enzymes of the consumer.

It has been ascertained that an overload of protein in the diet promotes the synthesis in the liver of those enzymes involved in catabolism of the excess amino acids produced. Thus, the rate of synthesis of rat liver arginase, which breaks down arginine into ornithine and urea, tripled when the level of casein of the diet fed to rats was increased from 8 to 70%. The rate of removal of arginine was unaffected.

As an example of increase in enzyme in response to a *dearth* of a food component, the rates of synthesis of enzymes of fatty acid formation climb steeply and remain at high levels when fatty acid is abruptly removed from an animal's diet.

Feedback repression/inhibition can also be observed. Thus, the level of the enzyme which catalyzes the first reaction unique to the pathway of cholesterol biosynthesis in mammals, hydroxymethylglutaryl CoA reductase:

$$\text{HOOC-CH}_2\text{-}\underset{\displaystyle \text{OH}}{\overset{\displaystyle \text{CH}_3}{\overset{|}{\underset{|}{\text{C}}}}}\text{-CH}_2\text{CO-CoA} + 2\ \text{NADH} + 2\ \text{H}^+ \rightleftharpoons$$

Hydroxy-β-methylglutaryl CoA

$$\text{HOOC-CH}_2\text{-}\underset{\displaystyle \text{OH}}{\overset{\displaystyle \text{OH}_3}{\overset{|}{\underset{|}{\text{C}}}}}\text{-CH}_2\text{-CH}_2\text{OH} + 2\ \text{NAD}^+ + \text{CoA}$$

Mevalonic Acid

is extremely sensitive to the presence of cholesterol in the diet amounting to as high as a 100-fold change in specific activity. Induction of serine dehydrase in livers of rats fed protein-containing food has been suggested as a measure of protein quality of the diet (Ogura 1970). In plants the most thoroughly studied example of enzyme induction is that of nitrate reductase (Chapters 34 and 37). Enzyme induction in higher plants was reviewed by Filner *et al.* (1969).

Animals, as mentioned, possess the ability to detoxify nonfood components in the diet by rapidly synthesizing enzymes which catalyze reactions converting these "xenobiotics," alien harmful substances, to other substances usually but not always less harmful. Older biochemistry textbooks refer to "detoxification mechanisms" involving the formation, presumably enzyme-induced, in response to benzene-like compounds, of mercapturic acids, hippuric acid and glucuronides. These compounds are derived from enzyme action involving cysteine, glycine and glucuronic acid as substrates, respectively. More recently a more encompassing liver enzyme system for oxidizing a variety of xenobiotics has been studied by pharmacologically-oriented investigators. Although much is known about this enzyme system (mixed-function or xenobiotic oxidase), interest in its relation to food quality has been awakened as evidenced by a study of its induction in rats fed varietal cauliflower leaves (Babish and Stoewsand 1975). The ingestion of almost all pharmaceuticals, as well as insecticides, herbicides, etc., leads to the derepression of a series of genes in the liver. This is manifested as a proliferation of smooth endoplasmic reticulum (see below) of liver cells. When liver tissue is ground up and centrifuged, this cellular membrane appears as the microsomes, the working source of the flavoprotein enzyme system, NAPH-cytochrome P-450 reductase. With the aid of this "P-450" system, substances can be oxidatively demethylated, dechlorinated, hydroxylated and decarboxylated. The response of the liver to such foreign substances is prompt and massive. Thus, as much as 30% of the membrane fraction can be made of cytochrome P-450. Activity of the enzyme system can increase five-fold in a few hours. The action of the enzyme system on a xenobiotic, X, can be depicted as follows:

$$X + Fe^{3+} \rightarrow Fe^{3+} \cdot X \xrightarrow{FADH_2} Fe^{2+} \cdot X \xrightarrow{O_2}$$

$$Fe^{2+} \cdot O_2 \cdot X \xrightarrow{FAD} Fe^{2+} \cdot O_2^- \cdot X \rightarrow XOH + Fe^{3+}$$

This system apparently also carries out normal metabolic oxidations such as the hydroxylation of steroids (Gunsalus *et al.* 1975).

Hormonal Regulation.—One way of viewing the effect of diet on enzymes is that the components constitute but one of the types of signals sensed by the genetic decoding machinery. Others include light, heat, time and, of course, hormones, which can act at the transcription level in higher animals and plants as well as at the substrate level. While epinephrine and the peptide hormones operate at the substrate level, the steroid hormones influence the level of enzyme by inducing the *de novo* biosynthesis of new enzymes at the level of transcription, probably by some direct interaction among the hormone, DNA, and a DNA and hormone receptor site in the target organ (Harris *et al.* 1976; O'Malley and Schrader 1976). Complexes of such receptor site interaction, which have been isolated, appear to go directly to the DNA and derepress the genes (Fig. 4.7).

While the nature of the hormonal regulation of enzyme synthesis at the transcription level in animals is just starting to be elucidated, even less is known about the role of plant growth regulators such as the kinetins and the gibberellins in this regard. Gibberellic acid can induce synthesis of α-amylase (Chapter 31), but at which level is not known. The level of our knowledge is typified by a report that addition of a gibberellin (GA_3) and cyclic AMP resulted in a doubling of RNA synthesis in isolated plant nuclei (Tarantowics-Marek *et al.* 1975).

Translation and Its Regulation

Translation, How Enzymes Are Synthesized.—Figures 4.6 through 4.8 depict how enzymes and other proteins are synthesized by the "decoding" machinery. This machinery is an exceedingly complex device indeed, comprising at least 50 distinct proteins and about the same number of different RNA molecules of which there are at least three distinct types: transfer (*t*RNA), ribosomal (*r*RNA) and messenger (*m*RNA). The proteins may also be tripartitioned into: enzymes in the classical sense, ribosomal proteins and special catalytic nonenzyme proteins involved in initiation, elongation and termination of the polypeptide chain. In eukaryotes, the sequence of steps by which enzymes are formed and "packaged" may be classified as follows: (1) activation of amino acid, (2) initiation of peptide synthesis, (3) elongation of peptide, (4) termination, (5) release of protein, (6) posttranslation modification-finishing, "editing," (7) packaging, and (8) shipment, delivery and compartmentation.

Figures 4.6 through 4.8 are three different versions of the translation, each of which emphasizes different aspects of this process.

Amino Acid Activation (Fig. 4.6).—This event occurs by specific enzyme-catalyzed interactions with *t*RNAs, each specific for each amino acid, resulting in the acylation of the 2′- or 3′-OH of the ribose of terminal *t*RNA nucleotide, always adenylic acid. However, the most salient feature of *t*RNA is the strategically positioned sequence of three nucleotides, the anticodon, each of whose bases pairs with that of the corresponding codon on *m*RNA. For instance, the anticodon for phenylalanine in Fig. 4.7, AAG, corresponds to codon UUC.

Initiation (Fig. 4.8).—With the cooperation of the catalytic proteins known as initiation factors (IF-1, -2, -3) the *m*RNA is attached to the 30 S ribosome (the "small" subunit in Fig. 4.8). The ribosome is a highly complex nucleoprotein containing 21 different protein molecules and an *r*RNA molecule. We are slowly learning just how each protein functions (Kurland 1977). Thus, ribosomal protein S binds to a unique sequence of nucleotides in *r*RNA so as to open up and stabilize this sector to allow for base pairing to that segment of *m*RNA involved in initiation (Dahlberg and Dahlberg 1975). Also involved in initiation is a unique amino-acid signal (codon AUG). For bacteria it is N-formyl methionine (the antibiotic streptomycin interferes with its binding to ribosome). Included in this complex is GTP. The next step is the combination of the 30 S with the 50 S ribosome to form the completed S-70 ribosomal unit (Garrett and Wittmann 1973). This reaction is powered by the hydrolysis of GTP. The assembled organelle is now triggered for the next step.

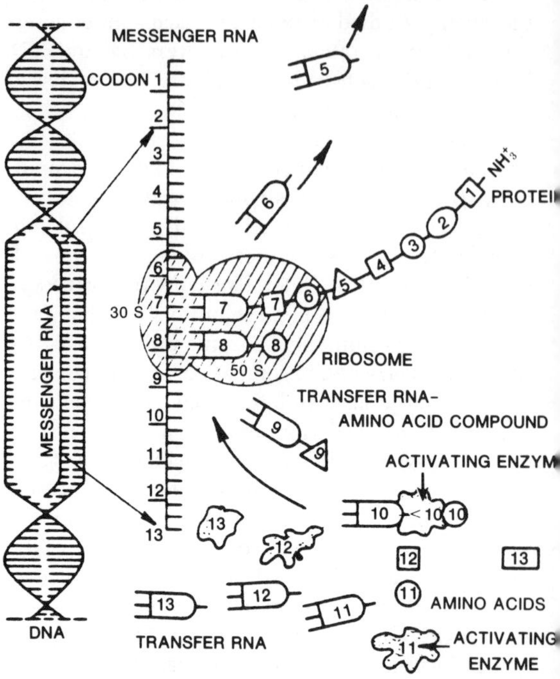

From Whitaker (1973)

FIG. 4.6. ENZYME BIOSYNTHESIS—AMINO ACID ACTIVATION

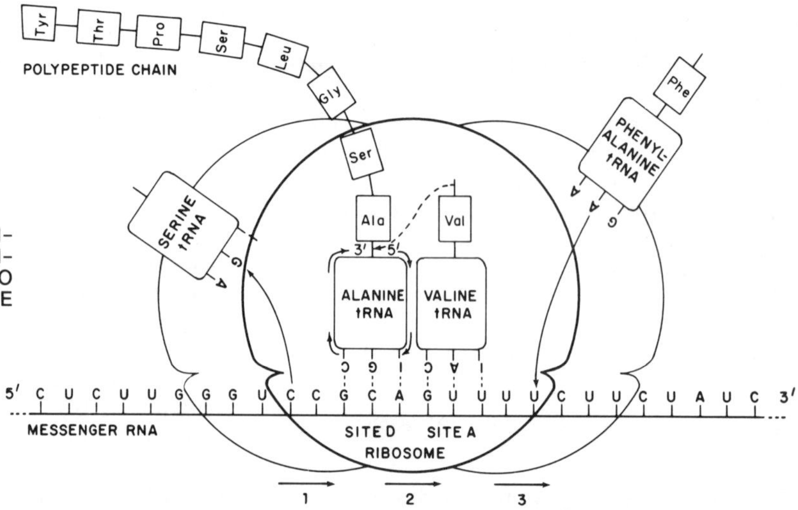

FIG. 4.7. ENZYME BIOSYNTHESIS—SEQUENTIAL ADDITION OF AMINO ACIDS TO GROWING POLYPEPTIDE CHAIN

Courtesy of Lipmann (1973). Reprinted with permission from Accounts Res. 6 (11) 362. Copyright by Am. Chem. Soc.

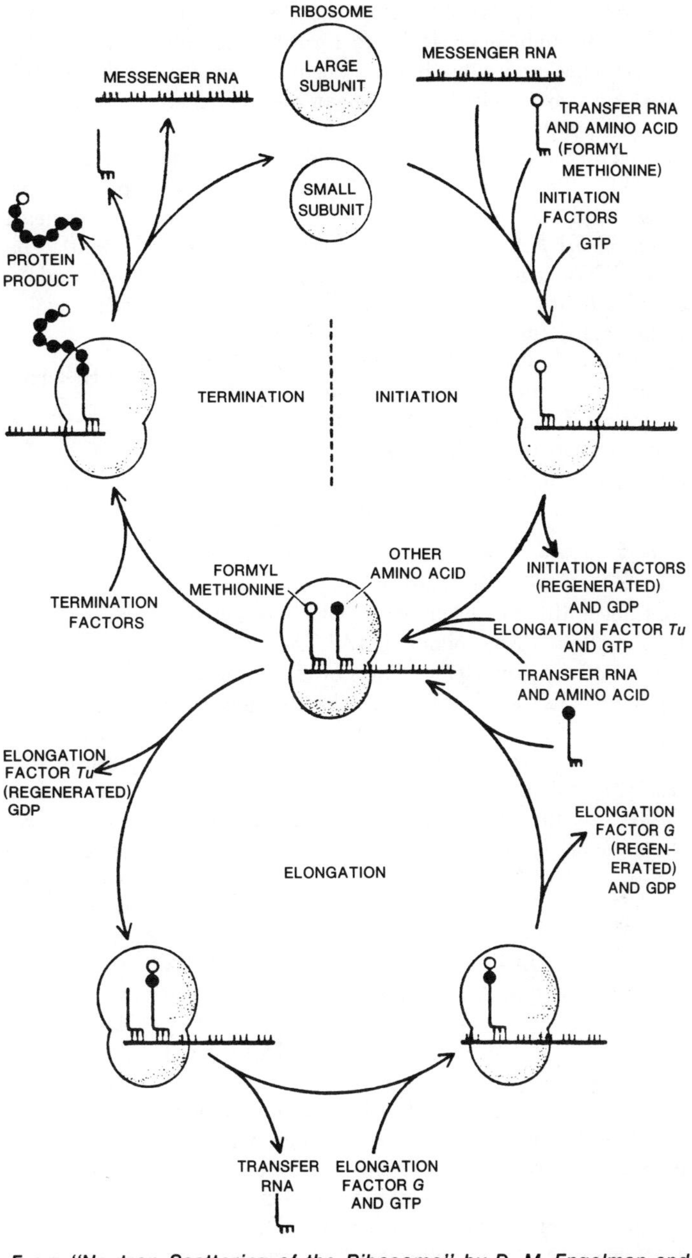

FIG. 4.8. ENZYME BIOSYNTHESIS—INITIATION AND TERMINATION FACTORS

From "Neutron Scattering of the Ribosome" by D. M. Engelman and P. Moore. Copyright © October 1976 by Sci. Am. All rights reserved.

Elongation.—There are two loci (sites) on the completed ribosomal complex which can accommodate the modified *t*RNA: The aminoacyl (A) and the peptidyl (P) shown as site A and site D in Fig. 4.7. In most representations these are located to the right and left halves of the ribosome when *m*RNA is displayed horizontally. In Fig. 4.6, amino acids 7 and 8 occupy sites A and P, respectively. The initiatory aminoacyl RNA binds to the ribosome at the P(D) site. The next aminoacyl RNA, the first amino acid at the COOH-end of the protein to be synthesized, binds *via* codon (*m*RNA)-anticodon (*t*RNA) base pairing at the A-site.

The next step is the transfer of initiator aminoacyl group from the *t*RNA on the P-site to the COOH-end-aminoacyl *t*RNA to form a dipeptide *t*RNA. This, and subsequent elongation, is catalyzed by "conventional" enzymes and is facilitated in a yet poorly understood manner by an elongation factor Tu (Fig. 4.8) and a peptidyl transferase *via* GTP. The "initiator" *t*RNA now drops out of its site and the whole ribosome shifts its position along the *m*RNA (translocation) so that the dipeptidyl *t*RNA now occupies the P-site; the message on *m*RNA has been translated. The empty A-site is now ready for the next codon-anticodon pairing and synthesis of the second peptide bond. Peptide chain growth proceeds from the amino- to the carboxyl-end. Diphtheria toxin inactivates these elongation factors by serving as an enzyme which catalyzes the transfer of ADP-ribosyl from NAD to elongation factor EF-2 (Pappenheimer 1977).

Termination.—This gliding of the ribosome along the *m*RNA and concomitant peptide synthesis continue until the termination, which is signaled by one of the three codons, UAA, UAG and UGA (Tate and Caskey 1974). The newly synthesized enzyme (NSE) is released from the P-site and its accompanying peptidyl transferase (PT) by interaction with "termination" or "release" factor(s) (RF) (Fig. 4.8) again with the aid of GTP. Tate and Caskey picture the release taking place on the ribosome as follows:

$$\text{PT-NSE} \ldots \text{UAA} \xrightarrow{\text{RF}} \text{PT-RF-NSE} \ldots \text{UAA} \longrightarrow \text{(NSE)} + \text{PT-RF-UAA} \xrightarrow{\text{GTP}} \text{PT} \ldots \text{UAA} + \text{(RF)}$$

where "..." indicates that the termination codon is not directly complexed on the ribosome with the preceding factors and "( )" indicates release from ribosome. The *t*RNA is released and the ribosome breaks up into subunits releasing *m*RNA. The synthesis of a second molecule of enzyme is ready to start.

Nonribosomal Synthesis—Synthesis of Oligopeptides.—Not all biologically formed peptide bonds are synthesized *via* the ribosomes. Small peptides such as glutathione, some of the peptide antibiotics (gramicidins) and many cell wall peptides which are parts of peptidoglycans appear to be synthesized on multienzyme-complex templates (Lipmann 1973) analogous to fatty acid biosynthesis.

In Vitro Organic Synthesis of Enzymes.—A discussion of the manner in which nature goes about synthesizing enzymes would not be complete without at least mentioning that enzymes can now be synthesized in the laboratory *ab initio* from constituent free amino acids. This was first accomplished for the low molecular weight pancreatic ribonuclease almost 20 years ago by Merrifield (1963). The stepwise synthesis is carried out on a polymeric support (PS) according to the following scheme:

$$\text{NH}_2\text{-CH(R}_1\text{)-COO (A)} \xrightarrow{\text{B-HN-CH(R}_2\text{)-COO-[PS]}}$$

$$\underset{\text{(dipeptide)}}{\text{B-HN-CH(R}_2\text{)-CO-HN-CH(R}_1\text{)COO-A}} \xrightarrow{\text{-B}}$$

$$\downarrow$$

$${}_2\text{HN-CH(R}_2\text{)-CO-HN-CH(R}_1\text{)-COOA} \xrightarrow{\text{B-HN-CH(R}_3\text{)-COO-[PS]}}$$

$$\downarrow \qquad \downarrow$$

$$\underset{\text{(tripeptide)}}{\text{B-HN-CH(R}_3\text{)-CO-NH-CH(R}_2\text{)-CO-NH-CH(R}_1\text{)COOA}} \xrightarrow{\text{etc.}}$$

where A and B are carboxy- and amino-protecting groups and [PS] is acylating agent attached to a polymeric support. Thus the enzyme is synthesized from the carboxyl end to the amino end. The method combines the advantages of using excess reagent to drive each reaction to near 100% yield without having the reagent present in the product. By the time this appears at least 25 enzymes and other proteins will have been synthesized *ab initio* and probably several-fold more by fragment condensation.

In addition to providing formal and complete verification of enzyme structure, this historically important advance in enzymology is useful both academically and perhaps practically. Thus, the three aspects of enzyme action, specificity, catalysis and regulatory properties, are studied by observing what happens when one amino acid or a peptide fragment is omitted or replaced. Unless a particular enzyme is difficult to obtain and has overwhelming medical importance, solid state peptide synthesis will probably be used mainly commercially for the synthesis of peptides for medical purposes. See Fridkin and Patchornik (1974), Erickson and Merrifield (1976) and Robson (1980), who discusses "scaffold" design, for reviews.

Aspects of Translational Regulation.—Lehninger (1975) pointed out that the regulation of enzyme synthesis operates mostly at the level of transcription rather than at that of translation, in keeping with the idea that it would be wasteful to make something and then suppress its activity (i.e., feedback inhibition works best on the first enzyme of a pathway). Some specialized mechanisms for regulation at this level have nevertheless been detected. Many are involved in suppressing the production of the "hardware" for translation. Thus, the formation of ribosomes is dependent on amino acids available by being indirectly tied to cyclic AMP mechanisms and especially to "magic spots," guanosine tetra- and pentaphosphate. These somehow turn off (stringent control) the message for the synthesis of ribosomes in a starving organism, one whose amino acid intake is limited. Not only do these nucleotides prevent the formation of ribosomes when the latter are not needed, but they also prevent the formation at site-of-synthesis of membrane phospholipids which are also not needed (Silbert 1975).

The plant growth regulators, the kinetins, may regulate enzyme synthesis by "inducing" and then modulating a specific enzyme pattern of amino acid—*t*RNA synthetase (Kahlem *et al.* 1975). There is also evidence for "latent" *m*RNA. Apparently control of enzyme synthesis may be dependent on the relative amounts of simple ribosome and polysomes, strings of closely spaced ribosomes. Enzyme synthesis proceeds more efficiently in polysomes. The proportion of polysomes in turn may be dependent on the presence of some plant growth regulators (Romani 1976). Controls at the translational level as well as the transcriptional level may be regulated by the types of enzymes made (Lodish 1976). The latter author stresses the role of viruses in elucidating this area.

POSTTRANSLATIONAL AND OTHER ASPECTS OF REGULATION

Posttranslational Modification

As the enzyme comes off the ribosome it spontaneously folds into its tertiary structure. Further modifications include: (1) the oxidation of SH of cysteine to form SS bridges, (2) oxidation of proline to hydroxyproline, (3) limited proteolysis, (4) assembly of subunits to form quaternary structures, (5) phosphorylation, and (6) combination with the prosthetic group—sugars and oligosaccharides. A host of covalent modifications of the 20 amino acids specified by the genetic code results in about 140 different amino acids as protein constituents (Uy and Wold 1977).

Cellular Aspects of Enzyme Synthesis

Site of Synthesis.—In bacteria (as indicated in Fig. 4.6) protein or enzyme biosynthesis occurs close to the site and synchronously with transcription. Thus, one end of the *m*RNA may be in the process of being formed *via* transcription from the DNA while the other end is being translated with enzyme on a ribosome. In multicellular eukaryotes, *m*RNA is manufactured in the nucleus and goes to the rough endoplasmic reticulum (ER), a vesicular membrane system which pervades the cytoplasm. Polypeptides synthesized on the ribosomes attached to the membrane can pass through it and are collected in closed sacs of the ER to become "free." Other enzymes are enclosed in sacs of the ER.

Packaging and Delivery.—Quite recently investigations have revealed the presence in the newly synthesized enzyme of a "zip code," a short peptide chain, which is eventually snipped off by a "signal" peptidase. This peptide serves as a guide which allows the enzyme to find its way to its final destination (Marx 1980B) in conjunction with special membrane receptor sites. For other enzymes, covalently attached carbohydrate and/or phosphate and transformations thereof serve, as expressed by Marx, as "address markers" for intracellular traffic.

As pointed out by Mercer (1968), in multicellular organisms the cells are specialists, i.e., they are highly differentiated and most of the decoding machinery is shut off. In some cells free ribosomes predominate and the enzymes collect in the cell sap to form specialized structures or accumulations on which the differentiated function depends. For example, the hemoglobin in red cells converts them into oxygen transports; the fibrous proteins in muscle cells are the basis of contractility; the toughened proteins of epidermal cells form a protective coat. In other cells a variety of functions are based on proteins originating from bound ribosomes and therefore enclosed in branches or vacuoles of the ER. These activities (summarized in Fig. 4.9) usually involve the participation of an important organelle, the Golgi apparatus or bodies, which may be regarded as a specialized region of the ER not associated with ribosomes. The various pathways of packaging and delivery involving the Golgi are summarized in Fig. 4.9. The diagram is a composite of several pathways not all seen in a differentiated cell.

In each case the enzyme or other protein is collected from the ribosomes in a sac of the ER

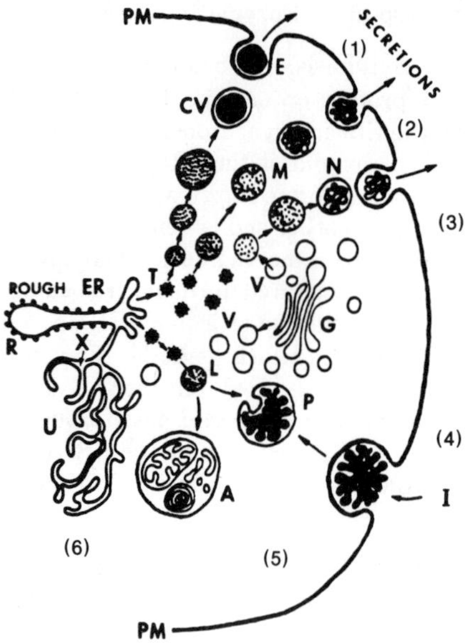

From Mercer (1968)

FIG. 4.9. ENZYME BIOSYNTHESIS—POSTTRANSLATIONAL FINISHING, PACKAGING AND DELIVERY

Depicted are events in a typical eukaryotic cell. See text for further details.

from where, according to its composition and role, it may follow any of the several courses: (1) Extracellular secretion of enzymes, etc. The protein is transferred by small vesicles to condensing vacuoles (CV) derived from the Golgi (G) and is passed ultimately through the plasma membrane PM at E. To facilitate delivery, microbial extracellular enzymes possess little or no disulfide crosslinks so that the enzymes can pass through the microbial envelopes in snake-like fashion and then assume their tertiary structure. (2 and 3) Enzymes polymerizing polysaccharides move to Golgi vesicles (M and N) where the monomers enter and are polymerized. In (2) polysaccharide only is formed; in (3) polysaccharides are linked to a protein to form mucopolysaccharides (mucins). (4 and 5) Hydrolytic, degradative enzymes collect in Golgi vesicles to form lysosomes (L). These may be used either (4) to digest the contents of food vacuoles (P) formed by phago- or pinocytosis or (5) to digest damaged parts of the cell itself which have been walled off in an autophagosome (A). (6) Various synthetic enzymes may find their substrates collected in vesicular or tubular extensions of the ER (V), e.g., steroids and lipids synthesis.

At least one important food-related enzyme, glucosinolase, is collected in such a vesicular extension of the ER (Chapter 21).

Lysosomes.—This compartmentation and organelle formation, as mentioned before, may be considered as one aspect of the control of enzyme activity at the posttranslational level by control of accessibility of enzyme to substrate. Perhaps of special interest to the food scientist is that 0.5 μ-diameter sac of acid hydrolyases, the lysosome. Unlike other organelles, such as the mitochondrion and the chloroplast which can make their own circular and catenated DNA which in turn can code for some of their own proteins (giving rise to the idea that a prokaryote cell is an evolved symbiotic collection of modified, interdependent unicellular microorganisms), all of the proteins in the lysosome are derived from transcription and translation of nuclear DNA. Medically these lysosomes are involved in the etiology of arthritis and at least 25 diseases in each of which a deficiency of a specific lysosomal hydrolase (degradation of complex lipids, mucopolysaccharide and glycoproteins) occurs (Neufeld *et al.* 1975; Dingle *et al.* 1979).

The convoluted interrelations between the functioning lysosomes and regulation of enzyme synthesis are illustrated by an example from gluconeogenesis (Pontremoli *et al.* 1975). During starvation or exposure to the cold not only is there a stimulation of the hormones involved in glygogen synthesis but also an increase in the number of lysosomes in the cells of animals so treated. These lysosomes contain specific proteolytic enzymes, which are under allosteric control and which as we mentioned earlier catalyze a limited hydrolysis of that key gluconeogenic enzyme, fructose 1,6-biphosphatase, rendering it now more sensitive to allosteric activation by histidine. Other aspects of the relation of lysosomes to hormone action and as regulators of the activity of other enzymes are discussed by Szego (1974) and Altman and Katz (1976). Corresponding organelles in plant cells may play a significant role in senescence and wound healing (Matile and Winkenbach 1971). Their roles as determinants of food quality will be frequently discussed in the ensuing chapters.

Isozymes and Enzyme Regulation

We have already explored multiple forms of enzyme or isoenzymes in relation to specificity. They also possess variable regulatory properties, the expression of which leads to the proper functioning of the cell and in turn the whole organism. The following provide a sampling of this particular aspect of isozymology.

An example of how isozymes ensure that an essential metabolic pathway be modulated but not completely cut out is that of the synthesis of the amino acid threonine in bacteria. Aspartate kinase, which catalyzes the first reaction unique to this pathway, exists in three multiple forms. However, only one of these is regulated *via* allosteric feedback inhibition by the end product, threonine.

Isozymes can also solve the problem of how to regulate the level of diverse end products whose synthetic pathways have in common the same chemical reaction. Thus, the first unique step in the synthesis of the aromatic amino acids tyrosine, tryptophan and phenylalanine is catalyzed by three different phospho-2-keto-3-deoxyheptanoate (PKDA) aldolases

$$\text{-OOC}\cdot\text{C}\cdot(\text{OPO}_3\text{H}_2^-)\text{=CH}_2 +$$
phosphoenol pyruvate
$$\text{OHC}\cdot(\text{CH}_2\text{OH})_2\cdot\text{CH}_2\text{O}\cdot\text{PO}_3\text{H}_2^- \rightarrow \text{PKDA}$$
D-erythrose-4-phosphate

Each aldolase is regulated both by allosteric feedback inhibition and by gene repression by only one of the three aromatic acids.

The now classical isozymes, lactate dehydrogenase (LDH), catalyzing the last step in glycolysis, provide another level of regulation of glycolysis/gluconeogenesis in mammalian organs, in addition to those already discussed. Now, each of the five isozymes of the tetrameric enzyme molecules is comprised of a combination of two distinct kinds of subunits or protomers, designated as M and H: M_4, M_3H, M_2H_2, MH_3, H_4. The latter, H_4, predominates in heart muscle and M_4 in skeletal muscle. M_4 has a high V_{max} and low K_m so that it can readily form lactic acid, and thus can rapidly supply energy anaerobically derived from glucose during muscular contraction.

In the heart, sudden energy is not needed and glycolysis proceeds steadily. In contrast to skeletal muscle, most of the pyruvate formed is funneled into the TCA cycle and practically no lactic acid piles up. This is ensured by kinetic properties of H_4 which are the reverse of M_4, a low V_{max} and a high K_m. Also, H_4, but not M_4, exhibits substrate inhibition, which thus acts as a "governor" on the small amount of pyruvate going to lactate. However, the fact that the system nevertheless is present suggests that it provides a safety valve. In emergencies the heart can draw on the energy reserve supplied by the lactate dehydrogenase system.

It was only relatively recently that a precise pinpointing of the principal difference between M and H isoenzymes of LDH in terms of the three-dimensional structure of this prototype of an oligomeric intracellular enzyme (Pfleiderer 1978) was possible as the result of the collaborative effort of Eventoff *et al.* (1977). Their detailed diagrammatic representation of the active site of LDH in its "ES" mode while binding to its cosubstrates, NAD^+ and lactate, is shown in Fig. 4.10 (compare with Fig. 4.3). They point out that the change of a relatively few amino acids within the active site, rather than a dramatic conformational alteration (the latter is effectuated by substrate-actuated "induced fit") accounts for the dramatic isoenzyme differentiation. For instance, the major catalytic difference is attributed to a replacement of Ala in the M-type subunit by Glu in the H subunit. The precision of our knowledge in this area is illustrated by the depiction in Fig. 4.10 of an interaction between residue 31 and the nicotinamide phosphate moiety of NAD^+ which occurs only with the H_4 isoenzyme. Food scientists have examined the roles of this enzyme in meat quality (Chapter 26).

Invertase, an important food-associated enzyme, exists in yeasts in multiple forms (Gascon and Ottolenghi 1967). The invertase isoenzymes in the interior are small and are unaffected by cell wall removal or by variations of growth conditions whereas the activity of the larger isozymes on the exterior of the yeast will change 1000-fold with changes in glucose concentration.

There is a vast and controversial literature on the isozymes of peroxidase in relation to growth, morphology, hormonal regulation, cellular differentiation and senescence in plants, including such foods as vegetables (Evans 1969), fruit (Frenkel 1971) and cereals (Tao and Khan 1975). Isozyme regulation is also of considerable interest to students of evolution and medically oriented investigators and other biologists (Markert 1975). The utility of isozymes in the food quality control laboratory will be explored in Chapter 37.

RELATION OF ENZYMES TO OTHER BIOLOGICALLY ACTIVE PROTEINS

We have conducted our discussion of the nature of enzyme action in Chapters 2, 3 and 4 by highlighting those salient features which distinguish this class of proteins from other biologically

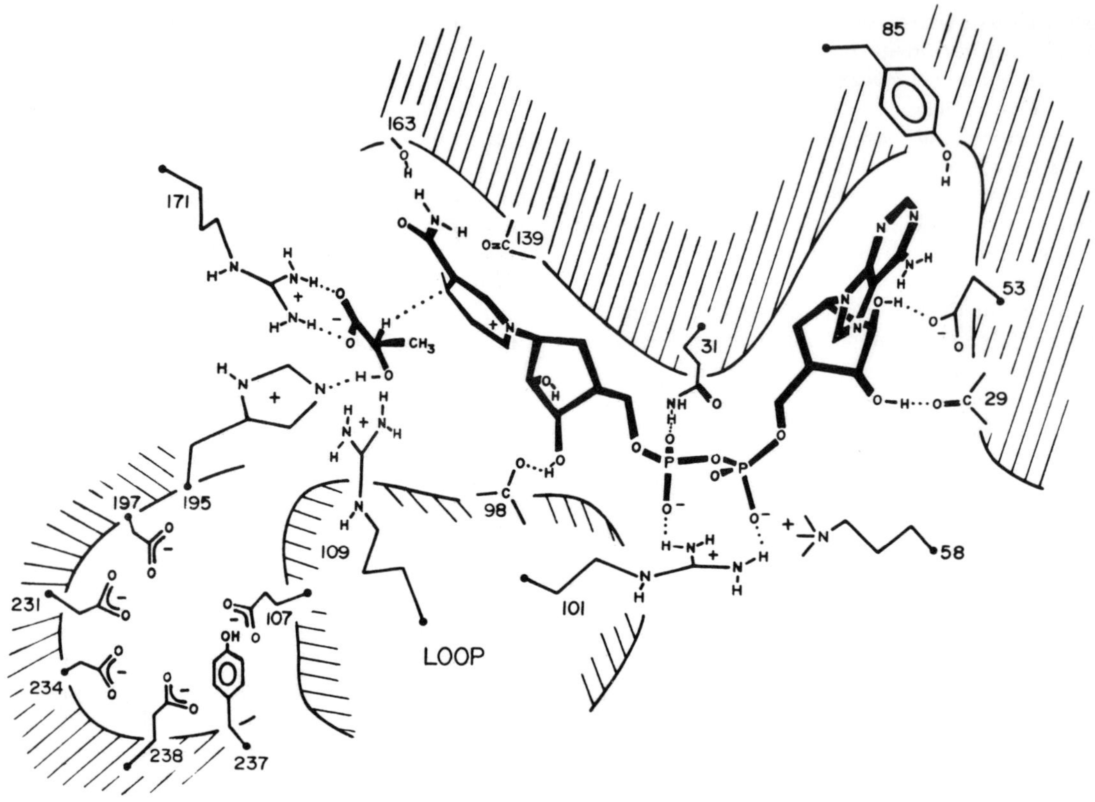

From Eventoff et al. (1977)

FIG. 4.10. ENZYME REGULATION BY ISOZYME DIFFERENCES AT THE ACTIVE SITE
Depicted are the changes in conformation of the active sites (shaded) of lactate dehydrogenase isozymes while binding substrate. Interaction between active site and residue 31 occurs only with H_4 isozyme. See text for further details.

active substances, especially other proteins. In this discussion we have repeatedly encountered such proteins which played essential roles in enzyme action but in themselves did not fit the definition of enzyme—repressors, hormones, specifiers, modifiers, proteins comprising the machinery of protein synthesis. Some of these "nonenzyme" proteins act in conjunction with other proteins to perform certain biological functions. Thus, ATPases and enzymes such as phosphoenol pyruvate transferase are involved in bacterial membrane transport of sugars (Peterkofsky and Gazdar 1975; Kundug 1976). We conclude this survey of the nature of enzyme action by pointing out and stressing the similarity of enzymes to these other proteins. For instance, the similarity in structure of the following five biologically active glycoproteins has been pointed out by Guidotti (1976): (1) Na^+/K^+-ATPase (transport), (2) Ca^{2+}-ATPase (muscle contraction, meat texture, see Chapter 26), (3) anion-exchange protein, (4) acetyl receptor protein (nerve impulse transmission) and (5) rhodopsin (vision). We have already seen how all proteins, enzymes or not, are assembled from amino acids by essentially the same decoding machinery.

Induced Fit

Perhaps the most unifying concept in this integration-oriented viewpoint is that of induced fit or ligand-induced conformation of proteins. Thus, Luria (1975) stated that "in both the study of enzyme action . . . every biological process becomes a problem of protein structural chemistry; how protein molecules alter their configuration . . ." Koshland (1973,1976) felt that the flexibility of proteins, the shape changes they undergo, is the key feature of the regulatory control of all enzymes in the physiological process. Lumry (1974) adopted the viewpoint that most proteins, enzymes or not, are useful because of their ability to undergo conformational changes and may be looked upon as devices for storage and transfer of free energy (see below).

The magnitude of the change in shape varies widely. As we mentioned before, changes in chymotrypsin, which are quite small, have been likened to the small rearrangement in the positions of various oranges in a bag when it is kicked (Lumry 1974). One possible advantage of such changes is that they occur several orders of magnitude more rapidly than protein-protein interaction, which can take minutes for completion.

Oxidative Phosphorylation.—Ligand-induced conformation is being evoked to elucidate the mechanism of one of the most elusive phenomena in biochemistry, oxidative phosphorylation. The first more-or-less plausible concrete explanation, the chemiosmotic hypothesis put forward by Mitchell in the early 1960s, proposed that an enzyme, mitochondrial ATPase, catalyzes the synthesis of ATP from ADP and P_i as consequence of being activated by a pH (proton) gradient generated across mitochondria, chloroplast and bacteria by the electron transport system (Mitchell 1967, 1979; Avron 1977; Fillingame 1980). The process according to this concept, taken from the excellent exposition of Hinkle and McCarty (1978), is shown in Fig. 4.11. For this concept and its subsequent strong experimental corroboration chronicled by Hinkle and McCarty, and for its broadening to include the utility of "vectorial metabolism" as yet another mode of *in vivo* regulation, Mitchell was awarded the 1978 Nobel Prize in Chemistry.

The question of the ultimate phosphorylation step remains, but it seems that it involves ATPase action in its synthesis mode. Boyer *et al.* (1975) suggested that phosphorylation was a reversal of ATP hydrolase action and that the oxidation in perhaps the abolishment of the gradient provides the energy for a conformational protein change which favors ATP synthesis. Racker (1976) proposed that a Ca^{2+}-dependent ATPase catalyzes ATP synthesis but that the energy derived from ion-protein interaction (ligand-protein induced conformation), more specifically from proton-protein interaction, drives the formation of ATP. These views are synopticized in a collection of reviews by the leading investigators including Mitchell (Boyer *et al.* 1977).

Before we lose site of our food-oriented theme, we should underscore the circumstance that the molecular bases of perception of the sensory attributes of food color, flavor and texture probably all involve ligand or energy-induced changes in the conformation of protein and that these changes are transmitted as signals to the brain. An important, relatively new emerging mechanism for the sensing and adaptation activities of many of these non-enzyme proteins is *via* methylation and demethylation: DNA recognition in recombinant technology

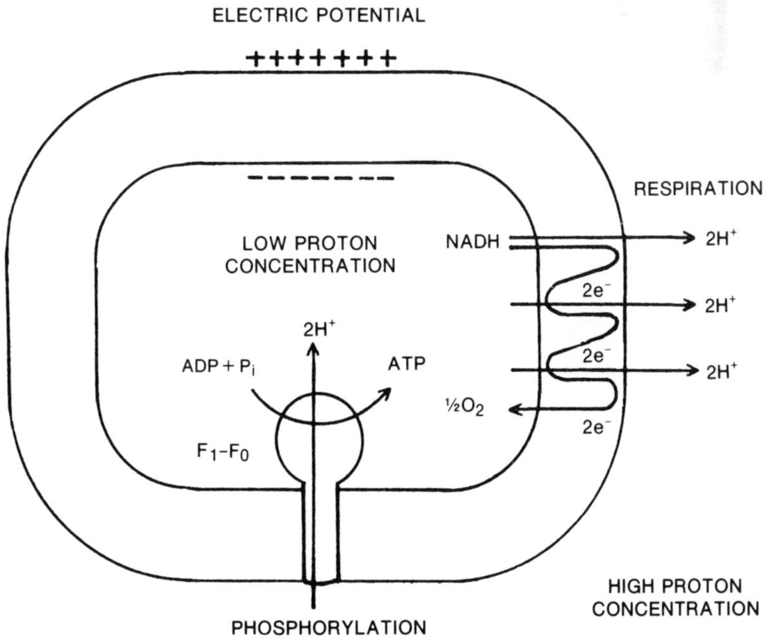

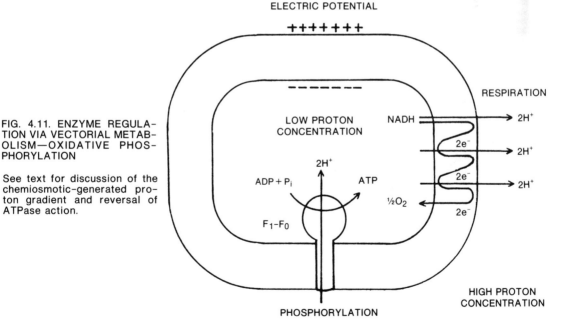

FIG. 4.11. ENZYME REGULATION VIA VECTORIAL METABOLISM—OXIDATIVE PHOSPHORYLATION

See text for discussion of the chemiosmotic-generated proton gradient and reversal of ATPase action.

From "How Cells Make ATP" by P. G. Hinkle and R. E. McCarty. Copyright © March 1978 by Sci. Am. All rights reserved.

(Chapter 5), histone function, chemotaxis (food-seeking "instinct") of bacteria and leucocytes, and neurological message transmission. The methylation and demethylation are themselves catalyzed by enzymes.

Bioenergetics.—Some investigators like to stress the role of the change of free energy in the function of biologically active proteins, including, of course, enzymes. We cited Lumry (1974) earlier. Knowles and Gutfreund (1974) consider proteins as devices for the transduction of energy.

Blondin and Green (1975) present a unifying model of bioenergetics in which enzyme catalysis, although a special manifestation of energy transfer, plays a central role. They are considered as electrical-charge separating catalysts in which the separation of charge in the enzyme (presumably in the active site) is coupled or paired with separation of charge in the substrate. What enzyme action has in common with the action of other proteins is that the latter are also involved in energy coupling and that in both cases protein facilitates charge substitution. According to the authors, in enzymes the role of metal prothetic groups or specially placed amino acids or even "thermally induced motions" are devices for initiating charge separation. Indeed, they consider enzyme catalysis and energy coupling as two sides of the same coin, the latter being a special form of enzyme catalysis. As previously mentioned these other biologically active proteins frequently act in conjunction with enzymes to perform a certain task. According to Blondin and Green, enzyme catalysis is a necessary ingredient whenever energy coupling involves the breaking or making of covalent bonds.

Part III

ENZYME PRODUCTION AND RELATED TOPICS

5

INDUSTRIAL FOOD ENZYMES—PRODUCTION, STABILIZATION, COSTS

Enzyme preparations as distinct food adjuncts—not as constituents of intact microbial cells or as substances secreted during food manufacture—have been used since man started to "engineer" the raw foods he/she gathered, killed or later cultivated and grew. Thus, South and Central American natives tenderized pieces of meat by folding papaya leaves around them or, as in Barbados, adding lumps of the green fruit to tough meat stew (Balls 1941). The same peoples discovered early that they could facilitate the processing of taro roots into fermented beverages with copious quantities of saliva. In the production of grain beverages, using only germinated barley in the process antedated the discovery that the malt could still be effective when added in relatively small amounts to unmalted barley previously cooked to gelatinize the starch. Crude preparations of what we now know to be amylases were used for the production of rice wines such as the Chinese yellow rice wines and Japanese saké long before microbes were being seriously considered as a source for the production of α-amylase and of other enzymes from microorganisms. This constitutes one of the first examples of the deliberate production of an "enzyme." We now know mammalian and fish organs to be rich sources of proteolytic enzymes, which were also used in tenderizing meat, clotting milk and in the preparation of fish sauces in Asia.

The first realization of the feasibility of producing commercial enzymes from microbial sources is usually attributed to Takamine—takadiastase from the *Aspergillus* fungi. Indeed, in spite of temporary set-back due to the adverse criticism occasioned by the detergent flap, microbes are now rapidly replacing plant and animal sources, with a complete replacement depending upon economics and the ability to replicate the nuances of alteration occasioned by the addition of the traditional enzymes and also, perhaps, by environmental considerations, i.e., use of waste products as enzyme sources.

Whatever the source, enzyme preparations are or will be purchased by food technologists and scientists first as catalysts for the preparation of food ingredients—sugars, syrups, nucleotide and glutamate flavor potentiators. Closely related but not within our purview is the exploitation of the enzyme systems of intact microorganisms as means for total synthesis or as a step in the manufacture of food additives such as amino, citric and glutamic acids (Crocco 1975; Böing 1976). Enzymes are, of course, bought for use in the food manufacturing process. The reasons for so doing may include the creation of identity, improvement of quality or simply improvement of efficiency and reducing operating costs such as in dough handling (Chapter 32) and juice and wine clarification (Chapter 30).

They are also purchased as reagents for analysis and testing the quality control process (Chapters 37–39). In the future we may see the use of enzyme preparations for solving environmental and energy problems and for the utilization of waste products of the food industry.

Probably the most important consideration determining whether or not commercial enzymes will be used in a food manufacturing process is simply whether economic value of the enzyme-induced improvement in quality or processing efficiency outweighs the cost of using the enzyme. Further mitigating circumstances impeding the expanded adoption of enzyme adjuncts include:

(1) natural inertia to innovation and modification of a time-honored process that is ostensibly satisfactory and certainly less risky than a relatively untested approach
(2) cost of technical services rendered by the enzyme supplier
(3) the optimum processing conditions—pH, temperature—may not coincide or may actually vary drastically from the optimum conditions for the action and stability of the added enzymes (such conditions may be optimal for undesirable microbial growth and contamination)
(4) correction of other safety and sanitation problems introduced, and
(5) legal and regulatory embroilments

Why Use Enzymes?

Alternatives to the application of commercial enzymes are either not to use them at all or use nonenzyme procedures which can accomplish roughly the same desirable goals. Nonenzyme approaches include chemical, physical and biological alternatives such as potentiation of the foods' own enzymes (Chapter 1) or using whole microorganisms, i.e., fermentative processes.

Underkofler (1972) lists the following idealized advantages of using enzymes (presumably as compared with analogous nonenzyme approaches) in food industrial products and processes. They are of natural origin and are, or should be, nontoxic. Because of their highly evolved specificity they catalyze reactions not easily carried out. These reactions proceed with fewer or without side reactions, working best under mild conditions of moderate temperatures and near neutral pH, thus not requiring drastically high temperatures, pressures and pH extremes. These latter conditions necessitate special expensive equipment and lead to undesirable side reactions. They act rapidly at relatively low concentrations. The rate of reaction can be readily controlled by adjusting pH, temperature, amount of enzyme employed and enzyme affectors. Their influence can be promptly removed by various modes of inactivation when the reaction has gone as far as desired.

A comparison with fermentative processes suggested to Edwards (1972) that use of enzymes, whether produced by microbial fermentation or by other means, circumvents some of the problems encountered in purely fermentative processes, at least for the manufacture of food ingredients. Some of the advantages of enzyme use which ameliorate some of these problems include: (1) possible minimization of side reactions which includes growth of the microbe, (2) a possible higher rate of production, (3) a simpler and faster product recovery, (4) prevention of unwanted microbial contamination with appropriate inhibitors which could not be used in a fermentation, and (5) reduced waste products as well as reduction of the accompanying headache of how to dispose of them. It is perhaps no fortuity that so far (1980) the progress in the successful nonanalytical application of commercial exploitation of immobilized enzymes (Chapter 8) has been confined to the manufacture of food ingredients, some of which are also produced by purely microbial fermentation processes.

Having listed the advantages of using enzymes in general, we can now delineate the rationale behind the ascendency of microbial enzymes produced from microbial sources by pinpointing problems posed by the use of nonmicrobial sources. These include: (1) competition for glandular tissues of animals with other more expensive products made from a limited supply of the same glandular tissue, (2) scarcity of such sources, and (3) irregularity and nonpredictability of supplies of enzyme sources which may be subject to seasonal, climatic and other agricultural-related uncontrollable variables. These enzyme sources come from widely diverse locations which leads to nonuniformity of quality, loss in enzyme potency and increased costs incurred in shipping, gathering and assembling, and higher likelihood of contamination from insects and other filth. Large quantities of material are required for the production of relatively small amounts of enzymes. There is a steady decrease in the availability of some of the raw materials such as the fourth stomach of the calf for the preparation of rennet (Chapter 33), mounting increases in cost, especially in labor, relative to those incurred in producing enzymes from microbial sources; they cannot, at least to a degree, be tailor-made to satisfy the varying requirement of the user or buyer. Production cannot, as in microbial production, be expanded at will in response to increased demand or ease of separation of the desired enzyme activities. Just about any enzyme found in a food can be induced in a microorganism, and usually at much higher levels of activity so that the industrial production of the enzyme becomes commercially feasible. Finally, microbial enzymes may be salvaged as useful by-products from other fermentations such as the recovery of glucose oxidase in penicillin manufacture or that of pectolytic enzymes in the manufacture of citric acid.

Figure 5.1 epitomizes the essential steps now in current practice for the commercial preparation of food enzyme adjuncts from plant, animal and microbial sources.

Production from Nonmicrobial Sources

Animal.—The traditional source of enzymes from animals is the meat packing industry. The animal organs and the type of enzyme activity utilized include stomachs as sources of pepsin and rennet to be used in cheesemaking.

The traditional source of rennet (Chapter 35) is the abomasum, the fourth stomach of the calf, but pepsin from the stomachs of older animals is also used and chicken stomachs are used as a source of a rennin-like pepsin in Israel.

Pregastric or oral tissues of the GI tract (mostly salivary of the lamb and kid) are sources of esterase and lipases (Nelson *et al.* 1977) for flavor production in dairy products (Chapter 33).

Pancreas, sold as pancreatin, is a source of enzymes which hydrolyze proteins, starch and fats. Although occasionally proposed to be used as a food adjunct, pancreatin is principally a pharmaceutical. Even diluted pancreatin preparations have objectionable flavors which may be eliminated in the future by the economical application of affinity chromatography (Chapter 7).

Liver is a source of catalase to remove excess hydrogen peroxide when the latter is used as sterilant in the cheese in manufacturing and also in conjunction with the use of glucose oxidase (Chapters 33 and 36).

Other proposed or potential animal sources of commercial enzymes include protease-rich fish processing waste (Chapters 28, 31) and the crystalline style of the clam, a potent source of a variety of glucanases (Jacober *et al.* 1980).

For most but not all commercial enzymes, the basic steps used to obtain the final product consist of dissolving the active enzyme in water or otherwise ensuring that the enzyme is soluble, removing

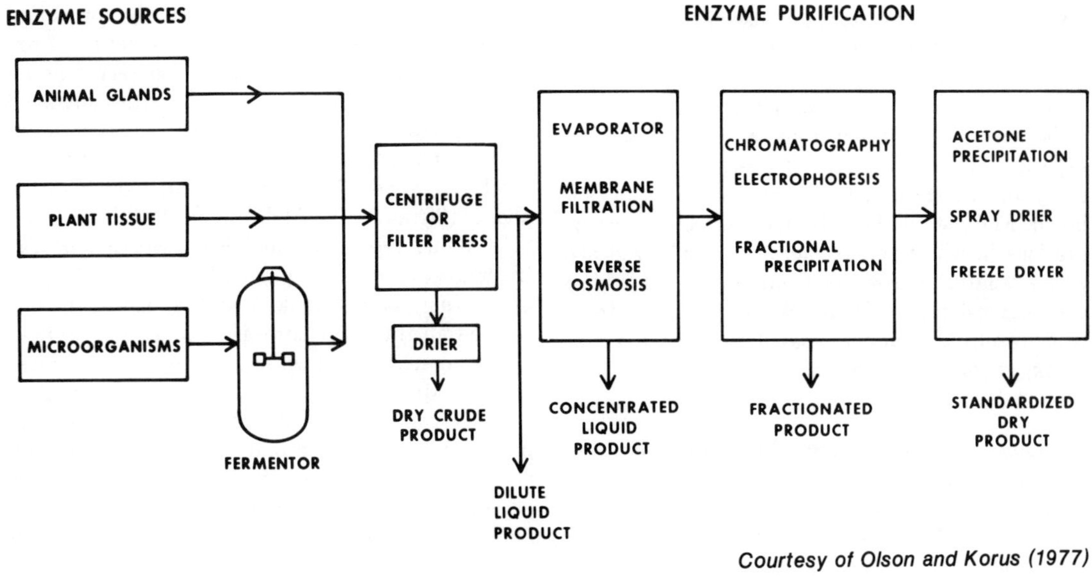

FIG. 5.1. PRODUCTION OF COMMERCIAL FOOD ENZYMES

the insoluble debris or whole cells usually by filtration or by centrifugation, and usually concentrating the enzyme activity by evaporation and/or precipitation. The production of animal enzyme preparations varies considerably with the type of tissue source. Thus, pancreatin is prepared by mincing hog pancreas in the presence of a small amount of duodenum to ensure conversion of the proenzymes (zymogens, inactive forms) to the active forms by means of the trypsin present. Usually the enzymes are not extracted but are simply concentrated by removing water in a vacuum drier and then removing lipids with a hydrocarbon solvent. For more active pancreatin, an extraction followed by acetone precipitation and drying is performed.

For the preparation of rennin, it is also necessary to ensure maximum conversion of zymogen to active enzyme. This is accomplished by allowing an extract of abomasal tissue—obtained by extraction of strips of mucosa at pH 5.2—to stand at pH 2.5, drying, grinding and reextracting in NaCl containing borate as a preservative (the optimum for the activating enzyme, pepsin). The extract is clarified by filtration and the enzyme is converted to a solid form by various procedures. Pepsin, prepared from adult hog stomach, is first activated and then extracted. In the preparation of pregastric lipase, a relative newcomer as a distinct industrial enzyme preparation (Farnham 1957), tissue from between the trachea and stomach of milk-fed animals is ground with an equal weight of milk powder and simply dried at not more than 43.3°C. The earlier history and use of this interesting commercial enzyme will be discussed in Chapter 33.

Catalase from liver is obtained by grinding beef or hog liver and purified from the fraction soluble between 35 and 50% (w/v) acetone. It is then sterilized and stabilized with glycerol and sold as a liquid. More stable but also more expensive preparations in which the extracts are lyophilized are also available. Lysozyme from waste egg white, used extensively in Japan, can be obtained from whey resulting from heat coagulation and removal of the denatured protein (Wilkinson and Dorrington 1975).

Commercial Enzymes from Flowering Plants.— By far the two major sources of enzymes from higher plants are the green papaya melon, for the preparation of products rich in proteolytic enzymes, and germinated barley, malt, which contains starch-digesting enzymes used in brewing and to some extent as an adjunct in breadmaking. Malt is used as such and is not further refined. It has functions in brewing besides serving as a source of amylases and other enzymes which contribute to the brewing process. As a means of converting starch to sugars in beermaking, malt is being rapidly replaced by microbial amylases in many countries. Production of malt will be further examined in Chapter 31. Plants which are added to foods primarily because of the action of enzymes in them, but are not refined further than making them into a flour after drying, include soybean flour (Chapters 17 and 32) and malted wheat kernel (Chapter 32).

Papain.—Although papain may some day also be replaced by microbial proteases for many food-related purposes, it is likely to be around for quite some time, perhaps because of socioeconomic reasons if nothing else; its source is confined entirely to the developing countries. As countries become affluent they lose interest in growing papaya for the sole purpose of manufacturing papain. Furthermore, papain will probably persist because of its widespread use for not only the food industry as we shall repeatedly have the opportunity to demonstrate in the ensuing chapters, but for other industrial and medically-related purposes enumerated in Chapter 1. These include uses, to mention only major industrial ones, for dry cleaning, as a preshrinking agent of textiles, for cleaning septic tanks and in leather tanning. It is still widely used as a digestive aid, for the prevention of postoperative adhesions and, until recently, for wound debridement. We delve into the production of papain in some detail because it also serves as an illustration of some of the complexities involved in nonmicrobial enzyme production.

Postwar imports of papain to the United States exhibit a steady 5-year cyclic variation between about 100 and 400 MT (Flynn 1975). In 1954 Schwimmer reported that Ceylon (Sri Lanka) was the largest producer of papain but that the Belgian Congo (now Zaire) would take the lead in the not too distant future. According to Jones and Mercier (1974), Zaire has indeed become by far the largest papain exporter. In 1971, however, according to Flynn, Uganda accounted for more than half of all papain exports from producing countries. Wolnak (1972) estimated that the 1971 dollar value of papain used in meat tenderizer formulations was about \$6/kg. The enzymes comprising commercial papain (Chapter 27) are present in a system of almost microscopic latex vessels directly beneath the surface of the green fruit. The specific activity of the fruit reaches a maximum in 3–5 months after planting (Skelton 1969; Madrigal *et al.* 1980). The latex is best tapped from fruits of young trees during warm weather in the early morning of a cloudy day, preferably after a rain. A sharp, nonmetallic object such as a sliver of glass or ivory bone is used to make three scratches since enzymes are sensitive to metals. The latex, as it issues from the open vessels, is a colorless fluid which soon turns milky and then coagulates after a few minutes. The fruit can then be tapped several times in a season before it ripens. Ripe fruit contains no papain. The drops of latex are collected in nonmetallic containers, formerly earthenware or porcelain pails, but probably now plastic and stainless steel pails. Latex coagulated on the fruit is scraped from the surface. It has been estimated that an average of 0.45 kg of latex per tree can be obtained (Fig. 5.2).

The latex is promptly dried. One kilogram of latex yields 200 g of crude papain. Drying procedures of varying degrees of complexity have been used and/or proposed. Sun-dried products, once the principal forms imported to the United States, are now legally barred. Improved methods include drying in moderate temperature ovens of various designs. It has been our personal observation that these white-to-deep-yellow products retain only a fraction of their original activities. Some products contain $NaHSO_3$ to activate and stabilize the enzymes. The enzyme activity can be stabilized with no loss in activity for over 20 years by vacuum drying in the presence of salt (Balls *et al.* 1941). The most successful commercial preparations obtained from Zaire are those in which the latex is refined immediately at the site rather than collected and shipped to a central processing area. The latex is stirred to liquefy the thixotropic gel, filtered, centrifuged and refiltered. It is then concentrated under vacuum, spray-dried and packaged. Although papain is generally recognized as safe (GRAS), it must now be partially purified by approved U. S. Food and Drug Administration procedures to be allowed to be shipped into the United States.

From Balls (1941)

FIG. 5.2. COLLECTION OF PAPAYA LATEX IN PAPAIN PRODUCTION

Source literature on papain as a commercial enzyme includes a series of papers edited by Tainter (1951), reviews by Schwimmer (1954) and Becker (1958) and more recently by Wolnak (1972). Jones and Mercier (1974) present more recent developments on commercial enzyme purification. The latest (to us) available discussion by Flynn (1975) is probably the most thorough assessment and analysis ever presented of production, import/export distribution and market potential of papain.

Other Plant Proteases and Nonbacterial Enzymes.—Many other plants besides papaya possess considerable proteolytic activity, but only pineapple (bromelin) and perhaps figs (ficin) are available as sources which can now be used with foods. Commercial food grade bromelin is produced in Taiwan from waste pineapple stems. The ash gourd and other plants have been proposed as sources of rennet (Sardinas 1976; Gupta and Eskin 1977). Table 5.1 lists some of these, as well as others which are not yet of economic importance (see also Chapter 33).

In the pages to follow we shall encounter occasional proposals for the use of other nonmicrobial sources of enzymes. Illustrative are orange peel as a source of pectin esterase for the manufacture of low-methoxyl pectin (Leo and Taylor 1962) and, as previously mentioned, waste egg white as a source of lysozyme. Lysozyme has been proposed for "hu-

TABLE 5.1

PROTEINASES IN FLOWERING PLANTS[1]

Scientific Name	Common Name	Enzyme	Parts of Plant	Sulfhydryl Enzyme
Carica papaya	Papaya	Papain	Latex	+
Ficus carica	Fig	Ficin	Latex	+
Anana sativa	Pineapple	Bromelin	Fruit, leaves	+
Asclepias mexicana	Milkweed	Asclepain	Latex	+
Euphorbia cerifera	Caper	Euphorbain	Latex	+
Soja hispidus	Soybean	Soyin	Sprouts	+
Pileus mexicanus	Cuaguayote	Mexicain	Leaves, fruit	–
Solanum eleagnifolium	Horse nettle	Solanain	Fruit	–
Hura crepitans	Jabillo	Hurain	Sap	–
Maclura pumifera	Osage orange	Pomiferin	Fruit	–
Arachis hypogea	Peanut	Arachain	Seed	–
Cucurbita pepo	Pumpkin	—	Fruit	–
Sarracenia purpurea	Pitcher plant	—	Flower	–
Drosera rotundifolia	Sundew	—	Flower	–
Cucumis melo	Melon	—	Fruit	–
Jarilla chocola	—	—	Fruit	+[2]
Benincasa cerifera	Ash gourd	—	Fruit	–[3]
Bromelia pinguin	Maya	Pinguinain	—	+
Cynera cardunculus	Prickly artichoke	Cardo rennet	Fruit	–[4]
Withania coagulens	—	—	—	+
Actinidia chinensis	Chinese gooseberry	Actinidin	Fruit	+[5]
Zingiber officale	Ginger	Zingibain	Rhizome	+[6,8]
Ulva lactuca	Sea lettuce	(Algae)	—	. .[7]

[1] Source: From either Schwimmer (1954) or Sardinas (1976) or as indicated; (–) indicates either not a sulfhydryl enzyme or unknown.
[2] Source: Tookey and Gentry (1969).
[3] Source: Gupta and Eskin (1977).
[4] Source: Barbosa *et al.* (1976).
[5] Source: Macdowall (1973).
[6] Source: Thompson *et al.* (1973).
[7] Source: Abdel-Fattah and Edrees (1973).
[8] Source: Ichikawa *et al.* (1973).

manizing" cows' milk and as a preservative in cheeses, saké and other foods (Chapter 36).

PRODUCTION OF MICROBIAL ENZYMES

Perhaps the most authoritative overviews of present-day commercial enzyme production are those of Underkofler (1966,1972,1976). A compendium of patents is available (Gutcho 1974). Other discussions and reviews representing different points of view and stressing different aspects include those of Beckhorn *et al.* (1965), Blain (1975), Melling and Philips (1975), Meyrath and Volavsek (1975) and Böing (1976). Books edited by Wiseman (1975), Spencer (1976) and Wang (1979) contain chapters on enzyme production engineering. Others will also be cited as they appear. However, as pointed out by Underkofler (1976), all major commercial enzyme manufacturers have kept and must continue to keep details of their processes entirely confidential in order to maintain a competitive position. Even in patents, only laboratory or relatively small-scale examples are usually cited and specific conditions for large-scale production may vary considerably. Even the specific procedures published by Underkofler of two detailed industrial processes, for bacterial α-amylase and fungal glucoamylase, were replaced. More recently an extensive Russian literature has appeared.

Selection of Microorganisms

According to Underkofler (1976), the most important single factor in the production of industrial enzymes is the selection and maintenance of high yielding and stable cultures. As seen in Table 5.2, commercial enzymes are produced aerobically from mold, bacteria and yeasts. In the United States for use in foods, enzymes derived from strains of *Aspergillus niger*, *A. oryzae* and *Bacillus subtilis* require no further clearance. Enzymes from certain *Aspergillus*, *Bacillus* and *Saccharomyces* groups are designated by the U. S. Food and Drug Ad-

TABLE 5.2

SOURCE ORGANISMS FOR SOME COMMERCIAL ENZYMES

Organism	Enzymes
Molds:	
Aspergillus oryzae	amylase, protease
Aspergillus niger	amylase, glucoamylase, cellulase, pectinase, glucose oxidase, catalase
Aspergillus saitoi	protease
Trichoderma reesei	cellulase
Rhizopus species	amylase, glucoamylase, pectinase, lipase
Penicillium species	pectinase, lipase
Mucor miehei, M. pusillus	microbial rennin
Endothia parasitica	microbial rennin
Bacteria:	
Bacillus subtilis	amylase, protease, penicillinase
Escherichia coli	asparaginase
Streptomyces species	glucose isomerase
Arthrobacter species	glucose isomerase
Streptomyces griseus	protease
Micrococcus lysodeikticus	catalase
Streptococcus hemolyticus	streptokinase-streptodornase
Yeasts:	
Saccharomyces cerevisiae	invertase
Saccharomyces fragilis	lactase

Source: Underkofler (1976).

ministration as GRAS. Production of industrial enzymes from *Bacillus* species has been reviewed by Fogarty and Griffin (1974) and Ingle and Boyer (1976) and of microbial alkaline enzymes by Kelly and Fogarty (1976). Milk clotting enzymes obtained from *Rhizopus, Mucor, Endothia* and *Micrococcus* have approval by food additive orders; to obtain the latter, petitions giving adequate information regarding safety must be supplied to the U. S. Food and Drug Administration. The necessary animal experimentation to demonstrate safety can be quite expensive.

Among other considerations which go into selection of a microorganism, Neubeck (1976) suggests that the following are of major importance: (1) ease of growing the organism, (2) concentration of the desired enzyme relative to the other products of growth, (3) presence or absence of undesirable factors in the organism such as pathogenicity, (4) toxic product formation, presence of undesirable enzymes, stability of the organism (it must be capable of consistently high production of the desired enzyme and does not easily mutate spontaneously), and (5) ease of removal of the desired activity from the cell mass.

This is not the place to present details of selection methods, but we would like to mention that screening methods constitute a most important procedure in the selection process. Microbes are isolated from sources rich in the substrate of the desired enzyme. Strains of the microorganism are grown as pure cultures and further selection is based on the assay of the enzyme activities of the enzyme produced. Further strains may be produced by mutation as discussed below.

Factors Influencing Yield

Once the organism is selected, a great deal of attention must be given on how to optimize yield. The classical environmental variables which have to be empirically and systematically varied are pH, temperature, medium composition (including the use of surfactants), aeration (oxygen transfer), stage of growth cycle and addition of inducers, especially the addition of those which are the substrates. Illustrative of the yield improvement is the study by Schwimmer and Garibaldi (1952) on *B. macerans* cyclodextrin transferase production. By optimizing the supply of air, ammonium ion and trace elements, they obtained a ten-fold increase in activity and reduction of fermentation time from 2 weeks to 12 hr. Specific recipes for the production of proteinases, amylases and pectinases are available (Meyrath and Volavsek 1975). It is quite probable that they should be useful for laboratory enzyme preparation but are not recommended as competitors to the closely guarded recipes of the larger enzyme manufacturers. The following catalogues further manipulations which may increase enzyme production (Demain 1973):

(1) Selection of organism.
(2) pH.
(3) Aeration (O_2 transfer).
(4) Temperature.
(5) Medium composition.
  (a) Surfactants.
  (b) Trace metals.
(6) Growth rate.
(7) Stage of growth cycle.
(8) Induction.
(9) End product inhibition = feedback repression.
(10) Catabolite repression.
(11) Gene dosage.
  (a) Episome transfer.
  (b) Phage escape synthesis.

It will be noted that many of them involve intervention in the regulatory processes discussed in the last chapter and some operate at the gene level. One of the most fruitful and probably the most highly guarded procedures is that used to obtain mutants.

Mutations.—Application of mutation to industrial microbiology goes back to over 30 years ago when X-rays were used to obtain strains of molds which gave improved yield of penicillin. Most mutants developed today are selected at random and developed empirically.

As pointed out by Outtrup (1976) in her cogent precis on the use of mutations in enzyme fermentations, commercial enzyme mutation involves problems which are entirely different from the ones often dealt with by university scientists. In addition to increasing yield, mutation is helpful in achieving the following which serve as the basic guidelines in the improvement of enzyme production: (1) safety to the manufacturing personnel and users of the products, (2) minimization of pollution, (3) economy of the process, and (4) optimization of product quality. By using mutations some of the major limitations and problems encountered in immobilized enzyme utilization might be partially overcome. Also, it has been possible to eliminate the accompanying production of undesirable side products such as potential carcinogens and antibiotics. Another effect helpful to maintain secrecy is to convert microorganisms which are sporulators to asporogenic strains. It should also be possible to influence the *in vivo* stability of the enzyme. The problem of "disappearing" enzymes *via* degradation and inactivation during microbial growth is discussed by Thurston (1972). We shall discuss *in vitro* stability shortly. Mutants can be obtained which produce less odor than the mother strain. Others thoroughly exhaust nutrients. This reduces the BOD and helps reduce pollution.

Mutations improve efficiency and lower production costs by reduction of medium requirements and purification costs. For instance, conversion of an induced to a constitutive enzyme eliminates the cost of adding expensive inducer. Further savings can be effected by producing mutations which result in facilitation of further steps in enzyme production such as the elimination of slime, more easily filterable mycelia, better flocculation and easier concentration. How mutants help "tailor-make" enzymes suitable for an adapted, particular end-use is shown in Table 5.3. In some cases as shown it is desirable to develop rather than suppress side activities.

Improvement of yield by mutation as observed in the laboratory is not always duplicated on a large scale. Furthermore, mutants which under laboratory conditions improve yield will, upon scale-up, give even better yields. This is at least in part due to control equipment that cannot be duplicated in the laboratory.

Both Outtrup and Underkofler stress that the method of mutation should result in a large number of mutants which then can be rapidly screened to obtain results quickly. The screening method preferred by Outtrup is the "top agar layer" method in which an active wild or parent strain does not exhibit any detectable activity under conditions of the assay. This eliminates all mutants which have activity equivalent to or less than the parent. Using this method she was able within a short interval to increase the activity/milliliter of an already very active parent amylase producer 20-fold in a seven step mutation. Alternatively, one may employ a stochastic "shotgun" gambling approach by selecting auxotrophs resistant to an-

TABLE 5.3

ADAPTING COMMERCIAL ENZYMES TO END-USE BY MUTATION: TAILOR-MADE ENZYMES

End Use	Enzyme Present in Preparation[1]			
	α-Amylase	Neutral Protease	Alkaline Protease	β-Glucanase
Desizing[2]	+	+	+	+
Brewing	+	+	–	+
Starch thinning	+	–	–	+
Baking	+	+	–	–
Barley brewing[3]	–	–	–	+

Source: Adapted from Outtrup (1976).
[1] (+) Indicates that the enzyme is present or that its activity is enhanced; (–) indicates absence or decrease.
[2] Parent or wild strain before mutation.
[3] For use in postmashing stage (Chapter 31).

tibiotics which inhibit the growth of the parent. Another roll-of-the-dice stratagem is to isolate enzyme-negative mutants and subsequently revert them, in the hope that the enzyme gene on the rebound has been duplicated.

Agents used for inducing mutations in industrial microorganisms include chemicals such as ethylene imine (EI) and N-methyl-N′-nitroso-guanidine (NTG) alone or in combination with either X-rays or UV radiation. In general, repeated reapplication of one mutagen will result in achievement of an activity plateau whereas alternate application of combinations can continue to boost the yield of enzymes considerably (Bailey and Markannen 1975). Successive treatment of an already high-yielding mutant strain of the GRAS organism *B. subtilis*, first with UV + EI, followed by selection of high-yielding clones and treatment of these clones with NTG + the antibiotic chloramphenicol (which blocks peptidyl transfer in ribosomal enzyme synthesis) resulted in a three- to five-fold boost in α-amylase production.

Molecular Biology of Mutation.—While the molecular basis underlying the successful application is far from being thoroughly understood, some investigators have successfully pinpointed the varying loci within the bacterium on the genes, the DNA, of some of the mutations. Most of this work has been done with the well-explored *E. coli* (Demain 1973; Pardee 1973). Thus, there are mutations which can increase both total number of enzyme molecules in the cell whereas others result in the increase in the activity of each enzyme molecule. In the first case this is accomplished by mutations in the regulatory genes—the repressor, operator and promoter segment of the DNA of a given genome (Chapter 4). As the result of such mutations which increase the affinity of RNA polymerase for the P-gene (promoter), rates of synthesis of the desired enzyme can be increased ten- to 50-fold and the mutation leads to the channeling of over 5% of all protein synthesis into β-galactosidase.

By mutating the I-gene which codes for the repressor protein of the β-galactosidase operon, 3% of the total protein produced can be diverted into biosynthesis of this enzyme. Mutation at an analogous I-gene locus was probably also responsible for the formation of a variant of the blue-green algae *Rhodopseudomonas spheroides*, 25% of whose protein is catalase (Clayton and Smith 1960).

Mutations in the operator or O-gene over which the RNA-polymerase must pass in order to start transcribing the coded message of the structural gene, can prevent the product of the I-gene, the repressor, from combining with the O-gene. This in effect converts an inducible into a constitutive enzyme. As mentioned below, this can also be done by recombinant technology. As previously mentioned, such a transformation obviates the necessity of adding expensive inducers to the growth medium. Finally, as outlined, means exist for increasing enzyme production by influencing the structural gene, usually by introducing several copies of the structural gene into the bacterium.

Recombinance and Genetic Engineering.—The approach which now is receiving the most attention in the press if not in applied molecular biology is that of introducing a structural gene into a bacterium *via* that aspect of genetic engineering known as recombinant DNA technology (Grobstein 1977; Sinsheimer 1977; an entire issue of *Science*, Sept. 19, 1980, is devoted to recombinant DNA). Not only can multiple copies of the gene be introduced but the genes from completely unrelated species, even those from higher organisms, can be introduced into microorganisms as shown in Fig. 5.3. To accomplish this, several interesting enzymes are needed. These are said to recognize, nick, create "sticky" ends, excise, mortise-cut, insert, patch, stitch, and repair segments of DNA. Recognition mechanisms of DNA-specific enzymes were reviewed by Jovin (1976). Among these enzymes are two which we wish to discuss briefly: the restriction endonucleases, needed for the enzyme used in recognition and snipping, and one needed for the "finishing" step, DNA ligase.

The former, whose discovery warranted the 1978 Nobel Prize in physiology and medicine, is a complex multifunctional enzyme whose function in a bacterial cell is to destroy foreign DNA by recognizing specific nucleotide sequences therein. Host DNA is protected by selective methylation of its purine and pyrimidine bases, thus rendering the host DNA refractory to enzyme hydrolysis. These enzymes require ATP, Mg^{2+} and S-adenosyl methionine, the latter for its methylation function (Meselson and Yuan 1968; Yuan *et al.* 1975; Wilkie 1976; Roberts 1976; Murray 1976; Hadi *et al.* 1980). First one strand of foreign duplex DNA is cleared followed by scission of the second strand right opposite the first to give double-stranded fragments of equal length.

DNA ligase, also known as DNA polynucleotide synthetase, catalyzes the formation of a phosphoester bond at the site of a single-strand break of double-stranded DNA (Olivera and Lehman 1967) as shown in Fig. 5.4. As mentioned in Chapter 2, one of the multiple forms of DNA ligase can utilize the hydrolysis of NAD rather than ATP to provide the free energy for the synthesis of the phosphoester bond in the strand.

Prospects and progress on the future of the microbial genetic engineering were presented from various points of view (including the aforementioned empirical presently-used approach) at an American Society of Microbiologist's conference coconvened by Halvorson and Demain (1976). Included were discussions of hybrid plasmids (recombinants) for the amplifications of DNA and hence gene products (Helinski 1976; see also Helinski *et al.* 1977) including, of course, enzymes. Plasmids are bits of circular DNA not associated with the rest of the DNA in the bacterial cell. One of the drawbacks in the introduction of this revolutionary technique is the unfortunate choice of *E. coli*, an intestinal microorganism which is not now accepted by the U. S. Food and Drug Administration for the manufacture of enzymes as food additives. Curtis (1976) proposed genetic manipulation of *E. coli* for use as a safe industrial microorganism. The application of these newer techniques for the production of food enzymes may be considered as a spin-off from more pressing applications (i.e., nitrogen fixation, insulin production). Dixon (1975) reported that the U. S. National Science Foundation rejected the financing of a proposal to insert the gene coding for calf stomach prorennin into *E.* coli. In the first successful improvement of protein production *via* recombinant engineering, incorporation into *E. coli* (K-12 strain) plasmids of the operator gene (O-gene, see below) of the *lac* operon (Chapter 4), resulted in a 40-fold increase in the uninduced (constitutive) production of the important food-associated enzyme, lactase (β-galactosidase) (Heyneker *et al.* 1976). By the time this appears, ways to transcribe and translate the codes of recombinant DNAs of structural genes into enzyme protein will probably have been accomplished. Looking to the future, recombinant technology combined with sophisticated, continuous fermentation-engineering processes should result in a drastic reduction in microbial production costs relative to both yield and specificity of industrial enzymes so that tailor-made enzymes will be used much more extensively in food processing than at present. It may even diminish the pressure for the search for immobilized systems (Chapter 8). Other possibilities in this area include production of animal (Nyiri 1974) and plant cells grown in pure culture, analogous to microbial culture for the purpose of using their enzymes. Still further in the future Wiseman (1971) envisioned, some decades hence,

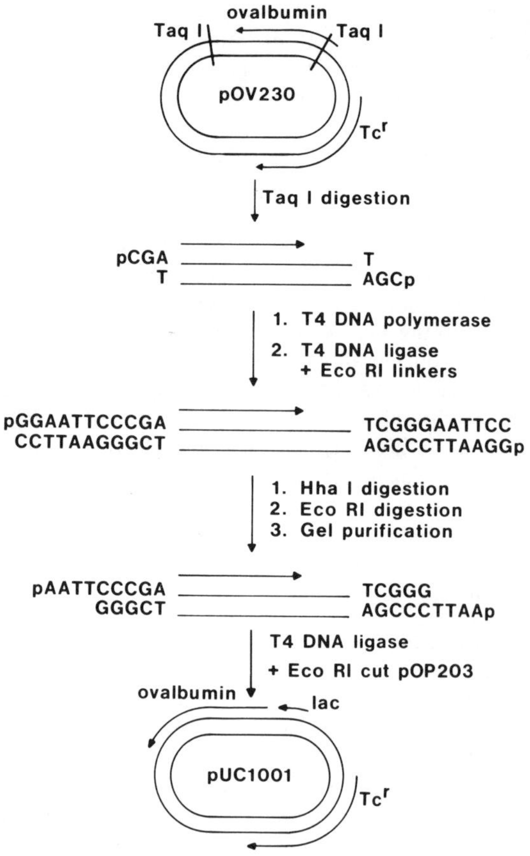

From Fraser and Bruice (1978)

FIG. 5.3. OUTLINE OF RECOMBINANT GENETIC ENGINEERING USING ENZYMES

In this particular application, the *lac* control elements are being fused to the ovalbumin gene. ECO RI, Taq I and Hha I are highly specific restriction endonucleases; pUC1001, *lac*-albumin fused plasmid; pOV230, plasmid containing all of the *m*RNA coding for the amino acid sequence of chicken ovalbumin gene. Other enzymes used include a DNA polymerase and DNA (nucleotide) ligase and alkaline phosphatase (not shown).

factory production of intracellular enzymes directly from ribosomes and other components of the machinery for protein biosynthesis. We believe that the suitable synzymes (Chapter 3) are more likely to be available before ribosomal factors. Continued food-related interest in these aspects of future progress is exemplified by a course in the Department of Nutritional Science, University of California at Berkeley by Chang (1976) on "Food Processing Enzymes: Prospects for Genetic Manipulation and Biochemical Engineering."

LARGE-SCALE OPERATIONS

Microbial Growth

Surface *vs* Submerged Culture.—The enzyme manufacturer must choose between these two principal general processes.

Surface Culture.—In this method, adopted from the Japanese for growth of fungi, wheat bran is mixed with *ca* 50% water, and the semisolid mass is steam-sterilized in rotating drums. Three growth configurations can be used: thin layer on trays, and thick layer ("high-heap" method) or drums in which the mold bran (koji) is slowly rotated. Adequate air is ensured by the porous nature of the soggy bran which also allows penetration of the fungal mycelia. Although the submerged process is considered more important, especially for growth of bacteria and yeast, the koji process is still used by most of food enzyme producers for obtaining glucoamylase for syrup production from *Rhizopus* strains, and proteolytic enzyme from *A. oryzae*. In addition to producing higher enzyme yields in these particular cases, surface culture still possesses some advantages over submerged culture in not having to take special precautions to prevent contamination, simplicity of operation and savings in energy. On the other hand, it does not lend itself to automation and continuous operation, resulting in high labor costs, and it requires much more space than do submerged set-ups.

Submerged Culture.—Techniques for large-scale microbial production of antibiotics some 30 years ago were rapidly adapted for the manufacture of enzymes. In modern, continuous fermentation configurations, previously sterilized liquid medium is introduced into sterile, closed fermentors whose capacity may reach 120,000 liters. Agitation and aeration as well as temperature and pH are carefully and automatically controlled. Compared to

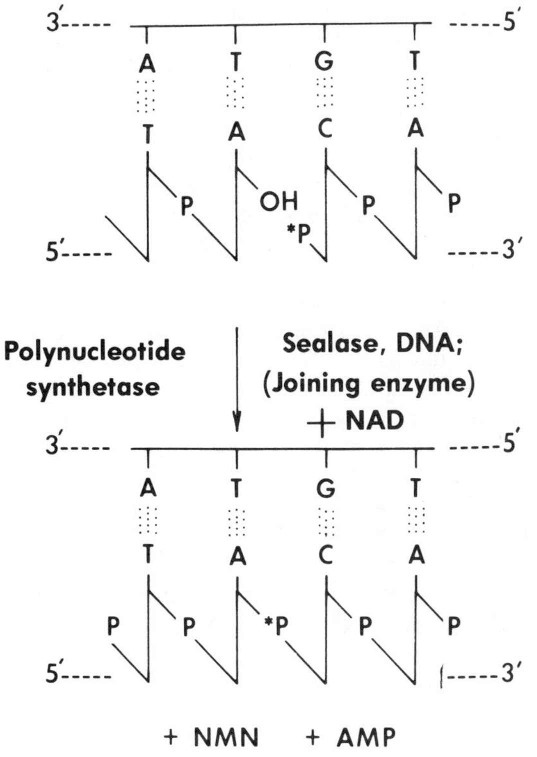

FIG. 5.4. ROLE OF NUCLEOTIDE LIGASE IN RECOMBINANT DNA GENETIC ENGINEERING

surface culture the advantages are: (1) opportunity for continuous operation, (2) diminished requirements for space, labor and energy, (3) ease of recovery offset to a degree by increased requirements for meticulous, continual monitoring for optimum yield, and (4) prevention of contamination. Both methods require sterilized nutrient media containing an adequate supply of carbohydrate and nitrogen. In both methods the media are inoculated with the selected microorganisms and both adequate air and rigid control of temperature are required to obtain optimum yield of enzyme.

Scale-up and Maintenance.—While we shall not go into the engineering aspects of the large-scale operations, we do wish to stress that scale-up of fermentations and isolation methods are the most critical phases in microbial enzyme production. As stated by Underkofler (1976) scaling up a submerged fermentation is one of the most difficult exercises in chemical engineering. The approach is empirical, each step requiring extensive trial-and-error testing. It is a hierarchy of experimentation from shake-flask to small glass-jar stirred reactor to pilot plant fermentors to partially full-scale plant fermentor to full volume-yield data which hopefully will give the optimum aeration and agitation rates, temperature and pH schedules and medium composition for maximization of enzyme yield. This procedure is followed through for each fermentation vessel and conditions, once established, are rigidly adhered to. As mentioned, the process has to be continually monitored to detect and correct, for instance, for incipient contamination by unwanted microbes, for evidence of caramelization during the sterilization in continuous processes or for loss of moisture in the koji fermentations. Finally, another major concern of a microbial enzyme production facility is the preparation, maintenance and stability of suitable inocula. Stability of the inoculum, that is, the preservation of the genetic information for high-yielding enzyme production through successive generations can be a problem, especially with mutants.

When to Harvest

Even after all the above-mentioned precautions are taken, the decision to stop the fermentation and harvest and recover the enzyme is based upon continual monitoring of the enzyme activity during microbial growth. Enzyme activity assays should not only be carried out as a rule under carefully standardized conditions (Chapter 37), but these conditions should simulate the particular transformations in the special application of the enzyme; i.e., an α-amylase to be used as a bread dough supplement should be assayed differently from similar enzymes to be used in beermaking.

For both surface and submerged cultures, enzyme stability may in part determine optimum harvesting time because in the plant, in contrast to the laboratory, subsequent operations may take hours instead of minutes.

In general, koji cultures are harvested in 1–5 weeks and submerged cultures in 1–7 days. For bacteria in submerged culture the point in the log growth serves as a suitable marker. Extracellular proteinases, in general, are elaborated during the logarithmic phase but the amylases of *B. subtilis* appear later, right before the microbes sporulate, in the stationary phase. Of course in the case of intracellular enzymes it may be advantageous, at times, to allow lysis (release of enzymes) to take place after the cells have gone through their maximum growth. But this is not always feasible so that means must be worked out for getting these enzymes, such as glucose isomerase, out of the cells (see Chapter 9).

RECOVERY, SEPARATION AND FURTHER HANDLING

Clarification

Once the microorganism has elaborated the maximum amount of enzyme, further steps in enzyme production do not differ greatly from those employed for plant and animal enzymes. In mold bran koji processes, the enzyme is extracted from the bran with water or buffer. In the koji and in the submerged culture processes the microbial cells, mycelia and other cellular debris are removed by filtration. Alternatively, both extraction and filtration are achieved in the koji process by percolation through the mass of bran by a countercurrent extraction system. In the rare case of intracellular commercial enzymes, it is necessary to disrupt the cells. This is done as a rule by allowing the cells to lyse. Even before filtration three classes of substances may be added: clarification agents, preservatives, and stabilizers (see below).

Clarification aids may either cause flocculation of unwanted substances which interfere with filtration or may simply aid in the filtration. Underkofler (1976) lists the following: ammonium phosphates, ascorbic acid, calcium salts, cellulose fiber, cysteine, diatomaceous earth, gelatin, gum arabic, hydrated lime, hydrochloric acid, phosphoric acid, sodium salts of phosphoric, sulfurous, citric and phosphoric acids. Preservatives include organic acids/salts, phenolics, quaternary ammonium compounds and fluoride—all employed within safe limits since the clarified solutions may be sold commercially as liquid enzyme.

Further Handling

Concentration.—The enzyme may also be sold in the liquid form after concentrating several-fold by vacuum evaporation. Alternatively the enzyme may be sold in the dry form. Although a few solid commercial enzymes, especially those which are unstable, are lyophilized products, most of them are obtained by precipitation with the appropriate precipitants. Among those used are volatile organic solvents which do not remain in the product, such as acetone, ethyl acetate, ethanol, 2-propanol and methanol, and salts, such as ammonium sulfate, sodium sulfate, disodium phosphate, along with some of the previously mentioned clarifying agents. In at least one plant salts were removed by ion exchange.

Spray-drying has been used in the preparation of enzymes employed as detergents (Kjaergaard 1974). According to Madsen (1974), membrane filtration is being used on a somewhat less-than-full industrial scale for concentration of most of the common commercial enzyme activities. Ultrafiltration, as of this writing, is tantalizingly close to the economical reach of enzyme producers because of several desirable characteristics. In the first place it is gentle and nondestructive. Furthermore one can, if needed, obtain simultaneous concentration and purification. No chemical or phase changes occur; concentration can be carried out at low temperature. At higher temperatures, microbial contamination may occur. Before being cast in final form, a purification step or two as discussed in Chapter 6 may have to be interposed.

Product Form.—Actually there is a gradation from crude (suspension, paste) → liquid (soluble) → concentrated liquid → standardized solid. Until 1973 almost all enzymes were sold in liquid form primarily because of the added expense incurred by further treatment; i.e., cost per unit of activity was, and is frequently, considerably less. Furthermore, liquid lends itself to blending equipment and automated metering. Also, drying to a powder, as was learned the hard way in the case of enzyme detergents, presents a potential (but avoidable) hazard to manufacturing personnel. Liquid enzymes are packaged in 5, 25 and 50 gal. (18.9, 94.6 and 189.2 liter) lined drums or in plastic containers which can be stored in refrigerated warehouses close to the areas where they are used.

Problems and disadvantages posed by commercial liquid enzymes include: (1) high shipping charges due to bulk caused by presence of water, (2) relative instability which may require refrigeration and/or stabilization, (3) liquid preparations subject to microbial contamination, (4) losses in activity which may occur due to mechanical agitation during shipping, and (5) storage facilities which may have to be provided. In the hands of unskilled operators much waste as well as incorrect dosage has been reported.

Conversely, powders incur lower shipping costs, have good storage stability, do not have to be refrigerated on the whole and can be sold in convenient, small packages. The problem of irritation of the powder can be circumvented by pelleting.

Stabilization and Standardization

In spite of the desirable characteristics of solid enzyme preparations, the advantages of using liquid are such that active search continues to find yet more effective agents which would stabilize the commercial preparations with respect to both enzyme activity and resistance to microbial deterioration. We are not concerned here with the latter nor with stability of enzyme at very high temperatures (Chapters 10 and 11) but only with temperatures and conditions that these enzymes will encounter without refrigeration before they are used or during their manufacture. This search is going on at both basic and applied levels. However, in general, an enzyme which is more stable at high temperatures is more likely to be stable at room temperature with all other factors such as susceptibility to oxidation (see below) being equal.

In general, to most enzyme preparations are added one or more substances like salts which may serve not only as an enzyme activity stabilizer but perform multiple duty as antimicrobial agents, as filter aids and, if the liquid is to be dried, as fillers, diluents, anticaking agents and for standardization of activity (see below).

Commercial preparations are quite likely to contain one or more of the hydrophilic substances such as glycerol, polyethylene glycol and sorbitol. The patent literature abounds with empirical formulations for stabilizing liquid enzymes. A few examples are shown in Table 5.4. From such patents, summaries of a symposium (Anon. 1976A), the reviews, interviews and discussions of Wiseman (1973), Elbein (1974), Wiseman and Woodward (1975) and Skinner (1975), the following armamentarium of stabilizers, procedures and circumstances along with comments on how stabilization is achieved are potentially available. Many of these are still in the exploratory stage and cannot yet be allowed because of safety considerations.

Substrate, Coenzymes: It is well known that coenzymes, cofactors and even enzyme reaction products stabilize enzymes against heat denaturation and probably also against attack by proteolytic enzymes.

Genetic Stabilization: Mutations resulting in the exchange of only one amino acid in a protein molecule, i.e., sickle-cell *vs* normal hemoglobin as well as certain lysozyme mutants can confer heat

TABLE 5.4

ENZYME STABILIZATION PATENTS

Enzyme	Stabilizing Agent or Procedure
Glycerokinase	Protein fraction from yeast.[1]
Catalase	Ascorbic acid[2]; sodium citrate + glycerol + thymol[3].
Papain	Lower alkanols added to incipient precipitation point[5]; partially hydrolyzed collagen + glycerol[6]; borates.[7]
Tomato "pectase"	Sorbic acid, probably acts as preservative.[8]
Amylases/proteases	Diamines, polyamines.[9]
Glucose oxidase	Methyl cellulose.[10]

[1] Brit. 1,488,988.
[2] Jpn. Kokai 74 62,687.
[3] U.S. 2,689,203.
[4] U.S. 3,133,001.
[5] U.S. 3,011,952.
[6] U.S. 3,296,094.
[7] U.S. 2,958,632.
[8] U.S. 2,982,697.
[9] Ger. 2,058,826.
[10] U.S. 3,006,815.

and presumably storage stability. The origin of the thermal stability of enzymes of thermophilic microorganisms is still the subject of much enigmatic debate and investigation (Zuber 1976).

Hydrophilic Agents: These include glycerol, polyethylene glycols, sorbitol, some sugars, polyvinyl alcohol and glycerol ethers which may affect hydrophobic interactions (Chapter 4), lower water activity at high concentrations (Chapter 13) or act as free radical scavengers (Chapters 16, 24).

Food Solutes: Salt and sugars in sufficiently high concentration lower water activity which contributes to enzyme stability (Chapter 13).

Metals: These include calcium in α-amylase (Schwimmer and Balls 1949), cobalt in glucose isomerase (Chapter 9) and zinc in phosphatase which may stabilize because they are needed for holding the tertiary structure together. Ca ions can form a network of electrostatic bonds.

Anions: Borate, acetate and other buffer anions may alter conformation or influence charge stabilization within the active site (Chapter 3). Since they do not, in general, exert their effect directly at the active site, Wiseman (1973) considers such substances as allosteric effectors, which, instead of directly affecting activity, affect stability.

Polyanions: Dicarboxylates, such as tartrates, and tricarboxylate, such as citrate, may be involved in chelation of destabilizing cations such as those of the heavy metals. Polyanions such as heparin, RNA, chondroitin sulfate and DNA are activators of a variety of enzymes (Elbein 1974) in part, at least, by virtue of their ability to stabilize conformation and in part due to removal of inhibitory proteins.

Polycations: These include mostly substances with free amino groups: diamines, lysine, polylysine, chitin, protamines and histones. They appear to stabilize the quaternary structure of oligomeric enzymes; i.e., they prevent dissociation of subunits (Chapter 4).

Proteins and Related Polymers: Bovine serum albumin is frequently used for prevention of inactivation of enzymes upon dilution during enzyme assay (Chapter 37) and purification (Chapter 6). This and proteins such as lysozyme, partially hydrolyzed collagen and casein, used in patented formulations, appear to stabilize enzyme activity as a result of protein-protein interaction. Elbein (1974) stresses similarity to polyelectrolyte interaction and cites instances in which polyelectrolyte-requiring enzymes stabilize themselves by forming aggregates so that the polyelectrolyte is no longer needed for stabilization. The amino group of the terminal amino acid of one molecule probably combines specifically with the non-active-site portion of a molecule of the same enzyme so that the catalytic conformation is retained. A more or less specific heat and acid stable enzyme stabilizer protein can be isolated from yeast by hot extraction followed by TCA precipitation (Beaucamp and Lilly 1977). Some neutral detergents such as Triton X-100 can also confer stability. It should be stressed that these and other stabilizing effects are highly dependent on the usual parameters of ionic strength and pH as well as stabilizer level and on the stabilizer/enzyme ratio.

Protease Inhibitors: Active enzymes in commercial preparation are more or less undenatured so that they are likely to be fairly resistant to action of accompanying proteolytic enzymes. Nevertheless the attrite action of these accompanying proteins during long periods of storage chew up the enzyme enough to cause considerable loss in activity. In the laboratory, protease-sensitive enzyme systems can be preserved by protease inhibitors such as benzamidine and polyphenylmethylsulfonyl chloride or by removal of the proteases (Chapters 6 and 7). Most natural plant antiproteases are not suitable.

Chelates: EDTA, citrate, pyro- and polyphosphates stabilize enzymes by tying up metal ion inhibitors and preventing interaction with SH groups and acids.

Antifoams: Substances such as octanol lower surface tension of liquids and prevent the surface denaturation and thus enzyme inactivation during handling and transport.

Reducing Agents, Antioxidants, Free Radical Scavengers: Cysteine, dithiothreitol and mercaptoethanol and cysteine are frequently used to stabilize enzymes in solution. They stabilize by protecting disulfide bridges required for tertiary structure. They also protect sulfhydryls. Disulfide bridges can be further oxidized to sulfoxides, which presumably would interfere with flexibility needed for ligand-induced conformational changes associated with enzyme activity. Also, oxygen can, *via* autoxidation or flavoprotein oxidase action, give rise to highly reactive species of oxygen which can destroy the enzymes. This suggests that catalase, glucose oxidase or superoxide dismutase as a side-enzyme or deliberately added may be beneficial in maintaining stability in liquid enzyme preparations. Other reducing agents, food additive antioxidants and specific free-radical quenchers and scavengers should also be helpful. Some of the polyhydroxy hydrophilic agents mentioned above may be protective because they are also excellent free-radical scavengers. Another approach to the stability of "disappearing enzymes" (Thurston 1972) is to substitute electron acceptors such as ferricyanide for oxygen. This improved the *in vivo* stability of a microbial dehydrogenase (Anon. 1976A).

Crosslinking-Protein Modification: This has frequently been suggested as a means of stabilization. Skinner cites the use of bifunctional agents, glutaraldehyde, dimethyl adipimate and photoreactive substances which stabilize by crosslinking at the active site. Protective groups can then be removed and the enzyme restored to full activity by simply exposing the dissolved enzyme to light. Some metal-stabilizing effects may be due to crosslinking.

Immobilization: Many but not all enzymes may be stabilized by immobilizing them, the effect being related to that of crosslinking. Reactions with dimethyl suberimidate with yeast invertase can either stabilize or labilize depending upon the yeast genus used as the source. Chymotrypsin is stabilized by reaction with citraconic acid.

Glycosylation: In Chapters 2 through 4, we called attention to the presence of sometimes large amounts of sugars attached to enzymes. Wiseman and Woodward (1975) attribute the stability of invertase to the presence of 50% phosphomannans.

Resistance to heat and to proteolytic degradation during presumably long-term storage stability can be conferred on enzymes by artificially conjugating them with polysaccharides, i.e., potato amylase linked to dextran (Marshall and Rabinowitz 1975) and lysozyme to dextrans or dextrins (Christensen *et al.* 1976). Undoubtedly numerous other examples will have appeared and perhaps have been applied to stabilizing commercial enzyme preparation by the time this appears. See also Chapter 13.

Standardization, Dilution.—As pointed out by Underkofler (1976), the concentrated enzyme products both liquid and solid are assayed for enzyme activity by procedures based upon what particular use to which they are put, or by a simpler test which serves as an index of such use (Chapter 37). The products are, of course, sold on the basis of activity per unit weight or volume. In the latter case, they are diluted to standard, uniform activities in order to ensure product uniformity. Dilution may also be necessary when the product is so potent that it becomes inconvenient to meter out the small volume or weights required.

Solid products are adjusted to standard specific activities (potencies) by addition of diluents such as starch, various sugars such as lactose, glucose and sucrose, flour, salts, gelatin and casein. Undoubtedly, at least some of these help to protect the activities of even solid formulations. Frequently, buffers and activator and stabilizing substances—citrate, phosphate, calcium sulfate, etc. (see below)—are used even in solid enzyme products. This is done to maintain favorable pH, activity and stability. Often an enzyme manufacturer will supply the same parent enzyme preparation at different activity levels or standardized with different diluents, depending upon the intended use. Further remarks on commercial enzyme standardization and some newer nonassay methods for assessing enzymes will be found in Chapter 37.

ECONOMIC CONSIDERATIONS

We have already briefly touched upon the economic aspects of a nonmicrobial enzyme, papain. Although there is not complete agreement on the value of commercial enzymes produced in the United States, figures in the middle tens of millions of dollars are most frequently quoted. Reliable estimates are difficult to obtain for a variety of reasons. For instance, for many years the papain used for chill-proofing beer was imported directly to the brewer. Wolnak (1972,1974) estimated and projected values of $30 million for 1971, $38 million for 1975 and $47 million for 1981. He predicted that the value of perhaps the most significant enzyme, glucose isomerase—entirely in the immobilized form—will amount to over $6 million. Samejima (1974) estimated the value of major industrial enzymes produced in Japan in 1972, when the yen/dollar was stable, at $13.3 million. It is of interest that the value of 5 MT of egg white lysozyme produced in Japan in 1972 (5×10^8 yen) is exceeded only by that of 10,000 MT of bacterial α-amylase (6×10^8 yen). Apparently some other enzymes not yet readily available in the United States for economic industrial use were also produced in Japan in 1972. These include naringinase and hesperidinase (Chapter 25) and hemicellulase (Chapters 29 through 32).

As these figures show, enzyme manufacture is not a major industry. To further put the economic impact of enzyme production in proper perspective we cite Edwards' (1972) estimate that the value of products of the fermentation industry grosses at least 20 times that of commercial enzymes.

Reed (1966) and Neubeck (1976) have estimated the supplemental cost adding commercially successful enzymes to various processes shown in Table 5.5. Singularly, inflation has not raised costs appreciably. The values do make it clear that a process in which a commercial enzyme is added must be inexpensive, indeed, in order to be adopted in the food industry.

It should be pointed out that a large share of the enzyme output goes to nonfood industries. It is estimated that about 90% of the "amylases" are used in the textile industry, a large share of the "proteinases" goes to the leather industry, a goodly amount to the cleaning industry and non-negligible quantities of enzymes are used for medicinal purposes. A timely survey of food enzyme economics is available (Daheny and Wolnak 1980).

TABLE 5.5

UNIT COSTS OF ADDING COMMERCIAL FOOD ENZYMES IN FOOD PROCESSING

Process or Food	Unit	Cost Per Unit From Reed (1966)	Cost Per Unit From Neubeck (1976)
Fruit juice clarification	100 gal. (378.54 liters)	10 –20¢	15– 30¢
Grape pressing	metric ton	—	25–100¢
Chillproofing beer	barrel	2 – 4¢	3– 6¢
Clotting milk in cheesemaking	1000 gal. (3785.4 liters)	20¢[1]	20– 50¢
Flavoring cheese with lipase	1000 gal. (3785.4 liters)	20¢	—
Meat tenderizing	100 lb (45.36 kg)	05– 2¢	*ca* 5¢
Improving bread flour			
with protease	100 lb (45.36 kg)	2 – 3¢	2– 5¢
with amylase		0.5– 1¢	2– 5¢
Syrup manufacturer, starch to glucose	100 lb (45.36 kg)	40–60	*ca* 50¢
Removing lactose from milk	1 qt (0.9 liters)	—	(15– 40¢)[4]

[1] Liquid rennet.
[2] Estimated cost for antemortem injection; home use, one to two orders of magnitude higher.
[3] Based on cost of starch at 10¢/lb (5¢/kg).
[4] Retail price in 1979 (Chapter 36). Interestingly, other prices have not appreciably increased due, in part, to increased competition among enzyme suppliers, in part to increased productivity and advances in enzyme production engineering, and perhaps in some instances, to comfortably absorbable markups.

6

ENZYME PURIFICATION, ENRICHMENT, ISOLATION

Don't waste clean thoughts on dirty enzymes.
—A motto of the enzyme purifier.

PRESENT PRACTICES FOR LARGE-SCALE PURIFICATION

Until quite recently and beyond crude fractionation of industrial enzymes, as discussed in the last chapter, further purification procedures have rendered the enzyme prohibitively expensive, at least as far as most uses are concerned. The aim of the enzyme manufacturer has been primarily concentration rather than purification. However, there have always been exceptions, and will be increasingly frequently in the future. Thus, glucose isomerase used for the production of fructose-containing corn syrups (Chapter 9) is extensively purified. Large-scale purification procedures with little regard to cost are used for obtaining enzyme preparations for scientific research in general, when used as pharmaceuticals and, as far as the food scientist is concerned, for making enzyme preparations available for analytical and test purposes, especially for the quality control laboratory. Furthermore, one or two steps beyond concentration are occasionally required to remove offending enzymes in enzyme preparations to be used as food additives or for the preparation of food ingredients.

Thus, these products are concentrated but not highly purified and the presence of enzyme contaminants is usually of little concern. But, as mentioned above, for certain applications these accompanying enzymes may be considered as contaminants because they adversely affect the product or process, and practical means have to be found to eliminate them. Of course such treatment does not necessarily result in a purification in the sense of enrichment of desired enzyme activity. Indeed the specific activity (Chapter 2) may actually decrease somewhat as the result of such treatment. For instance, proteinase contaminants in glucose oxidase preparations used for desugaring eggs (Chapter 19) would hydrolyze the egg protein and make it completely unacceptable by creating texture, taste, flavor as well as color problems, the latter of which the enzyme is added to solve in the first place. Proteinases used for antemortem injection (Chapter 27) have to be free of materials which might cause unwanted physiological reaction in the animals. The problems which had to be solved before the microbial rennets could be successfully applied in cheesemaking (Chapter 33) were the poor flavor of the resulting cheeses due to the presence of undesirable lipases and the tearing of cheesecloth due to the presence of cellulases. This flavor problem was solved for *Mucor pusillus* rennet by gel filtration through Sephadex G-75 (Somkuti 1974) but many other patents have been issued including batch adsorption on a variety of matrices.

As judged by the literature, especially by the number of patents issued, probably the most thoroughly investigated case is the elimination of contaminating transglucosylase activity from exo-1,4-α-glucanase (amyloglucosidase, glucoamylase). Transglucosylase (transglucosidase) prevented the attainment of close to 100% theoretical yields of glucose when early commercial batches were used in the manufacture of syrups (Chapter 9). This is because the transglucosidase transfers a glucosyl from the α-1,4-linkage in the ever-shortening dextrin chains to an α-1,6-linkage:

$$G_1 \xrightarrow{\alpha\text{-}1,4\text{-}} G_2 \ldots G_x \ldots G_n \xrightarrow{\text{I}}$$

$$xG + G_{(x+1)} \ldots G_n \xrightarrow{\text{II, GA}}$$

$$\underset{\text{Amylose dextrins}}{G_{(x+2)} \ldots G_n} + \underset{\text{Isomaltosyl dextrins}}{G_{(x+1)} \xrightarrow{\alpha\text{-}1,6\text{-}} GA}$$

where I is the glucoamylase, II the contaminating transglucosylase and GA the glucosyl acceptor, which can be the product itself, glucose. The presence of the resulting isomaltosyl dextrins, which can be detected by the use of appropriate reagents (Schwimmer and Bevenue 1956), not only is a manifestation of lower glucose yields but also interferes with glucose crystallization. The methods shown in Table 6.1, representative of the vast patent literature on the removal of this offending enzyme, are illustrative of measures taken to effect this type of "purification." However, it is not likely that the state-of-the-art details that spell the difference between success and failure in their application are always divulged by these patents. In general, though, the methods involve heat, acid or simple ion-exchange treatments plus a few special treatments discussed below.

Desirable Subsidiary Enzymes.—Going in the opposite direction, it frequently is advantageous to retain or even add subsidiary enzymes to commercial preparations. Thus, in the fruit juice and wine industry pectin-degrading enzyme action is probably the principal reason these commercial preparations are used, but other enzymes which are invariably present—proteinases, amylases and other glucanases—undoubtedly play an auxiliary role in clarification (Chapter 30). In the brewing industry, both plant (malt) and microbial sources of α-amylase may also possess proteinase, pullulanase, hemicellulase and tannase activities, each of which contributes to production, efficiency and quality of the final product (Chapter 31). In formulations for the enzyme candying of fruit (Hori and Fugono 1969) the *Trametes* culture broth used contains not only cellulases which help speed up the penetration of sugar, but also a whole host of other glycosidases as well as lipase which may contribute not only to acceleration of sugar penetration but also to flavor and color attributes of the candied products. Finally, when glucose oxidase is used for desugaring (Chapter 19), it may be necessary to supplement the catalase that is already present to prevent a build-up of hydrogen peroxide.

Large-scale *vs* Small-scale Purification

Before proceeding to outline principles and procedures in the purification and isolation of enzymes it may be worthwhile to contrast and compare small-scale as practiced in the research laboratory with requirements of large-scale mass isolation

TABLE 6.1

SELECTION OF U.S. PATENTS ON PREVENTION OF TRANSGLUCOSYLASE ACTION IN GLUCOAMYLASE[1]

Patent No.	Method of Prevention
3,042,584	Remove TG by adsorption onto clay.
3,047,471	Treat with acid (pH 2–5), precipitate TG with lignin or tannin.
3,067,108	Inactivate TG with acid and aryl-alkyl sulfonates.
3,108,928	Treat with alkali, pH 9–11, for 15 min to inactivate TG.
3,254,003	Adsorb TG on cation exchange resin.
3,268,417	Inactivate TG with proteolytic enzyme.
3,296,092	Precipitate TG with chloroform.
3,296,684	Conduct initial stage of starch hydrolysis below pH 3 where TG is inactive.
3,483,085	Precipitate TG with heteropolyacids.
3,483,090	Precipitate TG with maleic anhydride copolymer.

[1] TG = transglucosylase.

procedures (Charm and Matteo 1971; Dunnill and Lilly 1974; Underkofler 1976).

For most research purposes the aims of the experimentalist parallel that of the enzyme manufacturers in the need for highly active enzyme sources. An exception in food science research is the situation in which pronounced changes in some quality attribute of a food may be due to a suspected feeble enzyme activity. To understand how to either suppress or promote these changes, the food scientist must first assure himself or herself that the change is indeed due to enzyme action (see phytase, Chapter 36). To further characterize the change it may be necessary to purify and attempt to isolate the enzyme from a source which the "pure" enzymologist or an enzyme producer would shun (see below).

In contrast to commercial enzyme production, it is permissible for an investigator to take serious losses in activity during the isolation in order to obtain as pure a sample of enzyme as possible. These losses may be purely operational or physical ones, in going from step to step, or they may be due to enzyme inactivation. Depletion of product through the latter cause presents an extremely difficult problem in large-scale purification and isolation processes. It is important to minimize the latter by avoidance of those conditions which result in protein denaturation in general, i.e., pH and temperature extremes, surface denaturation due to foaming and high shear. In contrast to laboratory work, where glass is usually used in large-scale operations special attention has to be paid to the shape and composition of the vessels. Rather simple operations in the laboratory like transferring solutions become quite complex in the plant. They require complicated engineering decisions concerning pumping equipment, piping, flow rate, prevention of turbulence or introduction of air. In the laboratory, failure to completely remove gross insolubles such as connective tissue, seed coats and fat particles in the beginning is permissible but in a large-scale operation, accumulation of such material leads to clogged process lines.

Not all the equipment used in the laboratory is readily available, adaptable or economical on a large scale although there is much active research to adapt some of these on the one hand and on the other to adapt well established large-scale engineering and processing equipment for enzyme isolation needs. As an example, Charm and Matteo (1971) adapted a conveyor belt-drum device used in the food industry to separate fat, based on particle flexibility, for removing fat in the preliminary stages of large-scale enzyme purification. Such a device could not be used in the laboratory.

On the other hand, the centrifuge is a device used much more extensively in the laboratory than in the plant. In the early days of enzyme chemistry when enzymes were first crystallized, filtration was the method of choice in the laboratory. Northrop *et al.* (1948) frowned on the centrifuges available at that time. Now, a variety of high-speed centrifuge techniques are used which can accomplish separation in less than 2 hr. Although low-speed centrifuges are used in the large-scale operations, most separations are accomplished with rotary filters over long periods of time. Centrifugation would, even if possible, take too long not only because it would be uneconomical to use higher speeds but also because of the higher densities of the liquid needed for handling minimum volumes.

Thus, many of the contrasts boil down to the question of time. Short times for certain operations in the laboratory may lead to small but acceptable activity losses. With prolonged operating times which are almost unavoidable on a large scale these losses become unacceptable. In addition to the above-mentioned conditions of denaturation, other factors which only become important on a large scale are the previously mentioned action of proteinases (Chapter 5), and when the starting material is frozen, the rate and conditions of thawing.

In the remainder of this chapter we shall survey the principles and procedures now in use on a laboratory scale for the purification and isolation of enzymes. The prospects for and progress in scale-up will be given attention where applicable. Not included is an examination of related procedures clustered about the broad term "affinity chromatography." This approach, based upon the intrinsic properties of the enzyme as a catalyst and specifier rather than as a protein molecule, is such a promising and burgeoning field that we deem it worthwhile to devote a separate chapter to it. Also, in a sense, affinity chromatography also serves as a bridge between a consideration of enzymes in solution and the subject of immobilized enzymes. While these are undoubtedly the waves of the future, the following considerations will still serve as useful supplementary tools to be used in conjunction with affinity chromatography.

In addition to the references already cited, others which survey purification and isolation of enzymes include the first in English by Schwimmer and

Pardee (1953) and those of Dixon and Webb (1964), and Whitaker (1972A). References to most of the older illustrative results mentioned, in general those published before 1970, can be found in the above discussions. Volume 22 of *Methods in Enzymology* is devoted to enzyme purification and related techniques (Jakoby 1971A). Enzyme purification is a favorite exercise in biochemistry and chemistry classes (Foster 1979).

PRELIMINARY STEPS

Source of the Enzyme

The first problem confronting an investigator when he/she decides to purify an enzyme is choice of the proper starting material. The choice is a very important one and may mean the difference between success and failure in finally obtaining a purified product. Sources differ widely in the amounts of enzymes they contain; also, their enzymes, though catalyzing the same reaction, may otherwise differ greatly. For example, β-glucuronidase activities of three tissues of six species of animals were found to vary over a range of several hundred-fold. It was possible to crystallize yeast hexokinase which is fairly stable, but hexokinase from muscle which is difficult to solubilize has not been crystallized. One source may contain undesirable material that is absent in another. Varietal differences in enzyme content are also found in yeasts and plants. Stage of development or maturity and location within the tissue may have a profound influence on the enzyme content amounts of the enzymically active material. A single enzyme rarely constitutes more than 1% of the source, and much of this enzyme may be lost during purification. In order to prepare enough final product to be useful, kilograms of starting material may be required. As an example of a well worked-out procedure, 15 kg of mushrooms for the preparation of phenolase are required. There are exceptions; for example, L-amino acid oxidase from snake venom was isolated from a few grams of starting material. The many trials and assays that are made in the development of a new method of isolation require several times the amount of material needed for one isolation. Therefore, it is important to have a plentiful supply of inexpensive, reproducible material with a high enzyme content per unit weight. In many cases, particularly with animal tissues, the source must be very fresh because some enzymes rapidly deteriorate. Animal tissues should be iced as soon as obtained and worked up rapidly. Alternatively, the tissue may be frozen at dry ice temperature and kept below −20°C. Sometimes perfusion is necessary to remove interfering materials.

The various tissues of animals or parts of plants (seeds, fruit, embryos, tubers, etc.) may differ greatly in their enzyme content. Germination of seeds may greatly increase their content of certain enzymes. The enzyme content of microorganisms may be greatly increased by using proper conditions of growth, usually by addition of the substrate for the enzyme (Chapter 5). Bacteria were not readily grown in the laboratory by the pound, hence yeast which is commercially available was more often used for isolation of enzymes from microorganisms. Now microorganisms other than yeast such as *E. coli* are available in kilogram quantities.

Bringing the Enzyme into Solution

The general procedure of enzyme isolation consists of bringing the desired enzyme and also various impurities into solution and then either separating the enzyme from the solution or removing impurities from the enzyme solution. The first step is to prepare a solution containing as much enzyme and as little impurity as possible from the starting material. In doing this, care must be taken to avoid denaturing the enzyme by too high or low a pH, by heat, foam, organic solvents (under improper conditions) or oxidation, and to prevent action of destructive enzymes or microorganisms. In all cases, it is important to test the efficiency of the extraction by enzyme assays.

Concentration.—Sometimes the enzyme is found in solution (snake venom, gastric juice, bacterial culture medium). Usually the concentration is quite low and it is desirable to concentrate by removal of water. This may be accomplished by precipitation of the enzyme with salt or a cold organic solvent; however, if the protein is very dilute a colloidal suspension is formed rather than a precipitate. Also, very large amounts of precipitant may be required. Removal of water by distillation, even under vacuum, may not be satisfactory because of high temperature and formation of foam. Another technique is to freeze out the excess water by cooling the solution slightly below freezing, permitting about one-fourth to melt very slowly and discarding the remaining ice, or by placing the solution under the refrigerator cooling unit and removing the layer of ice that forms. These procedures also concentrate salts and may alter the pH so that denaturation is possible. Lyophilization (freeze-drying) may be used to remove water.

Other methods for water removal include placing the enzyme solution in a dialysis bag and blowing air over the bag (pervaporation). In pressure dialysis, the bag is opened on one end and placed in the neck of a dialysing medium-filled flask attached to a vacuum (Richardson and Kornberg 1964). Pressure dialysis was used in the purification of onion γ-glutamyl transferase (Schwimmer and Austin 1971A). The bag can also be placed in a solution or even in a powder of high molecular deliquescent weight material, like polyethyleneglycol (aquacides). Ultrafiltration and molecular sieves (see below) are also effective in water removal. In industry, as noted in Chapter 5, the most common method for concentration has been vacuum evaporation at relatively low temperatures. Another method of concentrating solutes, which we have explored briefly, would be to adapt the zone melting procedure, used extensively in the electronics industry and to some extent by organic chemists, for purification of organic compounds. A thin, moving, melted zone can be created along a cylinder of ice by regelation, the necessary pressure being supplied by centrifugal force.

Removal of Unwanted Tissues.—Frequently the removal of unwanted tissue not only results in enzyme enrichment but also makes subsequent purification much easier. This is especially true of fatty tissue. Fat can be removed by: (1) flotation, (2) filtration in the presence of a filter aid such as diatomaceous earth, (3) precipitation of the protein (including, of course, all of the enzyme) followed by removal of the fat by low-speed centrifugation, and (4) solvent extraction. Oil in oilseeds can be removed with a press. In the last case this is best done at low temperatures and preliminary tests should ensure that such treatment does not inactivate the enzyme. Gross debris can be removed by screening methods including the much-used cheesecloth, finer debris by such materials as miracloth and nylon. Occasionally the source is dry powder such as commercial digestive tract glands, takadiastase, other microbial commercial enzyme preparations, or an acetone powder (see below) purchased from a biochemical supplier or even a ground up seed material such as malt, mustard seed, etc. Sucrose solutions (Schwimmer and Bonner 1965) and trichlorethylene (Schwimmer and Kjaer 1960) have been used to effect a separation of seed coat and endosperm of seed sources of enzymes on the basis of density differences. Undoubtedly density-difference methods used in industry such as air grade classification would be helpful for preliminary separations.

Intracellular Enzymes and Cell Disruption

When enzymes are isolated from living cells, a great diversity of methods, each suited for different conditions, have been devised to bring them into solution.

Plant and Animal Cells.—The first step in preparing an extract of cells consists in breaking the cell wall so that the enzymes can escape. Plant and animal tissues are usually finely ground with a chilled meat grinder or tissue mincer. The Waring Blendor is used; however, it may not always be a desirable tool for this purpose because inactivation and changes have resulted from its use. Other mechanical methods of breaking plant and animal cells include grating, grinding with mortar and pestle, with or without sand, and use of a filter press which, as we have seen, is also useful for the removal of oil. Historically, it will be recalled, the press was the means of cell disruption of yeast cells by which Buchner was able to demonstrate the action of intracellular enzymes in alcoholic fermentation of sugars. Acetone also plays double duty in removing fat and disrupting cells, as in the preparation of "acetone powders." These are prepared by pouring cold, ground tissue of semisoup consistency into 5–10 vol of cool dry-ice cold acetone, filtering and drying. Cell disruption can also be accomplished by the use of frozen tissue which can be thawed and subject to mechanical treatment or applied, for instance, to an ordinary machine-shop grinding wheel. Of course, one must be sure that the enzyme is not sensitive to freezing (Chapter 12).

Microbial and Nonmechanical Cell Breakage.— The methods of cell disruption described so far are strictly mechanical. An even greater variety of techniques including chemical and enzymatic means have been devised to prepare extracts from microorganisms. One of the oldest and most effective is using the source's own enzymes, autolysis. One incubates a concentrated mass of organisms, usually with an immiscible organic liquid (toluene, ethyl acetate, $CHCl_3$), but sometimes with alkali, until liquefaction and release of CO_2 occur. As with other procedures, control of the pH, time and temperature can optimize yield as well as specific activity.

Exolysis, the disruption of cells by the application of external enzyme action, was first applied to microbial cells, but is now used extensively for plant cells as well. Thus, catalase was prepared by the aid of the exolytic enzyme lysozyme. More recently cellulases and pectic degrading enzymes and macerating enzymes have been used to remove plant cell walls for the preparation of protoplasts and "unicellular" foods discussed in later chapters.

Other methods, mostly mechanical, involve grinding with a large variety of hard particles such as alumina, sand, fine carborundum and, in the last 15 years or so, with fine glass beads used in reflector signs. Part of the effectiveness of these beads may lie in their ability to produce alkali during the grinding, so that care has to be exercised in controlling the time of grinding. Another very effective microbial cell disruption agent is ultrasonication. The consequences of cell disruption and autolysis during development, maturation and processing, and of enzyme action for food quality are explored throughout this book and in briefer discussions by Schwimmer (1969A,1972,1978) and by Wiley (1977). A list of the foregoing and other agents of cell disruption is shown in Table 6.2.

Cell Disruption in Scale-up.—As previously mentioned, autolysis in water is frequently allowed to proceed for the production of industrial/commercial microbial enzymes. Attempts to scale up the manufacture of pure asparaginase, at one time perhaps a hopeful cure for cancer, involved brief exposure of the microbial cells to alkali at pH 11.5–12.0. But for the most part not all the methods available in the laboratory for cell disruption can be adapted to large scale. The expense incurred, for instance, in transmitting adequate power to large volumes of suspensions of microbial cells eliminates the versatile use of ultrasonic energy. However, there is a trend towards the application of homogenizers and liquid shear devices common in chemical engineering. Such devices as the Dynomill and the Manton-Gaulin or APV-Manton Gaulin homogenizer, have been suggested or perhaps used for continuous operation (Dunnill and Lilly 1974; Underkofler 1976). The latter device seems to work partly by explosive decompression (Table 6.2). The former employs glass beads and can handle 5 kg of bacterial paste on a continuous laboratory basis and much more on an industrial continuous basis. In the production of food grade glucose oxidase from *A. niger*, a modified paint grinder has been used to disrupt and extract the mycelia in preparation for enzyme extraction.

TABLE 6.2

METHODS OF CELL DISRUPTION FOR EXTRACTION OF INTRACELLULAR ENZYMES

Freeze-thaw	Osmotic shock	Lysozyme
Solid shear	Cold shock	Alkali
Liquid shear	Irradiation	Mortar/Pestle (shear)
Grinding with abrasive	Agitation with abrasive	Filter press
Acetone powder	Explosive decompression (a liquid shear method)	

Extraction

Once the cells are broken, there are two general methods of extracting enzymes. One is to extract all possible material under mild conditions with some solvent like dilute salt solution in which most enzymes are stable. The other method is selective extraction using at first a solvent in which few enzymes are soluble and then repeating with other solvents in which other enzymes are soluble.

Quite different results may be obtained with different solvents. As examples: Several solvents were tested for extraction of a fungal lipase and water was found best. Pancreas enzymes are extracted with dilute H_2SO_4. Sometimes cysteine or protein or a substrate stabilizes an enzyme during extraction and preparation (Chapter 5). Glycerol solutions have been used for extraction. Treatment with trypsin and other proteinases is often used to release insoluble enzymes. Bile salts have also been used. Sometimes dilute buffer succeeds when water does not. Extraction with $(NH_4)_2SO_4$ solutions may prove useful. When enzymes are localized with or associated with a particular organelle or to other subcellular structures (provided they have survived the cell breakage treatment), considerable purification can be achieved by differential centrifugation and removal of the nonorganelle debris. Over 25 years ago procedures were developed for large-scale preparation of heart muscle mitochondria to be used for the isolation of enzymes in the research laboratory. With plants, purification can be frequently achieved by adsorption of the desired enzyme onto the insoluble debris followed by selective extraction (Schwimmer and Austin 1971A).

Cellular debris is removed usually first with a cheesecloth followed by some of the procedures cited under removal of unwanted tissues. Such treatment can be followed by clarification through large, soft-fluted filters which do not clog readily. Plant extracts frequently clarify themselves when the debris settles out in the cold overnight. Polyclar prevents browning and aids filtration.

The decision as to the method to be used should be based on determinations of enzyme activities of the various preparations compared to the original material or broken cell preparations. Enzyme activities of intact cells may be quite different from activities of broken cells, even though no enzyme is lost, because of impermeability of whole cells and loss of cofactors and activators on breaking the cells.

NONCHROMATOGRAPHIC ISOLATION PROCEDURES

General Considerations

Once a clear solution of crude enzyme solution is obtained the enzymologist has recourse to a virtual armamentarium of approaches from which to choose for further enrichment, purification and eventual isolation of the enzyme. However, enzyme isolation has been and is likely to remain a trial and error affair, but perhaps to a lesser degree with the further development of affinity chromatography (Chapter 7). There have been many failures involving extensive work for each published successful isolation. Even for following another's published procedure attention to small detail is extremely important. With the advent of affinity chromatography, however, failures should be less frequent.

In general, it is found that the repetition of a given step is usually advantageous in purification. This step may be repeated successfully or another series of fractionation procedures may be interpolated. Hence it is not always necessary to survey the whole range of salts, solvents, absorbents, etc. However, repetition of the same fractionation often may cause such great losses in total activity that it is not feasible. Furthermore, in most situations a radical change in procedure will probably result in the elimination of a different class of impurities.

Sometimes a choice must be made between two similar procedures, one of which gives a better enrichment and a poorer yield than the other. One way of making this choice is to imagine that the processes could be repeated many times, and to determine the overall enrichment, E, that would be achieved if the processes were repeated each time with the same enrichment, e, and yield, y, until the overall yield, Y, dropped to a certain figure, say 10%. At this point the procedures could be compared in terms of their enrichments. The process that achieved the greater overall enrichment would be superior. In a repeated purification of this sort, it can be shown that:

$$\log E = (\log e)\ (\log Y/\log y)$$

Figure 6.1 shows lines of constant overall enrichment for a 10% overall yield ($Y = 0.1$) on a plot of

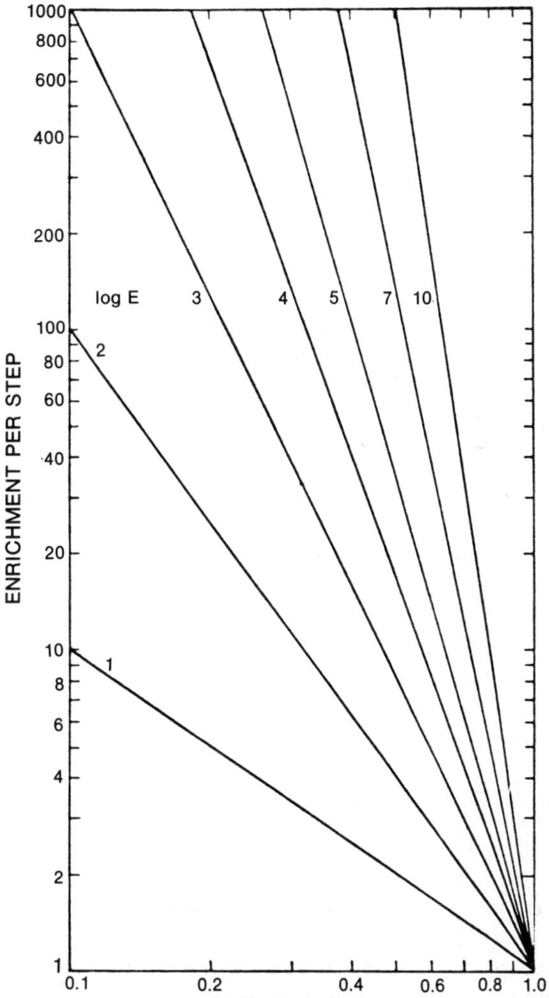

From Schwimmer and Pardee (1953)

FIG. 6.1. ENZYME PURIFICATION—YIELD AND ENRICHMENT

enrichment per step (e) *vs* yield per step (y). As an example of the use of this plot, suppose one compared two processes, one giving a three-fold enrichment ($e = 3$) and a 70% yield ($y = 0.7$) and the other giving a ten-fold enrichment ($e = 10$) and a 40% yield ($y = 0.4$). From the graph or equation one finds an overall enrichment of 1200× for the first process and 300× for the second, hence the first is the better process. The result would be the same if a different overall yield were used for the test, or if the overall yield for a given overall enrichment (say 1000-fold) were employed. Use of this graph also provides an indication of how much enzyme one can afford to lose for a given enrichment in a step.

As the enzyme becomes purer, the resulting solution becomes more dilute because of the removal of protein. Hence it is advisable to concentrate the purified enzyme so as to obtain solutions of at least 2% protein. Indeed, the failure of the early twentieth-century workers to achieve successful purification may be due, in part, to their insistence upon the maintenance of constant enzyme activity per unit volume rather than protein per unit volume throughout their purification.

Although theoretical considerations at first led to the conclusion that efficiency of fractionation increased with increasing dilution of the polymer solution, it was shown experimentally that the disadvantages of low concentration outweighed gains predicted by theory.

After a satisfactory step has been worked out, one should build up a sizable working amount of material before going on to a further purification step.

We have classified the general methods for isolating enzymes into the following five categories: (1) selective high capacity, (2) physical separation, (3) solubility methods, (4) crystallization, and (5) adsorption.

Selective High Capacity Operations

These preliminary high capacity steps which should precede those of low capacity (Whitaker 1972A) involve some sort of treatment or alteration which results in destruction of unwanted enzyme activity or removal of inert protein.

It is thus to some extent an unsorted array of procedures. However, many of these have in common the selective denaturation of inert protein. Thus, by careful heating of extracts, much of the inert protein material may be denatured, precipitated and removed without destruction of the desired enzyme activity. This is the first step in many enzyme purifications. Effectiveness is attested to by its widespread use in purifications, including α-amylase (Schwimmer and Balls 1949A). In each case emphasis is on controlled heating in the range of 35°–70°C for a limited number of minutes. Sometimes it is advantageous to add a substance which is known to protect or stabilize the enzyme activity (Chapter 5). In another class might be placed those few enzymes which are insensitive to relatively high temperature (above 90°C), examples of which are ribonuclease and adenylate kinase. Acid denaturation is also frequently used. Again the lower limits of pH stability have to be fairly accurately defined. Thus α-amylase is destroyed in malt extracts containing β-amylase at pH 3, but both activities are lost at a lower pH. It may be found, occasionally, that the inactivation of the enzyme is reversed upon allowing the heated solution to remain in the cold for a period (Chapter 11).

Often one finds that a heavy metal will precipitate inert protein or inactivate an unwanted enzyme. Also, metals are sometimes used for fractionation. Thus, lead salts have been used to purify the Q-enzyme from potato. Phenolase and β-amylase were prepared by fractional addition of lead salts. A more spectacular example is the destruction and denaturation of α-amylase when mercuric salts are added to a suspension of mixed crystals of α-amylase and limit dextrinase obtained from the mold *A. oryzae*. Furthermore, enolase could only be crystallized as the reversibly inactivated mercury salt. Lanthanum, copper and barium salts also have been used.

Denaturation by shaking the enzyme solution with an inert liquid such as chloroform or chloroform and alcohol has been used, for example, in preparing catalase, alcohol dehydrogenase and α-amylase. This treatment results in the formation of three layers: a lower chloroform layer, a second emulsion-like layer containing denatured protein and a top layer containing undenatured protein. The treatment, extended long enough to denature all of the protein, is used in the preparation of nucleic acids. To solubilize organelle enzymes associated with membranes, butanol or isobutanol, introduced by Morton (1950), is still effectively used, sometimes in conjunction with surface active agents (deoxycholate, detergents). Detergent extraction by itself is frequently used for effective enrichment and to remove substances which would otherwise give trouble later on.

Enzyme solutions can be purified and concentrated also by means of foaming (Charm and Matteo 1971). This process has been termed an adsorption on a gas, in which elution is unnecessary since the adsorbent automatically disappears. However, it is probably also an example of selective surface denaturation. In practice, a fine stream of nitrogen is passed through the enzyme solution and the foam is collected in a suitable apparatus. Liver phosphoribomutase is inactivated by foaming but not nucleoside phosphorylase. Pepsin and rennet were separated and purified by foaming, and more recently invertase and invertase inhibitor (Pressey 1972).

Occasionally, interfering material may be removed with the aid of an added enzyme, or by action of an enzyme already present (autolysis). Trypsin was used to remove protein in the purification of cytochrome *c* reductase and dehydropeptidase and amylase to remove glycogen in catalase purification. Tannic acid and other protein precipitants have found some use in enzyme purification.

It is advantageous to remove nucleic acids if they are present in large amounts, as in bacterial extracts which are much more readily fractionated after removal of nucleic acids, using protamines, $MnCl_2$ or streptomycin and also by adding nucleases. They can also be removed by adding two water soluble polymers such as dextran and polyethylene glycol (Albertsson 1962). This causes the formation of two liquid phases, one of which contains all the nucleic acid (Okasaki and Kornberg 1964). See below for further discussion of such systems.

Nucleic acids were used extensively and effectively as fractionation agents by German enzymologists. Ten-fold enrichments of lactic dehydrogenase and of phosphoglyceric kinase could be obtained by single nucleic acid fractionations. The nucleic acid is customarily removed by precipitation with a protamine such as salmine.

Mucin-like material interferes with crystallization in even trace concentration and can be most effectively removed by strong acids, copper hydroxide, filter-cell and acetone.

Such stratagems in large-scale purification schemes are, or have been, limited to relatively inexpensive procedures whose aim for the most part, as we have seen at the beginning of this chapter, is to remove offending enzyme activity rather than real enrichment. The procedures deplete the solution of protein and concentration means have to be used, among which are those enumerated previously and also the following methods based on physical separation.

Nonchromatographic Physical Separations

Procedures which effect a spatial separation of soluble enzyme from inert material without necessarily involving phase change may be grouped in three categories: membrane processes, electrophoresis and ultracentrifugation.

Membrane-mediated Separations.—Blatt (1971) considers membranes as one of three partition systems for enzyme concentration (the others being gel exclusion and adsorption chromatography). Membrane processes can further be divided into: (1) pervaporation, (2) dialysis against solutes of high molecular weights, (3) electrodialysis, (4) centrifugally accelerated ultrafiltration, and (5) pressure ultrafiltration. Others envision a filtration continuum with respect to cut-off size of excluded molecules or particles ranging from reverse osmosis (250) through filtration, ultrafiltration (*ca* 300,000), microparticle filtration up to true particle filtration. We have alluded to some of these such as pervaporation and pressure dialysis in concentrating dilute starting extracts and to ultrafiltration (Porter 1972) in industrial enzyme production (Chapter 5).

Dialysis.—Of the membrane-mediated separations, the simplest, oldest, most commonly and routinely used is dialysis. McPhie (1971) details techniques, membranes and apparatus. The principal function of dialysis is to remove small molecules such as salts from larger ones, especially, of course, the enzyme to be purified. The other variants of dialysis also remove water molecules, i.e., they achieve concentration. Aside from this, dialysis frequently results in purification by causing precipitation of globulin and other proteins. Because precipitation frequently occurs in the early stages of purification, it is unlikely that dialysis will be completely replaced by ultrafiltration devices. Where precipitation does not occur, undoubtedly some of the scores of ingenious variants of these techniques—among those mentioned by McPhie or used personally are marble ball, windshield wiper, rocking, rock-and-roll, rotatory, circulatory, the indeculator, countercurrent, arrangement for small samples such as the flat dialyzer—will be replaced by such convenient commercially-available ultrafiltration set-ups such as hollow fiber devices.

One of the principal problems in the use of dialysis and other membrane procedures is the loss of activity. Dialysis may cause loss of activity for several reasons. A cofactor or activator may be removed which is necessary for enzyme action; one can test this possibility by adding the known cofactor, or concentrated dialysis fluid, to the inactive enzyme and see if activity is restored. Another possibility is that the enzyme is denatured. A considerable drop in pH has been demonstrated on dialysis of $(NH_4)_2SO_4$. One should choose conditions to prevent destructive changes, perhaps using dialysis against buffers or dilute salt solutions. α-Amylase is irreversibly inactivated upon dialysis, as a result of removal of Ca^{++} which protects the enzyme. If dialysis is carried out against running tap water, heavy metal ions may diffuse in and inactivate the enzyme. Membranes may absorb some of the enzyme. Most bacterial and many plant extracts contain cellulase which will digest cellulose membranes. Animal membranes or collodion may be used for dialyzing such material.

Ultrafiltration.—Since ultrafiltration is a useful method which bridges the gap between laboratory scale and industrial scale set-ups, a few comments as presented by Underkofler (1976) may be in order. Although two types of ultrafiltration membranes have in the past been available—the microporous and the anisotropic diffusive—the former, which are subject to blockage because of the retention of rather large molecules within the structure of the membrane, have been largely superseded by membranes of the diffusive type because they are capable of more selective molecular discrimination.

The anisotropic diffusive ultrafilter is essentially a homogenous hydrogel membrane through which solvents and solutes are transported by molecular diffusion under the driving force of a concentration gradient. The anisotropic membrane consists of a highly consolidated but very thin skin ($0.1-5\ \mu m$) supported by a relatively thick ($20\ \mu m-1$ mm) porous substructure which acts as a support. Because the active layer is so thin, such membranes exhibit extremely high flux rates, while the porosity of the support is such that any molecule passing through the membrane is not retained by the support. The advantage of this type of membrane, which contains no pores in the conventional sense, is that it does not "plug" or block within the membrane and so there is no reduction in solvent permeability at constant pressure. At relatively low pressures the solvent activity is nearly directly proportional to hydrostatic pressure. Although high solvent flux rates are readily obtained using anisotropic membranes, the flow of solvent may be drastically reduced by a phenomenon known as concentration polarization. This arises as a result of the buildup of a layer of rejected solute at the membrane surface which impedes solvent flow. The design of ultrafiltration equipment has been aimed at keeping such polarization layers to a minimum. The laboratory stirrers increase the rate of removal of solute from the layer. In larger equipment, high flow rates through hollow channels have been employed. Flat membranes are sandwiched in a filter press arrangement. The material which is to be concentrated flows across the membrane through narrow channels designed to produce a laminar flow of liquid and thus cut down concentration polarization. In another arrangement the membrane is molded into a tube and then a number of such membrane tubes (hollow fibers) are incorporated into a cartridge. Liquid can be recirculated through the cartridge and sufficient pressure for ultrafiltration obtained by restricting the flow downstream of the cartridge. The hollow fiber cartridge system is used in both the laboratory and in various large-scale units with membrane areas of up to several square meters. Ultrafiltration is also speeded up when used in conjunction with ultra-centrifugation.

Since scale-up from laboratory ultrafiltration processes is relatively easy, ultrafiltration provides a powerful technique for both large-scale enzyme purification as well as concentration. Concentration followed by dilution with water or weak buffer is an efficient method for removal of salts prior to ion exchange chromatography. Other advantages and prospects for large-scale enzyme operation were discussed in Chapter 5. The use of ultrafiltration membrane devices for various kinds of enzyme immobilized reactors will be scrutinized in Chapter 8.

Centrifugation and Electrophoresis.—Although centrifugation is used extensively and routinely in enzyme purification for separating solids from liquids as part of a purification step, it was also used for purifying enzymes even without previous precipitation. More recently better separation has been achieved by performing the centrifugation in a density gradient composed of sucrose or glycerol. In this way not only the size but the density and shape of the enzyme and other protein molecules are utilized in the separation process.

In a simple version, a small hole is punched in the bottom of the plastic centrifuge tube at the end of the run and the liquid is collected drop by drop. Those drops with the major portion of the activity are collected. With a more sophisticated version, zonal centrifugation, whole cells, viruses and subcellular entities can be separated continuously (Cline and Ryel 1971).

Although electrophoresis with respect to enzymology is now largely used as a tool for separation of isozymes and other enzyme-related diagnostic analytical purposes, methods based on movement of ions in an electric field enjoyed some popularity for enzyme separation for quite a few years. Early large-scale variants such as electrodialysis (which is also membrane-mediated) go back over 50 years (Mintz 1970). Isoelectric focusing in a pH gradient (Vesterberg 1971) has been frequently used for laboratory enzyme purification and preparative gel-density gradient electrophoresis (Schuster 1971) appears promising. Another variant of electrodialysis, electrodecantation, was at least at one time considered a possible candidate for scale-up for proteins, including enzymes (Polson 1953; Mintz 1970).

Solubility Methods

The most commonly used techniques for purifying enzymes depend on separating more active from less active material by precipitation of one or the other. The procedures are based on solubility differences of proteins, and involve experimentally discovering the best possible conditions for enrichment and yield. Selective precipitation of the enzyme is usually superior to precipitation of impurities.

Salting Out.—One of the earliest successes in enzyme purification was achieved using separation with salts (Northrop *et al.* 1948) and a great many other enzymes have been subsequently isolated in this way. The most frequently used salt is $(NH_4)_2SO_4$. Its principal advantage is extreme solubility (*ca* 700 g/liter) which permits the salting out of practically any protein. The pH of concentrated $(NH_4)_2SO_4$ is about 4.5–5.5 and the solution should be adjusted to the desired pH before use. It is customary to make a ten-fold dilution of concentrated solutions before reading pH. There are at least ten publications exhibiting charts, tables and nomograms relating the weight of ammonium sulfate or the volume of a concentrated solution of the salt to be added to a protein solution to obtain a desired concentration. Earlier tables followed the lead of Kunitz (1952) in expressing salt concentration as the percentage of saturation of ammonium sulfate. While there are serious objections to the use of this ingrained convenient convention, as detailed by Wood (1976), the latter author presents a most complete and accurate compendium which retains the percentage of saturation but also gives the initial and final molarities. At 0°C, a saturated ammonium sulfate solution is 3.895*M*.

Salts do not seem to differ in effectiveness of separation in initial stages of purification; however, in later stages specific differences may appear. Some salts also serve as buffers and thus are advantageous. Ammonium sulfate concentration has also been expressed in terms of specific gravity.

The conditions under which a salt best separates enzyme from impurity must be determined experimentally. One method of doing this, described by Falconer and Taylor (1946), was devised as a criterion of purity. From these data one decides on the best concentration for precipitating inert material from enzyme and vice versa at a higher salt concentration. Another method of deciding on the best experimental conditions is to add successive increments of salt to one solution of enzyme, remove the precipitate after each addition and analyze to see what range of salt concentration is best. Still a third method based on the constant-solvent solubility test for purity as described by Northrop *et al.* (1948) has been used to fractionate a highly purified pepsin.

Solubility of a protein *(S)* depends on ionic strength *(I)* according to the equation $\log S = B - KI$, where B and K are constants. The relationship has been used in a number of studies, including enzyme purification. Temperature and pH strongly affect solubility by changing B but not K. Most proteins are more soluble in concentrated salt solution at low temperatures and in dilute salt at higher temperatures. Use has been made of this property for purification. Solubility is very dependent on pH, being the least near the isoelectric point. Thus, control and choice of pH in connection with salt fractionation are very important.

Probably the most complex factor in protein precipitation is the influence of other materials in the solution. Proteins combine loosely with a great variety of other materials, such as inorganic ions, organic molecules, lipids, nucleic acids and other proteins, and, like any other reaction, this depends upon pH, temperature, concentration of reactants and salt. Proteins combine with other proteins of opposite charge so that use of a pH above 6 or below 5, where most proteins have a net charge of the same sign, may reduce complex formation, as does high ionic strength or dielectric constant. The solubility of complexes may be quite different from each other and from that of the free protein. In a given solution part of the enzyme may precipitate as a complex with another protein, while other parts may exist as soluble complexes of free protein. The result is that the enzyme (activity) may precipitate over a wide range of salt concentrations, and that, as purification proceeds and impurities are removed, the solubility properties may change considerably. For example, the papain in crude latex is precipitated by 0.4 saturated $(NH_4)_2$-SO_4 (Lineweaver and Schwimmer 1941). Upon resolution, it can be reprecipitated by salt concentrations as low as 2% NaCl. Complex formation makes repeated fractionation procedures profitable. It is necessary constantly to apply preliminary tests and to analyze results at every step of the purification. For example, since it is good practice to work with a smaller volume upon repeating a precipitation, the enzyme will be more concentrated in the second step; therefore, it will start to come out of solution at a different salt concentration than it did the first time.

Experimental techniques of salting out are not complicated. Solid salt or concentrated solution is added to enzyme solution, usually slowly with stirring or from a dialysis bag, to avoid local high concentrations, under proper conditions of pH and temperature. Sometimes it is necessary to permit the material to stand several hours or days for complete precipitation. During this time the precipitate may settle, then much of the solution can be siphoned off. Remaining liquid is removed by filtration or centrifugation.

A point not always recognized is that considerable salt may remain in a precipitate, and the effect of this must be taken into account in the next purification step. Salt can be estimated roughly by assuming that the increase in volume when the precipitate is redissolved is caused by mother liquor in the precipitate, or salt can be estimated more precisely with a pycnometer to measure density, or the Nessler test for free ammonia. Salts can be removed by dialysis, ultrafiltration and various chromatographic columns.

When precipitates are dissolved, often a small amount of insoluble material remains in suspension. One should not try to dissolve every bit of material, because too large a dilution will result and inert material may be redissolved.

While centrifugation of precipitates has largely replaced filtration in the laboratory, the latter is, as previously mentioned, a most important step in industrial and other large-scale operations. Especially useful is rotary vacuum precoated filtration (Dunnill and Lilly 1974).

An important procedure is selective extraction of enzyme from inert material. Best conditions must be determined experimentally. The method has advantages over precipitation, especially in the sharper separations obtained. An effective Celite-column procedure which combines extraction with salt concentration gradients (Schwimmer 1953A) has been frequently used with effect. For instance, Doniger and Grossman (1976) incorporated this technique as a step in the purification of a DNA repair exonuclease (correxonuclease) from human placenta and Schwimmer (1971) employed it in the purification of the flavor-producing enzyme from onion. Selective extraction is also the method of choice in crystallizing enzymes (see below).

Organic Solvents.—A second group of precipitants are organic solvents miscible in water. The most commonly used of these are acetone and ethanol; others include methanol and dioxane. The principal difference in technique depends on the fact that most enzymes are very readily denatured by organic solvents at room temperature, but are much more stable in the cold. Therefore, one usually works at low temperatures, generally a few degrees below zero, adds the cold solvent very slowly, and uses a low salt concentration which improves specificity and reduces the rate of denaturation. Acetone is usually considered to be the best precipitant, ethanol fair, and methanol and propanol poor. Best pH is 6.7 and a low ionic strength is desirable. Organic solvents are less likely to split off prosthetic groups than are salts. Occasionally organic solvents are used for short times at room temperature for fractionation in order to take advantage of denaturation of unwanted protein.

A great amount of work was done by Cohn and his group (1950) on the fractionation of proteins from blood to obtain medically valuable blood protein fractions such as γ-globulins. This is an example of a commercially viable process in which the economic and social benefits of the products outweigh the considerable costs. Such costs could not be borne by a producer of enzymes as used in present-day applications of foods. The success of the approach depends upon the carefully controlled combination of organic solvents, pH, salts, protein concentration and metal ions such as Ba^{2+} and Zn^{2+} which form insoluble salts of the enzymes and blood proteins.

The use of so many controllable variables and the absence of high salt concentrations, which mask specific protein properties, permit choice of conditions to give cleaner separations than are obtainable with salts alone.

For the production of enzymes for the food industry, ethanol is used most frequently because of its acceptability. However, acetone and to a lesser extent isopropyl alcohol are also used, all at low temperatures. Recovery and reuse of solvent may offset the cost of refrigeration.

pH.—Another method uses differences in solubility at various pH values. The technique has been used extensively in enzyme purification, especially initially to remove large amounts of impurities. One adds acid or alkali gradually and determines the conditions under which best separation is obtained. Other conditions of course affect the results. Low salt concentration is usually used to permit more specific effects. Acid and alkali are protein denaturants, hence care must be taken in their use. One adds them slowly, preferably in the cold. Acetic acid, ammonium hydroxide or buffer solutions can be used to avoid local extremes of pH. It will be recalled that pH gradients are employed in isoelectric focusing.

Polyelectrolyte Polymers.—About 25 years ago Morawetz and Hughes (1952) found that several synthetic polyelectrolytes, including polymethacrylic acid, act as protein precipitants at pHs less than 6 and below certain polyelectrolyte concentrations. Above pH 6 and at higher concentrations, the precipitate redissolved. That this idea has promising commercial possibilities is attested to by the interest of one of the larger suppliers of industrial enzymes (Sternberg 1976C) who used polyacrylic acid for the separation of transglucosidase from glucoamylase, lipase from microbial rennet and β-amylase from lipoxygenase in soybean flour with some evidence of enrichment in the rennet. The procedure involves separation of the offending from the desired enzyme by precipitation of one or the other with polyacrylic acid between pH 3.0 and 5.8. If the offending enzyme is in the precipitate, as in the case of glucoamylase, the latter is ready for further handling as described in Chapter 5. If the enzyme is in the precipitate it can be solubilized and the precipitant removed by addition of a salt of a divalent cation such as Ca^{2+} or Mg^{2+} above pH 6 (see also Horvath *et al.* 1976). This insolubilizes the precipitant and solubilizes the enzyme. Since the precipitant is safe—it is an over-the-counter bulk laxative and inexpensive—a maximum of about $6/kg and recoverable and patents have been issued, it appears that this is a promising at-hand advance in commercial enzyme purification.

Nonionic Polymers and Multiphase Systems.—Thirty years ago Herbert and Pinsent (1948) purified catalase by partitioning the protein of the crude extract between two phases which were created by mixing proper proportions of water, ethanol and ammonium sulfate. A more effective and elegant method of separation of enzymes and other biological entities is the two-phase system of Albertsson. We previously referred to its use for the removal of nucleic acid from crude extracts. It has been used for several biological separations: bacteriophages, plant viruses, spores, single stranded from double stranded DNA, blood proteins and purification of the immunoglobulin. However, according to Fried and Chun (1971), it had hardly been used at all for enzyme enrichment and purification.

However, more recently Kroner and Kula (1978) advocated such a two-phase system for overcoming scaling-up problems in the production of commercial intracellular food enzymes. It is based on the principle that aqueous solutions of a mixture of two neutral polymers will form two phases, and each protein in this solution will have a unique distribution coefficient with respect to this system. Typically, the two-phase system will form in the presence of both a dextran and a polyethylene glycol (PEG). PEG is also used as a concentrating agent and as a fractionation precipitant, i.e., for separation of isozymes of glucoamylase (Yamasaki *et al.* 1977). It would appear that it might prove to be useful in large-scale continuous operations when used in a countercurrent distribution configuration. Such nonionic polymers may also serve as precipitants (Fried and Chun 1971).

Crystallization

The wave of enthusiasm which accompanied the spectacular success of Sumner and Northrop, Kunitz and coworkers in the late 1920s in crystallizing enzymes soon gave rise to a negative reaction usually stated in rather belligerent and skeptical terms, that after all, the appearance of crystals is by no means a rigorous criterion of purity. Several crystalline enzymes have been shown to be quite impure. A case in point is myogen A, a crystalline protein obtained from muscle extracts (Baranowski 1949). It was found later that these beautifully shaped bipyramids contained at least two consecutive glycolytic enzymes, glycerophosphate dehydrogenase and aldolase, and inert nonaldolase protein (Chapter 4). Although crystallization cannot be considered as an absolute criterion of purity, it can, however, be considered a definite and perhaps decisive step in characterizing an enzyme. The isolation of a crystalline enzyme from a large mass of inert protein is certainly an indication that chemical purity is at least being approached. Furthermore, the processes of crystallization and recrystallization are most effective procedures in actually purifying an enzyme. Thus, an eight-fold increase in specific activity was obtained upon crystallizing pyrophosphatase. The converse, removal of impurities by their crystallization, has frequently been used.

The state-of-the-art of enzyme crystallization developed during the quarter of a century following Sumner's crystallization of urease (Chapter 1) was detailed by Schwimmer and Pardee (1953). Since then, so effective have been developments in the appropriate, skilled hands that Jakoby (1971B) reported success with more than 100 proteins on which crystallization was attempted. The key to this success is the differential extraction of impurities at low temperatures by decreasing concentration of ammonium sulfate. As previously mentioned (Schwimmer 1953A,1971) this technique is far more effective than the more common increase in salt concentration used for fractionation. Crystallization occurs when the thus extracted protein is allowed to warm to room temperature. Other successful "tricks-of-the-trade" include seeding with crystals of enzymes from other sources (Eichele *et al.* 1980) and the "hanging drop" vapor diffusion technique (McPherson 1976).

In addition to being an effective purification measure, crystallization is an absolute requirement before X-ray and similar structural enzyme studies can be carried out. Methods for obtaining the relatively large macroscopic crystal—more than 2 mm long—needed for X-ray enzymology are detailed by Zeppezauer (1971), Davies and Segal (1971) and McPherson (1976).

CHROMATOGRAPHIC AND RELATED PROCEDURES

For the most part solid chromatographic column procedures used in enzyme purification employ four different kinds of column fillers: nonspecific absorbent, ion exchangers, hydrophilic gels and affinity matrices or adsorbents (Chapter 7). The following is a brief survey of the principles and use of the first three of these, with some comments concerning prospects and problems involved in scale-up and adaptation to commercial food enzyme production.

Adsorption Chromatography

Batch Adsorption.—Historically, adsorption has played an important role in the development of enzymology both positively and adversely. First used with newly formed uranyl phosphate as adsorbent by Jacoby (1900), it was rediscovered by Michaelis and Ehrenreich (1908) and developed by Willstätter and coworkers in the 1920s (Willstätter 1922). So effective was this approach that it outstripped methods for detection of protein and led some investigators to the conclusion that enzymes could not be proteins since adsorbents removed all detectable protein without removing enzyme activity from solution. The individual who first showed that enzymes were indeed proteins was able 20 years later to enrich yeast invertase 2000-fold by one adsorption step (Sumner and O'Kane 1948).

In classical batch adsorption, still used but infrequently, the enzyme solution is first acidified to pH 4−6. Adsorbent, usually in the form of a gel, is added and after a few minutes of stirring the suspension is centrifuged and the enzyme is eluted from the adsorbent either with salt or with a slightly alkaline solution. Alternatively, the amount of adsorbent can be varied so as to successfully adsorb impurities and then enzyme (Schwimmer 1947A). Right before the wholesale shift to chromatography, by far the most widely used adsorbent was calcium phosphate replacing the earlier popular alumina hydrate ("Alumina Cγ"). Others included bentonite, silicas and even ferric hydroxide. As can be seen from Table 6.1, batch adsorption is used industrially to remove enzyme impurities. On a large scale, batch adsorption is also used with ion-exchange cellulose as adsorbent (see below).

Adsorption Chromatography.—With the success of chromatography of small molecules in the 1950s, attempts were made to adopt the then widely-used batch adsorbent, calcium phosphate gel, to chromatography, for instance by deposition on a cellulose carrier (Koike and Hamada 1971). However, with the advent of more powerful alternatives (see below) nonspecific adsorbent chromatography (NAC), like crystallization, went out of fashion. With the systematic study of one such adsorbent by Bernardi (1971), NAC on hydroxyapatite has found a small, but useful, niche in the hierarchy of enzyme purification.

Hydrophobic Chromatography.—This relatively new separation procedure, some 6 or 7 years old, is a serendipitous offshoot of affinity chromatography (Chapter 7). It arose as a result of a control experiment in what was thought to be affinity chromatography (Shaltiel 1975). Some of the early successes of "affinity" chromatography of a food-associated enzyme, lactase (Steers *et al.* 1971), were according to Dean and Harvey (1975) due to the effectiveness of nonspecific hydrophobic interactions. The latter authors ascribe the "discovery" of hydrophobic chromatography (HC) to O'Carra *et al.* (1973).

HC, although usually included in discussions and expositions of affinity chromatography, is due to the action, as is much of NAC, of nonspecific van der Waal's forces. Furthermore, it lacks the hallmark of affinity chromatography—predictability. The "affinity" in HC arises through the hydrophobic interaction of the alkyl side chains attached to the chromatographic carrier (usually agarose, see below) and the hydrophobic regions of the protein to be adsorbed. The effectiveness of the side chains, that is, of a ligand in true affinity chromatography, also involves hydrophobic interaction but only with the active site of the enzyme (Morrow *et al.* 1975).

For the preparation of hydrophobic columns, insoluble matrices, usually Sephadex or agarose, are alkylated (i.e., coated with hydrocarbons) by methods briefly limned in the next chapter. What Shaltiel and coworkers found was that such columns retained glycogen phosphorylase B at least as tenaciously as did matrices on which glycogen was attached to the end of the "spacer" alkyl group. In one step the enzyme was enriched ten-fold from a crude muscle extract, purity approaching that of the crystalline enzyme, with only a 5% loss in activity. In general, to find the best conditions, one "coats" a series of columns with hydrocarbons of different chain lengths and finds which gives the best combination of adsorption and elution. The latter is accomplished with increasing salt gradients. Sometimes the best chain length (usually 4–7) for adsorption will result in an unelutable enzyme matrix complex (Broussard *et al.* 1976).

A further variant of hydrophobic chromatography—detergent chromatography—has been developed in which ionic charges are introduced into the spacer arm. It is especially useful in purification by selective exclusion. Aslam *et al.* (1976) removed albumin from blood serum, which combines avidly through hydrophobic interaction with fatty acids, by passing the serum through a column of agarose coupled to succinic anhydride through hydrocarbon spacer arms.

In addition to references already cited, a review by Miles Laboratories (1976A) is available. Although no large-scale applications have been discussed or are available as of this writing, the method has been patented by Shaltiel. Hydrophobic chromatography is starting to attract the attention of food enzymologists as exemplified by the pioneering purifications of lipoxygenase and phenolase by Jen and coworkers at Clemson University (Flurkey *et al.* 1978; Flurkey and Jen 1978).

Ion Exchange Chromatography

If one can select one event that brought chromatography as a routine enzyme purification step into the biochemistry laboratory, it is the introduction of DEAE cellulose as ion exchange resin by Sober and Peterson (1954). Previous attempts to use chromatographic columns in general and ion exchange resin in particular were sporadically successful (Schwimmer and Pardee 1953; Zechmeister 1950; Zittle 1953) especially with smaller protein molecules such as cytochrome *c* (Paléus and Neilands 1950) using strong traditional cation exchange resins, i.e., Dowex IRC 50. Presently DEAE and other mild ion exchangers are made by hooking ionizable groups to cellulose, Sephadex or polyacrylamide gel. Anion is more frequently used than cation exchangers. The stronger anion celluloses, such as TEAE- and guanidinoethyl- are used for operation in alkaline buffers greater than pH 9. Cation exchangers are especially useful for purification of basic enzymes which are adsorbed at or below their isoelectric point, but a given purification scheme may employ each in different steps. ECTEOLA cellulose is used mostly for isolation of supramolecular entities such as nucleohistones and ribosomes.

Like ammonium sulfate fractionation, the use of ion-exchange resins is somewhat of an empirical art. Experimental details have been discussed by Peterson and Sober (1962) and Himmelhoch (1971).

Use in Enzyme Production.—As in the laboratory, for large-scale use there are three types of ion exchangers available; resin, large-pore gels (see below) and cellulose derivatives, but the latter are used most frequently. As with other procedures, minor or easily-overcome problems encountered in the laboratory tend to become magnified on a large scale. For instance, the time factor and scale change make enzyme inactivation, especially on resins, because of higher charge density, quite troublesome.

For large-scale enzyme production, ion exchange celluloses are used batchwise in a very early step of purification. This is because in tall columns powdered cellulose derivatives tend to compress easily. Columns are used in the final step.

For large-scale work, in addition to giving consideration to the usual ion-exchange parameters—ion capacity, pH, ionic strength, limits which maintain enzyme activity and solubility, preparation of ion exchanger—shape and size of vessel, stirring and especially temperatures are quite important. Adsorption time can vary from 10 min to 24 hr. If conditions are satisfactory, considerable purification can be achieved in one step by binding all or most of the enzyme to the ion exchanger which is then allowed to settle. Microgranular suspensions of cellulose ion exchangers settle rapidly forming compact, stable masses from which the supernatant liquid containing nonbound protein impurities can be inexpensively decanted (siphoning or pumping). The enzyme is eluted, as in other batch adsorption procedures, by either change in pH or ionic strength or both. The ion exchange is reused. In large-scale processes a basket centrifuge or spin dryer is used to remove remaining supernatant.

Although Underkofler (1976) states that ion exchange celluloses are sometimes used in industrial enzyme purification, Meyrath and Volavsek (1975) argue that it is more logical to use ion exchange because of its low capacity for removal of impurities rather than of enzyme in order to avoid adsorption inactivation of enzyme.

Gel Filtration

Hard on the heels of ion exchange chromatography (Porath and Flodin 1959), the chromatographic separation and purification of enzyme molecules according to their size rather than ionic charge occupied an important niche in enzyme purification steps after the early 1960s. Unlike the ion exchangers the materials used serve triple duty as a means of concentrating protein, removing salts, as well as fractionating proteins during enzyme isolation. It should be pointed out that dialysis, ultrafiltration and centrifugation, either density gradient or zonal, also separate proteins according to size. Gel chromatography is also known as permeation, molecular sieve, molecular exclusion, restricted diffusional or steric chromatography. Technique is presented by Reiland (1971) and Determann (1968) and of course by the gel suppliers.

As in all chromatographic procedures, the sample is poured on the surface of a column of appropriate beads of the hydrated gel and solvent is percolated down the column. Those molecules too large to penetrate the porous structure of the beads are excluded and pass through. At the other end of the size scale, very small molecules are taken up by the beads at the top of the column and come out only when copious volumes of eluant are passed through. In between, molecules will sort out and would emerge from the column strictly according to size, if the gel were a perfect molecular sieve. However, nonsize effects such as ion exchange, adsorption and partition may also contribute to separation of enzymes (Reiland 1971). At present, three commercially available gel forming substances are in general use for enzyme purification:

Partially Crosslinked Dextrans (α-1,6-glucan). Sold under the tradename Sephadex, the crosslinking is accomplished with epichlorhydrin, the degree of crosslinking determining the molecular size exclusion limits. Thus, Sephadex G-25 will effectively retard those globular protein molecules with molecular weights between 100 and 5000 while the range for Sephadex G-100 is 4000–150,000 daltons.

Crosslinked Polyacrylamide Beads (Bio-gel P). These are also available with varying degrees of crosslinking. Thus, Bio-gel P-10 excludes molecules larger than 2000 daltons.

Agaroses. These are polymers consisting of alternate residues of D-galactose and 3,6-anhydrogalactose prepared from agar by removal of negatively charged pectin-like material. They are available commercially as the Sepharoses and Bio-gel A. One of their distinctive features is the ability to form gels at very low concentrations, somewhat below 1%. They are especially useful for separation of high molecular weight species such as the nucleic acids. Although the agaroses are not used as frequently as the other two molecular sieves in gel chromatography, they are used as matrix of choice, so far, in affinity chromatography.

Scale-up.—Gel chromatography as applied to industrial separation and concentration of protein (not necessarily enzymes) has been looked into

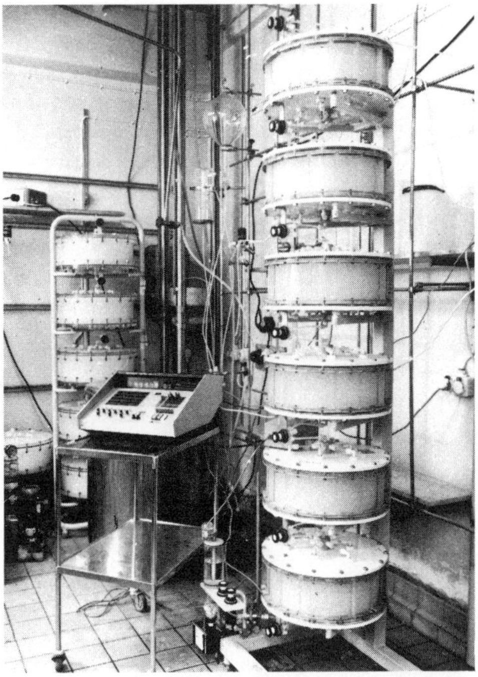

Courtesy of Pharmacia

FIG. 6.2. LARGE-SCALE GEL FILTRATION WITH SECTIONAL RADIAL COLUMNS

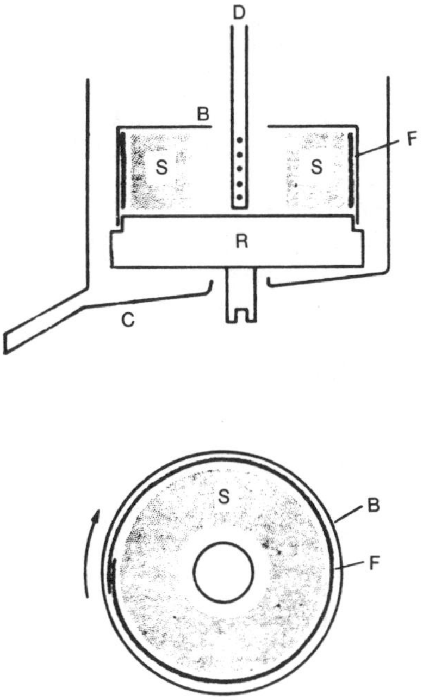

From Heftmann et al. (1972)

FIG. 6.3. THE CHROMATOFUGE
Schematic representation of apparatus for preparative rapid radial column chromatography: (B) basket, (C) cup, (D) tube, (F) filter, (R) rotor, (S) sorbent.

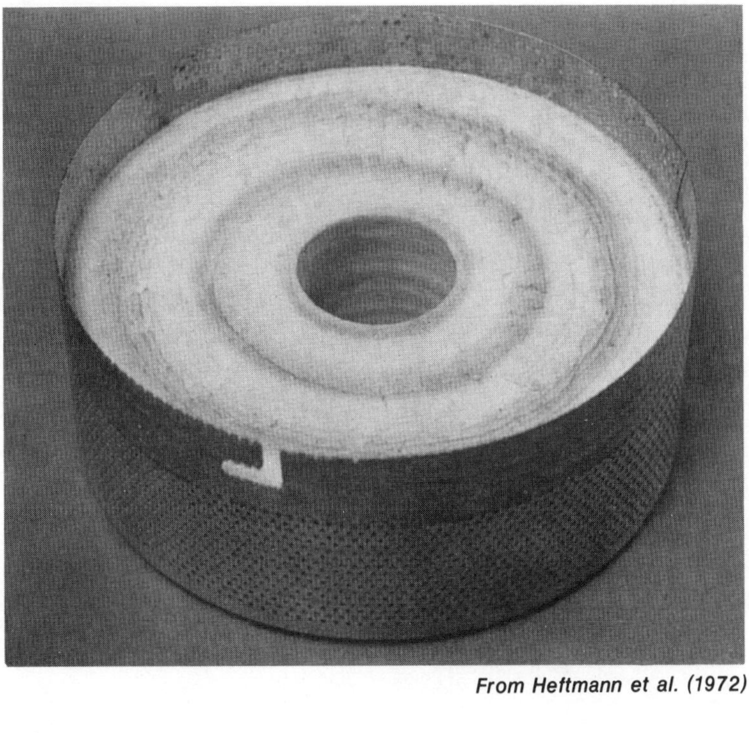

FIG. 6.4. A CHROMATOFUGE SEPARATION
Separation of dyes. The rotor has been detached to show migrating dyes.

From Heftmann et al. (1972)

fairly thoroughly and appears to be a good candidate for commercially successful scale-up. It is mild, fast flowing and reproducible. No enzyme is lost due to denaturation. Gradients are not needed for elution and no regeneration of the column bed is required. Reduction in flow due to the tall columns which would be required in a proportionate scale-up can be avoided by positioning a series of columns about the size and shape of a truck tire or large wheel of cheese in series as shown in Fig. 6.2. Gel filtration has been applied to the removal of lipase from microbial rennet (Somkuti 1974). Gravity feed is used to maintain eluant flow (Brewer 1971; Janson 1977; Curling 1977). One drawback is that no continuous or automated fraction collection has been developed. A potential continuous procedure that may have possibilities is rapid radial centrifugal chromatography (Heftmann *et al.* 1972). Efforts are now in progress to scale up this procedure (Finley *et al.* 1978). Figure 6.3 shows the set-up and Fig. 6.4 shows the separation of three dyes by this procedure. Commercial separators employing similar principles are now available.

7

AFFINITY CHROMATOGRAPHY OF ENZYMES

The methods of purification detailed in the preceding chapter have served (and will continue to assiduously serve) research biochemists, producers of fine biochemicals and to some extent commercial enzyme producers commendably well for over 50 years. However, the successful exploitation for purification purposes of those properties that impart to an enzyme its singularity, rather than those properties related to its being an ion, a macromolecule or a protein—affinity chromatography—is undoubtedly the most significant advance in enzymology—and for that matter other biological activity—purification since Sumner's crystallization of urease (Chapter 1). Its rising importance in enzymology is attested to by the devotion of a complete volume of *Methods of Enzymology* (Jakoby and Wilchek 1974) some six years after the cynosure paper of Cuatracasas *et al.* (1968). Only three years before (Jakoby 1971) affinity chromatography was included as another technique sandwiched between chromatography and electrophoresis, comprising two papers with 65 references cited. The 1974 issue contained 93 articles citing 1500 papers (references in these issues will not, for the most part, be cited in the ensuing discussion). According to Dean and Harvey (1975) applications of affinity chromatography grew from 21 in 1970 to 114 in 1974. This is about the same number of entries under "affinity chromatography" in 1 year of Volumes 85 and 86 of *Chemical Abstracts*. Of course the actual number of applications in this period were much greater, as it has become a routine procedure and is not reported prominently but appears in a paper as one of the steps in the isolation of an enzyme. This rise was heralded by the appearance of at least 30 reviews between 1970 and 1974 in English. Many of the more recent reviews are in other languages. In addition to the detailed compendium of Jakoby and Wilchek, the following selection of good reviews has since appeared: May and Zaborsky (1974), Porath and Kristiansen (1975), Baum and Wrobel (1975) and Turkova (1979), and discussions by Dean and Harvey (1975), Porath (1976), Wilchek and Hexter (1976), Lee and Kaplan (1976) and Gray (1980).

HISTORICAL BACKGROUND

Among the diverse observations which suggested to enzymologists of the early twentieth century the reality of the enzyme-substrate complex, among the most convincing was that enzymes could be desorbed from nonspecific adsorbents by elution with their substrates. The first observation in this regard was made by Hedin (1907), who eluted trypsin from charcoal by casein solutions. A year before, Bayliss (1906) had observed that if trypsin was added to a solution of the calcium salt of "caseinogen" and the solution was Berkefeld-filtered, the filtrate contained no trypsin. In the absence of caseinogen the trypsin was filterable. If the experiment was repeated with starch and malt amylase in the place of caseinogen and trypsin, a precipitate was formed which contained both the starch and amylase. In the 1920s and before, several investigators showed specific adsorption of enzymes onto their insoluble substrates such as lipase onto cream, pepsin onto coagulated hemoglobin and amylases onto starch.

The first successful attempt to use this phenomenon to purify and isolate an enzyme was that of Schwimmer and Balls (1949A). Based on the observation of Starkenstein (1910), we adsorbed barley malt α-amylase in ethanol solution onto a column of starch granules to obtain essentially pure enzyme. The attachment of the enzyme to the starch granules appears to be *via* the same site as that which attracts noncompetitive inhibitors rather than with the active site (Schwimmer and Balls 1949B). Adsorption isotherms are shown in Fig. 7.1. Other examples of affinity chromatography *via* attachment to polymeric (except for lipase) insoluble substrates include the following. Although some of such pre-Cuatrecasas procedures may be primarily of historical interest, it will be noted that their use extended into 1977:

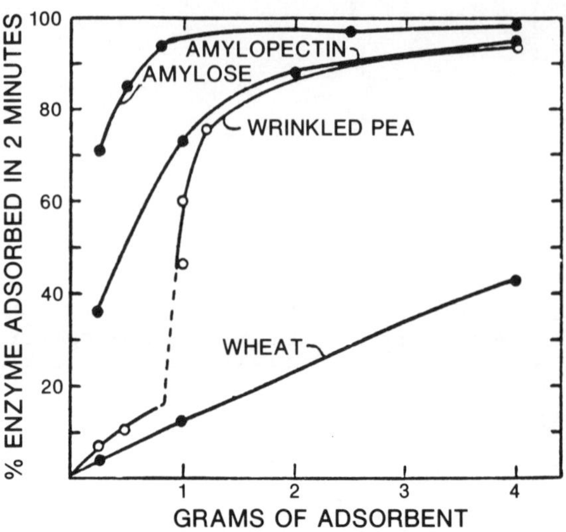

From Schwimmer and Balls (1949B)

FIG. 7.1. ADSORPTION ISOTHERMS FOR α-AMYLASE ON STARCHES

Isotherm expressed as Freundlich equation, log M/A = log (100−A) + a constant. M, adsorbent weight; A, % amylase absorbed.

α-Amylase on starch (Schwimmer and Balls 1949B; Thayer 1953; Schramm and Loyter 1966); crosslinked starch (Hostinová and Zelinka 1975) and on glycogen (Chapter 31).

β-Amylase, bacterial, on starch (Hoshino *et al.* 1975; Pasek *et al.* 1977).

Polygalacturonase on alginic acid = polymannuronic acid (Lineweaver *et al.* 1949); on crosslinked pectin (Rexová-Benková and Tibensky 1972).

Alginase on alginic acid (Favorov 1973).

Cellulases on cellulose (Gum and Brown 1976).

Xylanase on xylan (Nomura *et al.* 1969).

α-1,4-Glucan (starch) synthase on glycogen (Leloir and Goldenberg 1962).

Lysozyme on chitosan (Pryme *et al.* 1969; Weaver *et al.* 1977).

Galactose oxidase on agarose, a galactosyl glycan (Hatton and Regoeczi 1976).

Collagenase on collagen (Gallop *et al.* 1957; Nagai and Hori 1972).

Pepsin on ovalbumin (Northrop 1919).

Proteinase, wheat, on hemoglobin (Chua and Bushuk 1969).

RNA polymerase on chromatin (Bonner 1975).

Lipase on cream triglycerides (Willstätter and Waldschmidt-Leitz 1923).

Dickey (1949), working in Pauling's laboratory, prepared gels which preferentially adsorbed azo dyes which were present during the preparation of the gels. However, Lerman (1953) was the first to purify an enzyme by retarding its movement down a column *via* formation of a complex between the enzyme and substrate analog which was purposefully and covalently linked to the insoluble column matrix. The enzyme was mushroom phenolase, the substrate analog was an azophenol derivative (an inhibitor) and the matrix was cellulose. With this one step he obtained a more than 61-fold enrichment from a crude extract. Arsenis and McCormick (1964) obtained more than 20-fold enrichment of partially purified liver flavokinase, to apparently full purity, by chromatography on a column of cellulose linked to flavin derivatives, both substrate and inhibitors.

However, the modern era of affinity chromatography starts with the classical papers of Cuatrecasas *et al.* (1968) and Cuatrecasas (1970). Besides introducing the term "affinity chromatography" they introduced the use of spacer arms, stressed the importance of inhibitors as ligands and were able to achieve isolation of a bacterial ribonuclease preparation in one step, as shown in Fig. 7.2.

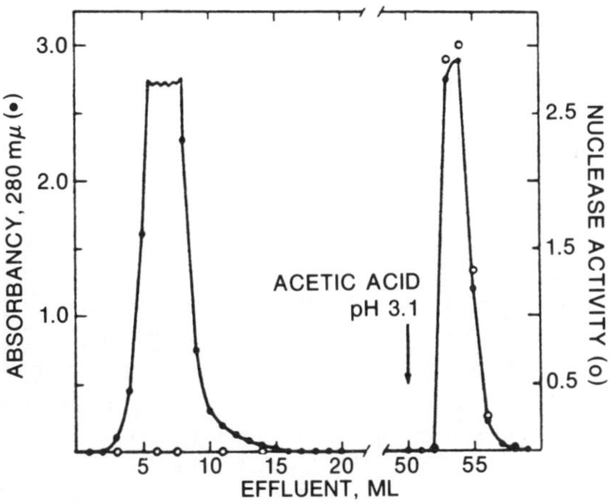

From Cuatrecasas et al. (1968)

FIG. 7.2. PURIFICATION OF A NUCLEASE BY AFFINITY CHROMATOGRAPHY

Depicting the first successful demonstration of affinity chromatography.

ADSORPTION

Kinds of Affinity

Although the term affinity chromatography is probably here to stay—it is listed in *Chemical Abstracts*—alternative terminology has been proposed to more accurately depict the distinctive basis of this approach. These alternative suggestions usually consist of one or two of the modifiers "liquid-solid," "bio," "specific," "ligand," "steric" and "affinity" coupled to one or two of the terms "adsorption" or "chromatography," or other special terms such as "elution," depending upon the particular adaptation. The objection to the word "affinity" is that, except for true gel permeation which might be considered to be the opposite of affinity since it is an exclusion (as are other partition-based separation processes), all other chromatography is based on some sort of "affinity" between the substances to be separated and the chromatographic matrix. As far as enzyme isolation is concerned, such techniques should be based on the exclusive stereospecific formation of a complex between a ligand (substrate, inhibitor or cofactor) and the catalytic site of the enzyme and with no other groups or sites. In actuality there exists a continuum of affinities with respect to specificities, but they can be divided into three groups: (1) those involving interaction with the catalytically active center, (2) specific interaction with other sites or areas on the enzyme usually involved in its regulatory properties, and (3) nonspecific interaction.

Interaction with the active site can be further broken down into: (a) stereospecific formation of a complex between the entirety of the active site, and (b) interaction with a particular component of the active site. In the former case the ligands are substrate analogs, substrates and competitive inhibitors. In a two-substrate reaction it may be advantageous to have the cosubstrate which is not immobilized on the column present in the enzyme solution. Examples of the second case are substances which combine with one component of the active site, especially with coenzymes or other cofactors. These substances and the matrices to which they are linked are usually referred to as group-specific ligands and matrices. Immobilized nucleotides such as AMP which is part of NAD^+ would be an example of a group-specific matrix for separation of the NAD-dependent dehydrogenases.

Interactions with sites other than the catalytically active site can be just as specific as those with the active sites. They may be further categorized as follows:

(1) Combination with antibodies which may perhaps be more specific than that with substrate analogs. The procedures involving this type of interaction are collectively referred to as immunoadsorption. There is a wide range of techniques by which purification of an enzyme can be achieved.
(2) Interaction of regulatory sites with allosteric affectors, noncompetitive inhibitors and other types of modifiers as ligands.
(3) Interaction of the carbohydrate moiety of the enzyme with lectins (plant haemagglutinins). Lectins, which are proteins, are considered to possess specificity with respect to individual sugars comprising carbohydrates of membranes and other biological entities including, of course, those enzymes that happen to be glycoproteins. In the mid months of 1977 this less than completely specific type of affinity chromatography was the most prevalently applied.
(4) Interaction between protomers (subunits) of the same enzyme.

In the third type of interaction, the specificity is much less of a factor than in the previous two.

Investigators who are trying to perfect biospecific affinity chromatography, as the term is used here, are interested in this category only in that they wish to suppress these interactions. This category may be further categorized as follows:

(1) Interactions based on the presence of free amino and carboxyl groups outside the above-mentioned sites and areas, i.e., those based on electric charge distribution. The techniques utilizing charge include ion exchange, detergent (Chapter 6), ion adsorption and perhaps charge transfer chromatography. Of course, electrophoresis is also based on unique charge distribution and is used in combination (affinity electrophoresis).

(2) Interaction with cystine residues by placing sulfhydryl groups on the matrix. Uniquely among chromatographic procedures the enzyme interacts covalently with ligand SH *via* SS-SH interchange. Alternatively, the ligand can be an organomercurial:

$$\text{Matrix-Hg-S-R} + \text{Enzyme-SH} \rightarrow \text{Matrix-Hg-S-Enzyme} + \text{RSH}$$

(3) Interaction with the hydrophobic regions *via* alkyl groups placed on the matrix. The technique associated with this type of interaction, hydrophobic chromatography, was discussed in Chapter 6.

(4) Interaction based on other nonspecific forces between matrix and enzyme and which results in the binding of the enzyme to the matrix. These may involve the making and breaking of hydrogen bonds and van der Waal's forces at sites other than those related to the biospecificity and regulatory properties of the enzyme (Chapter 3). Even the interplay of these forces must entail some "specificity" or else there would be no separation of molecules.

Not only enzymes but just about any substance which exhibits some sort of biological specificity or function can be separated by affinity chromatography. Use of ligands will be further elaborated on later.

Adsorbents

The most elaborate adsorbents as developed by Porath and others (Porath 1976) comprise the following sequence:

MATRIX - - - - - CONNECTOR, proximal - - - - -
(insoluble)

SPACER - - - - - CONNECTOR, distal - - - - -

LIGAND

The function of the spacer (arm) is supposed to be spatial separation of the ligand from the matrix so that the interaction with the enzyme, as mentioned above, will be more strictly biospecific by minimizing nonspecific adsorption forces. The connectors are derived from activating groups, placed on the matrix and the distal ends of the spacer, used to connect the matrix to the spacer and the spacer to the ligands, respectively. However, a survey of enzyme purification procedures in mid 1977 reveals that when affinity chromatography was included as one of the steps, in a majority of cases, the adsorbent did not contain a spacer. This may be because the most frequently used ligand was a protein, a lectin coupled to the matrix without the intervention of a spacer arm.

The Matrix.—Principally because of its approach to an ideal matrix but also because it is readily available—under the trade names Sepharose (Pharmacia) and Affigel (Bio-Rad)—by far the most widely used (about 90% of the enzymes) is beaded agarose, the nonionic fraction of agar.

Agarose approaches, but falls short of, the ideal matrix. It possesses the following properties of an ideal matrix: insolubility, uniformity of appropriate size and shape of beads, rigidity, mechanical stability at room temperature, permeability, lack of immunogenicity and lack of nonspecific adsorption capability. Agarose is also resistant to microbial deterioration, and microorganisms possessing agarose hydrolase activity are not widespread.

The properties which make agarose less than the ideal matrix as listed by Dean and Harvey (1975) are: mechanical instability upon freezing, chemical instability in the presence of activating agents and incompatibility with organic solvents, which restricts the range of activation procedures.

Some of the shortcomings can be circumvented by modifying agarose *via* crosslinking or by copolymerization with, for instance, acrylamide-Ultragel (Doley *et al.* 1976).

Indeed, polyacrylamide itself has been used as a matrix or support for affinity chromatography. Others which have been proposed include the following: polyacrylamide and at least ten derivatives thereof; porous glass (without success due to excessive general adsorption); Sephadex; agarose and derivatives; cellulose and derivatives; polyamides including nylons; polyesters and polyvinyl derivatives; ethylene-maleic acid copolymers; and copolymers of methacrylate and ethylene glycol derivatives and magnetic polymers, thus permitting separation of the matrix-bound enzymes by means of a magnet (Mosbach and Anderson 1977).

Matrix Activation.—Just as agar is the predominant matrix among many others studied, so is cyanogen bromide, CNBr, the favored reagent for activation of agarose although many other ingenious plans for activation have been devised. The complicated reaction of CNBr with vicinal hydroxyl groups of the agarose sugars ultimately yields iminocarbamates. They are especially suitable for attaching spacers or for direct linkage to the ligand. Other activated matrices which are available commercially and are being used are epoxy- and divinyl sulfone-activated agaroses. An application of the latter is exemplified by the binding of the insect hormones (Sage and O'Connor 1976). These can yield derivatives which are simultaneously connectors and activated spacers.

Activation of sugar-containing matrices can be accomplished by reaction with periodate which, as is well-known, produces highly reactive aldehyde groups. For polyacrylamide almost any reagent which reacts readily with free amino groups may be used—carbodiimides, isothiocyanates, as well as CNBr. Bifunctional reagents such as glutaraldehyde can form proximal connectors and activated spacers but the use of such bifunctional reagents introduces the possibility of crosslinking and other side reactions. This survey is merely suggestive. Others will occur to the organic chemistry-inclined reader. Other activators can be found in the abovementioned reviews.

Spacer Attachment.—The substance which reacts with the activated matrix should have an activatable group for what will become the distal end as well as a group which reacts with the activated matrix. For CNBr-activated matrices the substance of choice has been an α,ϵ-alkyl diamine:

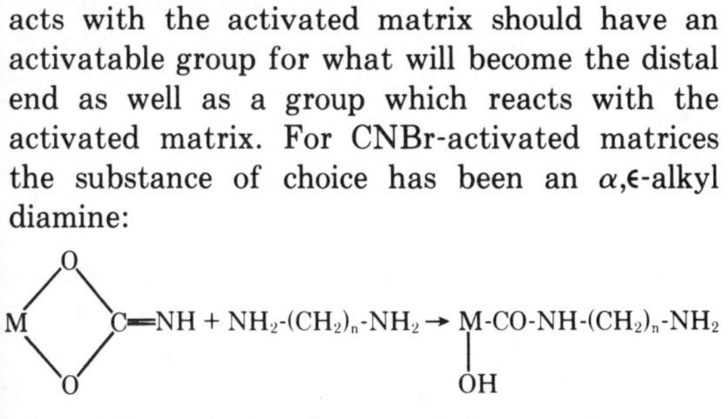

where M stands for the rest of the matrix.

As previously mentioned, one of the functions of a spacer arm is to minimize nonbiospecific interactions. Another function is allegedly to circumvent the steric hindrance, imposed by the looming matrix bulk, of the accessibility of the substrate to the active site of the enzyme. Insertion of the spacer-arm may localize the ligand in a milieu more favorable for normal enzyme-affector binding. In general, the weaker the affinity of the enzyme for the soluble ligand (for a substrate, the higher the K_m) the more likely the necessity for a spacer-arm. As a rule, macromolecular ligands, lectins, protease inhibitors, and protease substrates are attached directly to the matrix, activated or not.

Spacer-arm Length.—Early investigations established that the spacer length, that is, the number of carbon atoms in the alkyl chain, can be critical. Although many investigators used hexamethylene diamine, others later showed that for each enzyme there may be an optimum length. An example is afforded by the observations of Champagnol (1976) on effect of spacer-arm length on retention of plant carbonic anhydrase on agarose to which a sulfonamide inhibitor was attached through a spacer-arm. The enzyme was isolated when an 11-carbon atom was used as a spacer-arm but when a 6-carbon group was used the ligand-attached matrix did not retain the enzyme.

Improvements.—It was found that introduction of alkyl spacer-arms could involve nonspecific hydrophobic interaction (hydrophobic chromatography, Chapter 6) and that ionic groups inadvertently introduced into the matrix during its activation could involve nonspecific ion-exchange interaction with the enzyme. Furthermore, spacer-ligand leaching was frequently reported. Based on the work of Wilchek and coworkers, Miles Laboratories (1976B) made available poly-L-lysine agaroses and substituted hydrazide agaroses. The former reduces both leaching and hydrophobic interaction and makes possible the introduction of more than one ligand molecule per molecule of spacer. The latter provides, at appropriate pHs, charge-free spacer gels without introducing hydrophobic interactions. Hydrophobic interaction can also be suppressed with *N,N*-dimethyl formamide (Aukrust *et al.* 1976).

Spacer-arm Ligand.—It is necessary to either "activate" the distal end of the spacer-arm or groups in the ligand. Available commercially are matrix spacer gels in which the distal end is "activated" by the presence of NH_2COOH, SH, organomercurial (for covalent chromatography) or a succinyl group. The latter can condense with an NH_2 on the ligand *via* the well-known carbodiimide reaction.

Ligand activation *via* coupling of aromatic amines with phenol to form phenolase-inhibiting azodyes was used 25 years ago by Lerman. More recently azo and azo dyes have been used to activate the spacer-arm (Jakoby and Wilchek 1974; Fulton and Carlson 1980).

Finally, one can activate the ligand to permit reaction with the spacer-arm before attachment to matrix. The spacer-arm-bearing ligand can then be coupled directly to the activated matrix, i.e., the group specific ligand, AMP, bearing a spacer-arm is frequently coupled to CNBr-activated agarose.

The Ligands

As indicated before, there appears to be an almost inexhaustible diversity of ligands available for affinity chromatography. In the ensuing discussion we shall briefly elaborate on these "kinds" of affinity by presenting a sampling of investigations published in 1976 which are illustrative of different approaches for which the biospecific interaction between ligand and enzyme can be exploited for purposes of isolation and purification. For the most part, this discussion will be confined to the most frequently used categories: active site and nonactive site-directed ligands.

Active Site–directed Ligands

Although both Lerman and Cuatrecasas used inhibitors and ligands, most of the papers on affinity chromatography which appeared in the early 1970s presented procedures in which the substrates were used as ligands. This circumstance was due in part to a rather intensive investigation of the dehydrogenases especially with respect to the use of the nicotine adenine nucleotides (NAD) and one of its constituents, AMP, as affinity ligands. One of the peculiarities of these two-substrate reactions is that many or most of them are ordered and this circumstance can be utilized in affinity chromatography. When the nonNAD reaction partner is used as ligand, it is necessary to add NAD(P) along with the enzyme and to wash the column with NAD(P) because the NAD(P) has to first induce conformational change in the enzyme to allow the latter enzyme to combine with other immobilized substrate, the ligand. However, most dehydrogenases have been purified with either NAD(P), NADH (Fuller *et al.* 1980) or a fragment thereof as reaction ligand, especially AMP, which is also an effective ligand for the ATP-requiring kinases.

AMP as a Group Specific Ligand.—There are advantages in using a group specific ligand such as NAD(P), AMP or ADP since it is possible to separate and purify related enzymes in one purification procedure. Dean and Harvey (1975) cite the isolation of 21 distinct dehydrogenases using these group-specific ligands. It is also widely used for the separation of ATP-requiring kinases. Another reason for the widespread application of AMP and ADP as group-specific ligands is commercial availability. For instance ADP ligand, made by coupling N^6-(aminohexyl)-adenosine-2′,5′-biphosphate to activated agarose, is available as 2′,5′-ADP-Sepharose 4-B, and is now being used almost routinely in enzyme research laboratories. For instance, Mannervik *et al.* (1976) used it for the rapid and efficient purification of glutathione reductase. The isolation from red blood cells previously took 4 weeks and 10 purification steps. They were able to get a 9760-fold purification in a single step directly from the erythrocyte hemolysate (followed by three further steps) in 3 or 4 days. This illustrates the use of affinity chromatography to purify enzymes present in vanishingly small amounts in certain sources (Chapter 6), as contrasted to its use in speeding up the purification to avoid losses of inherently unstable enzymes.

The versatility of these adenosine-related group specific ligands is further illustrated by the use of 8-(aminohexyl)-amino-5′-AMP-Sepharose for separation and almost quantitative recovery of deoxyribonuclease and ribonuclease, and removal of contaminating chymotrypsin in commercial nuclease preparations (Lazarus *et al.* 1976). Other ligands with relatively broad specificity are the organophosphates which form covalent linkages with active-site serine. They have been particularly successful in the isolation of choline esterases.

Other coenzymes, cofactors and their congeners or other derivatives which have been used as ligands include: biotin, Coenzyme A, FMN, cyclic AMP, cobalamine, lipoic acid, pyridoxal phosphate and thiamin pyrophosphate. Cofactor metabolizing enzymes can also be purified with such ligands. Thus, thiamin pyrophosphokinase, enriched 200-fold in a 72% yield, was isolated by a one-step procedure (Wakabayashi *et al.* 1976). The ligand was the very potent product inhibitor, thiamin monophosphate.

ATP + thiamin → AMP + thiamin pyrophosphate

Macromolecules as Ligands.—In addition to the use of insoluble substrates as ligands without further purification, macromolecular substrates and inhibitors have been coupled to insoluble matrices, usually agarose without intervening spacers. Thus, chymotrypsin and other purified proteases attached to agarose have been used in the isolation of protease inhibitor proteins found in foods. Obversely, proteinase inhibitors have been effectively used for the purification of proteinases. Lima bean inhibitor was used as ligand to isolate modified chymotrypsin and trypsin in which one of the amino acids of the active site, serine, was converted into dehydroalanine (Ako *et al.* 1974). The modified enzymes could still bind to lima bean proteinase inhibitor through their active sites although they could no longer bind their substrates since they were completely catalytically inert.

An enzyme of major importance in food production is ferri-ferredoxin (FX)-nitrate reductase, one of the enzymes involved in nitrogen fixation:

$$NO_3^- + 3FX\ (oxidized) \rightarrow NO_2^- + 3FX\ (reduced)$$

Using ferredoxin coupled directly to agarose and a few nonaffinity preliminary steps, Ida *et al.* (1976) achieved isolation after a 1000-fold enrichment of the enzyme from spinach homogenates. The final affinity chromatography step alone resulted in 140-fold enrichment.

Another enzyme reaction involving small and large reaction partners and which may also be an important food-related enzyme, is the transglycosylation of collagen:

Uridinediphosphate glucose (or galactose) +
Hydroxylysine-Collagen → Uridinephosphate +
O-glycosyl (or galactosyl)-hydroxylysine-Collagen

The enzymes involved may have some hitherto unsuspected influence on the textural quality of meat (Chapter 27). Using denatured collagen bound directly to agarose, Ristelli *et al.* (1976) purified the glucosyl transferase 5000-fold and the galactosyl transferase 1000-fold. When they used a small molecule as ligand (an inhibitor, glucuronic acid) only the glucosyl enzyme was purified, thus eliminating any trace of one enzyme in the purified preparation of the other. Their work also illustrates the rule that the addition of a cofactor—in this case Mn^{2+}—enhances binding of the enzyme to the ligand.

One potential problem that has to be considered in the use of macromolecules as ligands is the effect of their heterogeneity on the ultimate resolution of an affinity system. Amneus *et al.* (1976) concluded that heterogeneity of soybean trypsin inhibitor when serving as a ligand arose not only from the natural ambience, but also as a result of alteration due to the procedure of immobilization on the matrix and to modication of the macromolecule by the proteinases (in pancreatin) to be fractionated. Such considerations would be particularly important in the separation of isoenzymes. However, as we shall forthwith discuss, the most effective macromolecules for use in affinity chromatography are those which combine at sites other than the catalytically active sites. The "blue dextrans," formed by attaching certain dyes to dextrans and similar macromolecules, appear to serve as ligands, when immobilized to agarose, for the affinity chromatographic purification of NAD(P)-dependent dehydrogenases and of some proteinases (Lowe *et al.* 1980; Fulton and Carlson 1980).

Nonactive Site-directed Ligands

Allosteric Modulators.—As previously pointed out, a high degree of specificity and resolution in affinity chromatography can be obtained when the ligand combines at sites on the enzyme molecule other than the catalytically active one. Among these sites are those involved in: regulation, subunit binding sites, antibody-binding sites and carbohydrate portion of glycoprotein enzyme molecules. An example of the use of an allosteric site affector as ligand is afforded by the affinity chromatography of the important allosteric regulatory glycolytic enzyme, phosphofructokinase (Chapter 4), using ATP bound to agarose through a 6-carbon spacer-arm as ligand. It will be recalled that at high concentrations, ATP is a powerful inhibitor (substrate inhibition).

Other examples include the use of tryptophan and tyrosine as affinity ligands for the purification of those enzymes for which they are feedback control inhibitors, i.e., enzymes uniquely involved in their biosynthesis, as discussed in Chapter 4. Purification has been successfully accomplished with tryptophan as ligand for anthranilate synthase, the first enzyme unique to tryptophan biosynthesis, and also for chorismate mutase, the first enzyme leading to phenylalanine synthesis. Further back along the metabolic pathway leading from carbohydrates to aromatic amino acids, both tryptophan and tyrosine each inhibit by feedback one of two difficult to separate isozymes of a special aldolase which forms the first 7-carbon metabolite, unique to the biosynthesis of all the aromatic amino acids. The tyrosine-sensitive can be separated effectively from the tryptophan-sensitive isozyme only by affinity chromatography using tyrosine as ligand. This whole area deserves more attention and further exploration, which it may well have had by the time this appears.

Subunits/Protomers as Ligands.—In addition to the modulation of their activities by allosteric effectors, the circumstance that these regulatory enzymes are oligomers (quaternary structure, Chapter 4) may also be exploited for purification by affinity chromatography. Thus, even if the subunits are identical, the enzyme will be retarded as it reversibly associates and dissociates with matrix-bound subunits as it travels down the column. This subunit exchange chromatography (Antonini *et al.* 1975) was first applied to hemoglobin but has also been used for the isolation of enzymes and active subunits of aldolase which have been bound to be covalently attached to a matrix and agarose matrix (Chan and Mawer 1972). Stockman *et al.* (1976) isolated the major seed protein of the common bean in essentially homogeneous form allowing its subunits, dissociated by pH adjustment, to complex with the same subunits covalently linked to an agarose matrix. Another food-associated protein, α-lactalbumin of milk, the regulatory subunit of the lactose synthase complex (Chapter 4), has been used as an agarose-linked ligand (no spacer) for the isolation of the galactosyltransferase moiety of the enzyme.

Lectins as Ligands.—So far we have confined ourselves to those ligands which, although they may or may not be part of the active site, are nevertheless involved in the catalytic properties of the enzyme. We shall now consider two sites or parts of an enzyme molecule which can be most effectively exploited for its isolation: its carbohydrates—when such are present—and its antibody binding site. Antibody may be considered as a ligand for binding one and only one enzyme, whereas the lectins are analogous to group specific ligands, like the nucleotides.

As previously mentioned, lectins are among the most widely used ligands. They are especially the ligands of choice if the enzyme happens to have appreciable carbohydrate. The lectins are plant proteins which combine fairly specifically with certain sugars attached to proteins and to other biological structures or entities such as cell walls (Sharon 1977). One of the reasons one of the lectins, concanavalin A, is enjoying current popularity is that it is readily commercially available attached directly to agarose. In one of those ironic twists of history, it turns out that this protein, first isolated from jack bean in 1919 by the same Sumner who 8 years later crystallized urease from the same source, should almost 60 years later play a much more prominent and central role than urease in enzymology and also be a key tool in the study of the role of the cell surface in control of tissue growth, mitogenesis and the cause of cancer.

Among the many enzymes purified with the aid of concanavalin-agarose not cited in the above reviews are a few food-related ones: refinement of commercial peroxidase by virtue of its mannose (Brattain *et al.* 1976), the purification of trehalase (Kelly and Catley 1976) and invertase. A food enzyme which should be a good candidate for purification through a lectin ligand is glucose oxidase (Chapter 17). Occasionally it may be advantageous to use different parts of the enzyme molecule in an enzyme isolation-double affinity chromatography. Thus, Miller *et al.* (1976) employed a sequence of concanavalin A-agarose followed by an active-site inhibitor (thiogalactose) bound to agarose in a two-step procedure resulting in a 20,000-fold enrichment and the isolation of liver lactase.

Antibodies as Ligands.—Almost 50 years ago soybean urease was purified by specific precipitation with jack bean antiurease (Kirk 1933). The modern approach is to use a column of the antibody, easily available immunoglobulin, attached to a matrix, usually agarose and—as with just about all macromolecules—without intervening spacer-arm.

Two examples with possible food-related potential will be cited to show how powerful this method can be. Owens and Stahl (1976) achieved a 1700-fold enrichment of liver lysosomal β-glucuronidase, which has been used as marker for other not easily distinguishable food-associated putatively lysosomal enzymes, i.e., cathepsins (Chapter 26). The use of immunoadsorption affinity chromatography to achieve purification of an unstable enzyme is exemplified by the work of Ferrante and Nicholas (1976). In this case they were fortunate to achieve a 54-fold enrichment of the oxygen-labile so-called Mo-Fe-S protein which binds N_2 and is thus the first enzyme in the nitrogen fixation cycle. This investigation also illustrates that immunoadsorption may be the only possible procedure in the absence of a substrate—in this case N_2—that can readily be attached to a matrix, at least by the usually available procedures.

Adsorption Steps

In the usual set-up for affinity chromatography a solution of the crude enzyme (about 10−15% of the total column volume) is placed at the top of a typical chromatographic column. The column is washed with 3-bed volumes of irrigant, frequently under conditions which are optimal for the action of the enzyme with the usual exception of temperature. In general, increase in temperature decreases the ligand-enzyme interaction, and indeed higher temperatures are sometimes used for elution so that, like most enzyme isolations, the procedure is carried out at 0°−5°C.

During this operation the enzyme is either completely adsorbed on the matrix if the interaction is strong (affinity constant in the range of 10^9 or greater) or is merely retarded on its way down the column if the interaction is weak (affinity constant in the range of 10^3). In the latter case the amount of irrigant should be minimized. In either case impurities pass through the column which is then washed with appropriate eluant (see below) and the enzyme-rich fractions collected and pooled.

To obtain maximum binding it may be advantageous, or even necessary, to add cofactor or activator or, as mentioned, cosubstrate to irrigant. Although ideally one should be able to start with a crude extract it is sometimes necessary to introduce preliminary nonaffinity steps in order to remove complexes, enzyme inhibitors (which would compete with the matrix-bound ligand), proteases, carbohydrates if the ligand is lectin, and enzymes which would otherwise degrade the matrix or the ligand.

ELUTION-DESORPTION

Just as there are many alternatives available for adsorption, so there are a multitude of approaches for removing the enzyme from the ligand-matrix complex. These can range from highly specific bioaffinity interactions between eluant and bound enzyme to nonspecific eluants.

Nonbiospecific Elution.—The simplest elution procedure is to continue to pass the irrigation buffer through the column until the enzyme comes off. Of course this will work best with enzymes that are bound weakly. Kiyohara *et al.* (1976) used this approach as part of a double affinity chromatographic separation of a microbial lactase from other glycosidases. One can also change conditions such as pH, ionic strength and temperature (as noted above) and composition of the eluant. In particular, solvents, including chaotropic salts (hydrophobic interaction interferers) are particularly effective when applied as gradients. For instance, Johnson and Travis (1976) describe a rapid one-step purification of human trypsin and chymotrypsin in which the two enzymes are separated during elution from a bovine pancreatic inhibitor column by a decreasing pH gradient. In general, these changes are interpreted in terms of "deformation" of either the enzyme or the ligand so that they no longer possess the same affinity for each other. Perhaps the most extreme deformation occurs with protein denaturants. Thus, 8*M* guanidinium salts were used to dissociate the avidin-biocytin complex and 8*M* urea dissociated a thiamin-binding protein from a column of thiamin pyrophosphate. When the binding is particularly strong, another drastic measure is to disrupt the covalent linkage between the ligand and the matrix and then find a method for separation of the latter two.

Biospecific Elution.—The other approach, to elute with a biospecific eluant by exploiting the enzymatic property of specificity, is probably more frequently used than nonspecific methods. Indeed, this elution by specific displacement is unavoidable if the ligand contains groups capable of binding considerable quantities of foreign material *via* nonspecific forces such as ion exchange (the latter capability can be countered by high ionic strength eluants).

Examples of different types of displacement eluants are shown in Table 7.1. In general, it is agreed that the most reliable, efficient and predictable eluant is the substrate of the enzyme being purified, competing with the matrix-bound ligand for the enzyme. Eluants include not only substrates, but also inhibitors and products (which, as a rule, are also inhibitors). Allosteric effectors may act as eluants by virtue of their ability to "deform" the enzyme rather than by competing with the ligand. According to Dean and Harvey, one can induce a two-substrate enzyme to come off the matrix by omitting a cosubstrate from the irrigation buffer.

TABLE 7.1

ENZYME PURIFICATION BY SPECIFIC DISPLACEMENT ELUTION

Enzyme	Ligand	Eluant	Type
PFK[1]	ATP	F-6-P + ADP	Dead-end ternary
LDH[2]	AMP	NAD + pyruvate	Ternary, general
Glyoxalase I[3]	GSH	GSH	Substrate competition
Pyruvate kinase[4]	None[9]	ATP; PEP + FDP	Mixed
β-Galactosidase[5]	Thiogalactoside	Galactonolactone	Inhibitor competition
β-Galactosidase[5]	Concanavalin	α-Methyl mannoside	Glycoside competition
Glutamine synthetase[6]	Anthranilic acid	AMP	Allosteric effector displaces inhibitor
Malic enzyme[7]	ADP	NAPD	Substrate competition
Thiamin pyrophosphokinase[8]	Thiamin monophosphate	Thiamin	Substrate replaces inhibitor

[1] Source: Ramadoss *et al.* (1976).
[2] Source: Bachman and Lee (1976).
[3] Source: Marmstal and Mannervik (1979).
[4] Source: Marie *et al.* (1976).
[5] Source: Miller *et al.* (1976).
[6] Source: Palacios (1976).
[7] Source: Yueng and Carrico (1976).
[8] Source: Wakabayashi *et al.* (1976).
[9] CM-cellulose used as nonspecific adsorbent.

For instance, one can dissociate an oxygenase by conducting the elution in nitrogen. Another approach for elution of two-substrate enzymes is *via* the formation of "abortive" or "dead-end" ternary complexes consisting of the enzyme, one of the two cosubstrates and "modified" cosubstrate, frequently the end-product derived from the other cosubstrate. Examples (Table 7.1) are use of a pyruvate-NAD adduct for elution of LDH, and a dead-end complex, FDP-ADP, for elution of PFK. Anderson and Wolfenden (1980) advocate elution with TSAIs (Chapters 3, 14) in what they term TSAI affinity jump chromatography.

In addition to substances which elute by virtue of their interaction with catalytic and regulatory sites, enzymes can also be eluted when complexed to a lectin with a sugar derivative, usually a methyl glycoside which competes with the matrix-bound lectin for the enzyme. Analogous reactions are carried out for the elution of enzymes in covalent chromatography *via* SS-SH or Hg-SH interchange.

Affinity Elution

As mentioned at the start of this chapter, what is now termed affinity elution, the biospecific desorption of enzymes from nonspecific adsorbents, was the first affinity phenomenon of this sort to be observed. Since then, it was used, until the last 10 years or so, sporadically, mostly through the capability of the substrate to stabilize the enzyme during elution. More recently and in parallel with the rise of affinity chromatography, it has been utilized in conjunction with ion exchange chromatography. Most frequently used are anion exchangers, especially phosphocellulose, carboxymethyl cellulose and carboxymethyl Sephadex. Affinity elution lends itself especially well to the successive elution of isozymes or other groups of closely related enzymes such as the amino acid-tRNA synthetase. It is also valuable where the ligand is highly charged or cannot be readily linked to an insoluble matrix because of the instability or expense of the ligand.

Much of the recent work has focused on two regulatory enzymes of glycolysis, pyruvate kinase and fructose diphosphate kinase as well as glucose dehydrogenase. It would appear that allosteric effectors are used more frequently than substrates. Kahn and coworkers (i.e., Marie *et al.* 1976) have emphasized the utility of what they call double affinity elution. The adsorbed enzyme is eluted first with substrate and then with a mixture of allosteric effectors which will form ternary dead-end complexes with the enzyme. The use of affinity elution for enzyme isolation is shown in Table 7.1.

ADVANTAGES AND ADAPTATIONS

In this section we briefly summarize and gather together some of the advantages of affinity chromatography, some of which have not been hitherto explicitly pointed out, and briefly allude to nonchromatographic adaptations—its use in the isolation of substances and biological entities other than enzymes, and its use for purposes other than purification.

Advantages and Utility

We feel it is worthwhile to recapitulate and collect the salient features of affinity chromatography that will ensure it a prominent if not dominating position in the armamentarium of enzyme purification procedures. In a few cases we shall document each advantage with a citation mostly from the 1976 literature. Examples of the others have already been given or they can be found in *Methods in Enzymology.*

(1) Drastic shortening of the time required to go through a purification procedure. In addition to the examples previously cited we should like to point out that the use of the TCA cycle enzyme, the malic enzyme for analytical purposes in recycling assays (Chapter 37), was made practicable by a one-step, 200-fold affinity purification from chicken liver which reduced the time of preparation from 2 weeks to 2 days.

(2) Decrease in the number of steps and manipulations so that the ultimate goal of isolation can be realized, as we have frequently alluded to, in one step.

(3) Increased yield as a result of the decreased time and the circumstance that the enzyme is immobilized during part of this time.

(4) The rapid purification also prevents alteration of the original *in vivo* enzymes which, like the phosphofructokinases, undergo rapid association-dissociation of their subunits so that there is a change in their subunit composition during purification (Ramadoss 1976). Of course, in some cases as mentioned, one can also utilize such interchange for purification purposes.

(5) Further stabilization of the activity by removal of enzymes, especially proteases, which might modify or inactivate the original enzyme (Abramovitz and Massey 1976). In the purification of proteases, autolytic self-destruction is prevented.

(6) Removal of inactive enzyme molecules. Sometimes the changes in the enzyme which result in loss of activity are so small that they cannot be separated from the active molecules by any other means, i.e., milk xanthine oxidase (Edmonson *et al.* 1972).

(7) In the *de novo* total chemical synthesis of enzymes from constituent amino acids only a small fraction of the peptides synthesized have enzyme activity. The synthesized enzyme can best be separated by affinity chromatography (see Chapter 4).

(8) Refinement of a fairly pure enzyme preparation. In addition to those already cited an interesting example is the removal of contaminating dehydrogenases from and increase in molar activity of what had hitherto been considered to be a fairly homogeneous cytochrome *c* oxidase (Holbrook *et al.* 1976).

(9) Purification and separation of deliberately chemically-modified enzymes, i.e., separation of inactive chymotrypsin with an altered serine residue. Wilkinson *et al.* (1976) also used affinity chromatography conditions to separate chymotrypsin from chemically-modified enzyme (phenacyl derivative) whose specificity was altered. Falkbring *et al.* (1972) readily separated red blood cell carbonic anhydrase from chemically-modified species in which the active-site histidine was chemically altered. Related to these capabilities are:

(10) Isolation of inactive enzymes resulting from "nonsense" mutations in parent organisms,

(11) Separation and purification of isozymes using dyes and phenyl borate ligands on agarose, and

(12) Separation and purification of zymogen (proenzymes) and active enzyme.

(13) Concentration of dilute enzyme solutions. This should be especially helpful in food enzyme studies where a trace of enzyme activity, as previously pointed out, may have a profound influence on food quality.

Nonchromatographic Configurations

In addition to the conventional use of columns, affinity purification can be carried out in a number of different configurations by adaptations of procedures examined in Chapter 6. The equivalent of "elution" in separating cells with fibers was to simply pluck the fiber. Details of these techniques can be found in *Methods in Enzymology.*

In addition to these methods, a promising adaptation of the Albertsson two-phase partition procedure discussed in the last chapter has been harnessed to affinity purification. In this, one application of affinity partitioning, one of the two soluble polymers which effect phase separation is covalently linked to a ligand to form a "macroligand." Thus, Hubert *et al.* (1976) linked a steroid derivative to polyethylene oxide to purify a steroid isomerase otherwise unpurifiable, even by conventional affinity chromatography. Undoubtedly by the time this appears other ingenious affinity variants will have appeared. One possibility which occurs to us is an adaptation of zone refining.

Purification of Substances Other Than Enzymes

In our discussion of affinity methods we have had occasion to allude to its use for purification of biological substances other than enzymes. To cite one example, Cuatrecasas (1972) purified the insulin receptor protein of liver cell membranes 250,000-fold from the liver homogenate by detergent extraction and affinity chromatography on an agarose-insulin adsorbent, followed by elution with urea-containing buffers in acid pH. Other substances, biological structures and entities which have been purified by biospecific affinity methods include binding sites for hormones, neurotransmitters, drugs, toxins, nucleic acids, ribosomes and ribosomal proteins (Ulbrich *et al.* 1980), complex polysaccharides, cell wall constituents, nonenzyme proteins such as contractile proteins (tubulin), vision receptors, taste and odor receptors and just about every conceivable biological activity involving specific molecular and/or energy transducing interactions. Isolation of the brain anti-pain receptor endorphin and its products of proteinase action, enkephalins as well as the anti-anxiety receptors, so prominent in the public news at the time of this writing, would not have been possible without the advent of modern affinity chromatography.

Nonpurification Applications

Perhaps the most fruitful scientific application of affinity methods to areas in enzymology not involving purification is in their capability of independently testing the validity of kinetically-derived enzyme mechanisms. For example, one can readily determine if a particular substance which inhibits an enzyme is competitive or a noncompetitive inhibitor. If the substance is a competitive inhibitor it will, when placed on an affinity column along with the enzyme, retard the progress of the enzyme down the column. This is because it will compete with the ligand for the active site of the enzyme. A noncompetitive inhibitor should have very little effect on the binding of the enzyme to an active site-ligand and one should obtain an insoluble ternary ligand/enzyme/noncompetitive inhibitor complex. In a more sophisticated application, Hollunger and Niklasson (1975), using affinity gels, were able to define the directional localization of a hydrophobic binding area in the vicinity of the anionic subsite in the active site of mammalian choline esterase. The active sites of the serine proteases have also been probed by affinity chromatography (Kumazaki *et al.* 1976; Chauvet and Archer 1976).

Other nonpurification applications include its use for analysis (Chapter 39) and as a probe for the elucidation of the structure, both chemical and morphological, of complicated molecules and biological entities.

PROSPECTS FOR SCALE-UP AND FOOD APPLICATIONS

The remarkable success of affinity chromatography at the workbench of the enzymologist may augur well for its large-scale industrial commercial application. Probably the sequence of adaptation to industry will be in the order: (1) availability from research biochemical supply houses to provide, hopefully, even yet purer and less expensive enzymes for research, for analysis in both medical and food laboratories (Hue *et al.* 1976), (2) for use in medicine as therapeutic agents [i.e., asparaginase—immunosorption is used to prepare antigens and carboxypeptidase G-1, a possible chemotherapeutic agent according to Cornell and Charm (1976)], (3) removal of enzyme impurities on a large scale from industrial food enzyme additives, and (4) purification of enzymes to be used for the preparation of food ingredients. Its use in the preparation of enzymes to be added directly to foods will develop slowly, if at all, unless there is a distinct economic advantage that compels its adoption. In this field it will undoubtedly first be used for removing impurities. Hsu *et al.* (1975) provide an interesting discussion of the economic feasibility of using affinity chromatography for the removal of enzyme contaminants. They point out that even for purposes of purification of enzymes presumably as research biochemicals, older methods may be more economical because of the stoichiometric relationship between binding sites of an enzyme and its immobilized matrix-bound ligand making affinity columns more expensive in terms of capacity. However, this limitation of capacity costs does not enter into the cost of removing enzyme contaminants. Even the use of antibodies to remove contaminants is considered economically feasible at least for the budget of a research biochemist.

Parenthetically, an interesting example of using affinity methods for removal of enzyme contaminants *via* adsorption of the enzyme is afforded by Hatton and Regoeczi (1976). Their studies on the structure and function of certain asialoproteins, important components of cell walls and membranes, were hampered by the presence of proteolytic enzymes in the galactose oxidase preparation used in these studies. Complete removal of proteolytic activity was accomplished by simply passing a solution of the oxidase through a column of unmodified agarose beads which served as both matrix and ligand. However, it is likely that the preferred mode of employment when used for food enzymes will be to remove the contaminating activity from the bulk of the enzyme extract. The expense of desorption, especially by the use of possible expensive and not readily removable specific ligands as eluants, precludes the use of affinity adsorbents for removing the desired enzyme.

Perhaps an advantageously economical mode of application of affinity methods in the food industry would be to provide the food processor him- or herself with a variety of packets of specific affinity adsorbents. The set-up using soldered net bags shown in Fig. 7.3 and presented by Porath and Kristiansen (1975) for laboratory use, allows for simultaneous adsorption of several enzymes. Depending upon the particular use, each adsorbent could be added immediately before use or even during processing, to batches of the same commercial enzyme sample. For instance, proteolytic activty would be removed when the glucose oxidase is to be used for egg desugaring whereas amylolytic activity would be removed when used for desugaring cold-stored potatoes as discussed in Chapter 19. The ratio of amylase to protease of crude commercial preparations could be altered at will for use in the baking industry depending upon the amount already present or what baked good would be made from the flour. Glucoamylase could be freed of transglucosylase by inserting a packet of agarose to which was attached α-1,6-glucosyl-containing ligand, perhaps dextran or panose. Offending proteinase inhibitor could be removed from high-protein foods such as soybeans by the use of packets of immobilized trypsin. Somewhat more difficult but feasible would be the removal of lipase and nonmilk-clotting proteolytic enzymes in the production of microbial rennets.

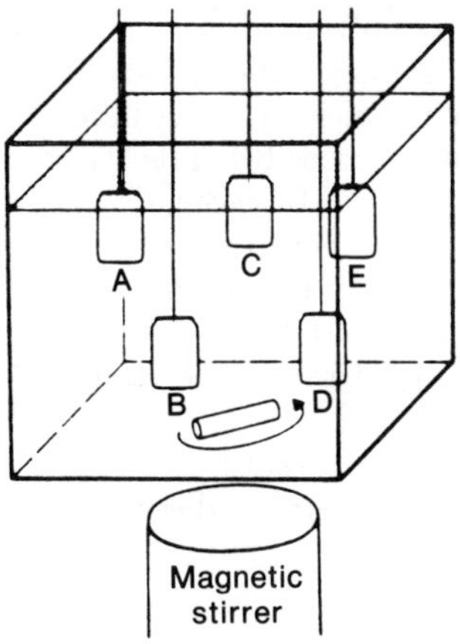

From Porath and Kristiansen (1975)

FIG. 7.3. AFFINITY PACKETS FOR REMOVAL OF NONTARGET ENZYMES

Since matrix-attached ligands for lipoxygenase have been prepared (Allen *et al.* 1977), it should be possible to remove this enzyme from foods in which its action impairs quality (Chapters 17 and 24). Polygalacturonases could be removed with packets of alginates in the production of low-methoxyl pectins through the use of pectin methyl esterase. Undoubtedly, other examples will occur to the reader and may have been in use by the time this appears.

Cost considerations would also result in the prior development of simple insoluble substrate-enzyme systems before matrix-ligand systems are used. Keay *et al.* (1972) describe sequential schemes for the large-scale commercially feasible separation of proteases and amylase from *B. subtilis* with adsorption of the latter on starch as one of the steps. Probably the first application for purification *via* adsorption-desorption processes will be one-step procedures when the enzyme is present in large amounts. Already a substantial literature has appeared on the separation and purification of the pancreatic proteinases in one step using mostly natural antitrypsin inhibitors as ligands. Thus, Porath (1976) starting with frozen dehydrated pancreas simultaneously isolated essentially pure chymotrypsin and trypsin in one step. One possible saving grace as far the economic feasibility of using such affinity gels is their stability since they lost only 33% of their capacity after 3 years of intensive use.

Another example of a proteinase purification study with food-application possibilities is the separation of collagenase, effective in tenderizing canners and cutters meat (Chapter 27), from SH-dependent proteinases (lysosomal cathepsins) which may be involved in aging-induced tenderization, by covalent affinity chromatography (Gillet *et al.* 1977). Also of food-related interest is the separation of mixtures of pepsin and rennin (Chernaya *et al.* 1976).

In the still more distant future, at least as far as food uses are concerned, these relatively simple matrix-ligands may eventually be exploited for purification of target enzymes. However, only with enzymes that will command high cost could the expense of the full regalia of activated matrix connected to the ligand through the spacer-arms be recovered. Robinson *et al.* (1974) describe an automated system with a 1.8 L-affinity column capable of putting out that much-investigated food-related enzyme, lactase (unfortunately obtained from a nonfood use organism, *E. coli*) at a rate equivalent to 60 g of pure enzyme and 100 g of crude enzyme per day.

8

IMMOBILIZATION OF ENZYMES

It would appear at first glance to be superfluous and redundant to devote a chapter to immobilization of enzymes after discussing how enzymes can be constrained in space for purposes of purification. As used in this chapter and in the context of the term as used today, the main differences between "affinity chromatography" and "enzyme immobilization" lie in a difference in aims. As we have seen, the principal purpose of immobilizing an enzyme in affinity chromatography is to accomplish some sort of purification. The main purpose of "enzyme immobilization" is to put the enzyme to work or as expressed by Zaborsky (1974) to perform a continuous catalytic process. Indeed it is better to use an already purified enzyme for the purpose of catalyzing the conversion of its substrate for industrial or analytical purposes or to learn how enzymes work in cells. True, the substrate may also be present in the performance of affinity chromatography but the aim is to allow it to combine with the enzyme without consummating the catalytic act.

Furthermore, the support or space-restricting parameter in affinity chromatography, be it a matrix-ligand or other means of enzyme confinement, requires a degree of complementarity to the enzyme. It is this complementarity which results in spatial restraints, whereas the means of confinement in enzyme immobilization is rarely through such biospecific complementarity. Indeed the principal means for enzyme localization may be some sort of physical or chemical spatial confinement based on solubility or molecular size differences between the enzyme and the instrumentality of confinement. The two fields are, however, different sectors of what has been termed solid-phase biochemical technology (another area being the solid-phase peptide and other macromolecule synthesis). Of course under special conditions the two are not necessarily mutually exclusive. It may be technically and economically feasible to immobilize an enzyme on, for instance, an insoluble matrix by affinity chromatography and then to "cement" it in place by one of the myriad available reactions. This can only be done if the coupling between matrix and ligand does not involve the catalytic active site of the enzyme in such a way that this site is no longer accessible to substrate added subsequently. Space and time since this chapter was written do not permit a full listing of the many reviews and books on enzyme immobilization. Key books designed as such and not as symposia proceedings are those edited by Wingard *et al.* (1976), Messing (1975), Mosbach (1976) and Brown *et al.* (1978). Immobilized food-related enzymes are dealt with in symposia books by Olson and Cooney (1974) and Pitcher (1981). Others can be found in association with specific chapters or papers cited, and in reviews by Shukla (1977), Mosbach (1980), Keyes (1980) and Sturgeon and Kennedy (1981).

Historical

The historical role of inorganic insoluble adsorbents in enzyme purification was briefly examined in the last two chapters. In the same year that Hedin showed that a substrate could elute an enzyme from nonspecific adsorbent, Michaelis (1908) observed that invertase lost none of its activity when adsorbed nonspecifically on a variety of solids. This observation was one of the cornerstones leading to the active-site concept (Haldane 1930). Furthermore, as enzymologists graduated from simple secreted extracellular enzymes to the more difficult to extract intracellular ones, the concept of desmoenzymes, enzymes tightly bound by the cells even after extraction, emerged in the early 1920s (Balls 1943). Willstätter in the 1920s set great store on trying to study the condition in which enzymes exist in living cells. He concluded that even such well known secretory pancreatic enzymes such as trypsin and amylases, as well as liver catalase, are largely bound *in vivo* ("desmos" is Greek for bond) to insoluble cell constituents (Willstätter and Rohdewald 1932). These ideas anticipated the modern concept that all intracellular enzymes are "immobilized," as shown, for example, by Kempner and Miller (1968) for the alga *Euglena*.

Soil, too, has been considered to be an immobilized enzyme system (McLaren 1974). In the 1950s and early 1960s some biologists and other scientists became interested in the creation of model immobilized enzyme systems to simulate and simplify *in vivo* systems in order to gain some insight as to how this binding might contribute to the maintenance of the living state. Among the principal advocates in the 1950s of this position who led groups which laid the theoretical basis for subsequent experimental development were McLaren and Katchalsky. These and other milestone investigations which heralded the literature explosion subsequent to 1965 in which it was demonstrated that what are now known as immobilized enzymes are listed in the following chronology covering the prior 45 years:

1921: Nelson and Griffin show that adsorbed invertase is active.
1925: Willstätter *et al.* show that adsorbed pancreatic amylase is active.
1929: Karrer and Mangelli show that cellulase is adsorbed to its substrate during the latter's hydrolysis.
1938: Langmuir and Schaeffer show that urease and pepsin monolayer films at air-water interfaces retain activity.
1945: Butler shows that enzymes adsorbed on solid nucleic acid are active.
1951: Zechmeister develops "paint brush" test in which substrate is streaked on an extruded column containing separated, adsorbed, active enzymes.
1953: Zittle and Schwimmer review enzyme adsorption and immobilization studies.
1954: McLaren finds that enzymes adsorbed on kaolinite are active but do not obey Michaelis kinetics.
1956: Mitz studies covalently linked active insoluble derivatives of enzymes as models of cellular metabolism.
1958: Reese and Mandels show that enzymes on partition chromatographic columns are active.
1960: McLaren develops theory of enzyme action in structurally restricted systems.
1962: Glazer *et al.* synthesize active polytyrosol trypsin.
1964: Manecke synthesizes active enzyme resin.

Although the technological potential of these experiments was pointed out by Grubhofer and Schleith in 1954, it was not until the latter half of the 1960s that a substantial literature on deliberately immobilizing enzymes on inert supports in order to exploit their catalytic capabilities started to develop. Much of the forefront of this activity has stemmed from grants to investigators in departments of chemical engineering by the RANN program of the U.S. National Science Foundation (research applied to national needs). Additional impetus was supplied by chemical supply houses, government laboratories and, for a while, by manufacturers of potential supports, matrices and membranes. By 1971 what was the first of many subsequent conferences and symposia, this one on "enzyme engineering," was held in which the principal investigators summarized and evaluated the progress resulting from about 350 papers published between 1966 and 1970. At least another 1000 papers on over 300 different immobilized enzymes were published in the 3-year span through the middle of 1975, as evidenced by a compilation of 1200 paper titles in this field by Weetal (1972–1975). The 100-odd entries listed in Volume 85 of *Chemical Abstracts* (Anon. 1976C) under "Enzymes, immobilization of" and "Enzymes, immobilized," are divided roughly equally among reviews, patents and reports of original investigations. The initiation of this drive and subsequent progress were and still are attended by quite a bit of fanfare, particularly in the specialized trade journals, about how immobilized enzymes will revolutionize or are revolutionizing that particular trade. It may be well to summarize the underlying causes of this rosy prognostication—at least as applied to food-related processing—of an approach which was hitherto confined to a rather academic search for one of the facets of biological activity and structure. While we cannot go into the myriad systems and variants proposed, we deem it appropriate to allude to those applications which appear to have relevance and some potential in control management and creation of foods and improvement of quality.

Why Use Immobilized Enzymes?

For the present we shall examine some of the reasons why enzyme immobilization has captured the imagination of the engineers and is currently looked upon by the food science community with more than passing interest—quite aside from the one spectacular success of its application to the manufacture of fructose-containing sugar syrups (Chapter 9).

In Chapter 1, we compared the advantages of enzyme-catalyzed reactions with noncatalyzed reactions for use in food processing. However, even this homogeneous catalysis by enzymes has its own shortcomings. Using economically and technologically feasible amounts of commercially available enzymes may result in changes which take place too slowly (sometimes hours) or else large amounts of enzymes have to be used. Almost invariably the enzyme is left in the product; this restricts the range of enzyme products which can be used in foods (Chapter 5). The reaction may be slowed down and the enzymes' potentials are not utilized to their fullest because of product inhibition and substrate deletion (Chapter 14). Precise control is often not feasible and cessation of the enzyme action can often only be accomplished by rapidly heating the food product. These shortcomings are overcome, at least in theory, by the use of immobilized enzymes.

Among the benefits and advantages forwarded by those who recite a litany of praise for this new technology are the following. Because the enzyme does not become part of the product, it can be repeatedly reused. In other words, full advantage is taken of that aspect of the catalysis which states that the catalyst is not used up in the reactions. Without immobilization only the specificity and acceleration features of enzymes are fully exploited. Also, repeated use allows economically feasible reclamation of enzymes which can only be recovered from their biological source in low yield. With appropriate reactor configuration more precise control of the action can be obtained, substrate is not depleted; hence, the reaction proceeds at a constant rate (zero-order kinetics) and can be easily terminated at any predetermined endpoint. One way not amenable to soluble enzyme action by which this can be accomplished is to simply vary the rate at which the substrate is continuously exposed to the substrate. This would be helpful, for instance, in the manufacture of a range of amylase-produced dextrins and sugar syrups with precisely controlled dextrose equivalent values (Chapter 9). Furthermore, by continuously exposing fresh substrate to the enzyme, it becomes possible to prevent appreciable buildup of product in the vicinity of the immobilized enzyme and thus obviate the above-mentioned product inhibition. Improved control can also be achieved *via* encapsulation which permits time release of immobilized enzyme when desired.

Another way to stop the reaction is simply to remove the immobilized enzyme from reaction vessel in the case of simple stirred tank reactors. For certain other engineering reaction configurations, like the fluidized bed reactor, enzyme can be removed with a magnet by immobilization of the enzyme on a magnetic support as, for instance, reported by Van Leemputten and Horisberger (1974) or Gelff and Boudrant (1974). Magnetic supports are particularly useful in systems of high viscosity and concentration of undissolved particles. Since, as in other heterogeneous catalysis in general, the effective substrate concentration is increased by an apparent lowering of the K_m (see below), immobilized enzyme action in a column configuration should make it possible to perform the task of practicably removing or transforming unwanted substances in foods such as toxins, antinutrients present in too dilute solution for batch homogeneous enzyme action.

In general it has been found that enzymes are frequently more stable when immobilized than when in solution, both under optimum conditions and also at extremes of pH and temperature. Stability by prevention of self-digestion of proteolytic enzymes is also achieved. Part of improved stability could be due to shifts in the pH and temperature optimum.

Economic Advantages.—The shift may be in a direction which is more compatible with other processing steps in the food production. In connection with industrial processing in general, it is frequently stressed that immobilization is compatible and lends itself more readily to continuous processing. This is supposed to result in a lower capital outlay.

This type of estimate leads to one of the prime alleged advantages of enzymes, namely, economics. Indeed Wolnak (1974) states that the use of immobilized enzymes offers a basis for overcoming this principal obstacle in the heretofore snail-like pace of expansion of industrial application of enzymes. Thus, in an analysis of the cost of milk-clotting enzymes in the manufacture of cheese, Wolnak arrived at a figure of about 0.5¢ per 500 kg of milk when immobilized rennet in a continuous process is reused the equivalent of 30 cycles. This compares with roughly 50¢ for present-day immobilized enzyme costs, according to Wolnak, a figure which is somewhat higher than the estimates shown in Table 5.6 for use of soluble enzyme. Perhaps this example may well serve to illustrate why immobilization is not being adopted so rapidly into existing processes.

Whatever significant economic benefits occur in such a shift-over, reduced enzyme costs may not be the aim (saving of a fraction of a cent per kilogram for a product which retails at about $4.00/kg) but rather the reduction of costs attendant upon enzyme addition by, for instance, automating the milk-clotting process. In any case, related economic benefits of immobilized enzyme use in industry include reduction of plant space requirements and disposal problems.

Safety Considerations.—Use of immobilized enzymes removes the necessity of direct contact between the enzyme and the food being processed when the substrate is small enough to penetrate the barrier imposed by the support, matrix or other restraint. The absence of the enzyme in the food should thus allow more latitude in the selection of enzyme source. This may help avoid problems confronted by the food processors in complying with federal regulations. A significant safety advantage is the reduced possibility of immunological reactions, especially in personnel who would otherwise be handling impalpable powders.

Use of Consecutive Two-step Reactions.—It is now well established that accumulation of product resulting from two or more consecutive enzyme-catalyzed reactions is higher when the enzymes are immobilized than when in solution (Mosbach 1974; Hultin 1974). This is especially advantageous when cofactors are involved. Other functions and presumably advantages in addition to acceleration of multienzyme aggregates as suggested by Hultin include: (1) modification of the nature of the reaction-production enhancement or, in some cases, inhibition by the presence of another protein as in the case of the lactase and tryptophan synthases complexes (Chapter 4); (2) stabilization of intermediates since they do not leave the site of the enzyme and are thus in a hydrophobic, water-poor environment; and (3) establishment of a unidirectional flow of reactants. Fundamental development of theories of multienzyme systems are usually directed to their utility in the living organism as touched upon briefly in Chapter 4. The kinetic behavior of artificial multienzyme membranes has been discussed and dissected by Katchalski (1970) and Lecoq *et al.* (1975). Probably the first application of such systems in food processing will be to utilize enzyme reactions in which a normally expensive cosubstrate can be regenerated. This would open the gates to the use of the NAD-dependent dehydrogenases proposed for such purposes as the removal of the off-flavor causing diacetyl from beer with diacetyl reductase (Chapter 24).

$$\text{NADH} + \text{Diacetyl} + \text{H}^+ \xrightleftharpoons{\text{Reductase}} \text{NAD}^+ + \text{Acetoin}$$

$$\text{NAD}^+ + \text{Glucose} \xrightleftharpoons{\text{Dehydrogenase}} \text{Gluconate} + \text{NADH} + \text{H}^+$$

Other Potential Benefits.—Other advantages and broadened horizons opened up in the application of immobilized enzymes include catalysis of the reverse of the conventional reaction as in the synthesis of plasteins (Chapter 34). See also Ingalls *et al.* (1975) on ester synthesis by the proteases chymotrypsin and subtilisin. Immobilized enzymes have already been proved to be of great benefit in analysis especially in connection with the development of enzyme electrodes and microcalorimetry as adumbrated in Chapter 38. The process of immobilization may remove soluble impurities from an otherwise crude preparation, thus eliminating the necessity for any further purification.

IMMOBILIZATION PROCEDURES

Modes of Immobilization

Reactor Modes.—A reactor, a term frequently encountered in chemical engineering discussions and introduced into the literature on immobilized enzymes, is simply any set-up for carrying out a chemical reaction. The term is more encompassing and has a broader connotation than "reaction vessel" since a reactor may comprise a multiplicity of components and especially includes operational procedures. Schematic representation of the more frequently used reactor modes for immobilized enzymes not mentioned elsewhere in the present discussion is shown in Fig. 8.1.

Although the many discussions of immobilized enzymes present a variety of schemata for classifying the modes of enzyme immobilization, they all agree on the separation into two broad classes, bound *vs* entrapped. Immobilization does not necessarily result in insolubilization of the enzymes; but most bound enzymes, especially those involving covalent attachments, are bound to water-insoluble supports. The classification shown in Table 8.1 also envisages two fairly distinct categories. The first involves chemical or physiochemical forces.

Entrapment.—In the second, the enzyme, by virtue of the relatively large size of its molecules, is entrapped or encaged within a mesh, lattice or other three-dimensional network of polymeric crosslinked support or matrix, or water-insoluble liquid phase as in some cases of microencapsulation. These materials include gels, cloth sheets, gel fibers, collagen or even biological cell envelopes of various types. Alternative terms include occlusion, physical restraint, spatial restriction or containment.

TABLE 8.1

MODES OF ENZYME IMMOBILIZATION

A. Known Physical/Chemical Interaction
  Adsorption
  Ionic Interaction with Ion Exchange Resins
  Covalent Linkage
    Between carrier-spacer and enzyme
    Crosslinking
      —intermolecular, of enzyme to itself via bifunctional reagents
      —cementing of adsorbed or ionically bound enzyme
      —incorporation into 3-dimensional matrix via copolymerization
    Chelation through metal ion complexes
    Within hollow fiber

B. Entrapment, Confinement, Containment of Soluble Enzyme
  —via "fixation" *in situ* in microbial cell or organelle
  —within polymer during polymerization and/or crosslinking
  —via separation of enzyme and substrate by a membrane (ultrafiltration)
  —via enclosure in a variety of devices: hollow fibers, cloth fibers, microcapsules, film

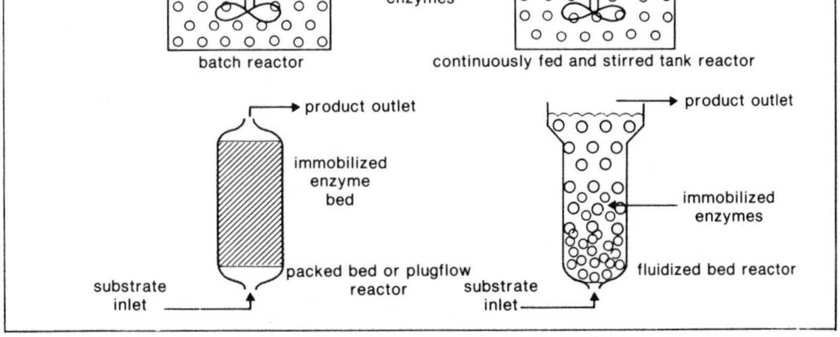

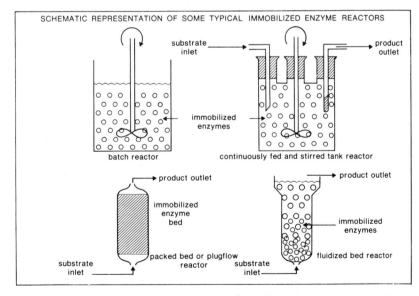

FIG. 8.1. REACTOR CONFIGURATION MODES

From Hupkes and van Tilberg (1976)

Two points of difference between entrapped and other immobilized enzymes are: (1) they usually act on relatively small molecules because substrates comparable in size and shape to that of the enzyme cannot penetrate the interstices of the restraint, and (2) the intrinsic properties of the enzyme are not changed as those of bound enzymes may be. However, restriction on activity due to substrate diffusion (mass transfer) and partition effects may be even more severe than those associated with covalently bound systems (see section on Kinetics).

Nevertheless, it is perhaps no accident that almost without exception in the first commercially successful immobilized enzyme systems the enzymes were entrapped within the cells in which they were elaborated or in ion exchange resins where the enzymes are probably held in place by a medley of factors, including physical entrapment. The most frequently used monomer for polymerization is acrylamide but as far as its use in medical and perhaps in food-related situations it is restricted by lack of biocompatability. 2-Hydroxyethyl methacrylate (HEMA) can be readily polymerized in the presence of enzyme to yield a satisfactory biocompatible immobilized enzyme gel (Korus 1976). Entrapment can also be effected by inducing polymerization with ionizing radiation (Maeda and Suzuki 1977).

Hollow Fiber Entrapment.—Space does not permit even a cursory examination of each of the other various kinds of entrapment shown in Table 8.1. We therefore shall briefly scrutinize what appears to us to be one of the more promising and flexible approaches with respect to food-related applications, the use of hollow fiber devices previously mentioned in connection with enzyme purification. A hollow fiber schematic is shown in Fig. 8.2. One of the great virtues of this device is flexibility. Not only can it be used as indicated in Fig. 8.2 with the enzyme entrapped within the membrane and the substrate circulating in the lumen but also in various other ways (Fig. 8.3). In each case the enzyme substrates and products are completely soluble but only the latter two can traverse the membrane. It will be noted that enzyme can be localized within the lumen or the fiber tube wall or can be restricted to the shell side (the region exterior to the hollow fiber). In addition to the variations in which the enzyme is soluble, most of the other means of immobilization can be used in conjunction with hollow fibers. Noncovalent containments which have been used include adsorption by various supports, confinement within liposomes and microcapsules.

Bound Enzymes

As indicated in Table 8.1, enzymes can be either physically or chemically bound to matrices or supports. In general only the binding by physical means is more or less reversible but by no means assured.

Crosslinking Agents and Covalent Bonding.—Frequently a physical binding can be made irreversible and the enzyme can be permanently

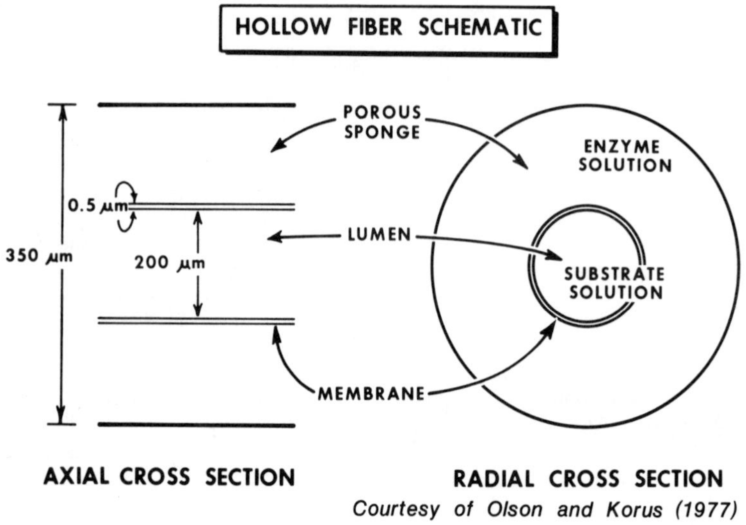

FIG. 8.2. IMMOBILIZATION OF AN ENZYME IN AN ASYMMETRIC HOLLOW FIBER

fixed onto an insoluble support by crosslinking with a bifunctional reagent as indicated in Fig. 8.4. The "cement" of choice is glutaraldehyde for binding to wide variety of supports including alumina, cheesecloth, DEAE cellulose, collodion, colloidal silica, chitin, filter paper and phenol-formaldehyde resin.

Quite a bit of labor and ingenuity have gone into devising interesting reactions for covalently linking enzymes to insoluble supports. Many can be found in Volume 44 of *Methods in Enzymology*. A sampling of such methods is presented to the reader to give a feeling of some of the activity in this field. Whether any of them will be adopted on an industrial scale is left for the future. Besides glutaraldehyde, three frequently used crosslinking agents are dialdehyde starch, diazobenzidine and di-isocyanate derivatives (Fig. 8.5). Unlike the procedure for linking enzyme to matrix in affinity chromatography, one can link the enzyme to the support with such bifunctional reagents without prior "ac-

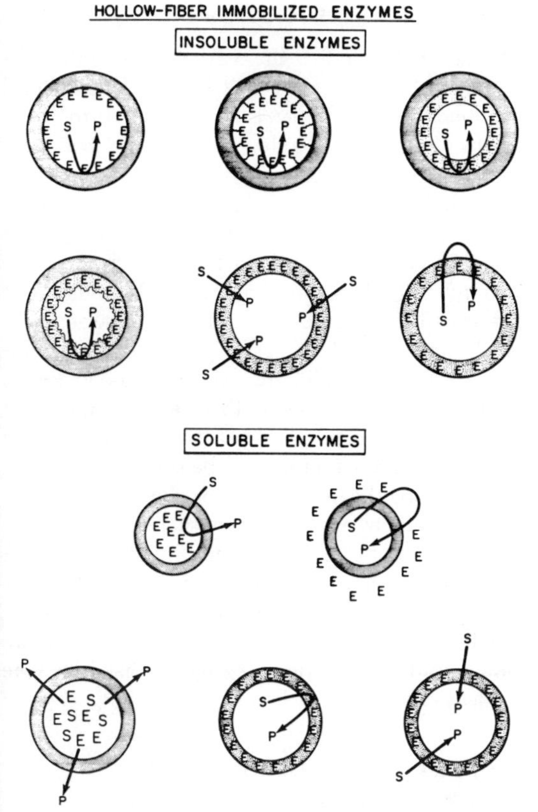

From Robertson (1976)

FIG. 8.3. ALTERNATIVE LOCALIZATION OF ENZYME IN HOLLOW FIBER DEVICES

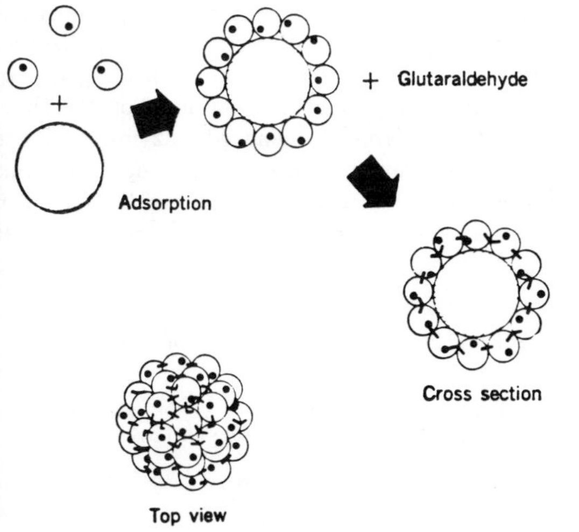

From Aiba et al. (1973)
Copyright Univ. of Tokyo Press.

FIG. 8.4. CONVERSION OF REVERSIBLE TO IRREVERSIBLE IMMOBILIZED ENZYME BY CROSSLINKING

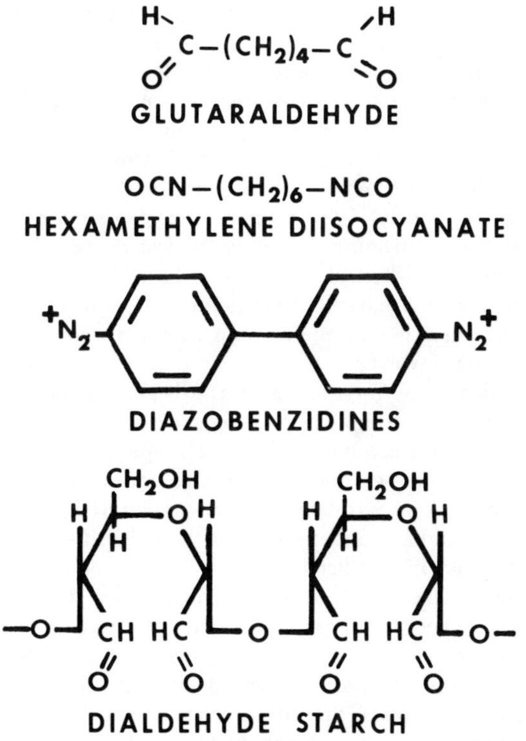

From Stanley and Olson (1974)

FIG. 8.5. CROSSLINKING AGENTS FOR ENZYME IMMOBILIZATION

tivation" of groups on the matrix supports, although this has also been resorted to. More frequently versatile reagents and reactions such as the carbodiimide or ingenious reactions such as that suggested by Zaborski (1973) are used:

$$\underset{\displaystyle\mathrm{O\text{-}R'}}{\mathrm{R\text{-}\underset{|}{C}{=}NH_2^+}} + \mathrm{E\text{-}NH_2} \rightarrow \underset{\displaystyle\mathrm{NH\text{-}E}}{\mathrm{R\text{-}\underset{|}{C}\text{-}\overset{+}{N}H_2}} + \mathrm{R'\text{-}OH}$$

Another imaginative covalent attachment of enzyme developed by Guire (1975) is based on the photochemical attachment of the enzyme to agarose through the photosensitive dye, 1-fluoro-2-nitro-4-azidobenzene. The photosensitive dye can be coupled in the dark to either support (carrier) or to enzyme, followed by exposure to light. According to Guire, a wide variety of enzymes bind facilely with little loss in activity.

Protein Nucleophiles.—In general, the support-matrix-carrier contains an electrophilic group, or one is inserted so that it can react with the nucleophilic group associated with one or more amino acids of the enzyme. While it is undoubtedly true that the most reactive of these groups is the ϵ-NH_2 of lysine, opportunity for reaction would include reactions with: Nitrogen—amino (lysine, N-terminal amino) acid, imidazole (histidine), indolyl (tryptophan); oxygen—hydroxyl (serine, threonine), phenolic hydroxyl (tyrosine); sulfur—sulfhydryl (cysteine), disulfide (cystine) and methylthioether (methionine). It is likely that the thioether of methionine or the threonine is seldom involved.

Ideally, as mentioned in the beginning of this chapter, the coupling resulting in a covalent linkage should not involve or disturb the spatial arrangement of the groups in the active site. In practice some active site modification undoubtedly does occur and accounts in part for the reduced activity of most immobilized enzymes. What makes coupling effective at all is the circumstance that the amino acid residues in active site as a rule are not as accessible to the coupling reagents, especially when the latter are associated with the support, as are nucleophilic groups on other amino acid residues. In any case it should be possible to reduce such interaction by carrying out the coupling in the presence of substrate or competitive inhibitor.

Composition of Insoluble Supports

Tables 8.2–8.5 list the various insoluble substances which have been tried as immobilized enzyme supports. They are classified into three groups: inorganic (Table 8.2), organic polymers (Table 8.3) and biologically derived materials (Table 8.4). An additional listing is made for cellulose (Table 8.5) although enzymes have probably been immobilized on perhaps as many derivatives of such much-used polymers as agarose. It would be of interest to see a list, if any, of insoluble substances which have failed to serve as supports for immobilized enzymes. There are, as far as we know, no hard and fast rules as to what will make the ideal support. In terms of practical use one of the main criteria must, of course, be cost. A social motive would be the use of waste products such as chitin, the crustacean and insect skeletal substance. Lignin might be another. Some investigators believe that the most suitable carriers should be hydrophilic. In this connection Super-Slurper, acrylonitrile-modified starch (Weaver *et al.* 1976), should find many interesting immobilized enzyme applications.

PROPERTIES OF IMMOBILIZED ENZYME SYSTEMS

As is to be expected the characteristics and enzymatic properties may be drastically altered as a result of immobilization. Of course the mode of immobilization may have a great influence on the behavior of the enzyme both positively and negatively. In spite of the diversity of immobilization effects, a considerable literature attempting to rationalize and predict data obtained on the characteristics of the immobilized enzyme and the behavior of immobilized systems in operation has developed. What are the fundamental chemical realities behind these observed changes? Rather

TABLE 8.2

SELECTED IMMOBILIZED ENZYME SUPPORT—INORGANIC

Alumina	Clay	Metalloglass
Aluminum hydroxide	Diatomaceous earth	Nickel oxide
Bentonite	Glass; beads, powder, porous, silylated	Sand
Brick		Silica
Calcium phosphate	Hydroxylapatite	Titania
Calcium carbonate	Kaolin	Zirconium
Celite	Macadam	Stainless steel

TABLE 8.3

SELECTED IMMOBILIZED ENZYME SUPPORTS—SYNTHETIC ORGANIC POLYMERS

Acrylamide	Ethylene-maleic anhydride	Iodoalkyl methacrylate
Acrylonitrile	HEMA	2-Hydroxymethyl ethyl methacrylate
Acrylamide	Ion exchange resins	Nylon and derivatives
N,N methyl-bis-allylcarbonate	Isocyanate polymers	Phenol formaldehyde resins
		Triazynyl duolite
Aminostyrene	Maleic anhydride	Vinyl alcohol
Epidex, derivatives	4-Methacryloxybenzoic acid esters	Silastic resin
		Styrene
Ethylene polymers		Tygon (tubing)

TABLE 8.4

IMMOBILIZED ENZYME SUPPORTS—
BIOLOGICALLY DERIVED

Agarose	Erythrocytes	Rubber latex
Alginate	Liposomes	Starch, derivatives
Cells, animal	Pectin	Starch
Chitin, chitosan	Protein,	Tannic acid
Dextran, Sephadex	denatured	Wool
Silk	Feathers	Cellulose

TABLE 8.5

IMMOBILIZED ENZYME SUPPORTS—CELLULOSE DERIVATIVES

Acetate	CMC-hydrazide	Epichlorhydrin-triethanolamine
Acetate-butyrate	CMC-azide, -ether	ECTEOLA
Amine-aryl derivatives	Cellophane	Halogen acetyl
2-Amino anisole	3-Cl-2-OH-propyl	Methyl
p-Amino benzoate	Cotton	Nitrate
Carbonate	Cyanoethylated	Oxidized (partially)
Carbonate—2,3-*cyclic*	Dialdehyde	Triazine, derivatives
Carboxymethyl (CMC)	Diethylaminoethyl	

than go into a detailed critique of the present status of immobilized enzyme kinetics, it may be well to examine how authorities have classified the forces and pressures at work at the molecular level, present some of the simpler concepts derived for immobilized enzyme kinetics and proffer a comment or two concerning their general applicability.

Engasser and Horvath (1976) cite the following hierarchy in the characteristic complexity of kinetic or rate parameters:

(1) *Inherent*—those associated with the soluble unmodified enzyme;
(2) *Intrinsic*—those imposed upon the enzyme by virtue of its being immobilized except for the effect of diffusion of substrate and other ligands;
(3) *Effective*—those observed experimentally in the presence of diffusional effects.

Goldstein (1976) distinguishes among the following effects of immobilization:

(1) *Conformational and steric*—as mentioned, except for soluble enzyme held in the lumen or shell of a hollow fiber device, the act of immobilization influences the active site and the conformation of the enzyme.
(2) *Partitioning*—the concentration of substrate or other affector of enzyme activity may be distributed, even if and after equilibrium has been reached, unevenly between the bulk of the solution and that in the vicinity or domain of the support. The resultant effects (i.e., changes in K_m as affected by ions, ionic strength, pH-activity profiles and hydrophobic interactions) will be greatly dependent upon the chemical nature of the substrate, and the support.
(3) *Microenvironmental*—the catalytic events occur in an environment different from that in solution.
(4) *Diffusion or mass transfer restriction*—investigators distinguish between external, which applies to surface bound enzymes, and

internal, which applies to enzymes (mass transfer) held within a support.

Some Kinetic Considerations

Partition Effects.—Swaisgood (1977) classifies these effects as specific interactions between substrates as well as small ligands within the surface of the support and goes on further to divide the effects between ionic and nonionic interactions.

From simple electrostatic theory, when the support is charged the pH optimum is supposed to be shifted according to the following equation:

$$pH^* = pH + 0.43\ \epsilon\Psi/kT$$

where pH* and pH are the localized and bulk pH optima, ϵ the electronic charge, Ψ the electrostatic potential of the support and k the Boltzmann constant. What this equation predicts is that if an enzyme is bound to an anionic surface, the pH optimum will be more alkaline; if bound to a cationic support it will shift to a more acid pH. The classical example of the verification of this prediction is shown in Fig. 8.6.

The open circles represent the pH-activity profile of chymotrypsin in solution and the closed circles that when bound to a polyanionic support (actually copolymerized with a polyanionic matrix) and the triangles that when bound to a copolymerized with a polycationic support. However, this relationship does not always hold, as shown in Table 8.6 (Charles 1976).

With maleic anhydride copolymers as supports, the shift in the pH optimum observed was always towards the alkaline direction for four proteinases, independent of the ionic character of the support. Thus, in addition to electrostatic interaction and partitioning effects the pH-activity profile will be influenced by the other parameters such as changes in enzyme configuration, charge effects associated with ligands and even, as suggested by Charles, with reaction-generated pH and concentration gradients.

Electrostatic partition theory predicts that the K_m will change in an analogous manner:

$$v = \frac{V[S]}{K_m' + [S]}; \quad K_m' = K_m \exp(Z\epsilon\Psi/kT)$$

where v, V, K_m and [S] have the same meaning as in Chapter 2, and Z is the charge on the substrate at the pH of operation. If both the substrate and the support have the same charge the observed concentration of substrate at half maximum rate ($S_{0.5}$ = observed K_m) will be greater than that in bulk solution, but if they have opposite charges the apparent K_m will be lower. Theory also predicts that the apparent K_m will be strongly affected by the electrochemical environment, i.e., the ionic strength. In general $S_{0.5}$ decreases. The same kind of interpolation of electrostatic theory into the rate equations for enzyme inhibition (Chapter 2) leads to some rather interesting predictions. Suppose the support surface possesses negative charges which generate an electrostatic potential of 100 mV, the substrate is charged positively (+1) and the inhibitor negatively (−1). The effectiveness of the inhibitor will then be reduced to *ca* one part in 2000 of that in solution. Similarly, substrates which are hydrophobic will exhibit an increasingly lower $S_{0.5}$ with increasing hydrophobicity of the support, i.e., $v = V[S]/(K_m' + [S])$. K_m' is the $S_{0.5}$ and is equal to K_m/P where P is the partition coefficient for the distribution of the substrate between bulk solution and matrix.

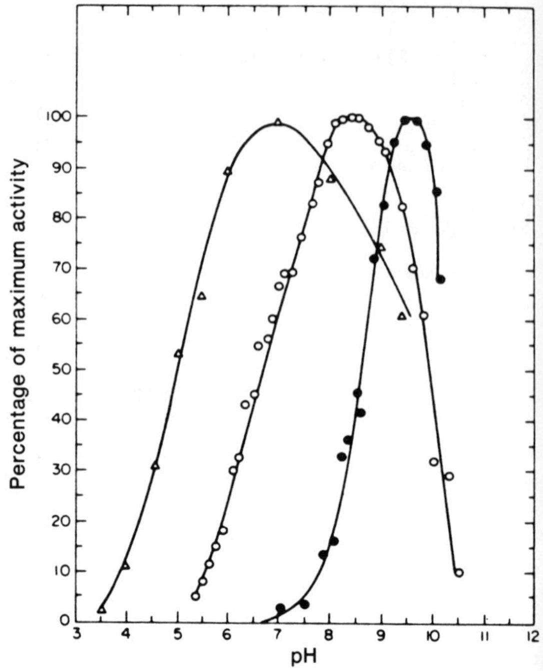

From Goldstein (1976). Copyright Academic Press.

FIG. 8.6. EFFECT OF SUPPORT CHARGE ON pH-ACTIVITY PROFILE OF AN IMMOBILIZED ENZYME

Open circles, no support, soluble chymotrypsin; closed circles, a polyanionic support; triangles, a polycationic support.

TABLE 8.6

EFFECT OF SURFACE CHARGE ON pH OPTIMUM[1]

	Trypsin	Chymotrypsin	Subtilisin Novo	Papain
Native	8.0	8.3	8.2	6.3
EMA-MDA[2]				
Anionic	9.0	9.5	9.2	7.5
Cationic	9.0	9.2	9.0	7.5
EMA-hydrazide[2]				
Anionic	9.2	9.7	9.5	7.5
Cationic	9.2	9.7	9.5	—

[1] Source: Charles (1976).
[2] MDA = *p,p'*-diaminodiphenylmethane; EMA = ethylene maleic anhydride copolymer.

Mass Transfer and Diffusion.—Once an enzyme is immobilized, the substrate may not be as readily available as in solution and the rate of the reaction is now governed by the rate at which the substrate arrives at the support site, as shown in Fig. 8.7. As mentioned, one must distinguish between external mass transfer and internal mass transfer. For solid supports in which enzyme is strictly bound to the surface only the former enters into consideration, whereas both must be considered when dealing with porous support particles and other situations where the enzyme is restrained within the support.

This is not the place to go into detail into the wedding of Michaelis kinetics and heterogeneous catalysis terminology adopted by most engineers. Goldstein (1976) cites some 15 general references to this branch of enzyme-immobilized enzyme kinetics. Such expositions usually point out that the rate of diffusion of the substrate from the bulk of solution to the localized enzyme is proportional to the difference in concentration (whose dimensions are cm^2 per sec) between these two "phases." The constant relating flow to this difference is the diffusivity constant D which compares the ten-

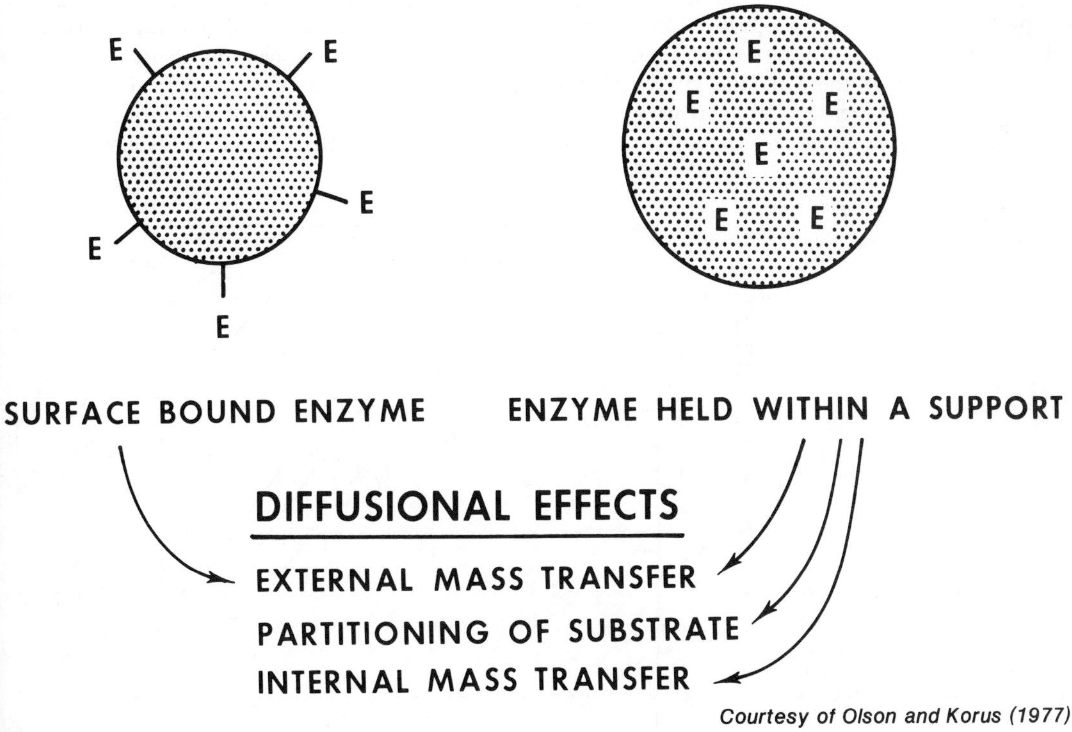

Courtesy of Olson and Korus (1977)

FIG. 8.7. DIFFUSION EFFECTS OF ENZYMES IMMOBILIZED AT SURFACES dency of different substrates at the same concentration differential to diffuse. Whereas in solution kinetics one assumes a steady state with respect to the concentration of the enzyme-substrate complex (Chapter 2), in diffusion controlled kinetics it is assumed that steady state is reached when the rate at which the substrate gets to the enzyme is equal to the rate at which the enzyme transforms the substrate:

$$v_{transfer} = v_{enzyme}$$

or

$$D([S]_b - [S]_s) = V[S]_s/(K_m + [S]_s)$$

where the subscripts b and s refer to bulk and surface.

Most mass transfer kinetics introduce one of two parameters which are related, not to the observed rate of the enzyme reaction, but to the effectiveness η, the ratio of the rate of action of the immobilized enzyme when diffusion is restricting to that when diffusional effects are not present. Thus, this dimensionless factor always has to be less than unity. Two parameters frequently used in describing such systems are the Thiele modulus = r [or l] times $(V'/K'_m \times D)1/2$, and the Damköhler number, $r^2(v'/K_m \cdot D)$ where r is the radius of the support particle, l thickness if support is a membrane and V′ the activity of the immobilized enzyme. The use of either of these moduli vividly shows how effectiveness decreases with increase in either of these parameters to yield curves similar to those in Fig. 8.8. Unlike solution kinetics, increasing enzyme concentration decreases effectiveness whereas increasing the substrate increases effectiveness. Apparent K_m, as well as effectiveness varies with the Damköhler number, the enzyme concentration, layer thickness and flow rate in column reactors. External diffusion enters in the latter case only if the apparent activity increases with increasing flow rate.

In terms of the more familiar Michaelis kinetics (Chapter 2)

$$v = \eta \frac{V[S]_s}{K_m + [S]_s} \text{ and } 1/v = \frac{K_m}{\eta \cdot V} \cdot \frac{1}{[S]} + \frac{1}{\eta \cdot V_m}$$

Since both [S] and the effectiveness factor η can vary with the directly observed $[S]_b$, the double reciprocal Lineweaver-Burk curve is no longer linear but can take on a variety of shapes (Fig. 8.9).

Temperature.—The effect of temperature on the rate of immobilized systems can be used to determine if the reaction is diffusion controlled. The more mass transfer controls the reaction and the higher the temperature, the less effect temperature will have on the measured rate, so that at a high enough temperature, the rate will be independent of the latter and will not vary with temperature at all (Buchholz and Reuss 1977). This may have some benefits in practical application. For example, the energy of activation of barley malt β-amylase immobilized by covalent attach-

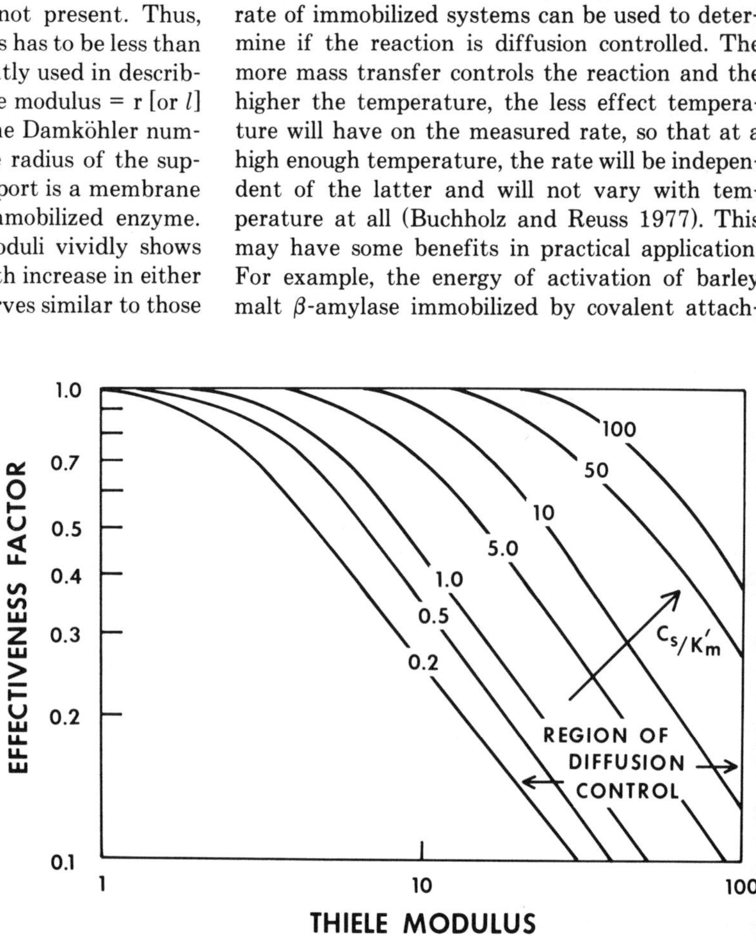

FIG. 8.8. THE THIELE MODULUS

Depicting the use of this dimensionless parameter for estimation and predicting effectiveness of enzyme immobilization.

Courtesy of Olson and Korus (1977)

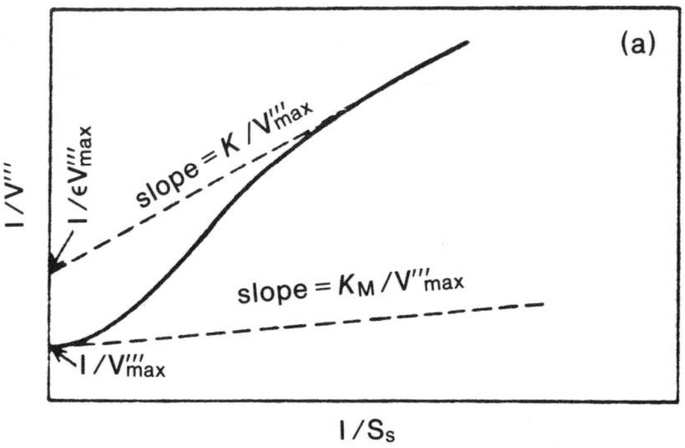

From Engasser and Horvath (1973). Reprinted with permission of Academic Press, London.

FIG. 8.9. EFFECT OF DIFFUSIONAL LIMITATIONS IN IMMOBILIZED ENZYME SYSTEMS ON THE LINEWEAVER-BURK PLOT

V''', initial reaction rate per unit volume of porous support. S_s, S/unit of support surface.

ment to porous silica was calculated as 7.6 kcal/mole, which is about one-half the normal value (Lee and Tsao 1974).

Internal Diffusion and Pore Size.—When the enzyme is embedded within a particle the kinetics become more complicated. The substrates, products and other ligands which might affect activity have to find their way *via* a labyrinthine maze through the surface layer and then through the pores before forming the enzyme-substrate complex. However, some understanding of this progression is necessary for optimal reactor design. It is of interest to note that under conditions of severe restriction of rate due to diffusion in such systems, the rate is proportional to the square root of the enzyme concentration, somewhat reminiscent of the now almost-forgotten Schutz law for the enzymatic hydrolysis of polymers (Chapters 2 and 9). The now classical paper on internal and external diffusion-controlled rate of enzyme action is that of Engasser and Horvath (1973,1976). Charles (1976) cogently discussed effects of pore size. Pore size exerts control over several parameters related to the final rate at which the immobilized enzyme system will act. These include extent of immobilization, surface area, charge effects, as well as intrinsic reaction behavior.

For the most part the larger the pore size, the less the resistance to diffusion and total but not necessarily available or effective surface area. On the other hand, large pore size usually means that more of the enzyme can be immobilized and that large polymeric substrate molecules can get to the enzyme. In general it would appear that the maximum immobilization occurs in porous supports when the pore diameter is about twice that of the major axis dimension of the enzyme so that for a given pore size there is an optimum enzyme size. Conversely in a mixture of polymeric substrates, such as partially hydrolyzed cornstarch, there is an optimum degree of polymerization which corresponds to maximum enzyme activity. These effects as well as problems of product inhibitor accumulation are peculiar to porous supports which in turn are related to the diffusion controlled effects.

It should be emphasized that while these elegant kinetics of immobilized systems have been developed in some detail and are useful under restricted conditions, some investigators (Olson and Korus 1977) feel that the assumptions needed to obtain solutions to simplify the already complicated differential equations are of virtually no use in most practical food-related situations. Thus, the assumption that the substrate concentration is such that first-order kinetics obtain is not realistic for most immobilized enzyme systems likely to be encountered or adopted in food processing.

Stability of Immobilized Enzymes

One of the most important properties of an immobilized enzyme is its ability to be constantly reused, so stability has been studied in detail. According to Mosbach (1974,1980), one of the two more common questions asked about immobilized enzymes relates to their stability. In our enumeration of the benefits of immobilization we also mentioned increased stability not only towards heat and other denaturing conditions but also to self-hydrolysis in the case of proteolytic enzymes. However, in spite of the many claims for increased thermostability, according to Mosbach (1976), few reports have yet appeared which prove beyond a doubt the putative increased stability for immobilized enzymes. Mosbach does state that stabilization can be accomplished by either covalent multipoint attachment of the enzyme to the matrix or by embedding the enzyme in a stabilizing protective microenvironment. Some of the previous reports on the increased stability may be caused by overloading the support with an enzyme. Thus, Korus and O'Driscoll (1975) showed that diffusion effects cause an apparent increase in stability. They found on theoretical grounds that the apparent half-life of diffusion-controlled immobilized enzyme increases with increasing enzyme concentrations and support dimension and decreases at low substrate concentrations. At high enough enzyme concentrations, when the rate is completely diffusion controlled, the apparent or pseudo half-life will be twice as long as the true half-life. This could lead to misjudgment in the choice of a suitable support.

How long an enzyme will retain activity in immobilized form will depend upon several factors:

(1) The intrinsic stability of the enzyme as a protein molecule, i.e., its "natural" resistance to denaturation while it is, so to speak, idling.
(2) What happens during the action of the enzyme. Heat, hydroxyl or hydrogen ions may be liberated; the enzyme may be subject to reaction inactivation (Chapters 15 and 24).
(3) Accumulation of injurious products such as hydrogen peroxide in diffusion controlled reactions. In solution these would diffuse away from the support before they had a chance to inactivate the enzyme, i.e., superoxide anion-producing enzymes (Chapter 16).
(4) How tightly the enzyme is bound to the support. An equilibrium between the bulk solution and the support, even if overwhelmingly in favor of the support could, after appreciable operation when water solution is continuously replaced, result in serious loss of enzyme from the support.
(5) The support material itself may slowly be dissolved or be removed by other means of attrition. The slow disappearance of the support material by dissolving into the bulk solution or otherwise disintegrating and taking the enzyme along with it. Thus, invertase immobilized on glass was found to be perfectly stable, but in column operation the glass particles slowly dissolved.

APPLICATIONS

Food Ingredients (Table 8.7)

At the time of this writing the available information indicates that the controlled action of immobilized enzymes in food-related manufacturing processes on a large scale have been commercially successful in the following instances, the first two of which are in use in Japan and the others in Japan and also in the United States. [The fate of earlier patents on the use of invertase to prepare invert sugar (Chapter 9) is not known.]

Production of L-Amino Acids from Racemic Mixtures of DL-Amino Acids.—Although at first restricted to L-methionine to provide a supplement to methionine-deficient animal feeds when such supplementation was restricted by law to the L-form (now no longer required) the Tanabe Seigaku Company of Japan now supplies, in addition, the L-form of at least the essential amino acids: valine, tryptophan and phenylalanine, as well as alanine and others. This is the first successful commercial scale process, in use since 1969. A flow diagram of the process is shown in Fig. 8.10.

Over 250 MT of L-methionine alone, with an estimated cost of 40% of the previously-used non-continuous process, are produced each year.

Production of Inosinic Acid and Other 5′-Nucleotides.—This is a more recent adaptation to im-

TABLE 8.7

IMMOBILIZED ENZYME ACTION IN THE MANUFACTURE OF FOOD INGREDIENTS

Food Ingredient	Enzyme	Function or Purpose	Immobilization Mode
L-Amino acids	Amino acid N-acyl hydrolase (Deacylase)	Product easily separable from unreacted D-acyl amino acid remaining from starting racemic mixture.	Adsorption on a column of DEAE Sephadex.
Sucrose (beet sugar)	α-Galactosidase	Removes raffinose which interferes with sucrose crystallization.	Enzyme fixed *in situ* in pellet of fungus.
5′-Nucleotides	5′-Ribonuclease	Production of Flavor Potentiators.	Adsorption on porous ceramic support columns.
Inosinic acid	AMP deaminase	Same	Same
High fructose corn syrup	Glucose isomerase	Isomerization of corn-derived glucose.	See Chapter 9.

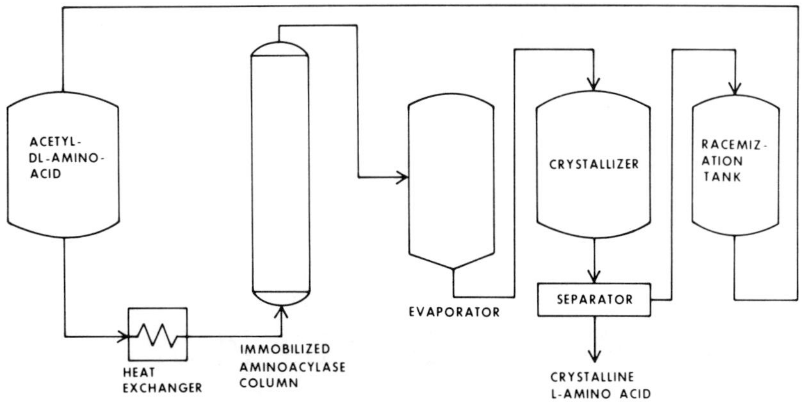

Courtesy of Olson and Korus (1977)

FIG. 8.10. FIRST FOOD USE OF IMMOBILIZED ENZYME TECHNOLOGY

Process flow sheet for continuous production of L-amino acids from racemic mixtures of DL-amino acids.

mobilization of a process which goes back to 1961. In this process an enzyme from the fungus *Penicillium citrinum*, an RNAse designated as a "5′-phosphodiesterase" or "nuclease P_1," produces 5′-nucleotides from commercial grade yeast nucleic acid. Of the four 5′-nucleotides obtained from this hydrolysis only 5′-guanosine monophosphate (also designated as guanylic acid or 5′-GMP) and 5′-AMP are converted by microbial AMP deaminase into inosinic acid (5′-IMP) which is also used in flavor potentiation formulations along with glutamate and/or GMP ("Ribotide").

Other enzymatic aspects of glutamate and 5′-nucleotides as gustators will be discussed in Chapter 20.

Improvement in Production of Sucrose.—Raffinose, whose presence in beet sugar juice interferes with crystallization of sucrose, is removed by the action of α-galactosidase immobilized *in situ* within pellets of the fungus *Mortierella vinacea*. The unrefined molasses is passed through a succession of pellet-containing chambers, separated by screens, within a vat, being in effect a stirred tank reactor (Shimizu and Kaga 1972). A favorite source of a heat-stable α-galactosidase for immobilization studies is that from *Bacillus stearothermophilus* (Reynolds 1974).

As much as 4–5 MT of raffinose, present in sugar beet juice extracted from 3000 MT of beets containing 600 MT of sucrose, can be removed in a day. The loss of raffinose is more than compensated by increased crystalline sucrose yield and speed-up of operations (Wolnak 1974).

Production of High Fructose Corn Starch Syrups.—Enzymatic isomerization of glucose to fructose constitutes by far the most revolutionary and important application, in terms of political, social, economic and technological impact, of large-scale industrial use of immobilized enzymology as of this writing. To place this breakthrough in proper perspective and to serve as a paradigm of applied enzymology in this, perhaps the most enzyme-intensive food- related industry, we shall devote the next chapter to the role of enzymes and their immobilization in the complete transformation of its technology. This discussion also provides an opportunity to survey a topic which does not readily fit into other chapters but which certainly should be covered in a food enzymology book. As previously emphasized, we are dealing with the use of enzymes to prepare food ingredients rather than foods per se.

Also, we shall take the pedagogic license of proceeding biochemically from large to small molecules followed by the rearrangement of these small molecules, rather than on a strictly historical or technological basis. In addition we shall look into other enzymes which may in the future be used to produce carbohydrate sweeteners commercially.

9

ENZYMOLOGY OF SWEETENER PRODUCTION

STARCH-DERIVED SUGAR SYRUPS

The ensuing discussion takes up the role of enzymes in the hydrolysis of starch to produce so-called "modified" starches, (corn) starch syrups and glucose as well as the manufacture of mixtures of glucose and fructose from either the ultimate product of starch, glucose, or from sucrose (invert sugar). The products from the action of these enzymes are used extensively in the food industry as ingredients for a wide variety of processed and fabricated foods. This chapter also examines enzymes which hydrolyze cellulose and xylans as possible future agents for the production of sweeteners and energy sources.

Before proceeding to the purely enzymatic aspects of the production of these food ingredients, we list their food-related properties and quality attributes which they impart to foods, their present and potential use as food ingredients and their production by nonenzymatic antecedents.

The "modified" starches, most of which are chemically altered without concomitant hydrolysis, conserve or maintain identity as "starches" rather than "syrup solids." Most of the modified starches are used for nonfood purposes such as adhesives, sizings, laundry stiffeners, etc. However, some of them, including those produced by a very limited cleavage (enzyme- or nonenzyme-catalyzed), are finding increasing use for imparting special texture-related properties, as do modified proteins (Chapter 34), to a variety of fabricated foods. Among these properties are, for instance, the facilitation of dehydrated foods such as soups, purees and coffee whiteners to rehydrate smoothly in hot water, the capacity for controlling water content, setting time, rigidity and stability of a variety of starch-based foods such as gumdrop candies, puddings, etc., and as binders, texturizer thickeners in extruded new protein foods (Chapter 34). With the new commercialization of pullulanase (see below), one of these "modified starches," amylose, will find use in the food industry as a film barrier to oxygen and fats (Allen and Dawson 1975).

The starch syrups and dextrose as well as invert sugar products can and have been used wherever sweetness is desirable. But they also impart to food many nonflavor attributes as well. Mostly, however, they are used along with or as replacement for sucrose. About 50% of production of several billions of kilograms per year goes into confections but they are also used in the canning, baking, dairy and beverage industries. Foods to which they are added either as ingredients or as part of the production process include such diverse items as applesauce, beer, baby foods, sweet beverages, confections (especially malt), high maltose syrups in hard candies, cookies, cookie filling, foods in which they are used as fermentation media (brewing, baking), fountain syrups, ice cream, icing for cakes, jams and jellies, malt liquor, maraschino cherries, marshmallows (high maltose), salad dressings, sherbet, table syrup, whipped cream and wine and perry. They are also used extensively in the pharmaceutical, tobacco and paper industries.

Desirable properties that they impart to foods of which they are ingredients include the following (some indications of the role of the degree of polymerization on these attributes are also shown): M, maltose; S, syrup (intermediate); A, modified starches, high amylose; DDE and IDE indicate enhancement of attribute with decreasing and increasing dextrose equivalent, respectively):

—Impart sweetness; accentuates true flavor (M, S, IDE); promote flavor transfer.
—Contribute to color stability (M); afford opportunity for nonenzymatic browning (IDE); M improves color, resistance to off-color and brilliance in jams and jellies.
—Prevent sandiness in ice cream due to sucrose crystallization.
—Modification and control of other frozen dessert characteristics such as melt-down.
—Control of humectancy, hygroscopicity, stickiness in candies (M, S).
—Enhance resistance to microbial spoilage by virtue of high osmolality (IDE).
—Make fried food feel less greasy (DDE, A).
—Impart texture-associated attributes such as body, cohesiveness, smoothness, mouth-feel (DDE).
—Form quick-setting gels, thickeners, improve foam stability (DDE).

Nonenzymatic Transformations

Starch Syrups and Glucose.—A year after Nicholas Appert published and received an award from the government of Napoleon for his successful invention of canning technology, the St. Petersburg chemist Kirchoff published in 1811 his process for the sulfuric acid-catalyzed hydrolysis of potato starch for which he, like Appert, was rewarded handsomely, in this case by the Tsar. In a crude form, the Kirchoff process—a plant for which was erected in France in 1814 and in the United States in 1842—resulted in what until about 10 years ago were the three principal classes of products obtained from starch hydrolysis: sugar syrup, sugar syrup solids and D-glucose, which is still referred to in this industry by its older name, dextrose.

Until the economic crystallization of glucose hydrate was achieved in the early 1920s, only the former two products were sold and produced industrially, however.

Dextrose Equivalents.—The conventional measure of the degree of hydrolysis (from the viewpoint of the product, the degree of saccharification) is the Dextrose Equivalent or DE, the percentage of the glucosidic bonds hydrolyzed. One measures the total reducing groups (officially by the venerable Fehling's Solution) and assumes that the value obtained is all due to glucose. The DE value determines or defines the distribution of the hydrolysis products in the acid-hydrolyzed, but not in enzyme-hydrolyzed starch syrups. According to the Corn Refiners' Association, the distribution, to the nearest percentage, of the products in 20 and 60 DE syrups are as follows:

Polym.	*1*	*2*	*3*	*4*	*5*	*6*	*7*	*"Dextrins"*
%, 20 DE	6	6	6	6	6	4	4	62
%, 60 DE	43	20	13	8	5	4	2	5

where "Polym." 1 and 2 are glucose and maltose, respectively, etc.

Several drawbacks and circumstances prompted the curtailment but not abandonment—some 700,000 MT were recently produced in the United States in 1 year according to Barfoed (1976)—of acid and switch-over to enzymes for the production of both syrups and glucose. For syrups, the DE limit, set at 60, was dictated by the development of bitter and colored compounds above this limit as well as by the tendency for glucose to crystallize out. Also, as mentioned, the variety of syrups was restricted by the invariant proportions of glucose and maltosaccharides for a given DE. For new plants there was the incentive of not having to install expensive corrosion-free equipment.

While glucose produced from starch has been available as a food ingredient for many years, the price that has had to be charged was too high because of the added expense involved in removal of these bitter, colored and other types of impurities, and the failure to achieve theoretical 100% splitting of the glucosidic bonds. Much of this failure was due to reversion reactions in which the glucose already formed repolymerized to di- and higher saccharides.

Isomerization of Glucose to Fructose.—The possibility of converting glucose to fructose as an industrial process has been explored for over 150 years since the classic investigations of Lobry de Bruyn and von Eckstein on the alkaline isomerization of glucose. A small amount was produced before the enzyme process was developed. In addition to the use of inorganic catalysts (synzymes) discussed in Chapter 3, other selected attempts at industrial isomerization are listed by MacAllister (1979). The nonenzymatic isomerization of glucose illustrates in a particularly striking manner the role and advantage of specificity of action that enzyme use entails. Thus, alkaline isomerization involves not only an interchange of keto and hydroxyl at carbons -1 and -2 but also epimerization at carbon -2 to yield mannose and at carbon -3 to yield D-allose (psicose). Shown here are the first 3 carbons:

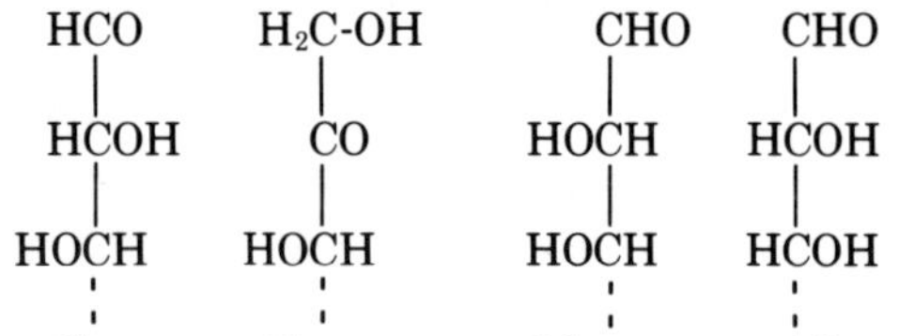

D-Glucose D-Fructose D-Mannose D-Allose

Starch Hydrolyzing Enzymes

We discuss the various actions of the amylases and other starch-transforming enzymes throughout this book. It may be appropriate at this time to review the action of hydrolytic enzymes that act on the two polymeric components of starch, amylose and amylopectin. It will be recalled that amylose is (ideally) a straight-chain polymer of glucose residues linked to each other *via* α-1,4 linkages. Amylopectin is a branched polymer consisting of α-1, 4-glucosyl (glucopyranosyl) oligomers with branching due to the formation of α-1,6-glucosyl linkages. Amylopectin constitutes 75–80% of most starches. The ensuing tabulation classifies the known starch hydrolyzing enzymes.

THE STARCH HYDROLASES

α-1,4-Glucan Hydrolases

Exo Enzymes.—Glucoamylase (amyloglucosidase, exo-1,4-α-D-glucosidase): Splits successive linkages from nonreducing end to form glucose but can also split α-1,6-linkages (see below).

β-Amylase (exo-maltohydrolase): Splits every other linkage from nonreducing end to form β-maltose and β-limit dextrin from amylopectin; highly specific for α-1,4 linkage.

Classical sources include malt and sweet potatoes, and newer sources include bacteria such as *Bacillus polymyxa, Bacillus cereus* var *mycoides* and *Streptomyces* spp.

Exo-maltotetraohydrolase: Splits every fourth linkage from nonreducing end to form maltotetraose: A source, *Pseudomonas stuzerei.*

Exo-maltohexaohydrolase: Splits every sixth linkage to form maltohexaose. Source: *Aerobacter aerogenes.*

Iso-pullulanase: Does not act on intact starch components but does act as an exoenzyme synergistically by rapidly removing otherwise slowly hydrolyzed terminal isomaltosyl groups as they are produced by the action of amylases. Source: *A. niger.*

Endo Enzymes.—α-Amylases: Hydrolyze internal linkages more or less at random to form straight and branched chain dextrins.

Classical or Historical Amylases: Classical sources, higher plants, animals, fungi especially *Aspergillus* spp.

Acid α-Amylases: From strains of *Aspergillus* via mutation (Chapter 6).

Alkali α-Amylases: From bacteria, especially from strains of *Bacillus subtilis* and mutant strains of related species.

Heat Stable α-Amylases: From bacteria, *B. stearothermophilus, B. coagulans, B. licheniformis.*

Debranching Enzymes, α-1,6-Glucan Hydrolases

Exo Enzymes.—Exo-pullulanase: As does glucoamylase produces glucose, hydrolyzing both α-1,6- and α-1,4 linkages but, in contrast, the former much faster than the latter; can act on amylopectin and maltose, but preferred substrate is pullulan; source: *Cladosporium resinae.*

Endo Enzymes.—Pullulanase (R-enzyme, β-limit dextrinase): Hydrolyzes α-1,6-linkage in pullulan, an α-1,4-, α-1,6-linear glucan in which every third linkage is α-1,6; and less efficiently in amylopectin, α- and β-amylase limit dextrins and hardly at all in glycogen; "minimum" substrate is I:

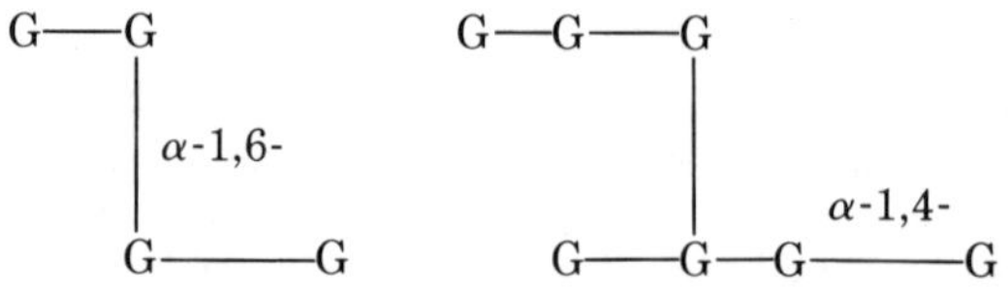

I II

where horizontal lines are α-1,4- and vertical lines are α-1,6-linkages. Sources: plants, animals, more recently found in bacteria.

Isoamylase: Completely debranches glycogen but cannot utilize pullulan as substrate; minimum substrate, II, above; source, bacteria such as *Pseudomonas* and *Cytophaga.*

References to most information just listed can be found in the review of Marshall (1975A).

From the viewpoint of industrial starch transformation as practiced today, Barfoed (1976) distinguishes among malt amylases, bacterial α-amylases, fungal α-amylases, glucoamylase and α-1,6-glucosidases (pullulanase and isoamylase).

Malt β-amylase, as mentioned, keeps snipping off maltose from the nonreducing ends of the outer branches of the amylopectin until it encounters a branch point leaving the so-called β-limit dextrin. This β-limit dextrin contains all of the α-1,6-linkages of the original starch and constitutes about 60% of the original amylopectin molecule. It can, of course, be further hydrolyzed and β-amylase can continue to act in the presence of endo-α-amylases and α-1,6-glucosidases. In terms of its use in modifying starch, malt α-amylase action causes a very rapid and highly striking thinning of heated starch suspensions, leading to the rapid formation of dextrins (glucose oligomers) whose chain lengths range from 6 to 10 glucose units ending up eventually in the formation of glucose, maltose, maltotrioses and isomaltotrioses (Schwimmer and Balls 1949A).

The malt α-enzyme is relatively heat stable and acid labile whereas the reverse is true for the β-enzyme. The bacterial α-amylases are characterized by very high resistance to thermal inactivation, some commercial enzymes acting at temperature as high as 110°C.

The original microbial commercial food-grade enzymes were of fungal origin, Takamine's takadiastase from *Aspergillus oryzae.* In a sense, action of these fungal α-amylases may be considered as an eclectic blend of α- and β-action because they form glucose, maltose and maltotriose much more rapidly and in larger amounts than do malt or bacterial enzymes, but still get around branch points. The enhancement of glucose levels is at least in part due to the presence of glucoamylases in all commercial preparations. As indicated, glucoamylases and exopullulanase are exoenzymes splitting off one glucose unit at a time from the nonreducing ends of both amylose and amylopectin. They appear to be specific for the hydrolysis of both α-1,4- and α-1,6-linkages although exopullulanases act much faster on the latter. *Aspergillus* glucoamylase in present commercial enzymes requires the presence of some α-amylase to act efficiently in the production of glucose and syrups (Marshall 1975A). The α-1,6-glucosidases, which as far as starch is concerned are debranching enzymes, especially microbial pullulanase, show greater promise in the production of modified starches and of maltose. The salient steps in the present and perhaps future production of glucose, maltose and fructose-containing syrups are shown in Fig. 9.1.

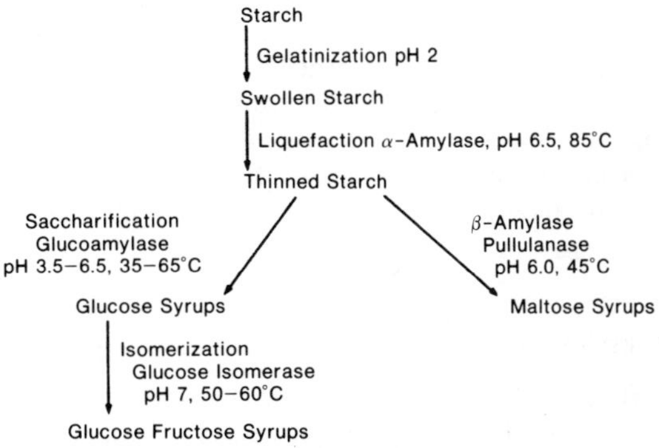

Courtesy of Olson and Korus (1977)

FIG. 9.1. PRODUCTION OF CORN SYRUP SWEETENER

Modified Starches and Nonsweet Dextrins

The properties and food uses of such products were briefly mentioned. As practiced today the production of enzyme modified-starches and the related low-DE dextrins is basically the first step in the production of starch sugar syrups and solids. Gelatinized starch is subjected to a very brief exposure of the thinning and liquefying action of an α-amylase preparation, usually of bacterial origin but of long enough duration to prevent retrogradation. In a typical process, a suspension of starch at pH 7 is continuously heated, at first slowly but later faster, in the presence of bacterial α-amylase until the DE of the product is about 5–6. The mixture may be kept at 100°C to precipitate small amounts of protein and inactivate the enzyme.

In a special process ultrafiltration was used to separate a branched dextrin (essentially β-limit dextrin) to be used as food-base material by treatment with either glucoamylase or β-amylase (Murayama *et al.* 1976). Reviews of this subject with relation to foods are sparse because, perhaps, of the overwhelming preponderance of the use of modified starches in nonfood industries and also because most modified starches are made by means other than the action of enzymes. Two reviews which include enzyme modified starches for food use are those of Marrs (1975) and Zobel (1975), the latter in German.

Another class of modified starches for use in the food industry which will probably be produced by enzyme action are the straight-chain maltodextrins or amyloses by the debranching action of the α-1,6-endoglucosidase pullulanase on amylopectin (Allen and Dawson 1975). The resulting straight chain maltodextrins together with the amylose already in the starch may find some use, not only as ingredients of fabricated foods for their functional texture-associated properties, but also perhaps as edible, biodegradable replacements of plastics in packaging of foods.

Starch Syrups

Because of the flexibility in composition introduced by the use of fungal and other microbial enzymes, a wide range of syrups is now available for a variety of purposes, ranging, in the case of corn conversion syrups, from "low" (DE = 28–38), "regular" (38–48), "intermediate" (48–58), "high" (58–68) and "extra high" (68 or higher). At the low end of the range, bland products with only a small percentage of the sweet-tasting oligomers and sugars (basically a continuation of the above-discussed modified starches) are used where low hygroscopicity and solubility are desired in bulking agents. They are made, as with modified starches, by treatment with controlled amounts of microbial α-amylases. As pointed out by Murray and Luft (1973) they replace the so-called acid-produced maltodextrins because the latter are not completely soluble, they contain native starch, and have a cereal or acid-hydrolyzed taste. Murray and Luft list foods where the special properties of the low-DE enzymatically produced starch hydrolysates are particularly desirable: dehydrated foods as carrier and drying aids; frozen sauces and fillings to impart more sheen, brighter color and appropriate body and texture; frozen desserts for slower melt-down and better texture; and for coating applications.

Progressing to higher DE syrups, it should be pointed out that malt was used to a limited extent to prepare sweet syrup from corn before the advent of microbial amylase. Recently, Goering and Eslik (1976) proposed that cornstarch syrup used in brewing application be replaced by a self-liquefying waxy barley starch. These low-amylose starch granules retain sufficient α-amylase-like activity to liquefy the starch in them plus four times their weight of normal cornstarch. Historically, use of enzymes to really control the composition of such syrups stems from pretreatment with acid to thin the starch followed by treatment with fungal enzyme preparations, largely α-amylase (Dale and Langlois 1940). This was followed by (1) substitution of bacterial α-amylases for the thinning operation (enzyme-enzyme process), and (2) a host of patents describing different combinations of starch-hydrolyzing enzymes with or without differing degrees of preliminary acid thinning to get differing ratios of glucose to maltose, plus small amounts of tri- and higher saccharides. In a comparison of acid-enzyme with enzyme-enzyme syrup production processes, Bodnar *et al.* (1972) concluded that higher levels of maltose plus glucose were attainable with the latter process. The ultrathermostable α-amylase from *B. licheniformis*, which is also less dependent on Ca^{2+} than are other α-amylases and is now available commercially for food use, has been suggested as a particularly good thinning enzyme for the production of syrups (Barfoed 1976; Olson and Korus 1977).

An example of the utility of being able to vary the proportions of glucose and maltose is afforded by the circumstance that although sweetness may be the main aim of a syrup so that one wants as much glucose as possible because glucose is the sweetest of the starch hydrolysis products, it tends to crystallize out of such syrups when it constitutes more than 40% of the total dry weight. With a suitable combination of hydrolases, this limit is achieved in a 68–70 DE syrup (almost equal concentration of glucose and maltose) whereas, as mentioned, one cannot go beyond a DE of 60 for acid hydrolyzed syrups. Among the enzymes used to obtain desired ratios of the sugars are, in addition to the α-amylases of malt and *Aspergillus oryzae*, the β-amylases (presumably also from malt grain, but in the future perhaps also from microbial sources as tabulated on p. 154) and of course glucoamylases. As indicated, most *Aspergillus*-derived commercial amylases contain some glucoamylase. The latter enzyme is also frequently added along with the other amylases. One problem in its use for this purpose is that, unlike fungal amylase, its action cannot be economically terminated by adding acid. Whatever the source, MacAllister *et al.* (1975) point out that in contrast to the purely acid-produced products these syrups are or can be sweeter but possess no flavor other than sweetness. For a synopsis of the progress of immobilization of the α-amylases, see Olson and Korus (1977).

Maltose and High Maltose Syrups

For many reasons inexpensive maltose syrup and pure maltose would be an advantageous food ingredient. According to Allen and Dawson (1975) maltose is the least hygroscopic of the maltosaccharides. It is not as sweet as glucose but has a more acceptable sweetness. It is fermentable and nonviscous, and does not crystallize readily. Its preponderance in syrups facilitates boiling, prevents stickiness in hard candies and allows increase of the ratio of syrup to sucrose where these two are used together.

As an example of the use of a judicious combination of enzyme treatments as practiced now to attain a desired syrup composition, in this case a high maltose syrup, the recipe of Barfoed is instructive. The starch is hydrolyzed to a DE of 10−12 by acid or enzyme, followed by further degradation with fungal α-amylase with a low glucoamylase activity to give a syrup with 60% maltose and less than 5% glucose. According to Nakajima *et al.* (1974) a still higher yield, 86%, can be achieved by removing glucose during action of *Actinomyces* amylase (high temperature for thinning, lower temperature for conversion) of potato starch. However, the key enzyme in production of maltose is β-amylase, presumably of cereal origin, as practiced today. MacAllister *et al.* (1975) state that limited prehydrolysis (the more limited the higher the yield of maltose but not so low as to release retrogradable amylose molecules) at high temperature followed by β-amylase at lower temperature (where microbial contamination can become a problem) results in syrups with 70% maltose. In the future when the enzymes are inexpensive and allowable, maltose may be produced by the successive action of the amylopectin debranching enzyme pullulanase and β-amylase.

We can expect both pullulanase and the β-amylase to be of microbial origin. Thus, Takasaki (1976) produced maltose from starch in 95% yield from the action of β-amylase and pullulanases elaborated by the same organism, a *B. cereus* from soil. In the still more distant future it may be possible to produce maltose using immobilized enzymes. Both β-amylase and pullulanase have been immobilized separately and together (Lee *et al.* 1978; Martensson 1974). High maltose syrups can also be produced with immobilized β-amylase/glucoamylase mixtures (Bohnenkamp 1979). Close-to-pure maltotriose can be obtained with a rice debranching enzyme (Yamada and Izawa 1979).

Glucose, Glucose Syrups and Glucoamylase

The manufacture of these products with the aid of enzymes, which began about 20 years ago, became possible only with the availability of a commercial source of glucoamylase.

Probably the first patent in which this enzyme was used in glucose production was issued to an enzyme supplier (Wallerstein 1950). According to Wolnak (1972) a patent issued in 1948 formed the basis of a legal action against a company in the corn wet milling industry in which Baxter-Wallerstein received damages of $5 million. This brings up a point worth mentioning at this time: There is constant, fierce competition between the enzyme producers and those companies that prefer to produce the enzymes they use themselves.

The problem posed by the presence of transglucosidase and its solution was referred to in Chapter 5. Presence of proteases in glucoamylase preparations may also pose potential problems in light of the demonstration that their action can result in some loss in activity as well as lowering the extent to which glucoamylase can break down branched glucans. Another interesting modification is the loss of ability to attack raw starch granules (Hayashida *et al.* 1976), a property which may have some practical implications in the future.

However, it is probably not desirable to remove other starch hydrolyzing enzymes from glucoamylase preparations. In a sense the production of glucose from starch can be looked on as merely an extension of sugar syrup production by application of special contributions of the same starch hydrolyzing enzymes used in the production of other starch syrups. As for the other syrups, the first processes used acid for starch thinning (acid-enzyme). Acid was then partially replaced by bacterial α-amylase (enzyme-enzyme). In the switch-over to enzymes, some corn wet milling companies produced their own enzymes and this has continued in the case of fructose production. However, the aim is to approach, as close as is compatible with economic realities, 100% cleavage of all of the glucosidic linkages in the starch for the production of: (1) crystalline glucose (dextrose) hydrate, (2) high glucose syrups containing more than 95% of this sugar, and (3) substrate for glucose isomerization, i.e., as an intermediate for the production of fructose-containing corn syrups. In 1976 the market for glucose alone was close to 1 MMT (million metric tons). The key enzyme is glucoamylase obtained mostly from strains of *A. niger* and *Rhizopus* fungi.

Petitions for glucoamylase use in foods continue to be issued. Thus, in 1975 the U.S. Department of Agriculture permitted the use of *Rhizopus niveus* "amyloglucosidase" to degrade gelatinized starch for the specific purpose of production of distilled vinegar, provided that the maximum level of the enzyme does not exceed 0.1% of the weight of the starch (Food and Drug Administration 1975). While acting on both α-1,4- and α-1,6-linkages of amylopectin, glucoamylase hydrolyzes the latter linkage much more slowly (Marshall 1975A). Marshall also showed that α-amylase, present at adequate levels in most (but not all) industrial glucoamylases, is absolutely essential for the action of this enzyme (Table 9.1).

In practice, 30% starch suspensions are thinned with either acid or bacterial α-amylase in the presence of Ca at pH 7 (unless a newer noncalcium requiring enzyme is used) to a DE of about 15–20 followed by treatment with glucoamylase at pH 4 until at least 95% of the theoretical amount of glucose is formed. The glucose is crystallized by standard chemical engineering practice (MacAllister *et al.* 1975). Unfortunately one of the problems encountered with acid, namely repolymerization, also persists in the enzyme processes so it is important to use an empirically determined optimum combination of starch concentration and of enzyme:starch ratio to maximize glucose yields. According to Madsen and Norman (1973) glucose syrups with DEs ranging from 90 to 98 have been manufactured. They have to be kept at elevated temperatures to prevent crystallization.

The Future Outlook

Judging from the tremendous outpouring of papers on immobilized glucoamylase—over 50 had been published by mid 1976—it would appear that this enzyme will be next after the now well-established glucose isomerase to be immobilized for use in a food-related industry. The supports run the gamut of the various types discussed in Chapter 8. As selected examples of the problems, progress and promise we take note of its immobilization onto carboxymethyl hydrazide *via* covalent attachment to the enzyme's carbohydrate, a technique borrowed from affinity chromatography (Christisen 1972) and the extremely high yields, better than 95%, which were obtained using a stabilized glucoamylase immobilized on chitin with glutaraldehyde (Stanley *et al.* 1977). On the other hand pilot plant experience with porous silica as support (Lee *et al.* 1976) resulted in consistently lower (1–15%) yields than when using the free form of the enzyme. Apparently this lowered yield is due to an exacerbation of the above-mentioned repolymerization brought about by the circumstance that the glucose [product] is present at higher concentrations in the pores of the support than in solution. If and when immobilized glucoamylase is used industrially, their data indicate that the enzyme-enzyme process will be used to the exclusion of the acid-enzyme procedure.

It should be pointed out that there is a fundamental difference between glucose isomerase and glucoamylase with respect to the urgency and economic exigency for immobilization. The latter enzyme is extracellular and can be prepared readily and with high activity whereas the former is intracellular and has to be solubilized. Thus, an estimate of the cost of the enzyme by Wolnak (1972) suggests that its use may not contribute more than 1% to the retail cost of glucose hydrate in 1977. In contrast, Wolnak also opines that if glucose isomerase were to be used only once in production of HFCS, product price would be prohibitive.

Basic research on starch hydrolyzing enzymes suggest that improvement in glucose production may come about from a better understanding of the specificities of the more newly discovered enzymes. Table 9.1 is a selection of data which shows how the extent of glucose production from amylopectin and raw starch may be stimulated in light of this new knowledge. The enhancement of mammalian α-amylase-induced digestion of raw wheat starch by *A. oryzae* "maltose," observed 33 years ago (Schwimmer 1945), may have been due in part to a pullulanase-like stimulation of raw starch digestion (Ueda and Ohba 1976). The action of such "supplemental" enzymes may also explain the commercially successful production of glucose using, putatively, one enzyme, a bacterial α-amylase preparation for both thinning and saccharification (van Twisk *et al.* 1976). The general significance of the stimulation by glycosidases of enzyme hydrolysis of insoluble polysaccharide substrates will be discussed in connection with cellulolytic degrading systems further on in this chapter.

TABLE 9.1

RAW STARCH AND AMYLOPECTIN DIGESTION

Substrate	Enzyme Principal	Enzyme Supplementary	Time (min)	% Hydrolysis to Glucose Principal	% Hydrolysis to Glucose + Supplement
Wheat starch[1] granules	α-Amylase	Maltase	120	40–50	70–80
Waxy maize[2] granules	Glucoamylase I *Rhizopus* sp.	Pullulanase *Aerobacter*	65	28	63
Waxy barley[3] granules	α-Amylase, endogenous, absorbed to granules		Liquefies 4 times its weight of cornstarch when heated in water.		
Amylopectin[4]	Glucoamylase[5]	α-Amylase	60–90	60	98
Amylopectin	Glucoamylase *A. oryzae*	Exo-pullulanase[6] *Cladosporium*	10	59	94

[1] Source: Schwimmer (1945).
[2] Source: Ueda and Ohba (1976).
[3] Source: Goering and Eslik (1976).
[4] Source: Marshall (1975A).
[5] Comparison of low- ("principal") with a high (+"supplementary") α-amylase-containing commercial fungal glucoamylase preparation. The "high" sample contained *ca* 180-fold more α-amylase activity *per unit* of glucoamylase than did the "low" sample.
[6] Exo-pullulanase, an alternative rather than a supplementary enzyme.

Invert Sugar

Isomerases and Invertase.—The two enzymes which catalyze reactions which result in the production of fructose for food use, invertase (β-fructofuranosidase) and glucose isomerases, trace their historic roots to the same area of biochemistry and enzymology. Yeast invertase was considered to be a prime example of an "unorganized ferment" whereas, in pre-Buchner terminology, the production of alcohol from sugar was supposed to be carried out by "organized" ferments, which we know today as the enzymes of the glycolysis. One of these enzymes is glucose-6-phosphate isomerase. Its discovery by Lohmann (1933) laid the groundwork for the gradual realization that glucose as well could be enzymatically isomerized and, more than 30 years later, also for the eventual replacement of invertase by glucose isomerase for the production of glucose-fructose syrups. Invertase also played an important role in the development of enzymology because it could be conveniently assayed by polarimetric methods (from whence the name invertase derives). Much of classical kinetics grew out of studies of invertase behavior in spite of the fact that it turned out to be one of the more difficult enzymes to purify owing, in part, to its high content (50%) of the polysaccharide mannan.

No doubt this early attention and its association with well-established industrial processes led to its adoption for industrial use in the preparation of invert sugar. The virtues of invert sugar syrup as a food ingredient compared with unhydrolyzed sucrose or starch syrups are its uncrystallizability in extremely high concentrations, stability of sweetness level in low-pH foods (no hydrolysis of sucrose) and the enhancement of fruit flavors by the fructose component. For these reasons it has been used extensively until recently in the confection manufacturing industry. In Europe it was fairly extensively used as a supplemental yeast food in brewing. In 1971 invert sugar accounted for 10% of the sucrose used in the United States and Europe (Wolnak 1972). In the United States invert sugar has been produced solely by acid hydrolysis (HCl) followed by neutralization with Na_2CO_3. As pointed out in Chapter 1, the circumstance that sucrose is quite readily subject to hydrolysis has made this nonenzymatic option still viable. That invertase which is produced in the United States for food additive purposes goes into making high test molasses (and perhaps sorghum syrups), in which complete hydrolysis of the sucrose is attained; ordinary invert sugar from purified sucrose conventionally contains 20% unconverted sucrose as well as 40% each of glucose and fructose. It is probably also used for recovery of candy scrap (Anon. 1973B). Not much is needed for its other obligatory application; it is injected into chocolate coated sucrose-containing candies whose interior is transformed into the creamy consistency typical of these fondant-based confections. Undoubtedly a small but sustained market will continue to demand the production equivalent to $100,000 worth produced in the United States in 1971 (Wolnak 1972). By contrast, quite a bit of invertase was used to make most of the invert sugar which has been produced in Europe at a cost, in 1971, of about 40¢ per 100 kg of sugar.

Production of Invertase.—Part of the reason for the relative low cost of invertase, in contrast to glucose isomerase, is the ease of production and ready availability of enzyme source, brewers' yeast. The enzyme is induced by growing the yeast in sucrose and is extracted from the cake obtained by centrifugation *via* autolysis at 30°C and pH 5.8 for 4 hr. According to Wolnak, this renders the enzyme, which is localized between the membrane and the wall (periplasma) of the yeast cell, soluble so that it can be precipitated out of solution with typical organic solvents (Chapter 5). Alternatively, the solution can be vacuum concentrated and sold as liquid. In older methods, such as those which used ethyl acetate as autolyzing agent, the enzyme remained insoluble and had to be solubilized by the addition of proteolytic enzymes such as papain (Tauber 1949).

Immobilization. Although as a food enzyme adjunct invert sugar faces a bleak future as the glucose isomerase-converted glucose-fructose syrups take over, invertase is increasingly helpful in a different food-related context, as a specific analytical reagent in the determination of sucrose and other sugars (Chapter 38). Furthermore, in analogy to its role as a model enzyme for study in the formative years of enzyme kinetics and mechanism elucidation, it has played an important role in unearthing the distinctive behavior of enzymes in the immobilized state. It has been immobilized in just about every conceivable fashion onto just about every conceivable support. Probably the first demonstration that an enzyme can act while adsorbed involved invertase (Nelson and Griffin 1916). The enzyme was used in the early formulation of enzyme kinetics in restricted systems (McLaren 1960) and in demonstrating covalent attachment of an enzyme to a synthetic polymeric support (Manecke *et al.* 1960).

Probably the first practical proposal to use immobilized enzymes in an industrial process involved invertase. Tate and Lyle Ltd. and De Whalley (1944) were issued a patent for the simultaneous inversion and decolorization of sugar liquors by passing them through a body of bone charcoal to which invertase had been adsorbed. We do not know if this method was actually put into practice, but perhaps part of the estimated $500,000 worth of invertase produced in the United States in 1971 may have been immobilized for this purpose.

An interesting proposal which might put invertase back into the food ingredient picture involves the use of immobilized invertase in conjunction with glucose isomerase (see below) to produce, in a continuous cyclic process, pure fructose from sucrose (Hamilton *et al.* 1974). Each 1 kg of sucrose should yield more than the estimated 1 kg fructose since the invertase step would add a weight equivalent to 1 molecule of water.

$$\text{Sucrose} + H_2O \rightarrow \text{Glucose} + (\rightarrow) \text{ Fructose}$$

FRUCTOSE SYRUPS AND IMMOBILIZED GLUCOSE ISOMERASE

Impact as Food Ingredients

Perhaps it was in a slightly ironical mood that Clio, the muse of history, arranged events so that the first patent on use of an immobilized enzyme, invertase, should be concerned with the production of a food ingredient closely resembling that obtained by a process which for the first time uses an immobilized enzyme on a truly large scale and which not only almost completely replaces invertase, but also makes inroads or threatens to replace a significant portion of the invertase substrate, sucrose. By the time this appears, it is likely that the annual production of this replacement (commonly and hereafter referred to as HFCS (for high fructose corn syrups, with the understanding that, not only corn but other high-starch plants may be used in their production) in the United States alone has now exceeded 4.5 billion kg/year, some 10 years after it was first made available commercially. As of midsummer of 1979 it was selling wholesale out of Decatur, Illinois for 24¢/kg as compared with 38¢/kg for beet sugar, sucrose. In 1968, when HFCS with a somewhat lower fructose content was first placed on the market, 0.09 billion kg were produced and it was sold for about 2¢/kg. In an economic analysis of sugar policy option for the United States, Jesse and Zepp (1977) made a projection in which HFCS in 1980 will constitute some 35% of the total United States production, if the world raw sugar price is 15¢/kg and 20% at a world sucrose price of 15¢, close to the present one at which point, he predicted, the United States would be exporting sugar as compared with the present necessity of importing about half of our total consumption. A confidential survey of the then eight HFCS American producers (Andres 1976) indicated that the United States output would reach 2.5 billion kg in 1978. Andres also points out that replacement of all the imported sugar by HFCS would use up only 6% of the 1975 corn crop and at that time the production of HFCS was already contributing at least $1 billion to the United States trade balance. However, in contrast to the simplifying assumption of Jesse and Zepp, rapid expansion of this immobilized enzyme-based process is not confined to the United States. Political repercussions are typified by the protest of beet growers in the Common Market countries who are protected by tariffs on imports of sugar but not on starch or corn. In 1980 the United States per capita consumptions of HFCS and refined sucrose were 9.1 and 39.5 kg, respectively. This figure for HFCS is bound to rise since the announcement, in 1980, by the world's leading beverage manufacturer of an additional 0.5 billion kg/year, more than enough to take up the slack due to "overbuilding" of enzyme conversion plants.

There does, however, appear to be a difference in source of supply of immobilized enzyme. In the United States many of the HFCS producers and/or suppliers are either from the corn wet-milling industry or the sugar cane industry. They have made their own immobilized enzymes or else they make some sort of arrangement with these carbohydrate processors (Table 9.2). In Europe and other parts of the world it seems that the HFCS producers are purchasing their immobilized enzymes from well-established enzyme producers. In other instances the manufacturers are under license to American countries for using a particular microorganism to manufacture their enzyme, or are using the know-how accumulated by other companies. This cross-fertilization can become somewhat involved; apparently the first HFCS production in the United Kingdom is being carried out in a plant in Tilbury affiliated with a sugar company (Albion Sugar), which is a subsidiary of a Netherlands based firm, KHS, which in turn is using the know-how under a license or by agreement with R.J. Reynolds, an American-based food production firm. Subsequently KHS became affiliated with Avebe GZ.

One can gain some idea of the scale of these operations from the size of single plants as reported by research-oriented personnel associated with them. The output of HFCS by one Corn Sweeteners Plant in Cedar Rapids, Iowa, was anticipated to run as high as 2 million kg per day from 130,000 bu of corn by 1978 (Langhurst *et al.* 1976) and the plants using Novo's immobilized glucose isomerase (hereafter referred to as GI) have design capacities of up to 1.1 million kg per day (Oestergaard and Knudsen 1976). In late 1978 and early 1979 there was some misplaced apprehension of overly rapid expanding of production facilities.

Composition and Uses.—What has become the standard HFCS, equivalent in composition to Clinton-Standard Brand's Isomerose 100, is a syrup

TABLE 9.2

SOME PROCESSORS/DISTRIBUTORS OF HFCS AND GLUCOSE ISOMERASE

Trade Name	Holding Company and Licensee or Subsidiary	Source of Enzyme
Amerose	Amstar	—
Corn Sweet	ADM	—
Invertose	CPC	*Streptomyces chromogenes*
Isomerose	Standard Clinton	*S. alba*, other *S.* spp.
Isosweet	Staley	*Streptomyces* spp.
Iso-syrup	Miles	*S. olivaceus*, whole cells
Maxazyme	Anheuser-Busch, Gist-Brocade	*Actiplanes missouriensis*
Sweetzyme	Novo	*Bacillus coagulans*
TruSweet	American Maize	—
Various	Reynolds, KHS	*Arthrobacter* spp.

[1] As of 1976, 1977; dashes indicate not known to author.

containing 71% solids, of which 42% is fructose, 50–52% glucose and the remainder higher maltosaccharides. It has about the same sweetness as sucrose. Dingwall and Campbell (1975) stress the enhanced flavor that free fructose imparts synergistically to many of these foods. It is also supposed to suppress the aftertaste of saccharine. Other benefits are increased osmotic pressure and hence greater resistance to microbial deterioration and, for instance, enhanced tenderness in marshmallows as compared with either corn syrups or sucrose. On the negative side, the reducing sugars comprising HFCS may produce unwanted nonenzymatic browning in protein-rich foods. Honey and fresh fruit juices are easily subjected to HFCS adulteration, which, however, is detectable by estimating $^{13}C/^{12}C$ ratios (see "Honey Enzymes," Chapter 37). Also, fructose allegedly may contribute to heart disease.

By 1977, Clinton, the first producer of HFCS, had made a 90% fructose syrup (Isomerose 900). Several firms are looking into a 95% fructose syrup. By 1980, Cetus, one of the new gene-splicing firms (Chapter 5) had announced a new process for almost pure fructose production. Fructose can accumulate in a GI reaction mixture by adding borate, which complexes readily with glucose so that it is no longer available for the reverse reaction (Takasaki 1972). However, it is more likely that the fructose is separated from the glucose on anion exchange columns in the bisulfite form (Takasaki 1974). The 90%-fructose syrup costs about one-third more than sucrose but is 20–60% sweeter so that on sweetness basis the cost is about the same, but at the most is only one-fourth of that of pure fructose. In addition to other advantages of fructose sweeteners, this 90% HFCS results in reduction of the caloric intake for the same sweetness; hence, its use diet formulations. Also, fructose is more readily tolerated by diabetics than is sucrose. On the other hand, fructose has been recommended as bulking material for some high protein food systems because it melts at much lower temperatures than does sucrose, about the same as the boiling point of water.

Glucose Isomerase

The specificity of GI illustrates how patterns of enzyme specificity may contravene stereochemical configuration classification. Conventionally, as shown in elementary textbooks, all the D-sugars are built up from D-glyceraldehyde (Fischer formula) by conceptually inserting a HCOH or a HOCH group between the aldehyde (CHO) and the proximal or number 2 carbon atom of the monose to obtain the next higher sugars. In this scheme the direct pentose precursor to D-glucose and D-mannose is D-arabinose and not D-xylose. However, what was used eventually as the first commercial glucose isomerizing enzyme is basically a xylose isomerase, obtained initially by Hochster and Watson (1954) from *Pseudomonas hydrophilia.* Marshall and Kooi (1957) found that glucose but not arabinose is a substrate, but that only xylose could induce the biosynthesis of this enzyme. This means that the enzyme "looks" at the substrate from the CHO down rather than from the primary CH_2OH up, as in the Fischer scheme. Of course in nonalkaline solutions most of the sugars are in the ring form but it may be significant that most of the GIs have alkaline pH optima. In other words it is the chirality of the carbons 2–4 in a sugar which accommodate the conformation of the active site of aldose-ketol isomerases. Thus this specificity pattern holds for arabinose isomerase which can utilize the 6-carbon sugar L-fucose as well as D-arabinose, but not xylose (Cohen 1953). However, the evidence seems to indicate that the α, but not the β-, anomer of D-glucopyranose is the substrate.

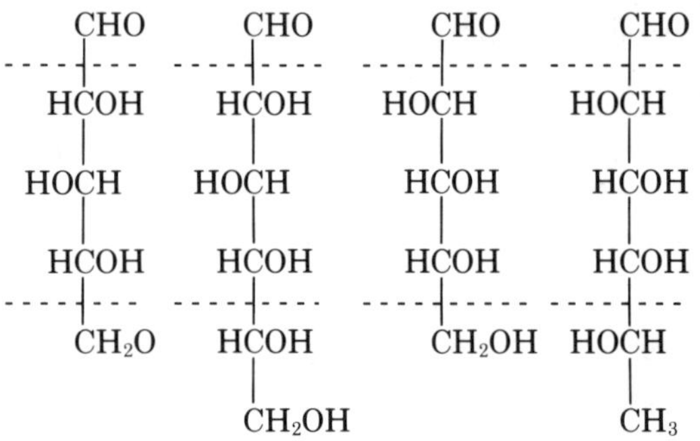

D-Xylose D-Glucose D-Arabinose L-Fucose

Glucose isomerase and discovery of various aldose-ketol isomerases were reviewed by Chen (1980) and Casey (1977), the latter as background material leading to the practical application of GI. Food-related technological possibilities of this type of enzyme specificity were first sensed and acted upon by A. L. Elder, director of research for one of the larger starch processors, who directed the attention of his staff to the Hochster-Watson xylose isomerase. This led to the finding that the *Pseudomonas* enzyme could use glucose as well as xylose as substrate (Marshall and Kooi 1957) and to the first patent on its use to prepare fructose-containing syrups. However, the activity of the enzyme was relatively low, equilibrium was established at 35% fructose, expensive xylose was required as inducer of the enzyme and arsenate was needed to obtain maximum isomerization rates. In these respects this particular enzyme seems to be halfway between simple aldose and aldose phosphate isomerases. In the 1940s it was found that arsenate could replace phosphate in most enzyme reactions in which phosphate is a cosubstrate. Thus, phosphorylase and glyceraldehyde phosphate dehydrogenase yield glucose and 3-phosphoglyceric acid instead of glucose-1-phosphate and 1,3-diphosphoglyceric acid.

It remained for a limited number of groups of Japanese investigators to more thoroughly explore the enzymology of glucose isomerization in the 1960s. This has resulted in more than 20 publications in the Japanese journal, *Agricultural and Biological Chemistry*. Much of the work was performed at two government laboratories, the Food Research and Fermentation Institutes in Tokyo and Chiba City, respectively. Key and representative papers are those of Takasaki *et al.* (1969) and Tsumura and Sato (1965). According to Casey (1977), Mermelstein (1975) and others who have reviewed this story, the know-how for practical application was purchased from the Japanese government in the mid 1960s by Clinton-Standard Brands and that the basic U.S. patent relating to this transaction is that of Takasaki and Tanabe (1971). Clinton was producing HFCS in 1968.

One of the key innovations that led to this practical application was finding a microorganism which had xylanase activity in order to provide the xylose *via* hydrolysis of inexpensive xylan for induction of the enzyme. *Streptomyces* GIs have an absolute requirement for Mg, also require Co for stimulation and stabilization, and need some sort of reducing agent to protect them from oxygen. The latter requirement has usually been fulfilled by SO_2 and its soluble relatives as indicated by several patents (i.e., Cotter *et al.* 1971) and by injecting inert gas (Cory 1975). Oxygen is also undesirable because HFCS tends to be colored in its presence.

Although the first GI used commercially was essentially a xylose isomerase obtained from microorganisms which required xylose to induce the enzyme, it was not long after the initial application that revelations of true "glucose isomerases," obtained from microorganisms that produced GI without xylose, appeared in the patent literature and elsewhere, as indicated in Table 9.2. Other glucose isomerases with a diversity of characteristics have so far been found in at least 60 distinct microorganisms distributed among 16 genera, including 17 species of *Streptomyces* (Gams 1976). It is most likely that the industrial GIs were obtained from mutants of these species (Chapter 5). A CPC-patented microorganism which produces a constitutive GI is a mutant of *S. olivochromogens* (Armbruster *et al.* 1974). Those used industrially as far as we can ascertain are listed in Table 9.2. That from *S. alba* has been well characterized. It possesses, as probably do all GIs, quaternary structure, its molecule consisting of four identical subunits. The molecular weight of this GI and others which have been determined lie around 150,000 (Berman *et al.* 1974). The pH optima of most GIs lie in the slightly alkaline range. Temperature optima tend to be on the high side as compared with most enzymes, going to about 90°C for the enzyme from *Actinoplanes*. As a rule, the higher the temperature optimum the more stable the enzyme. One of the beneficial effects of Co is to stabilize the enzyme against heat inactivation. As mentioned, all of the GIs require Mg for activity. However, in *S. alba* the requirement for Mg becomes less stringent with increasing substrate concentration.

Not surprisingly sugar alcohols are competitive inhibitors as are the heavy metals. For some but not all GIs, inhibition by Ca conflicts with the requirement of this metal for the stability and functioning of most bacterial amylases used in starch thinning operations at the beginning of the process leading to HFCS. Some bacterial amylases, such as that from *B. licheniformis*, are much less metal ion dependent (Barfoed 1976).

As mentioned, most of the GIs are sensitive to oxygen, but Tsumura and Sato (1965) found one species of *Streptomyces* that is not. In general, the isomerase reaction reaches an equilibrium from either direction when the ratio of fructose to glucose is 1:1, but this ratio can vary somewhat, depending upon pH and temperature, i.e., from 0.87 at 30°C to 1.10 at 70°C. The ratio may be shifted to 9:1 by addition of borate which forms a complex with glucose so that the latter cannot readily participate in the reverse ketose-aldose reaction.

Most but not all GIs are intracellular enzymes. One can allow the enzyme to remain in the cell or in the mycelium of *Streptomyces* species or the enzyme can be solubilized. Takasaki described several methods for solubilizing the enzyme: ultrasonics, lysozyme, surfactants "potentiation" of lytic enzymes by heating the mycelia below 50°C (higher temperatures inactivate the autolytic enzymes) (see below). Korus and Olson (1977) observed complete solubilization of *S. phaechromogenes* GI by suspension of the cells in water at 5°C for 2 weeks. CPC's *S. olivochromogens* appears to be extracellular (Iizuka *et al.* 1971).

IMMOBILIZATION AND PRODUCTION

Modes of Immobilization

Although the first HFCS produced in 1968 involved the use of soluble enzyme and were produced batchwise, it is probably true that not only Clinton but all the manufacturers now use immobilized GI in highly sophisticated automated continuous operations. Of course one of the prime concerns is safety, so great precaution is taken to avoid the introduction of unnecessary materials. It is probably for this reason that so far no system in which the enzyme is covalently bonded to the immobilization support is used. It is probably also true that in no instance is the enzyme purified to any great extent except, perhaps, to remove proteases. However, the producers of HFCS may be missing a bet in this regard. This is because an analysis of the economic implications of even a rough partial purification before immobilization (fractionation with acetone and ammonium sulfate) led Ladisch *et al.* (1977) to the conclusion that the total process cost for 1 million kg/year production facility fell dramatically as enzyme purity increased, and that economic gain may accompany purification prior to immobilization. The following is an attempt to impart some idea as to what immobilization procedures may be actually applied, as gleaned from the patent literature and trade journal write-ups, along with a few comments on variations and future possibilities.

Entrapment.—The Japanese investigators at first described laboratory production of fructose by the action of GI retained within the whole live organism. However, the glucose-fructose mixtures obtained were contaminated with protein because of the action of autolytic enzymes. By first exposing the cells to temperatures between 60° and 80°C, the enzyme becomes fixed within the cell. This method of immobilization is covered in U.S. patents issued to the Japanese government (Takasaki 1971) and to Clinton-Standard Brands (Lloyd *et al.* 1972). In the latter procedure, the finished fermentor broth is first pH-adjusted and then heated to fix the enzyme. Park *et al.* (1976) isomerized glucose at 70°C with cell-bound isomerase of *S. bikiniensis*. The enzyme is thus entrapped within the confines of the cell. Dinelli *et al.* (1973) entrapped the enzyme in fibers of cellulose triacetate after a solution of the former in the latter was passed through a spinneret.

As an example of other methods which have been proposed, that of Korus and Olson (1977) provides an interesting example of entrapment using a hollow fiber device (Chapter 8). They point out that the use of such devices is well established in food processing, no chemicals are involved in the immobilization, the device is easily cleaned and the drop in rate which may occur due to pressure drop in column type immobilization is avoided. They found that the time course of the reaction was not affected by immobilization. However, they had to coat the fiber with protein to prevent activity loss which means they would have to ensure the absence of proteolytic activity.

Glutaraldehyde Procedures.—In at least three cases the enzyme which is supplied to HFCS producers who do not use their own is immobilized by procedures involving crosslinking of the GI with glutaraldehyde as described in Chapter 8. Patents issued by Miles (Zienty 1971) and Novo (Amotz *et al.* 1976) both involve treating the cells, which may be at least partially disrupted or lysed, with rather strong (25 and 50%) solutions of glutaraldehyde. In a variation of this procedure for the preparation of GI offered by Gist-Brocades which combines entrapment with glutaraldehyde, fixation results in the formation of microcapsules. As described by Hupkes and van Tilburg (1976), the enzyme-occluded microcapsules are made by pouring a suspension of heat-treated crude *Actinoplanes* mycelia in 8% gelatin into an immiscible solvent such as butanol. The microcapsules (*ca* 0.5 mm in diameter) are then stabilized by treatment with glutaraldehyde. Like the production of the HFCS, the production of the enzyme is continuous and automated (Fig. 9.2). Absence of proteinase prevents hydrolysis of the gelatin as well as stabilizes the GI.

Adsorption.—In contrast to the enzyme suppliers it is most likely, from a perusal of the patent literature, that the HFCS producers who use their own enzyme employ adsorption as the prevalent mode of immobilization. Although the Takasaki procedure initially adopted by Clinton employs entrapment *via* fixation of the enzyme in the mycelia, this producer switched over to adsorption in 1972, using immobilization procedures akin to those detailed in their patent authored by Thompson *et al.* (1974). The enzyme is solubilized (perhaps by ultrasonication) and then mixed with an anion exchange column, most likely DEAE cellulose, so that the enzyme is held to the ion exchanger as a counter-ion. Other producers have also issued patents in which the use of ion exchangers figure prominently. For instance, one of several of such patents issued to CPC, the immobilized enzyme support, is a porous articulated strongly basic ion exchange resin having a quaternary ammonium, $-N(CH_3)_4^+$, as the ion exchange group (Tamura *et al.* 1976).

CPC has also patented the use of inorganic immobilization supports to which GI is apparently held by purely physical adsorption. These include basic magnesium carbonate (Heady and Jacaway 1974A) and silicate-based ceramic materials such as titania-alumina which is treated in a manner such that the enzyme is adsorbed only to inner surface (CPC Int. Inc. 1976). Similar types of inorganic support, silica coated with zirconia or titania, have also been used to immobilize GI *via* covalent attachment in a patent issued to Corning, a supplier of immobilized enzyme supports (Tomb and Weetall 1977).

Reactor Types

In the Clinton process, as revealed by patents and discussed by Mermelstein (1975), the slurry of the enzyme is pumped through a pressure leaf filter until each leaf is coated with a 2.54 to 3.8 cm layer. The Lloyd patent mentions a depth to width ratio of less than 2. In other Standard Brand patents shallow beds, 10–100 times wider than longer, are used (Barker 1975).

It would appear that most HFCS processors use more conventional reactors in the form of columns which can be operated in series or in parallel. Langhurst *et al.* (1976), describing a 140,000 bu/day corn processing plant, mention that more than 20 isomerizing columns were in simultaneous operation and that many will eventually be used. Use of bedform tower columns was considered a key innovation needed for successful continuous isomerization.

From discussions by investigators associated with food enzyme producers we may assume that commercially available immobilized GI is also used in conventional column type reactors. These include downward flow fixed bed reactors and upflow in what Oestergaard and Knudsen of Novo (1976) term expanded bed reactors. Hupkes and van Tilburg discuss the use of their immobilized enzyme in packed bed plugged flow and fluidized bed reactors as shown in Fig. 9.3.

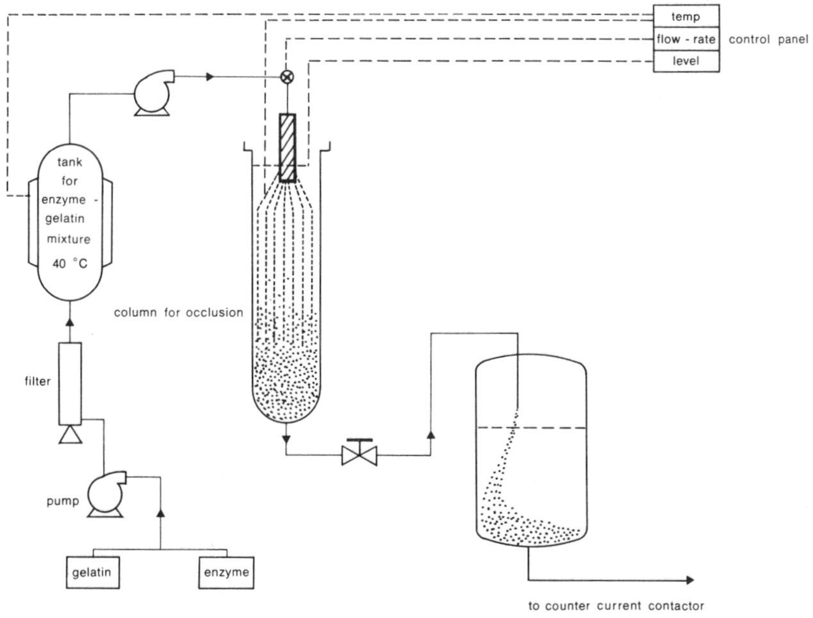

FIG. 9.2. PRODUCTION OF IMMOBILIZED GLUCOSE ISOMERASE

Courtesy of Hupkes and van Tilburg (1976)

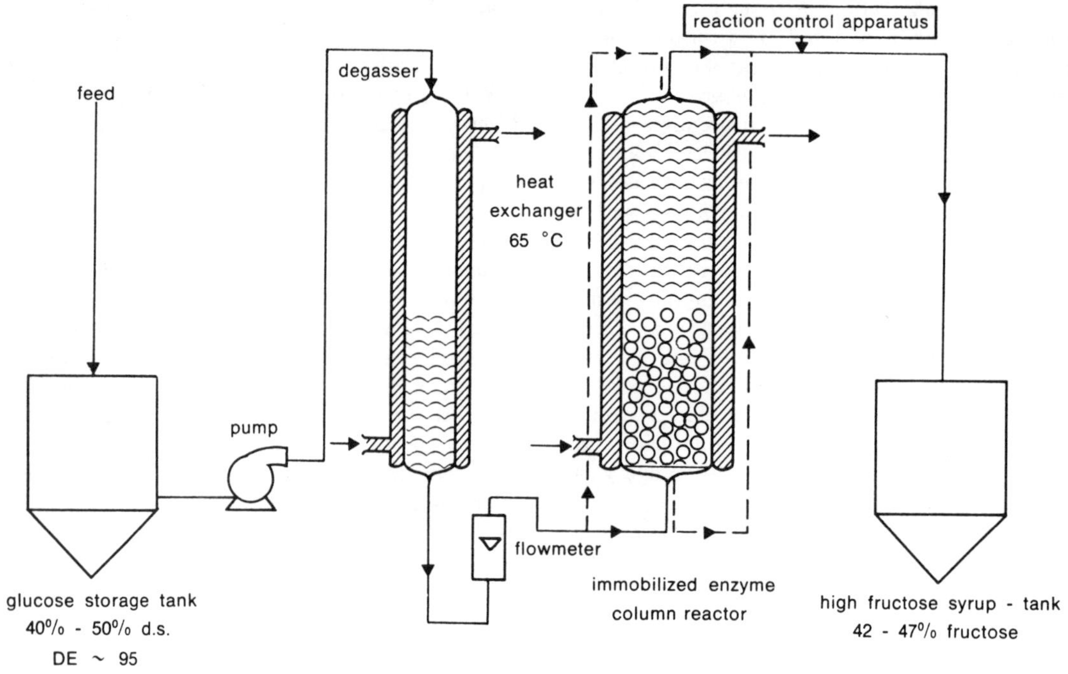

Courtesy of Hupkes and van Tilburg (1976)

FIG. 9.3. CONTINUOUS GLUCOSE ISOMERIZATION FOR HFCS PRODUCTION

Schematic shows immobilized glucose isomerase packed in a plugged flow or fluidized bed reactor.

Enzyme-related Operations

While the detailed conditions and parameters relating to the operation of an HFCS plant may vary considerably, there does appear to be a sort of common denominator for many of these operations. Thus, in the charging of the reactor with the immobilized GI, it is usually recommended that the glucose syrup be mixed with the immobilized enzyme *before* the latter is placed into the reactor. This is especially important for continuous operation. As an example Hurst (1977) in a patent issued to one of the HFCS pioneer companies, Staley, describes the treatment of spray-dried crude isomerase from *Bacillus coagulans* with glucose syrup (94% dextrose) to which are added the activators and stabilizers—0.02*M* $MgSO_4$, 0.001 $CoNO_3$ and 0.01 Na_2SO_3. This stabilized enzyme is then mixed for 30 min at 23°C and pH 6.4. The claim is made that this isomerase is capable of producing higher yields of fructose than untreated dry isomerase.

In continuous operation it is usually recommended that the input syrup solids contain no less than 92% glucose substrate and that the concentration of solids lie in the range of 35–45%. The conversion is always operated at elevated temperatures, as low as 60°C but also as high as 80°C. Programmed temperature control in which the temperature starts at 50°C and then increases has been recommended (Heady and Jacaway 1974B). One has to balance the benefits of decreased residence time of the enzyme (due to speed-up of isomerization which increased temperature confers) with increased chances for nonenzymatic browning and its attendant increased refining costs.

The reaction is usually run at a pH near neutrality. The exact pH depends on the source and purity of the enzyme, the effect of immobilization, the actual pH optimum of the enzyme, and conditions which are consonant with activation and stability of the enzyme. Thus, *Streptomyces* GI has to be protected against O_2 and needs Co ions for maximum stability. On the other hand the GIs from other organisms are not quite so fussy. Thus, Novo's Sweetzyme from a strain of *B. coagulans* is used at pH 8.5 where it does not require Co addition. At 60°–65°C it requires neither Co nor oxygen-protective agent (Oestergaard and Knudsen 1976; Poulsen and Zittan 1976). Eventually the activity of the immobilized GI starts to decline and the column either has to be replaced, which can be accomplished during continuous operation (Thompson *et al.* 1974), or the column can be recharged with fresh enzyme (Tamura *et al.* 1976). In general it would appear that such renewal operations have to be done every 1 or 2 weeks.

PROPOSED AND POTENTIAL ENZYMATIC SUGAR SWEETENER PRODUCTION

The remainder of this chapter is devoted to enzymes which, unlike the successful application of the starch and glucose transforming enzymes, have not at the time of this writing found their way into the marketplace, but for which there is hopefully great potential. A great deal of effort is going into bringing to practical fruition at least one of these enzyme systems, cellulase, because of the high stakes and pluralistic nature of the problems which would be solved by its successful application. The other two enzyme systems are those involved in the hydrolysis of the xylans to produce xylose to make xylitol, and a small but significant effort to make sugar syrups from lactose by action of lactase. While there is some disparity in efforts to use these enzyme systems for the production of sweeteners, what they do all have in common is that their substrates, unlike starch, are or can be under certain circumstances waste products. The incentive to make these enzyme systems work practically comes from the possible secondary transformations of the products of their action. We shall have occasion to discuss these enzymes in contexts other than their use as catalysts for the production of sweetener ingredients for foods. Thus, both cellulase and xylanase (as well as other hemicellulase) action influences the texture of plant foods, both as endogenous enzymes and as commercial enzyme preparations (Chapters 29 and 30). Lactase has been proposed and is being used for a variety of health and texture-related purposes in addition to its use in the production of sweetness and sweeteners (Chapters 34 and 35). However, the task being asked of the cellulase system and consequences of successful application are of monumental proportions as compared with xylanase and lactase.

Cellulase

Cellulose.—This most abundant organic substance on the earth is the main constituent of most plant cell walls in which it is present as micelles, crystalline aggregates and/or microfibrils of well oriented, highly ordered hierarchy of hydrogen-bonded chains. These chains are polymers of D-glucose joined by β-D-(1→4) linkages. Unlike starch molecules there is no branching as in amylopectin or helical conformation as in amylose.

In the case of plant fibers such as cotton and ramie the cellulose is present in almost pure form as the remains of elongated cells with heavily thickened walls. However, as a substrate for the proposed uses of cellulase, it probably will be frequently accompanied and interpenetrated by other cell wall constituents such as the hemicelluloses (including xylan) and frequently lignin. Since it is not a human food it is a by-product or waste product in some food processing operations. However, the ramifications go beyond food processing and include by-products made from trees such as newsprint, municipal solid waste and agricultural residues. Thus it has been estimated that 400–500 MMT of crop wastes are produced in the United States each year. Cellulose by-products from both harvesting and processing operations include corncobs, cottonseed hulls, flax shives, sugar cane bagasse, oat hulls, rice straw and smaller but by no means negligible amounts from the processing of fruits and vegetables (Ben Gera and Kramer 1969). Of course many of these by-products are used for feed. Nevertheless those who advocate and are trying to develop enzymatic conversion of cellulose stress that it is not only the sweetening properties of the resulting glucose which can be utilized, but that glucose has the potential of being the starting material for further conversion into such diverse products as alcohol, methane and single cell protein (Chapter 34). In the more distant future some of these advocates envision a process of multiple immobilized enzyme technology in the first step of which cellulose, converted eventually into starch, is converted into glucose by a cellulase system (Adams *et al.* 1974). Cellulose utilization may be considered as a source of solar energy to hopefully replace less abundant energy sources. The importance attached to the success of this inchoate cellulase technology is attested to by the devotion of at least two Biotechnology and Bioengineering Symposium series to cellulase and its application to the production of glucose (Wilke 1975; Gaden *et al.* 1976). These publications provide central reference points and keys to the literature.

Discovery of Cellulase.—Unlike the amylases whose investigation as a scientific activity goes back to at least 1811 (Chapter 1), the cellulase story is about 55 years old although the vague term "cytase" was applied to plant-wall digesting enzymes in 1890. Karrer and associates in a series of ten papers published between 1923 and 1929 describe the presence, in the hepatopancreas and gastric juice of snails, of an enzyme system capable of producing glucose from a variety of cellulosic substrates. The first substrate investigated was the "reserve cellulose" or lichenan of lichens which we now know is a β-D-linked glucose polymer in which the linkages are both $(1\rightarrow3)$ and $(1\rightarrow4)$. This was followed by use of the then so-called artificial silks (rayons), viscose cotton and eventually native crystalline cellulose including filter paper (Fig. 9.4). Some of the characterizations of modern cellulase research were then established. Thus, the presence of a cellobiase and a cellulase proper was established as was the estimation of the extent of enzyme action by determining the percentage of the theoretical reducing power of enzyme digests as glucose. The Zurich group also compared the susceptibility of cellulose treated to alter its solubility with native crystalline cellulose (the latter experiments took weeks to complete), carried out an X-ray crystallographic study of the effect of cellulases on cellulose structure (Faust and Karrer 1929) and then, as now, deplored the lack of a "standard" substrate. The Swiss group also laid the foundation for the endogenous effects of cellulase action in foods by demonstrating activity in malt, spinach and corn kernels. Kinetics followed Schutz's law (rate proportional to the square root of the enzyme concentration) which, as mentioned in the last chapter, holds for immobilized enzyme systems which are diffusion-limited. Incidentally, since mention of cellulase activity occasionally conjures up the image of termites eating away on wood, we digress to mention that the long-held belief that the termite's ability to utilize wood is due exclusively to the cellulase secreted by symbiotic gut microorganisms, especially protozoa (which may constitute almost 25% of some species' fresh weight) has been modified to include other sources. These comprise diverse tissues of the insects' gastrointestinal system such as ventrical cells, gastric juice, midgut gland (Yokoe 1963), and salivary glands of wood-eating cockroaches; and for one species of termite, the fungus which it cultivates for food (Martin and Martin 1979).

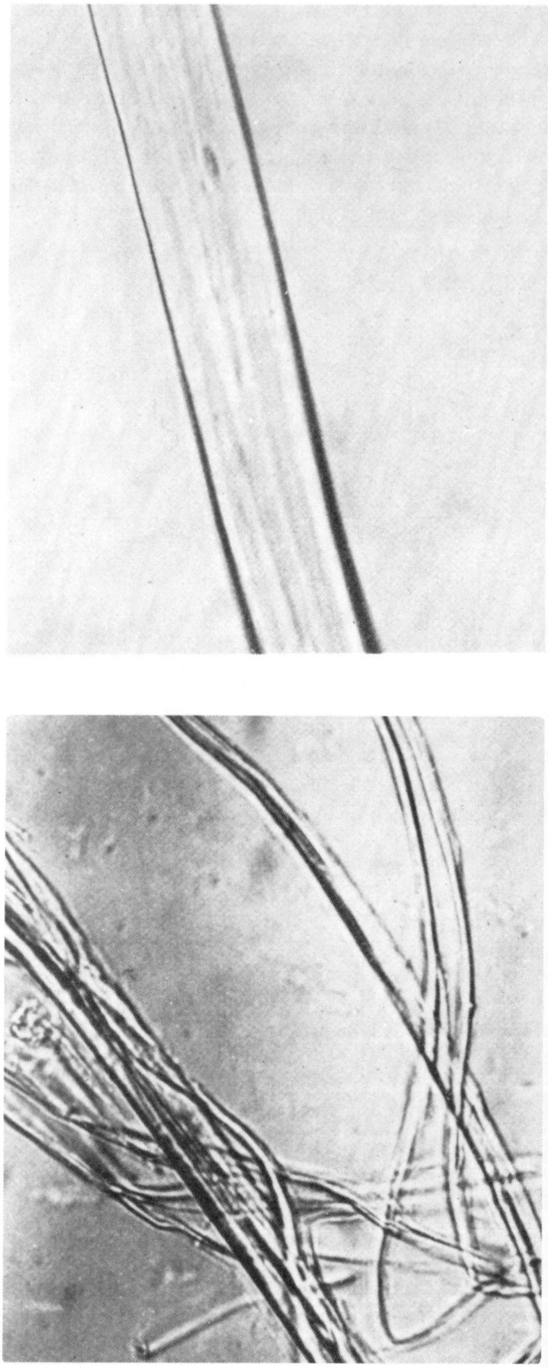

From Karrer et al. (1925)

FIG. 9.4. EFFECT OF SNAIL CELLULASE ON "COPPER SILK"

Magnification, 700×. Top—before; bottom—after treatment.

Cellulase Production from Microorganisms.— Subsequent to Karrer's pioneering investigations, cellulase studies were pursued by a motley assortment of investigators interested in rumen digestion, wood-rotting fungi, plant pathogens and later by the military, who were interested in the rotting of cotton goods in the tropics and other warm, humid climates, These studies, especially the latter and more specifically the group at the U.S. Army's Natick, Massachusetts, Laboratories, provided the background for the development of the feasibility of putting the cellulase to work to produce glucose.

As with other enzymes, the cellulases were found distributed among a vast number of microorganisms and could be induced and enhanced by mutation. For the production of glucose, however, the list of suitable microorganisms was limited because it is necessary to have the right combination of the full complement of the enzymes constituting the cellulose-to-glucose enzyme system. Fungi seem more likely to contain such systems at higher levels than other microorganisms. Young *et al.* (1977) found that a thermotolerant fungus, *Chaetomium cellulolyticum*, which utilized cellulose very efficiently, did a poorer job of producing glucose than the now well-established *Trichoderma viride* (now *T. reesei*) of the Natick group, but did grow faster than did the latter fungus. Also, Bellamy (1976) reported on a thermophilic cellulolytic *Cytophaga* species (grows at 65°C) which digests cellulose more rapidly than any organism so far tested but which secretes no cellulase. Wicklow *et al.* (1980) report that *Cyathus stercoreus*, a basidiomycete that grows on the surface of decomposed manure, degrades lignin (45% in wheat straw) and at the same time frees greater amounts of cellulose for cellulase-directed hydrolysis than any other fungus previously reported.

As mentioned, by far the most thoroughly investigated cellulase system is that of *T. reesei*. Early work established that the enzyme was induced rather than constitutive and that among the most effective inducers were cellobiose (4-O-β-D-glucopyranosyl-D-glucose), sophorose (2-O-β-D-glucopyranosyl-D-glucose), as well as cellulose, and that nonionic surfactants such as digitonin and sucrose monpalmitate stimulated synthesis of the enzyme (Reese 1976). Later work concentrated on cellulose as the most practical and efficient inducer, when used at the correct levels for large-scale production (Sternberg 1976A; Breuil and Kushner 1976). Glucose in the culture medium suppresses cellulase synthesis (catabolite repression, Chapter 4) and the metabolic products of glucose metabolism in the fungus actually lead to enzyme inactivation (Mandels 1975). The latter investigator stated that constitutive mutants of *T. reesei* have been looked for, but at that time had not yet been found. Enzyme production as part of systems for conversion of cellulose to glucose on an industrial scale will be outlined below. In addition to those authors already cited, Brown (1976) and Enari and Markkanen (1977) also provide discussions of microbial cellulase production.

The Cellulolytic Enzymes.—The magnitude of the problems involved in the selection of the appropriate microorganism as a source of cellulase for industrial glucose manufacture cannot be appreciated without having some idea of the complexity of the system responsible for the degradation of native crystalline cellulose. This complexity stems in part from the repeatedly stated lack of standard substrates. Excluded from most cellulase studies as substrates are natural substrates, in which cellulose is mixed and integrated with other cell wall constituents such as the xylans, other hemicelluloses and lignin although knowledge of how these substituents affect cellulase action is of utmost importance before commercial glucose production becomes a reality. At the present stage of cellulase enzymology, with the exception of deliberate derivatization, only as pure as possible polymers constituted of β-1,4-linked glucose units are used as substrates. In general we may distinguish among:

(1) Insoluble substrates—insoluble highly-ordered native crystalline cellulose such as cotton fibers (dewaxed), cotton linters and filter paper (the latter is derived from wood treated to remove noncellulosics). Included are somewhat degraded forms obtained by mechanical treatment such as microcrystalline and brushheap cellulose.
(2) Treated but still highly ordered and insoluble substrates such as Avicel. The cellulose chains have been degraded from DPs in the tens of thousands to the low hundreds.
(3) Insoluble but amorphous highly hydrated forms of cellulose obtained after treatment with alkali (mercerization) or with strong 85% H_3PO_4 (Walseth cellulose) and/or copper-ammonium salts (see Fig. 9.4).

Among the soluble substrates the most prevalent are carboxymethyl celluloses (CMC) derivatized to the extent that the cellulose becomes soluble. According to Wood (1975), the hydroxyethyl derivative would be more suitable. Introduction of too many groups prevents action of some of the components, especially the exo-acting enzymes of the cellulase system. Also used as substrates are the cellodextrins or cellooligosaccharides, relatively short chains of β-1,4-linked glucose molecules. Brown (1976) restricts the group to cellobiose, DP = 2 to cellohexaose, DP = 6.

Endocellulases or Endo-β-(1→4)Glucanases.— In analogy to α-amylase acting on the amylose component of starch, by far the most thoroughly investigated and perhaps most widely distributed components of the cellulase system (Comtat and Barnoud 1976) are those that randomly split the internal β-1,4-linkages of CMC to form cellodextrins and eventually cellobiose and cellotriaose. For the most part the substrates for these endocellulases are the CMCs and cellodextrins. By themselves the endocellulases exhibit vanishingly little activity towards crystalline cellulosic substrates. Again, in analogy to α-amylase which was named "α" because the products of the reaction it catalyzed retained α-configurations at C_1 of the newly formed glucose residue end-groups at point of scission, the products of some but not all endocellulases also retain the original β-configuration. The endocellulases are usually present as multiple forms. Several of these isoenzymes have been purified and apparently isolated. Most investigators include an adsorption on some form of insoluble cellulose (Table 7.1) in their purification schemes but this is usually followed by fractionation on a Sephadex column. For example, Berghem *et al.* (1976) isolated two *T. reesei* endocellulases, glycoproteins with molecular weights of 12,500 and 50,000. Similar results have been reported by Japanese workers (Okada and Nishizawa 1975). Most molecular weights range below 40,000. Other references to endocellulases can be found in discussions by Wood (1975) and Comtat and Barnoud (1976).

Exocellulases.—Although there is limited evidence that some microorganisms produced enzymes, in analogy to glucoamylase, that hydrolyze cellulose to glucose by removing one glucose at a time from the nonreducing end of the molecule, attention has been centered on evidence for the presence of an exocellulase which, starting at the nonreducing end of the cellulose chain, hydrolyzes every other linkage to form cellobiose exclusively. In contrast to the endocellulases these enzymes, cellobiohydrolases (cellobiosylhydrolases) act better but still rather feebly on insoluble cellulose than they do on CMC, probably because of the steric interference of the carboxymethyl groups of CMC in the progress of the enzyme down the cellulose chain. They act more readily than do the exoenzymes on swollen and partially degraded but still crystalline forms (Avicel, hydrocellulose). In analogy to the corresponding starch-hydrolyzing enzymes, the β-amylases, exocellulase action causes inversion of the β-configuration so that the cellobiose resulting is the α-anomer. Unlike the β-amylases, however, they are present in a wide spectrum of microorganisms and almost invariably in association with endocellulases. Their molecular weights range in the 40–60 thousands, somewhat higher than those of the endo-enzymes.

With a few exceptions the exocellulases are glycoproteins with a moderate amount of carbohydrate (0–13%). One of the cellobiohydrolases (C) has been examined in detail with respect to the nature of and linkages between the sugars of its carbohydrate (Gum and Brown 1976). This enzyme contains 11.3% total neutral carbohydrate, present for the most part as 1,6-linked disaccharides, some 15 per molecule. One of the difficulties in sorting out the cellulase complex is that accompanying glycosidases cleave these carbohydrates to variable extents resulting in heterogeneity. Most of the purified cellulases are rich in serine, threonine and glutamic and aspartic acids.

Less-random Cellulases.—The analogy with starch-digesting enzymes is not adhered to rigorously because of the presence of at least one enzyme (and probably others once they are looked for) in a cellulase system exhibiting some characteristics of the both exo- and endo-enzymes. Such an enzyme was found in *T. reesei* by Okada (1976). This enzyme, obtained as a subfraction of a rather complicated mixture of cellulases, could readily utilize CMC as a substrate and when acting on cellohexaose produced β-anomers (like a random-acting endo-enzyme) of glucose and cellobiose. Interestingly, the enzyme readily catalyzed the condensation of cellobiose to cellotetraose. However, like cellobiohydrolases it could act on more ordered insoluble substrates such as Avicel, filter paper and cotton. As shown in Fig. 9.5 the enzyme exhibited less randomness of action than did a more "more" random cellulase, i.e., the change in fluidity of CMC per unit amount of reducing power gener-

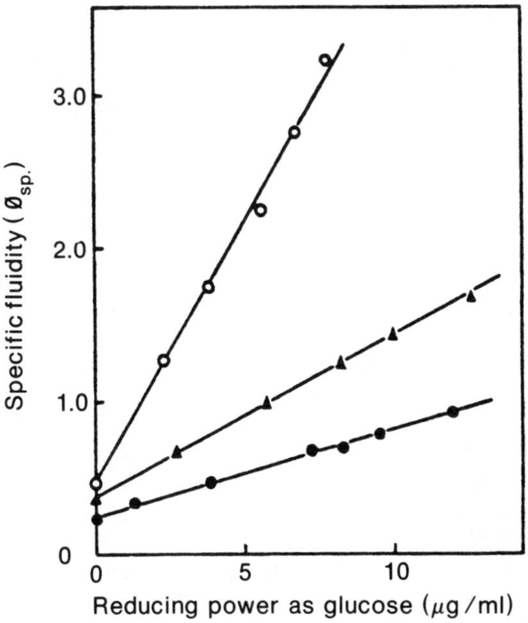

From Okada and Nisizawa (1976)

FIG. 9.5. ESTIMATION OF RANDOMNESS OF CELLULASE ACTION

From the slope of the fluidity *vs* reducing power (covalent bonds split) curve. Shown are the curves for three isozymes from a strain of *Trichoderma reesei*. Increased randomness is reflected in increased slope.

ated was intermediate between that of a "true" endoenzyme and an exoenzyme. Indeed Okada suggests that cellulase systems, rather than containing distinct endo- and exo-acting β-glucanases, are mixtures of "more" random and "less" random types. It is obvious that the whole story on the nature of the exo-β-1,4-glucanases of *T. reesei* is not in yet, especially since each group of investigators comes up with a different set of isoenzymes—this in spite of what appears to be fairly good evidence for homogeneity of each fraction. It is tempting to suggest that interchange of enzyme subunits of protomers (Chapter 4) may occur during the purification. This could give rise to distinct and apparently pure proteins each with its own specificity and other characteristics, and that the type of activity (exo/endo more or less random) depends upon what subunits combined.

Cellobiase or β-Glucosidases.—Each cellulolytic system comprises not only β-glucanases but also what should be the final enzyme needed for conversion to glucose, the cellobiases, cellobiose + H_2O → glucose. Because many of these enzymes act on other β-glucosides, they have been considered to be β-glucosidases. However, not all β-glucosidases can attack or hydrolyze cellobiose. Furthermore, cellobiases can also readily hydrolyze cellooligosaccharides up to at least cellohexaose but acting alone have no action whatsoever on the more complex cellulase substrates. Indeed cellotriaose is a better substrate than cellobiose for at least one cellobiase from *T. reesei* (Emert *et al.* 1974). It is of interest to note that cellohexaose is a substrate of all of the components of the cellulose system. Like the β-glucanases proper, amino acid analysis of cellobiase exhibits predominance of the acidic and hydroxyamino acids. Gong *et al.* (1977) isolated three isoenzymes of *T. reesei* cellobiase with molecular weights of about 76,000.

Characteristics of the enzyme which have to be considered in maximizing glucose production are its relative overall instability and its inhibition by high concentration (K_i = *ca* 30 m*M*) of substrates including both cellobiose and cellotriose (Emert *et al.* 1974). It is also important in designing efficient glucose production processes that this range of concentration is more than adequate to completely suppress (K_i = 0.07μ*M*) the solubilization of crystalline cellulose by a cellobiohydrolase (Halliwell and Griffin 1973). Its crucial role in efficient glucose production (from crystalline cellulose) will be discussed below. Cellobiases are also product-inhibited with a K_i for the noncompetitive formation of enzyme-glucose complex of 1.22m*M*, an inhibition which is stronger than that exerted by excess substrate but still much weaker than that of the inhibition of cellobiohydrolase by cellobiose.

C_1, C_X, Initiation Factors and Synergy

Observations accumulated by the Natick group and others in the 1940s that: (1) the slow but fairly extensive action of unfractionated fungal culture filtrates on native crystalline cellulose disappeared upon fractionation, (2) culture filtrates from different microorganisms exhibited different ratios of enzyme activities as measured on soluble substrates especially CMC (C_x) as compared to insoluble substrates or crystalline cellulose, (3) the activity towards crystalline cellulose almost disappeared except for a fraction (C_1) which showed feeble activity, and (4) the original activity was restored when the fractions were recombined. This, together with the persistent presence of cellobiase in a wide variety of cellulolytic sources, led Reese *et al.* (1950) to propose that the degradation of cellulose to glucose followed the pathway:

$$\text{Crystalline Cellulose} \xrightarrow{C_1} \text{Amorphous Cellulose} \xrightarrow{C_x}$$

$$\text{Soluble Intermediates} \xrightarrow{C_x} \text{Cellobiose} \xrightarrow{\text{Cellobiase}} \text{Glucose}$$

To C_x was assigned the endo- and exo-acting components of the cellulase system. Adding C_1 to C_x, as mentioned, restored to a large extent the original ability of the culture filtrate to degrade crystalline cellulose, the activity increasing as much as 20–30 fold (Wood 1975). While all investigators agree that several enzyme types and/or factors are involved in the degradation of crystalline cellulose, there is little agreement as to what the initiators of this action are. Since C_1 acted, albeit sparingly, on native cellulose Reese and colleagues assigned C_1 the task of initiation. C_1 was supposed to somehow or other break the complex hierarchy of hydrogen bonds keeping the cellulose inpenetrable to the C_x enzymes. This problem of initiation of enzyme action is not unique to cellulase but recurs in attempts to explain just about all biological polymers in which the hydrogen bonds play key roles in ordering the conformation and association of polymer chains—DNA, starch, chitin, collagen. In the case of cellulase C_1, perhaps a "hydrogen bondase" induces a modification (possibly a distortion) of the crystalline lattice so that the cellulose chains now accept water, swell and thus become susceptible to cellulolytic degradation. The controversy surrounding C_1, its nature and its role in initiation continue. As of this writing the following mechanisms for initiation of the degradation of crystalline cellulose have been proposed.

Reminiscent of lipase's interfacial recognition site (Fig. 25.2), C_1 is a specific form of any enzyme which can split a covalent bond of cellulose, but whose action is *restricted to the crystalline surface.* Being specific for such surfaces, it does not participate in further degradation of cellulase chains, this task being taken over by the exo- and endo-cellulases (Reese 1976).

Action is initiated by randomly-acting C_x endocellulases. This creates nonreducing group ends which are substrates for cellobiohydrolases, identified as C_1. Thus, C_1 no longer has a "C_1" function. The further combined action of C_1 and C_x enzymes produces cellobiose which is then hydrolyzed to glucose by cellobiase (Wood 1975).

A small wedge-like endoglucanase enters the less ordered, crystalline regions of the native cellulose fiber and opens up the structure to attack by other endonucleases whose action is reenforced by the exoglucanases acting on the nonreducing ends of the newly created chains (Comtat and Barnoud 1976).

The hydrogen-bonded superstructure imposed on cellulose is disturbed by glucuronic residues created in the cellulose chain by the action of a "cellulose oxidase." The evidence is based on the observation that fungal culture filtrates break down cellulase in the presence of air faster than its absence and that oxygen is consumed in the process (Eriksson *et al.* 1974). Unfortunately, the novel cellobiose: quinone oxidoreductase, which connects the degradation of lignin with that of cellulose while oxidizing cellodextrins as long as cellopentaose, does not act on cellulose (Westermark and Eriksson 1975). We suggest that the enzyme may serve a "windshield wiper" role in high lignin-low cellulose-degrading systems analogous to that played by cellobiase in high cellulose systems.

The hydrogen bonds are disturbed by oxidative disruption of the β-1,4 linkage. This disruption is caused by the combined action of iron salts and hydrogen peroxide (Koenigs 1975). While Koenigs suggests that the initiation of the decomposition in brown rot fungi which lack cellobiohydrolase activity is due to H_2O_2 produced from glucose by a glucose oxidase in the fungus, the initiating disruption may be due to a more active oxygen species such as singlet oxygen, superoxide anion or hydroxy free radical. H_2O_2 + Fe is Fenton's reagent, a good source of these active forms of oxygen. Furthermore there are oxidases in most cells which can produce superoxide anion, and eventually the other active oxygen species. Methyl cellulose can be so depolymerized by the action of xanthine oxidase (Fig. 9.6). Mannitol and DABCO quench $\cdot OH$ and O_2^*, respectively (Chapter 16).

Role of Cellobiase.—Whatever the initiation factor is, it is becoming increasingly clear that to obtain not only glucose but to also complete solubilization of crystalline cellulose, it is necessary to have the cooperative participation of all of the components of the cellulolytic degradation system. In particular the participation of cellobiase in this synergy appears to be crucial, as an inspection of the collation of data in Table 9.3 from various investigations clearly shows. As mentioned, this

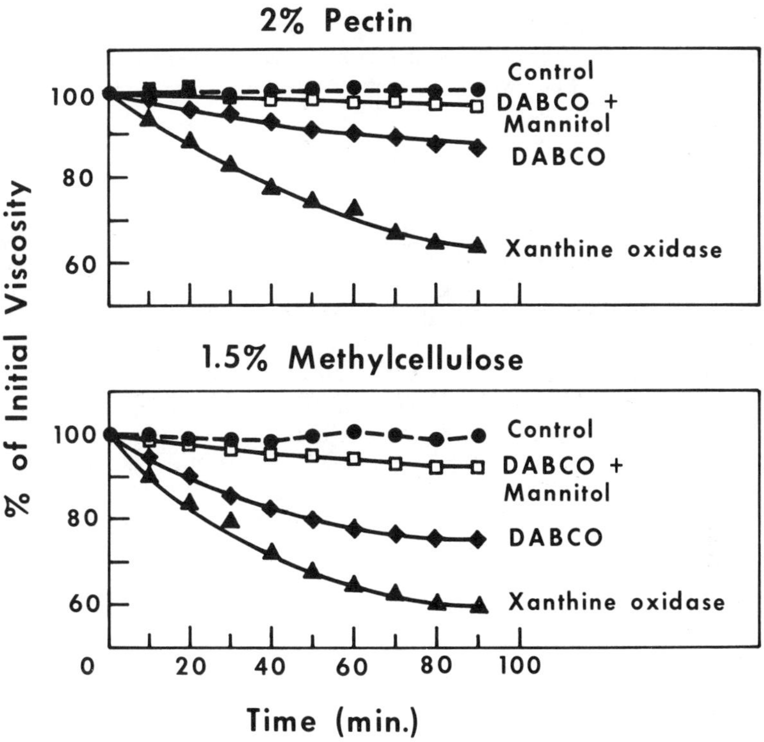

FIG. 9.6. OXIDATIVE DEGRADATION OF PECTIN AND METHYL CELLULOSE

From Kon and Schwimmer (1977)

TABLE 9.3
WINDSHIELD WIPER EFFECT OF CELLOBIASE AND OTHER CELLULOSE HYDROLYSIS STIMULANTS

Substrate	Cellulase-source, Component	Time (hr)	Measurement (%)	Effect of Treatment Before	After
Cotton, dewaxed[2]	Cellobiohydrolase (C_1)	336	Solubility	22	50
Cotton, dewaxed[2]	$C_1 + C_2$	168	Solubility	23	41
Various[3]	*T. reesei* culture filtrate	50	Total sugar (TS)	(5–42% stimulation)	
Solka-floc,[4] ball milled	*T. reesei* system	40–50	TS	60–65	65–75
Newsprint,[4] ball milled	*T. reesei* system	40–50	TS	10	16
Cotton[5]	*T. koningii* $C_1 + C_x$	—	Swelling	90	99
Other treatments:					
Newspaper	± Ball milling[6]	25	TS	9	18
Newspaper	± Two-roll milling[6]	25	TS	9	27
Agricultural residues	± Cadoxin[7]	20	Glucose	10–27	86–100

[1] Sources of cellobiase include separated enzymes from *T. reesei*, almond emulsin (Chapters 22 and 35), *B. theabromae*, *F. solanii* as well as cellulolytic systems with high and low ratios of "cellulase" to cellobiase activity.
[2] Source: Halliwell and Griffin (1973,1978).
[3] Source: Sternberg (1976B).
[4] Source: Yamanaka (1975).
[5] Source: Wood (1975).
[6] Source: Tassinari and Macy (1977).
[7] Source: Ladisch *et al.* (1978); Cadoxin is 5% cadmium oxide dissolved in 28% aqueous ethylene diamine.

need for cellobiase activity is dictated not only for full glucose production but by the strong inhibition by its substrate, cellobiose, of the exoglucanase cellobiohydrolase. This "windshield wiper" effect is analogous to the previously mentioned strong stimulation of the α-amylase-catalyzed digestion of raw starch by fungal β-glucosidase, maltase. We suggest that the inhibitory effect may be more crucial in such heterogeneous systems than it would be in solution because glucanases, in order to effectively "unzip" the cellulase chain, must remain uninterruptedly associated with the chain during their progression down the polymer (exoenzyme action). Furthermore, it will be recalled from Chapter 8 that two or more enzymes which catalyze consecutive reactions do so more effectively when they are immobilized than when in solution. Cellobiose may very well prevent the continuity of this vital association or immobilization periodically by, in effect, affinity eluting (Chapter 7) glucanases from the matrix-substrate. A more speculative suggestion, as mentioned in connection with C_1, is that glucose formed from cellobiase may be oxidized by an enzyme which produces, as do many oxygen-utilizing oxidases, active forms of oxygen. As we have seen, these can make crystalline cellulose more susceptible to the action of the cellulase system. Perhaps this would contribute to the oxygen-mediated stimulation of cellulose degradation observed by Eriksson *et al.* (1974). A newly described cellobiase may participate in the regulation of cellulase induction (Inglin *et al.* 1980).

It would seem that those interested in glucose production have only recently taken cognizance of the importance of cellobiase. Sternberg (1976A,B) was able to consistently increase the yield of cellobiase 5- to 10-fold during *T. reesei* culture by not allowing the pH to go below 5.0 because the enzyme is quite sensitive to acid. The time of saccharification was cut in half by this enhancement of cellobiase activity. He also showed that most of the glucose produced from crystalline cellulose is the result of cellobiase activity but that glucose from pretreated celluloses stems largely from the action of the glucanases. Further evidence of the awareness of the crucial role of cellobiase is attested to by its use in supplementing cellulolysis for glucose production (Yamanaka *et al.* 1977), and its immobilization *via* covalent attachment to cellulose by investigators interested in cellulose degradation (Srinivasan and Bumm 1974). More recent immobilization studies, not directed towards cellulase as such, but which may have some use in this respect, include that of Van Dongen and Cooney (1977) who examined hollow fibers as a means of immobilizing β-glucosidase.

Large-scale Glucose Production

Large-scale trials are being conducted by the Natick group and by a group at the Department of Chemical Engineering, University of California (Wilke 1975). Investigations at Louisiana (Srinivasan 1976) and General Electric (Bellamy 1976) have also looked into the economic aspects of such processes. The prospect for using waste from sugar cane processing has been explored by groups in Louisiana (Srinivasan 1976), Japan (Toyama 1976) and Taiwan (Wang and Kuo 1976).

In essence an industrial glucose facility for the conversion of cellulosic materials to glucose would consist of three sections: a pretreatment plant, a fermentor and a cellulose-to-glucose reactor. In the Natick set-up (Fig. 9.7) the *T. reesei* culture filtrate is pumped from the fermentor into the reactor previously charged with cellulosic substrate (Nystrom and Andren 1976). After a suitable period of several hours to several days under proper conditions of pH, temperature (preferably above 50°C) and stirring, sugary syrup is removed and the unreacted solids (highly crystalline cellulose, lignin) filtered off. Quite a bit of effort has gone into how to most economically recover the enzyme system. In both the Natick and Berkeley schemes, the enzyme in the filtrate is adsorbed onto a fresh charge of substrate. Earlier, the Natick group had demonstrated that the enzyme could also be removed by ultrafiltration (Ghose and Kostick 1970). One of the first lessons learned in going from bench to small pilot-plant scale was that it was necessary to lower the reactor temperature from 50° to 40°C to prevent excessive heat inactivation of the enzymes.

In the Berkeley process (Wilke and Mitra 1976) as shown in Fig. 9.8 part of the cellulose is used as food and carbon source for the growth of the cellulase-producing microorganism, the enzyme system is precipitated with acetone before being added to the reactor, and the reaction is carried out at around 50°–60°C for 40 hr. After removal of the enzyme from the spent solids by adsorption, washing and filtration in a multistage countercurrent thickening process, the spent solids are dried and used as fuel to produce process heat. In the fermentation stage, provision is also made for using the microorganism alternatingly for production of solu-

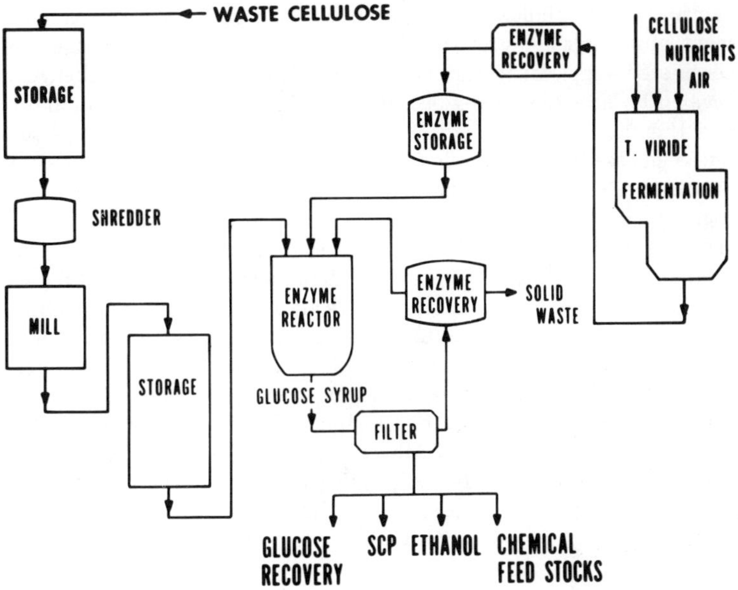

FIG. 9.7. CELLULOSE SACCHARIFICATION—THE NATICK PROCESS

A schematic for the conversion of waste paper products to glucose.

Courtesy of U.S. Army Natick Research and Development Command

ble cellulase and biomass. The latter can be used as feed. A theoretical model for such a continuous multistage enzyme-biomass production system has been developed (Von Stockar *et al.* 1977). We previously alluded to the Berkeley group's study of the feasibility of supplementation of the cellulase system with cellobiase. Not only the concentration of enzyme and substrate have to be considered but also the ratio of enzyme to substrate.

The Louisiana group used cellulose as sole carbon source for the growth of *Cellulomonas* and *Aspergillus* species. Further cellulose hydrolysis by or-

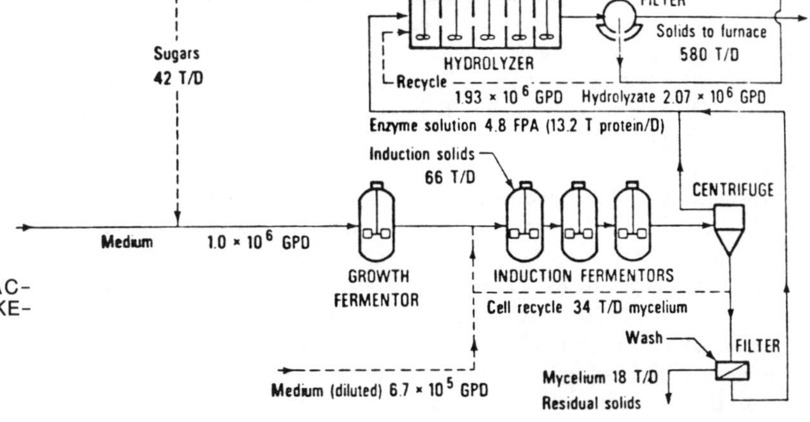

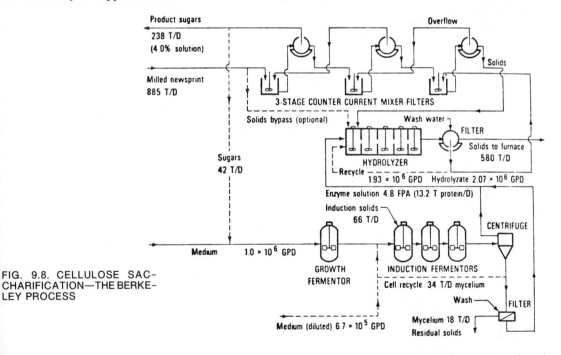

FIG. 9.8. CELLULOSE SACCHARIFICATION—THE BERKELEY PROCESS

From Wilke and Yang (1975)

ganism-bound cellulase occurred in an environment unsuitable for growth. The group has also considered supplementation of their process with covalently immobilized β-glucosidases (Srinivasan and Bumm 1974). In Japan, Toyama (1976) produced glucose from cellulases from *T. reesei* grown on wheat bran in a Koji-like process analogous to sugar production by amylases from *Aspergillus oryzae* grown on steamed rice (autosaccharification). In Taiwan the problem of disposal of bagasse pith has led to attempts at saccharification with cellulase (Wang and Kuo 1976). After ascertaining the optimum substrate concentration (5% bagasse pith) and pith to enzyme ratio (12:1) as well as suitable pretreatment regimes (hammermill plus treatment with NaOH) they achieved some pilot-plant success in completely hydrolyzing the cellulose, comprising 60% of the pith, in 12 hr at 50°C. The resulting glucose was converted into fodder yeast.

The Pretreatment Problem.—The principal factor which may limit or prevent the establishment and growth of a viable industry based upon the production of glucose from cellulosic wastes appears to be the refractoriness to enzyme action of the substrates present in this waste material. In spite of the synergy exhibited by the components of the cellulase systems on the relatively pure, crystalline cellulose employed in cellulose research, the degradation of unmodified cellulose as it exists in these wastes *in situ* is at the present time not economically feasible, according to just about all writers and investigators of the subject. This recalcitrance stems not only from the aforementioned lack of water in and crystallinity of the cellulose but also to the intimate association, interpenetration and encrustration of the cellulose with cell wall materials such as xylans (see below) and lignin, whose enzymatic transformation (what little was known) was recently reviewed by Hall (1980). Cowling (1975) describes in detail these physical and chemical constraints.

The problem does not lie in a shortage of methods of pretreatment. (Table 9.3 displays a sampling of data): chemicals and heat alone or in combination—NaOH, CaO, Na_2CO_3, ammonia, zinc and, promisingly, cadmium compounds (Ladisch *et al.* 1978; Hsu *et al.* 1980); solvents such as dimethyl sulfoxide/paraformaldehyde (Johnson *et al.* 1976) combined with mechanical ball-mill type treatment and other grinding and shearing modes; ionizing radiation, which, like the Fenton reagent, produces free radicals which oxidatively disrupt glucosidic linkages and thus break the hydrogen bonds and weaken crystallinity. Unfortunately, so far, most estimates of the costs of these pretreatments boost the price of the subsequently produced glucose to noncompetitive levels. Of course experiments are continuing to develop more effective and economical pretreatment. Thus, Natick investigators Tassinari and Macy (1977) reported that while ball milling of newspaper for 24 hr increases its susceptibility to cellulase degradation 64%, pretreatment for 10 min with a "differential speed two roll milling," used in the plastics and rubber industry, increased susceptibility two-fold to 125%. Susceptibility of a pure crystalline cellulose, cotton, was increased 15-fold. A cogent review on pretreatment is that of Millet *et al.* (1976).

Immobilization.—If and when glucose production is adopted commercially, it will constitute another instance of the successful use of immobilized enzyme technology. The enzyme is adsorbed onto a carrier, in this case its substrate, and one of the features of the proposed processes which may make them viable is the provision, as we have just seen, of repeated reuse of the enzyme system. Other investigators feel that cellulase reutilization can be improved by means of immobilization other than simple as adsorption onto substrate. Thus, Karube *et al.* (1977) describe complete hydrolysis of cellulose in a fluidized bed reactor (column with upward flow) using cellulase immobilized onto collagen which in turn served as a coating on stainless steel beads or macadam.

Xylanase and Xylitol

When noncotton sources of cellulose are hydrolyzed by crude culture filtrates of potent cellulase-producing microorganisms, the sugar xylose invariably accompanies glucose, as high as 20% of the total sugar produced from corn processing waste, but running in the range of 5—10% for most other sources (Andren *et al.* 1976; Nystrom and Andren 1976). This xylose is formed as the result of the action of xylan-degrading enzyme systems which almost invariably accompany (if are not identifiable with) cellulase just as xylans almost always accompany cellulose (except for such fibers as cotton and ramie) in woody plants and agricultural wastes; 50% of corncob polysaccharide is xylan. As a result of the cornstarch-fructose revolution, we may expect to see an increase in this particular agricultural by-product, corresponding to the expected diminished production of bagasse from sugar cane processing. It would be ironic indeed if fructose in turn were to be partially replaced by xylitol, derived from a by-product of the same wet cornstarch processing industry.

Although it has long been recognized that the sweetening power of xylitol, the sugar alcohol derived from D-xylose by a chemical reduction, is roughly the same as that of sucrose, it is only recently that interest in this sweetener has revived. This has come about as the result of the discovery by a group of Finnish dental scientists that xylitol can be anticariogenic in concentrations above 50% (Mäkinen 1979; Sharman 1976; Moore 1977). Other properties which suggest that it would be a suitable food ingredient as compared with other sweeteners include the following:

(1) the cooling effect of solid xylitol—its endothermic heat of solution is 10 times that of sucrose, thus creating a pleasing cooling sensation
(2) flavor enhancement
(3) reduction of viscosity—a 60% solution is one-third as viscous as sucrose
(4) lower hygroscopicity than sucrose or fructose, and
(5) like sucrose, it cannot cause nonenzymatic browning; unlike sucrose it does not caramelize.

In addition xylitol possesses several nondental health related properties which have attracted some interest. It has the same caloric values as the 6-carbon sugars but is not as readily converted into fat, mainly because it does not go through the same metabolic pathways and therefore has been suggested as treatment for ketogenesis and diabetes. At the time of this writing inconclusive preliminary work has suggested it may be carcinogenic in larger doses. However as mentioned, its principal virtues are its anticaries (above 50%) and noncaries effects (below 50%). At concentrations over 50% it has been recommended wherever sucrose may be suspected as contributing to dental caries in sticky sweet foods such as pies, ice cream, jams, sticky confections and syrups. As of this writing it is widely used throughout the world in chewing gums at noncariogenic levels and in chocolate candies in the USSR. Parenthetically, among the proposals for mechanism of the anticariogenic effect, is the suggestion that it induces increased production of salivary lactoperoxidase which, as discussed in Chapter 11, can act as an effective germicide in the presence of appropriate acceptors such as Cl^- and of course H_2O_2.

At the present time xylitol is manufactured by the catalytic (inorganic) hydrogenation of xylose obtained from the acid hydrolysis of woody material such as birch chips. In view of the low cost of hydrogenation it is unlikely that it will be replaced by enzymes although an NADPH-dependent reductase is associated with glucose isomerase in *Enterobacter liquefaciens* (Yoshitake *et al.* 1976).

Xylans (Timell 1965; Wilkie 1979).—Xylanase substrates comprise one of the group of plant wall polysaccharides classified as hemicelluloses. They are present at especially high levels in corncobs (up to 50%), in hardwoods and some fibers such as sisal (about 25%), and low in softwood and other fibers (12% and less). Xylan is localized in close association with cellulose fibers and may actually be encrusting them in some woody tissue (Sinner 1975). Like other hemicelluloses xylans are polydisperse mixtures of sugar chains, each consisting of repeating xylopyranosyls linked together, as in cellulose, by (1→4)-β-D linkages (DP = 150–250). Off from this backbone are periodic branchings of either short runs of xylose, "true" xylans which are rather rare, or of other sugars such as glucuronic acid derivatives and/or arabinose. A composite of such a xylan chain may be represented thusly:

```
X-X- - -X-X- - -X- - -X- - -X- - -X-X- - -X-X
           |       |     |     |      |        |  |
  X-X-X-X       I     I    II   III      IV V
```

I is (1→2)-(4-O-methyl)-α-D-glucuronic acid
II is (1→3)-α-L-arabinofuranose
III is (1→2)-α-D-glucuronic acid
IV is (1→2)-O-acetyl-L-arabino-(4-O-methyl)-α-D-glucuronic acid
V is (1→2)-O-methyl

The side chain designated by the Roman numerals is listed in order of abundance and frequency of occurrence. Xylans which have glucuronic acid side chains are found in woods. Wheat xylan is an arabinoxylan. In addition to the latter, Dekker and Richards (1976) list the following as substrates of the "xylanases": arabinoglucuronoxylan, arabino-4-O-methyl-D-glucuronoxylan and glucuronoxylan. Examples of recent studies on xylan structure show, for instance, that the pentosan from jute bark is glucuronoxylan (Haq *et al.* 1975) whereas those from grape vines are exclusively 4-O-methyl-glucurono- and glucuronoxylans (Dudkin *et al.* 1976).

The backbone chain differs from cellulose only in the absence of the nonring CH_2OH projecting out from the xylanopyranose ring. Incidentally in algae the xylans are composed of β-D-(1→3)-linked xylose residues and, not surprisingly, the corresponding hydrolases are found in marine microorganisms in association with seaweed (algae) and bottom sediments as well as in seawater. As substrates for enzymatic hydrolysis, the xylans are differentially alkali-extracted from wood usually after the removal of lignin. After extraction they exhibit variable solubilities in typical enzyme reaction mixture solutions. The arabinoxylans are likely to be more soluble than acidic xylans which are solubilized by further treatment with alkali. Like starch, xylans have a tendency to retrograde from solution. Eda *et al.* (1976) isolated a completely water-soluble unbranched straight chain xylan from tobacco stalk (PD = 1000) which should serve as suitable xylanase substrate. Some investigators use carboxymethylxylan as substrate. Considerable insight into the details of xylan structure have been gleaned from analysis of products obtained from xylanase action.

Xylanases.—References to most investigations and results mentioned in the following discussion can be found in the valuable reviews of Dekker and Richards (1976) (English), Comtat and Barnoud (1976) (French) and Sinner (1975) (German). Just as their substrates are found in close association, so are the enzymes that degrade xylose and cellulose. They are found along with cellulases in the hepatogastric organ of the snail and in other invertebrates such as lobster, in marine plants (algae), in wood-eating insects and in higher plants. Thus, xylanase activity is greater than that of any other carbohydrase during germination of the tropical legume *Stylosanthes humilis* whose reserve polysaccharide is arabinoxylan rather than starch (Dekker and Richards 1976) and is present in barley, corn and other grains. It is also present in many genera of microorganisms, but is especially abundant and widespread among the fungi. Dekker and Richards list over 100 species.

Our knowledge of this class of enzymes is still in the formative stage. Thus it is not clear if the enzyme is constitutive or is induced in the species where this has been investigated. Early workers were convinced that the xylanases from wood-rotting fungi were constitutive, but they could have been induced by the xylan present in the fungi mycelia (self-induction). However, there seems to be no doubt that the addition of xylan or xylan-containing plant parts stimulates xylanase production. There even appears to be specificity with respect to the source of the xylan and microorganism (*Trichoderma* and *Fusarium* spp.) for maximum enzyme production (Strizhevskaya 1975). Nakanishi *et al.* (1976) found that *Streptomyces* xylanase was produced in the presence of xylan and especially methyl-β-xyloside as sole C source but not with glucose or starch. On the other hand, Lee and Lee (1975) found that high-producing strains of *A. niger* could be stimulated to produce more xylanase in the presence of carboxymethyl cellulose. Most if not all xylanases are extracellular. Plant saprophytic and parasitic pathogens excrete the same xylanases when grown on xylan as a sole carbon source as they do when acting as pathogens (Doux-Gayat and Auriol 1976).

Xylan-degrading systems appear to comprise true endoxylanases and a xylobiase (β-xylosidase). Several of the xylanases have been purified and crystallized (Hashimoto *et al.* 1971) by the conventional nonaffinity techniques detailed in Chapter 6 so far without the aid of affinity chromatography. Adsorption onto insoluble xylans was used in one partial purification (Table 7.1). Xylanases have molecular weights in the range from 6000 to as high as 35,000 suggesting that they do not possess quaternary structure. Yet, in spite of the low molecular weights reported for *T. reesei* isozymes of less than 20,000, what appear to be subunits of such isozymes have been prepared (Hashimoto and Funatsu 1976). In the few instances investigated, the xylanases appear to contain carbohydrate. Their pH and temperature characteristics run the usual gamut of extremes with most of them acting optimally between pH 4 and 6 and at temperatures in the vicinity of 40°–50°C.

What is of special interest about them are the singularities in specificities which they exhibit, singularities which may be pertinent to their eventual industrial application. When acting on straight chain unbranched β-D-(1→4)-linked xylans, most xylanases behave as "true" endoglucanases, attacking the linkages in the polymer randomly to produce mostly xylobiose and xylotriose. Their behavior towards other xylans and oligooxylosides may be summarized as follows: Some but not all xylanases split the α-1,2 linkage at the branch point between the 2-position of the xylose residue in the backbone and the 1-position in the side chain arabinose in arabinoxylans to liberate free arabinose. An analogous reaction does not occur with the glucuronic acid-containing xylans.

In the algal xylan consisting of both β-D-(1→4) and β-D-(1→3) linkages, a partially purified xylanase preparation from *A. niger* splits both linkages with a preference for the β-D-(1→3). It will split the middle of a run of three or more β-D-(1→4) linkages.

In branched xylans most xylanases no longer split β-D-(1→4) linkages randomly but seem to have a preference for those near the branch points. This may arise from the change in conformation in the xylose residue brought about by the presence of a branch point and it may be a fortuity that this new conformation favors increased catalytic activity. This may also ultimately be a problem in attaining complete hydrolysis in xylose manufacture because it leaves less substituted, shorter-chain xylans which, like partially degraded amylose, tend to retrograde (come out of solution) and lose susceptibility to further enzyme attack.

As far as the mechanism of action at the active site has been ascertained, Dekker and Richards suggest that the binding site can accommodate four consecutive xylose residues (three linkages) but that only one of the outside linkages accommodates the catalytic site, i.e., - - -X-X-X↓-X- - -. Apparently adequate binding is also achieved when arabinose is linked to the 4-xylose fragment because the enzyme acts even more readily on such a combination.

At least some fungal enzymes can liberate xylitol from the trisaccharide consisting of two xyloses plus xylitol in the position of what would have been the reducing end of xylotiose (xylotriitol). This behavior has been observed with two well-investigated cellulolytic cellulase-containing fungi, *A. niger* and *T. reesei*. Exploitation of this specificity may be possible by chemically hydrogenating the xylose to xylitol before enzyme hydrolysis.

Some authors claim to have obtained highly purified xylanase preparations whose cellulase activity persisted throughout the purification and that the two activities could not be separated by a variety of procedures. Such preparations have been obtained not only from *Trichoderma* species but also from a strain of *B. subtilis* (Ohtsuki *et al.* 1976), *Irpex lacteus* (Kanda *et al.* 1976) and *Verticillium albo-altrum* (Nicholson *et al.* 1975).

These investigations also present rather convincing kinetic evidence for the coexistence of the two activities in one molecule, but none has tried affinity chromatography. If both activities do indeed reside in the same molecule of some xylanases, this would have to be taken into account in developing industrial processes for both cellulose and xylan conversion.

As do all glycanases, the xylanases can catalyze the synthesis of higher oligosaccharides from xylose and the lower ones. However, some of the xylanases such as two of the three isozymes isolated from *A. niger* van Tieghem by Takenishi and Tsujisaka (1975) are particularly adept in producing highly polymerized xylooligosaccharides from xylotetraose or xylopentaose. Such behavior would also have to be taken into account in a maximizing enzymatic xylose production. Although demonstration of the flexible specificity of xylanases seems, on the whole, to be based on fairly convincing experimental evidence, the final verdict on some of them, as pointed out by Comtat and Barnoud, must be held in abeyance. It is a question of incomplete purification or an absence of specificity which can not be answered at the present time. Certainly, application of the latest advances in affinity chromatography could contribute to the answer.

The question of purity also beclouds the establishment of the existence of exoxylanases. Production of large quantities of xylose could be due to the combined action of endoxylanase and β-xylosidase. The latter term was applied by Rodniova *et al.* (1975) to an enzyme they isolated which, more in analogy to glucoamylase than to α-glucosidase, hydrolyzed xylans, xylooligosaccharides and β-methyl-D-xyloside to xylose. As expected of a true exoglycanase, it did not decrease the viscosity of carboxymethylxylan solutions. On the other hand, Takenishi and Tsujisana (1975) described a more specific β-xylosidase which liberated xylose mainly from xylobiose and a few higher oligomers and selectively added a xylose residue to position 4 of the terminal xyloses in xylooligosaccharides.

Evidence for the existence of an exo-acting "xylobiosylhydrolase," in analogy to β-amylase and cellobiosylhydrolase action on amylose and cellulose, appears to be rather meager.

Enzymatic Xylose Production.—We alluded to some of the problems future production-oriented investigators will have to face in light of some of the special properties of the xylanase systems. Probably the first attempt to produce xylanase breakdown products in a continuous process using immobilization technology is that of Puls *et al.* (1974). The 30% yield obtained consisted of xylobiose, xylotriose and 4-O-methylglucuronoxylotetraose. The enzyme acting in solution on the same substrate yielded the corresponding glucuronic acid derivative of xylobiose rather than xylotetraose.

As far as we are aware the only group investigating the xylanase systems from an engineering point of view has been that of Reilly at Iowa State University (Reilly 1977). The xylanase system this group is investigating comprises over 10 enzymes. Undoubtedly by the time this appears more definitive information on the feasibility of using an enzyme as a step in the production of the sweetener xylitol will have been accumulated.

Chitinase is now in the process of being examined as another enzyme to be exploited in the bioconversion of yet another waste polysaccharide of food processing, chitin (a polymer of N-acetylglucosamine), for eventual production of single cell protein (Tom and Carroad 1981).

Lactase Generated Sweeteners

The sweetness of lactose can be enhanced by isomerization to lactulose or by hydrolysis to glucose and galactose using lactase (β-galactosidase) as suggested and explored by Wierzbicki and Kosikowski (1973A). They produced a high solids syrup from whey in which the lactose had been hydrolyzed with *A. niger* lactase and suggested that their product be used in imitation maple syrup, fruit juice and puddings. One of the benefits claimed for a yeast lactase-treated dairy product is increased sweetness (Guy 1973). When and if immobilized lactase, which has been worked on quite extensively, is a commercial reality (Chapter 36), one of the uses to which it can be put, perhaps, as a by-product of more important health-related applications, would be to produce sweetener.

Increased sweetness (as well as obviation of sandiness due to crystallization of the lactose—see also Table 33.5) is also the aim of adding lactase to milk destined to be concentrated to a sweetened condensed milk product (Kisza *et al.* 1973). Other potential applications of lactase to dairy products, especially whey, were discussed by Bouvy (1975). See Moore (1980) and Finocchiaro *et al.* (1980) on the use of immobilized lactase for removal of lactose from whey. Health-related aspects of lactase utilization are dealt with in Chapter 36 and enzymatic aspects of sweetener-substitute production in Chapter 20.

Part IV

NONBIOLOGICAL CONTROL AND MANAGEMENT OF ENZYME ACTION

10

CONTROL OF ENZYMES BY ENERGY INPUT

CONTROL BY THERMAL ENERGY

In Chapter 3 we briefly described how knowledge of the influence of temperature on the rate of enzyme-catalyzed reactions led to insight into the nature of the catalytic event. In this chapter we broaden the historical and experimental base of temperature-enzyme relations and describe how this information is put to use in relation to food quality and processing, especially with respect to the prevention of undesirable enzyme action—blanching.

Temperature Optimum and Thermal Stability

It is necessary to distinguish between the heat stability curve and the temperature-activity profile. Only the latter exhibits an optimum corresponding to maximum activity. Heat stability is measured by incubating the enzyme, usually in the absence of substrate, at different temperatures for an arbitrary but constant period of time and then determining (assaying for) the enzyme activity under standard assay conditions. The heating period is usually far greater than the assay time.

The temperature-activity profile is obtained by conducting a series of enzyme reactions under standard conditions except that the temperature is changed. The activity-temperature profile always contains an element of "stability" since it is practically impossible to measure instantaneous initial rates (Chapter 3). Figure 10.1 shows typical heat stability curves for one of the papain proteinases and Fig. 10.2 displays a temperature-activity profile for invertase, free and immobilized to glass beads. These temperature-activity relations were already well characterized at the turn of the century and theories had been developed to account for them. It was understood that the temperature optimum was due to or created by the counteract-

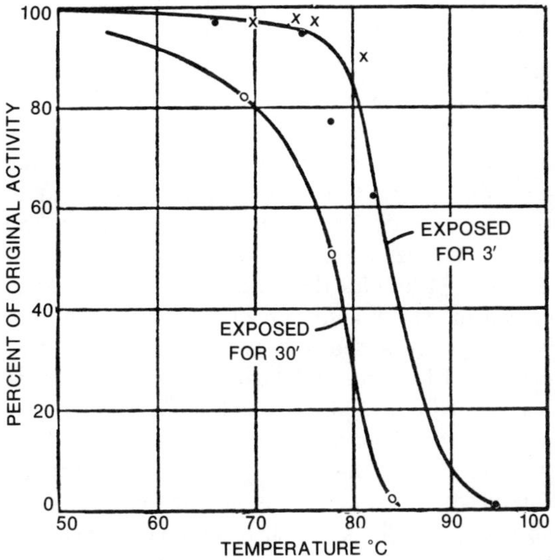

From Lineweaver and Schwimmer (1941)

FIG. 10.1. HEAT STABILITY CURVES FOR PAPAIN GLUCOSE ISOMERASE

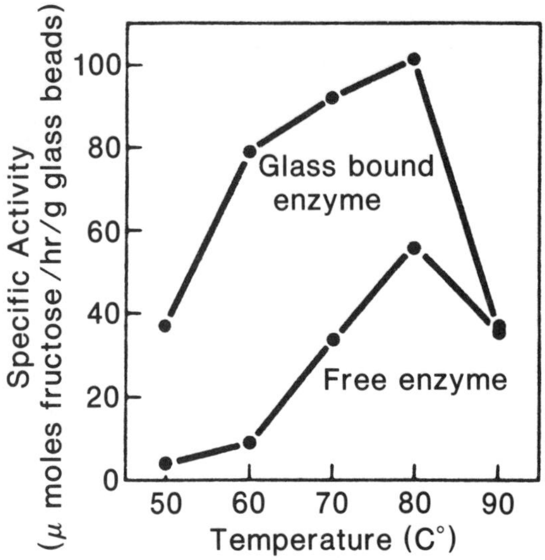

From Strandberg and Smiley (1972)

FIG. 10.2. TEMPERATURE-ACTIVITY PROFILE FOR SOLUBLE AND IMMOBILIZED GLUCOSE ISOMERASE

ing effects of increasing temperature and the destruction of the enzyme at the higher temperatures, which we now know is the result of protein denaturation. Furthermore, it was clear more than 80 years ago that the temperature coefficient of the "deteriorative" reaction was much higher than that of the enzyme-catalyzed reaction. The data of Tamman (1895) for the catalysis of the hydrolysis of salicin, the β-glucoside of benzaldehyde, was widely quoted in the early twentieth century textbooks. For instance, in a book on the physiology of bacteria, Rahn (1932) presented Tamman's data as the three-dimensional figure shown in Fig. 10.3. Assuming that the rates of both the enzyme-catalyzed reaction and the destruction of enzyme obeyed first-order kinetics, Tamman arrived at the following equation:

$$X = S\,[1\text{-exp}\,(k_oq/K_oQ)\,P\,(1\text{-exp}(-\,KQt)]$$

(Equ. 10.1)

where S and P are the initial substrate and enzyme concentrations, X = [S] at time t, k_o and K_o are rate constants at 20°C for the catalyzed and deteriorative reactions, and q and Q are the van't Hoff rule dependencies of rate constants on temperature, i.e., $K = K_oQ$.

From this equation Tamman found that the coefficient (Q_{10}, see below) for the enzyme reaction was 1.26 and for the deteriorative reaction, 6.36. Thus, when the temperature was raised 40° above 20°C, the rate of the enzyme-catalyzed reaction increased $(1.26)^4$ = 2.52-fold, whereas the rate of enzyme destruction increased $(6.36)^4$ = 1636-fold, leaving some 10.5% of the original enzyme after 30 min. Actually Tamman, anticipating more modern information, found somewhat higher survival rates which he attributed to the protective action of the substrate. Rahn also lists the following enzyme action/enzyme destruction Q_{10}'s: "rennet" 2.61/65.8, trypsin 1.61/17.8 and pepsin 2.06/33.7.

As clearly shown in Fig. 10.3 the optimum decreases with increasing temperature from 55°C (confining our attention to 30 min) to 25°C after 40 hr. Of course in constructing an authentic temperature-activity profile one should use initial rates as accurately as is feasible. The old data do illustrate the lack of sharp definitiveness in the temperature optima.

We briefly digress here in order to review the concept of temperature coefficients, usually designated as Q_{10}, the ratio of the rate of a chemical reaction at one temperature v_T to the rate at a temperature 10°C below $V_{(T-10)}$, the latter temperature usually taken at 25°C. On the average, it will be recalled, the Q_{10} of most chemical reactions is *ca* 2; those of enzyme-catalyzed reactions somewhat under 2. Since the rate of reactions, including those catalyzed by enzymes, follows the Arrhenius rule (Chapter 3), the Q_{10} will, of course, vary with temperature. From equation (3.1) it can be shown that

$$\log \frac{v_T}{v_{(T-10)}} = \frac{E_a}{2.30\,R} \times \left(\frac{1}{T-10} - \frac{1}{T} \right)$$

(Equ. 10.2)

It will be recalled that small differences in Q_{10} or activation energies can result in differences of several orders of magnitude of rates at appropriate temperatures. The relation between Q_{10} and Q_A where A represents temperature differences from 1° to 40°C is shown in Fig. 10.4.

For most nonenzymatic reactions with positive temperature coefficients, Q_{10}'s decrease somewhat with increasing temperature. For enzyme reactions, the decrease is more precipitous than predicted from equation (10.2) because of the contribution of enzyme inactivation. The situation can become more complicated when, as in the case of "malt amylase," more than one enzyme, each of which has its own distinct temperature response, acts on the same substrate. Thus, as reported by

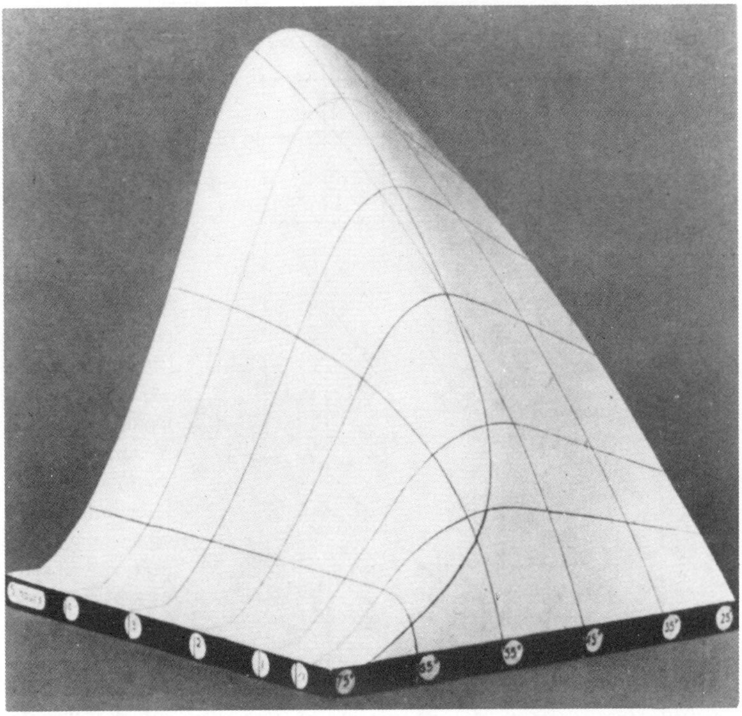

FIG. 10.3. INFLUENCE OF TEMPERATURE AND TIME ON ENZYME ACTION

The three-dimensional representation of Tamman's equation (see text) for the action of β-glucosidase (emulsin) on salicin.

From Rahn (1932). Copyright © 1932, P. Blakiston's Son & Co., Inc. Used with permission of McGraw-Hill Book Company.

Sandstrom (1946), the Q_{10} of malt amylase—really a mixture of heat-labile β-amylase and heat-stable iso-α-amylases—dropped from 1.67 at 10°–20°C to 0.62 at 40°–50°C, a 63% decrease. The Arrhenius relation predicts a drop of about 11%. In the next temperature interval, 50°–60°C, the Q_{10} decreased to 0.59, about what equation (10.1) predicts. An important experimentally practical consequence of the drop in Q_{10}'s is the necessity, for the most part, to perform enzyme assays at constant temperature. Thus, Kempner (1972) reported that an error of 1°C in the assay of aldolase will create an error of 12% in assay results.

Heat Stability of Food-related Enzymes.— There is of course, as illustrated by the example just cited, a considerable variation in the thermal stability of different enzymes. Even for a particular enzyme it is dependent upon a host of temperature-related parameters and especially on the environment, as we have had and will have occasion to discuss. For instance, we have alluded to immobilization (Fig. 10.2) and to such stabilizing substances as substrates, special ions and others mentioned in Chapter 5 on methods used and proposed to stabilize commercial enzymes. In Chapter 9 we showed how important it was, from the viewpoint of achieving successful commercialization, to maximize the stability of glucose isomerase. In most of this and the next chapter we shall be concerned with means of destroying enzymes by heat rather than with maximizing their action.

As a general rule, enzymes start to lose their activity in minutes at about 60°C but most are stable at low temperatures (Chapter 12). There are exceptions. The classical heat resistant enzyme adenylate kinase (AMP + AMP → ADP + ATP) withstands 100°C at pH 1 for prolonged periods. As pointed out in Chapters 5 and 9, microbial enzymes, particularly the bacterial amylases, can be made almost as heat resistant as desired by appropriate genetic manipulation. Utilization of information concerning heat stability of the heat stable peroxidases in conjunction with blanching will be taken up in Chapter 11.

Intermediate in stability are those food-related enzymes whose action is potentiated when the unprocessed food is subjected to intermediate temperatures. As pointed out in Chapter 1, these enzymes survive and act in the range of 40°–80°C

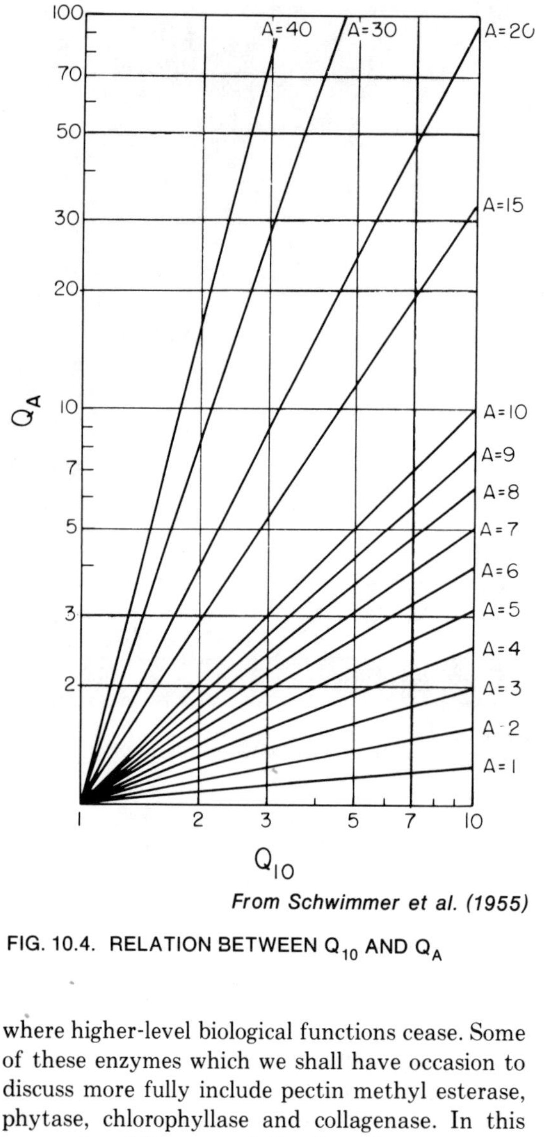

From Schwimmer et al. (1955)

FIG. 10.4. RELATION BETWEEN Q_{10} AND Q_A

where higher-level biological functions cease. Some of these enzymes which we shall have occasion to discuss more fully include pectin methyl esterase, phytase, chlorophyllase and collagenase. In this range also fall many of the fungal enzymes, especially the amylases. Somewhat more sensitive to heat are the grain amylases, especially, as mentioned, β-amylase. In this intermediate range also fall the group-specific alkaline phosphatases, used as an index of pasteurization of milk and other dairy products (Chapter 11).

More normal temperature stability exhibited by the vast majority of enzymes is represented in the food field, for instance by the lipases and by garlic alliinase which act over the range of 0°–60°C, have an activity optimum at 37°C by most assays and cannot withstand temperatures higher than 65°C.

Protein Denaturation and Enzyme Inactivation

Historically the term "denaturation" was applied to the irreversible insolubilization (coagulation) of a "native" protein brought about by heat, acid and other agents. Probably the earliest systematic investigation of protein denaturation is that of Chick and Martin (1912) some 70 years ago. The evolvement of technology and theories on the mechanism of protein denaturation may be assessed from reviews spanning the last 35 years (Neurath *et al.* 1944; Johnson *et al.* 1954; Kauzman 1959; Joly 1965).

We now know, of course, that denaturation can occur in solution (in the presence of urea, guanidine or salicylate) and can be reversed, as we shall have an opportunity to discuss further in Chapter 11. Thermal protein denaturation usually occurs without change in the molecular weight. Proteins with quaternary structure will dissociate into subunits in denaturing solvents such as urea.

During denaturation, the quaternary, tertiary and secondary structures of the protein successively disappear, mainly due to the progressive disruption of hydrogen bonds. The energy of a strong hydrogen bond amounts to between 5 and 8 kcal and the high energy of activation of the denaturation reaction can be largely accounted for in terms of the disruption of a large number of these bonds. Of course disulfide and polar COO^-/NH_3^+ bonds are also disrupted and weaker hydrogen bonds may be reformed during denaturation. While the high entropy of activation of the denaturation process known for many years bespoke of an enormous number of accessible configurations, it is only in the last 10 years or so that direct evidence has been forthcoming which concludes that during denaturation, the protein goes through a precise series of unfoldings operating *via* cooperativity mechanisms (Chapter 4) and that these are accompanied by characteristic conformational changes (Baldwin 1975).

A sharp rise, almost a quantum jump, in the understanding of the details of the folding and unfolding of proteins has come about by newer experimental techniques: optical rotary dispersion (ORD), circular dichroism and especially differential scanning microcalorimetry. This latter technique has enabled the acquisition of precise "melting point" values which can be compared with values calculated from a particular theory. The melting points of some food-related enzymes are:

lysozyme, 72°C; ribonuclease, 62°C; pepsinogen, 64°C; trypsinogen, 58°C; and chymotrypsin, 57°C (Bull and Reese 1973). Other information and discussions of folding and unfolding of proteins and enzymes and their relation to the role of water and hydrophobicity (Chapter 3) are provided by Privalov (1974), Lumry (1973), Wetlaufer and Ristow (1973) and Lapanje (1978).

Differential scanning calorimetry is now being applied to food science in order to obtain a better understanding of the molecular forces involved in time-honored but little understood food processing practices (Chapter 33; Donovan 1977,1979; Silano and Zahnley 1978). In relation to protein conformation and heat stability, it is of interest to note that Alexandrov (1977) in his biologically-oriented novel approach to the connection between conformational flexibility of macromolecules and ecological adaptation, marshals impressive documentation that a parallelism exists between the heat stability of proteins and their susceptibility to attack by proteolytic enzymes. This certainly would be in accordance with the idea that hydrogen bonds have to be disrupted in each case. The role of the hydrogen bond in susceptibility to enzyme action was explored in Chapter 9.

BLANCHING

Terminology and Parameters

Definitions, significance and application of two of these parameters, the temperature coefficient Q_A, and the energy of activation E_a, frequently used in connection with blanching, have already been subject to examination in this chapter and in Chapter 3. Food sterilization terminology in the canning industry, delineating the thermal characteristics of microbial death and thus providing for a rational time-temperature scheduling, are also applied to enzyme inactivation in the blanching process. The three most frequently used parameters are the "D-," "z-" and "F-values." As applied to enzyme activity they may be defined as follows.

The D-value, decimal reduction dose, is the *time* (usually expressed in minutes) it takes to inactivate 90% or 0.9 of the original enzyme activity at definite constant temperature. In sterilization the temperature is that to which the food is subject, usually at or around 120°C. For first-order inactivation (constant = k) the D-value is the time for the enzyme activity to decrease by one log cycle; D = 2.303/k.

The z-value is the *temperature* interval (°F) required to reduce the enzyme activity by 90% at a constant time or that necessary to change the reaction rate by one order of magnitude. When the enzyme inactivation follows first order kinetics, a plot of log time vs temperature yields a straight line. The z-value is the temperature interval corresponding to one log cycle. The z-value is the reciprocal of the slope. The z-value is related to the energy of activation E_a by the following equation (Lund 1977):

$$z = \frac{2.303\, R \cdot T \cdot T_r}{E_a} \cdot \frac{9}{5}$$

where T and T_r define the temperature interval.

The F-value is the time required to inactivate a fixed amount of enzyme. F_0 is the time (usually expressed in minutes) to inactivate a given amount of enzyme at a definite temperature (for sterilization, during canning 250°F) when the z-value is 10°C (18°F). As originally applied in the canning industry these relationships are expressed as thermal death curves. A typical death curve for bacterial spores is shown in Fig. 10.5. In canning practice, typical F and z values are 2–3 min and 18°–20°F, respectively, values which presumably ensure destruction of heat resistant spores as well as vegetative bacteria. Corresponding values for enzyme inactivation will be shown in Chapter 11.

Types of Heat Treatment of Foods

Thermal energy input into foods encompasses a variety of purposes, quite aside from the usual "cooking" requisites. While they overlap, these purposes have different temperature ranges. We may conveniently distinguish among the following.

(1) Low temperature treatment, 40°–80°C, to activate or potentiate desirable enzyme action.

(2) Pasteurization, 60°–85°C, to remove a selected range of possible offending microorganisms. Some of the more heat stable enzymes survive.

(3) Blanching, 70°–105°C, to inactivate most or preferably all enzymes in order to ensure absence of offending enzymes. In general, these enzymes cause off-flavors in vegetables and off-color in fruits.

(4) Sterilization, 100°–130°C, usually under pressure, to destroy vegetative microorganisms and spores.

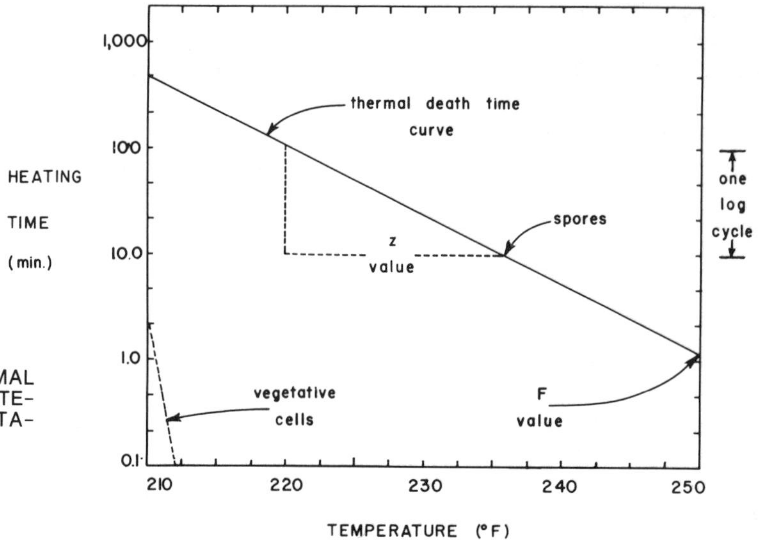

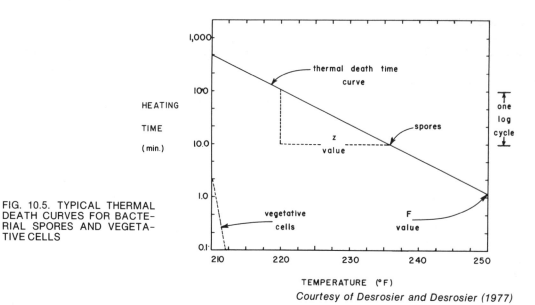

FIG. 10.5. TYPICAL THERMAL DEATH CURVES FOR BACTERIAL SPORES AND VEGETATIVE CELLS

Courtesy of Desrosier and Desrosier (1977)

(5) Steam blanch, greater than 100°C, shortens time required to destroy enzymes.
(6) Dehydration, covers the whole range of temperatures from those at freezing (freeze-drying) to those of superheated steam.

The term "blanching" or "scalding" was originally coined to designate those operations in the processing of frozen foods which prevent deteriorative changes in these foods not so treated during subsequent storage. Just as pasteurization is associated with removal of microorganisms, so blanching is now associated with removal of enzyme activity. The question of what enzymes have to be removed is a moot one which we shall examine in Chapter 11. Today the term is used not only in frozen food processing but also in the dehydration and canning. Whatever the process, the general rationale is to apply just enough, that is, the minimum heat consonant with adequate removal of undesirable enzyme activity. Any further heating may cause undesirable alterations in the food which could cancel out the beneficial effects of blanching (see "Disadvantages").

Probably the most comprehensive and useful review devoted solely to blanching, although somewhat dated, is that of Lee (1958). Other generally informative sources are reviews by Joslyn (1949, 1961A,1966), Olson and Dietrich (1968) and Voirol (1972). Innovations in blanching engineering are succinctly discussed by Lee (1975). Blanching in the context of food preservation was cogently discussed by Desrosier and Desrosier (1977).

The blanching step in a food processing line usually follows cleaning of the commodity. In water blanching, the temperature of the heated blanching water starts to decrease when it contacts the vegetables or fruits and then starts to fall again upon subsequent cooling. A special cooling step is needed for frozen foods prior to blast freezing. It is the general convention to consider the "blanching time" as being that time measured from the moment the blanch water regains its input temperature to the time when cooling is initiated, 3–5 min. In some operations the blanch and sterilization steps are combined so that some elements of enzyme inactivation as well as complete removal of microorganisms are involved.

Trade-offs.—Blanching constitutes an excellent case-in-point of the trade-offs and benefits-to-cost considerations which have to be taken into account even for such a well-established process. For didactic reasons we shall be confining this discussion mainly to water blanching.

Advantages and Accrual of Entrained Benefits.—If the only positive virtue of blanching as practiced today would be that it inactivates enzymes, it probably would not be used so extensively. Like most processes and unit operations, other positive benefits accrue which outweigh the accompanying disadvantages enumerated below.

In the processing of most fruits and vegetables the main virtue of blanching is, as previously stated, to prevent enzyme action leading to off-color and off-flavor which develop during subsequent storage or in holding periods between processing steps in vegetables and fruits which are not blanched. It is also used in processing of shrimp and less extensively in some meat-processing (see below) and especially in the USSR, at least as judged by the extensive literature on meat blanching emanating therefrom. In the manufacture of tomato juice and related products, the main purpose of blanching, the "hot break," is the retention of consistency. These are the principal goals of blanching. Other benefits and advantages are as follows: Foreign materials such as small berries, pebbles and grit particles which escape previous cleaning operations are removed.

Partial asepsis is achieved by the treatment with water and heat. This is particularly important, as in the case of apricots when incipient fungal infection has already been initiated, or if there is an extensive holding period before final processing (Chapter 29).

Shrinkage can be prevented by expellation of air and wilting of leafy vegetables. Without the latter, the container, in the case of canning, would not be full after retorting and sealing, contrary to regulations. Even with nonleafy vegetables such a benefit may be significant. Thus, Ramaswamy and Rege (1975) reported that blanched pieces of aubergines (brinjals, eggplant, *Solanum melongena*) destined for both dehydration and canning gave 67% can fill *vs* 63% for unblanched ones, although blanching had no effect on color which seemed to be due to nonenzymatic browning. In general, however, the expelling of gases improves the apparent color and appearance altered by the presence of such gases.

One advantage which pertains uniquely to dehydration is the reported enhanced facility for rehydration. The effect has been ascribed to the removal of enzymes which alter the structure of macromolecules involved in texture and especially in ease of penetration of water or, as stated by von Loesecke (1955), to their "colloidal" state.

Removal of undesirable food constituents other than enzymes, usually performed by leaching, can also be accomplished by denaturing proteins with undesirable biological activities. These include inhibitors of the consumers' digestive enzymes and hemagglutinins (Chapter 35). Off-odors, both natural (i.e., pea vine juice) and adventitious, may be expelled. More recently interest has shifted to removal of adventitious chemicals such as herbicides, pesticides (Kleinschmidt 1971; Fair *et al.* 1973; Davis *et al.* 1975) and undesirable natural chemicals, such as nitrate and oxalates (Sistrunk and Cash 1975; Heintze *et al.* 1975). Although some investigators report simultaneous removal of these constituents, others could not confirm this (Paulus *et al.* 1975). Suggested alternative means of removal of oxalate include special treatment such as heating with ammonia (Bengtsson and Bosund 1975). However, this was offset by loss of green color or conversion to physiologically inactive Ca oxalate (Richter and Handke 1973). Lee (1958) suggests that the leaching occurring during blanching removes substances which react with chlorophyll; blanching thus aids in chlorophyll retention (Chapter 17). In a more recent confirmation of this chlorophyll-retention effect of blanching, Sistrunk and Bradley (1975) reported that water-blanching was superior to steam-blanching of turnip greens for the preservation of green color and the retention of carotene as well as riboflavin. Other advantages will be mentioned in later chapters.

Disadvantages and Problems.—Accrual of nonenzymatic-associated benefits occasioned by blanching is counterbalanced by the entrainment of some rather serious negative aspects. While there are, of course, countermeasures for ameliorating at least some of these problems, we discuss the more important ones. As previously mentioned, heat application must be just enough but no more than is necessary to accomplish prevention of enzyme-induced quality deterioration. However, it is not always possible to fully accomplish these two objectives and blanching usually results in loss of texture and frequently color. For instance, Crivelli and Boncore (1975) concluded that in the freezing of asparagus tips blanching exacerbated histological lesions (reflected in a weakening of the tissue structure) which occurred during freezing after 2 min blanch at 98°C. Textural deterioration is a particularly vexatious problem in blanching of fruit. As far as color is concerned, we mentioned that blanching improves color associated with the presence of gas in vegetables and by removal of enzymes in fruit. But in some special cases blanching will promote certain nonenzymatic color-producing reactions such as that resulting in pinking of pears and peaches.

While blanching drives off undesirable flavors already present in the food and prevents formation of new ones during subsequent storage, it may also result in driving off desirable flavors as discussed in Chapter 23. A relatively recent example is afforded by the observation of Gormely and O'Riordan (1976) that unbalanced oyster mushrooms develop a strong off-flavor after 3 months storage at −30°C but that a considerable amount of desirable flavor loss was unavoidable during blanching. An example of an ostensibly nonenzyme-related undesirable flavor change induced by blanching is afforded by the observation that blanching (along with sulfiting) of banana contributes to leakage of tannin-like compounds from latex cells, to impart an undesirable astringency to the resulting intermediate moisture product (Ramirez-Martinez *et al.* 1977).

Blanching usually results in losses of soluble solids in general and in desirable nutrients in particular. Solids-loss problems also exert economic and environmental impacts. Compilations of such losses have been made for peas, carrots, broccoli, cauliflower, green and lima beans and brussels sprouts (Bomben *et al.* 1975).

Solids losses, as well as other results of blanching, are approached in an integrated way in a series of related papers on spinach blanching put out by Paulus and associates at Karlsruh, Germany. The last paper deals with differential heat inactivation of isozymes of peroxidase (Delincée and Schaefer 1975). Some representative data obtained in these studies is shown in Table 10.1. The researchers observed a considerable loss in total (drip) weight but not in the percentage of solids and an actual enrichment of insoluble substances, such as precipitated protein. The investigators concluded that on the whole, blanching confers beneficial effects. As might be expected, the calcium present in hard blanch water tended to be enhanced in vegetables containing good sequesterers such as oxalate and phytate. Presence of excess calcium in blanch water can result in undesirable overtoughening of the texture rather than a loss in texture (Chapter 21). While losses have been reported for other nutrients such as thiamin, carotene and niacin during the preparation of processed foods, it is by no means clear that these losses occur during blanching (Chapter 35). Sistrunk and Bradley (1975) found that water is better than steam blanching for riboflavin retention and that niacin losses seem to depend more upon the duration than on temperature of blanching, indicating the importance of leaching of this nutrient.

Another point which a survey of the literature on processing losses seems to bring out is that a considerable part of these losses occurs in the postblanch period, especially during cooling. These cooling losses can be exceptionally severe in steam blanching, accounting in some cases, according to Bomben *et al.* (1975), for two-thirds of the total solids loss.

Blanching as conventionally practiced has had a considerable adverse environmental impact because of high inputs of water and energy and output of solids-containing effluent (especially for water blanch) which can have a high BOD. It is anticipated that effluent limitations set by the Environmental Protection Agency may be set at zero discharge by 1983 (Anon. 1974).

While this is not the place to go into engineering detail it is of interest to note that the bulk of energy loss with the steam blanching equipment hitherto available appears to arise from the relatively slow rate of heat penetration of food pieces in comparison with their rate of travel through the blancher. Because the steam is wasted at the inlet and outlet, the heat does not have the opportunity

TABLE 10.1

BLANCHING LOSSES AND GAINS IN SPINACH [1]

Measurement	Value	Units
Total weight, loss	25.7	%
Solids, gain	5.8	g/kg FW
Vitamin C, loss	D-value, 21	min
Vitamin B_1, loss	D-value, 33	min
Protein, gain	From 360 to 465	g/kg DW
Nitrate, loss	From 0.42 to 0.22	g/kg DW
Oxalate, loss	From 96 to 88	mg/kg DW
Peroxidase, loss	D-value, 1.6;	min
	z-value, 3.3	°C

Source: Selected from series of papers starting with that of Paulus *et al.* (1975).
[1] Blanch data at 90°C, 3 min unless otherwise stated; FW, fresh weight; DW, dry weight.

to adequately penetrate the pieces. The process is also made less efficient because of air infiltration and increased condensation on the walls, which results in increased effluent by entrapped volatiles and by a stagnant layer of air accumulating on the heater. Finally steam-blanched peas are more likely to possess a vinyl or grassy off-flavor than water-blanched ones.

Several factors and circumstances tend to counteract the effect of heat in inactivating enzymes. These include: (1) presence of substrate, (2) a pH at the optimum of the enzyme, (3) low moisture level, and (4) immobilization onto the cellular debris arising as the result of processing.

In the canning process some enzymes may be "activated" if the food is underblanched. This could happen during sterilization (Winter 1976). This "leaky" effect of blanching may explain the decrease of digalactosyl glycerides and the concomitant increase of monoglycerides observed by Fricker *et al.* (1975) during the blanching of spinach. The presence of air also protects the enzyme from inactivation by the above-mentioned poor heat transfer. Finally some heat-inactivated enzymes can regenerate their activity, as discussed in the next chapter.

An example of how many of these countervening effects came to play in a fortuitous conjunction of events is illustrated by a report of Haisman and Knight (1967). Cans containing plums are notoriously subject to corrosion, apparently due to enzymes since by and large blanching prevents this corrosion. The investigators found that processing conditions which resulted in complete inactivation of the selected indicator enzyme in the pulp of the fruit (and putatively the causative factor), β-glucosidase, did not destroy the enzyme activity within the kernel. The β-glucosidase in plums is not a particularly heat-resistant enzyme like peroxidase (D-value of 1.6 min at 80°C as compared with at least 30 min for peroxidase). However, the β-glucosidase in the kernel, unlike that in fleshy part of the plum (pulp), was at its pH optimum of about 6 (as compared with pH 3.0 in the pulp), in a low moisture environment of 50–60% (95% in the pulp) and the entire substrate of the enzyme was localized in the kernel. Slow heat penetration to the center of the fruit, exacerbated by a layer of protective air around the kernel, destroyed the integrity of the cells without destroying the enzymes. This allowed the enzyme to act on the substrate—hence, the apparent "activation" previously referred to. Whatever enzyme was responsible for the release of the "corrosivity" factor, the latter leaked from the kernel to the inner surface of the can.

Solution, Improvements, Trade-offs

Attempts to overcome some of the above-listed shortcomings of the blanching process come under two general headings: (1) manipulation and control of process parameters and associated engineering developments and innovations, and (2) combinations of heat with other means including the use of nonthermal forms of energy to prevent enzyme action.

Process Optimization.—The traditional route for minimizing these problems is through the empirically-oriented approach of optimization of processing parameters. Paulus (1975) presents a valuable and timely discussion of process optimization in food manufacturing, using blanching of spinach as a model.

Aspects taken up by Paulus include optimization of individual stages. In addition to pinpointing some of the above-mentioned points and stages in the blanching process which can be optimized, Paulus stresses the necessity of compromise between high quality of the processed food and the dictates of economic realities in order to establish maximum and minimum requirements.

Process optimization of blanching (enzyme destruction) *vis-à-vis* sterilization (microbial destruction) is alluded to by Lund (1977) in a discussion of maximizing nutrient retention. Lund points out that, in general, since the rate of enzyme inactivation equals the rate of microbial death in the range of 132.2°–143.3°C, products containing heat-resistant enzymes must be processed on a blanching basis. This was once a problem in aseptic canning.

To simplify the discussion of disadvantages we largely confined the discussion to water blanching. Some of these problems, such as excess solids loss, use of large amounts of water, etc., are immediately reduced by shifting over to steam (pipe) blanching. For instance, the volume of liquid waste is reduced by 10–20-fold. However, steam blanching in turn presents new problems and has only incompletely solved the old problems.

We wish to stress that no pretense is made here to present a rigorous engineering analysis and review but we do feel that brief synoptic descriptions of the proposals, some of which are being adopted by the food industry, graphically illustrate the close association of enzyme action considerations to food processing in general and to the solutions of blanching problems in particular. Assaults on these problems can in general be segregated into those which are improvements of existing blanching procedures (mostly improvements in steam blanch) and those which depart radically from the traditional approaches.

Water Blanching.—Up until the last 10 years or so water blanching was probably the most widely used blanching procedure in food processing, followed closely in more recent times by steam blanching. Most heating innovations bypass water blanching as ordinarily practiced. An exception is in-can blanching in which hot water is injected into cans filled with peas. After 20–40 sec at 97°C, the peas are subject to routine sterilizing temperature (Mitchell 1972). By not having to move the peas from the blancher into the can, the proposed system would reduce cell disruption (Chapter 1) and tissue damage, thereby preventing the undesirable enzyme action, improving quality of the final canned product. On the other hand, the amount of water used, some 6700 liters per MT of peas, could create environmental problems unless in-can blanching is combined with a scheme for recirculating the blanch water.

Innovations in Steam Blanching.—*HTST.*— First developed for milk pasteurization, one of the earliest and most successful innovations in blanching was the use of higher temperature and shorter times (HTST) than had been standard practice before its adoption. For instance, instead of batch pasteurizing milk at 63°C for 30 min, the milk is heated for 15 sec at 72°C in a continuous process. With the proper equipment, which could be costly, fruit juices can be HTST pasteurized in 2 sec at 120°C (flash pasteurization) instead of the standard 145°C for 30 min. Applied to blanching, HTST is especially suited for the processing of acid foods. An alleged drawback to HTST is that it may intensify the regeneration of heat-inactivated enzymes as discussed below and in Chapter 11.

Individual Quick Blanching, IQB.—In this advance in steam blanching the food is diced and placed on a moving tray as a single layer so that each individual piece receives unobstructed but short exposure to high temperature steam (Lazar *et al.* 1971; Bomben *et al.* 1973; Lund 1974). As shown in Fig. 10.6 after the relatively short exposure to live steam and before the interior of the pieces have reached maximum temperature, they are transferred *via* moving belt into a second insulated chamber where they form a deep bed. The food pieces are held in this chamber adiabatically (no heat exchange with the surroundings) until the centers of the pieces achieve a mass average temperature sufficient to inactivate the enzymes. Typical schedules are 15 sec exposure to live steam followed by a 50 sec adiabatic holding time. The method was improved by interpositioning of a partially dehydrating "preconditioning" of 5–8 min at 65.5°C. Previously, Dietrich and Neumann

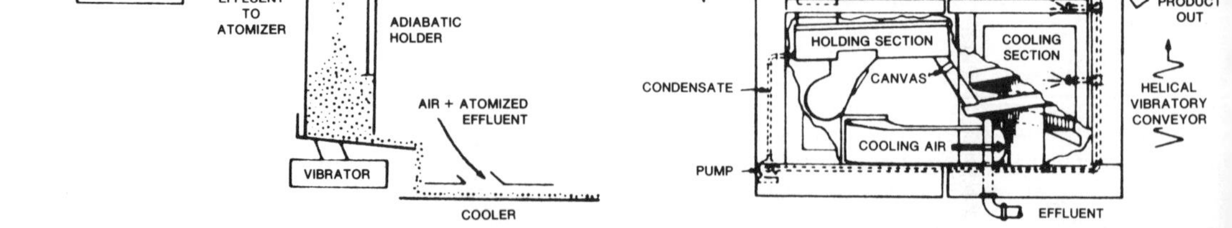

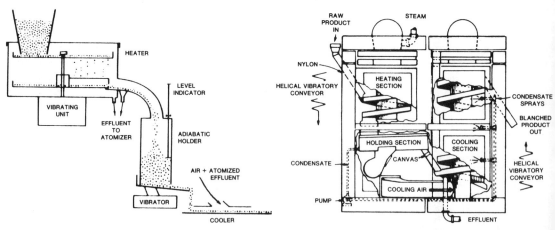

From Bomben et al. (1978) and Brown et al. (1974)

FIG. 10.6. THE VIBRATING INDIVIDUAL QUICK BLANCHER-COOLER

Left—prototype experimental, right—schematic of commercial blancher (Rexnard Co.; Louisville, Ky).

(1965) had found that such a preconditioning step in ordinary deep-bed blanching decreased time required for enzyme inactivation and reduced chlorophyll degradation. Preconditioning removes gas and may also allow desirable enzyme action (potentiation) to occur.

The principal improvement brought about by IQB is environmental in that it drastically reduces effluent volume by as much as 75%, solids and BOD by 40–70%, and product yield is increased by as much as 4%. Serendipitously, IQB also utilizes steam energy more efficiently and improves frozen food quality as measured by such standard criteria as chlorophyll loss and nutrient retention. Even texture is improved, perhaps as a result of the potentiation of pectin esterase during preconditioning (Chapter 21). Lund (1974), in an EPA assessment, concluded that IQB fulfills the objectives of blanching without quality loss while significantly reducing effluent volume in canning as well as in the freezing of vegetables. Lund also called attention to a lowering of the quality of canned but not frozen food which had been subject to preconditioning.

Vibrating IQB.—Brown *et al.* (1974) and Bomben *et al.* (1978) introduced modifications into IQB which almost doubled steam utilization efficiency and reduced space occupied by first generation IQB equipment while still maintaining product quality through adequate enzyme inactivation and reduction of stream pollution (Fig. 10.6).

Perhaps the most distinctive changes were elimination of the preblanching step and use of a solid-surface vibrating conveyor and of product seals. Also, as will be mentioned later, Brown *et al.* (1974) made some improvements in cooling treatment.

Hydrostatic Sealed Steam Blanching.—(Lee 1975; Ray 1975; Layhee 1975). In hydrostatic sealed steam blanching, blancher and cooler comprise an integrated system in which the only water added is cooling water which runs in gutters past the bottom of the blancher on its way to recirculation to the washer. The presence of this flowing water makes a seal with the lid of the blancher which in turn creates a hydrostatic head in the blancher of about 5 cm, thus raising the temperature of the steam. This system results in the following improvements over conventional blanching: It reduces steam consumption in half, lowers effluent volume (in 1975 it met OSHA requirements) and eliminates live steam exit to the atmosphere; blanch time is reduced because of higher blanch temperature [e.g., from 3.5 to 2.5 min for broccoli (frozen)]; it doubles output in the same space and results in improved sanitation. Layhee (1975) reported that Seabrook processed spinach and broccoli for freezing at the rates of 13,608 and 2948 kg/hr. Other products that have been blanched include cauliflower, celery hearts and pieces, artichoke hearts, sweet potatoes, green beans, carrots and meat balls.

Blanching Without Water.—The urgency of the environmental considerations has led to proposals and considerable experimentation on the complete elimination of water. The leading contenders are Microwave and Hotgas (Dry) blanching.

Microwave Blanching.—(Decreau 1972; Lee 1975; Lorenz 1976). Experiments in microwave or "electronic" blanching were reported almost 35 years ago (Moyer and Stotz 1945). Since no water is added, microwaves would appear to be ideal for the solution of some of the problems discussed, such as effluent, solids loss, etc., and indeed the limited literature on the subject suggests that it is good in these respects. It also appears to be potentially useful in the blanching of large objects such as whole potatoes (Collins and McCarty 1969) and corn-on-the-cob (Huxoll *et al.* 1970) since it heats an object uniformly independent of size.

A long-standing controversy as to whether microwaves affect enzymes directly or cause inactivation by raising the temperature appears to have been settled by the definitive experiments of Lopez and Baganis (1971). By evaporating water from their systems (consisting of both model purified enzymes and enzymes in foods) to keep the temperature constant, they conclusively demonstrated retention of enzyme activity (peroxidase, phenolase, pectin esterase and catalase) at 20°C upon prolonged exposure of the enzymes to microwaves at a radiofrequency of 60MHz.

Unfortunately microwave blanching introduces a new set of problems, most of them economic. Henderson *et al.* (1975), also using purified "model" enzyme-substrate systems, showed that peroxidase activity can be reduced by more than 99%, but that a very high microwave power density (as high as 2450 MHz or 375 W/cm^3) at 25°C and relatively long period, up to 40 min, are required.

Another problem is the dehydration of the exterior of the food pieces. This was solved by a combination of conventional and microwave blanching. For instance, the blanching time of brussels sprouts was reduced from 20 to 5 min by a combination of steam and microwaves. It has been suggested that microwaves be used in special enzyme-caused problems such as inactivating PE to prevent gel formation in frozen orange concentrate (Copson 1954) and the modulation of α-amylase in wheat flours (Edwards 1964). Excess α-amylase results in poor doughs and unacceptable baked products (Chapter 32) but microwaves denature the gluten.

Perhaps the most telling barrier to its widespread use is operation cost. Ralls and Mercer (1973) estimated that it cost eight-fold more than water or steam blanching. Perhaps it will find some specialized use where the benefit derived by enzyme inactivation outweighs the cost, especially when combined with conventional heating (Sale 1976).

Hot-gas or Dry Blanching.—(Ralls and Mercer 1973, 1974; Smith 1975; Lee 1975). In this variant, a mixture of heated air, nitrogen, carbon dioxide and water (supplemented on occasion by injected steam) is passed over vegetable pieces as they are conveyed through a high temperature zone. The temperature of the pieces is raised by transfer of heat from the hot gas mixture until sufficient to inactivate enzymes as well as to drive off occluded gases. It is important to have enough water in the form of steam to prevent drying. As compared with conventional blanching of spinach, hot-gas blanching resulted in an over 99% reduction in volume of waste water, BOD and weight of soluble solids. In a direct comparison with steam, water and microwave blanching, Ralls and Mercer found that in most other respects—flavor, texture, appearance, retention of nutrients (except for a higher retention of ascorbic acid in spinach and peas)—the hot-gas produced foods were at least as acceptable as those processed by the aforementioned blanching procedures.

The canning counterpart of hot gas blanching, Steriflamme, has been shown to contribute to the prevention of enzyme action during the sterilization step of canning. As mentioned, one of the problems of sterilization is the so-called "activation" during the heating-up before the can reaches maximum temperatures. According to Winter (1976) Steriflamme-processed fruit cocktail showed improvements, such as better color, increased firmness and improved syrup clarity, which could be attributed to prevention of enzyme action.

An ancient nonwater method of cooking which might be adaptable to blanching is the Chinese technique of cooking in hot oil which, as is well known to any Chinese food aficionado, results in superior color and flavor. In this treatment a blanket of steam forms which protects the vegetable against oil absorption and ostensibly inactivates enzymes. Relatively nondestructive synthetic water-repellent liquids of high heat capacity, such as the silicones, might serve the same function. How such foods would fare upon freezing and storage would be of considerable interest. Removal of oil might present a problem. Other innovations in blanching discussed by Lee (1975) include: Thermocycle Blanching based on the Venturi effect said to save energy, lower effluent volume and eliminate steam clouds in the working area, and Fluidized Bed Blanching in which an updraft of saturated steam and air is blown into a sloped bed of peas to inactivate peroxidase at somewhat lower temperatures (90°–95°C) than required in a control water-blanch (98°–100°C).

Cooling Innovations.—As discussed, the necessity of cooling a product before freezing in order to prevent enzyme action creates both environmental and food quality problems. As a rule it is not necessary to cool food in canning and dehydration operations although Sistrunk and Bradley (1975) state that turnip greens are normally cooled in tap water before filling the cans. Advocated as an amelioration of some of the adverse effects created by water cooling (increased effluent volume, nutrients loss and BOD), air cooling, perhaps one of the earliest alternatives, is now in commercial use in frozen food plants (Smith and Robe 1973). Perhaps one of the reasons for the successful commercial adoption of air cooling by some frozen food processors is that it contributes to the recovery of the *total* weight lost during heat treatment although it may impose an additional gratuitous economic penalty in that water but not air cooling can contribute to the regain of some of the total solids lost during blanching.

Some of the loss during cooling may be overcome by incorporation of water mist into the air spray. In closed loop steam blanching in which the condensate from the steam blanch is incorporated into the cooling spray, solids and nutrients are retrieved so that chemical oxygen demand (COD) and effluent are further reduced (Bomben *et al.* 1975).

Cain (1974) recommends closed loop blanching with hot water. We mentioned the use of blanch water for cooling in Hydrostatic blanching. In this process still more efficiency is gained by reuse of cooling water and condensate as make-up water in the wash tank. Since this wash water is already somewhat hot (43.3°C), less energy has to be expended in blanching.

Combination of Blanching and Chemical Treatment

Another approach to meeting and overcoming some of the problems and disadvantages associated with blanching is to combine this step with the addition of a safe edible substance. While we shall examine inhibitors as a means of preventing enzyme action in somewhat more detail (Chapter 14), we deem it appropriate to present a summary at this time of how some of these substances have been proposed to be used in conjunction with heating to further enforce the suppression and elimination of enzyme action when this is advisable in food processing.

Lowered pH.—Since in general enzymes are more vulnerable in acid media, it should be possible to decrease the blanch time and/or temperature by lowering the pH during blanching, if this is compatible with other requirements of the process and product. Some organic acid (usually citric) is incorporated in the blanch water during potato processing. While these acids in general prevent discoloration (Chapter 15) by acting as sequestering agents, part of their effectiveness may be due to lowering of the pH resulting in partial inactivation of phenolase. In the case of eggplant, citric acid also suppresses bitterness and astringency (Constantin *et al.* 1974; Voirol 1972). Although acid was combined with heat to more effectively inactivate polygalacturonase in the "hot break" processing of tomatoes (Chapter 21), improved consistency was due to the formation of cellulose microfibrils (Chapter 29).

Sequestering Agents and Detergents.—While citric acid is, as mentioned, an effective supplement to heat in preventing enzymatic darkening of some fruits and vegetables it is even more effective when combined with EDTA or other sequestering agents (Voirol 1972). Hexametaphosphate has also been used in the water blanching of peas mostly to prevent toughening when the water is exceptionally hard. In this connection it has been proposed to incorporate detergents in a cold wash prior to steam blanching to prevent the development of a hay-like off-flavor occasionally associated with steam blanching. These and other proposals can be found in the excellent historical review of Lee (1958).

Increased pH.—Increased stability of chlorophyll at higher pH's (Chapter 17) is probably the reason for the retention of green color of spinach steam-blanched with ammonia. Odland and Earhart (1975) also found improved overall acceptibility as well as higher solids and greater mineral and ascorbate retention as compared with steam-blanched broccoli in the absence of ammonia. Addition of 5% Na_2CO_3 or 10% NaCl to the blanch water shortened the blanch time for green beans (Thomopoulos 1975) and soaking of dried peas in a solution of Na_2CO_3 reduced the subsequent blanch time (Daglety Franklin Ltd. 1974).

Ascorbate.—Ascorbic acid is used in conjunction with blanching to prevent discoloration of cut potatoes, mushrooms, cauliflower and also cabbage destined for conversion into sauerkraut. In addition to preventing discoloration, supplementation of blanching with ascorbate can also aid in the removal of nitrite from spinach (Voirol 1972). The enzyme-related roles of ascorbate and sulfite, which also has been advocated as an addendum to blanch water to protect ascorbic acid, will be dealt with in more detail elsewhere.

Other Approaches.—Alum or aluminum sulfate was used in conjunction with HTST blanching to prevent the discoloration of black-eyed and purple peas (Culver and Cain 1952). However, this discoloration was not due to enzyme action but to the leakage of colored anthocyanin pigment from the seed coat. The action of the alum was thus a physical phenomenon. Another unexplored approach to energy conservation would be to combine blanching with enzyme treatment. Enzymes are available which denature proteins and inactivate enzymes, in addition to proteases, by catalyzing disulfide-sulfhydryl (thiol) interchanges (Francis and Ballard 1980).

Another way to solve the environmental problem posed by effluent discharge from blanching is *via* anaerobic microbial treatment of effluents. This approach proposed by and developed on a laboratory scale by Schlegel (1975) would probably increase the problem of energy conservation since the procedure involves prolonged cooking and storage.

We mentioned that blanching removes desirable flavors as well as desirable flavor-producing enzymes but not flavor precursors. We discuss the use of "flavorese" to regenerate flavor in blanched foods in Chapter 23. There has been some experimentation with combining blanching with ionizing radiation as means of food preservation and prevention of enzyme action. This will be discussed shortly.

CONTROL BY NONTHERMAL ENERGY

Enzyme inactivation or activity control can be achieved by input of forms of energy which usually exert their effect by protein denaturation. This includes energy in broad sectors of the electromagnetic spectrum, ionizing radiation, and ultrasonic radiation and related surface energy effects. These other forms of energy are used in the food industry primarily for purposes other than those involved directly with enzyme considerations, i.e., drying, cooling, sterilization and chemical analysis. Just as in the case of the input of thermal energy into living or *in situ* systems, at certain ranges of dosage the result of supplying the system with these other forms of energy is to exacerbate rather than prevent enzyme action. However, the mechanisms by which this type of potentiation occurs may be radically different from that by which thermal energy exerts its potentiating effects. We shall discuss these different forms of energy in decreasing order of dosage intensity.

Ionizing Radiation

Radiation consists of γ-rays, electrons and α-rays given off by radioactive isotopes or generated by appropriate devices such as the Van de Graaf machine. Ionizing radiation is defined legally as a food additive and is measured in rads. A rad is an adsorbed dose of ionizing radiation of 100 ergs per gram of absorbing material. Figure 10.7 illustrates some of the varied objectives of ionizing radiation as applied to foods and in order of decreasing dosage required to achieve these objectives. The terms radicide, radurize and radappertize are frequently used to describe the analogous objectives of pesticides, pasteurization and canning sterilization. Progress or lack of it in achieving these objectives, especially in the United States as compared to other countries, was discussed by Josephson (1976) and Hall and Eiss (1976). Other proposed applications related to foods include the medical objective of feeding highly radappertized foods to individuals whose immune systems are missing or deficient, and sterilization of sewage and sludge for animal feed.

Target Theory.—The amount of ionizing energy required to cause discernible damage in a living organism or biologically active polymer is inversely proportional to the biomass of the biological entity (Fig. 10.7), going roughly from man to enzyme. Since an enzyme molecule is incomparably smaller than a vegetative microbe or even a spore, most enzymes survive radappertization. Other targets within an organism whose perturbation by irradiation could be responsible for ultimate death include cell envelopes, nuclei, other cell organelles and ribosomes needed for enzyme synthesis.

As in the case of heat energy, enzymes are less sensitive to ionizing radiation in the dry state than in solution. However, the mechanism by which enzymes become inactivated in solution is considerably different. Yamamoto (1977), as well as providing a key to the literature, also provides an excellent assessment and analysis of the role which individual amino acid components play in absorbing ionizing radiation and the subsequent inactivation, especially that caused by crosslinking of individual protein molecules to form biologically artifactual oligomers. Furthermore, enzymes are much more stable in the absence of oxygen than in its presence. Inactivation of pure enzymes in the dry state can, by and large, be adequately accounted for by target theory which assumes that a direct hit anywhere in a structure, be it simple molecule, enzyme or larger biological structure, will result in inactivation of this entity. If D_0 is the amount of incident radiation, then

$$N/N_o = -\exp\ (D/D_o)$$

where N_o and N are the number of original and surviving targets and D is the amount of energy absorbed by the target. When $D = D_o$, $\log N/N_o = -\log(\exp) = -0.43$ so that $N/N_o = 0.37$.

Sensitivity of an enzyme to ionizing radiation is measured by D_{37} which is the dose required to inactivate 63% of the original activity. A plot of dose against log survival yields a straight line from which D_{37} and target size can be calculated as the reciprocal of the slope. In the food field the effect of ionizing radiation is frequently expressed in

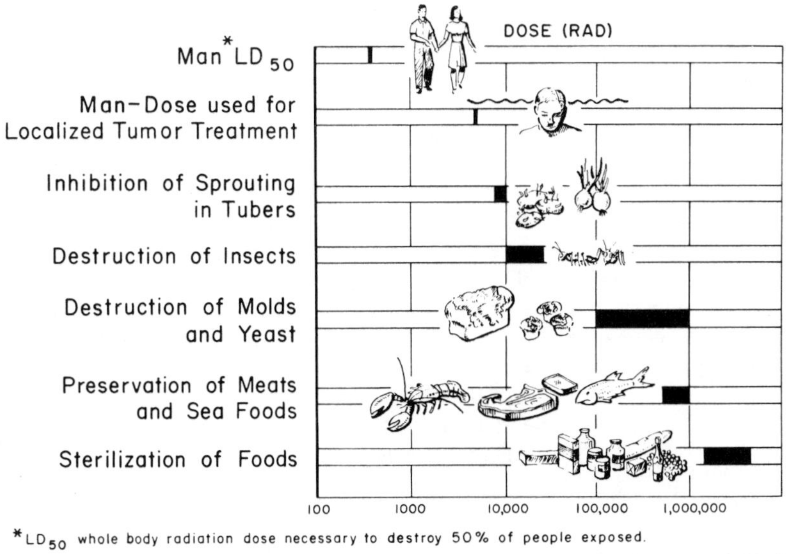

FIG. 10.7. EXPLOITATION OF IONIZING RADIATION

From Goldblith (1963)

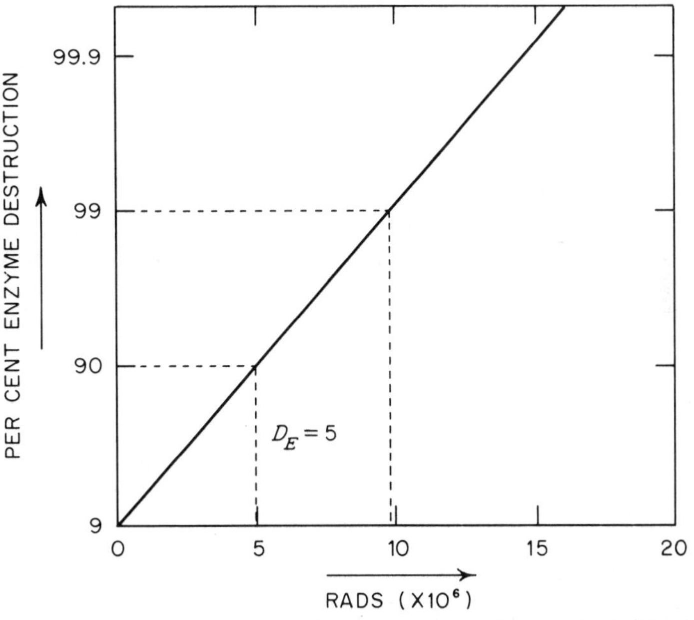

FIG. 10.8. IRRADIATION INACTIVATION CURVE FOR AN ENZYME—D_E VALUE

From Desrosier and Rosenstock (1960)

terms more analogous to that used for thermal inactivation. Thus Desrosier and Rosenstock (1960) define D_E as that dose of radiation resulting in 90% inactivation (Fig. 10.8). D_{37} values were used by Roozen and Pilnik (1971) to determine the effect of water activity on the sensitivity to electron bombardment of some organic polymer-containing enzymes in model systems (Chapter 13). Values, in megarads typical of enzyme in the dry state, were: alkaline phosphatase, 4–5; pectin esterase, 6; peroxidase, 3–7, depending upon pH, water activity and presence of polymers. This type of information has suggested to several investigators that ionizing radiation might be useful in assuring the safety of food enzymes (Chapter 35). Doses in the range below 1 Mrad (megarad, rad × 10^6, 1 million) are apparently quite sufficient for eliminating bacteria from commercial preparations of proteolytic enzymes, amylases, pectinases, naringinase and hesperidinase (Delincée and Schaefer 1975; Kawashima *et al.* 1975). Since the most sensitive locus in the enzyme molecule is most likely to be the active site (this is somewhat contrary to target theory), protection against ionizing radiation may be afforded by the presence of substrate, inhibitor or similar ligand. In this way one could reduce undesired activities in commercial enzyme preparation (Sanner *et al.* 1974).

Sensitivity in Solution.—Related to these suggestions are the observations of Lynn (1977) that serine proteinases and ribonuclease in solution were protected against inactivation by the formation of noncovalently bound complexes with DNA and silica. D_{37}'s fell in the range 0.01 Mrad with 0.5 or 0.006 Mrads in the presence of silica or DNA, respectively. In the solid state and in concentrated solution, protein and enzyme molecules crosslink with each other, i.e., "intercrosslink," to form dimers. Yamamoto (1977) lists these dimers as being blood albumin, lysozyme, peroxidase, trypsin and ribonuclease. This will not occur, however, with enzymes in very dilute solutions, according to Lynn (1977). For smaller biologically active molecules, which do not possess secondary or tertiary structures, the inactivation can also involve crosslinking resulting in dimerization.

In solution, enzymes as well as other macromolecules are more sensitive to ionizing radiation because of the magnifying effect of water occasioned by the formation of a variety of free radicals resulting from the radiolysis of water. Also known as the "indirect" effect, radiolysis results in the formation of: $\cdot OH$, e_{aq}, H, HO_2 and, of course, H_2O_2. As pointed out elsewhere, some of these can be generated by the action of oxidative enzymes. Further sensitization occurs in the presence of O_2, presumably because of the formation of superoxide anion (O_2^-) which can then interact to raise the energy level of other active molecules and residues. While single hits contribute to the death of microorganisms in aqueous suspension, enzymes in solution are inactivated exclusively by indirect action. Thus, in the relatively dry "model" system of peroxidase of Roozen and Pilnik cited above, it took about 7 Mrad to inactivate 63% of the enzyme. In solution 85% of peroxidase activity was lost upon exposure to 0.24 Mrad (Macris and Markakis 1971).

Postirradiation Inactivation.—In general most enzymes continue to lose activity after removal of the source of ionizing radiation. Thus, Macris and Markakis observed a 90% diminution of activity after irradiation of their model peroxidase solutions, 20% of which occurred in 1 day after irradiation absorption of 24 Krads.

Other enzymes whose activity has been observed to decrease after irradiation include lipoxygenase (Farkas and Goldblith 1962) and the milk-clotting rennet-like protease from *Mucor pusillus* (Sanner *et al.* 1974). Roozen and Pilnik observed that the effect is much slower with dry than with dissolved enzymes. Macris and Markakis suggested that irradiation destabilizes some of the molecules without inactivating them outright.

Effect in Foods.—In foods the gap in sensitivity between microorganisms and enzymes persists. Thus meat, after being radappertized by about 5 Mrad still goes "bad" due to survival of protease activity (Losty *et al.* 1973). Rhodes and Meegungwam (1962) report a D_{37} of 10 Mrad for the protease system responsible for liver autolysis. Lipase, which retained full activity at 10 Mrad, exhibited a D_{37} of about 30 Mrad. However, lipase in living tissue is not always so insensitive to ionizing radiation. Indeed it illustrates high sensitivity at the other end of the scale, as shown in Table 10.2, which also serves to illustrate the protective effects of drying, presence of substrate and the incorporation of the enzymes in a macromolecular biological structure, in this case the casein micelle (Chapter 33). The latter effect seems to be related to the observed strengthening of the micelle due to irradiation-induced polymerization. Interestingly, this crosslinking also decreased the apparent milk clotting activity of rennets, thus providing an example of how ionizing radiation can reduce the activity of an added enzyme by making the substrate less susceptible to the action of the enzyme. Another enzyme which is easily inactivated is plum pectin esterase. However, the benefits derived therefrom in the preparation of slivovitz (plum brandy) may be nullified by the radiation-induced release of methanol (Gitenshtein 1974) as discussed in Chapter 35.

Prevention of Enzyme Action After Radappertization.—In spite of these interesting effects, by and large there is a wide disparity between germicidal and enzyme inactivation dosages. Thus, according to Desrosier and Rosenstock (1963) 90% of all spoilage microorganisms are destroyed at a dose

TABLE 10.2

GAMMA IRRADIATION SENSITIVITY OF LIPASE IN MILK [1]

Material Irradiated	Dose Rate (Rad/hr)	D_{37}, Rads
Skim milk (SM)	100×10^5	3×10^5
SM + substrate[2]	100×10^5	25×10^5
Freeze-dried milk	23×10^5	32×10^5
Casein micelles (Chapter 33)	8×10^5	24×10^5

[1] Source: Saito (1974).
[2] Substrate, milk fat from homogenized milk.

level of 0.2 Mrad (D-value) and complete destruction is ensured at 2.4 Mrad. Hence, enzyme activity continues after complete sterilization. The most obvious solution is to combine ionizing radiation with thermal energy input, heating. In the above-mentioned study of Losty *et al.* blanching the meat for 5 min followed by irradiation completely eliminated the proteolytic activity.

Some controversy exists as to whether the combination of these forms of energy exhibits any synergy. A reading of the literature suggests that it depends upon the system being treated. Thus, Goldblith (1963), from his observation on the survival of peroxidase activity in peas, concluded that synergy was absent. In a model enzyme system Sanner *et al.* (1974) report a reduction of heat sensitivity after irradiation. According to Radola and Delincée (1972) irradiation followed by blanching is more effective than the reverse, which simply yields the additive effects of the two energy sources. Previously, Farkas and Goldblith (1962) also observed a similar effect but only when pea slurry was added to their model system.

Behavior of Enzymes in Foods Subject to Nonsterilizing Dosages.—The fact that enzymes survive radappertizing doses of ionizing radiation does not mean that the enzyme population remains static at all levels of irradiation. An entire spectrum of effects is observed, including an apparent increase of enzyme activity in the postirradiation period, especially if the food remains viable. Those changes associated with irradiation of potatoes will be discussed in Chapter 19. Indeed the effects observed at relatively low dose in the Krad range and below are illustrative of the perturbation of normal metabolic process which results in the accumulation of "nonphysiological" food components (Chapter 1). This accumulation may in turn be due to changes in the levels and ratios of enzyme activities. These effects appear to originate, not from the action of ionizing radiation on the enzymes themselves, but on those target structures within the cell, discussed in Chapter 4, involved in the control of the level of enzyme activity. These targets comprise a hierarchy of supermolecular entities: whole tissue, cell, cell membrane, organelle, organelle membrane, the complement of software and hardware of the machinery for synthesizing enzymes—nucleus, nucleic acids, nucleoproteins, ribosomes and associated enzymes. By changing the level of one enzyme in a metabolic pathway, the cell metabolism may manifest itself by a shift to other pathways.

While it is not possible to review the whole literature, a few examples serve to illustrate the diversity and ubiquity of this phenomenon more or less in order of increasing dosage required to elicit the observed enhancement:

—Chicken liver, 50 to 1000 Krad; total catheptic activity was higher (*ca* 20%) and free catheptic activity was lower (Ali and Richards 1975).

—Milk, 500 Krad (radurization dose); increases in catalase (2%), lipase (11%) and alkaline phosphatase (9%) activities but comparable decreases in activities of amylase, acid phosphatase and peroxidase (Ismail *et al.* 1975).

—Navy beans, 200 Krad; for reduction of flatulence; α-galactosidase action potentiated as evidenced by removal of 40–80% of α-galactosides (Snauert and Markakis 1976).

—Bananas, 35 Krad; for improved texture; stimulation of gluconeogenesis and especially the glyoxylate pathway (Chapter 20) the activities of two of whose enzymes, malate synthase and isocitrate lyase, were enhanced 3-fold (Surendranathan and Nair 1976).

—Potatoes, 10–20 Krad; sprout suppression; enhancement of phenolase, peroxidase, lactase dehydrogenase and chlorophyll synthesis during storage (Schwimmer *et al.* 1957A,1958); Schwimmer and Weston 1958; Tatsumi *et al.* 1972; Berset and Sandret 1976). See Chapter 19.

Apparent stimulation of enzyme activity in the postirradiation phase has been attributed to the following:

(1) Stimulation of cell growth (and therefore of enzyme level) by the debris of destroyed cells which provide nutrients for growth.
(2) Greater radiosensitivity of those cells with lower enzyme levels.
(3) Artifacts created by change in the extractability of the enzyme.
(4) Differential destruction of enzyme inhibitor.
(5) Artifacts due to the way the activity is calculated and expressed.
(6) Change in the physical state of the enzyme.
(7) Delocalization of the substrate when measuring autolytic effects.
(8) Destruction of repressor apparatus which codes for the synthesis of repressor protein which normally would prevent uncontrolled synthesis of the enzyme.
(9) Differential sensitivity (target size) of the apparatus responsible for elaborating factors involved in the degradation of the particular enzyme as compared to that involved in its synthesis (Schwimmer *et al.* 1958). Enzyme activity could indeed decrease. However, many enzyme molecules would accumulate if the former (proteinase synthesis) is more sensitive than the latter, proteinase degradation.

Occasionally this stimulated enzyme activity can result in unwanted food attributes, such as changes in texture and color. For instance, the stimulation of after-cooking darkening of potatoes by radiation may possibly be owing to stimulation of enzymes in the pathway of the synthesis of phenolics. This stimulation can be avoided by the use of less penetrating electrons (β-rays), for the purpose of sprouting by killing the surface "eyes," instead of γ-rays, which penetrate the whole tuber (Berset and Sandret 1976).

Ionizing radiation, just as any other form of energy, can be used for analysis and measurement. A promising application is the use of nitrogen γ capture for the determination of total nitrogen in foods (Roberts and Eckhoff 1973). Irradiation can also be used for thickness and density determination measurements.

Nonionizing Nonthermal Energy

Visible and Ultraviolet Radiation.—Visible radiation, light, in the presence of suitable sensitizers such as methylene blue, can inactivate enzymes *via* photooxidation. Historically, this phenomenon was important in establishing histidine imidazole as playing a prominent role in the act of enzyme catalysis (Weil and Seibles 1955). Ultraviolet radiation interacts directly with enzymes to inactivate them. Involved are the aromatic nuclei of tyrosine and tryptophan as well as cysteine sulfur (Luse and McLaren 1963). However, in the range of these forms of energy the quantum yield is too low to serve as a practical means of controlling or eliminating enzyme action in food processing.

Infrared.—Of course radiation in the range of micrometers is widely used in food processing and ends up essentially as another form of thermal energy. Pasteurization by micronization, in which the infrared is generated from heated tile, is particularly useful. Other specific applications include lye peeling of potatoes and other vegetables and fruits, drying sucrose, sterilizing containers, shucking oysters and *in situ* gelatinization of the starch in wheat kernels. Possibilities with respect to enzyme inactivation have not been explored extensively although its kinship to hot gas blanching is obvious. Russian literature abounds in studies on the physical and engineering aspects of infrared processing of foods (Rogov *et al.* 1975). Descending the energy scale in the electromagnetic spectrum, we come next to microwaves. As we have seen, the effect of microwaves on enzymes is largely through the heating of the water in which they are dissolved. However, the rate of inactivation in a food subject to microwaves is probably influenced markedly by the variable distribution of the heat generated, especially in large pieces.

Ultrasonics and Surface Denaturation.—Ultrasonic energy has been used in food processing for a supplement in the evaporation of sugars, in flour milling, and analytically for the nondestructive determination of the fat content of meat and has been proposed for the pasteurization of wines. However, this form of energy can also result in protein denaturation and hence enzyme inactivation. The denaturation is primarily a surface phe-

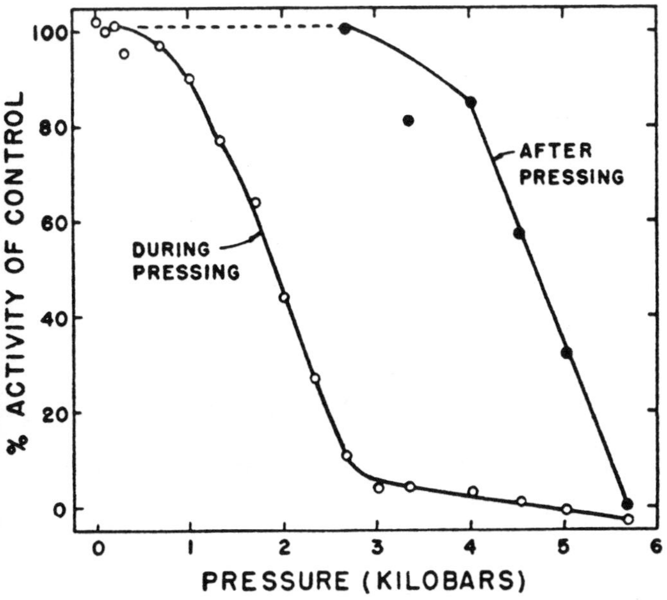

From Curl and Jansen (1950)

FIG. 10.9. PRESSURE INACTIVATION OF AN ENZYME
The enzyme reaction is the trypsin-catalyzed conversion of chymotrypsinogen to chymotrypsin.

nomenon occasioned by the cavitation which occurs in aqueous solutions subject to ultrasonication and perhaps secondarily to oxidation. Surface denaturation of proteins *via* foaming has been proposed and used in the laboratory for purifying enzymes, as described in Chapter 6. In analogy to the biological effect of ionizing radiation, ultrasonication of living systems can also result in an apparent increase in enzyme activity, i.e., various enzymes in the milk of cows whose udders had been subject to ultrasonication (Akatov and Khodakov 1974).

Enzymes can also be inactivated and inhibited by application of extremely high pressure in the range of 10^3-10^4 kg/cm^2 (Miyagawa and Suzuki 1964) and as high as 6100 bars (Curl and Jansen 1950). High pressure was considered as an alternative to heat for inactivating enzymes. Pressure causes protein denaturation and also may prevent some conformational changes obligatory for the catalytic act (Chapter 3) when the pressure is applied during the enzyme reaction. The effect of 6100 bars of pressure on the activation of chymotrypsinogen (pressing time, 60 min) is shown in Fig. 10.9.

A unique combination of heat and sonic vibrations has been proposed for drying of foods and related biological material (Lockwood 1970). Using a sonic jet airplane engine for drying, the method provides heat, airflow and air vibration. The combination is supposed to be particularly faster and better than just heat alone for sticky slurries and particulates. It would be quite interesting and instructive to learn how enzyme inactivation patterns are altered in this system.

11

PEROXIDASES, OTHER ENZYMES AND ADEQUACY OF HEAT TREATMENT AND REGENERATION

At this juncture we shall raise the question of the adequacy of heat treatment of foods as measured by the extent of enzyme inactivation. Other aspects of the use of enzymes as indicators or markers and indices related to food quality control will be described in Chapter 37.

In this chapter we shall not attempt to discuss the multiplicity of roles played by enzymes in causing the diverse aspects of food deterioration which blanching ostensibly circumvents or prevents. These will appear at the appropriate occasion in connection with the particular quality attribute being discussed.

The two heat treatments for which enzymes are extensively used as markers are pasteurization and blanching. The most widely applied enzyme with respect to blanching is peroxidase. We shall assess the available information concerning its putative role in quality deterioration of processed foods as well as its use as a marker.

Biochemistry of the Peroxidases

Before discussing their food-related roles we shall briefly survey the status and intriguing functions of the peroxidases in biochemistry, biology and medicine.

The peroxidases comprise the subclass of oxidoreductases acting on hydrogen peroxide as acceptor (EC 1.1.1.):

$$\text{Donor} + H_2O_2 \rightarrow \text{Oxidized Donor} + H_2O$$

The particular subclass as represented by the first to be isolated and crystallized, horseradish peroxidase, has been and continues to be of considerable interest since its description by Schoenbein (1863). In 1976 it appeared in over 1100 entries in *Chemical Abstracts.* In addition to being the first enzyme to be used as a specific analytical reagent (Chapter 38), this enzyme marks a turning point in the history of enzymology in that the first enzyme-substrate-like complex to be directly demonstrated was the so-called peroxidase-H_2O_2 complex by Chance and coworkers (Chapter 3). However, this turns out to have been a poor choice didactically since there is more than one complex formed and the nature of the complexes is still open to question. The first observed compound is no longer considered to be the initial H_2O_2-enzyme complex. Research is still being conducted on the sequence of events involving transformation of the various valencies of iron in the peroxidase hemin. One of the versions being currently proposed, somewhat simplified, is as follows:

$$Fe(+III) + H_2O_2 \rightarrow [Fe(+III) \cdot H_2O_2] \rightarrow Fe(+V)O + H_2O$$
Compound I

It was only relatively recently that compelling evidence was obtained showing that there is an extra oxygen in Compound I and that this oxygen was once part of H_2O_2 (Jones and Middlemiss 1974). It appears that the reaction is a two-electron equivalent oxidation of the enzyme in which the formal oxidation state of the Fe in the oxy-cation product has increased from +III to +V, i.e., Compound I retains both oxidizing equivalents of its parent peroxide substrate. Only in this sense can Compound I be considered to be an enzyme-substrate complex.

What subsequently happens to Compound I depends upon the nature and source of the peroxidase and the hydrogen donor (co-substrate). When the donor is another molecule of H_2O_2, as it is when the enzyme is catalase, or a peracid when the enzyme is chloroperoxidase (Hager *et al.* 1975), Compound I reverts back to peroxidase, Fe(+III), and free oxygen is released

$$\text{Compound I} + H_2O_2 \rightarrow O_2 + \text{Peroxidase}$$

When the peroxidase is of plant (vegetable) origin such as horseradish, Compound I is supposed to react with the hydrogen donor to give Compound II in which the Fe is tetravalent:

$$\underset{\text{Compound I}}{Fe(+V)O} + AH_2 \rightarrow \underset{\text{Compound II}}{Fe(+IV)OH} + AH$$

$$Fe(+IV)OH + AH_2 \rightarrow \underset{\text{Peroxidase}}{Fe(+III)} + AH + H_2O$$

Only the second reaction, the re-formation of "ground level" trivalent peroxidase from Compound II, is accompanied by large conformational changes (Chapters 2 and 3).

Kinds of Peroxidases.—Most but not all of the peroxidases which can be distinguished by differences in specificities are heme proteins containing protohemin, ferriprotoporphyrin III (red peroxidases) or other Fe-protoporphyrins [green (verdo) peroxidases] as prosthetic groups. Exceptions are the NAD(P) peroxidases (Florkin and Stotz 1973) and the selenium-containing glutathione peroxidase (Ladenstein and Wendel 1976). In addition to distinctly specific enzymes shown in Table 11.1, each source of peroxidase comprises a veritable plethora of isoenzymes. Their role in food science and technology will be discussed at the appropriate times.

Although these distinct peroxidases are differentiated according to the most efficient reaction catalyzed, there are overlaps. Thus, lactoperoxidase and myeloperoxidase as well as chloroperoxidase and the thyroid enzyme efficiently incorporate Cl^- (probably *via* ClO^-) and I^- into organic molecules. Cytochrome c peroxidase and chloroperoxidases as well as catalase catalyze the release of O_2 (Erman and Yonetani 1975). NAD peroxidase uses ferricyanide and other H acceptors as well as H_2O_2. Myeloperoxidase as well as clam luciferase catalyze bioluminescent reactions (Cormier *et al.* 1976). Possibly because of a similar reaction, leucocytes (white blood cells) give off light when they kill bacteria (Allen 1975). Furthermore, in the presence of dihydroxyfumaric acid, plant peroxidase utilizes O_2 instead of H_2O_2 as a substrate but oxidizes I^- to I_2 rather than incorporating it into organic molecules. Classical horseradish peroxidase and probably other plant peroxidases have molecular weights of about 40,000. That of verdoperoxidase is at least 80,000. The Fe is bound to four of the nitrogens of protohemin through four of the former's six coordination valences. The fifth valence is bound to the protein and the sixth is involved in catalysis. The enzymes have broad pH optima whose values and range depend largely upon the substrate, but which in general lie on the acid side in the range of 3–7. They all contain some carbohydrate: plant peroxidase about 15% (Morita and Yoshida 1970) and chloroperoxidase 30%. Plant peroxidases appear to be metabolically stable and survive autolysis (Axelrod and Jagendorf 1951).

Biological Functions.—The question of why the peroxidases are present in almost all living things and what they are doing there has intrigued generations of biochemists. A rather vague answer was afforded by the discovery of the H_2O_2 generating flavoprotein oxidases whose substrates are O_2 and oxidizable substances such as glucose, D- and L-amino acids, and xanthine and aldehydes. However, it does not make much sense for an organism to make H_2O_2, only to have it broken down again to re-release oxygen by catalase. According to Fridovich (1974), a principal function of the peroxidases (including catalase) as well as superoxide dismutase (Chapter 16 and elsewhere) is to scavenge H_2O_2 so that it does not have the opportunity to react with O_2^- to yield the highly reactive singlet oxygen (Schwimmer 1978). It has also been implicated in the production of ethylene in plants. Haard (1977) provides a broad discussion of its possible role in: (1) biosynthesis of lignin, (2) formation of certain stress metabolites, especially the furanoterpenoids, (3) disorders of texture-related attributes of fruits (pears) and vegetables (asparagus), and (4) ripening through its involvement in the degradation of the plant growth regulator indole acetic acid (IAA). Some investigators have suggested that in higher plants, including most of the vegetables which are blanched, the enzyme may be an atavist with no vital function other than testimony to the plants' evolutionary forebears. At least one enzyme, galactose oxidase, is activated upon being acted upon by peroxidase. The activation is accompanied by the formation of intramolecular O, O-dityrosine crosslinks within the oxidase molecule (Tressl and Kosman 1980).

Role in Mammals.—The verdoperoxidases appear to be the first line of defense against invading microorganisms (Klebanoff 1968). In the presence of I^-, Cl^- or NCS^-, they kill while simultaneously incorporating the halogen into the invading bacteria. Their contribution and relationship to other germicidal capabilities of the white blood cell are now under intensive scrutiny, as we shall have occasion to discuss in later chapters. A sampling of the extensive literature of the involvement of active oxygen includes papers by Badwey and Karnovsky (1980), Takanaka and O'Brien (1975), Krinsky (1974) and Klebanoff (1974).

Lactoperoxidase has also been assigned antibiotic roles. Its presence in milk has been ascribed to its remarkable antibacterial action, similar to that of leucocyte myeloperoxidase. The hydrogen peroxide needed could come from the action of xanthine oxidase present in milk of or from *Lactobacillus* spp. The alleged benefits of these microorganisms in milk products such as yogurt and buttermilk may be derived from this synergism. Quite apart from the artifactual introduction of lactobacilli

TABLE 11.1

THE PEROXIDASES

Prefix (P)	Prosthetic Group	Remarks
Plant, horseradish (P)	Heme, ferriprotoporphyrin IX	H-donor broad specificity
Lacto, verdo- (P)	Heme different from protoheme	Green: used for iodination
Iodide, thyro- (P)	A heme, not covalent	May be same as myelo-(P)
Myelo-, leucocyte (P)	A heme	Green: with Cl^-, I^-, NCS^-, a potent bactericide; incorporates halogen into bacteria
Chloro-, chloride (P)	A heme	A glycoprotein of *Caldariomyces* fungi, converts I to IO_3^-, H_2O_2-compound I to O_2
Cytochrome (P)	A heme	Cytochrome c best H donor
Catalase	A heme	H_2O_2 best but not exclusive H-donor
NAD (P) NADP (P)	FAD	Ferricyanide can replace H_2O_2
Luciferase	Cu plus	Found in clams; involved in bioluminescence
Glutathione (P)	Se plus	Removes lipohydroperoxides, may be involved in aging and senescence
Fatty acid (P)	Fe plus	Involved in α-oxidation of fatty acid in germinating seeds; wax biosynthesis (Chapters 21, 25)
Tryptophan pyrollase	Heme	Appears to be an oxidase but has many properties in common with classical peroxidase
Indole acetate oxidase	—	May be a plant peroxidase isozyme
Lipohydroperoxide (P)	—	See Chapter 25

[1] Animal peroxidases contain covalently bound prosthetic groups other than protoheme (Yamasaki 1974).

into milk, Schanbacher and Smith (1975) point out that the presence of lactoperoxidase in milk may be important to the survival of the newborn since both milk and colostrum contain growth factors for these microorganisms which appear to ensure their survival in the gut of the newborn. As mentioned in Chapter 9, the antibacterial action of this enzyme induced in saliva may be responsible for the non- and anticaries effect of xylitol. Aune *et al.* (1977) stress the role of thiocyanate ion, SCN^-, which yields cyanogen, C_2N_2, in producing sterilizing effects in human fluids.

Synthesis of the hormone thyroxin and iodinated proteins such as thyroglobulin arises as the result of the action of a special thyroid peroxidase (Pommier *et al.* 1975).

The metabolism of tryptophan as well as that of the closely related IAA appears to involve peroxidase (Gove and Hoyle 1975). Indeed, tryptophan pyrollase was originally placed in the peroxidase subclass (EC 1.11.1.4).

Perhaps the most unequivocal and spectacular function of a peroxidase as well as catalase is afforded by the weapon possessed by the bombardier beetle (Aneshansley *et al.* 1969). The insect's artillery is a fine mist of steam at 100°C containing oxygen, quinone and methyl quinone. The oxygen and steam arise as the result of catalase action and the toxic quinones from the action of peroxidase, both functioning by being ejected into a special gland containing 25% H_2O_2 and 10% hydroquinones.

Applications

Quite aside from their biological functions, the peroxidases have been broadly applied for a variety of research and analytical purposes, many of which have important relations to foods. Scientifically one of the most interesting is as an aid in probing the topography of the surface and molecular architecture of supramolecular biological edifices such as ribosomes, the myelin sheath of nerve cells, antibodies and erythrocyte membranes. The general principle is to incorporate radioactive iodine into these arrangements with the aid of lactoperoxidase (supplied commercially in both free and immobilized forms) and then to take them apart, both physically and chemically *via* the aid of proteolytic enzymes, and then to examine the distribution patterns of the iodine label in the resulting products (Litman and Cantor 1974; Mueller and Morrison 1975; Poduslo and Braun 1975; Sikes *et al.* 1977).

Peroxidase has also been frequently used as a marker for the immunological estimation of biological activities or of relatively small molecules such as the estrogens as alternatives to radioimmunoassay, enzyme immunoassay (Chapter 39).

In quantitative analysis, peroxidase, both free and immobilized, has largely replaced catalase in the determination of glucose (Chapter 38) and may be used in general for the determination of any substance which upon oxidation produces H_2O_2. Immobilized peroxidase has been suggested for the cold sterilization of fluids including those which happen to be foods (Henry *et al.* 1974).

The use of peroxidases as markers or indicators of various biological parameters will be examined in Chapter 37. Finally, peroxidases are used as indices of the adequacy of heat treatments. Properties related to this area are subjects of the following discussion.

THERMAL PROPERTIES OF PEROXIDASE

The universal adoption of peroxidase tests as indicators of the adequacy of blanching is predicated on the following, although not necessarily always valid assumptions. In general it is preferable to get rid of all enzyme activity in most foods, especially frozen foods. Since peroxidase is supposed to be the most heat stable enzyme, its destruction ensures that of all other enzymes in the food exposed to the same heating conditions. Peroxidase was selected because it is easy to test for and is practically ubiquitously distributed (Joslyn and Cruess 1929; Joslyn 1949,1961A,1966). As stressed in Chapter 10, one wants to blanch for minimal times at temperatures consonant with the prevention of quality deterioration. It is therefore important to have some detailed knowledge of the peroxidases' properties involved in denaturation.

The Energy of Activation of the Denaturation of Vegetable Peroxidases

Values reported for the energy of activation of the denaturation of peroxidase are unusually low as compared with those of less heat stable enzymes. Thus, values for E_a have been reported at 15,700 cal/mole, sweet corn (Yamamoto *et al.* 1962); 19,000, green beans (Zoueil and Esselen 1959); 32,000, horseradish (Lu and Whitaker 1974), as compared to 61,000 for catalase (Sapers and Nickerson 1962) and about 100,000 for lipoxygenase (Farkas and Goldblith 1962) and phenolase (Dimick *et al.* 1951). In contrast to typical protein denaturation, the entropy of activation for peroxidase denaturation (ΔS) is negative or else has a very low positive value. This indicates that the peroxidase molecules are more ordered in the activated or transition (Chapter 3) than in the "ground" or resting state. Lu and Whitaker attribute this to the events leading to the release of the hematin group, and it may explain why the peroxidases are so easily regenerated, as will be discussed shortly in more detail.

Another special feature characterizing the thermal properties of vegetable peroxidase, which it also appears to share with catalase, is the deviation from linearity (first order kinetics) of the log survival *vs* time curves at temperatures below 90°C. Although most investigators show a curve of continuously decreasing slope (Yamamoto *et al.* 1962; Resende *et al.* 1969), Duden *et al.* (1975) conceive of the deviation as a sharp break. More sophisticated studies on the intimate details of the cooperative thermal unfolding of other enzymes follow biphasic (ribonuclease, chymotrypsinogen) and even triphasic (cytochrome *c*) kinetics (Tsong 1973). Further insight into the peculiarities of the heat stability of the peroxidases will be examined in association with regeneration of activity.

Data through 1955 on the stability characteristics of peroxidase in foods were collected and classified by McConnell (1956). Abbreviated updated versions of his compilations dealing mainly with canning are shown in Tables 11.2 and 11.3. Although values are not really directly comparable since they are based on different assays and different criteria of inactivation, it will be noted that z-values of low acid fruits and vegetables range from 8.8° to 71.9°C. Higher z-values may be more reflective of the true situation since, as pointed out by Duden *et al.*, denaturation curves should be determined with suspensions to which peroxidase is adsorbed rather than with soluble filtered or centrifuged extracts because more than 99% of plant peroxidase is present in bound form in the cell. Furthermore, the higher values are more in line with what appears to be the only attempt to assign a z-value (116°F) to off-flavor in frozen foods (Diehl *et al.* 1933). The average value of 62°F is still above the 18°–20°F normally used in pro-

TABLE 11.2

HEAT RESISTANCE OF PEROXIDASE IN LOW ACID FOODS[1]

Food	z-Values[2] °F	F_T-value (min)[3] at 180°F	300°F
Artichokes	105	2	0.14
Asparagus	55–161	2–27	0.2–1.3
Beans, fava	105	2	0.14
Beans, green	86–88	3.8	0.16
Beans, lima	67	30	<0.1
Corn[4]	70	—	0.5
Peas	48	60	<0.1
Turnips	58–73	2–3	0.2–0.3
Spinach[5]	59,81	6–10	<3
Milk	0–4	<0–1	—

[1] Source: Modified from McConnel (1956) where references not cited can be found; some values shown were obtained by extrapolation and interpolation.
[2] Slope of the logarithmic thermal inactivation curve.
[3] Time at indicated temperature to inactivate enzyme (<1% survival).
[4] Source: Vetter *et al.* (1959).
[5] Source: Duden *et al.* (1975).

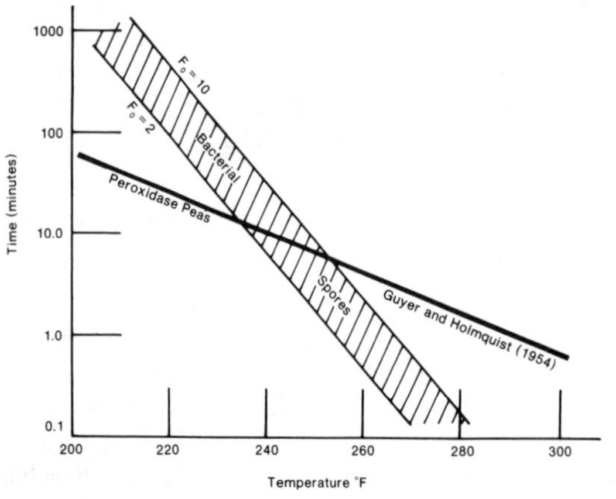

From McConnel (1956)

FIG. 11.1. THERMAL DESTRUCTION CURVES OF SPORES AND OF PEROXIDASE IN LOW ACID FOODS

TABLE 11.3

HEAT RESISTANCE OF PEROXIDASE IN ACID FOODS[1]

Food	z-Values	F_T-values (min) at 180°F	212°F
Peaches	20–31	0.3–0.9	0.1
Peaches in sugar	37	2–4	0.3–0.4
Pears	20–32	0.5–8.2	0.1
Pears in sugar	12–28	0.6–11.2	0.1–0.7
Cucumbers, fresh	22	48	1.7
Pickles, salt pack	20–33	5–42	5
Pickles, fresh pack (after acidification)	10–20	0.2–4	0.1–0.4
Tomato juice	18	0.9	0.1
Apples, fresh juice, cider	11–34	5–7	0.1–0.7
Apples in sugar	27–28	10–11	0.8

[1] Source: Modified from McConnel (1956); see Table 10.2.

cessing canned foods. This is graphically illustrated in Fig. 11.1. This shows that processing above 250°F necessitates heating for longer than is necessary to destroy spores, assuming of course that almost complete destruction of peroxidase is necessary to prevent subsequent quality deterioration.

In acid foods peroxidase is probably not the enzyme of choice. The effect of low pH of foods in destabilizing peroxidase is clearly shown in Table 11.3 where z-values range from 10° to 34°F, well within the normal processing range. The stabilizing effect of concentrating the enzyme, and hence insoluble impurities to which the enzyme can adsorb as well, is aptly illustrated in Table 11.3. A study of the influence of heat treatment *per se* on organoleptic quality of vegetables yielded the following z-values: green beans, 52°F; corn, 57°F; peas, 51°F (Hayakawa *et al.* 1977).

PEROXIDASE ASSAY AND ADEQUACY OF BLANCHING

What Is the Best End-point?

While the use of the peroxidase assay—aside from the problem of activity regeneration—is accepted, and rightfully so, as good measure of the adequacy of blanching (and also quite aside from the question as to whether it contributes directly to subsequent deterioration), the question remains as to how much has to be lost to have an adequately blanched product. Undoubtedly the end-point depends upon the product, the process and assay procedure.

However, there seems to be accumulating evidence that complete destruction, i.e., no enzyme detected with standard assay methods, may be going too far. Thus, Campbell (1940) showed qualitatively that after "scalding" of cut corn for 30 but not for 15 sec, the qualitative test for peroxidase was slightly positive but definitely detectable and the beans did not develop off-flavor after storage at −15°C for 11 months. This result is echoed in a more rigorous fashion by Böttcher (1975) who concluded that complete absence of peroxidase indicated overblanching. As previously stressed, blanching more than is necessary results in needless lessening of quality of the finished food product. Böttcher recommends the following residual activities (as % of original activity): peas, 2−6.3, depending upon the variety; green beans, 0.7−3.2; cauliflower, 2.9−8.2; brussels sprouts, 7.5−11.5. In most cases a 2 min blanch at 90°−100°C is considered suitable. Another reason for not trying to get rid of the last trace of peroxidase activity relates to the varying heat resistance of multiple forms of peroxidase inevitably comprising the "peroxidase" of any given vegetable. Thus, Winter (1969) found that between 1 and 10% of the activity in 10 different commonly used vegetables could be accounted for by ultraheat-resistant peroxidase isozymes. This has been confirmed by Delincée and Schaefer (1975).

Peroxidases Assay

The question of how much peroxidase to get rid of is closely linked to the actual methods used to detect and determine the activity of the enzyme. In general, peroxidases are assayed by measuring the rate of conversion of colorless H donors to colored compounds (chromogenic assay). In the 1920s and 1930s the standard peroxidase unit was the "P.Z." or Purpurogallin Zahl of Willstätter and colleagues based upon the oxidation of pyrogallol to the highly colored purpurogallin. Since then, many such chromogenic hydrogen donors have been used (Maehly and Chance 1954; Hunting *et al.* 1959). In one interesting variation, two hydrogen donor cosubstrates, iodide ion and either ascorbate or thiosulfate along with starch, are added to the enzyme reaction mixture. The time required for the sudden appearance of a blue color (starch-iodine) due to the liberation of iodine after exhaustion of the ascorbate is noted (Schwimmer 1944A). Ascorbic acid is also used in the AOAC 1975 method.

For the purpose of adequacy of heat treatment these methods have been largely replaced by the guaiacol assay in which the conversion of the latter compound to the highly colored oxidation product, tetraguaiacol, can be followed continuously with a recording spectrophotometer or intermittently with an ordinary colorimeter. Alternatively, for fast results the Morris test paper now used worldwide, which is based upon the use of urea peroxide, gives adequate semiquantitative results in many cases (Morris 1959; Bakowski and Maleski 1968). Continuous surveillance during blanching operation is made practicable with the Morris test. Varoquaux *et al.* (1975) developed a rapid, reproducible and automated guaiacol assay.

As an example of the details of simple enzyme test which has official and forensic status, we describe a peroxidase assay, essentially that of Masure and Campbell (1944). It is one of the technical inspection procedures for the use of U.S.

Department of Agriculture processed food inspectors (U.S. Dep. Agric., Agric. Mark. Serv. 1975). In this test which is completed in 30 min, a 200 g sample taken from the processing line is cut into small pieces and blended at full speed with 600 ml of water for 1 min. The macerate is filtered through a cotton milk filter. To 2 ml filtrate in a test tube is added successively without stirring 1 ml each of 0.5% guaiacol and 0.08% H_2O_2. The tube is then mixed exactly three times. Any color difference between the enzyme reaction mixture and a blank containing all of the ingredients except H_2O_2, which appears within 3½ min, is indicative of a positive peroxidase test and the sample is considered underblanched.

This is not the usual type of assay in which the pH and temperature are under careful control (Chapter 37). In the AOAC (1975) method (Joslyn and Neumann 1963) both ascorbic acid and the traditional ascorbate oxidizing dye 2,6-dichlorophenolindiphenol are added to the reaction mixture. The hydrogen donor is considered to be the reduced dye added at the start of the reaction. This dye is reduced by the ascorbic acid present in excess. The first-order constant of the rate of decrease of ascorbic acid, as measured by titration with the dye, is used as a measure of peroxidase activity.

While serving well in early days of the frozen foods industry, this method would appear to be somewhat dated. It is cumbersome, using unnecessarily large volumes which introduces timing errors. There are now available very inexpensive colorimetric devices to replace time-consuming indirect titration. As stated in Chapter 37 in this type of assay, first-order reactions are to be avoided if zero-order reaction or initial rates can just as easily be measured. It is rather ironic that the same edition of AOAC does contain a method for the determination of glucose in which peroxidase is used to oxidize a chromogen, *o*-dianisinidine. This dye is widely used by biochemists in a simple accurate and rapid zero-order assay for peroxidase (Worthington 1972). An improved assay has been adopted by Worthington (1976) using 4-amino antipyrine as an H-donor.

Whatever method is adopted, there are several neglected areas which should be investigated. Dietrich (1975) has called the attention of frozen food technologists to the following problems, comments and suggested solutions centering around the Masure-Campbell Guaiacol test. Many of the points raised would apply to other hydrogen donor peroxidase cosubstrates.

—Should pH, temperature, salt concentration as well as time be controlled in the extraction of the enzyme? Nobody knows the effect of these variables on the reproducibility and validity of the test.

—After extraction, it is traditional to use a cotton-backed milk filter. This can result in turbid filtrates which obscure the peroxidase-induced color generation.

—Should temperature, pH and salt concentration be controlled as is usually done in enzyme assay? Dietrich reports that buffering vegetable extracts at pH 4.5 resulted in apparent increased enzyme activity. Should the time, then, be reduced from the standard 3½ min? The temperature at which the assay is run may vary from 18° to over 30°C, which could give more than a two-fold difference in activity.

—The test is characterized by a lag period before the appearance of the color. Very little is known about the cause and the effect on reproducibility of this lag. Dietrich suggests that the following variables should be examined: (1) thawing, if any, before blending, (2) time delay between extraction and assay proper, (3) particle size, and (4) presence of changes in endogenous peroxidase cosubstrates (hydrogen donors), especially ascorbic acid.

—Should visual judgment be replaced by a quantitative colorimetric measurement? Dietrich found measurement at 420 nm in a simple colorimeter just as convenient and indicative of adequate blanch as the visual method.

—The question of the variability of the enzyme content of the raw material and the relation of this to how long to blanch and what end-point (see above) to use have not been adequately examined. Dietrich states that size variation probably has more effect on blanch time than does variation in the level of peroxidase in the raw material. Interrelations among size of the sample, heat penetration, uniformity of heat distribution within the blancher, vis-à-vis the peroxidase test, as well as the previously mentioned factors have not, with some exceptions, received adequate attention, especially in freezing as compared with canning. Several questions relating to regeneration of enzymes will be examined shortly.

Miscellaneous Uses of the Peroxidase Assays in Relation to Blanching

In addition to its use in the preparation of frozen, canned and dehydrated foods, peroxidase has been used as a measure of adequacy of heat treatment with other foods and food products. Thus, Gardner *et al.* (1971) found that inactivation of peroxidase in corn as a result of heating the corn correlated well with the consistency and water absorbing properties of grits prepared from the corn. Parenthetically (since this should really belong in Chapter 37), the *original* corn peroxidase level correlates with the percentage of crude oil in dry-milled high-lysine corn. As previously mentioned, the USSR has published extensively on blanching of meat. Here, too, peroxidase appears to be a useful guide for the effectiveness of heat treatment (Kuchinskii and Yasskaya 1976).

OTHER ENZYMES AS INDICATORS OF ADEQUACY OF HEAT TREATMENT

When one is reasonably sure that a particular kind of loss in quality is due to the action of a specific enzyme type, it is preferable not to heat beyond the point necessary to ensure complete inactivation of the enzyme. Such enzyme assays include phenolase assays for prevention of off-colors in fruits, polygalacturonase for prevention of changes in consistency in such vegetables as tomatoes, eggplant and potatoes, lipoxygenase in soybean and lipase in cereal products. As shown in Table 11.4 these enzymes are much less heat-stable. Another enzyme whose action is not particularly associated with a particular change in food quality but whose inactivation is apparently still being used as an index of blanching is catalase.

Catalase

Enzymology.—Historically catalase is probably the first enzyme whose action was observed to be universally distributed. As mentioned previously, catalase is really a peroxidase whose preferred H-donor as well as acceptor is H_2O_2.

$$\text{Compound I (FeV-O)} + H_2O_2 \xrightarrow{k_2} \text{Compound II} + O_2$$

Other complexes (II, III, IV) are formed as a result of reaction with other H-donors. These donors react rather feebly in comparison with H_2O_2 (i.e., ethyl hydrogen peroxide). An exception is the catalysis of the decomposition of linoleate hydroperoxide (Chapter 24), a product of the lipoxygenase reaction. The rate constant for the catalase-catalyzed decomposition is 160 times that of peroxidase. The k_2 for H_2O_2 is at least 10,000 times that for methyl hydrogen peroxide.

In contrast to peroxidases, the catalases which have thus far been isolated all possess quaternary structure. Each consists of a tetramer so that there are four molecules of ferriprotoporphyrin III per molecule of enzyme. Molecular weights run around 225,000 daltons. Until relatively recently they were considered to be the most efficient catalysts known, with a molar activity of about 10^5 (Table 3.1). Now several enzymes, including carbonic anhydrase, have higher molar activities with

TABLE 11.4

HEAT RESISTANCE OF OTHER ENZYMES IN FOODS[1]

Enzyme and Food	z-value, °F Avg.	z-value, °F Range	F-value at 180°F (min) Avg.	F-value at 180°F (min) Range
Pectin esterase in citrus juice	14	7–17	43	2–94
Polygalacturonase in citrus juice	16	5–60	12	10–84
Polygalacturonase in papaya[4]	11	—	23	—
Phenolase in several fruits	12	7–17	1.1	0.05–4.3
Phenolase (?) in cranberry[5]	24	—	—	1.6
Catalase in vegetables[2]	28	15–36	6	0.03–2.5
Ascorbate oxidase in peach and veg.	59	34–70	2	1–4
Phosphatase in orange juice	9	—	—	—
Phosphatase in milk	13	—	—	—
Chlorophyllase in spinach[3]	22	—	2	—
Lipoxygenase + peas solids[4]	16	—	<0.1	(176°F)

[1] Source: Adapted from McConnel (1956).
[2] Source: Includes data of Sapers and Nickerson (1962).
[3] Source: Resende *et al.* (1969).
[4] Source: Aylward and Haisman (1969).
[5] As evidenced by destruction of anthocyanin pigment (Chapter 16).

values running to 600,000. Catalases are characterized by an exceptionally low energy of activation, about 5600 cal/mole. They are inactivated by their own substrate, H_2O_2, but reactivate when immobilized (Wang *et al.* 1974).

Catalase, as mentioned, is one of the most widely distributed and easily detected enzymes—hence, it is rather all-inclusively named the "universal catalyst." It is probably this circumstance that prompted its extensive use in the early days of the frozen food industry. Its physiological function, perhaps among others, is to scavenge H_2O_2 as discussed elsewhere. It is localized in animal cells in microbodies known as peroxisomes where it apparently decomposed H_2O_2 produced by the high level of peroxide-generating flavoproteins which are mainly involved in the conversion of noncarbohydrate metabolites into carbohydrates (De Duve 1969).

Heat Stability.—Ubiquitous distribution of catalase and ease of demonstrating its presence, rather than its intrinsic worth as an indicator of adequate blanching, led to its use for this purpose in the early days of the frozen food industry. The stability of catalase and other quality-related oxidoreductases expressed in food processing terminology are shown in Table 11.4. Sapers and Nickerson (1962) made a rather thorough study of the stability of partially purified spinach catalase and found that such preparations were quite stable below 36°C but were completely inactivated at temperatures greater than 80°C. Thermal stability was found to be highly dependent upon pH, dropping off rapidly in the acid region. Undoubtedly the fact that catalase has a quaternary structure while peroxidase does not accounts significantly for the profound differences in the heat stability of these two similar-acting enzymes. As a rule, the simpler and smaller a protein is the more heat stable it is. However, sometimes very subtle alterations in a protein, such as the change of one amino acid in a bacteriophage lysozyme, can confer heat stability (Feeney 1976). A direct comparison of the heat stabilities of broccoli enzymes is shown in Fig. 11.2.

There was at one time controversy as to which enzyme test should be used in the determination of the adequacy of blanching. Thus, early in the history of the frozen food industry catalase was extensively used (Diehl 1932; Diehl *et al.* 1933; Campbell 1940), especially for peas. The literature of the intervening years recommends catalase for peas and certain other vegetables and peroxidase for others. Joslyn (1949) concluded, from his work and that of his colleagues from 1933 to 1940, that peroxidase was found to parallel the enzymes responsible for off-flavor more closely than did catalase.

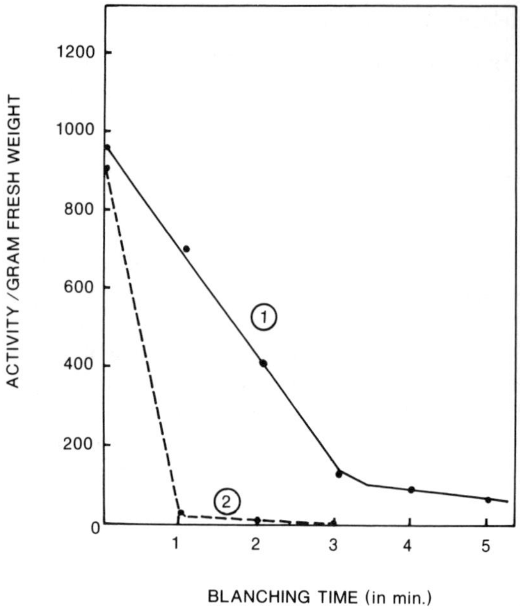

From Buck and Joslyn (1953)

FIG. 11.2. HEAT RESISTANCE OF PEROXIDASE (1) AND CATALASE (2) IN BLANCHING BROCCOLI

In a direct comparison of the thermal stability of catalase and peroxidase in southern peas, Lopez *et al.* (1959) confirmed the general instability of catalase as compared with peroxidase. The U.S. Department of Agriculture Technical Procedures Manual (U.S. Dep. Agric., Agric. Mark. Serv. 1975) states that for the majority of vegetables, inactivation of catalase alone is not an indication of adequate blanch and that inactivation of peroxidase is considered necessary to minimize the possibility of future deterioration in quality. Thirty to fifty percent more blanch time is required to inactivate peroxidase than would be required to inactivate catalase alone. Catalase is also a poor indication of blanch in products after they have been in storage for several months, since the catalase activity gradually disappears.

Nevertheless, catalase is still used as an official test in the specifications set up by the U.S. Department of Agriculture and also in the 1975 edition of the AOAC Methods of Analysis. It would appear that the disappearance of catalase is on the borderline of adequate blanching. Its successful use depends upon the assay used and the uniformity of the raw material. Its continued use bespeaks the desire to keep blanch time down to a minimum. It depends upon the trade-off between heat-caused deterioration in quality and the gamble that the catalase test will be adequate if not infallible, as it probably is for peas and some other vegetables. The use of a peroxidase test is frequently described as a margin of safety.

For the most part there are four catalase tests in use. The simplest is that described by Joslyn (1949) in which one adds H_2O_2 to crushed vegetables in the bottom of a test tube. The absence of bubbles is an indication of the adequacy of blanching. The 1975 U.S. Department of Agriculture inspection test similar to that described by Dietrich and Neumann (1956) involves preparation of an extract. It also depends upon the observation of gas evolved, but in a Smith fermentation tube used by microbiologists. Evolution of less than 0.1 ml of gas is an indication of adequate blanching. In a special test for brussels sprouts, H_2O_2 is added directly to the cut surface. A pink color develops in *ca* 15 sec if the vegetable is underblanched. The AOAC method is essentially the classical "Katalase Fahigkeit" as modified by Balls and Hale (1932). This assay, as in the peroxidase assay, a manual titrimetric macromethod, is performed at 0°C and is run under favorable conditions to give a first-order reaction course. While these methods were quite quantitative and useful in their day, they were developed before the widespread availability of inexpensive colorimeters and spectrophotometers. Thus, catalase has been assayed rapidly, using zero-order kinetics at ambience by enzymologists for over 25 years by simply following the linear rate of decrease of H_2O_2 as measured by its UV absorbance (Worthington Biochemical Corp. 1976). These and other assays were reviewed some time ago by Chance and Maehly (1955). Such a catalase assay would be appropriate to meet the sensitive requirements of a suitable indicator of the adequacy of blanching of most vegetables except leafy vegetables, carrots and beans (Winter 1969).

A Note on the Adequacy of Pasteurization

Although we are not intimately concerned in this book with food microbiology, the fact is that a knowledge of food enzyme properties is needed to serve as indicators of microbiologically associated heat treatment of foods (Chapters 33 and 36), in the present case to ensure the destruction of a particular group of microorganisms. The classical example is the assay for the alkaline phosphatase of milk whose reported z-values in milk are less than 10 (Hetrick and Tracy 1948), lower than those of the enzymes used for adequacy of blanching. The rationale behind the use of this enzyme is similar to that of other such tests, that is, to use a minimum amount of heat, just enough to ensure adequate removal of tuberculosis-causing microorganisms. This happens also to be the amount of heat necessary to just inactivate the alkaline phosphatase in the milk. Any further heating results in the change of milk flavor, waste of energy and additions to the cost of the final product.

Milk and Dairy Products.—The use of phosphatase as an indicator of adequate pasteurization was established over 40 years ago by Kay and Graham (1935). In the method, accepted as official in the early editions of the AOAC Methods of Analysis, phenyl phosphate was the substrate and the liberated phenol was determined with the Folin reagent over a 24 hr period. This substrate was superseded by phenolphthalein phosphate because of the extreme simplicity in measuring the color of the liberated red phenolphthalein at the pH of the assay, 9–10, after dialysis. Although this method is included in the 1975 AOAC edition, the preferred assay again goes back to phenyl phosphate as substrate, but the liberated phenol is determined by reaction with BCQ, 2,6-dibromoquinone-4-chloridate, to yield an intensely colored compound which can be readily measured spectrophotometrically at 610 nm. Continuing improvements in this method (Schiemann and Brodsky 1976) in which as little as 0.1% of raw milk in pasteurized milk can be detected, may be supplemented by the use of *p*-nitrophenyl phosphate as substrate in automated analysis (Schiemann 1976).

The milk enzyme has been purified and thoroughly studied. Like other alkaline phosphatases, the enzyme is activated by Mg^{2+} and displays half-site reactivity (Chapter 4). The paper by Linden *et al.* (1977) provides a key to literature on this enzyme.

Regeneration of phosphatase will be discussed below.

Of course many enzymes survive pasteurization and their subsequent action could lead to quality control problems especially when the milk is incor-

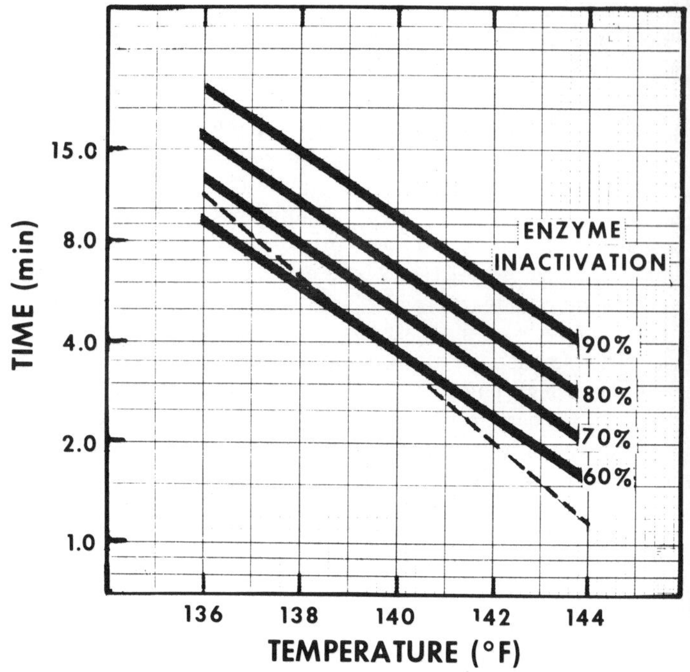

FIG. 11.3. THERMAL DESTRUCTION CURVES FOR *SALMONELLA* (DASHED CURVE) AND *N*-ACETYL-β-D-GLUCOSAMINIDASE (SOLID CURVES)

Courtesy of Hansen (1971)

porated into other foods. For instance, bacterial proteinase, responsible for a source of off-flavor in dairy products, which frequently survives UHT pasteurization (up to 150°C for 30 sec or less), can be eliminated by a quasi-potentiation pretreatment of exposure of the milk to 55°C for 5–60 min (Adams *et al.* 1979). The presence of *acid* phosphatase in milk is indicative of microbial contamination.

Pasteurization of Nondairy Foods.—Phosphatase assays have been looked into as indicators of pasteurization of a variety of foods such as orange juice (Axelrod 1947), canned hams and eggs (Guenther and Burkhart 1967). There is no *a priori* reason why phosphatase should meet the criteria for adequacy of pasteurization in such a wide variety of foods. Indeed, Hansen (1971) determined that the thermal characteristics of egg phosphatase did not meet these criteria, which are 3.5 min at 65°C, in order to ensure removal of *Salmonella*. She did find, however, that *N*-acetyl-β-D-glucosaminidase did (Fig. 11.3). Figure 11.4 shows the reactions used as the basis for the assay.

REGENERATION OF ENZYME ACTIVITY

Regeneration of peroxidase and catalase activity in extracts of blanched vegetables was observed at the inception of food freezing in the early 1930s (Diehl 1932; Davis 1942; Balls 1942; Joslyn 1949). Quite aside from being suspected of contributing to quality deterioration, regeneration poses a problem in that quality specifications for frozen foods usually do not allow for it.

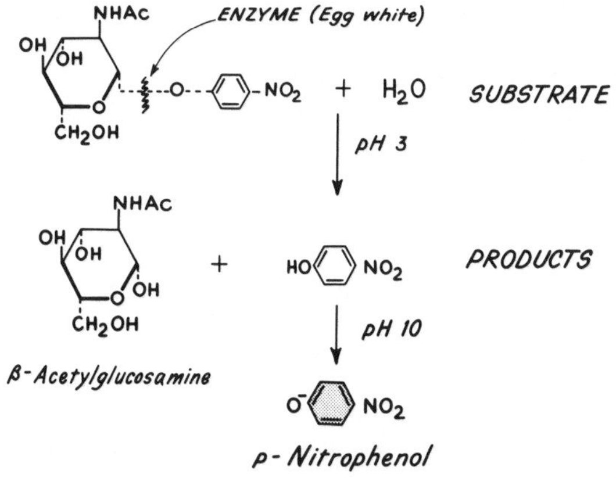

Courtesy of Hansen (1971)

FIG. 11.4. BASIS FOR ASSAY OF *N*-ACETYL-β-D-GLUCOSAMINIDASE

Categories of Regeneration

Recovery of peroxidase activity after being heated is but one of many examples of the reversal of protein denaturation. For instance, several proteinases are also capable of recovering activity after heating (Balls 1942). This capacity to recover activity after inactivation is not only confined to enzymes which have lost activity because of thermal denaturation.

Thus, bromelin, trypsin, polygalacturonase and RNase have been reported to recover activity after ostensible acid or alkaline inactivation by adjusting the pH to near neutrality. Urea and guanidine are general denaturants which reversibly inactivate by unfolding the tertiary structure. Similarly ribonuclease can recover activity after being inactivated by agents which reduce the S-S linkages crucial to tertiary folding and active-site formation. Obversely, papain loses activity when its essential SH is oxidized but can regain activity upon treatment with appropriate reducing agents or HCN (HCN + S-S → -CNS + SH). Inactivation by active-site reagents (Chapter 3) such as the alkyl fluorophosphates can be reversed with nucleophiles. Finally, frequently apparent inactivation by substrate such as that of immobilized catalase by H_2O_2 can be reversed by adjustment of pH or slowly by removal of substrate. It is evident from a glance at the literature relating to this phenomenon that the principle of self-assembly *via* cooperativity (Chapter 4) is manifestly operating here.

Mechanism of Peroxidase Inactivation and Regeneration

Within the heme proteins, the reversible denaturation of the nonenzyme proteins cytochrome c and myoglobin have been the most thoroughly studied and understood so that these proteins provide valuable background for understanding peroxidase regeneration. See Lu and Whitaker (1974), Tamura and Morita (1975) and Gibriel *et al.* (1978) for pertinent literature.

Regeneration of peroxidase was first observed in a U.S. Department of Agriculture laboratory by Woods (1902). Investigations during the next 40 years amounting to about 12 papers were reviewed by Schwimmer (1944A). In the latter investigation it was suggested that heat inactivation of peroxidase in turnip extracts proceeded *via* denaturation of the protein followed by dissociation of the hematin from the apoenzyme and that regeneration occurred by recombination of the renatured apoenzyme (previously adsorbed and protected from irreversible denaturation) with the soluble hematin. Presence of isozymes with differing heat stabilities was utilized in establishing this mechanism. In the same year, Herrlinger and Kiermeier (1944) described the dependence of the extent of regeneration on such factors as pH, buffer salt, and enzyme concentration as well as on time and temperature of heating and regeneration. They concluded that the regeneration was due to disaggregation of heat-induced aggregation of "carrier" protein in analogy to "cryolysis," a term used by some German colloid chemists in the 1930s and 1940s. Literature cited by Tamura and Morita demonstrate a heat-induced aggregation of myoglobin and cytochrome c. The earlier suggestion that the regeneration was somehow or other tied up with cytochrome c (Purr 1950) has not been substantiated. Joffe and Balls (1962) stress that contrary to *in situ* (see below), completely inactivated purified peroxidase did not regenerate. They also observed a lag period which they attributed to a finite relaxation time which is considered to be critical to the partial renaturation of the apoenzyme. As mentioned, this is necessary for recombination with hematin. Wilder (1962) separated the isoenzymes of peroxidase before regeneration. The role of isoenzymes in the regeneration phenomenon was further explored by Wang and DiMarco (1972). Using UV spectrometry and sedimentation analysis, they concluded that the above-mentioned reversal of heat-induced aggregation may contribute only partially to the regeneration of activity, much of the latter being due to conformation changes (refolding). Lu and Whitaker (1974) added the peroxidase prosthetic group, hematin, at various stages of their inactivation-regeneration systems. They reported that the regeneration was complex, following neither first nor second order kinetics. Preliminary plotting of their data suggests that they follow third order kinetics, previously found by Schwimmer (1943), Fig. 11.5. Their data convincingly confirm earlier intimations that some sort of conformation change occurs in the protein before dissociation of the prosthetic group takes place upon heating. As previously mentioned, the rather low energy of activation of denaturation (32 kcal/mole) that they found reflects conformational changes in the region of the prosthetic group, allowing the latter to be released from the apoenzyme with comparative ease.

An eclectic synthesis of the previous ideas on the mechanism of peroxidase activity regeneration is

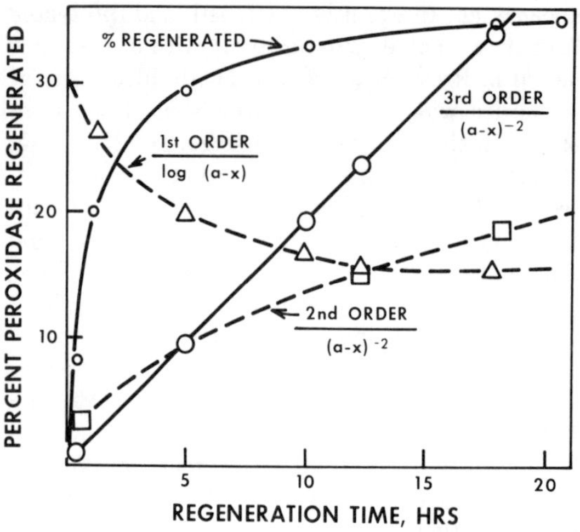

From Schwimmer (1943)

FIG. 11.5. PEROXIDASE REGENERATION FOLLOWS THIRD ORDER KINETICS

afforded by the studies and conclusions of Tamura and Morita (1975) as shown in Fig. 11.6. Further refining the measurement of activity and by means of applying rapid scan visible and fluorospectrophotometry and circular dichroism, they conclude that, as with myoglobin and cytochrome c, the first step in heat denaturation consists of a conformational change in the vicinity of the hematin. This in turn induces changes in the ligand groups coordinated with the heme. The next step, reversible denaturation of the protein moiety, involves changes in the tertiary structure which simultaneously accompany dissociation of hematin. They point out that no aggregation or precipitation occurred (with the particular isoenzyme of the radish peroxidase used) because of the presence of carbohydrates in the enzyme molecule. Indeed they found that the apoenzyme was more stable than the holoenzyme at pH 7. In alkaline solution, regeneration fails because of alteration of the prosthetic group, whereas in acid media no regeneration takes place because irreversible changes in the protein occur, largely the breakdown of β-pleated sheets (secondary structure). The well-recognized heat stability of plant peroxidases and increased stability of the apoenzyme are attributed to the stabilizing influence of the indicated four disulfide bridges. This fine paper also serves as a gateway to literature on both peroxidase and on modern concepts of protein denaturation. An excellent exposition of the sequences of folding and unfolding of proteins and on the detection of intermediates is that of Baldwin (1975).

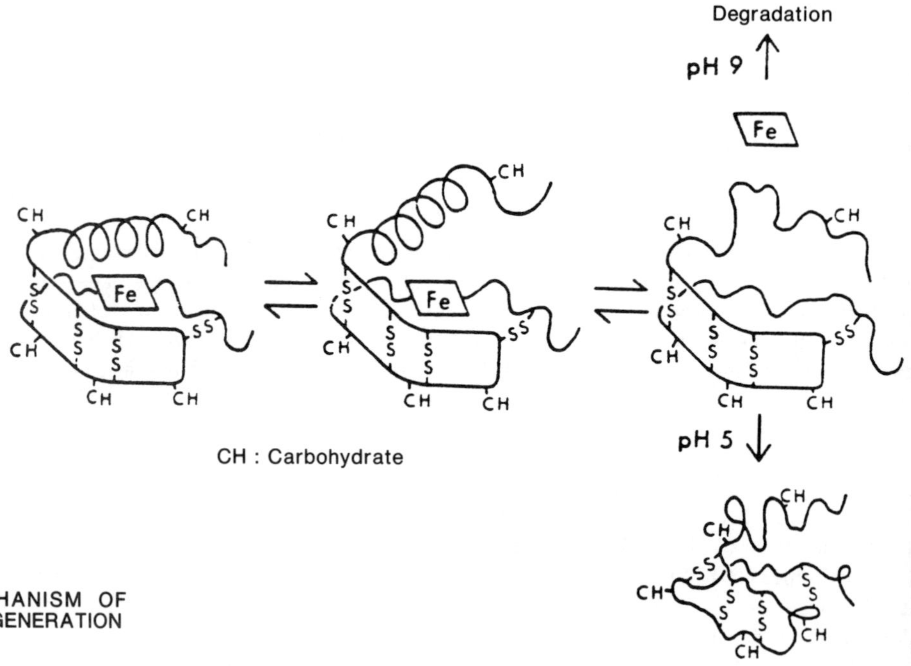

FIG. 11.6. MECHANISM OF PEROXIDASE REGENERATION

From Tamura and Morita (1975)

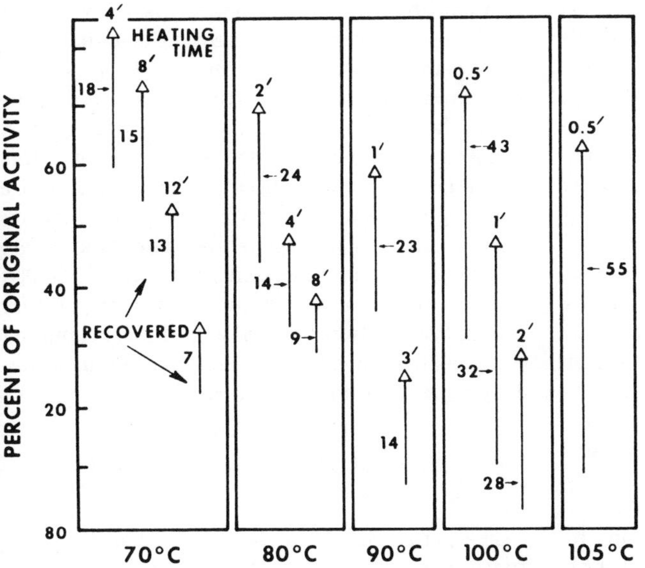

From Schwimmer (1944A)

FIG. 11.7. PEROXIDASE REGENERATION IN HEATED TURNIPS

Regeneration in Blanched Foods

Although regeneration of peroxidase in extracts of blanched vegetables was observed quite early in the food freezing, systematic studies on *in situ* regeneration, allowing the activity to recover before making extracts, goes back some 30 years (Herrlinger and Kiermeier 1948). In the freezing of foods, regeneration was more apparent and considered to be more of a problem with steam- than with water-blanched vegetables. Interest in regeneration in the 1950s extended to the canning industry upon the adoption of HTST processes. As pointed out by Schwimmer (1944A,1969A,1972), Farkas *et al.* (1956) and others, regeneration frequently is highly dependent upon the rate of heating rather than upon the actual temperature and time. Figure 11.7 shows some data on inactivation and regeneration of peroxidase activity in turnip under varying schedules of time and temperature. Using this basic data, Fig. 11.8 shows the regeneration as a function of the rate of heating. HTST characterized by extremely high heating rates is usually associated with regeneration.

Is Regeneration Really Undesirable?—The problem posed by regeneration, as evidenced from the literature and informal communications, may be largely artifactual, stemming, as mentioned, from the circumstances that most of the specifica-

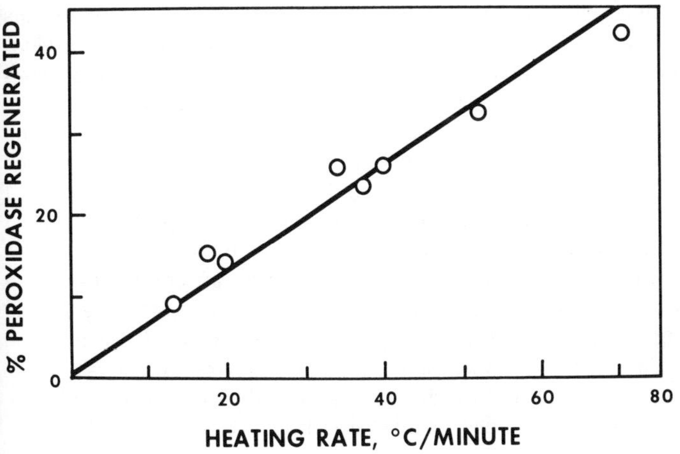

From Schwimmer (1972)

FIG. 11.8. PEROXIDASE REGENERATION AS FUNCTION OF HEATING RATE tions for quality control stipulate practically complete absence of peroxidase activity. Thus, according to Dietrich (1975), those charged with quality control responsibilities in the frozen food industry, in using the official U.S. Department of Agriculture methods, are concerned with a change to a positive reading after a few days to weeks in frozen storage for an adequate blanch from "adequate" (3.5 min or more) *directly* after blanching to "inadequate" blanch (1−1.5 min). Off-hand this change would appear to be due to regeneration, but could just as well be due to nonuniform blanch in large lots, or to an artifact such as regeneration during the test itself.

The possibility that some regeneration may not be harmful is corroborated by the previously mentioned demonstration that the maximum quality corresponds to residual but not unmeasurable activity in many vegetables. As a rule, according to Joslyn and Neumann (1963) and Joslyn (1966), regeneration only occurs at storage temperatures higher than 10°C and in previously frozen vegetables thawed at temperatures and times too short for complete inactivation. Furthermore where it does occur, it does not appear to be accompanied by off-flavor impairment in organoleptic quality (Wagenknecht and Lee 1958). The question as to whether peroxidase itself participated in the generation of these off-flavors will be discussed in Chapter 24. Indeed, as mentioned in Chapter 1, samples in which regeneration has taken place appear to be superior to those in which it has not (Winter 1976).

At the present time, the reasons as to why such regenerated samples are better is speculative. Perhaps the activity of "flavorese" enzymes is preferably regenerated along with peroxidase. Another possibility is the opportunity for regenerated or surviving proteinases to act on heat-denatured proteins now rendered susceptible to such action. This could change flavor and liberate "protective" peptides and amino acids.

In canning, the effect of regeneration of peroxidase on quality is also far from clear (Adams and Yawger 1961). Nebesky *et al.* (1950) found an association between regenerated residual activity and decrease in quality. This may reflect the circumstance that canned foods are stored at much higher shelf temperature than are frozen foods before they are consumed, thus changing the regeneration pattern as well as resulting in more enzyme action. It is obvious that further investigations are needed if further reliance on blanching is to diminish. For instance, nothing is known about the fate or influence of superoxide dismutase vis-à-vis peroxidase on quality during the processing and storage of fruits and vegetables. It may be desirable to keep the activity of this enzyme at a maximum. It would also be of interest to determine if animal peroxidase, whose prosthetic groups are more tightly bound than those of plant enzymes (Yamasaki 1974), is regenerated and what effect this would have on processed meat quality. Finally, the fact that foods are subject to temperature fluctuations between processing and consumption could, as mentioned in the next chapter, influence enzyme regeneration patterns.

Regeneration of Other Food Quality Indicator Enzymes

Like peroxidase, catalase can regenerate its activity after heat inactivation. Over 40 years ago Diehl *et al.* (1933) reported that catalase, inactivated for the moment, may be active in blanched peas after cooling and standing for 30 min. Sapers and Nickerson (1962) investigated the reappearance in spinach of catalase in some detail. As expected, the regeneration is affected by pH and regeneration temperature as well as by heating conditions. As much as 30% of the original activity could be recovered at pH 7 within 2 hr at 30°C. However, in vivid contrast to peroxidase, the regeneration is transitory and no activity is detectable within 24 hr.

After pasteurization of milk, alkaline phosphatase activity can be regenerated. Maximum reactivation occurs in products previously heated at 104.4°C, incubated during regeneration at 34°C and adjusted to pH 6.5 (Murthy *et al.* 1976). Of course regeneration of the enzyme has nothing to do causatively with the presence of microorganisms that pasteurization removes. However, a product made from milk in which regeneration had taken place would be unjustly considered suspect. However, thanks to change in certain properties of the regenerated enzymes, it is relatively simple, if not always certain, to distinguish between original and reactivated (as the investigators prefer to call it) phosphatase. In the AOAC method (1975) results of comparing the activity of an undiluted sample *not* containing Mg (A) is compared with the sample containing Mg diluted five-fold (B). If the activity of B is greater than A, then the sample is considered to contain reactivated phosphatase, and thus is adequately pasteurized; if less, the sample is con-

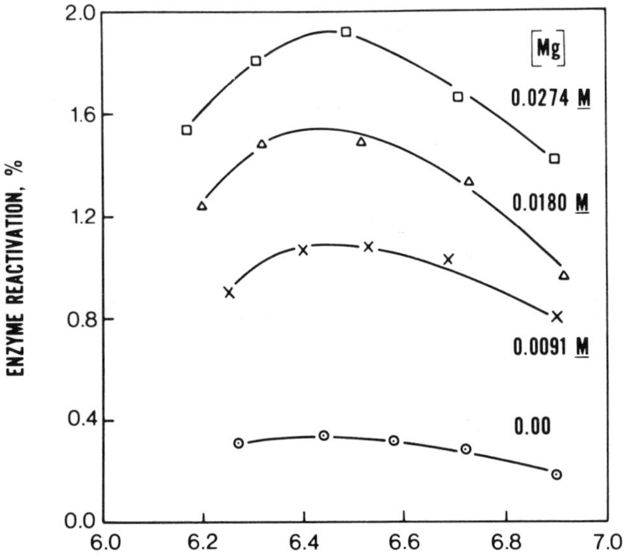

FIG. 11.9. REGENERATION OF ALKALINE PHOSPHATASE IN MILK

Shown are the effects of pH and Mg^{2+} concentration which form the basis for ascertaining if phosphatase in dairy products is due to not pasteurizing or to activity regeneration.

From Murthy et al. (1976)

sidered to contain residual phosphatase (inadequate pasteurization). Apparently there is some conformational change in at least one of the reactivated phosphatase isoenzymes which makes it more activatable by Mg than the original. In addition to detecting changes in such isoenzymes, Murthy *et al.* (1976) also provide a critique of the presently not infallible method of detecting reactivation, a thorough study of the factors affecting the degree of activation, and a key to the pertinent literature. A typical experiment is shown in Fig. 11.9. Note how the percentage of enzyme regenerated increases with increasing Mg level. Improvements and speedup of the AOAC test continues (Kleyn and Ho 1976).

Purified banana phenolase shows considerable regeneration of activity at 0°C after heat inactivation (Galeazzi *et al.* 1976).

12

ACTION AND CONTROL OF ENZYMES AT LOW TEMPERATURES

As in the case of high temperature in food processing and preservation, exposing foods to low temperatures involves a pluralism of aims, purposes and action. Thus, the primary reason for exposing foods to low temperatures, especially frozen foods, is to prevent microbial growth while maintaining the food as close to the fresh as possible in form, appearance and quality over extended storage periods. Somewhat higher temperatures, but still considerably less than ambience, are also used in prolonging the storage life of fresh produce as we shall briefly elaborate on later in this and other chapters. However, the main thrust of the ensuing discussion is geared to enzyme action in processed foods. As in the case of microbial growth and, for that matter, most chemical reactions, enzyme action by and large decreases with decreasing temperatures. The decrease in reaction rate, to a first approximation, follows the Arrhenius relation (Chapters 3 and 10).

However, low temperatures encountered by frozen foods do not prevent some of the deleterious consequences of enzyme action persuant to long-term storage, as Joslyn and Sherril (1933) pointed out almost 50 years ago. Although great strides have been made in the retention of high quality frozen foods, there still remain some less-than-gourmet-quality products consumed by the public. We address ourselves in this chapter to a consideration of enzyme-related factors which may lead to this deterioration in quality (or in some instances improvement in quality of long-term stored processed foods). We divide the ensuing discussion into the exploration and explanation of phenomena relating to factors which tend to reduce or enhance enzyme action in foods (and model systems) when the food encounters temperatures below ambience. Earlier reviews on the effect of cold temperatures on food enzymes include those of Balls and Lineweaver (1938), Joslyn (1949, 1966), Chilson *et al.* (1965) and Tappel (1966A). Especially valuable are more recent discussions by Fennema (1973,1975A,B,1976). Partmann (1975) also touches on some enzymatic aspects. Books which deal in part with the same general subject relating water, temperature and biological phenomena include those of Duckworth (1975), Meryman (1966), Fennema (1976) and Alexandrov (1977). For the practical side of freezing preservation of foods see Tressler *et al.* (1968) and the textbook by Desrosier and Tressler (1977). Other relevant reviews include those of Aylward and Haisman (1969), McWeeny (1968) and Singh and Wang (1977). The journal *Cryobiology* is the principal repository for current research on low temperature effects in biological systems.

FACTORS TENDING TO MINIMIZE ENZYME ACTION

Most enzymes obey the Arrhenius relation over a range of temperatures ranging somewhere between optimum and some lower temperature usually above 0°C. However, below this temperature supplemental suppression of activity of most enzymes beyond that accounted for by extrapolation of the Arrhenius equation is frequently observed. Experimentally we may distinguish among the following:

(1) Moderate deviation to a new straight-line Arrhenius relation but with reduced slope.
(2) Complete but usually reversible inactivation.
(3) Further drop of activity below that predicted by extrapolations of the Arrhenius relation in (1) achieved by freezing.
(4) Further loss of activity upon thawing.

The above observations have been established on the basis of both model enzyme-substrate solvent systems and in *in situ* or *in vivo* biological systems including of course many which happen to be foods.

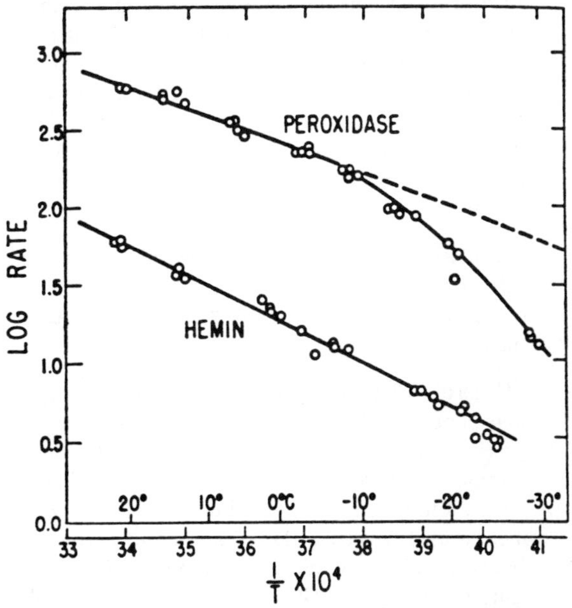

From Tappel (1966A). With permission from Cryobiology, H. Y. Meryman, Editor, Copyright by Academic Press, London.

FIG. 12.1. ARRHENIUS PLOTS OF OXIDATION OF GUAIACOL BY HEME AND PEROXIDASE

Deviation from the Arrhenius Relation

Reviews, discussions and key experimental results include papers by Lumry (1959), Tappel (1966A), Talsky (1971) and Douzou (1973,1975). As mentioned, almost all enzymes exhibit this temperature anomaly. In general the Arrhenius plots fall into two categories: A continuous convex curve as exemplified by acetylcholine esterase, or two intersecting lines (or at least they are so interpreted). Enzymes which have been reported to exhibit this behavior include trypsin, invertase, lipase and peroxidase. Lumry states that the entropy changes are the greatest in which the break is the sharpest. Figure 12.1 shows the "anomaly" for horseradish peroxidase. In this particular case the discontinuity break occurs below 0°C. For most other enzymes, especially the proteinases, the break occurs in the region of 10°–20°C. Figure 12.1 also shows that the anomaly is due to the protein and not to the prosthetic group, hemin. On closer examination, using much smaller temperature intervals than hitherto attempted, Talsky (1971) claimed to have detected a temperature anomaly over the entire range in which the enzyme was active up to the optimum. The anomaly was characterized by sinusoidal descent of the Arrhenius curve (log rate *vs* 1/T) with increasing values of 1/T. Previous to this finding a controversy existed as to whether the breaks were real or whether all such curves belonged to the first category (Kistiakowsky and Lumry 1949).

Explanations.—A critical analysis of these anomalies is beyond the scope of our discussion. Some of the hypotheses which have been put forward to explain these deviations are exhibited below. Most of the data supporting these conclusions were obtained in model systems in solution, frequently at temperatures lower than 0°C. If "super" hydrogen bonding does play a role, then the enzymes should also be supplementarily protected from proteolytic action, according to the Alexandrov hypothesis (Chapter 10).

Urease.—Shift of equilibrium between enzyme and a reversible inhibitor (Kistiakowsky and Lumry 1949).

Lipase.—Increased rigidity of conformation decreases accessibility to and correct orientation of substrate with respect to active site (Belehradek 1954).

Phosphatase, Peroxidase.—Increase in intramolecular hydrogen bonding (Maier *et al.* 1955).

Trypsin and Papain.—Change of rate-determining step from acetylation at high temperatures to deacetylation at low temperatures (Wilson and Cabib 1956; Angelides and Fink 1978). The latter investigators consider the conformational shifts of active site imidazole as the key feature of the mechanism of action of papain at low temperatures.

Catalase.—Change in rate determining step from diffusion-controlled formation of ES to formation of Compound I (Chapter 11) (Strother and Ackerman 1961).

General.—Change in conformation of the enzyme and hence of ES in vicinity of active site; change proceeds with reversible phase transitional conformation involving very few amino acids (Talski 1971). Superhydrogen bonding between enzyme molecules, between water clusters and active site, between substrate and H_2O; change in ionization, shift in pH optima (Kavanau 1950; Hultin 1955; Tappel 1966A; Fennema 1973; Douzou 1975).

Myosin ATPase.—Rate at low temperature associated with temperature insensitive isomerization of complex between product ADP and subfragment of myosin; ADP dissociation is markedly temperature-dependent (Bagshaw and Trentham 1974); change in the rate-limiting step of the hydrolytic cycle due to change in essential SH-dependent change of conformation of certain intermediates (Watterson *et al.* 1975).

Starch Synthase.—Anomaly (discontinuity) observed with enzyme from chill sensitive (corn, sweet potato) but not in chill resistant plants (potato) due to phase change in lipids which melt at 12°C (Chapters 1 and 19) (Downton and Hawker 1975).

Chymotrypsin.—This and many other enzymes have maximum structural folding stability at 10°–15°C; lower temperature gives rise to more unstable structures which cause changes in active site (Baldwin 1975).

Enzymes Acting on Insoluble Substrates.—Berge *et al.* (1980) demonstrated that such discontinuities for enzymes, such as long chain (Palmitoyl) CoA hydrolase, reflect decrease of substrate available to the enzyme. This decrease is due to the formation of mixed micelles of substrate molecules.

Palmitoyl or Long Chain Acyl CoA Hydrolase.—Discontinuity is due to formation of mixed micelles, leading to a decrease in availability of substrate (Berge *et al.* 1980).

Reversible Inactivation

In addition to enzymes whose activities drop faster than expected as the temperature is lowered, several enzymes apparently *completely* lose their activities, at least as measured by removing them from the cold and assaying immediately. Such enzymes are shown in Table 12.1. Apparently in most cases the reason for this inactivation can be traced to the dissociation of subunits and to some extent to a change in conformation that makes the active site unavailable to the substrate. It is of interest that these enzymes, besides having quaternary structure, are all allosteric (Chapter 4) and the dissociation has to be preceded by a conformational change brought about by the cold. Frequently present are substances such as sucrose and glycerol which prevent inactivation, initiate reactivation and/or stabilize the enzyme.

Decrease in Enzyme Action During Freezing and Thawing and While Frozen

In this section we examine those reports which

TABLE 12.1

REVERSIBLE COLD INACTIVATION OF ENZYMES

Enzyme	Mechanism, Remarks
Arginosuccinase[1], pyruvate carboxylase[2]	Dissociation of subunits preceded by conformational changes of oligomer.
*t*RNA synthetase[3,4]	One of several functions of the enzyme is inactivated; active and inactive forms are separable.
Glycogen phosphorylase[5]	Cold induced conformational change of enzyme structure; no loss of enzyme-bound pyridoxal phosphate.
Glutamate dehydrogenase[6]	Enzyme from mutant of *Neurospora crassa*, a fungus; has little activity at 20°, but becomes active after a few minutes at 35°.

[1] Source: Havir *et al.* (1965).
[2] Source: Scrutton and Utter (1965).
[3] Source: Papas and Mehler (1968).
[4] Source: Lee and Muench (1969).
[5] Source: Graves *et al.* (1965).
[6] Source: Fincham (1957).

indicate further loss of activity during freezing and thawing, and in the frozen state. We present what are considered to be the underlying reasons for these losses.

Enzyme Action in the Frozen State.—During freezing the removal of water results in increasingly higher concentration of solutes in the remaining liquid. At the usual temperature which frozen foods experience, there may always be enough liquid water around for the enzymes to continue acting. Hence the terms "solid" or "frozen" refer to the appearance of the whole system and not necessarily to the state in which the enzyme finds itself. However, the action of most enzymes is considerably less than that predicted by extrapolation of the Arrhenius plot, even taking into account the deviation from the above-mentioned anomaly. Examples of the lowered rate in this quasifrozen state as compared to that in the liquid state at the same temperature are shown in Table 12.2. Lund *et al.* (1969) showed that for an invertase-sucrose-buffer-water system, rates of sucrose hydrolysis in frozen reaction mixtures at given temperatures are the same as that of an unfrozen but more concentrated system in which all the components are present in the same proportion. Thus, the rates in five- and 15-fold concentrated systems were found to be the same as the rate of the *unconcentrated* system in the frozen state at −3.7° and −16°C, respectively. Hiranpradit and Lopez (1976), from measuring the rate of maltose liberation from starch by α- and β-amylases at temperatures as low as −23°C, concluded that enzyme concentration is the limiting factor. Our calculations indicate that maltose accumulated at −13°C at about 1/40th the rate of that at 4°C, about 25% of the Arrhenius extrapolated value.

TABLE 12.2

EFFECT OF FREEZING ON ENZYME REACTION RATES[1]

Enzyme	Temperature (°C)	% of Activity, Frozen, to That at Same Temperature, Unfrozen
Chymotrypsin[1]	−5	5
Cathepsin[1]	−2.8	33
Lipase[1]	−18.0	16−20
Lipoxygenase[2]	−5	1
Peroxidase[3]	−5.6	48

[1] Source: Lineweaver (1939).
[2] Source: Tappel *et al.* (1953).
[3] Source: Tappel (1966A).

Activity in the frozen state is also affected by the rate of freezing and by the nadir (lowest temperature reached). Both acceleration and suppression of reaction rates have been observed as functions of these variables (Behnke *et al.* 1973; Fennema 1975A). Thus, minimal invertase action was achieved by either freezing slowly to an extremely low temperature (−60°C) or by rapid freezing to a nadir no higher than −40°C.

Several investigators have observed a sharp break in Arrhenius curves in going from the liquid to the solid state. Enzymes tested include lipase and several proteinases (Lineweaver 1939; Sizer 1942). Lund *et al.* (1969) found that invertase still gave a straight-line Arrhenius plot in the frozen state with the rather high (for enzyme) energy of activation of reaction rates (not denaturation) of 39,000 cal/mole as compared with a value of 10,000 for unfrozen invertase in the vicinity of 0°C. This finding is in contrast to the HCl-catalyzed hydrolysis of sucrose which was marked by a flattening out of the Arrhenius curve incident upon freezing.

Frozen Foods.—The rate of enzyme action, important as a determinant of frozen food quality, may decrease with decreasing temperature (but not always) in frozen foods. Thus, lipase action in frozen peas, as measured by the rate of accumulation of free fatty acids decreased five-fold in going from −5° to −20°C and ceased completely at −26°C (Bengtsson and Bosund 1966). The first-order rate constant for phospholipase in frozen fish (Chapter 28), again as measured by accumulation of free fatty acid, dropped 80% in going from −7° to −29°C (Olley *et al.* 1969). In contrast to these model systems, Hirandpradit and Lopez (1976) noted that the α- and β-amylases of frozen sweet potato purees were inactive (but not inactivated) at temperatures where their model systems were producing maltose.

Another phenomenon which may tend to contribute to apparent suppression of activity in foods is the frequent observation made in the above-cited investigations that enzyme action in the frozen state for both model systems and foods almost invariably stops short of completion, although the enzymes are not inactivated.

Some Explanations.—While the above-cited experimental evidence incontrovertibly demonstrates the imposition of additional barriers to enzyme action when a food is frozen, the molecular reality behind this data is far from being elucidated. Presumably the reasons for the liquid-state anomalies (Table 12.1) are still valid in the frozen state, but these are not enough to explain the further suppression of activity. This is true for model systems. In foods the situation is enormously more complex.

Older theories and discussions invoked ice crystals as barriers to interaction of enzyme and substrate and also to the removal of the inhibitory products of the enzyme reaction. The influence of increased viscosity which should also create diffusion barriers is still a much-debated question. It is generally conceded that it must make a significant contribution to the failure of enzyme reactions in the frozen state to go to completion. Undoubtedly, reversal of enzyme action plays an equally important role. Most recent discussions hold that the important event is the removal of water during freezing with its attendant concentration of system components, resulting in a poised alteration of the microenvironment. Importantly, there is increasing awareness of the relation of the effects observed to those observed in dehydration (Chapter 13), and that the effects may indeed be due to dehydration.

Tappel (1966A) stresses the point that many of the common solutes such as salt buffers, sugars and other carbohydrates are effective inhibitors in higher concentrations (Chapter 14). Increased solute concentration would also bring into play excess substrate and product inhibition.

Changes in buffer concentration and composition (eutectic phenomena) can shift pH as well as the pH optimum of the enzyme. Fennema (1975A) cites evidence that pH increases in high protein foods such as chicken and fish and decreases in low protein foods with a relatively low starting pH such as in fruits. Somewhat more fundamental, the "effective" acidity may also be increased by the increased mobility of the hydrogen ion (proton) in ice as compared to water (Eigen and de Maeyer 1958).

Dissociation and association of protein subunits may also be involved in lowering the activity without complete inactivation as previously mentioned. Fennema has invoked the persistence or slow reassociation of such action to explain the dependence of the suppression of enzyme activity on both the nadir and on the rate of freezing.

In complex food systems containing inhibitors, the inhibitory effect will, from simple Michaelis kinetics considerations, rise faster than will the accelerating effect due to increased substrate concentration (Chapter 14) for the opposite effect of dilution. This type of effect could account for the difference reported among pure enzymes, impure crude preparations, extracts and enzyme in foods. Which of these factors contribute most significantly to deterioration of quality in frozen foods and how to examine their effects in this regard await investigation.

Irreversible Inactivation

In Model Systems.—An additional factor leading to the further suppression of the overt manifestation of enzyme action is abundant evidence that some enzymes under some conditions are simply irreversibly inactivated. Experimentally, such evidence is obtained by assaying for enzyme activity under standard (and presumably optimal) conditions after a rapid thawing. Although this approach leads to some inaccuracy and quantitatively valid data are scarce, some definitive conclusions have been reported mostly in model systems but also in foods and related biological systems.

In general, it appears that most enzymes become inactivated during freezing *and* thawing. However, this does not mean that once frozen, further progressive inactivation does not occur. We remarked on the instability of catalase in the last chapter. Kiermeier (1949), who first noted this, also observed that slow freezing was more conducive to catalase inactivation than was fast freezing. More than 40 years ago Balls and Lineweaver (1938) observed inactivation of the then new crystalline enzymes, trypsin, chymotrypsin and carboxypeptidase, after exposure of a solution of the enzymes to −180°C for 44 hr.

Using low concentrations of several oligomeric dehydrogenases, Chilson *et al.* (1965) found that slow freezing and thawing caused more loss of activity and rearrangement of the enzyme subunits than did fast freezing. At high enzyme levels this effect became less pronounced but the enzyme became more unstable in the absence of coenzyme. Complete recovery of enzyme activity after freezing and thawing could not be achieved, irreversible inactivation ranging from 10 to 85%.

In a study by Darbyshire (1975), comparing destruction of freezing- and drying-induced inactivation of peroxidase in model systems, 38% of the activity in water and 53% in phosphate buffer were recovered after freezing. Lability increased with decreasing enzyme concentration. Working with a myosin B (ATPase) model system, Hanafusa (1972) observed that enzyme inactivation was associated not only with slow freezing, but also with low nadirs. However, according to literature cited by Fennema (1975A), once a system is frozen, high subfreezing temperatures (−2° to −10°C) are more frequently conducive to enzyme inactivation than are lower temperatures.

In Foods and Other *in Situ* Systems.—*Catalase.*—We alluded to the observation that catalase tends to disappear in frozen food packs. The following documents this conclusion. About 50 years ago Carrick (1929) found that that loss of catalase activity in apple tissue seemed to be dependent upon whether the cells remained viable, but that the extent of inactivation was variety-dependent. In contrast to most results obtained in *model* systems, Kaloyereas (1947) reported greater destruction of catalase in spinach frozen at −65° than at −17°C. Catalase added as a crystalline preparation to pea puree lost some but not all of its activity after storage (Wagenknecht and Lee 1958). The findings of Sapers and Nickerson (1962) were cited in the last chapter.

Other Enzymes.—Added lipoxygenase also lost activity when added to pea puree in the above-cited experiments. Other evidence for irreversible enzyme inactivation in frozen tissue comes from medically-oriented studies whose mission is preservation of viable human tissue in the frozen state. This evidence includes the observation of Lehmann (1965) that 50% of the aldolase disappeared from blood stored for 3 years at −196°C, and that of Fishbein and Stowell (1969) of substantial reduction in the activity of cytochrome reductase of frozen mouse liver. Johnson and Daniels (1974) concluded from a study of human skin enzymes that although some enzyme inactivation does occur during freezing, there is little evidence that denaturation occurs while the cells are still alive.

Explanations.—Inactivation and irreversible inhibitions of enzymes during freezing and in the frozen state can be traced back to the same primary reasons for the reversible inhibition or suppression of enzyme activity, viz., concentration of solutes and the attendant radical changes in the microenvironment of the enzyme. Thus, pH depression can denature and inactivate an enzyme as well as slow it down. The pH of phosphate buffer can change from 7 to 3.5 upon freezing, to irreversibly inactivate many enzymes. Denaturation and precipitation as results of such changes were observed about 75 years ago by Gorke (1906).

Changes in ionic strength as well as pH can result, as shown by Chilson *et al.* (1965), in dissociation of quaternary proteins (i.e., lactic dehydrogenases) into subunits which reassociate very slowly during thawing. Reassociation was hindered by the denaturing effect of increased concentration of halides, to which the subunits are more denaturation-sensitive than are the oligomers.

Another factor contributing to inactivation in cells is the perturbation of sulfhydryl groups essential to the activity of many enzymes. This perturbation may be brought about by increased opportunity for SS-SH interchange afforded by the increased concentration of disulfide groups already present, conformational changes, and accelerated oxidation of sulfhydryls because of increased oxygen concentration in ice. Thus, Fennema (1975A) states that the oxygen concentration in a "partially" frozen system at −3°C is 1150 × that in a solution at 0°C. Injury to enzymes through oxygen could also come about from the increase in concentration of active oxygen and free radicals which would be stabilized by and at low temperatures and react with the protein (Chapter 16). We again emphasize that it is important to assume, and there is good reason to believe, that even in the most deeply frozen foods microdomains of liquid phase persist. Incidentally, this sort of explanation has been evoked for frost injury in plants (Levitt 1962). Finally, as mentioned before, extreme desiccation has been advanced to explain irreversible inactivation as will be expanded upon in the next chapter. A minimum amount of structured water is necessary to maintain integrity of protein structure.

PERSISTENCE OF ENZYME ACTION AT LOW TEMPERATURES

Although there undoubtedly are situations where, as we have just shown, enzyme action can be depressed quite below that predicted from extrapolation of the temperature-activity relation at ambience, the fact is that enzyme action stubbornly persists at the lowest temperatures to which enzymes in foods are likely to be subjected. Perhaps this action is more relevant to food quality than lack of action because we know after all that this action must be intense and its consequences for food quality severe enough to require some means such as blanching to destroy the offending enzymes. In the following discussion we attempt to pinpoint not only why enzyme action persists, but also why the action of the enzymes (even when not intrinsically enhanced) at these low temperatures can result in the exacerbation of deterioration of quality attributes, especially when the food is restored to ambient temperature or cooked.

In general the following may account largely for many enzyme-related problems of food quality which arise during low temperature storage:

(1) Concentration of solutes attendant upon removal of water by freezing.
(2) Relative predominance of enzyme action compared to other biological, physical and chemical changes because of the relatively low energy of activation of enzyme reactions.
(3) Fluctuation of temperatures during storage and distribution.
(4) Cryptic accumulation of intermediates, side products and secondary products.
(5) Cellular disruption and attendant decompartmentalization of enzymes with its resultant formation of nonphysiological food components.
(6) Shift in physiological function upon lowering of temperature, i.e., as response to stress in living tissue.
(7) Presence of cryoprotective agents in biological tissues.
(8) Regeneration of enzyme activity, dealt with at length in the previous chapter.

Effects Related to Solute Concentration

Removal of water, which may result in suppression of enzyme activity, can also result in increased enzyme activity under the proper circumstances. In model systems, the now classical example is the two- to three-fold freezing-induced accelerated rate of hydroxylaminolysis (transfer or exchange between amino acid esters and hydroxylamine) catalyzed by trypsin (Grant and Alburn 1967). Previously, Kiermeier (1949) reported a 12-fold increase in peroxidase activity in kohlrabi *during* freezing. Tappel observed that during slow freezing the initial rate of the oxidation of guaiacol by peroxidase in a model system was doubled whereas during rapid freezing, suppression of activity occurred. Tong and Pincock (1966) made similar observations with an invertase model system. According to Fennema such behavior occurs only with very dilute enzyme solutions.

Once the solution has been frozen, many of the same circumstances and factors which can lead to inhibition and inactivation may also lead to activation. These include: increase in both enzyme and substrate concentrations, pH shifts favorable to enzyme action, increase in ionic strength favorable to some enzymes, and increase in salts and other solutes which activate and stabilize certain enzymes. Even the hazards of extreme desiccation may be mitigated or countered by simultaneous concentration of proteins and other substances which afford cryoprotection through their ability to mobilize water (Darbyshire 1975). Furthermore, the increased solute concentration lowers the melting point and thus maintains a liquid microenvironment with attendant increased opportunity for enzyme action. Thus, Scrutton and Utter (1965) describe protection by added sucrose, coenzyme and cosubstrate against the reversible cold inactivation of the pyruvate carboxylase complex (Chapter 4). Certainly, if freezing can indeed be considered as a special case of dehydration, then the removal of water should stabilize the enzymes, at least before extreme desiccation occurs. While increased oxygen in the reaction space in frozen systems could contribute to the inactivation of sulfhydryl enzymes, it could also lead to increased activity of oxygenases. With respect to food quality the action of the oxygenases further exacerbates deterioration *via* the activation of oxygen, as we shall have ample opportunity to show.

In the last chapter we drew attention to the remarkable change in the activability by Mg^{2+} of milk alkaline phosphatase upon regeneration, indicating that temperature extremes could permanently change the conformation of an enzyme without inactivating it. Promotion of enzyme action after thawing could be elicited by rendering the enzyme insensitive to certain inhibitors as suggested by Kiermeyer (1949) to account for the eight-fold transitory increase of catalase activity in frozen vegetables which he observed. A paradigm for such "activation" is afforded by the observation of Westick *et al.* (1974) that the enzyme adenine phosphoribosyltransferase is still active but no longer inhibited by certain gold salts (sodium aurothiomalate) after freezing and thawing. The effect appears to be due to a change in availability of enzyme sulfhydryl groups needed for inhibition. Again the cold induced a permanent conformational change.

Predominance of Enzyme-catalyzed Reactions

A consequence of the generalization that the energy of activation of enzyme-catalyzed reactions is lower than that of nonenzymatic reactions is that, upon extrapolation of the Arrhenius equation to low temperatures, enzyme reaction will predominate although both types of reaction may have contributed equally to overall change at ambience as shown in Fig. 12.2 (Lineweaver 1939). For instance, at -15°C the effective ratio of the activity of trypsin (E_a = 11,000 cal/mole) to that of catalase (E_a = 6400 cal/mole) will be one-sixth of that at 30°C. Thus, the action of the breakdown of proteins would tend to be less important over the long run than that of catalase and perhaps of other oxidoreductases. The values for energies of activation for microbial growth run in the range of 20–50 kcal/mole. This is probably the basis for the empirically established knowledge that enzyme action in inadequately blanched frozen foods is more of a problem than is microbial action.

Nonenzymatic Reactions.—While such considerations probably have some relevance to the action of enzymes in foods, we have already seen that even in model unfrozen systems a departure from Arrhenius relation occurs and that this departure is deepened upon freezing. From the observations summarized by McWeeny (1968), Fennema (1975A,B) and others, it would appear that there are factors that tend to make a nonenzymatic reaction go faster than one would at first glance expect them to in the frozen state. Indeed some have negative temperature coefficients. Nonfood reactions which have exhibited such reversals include: (1) spontaneous hydrolysis of many lactones and anhydrides, (2) base-catalyzed hydrolysis, (3) imidazole-catalyzed reactions, (4) double decomposition of ethylene chlorohydrin, (5) reduction of potassium ferricyanide, (6) UV-catalyzed purine dimerization, and (7) decomposition of $FeCl_3$. Of particular relevance to foods is the acid or H^+-hydrolysis of sucrose, the conversion of nitrites to nitrosamines and oxidation of ascorbic acid.

While Fennema (1975B) concludes that the acceleration of nonenzymatic reactions in model nonbiological systems can be satisfactorily explained by concentration effects, earlier investigators such as McWeeny invoke supplementary factors such as increased acidity brought about by the abovementioned increased mobility of protons, catalysis by ice surfaces and changes in dielectric constant of the medium which can promote nucleophilic reactions. Certain reactions listed by McWeeny as nonenzymatic, in our opinion, may have some enzymatic component or may be influenced by enzymes indirectly.

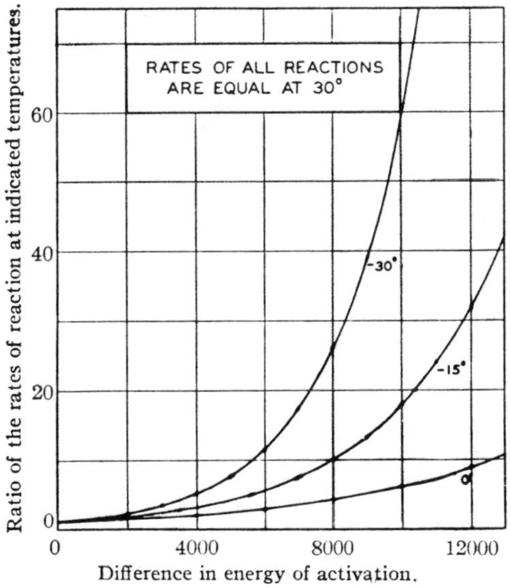

From Lineweaver (1939)

FIG. 12.2. RATIO OF REACTION RATES DEPENDS ON DIFFERENCE IN ENERGY OF ACTIVATION

Cryptic Accumulation.—As pointed out in Chapter 1 many so-called enzyme reactions causing change in food quality arise from secondary nonenzymatic reactions. If, as seems likely, these secondary reactions have higher energies of activation than do the primary catalyzed reaction, then the intermediates formed in these secondary reactions will tend to accumulate. If they do, we conceive of two possibilities: First, an observed reaction will not manifest itself until the food is brought back to room temperature, somewhat of a cryptic accumulation (see below). Second, these intermediates will react through pathways other than those which prevail so that overall reaction as measured by the rate of appearance of the final product will appear to have a negative temperature coefficient. Also the apparent specificity of the enzyme will appear to have been altered.

The first type of effect is illustrated by the clotting of milk "catalyzed" by the rennet enzymes (Fig. 12.3). What the rennets really catalyze is the hydrolysis of a particular peptide bond in casein (Chapter 33). This is followed by aggregation of

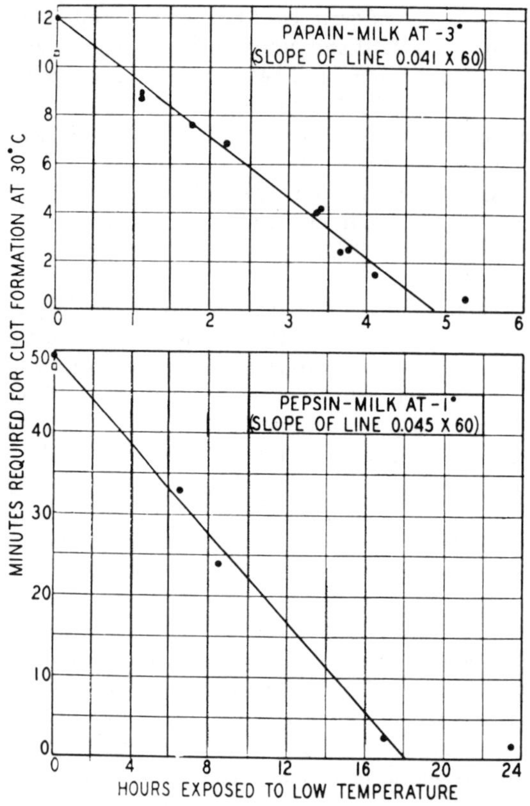

From Balls (1938)

FIG. 12.3. RENNET ACTION: EFFECT OF EXPOSURE TO LOW TEMPERATURE ON CLOTTING TIME WHEN THE RENNET-MILK SYSTEM IS BROUGHT TO 30°C

hydrolysis product and appearance of the clot. In the vicinity of 0°C, no clotting takes place. However, the longer the time (t_L) the milk is exposed to the enzyme, the shorter the time required for clotting when the reaction mixture is restored to the higher temperature. This latter time, t_x, can be predicted from a knowledge of the ratio, Q_{hyd}, of the rates of peptide bond cleavage at these two temperatures.

$$t_x = -1/Q_{hyd} \cdot t_L + t_H$$

where t_H is the clotting time at the higher temperature *without* prior low-temperature incubation. The comparatively cold-stable lipoxygenase-generated lipohydroperoxides (Chapter 21) could also contribute to the acceleration of food quality change upon thawing. Stable enzyme-substrates may comprise another class of apparently hitherto unsuspected class of "intermediates" which could conceivably affect food quality especially when they decompose upon thawing. Fink and Angelides (1976) present convincing evidence for the accumulation of at least three acyl-papain intermediates in the presence of cryosolvents such as dimethyl sulfoxide. Other enzyme-substrate or intermediate complexes, which accumulate in low temperature systems cited by the above authors and by Fink (1976) in a theoretical exposition of "cryoenzymology" include peroxidase, chymotrypsin elastase, lactase, β-glucosidase—all enzymes we encounter in our exploration of food enzymology, as well as bacterial luciferase.

Evidence that the specificities of enzymes in frozen foods can, to all intents and purposes, be altered by shifting to lower temperature is afforded by the observations of Bengtsson and Bosund (1966) that the composition of enzymatically produced fatty acids in frozen peas stored in the vicinity of 0°C was function of temperature. The lower the temperature, the higher the proportion of unsaturated fatty acids. This is not a true change in specificity but rather a change in substrate availability; the unsaturated fats are the last to be solidified upon decreasing the temperature. This in turn might make the action of lipoxygenase relatively more prominent since this enzyme can act only on certain polyunsaturated fatty acids.

Cellular Disruption and Decompartmentalization

We have discussed the role of the maintenance on one hand and dissolution on the other of cellular integrity in prevention of undesirable and promotion of desirable enzyme action (Chapter 1; Schwimmer 1969A,1972,1978). One of the agents of cellular disruption mentioned is temperature change. Considerable evidence has accumulated showing that the end effect of certain alterations of food quality which occur more rapidly at lower temperatures including the frozen state are due to enzyme action as a consequence of cell disruption. Perhaps the most well-documented organelle involved in these phenomena is the lysosome (Tappel 1966A,B). One of the experimental criteria for the existence of lysosomes is the test of *latency*; lysosomal enzyme is inactive until the suspected lysosome preparation is subject to a freeze-thaw cycle or to other treatment which disrupts the membrane enclosing the enzymes. This disruption not only allows for enzyme action in the frozen state but also for continued and even further acceleration upon thawing and warming on the way to being cooked.

Evidence for cellular and internal membrane disruption has been obtained for the following changes in foods: (1) potentiation of phospholipase in fish (Chapter 28) and attendant changes in texture (Lovern and Olley 1962), (2) formation of metmyoglobin in meat (Chapter 18), a color problem (Bito 1976), (3) ATPase (Chapter 26) attributable to release of Ca^{2+} occasioned by disruption of membranes which control this release (Bendall 1973), and (4) cold shortening of meat muscle (Chapter 27).

Changes in the level of glycogen in meat (Behnke *et al.* 1973) appear also to be due to cell disruption with the additional circumstance that while there is no direct inactivation of phosphorylase (the enzyme responsible for breakdown of glycogen) the rate of buildup exceeds or lags behind but does not keep pace with the rate of breakdown of glycogen.

In fruits it is conceivable that polygalacturonase cannot act or be effective in manipulating texture until pectin esterase acts. The latter enzyme has to be activated by salts which are released upon cellular disruption as detailed in Chapter 21. Other examples of where there is an apparent increase in enzyme activity related to a specific change in food quality will be dealt with in appropriate chapters.

Shifts in Physiological Function.—Effects of cooling but not freezing of fruits and vegetables in response to stress has been ascribed to an imbalance of enzyme rates leading to accumulation of intermediates of intermediary metabolism. This includes the sweetening and internal browning of potatoes (Chapters 15 and 19), the purpling of cauliflower and the pitting of grapefruit. These and similar effects will be treated elsewhere in this volume.

Fluctuating Temperatures

Many foods, both before and after processing, are subject to widely diverse temperatures. Deterioration appears to occur more rapidly in foods subject to such fluctuations than in foods stored at relatively constant temperatures. This is true whether the product is a canned food stored in warm markets or frozen and stored in special refrigerated warehouses.

Of course there has been considerable interest in the final outcome of such fluctuations and why they affect foods as they do. Hicks (1944) and Schwimmer *et al.* (1955) developed equations in which a sinusoidal function of time replaced temperature in the Arrhenius equation. Schwimmer *et al.* extended this theory of prediction of reaction rates fluctuating temperature in systems by developing equations for additional modes of fluctuations and expressed the effects of the fluctuations in terms of effective temperatures at which the reactions would have occurred at the same rates. Labuza (1979) showed that losses of less than 50% degradation are the same for zero and first-order so that determination of reaction rate order is not critical but that the temperature sensitivity *is* critical in making predictions concerning shelf life. Labuza states that Hicks and Schwimmer *et al.* assumed zero-order reaction kinetics, and that Schwimmer *et al.* were incorrect in their conclusion that the relationship between a function of the effective temperature and Q_A, the ratio of rates at fluctuation extremes of amplitude A is independent of the order of the reaction. As far as we can ascertain, neither Hicks nor Schwimmer *et al.* assumed any order of reaction. Schwimmer and Ingraham (1980) explicitly demonstrated that in principle as well as in approximation the effective temperature relationship is indeed independent of reaction order. Of course, as Labuza clearly demonstrated, the *amount* of product *is* dependent upon reaction order, once the effective temperature has been calculated.

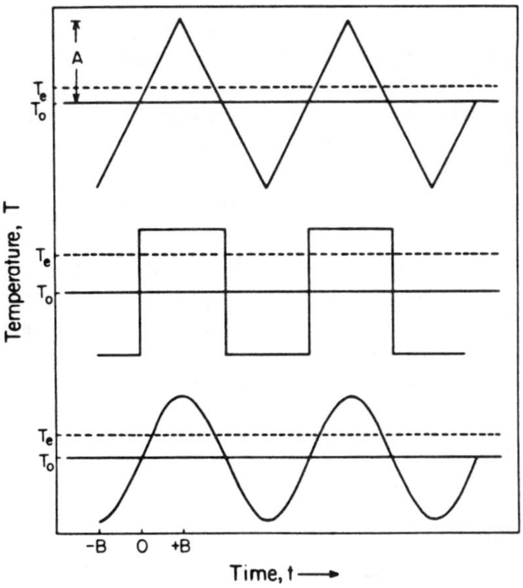

From Schwimmer et al. (1955)

FIG. 12.4. EFFECTIVE TEMPERATURE IN FLUCTUATING TEMPERATURE REACTION SYSTEMS

For sinusoidal, square wave and sawtooth fluctuations: t_e = effective temperature; A, amplitude; B, period.

As shown in Fig. 12.4, the effective temperature is always greater than the mean temperature for any reaction with a positive temperature coefficient. The more abrupt the change, as would be likely to happen in food handling, the higher the effective temperature and therefore the greater the expected deterioration. The theory also predicts that the effective temperature will increase with increasing amplitude but will be independent of frequency. That this was actually the case in frozen foods is borne out in a study of the loss of ascorbic acid in frozen strawberries as shown in Fig. 12.5. This study was published in one of more than 24 papers on experimental time-tolerance of frozen foods; the first paper is that of Van Arsdel (1957), the last that of Dietrich *et al.* (1962). A comparison of observed *effective* temperatures with those predicted from reaction rates at the nadir and a zenith of the fluctuations is shown in Table 12.3.

As briefly alluded to in Chapter 1 in connection with sources of diversity of food constituents, change in the quality of such foods comes about not only because the effective is higher than the mean temperature, to give increased yields of undesirable products, but is also due to a change in the ratio of possible products. These ratios would not be possible at the constant effective temperature. Temperature fluctuation thus effects a qualitative as well as quantitative change in food constituents.

Enzyme Systems.—Perhaps the only investigations dealing experimentally and directly with the effect on enzyme systems have come from the laboratory of Powers and coworkers at the University of Georgia. In a series of papers, the most recent being that of Wu *et al.* (1975), they used lipase, invertase and chymotrypsin and lysozyme. Some typical results are shown in Fig. 12.6. They found that at low but not freezing temperatures and well below the temperature optima of these enzymes, the yield of enzymatically-produced product was as dependent upon the amplitude but independent of frequency of the temperature fluctuations, as predicted by Schwimmer *et al.* Deviation from theory which did occur was dependent upon the temperature value with respect to its position in the cycle, i.e., how close the starting temperature was to the zenith or nadir. In general, they observed "overshoot" at temperatures below the optimum and more inactivation than expected, and undershoot at temperatures above the optimum. Near the optimum, for some of the enzymes such as chymotrypsin, product yield did depend upon frequency and values for apparent energy of activation and temperature optima were shifted

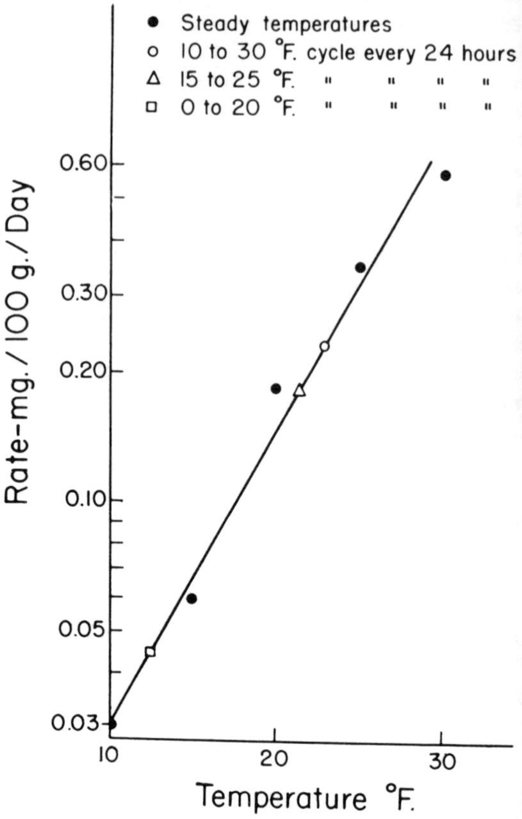

From Guadagni and Nimmo (1958)

FIG. 12.5. ASCORBIC ACID LOSS IN STRAWBERRIES DURING FROZEN STORAGE WHILE BEING SUBJECT TO FLUCTUATING TEMPERATURE

Points for fluctuating temperature are plotted at the calculated effective temperature.

TABLE 12.3

EFFECTIVE TEMPERATURES IN FOODS STORED IN FLUCTUATING TEMPERATURES

Food	Measurement	Effective Temperature, °F Found[1]	Effective Temperature, °F Predicted[2]
Green beans	Chlorophyll	14.1	14.1
Peas	Chlorophyll	13.4	13.0
	Color	14.3	13.2
Cauliflower	Color	12.5	12.9
	Ascorbic acid	12.3	12.4
Strawberries	Color	11.7	12.6
	Ascorbic acid	12.1	12.9
Raspberries	Color	12.4	12.6
	Ascorbic acid	12.7	12.7
Chicken	Moisture	13	12

[1] From time-temperature-tolerance-in-frozen-food studies (see text).
[2] As predicted from theory of Schwimmer *et al.* (1955).

downwards. Interestingly, the investigators ascribed some overshoot to a "reactivation" of enzyme (Chapter 11) as it cooled down after reaching the zenith where it had been subject to partial inactivation.

Much remains to be elucidated concerning the effect of temperature fluctuations encountered by foods on their way to the consumer, especially in regard to the consequences of enzyme action on the quality of the foods. As far as we know, no research specifically designed to directly measure enzyme action in foods under such conditions has been published. However, some of the results obtained in the above-cited time-temperature-tolerance studies were probably due to enzyme action. Available evidence tentatively suggests that such fluctuations would tend to magnify the deleterious consequences of enzyme action in foods especially at temperatures well below the optimum. For instance, the accumulation of phospholipase-produced free fatty acids decreased with decreasing subfreezing temperatures (Olley *et al.* 1969). However, raising the subfreezing temperature allowed the stopped reaction to resume until the product accumulated to a new, higher plateau characteristic of the new temperature. In a fluctuating system this particular behavior would be manifested eventually in a level of product characteristic of the zenith temperature, somewhat higher than that predicted from the Hicks-Schwimmer-Labuza analysis (Schwimmer and Ingraham 1980).

Another synergy resulting from a fluctuating system is related to the previously discussed accumulation of secondary products of enzyme action. The low temperature at the nadir would tend to stabilize such secondary products which would then accumulate slower than in a steady temperature situation but would then react more readily at a higher concentration to give more and different products as the temperature rose and approached the zenith. Another consequence of such fluctuations would be the reactivation of the aforementioned cold-labile enzymes, such as lactic dehydrogenases and muscle glycogen phosphorylase, near the zenith of the cycle. Since these glycolytic enzymes appear to play a role on meat quality (Chapter 26) it is not hard to imagine that fluctuation of meat temperature could affect such quality through such a mechanism.

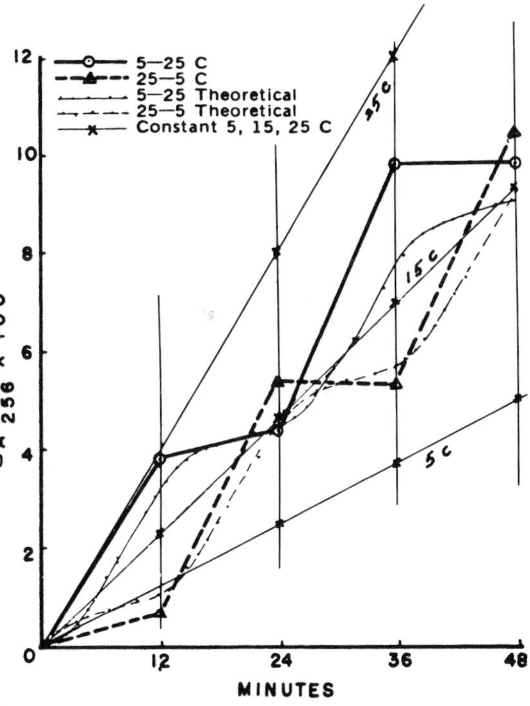

From Wu et al. (1975)

FIG. 12.6. PRODUCT YIELD FROM ACTION OF CHYMOTRYPSIN AT FLUCTUATING TEMPERATURES

Cryoprotective Cryoadaptive Effects

Further progress in understanding and controlling effects of low temperatures on enzyme action in foods may come about in the future by exploiting and applying information gleaned as a spin-off of studies in two related areas dealing with cryoprotective and cryoadaptive enzymes. Cryoprotective substances are present in cold-adapted fish (Feeney *et al.* 1972) and in plants (Volger and Heber 1975). We already alluded to the cryoprotective effects of solutes during freezing-induced dehydration of model enzyme systems. In both fish and plants these cryoprotective substances are carbohydrate-containing proteins. We have seen (Chapter 11) how the carbohydrate moieties of enzymes may be involved in enzyme regeneration phenomena. Cryoprotective proteins in cold-adapted fish accomplish their mission by lowering the freezing point of the fish's blood whereas those of plants prevent damage to the cell while allowing freezing to proceed. Protection of enzymes and other proteins against the proteolytic enzymes by virtue of "super" hydrogen bonding was alluded to earlier in the chapter.

A field of biology related to the study of the enzymes and which may have some relevance to understanding food quality changes in cold temperatures is the study of cryophilic organisms, which prefer to live at low temperatures (Feeney *et al.* 1972; Feeney 1976; Alexandrov 1977). Feeney describes experiments on the inactivation by nucleotides of glyceraldehyde-3-phosphate dehydrogenase of rabbit but not that of cold-adapted Antarctic fish. The reason for this resistance to inactivation is that the latter enzyme is not dissociated as the former is by the nucleotides. Cold-adapted fish also have a lower energy of activation of catalysis. It is of interest to note that at least one fish enzyme adapted to low temperature, aldolase, is quite unstable at what we would consider to be normal, ambient temperature, apparently due to the exposure, as the result of conformational changes, of previously hidden sulfhydryl groups.

13

ACTION AND CONTROL OF ENZYMES AT LOW WATER ACTIVITIES

As in cooling and heating, reasons for drying foods and keeping them dry are pluralistic. Perhaps the principal goal, especially for deliberate dehydration, is the reduction or elimination of microorganisms. In addition, drying reduces weight and bulk, achieves compactness and can create new identity such as producing prunes and raisins from plums and grapes. Drying to intermediate moisture allows for storage stability without the necessity of rehydration so that such intermediate moisture foods, IMF, can be eaten "out-of-hand." Enzyme action, in general, proceeds slower in such foods than in high moisture foods. As we pointed out in Chapter 12, many aspects of the special behavior of enzymes in frozen systems are echoed in low moisture conditions. This is probably not a coincidence but reflects the probability that dehydration occurs in frozen as well as in dried foods. Hence clues which point to methods of control of enzyme activity in frozen foods might be helpful in preventing such action in dried foods and, of course, vice versa, as we shall have ample opportunity to demonstrate in this and other chapters.

PHYSICS AND ENGINEERING OF DEHYDRATED FOODS

Before we examine the evidence for action of enzymes in water-deficient foods, we shall adumbrate some of the elementary physical and engineering information and concepts involved in dehydrated food technology. For a more formal, in-depth treatment, any good book on food technology has a discussion of dehydrated foods such as Potter (1978) and Joslyn and Heid (1964). Among books devoted mainly to general principles of drying physics in the chemical industry is that of Nonhebel and Moss (1971). The overall food technology of dehydrated foods is dealt with in detail in the two-volume treatise edited by Van Arsdel *et al.* (1973). Collections of reviews and discussions presented at symposia include those edited by Spicer (1974), Duckworth (1975) and Rockland and Stewart (1981). A compendium of patents on food dehydration is also available (Torrey 1974). Perhaps the most comprehensive review of research and development in this general field is that of Karel (1973A). For a more popular, shorter, trade-journal approach see Konigsbacher (1974). Monographs by Troller and Christian (1978) and by Ginzburg (1973) in Russian provide a good academic background from the viewpoint of the researcher.

Water Activity and Potential

Water Activity, a_w.—Central to all work and discussions, application and understanding of dehydration processes and food stability (including that of enzyme) is the concept of water activity, a_w, numerically equal to the relative humidity divided by 100 and is defined as $a_w = P_f/P_w$, where P_f and P_w are the partial pressures of water vapor above a substance, in this case a food, and that above pure water at a given temperature. Water activity is usually displayed as a water sorption isotherm which relates a_w to the water content of the food. Karel lists empirical and semiempirical equations for water sorption isotherms proposed for foods. More recently Iglesias and Chirife (1976) developed the following equation which describes reasonably well the relation among a_w, moisture content, X, and temperature, T, for many foods:

$$a_w = \exp(-\exp(bT + c) \cdot X^{-r})$$

where b, c and r are temperature-independent parameters whose values depend upon the type of food and upon whether the water activity measurement involves progressive water take-up (absorption) or removal (desorption). For most foods the ranges of values for r, b and c are, respectively: 1.5–2.6, −0.011 to −0.23 and 2–5. Rockland (1969) detected discontinuities in sorption isotherms by using a function of a_w more directly related to chemical potential (see below) than is a_w itself. Sorption isotherms for representative foods are shown in Fig. 13.1.

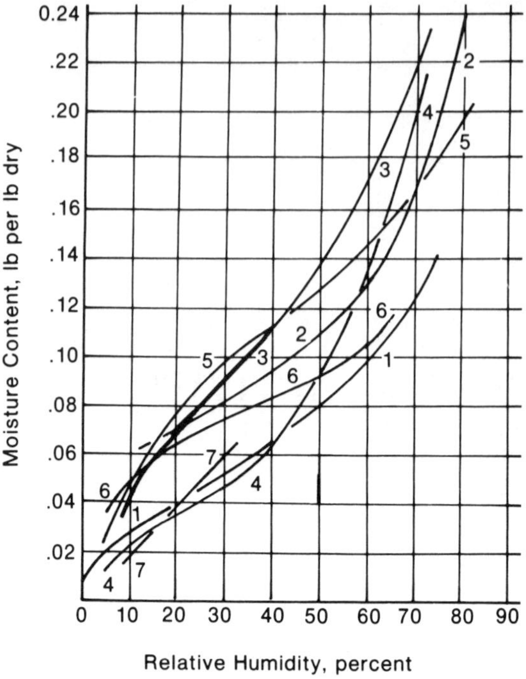

From Van Arsdel et al. (1973)

FIG. 13.1. WATER SORPTION ISOTHERMS OF SELECTED FOODS

Shown are: (1) egg solids, (2) beef, (3) codfish, (4) coffee, (5) starch gel, (6) potato, (7) orange juice; all at various temperatures.

For all foods, a_w is less than unity. Although we will not discuss the molecular basis of depression of a_w in foods here, a nodding acquaintance with some of the underlying factors is necessary in order to understand the nature of enzyme action or lack thereof in these water-deficient systems. Karel (1973A) succinctly epitomized these factors.

Factors related to Raoult's law and deviation therefrom: Raoult's law states that equal molar concentrations of solutes in ideal solution will depress the vapor pressure of the solvent to the same extent. For the purpose of the present discussion

$$a_w = \frac{\text{molecules of water}}{\text{molecules of water + molecules of solute}}$$
$$= \text{mole fraction of water}$$

Deviations from Raoult's law, which invariably occur in food systems, are attributed to water and solute bound to food insolubles and solute dissociation. Scherrer and Eaton (1978) stress the importance of charge and elastic properties whose contributions to deviations of Raoult's law in synthetic polymer solutions have been well appreciated by organic chemists. As viewed by Ross (1979) such deviations occur because attractive or repulsive forces allow formation or disruption of molecular clusters between solvent molecules or between solvent and solute molecules. Salting-in-and-out effects, discussed later in this chapter, are manifestations of such deviation.

Water trapped in capillaries cannot contribute fully to the ideal vapor pressure. This results in a lowering of a_w. The role of capillaries in highly porous food constituents such as cellulose in modulation of enzyme action will be examined later in this chapter.

Water bound to specific polar sites in the foods creates a monolayer the extent of which can be estimated from the Brunauer-Emmet-Teller (BET) equation (Brunauer *et al.* 1938). The theory from which the equation was developed enables one to estimate the total number of polar groups in the food which bind water. For most foods monolayer values fall in the range of 0.05–0.15 g of water per gram of solids. BET theory provides a starting point for consideration of the effect of mobility of small molecules as substrates or as affectors of enzymes.

Water Potential.—Water relations in foods may also be expressed as the chemical potential, $\Delta\mu_w$, measured in bars. The chemical potential of water is the difference between the partial specific Gibbs free energy of water in a system and that of pure water and is related to a_w by the following equation (Slayter and Taylor 1960):

$$\Delta\mu_w = 2.303 \cdot R_w \cdot T \cdot \log a_w$$

where R_w is the gas constant per mole of water and T the absolute temperature. Since a_w is less than unity, values for water potential are negative.

Although not frequently used in food-related investigations, the chemical potential is preferred to water activity by agronomists, plant physiologists and soil scientists and has been interchanged with terms such as "diffusion pressure deficit" and "suction force." It is finding its way into literature on the behavior of enzymes in water deficient systems which are, of course, of vital interest to plant scientists studying plant and soil water relations. Relationships between water potential and water content of two intact vegetables and a fruit, before processing, are shown in Fig. 13.2. This graphically illustrates how the water potential (and thus a_w) can vary widely without an appreciable change of water content and also shows that plant food water potential is controlled by changing solute. Since enzyme action parallels a_w rather than water content (see below) such action can also exhibit wide disparity in values at the same water content in a given food. Food-oriented studies in which it is advocated that log a_w, more closely related to chemical potential than a_w, be used include those of Rockland (1969) and Strolle (1977). On the other hand, some plant scientists feel that there is no logical reason for introducing and using this term (Zimmermann and Steudle 1978).

Low Moisture Foods

At some point in their history most water-deficient foods have had their water removed either *in vivo* mostly through the agency of solar energy, as for instance grain and nut ripening, or *in situ*, as in raisin, dried prune and apricot production, also through both the agency of solar energy and *via* processing, using artifactual energy sources. In some foods, such as dried soups, dehydration takes place before the ingredients of the food are mixed. Authorities classify water-deficient foods on the basis of their water activity, and indirectly on the moisture left after the dehydration process. This classification also is probably the most meaningful from the viewpoint of enzyme action. Going from

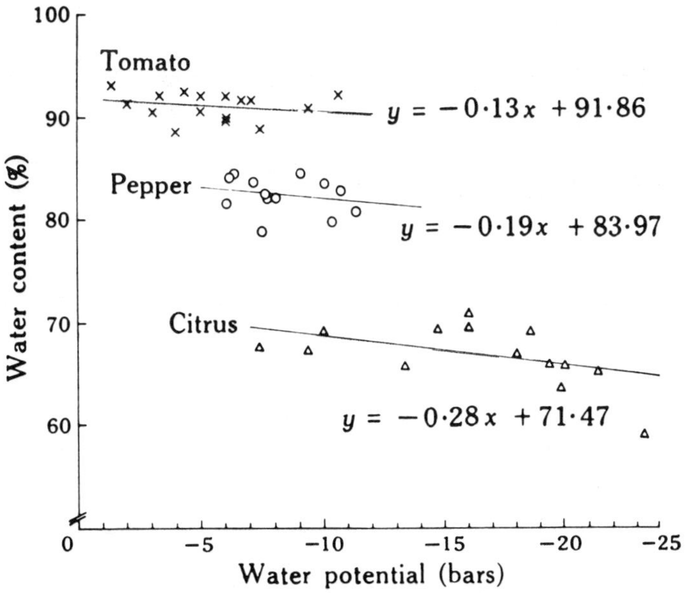

FIG. 13.2. WATER POTENTIAL OF WATER CONTENT FOR FOOD PLANTS

From Darbyshire and Steer (1973)

high to low a values of a_w, it is customary to distinguish among the following:

Concentrates, foods with moisture contents usually more than 25% and a_w's usually above 0.9. Foods in this category in which enzyme action may contribute to quality include honey, sugar syrups and frozen juice concentrates.

Intermediate Moisture Foods (IMF), with a moisture range of 15–60% and a_w's between 0.6 and 0.9. Included in this category are the traditionally dried fruits as well as jams, dressings and pie fillings and perhaps most prominently from a strictly economically lucrative viewpoint, pet foods. Much effort is being expended to extend the IMF range, mostly by incorporating polymer humectants which provide supplementary sites for the reordering of bound water (Gee *et al.* 1977). Another approach is to alter the polymers in the food, sometimes by enzymatic means. For instance, Borochoff *et al.* (1971) generated such humectants in pet foods *in situ* by adding glucoamylase and α-amylase. These manipulations result in a lowering of a_w values at relatively high moisture contents, thus imparting desired texture-associated attributes such as optimum mouth-feel and yet without providing a medium favorable for bacterial growth.

Dry-to-the touch foods, mostly in the moisture range between 5 and 14%, with a_w values falling between 0.1 and 0.4. These include conventionally "dehydrated" foods. The drying of such liquids, i.e., coffee, have no enzymatic impact whereas in the drying and storage of fruits and vegetables enzymes may play an important role in quality maintenance and deterioration.

Vacuum and Freeze-dried Foods with a moisture content between 1 and 5% and a_w values less than 0.2.

Dehydration Engineering.—As previously mentioned, many foods are dried without intervention of man, mainly by a type of solar heating. Time-honored practices which are still used in which water is deliberately removed from biological entities to be converted into foods include baking, frying (heating with a water-immiscible solvent), salting (osmotic drying) as in the preservation of fish, and wilting of vegetables by low temperature forced air draft. Moisture content of milk is lowered in cheese-making in conjunction with the use of rennet enzymes. Most of these processes of prehistoric origin have been and are being supplemented in modern food manufacturing by increasingly sophisticated automated engineering developments. We briefly limn some of the more prominent and promising modes and techniques of dehydration used or proposed for food manufacturing because they may very well dictate the effect and extent of enzyme action during and subsequent to processing.

It is convenient and customary in considering these techniques of water removal to distinguish between concentration and dehydration. However, there is no immutable barrier which precludes a continuity of the concentration processes from the area of concentration to the area of dehydration. Concentration, resulting in liquid foods containing a minimum of 30% moisture, is usually achieved in steady-state processes by molecular and eddy transfer of the liquid. True dehydration, resulting in palpably dry products with moisture contents of less than 13%, is achieved by quasiequilibrium processes such as molecular diffusion from solid particulates or from liquid droplets in sprays.

King (1974) lists 44 separation processes of which 20 are considered as feasible alternatives for achieving separation of water. Other classifications of processes for the removal of water from foods distinguish among: (1) the mechanism by which energy is supplied to the food, (2) the machinery or container configuration, and (3) the movement of the object being dried.

Concentration.—For most foods the most economically feasible and prevalently used concentration techniques are evaporation processes. Honey is, as mentioned, a concentration process by evaporation accomplished without human intervention. Evaporation is essentially an equilibrium process in which efficient separation of water is obtained at phase equilibrium of all components between the concentrating and extracting phases (Thijssen 1974). Spicer (1974) lists 20 modes of evaporation for concentrating liquid foods. Other equilibrium processes which have been considered for the removal of water from foods for concentration purposes include freeze-concentration and clathration in both of which the liquid food is partially frozen to form a slush and the relatively pure crystals of ice containing little or no solute are removed, leaving behind a concentrate. Clathration, involving the formation of special complexes of the solutes in partially frozen systems with nonpolar clathration agents such as freons and other hydrocarbons, is not considered practical for food use. However, freeze-concentration has been at least considered for food use because of the low temperatures used; the latter would tend to preserve the enzymes in the foods and allow enzyme action to proceed, subject to the environmental influences discussed in this and in the previous chapter.

Nonevaporative Processes.—Much effort has been expended in the development of nonequilibrium approaches to water removal in foods, almost all of which involve membranes, such as direct osmosis, reverse osmosis, pervaporation, ultrafiltration and related techniques involving electrophoretic mobility of the food constituents, and electrodialysis and electrodecantation. While some of these techniques have proved to be valuable laboratory aids for enzyme purification (Chapter 6) and have been used industrially for the preparation of costly products such as pharmaceuticals, only reverse osmosis has so far been seriously considered for the removal of water from foods.

Dehydration.—There are many classifications of dehydration processes according to: (1) the initial phase of the water, liquid or solids; (2) receiving phase of the water, vapor or immiscible liquid or solid; and (3) means of lowering the chemical potential of the water in the receiving phase. The latter includes use of vacuum, freezing, screening, precipitation, adding carrier and use of temperature and density differences (centrifugation) as outlined by King (1974).

Most discussions on true dehydration distinguish between air drying and nonair drying procedures. One of the most widely adopted techniques is spray drying. Confined at first to liquids such as milk, coffee, fruits and vegetable juices, and corn hydrolysates, this versatile technique has spread to the drying of cake mixes, cheeses, potatoes, fish concentrates and other high protein foods (Karel 1937A).

However, the most commonly used approach to drying involves some sort of heated chamber, frequently a tunnel-like object, into which the food enters by a belt, or else the food is rotated in a drum—hence the terms used in describing such arrangements, such as roller, belt, trough, pan, rotary, drum. Depending upon how the air is delivered, the drying may occur *via* air blast, air lift, convection/forced air (centrifugal/stirred), fluidized bed, turbulent film, gas injection spray and foam mat. Other means of water removal, some of which we shall encounter in our subsequent discussions, include freeze-drying and its variations (King 1974), osmotic drying with sugars (especially useful for fruits), microwave, puff and slush drying, solvent extraction (fish), in-package desiccation and drying with superheated steam (Lazar 1972). Some of these innovations have yet to be developed commercially.

One of the problems addressed in the design of these drying processes is that of product quality and achievement and subsequent maintenance during storage. To this end there have been numerous studies directed toward improvement of reconstitutability (rehydration) of the dried food, by additions of either suitable substances including enzymes (Morgan and Schwimmer 1965) or other substances such as enzyme modified starch (which amylase in the food could further degrade), glycerol (Shipman *et al.* 1972; Neumann *et al.* 1980), monomers to be polymerized in the food during drying (Schwimmer 1969C) or by physiochemical manipulations such as "sublihydration" (McCormick 1976A).

ASSESSMENT OF ENZYME AND NONENZYME ACTION IN DRIED FOODS

As mentioned, in addition to economic feasibility, one of the factors to be considered in the adoption of a particular drying procedure is the quality of the product. While, as will be subsequently developed, enzyme action can indeed affect the final quality and storage-stability of water-deficient foods, traditionally nonenzymatic alterations or changes in such foods have been under much more penetrating scrutiny. These nonenzymatic changes may be initiated biologically through microbial growth or nonbiologically as exemplified by nonenzymatic browning and lipid oxidation.

Microbial Growth

A graphic illustration of the range of enzyme action with respect to a_w as compared with nonenzyme action is shown in Fig. 13.3. It will be noted that most microbial growth ceases at values of a_w at which enzyme action could but does not necessarily continue. Bacterial growth ceases below $a_w = 0.9$, that of yeast below 0.8 and that of molds below 0.7. Therefore, as a rule in the absence of rarely encountered xerophilic fungi and yeast, foods with a_w values below 0.70 are not subject to microbial spoilage. Since IMF can approach or even occasionally exceed this value, some attention has been given as to how to stop microbial growth in IMF (Duckworth 1975; Bone 1973; Haas *et al.* 1975; Mossel and Shennan 1976). One general rule

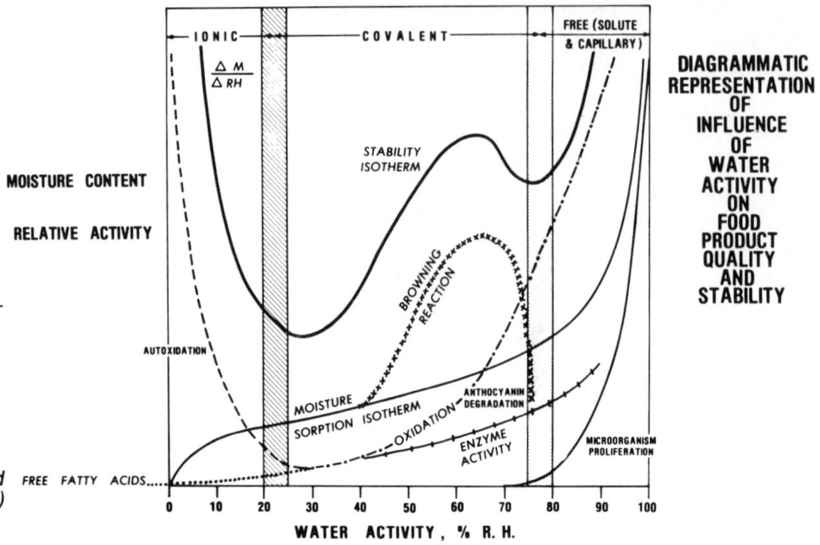

FIG. 13.3. WATER ACTIVITY AND FOOD STABILITY

Courtesy of Rockland and Nishi (1980)

established in this regard is that IMF systems prepared by absorption of water present a more hostile environment to microbial growth than do corresponding foods prepared by desorption (Labuza *et al.* 1972). Most dehydration processing procedures, especially those involving movement of hot gas, cause complete destruction of vegetative microorganisms. This, however, is not true of commercial onion and garlic dehydration where drying temperatures are relatively low (81°–87°C in the brief first stage of drying followed by a longer period at 55°–60°C). This schedule is necessary to prevent destruction of the flavor generating enzyme system, including both enzyme and substrate (Chapter 21). For such products, monitoring microbial stability has to be conducted somewhat more closely and stringently than for other dehydrated foods. Although the a_w of such products may be considerably below 0.7 and microorganisms cannot grow in them, the latter could start to grow when the products are reconstituted. In the low-temperature method of Gee *et al.* (1977) modeled on home dehydration, temperatures as low as 48°C result in products with minimum contamination.

One more point concerning the relation of microbial growth to enzyme action in low-moisture foods is the circumstance that some of the enzymes in the dried food may arise, not from the cells of the food itself (endogenous enzymes) but may be released by contaminating organisms. These exogenous enzymes may continue to act even more efficiently after release into the food, even after microbial growth has ceased. An example is the continued action of bacterial lecithinase, resulting in off-flavors and off-colors in improperly dried eggs (Chapter 19).

Nonbiological Changes in Dehydrated Foods

Both nonenzymatic browning and nonenzymatic lipid oxidation (rancidity) can be distinguished from the corresponding quality changes generated by enzyme action by examination of their dependence on water activity. As shown in Fig. 13.3 only nonenzymatic reactions exhibit maxima in the IMF range. In addition nonenzymatic lipid oxidation starts to rise again as further water is removed from the system. Furthermore, nonenzymatic browning is more highly exacerbated by elevated tropical temperatures than is enzymatic browning as evidenced, for instance, by the increased nonenzymatic browning of IMF meat products stored at 38°C (Obanu *et al.* 1976). It is of interest to parenthetically note that there might be a connection between these two major modes of nonenzymatic changes in that the nonenzymatic browning has been reported to produce antioxidants which stabilize foods and fats against oxidative rancidity. It is not unreasonable to postulate intermediates formed in these nonenzymatic reactions can affect the action and survival of enzymes in the food. This is especially true of free radicals which tend to be stabilized in low moisture systems (Karel 1975). Conversely, continued enzyme action in dried foods could provide reactants for these nonenzymatic reactions as emphasized in

Chapter 1 and illustrated throughout the book. To cite one example, carbohydrases and proteinases yield the reactants for nonenzymatic browning.

In addition to lipid oxidation and nonenzymatic browning, other nonenzymatic quality changes and reactions occurring in dehydrated foods which have enzymatic counterparts include carotene degradation, oxidation of ascorbic acid and myoglobin—reactions which result in loss of protein solubility and nutritional availability (Chapter 34) and change in fish texture (Chapter 28). Pertinent discussions not previously cited are those of Lee and Labuza (1975), Labuza (1980) and Labuza *et al.* (1977). One deteriorative change in a dried food which appears to proceed strictly nonenzymatically is the degradation of anthocyanins in freeze-dried strawberries (Erlandson and Wrolstad 1972). While extensive information is available on nonenzymatic oxidative rancidity, very little is known about the contribution of lipoxygenase type enzymes to lipid instability in the IMF moisture range.

ENZYME ACTION IN DRIED FOODS

Evidence from Blanching

The principal although circumstantial evidence that enzyme action does indeed occur in water-deficient biological systems including foods, is the necessity for blanching before dehydration for most such foods. According to Konigsbacher (1974), the first patent ever granted for a food dehydration process about 200 years ago incorporated a "scalding" of vegetables prior to drying them in a heated room. Just 30 years ago Joslyn (1951) drew attention to the hay-like odor and off-taste which developed in dried unblanched peas, string beans and corn, quite similar to that encountered in unblanched frozen vegetables. Joslyn also pointed out that elimination of oxygen retarded off-flavor development but that the rate of inactivation of neither peroxidase nor catalase paralleled the thermolability of the system responsible for the off-flavor. Improvement in ascorbic acid retention in dehydrated foods upon blanching occurs only in foods containing ascorbate oxidase (Chapter 16). Underblanched prunes may develop phenolase-associated tobacco-like flavor (Chapter 25). Other examples of changes due to enzyme action which occur slowly during prolonged storage include the development of "tallowing" in dairy products which Acker (1962) attributes to peroxidase action but could arise as the consequences of superoxide anion generation arising from the action of xanthine oxidase, an enzyme which is quite active in such products (Chapters 16 and 25).

Cell Disruption.—Cell disruption tends to exacerbate enzyme action in dried foods. Thus, suitably dried wheat kernel grains can be stored at ambience indefinitely whereas ground wheat deteriorates in a few weeks. Balls (1947) pointed out that insufficient blanching is worse than no blanching at all, precisely because blanching kills and disrupts the cells and allows surviving enzymes to act much more rapidly than they would have in the unblanched food. Where drying is conducted in a manner such that there is no gross cellular disruption, as in the aforementioned controlled low-temperature forced air drying of Gee *et al.* (1977), blanching may not be necessary.

Other evidence from the older literature that enzymes potentiated by cell disruption cause undesirable changes in dehydrated or dry food was reviewed by Acker (1962). Examples include: off-flavor in dried cereals, milled cereals, cake mixes and bread; bitterness in oats, corn and corn products; loss of lecithin in noodle, macaroni and other pastas; off-flavors in dried apricots; and lipase action in oilseeds, dried meats and eggs. With regard to the latter, considerable effort was required to establish that undesirable changes which dried eggs undergo are in fact mostly due to the action of enzymes (Chapter 19).

Beneficial Effects.—Not all the long-term changes which occur in dehydrated foods are deteriorative: for instance, long-stored rice grains which have become less sticky (after cooking) because of attrite action of amylases on amylose chains; also, interaction of enzyme-liberated fatty acids and phospholipids with amylose could improve texture of rice as well as bread (Chapter 32). Continued action of added bacterial amylase improves the mouth feel of certain baked goods. In the dehydration of sprouted onions, flavor is enhanced during storage after drying by the slow action of the transpeptidase which releases amino acids from γ-glutamyl peptides. These liberated amino acids serve as substrates of the flavor-generating lyases (Chapters 21 and 23).

Behavior of Enzymes in Model Dry Food Systems

We trust that the above brief survey demonstrates that enzymes do act in water-restricted food systems. It will be noted that the evidence is strictly empirical. We shall now turn to an examination of the mechanisms which govern such action and try to develop some idea as to how they resemble or differ from action of enzymes in solution and in the frozen state. We also have something to say about stability under these conditions and conduct a brief survey of means of and proposals for counteracting undesirable enzyme action in dried foods.

Most of the basic information and proposed mechanisms arise from the construction and use of model systems and observation of their behavior. Two of the most prominent groups doing research in this field are those of Acker and coworkers at Munster and Drapron and colleagues at IRNA, Massy. Key papers from these laboratories include those of Acker (1969), Potthast *et al.* (1975), Tome *et al.* (1978) and Drapron (1979).

Concentration Effects.—Many of the remarks concerning the concentration of solutes during freezing in the last chapter are quite relevant to removal of water at higher temperatures to yield concentrated solutions. For instance, with regard to pH effects Tome *et al.* (1978) found that in a Type 2 model enzyme system (see below) with glycerol as solvent, the pH optimum of phenolase shifted from 7.0 at $a_w = 1.0$ to 7.6 in a glycerol-water mixture corresponding to an $a_w = 0.85$. Although the activity of the latter at its pH optimum amounted to about 90% of that of the $a_w = 1$ system at its pH optimum, the glycerol system was, as the result of this shift, considerably more active in the higher pH range. This could have considerable significance for storage stability of dehydrated foods, especially those with high protein contents as discussed in Chapter 12. Such changes have been interpreted as being due to perturbation of the structure and subsequent alteration of loose clusters of water molecules.

In particular there are two special effects which may play dominant roles during the concentration phase stage. These are decrease in activity of the enzyme with increasing substrate concentration (substrate inhibition) and catalysis by the enzymes of reverse reactions. In the case of hydrolases this would result in the further water depletion of the already dry system. These effects will be taken up in further detail in the next chapter.

Invertase.—These effects have been invoked to explain the kinetics of invertase. Although, perhaps, a minor note in the general advance of enzymology, substrate inhibition of yeast invertase has intrigued three generations of enzymologists ever since it was first described 65 years ago by Michaelis and Menten (1913). They reported that the rate of hydrolysis of sucrose reached a maximum with increasing substrate concentration at about 0.192*M* (6.55% w/v). Two key investigations in the intervening years supporting diametrically opposite mechanisms are represented by the papers of Nelson and Schubert (1928) and Ruchti and McLaren (1964). The former concluded that substrate inhibition was largely due to a decrease in the concentration of water. One mole of sucrose can take up 7 moles of water and the sugar behaves with respect to Raoult's law as if it were twice as concentrated than it actually is. Ruchti and McLaren concluded that the inhibition was typical excess substrate inhibition frequently observed for many enzymes and thus probably due mainly to the combination of the enzyme with more than one molecule of sucrose rather than water, with perhaps a small contribution due to increased viscosity. Bowski *et al.* (1971), in an elegant exercise in kinetic modeling, found that no one model hitherto proposed would fit their extensive experimental data, an example of which (the effect of water concentration) is shown in Fig. 13.4, but that a model incorporating both substrate inhibition and change in water concentration did. Parenthetically they observed a maximum velocity at a sucrose concentration about one-third higher than did previous workers, who did not take into account the fact that invertase catalyzes the formation of oligosaccharides by means of glucosyl transfer as well as hydrolysis of sucrose. They also found that sucrose did not take up the theoretical 7 molecules of water and suggested that sucrose molecules form intermolecular hydrogen bonds which result in clusters which are not available as invertase substrate because they are inaccessible to the active site, as shown in Fig. 13.5.

Catalysis of Reverse Reactions.—This is exemplified by the plastein reaction, the reversal of proteolysis, discussed at some length in Chapter 34, which takes place in 30% solutions of substrate. Citations of apparent reversal of glycosylase, lipase and proteinase action in the 1920s and 1930s were frequently due to transfer reactions. In a modern echo of these older investigations Ingalls *et al.* (1975) were able to readily demonstrate ester (not peptide) synthesis from alcohols and *N*-acetyl tyrosine, catalyzed by immobilized proteinases in glycerol-containing media containing low concentrations of water (10% v/v). It is this low water concentration which supplies the free energy required for driving the synthesis. It is not difficult to visualize how "nonphysiological" products, some perhaps undesirable, could be formed in dehydrated foods due to such reactions. Reaction reversals, such as that catalyzed by lipase, for instance (Chapter 32), could beneficially affect food quality.

Endogenous Inhibitors.—An additional factor which has not hitherto been forwarded is that of increased inhibition of enzyme activity due the presence of endogenous inhibitor in the food (Chapters 19, 37).

In the presence of a noncompetitive inhibitor, in a Type I system, the percentage inhibition of the enzyme increases with increasing water removal and subsequent concentration:

$$\text{Percentage Inhibition} = \frac{\text{A constant} \times \text{Concentration Factor (fold-concentration)}}{K_i + \text{Concentration Factor}}$$

where K_i is the enzyme-inhibitor dissociation constant. Note the formal resemblance to the Michaelis-Menten equation. If for instance the constant is set at unity ($K_i = 9$), then, if an endogenous inhibitor in a food inhibits to the extent of 10% before removal of water, it will inhibit to the extent of more than 50% upon the removal of 90% of the water from the food.

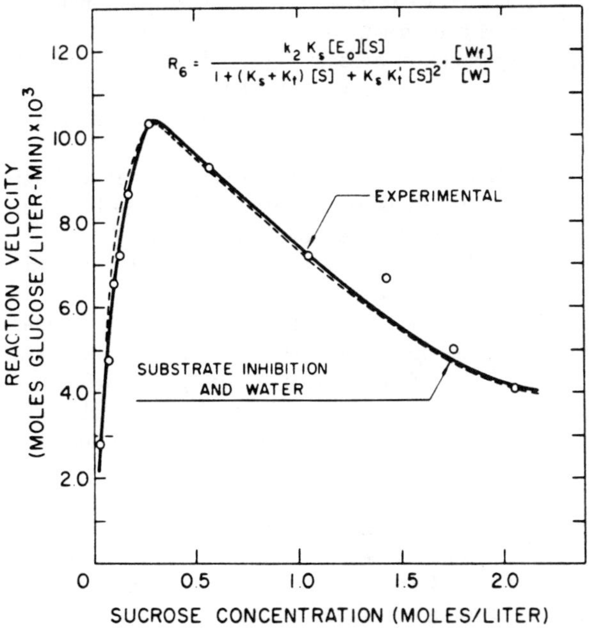

From Bowski et al. (1971)

FIG. 13.4. EFFECT OF SUBSTRATE CONCENTRATION ON INVERTASE ACTIVITY

Theoretical smooth curve takes into account substrate inhibition *and* water concentration effects.

Solid Food Systems Models

In order to obtain clear insight into the nature of enzyme action in water-deficient foods and other dried biological systems, model experiments have been conducted in which enzyme and substrate are allowed to interact under rigorously controlled conditions of temperature, water activity and medium composition. In this way changes that are unequivocally enzymatic and cannot be masked by other reactions are more clearly defined. Usually the main variable changed is the water activity, either by absorption or desorption. Hysteresis ef-

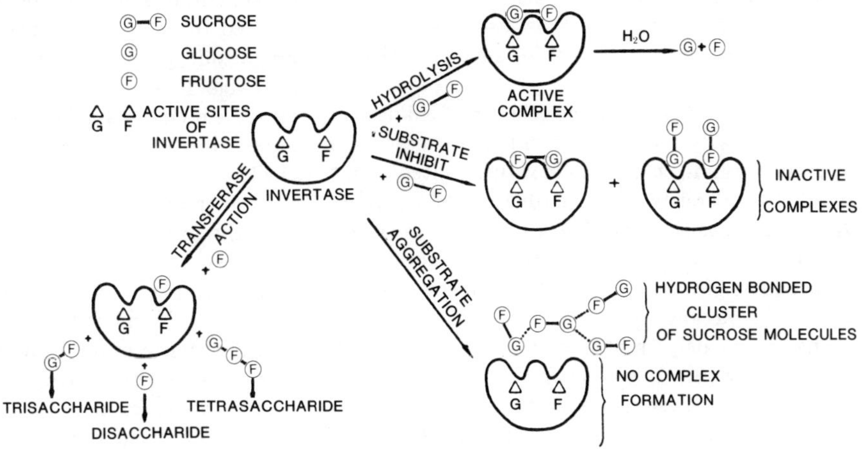

FIG. 13.5. A MODEL FOR HYDROLYSIS OF SUCROSE BY INVERTASE

From Bowski et al. (1971)

fects are frequently observed when both modes of change are used, although some hysteresis may be experimental artifact. The following systems are most frequently assembled for the study of the action of enzymes in water restricted systems:

(1) Only enzyme, substrate and buffer, and limited amounts of water are used, as in the invertase system of Bowski *et al.*
(2) Water is replaced, in part, by a water-miscible solvent such as glycerol.
(3) As in (1) or (2) with the addition of a polymer, usually cellulose, which can take up water and provide capillarity or adsorptive surfaces.
(4) A dried food, usually freeze-dried, plus substrate and varying amounts of water are used.
(5) As in (4) but no substrate is added and the change in level of particular components is followed as a function of the water content or a_w. Here it is assumed that what is measured is the resultant of the action of endogenous enzyme on endogenous substrate.

In addition to the previously cited reviews and contributors, the following have used model enzyme systems for a better understanding of dried food behavior. Blain (1962) worked with lipoxygenase and peroxidase in water-restricted systems containing glycerol (system 2). Using a system 1, Skujins and McLaren (1967) found that measured enzyme activity in a urease-urea system paralleled the sorption isotherm of the enzyme and not that of the substrate. Drapron (1972) conducted extensive studies on the behavior and change in specificities of carbohydrases in such a system. The following phenomenology has been observed and conclusions have been reached.

No matter what system is used, a minimal amount of water is necessary for enzyme action to occur.

Acker and others stress the relation of changes in enzyme activity (or water potential) to changes in a_w rather than to changes in moisture content. This parallelism persists when a function of a_w, $\Delta M/\Delta a_w$, is used, as shown in Fig. 13.6. Such a relationship can form the basis for predicting storage life of products, where deterioration depends largely on enzyme action.

The previously mentioned deviations from Raoult's law will thus be mirrored in the behavior of enzymes in these model systems.

The parallelism to deviations from the Arrhenius relation as discussed in Chapter 12 is striking. One

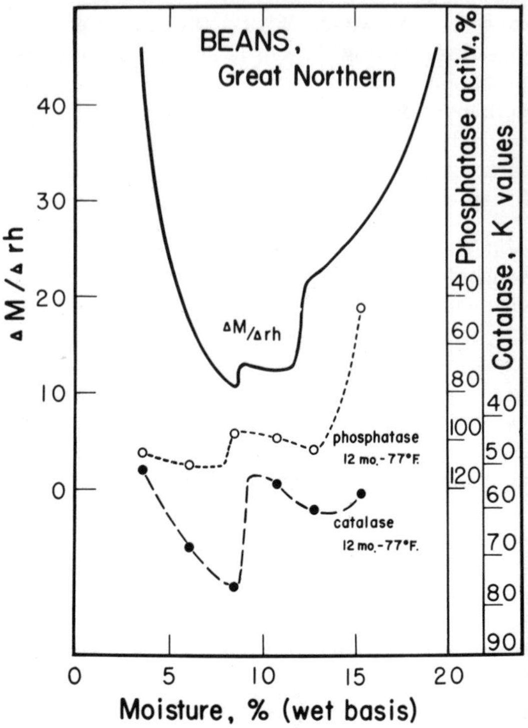

From Rockland (1957)

FIG. 13.6. PARALLELISM AMONG ENZYME ACTIVITY, WATER ACTIVITY AND FOOD QUALITY AS AFFECTED BY WATER CONTENT

of the manifestations of this deviation, important in the structuring of IMF, is the "salting in" and "salting out" effect of two or more solutes (Bone 1970). For instance, the effective concentrations of both urea and NaCl are decreased in the presence of each other (salting out) whereas the effective concentrations of sucrose and mannitol are increased in each other's presence. An idea of the magnitude of effects is afforded by the data shown in Table 13.1. Undoubtedly such aberrations can effect enzyme behavior as shown in Table 13.2, adapted from the investigation of Tome *et al.* (1978).

It can readily be seen from the data presented that the greater the deviation from Raoult's law, the deeper the phenolase inhibition. The same investigators also observed, also in analogy to what is observed in model frozen food systems (Chapter 12), that quite aside from inhibition of the initial rate, the *extent* of the reaction decreases with decreasing water activity.

With few exceptions no enzyme action takes place in that part of the sorption isotherm cor-

TABLE 13.1

MUTUAL SALTING-IN OR -OUT REDUCES OR INCREASES EFFECTIVE CONCENTRATIONS

Solutes	Concentration	Molal Activity Coefficient Alone	Together
Salting In:			
NaCl	1 *M*	0.99	0.82
Urea	20 *M*	0.58	0.42
Salting Out:			
Mannitol	1 *M*	1.013	1.259
Sucrose	2 *M*	1.422	1.595

[1] Source: Bone (1970).

responding roughly to the completion of a monolayer of molecules of absorbed water, the BET layer. For most foods this occurs in the neighborhood of $a_w = 0.1-0.2$, corresponding to a moisture content for most foods of 5% and over. Skujins and McClaren (1967) found that only after 1.3 molecules of water/amino acid were taken up by urease, was enzyme activity detectable. Thus, enzymes can already act in the presence of loosely bound water; completely "free" water is not absolutely necessary.

There are exceptions, as mentioned. Where the substrate itself has sufficient mobility, i.e., an unsaturated fat, enzyme action, mainly that of lipase and lecithinase have been observed below a_w's corresponding to the apparent BET layer. In a special case, Duden (1971) reported the hydrolysis of indoxyl acetate at a_w's as low as 0.01 over an extended period of time measured in months in a type 3 system. In this case the enzyme combined with gaseous substrate to form an ES complex which then broke down into products. Only in reactions not requiring water can glycerol replace water as a medium of transport.

Another lesson learned from these investigations is that the primary function of water in these water-deficient systems is to serve as a medium of transport and secondarily as a reaction medium. Thus, oxygen-dependent enzymes such as phenolase and lipoxygenase respond to changes in water in a manner similar to the hydrolases, except for the substitution of glycerol.

Since transport is of paramount importance, it appears that the capability for enzyme action would depend on the molecular weight or size of the substrate, its physical state, its state of aggregation and its solubility. These conclusions are not contradicted by the observation of Potthast *et al.* (1975) that the high molecular weight glycogen is not as readily broken down as are nucleotides in such water poor systems.

The effect of the physical state of the substrate on reaction rate in such media is illustrated by the observation of Acker and Wiese (1972) that the insoluble saturated triglyceride trilaurin *alone* was hardly hydrolyzed in a type 3 system, but because it is soluble in unsaturated triglycerides it was readily hydrolyzed when triolein was added to the system. The effect of substrate solubility is further illustrated by evidence that the relatively insoluble inosine is not degraded in dehydrated meat at a_w's below 0.55, whereas ATP, the ultimate precursor of inosine, is readily degraded at $a_w = 0.4$ (Potthast *et al.* 1975). In general, because mobility and transport are limiting and decisive, the step which paces enzyme action at low moistures is the speed with which the substrate can get to the enzyme.

TABLE 13.2

INFLUENCE OF WATER ACTIVITY AND DEVIATION FROM RAOULT'S LAW ON ENZYME ACTIVITY AND STABILITY: MODEL SYSTEMS

System		Percentage of Original Activity		
Activity[1]	a_w's ⟶	0.8	0.9	0.95
Phenolase plus:				
Ethanol (large (+) deviation)		2	27	54
Diethylene glycol (large (−) dev.)		34	65	83
Glycerol (very small (−) dev.)		63	95	107
Stability[2]	a_w's ⟶	0.11	0.43	0.75
Peroxidase plus:				
Carboxymethyl cellulose (CMC)		83	32	100
Do + substrate (guaiacol)		89	89	100
Polyvinyl pyrrolidone (PVP)		56	74	7
Starch		100	23	—

[1] Source: Interpolated values from data of Tome *et al.* (1978). a_w's estimated from mole fraction of water in enzyme reaction system at 30°C containing indicated solvents.
[2] Source: Roozen and Pilnik (1971). 25°C; 9–15 days.

Rockland (1977) has suggested that water molecules may be available for the support of enzyme action even at very low pseudo-monolayer water potentials because of the dynamic interchange of the molecules among the various states it can assume (bound, ionized, "loosely" bound, hydrogen-bonded, free, etc.) so that statistically there is a finite chance that some of these molecules for a finite time will be in a state available for supporting enzyme action, i.e., as a medium of transport.

In this connection *capillarity* may play a decisive role in that the manner in which the substrate is distributed is also important. According to Bone (1973) formulated foods such as dry soups and cake mixes are best simulated in model systems by impregnation of cellulose with enzyme solution, freeze-drying and *then* adding substrate. Although some investigators claim that the sequence of addition of components is important, especially in simulating IMF systems, Sloan *et al.* (1977) concluded that neither method nor sequence of addition of humectants such as polyethylene glycol to a collagen-sucrose system affected the final a_w. This does not exclude the possibility of enzyme action being so affected in Bone's systems or in dehydrated foods.

Such effects could arise as the result of capillarity contributed to such systems by collagen or cellulose. Thus Drapron (1972) and earlier Kiermeier (1949) could not detect hydrolysis of starch by amylases in a type 1 system at $a_w = 0.7$ but in a type 3 system in the presence of cellulose, starch hydrolysis could be readily detected at $a_w = 0.46$. Again, in striking analogy to the effect of varying temperatures in the frozen state, it was found that the extent as well as the rate of enzyme action depends on the value of a_w. After the reaction levels off at a particular a_w, it will resume again upon adding more water to achieve a new, higher a_w. The microdroplets of the quasifrozen state are analogous to capillaries in the dry state.

Enzyme Reaction Pathways Altered in Water-restricted Food Systems

Probably because of the slowness of diffusion of intermediates from the surface of the enzyme, enzyme action at low a_w tends to prevent intermediate accumulation or to channel reaction pathways. Some examples are shown in Table 13.3.

Drapron and Guilbot (1962) observed that at a_w's below 0.85 in a system containing α-amylase and soluble starch only glucose and maltose, and a trace of maltotriose were detected, whereas above this value intermediate oligosaccharides and dextrins were produced as well in the same time interval, as shown in Fig. 13.7.

TABLE 13.3

ENZYME ACTION AT LOW a_w TENDS TO PREVENT INTERMEDIATE ACCUMULATION OR CHANNELS REACTION PATHWAYS

Enzyme	End-product Observed	Not Observed
α-Amylase	Glucose	Intermediate dextrins, maltose
Lipoxygenase	Oxygenated polymers	Lipohydroperoxide decompos. prod.
Phytase	P_i + myo-inositol	Phosphoinositol intermediates

Similarly, Brockmann and Acker (1977) reported that for lipoxygenase action at low moisture levels the primary lipohydroperoxide products were transformed almost quantitatively into polymerized polar products with increased oxygen content, in contrast to the diversity of products formed in conventional aqueous enzyme reaction systems (Chapters 21 and 24). It would be of considerable interest to ascertain if this product, consisting largely of polymers of ricinoleic acid (12-OH,$\Delta^{9,10}$-octadecenoic acid), is an activator of castor bean lipase.

As in wheat flour suspensions, the ultimate preponderant products of lipoxygenase action in stored wheat flour are the mono- and dihydroxy fatty acids. In contrast to the aqueous flour suspensions where the mono-OH derivative predominates and other secondary products can be detected (Chapters 24 and 32), the low water activity prevailing during flour storage channels the reactions almost exclusively to fatty acids with di-OH derivatives predominating (Warwick and Shearer 1980), as shown in Table 13.4.

Also, endogenously enzyme-catalyzed hydrolysis of phytic acid *in* intact beans *in situ* does not permit buildup of intermediate phosphate esters, whereas these esters do accumulate in autolyzing slurries of bean powder (Chapter 36). This illustrates how the products of enzyme action may change at low a_w's. Similarly, glycogen in meat is not degraded in type 5 systems at a_w's at which glycogen in a type 3 system is readily degraded (Matheson 1962; Potthast *et al.* 1977). Undoubtedly, compartmentation as well as lack of capillarity may have been influential here.

Illustrative of the manner in which drying can alter the consequence of enzyme action is the ob-

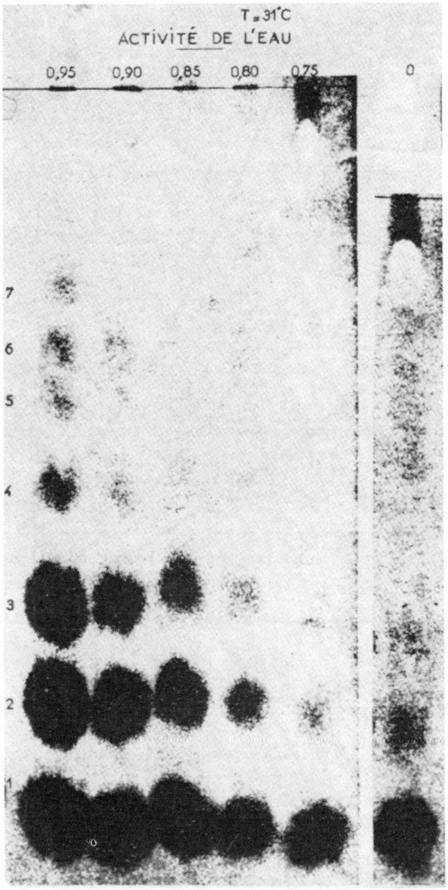

From Drapron and Guilbot (1962)

FIG. 13.7. QUALITATIVE CHANGE IN ENZYME REACTION PRODUCTS AT LOW WATER ACTIVITIES

The numbers 1–7 refer to the dextrin chain length resulting from the action of α-amylase at various water activities.

servation that uncooked dried beef tends to brown more rapidly than do similarly treated cooked samples. Aldolase and other glycolytic enzymes are presumably still acting in the first case to give rise to reducing sugars which can participate in the subsequent nonenzymatic browning favored at low moisture levels.

Enzyme Stability

Up to now we have been concerned with the action of the enzyme in presence of substrate at low water activities. One factor not hitherto mentioned which contributes to such action or lack of it is how much enzyme survives the drying process and subsequent storage. In this section we examine the consequences of water removal in foods and model food systems on enzyme stability. It is generally recognized that enzymes tend to be more stable in the dry state at ambient a_w than in solution. Thus, as shown in Fig. 13.8, inactivation of lipase in ground oat kernels becomes detectable at 30°C in the presence of 23% moisture but is not discernible in the presence of 10% moisture until a temperature of 60°C is reached. Although few systematic investigations have been conducted on the molecular factors which tend to preserve and protect enzymes in the dry state, it is probably true that some of the factors discussed in Chapters 5 and 6 on the stability of enzyme preparations are probably applicable here to enzyme stability in dried foods.

More attention has been specifically devoted to the question of why enzymes *in situ* or *in vivo* are unstable at sufficiently low moisture levels than what stabilizes them. Much of this interest stems from an attempt to understand what happens during freezing.

Central to such an understanding is a knowledge of the nature of the interaction of water with proteins, briefly inspected in Chapters 3 and 5. The analogy to freezing carries over to this area since some investigators envision the water bound to surfaces of dried protein as assuming an ice-like configuration. However, this ice-like configuration occurs only after sorption of a critical quantity of water (Berlin *et al.* 1970), below which the protein is no more than a randomly crosslinked "dust" through which water molecules move, according to Lumry (1973).

TABLE 13.4. LOX-GENERATED HYDROXY FATTY ACIDS IN WHEAT FLOUR[1]

			mg/100 g Dry Flour		
a_w		Time	Mono-OH	Di-OH	Tri-OH
1.0	10% suspension[1]	60 min	25	5	75
0.6	Dry flour, weak[2]	60 months	11	23	0
0.6	Dry flour, strong[2]	60 months	6	13	0

[1]Source: Graveland (1970).
[2]Source: Warwick and Shearer (1980).

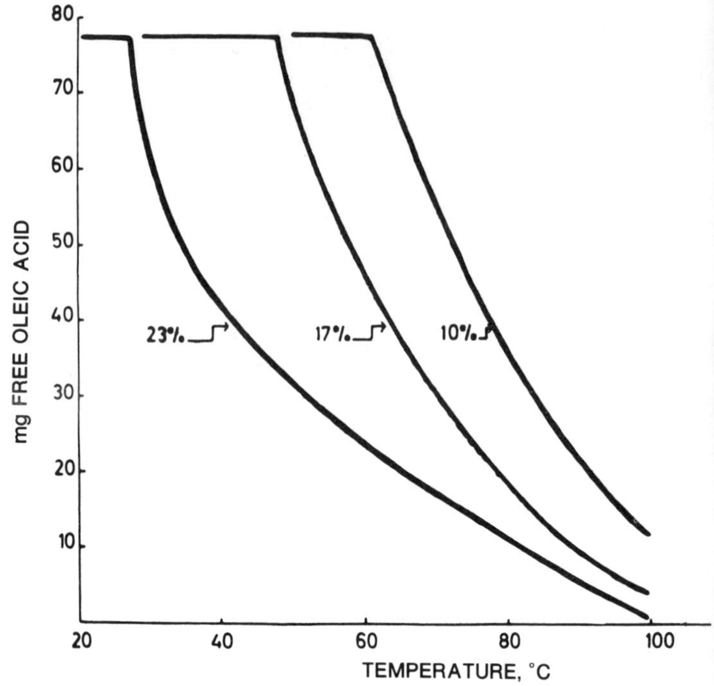

FIG. 13.8. THERMAL STABILITY OF LIPASE IN OATS AS A FUNCTION OF WATER ACTIVITY

From Acker (1969)

Model Studies.—As in the case of interaction of substrate and enzyme, models representing what might happen to enzymes in foods have been assembled and observed. Earlier investigations in this area are cited by Roozen and Pilnik (1971) and in the field of plant physiology by Darbyshire (1974A,B, 1975). Representative results obtained by Roozen and Pilnik are summarized in Table 13.2. As mentioned in the last chapter, Darbyshire and coworkers examined and compared inactivation effects of dehydration and freezing. For instance, at a water potential of −14 ($a_w = 0.55$) both aldolase and RDP carboxylase lost about 80% activity presumably due to fragmentation of enzyme molecules. From similar studies with and without additives and enzymes including glucose oxidase, catalase, peroxidase and glucoamylase, in general it would appear that low moistures and protein-protein interactions exert a xeroprotective effect analogous to and perhaps at least partially identical with cryoprotection in frozen systems and tend to stabilize enzymes. As in the case of freezing, large molecules, some buffers and some salts also exert protective effects when they are concentrated by water removal. In general, carbohydrates both as part of the enzyme molecule or added during drying afford xeroprotection.

In a food-related study on the stability of myosin (ATPase, see Chapter 26) in a meat model system, Nakano and Yasui (1976) found that myosin stability decreased with decreasing a_w until the BET region was reached and then increased upon further drying. Some inactivation also occurred during the drying process. They found that sucrose was an effective xeroprotectant.

A study which lies on the borderline of drying and heating effects is that of Multon and Guilbot (1975) on the stability of ribonuclease in wheat whose water content was varied from 4.4 to 45% and temperatures from 45° to 130°C. From the calculated energies of activation of the denaturation of this enzyme and application of absolute reaction rate theory, they concluded that the role of water in thermal denaturation in their samples was basically the same as its role in solution. However, they do qualify this conclusion by postulation that enzyme inactivation of the dryish enzyme is a two-stage process consisting first of a reorganization of bonding energies between water and polar sites followed by hydrogen bond breaking in the enzyme in which water participates. Parallelism between inactivation of ribonuclease in wheat kernels and change in a_w when the wheat is heated is shown in Fig. 13.9.

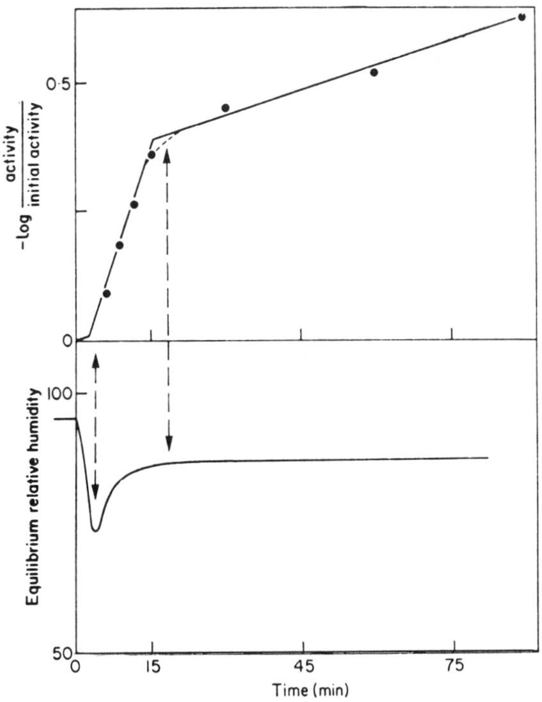

From Multon and Guilbot (1975). Copyright by Academic Press, London.

FIG. 13.9. PARALLELISM BETWEEN CHANGE IN RELATIVE HUMIDITY AND INACTIVATION OF NUCLEASE DURING THE HEATING OF WHEAT

Enzyme activity may also be lost by intermediation of active intermediates formed nonenzymatically by reactions which proceed more readily in the dry state. Thus, Kanner and Karel (1976) showed that peroxidizing methyl linoleate (which could also arise from enzyme action) inactivated the enzyme lysozyme in a freeze-dried model system. Loss of activity of the enzyme increased with increasing values of a_w.

APPROACHES TO ENZYME MANAGEMENT IN DRIED FOODS

While we shall have more than one occasion to illustrate modes of control of enzymes in dehydrated foods in subsequent chapters it may be apropos to briefly outline and classify in this chapter some of the approaches used and proposed for control or promotion of enzyme action in relatively dry foods and concentrates.

We conceive of three different ways the enzymes in dried foods can be treated. We may wish to: (1) completely eliminate all enzymes so that upon rehydration they will also be absent; (2) prevent them from acting during storage in the dry state, but preserve them so that they will act when the food is dehydrated; and (3) promote their action while the food is dry.

Blanching

For commercial dehydration of most foods, with certain exceptions outlined below, hopefully complete enzyme inactivation is accomplished by blanching, as first pointed out in the start of this chapter. Blanching of foods destined to be dried serves an additional function of partially cooking the tissue, resulting in an increased rate and extent of drying. This in turn improves texture and saves energy and money (Barta *et al.* 1973). Unlike freezing or canning, as mentioned, no correlation has been established between adequacy of blanching and the familiar standard enzyme tests.

Illustrative of the extension of the blanching approach beyond fruit and vegetables is the inclusion of blanching (moisture cooking) in a patent for the manufacture of a wheat kernel-based meat extender (Desrosier 1976). On the other hand, dehydration at 46°C of protein destined for use as a supplementary foodstuff (Lyall 1976) ensures the preservation of most enzymes as well as vitamins.

Other Approaches

Enzyme action in dry foods can be prevented by further decreasing their water content. One way to accomplish this is to freeze-dry. However, with the exception of specialty foods such as chives, most of the successfully freeze-dried items such as coffee are already devoid of enzymes.

Occasionally it is possible to prevent enzyme action using relatively economic means by simply lowering the moisture level by a rather trivial amount. A case in point is the successful prevention of hydrolytic rancidity due to lipase action (Chapter 25) in a fortified wheat blend by simply removing water to below a critical 12% at which lipase action ceased, and then providing waterproof packaging (Fellers and Bean 1980). In some instances this entailed a drop of only a few tenths of 1%. A rather novel principle was used to prevent enzymatic oxidative rancidity due to lipoxygenase action in such flour blends. Instead of lowering the moisture, content of the latter as well as the temperature was transiently raised to permit the lipoxygenase to self-destruct *via* reaction inactivation (Chapter 15). As expected, removal of lipid stabilized the enzyme (Wallace and Wheeler 1972).

Drying processes in themselves may be effective in eliminating unwanted enzymes. Limited empirical data on the drying of some sausages (Mihalyi and Koermendy 1967) and macadamia nuts (Prichavudhi and Yamamoto 1965) suggest that slow is more effective than rapid drying for preventing undesirable enzyme action (Schwimmer 1969A). While this generalization appears to be confirmed by the low temperature wilt-drying of Gee *et al.* (1977), the mechanisms by which the enzymes are prevented from acting are probably not the same.

As an example of allowing the enzymes to remain in a latent state, we point out that several dried fruits are treated with sulfite to prevent oxidative enzyme action (Chapter 15).

Another approach to preventing enzyme action in dried foods without destroying the enzymes is to ensure cellular integrity. As previously mentioned, wheat kernels can be kept indefinitely.

Utilization of Enzymes upon Rehydration.— The objective of dehydration of alliaceous foods such as onion, garlic, chives, leeks, shallots, as well as some seasonings and spices, is to remove water without inactivating the enzyme responsible for flavor production (Chapters 21 and 23). In addition the substrate must also be preserved so that one must prevent the two from getting together during the drying. Such flavor-precursor enzyme systems exist in many foods and drying may remove the enzyme and flavor without removing the precursor. Morgan and Schwimmer (1965) added enzyme-containing freeze-dried foods back to the foam-mat-dried foods to restore flavor. Pectin-transforming enzymes, PG and PE, may aid the rehydration process.

Finally, in a limited number of moisture-deficient foods it is desirable to allow enzyme action to continue during storage. We mentioned prevention of rice stickiness. A case where continued enzyme action during drying is absolutely essential is the development of high quality dates. We already mentioned that prevention of retrogradation of starch and its attendant loss of desirable texture in baked goods can be accomplished by the use of heat-stable bacterial amylase during partial drying taking place in baking. In the preparation of dehydrated sweet potatoes it is advantageous to activate its α-amylase prior to drying. These and examples discussed in more detail elsewhere in this book are illustrative of the approaches which are used to manage enzyme activity in dried foods.

14

CONTROL OF ENZYMES BY PERTURBATION OF MOLECULAR ENVIRONMENT

FOOD ADDITIVES AS ENZYME INHIBITORS

While manipulations of the temperature and water activity of foods are the principal modes for keeping unwanted enzyme action in check, the use of specific food additives can provide an ancillary approach. However, the opprobrium associated with the word "chemical" in the food additive context makes the future of this alternative route somewhat uncertain. In our previous discussion, consideration of changes and manipulation of the molecular environment as it affects enzyme action have been unavoidable, especially with regard to inhibition of enzymes and its different modes. The pluralism of purpose noted for these other means of stopping enzyme action also holds for enzyme inhibitors. Many of the commonly used food additives and adjuncts designed for specific nonenzyme purposes will also inhibit food enzymes as a glance at the GRAS (generally recognized as safe) list or a listing according to function (Fig. 14.1) reveals. Thus, antioxidants added to foods to prevent autoxidation are effective inhibitors, of several oxidative enzymes, and some of the preservatives are effective because they inhibit the enzymes of invading microorganisms and could also inhibit enzymes present in the food.

In assessing the suitability and selection of an additive as an effective enzyme inhibitor, there is available to the ingredient specialist an imposing array of substances amassed by investigators representing a wide spectrum of chemical, industrial and biological disciplines. Examples are: Physiologists using inhibitors to trace cellular function by inhibition of particular enzymes, toxicologists deal-

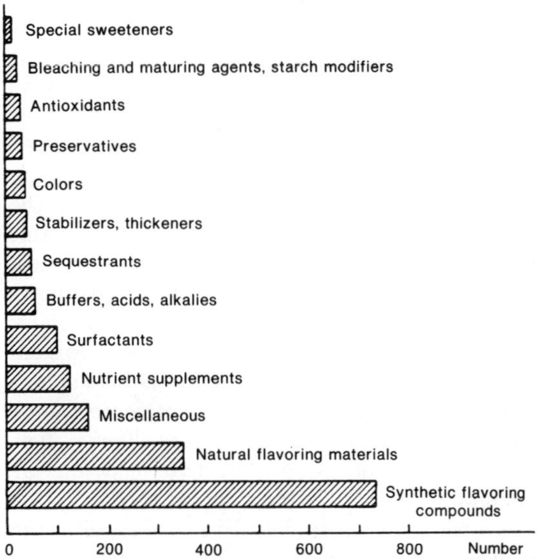

From Siu et al. (1977). Reprinted from Fed. Proc. 36, 2519–2562.

FIG. 14.1. FUNCTIONAL DISTRIBUTION OF FOOD ADDITIVES

ing with tracing the effect of poisons to enzyme inhibition, pharmacologists and medical scientists developing new medicines on the basis of their specific enzyme-inhibitory properties which are curative by virtue of their ability to inhibit certain key enzymes, and, of course, biochemists and enzymologists who use enzyme inhibitors as highly effective tools for elucidation of the nature of enzyme action. Historically, enzyme inhibitors were associated with toxins, as attested to by a chapter on "Enzyme Poisons" in Haldane's 1930 classic.

Design of Food Enzyme Inhibitors

One frequently encounters the designation "design" in enzyme inhibitor literature (Webb 1963; Lindquist 1975; Wolfenden 1978). The inhibitor may be designed as a specific medicine or designed to gain a more rational perspective of physiological processes such as ion transport or nerve signal transmission. How would one design effective enzyme inhibitors to be used in a food? Ideally, if we knew exactly which enzymes were causing which particular food quality defect, we could design an inhibitor which specifically inhibits that enzyme and no other. However, we would have to add the two additional criteria of safety and cost. We shall discuss safety later. With regard to cost, we want to use as little of an inhibitor as possible, but an amount which is extremely effective at low levels and is commercially available. Unfortunately the most potent, inexpensive and readily available inhibitors (for nonfood purposes) such as cyanide, Hg and other heavy metals are not particularly specific and most important, do a particularly excellent job of inhibiting the consumers' own enzymes, i.e., they are violent poisons. Others in lower ranges of concentration are therapeutic agents. In either case, these substances are eliminated as potential candidates for effective food enzyme inhibition, at least for the time being.

In some instances it should be possible either to remove or to attenuate untoward effects of the enzyme inhibitor in the food once it has accomplished its assigned task, so that its addition to the food would no longer be objectionable. One way to ensure such safety would be to attach the inhibitor to a large molecule in a setup akin to affinity chromatography (Chapter 7) so that while it still inhibits the particular enzyme whose action we want to prevent, it cannot be absorbed into the food consumer's bloodstream.

It would be preferable that the inhibitor be a natural product, especially one commonly present in foods. If not on the GRAS list, it should beyond reasonable doubt be safe for consumption at the level at which it is an effective inhibitor. Since there exists such a wide variety of quality attributes and enzymes which affect them, and since we as yet have not acquired the insight to make completely rational predictions, we may have recourse to "shotgun" inhibitors which, as we try to do in blanching, are aimed at irreversible inhibition or inactivation of all of the enzymes in the food and thus assure that the offending ones are no longer functional.

VARIETIES OF INHIBITION

With the above criteria in mind, we shall attempt to use classification schemata of enzyme inhibitors as a framework for the exploration of chemicals which might be used to control enzyme action, especially as an alternative to blanching. There are several approaches to categorization of enzyme inhibition. Haldane (1930) pointed out over 45 years ago that it would be theoretically advantageous to divide the effects of enzyme "poisons," a term he used interchangeably with inhibitors, into those which cause reversible and irreversible changes, i.e., inhibitors and inactivators. Enzymology has progressed to the point where enough is known about reversibility of enzyme activity to suggest that this is not the most logical division. Newer knowledge of the details of protein structure and the nature of enzyme action suggests a classification based upon:

(1) what part of the enzyme molecule binds the inhibitor
(2) the nature of the linkage between inhibitor and enzyme molecule, and
(3) the mechanism by which diminution of the enzyme activity is attained.

In a few cases, some inhibitors act so broadly or uniquely that special categories have to be provided for them. Accordingly we suggest the following classification:

A. Molecules which combine, for the most part noncovalently, with or at the active site of the enzyme by virtue of their topography vis-à-vis this site. The geometry of the inhibitor is thus of seminal importance for this class of inhibitors.
B. Small molecules each of which combines at a definite locus in the enzyme molecule other

than a locus in or within the domain of the active site.

C. Molecules, essentially chemical reagents, which react to form covalent bonds with specific active-site amino acids via their side chains. Such reactions may be catalyzed by other nontarget enzymes.
D. Large molecules, which interact with enzyme molecules via protein-polymer interaction, including other proteins.
E. Reagents which inactivate the enzyme by denaturation of the enzyme molecule. Such reagents destroy irreversibly the secondary, tertiary and sometimes the quaternary structure, usually by virtue of their being hydrogen bond breakers.
F. Substances which perturb those water-protein relationships which may be involved in catalysis, as limned in Chapter 3.
G. Highly active intermediates of the enzyme reaction which inactivate the enzyme.
H. The hydrogen ion at optimal concentrations.
I. Molecules which react with nonamino acid moieties of the enzyme molecule such as metals, coenzymes, carbohydrates and prosthetic groups.

Some of the more effective general inhibitors such as cyanide and heavy metal ions act through two or more of the above categories, which is probably one of the reasons that they are so effective. Each category is examined in terms of: What types of substances belong to what category, how they exert their inhibitory action beyond the mechanism implied in their categorization, which are used at present as practical enzyme inhibitors, and a prognosis and assessment of the potential of substances and categories hitherto not used or suggested as inhibitors of food enzymes.

ACTIVE-SITE DIRECTED INHIBITORS

Into this category fall most of the substances which have been shown experimentally to be fully competitive inhibitors. As sketched in previous chapters, the usual explanation for the kinetic behavior of this type of inhibition is that the inhibitor competes with the substrate for the active site of the enzyme. For the most part, a competitive inhibitor possesses a structure with a strong resemblance to that of the substrate.

Analog Inhibitors

The classical example is the inhibition of the succinate ($HOOC\text{-}CH_2\text{-}CH_2\text{-}COOH$) dehydrogenase complex by malonate ($HOOC\text{-}CH_2\text{-}COOH$). Such substances are frequently called analog inhibitors, substrate analogs and/or antimetabolites. There are as many different kinds of analogs as there are variations in structure: the just illustrated homolog analog and analogs based on chirality; positional (syn-anti), geometric (cis-trans), enantiomorphs and anomers. The latter are especially abundant in studies on mechanism of action of the glycosidases. As a recent example, the enzyme phosphoribosylaminoimidazole carboxylase, a central intermediate in the purine nucleotide biosynthetic pathway in most organisms (Chapter 20), is inhibited by β-D-5-carbon analogs and also by β-2-bromoderivatives synthesized for this study (Mackenzie *et al.* 1976). Occasionally the D-isomer will activate the enzyme acting on the L-isomer as substrate, i.e., trypsin hydrolyzing an arginine ester (Trowbridge *et al.* 1963).

Medical Applications and Significance.—The first of the "miracle drugs," the sulfonamides, are supposed to be effective because they inhibit an enzyme involved in the incorporation of benzoic acid, to which they are considered to be analogs, into folic acid. Antibiotics are frequently analog inhibitors of specific microbial enzymes. The chance for the successful medical application of an antibiotic often depends upon whether the particular enzyme being inhibited is peculiar to the microorganism being attacked by the antibiotic or is a "universal" enzyme found in microbes and man. For instance, penicillin inhibits an enzyme which catalyzes a crosslinking reaction involved in the incorporation of peptidoglycans into the bacterial cell wall. No equivalent reaction takes place in human cells. Similarly the classical example of an antibiotic, cycloserine, inhibits the incorporation of D-alanine into such cell walls. On the other hand, the antibiotic azaserine cannot be used therapeutically because it is an analog substrate of glutamine, which, as discussed in Chapter 4, is seminal to the regulation of all metabolism in all forms of life. The principal enzyme inhibited is involved in purine biosynthesis.

$$HOOC\text{-}HCNH_2\text{-}CH_2\text{-}CH_2\text{-}CONH_2$$
Glutamine

$HOOC\text{-}HCNH_2\text{-}CH_2\text{-}O\text{-}CO\text{-}HC{=}\overset{+}{N}{=}\overset{-}{N}$
Azaserine

Substrate Analogs of Food Enzymes.—An example in the food field of inhibition by a seeming substrate analog, the corresponding sulfoxide-less amino acid, is that of alliinase (Chapter 21), the flavor-generating enzyme in onions and other alliaceous foods (Schwimmer *et al.* 1964B; Schwimmer 1978).

$R\text{-}SO\text{-}CH_2\text{-}HCNH_2\text{-}COOH$
Substrate

$R\text{-}S\text{-}CH_2\text{-}HCNH_2\text{-}COOH$
Substrate Analog

In this case the inhibition is only partially competitive. Apparently both the inhibitor and substrate can occupy the active site or rather the *domain* of the active site simultaneously, but the presence of the analog diminishes the affinity of the enzyme for the substrate. This is manifested experimentally as a decrease in K_m but not V_{max}. The different modes of partial and complete inhibition are depicted in Fig. 14.2. The effect of homology on the inhibitor is shown in Fig. 14.3. The observed linear increment of effectiveness, related to a free energy function of homologous competitive inhibitors, has been frequently reported and seems to be independent of the type of reaction catalyzed, i.e., the lyase alliinase and the hydrolase choline esterase exhibit essentially the same behavior towards their homologous inhibitors; due, perhaps, to hydrophobic interaction between the nonpolar alkyl chains of the inhibitor and the hydrophobic domains near or at the active site of the enzyme molecules.

It is unlikely that the practical large-scale applications of the adaptation of substrate analog inhibition for the control of enzyme action in foods constitutes a fruitful avenue of exploration except for those listed in Table 14.1. Many are not natural products and thus would have to be tested extensively for safety and are likely to be expensive. This is especially true of nonsubstrate analogs (category I) since many of these are closely related to vitamins.

Product Inhibition

We have had occasion to refer to inhibition of an enzyme by the stable products of the reaction it catalyzed in sundry contexts. If the reaction is reversible, such products are themselves substrates

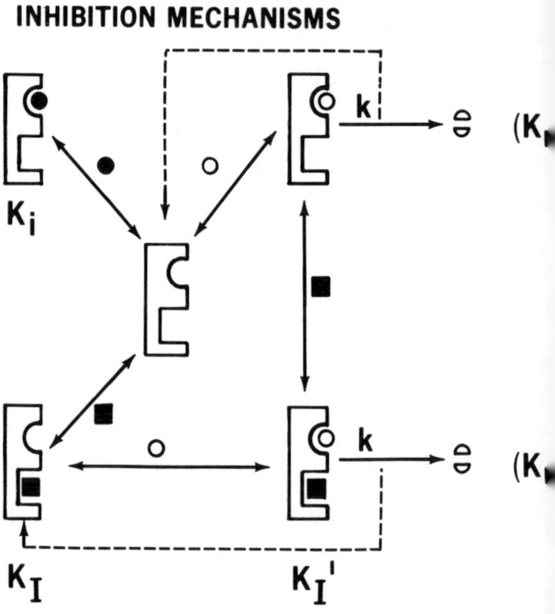

FIG. 14.2. MODES OF ENZYME INHIBITION

and are most likely to combine with the active site. Many product inhibitors of irreversible enzyme-catalyzed reactions also have such capabilities as, for instance, the competitive inhibition of phytases by phosphate as shown in Fig. 14.4. Product inhibition can thus be another form of substrate analog inhibition. While, as pointed out in Chapter 3, this product inhibition is representative of a relatively crude, imprecise mode of metabolic regulation, Frieden and Walter (1963), in a discussion of the role of product inhibition in evolution, suggested that product inhibition could form the basis for the operation of biological clocks.

While many enzyme reaction products probably are competitive inhibitors, this is not always the case. This is especially true of the depolymerases, where the intermediate products of breakdown are themselves substrates and hence competitive inhibitors of the breakdown of the parent substrate whereas the final product of the reaction may not combine with the active site of the enzyme as in the case of α-amylase acting on amylose (Schwimmer 1950). Other possible modes of interaction of the product with the enzyme to give different types of inhibition is shown in Fig. 14.5 and the corresponding chronometric integrals in Fig. 14.6.

Hsu and Tsao (1979) introduced a useful, convenient transformation of these equations. By plotting c against $(1/c)\ln[1/(1-c)]$, where $c=(S-P)$ of Fig.

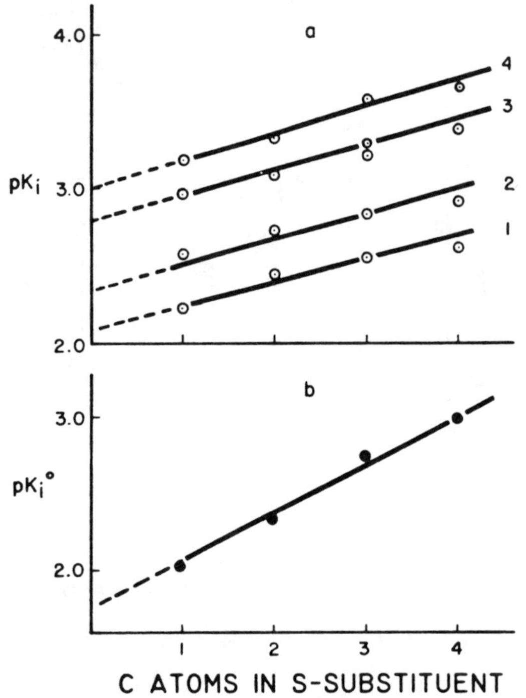

From Schwimmer et al. (1964A)

FIG. 14.3. EFFECTS OF HOMOLOGY OF SUBSTRATE AND INHIBITOR ON ENZYME KINETICS

The enzyme is onion alliinase, the substrates the homologous series of S-alkyl L-cysteine sulfoxides and the inhibitors are the corresponding S-alkyl L-cysteines. Abscissa, number of C atoms in S-substituent of the inhibitor. Numbers next to curves in a, C atoms in S-substituent of substrate; $pK_i^\circ = pK_i$ extrapolated to no C atoms.

14.6, one can ascertain, with minimum experimentation, if and which type of product inhibition of Fig. 14.6 is operational. Sometimes the final stable product will not affect the enzyme. Thus, pyruvate does not inhibit alliinase (Schwimmer *et al.* 1960). (See below for further comments on modes of product inhibition which are not strictly competitive.)

Relation to Foods.—Product inhibition can be a problem in application of enzyme adjuncts to foods. By both preventing the reaction from going to completion and slowing it down, such inhibition can spell the difference between practical application and laboratory curiosity. For example, as discussed in Chapter 36, the development of the use of lactase in foods, especially in immobilized systems, was at first hampered by intense product inhibition. According to Engasser and Horvath (1974), however, such inhibition tends to be less effective in immobilized enzyme systems where diffusion limitations exist (Chapter 8). This effect arises from the "antienergistic" interaction between chemical and diffusional inhibition of the activity. The consequence for product inhibition is that such antienergistic interactions make the immobilized enzymes less sensitive to changing product concentration in the macroenvironment. Conversely, an inhibitor product attenuates the magnitude of diffusional inhibition.

Product inhibition can also interfere with the desirable potentiation of desirable endogenous enzyme action. Thus, product inhibition by phytase-generated phosphate in beans apparently contributes to the failure to completely remove this sub-

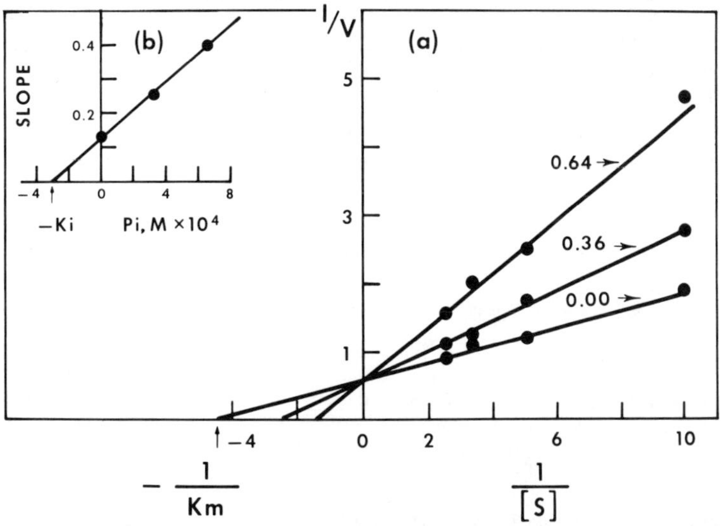

From Chang and Schwimmer (1977)

FIG. 14.4. PRODUCT INHIBITION OF BEAN PHYTASE

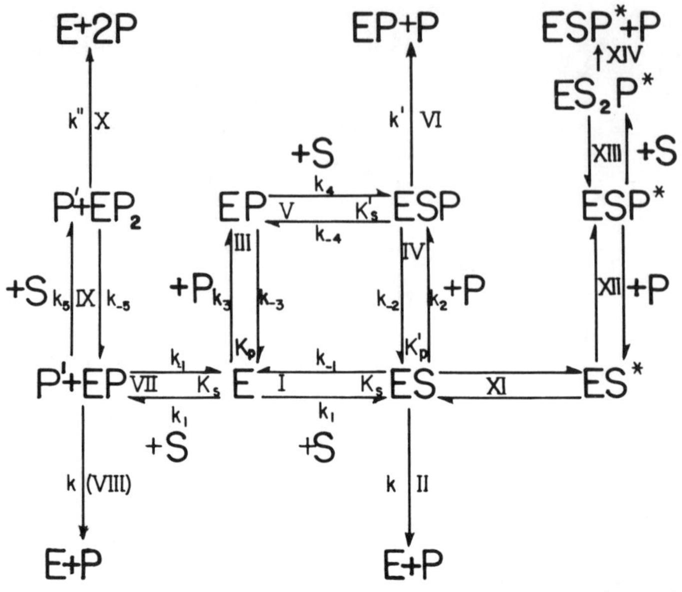

FIG. 14.5. MODES OF PRODUCT INHIBITION

From Schwimmer (1961A)

stance from beans exposed to high water activities (Chang *et al.* 1977).

This does not mean that product inhibition is always undesirable. Indeed it can be exploited to improve food product quality. For instance, it should be possible to suppress the action of unwanted enzymes present as impurities in commercial food enzyme adjuncts by adding the product of the reaction catalyzed by the unwanted enzyme impurity. Thus, lipases which hydrolyze the ester bonds in the α (1′) but not β (2′)-positions of triglycerides and which uninvitedly accompany many microbial food enzymes, i.e., lactase, can be controlled by adding 2′-monoglycerides. This would prevent development of a soapy taste in dairy products treated with lipase-containing lactase.

In vivo product inhibition has also been exploited to improve the quality of sweet corn, although as far as we know, this method has not been used commercially. In this somewhat futuristic application based upon an understanding of plant intermediary metabolism, pyrophosphate salts were

$$ket = \alpha \ln (S/S - P) + \beta P + \gamma f(P) \qquad (3)$$

Type of Inhibition[1], [2]	Reactions, Fig. 14.5	Alpha	Beta	Gamma	f(P)	Equation No.
No Inhibition	I, II	K_M	1	—	—	(4)
Fully Competitive (7)[2]	I, II, III	$K_M (1 + S')$	$1 - K_M/K_P$	—	—	(5)
Sequential Formation (7)[2]	VII, VIII	$(k/k_1)(1 + (k_{-1}/k_1)S)$	$1 - K_S$	—	—	(6)
Uncompetitive (7)[2]	I, II, IV	K_M	1	$\frac{1}{2K_P}$	P^2	(7)
Non-Competitive (11)[1]	I, II, III, IV, V, $K_S'=K_S$	$K_S (1 + S')$	$1 - K_S/K_P$	$\frac{1}{2K_P}$	P^2	(8)
Mixed Type (11, 12, 13)[1]	I, II, III, IV, V	$K_S (1 + K_r S')$	$1 - K_r(K_S/K_P)$	$\frac{K_r}{2K_P}$	P^2	(9)
Partially Competitive (11)[1]	I, II, III, IV, V, VI, $k=k'$	$K_S \frac{(1 + S')}{(1 + K_r S')}$	1	$\frac{K_S' - K_S}{1 - K_r S'}$	$\ln(1 + (K_r/K_P)P)$	(10)
Partially Non-Competitive (11)[1]	I, II, III, IV, V, VI, $K_S=K_S'$	$K_S \frac{(1 + S)}{(1 + S'/k_r)}$	k_r	$k_r K_P(1 - k_r)(1 + K_S(S - k_r K_P))$	$\ln(1 + (1/K_P K_r)P)$	(11)
Generalized Inhibition (12)[1]	I, II, III, IV, V, VI	$K_S \frac{(1 + S')}{(1 + (S/K_r k_r))}$	k_r	$k_r K_S \left[1 - k_r + K_S \frac{(1 - K_S'/K_P)}{(S + k_r K_S')} \right]$	$\ln(1 + (1/K_S' k_r)P)$	(12)
Formation of EP_2 (6)[2]	VII, VIII, IX, X	$Z - Y + S(Q - ZR)$	$R - Q$	$S (Y + RS)$	$[(1/S(S - P)]P$	(13)

From Schwimmer (1961A)

FIG. 14.6. CHRONOMETRIC INTEGRALS FOR PRODUCT INHIBITION

used to increase the sweetness of sweet corn (Amir *et al.* 1971). Pyrophosphate, a GRAS substance, is an end-product of the key reaction which is concerned with the conversion of sucrose to starch (Chapters 19 and 20) responsible for the rapid decrease in the quality of freshly harvested sweet corn.

$$\text{ATP} + \text{Glucose-1-phosphate} \rightleftharpoons \text{Pyrophosphate} + \text{ADPglucose}$$

By injecting pyrophosphate into the inner parenchymatic tissue of the ear of the corn in the field, Amir *et al.* were able to maintain the initial high sugar content and therefore desirable flavor. By inhibiting the reaction catalyzed by ADPglucose pyrophosphorylase shown above, the amount of ADPglucose was diminished and thus was less available for the formation of starch, catalyzed by starch synthase, one of whose substrates is ADPglucose (Chapter 19).

Active-site Irreversible Inhibition: Suicide Substrates

These constitute a special class of substrate analog inhibitors which react covalently with one or more groups within the active site. They are distinguished from general amino acid side group reagents by virtue of their shape vis-à-vis the active site. They are essentially substrates bearing highly reactive functional groups which react to form covalent bonds with groups within the active site (Chapters 3 and 37). It is sufficient to point out at this juncture that an excellent overall introduction to this type of inhibition is the monograph of Baker (1967).

Although Baker pays tribute to the organophosphates (Balls and Jansen 1952) as historically important, partially specific active-site irreversible inhibitors, he considers azaserine (see below) to be the first case of what he considers a true active-site-directed irreversible inhibitor. It indeed combines the geometrical restriction of the classical substrate analog. In the initial stages of its interaction with glutamine metabolizing enzymes it behaves as a competitive inhibitor with the special feature of reacting irreversibly with a specific group in the active site.

Prognosis for the application of such reagents to prevent enzyme action in foods is rather beclouded. Thus, considerable time and effort have gone into detection of even the faintest trace in foods of the grandparent of these inhibitors, the organophosphates, in spray residues of foods (Chapter 39). Of course the latter are effective insecticides because they inhibit the neurologically associated esterases of insects which attack crops. Those suicide substrates that are antibiotics would also have a difficult time passing the rigorous requirements of a food additive. Avenues which have to be explored to develop an approach based upon the use of active-site directed inhibitors should include, as mentioned, discovery of much more specific active-site inhibitors of this class which are completely specific for the particular enzyme whose action is not wanted, and attachment to matrices which prevent them from being assimilated into the bloodstream.

Illustrative of the difficulties which lie ahead is the interesting observation that aspirin but not salicylate is such an active-site directed inhibitor of prostaglandin synthase (Chapter 1), an enzyme which performs a similar but not identical task as that of the intensely food-related enzyme, lipoxygenase (Chapters 21 and 24). While aspirin itself apparently inhibits neither plant (Wardle 1978) nor animal (Siegel *et al.* 1980) lipoxygenases, the latter investigators did find that aspirin inhibits (as does salicylate) a platelet enzyme which converts the product of lipoxygenase, lipohydroperoxide, to hydroxyfatty acids, homologous to those produced by similar enzymes in many plant foods (Chapter 24). It would indeed be ironic if headaches in food processing created by lipoxygenase action could be cured by inhibiting this enzyme with aspirin, only to have this solution contravened by food and drug laws. Hopefully, if aspirin does indeed inhibit enzymes in foods, it may be effective at levels far below therapeutic dosages.

Transition State Analog Inhibitors (TSAI)

In Chapter 3 we briefly alluded to the importance of this class of inhibitors in contributing to an understanding of the nature of the efficiency of enzyme catalysis. Basically, the main idea is that the transition state of the substrates binds several orders of magnitude more tightly to the enzyme than does the ground-state substrate. To paraphrase Pauling (1948,1976), the conformation or geometry/morphology of active site is more complementary to a molecule with an only transient existence—the activated ES-complex—than to the substrate itself. Thus, catalysis by enzyme is likened by Wolfenden (1976) in his excellent review to midwives, in their easing the difficult passage of substrate from one metastable state to another, and the ease with which this is done depending upon extremely tight binding forces between the enzyme and the transition state. Therefore, substances whose structures resemble or are analogs of the transition state should be much better inhibitors than classical substrate analogs. The main criterion of the tightness of binding is comparison with the K_m value, although as pointed out previously K_m is made up of more than just the rate constants that go into a true dissociation or affinity constant measurement. Nevertheless the first test of a TSAI-like inhibitor is that its K_i be much smaller than the K_m of the substrate of which it is an analog. Another important criterion and one that is especially relevant to possible food use is that ordinarily the inhibitor should be completely inert. A further requirement as expressed by Wolfenden is, of course, that the substance bear some spatial, mechanistically understandable relationship to substrate and products. Lindquist (1975) points out that one should consider the metastable state of probable intermediates: carbonium ion, oxonium ion (lysozyme), tetrahedral adduct (chymotrypsin), and trigonal pyramid (ribonuclease).

Indeed some of the most successful transition state analogs consist of product and substrate covalently linked to each other. Thus, NAD-pyruvate adduct inhibits lactate dehydrogenase (Trommer *et al.* 1976) and NAD-acetaldehyde adduct inhibits alcohol dehydrogenase (Everse *et al.* 1971). Perhaps some of the early reports on competitive inhibition by substances whose structure bears only a remote resemblance to that of the substrate may owe their inhibitory effects at least in part to their resemblance to transition states. Included may be several amines referred to by Haldane as aldehyde reagents, as well as aldehydes. Wolfenden points out the inhibitory effectiveness of several aldehydes compared to the corresponding alcohols. The powerful inhibition of chymotrypsin by one of the most effective of such substances, acetyl-L-phenylalanylaminoacetaldehyde, with a K_i of 5×10^{-8}, could only be understood in terms of transition state theory. Leupeptines and antipain are extremely potent naturally-occurring peptide aldehyde inhibitors of several proteases (Lindquist 1975). Perhaps the most spectacular example which has come to our attention is the more than million-fold increase in affinity of thiazoline pyrophosphate for pyruvate dehydrogenases complex, in comparison with the K_m of the natural substrate, thiamin pyrophosphate (Gutowski and Lienhard 1976). However, according to some investigators an increase in affinity of this order of magnitude is still far from that required for a true or ideal transition state analog inhibitor. These and other substances which have been designed or proposed as TSAI and which are food-related in the context of this book are shown in Table 14.2. Not included are the aldonolactones, suggested as transition state analogs (Levvy and Snaith 1972), because, as pointed out by Pokorny *et al.* (1975), their inhibition constants are of the same order of magnitude as the K_m's of normal substrates.

The antibiotic nojirimycin (5-amino-5-deoxy-D-glucose) gives a 50% inhibition of β-glucosidase at $6 \times 10^{-7}M$ and is some 300 times more effective than D-gluconolactone. The α-glucosidase inhibitor shown in Table 14.1 is effective against enzymes hydrolyzing α-glucoside linkages as diverse as those present in amylose starch and sucrose (Schmidt *et al.* 1977). This inhibitor has been demonstrated to be effective in weight control of human subjects. Other antibiotics such as penicillin and cephalophorin may well be TSAIs rather than substrate analogs as are many of the naturally occurring proteins which inhibit proteinases such as the antitrypsins of egg white and soybean (see below).

TABLE 14.1

TRANSITION STATE ANALOG INHIBITORS OF FOOD-RELATED ENZYMES

Enzyme	Inhibitor	Inhibitory Effectiveness[1]
Phosphatase, acid	Tungstate	7500
α-Glucosidase	Oligosaccharide[2]	100,000
β-Glucosidase	1-Aminoglucose	70
Lysozyme	(N-acetylglucosamine)-4-lactone	70
Papain	Acetyl phenylalanyl glycinal	700
Subtilisin[3]	Benzeneboronic acid	110
Phenolase[4]	α-Hydrazinophloretic acid	100
Lactate dehydrogenase	Oxalate (*vs* pyruvate)	4000
Lactate dehydrogenase	NADH-pyruvate adduct	25

Source: Wolfenden (1976,1978).
[1] Ratio, K_m:K_i.
[2] Source: Schmidt *et al.* (1977). The inhibitor may be represented as $(Glu)_m$-Cyc-amino sugar-$(Glu)_n$ where Glu = D-glucopyranose unit, Cyc = an unsaturated cyclitol unit and linked through N of the amino sugar 4,6-dideoxy-4-amino-glucopyranose; $m + n = 1-8$.
[3] A microbial protease used in foods (Chapters 5 and 34).
[4] Source: Fourche *et al.* (1977).

TABLE 14.2

PRODUCT INHIBITION: NONCOMPETITIVE IS MORE EFFECTIVE THAN IS COMPETITIVE

% Reaction Completed	Time to Reach Indicated Completion (Relative) No Inhibition	Competitive	Noncompetitive
50	1.0	1.4	4.7
60	1.2	1.6	5.0
70	1.5	1.75	5.6
80	1.7	2.00	6.8
90	2.1	2.35	14.6

For an enzyme obeying Michaelis kinetics with $K_m = K_i$ (competitive) $= K_i$ (noncompetitive) $= 10^{-3}M$; substrate concentration, $10^{-2}M$; see Fig. 14.5 and 14.6 and Frieden and Walter (1963).

Application to and Prognosis for Foods.—It will be noted from Table 14.1 that most of the food-related enzymes listed are those whose action in foods is usually considered desirable. It will also be noted that some of these inhibitors such as oxalate are themselves naturally occurring food constituents and that they are effective at extremely low levels. We predict that useful practical inhibitors or at least practical information gained will arise from this branch of enzymology. It may be possible to control the acidity of fruits with analogs of phosphoenolpyruvate and ribulose diphosphate metabolizing enzymes (Chapter 20). Perhaps the efficacy of cinnamic acids, proposed for suppression of browning in fruits (Walker 1976) may be due to the role conferred upon them as TSAI-like substances. Binding of a TSAI to an enzyme may also result in the latter's inactivation (Gawon and Jones 1977).

NON-ACTIVE-SITE DIRECTED INHIBITORS

Classical Noncompetitive and Substrate Inhibition

To recapitulate (Chapter 2), experimentally a noncompetitive inhibitor (NCI) is basically one which reversibly decreases the molecular activity of an enzyme supposedly by combining at a site other than the active site. Unlike the competitive mode the degree of noncompetitive inhibition is independent of the substrate concentration, behavior which should be advantageous in selection of a food enzyme inhibitor. Kinetically this is manifested as constant K_m and decreasing V_{max} with increasing concentration of inhibitor. Actually, there are other modes of combination which obey some of these criteria. Thus, many irreversible inhibitors can prevent the substrate from combining at the active site but adding substrate also does not restore activity. However, as mentioned, NCI usually refers to reversible inhibition and involves non-active-site functional groups in the enzyme molecule, such as metal ions or sulfhydryls. For enzymes where the metal ion is not tightly bound to the protein but is essential for activity, most metal chelators and sequestering agents are NCIs. The anion fluoride, a broadly-acting inhibitor of lipases, phosphatases and other enzymes, usually acts noncompetitively. Salicylate is an NCI of most β-glucosidases whereas aspirin, N-acetyl salicylate is, as mentioned, a TSAI of an oxygenase. Other examples will be mentioned at the appropriate place.

Product Inhibition.—While most products are competitive inhibitors, there are enough exceptions and peculiar behavior in this regard to warrant a special discussion. Thus, α-glucose, one of the products of invertase action, is a NCI whereas β-glucose and α-fructose, the other products, inhibit competitively. The mixed type of inhibition observed for galactose, one of the products of lactase action (Kobayashi and Susuki 1975), may be due to the net effect of mixtures of α- and β-galactose.

Perhaps the one characteristic which earns such product inhibitors a special discussion is that in terms of persistency of inhibition, noncompetitive inhibitors are in a sense more effective than competitive product inhibitors. Kinetic analyses of chronometric integrals (Schwimmer 1961A; Frieden and Walter 1963) reveals that noncompetitive products slow up the reaction at later stages much more than do competitive inhibitors, although the degree of inhibition of initial rates is the same. This is dramatically illustrated in the data of Table 14.2.

We previously mentioned that the final product of depolymerase action inhibits noncompetitively whereas intermediate product oligomers, in themselves substrates, inhibit competitively. This behavior is not relegated to polymer hydrolases only. Some of the NAD-dependent dehydrogenases exhibit rather peculiar product inhibition behavior in this regard. For both lactate and alcohol dehydrogenases

$$RH_2 + NAD^+ \rightleftharpoons R + NADH + H^+$$

where RH_2 can be ethanol or lactate, and R acetaldehyde or pyruvate, R inhibits the forward reaction noncompetitively and RH_2 inhibits the reverse reaction noncompetitively. Conversely NADH inhibits the forward reaction competitively as does NAD^+, the reverse reaction. On the other hand all corresponding products of the glyceraldehyde phosphate dehydrogenase reaction inhibit noncompetitively in both directions, as does the apparent substrate analog, threose-2,4-diphosphate.

Not restricted to product inhibition, there is some evidence that such paradoxical behavior may in part be due to the ability of the NCI to enter the domain of the active site or on the fringes thereof and not completely replace the substrate. This is illustrated in Fig. 3.5 which depicts the noncompetitive inhibition by most anions of the Zn-containing carbonic anhydrase. The Zn atom is located at the bottom of a cleft comprising the active domain. The substrate or competitive inhibitor is initially bound to hydrophobic areas of the active site but then replaces the OH of the Zn-coordinated H_2O (involving the imidazole side group of histidine Im). Since the inhibitor is bound directly to the metal, it does not interfere with the complexing of the substrate to the active site but it does prevent catalysis. This is manifested as NCI.

We have elaborated on the foregoing illustrations of NCI to point out that the molecular reality behind the experimentally observed NCI which appears to be simple can be quite complex and is still not completely understood. As mentioned in Chapter 3, NCI was one of the peculiarities of enzyme behavior that led Koshland and colleagues to the concept of induced fit.

Substrate Inhibition.—We have had the opportunity to examine inhibition by excess substrate in Chapters 2, 4 and 13. It is most likely that in most instances an additional substrate molecule which combines with the enzyme to give rise to these types of inhibition does so at loci other than the active site.

Feedback Inhibitors and Allosteric Modulators

These constitute classes of enzyme inhibitors which have been largely unexplored and unexploited from the viewpoint of preventing unwanted enzyme action in food. This benign neglect may not be entirely accidental since these substances are usually associated with the regulatory enzymes of intermediary metabolism. Traditionally most of the food-related enzymes are not under *in vivo* allosteric control. This situation does not have to persist into the future. For instance one of the key enzymes catalyzing a reaction which eventually leads to C_6-C_3 phenolic compounds in fruits and vegetables (flavonoids, anthocyanins, etc.), phenylalanine lyase, is inhibited by certain flavonoids (Attridge *et al.* 1971) which may be considered as end-products of the pathway of which this enzyme is a key participant. Conceivably, application of appropriate flavonoids could attenuate enzymatic browning where it is not wanted by preventing the formation of phenolase substrates (Chapter 15). The circumstance that such inhibitors may have to be applied *in vivo* may present some difficulties. Perhaps the previously discussed encouraging work on the use of pyrophosphate to maintain the fresh flavor of sweet corn points the way. In any event this type of inhibition would probably have to involve "natural" food ingredients to be acceptable and thus appears to offer promising ground for future research and application to foods.

Side Chain and Metal-directed Reagents as Enzyme Inhibitors

For the most part the effects of these types of reagents are manifested kinetically as noncompetitive inhibitors. They can react with sulfhydryl, free amino and carboxyl groups or with the carbohydrate moieties of the enzyme as well as with individual amino acid side chains and metals. The more generally they react, however, the more poisonous they are. They include such violent toxins as mercury compounds, cyanide, azide arsenicals and hydrogen sulfide. Hence most members of this class of inhibitors are not suitable for food use. However, there are some exceptions. Thus, sorbic acid may act as a good food preservative by inhibiting microorganisms' sulfhydryl enzymes. There are also the numerous GRAS metal chelators which may inhibit metal-requiring enzymes in foods, such as citrate, pyrophosphate and EDTA, although such substances can also activate other enzymes by removing offending metal ions.

Members of this subclass of inhibitors which have been proposed specifically for food use are the phosphatides, found abundantly in animal brain tissue blood clots, membranes, etc. (Fullington 1971). Used in conjunction with the divalent cations such as Ca^{2+}, they apparently form chelates with the aid of the amino and hydroxy groups of the enzyme amino acids. Although not presently in commercial use to our knowledge, this general approach deserves further development especially because the substances are themselves food constituents.

Some investigators emphasize the difference between reversible interaction with these groups and more drastic irreversible "protein" modification (Chapter 34).

Sulfite.—In the form of salts of sulfite (SO_3^{-2}), bisulfite (HSO_3^{-1}), and as the gas sulfur dioxide, SO_2 is one of the few chemical additives still allowed for the suppression of off-color in a wide variety of fruits and fruit products. Like other means of preventing enzyme action it also suppresses microbial growth. Its mode of action encompasses a protean collection of mechanisms, not all of which are thoroughly understood. Reviews on how it prevents browning and affects enzymes include the earlier ones of Joslyn and Ponting (1951), Joslyn and Braverman (1954) and more recently that of Haisman (1974). Speculation and experiments on the nature of its action go back over 50 years to the investigation of Overholser and Cruess (1923) who concluded that SO_2 prevented browning by destroying "organic peroxides." Since then several enzymes have been found to be inhibited by sulfite. Thus the NAD-dependent dehydrogenases are inhibited by forming a complex between sulfite and NAD. It inhibits thiamin-requiring carboxylases by cleaving the thiamin moiety of cocarboxylase, thiamin pyrophosphate. Perhaps the most intense inhibition by sulfite is that of aryl sulfatases, apparently by competitive inhibition. It inhibits urease and probably other sulfhydryl-requiring enzymes by oxidation. Interestingly the latter inhibition is more effective at lower temperatures (Kistiakowski and Lumry 1949). Also inhibited are lipoxygenase and ascorbate oxidase. Its inhibition of phenolase is due to at least two factors: A truly irreversible inactivation and the formation of colorless adduct with quinones, products of phenolase action, thus prevent them from being transformed into colored melanins. The frequently observed softening of fruits and vegetables by sulfite may be due to inhibition of superoxide dismutase and perhaps catalase (Kon and Schwimmer 1977). Its use in conjunction with other food enzyme inhibitors will be discussed in the appropriate chapters. Like so many other food additives, a renewed look at safety aspects has suggested some alleged hazard associated with its use in some foods. Most countries do not permit its use in foods at any level. Unless some inexpensive means are found to remove SO_2, perhaps by enzymatic means (i.e., sulfite oxidase), its future as an enzyme inhibitor in food looks dim.

Polymer-Enzyme Interactions

Protein-Enzyme.—Many hydrolases are inhibited by substances consisting of polymeric molecules some of which are specific inhibitors elaborated by foodstuffs. The most intensively studied are the antiproteases, which as mentioned earlier in this chapter may exert their inhibitory action by being transition state analog inhibitors. Their study from both plants and animals has become a subdiscipline in itself, with some 30 to 40 entries in each issue of *Chemical Abstracts.* Background reviews include those of Tschesche (1974) and Means *et al.* (1974). Although under proper conditions they might be used to advantage in foods, in general those found in plant foods, especially in soybean, present a safety or health problem. Analogously, many plants, especially the cereals, elaborate proteins which inhibit animal but not plant α-amylases. Marshall *et al.* (1976) suggest that inhibitors of pancreatic α-amylase might be used as novel dietetic or weight-reducing agents. However, one wonders what would happen when the undigested starch was exposed to the flora of the lower gastrointestinal tract where the effectiveness of the inhibitor on microbial amylases might be attenuated.

By isolating pancreatic lipase inhibitor from soybean, Satouchi *et al.* (1974) completed the list of polypeptide inhibitors from legumes which inhibit pancreatic enzymes responsible for the digesting of the three major foodstuffs: proteins, carbohydrates and fats. Other plant proteins which are specific hydrolase inhibitors, reviewed by Pressey (1972), will be mentioned in appropriate chapters. Microorganisms can also elaborate similar inhibitors.

Some *Streptomyces* elaborate amylase inhibitors of unusual composition. The ratio of amino acids to natural sugar in what are really peptidoglycans rather than glycoproteins was found to be as low as 0.09. *Streptomyces* also produce a proteinaceous alkali proteinase inhibitor (Uyeda *et al.* 1976) and are also the source of an interesting acid-protease inhibitor, pepstatin (inhibits pepsin), isovaleryl-L-valyl-L-valyl-4-amino-3-hydroxy-6-methyl hepatanoyl-L-alanyl-4-aminoheptanoic acid (Barrett and Dingle 1972). It is thus similar in structure (except for being an acid rather than an aldehyde) to the previously mentioned naturally-occurring peptide aldehydes, but is apparently not considered to be a transition state analog inhibitor.

Another kind of enzyme-protein interaction which could result in inhibition is that between the enzyme as antigen and its antibody. While this is of prime interest and importance in analytically related applications of enzymes (Chapters 37 and 39) antibodies are not likely to prevent unwanted food enzyme action in the near future.

Polyelectrolytes and Other Macroions.—We alluded to these substances in our discussion of enzyme stabilization in Chapter 5. Table 14.3 shows some of the polymers that might be of interest to the food scientist. For the most part they combine with groups on the enzyme which bear the opposite charge to that of the polyelectrolyte and thus interact with enzyme *via* electrostatic forces. The resulting complex may result in an enzyme with higher or lower activity or simply greater stability. Like most effective, broadly-acting enzyme inhibitors, the mechanism of inhibition (or of activation) is pluralistic. Elbein (1974) refers to the following factors which might lead to inhibition: induction of conformational changes in the enzyme, covering up of the active site, bringing enzyme and substrate into improper alignment, and removal of an activator. Nonionic polymers are said to cause an increase in affinity.

With regard to their possible use as food enzyme inhibitors, one advantage macroions have is that they are not likely to be digested and absorbed. Heparin, a mixture of sulfated mucopolysaccharides found in mammals, appears to be a possible candidate. Nucleic acids may be too expensive and could cause gout. Gatfield and Stute (1975), recognizing the potential application of such macroions as effective food inhibitors, investigated the inhibition of horseradish peroxidase by the natural macroions, pectin and carrageenan. The latter, like heparin, is a mixture of sulfated glycans, obtained from Irish moss. With pectin the net result was a shift in the pH optimum from about 5 to about 8 so that at neutrality and above, pectin activated, the maximum absolute rate being about 10% above that of peroxidase alone. At pH 5, close to the pH optimum of the enzyme alone and to that of many vegetables, a profound inhibition—almost 90%—was observed. Gatfield and Stute believe that because of the complexity of such systems, a general application of the inhibitory effect of macroanions appears at the present time to be limited.

Thus, they might cause the activation of some undesirable enzymes while inhibiting others. However, we believe that they may show some promise in conjunction with blanching and some of the other means of preventing enzyme action. In this connection Muller (1973) was issued a patent in which raw fruits and vegetables were mixed with an "enzyme blocking" agent having negatively charged portions on its molecule, a suitable agent consisting of pectin and carrageenan added to a food below blanching temperatures. Most if not all extracellular biological macroions are polyanions (Scott 1981).

TABLE 14.3

POLYELECTROLYTES AS FOOD ENZYME INHIBITORS

Inhibitor	Enzyme Inhibited[1]
Heparin and related sulfated polymers such as chondroitin, chitin	I–VII
Sulfated starch, amylose and amylopectin	III, VII
Pectin, and sulfated pectin and related polymers	III, IV, VIII
Sulfated cellulose, dextrans and related polymers	V, VII
Polyacrylic acid and other synthetic polymers	IX
Cations, protamine, polylysine and histone	VI, VII, X

Source: Adapted from Elbein (1974).
[1] I, alkaline phosphatase; II, acid phosphatase; III, the amylases; IV, ribonuclease; V, lysozyme; VI, trypsin; VII, pepsin; VIII, peroxidase; IX, catalase; X, lipoprotein lipase (Chap. 33).

Tannins.—These are substances which showed great promise as possible general-acting enzyme inhibitors. Tannins may inhibit depolymerases in general by virtue of their entropy change-associated interference with the polymeric substrate. Thus it is frequently observed that tannins do not inhibit the action of such depolymerases acting on homologous oligomers as substrates. Although tannins were used as enzyme adsorbents in the early days of enzymology, it was not much more than 35 years ago that measurement of the inhibition of amylase was suggested by Barnell and Barnell (1945) as an index of astringency in bananas (Chapter 37). Inhibition of enzymes by tannins used to be reviewed often. A comparatively recent review which cites earlier work and reviews is that of Tamir and Alumot (1969). This inhibition presents a problem to plant biochemists who are interested in extracting and determining the level of enzymes. The object here is to remove the tannins so that they do not interfere with extraction and determination of the target enzyme. To this end, of the several tannin complexing agents which have been investigated, polyvinylpyrrolidone (PVP) has emerged as the complexing agent of choice (Loomis and Battaile 1966). Tannins are also of interest to plant pathologists because they appear to be part of the defense mechanism of the plant, involving inhibition of enzymes of insect and microbial, especially fungal, predators. Thus, Chan *et al.* (1978) found that a tannin is the specific anti-boll weevil factor in resistant cottons. Also microbial cellulase inhibition has been exploited for nonfood uses such as protection and preservation of fishnets and other cellulosic fiber artifacts such as sandbags, ropes and tents.

Foods.—One of the earliest reports on the use of tannins for prevention of deterioration of food quality was that of Joslyn (1929) who observed that the softening of dill pickles was inhibited by grape leaves. In a series of papers and patents by Bell and Etchells and coworkers, the last of which is that of Bell *et al.* (1965), it was shown that grape leaves do indeed contain tannins which inhibit cellulase and pectinases which cause softness in pickles. As expected, astringent fruits such as persimmon and some sorghums are also good sources of inhibitory tannins. That tannins can inhibit enzymes other than hydrolases is attested to by the inhibition of NAD-dependent dehydrogenases by wattle tannin (Goldstein and Swain 1965) and the apparently specific phenolic, possibly tannin-like, noncompetitive endogenous inhibitor of nitrate reductase in radish cotyledons (Stulen *et al.* 1971), using our endogenous inhibition kinetic analysis (described in Chapter 37) to estimate its level.

Unfortunately tannins as food additives are now looked upon with disfavor from a health viewpoint (Chapter 35) and it is unlikely that they will join the armamentarium of effective, inexpensive food enzyme inhibitors.

Inhibition of Enzymes by Disturbance of Protein-Water Relations

We have sallied into this region in force in Chapters 12 and 13 in order to explain the behavior of enzymes at low temperature and moisture levels. We shall now elaborate on some of the more significant contributions of perturbation of water-protein relations from the viewpoint of the resultant modification of the chemical environment and its consequences for enzyme action in foods, especially with regard to deliberate rational control of such action.

Salts and Other Solutes.—Even in dilute solution, inhibiting effects of some salts on some enzymes have been ascribed to perturbation of poised water-protein relations. According to Low and Somero (1975), surface-exposed amino acid side chains and peptide bonds in an enzyme are involved in water transfer in conjunction with conformational changes (induced fit, Chapters 2–4) occurring during enzyme catalysis. Salt, by influencing the degree to which water can organize around these shifted enzyme groups, modifies the thermodynamics of the reaction. The extent and direction of such modifications determine whether the salt inhibits or activates. High concentration of salts disturbs water-protein relations in a much less subtle manner, simply by removing water from the reaction medium. Inhibition by high concentrations of sucrose and other sugars probably involves similar water-protein changes as does reversible inhibition by organic solvents (Chapter 13).

Hydrogen Bond Breakers.—Into this category fall substances which completely destroy the tertiary and secondary structure of enzymes and are essentially denaturants which cause irreversible inhibition, i.e., inactivation. As discussed in previous chapters, denaturation entails almost total disruption of the bonds which hold the enzyme molecule more or less locked into its unique configuration which, as previously mentioned, is highly dependent upon the state of the water molecules. This also occurs when enzymes are heated. Chemicals such as urea, guanidine, organic solvents and hydrogen ions can do the same thing. It is not likely that such chemicals in the high concentration required (other than hydrogen ion as discussed below) will have an impact as food enzyme inhibitors.

Surfactants and Disturbance of Hydrophobic Interaction.—Many but not all detergents inhibit enzyme activity, probably by virtue of their ability to cause unfolding of the enzyme molecule through disruption of hydrogen bonds, but especially by disturbance of hydrophobic interactions. In the proper detergent, i.e., sodium dodecyl sulfate, most quaternary enzyme molecules not only dissociate into protomers (subunits, Chapter 4) but also completely unfold. Detergent molecules completely surround the subunits and lend their charge to it. This is the basis for the widespread use of zone electrophoresis for the determination of the molecular weight of proteins, including of course, enzymes (Weber and Osborn 1969). In food applications *in situ* detergents may influence enzyme action indirectly by changing cell permeability and disrupting cellular membranes. Their proposed use in acid and organic solvent, and as an alternative to blanching, will be discussed below.

Inactivation of Enzymes by Enzyme Action: Suicide Products

Enzyme action itself can be exploited to prevent by chemical means undesirable enzyme action. This seemingly paradoxical circumstance can come about in three ways: (1) modification of products and substrate by other than the target enzyme, (b) inactivation of the target enzyme by other enzyme, i.e., proteinase action on unwanted enzyme, and (3) inactivation by highly reactive products (reaction inactivation, suicide products). These were briefly discussed in Chapter 1. We also pointed out that some modifications following enzyme action in foods were due to secondary reactions resulting from these highly reactive secondary products. As discussed in Chapter 1 some of these primary and occasionally secondary products can react with enzymes in such a manner as to inactivate them. Its use in preventing the undesirable consequences of lipoxygenase action is covered elsewhere, as is reaction inactivation of phenolase and other enzymes (Van Sumere *et al.* 1975). Phenolases can also give rise to somewhat more stable quinones which can then inactivate other enzymes such as phosphorylase (Chapter 15). The products of the flavor-producing enzyme reactions of radish, onion and garlic and other crucifers and alliaceous foods are general enzyme inhibitors. Hydrogen peroxide, superoxide ion, products of oxygenase action and other active forms of active oxygen generated by their interaction can also inhibit and inactivate many enzymes if present in the food long enough. As an example, as mentioned in Chapter 13, peroxidizing methyl linoleate (which could be formed through the action of lipoxygenase) reacts with muramidase (lysozyme) in low moisture systems to inactivate it, presumably by inducing free radical formation in the enzyme.

A rather specialized inhibiting effect of one free radical, the superoxide anion, on an enzyme is afforded by its involvement in the mechanism of oxygen inhibition of the NADPH-dependent nitrate reductase which converts aromatic nitro groups to the corresponding amines. Normally,

$$\text{Ph–NO}_2 \longrightarrow \text{Ph–NO}_2^- \longrightarrow\text{'} \text{Ph–NH}_2$$

In the presence of O_2, the nitroaromatic anion radical reacts with oxygen to form superoxide anion and reverts back to the original nitro compound:

$$\text{Ph–NO}_2^- + O_2 \longrightarrow \text{Ph–NO}_2 + O_2^-$$

The apparent prevention of enzyme action is thus somewhat reminiscent of the re-reduction of phenolase-produced quinones back to the original phenolic substrates in the presence of reducing substances such as ascorbic acid (Chapter 15). The suggested use of the disulfide/thiol interchange enzyme to inactivate other enzymes was discussed in Chapter 10. There are probably many such special cases which might be advantageously exploited for the prevention of unwanted enzyme action in foods.

HYDROGEN ION AS ENZYME INHIBITOR

The observation that a substance such as a macroion salt can act as both an activator and an inhibitor of an enzyme has been alluded to several times. The hydrogen ion is an example par excellence of such a schizophrenic enzyme affector. As a result of this ambivalent behavior, when one investigates the effect of pH on a particular enzyme one in general observes that the enzyme is active within a restricted range of pH, that the enzyme acts maximally over a much narrower range of pH, and finally that at extreme pHs on both the acid and alkaline side, the enzyme is irreversibly inhibited (inactivated) if left at such an extreme for a sufficiently long period. For many enzymes, the activity as measured by the initial rate of reaction as well as by the extent of the reaction will exhibit a typical Gauss-like response to pH, analogous to that observed with temperature (Chapter 10). The final shape and location of the optimum of such pH-activity profiles may depend not only upon the intrinsic activity but also upon such factors as method of assay, degree of inactivation (especially at extremes), buffer anion, stabilizers, temperature and on substrate and product inhibition. It is in practice necessary to distinguish between pH-activity profile from which the pH optimum is estimated (the activity corresponding to maximum activity) and the pH-stability curve. For such a curve the enzyme is exposed to extremes of pH; the higher the temperature the more restricted the range over which the enzyme is stable. The pH-activity profile for onion alliinase, a C-S lyase, is shown in Fig. 14.7 and pH stability for papain in Fig. 14.8. In this case the optimum is rather sharp and the curve is rather skewed. In many of the old textbooks one finds lists of pH optima. We now know that enzymes catalyzing the same reaction can have pH optima which vary over a wide range. This is especially true of some of the tailor-made food enzymes produced from microbial sources. Thus, the need for an alkali-stable amylase to be used in conjunction with detergent prompted the development of alkali active and stable microbial enzymes. Even pepsins, which traditionally are considered to act optimally in a pH range of 1.8–2 can have optima as high as 4, as in the case of the chicken enzyme (Chapter 33). It is traditional to place the designation "acid" or "alkaline" before an enzyme, to indicate the pH range over which it is active. There may be exceptions to the general rule that the lysomal hydrolases, such as the cathepsins, all act optimally in slightly acid media (Chapter 27).

pH rate profiles occasionally exhibit multiple or double optima by apparently the same enzyme activity. Their appearance may arise from a variety of cases in addition to the presence of two distinct isozymes. These include the presence of inhibitor or peculiarities in the ionization of the substrate. Examples include invertase (Chapter 19), β-glucuronidase (Levvy and Snaith 1972) and polygalacturonase (Patel and Phaff 1960).

Theory of pH Control of Activity

The appearance of pH optima is causally related to the circumstance that enzymes like all proteins are prototropic ampholytes. That is, they contain prototropic groups which can both accept or give off protons. It is clear that, by and large, the change in activity is guided by the degree of the ionization of the prototropic side groups of the amino acids of the active site. Such change affects the conformation of the active site and dictates whether the active site will receive the substrate.

This state of affairs was anticipated some 65 years ago by Michaelis and Davidsohn (1911) and elaborated upon by Michaelis (1922). The basis of

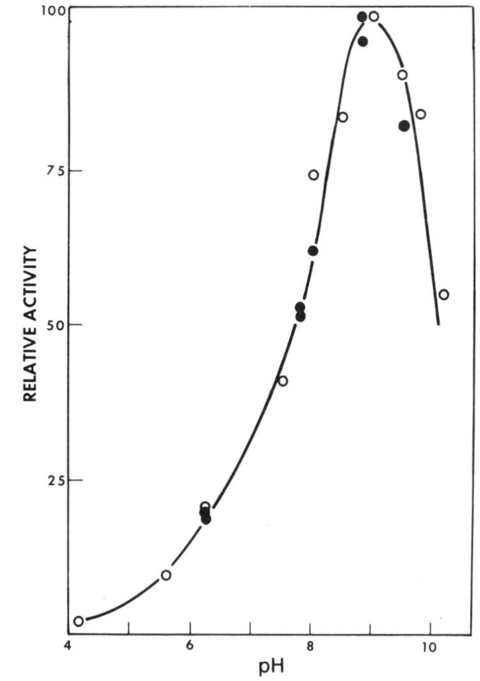

From Schwimmer and Mazelis (1963)

FIG. 14.7. pH-ACTIVITY PROFILE OF ONION ALLIINASE

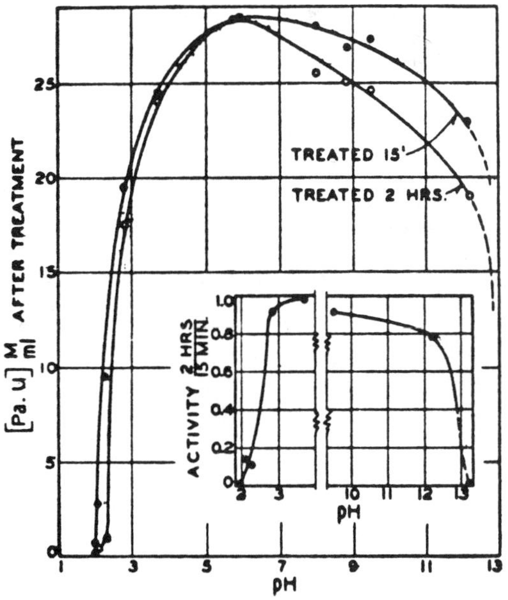

From Lineweaver and Schwimmer (1941)

FIG. 14.8. pH STABILITY CURVE FOR PAPAIN

this idea is that only one of the prototropic forms of the enzyme $EH^{+n} = E$ possesses catalytic activity. Letting $D = EH^{+(n-1)}$ and $F = H^{+(n+1)}$, the theory in its simplest steady state form is as follows:

$$E + S \rightleftharpoons ES \xrightarrow{k} E + P$$

$$D + H^+ \rightleftharpoons E \qquad K_1 = ([D][H^+])/[E]$$

$$E + H^+ \rightleftharpoons F \qquad K_2 = ([E][H^+])/[F]$$

By replenishing or depleting the E reservoir H^+ can act as an activator or inhibitor. At steady state, the rate of formation of ES is equal to the rate of its disappearance (Chapter 2) and the velocity of the reaction v = k [ES]. The total enzyme concentration can be expressed as the sum of its various forms:

$$e = [E] + [ES] + [D] + [F]$$

Each of the terms can be expressed as a function of ES:

$$[E] = \frac{K_m}{[S]}[ES]; \quad [D] = \frac{K_1 K_m}{[H^+][S]}[ES];$$

$$[F] = \frac{K_m [H^+]}{K_2 [S]}[ES], \text{ and}$$

$$\frac{e}{[ES]} = \frac{K_m}{[S]} + 1 + \frac{K_1 K_m}{[H^+][S]} + \frac{[H^+]K_m}{K_2[S]} = \frac{ke}{v}$$

Casting the equation into the standard Michaelis form (Chapter 2) we obtain:

$$v = \frac{ke[S]}{K_m(1 + K_1/[H^+] + [H^+]/K_2) + [S]}$$

The term which modifies K_m is the Michaelis pH or f function. Of course there are much more elaborate constructions which take into account the formation of unproductive or partially suboptimally productive enzyme substrate complexes with more than one prototropic form of the enzyme. Others include union with only one ionized form of the substrate. The formulations all contain Michaelis f functions. These equations are succinctly summarized by Dixon and Webb (1964) and Whitaker (1972A).

In the early days of enzymology, it was suggested and widely assumed that the catalytically active prototroph, E^{+n}, was that which was maximally present at the isoelectric point of the enzyme. However, further kinetic analysis, especially that involving the effect of pH on apparent K_m and V_{max}, revealed the presence of a prototropic group in most enzymes with a pK_a value of about 6–7. pK_a, it will be recalled, is the pH at which half of groups are ionized. The observed pK_a range corresponds to that of the imidazole group of histidine, which as we have seen, is usually part of the active site and is seminal for the consummation of the catalytic act. Other side groups for which there is evidence of participation in determining the shape of the pH-activity profile include, at least for some enzymes, the ε-amino group of lysine (pK = 9–11), aspartate and glutamate carboxy (pK_a = 3–5), guanidinium in arginine (pK_a = 11.5–12.5) and sulfhydryl in cysteine (pK_a = 8.0–8.5). Parkash and Bhatia (1980) have provided a convenient procedure for the graphical determination of pK_a values for such groups within the active site of the enzyme.

pH-associated Control of Enzyme Action in Foods

Control Without Inactivation.—While it is quite a simple matter to control pH in model experimental enzyme systems and in some liquid foods such as beverages, it may be difficult to alter at will the pH of muscle tissue, fruits or vegetables by exogenous addition of acid, alkali or buffer without deleterious side effects. This is especially true if the aim is to control rather than inactivate enzymes. There are a few exceptions, such as the transitory lowering of pH in onion processing to prevent bitterness (Chapter 21). While it is usually not feasible to control endogenous food enzyme by modulation of pH, it has become possible to select an exogenous food enzyme which has the proper pH. With the advance of enzyme production (Chapter 5) it has now become routine practice to select a commercial enzyme adjunct having a pH optimum close to that of the food product. Also in the production of food ingredients, as exemplified by the corn sweeteners, we have seen how important and relatively easy it is to adjust the pH to obtain maximum yield of the desired mixture of sugars (Chapter 9). One of the problems to be overcome in the successful use of lactase in dairy products was to find an enzyme which acted optimally at the pH of milk. With regard to "natural" control of enzyme action, it has been suggested that one of the reasons that some green, immature fruit undergo browning more rapidly than ripe fruit is because the pH is more favorable for phenolase action. We believe the answer lies elsewhere (Chapters 1 and 15).

Stability to and Inactivation by pH Extremes.— All enzymes are eventually destroyed by pH extremes if the extreme and time are great enough. Among the more acid-stable enzymes are the classical muscle adenylate kinase (myokinase) and pancreatic ribonuclease. The former survives 0.1N HCl at 100°C for 30 min and even the protein precipitant trichloroacetic acid at 5°C. Ribonuclease can also be boiled but in less acid media pH 2 with little loss in activity. These data also illustrate that pH optimum has little to do with pH stability since optima of these two enzymes lie between pH 7 and 8.

Mechanism of Acid Inactivation.—At the molecular level, inactivation follows denaturation of the enzyme molecule which occurs in a stepwise fashion. First the quaternary structure, if there is one, of an oligomeric enzyme is disrupted, leading to dissociation into subunits. The next event is dissociation of the prosthetic groups and coenzymes, if any. Tertiary structure is loosened by the progressive interference with hydrophobic interaction followed by the disordering of enzyme-associated water molecules and release of lightly linked metal ions. Finally the complete unfolding of the molecule ensues upon disruption of the hydrogen bonds, especially those involving the N, O and H of the peptide bond, N . . . O and N . . . H, which hold secondary structure together.

As shown in Table 11.5 some enzymes regain activity after inactivation at low pH. In an elegant study of the rate of change of the environment of Trp 140 in the active site of a staphylococcal nuclease, Epstein *et al.* (1971) reconstructed the following regeneration model. First, there is a "nucleation" of the α-helices that form the pocket in which the indole ring of the tryptophan lies. This is followed by the positioning of side chains around the indole ring via reestablishment of hydrophobic interactions. Aggregation and disaggregation are not involved.

Food-related Considerations.—Extremes of pH can be critical in the processing and manufacture of food products. Many foods are treated with acid—soft beverages, sweet-and-sour meats, or are allowed to become acid via fermentation—cucumbers, sauerkraut, yogurt. Acid treatment of crude soybean flour is an essential step in the preparation of isolates. Addition of ascorbic or citric acid serves several purposes including prevention of enzyme action partially due to lowering of pH (Chapter 16).

Alkali is used for debittering olives and in the preparation of some corn products such as Mexican tortillas and hominy, and recently has been extensively used for peeling potatoes and other vegetables and fruits.

The primary purpose of these treatments is not to inactivate enzymes. The use of pH extremes as alternatives to blanching has been somewhat neglected. Use of ammonia in the Pro-Xan process for alfalfa not only aids in the separation of nutritious high-protein fraction but also inactivates lipoxygenase and thus preserves xanthophyll and carotenes and prevents off-flavor development. The pH adjustment occasioned by spraying fish with ammonia (Mandal and Mukerjee 1974) designed primarily as a preservation measure also entrains desirable enzyme inactivation as a spinoff. On the other hand, exposure of high protein foods to elevated pH may result in the formation of nonnutritional amino acids derived from the interaction of the dehydroalanine formed with lysine, such as (lysinoalanine), cysteine (lanthionine) and probably other amino acids.

One of the paradoxical effects which ensues from treatment of olives and other phenolase-containing foods with alkali is that during the treatment, before the food has reached its ultimate uniformly high pH, the alkali promotes cellular disruption, thus allowing the phenolase and probably other enzymes to be potentiated and causing enzymatic browning to occur. This browning is accelerated by the neutralization of naturally occurring acids if any phenolase survives so that the pH of the tissue may transitorily pass through the optimum of the discoloring enzyme. To avoid or prevent phenolase action in this particular case, avoidance of exposure to oxygen may be combined with heat inactivation and expeditious readjustment of the pH of the tissues to neutrality.

As we have had occasion to point out, not all enzymes have the same stability so it has been possible to destroy unwanted enzymes in commercial food enzyme adjuncts (Chapter 5). The now classical example is the differential inactivation of either the amylase or protease activity in the same fungal enzyme preparation (Miller and Johnson 1954) depending upon what bread-related property of dough is being adjusted (Chapter 32). Other examples are mentioned in Chapter 5.

Blanching Alternatives.—As mentioned, not many studies have been deliberately designed to use pH extremes primarily for preventing enzyme action as an alternative to blanching. In one such study by Makower and Boggs (1960) treatment of cabbage, celery and carrot with acid was combined with vacuum infiltration, surfactants and ethanol. While the results were mixed, they do give some clues to future possibilities of such an approach. Of the many detergents (surfactants) used they found that anionic detergents were more effective than were cationic detergents, with dioctyl esters of sodium sulfosuccinic acid (aerosol OT) being especially effective in inactivating peroxidase. Analogous to its heat stability, peroxidase was the most resistant enzyme of those tested (catalase, esterase and phosphatase). Flavor could be improved by using 0.1N HCl instead of H_2SO_4. Attesting to the high buffering capacity of vegetables, they observed that even at the high acidity of the infiltrating medium, 1.2, the final pH of ground up cabbage after treatment was about 5.4, not too far from that of the untreated cabbage. This chemical inactivation resulted in quality retention and stabilization about equivalent to that observed after steam blanching with respect to taste, color and ascorbic acid level, but left quite a bit to be desired with respect to odor.

Other explorations in the realm of pH extremes for the express purpose of preventing enzyme action in foods include the acidified hot break in tomato processing to prevent polygalacturonase action (Chapter 21), the grinding of beans in acid to prevent lipoxygenase action (Chapter 34), exposure of vegetables to propionic acid vapors (Matyas *et al.* 1974), and in modulating α-amylase action in cereals (Chapter 32). Other combinations of heat blanching and chemical treatment in which pH extremes may have played a role were mentioned in Chapter 10.

Part V

ENZYME ACTION AND FOOD COLOR QUALITY

15

FOOD COLOR, THE PHENOLASES AND UNDESIRABLE ENZYMATIC BROWNING

COLOR AS A FOOD QUALITY ATTRIBUTE

In the next several chapters we shall devote our attention to an evaluation and assessment of how enzymes, both in food or deliberately added, eventually influence the consumer's conception of the food. In other words, what is the net result of the action of these enzymes on not only the traditional organoleptic quality attributes of foods, but also how enzymes either aid in or are the sole instrumentation of changing the identity of an agricultural commodity into a finished food product? It should be pointed out at the outset of this discussion as mentioned in Chapter 1 that very seldom, if ever, is the consequence of the action of a particular enzyme confined solely to changing one quality attribute in any given food in which the enzyme acts. One way of dividing the ensuing discussion would be to take up the stimuli each sensory organ receives. However, if this were done, we would have to place the lachrymatory properties of onions in juxtaposition to appearance, color, taste and some aspects of texture, all perceived by the eye, into one category. Instead, we shall divide the discussion into the traditional classification of color, texture and flavor plus special discussions in which the enzyme action is characterized by gross noncolor changes in appearance and identity. These include those enzyme-related change-of-state modifications involved in production of dairy products (mostly cheese), alcoholic beverages, bread and other baked goods. It will be noted that these three food products are all distinguished by and have in common the circumstance that in each case desirable enzyme action is followed by microbial fermentation. Finally, we shall take up changes in appearance of beverages, especially with regard to control of haze and turbidity. This includes both fermented beverages such as wine and beer, as well as nonfermented fruit and vegetable juices. Frequently, the motives for consideration of enzyme action in these cases are dictated by concern for efficiency of processing and economy more than for optimization of product quality.

Appearance and Color

In this and the next four chapters, we focus our attention on enzyme induced changes of what our eyes perceive as color. The integrated experience of perception of the eye is, of course, appearance, of which color is only one component. As pointed out by Clydsedale (1976), appearance includes form, texture and markings. Before going into how enzymes affect the color of food, one should have something more than an intuitive feeling of what is meant by color. A physicist would look upon color as did Newton some 300 years ago in his "Optiks" as cited by Little and Mackinney (1972): "Colours in the object are nothing but a Disposition to reflect this or that sort of Rays more copiously than the rest." A biochemist or molecular biologist would define color as a response of an organism to the quanta brought about in the first instance by a transduction of light energy into conformational changes in protein intermediated through pigment linked to the protein. In this sense, the first act in the perception of color is not unlike that of the primary event in enzyme catalysis and a host of other biological phenomena as explained in Chapter 3. The difference is that this conformational accommodation of protein-pigment (rhodopsin) leads to an alteration of membrane permeability to Ca^{2+} which eventually enables a message coded as a difference in electric potential to be transmitted to the brain, whereas the message signaled to the enzyme by the substrate is to split a chemical bond. A physiologist-anatomist would define color in terms of the various structures of the eye and especially the rods and cones and the eyes' afferent nerves. A definition related to color of a *food*, however, is essentially and in the final analysis a subjective psychological response. A psychophysical definition is that proffered by the optical and illuminating engineering societies as cited by Mackinney and Little as "a noninhomogeneity consisting of characteristics of light ... [that] aspect of radiant energy of which a human observer is aware through visual sensations which arise from the stimulation of the retina of the eye." As an aspect of visual experience, Little and Mackinney define light psychologically as "an attribute of visual experience that can be described ... in the quantitatively specificable three dimensions of hue, saturation, and brightness (lightness). Hue is the attribute by which the color is identified (red, green, etc.); saturation is the proportion of chromatic content in the total perception; and lightness relates to the fraction of total incident light reflected by or transmitted through the food." Another parameter of food color as with all other food quality attributes is the purely subjective one of preference. In general, color is a guide and indicator of maturity, quality and wholesomeness.

Much effort has been expended for development of the instrumentation and analysis to obtain an objective description of food color (Watada 1975; Little 1976). This is not the place to go into a description of this instrumentation and expression of results obtained therefrom which have been designed to translate these subjective experiences into objective measurements including exegetic books (Mackinney and Little 1962; Francis and Clydsedale 1975; Hunter 1975) and shorter reviews and explications (Little and Mackinney 1972; Chichester 1972; Little 1976; Clydsedale 1976; Kramer 1976). Such instrumentation is serving as a catalyst for the mechanization of agriculture (Schatzki *et al.* 1980).

Food Colorants and Pigments.—Perception of food color, be it via instrumentation or visually, is a function of how the particular food is presented and inspected with regard to conditions and quality of illumination. Especially important is the degree of light scattering, absorption and reflection during examination. Although to say that food color is determined by the pigments present would appear to be a circular statement, the facts are that color in the present context depends not only on the substances which are colorants or pigments but also on their physical state and on the presence and state of nonpigmented constituents of the food. For instance, the color of green vegetables changes from a dull to a bright green upon blanching. This is not, for the most part, caused by a change in composition of the chlorophyll pigments but is due to change in their physical state within the chloroplasts, to expulsion of air and to increase in translucency of intercellular components as well as that of surrounding tissue. The perception of carotene color in marine food depends on the amount and disposition of colorless lipid material (Little 1976). Some parameters of food appearance which can also have some influence on the color perception include gloss, distinctness-of-image, haze, reflection, luster and turbidity. On the other hand, color in food is seldom perceived in the complete absence of colorants as for instance that elicited by refraction, i.e., prism effects, such as the color of butterfly wings. Exceptions are the perception of opaque whiteness when transparent or translucent foods such as sugar and egg white are subdivided by grinding, powdering or beating into small particles separated by air bubbles so as to scatter the incident light.

Food Pigments.—The colorants present in and not deliberately added to biological material are in general referred to as pigments. From the viewpoint of enzyme action, we divide such pigments into those which are present *in situ* or *in vivo* formed via the natural metabolic physiological processes in the living organism and those formed as a result of injury or in subsequent processing.

To refresh the reader's memory (they are mentioned in all food science texts), Table 15.1 lists

TABLE 15.1

BIOLOGICAL PIGMENTS

Class	Colors
Tetrapyrrole compounds, Mg	green → brown (chlorophylls)
Tetrapyrrole compounds, Fe	red → brown (hemes)
Carotenoids	red → orange → yellow
Flavonoids; anthocyanins	all colors
Flavins	yellow
Copper proteins	blue → green
Melanins	brown, black, blue
Melanoidins	brown, black, blue
Caramel	brown → tan
Physical effects, particle size, air bubbles, translucency, opalescence, prism effects, luster, etc.	all colors

naturally occurring pigments and Table 15.2 some examples of nonphysiological colors in foods and food products formed as in the postharvest and postslaughter stages of the food processing chain. Strictly speaking, many of the latter are due to nonenzymatic influences such as heat. However, as exposited in Chapter 1 and expanded upon in the following chapters, even some of these may be considered as secondary reactions arising from the interactions of products of enzyme action. Of course, many of the pigments formed are responsible for the marked changes during the life of plants as they go through maturation and senescence. The coloring of fruit while ripening on the tree is due largely to destruction of the original chlorophyll as well as biosynthesis of anthocyanin and carotenoid pigments. With oranges, this change requires cool nights perhaps as a response to cold stress (Chapter 1) as well as warm days. When oranges and similar citrus varieties are grown in warm climates, the green color may persist in the peel though the inner fruit tissue may be fully colored.

TABLE 15.2

A RAINBOW OF NONPHYSIOLOGICAL PIGMENTS: FOOD OFF-COLORS

Color	Food	Substance/Precursor/ Correlative
Violet-blue	cauliflower	Flavonoid + sulfur
Greens	tuna	Metmyoglobin-S-S-R
	garlic	Unknown
	green vegetables	Chlorophyllins
Yellow	vegetables, fruit	Pheophytin
Brown	vegetables, fruit	Melanins from polyphenols
	asparagus	Fe^{2+} + rutin → colored complex
	meat	Metmyoglobin
Black	apple scald	Farnesene
Pink	onion	Lachrymator = thio propanal S-oxide
	pears, beans, Brussels sprouts	Proanthocyanidin, flavan-3,4-diols
Green → red	lobster	Astaxanthin-protein

Yokoyama and coworkers (Yokoyama 1975; Poling *et al.* 1975) have discovered a class of aryl-substituted alkyl amines which serve as regulators, perhaps as derepressors, for the genes controlling carotenoid biosynthesis. Alternatively, these amines repress genes coding for enzymes responsible for cyclization during carotenoid biosynthesis. This would result in the pile-up of colored alicyclic intermediates (Chang *et al.* 1977; Hayman *et al.* 1977). A typical compound of this class is 2-(4-chlorophenylthio)triethylamine. It can stimulate synthesis in citrus of lycopene, the natural carotenoid responsible for the color of red tomatoes. Similar compounds stimulate the synthesis of the carotenes. The color and vitamin potency of a wide assortment of plant foods such as sweet potatoes, apricots, carrots and peaches could be controlled once U.S. Food and Drug Aministration approval is obtained.

Although somewhat incidental to our main theme, it may be of interest to mention that the beautiful leaf colors associated with autumn's foliage is, as in fruits, due to a programmed senescence which is signaled by derepression of those segments of DNA or by unmasking preformed latent messenger RNA coding for the formation of enzymes which cause the breakdown of chlorophyll. These genes are linked to those that participate in the biosynthesis of anthocyanins responsible for the variegated hues of autumn foliage. Creasy (1974) elucidated the sequence of the development of autumn coloration in strawberry leaves. First chlorophyll starts to disappear. This may release an inducer of the frequently alluded to phenylalanine ammonia lyase (PAL), the first enzyme in the pathway to flavonoids (Billet and Smith 1980), including colored anthocyanins which accumulate. The influence of temperature stress on anthocyanins as an "off-color" in a food is exemplified by their accumulation in normally white cauliflower grown during an abnormally cool summer.

UNDESIRABLE FOOD COLOR AND PHENOLASE ACTION

By and large, the brown color appearing in senescent and injured plant foods is due to the melanins formed as a consequence of the action of phenolase. In general, this action is considered to be undesirable. Its action contributes at least part of the color to such intermediate moisture fruits as dates, figs and raisins. While it is usually involved in imparting color to foods, we shall see that it is also involved in decoloration of the anthocyanins and perhaps other natural food pigments. As we shall learn in later chapters, it is also involved, at least indirectly, in the generation of desirable flavor in cider, dates, tea, cocoa and probably other foods and in the bitter taste accompanying infected browned fruit tissue. It may even play a role in haze formation in juices and in some textural attributes. In the present chapter, we shall confine our attention to its role in undesirable oxidative enzymatic browning. But first we survey the "present" state of knowledge of the nature of the phenolases.

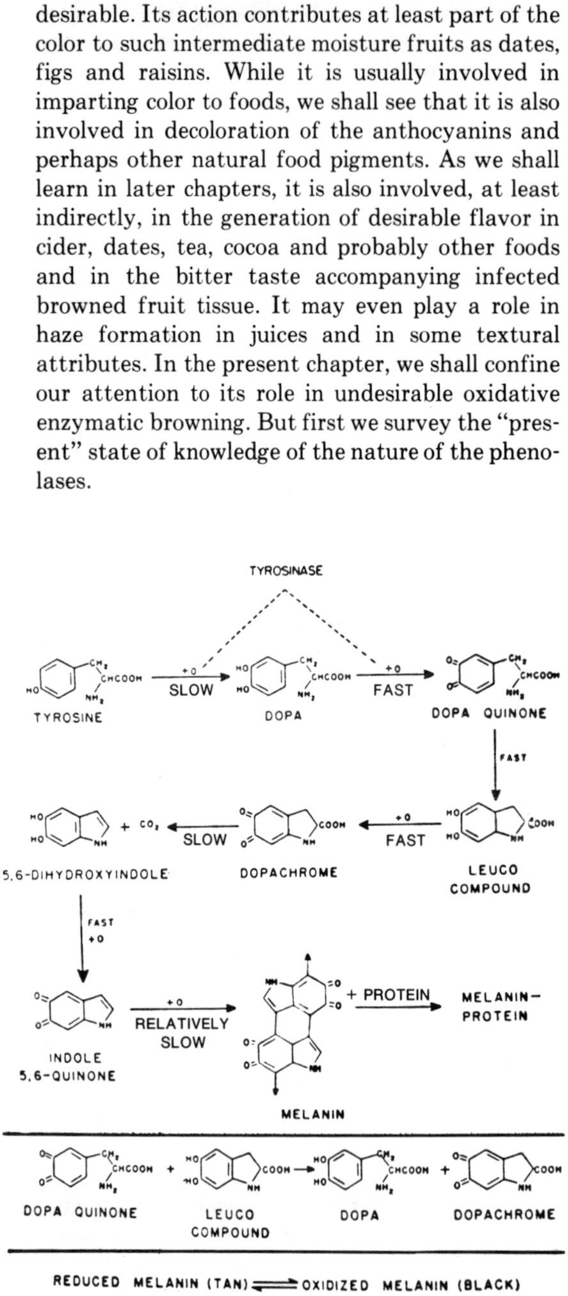

From Lerner (1953)

FIG. 15.1. FORMATION OF BROWN MELANIN VIA PHENOLASE ACTION ON TYROSINE

Phenolase Biochemistry

Historical.—The origins of the discovery of phenolase are somewhat clouded. According to Whitaker (1972), Schoenbein in 1856 suggested that the oxidation of certain compounds in plants was due to a widespread organic catalyst. Incidentally this same Schoenbein (1861) used paper chromatography as an example of the role of "absorption" forces in enzyme catalysis. Eskin *et al.* (1972) ascribes to Lindet's recognition in 1895 of the enzymatic nature of the browning in cider and other foods and to Bourquelet and Bertrand's [also cited by Sumner and Somers (1947)] in the same year, recognition of tyrosine as the substrate for the oxidase in mushrooms, the classical source of this enzyme. However, it is the work of Onslow over a period of at least 16 years starting over 60 years ago (Onslow 1915) that what is herein called phenolase—enzymes acting on dihydric phenols—was established as definite enzyme entities responsible for enzymatic browning in most plant tissues.

Phenolase has been the subject of numerous reviews as an enzyme, as metalloprotein and as food browning participant starting with Onslow (1931) and including Raper (1932), Sutter (1936), Nelson and Dawson (1944), Dawson and Tarpley (1951), Joslyn and Ponting (1951), Sizer (1952), Lerner (1953), Mason (1955,1957), Pridham (1963), Hayashi (1962), Malkin and Malmstrom (1970), Hamilton (1969,1974), Mathew and Parpia (1971), Scandalios (1974), Coleman (1974), Van Sumere *et al.* (1975), Holwerda *et al.* (1976) and one in German (Hermann 1976) and one in Russian (Rubin *et al.* 1976). The need for reviews is attested to by the appearance of 130 entries in the July-December 1977 substance index of *Chemical Abstracts.*

Classification and Specificity.—This group of enzymes which catalyze the oxidation of phenolic substances was first referred to as oxygenase and later variously as tyrosinase, cresolase, monophenoloxidase, monophenol monoxygenase, catecholase, *o*-diphenol oxidase, *o*-diphenolase, laccase, *p*-diphenolase, polyphenol oxidase, polyphenolase and phenolase, each indicating a somewhat different specificity. With the possible exception of certain fungal and *Rhus* laccases these enzymes may differ only in relative and not absolute specificity and may even form a continuum of susceptibility to various substrates. Specificity may also depend may even form a continuum of susceptibility to various substrates. Specificity may also depend upon what else is present in the enzyme reaction system. Although many recent papers refer to such enzymes as polyphenol oxidases or PPO, for consonance and convenience we have and shall, for the most part, refer to this enzyme type as simply "phenolase." It is not to be confused with phenol oxygenase, also known as phenolhydroxylase, which performs only hydroxylations of phenol and other phenolics and for which NADPH is a cosubstrate along with the oxygen (Massey and Hemmerich 1975).

However, the continuing confusion in specificity and classification is attested to by the continuing alteration of EC classification schemata with each revised edition of the official enzyme nomenclature compendium (in addition to interim recommendations issued from time to time in biochemical journals) as detailed in Chapter 2. Laccase, traditionally considered to be a *p*-diphenolase, along with *o*-diphenolase activity has been reported not only in the classical fungal *(Polyporus)* and plant (lacquer tree and other *Rhus*) species, but also in various fractions, after special treatment or in certain stages of development of extracts and some plants including tobacco and foods such as grains, apples and peaches. However, probably more frequently, no *p*-diphenolase activity is reported especially after purification, i.e., apple (Joslyn and Ponting 1948), avocado (Kahn 1977), olives (Ben-Shalom *et al.* 1977), mango (Joshi and Shiralkar 1977), eggplant (Ramaswamy and Rege 1975B) and dates (Hasegawa and Maier 1980). The latter three phenolases were also inactive towards tyrosine and other monophenols (see below). Also, the EC classification had changed from an emphasis on phenols (i.e., an oxidoreductase acting on diphenols with oxygen as acceptor) to an emphasis on oxygen (an oxidoreductase acting on paired donors with incorporation of oxygen). Yet, as will be brought out later, the state of Cu and the mechanism of its involvement with fungal and *Rhus* laccases appears to differ fundamentally from that of the catecholases or tyrosinases of mushroom.

As stated in a correction and addition supplement (Anon. 1979) the classification of this most thoroughly studied "group of enzymes . . . is controversial."

Phenolases as Metalloproteins.—The phenolases which have been isolated from food sources are oligomers with one copper per subunit.

Copper and the Active Site.—The presence of copper in phenolase was established over 40 years ago by Kubowitz (1937). In the intervening years it has become clear that there is one copper atom per subunit regardless of source (Jolley *et al.* 1974; Lerch 1976). What has not been clear is how the copper is distributed among the subunits. Until quite recently the status of the oxidation states in the resting enzyme has been a bone of contention ever since Kertesz (1957) questioned the presence of Cu^{2+} (Cu II) in mushroom phenolase. As late as 1974, Coleman (1974) concluded that the resting enzyme contains either Cu(I) or magnetic dipole-dipole coupled Cu(II) pairs, similar, in the latter respect, to that of the hemocyanins, the oxygen-carrying protein of arthropods and molluscs. Indeed, according to Schoot Uiterkamp *et al.* (1976), the active site of the H_2O_2-treated enzyme is structurally related to that of hemocyanin. Makino *et al.* (1974) present convincing evidence that the active site of the enzyme does indeed contain a pair of antiferro-magnetically (diamagnetic) coupled Cu^{2+} ions quite close to each other. They speculate that the active site consists of a disulfide-copper pair complex. Such a pair has also been detected in the *Neurospora* enzyme (Deinum *et al.* 1976), but earlier work (Fling *et al.* 1963) indicates that this enzyme does not have disulfide bridges. Such bridges are present in *Polyporus* enzyme, but the copper is present in three distinct forms, one of them (Cu^{2+}) contributes the blue color and one is "non-blue cupric" and corresponds roughly to the oxidation state Cu^{2+}. Two others, apparently like the pairs in the mushroom enzyme, constitute a spin-paired Cu^{2+}-Cu^{2+} unit (Malkin and Malmström 1970). The state of Cu in the *Rhus* laccase enzyme is similar (Howlerda *et al.* 1976).

One of the reasons for confusion about the state of Cu in the phenolases is that, with the exception of the two Cu atoms in fungal laccase, the active-site Cu appears to be "silent" with respect to electron paramagnetic resonance (spin) measurements (EPR, ESR), so that it has been necessary to change the enzyme from its resting state or to reason from aberrations in optical, UV circular dichroism spectroscopy measurements as was done in all of the above cited work. Hopefully, the newer technique of X-ray photoelectron spectroscopy which apparently can distinguish among the various forms of Cu including a signal from EPR—and optically-silent Cu^{2+} (Rupp and Weser 1976) will settle the true state of the copper in the phenolases.

Mechanism of Action: Quinone Formation and Oxygen Reduction.—Until the tertiary structure and a better understanding of the nature of copper in phenolase is forthcoming, intimate detailed knowledge of the sequence of events that lead to the formation of the relatively stable products, quinone and water, will be partly conjectural. This does not mean that a great amount has not been learned by shrewd guesses, the use of inhibition and other kinetic isotopes, and the above-mentioned spectroscopies. Whatever else they are, most phenolases may be considered to be mixed-function oxidase isozymes which can catalyze both hydroxylations of the benzene ring of monophenols and the removal of hydrogen from diphenols, with each isozyme acting within its own domain of specificity. Since the course of the monophenolase reaction exhibits a lag period, and with the aid of ^{18}O and ^3H isotopes, it now seems clear that the oxygen of the new phenolic OH comes from O_2 (Mason 1957; Pomerantz and Warner 1966). It appears that an infinitesimal amount of an *o*-diphenol has to be present to "spark" or prime the hydroxylation. The latter is coupled to an electron source coming from the formation of the quinone. As far as the mechanism of oxidation of monophenols is concerned, the evidence so far as summarized by Coleman (1974) and Ullrich and Duppel (1975) suggests the following:

Monophenol (A) + *o*-Diphenol (B) + $O_2 \rightarrow$
   o-Diphenol (A) + *o*-Quinone (B) + H_2O

Concerning the mechanism of action of the phenolases on *o*-diphenols, while we have come a long distance from the postulated formation of free H_2O_2 (Onslow 1931), all the evidence so far is not in. Interpretation of the above-mentioned measurements seems to clearly indicate several aspects of the events that do occur. Thus, it is fairly clear that the oxygen is bound by the Cu and that the Cu is a good electron acceptor. The binding is ordered and oxygen probably binds first. Whitaker (1972A) suggests an ordered bi-bi mechanism involving two substrates and two products. However, there actually appear to be three substrate molecules involved: one O_2 and two *o*-diphenols, or one *o*-diphenol and one monophenol to form a quaternary (in the kinetic, not protein structural sense) complex for both the hydroxylation and *o*-diphenol oxidations. Kinetics clearly reveals two separate but independent binding sites: one for the two phenolic substrate molecules and the other, the Cu locus, definitely the binding site for oxygen.

Hamilton (1969) speculated that oxygen forms a perhydroxyl (OOH) complex from the H transferred from the phenols through on Cu and that this complex then combines with the second phenolic substrate. A modified version (state of the Cu is not indicated) follows:

$$\begin{array}{c} E \\ | \\ Cu \end{array} \xrightarrow{O_2} \begin{array}{c} E \\ | \\ Cu \cdot O=O \end{array} \xrightarrow{2R(OH)_2} \begin{array}{c} R(OH)_2-E \\ | \\ R(OH)_2 \cdot Cu \cdot O=O \end{array}$$

$$\xrightarrow{-H} \begin{array}{c} R(OH)_2-E \\ | \\ HORO-Cu \cdot O=O \end{array} \xrightarrow[-H]{-RO_2}$$

$$\begin{array}{c} E \\ | \\ HORO-Cu-OOH \end{array} \xrightarrow{-H_2O} \begin{array}{c} E \\ | \\ Cu-H \end{array} \xrightarrow{+H_2O}$$

$$\begin{array}{c} E \\ | \\ HO-Cu-OH \end{array} \xrightarrow{-2OH^-} \begin{array}{c} E \\ | \\ Cu \end{array}$$

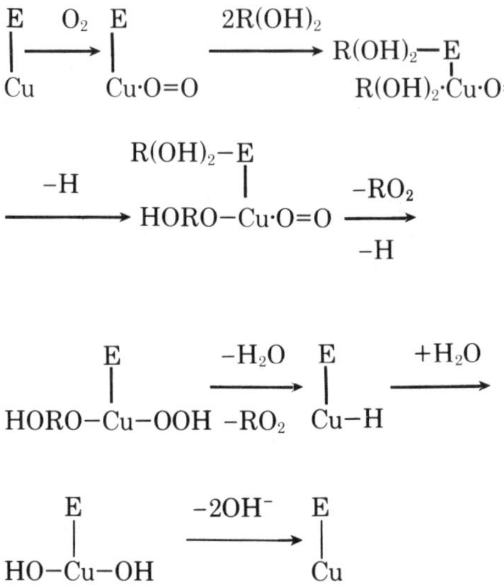

Role of Cu in Catalysis.—The state of our understanding of what occurs is exemplified by the speculation of Makino *et al.* (1974) the oxidation-reduction might go through the following cycle:

$$\begin{array}{cc} -S\!-\!\!-\!\!-\!S- \\ | \quad\quad | \\ | \quad\quad | \\ Cu(II) \; Cu(II) \\ \text{Fully Oxidized} \end{array} \xrightarrow{2e} \begin{array}{cc} -S\!-\!\!-\!\!-\!S- \\ | \quad\quad | \\ | \quad\quad | \\ Cu(I) \; Cu(I) \\ \text{Half Reduced} \end{array} \xrightarrow{2e} \begin{array}{cc} -S\!:^- \quad ^-\!:S- \\ | \quad\quad | \\ | \quad\quad | \\ Cu(I) \; Cu(I) \\ \text{Fully Reduced} \end{array}$$

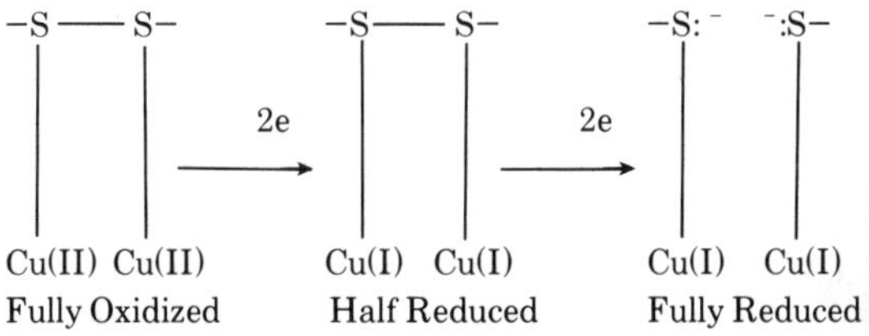

but point out that because the Cu atoms are optically and magnetically silent, there is no certainty as to whether this structure represents the resting form of copper. Furthermore, Coleman points out that a mechanism must solve the problem of how to couple one- or two-electron transfers in the oxidation of the substrate to the four-electron transfer required to reduce oxygen to water. By the time this appears, it is likely with the application of X-ray photoelectron spectroscopy and the elucidation of the shape of the active sites from sequencing and X-ray crystallography we shall have a clearer picture of what happens first at a submolecular level when foods turn brown.

Laccase.—For the *Polyporus* and *Rhus* laccases, in contrast to the mushroom enzyme, kinetics indicate that the substrate molecules combine one at a time with the enzyme (ping-pong kinetics). The role of the three types of Cu is fairly well delineated. Oxygen first combines at the "blue" Cu site where oxidation (H-removal) of substrate occurs. A second type, the "nonblue" Cu, stabilizes an intermediate, Hamilton's perhydroxyl. The third, occurring as the silent Cu^{2+}-Cu^{2+} pair which seems to be the only form in the food phenolases, acts as a two-electron accepting unit. For *Rhus* laccase, the substrate (in this particular case, hydroquinone QH_2, quinol) attacks the "colorless" Cu located in a pocket constituting the active site of the enzyme. This attack by substrate involves reduction of the Cu^{2+} and formation of QH^-, coordinated with a conformational "protein activation" which opens up the pocket to make the other two types of Cu, "silent" and "blue," available for reduction. Based on a penetrating circular dichroism study, details of *Rhus* laccase mechanism were further elucidated by Farver *et al.* (1980).

Reaction Inactivation.—Phenolase action preeminently illustrates the general principle elaborated in Chapter 1 that many of the observed effects of enzyme action are not due to the primary stable product but to secondary reactions. For phenolase this occurs even before the product is released from the enzyme. Phenolase investigators knew early in the 20th century that the fruit enzyme was remarkably more unstable than that of *Rhus* laccase. Apple phenolase activity is lost by lyophilizing and even upon storage at −35°C. Phenolases were also shown to lose activity while they are acting, as probably first described by Miller *et al.* (1944). In their system, the inactivation could not have been due to accumulating free product quinones since these were re-reduced back to substrate (catechol) by the ascorbic acid present. This type of inactivation which could be accounted for by neither free stable reaction product nor by inherent enzyme instability was termed "reaction" and later "syncatalytic" inactivation.

What we do know about the nature of this inactivation is due to Ingraham and coworkers (Ingraham 1954; Wood and Ingraham 1965). From a close look at the kinetics of the inactivation (it follows first-order kinetics) and an observed difference between the energy of activation of the enzyme catalysis and that of the inactivation, it was at first assumed that a fleeting intermediate semiquinone reacted at the active site. However, later work, in which it was shown by tracer technique that transformed substrate was covalently attached to the enzyme protein, led Wood and Ingraham to suggest that essentially the same product of reaction leading to melanin (see below) also leads to inactivation. The principal difference in the case of tyrosine oxidation is that instead of the cyclization of the quinone product (DOPA quinone) through its own amino group (Fig. 15.1), the latter is supplied by the protein as an epsilon amino of an active-site lysine. The resulting covalently linked product then blocks entry of substrate. Another difference is, of course, that the quinone product never gets very far, if at all, from the surface of the enzyme molecule. Free quinones also react with lysine ε-amino groups of other proteins (Chapter 34) but not quite so facilely. If this turns out to be a valid mechanism, then reaction inactivation should be a powerful tool for labeling the active site of the enzyme. This mechanism has to be reconciled with the report of Ponting (1954) that reaction inactivation of fruit phenolases as well as an ascorbic acid-reversible inactivation could be reversed by acetone precipitation and recovery of the protein. Also, Jolley *et al.* (1969) observed that iodinated mushroom phenolase was more easily reaction inactivated after iodination, presumably of tyrosine in the enzyme.

In an application of the theory of Wood and Ingraham, Letts and Chase (1973) tried to prevent reaction inactivation by blocking the lysine groups. Table 15.3 shows the inactivation-enhancing effect of ascorbate, its partial reversal by lysine modification. The authors hope eventually to immobilize such modified phenolase for the preparation of DOPA used in the treatment of Parkinson's disease. Reaction inactivation might also be useful, as discussed below, in prevention or control of enzymatic browning just as lipoxygenase reaction inactivation has been used to prevent off-flavor development (Chapter 14). As mentioned later, reaction inactivation may be a limiting factor in such browning.

Phenolases as Oligomeric Proteins.—By the time this appears, the amino acid sequence of at least one of the phenolases will have been determined. Interestingly, the N-terminal serine of one of the subunits of the *Neurospora* enzyme is N-acetylated (Nau *et al.* 1977). So far, phenolases from mammals but not from plants appear to be

TABLE 15.3

REACTION INACTIVATION OF PHENOLASE AND ITS CONTROL[1]

Time (hr)	Before Lysine Modification			After Lysine Modification		
	No Subst.	Subst.	Subst. + As. Ac.	No Subst.	Subst.	Subst. + As. Ac.
1	100	55	100	98	98	98
2	95	50	80	76	50	90
8	90	40	10	50	40	80

Source: From data of Letts and Chase (1973).
[1] Subst., substrate; As. Ac., ascorbic acid; values are percentage of untreated, unincubated control.

glycoproteins. At first the enzyme was thought to consist of four equivalent 30,000 dalton monomers. The principal isozyme of the classical mushroom phenolase (tyrosinase) has now been shown to consist of the tetramer L_2H_2, the molecular weights of L and H being 13,400 and 43,000, respectively (Strothkamp *et al.* 1976). The functional oligomer may be smaller than L_2H_2. Similarly, the phenolase of the fungus *Neurospora*, at first reported as a 30,000 dalton monomer, has now been found to be a dimer consisting of two unequal monomers (Lerch 1976). The phenolase (laccase) from the fungus *Polyporus* also consists of four monomers held together by disulfide bonds (Butzow 1968). Preliminary evidence indicates that banana phenolase appears to be a dimer consisting of two identical 30,000 dalton subunits (Galeazzi *et al.* 1976).

The spinach chloroplast enzyme appears to be a 36,000 dalton monomer which can be induced to associate to a dimer in the presence of 2,3-dihydroxybenzaldehyde (Sato 1976). In general, the quaternary structures of phenolases are quite responsive to their environment, the oligomers mostly dissociating under the influence of temperature, ionic strength, detergent and chelators, by simply changing enzyme concentration and by chemical modification. The dissociation may or may not be accompanied by activity changes. They are obviously related to observed isozyme patterns. The role of the quaternary structure in regulatory aspects of this enzyme remains to be elucidated.

Isozymes.—As indicated in our preliminary discussion of phenolase nomenclature, multiple forms of the enzyme exist with varying specificities from the same source. In addition to mushroom (Jolley *et al.* 1969,1974) and previously mentioned examples of phenolase, isozyme sources in foods include fava bean (Robb *et al.* 1965), grape (Durmishdze *et al.* 1975), avocado (Kahn 1976), cherry (Benjamin and Montgomery 1973; Pifferi *et al.* 1974A), and mango (Joshi and Shiralkar 1977). As an example of the type of results obtained, Matheis and Belitz (1975) report the detection of 17 multiple forms of potato phenolase by conventional gel electrophoresis. All of them utilized *o*-diphenols and five also utilized monophenols as substrates. They report separating by gel chromatography a 30,000 dalton monomer, several oligomers ranging up through tetramer and isozymes reaching into the polymer range, all of them active. This fits in with the possibility that some phenolase monomers are disulfide-linked. Some isozymes probably also arise as the result of interchange of subunits (Fling *et al.* 1963; Lerner *et al.* 1972). The distinction between "true" isozymes and differing degrees of subunit association is blurred.

Regulation via Activation, Synthesis and Translation.—While the regulatory significance of the subunit structure is not clearly understood, the state of the subunits may be involved in the activation of inactive proenzyme which is found in extracts of some, but not all, tissues. Historically, these include grasshopper eggs (Allen and Bodine 1941), mealworm and crayfish as well as broadbean leaf (Kenten 1957; Robb *et al.* 1964), spinach chloroplast (Sato and Hasegewa 1976), grain (Szarkowski 1957) and grapes (Lerner *et al.* 1972). Treatment with a variety of substances and conditions can, within 5 min, increase the activity in crude extracts more than 50-fold. These treatments include pH extremes, proteolytic enzymes, surfactants such as sodium dodecyl sulfate, high salt concentration and some protein denaturants such as urea and guanidine as well as macroanions such as carboxymethyl cellulose. Treatment of grain extracts with trypsin whose untreated phenolase acted only on *o*-diphenols elicited monophenolase (cresolase) activity as well. Cresolase also developed upon prolonged storage of untreated grains. Trypsin treatment was also responsible for a four- to 10-fold activation of grape phenolase.

The activation phenomenon has been ascribed to rearrangements of both a quaternary (subunits) and tertiary (conformation) structure.

Increase in phenolase activity along with increase in its substrates and PAL, the key enzyme involved in their synthesis (Yoshida 1969) is almost invariably observed in phenolase-containing plants in response to such stresses as wound injury and infection (Kuc *et al.* 1956; Anisimov *et al.* 1975; Van Sumere *et al.* 1975; Uritani 1978). In at least one plant, the sweet potato, the newly formed active enzymes are isozymes distinctly different from those present in the uninfected root (Hyodo and Uritani 1967). Conversely, partial blockage of *de novo* phenolase synthesis also results in a rapid change in isozyme pattern. These prompt responses are due to regulation of phenolase (and PAL) synthesis at the translational level (Chapter 4). The RNA which translates the DNA code into an enzyme is present in germinating embryos as latent or "conserved" messengers (Taneja and Sachar 1976). The presence of such masked messenger RNA may be the reason why certain dosages of ionizing radiation, as discussed in Chapter 10, induce *de novo* synthesis of certain enzymes, including phenolase (Schwimmer *et al.* 1958; Thomas and Janave 1973).

Occurrence, Distribution, Localization and Function.—The phenolases are ubiquitously distributed not only in plants but in other forms of life, including insects, especially in the eggs and larvae where they may participate in the sclerotinization signalling pupation; in crustacea, mollusca and mammals where they are involved in skin pigmentation via their action on tyrosine. Phenolases are particularly abundant and readily detectable in tissues of plants which brown or otherwise discolor rapidly upon injury. They have been purified and their role in browning has been substantiated in, at least, the following fruits: apple, apricot, avocado, banana, cherry, cranberry, fig, grape, mango, peach, plum, pear and quince. Fruits in which they are not present, or present no problem as to unwanted browning, include papaya (no enzyme), the kiwi (no substrates) and citrus and other highly acidic foods (no enzyme and/or inactivation/inhibition due to low pH—they act optimally between pH 6 and 8). A similar list of vegetables includes bean and other legumes, carrots, celery, eggplant (aubergine, brinjal), lettuce, mushroom, parsnips, potato, sweet potato and perhaps tomato. Phenolases also occur as undesirable enzymes in sugar beet, nuts, wheat bran and germ, and at heightened levels in the flour of "miracle" wheat of the Green Revolution.

As mentioned, phenolase-induced browning may be used to impart desirable color to some foods such as the beverages, cider, tea, cocoa and perhaps coffee and olives (Chapter 16). At least part of the color of intermediate moisture fruit products such as raisins, prunes, dates and figs is due to phenolase action. Their presence in commercial preparations of fungal pectin-degrading enzymes has posed a problem in the use of such preparations for improving filterability and clarity of white wines. Some of the phenolase activity of grapes has been attributed to accompanying contaminating fungi.

In many plant foods, the phenolases are localized or at least concentrated in specific organs. The vascular bundles of fruit are particularly rich sources; however, this is not always true. Thus, the phenolase activity of parsnips is confined almost exclusively in surface tissue where it creates a browning problem and is almost undetectable in the vascular cylinder (Chubey and Dorel 1972). Ever since Arnon (1949) demonstrated the presence of the enzyme in green tissues of the beet, phenolases of all green plant tissues so far examined appear to be localized in the chloroplast (Harel and Mayer 1968; Parish 1972; Tolbert 1973; Scandalios 1974; Sato 1976; Ben-Shalom *et al.* 1977). The phenolases are strongly attached to the chloroplast membrane and are usually present as two main isozymes. From the just-cited work and other reports, it appears that the total phenolase activity decreases with increasing development and maturity of the tissue. In at least the cases of olive and apple, phenolase activity becomes increasingly soluble with little change in properties as the chloroplast and green color disappear and anthocyanins appear (Ben-Shalom 1977). In tissues which never were green, the enzyme has also been reported to be associated with particulates, but this has not been fully explored.

Biological Function in Plants.—Besides giving rise to colored melanins, it appears that phenolases in plants may be involved in terminal respiration, in resistance to infection and in biosynthesis of plant constituents. In this latter capacity, phenolase may be an enzyme of the pathway leading from cinnamic acid to the host of C_6:C_3 plant constituents such as flavonoids, coumarins, etc.

Investigations have shown, for instance, that the monohydric *p*-coumaric acid, the *p*-hydroxy derivative of cinnamic acid, is converted into the *o*-diphenol, *cis*-caffeic acid in the presence of ascorbic acid by chloroplast phenolase (Sato 1966; Parish 1972). Although it appears that this may not be physiologically significant (Gestetner and Conn 1974), analogous nonphysiological action at the site of browning fruits due to senescence and infection may account for the associated bitter flavor due to formation of esculetin, scopoletin and other bitter coumarins as mentioned (Swain 1962; Sato 1967).

Phenolase action has also been implicated in the biosynthesis of tannins and *p*-diphenolase (laccases) actions in the biosynthesis and degradation of lignins (Mason 1955; Gierer and Opara 1973). It will be recalled that the degradation of cellulose may be linked to that of lignin via the cellobiose-quinone oxidoreductase in certain fungi (Chapter 9). The quinone could be supplied to this system *via* phenolase action. In its role as a defense mechanism, Van Sumere *et al.* (1975) suggest that quinones formed as a result of phenolase action may be toxic to the pathogen, or the quinone could undergo several secondary reactions with proteins (Chapter 34) to reduce the nutritive capacity of the host plant, inhibit and inactivate its own enzymes, eventually resulting in the death of surrounding cells. This dead tissue acts as a physical barrier which seals off the infected area, thus preventing further infection.

TABLE 15.4

ENDOGENOUS PHENOLASE SUBSTRATES IN FOODS

Food	Reported Substrates
Banana[1]	3,4-Dihydroxyphenylethylamine (Dopamine)
Apples[2]	Chlorogenic acid (flesh); *o*-catechin (peel)
Cocoa	*epi*-Catechins
Coffee beans[3]	Chlorogenic and caffeic acids
Dates[4]	Caffeoyl shikimic acid
Eggplant[5]	Caffeic, coumaric, cinnamic acid derivatives
Fava beans	Dihydroxyphenylalanine (DOPA)
Lettuce	Tyrosine
Mushrooms	Tyrosine
Olives	Urushiol
Pears[6]	Chlorogenic acid
Potatoes[7]	Tyrosine, chlorogenic acids, flavonols
Quinces[8]	Chlorogenic acid, catechins, flavonols leucoanthocyanidins
Sweet potatoes	Chlorogenic acid
Tea[9]	Flavanols, catechins, tannins

[1] Source: Weaver and Charley (1974).
[2] Source: Weurman and Swain (1953) and Siegelman (1955).
[3] Chapter 16.
[4] Source: Maier and Metzler (1965).
[5] Source: Ramaswamy and Rege (1975).
[6] Source: Weurman and Swain (1953).
[7] Source: Mapson *et al.* (1963), Alberghina (1964), Schwimmer and Burr (1967), and Schaller and Amberger (1974).
[8] Source: Gumbaridze (1973).
[9] Chapter 16.

Phenolase Action in Foods

Natural Substrates.—As shown in Table 15.4 the two most prevalent naturally occurring substances which could potentially be phenolase substrate in plant foods are tyrosine and the chlorogenic acids (Fig. 15.2). It is unlikely that monophenols other than tyrosine give rise to enzymatic browning products. Simple phenols such as catechol, commonly used in phenolase assays, are usually not found in association with the enzyme, although catechol is present in onion skin (Link and Walker 1933). The presence of the substrate does not ensure its role as substrate in phenolase action which leads to enzymatic browning as discussed below. The substrates for tea, coffee and cocoa will be discussed in the next chapter.

Pigment Formation.—As mentioned (Chapter 1), pigment formation is due to secondary reactions resulting from the interaction of the primary enzyme products, quinones. Because it occurs in human skin, we probably have a more thorough comprehension of the events leading to the formation of melanin pigments of phenolase action from tyrosine than those formed in other plant substrates. When other constituents are not involved, it is likely that when tyrosine is the primary substrate in plants, it undergoes changes similar to the by now classical mammalian tyrosinase-associated reactions leading to the red pigment DOPAchrome, previously called hallochrome, as shown in Fig. 15.1. DOPAchrome is the gateway to melanin formation through a series of rearrangements, decarboxylation and spontaneous oxidations, to give successively 5,6-dihydroxyindole and 5,6-indole quinone. The latter may be considered as the simplest monomeric form to undergo spontaneous polymerization to the high molecular weight, colored melanins.

In the disrupted plant cell, the melanin can undergo further oxidation and interaction with other plant constituents. The indole quinone as well as DOPA quinone, and, in plants, chlorogenate-derived quinone, can interact with plant constituents such as metals (Breunger *et al.* 1967), amino acid and proteins (Van Sumere *et al.* 1975) to form highly complex, colored substances. If one

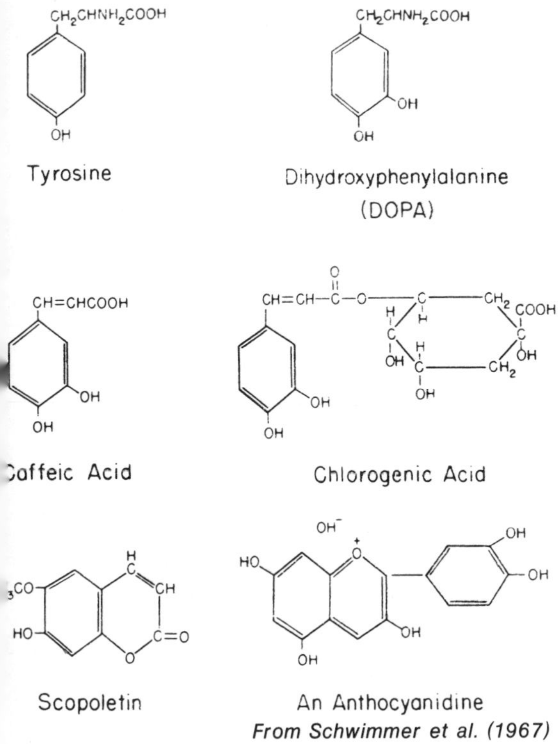

FIG. 15.2. SUBSTRATES OF PHENOLASE PRESENT IN FOODS

adds to this variation of the substrate, the nature of the products can be varied and complex indeed. These melanin pigments can thus be blue, pink, tan, green, brown, purple, black, depending upon such circumstances. Incidentally, tyrosine-derived pigments contribute to the color of the skin and similar integuments not only of man but also the eyes, hair and even feathers of birds, crocodiles, chordates, frogs, fish, sea squirts, salamanders, lizards, snakes and squid ink, as well as mammals other than man. With regard to food plants, melanin formation studies are exemplified by that of Andrews and Pridham (1967), who examined the melanins produced via phenolase catalyzed oxidation of dopamine and tyrosine and as well as that produced naturally from the pods and flowers of the fava bean. From infrared and degradation data, they concluded that the melanins produced from DOPA-containing plants such as the fava bean are largely composed of catechol type pigment.

Very little has been done on the nature of the pigments from chlorogenic acid. Horikawa *et al.* (1971) separated brown, green and blue-violet pigments from the action of apple phenolase on chlorogenic acid. Pink center in brussels sprouts may also be due to phenolase action since it is abolished by blanching (Chapter 10). Since the interaction of quinones with proteins may have nutritional consequences, it will be dealt with in Chapter 34.

Phenolase-induced Changes in the Pigmentation of Food Plant Tissue: Browning.—The most frequently encountered off-color pigmentation in plant foods is the brown discoloration (browning) due to the formation of melanins. These browning products vary in hue from yellowish to reddish brown through dark brown. Less frequently encountered are pink to red and blue to black pigmentations. Natural pigments may be masked or, as discussed below, removal of desirable anthocyanin pigment concomitant with the appearance of melanin may occur as the result of phenolase activity.

A variety of external agents evoke the interaction of phenolase and endogenous substrates to usually yield unwanted pigment in fruits and vegetables. One of the causative agents is mechanical damage due to bruising, excessive pressure and shear forces.

The advent of widespread use of insecticides has resulted in almost complete elimination of damage due to predators such as insects and, much to the dismay of at least some naturalists, birds. More frequently encountered is browning damage caused by mechanical harvesting, although future generations of harvesters may solve this problem. Other examples of the cause of browning are: in the orchard, windburn and bruising due to windfall; in the processing plant, slicing, inspection and pressing (Traverso-Rueda and Singleton 1973); lye peeling (Burkhardt *et al.* 1973); mechanical disruptions causing discoloration of cherry stems (Siegelman 1953); reddening of the butts of head lettuce; blueing of injured horse beans. Fresh produce can brown due to bruising if the container in which it is placed is not rigid enough to prevent transmission of pressure. In the harvesting of potatoes, browning may occur in potatoes placed in the bottom of the gathering conveyance, not only because they fall a greater distance, but also because of the greater continuous pressure to which they are subjected.

Phenolase-induced browning will also occur as the result of infection due to molds, bacteria, viruses, as a result of other diseases, and by exposure to high oxygen tension and toxic vapors. In all of these cases, cell disruption allows phenolic substrates to become accessible to the action of the phenolases. This usually is due to cell disruption. This common browning response to such a variety of stresses may be a defense mechanism of the whole organism. Also associated with this type of defense is the healing process as evidenced by formation of suberin in root vegetables. Thus, Dorell and Chubey (1972) concluded that any damage to the root vegetables carrots and parsnips which stimulates the suberization process also accelerates surface browning. While this ensures survival of the uninjured tissue, it also adversely affects market quality.

In general, the browning occurs at the site of the injury. An exception is the development of Black Spot in potatoes bruised at restricted sites. Blackening frequently occurs underneath uninjured skin in sites adjacent to the injuries (Weaver *et al.* 1971).

Internal browning, apparently due to phenolase action, can occur deep within the food plant tissue upon senescence of many fruits. It may also occur in superficially sound, relatively fresh tissue to give rise to such disorders as Black Heart in potatoes. The internal browning of pineapple was attributed by Teisson (1972) to a deficiency of ascorbic acid. Browning may also occur in vegetables stored in the cold or under controlled atmosphere. Thus potatoes appear to be more susceptible to browning when stored in the cold (Amberger and Schaller 1974). Intensity of browning of polyvinyl-wrapped mushrooms increased with decreased respiration occasioned by not having the optimal number of perforations per package (Nichols and Hammond 1975).

Internal browning may also occur as the result of ionizing radiation and constitutes an obstacle in successful application of such radiation to the preservation and prolonging the shelf life of some fresh fruit. Such treatment actually apparently induces *de novo* synthesis, perhaps *via* the unmasking of messenger RNA, of phenolase in mangoes (Thomas and Janave 1973). This constitutes another instance of heightened levels of enzyme action (including potato browning) in response to nonlethal dosages of ionizing radiation (Chapters 10 and 19).

One of the RNA unmasking agents may be ethylene which, when not judiciously applied, may result in discoloration of vegetables such as sweet potatoes (Buescher *et al.* 1975) or lettuce where it appears to be responsible for russet spotting (Klaustermeyer and Morris 1975). The browning which occurs in advanced stages of senescence may be traced, perhaps, in part, to endogenous ethylene action. Pink rib, a symptom of lettuce senescence (Morris *et al.* 1975), can also be induced by exogenous ethylene.

In addition to some major problems accompanying the Green Revolution, there is a relative minor but still troublesome problem relating to phenolase action which has impeded its adoption in at least one country, India. Many of the high yielding dwarf wheat varieties contain increased levels of phenolase and phenolase substrates (Abrol *et al.* 1971; Singh and Sheoran 1972). This enzyme action is especially objectionable in India because it results in discolored chapaties, a kind of unleavened pan-baked bread made of whole wheat meal. This discoloration is unacceptable to consumers (Dhonukshe and Bhowal 1974).

Browning due to oxidation of phenols by phenolase may also involve the participation of other enzymes, may be nonenzymatic or browning may not involve phenols at all. These cases will be explored after discussion of the means of preventing induced darkening when and where it is not wanted. Before this discussion, we would again like to stress that phenolase-induced browning is deliberately allowed to proceed to improve the color quality of semidry fruits such as dates and raisins as well as the beverages cider, cocoa, tea and probably coffee (Chapter 16). Immobilized phenolase was proposed for preparation of DOPA as a valuable pharmaceutical *via* hydroxylation of tyrosine by Wykes *et al.* (1971) and Letts and Chase (1973).

Pacers of Plant Food Browning.—The ensuing discussion revolves about the control of the rate and extent of browning of phenolase-induced browning in *in situ* circumstances. Of course, the past history of the crop and variety are quite important. Thus, for instance, it has been known for years that high K-containing soils are associated with heightened phenolase and browning levels in potatoes (Schwimmer and Burr 1967; Birecki *et al.* 1971). However, here we shall narrow our concern to the enzyme and substrate, and what influences their getting together.

Potatoes.—One of the main ways of approaching this question is to make a statistical evaluation among suspected influences and browning. The statistical approach is exemplified by the analysis of Mapson *et al.* (1963) which revealed a direct correlation between browning and tyrosine content. In keeping with the general experience that this amino acid is frequently the most abundant phenolase substrate in potatoes (Schwimmer and Burr 1967), their work revealed no direct correlation between browning and either phenolase activity or chlorogenic acid level. More recently, Schaller and Amberger (1974), using factorial and multiple regression analysis, reported that browning of potatoes depends not only on tyrosine and phenolase but also on total phenol content, basic amino acids, dry matter content, chlorogenic acid and flavonoids. Of course, this type of approach gives little insight as to what the actual pacemaking steps or limiting factors in browning susceptibility actually are. The same authors found that the total phenol, but not phenolase, activity increased with decreasing temperature. At 2°C, total substrate was more of a controlling factor than was enzyme level. Undoubtedly, the level of substances, such as ascorbic acid, would be important since they are capable of reducing the quinones formed as a result of the enzyme reaction.

Another factor operating not only in potatoes but also in other plant foods would be the ease of accessibility of substrate. As pointed out in Chapter 1, this could depend upon the fragility of the membranes which keep them apart. The more membrane, the greater would be the resistance to this type of cellular disruption. Mondy and coworkers (Mondy and Mueller 1977) have amassed evidence that the susceptibility of potatoes to enzymatic darkening was greatest in those tissues of a single potato tuber with the least lipid. On a rational basis it would appear that the phospholipid content should be a better indicator than total lipid, since membranes contain only the former. However, Mondy and Mueller found that the total lipid content followed the phospholipid level.

Other Foods.—In corroboration of the idea that the separation of enzyme and substrate may be limiting is the general observation that some plant foods, such as olives, become more susceptible to browning when the phenolase becomes detached from organelles such as the chloroplasts (Ben-Shalom *et al.* 1977).

Another approach to the question of what are the important factors involved is to see how each factor changes upon browning. Frequently, this amounts to finding which substrates are depleted. To take a relatively early example, Ingle and Hyde (1968) report that bruising of apples reduced the total concentration of phenols, flavonols and chlorogenic acid by 10–29, 60–90 and 20–40%, respectively. Similarly, Weaver and Charley (1974) found that the best of several correlates of banana browning was the decrease in its natural substrate, DOPAmine. In a more recent investigation of the effect of deliberately introducing air into apple pulp ordinarily used for apple juice preparation, Van Buren *et al.* (1976) reported almost complete loss of chlorogenic acid and free flavonoids but substantial retention of flavonol glycosides and nonflavonoid-nonchlorogenic acid phenolic substances.

Substrate depletion was also used to obtain some information on limiting factors in the browning of mushrooms (Murr and Morris 1975) and apricot (Gajzago *et al.* 1977). Increases in discoloration of mushrooms during storage correlated with decreases in total phenols as well as with increases in total phenolase activity. The latter authors concluded that in certain apricot cultivars the substrate depletion was the limiting factor (phenolase activity with respect to substrate level was at least five times that found in apple), whereas in other cultivars the limiting factor was reaction inactivation.

Another approach is to ascertain whether observed correlation between activity and browning tendency among varieties of cultivars is due to varying levels of enzyme or to nonenzyme factors (i.e., inhibitors) which affect the activity of the same isoenzyme and then to ascertain if the observed differences are due to intrinsically different enzymes or simply to differing levels of the same enzyme or isoenzyme complex. In the case of three avocado varieties investigated by Kahn (1976, 1977), it is clear that the variation in activity is due to varying levels of the same isoenzyme complex. As mentioned, kiwis, sunbeam peaches, and perhaps golden delicious apples, which do not readily undergo enzymatic browning, it is the dearth of substrate which is the controlling factor and in the gourds, cantaloupe, tomato and papaya, it is the almost complete absence of enzyme. It may be coincidental, but most of these plant foods happen to possess considerable proteolytic activity.

Another way of finding out if certain correlates are causally related to browning is to vary a parameter which causes a divergence in the parallelism and observe which factor changes or remains the same in keeping with the change, if any, in browning. As shown in Fig. 15.3, Buescher *et al.* (1974) observed association of brown-end discoloration of snap beans (BED) with not only phenolic content and phenolase activity but also with perox-

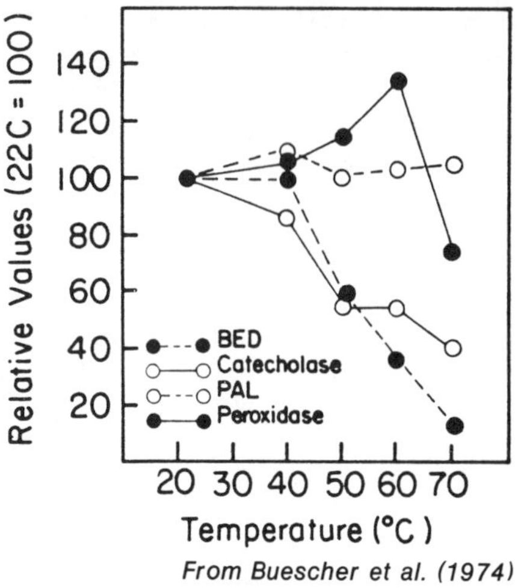

FIG. 15.3. ENZYME CORRELATES OF FOOD BROWNING

Shown is brown-end discoloration (BED) of snap bean pods; PAL, phenylalanine lyase.

idase and phenylalanine ammonia lyase (PAL), which as we mentioned is a key enzyme leading to the formation of chlorogenic acid and other diphenolics. The loss of BED was most closely paralleled by phenolase (catecholase) activity up through 70°C. Obviously, a peroxidase test for the adequacy of blanching would not have been meaningful in this case.

PREVENTION AND RETARDATION OF UNDESIRABLE PHENOLASE-INDUCED DISCOLORATION

A diversity of approaches to prevent or delay phenolase-induced discoloration and to retain the original, desirable color of plant foods have been explored. Applications of each of the approaches categorized in Chapter 1 are available or have been proposed. Here are all treatments classified according to what happens to enzymes or products or to the interaction of the two.

(1) Preharvest control and treatment: Agronomic practices, breeding and mutation.
(2) Preventing unregulated access of the enzyme to the substrate.
(3) Irreversible destruction of the enzyme.
(4) Optimization of processing and storage parameters to minimize enzyme action.
(5) Inhibition of enzyme activity by chemicals.
(6) Modification or removal of the cosubstrates, phenols and/or oxygen.
(7) Modification of the primary products, quinones.
(8) Superoxide dismutase? (see Chapter 16).
(9) Other treatments including those whose mechanism action, with respect to categories 1 to 6, is uncertain.

Practical strategies usually demand a combination of these approaches, since empirically such combinations frequently act synergistically. Even the order of treatment may be crucial. They also include measures designed to counteract possible adverse effects on noncolor food attributes such as texture and flavor.

Preharvest Control

It has long been known that agronomical variables such as fertilizer, variety and climate can affect the rate of browning. As mentioned, heavy application of potassium increases the phenolase of potatoes as cited by Schwimmer and Burr (1967). Foliar sprays of gibberellic acid (Chapter 31) or ethepon reduce postharvest browning by decreases in biosynthesis of the enzyme phenolase (Paulson *et al.* 1980).

We already pointed out that natural control may exist because of the absence of substrate or enzyme or both, or because the pH of the food as processed or as eaten is unfavorable to phenolase action.

An example of how mutation may solve a browning problem is afforded by investigation of Dhonukshe and Bhowal (1974) on the browning tendency of Green Revolution dwarf wheat. Using a simplified semiquantitative phenolase assay, they screened more than 9000 plants grown from seeds treated with 20 and 25 kilorad of acute doses of X-rays. They obtained one mutant which was practically phenolase-free. From the progenies of other low-phenolase mutants, lines were established in three varieties which could be made into acceptable chapaties, staples of the Indian diet.

Postharvest Prevention of Access to Substrate

The most widespread way to attain this goal has been to maintain and prolong the viability and hence the cellular integrity of the plant food tissue. This can be done by providing conditions for prolongation of viability of the food. In addition to the well established low temperature and controlled atmosphere storage and more recently subatmospheric (hypobaric) storage (Haard and Salunkhe 1975; Mermelstein 1979), the future may see the use of safe plant senescence inhibitors.

Measures can also be taken to prevent cellular disruption brought about by bruising or pressure. Thus, Weaver *et al.* (1965) recommended bagging potatoes as an alternative to shipping in bulk, or to soften their fall during transfer to trucks by partially filling the latter with water. More recently, Krochta *et al.* (1977,1978) were able to accomplish similar results with tomatoes (Chapter 29) by harvesting them into foam and/or by air-cushioning harvesting. Browning of potatoes and fruit can also be prevented by the use of relatively small and rigid corrugated containers to allow ventilation and also to prevent bruising due to pressure transmitted from the top to the bottom of the lot of produce (Hudson 1975).

Although sucrose plays other important roles in preventing enzymatic browning, at least one of its beneficial effects in prefreezing processing of Golden Delicious apples appears to be protection of the parenchyma cells (Ponting and Jackson 1972). Paradoxically, removal of peel may aid in the prevention of browning during the thawing of frozen peaches when this is done with ascorbic acid (Philippon and Rouet-Mayer 1973). This enhanced browning due to the skin appears to be related to its high phenolase content and to its acting as a barrier to ascorbic acid penetration during immersion.

Another means of separating enzyme from substrate is to adsorb the enzyme onto an inert insoluble carrier which is then removed by filtration. Thus, bentonite, a "fining" adsorbent in wine (Chapter 30), also adsorbs phenolase (White and Ough 1973) but was found to be less effective than low levels of SO_2 ($< 0.01\%$). It will be recalled that bentonite was a choice selective adsorbent in enzyme purification (Chapter 6).

Inactivation

Chelators.—Irreversible destruction of phenolase can be effected by heat, SO_2, organic solvents and more specifically by substances which remove Cu from the active site—diethyldithiocarbamate, 8-hydroxyquinoline, thiourea and other thiol compounds. Although the mechanism is not fully elucidated, citric acid in addition to providing high acidity also contributes to phenolase destruction by combining with Cu.

Citric acid is commonly used in conjunction with sulfite to delay or prevent browning in the commercial preparation of prepeeled potatoes (Feinberg *et al.* 1964). Malic acid has also been proposed as a phenolase inactivator.

CO and SO_2.—Similarly, phenolase inactivation of phenolase by CO (see below), recommended as a practical means of preventing discoloration of shiitake mushrooms (Fujimoto *et al.* 1972), may also involve active-site Cu, as does the slow irreversible inactivation by sulfite (Markakis and Embs 1966; Haisman 1974).

An example of how this slow irreversible inactivation can be exploited is afforded by its application to the bulk storage preservation of cling peach halves for several months (Ponting *et al.* 1975). This allows processors to extend plant operations from 8 to 20 weeks per season. The sulfite is removed by rapidly extracting with hot water or by oxidation with H_2O_2 without loss of flavor after sugar and acid are restored. Other roles of sulfite in browning prevention will be discussed shortly. Sulfite is frequently used together with blanching (Chapter 10) and other treatments such as heat to more effectively remove phenolase activity, especially in the production of frozen and canned fruits.

Heat.—However, the amount of thermal energy required to completely inactivate phenolase frequently results in serious flavor and solid (leaching) losses. A judicious, commonly applied heat treatment of peaches, used to prevent phenolase-induced browning at cut fruit surfaces during processing, consists of a brief exposure to steam.

Results of some investigations do suggest that other carefully controlled heat treatments can be useful while others entrain, as mentioned, serious side effects. Thus, Mapson and Tomalin (1961) found that blanching of potatoes (hot water, microwave) can prevent browning of prepeeled potatoes but results in weakened texture, off-flavor and microbial spoilage. On the other hand, cherries can be successfully blanched by microwaves. Adequacy of blanching to inactivate phenolase may be ascertained qualitatively but rapidly by applying dilute catechol to the cut surface of the fruit (Ponting 1944). Somewhat more quantitatively, Jankov and Kirov (1972) have ascertained that the inactivation temperature T of phenolase in grapes obeys the equation, $T = a \log t + c$, where t is the time in seconds and the intercept c is the temperature at which inactivation occurs within 6 sec.

Values of c varied from 80°C for Thompson Seedless to 89.5°C for White Muscat.

Finally, we suggest that it may be possible to exploit reaction inactivation of phenolase to prevent enzymatic browning in foods. To do this one needs to add a substrate which yields colorless, stable products. The ascorbate dependent formation of caffeic acid from *p*-coumaric acid, mentioned earlier in this chapter, might be such a suitable phenolase-catalyzed reaction, provided the caffeic acid itself is not bitter.

Optimization of Processing Parameters

It is possible to slow down enzyme action and to at least delay browning without the sometimes Draconian measures required to inactivate the enzyme. As an example, Luh and Phithakpol (1972) suggested the following for control of the browning problem in peach canning: Wash the peaches thoroughly, cool with water spray and pass through 1% citric or malic acid bath or spray with these acids so that the pH is always less than 4. Design and arrangement of equipment should allow washing, cooling and retorting within 10 to 15 min after peeling.

An interesting investigation of one of the parameters which control enzyme activity as applied to phenolase browning is that of Acker and Huber (1969). In a model water-deficient system (Chapter 13) ascorbic acid influenced the already diminished browning only when dissolved together simultaneously with other components of the system.

Reversible Inhibition

Reversible inhibition of phenolase by high concentration of sugars, especially sucrose, may be due to their bonding of water to an extent such that the reaction is slowed down due to lack of solvent (Chapter 13). Phenolase is also inhibited at high concentration of various salts. NaCl is perhaps the most effective; it inhibits noncompetitively (Knapp 1965; Sharon and Mayer 1967; Luh and Phithakpol 1972). The use of 2% NaCl brine to inhibit discoloration of sliced mushrooms in preparation for freeze-drying was not as effective as either sulfiting or blanching (Fang *et al.* 1971). Carboxylic acids, especially aromatic carboxylic acids (Pifferi *et al.* 1974; Walker 1976), are potent inhibitors but not many are acceptable as food additives. Aromatic carboxylic acids exhibit competitive inhibition with respect to the phenol cosubstrate but inhibit noncompetitively with respect to O_2. Conversely, cyanide replaces O_2 competitively but exhibits noncompetitive inhibition with respect to phenols. Walker (1976) proposed adding cinnamic acid and the monophenolic and monomethylated diphenols such as *p*-coumaric and ferulic acids to fruit juices at 0.5 *M* to prevent phenolase browning. These may resemble transition-state analogs, as does α-hydrazinophlorate (Table 14.1). Inhibitors of this enzyme have also proven to be useful for distinguishing between enzymatic and nonenzymatic browning (Wong 1975). This can be an important decision to make in the wine industry.

The aforementioned inhibition by CO which is effective *in vivo* in reducing mushroom browning could, according to Murr and Morris (1975), be due to competition with O_2 or to a delaying effect on maturation as well as chelating with Cu.

The requirements for the use of a good enzyme inhibitor applied to foods (Chapter 14) are that it be inexpensive, wholesome and preferably a food or food ingredient. Sucrose and related sugars are the most suitable inhibitors of phenolase for many foods, especially those that are naturally sweet.

The most successful application of these requirements is the use of sucrose in the Osmovac process, the osmotic dehydration of fruit (Ponting 1973). Basically, the process consists of removing water to the extent of 30–50% by placing fruit pieces in contact with dry sugar or heavy syrup, draining to remove excess sucrose and drying to 0.5–3% H_2O at a pressure of 10 mm or less at 60°C. Optional alternatives are not to remove excess syrup or inactivate the phenolase by a 2–5 min blanch at 100°C. In addition to being an effective inhibitor of phenolase, sucrose prevents the loss of volatile flavoring constituents.

Modification, Removal of Substrates

Several proposals have been forwarded whose aim is to render one or both of the phenolase cosubstrates, phenol and/or oxygen, unamenable to phenolase action. We briefly alluded to the possibility of breeding fruit devoid of substrate and/or enzyme and the use of bentonite in winemaking. Potential browning substrates may also be removed with nylon or PVP (Wong 1975). Another possibility is to remove phenolic compounds with organic solvents, although practical means of doing so have not been developed.

Another approach is to modify the substrate chemically so that it is no longer susceptible to enzyme action. The borate and titanate complexes of *o*-diphenols are refractory to phenolase action (Schwimmer and Burr 1967) but at the present are not food additives. Another way to block the phenolic hydroxyl groups is to methylate them (Chapter 16).

Browning may also be prevented by keeping the other substrate, oxygen, away from phenolase. Thus, storage of mushrooms in 100% CO_2 reduced both browning and phenolase action for as long as 7 days (Murr and Morris 1974). Other means of preventing access of oxygen which might be applicable in preparation of fruits for processing and in preserving the light color of white wines include replacement—purging—of air with N_2 (Rankine and Pockock 1970), vacuum under appropriate conditions, simple immersion in water or in thick syrups where O_2 is less soluble. Application of glucose oxidase will be discussed in Chapter 16. Limited access of O_2 may account for reduced browning of raisins prepared from methyl oleate-treated grapes-on-the-vine (Bolin *et al.* 1975). In general, as pointed out by Henderson *et al.* (1977), systems which inhibit the biosynthesis of phenolics in response to injury, such as CO_2, tend to ameliorate browning problems in fruits and vegetables.

Modification, Removal of Primary Products—Quinones

The first relatively stable but reactive products of phenolase action in foods are, for the most part, *o*-quinones, which in themselves are colorless. Means to prevent their conversion to colored melanin pigments have been used to prevent browning. This is accomplished in two ways: The quinone may be re-reduced to original substrate or the quinone oxygen atoms can form complexes, thus effectively preventing conversion of the quinones to melanins. Ascorbic acid (Chapter 16) is the primary practical agent for the first and SO_2 or sulfite for the second mode of action. As previously indicated, empirically, there does not seem to be any added advantage of using them together (Ponting and Jackson 1972). According to Embs and Markakis (1965), low levels of SO_2 have little protective effect on ascorbic acid, but ascorbic acid can under certain conditions counteract the effect of sulfite and result in accelerated browning (Markakis and Embs 1966).

At low concentrations, sulfite forms stable additional compounds with quinones. The evidence for this mode of browning control, elicited by Embs and Markakis, was elegantly adumbrated by Haisman (1974) in his overview of the kaleidoscopic effects of sulfite on food enzymes. Haisman also pointed out that lower levels of sulfite are required to prevent browning of tyrosine-containing foods than foods containing preponderately polyphenols like chlorogenic acid. Like ascorbate (Chapter 16), sulfite can be aerobically oxidized by an O_2-initiated free-radical chain reaction (Chapter 16). The latter can, as we shall see, arise via flavoprotein oxidase action. Application of some of these new findings to controlled stabilization or removal of sulfite will probably be forthcoming.

Cysteine, proposed 45 years ago as a browning inhibitor by Balls and Hale (1935), also appears to act by combining with quinone rather than reducing it as ascorbic acid does (Muneta and Walradt 1968). Binding to quinones appears to be partially responsible for retardation of postharvest discoloration of cultivated mushrooms by succinic acid-2,2-dimethylhydrazide (Murr and Morris 1974). In addition, this compound results in a heightened level of proteases which, the authors believe, degrades phenolase, thus contributing to the prevention of browning.

Finally, there are substances such as glutathione and *m*-hydroxyphenol derivatives which modify the melanin to yield light color polymers (Waggoner and Dimond 1957). This area should be looked into further.

Of course, not all the approaches to prevention of phenolase-induced browning are amenable to reduction to practice and the magnitude and relative contributions of these parameters are not absolute.

16

OTHER ROLES OF ENZYME ACTION ON PHENOLS AS DETERMINANTS OF FOOD COLOR QUALITY

In the last chapter we emphasized the role of phenolases in converting colorless phenolic substances to undesirable high molecular weight colored melanins and we also discussed means used and proposed for minimizing such conversions. In this chapter we shall expand and comment on suggestions and evidence that other enzymes may be involved in control of enzymatic browning. We also will survey the role of phenolases in the formation of desirable color in hot beverages such as cocoa, coffee and especially tea. Finally, we shall discuss enzymatic decolorization of anthocyanins in some fruits and during cacao seed curing by both phenolase and nonoxidase action.

SUGGESTED ROLES OF NONPHENOLASE OXIDOREDUCTASES

Ascorbic Acid Involvement

Like sulfite, ascorbate helps prevent the formation of colored polymers in more than one fashion. These include: the reduction of the primary oxidation products, the quinones; direct inhibition of the enzyme (Fig. 16.1) and by competing with the substrate; and inactivation, due perhaps to the interaction of the enzyme with partially oxidized highly reactive free radical form of ascorbic acid, although Varoquaux and Sarris (1979) present evidence that ascorbic acid neither inhibits nor activates the enzyme. In terms of preventing browning, any measures taken to prevent disappearance of ascorbic acid would also delay browning. These might include sequestration of trace metal ions

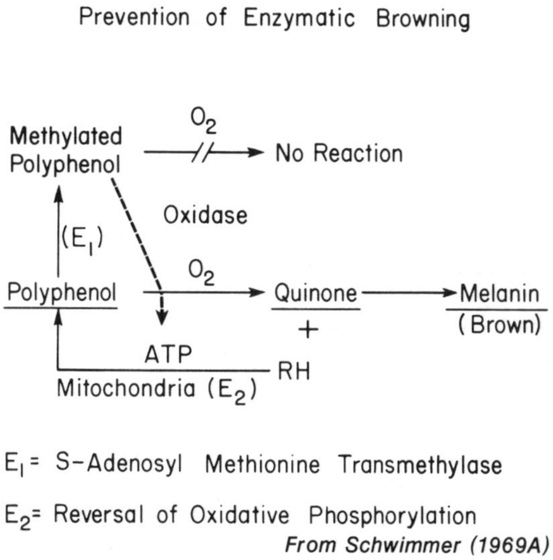

FIG. 16.1. PREVENTION OF PHENOLASE ACTION BY ENZYMATIC MODIFICATION OF SUBSTRATE OR PRODUCT

such as Cu^{2+} by citrate and other sequestering agents, inhibition of nonphenol oxidases, perhaps ascorbate oxidase and peroxidase, and those enzymes which produce superoxide anion as well as promotion of superoxide dismutase action, and creation of a reducing milieu by potentiation of enzymes involved in the reverse of oxidative phosphorylation. Contributing to the increased enzymatic browning of irradiated potatoes is a decrease in ascorbic acid content. Berset and Sandret (1976) suggest application of β-rays instead of γ-rays which penetrate surface tissue only deeply enough to affect the eyes. Although designed to prevent after-cooking discoloration (blackening), a phenol-associated defect of nonenzymatic origin (Schwimmer and Burr 1967), maintenance of a high level of ascorbate would also tend to delay enzymatic darkening.

Production of a Reducing Milieu with ATP.— Not yet feasible for commercial exploitation for control of browning is the observation that ATP, although itself not a reducing agent, can prevent or delay the darkening at the surfaces of raw potato, quite similar in its effect to that of ascorbic acid as is shown in Fig. 16.1. This mitochondrial-dependent effect (Table 16.1) appears to be due to conservation or formation of new NAD(P)H (Makower and Schwimmer 1956,1957; Makower 1964). The postulation that this is due to the reversal of oxidative phosphorylation (reverse electron transport) is strengthened by newer insight into the latter process since achieved. Rienits *et al.* (1974) demonstrated an ATP-driven reverse electron transport in lettuce chloroplasts. Also there is the growing realization that there is no true free intermediate in oxidative phosphorylation, as discussed in Chapter 4. The formation of ATP is simply a reversal of ATPase-catalyzed hydrolysis of ATP. The free energy evolved in electron transport attending mitochondrial dependent oxidation stored in proton (pH, H^+) gradient is transduced to the mechanical energy involved in a H^+-induced protein conformational modification which results in the release of the ATP from the ATP generation apparatus. The analogy and similarity of the role of ATP in muscle contraction (Chapter 26) is striking. The action of ATP illustrates a rather special instance of cellular disruption in which at least one of the subcellular organelles survives and potentially, at least, may participate in the control of enzyme food quality. Whether development to practical feasibility is ever achieved may depend upon the progress toward, and would occur as a spin-off of, ongoing immobilized enzyme engineering technology efforts aimed at economical regeneration of both ATP (Whitesides *et al.* 1976) and NADH (Mosbach 1974, 1980).

TABLE 16.1

INHIBITION OF COLOR FORMATION BY MITOCHONDRIA AND ATP (POTATOES)

	Color	
	−ATP	+ATP
A. Potato Extract	100	66
B. Mitochondria from A	66	5
C. B + "Uncoupler"	43	42

Source: Makower and Schwimmer (1954).

Superoxide Anion (O_2^-) and Superoxide Dismutase (SOD).— The relation among ascorbate, superoxide dismutase, the superoxide anion, O_2^-, (also designated as $O_2^{\cdot}$) and enzymatic browning as well as other undesirable changes in food quality was first brought to light in the patent of Michelson and Monod (1975). In particular this work was directed to stabilization of antioxidants, including ascorbic acid, in slightly alkaline foods where ascorbate ordinarily is rapidly oxidized.

Nishikimi (1975) in a somewhat more rigorous manner demonstrated that the oxidation of ascorbic acid by O_2 generated by a xanthine-xanthine oxidase system was prevented from occurring in the presence of SOD.

$$\text{Xanthine} + H_2O + O_2 \xrightarrow[\text{Oxidase}]{\text{Xanthine}} \text{Urate}^- + 2H^+ + O_2^-$$

$$O_2^- + AH_2 + H^+ \longrightarrow H_2O_2 + AH\cdot$$

$$O_2^- + AH\cdot + H^+ \longrightarrow H_2O_2 + A$$

where AH_2 is fully reduced ascorbate, $AH\cdot$ is the fleeting monohydrogen ascorbate free radical and A is dehydroascorbate.

Some of the other effects of SOD reported by Michelson and Monod are shown in Table 16.2. This enzyme not only stabilized ascorbic acid but also markedly decreased the enzymatic darkening of mushrooms, apples and potatoes. Since this may be a key patent which may constitute a departure from the traditional concepts of enzymatic plant food discoloration, we cite one of their experiments

TABLE 16.2

SUPEROXIDE DISMUTASE (SOD) AS FOOD ANTIOXIDANT

Food/Constituent	Effect of SOD
Ascorbic acid	Suppresses oxidation (1 week)
Antioxidant	Prolongs effectiveness
Anchovy fat	Inhibits autoxidation (78%) for 44 hr
Mushroom; slice in H_2O	Suppresses color in supernatant
Apple; slice, dry[1]	Suppresses color
Potato; slice, dry[1]	Color, odor improved
Potato, carrot; cook, freeze	Flavor, odor improved
Reduced ribonuclease (SH)	Prevents reactivation of enzyme by preventing SH→SS
Salmonella medium	Prolongs effectiveness of medium by inhibiting oxidation of sodium tetrathionate

Source: Michelson and Monod (1975).
[1] However, Kon (1980), upon repeating these experiments using mammalian SOD, could not observe a discernible, significant difference between treated and untreated slices.

in detail. A 10% suspension of thinly cut mushroom slices (previously freed of their lamellae) containing 0.02% ascorbic acid at pH 7.8 was allowed to incubate at ambience for 40 hr with and without SOD prepared from a sea bacterium, *Photobacterium leignathi.* Absorbances at 400 nm of the drained liquids with and without SOD were 0.002 and 0.173. Experiments with apples and potatoes were less quantitative but still showed a striking retardation of darkening.

As indicated in Fig. 16.2, active forms of oxygen may be generated in organisms which need oxygen in order to function. These active forms of oxygen may be involved in many of the forms ouf quality changes observed in fruits and vegetables, in milk (Chapter 23) and meat (Chapter 29).

If not phenolase-induced color (Kon 1980), then certainly flavor (Table 16.2) and texture (Kon and Schwimmer 1977) (Fig. 9.6, Chapters 29 and 30) are affected. Thus, O_2^- like quinones may also be considered as an unstable primary product of enzyme action which can give rise to a host of secondary food quality-affecting secondary products. Perhaps O_2^-, used in a process to dispose of toxic wastes (Sawyer and Roberts 1981), could be generated by immobilized xanthine or other flavoprotein oxidases (Fig. 16.2).

Ethylene Biogenesis.—Although it was first believed that the C atoms of ethylene originate from unsaturated fatty acids (Chapters 17, 21 and 24) and some investigators continue to believe so, most consider that ethylene is ultimately derived from methionine, or more precisely from S-adenosyl methionine (I) via 1-aminocyclopropanecarboxylic acid (the enzyme doing this is a phosphopyridoxal synthase) (Yang 1980):

$$CH_3\cdot\overset{+}{\underset{\underset{R\ \ (I)}{|}}{S}}\cdot CH_2\cdot CHNH_2\cdot COOH \xrightarrow[\text{[O]}]{[\cdot OH]}$$

$$\underset{(II)}{CH_3}\cdot\overset{+}{\underset{\underset{R}{|}}{SH}} + \underset{\text{Ethylene}}{CH_2CH_2} + \uparrow CO_2 + [NH_3]$$

I is regenerated by action of an enzyme which couples II, S-adenosyl methyl mercaptan, to homoserine, III.

$$CH\cdot\overset{+}{\underset{\underset{R}{|}}{SH}} + \underset{(III)}{HOCH_2\cdot CH_2\cdot CH_2\cdot CHNH_2\cdot COOH} \rightarrow I + H_2O$$

The [O] in plant food chloroplasts may arise from a ferredoxin-catalyzed reaction which, with appropriate donors such as NADH or light plus pyridoxal phosphate, produces ethylene from methionine (Konze and Elstner 1976).

Other rather compelling evidence suggests that some of the hyperactive oxygen is generated during the action of lipoxygenase and perhaps also in the presence of peroxidase (Mapson and Wardale 1972; Frenkel 1978):

Superoxide Dismutase.—The reason all oxygen-requiring organisms have SOD, as pointed out by Fridovitch, one of the pioneers in the discovery of this enzyme, is that the adoption of oxygen during evolution as the ultimate electron acceptor in respiration brought with it the danger of these reactive oxygen types. The importance of SOD in microorganisms and other aerobic life forms has been thoroughly reviewed (Fridovich 1974; Malmström *et al.* 1975; Michelson 1976). A key to longevity may be a sustained high level of specific SOD activity since the ratio of the latter to specific metabolic rate of primate tissue, or of whole adult organism, correlates very strongly with the maximum life span of at least primate species (Tolmasoff *et al.* 1980).

At first the sole function of this enzyme was, as indicated in Fig. 16.2, considered to be the acceleration of the spontaneous dismutation of superoxide

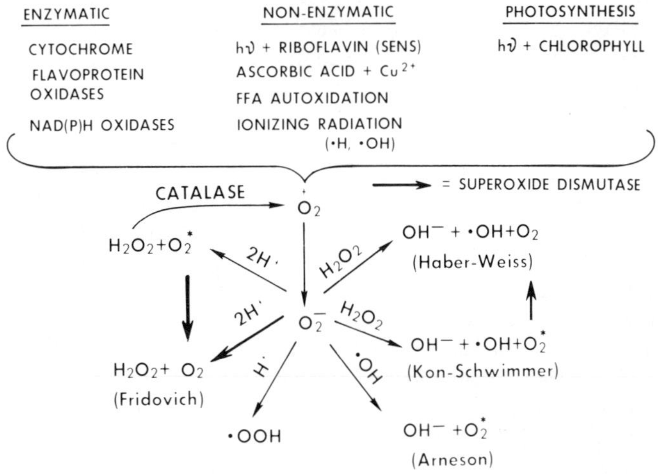

FIG. 16.2. GENERATION OF ACTIVE OXYGEN IN FOODS

Superoxide anion as unstable primary product of enzyme action leads to other active oxygen species putatively responsible for altered food quality.

From Schwimmer (1978). References to the indicated reactions are: Haber and Weiss (1934), Kon and Schwimmer (1977), Arneson (1970) and Fridovich (1974).

anion into hydrogen peroxide and an ordinary (ground, triplet state) oxygen molecule. This prevents interaction of O_2 with H_2O_2 to produce the hydroxy free radical, ·OH. However, its primary purpose may be to prevent the formation of singlet oxygen [O_2^*, $^1O_2$, $O_2(^1\Delta_g)$]. Singlet oxygen, it will be recalled, is a highly reactive form of oxygen which is characterized by the localization of the two highest energy electrons in the same orbital state and whose spins are opposed. In triplet state oxygen each of these electrons occupy separate degenerate pi molecular orbitals. The respective orbital occupancies may be represented as ↑↓ - and ↓↓ (Bland 1976).

SODs from eukaryotes (higher forms of life possessing true cell nuclei) contain Cu and Zn whereas bacterial SOD may contain Mn or Fe. The SOD from red blood cells has been thoroughly characterized. It is a single 151-amino acid polypeptide of known sequence possessing one atom each of Zn and Cu. It is of interest to note that the SOD of mitochondria, which may have evolved from microbial symbionts, contains Mn. Literature on the presence and isolation of SOD in plant foods is rapidly expanding and includes its isolation as isozymes from spinach chloroplasts (Asada *et al.* 1973), peas (Sawada *et al.* 1973), wheat germ (Beauchamp and Fridovich 1973) and other seeds (Hai *et al.* 1975). Baker (1976) found about the same levels of SOD activity in extracts pre- and postclimacteric of apple, avocado, tomato and banana. So far there has been no exception to the idea that it is present in any organism or tissue which utilizes oxygen including those which are foods and is absent from those which do not. A seeming exception is milk where it has been detected in trace amounts probably derived from red blood cells (Chapter 25). Undoubtedly by the time this appears, a respectable food science literature on this enzyme and its role in controls of active forms of oxygens will have been developed. It should be pointed out that one possible drawback to the application of SOD as a food adjunct is that it also protects anaerobic bacteria and phages and thus might encourage their growth if proper precautions are not taken to ensure against this possibility.

Putative Role of Peroxidase.—This enzyme has been forwarded as a prime candidate involved in color changes in foods brought about as the result of the action of oxidoreductases on phenolic substrates. Its ubiquitous distribution, heat stability and the confirmed observation that it can indeed catalyze the *in vitro* oxidation of phenols to form colored products in the presence of H_2O_2 has prompted some consideration of its involvement in not only the browning of foods but also in the decolorization of anthocyanin-containing foods as mentioned later in this chapter. It could cause browning *in situ*, assuming that H_2O_2 is available (Frenkel 1978), acting directly on phenols or indirectly via ascorbic acid oxidation.

The possible involvement of peroxidase as well as

catalase in black spot susceptibility of potato tubers was investigated by Weaver and Hautala (Weaver and Hautala 1970; Weaver *et al.* 1971). They showed that, while confirming that peroxidase-mediated darkening occurred *in vitro*, the act of bruising the tuber altered the level of activity of neither peroxidase nor catalase. The change in the activity ratios between each of these enzymes and that of phenolase could not be attributed to or associated with susceptibility or resistance to black spot. As a prelude to their possible involvement in starch-sugar interconversions (Chapter 19), rather than in enzymatic browning, Kahn *et al.* (1981) have rather thoroughly characterized the bound and soluble isoperoxidases of the potato tuber. On the other hand, Schaller (1974) reported that the frequency of occurrence of black spot correlated strongly with the activities of phenylalanine lyase and invertase but only weakly with that of phenolase. Of course the latter perhaps could have been due to reaction inactivation at the site of action.

The idea that peroxidase may be involved in such browning stems in part from older literature which indicated a drastic change in the activity and isoenzyme pattern at the time of respiratory upsurge of climacteric fruit. However, Haard and Tobin (1971) showed that the spectrum of peroxidase isoenzymes in developing and ripening bananas was fairly constant if phenols were removed first with PVP (Chapter 15). The total activity, most of it bound, increased upon yellowing of the peel and started to decrease when the peel started to turn brown. Also arguing against direct participation of peroxidase in browning are well-established blanching procedures which show that application of sufficient heat to destroy phenolase prevents browning, although heat stable peroxidase remains. This circumstance does not preclude its participation in off-flavor development in frozen foods (Chapter 24).

Peroxidase could play an indirect role in browning if some of the plant physiological roles attributed to its action (indole acetate oxidase, respiratory control, control of other growth regulator level, ethylene biogenesis, membrane integrity) prove to be valid. Especially in the latter case, the level of peroxidase activity could be a determinant of the availability of phenolase substrate. Of course one might argue that peroxidase (as well as catalase) could counter phenol-induced browning by prevention of the formation of these the above-mentioned destructive forms of oxygen. This would, for instance, tend to preserve ascorbate in complicated food systems. Although the resultant H_2O_2 and ascorbate are perfectly good substrate pairs for peroxidase action, its presence could delay or retard ascorbic acid loss, brought about by other means which involve these other active forms of oxygen.

Ascorbate Oxidase.—In principle such an enzyme should have some influence on enzymatic browning since it would remove one of the major factors delaying such browning. However, as will be brought out in the ensuing discussion, it is questionable if its action plays a significant role since accumulated evidence indicates that it is usually present in plant food which does not readily turn brown. Indeed it might be more appropriate to describe its action in connection with health-related food quality (Chapter 35). However, since it appears to act in *some* phenolase-containing plants, some of which are foods, has a historical enzymatic browning connotation and as an enzyme has many properties in common with the phenolases, we include it in our present overview of non-phenolase enzymatic involvement in pigment formation from colorless phenols. Much of our understanding of this enzyme is due mainly to the many years of investigation by Dawson and coworkers (Stark and Dawson 1963; Strothkamp and Dawson 1974; Dawson *et al.* 1975). Other reviews have focused on the role of its copper in catalysis (Malkin and Malmström 1970; Deinum *et al.* 1974; Howlerda *et al.* 1976). This enzyme appears to be more related to fungal laccases in structure and action than to the majority of phenolases. Dawson and coworkers succeeded in obtaining this enzyme as a pure crystalline protein from zucchini peel. It is a blue protein of MW = 146,000 daltons containing 8–10 Cu atoms of which there are the same three types as in laccase (Chapter 15). The molecule consists of four subunits comprising two subunits each consisting of two distinct chains $(\alpha\beta)_2$. Each $\alpha\beta$ pair is formed by disulfide crosslinking. The two protomers are held together by non-covalent forces.

The active form of the reductant substrate is the monoanion

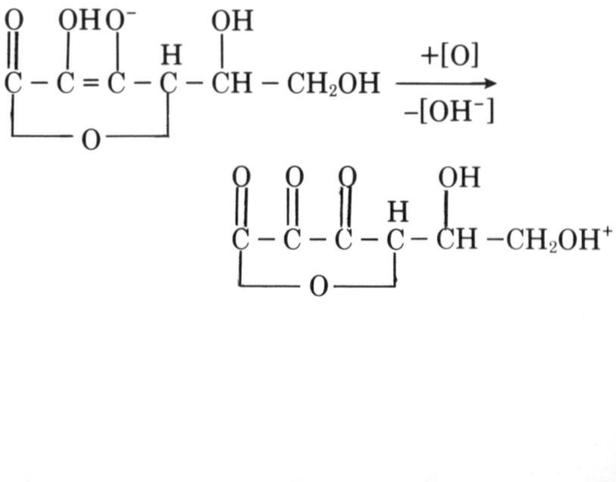

Interestingly, the V_{max} is increased, i.e., the enzyme is activated severalfold, by prior exposure to substrate in the absence of oxygen (Gerwin *et al.* 1974). This may be related to the above-mentioned disulfide bond holding together the two dimers comprising the enzyme molecule. Ascorbate oxidase is quite responsive to its environment; for instance, the enzyme is very sensitive to sulfite inhibition (Ponting and Joslyn 1948; Haisman 1974) and is also inhibited by citrate (Gerwin *et al.* 1974). Certain citrus flavonoids and anthocyanins affect the activity of this enzyme in a complex manner (Stenlid and Samorodova-Bianki 1966; Sistrunk and Cash 1974; Shrikhande and Francis 1974). For instance quercitin, a flavonoid constituent of many plant foods, inhibits ascorbate oxidase but activates the phenolase-mediated oxidation of ascorbic acid (Stenlid and Samorodova-Bianki 1966).

In addition to squash, other members of *Cucumis*, especially cucumber, are favorite sources for the isolation and study of this enzyme. While many foods have been reported to possess ascorbate oxidase activity, the possibility of an artifact exists due to:

(1) Oxidation by phenolase-generated quinones (LeGrand 1967), i.e., in the presence of a trace of an *o*-quinone (or *o*-diphenol), phenolase behaves as an ascorbate oxidase:

$$\text{Ascorbate} + o\text{-Quinone} \xrightarrow{\text{Spont.}} o\text{-Diphenol} + \text{Dehydroascorbate}$$

$$o\text{-Diphenol} + \tfrac{1}{2}O_2 \xrightarrow{\text{Phenolase}} o\text{-Quinone} + H_2O$$

the overall reaction being:

$$\text{Ascorbate} + \tfrac{1}{2}O_2 \xrightarrow{\text{Phenolase}} \text{Dehydroascorbate} + H_2O$$

(2) Oxidation by flavoprotein-generated superoxide anion as discussed earlier in this chapter.

(3) Nonenzymatic oxidation *via* metal-ion catalysis *via* lipohydroperoxide formation or via other nonenzymatic oxygen-activating mechanisms.

Indeed it has been suggested that ascorbate oxidase and phenolase activity are rarely, if ever, found in the same plant. If this is true, then the role of ascorbate oxidase in color quality of non-green vegetables is problematical since phenolase-induced darkening would not be influenced by the action of this enzyme. With this caveat in mind, the following foods have been reported to possess ascorbate oxidase activity.

Vegetables and Related Plants.—All *Cucumis* species, like squashes, pumpkin, cucumber, cauliflower, and all green vegetables tested, such as spinach, kale, grape leaves, cabbage, parsley, green bean, have it localized in the cell walls (Hallaway *et al.* 1970). Attesting to the difficulty of ascertaining the presence or absence of the enzymes in vegetables it is only recently that the absence of ascorbate oxidase in potatoes has been confirmed. This was accomplished by comparing the ample activity in a protein fraction isolated from squash with the total absence of activity in a comparable fraction from potatoes using strong anion exchange resin instead of PVP (Chapter 15) to remove interfering phenols. It is of interest that although the enzyme has been reported in tomato (Joslyn 1961B) and the interiors of bruised tomatoes were reported by Krochta *et al.* (1975) to be darker than undamaged controls, little loss of ascorbic acid could be detected as a result of bruising. (On the other hand complete homogenization in a blender followed by holding at 37.7°C and 100% R.H. for 2 hr resulted in almost complete disappearance of ascorbic acid.) Perhaps ascorbic acid is protected in the bruised tissue by superoxide dismutase (Baker 1967).

Fruits.—Citrus rind, watermelon, quince, apple, cherry, black currant and berries—The enzyme appears to be so active in saskatoon berry that in juices prepared from it only the oxidized form of ascorbic acid can be detected (Panther and Wolfe 1972).

Seeds and Grains.—Cereals such as wheat, rice and corn; mustard (seedlings); bean (leaves)—It is of interest to note that in mustard and other crucifers, ascorbic acid activates the enzyme glucosinolase, whose action contributes to the flavor of these crucifer foods (Chapter 21).

While a great deal of attention has been given to the role of ascorbic acid as a vitamin and its function at the metabolic level, we deem it of interest to outline some of the rather interesting biological effects to which this enzyme has been linked. Its activity in intact plants where it is located at the cell wall has been reported to vary diurnally (Okuntsov and Plotnikova 1970). Its biosynthesis appears to be closely associated with the action of the important photobiological protein, phytochrome (Drumm *et al.* 1972; Attridge 1974). According to Omura *et al.* (1974), ascorbate oxidase (as well as Shiitake mushroom phenolase), by virtue of its ability to induce depolymerization of DNA, possesses antitumor activity. It has also been implicated in collagen biosynthesis via lysyl oxidase, prolyl hydroxylase and in cytochrome P450 related oxidation via xenobiotic oxidases.

Enzymatic Prevention of Phenolase-induced Discoloration

In Chapter 15, we pointed out that undesirable enzyme action might be prevented by modification or removal of substrate. In this section we discuss proposed involvement and use of nonphenolase enzymes to render phenolase substrates inaccessible to their enzymes.

Transformation of Phenols.—*O-Methyl Transferase.*—One way to block phenolase action is to block the phenolic hydroxyl groups of substrates by methylating them. Finkle (1964) and Finkle and Nelson (1963) invoked transmethylation *via* O-methyl transferase to explain the delay in darkening of apple slices placed in slightly alkaline (pH 8) buffers. Such a reaction:

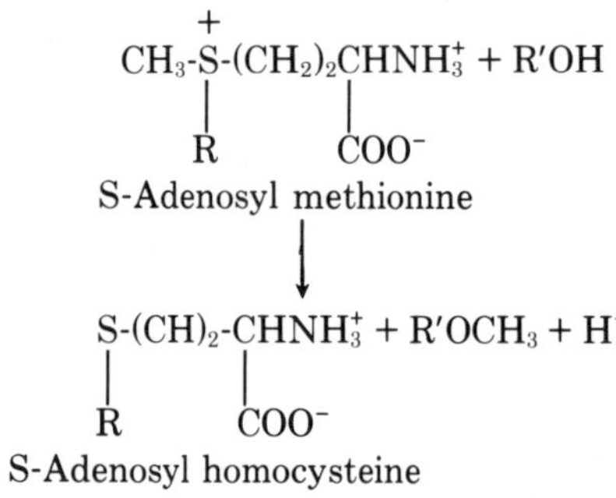

would require active participation of the fruits' living normal metabolic systems. This implies that oxygen is needed in order to supply the ATP required to continually synthesize the active methyl donor, S-adenosyl methionine. Finkle also suggested that the enzyme could be added to the fruit, if it were not present in adequate levels. Its presence in plant foods and participation in enzymatic prevention of enzymatic browning would constitute another striking example of potentiation of the food's own enzyme to improve quality. Walker (1976) pointed out that the effectiveness of such a system may be reinforced by the inhibition of phenolase action on surviving phenols by their methylated derivatives as indicated in Fig. 16.1. Another interpretation would be that superoxide dismutase is potentiated at this pH. This would be in accordance with the observations of Michelson and Monod (1975) (Table 16.2) on apples, potatoes and ascorbic acid-treated mushrooms dipped in SOD and also with the known properties of SOD. However, since Kon (1980) did not observe such effects using mammalian instead of bacterial SOD, we suggest, alternatively, it would not be far-fetched to invoke a chemiosmotically-produced production of ATP (Chapter 4) which would then create a reducing milieu as discussed earlier in this chapter. The requisite proton gradient would occur during adjustment of pH.

Practical application of the principle of pH adjustment, first suggested by Finkle, was developed by Bolin and Stafford (1964) for preservation of the color of refrigerated apple slices destined for the baking trade. Their investigation constitutes an excellent example of the principle of the synergistic effect of two or more treatments. In a later refinement Ponting *et al.* (1972) by combining pH adjustment with low-level sulfite treatment (0.01%) or with ascorbic acid and calcium salts, improved both color and texture retention (Fig. 16.3).

Benzene Ring-splitting Enzymes.—In principle it should be possible to remove phenolic substrates by splitting their benzene rings. Indeed the normal pathways of catabolism of the aromatic amino acids of most organisms proceed through the appropriate ring-splitting enzymes. This has been accomplished in apple juice, again in principle, by Kelly and Finkle (1969) through the addition of a highly purified preparation of ring-splitting ring-cleaving bacterial procatechuate-3,4-dioxygenase, albeit at an uneconomically feasible and noncatalytic concentration of 17.4 mg of pure enzyme protein per ml of apple juice. This is one place where enzyme immobilization might be profitably applied.

Removal of Oxygen.—In Chapter 15 we reviewed physical methods for removal of the other substrate of phenolase, oxygen. This can also be removed by means of added enzymes; the principal enzyme proposed for this purpose is glucose oxidase, discussed in more detail in Chapter 19. Ac-

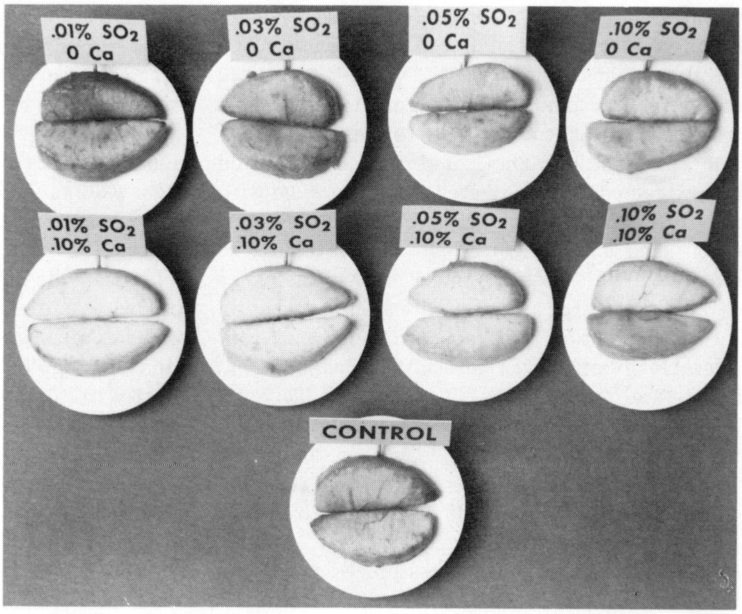

FIG. 16.3. CONTROL OF ENZYMATIC BROWNING IN PEACHES

Using a combination of pH adjustment, SO_2 and Ca salts.

Courtesy of Ponting et al. (1972)

cording to Scott (1975) oxygen may be removed by this means for preventing enzymatic browning of fresh frozen fruits and is particularly effective on cherries. The glucose oxidase preparation is incorporated into the sugar cap, resulting in the deoxygenation of the surface layer. Glucose oxidase action also results, according to Scott, in depigmentation of bruised areas, suggesting competition with phenolase for O_2.

Removal of O_2 with glucose oxidase during winemaking has been attempted with promising preliminary results being reported (Scott 1975; Ough 1975). Some potential problems attending this particular application include: conversion of ethanol to acetaldehyde, requirement for adding glucose to very dry wines and inhibition of glucose oxidase by sulfite frequently used in the processing of white wines.

Removal of O_2 by phenolase action itself has been suggested for deoxygenation of orange juice. This can be accomplished economically by adding small amounts of phenolase-containing apple juice to bottles of orange juice before pasteurization. This is especially effective in reducing subsequent nonenzymatic browning of the juice during storage (Joslyn 1961B).

DESIRABLE ENZYMATIC BROWNING

Until now we have considered the formation of pigments formed via phenolase and other oxidases as undesirable. As pointed out previously, accumulation of such quinone-derived pigments imparts desirable characteristic visual-quality attributes to several beverages, i.e., cider, teas, cocoa and perhaps to coffee and to such partially dehydrated fruits such as figs, dates and raisins, but probably not to prunes. The substrates of those phenolases involved in the development of these desirable pigments are usually tyrosine, chlorogenic acids or related compounds so frequently the substrates for undesirable enzymatic browning, but also flavonoids and other less widely distributed phenol derivatives. Although it is a rather complicated enzyme substrate system, perhaps far more is known about the phenolase action and its role in tea color than in any of the other foods.

Tea Color and Phenolase Action

Unlike many of the processed foods whose preparation frequently involves simultaneous (fish sauces, pickled vegetables, meat sausage, some fermented dairy products, coffee, wine) or more or less sequential (beer, cheese, cocoa, bread) action of enzyme systems of intact microorganisms (fermentation) and free enzymes (either added or present in the food), involvement of enzyme action in the so-called "fermentation" step in tea manufacture involves only the tea leaves' own enzymes.

From the viewpoint of intensity of endogenous enzyme action or "fermentation" the three principal tea categories are: black (fermented), oolong (semifermented) and green (unfermented). Phenolase action in the latter is largely attenuated by suitable means. For instance, instant green tea powder may be made by heating the fresh leaf to a temperature sufficiently high to cause total enzyme inactivation. The leaf is then comminuted and extracted with hot water. However, even with fully fermented tea the action of phenolase, while encouraged and promoted, must nevertheless be controlled. Since both the flavor and especially the color of black tea infusions are so important in their consumption, it is not surprising that considerable investigative effort has been expended to understand the complex chemistry and biochemistry behind the age-old well established technology by which fresh young leaves of the shrub *Camellia sinensis* are converted to tea. A selection of reviews over the last 35 years constituting a perspective of the gradually emerging understanding and rationalization of these processes follows: Roberts (1942,1952,1962), Eyton (1972), Sanderson *et al.* (1972) and Jurd (1972). A most interesting, somewhat unorthodox viewpoint stressing the health-related aspects of the beverage by pioneers in the field is marred by incorrect structures of tea flavonoids (Bokuchava and Skobeleva 1980). Other keys to the vast literature of the enzymology involved may be found in relatively recent investigations by Coggon *et al.* (1973A,B) and Sanderson and Graham (1973). Sanderson and Coggon (1977) provide an excellent, concise, overall view of the role and use of enzymes in the manufacture of black tea and instant tea. Other reviews will be cited in connection with the role of phenolase in tea flavor development (Chapter 22).

The Fermentation Step in Tea Production.— The color and flavor qualities and processing parameter control are in good measure determined by preharvesting conditions, as in other foods. Enzymatic aspects of the role of water stress in achieving desirable flavor were alluded to in Chapter 1. In considering the usual horticultural variables—climate, soil, cultivation, altitude (mountain-grown yield better teas than lowland-grown bushes)—special attention is given to ensure fast growth of shoot tips.

Considerable care is also given to selection of which parts of the immature, rapidly growing tea bush to pick ("pluck"). In general plucking is confined to the shoot tips—the flush—consisting of the terminal bud, the first three leaves and associated stem. Grades of tea are, in part, determined by age, size and position of the leaf on the shoot. Thus, Orange Pekoe comprises largely the first and second leaves.

The leaves are then "withered" by exposure to the heat of the sun or hot oven. Withering removes about one-fourth of the water in two separate steps and undoubtedly determines the course of enzyme action during the drying process and in subsequent steps.

Thus, in the next step, maceration, the leaf can be more easily twisted and curled or "rolled" without causing excess tearing and its attendant increased opportunity for phenolase and its substrates to interact. In point of fact, the procedure up to this point is the same for most green and black teas. If dried immediately, the dull green leaves become green teas. When rolled into pellets, the product is known as green "gunpowder" tea.

Fermentation.—The relatively short and simple step which distinguishes black tea manufacture is the interpolation of the so-called fermentation already begun with first cellular disruption occasioned by the first mechanical damage. The rolled tea is spread out on a flat surface delaying the subsequent drying (firing) for as little as 1 hr. This allows potentiation of the phenolase, whose action during this short period results in conversion of the leaves from green to copper red, in flavor precursors, and in substances involved in haze in instant tea (Chapter 30). This fermentation step also ensures uniform color development.

This action is initiated and controlled by the movement of the substrates to the enzyme rather than the reverse since cell disruption occasioned by withering and rolling promotes an "affinity chromatography" in which the enzyme interacts with a small amount of the flavonol substrate to form insoluble complexes which are nevertheless capable of carrying out "immobilized" enzyme reactions during fermentation.

The firing step will determine the subsequent pathway of secondary reactions. An important step, but perhaps not relevant to enzyme consideration, is sorting, in which the black tea is sifted for size; i.e., only certain sized particles are used for Orange Pekoe.

The Substrates and Colored Products.—The ultimate precursors to black tea color substrates for tea phenolase comprise the colorless flavonols, epicatechins; and their galloylated esters, the tannins, constitute 20–30% of the dry weight of fresh tea leaves. The gallocatechins are reduction products of the colored anthocyanidins.

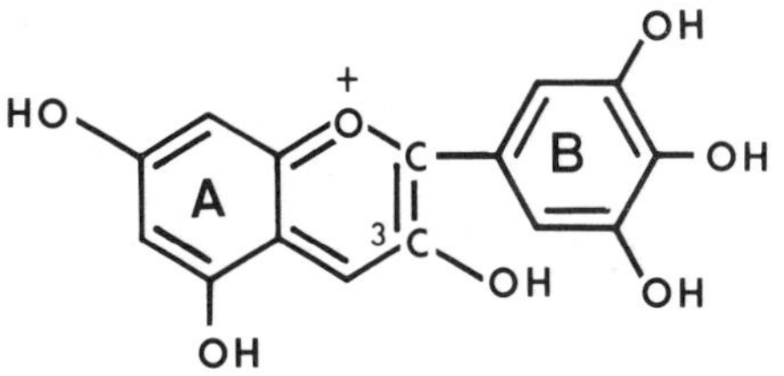

AN ANTHOCYANIDIN
(DELPHINIDIN)

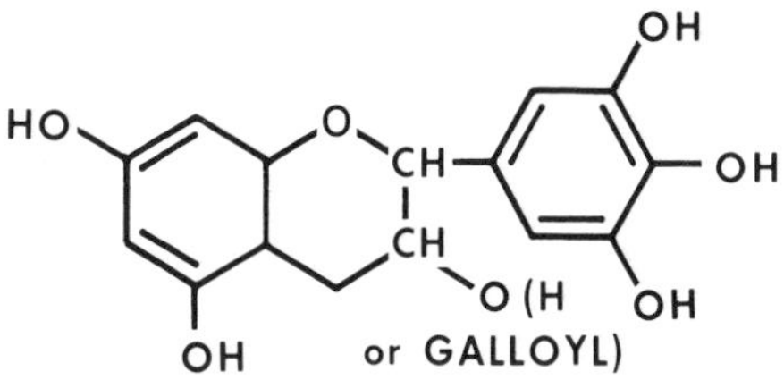

A GALLOCATECHIN (EPI)
(TEA CATECHIN)

The distinction between catechins and their diastereomers, the epicatechins, involves the stereochemistry of the B ring with respect to the hydrogen in position 3 whose OH is involved in galloylation. A catechin has two and a gallocatechin three B-ring hydroxyls. The epicatechins, comprising (–) epigallocatechin gallate and (–) epicatechin gallate, account for about 10% of all the flavonols in fresh tea leaves (Bhatia and Ullah 1968; Coggon *et al.* 1973B). There are at least six such substrates for phenolase action present in most teas (Fig. 16.4). The sequence of reactions leading to black teas may be represented as follows: Epicatechins and their gallates → corresponding quinones → theaflavins → thearubigins.

The colored substances of tea are the theaflavins and thearubigins (Fig. 16.5). Theaflavins appear to be dimers of the oxidation products (quinones) of epicatechins and have been considered to be relatively unstable bisflavonol or benzotropolone derivatives. The relatively unexplored thearubigins comprise some 10–20% by weight of black tea solids. At first considered to be polymeric proanthocyanidins derived from theaflavins by further oxidative condensation (Brown *et al.* 1969; Eyton 1972; Sanderson *et al.* 1972), at least some appear to be pentameric flavon-3-ols and their gallates, containing interflavonoid links as well as benzotropolone units. The overall quality of teas has been correlated with what was supposed to be thearubigin optical properties (absorbance ratio, 380 nm to 460 nm). More recently this ratio was shown to involve contributions by flavonol glycosides (Cattell and Nurnsten 1977).

Tea Phenolases and Their Action.—Historically, several oxidative enzymes have been implicated as being involved in tea color. These include enzymes from the putative microorganisms growing during misnamed fermentation, as well as tea leaf derived cytochrome oxidase, ascorbate oxidase and peroxidase (Eyton 1972). Because of its heat stability, peroxidase may contribute to decline in color quality by removing theaflavins (Cloughley 1980). Much of the history of tea enzymology consists of disproving these ideas and showing that catechol oxidase, undoubtedly, as expressed by Sanderson and Coggon (1977), is not only the most important single enzyme in tea manufacture but probably the only one involved directly in color change.

Establishment of phenolase as a distinct enzyme entity goes back some 35 years as a result of the investigations of Sreerangachar (1943). Although the enzyme is apparently a typical *o*-diphenolase, utilizing catechol as well as tea flavonols as substrates, chlorogenic acid, an important substrate for other food phenolases (Chapter 15), cannot serve in this capacity. Relatively recent investigations on the nature of tea phenolase are exemplified by those of Roberts (1972) and of Coggon *et al.* (1973A,B). Roberts confirmed the presence in tea of a readily available phenolase of the catecholase type which is active *in situ* in an insoluble state as mentioned during tea fermentation. The latter investigators purified tea phenolase 200-fold and confirmed that tea phenolase is comprised of three copper containing isozymes. These act optimally at pH 5.7 and specifically on the tea flavonols (Fig. 16.4) by oxidizing the 3′,4′-*o*-dihydroxyls to form the respective quinones. Closely associated with the enzymes were a peroxidase as well as a weak tannase (gallate esterase) activity. Reaction systems containing the enzyme (devoid of peroxidase) and tea flavonol substrates were capable of forming the components of tea color, theaflavins and thearubigins. The substrates, after being oxidized, were also epimerized and degallated in side reactions (Fig. 16.6). Phenolases can thus, with the

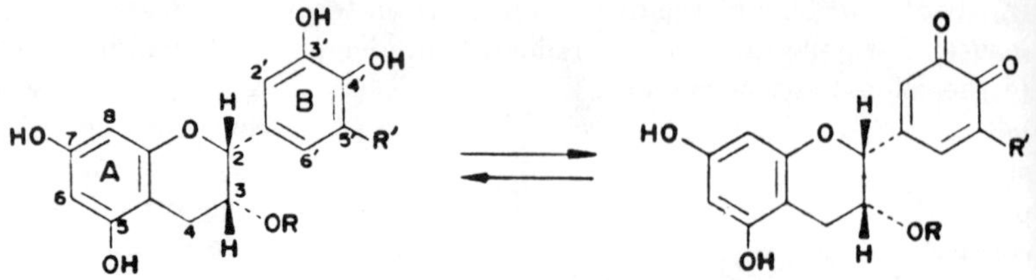

(I) L-(−)-epicatechin R = R' = H
(II) L-(−)-epicatechin gallate R = G; R' = H
(III) L-(−)-epigallocatechin R = H; R' = OH
(IV) L-(−)-epigallocatechin gallate R = G; R' = OH

(V) R = R' = H
(VI) R = G; R' = H
(VII) R = H; R' = OH
(VIII) R = G; R' = OH

G = Galloyl = 3, 4, 5-trihydroxybenzoyl

FIG. 16.4. TEA PHENOLASE SUBSTRATES

From Coggon et al. (1973A). Reprinted with permission from Phytochemistry 12, 578. Copyright 1973 Pergamon Press, Ltd.

proper substrates, act as "tannases." This oxidative degallation, rather than the feeble tannin esterase activity of tea, accounts for the degallation which occurs during the commercial fermentation of tea leaves. Thus, development of tea color illustrates indispensable potentiation of enzyme activity *via* extensive cellular disruption as well as the production by enzyme action of highly reactive intermediates which undergo changes resulting in multiplicity of food attributes. Since phenolases are central to the development of tea quality, patents have been issued on the addition of both endogenous tea (Seltzer *et al.* 1961; Millin 1972) and nontea phenolase preparations (Fairley and Swaine 1973) for the manufacture of soluble tea. If not already in the literature, we predict that someone will produce tea experimentally using deliberately immobilized phenolase, thus mimicking the traditional fermentation.

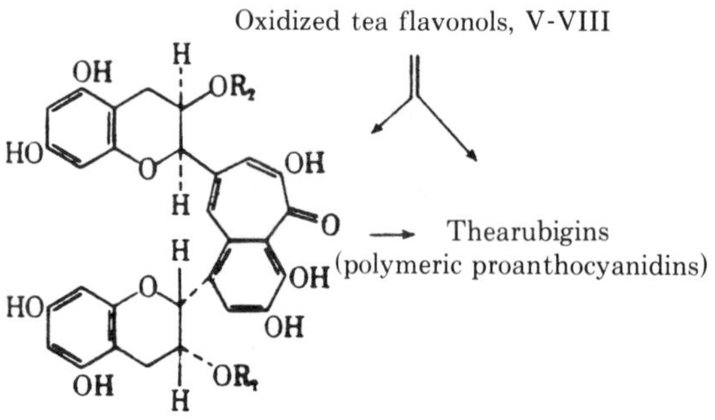

Theaflavins, IX-XII

From Sanderson and Graham (1973). Reprinted with permission from J. Agric. Food Chem. 21, 578, Fig. 4. Copyright by Am. Chem. Soc.

FIG. 16.5. TEA COLOR PRODUCTS OF PHENOLASE ACTION

Color Quality of Cacao Seed Products.—In contrast to so-called tea "fermentation," the processing of cacao bean for the production of cocoa and chocolate "curing" comprises a true microbial fermentation phase followed by a drying phase. The process and the mechanism of curing were reviewed by Forsyth and Quesnel (1957) and Forsyth (1963). Although the principal aim of curing and the main thrust of recent investigations of the curing process are concerned with and directed to flavor development, color changes of course do occur.

These color changes due to the action of the cacao seed's endogenous enzymes occur only after microbial growth is well advanced. Polysaccharidases released by the growing organisms remove mucilage in which the seeds are embedded. Microbial growth also kills the seed cells. Cessation of seed viability allows decompartmentalization of enzyme and substrate to occur. Cell death due to lowering of pH and increase in temperature to 45°−50°C can occur within 2 days but usually lasts from 4 to 8 days. As in the case of tea fermentation, the substrates in the disrupted dead pigmented cells seek their way to the enzymes which remain immobilized, probably in nonpigmented cells, into which colored substrate released from pigment cells must migrate or to insoluble cell fractions within the same cells. In the latter situation the substrate released from vacuoles, mostly colorless phenolics, find their way more readily to the enzyme.

Enzyme action affecting color in the curing process is of two kinds: glycosidase action which removes anthocyanins responsible for the undesirable purple color of the unprocessed bean, discussed below, and phenolase which occurs during the subsequent drying phase when oxygen is accessible to the enzyme. Development of typical cocoa color is dependent on the phenolase action during the drying phase. The main substrate for the phenolase action is probably epicatechin but leucocyanidins may also participate. These substrates are present at the end of the fermentative phase but disappear after drying. A more recent corroboration of this pattern of events and its relevance to chocolate quality is afforded by the investigation of Kharlamova (1972), who found that inferior chocolate produced from incompletely cured cacao beans still contained (−)epicatechin, catechol, leucocyanidins and neochlorogenic acids.

In their review, Forsyth and Quesnel array the experimental evidence which shows that the disappearance of these phenols and the appearance of color is due to the action of endogenous phenolase. For instance, fermented beans dried in the absence of oxygen failed to turn brown. More recently, Quesnel and Jugmohunsingh (1970) characterized the phenolase activities of both fermented and unfermented cacao. Their study was designed to find out why the phenolase did not work during the fermentative phase and to eliminate the possibility that phenol oxidation during the drying phase was nonenzymatic. The study demonstrated the action of phenolase and defined limits corresponding to inactivation of the phenolase. As with tea, the action of phenolase contributes positively to the final flavor as well as to the color of the cocoa and chocolate. It may also aid in the removal of astringency.

Investigations of the role of phenolase action in coffee quality are rather sparse and of recent origin (Amorim and Amorim 1977). Preliminary results indicate that better coffees have higher phenolase activity and that more chlorogenic acid is bound to protein. Since the color of coffee develops during roasting and is thus of nonenzymatic origin, it is unlikely that phenolase plays a significant role directly in color development. However, it may play some role in flavor (Chapter 22) by providing reducible colorless quinones.

DECOLORIZATION OF ANTHOCYANINS

The class of flavonoid phenolic pigments known as anthocyanins, glycosides of anthocyanidins, may be destroyed and each lose its characteristic color *via* nonenzymatic reactions, by action of phenolase and by the action of glycosidases. Reviews emphasizing factors influencing stability of anthocyanins in foods include those of Hrazdina (1974) and Markakis (1974,1975), who cite the brief enzyme-related literature through 1964. Subsequent literature citations are provided by Pifferi and Cultrera (1974).

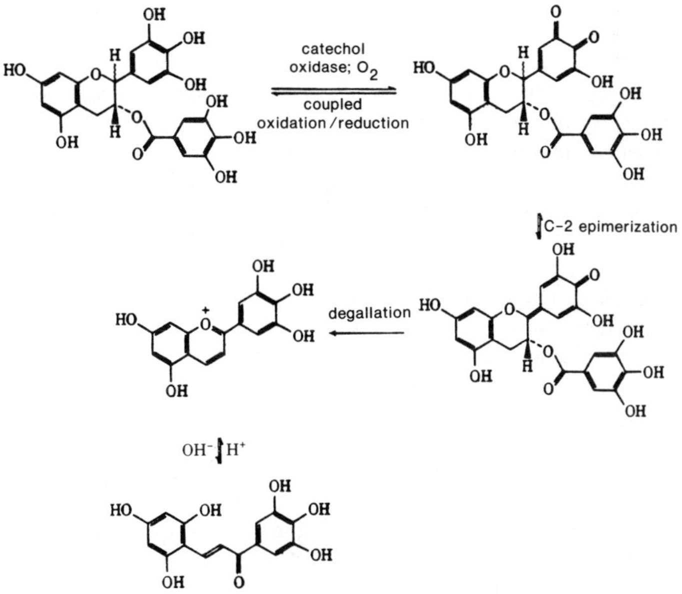

FIG. 16.6. ACTION OF TEA PHENOLASE ON TEA PHENOLS (GALLOCATECHINS)

From Coggon et al. (1973B). Reprinted with permission of the J. Agric. Food Chem. 21, Scheme I, 730. Copyright by Am. Chem. Soc.

Decolorization *via* Glycosidase (Anthocyanase)

In Cocoa Manufacture.—As previously mentioned, an important aim of cocoa seed curing is to eliminate the purple color due to the presence of the anthocyanins, cyanidin 3-β-D- and 3-α-L-arabinosides. The latter is the major anthocyanin in Forestero, the most important variety for the manufacture of cocoa. Such action not only removes undesirable color but may also provide precursors to flavor and color *via* phenolase action on the liberated aglycones. These pigments are present in specific storage cells in the cotyledon. As previously mentioned these cells have to be disrupted via death caused by the changed conditions brought about by the microbial fermentation. Thanks to the work of Forsyth and Quesnel, it is clear that highly specific β-D-galactosidases and β-L-arabinosidases (they will not attack glucosides or xylosides) located in unpigmented cacao seed cotyledon cells are responsible for the hydrolysis of these colored anthocyanins and thus for removal of the purple cast from the seeds; cyanidin-3-α-L-arabinoside + H_2O → cyanidin + L-arabinose. The liberated cyanidin appears to be converted into a colorless acid-stable "pseudo" base which may serve as a phenolase substrate during the subsequent drying period. The glycosidases have to perform their task before their inactivation by the complex flavolan tannins which accompany the pigments in the purple cells.

Fruits.—Anthocyanase or glycosidase action on such pigments in fruits or their processed products is largely derived from fungi which may be adventitiously present. Thus, addition of an enzyme preparation from a strain of *A. niger* to berry juices resulting in losses amounting to as high as 67% was shown by Huang (1955) to be due to the action of glycosidase(s) on the anthocyanins. The aglycone specificity remains unclear as does the subsequent transformation of the liberated aglycone moiety of the anthocyanidins. Apparently the newly liberated but still colored aglycone anthocyanidins undergo spontaneous transformations to colorless and insoluble substances.

At first suggested for use in removing excess red color due to anthocyanin from grape juice or red wines or for decolorizing pigmented grapes for white wine manufacture, action of glycosidases present in commercial pectinase preparations used in juice and winemaking (Chapter 30) may have been responsible for the occasional loss in desirable wine color (Wong 1975). However, most pectic enzyme preparations, when used properly, apparently are relatively anthocyanase-free since their use has been reported to result in an increase in anthocyanin-associated desirable color rather than in a loss (Ough *et al.* 1975). This increase is due to better extraction of these pigments brought about by the breakdown of pectin and not to enzymatic or any other type of browning. However, that anthocyanase may still be present in some pectinase preparations is suggested by their finding that this color-related advantage of using pectic enzymes is diminished if the enzyme incubation time is increased from 48 to 72 hr or longer (Table 16.3).

TABLE 16.3

ADDED ENZYME INDUCES COLOR CHANGES DURING WINE MAKING[1]

Variety	Skin Contact Time (hr) 24	48	72
Grenache	1.00	4.00	1.39
Zinfandel	2.10	3.93	1.69

Source: Adapted from Ough *et al.* (1975).
[1] Values, relative to that of "Grenache" at 24 hr, represent color differences (color calculated from transreflectance measurements) between untreated control and enzyme (commercial pectinase) treated grapes.

This illustrates the necessity of painstaking attention to control process variables in order to achieve maximal usefulness of commercial enzyme preparations in food processing.

An instance of the presence in fruit of an anthocyanin glycosidase in association with its natural substrate is that of β-galactosidase present in apple peel (Schmid 1971). As part of a carbohydrase-oxidoreductase system, the glycosidase decomposes the endogenous red cyanidin-3-galactoside (idaein) as well as the hyperoside, quercetin-3-galactoside. As with other anthocyanidins, the aglycone is unstable and is oxidized. The enzymes are apparently present in a latent form in the peel since, even after homogenization, they were fully active only after detergent treatment.

Oxidative Enzymatic Decolorizing Degradation of Anthocyanins

In contrast to the consequence of its action on colorless phenolic substrates in foods, phenolases act on colored anthocyanins to decolorize them. In this regard, most investigators had assigned a role to phenolase analogous to its role in the oxidation of ascorbic acid; anthocyanins are degraded as a result of the cyclic oxidation-reduction of trace amounts of colorless phenolics such as chlorogenic acid.

$$\text{Colored Anthocyanin} + \text{Colorless Quinone} \xrightarrow{\text{Spontaneous}} \text{Colorless Phenol} + \text{Anthocyanin Oxidation Product}$$

$$\text{Colorless Phenol} + \tfrac{1}{2}O_2 \xrightarrow{\text{Phenolase}} \text{Colorless Quinone} + H_2O$$

$$\text{Colored Anthocyanin} + \tfrac{1}{2}O_2 \rightarrow \text{Colorless Anthocyanin Oxidation Product} + H_2O$$

However, Sakamura *et al.* (1966) separated two oxidases from eggplant, only one of which appeared to catalyze the colorless phenol-dependent reaction while the other appeared to oxidize anthocyanins directly. Pifferi and Cultura (1974), taking special precautions to remove traces of phenols from their enzyme preparations, demonstrated feeble cyanidin-3-glucoside decolorizing activity by an enzyme present in the skin of sweet cherries. The pulp contained a phenolase which vigorously degraded anthocyanins *via* the phenol-mediated reaction. The pulp isoenzymes were studied in detail. It is not clear if the skin enzyme is indeed an oxidase or a glucosidase.

That such oxidations, whether direct or indirect, may influence food quality is evidenced, for example, by the studies of Sistrunk and coworkers such as that by Cash *et al.* (1976) on involvement of phenolase action in color loss of Concord grape, as shown in Fig. 16.7.

The end-products of this anthocyanin degradation settled out as polymerized brown insoluble

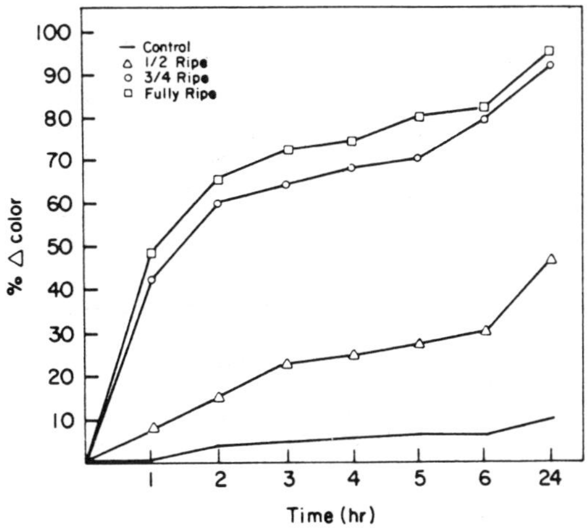

From Cash et al. (1976)

FIG. 16.7. LOSS OF GRAPE PIGMENTATION DUE TO PHENOLASE ACTION ON COLORED ANTHOCYANINS

oxidation products. The investigators concluded that the activity of phenolase is one of the most important factors involved in loss of color in such juices. Phenolase action may also contribute to color loss in cranberry juice (Chan and Yang 1971). Methods proposed for anthocyanin stabilization in foods are included in Table 15.5.

In the presence of H_2O_2, peroxidase can also decolorize anthocyanins (Grommeck and Markakis 1964). The H_2O_2 could conceivably arise from a metal-catalyzed oxidation of ascorbic acid (Shrikhande and Francis 1974; Tanchev 1974), although it is more likely that the more active forms of oxygen discussed earlier in this chapter would be involved in ascorbate-mediated anthocyanin destruction.

17

COLOR CHANGES INDUCED BY ENZYME ACTION ON NONPHENOLIC SUBSTRATES IN PLANT FOODS

ENZYMATIC DESTRUCTION OF CAROTENOID PIGMENTS

In communications to Sumner and Somers (1947), Bohn and Haas reported in 1927 that they found the yellow color of wheat flour and egg yolk was destroyed by an oxygen-requiring enzyme present in various beans, and that soybeans were a specially active source of this enzyme. This bleaching action was ascribed to a "carotene oxidase." Sumner and Sumner (1940) and Tauber (1940) independently reported that the bleaching of carotene was due to a coupled oxidation *via* unsaturated fatty acid "peroxide." The peroxide is formed, as we now know, as the result of the catalysis of the hydroperoxidation of *cis-cis* pentadienes, including unsaturated lipids such as linoleic and linolenic acid, catalyzed by lipoxygenase.

Since perhaps a more predominant attribute imparted to foods by the action of this enzyme is flavor, lipoxygenase will be discussed in somewhat more detail in Chapters 21 and 24 and in other discussions of food quality attributes affected by its action. Like many of the enzymes we have and will be discussing, lipoxygenase produces a class of highly reactive primary products which can then react in a multiplicity of diverse secondary reactions with food constituents to produce dramatic changes in almost all of the common quality attributes of foods, including not only color and flavor but also physical and textural properties, as well as those related to health and nutrition (Chapter 34). Lipoxygenase may also be important by virtue of its possible participation in the genesis of ethylene in the preprocessing history of plant foods. The protean nature of the influence of lipoxygenase action on food quality is illustrated by the following reported effects and alleged functions of this widespread enzyme. Its action, either directly or through secondary reactions, may lead to:

Color Changes:

Bleaching of hard wheat flour *via* the carotene destruction (desirable).

Bleaching of pasta products, also by carotenoid destruction (undesirable).

Participation in loss of desirable green color due to chlorophyll destruction in frozen and perhaps other processed vegetables.

Destruction of xanthophyll and other colored carotenoids important in assessing the quality of alfalfa-derived animal feeds (undesirable).

Destruction of added food colorants (undesirable).

Participation in meat color as prostaglandin synthase as well as lipoxygenase (?).

Destruction of skin pigmentation in some food fishes.

Superficial scald in stored apples.

Flavor Changes:

Production of components of volatiles which are responsible for desirable aroma in fruits and fresh vegetables such as tomatoes, string, green, or snap beans, bananas and cucumbers.

Production of off-flavors in frozen vegetables and perhaps other processed foods.

Production of off-flavors in stored cereals such as the "cardboard" flavor of barley.

Production of off-flavors in high protein foods, especially the legume seeds.

Participation in rancidity in meats (as prostaglandin synthase and lipoxygenase).

Texture Changes:

Production of favorable effects on the rheological properties of wheat flour doughs and eventually on baked goods texture *via* control of SS-SH balance and hydrophobic bonding of lipids to glutens.

Nutritional Quality Changes:

Destruction of vitamin A and provitamin A.

Destruction of nutritionally essential polyunsaturated fatty acids.

Interaction of product of action with some essential amino acids of proteins so as to lower the protein's nutritional quality and functionality.

In Vivo Functions:

Participation in the biogenesis of ethylene by creating active oxygen needed to convert methionine to ethylene (Chapter 16).

Conversion of carotenoids to plant growth affectors such as abscisins and growth inhibitors.

Conversion of the unsaturated mammalian fatty acid, arachidonic, to hydroxyfatty acids.

Destruction, anaerobic, of peroxides arising from wound tissue.

Excellent reviews of this enzyme include those of Axelrod (1974) which deals more with the basic biochemistry, of Grosch (1972) in German which also delineates the many-faceted aspects of its action in foods, and the comprehensive discussion of Eskin *et al.* (1977). At the present, we shall examine the role of lipoxygenase and other putative enzyme action as determinants of the color quality of carotenoid containing foods.

Enzymology of Carotenoid Bleaching

Action of Lipoxygenase.—The pentadiene hydroperoxide formed *via* oxidative addition of O_2 to the leftmost or penultimate carbon of the pentadiene moiety of polyunsaturated fatty acids such as linoleic acid oxidizes carotenoids to colorless compounds. Few investigations have been carried out on the fate of the hydroperoxide or of the carotene in this particular reaction. As colorfully phrased by Sumner and Somers (1947), the actor, the inductor and the acceptor are oxygen, unsaturated fat and carotenoids, respectively. Among the colored carotenoids which are bleached as a result of the action of this enzyme are: the carotenes, vitamin A, xanthophyll, phytofluene and bixin. These and other pigments such as chlorophyll, (see below), violoxanthins and synthetic dyes, as well as nonpigments such as cholesterol and thyroxine, are destroyed in the "flame" of the lipoxygenase-fatty acid reaction.

Although the action of lipoxygenase had long been equated with carotene bleaching, this statement has to be qualified somewhat. What was considered to be *the* lipoxygenase of soybean was crystallized in 1947 by Theorell *et al.* In the same year Kies (1947) questioned the identity of "lipoxidase" with "carotene oxidase." She showed that controlled heating of partially purified soybean enzyme resulted in complete loss of carotene destroying activity without eliminating the catalysis of linoleate peroxidation. Some 22 years were to elapse before Kies *et al.* (1969) demonstrated that the Theorell isoenzyme (now termed L-1) is very inefficient in bleaching and pointed to the then preliminary literature on the existence of lipoxygenase isoenzymes in soybeans. It remained for Weber *et al.* (1973A,B) to isolate a class of isoenzymes from soybeans as well as from peas and wheat, which not only are carotene bleachers in the presence of linoleic acid but also act optimally at pH 6.5, the same pH at which soyflour is used as a flour bleaching agent and is indeed distinct from pH 8–9 at which L-1 acts optimally. These observations were corroborated and elaborated upon by Grosch *et al.* (1977) and would appear to be in qualitative agreement, at least for one nonseed tissue, with those of Grossman *et al.* (1969). They isolated a fraction from alfalfa extracts which possessed a double pH optimum with respect to carotene destruction, one at pH 4.0 and one at pH 6.5, the latter corresponding to the optimum for lipoxygenase action. Nevertheless, more recent findings (Ramadoss *et al.* 1976; Ikediobi and Snyder 1977) strongly implicate L-1 participation in coupled carotene oxidation after all because as well as being overwhelmingly the most preponderant and stable isoenzyme, it acts synergistically with other soybean isolipoxygenases. The particular isomer of lipohydroperoxide formed (Chapter 24) appears to be crucial for bleaching. This problem illustrates both the roles of isozymes and of alternative secondary nonenzymatic pathways on food quality attributes. It is also apparent that as of this writing, the question of just how carotene is bleached in wheat flour is far from settled.

Nonlipoxygenase Carotene Destruction.— The report of Weber *et al.* (1973B) states that the lipoxygenase:carotene bleaching activity ratio (4.2 moles of linoleic acid reacted for each mole of carotene destroyed) remained constant during purification of their isoenzymes, apparently leaving little room for nonlipoxygenase participation. However, significant, earlier literature on the existence of such factors (including enzymes) does, nevertheless, exist. Thus, Maier and Tappel (1959) demonstrated that such heme-proteins as peroxidase, cytochrome c and hemoglobin catalyzed nonspecific autoxidation of unsaturated fatty acids, and a variety of hemoproteins as well as ferrous salts plus preformed lipohydroperoxide can replace one of the soybean isolipoxygenase (L-3) needed for the above-mentioned synergy observed by Ramadoss *et al.* (1976).

Experiments on carotene bleaching by plant extracts which seemed to eliminate lipoxygenase action as the principal enzyme involved were interpreted by some investigators to be due to: (1) participation of "lipoperoxidase factors" (Gini and Koch 1961; Dillard *et al.* 1961; Blain 1970; Grosch 1972) which destroy linoleate peroxides (i.e., soybean extracts would destroy carotene only when the lipid was already slightly oxidized); (2) hydroperoxide isomerases, identified in Chapter 21 (Zimmerman and Vick 1970; Eskin *et al.* 1977), which might be used to slow down bleaching since they could conceivably compete with carotene-bleaching systems (Moutounet 1979). It would appear that wherever there is a possibility that oxygen can be activated, the possibility also exists that this active oxygen, whether in the form of lipohydroperoxide, hydroperoxide or its free ionic form, the superoxide anion O_2^- and its transforms (Chapter 16), can destroy the color of many food pigments including carotenoids. Apparently the oxygen of hydrogen peroxide itself is not sufficiently active since even in the presence of peroxidases from beans or wheat it will not, in moderate concentrations, bleach carotene (Weber *et al.* 1973B). From the foregoing discussion it is obvious that the final word on the diverse contributions of lipoxygenase, nonlipoxygenase enzymes and nonenzyme action is not yet in. Even pinpointing these contributions will require, in addition, assessment and quantification of these various contributions in specific cases. In the meantime, carotene destruction—whether desirable or undesirable—occurs in foods, and measures are taken (in many cases empirical) to promote or prevent the action of enzymes involved.

Influence on Color Quality of Food.—*Beneficial Effects.*—In 1934 Haas and Bonn patented the use of soybean flour for bleaching of wheat bread flour as an alternative to the chemical bleaching agents used at that time, such as the toxic nitrogen trichloride, chlorine, nitrogen oxide, etc. In their first patent, they describe washing the soybeans, grinding them, removing the hulls and mixing the resulting flour—still containing the lipid—with four parts of corn flour. About 1% of the latter mixture was adequate for bleaching which was found to occur most rapidly during dough mixing at 40.5°–48.8°C. Haas and Renner (1935) patented the addition of suitable fats and oils to serve not only as shortening, but also in conjunction with soybean flour as bleaching agent. Appropriately, it has been sold by the name of "Whytase." Reed and Thorn (1971) recommended that it be added at the bake shop. Although up to 3% of denatured soyflour can be added in bread by the Standard of Identity, only up to 0.5% of enzyme-active flour, based on the weight of the flour, is permitted. Its use may be increasing since it lends itself well to continuous mix processes which are finding wider acceptance, as in the Chorleywood no-time process (Mecham 1975). In a continuous modification of the Blanchard two-step batter process in Great Britain, 0.7% soyflour is added in the first step along with 75% of the total flour to be added and most of the water. Ascorbic acid, which could possibly inhibit the lipoxygenase activity, is added with other ingredients in the second step after the bleaching is presumably completed. Indeed, such an inhibition might prevent the development of lipoxygenase-induced off-flavors. According to Wolf (1975), the levels of soybean flour used (usually defatted "enzyme-active") for bleaching give borderline adverse taste effects. As an example of a typical application, Brown and Church (1972) found an optimum level of 1.4% of full fat soy flour gave better results than 0.8 or 1.0% for the production of Italian white breads. Other legume seed flours, especially that of fava bean, are used for bleaching wheat flour in Europe. Perhaps in the future microbial lipoxygenase, a cobalt-requiring heme protein (Matsuda *et al.* 1976), will be adopted.

Prevention of Unwanted Color Changes.—The removal of carotenoids from wheat flour in breadmaking is a rather special case. In general it is desirable to maintain the color and nutritional qualities of the carotenoids and vitamin A together. However, there are instances in which the main and probably sole consideration in preventing carotenoid loss is to retain or improve the color and consumer acceptability of the food. These foods are: (1) the pastas made from the yellow-colored semolina flours from durum wheat; (2) chicken skin and egg yolk whose color is augmented by the incorporation of xanthophyll-containing alfalfa into the diets of the chickens; the deeper the color the more attractive is the color quality and the economic rosiness of these foods; (3) paprika, whose discoloration due to carotene loss is a perennial problem; and (4) some fish and fish roe, especially salmon, whose color is in part due to carotenoids.

Although wheat flours have rather weak lipoxygenase activity as compared with soybean, the presence of even low activity in flour made from durum varieties of wheat (the enzyme is probably located in the germ and some of this gets through and survives the screening and separation processes in milling) has been shown to result in the bleaching of the carotenoids (mostly lutein) of some varieties, to yield pale, unattractive pasta products such as macaroni and spaghetti (Irvine and Winkler 1950; Irvine and Anderson 1953; Irvine 1955; McDonald 1976; Matsuo *et al.* 1968). Catalase has also been implicated in pigment bleaching in flour (Kruger 1976).

Probably the most practical and most effective method of preventing this color loss, to date, is to incorporate into the dough ascorbic acid, which is being used as a nutritional additive anyway. Walsh *et al.* (1970) found that ascorbic acid in the range of 10 to 100 ppm added to high lipoxygenase-activity durum flour resulted in a spaghetti that was a bright yellow color instead of tan-grey of the control. It is of interest to note that the same authors found ascorbic acid to be a very potent fully competitive inhibitor of lipoxygenase, $K_i = 4 \times 10^{-6}M$. This indicates considerable inhibition at a concentration equal to about 0.1% of that used by Blain (1970) in agar gel assays of carotene bleaching activities. Also, these results are not in accord with the citation of Scott (1975) that ascorbic acid as well as tocopherol tends to enhance the oxidative reaction.

At any rate, the use of other inhibitors such as the standard fatty acid antioxidants which are also considered to be somewhat less effective as lipoxygenase inhibitors, would not be as advantageous as ascorbic acid. It is of interest to note that polyacetylenic analogs of the polyunsaturated naturally-occurring fatty acids are potent inhibitors. According to Blain and Shearer (1965) 5 ppm (*ca* $10^{-5}M$) can suppress more than half the activity.

Since most of the lipoxygenase comes from the germ, this suggests careful exclusion of the germ fraction during milling may be effective in color retention. Other attempts to reduce pigmentation loss in pasta products, which have met with varying success according to Walsh *et al.* (1970), include selection and breeding of durum wheats for low lipoxygenase activity and exclusion of air.

Other possibilities which food scientists of the future might look into are to use other enzymes which would compete with the lipoxygenase for the substrates or with carotene for the lipohydroperoxide formed by lipoxygenase. Thus, linoleate isomerase, a bacterial enzyme, converts linoleate to 9-*cis*-11-*trans*-octadecadienoate (Kepler and Tove 1967). Since the double bonds in the latter compound are not conjugated, it cannot be attacked by a true lipoxygenase. Other alternatives will be explored in Chapters 21 and 24.

Xanthophyll.—Haas and Bohn (1934) observed that soybean flour bleached egg yolk. The color of egg yolk as well as of chicken skin is at least in part derived from carotenoid xanthophyll present in the alfalfa fed to the chickens. Lipoxygenase activity results in bleaching of this desirable pigment (Grossman *et al.* 1969) and steps must be taken to minimize its loss during drying. The use of antioxidants such as BHA and NDGA aid not only by preventing coupled nonenzymatic autoxidation but also by inhibiting or at least partially blocking lipoxygenase activity. Scalding or blanching the alfalfa has also been used to minimize enzyme activity. In the PRO-XAN process for alfalfa processing, early application of high pH (ammonia) results in the rapid inactivation of lipoxygenase activity (Knuckles *et al.* 1970).

Another potential color problem posed by lipoxygenase and by similar oxidative enzymes is their ability to couple the oxidation of lipid to the bleaching of synthetic dyes, some of which could be food additives (Fritz and Beevers 1955). On the other hand lipoxygenase is said to be inhibited by a food coloring product of Maillard-type nonenzymatic browning reaction discussed in Chapter 19 and elsewhere (Samejima *et al.* 1972). Some attention has been given to a significant loss of desirable color of sweet red bell pepper, especially in the form of its dehydrated product, paprika *(Capsicum annuum)* which turns from red to brown upon exposure of paprika powder to light. There is no direct evidence to date that this color degradation, shown to involve a light-sensitized oxidation of carotenoids by De La Mar and Francis (1969), involves enzyme action. On the other hand, Kanner *et al.* (1976) did show that the fresh pepper contained a protein, presumably an enzyme, which was in part responsible for color loss in an *in vitro* carotene-linoleate model system. The remainder of the carotenoid-oxidizing activity was due to ascorbic acid, a historically prominent constituent of this food.

A beneficial effect of enzyme action with regard to carotenoid availability is that brought about by the lytic β-glucanase (Chapters 31, 34) complex from *Bacillus circulans* WL-12. From certain yeasts it liberates astaxanthin (Johnson *et al.* 1979). This natural fish pigment supplements salmonid and, as does xanthophyll, poultry feeds.

Transformation of food fish carotenoids will be briefly highlighted in Chapter 18, and their controlled transformations by the use of bioregulators was alluded to in Chapter 15.

COLOR CHANGES ASSOCIATED WITH CHLOROPHYLL TRANSFORMATIONS

Retention of the freshly-harvested green color of many vegetables during processing and subsequent storage is a goal towards which the frozen food industry has strived since its inception. As far as the consumer is aware, by and large this goal has been achieved for frozen legume fruit vegetables such as peas, green lima beans and pole or string beans, for the herbage vegetable, green spinach and other leafy vegetables as well as for the more specialized freeze drying of chives. This degree of success has not been reached in the canning and in conventional drying of these vegetables. The brining of pickled green vegetables (such as the vine fruit vegetable, cucumber and the berry fruit vegetable, green tomato) imparts off-green color which is usually acceptable. In the case of unprocessed plant foods, desirable loss of green color is associated with most fruits (oranges, bananas, some pome varieties) except limes. Means of preventing green color are associated with potatoes and etiolated vegetables such as celery and asparagus. The latter are examples of consumer acceptance of both green and nongreen forms of the same vegetable. The apparently divergent enzyme and nonenzyme mechanisms involved suggest that we separate the ensuing discussion of the role of enzyme action into three categories:

(1) Chlorophyll degradation in frozen vegetables.
(2) Chlorophyll degradation in canning, drying and other thermal processing.
(3) Chlorophyll changes in the postharvest preprocessing stages.

Class One changes apparently involve enzyme action to some extent although contribution of direct enzyme action on chlorophyll and that due to indirect secondary action of lipase and lipoxygenase as well as that due to nonenzyme actuated autocatalysis is by no means solved.

In the second class of changes it is well established that chlorophyll degradation is due mainly to a nonenzymic reaction. The putative involvement of enzyme action comes about in attempts to modulate these changes by enzymatic conversion of chlorophyll to more heat-stable green products *via* potentiation of endogenous enzymes.

The third class of changes involves *in vivo* transformation. We include some examples and interactions of the *in vivo* enzymes involved because the control of the green color can be manipulated.

Since the natural green color of plant foods is due to the chlorophylls "a" and "b," a brief summary of these forms of chlorophyll and their breakdown products is in order. Figure 17.1 shows the Mg-tetrapyrroles comprising the chlorophylls. Table 17.1 defines the structures of the derivatives of interest in terms of Fig. 17.1. The conventional view of chlorophyll transformations envisions the ensuing reactions and conditions:

$$\text{Chlorophyll} \xrightarrow{\text{Acid, Heat}} \text{Pheophytin} + \text{Mg}$$

$$\text{Chlorophyll} \xrightarrow{\text{Chlorophyllase}} \text{Chlorophyllide} + \text{Phytol}$$

$$\text{Chlorophyll} \xrightarrow{\text{Acid + Heat}} \text{Pheophorbide} + \text{Mg} + \text{Phytol}$$

$$\text{Chlorophyll} \xrightarrow{\text{Lipoxygenase, hydroperoxide isomerase}} \text{Bleached Chlorophyll}$$

TABLE 17.1

STRUCTURE AND COLOR OF CHLOROPHYLL AND DERIVATIVES[1]

Substance	R^2	R_4	Metal	Color
Chlorophyll	CH_3, CHO	Phytyl	Mg (Zn, Cu)	Green
Pheophytin	CH_3, CHO	Phytyl	None	Brown-Olive
Chlorophyllide	CH_3, CHO	H	Mg	Green
Pheophorbide	CH_3, CHO	H	None	Brown-Olive

[1] See Fig. 17.1 for tetrapyrrole nucleus and positions of R and R_4.
[2] Chlorophyll a, R = CH_3; chlorophyll b, R = CHO.

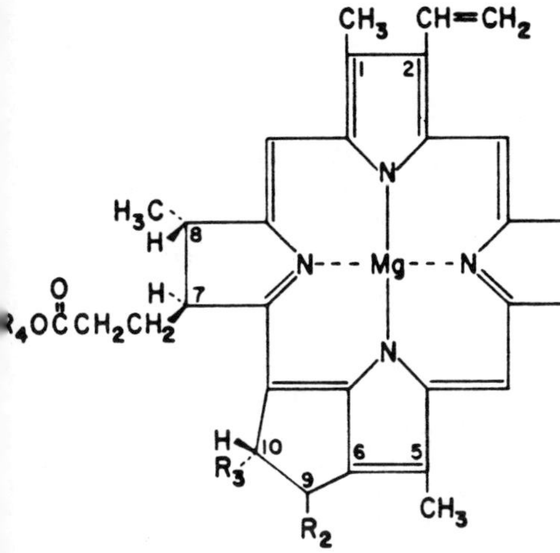

From McFeeters (1975)

FIG. 17.1. STRUCTURE OF CHLOROPHYLL

See Table 17.1 for terminology.

Frozen Foods

Destruction.—*Pheophytinization (Browning)*.—As annotated by Joslyn (1949), vegetables undergo changes during freezing which cause the green to be transformed into unattractive yellow or olive-brown hues during subsequent low-temperature storage. Other reviews on color changes in frozen foods include those by Aylward and Haisman (1969), Joslyn (1961A,1966) and Lee (1958). Early work such as that of Campbell (1937), Wagenknecht *et al.* (1952) and Dietrich *et al.* (1957) attributed these color changes to the conversion of chlorophyll to pheophytin (pheophytinization) brought about by acidity (H^+) due to lipase action, or to H^+ already present. However, Wagenknecht and Lee (1958A,B) attributed this conversion to the preferential action of pea lipoxygenase on free fatty acid (formed *via* lipase action) and that the action of lipoxygenase on the liberated free fatty acid was somehow responsible for Mg removal. Their conclusions were based upon the stimulation obtained by adding preparations of lipoxygenase, and especially of lipase (as well as the hematin-containing peroxidase and catalase) to blanched pea purees. Most of the enzymes in the purees survived the long storage at freezing temperatures. By contrast, Buckle and Edwards (1970) detected neither *endogenous* lipoxygenase nor chlorophyll-bleaching activities in unblanched pea purees after prolonged frozen storage as indicated in Table 17.2.

So far as is known, this color transformation to pheophytin is brought about by chelating agents, by heat or by replacement of Mg^+ with H^+. A more definitive study of lipid hydrolysis in unbleached frozen peas by Bengtsson and Bosund (1966) showed preference for hydrolysis of unsaturated fatty acids, both linoleic and linolenic. Another advocate for lipid oxidation-associated pheophytin formation was Walker (1964) who speculated that it occurred during the anaerobic initiation stage of fat peroxidation in frozen french beans, followed by pigment destruction during the aerobic phase of lipohydroperoxide formation. A naturally-occurring antioxidant in Walker's beans delayed onset of fat peroxidation and chlorophyll destruction. This implied participation of lipoxygenase, as well as nonenzyme autoxidation. In order to provide a rational basis for explaining the loss of chlorophyll in frozen foods, a balance sheet, showing just what happens to the chlorophyll as well as other components and which theory tells us what might be associated with it, is needed. Perhaps the most thorough analysis to date (1980) is that of Buckle and Edwards (1970). Indeed this analysis (Table 17.2) does put some of the loose ends together. As expected, these changes were slowed down by lowering the temperature from −9.4° to −23.3°C and by storage in N_2 instead of O_2. There is no question that chlorophyll is lost. Also confirmed is the conversion of chlorophyll to pheo-

TABLE 17.2

GREEN COLOR-ASSOCIATED CHANGES IN UNBLANCHED FROZEN PEAS[1]

Changes Measured	After Storage at -9.4°C in O_2 for: 0 Weeks	8 Weeks	20 Weeks
Total chlorophyll pigment (P) lost (%)	0	24	36
Conversion to Mg-free P (%)	6	57	87
Conversion to phytol-free, P (%)	0	23	82
TBA value (% max. value)	14	53	100
Hydroperoxide (% max. value)	9	100	94
Lipoxgenase activity destroyed (%)	0	77	99
P-bleaching "enzyme" destroyed (%)	0	34	90

Source: Adapted from Buckle and Edwards (1970).
[1] Some of the values were calculated from complete analysis of "a" and "b" forms of chlorophyll, pheophytin, chlorophyllide and pheophorbide.

phytins and loss in total pigment. What is surprising is the hitherto unsuspected formation of the phytyl-shy derivatives, chlorophyllide and pheophorbides plus free phytol in the absence of detectable chlorophyllase. The latter result may be related to the observation of Arckoll and Holden (1973) on the apparent potentiation of chlorophyllase in green leaves at 80°C and its destruction at 100°C. By contrast, pheophytinization was pinpointed as the cause of the marked color change occasioned by washing the leaves with acid.

"Bleaching" of Chlorophyll.—Parallel to the loss of the green color upon prolonged storage of frozen green vegetables has been a series of *in vitro* experiments in which crude and/or purified enzymes have been added to chlorophyll. Under controlled conditions "bleaching" of chlorophyll occurs without concomitant formation of pheophytin. All subsequent conclusions confirm those of Strain (1941) who observed that crude soybean extracts converted chlorophyll to colorless substances in the presence of fat and oxygen, thus implicating, at least in part, the action of a lipoxygenase. Corroboration of this bleaching effect in model systems includes investigations of Wagenknecht *et al.* (1956), Walker (1964), Holden (1965) and Buckle and Edwards (1970). The analogy to carotene bleaching is striking and is cited by several investigators. Thus, in many of the extracts (soybeans, peas, clover) the maximum bleaching effect occurs at pH 6 whereas linoleate hydroperoxide formation is optimal at pH 8. There is some disagreement concerning the formation of pheophytin *vs* a true bleaching effect. With pure chlorophyll, apparently all agree that bleaching occurs. Little is known about the colorless chlorophyll derivatives which are formed. Holden (1965) conjectured that the conjugated double-bond system of the porphyrin ring of chlorophyll may be interrupted due to oxidation by lipohydroperoxides or by their isomers, the ketohydroxy fatty acids which are products of hydroperoxide isomerase action. Pyrrole ring scission may also occur.

Older workers cited carotene bleaching as a precedent for nonlipoxygenase involvement. But since then, some of the so-called secondary effects in model systems can be explained by lipoxygenase-mediated interaction between substrate and product as discussed in Chapters 21 and 24, or to synergy effects resulting from the cooperative action of isolipoxygenases. Indeed the results of Sakai-Imamura (1975), who definitely ruled out the participation of hydroperoxide isomerase in beans, peas and cereals, can be most facilely so interpreted. However, in stored foods, in contrast to model systems, the bulk of evidence does indicate that some lipoxygenase isozymes plus other heat-labile factor(s) do contribute to the loss in green pigment during frozen storage of vegetables. Loss and eventual disappearance of lipoxygenase activity during frozen storage (Buckle and Edwards 1970) is in accord with both action of the enzyme in frozen media—reaction inactivation (Wallace and Wheeler 1972)—and production of inactivating concentration of H^+ resulting from lipase action and possibly other acid-producing enzyme actions. Nonenzymatic oxidation may be catalyzed by food additives such as sulfite (Peiser and Yang 1975) or by linoleate plus light, similar in this respect to nonenzymatic oxidation of carotene.

In summary, possible enzymatic factors leading to deterioration of color quality of frozen green vegetables include:

(1) Lipase action, leading to the formation of pheophytin responsible for the disagreeable olive-brown color of unblanched vegetables.

(2) Action of an isoenzyme of lipoxygenase leading to the formation of unsaturated hydroperoxide.
(3) Action of a second-heat labile factor which may couple the oxidation of chlorophyll to the product of lipoxygenase action leading to a loss in pigment.
(4) Unspecified nonlipoxygenase mechanisms in which oxygen is activated to the extent that chlorophyll is oxidized and/or bleached.

These may include nonspecific hematin-catalyzed reactions and possibly other reactions which can give rise to active oxygen such as O_2^- and $\cdot OH$. Indeed there is evidence (Elstner *et al.* 1975) that O_2^- is involved in photosynthetic O_2 reduction (Fig. 16.2). It is of special interest that older literature refers to enzymatic oxygen activation by glycolate oxidase (Wagenknecht *et al.* 1952). This enzyme, also involved in flavor production (Chapter 20), is a flavoprotein which plays a key role in plant photorespiration and may also be a source of O_2^-.

Finally the appearance and possible role in color quality deterioration of the products of the chlorophyllase reaction, pheophorbide/chlorophyllide and phytol in frozen vegetables in apparent absence of an easily detectable chlorophyllase action needs further exploration. Model studies have been conducted at high temperatures over very short periods. Some of the answers must await development of the special enzymology and biology of the frozen state (Chapter 12).

Retention.—The preceding survey indicates that recent years have not witnessed a great effort to further our understanding of the underlying causes of the color quality deterioration of frozen green vegetables. One reason for this state of affairs may be the by-and-large empirically successful measures for combating enzyme-mediated quality deterioration by blanching. Although one may assign processing parameters and predict how a given lot of fresh produce will behave, there still is a tendency to underblanch to avoid the disadvantages associated with blanching. Fluctuations in temperature can lead to undesirable color changes (Boggs *et al.* 1960). Overblanching, also for some unknown reason, apparently leads to an accelerated conversion of chlorophyll to pheophytin during subsequent frozen storage (Walker 1964; Dietrich *et al.* 1955). Since blanching itself is being increasingly subject to environmental and energy restraints (Chapter 10), a more thorough understanding of the contribution of enzyme action in the frozen vegetable to chlorophyll disappearance is still a timely and relevant problem worthy of further investigation.

Thermally Processed (Canned) Foods

Unlike processed frozen foods, the pristine green of freshly harvested green vegetables is retained in neither heat sterilization (canning) nor dehydration. For the most part the causes of this color change and attempts to prevent it are largely nonenzymatic in nature. To a lesser extent putative enzyme action has been evoked to explain some beneficial effects obtained by potentiation of the plant's own chlorophyllase.

It is quite clear that the color of canned green vegetables is largely and perhaps entirely due to the nonenzymatic conversion of chlorophyll to pheophytin, which occurs mostly during sterilization. The basis for this knowledge rests on the model experiments of Mackinney and Joslyn (1938,1940,1941) defining the effect of time, temperature and acidity on the conversion of the chlorophylls to pheophytins. Over a wide range of these variables, especially within the range foods might encounter from harvest to consumption, very few and only traces of side products were found. Chlorophyll a was found to be about eight times more reactive than chlorophyll b. In somewhat more refined measurements, Schanderl *et al.* (1962) determined the rate constants of pheophytinization of chlorophyll in 10^{-4} *N* HCl. They also investigated the conversion of chlorophyllides to pheophorbides, the results of which are germane to the putative role of chlorophyllase discussed below.

Of course there is a vast difference between such model systems and commercial canning retorts. Nevertheless data accumulated over the years tend to bear out the findings with model systems. Thus, Gold and Weckel (1959) found 50% conversion during the retort "come-up" period of conventional canning procedure. Even blanching destroys some chlorophyll. The pseudo first-order rate constant for a puree is linearly related to the titratable acidity of the puree as measured before heat processing. As if to emphasize the nonenzymatic nature of this conversion, at retorting temperature (*ca* 120°C) it went faster in blanched peas than in unblanched peas. No evidence of breakdown products other than pheophytin was obtained.

The above experiments were performed in *in vitro* mixed systems. Only recently has pheophytinization during the *in vivo* to *in situ* phase of heating of vegetables been studied in depth in a cogent investigation by Haisman and Clarke (1975). Pheophytinization appears to be analogous to the potentiation of enzyme action by temperature-induced intracellular disruption (Chapter 1). The difference is that instead of *enzymes* being potentiated, it is the migration of hydrogen ions, or rather H_3O^+ (hydronium ion) to the negatively charged lipid/water interface of the chloroplast membrane which is potentiated. This comes about from disruptive partial collapse and reorientation of the chloroplast structure (Fig. 17.2) occasioned by heating vegetables to 40°–60°C, the same range of temperature for many enzyme potentiations (Schwimmer 1978). These newer insights provide both a rationale for the above-mentioned empirical procedures and a framework for further progress in this area of food processing enzymology.

Coming to actual processing conditions such as those for the preparation of spinach puree the principal color change is attributed to nonenzymatic conversion of chlorophyll a to pheophytin and secondly to an analogous degradation of chlorophyll b (Tan and Francis 1962). Nonpheophytin chlorophyll degradation products were detected by Buckle and Edwards, but only after 18 months storage of HTST processed pea puree at room temperature.

In thin layer chromatographic studies on color changes in green peas during canning, Attilla (1971) reported that the chlorophyll isomers were converted to pheophytins only, during both blanching (4–8 min at 95°C) and sterilization at 120°C. The changes at the latter stages of processing were due to heat alone. Not even traces of the original chlorophyll were detected in the final product; the nonenzymatic transformation by heat to pheophytin was found to be irreversible.

Approaches to Chlorophyll Retention.—Historically, attempts to mitigate the severe loss in green color of vegetables during canning operations have been channelled mainly into the following four

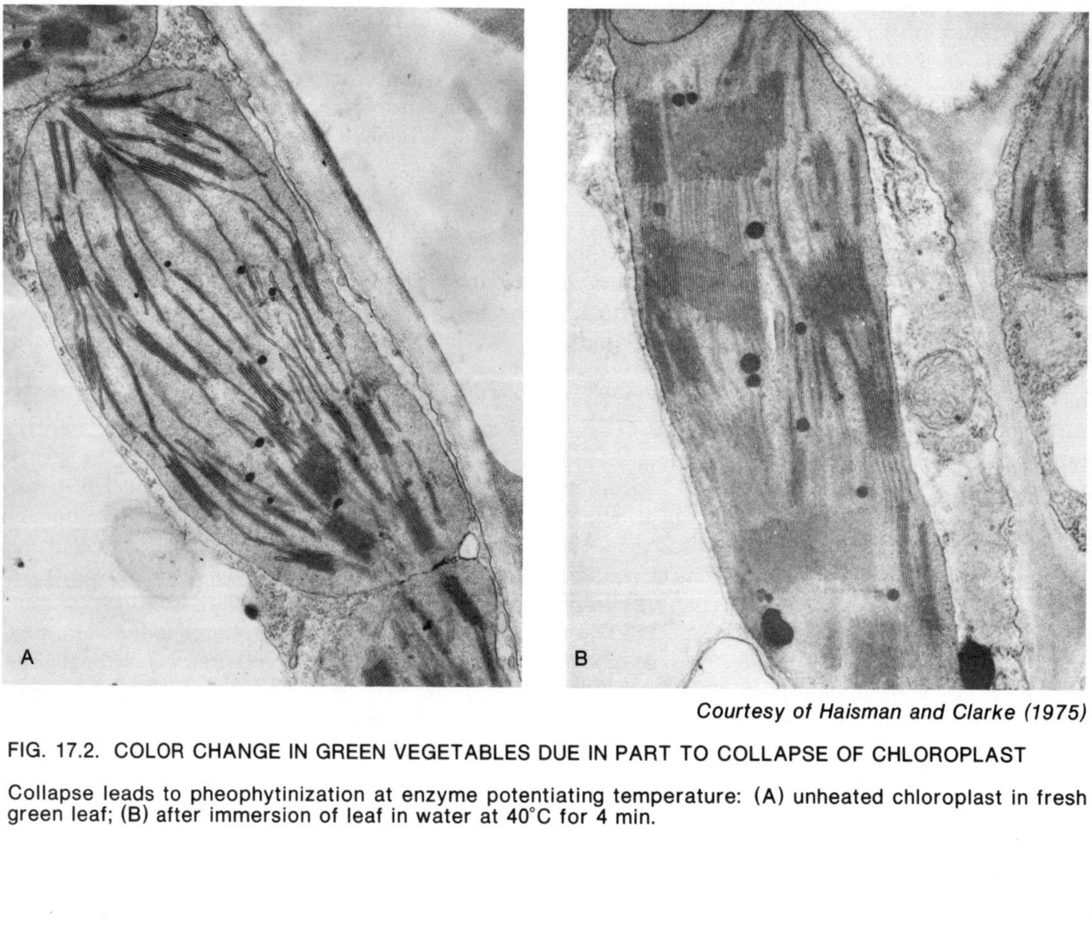

Courtesy of Haisman and Clarke (1975)

FIG. 17.2. COLOR CHANGE IN GREEN VEGETABLES DUE IN PART TO COLLAPSE OF CHLOROPLAST

Collapse leads to pheophytinization at enzyme potentiating temperature: (A) unheated chloroplast in fresh green leaf; (B) after immersion of leaf in water at 40°C for 4 min.

approaches: (1) High Temperature-Short Time blanching discussed in Chapter 10 (HTST), (2) pH adjustment (alkalization), (3) color reversion, and (4) potentiation of action of the vegetable's own chlorophyllase.

Adams and Yawger (1961) found that HTST peas HTST processed at 122°C for 9 min (optimum for both stabilization of color and sterilization) exhibited a temporary (2 weeks) improvement in color. This instability was partially due to the slow pheophytinization of the chlorophyll in the cans. Voirol (1972) recommends minimum heating periods, especially for green vegetables.

Alkalization of thermally processed vegetables to prevent discoloration used by the consumer when he/she adds a pinch of baking soda to spinach while cooking is based on the retention of Mg in the porphyrin nucleus by preventing its exchange with H. Alkalization also neutralizes organic acids which, as noted, participate in chlorophyll loss. These acids are liberated when cut vegetables are held for 2 or 3 min in a blancher. Alkalization patents include those of Blair and Ayers (1943), Stevenson and Swartz (1942), Bendix *et al.* (1952) and Malecki (1965). A prominent caveat in the patents is to not exceed an upper limit of pH 8, beyond which adverse texture and flavor effects are noted. That nonenzymatic oxidative processes may play a role is supported by the observation that frozen storage of blanched peas in N_2 is even more effective in preventing chlorophyll loss than is storage in oxygen. As an example from the patent literature in which both HTST and alkalization are used (Malecki 1965), peas destined for canning are blanched for 5 min at 93.4°C in a $CaCO_3$ containing medium adjusted to pH 9.0 to 10 with $Ca(OH)_2$—or glutamic acid. To the canning medium is added 2.7 ml of NaOH (10% by weight) per 16 oz can. After sterilization at 124.5°C for 8 min and cooling, the pH of the peas is 7.6−7.9. According to Malecki, there is no change in color after 10 months' storage without refrigeration. Gupte and Francis (1964) noted improved retention of chlorophyll pigments in spinach purees immediately after processing (but not during subsequent storage) using HTST and/or alkalization with $MgCO_3$. Conversion during subsequent storage is probably nonenzymatic (Buckle and Edwards 1969). Hudson *et al.* (1974) obtained maximum chlorophyll retention without the undesirable effect of high pH by use of NaCl solutions, the concentration depending upon the hardness of the cooking water. Haisman and Clarke (1975) observed less pheophytinization in the presence of detergents (see above).

The third nonenzymatic mode of color retention, color reversion, is based on attempts to replace the Mg lost from chlorophyll by other metallic ions whose porphyrin complexes are also green. Thus, the chlorophyll, highly touted a generation ago as a deodorant, is mostly cupric chlorophyllide. Sweeney and Martin (1958), in a paper devoted principally to problems involved in the determination of chlorophyll and pheophytin and their metallic derivatives, suggested using Zn to replace Mg eliminated from chlorophyll by cooking. Stable Cu and Zn complexes of pheophytin a are responsible for the bright green color which may develop during the storage of canned green vegetable purees (Schanderl *et al.* 1965). There is enough Cu normally present in green vegetables to account for this. Voirol (1972) points out the potential health hazard posed by *addition* of copper salts. Probably the only enzyme-related aspect of their reversion is that both Cu and Zn are probably released from various Cu- and Zn-containing denatured enzymes during thermal processing and diffuse the chlorophyll. Wilson *et al.* (1973) report that greenness as well as clarity, flavor and aroma of green beans are retained by addition of 300 mg ascorbic acid to process water containing Ca and Mg salts and the sequestering EDTA.

Putative Role of Chlorophyllase.—We have previously mentioned that products of chlorophyllase action are present in frozen green vegetables (Buckle and Edwards 1970). The concept that its action contributes to color stability in green vegetables is 100 years old, being customarily attributed to J. Borodin in 1882 by, for instance, Sumner and Somers (1947). Chlorophyllase is a highly specific enzyme reversibly catalyzing the following action (McFeeters 1975).

$$M^{+x} \cdot \text{pheophorbide}_{a,b}\, R_1 + R_2OH \rightleftharpoons M^{+x} \cdot \text{pheophorbide}_{a,b}\, R_2 + R_1OH$$

where M^{+x} may be H or a Mg. In chlorophyll, R_1 is the naturally occurring phytyl group and R_1OH is phytol and R_2 is H or various alkyl groups such as methyl or ethyl. Its natural *in vivo* function appears to act as participant in the pathway of chlorophyll biosynthesis catalyzing perhaps, the final reaction, rather than in its degradation (Holden 1963; Sudyina, 1963; McFeeters 1975). In older investigations the enzyme assay was carried out in concentrations of alcohol and acetone as high as 40%. Using this type of assay to solubilize the substrate, Weast and MacKinney (1940) reported rather bizarre properties such as the dependence of its heat stability on the season in which it was harvested. More recently purified, water-soluble enzyme preparations were used by McFeeters (1975) in an aqueous reaction system at pH 7.8 incorporating detergents such as Triton X-100. A thorough analysis by McFeeters of chlorophyllase specificity revealed that the presence of a 9-keto group and a methyl group at the 7-position (chlorophyll$_a$) results in maximum binding affinity. Thus, the K_m for hydrolysis of 7-methyl pheophorbide$_a$ was found to be about 1/30 that of chlorophyll a or methyl-chlorophyllide a.

On the basis of the idea that chlorophyllide is more stable than chlorophyll, proposals in the form of patents have been made as far back as 50 years ago to potentiate the *in situ* chlorophyllase activity by low temperature blanching (Thomas 1928; Lesley and Shumate 1937). Loef and Thung (1965) reported that a long blanching time (25 to 45 min) within a relatively low temperature range (60°–65°C) improved the color of sterilized spinach. They attributed this improvement to the enzymatic formation of green chlorophyllide, presumably able to withstand subsequent retorting better than does chlorophyll.

This question has not been satisfactorily resolved. Certainly the finding of Schanderl *et al.* (1962) that in 10^{-4} *N* HCl chlorophyllide a loses its Mg faster than does chlorophyll a (k = 0.0171 *vs* 0.0103 at 25°C) and that at higher temperatures (i.e., 55°C) the ratio of rates is about the same—does not bespeak an increased stability of chlorophyllide. Clydesdale and Francis (1968) find chlorophyllide *in vitro* as stable as chlorophyll. Of course there is still the possibility that at higher pH's the situation may be reversed. To prove this would be difficult since one has to separate the aforementioned beneficial effects of raising the pH which are due to nonenzymatic transformations from those which can be attributed to enzyme action. Just such an attempt was the goal of Clydesdale and Francis (1968). Some of the results obtained, adapted from the original data, are shown in Table 17.3. As can be seen, there does seem to be a real, if temporary, improvement brought about by pH, but only a slight improvement could be attributed to enzyme potentiation. The authors state that trends of the data suggested, albeit inconclusively, that chlorophyllide b was more stable than was chlorophyll b at normal but not at alkaline pH. But this still does not answer the question: Are the phytyl-free derivatives, the chlorophyllides and pheophorbides more stable than are the chlorophylls in canned vegetables? Combination of enzyme potentiation plus alkali treatment retained a significantly higher level of green pigment. However, this advantage was abolished upon subsequent storage, apparently due to acid production. Due to the small amounts of chlorophyllides present, the authors felt that conclusive statements could not be drawn. However, as they also point out, they did not adjust the pH *before* enzyme potentiation.

Also they did not achieve close to 100% hydrolysis of the chlorophyll. Lack of agreement concerning previously reported success may very well be due to variable interaction of *in situ* chlorophyll with chlorophyllase (Braverman 1963) plus yet unknown factors (Thung 1963). We feel that at the time of this writing, the final verdict as to whether enzyme potentiation contributes significantly to green color retention in canned vegetables is not

TABLE 17.3

INFLUENCE OF ALKALIZATION AND CHLOROPHYLLASE ON COLOR RETENTION OF PEA PUREE[1]

Blanching Treatment[1]	Visual Appearance	Chlorophyllase[2]	% Chlorophyll Remaining After: Blanching	Retorting	Storage
Conventional	Worst	0	32	3.6	0.2
Chlorophyllase potentiation (I)	Some improvement	33	38	8.9	0.2
Alkalization (II)	Very good	0	67	8.7	4.3
I + II	Best	26	69	14.9	0.3

Source: Adapted from data of Clydesdale and Francis (1968).
[1] Conventional, 5 min at 190°F; I, 30 min at 155°F; II, 5 min at 190°F; pH adjusted to 8.5 with $MgCO_3$. Storage for 6 months at room temperature.
[2] Percentage of unprocessed.

in. Indeed a more significant food quality-related effect of chlorophyllase action may reside in its effect on health, at least to farm animals (Chapter 35).

Dehydrated Food

Putative Nonenzyme Involvement.—The recent upsurge of interest in dehydrated foods as investments by consumers against future famine should contribute to a renewed interest in means of retaining their fresh-like qualities, including color. The impetus in the early 1940s for such efforts was an urgency, imposed by World War II, to develop palatable and otherwise attractive dehydrated foods for the Armed Forces. Among such efforts was that of Dutton *et al.* (1943) on the changes in color and pigments of dehydrated spinach during processing. They found that the destruction of chlorophyll, almost entirely due to loss of Mg to form pheophytin, was positively correlated with the moisture content of the unblanched spinach powders. Thus, after 2 weeks storage in air, 5% and 93% of the original chlorophyll was converted to pheophytin in powders containing 2.0% and 16.3% H_2O, respectively. As one would expect from our previous discussion, blanching in steam followed by drying resulted in as high as 50% pheophytinization even before storage. The results of Dutton *et al.* were confirmed and extended by LaJollo *et al.* (1971). Observing the rapid change of color from bright green of blanched (1 min boiling H_2O) freeze-dried spinach to the olive-brown of the pheophytins, their results showed that the Mg-removal reaction could be catalyzed in systems where all the water is tightly bound. In this sense the pheophytinization is analogous to that in the frozen state where most of the water is tightly bound. At a_w's lower than 0.32 (5.4% H_2O) products are formed which resembled those observed by Buckle and Edwards (1969) for canned pea puree. In model systems containing chlorophyll, cellulose, inert oil and H_2O, Mg was removed at a water concentration below that corresponding to enclosure of a particle by a monomolecular layer of water. It is of interest to note that in the presence of methyl linoleate, a lipoxygenase substrate, loss in total pigment as well as pheophytinization occurred. Presumably autoxidation of methyl linoleate resulted in lipohydroperoxide formation coupled to chlorophyll oxidation, but enzymatic oxidation could also have contributed to the situation. This is in accord with our conclusion that chlorophyll loss in frozen foods may occur via several routes, both enzymatic and nonenzymatic. In well-dried foods, the loss of color appears to be largely nonenzymatic. However, the study of LaJollo *et al.* (1971) does provide insight into certain obscure aspects of enzyme action in food. Thus, the latter study unearthed evidence for compartmentalization of chlorophylls at low water activities. Some of the more firmly bound chlorophyll, both in the processed food and in model systems, was less reactive. Analogous compartmentalization probably influences enzyme action in many foods under diverse conditions, as indicated elsewhere in this book.

Postharvest *in Vivo* Chlorophyll Changes

Although our subject matter emphasizes enzymes and their action in nonliving food systems we deem it appropriate to digress and briefly elaborate on the circumstance that the green color of plant foods may change between the time they are harvested and the time they are either eaten as fresh produce or subjected to the rigors of processing.

Enzymes of Chlorophyll Metabolism *in Vivo*.— During the ripening process of many fruits and vegetables chlorophyll disappears. Bananas and oranges are harvested green and then treated with ethylene to hasten the ripening and thus destroy the chlorophyll. Ethylene will also hasten the ripening and drying of immature wheat but the ignition hazards of this gas have prevented it from being used in grain elevators (Hale *et al.* 1943). Ethylene is produced by the live fruit itself at the onset of the climacteric and ripening. At one time it was conjectured that the source of ethylene was chlorophyll since the disappearance of the latter almost invariably signals the appearance in the former. However, as adumbrated in Chapter 16, we now know that this is most unlikely and that the precursor to ethylene is in all probability methionine. Ethylene, however, is probably involved in the expression of the genome coding for the enzymes of chlorophyll degradation either through derepression of the gene proper (DNA) or *via* unmasking of latent or masked messenger RNA (Chapters 4 and 15).

Although the enzymes of the biosynthesis of the pyrrole nucleus are well known (Lascelles 1965), it is only recently that insight into the regulation of chlorophyll synthesis is accumulating *vis-à-vis* that of the structurally related protohemin (Castelfranco and Jones 1975; Harel 1978). A scheme for such regulation is shown in Fig. 17.3. *In vivo*, the chlorophyllase does not seem to be involved in chlorophyll degradation. Thus, the product of such a hydrolysis, chlorophyllide, cannot be detected. Indeed, as we mentioned previously, chlorophyllase is more likely to be involved in synthesis, rather than in the breakdown of chlorophyll (Holden 1963; Sudyina 1963; Liljenberg 1977). Lynn Co and Schanderl (1967) and Schanderl and Lynn (1966) separated compounds which may be intermediates in the breakdown of chlorophyll in ripening bell peppers, banana peel and cucumber peel.

Controlled Atmosphere (CA) Storage.—This is a successful means of retaining the green color due to chlorophyll. However, upon removal from prolonged CA storage, the produce loses its green color very rapidly. Thus, after storage in various controlled atmospheres (0–8% CO_2; 2–13% O_2) for 9 months at 1.7° or 3.3°C, Golden Delicious apples lost about one-half of their chlorophyll in 1 week at 20°C and about 85% after 1 month (Patterson *et al.* 1974). Similarly after subatmospheric storage at 102 torr, 12.5°C for 100 days, green tomatoes lost over 90% of their chlorophyll in 4 days after return to 640 torr (Salunkhe and Wu 1974). This promising variation of CA has apparently progressed beyond the stage of economic analysis and engineering developmental problems such as how to build inexpensive containers which can withstand external atmospheric pressure (Burg 1975; Mermelstein 1979).

The post-CA behavior of produce just described is somewhat reminiscent of the cryptic reactions occurring at low or freezing temperature but whose deleterious consequences in terms of food quality are manifested only after return to ambience (Chapter 12). The food has been subject to stress and responds abnormally biochemically. That accelerated chlorophyll catabolism is a response to stress in CA as well as in subatmospheric storage is strikingly illustrated by results of the investigations of Wang *et al.* (1971) on the fate of chlorophyll during and after CA storage of asparagus (5% CO_2; 1.6°C; a_w, 0.95). More chlorophyll was lost in the air-stored controls. What did disappear showed up as pheophytin probably formed as the result of the accumulation of acid; the higher the CO_2 level, the higher the pH (less acid produced) and chlorophyll retained. However, there were unexpected anomalies in terms of a straightforward explanation of ease of removal of Mg with increasing H^+ concentration. First of all they found better chlorophyll retention in the second inch-long section from the tip than in the first section although pH of the former was lower. Secondly, the *in vivo* reaction rates were much lower than those observed by Schanderl *et al.* (1962) in their organic solvent system. Hence, although CA storage may re-

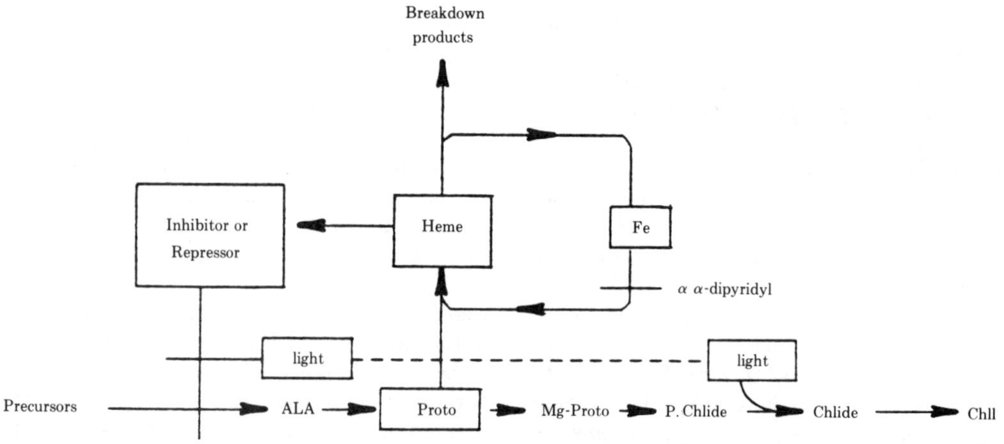

From Castelfranco and Jones (1975)

FIG. 17.3. PROPOSED LIGHT-DEPENDENT REGULATION OF CHLOROPHYLL BIOSYNTHESIS

Light abolishes interference with 5-aminolevulinic acid (ALA) biosynthesis, inhibits synthesis of heme from protoporphyrin X (Proto) and promotes conversion of protochlorophyllide (P. Chlide) into chlorophyllide (Chlide); Chll, chlorophyll.

sult in a nonphysiological stressful environment leading to increased accessibility of acid to partially decompartmentalized chlorophyll, the structure of the cell remains intact to the extent that compartmentalization and other protective factors such as protein and phospholipid binding to chlorophyll slow down pheophytinization. That pheophytin formation is indeed a response not only to a nonphysiological situation but also to pathological stress is further corroborated by the investigation of Kato and Misawa (1974) who observed that the chlorophyll of virus-infected tobacco leaves was rapidly converted to pheophytin without accumulation of the chlorophyllase reaction products, chlorophyllide and pheophorbide. Indeed chlorophyllase activity decreased in response to the infection.

In the future we may, as mentioned previously, yet be using consumer-safe plant regulators such as senescence inhibitors, perhaps akin to kinetin, to prolong storage life and maintain the color quality of fresh green product (Vreman and Corse 1976). Cytokinins may not be applicable in all cases. Thus, Bata and Neskovic (1974) found that gibberellic acid at 1 ppm, but not kinetins, prevented chlorophyll destruction of duckweed cultures so that they remained green 2–3 weeks longer than did controls. Indeed kinetin when used in combination with CA hastened loss of the green color of lettuce (Salunkhe and Wu 1974).

Prevention of Chlorophyll Biosynthesis.—In some foods it is desirable to prevent chlorophyll synthesis and the accompanying greening. The simplest way to do this is to keep the food in the dark, i.e., etiolation of celery, asparagus and bean sprouts. Greening poses a special problem in potato storage because of the economics and the concomitant buildup of the poisonous steroidal alkaloid, solanine. The greening of potatoes, its prevention and enzymes involved have received some attention. Thus, Ramaswamy and Nair (1974) measured maximum chlorophyll synthesis in cold-stored tubers at 0 and 30–40 lux illumination. Chlorophyll a content exceeded chlorophyll b in the first 4 weeks. After 4 weeks some of chlorophyll a was converted to the b form by a dehydrogenase type enzyme system, presumably according to the following reaction:

$$\underset{\text{Chlorophyll a}}{RCH_3} + NADP^+ \xrightarrow{+(O)} \underset{\text{Chlorophyll b}}{RCHO} + NADPH + H^+$$

Alternatives to dark storage as means of preventing greening have been sought. Some successes in retarding greening without hopefully otherwise affecting the quality of the tubers include:

(1) Enclosure in colored cellophane but not polyethylene bags (Yamaguchi *et al.* 1960).
(2) Crating potatoes with certain waxes (Yamaguchi *et al.* 1960; Wu and Salunkhe 1972A).
(3) Dipping in acetone-diluted solutions of vegetable oils, in mineral oil and in glycerin (Wu and Salunkhe 1972B; Jadhav and Salunkhe 1974). The latter two treatments prevent access to air and may result in internal breakdown caused by anaerobic respiration, another case of a nonphysiological metabolic aberration in attempts to improve food quality.
(4) Treatments with special chemicals such as 3-amino-1,2,4-triazole (Schwimmer and Weston 1958); other synthetics are also effective (Jadhav and Salunkhe 1974).
(5) Carbon dioxide enriched packaging (Patil *et al.* 1971).
(6) Subatmospheric storage (Salunkhe and Wu 1974,1975).
(7) Ionizing radiation (Schwimmer and Weston 1958). Figure 17.4 shows a reduction in chlorophyll but not carotenoid synthesis.
(8) Treatment with household detergent (Sinden 1971).

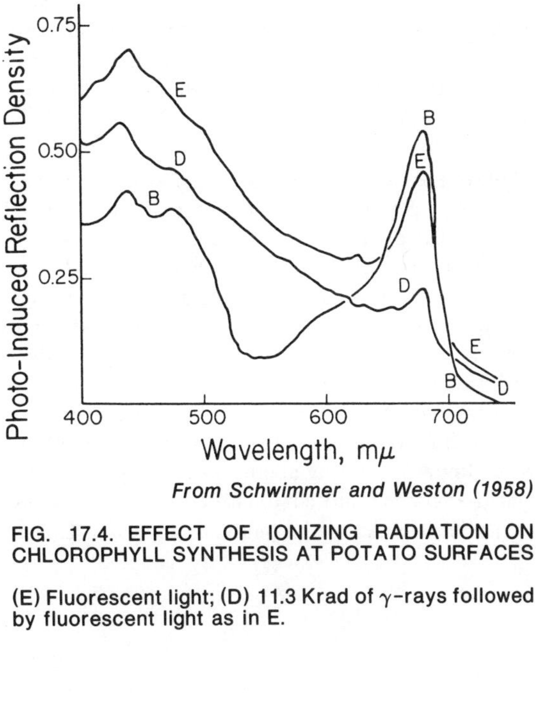

From Schwimmer and Weston (1958)

FIG. 17.4. EFFECT OF IONIZING RADIATION ON CHLOROPHYLL SYNTHESIS AT POTATO SURFACES

(E) Fluorescent light; (D) 11.3 Krad of γ-rays followed by fluorescent light as in E.

OTHER ENZYME-RELATED COLOR CHANGES

Pigmentation of Allium Foods—Lyase Action

Up to this point in our discussion of enzymatic aspects of the formation of pigments in foods, we have dealt mainly with the oxidoreductases and to some extent with the action of hydrolases. In this section we show how the action of an enzyme belonging to the fourth class, the lyases, results in the formation of unwanted pigments in vegetables of the *Allium* genus, onion *(A. cepa),* leeks *(A. porrum)* and garlic *(A. sativum).* Although the chemistry and mode of enzyme action are apparently different (we shall discuss these aspects in somewhat more detail in Chapter 21 on flavor), they do have in common, along with other enzymes which remain active after harvest and during processing, several similarities. In addition to contributing to color changes, the enzymes manifest their action only after cell disruption and the subsequent decompartmentalization of enzyme and substrate so that they can interact. Each catalyzes the formation of a highly reactive primary product. This primary product can interact with itself, the substrate or can react with other food constituents to cause changes in a quality attribute of a food. Also, more than one quality may change depending upon the components of the foods, although we traditionally associate the action with one type of quality with one enzyme. So it is in the case on the lyases (alliinase) of onion and garlic. What has been traditionally considered to be an enzyme involved exclusively in flavor development, especially odor attributes, is also involved in color formation.

Pinking of Onions.—A problem which has occasionally plagued the commercial onion dehydrator is the development of what he/she considers to be an undesirable nonuniformly distributed pink discoloration in some of the dried pieces. Joslyn was probably the first, in 1931, to investigate this pinking systematically in the laboratory (Joslyn and Peterson 1958). Although structure of this pigment (which is a deep red-purple when concentrated) has to our knowledge not been elucidated, we do know something about one of the precursors, and the enzyme and circumstances involved in its formation thanks to the pioneering work of Lukes (1959) (only one of several papers is cited), Joslyn and Peterson (1960), Yamaguchi *et al.* (1965), who studied agronomical parameters, Shannon *et al.* (1967) and more recently by Bandyopadhyay and Tewari (1973). On the basis of the accumulated observations to 1967 of these investigators, Shannon *et al.* proposed the following:

(1) A Precursor $\xrightarrow[\text{very fast}]{\text{enzyme}}$ Color developer (I) (colorless, ether soluble)

(2) I + Amino acid $\xrightarrow[\text{fast}]{}$ Pigment precursor (II)

(3) II + A carbonyl $\xrightarrow[\text{slow}]{}$ Pink Pigment

Based on experiments which strongly indicated that the progenitors of odor-bearing components formed enzymatically in onions may also give rise to the Luke's color developer, Shannon *et al.* surmised that the unknown precursor in this equation was propenyl-S-1-cysteine sulfoxide, the principal substrate of the lyase in onion whose action gives rise to most of the other sensory attributes of onions when eaten raw or when freshly minced or comminuted as will be discussed in Chapter 21.

Strong confirmatory evidence of the cumulative views of the above-mentioned investigators was provided by Bandyopadhyay and Tewari (1973) who isolated and characterized Lukes' color developer. The major component of the three sulfur-containing compounds they isolated was identical to the lachrymator in onion, thiopropanal S-oxide (Chapter 21).

In confirmation of the conclusions of Joslyn and Peterson, Shannon *et al.* found that the rate determining step of color development in their model systems (as well as in onion purees and *in situ* during onion dehydration) was probably the concentration of carbonyls. This explains why Yamaguchi *et al.* could not find a relationship between extent of "pinking" of different onion varieties and objective indices of pungency and odor such as volatile reducing substances (Farber 1957) or enzyme-produced pyruvic acid (Schwimmer and Weston 1961; Schwimmer and Guadagni 1962; Schwimmer 1971B). Since the final step is rate limiting we may formulate the sequence of reactions occurring in onion which lead to red pigment formation as follows:

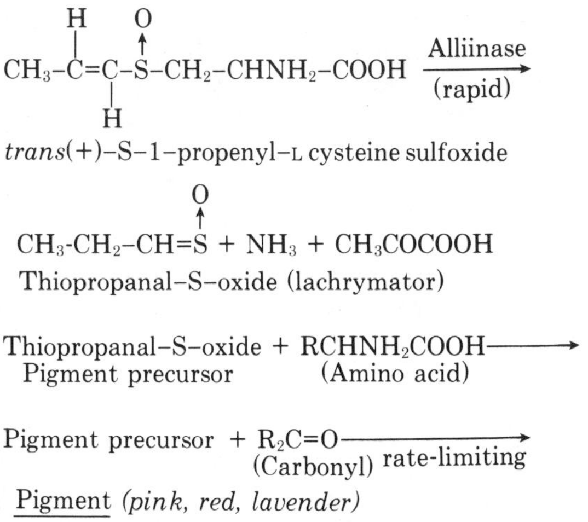

As of this writing there are still several loose ends to tie up. The qualities of the natural occurring carbonyl compounds in onions and whether or not they are produced enzymatically are not known. The general structure of the pigment has not as yet been ascertained. It will probably consist of a class of compounds since there are many carbonyls and many amino acids. These pigments contain carbon, nitrogen, sulfur and methoxy groups and absorb maximally between 490 and 540 nm, depending upon the source of onion and on experimental conditions. They are similar in some respects to the betacyanins of beets. An analytical method for the determination for the lachrymator is based on this pinking (Tewari and Bandyopadhyay 1975).

Prevention of Pinking or Reddening.—Since red color is not associated in the consumer's mind with what is desirable in onion color (at least after it has been peeled), onion and leek processors would like to prevent this pigmentation. They cannot use blanching or acid treatments since these would destroy the labile lyase responsible for the catalyzing flavor-producing reaction.

One solution would be to apply and extend the findings concerning the relation of pinking to variety, maturity, horticultural practices and storage. Of course since the unique causative factor is the precursor to flavor, it is not surprising that agronomical factors affect both pinking and pungency similarly. For dehydrated products, it is desirable during the processing to minimize cell disruption and maximize maintenance of compartmentalization of enzyme and substrate. Joslyn and Peterson observed that avoiding bruising and very careful slicing before dehydration tended to prevent pinking (or reddening). For purees, disruption of the onion tissue is unavoidable and for some purposes the preservation of maximum enzyme-produced fresh flavor is not so important (Schwimmer 1968A,1969D). Ascorbic acid is effective in reducing rate and extent of pinking in onion purees (Joslyn and Peterson 1958) and ascorbic-citric acid prevents its formation in both onion and leek purees (Korner and Berk 1967). A patent was issued to Li *et al.* (1967) on the use of cysteine to prevent pinking in onion purees. Concentrations of cysteine in the range of 0.01 to 0.3% are adequate for this purpose. However, cysteine, even at these low concentrations, intensifies and modifies the odor of such macerates (Schwimmer and Guadagni 1967).

Leek.—Leek has also been reported to develop a pink pigmentation during processing (Korner and Berk 1967). The properties of the pigment and the circumstances associated with its formation are similar to those prevailing in purees of onion which undergo pigmentation. For instance, amino acids in the ether-insoluble fraction are necessary for pigment formation. However, since they found phenolic substances in the ether extract, they concluded that phenolase action produced quinones which then reacted with the amino acids. In light of what we know now about pigment formation in *Allium* it is likely, but by no means proven, that a lyase-linked rather than a phenolase-linked enzyme action is involved.

Green Pigment Formation in Garlic Purees.—Like most plant foods, off-color of garlic products such as purees and juices develops unless a blanching step is interposed at the appropriate step in their production. Since lyase action is required for development of the requisite flavor (Chapter 21), the blanching step *follows* and may also be accompanied by acidification (Sugihara and Cruess 1945). In this and in similar processes, a green color was occasionally observed. This greening was investigated by Joslyn and Sano (1956) and Yamaguchi *et al.* (1965). Taking a clue from onion-pinking investigations, they were able to induce a variety of colors and hues in garlic purees by the addition of various carbonyl-containing compounds and/or mixing the garlic with onion puree.

Not enough is known about this type of pigmentation development in order to ensure means of preventing it. Pulverizing the garlic while frozen, followed by cellulase treatment, centrifugation and spray drying (Noznick and Bundus 1966) might circumvent such a color problem.

The foregoing survey of the enzymatic aspects of color change and retention in plant foods probably does not encompass all such changes which occur, only those for which we have evidence that enzymes are involved. Undoubtedly other color changes whose enzymatic credentials have not been established do occur, for instance, the statistical and apparently causative relation between the occurrence of apple and (perhaps banana) scald and accumulation of farnesene (Meigh 1969). Farnesene is a 15-carbon unsaturated hydrocarbon, $(CH_3)_2C{:}CH(CH_2)_2C(CH_2){:}CH(CH_2)_2C(CH_2){:}CHCH_3$ whose accumulation may be an aberration in the regulation of the enzymes on the pathway of steroid and flavorant biosynthesis (Chapter 22). More recently Feys *et al.* (1980) presented rather cogent evidence implicating lipoxygenase action in the etiology of this physiological disorder.

18

ENZYMATIC ASPECTS OF COLOR CHANGE AND RETENTION IN MEAT AND FISH

MEAT PIGMENTS

Although the switch from the green color of plants to the red color of meat at this point appears to be a rather sharp deviation in our discussion of enzymes and color quality, the enzymatic determinants of color of these foods have much in common. Thus a metal porphyrin ring is common to the pigments involved. In each case an oxidation may occur which can give rise to undesirable brown pigments which may, in part, be associated with lipid hydroperoxide formation. Furthermore, the reactions involved also cause other food quality changes. For meat, flavor deterioration is inextricably associated with discoloration. For cured meats color changes are accompanied by health-associated quality changes. On one hand the nitrites act as antibacterial agents but on the other are implicated as precursors to the carcinogenic nitrosamines as alluded to in Chapter 35 (Gray and Dugan 1975).

The many excellent reviews on meat color include those of Fox (1966) on curing chemistry, of Solberg (1970) who linked the intrinsic properties of myoglobin to meat color quality, and of Giddings (1977A,B) on the molecular basis of color in muscle foods. Lipid involvement and the role of oxidation generators are discussed by Greene (1971) and Benedict *et al.* (1975). Means of preventing discoloration are reviewed by Govindarajan (1973) and Greene and Price (1975). Other reviews on factors affecting color quality, especially from the more food technological viewpoint, are those of Lawrie (1966), Briskey *et al.* (1966, 1967), Solberg (1968) and Giddings (1977B).

The colors associated with good meat quality are dependent, of course, on the type of meat; for most fresh beef products, both as slices, roasts, steaks on one hand, and as ground meat on the other a bright red color is wanted; for lamb, chicken, turkey and pork, and for cured meats in general, pink to "flesh-" like colors are considered to be normal. In the case of some of the newer comminuted products such as turkey hamburger as well as mixtures of meat and nonmeat protein (Chapter 34), the product may be treated to optimize color quality. Solberg pointed out that the instability of these colors in red meat constitutes an obstacle to centralization of cutting and packaging facilities. In view of the double-digit rise in prices, the savings achieved by effective stabilization of raw meat color would be more welcome now than in 1970.

The pigments associated with the bright red color on the surface of uncooked meat are oxymyoglobin, MbO_2, and oxyhemoglobin, HbO_2, for fresh meat and the corresponding nitrosyl derivatives (MbNO and HbNO) for cured meats (Table 18.1). Although about 20 to 30% of the total pigment of fresh meat is present as the hemoglobins, attention has been focused on the role of myoglobins in most

TABLE 18.1

MEAT PIGMENTS AFFECTED BY ENZYME ACTION

Pigment	Ligand/Feature	Color	Symbol
Myoglobin	Fe^{2+} Fe (II)	Purple	Mb
Hemoglobin			Hb
Oxymyoglobin	$Fe^{2}\ O_2$; $Fe^{3+}O_2^-$	Red	MbO_2
Oxyhemoglobin	Fe–O–O≃120°		HbO_2
	Fe...O... (120°)		
Metmyoglobin (Ferri)	Fe^{3+}	Brown	MMb
Methemoglobin (Ferri)		Brown	MHb
Nitrosyl Myoglobin	$:(Fe–NO)^{2+}$ or	Red	MbNO
Nitrosyl Hemoglobin	$:Fe{=}N^{+\cdot}–O^-$	Red	HbNO
Nitrosyl Hemochrome	After cooking	Red	—
Nitrosyl Myochrome	$^+ON–Fe^{2+}–NO$	—	—
Ferryl hemoglobin	FeO (IV)	—	—
Choleglobin	Porphyrin ring oxidized	Green	—
Cystine-globin	R-S-S-Protein (denatured)	Green (tuna)	—
Nitrimetmyoglobin	Excess nitrite	Green	—
Irradiated, modified	—	Red	—

meat color studies (Fox 1966). We shall henceforth discuss and use the terminology of myoglobin and derivatives with the caveat that the rate of certain reactions involved in color transformations of Hb and Mb may be decidedly different. The intensity of color quality depends upon the concentration of these pigments, a high concentration being sought for in fresh uncooked beef and relatively low levels in other meats such as pork, rabbit, lamb as well as in fish such as tuna. The undesirable tan-brown pigment is principally the ferric form of myoglobin, metmyoglobin (MMb). A maximum of 30% of this oxidized pigment at the surface of the fresh beef is consonant with acceptable color quality and over 60% is considered to be unacceptable.

In unspoiled meat the interior pigment consists principally of purple unoxygenated myoglobin, Mb. This color at the surface exposed to air is considered to be quite undesirable in the so-called dark-cutting meat whose surface stays purple (Ashmore *et al.* 1973).

Although the conventional point of view considers the iron in MbO_2 to be in the ferrous form, $Fe(II)O_2$, Raman spectroscopic analysis of oxyhemoproteins has revealed a low spin state indicative of the presence of Fe(III) (Spiro and Burke 1976). According to this formulation O_2^- is antiferromagnetically coupled to low spin Fe(III) and the structure is denoted as $Mb{\cdot}Fe^{+3}{\cdot}O_2^-$. That the last word on the nature of the linkage of oxygen to iron in the oxyglobins is not yet in is attested to by alternative formulations, which may perhaps be considered as resonance forms of $Fe\text{-}O_2^-$, iron dioxygen, Fe ···· O ···· O (Caughey *et al.* 1975). A structure first proposed by Pauling (1975) and symbolized as Fe-O-O, consists of a low spin ferrous iron cation and neutral dioxygen ligand to best account for the pattern of charge transfer transition deduced from the polarized absorption spectra by Churg and Makinen (1978). In all these formulations the two O atoms form a 110°–120° angle. A more lucid understanding is important to the rationalization of measures to maintain optimum color.

Misra and Fridovitch (1972) postulated the $Fe^{3+}O_2^-$ dissociate to yield superoxide anion, O_2^-.

$$\text{Oxyglobins} \rightarrow \text{Metglobins} + \text{superoxide anion}$$

O_2^- and O_2^* appear with ever-increasing frequency in the literature of globin transformations. An even higher oxidation state of iron (MbO^{IV}) present in ferryl myoglobin, which is formed when H_2O_2 is added to ferrimyoglobin, MbO_2, has been implicated in mitochondria-myoglobin interactions (Uyeda and Peisach 1975).

Proposed structures for the complexes of NO with Mb are analogous. These and other forms of the globin pigments, including the interesting but undesirable pink pigments of irradiated meat (Satterlee *et al.* 1971), are assembled in Table 18.1.

Myoglobin, *in Vivo* Structure, Function and Enzymatic Reduction

Myoglobin (Mb) is specifically designed and situated in muscle cells to facilitate access of oxygen to these cells' mitochondria (Kagen 1973; Wittenberg *et al.* 1975) so that the transduction of chemical to mechanical energy *via* ATP hydrolysis catalyzed by the contractile structural muscle proteins whose disposition determines meat texture (Chapter 26) can proceed *in vivo* with optimum efficiency. Myoglobin is particularly prominent in the history of biochemistry and molecular biology because it is the first complex protein whose tertiary structure was virtually completely elucidated through X-ray analysis by Kendrew (1963) and Perutz (1978).

Our knowledge of this protein *in vivo* might be expected to shed some light on meat myoglobin *in situ*, that is, in meat. In the analogous area of human hemoglobin study, impetus has been provided by the occurrence of the disease methemoglobinemia (Jaffe and Neumann 1968). In nondiseased individuals whatever MHb is formed is immediately reduced to Hb. This disease, which causes mental retardation, is accompanied by a decrease of what at first were termed "diaphorases" but later pinpointed as cytochrome b_5 reductase (see below) by Leroux *et al.* (1975). "Diaphorases" were considered to be flavoproteins oxidizing NAD(P)H in the presence of a mixture of acceptors. The term diaphorase is no longer accepted as an official enzyme. Indeed the EC list of 1973 does not even recognize the existence of a specific hemo- or myoglobin reductase. However, the term "DT diaphorase" is considered acceptable as an alternative to the recommended "NAD(P)H dehydrogenase (quinone)" because a quinone can intermediate its action (Crane 1977). Although the natural acceptor of this enzyme, prevalent in all mammalian meat tissues, is still unknown (West *et al.* 1977), its salient characteristic is its high sensitivity to dicoumarol and to all similar anticoagulant drugs. This inhibition, although formally competitive with respect to NADPH, involves cooperative interaction between separate substrate and inhibitor binding sites through a Ping Pong Bi Bi sequence (Hollander and Ernster 1975). However it is believed that the NAD(P)H serves as reductant and that Mb reduction is catalyzed by enzymes. There are two conflicting claims as to what the actual enzyme mechanism is.

Cytochrome b_5 Participation.—One claim invokes cytochrome b_5 *via* cytochrome b_5 reductase. Cytochrome b_5 plays a protean role in facilitating reactions outside of mitochondria, whereas other cytochromes *a, b* and *c* are constituents and participate in the free energy producing electron-transport system within the mitochrondion. Extrapolating to muscle myoglobin, cytochrome b_5 may participate in the maintenance of myoglobin in a form where it is useful in shuttling oxygen from extracellular space into muscle cell mitochondria. The following reactions may occur (Passon and Hultquist 1972).

$$H^+ + NADH + \text{Ferrocytochrome } b_5\ (Fe^{3+}) + A \longrightarrow NAD^+ + (Fe^{2+}) + AH_2$$

$$(Fe^{2+}) + MMb(Fe^{+3}) \longrightarrow \text{Ferricytochrome } b_5(Fe^{+3}) + Mb(Fe^{2+})$$

$$Mb(Fe^{2+}) + O_2 \longrightarrow Mb(Fe^{3+} \cdot O_2^-) = MbO_2$$

$$MbO_2 + \text{Mitochondrion} \longrightarrow MbO_2 - \text{Mitochondrion complex } (Mb \cdot O_2 \cdot C)$$

$$Mb \cdot O_2 \cdot C \longrightarrow Mb + O_2\text{-enriched mitochondrion}$$

The production of O_2^- is ruled out because neither SOD nor catalase influences this reduction.

Ferriglobin Reductase.—The other point of view is that specific enzymes are responsible for the reduction of MMb and MHb. An MHb reductase has been obtained from erythrocytes and MMb reductases from muscle tissue of food animals (Shimizu and Matsuura 1968,1971) and fish (Yamanaka *et al.* 1973; Al-Shaibani *et al.* 1977). The latter found the highly purified bluefin tuna enzyme activated by Mn^{2+}, Fe^{3+} and I^- whereas the enzyme is inhibited by Cl^-.

Unexpectedly, perhaps, the purified enzymes are not "diaphorases" in the sense that they contain neither of the flavin nucleotides, FAD and FMN, nor are they cytochromes. It may very well be that this type of globin reductase activity is very similar if not identical with that present in a variety of other muscle tissue (Leroux and Kaplan 1972).

In addition to the action of these enzymes, reduction of MMb *in vivo* has been ascribed to reductants present in the same tissues as the globins. Especially prominent in this area are studies on ascorbate (Carra and Colleran 1969; Fox *et al.* 1975; Halliwell and Foyer 1976). The presence of ascorbate can cause both reduction and oxidation of sulfhydryl molecules such as glutathione, which is reversibly oxidized by the selenoprotein glutathione peroxidase (McCay *et al.* 1976) and an NAD(P)H-dependent FAD-requiring reductase (Asghar *et al.* 1975). Giddings does not believe that the latter can have more than a minor role to play in this particular phenomenon, particularly with regard to Mb since mammalian Mb unlike Hb and perhaps tuna Mb (see below in this chapter) contains no cysteine sulfur and is not an allosteric protein (Perutz 1978).

ENZYME ACTION AND RED MEAT DISCOLORATION

Early workers had thus made uniquivocally clear that maintenance of myoglobin in the MbO_2 state is the key to prolonging the color quality of fresh meat. What are the factors that prevent this optimization? Although we are concerned primarily with the influence of enzyme action, understanding of the intrinsic properties of Mb is of seminal importance for achieving this color optimization. Thus the influence of oxygen, temperature, pH and ionic strength on the nonenzyme reactions:

(1) $Mb^{2+} + O_2 \rightarrow MbO_2^{2+}$ (stable), an *oxygenation,* and
(2) $Mb^{2+}O_2 \rightarrow Mb^{3+} + [O_2^-]$, an *oxidation,* are quite profound.

Nonenzyme Factors and Model Systems

The opportunity for MMb formation is dependent on both free Mb and on O_2 concentrations. Mass action would tend to make more Mb available at lower O_2 levels. The rate of reaction *in vitro* is maximal at a pO_2 of about 14 torr (mm Hg) and levels off to a constant value at about 30% of the maximum with increasing oxygen tension.

It is at the low oxygen levels short of anaerobiosis that off-colors in meat are developed. This effect can be observed in a freshly cut cross-section of a beef slice from meat which has been held in the refrigerator for several days. The pigment at the periphery of the slice in the vicinity of what was the surface before slicing the bright red is MbO_2 resulting from the mass action effect of O_2 which has diffused into this region. As the amount of available O_2 diminishes, Mb available for conversion to MMb increases. This is manifested as a thin brown layer a few centimeters below the original surface. The O_2 is exhausted in the region extending further toward the center of the slice so that Mb preponderates, imparting a purplish hue to this region.

The rate of MbO_2 oxidation is also profoundly affected by both pH and temperature. The rate at pH 8 can be three to four times that at pH 5 to 7. With regard to temperature, the Q_{10} (oxidation) of Mb in model systems in the pH region of most meats is 4.8, high as compared with 2–3 for most chemical and enzyme-catalyzed reactions.

Another factor which may influence the rate of MHb formation, but which does not appear to have been hitherto considered, is the influence of certain phosphate esters on the $Hb + O_2 \rightleftharpoons Hb{\cdot}O_2$ equilibrium. Phytic acid in birds and 2,3-diphosphoglyceric acid in mammals induces conformational changes in the Hb but not in the Mb molecule. This shifts the above equilibrium to the left, i.e., O_2 is replaced by these organophosphates. One would expect, therefore, an enhanced opportunity for the Hb → MHb conversion to occur at a given oxygen tension for Hb but not for Mb (Perutz 1978; Benesch *et al.* 1968; Nelson *et al.* 1974; Hedlund *et al.* 1975). According to the latter authors, the function of such organophosphates is largely one of pH control. Phytic acid can induce cleavage of the bond-linking hemoglobin histidine to iron in nitrosyl hemoglobin (Maxwell and Caughey 1976).

Enzyme-related Factors

Having briefly outlined what may happen *in vivo* to maintain MbO_2 on one extreme and what happens in relatively pure myoglobin model systems, we now turn our attention to the question of what factors tend to cause accumulation of MMb at surface of displayed refrigerated meat. Progress in understanding color changes is highly dependent upon convenient valid measurements of these forms of heme pigments at the surface of the meat. Quantitative reflectance spectroscopy (Fig. 18.1) plays a vital role in these investigations. We may conveniently divide the effects as follows: bacterial, lipid-heme mutually reciprocal autoxidations and other oxygen-activation reactions, enzyme reactions which probably counteract the active M·Mb reducing enzyme systems in beef tissue, and other correlates of deleterious color formation which may possess enzymatic components.

Bacterial.—According to Benedict *et al.* (1975) most browning of fresh beef could possibly be traced to bacterial reaction. On the other hand the same authors point out that at the unrefrigerated temperature of 23°C, the color of ground beef turns rufous in 15 min and to brown (all MMb) in 2 hr. Such behavior cannot be accounted for by bacterial action. Color problems in dry sausages may be caused by microorganisms which produce hydrogen peroxide which oxidizes Mb to MMb (Rozier 1971; Devore and Solberg 1974). Benedict *et al.* (1975)

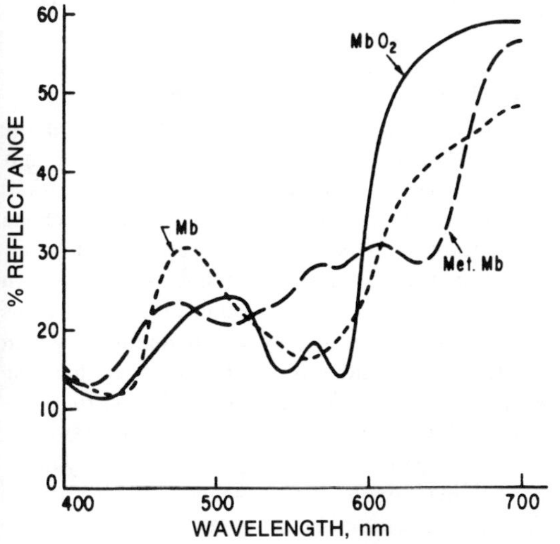

From Strange et al. (1974)

FIG. 18.1. REFLECTANCE SPECTRA OF BEEF MYOGLOBINS

argue that the aerobic bacteria on the meat surface and the unavailability of unsaturated lipids tend to diminish oxygen available for tissue lipid oxidation at the surface of cut meat. Devore and Solberg found little if any effect of lipids, but they also eliminated contaminating bacteria as a source of oxygen uptake at the surface of fresh cut meat. Long-term storage does afford opportunity for a microbial enzyme contribution to color change. Thus, Dedyukhina *et al.* (1975) attribute Mb-associated changes in whale meat color during frozen storage to bacterial action.

Lipid Involvement.—The situation is quite different in ground meat where several factors make unsaturated lipid available for mutually reciprocal autoxidation of lipid and heme pigments. The oxidation of the lipid results in a particularly objectionable deteriorative odor (especially important in pork) while that of the heme pigments results in poor color quality. As pointed out and shown experimentally by Benedict *et al.* (1975) the deliberate addition of polyunsaturated fat to new high protein products (Chapter 34) containing meat and vegetable protein (Dinius *et al.* 1974) would tend to exacerbate the autoxidation of heme pigments and the development of off-color. Even in unground meat this may be a problem. As previously mentioned the development of centralized operations in the fresh meat industry awaits effective means of preventing the oxidation of Mb to MMb and the retention of MbO_2 at cut surfaces for extended periods. This, in turn, awaits a clearer understanding of the nonbacterial-associated changes.

The chief proponents of the concept of mutual oxidation of unsaturated fats and heme are Greene and Price (1975), who stress autoxidation, and Hultin and coworkers (Govindarajan *et al.* 1977; Lin and Hultin 1977), who stress the role of intracellular enzymes.

The oxidation may occur *via* nonenzymatic self-propagating free radical reactions which may be depicted as:

$$R{\cdot}H \xrightarrow{Fe^{+3}\ (MMb);\ h\nu} R{\cdot} + H{\cdot}$$

$$R{\cdot} + O_2 \longrightarrow ROO{\cdot}$$

$$ROO{\cdot} + R{\cdot}H \longrightarrow ROOH + R{\cdot}\ \text{(fatty acid radical)}$$

$$H^+ + ROOH + Fe^{+2}\ (Mb) \longrightarrow Fe^{+3}\ (MMb) + H_2O + R'\ \text{(decomposition of R)}$$

Thus, both unsaturated lipid (R·H) and reduced myoglobin are oxidized.

The free radicals appear to attack heme directly rather than to denature the protein. The oxidized lipids create off-odors and the oxidized globins off-color. Participation of enzymes is, as of this writing, still speculative. In intact tissue, fatty acids appear to be in part present as complexes with myoglobin, which thus functions as a transport protein for fatty acids. Formation of such complexes in turn ensures propinquity between heme and fatty acid. This would lead to nonphysiological oxidation upon cell disruption occasioned by grinding (Gloster and Harris 1977).

Grinding potentiates the action of several enzymes. These include lipases and phospholipases acting on substrates containing unsaturated fatty acids. The polyunsaturated fatty acids come from cellular and subcellular membranous envelopes. At least as important is the physical association and propinquity of the released fatty acids to the heme pigments. The phospholipids, by virtue of being a supplier of polyunsaturated acids, provide substrates for the formation of active oxygen. Hultin and coworkers point out that it is probably the intracellular lipid which is responsible for deteriorative color and flavor changes rather than that from the highly visible adipose tissue in meat.

Although investigations in meat color research have not invoked lipoxygenases *per se*, a closely related enzyme, prostaglandin synthase, and lipoxygenase (Chapter 24) are present in mammalian tissue and may influence meat color.

Other Sources of Free Radicals and Excited Oxygen.—As frequently pointed out throughout this book, highly reactive forms of oxygen and free radicals can be generated in the absence of lipids and/or lipoxygenase action. Benedict *et al.* (1975) cite evidence from this type of autoxidation of Mb to MMb for the creation of O_2^- in the discoloration of meat due to the autoxidative conversion of Mb to MMb. In addition to the previously cited work of Misra and Fridovitch (1972), other pertinent investigations in which it is concluded that O_2^- is generated in the oxidation of either Hb or Mb include those of Gotoh and Shikima (1976) on its generation by autoxidation of Hb and of Goldberg *et al.* (1976) on its enzymatic generation. Analogous to lipids, epinephrine also engages in complementary oxidation with hemoproteins which generate O_2^- (Fridovitch 1974).

Light.—There is some evidence, which some consider marginal, that light itself may contribute to the initiation of autoxidation (Setser *et al.* 1973; Hunt *et al.* 1975; Owen *et al.* 1976). Light is also involved in the release of nitrous oxide from MbNO. The plastic film used to wrap fresh raw meat has to be permeable to oxygen so that the desirable red color can be maintained. However, with cured meats the plastic packaging film has to be oxygen-impermeable because light initiates the release of NO from Mb·NO resulting in an undesirable discoloration.

The focal excited oxygen species involved in the light-induced formation of MMb in both meat and in model experiments is not O_2^- but O_2^*, generated from both free and globin-bound oxygen. This is also an autoxidation in the sense that no direct enzyme action is involved but rather a photocatalysis—either indirectly through riboflavin-associated compounds or directly through photosensitization of heme or alteration of an amino acid in Mb (Possani *et al.* 1970; Lynch *et al.* 1976; Koppenol *et al.* 1976; Giddings 1977A).

Energy for the photocatalyzed destruction of Mb may also reside in nonoxygen electronically excited molecules such as that produced in the complementary oxidation of Mb and acetoacetate, an intermediate in lipid metabolism:

$$\underset{\text{Acetoacetic Acid}}{CH_3\text{–}CO\text{–}CH_2\text{–}COOH} + O_2 + Mb \longrightarrow \underset{\text{Methyl Glyoxal}}{CH_3\text{–}CO\text{–}CHO} + CO_2 + H_2O + MMb$$

The electronically excited methyl glyoxal according to Vidigal and Cilento (1975) sensitizes the photodestruction of Mb. It is conceivable that similar processes occur in meat. The color of cooked meat products such as sausages and luncheon meats may be affected by fluorescent light.

Ionizing radiation, although not yet (if ever) in use for meat preservation (Chapter 10), also gives rise to an abundant supply of superoxide ion which can have profound effects on color including the peculiar off-pink color previously alluded to. Some light on the nature of what is happening is afforded by study of Haristoy (1977) on the reactivity of O_2^- in an HbO_2-MMb system. He concludes that *in vivo* (and presumably in meat), reactions occur which give rise to a steady-state concentration of O_2^- and O* which facilitate or even initiate reactions favoring oxidation by oxygen.

Other Enzyme Systems.—Liver, whose red color rapidly changes to brown with a corresponding development of off-flavor, contains the previously mentioned cytochrome P-450 system (Chapter 4) which apparently is involved in the formation of lipohydroperoxides (O'Brien and Rahimtula 1975; see also Chapter 5). Moreover there is some evidence that other forms of activated oxygen act at the site of this powerful oxidation (Rikans *et al.* 1979). In a structurally disintegrating system, as meat is, one would expect the selenium-containing glutathione peroxidase to participate as a determinant of meat color (Prohaska 1979). Undoubtedly other flavoprotein oxidases and even heme itself could produce O_2^-. Another factor which would tend to enhance activated oxygen production derived from autoxidation would be the lowering of the pH in the muscle-to-meat conversion process (Chapter 26) from 7.4 to 4.7 as discussed shortly.

The interesting nonheme iron-H_2S protein ferredoxin whose easily detachable H_2S may contribute to food flavor (Schwimmer and Friedman 1972) has also been implicated in lipid peroxidation and thus in meat color formation (Liu and Watts 1970).

Dehydrogenase Dysfunction via Oxygen and pH.—The rate and extent of O_2 uptake by meat profoundly influence MMb formation therein. As expected from its behavior in model systems, the rate is greater at lower oxygen pressures than at ambience (Ledward 1972). Autoxidation is most pronounced at a pO_2 range of 1 to 4 torr. As shown in Fig. 18.2, a negative linear relation obtains between log oxygen uptake into unground beef and color stability defined as retail display life (Atkinson and Follett 1973). This effect of oxygen appears to relate to the malfunctioning of MMb reducing enzyme systems. After all, ambient oxygen pressure is a rather unphysiological condition for semimembranous muscle to find itself in. Thus NAD, involved in one of these Mb conversion systems, also disappeared in this system. Other investigators (Bendall and Taylor 1972; DeVore and Solberg 1974,1975) attempted to get at the primary reasons for this O_2-dependent NAD depletion. It would appear that substrate depletion, i.e., NAD, enzyme (cytochrome c reductase) and malfunction of mitochondria all play roles in limiting respiratory oxygen conversion. We have already referred to the role of pH in increasing the formation of MMb. In the present context lower pH would also tend to denature some of the active reducing enzymes and damage the mitochondrion itself. Furthermore, upon prolonged storage disrupted lysosomes may release cathepsins which would attack mitochondrial and perhaps the sensitive enzymes involved in the MMb reducing system. In fact, mitochondria could participate in meat MMb reducing systems. Other enzyme actions include those more commonly associated with textural quality of meat such as the activation of ATPase (Bendall 1972) due to the leakage of Ca^{2+} (Chapter 26) by decay and disruption of intracellular membranes and the recently discovered Ca-activated CAF, nonlysosomal protease. It is possible that this and other Ca-related protease action can affect color or at least appearance of meat through the enzymes of the blood-clotting cascade (Chapter 4). Even the buildup of the previously mentioned organic phosphates which affect the Hb/HbO_2 equilibrium by competing with O_2 for the Hb could contribute to the loss of color quality. If this turns out to be the case, then if the action of phosphatase proceeds faster than the accumulation of these organic phosphates, one might expect an improvement in color. Another effect of these organic phosphates, in addition to competing with oxygen, is their ability to enhance a peroxidase-like activity of MHb (Mauk *et al.* 1973). It remains to be determined if such an effect would protect meat color quality by preventing H_2O_2 and O_2^* access to Mb and Hb or would accelerate MMb formation via autoxidation.

In connection with ATPase action and color, the occurrence of freezer burn at the surface of frozen meat during prolonged storage is accompanied by decrease in activity of ATPase but not that of other enzymes in the affected area (Kaess and Weidmann 1967). Other useful empirical correlations between enzyme activity and tendency to brown have been reported. For instance using a histochemical approach, Franke (1974) disclosed a very close correlation between metmyoglobin formation in intact beef muscle tissue and the proportion of succinic dehydrogenase positive fibers after less than 1 week's storage at 4°C.

ENZYME ACTION AND FRESH MEAT COLOR RETENTION AND CREATION

Metmyoglobin Reducing Activity (MRA)

Just as color retention which occurs in model meat systems and in meats at very low pO_2 can actually take place in meat on the display shelf, so the beneficial effects of very high oxygen tension can also be demonstrated. Zimmerman and Snyder (1969) appear to be the last of several investigators who showed that pO_2 above ambience (i.e., 80 psig) prolonged the display life of red beef, apparently by mass action effect on the $Mb + O_2 \rightarrow MbO_2$ reaction. Since little Mb is available for oxidation and since oxygenation of Mb increases its stability, the possible involvement of increased lipid induced oxidation is cancelled out. However, the beneficial effects of high O_2 are partly counterbalanced by the destructive effects of high O_2 on the metmyoglobin reducing activity, MRA (Zimmerman and Snyder 1969) and as previously noted by the increased opportunity for activation of oxygen via lipids.

Just as higher pH favors MbO_2 formation in the test tube, so it does in meat. In dark cutting meat or meat from animals injected with adrenalin (Chapter 26), the pH remains elevated during storage and myoglobin remains in the reduced state (Ashmore *et al.* 1973) leading in some meats to the aforementioned "dark cutting." This can be prevented by preslaughter injection of drugs such as the β-blocker propranolol acting as antagonists to adrenalin. However, it is not high pH *per se* which keeps the meat from turning brown. High pH stabilizes and potentiates enzymes and organelles, especially mitochondria, which enclose these enzymes, especially those responsible for the maintenance of MRA (Cheah and Cheah 1971).

MRA can be easily observed by placing an oxygen-impermeable wrapper around a patty of brown gound beef. The bearer of the brown color, MMb, will be gradually converted to the purple Mb. Upon reexposure to air the surface Mb will combine with O_2 to yield the desirable "bloom." MRA has been investigated in detail by Watts and coworkers (Watts *et al.* 1966; Liu and Watts 1970).

Using anaerobic reduction of ferricyanide as a measure of MRA in beef, Ledward (1972) failed to find a correlation between MRA and MMb formation after exposing the meat to decidedly low followed by ambient pO_2. He did note, however, a strong negative correlation between MRA under aerobic conditions and MMb level. This led him to suggest that the mode of MMb reduction may be different in air than under completely anaerobic conditions. Giddings (1974,1977A,B) correlates and marshals the facts and theories on this aspect thoroughly and forwards his own thesis that mitochondria are involved in this reduction by supplying the meat tissue with that key reducing cofactor, NADH, generated via reversal of electron transport. This is reminiscent of ATP-driven reverse electron transport of chloroplast and it will be recalled that reversal of mitochondrial oxidative phosphorylation to produce a reductant, probably NADH, is the mechanism by which ATP prevents potato slices and other fruits and vegetables from phenolase-induced blackening as discussed in Chapter 19.

This should all be reexamined, and it makes even more sense, in light of the Boyer hypothesis (Boyer *et al.* 1977) that the free energy in the proton gradient produced as a result of electron transport is used to bring about a conformational change in a muscle-like protein in the mitochondria. This change then releases ATP formed via reversal of ATPase action. The idea that ATP-linked energy transductions occur in both muscle (Chapter 26) and mitochondria is in agreement with the above concepts. Since the synthesis requires the presence of ADP, the following sequence is envisioned.

$$\text{ATP} \xrightarrow{+\text{E}} \text{ATP–E} \xrightarrow{\text{ADP+Pi}} \text{ATP–E}\begin{matrix}\cdot\text{Pi}\\ \cdot\text{ADP}\end{matrix}$$

$$\xrightarrow[\text{conformation change}]{\text{energy-linked}} \text{ATP–E*}\begin{matrix}\cdot\text{Pi}\\ \cdot\text{ADP}\end{matrix}$$

$$\xrightarrow{-\text{ATP}} \text{E*}\begin{matrix}\cdot\text{P}\\ \cdot\text{ADP}\end{matrix} \xrightarrow{-\text{H}_2\text{O}} \text{E–ATP} \longrightarrow \text{E + ATP}$$

Again we return to a link between the ATP-muscle system involved in texture and color and textural quality of meat through the ATPase system.

Color Retention

The presence of an enzymatic reducing system in meat suggests that anaerobic packaging can be a practical approach to prevention of off-color browning due to MMb at the surface of the meat. Of course the reduced myoglobin, Mb, which would be formed is purple. However, upon opening the package the air should turn the surface to the right shade of bright red (MbO_2) associated with good color quality. However, according to Greene (1971) anaerobic packaging prevents Mb oxidation only if the meat possesses adequate MRA. This may be related to the finding that the effectiveness of anaerobiosis, both with respect to reduction of brown MMb to Mb and the subsequent regeneration of MbO_2, the "bloom" of fresh beefsteak, are predicated on minimization of the period between slicing and vacuum packaging.

Additives.—At any rate, substances which maintain and/or stimulate the MRA system should be effective. One of these which supplies the metabolites for NADH and which is inexpensive enough is glutamate (Saleh and Watts 1968). Niacin (nicotinamide) is also effective because it lessens the activity of NAD nucleosidase *via* product inhibition. For that matter, the addition of any safe inhibitor of phosphatase action on the Hb-binding organic phosphates (i.e., 2,3-diphosphoglycerate or more practically, phytic acid) should contribute to maintaining desirable fresh meat color. Of course, any additive must not only be safe and allowable but should not mask any concomitant deterioration in health-related quality. Based upon the previously mentioned correlation between O_2 uptake and color deterioration (Fig. 18.2), Atkinson and Follett (1973) suggested that any treatment which slows down O_2 uptake as well as maintains endogenous NAD levels of unground cuts of lamb should also be effective in retention of the meat's red color. Both propyl gallate and BTH as well as ascorbate were found to be effective in color retention of ground meat (Greene 1971; Govindarajan 1973) but not when added to whole muscle section (Devore and Solberg 1974) probably because the complementary oxidation of polyunsaturated fatty acids with the globin hemes is re-

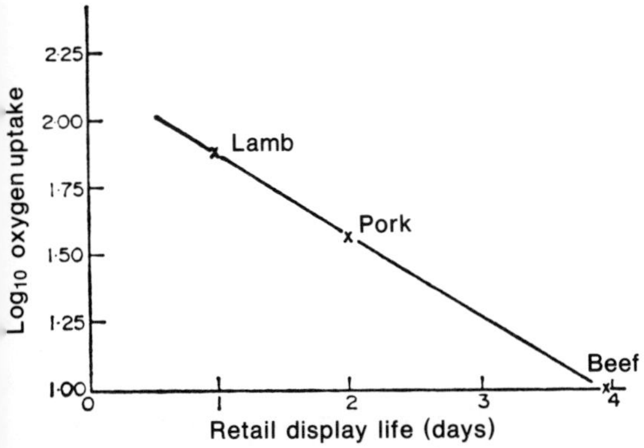

FIG. 18.2. UTILIZATION OF OXYGEN BY MEAT ENZYMES LEADS TO REDUCED SHELF LIFE

stricted by maintenance of structural and cellular integrity with its accompanying compartmentalization (Liu and Watts 1970; Schwimmer 1978). However, this does not explain the ineffectiveness of such natural antioxidants as α-tocopherol and ascorbyl stearate reported by Benedict *et al.* (1975), who also observed an acceleration of MMb formation in ground beef upon addition of ascorbic acid or sodium bicarbonate, the latter presumably to utilize the beneficial effects of higher pH. However, added fat (33% w/w) may have contributed to this acceleration since Greene (1971) suggested that actual maintenance of a high heme:lipid ratio prevents catalysis of lipid oxidation. Also ascorbic acid, as previously discussed, during oxidation in alkaline solution can give rise to H_2O_2 and excited oxygen species. On the other hand, the reducing properties of ascorbate if properly deployed, i.e., *via* preslaughter injection of beef as carried out by Hood (1975), can be exploited to maintain the proper color of unground meat by retention of MbO_2. In a similar vein the proper color of fresh pork can be stabilized by the addition of the reduced form of glutathione (Kortz 1973). It will be recalled that both of these naturally occurring additives have been implicated in maintaining the proper *in vivo* redox balance of at least hemoglobin.

Benedict *et al.* also reported a marginal effect of citric acid in beef color retention. Adequate concentration of this acid as well as oxalic acid and other chelating agents slowing Mb oxidation, presumably by preventing initiation of metal ion-actuated free radical autoxidation, were reported by Govindarajan (1973) to be effective.

Several polyphenols such as pyrogallol and quinones, whose action may be intermediated by the DT diaphorase of the MRA system, by lipid-heme interaction, by accelerating a globin reductase or simply through removal of oxygen, may also be effective in maintenance of optimal meat color (Greene 1971; Saleh and Watts 1968; Roy and Banaschak 1971; Greene and Price 1975).

Addition of Enzyme

So far we have been concerned exclusively with ways of utilizing enzymes and enzyme systems of the meat to maintain color quality. We now wish to briefly discuss the possibility of adding commercial enzymes for this purpose. One of these is papain. Hutchins *et al.* (1967) pointed to the observation that meats treated with papain appear to have a color which suggests that most of the heme pigment is in the reduced form; indeed they found that papain increased and prolonged MRA activity. Part of this may have been due to reducing activity in some samples of commercial papain, but the authors attribute the effect to enzyme potentiation via decompartmentalization of participants in MRA. Catalase was added to sausage to prevent the buildup of H_2O_2 which according to Rozier (1971) is produced by bacterial growth. Glucose oxidase has been reported to help keep shrimp in the pink (Scott 1975). Scott also reported on experimental wrapping of luncheon meats with glucose oxidase + catalase-treated film to prevent discoloration which, he reports, occurs when the meats are exposed to fluorescent light. The resulting gray discoloration can be prevented by ensuring adequate vacuum packing and use of ascorbic acid. Scott found that in the absence of adequate catalase, the meat turned green upon opening of the package. This was probably due to the continued production of H_2O_2. Superoxide dismutase (Chapter 17) may yet be practicably useful in keeping color retention of meat and, as has already been shown (Table 18.2), inhibits the formation of Hb from HbO_2 (Lynch *et al.* 1976). Related are the findings of Lin and Hultin (1977) that the protective effect of glutathione on retention of red Mb is potentiated by glutathione peroxidase, as shown in Table 18.2. Like superoxide dismutase, the peroxidase scavenges a reactive oxygen species (H_2O_2) formed as a result of the autoxidation of MbO_2.

Enzymology of Ham Curing

Current discussion suggests that the future of this ancient use of nitrite in ham curing is cloudy

TABLE 18.2

RETENTION OF RED MEAT PIGMENTS BY ACTIVE OXYGEN-SCAVENGER ENZYMES

Pigment Oxidation System	"Retention" Enzyme	Pigment Loss[C]
MbO_2 + muscle microsomal peroxidation system + O_2[A]	Glutathione peroxidase	32[D]
HbO_2 + photoreduced HbO_2[B] + O_2 + mannitol	Catalase (CAT)	70
	Superoxide dismutase (SOD)	60
	CAT + SOD	35

[A] Source: Calculated from data of Lin and Hultin (1977).
[B] Source: Lynch *et al.* (1976).
[C] Rate of pigment loss, percentage of that without addition of "retention" enzyme; duration of experiments, 9 and 15 min, respectively.
[D] Glutathione without enzyme, 88%.

from a chemical as well as from a quality *vs* health point of view; the process has presented a dilemma. Although seemingly academic, the lessons learned from its resolution may provide guidelines for alternative curing processes. How can the pigment in cured ham in the presence of excess of nitrite salt, a powerful oxidizing agent—it is used for quantitative preparation of oxidized MMb (*cf* Mauk *et al.* 1973)—be the *reduced* nitrosyl myoglobin? The investigations of Walters and coworkers over the last 10 years or so (Walters *et al.* 1975) have resolved much of this dilemma and have greatly clarified the nature of the enzymes involved. Based primarily on observation of enzymes of pig muscle mitochondria, they established that in the presence of nitrite, CN^- and CO, the driving force is the production by the mitochondria of reducing power *via* NADH, to reduce MMb·NO to Mb·NO. The two key enzymes of the mitochondrial respiratory chain involved are probably the Cu enzyme cytochrome oxidase which normally reduces O_2 to H_2O as the terminal step in the respiratory chain of mitochondria:

$$4 \text{ Cytochrome } c\ (Fe^{2+}) + O_2 + 4H^+ \longrightarrow 4 \text{ Cytochrome } c\ (Fe^{3+}) + 2H_2O$$

and cytochrome *c* reductase (NADH dehydrogenase) with support from $NADP^+$ transhydrogenase to provide an ancillary supply of NADH:

$$NADPH + NAD^+ \rightleftharpoons NADP^+ + NADH$$

$$NADH + \text{Cytochrome } c\ (Fe^{3+}) \longrightarrow NAD^+ + \text{Cytochrome } c\ (Fe^{2+})$$

According to this scheme, nitrite nonenzymatically oxidizes all the Mb to MMb.

$$MbO_2(Fe^{2+}) \longrightarrow Mb(Fe^{2+}) + O_2$$

$$HNO_2 + Mb(Fe^{2+}) \longrightarrow MMb(Fe^{3+}) + NO + OH^-$$

When nitrite arrives at the mitochondria, it acts as a substrate substitute for oxygen in the cytochrome *c* oxidase system:

$$\text{Cytochrome } c\ (Fe^{2+}) + HNO_2 + H^+ \xrightarrow{\text{Cytochrome Oxidase}} \text{Cytochrome } (Fe^{3+})\ NO + H_2O$$

The rate limiting factor appears to be the transfer of the nitric oxide (NO) from the nitrosyl ferricytochrome c (Fe^{3+}) to MMb in a coupled cooxidation of $NADH^+$:

$$\text{Cytochrome } c\ (Fe^{3+})\ NO + MMb(Fe^{3+}) + NADH \longrightarrow \text{Cytochrome } c\ (Fe^{2+}) + MMb{\cdot}NO + NAD^+$$

More recently Cheah (1976) presented data which supports the involvement of lactate dehydrogenase (presumably from mitochondria) in the formation of MbNO from MMb in bacon as shown in Fig. 18.3.

Much of this information accumulating on the enzyme systems in cured meats may be helpful in the ongoing efforts to find substitutes for nitrites or to find treatments and supplements which will retain the color the consumer associates with cured meats and prevent the formation of nitrosamines. Upon cooking, the globin is denatured and presumably the liberated heme, nitrosyl hemochrome, is formed. It is this pigment which is responsible for pink-red color of cured cooked meat. According to Tarlagdis (1962) the pigment may contain two molecules of NO. For a recent review, see Cassens *et al.* (1979).

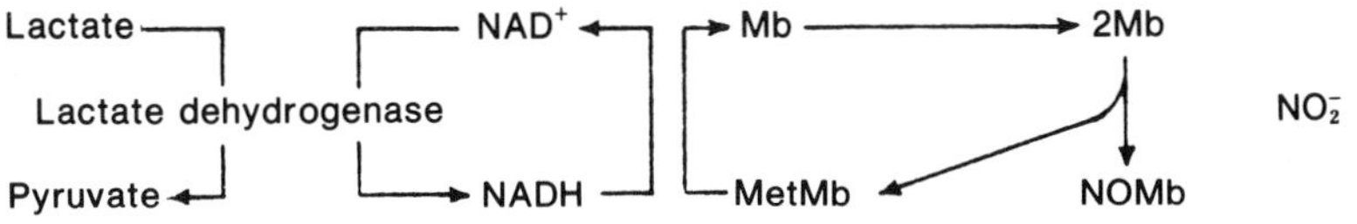

From Cheah (1976)

FIG. 18.3. ENZYME INVOLVEMENT IN FORMATION OF RED PIGMENT ASSOCIATED WITH NITRITE

Off-color metmyoglobin is converted to the bright red nitrosomyoglobin.

Nonbrown Discolorations

Green pigment occasionally forms in cured meats and more rarely in cooked uncured meat and seldom in raw sterile uncooked meat. For the most part, most of the green color appears to arise as the result of nonenzymatic reactions unless the meat has been contaminated by microorganisms such as might happen in the interior of fermented sausages (Fox 1966). In the latter case the heme of the pigment produces a nitrited porphyrin ring in the presence of excess nitrite, nitrimetmyoglobin. Fox also points out that some reductants such as ascorbic acid can produce H_2O_2 which can cleave the porphyrin ring to yield highly colored bile pigments. It is conceivable, but not yet demonstrated, that some of the oxygen-activating enzymes in meat can, under the appropriate conditions, induce such changes. In the presence of reducing agents such as cysteine or other sulfhydryl compounds, a green pigment, cholemyoglobin, in which the porphyrin ring is oxidized and sulfur is added to the heme ring, is formed.

Cysteine plus a mild oxidizer is involved in the generation of the green color which may appear in cooked tuna. However, in this case the oxidant is, according to Grosjean *et al.* (1969), trimethylamine oxide; and the sulfur is attached to the denatured protein moiety of tuna myoglobin through S-S bonds. The latter globin, in contrast to mammalian myoglobin, contains cysteine:

$$\text{MMb (native)} \xrightarrow{70^\circ\text{C}} \text{MMb·SH (denatured, I)}$$

$$\text{I} + \text{RSH} + [\text{O}] \xrightarrow{Fe^{3+}} \text{MMb·SSR} + H_2O$$

Actually there is very little, if any, direct involvement of enzyme action except for the enzyme action whch leads to formation of trimethylamine oxide, the source of [O] in the above equation and apparently the limiting factor in the reaction (Yamagata *et al.* 1971).

Enzymatic transformation of trimethylamine oxide in fish may also create problems in flavor (Chapter 25) and texture (Chapter 28) as well as in color quality. Because trimethylamine oxide is the limiting factor, Yamagata *et al.* have used the level of this oxide as a predictor of the green color of the cooked fish. The green color is abolished by sodium sulfite but not by other reductants.

Processing of crab meat can also give rise to several color problems, some of which may be enzymatic components (Boon 1975). The role of a phenolase-like action perhaps inherent in the Cu-containing blood oxygen carrier, hemocyanin, is under active consideration as contributing to the "bluing" of frozen crab meat. This point is still controversial as are the proposed means of prevention, such as treatment with acid, bisulfite, reducing agents or chelators, followed by rinsing. Occasionally frozen raw shrimp turn green. The greening appears to be associated with airborne iron particles (Thompson and Farragut 1969). The circumstances leading to exacerbation of these problems, such as elimination of the possibility of microbial contamination, high pH and alternate freezing and thawing, suggest involvement of the action of endogenous oxidoreductases.

Fish Carotenoids.—Although there is little if any evidence that lipoxygenase or lipoxygenase-like action influences the desirable deep orange color of salmon flesh and roe, a "discoloration" of skins of red fish, including some foods (rockfish, skipjack and a variety of red snapper), has been attributed to a lipoxygenase-like action in the skin of such fish on the carotenoid pigments responsible for the color. Tsukuda and coworkers have published extensively and practically exclusively on this subject (Tsukuda 1970). While it is not clear from available abstracts if this "discoloration" is a bleaching or a darkening, Tsukuda and Amano (1972) treated prawns with $NaHSO_3$ in brine to prevent "blackening." Since the carotenoid pigments (astaxanthin, violaxanthin, tunaxanthin) are present as esters of fatty acids, lipase (as in plants) may also participate in this discoloration (Tsukuda and Kitahara 1974). The same laboratory has investigated the intensification of salmon roe color with nitrite.

19

ENZYMATIC ASPECTS OF NONENZYMATIC BROWNING

NONENZYMATIC BROWNING

Relation to Enzyme Action

It might appear to be somewhat paradoxical to include the topic of nonenzymatic browning (NEB) in a book on enzymes. Nevertheless enzymes are involved in the formation of food color resulting from NEB in several ways.

First, endogenous *in vivo* enzyme action in the live food can produce effective levels of participants in NEB. The question we explore forthwith is how the enzyme action is or might be practically controlled so that NEB does not occur during subsequent processing and storage in, mainly, processed potato products and in dehydrated eggs.

Second, the action of both *in situ* endogenous and added enzymes may produce such NEB precursors. For instance both endogenous and added amylase action can improve bread crust color by providing the requisite sugars needed for NEB (Johnson and Miller 1961; Benedickt 1972). Especially useful are heat-stable bacterial α-amylase preparations which not only improve crust color but, by virtue of their partial survival during baking, prevent staling of baked goods. Their continuing action offsets delays or reduces factors responsible for or related to staling: starch retrogradation, increasing crumb firmness, moisture loss, increased tendency to crumble, hard slicing and decreased shelf life (D'Appolonia 1971).

Third, enzyme preparations containing glucose oxidase and catalase are added to foods to remove glucose, one of the participants in NEB reactions.

Fourth, steps then to prevent enzyme action, i.e., blanching, can influence subsequent NEB (Engl 1967).

A somewhat more philosophical reason for including NEB here is that it can be looked upon as a protomorphic paradigm of enzyme action. The classical Maillard NEB, it will be recalled, involves the reaction of an NH_2 group, traditionally in an amino acid, with the aldehyde (carbonyl) group of a reducing sugar. The amine may be considered a primitive biological catalyst and the carbonyl as the substrate. During the ensuing reaction, at some point amine is liberated and polymerization occurs. Of course Maillard-type NEB differs from enzyme action in that the former is favored by ever-increasing alkalinity and temperature (eventually true caramelization supersedes NEB).

As pointed out in Chapter 1 and copiously illustrated in the ensuing discussion many reactions considered to be enzymatic in nature are in reality secondary ones following primary enzymatic events. Thus, what we call enzymatic browning (Chapter 15) may also be considered as "nonenzymatic browning" in which one of the participants, a quinone, is the primary product of prior enzyme reaction. Conversely, one of the partners of NEB in many foods may be considered as an intermediate in a type of "enzymatic" browning in which invertase is the primary causative enzyme. This is roughly what happens in potatoes. This type of enzyme action differs from the traditional enzymatic browning in that the specific reactants in NEB in potatoes, the invertase-generated glucose and fructose, are far less reactive than are the phenolase-generated quinones. Only when heated at low a_w do the invertase-produced primary products become unstable and yield an abundance of NEB products which confer flavor and antioxidant as well as color properties on the food. Another difference is that invertase action takes place in an uninjured although stressed living cell, whereas phenolase action occurs as a rule after death and certainly only after some cellular disruption has occurred.

Types of Nonenzymatic Browning

Both current and accumulated evidence suggests five types of NEB occurring in foods.

Caramelization.—This involves pigment formation from carbohydrates without participation of amino acids, proteins or other NH_2-containing food constituents. The term is frequently used interchangeably with NEB. At times it is used in the broader sense as any heat-induced browning. It is very important as a desirable reaction in the confectionery industry (Sugisawa and Edo 1964; Heyns and Klier 1968; Orsi 1972; Greenshields and MacGillivray 1972; Arnestard and Bordalen 1972).

Maillard NEB.—Maillard (1912) first described in detail the now classical type of NEB involving reactions between reducing sugars (or other active carbonyl carriers) and amino acids and proteins (or other NH_2-containing compounds). The classic review in this field is that of Hodge (1953). More recent discussions (Hodge 1967; Hodge *et al.* 1972) stress the flavor-associated aspects of the Maillard reaction. Other discussions and reviews include those of Ellis (1959), Braverman (1963), Hurst (1972), Greenshields and MacGillivray (1972), McWeeny *et al.* (1974) and Berk (1976). Papers which bring out more specific aspects of NEB are those of Stadtman (1948) on NEB of fruits, Mohammed *et al.* (1949), Chichester *et al.* (1952) and Schwimmer and Olcott (1953).

Quinone-induced Browning.—This may be considered to be a special case of NEB. In addition to the interactions previously mentioned, a model system emphasizing the Maillard-like interactions between quinone and amino acids has been investigated (Hatanaka and Omura 1973).

Ascorbic Acid Browning.—This occurs in some processed orange juice without necessarily involving amino groups (Joslyn and Marsh 1935; Joslyn 1957B; Braverman 1963). Ascorbate does, of course, react readily with amino acids to yield brown pigment (Yu *et al.* 1974). The analogy between quinone and ascorbate as browning agents is quite striking. In each case an enzyme (phenolase or ascorbate oxidase) or a nonenzymatic catalyst transforms a reduced, relatively inert precursor (phenol/ascorbate) to a highly active primary product (quinone/2,3-diketogulonate) which can then either form pigment without the intervention of amines (melanins, caramels) or can react with amines to eventually yield N-containing melanins or melanoidins, the brown polymers of NEB.

Lipids.—Those containing unsaturated fatty acids can react with proteins to form brown pigments. The reaction has been considered to involve a typical Maillard reaction between carbonyl-containing breakdown products of lipohydroperoxides such as malonaldehyde ($COOH \cdot CH_2 \cdot CHO$) and protein (Kwon *et al.* 1965; Buttkus 1967). More recently Pokorny *et al.* (1974) provided evidence that the hydroperoxides themselves can and do react directly with protein (Chapter 34). Again we see the involvement of enzyme action—in this case that of lipase followed by lipoxygenase to form lipohydroperoxides in providing reactants for NEB. Also, as we shall see in our discussion of fish texture (Chapter 20) the action of lipase and phospholipases leads to an interaction between fatty acid and protein which may affect the textural qualities of the cooked fish. Although historically the Maillard reaction has been associated with color changes, increasing attention as mentioned is being directed to its effect on other parameters of food quality: formation of flavor compounds (Hodge 1967; Hodge *et al.* 1972; Adrian 1973), formation of antioxidants (Hatanaka and Omura 1973; Yamaguchi and Fujimaki 1973), deterioration of nutritional quality (Lea *et al.* 1960; Buttkus 1967; Adrian 1973; Lee *et al.* 1974; Tanaka *et al.* 1974; Schwimmer 1975), and to toxic effects (Adrian 1973), especially to what is at the time of this writing a controversy concerning the formation of nitrosamines (potentially carcinogenic). Apparently nitrosamines do not appear in absence of nitrite (Kawabata and Shazuki 1972) but can be detected if NEB is carried out in the presence of nitrite (Heyns *et al.* 1974; Samejima *et al.* 1972). Finally we wish to point out that the NEB products may themselves be effective activators or inhibitors of enzymes. Although most NEB-producing processing conditions would preclude survival of enzymes in such processed foods, there are some foods in which they can coexist; of course, added enzymes can also be affected.

ENZYME ACTION AND COLOR OF PROCESSED POTATO PRODUCTS

In a favorable environment, mature potato tubers do not contain more than traces of sugars. In response to such stressful situations as low or high temperatures, ionizing radiation and senescence, some of the starch is converted to sugar or perhaps, more precisely, sugar accumulates in the course of starch catabolism (Schwimmer *et al.* 1954; Schwimmer and Burr 1967; Schwimmer 1978; Isherwood 1973, 1976; Isherwood and Burton 1975). Total sugar (glucose + fructose + sucrose) may during storage in the vicinity of 0°C account for as much as 10% of the dry weight and hence one-third to one-half the nonstarch solids. Innumerable studies have confirmed this phenomenon for almost all varieties. Several other starch containing foods, notably bananas and peas, exhibit similar temperature-dependent starch-to-sugar conversions. This is just one particular aspect of the sugar-starch interconversions which in turn constitute a specialized manifestation of the regulation of carbohydrate metabolism (Chapter 20).

The significance for food quality is that it presents a Scylla/Charybdis-like dilemma for potato storage and processing. If it did not occur it would be possible to keep potatoes in a relatively fresh state suitable for processing and distribution for extended periods of time at low temperature (<10°C) and thus delay or prevent deteriorative changes such as excessive weight and starch loss, microbial attack and sprouting. However, the presence of excess reducing sugar in cold-stored potatoes results in NEB during processing. It is therefore necessary to subject cold-stored potatoes to "conditioning" which consists of storage for a few weeks at ambient temperatures above 15°C. Conditioning results in sugar loss apparently by reconversion to starch. Furthermore, of the components of this accumulated sugar, only the reducing sugar content correlates strongly and consistently ($r = 0.95 - 0.99$) with the color intensity of various processed products made from those potatoes (Schwimmer *et al.* 1957B). The latter citation is but one from a sea of literature demonstrating a probable causal as well as a statistically significant relationship between reducing sugar content of potatoes and the intensity of NEB in high temperature-processed potato products. Figure 19.1 depicts the interrelations among cold storage, conditioning, reducing sugar content and potato chip color.

Schwimmer *et al.* (1957B) suggested that phosphate ion (Pi) (Schwimmer and Olcott 1953) influences browning and that this may account for some scatter in their observed correlation. The Pi content of potato increases during cold storage (Samotus and Schwimmer 1962B). Hart and Smith (1966) found that ^{32}P added to the nutrient during the growth of the potato plant was more or less concentrated in the browner areas of individual potato chips. While color is probably the most important quality-related consequence of the starch to sugar conversion, subsidiary undesirable effects on texture and flavor also result, especially in cooked products.

Enzymes of the Starch–Sugar Interconversion Cycle in Potatoes

On the basis of known enzyme reactions in potatoes the sequence of reactions shown in Fig. 19.2 to account for the interconversion of starch (G_n) to sucrose and the reducing sugars fructose (F) and glucose (G) (Schwimmer 1953; Samotus and Schwimmer 1963; Schwimmer and Burr 1967; Schwimmer 1972; Isherwood 1973). The latter investigator found that the sum of changes in starch content, sugar content and respired CO_2 was invariant with respect to temperatures of storage. However, he calculated that one equivalent of ATP is used in going from starch to sugars whereas two equivalents, as shown in Fig. 19.2, are used in reconversion of sucrose to starch if one assumes that sucrose is first hydrolyzed to free fructose and glucose.

Reducing Sugar Accumulation; Invertase and Invertase Inhibitor.—While an unequivocal understanding of the nature of the trigger which sets off sugar accumulation at low temperature and starch

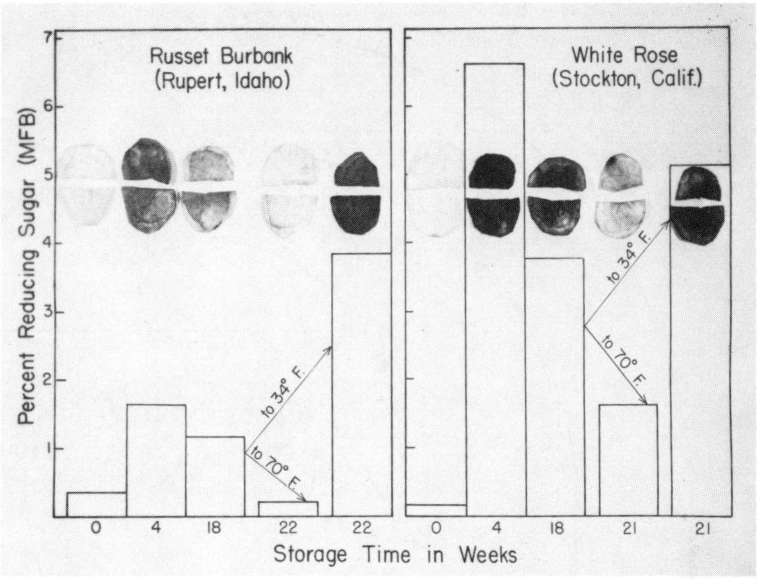

FIG. 19.1. COLD STORED POTATOES PRODUCE REDUCING SUGARS WHICH PARTICIPATE IN PRODUCING OFF-COLOR CHIPS

From Schwimmer (1953B)

synthesis at conditioning temperature still eludes us, control of the action of the enzymatic formation of the reducing sugar has been fairly well clarified. In Fig. 19.2 this reaction is denoted as:

$$\text{Sucrose} \rightarrow \text{F, G}$$

and the enzyme responsible for the hydrolysis is invertase, β-D-fructofuranoside fructohydrolase. Early conflicting reports on the presence and/or feeble invertase activity in potato tubers was shown by kinetic analysis (Chapters 13 and 37) to be in part due to the presence of an inhibitor in potato extract (Schwimmer *et al.* 1961) and also to a peculiar pH activity profile which pointed to the existence of two pH optima (Chapter 14) due to one enzyme (Rorem and Schwimmer 1963) and that the invertase activity was more easily demonstrable in extracts of cold-stored potato than in those of potatoes stored at room temperature (Schwimmer and Rorem 1960; Hoover and Xander 1963; Tishel and Mazelis 1966). The data suggested that the two phenomena were related. Theoretical analysis revealed that the invertase inhibitor is one of the ionized forms of a high molecular weight ampholyte whose pK values lie within the range of activity of the inhibited enzyme (Schwimmer 1962). The maximum inhibitor level coincided with the invertase's pH optimum. This analysis predicted that the uninhibited enzymes would act optimally at about pH 4.5–5.0 (see Fig. 19.3). Pressey (1966) separated the invertase from the inhibitor; the latter proved to be a protein which inhibited noncompetitively. The effect of adding the inhibitor back to inhibitor-free invertase is shown in Fig. 19.4. The pH maxima and minima coincided with those found in crude extracts containing both inhibitor and invertase (Rorem and Schwimmer 1963) and those predicted from Schwimmer's theory. During development of the tuber no net or basal activity can be detected because immature tubers contain excess inhibitor. Sucrose translocated from the leaves to the potato tuber during growth is thus promptly converted to starch. When mature tubers are placed at low temperature, there appears to be a simultaneous disappearance of inhibitor and a net *de novo* synthesis of invertase, so that sucrose is converted into reducing sugars (Pressey and Shaw 1966; Pressey 1972). The invertase-inhibitor complex forms rather slowly *in vitro* but is essentially undissociable (Ewing and McAdoo 1971).

Tishel and Mazelis (1966) suggested that the accumulation of sucrose may induce invertase activity via derepression of relevant genes. The molecular biological basis of enzyme induction which has yielded insights into other areas of biology (Chapter 4) has not been explored with respect to the cold sweetening phenomenon.

Regulation of Enzymes in Starch-Sucrose Conversion.—*Phosphorylase.*—In contrast to the elucidation of the origin of the reducing sugars, the

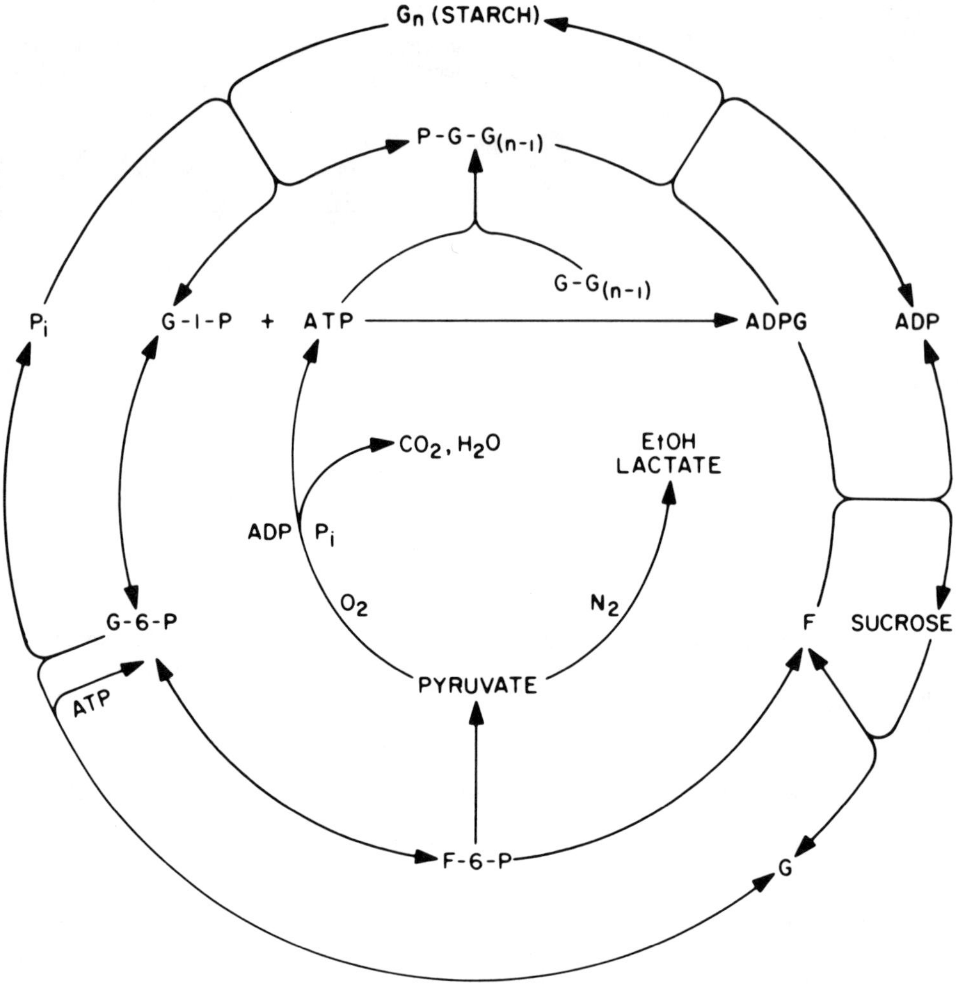

From Schwimmer (1972)

FIG. 19.2. THE STARCH–SUGAR INTERCONVERSION CYCLE IN POTATOES

nature of the events which lead to net synthesis of sucrose from starch is still under investigation, although all the enzymes probably involved have been fairly well characterized. Theories have been forwarded which invoke either temperature-dependent effects on key regulatory enzymes or on intracellular membrane sensitivity.

One of the most widely investigated enzymes of potato is the glycolytic enzyme (Chapter 4), α-1,4-glucan phosphorylase, usually referred to simply as "phosphorylase."

$$G_n\ (\alpha\text{-1,4-Glucan, Amylose}) + Pi \rightleftarrows \text{G-1-P} + G_{n-1}$$

where G-1-P is glucose-1-phosphate. The history with respect to its role in controlling the starch-sugar conversion is rather interesting. In the 1940s when it was believed to be involved in the synthesis of starch in plants and glycogen in animals, theories were tendered on the development of phosphorylase inhibitors in the cold. Schwimmer and Weston (1956), Schwimmer (1958) and Nakamura (1960) showed that apparent inhibition by alcoholic extracts from cold-stored tubers of the starch-synthesizing activity of partially purified phosphorylase was due to a complex set of artifacts. One of these was the increased inorganic phosphate content of cold-stored tubers. This induced shortening of the amylose synthesized without affecting the activity of the enzyme. Separation of components of this extract did reveal the presence of one true inhibitor precursor, chlorogenic acid. This was converted to quinone inhibitors of phosphorylase by the phenolases present (Blank and Sondheimer 1969). However, the level of inhibitor is not dependent upon storage temperature of the potatoes (Schwimmer 1958; Henderson 1968).

Having established that the role of phosphorylase in animals is to initiate the degradation rather than the synthesis of α-1,4-glucans, most investigators have believed that it plays a similar role in the cold sweetening process (Ioannou *et al.* 1974; Chism *et al.* 1975; Kennedy and Isherwood 1975). However, the results of diverse investigations do not support a role for this enzyme in triggering cold sweetening (Schwimmer *et al.* 1958; Kennedy and Isherwood 1975; Ioannu *et al.* 1974).

The principal isozyme of potato phosphorylase has been isolated as a sulfhydryl and PLP-containing 216,000-dalton dimeric protein (Lee 1966; Kamogawa *et al.* 1968; Iwata and Fukui 1973; Gerbrandy 1974). It can be cleaved in two by endogenous protease without much loss of activity (Ariki and Fukui 1975; Gerbrandy *et al.* 1975). This protease action, in part responsible for the isozyme pattern found in extracts of frozen and thawed potatoes, does not take place in freshly harvested tubers, perhaps because of the high levels of protease inhibitor present.

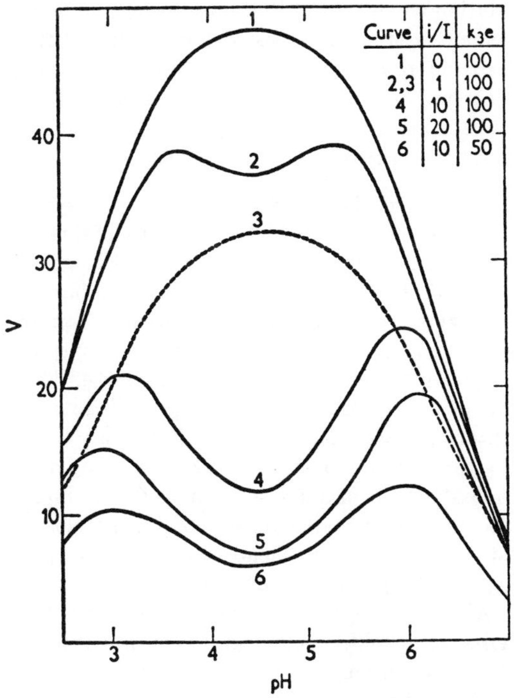

From Schwimmer (1962)

FIG. 19.3. ENDOGENOUS INHIBITION BY AMPHOLYTE CREATES DOUBLE pH-OPTIMA

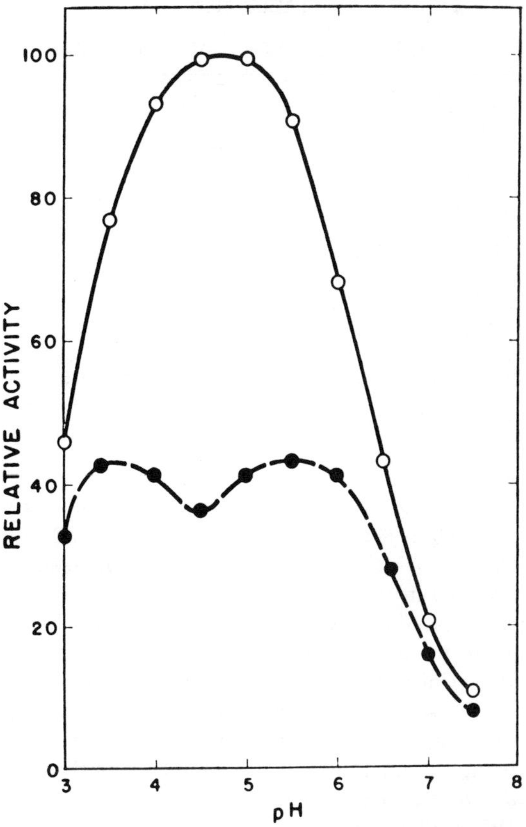

From Pressey (1966)

FIG. 19.4. EFFECT OF ENDOGENOUS INHIBITOR ON pH-ACTIVITY PROFILE OF POTATO INVERTASE

Like most phosphorylases it requires a trace of a starch-like primer to initiate starch synthesis (Kamogawa *et al.* 1968). Based primarily upon primer specificity and secondarily on differences in kinetic behavior of two "slow" and "fast" isozymes (Table 19.1), Shivaram (1976), a colleague of Gerbrandy, suggested that polyglucan (starch) synthesis and degradation may be regulated by the change in proportion of slow and fast enzymes. One would expect their to be more fast enzyme in low temperature-stored potatoes. Multiple forms of phosphorylase have been isolated from other temperature-dependent starch to sugar foods including the previously mentioned banana (Singh and Sanwahl 1976) and peas (Matheson and Richardson 1976).

TABLE 19.1

POTATO PHOSPHORYLASE ISOZYMES

Nomenclature	Fast	Slow
Source	Fresh tubers	Sprouted tubers
pH optima (dependent upon primer)	5.2–6.0	5.2–6.0
Temperature optima (°C)	30°	35°
Isoelectric point (pH)	5.0	5.5
Molecular weights (daltons)	209,000	165,000
Subunits/molecule	2	4
K_m, (glucose-1-phosphate), mM	6.7	8.0
K_m, (amylose)	3.1	1.1
More active in starch	Synthesis	Phosphorolysis

Source: Shivaram (1976).

Unlike mammalian phosphorylase the potato and other plant enzymes are not activated by AMP or by cyclic AMP, the ubiquitous "second messenger" involved in transmitting hormonal effects by specific cascade activation of key enzymes (Chapter 4). In the connection it is of interest to note that while Chism *et al.* (1975) added theophylline, routinely used to inhibit any cyclic AMP phosphodiesterase (PDE), Pi, which increases in cold-stored potatoes (Samotus and Schwimmer 1962B) is also a potent inhibitor of this PDE (Shimoyama *et al.* 1972). Pi is also the product of action of a 5′-nucleotidase which is, in turn, inhibited by cyclic AMP (Polya 1975). Pi, when present in excess as one of the substrates of the starch-degrading action of phosphorylase, limits the size of the starch (amylose) formed when the reaction is run in the direction of synthesis but does not inhibit the enzyme.

Starch Synth(et)ase.—Whereas phosphorylase has until quite recently been considered to be the agent of starch degradation, the synthesis of the amylose moiety of starch has been assigned to this enzyme which is closely associated with the starch granule. In Fig. 19.2 the reaction is depicted with adenosine diphosphate glucose (ADPG) as the substrate of ADPG-starch glucosyl transferase:

$$\text{Pr-(G)}_n + \text{G}_m \xrightarrow[\text{G-1-P}]{\text{ADPG (or)}} \text{Pr-(G)}_{(n+m)}$$

Lavintman *et al.* (1974) and Manners (1974) have discussed the roles of the particulate transferases (transglucosylases) in starch metabolism. Enzymes utilizing ADPG and the uridine analog, UDPG, are stimulated by KCl, the significance of which for regulating starch synthesis was pointed out by Davies *et al.* (1974). Although it should be termed starch "synthase" rather than "synthetase," usage has overridden EC rules (Chapter 2). In contrast to most other enzymes, the activity of the starch synthase adhering to potato starch granules is slightly higher at 0°–4°C than at 45°C (Frydman and Cardini 1967). Perhaps one of the penetrating advances in this complex intractable problem of exactly how starch is synthesized is the paper by Lavintman *et al.* (1974) which is also a key to previous literature. They isolated from potatoes a protein, Pr, which transfers glucose from UDPG to form a glucoprotein, the "glucoproteic acceptor," Pr(G)_n, in the above reaction, a short chain of α-1,4-glucosyl residues. The subsequent elongation of this chain can be catalyzed by a G-1-P glucosyl transferase which appears to be distinct from the traditional primer-requiring amylose phosphorylase. Indeed, it may be a "true" starch synthase. Similar "primerless" activities were reported by Slabnik and Frydman (1970) in potatoes and in rabbit muscle by Tandecarz *et al.* (1977).

Classical phosphorylases can catalyze starch synthesis in plant plastids during oxidative phosphorylation when the ATP/Pi ratio is high. Low ratios which exist in potatoes at low temperatures are associated with starch degradation (Samotus and Schwimmer 1962B,C; DeFekete 1968). We have thus almost come full circle and again assign a putative synthetic role to phosphorylase.

Q-Enzyme.—No discussion of starch synthesis would be adequate without at least alluding to the formation of the larger portion of most starches, amylopectin. It is synthesized via the action of the Q-enzyme, a term which in its pristine simplicity is in vivid contrast to its systematic name, 1,4-α-D-glucan 6-α-(1,4-α-glucano)-transferase. The enzyme can form amylopectin via both intra- and inter-amylose chain transfer. The latter seems to be the *in vivo* mode of synthesis involving perhaps a double helix of amylose as illustrated in Fig. 19.5. R denotes the reducing end of the amylose substrate and the arrowhead denotes the α-1,6-linkage at the point of branching.

Although no role has been attributed to this enzyme in cold sweetening, it is of interest to note that the minimum degree of polymerization required for an amylose chain to be a Q-enzyme substrate decreases with decreasing temperature. The resulting hyperbranched structures might then serve as efficient primers for phosphorylase.

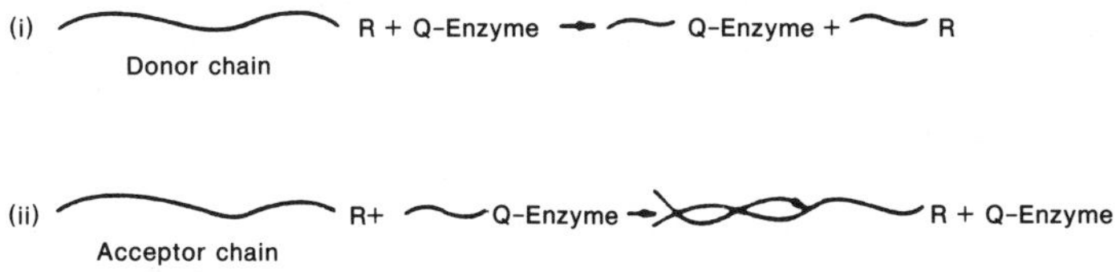

From Borovsky et al. (1976)

FIG. 19.5. Q-ENZYME ACTION AT INITIAL STAGE OF AMYLOPECTIN SYNTHESIS

Sucrose Synth(et)ases.—It will be noted (Fig. 19.2) that the enzymes leading to synthesis of both starch and sucrose share in common the nucleoside diphosphate glucoses. Potatoes possess both well-defined sucrose synthesizing enzymes UDPG:D-fructose-2-glucosyl transferase or sucrose synthetase,

$$\text{U-R-P-P-G (UDPG)} + \text{F} \rightleftharpoons \text{U-R-P-P (UDP)} + \text{G-F (sucrose)}$$

and sucrose phosphate synthetase

$$\text{UDPG} + \text{F-6-P} \longrightarrow \text{UDP} + \text{sucrose 6-phosphate}$$

No significantly great differences have been observed between activities of the sucrose synthase in cold and in warm-stored tubers (Schwimmer and Rorem 1960; Pressey 1970; Sowokinos 1973; Murata 1974). Its main function in most food plants is to cleave sucrose translocated from the leaves to the tuber or analogous organs during the plant development in order to provide ADPG in preparation for incorporation of glucose into starch. The reaction is readily reversible. Indeed, the dramatic drop in the activity of this enzyme upon maturity, depicted in Fig. 19.6, makes it useful as an indicator of maturity (Chapter 37).

In contrast, sucrose phosphate synthetase activity does respond to low temperature storage by increasing (Pressey 1970; Pollock and Ap Reese 1975A). The activity per unit weight of fresh tissue is much higher than that of sucrose synthetase at all temperatures, and the increase upon cold storage matches the corresponding cold-induced sugar accumulation.

Also, in contrast to sucrose synthetase, the reaction catalyzed by sucrose phosphate synthetase is practically irreversible, so that it is quite likely that this enzyme is ultimately responsible for sucrose formation in cold-stored potatoes. By sequestering the nucleoside diphosphate glucose, it may also signal starch degradation.

Glycolytic Enzymes.—Another school of thought concerning the triggering of the cold-induced starch-to-sugar conversion holds that certain key enzymes of the glycolysis (Pollock and Ap Reese 1975A) or of the hexose/pentose phosphate shunt (Tishel and Mazelis 1966) act disproportionately slowly or actually inactivate during cold storage. Among the glycolytic enzymes in this regard are PFK, GAD and pyruvate kinase (Chapter 4). Since sucrose phosphate synthetase follows allosteric kinetics (Chapter 4), sucrose synthesis will not proceed appreciably until a threshold concentration of F-6-P is exceeded. The behavior of the above-mentioned glycolytic enzymes at low temperature favors such a buildup.

In both low and room temperature-stored potato halves, the specific activities of G-6-P dehydrogenase, the gateway to the above-mentioned pentose shunt, were observed to fall precipitously. The change in the activity and lability of this enzyme

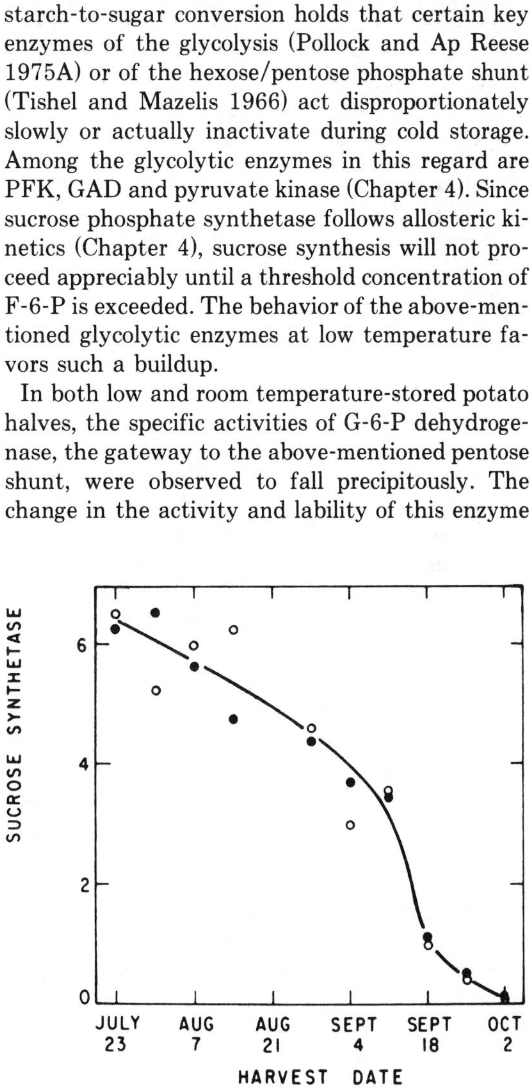

From Pressey (1970)

FIG. 19.6. DISAPPEARANCE OF SUCROSE SYNTHASE FROM MATURING POTATO TUBERS as well as that of glycolytic enzymes such as phosphoglucomutase and FPK may be due to increased attack by proteinases (Sasaki *et al.* 1973; Kahl and Gaul 1975; Nowak and Skiera 1975). It will be recalled that potato phosphorylase (and sweet potato phosphorylase) remain active after such proteolytic cleavage.

Another set of observations which implicates both glycolysis and sucrose synthesis is that of Sowokinos (1976) who found that F-6-P and other glycolytic intermediates activate the enzyme which splits (pyrophosphorylizes) the most favored substrate for starch synthesis, ADPG, but not that for sucrose synthesis, UDPG:

$$\text{Nucleoside-P-P-glucose} + \text{PP} \rightleftharpoons \text{Nucleoside-P-P-P} + \text{Glucose-1-P}$$

It is of interest to note that pyrophosphate (PP), a product of the reverse action, inhibits the action of starch synthetase when UDPG is a substrate but stimulates starch synthesis in the presence of ADPG.

Another way by which glycolysis could influence the NEB of potato products might be *via* participation of hexose phosphates which are glycolytic intermediates, especially G-6-P. The latter is a more intense NEB reagent than is glucose (Schwimmer and Olcott 1953). Incidentally an interaction between the NH_2 of valine at the terminal end of hemoglobin (and possibly other proteins) with excess G-6-P which accumulates in the blood of diabetics appears to be the cause of the traumatic sequelae of diabetes (Cerami and Koenig 1978). The resulting protein-sugar complex is formed via the first two steps in NEB: Schiff-base formation between the reducing aldehyde of the sugar and the terminal N of the hemoglobin, followed by an Amadori rearrangement. This results in a fructose derivative of the protein which may undergo further symptom-causing aggregation akin to that which occurs in NEB.

Cell Membrane Sensitivity and Enzyme Decompartmentalization.—With increased emphasis on the role of membranes on biochemical control and regulation systems of food tissue (Haard 1972; Romani 1972; Hultin 1972; Schwimmer 1969A,1972,1978) it is not surprising that these cellular envelopes have been considered as participants in cold sweetening of potatoes. Explanations of cold hardiness on the basis of the physical properties of membrane phospholipids have met with considerable success (Lyons *et al.* 1980; Sharpe and Goeschl 1975). Chill-sensitive plants show a discontinuity at about 12°C in the Arrhenius plot for their starch synthetases (Chapter 12) whereas the synthetases from potatoes do not (Downton and Hawker 1975). The absence of this break is attributed to the absence of lysolecithin present in the starch of chill-sensitive plants but not in that of the potato. Alternatively, the break may be due to the lower transition (melting) temperature of the more highly unsaturated phospholipid in potatoes which thus play a role in chill-resistance.

Paez and Hultin (1970) found that potato mitochondrial respiration displayed a response which was typical of that of other plant mitochondria. Their work suggests that injury to mitochondria cannot account directly for cold sweetening. On the other hand, limited action of added protease on potato mitochondria elicits latent ATPase activity (Jung and Laties 1975).

Another phenomenon which may merit some looking into is the stimulation of membrane-associated phospholipid exchange between mitochondrial and microsomal potato fractions by a protein isolated from potatoes (Kader 1975). Direct involvement of membrane transformation as the trigger which initiates cold sweetening has been proposed by Ohad *et al.* (1971) based upon electron micrographs similar to those displayed in Fig. 19.7 which show: (A) a section through the cell of an immature potato tuber. Within the cytoplasm is a protoplast (PP) surrounded by a double membrane (27,800×); (B) a section from a mature tuber immediately after harvest; a starch granule is seen surrounded by the intact plastids membrane, pm (64,000×); (C) a section from a tuber stored at 25°C for 31 days (13,900×); (D) a section through a tuber stored at 4.4°C for 12 days. The membrane has disintegrated and moved away from the starch granule, whereas the vacuolar and cytoplasmic membranes remain intact (25,200×).

Lowering of the temperature appears to cause disruption and/or modification of the double membrane surrounding the plastids in which the starch granule is embedded in freshly harvested and room-stored tubers. At low temperatures the freshly exposed starch now becomes susceptible to starch-degrading enzymes. In contrast to their further reports of alteration of some of the physical properties of membrane material isolated from cold-stored tubers, Jarvis *et al.* (1974) concluded that the chemical changes in glycolipids and phospholipids at 4°C were not sufficient to account for

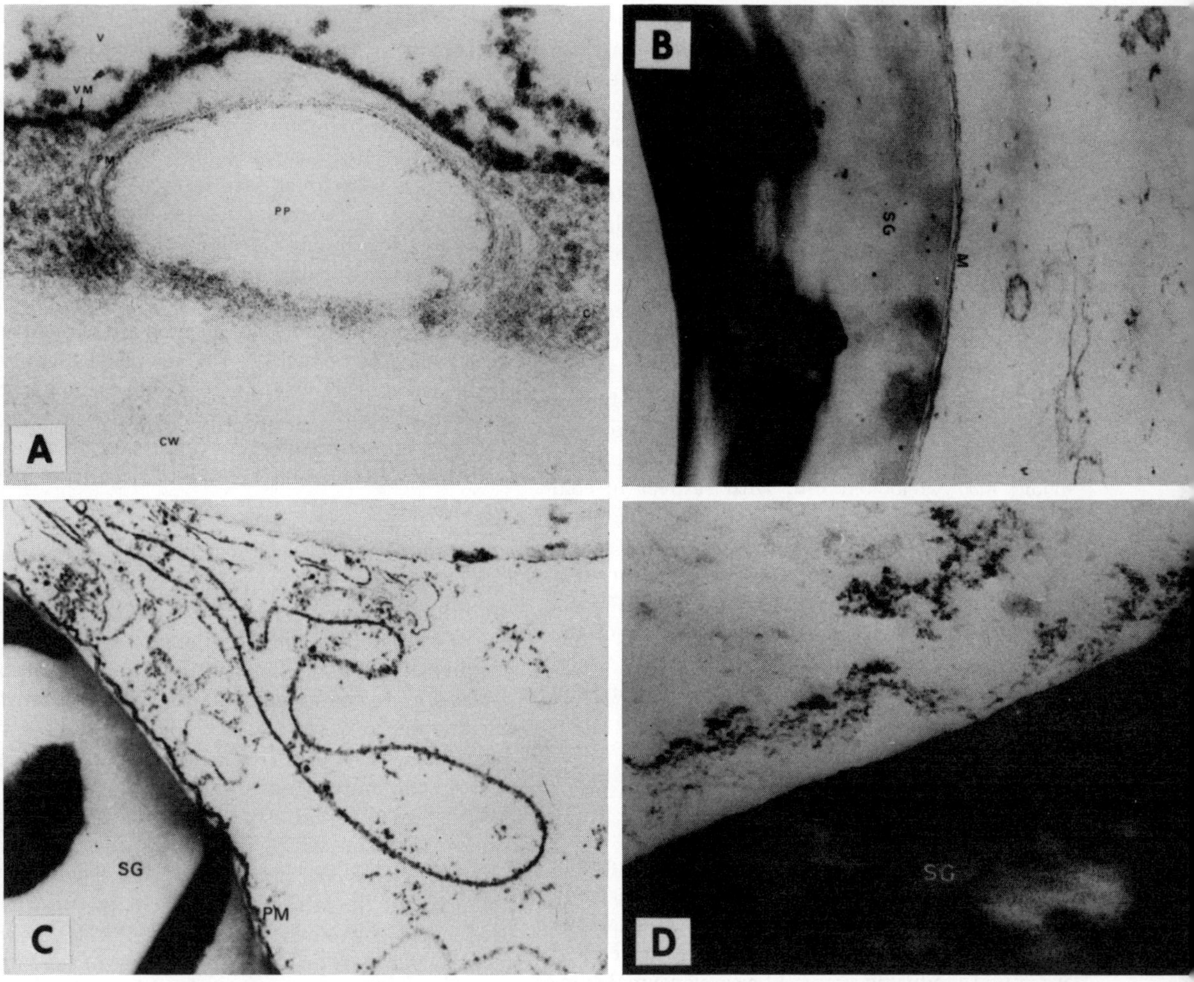

From Schwimmer (1972)

FIG. 19.7. DEVELOPMENT AND DISINTEGRATION OF AMYLOPLAST MEMBRANES

See text for detailed discussion. Based on work of Ohad and coworkers.

severe membrane degradation. Although Isherwood (1976) could not confirm the presence of disintegrated amyloplast membranes in cold-stored tubers, he did find them in association with sugar accumulation due to tuber senescence and further implicated these same membranes as the site of an electron-transfer mediated starch-sugar interconversion. Electron transport "poisons," not including dinitrophenol, cause immediate sweetening (Amir *et al.* 1977).

Amylase.—Finally we wish to point out that although amylase has rarely been implicated in cold sweetening, Dunn (1974) assembled data including the early work of the author (Balls and Schwimmer 1944; Schwimmer 1945; Schwimmer and Balls 1949A,B) which point to the uniqueness of this type of enzyme action in attacking native starch granules (Chapter 9), especially in sprouting tubers. The α-amylase of potatoes has only comparatively recently been isolated and characterized (Fan 1975). Its optimum temperature is 42°C and the reaction it catalyzes has a rather low energy of activation for an enzyme of 8000 kcal/mole. If this trend continues to sweetening temperatures, the enzyme may be active enough to participate in, if not trigger, the starch-sugar conversion.

A fundamental construct in Dunn's model of starch granule disintegration is that α-amylase (and presumably not phosphorylase) is the only degradative enzyme which can act directly. The kinetics of the granule digestion follow those that were developed by McLaren (1963), one of the pioneers in immobilized enzyme technology (Chapter 8), on the basis that adsorption of α-amylase on starch obeys the Freundlich adsorption isotherm (Schwimmer and Balls 1949B).

If indeed amylases do turn out to have some significance in cold sweetening, then the action of both isoamylase, the R-enzyme (Chapter 9), which hydrolyzes α-1,6-branchpoint linkages in amylopectin, and of α-glucosidases (both of whose roles in raw starch digestion were examined in Chapter 9) should also be looked into with respect to cold sweetening. Perhaps the demonstration of Pollock and Ap Reese (1975B) on the cold-induced sweetening of potato tissue cultures and the use of chemically debudded tubers (Amir *et al.* 1977) will provide new valuable tools for elucidation of the complex metabolism whose understanding would lead to more rational means for its prevention than are now at hand.

Prevention of Browning of Processed Potato Products

As previously pointed out, potatoes are stored in the cold to prevent sprouting, weight loss and general deterioration before being processed. However, during such cold storage they lose starch and accumulate reducing sugar. The conventional standard means of preventing the resulting NEB of cold-stored potatoes is to condition them—to store them at room temperature from 1 to 3 weeks after removing them from cold storage, depending upon variety and cultural history. Conditioning removes the reducing sugars which participate and control the extent and intensity of NEB. An extensive study of the influence of storage on potato quality, illustrative of research in this area and using 70 varieties, led to the suggestion that tubers intended for processing into dehydrated products should be stored at low temperatures (< 5°C) and high a_w (Chapter 13) with preremoval warming. Potatoes intended for baking should be stored aseptically at 7°−10°C (Pätzold 1974).

In general, other efforts to prevent sugar NEB in potatoes fall into four principal categories: leaching, horticultural practices, storage in low oxygen atmospheres and prevention of sprouting at room temperature by the use of chemical sprout inhibitors or by the use of ionizing radiation.

Leaching.—This is limited to products such as french fries where only the surface is exposed to NEB conditions. Leaching may be combined or the term is sometimes used interchangeably with blanching, as in the manufacture of frozen parfried potatoes (Brown and Morales 1970; Weaver *et al.* 1975). While leaching-blanching can be used to remove reducing sugar at the surfaces of potato slices, the introduction of this step entrains other beneficial effects such as the cessation of undesirable enzyme activity, a more uniform final product, reduction of fat absorption, reduced frying time and improved color and texture, but may also lead to loss of desirable flavor constituents (Chapter 21).

Horticulture.—Disposition towards accumulation of reducing sugar and the propensity to proper conditioning can be controlled more or less by maturity (Samotus and Schwimmer 1962A; Yamaguchi *et al.* 1960), by horticultural practice and by varietal and genetic manipulation. While confirming the influence of both variety and year of vegetation, Mica (1977) concluded that the latter exerted the more significant effect.

With regard to variety, it is well known, for instance, that the White Rose variety is an active whereas Russet Burbank is a modest sugar accumulator, but the latter can still yield overbrowned products (Schwimmer *et al.* 1954,1957B). More recently Jarvis *et al.* (1974) found that of ten varieties grown in Scotland, nine accumulated too much reducing sugar to be acceptable for processing immediately after cold storage. The remaining variety, Pentland Ivory, whose low sugar content (0.1% fresh weight) was not affected by cold storage, appeared to be suitable for chip (crisp) processing except for its extremely pale flesh. By genetic manipulation Lauer and Shaw (1970) grew potatoes which produced acceptable potato chips when processed directly from storage at 4.5°C. Geneticists and breeders might advantageously exploit some of the findings of the enzymology of cold sweetening by breeding for increased levels of invertase inhibitor, for instance.

Exclusion of Oxygen.—Figure 19.2 shows that potatoes in air lose starch and respire normally providing the ATP needed for sucrose synthesis is present. In nitrogen and in the absence of oxygen starch is also lost but does not appear, as in cold sweetening, as sugars. The pyruvate formed via glycolysis, instead of going through the TCA cycle (Chapter 4) and its attendant oxidative phosphorylation, yields lactate, ethanol and other reduced end-products of anaerobic metabolism (Samotus and Schwimmer 1963). Since the use of N_2 may be prohibitively expensive on a large scale, Samotus (1971) investigated the utility of storing tubers in water, where the rate of diffusion of O_2 is infinitesimally slower than in air. He found that the water-stored tubers did not accumulate sugars at 1°C and were acceptable as food after 8 weeks. Addition of safe aseptic agents such as thymol and H_2O_2 prevented microbial contamination.

Sprout Inhibitors.—In contrast to the last two approaches, it would not be necessary to store tubers in the cold if they could be prevented from sprouting at ambience. Considerable effort has gone into finding safe, inexpensive sprout-inhibiting chemicals. Their use and acceptance has been reviewed by Sawyer (1975). The four chemicals being used in 1975 in the United States were maleic hydrazide (MH), isopropyl-N-chlorophenyl carbamate (CIPC), tetrachloronitrobenzene (TCNB) and the methyl ester of naphthalene acetic acid (MENA). MH is applied as a spray to leaves of the the potato plant before harvest. The others are applied to tubers as dips or dusts distributed via the ventilating system of the storage facility. Problems (other than those which are strictly economic or engineering in nature) include inhibition of wound periderm formation, development of internal sprouts and a disputed increased susceptibility to infection. In England and Denmark the vapors of higher aliphatic alcohols such as octanol and nonanol have been widely adopted (Burton 1958; Sawyer 1975). Dipping in a solution of 0.05% H_2SO_4 for 24 hr is effective for local consumption of the tubers (Said *et al.* 1973).

Ionizing Radiation.—Interest in the use of ionizing radiation (Chapters 1 and 10) continues some 25 years after the initial observation of Sparrow and Christensen (1954) that γ-rays inhibit sprouting. Since then a considerable literature has developed including that of the author and associates (Schwimmer *et al.* 1957A,B,C; Schwimmer and Weston 1958; Schwimmer *et al.* 1958). The marked increase in sucrose at room temperatures observed by us in the sprout-inhibiting range of 8–15 kilorad has been repeatedly confirmed as exemplified by the publications of Jaarma (1958) and Burton (1975). Like most of the sprout-inhibiting chemicals, ionizing radiation abolishes wound periderm formation, may increase susceptibility to microbial attack and hastens "senescent" sweetening (Burton 1975). It also inhibits chlorophyll synthesis (Chapter 17) and results in an increase in the apparent activity of several enzymes related to the starch-sugar interconversion process, usually in synchrony with the rise in sucrose (Schwimmer *et al.* 1958; Kodenchery and Nair 1972). More recent observations are in agreement with our earlier suggestion that the breakdown of enzymes in potatoes is more sensitive to such radiation synthesis than is protein (enzyme) synthesis.

Relatively recent reports indicate that several countries, excluding the United States, have or are still considering potato irradiation: Yugoslavia, France and other EEC countries (Pomarola and Sandret 1973), Canada (Sawyer 1975) and Japan (Sato 1973). Sato, in Japanese, covers key aspects of the commercialization of the use of ionizing radiation such as the necessity for sprout inhibition, comparison with other methods, irradiation effects, safety of the irradiated tubers, economic and environmental considerations and irradiation apparatus. The use of ionizing radiation to prolong storage life of potatoes and commercial applications of the process are considered in a concise discussion (Anon. 1973A). Persistence of such literature suggests that it will finally be in wide use throughout the world by the time this appears. Indeed, a literature is developing on means of recognizing *a posteriori* potatoes which have been subject to irradiation (Magaudda 1973; Sandret *et al.* 1974). From these and other studies it would appear that at sprout-inhibiting dosages of ionizing radiation the replication mode of DNA is completely destroyed but not necessarily its transcriptional function or the translational capacity of messenger RNA (Chapter 4).

Another approach to prevention of NEB in potato products would be to employ extraneous enzymes. Schwimmer (1953C) found that a commercial preparation of glucose oxidase also contained considerable amylase, phosphorylase and G-6-Pase activities so that the action of these enzymes provided a steady supply of reducing sugars faster than the oxidase was removing them. Perhaps more recent preparations are more suitable or can be made so by application of appropriate affinity chromatography packets as discussed in Chapter 7. In any case, the remarkably versatile glucose oxidase as it exists today is indeed effective in removing glucose from eggs.

GLUCOSE OXIDASE AND EGG DESUGARING

Enzymology of Glucose Oxidase

Glucose oxidase, β-D-glucose:oxygen oxidoreductase (GOX) catalyzes the oxidation of β-D-glucose by oxygen to β-D-glucono-δ-lactone and hydrogen peroxide; β-D-glucose + O_2 → β-D-glucono-δ-lactone + H_2O_2. The lactone hydrolyzes spontaneously, but more rapidly via the action of D-glucono-δ-lactone hydrolase, present in commercial preparations: D-glucono-δ-lactone + H_2O → D-gluconic acid.

Although GOX is one of the few nonhydrolytic enzymes to be successfully exploited in the food industry, the practicability of its use may well hinge on an adequate level of lactonase activity.

Another enzyme usually present in commercial preparations which increases the efficiency of oxidation is mutarotase, an anomer isomerase which speeds up the spontaneous conversion of α-D-glucose to β-D-glucose. GOX acts almost 200 times faster on β-D-glucose than on its anomer, α-D-glucose. Bright (1974) states that activity with β-D-glucose is at least 100 times greater than with any other naturally occurring monose. However, the V_{max} for oxidation of deoxyfluoroglucose by GOX from *P. fluorescens* was about the same as that for glucose (Taylor *et al.* 1975). GOX activity was first obtained by Maximow (1904) as acetone powder of *Aspergillus niger* mycelia. It was first designated as an oxidative enzyme by Müller (1928). Evidence slowly accumulated that the enzyme, yellow in color, contains a prosthetic group, flavin adenine dinucleotide (FAD), and that the oxidation is mediated through FAD (Keilin and Hartree 1948). In the early 1940s it was temporarily confused with penicillin-like substances and was actually considered to be an antibiotic ("notatin," penicillin B). This was due to its ability to produce the germicidal H_2O_2. Reviews include those of Neims and Hellerman (1970), Whitaker (1972A) and Bright (1974).

The enzyme is usually studied and isolated from fungi, especially from *Aspergillus* (Haupt *et al.* 1974) and from *Penicillium* (Gorniak and Kaczkowski 1974). It has also been purified from bacteria such as *Pseudomonas* (Taylor *et al.* 1975). As a commercial food additive it is prepared from *A. niger* in the United States, from *P. amagaskinense* in Japan and from *P. vitale* in the USSR. Depending upon its intended use, commercial GOX preparations may, in addition to the enzymes mentioned, contain catalase. The catalase may be present in these preparations or may be added to destroy unwanted H_2O_2 so that the overall reaction is:

$$\text{D-Glucose} + \tfrac{1}{2}O_2 \rightarrow \text{D-Gluconic acid}$$

Both catalase and peroxidase (separately) are added to GOX for use as an analytical reagent (Chapter 38).

The enzyme from *A. niger* consists of two subunits of about 80,000–90,000 MW, each linked to one tightly associated noncovalently bound FAD (Swoboda and Massey 1965). Each subunit, at least in the enzyme from *A. vitale*, can be further subdivided into polypeptide chains each of about 33,000 daltons, by disruption of SS bonds (Bogdanov *et al.* 1974). There are two masked free cysteine-SH groups in the intact protein which were unmasked only after denaturation.

Glucose oxidases from different sources all contain a considerable amount of carbohydrate. The enzyme from *A. niger* contains about 16% carbohydrate (Wellner 1967). The purified, crystallized Soviet enzyme contains 8–10% of neutral carbohydrate (Bogdanov *et al.* 1974). Each enzyme molecule contains six oligosaccharides, each of which in turn comprises 18 monoses. Each oligosaccharide is bound to the peptide chain of the enzyme through a glucosyl-amide bond between β-acetyl-glucosamine and asparagine of the peptide chain.

In two independent investigations, it was found that removal of the carbohydrate from the enzyme by periodate oxidation renders it heat-labile, both in the dry state (Chapter 13) and in solution in the presence of Na dodecyl sulfate (Nakamura and Hyashi 1974). Slightly enhanced thermal stability of *A. niger* glucose oxidase was achieved by coupling the carbohydrate to water insoluble *p*-aminostyrene, resulting in immobilization of the enzyme (Zaborsky and Ogletree 1974).

The enzyme is relatively stable over a wide range of pH's and temperatures. Thermal stability is a particularly important property from a practical point of view. Scott (1975) was able to improve the heat stability by supplying the commercial enzyme in coated tablet form. In this form, both glucose oxidase and catalase survived pasteurizing temperature (87°C, 3–6 min) in buffers and in a wide variety of fruit and vegetable juices. Proposed mechanisms of action (elegantly evaluated by Bright 1974) involve, as for other flavin nucleotide-containing enzymes, alternate oxidation and reduction of the prosthetic group, FAD. Thus, the enzyme reaction consists of a turnover cycle for FAD comprising two half reactions:

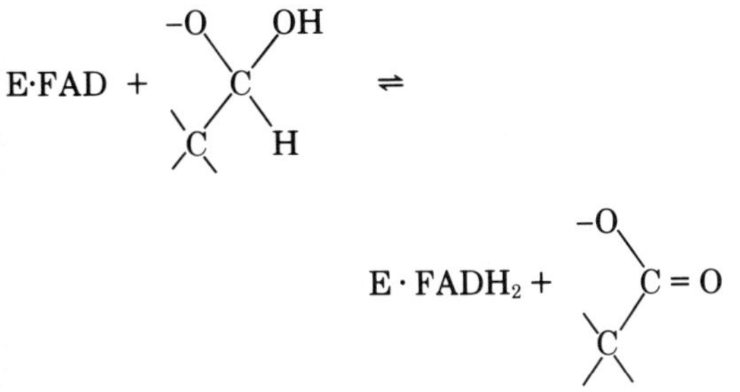

$$E \cdot FADH_2 + O_2 \rightarrow H_2O_2 + E \cdot FAD$$

Substrate-rate data obey "ping-pong" kinetics; that is, double reciprocal plots of 1/rate *vs* $1/[O_2]$ for various glucose concentrations yield a series of parallel straight lines, as did a plot of 1/rate *vs* 1/[glucose] for various O_2 concentrations (Chapter 2). This means that the enzyme and substrates interact in a manner such that only binary complexes are formed and that the reaction is ordered, i.e., first glucose combines and then O_2. Such kinetics cannot reveal whether the product (gluconolactone) is released from the enzyme. Bright, using stopped flow rapid reaction techniques which record transient state kinetics (Chapter 3) and others cited by Bright, considered the detailed chemical mechanism and concluded that free radical (homolytic) reactions are not involved, but that adduct formation with FAD proceeds through the C-1 of glucose. The hydrogen bound to this carbon is removed as a proton but the resulting carbanion, $(\text{-O-}\underset{|}{\overset{|}{\mathrm{C}}}\text{-OH}^-)$, is very short-lived. The half reaction involving reduction of FAD is probably a general base-assisted proton removal from C-1 of glucose preceding or synchronous with flavin reduction.

In the second half reaction of the cycle, the protein moiety, the "apoenzyme," is probably involved in the interaction of O_2 with FAD to form H_2O_2. The intermediate, $E \cdot H_2O_2$, is very short-lived if it exists at all. However, a free radical is formed since oxidized cytochrome *c*, reducible by O_2^- (Chapter 16), is reduced during the action of GOX. O_2^- probably is not involved, however, since this reduction is not inhibited by superoxide dismutase (Kovacs *et al.* 1975). Bright states that all nonmetal-containing flavoprotein oxidases produce H_2O_2 directly, without intermediation of O_2^- (superoxide anion).

In this connection it may be of interest to note that the Cu-containing galactose oxidase *is* inhibited by superoxide dismutase. The redox cycle consists of oscillation between Cu (I) and Cu (III) (which converts the primary alcohol at C-6 of galactose to the aldehyde), with O_2^- bound fleetingly to (Cu II) in the enzyme (Hamilton 1974). This interesting enzyme may itself possess strong superoxide dismutase activity (Cleveland and Davis 1974). Unlike glucose oxidase, galactose oxidase acts on di- and oligosaccharides.

Glucose Removal in Dry Egg Processing

Production of dried eggs to be used in bakeries dates back to the 1880s (Bergquist 1973). Twenty years later a much more inexpensive, light-colored, granular product imported from China which had been subjected to an arcane fermentation replaced the local product. Later, high import duties, civil strife in China and the approach of World War II dried up imports and gave impetus to the reestablishment of a dried-egg industry in the United States. One of the first problems encountered in reestablishing the industry was the browning which occurred during the drying process and upon subsequent storage. Balls and Swenson (1936) were the first to show that the browning was probably due to Maillard-type reactions, with accompanying fluorescence, between sugar and NH_2 groups of protein. More definitive proof of the first step in the interaction between glucose and egg white protein, even in dilute solution, was provided by Feeney *et al.* (1964), who used electrophoretic probes. This was substantiated by several groups whose work has been reviewed by Hill and Sebring (1973) and Kilara and Shahani (1973). Details of early work were covered by Lightbody and Fevold (1948). It is now clear from the work of Kline *et al.* (1954) that two principal types of nonenzymatic browning take place in drying and in dried eggs.

One is due to reaction of glucose with amino groups of proteins and amino acids in the whites to produce undesirable brown color as well as to interfere with desirable baking and related functional properties such as in whippability, solubility, cake volume, etc.

The other is due to a reaction between glucose and cephalin, 3-*sn*-phosphatidyl aminoethanol, present in yolk, which not only yields brown products but also results in the development of off-flavors. The effect of desugaring of egg white (albumen) and yolk on properties of the product is shown in Table 19.2. Parenthetically, GOX, because of the H_2O_2 released, probably causes oxidation of SH to SS in pasta proteins. Cantagalli and Tassi-Micco (1974) suggested that protein denaturation might be used as evidence of the prior action of GOX.

From such investigations it became clear that the only practical means of preventing browning in eggs was to remove the reducing sugar, most of which is glucose, present at a level of about 1%. Unlike the potato story, no extensive parallel investigation on the source and enzymology of this glucose has developed. Egg browning and its prevention have received some attention. As mentioned, the subject is dealt with briefly by Vahedra and Nath (1973), Kilara and Shahani (1973) and more comprehensively by Hill and Sebring (1973) and by Scott (1975). Hill and Sebring (1973) emphasized the microbial approach (see below), whereas Scott discussed the application of glucose oxidase. It is of interest to note that Hill and Sebring are listed as being associated with Armour Food Company, an early developer of yeast fermentation methods, whereas Scott has been closely associated with the entire history of the application of GOX in the food industry (Scott 1953, 1975). Approaches to glucose removal (desugaring) may be classified as follows:

(1) Microbial fermentations
  (a) Spontaneous
  (b) Controlled bacterial
  (c) Yeast
(2) Enzymatic, glucose oxidase
(3) Other proposed experimental approaches
  (a) Reverse osmosis (Kearsley 1974)
  (b) Cysteine (Kato *et al.* 1974)
  (c) Glucose reduction

Of these approaches, according to Hill and Sebring, only two are in widespread use in the United States egg industry: controlled bacterial fermentation and addition of glucose oxidase. Spontaneous fermentation was used for about 15 years to the mid-1940s. Hill and Sebring also state that bacterial fermentation is used almost exclusively to desugar egg white (albumen), while glucose oxidase and to some extent yeast are used to desugar whole eggs and egg yolk. Economic consideration and improved protein functionality have preempted bacterial fermentation of egg white, whereas microbial fermentation of yolk-containing raw material may yield products with unacceptable organoleptic attributes. This illustrates how the final choice in the use of enzyme is governed by considerations not directly dependent upon its primary function, improvement of color, but upon assessment of entrained considerations. Kilara and Shahani (1973) conclude from their review of the nonpatent literature that egg powders prepared after removal of glucose are uniformly improved with respect to storage ability and functional properties. The desugaring of eggs, especially by microbial fermentation, exemplifies the difficulty of assessing the present state-of-the-art, because, as stressed by Hill and Sebring, the pertinent information is considered quite confidential. It is of interest to note that the nonpatent United States literature practically ceased in the late 1950s. In the last 15 to 20 years almost all investigations reported have been from outside the United States (*cf* Kiss 1967; Sigmund 1968; Lobzov and Volik 1972). Scott and Klis (1962) and Scott (1975) described the production of *Salmonella*-free eggs by a process developed in the early 1950s and patented by Scott (1964).

As applied to whole egg and to egg yolk about 100,000 Sarett units of glucose oxidase per 1000 are added to approximately 450 kg egg after having added, with adequate stirring, 600 ml of 35% H_2O_2. One Sarrett unit is that amount of enzyme in the presence of excess catalase which catalyzes the conversion at 30°C of 10 mm O_2 per minute per ml reaction mixture containing glucose and an acetate-Pi buffer at pH 5.9. H_2O_2 is added, along with

TABLE 19.2

GLUCOSE OXIDASE TREATMENT OF EGG AND YOLK IMPROVES PRODUCT QUALITY[1]

Eggs Treated with Glucose Oxidase (?)	Yes	No
Whole Egg Properties		
Browning, whole egg powder % Reflectance	32.0	22.0
Browning, phospholipid, absorbance	0.03	0.23
Protein, % solubility	85.0	60.0
Egg-containing Foods		
Scrambled eggs, taste score (1–10)	7.4	2.8
Sponge cake volume (ml)	468.0	172.0
Doughnuts (% fat)[2]	32.12	25.77

Source: All data except for doughnuts from Kline *et al.* (1954).
[1] Dried whole eggs (5% H_2O), products stored for 5 weeks at 30°C in N_2.
[2] Source: Paul *et al.* (1957); doughnuts contained dried yolk stored for 90 days at 35°C.

catalase, to provide more oxygen and thus faster oxidation of glucose than is available by slow diffusion of oxygen in the air into solution and thence to the site of enzyme action. However, one of the problems in the use of this method is that the added H_2O_2 apparently oxidizes some of the lipids, resulting in some off-flavor. Catalase accompanying the glucose oxidase in commercial preparations decomposes the peroxide so that foaming occurs without actual accumulation of foam on the surface. The peroxide is continuously replenished. Foaming helps maintain the emulsion and also helps to exclude bacteria from the emulsion by floating them off, thus contributing to the maintenance of sterility. Sterility is also promoted by conducting the desugaring at the relatively low temperature of 13°C. The glucose content is lowered in whole egg from about 1% to 0.1% in whole egg in about 4 hr and in yolk in about 6 hr. Scott states that the choice of desugaring time of yolks and fresh whole eggs is largely determined by effect on quality. Table 19.3 compares conditions in the above-described methods (Scott 1975) with a previous United States process (Scott 1953) and with that described in representative European publications (Kiss 1967; Sigmund 1968). Further details of some of Scott's patents can be found in Noyes (1969) and in Wieland (1972).

Before leaving the subject of egg color we wish to mention that enzymes have been used empirically to help solve other color problems in this food. Thus, a patent was issued to Ng (1971) for a method of reducing the disagreeable characteristic gray-green discoloration at the egg-white interface of hard-boiled eggs. This was accomplished by mixing liquid yolk with 0.01–0.03% of proteolytic enzyme followed by heat coagulation and further heating in the presence of 1–5% H_2O_2.

Other Applications of Glucose Oxidase

Applications of glucose oxidase straddle the various food quality attributes so it is difficult to disentangle them. Their applications in other color problems were alluded to in previous chapters. We have just seen how prevention of browning in dried eggs prevents deterioration of the functional properties of the egg proteins and improves baked goods made from them. Scott (1975) pointed out that the versatility of glucose oxidase arises from its specificity and from its unique products. Each of these, the oxygen and glucose removed and the hydrogen peroxide and gluconic acid formed, can be a determinant of food quality. Indeed, the first patent on the use of glucose oxidase in foods issued to Baker (1950) is concerned mainly with the removal of oxygen from foods. The first publications on its use for the removal of glucose from foods appeared in 1953 [Baldwin *et al.*; Carlin and Ayers; Schwimmer (C); Scott]. Since this chapter is concerned with color, it may be well to enumerate the various effects of glucose oxidase on the color quality of foods other than eggs.

We have in this chapter briefly alluded to the potential use of GOX in the prevention of potato nonenzymatic browning by removal of glucose; improvement of wine color quality by removal of one of the substrates of enzymatic browning, oxygen; experiments on its use in the prevention of phenolase-induced browning in frozen fruits; and its use in the prevention of nonenzymatic browning of canned orange juice. Scott (1975) suggested that it

TABLE 19.3

DESUGARING WHOLE EGGS AND YOLKS WITH GLUCOSE OXIDASE-CATALASE

Raw Material	Glucose Oxidase (Units/kg Egg)[1]	Temp. (°C)	Time (hr)	H_2O_2 (%)
Whole eggs[3]	330	35–38	3.5	0.3
Whole eggs[3]	495	35–38	2	0.3
Whole eggs[4]	220	16	6	2
Yolks[5]	165	16	4	2
Whole, yolk[6]	400	25–30	3	0.25
Whole, yolks[7]	1000	30	2	0.02

[1] Sarett units, as defined in text. Standard glucose oxidase, 75,000 units/lb or 165 units/g.
[2] Introduced into a 450 kg batch of eggs at the rate of 5 ml of 35% H_2O_2 per min. In each example, except that of Kiss (1967), further H_2O_2 added to keep the level constant until near the end of the run.
[3] Source: Scott (1953).
[4] Source: Scott (1964).
[5] Source: Scott (1975).
[6] Source: Sigmund (1968).
[7] Source: Kiss (1967).

may be useful in removing glucose from meat to yield an improved dried product. By virtue of its oxygen-scavenging ability, it has been used to improve and retain the clarity and brilliance of beer (Ohlmeyer 1957) and to retain color as well as prevent corrosion in canned fruit beverages (Joslyn 1961B; Scott and Hammer 1962).

Some applications are not directly related to improvement of food quality. The most important of these is as immobilized enzyme in analysis for the routine determination of glucose (Chapter 38). GOX, both free and immobilized, has been incorporated into packaging materials in several ingenious ways (Scott 1975; Utsumi 1974). It has been used, in one case at least, to prevent the oxidative deterioration of the packaging materials themselves (Higuchi 1973). It may increase the yield of must in winemaking (Zinchenko *et al.* 1973). As a therapeutic agent, this remarkably versatile enzyme has been experimentally reported to help in the decrease of plaque formation when incorporated into mouthwashes along with amyloglucosidase (Koch *et al.* 1973) and, when immobilized on glass beads, was part of an extracorporeal shunt in reducing blood glucose levels in dogs (Venter *et al.* 1975).

Immobilized Glucose Oxidase.—We have alluded to immobilized glucose oxidase in the preceding discussion in connection with its use in foods, in determination of glucose and in medicine. About 300 papers as of late 1979 on immobilized glucose oxidase have appeared since it was first accomplished by Avrameas (1969). About 80 of these are concerned with glucose analysis. The following are some of the immobilization supports, reagents and techniques which have been used: alginic acid, Celite, chitin, clays, collagen, adsorption, attapulgites, DEAE-cellulose, dialdehyde cellulose, dialdehyde Epidex, erythrocytes, epoxy-copolymer, gel entrapment, glutaraldehyde, gamma-aminopropyl-tri-ethoxyl-silane (silanization), 2-hydroxymethacrylate, kieselguhr, microbe cells, nickel oxide, nylon tubing (activated), polyacrylamide, polystyrene, Sepharose 4B, silastic entrapment, silica, alumina, titania and vanacryl. Representative papers by Greenfield and coworkers (Greenfield and Laurence 1975; Markey *et al.* 1975) report studies aimed towards the eventual economical use of immobilized glucose oxidase in the food industry. They found that glucose oxidase is inactivated, when immobilized, by the hydrogen peroxide evolved. This could be prevented by immobilizing catalase on the same support. Since for most purposes catalase is needed in any case, they determined that the fungal catalase accompanying commercial glucose oxidase was the most stable and economical to use. Although they found that adsorption was the more convenient immobilization technique, it was necessary to use a chemical coupling reagent such as glutaraldehyde to keep the enzyme bound to the support. In another approach, Atallah and Hultin (1977) stabilized both glucose oxidase and catalase and increased their binding to inorganic supports by conjugating them without insolubilizing them with glutaraldehyde. Since then experiments on the use of immobilized glucose oxidase to desugar eggs have been reported.

Part VI

ENZYME ACTION AND FOOD FLAVOR QUALITY

20

ENZYME ACTION AND ACCUMULATION AND RETENTION OF NONVOLATILE FLAVORANTS

THE NATURE OF FOOD FLAVOR

Although flavor substances are usually defined as those which are perceived by the chemical senses, taste (gustation) and odor (aroma), other sensations such as tactual perception, response to temperature, astringency, mouth feel and even pain may contribute to flavor awareness and especially to preference and/or rejection of food. In this regard, flavor, perhaps more than any other food quality attribute, is closely associated with health and nutrition since a perfectly wholesome food is more likely to be rejected on the basis of odor than on objectional color or texture. Off-odor is frequently the first clue of unwholesomeness. Conversely, completely valueless food from a nutritional standpoint may be unadvisedly substituted for wholesome food on the basis of flavor. Indeed Maier (1970) considers flavor to be an auxiliary dietary factor. For these reasons some topics which might well fit into our ensuing discussion will be deferred or reemphasized and integrated into discussion of nutritional quality, i.e., bitter peptides and off-flavors in unconventional high-protein foods (Chapter 34).

Another consideration which imputes prime importance to the perception of flavor is the circumstance that the odorant receptor tissue in the nose is closer to the brain than is any other sensory receptor. Recent research indicates very strongly that flavorant receptors, like enzymes, are proteins which undergo conformational change upon association with a flavor molecule. Of course, the visual receptor rhodopsins (Chapter 15) which also undergo conformational changes are involved in the transduction of photon energy into electrical energy and information. Indeed, there is some evidence that vitamin A is also involved in odor perception; vitamin A deficiency is frequently accompanied by anosmia. The chemical senses involve transduction of structural parameters and chemical energy into the electrical energy of nerve impulses and convey information to the brain.

As in enzyme theory with its lock-and-key concept which emphasizes the shape of the substrate molecule with respect to the complementarily-shaped active site of the enzyme, at least one theory of odor perception stresses the shape of the odor-bearing molecules (Amoore 1970). Other theories can be found in the references cited below. Klopping (1971), in reviewing this and other theories, concluded that odor quality of a pure compound is determined by a complicated interplay of (1) size-shape, (2) orientation with respect to surface of the receptor in the olfactory cells in the nose, and (3) configuration and properties of the electron orbitals of functional groups in the molecule. Theories of enzyme action (Chapters 2 through 4), although more sophisticated and advanced, also contain these elements of interplay.

More recently Price (1978), using the powerful technology of affinity chromatography (Chapter 7), reported the purification from mammalian olfactory tissue of a protein which may be the odor receptor for anisole.

The role of molecular structure relationships in taste perception is exemplified by the theory of Shallenberger (1971) and confirmed for fructose by, among others, Lindley and Birch (1975). Briefly, Schallenberger postulated that each sweet compound consists of two moieties: (1) AH, where H is an acidic proton; and (2) an electronegative center, B. The compound is sweet when the distance between AH and B is about three Å with formation of a hydrogen bond AH ··· B. Such studies as well as more intimate knowledge of the relation between taste and structure of phenolic glycosides (Horowitz 1964) will eventually be helpful in the selection of acceptable sweeteners (Inglett 1974). Dastoli (1974), Lee *et al.* (1975) and Vanderheijden *et al.* (1978) have marshalled experimental evidence which strongly points to specific separate sweet and bitter taste receptor proteins in mammalian tongue.

Still earlier physiological studies led some investigators to the conclusion that although a strong formal analogy between enzyme action mechanism and chemical sense perception (i.e., hyperbolic relation between flavorant concentration and response intensity, equivalent to substrate concentration and rate of enzyme action) exists, the forces holding together the "ligand-receptor" complexes in flavor perception are much weaker than those involved in enzyme action. Hornstein and Teranishi (1967) point out that the latter binding strength range agrees with strictly physiological-sensory experiments of Beidler (1966) and corresponds to adsorptive forces no greater than those provided by hydrogen bonding.

The biochemical basis of the effect of certain substances on acuity appears to lie in their chelating properties according to Beidler (1966) who cites observations on the strong cation-chelating abilities of the flavor potentiators such as the 5′-nucleotides and presumably of monosodium glutamate.

On the other hand, action of miraculin, the protein of miracle berry which makes sour-tasting foods taste sweet (Inglett 1974; Harvey 1974), is cited by Beidler as an example of a taste inhibitor. Another class of taste inhibitors is certain thiol-containing drugs, in particular D-penicillamine, which decrease taste acuity in man and animals. Return of taste to normal occurs when transition metal ions such as copper (II) and zinc (II) are administered. To rationalize these observations Henkin and Bradley (1969) suggested that taste depends upon the state of a regulatory protein that is controlled reversibly by the local balance of interacting thiols, disulfides, metal ions and metal complexes, all of which would approach dynamic equilibrium locally as suggested in Fig. 20.1. The regulatory protein may be an enzyme whose activity is controlled or may be part of a membrane whose permeability is modified. They suggest three ways in which the thiol-metal balance may regulate taste acuity:

(1) It may control the degree of the sulfhydryl-disulfide interchange, adjusting the concentration of the active protein.
(2) Ions of a particular metal (e.g., zinc or copper) may be specific enzyme activators or may bring about specific aggregation-favoring activity. The free thiol would control the activity of the critical metal and other metals would act indirectly through effects on thiol activity.
(3) A particular thiol and metal may act jointly to activate a regulatory protein. Other thiols and metals would inhibit by competing with the specific active agent. It may be possible to obtain additional evidence of flavor genesis processes at the molecular level. More recently this theory has become the subject of some controversy (Cullitan 1975).

In order to place the role of enzymes in flavor development in proper perspective, we briefly review schemata proposed for the classification of origin of food flavors such as that presented in Table 20.1 (Schwimmer and Friedman 1972). This classification distinguishes "physiological" flavor generation from that arising as the result of processing. Physiologically associated flavors refer to those flavors which are present at time of harvest or slaughter. The flavors are further classified as ubiquitous, such as the lower aldehydes and alcohols (odor) and the sugars and organic acids (taste), or specialized, such as those which impart to a given food its special unique aroma or taste. Such special flavors most frequently arise as the result of the perception of a spectrum of volatiles rather than that of gustatory nonvolatiles. The character of the aroma of many foods, especially beverages such as tea (Sanderson 1975) is due to the blending of several key aroma constituents. On the other

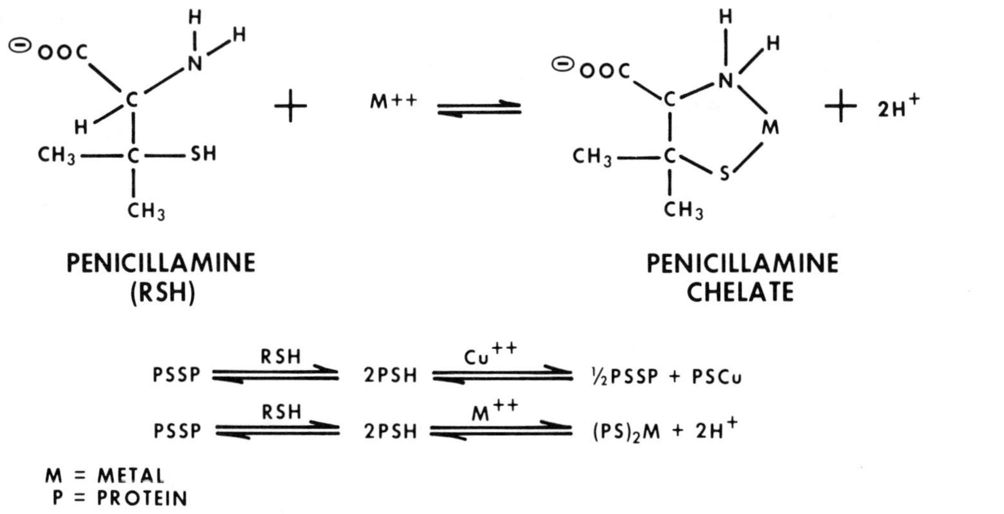

From Henkin and Bradley (1969)

FIG. 20.1. MODEL FOR CONTROL OF TASTE ACUITY

The control is established through balance of thiol groups and metal ions, M^{++}.

hand, the characteristic odor of other foods such as garlic and many of the essential oils arises from a limited number of related substances with similar odors. Exceptions are those substances which are more likely to convey a pain component of flavor such as the astringency of persimmons and tea and the piquancy of pepper.

In the category labeled as flavor arising after harvest, we distinguish between food in which the cells are still alive, new flavors originating during the storage and, alternatively, foods in which the cells are disrupted (Schwimmer 1972). As a result of the disorganization of the cell, many new compounds may be formed among which are those which have flavor. Usually, undesirable flavors develop in plant foods which are senescing. Alternatively the flavor, usually desirable, may arise as a result of the catalysis of a reaction by a specific enzyme acting on a limited class of substrates. This is due to the decompartmentization of enzyme and substrate accompanying disruption of the cell. The enzyme and substrate may be in separate cells or in different organelles within the same cell.

As shown in Table 20.1, components contributing to flavor of foods may also arise during processing. The agencies responsible for this transformation may, as shown, be enzymatic or nonenzymatic. Enzymatically developed flavors may also arise as the result of the introduction of microorganisms, i.e., fermentation. Margalith and Schwartz (1970) reviewed the contribution of fermentative processes to flavor in baked products, fermented beverages, dairy products, pickles and oriental foods.

TABLE 20.1

GENESIS OF FOOD FLAVORS

I. PHYSIOLOGICAL
  A. Present at harvest or slaughter
    1. Ubiquitous flavors
    2. Specialized flavors
  B. Postharvest or Postmortem
    1. Food source still alive—postharvest
    2. Cells of food disrupted
      a. Autolysis, general enzyme catalyzed macromolecular decomposition
      b. Specific catalysis of conversion of flavor precursors

II. PROCESSING
  A. Nonenzymatic
    1. Heating
    2. Chemical reactions
    3. Storage
    4. Addition of flavorings
  B. Enzymatic
    1. Fermentation
    2. Deliberate potentiation
    3. Adding enzymes
    4. Storage

In a discussion of the role of genesis of volatile sulfur-containing food flavors, Schwimmer and Friedman (1972) discerned four distinct modes of flavor genesis:

(1) Flavors preformed in the food via intracellular biogenetic processes before harvest, slaughter and/or processing. The presence of preformed volatile sulfur compounds in foods is probably the exception rather than the rule but may include sulfur-containing esters in Concord grapes (Stern *et al.* 1967).
(2) Compounds formed by enzyme action when the integrity of the cell is destroyed, so that the compartmentalized flavor precursor (substrate) intermingles with endogenous enzyme to yield volatile odoriferous products.
(3) Compounds formed by the action of microorganisms on food constituents.
(4) Compounds formed nonbiologically by the agencies of heat, light or ionizing radiation during food processing and storage of the food product.

From a penetrating analysis of the origin of tea flavor, Sanderson and Graham (1973) classified flavors according to whether they are:

(1) *Biosynthetic*, similar to category A-1 Physiological, Table 20.1.
(2) *Direct enzymic*, in which the flavors are produced by enzymes which catalyze the transformation of specific flavorless precursors to flavor-bearing components.
(3) *Oxidative or indirect enzymic*, in which flavor precursors are transformed into flavor-bearing compounds by enzymatically-formed oxidizing agents.
(4) *Pyrolytic flavor-bearing chemicals*, arising as a result of baking, roasting or otherwise heating food during the course of its processing or other preparation for consumption.

As briefly mentioned in Chapter 19, one of the principal reaction types responsible for these pyrolytic flavors is the Maillard nonenzymatic browning reaction (Hodge *et al.* 1972; Tressl 1975).

Categories 2, 3 and 4 of Sanderson and Graham may be looked upon as involving the same type of reaction sequence except that the time scale is expanded in going from "direct enzymic" to pyrolytic. As previously discussed, in each category the primary event is an enzyme-catalyzed reaction which gives rise to a primary product. This primary product then may be considered as an intermediate which can further interact with food constituents to give rise to stable flavor-bearing components. In direct enzymic reactions, the formation of this intermediate is very rapid, the intermediate is highly unstable and either decomposes, interacts with itself and/or with food constituents very rapidly at comparatively low temperatures. At extremely low temperatures, i.e., frozen food storage of unblanched foods, the end-products are formed quite slowly and usually give rise to objectionable flavors. At the other extreme, pyrolytically-derived flavors arise from relatively stable intermediates and are formed only at high temperatures.

Chase (1974) distinguishes the following modes of flavor biogenesis by:

(1) biosynthetic enzymes in living tissues,
(2) degradative enzymes which begin to act when living tissue is subject to shear forces,
(3) enzymes in microorganisms grown in cultured or fermented foods, and
(4) enzymes which generate compounds which react nonenzymatically to form flavor components.

The viewpoint of the flavor ingredient supplier is represented by the classification scheme of the following areas (Wiener 1974):

(1) Biochemical physiology which is the molecular basis of flavor perception.
(2) Biochemical production of raw materials which includes the use of enzymes added to food either as commercial preparations, fermentations or by isolation of flavor-bearing compounds using "biochemical methods," synthesis, etc.
(3) Biochemistry of flavor formation in natural products.
(4) Flavor formation during food processing.
(5) Biochemical rationalization of flavor creation by application of the knowledge of enzymology and other branches of biochemistry.

Since the emphasis in our discussion has been on the effects of enzyme action on the quality attributes of foods, we find it convenient and useful to classify flavors as being desirable or undesirable. Further division of the subject matter will more or less follow the classifications outlined above. Of course the judgement of a given substance as desirable or objectionable will depend upon the food, the substance in the food, its association with other flavor components and its contribution to the overall flavor.

A valuable brief account which serves as an excellent introduction to the general subject of flavor is that of Hornstein and Teranishi (1967). Several reviews on this, as in other fields, are available in books detailing the proceedings of symposia. Books on general aspects of flavor include those of Campbell Soup Company (1961), Gould (1966), Schultz *et al.* (1967), Ohlhoff and Thomas (1971), Drawert (1975) and Charalambous and Inglett (1978). As a result of the spectacular successes of gas-liquid chromatography/mass spectrometry, most of these books are oriented to evaluation of the volatiles as bearers of odor. One of the most definitive books in this area stressing instrumentation and preparation of sample for GLC is that of Teranishi *et al.* (1971). However, GLC has an important limitation—its relatively high temperatures which destroy thermolabile volatiles that give distinctive odors at ordinary temperatures. Newer techniques such as high pressure (performance) liquid chromatography and X-ray fluorescence spectroscopy will help fill this gap. These methods, we predict, will play a key role in future studies of the chemistry of flavor compounds because they make it possible to determine such compounds under mild, nondestructive conditions. Moreover, the task of elucidating the original genesis of flavors based upon structural studies will require not only advanced analytical techniques, but hopefully a judicious eclectic blend of biochemical and organic chemical theory.

Viewpoints on the role of enzymes in flavor development in the segment of the food industry which supplies flavorants as food ingredients can be found in the brief editorial-like comments of MacLeod (1970), Wiener (1974) and Nursten (1975). More detailed discussions of the role of enzyme action in food flavor development are those of Hewitt *et al.* (1956), Reed (1966), Schwimmer (1969A), Chase (1974) and Schwimmer (1978). Several reviews on the role of sulfur compounds in food flavor have appeared since that of Schwimmer and Friedman (1972) on both enzymic and nonenzymic genesis of volatile sulfur-containing food flavors. Including comments on yeast enzymes involved in the generation of sulfur volatiles are reviews by Shankaranarayana *et al.* (1974), Schutte (1974), Maga (1976A) and Schrier (1975). A review of the volatile flavors of crucifer vegetables by Macleod (1976) is also interesting. In the present chapter we direct our attention to enzymes involved in the formation and removal of flavor bearing solids whose organoleptic properties are perceived primarily as taste.

ENZYMES OF FOOD ACID ACCUMULATION

Action and Regulation of Biosynthetic Enzymes

In the following examination of enzymes and enzyme reactions, which may be involved in the production of flavor-bearing compounds present in the living cell before harvest or slaughter, we shall see that most of the enzymes are ubiquitously distributed in almost all organisms and are usually part of or closely associated with universally used metabolic pathways. We therefore wish not only to ask what enzymes form these flavorants, but what are the enzymatically-related circumstances that lead to their accumulation in concentrations above taste threshold levels. While we cannot detail the vast literature on their biosynthesis using the "tracer" approach, we deem it timely to outline and assess (1) what some of these enzymes might be, (2) how their properties might relate to flavor accumulation, (3) the meager information on the regulatory mechanisms which might lead to accumulation, (4) enzymatic aspects of agronomically-related and other measures taken to optimize levels of these flavors, (5) the preservation of these levels by preventing enzyme action which would catalyze transformations of these substances, and (6) notes on how these substances are produced for uses as food additives.

Nonvolatile, desirable, taste-bearing substances comprise, to a large extent, acids (sour/acerb), sugars (sweet), amino acids, peptides and the 5′-nucleotides (potentiators, gustators), astringents, and biting, peppery compounds. The ensuing discussion is devoted to sourness and acidulants followed by gustators.

In a wide variety of foods, sourness is the predominant desirable sensory attribute. In foods such as sauerkraut, pickles, some salads, buttermilk and sourdough french bread, the sourness can be attained by either adding acid (usually citric, phosphoric or acetic acid) or may develop as lactic and acetic acid during fermentation. The degree of "sourness" is governed by the acidity, the pH and thus the concentration and strength (pK_a) of the acids. The optimal acidity is very high for citrus, sorrel and rhubarb, somewhat less for pome fruits, and even less for tomatoes and spinach. Especially for fruit, the plant food accumulates excess acid in the preripe stage of development. Upon ripening the acidity decreases and the sugars increase. Each fruit has its own special range of acid-to-sugar ratio corresponding to optimum taste. This ratio, relatively easy to determine, may also serve as an

TABLE 20.2

FOOD ORGANIC ACIDS AND INTERMEDIATES—ACERBITY AND SOURNESS

I COOH CH_3 Acetic	II COOH CH_2OH Glycolic	III COOH CHO Glyoxylic	IV COOH COOH Oxalic	
V COOH CO CH_3 Pyruvic	VI COOH HOCH CH_3 L-Lactic	VII COOH $COPO_3H_2$ CH_2 Phosphoenol-pyruvic (PEP)	VIII COOH CHOH $HC(OH)OPO_3H_2$ 3-Phosphoglyceric (3-PGA)	
IX COOH CH_2 CH_2 COOH Succinic	X COOH CH_2 OHCH COOH L-Malic	XI COOH HCOH HCOH COOH L-Tartaric	XII COOH CH_2 C=O COOH Oxalacetic (OAA)	XIII COOH CH_2 CH_2 O=C-COOH α-Ketoglutaric (2-Oxoglutaric)
XIV CH_2 COOH HO C COOH CH_2 COOH Citric	XV CH_2COOH CH COOH OH CH_2 COOH *iso*-Citric	XVI CH_2 COOH HO C COOH CH_3 Citramalic	XVII COOH CH HCR Caffeic R=3,5-dihydroxybenzyl	

XVIII = Quinic; R = carboxy - 1,3,4,5-tetrahydroxy cyclohexane
XIX = Shikimic, R = carboxy - 3,4,5-trihydroxy-5-cyclohexene
XX = Chlorogenic; caffeoyl quinic (several isomers)

objective criterion of maturity. Citric and malic acid are by far the most predominant contributors to the desired degree of sourness in most fruits.

In the case of fruits there appears to be some uncertainty as to the anatomical origin of organic acids. Ulrich (1970) summarizes evidence pro and con on translocation of acids from roots and leaves *vs* synthesis in the fruit. In addition to the results of tracer studies, the existence in fruit of enzymes which are capable of forming most of the acids strongly suggests that acids are synthesized within the fruit. The structure of these acids is shown in Table 20.2. Also shown are acids which are probably intermediates in their interconversion and formation, or which may play a role in accumulation but do not usually contribute to sour taste. Some of the more prominent enzymes which catalyze transformation and, potentially, accumulation of the acids of Table 20.2 are shown in Table 20.3.

TABLE 20.3

ENZYMES WHICH MAY BE INVOLVED IN ACID ACCUMULATION[1]

Name (Trivial)	Reaction Catalyzed[2]
(1) Malate dehydrogenase	$XII + NADH + H^+ \rightleftharpoons X + NAD^+$
(2) Malic enzyme	$CO_2 + V + NADH + H^+ \rightleftharpoons X + NAD^+$
(3) Pyruvate decarboxylase	$V \rightarrow$ Acetaldehyde $+ CO_2$
(4) Pyruvate carboxylase	$CO_2 + V + ATP + H_2O \rightleftharpoons XII + ADP + Pi$
(5) PEP carboxylase	$CO_2 + PEP + H_2O \rightleftharpoons XII + Pi$
(6) PEP carboxykinase	$CO_2 + PEP + ADP \rightleftharpoons XII + ATP$
(7) Citrate synthase	$I\text{-}CoA + XII + H_2O \rightleftharpoons XIV + CoA$
(8) Citrate-cleavage enzyme	$XIV + ATP + CoA \rightleftharpoons XII + I\ CoA + ADP + Pi$
(9) Aconitase	$XIV \rightarrow XV$ (overall)
(10) Glycolate oxidase	$II + O_2 \rightarrow III + H_2O_2$

[1] The roman numerals refer to the acids of Table 20.2.
[2] The direction of enzyme reaction during ripening of fruits.

Enzymes Catalyzing Acid Interconversions

Origins.—From our present knowledge, the accumulation of organic acids appears to involve the operation and regulation of several alternative metabolic sequences and enzyme reaction types. Acids may arise from carbohydrates *via* glycolysis and the TCA cycle. Malic and citric, two major fruit acids, and succinic acids are members of the TCA cycle, and in some plant foods of the glyoxylate cycle. The relation between these two cycles is sketched in the following scheme:

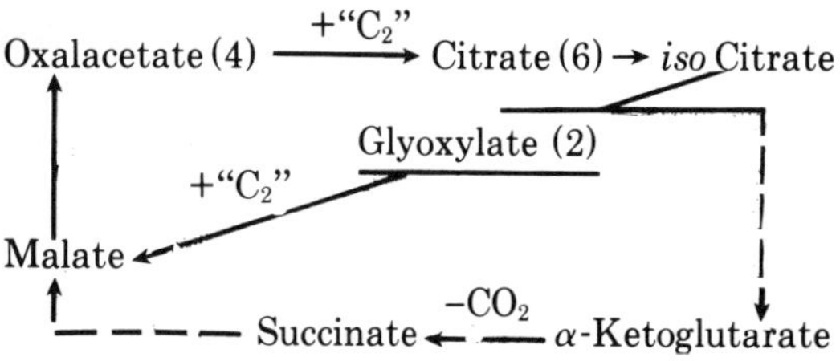

where the dashed arrows indicate steps of the TCA cycle, the number of Cs in parentheses. Acetyl CoA ("C_2") can arise from both fatty acids and from sugars. The importance of the glyoxylate cycle in acid accumulation in many fruits is still uncertain. Thus, Bogin and Wallace (1966A) could not obtain clear-cut evidence for the presence in grapes of malate synthase, one of the key enzymes of the glyoxylate cycle (Higgins *et al.* 1972; Doig *et al.* 1975).

This cycle is found in the glyoxosomes of food plants such as watermelon, cucumber and peanuts, as well as in castor beans, a favorite source material for the study of this cycle. This is because the glyoxosome enzymes make possible the conversion of lipid into carbohydrate by bypassing the decarboxylation steps of the TCA cycle. The glyoxylate cycle does this by making possible the utilization of acetate for the manufacture of fatty acids, carbohydrates and amino acids. Levels of enzymes leading to and away from glyoxylate, whose concentration parallels fresh weight during development of mango fruit, are shown in Fig. 20.2.

Acid may arise from amino acids *via* aminotransferases

$$R_1CHNH_2COOH + R_2COCOOH \rightleftharpoons R_1COCOOH + R_2CHNH_2COOH$$

Donor amino acid — Acceptor ketoacid — New ketoacid — New amino acid

The earliest reported and most frequently encountered acceptor ketoacid is α-ketoglutaric and the "new" amino acid is glutamic acid in over half of the 50 odd distinct amino transferases so far reported. Note that α-ketoglutaric acid is also a member of the TCA cycle.

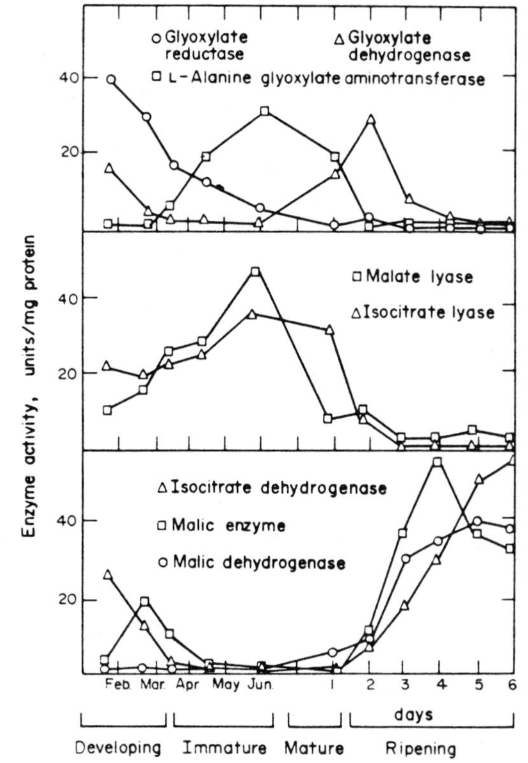

From Banqui et al. (1977). Reprinted with permission from Phytochemistry 12, 52. Copyright 1973 Pergamon Press, Ltd.

FIG. 20.2. FRUIT TASTE AND THE GLYOXYLATE PATHWAY

Shown are the activities of enzymes leading to and from glyoxylate during development and ripening of mango fruit.

Enzyme reactions involving the uptake (carboxylation) or removal (decarboxylation) of CO_2 interconvert one acid to another. In addition to serving as important enzymes of metabolic schemes (i.e., PEP carboxylase in "C4" photosynthesis) these anaplerotic (filling up) reactions have been considered as agents of replenishment of the compounds of the TCA cycle which are removed for the synthesis of amino acids. In addition to ascorbic acid itself arising from carbohydrates, two other special acids which contribute predominantly to the sources of a few foods, oxalic and tartaric acid, may arise directly from ascorbic acid as discussed below. Finally, they may arise from the oxidation of long chain fatty acids, about which we shall have more to say in connection with volatiles. It is sufficient here to mention that one of the key acids involved in enzymatic development of sour taste, glyoxylic acid, inhibits the fatty acid oxidizing capacity of mango mitochondria (Banqui *et al.* 1977).

Acids may also arise from carbohydrate via the hexose/pentose phosphate shunt and the 7-carbon sugar-shikimic acid pathway to phenolic acids such as quinic, caffeic and especially the chlorogenic acids, as well as shikimic acid. The latter, normally the first distinct phenolic-like but fleeting intermediate, accumulates to become the predominant organic acid in the gooseberry. Another carbohydrate-derived acid is galacturonic present in pomes as the result of the action of polygalacturonase and pectin esterase (Chapter 31).

Enzymes.—*Malate Dehydrogenase.*—This TCA and glyoxylate cycle enzyme has been detected in several fruits and vegetables. Its activity *decreases* in the ripening tomato, closely paralleling the decrease in acidity (Swardt and Duvenage 1971). This decrease in enzyme activity is considered to be a significant process signaling the onset of senescence in this climacteric plant food. Hawker (1969) reported a dramatic drop in activity of this enzyme between the third and fourth week after flowering in grapes (Table 20.4). As with all other enzymes it is present as isozymes (Curry and Ting 1975). There is special interest in the isozymes of malate dehydrogenase since they should be present in mitochondria, site of the TCA cycle, in glyoxosomes, site of the glyoxylate cycle, as well as in the chloroplasts of "C3" plants (Wolpert and Ernst-Fonberg 1975). Vines (1968) isolated and characterized malate dehydrogenase of grapefruit vesicle mitochondria.

The Malic Enzyme.—This anaplerotic enzyme may be involved in the accumulation of malic acid in fruits. Ulrich (1970) summarized evidence for its existence in a variety of fruits and that, in contrast to malate dehydrogenase, it increases during ripening in pome fruits and dramatically during the respiratory climacteric. In grapes (a nonclimacteric fruit) no consistent trend in the activity of this enzyme could be detected, in contrast to apples. This might indicate that the enzyme is acting in the direction of decarboxylation in grapes (Lakso and Kliewer 1975). In green plants it is a very important enzyme in Hatch-Slack "C4" photosynthesis (Hatch *et al.* 1974). A clue to the role of this enzyme in accumulation of acid in fruits is afforded by the finding of Do Nascimento *et al.* (1975) that tartrate ion, which accumulates in grapes (see below), *activates* malic enzyme when measured by the ability of pyruvate to take up CO_2 but *inhibits* the enzyme in its decarboxylation mode. This observation explains how malic acid stimulates its own synthesis. The exceedingly complex regulatory properties of this enzyme in vegetable mitochondria was explored in some detail by Davies and Patil (1975).

Phosphoenolpyruvate (PEP) Carboxylase.—In addition to being a ubiquitous anaplerotic enzyme, it has recently been studied in some detail because of its key role in "C4" photosynthesis (Zelitch 1975). It is the enzyme which catalyzes the primary uptake of CO_2 in the chloroplast of mesophyll cells.

As shown in Fig. 20.3 it plays a role analogous to that of ribulose-biphosphate carboxylase (dismutase) in the classical "C3" Calvin pathway present in the bundle-sheath of "C4" plants such as corn and sugar cane. Investigations have established the allosteric regulatory nature of this enzyme in its photosynthetic function (Mukerji 1974; Coombs *et al.* 1974; Shio 1979). As such, one might anticipate that it would play a key role in acid accumulation in plants. As shown in Table 20.4, Hawker (1969) observed relatively high activity in

TABLE 20.4

ACTIVITIES OF ACID METABOLIZING ENZYMES IN GRAPE EXTRACTS

	Weeks After Anthesis (Flowering)				
Activity Measured	2–3	4	6–7	9	11–12
PEP Carboxylase[1]	100	66	10	5	2
PEP Carboxylase[2]	95	100	86	59	90
PEP Carboxykinase[2]	44	100	41	44	44
Malate Dehydrogenase[1]	100	31	48	27	31
Pyruvate Decarboxylase[1]	100	35	19	2	14

[1] Relative activity per gram of grape tissue; adapted from the data of Hawker (1969).
[2] Source: Adapted from the data of Ruffner and Kliewer (1975).

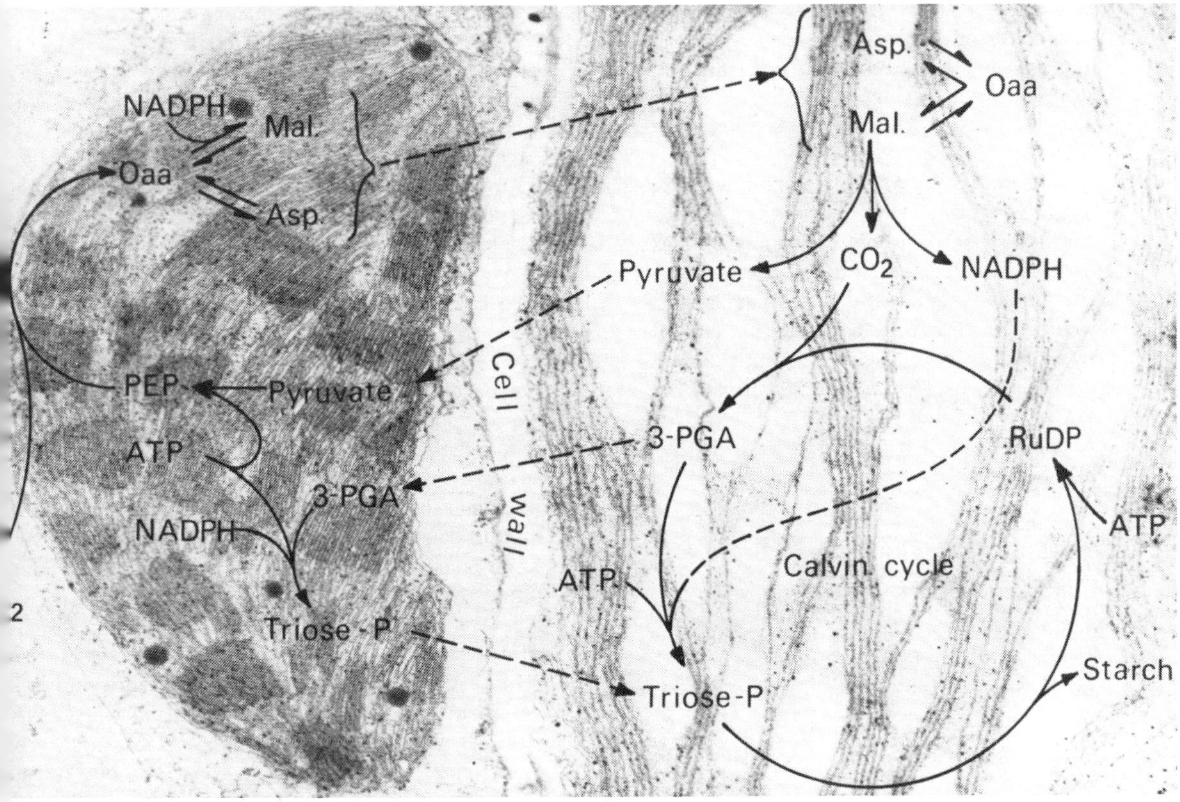

From Berry (1971)

FIG. 20.3. COMPARTMENTATION OF ENZYMES OF THE C3 AND C4 PHOTOSYNTHESIS ROUTES IN SEPARATE CELLS

Solid lines represent chemical reactions, dotted lines movement of the chemicals. The background is an electron photomicrograph showing bundle sheath, site of C4, and adjacent mesophyll cells, site of C3 photosynthesis.

very young grapes and a decrease to a low value at about the same time that the level of malic acid reached a maximum. They concluded that this enzyme, along with the malic enzyme, was primarily responsible for fixing CO_2 in grapes. Among the properties of the enzyme in extracts of 2 to 3 week old grape berries studied by Lakso and Kliewer (1975) is its inhibition *in vitro* by relatively high concentrations of malate (50% inhibition at 35 m*M*). This would indicate that this enzyme is important in controlling the level of sourness in grapes. By comparison, the K_m value with PEP is 90 μM. As an illustration of how the presence of other enzymes can influence enzyme efficiency (Chapter 8) Wolpert and Ernst-Fonberg (1975) found that the affinity of PEP carboxylase for CO_2 as bicarbonate ion is ten times greater when the enzyme is part of a complex (isolated as such from algae) containing malate dehydrogenase as well as acetyl CoA carboxylase, than that of the free enzyme.

As a clue to how this enzyme might function in terms of accumulation of enough acid to detect it as a sour taste, we turn to a salt tolerant nonfood plant. Von Willert (1975) found that Pi counteracted the inhibition of this crassulacean enzyme by malate. Furthermore this effect of Pi increased with decreasing pH (increasing sourness).

PEP Carboxykinase.—In addition to being considered as playing a role in anaplerotic enzyme reactions this enzyme has also been established as the one catalyzing the first reaction in the reversal of glycolysis in both plants and animals. Thus it should play a key role in setting the stage for the shift from acid to sugar in ripening fruits (veraison) (Ulrich 1970; Ruffner and Kliewer 1975). The latter investigators observed that the activity in grape extracts reached a maximum at 4 weeks after flowering; in contrast, PEP carboxylase changed little except for a sharp reduction at the onset of ripening; also compare with Hawker's results (Table 20.4) (1969). This was apparently the first time that this enzyme had been characterized in woody plants. Changes in the levels of neither PEP carboxykinase nor those of the malic enzyme and PEP carboxylase which would lead to malate consumption can, according to Ruffner *et al.* (1977), account for the lower acidity levels occurring at maturity in grapes grown at relatively high temperature. Much remains to be learned of its role in the level of acids in plant foods.

Citrate Synthase.—In citrus fruits citric acid is the principal source of sourness. From the evidence in the literature marshalled by Ting and Attaway (1971) it is synthesized in the fruit vesicles rather than translocated from the leaves. Since the citrate synthase, a well-established enzyme of the TCA cycle, has been reported to be present in the mitochondria of citrus fruits, it is not unreasonable to assume that this enzyme is exclusively responsible for the formation of citric acid.

Thanks to the work of Bogin and Wallace (1966B) and of Srere (1974) some insight into the manner in which citric acid accumulates is available. This is depicted in the following scheme:

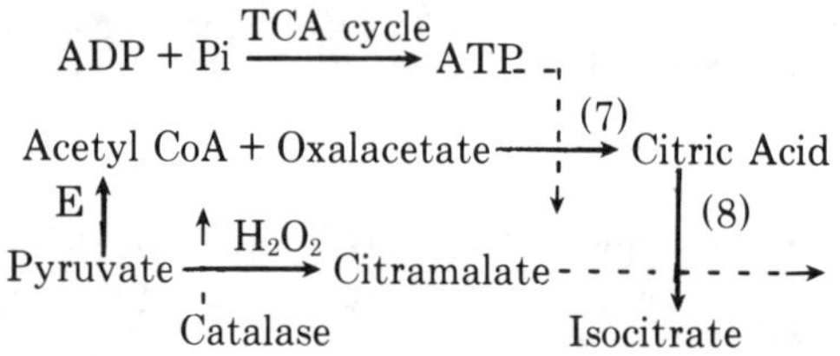

The broken arrows denote inhibition. The numbers indicate the enzymes of Table 20.3 and E denotes enzymes involved in the entry of pyruvate into the TCA cycle. Somewhat more speculative is the suggestion that the higher activity of catalase in sweet lemon interferes with the conversion of pyruvate to citramalate by removing H_2O_2 putatively required for one of the steps in the conversion. ATP formed as the result of the operation of the TCA cycle inhibits citrate synthase by increasing the K_m (lowering the affinity) for acetyl CoA. Therefore, weakening of TCA cycle activity would tend to increase citric acid accumulation. In agreement with this idea, Bogin and Wallace found that the O_2 uptake and ATP forming capacity of mitochondria of sour lemon was about one-half that of sweet lemon. Of course, this might have been due to damage of the mitochondria at the low pH of sour lemons. These investigators also found that citramalic acid, a minor acid in citrus (XVI, Table 9.2), competitively inhibited aconitase, the TCA enzyme catalyzing removal of citric acid. Sour lemon formed more citramalate from pyruvate than did sweet lemon. As far as can be ascertained, the only confirmation of the presence of citrate synthase in citrus is unpublished work cited by Ting and Attaway (1971). The evidence that citrate synthase decreases with ripening (Table 20.3) is based upon the use of very sour lemons. For general discussions of the regulation of this key TCA enzyme, see Goodwin (1968) and Srere (1974).

Further insight into regulation of citric acid levels in fruits, especially during ripening, is afforded by work on mango by Matoo and Modi (1970) who found that the decrease in citric acid during ripening was paralleled by a five-fold increase of specific activity in extracts of citrate cleavage enzyme. This enzyme may, like PEP carboxykinase, be a bridge to sugar buildup in fruits because it is stimulated by glucose and fructose. The authors also found that fatty acids also stimulate the citrate cleavage enzyme and suggest that this enzyme reaction provides the building block for many of the new compounds found in the ripe mango.

Accumulation of Other Acids.—Other organic acids which contribute to the sourness of fruits are ascorbic, tartaric and oxalic. Accumulation of tartrate is characteristic of grapes. Oxalic acid is largely responsible for the sourness and acerbity of rhubarb and wood sorrel sometimes used as a food and is present in spinach. The synthesis of ascorbic acid in plants has been fairly well clarified and the pertinent enzymes have been detected. The situation with respect to the other two acids is by no means settled. The enzymes in foods responsible for their formation are virtually unknown, although several microbial enzymes catalyzing the formation of these substances have been discovered. Penoud and Ribereau-Gayon (1971) reviewed evidence for the origin of tartaric acid from glucose via gluconic acid. Since then several investigators have suggested that ascorbic acid was the direct precursor to tartaric acid while others have claimed that this conclusion is based on artifacts. Wagner *et al.* (1975) convincingly showed, by feeding C^{14} labeled ascorbic acid to grapes, that tartaric acid is a major breakdown product of ascorbic acid and that (+)-tartaric acid is the *only* form produced. Immature tamarind fruits contain (+)-tartaric acid while the ripe fruits accumulate the *meso*-isomer. The enzyme presumably responsible for this isomerization is tartrate epimerase (Ranjan *et al.* 1961). Although tartrate-metabolizing enzymes have been detected in plants and purified from microorganisms, the apparently unconfirmed brief of Ranjan *et al.* appears to be the only report of tartaric acid metabolizing enzyme in fruits which accumulate this acid. A similar type of investigation from Loewus' laboratory led the authors (Yang and Loewus 1975) to cautiously suggest that ascorbic acid decomposition could be responsible for oxalic acid accumulation in the foods spinach and wood sorrel (oxalis), among other plants. Among several enzymes from nonfood sources which catalyze the formation of oxalic acid, the most direct is the hydrolase oxalacetase (Hayaishi *et al.* 1956):

$$\text{Oxalacetate} + H_2O \longrightarrow \text{Oxalate} + \text{Acetate}$$

Another possibility is that oxalate may arise as a result of glycolic acid associated enzymes present in the subcellular organelles of green plants known as peroxisomes, which only recently have been sharply distinguished from glyoxosomes. Parenthetically the peroxisomes are the site of photorespiration, light-induced O_2 uptake which can amount to more than 50% of the CO_2 fixed by photosynthesis (Zelitch 1975).

An important source of glycolate is that which arises as the result of oxygenase action of the same enzyme which fixes CO_2 in photosynthesis, ribulose-biphosphate carboxylase. In its oxygenase mode of action (which appears to involve SH groups) it forms phosphoglycolate from the same substrate and O_2 instead of CO_2 which rapidly breaks down into glycolate (Jensen and Bahr 1977). The rate of glycolate biosynthesis is probably the most important factor controlling photorespiration. Glycolate is oxidized by the flavoprotein glycolate oxidase (Table 20.3) present in peroxisomes of green leaves, including spinach (Halliwell and Butt 1974).

Being a flavoprotein, this enzyme may produce superoxide anion, O_2^-, whose effects on food quality have been and will be discussed elsewhere. Further oxidation of glycolate would give rise to oxalic acid. A much more important facet of glycolate metabolism is that a complete understanding of its role in photorespiration could lead to ways to suppress photorespiration without suppressing photosynthesis. This suppression could lead to greater yields of foods than now possible as discussed in this chapter in conjunction with carbohydrate accumulation (see below).

A microbial enzyme has been described which can oxidize glyoxylic acid to oxalic acid. Another possibility is that glyoxate-oxalate conversion takes place *via* the ubiquitous lactate dehydrogenase. In addition to lactic oxidation, LDH catalyzes the simultaneous oxidation and reduction of glyoxylic to oxalic and glycolic acids (Duncan and Tipton 1969).

$$\underset{\text{Glyoxylic}}{\begin{matrix} COOH \\ | \\ CHO \end{matrix}} + NADH + H^+ \rightleftharpoons \underset{\text{Glycolic}}{\begin{matrix} COOH \\ | \\ CH_2OH \end{matrix}} + NAD^+$$

$$\begin{matrix} COOH \\ | \\ HCOH \\ | \\ H \end{matrix} + NAD^+ + H_2O \rightleftharpoons \underset{\text{Oxalic}}{\begin{matrix} COOH \\ | \\ COOH \end{matrix}} + NADH + H^+$$

While oxalic acid itself contributes to the acerbity of only a few foods, it may, as oxalate ion, participate in controlling the level of sourness in other foods by its ability to inhibit the activity of many enzymes mentioned above at levels far below its taste threshold. For instance the K_i value of oxalate for malate synthase is about 2×10^{-5} *M*, i.e., 50% inhibition at less than 2 ppm (Higgins *et al.* 1972).

Accumulation of acids in foods and the biosynthesis of related enzymes are, like all enzymes, under genetic control. Their formation is dependent upon gene derepression. Few studies have been made on the effect of gene derepressants such as plant growth regulators on their level. For instance, some enzymes of the glyoxylate cycle and formation of glyoxosomes are dependent upon gibberellic acid (Doig *et al.* 1975). We do know empirically in some measure how to maintain a steady state of flavor quality with respect to acids as well as other constituents as annotated in a recent symposium on postharvest biology and handling of fruits and vegetables (Haard and Salunkhe 1975). In connection with one of these measures, CO_2-controlled atmosphere storage, it has been suggested that the increased CO_2 level results in slowing down respiration and in the onset of senescence by the CO_2 inhibition of one of the key enzymes of the TCA cycle, succinate dehydrogenase. Succinic acid accumulates in CO_2-stored apples (Hulme 1956) at 20% (but not at 10%) CO_2. At 3°C but not at 10°C the apples showed signs of injury and the levels of accumulated succinic acid (21 mg/100 g) were less than 2% of the malic acid content.

ENZYMES OF SUGAR SWEETENER ACCUMULATION

In this section we attempt for the most part to provide insight and organize available information on the role of natural or endogenous enzyme action involved in the formation and accumulation of sugars in foods (mostly plants) to levels above which they are perceived as desirable sweet substances. We shall also be concerned with enzymes that remove these sweeteners and the means by which such enzyme action can be prevented. Unprocessed foods which accumulate sugars range from intensely sweet honey through beets and sugar cane used for sucrose manufacture to fruits to sweet vegetables such as yams to those vegetables and other foods such as peas in which a modicum of sweetness contributes to the overall flavor but is not the dominant note.

The principal naturally-occurring sugars present in most plant foods are glucose, fructose and sucrose. Sorbitol is a substantial contributor to sweetness of pomes and maltose to that of such starch vegetables as yams. In Chapter 9 we dealt in some detail with the use of enzymes in the industrial manufacture of food ingredient sweeteners, and in Chapter 19 with sugar accumulation in potatoes. To understand how these sugars accumulate in these food plants and what enzymes are involved requires a knowledge of their carbohydrate metabolism. Good reviews and source references in this intricate field include Turner and Turner (1975), Kelly *et al.* (1976) and Coombe (1976). Sugars are formed and may accumulate via following pathways:

(1) Photosynthesis via the Calvin or "C3" pathway.
(2) Photosynthesis via the Hatch-Slack or "C4" dicarboxylic acid pathway.
(3) Gluconeogenesis from carboxylic acids.
(4) Interconversion of sugars.
(5) Degradation, usually hydrolytic, of di-, oligo- and polysaccharides.

When still on the vine, bush or limb, the enzymes whose action results in acid formation and accumulation are also involved in sugar accumulation *via* photosynthesis and gluconeogenesis. A signal, be it period of the day, stress situation such as change in available moisture or plant growth regulators, leads to derepression of the genes coding for the synthesis of those enzymes which are directly involved in sugar formation as well as those indirectly involved in regulation in such a way as to lead to accumulation of these sugars.

Enzymes of Gluconeogenesis

The glyoxylate cycle is the pathway to sugar accumulation from the carboxylic acids. The gateway is a pyrophosphate-requiring PEPcarboxykinase. This enzyme pushes the product of the glyoxylate cycle, oxalacetate, to PEP, one of the final products of the glycolysis. Proceeding in the opposite direction from that leading to acid accumulation, this enzyme catalyzes the following reaction:

Pyrophosphate (PP) + oxalacetate (OAA) ⟶ orthophosphate (Pi) + phosphoenolpyruvate (PEP)

In sugar cane, which photosynthesizes sugar via the C4-dicarboxylic acid pathway, sucrose accumulates through PEP carboxykinase and the intact gluconeogenesis machinery (Kelly *et al.* 1976). This means that the same type of enzymes, if not identical, which catalyze glycolytic reactions from FDP onward is also involved in the formation of sugars but not in the same cell or subcellular site. As pointed out in Chapter 4, reversal of the phosphofructokinase (PFK) reaction is thermodynamically and metabolically unfavorable. It will be recalled that the key enzyme which allows the continuation of gluconeogenesis is the highly allosterically regulated fructose diphosphate phosphatase (FDPase). Without exception, the activity of this enzyme appears to rise along with sugar accumulation in all of the plant foods thus far studied. Attention has been focused on influence of one of the prime allosteric affectors of this enzyme, AMP. For instance, the FDPase activity of mango comprises an alkaline and an acid phosphatase both of which are allosterically inhibited by AMP although not to the same extent (Rao and Modi 1976). AMP does not appear to play as critical a role in regulation during germination as it does in other situations in both plants and animals. Turner and Turner (1975) suggest that the key point of regulation is at the glucose phosphate level *in vivo* and that Pi may play a more crucial role than does AMP.

If gluconeogenesis and glycolysis were not compartmentalized, the effect of FDP would be nullified by the stimulation of PFK resulting in a "futile" cycle which would tend to favor glycolysis. Ap Rees *et al.* (1975) working with squash cotyledons suggest that gluconeogenesis occurs in the cytoplasm rather than in the glyoxosomes. It should be emphasized that a too rigid compartmentalization might hinder sugar accumulation. It will be recalled that sugar accumulation frequently occurs in response to stress during senescence. In each instance some cell disorganization may occur leading to decompartmentalization. In the last chapter we showed how this could lead to sugar accumulation in a starchy food, potato. One can readily envision how such loosening of the cell structural characteristics could lead to gluconeogenesis. Thus citrate which accumulates in the vesicles of ripe fruit could, after decompartmentalization, be transferred to the site of PFK action, inhibit this key glycolytic enzyme, thus in effect promoting FDPase and gluconeogenesis and as a consequence ultimately sugar accumulation.

Sucrose Accumulation in Nonstarchy Food Plants

Once the monoses are formed via gluconeogenesis, they are usually converted into sucrose which can be transformed into reserve polysaccharide in a process which may or may not involve translocation of the sucrose from the leaves into the more edible tissues.

As we mentioned in the last chapter, the enzymes of sucrose synthesis catalyze the reactions between UDPG and fructose or F-6-P to form UDP and sucrose or sucrose phosphate. Regulation of sucrose synthesis in such plants has been reviewed by Davies (1974). Two investigations, typical of the type of results obtained, are one of the earlier ones in this relatively new field (Hawker 1969) and that of Meynhardt *et al.* (1974). The former investigator noted that sugar accumulation in grapes was paralleled by increases in the following enzymes: sucrose synthase, sucrose phosphate synthase and sucrose phosphate monoesterase (phosphatase). Prior disappearance of glucose-6-phosphate dehydrogenase, the first enzyme of the pentose phosphate shunt, illustrates one aspect of carbohydrate regulation and accumulation—the shutting off of "escape" routes leading to transformation of the sugars through which the monoses would otherwise pass in the performance of their normal metabolic functions.

Accumulation via Translocation and the Roles of Invertase.—Meynhardt *et al.* showed that also involved in sugar accumulation are the hexokinases and hexose isomerases needed for the interconversion of glucose, fructose and their phosphates. Because of the formation of sucrose phosphate from F-6-P, these investigators suggested that these enzymes as well as sucrose phosphate phosphatase may be essential for the movement of sugars from leaves to fruit against a concentration gradient. However, the authors point to the apparently paradoxical role of the hydrolase invertase as a key enzyme in accumulation rather than in the breakdown of sucrose in ripening fruits. Invertase isozymes serve as a paradigm for the regulatory function of isozymes in general.

In addition to fixing the point of entry of the acids into the pathway of sugar accumulation *via* studies on PEPcarboxykinase action, the sugar cane plant also affords a most penetrating analysis of the role of isoinvertases in sucrose accumulation, an analysis which has been reviewed, epitomized and extended to all fleshy fruits by Coombe (1976). The following sequence of events takes place in three distinct cellular locations.

(1) *The cell wall*: Sucrose present at the cell surface is hydrolyzed by an acid invertase into glucose and fructose. This allows the sugar to enter the cytoplasm of the cell. The resulting glucose and fructose enter the cell via an ATP-energized active transfer process which does not involve invertase. Undoubtedly the existence of multiple forms of acid sugar phosphatases bespeaks of yet another layer of control of sugar accumulation at the active transport level.
(2) *The cytoplasm*: Sucrose phosphate, synthesized via the action of appropriate enzymes, is then expelled into the vacuole against a gradient in yet another energy-requiring process. Again phosphatase and invertase action contribute to regulation.
(3) *The vacuole*: In immature plants in which the cells are still elongating, action of a neutral invertase results in the accumulation of reducing sugars. But with ripening this invertase disappears and sucrose accumulates.

The message for the synthesis of the cell-surface free acid invertase is already present in the cell in a latent or masked RNA. A function of the plant hormone abscisin is apparently to unmask this messenger RNA. Thus, the conclusion by Alleweldt *et al.* (1975) that there is a cause-and-effect relation between abscisic acid level and sugar accumulation in grapes suggests a similar mechanism for other fruits and vegetables which do not possess a starch-sugar interconversion system.

This understanding of the hitherto recondite roles of invertase may have some practical applications. Thus, invertase action appears to play a role in the prolongation of cut flower life which occurs when the stems are soaked in sucrose solutions instead of water (Hawker *et al.* 1976; Chin and Sacalis 1977). Apparently the cell wall-associated acid invertase is involved, along with sucrose synthase, in the translocation of the sugars to the flower petals. These are kept fresh and viable in part by the ensuing vigorous metabolism afforded by an abundance of metabolizable food. We suggest that quenching of O_2^* and other active oxygen species by sugars may also play a role. Perhaps some of these ideas can be applied to prolong the shelf life of fresh produce.

Sugar Accumulation in Honey.—Honey may be considered as processed food in which the processing occurs in the absence of human intervention. Most of the sugar in flower nectar gathered by the honeybee is sucrose, while the honey itself is largely invert sugar plus some trisaccharides, such as erlose, which may be looked upon as glucosyl-1,4-sucrose (i.e., fructose-glucose-glucose). As expected the enzyme responsible for its formation is invertase. Also present are amylase and an interesting α-galactosidase with maltase (α-glucosidase) activity (Echigo and Takanaka 1973).

The invertase is, along with α-amylase, secreted from food glands in the thorax of the honeybee, rather than from the hitherto suspected nectar or pollen (Rinaudo *et al.* 1973) and appears to be an α-glucosidase rather than a furanosidase (Barker and Lehner 1972). This enzyme also appears to be involved in sex differentiation; worker but not queen bees contain glucose (Ishay *et al.* 1976). The enzyme is extremely thermostable as it exists in the honey, due undoubtedly to substrate stabilization (Chapter 5). The enzyme has not been thoroughly characterized but some progress is being made towards its purification. It is also present in the hemolymph of workers but not in larva (Bounias 1976). Its activity along with that of amylase has long been used as a mark of quality and prior treatment, as discussed in Chapter 37. Honeybee invertase has been used to convert sucrose syrups to bee food.

Postharvest Action of Invertase in Dates.—A ripe date suitable for eating is, physiologically speaking, in a much more advanced stage of senescence than other fruits when the latter are considered ripe. The *sine qua non* for this circumstance is intense invertase action on abundant substrate (Hasegawa and Smolensky 1970). This action not only produces sweetness but also contributes to color (via NEB, Chapter 19) and also to textural qualities. Poor quality can be converted into high-quality dates either by potentiating the dates' own invertase via cell disruption or by adding an external source of the enzyme. Purification and characterization of this 2-subunit glycoprotein enzyme was accomplished by Al-Bakir and Whitaker (1978).

Enzymes of Sugar Accumulation in Starch–containing Foods

We have previously elaborated schemata relating to enzymes of the starch-sugar interconversion cycle in connection with the undesirable development of reducing sugars in cold-stored potatoes. In general, of course, the accumulation of sugar in fruits is desirable and the enzymes which are expected to be involved are being actively investigated. Singh and Sanwal (1976) report that one of the isoenzymes of banana, α-glucan phosphorylase, is under allosteric regulation and is most probably involved in the regulation of starch metabolism since it is variously affected by nucleotides and glycolytic intermediates.

We also have had occasion to refer to the investigations of Amir *et al.* (1971) on adding pyrophosphate to ears of corn to retain their fresh sweetness (Chapter 14). Amir ascribed this effect to the product inhibition of ADPglucose phosphorylase and not to the chelating properties of PP, as was suggested by Marshall (1976) in connection with translocation of the sucrose from endosperm tissue through the membrane at the edge of the kernel. Although Amir could not find any increase in sweetness as did Marshall with other chelators such as EDTA, chelates may still play a role in promoting the translocation via the PEP-HPr phosphotransferase system (Chapter 4) discussed below in connection with sorbitol. PEP and PP could also be involved in this increase in sweetness by virtue of their being cosubstrates for the sugar accumulation "gateway" enzyme of gluconeogenesis, PEPcarboxykinase, as discussed above.

The general topic of starch-sugar interconversion, especially during development, germination and senescence, was reviewed by Turner and Turner (1975) and Meisel and Ulmann (1974). In contrast to most enzymes involved in sugar accumulation in low temperature of stored potatoes, a true *de novo* synthesis of sugar accumulation enzymes via genome derepression (transcription) or mRNA unmasking (translation) occurs in germinating plants such as barley, malt and wheat used for brewing and baking (Chapters 31 and 32).

Amylase.—Another sugar forming enzyme made during germination is α-amylase. Its putative role in the starch-sugar interconversion was discussed in Chapter 19. It only appears to be involved when excessive amounts of rapidly utilized starch constitute the sole source of carbon for rapidly growing, differing plant tissue.

Sweet Potatoes.—The action of α-amylase is exploited in many different ways as we have already and shall have more opportunity of demonstrating throughout this book. One of the more interesting applications of the exploitation of α-amylase action is in the processing of dehydrated sweet potato flakes.

The references cited in Walter *et al.* (1976) encompass a rather intriguing case history of the use of enzyme potentiation with minor excursions (as with dates) into the application of commercial enzyme preparations and *in situ* potentiation *via* "curing" process (interval of storage of the whole roots at elevated temperatures). In the procedure finally adopted, the action of the enzyme in slurries is potentiated by brief exposure (5 to 10 min) to temperatures from 75° to 80°C as shown in Fig. 20.4. In contrast to other heat-induced enzyme potentiation, in this case the substrate, starch, is made accessible to the enzyme by heating. Heat causes the starch to gel and thus become susceptible to the relatively low α-amylase activity present. Most of the work of breaking down the starch falls to the potent β-amylase present at relatively high levels (Schwimmer 1947C). This sequence of action accounts for the persistent presence of both relatively highly polymerized dextrins on the one hand, and maltose as the sole oligo- or disaccharide on the other. It will be recalled that β-amylase attacks both amylose and amylopectin from the nonreducing end to form maltose only, and, in contrast to α-amylase, cannot attack raw uncooked starch even at high levels (Chapter 9). Postharvest action of amylase is responsible for increase in sweetness due to the accumulation of maltose in many other starch-containing foods. Thus Park *et al.* (1971) reported that both amylase activity and maltose content as well as starch content were higher in southern peas held at 7°−10°C than those held at

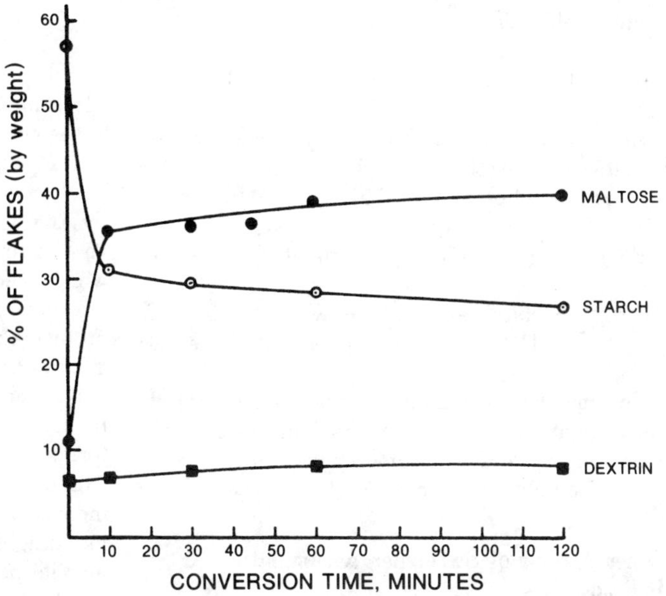

FIG. 20.4. CHANGES IN AMYLOID CARBOHYDRATES DURING DEHYDRATED SWEET POTATO FLAKE MANUFACTURE

From Walter et al. (1976)

24°–27°C. A metabolic block at the low temperature channelled the smaller amount of starch degraded into accumulated sugar instead of into respired metabolites.

Like the action of many of the enzymes we have met, that of α-amylase results not only in flavor improvement of dehydrated sweet potatoes and other foods but also in color (*via* NEB) and several texture-related attributes such as "mouth feel" and feeling of the appropriate "moistness" due to the size redistribution of the remaining starch, the dextrins, and of course the sugars.

Enzymes of Sorbitol Accumulation

Another substance closely related to the sugars which probably contributes to the sweet taste of rosaceous fruits—plums, apples, pears—is the polyol sorbitol. The closely related mannitol is also present in fruits but usually in amounts too small to be tasted; however, it is the key to carbohydrate metabolism in some of these species (Stacey 1974). Several studies have shown, for instance, that CO_2 is photosynthesized principally to sorbitol and that it is readily convertible to and translocated from leaves to fruit along with fructose (DeVilliers *et al.* 1974). However, almost all knowledge of the pathways and the enzymes involved is conjectural and we must fall back on studies with microorganisms and mammalian tissue.

Sorbitol can be formed from and converted to fructose by sorbitol dehydrogenase, also known as L-iditol dehydrogenase.

$$\text{D-Sorbitol} + NAD^+ \rightleftharpoons \text{L-Fructose} + NADH + H^+$$

At least in bacteria sorbitol is translocated across membranes *via* the PEP HPr protein acceptor system (DeLobbe *et al.* 1975).

$$\text{PEP} + \text{HPr} \longrightarrow \text{HPr-P} + \text{Pyruvate}$$

$$\text{HPr-P} + \text{Sorbitol (external)} \longrightarrow \text{HPr} + \text{Sorbitol-6-phosphate (internal)}$$

The membrane is impermeable to the sorbitol-6-phosphate which, once formed inside the cell, cannot leave and then after phosphatase action may accumulate as free sorbitol.

Other Aspects of Sweetener Accumulation in Foods

In this section we briefly give some attention to other roles of enzyme action in the production of carbohydrate sweeteners both *in vivo* and in postharvest and processed foods not previously alluded to in this chapter. The enzymatic production of sugar as food sweeteners was dealt with in detail in Chapter 9. We also pointed out that invertase as a commercial food enzyme is assuming less importance in cane sugar manufacture. In refining of cane sugar endogenous invertase action is unwanted and considerable pains are taken to see that it does not occur. Unwanted invertase action is avoided by painstakingly separating the cane stems from the tassels, which are a rich source of invertase.

Three other enzymes merit the attention of cane sugar refiners: raffinase or α-galactosidase, dextran sucrase and dextranase. In Chapter 8 we mentioned the use of immobilized α-galactosidase "fixed" within certain fungi to remove raffinose, the trisaccharide which interferes with the subsequent sucrose crystallization. In this connection some plant foods such as the coconut kernel are fairly good sources of α-galactosidase. Balasubraniam *et al.* (1976) describe the interconversion of two forms of this enzyme catalyzed by a protein which itself has no α-galactosidase activity.

A significant portion of the sucrose in stale sugar cane juice can be lost by conversion into the ropy polysaccharide dextran, α-1,6-glucan. This is accomplished by the action of dextran sucrase elaborated by the microbes which develop in such juices.

$$\text{Sucrose (nF-G)} \longrightarrow \text{Dextran } (G_n) + \text{Fructose (nF)}$$

In some sugar cane mills shutdown-related sugar losses can amount to as high as 14% and the resultant ropy mixture is almost useless unless the dextran is first removed. This can be accomplished by the addition of commercial preparations of an enzyme (termed Glucanase D-1, Talozyme) which specifically hydrolyzes the α-1,6- linkage in dextran, dextranase. This enzymatically caused problem and its enzymatic solution are reviewed in a series of symposium papers introduced by Wells and James (1976). Formation of dextran can also be a problem in beer manufacture (as well as in formation of dental plaque).

Inadvertent glycosidase enzyme action leading to increased sweetness may occur in wine and juice manufacture. The source of such action is commercial pectin degrading enzyme used for nonflavor related purposes (Chapter 30). While such action may be desirable, at least in the case of juice manufacture, the yeast-derived action of invertase on sucrose added to certain wines is considered undesirable (Burkhardt 1973).

Lactose Synthase.—A discussion of the biological sources of sweetness in foods would not be complete without at least a nodding reference to that most singular of biosynthetic pathways responsible for the one sugar capable of being transiently accumulated by mammals, lactose. As briefly limned in Chapter 4, this is one of the textbook illustrations of the regulatory properties of enzymes. The galactosyl transferase function of lactase synthase complex is utilized for synthesizing lactose only during the brief period when the mammal is producing milk. This milk contains the unique protein α-lactalbumin whose regulatory function, as pointed out by Hill and Brew (1975)—that of binding monosaccharides, unlike that of any other regulatory protein—resembles more closely that of coenzymes, metals and other cofactors. For most of the life of the mammal, the main function of the transferase, which appears to have evolved from lysozyme, is to catalyze the formation of a β-1,4-linkage of galactose to *N*-acetylglucosamine present in certain proteins. The consequence of this unique combination of two completely diverse proteins—lactose—has provided opportunities, problems and challenges to food scientists and technologists, as detailed throughout this book.

Short-circuiting of Photorespiration

In the future a sophisticated and rational approach to increased *in vivo* sugar accumulation via manipulation of enzymes is that promised by the investigations of Ogren and colleagues (Ogren and Hunt 1978; Somerville and Ogren 1980). They produced mutants of a crucifer, mouse-ear cress, in which photorespiration, but not photosynthesis, was strongly repressed. Although mainly for boosting food crop productivity, such an approach might be also be useful in manipulating the levels of sugar and organic acids, and thus the flavor of many foods. Due to photorespiration most major C3 photosynthesizing food plants lose about 25% of the potential sugar produced photosynthetically via CO_2 fixation. The target reaction for suppression of photorespiration is the initial one, oxidation of ribulose-1,5-bisphosphate. The Illinois investigators succeeded in producing cress mutants in which the oxygenase function, but not that of carboxylase, of the key enzyme ribulose-1,5-bisphosphate carboxylase-oxygenase is severely repressed. They have also produced mutants which are deficient in the photorespiration-associated enzymes serine-glyoxylate aminotransferase and phosphoglycolate phosphatase. Suppression of photorespiration may become especially urgent in conjunction with the anticipated introduction of the free energy-costly nitrogen fixing enzyme systems via genetic engineering into hitherto N-fertilizer requiring food plants (Andersen *et al.* 1980). Of course the crux of the success of this approach depends on the non-essentiality of this energy-extravagant process for plant survival. If, as has been suggested (Heber and Krause 1980), photorespiration plays a role, as does superoxide dismutase (Chapter 16), in ameliorating the potential baneful effects of oxygen during photosynthesis, especially under the stress of water dearth and perhaps salinity, then it may be cogently relevant to look for mutants in which the *carboxylase* function of the enzyme is preferentially suppressed and the oxygen function is enhanced. Such mutants of temperate-adapted food plants might be able to survive in the desert and thus open up much more area to food production which may more than compensate for lowered yields.

Finally, we point out that enzymes may, in the future, play some role in the biochemistry and applications of some of the alternatives to carbohydrate sweeteners, such as the dihydrochalcones (Horowitz and Gentili 1963), stevioside derivatives produced by α-glucosyl transferase (Miyake 1980), and others.

ENZYMES OF FLAVOR POTENTIATOR ACCUMULATION

Although the lion's share of the gustators, glutamate and 5′-nucleotides, are consumed in foods as additives in industrialized countries with highly developed food processing industries, these substances nevertheless do accumulate naturally *in vivo,* postharvest and postmortem in many foods to reach levels at which they contribute to the enhancement of overall flavor.

Enzymes of Glutamate Accumulation

This important metabolically active amino acid which, through its amide, glutamine, plays such an important role in the regulation of all biological life processes is normally present in substantial quantities in foods as the glutamyl moieties of γ-peptides, proteins and their hydrolytic products, in γ-glutamyl peptides such as glutathione, glutamine and free glutamate. It is also present in several intermediate metabolites in trace amounts which however taken all together may constitute an appreciable fraction of the total. In the free form it may constitute at least half of the normal amino acid pool. In general the level of glutamate tends to remain constant while that of other amino acids changes. Glutamate would thus tend to contribute to food flavor under conditions of rapid amino acid utilization such as takes place during accelerated protein synthesis, i.e., when the tissues are differentiating or laying down a reserve protein. This would include germinating and senescent food tissues.

Upon death of the food tissue occasioned in the food processing and both postharvest and postmortem, there would probably be a tendency for accumulation due to indiscriminate action of glutamate-metabolizing enzymes, as pointed out in Chapter 1. Of course there would be an acceleration of both formation and degradation (turnover) of glutamate due to decompartmentalization of enzymes and substrates, but the net result is likely to be accumulation. A survey of the possible enzyme reactions involved suggests that those enzymes leading to glutamate formation would tend to more than offset those causing its removal. Because few studies have been directed towards factors (especially enzymatic) leading to changes in glutamate as a flavoring in foods, we deem it appropriate to survey known enzyme reactions which could lead to glutamate accumulation.

Enzymes Which Produce Free Glutamate.— These may be classified into the following categories: (1) α-peptide hydrolases, (2) γ-peptide hydrolases, (3) enzymes using α-oxo(keto)glutarate as cosubstrate, (4) enzymes using glutamine as a substrate usually involved in anabolic pathways, and (5) enzymes catalyzing reactions in the catabolism of several amino acids.

α-Peptide Hydrolases.—Proteinase action in foods undoubtedly releases glutamate in amounts capable of contributing to food flavor. Only when autolysis is a major step or where an exogenous source of proteolytic enzyme is used would glutamate from protein make a *major* contribution to the flavor of a food. Undoubtedly the action of proteinases from the milk, the rennet starter, ripening and other desirable microorganisms which contribute to the flavor of cheese during their ripening liberates glutamate as well as some glutamyl peptides (Chapter 33). Similarly, the copious amounts of glutamate formed in soy and fish sauces owe their origins to a mix of autolytic and microbial proteinase action (Chapters 28 and 34). The use of proteinases for the purpose of preparing monosodium glutamate, MSG, as a food additive will be touched upon below. Noguchi *et al.* (1975) isolated broth-like glutamic acid-rich oligopeptides from fish protein hydrolyates (Chapter 34).

Another α-peptide, pteroyl glutamic acid (folic acid), is hydrolyzed by a special peptidase, carboxypeptidase G_1. This enzyme will also hydrolyze the terminal glutamate from any peptidyl-L-glutamate.

γ-Glutamyl Peptidases.—Since folic acid exists as γ-glutamyl conjugates, it is not surprising that an enzyme exists capable of removing glutamate stepwise. This enzyme, γ-glutamylglutamate carboxypeptidase specifically catalyzes the following reaction: Pteroyl-(-L-glutamate)$_n$ + H_2O → pteroyl-(-L-glutamate)$_{n-1}$ + L-glutamate.

Enzymes capable of hydrolyzing the γ-glutamyl peptide linkage in glutathione and other γ-glutamyl peptides, as well as γ-glutamyl transpeptidases (γ-glutamyl transferase) which can function hydrolytically (Austin and Schwimmer 1971) undoubtedly contribute to the glutamate pool in many foods.

α-Ketoglutarate Converting Enzymes.—At least three enzymes utilize this keto(oxo)acid, HOOC-CO-CH_2-CH_2-COOH, to produce glutamate. In addition to the traditional and well-known transaminases or aminotransferases

$$\alpha\text{-Ketoglutarate} + RCHNH_2COOH \rightarrow RCOCOOH + \text{Glutamate}$$

and the reversal of glutamate dehydrogenase

$$\alpha\text{-Ketoglutarate} + NADH + H^+ \rightarrow \text{L-Glutamate} + NAD^+$$

α-ketoglutarate is utilized in the relatively recently discovered reactions catalyzed by glutamate synthase (Miller 1976). The Enzyme Commission distinguishes between an NADP-dependent and a ferredoxin-dependent enzyme although the former is stated to be, like ferredoxin, an iron-sulfur protein:

$$\text{Glutamine} + \alpha\text{-ketoglutarate} + NADPH + H^+ \text{ (or 2 reduced ferredoxin)} \rightarrow 2\ \text{Glutamate} + NADP^+ \text{ (or 2 oxidized ferredoxin)}$$

TABLE 20.5

GLUTAMATE FROM GLUTAMINE VIA BIOSYNTHETIC PATHWAYS:
GLUTAMINE + A + B → GLUTAMATE + ANH_2 + C

A	B	ANH_2	C
α-Ketoglutarate[1]	NADPH	Glutamate	$NADP^+$
α-Ketoglutarate[2]	Ferredoxin (red)	Glutamate	Ferredoxin (ox)
Fructose-6-P[3]	—	Glucosamine-6-P	—
Ribose-1,6-diP[4]	—	5-P-ribosylamine	—
Xanthosine-5′-P[5]	ATP	GMP	AMP

[1] Glutamate synthase (NADPH) once considered to be a transferase, now an oxidoreductase.
[2] Glutamate synthase (ferredoxin).
[3] Glucosamine isomerase.
[4] Amidophosphoribosyltransferase.
[5] GMP synthetase.

This light-dependent ferredoxin chloroplast enzyme plays a central role in regulating the nitrogen metabolism of green plants including, of course, many green vegetables. Because glutamate derepresses the gene coding for a closely associated ferredoxin enzyme, nitrate reductase, this glutamate synthase is probably also involved in protein accumulation (Nicklish *et al.* 1976). The bacterial enzyme has been used for the enzymatic determination of glutamate (Miller 1976).

Enzymes Utilizing Glutamine.—Glutamate synthase is but one of several glutamine-utilizing enzymes which catalyze one particular step in the biosynthetic pathway of individual amino acids and several other food constituents. The reactions follow the general pattern

$$\text{Glutamine} + A + B \rightarrow \text{Glutamate} + ANH_2 + C$$

The well-established and related ones, their enzymes and the biosynthetic pathways to which they belong are shown in Table 20.5. Glutamate is also derived from glutamine by the ubiquitous hydrolase glutaminase. Other glutamate forming enzymes catalyze steps in catabolic pathways usually *via* oxidative degradation. Not all foods contain these enzymes nor would all be effective in influencing the level of glutamate, but they can form a rational basis for a study of what factors affect the level of glutamate and hence flavor in the absence of MSG additive in many foods. While it is unlikely that any one of these 100 odd glutamate-producing reactions would contribute significantly to the accumulation of glutamate and flavor, the summation of their effects in food tissues could very well contribute to the free glutamate pool, thus flavor, especially under conditions of severe cell disruption.

Production of Glutamate as a Food Additive.—The manufacture of this gustator provides an interesting case in which fermentation has apparently won out over both enzymes and acid. In the early days glutamate was made by acid hydrolysis from wheat or from beet sugar waste in the United States and in Japan from tangleweed and dried bonito and later from soy and similar foods in which proteinases have acted extensively (Nasuno and Nakadai 1971). Efforts have been made to increase the glutamate content in protein hydrolysates by combining a microbial proteinase action with that of glutaminase (Yokotsuka *et al.* 1974). However, recent developments in fermentation technology have resulted in several patents, mostly from Japan, which have been reviewed by Okada (1977), in which yields as high as 8% (w/v) of the culture broth (Arima *et al.* 1977) and recoveries accounting for as much as 60% of the carbon of the added nutrients have been reported. Success as judged from these patents seems to have depended upon proper organism selection (mostly strains of *Brevibacterium* and *Corynebacterium*), use of surfactants and penicillin, which actually stimulates the production, and special care to prevent phage contamination. With such success it is unlikely that use of enzymes even upon immobilization (Rokugawa *et al.* 1979) will in the near future be competitive. A kindred area, production of soy sauces and related seasoning by protease action, will be discussed in Chapter 34.

Enzymatic Aspects of 5′-Nucleotide Accumulation and Retention

To refresh the reader's memory, it will be recalled that adenine, guanine and hypoxanthine, the purines most directly involved in food flavor, are, respectively: 6-amino-, 2-amino-6-oxy- and 6-oxypurine. The corresponding nucleosides or more correctly ribonucleosides (position 9 of the purine covalently linked to the 1′- position of the pyranosyl group of ribose) are adenosine, guanine and inosine; the corresponding 5′-phosphates of the nucleosides are adenosine 5′-phosphate (AMP), guanosine 5′-phosphate (GMP) and inosine 5′-phosphate (IMP) or, alternatively, adenylic acid (adenylate), guanylate and inosinate. The distinction between 3′- and 5′-nucleotides resides in the position on the ribose to which the phosphate is attached, as symbolized in Fig. 2.1. Of the nucleotides only GMP and IMP bear the proper gustatory quality, but AMP is intimately involved in the formation of IMP.

5′-Nucleotides as Natural Flavorants.—Like glutamate, the 5′-nucleotides are present in foods *in vivo* and even more *in situ* after harvest, processing, storage and cooking at levels adequate to contribute to gustatory attributes. Indeed they frequently occur together and contribute synergistically with glutamate to overall flavor, a circumstance applied by manufacturers of flavor potentiators. The bonita (tuna) which is considered to be the historical source of nucleotides contains 500–600 mg/100 g. However, the purple seaweed, long considered to be one of the ancient sources of glutamate, owes its organoleptic properties as much to 5′-nucleotides as to glutamate (and alanine) (Shimazono 1964; Kuninaka *et al.* 1964; Kojima 1974; Noda *et al.* 1975).

The principal 5′-nucleotide contributing to flavor in all mammalian muscle foods, meats and in most vertebrate fish is IMP. Thus, Kojima reported that mammalian muscle tissue contains *in vivo* 0–3 mg whereas nonmuscle tissue contains as high as 30 mg per 100 g of tissue. The threshold of effectiveness as a gustator lies in the order of magnitude of 1–10 mg/100 g. Most invertebrate seafoods accumulate IMP whereas prawns occupy an intermediate position, accumulating both IMP and AMP (Hiltz *et al.* 1974).

In plant food, GMP is the principal gustatory nucleotide. It, along with other nucleic acid-derived nucleotides, has been reported in potatoes, peas, asparagus, tomatoes and orange juice. It is particularly abundant in the shiitake mushroom. References to these reports include those of Shimazano (1964), Mouri *et al.* (1970), Buri and Solms (1971), Davidek *et al.* (1972) and Barmore and Biggs (1972). Its effectiveness in potentiating aroma as well as taste is indicated by the report of Schinneler *et al.* (1972) that GMP lowers the odor threshold for octanol. One food whose flavor is not affected by endogenous 5′-nucleotides is beer because of the low but detectable levels present (Pickett *et al.* 1974).

Enzymes of Nucleotide Accumulation.—The three principal sources of nucleotides as natural constituents of foods are: through biosynthesis starting with glycine; from ribonucleic acids via ribonuclease (phosphodiesterase) action; and from ATP through the action of ATPase adenylate kinase (myokinase) and AMP deaminase.

From ATP.—The IMP found in meat and fish is derived from ATP present in these muscle foods to power the transduction of its chemical energy to the mechanical energy of muscle motion. Hence the enzymes involved also participate as determinants of the texture of these foods as discussed in Chapter 28. Three major ones are ATPase or apyrase (I), nucleoside kinase (II) and AMP deaminase. The latter is considered in most organisms to be the first enzyme on the pathway of purine degradation. It appears to be an allosteric enzyme and is thus inhibited by 2,3-biphosphoglycerate (BPG) (Sasaki *et al.* 1976) which also is an allosteric regulator of the uptake of oxygen hemoglobin as briefly discussed in Chapter 18. This suggests that naturally occurring phosphates such as phytic acid, which takes the place of BPG in avian blood (Chapter 30), could conceivably hinder accumulation of desired IMP in poultry meat, and that Ca salts could promote accumulation by tying up these organic phosphates. Undoubtedly postmortem cell disruption would tend to favor the catabolic degradation of the purines rather than their biosynthesis. AMP deaminase has been used to prepare seasoning by adding (along with cell wall-destroying enzymes) to 5′-nucleotides obtained from the autolysis of suitable microorganisms (Takeda 1966).

From Biosynthetic Pathways.—The first distinct compound containing an authentic purine in purine biosynthesis is IMP which, as mentioned, is also the first product of purine catabolism. Hence it might be possible to favor its accumulation to improve overall flavor rationally by intervening in this biosynthesis (Fig. 20.5).

Starting with glycine, nine enzyme reactions take place to reach IMP; each enzyme in effect adds alternatively a carbon and a nitrogen atom. The IMP can then be converted to either AMP which from the flavor point of view is undesirable or the

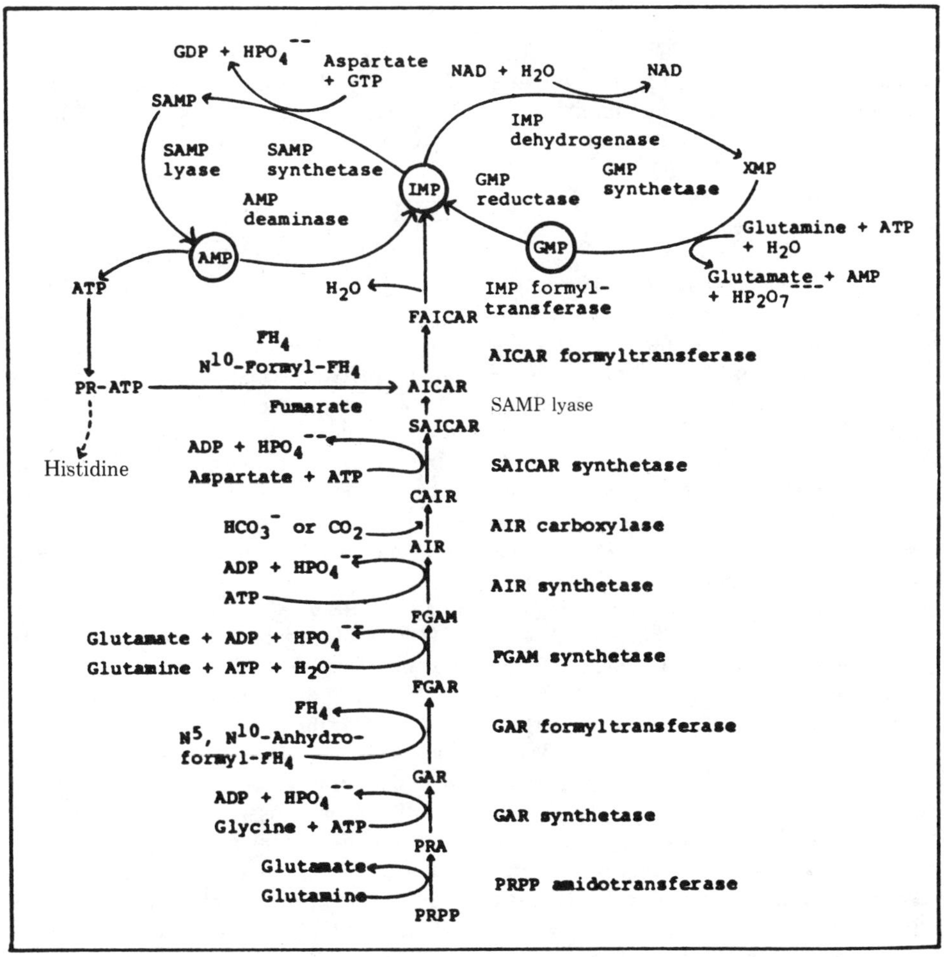

PRPP : 5-phosphoribosylpyrophosphate
PRA : 5-phosphoribosylamine
GAR : glycineamide ribotide
FGAR : formylglycineamide ribotide
FGAM : formylglycineamidine ribotide
AIR : aminoimidazole ribotide
CAIR : 5-amino-4-imidazole carboxylic acid ribotide
SAICAR : 5-amino-4-imidazole-N-succino-carboxamide ribotide
AICAR : 5-amino-4-imidazole carboxamide ribotide
FAICAR : 5-formamido-4-imidazole carboxamide ribotide
SAMP : adenylosuccinic acid
PR-ATP : 1-(5′-phosphoribosyl) adenosine triphosphate

From Shibai et al. (1978)

FIG. 20.5. BIOSYNTHESIS OF NUCLEOTIDE GUSTATORS, IMP AND GMP

desirable GMP by each of distinct pairs of enzymes. One might, for instance, find a highly specific nonpoisonous inhibitor of one of the enzymes responsible for conversion of IMP to AMP, adenylosuccinate (SAMP) synthetase:

$$\text{IMP} + \text{GTP} + \text{L-aspartate} \longrightarrow \text{GDP} + \text{Pi} + \text{adenylosuccinate}$$

The product, adenylosuccinate, is a fairly good competitive inhibitor (K_i + *ca* ^{-4}M) (Fromm 1958), is fairly inexpensive to synthesize and easily degradable after accomplishing its task. However, a transition state analog would probably be even more suitable.

The enzyme catalyzing the final reaction leading to AMP, adenylosuccinate lvase

$$\text{Adenylosuccinate} \longrightarrow \text{fumarate} + \text{AMP}$$

is inhibited by one of the intermediate metabolites of the purine biosynthesis, 5′-phosphoribosyl-4-

carboxamide-5-aminoimidazole (AICAR). It is conceivable that in the disrupted cell this inhibition would be enhanced beyond its normal regulatory function and lead to enhanced accumulation of IMP. We bring this up to point out that there may be unsuspected sources of quality-affecting substances brought about by nonphysiological enzyme action in foods after harvest and slaughter.

From Nucleic Acids and Nucleotide Coenzymes.—In plant foods the major enzyme action leading to the gustatory nucleotides is the ribonucleases (RNAase, RNase, phosphodiesterase) which hydrolyze the ester linkage between the orthophosphate and the 3-position of the ribose to liberate 5′-nucleotides, with the linkage to the 5-position intact. Wilson (1975) has provided an excellent survey of the plant nucleases. Unlike the traditional distinction between DNases and RNases found to hold for many of the microbial and animal nucleases, a class of the plant nucleases which liberate 5′-nucleotides is equally at ease in degrading both RNA and DNA. They do, however, maintain rigid specificity with regard to the ester bond hydrolyzed (5′ or 3′) and also with regard to endo- and exo- action. Endonucleases are most likely involved in flavor nucleotide liberation since they attack the 3′-ribose-phosphate linkage optimally at pH 5–6.5, close to that of most plant foods. The exonucleases act and exhibit pH optima in the range of 7–9. Also widely distributed in plants are certain RNases which may not be as important in flavor development because they release the 2′:3′-cyclic nucleotides. They are of significance in flavor development only in that they could compete with the flavor-releasing nucleases. Nucleases have been detected in the following plant foods: carrot, sweet corn, peas, potatoes, spinach, wheat, rye, barley, mung bean and canteloupe.

Judging from reports on the increase in nucleotides upon heating certain raw foods, it would appear that the action of the nucleases is potentiated at least in the early stages of cooking of some foods. These include potatoes (Buri and Solms 1971), mushrooms (Shimazano 1964; Mouri *et al.* 1970; Davidek *et al.* 1975), sweet corn and asparagus (Mouri *et al.* 1968) and probably many other vegetables. The latter authors showed that the action of RNases resulting in improved flavor can be promoted in asparagus by adjustment of pH from 6 to 7.5 (which probably allows exonucleases to participate) and in mushrooms and shrimps by freezing. In this connection we should point out that although much of the nucleotide flavorant in vertebrate food comes from the action of ATP and the action of AMP deaminase, under some conditions it may also arise as the result of the action of RNase. During cooking, the nucleotides indeed could be released by nonenzymatic hydrolysis of the phosphate ester linkages. Even the most stable of these bonds is amenable to hydrolysis at 90°C (Davidek *et al.* 1972). The pyrophosphate bonds in ATP and other nucleoside triphosphates and similar bonds in coenzymes such as ADPG, NAD and FAD are even more unstable. These cofactors are undoubtedly attacked by phosphatase as well as by the phosphodiesterase, which acts on cyclic AMP and pyrophosphates present in all foods.

In normal catabolism of AMP the nucleotide substrate is, as mentioned, IMP. The nucleotidases differ from previously discussed phosphate associated hydrolases in that they are true phosphomonoesterases, which act on a C-O-P bond and not, like the nucleases, acting on one of the bonds in C-O-P-O-C (phosphodiesterases), nor like the pyrophosphatases and ATPases, capable of splitting the P-O-P-bonds. As isolated, and probably *in vivo*, nucleotidases are highly specific for different nucleotides. However, nucleotides in postmortem food tissue are probably also acted upon by nonspecific phosphatases due to the decompartmentalization ensuing upon cell death.

The actions of these enzymes are frequently used as indices of freshness of foods, especially fish. An extensive literature has developed which shows that the amount of such action seems to correlate highly with extent of cell damage in frozen fish. Kemp and Spinelli (1969) concluded that the flavor-contributing effect of frozen-and-thawed (slacked) fish is the same as in fresh fish only if the former was frozen and stored at or near a temperature of −29°C. In shrimp AMP degrades somewhat faster than does IMP, although the latter was found to be present initially at higher levels than in mammalian muscle. Similarly, rates of AMP degradation and of hypoxanthine accumulation in scallops at −5°C were similar to those in the unfrozen food (Hiltz *et al.* 1974). AMP seemed to go directly to hypoxanthine with no buildup of IMP. These results suggest that in shellfish the AMP is hydrolyzed directly to adenosine and then to adenine which is then deaminated to hypoxanthine by a purine deaminase. Hypoxanthine is tasteless.

Very little recent progress has been made on the characterization of the specific nucleotidases and even less on nucleosidases actually responsible for removal of flavor nucleotides or even in demonstrating that the removal is really enzymatic. Yamamoto *et al.* (1967) described an alkaline carp phosphatase with a strong affinity ($K_m=10^{-5}$) for IMP. Mandal and Mukherjee (1974) found substantial differences in the rate of IMP dephosphorylation among extracts of various fish species. These differences were echoed in the postmortem rate of the disappearance of IMP in whole fish during storage. Similar nucleotide and nucleoside degrading enzymes have been detected in potatoes (Reese and Duncan 1972) and in milk (Caulini *et al.* 1973).

Retention of Flavor by Prevention of Nucleotidase Action.—These nucleotide degrading enzymes act, of course, just as readily on added substrates in the form of commercial flavor potentiators as well as on endogenous substances. Mainly because of the former, considerable effort has gone into finding ways of preventing these enzymes from acting. For the most part such measures, compiled in Table 20.6, consist of using nucleotidase inhibitors but also include illustrations of the use of low moisture and pH extremes (Chapters 13 and 14). Many of them consist of the purely empirical addition of plant tissue extracts.

TABLE 20.6

PROPOSALS FOR PREVENTING NUCLEOTIDASE ACTION IN FOOD[1]

Inhibition by:
EDTA[2]; histidine (5168–67); diethyl pyrocarbonate (17981–69); phosphate esters (8694–68); extracts of animal tissue (5708–68); *Uncaria gambier* leaves (24817–70); *Areca catechu*—betel nuts (20547–70); seeds of *Cinnamomum cassia* (20545–70); *Vaccinium vitis*—blueberry (20544–70); *Pterocarpus marsupium*—gum kinos (20543–70); bloodwort (20542–70). Treat with ammonia, used primarily as preservation measure.[3] Lower moisture in spices, wheat and soy flour to less than 7%.[4]

[1] Numbers in parentheses refer to Japanese patents issued to Takeda Chemical Industries, Ltd.
[2] Source: Groninger and Spinelli (1968).
[3] Source: Mandal and Mukherjee (1974).
[4] Source: Ishii *et al.* (1968).

Some of the suggested chemicals, such as diethylpyrocarbonate, can no longer be used. One of the suggested inhibitors, histidine, was first reported to be complexed to IMP as the flavor component of bonito by Kodama in 1913 but this apparently was later retracted (Kojima 1974). EDTA was specifically used to inhibit nucleotidase in refrigerated fishery products (Groninger and Spinelli 1968) presumably by the binding of Mg^{2+}, an obligatory cofactor for many but not all nucleotidases. Because of a lag period between EDTA application and its effective inhibition of the enzyme, this chemical worked only in fish whose extracts exhibited low nucleotidase activity.

The purine-degrading enzymes, especially nucleosidases and nucleoside phosphorylase, may be considered as determinants of other aspects of food flavor and color in the same sense that other glycosidases which produce reducing sugars provide reactants for nonenzymatic browning. Both ribose and ribose phosphate have been implicated as important precursors or reactants of flavor components which develop in cooked meat. Furthermore, hypoxanthine is a good substrate for xanthine oxidase, which as we have discussed is a powerful generator of superoxide anion, and other forms of excited oxygen which in turn can affect the properties of foods in so many diverse ways.

Production of 5′-Nucleotides.—In Chapter 8 we pointed out that these gustators were being commercially produced via the agency of immobilized enzyme action on yeast ribonucleic acid plus chemical synthesis to convert all of the purine into flavorants (Shibai *et al.* 1978). As in the case of glutamate production, improved fermentation technology threatens to replace prepared enzymes as the principal industrial source of 5′-nucleotides. Interestingly, the same *Bacterium brevis* which has proved so useful in glutamate production has also been used to obtain increased yields of 5′-nucleotides. As in glutamate production, the presence of certain antibiotics as well as drugs, such as the barbiturates, enhances yield (Schwartz and Margolith 1972; Okada 1977).

21

ENZYME ACTION AND AROMA GENESIS IN VEGETABLES

ORIGIN OF VOLATILE FLAVORANTS IN FOODS

The ensuing five chapters are almost entirely devoted to the enzymology of volatile flavor development in foods. In this chapter we shall briefly inspect reactions leading to desirable odorants in foods in general, assess the overall role of enzymes in such development and then direct our attention to enzymatic aspects of desirable volatile flavor genesis in vegetables. Inevitably there will be overlapping, which we hope to minimize, with odor development in other foods.

In an opening address to a symposium on aroma research, Drawert (1975) drew a sharp distinction between "primary" volatiles present in the cells of the living food tissues and "secondary" volatiles formed as the result of processing. Similarly, Schutte (1974) distinguished among "bioformation," "enzymes" and "heat" as the three main modes of formation of volatile flavor compounds from nonvolatile precursors. To this we might add "light" which can form undesirable flavors such as that in "sunstruck" beer (prenylmercaptan, $(CH_3)_2C{=}CHCH_2SH$) and perhaps in milk. Indeed one might include all forms of energy input discussed in Chapter 10. Chase (1974) divided fresh food flavors into those that are produced by biosynthetic enzymes in living tissue and those produced by degradative enzymes which begin acting only when the tissues are cut or crushed. We believe that a more realistic approach is lookıng upon the process of food volatile generation as a continuum starting at one extreme from true biosynthetic accumulation of the substance in intact cells in the food tissue as part of the normal metabolism, i.e., the presence of terpenoid in special cells of fruits and essential oil-bearing plants. Closely allied events leading to volatile aroma production would be those involving intracellular membrane destabilization which might occur during stress and "curing" in postharvest plants and "aging" in postmortem meat, processes which accelerate with increasing maturity and eventual senescence in plants. The volatiles accumulate without gross macroscopic change in appearance of the foods. Further down the line, flavors arise from some injured cells or from altered membrane properties leading to altered metabolism (Chapter 1) in the intact cell under stress.

In many fresh vegetables the aroma develops almost instantaneously upon mastication. For the most part the perceived odor is the result of production of volatiles *via* enzyme action although in some instances the preformed volatile is merely liberated mechanically from the disrupted cell by the chewing. Further along in this continuum we come to tissue comminution and accompanying cell disruption in processing situations designed for the purpose of liberating and producing the flavorants such as in the manufacture of essential oils and beverages such as tea and cocoa. In fermented foods both endogenous enzymes from the plants or animals and enzymes derived from microorganisms cooperate in creating the proper proportion of flavorants which give the characteristic desirable flavor we associate with foods such as cheese, other dairy products, alcoholic beverages, fish-derived sauces, herring, meat sausages and many pickled foods. Surprisingly some of the newly found, highly intense odorants such as the 2-methoxy-3-pyrazines and the fused ring volatile terpenoid, geosmin (see below), appear to be formed in trace amounts from microorganisms in vegetables, the former compounds from *Streptomyces* and the latter from *Pseudomonas* (Buttery 1977, 1980). It may very well turn out that other odor-bearing volatiles presumed to be "preformed" arise as the result of such subtle microbial metabolism. The microorganism and the plant may form a symbiotic flavor generating system.

The opposite end of the continuum is occupied by added commercial food enzymes, frequently of microbial origin (Chapter 5), which produce flavorful volatile compounds either in the food directly or, as in the case of sugars and gustators, used in the manufacture of food ingredients as discussed in Chapters 9 and 20. Although of promising potential, the former application so far finds practical outlets in the use of lipase for cheese and candy manufacturing (Chapter 23).

The production of volatile flavors serves to illustrate, perhaps more than that of other quality-bearing food constituents, how a network of myriad secondary compounds is formed from a limited number of primary products of enzyme reactions. The time scale for development of this secondary network extends from milliseconds in the case of the odors perceived when fresh salad vegetables such as cucumbers, onions and radishes are chewed, to hours in cooking and baking giving rise to the seemingly limitless volatile components resulting from pyrolytic decomposition and Maillard reactions among substances present at ambience. The task of elucidating the nature of the enzymes ultimately responsible for the formation of these natural flavor components which have been almost completely revealed by GLC in the past 15 years presents formidable challenge to food enzymologists. In the ensuing discussion we shall attempt to inform the reader of the status of the progress in meeting this challenge.

In this endeavor the enzymologist has the accumulated knowledge of intermediary metabolism as a backdrop. This repository of biochemical lore suggested to Tressl and Renner (1976) that volatiles arise from: (1) mevalonic acid which gives rise to the highly aromatic terpenes, (2) branched chain amino acids (valine, leucine) to yield the branched chain aliphatic aldehydes and some of the allyl and branched alkyl volatile sulfur compounds, and (3) benzenoid compounds to yield the aromatic aldehydes. Another source would be the lipids leading to the unsaturated alk(en)yl aldehydes. The role of intermediary metabolism as exposited by Salunkhe and Do (1976) in their excellent compendium of the aroma constituents of each of 20 fruits and vegetables suggests that just about every major nonvolatile constituent of foods which is present in substantial quantities can contribute to creation of aroma. We largely confine our remarks to those reactions in which an enzyme is at least putatively involved.

In general the development of solid information on the biogenesis of volatile aroma-bearing substances in fresh foods falls into a general pattern comprising the following idealized sequence. Papers which establish the presence of individual components of the aroma profile as revealed by GLC contain discussions which frequently call upon the above-mentioned repository of knowledge of intermediary metabolism to speculate on the nature of the precursors and reactions leading to the observed volatiles. The first valid data testing these hypotheses are often obtained through experiments in which radioactivity-labeled suspected precursors are fed to intact tissue and the volatiles then reisolated. This eliminates many possibilities and strengthens others.

At this juncture studies may appear specifically designed to ascertain if the substance is preformed in the intact tissue. To do this all enzyme action has to be prevented during isolation and preparation of the volatiles for analysis. The GLC pattern of the resulting aromagram is compared with that prepared in such a way as to promote enzyme action. The latter is usually not as much of a problem as is the former. Methods for inhibiting enzyme action include the use of very low temperatures combined with "inhibitors," usually inactivators, especially protein denaturants and precipitants such as trichloracetic acid. Heat inactivation is proscribed not only because the delicate volatiles are frequently heat sensitive but also because heating often complicates subsequent separation procedures. Such an "inhibitor" approach was used to establish that monoterpenes comprising the flavor of parsley leaves (Freeman *et al.* 1975), dithiolanes contributing to asparagus flavor (Tressl *et al.* 1977) and the 2-thio 2-isobutylthiazole of tomatoes are preformed *in vivo* and are not formed as the result of enzyme action ensuing from tissue and cell disintegration. However, as mentioned, such compounds may also result from symbiotic microbial activity. If such experiments indicate enzyme action, they may be followed by studies showing that tissue extracts, and hopefully eventually purified enzyme and substrate *in vitro*, can give rise to the selected volatiles. Conclusions may be followed up and secured by establishing high correlation among enzyme action, precursor disappearance, volatile or accompanying product production and production of aroma as measured subjectively by a flavor panel.

Origins of Enzymatically Produced Flavors in Vegetables

In this discussion we shall confine ourselves largely to those desirable substances whose enzymological origins are fairly well understood and arise as a result of cell disruption of vegetable tissue. This does not preclude the possibility that the action of these enzymes and the volatiles they form do not contribute to the flavor, both desirable and undesirable, of other foods or that enzymes and flavorants not mentioned here do not contribute to vegetable flavor and its genesis.

The principal volatiles in vegetables from the viewpoint of their contribution to fresh flavor and also with regard to the role of enzyme action in their creation may be categorized as follows:

(1) The medium length straight chain C_5-C_{10} aliphatic aldehydes and ketones.
(2) Branched chain aldehydes.
(3) Aliphatic alcohols.
(4) Sulfur volatiles ultimately derived from the S-substituted cysteines found mostly in the alliums.
(5) Sulfur volatiles ultimately derived from the mustard oil thioglucosides (glucosinolates) present mainly in crucifers.
(6) Other odor-bearing volatiles such as esters, aromatic aldehydes, terpenes present in fresh vegetables which have the sweet aromatic aroma usually associated with nonvegetable foods such as fruits, many essential oils and some beverages.
(7) Other odorants which are quite important but whose biological metabolic origin and enzymological credentials remain at least one order of magnitude more obscure than the substances listed above, such as the potent methoxypyrazines and the sedanolides (phthalide derivatives responsible for celery flavor) will be briefly scanned at the end of this chapter.
(8) Finally there are the vast number of volatiles formed upon cooking. Many but not all are cyclic compounds, i.e., the ubiquitous methional, $CH_3{\cdot}S{-}CH_2CH_2{\cdot}CHO$ (threshold two parts per 10^{10}) and H_2S. Some aromas of cooked vegetables, as previously noted, arise from amino acids via nonenzymatic browning Maillard-type reactions and Strecker degradations (Chapter 19).

ALDEHYDES AND KETONES

Aldehydes

While the development of the lower aldehydes such as acetaldehyde has long been associated with the development of off-flavors in frozen and other processed vegetables, many of the unsaturated alkenals in the C_5 and C_{10} range at appropriate levels and in optimal proportions impart desirable flavor attributes to fresh vegetables. In particular, much attention has been paid to the 6-carbon "green leaf" aldehyde. It appears under various names such as hexenal and more precisely as hex-2-enal, hex-*trans*-2-enal, hex-*trans*-2-en-1-al, 2E-hexenal (E and Z replacing *trans* and *cis*). The true "fresh leaf" odor may be due to its more potent precursor hex-*cis*-3-enal which is rapidly isomerized to the *trans*-2 isomer. Mouthfeel and "blending" as well as aroma properties have been attributed to hexenal at least when present in canned tomato juice (Kazeniac and Hall 1970). Other alkenals and alkenones present in the volatiles of vegetables are shown in Table 21.1. It should be pointed out that the contribution to overall perception and acceptability of a given aldehyde is highly dependent upon its absolute and relative concentration with respect to other aldehydes and even other volatiles. Typical cucumber and melon-like odors are due to 9-carbon nonenals. Like other natural aromas, some flavors are used as flavor additives in nonvegetable foods, i.e., the patenting of Z-2-nonenal as coffee flavor enhancer (Parliment *et al.* 1976).

TABLE 21.1

OCCURRENCE OF C_4 TO C_9 ALIPHATICS IN FOOD VOLATILES

C Atoms, Main Chain	Side Chain CH_3	Functional Group in Main Chain -ene	-al	-ol	Chirality E/Z	Food
		Position in Chain				
4	—	—	1	—	—	Beans, oranges
4	—	—	—	2,3[2]	—	Peas, dairy
4	3	—	1	—	—	Tomatoes
4	3	—	—	1(2)[3]	—	Grapes (raspberry)
4	2	—	—	—	—	Lima beans
4	—	2	1	—	—	Onions
5,6	—	—	—	—	—	Peanuts, soybean
5	—	—	—	1	—	Tomato
6	—	—	1	(1)	—	Most fruits, vegetables
6	—	—	—	1	—	Apricot, most fruits
6	—	3	1	—	Z	Freshly cut vegetables
6	—	—	2	—	E	Raw vegetables
6	—	—	3	—	—	Many vegetables
6	2	—	3	—	—	Tomato
7	—	2	1	—	E	Potatoes, peas
7	—	—	—	1	—	Grapes
8	—	—	1	(1)	—	Peanuts
8	—	3,5	—	—	—	Beans
9	—	6	1	—	E	Melon
9	—	2	1	—	—	Carrots, cooked
9	—	2,6	1	—	E,Z	Cucumber, freshly cut
9	—	3,6	1	(1)	Z,Z	Cucumber

[1] E = *trans*, Z = *cis*.
[2] An "-one" function at position 2.
[3] Indicates alternate function or position.

Origin.—Aliphatic aldehydes may arise and accumulate in vegetables as a result of enzyme action from the following precursors:

(1) From esters via hydrolysis by appropriate esterase followed by oxidation, presumably by NAD(P) dependent dehydrogenases, of the liberated alcohol to the corresponding aldehyde:

$$R_1COOR_2 \rightarrow R_2CH_2OH \xrightarrow{NAD(P)} R_2CHO$$

(2) From amino acids via the combined sequential action of transaminases (I) and appropriate α-ketodecarboxylases (II).

$$\text{Amino Acid} + \alpha\text{-ketoglutarate} \xrightarrow{I}$$

$$\alpha\text{-ketoacid} \xrightarrow{II} \text{aldehyde}$$

(3) From interruption of the classical β-oxidation step of the spiral catabolic pathway for fatty acids.

$$RCH{=}CHCOSCoA \longrightarrow$$

$$RCH{=}CHCHO\ (+\ CoASH)$$

A similar abortion of β-oxidation has been invoked to explain the production of β-methyl ketones, 2-alkanones (Chapter 22).

(4) From nonenzymatic oxidation of alcohols via the indirect action of phenolase (I) (Chapter 15):

$$\text{Phenols} \xrightarrow{I} \text{quinones}$$

$$\text{Quinones} + \text{alcohols} \longrightarrow \text{phenols} + \text{aldehydes}$$

(5) From fatty acid via lipase followed by lipoxygenase action, probably the most important single source responsible for the desirable fresh flavor of tomatoes, cucumbers, peanuts, string beans and other leafy salad vegetables. Lipids as the origin of leaf aldehyde were first suggested by Nye and Spoer (1943).

From Fatty Acids: Evidence for the Participation of Lipoxygenase.—The development of definitive information on the biogenesis of the unsaturated aldehydes in vegetables has provided an apposite paradigm for the application of the idealized pathway of research outlined earlier for aroma research in general. The pertinent investigations include one or more of the following types of experiments.

(1) establishment of the presence of the alkenal, usually the "leaf" aldehyde or one of the nonenals;
(2) showing that the volatile is generated enzymatically by the general methods outlined above;
(3) establishment of linoleic and/or linolenic acid as the precursor/substrate by the use of ^{14}C-labeled acids injected into the intact vegetable before comminution;
(4) partial purification of the associated enzyme which usually has typical lipoxygenase properties;
(5) repetition of the tracer experiment with purified enzyme or a subcellular organelle fraction;
(6) showing that the 2E-hexenal arose *via* isomerization of 3Z-hexenal. An analogous isomerization of nonenal occurs;
(7) showing that the 3Z-hexenal came directly from the 13-hydroperoxide of linolenic acid;
(8) demonstrating that there is adequate linoleic or linolenic acid to account for the aldehydes and that these acids decrease under condition of aldehyde formation;
(9) frequently demonstrating that the formation of the aldehyde is dependent on the prior action of lipase to liberate the linolenic acid from its linkage to glycerol;
(10) occasionally discussing the relative merits of invoking lipoxygenase action as compared to nonlipoxygenase pathways.

Based upon this information, with each group of investigators placing special emphasis on their own work, pathways have been proposed for the formation of 2E-hexenal in tomatoes (Stone *et al.* 1975, Galliard *et al.* 1977) and tea (Sekiya *et al.* 1976) and the formation of nonenals in cucumber extracts by Hatanaka *et al.* (1975) and especially by Galliard and coworkers (1976); for enzyme studies see Wardale and Lambert (1980). The principal aroma-bearing constituent is believed to be 2E-6Z-nonadienal—"violet leaf perfume" (Sondheimer 1952)—derived from linolenic acid (18:3) via isomerization of 3Z-6Z nonadienal. 2E-Nonenal from linoleic acid and other C_9 aldehydes also imparts character impact to the overall flavor. The circumstance that determines whether the flavor-bearing components are six or nine carbon atoms long—"leaf" *vs* "cucumber/melon" aroma—is the positioning of the hydroperoxide resulting from the lipoxygenase catalyzed insertion of oxygen. Mechanistically, the decision to form 13- or 9-hydroperoxide depends upon which of the two hydrogens attached to the asymmetric carbon-11 is removed. Removal of L_S or D_R hydrogen may yield a fleeting free radical which forms the 13- or 9-isomer, respectively. When the resulting hydroperoxide is in the 13-position, subsequent cleavage gives rise to hexenals whereas insertion into the 9-position yields nonenals. Thus, the degree of unsaturation of the resulting long-chain aldehyde depends upon whether linoleic (18:2) or linolenic acid (18:3) is acted upon. In general the more fully saturated aldehydes such as *n*-hexanal are less likely to have as strong or pleasant aromas under most circumstances than the more unsaturated. This again illustrates the crucial role of enzyme specificity as a determinant of food quality. The alternative pathways with α-linolenic acid as lipoxygenase substrate may be depicted as follows:

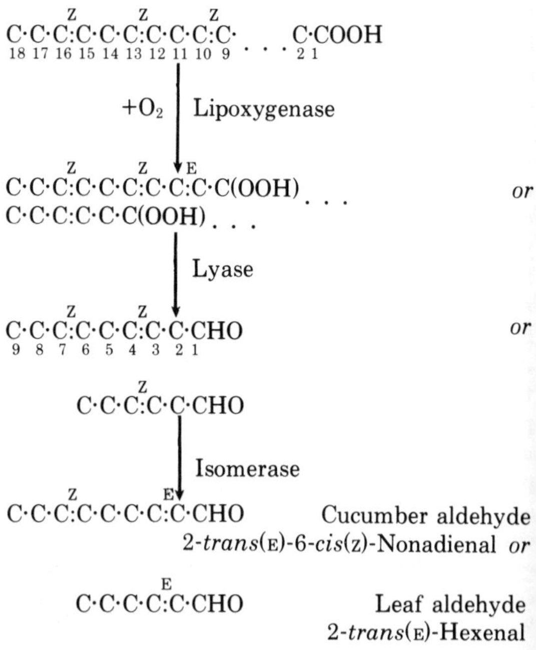

It would appear that although action of the tomato lipoxygenases results in a preponderance of the 9-lipohydroperoxide, only the 13-isomer is cleaved to form C_6 aldehydes (Galliard and Matthew 1976). Lyase activity has been detected in watermelon

(Vick and Zimmerman 1976), in cucumber (Galliard *et al.* 1976) and was proposed earlier for banana by Tressl and Drawert (1973). The specificity with regard to geometric isomerism of LOOHs (Chapter 24) should be ascertained.

One of the difficulties in establishing the validity of a lyase is that such a preparation may be accompanied by lipoxygenase which itself may participate in lipohydroperoxide cleavage. Nevertheless, Phillips and Galliard (1978) localized and partially purified a membrane-bound cleavage enzyme. In the fruit itself such activity was distributed among plasma membrane, Golgi apparatus and endoplasmic reticulum (see discussion of glucosinolase in this chapter). Even the participation of lipoxygenase or at least its hydroperoxide forming function has been questioned. Thus, Major and Thomas (1972) reported that the formation of 2-hexenal in macerated *Gingko* leaves is catalyzed by a "water-soluble substance" that is not lipoxygenase. Grosch *et al.* (1974) were able to generate typical cucumber odor from linolenic acid and singlet oxygen. They suggest that the lipoxygenase may act simply by cleavage of the double bonds of the unsaturated fatty acids, presumably through fleeting dioxetane intermediates proposed by Cilento (1975) for bioluminescent and dioxygenase systems. It is of interest in this connection that spinach, a rich source of superoxide dismutase which, it will be recalled, prevents O_2^* formation and inhibits lipoxygenase (Richter *et al.* 1975), does not produce 2E-hexenal. Although Grosch and coworkers do not eliminate the possibility of lipoxygenase action, such a presumption is not too farfetched. Other aspects of odorous volatiles derived from aldehydes via lipoxygenase, including lipoxygenase enzymology, action in food and conditions under which it imparts undesirable flavors, the importance of isoenzyme specificity of hydroperoxide isomer formation, and auxiliary secondary hydroperoxide transformations which would prevent aldehyde formation, as well as nonlipoxygenase transformation of fatty acids, will be discussed in Chapter 24. The unsaturated aldehydes can be removed enzymatically via reduction to alcohols (Chapter 22), or presumably by oxidation by "aldehyde" oxidases such as that present in potatoes which can utilize at least one alkenal, crotonaldehyde, as substrate (Rothe 1975).

Branched Chain Aldehydes.—These have been reported to be present in some vegetables such as tomatoes after tissue disruption. While their contribution to the flavor of such vegetables is uncertain, they are more likely to play a significant role in fruits as precursors to branched chain alcohols, which in turn give rise to the sweet-smelling esters associated with fruit aroma. 3-Methylbutanal (IV) in tomato volatiles probably arises from leucine (I) via the sequential action of a transaminase (II) and an α-ketocarboxylase (III)

$$(CH_3)_2\text{-}CH_2\text{-}CH_2\text{-}CHNH_2\text{-}COOH\ (I) + HOOC\text{-}CH_2\text{-}CH_2\text{-}CO\text{-}COOH \xrightarrow{II}$$

$$(CH_3)_2\text{-}CH_2\text{-}CH_2\text{-}COCOOH + HOOC\text{-}CH_2\text{-}CH_2\text{-}CH\text{-}NH_2\text{-}COOH$$

$$\downarrow III$$

$$(CH_3)_2\text{-}CH_2\text{-}CH_2\text{-}CHO\ (IV)$$

Of course the α-ketoacid could also arise from the action of amino acid dehydrogenase

$$\text{L-Amino acid} + H_2O + NAD^+ \longrightarrow \alpha\text{-Ketoacid} + NH_4^+ + NADH$$

Although prominently and widely distributed in vegetables as well as in fruit volatiles, for the sake of unity the alcohols and associated enzymes will be discussed in Chapter 22 in connection with ester formation.

SULFUR VOLATILES—THE DISULFIDES AND RELATED VOLATILES

The Sulfur Volatiles and Their Genesis

Representative of earlier reviews on plant sulfur chemistry and metabolism are those of Kjaer (1958) and Ettlinger and Kjaer (1968). The first integrated discussion encompassing the origin and genesis of the sulfur-containing odor-bearing volatile sulfur compounds in foods is that of Schwimmer and Friedman (1972). Further complementary and even more comprehensive reviews stressing the chemistry of sulfur volatiles and their nonvolatile precursors include those of Shankaranarayana *et al.* (1973) and Schutte (1974). Other reviews which cover broad aspects of the enzymology of sulfur volatile flavor production include those of Salunkhe and Do (1976) and Chase (1974). Other thorough and more specialized reviews will be mentioned later.

This upsurge of interest in sulfur flavors is the result of a multidisciplinary convergence of organic chemistry, analysis and enzymology which has led to a fairly complete understanding of two of the major sulfur volatiles formed in alliaceous and cruciferous plants which comprise some of the more strongly flavored vegetables. Some are so strong that this subject might just as well have been put under flavorants (Chapter 22) since the substances are commercially available as essential oils but not as fragrances. Perhaps more important has been the realization that, rather than the usual pejorative flavor connotation frequently associated with volatile sulfur compounds, they contribute to the flavor profile of almost every food, albeit at a somewhat lower level of odor intensity and in smaller concentrations and much more subtly than, for instance, onion flavor. Their role in undesirable flavor is probably largely confined to their development during microbial deterioration of food. Indeed their sensory effects in this regard serve, as mentioned, as an early warning signal of food unwholesomeness. The strong sulfur-like odors associated with this element are actually restricted to divalent noncyclic compounds. Their flavor properties are ultimately conferred by the unique position of S in the periodic table and its electronic orbitals; as many as ten electrons can occupy the valence shell of its third orbital when it is in the divalent state. This peculiarity also leads to its relative polarity, volatility of its divalent compounds and its diminished capacity for forming hydrogen bonds in its divalent compounds. Such divalent sulfur volatiles are more likely to be more direct products of enzyme action than those compounds in which sulfur is part of a ring. The harsher the treatment to which a food is subjected (i.e., roasting) the more the sulfur is incorporated into heterocyclic ring compounds such as thiophenes, thiazoles, thiazolines, polythiolanes and cycloalliin. In such compounds which have a degree of aromaticity (in the chemical sense, not the flavor sense), the electrons of the outer orbitals of the sulfur atom are delocalized, thus depriving the sulfur atoms of those distinctive properties conjectured to be the basis of odor quality. The odor quality of these less reactive heterocyclic compounds would appear to be determined more by molecular shape than by the presence of the sulfur atom. On the other hand, the only two sulfur volatiles in vegetables reported to have been present in vegetables before crushing of the tissue are 2-*iso*butylthiazole in tomatoes and the dithiolanes in asparagus. However, both of these were detected by GLC which could easily have converted unstable endogenous compounds into volatile heterocyclics.

Our most advanced knowledge concerning the role of enzyme action in sulfur volatile generation is largely associated with two principal classes of aroma precursors, the alkyl S-alk(en)yl-L-cysteine sulfoxides (alliins) and the glucosinolates (mustard oil thioglucosides). They both give rise via enzyme action to volatile divalent sulfur odor-bearing volatile compounds. Similarities between mustard and garlic oils were observed over 130 years ago by Wertheim, according to Liebig (1859).

The Cysteine Sulfoxides in Alliums

Garlic.—The chemistry, enzymology and some aspects of the food technology of the development of flavor in alliaceous vegetables have been reviewed by Stoll and Seebeck (1951), Carson (1967), Granroth (1970), Schwimmer and Friedman (1972), Schutte (1974), Freeman and Whenham (1976B), and, perhaps most comprehensively for onion, by Whitaker (1976). The reaction leading to odor-bearing sulfur volatiles may be formulated as

$$\underset{\text{(+)-S-Alk(en)yl-L-cysteine sulfoxide}}{R\overset{O\uparrow}{S}CH_2CH_2CHNH_2COO^-} \longrightarrow S\text{ (volatile)} + NH_3 + \underset{\text{pyruvate}}{CH_3COCOO^-}$$

The (+) denotes that the naturally occurring L-cysteine sulfoxides are dextrorotatory by virtue of the chirality of the sulfur atom. The latter is asymmetric because of the presence of the sulfoxide oxygen (four isomers of cysteine sulfoxide are possible). From the pioneering work of Stoll and Seebeck we have known for 40 years that the R in the flavor precursor in garlic *(Allium sativum)* is (+)-allyl [(+)-2-propenyl]-L-cysteine sulfoxide which the Swiss investigators termed alliin. The relatively stable isolatable sulfur-containing product of this reaction is allicin, $CH_2{=}CH\,CH_2\text{-}S\text{-}\overset{O\uparrow}{S}\text{-}CH_2CH{=}CH_2$, the allyl ester of allene sulfenic acid, or allyl allyl (diallyl) thiosulfinate. Allicin is the principal component responsible for the fresh odor of garlic, while that of cooked garlic is largely due to diallyl disulfide, $CH_2{=}CH\text{-}CH_2\text{-}S\text{-}S\text{-}CH_2\text{-}CH{=}CH_2$. The primary product of the reaction may be allyl sulfenic acid (I)

$$CH_2{=}CH\text{-}CH_2\text{-}SO\text{-}CH_2CHNH_2COOH \longrightarrow CH_2{=}CH\text{-}CH_2\text{-}SOH + \ldots$$

According to Seebeck and Stoll, two molecules of this primary product condense with the elimination of one molecule of water to yield allicin

$$2\ CH_2{=}CH\text{-}CH_2\text{-}SOH \rightarrow \text{Allicin} + H_2O$$

However, this primary product has not been detected and alternative mechanisms have been proposed (Granroth 1970). The enzyme in garlic which catalyzes the formation of allicin was called alliinase by Stoll and Seebeck. The latest official EC recommended designation is alliin lyase. Uniquely, the classification is based upon a yet unidentified reaction product: S-alkyl-L-cysteine sulfoxide→2 *aminoacrylate* + an alkyl sulfenate. Other designations for this and similar enzymes are 3-alkylsulfinyl alanine alkylsulfinatelyase (deaminating), S-alkylcysteinesulfoxide lyase, S-alkylcysteine lyase. The latter term has been reinstated (1976) to designate enzymes which act on both cysteine and cysteine sulfoxide derivates.

Onion.—The situation is a bit more complicated in onion as illustrated in Fig. 21.1. Fresh onion odor is due to thiosulfinates (RSSOR) which can undergo disproportionation to disulfides

$$R\text{-}S\text{-}SO\text{-}R + R\text{-}S\text{-}SO\text{-}R \longrightarrow R\text{-}S\text{-}SO_2\text{-}R + R\text{-}S\text{-}S\text{-}R$$

When comminuted onions are heated, the unstable thiosulfinates rapidly disappear and one finds mainly disulfides, trisulfides and cyclic thiophenes (Schutte 1974). The latter, but not thiosulfinates, survive GLC treatment.

Although the propyl and methyl derivates of L-cysteine sulfoxide are also present in onion, the odor of freshly sliced or comminuted onion is largely due to *trans*-1-propenyl derivatives (Schwimmer 1968A,1969D; Freeman and Whenham 1976A). However, it shows up as propyl propenyl disulfide and perhaps as propyl disulfide in GLC analyses. Just how the original 1-propenyl shows up as propyl in GLC awaits clarification. Not only is the propenyl derivative the principal source of odor, it is also responsible for the sensory attributes evoked when the structural integrity of the onion cell is destroyed by mastication, slicing or other means of comminution. When one bites into an onion, almost immediately the tongue burns and the eyes water. Within minutes a freshly prepared comminuted onion suspension will develop a bitter alkaloid-like taste and after several hours the suspension will turn pink as discussed in Chapter 17. All of these attributes have been shown to occur with the propenyl derivative as sole substrate as depicted in Fig. 21.1. Presumably, the primary product of the reaction is the propenyl sulfenic acid which rearranges very rapidly to form the still unstable but detectable (Wilkins 1961; Brodnitz and Pascale 1971) and analytically measurable (Tewari and Bandyopadhay 1977; Freeman and Whenham 1975; Lukes 1971) lachrymator, thiopropanal S-oxide, 95% of which is, when isolated, in the *syn*

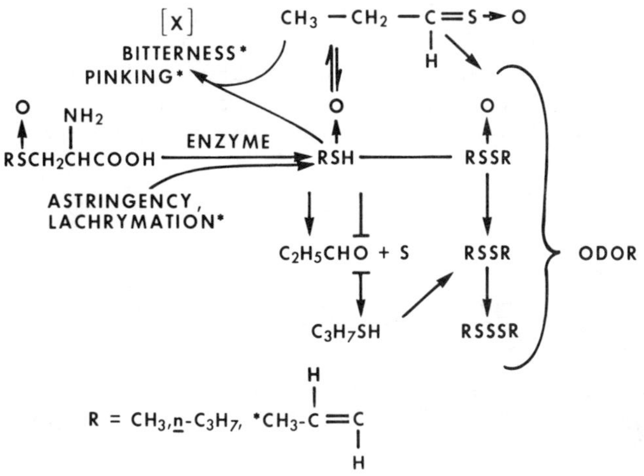

FIG. 21.1. DEVELOPMENT OF FLAVOR AND COLOR IN ONION VIA ALLIINASE ACTION

form (Block *et al.* 1979), thus echoing the chirality of the glucosinolates (see below), except that here we are dealing with a product and not a substrate of enzyme action.

Parenthetically, health-associated properties have also been attributed to garlic and onions. These include antibiotic activity (Mantis *et al.* 1978), goitrogenic activity due to the disulfides (Saghir *et al.* 1966), fibrinolytic (anti-blood clotting) activity (Menon *et al.* 1968) and prevention of heart disease, perhaps in part due to the reported ability of some cysteine derivatives to lower blood cholesterol and perhaps, in part, to prostaglandin A_1 (Attrep *et al.* 1973). Indeed, most food volatiles possess some sort of physiological activity. Of course, onions are not eaten for this purpose. In addition to their flavor function, they also impart texture as betokened by their use in hamburgers. About the only food attribute not associated with onions, as pointed out by Freeman and Whenham (1976), is any nutritional effect (as distinct from the above-mentioned other health-associated properties).

The Enzymes.—Further evidence of the primacy of the S-1-propenyl derivative of cysteine sulfoxide in flavor production is afforded by its kinetic behavior as substrate of onion alliinase. Thus, its affinity for the enzyme (as measured by K_m) is six times that of the methyl derivative and its V_{max} is at least four times that of the alkyl derivatives. The relative contribution of the propenyl derivative to the overall reaction rate in the presence of the other two cysteine sulfoxides, for a strong onion, amounts to at least 98%.

In common with other alliinases, only S-substituted (both alkyl and aryl) derivatives of L-cysteine sulfoxide serve as substrates. Neither the closely related cycloalliin (Fig. 21.2) (probably formed during the processing of dehydrated onions) nor the corresponding S-alkyl-L-cysteines are substrates (Schwimmer *et al.* 1964A). The latter, however, do inhibit by a partially competitive mechanism (Chapter 14). In general the naturally occurring (+)-diastereomer is a better substrate than the levorotatory synthetic isomer (because of the three asymmetric centers, eight isomers of S-1-propenyl cysteine sulfoxides are possible). The alliinases (as we shall refer to them for the sake of simplicity) are phosphopyridoxal enzymes and presumably catalyze α,β-elimination reactions in a mechanism similar to that depicted in Fig. 21.3. When partially pure, they are stimulated by addition of their coenzyme pyridoxal phosphate (Schwimmer 1964B). The purified onion enzyme, not so affected, is a 50,000 dalton monomeric phosphopyridoxal glycoprotein (Tobkin and Mazelis 1979). The enzyme and substrate are most concentrated in the inner part of the bulb.

Biosynthesis of the Flavor Precursor.—Thanks largely to the elegant tracer work of Granroth (1970), we are now fairly certain that the propenyl side chain is biosynthesized from valine by a series of reactions following roughly that depicted in Fig. 21.4 (coenzymes and cofactors not shown). Only beginnings have been made in showing the enzymes involved. With the aid of *Albizzia* cysteine lyase (see below), Schwimmer and Granroth (1975) demonstrated conversion of S-(2-carboxyethyl)-L-cysteine to the S-1-propenyl derivative by injecting the former into the onions. The cysteine moiety appears to arise from the action of cysteine synthase from O-acetyl serine (Granroth 1974).

$$CH_3COCH_2CHNH_2COO^- + RSH \longrightarrow RS\text{-L-Cysteine} + \text{Acetate}$$

It is to be expected that such information will eventually be applied in controlling or maximizing the level of the precursor.

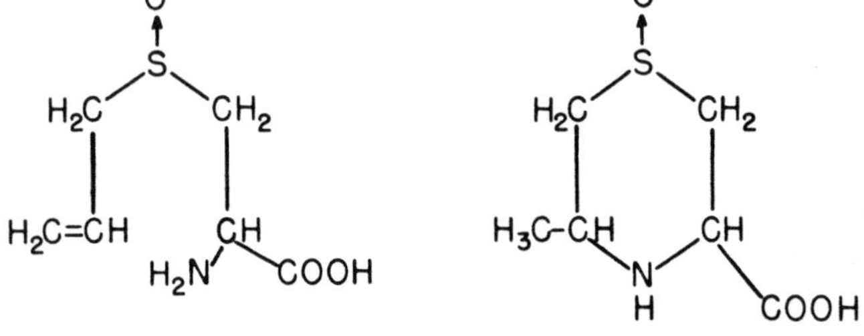

Alliin　　Cycloalliin

FIG. 21.2. STRUCTURE OF ALLIIN, PRECURSOR TO GARLIC FLAVORING, AND CYCLOALLIIN

Cycloalliin is an easily formed flavorless transform of *onion* flavor precursor (Fig. 21.1)

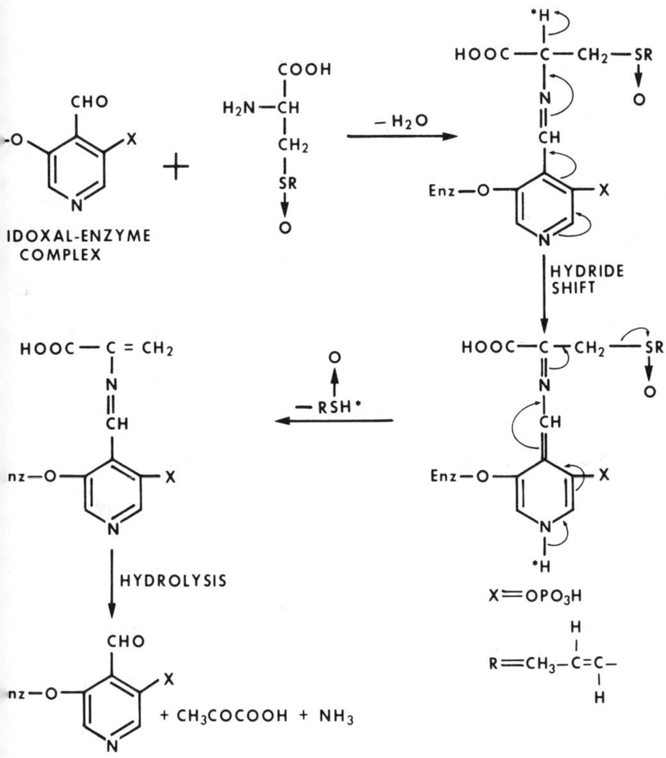

FIG. 21.3. PROPOSED MECHANISM OF ALLIINASE CATALYSIS INVOLVING PYRIDOXAL PHOSPHATE

From Schwimmer and Friedman (1972)

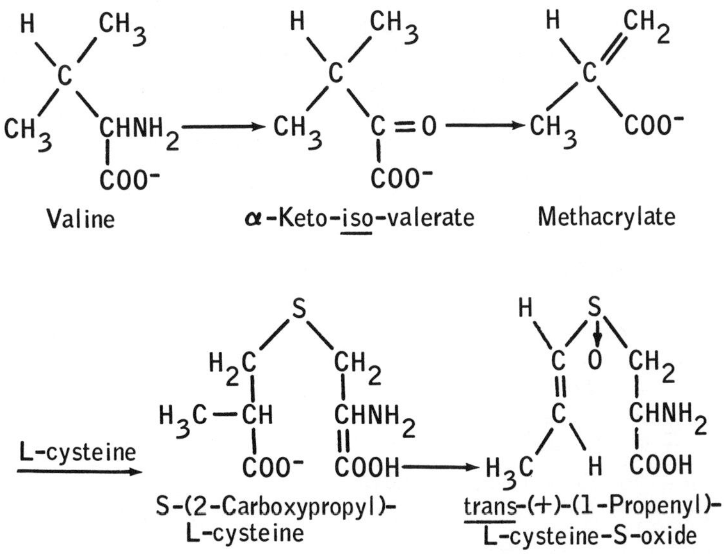

FIG. 21.4. PROPOSED BIOSYNTHETIC PATHWAY FOR ONION FLAVOR PROGENITOR

From Granroth (1970)

Control and Retention of Enzyme-produced Flavor Potential.—While it is true that it is not always desirable to retain the full enzymatic flavor potential of the fresh onion in many of the uses of this vegetable (pickled, browned or even when eaten raw), in dehydration it is desirable to do so. Indeed the commercial onion dehydration process is a compromise between rapid drying and retention of enzymes and substrate. Nevertheless measurement of the flavor potential indicates that a large proportion of this potential is lost (Schwimmer *et al.* 1964B; Bernhard 1968; Freeman and Whenham 1974). As shown in Fig. 21.5 and Table 21.2, most of the loss is due to disappearance of substrate rather than to enzyme. Freeman and Whenham (1976A,B) have examined in detail the various factors which go into determining the final level of substrate, starting from preharvest through postharvest and processing. For instance, the importance of an adequate supply of sulfur in the soil is illustrated by the complete absence of flavor precursor in soils extremely deficient in sulfur. Just as sugars accumulate in potatoes in response to stress, so are the onion flavor precursor stress metabolites accumulating in particular in response to meager water regime. During the postharvest storage the flavor potential increases for a few months due to liberation of the flavor precursor by γ-glutamyl transferase action (see below) and is followed by a decrease probably due to the entry of the flavor precursors into the common amino acid-nitrogen pool to be used as nutrients for sprouting. Another possibility is that the substrates are attacked and degraded by active oxygen such as ·OH, O_2^*, etc. (Nishimura and Mizutani 1973). On the other hand, loss of flavor potential in irradiated onions during storage has been attributed to an effect on the enzyme rather than on the substrate (Kawkishi *et al.* 1971; Nishimura and Mizutani 1975).

Other Alliums and Vegetables

Substrates.—Fujiwara *et al.* (1955) were probably the first to survey plant species for the disulfide flavor-producing system by allowing the enzymically generated thiosulfinates to react in slightly alkaline solutions with thiamin (Chapter 36) or cysteine to form crystallizable, easily quantifiable compounds. Because they used alkali, they did not detect the most important product of the action of the enzyme on the S-1-propenyl derivative. Similarly, earlier surveys of *Allium* species by means of GLC of head space confused 1-propenyl-containing disulfides with allyl disulfide (Bernhard 1968). Perhaps the most complete survey to date is that of Freeman and Whenham (1975) using a combination of analytical approaches. From a survey of 27 *Allium* species and cultivars, these investigators discerned three classes based upon the S-substituent in principal alliinase substrate: (1) S-1-propenyl-, (2) S-2-propenyl-(allyl), and (3) S-ethyl-L-cysteine sulfoxides. The flavor precursor is the principal substrate of alliinase action. Class (3) species were mainly ornamental. To class (1) belong not only onion but also shallots, chives, leek and rakkyo, which is usually used in pickled and blanched products. In addition to garlic, class (3) includes Chinese chives and wild onion. To this class we may add the Japanese vegetable, ainunegi (caucus, *A. victorialis*) which produces relatively large amounts of methyl allyl disulfide probably responsible for the distinctive odor of this vegetable (Akashi *et al.* 1975). As expected, class (2) members have little or no lachrymatory potency. Besides the historical isolation of the garlic ami-

TABLE 21.2

EFFECT OF PROCESSING OF ONIONS ON ENZYMATIC FLAVOR PRODUCTION SYSTEM[1]

Substrate Transformed by:	Freeze-dried[2]	Frozen[2]	Frozen[3]	Dried[3]	Boiled[2]	Pickled[3]
Endogenous alliinase	67	59	3	19	4	18
" + added alliinase	75	91	46	19	73	23
Nonenzymatic means	2	5	48	78	24	41
Alliinase activity	45	18	7	10	5	0.01

Source: Freeman and Whenham (1976B).
[1] Results expressed as percentage of that of fresh onion as estimated from pyruvate production (Schwimmer and Friedman 1972).
[2] Laboratory prepared samples.
[3] Commercially prepared samples.

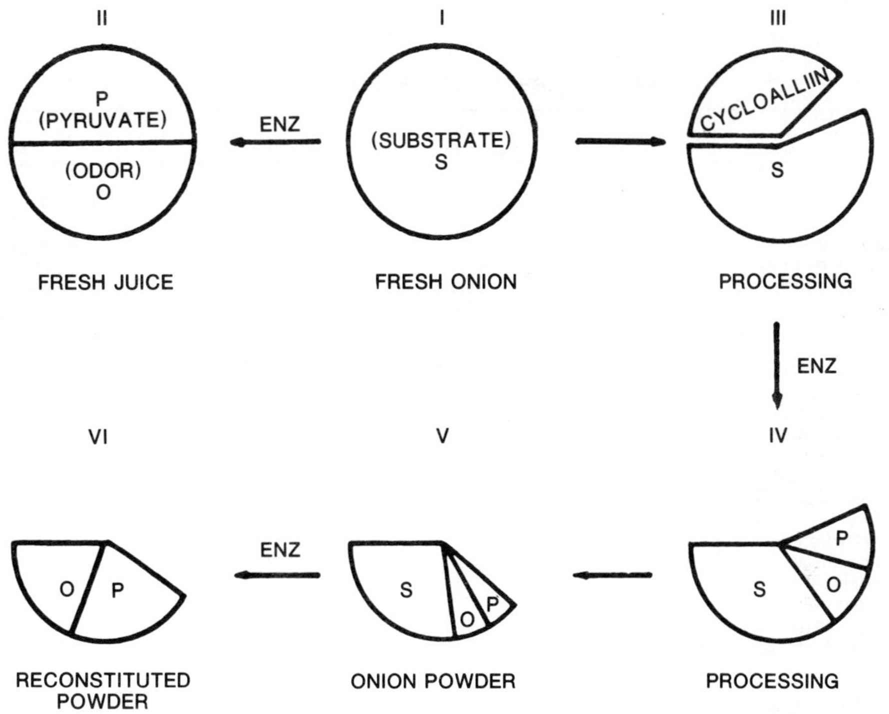

From Schwimmer et al. (1964B)

FIG. 21.5. FATE OF FLAVOR PRECURSORS AND FLAVOR DURING AND AFTER ONION DEHYDRATION

no acid, so far only onion has been subject to a thorough direct free amino acid analysis by Matikkala and Virtanen (1967), who found an overwhelming preponderance of the S-1-propenyl derivative. Other studies of other alliums have relied on identification of secondary volatile products using improved GLC. Thus, Dembele and Dubois (1973) reported that shallots vapor contains more 1-propenyl than does onions, and Schreyen *et al.* (1976) showed that carefully prepared headspace condensation of leek volatiles contains several 1-propenyl and propyl-containing disulfides.

Other Cysteine Lyases.—Enzymes of food-related interest which cleave the C-S bonds are shown in Table 21.3. As far as is known they are all phosphopyridoxal enzymes. Some will act only on the sulfoxide of cysteine while others will act on S-substituted derivatives of cysteine as well as on sulfoxide derivatives. Those whose specificity restricts their action to the sulfoxides fall into two distinct subgroups. The enzymes from onion and from *Brassica* foods such as cauliflower and cabbage act optimally at pH 8−9 and the reaction can be readily made to go to completion at these pH's. The enzymes from garlic and from a related member of the Liliaceae, *Tulbaghia violacea*, undergo reaction inactivation at their pH optima in a slightly acid medium. In this pH range, the onion enzyme also undergoes reaction inactivation, suggesting that a protonated intermediate, still enzyme-bound, apparently causes inactivation. This intermediate may be a reaction of the primary product, sulfenic acid, with the coenzyme pyridoxal phosphate. The best substrate for each enzyme is its natural substrate insofar as this has been ascertained and where investigated isoenzymes have been detected.

Of special interest is the shiitake mushroom responsible for the formation of the highly aromatic 1,2,3,5,6-pentathiepane from the γ-glutamyl-L-cysteine sulfoxide lenthionine. Like the *Albizzia* enzyme, the shiitake cysteine lyase will not utilize the peptide in which it is present in the mushroom but does act on S-substituted cysteine derivative with and without an oxygen on the sulfur.

The *Albizzia* enzyme was found to be useful as a "flavorese" in study of the kinetics of the development of onion odor in frozen onion (Schwimmer and Guadagni 1967), as an adjunct in the study of

TABLE 21.3

CYSTEINE CARBON-SULFUR LYASES IN PLANTS

Source	Sulfur Specificity	pH Optimum	Natural Substrate
Allium sativum, garlic[1]	-SO-	5.6–6.4	Allyl[10]
Allium cepa, onion[2]	-SO-	7.4–8.8	1-Propenyl[10]
Tulbaghia violacea[3]	-SO-	6.6	Ethyl[10]
Brassica, cabbage, etc.[4]	-SO-	9.4–9.7	Methyl[10]
B. napobrassica, rutabaga[5]	-S-S-	9.10	Cystine
Lentinus edodis, mushroom[6]	-S-, -SO-	8.9	DGLA[11]
Albizzia lophanta (seeds)[7]	-S-, -SO-	8.4	Djenkolic acid
Penicillium corymbiferum[8]	-SO-	6.5	Allyl[10]
Pseudomonas spp.[9]	-S-	8.8	—
Acacia farnesiana[5]	-S-, -SO-[12]	8	—

[1] Source: Stoll and Seebeck (1951); Mazelis and Crews (1968).
[2] Source: Schwimmer (1971A); Tobkin and Mazelis (1980).
[3] Source: Jacobsen *et al.* (1968).
[4] Source: Mazelis (1963).
[5] Source: Mazelis (1975).
[6] Source: Iwami *et al.* (1975).
[7] Source: Schwimmer and Kjaer (1960).
[8] Source: Durbin and Uchytil (1971).
[9] Source: Nishizuka (1971).
[10] S-Substituent of L-cysteine.
[11] Desglutamyl lentinic acid.
[12] Also cleaves ether linkage of O-methyl serine.

onion γ-glutamyl transferase (see below), in establishing the nature of the heme-protein linkage in cytochrome *c* (Schwimmer and Kjaer 1960), in elucidating the structure of an intermediate in the plant metabolism of the herbicide atrazine (Lamoureux *et al.* 1973) and to establish configuration of naturally occurring β-methyl lanthionine (not lenthionine) from the antibiotic subtilin (Schwimmer and Alderton 1960). The enzyme from the closely related *Acacia farnesiana* has the distinction of being the first plant C-S lyase to be purified to homogeneity. Each molecule consists of two subunits and one pyridoxal phosphate. Another interesting lyase is that obtained from a fungus growing on rotting onion bulbs. It seems to be similar to the host alliinase except for its elevated pH optimum and lack of stimulation by divalent cations or EDTA (Mazelis 1978).

γ-Glutamyl Transferase and Food Flavor

All alliaceous foods so far examined contain substantial quantities of γ-glutamyl peptides, the amino acids of some of which happen to be flavor precursors, that is, alliinase substrates. Unfortunately, alliinases will not act on these precursors when they are so tied up, and the free allium vegetables do not have enzymes which can liberate them so that they become available to the flavor production (Austin and Schwimmer 1971). However, the breaking of dormancy during postharvest overwintering and upon sprouting is associated with increase in flavor (Schwimmer and Austin 1971A) and lachrymator (Bandyopadhyay and Tewari 1976) in internal and external sprouts. The available evidence suggests that this increase is due to the appearance of a γ-glutamyl transpeptidase (transferase) capable of liberating the flavor precursor from γ-glutamyl bondage; γ-glutamyl-*trans*-S-1-propenyl L-cysteine sulfoxide (γ-glutamyl I) + amino acid $\rightarrow$ γ-glutamyl-amino acid + I $\xrightarrow{\text{Alliinase}}$ CH_3CH_2 - $CH{=}S{=}O$, etc. This or a similar enzyme is an important part of the γ-glutamyl cycle (Chapter 19) and may be involved in creating *in vivo* crosslinks in certain mammalian proteins such as that in the lens of the eye (Zigman 1977). Its endogenous action is also responsible for the liberation of the precursors to shiitake mushroom flavor, des-glutamyl lentinic acid, from its γ-glutamyl peptide, lentinic acid.

The presence of such an enzyme in older onions is in agreement with the observation that the ratio of free flavor precursors to that bound to glutamic acid was much higher in commercial powders dehydrated from old onion (2.3) than that obtained from fresh onion (0.5). In old onions the sum total of flavor precursors plus precursors to the precursors amounted to more than 2%. In commercially dehydrated garlic they can account for 5% of the weight.

Of course the possibility exists that the S-1-propenyl-L-cysteine sulfoxide could accumulate due to the very slow action of undetectable amounts of the γ-glutamyl transferase present even in fresh onions. However, even in sprouted onions, the activity, although detectable, is still quite feeble compared to rich sources of this enzyme such as mammalian kidney (1000-fold greater specific activity) or even asparagus (50- to 100-fold). The enzyme isolated by Schwimmer and Austin, although a true transferase, also possessed hydrolytic activity. There also appears to be a specific γ-glutamyl hydrolyzing enzyme in onions and in chive seeds, which may also possess glutaminase activity, but such an activity has not gained official EC recognition. In both model experiments (Schwimmer 1971B) and in a patented process (Schwimmer 1973) the extent of the flavor potential of onion was shown to be extended by addition of transferase preparations. The γ-glutamyl transferase-induced increase in flavor in a sample of commercial dehydrated onion is shown in Fig. 21.6. Increase in pyruvate has repeatedly been shown to correlate with onion flavor and with other indices of onion quality. A particularly rich and suitable source for food use would be milk (Baumrucker 1980).

ISOTHIOCYANATES AND GLUCOSINOLASE

The Substrate and Principal Product

It is well known to every one that mustard-seed powdered, and formed into a paste with water, yields in the course of a few minutes a mixture which, placed upon the skin, produces excessive irritation, so much so as even to raise blisters. This action is caused by a volatile oil containing sulphur, but free from oxygen. This oil may be obtained from mustard by distillation with water, in the same manner as the oil of bitter almonds is obtained from bitter almonds. It is to this volatile oil that the mustard eaten at table owes its smell and taste. In its purest state it is frightfully acrid and pungent. Now, in the mustard-seed there exists no trace of this oil; the fixed oil expressed from mustard-seeds is bland and destitute of any pungency. The volatile oil is formed from a substance rich in sulphur and nitrogen, and possessing itself no pungency. This substance, by the action of the vegetable caseine contained in the mustard-seeds, undergoes decomposition immediately upon the addition of a sufficient amount of water, and the volatile pungent oil is one of the new products originating from the transposition of the elements. As vegetable caseine, in the seeds of mustard and in almonds, exercises a decomposing influence upon the other constituents of the same seeds, in consequence of the state of transformation into which it passes immediately upon coming into contact with water,

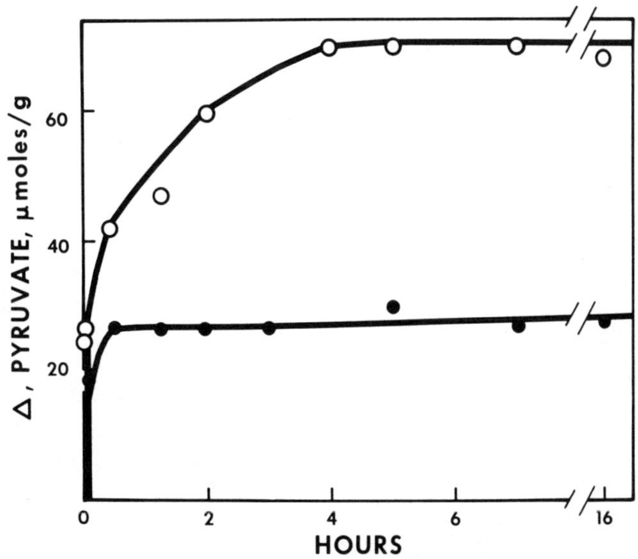

FIG. 21.6. γ-GLUTAMYL TRANSFERASE EVOKES FULL FLAVOR POTENTIAL OF ONION

From Schwimmer and Austin (1971B)

so also the similarly constituted sulphur and nitrogen compounds of nearly all seeds comport themselves. . . .

The above remarks were made by Liebig (1859) some 25 years after Faure (1835) first described the action of the "vegetable casein" now known as glucosinolase and 40 years before Gadamer described "myronate," the potassium salt of sinigrin, the first substrate of "sinigrinase" to be isolated, but some 100 years before the *correct* structure of this compound was elucidated.

Isothiocyanates.—The traditional designation "mustard oils" usually refers to distillates of the isothiocyanates, esters of thiocyanic acid, R-N=C=S. They are present in seeds and other tissues of all members of the Cruciferae and are especially abundant in foods of the *Brassica* genus. For instance Van Etten *et al.* (1976) quantitatively determined 11 isothiocyanates in cabbage. Although the pungency of horseradish is largely attributable to allyl- and 2-phenylethyl isothiocyanates, six others have been identified, some of which had not been previously detected in food plants (Gilbert and Nursten 1972).

The presence of isothiocyanates in food does not always ensure a desirable aroma or any aroma at all. Thus the *p*-hydroxybenzyl derivative present in yellow mustard seed produces a sharp taste without a pungent aroma. In the wrong context such as their presence imparting a burnt flavor to milk of cows grazing on cruciferous weeds and in feed they are undesirable. The aroma is not always associated with vegetables. Thus, the benzyl derivative present in the volatiles of papaya undoubtedly contributes to its distinctive mild fruity aroma. Similarly, they may, in rare instances, be preformed. Thus, a relative of the caper, known in India as "dhuti," contains 4,5,6,7-tetrahydroxydecyl isothiocyanate, apparently preformed as in the case of the papaya (Gaind *et al.* 1975).

Isothiocyanates can also directly and indirectly influence the health-associated attributes of foods (Chapter 35). Like the alliinases products, they appear to possess antibiotic activity. One of them, 2-hydroxy-butenyl, is the precursor to the potent goitrin (see below) present in large amounts in rutabagas, used as a major staple of the diet of destitute farmers during the depression and immortalized in Erskin Caldwell's "Tobacco Road." Also, an outbreak of goiter in New Zealand was traced to milk from cows fed rutabagas. They are also of concern because of their undesirable presence in some oilseeds which, but for their presence, would be more widely adapted as rich protein sources (Chapter 34). The heightened awareness of toxic substances in foods, as occasioned for instance by the introduction of new plant food varieties (such as the solanine-containing Lenape potato) has resulted in reexamination of the presence of undesirable isothiocyanates (Chapter 34) in traditional vegetables such as cabbage (Hanson 1974; Van Etten *et al.* 1976; Tookey *et al.* 1977).

Glucosinolates.—The isothiocyanates are formed via the action of enzymes on the mustard oil thioglucosides or glucosinolates, a name suggested by the investigators largely responsible for their discovery and structure elucidation and presence in plants as more than 50 distinct derivatives (Ettlinger and Kjaer 1968). These can be classified as follows:

(1) *Aliphatic.* This includes the traditional allyl derivatives present and responsible for the flavor of black mustard and horseradish and contributes to the flavor of other crucifers such as cabbage representative of the subclassification alkylalkenyl. Other related subclasses are acyloxyacyl ketoalkyl, hydroxy ethyl and diacid-derived derivatives.
(2) *Methyl thioalk(en)yl.*
(3) *Arylmethyl.*
(4) *2-Aryl ethyl.*

Not all of these are found in foods.

Biogenesis.—The broad outline for the biosynthetic pathway of the glucosinolates was worked out in the 1960s in large part by Canadian investigators at the Prairie Regional Laboratory as exemplified in one of their more recent and perhaps final papers (Chisholm 1973). Further references can be found in this paper and in the review of Chase (1974). It is clear that the side group R (Fig. 21.7) arises from amino acids. Thus, the isopropyl derivative arises from valine. The allyl group of the classical glucosinolate sinigrin, at first thought to arise from glycine, apparently comes from methionine as do the extensive series of S-methyl alk(en)yl derivatives.

In general, the α-C ($\overset{*}{C}$ in the ensuing equations) of the amino acid (I) in the above scheme becomes part of the side group R directly attached to the C in the functional group ($\overset{**}{C}$) but does not furnish the latter or the N of the functional group. The $\overset{**}{C}$ arises from the methyl group in acetate via a convergence of amino and fatty acid metabolic pathways.

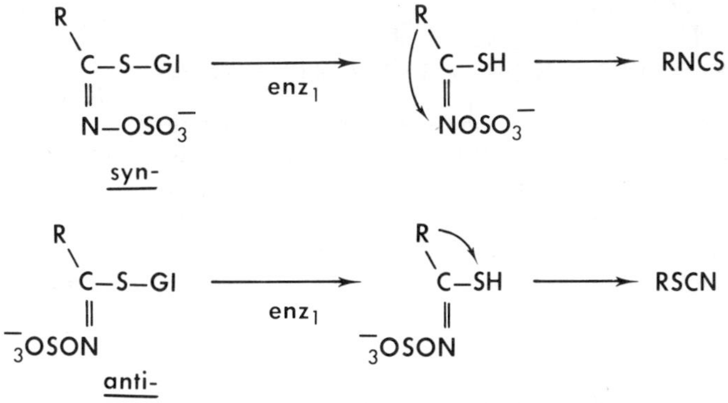

FIG. 21.7. GLUCOSINOLATE STRUCTURE

Depicting the putative existence of both *syn* and *anti* stereoisomers

From Schwimmer (1978)

$$\underset{\text{I}}{\overset{*}{R}CHNH_2(HOOC)} \xrightarrow{\overset{**}{C}H_3[(CH_2)_n]CO-}$$

$$\underset{\text{II}}{R\overset{*}{C}H(COOH)\text{-}\overset{**}{C}O(COOH)} \xrightarrow[+[NH_2]+[O]]{-2CO_2} \underset{\text{III}}{R\overset{*}{C}\text{—}\overset{**}{C}{=}N\text{-}OH}$$

The original amino acid I is transformed into II via the action of aminotransferase and acyl CoA transferase. This explains the existence of S-alk(en)yl derivatives longer than three carbons. Other alterations in R may occur at various steps in the pathway. The dicarboxylic acid (II) is transformed into an aldoxime (III) via known aminotransferase action to put on NH_2 on $\overset{**}{C}$ followed by a series of decarboxylations (enzymes known) and oxidations of the NH_2 (enzymes unknown). Sulfur is probably introduced from cysteine [enzyme(s) unknown] and the resulting thiohydroxamate is glucosylated by a known UDP glucose enzyme:

$$\text{UDP glucose} + \underset{\text{IV}}{R\text{-}\overset{**}{C}(=NOH)\text{-}S^-} \rightarrow \underset{\text{V}}{R\text{-}\overset{**}{C}(=NOH)\text{-S-glucose}} + \text{UDP}$$

The glucosyl thiohydroxamate is probably converted into the finished glucosinolate via 3′-phosphoadenosyl-5′-phosphosulfate (PAPS)

$$\text{V} + \text{PAPS} \rightarrow \text{glucosinolate} + \text{adenosine-3}',5'\text{-diphosphate}$$

Eventually such information may reduce reliance on breeding or at least speed up the approach to enhance flavor in crucifer vegetables or remove unwanted glucosinolates and their reaction products from seed proteins (Chapters 34 and 35).

Glucosinolase-induced Transformations

Isothiocyanate Formation.—The action of glucosinolase serves as a model of the origin of the diversity of food constituents as the result of the reaction of primary unstable products of enzyme action. The isothiocyanates are formed from the glucosinolates by an enzyme that catalyzes the following overall reaction.

$$R\text{-}C(\text{S-glucose}){=}NOSO_3^- + H_2O \xrightarrow{\text{Glucosinolase}} R\text{-}N{=}C{=}S + HSO_4^- + \text{Glucose}$$

The enzyme, which is basically a thioglucosidase and is officially classified as such, has been and is still being called by a variety of names: myrosin (as in pep*sin*) myrosinase, myrosulfatase and sinigrinase. We prefer the term glucosinolase as suggested by Ettlinger and Kjaer.

It will be noted that there is a formal similarity between the above reaction and that which produces *Allium* flavors. Thus, three main products are formed, only one of which is a sulfur volatile. The analogy does not end here. Both plant enzymes coexist in the intact cell without interacting with their respective substrates; they interact only after cellular disruption. Neither enzyme is inhibited at moderate concentrations by the final stable products. In both cases the primary enzymic event appears to produce a relatively unstable intermediate. In the case of glucosinolase action, this is a thiohydroxamic O-sulphonate (Schwimmer 1960, 1961B; Ettlinger *et al.* 1961).

$$R\text{-}\underset{|}{\overset{\text{S-Glucose}}{C}}{=}NOSO_3^- + H_2O \longrightarrow R\text{-}\overset{S}{\overset{\|}{C}}NH\text{-}OSO_3^- + \text{Glucose}$$

This does not preclude the existence of competent sulfatases as that from the snail (Thies 1979). The release of sulfate is considered to be spontaneous, accompanying the subsequent Lossen rearrangement depicted in Fig. 21.7 (Ettlinger *et al.* 1961).

$$R\text{-}\overset{S}{\overset{\|}{C}}NH\text{-}OSO_3^- \longrightarrow RNCS + HSO_4^-$$

Transformation of the Isothiocyanates.—That derived from glucobrassicin, 3-indolylmethyl glucosinolate, may decompose into indolylmethyl alcohol and thiocyanate ion which has goitrogenic activity. The alcohol can condense with the liberation of formaldehyde or, in the presence of ascorbic acid, from an adduct, ascorbigen. 2-Hydroxyalkylalkenyl glucosinolates cyclize to form oxazolidine-thione derivates, the most notorious of which is goitrin, (S)-vinyl-oxazolidine-2-thione, derived from the glucosinolate goitrin via 1-hydroxy-2-butenyl isothiocyanate (Fig. 21.8). Thanks to the work of the investigators at the Northern Regional Research Laboratory summarized by Van Etten *et al.* (1969), we now know that an even greater variety of products can be derived from this glucosinolate. Some of these, such as the epithiobutane derivatives (Fig. 21.8), require a special enzyme (epithio specifier protein) present in crambe seeds (Tookey *et al.* 1977). 3-Hydroxyalkyl isothiocyanates cyclize to the corresponding 6-membered ringed oxazine derivatives.

Nitrile Formation.—At low pH's the allyl primary product decomposes to vinylacetonitrile and sulfur (Schwimmer 1960).

$$CH_2{=}CH\text{-}CH_2\text{-}\overset{S}{\overset{\|}{C}}\text{-}NH\text{-}OSO_3^- \longrightarrow$$

$$CH_2{=}CH\text{-}CH_2\text{-}C{\equiv}N + HSO_4^-$$

Under the proper conditions in cabbage the glucobrassicin-derived isothiocyanate can also give rise to nitriles as can progoitrin under the influence of Fe^{2+} (Tookey *et al.* 1977). Twelve nitriles derived from glucosinolase action have been identified in autolyzing cabbage (Daxenbichler *et al.* 1977).

Thiocyanate Formation.—A rather novel source of diversity of products resulting from glucosinolase action arises from a rather subtle difference in the chirality of the substrate. Gmelin and Virtanen (1959) found that glucosinolase acting on benzyl glucosinolate in plants, such as garlic mustard and pennycress, resulted in the formation of benzylthiocyanate, which has a garlic-like odor, instead of the isothiocyanate derivative. They suggested that either the specificity of the pennycress glucosinolase is different or that a second enzyme catalyzes the isomerization of benzyl isothiocyanate to thiocyanate. This hypothetical enzyme has been raised to the dignity of being assigned an EC number. More recently Schlüter and Gmelin (1972) suggested that such glucosinolates (i.e., the 4-methyl-

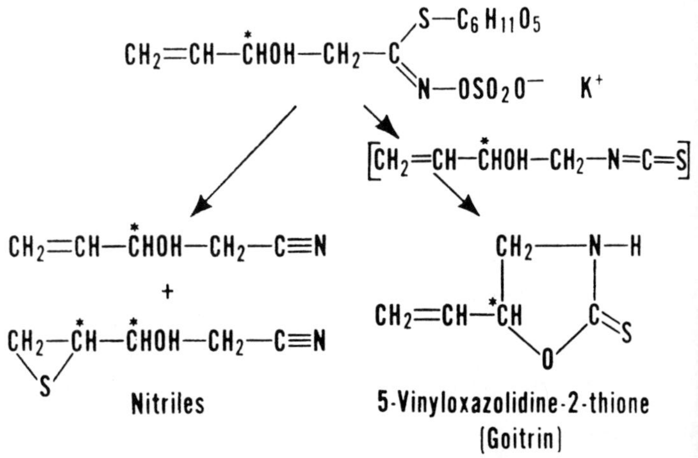

FIG. 21.8. TRANSFORMATIONS OF GOITRIN IN AUTOLYZING CRAMBE SEED

Courtesy of Van Etten et al. (1969)

thio-derivative) exist in the plant in a slightly different geometric from that obtained when the glucosinolate is isolated. They tentatively suggest that *syn-anti* isomerism and not the specificity of the glucosinolase may be involved, as depicted in Fig. 21.7. The corresponding products possess a "scorched-pork" odor. The *syn* form is found in most foods. In the *anti* form, the sulfonate sterically hinders the spontaneous Lossen rearrangement.

The Glucosinolases

General Properties.—As a result of investigations over the past 20 years, we now definitely know that the glucosinolases exist as a bewildering array of isozymes which differ according to genus, species, tissue and stage of development, both as evidenced by the now classical gel electrophoretic procedures (MacGibbon and Allison 1970; Henderson and McEwen 1972; Vose 1972) and by actual isolation (Ohtsuru and Hata 1972; Björkman and Lönnerdal 1973). Some of the enzymes have been purified to the extent that we can say with some confidence that they are oligomeric glycoproteins with two or four subunits per molecule whose molecular weights lie in the range of $30-40 \times 10^3$.

Ascorbate Activation.—Perhaps the most interesting characteristic of the glucosinolases is their activation by ascorbic acid. Just as the full activity of the alliinases is evoked by the vitamin-related pyridoxal phosphate, so some of the glucosinolases are activated by, but are not completely dependent upon, ascorbic acid (Nagashima and Uchiyama 1959; Schwimmer 1961B; Ettlinger *et al.* 1961). Perhaps in no other enzyme system yet unearthed does ascorbate act so specifically. Unlike most activators both V_{max} and K_m are increased. The enzyme activity-ascorbate concentration profile exhibits an optimum at about $10^{-3}M$ (Ohtsuru and Hata 1973). This activation does not depend upon the reducing function of the ascorbate nor is ascorbate consumed during the enzyme action. Instead, it is protected from oxidation. Extensive kinetic and preparative studies by Tsuruo and coworkers led to the model of glucosinolase action shown in Fig. 21.9. As with other glycosidases, the active site of glucosinolases is considered to comprise aglycone and glycone sites. In addition, an ascorbate-binding site exists near the active site. Ascorbate is an allosteric effector, i.e., it induces a change in the conformation of the aglycone-binding site, so that the nonglucose moiety of the glucosinolate fits better in a way that enables the enzyme-substrate complex to dissociate more quickly into product and free enzyme. *p*-Nitrophenyl glucoside is also a substrate for glucosinolase, albeit a poor one. Its hydrolysis is not activated by ascorbate because the aglycone, *p*-nitrophenol, does not extend into the ascorbate-sensitive part of the aglycone-binding site. More recent work of Ohtsuru and colleagues seems to corroborate this theory (Ohtsuru and Hata 1975). At maximum activation one ascorbate molecule binds to each of the four subunits of their particular yellow mustard isozyme. There is still some uncertainty as to whether all or which isozymes are activated by ascorbate. The latter should serve as a good ligand in affinity chromatography of the glucosinolases.

Microbial Thioglucosidases.—In a search for "flavoreses," discussed in Chapter 23, Reese and colleagues at Natick (Reese *et al.* 1958) were able to obtain a highly specific extracellular thioglucosidase from the fungus *Aspergillus sydowi* grown in glucosinolate-containing media. Further investigation, conducted by workers in Kyoto, is summed up and the properties of microbial and plant thioglucosidases are compared in a paper by Tani *et al.* (1974). In addition to confirming the Natick group's results, the Kyoto group isolated an intracellular thioglucosidase from *A. niger* and from certain bacteria. In general the microbial enzymes differ from the plant enzymes in the following respects: They are less heat-stable and, with one or two possible exceptions, do not appear in multiple forms as isozymes in the same culture. Some are activated by Co^{2+}. Perhaps most interestingly of all, while ascorbate does not activate them, it may have no effect *(A. sydowi)*, may stabilize the enzyme *(A. niger)* or may inhibit the bacterial enzyme (*Enterobacter cloacae*).

Localization of the Enzyme in the Endoplasmic Reticulum.—Most enzymes of the type discussed are typically found or presumed to be associated with conventional intracellular structures such as the mitochondrion or within the lysosome. The glucosinolases and certain other enzymes appear to constitute exceptions. Iversen and coworkers found that the glucosinolases seem to be localized in structures unique to crucifers, namely, dilated cysternae of the rough endoplasmic reticulum (Iversen 1973; Pihakaski and Iversen 1976). The latter is the site of synthesis of all enzymes and other proteins in all eukaryotes (Chapter 4). Unlike the con-

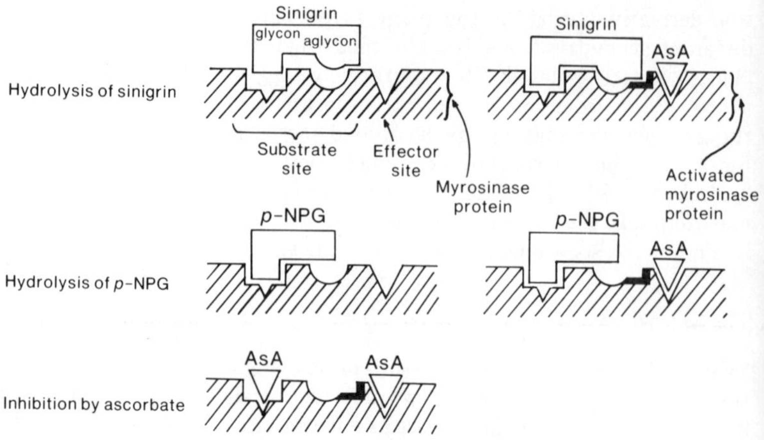

FIG. 21.9. MODEL FOR ASCORBATE ACTIVATION OF GLUCOSINOLASE

From Tsuruo and Hata (1968)

ventional cell arrangement where some enzymes are packaged in Golgi bodies which evolve into enzyme-bearing packets, it appears that the glucosinolases remain associated with the site of their synthesis. Even more surprising is the finding of Hoefert (1975) that these dilated cisternae contain tubules as shown in Fig. 21.10. These tubules are about 30 nm wide and the cisternae that contain them are about 10 mμ long. The ribosomes lining the endoplasmic reticulum are clearly visible as black dots. Hoefert suggests these tubular inclusions may actually be built up of globular glucosinolase isozymes. Answers to questions of just why they are present in such a complicated and unusual array and what their function might be must await further investigation. It would also be interesting to ascertain the localization of the substrate. Perhaps the elucidation of this problem will yield a more intimate understanding of how and why enzyme action is initiated and potentiated when the living cell is subject to the kinds of perturbations discussed, and will lead to rational and purposeful control of the modification of the cellular integrity in fruits and vegetables.

As an example of the application of glucosinolase enzymology, we cite a Soviet Union patent in which a regime of vinegar addition, concentration and temperature is instituted so as to maximize sugar (glucose) yield, as well as that of flavor, in order to cut down on the amount of sugar that has to be added in mustard manufacture (Khantsin 1976).

OTHER VEGETABLE AROMAS

Of course there are many other substances and enzymes involved in generation of vegetable aroma not discussed in detail in this chapter because they will be discussed elsewhere, or because there is not enough known about their genesis nor about the enzymes involved in their formation in the disrupted cell to develop a meaningful discussion. Some of these volatile aromas are nevertheless quite important and constitute an impressive list. Some of the important flavors are formed only during heating and may be enzymatic only in the broadest sense in that their formation proceeds via "secondary" reactions quite removed in both time and conditions from the primary enzyme reaction. What the overwhelming majority do have in common is their amino acid heritage. Furthermore many of the reactions take place, even when enzymatic, during the actual heating up (thermal potentiation) of the vegetable. According to Tressl *et al.* (1977) most of the flavor compounds in cooked vegetables are formed during heating.

Some of these as well as the preformed microbially-derived volatiles constitute some of the most specifically characteristic potent odors we associate with vegetables. Thus, one drop of 2-methoxy-3-propylpyrazine in a large city swimming pool would be more than enough to give the whole pool the odor of raw potatoes (Buttery 1977). The 2-methoxy-pyrazines and their homologs are responsible for or contribute to the characteristic aroma of fresh and cooked bell peppers, lettuce, beans, pea and bean pods, carrots, lettuce and watercress. Work on their biogenesis via microbial symbiosis is in the speculative stage. Murray and Whitfield (1975) postulated a reaction between branched amino acid amides and glyoxal. The experimental difficulties in ascertaining their biogenetic pathway are attested to by the presence in

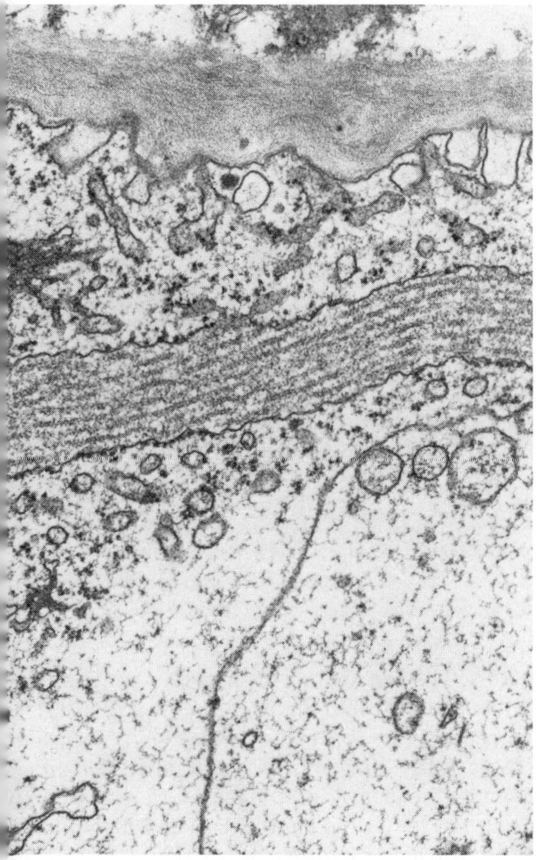

Courtesy of Hoefert (1975)

FIG. 21.10. LOCALIZATION OF GLUCOSINOLASE N CYSTERNAE OF ENDOPLASMIC RETICULUM OF PENNYCRESS

›ell peppers, for instance, of five parts per billion ›f 2-methoxy-3-isobutylpyrazine.

The other microbially-derived trace volatile, re- ›ponsible for the characteristic aroma of red beets, s geosmin *trans*-1,10,-dimethyl-*trans*-9-decalol; ›dor threshold, two parts per 10^{11}. It is related to he terpenes. Like the latter, such as that respon- ›ible for and presumably preformed in parsley, t is derived from acetate via mevalonate (Chap- er 22). The geosmin odor is particularly strong n cooked beets due to increased volatility. In oth- er vegetables such as carrots, the terpenoids are driven off during cooking and hence contribute very little to cooked vegetable aromas.

The 2-methyl-isobutylthiazole in tomato volatiles has been conjectured to arise from methionine (Kazeniac and Hall 1970) or from leucine through isovaleraldehyde (Schutte 1974). It is not too much to conjecture that the apparently preformed dithiolane carboxylic acid present in asparagus arises from 2,2′-dithiolisobutyric acid, which was isolated by Jansen (1948) in search for the precursor of the substance responsible for the characteristic odor of an asparagus consumer's postprandial urine. Retracing its metabolic routes, eventually the source may be the important cofactor lipoic (thioctic) acid, 2,4-dithioloctanoic acid.

Another preformed highly characteristic aroma, that of celery, is borne by phthalide derivatives, the sedanolides which probably arise from amino acids. Another singular celery component, the ester *cis*-3-hexen-1-yl pyruvate, may have considerable flavor impact, as do most esters in fruit (Chapter 22).

Many of the volatiles of cooked vegetables and other foods as mentioned can be traced back to amino acids. Thus, the widespread dimethyl sulfide and methional (Schwimmer and Friedman 1972) come from methionine. Although both could arise from enzyme action, the former from methyl methionine sulfonium salts $CH_3\overset{+}{S}CH_2CH_2CHNH_2COOH$ (Hattula and Granroth 1974), and the latter probably does arise from the action of microbial amino acid decarboxylase in cheese (Chapter 33), in most foods they arise spontaneously or upon cooking. Nonenzymatic browning reactions, which give rise to a wide variety of desirable odor-bearing cyclic volatiles such as dimethyl pyrazine derivatives, largely responsible for the odor of baked potatoes. Related compounds found in baked and roasted foods whenever the temperature of at least part of the food rises about 150°C (Buttery 1977) are derived from amino acid. Cooked bean volatiles include a gamut of N and S containing heterocyclics including the 6-membered thialdine (3C,2S, N) the 5-membered trithiolane (2C,3S) and thiazol(in)es (3C,1S,1N) which form spontaneously from NH_3, H_2S and HCHO.

22

ENZYMATIC AROMA GENESIS IN FRUITS, FLAVORANTS AND BEVERAGES

In this chapter we shall direct our attention to enzymes responsible for the formation of the sweet smelling volatiles usually associated with foods ranging from hot beverages such as cocoa and from tea with its delicate nuances of aroma to the fruits with their distinctive ester-like aromas to wines with their subtle bouquets to more pungent strong smelling foods and food ingredients such as spices and many of the essential oils with their pungent, fragrant redolences. The enzymology of the essential oils used for food ingredients overlaps their use in creating the fragrances from flowers of the perfumery industry and has some medical and health ramifications, as in the case of essential oils made from the cyanogenic glycosides (Chapter 35). In large measure we shall discuss alleged enzyme involvement in preformed or "bioformed" esters and terpenoids, and enzymes responsible for the release of flavor from the cyanogenic glycosides in crushed tissues.

FRUIT AROMAS AND ESTER BIOSYNTHESIS

On the average the volatiles of fruit consist of about 2000 distinct compounds which, from the viewpoint of flavor impact, may be roughly divided into: esters, aldehydes, lactones, terpenoids, carbonyls other than aldehydes, nonterpene hydrocarbons (toluene, xylene), alcohols and some thiaesters. The distribution of these classes in common fruits and wine in terms of the number of different compounds in the major categories is shown in Table 22.1. Unlike fresh vegetables, the aroma is more frequently preformed and arises directly from the intact food and tissue, and disruption is not always required. As to whether the complete integrity of the cells is maintained or whether there is some loosening of cellular structure or even complete disruption of a fraction of some cells while maintaining the appearance of the fruit remains, in most instances, still to be answered. In many of the investigations from which the information displayed in Table 22.1 was collated, it is not always clear or explicitly stated whether or not the volatiles were isolated under the conditions of "enzyme inhibition" as discussed in Chapter 21. An exception, serving as a model of hopefully how to obtain preformed flavors as close as possible to the original are the data of Schreier *et al.* (1976) displayed in Table 22.1 for grapes. Even in this case, some of the volatiles could have arisen from symbiotic microbial activity in the intact fruit as previously discussed. However since much of the flavor does occur in the postclimacteric phase of fruit development, one may assume that loosening of regulatory controls and some decompartmentalization of enzyme and substrate needed to form these volatiles has occurred. Of course during crushing many of the aldehydes and other volatiles

TABLE 22.1

DIVERSITY OF AROMA BEARING FRUIT AND WINE CONSTITUENTS[1]

Fruit	Chemical Class				
			Number of Constituents		
Fruit	Esters	Alcohols	Acids	Carbonyls[2]	Terpenes + Others
Apple[3]	72	30	21	37	5
Cider[4]	50–60	12	1	15	3
Banana[3]	18	20	7	8	3
Canteloupe[5]	37	7	1	—	—
Grape[6]	31	31	48	40	—
Wine, sherry[7]	42	13	23	—	—
Jack fruit[8]	66	4	—	—	—
Orange[9]	32	12	10	38	66
Peach, cling[10]	6	—	—	10	—
Plum[11]	18	18	—	4	9
Strawberry[3]	57	27	21	37	5

[1] Does not necessarily include all investigated, nor is each study all-inclusive; but the data are representative enough to show predominance of esters in most fruits.
[2] Mostly aldehyde and ketone, but also lactone, ether, acetal, dioxalane.
[3] Source: Nursten (1970); for a more recent review see Lemperle (1977).
[4] Source: Williams and Tucknott (1978).
[5] Source: Yabumboto *et al.* (1977).
[6] Source: Schreier *et al.* (1976).
[7] Source: Webb and Noble (1976); see also Williams and Strauss (1978).
[8] Source: Swords *et al.* (1978).
[9] Source: Bruemmer (1975).
[10] Source: Spencer *et al.* (1978).
[11] Source: Forrey and Flath (1974).

discussed in Chapter 21 are released and contribute to the overall flavor.

The Esters

If there is one class of compounds that is associated principally and makes the most striking impact and contribution to what we usually perceive as fruit flavors, it is the esters of carboxylic acids. Not only are they present in fruit volatiles at higher levels than in other foods, but there is, as indicated in Table 22.1, a greater variety of them than any other class of compounds. That wine volatiles contain even more esters than do grapes—some 40% of all the compounds detected in sherry wine aroma—probably reflects the imposition of ester synthesizing enzyme systems present in the yeasts. While all these esters may contribute to the character impact of the resultant aroma, certain individual esters have been associated with the aroma of specific fruits. In some cases this association may be a historical accident occasioned by the first discovery and isolation of this particular ester in a given fruit before its detection in other fruits. Nevertheless each ester does have an aroma reminiscent of the fruit in question. These include: 2-methyl butyrate, apples; isopentyl acetate, bananas; methyl and ethyl esters of *trans*-2-*cis*-4-decadienoate, the Bartlett pear; ethyl 2-methyl butanoate, canteloupe; methyl anthranilate, Concord grapes; ethyl butyrate, orange; ethyl 3-methyl butyrate, blueberry; 3-methyl-1-butyl acetate, bilberry; benzoyl benzoate, cranberry; and methyl β-methiopropionate, pineapple. There are, of course, exceptions. The key compound in peach aroma appears to be due to α-lactones (although as in other *Prunus* species benzaldehyde plays a role), in raspberries to "raspberry ketone," 1-(*p*-hydroxyphenyl)-3-butanone. Until quite recently in other fruits, notably the strawberry, no one ester compound or even class of compounds has been singled out as reminiscent of or providing the requisite character impact to the fruit's aroma. Certain furanones, found first in pineapple and more recently in strawberry, do possess typical character odor reminiscent of strawberries. Of course the contribution of the terpenoids to aroma of fruits, to be discussed later in this chapter, cannot be or should not be underestimated.

Biosynthesis.—Surprisingly what would seem to be a straightforward problem amenable to ready solution, the origin of the esters has not at the time of this writing been satisfactorily explained. The

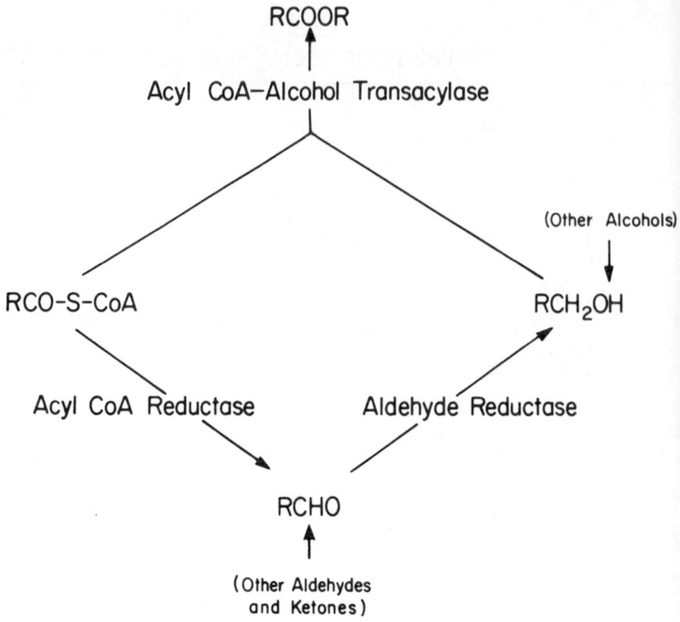

FIG. 22.1. FLAVOR-BEARING ESTER BIOSYNTHESIS IN FRUIT

Proposal involves acyl CoA as source of both acid and alcohol moieties.

Courtesy of Croteau (1978)

lore of intermediate metabolism would suggest a scheme such as that proposed in Fig. 22.1 by Croteau (1978) parts of which had been suggested by others (Nordstrom 1964; Nursten 1970; and Tressl and Drawert 1973). The enzyme directly responsible for ester formation is assumed to be an acyl CoA-alcohol transacetylase. Although a series of specific acetyl CoA transferases have been found which acetylate such disparate but important metabolites such as glucosamine, imidazole, glutamate, the lysine of histone and even terpenoids, and the acyl CoA-long chain alcohol transferase activity apparently responsible for the synthesis of the wax and cutin on fruit surfaces has been detected, many attempts to find this enzyme type in fruits have not met with success as of this writing. Perhaps the long sought ester synthase may simply be the transferase activity of the well-known acyl CoA thiohydrolase. If so, then water must compete with alcohols and alcohols must compete with each other for the acyl CoA. In general, in this connection it is found that for many fruits the most abundant alcohol in the volatile fraction is usually present in the most abundant ester. Some information is available from tracer studies about the precursors to the esters. These studies point to two classes of precursors: amino acids, especially as expected the ones containing branched chain and benzyl groups as are found in melons (Yabumoto *et al.* 1977) and cranberries (Croteau 1978), and from fatty acids, presumably through the lipase-lipoxygenase pathway found in bananas (Tressl and Drawert 1973) but perhaps also through fatty acid oxidation (α-oxidative metabolism) and as suggested by Croteau (1978) and depicted in Fig. 22.2 via abortive shunting of the fatty acid β-oxidation spiral alluded to in Chapter 21. In a somewhat different experimental design, Yamashita *et al.* (1975) detected more than 70 C_2–C_6 carboxylic acid esters after exposure of intact strawberries to the vapors of one or more of nine C_1 to C_6 alcohols. One is almost tempted to suggest that the esters may have been synthesized nonenzymatically by simple spontaneous or esterase-catalyzed interaction of acid and alcohol in a water-poor environment (Chapter 13). However, if we are to invoke a mechanism involving CoA such as that discussed above and for which indirect evidence has been obtained, then an abundant supply of CoA through the intermediation of mitochondria and ATP would be available in the intact fruits. Indeed the requisite enzyme acetyl CoA synthetase has been detected in fruit and is particularly active in the vesicles of the orange (Potty 1969). In particular it may be associated with the synthesis of esters in the same way as the fatty acid synthesizing system is associated with building up of the side chains of the glucosinolates (Chapter 21). This hypothesis

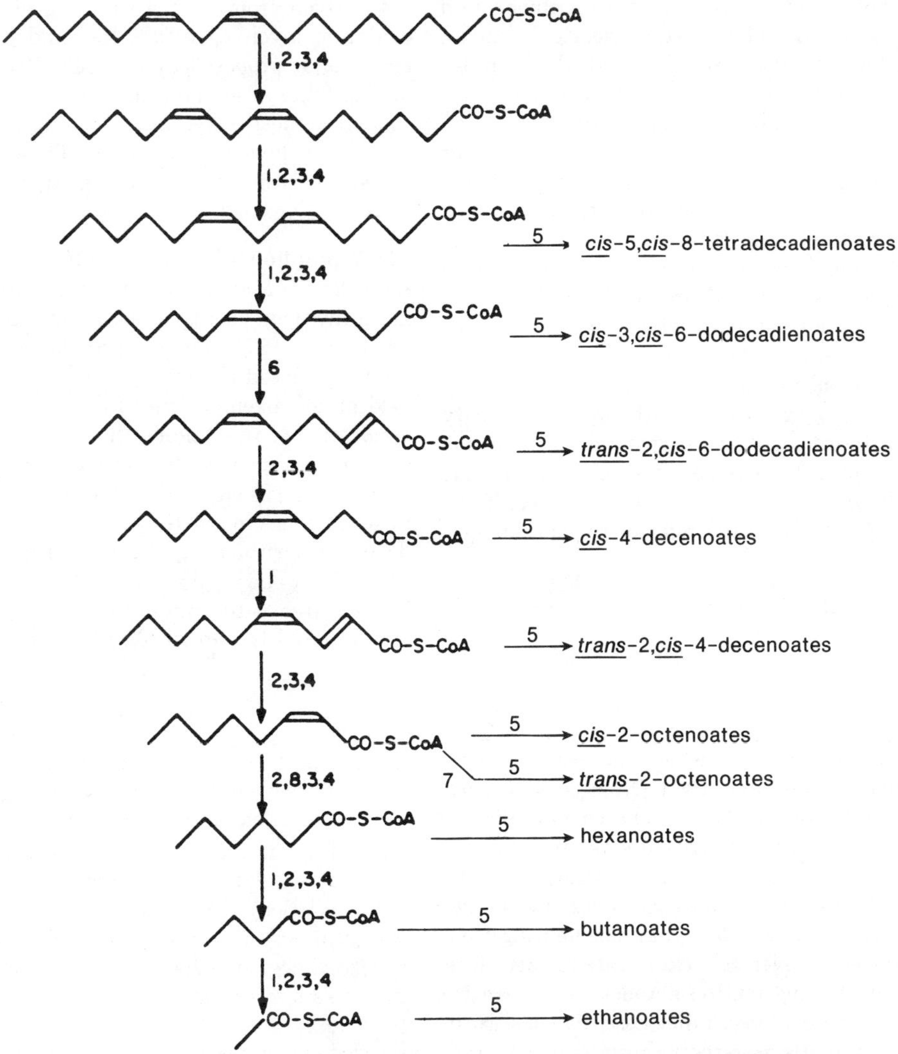

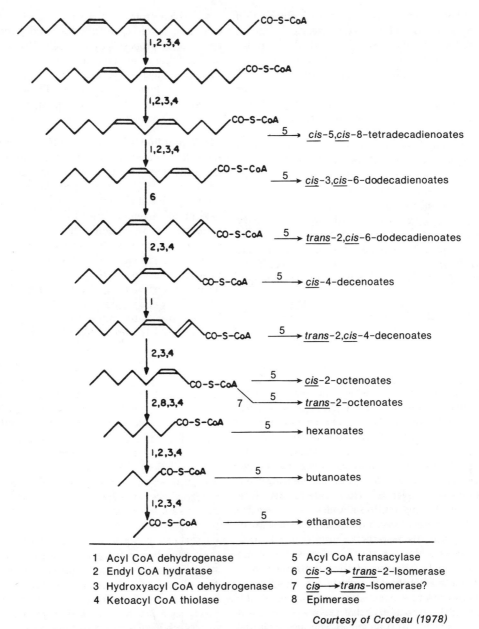

1 Acyl CoA dehydrogenase	5 Acyl CoA transacylase
2 Endyl CoA hydratase	6 *cis*-3 ⟶ *trans*-2-Isomerase
3 Hydroxyacyl CoA dehydrogenase	7 *cis* ⟶ *trans*-Isomerase?
4 Ketoacyl CoA thiolase	8 Epimerase

Courtesy of Croteau (1978)

FIG. 22.2. ESTER BIOSYNTHESIS VIA ABORTIVE SHUNTING OF THE β-OXIDATION SPIRAL

would be in accord with the finding that the acyl moieties of the fruit esters are less likely to be branched chains than are the alcohol moieties. Where branched chains do occur occasionally as the ethyl 2-methylbutanoate in cantaloupe (Yabumoto *et al.* 1977), they probably are derived, as mentioned, from amino acids (although there do appear to be some anomalies in labeling patterns after administration of ^{14}C-labeled leucine and isoleucine). At any rate it would appear that the branched-chain alcohol moieties come exclusively from the branched chain amino acids through the action of transaminases (aminotransferases) and α-decarboxylases to give aldehydes if the biosynthesis does indeed proceed as indicated in Fig. 22.1. Not only has direct verification of esterification via

CoA-linked transacylase eluded experimentation but so has the postulated reductase. Again, analogous reductases have been detected which are involved in the biosynthesis of plant waxes and cutin (Croteau 1978) and lignin (Wengenmayer *et al.* 1976). It should be emphasized, however, that most precursor labeled metabolism experiments have not *ruled out* the scheme depicted in Fig. 22.1. A completely alternative mode of ester biosynthesis has been observed in a microorganism (*Pseudomonas*) which, if present in fruit, may be the key to the mystery of why it has been so difficult to observe ester biosynthesis via the CoA route. Britton and Markovetz (1977) purified a ketone monooxygenase (I) with broad specificity capable of inserting an oxygen into the carbon chain of some ketones on either side of the carbonyl group to form an ester; for instance, if the oxygen is inserted to the right of the carbonyl,

$$CH_3(CH_2)_mCH_2CO\overset{\downarrow}{}(CH_2)_nCH_3 \xrightarrow{I,\ O_2+NADPH}$$

$$CH_3(CH_2)_mCH_2COO(CH_2)_nCH_3$$

and

$$CH_3(CH_2)_mCH_2COO(CH_2)_nCH_3 \xrightarrow{esterase,\ H_2O}$$

$$CH_3(CH_2)_mCH_2COOH + HOCH_2(CH_2)_{n-1}CH_3$$

They found that the two-subunit flavoprotein (FAD) enzyme acted on 2-alkanones from C_7 through C_{14} as well as on the 13-carbon ketones (tridecanones) with carbonyl groups varying from position two to seven to yield a wide variety of esters. Odor-bearing methyl ketones have been reported in oil of cloves, other essential oils, coconuts, peanuts, palm kernels (Forney and Markovetz 1971) and especially in cheeses and other dairy products. Enzymes of their synthesis will be discussed in Chapter 23. A character impact volatile of mushrooms is oct-en-3-one (CH_2=CH-CO-$(CH_2)_4CH_3$). The action of this enzyme is somewhat reminiscent of insertion of oxygen between the eight and nine positions of lipoxygenase-formed linoleic-9-hydroperoxides to form vinyl ethers (Chapter 24). The participation of NADPH in an oxidation situation suggests that singlet oxygen (Chapter 23) may be involved in the formation of these esters.

In the last chapter we mentioned abortive catabolic β-oxidation spiral of fatty acid as a source of aroma-bearing aldehydes. A scheme for the appearance of esters in fruit such as the pear which are derived from such a shunting of classical β-oxidation involving established and hypothetical enzymes as proposed by Croteau (1978) is shown in Fig. 22.2. A similar pathway could also be the source of methyl ketones (Chapter 23) which in turn could give rise to esters. These latter suggestions may merit some looking into in future food flavor enzymology.

Manipulation of Ester Levels in Fruits.—The ester level of apples in the peel increases dramatically upon peeling as shown in Fig. 22.3 (Guadagni *et al.* 1971). Under optimum conditions the esters increased as much as 30-fold in apple peels held at room temperature for 1 to 2 days. In an application of this finding the investigators developed a pilot plant scale process in which the conditioned peels gave aroma solutions with from two to seven times more "flavoring capacity" than no-delay peels without significant sacrifice in quality.

In at least one instance the development of esters is associated with undesirable flavor. Chan *et al.* (1973) noted the presence of the methyl esters of

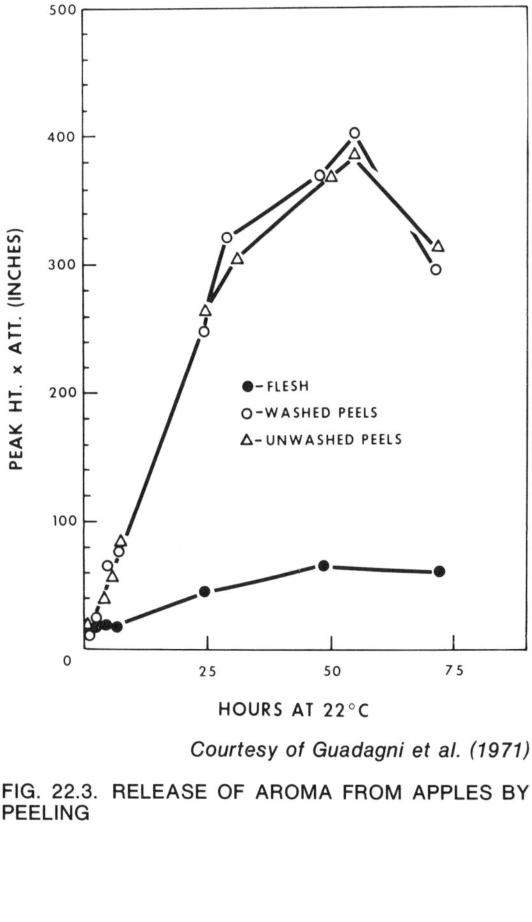

Courtesy of Guadagni et al. (1971)

FIG. 22.3. RELEASE OF AROMA FROM APPLES BY PEELING

butyric, hexanoic and decanoic acids in evil-smelling, improperly processed papaya purees. These esters were not present in the fresh fruit nor in adequately blanched and otherwise improved puree.

In both of the above cases apparently enough cellular integrity remained after treatment to allow continuation of synthesis of whatever coenzymes were necessary to fuel the ester synthesis. In the latter case the decompartmentalization was extensive enough to potentiate or evoke esterase action which gave rise to the malodorous free lower fatty acids. Potentiation of carboxyesterase activity in orange juice has also been implicated in decreasing the flavor and quality of orange juice (Bruemmer and Roe 1976). Inadvertent esterase action in added commercial pectinases (Chapters 5 and 30) may be responsible for reported ester hydrolysis in wines and grape juice. This hydrolysis is occasioned by a decrease in aroma intensity (Jakob *et al.* 1973; Kwasniewski 1975).

Enzymes of Alcohol Production

The volatile esters and aldehydes of most plant foods are almost invariably accompanied by their correspondingly less intensely flavorful reduction products, saturated alcohols. In general the alcohols are much weaker odorants than the aldehydes. According to Eriksson *et al.* (1977), the odor detection thresholds of the alcohols are one to two orders of magnitude greater than those of the aldehydes, but that for hex-*trans*-2-ene compounds there is very little difference in odor quality. Alcohols could conceivably accumulate in foods as the result of the action of the following enzymes: esterases, hydratases adding water to double bonds, epoxide hydratases, oxidases acting on paraffins such as the flavoprotein cytochrome P-450 systems found in both plants and animals, oxidases which produce superoxide anion under conditions in which it can act as reductant, and lastly but perhaps most importantly of all through dehydrogenases acting as reductases on aldehydes with NAD (P)H as cosubstrate. Under appropriate but altered redox conditions in a food, the dehydrogenases can also account for the disappearance of the alcohols. They may also be removed from food systems via the action of the following: as yet undetected ester synthases with acyl CoA as cosubstrate; dehydratases, most of which contain pyridoxal phosphate as coenzyme; enzymes which liberate active oxygen species; phenolases in a secondary reaction involving the reduction of the primary phenolase reaction product, quinone, with the concomitant oxidation of the alcohol. Alcohols are also formed nonenzymatically in foods, especially at high temperatures which convert some of the amino acids present into the corresponding lower aldehydes via the Strecker degradation and thence to alcohols as discussed in connection with tea flavor.

Alcohol Dehydrogenase.—Evidence from tracer experiments in fresh vegetables and fruits points to the unsaturated fats as the primary precursors of the middle alcohols via the lipoxygenase-generated aldehydes. It is not surprising, therefore, that alcohol dehydrogenases have been looked for and found. However, interest in this enzyme in alcohol formation is usually associated with off-flavors and their prevention (Eriksson *et al.* 1977) or with its role in producing precursors to esters in fruits. A synopsis of our knowledge of alcohol dehydrogenases in plant foods, both vegetables and fruits, ensues.

Bean.—The enzyme in NAD-dependent string beans is less heat stable than lipoxygenase; blanching favors a higher aldehyde/alcohol ratio (Stone *et al.* 1976).

Pea.—Unsaturated alcohols serve as more efficacious substrates than do the corresponding alkanols for this reversible NAD-dependent enzyme which is inhibited by a variety of substances including fatty acids, methanol and histidine, and whose action can profoundly affect the alcohol/aldehyde ratio and thus the flavor of peas (Eriksson 1968).

Peanut.—The action of the zinc containing NAD-dependent enzyme isolated and characterized by Swaisgood and Pattee (1968) is closely associated with off-flavor which may develop in high-temperature cured peanuts.

Potato.—Potatoes contain three specific NAD-dependent alkanol dehydrogenases whose K_m decreases (affinity increases) with increasing chain length, two NADP-dependent aromatic alcohol dehydrogenases whose role in flavor has not been explored, and a terpene alcohol dehydrogenase (Davies *et al.* 1973).

Orange.—NAD-dependent reduction of aldehydes, catalyzed by alcohol dehydrogenase, is faster than oxidation of alcohols; unsaturated oxidized faster than saturated alkanols (C_2—C_8); activity determines prevalent form alcohol-aldehyde pair in fruit (Bruemmer and Roe 1971). Alcohol/aldehyde ratios may be of particular importance in flavor threshold and ultimate flavor quality of oranges (Ahmed *et al.* 1978).

Melon.—According to Rhodes (1973) melons possess an unusual enzyme which can utilize both NADH and NADPH as cosubstrates for reduction of aldehydes.

Strawberry.—Seeds contain an NAD-dependent enzyme specific for ethanol and presumably other alkanols, and an NADP-dependent enzyme acting on benzyl alcohol and geraniol; may be involved in ester biosynthesis (Yamashita *et al.* 1977; Croteau 1978).

Tea.—Leaves also contain NAD- and NADP-dependent dehydrogenases which act on long chain (C_4—C_5) alcohols, both saturated and unsaturated (Sekiya *et al.* 1976).

Other Plants.—The cambial regions of most higher plants possess NADP-dependent dehydrogenases which appear to be highly specific for derivatives of cinnamyl alcohols especially coniferyl alcohol (Mansell *et al.* 1976). This suggests that some of these enzymes may be involved in lignin biosynthesis.

Noteworthy is the presence in these foods of both NAD- and NADP-dependent enzymes. In general the higher alcohols and aldehydes can serve as efficient substrates. It should be emphasized that the aldehyde/alcohol ratio will be highly dependent upon the environmental conditions and especially on the NAD(P)/NAD(P)H ratio which in turn is highly pH dependent. Thus the 6-carbon alcohol but not aldehyde is formed when inner cabbage leaves are crushed, whereas no aldehyde but a substantial quantity of the alcohol is generated in crushed outer leaves. The effect of pH on the equilibrium of NAD-linked dehydrogenase reactions arises from the circumstance that the hydrogen ion participates:

$$RCH_2OH + NAD(P)^+ \rightleftharpoons RCHO + NAD(P)H + H^+$$

High concentration of H^+, low pH, would tend to drive the reaction towards alcohol formation. In general highly acid foods tend to accumulate alcohol at the expense of aldehyde.

Also of relevance is the observation of Sieso *et al.* (1977) that the same tomato dehydrogenase which converts hex-*trans*-2-enal to *trans*-2-enol can also reduce the double bond of the former to form hexanal which could then be further reduced to hexanol. The dehydrogenase(s) present thus catalyze at least three reactions starting with hex-*trans*-2-enal

$$2CH_3CH_2CH_2CH{=}CHCHO + 2NADH + 2H^+ \rightleftharpoons 2NAD^+ + CH_3CH_2CH_2CH{=}CHCH_2OH +$$

$$CH_3CH_2CH_2CH{=}CHCH_2OH \xrightleftharpoons{NADH + H^+} NAD^+ + CH_3CH_2CH_2CH_2CH_2CH_2OH$$

Further roles played by alcohol dehydrogenases will be found in our treatment of other aspects of flavor and food quality.

Other Fruit Flavorants

The various possibilities for the production in plants of the aldehydes and alcohols was broached in Chapter 21. We have also just summarized evidence for the participation of alcohol dehydrogenases. In addition to the NAD-linked ubiquitous dehydrogenases specific for straight chain alcohols, fruits may also contain NADP-linked dehydrogenases specific for monoterpene alcohols. As in the case of vegetables, it appears reasonable to assume (but by no means proves) that the alcohols in the volatile fraction of fruits come from aldehydes via these dehydrogenases. Also as mentioned, the aldehydes may also come from the amino acids via transaminase action followed by the action of an α-decarboxylase to yield the next lower aldehyde. Bruemmer *et al.* (1977) detected a broadly acting α-decarboxylase in orange juice. Of special interest is pyruvate decarboxylase which yields acetaldehyde. Although not in itself a unique intense odorant, it is a universal accompaniment of plant food aroma volatiles and is present in both desirable and undesirable odors. Its nonflavor role in the control of desirable food attributes is exemplified by its measurement as an index of apple maturity and ripeness (Sapers *et al.* 1977). The more complicated lipoic acid-requiring pyruvate dehydrogenase complex (Chapter 4) which provides the bridge between glycolysis and the tricarboxylic acid cycle, is said to be responsible for microbially-generated development of a malty off-flavor in orange juice (Chase 1974).

Evidence for the participation of unsaturated fatty acid via lipoxygenase action in fruits is largely confined to the banana (Tressl and Drawert 1973) and oranges (Bruemmer *et al.* 1977). We discussed earlier the putative formation of aldehydes (and eventually esters) *via* the abortive shunting of the fatty acid β-oxidation spiral. Tressl and Drawert (1973) suggest and present some evidence that aldehydes in bananas may be formed, in addition to the lipoxygenase pathway, via two enzyme reactions analogous to those which occur in the fatty acid synthesizing (anabolic) system. One is a fatty acyl NADPH-linked oxidoreductase (analogous to β-ketoacyl ACP reductase) which forms unsaturated fatty acids. Lactones contribute to the flavor of most foods and may even have character impact on some fruits such as peaches. The production of α-hydroxyfatty acids is mediated by the same enzyme system in plants which was previously shown to convert unsaturated fatty acids (C_{12} to C_{18}) to the next lower aldehyde, thus accomplishing an α- rather than the classical β-oxidation of fatty acids (Chapter 24). The system is a fatty acid peroxidase (Chapter 11)

$$RCH_2COOH + 2H_2O_2 + H_2O \rightarrow RCHO + 3H_2O + CO_2$$

Recent experimentation and speculation have implicated active oxygen in the form of O_2^- and O_2^* (Chapter 16) in creation of and altering aldehydes.

Very little work has been done on the distribution of the enzymes of fruit aroma biogenesis probably because we still do not know what many of them are. Bruemmer *et al.* (1977) looked into the distribution of several enzymes among citrus fruit tissues. Tang (1973) studied the distribution of glucosinolate and glucosinolase in the papaya. As mentioned in the last chapter this fruit is unique in being a noncrucifer food plant which contains the glucosinolate-glucosinolase system and appears to be one of the few plants in which this system is ostensibly operating in the intact tissue. The product, benzyl isothiocyanate, seems to play a role in defense against infection and as an inhibitor of papain. The substrate benzyl glucosinolate (glucotropaeolin) is localized in the immature fruit in the papain-containing latex at the surface. In mature papaya seeds the enzyme is localized in the sarcotestae (the soft gelatinous outer covering of the seed) whereas the substrate is confined exclusively in the endosperm of the seed proper.

Much remains to be done in the field of the biosynthesis and enzymology of fruit flavor development. We shall have the opportunity of examining the origin of other important aroma constituents of fruit in the ensuing discussion of food flavorings.

ENZYMES OF FOOD FLAVORINGS—CYANOGENIC GLYCOSIDES AND EMULSIN

The "flavor" of alchemy still lingers in the continued separation of the "Essential Oils," with their evocation of retorts and alembics, as a distinct category of ingredients—essences—obtained from plants (Guenther 1948) which possess desirable aromas. They are prepared by the distillation of press extracts of plant tissue so that all of the components of a properly prepared essential oil are, at least when fresh, all volatile at room temperature (of course polymerization may occur). Some are used in food for additives and others as perfumes or as both. As far as the enzymology of their origin is concerned we categorize them into those in which the flavor components are preformed in the intact plant and those produced by tissue and cellular disruption, allowing the enzyme to have access to the substrate. In the first case the essential oil components comprise relatively heat stable molecules which persist through the preparation of the oil. With these, the enzyme associated relevance is directed to what enzymes are involved in their biosynthesis.

In the second category the fresh crushed tissue aroma is frequently distinctly different from that of the *essential oil* prepared from this crushed tissue. This, as we have seen in the case of the alliums, is because these fresh aromas are heat-labile. The change is particularly dramatic when the processing causes cyclization to occur such as that in onion oil preparation, where the original lachrymator is converted into thiophenes, as mentioned in Chapter 21.

In both categories, the components of essential oils have properties which intrude into the subject of and are present in the aromas of fruits and vegetables. They also possess properties which intrude into the areas of health (Chapter 35), both beneficially and pejoratively. We have already dealt with the enzymes responsible for the formation of major sulfur volatiles, the esters, and to some extent the carbonyl-bearing volatiles. So in essence what remains for scrutiny are the enzymes responsible for release of odor-bearing volatiles from glucosides, mostly cyanogenic in the "crushed tissue" category, and the enzymes of terpenoid biosynthesis and interconversion in the "preformed" category.

The Cyanogenic Glycosides

Amygdalin, the precursor of benzaldehyde in bitter almond oil, occupies a prominent niche in the history of enzymology in that Wöhler and Liebig collaborated on the study of the breakdown of amygdalin in 1832, some 4 years after Wöhler's synthesis of urea (Liebig 1859). They were among the first to give a name to what we now know as enzymes. They called the active principle emulsin. As in the case of mustard oils, Liebig considered his and Wöhler's emulsin to be a "vegetable casein . . . which, brought into a solution of amygdalin . . . decomposition takes place in the course of a few seconds and the amygdaline atom [sic], in consequence of a new mode of molecular rearrangement, resolves itself into hydrocyanic acid, volatile oil of bitter almonds, and sugar, the atoms of which, ninety in number, were, with exception of . . . water, . . . aggregated into groups in the amygdaline atom . . . . The white constituent of bitter almonds is absolutely identical with the vegetable casein of sweet almonds." The amygdalin-emulsin system is also of historical interest because it is probably the first reversible enzyme system to be found and also because this synthesis is probably the first asymmetric synthesis accomplished with the aid of an enzyme (McKenzie 1936). The cyanogenic glucosides continue to hold a special fascination for those interested in the chiral interrelations among naturally occurring molecules.

Present keys to the modestly burgeoning literature of these and related cyano compounds can be found in the reviews of Conn (1969,1979) and Seigler (1975). The number of naturally occurring cyanogenic glycosides is growing, from 12 listed by Conn in 1969 to 26 by Seigler 6 years later. A cyanogenic glycoside is a β-glycosyl-1-hydroxyacylonitrile:

$$\begin{array}{c} \text{O - glucosyl-glucosyl (usually)} \\ | \\ R_1\text{-C-C}\equiv\text{N} \\ | \\ R_2 \end{array}$$

where R_1 is an organic radical and R_2 can be either an organic radical or hydrogen. Thus the classical amygdalin is gentiobiosyl mandelonitrile (Fig. 22.4). The asymmetric carbon to which the sugar is attached makes for the existence of the diastereomer. The diastereomer of prunasin, the β-glucoside of D(-)mandelonitrile is sambunigrin. Although some of the cyanogenic glycosides and their glycones are found in food plants, their presence has been established in more than 800 plant species, most of which are not foods.

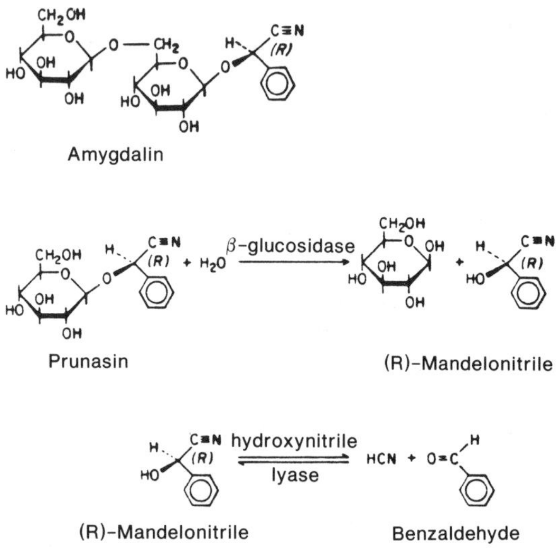

Courtesy of Conn (1978)

FIG. 22.4. AMYGDALIN AND EMULSIN ACTION

Depicted is the production of HCN from prunasin derived from amygdalin via β-glucosidase action.

Biosynthesis.—Thanks primarily to the investigations of Conn and coworkers, the pathway of biosynthesis of the cyanogenic glycosides is known in considerable detail, including the isolation and characterization of the associated enzymes. The pathway of biosynthesis follows roughly that of the glucosinolates (Chapter 21). With few exceptions the side groups arise from amino acids and give rise to an aldoxime, a β-hydroxynitrile and eventually to the glycoside, as shown in Fig. 22.5 and 22.6.

Thus, the parent compound of amygdalin and prunasin is phenylalanine, that of dhurrin is tyrosine and that of linamarin is valine. Other precursors are leucine and isoleucine. These amino acids recur as precursors to the "side groups" in most of the volatiles we have been discussing.

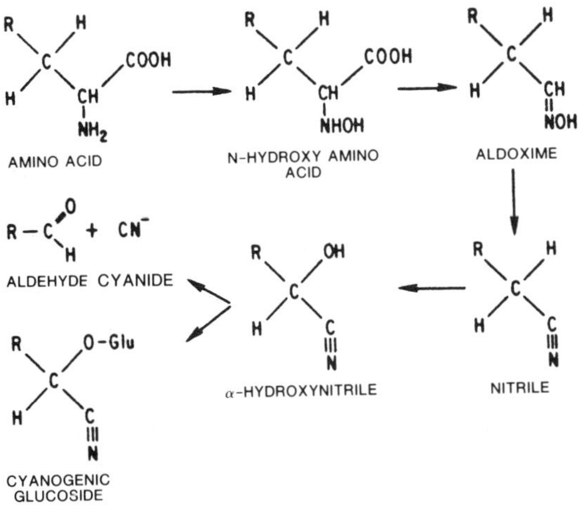

Courtesy of Conn (1979)

FIG. 22.5. THE BIOSYNTHETIC PATHWAY FOR CYANOGENIC GLUCOSIDES

Unlike the glucosinolates, the original amino acid N in the α-amino group is retained as the N in the glycoside's functional group. The enzymes needed to get to the α-hydroxynitrile are present in the microsomal fraction of the plant cell and only NADPH needs to be added *in vitro* to such a fraction. Since the microsome fraction usually contains the remains of the protein synthesis machinery, it would appear that these enzymes, like glucosinolases, are closely associated with the endoplasmic reticulum and their accompanying ribosomes.

In contrast, the finishing step in the biosynthesis is catalyzed by an enzyme, UDP glucosyl transferase, present in the soluble cytoplasm fraction (McFarlane *et al.* 1975). Undoubtedly by the time this appears the microsomal enzymes will have been separated.

Although we are discussing this class of compounds in connection with desirable flavors, it should be pointed out that amygdalin and prunasin are the only true flavor precursors in this family of compounds. These glycosides are found not only in bitter almonds but also in many species of *Sorbus, Prunus* and other Rosaceae. This includes fruits such as plums, peaches and cherries. The consequences of enzyme action on other cyanogenic glycosides in foods, i.e., lima beans, usually results in diminished quality and the problem is how to prevent enzyme action, (Chapter 35).

Emulsin; β-Glucosidase and Nitrile Lyase.— We learned earlier that onion and garlic flavors were formed as the result of the action of lyases and those of mustard oils from the action of (thio)-β-glucosidase. To produce oil of bitter almonds, both lyase and β-glucosidase action are required. Almost invariably these enzymes accompany their substrates. As with these other flavorants, as a result of the sequential action of these two enzyme types, three products—glucose, HCN and benzaldehyde—are formed, only one of which is aroma-bearing. Unlike the above enzyme reactions all three primary products—glucose, hydrogen cyanide and benzaldehyde—are relatively stable. One possible function of HCN in plants may be in nitrogen utilization by regulating a key enzyme in protein synthesis. At least in algae, nitrate reductase is present in an inactive form bound to HCN (Lorimer *et al.* 1974). From the older (Conn 1969) and more recent verifying investigations (Lalegerie 1974; Hösel and Narstedt 1975) it is clear that these three compounds are liberated as the result of three reactions

(R)-Amygdalin + H_2O ⟶ (R)-Prunasin
(R)-Prunasin + H_2O ⟶ (R)-Mandelonitrile
(R)-Mandelonitrile ⟶ Benzaldehyde + HCN

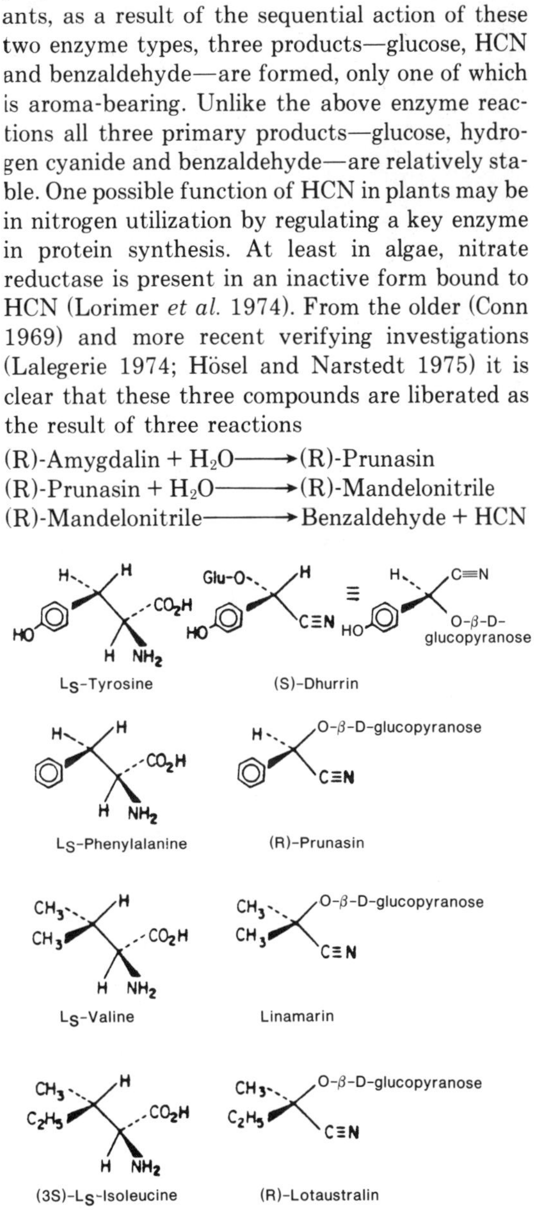

Courtesy of Conn (1979)

FIG. 22.6. CYANOGENIC GLUCOSIDES: PRECURSOR-PRODUCT RELATIONSHIP

The first two reactions are catalyzed by β-glucosidases and the last by nitrile lyase earlier termed oxynitrilase by Krieble (1921). The complex is the classical emulsin, the "vegetable casein" of Liebig and Wöhler. For the sake of historical perspective we briefly outline the preparation of "emulsin," not too different from that devised some 150 years ago. After expressing the oil from bitter almonds, the resulting white, dry residue is extracted with ether and dried further. Emulsin is obtained from an alkaline extract of this powder by acid precipitation, filtration, redissolving and reprecipitation with ethanol, and finally drying with ether.

In spite of the considerable work done on the hydrolysis steps for the removal of the sugars—older work was reviewed by Gottshalk (1950) and more recently by Conn (1969)—there is still no unanimity of opinion as to whether the removal of the two glucose molecules requires two distinct enzymes, one for the hydrolysis of amygdalin to prunasin and the other for release of the nitrile. From the above-cited references of Lalegerie, Hoesel and Darmstadt who used the enzyme system from arrow grass (the accompanying cyanogenic glucoside is triglochinin, O-β-D-glucopyranosyl-1-cyano-1-hydroxy-2-methylcarboxy-2E,4Z-butadiene), we can now confidently say that there are several isozymes of β-glucosidase in emulsin and in other enzyme sources. When one of these is purified it always turns out to be an oligomer of three to five subunits. While the purified enzymes can act on a variety of noncyanogenic β-glucosides (and even some β-galactosides) they all show much greater specificity for their natural substrates. For instance the K_m of the enzyme which accompanies triglochinin is about 1% of that from any other of many glucosides used. For this particular enzyme, only those hydroxynitrile glucosides with an (S)-configuration at the asymetric carbon were hydrolyzed and no other sugars other than glucose could be used. The question of whether the original emulsin requires one or two β-glucosidases remains unanswered.

Nitrile Lyase.—The lyase which accompanies the glucosidase(s) is essentially an aldolase and it exhibits typical aldolase activity towards certain derivatives of 4-carbon sugars. Its action on nitriles may be depicted as follows:

$$\text{R-}\overset{\displaystyle \text{OH}}{\underset{\displaystyle \text{H}}{\text{C}}}\text{-CN} \rightleftharpoons \text{R-}\underset{\displaystyle \text{H}}{\text{C}}\text{=O} + \text{HCN}$$

The enzyme from almond emulsin has been crystallized, thoroughly characterized and immobilized by Becker and coworkers (Becker and Pfeil 1966). Surprisingly, it contains FAD with the same spectral properties as that in glucose oxidase (Ghisla *et al.* 1974). The ability of nitrile lyases to readily catalyze the reverse reaction stereospecifically has been exploited in the synthesis of certain pesticides (Elliot *et al.* 1974) and for tracing the metabolism of growth regulators (Dierickx *et al.* 1975). We suggest that this nitrile lyase when injected with a suitable cosubstrate might be effective in combatting cyanide poisoning (Chapter 35).

Localization.—Quite appropriately this classical enzyme-substrate system whose potentiation depends upon cellular disruption is one of, if not the first, whose anatomical and subcellular localization has been clearly delineated. Thanks to the strides made primarily by Conn and colleagues (Saunders and Conn 1978; Kojima *et al.* 1979), it is quite clear that at least one emulsin substrate, dhurrin (Chapter 35), and presumably the flavor-producing cyanogenic glycosides are anatomically localized entirely in cells of the epidermal tissue whereas the emulsin enzyme system is confined to actively growing internal mesophyll tissue. Successful separation of the cells of these tissues relies on the efficient action of cell wall-decomposing enzyme preparations such as Macerase and Cellulysin (Chapter 29) for the preparation of wall-less plant cells, the protoplasts. In what may well be the archetype experiment in this field, Kojima *et al.* clearly demonstrated as shown in Table 22.2 that the end-product of emulsin action, HCN, is released in maximal amounts only in the presence of protoplasts prepared from both mesophyll and epidermal cells.

The investigators also went one step further and ascertained, by actual isolation of the organelles, the intracellular distribution of both substrate and enzymes. Unlike the glucosinolates (and perhaps the enzymes for synthesis of the substrates) which appear to be associated with the endoplasmic reticulum, dhurrin and its emulsin accumulate, in the classical mode, in the vacuoles of the sorghum cells, albeit in different kinds of cells. Thus contrary to earlier speculations that enzyme and substrate are each located in separate organelles within the same cell, the emerging picture strongly suggests that enzyme and substrate may be generally located in separate cells (reminiscent of the enzymes of the C3/C4 photosynthetic pathways, Chapter 20).

TABLE 22.2

EMULSIN ACTION BY COMBINING SUBSTRATE- WITH ENZYME-CONTAINING PROTOPLASTS[1]

No. of Disrupted Protoplasts Containing: Substrate Alone[2]	Enzyme Alone[3]	Enzyme Activity % of Highest[4]
7.5×10^3	—	2.5
7.5×10^3	7.5×10^3	18.9
7.5×10^3	7.5×10^4	100.0
—	7.5×10^3	2.5
—	7.5×10^4	8.2

Source: Modified from Kojima *et al.* (1979).
[1] No exogenous emulsin used.
[2] Present in epidermal cells of *Sorghum* leaf blade.
[3] Present in mesophyll cells of *Sorghum* leaf blade.
[4] As measured by release of HCN during incubation at 37°C for 4 hr. The enzyme is the emulsin complex and the substrate is dhurrin.

Flavor-related Aspects

The bitter almond tree, *Prunus amygdalus* var. *amara* is grown in southern Europe as a source of essential oil. In the United States only the sweet almond, var. *dulcis*, is grown as a source of nuts, but the bitter variety is used as a graft stock. According to Conn (1969) only the bitter variety contains significant quantities of amygdalin and according to Liebig (1859), even this variety can be induced not to produce amygdalin. The sweet variety does as mentioned by Liebig possess considerable β-glucosidase activity.

In the manufacture of the essential oil, the kernels are ground or crushed, macerated for 1 hr or so with warm water and benzaldehyde, constituting the principal constituent flavor, is isolated by distillation. Undoubtedly benzaldehyde, present in the volatiles of peaches, apricots, cherries and plums (Forrey and Flath 1974; Salunhke and Do 1976) and other *Sorbus* and *Prunus* foods, originates from the action of the β-glucosidase-nitrile lyase system and contributes to the flavor of these fruits. Perhaps surprisingly, the flavor of marzipan, derived from apricot and almond pits and used in confections and baked goods, does not seem to depend upon the products of emulsin action. Thus, Schab and Yannai (1973) ostensibly removed such flavors as well as the bitter amygdalin by steam distillation of the volatile flavorant, benzaldehyde, after thermally potentiating emulsin action at 55°C for 2 hr. Perhaps enough benzaldehyde and a trace of amygdalin adhered to the insoluble coarsely ground kernel meat to contribute to the typical marzipan flavor.

Other Glycosidase-generated Flavors.—Methyl salicylate, *o*-hydroxybenzoic acid methyl ester, the principal flavoring constituent of both wintergreen leaves and sweet birch bark, is also liberated from a glycoside by the action of endogeneous glycosidases. To prepare the essential oil, the leaves are chopped with water and allowed to stand overnight at about 50°C before distillation. Wintergreen leaves have been largely replaced by young twigs of the sweet birch. Most essential oils used in foods and perfumes consist of preexisting terpenoids (see below). However, the aromas from at least one source of essential oil, rose petals, exist as glycosides which have to be enzymatically hydrolyzed before the full attar aroma develops. (Chapter 23).

Vanillin.—A discussion of the liberation of food aromas by native glycosidases would not be complete without mentioning that of vanillin, *m*-methoxy-*p*-hydroxybenzaldehyde. In contrast to the above-mentioned aromas, the vanilla fruit pod is not deliberately crushed in order to potentiate the required β-glucosidase action which liberates the vanillin. To do this the vanilla beans are "cured." Although an ancient practice, as recently as 15 or so years ago a patent on the curing of vanilla beans (Kaul 1967) serves to illustrate improvement of a traditional process. Loosely packed green unripe beans are enclosed in a water-impermeable container and subjected to elevated temperatures and humidities for a few days. Undoubtedly some cellular disruption and cell death ensues, allowing for potentiation of the requisite glycosidase and other enzymes.

The traditional procedures, the Mexican and Bourbon processes for vanilla bean curing, were described by Balls and Arana (1941). In both methods the curing process consists of four stages.

The beans are wilted and then "sweated," i.e., kept at 45° to 65°C for several days. The beans are next dried in air to the desired moisture content (until about two-thirds of the original water has been lost) and finally conditioned by storage in a nearly airtight container for 6 months or longer.

The Mexican method of curing, first practiced by the Aztecs, consists of wilting the beans in the sun and then wrapping them in blankets where they are kept warm by intermittent exposure to more sunlight. After this treatment the beans are dried and conditioned. The Bourbon process, used in the French islands of the Indian Ocean whence most of the world's vanilla comes, achieves the wilting by dipping the beans briefly into hot water. They are then sweated in warm ovens, dried on racks and placed in a zinc-lined conditioning "trunk." Many variations of the wilting process are known. Sometimes the beans are rubbed with alcohol, at other times they are scratched with a needle.

As the curing process progresses, the aroma of vanillin develops in the heretofore odorless beans. During the conditioning period, crystals of vanillin frequently appear and the beans take on the characteristic fruity flavor of natural vanilla. Initial treatment of the beans with dilute ethylene was found by Balls to hasten the curing process and to improve the quality of the final product.

Although the action of β-glucosidase is undoubtedly a prerequisite to aroma development, Balls and Arana suggested that the recognized superior quality of true vanilla extract over the synthetic tincture (obtained via chemical conversion of the more abundant eugenol, in which the latter's allyl is converted into aldehyde, $R\text{-}CH_2\text{-}CH{=}CH_2 \longrightarrow RCHO$) is due to action of other enzymes, probably phenolases and perhaps other oxidoreductases including peroxidases, as well as nonoxidative enzymes, such as proteinases (Wild-Altamirano 1969). In this sense, the enzymological origin of these supplementary components which lend to vanilla extracts their superior aroma, is not unlike that of tea and cocoa, discussed below.

ENZYMATIC ASPECTS OF HOT BEVERAGE AROMA GENESIS

Tea Aroma

The role of enzyme action in the development of tea aroma as well as other aspects of tea manufacture has been thoroughly discussed by Sanderson (1975) and Sanderson and Coggon (1977). Of the 300-odd compounds present in the volatiles of hot tea, the most conspicuous are carbonyl compounds, aldehydes, alcohols, ketones and lactones. A clue to the relative contributions of types of volatile components to tea aroma may be gleaned from the synthetic flavorings simulating tea flavor. For instance geranyl acetone and δ-decalactone intensify tea flavor (Horman and Cazenave 1976). Jasmin, a 10-carbon lactone, is considered to be an important constituent of highly flavored teas (Yamanishi *et al.* 1973). One class of plant food volatiles missing from black teas are the esters.

The role of enzymes in the development of tea flavor has been examined at all stages of tea manufacture going all the way back to planting. Especially prominent in the search for enzyme involvement at the cellular level is the question of the degree of decompartmentalization of enzymes and substrates. The role of climatic conditions during the growth of the tea leaf was alluded to in Chapter 1 as an example of potentiation of enzyme pathways via environmental stress. Such stress results, in this case, in fragility of the chloroplast membrane according to Wickremasinghe (1974), who suggested that the important precursors to the terpenoids (see below) are the amino acids.

Tea manufacture was outlined in Chapter 16 in connection with tea color enzymes. The withering step results in partial disruption of cells and cell membranes, allowing phenolases, proteinases, carbohydrases and perhaps lipases to act on their respective substrates. This evidence is experimentally in agreement with the observed increase in amino acids, sugars and organic acids during withering. It is probably at this juncture that the esters are lost via prolonged esterase action. The injured cell cannot resynthesize these esters. They may thus, together with other preformed and firing-generated volatiles, contribute to green tea aroma (see below).

As previously pointed out—and this is even more important in aroma genesis than in color development—the products of enzyme action formed during the withering provide the precursors/reactants for production of flavor during continued fermentation and especially during firing operations. From the work of Sanderson and colleagues at Lipton, as reviewed in the above-cited references, it appears that much of the aroma arises from reactions of phenolase-produced quinones (flavonones) with amino acids, fatty acids, aldehydes and especially carotenoids. This does not completely eliminate the role of preformed flavors, of course. Nevertheless, because of the intense phenolase action promoted by the rolling process as discussed in Chapter 16 and during the subsequent fermentation, and because of the singular nature of the phenolase substrates, some of the same volatiles present in other foods which arise from reactions closer to primary enzyme action are produced in tea via alternative pathways. In each case the flavor precursor is oxidatively degraded with the aid of the phenolase-produced quinones.

$$\text{Flavonoids} \xrightarrow{\text{Fermentation—Phenolase}} \text{Quinones (+ Precursors)} \xrightarrow{\text{Firing—Heat}} \text{Flavonoids + Flavorants}$$

Note that the quinones are reduced back to flavonols which thus perform catalytically.

Other Roles of Phenolase-produced Quinones in Other Foods and Beverages.—We have already alluded to such a role as an auxiliary mechanism responsible for the enhanced flavor of natural vanilla extracts. Phenolase may also play a similar role in cocoa (see below) and even in wine flavor. Wildenradt and Singleton (1974) postulate that the quinones oxidize alcohols to aldehydes which develop during the aging of some superior wines. They also suggest that autoxidation of phenols occurs and coproduces a strong oxidation—they postulate H_2O_2 but it could also be an active oxygen species such as -OH, O_2^- or O_2^*—which forms the aldehydes. As previously discussed, aldehydes are in general more potent and pleasant flavorants than the corresponding alcohols. It is these aldehydes which are considered to be responsible for the improved flavors of these fine wines.

Amino Acids.—As we have seen, the branched chain and some of the straight chain aldehydes in many fruits and vegetables arise from the sequential action of aminotransferases and α-ketodecarboxylases. In tea, the amino acids are converted into aldehydes through a quinone (and hence phenolase) mediated Strecker degradation:

$$R\text{-}CHNH_2\text{-}COOH + \text{Quinone } (R_1O_2) + H_2O \longrightarrow RCHO + NH_3 + CO_2 + R_1(OH)_2$$

In other foods the Strecker degradation of amino acids may be induced via thermally produced α-dicarbonyl compounds.

Carotene and Terpenoids.—Similarly, some of the terpenoids of tea volatiles, rather than being preformed via biosynthesis from mevalonic acid (see below) arise as the result of quinone-aided oxidative fragmentation during the firing process of carotenes and other terpenoids. Sesquiterpenes may degrade to monoterpenes. Thus, one of the fragments, limonene, normally found in large amounts in citrus volatiles, is converted to the pleasant smelling linalool (see below). Other aromatic fragments, β-irone, theaspirone, 2,2,8-trimethyl cyclohexanone, and the tea ketone, 6-methyl-E3,5-heptadiene-2-one, are all derived from the tetrahydroxybenzenoid ends of the carotene molecule.

Fatty Acids and Unsaturated Aldehydes.—The evidence that the ubiquitous hexanal and 2E-hexanal arise from analogous reactions of unsaturated fatty acids with quinones is not unequivocal, as judged from Sanderson's 1975 discussion. Coggon *et al.* (1977) report that oxidation, or perhaps peroxidation, of fatty acids in tea is mediated through a "nonenzyme metalloprotein" or an exceedingly heat-stable enzyme, which also catalyzes the lipolyis of these fatty acids. If the quinones are indeed responsible for or involved in lipid oxidation in tea, then they may also be involved in the formation of aroma-bearing lactones.

Further interaction of primary products of enzyme action other than phenolase-induced quinones may occur to produce undesirable volatiles, and part of the function of firing may be to drive off volatile excess which would unbalance the poised system said to be critical in creating desirable tea aroma. C_6-aldehydes could also arise from the operation of a lipoxygenase system present in the chloroplasts (Hatanaka *et al.* 1978; Selvendran *et al.* 1978).

Green Tea Aroma.—In the manufacture of green tea, phenolase action is suppressed by steaming the tea flush before rolling. However, there is some evidence that enzyme activity persists since green tea volatiles contain some of the enzyme-derived aromas found in black teas. These include the pyrazines which are probably formed nonenzymatically from flavor precursors which in turn may have arisen from heat-stable enzyme action during or even after the steaming (Yamanashi *et al.* 1973).

In contrast to black tea the volatiles of green tea contain a few, albeit unusual esters, Z3-hexenyl hexanoate and Z3-hexenyl E2-hexenoate. Whether they arose from 6-carbon aldehydes via metalloprotein catalysis or perhaps via nonenzymatic oxidation of the long chain alkenones as mentioned previously in our discussion of ester formation has not been ascertained.

Cocoa Aroma

Although a fermentation takes place during the manufacture of cocoa and hence might be more properly discussed in Chapter 23, we believe that most of the important aroma constituents may arise from reactions similar enough to those in tea to discusss them at this juncture. The role of phenolase and glycosidase derived from the cocoa bean in color quality were covered in Chapter 16. For a concise description of the process of cocoa manufacture see Potter (1978). Some 400 compounds have been identified in cocoa volatiles. A key to the literature on these volatiles is that of Vitzhum *et al.* (1975), who found that liquid CO_2 has been particularly useful in separation of the volatile fraction.

The following enzymes have been implicated in aroma development and as markers of quality of cocoa aroma:

Phenolase—As in tea, the quinones resulting from the action of this enzyme are strong oxidants which catalyze the Strecker amino acid degradation reaction (Purr 1972). The quinones also oxidize sulfhydryl to disulfide in certain key compounds, reactions which are considered to be quite essential in the development of high quality chocolate aroma (Lopez and Quesnel 1974,1976).

Pectin esterase—In order for polygalacturonase to act effectively during fermentation, the pectin has to be first converted into pectic acid or low-methoxyl pectin. The enzyme is not present in immature beans and increases progressively with increasing ripeness (Gamble 1973).

Polygalacturonase and other glycanases—The functions of these enzymes are to remove mucilage, loosen skin from the "nibs," and produce reducing sugars, which are precursors to Maillard reaction-produced flavors formed during the heating steps. While some of these enzymes are produced by microorganisms during sweating and fermentation, the "resident" enzymes (as evidenced by electron microscope studies of the cell membrane changes during sweating) contribute to the breakdown of the tissue during fermentation (Biehl 1972).

Invertase—Undoubtedly the action of this enzyme also contributes to the formation of requisite reducing sugars. However, it should be pointed out that for the best cocoa and chocolate, more is not better, as in the case of sweetener manufacture (Chapter 9). Most of the sugar is fermented to alcohol and the fraction of total carbohydrate participating in subsequent flavor-producing thermal reactions is quite small in comparison with, say, that in cold-sweetened potato during frying. Only limited and controlled enzyme action leads to acceptable and optimal aroma quality.

Proteinases—These provide the precursors to be the important impact character volatiles such as isobutyryl and valeryl aldehydes and mercaptans such as diisoamyl disulfide which develop during heating and drying steps. The amino acid aroma precursors, valine and leucine, seem to be preferentially released. As in the case of other enzyme action, too much protease action may be harmful because not only free single amino acids, but also some peptides have been reported to be crucial precursors for subsequent aroma development. Too much proteolysis would result in the hydrolysis of such peptides. Protease-liberated amino acids give rise to pyrazine and pyridine derivatives which contribute to chocolate aroma (and for that matter, the aroma of other cooked food).

Lipase—Fatty acids in proper proportions and levels constitute an important part of chocolate aroma. They apparently result from the action of the microbial rather than cocoa seed lipases during fermentation (Guenot *et al.* 1976; Lopez and Quesnel 1976). Interestingly some of the lipase comes from the fungal rennet producing *Mucor pusillus*, specially grown as the principal microorganism which survives the cocoa fermentation. As with other enzymes, too much lipase action results in off-quality; in this case fatty acids, responsible for the "dull, insipid" off-flavors occasionally encountered. The use of commercial lipase in the confection industry will be alluded to in Chapter 23.

Other hot beverages whose aroma constituents may have some enzyme-related antecedents are coffee and soups. Serious investigation into this phase for coffee is still in the embryonic stage. Amorim and Amorim (1977) noted a close correlation between phenolase activity of the berries and overall flavor quality of the roasted berry infusions. As in cocoa, microbial and endogenous glucanase action, whose primary function is to loosen the skins to facilitate their removal, produce sugars which undergo thermal degradation to produce volatiles which contribute to the aroma of this hot beverage.

GENESIS OF THE TERPENOIDS

While many of the terpenoids in processed tea and other beverages may arise as during heating via quinone-induced oxidations, in general terpenoids in most foods and in essential oils are preformed and arise via biosynthetic enzyme action. Terpenoids, although widely distributed in tea and fruits, especially citrus, are especially prominent constituents of redolent essential oils. Since most of them are preformed, some food research-oriented investigators have started to look into their biosynthetic pathways and the enzymes of these pathways. Much of the progress has been the result of the exploitation of the aforementioned background in intermediary metabolism, specifically in this case as a spin-off within the last 10 years of the Nobel Prize winning work of Bloch (United States), Popjack (Great Britain) and Lynen (Germany) on the 25-step biosynthesis (each step being catalyzed by a different specific enzyme, starting with acetate) of cholesterol in mammals. As every biochemistry textbook published since then explains, the key repeating unit, not only of the steroids but also of the terpenoids (be they mono-, sesqui- or tri-), rubber, carotenoids, vitamin A and the phytyl group of chlorophyll, is the five-carbon isoprene unit, CH_2=C(CH_3)-CH=CH-. The key metabolite leading to the polymerization of these isoprene units is mevalonic acid HOOC-CH_2-COH(CH_3)-CH_2-CH_2OH. The latter is ultimately derived via enzyme action from that deceivingly simple but protean fleeting metabolite, acetyl CoA.

Among the terpenoids which contribute to aroma of foods, spices and flavoring, the 10-carbon monoterpenoids are probably the most predominant. The structures of some of these are shown in Fig. 22.7. The 15-carbon sesquiterpenes are less so but by no means negligible aroma contributors. Terpenoid biosynthesis in citrus fruits has been reviewed by Bruemmer (1975). More extensive reviews on their biosynthesis in plants used primarily for the preparation of food flavorings include those of Loomis and Croteau (1973), Croteau (1975,1978) and Charlwood (1978). Since terpenoid-related aromas in these plants used for flavorings are much stronger than those from citrus, much more progress has been made in the understanding and isolation of the enzymes from essential oil plants such as those from peppermint *(Mentha piperita)* and spices such as sage *(Salvia officinalis)*, and more progress can be expected from workers investigating them. The enzymes catalyzing the conversion of acetyl CoA into mevalonic acid in higher organisms, hydroxymethylglutaryl-CoA synthase and hydroxymethylglutaryl-CoA reductase have been looked for but not yet detected in citrus. It is possible that an alternate pathway forms mevalonic acid in these fruits. Both in citrus and other terpenoid-containing plants the enzymes catalyzing the reactions leading from mevalonic acid to geraniol have been detected. Geraniol—a flavorant which in the form of pyrophosphate is a normal intermediate in sterol biosynthesis—is apparently the first aroma-bearing monoterpenoid from which the others are formed in plants. Apparently it is at the geraniol formation step that the pathway common to the biosynthesis of the monoterpenes and the sterols branches off. The last enzyme these two pathways have in common is dimethyl allyl transferase which catalyzes the transfer of dimethyl allyl pyrophosphate (I) from its pyrophosphate to isopentyl pyrophosphate (II) in a head-to-tail fusion of C_5 isoprene units:

I + II → *trans*-geranyl pyrophosphate (III) + PP

In sterol and sesquiterpene biosynthesis the same enzyme continues to act and adds on another 5-carbon unit, whereas the first unique step in the pathway to the monoterpenes probably consists of the removal of the pyrophosphate from (III) by a pyrophosphatase to form geraniol, followed by oxidation to the flavorant geranial by the action of the above mentioned NADP dependent geraniol dehydrogenase which has been detected in strawberries, oranges, and potatoes (Table 21.3). In citrus this geranial ends up primarily as limonene (not to be confused with limonin discussed in Chapter 25). As pointed out by Bruemmer (1975), the specificity and activity of the enzymes which give rise to oxygenated monoterpenes such as geranial and α-terpineal (*ca* 5% of orange oil) are ultimately responsible for imparting to each citrus species its distinctive aroma. Linalool appears to be

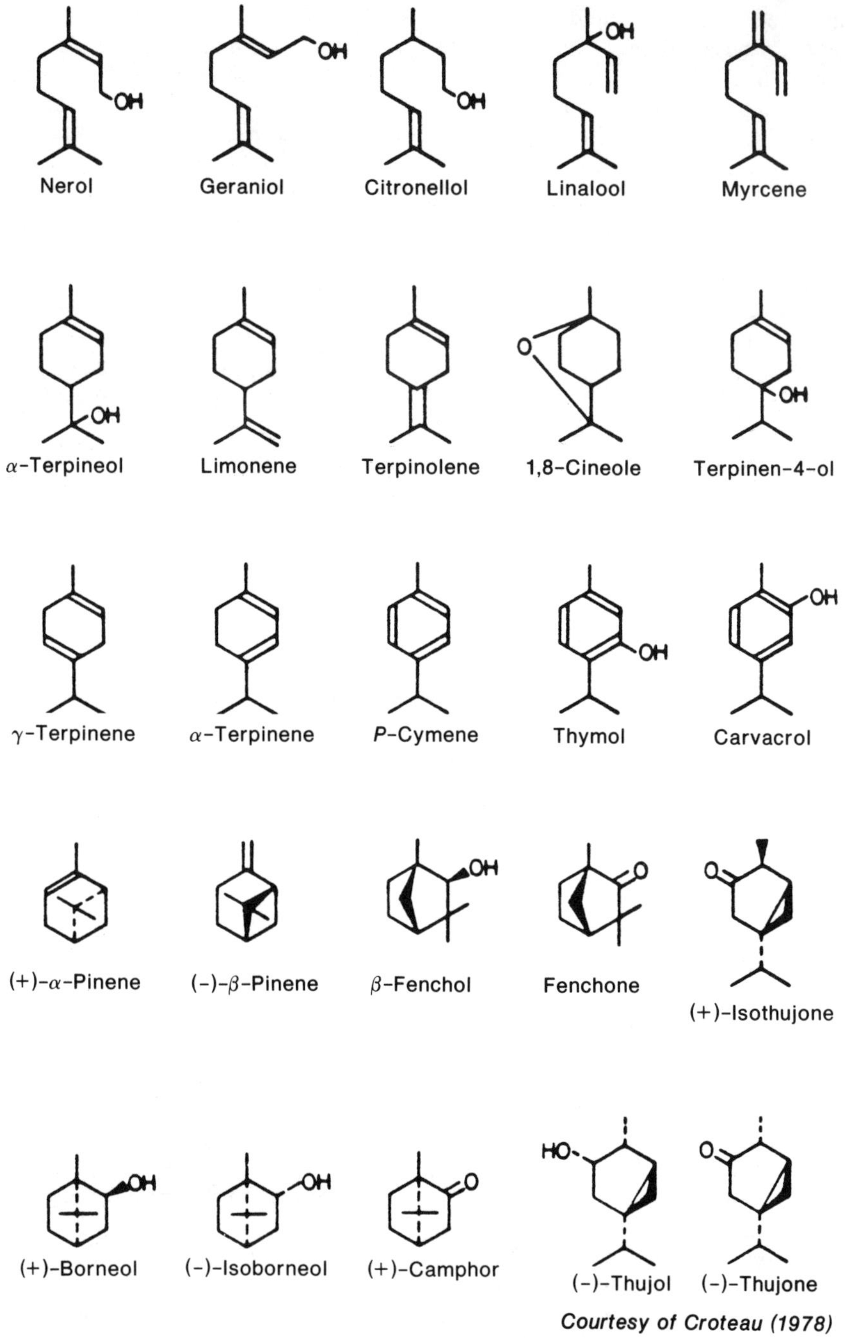

FIG. 22.7. FLAVOR RELATED MONOTERPENES OF FRUITS

a key intermediate in monoterpenoid interconversion and metabolism in fruits.

In the further metabolism of the monoterpenes from *trans*-geranyl pyrophosphate, the work of Croteau and colleagues indicates, the pyrophosphate stays on the geraniol and the geranyl pyrophosphate is isomerized to *cis* isomers of the pyrophosphates of nerol, linalool and geraniol via an enzyme which catalyzes a reversible *trans-cis* isomerization. For steric reasons only *cis* but not *trans* isomers of the monoterpenoids can cyclize.

Recently Croteau (1978) has found that the pyrophosphate group ionizes during interconversion. The next step in the interconversion of the terpenoids appears to be a cyclization of the *cis* isomers of geranyl, neryl or linaloyl pyrophosphates or all of them to form what was thought at first to be a carbonium ion, a common intermediate which could then spontaneously give rise to α-terpineol or limonene, cyclic monoterpenoids. Croteau and Karp (1977), from a study of terpenoid biosynthesis in cell-free systems from sage, concluded that these cyclic monoterpenoids are derived independently from neryl pyrophosphate to form hitherto unknown cyclic monoterpenoid pyrophosphate intermediates rather than as free intermediates of a common reaction sequence. These cyclic pyrophosphates are rapidly cleaved to give the free cyclic flavorants.

Just as the synthesis of the monoterpene flavorings are an offshoot from the main cholesterol biosynthetic pathway at the level of geranyl pyrophosphate, so the formation of flavor-bearing 15-carbon sesquiterpenoids such as nootkatone in grapefruit aroma (Stevens *et al.* 1970) may arise as offshoots from the classical three-isoprenoid metabolite on the same cholesterol biosynthesis pathway, farnesyl pyrophosphate. The enzymes for the synthesis of terpenoid flavorants appear to be compartmentalized from those involved in the synthesis of 30-carbon triterpenoids and cholesterol. According to Loomis and Croteau this is accomplished by the differentiation of the plant tissue during their development into special oil gland tissue.

To summarize, a broad outline of the biosynthesis of the terpenoid flavorants is depicted at the end of the chapter with covalently bound activity groups CoA and PP shown in parentheses.

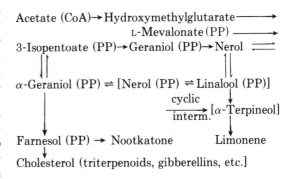

Acetate (CoA)→Hydroxymethylglutarate→
L-Mevalonate (PP)→
3-Isopentoate (PP)→Geraniol (PP)→Nerol ⇌
⇅ ⇅
α-Geraniol (PP) ⇌ [Nerol (PP) ⇌ Linalool (PP)]
↓ cyclic interm. → [α-Terpineol] ↓
Farnesol (PP) → Nootkatone Limonene
↓
Cholesterol (triterpenoids, gibberellins, etc.]

In hops used in flavoring beer, the main-chain atoms of the constituents' molecules are terpenoids whereas the side chains come from leucine (Nursten 1975). Increase in the release of both aroma and bitter substances in hops can be accomplished with the aid of proteolytic enzymes. The latter illustrates how enzymes can be involved in production of food flavorants in modes other than biosynthesic. Pectolytic enzymes are also involved in the retting operation which releases flavor in the preparation of white pepper (Lewis *et al.* 1969). Fatty acid mixtures used in the preparation of confection and dairy products are made by the application of food grade commercial lipase additives (Chapter 23) to both animal and plant fats and oils for the controlled hydrolysis and modification of their esters (Seitz 1974).

23

ENZYMATIC AROMA GENESIS IN MEATS, FERMENTED FOODS AND VIA FLAVORESE

In the last chapter the thrust of our discussion on the desirable odor and aroma genesis in food was largely directed toward the range of the food volatile aroma generation continuum or spectrum (Chapter 21) in which the action of the food's enzymes, whether from crushed tissue and dead cells or preformed in intact tissue and live cells, immediately and directly or eventually and indirectly results in aroma release. Furthermore we have dealt mainly with plants which may be consumed as foods immediately after harvest. We did, nevertheless, have occasion to make some incursions into the other end of the flavor genesis continuum in connection with the following: Further treatment and processing of the plant tissue as in tea and cocoa manufacture; the role of microorganisms in producing enzymes whose actions result in flavor, as in the suspected symbiotic microbial origin of the 3-methoxypyrazines in fresh vegetables; microbial enzymes of supplemental ester formation in wine; the putative aroma-contributing enzyme action during cocoa fermentation. Illustrations of added commercial food enzymes' influence on the aroma of some of the above-mentioned foods were also presented.

In this chapter this latter area of the role of enzymes in aroma genesis will be explored in depth, especially in connection with foods in which processing, microbial enzymes and commercial enzyme addition either assume predominant or parallel importance, or have been proposed to do so as in the "flavorese" concept. As mentioned in Chapter 1, parallel enzyme action from the original nonmicrobial portion of the food and from the microorganisms which have been acting and growing frequently occur in such "fermented" foods as cocoa, dairy products (cheese, yogurt, buttermilk), some pickled vegetables (cucumbers, cabbage, tomatoes, etc.), wine, beer, cider, sausages, herring and fish sauces of the Far East. Although not considered a "fermented" food, bread has also in its manufacture been subject to the action of enzymes from both wheat and yeast.

For each food, aroma is a component of the quality attributes acquired by the parallel action of enzymes from these two sources. It is to the aroma-generating action of enzymes in these foods that we now direct our attention. The role of microbial enzymes, both when present in the living microorganism and when added as commercial foods enzymes, was overviewed by Shahani *et al.* (1974) in an introductory paper to a symposium devoted to microbially induced flavors. Other papers in the symposium include one on sherry wine mentioned in Chapter 22, but most are devoted to flavor of dairy products, especially cheese.

ENZYMES OF CHEESE AROMA GENESIS

Aroma Components and Sources

Ever since the pioneering investigation of dairy products aroma by Day (1967), a rising tide of literature has developed. This literature not only delves into the enzymatic origin of these aromas and their components, but also on the application of the findings, not so much to improve cheese flavor (although it is certainly one of the goals) but to impose quality control and render the cheese manufacturing process more efficient. Optimum flavors achieved in a short time are comparable to those present, hitherto, in only the finest aged cheeses. Some of these applications extend beyond cheese-making to confections and even baked goods. Key reviews exclusively devoted to flavor of cheese and other dairy products include an exhaustive discussion by Dwivedi (1973) and a brief but lucid synopsis by Horwood (1975). Although restricted to flavor biogenesis in blue and Roquefort cheeses, the reviews of Kinsella and Huang (1976A,B) serve as excellent background for the assessment of the role of enzymes in cheese aroma development in general. Other views which zero in on enzymatic aspects of cheese aroma genesis and their use in accelerating cheese manufacture include those of Shipe (1975) and Shahani *et al.* (1976).

The enzymes which participate in flavor genesis during cheese-making arise from three sources: the endogenous enzymes of the uncontaminated milk; added enzymes, mainly those present in milk clotting rennets (Chapter 33) and in lipases used for flavor enhancement (see below); and those present in the microorganisms. The latter include the starters, mostly selected *Leuconostoc* and *Streptococcus* strains or species, ripening microorganisms and organisms adventitiously introduced into the milk before processing (Chapter 33).

Enzymes from Milk.—The verdict of the literature would seem to eliminate the action of milk enzymes as positive agencies of flavor biogenesis in cheeses. This appears to be true even though the same type (i.e., lipase) of enzymes is present in milk as in the other sources. Thus, milk lipase tends to produce a combination of fatty acids which impart an undesirable aroma; the action of milk proteases produces bitter peptides (a taste rather than an odor but undesirable nevertheless) and branched-chain aldehyde blends which do not contribute to desirable aroma. Of course many of these enzymes are inactivated during pasteurization. Perhaps one milk enzyme which could conceivably contribute to desirable cheese aroma is the superoxide dismutase which appears to be stable enough to survive pasteurization (Chapter 25). It is perhaps owing in part to the circumstance that the enzymes of milk do not contribute to desirable aroma that in cheese, perhaps even more than in other foods, it is the right combination of volatiles in the right proportion which is critical to proper flavor perception.

The strict purviews of our aims in this book would thus leave us with only a discussion of the roles of added commercial enzymes on aroma genesis in cheese. However, the microbial-wrought changes in flavor are so important that any omission of the role of their enzymes would lead to a rather warped and unbalanced view of just how cheese flavor is generated. We therefore include a synopsis of what is known about the role of these microorganisms.

Now while the role of the enzymes of milk in flavor biogenesis may be minimal, the milk does nevertheless supply the substrates, or, perhaps more precisely, the flavor precursors analogous to those supplied by tea leaves, cocoa, wheat, meat and other foods. The main difference is that in cheese the subsequent transformation of these flavor progenitors is through the agency of microbial growth whereas the agency in the other foods is the heat required to effect the requisite pyrolytic transformations.

Volatile Aroma Components of Cheese

Many of the volatiles of cheese are universally distributed: aldehydes, (especially acetaldehyde) alcohols, sometimes esters (in cheddar cheeses), ketones and sulfur containing compounds, not the least of which as an aroma contributor is hydrogen sulfide, again in the proper range of concentration.

In general, for any one cheese aroma, the total number of compounds identified is far less than in those foods in which pyrolytic decomposition has occurred. The hundreds of heterocyclics present in such foods are missing in cheeses and in other fermented but heated dairy products. Kristoffersen (1973) stated that characteristic cheese flavor is related to the concentration of relatively few key components. Perhaps not more than 100 volatiles themselves have been identified.

Besides the universal classes of compounds many cheeses do have in common certain volatiles not found so frequently and so widespread in other foods. These include the lower fatty acids (C_4–C_{10})

which impart the well known pungent notes; diacetyl, CH_3-CO-CO-CH_3, which imparts the characteristic creamy note to most cheese and other dairy products; ammonia and amines; and for special notes, the lactones and the 2-methyl ketones. At least one of the alkenals, 4Z-heptenal, is said to possess a distinct creamy, dairy-associated aroma. From our purview, we wish to ascertain just what are the reactions and the corresponding enzymes which transform some of the bulk components of milk—especially triglycerides, the proteins (mostly casein) and the principal carbohydrate, lactose—into these aromas.

Carbohydrate and Diacetyl.—Lactose, in addition to serving as nutrient for the growth of starter microorganisms, is probably the principal flavor precursor to diacetyl. Assimilation of lactose by starter bacteria is probably accomplished with the aid of lactase. Although the pathways to diacetyl have been established in other microorganisms, that in cheese starter microorganisms has yet to be clearly demarcated.

Several pathways have been proposed, some going back to 60 years ago. What the more recent ones have in common is an association with thiamin-dependent enzymes or enzyme systems, especially with some regulatory aspect of the function of the pyruvate dehydrogenase complex operating at the junction of the glycolytic pathway with the TCA cycle (Chapter 4). Apparently, at least for homofermentative bacteria (i.e., leading to one major product), in this case lactic acid starters such as the streptococci, the abundant supply of lactose leads to a potential building-up of excess pyruvate or acetaldehyde because the enzymes of the TCA cycle cannot handle the influx coming from glycolysis. To prevent this buildup, these microorganisms have developed safety valves in the form of pathways leading to diacetyl evolution. Four enzymes producing diacetyl or its reduction product, acetoin, are known.

Diacetyl is probably generated through acetoin from pyruvate via intermediation of the TPP-requiring α-acetolactate synthase which catalyzes the overall reaction:

$$2\ CH_3{\cdot}CO{\cdot}COOH \longrightarrow \underset{\alpha\text{-Acetolactate}}{CH_3{\cdot}CO{\cdot}C(CH_3)OH{\cdot}COOH} + CO_2$$

Undoubtedly active acetaldehyde, $[CH_3CO]{\cdot}TPP$, is involved. Thus, acetaldehyde itself can react with the active aldehyde to form acetoin directly:

$$[CH_3CO]{\cdot}TPP + CH_3{\cdot}CHO \longrightarrow \underset{\text{Acetoin}}{CH_3{\cdot}CO{\cdot}CHOH{\cdot}CH_3}$$

Alternatively, acetoin is formed from α-acetolactate via the action of α-acetolactate decarboxylase, also a TTP-enzyme:

$$(-)\alpha\text{-Acetolactate} \longrightarrow (-)\text{-2-acetoin} + CO_2$$

Note that acetoin C_2 is asymmetric. The acetoins can be formed by two distinct stereospecific enzymes, (+)- and (−)-2,3-butanedioldehydrogenases, e.g.:

$$(+)\text{-}CH_3{\cdot}CHOH{\cdot}CHOH{\cdot}CH_3 + NAD^+ \longrightarrow (+)\text{-acetoin} + NADH + H^+$$

(+)-acetoin can be converted into (−)-acetoin via a specific acetoin racemase. Diacetyl is then formed via the action of acetoin dehydrogenase (diacetyl reductase):

$$CH_3CO{\cdot}CHOH{\cdot}CH_3 + NAD^+ \longrightarrow \underset{\text{Diacetyl}}{CH_3{\cdot}CO{\cdot}CO{\cdot}CH_3} + NADH + H^+$$

In some microorganisms the accumulation of diacetyl may be modulated by the action of acetolactate mutase:

$$CH_3{\cdot}CO{\cdot}C(CH_3)OH{\cdot}COOH \longrightarrow CH_3{\cdot}C(CH_3){\cdot}OH{\cdot}CO{\cdot}COOH$$

In this case excess glycolysis might be funneled into the amino acid valine:

$$CH_3{\cdot}C(CH_3)H{\cdot}CHNH_2COOH)$$

An older suggestion that diacetyl is formed via the action of an α-acetolactase "oxidase" has to our knowledge not been substantiated. Furthermore not all of the above-discussed enzymes have been established as being present in starter microorganisms. It is likely that in starters the level of diacetyl is most directly controlled by acetoin dehydrogenase, proposed for *removal* of diacetyl from beer and other foods where its contribution to flavor is considered objectionable (Chapter 25).

The addition of cyclic AMP in a spray during the manufacture of cheddar cheese in order to hasten ripening (Sullivan and Infantino 1975) may be associated with derepression of genomes of ripening microorganisms which code for enzymes of carbohydrate metabolism which lead to many flavorants of cheese volatiles (Chapter 4).

Proteins as Flavor Precursors

Intact milk proteins have been assigned a role in flavor development in cheeses by virtue of their cysteine-bound sulfur. Kristoffersen (1973) suggested that ultimately, the development of the characteristic flavors is determined and controlled by the disulfide-sulfhydryl ratio: the higher this ratio the better the flavor because of the capacity of the protein for accepting hydrogen generated anaerobically during the ripening process. The more conventional view of the role of proteinase involvement in aroma genesis is as a supplier of flavor precursor amino acids as auxiliary to their more vital roles in texture, milk clotting and softness of the final cheese (Chapter 33) and also in taste perception (Chapter 34). While it has been empirically established that vigorous proteolysis is instigated by microbial enzymes in the starter, to some extent in resident and in ripening organisms and by added rennets, it is not at all clear what specific enzyme events conspire to give rise to aroma constituents which trace their origin to these amino acids. Some factors and parent amino acids which either have been fairly well documented or have been proposed include the following. Alanine is probably the direct precursor of propionic acid, responsible for part of the aroma as well as taste of Swiss cheeses. Threonine is a potential source of diacetyl through acetaldehyde via the enzyme threonine aldolase:

$$CH_2OH{\cdot}CH_2{\cdot}CHNH_2{\cdot}COOH \rightarrow CH_2NH_2{\cdot}COOH + CH_3{\cdot}CHO$$

The natural function of this phosphopyridoxal enzyme is to catalyze the synthesis of serine from glycine by transferring a methyl group from 5,10-methylenetetrahydrofolate (I) to glycine.

$$I + CH_2NH_2{\cdot}COOH \rightarrow HOCH_2{\cdot}CHNH_2{\cdot}COOH$$

It functions *in vivo* as a transferase rather than a lyase. Valine represses expression of the genome coding for enzymes of valine biosynthesis (feedback repression). This makes α-acetolactate, normally an intermediate in the synthesis pathway of valine, available for diacetyl formation and accumulation. Branched amino acids valine and leucine yield branched aldehydes which according to Diwidi (1973) play an important role in the flavor of cheddar type cheeses. On the other hand the leucine-derived 3-methyl butanal formed via aminotransferase and α-ketodecarboxylase action has a malty-like odor which is considered to be a flavor defect in dairy products (MacLeod and Morgan 1956).

In addition to these individual amino acids there are several flavor associated effects produced (Chapter 35) by amino acids in general. Thus they provide the brothy background characteristic of most cheeses. Furthermore amino acids are precursors to amines (Chapter 35). These amines are produced by the action of individual specific amino acid decarboxylases which act on the amino acids directly without the intervening action of aminotransferase which, as we have seen, results in the formation of aldehydes:

$$RCHNH_2COOH \rightarrow RCH_2NH_2 + CO_2$$

vs

$$RCHNH_2COOH \rightarrow RCOCOOH \rightarrow RCHO + CO_2$$

Another way that proteins are involved in flavor is through the ability of protease-generated amino acids to stimulate production of free fatty acids. Reciprocally, the addition of fatty acids (along with salts) can be used to control proteinase action, thus serving as a regulator of the production of amino acid-derived aroma constituents. While the actual enzymes involved in these protease-lipase interactions are unknown, that they are effective is evidenced, for instance, by the observation of Nakanishi and Itoh (1974) that adding a purified *A. oryzae* proteinase to milk doubled the fatty acid (C_4–C_8) level of the resulting cheeses. Previously Lawrence (1965) reported a stimulation by amino acids of production of fatty acid-derived 2-methyl ketones.

Applications.—Whatever the mechanisms, several applications of these interactions have not been long in coming. Proteinase preparations, sometimes in conjunction with lipases (see below), have been used to accelerate and enhance flavor development in ripening cheeses. In addition to the above-mentioned work, Kosikowski (1976) used fungal neutral proteinase, along with commercial lipase preparations, to hasten the development of cheddar type cheeses.

Lipids as Flavor Precursors

While carbohydrates and proteins and their attendant enzymes undoubtedly play important roles as precursors to cheese aroma, the fatty acids are unquestionably the key to both desirable aroma and undesirable odors (Chapter 25) in most cheeses both as precursors and as aroma-bearing constituents in their own right. In addition to being important aroma-bearing constituents themselves they perform the following precursor functions:

(1) They are the precursors to the 2-methyl ketones in all cheeses but are especially important in blue and Roquefort cheeses.
(2) They are the precursors to the fatty acid lactones.
(3) They are the precursors to 4-*cis*-heptenal which has a dairy-associated "creamy" aroma.
(4) They are probably precursors to the few esters found in cheese, perhaps through action of the ketone monooxygenase discussed in Chapter 22.
(5) The acids modulate proteinase activity, as previously noted.

A completely different role played by fatty acids in aroma perception is that they provide the appropriate physical milieu for the action of the enzymes responsible for the formation of the aroma. They also serve as a repository for almost all of the flavorants. As such, they determine the relative composition of the vapor or headspace above. It is this vapor and not the concentrated volatile fraction which is perceived by the consumer.

Lipase.—The keystone enzyme activity for flavor development in cheeses is that of lipase. This includes development of both desirable and undesirable odors (Chapter 24). Lipase enzymology will be outlined also in Chapter 24. A symposium proceedings devoted to lipases (mostly milk lipase) has been published (Downey 1976). The lipases which act to produce desirable aromas come principally from the ripening microorganisms and to a much smaller extent from the starter but probably not from milk. For instance, the free fatty acid content of blue cheese increased ten-fold to 1000 meq/kg after 24 weeks of ripening of blue cheese (Kinsella and Hwang 1976A). Of course the free fatty acid level in each cheese at any given moment is the resultant of the rate of its formation and removal. Since the milk lipase action generally results in undesirable flavors, there must obviously be some major differences between the specificities of milk and ripener lipases. However, not all microbial lipases give rise to desirable flavor, viz, the problem now solved, of removing undesirable lipases in microbial rennets (Chapter 33) as described, for instance, by Somkuti (1974). Again, as in the case of the other foods, the proportions of the aroma constituents are at least as important as the individual level of any one of them. Thus, acetic acid is considered essential for good cheese aroma because it modulates the otherwise harsh aroma of the higher fatty acids (C_4–C_{10}). Presumably the acetic acid is derived via esterase action on the esters present in most cheeses.

In addition to the documentation of the preceding discussion provided by the above reviews, a further example of recent work on the importance of lipolytic microorganisms in ripening is afforded by the investigation of Singh *et al.* (1976). Cheddar cheese ripened with lipolytic bacteria (*Serratia* spp.) alone contained more free fatty acid than cheese which was treated with both starter and ripening organisms. In the case of blue cheese thermolabile lipase elaborated in *P. roqueforti* is present in both the mycelia and the medium into which it is excreted (Kinsella and Hwang 1976B). A clue to its effectiveness is its predilection for the short chain fatty acids which contribute more effectively to aroma. See below for roles of added lipase.

2-Methyl Ketones.—As mentioned, one important character impact in cheese flavor is that of 2-methyl ketones derived from fatty acids. The important ketone in blue and Roquefort cheeses is 2-methyl heptanone. Because the capacity of blue cheese fungi to form ketones exceeds the capacity of the lipase-actuated liberation of fatty acids by starter organisms, the rate of lipase action may be the limiting factor in methyl ketone production. Although one might anticipate some sort of feedback regulation to limit the level, it turns out that while high fatty acid levels in reconstituted model systems containing cheese microorganisms do indeed inhibit methyl ketone formation, in cheese itself the free fatty acids are compartmentalized in the fat. They are thus not available for the purpose of inhibiting or repressing the expression of enzymes of the genomes coding for these enzymes.

Enzymes of 2-Methyl Ketone Synthesis.—Just as the enzymes involved in diacetyl formation are induced in starter organisms to provide a safety valve pathway regulating the flow of pyruvate into lactic acid and the TCA cycle at the junction of the latter and glycolysis, so are the enzymes involved in methyl ketone formation elaborated in ripening organisms in response to the high flux of metabolites through the fatty acid β-oxidation spiral pathway. Once again a shunting or abortion of this pathway (Fig. 22.2) results in the accumulation of odor-associated food components such as the glucosinolates (Chapter 21) and aroma constituents in fruits (Chapter 22). In order to be shunted from this spiral, the β-keto acyl CoA's are probably hydrolyzed to free CoA and α-keto acid by an appropriate thiohydrolase (deacylase) or by a CoA transferase with a suitable CoA acceptor such as succinic acid.

$$(1)\ CH_3(CH_2)_nCOCH_2COSCoA + H_2O \rightarrow CH_3(CH_2)_nCOCH_2COOH + CoA\text{-}SH$$

or

$$(2)\ \beta\text{-Ketoacyl CoA} + \text{succinate} \rightarrow \beta\text{-keto acid} + \text{succinyl CoA}$$

Documentation for the existence of such enzymes in cheese organisms is not extensive. The enzyme directly responsible for methyl ketone formation, β-ketoacyl decarboxylase (2-oxo-carboxylase) has been demonstrated but only after a long series of papers showing that the ketones came from fatty acids via β-oxidation; i.e., showing their presence in mold-included rancid coconut oil (Starkle 1924) and in cheeses (Thaler and Geist 1939) and culminating in partial purification of enzymes from the responsible fungi (Hwang *et al.* 1977).

$$RCH_2{\cdot}CO{\cdot}CH_2{\cdot}COOH \rightarrow RCH_2{\cdot}CO{\cdot}CH_3 + CO_2$$

The fungal isozymes act optimally on the 12-carbon lauric acid to form the 11-carbon 2-undecanone. Indeed one of the strongest arguments for the involvement of β-oxidation spiral in methyl ketone formation is the circumstance that each ketone formed by the blue cheese and similar fungi always contains an odd number of carbon atoms. The presence of the corresponding alcohols in cheese aromas can be accounted for by the action of demonstrable NAD-dependent derhydrogenases in these fungi, as documented by Diwidi, Kinsella and Hwang.

Lactones.—Their ubiquitous distribution among some 30-odd food classes, pleasing and extensive aromas, and their frequent use in imitation flavor formulations bespeaks the contribution of the lactones, dehydrohydroxyfatty acids, to the flavor of cheeses as well as of other foods. Depending upon the particular lactone (over 60 are naturally occurring) and the concentration at which they are sensed, their aromas have been described as fruity sweet, peach, plum, coconut-butter, caramel, malty, herbaceous and cocoa-like. Dairy-associated aromas have been assigned to lactones of the following fatty acids; δ-nonaoic, 4,4,-dibutyl-γ-butyric and γ-butyric. They are usually derived from straight-chain hydroxyfatty acids with chain length varying from C_4 (γ-butyrolactone) to C_{12} (γ- and δ-dodecanolactones). A few are branched. A rare one is ambrettolide, or ω-6-hexadecenolactone, with a strong musk odor. Various aspects of their presence in foods, structure and contribution to aroma have been reviewed by Maga (1976B). Perhaps more is known about how they form nonenzymatically than about the enzymes involved in their biosynthesis. Explanations of their formation in any case are still largely conjectural. The following systems or reactions have been implicated: Formation from ketoacids via NAD-linked reductases; from unsaturated fats via hydrations or indirectly from the lipohydroperoxide formed as the result of lipoxygenase action; from esters; in plants via an enzyme (system) which splits long chain hydroxyfatty acids into fragments, including γ- and δ-derivatives capable of forming lactones, or even from carbohydrate degradation. Some of these possibilities have been examined without any conclusive results.

One of the problems in establishing synthetic pathways for the lactones is that they are formed from γ and δ hydroxyfatty acids:

$$\gamma\text{-lactone: } CH_3{\cdot}(CH_2)_n{\cdot}CHOH{\cdot}(CH_2)_2{\cdot}COOH \xrightarrow{-H_2O} CH_3{\cdot}(CH_2)_n{\cdot}\underbrace{C{\cdot}(CH_2)_2{\cdot}CO}_{O}$$

$$\delta\text{-lactone: } CH_3{\cdot}(CH_2)_n{\cdot}CHOH{\cdot}(CH_2)_3{\cdot}COOH \xrightarrow{-H_2O} CH_3{\cdot}(CH_2)_n\ \underbrace{C{\cdot}(CH_2)_3{\cdot}CO}_{O}$$

The enzymes for the formation of other hydroxyfatty acids are well established. They include: enzyme- and nonenzymic conversion from lipoxygenase-generated lipohydroperoxides (Chapters 14, 24, 32), the fatty acid peroxidase of Shine and Stumpf (1974) which also forms the next lower aldehyde, the now classical enoyl-CoA hydratase of the β-oxidation spiral

$$R\text{-}C{=}C\text{-}COOH + H_2O \rightarrow R\text{-}CH\text{-}OH\text{-}CH_2\text{-}COOH;$$

and the enzymes in mammalian liver which convert the last methyl in fatty acids into ω-hydroxy acids via a cytochrome P-450-NADPH flavoprotein oxidizing system (Lu and Coon 1968). Similar P-450 oxidase systems present in plants are involved in the metabolism of herbicides (Frear *et al.* 1969), cinnamic acid (Russell 1971) and alkaloids (Ambike *et al.* 1970). They are also present in microorganisms, such as *Pseudomonas putida* (Ishimura *et al.* 1971) where they catalyze the oxidation of long chain hydrocarbons into alcohols and fatty acids and presumably ω-hydroxyfatty acids (Cardini and Jurtshuk 1968).

Applications of Lipase.—Whatever the mechanism, it is most likely that lactones are derived from free fatty acids so that the action of lipase would be the first and conceivably a critical step in their formation. For instance Jolly and Kosikowski (1975) found that the concentrations of δ-decalactone and dodecalactone were five times higher in lipase-treated blue cheese than in untreated controls. In view of the central role played by lipase in aroma genesis, it is not surprising that its use as an enzyme adjunct in the dairy industry is now well-established. Reviews on lipase application in addition to the general reviews on cheese aroma already cited include those of Seitz (1974) and Shahani *et al.* (1976), who have provided synoptic overviews of the varied application to cheeses and other foods, and a review by Nelson *et al.* (1977) covering the enzymology, discovery, anatomy, histology and dairy application of the principal lipases used as commercial enzymes.

The addition of lipases to cheeses originated from an investigation of the nature of the agents which gave to Italian cheeses such as Romano and Provolone their characteristic bouquets. This causative factor was traced to the "rennet" used in their manufacture. These rennet pastes were prepared traditionally from the contents of the milk-filled stomachs of lambs, kids or calves which had just suckled. After much trial and error, principally at the Dairyland Farms in Wisconsin, to find an enzyme whose addition to milk in cheese-making would replicate the distinctive nuances of piquancy associated with Italian cheeses, success was attained with a lipase obtained from special glands located at the base of the tongue of the above-mentioned young animals (Farnham 1950). These preparations have been referred to as pregastric, oral lipase or esterase.

These lipases have been used not only for the preparation of Italian-type cheeses but also for intensifying the flavor of almost any cheese, to improve and hasten the ripening process and to prepare flavoring which is then added to cheese during its manufacture. These lipases are also used as such for a wide range of other products including confectionery products, especially milk chocolate; toffees, caramels and cream centers; cereal-derived foods such as breakfast cereals; dry mixes of various kinds (cheesecake, cookie, pancake, sweet cake); and nonyeast leavened products.

Lipases as food adjuncts have also been suggested for the improvement of a diverse range of foods including soybean milk, smoked carp, rice alcoholic beverages, coffee whiteners and dog foods for which pancreatic lipase is most suitable (Marshall 1976). References to these uses can be found in the above-mentioned reviews on lipase application. It is also used, as previously mentioned, in connection with commercial proteinases and in conjunction with other microorganisms in cheese manufacture. Illustrative of the success of such enzyme action on a practical scale is attested to by the report of Andres (1977) on a new semifluid cheddar cheese-like dairy product with five times the aroma intensity of regular cheddar cheese. It can be blended with the latter, especially as an ingredient to "bring up" the flavor of less-aged cheddar cheese with substantial economic advantages and without sacrifice of quality, "authenticity" or the use of "chemicals." Other uses include dips, spaghetti sauces, soups, crackers and, in general, for production of intensified butter flavoring to be used in the above-mentioned products (Andres 1980).

Specificity plays an important role in aroma development as illustrated by the selective liberation of fatty acids from cream by various sources of lipase as shown in Table 23.1. This specificity results, when oral lipase is allowed to act on butterfat, in a product relatively free of "soapy" or "bitter" notes which develop with lipases from other sources (Nelson *et al.* 1977). Even lipase concentration and extent of its action can influence not only the intensity but the quality of the aroma. Thus, Nelson (1972) noted that with increasing levels of the enzyme, there is first a sensation of richness without any detectable free fatty acid flavor, then creamy or butter flavors and a cheese-like flavor at relatively high levels. By contrast the importance of the substrates is further illustrated by the circumstance that when milk lipase acts on milk fats it produces a mixture of fatty acids which we perceive as undesirable, yet when acting on cocoa fat as it does in chocolate a combination of fatty acids is produced which contribute to the desired aroma and texture of this confection. In this connection, Nelson *et al.* (1977) point out that

TABLE 23.1

SPECIFICITY OF LIPASE-ESTERASES TOWARDS CREAM LIPIDS

Source of Enzyme	No. of C Atoms in Liberated Fatty Acids[1] C = 4	C = 6−10	C = 12
	Moles Liberated Relative to Butyric (C = 4)[2]		
Pregastric[3]	1.0[2]	0.7−0.8	0.4−1.1
Milk	1.0[2]	2.0	4.4
Pancreas	1.0[2]	0.3	10.6
Fungal *(A. niger)*	1.0[2]	1.3	0.0

Source: Adapted from data of Huang and Dooley (1976).
[1] Even numbers only.
[2] Actual mol % liberated, going down the column; 36.7−48.1, 13.5, 8.4 and 4.3.
[3] Includes three preparations: esterases from kid, lamb and calf and an imported kid rennet.

lipase-modified butterfat added to cocoa improves the overall flavor and also mitigates the sensation of intense sweetness known as "sugar burn." Lipase can also affect texture of foods by producing monoglycerides which have surface-altering properties and are widely used in baking.

In connection with the undesirable flavor produced by the action of lipase on milk lipids, a perusal of Table 23.1 reveals the molecular basis of the particularly objectionable odor associated with hydrolytic rancidity of milk (Chapter 25). Thus while the pancreatic enzyme liberates too much of the higher fatty acids and too little of the fungal enzyme of the lower fatty acids, the milk enzyme produces a superabundance of *both* lower and higher fatty acids.

Lipase action, presumably desirable, endogenous or from microorganisms, also occurs in other fermented foods such as pickled cucumbers, sauerkraut and other fermented vegetables. It also plays a role in flavor generation in smoked hams and sausages (Dwivedi 1975) and in fermented fish products (see below).

Other foods in which fermentation takes place and in which the fermentative organisms may well provide enzymes in flavor genesis include wine, cocoa and coffee discussed in Chapter 22, and oriental foods such as tempeh. The role of enzymes in the genesis of H_2S in many foods was reviewed by Schwimmer and Friedman (1972).

Enzymes Involved in Bread Aroma Genesis

From the viewpoint of flavor precurors' origin, bread represents a hybrid food in which microbial enzymes, enzymes from the wheat flour and added enzymes, as in cheese, all contribute to the final aroma of the bread. Bread aroma arises from application of heat to enzyme-produced aroma precursors comprising over 200 components distributed among several chemical classes.

Compounds include one terpenoid (limonene), maltol and two ether derivatives. This distribution is to be contrasted with that present in previously discussed foods. While aldehydes and alcohols among the nonpyrolytically-derived volatiles are varied and abundant, the number of esters is quite low and terpenoids and hydrocarbons are practically nonexistent. Noteworthy is the absence of the all-pervasive strong odorant, methional, among the nonenzymatically produced sulfur compounds. Maga (1974,1978) has provided in-depth reviews of bread flavor components and factors affecting the sensory properties of bread. The role of enzymes in the production of these volatile compounds is not unlike their role in cocoa aroma production, with the glaring exception of the absence and lack of a central role in most wheats of phenolase. The lipases and other esterases release fatty acids which in themselves may be odorants or may provide substrates for appropriate desirable lipoxygenase action. On the other hand, it is likely that aminotransferases of yeast account for volatile α-keto acids, especially the methionine-derived S-methyl-thio-α-keto butyric acid, which tends to accumulate and then disappear in doughs during fermentation. The optimum level appears to be dictated by the relative activity of the yeast aminotransferases (Suomalainen and Ronkainen 1963). The few reported esters in rye bread aroma, about twice the number present in wheat, comprise mostly two or three carbon alcohols and acids. They probably arise from the action of ester-synthesizing enzymes in the yeast. Lipoxygenase action from wheat flour and/or added as soybean flour can, when properly controlled, contribute to desirable bread aroma. The exact compounds which arise, their level and proportion to other compounds, have not been ascertained. We do know, however, that wheat possesses some interesting lipohydroperoxide transforming enzymes discussed in Chapter 24 which must have some influence on the degradation pathway of these intermediates and eventually on bread aroma. Indeed, soybean flour supplied to the baking industry as an enzyme additive is used not only for bleaching (Chapter 17) and improving texture (Chapter 32) but also for improving flavor (Kleinschmidt 1963). The first clue that lipoxygenases are involved in bread flavor was the observation of King *et al.* (1937) that the level of shortening can determine to a degree contribution of added lipoxygenase to bread flavor. When used to improve flavor, the soybean flour is not, as it is when used for bleaching, added directly to the wheat flour, but is first mixed with a source of unsaturated fatty acid substrate such as cottonseed or soybean oil until 0.03 to 0.04% hydroperoxide is present (Kleinschmidt 1963). Less than this level is ineffective (probably due to reaction inactivation) and very little more (one part of peroxidized oil to 200 parts of dough) imparts undesirable flavor. Although this use of lipoxygenase was introduced in the baking industry over 15 years ago, enzyme-related questions concerning just how this procedure results in improved flavor still need answering. Do the peroxides split to form the 6- and 9-carbon aliphatic straight chain aldehydes? If they do, is there a lyase present similar to that reportedly present in bananas, cucumbers and tomatoes (Chapter 22)? It is conceivable that lipohydroperoxide can function as a general-acting strong oxidizing agent in subsequent baking operations, a manner analogous to that of the phenolase-produced quinones in tea, cocoa and perhaps coffee.

The color problem presented by the presence of heightened levels of phenolase in the new varieties of "miracle" wheat (Chapter 16) may have a counterpart in bread flavor aberrations. Increased phenolase activity may upset the delicate balance of volatile components corresponding to the aroma associated with good fresh bread. One might expect, for instance, some notes of tea flavor due to excessive conversion of carotenoids to terpenoid-like substances. This would be especially prominent in the high-carotene semolina flours.

Undoubtedly some of the flavor of leavened baked goods arises from the action of enzymes in the yeast. One such enzyme which could conceivably be so involved is a recently delineated NAD-dependent secondary alcohol dehydrogenase. This enzyme catalyzes the conversion of such alcohols to 2-methyl ketones (Patel *et al.* 1979). The latter, it will be recalled, are important constituents of cheese flavor.

Proteinase.—Action of wheat proteins of both endogenous and added commercial proteinases provide amino acids which serve as precursors to many of the aroma-bearing volatile components of bread flavor. These include: aldehydes, produced via the Strecker degradation; pyrazines, intermediates in or side-products of the Strecker degradation; and ketones, created via the Maillard nonenzymatic browning pathway. The qualitative difference between the aromas of wheat and rye breads has been attributed to, ultimately, a quantitative difference in amino acid levels between the two grains with the level of rye being the higher.

Further evidence for the participation and importance of the amino acids in flavor development and the application of this knowledge is provided by the noted improvement of aroma attendant on the addition of commercial proteases such as the Rhozyme preparations (El-Dash 1971) and undoubtedly others in the United States and Amylorim PX in the USSR.

The degree of proteolysis is critical, due, perhaps, to the central role of proline as flavor precursor amino acid. Baking-generated proline-derived pyridine volatiles lend a strong character impact to fresh bread flavor (Hunter *et al.* 1969). Hence an examination of the proline-releasing properties of the concerned enzymes might be germane to further progress in this area. Proline, it will be recalled, is an amino acid which does not possess a true α-amino group, so that it is not surprising that organisms have elaborated specific peptide hydrolases as indicated in Table 23.2. Balanced judicious additions to wheat flour would result in the selective, controlled accumulation of proline. Indeed the level of the product of proline liberating peptidases' action is used as an index of the malting and brewing quality of barley (Chapters 31 and 37).

Carbohydrases.—Undoubtedly the commercial amylases together with the action of those from the flour and invertase from the yeast provide reducing sugars which participate along with amino acids in Maillard reactions during baking to produce some of the flavors associated with bread flavor. In addition, furan derivative products of the pyrolytic decomposition of sugars also bear odors which contribute to bread aroma.

MEAT FLAVOR

Except for the fermented sausages, meats and meat products are foods whose desirable odors develop almost entirely during thermal treatment,

TABLE 23.2

PROLINE-ASSOCIATED PEPTIDASES

Recommended EC Name	Action, Comments
Proline amino-peptidase	HN·[prolyl]·CO ·NHCH (R) CO...
Iminopeptidase	Acts on polypeptides found in bacteria; usually requires Mn^{2+}
Proline carboxypeptidase	..CO·N·[prolyl]CO ·NHCH(R) COOH Proline is penultimate amino acid from the carboxy end of the peptide chain, involved in blood pressure regulation.
Prolyl dipeptidase Prolinase Iminodipeptidase	HN·[Prolyl]CO ·NH·CH(R)COOH Ubiquitously distributed, this enzyme also acts on hydroxyprolyl dipeptides
Proline dipeptidase Prolidase Iminodipeptidase	HOOC·CH(R)·CO ·N·[Prolyl]COOH High activity in mammalian kidney, intestinal mucosa, strictly specific for *trans* form of peptide bond of amino acyl proline dipeptides (Lin and Brandts 1979).

be it cooking, frying, broiling, roasting or baking. Thus, very seldom is an aroma constituent of meat the immediate product of an enzyme reaction. This is reflected in the high proportion and wide variety of heterocyclics comprising some 100 of the 300-plus volatiles isolated from the cooked beef (Dwivedi 1975). Of course at the time of cooking almost all meat constituents have accumulated as the result of enzyme action. However, at time of slaughter the combination of fats, carbohydrate and amino acids, which serve as precursors to the aroma constituents which develop during cooking, are not conducive to desirable aroma (or taste) development. The postmortem aging process (Chapter 26) not only leads to development of desirable texture, but by allowing potentiation of certain enzymes via decompartmentalization of substrate attendant upon cell disruption, also results ultimately in desirable cooked meat odor.

Although there are no major formal investigations of these enzymes specifically directed to their roles as generators of precursors to meat flavor, it is not hard to surmise what they might be from a knowledge of components present in meat extracts, and of what enzymes are active in muscle tissue. An even more salient clue is a comparison of change in composition of such extracts due to aging and cooking. In one of the few such investigations, for instance, Wood (1961) observed that phosphorylated sugars and NAD disappear and free sugars appear upon aging indicating that phosphatase action may be helpful in flavor generation. There is, of course, ample evidence that free amino acids increase during aging of beef and thus the endogenous proteases play a role in flavor development. Besides releasing the 5′-nucleotide flavor potentiators, ribonuclease also, along with nucleotidases, ATPases, coenzyme hydrolases and nucleosidases, undoubtedly gives rise to ribose phosphate and ribose, both of which when heated give rise to meat-flavored aromas.

Although accounting for not more than 2% of the total carbohydrate (glucose may account for 98%, fructose up to 7%) in diffusates of various meats before heating, only *ribose* disappeared completely upon heating of the diffusate (Macy *et al.* 1964). Lipase and perhaps animal lipoxygenases and prostaglandin synthases may give rise to precursors of cooked meat aroma but also may be involved in off-flavor development (Chapter 24).

Somewhat but not quite analogous to the use of lipase to prepare dairy product flavorants, proteolytic enzymes have been used to prepare meat flavorants, at least as judged by a sizeable patent literature. An interesting illustration in which dairy products are connected to a meat flavorant is afforded by a patent issued to Givaudan (Jaeggi and Krasnobajew 1977) in which yoghurt or other fermented dairy products are subject to proteinase action. The analogy with lipase breaks down here in that the resulting amino acid-containing proteolysate is subjected to further treatment, viz, it is dried and heated in the presence of cysteine at a temperature which accelerates nonenzymatic browning. Their use of enzymes has health-associated overtones in that the intensive meat aroma-bearing flavorant may be especially suitable for use in bland high-protein foods (Chapter 34).

Proteolytic enzymes from microorganisms and not from meat are probably responsible for liberation of amino acids which in turn are major contributors to amine-associated sausage flavor via α-decarboxylase action (Mihalyi and Kormendi 1967; De Ketelaere *et al.* 1974). Inasmuch as we have already outlined the nature of some of the nonenzymatic pyrolytic reactions both in this chapter and elsewhere, and this is not a food chemistry compendium, we refer the reader to the above-mentioned review of Dwivedi for further discussion of thermally-generated flavors. The distinctive flavor of eel may be due to an abundance of the constituent amino acids of carnosine, β-alanyl histidine, also present in other fish and mammalian muscle tissue. These amino acids are formed during storage of eel meat due to the continuous action of a highly specific β-alanyl peptidase, carnosinase (Delincee and Partmann 1979). By the same token, hydrolysis of anserine, β-alanyl-1-methyl histidine, by anserinase may contribute to the flavor of pike, cod and trout.

Very few investigations have been directed to the enzymatic aspects of the generation of fish aroma. Undoubtedly nitrogen compound-metabolizing enzymes dominate in this regard. Dougan and Howard (1975) state that the aroma of fermented fish sauce comprises three distinct notes: cheesy, due to the lower fatty acids; ammoniacal, due to ammonia and amines; and meaty, produced by atmospheric oxidation of flavor precursors. Beddows *et al.* (1976) concluded that with mackerel, microorganisms play a significant role in aroma development, as they undoubtedly do in fermented sausage. Dimethyl sulfide, an important constituent of oyster and clam aroma, probably arises from β-propriothetin (Schutte 1974). Proteinase action as an agency of flavor development in barrel-salted herring was studied by Kiesvaara (1975).

FLAVORESE

We conclude our survey of the enzymatic antecedents of food aroma and related and desirable food flavors with a brief examination of the origin and present status of the "flavorese" concept. About 25 years have passed since this concept was promulgated by Hewitt *et al.* (1956) and a patent was issued (Hewitt *et al.* 1960). In a discussion of why the idea has never been completely accepted, Konigsbacher (1976), one of the pioneers of this field, felt that his experience showed that the food processing industry was not until then inclined or able to support the prolonged and sophisticated research efforts required to the point of practical application. Another possibility is that judging from the paucity of research published from industry, pertinent proprietary information is not being published or patented. The literature which has been published has time and again demonstrated improvement and/or restoration of flavor in both processed fruits and vegetables and perhaps in meats and nonplant foods as well.

The term "flavorese" was chosen because in the early history of enzymology it was suggested that names for enzymes which catalyzed synthetic and degradative reactions should end with "ese" and "ase," respectively. Thus, food flavors are being synthesized. This nomenclature still lingers on (aside from flavorese) in the enzyme rhodanese, thiosulfate cyanide sulfur transferase

$$S_2O_3^{=} + CN^{-} \rightarrow SCN^{-} + SO_3^{=}$$

This enzyme is widely distributed in mammalian tissue and may function as a detoxification mechanism by protecting mammals from cyanide poisoning resulting from the action of the emulsin on the cyanogenic glycoside (Chapters 21 and 35). Administration of thiosulfate is extremely effective in preventing death due to acute cyanide poisoning. The general idea behind the flavorese concept depicted in Fig. 23.1 is that flavors—whether produced via normal metabolic processes (preformed), accumulated in response to stress, or arising via tissue comminution and cell disruption—are lost during the processing of the food (usually through heating). Simultaneously, flavor-producing enzymes but not flavor-producing substrates (flavor precursors) are eliminated from the food. One should be able, in principle, to restore the fresh flavor by adding to the processed food those enzymes involved in flavor generation. Ideally, the best source of these enzymes is the same fresh unprocessed food. The original investigators used air-dried acetone-precipitated proteins (acetone powders, Chapter 6) derived from water extract of such foods, especially fruits and vegetables. With these "flavorese" preparations, they reported successful restoration of fresh-like flavors to a wide variety of dehydrated and frozen foods including celery, spinach, cabbage, carrots, milk, onions, or-

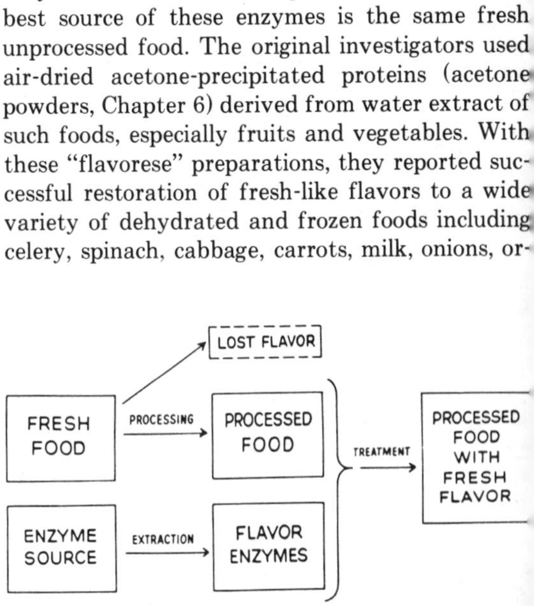

From Hewitt (1963). Reprinted with permission of J. Agric. Food Chem. 11, 155, Fig. 1. Copyright by Am. Chem. Soc.

FIG. 23.1. THE FLAVORESE CONCEPT

anges, pineapples and strawberries. Even parsley, whose flavor appears to be preformed (Chapter 21) improved in flavor upon addition of the appropriate parsley enzyme fraction. In addition to the aforementioned and further publications from the inventors' laboratories and those of their collaborators at Natick (Mackay and Hewitt 1963; Konigsbacher and Hewitt 1964; Mackay and Hewitt 1959), further experimentation has tended to confirm, elaborate and extend the fundamental basic premises.

Flavorese Experiments with Vegetables

Cabbage.—Bailey *et al.* (1969) reported the restoration of some but not all of the components of the various classes of compounds present in fresh cabbage volatiles, as measured by GLC. Components measured were isothiocyantese and di- and trisulfides, thus bespeaking glucosinolase and alliinase action. Schwimmer (1963) confirmed previous results, using mainly cabbage but also dehydrated and frozen broccoli, peas, beans and carrots. The flavor of each enzyme-treated processed food approached that of the fresh vegetable but never was considered to be identical, since the fresh could always be distinguished from the treated samples. Addition of the acetone powders tended to over-emphasize certain notes, indicating that some flavor-generating systems were operative and others were not. This state of affairs could be due to change in substrate, loss of some enzymes during the preparation of the flavorese, or the hydrolysis of cofactors such as NAD, acetyl CoA or CoA itself, usually via nonspecific hydrolases. The effect of substrate and enzyme sources on the flavor effect of rehydrated food is shown in Table 23.3.

Mustard.—The isothiocyanates in commercial mustard produced via the action of the glucosinolases (Chapter 21) are slowly decomposed or transformed into flavorless thioureas. Konigsbacher (1976) reported that he was able to extend the shelf-life of mustards by slowly releasing glucosinolase from microcapsules added to the aging condiments. This particular application extends the flavorese concept from restoration of flavor in foods to maintenance of flavor during storage. Glucosinolase preparations have been used to prepare "flavor extracts of vegetable materials" (Kinjirushi Wasabi 1974).

Onions.—The application of enzymes to enhance onion flavor was discussed in Chapter 21. More specifically, the use of γ-glutamyl transferase extends the flavorese concept to include enzymatic transformations of components which normally are not flavor precursors.

Carrots.—The investigations of Heatherbell and Wrolstad (1971) illustrate some of the difficulties in this field. Even when improved flavor was noted in a series of experiments which yielded erratic results, they could find no correlation between improved flavor and appearance of new GLC peaks.

Beans.—Schwimmer (1963) found that oxygen was necessary to obtain improved flavor in accord with the role of lipid peroxidation in fresh vegetable aroma development discussed in Chapter 21.

Tomatoes.—Using spray-dried tomato powder as a source of substrates (flavor progenitors) and crude or density-fractionated extracts of various tissues of the tomato plants as enzyme sources,

TABLE 23.3

ENZYMATIC ALTERATION OF FLAVOR

Source of Flavor Precursor (Substrate)	Source of Enzyme	Description of Flavor Difference[1]
Cabbage dehydrated, blanched	Cabbage, dehydrated, unblanched	Slight pungency, more like fresh
blanched	Cabbage, fresh —Supernatant (Sup.)	Biting, isothiocyanate
blanched	Mitochondria (Mit.)	Pungent, more like fresh cabbage
blanched	—Sup. + Mit.	Cabbage heart, turnip, chestnut
blanched	Mustard	Biting, horseradish
blanched	Onion	Pungent, onion, lachrymatory
Cabbage, dehydrated, unblanched	Mustard	No difference
Mustard:Sinigrin	Cabbage, dehydrated, unblanched	Biting, pungent, horseradish, isothiocyanate

Source: Schwimmer (1963) and Morgan and Schwimmer (1965).
[1] After addition of enzyme or flavor precursor.

Gremli and Wild (1974) observed that, as for cabbage (Table 23.3), authentic regenerated aroma could be more closely approximated by combining soluble and particulate sedimented fractions of the extracts than by using either separately. Even with uncentrifuged, unfractionated, whole extracts as enzyme the regenerated aromas, some of which could not be described as desirable, depended upon the flavorese source. In addition to a "green" note present in all of the regenerated aromas, additional flavor nuances, each characteristic of each enzyme source, were sensed.

The closest to fresh aroma was attained with enzymes from both ripe and green fruits, the latter having a heavier overlay, appropriately, of the "green" note. Enzymes from tomatoes at the yellow stage of maturity generated a "fruity" note while those from leaf and stem tissues developed "sweet" and "earthy-woody" flavor notes, respectively.

Several identified volatile components obviously arose from the action of aminotransferase, lipoxygenase and alcohol dehydrogenases shown to be present in the extracts of ripe tomatoes (also present were phenolase and peroxidase). The action of the dehydrogenase at the low pH of the reconstituted tomato powder probably accounts for the preponderance of alcohols over aldehydes in the accompanying volatiles. It will be recalled (Chapter 22) that low pH's favor the reductase activity of NAD(P)-dependent dehydrogenases. It is perhaps surprising that enough NAD-associated cosubstrates presumably present in the tomato powder survived the spray-drying processing or that there were enough present in the fresh fruit, unless a coenzyme-regenerating system was also present.

Flavorese Experiments with Fruits

Bananas.—Hultin and Proctor (1961) found that fresh banana aroma was regenerated in heat-processed purees to which crude banana protein was added as an enzyme source. Comparable results were obtained with banana extract as flavorese source (de Menezes and Boas 1973). Further stimulation of aroma production was attained by adding substrates of some of the enzymes we have been discussing. These substrates included pyruvate, valine and oleic acid. Other compounds, leucine and acetate, did not afford stimulation of aroma development. While the participation of oleic acid is readily comprehensible, just why valine and pyruvate were effective in absence of added biosynthetic cofactors is puzzling. We suggest that contrary to the prevailing ideas, some of these cofactors, i.e., for the dehydrogenases, may have survived the rigors of cellular disruption.

While there may have been, as for tomatoes, an increase in alcohols upon standing in these flavorese puree reaction mixtures, they probably arose from esterase action rather than that of dehydrogenase. At least esterase action also produces acids which may account for the "rotten" banana smell which developed in these mixtures. Alternatively, the smell could have arisen from the action of enzymes in incipiently developing microorganisms.

Raspberries.—Weurman (1961) observed that enzymes prepared from the center cores of this fruit, when added to a flavorless substrate source consisting of thick syrupy residue from the steam distillation of fresh raspberries (distillers wash), caused coordinate development of a distinct fresh raspberry aroma and of new volatile compounds as measured by GLC. Of a variety of commercial food enzymes and of highly purified biochemical grade enzyme preparations, only the extremely crude commercial mixtures and purified alcohol dehydrogenase resulted in the formation of new volatiles. None of these preparations could, as did the raspberry enzyme, restore fresh flavor. The author concludes, not surprisingly, that more than one enzyme is involved in raspberry aroma formation.

Bruchmann and Kolb (1973A) showed that an insoluble fraction from raspberry sepals (calyx cone) of the ripe fruit served as a potent enzyme source for the development of fresh raspberry flavor and associated volatiles as shown in Fig. 23.2. The substrate sources were either homogenates of unripe green raspberries, or the abovementioned distillers wash. By the use of what they considered specific inhibitors, i.e., copper chelators (gluconolactones), each of which at low levels prevented appearance of new peaks in the GLC aromagrams, they hypothesized that several membrane-associated glycosidases and heavy metal-containing enzymes constitute a sequential synergistic aroma system. When one enzyme of the system is inhibited, the reaction chain collapses, no aromagram develops and the remaining enzymes act fruitlessly in parallel instead of in a series. Thus, these flavorless precursors may exist in the distillers wash as glycosides of such substances as, say, raspberry ketone, *p*-hydroxy-1-(*p*-hydroxyphenyl)-3-butanone (Chapter 22).

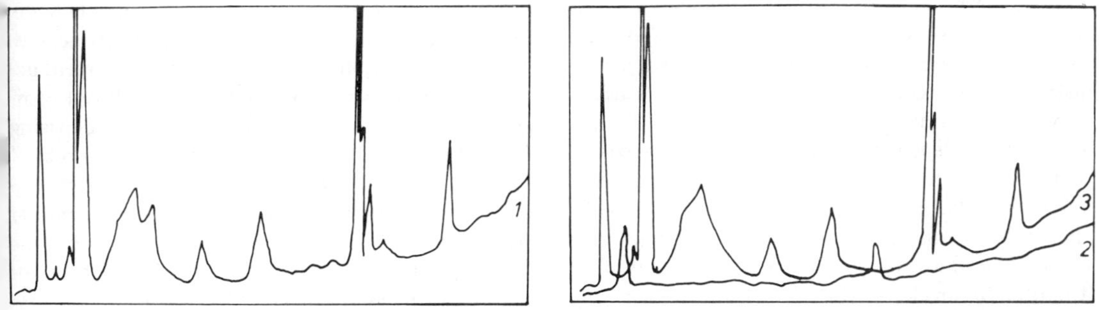

From Bruchmann and Kolb (1973A)

FIG. 23.2. RESTORATION OF RASPBERRY FLAVOR WITH FLAVORESE

Depicted are GLC aromagrams of: ripe fruit (1), unripe fruit without (2), and with (3) added flavorese from calyx cone sepals.

Peanuts and Flowers

Peanuts.—Raw peanuts develop what some consumers consider to be a desirable aroma when chewed or otherwise crushed or comminuted. Although they were not primarily interested in application of flavorese systems, investigators at North Carolina State University did demonstrate flavorese in action (Singleton *et al.* 1976). Reaction mixtures of peanut acetone powders (enzyme) and enzyme-inactivated peanut homogenates (substrate) developed typical raw peanut aroma as well as volatiles whose GLC aromagram was almost qualitatively identical with that of fresh uninhibited homogenates and also with that of a model system consisting of purified peanut lipoxygenases and linoleic acid as shown in Fig. 23.3. Also, exogenous aliphatic alcohols added to peanut homogenates before blending were rapidly converted to the corresponding aldehydes, thus demonstrating that alcohol dehydrogenase action at pH's higher than in tomato favors aldehyde production. NAD must have survived the homogenization. Pentane, pentanal and hexanal arose directly from hydroperoxide via lipoxygenase action. To ensure that this enzyme was, in fact, the prime causative factor in the development of desirable peanut aroma, the investigators ran controls with inhibitors and cosubstrates.

The applied aspect of this particular study to flavorese development is the demonstration that by controlled enzyme action, the judicious use of food antioxidants and application of information on substrate specificity, it is possible to alter the comparative amounts of volatile components constituting aroma and thus maximize or optimize desirable flavor in newly developed peanut products.

Flower Aroma.—Although the redolences and fragrances of flowers are usually not employed as food flavorants, some of them such as essential oils like jasmine are used to flavor teas.

The flavorese concept has been used to improve the yield of fragrance obtained from essential oils of flowers. Konigsbacher (1976) cited renewed

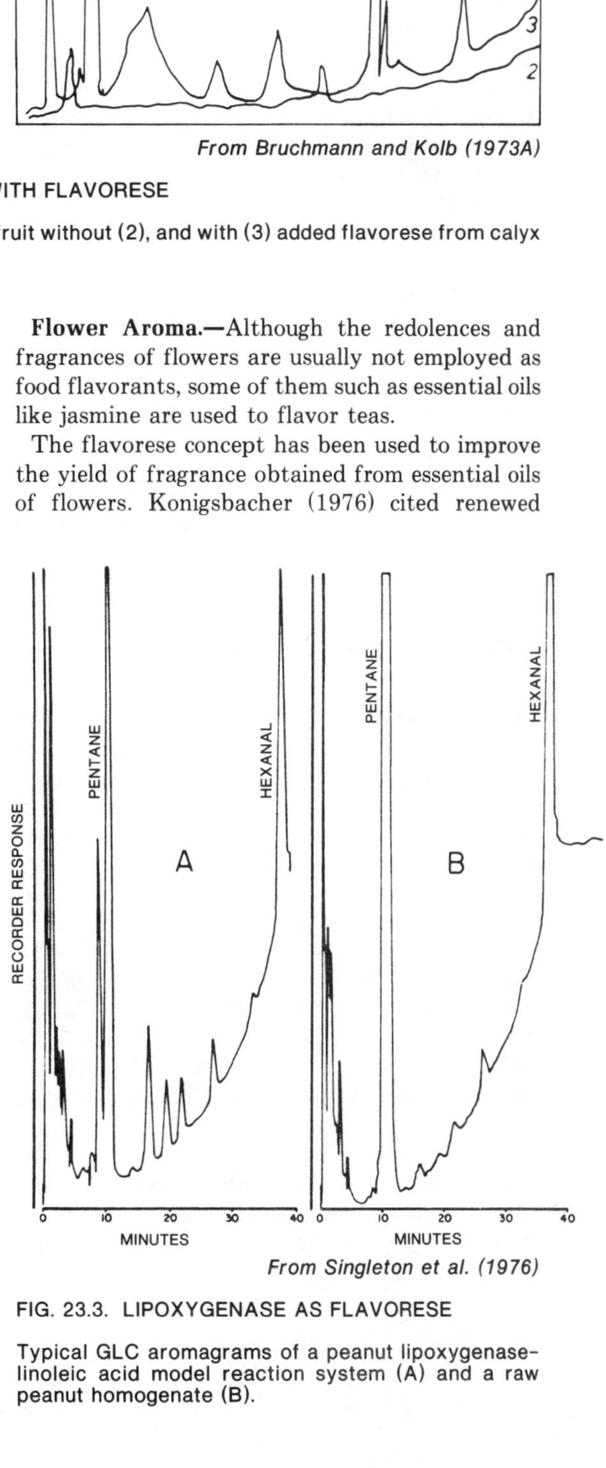

From Singleton et al. (1976)

FIG. 23.3. LIPOXYGENASE AS FLAVORESE

Typical GLC aromagrams of a peanut lipoxygenase-linoleic acid model reaction system (A) and a raw peanut homogenate (B).

aroma generation from seemingly exhausted petals of mimosa, rose and jasmine by adding flavorese enzymes prepared from fresh petals not subject to distillation or extraction. Again, as in other studies, they found that nuances of redolence of the evoked aroma were dependent upon a variety of conditions. Thus, a commercial β-glucosidase applied to a mimosa substrate source evoked an undesirable, almost bacon-like odor. Rose petal enzyme fractions isolated by precipitation at different pH's (Chapter 6) developed different floral notes. Only the pH 4-insoluble fraction yielded rose aroma, probably from geraniol-β-D-glucoside (Francis and Allcock 1969).

Enzymes acting on β-D-glucosides and similar glycosidases were also believed to be responsible for the supplementary pleasant but not identical aroma developed from jasmine flowers previously exhaustively solvent-extracted. At $200 to $250 per kilogram, this could turn out to be one of the earliest economically feasible applications of the flavorese procedure.

In addition to the above, we have discussed several instances of the use of enzyme preparations such as the lipases and proteinases for the preparation of flavorants.

Rationale, Prognosis and Progress

These illustrations of experimentation with and application of the flavorese concept would appear to be indicative of some promise of its wide adoption. One of the problems in the way of successful economic development is the expense of even partially purifying enzymes (Chapters 5 and 6). To avoid having to purify the requisite enzymes, Morgan and Schwimmer (1965) found that flavor of foam mat-dried foods was improved by adding to such processed food about 1% by weight of the freeze-dried fresh fruit or vegetable from which the food was processed. Of course, such foods probably contain all of the most unstable enzymes, both desirable and undesirable. The demonstrated improvement in flavor by the application of flavorese preparations is thus somewhat of a paradox in view of the traditional ideas that enzymes must in general be destroyed, as in blanching, to maintain quality and prolong shelf-life (Joslyn 1963). From the evidence so far accumulated, as outlined above, it would appear that the answer is multifaceted. Thus, most obviously the relatively short-term action required to regenerate flavor results in compounds that differ qualitatively from those arising from the attrite action over extended periods of storage. Another factor might be the relative stability of the involved enzymes. Thus, Schwimmer (1963) found that old flavorese preparations no longer produced pleasant flavors. Another factor may be differences in volatility. The "good" aromas appear to disappear more readily then do the "poor" ones. Also one gets the overall impression that at least some of the flavorese enzymes seem to be associated with particulates in the cell, whereas the undesirable enzymes when they act appear to be more soluble or more readily extractible.

Microbial Enzymes as Flavoreses.—Several authors (Bruchmann and Kolb 1973A; Drawert 1975; Konigsbacher 1976) have suggested that the solution of flavorese adoption may lie in the development of microbes and microbial enzymes. We have had occasion to discuss their use in connection with onion alliinase and glucosinolase. That limited work is underway to further explore this approach is evidenced by German and Japanese investigations reviewed in Japanese (Kanisawa 1975) results of which are available only as abstracts. That there are problems to solve is illustrated by the seemingly conflicting reports on the effect of a particular microorganism *(Oospora suaveolens)*. The Japanese investigators (Hattori *et al.* 1974) reported that this organism, which they isolated from preripe green strawberries, when inoculated into sterilized strawberry puree, produced aroma suggestive of overripe rather than fresh ripe fruit. The same organism also produced overripe aromas when inoculated into purees or juices of grape, banana, peach and pineapple.

The German investigators (Askar and Bielig 1976) report that they detected a decidedly improved flavor upon addition of "enzyme" prepared from a strain of the same organism isolated from overripe strawberries. Fruit-like aromas in dairy and other products where such flavors are unwanted arise from *in vivo* enzyme in adventitious organisms which may be a good place to look for "flavoreses."

24

ENZYME INVOLVEMENT IN OFF-FLAVOR FROM THE OXIDATION OF UNSATURATED LIPIDS IN NONDAIRY FOODS

CONTRIBUTION OF ENZYME ACTION TO FOOD OFF-FLAVORS

Nonbiological Contributions

To keep a proper perspective on the role of enzyme action in causing or preventing flavor defects in foods, it should be stressed that nonbiological agents such as adventitious contaminants with nonfood substances carried by air and water are probably the major sources of flavor defects (Goldenbenger and Matheson 1975). These contaminants, which can enter the food processing chain at almost any stage (Fig. 1.2) include packaging material, paint ingredients, disinfectants, pesticides, herbicides, petroleum-associated products and detergents. Of course it is conceivable that substances in the aforementioned categories could produce off-flavors by interaction with some components of the food's biological system and give rise to untimely interaction of enzyme and substrate. However, for the most part, where the cause of the defect is known these substances themselves bear the objectionable flavor. Occasionally the off-flavor may arise as a result of a simple chemical reaction with some food constituent. Thus, mesityl oxide interacts with the sulfhydryl groups present in food constituents (including enzymes) to produce catty off-flavors (Maarse and Ten Noever de Brauw 1974).

$$(CH_3)_2{=}C{=}CH\text{-}CO\text{-}CH_3 + RSH \rightarrow (CH_3)_2{=}C(SH)\text{-}CHR\text{-}CO\text{-}CH_3$$

Blanching itself may induce off-flavor, as in the avocado, due to the heat-induced production of 1-acetoxy-2,4-dihydroxy-n-heptadec-16-ene and similar compounds (Ben-Et *et al.* 1973). A substance which contributes to desirable aroma or taste of one food may be considered objectionable when present in other foods, i.e., the clover-derived cinnamon-like "off-odor" present in some wheat flours and bread (Buttery *et al.* 1978). We previously mentioned the light-induced flavor defect of "sunstruck" beer due to the nonenzymatic production of prenyl mercaptan (Schwimmer and Friedman 1972). Light also induces undesirable odor derived from decomposition of thiamin (Seifert *et al.* 1978). Thiamin itself is responsible for the objectionable flavor of yeast (Höhn and Solms 1975). As mentioned in Chapters 23 and 25, objectionable fruit-like aromas can develop in dairy products; diacetyl, which imparts characteristic aromas to many dairy products, is objectionable in beer.

Enzymatic Involvement.—Aside from ancillary and largely nonbiological considerations, further delineation of the reticulations of enzyme action into the problems and solutions of undesirable food flavors may be separated into the following five divisions.

Off-flavor may be due to microbiological deterioration of the food. The enzymes of the offending organisms convert food constituents into objectionable substances. However, as adumbrated in Chapter 1, occasionally what would appear to be a clear-cut case of off-flavor due to microbial spoilage such as holding coleslaw at 7°C for almost 2 weeks turns out to be a problem of endogenous enzyme action.

Off-flavor may be due to the presence of substances accumulated normally or "physiologically" during the normal unstressful life cycle of the food organism.

Flavor defects may arise in plant foods due to accumulation of stress metabolites (Chapter 1) although stress metabolites are, as we have seen, at least as likely to contribute to desirable flavor.

Another source of off-flavor with enzymatic credentials in animal-derived foods is that originating in the animal feed. Indeed, a substantial measure of progress in our understanding of the enzymology of flavor production in the crucifers and alliums (Chapter 21) is due to the funding of projects whose principal aims were in finding how enzymatically-caused off-flavors in the milk of cows, which had been grazing on plants containing these enzymes, could be gotten rid of. Also, the teleology and ecology of these secondary plant metabolites and their attendant enzymes is relevant in this respect. Brighton and Horne (1977) observed that the presence of the cyanogenic glycosides in white clover seemed to reduce the resistance to stress which presumably led to a decreasing frequency of distribution of these compounds with increasing altitude. However, the glycosides do have a survival value in that they are fairly impalatable to grazing animals. The authors conclude that the interplay of temperature effects with differential grazing may maintain the polymorphism (persistence of both cyanogenic and acyanogenic strains of plants in a given location) and govern the ecology of the different forms of white clover in at least some locations.

We are especially interested in off-flavor which may arise as the result of cellular disruption during processing and during subsequent storage of the food. Progress in this field tends to follow a pattern. First, the flavor defect is found to be associated with parameters or variables such as time, temperature and treatment of the food, and its general ambience. Next, the defect and its intensity are correlated with the presence and level of definite food constituents. Usually, but not always, these substances may turn out to be the flavor-bearing agents. Next ensues the development of a connection between an enzyme (or enzymes) which produces these substances, usually first by demonstrating a correlation between the activity of the suspected enzyme and the intensity of the odor. This may be followed by introduction of measures to prevent the enzyme from acting, measures which are based upon intimate knowledge of its properties and behavior. A survey of some of the off-flavors in foods which thus have been shown to have enzymatic origins is shown in Table 24.1.

In practice, economics may dictate an empirical approach to preventing off-flavor even before the enzymatic nature of its genesis has been established; indeed, frequently the means of preventing off-flavor pertains universally to the prevention of other enzymatically-caused food quality defects. Furthermore, as our discourse has repeatedly brought out, more than one quality attribute can be imparted by the action of the same enzyme; the same enzyme can improve as well as cause deterioration in food quality.

A fifth aspect of the role of enzymes in off-flavor is their use as additives to remove the offending substance (Table 24.2). The addition of enzymes for nonflavor purposes may inadvertently cause the development of objectionable flavors. Unwanted generation of the bitter peptides due to action of proteolytic enzymes, one of the main examples, will be dealt with in Chapter 34.

While we have categorized some of the above-outlined aspects of enzymatic involvement in off-flavors, the distinction cannot always be clearly separated, as we shall see in the ensuing discussion. The fact is that in far too many instances we are abysmally ignorant of the causative factors determining the direction towards which flavor-generating systems will develop.

Lipid Involvement

By and large the enzymes involved in the development of a majority of off-flavors act on lipids and transforms thereof. The two major enzyme types directly involved are the lipases and the lipoxygenases, or closely-related enzymes. Involve-

TABLE 24.1

SURVEY OF SOME FOOD FLAVOR DEFECTS CAUSED BY ENZYME ACTION[1]

Flavor Defect	Food	Enzyme Implicated
Diacetyl	Orange juice, beer	Peroxidase[2]
Tobacco-like	Prune	Phenolase[3]
Bitter	Eggplant, senescent fruit	Phenolase[4]
H_2S, excess	Wine, beer	Yeast C-S lyase ATP-Sulfurase
	Eggs	Not C-S lyases
	Fish	Proteases
Overly fishy (trimethylamine)	Fish	Trimethylamine-N-oxide reductase (Chap. 28)
Bitter	Cheese, FPC, SCP and other high protein sources	Protease (Chap. 34)
Bitter	Onion	Alliinase, associated with pinking (Chaps. 17, 21)
Sour, musty (nitriles)	Cabbage	Glucosinolase, secondary transformations
Metallic (a vinyl ketone)	Butterfat	Lipoxygenase, secondary transformations
"Oxidized" flavors	Beer, orange juice	—

[1] Includes enzymes not discussed in this and the next chapter.
[2] Source: Bruemmer (1975).
[3] Source: Chari *et al.* (1948).
[4] Source: Flick *et al.* (1977).
[5] Source: Lawrence and Cole (1968) and Eschenbruch and Bonish (1976).
[6] Source: Swoboda and Peers (1977).

TABLE 24.2

PROPOSED USES OF ENZYMES AS ADDITIVES FOR REMOVAL OR PREVENTION OF FOOD OFF-FLAVORS

Enzyme	Food	Function
Dehydrogenases	Citrus	Remove limonene
Dehydrogenase	Beer	Remove diacetyl, an off-flavor
Glucose oxidase	General	Remove oxygen
Sulfhydryl oxidase	Dairy	Prevent "oxidized flavor"
Aldehyde oxidase	Soybean	Removes objectionable odor-bearing aldehydes
Superoxide dismutase	Herring oil	Prevent rancidity, oxidative off-flavors
β-Glycosidase	Citrus	Remove bitter naringin (excess)
β-Glycosidase	Safflower	Remove bitter glycosides
Phospholipases C, D	Milk	Inhibit oxidative off-flavors
Proteases	High protein	Remove off-flavor, bitterness
Carboxypeptidase	High protein	Peptides (Chap. 34)
Lysozyme	Several	Prevent off-flavors due to microbial contamination
Alcohol dehydrogenase	Apple juice	Adjust alcohol/aldehyde ratio for optimal flavor
Ribonuclease	Fish	Reduce odor formation
Emulsin	Bakery products	Debitter apricot kernels
Urease	Shark meat	Remove urea which makes meat bitter

ment of these two enzymes shows some striking contrasts. Thus, the immediate, stable end-products of lipase action, the free fatty acids (FFA), are the actual odor-bearing constituents, whereas the major proportion of the contribution of lipoxygenase action to off-flavor development probably arises from secondary reactions of the primary products (the lipohydroperoxides) although the latter appear to possess a bitter taste and some off-odor (see below). Furthermore, lipase action can occur in the absence of lipoxygenase whereas lipoxygenase frequently acts only after the lipases have liberated FFA from triglycerides and phospholipids. Of course, lipids can give rise to off-odor via nonenzymatic processes such as autoxidation as well as via nonlipoxygenase-mediated enzyme action. We shall delay further examination of lipase action until the next chapter and shall devote the remainder of this chapter to the genesis of nondairy-related off-flavors (mostly odors) which appear to arise from secondary reactions following action of lipoxygenase and other lipid-transforming enzymes in foods. We shall also glance at the "pure" enzymology involved. Other biological and food quality associated effects of this enzyme were summarized in Chapter 17.

LIPOXYGENASE

Soybean

Off-flavor.—When bean flour is mixed with water, rather repugnant flavors, characterized by bitter tastes and grassy, beany odors with overtones of rancidity, are rapidly perceived (Goossens 1975; Wolf 1975; Rackis *et al.* 1975; Rackis 1977).

These tastes and odors create flavor problems in soy products and in foods to which soy is added as an enzyme adjunct unless appropriate precautions are taken to prevent their formation. When used as a source of lipoxygenase for bleaching flour (Chapter 17) the problem is not acute. Off-flavors more frequently have posed problems in development of soy products such as soy milk and full-fat soybean products and they linger in protein concentrates and isolates prepared from soybean flour (Chapter 34).

According to Rackis (1977), some of the beany flavor constituents are preformed. However, extensive enzyme action leading to odor intensification may arise during "tempering" in which the soybeans are slightly wetted in order to improve oil yield during the subsequent solvent extraction.

The precise nature of the components responsible for the off-flavor has not been established. Thus, Wolf (1975) states that the flavor is caused by the presence of a large number of odorants rather than being due to the impact of a limited number of substances. From time to time certain compounds have been implicated as key contributors to these off-flavors. These include the alcohols isopentanol, hexanal, heptanol and 1-octenol, and the aldehydes hexanal, 2-pentenal and 2,4-decadienal as well as 2-pentylfuran. Vinyl ethyl ketone, CH_3-CH_2-CO-CH=CH_2, has also been implicated as a principal contributor (Pagington 1975). Kalbrenner *et al.* (1974) found that ostensibly pure samples of both linoleic and linolenic acid hydroperoxides in water possessed odors and tastes similar (except for an additional astringency) to those of soybean. Baur *et al.* (1977) describe enzymatically-produced "bitter tasting" fatty acids. Another candidate for the objectional taste is the dimer formed upon oxidation of phosphatidyl choline (Sessa *et al.* 1974). Takahashi *et al.* (1979) attributed the objectionable beany odor to medium straight chain aldehydes and developed a process for their removal with beef liver aldehyde oxidase. In addition to lipid-derived off-flavors, several other substances have been implicated. These include 4-vinyl guaiacol and substances derived, according to Goossens (1975), by "oxidative" deterioration of amino acids and by "enzymolysis"; presumably the latter refers to formation of bitter peptides through the action of endogenous peptidases (Chapter 34).

Furthermore, the concentration of these odorants may be quite important and the noxious odor may in part be due to their high levels, which in turn may be due to the saturating levels of substrates available to the pertinent enzymes. Although tomato juice rather than soy was used by Kazeniac and Hall (1970), their experiments are instructive in this regard. The three common C_6 aldehydes, *cis*-3-hexenal, *trans*-2-hexenal and hexanal, impart pleasant qualities below levels of 1, 2 and 0.5 ppm, respectively, but are perceived as rancid off-flavor above these levels. At high concentrations *trans*-2-hexenol has been described as "rubber" and "candle-like" whereas that of the corresponding aldehyde as "fruity, green, leafy" according to a detailed organoleptic analysis by Eriksson *et al.* (1977) who nevertheless conclude that interconversion will result in a change in odor intensity but not quality (see also Chapter 21). Thus, in soybean, with its high oil content, the high levels of such substances may arise from the abundance of substrate available to the pertinent enzymes, the most important of which, at least for many plant food off-flavors, is probably lipoxygenase.

Soybean Lipoxygenase.—Although far from constituting a completed chapter in the history of enzymology, knowledge of soybean lipoxygenase is probably more advanced than that from any other source. Two excellent comprehensive reviews of lipoxygenase enzymology are those of Axelrod (1974) and Eskin *et al.* (1977). For more specialized discussion of the chemistry, reviews of the Peoria group are especially valuable (Gardner 1975; Gardner and Sessa 1977). Reviews including contributions of the groups at the University of Utrecht and at Colney, England, are those of Boldingh (1975) and Galliard (1975). By the time this appears some of the vexing mystery surrounding this enzyme will undoubtedly have dissipated. To reiterate, the action of all plant lipoxygenases known to date is restricted to lipid substrates which possess the *cis,cis*-penta-1,4-diene moiety. Although this limits the number of substrates available to this enzyme, in plants usually only linoleic (18:2) and linolenic (18:3) acids and in animal, arachidonic (20:4) (numbers in parentheses refer to the number of carbon atoms and double bonds per molecule), the number of primary products, the lipohydroperoxides (LOOHs), is greater because of the opportunity for both positional (9- and 13-LOOHs) and geometric *(cis-trans)* isomers. The number of compounds resulting from transformations—much larger as we shall see—is the prime factor responsible for the origin of a diversity of products (Chapter 1) rather than, as in the case of phenolases, the availability of diverse substrates.

It is fairly well established that soybean contains three principal isozymes of lipoxygenase with fairly distinct properties, designated as L-1, L-2 and L-3. L-3 comprises two closely related proteins which so far are only electrophoretically distinguishable. L-1, L-2 and L-3 differ from each other with regard to (1) influence of pH on activity (Chapter 17), (2) availability of triglycerides as substrate (specificity) which may be related to the charge and polarity of the substrate (Bild *et al.* 1977), (3) pigment-bleaching ability, that is, the specificity with respect to LOOH isomer formed although this can also depend upon a host of other factors, (4) specificity with regard to the breakdown products of the LOOH, and (5) activation (or inhibition) by Ca^{++}.

Under conditions usually used to assay lipoxygenase activity, oxygen uptake exhibits a lag period. The enzyme furthermore usually does not obey simple hyperbolic Michaelis kinetics, probably due to the increasingly micellar character of the fatty acid substrates with their increasing concentration. This in turn influences the measured enzyme activity. Apparently the controversial effects of Ca salts (Restrepo *et al.* 1973) which can normalize the v *vs* S relationship is due to their capacity to affect the critical concentration of lipid substrate associated with the formation of these micelles. The entire question of micelle formation vis-à-vis lipoxygenase kinetics was explored in depth by Galpin and Allen (1977).

The task of sorting out and isolating these isozymes would have been incomparably simpler had affinity chromatography been made available as it now has been by Grossman *et al.* (1972) using aminoethyl derivatives of linoleic acid attached to agarose. However, they could not separate the isoenzymes. More recently, Allen *et al.* (1977) did accomplish isolation of L-1 in a one-step procedure (16-fold purification) using aminohexyl derivatives of highly unsaturated fatty acids as ligands. The greater the degree of unsaturation, the more specific and sharper the separation.

Lipoxygenase contains one atom of iron and may be an oligomer. Proposed biological functions of this enzyme are listed in Chapter 17. In addition to the evidence marshalled by Axelrod (1974) that lipoxygenase is a true dioxygenase in that it catalyzes the incorporation of the entire oxygen molecule into the cosubstrate, more evidence is afforded by the observation that nitrocatechol is a potent inhibitor (Galpin *et al.* 1976), a behavior suggested previously as a benchmark for such iron nonheme dioxygenases (Tyson 1975).

Mechanisms.—Development of an intimate understanding of just how lipoxygenase works is still in the formative stages. In all postulations Fe plays a central seminal redox role, although by 1979 the nature of the ligand to which the Fe is bound had not yet to our knowledge been elucidated. A rather elaborate scheme proposed by the Utrecht group which accounts for many of the known properties of L-1 at its pH optimum for common substrates is shown in Fig. 24.1. According to this concept, the enzyme Fe in the resting state is in the reduced ferrous form (see also Aoshima *et al.* 1975). The enzyme assumes its active (ferric) form by reaction with a trace of 13-LOOH. As alluded to in Chapter 21, the first and pace-setting reaction of the activated enzyme is the sterospecific abstraction of one of the hydrogens (pro-L_S) from carbon 11 of the substrate, in this case, linoleic acid (as previously mentioned, this leads to the formation of the 13-LOOH) to yield a yellow-colored fatty acid free radical-enzyme complex (purple complexes have also been observed). The H abstraction is followed by O_2 insertion. This may be accompanied by activation of the oxygen to O_2^*. The O_2 is inserted on the opposite side via an "antarafacial" (backside) mechanism. This neatly accounts for the formation of the corresponding 13-LOOH, the conversion *cis* to *trans* and the rigid requirement for the *cis, cis*-pentadiene configuration. The latter is the only configuration that allows the substrate to be oriented on the enzyme in the planar form, necessary for an antarafacial mechanism to proceed. The complex spatial configurations, in part, account for the aforementioned diversity of products.

The ferric form of the enzyme is regenerated by reduction of the fatty acid free radical to form the hydroperoxide and the enzyme is now ready to start another cycle.

Heterogeneity of Hydroperoxide Formation; Positional Specificity.—The soybean L-1 isoenzyme

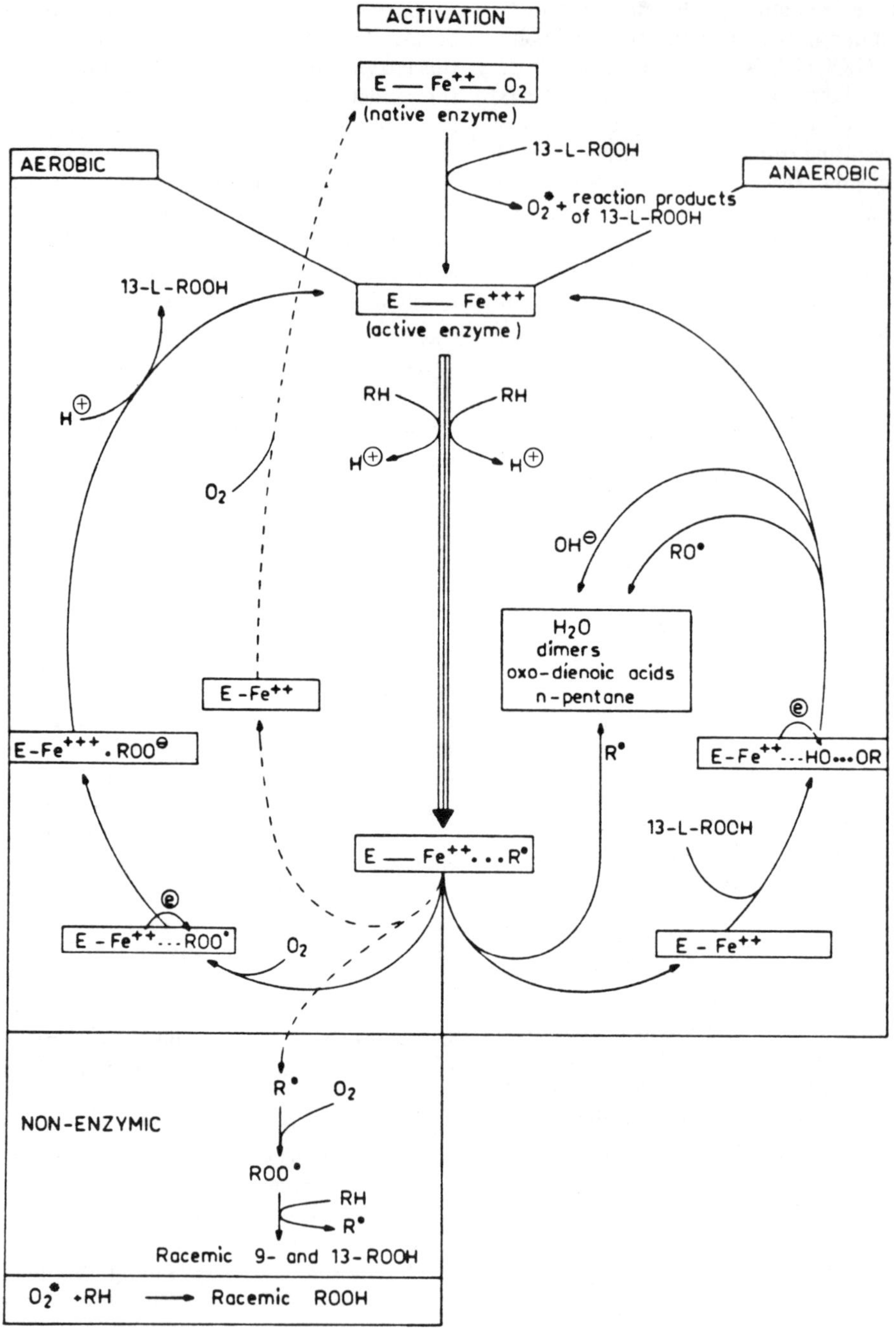

From de Groot et al. (1975)

FIG. 24.1. PROPOSED MECHANISM OF LIPOXYGENASE ACTION

forms predominantly the 13-LOOH at pH 9 with linoleic acid as substrate. Contrary to the earlier notion that only *cis,trans* are formed enzymatically, at least three other LOOH isomers can be formed as a result of lipoxygenase action. The four isomers are: 13-hydroperoxy-*cis*-9,*trans*-11-(L-13-LOOH); 13-hydroperoxy-*trans*-9,*trans*-11-(D-13-LOOH); 9-hydroperoxy-*cis*-10,*trans*-12-(D-9-LOOH); and 9-hydroperoxy-*trans*-10,*trans*-12-(L-9-LOOH). This recent finding was made possible by the advent of high pressure liquid chromatography (Pattee and Singleton 1977). Further evidence of the utility and high performance of this technique in this field is afforded by its use in monitoring the purification of lipoxygenase-generated LOOHs from arachidonic and γ-linolenic acid, both precursors to prostaglandins (Funk *et al.* 1976). It would be interesting to compare the organoleptic properties of these highly pure samples with those of the corresponding LOOHs generated in plants. A large number of factors other than enzyme specificity determine which and how many of the hydroperoxides are formed. A small percentage of the racemic mixture is probably formed nonspecifically, perhaps via the presence of O_2. Other variables include substrate concentration, temperature and Ca^{2+} concentration.

Transformation of the Primary Products.—Diversity of products already arises if we consider the free radical (·OOL) formed as the result of extraction of hydrogen as from C-11 as the primary product as evidenced by the anaerobic reactions of lipoxygenase (right hand side of Fig. 24.1). However, we prefer to consider the LOOHs as the primary products, since they can give rise to an even wider diversity of secondary products which have great impact on the flavor of the food in which they are formed. Gardner (1975) has provided a good overview of these pathways, not all of which may occur in soybean, as shown in Fig. 24.2. They include: (1) reduction or nucleophilic reactions involving peroxidase systems including glutathione peroxidase, (2) enzymatic isomerization to polyhydroxy derivatives and ketols, (3) epoxidations by flour-water suspensions, (4) vinyl ether formation by a potato enzyme, (5) anaerobic lipoxygenase catalyzed reactions including dimerization due to a reaction between linoleic acid and its hydroperoxide and production of pentane (Fig. 24.2) (peanut enzymes also produce pentane anaerobically) (Singleton *et al.* 1975), and (6) production of volatile aldehydes and ketones. As mentioned in Chapter 21 there has been some controversy as to whether a special "lyase" is needed to accomplish the scission of LOOH. Consensus had appeared to favor participation of lipoxygenase itself, but a lyase has been isolated.

These various modes of transformation yield literally hundreds of different products so that it is not surprising that the same enzyme can influence food quality in both desirable and undesirable ways. In many instances, however, our knowledge of the chemical transformations outpaces our understanding of just how these transformations affect the relevant food properties and how they can be applied practically. For instance, as mentioned, not many data are available on the organoleptic properties of the C-18 transforms. If they turn out to be innocuous, their formation might be used to prevent via competition some of the undesirable reactions such as the formation of some of the unsaturated aldehydes and ketones (see below). These latter may, in the proper proportions, contribute to the unacceptable flavor of bean products and unblanched frozen foods as discussed forthwith.

A seventh category might be those reactions of the primary product with nonlipid food constituents. Among these are other enzymes such as proteinases (Matsushita *et al.* 1970) and pea alcohol dehydrogenase (Grosch 1969) and lipoxygenase itself as discussed later in this chapter. Inactivation of alcohol dehydrogenase would lead to a redistribution of alcohols and aldehydes, and thus influence odor.

Several comparisons have been made of the volatiles evolved from the action of the various isozymes of soybean both among themselves and in comparison with those from other lipoxygenases of plant foods. Thus, Grosch and Laskawy (1975) reported that both L-3 and L-2 formed wider ranges of volatiles (C_7-, C_8- and C_9-carbonyls as well as propanal, 2E-pentenal and 2E-hexenal) than did L-1, which formed the latter but not the former three compounds, at least under the conditions of their experiment. Some investigators believe that many of these volatiles come from both LOOH and directly from the fatty acid free radical (Heimann *et al.* 1976). As indicated in Fig. 24.1 and 24.2 not all of the LOOH transformations degrade the carbon chains. Persuant to our interest in why some foods such as soy and other legumes give rise to objectionable flavors while others such as fresh peas give rise to desirable flavors, it is instructive to compare the nature of some of these LOOH transforms in homogenates of these two legumes.

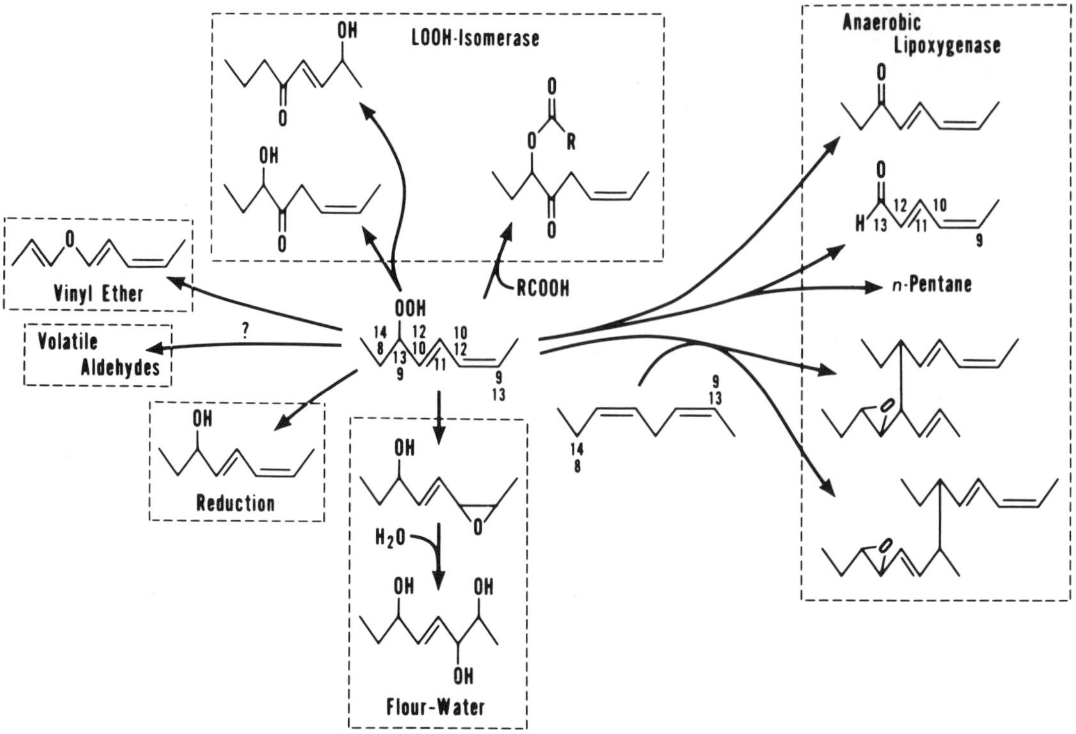

Courtesy of Gardner (1975)

FIG. 24.2. ENZYMATIC TRANSFORMATIONS OF LINOLEIC ACID HYDROPEROXIDE

This opportunity is afforded by the comparison of Gardner and Sessa (1977) of products from the transformation of LOOH by various chemical and biological systems. Table 24.3, adapted from their paper, shows the sequence of various substituents along the carbon chain of the transformed C_{18} LOOHs. The absence of epoxy derivatives from the volatile of fresh pea homogenates suggests that these may contribute to the off-flavor of soybean. This abundance and variety of hydroxyfatty acids in peas may be due, in part, to participation of a rather special lipohydroperoxidase, termed "peroxygenase" by Ishimaru and Yamazaki (1977). This enzyme utilizes LOOH to hydroxylate benzenoid substrates, reminiscent in this respect of the cellobiose:quinone oxidoreductase of Chapter 9. Another difference may be related to the circumstance that at least one of the soybean isolipoxygenases, L-2, can act on triglycerides as well as on free fatty acids. Hence, prior lipase action is not required for this particular isoenzyme, although lipase is present in soybeans.

Other Plant Foods

Other Legumes.—Lipoxygenase action creates flavor problems with other legumes such as dry beans, scarlet runner bean, fava beans and in unblanched fresh, frozen or dehydrated beans and peas after extended storage. Literature on dry bean lipoxygenases was reviewed by Hale *et al.* (1969). Klein (1976) and Borisova *et al.* (1977) have contributed to the purification and isozymology of the pea enzyme. Fava bean flour, used instead of soybean as a wheat flour dough improver in Europe, has presented lipoxygenase-induced problems due to a "dried pea" flavor and bitter aftertaste sometimes encountered in products to which it is added (Drapron *et al.* 1974; Hinchcliffe *et al.* 1974). It takes at least 15 min at 160°C to inactivate the enzyme. Beaux and Drapron (1975) have crystallized the principal isozyme. According to Grosch *et al.* (1976), the main isozyme of the scarlet runner bean *(Phaseolus coccinea)* resembles soybean isozymes L-2 and L-3 in hydro-

TABLE 24.3

SECONDARY LIPOXYGENASE PRODUCTS IN SOYBEAN AND PEAS

Sequence of Functions[1]	Presence (+) in Homogenates of[2] Soybeans	Peas
CO ⟶ C=C ⟶ C=C	+	+
OH ⟶ C=C ⟶ C=C	+	+
C(O)C ⟶ C=C ⟶ CO	+	+
C(O)C ⟶ OH ⟶ C=C	+	−
C(O)C ⟶ C=C ⟶ OOH	+	−
C(O)C ⟶ C=C ⟶ OH	+	−
OH ⟶ OH ⟶ C=C ⟶ OH	−	+

Source: Gardner and Sessa (1977).
[1] In sequence from ω-end of chain: CO, keto; C=C, -ene; OH, hydroxy; OOH, hydroperoxy; and C(O)C, epoxy derivative functions of octadecanoic acid.
[2] Homogenates of soybean give rise to off-odor; those of peas to a pleasant odor.

peroxide-forming specificity and possesses a specificity intermediate between that of L-1 and L-2 with respect to volatile carbonyl-forming pattern.

A LOOH-producing enzyme system which, as just mentioned, when coupled to lipoxygenase, may be involved in color as well as flavor is the LOOH-dependent hydroxylation of monophenols to diphenols of Ishimaru and Yamazaki, in the microsomal fraction of peas. This hydroperoxidase, or peroxygenase, reduces the LOOH to involve a hemoprotein similar to but not identical with cytochrome P-450. Peas will be examined further in connection with frozen foods.

Cereals.—Long-term storage of whole ground and whole dry cereal grains can result in rancidity due to the consecutive action of lipase and lipoxygenase, particularly if the water activity rises above 0.1, as discussed in Chapter 13. Raw cereal products tend to show some resistance to oxidative rancidity so that the lipolytic enzymes which are relatively more active at low a_w's play a relatively greater role than lipoxygenases in off-flavor development.

Barley.—Stored barley may develop a "cardboard" flavor due, perhaps, to lipohydroperoxide isomerase (Esterbauer and Schauenstein 1977). Stale flavor in beer may be due to barley lipoxygenase (Lulai and Baker 1976). According to Schreier and Heimann (1971), degradation of LOOH leads to rancid carbonyl-associated flavors in cereals whereas polymerization of LOOH (mediated through antioxidant, moisture and peroxidase) leads to a bitter taste in stored cerals.

Wheat.—The mischief generated by the lipase-lipoxygenase system in fortified wheat flours has been thoroughly chronicled by Fellers and Bean (1977). Wheat lipoxygenase has also been alluded to in our discussion of pasta color (Chapter 17). According to Grosch the wheat isozymes form the 9-LOOH but otherwise resemble soybean L-1. Wallace and Wheeler (1979) separated two isozymes and obtained evidence for a protein "activator" whose mode of activation resembles the previously mentioned effect of Ca^{2+} in that it seems to affect the micellar association of the fatty acid substrates. From the findings of Morrison and Panpaprai (1975) on the oxidizability of ^{14}C-labeled unsaturated lipids added to wheat flour in the presence and absence of soybean flour, it would appear that not only does the latter raise the level of total lipoxygenase activity in the wheat flour, but it also broadens the lipoxygenase specificity so that more than just FFA and monoglycerides are amenable to enzyme action.

A start has been made towards characterizing the lipoxygenase systems of rice bran and oats where it can cause serious quality defects in rice oil and in grains during storage (Shastry and Rao 1975; Heimann and Klaiber 1977).

Various investigators have found evidence for the operation in wheat of the anaerobic mode of lipoxygenase action as well as other modes of LOOH transformation. Historically, flour-water suspensions were found to be the source of enzymes responsible for the formation of the epoxy and hydroxy derivatives of the unsplit carbon chains of the fatty acids (Graveland 1973) which were alluded to previously. It is as yet unclear just how these hydroxy and epoxide derivatives are formed and it would appear that they arise from the action of special "glutenin" enzymes *via* lipoxygenase through the LOOHs and nonenzymatically via a Fe^{2+}-catalyzed reaction (Gardner and Sessa 1977). The role of flour thiols in the formation of hydroxy acids will be explored in Chapter 32.

Corn.—Quality-related interest in corn centers around a concern for the stability of the oil (Gardner and Inglett 1971), although lipoxygenase may play a role in the deterioration of underblanched frozen corn. Its enzymology is still in the formative stages. As with other lipoxygenases the LOOH-forming specificity is pH dependent, the 13-LOOH predominating at pH 9. Corn appears to be protected from the undesirable consequences of lipoxygenase action such as rancidity by the presence of an active lipohydroperoxide isomerase. The organoleptic characteristics of the resulting α-and γ-ketols (Fig. 24.2) have not been explicitly described, but presumably are not objectionable.

Isomerases from different sources, as documented by Gardner and coworkers, from corn and alfalfa yield quite different isomers (Gerritson *et al.* 1976). Corn enzymes form both α- and γ-ketols; those from alfalfa form only an isomer of the corn γ-ketol, different from that from corn.

Other Plant Foods

Peanuts.—In our discussion of desirable flavors and flavorese in Chapter 23, we briefly examined the role of lipoxygenase in creating desirable peanut aroma associated with cell disruption upon mastication of the fresh food. By contrast, in raw peanuts stored for long periods before roasting, some cellular disruption occurs which is accompanied by lipoxygenase action and the resultant accumulation of off-flavors. As with other foods, this lipoxygenase action has to be preceded by that of lipase. Much of our information concerning peanut lipoxygenase comes from North Carolina State; pertinent references may be found in Pattee and Singleton (1977). An alkaline and at least two acid isolipoxygenases have been detected and purified. One of the acid isozymes shows preference for the formation of the 13-LOOH and all of the latter is the L-13-LOOH isomer, but all isomers are formed. As previously mentioned, pentane is a major volatile which must have considerable flavor impact. Contrary to its depicted genesis via anaerobic action of lipoxygenase (Fig. 24.1), the pentane production by the peanut enzyme is stimulated by O_2, apparently at the expense of the carbonyl pathway. Peanuts also possess lipohydroperoxide isomerase activity (Mitchell 1972).

Potatoes.—There has been no extensive discussion of the role of lipid-degrading enzymes on potato flavor, desirable or otherwise. Most potatoes are heat-treated so that enzyme activity presumably is not an important factor in storage stability of processed products. Some enzymes may survive in parfried products used for frozen french fries. Nevertheless, we do know quite a bit about lipid transforming enzymes in this vegetable (Grosch *et al.* 1976; Berkeley and Galliard 1976; Sekiya *et al.* 1977). Lipoxygenases in such nonseed food plants seem to be localized in specific fragile organelles. The potato enzyme (or principal isozyme before fractionation into isozymes) prefers but does not act exclusively on free fatty acids. It forms the 9-LOOH predominantly at its pH optimum of 5.5; its carbonyl-forming pattern and carotene oxidizing ability resemble those of the lipoxygenases of wheat, according to Grosch *et al.* (1976). It is accompanied by an interesting enzyme converting 9- but not 13-LOOHs into a vinyl ether derivative, colneleic acid CH_3-CH_2-CH=CH-CH_2-CH=CH-CH=CH-O-CH=CH-$(CH_2)_6$-COOH. Unlike O-containing groups in other LOOH transforms (Fig. 24.2), the ether oxygen atom is not derived from the "O" of the hydroperoxide group (Galliard and Matthew 1975). Other fruits, vegetables and cereals in which lipoxygenase has been more or less demonstrated and characterized include lentil beans, garbanzos, cabbage, cauliflower, eggplant, tomatoes, bananas, oats, alfalfa, lupines turnip and kohlrabi (Eskin *et al.* 1977). It is undoubtedly a companion of perhaps most if not all plant foods but does not create off-flavors in all of them as we have discussed in Chapter 21.

ENZYMATIC ORIGINS OF OFF-FLAVOR IN FROZEN FOODS

The Flavor Defects

Although enzymes undoubtedly play important roles in the development of off-flavors in other processed foods, we shall discuss frozen foods and more specifically peas as a paradigm of foods and food products which develop off-flavors (mostly odors) upon subsequent storage. While the most common off-flavor developed in frozen vegetables is usually described as "hay-like," Joslyn (1949) pointed out that many vegetables develop their own peculiar overtones. Thus, "alfalfa" has been used for peas, spinach and green beans. The related terms "silage," "grass" or "composted" have been used to describe the odor of spinach, asparagus and artichoke hearts. The adjective "sharp" has been applied to artichoke hearts, squash, asparagus, spinach and beans; the connotation "sweet" applies to green beans, horse beans and peas. The stale cabbage odor of brussels sprouts and the oil-of-bitter-almond odor developed in lima beans bespeaks the undesirable action of glucosinolase (Chapter 21) and the cyanogenic glycoside (Chapter 22) degrading enzymes, respectively. (Fava beans should also give rise to a cyanogenic glycoside-derived odor.) Other adjectives which have been used to describe frozen (and also dehydrated) vegetable off-odors include "acrid" (squash), "fruity" (peas), "fishy" (artichoke hearts) and "oxidized linseed oil" which replaces the almond-like odor in lima beans after long periods of storage. In addition to these odors frozen vegetables in general develop off-tastes described as "sour-stale," "acid cooked" and "acid oxidized" due in part to simultaneous perception of the effect of lowered pH and off-odors.

While some of these characteristic odors could be due to the action of special enzymes, the off-flavors which most frozen, unblanched vegetables have in common would appear to be due to the transformations of lipids involving the action of lipases and lipoxygenases (Lee and Mattick 1961; Bengtsson and Bosund 1966). For dehydrated vegetables, nonenzymatic oxidative degradation of unsaturated fatty acid appears to be at least as important as enzymatic degradation modes. For instance, Buttery *et al.* (1970) demonstrated a high correlation among losses of the unsaturated FFA, O_2 uptake and off-flavor after reconstitution of dehydrated potatoes. Thus, the off-flavor may arise: immediately upon crushing of the tissue of leafy foods; as the result of water addition to seedfood powders such as beans; in foods in which the off-flavor develops only on long storage; and in food plants such as peas. Upon tissue comminution they develop fresh flavor at first (hence their use in fresh salads) but the odor which develops after storage is considered undesirable.

Off-flavor Constituents

Empirical Correlations.—The observation that the acetaldehyde content of blanched unfrozen vegetables rose consistently and that there frequently existed a strong correlation between off-flavor and ethanal content was well-established in the 1930s and used in the 1940s by Joslyn and coworkers (Joslyn 1949,1966). They also showed that heating vegetables such as green snap beans, brussels sprouts, asparagus and lima beans prevented this accumulation. Later Gutterman *et al.* (1951) clearly demonstrated that whenever the acetaldehyde level in frozen peas or asparagus reached a certain range of concentration, taste panels invariably considered the product to be of impaired quality and frequently inedible. However, it soon became clear that the acetaldehyde itself does not contribute to off-flavor. Its accumulation is in large extent due to the continued operation of the glycolysis in the absence of operable mitochondria (and associated TCA cycle) and sufficient oxygen. Samotus and Schwimmer (1963) observed that potatoes held in N_2 developed a cucumber-like odor, presumably a nonenal from 9-LOOH, an indication of some lipoxygenase action utilizing the limited available O_2 (Chapter 21). Hexanal levels which occur simultaneous with off-flavor development have limited value when applied to samples of unknown history (Bengtsson *et al.* 1967).

That enzymes were largely responsible for many of these flavor aberrations was well documented by Joslyn and coworkers. Their work and evaluation of the work of others pointed to the participation of lipase, lipoxygenase and peroxidase in vegetables. Of course, phenolase action was implicated in color development in fruits (Chapter 15). Most of this work was based on the behavior of the frozen food systems after these enzymes had been inactivated, usually by blanching (Chapters 10 and 11). Another approach, adopted by few investigators, was to observe what happened when these suspect enzymes were added to the vegetables. There are all sorts of experimental pitfalls and control problems in the way of arriving at valid inferences from such experiments. Nevertheless, we deem it well worthwhile to briefly review this experience.

As one might expect from our discussion on Chapters 10 and 11, special attention has been paid to peroxidase. Wagenknecht and Lee (1958) reported only a moderate acceleration in flavor deterioration of pea purees stored at −17.8°C for between 3 and 17 months after treatment with peroxidase (horseradish), catalase (liver) or lipoxygenase (soy). Flavor deterioration in the presence of lipase was at least in part due to the flavor of the lipase itself. In a related study Lee and Wagenknecht (1958), using as enzymes more or less purified preparations from the same batch of peas used to prepare the purees, reported pronounced off-flavors with or in all of the enzyme-treated samples except peroxidase.

Shannon (1966), in unpublished work, could detect no significant flavor (and other quality) differences between blanched, salt-treated vegetable purees with and without added biochemical supply house-grade horseradish peroxidase, after storage for 8 months at −24.8°C. Vegetables included purees of peas, lima beans, soybeans, snap beans, spinach and broccoli. On the other hand, Zoueil and Esselen (1959) reported significant flavor differences between blanched pea purees stored at ambience and above for 1−12 weeks and blanched purees to which had been added peroxidase equivalent in activity to ten times that in the peas before blanching. The blanched peroxidase-treated pea control flavor was described as "acid cooked" and that of the unblanched peroxidase-treated samples as "stale." As mentioned below, heating peroxidase "activates" its LOOH-forming capacity. The change in flavor was accompanied by enhanced accumulation of acetaldehyde, indicating that the acetaldehyde does not have to arise from glycolysis (presumably blanching destroyed the glycolytic pathway). This is a field which needs further renewal and further development, esepcially in light of the problems of heat blanching now being used to destroy enzymes as discussed in Chapter 10.

What Are the Responsible Constituents?.— While some of the special off-flavors such as that of the oil of bitter almonds developed in lima beans may be due to specific substances, just what is responsible for the general hay-like odor developed in many frozen vegetables is still an open question. In terms of ascribing the odor to one or a limited number of volatile constituents, the question may never be answered. We can say this with a certain degree of surety thanks to the investigations of Murray and colleagues (1976) on ascertaining the volatiles of off-flavor unblanched peas. In the latest publication on this subject available to us, which also serves as a key to the pertinent literature, they suggest that the off-flavor is caused (as for soybeans) by a complex mixture of mono- and disaturated carbonyls (aldehydes and ketones) and saturated alcohols which accumulate as the result of the action of the lipoxygenases and alcohol dehydrogenases. Distribution according to chemical type of these 90 to 100 odd volatiles (all but five of which were secondary products of lipoxygenase- or other enzyme-induced decomposition of unsaturated fatty acids) of unblanched frozen peas stored at −30°C as compared with fresh pea shells showed some significant differences. Their findings afford an instructive example of how the same general kind of enzyme action can give rise to both desirable and/or undesirable food quality depending upon the treatment of the food as discussed in Chapter 23 in conjunction with flavorese. Thus, the Australian investigators reported that the flavor panel gave a higher rating to fresh than unblanched frozen peas. There existed an optimum time delay (1 to 2 hr) between harvesting and processing for maximum flavor rating by a panel. Presumably this period corresponded to potentiation of desirable enzyme action as a result of moderate tissue damage sustained during harvesting. If pea shell volatiles are to be considered as representative of a desirable fresh aroma, then it is of interest to note the increase in the numbers and levels of odors which may account for the undesirable (for peas) fruity off-flavor. Unfortunately, the experiment in which volatile composition of fresh panel-preferred peas are compared with peas which have developed an off-flavor remains to be conducted.

Among the factors and parameters determining in which direction, desirable or undesirable, such complex poised enzyme flavor systems will go are the following:

(1) Change in absolute concentration. Thus as previously mentioned, the C_6 compounds tend to acquire a rancid note to their odors above certain levels. In the Australian investigation context, hexanal was found to be present at a level 2000 times that of the hexenals.
(2) The diversity of volatiles may be a contributory factor. Thus, the volatile fraction of the unblanched peas contained a greater variety of compounds suggesting that the farther the products are from the primary products in "reaction time space," the more likely they are to contribute to off-flavor.
(3) The relative amounts of the responsible constituents changing during storage may contribute to the perception of the resulting odor as objectionable.
(4) Another important factor may be the relative rate of formation of products of their degradation. Thus, enzymatic reduction of the aldehydes is retarded more at the temperature of stored frozen foods (partly due to super-cooling effects) than is the preceding (enzymatic and nonenzymatic) formation of these aldehydes from unsaturated fatty acids (Eriksson and Svensson 1973).
(5) Further complexity could be introduced due to the differing stabilities and specificities of the involved enzymes as affected by low temperatures (Chapter 12) and low water activity (Chapter 13).
(6) A radically different set of enzymes catalyzing transformation of lipids discussed later in this chapter may come into play during prolonged storage. However, it is likely that lip-

oxygenases still play a significant role in the formation of off-flavor and one of the more thoroughly studied lipoxygenase systems in this regard is that of peas.

(7) In reviewing the literature, one receives the general impression that membrane phospholipids other than triglycerides are major substrates for lipid-associated off-flavor in stored, processed foods. The higher proportion of polyunsaturated acids in phospholipids could account in part for this.

Pea Lipoxygenase

In addition to evidence gleaned from lipid changes during the storage of frozen foods, literature is, as mentioned, available on pea lipoxygenase (Eriksson and Svensson 1970; Borisova *et al.* 1977; Haydar *et al.* 1975; Klein 1976). The crude enzyme possesses a broad pH activity profile acting optimally around pH 6. The activity seems to be largely confined to the outer portion of the cotyledon. From the viewpoint of flavor deterioration perhaps one of the most salient observations is that all but one isozyme (resembling L-3 of soybean) disappears during long-term storage at low temperatures. This storage-resistant form appears to oxidize polar substrates as Bild *et al.* (1977) found for L-3 soy isozyme. On the other hand, pea lipoxygenase has been reported to discriminate against those substrates bound in polar lipids but not against those in triglycerides (Haydar *et al.* 1975). Similarly, Klein (1976) reported that the behavior of partially purified enzyme preparations from fresh peas and snap beans was qualitatively distinct from that from more mature dried peas. According to Leu (1976), pea lipoxygenases form equal amounts of the 9- and 13-LOOHs from linoleic acid. Unlike soy L-1 and potato enzyme, they reported that LOOH specificity was not influenced by O_2, pH and temperature.

Eriksson (1975) reported the usual range of C_3–C_{10} aldehydes and alcohols in model pea lipoxygenase-linoleic systems. However, missing were some of the products which contribute (at least at low levels) to pleasant flavors such as pentane and the hexenals. The pattern of volatiles resembles those of soybean L-1. Means of preventing lipoxygenase action in frozen peas and other foods will be discussed below.

NONLIPOXYGENASE, NONDAIRY-MEDIATED LIPID TRANSFORMATIONS WHICH MAY LEAD TO OFF-FLAVOR IN MEATS, VEGETABLES AND FISH

We have frequently alluded to the possibility that lipids can undergo peroxidation or otherwise experience flavor-generating transformations via action of enzymes other than lipoxygenase. Although such enzyme-associated reactions could occur in plant foods, perhaps preoccupation with the role of lipoxygenase has confined the experimental evidence largely to such systems in meat and fish. Although quite diverse in mechanistic detail, what most of these systems do have in common are their capacities for generating active excited oxygen and the participation of these oxygen species in lipid peroxidation and other lipid transformations (Fig. 16.2). Indeed, even lipoxygenase may be considered as a vehicle for oxygen activation in this mode, especially in light of its at least partial inhibition by superoxide dismutase (see below), suggesting that singlet oxygen is being brought back to the ground or triplet state by the SOD.

Modes of Fatty Acid Transformation

Mutual Oxidation of Myoglobin and Lipid in Meats.—This results in development of off-flavor as well as off-color as discussed in Chapter 18. This system is perhaps essentially nonenzymatically catalyzed but its initiation requires the myoglobin iron to be in the ferric form. To achieve this state, evidence for some enzyme intervention has been postulated as adumbrated in Chapter 18. Some of this experimental evidence has been extended to include the relatively recent, commercially successful development of mechanically deboned meat. Lee *et al.* (1975) conclude that hemoproteins are the predominant catalyst of lipid oxidation in this product. The relative ratio of polyunsaturated fatty acid to hemoprotein is in the range where rancidity would proceed at a maximum rate. An example of an enzymatic factor which might contribute to rancidity in meats is the interaction of globin-associated pigments with a hemolytic agent such as phenyl hydrazine (Goldberg and Stern 1975). It is not likely that this particular agent would be normally present in beef but is conceivable that adventitious contamination with such a substance (as mentioned in the introduction) could, via secondary effects, lead to off-flavors.

Other Nonenzyme Fe-dependent Peroxidations.—These include autoxidation by Fe^{2+}, free heme iron-amino acid complexes groups liberated by cooking, nonspecific heme proteins and non-heme iron-containing proteins, especially ferredoxin. Although we are mainly concerned with enzyme-catalyzed reactions, it might be well to illustrate nonenzymatic but enzyme model-like options by showing in Fig. 24.3 a proposed mechanism via Fe (III)-cysteine complex for the homolytic decomposition of LOOH to form ketohydroxyoctadecenoic acid (Gardner 1975).

Ferredoxin-mediated oxidation of lipid has been implicated in subcellular membrane destabilization (Kaschnitz and Hatefi 1975). Such destabilization, taking place during the heat-up period of foods which are to be cooked, could lead to enzyme-substrate decompartmentalization and thus affect flavor.

That such interactions may be important in plant foods as well as in meats is suggested by the observation of Stone *et al.* (1976) that blanched green beans retained their capacity to convert ^{14}C-labeled unsaturated fatty acids to aldehydes. This observation may also be related to that of Eriksson and Vallentin (1973) that heating actually activates the capacity for peroxidase to serve as a lipid oxidation catalyst (see below). However, in this case membranes are not directly involved but rather are the conversion of the peroxidase molecule to a polymolecular aggregated form induced by heating (Chapter 11).

NADPH.—A fast-moving literature has sprung up on the highly efficient peroxidation of lipids by NADPH-cytochrome-linked reductase systems usually found in association with the microsomal fractions of animal cells. This area of enzyme-generated free radicals as initiators of lipid peroxidation in mammalian tissues has far-reaching interrelated implications in medicine and in biological membrane research (Tappel 1973; McCay and Poyer 1976). In some tissues, this oxidation is further stimulated by the addition of ferric salts.

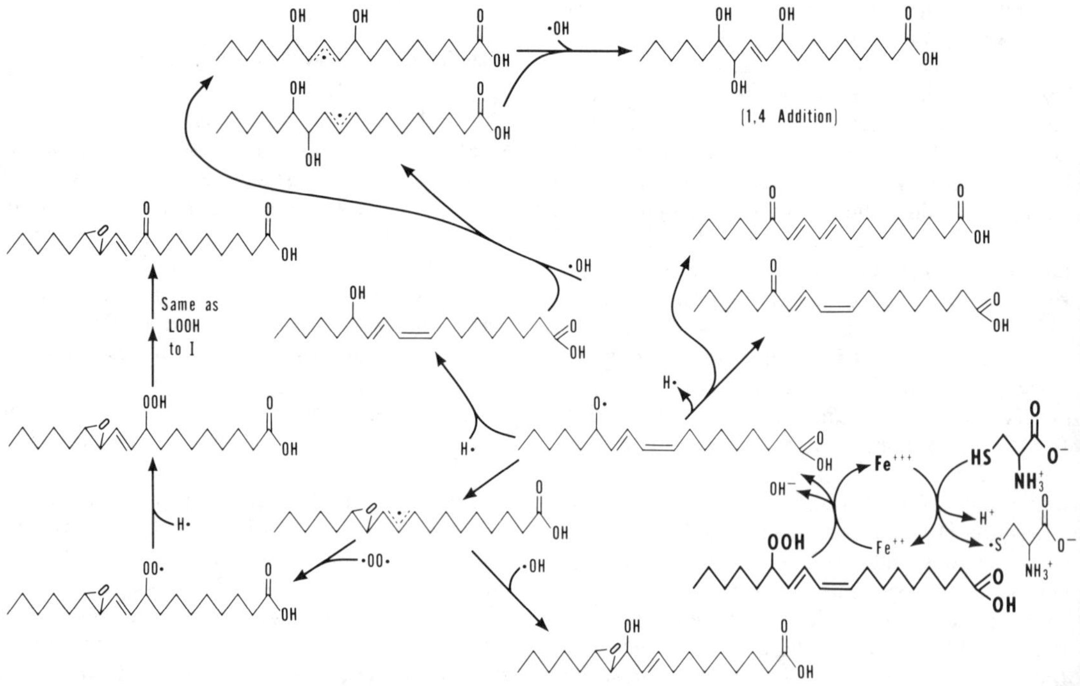

Courtesy of Gardner (1975)

FIG. 24.3. HOMOLYTIC (FREE RADICAL) DECOMPOSITION OF LINOLEIC ACID HYDROPEROXIDE

Depicted is a model enzyme system (Chapter 3) consisting of Fe(III) and cysteine.

Investigations in this area include those of Bidlack (1973) with cytochrome b_5, Högberg *et al.* (1975) with cytochrome P-450, King *et al.* (1975) on the role of singlet oxygen and Roders *et al.* (1976) on the role of Ca^{2+}. Such systems are especially active in liver and may account in large part for the rapid development of off-flavors in this food. That they probably also play a major role in the flavor of muscle meats is now fairly well-established (Lin and Hultin 1977; Player and Hultin 1977).

Superoxide Generation.—The action of xanthine oxidase in the presence of unsaturated lipids results in their hydroperoxidation through the agency of singlet oxygen generated from O_2^- (Pederson and Aust 1973; Kellogg and Fridovich 1975). While this is undoubtedly important in the flavor of milk and dairy products (Chapter 25), it is not far-fetched to assume that other flavoprotein oxidases which produce superoxide can participate in the lipid oxidation and off-flavor in both plant and animal foods. We referred previously to the possible role of glycolate oxidase in this regard. These enzymes would be especially effective in the disorganized metabolic systems attendant upon cell disruption which accompanies the postharvest and post-slaughter interval before and during processing.

Oxidation of Ascorbate.—As previously mentioned, this is accompanied by the generation of active oxygen which could, in a disorganized biological system such as pertains to many foods, peroxidize lipids (Epel *et al.* 1973). Such oxidation in foods, as measured by bleaching of β-carotene, was also demonstrated by Kanner *et al.* (1977). All this takes place at relatively low ascorbate concentrations. Above $10^{-3}M$ ascorbate usually behaves as an antioxidant.

Photosynthesis.—Superoxide anion is generated during and perhaps as part of the process of photosynthesis according to several authors (Elstner *et al.* 1975; Allen 1975). It is conceivable that in a slightly disorganized protoplast which can still carry out at least part of photosynthesis, this O_2^- and its product, O_2^*, could survive long enough to affect peroxidation of lipid and thus contribute to food flavor.

Prostaglandin Synthase and Mammalian Lipoxygenases.—This fatty acid cyclooxygenase is present in meats, and could contribute to flavor development in meats. The main differences between the plant lipoxygenases and animal prostaglandin synthases are that the former form hydroperoxide, while the latter require free heme and form cyclodioxygen derivatives. They do, however, share in common several mechanistic features (Samuelsson *et al.* 1978) and plant lipoxygenases have been frequently used in prostaglandin studies (Boldingh 1975; Skinner 1975). Formation of a prostaglandin analog by the action of soybean lipoxygenase on γ-linolenic acid is shown in Fig. 24.4. Although largely confined to vesicular and glandular tissue, it is conceivable that its action could contribute to meat flavor. Mammalian blood platelets contain a "true" hydroperoxide-forming lipoxygenase which catalyzes the formation of LOOH, and converts prostaglandin precursors into non-prostaglandin derivatives (Nutgeren 1975; Siegel *et al.* 1980). Together with other LOOH-transforming enzymes (Jakschik and Lee 1980), lipoxygenase could cause off-flavor in blood-rich meats.

β-Oxidation.—In addition to the aforementioned modes of oxidations, lipids can, as we have already seen, be transformed via the classical β-oxidation spiral. As we have seen in Chapter 23, this may be more likely to lead to desirable (esters) than to undesirable flavors. A fruity note, as noted in off-flavor peas, is probably due to undesirable formation of esters through this pathway.

α-Oxidation.—What may be just as likely in such enzymes is that the aforementioned α-oxidation of fatty acids (Chapter 22) could contribute to off-flavor. Since the system requires NAD, it is likely [and more recently detected in cucumber by Galliard and Matthew (1976)] that here we are again dealing with active oxygen.

Decarboxylation to Hydrocarbons.—Some food plants have been reported to contain enzyme systems capable of decarboxylating higher fatty acids to straight chain hydrocarbons such as $C_{31}H_{64}$ which could contribute to flavor (Khan and Kolattukudy 1974).

Fish.—Oxidative rancidity can be a problem in unprocessed fish such as mullet in which after a few days storage at 1°C it was detectable (Mendenhall 1972). Somewhat reminiscent of off-flavor development in coleslaw on display shelves (Chapter 1), this rancidity is caused by the enzyme system in the mullet rather than by invading microbial enzymes, since rancidity could not be prevented by icing of the carcasses before filleting, whereas icing did delay bacterial contamination. From the studies of Shono and coworkers (Toyo-

FIG. 24.4. LIPOXYGENASE AS A PROSTAGLANDIN SYNTHASE

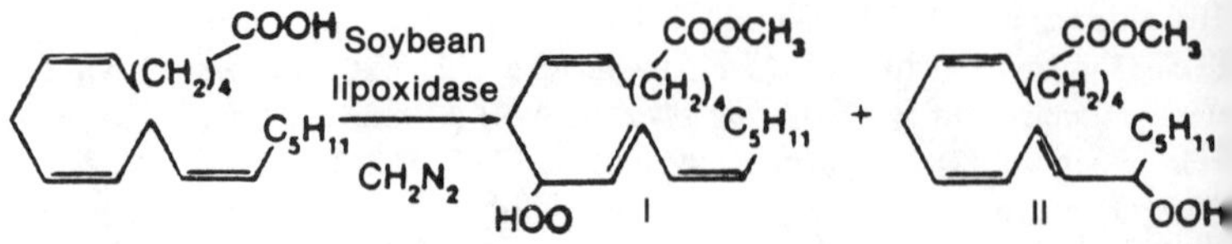

All-*cis*-6,9,12-octadecatrienoic acid   ω-10 and ω-6 hydroperoxides

From Skinner (1975). Reprinted with permission from Chem. and Eng. News, Dec. 1
Copyright by Am. Chem. Soc

mizu *et al.* 1976) we know that in the jack mackerel the principal substrate for the enzymes whose action leads to this oxidative rancidity is a highly polyunsaturated C_{22} fatty acid (22:6). Off-flavor which develops in fish is also frequently due to lipase-induced hydrolytic rancidity without oxidation of the liberated fatty acids (Chapter 25) as well as to volatile nitrogen derivatives. Especially prominent in this regard is trimethyl amine formed via reductive enzymes of adventitious organisms acting on the endogenous trimethyl amine oxide which in itself has no objectionable odor. In this connection, peroxidation of lipid participates in control of this type of off-flavor in fish. The LOOH formed reoxidizes trimethyl amine to its oxide:

$$(CH_3)_3N + LOOH \longrightarrow (CH)_3N{\rightarrow}O + LOH$$

Both FFA and trimethyl amine have been implicated in enzyme-mediated undesirable textural changes in fish (Chapter 28).

The off-flavor described by Love (1980) as "wet cardboard" or "boiled clothes" which may develop in cold-stored fish has been ascribed to *cis*-hept-4-enal, presumably a secondary product of action of a lipoxygenase-like enzyme. One way to prevent this off-flavor from developing is to deprive the enzymes involved of their substrates, the FFAs, by starving the fish. The level of the C_7 aldehyde can be reduced by 90% by this means. The subject of prevention of these undesirable enzyme actions in other foods is taken up in detail forthwith.

PREVENTION OF NONDAIRY LIPOHYDROPEROXIDE ASSOCIATED OFF-FLAVORS

Many of the measures instituted to prevent enzyme action which results in deterioration of food quality attributes such as color, texture and wholesomeness can be and have been used to prevent off-flavor in foods. We have had occasion to refer to several of these repeatedly throughout this book. Methods for preventing lipoxygenase action in soybean protein products, reviewed by Wolf (1975), include thermal inactivation methods such as dry heating-extrusion cooking, blanching of the whole bean and grinding with hot water, a standard method which affords a more effective destruction of lipoxygenase than does microwave heating (Chapter 10) (Armstrong and Stanley 1975). Heating may be objectionable when it is desirable to maintain the native protein, as in the preparation of isolates.

Attempts to inactivate lipoxygenase in legumes have brought into serious consideration some of the less widely used approaches to enzyme inactivation such as the use of organic solvents and pH extremes (Chapters 1 and 14). Thus, Eldridge *et al.* (1977) were able to considerably reduce the off-flavor of soy protein and its putative cause, lipoxygenase activity, by steeping whole soybean ambiently in 40–60% ethanol for 24 hr. They recommend further conventional steam treatment to inactivate the endogenous trypsin inhibitors (Chapters 34 and 35). Ethanol and other organic solvents are used in the production of soybean concentrates (Wolf 1975; Daftary 1975). Inactivation of legume seed lipoxygenase by acid treatment of the whole bean has been recommended and applied to the development of new products for soybean and especially other *Phaseolus* beans (Kon and Dunlap 1977). This development serves as an apposite illustration of accrual of serendipitous benefits without which the processes would not be economically viable (Chapter 1). What tips the scale towards favorable prognosis for practical application are the following observations: (1) increased digestibility of nutrients occasioned by cell wall breakage, (2) increased extractability and availability of proteins (Chapter 34), (3) not having to retort and thus expend energy, (4) inactivation of trypsin inhibitors and (5) ending up with an already salted product where one is needed.

Recent data on the inactivation of lipoxygenase in peas *in situ*, in purees, and as purified enzyme in direct comparison with peroxidase permitted calcuation of energy of activation of the denaturation to lie in the range of 600 kJ mol^{-1} as compared with 142 for peroxidase (Svensson and Eriksson 1974). Thus if, as indicated, lipoxygenase is a major contributor to off-flavor in frozen peas, heating to the point of complete removal of peroxidase activity may not be necessary.

As with other quality-associated enzymes, consideration has been given to the possibility of breeding foods with low enzyme activity. From the work of Chapman *et al.* (1976) we know the level of lipoxygenase cannot be influenced by horticultural practice and that the levels are genetically determined. This suggests that soybean genotypes could be genetically selected for low lipoxygenase activity as well as for high oil content and quality. Chapman and coworkers also reported that lipoxygenase is quite heat sensitive at low water activities. Off-flavor in peas can be minimized by chilling and the avoidance of bruising prior to freezing according to Dupuy and Rigaud (1970), who suggest that a better alternative to chilling may be to develop new varieties with low enzyme activities. Interestingly, they found that wrinkled pea varieties, whose starch is higher than normal in amylose, always possessed more lipoxygenase activity than did the smooth varieties.

Enzyme Inhibitors

The use of food additives to prevent lipid-associated off-flavor is fairly well established and in widespread use. This is especially true of additives designed to prevent or delay the onset of oxidative rancidity. The most common of these are the antioxidants which act by destroying hydroperoxides (Hiatt 1975), by interrupting free radical propagation chains or by quenching excited activated oxygen. Among the latter are mannitol and ethanol, both good ·OH scavengers. Indeed, ethanol may be especially effective in the above-mentioned process proposed for preventing soybean off-flavor by both inactivating lipoxygenase and scavenging free radicals as well as by extracting substrates and products.

In addition to the well-known and almost universally used NDGA, BHT, BHA and propyl gallate, other antioxidants mentioned by Hiatt include metal salts of the dithiocarbamates, sulfite (Haisman 1974) and certain phosphates. Interestingly, one class of potent lipoxygenase inhibitors, hydroquinone derivatives, were found not to be effective in preventing off-flavors in bread baked from flour treated with excessive broad-bean lipoxygenase (Palla and Verrier 1974). Cysteine also appears to be an effective antioxidant whose diverse use in foods was reviewed by Diamalt-Aktiengesellschaft (1974). Undoubtedly, many of these compounds inhibit lipoxygenase by decomposing LOOH formed by this enzyme to innocuous compounds. The use of ascorbate as a lipoxygenase inhibitor in pastas was alluded to in Chapter 17. Some legumes and many other vegetables appear to contain antioxidants which apparently are natural endogenous lipoxygenase inhibitors (Pratt 1972; Pinsky *et al.* 1971). As mentioned in Chapter 19, the Maillard nonenzymatic browning reactions yield not only color and odor-bearing substances but also antioxidants. They can arise in vegetables as evidenced by their formation by heating xylose and arginine (Eriksson 1975) and in roasting meat (Sato *et al.* 1973).

In the future one can anticipate the development and application of safe lipoxygenase inhibitors based upon the principles laid forth in Chapter 14. Transition-state analog inhibitors specific for plant but not mammalian lipoxygenases would be most helpful. A clue as to what to look for in the area of active-site reagents is afforded, as previously pointed out, by the finding that the reason that aspirin is such a helpful medical agent is related to its ability to inhibit at very low concentrations the lipoxygenase-like prostaglandin synthase (Samuelson *et al.* 1978) by acting as a specific active-site acetylating agent. The enzyme is accompanied by an endogenous heat-labile nondialyzable inhibitor. Several other nonsteroidal drugs and other lipoxygenase substrates such as the triple bond analog of arachidonic acid, 5,8,11,14-eicosatetraynoic acid, also inhibit prostaglandin synthase (Downing *et al.* 1970). These nonsteroidal drugs, including aspirin, while not, apparently, effective LOX inhibitors (Chapter 14), do inhibit a blood platelet enzyme, somewhat confusingly if not erroneously referred to as a hydroperoxyeicosatetraenoic acid peroxidase (Siegel *et al.* 1980). This enzyme catalyzes the extraction of one oxygen atom from the 12-LOOH which is the product of LOX action on arachidonic acid to form the corresponding 12-hydroxytetraenoic acid. It would be of interest to ascertain the effects of these inhibitors on food enzymes and quality.

Enzyme Action

It should be possible to mitigate the deleterious

effect of lipid-mediated off-flavor production by the effective exploitation of enzymes (Table 24.2) which (1) remove oxygen, (2) deactivate oxygen, (3) remove fatty acids, (4) act on LOOHs so as to bypass the formation of undesirable constituents, (5) transform the odor-bearing components, and (6) by using lipoxygenase action, itself. Not all of these enzyme possibilities have actually been deliberately used or investigated for the express purpose of preventing off-flavor. Knowledge of the presence of enzymes, monitoring of their level in the future quality control laboratory so as to minimize blanching and rationalization of other steps in the processing of the foods optimally to potentiate enzyme action, if present, may be of use.

Oxygen Removal.—We have discussed the use of glucose oxidase for maintaining the quality of foods in Chapter 19. Apparently serious attempts to use it to prevent lipid oxidation by removing oxygen (not glucose) have been largely confined to dairy products (Chapter 25).

Oxygen Deactivation.—Superoxide dismutase inhibits lipoxygenase directly (Richter *et al.* 1975) and can also prevent the NADPH-catalyzed peroxidation of lipids (Kellogg and Fridovitch 1975). Michelson and Monod (1975) reported complete suppression by SOD of off-odor in a 10% suspension of anchovy oil accompanying a 70% reduction in oxygen uptake. SOD also prevented off-odor as well as delay of off-color (Chapter 16) in potatoes (Table 16.2). Undoubtedly, the presence of this enzyme in all vegetables where it has been sought should influence the subsequent flavor developments. Another enzyme, glutathione peroxidase (Chapter 11), effective in shunting LOOH to reductive pathways (Fig. 24.2), might also be as effective as SOD in preventing free radical and off-flavor formation (Halliwell 1974).

Removal of Fatty Acid Substrate.—Such enzymes prevent the formation of LOOHs. Among the enzymes which could so function are those discussed earlier in this chapter and in Chapter 23 such as the fatty acid peroxidase system. While they were discussed in connection with the formation of off-flavor, it is conceivable that LOOH formation could under certain circumstances contribute to the prevention of off-flavor. Another possible enzyme reaction in this regard is the linoleate isomerase of Kepler and Tove (1967) which converts linoleate, 9-*cis*-12-*cis*-octadecadienoate, to 9-*cis*-11-*trans*-octadienoate, containing conjugated double bonds and hence no longer a substrate for lipoxygenase.

Transformation of the Primary Product LOOHs.—Theoretically, it should be possible to avoid the formation of some of the lower carbonyl-containing volatiles contributing to off-odor by bypassing the degradation of the LOOHs via conversion into one or more of the five enzymatically-mediated types of transformation of these primary products of lipoxygenase action. The problem is that the literature has not reported what the flavor of most of these transformed products is. If they are innocuous, then adding them or potentiating their action may prove to be useful. At least one of these modes of conversion, that through glutathione peroxidase, has been used to prevent enzymatic off-color and hence presumably off-odor in chicken leg muscle (Lin and Hultin 1977). Other possibly useful enzymes are the LOOH isomerases and the potato enzyme which catalyzes the insertion of "O" to form the vinyl ether colneleic acid of the Colney, England, investigators. The fact that bad odors are nevertheless formed in foods in which they are present might merely indicate that they are not present in effective levels.

Action of Lipoxygenase.—The fifth type of enzyme action on the LOOHs which would divert them away from forming lower carbonyls is that catalyzed by lipoxygenase itself in the absence of an oxygen (Fig. 24.1 and 24.2). While we know that one of the products of this reaction, pentane, can contribute to flavor, again it is uncertain how the others, the dimers and oxo-acids, affect flavor. Another mode of action of lipoxygenase has been utilized to mitigate its deleterious effects on flavor by allowing it to catalyze its own destruction-reaction inactivation. This reaction inactivation has been characterized for soybean L-1 isoenzyme by Smith and Lands (1972) and studied in detail by Galaway and Gibian (1975), and in wheat by Wallace and Wheeler (1972) and Wheeler and Wallace (1978). Singlet oxygen is not the inactivating substance. As mentioned in Chapter 13, this property of lipoxygenase has been used to stabilize fortified wheat blends.

Transformations of Secondary Products (Odorants).—We pointed out in Chapter 1 that one can avoid the deleterious consequences of enzyme action by removing the products which are the cause of the deterioration in quality. This approach has been used to improve soybean flavor. Specifically, employment of solvent extraction using mixtures of hexane with either ethanol or isopropyl alcohol is effective without the protein denaturation and loss of functional properties accompanying thermal inactivation of the offending enzymes (Wolf 1975).

One can also exploit enzyme action instead of using solvents to remove volatiles. One of the problems in the use of solvents or vacuum to remove "volatiles" is that the latter are not so volatile as when isolated because they complex tightly with food constituents, especially proteins, hence, advocacy of the use of proteinase action (taken up in Chapter 34) to liberate these flavors. It is also possible to use enzymes either by adding them or potentiating their action *in situ* to transform the volatiles themselves. One such candidate is alcohol dehydrogenase (Chapters 22 and 23). So far, only apple juice has been suggested as a food whose flavor might be improved by the addition of this enzyme (Eriksson *et al.* 1977). Presumably the use of such enzymes would be effective because, as mentioned before, in high concentrations aldehydes tend to have a less pleasant flavor in their own right. Furthermore, the change in the aroma spectrum wrought by the dehydrogenase action, while changing the overall flavor, does not necessarily guarantee a resulting improvement in acceptability. At any rate, the application of outside sources of dehydrogenases awaits immobilization of the enzyme and an inexpensive source of NAD, probably via regeneration, as discussed in Chapter 8. Other suggested applications of alcohol dehydrogenase to improve flavor will be given attention in the next chapter. As previously mentioned in this chapter, Takahashi *et al.* (1979) circumvented the problem of coenzyme regeneration by switching from an NAD-dependent dehydrogenase to aldehyde oxidase to remove soybean off-odors. Other suggested applications of dehydrogenases to improve flavor will be assessed in the ensuing chapter.

25

LIPASE AND OTHER ENZYME INVOLVEMENT IN DAIRY OFF-FLAVORS, CITRUS BITTERNESS AND OTHER FLAVOR DEFECTS

With this chapter we conclude the main part of our survey of the role of enzyme action in the genesis and control of flavor attributes in foods. Further consideration of flavor will arise from time to time in conjunction with other almost inextricably associated aspects of food quality. Here we discuss: lipase, its characteristics as a class of enzymes and as cause of flavor defects in dairy and nondairy foods; other enzymatically-caused undesirable flavor in milk and dairy products; citrus bitterness; and a potpourri of other undesirable flavors whose origins are likely to be enzymatic and/or which can be controlled or removed by enzymes.

THE LIPASES

General Aspects of Lipase Enzymology

We have had occasion to comment on lipases in connection with development of aroma, as an example of a group-specific enzyme type, as the precursor enzyme to sources of unsaturated fatty acids which serve as substrates for lipoxygenases and other modes of lipid oxidation. The lipases possess rather singular properties which they do not share with other enzymes. One of these characteristics is their specificity.

Specificity.—Since 130 years ago when the action of digestive jucies on fats was first described (variously attributed to Claude Bernard, Eberle and Marcet) our concepts of what lipase is and how it differs from other carboxyesterases have and are still continuing to evolve. One of the bitterest disputes had to do with the classical animal source of this enzyme, pancreas, which was supposed to contain both a true lipase which acted only on neutral fats (long chain triacyl glycerols) and an "esterase" which acted on short chain glycerides. We now know that this lipase, although present in multiple forms, will readily hydrolyze both long and short chains, provided the latter are present in sufficiently high concentrations so as to be present in the solid phase. Figure 25.1 shows the hydrolysis of triacetin by an esterase and by a lipase as a function of total substrate added. By 1953 one of the principal developers of the EC system, Hoffman-Ostenhoff, reflecting the generally accepted view at that time, considered lipase as a distinct class of group-specific "carbonic" (carboxylic) esterases distinct from "simple" esterases. He also placed in special subclasses the two other enzyme types known at that time to be capable of catalyzing the liberation of fatty acids from glycerol in phosphatides: phospholipase A (or lecithinase A) and phospholipase B (or lecithinase B). That our ideas of lipase specificity continue to change is evidenced from the following synopsis.

For EC 3.1.1.3, the recommended name, originally glycerol ester hydrolase (lipase) in 1964, was changed to triacyl lipase in 1972, and more recently Kinnunen *et al.* (1977) strongly recommended that this number be reserved for lipoprotein lipase.

For EC 3.1.1.4, the names of phosphatide acyl hydrolase or phospholipase A (1964) were changed to phospholipase A_2 or lecithinase A (1972). In 1976 we saw a minor change in reaction terminology.

For EC 3.1.1.5, "lysolecithin acyl hydrolase" or "phospholipase B" (1964) was changed to "lysophospholipase" or "lecithinase B" with a minor change of the reaction terminology in 1976.

For EC 3.1.1.23, the term "mono acyl glycerol lipase" (1964) has not been revised since the 1964 designation for this number.

For EC 3.1.1.32, the designation phospholipase A_1 was absent in the 1964 classification; it first appeared in 1972 and did not change in 1979.

For EC 3.1.1.34, the designations "diacylglycerol lipase" or "clearing factor" first appeared in 1972 only to be replaced by lipoprotein lipase in 1979.

New EC numbers have been added and dropped, and even the delimitation of the specificity of the enzymes which have retained their EC number has changed.

The first classification scheme in 1964 retained the suggestions of Hoffman-Ostenhoff. There were the "true" lipases which hydrolyzed all triacylglycerols without distinction as to how many and which (α or β) fatty acids came off. Of course differences were considered to exist within this subclass. For instance, one of the distinct differences between the two classical animal and plant sources, pancreas and castor bean, was that the pancreas lipase prefers to attack the α or 1- and 3- positions of the glycerol molecule whereas the castor enzyme was supposed to expeditiously liberate all three of the fatty acids. Later it was learned that, if anything, the castor lipase is even more stringent in its specificity with regard to hydrolyzing only α-linkages and that the appearance of stoichiometric amounts of fatty acid in castor bean lipase hydrolysates is due to migration of the fatty acid from the 2 or β-position to the 1 or α-position (Borgstrom and Ory 1970). It turns out that the pH optimum (*ca* 6) at which these castor bean studies were conducted is conducive to such migration, whereas the optimum pH of the pancreas enzyme (*ca* 8) is not:

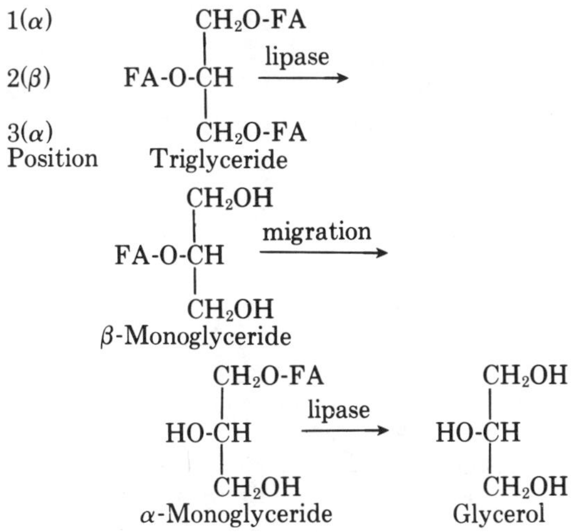

where FA is fatty acyl. The β-linked fatty acid can also migrate to the equivalent 3 position and also at the diglyceride stage (not shown). For any given enzyme the optimum fatty acid chain length attached to the glycerol or for that matter to most other alcohols (provided the resulting ester is insoluble) corresponds for most lipases to the C_4 butyryl esters. Microbial and other lipases which do not appear to follow these specificity guidelines include that obtained from the mold *Geotrichum candidum* which preferably liberates certain unsaturated fatty acids such as oleic acid even if attached to the 2- position of the glycerol (Franzke *et al.* 1973) and lipases from *Staphylococcus aureus* (Alford and Smith 1965) and human milk (see below). These two lipases thus do not exhibit positional specificity.

By 1972 the number of EC entries corresponding to enzymes which liberated fatty acids from nonphosphorylated glycerol had risen from one to three and the preferred name for the "original" lipase had changed from glycerol ester hydrolase to triacyl lipase. In 1976 the so-called diacylglycerol lipase was changed to what most investigators had been calling it anyway, lipoprotein lipase, simply because it was learned that in its natural habitat the lipoprotein lipase does act on triglyceride or, preferably, triacylglycerol. That these two new EC numbers may indeed be short-lived is betokened by the assignation by leading investigators of the EC number of simple lipase 3.1.1.3 to lipoprotein lipase (Kinnunen *et al.* 1977). The other "new" glyceride-splitting enzyme, monoacylglycerolipase, is considered by Brockerhoff and Jensen (1974) in their review of lipases to belong to the subclass of nonspecific esterases. On the other hand the apparently distinct hormone-sensitive mammalian lipase which is regulated by the cyclic nucleotide cascade mechanism limned in Chapter 4 may earn a special EC number. Incidentally, the reason that coffee drinking results in an increase of blood FFA is that the caffeine in the coffee inhibits the breakdown of the cyclic nucleotides by phosphodiesterase in the bloodstream. This leads to the eventual activation of this lipase and release of FFA. Contrary to the suggestion that cAMP and cGMP regulate metabolic processes in opposite directions (the *yin-yang* hypothesis), Khoo *et al.* (1977) showed that this hormone-sensitive lipase is equally efficiently activated by both cascade members, cGMP- and cAMP-dependent protein kinases: Lipase (inactive) + ATP → Lipase-P (active) + ADP.

The Phospholipases.—Our concepts of the specificities of enzymes which split off FFA from phosphoglycerides (phosphatides) which are, metabolically speaking, the parents of triacylglycerols, have also undergone some revision. Originally there were two acyl hydrolases which were at first called phospholipase A and B and later lecithinase A and B. (Phospholipases C and D hydrolyze phosphodiester linkages and thus are not really lipases in the sense used here.) The A enzyme liberated the fatty acid at the 2-position of the glycerol to form lysolecithin, 1-acylphosphatide, which was then attacked by what was considered to be a highly specific phospholipase B or lysolecithinase. Incidentally one of the best sources of phospholipase is snake venom which may owe part of its toxicity to this phospholipase because a product of its action, lysolecithin, lyses red blood cells.

Two distinct phospholipases A are now recognized. A_2 has the same specificity of the old "A." The food quality-related significance of phospholipase A_2 extends beyond flavor defects because (1) the fatty acid liberated is almost always polyunsaturated and thus a substrate for lipoxygenase action leading to aroma in fresh vegetables (Chapter 21), and (2) its action in frozen fish may lead to undesirable texture changes (Chapter 28). A_1 attacks, as indicated by the subscript, the 1 position to form 2 acyl-lysolecithin. Apparently phospholipase B, now lysolecithinase, is not so specific as once believed since most preparations have been reported to attack both lysolecithins:

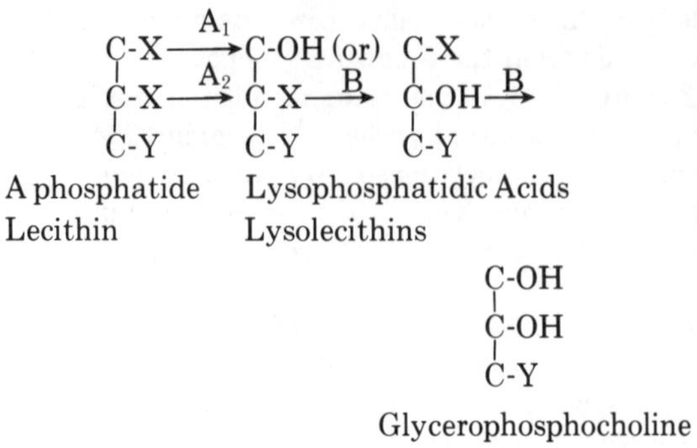

where Y is a phosphodiester. Indeed it may turn out that the phospholipase Bs are, as are the monoacyl glycerol hydrolases, simple nonspecific carboxylesterases, EC 3.1.1.1. As pointed out by Egelrud and Olivecrona (1973) specificity differences of the true lipases reside in the physicochemical nature of the substrates as much as in their chemical structure.

In addition, there are some enzymes which appear to be quite specific and which can liberate fatty acids from naturally occurring esters. These include cholesterol esters formed during digestion of food, the galactosyl diglycerides and certain sulfolipids, the latter two of which are widely distributed among food and other plants (Hitchcock and Nichols 1971).

Activation.—Perhaps more than other enzymes the phenomenon of activation has been readily observed when assaying for lipase from whatever source. Some of this activation is simply a matter of making more substrate available to the enzyme by more thorough emulsification, hence the activation by surface active agents. Indeed lipase assays require such surfactants, be it one of the Tweens or the previously widely used bile salts. Especially effective among the newer emulsifiers are siliconized glass beads which speed up pancreatic lipase-catalyzed hydrolysis of triproprionin 1000-fold (Brockerhoff and Jensen 1974).

Another widely reported activator is Ca^{2+}. Its mode of action for activation, at least for the acylglycerol lipases, is simply to remove liberated FFA as insoluble calcium salt. On the other hand the activity of pancreatic lipase is considerably enhanced by a Ca^{2+}-requiring glycoprotein activator present in human pancreatic juice (Kimura *et al.* 1974).

Ca^{2+} is also an authentic obligatory cofactor for pancreatic phospholipase A_2 (Verger and de Haas 1976).

As previously mentioned, in common with phosphorylase and other metabolically important enzymes, the hormone-sensitive lipase of the mammalian blood and adipose tissue is activated by a kinase-catalyzed covalent attachment of orthophosphate. An interesting naturally-occurring activator for castor bean lipase was reported by Ory *et al.* (1964) to be a cyclic tetramer of its natural substrate, ricinoleic acid (see Chapter 13).

$$\left[O{=}\overset{O-}{\overset{|}{C}}(CH_2)_7CH{=}CHCH_2\overset{-O}{\overset{|}{C}}H(CH_2)_5CH_3\right]_4$$

Another biologically meaningful activation (as well as inhibition) is that of heparin-stabilized lipoprotein lipase, discussed below, by separate, specific proteins which are part of the lipoprotein complex of the "very low density" lipoproteins. The latter are vehicles for transport of triglycerides from blood to tissue. Phospholipids have also been reported to be involved in lipoprotein activation (Chung *et al.* 1973). Some of the above may activate by stabilizing, others by being true cofactors.

Activation and inhibition of phospholipase A_2 have deep health ramifications since the enzyme in leucocytes is activated by specific peptides such as Met-Leu-Phe which also elicit the chemotactic response by which white blood cells fulfill their role as a defense against illness. Furthermore this activation of phospholipase A_2 is specificially inhibited by a protein elaborated in leucocytes treated with anti-inflammatory agents such as hydrocortisone (Hirata *et al.* 1980). Undoubtedly this enzyme regulatory control has food quality implications.

Like other pancreatic enzymes, phospholipase A_2 is secreted as a proenzyme which is activated by the splitting off of a hexpeptide (Pieterson *et al.* 1974). Unlike these other activations a new catalytically active substrate site is not formed as a result of this limited proteolysis because it is apparently already present in the proenzyme, i.e., the proenzyme hydrolyzes short chain soluble substrates but not insoluble long chain phosphates. What is induced by this limited proteolysis is the formation of the "Interfacial Recognition Site."

The Interfacial Recognition Site of Lipases.— The hallmark of authenticity of true lipase action is that the substrate be insoluble. Almost 20 years ago Desnuelle (1961) defined lipases as esterases which act at interfaces (Fig. 25.1).

Thus, lipase action can be looked upon as the reverse of immobilized enzyme action in which soluble enzyme is acting on insoluble substrate. The insoluble substrate may be in the form of emulsion, as micelles (*cf* lipoxygenases, Chapter 24) or as surface monolayers. The latter is of particular importance to the cell molecular biologist whereas emulsions may be more relevant in food situations. This activation by the surface is thus the most salient feature of its action.

Early suggestions forwarded to explain this surface dependence included increase in local concen-

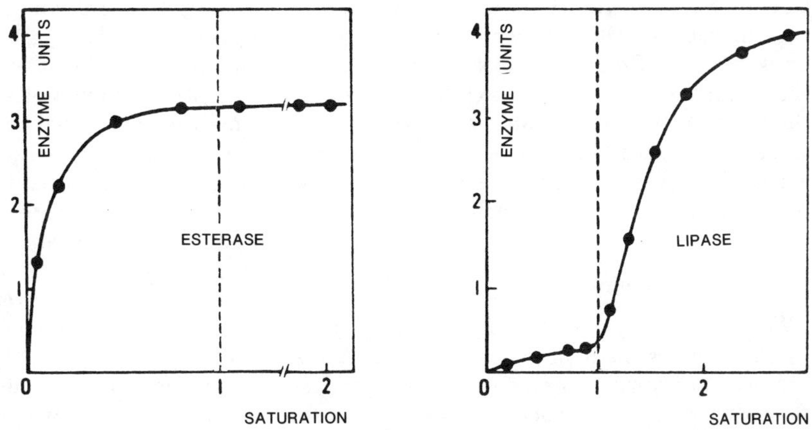

FIG. 25.1. LIPASES REQUIRE INSOLUBLE SUBSTRATES

Triacetin is hydrolyzed appreciably by lipase only after solubility limit (vertical line) is exceeded.

From Verger and De Haas (1976). Adapted from Desnuelle (1961).

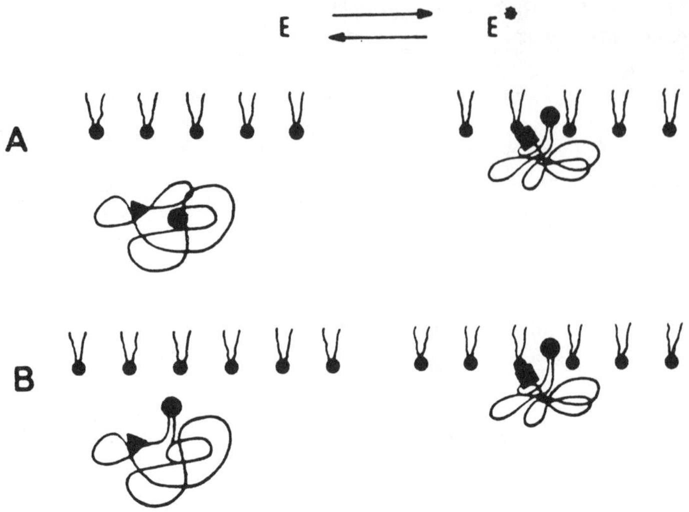

FIG. 25.2. LIPASE INTERFACIAL RECOGNITION SITE

The site, denoted by a large closed circle, when in contact with substrate (hair-pin structures) induces conformational alteration in catalytic site to make the latter catalytically competent (triangle to square).

From Verger and De Haas (1976)

tration of substrate, increased hydration of substrate and surface pressure.

However, the consensus now ascribes (not surprisingly in view of its widespread applicability) this activation to a change in conformation of the active site (triangle to square in Fig. 25.2) so that it can latch onto the substrate. This conformational change takes place when a portion of the substrate or the interface itself, by virtue of steric charge and other factors, makes contact with a special site (circle) different from the catalytic site, as speculatively illustrated by Verger and de Haas (1976) in Fig. 25.2. This is termed the Interfacial Recognition Site by Verger and De Haas and by Desnuelle and the Supersubtrate Binding Site by Brockerhoff and Jensen. Other terms are penetration or anchoring site. Brockerhoff looks upon lipases as simple carboxylic esterases which have developed these special sites, presumably evolutionarily. He suggests that such binding sites are very common among intracellular enzymes and play decisive regulatory roles in intracellular metabolism.

Lipases of Vegetables

We have already seen how lipoxygenase action usually requires the prior action of lipases. A special role in creating flavor defects in food assigned to the action of phospholipase A_2 stems from the circumstance that in most phospholipids the secondary hydroxyl of glycerol (position 2 or β) is esterified with an unsaturated fatty acid. Physiological interest in lipase action in vegetables centers around its possible role in the control of dormancy, germination and senescence. Thus, the hydrolysis of reserve fat in the apple seed embryo was reported to be catalyzed by two different lipases, one of them an acid lipase acting during the cold-mediated breaking of embryo dormancy. The other enzyme is an alkaline lipase which acts during germination (Zarska and Lewak 1976). Evidence that desirable flavor aromas associated with lipid transformations which occur immediately following tissue disruption can be traced back to action of phospholipase is afforded by the investigation of Galliard and colleagues (Chapters 1, 21 and 24). According to Wardale and Galliard (1975) vegetable lipases, unlike lipoxygenase, appear to be localized within the cell in classical lysosomal bodies along with other acid hydrolases. That lipase action continues after an initial burst in FFA liberation is attested to by demonstration of increasing FFA levels in unblanched frozen foods, especially peas (Lee and Mattick 1961; Bengtsson and Bosund 1966). Again, as for fresh vegetables, a preferential liberation of unsaturated over saturated FFA acid is observed and the preference becomes more pronounced at lower temperatures of frozen storage. Indeed, as mentioned in Chapter 12, at certain critical temperatures the trend towards lower rates of enzyme activity with lowering temperature may be reversed. A further example is afforded by the observations of Mullenax and Lopez (1975), who reported an enhanced rate of FFA release from olive oil at −23°C than at higher temperatures over a period of 90 days.

Lipoxygenase action preceded by that of lipase is mainly responsible for the development of off-flavor in soybeans which provides ample lipase activity for potentiation of the isolipoxygenases for which only FFA but not triglycerides are substrates (Perl and Diamant 1963). By contrast in other seeds the action of lipase itself can lead directly to off-flavor released during grinding. Thus, lipase action results in the development of a soapy taste in stored ground seeds of *Cucuropsis edulis*, a high protein seed used in West Africa as a soup thickener (Idiem Opute 1975).

Although not due to lipase action, other off-flavors in high protein seeds potentially useful as human food develop as the result of the action of the glucosinolases and associated enzymes present in rapeseed and mustard (Chapters 21 and 24).

While lipase-induced off-flavors in fruits are rarely reported, the observation of Chan *et al.* (1973) that butyric, hexanoic and octanoic acids are present in improperly processed papaya purees suggests that lipase may play a pejorative role in creating the resulting flavor defect. While the unsaturated fatty acid linoleic was located, as is usually found, at the 2-position of the phosphatides of orange juice, Braddock and Kesterson (1973) question occurrence of enzymic hydrolysis of lipids during storage of commercial orange juice. The presence of lipase in spices may create a flavor problem when the spice is used in combination with a food oil (Gross and Ellis 1969). The off-flavor described as "soapy" was traced to the liberation of lauric acid (12:0).

Peanuts.—We briefly examined the role of lipoxygenase in creating desirable aromas in fresh peanuts (Chapter 23). However, in peanuts stored for long periods before roasting, cell disruption with attendant lipoxygenase action and consequent accumulation of undesirable volatiles may occur. At least part of the off-flavor developing in these peanuts comes directly from the lipase which is localized in the mitochondria. Uniquely, lipase activity in germinating peanuts is inhibited by phosphate, whereas that from germinating castor bean, cotton and pumpkin seeds is not. Jacks and Yatsu (1974) suggest that lipase action can be suppressed in raw peanuts, in comminuted meats (peanut butter) during storage and during processing of peanuts into oil and edible protein (Chapter 34) with aqueous solvent by addition of phosphate.

Cereals.—After cereals are processed into baked goods, oil, etc., presumably no endogenous enzyme action occurs. About the only enzymes which might act are heat-stable added bacterial α-amylases which are designed to survive baking. Such action is not likely to cause off-flavor problems. In such cereals, including rice as well as wheat, such problems, characterized by both hydrolytic and oxidative rancidities, may occur when the a_w value rises significantly during storage (Chapter 13). In contrast to legume seeds, considerable attention has been paid to the role of lipases as agents of hydrolytic rancidity. However, it should be pointed out that enzyme-related deterioration of grains is but one aspect of their storage stability.

Rice.—Particular attention has been directed to rice bran lipase since its action during storage of unpolished rice results in hydrolytic rancidity. This not only may affect the quality of the subsequently polished rice, but also the oil prepared therefrom. Keys to literature on rice lipase and rancidity may be found in papers by Shastry and Rao (1976) and by Aizono *et al.* (1976), who have isolated and characterized two isolipases from rice bran. Like most other lipases, although they hydrolyze the long chain fatty acyl glycerols present in rice oil quite efficiently, they do show a preference for short chain triglycerides. Among the known properties of rice lipase which might be helpful in controlling its action is the finding that Ca^{2+} apparently activates at low but, as does EDTA, inhibits at high concentrations.

Corn.—While hydrolytic rancidity caused by lipase action in corn has usually been a problem, much of the concern with this enzyme has been focused of late on the discovery that the infamous southern corn leaf blight is accompanied by increased FFA liberated by the lipases of the infecting organism, *Helminthosporum maydis* Race T (Bennet *et al.* 1976).

Wheat.—The deteriorative mischief that the lipase-lipoxygenase systems of wheat-based foods can perpetrate during their storage has been chronicled by Fellers and Bean (1977). As mentioned, barley may develop a "cardboard" odor. Sprouted barley (malt) may also contribute to the off-flavors in bread, beer and other foods to which it is added, if proper programming of temperature and other parameters are neglected (Narziss and Sekin 1974). As mentioned, lipase action can occur at rather low a_w values as shown in Table 25.1

TABLE 25.1

WHEAT LIPASE ACTIVITY AS A FUNCTION OF WATER ACTIVITY[1]

Water Activity	Dry Weight (%)	Relative Lipolysis at 20°C	30°C	40°C
0.2	5.7	6	8	12
0.5	11.0	20	40	55
0.8	25.0	40	90	100
1.0	—	30	80	65

Source: Caillat and Dapron (1974).

[1] Normalized data expressed as percentage of activity at 40°C and $a_w = 0.8$ at which the samples became visibly wet.

(Caillat and Drapron 1974). Indeed, inhibition was initiated upon appearance of solvent water. Like most other lipases, only the 1 and 3 positions of the triglycerides are attacked. A mixture of $Na^+ + Mg^+ + K^+$ proved to be an effective activator. Pancholy and Lynd (1972) report that the fungal toxin, aflatoxin, is a potent inhibitor of wheat germ lipase activity. Information on the distribution among milling fractions and the effect of horticultural practice on wheat lipase activity is available (Colas and Charlegegue 1974). The dispute concerning the role of lipase in bread texture will be examined in Chapter 32.

Meat and Fish.—Lipase action of fish will be discussed in more detail in connection with fish texture. As an agent of off-flavor, the action of this enzyme figures prominently in hydrolytic rancidity problems in both meat and in fish (Potthast and Hamm 1973; Reineccius 1979; Love 1980).

Plasmalogens.—An interesting and so far singular instance of lipase action indirectly causing the formation of off-flavor (and off-color) in meat and meat products is afforded by reports so far from Japanese laboratories on the presence in meat (especially pork, some in fish) but not in plant foods of long chain ($C_{24}-C_{28}$) 2,3-dialkylacroleins, R_1-CH=CR_2-CHO (Waku and Nakazawa 1972; Nakashine and Suyama 1974; Kuroda and Negishi 1975). Phospholipase action liberates these aldehydes, corresponding to the more conventional FFAs, from plasmalogens. Plasmalogens are acetal phosphatides analogous to cephalin, phosphatidylethanolamine:

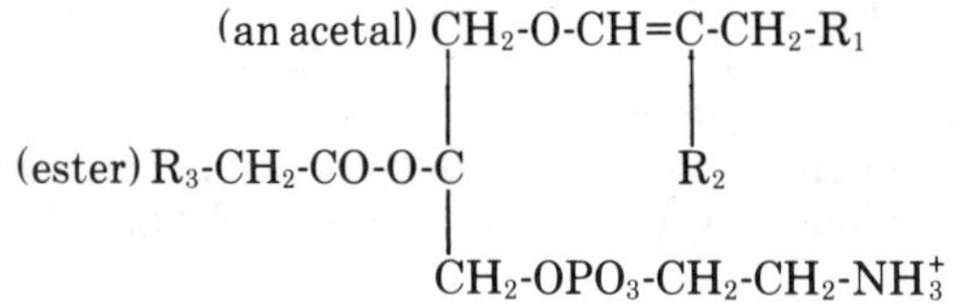

These aldehydes then undergo aldol condensation (probably nonenzymatic) and dehydration reactions. Their distribution and relative concentration may contribute to both aroma and off-flavor in ham, bacon and other pork products.

Phospholipase-generated plasmalogens may well contribute to the off-flavor in venison, hare and other wild game meat prepared from animals whose hind leg nervous tissue has not been removed. Plasmalogens are found in such tissue as are other lipoid substances such as the cerebrosides, sphingomyelin and gangliosides whose products of enzymatic hydrolysis could lead to off-flavors.

MILK LIPASE AND HYDROLYTIC RANCIDITY

Lipases, which, as we have seen (Chapter 23), play a seminal role in the development of desirable flavors in dairy products, are not endogenous in origin but are deliberately added either in the form of commercial oral or pregastric lipases, or are present in deliberately selected starter and ripening microorganisms. Lipase action of the milk itself is the source of off-flavors. Lipase action in milk can lead to off-odor directly by virtue of the presence of the products of hydrolysis (hydrolytic rancidity), or can set the stage for the development of oxidative rancidity. However, not all flavor defects in milk due to lipid transformation are necessarily funneled through lipase action.

A good overall review containing a cogent précis of circumstances leading to and consequences of lipolysis in dairy products and how to avoid them is that of Deeth and Fitz-Gerald (1976). Other reviews covering various aspects of dairy-related hydrolytic rancidity and the enzymes involved include those of Shahani *et al.* (1974), Downey (1975), Olivecrona *et al.* (1975) and Olivecrona (1980). Undoubtedly many others will be available by the time this book appears.

The Problem

Although we have referred to hydrolytic rancidity before, it may be apropos to reiterate that the perceived flavors, both taste and odor, are due principally to the lower FFAs, especially butyric acid but also caproic, capric and caprylic acids (C_4-C_{10}). At low enough levels and in the presence of other odorants, these acids, as we have seen, contribute to the aroma of milk and cheeses (ADV = 1.5 for milk, 2.5 for cheese; ADV = acid degree value, milliequivalents of alkali needed to neutralize the FFA per gram of fat × 100). With increasing levels of acid (ADV > 5) these miasmas are perceived as cowy, stale, old, unclean, bitter, goaty, soapy, butyric and, of course, rancid.

Analogous to accrual of added benefits or disadvantages occasioned by the potentiation of desirable enzyme action (Chapter 1 and elsewhere) the action of lipase in milk and other dairy products entrains a host of consequences, mostly undesirable, besides producing flavor defects. These include increased churning times and foaming of cream in butter manufacture, inhibition of some starter bacteria in cheese manufacture, fat losses because of separation difficulties, and the prevention of proper, adequate foaming of milk in espresso coffee machines. A benefit of this antifoaming capacity is to prevent foam formation during concentration processes.

Quite apart from any knowledge of the causes, it was early recognized that there are three distinct causes of lipolysis in milk—induced, spontaneous and microbiological. In each case the action of lipase is potentiated at the wrong time at the wrong place in the food chain. Induced lipolysis as defined by Deeth and Fitz-Gerald is brought about when raw milk is subjected to a variety of physical treatments during or after milking. The more infrequently encountered spontaneous lipolysis occurs in milk from certain cows and is initiated without further physical treatment when the milk is simply cooled.

Induced lipolysis is manifested when the milk is subject to physical manipulations such as agitation, foaming, rapid up-and-down temperature changes, homogenation, freezing and thawing. Higher temperatures favor susceptibility to induced lipolysis. Thus, such changes could be caused by faulty milking machines and practices, faulty pipelines and pipeline design, pumping practice and storage. In the factory, hydrolytic rancidity may be induced during pumping and separating operations and during storage all of which may involve both physical agitation and temperature fluctuation.

Investigations have established several parameters associated with the occurrence of spontaneous hydrolytic rancidity of milk from about 3–5% of a given herd which are likely to yield such milk. Among the more important are stage of lactation, feed and nutrition. Indicative of the quest for additional factors is the report of Bachmann (1973) that altitude (as simulated in an altitude chamber) had no effect on the lipase activity per milliliter of milk. To induce spontaneous rancidity, the obligatory cooling to below 15°C must occur soon after milking; the sooner the milking and the lowering of temperature of subsequent storage, the more intense the flavor defect. The problem is somewhat mollified by the circumstance that mixing a minimum of four parts of normal milk to one part of spontaneous milk suppresses the spontaneity of the latter.

Microbially induced rancidity is most likely to occur when the lipolytic psychrophilic bacteria, normally present in milk but held in check, exceed a count of about 1 million/ml.

The Enzymes.—Although lipolysis by enzymes in milk is easily demonstrated, studies of the lipases are beset by difficulties, partly because of their reputed heterogeneity and partly because of their inherent instability; they are readily inactivated by pasteurization. Early workers had to resort to deaeration with nitrogen to preserve the original activity (Forster *et al.* 1956). Furthermore, the correlation of lipase activity, as measured by most assays, with degree of hydrolytic rancidity has yielded ambiguous results (Driessen and Stadhouders 1975). At first it was generally believed there were at least two kinds of lipase. It is still not clear if there are several isoenzymes. Molecular weights have been reported from under 10,000 to over 200,000. As pointed out by Shahani *et al.* (1974) these differences might be due to the propensity of the lipases to combine strongly with casein micelles (perhaps through their interfacial recognition sites). There is even some question if the active lipase found in the sediment from clarification (clarifier slime) containing nonmilk udder cells and some bacteria is the same as that found in milk (Richter and Randolph 1971).

Some investigators believe that milk lipases and pancreatic lipase have a common origin (Shahani 1975). At least one of the lipases, if there are more than one, has been thoroughly purified and characterized as a lipoprotein lipase by a group of Swedish workers who have effectively applied affinity chromatography (Chapter 8) of this enzyme (Bengtsson and Olivecrona 1977).

Lipoprotein Lipase.—We referred to lipoprotein lipase in our discussion of lipase specificity, classification and nomenclature. Stemming from the original observation of Hahn (1943) that injection of the polysaccharide heparin (Chapter 5) into fat-fed animals resulted in disappearance of lipemia as evidenced by the clearing of the blood plasma, establishment of the "clearing factor" as a distinct lipolytic enzyme by investigations during the ensuing years was attested to by the review of Schwimmer (1957). This physiologically and medically important enzyme is specifically activated by a 78-amino acid protein (formed via solid phase peptide synthesis) present in one of the lipoprotein blood fractions (Kinnunen *et al.* 1977). Among the enzyme's characteristic properties are its stabilization by heparin and inhibition by salts at neutral pH's.

Cause of Spontaneous Hydrolytic Rancidity.—Now according to the Swedish group (Olivecrona *et al.* 1975) there is essentially one lipase in milk and it is such a lipoprotein lipase. It possesses the abovementioned special characteristics of lipoprotein lipase, and the purified enzyme crossreacts immunologically with that from postheparin plasma. They believe that a specific protein termed apolipoprotein or similar activator from the cow's blood activates or stimulates the lipoprotein lipase already present in the cow's milk to create the phenomenon of spontaneous rancidity. Other evidence led Downey (1975) to a similar conclusion, i.e., the same enzyme produces the activities attributed to lipase and to lipoprotein lipase. The Irish investigator does, however, stress the variation of the structural integrity of the fat globule membrane (see below) as a key element in susceptibility to lipolysis in spontaneous milk. Similarly, Driessen and Stadhouders (1975) concluded that spontaneous milk contains thermostable factors, in this case small phospholipoproteins, which form complexes with lipoprotein lipase which then adsorbs to the fat globules and hydrolyzes the triglycerides in the globule. Normally, according to this concept and corroborated by others, lipoprotein lipase in the milk serum probably is bound to casein—a situation which, as mentioned, has in the past created difficulties in the study of lipase. According to Posner and Bermudez (1977) this milk lipoprotein lipase in milk is bound to a stabilizing factor which appears to be a large nonprotein molecule, perhaps akin to heparin. In addition to the lipoprotein lipase, which can also cause rancidity problems in stored human milk, the latter contains a second, inactive lipase activated by bile salts but not by blood lipoprotein peptides. This second lipase is responsible for the digestion of fats in the small intestine of the human infant (Olivecrona and Hernell 1976). Interestingly, this is said to be one of those rare lipases which readily hydrolyze all three ester bonds in triglycerides. Similarly, the Dutch workers Driessen and Stadhouders (1975) detected a lipase present in colostrum which is not bound to casein micelles nor is it activatable by their phospholipoproteins.

It should be pointed out that this view of spontaneous lipolysis is not universally held. Deeth and Fitz-Gerald state that an inhibiting factor has been invoked to explain the prevention of spontaneous lipolysis. According to this concept the inhibiting factor missing in spontaneous milk interferes with cold-induced adsorption of the enzyme to the fat globule. The reviewers believe that susceptibility depends upon a combination of some or all of the lipase-associated factors just discussed.

The Substrate and Induced Lipolysis.—We hitherto have concentrated on the enzyme and not on the substrate. Of course with lipase as the enzyme, the physical state of the substrate is a most important factor in hydrolytic rancidity, whether spontaneous or induced. Indeed, contrary to the situation in other foods, where the enzyme is confined to a particulate and the substrate may be relatively free, in milk the substrates are confined to microscopic fat globules which are surrounded by membranes as are all subcellular organelles. This globule does indeed represent a delicately poised system which is easily disoriented and thus subject to attack by the lipoprotein lipase present in the skim milk. It is this fragility which is probably responsible for induced lipolysis. It should be mentioned, however, that the properties of the globule (substrate) have been said to play a role in spontaneous lipolysis also. Thus, globule "weakness" of size, shape and especially hardness have been considered. Typical of the evidence that changes in the fat globule membrane are involved in induced hydrolytic rancidity is the finding of Anderson *et al.* (1972) that the aging of cooled milk resulted in a greater loss of phospholipid (probably associated with membrane) during the separation of such membranes than that observed with fresh milk. Other investigators stress protection by protein components of the membrane. Sustek *et al.* (1975) detected considerable lipase activity associated with the membrane under conditions which cause lipolysis. Thus, according to one idea, during homogenization the natural membrane is replaced by a coating of casein micelles which carries along with it associated lipase. The latter is now in close proximity and available to its substrates. Freezing causes complete disruption of the membrane allowing decompartmentalization of enzyme and substrate. One point that should be stressed is that the milk enzyme, although a lipoprotein lipase, can still attack insoluble emulsified carboxylic esters rather indiscriminately (Egelrud and Olivecrona 1973). In this diversified sweeping specificity may lie a clue as to why milk lipase action results in off-flavors whereas the more selective, controlled action of the pregastric lipases results in desirable flavor (Chapter 23, Table 23.1).

Microbial Lipases.—As mentioned, many of the psychrotrophic microorganisms which tend to thrive at refrigerator temperatures in milk, if allowed to grow beyond permissible levels, produce lipases in sufficient amounts such that their action results in flavor defects. Unlike milk lipase, which is completely inactivated during pasteurization, some of the microbial lipases are quite heat stable and can cause off-flavor problems in pasteurized milk and in products made therefrom. Kishonti (1975) reported that treatment of milk at 150°C was necessary to inactivate such lipases although the organisms which contained them were quite heat-sensitive and were destroyed under normal pasteurizing conditions. The contribution of these lipases to off-flavor may be as great as those of the milk. Thus, Suhren *et al.* (1975) reported that of almost 200 samples of pasteurized milk, 20% developed FFA to levels perceived as rancid.

The mechanism by which the fat globule is attacked by microbial differs from that by milk lipase. Since such microorganisms contain phospholipases as well as triacylglycerol lipase, the action of the former on the phosphatides of the fat globule membranes may blaze a pathway for the action of the latter (Fox *et al.* 1976). At least one flavor defect, broken cream, has been attributed to phospholipase action. In this particular instance the responsible enzyme is phospholipase C, a phosphodiesterase and not a carboxylesterase, which splits off phosphoryl choline from lecithin and phosphoryl ethanolamine from cephalin. Not only do phospholipids produce undesirable fatty acids, but they also activate the milk's own lipase. The activation is accomplished as a result of modification of the fat globule membrane, thus making it susceptible to lipase action.

In another aspect of undesirable lipase action vis-à-vis dairy products, its presence threatened to delay or prevent the successful application of microbial rennet (Chapter 33). These rennets are by no means pure milk-clotting enzymes and contain most of the proteins of the culture medium or are secreted into the medium. Fortunately, lipases can be inexpensively removed, as mentioned in Chapter 6.

Prevention and Control

If milk lipase action could be suppressed before pasteurization, this endogenous enzyme would pose no flavor problem in milk products since, as mentioned, the milk enzyme but not that from adventitious organisms is completely inactivated by pasteurization. Unfortunately this cannot be done as yet so that the best thing to do is to carry out practices in the farm, the dairy and factory which minimize such action. As outlined in more detail by Deeth and Fitz-Gerald, on the farm these include: (1) avoidance of excessive air intake into the vacuum section of the milking equipment; (2) minimization of turbulence caused, for instance, by overmilking cows to the "bitter end" (a case where the beverage is not good to the last drop); (3) controlling temperature by keeping the latter as low as possible without freezing and avoiding temperature fluctuation; (4) maintaining cleanliness approaching adequate asepsis; and (5) treating the cows with respect to feeding, breeding, lactating and culling so as to minimize the number giving spontaneous milk. In the factory the authors suggest: (1) avoidance of excessive agitation, turbulence and foaming; (2) paying attention to such plumbing and engineering details as avoidance of leaks, mixing and flow speeds, and pump operation; (3) attention to the regulation of storage conditions; (4) pasteurization before homogenization; (5) never mixing homogenized and raw milk; (6) separation of cream should not be conducted between 25° and 35°C; and (7) of course, very strict hygiene practice.

With the accumulation of fundamental knowledge on milk lipase, it should be possible to apply it in the future to minimize its action. For instance the use of bags of affinity matrices as discussed in Chapter 7 might be practicable. In this case the ligand would probably be heparin or similar mucopolysaccharide. Lipoproteins of blood contain not only the specific inhibitor, the above-mentioned apolipoprotein C-II (apoC-II), but also at least one very potent lipoprotein lipase inhibitor protein, apoC-III (Morisett *et al.* 1975).

Still further in the future, avoidance of lipase action might be accomplished by stabilization of fat globule membrane and by other applications of our knowledge of the control of enzyme action as detailed in Chapter 14.

ENZYME INVOLVEMENT IN OTHER FLAVOR DEFECTS IN DAIRY PRODUCTS

Oxidative Rancidity

Lipolysis is just one of the lipid transformations leading to off-flavor. Most frequently, but not exclusively, these other flavor defects are due to lipid oxidation. The resulting oxidative rancidity has been variously perceived as "tallowy," "coconut," "oily cardboard," "cappy" and, of course, "oxidized." Lipolysis and hydrolytic rancidity are linked to oxidative rancidity not only by virtue of the well-known increased propensity of unsaturated FFA to undergo such oxidation whether enzymatic or not, but also because FFAs catalyze decomposition of the lipohydroperoxides and retard the action of endogenous antioxidants.

Although a rigorous cause-and-effect relation between enzyme presence and the development of such oxidative rancidity has not been firmly established, some of the oxidative enzymes of milk have been implicated. Older literature refers to "oleinase" activity. More than 20 years ago Aurand and Woods (1959) presented convincing evidence, which was followed up by subsequent investigations, that xanthine oxidase is involved via unsaturated fatty acids in the evolution of acetaldehyde in dairy products. Arguments put forward for the "peroxidatic" activity of this enzyme are particularly convincing in light of the subsequent finding (Chapter 16) that this enzyme is a rich source of active oxygen species. According to Aurand *et al.* (1977) O_2^* is the immediate source of hydroperoxides which initiate fatty acid oxidation. Less convincing is the role of milk peroxidase action in causing the "tallowy" flavor. Of course there are just as likely to be nonenzymatic as enzymatic sources of oxidative rancidity. These are reviewed by Sattar and De Man (1975). They include light, metal ions and their chelates. Among the latter are the hemoproteins, especially peroxidase acting in its nonenzymatic mode (Eriksson 1970). As with vegetable peroxidase (Chapter 21) the lipid-oxidizing capacity of lactoperoxidase is activated by heating and thus during pasteurization. Furthermore, xanthine oxidase largely survives pasteurization. Thus, pasteurization, unlike its ability to prevent milk-lipase induced hydrolytic rancidity, actually may increase the propensity towards oxidative rancidity.

Prevention and Removal.—Enzymes preventing the onset of oxidative rancidity include glucose oxidase, superoxide dismutase, proteinases and phospholipases C and D. The rationale behind the use of the oxidoreductases is quite straightforward, while the reasons for the latter work are not completely clear. In addition to these enzymes, another enzyme, sulfhydryl oxidase, has been proposed for preventing and removing the abovementioned "cooked" and related off-flavors.

Glucose Oxidase.—In this case one of the cosubstrates of autoxidation, O_2, is removed before interacting with the oxidizable lipid. Application of glucose oxidase to prevent oxidative deterioration has been confined largely to concentrated milk products, especially dry milks. Thus, Meyer *et al.* (1960) demonstrated that repeated nitrogen flushing of cans could not remove oxygen as well as could packets of glucose oxidase-catalase and glucose designed as oxygen scavengers. Glucose oxidase (along with lactase to produce glucose) has been proposed for use in dairy products for nonflavor purposes such as production of acid as a substitute for rennet (Rand and Hourigan 1975) and also to prevent delayed browning in evaporated milk.

Superoxide Dismutase.—Perhaps one of the reasons that milk is not more prone to oxidative rancidity than it has actually been found to be is because of the presence of a heat-stable pasteurization-survivable superoxide dismutase (Hill 1975; Hicks 1980). In contrast to glucose oxidase, which removes the oxygen outright before it has a chance to get to the unsaturated fatty acids responsible for oxidative off-flavor, SOD deactivates the oxygen after the latter has been brought into an excited state via a variety of both enzymatic and nonenzymatic pathways including interaction with lipids (Fig. 16.2). Experimental evidence for the quenching of both O_2^- and O_2^* generated in milk or by milk xanthine oxidase preparations by both endogenous and exogenous SOD adapted from the results of Asada (1976) and Aurand *et al.* (1977) is shown in Table 25.2. Perhaps in the future it might be helpful to breed cows whose milk has high levels of SOD and low levels of xanthine oxidase. However, Hill (1975) suggests that what he considers to be trace levels of SOD in milk are due to leakage of the enzyme from red blood cells, reminiscent of the purported leakage of the lipase-activating lipoproteins previously discussed. This suggests that the SOD levels should be raised in "spontaneous" milk.

Proteolytic Enzymes.—In 1939 Anderson reported that trypsin added to milk delays the

TABLE 25.2

QUENCHING OF MILK LIPID PEROXIDATION AND OTHER OXIDATIONS BY SUPEROXIDE DISMUTASE

Active Oxygen		Superoxide Dismutase Effect	
Generator	Acceptor	Source	Quench Quotient, %
Xanthine Oxidase in cream (XO)	Cytochrome *c*	Milk, skim, *in situ,* 1 ×	40[1]
Xanthine Oxidase in cream (XO)	Cytochrome *c*	Milk skim, *in situ,* 2 ×	70[1]
FMN + hν	Nitrotetrazolium blue	Cream, *in situ*	12[1]
FMN + hν	Nitrotetrazolium blue	Spinach[3]	100[1]
FMN + hν	Nitrotetrazolium blue	Milk, *in situ*	90[1]
Milk, whole (FMN) + hν	Milk, whole, *in situ* (lipid)	Mammalian[3], 1 ×	28[2]
+ Cu	Milk, whole, *in situ* (lipid)	Mammalian[3], 167 ×	75[2]
+ XO, added	Milk, whole, *in situ* (lipid)	Mammalian[3], 167 ×	91[2]

[1] Source: Asada (1976). Effects ascribed primarily to superoxide anion, O_2^-, except perhaps that from FMN + hν.
[2] Source: Aurand *et al.* (1977). Effects ascribed to singlet oxygen, O_2^*.
[3] Isolated crystalline enzyme used.

development of an oxidized flavor and Placek *et al.* (1960) suppressed the development of a cardboard flavor in milk by the addition of a "trace" of pepsin. Too much enzyme resulted in the formation of bitterness which we now know is due to the "bitter" peptides (Chapter 34) and, of course, in the clotting of milk. This phenomenon has been further investigated and shown to occur with immobilized trypsin and was reviewed by Shipe *et al.* (1975). The most likely reason for the effectiveness of trypsin is that the amino acids and peptides released from milk protein tie up copper ion, so that the latter cannot serve as initiator of lipid peroxidation.

Perhaps contributing to the observed antioxidant effect of trypsin is its destabilization of xanthine oxidase so that it does not survive pasteurization.

It may be well to mention at this juncture that the enzymes responsible for splitting and transferring of peptide bonds in milk products may be involved in the dairy flavors in other ways, both desirable and undesirable. We already mentioned their role in creating taste-desirable substances in Chapter 20. They can also produce bitter peptides as will be discussed in Chapter 34. They may also be looked upon as the ultimate progenitor enzymes for the production of microbially induced off-flavors discussed below. Perhaps endogenous glutamyl transferase (Wiesemann and Brinkley 1977) whose role in vegetable flavor was scanned in Chapter 21, may affect flavor by shifting the balance of free amino acids. Finally, one of the enzymes of the lactose synthase complex, galactosyl transferase (Chapters 4 and 20), is present in milk as two isozymes, one of which can be converted into the other by a trypsin-like enzyme in milk (Magee and Ebner 1973). Too much proteolysis results in inactivation and thus could affect the taste of milk.

The Phospholipase C and D.—These are phosphodiesterases and therefore, as mentioned, are not lipases in the sense of hydrolyzing a carboxyl ester bond of an insoluble substrate. Their capacity for prevention of oxidative off-flavors when added to milk, first described by Shipe and coworkers, has been ascribed to rearrangement of the components of the fat globule membrane. However, both trypsin and phospholipase C have been reported to increase lipolysis, so that both of these enzymes would have to be added after pasteurization. As yet, apparently neither of these hydrolases has been used in industry. As mentioned in Chapter 23, treating milk with pregastric lipase tends to mitigate the effects of lipid oxidation.

Other Flavor Defects

In contrast to these oxidized flavors, flavors produced by excess reducing capacity in the milk are referred to as "cooked" flavors and do not arise directly from lipid transformations. This cooked off-flavor can be effectively prevented by treatment with sulfhydryl oxidase, present in raw milk (Swaisgood and Horton 1975). Like trypsin, the enzyme works effectively while immobilized. Perhaps one problem in the application of its use would be to take precautions that any possible excited oxygen generated by the action of this iron-containing enzyme does not cause oxidative rancidity

$$2RSH + 2O_2 \longrightarrow 2RS\cdot + 2O_2^- + 2H^+$$

$$2RS\cdot \longrightarrow RSSR$$

$$2O_2^- + 2H^+ \longrightarrow H_2O_2 + O_2^* \xrightarrow{\text{(Lipids)}} \text{Lipohydroperoxides}$$

Other off-flavors are caused by substances which, in their proper setting, would contribute to desirable flavors. For the most part they are produced by enzymes contained within psychrotrophic microorganisms which as already seen are responsible for the active heat-stable lipases in milk. Some of these flavors, the substances probably responsible for them, the responsible microorganism, the cause, the putative flavorless precursors and the enzymes which could reasonably be assumed to be involved in their genesis are shown in Table 25.3. Shipe *et al.* (1975) suggest that offending carbonyls might be removed from off-flavor milk and products with certain specific microorganisms by virtue of the action of enzymes contained within these organisms. Their potential use as sources of "flavoreses" was mentioned in Chapter 23.

ENZYME INVOLVEMENT IN CITRUS BITTERNESS

In contrast to most aspects of enzyme involvement in flavor defects which we have heretofore considered, the major enzyme "connection" in the area of citrus bitterness arises from attempts to use exogenous enzymes to remove the responsible bitter substances. They comprise two different type of compounds: the limonoids, as represented by limonin, and the flavonone glycosides, the most offensive of which is naringin (Fig. 25.3). The CSIRO Australian investigators, Chandler and Nicol (1975), have provided a succinct yet comprehensive and critical assessment and access to the entire literature through 1974 of enzyme debittering of citrus products. With few key exceptions we shall cite only more recent references not cited by these authors.

Limonin

As pointed out by the Pasadena group, also very active in this field (Hasegawa 1976; Brewster *et al.* 1976) and others, this bitter triterpenoid dilactone does not exist as such in the intact citrus fruit. It is formed in navel and related varieties of orange juice as a result of the acid-and-enzyme catalyzed conversion of its flavorless, highly soluble, naturally-occurring precursor, limonoic acid A-ring (mono) lactone. It will be recalled that lactone formation is favored in acid milieu. There is thus a delay in perceived flavor, referred to by Joslyn and Pilnik (1961) as "delayed bitterness." The delay, which is shortened by heating, is partly due to the finite time for the H^+- and enzyme-catalyzed reactions to proceed, but also reflects the relative insolubility

TABLE 25.3

OFF-FLAVORS IN DAIRY PRODUCTS[1]

Description	Organism	Principal Compounds	Precursor	Enzyme
Malty	*S. lactis* var. *maltigenes*	2-Methyl butanal	Leucine, valine	Aminotransferase, α-ketodecarboxylase
Strawberry, fruity	*Pseudomonas fragi*	Ethyl isovalerate	Fatty acid	Lipase, ester synthase (Chap. 22), dehydrogenase
Geranium, mushroom	—	Vinyl ketones	Polyenoic acids	Lipoxygenase, lipohydroperoxide transformation
Pepper, potato	*Pseudomonas* spp.	2-Methoxy-3 alkyl-pyrazines	Amino acids	(See Chap. 20)

[1] Off-flavor which might be considered as desirable aroma constituent of nondairy foods or other kinds of dairy food. See Morgan (1976) and Swoboda and Peers (1977).

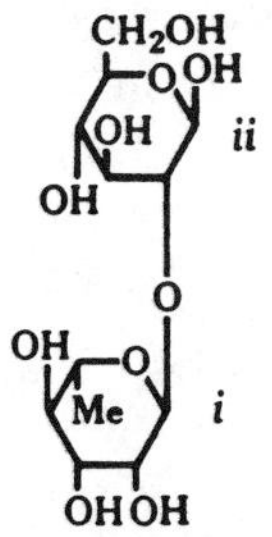

I Neohesperidose
i Rhamnose unit; Me = CH_3
ii Glucose unit

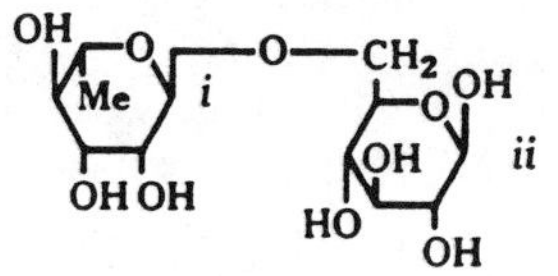

II Rutinose
i Rhamnose unit; Me = CH_3
ii Glucose unit

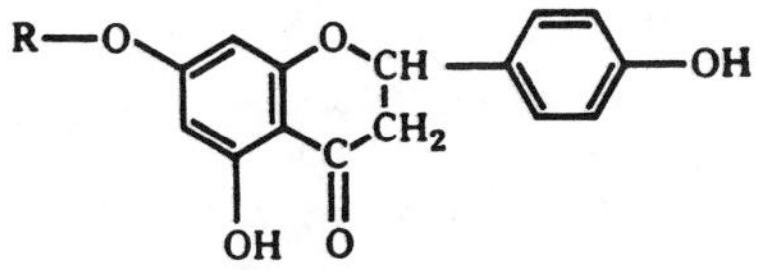

III*a* Naringin; R = Neohesperidose
III*b* Narirutin; R = Rutinose

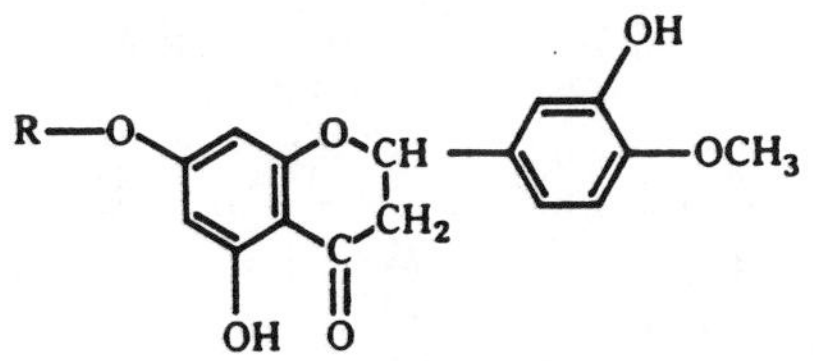

IV*a* Hesperidin; R = Rutinose
IV*b* Neohesperidin; R = Neohesperidose

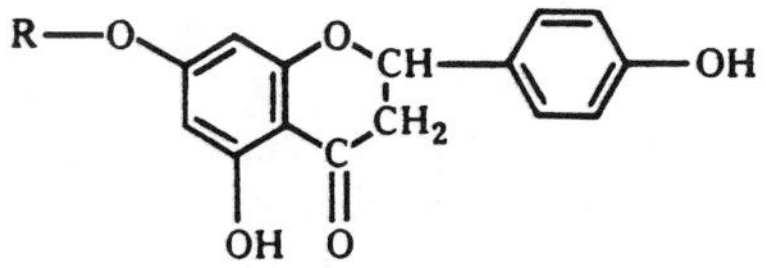

X*a* Prunin; R = Glucose
X*b* Naringenin; R = Hydrogen

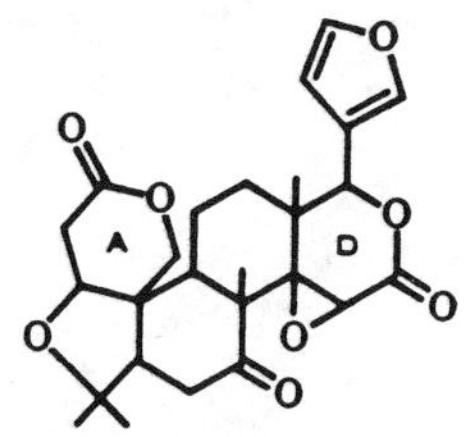

V Limonin

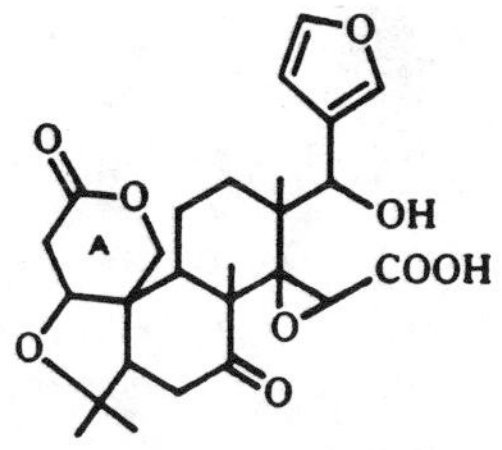

VI Limonoic acid
A - ring lactone

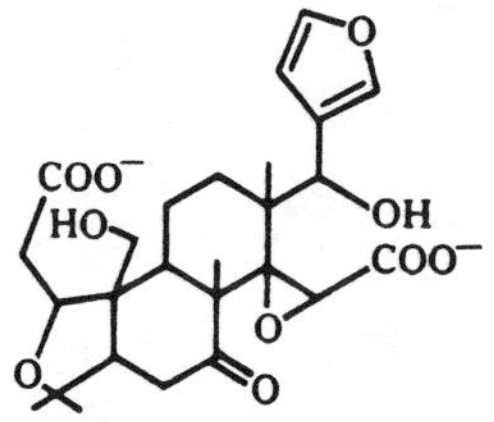

XI Limonoate ion

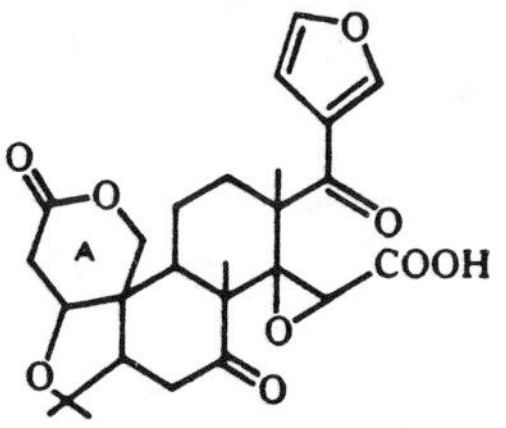

XII 17 - Dehydrolimonoic
acid A - ring lactone

Modified from Chandler and Nicol (1975)

FIG. 25.3. BITTER AND ENZYMATICALLY TRANSFORMED NONBITTER CITRUS CONSTITUENTS of the newly formed dilactone and the time required for it to pass into solution. The lactonizing enzyme responsible in large measure for intensification of bitterness which may occur during the processing of grapefruit, limonoate D-ring lactone hydrolase, can catalyze hydrolysis equally as well as lactonization. Limonoic acid A-ring lactone accumulates in the white of the peel (albedo) at the blossom end of the fruit (Fig. 25.4), reaching levels as high as 0.06%, but is not present in the juice sacs. The bitterness of limonin, unlike that of the flavonones, is wholly objectionable under all circumstances (threshold concentration, *ca* 9 ppm in orange juice) and increases sharply with increasing concentration. Differences in bitterness of different samples is apparently entirely due to differences in limonin levels and not to different types of limonoids. Like some of the more pleasantly flavored terpenoids (Chapter 22) limonoate, probably synthesized via the mevalonate pathway, is considered to be an intermediate on the metabolic pathway to the steroids. This is probably the reason it disappears upon maturity (Flavian and Levi 1970) and why, pehaps, as a stress metabolite it persists (at least in Shamouti oranges) after a season of abnormal weather (Gutter 1973).

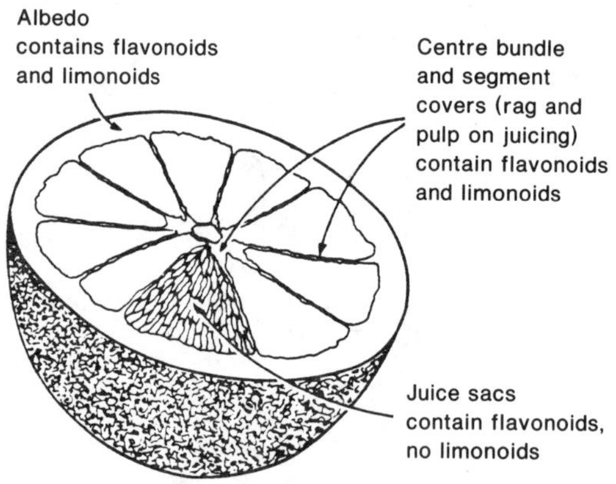

From Chandler and Nicol (1975)

FIG. 25.4. LOCATION OF BITTER SUBSTANCES IN CITRUS FRUIT

Prevention.—Chandler and Nicol point out that there are nonenzymatic nonbiological processing approaches to preventing limonin-caused bitterness. For instance, in the manufacture of juice the rag and pulp should be removed from freshly expressed juice as rapidly as possible. It is possible at least in the laboratory to obtain after centrifugation a juice completely devoid of bitterness.

Adding Enzymes.—Still in the developmental stage as of this writing after 30 years of research is the application of and treatment with enzymes. Researchers have followed two lines of investigations: (1) addition to citrus products during processing of external exogenous enzymes usually from microorganisms, and (2) potentiation of the enzymes of the citrus fruit's own limonoid-metabolizing systems. Far more work has gone into the former than the latter approach. Early favorable results using very crude fungal preparations putatively containing the requisite enzymes were due to the action of pectin-degrading enzymes. The result of this action was to drag out of solution desirable cloud (Chapter 30) as well as limonin.

A more rational approach was instituted by the Pasadena group by growing specific microorganisms in enrichment culture with limonin as a sole source of carbon and then looking for the first enzyme on the pathway to its utilization. Several organisms and corresponding enzymes have been unearthed. The latest enzyme, from a *Pseudomonas* strain, is an NADP-dependent dehydrogenase which forms dehydrolimonoic acid A-ring lactone from the corresponding limonoate (Fig. 25.3). This dehydrogenase acts more efficiently than that previously obtained from other organisms, by virtue of being more stable at the low pH (3.5–4.0) of orange juice (Brewster *et al.* 1976). Using sufficiently high levels of partially purified preparations, they were able to reduce the limonin content of navel orange juice from 21 to 3 ppm. However, the pH optimum is still not that of orange juice. Perhaps even more promising, if not altogether clear, is the report of Vaks and Lifshitz (1975) of an enzyme from *Acinetobacter* sp. of soil bacteria which catalyzes in a yet undetermined fashion the transformation of limonin (and not limonoate) at the pH of orange juice, but acts optimally at pH 6. Apparently unlike the Pasadena group's dehydrogenase, NAD is the cosubstrate.

Potentiation.—The second approach, that of using the orange's own enzymes, stems from the previously mentioned observation of Flavian and Levi (1970) and of Chandler (1971) that citrus peel possesses an enzyme system capable of degrading limonoate and from the detection of limonoate dehydrogenase activity in navel orange albedo. Chandler and Nicol conclude from experiments with this enzyme that its use has several advantages: it retained 50% of its activity at pH 4, possible activation, pectin in added albedo ensures or improves cloud stability (Chapter 30), and there are no food regulatory hassles. Disadvantages include the occurrence of high activity over a narrow period of maturity and the expense incurred because of the further purification probably required.

An alternative to isolating enzymes from citrus via a "flavorese" approach (Chapter 23) is to potentiate the requisite enzyme system *in vivo*. Earlier studies such as that of Rockland and Beavens (1957) have been reexamined and modified by the Pasadena group (Maier *et al.* 1973). They were able to halve the limonin content of navel oranges in 3 hr by exposing them to 20 ppm of ethylene followed by 5 days storage at ambience. In actual practice, oranges can be treated while being trucked to the processing plant. Chandler and Nicol point out that a major problem is the development of off-flavors in oranges exposed to conditions even slightly outside the narrow range of effective conditions. Thus, as of this writing, the full development of a practical process of limonin removal based on enzymes still has some way to go.

Flavonones

Most of the flavonone-induced bitterness in citrus fruits is due to naringin, the 7-β-neohesperidoside of naringenin. Neohesperidose is L-rhamnosyl-α-1,2-glucose and naringenin is 5,7,4′-trihydroxyflavonone (Fig. 25.3). The aglycone is synthesized via a pathway not too unlike that of fatty acids; successive additions of malonyl CoA to a C_6-C_3 cinnamic acid precursor CoA. The enzyme complex, flavonone synthase, was purified from parsley and from plant cell suspension cultures (Kreuzaler and Hahlbrock 1975; Saleh *et al.* 1978).

$$\underset{p\text{-Coumaryl CoA}}{p\text{OH-}\phi\text{-CH=CHCO-SCoA}} + \underset{\text{Malonyl CoA}}{3\text{HOOC-CH}_2\text{-COSCoA}} \rightarrow \text{Naringenin} + 4\text{CoASH} + 3\text{CO}_2$$

In addition to belonging to a different class of organic compounds from that of limonoids, the flavonones differ from the latter in being quite soluble, in being found in the juice sacs as well as in albedo, and in being required at low levels as an essential contributor to the taste of grapefruit juice. Furthermore, because of its greater solubility, its bitterness is instantly perceived and then disappears due to dilution. Thus, a 1mM solution of naringin is about as bitter as a 0.01mM solution of limonin. The desirable modicum of bitterness imparted by naringin has health-associated overtones in that, by inducing alleged enhancement of taste perception, food eaten immediately afterwards needs less sugar—hence the popularity of grapefruit in slimming diets. Another difference from limonin is that the intense bitterness is related not only to the concentration but to the type of flavonone as well.

Prevention.—Like limonin bitterness, that of the flavonones can be readily eliminated by complete and rapid centrifugation. The enzyme approach was initiated by Japanese investigators in the early 1950s. Almost all such proposals are concerned with removal of the sugar from the aglycones. The crude preparation, now available from some biochemical supplyhouses in the United States and Europe (Versteeg *et al.* 1977), is called "naringinase." These partially purified fungal culture fluids contain two distinct glycosidases, β-L-1-rhamnosidase which hydrolyzes off the terminal rhamnose of naringin to yield prunin (Fig. 25.3), and a β-glucosidase which converts the mildly bitter prunin into naringenin aglycone and glucose. Such commercial preparations may be richer in one or the other enzyme activity, but they do usually succeed in debittering orange juice as well as removing the related narirutin, according to Versteeg *et al.* (1977).

Such enzyme preparations should be devoid of pectin-degrading enzymes because, as with limonin, these enzymes carry down debris resulting from disruption of the unstable hydrocolloid system constituting the "cloud." A good example of partial purification of commercial food enzymes to remove unwanted enzymes (Chapter 6) is that the removal of pectinase involves either alcohol fractionation or heat inactivation.

Naringinase has to be inactivated after treatment in order to avoid complete disappearance of the desired trace bitterness. For instance, a 10 min, 80°C in-can treatment of grapefruit segments to allow naringinase but not pectinase to act, recommended by Japanese investigators, would have to be followed by a more intense and not particularly otherwise desirable heat treatment to prevent complete disappearance of the naringin.

Immobilization of Naringinase.—Hence the search at least for juices to find an immobilized enzyme system which would obviate the necessity of heating the citrus products beyond pasteurization. First accomplished by Goldstein *et al.* (1971), several patents based upon the immobilized-enzyme principle using a variety of supports have been issued. However, in the considered opinion of Chandler and Nicol, the matrices themselves probably adsorb the bitter substances and they are altogether too expensive. Parenthetically, naringinase-like preparations (hesperidinase) have been proposed for hydrolyzing hesperidin as a step in the preparation of dihydrochalcone sweeteners (Horowitz and Gentili 1963; Eisenstadt 1978). Since 1965, interest in the field lagged until recently as witnessed by the above-mentioned paper of Versteeg *et al.* and especially by the successful use of immobilized naringinase in hollow fiber arrangements discussed in Chapter 8 (Olson *et al.* 1979). Some results are shown in Fig. 25.5. Using such setups obviates the necessity for heating, precludes adsorption on a matrix and allows complete control of the degree of hydrolysis. As we have seen, this is important in permitting some bitterness to remain in grapefruit products. The investigators report preference for their product by a trained panel over untreated controls.

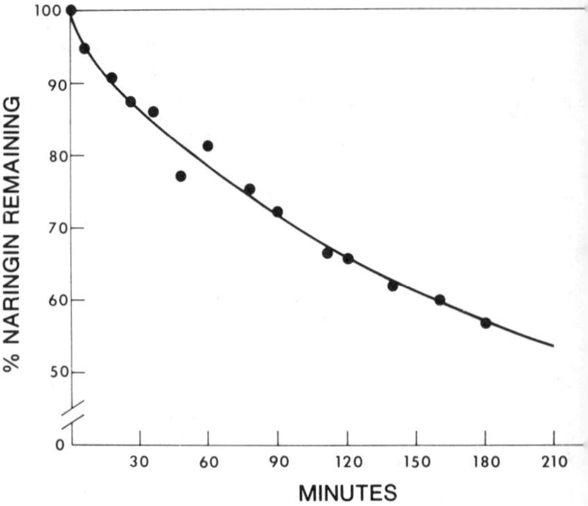

Courtesy of Olson et al. (1979)

FIG. 25.5. REMOVAL OF EXCESS BITTERNESS FROM GRAPEFRUIT JUICE

By hydrolysis of naringin with naringinase using a hollow fiber immobilized enzyme system.

Looking to the future, the results of recent studies of flavonoid biotransformations might be profitably exploited. One may be to look into specific inhibitors of flavonone synthase such as the antibiotic cerulenin (Kreuzaler and Hahlbrock 1975). Another approach is to look for alternatives to glycosidase action. For instance naringenin (and perhaps naringin) is readily oxidized to the corresponding flavone apigenin (5,7,4′-trihydroxyflavone) by an enzyme in parsley, also a rich source of apigenin. Specifically, a double bond is introduced between positions 2 and 3 of the nonphenolic ring of naringenin.

The bitter principle in squashes, cucurbitacin, can be removed by the action of cucurbitacin 19-hydroxylase present in the unripe fruit of this vegetable (Schabort 1978).

OTHER ENZYMATIC ASPECTS OF OFF-FLAVORS

Genesis of Off-flavors

Hydrogen Sulfide.—Normally a volatile component of desirable aromas at appropriately low levels and in combination with other volatiles (Schwimmer and Friedman 1972), formation of H_2S as a primary product of enzyme action is the exception rather than the rule. Even its production as result of alliinase or glucosinolase action is the result of secondary reactions fairly far removed from the enzymatic event (Chapter 21). Far more frequently H_2S and other volatile sulfides are formed as the result of cooking the food, *in vivo* from the action of enzymes of microorganisms in the foods, or less frequently decomposition of unstable sulfur compounds such as the ferredoxins (Chapter 23). It is only recently that such distinctions and the precursors to these volatiles are being firmly established on an experimental basis and for individual foods. We cited a few instances such as the development of off-flavors in coleslaw (Chapter 1) in which what appeared to be a microbially induced off-flavor turned out to be due to cabbage enzymes. Table 25.4 lists foods in which off-flavor due to excess volatile sulfides may occur, along with the precursor and mode of production of the off-flavor. It will be noted that the foods' own enzymes, even when involved, serve as suppliers of the substrates for microbial enzymes or as reagents for pyrolytic reactions leading to the formation of off-flavor-bearing compounds.

Amines.—On the other hand, both bacterial and fish enzymes contribute to the formation of the unpleasantly odoriferous methyl amines. Older investigations established that trimethylamine arises in stored raw fish as the result of the reduction of trimethylamine oxide (TMAO) by an NADH-dependent reductase (Chapter 28).

More recently, as mentioned in Chapter 23, lipoxygenase-like generation of lipohydroperoxides can reverse this effect via reoxidation of this

TABLE 25.4

GENESIS OF OFF-ODOR-BEARING SULFUR COMPOUNDS

Compounds	Food	Precursors	Causes
Dimethyl sulfide	Cabbage; milk, overcooked	Methyl methionione sulfonium	Thermal
	Salmon	Dimethyl propiothetin	Contaminant enzyme
Mercaptans	Cheeses, unripe	Cysteine	Microbial enzymes
	Codfish	Sulfur amino acids	Fish proteases + contaminant enzymes
H_2S, SO_2	Eggs	Protein	Thermal + contaminant enzymes
	Beer, wine		Yeast enzymes

amine, in a reaction reminiscent but the obverse of the re-reduction of phenolase-produced quinone by appropriate reductants (Chapter 16). Similar off-flavors in cold-stored sausage meat have been attributed to uncontrolled activity of both microflora and meat enzymes (Halliday 1972). Presence of other primary amines in foods, which in excess may contribute to off-flavor but perhaps create more of a health-related problem (Chapter 35), arises as a result of the action of microbial amino acid decarboxylases.

Phenolic-derived and Other Off-flavors.—We have explored the role of phenolase action and other phenolic transformations in the creation of color changes, both desirable and undesirable (Chapters 15, 16, 17 and 22). Phenolase and other metallo-oxidoreductases have also been implicated in the production of off-flavors. Thus their action has been implicated in the development of the tobacco-like flavor developed in some prunes during prolonged storage (Chari *et al.* 1948); the bitter flavor of cooked eggplant (Flick *et al.* 1977); off-flavors in orange juice (Bruemmer *et al.* 1976); the harsh, lingering aftertaste and subsequent stale flavor imparted to some beers by the highly oxidized polyphenols (Engan 1975; Gheluwe and Valyi 1974)—off-flavors which can be prevented by avoiding air accumulation at the end of the mashing step; and finally the accumulation of bitter substances, perhaps similar oxidized polyphenols, accompanying the melanins in phenolase-induced discolored brown areas of injured tissue of fruits and vegetables. Part of the bitter taste may also be due to scopoletin (Fig. 15.2), esculetin and coumarins formed via action of phenolases and methyl transferases present in the food and invading microorganisms. Related may be the accumulation of the substance responsible for the bitterness in carrots, 3-methyl-6-methoxy-3,4-dihydroxyisocoumarin.

We referred to the development of off-flavor in refrigerated coleslaw in Chapter 1. A discussion of the role of phenolics in taste perception would not be complete without mentioning that some of them, especially the tannins, are causes of astringency in many plant foods and derived products.

The bitter taste of yeast is due to thiamin, otherwise a most desirable food constituent as an essential nutrient with vitamin B_1 activity. Its accumulation in yeast has created a challenge to those who see yeast as a major source of protein as SCP (Chapter 34).

Prevention and Removal

These and other examples of off-flavor development which may occur to the reader may have enzymatic antecedents and the problems they create may have enzymatic answers. For instance, we suggest that excess thiamin in yeast might be efficaciously removed by the judicious controlled action of the heretofore totally unwanted thiaminase, discussed in Chapter 35. Like many of the other undesirable changes in food quality, some of the above-mentioned flavor defects can also be prevented by the applications of the principles outlined in Chapter 1 and applied to specific instances in the ensuing chapters. Among these measures are the treatment of the food with more or less specific enzyme preparations to remove the off-flavor or to prevent its formation. A collection of proposals for such exploitation of enzyme action is shown in Table 24.2. Further discussion of flavor-related problems closely associated with other quality attributes will be alluded to in other chapters, such as trimethyl amine development in fish (Chapter 28).

Part VII

ENZYME ACTION AND THE TEXTURAL QUALITY OF FOODS

26

ENZYME ACTION AND FOOD TEXTURE: MUSCLE TO MEAT

THE TEXTURE OF TEXTURE

Authorities have presented the following definitions of food texture: ". . . those perceptions which contribute to the evaluation of a food's physical characteristics by the skin or muscle senses of the buccal cavity, excepting the senses of temperature or pain" (Matz 1962); ". . . composite structural element of foods and the manner in which it registers with the physiological senses" (Szczesniak 1963); ". . . that one of the primary sensory properties of foods which relates . . . to the sense of touch or feel and . . . potentially capable of precise measurement objectively by mechanical devices in units of mass and force" (Kramer 1972).

A somewhat less formal description of food texture which may impart to the reader a feeling of food texture consists of listing texture-associated adjectives which have appeared in the technical literature (Table 26.1). Some of these terms such as "soft" or "firm" are of broad connotation whereas others such as "soupy" are more restricted. Many of these adjectives are also used in daily conversation. As such they have been collected and classified in conjunction with controlled psychological association tests (Szczesniak 1963; Yoshikawa *et al.* 1970) and as elements of consumer texture profile analysis of foods (Szczesniak *et al.* 1975). Of course, food texture does not exist "in a vacuum" so to speak, nor is it confined to reactions of the tongue, but interacts and is closely associated with perception by the eye in an interplay with color and especially appearance, as pointed out in Chapter 15.

The increasing interest in food texture is attested to by the continued issuance of the *Journal of Textural Studies* started in 1969. A convenient point of departure on the basic philosophy and experimental approaches to modern studies on food texture is afforded by the translation from Japanese of the classic by Sone (1972) and by the International Food Technologists symposium on Texture Measurement published in *Food Technology* in 1972. The very definition of food texture as a significant parameter of food quality constitutes a subset of problems to which several investigators have addressed themselves. These problems were reviewed by Kramer (1972), who concluded that the accuracy or predictability via maximization of the calibratability of an objective (and therefore indirect) method is predicated only by its correlation with a subjective (direct) method involving sensory panel response. Abbot (1972) in reviewing the sensory assessment of food texture points out that the highly specific approach required for each specific purpose, i.e., preference/acceptance, discrimination and descriptive test methods, must be assessed for each food tested. Szczesniak (1972) discussed applications, advantages and short-

TABLE 26.1

ADJECTIVES ASSOCIATED WITH TEXTURAL QUALITY OF FOOD[1]

Aerated	Gooey, grainy	Sandy
Adhesive	Granular	Silky
	Greasy	Slimy
Cellular	Gritty, gummy	Slippery
Chewy		Sloppy
Coarse	Hard	Sloughy
Cohesive		Smooth
Cookable	Impalpable	Soapy
Crackly		Softy
Creamy	Leathery	Solid
Crisp	Light	Slushy
Crunchy	Limp	Soggy
Crumbly	Loose	Soupy
Crystalline	Lumpy	Springy
		Sticky
Dense	Mealy	Stiff
Doughy	Moist	Stretchy
Dry	Mucilaginous	Stringy
	Mushy	Succulent
Elastic		Swollen
	Oily	
Fibroid		Tacky
Firm	Palpable	Tender
Flabby	Pasty	Thin, thick
Flaccid	Plastic	Tough
Flexible	Pliable	Turgid
Flaky	Plump	
Full	Powdery	Unctuous
Fracturable		
Friable	Rigid	Viscous
Full	Rigor-like	
	Rough	Waxy
Gelatinous	Rubbery	Watery
		Wooly

[1] As described in the food science and technology literature.

comings of diverse devices, each of which measures only a portion of the spectrum of physical properties which contribute to food texture. In general, texture measuring devices comprise penetrometers, compressimeters, shearing devices, cutting devices and masticometers. For liquid foods consistometers and viscometers are most commonly used. Finney (1972) in reviewing elementary concepts of rheology relevant to food texture studies covers the physical state of the food, force and deformation, stress and strain, elasticity, plasticity and viscoelasticity. Table 26.2 shows values of the modulus of elasticity of some food products. These are arranged in order of increasing firmness as evaluated subjectively.

Sherman (1972) points out that the textural properties of foods are greatly influenced by internal structure of the food and that in the case of natural foods such as meat, fruit and vegetables, an understanding of food texture must await a greater knowledge of the complex of natural factors which induce changes in the microstructure. Among these factors are the action of enzymes on the components of this microstructure. Physiological and psychophysiological (kinaesthesic) aspects of food texture are adumbrated by Yeatman (1972) and Morrow (1972). Among the themes covered by the latter authors is the development of measurement of textural qualities of textured and intermediate moisture foods. These measurements should be made amenable to fabrication of any combination of desired textural qualities. The application of enzymes may contribute to the attainment of such texture, especially in the field of soybean protein technology (Chapter 34). For each food there is a range of texture which is acceptable to the consumer.

In the following chapters we shall define the role of enzymes in either achieving the desired texture or in causing the texture to be altered beyond the limits of this acceptability. The enzymes may be present in the food as natural constituents or they may be added in the form of commercial preparations. Chapters 26 through 29 cover solid foods which more or less retain their original form. Chapters 30 through 33 present aspects of the changes of transformed foods which during processing have lost their original appearance, form or identity and may have undergone a change in state, i.e., bread, beverages, cheese, juices, purees, etc.

Since texture involves structural elements of the food, the enzymes involved in food texture are primarily those which act on biological polymers such as proteins in meats and polysaccharides in fruits and vegetables. However, small molecules and the enzymes associated with their transformation play ancillary regulatory and control roles. In some instances, such as the action of pectin esterase, an enzyme can cause a desirable increase in firmness. On the other hand, the action of lipases in cold-stored fish may result indirectly in an undesirable increase of textural toughness.

TABLE 26.2

VALUES FOR THE APPARENT MODULUS OF ELASTICITY OF SOME FOODS[1]

Food	Young Modulus[2]	Food	Young Modulus[2]
Bread	1–3	Apples	600–1400
Gelatin	20	Potatoes	600–1400
Bananas	80–100	Pears	1200–3000
Peaches	200–2000	Carrots	2000–4000

Source: Adapted from Finney (1972).
[1] Raw fruits and vegetables.
[2] $\times 10^5$ dynes/cm^2.

Meat Texture

Of the myriad ways in which enzyme action can change the texture of food, probably the most thoroughly researched and most extensively applied are those involved as determinants of meat texture. In addition to the many reviews and books on the subject the following are of special relevance and serve as keys to other reviews and meat science literature: Briskey *et al.* (1966,1967), Price and Schwiegert (1971), Forrest *et al.* (1975) and a symposium on the basis of tenderness in muscle foods introduced by Marsh (1977). The strictly biochemical aspects of muscle biology devoid of food implications are usually explored in standard biochemistry textbooks such as that of Lehninger (1975). Three articles in *Scientific American*, spanning ten years, furnish a lucid, authoritative, yet not too technical panorama of progress and insight into the nature and mechanisms of muscle functions (Huxley 1965; Murray and Weber 1974; Cohen 1975).

By and large most of the research on the tenderness of meat has been devoted to the skeletal muscle tissue of beef, pork, poultry and, to a lesser extent, lamb and mutton, as contrasted to organ tissue such as heart and lung, pancreas, spleen, kidney, liver and brain, each of which have been utilized as food. The introduction to a substantial proportion of papers relating to meat texture serves as testimonial to the primacy of tenderness as the determining factor in the assessment of meat quality and acceptability. Thus we quote Dikeman *et al.* (1971): "Tenderness is the most desirable eating characteristic of meat. . . ." Joseph (1970) states that "texture is above all the quality of beef desired by the consumer." Some of the parameters which have been investigated as determinants of texture quality are shown in Table 26.3.

TABLE 26.3

FACTORS AFFECTING SKELETAL MUSCLE TEXTURE

Nutrition, diet	Enzyme injection
Exercise (Pork)	Positioning of carcass
Age, breed	Bleeding
Struggling	Muscle restraint
Tension	Temperature, freezing
Shivering	Ionizing radiation (radurization)
Emotional stress	Feather plucking (poultry)
Tranquilizers	Cooking method
Hormone injection	Pressure, massaging
Other drugs (curare, etc.)	Koshering

MUSCLE—COMPOSITION, STRUCTURE AND FUNCTION

Proteins

The normal function of skeletal muscle is the transduction of chemical energy to mechanical energy. Both muscle function and meat texture are largely determined by the protein moieties of muscle tissue which can be conveniently categorized according to their three sources: (1) the sarcosomal proteins which are mostly enzymes present in the cytoplasm and nucleus, (2) the scleroproteins of the connective tissues most of which are extracellular and consist mainly of collagen and elastin and to some extent resilin, and (3) the proteins of the myofibril, which is the site of the contractile apparatus. The last is most important both quantitatively and functionally for it is that fraction by which the muscle cell attains its specialized function. Although each of these protein classes has been studied with respect to contribution to meat texture, the myofibrillar and connective tissue proteins are the principal and independent contributors to texture (Jones 1977; Marsh 1977). The collagen of connective tissue tends to gelatinize during cooking and is probably the principal substrate for both added and thermally potentiated endogenous enzymes. The scleroproteins contribute to texture by virtue of age-dependent crosslinks which render them increasingly resistant to gelatinization and, as we shall see, to the action of proteinases (Friedman 1977) and, to a lesser extent, by their level. Their contribution to texture is fairly well set before slaughter and processing, and under most circumstances takes precedence to that of the myofibrillar proteins.

Nevertheless the latter have received more attention as determinants of meat texture because they are the substrates for enzymes acting in the cold during normal aging of beef and are responsible for most the textural aberrations which develop postmortem (Marsh 1977). These proteins are also involved in texture problems encountered with frozen and dehydrated meats, problems which have only recently received serious attention (Yamamoto *et al.* 1977; Nesterov *et al.* 1977).

Protein Interactions.—Based on what happens to the proteins of muscle postmortem, we may discern four reaction modes: (1) protein-protein interactions; (2) small molecule-protein interactions, i.e., substrates, inhibitors, activators, hormones and solvents, including changes in solubility; (3) protein conformational reactions brought about by (1) and

(2)—the most extreme form of which is denaturation; and (4) changes in covalent bonds in proteins, including disulfide and peptide bonds. As will become evident, all of these interactions are considered to play a role in the muscle-to-meat conversion process.

From the viewpoint of the enzymes involved, with special reference to determinants of texture, one can divide the process of the conversion of skeletal muscle to meat into two phases: (1) that leading up to and including prerigor mortis, involving the ATPase activity of the myofibrillar proteins, the sarcoplasmic enzymes of the glycolytic sequence and enzymes of the sarcoplasmic reticulum; and (2) the postrigor aging process which may or may not involve the hydrolytic enzymes of the lysosomes. The textural aspects of these two phases are linked by the general rule that the tougher the muscle tissue is at rigor, the more likely the meat will be tough at the point of consumption or utilization (Smith *et al.* 1971). It should also be pointed out that poultry muscle goes through the conversion of muscle to meat in a much shorter period than most mammalian muscle, about 1 day or less as compared to other meats which take more than 1 week (de Fremery 1966). A synopsis of the progression of postmortem events is shown in Fig. 26.1.

The Myofibrillar Proteins and the Contractile Apparatus.—The myofibrillar fraction is present as long filiary entities extending the entire length of the muscle cell each of which is enormously extended fiber which traverses the length of the muscle as a myofibril. The light, the transmission and, more recently, the scanning electron microscopes reveal both horizontal and longitudinal fibers. The most prominent of the longitudinal fibers—and perhaps the most pertinent to a discussion of the role of enzymes in meat texture—is the Z-line (Z-disc, Fig. 26.2 and 26.3). The part of the muscle cell between two Z-lines is the sarcomere *ca* 1 μ in length (Fig. 26.2). The horizontal striations, Fig. 26.3, delineate the myofilaments.

On a supramolecular level, as sketched in Fig. 26.3, the myofilaments consist of interdigitating filaments, the stationary thick filaments, of the protein myosin (*ca* 25%), and the thin filaments containing actin (>50%) as well as the actinins, of which the Z-disc is made and to which adheres the elongated double helical coiled coil tropomyosin (10%) which is considered to be nesting in one of the grooves made by the aligned globular actin molecules and spanning seven of the latter; and troponin occurring at 400Å intervals along the thin filaments about one-third from the end of a tropomyosin molecule. The troponin molecule com-

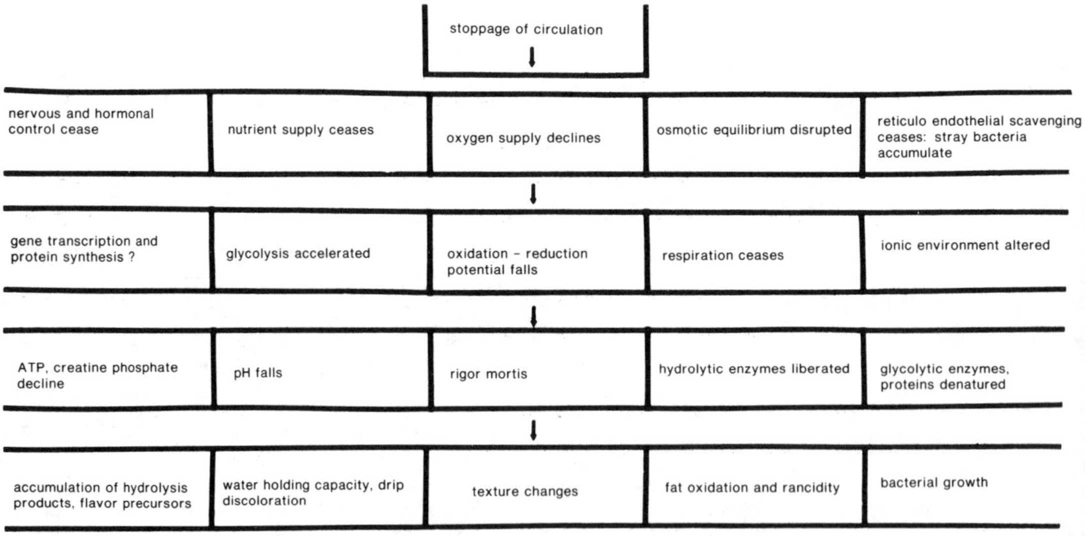

From Haard (1971). Adapted from Lawrie (1966).

FIG. 26.1. CONVERSION OF MUSCLE TO MEAT

Progression of events in muscle tissue leading to development of muscle tissue as a food.

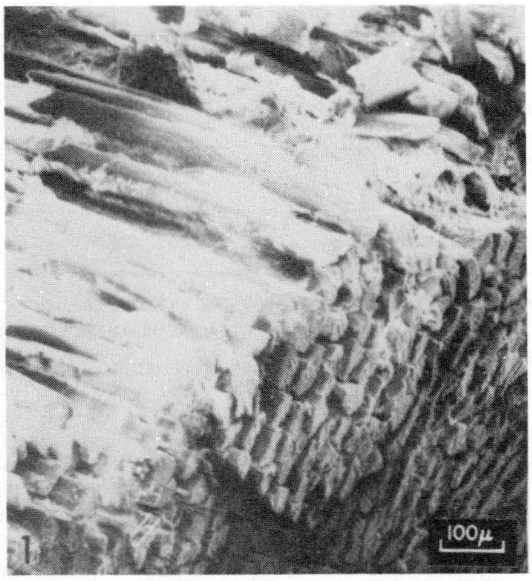

A Muscle fibers

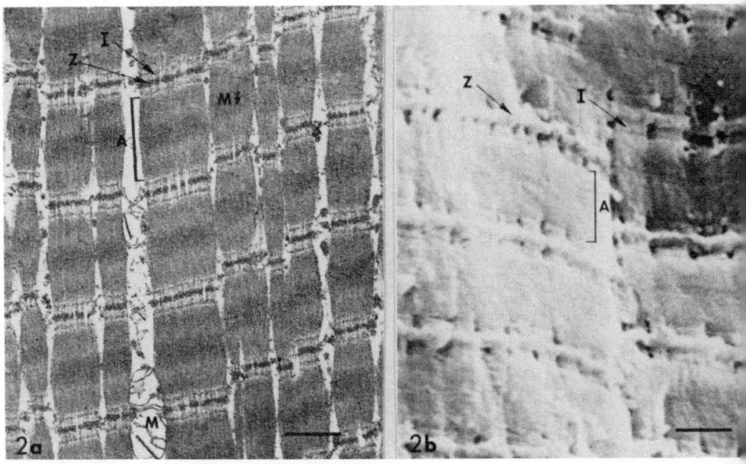

B Aged myofibrils (transmission electron microscope)

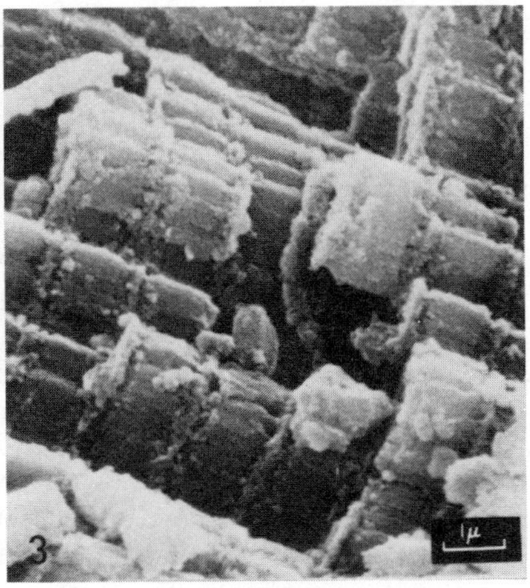

C After heating at 90°C for 45 min

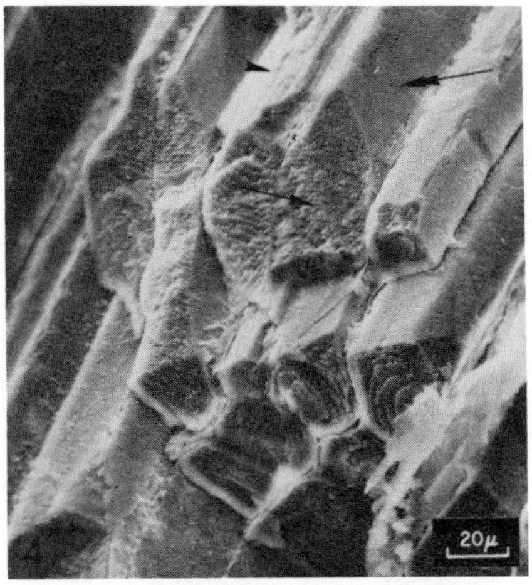

D As in C with surface exposed by freeze etching

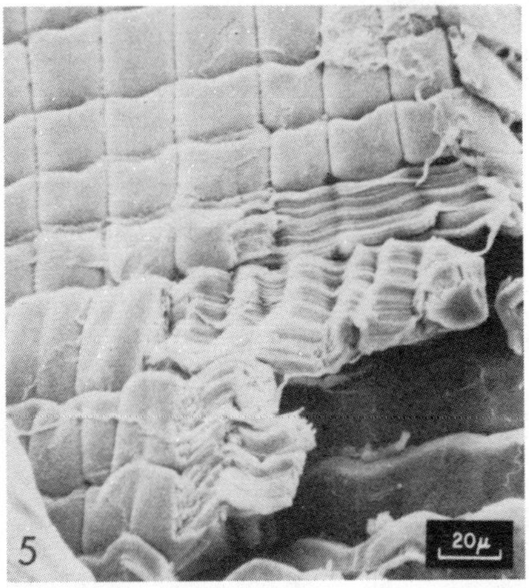

E Cold shortened fibers

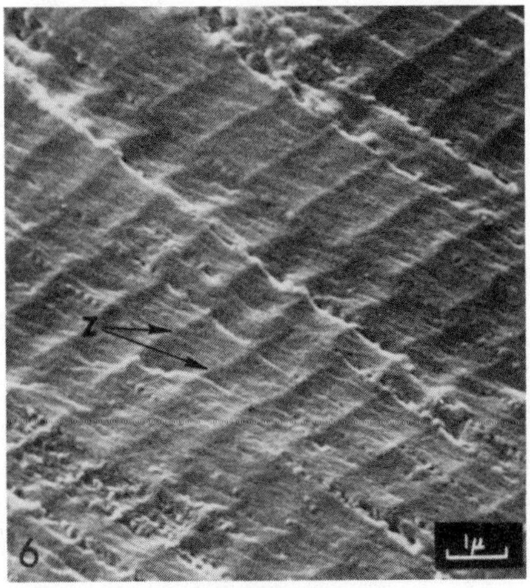

F Cold shortened myofibrils

FIG. 26.2. MUSCLE / MEAT STRUCTURE AS REVEALED BY THE SCANNING ELECTRON MICROSCOPE

Courtesy of Jones (1977)

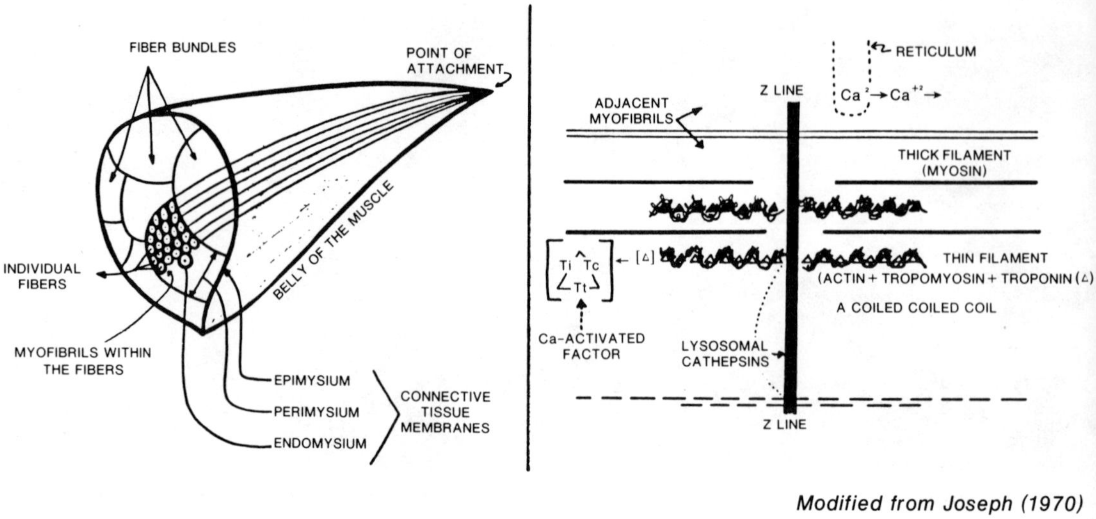

FIG. 26.3. ELEMENTS OF MUSCLE STRUCTURE

Showing possible points of attack of tenderizing enzymes during aging (right); morphology (cross section) of a typical muscle (left).

prises three distinct subunits, each with a specific function: (1) Tn-C (also T-C, Tc or TN-C) contains the Ca-binding site (by "Ca" in this chapter and elsewhere we refer to the calcium ion, Ca^{2+}, Ca^{++}, Ca (II), or to calcium salts and not to the free metal); (2) Tn-T (Tt) contains the tropomyosin binding site—the locus of the latter has been, so far, if not pinpointed, at least narrowed down with the aid of affinity chromatography (Pearlstone and Smillie 1977); and (3) Tn-I (also Ti) contains an actin-binding site that regulates and prevents the binding of myosin to actin and hence the thick to the thin filament, but only in the absence of Ca.

The myosin molecule in the thick filaments combines both globular enzyme and fibrous structural protein in one functional, covalently linked unit. The rod-like portion of the molecule (tail or light meromyosin) is responsible for the formation of filaments at low ionic strength while the globular region (head, heavy meromyosin) interacts with actin via the swiveling movement of a hinge between tail and head. The latter is the locus of ATPase activity.

In normal resting, relaxed muscle the interdigitating myosin and actin are dissociated due to the presence of ATP at low levels of Ca and can slide passively past each other. The texture of living muscle tissue is thereby made soft and pliable and highly extensible. This is known as the plasticing effect of ATP. Thus, ATP is acting as an allosteric effector by holding the myosin in a conformation such that the latter, in conjunction with the just-mentioned Tn-T positioning on actin, cannot couple to the actin. The ATP associated with the myosin may be present as ADP and Pi tightly bound to an energized conformation mode of myosin which cannot achieve union because its attachment site on the actin is blocked. In this state as envisioned by some investigators, at least as of this writing (the complete story is not yet in), troponin does not cover up the myosin binding site on the actin by actually competing with myosin for this site. Rather, a steric hindrance effected by the interposition of tropomyosin may be visualized in the following stylized version of the fate of one set of the major thin filament molecules which are involved in the contraction or in its regulation. It also depicts our own impression of how these components may contribute to contraction.

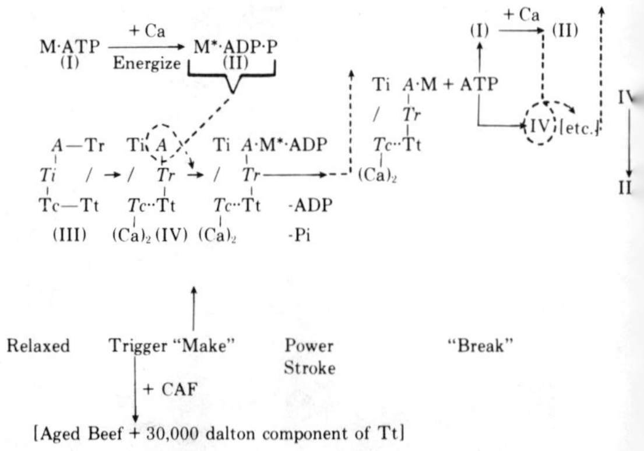

where M is myosin; M*, energized myosin; *A*, actin; Ti, Tc and Tt, the three subunits of troponin; Tr, tropomyosin; and CAF the calcium-activated protein. The symbols |, —, ··, · and / indicate different modes of linkage between components. Contraction is denoted by the vertical dashed arrows and by the upward shift of the protein-complex IV.

The Twitch Switch.—In living muscle the trigger which sets a muscle into contraction is a neuronally-derived electrical impulse arriving at a muscle membrane, thus depolarizing it. This depolarization leads to increased ion permeability, thus allowing the level of Ca in the vicinity of the myofilaments within the sarcoplasm to reach a concentration ($5 \times 10^{-6}M$) sufficient to activate the ATPase function of the actomyosin complex. The initial triggering action of Ca appears to be associated with the Tn-C subunit of troponin, Tc, the twitch switch. Release of Ca sets into motion a chain of successive events which amplify these conformations into movements and whose ultimate purpose is to create mechanical energy and movement on a macroscopic visible-to-the-eye rather than only a macromolecular scale. Two molecules of Ca which arrive from the sarcoplasmic reticulum membrane combine with one molecule of the Tn-C subunit of troponin to induce a conformational change in this subunit. This induces conformational change in the Tn-I (Ti) subunit such that the troponin molecule is no longer in direct contact with the actin. This in turn creates a space into which the tropomyosin (Tr) coils "roll" so that their position in the filament groove is shifted thus bearing the myosin-binding site on actin. Adelstein and Eisenberg (1980) propose an allosteric alternative to this "all-or-none" scenario.

While all this is going on in the thin filaments, Ca also starts a parallel train of events in the thick filaments by combining with a section of the myosin tail which in turn induces conformational changes in the head leading eventually to binding of the head to the actin to form the actomyosin complex. Ca also activates the ATPase function of the resulting actomyosin complex. It will be recalled (Chapters 4, 16 and 19) that the energy put into ATP is also ultimately associated with ATPase. It is the energy released in this reaction, still not fully understood, that causes the head to shift angles (swivel) and the actin (and along with it the whole thin filament) to shift in space. This swivel breaks the contact between actin and myosin and the cycle starts anew. Ultimately, this shift is, of course, amplified along the whole filament and is manifested as a twitch, a muscular contraction.

Incidentally, not only the function of the Ca-actuated switching subunits of troponin and myosin but also their amino acids sequence (homology) are quite similar (Wu and Yang 1975). At least two other Ca-binding proteins, the fish parvalbumins (Collins 1976) (see Chapter 28) and the protein activator of cyclic AMP phosphodiesterase (Drabikowski *et al.* 1977), possess structured similarities to Tn-C), as does the ubiquitous and versatile, yet highly specific, intracellular calcium receptor, calmodulin (Means and Dedman 1980).

Denouement.—After contraction, Ca is pumped away from the vicinity of the myofibril into the sarcoplasmic reticulum, a membranous structure pervading and encasing the muscle cell. It is the site for protein synthesis and (more to the point of the present discussion) its cisternae are also the site of the "relaxing factor." The latter, also called the calcium transport ATPase, catalyzes the following reaction (Marsh 1966; Inesi and Scarpa 1972; MacLennan and Holland 1976):

$$\text{ATP} + \text{Ca (out)} \xrightarrow{\text{Relaxing Factor}} \text{ADP} + \text{Pi} + \text{Ca (in)}$$

where "out" and "in" refer to the location of Ca with respect to the cisternae of the sarcoplasmic reticulum which also contains other Ca-binding proteins including the aptly named calsequestrin. This ATPase is regulated by two enzymes of the cyclic AMP cascade, protein kinase and phosphodiesterase (Hörl *et al.* 1978). It would appear that more and more enzymes are being found in cisternae of the subcellular sites associated with protein synthesis, i.e., glucosinolase, LOOH lyase (Chapter 21). As we shall soon see, the integrity of this entity is closely linked to postmortem textural changes.

ATP Replenishment and Glycolysis

The ATP reservoir in the live, functioning muscle is replenished instantaneously by a phosphogen, phosphocreatine, present in concentrations about four to five times that of the average ATP level via the action of the appropriate kinase:

$$\text{ADP} + \text{Phosphocreatine} \longrightarrow \text{Creatine} + \text{ATP}$$

The source of this energy required for the replenishment of ATP comes from the chemical energy of the food through the operation of the glycolytic and the respiratory cycles (mitochondrial respiration). For the present discussion, we shall focus on the role of the glycolytic system because it appears to play more of a central role in texture in the initial stages of the conversion of skeletal muscle to meat. Furthermore, as Lardy (1966) has pointed out, respiratory systems are not as amenable, if at all, to controlled management of tenderness as is glycolysis. This is because we now have much better understanding of the regulation of the glycolytic cycle enzymes in general, the fate of ATP in particular and because the muscle-to-meat conversion occurs in the absence of respiration. However, this does not mean that the apparatus for respiratory ATP production, the mitochondrion, should be neglected since it appears that at least in avian muscle the glycolytic system is localized within the mitochondrion (Southard and Hultin 1969); also, lactic dehydrogenase, a glycolytic enzyme may, under some circumstances, be bound to myofibrils. Hultin *et al.* (1972) pointed out that this enzyme would become solubilized postmortem and thus provide a mechanism for maintaining glycolysis after death in muscle tissue under anaerobic conditions.

The rate of operation of glycolysis is under control of intermediates and enzymic cofactors as participants, substrates, end-product inhibitors and as allosteric effectors and by hormonal controls as discussed further in connection with texture aberrations below and in Chapter 4.

NONPROTEOLYTIC POSTMORTEM EVENTS

The Role of Glycolysis in the Muscle-to-Meat Conversion

Living tissue glycolysis produces pyruvic acid which enters the tricarboxylic acid respiratory cycle or in the case of heavy exercise (oxygen deficit) is converted into lactic acid which can then be reconverted into glycogen. But upon death the contractile apparatus and associated enzymes continue to function, albeit aberrantly. Accessibility to aerobic pathways is dramatically reduced or nonexistent—and lactic acid accumulates. Useful metabolic energy is no longer generated via respiration and the remaining glycogen glycolizes with the resultant accumulation of lactic acid. Inhibition of glycolysis by oxygen (Pasteur effect) is minimized. Furthermore, no or very little ATP is formed. What ATP is formed as the result of glycolysis (two molecules per six carbons as compared with 30 molecules formed with each turn of the respiratory cycle) is destroyed via myosin myofibrillar ATPase and perhaps also by the so-called apyrases of the sarcoplasm. Furthermore, the lowered ATP concentration probably also diminishes the operation of the relaxing factor so that the accumulation of Ca may also contribute to the loss of ATP.

Rigor.—In the absence of ATP myofilaments contract, as just discussed. This results, especially at the somewhat acid pH due to the accumulation of lactic acid resulting from glycolysis, in a stiffening of the muscle known as rigor mortis. Locker (1960) was one of the first investigators to put the connection between muscle contraction state and meat tenderness on a firm experimental basis.

Rigor mortis is measured objectively by a variety of parameters such as loss in extensibility, elasticity, compaction and shortening (Szczesniak 1972). The difference between shortening (directly related to ATP) and extensibility in molecular terms has been discussed by Goll *et al.* (1970). The following further changes usually accompany rigor and are frequently used as indices or predictors of muscle tenderness in the early phases of the muscle-meat conversion: sarcomere shortening solubilities of the myofibril proteins in carefully defined solvents, myofibril fragility and appearance of certain protein fragments (see below), water holding capacity of the muscle and a lowering of the pH from about 7 to as low as 5.5. The fact that this pH is also the isoelectric point of the actomyosin complex is probably related to the aforesaid solubility and water holding changes.

While the main body of our present discussion is concerned with enzymic factors influencing tenderness, nonenzymic factors (Table 26.3) can influence tenderness in terms of the conformation of the myofibrillar protein during post-slaughter. Of special importance in this regard is the position of the carcass and the degree of restraint upon the muscle (Smith *et al.* 1971; Klose *et al.* 1970). It has been suggested that the tenderness of expensive cuts of steaks is owing to the manner in which the carcass is hung in that the actin and myosin in these tissues tend to separate because of the carcass orientation (Chapter 27). Stretching of the scleroproteins in the connective tissue has also been implicated in this regard.

From the above discussion it would be appropriate to deduce that postmortem tenderness or texture in the rigor phase would be influenced in large part by factors that control the level of ATP and especially the postmortem rate of decrease of ATP in the sarcoplasm bathing the muscle. That this surmise is based upon a firm foundation of experimental evidence and detail is clear from diverse sources. Thus, it has been repeatedly and consistently demonstrated that muscle tissue undergoing rigor at temperatures in the range of 15°C are more tender than at ambient body temperatures of 35°−40°C. Lower temperatures decrease the rate of glycolysis and thus the rate of hydrolysis of ATP and lactic acid accumulation prevents the intense actin-myosin interaction which would occur in the complete absence of ATP at the low pH (Khan 1971). Also lower temperatures tend to favor the recycling of glycolysis and retention of the activity of the relaxing factor, thus tending to diminish the rate of ATP disappearance. The parallelism between toughness of meat and relative rate of ATP disappearance in muscles held at various temperature is shown in Fig. 26.4.

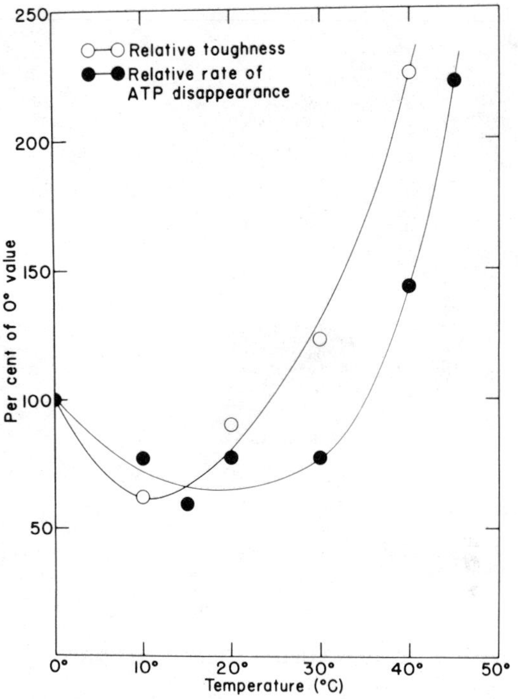

From de Fremery (1963)

FIG. 26.4. CHICKEN MEAT TEXTURE AND ATP LOSS

Texture Aberrations

Cold Shortening.—Augmentation of tenderness with lowered temperature does not, however, proceed for most meats with decreasing temperature below about 10°C. Thus freshly excised beef exposed to temperatures in the region of 5°C exhibits as much as a 60% decrease in length as compared to 20% when held at 15°−20°C, an effect on tenderness which is decidedly different from that which would have occurred had the cooling been delayed several hours. This decrease in length is accompanied by an enhanced, totally unacceptable toughness. The force required to bite through the meat may increase about five-fold. Increasing incidence of cold shortening is closely associated with speeding up the meat-to-muscle conversion process to make it more efficient and economic. Two scanning electron microscope views of this defect are shown in Fig. 26.2. Cold shortening occurs only in red meat and not, for instance, in poultry or lamb. As pointed out by Jones (1977), there is probably a complete overlap of actin and myosin filaments so that the latter are buckled against the Z-disc.

Causes.—Marsh (1966) suggested that cold-shortening was due to inactivation of the calcium transport ATPase, thus releasing Ca and resulting in activation of contractile ATPase. Davey and Gilbert (1974) measured a 40-fold rise in myofibril Ca as the temperature of prerigor muscle was reduced from 15° to 0°C. The absence of cold shortening in white meat has been attributed to the more highly developed (than in red meat) sarcoplasmic reticulum and consequently to a better ability to sequester Ca, and also to a sparsity of mitochondria which, when damaged by the sudden change in temperature, could leak Ca. Figure 26.2A shows a damaged mitochondrion in aged beef muscle.

While the most obvious way of preventing cold shortening is not to cool the carcass too rapidly before rigor sets in, the time spent in doing so is incompatible with modern processing schedules. Perhaps in the future, safe ATPase inhibitors may be feasible. For instance, Yount *et al.* (1973) found that certain purine analogs of ATP were very potent inhibitors of contractile ATPase. Such analogs would have to be designed to be completely eliminated perhaps by irreversible reaction with active site (Chapter 14) or through thermal decomposition during cooking by the time the meat is eaten by the consumer. Another approach is to raise the ATP input by accelerating glycolysis via electrical stimulation or high pressures (Bouton *et al.* 1977). Acceleration via electrical stimulation is due to increased binding of several glycolytic enzymes to actin (Clarke *et al.* 1976).

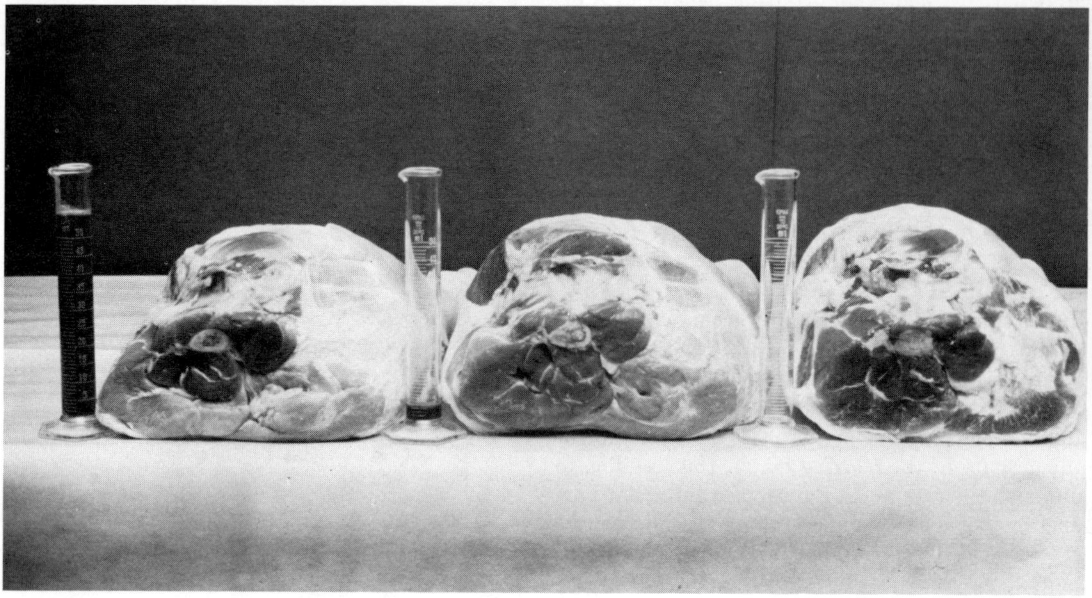

Courtesy of Davis (1975)

FIG. 26.5. PSE AND DFD—DEFECTS IN PORK TEXTURE

The flabby texture of pale, soft exudation, PSE (left), is evidenced by flattened oval shape of hams and increase in exudate volume. The ham on the right is dark, firm and dry, DFD; the one in the center is normal. Two hour fluid loss also shown.

Thaw Rigor.—A closely related texture problem is that of "thaw rigor" in which there is an enhanced toughening of thawed meat frozen prior to the onset of rigor. In thaw rigor the myofilaments appear to be in a supercontracted state to the extent that they can be observed buckling against the Z-line. As in the case of cold shortening, inactivation of the relaxing factor along with further release of Ca^{++} due to salt flux (upon thawing solutes will tend to concentrate in liquid phase) are probably involved (Okubanjo and Stouffer 1975).

Pale Soft Exudates.—While cold shortening appears to be associated with excessively rapid loss of ATP, not compensated for by glycolysis, the erratic appearance of pale soft exudate, PSE, in meat from certain strains of pigs exhibiting the "porcine stress syndrome" (PSS) appears to be associated with an unregulated rapid glycolysis. The extent and especially rate of glycolysis-produced lactic acid (as measured by a drop in pH to as low as 4.8 instead of the normal pH 5.5 within minutes after slaughter) is just one of the dramatically altered features of this rather spectacular change in postmortem texture (Fig. 26.5). Such cuts of meat, besides leaking liquid and appearing to be pale and "soft," assume a flattened, oval shape and cannot contract. This meat, although "soft" looking, is tougher than normal meat. This is a case in which the "appearance" of texture (Chapter 15) belies the texture perceived by chewing. In contrast to the buckling of the myosin up against the Z-disc observed in cold shortening, the actomyosin bridges between thick and thin appear to be "locked-in" or "frozen" with the contractile apparatus in a stretched configuration.

Causes.—The rapid production and abnormally high levels of lactic acid strongly suggested to earlier investigators that the ultimate cause of PSE is enhanced glycolysis brought about by depletion of glycogen reserves in the nervous animal. Since the animal is under stress it undoubtedly secretes epinephrine which in turn activates the formation of the second messenger, cyclic AMP, which, as adumbrated in Chapter 4, through the cascade regulation mechanism results in the activation of glycogen phosphorylase and thus acceleration of glycolysis. Hence, we have the finding of Ono *et al.* (1976) of a transient rise in this second messenger immediately after slaughter.

With regard to the immediate cause of the altered texture, most investigators have suggested a rapid deposition of acid-denatured sarcoplasmic proteins on the myofibrils. Delving into the enzymatic molecular aspects more deeply, Sung *et al.* (1976) ruled out inactivation of ATPase as a causative factor. Another suggestion which has been tentatively forwarded is that the lower pH defunctionalizes mitochondria even faster than they are inactivated in normal postmortem meat so that the metabolites from glycolysis have nowhere to go but into lactic acid even faster than usual. Undoubtedly the state of LDH, stressed by Hultin and colleagues, is profoundly affected. Cheah and Cheah (1976), however, assign to mitochondria a somewhat different role in development of the PSE phenomenon; they are assumed to be the source of increased Ca. Indubitably, the recent finding of Cooke and Franks (1980) that all myosin heads form bonds in rigor muscle will contribute to the clarification of this textural aberration. This heightened efflux of Ca from PSE pig mitochondria is considered to be the trigger for PSE and other porcine aberrations.

Another suggested and plausible immediate cause of PSE, at least that aspect of it which has to do with loss of contractility, is the irreversible binding of actin to myosin (Izumi *et al.* 1977). Presumably the myosin "hinge-swivel" mechanism is impaired so that the "break" phase of the scheme on p. 466 is inoperative—Park *et al.* (1975) describe an "arrowhead" myosin obtainable only from PSE pigs—or perhaps a specific inhibitor prevents the break reactions (Lynn and Taylor 1971).

As far as preventing this phenomenon is concerned, it is known, for instance, that struggling and poor nutrition make the meat more tough whereas sedation and a nutritious diet have the opposite effect. It would thus appear that humane practices both before and at slaughter fulfill the requirements for improved meat quality and reduce economic losses; quite aside from textural considerations, PSE hams lose up to one and a half times more weight than do normal hams (Davis 1975).

Postmortem Modulation of Regulation of Glycolysis

If adrenalin is injected into an animal before slaughter, glycogen reserves are depleted but in the living animal, of course, the mitochondria handle the increased efflux of metabolites aerobically. Hence, when the animal is slaughtered, glycolysis cannot continue in absence of glycogen so that the pH of the treated meat is actually elevated to as high as 6.8 (Lawrie 1966; de Fremery 1966). The resulting meat is more tender than untreated controls. Poultry muscle under these conditions goes through rigor almost instantaneously and maximum tenderness is achieved within 1 hour of postmortem slaughter. A summary of the influence of hormones on glycolytic enzymes is shown in Table 26.4.

TABLE 26.4

GLYCOLYTIC ENZYMES INFLUENCED BY HORMONES AND RESULTANT EFFECT ON TEXTURE

Enzyme	Hormone	Effect on Enzyme	Effect on Texture
Phosphorylase	Epinephrine	Stimulates	More tender
Hexokinase	Insulin	Induces	More tender
Phosphofructo-kinase (PFK)	17-OH-corti-costeroids	Stimulates	Develops PSE
Aldolase	17-OH-corti-costeroids	Stimulates	Develops PSE
Pyruvate kinase	17-OH-corti-costeroids	Stimulates	Develops PSE
Pyruvate kinase	Aldosterone	Inhibition	Develops PSE
LDH	Sex hormones	Isozyme pattern Changes	Indices of changes (Chapter 37)
LDH	Thyroxine	(as above)	(as above)

In addition to the influence of and modulation by hormones on the individual enzymes involved in control of meat texture up to and through rigor, information is accumulating on the control and state of activity of these postmortem enzymes by metabolites (Chapter 4). As an example of the influence of metabolite in a texture context, we point out that some of the ADP formed as a result of contractile ATPase activity is converted via nucleotide kinase (myokinase).

$$2\ \text{ADP} \rightleftharpoons \text{AMP} + \text{ATP}$$

Both AMP and Pi stimulate the activities of both phosphofructokinase (PFK) and of glycogen phosphorylase (Chapter 4). Pi also increases the K_m of aldolase, the next enzyme after PFK in the glycolytic sequence (thus in effect inhibiting its action). This results in the raising of the level of fructose diphosphate which—contrary to the general rule that products of an enzyme reaction inhibit the reaction—activates the PFK and thus again leads to an accelerated glycolysis.

In order to identify the determinants of the rate of glycolysis, several investigators have looked into the level of the intermediates and activities of the glycolytic enzymes. Thus, Kastenschmidt (1970) has summarized the buildup of such intermediates in a "crossover" plot (Fig. 26.6) in which the intermediates (in sequence of their appearance in glycolysis) are displayed on the abscissa and the ratios of the levels of intermediates in "fast" to those in "slow" glycolyzing muscles postmortem on the ordinate. The occurrence of a crossover between fructose-6- and fructose-1,6-diphosphate emphasizes the sensitive role of PFK in the metabolic control of glycolysis in muscle tissue after as well as before slaughter. Kastenschmidt concluded that inhibition at the PFK step is the primary cause of cessation of postmortem glycolysis. Many of the factors involved in raising the level of PFK activity in the living muscle such as AMP and Pi may be even more effective in activating PFK in the postmortem tissue. Futher involvement of this key enzyme in the course of postmortem glycolysis is afforded by the studies of muscle tissue mince (Newbold and Scopes 1971) in which it was concluded that the cessation of metabolism in these minces—which paralleled a rise in fructose diphosphate levels—was a consequence of the destruction of NAD, the adenine moiety of which was converted to inosinic acid via the enzyme adenylate deaminase.

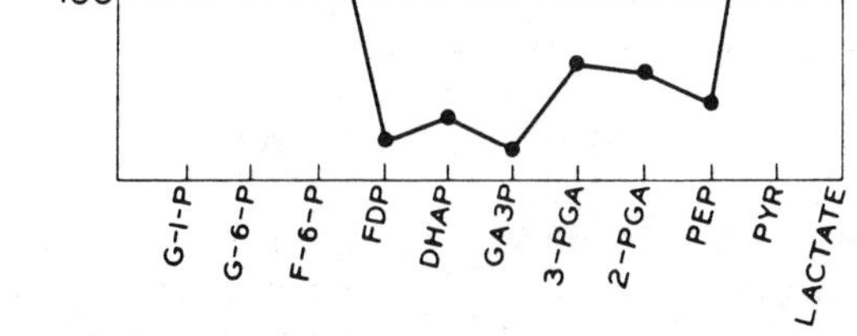

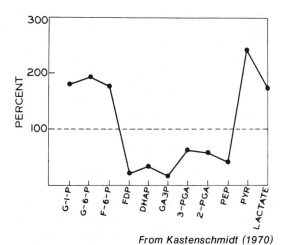

From Kastenschmidt (1970)

FIG. 26.6. CROSSOVER PLOT OF GLYCOLYZING INTERMEDIATES

Ordinate, ratio of fast- to slow-glycolyzing beef muscle postmortem times 100; abscissa, level of the intermediates in glycolysis. See text for significance for texture.

$$\text{Adenylate} \longrightarrow \text{Inosinate} + NH_3$$

Although this enzyme reaction is usually considered to be involved in meat flavor rather than texture (Chapter 23), it may participate in textural quality of meat since it influences the AMP level. On the other hand, its action may be too late postmortem; Aberle and Merkel (1968) concluded that its activity is not related to PSE.

In further investigation, Newbold and Scopes (1971) showed that addition of NAD prevented the accumulation of fructose diphosphate, due to a reduction in effective PFK activity. As a final note on the apparent central role of PFK are the observations of Southerd and Hultin (1969) that the ADP produced by PFK activity associated with mitochondria is more available for respiration than is exogenous ADP; and of Chism and Hultin (1977) that ATP solubilizes PFK bound to intracellular particulates in chicken muscle cells. This suggests a special structural relationship of PFK to the apparatus of the respiratory chain.

Glycogen phosphorylase has also been examined with regard to its role in controlling postmortem glycolysis. Within 10 min after slaughter this enzyme is primarily in the b form (Sayre *et al.* 1963). Phosphorylase b in the absence of AMP is much less active than phosphorylase a. The b enzyme has two subunits (instead of four in the a form), lacks phosphate, and is activatable by AMP. Furthermore, AMP binds tighter to the enzyme in the presence of Ca, again illustrating the complex and varying role of Ca in postmortem glycolysis. On the basis of studies with muscle mince, Newbold and Scopes (1971) concluded that inactivation of phosphorylase was the key phenomenon in the cessation of glycolysis. Binding of glycolytic enzymes to actin also controls glycolysis. This binding, as mentioned previously, is increased by electric stimulation of the muscle tissue.

Finally, we wish to point out that the state of sulfhydryl groups can profoundly influence the rates of both glycolysis and the ATP breakdown and thereby texture. The activity of phosphoglyceraldehyde (PGA) dehydrogenase—PGA + Pi + $NAD^+ \rightarrow$ 1,3-diphosphoglycerate + NADH + H^+—is dependent upon its content of free sulfhydryl groups (SHs). Blocking of these groups by an appropriate sulfhydryl reagent such as iodoacetate results in complete cessation of activity and thus cessation of glycolysis. The level of activity of myosin ATPase is also uniquely dependent on the presence of two different types of SH in the myosin molecule. They differ in their reactivity to blocking agents. Blocking of the more reactive SHs results in enhancement of ATPase activity. Further blocking reduces the activity. De Fremery (1963) demonstrated that injection of poultry antemortem with iodoacetate resulted in a postmortem increase in both tenderness and rate of loss of ATP. Sulfhydryl groups may also play a role in activation of proteinase, both endogenous (cathepsin) and added (Chapter 27), and thus play a role in the postrigor aspects of meat tenderness. In traditionally cured meats employing nitrites, the latter may influence texture by converting SH to disulfide (S-S):

$$NO_2^- + 2\ SH + O_2 \rightarrow S\text{-}S + NO_3^- + H_2O$$

High levels of SH in beef seem to be associated with toughness. Dube *et al.* (1972) found that beef muscles which had highest initial SH contents and the highest percentage of reversible oxidation of these groups also had the highest shear values and the shortest sarcomeres.

ROLES OF NONGLYCOLYTIC ENZYMES IN AGING

The Case for Protease Participation in 1972

Postmortem rigor is followed by a slow relaxation period (about 1 day for poultry to about 1 to 2 weeks for beef) during which the meat attains the texture commensurate with consumer acceptability. This "aging" is usually conducted at cold temperatures to minimize attacks by microorganisms. In contrast to a rather substantial consensus concerning the enzymatic aspects of texture up through the development of rigor mortis, as outlined above, the assignment of a role to enzymes, especially proteolytic enzymes, in the postrigor production of tenderness was by no means universally accepted prior to 1972. Thus in 1969, de Fremery and Streeter stated that "for the present the biochemical mechanism of postmortem tenderization of meat must remain obscure." In the ensuing discussion we shall assume advocacy of a significant role of proteolytic enzymes in the aging process, with the understanding that this is not an inflexible stance but rather a heuristic approach. First we shall present positive evidence, albeit mostly circumstantial, and then evidence against such a role, with an appropriate rebuttal.

Historically, the putative role of proteolysis in the aging process can be traced back to the studies of meat "autolysis" by Hoagland and coworkers (1917) at the U.S. Department of Agriculture. Autolysis is self-digestion due to the inherent proteolytic enzymes of the meat producing amino acids and presumably accompanying tenderization. The phenomenon of autolysis has been repeatedly demonstrated during the ensuing 50-odd years. A relatively recent example, also the result of a U.S. Department of Agriculture investigation, is shown in Fig. 26.7.

Further circumstantial evidence of the potential capability possessed by proteases of meat to effect gross textural changes is afforded by the increased amino acid production, intensified autolysis, and the overtenderization which ensue upon exposure of meat to sterilizing doses (4–5 Mrads) of ionizing radiations (*cf* El-Badawi *et al.* 1964). These workers also showed that heating beef prior to irradiation resulted in a firmer meat, as did presoaking in distilled water. In both cases protein denaturation (resulting, presumably, in proteinase inactivation) was suggested as the cause of this partial reversal of the over-tenderizing effect of ionizing radiation. Even higher doses of radiation do not completely inactivate the proteinases (Rhodes and Meegungwan 1962) but may destroy lysosomal and other membranes which would release proteinases and thus enhance their action (see below).

Cathepsins.—As far back as 1938 Balls suggested that meat autolysis was due to the action of the intracellular cathepsins. Like Mt. Everest, they are there. There are now six well-established classes of cathepsins. The properties of the four classical categories of cathepsins (A, B, C and D) are summarized in Table 26.5. Each of these (mostly from nonskeletal muscle tissue) possess specificities analogous to those of the more thoroughly characterized extracellular pancreatic proteases. Since the latter can attack the myofibrillar proteins, one would expect that the cathepsins would also be capable of degrading them.

Some of the individual meat cathepsins have been purified and implicated in muscle autolysis. Thus, by a careful survey of the effect on autolysis of factors and substances which influence cathepsins, Caldwell (1970) concluded that cathepsins B and D were responsible for the observed autolysis of chicken muscle. Although most of the research on cathepsins has been accomplished with nonmeat tissues, several papers have dealt with purification of poultry and beef muscle cathepsins (Landmann 1963; Bodwell and Pearson 1964; Randall and

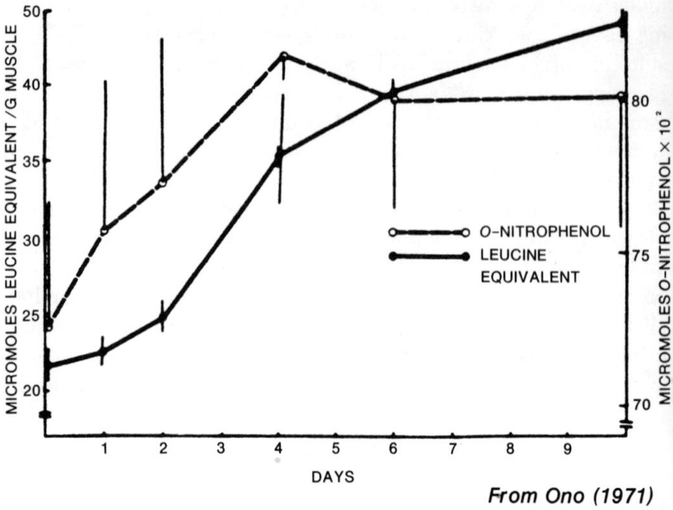

From Ono (1971)

FIG. 26.7. AUTOLYSIS PARALLELS LYSOSOMAL DISRUPTION

During postmortem aging of beef muscle: Autolysis measured by liberation of amino acids, lysosomal disruption by solubilization of β-galactosidase.

TABLE 26.5

PROPERTIES OF THE CATHEPSINS

General Class	Endopeptidases (B, D, E, G, L)		Exopeptidases (C, A, Carboxypeptidases)	
Designation; EC #	D; 3.4.23.5	B(B_1B_2); 3.4.22.1	C; 3.4.14.1	A; not listed.
Typical synthetic substrate	None	Benzoyl arginine-amide	Glycyl phenylalanine amide	Carbobenzoxy glutamyl tyrosine
Action resembles	Pepsin	Trypsin	Chymotrypsin	Carboxypeptidase
Activators	None	Thiols, EDTA	Thiols, chloride, cyanide	None (?)
Specific inhibitors	Pepstatin (Chap. 14), 3-phenyl-pyruvic acid	Leupeptin, 6-NH_2-caproic acid	SH reagents, oxytocin	None
Nonaffectors	Organophosphates, thiols	Cyanide, DFP, N-ethyl maleimide, trypsin, soybean trypsin inhibitor	DFP, 2,4-dinitrophenol, p-Hg-chlorobenzoate	SH reagents
Molecular weight	58,000	52,000 (B2); 25,000 (131)	210,000	
pH optima, range	3–4.5	4–6.5	5–6	5
Other salient characteristics	Attacks collagen fragments, other proteins	Inactivates most other enzymes; activates trypsinogen	Dipeptidyl aminopeptidase; dipeptidyl transferase; polymerizes peptides; Cl^- required for action on proteins	Acts synergistically with Cathepsin D, attacks glucagon

MacRae 1967; Iodice 1967; Caldwell and Grosjean 1971). In all cases the properties of the purified enzymes appear to be compatible with a role in post-rigor aging. For instance, as shown in Fig. 26.8, the cathepsins present in the postnuclear homogenates of chicken muscle act optimally in the pH of postmortem meat.

The locus of the catheptic activity within the cells of the tissues of animals appears to be the lysosome, the intracellular, membrane-encased organelle containing a variety of degradative hydrolytic enzymes (Chapter 4). Although not all the evidence is in, it would appear that muscle is no exception. Problems in assessing cathepsin "lysosomality" has prompted the use of marker enzymes such as β-galactosidase in the lysosomes of aging beef muscle which decreases whereas that of the soluble fraction increases roughly paralleling autolysis, as shown in Fig. 26.7.

Lysosomal Heterogeneity.—Caldwell and Grosjean (1971) found evidence for heterogeneity of their chicken lysosomes as did Stagni and deBernard (1968) for the lysosomes of beef skeletal muscle. The evidence was based on variation of "latency" (the lysosomal enzymes do not manifest their activity until the lysosome membrane is removed or damaged). These observations are but part of mounting evidence that catheptic activity of meat may come from several sources besides the muscle fibril alone. The specific proteolytic activity of sheep lung macrophage amounts to 120 times that of rabbit leg muscle. Indeed Canonico and Bird (1970) presented completely independent evidence for the existence in rat muscle tissue of lysosomes originating in both macrophages and in connective tissue. Ali and Evans (1969) implicated cathepsin B in the autolytic degradation of cartilage matrix. Davey and Gilbert (1968) suggested proteolysis of bacterial origin as a factor in enhancing the changes of aging. Perhaps immunodiffusion techniques (Stoller and Levine 1963) using antibodies to purified nonskeletal muscle proteinases with skeletal muscle proteins as antigens might be helpful in answering the question of the participation of these sources in aging.

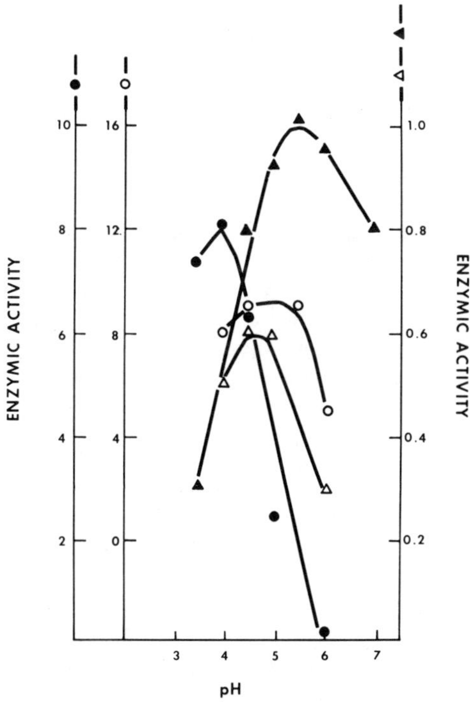

Courtesy of Caldwell and Grosjean (1971)

FIG. 26.8. ACTIVITY-pH PROFILES OF CHICKEN CATHEPSINS

Noncatheptic Proteases.—In addition to the cathepsins, which act at the slightly acid pH of meat tissue, protein-digesting activity at slightly alkaline pH's has also been reported. Thus Landmann (1963) observed a peak at pH 8 in the activity-pH profile of the proteolytic enzymes present in a preparation of beef muscle protein. An apparently distinct and now well-characterized alkaline proteinase, exclusively localized in the myofibrils of rat muscle, has been implicated in the autolytic breakdown of rat and skeletal protein (Noguchi and Kandatsu 1971). Examined in more detail in Chapter 27 is the beef muscle neutral "true" collagenase of Laakonen (1973B). It attacks native triple-helixed as well as denatured collagen, retains half of its optimal activity at pH 5.5 of beef and is stimulated by Ca. Its action is potentiated by the heightened temperature at the start of the cooking process as discussed in Chapter 27. Once this collagenase has acted the connective tissue can presumably be further degraded by the lysosomal cathepsins. It may be apropos to mention that bones are also a source of collagenase (Evanson 1971) and might, in special cases, contribute to the autolysis of some meat tissue perhaps during the early stages of cooking.

It is also possible that in addition to the action of the proteases, autolysis might create localized regions of high pH. The production of ammonia from the action of adenylic deaminase and glutaminase comes to mind. Experimental techniques for detecting such micro-environmental aberrations have been developed (Chance and Nishimura 1967). The aforementioned nonacid endopeptidases could presumably act in concert with alkaline exopeptidases such as muscle leucine aminopeptidase (Schwimmer 1944B; Randall and MacCrae 1967; Otsuka *et al.* 1976).

The Case Against Enzymatic Participation in 1972

An impressive array of evidence through 1969 against the participation of enzymes, especially proteinase, in postrigor aging was marshalled by Goll *et al.* (1970). The arguments may be condensed to the following main points.

In the first place the evidence for this participation is strictly circumstantial. This argument has much validity. However, the definitive experiment one way or the other in this field had yet not been performed or at least published.

Secondly, the catheptic activity of muscle tissue is very low as compared with other tissue. Thus, as far back as 1958 Zender *et al.* reported that the proteolytic activity of rabbit heart, lung, liver and kidney were 13, 60, 64 and 76 times that of the psoas muscle. Such earlier reports of low catheptic activity may reflect incomplete separation and solubilization and failure to ensure adequate concentrations of critical cofactors such as sulfhydryls (Caldwell 1970). On the other hand, it may be precisely because of the low proteolytic activity of meat muscle, resulting in a slow, modulated and controlled resolution of rigor during aging, that the resultant meat has just the right texture. Vigorous proteinase action, if it were present in muscle, would carry the aging process far beyond the desired texture.

Thirdly, it is difficult to relate this small amount of proteolysis which does occur to the corresponding changes in texture during the aging process. For instance, Parrish *et al.* (1969) reported that most of the autolysis of beef muscle occurs *after* most of the tenderness has been achieved. This, however, does not negate the findings of McCain *et al.* (1968) on the highly significant correlation coefficients among ninhydrin-positive material (amino acid levels) and organoleptic measurements of elasticity and other texture parameters. Furthermore, coincidence in time between the resultant of proteolytic activity and texture would hold only if such activity were the rate-limiting step in what is admittedly a complex phenomenon involving nonproteolytic events. This again points to a controlled release of proteinase action in texture change in meat muscle during aging.

Fourthly, muscle proteinases do not attack the isolated proteins of the muscle at pH 5.5, that of normal meat, for instance, Martins and Whitaker (1968) detected no action of purified chicken cathepsin D on actomyosin. However, the authors did not ascertain whether other meat proteinases acting either alone or in conjunction with cathepsin D were effective.

Fifthly, little or no proteinase-induced change in muscle tissue components can be detected during the aging process. Thus, Goll *et al.* (1970) stated that no new bands indicative of proteolytic action occur in the gel electrophoretic patterns of sarcoplasmic proteins. Zender (1958) did observe some changes in electrophoretic patterns. Furthermore, the sarcoplasm contains a mixture of many proteins. As the result of a study of the stability of the alkali-insoluble connective tissue of poultry, de Fremery and Streeter (1969) concluded that concomitant textural alterations during aging are not caused by the breakdown or dissolution of this class of protein. However, Kruggel and Field (1971) reported that the aging of bovine muscle is accompanied by changes in the molecular structure of intramuscular collagen—as indicated by sucrose-gradient density analysis of denatured collagen—attributable to the cleavage of crosslinks between polypeptide chains of this molecule. This again points to a slow, subtle, but nevertheless texture-affecting action of proteinase. Wu and Sayre (1971) found that the myosin from aging chicken muscle was accompanied by a myosin-like fraction which increased with aging. It might have been a product of the transpeptidase action of cathepsin C. Another myofibrillar protein, actin, was reported by de Fremery to be invariant with respect to amount and to the nature of its amino end, phenylalanine, during the aging. This shows that detachment of actin from the Z-line, discussed below, does not involve carboxypeptidase-like action on actin. Perhaps the most telling argument involved the principle of Occam's Razor, i.e., there exists a body of experimental observations on changes that occur during aging—mostly based upon alterations in the extractibility and solubility of myofibrillar proteins and especially on microscopic changes—for which it does not appear to be necessary to invoke peptide bond hydrolysis. Some of these changes are: (1) the actomyosin bond is not broken during aging (Sayre 1968), (2) an increase in the ATPase activity of the myosin apparently due to alteration of the state of its SH groups, (3) increasing fragility of the myofibrils (Fig. 26.9) (Sayre 1970), (4) loss of adhesion between adjacent myofibrils, and (5) perhaps most pertinent of all, progressive weakening and eventual disintegration of the Z-lines concomitant with the rupture of their junction with actin filaments (Goll *et al.* 1970; Davey and Gilbert 1968). Figure 26.2 shows such a weakened Z-line. Since this phenomenon can be prevented by EDTA and accelerated by Ca, it has been suggested that the progressive loss of the capacity of the sarcoplasmic reticulum (relaxing

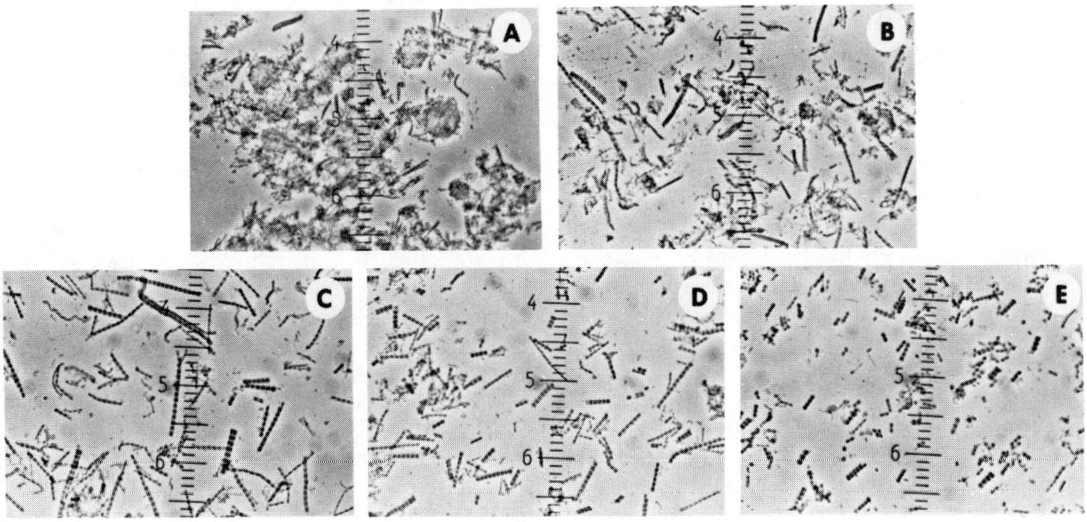

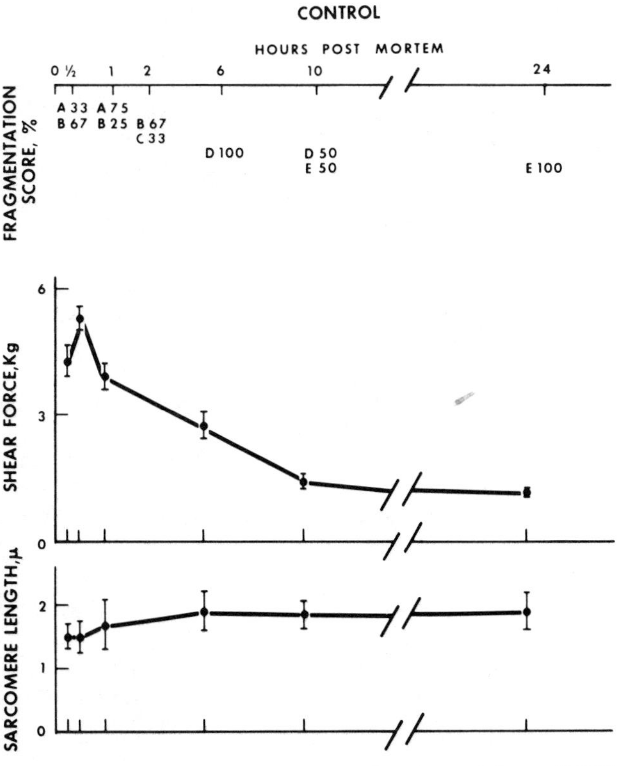

Courtesy of Sayre (1970)

FIG. 26.9. MYOFIBRILLAR FRAGMENTATION AND TENDERNESS

factor, see above) to accumulate Ca results in a localized accumulation of this ion at the site of the Z-line. This is considered to be the determining event in modulating the textural changes which accompany postrigor aging. Since aging is probably a much slower process than ionic diffusion, we suggest that enzymic destruction of the sarcoplasmic reticulum and lysosomal membrane resulting in potentiation, due to removal of latency and activation, of hitherto Ca-dependent lysosomal cathepsin, may play a role in aging. Perhaps the role of enzyme action, not necessarily proteolytic, on membranes and on the substance(s) linking actin to the Z-line should be looked into further. Goll *et al.* (1970) as well as Parrish *et al.* (1969) and Davey and Gilbert (1968) suggested participation of catheptic or bacterial enzymes in dissolution of the Z-line.

Status of Proteinase Action in Meat Aging After 1972; The Calcium Activated Factor (CAF)

In a series of papers starting in 1972, investigators at the Universities of Wisconsin and Missouri mounted a series of experimental attacks on this problem which has apparently pinpointed the enzyme link to postmortem aging of beef to the action of a noncatheptic "CAF," most likely a protease, originating in and localized within the sarcoplasm. The properties of the purified CAF and its effect on the contractile apparatus at molecular, supramolecular microscopic, sensory and macroscopic levels, when deliberately added to myofibrils, are almost precisely what is observed in beef muscle during normal aging of beef (Busch *et al.* 1972; Dayton *et al.* 1976; Olson *et al.* 1977; Olson and Parrish 1977; MacBride and Parrish 1977; Cheng and Parrish 1978). This rather high molecular weight (for a proteinase) enzyme consisting of two protomers (30,000 and 80,000 dalton subunits) acts optimally at pH 5.5, the precise pH of aged beef.

Perhaps the most potent tool in establishing the parallelism between what happens during aging and what happens when the enzyme is deliberately added was improved gel electropheresis. With this technique the investigators demonstrated appearance of a 30,000 dalton fragment, apparently derived from the highly specific and limited action of CAF on one and only one of the subunits of troponin, Tn-T. Tn-T (Tt in Scheme I), it will be recalled, binds this regulatory protein to its regulatory partner, tropomyosin. Its restricted action calls to mind another important food-related proteinase, rennin, Chapter 33. The 30,000 dalton fragment, as shown in Fig. 26.10, appears in meat which becomes tender but not in meat which does not age well and ends up as tough meat.

Other parallelisms include disassembly of contractile apparatus without disappearance of any other myofibrillar component other than Tn-T; removal but not proteolytic breakdown of the α-actinin, the main component of the Z-disc; disintegration of the Z-disc; increase in sarcomere fragility; changes in shear forces required to cut the meat; dependence of myofibrillar changes on Ca availability (Cheng and Parrish 1978); and finally changes in meat texture as judged by a panel. Penny *et al.* (1974) were able to tenderize freeze-dried meat by adding CAF to the point of disintegration of the Z-line.

Problems.—Some unanswered questions remain. Thus, although the Z-disc is readily disintegrated and α-actinin released, the latter is not readily hydrolyzed. The sequence of events leading from (hydrolytical) release of the Tn-T fragment to the disassembly of the contractile apparatus remains to be elucidated. Penny (1974), using the then admittedly impure CAF, did observe extensive enzyme action on α-actinin. In any case, just as before death, Ca activates the trigger which sets into motion the chain of events leading to muscle contraction, so the same ion after death sets into action a train of events which converts muscle into meat. This may be more than poetic license, since as mentioned, Ca-activated proteins associated with muscle enzymology appears to have quite similar amino acid sequences.

Non-CAF Proteinase Since 1972.—The search for other myofibril associated proteinases and assignment of them to roles in postmortem aging and meat texture has continued. Furthermore, evidence continues to be advanced, in apparent contradiction to the above-mentioned results, that sarcoplasma protein molecules, *other* than a particular, unique troponin subunit, undergo changes during aging (probably due to the action of proteolytic enzymes), whose ultimate consequence is the modification of meat texture resulting in aging.

Perhaps the most serious challenge to the exclusive supremacy of CAF comes out of the University of Tokyo by investigators who have briefly reported the presence in beef muscle cells of a cathepsin which acts maximally at pH 5.6−5.8, close to that of meat (Arakawa *et al.* 1976A). This cathepsin attacks both myosin and α-actinin *in*

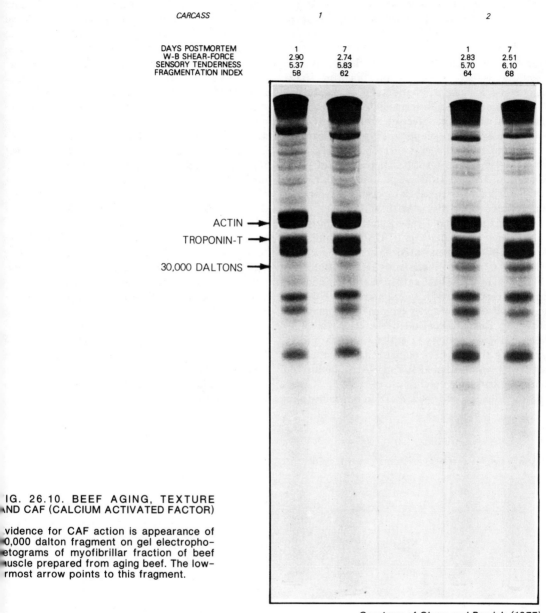

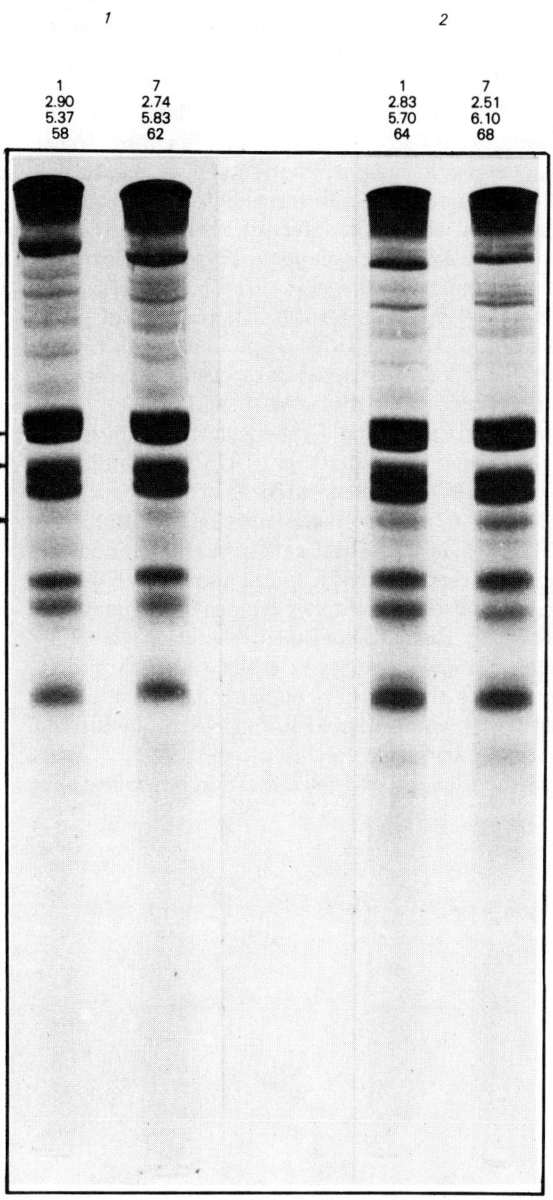

IG. 26.10. BEEF AGING, TEXTURE
ND CAF (CALCIUM ACTIVATED FACTOR)

vidence for CAF action is appearance of
0,000 dalton fragment on gel electropho-
etograms of myofibrillar fraction of beef
uscle prepared from aging beef. The low-
rmost arrow points to this fragment.

Courtesy of Olson and Parrish (1977)

itu, as revealed by gel electrophoresis. Further-
ıore, this behavior is paralleled in intact myofi-
rils exposed to relatively high temperatures in
bsence of the added cathepsin (Arakawa *et al.*
976B). The alkaline proteases associated with
ıammalian skeletal muscle tissue, and their origin, has attracted the attention of medically-oriented investigators, and is shrouded by controversy. Some believe that the Ca-activated protease and cathepsin D are the only proteases originating in muscle tissue which are capable of degrading myofibrillar protein. These and other view-

points and evidence for even newer proteases can be found in a recently published symposium on protein turnover in muscles (Anon. 1980). A cursory review of selected reports on non-CAF proteinases associated with muscle which have appeared since 1972 suggest that if they do not participate in aging they play a role in flavor development and as texture determinants in curing, freezing, drying, ionizing irradiation and, of course, enzyme potentiation at intermediate temperatures. Thus, a study on the effect of curing agents used in ham manufacture on porcine lysosomal cathepsins reconfirmed the existence of a cathepsin D acting optimally on myofibrillar proteins at pH 9 (Deng and Lilliard 1973). A perhaps similar myofibrillar alkaline protease is especially effective in autolyzing muscle tissue with the release of 3-methyl histidine and is particularly inhibited by tryptophan (Hansen *et al.* 1978). Robbins and Cohen (1976) demonstrated that ultrastructural changes in muscle fibers were restricted to the Z-line region when beef cathepsins were added to beef muscles. Pavlovskii and Simbireva (1976) who have published extensively on meat lysosomal proteases in Russian, concluded that the lysosome-liberated (free) activity of cathepsins C, A and D, as well as that of a lysosomal collagenase associated with connective tissue, persisted under the normal meat aging conditions.

Cathepsins are also released from lysosomes as a result of newer innovations of beef muscle processing, such as high temperature conditioning (Moeller *et al.* 1977), ionizing radiation (Ali and Richards 1975) and freeze drying, thawing and rehydration (Nesterov *et al.* 1977). However, in the latter case, freezing the tissue while hot preserves the lysosomal membrane. While these classical cathepsins are more or less readily released from the lysosome, an insoluble carboxypeptidase A-like activity in skeletal muscle could not be solubilized after repeated freeze-thaw cycles (Meyer and Reed 1975). A hint that cathepsins may be involved indirectly in the determination of meat texture through their effect on glycolysis is afforded by reports such as that of Bond and Barret (1979) that cathepsin B (as well as the recently described G and L) readily inactivates the key glycolytic enzyme aldolase. Undoubtedly the many ramifications of the action of these lysosomal proteases, not so much in meat quality, but much more significantly in all aspects of biology and medicine—protein turnover, tissue regeneration, starvation, programmed death, human aging (senescence), diseases such as arthritis, muscle-wasting diseases and even emphysema (tobacco smoke constituents may oxidize essential SHs), to name a few—will ensure a steady stream of information and its application to food science and technology, as we shall have the opportunity to explore in the ensuing chapter.

27

APPLIED ENZYMOLOGY OF MEAT TEXTURE OPTIMIZATION

In the last chapter we directed our attention to the role of enzymes as determinants of meat texture during normal operations and under standard postmortem conditions. Except for a brief allusion to ways of preventing the textural aberrations of beef cold shortening and pork PSE, we did not address ourselves to the deliberate intervention in this process in order to optimize and improve meat texture. In this chapter we examine the methods and approaches—some quite empirical, others based upon information gleaned from fundamental muscle biochemistry, either actually used or experimentally proposed—to exploit enzyme action towards these ends. As with other foods and food processes, it is convenient to distinguish between the meat's own enzymes on one hand and the use of external sources of enzymes on the other. In each situation we shall endeavor to systematize, on the basis of experimental evidence, just how the enzymes act to influence texture.

Before we embark upon this analysis we should like to point out that these procedures and efforts are usually, but not always, directed towards modification of the collagen and other scleroproteins rather than of the structure of the contractile apparatus and its constituent myofibrillar proteins. There is an economic incentive at work in propelling research in this direction. As pointed out by many investigators, such as Davey and Gilbert (1975), lower grades of beef meat and meat from older animals, whose toughness is primarily due to the scleroproteins of the connective tissue, can be upgraded and utilized by use of these approaches. Furthermore, even when the target for enzyme action is initially not known, further basic research frequently reveals that it is this fraction which is being acted on. Nevertheless there is a continuing school of thought and line of research going back at least half a century which considers that beneficial results of enzyme action on meat can be best achieved in terms of modification of the contractile apparatus. One of the reasons for this effort is that perhaps tenderness achieved by modification of the contractile protein is closer to what happens during postmortem aging and also to what the consumer desires than that achieved by action on the connective tissue.

TENDERNESS THROUGH ENDOGENOUS ENZYME ACTION

Antemortem.—In addition to manipulation of the enzymes associated with transformations or modifications of the contractile apparatus and potentiation of endogenous proteinases which probably act on both the sclero- and myofrillar proteins, an alternate means of influencing texture (Table 26.3) is by antemortem control of the animal's environment and diet. As briefly discussed in Chapter 4, the presence or absence of specific dietary components can cause induction of many enzymes in the animals. Proposals have been forwarded to administer specific substances which are conjectured to affect the texture of the meat.

These substances include hormones, vitamins and collagen diminution agents.

Hormones.—We have discussed the role of some hormones as determinants of the activity levels of glycolytic enzymes which in term may influence meat texture (Table 26.4). One of these, adrenalin (epinephrine), has been proposed and its use patented as means of producing an "antiautolytic" effect which purportedly results in increased tenderness and also (along with the serotonin and the tryptophan and tryosine derivatives other than adrenalin) prevents pork PSE (Radouco-Thomas 1962).

Lawrie (1966) suggested administration of cortisone to meat animals. This hormone is presumed to be effective because of its suppression of the genes coding for the biosynthesis of enzymes involved in the biosynthesis of mucopolysaccharide "ground substance" associated with connective tissue in which the scleroproteins are embedded and with which they are covalently linked. As mentioned below, these mucopolysaccharides have been associated with toughness ascribed to connective tissue. Although not directly implicated in enzymatic aspects of tender texture, the use of diethylstilbestrol as a fattening agent for cows probably involves alteration of enzyme activities.

Vitamins and Other Substances.—Manipulation of vitamin levels of meat animals may be of benefit for influencing texture. Thus vitamin A causes decomposition of liver lysosomal membranes (Ogasawara 1967) and of porcine cartilage (Fell and Dingle 1963) involving in each instance release of catheptic proteolytic enzymes. The poor keeping qualities of liver, rich in both lysosomes and in vitamin A, may be due to extensive lysosomal disintegration. Perhaps carefully controlled application of this vitamin to muscle-derived meat could be used to increase tenderness by controlled release of endogenous cathepsins from lysosomes. However, Meyer *et al.* (1967) reported a decrease in the juiciness of steaks from cows fed vitamin A supplement. Deficiency of vitamin C, by virtue of the resulting diminished ground substance (Stone and Meister 1962), and vitamin E by virtue of the resulting enhancement of lysosomal enzyme activities (Tappel 1970) have been proposed as means of artificially tenderizing meat (Lawrie 1966).

Jenevein (1971) proposed antemortem injection of collagen-diminution agents (e.g., β-amino-proprionitrile or cysteine antagonists) for enhancing tenderness. Repression of the biosynthesis of lysyl oxidase, an enzyme involved in formation of collagen and elastin crosslinks (see end of this chapter) might be worthwhile looking into. Increased tenderization *via* antemortem injection, 5 to 20 min prior to slaughter, of sulfhydryl-containing compounds such as cysteine and glutathione (20 to 40 ppm of live weight) has been patented (Weber 1971). Such compounds are considered to exert their tenderizing effect by activation of naturally occurring proteinases, presumably cathepsins B and C (Table 26.5). The activation takes place during the shortened aging period, 4 to 7 days.

Modification of the Contractile Apparatus.—As mentioned, procedures have been proposed whose principal effect appears to be associated with the state of tension of the myofibrils. More precisely, these procedures which are said to interfere with the attainment of rigor mortis (Streitel *et al.* 1977) consist of either cooking meat before rigor or of postmortem administration via injection, perfusion or infusion of certain metal chelators, principally the polyphosphates and citrate.

The Chelators.—As examples of the effects of the polyphosphates, Bendall (1954) observed that both ortho- and polyphosphate salts caused increases in minced muscle volume. Carpenter *et al.* (1961) reported that perfusion of muscles of a freshly slaughtered animal with *ca* 15% solutions of hexametaphosphate salts resulted in demonstrable tenderization of the resulting cooked (beef) and cured (ham) products. Results of a more recent investigation are shown in Table 27.1 (Streitel *et al.* 1977). Phosphates and magnesium salts are muscle relaxants while the target of sodium citrate is considered to be glycolysis by virtue, perhaps, of its inhibition of the key regulatory enzyme, PFK. On the other hand, Mg, really part of the ATPase substrate (the Mg salt of ATP), has been considered to be the target of added pyrophosphate with which it chelates. Of course these three chelators can also bind Ca. Binding of Ca leads to prevention of the chemical and mechanical events in the muscle upon which it is dependent. As expected, mixtures of these three chelators produce profound alteration in texture (Table 27.1). The polyphosphates are readily hydrolyzed by active beef muscle polyphosphatases. The latter are of particular significance in comminuted meat to which polyphosphates have been added for improved protein functional properties (Chapter 34). They are the subject of a series of at least 11 papers by German investigators (Neraal and Hamm

TABLE 27.1

EFFECT OF CHELATING AGENTS ON BEEF TEXTURE

Treatment	Percentage of Control Shear[1]	Stretch[1]
Sodium pyrophosphate (P)	94	54
Sodium hexametaphosphate (H)	75[3]	51
Magnesium chloride (M)	78[2]	24[3]
P + H	74[3]	
M + H	84[2]	36
M + P	85[2]	57[3]
M + P + H	97	51
Sodium citrate	76[3]	33

Source: Adapted from Streitel *et al.* (1977).
[1] As compared to controls whose mean values ranged from 38.55 to 45.38 (shear) and from 13.66 to 30.83 (stretch); both shear and stretch measured with Warner-Bratzler Shear apparatus.
[2] Significant difference at 5% level ($p < 0.05$) from T-test statistics.
[3] $p < 0.01$.

1977). Tripolyphosphatase is activated by $MgCl_2$, EDTA, and NaCl and inhibited by $CaCl_2$ and diphosphate. Diphosphatase (pyrophosphatase) is activated by $CaCl_2$, EDTA and high concentrations of $MgCl_2$, and is inhibited by NaCl and low Mg concentrations. Freezing and thawing result in regeneratable inactivation. The effect of polyphosphate chelators on meat texture may thus depend upon the presence of other important meat-associated factors.

Prerigor Inactivation by Cooking.—De Fremery (1963) suggested that inactivation of the glycolytic enzymes before rigor would prevent its achievement and demonstrated the validity of this hypothesis by producing more tender chicken meat using boiling as the means of inhibiting these enzymes. Increased tenderness has also been achieved in beef by broiling, oven roasting, deep fat frying (Paul *et al.* 1952) and exposure to microwaves (see also Chapter 10). Weidemann *et al.* (1967) attributed the improved texture not so much to enzyme inactivation as to disruption of the structure of myofibrils by what they called "supercontraction" clots of heat-coagulated protein. It is likely that during the attainment of the enzyme-inactivating temperatures, some potentiation of proteinase action occurs which could contribute to the tenderness as discussed below.

Potentiation of Proteolytic Enzymes.—*In Frozen Meat.*—While the role of lysosomal enzymes in normal postmortem aging may be debatable, they definitely do appear to participate in improvement of meat texture subsequent to appropriate cool-freeze-thaw programming of muscle. Winger *et al.* (1976) recommend, for instance, completion of rigor at ambient temperatures followed by freezing at −3°C and subsequent aging at 15°C for 24 hr. The freezing step undoubtedly releases lysosomal proteinases which speed up the aging process.

Potentiation of Collagenase Activity.—Temperature manipulation going in the opposite direction can also result in improved meat texture. Although the tenderizing effect of long time-low temperature cooking had been recorded in the science literature over 40 years ago and was probably known to cooks before that, it is only in the last 10 years or so that this effect has been traced to the potentiation of a muscle collagenase briefly referred to in Chapter 26 (Laakonen 1973A); the effect has been developed into a patentable process (Laakonen 1973B). Significant increase in tenderness can be achieved by exposing beef muscle to temperatures in the range of 50°−60°C for 6 to 10 hr (Fig. 27.1). Measures have to be taken to reduce and maintain low bacterial counts during this treatment. Since at these temperatures other proteinases are likely to become activated through potentiation—a collagenase acts on native undenatured collagen whereas just about any protease will act on heat-denatured collagen—one would expect that these other enzymes might also contribute to the resulting tenderness. Furthermore, such enzymes would also be expected to act on the myofibrillar proteins. This would explain the finding of Bouton *et al.* (1976) that normal aging of veal is enhanced and accelerated at such temperatures.

Tenderay.—A discussion of the means of accelerating aging would not be complete without mentioning the "Tenderay" process proposed over 25 years ago. In this process, aging is accelerated by exposing meat to 15°C and UV rays for 3 days instead of the usual 2 weeks at 2°C. According to Wilson (1960) aging temperatures as high as 43°C are compatible with acceptable meat in the presence of adequate bacteriostatic agents. In this connection Schreiner (1969) claimed adequate tenderization of beef within 24 hr after slaughter of beef exposed to 45 p.s.i. of oxygen at 35°C. Also similar to this approach is the glazing of fresh meat with a microbial-impermeable coating such as polyacrylate. Of course certain choice cuts of beef are deliberately aged longer than the normal period to further the protein disintegration process not only by the meat's own enzymes, but perhaps *via* the

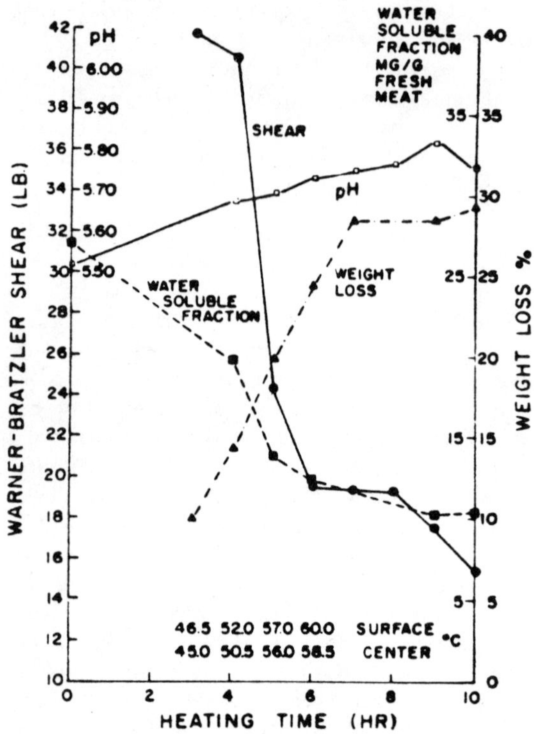

From Laakonen et al. (1970)

FIG. 27.1. THERMAL POTENTIATION OF BEEF MUSCLE COLLAGENASE

Results in meat tenderization.

enzymes supplied by surface microbes, especially the mold *Thamnidium*, which appear during such prolonged aging.

Other Procedures.—Several procedures, mostly patented, for tenderization of meat appear to be due to the action of meat proteases. We have already mentioned the antemortem injection of sulfhydryl-containing compounds and that they probably activate some of the cathepsins. Postmortem physical treatment with chemicals has also been reported, mostly in the patent literature. Thus the success of exposing postrigor beef muscle heated to 40°–60°C to pressures of the order of 100 MNm^{-2} applied for 2.5 min was attributed to more effective alteration, through denaturation, of the myofibrillar component of toughness (Bouton *et al.* 1977). We believe that the treatment would also increase accessibility of potential substrates to the native proteinases. Injection of water (3% of the live weight of the animal) and gas (Williams 1964A), or solutions of carboxymethyl cellulose (Williams 1964B) presumably tenderize the meat by expanding the muscle fiber bundles, which are penetrated, separated and saturated by the water, allegedly resulting in the enhancement of the activity of hydrolytic enzymes. Enzyme involvement in this manner may lead to increased tenderness through massaging (Siegel *et al.* 1978).

Sodium chloride, about 1%, added to rabbit muscle mince (Bendall 1954), to freeze-dried muscle (Wang *et al.* 1958) or as a result of "koshering" in which the meat is soaked, sprinkled with salt crystals and the latter rinsed away (Biran 1974), has a definite tenderizing effect perhaps in part due to the specific activation of cathepsin C (Table 26.5). NaCl and the other salts mentioned in the foregoing discussion are additives in proteolytic enzyme preparations used as meat tenderizers (Chapter 5).

In a sense the new hot processing of beef may be considered as an instance of exploiting endogenous enzyme action. Muscle tissue is removed from the warm carcass of electrically stimulated muscle. The resulting meat is tender because the electric stimulation speeds the onset of rigor, presumably by affecting the activities of glycolytic enzymes in the region of potentiation temperatures (Cross and Berry 1980).

ACTION OF THE MEAT TENDERIZING ENZYMES

Biochemistry of the Meat Tenderizers

Papain.—Although by far the most widely used of meat tenderizers, the term "papain" is ambiguous not only with respect to pronunciation (*pa*·pa'in, pa'*pa*·in, pa·pa in'), but also to meaning. It can refer to the enzyme preparations of commerce or to one of the two minor "papainases" present in the latex of the papaya tree (Chapter 5). Papainases are plant enzymes which are activated by reducing agents and inactivated by oxidizing agents by virtue of the presence of a sulfhydryl group in their active sites. Papain itself was the first of the papainases to be isolated and characterized (Balls *et al.* 1940; Balls 1941; Lineweaver and Schwimmer 1941) and is probably the most well-characterized of proteinases not present in mammalian pancreas (Drenth *et al.* 1971; Glazer and Smith 1971). The sequence of the 102 amino acids comprising the 21,000 dalton molecule is shown in Fig. 27.2. The active site of the enzyme, which can accommodate seven amino acid residues of a protein substrate, is in the shape of a cleft made up of two distinct parts of the chain involving histidine 159 and cysteine 25. The presence of cysteine in the active site explains the well-known requirement of papain for the presence of an appropriate reducing agent to assure maximal activity. This characteristic has rather important practical consequences when used as a meat tenderizer. The involvement of this cysteine sulfhydryl group in the mechanism of action is usually depicted as an exchange of an acyl group of a peptide for a hydrogen:

$$\text{Papain-SH} + \text{R-CO-NHR}' \rightarrow \text{Papain-S-CO-R} + \text{R}'\text{NH}_2$$

The reaction is completed by exchange with water in a second analogous reaction:

$$\text{Papain-S-CO-R} + \text{H}_2\text{O} \rightarrow \text{Papain-SH} + \text{RCOOH}$$

Papain exhibits appreciable activity at the pH of meat (see below). The specificity of papain in terms of preference for peptide bonds in protein substrates is shown in Table 27.2. Noteworthy is the susceptibility of the bond once removed in the direction of the carboxyl end of the protein substrate. Apparently this type of specificity extends to other amino acids including phenylalanine. However, it can split almost any peptide bond.

Other Meat Tenderizing Enzymes.—In addition to "papain" the latex of the papaya fruit contains at least two distinct papainases, papaya peptidase A and chymopapain present as distinct isozymes. The latter is a more active rennet than papain is (chymo = milk) and is characterized by its extreme acid stability. It is probably the enzyme that survives the acidity of the stomach and is thus the active ingredient in the use of crude papain as a digestive aid (Chapter 36). The role of chymopapain in meat tenderization has been neglected until relatively recently (Kang and Warner 1974). Thanks to these and related investigations it is now fairly certain that chymopapain, and not the

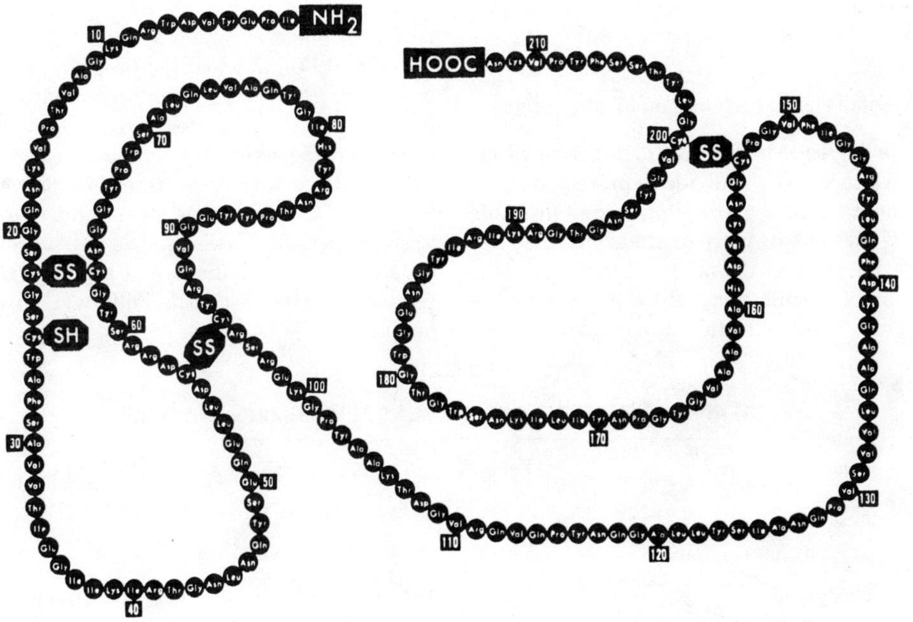

From Drenth et al. (1971)

FIG. 27.2. PRIMARY STRUCTURE OF PAPAIN

Showing the sequence of the 212 amino acid residues, the location of the essential sulfhydryl group (SH) and that aspect of the tertiary structure determined by the intramolecular crosslinking of distant amino acids through cystine disulfide bridges (SS).

first of the papainases to be isolated, papain, is the major contributor to meat tenderizing action of crude papain sold as meat tenderizer. Its pH optimum for meat proteins is about 5, close to that of meat, as compared to 7–8 for the other papaya proteinases. Its greater heat stability allows it to act during the heating-up period in cooking when SH groups, exposed as a result of protein denaturation, become available for activation of the meat tenderizing function.

Also available commercially as GRAS (by the United States Department of Agriculture) tenderizers of not only beef (prior to 1981) but now also *all* red meats (pork, lamb) as well as poultry are proteinase preparations from figs (ficin), from pineapple (bromelin, Table 5.1) and from the *Aspergillus* fungi, *A. oryzae* and *A. flavus.* As discussed later, tenderizing formulations may contain one or more of these proteinases.

The pancreas is another source of enzymes proposed as meat tenderizers. The action on meat of the commercial preparations used for this purpose are probably due to trypsin, chymotrypsin and, perhaps more important, elastase and a reported collagenase activity. Finally, specific acting collagenases have been proposed as components of meat tenderizers. Peptide bond specificities are shown in Table 27.2.

Action of Tenderizing Enzymes on Meat Proteins

Action *in Vitro*.—*Myofibrillar and Sarcoplasmic Proteins.*—Data on the individual protein components of meat tissue *in vitro*, summarized in Table 27.3, show that one or more of these enzymes can hydrolyze one or all of the main classes of proteins in meat—sarcoplasmic, myofibrillar and stromal—isolated from meat skeletal muscle tissue. Thus McIntosh and Carlin (1963) reported significant hydrolysis of the beef myogen [easily water soluble muscle sarcoplasmic proteins (Chapter 5)] by huge dosages of crystalline papain. Other sarcoplasmic proteins, especially purified enzymes of the sarcoplasm, have been subjected to the action of meat tenderizing enzymes for the purpose of determining amino acid sequence.

McIntosh and Carlin also found appreciable digestion of the myofibrillar proteins of beef, myosin and actomyosin, by the crystalline enzyme. Using a myosin:papain ratio of 8:1 they found 10.2% of the myosin nitrogen converted into an ultrafilterable form in 20 hr at 45°C and pH 7. By contrast, purified enzymes which are present in meat tenderizers are employed in extremely small amounts to elucidate the structure of the myofibrillar proteins. A complex of the regulatory myofibrillar proteins, tropomyosin-troponin, is readily split by trypsin (Ebashi and Kodoma 1966) at a protein:enzyme ratio of less than 1000:1. Similar ratios were used by Rattrie and Regenstein (1977) in a study designed to ascertain the short-time action of crude papain, presumably of meat tenderizing grade, on the major isolated myofibrillar proteins. They too observed a fast but very limited proteolysis, as measured by appearance of new bands in gel electrophorograms. The action was particularly rapid when the proteins of the myofibrils were contracted. The accumulated evidence based upon such model experiments, especially those showing action at very low levels of enzymes, is compatible with the suggestion that the myofibril is a site of action meat tenderization over a short period. However, the scleroproteins are usually considered the prime target of the meat tenderizing enzymes, especially over an extended period.

TABLE 27.2

PEPTIDE BOND PREFERENCES OF MEAT TENDERIZING ENZYMES[1]

Papain	Val-X-Y	Lys-X	Arg-X	X-Y	X $\not-$ Pro
Trypsin	—	Lys-X	Arg-X	—	—
Chymotrypsin	Trp-X	Phe-X	Leu-X	Met-X	—
Elastase	Ala-X	X_n-Y	—	—	—
Collagenase	X-↓Gly-Pro	X-Gly	—	—	—
Mold acid proteases	X-Y except	Gly $\not-$ X &	Ala $\not-$ X	—	—

Source: Constructed from information in Boyer (1970–1976) and Drenth *et al.* (1971).
[1] X-Y denotes general specificity, the arrow which of two bonds is split and X_n a neutral amino acid. Left to right indicates progression from amino to carboxyl end of the protein chain. The symbol $\not-$ denotes bond not split.

TABLE 27.3

ACTION OF PROTEINASES ON MUSCLE PROTEINS *IN VITRO*

Authors[1]	A	B	C	D	E	F	G
Proteinase (E)	Trypsin, (X)[2] Papain	Collagenase	Papain Bromelin	Papain, (X)	Papain Ficin	Papain, (X) "Adolphs"	Pancreatin "Viokase"
Substrate (S)	Myosin	Collagen	Various	Various	Collagen	Mucoprotein	Collagen, "acid sol."
E/S, mg/g	0.33 / 0.1	67	1.6	50	20	42 / 833	98
Time	10 min	Variable	3 hr	20 hr	30 min	20 hr	24 hr
Temperature	Ambient	37°C	38°C	45°C	Variable	45°C	37°C
Measure	Single-scission	OH-Proline soluble	Soluble $N,\text{-}NH_2$	Viscosity, Soluble N	TCA-sol. N	Ultra filt., Hexos-amine	OH-proline, soluble

[1] Sources: (A) Lowey *et al.* (1969); (B) Goll *et al.* (1964); (C) Kang and Rice (1970); (D) McIntosh and Carlin (1963); (E) Hinrichs and Whitaker (1962); (F) McIntosh (1967); (G) Satterlee (1971).
[2] Indicates that crystalline enzyme was used.

Stromal or Scleroprotein.—The morphological relation of the connective tissue, site of the scleroproteins, to the myofibrils is shown in Fig. 26.3. One of the few earlier studies of meat tenderizing enzyme action on isolated scleroproteins is that of Hinrichs and Whitaker (1962), who reported that neither ficin, papain, bromelin nor trypsin attacked native beef Achilles tendon collagen. On the other hand McIntosh and Carlin (1963) reported considerable action of crystalline papain on the collagen from commercial Achilles tendon and from laboratory prepared connective tissue. The apparent discrepancy may be due to differences in the preparation of the Achilles tendon collagen which is complexed *in vivo* to a glycoprotein (Anderson and Jackson 1972). However, the data of McIntosh and Carlin are at variance with those of Satterlee (1971), who reported complete refractoriness of the "acid soluble" collagen and connective tissue to even more massive doses of papain than used by McIntosh and Carlin.

However, Satterlee did observe considerable action of porcine pancreatin on various fractions of connective tissue collagen and succeeded in partially purifying this pancreatic "collagenase." He suggested that this organ might yield an inexpensive source of a specific collagenase for meat tenderizing formulations. Other possible sources of useful collagenases include the previously mentioned heat-activatable collagenase of beef tissue (Laakonen *et al.* 1970), a highly activatable collagenase in cultures of bone explants (Vaes 1972) and the well-characterized classical collagenases of bacteria especially those of *Clostridium* (Seifter and Harper 1971). The specificity of one of these "true" collagenases is shown in Table 27.2. About one out of every three amino acids in collagen is glycine and one out of every four is proline. Both of these amino acids are involved in the specificity of this enzyme. However, "true" collagenases must also possess some special properties which allow them to act on the highly complex native collagen.

One of these collagenases (from *C. histolyticum*) was used to probe age-associated changes in bovine muscle connective tissue (Goll *et al.* 1964). The observed decreased digestibility of the connective tissues with advancing age (except for steer collagen which appeared to be protected from enzyme action by an excess of lipid) was attributed to stronger and more extensive crosslinking in the collagen from older animals. The implications of these observations for effectiveness of meat tenderizing will be brought out later. Denatured collagen, i.e., heated during cooking of meat, is probably susceptible to plant, animal and microbial tenderizing enzymes. Although no data are available on the action of purified preparations of the enzymes in fungal commercial meat tenderizers on isolated collagen, it is of interest to note in Table 27.2 that fungal acid proteases do not as a rule readily attack glycyl peptides. This may explain the oft-cited resistance of connective tissue to this class of enzymes.

Discussions of the antemortem aging and health implications as well as the chemistry of collagen crosslinking are available in one volume (Friedman 1977B).

Elastin, the other major scleroprotein of meat tissue, is readily attacked by the plant-meat tenderizing enzymes (McIntosh and Carlin 1963) and of course by meat tenderizers containing the pancreatic enzymes because of the elastase content of the latter. Pancreatic elastase will split almost all peptide bonds in which a neutral amino acid is one of the partners (Hartley 1971). As mentioned, microbial enzymes act very sluggishly, if at all, on alanyl and glycyl peptides which constitute a large proportion of elastin. Partridge (1966), in commenting on reports (not cited) that the rate of elastase digestion of elastin increases with the chronological age of the elastin source, suggested that this effect could be due to more effective screening by ground substance mucopolysaccharides in the *young* tissue. This effect would be in the opposite direction to that observed with collagenase acting on collagen.

Kang and Rice (1970) compared proteolytic action on the sarcoplasmic fraction protein. Collagenase (presumably bacterial), bromelin and trypsin acted preferentially on the stromal (connective tissue) protein whereas papain, ficin and Rhozyme P-11 acted more extensively on the myofibrillar protein. In further investigations, Kang and Warner (1974) delineated the action of the papaya proteases on the same meat protein fractions.

On the premise that alteration in the "ground substance" of muscle tissue might be responsible for textural changes during both postmortem aging and tenderization by added enzymes, McIntosh (1967) isolated the mucoprotein fraction from bovine skeletal muscle and showed that it was readily degradable by papain.

Action *in Situ*.—The alternative approach to an understanding of the action of meat tenderizing enzymes is to measure the changes induced in meat proteins while they are still in the meat, as summarized in Table 27.4. One of the earlier examples is illustrated in Fig. 27.3 which depicts the effect of salted, partially dried papaya latex on a beefsteak (Balls *et al.* 1940). Instead of measuring the products of the action of the meat tenderizers, the decrease and change in character of the remaining protein may be measured. Using this approach, Miyada and Tappel (1956) reported "good" hydrolysis of the soluble beef proteins and of collagen by papain, bromelin, ficin, trypsin and by the fungal enzyme preparation, Rhozyme P-11. Papain and ficin could attack elastin. Certain other microbial commercial preparations and pepsin showed very little action. The effectiveness of these enzyme preparations on collagen was probably owing to action on the denatured rather than native form of this protein which is attacked only by true collagenases. Thus, in a similar study, El-Garbawi and Whitaker (1963) concluded that collagen must be denatured before ficin and bromelin (both crude) could render this protein soluble. These authors also reported that all of the protein fractions of beef are solubilized maximally at pH 7 and 80°C and that ficin is particularly effective in the solubilization of elastin at temperatures as low as 20°C.

Histological Changes.—This general pattern of action is in agreement with the relative effectiveness of commercial enzyme preparations on the structural integrity of meat muscle tissue components as observed under the microscope (Wang *et al.* 1958). These authors could detect little or no action of microbial enzymes on the collagen of isolated perimysium, but did demonstrate collagenase activity of plant and pancreatic meat tenderizers.

TABLE 27.4

ACTION OF PROTEINASES ON BEEF MUSCLE PROTEINS *IN SITU*

Reference	Gottschall and Kies (1942)	Wang and Maynard (1955)	Miyada and Tappel (1956)	Tappel *et al.* (1956)	El-Garbawi and Whitaker (1963)	Stromer *et al.* (1967)	Dubois *et al.* (1972)
Proteinase	Papain, salted latex	Papain "Adolphs"	Papain	Papain "Nutritional"	Ficin "Merck"	Trypsin, purified "Sigma"	Papain "Miles"
Substrate	Beef, ground	Beef, cubes, 1.5 ml	Beef, freeze-dried	Beef, freeze-dried	Beef, freeze-dried (biceps)	Myofibrillar suspension	Beef-extract, homogenate
E/S[1]	0.6	3–30	30	14	7	10	1
Time	30 min	various	1 hr	1 hr	1 hr	24 hr	various (to 24 hr)
Temperature	various	various	60°C	60°C	60°C	2°C	various
Measure	TCA-sol. N	Histology	Collagen, Sol. N.	Histology	Collagen, various	Histol. Z-line	Oil emulsification

[1] Based upon average protein content (20%) of fresh beef.

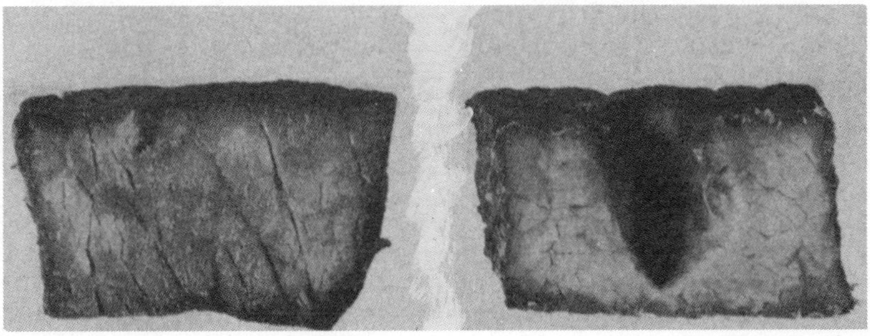

FIG. 27.3. (RIGHT) DIGESTION OF BEEFSTEAK WITH PAPAIN AT 5°C. (LEFT) UNTREATED

From Balls (1941)

Indeed to Wang and coworkers we owe the basis of the present knowledge on the structural changes induced by the action of enzymes in commercial meat tenderizers. Within the muscle fiber cell they found the following sequence of disintegration of the structural elements: sarcolemma, muscle fiber envelope, muclei, fading of the structure of the actomyosin complex and merging of the fibers. These and other results suggest, at first hand, that one difference between postmortem aging and enzyme addition is that in aging no change in the outer membrane of the muscle fiber can be detected in the light microscope. However, this is not a valid comparison since the levels of added enzyme required to effect these changes are much higher than those required for optimal tenderization of the meat.

Traditionally, plant and animal enzymes act only on connective tissue proteins whereas the fungal proteases do not. Sato *et al.* (1967) described a meat tenderizing *Aspergillus* protease which, based upon microscopic observations, attacked collagen more strongly than other muscle proteins. Tappel *et al.* (1956) used a histological approach to study the attack of papain at the surface of cooked beef. They conclude that the added enzyme did not affect the sarcolemma membrane nuclei nor muscle fiber at distances greater than 0.5 mm. In a somewhat more defined system Stromer *et al.* (1967) found that treatment of myofibrils with trypsin resulted in a rapid loss of the Z-line. Treatment of supercontracted myofibrils caused a return to the banding pattern of resting muscle.

Protein Functional Properties.—The effect of enzymes on protein functional properties will be explored in more detail in Chapter 34. Here we wish to point out that in addition to chemical and histological changes, several physical alterations induced by proteases have been observed. Thus, from a study of protease-induced viscosity changes in muscle extracts, DuBois *et al.* (1972) suggested that some of the resulting fragments perhaps similar to that released from troponin T by CAF (Chapter 26) undergo physical aggregation during proteolysis. The same authors also found that oil emulsification capacity is increased by papain; after 4 hr the treated samples took up 140 ml of oil per 10 ml of muscle homogenate as compared to about 100 ml for the control. Meat tenderizers also decrease the water-holding capacity (Jay 1966; Sato *et al.* 1967; Hasegawa *et al.* 1970) and the extract release volume of a filtrate of meat homogenate under standard conditions. The latter two measurements are reliable indices of the microbial quality of beef, probably because of the action of proteases present in the contaminating microorganisms. In this connection Hasegawa *et al.* (1970) observed extensive inactivation of a diverse array of sarcoplasmic enzymes by the proteases of meat spoilage microorganisms.

Tenderizing Action

Effectiveness.—In addition to the above biological, chemical and physical parameters delineating the action of meat tenderizing enzymes on meat, there is, of course, "proof-of-the-pudding" evidence which demonstrates directly—by both sensory and by such physical measurements such as shear force, penetrability, etc.—that increase in the tenderness of meat does occur (see Tables 27.5 and 27.6). The data show that rather low enzyme:substrate ratios and short times are involved in these experiments as compared to those previously discussed. For the most part the specific activities of commercial tenderizers used in these experiments lie in the range of one to two orders of magnitude less than those of purified proteases, especially

TABLE 27.5

TENDERIZING ACTION OF PROTEINASES: CONDITIONS

Authors	"Substrate"	"Enzyme"	E/S	Time	Temperature	How Applied
Hay *et al.* (1953) (A)	Beef, raw steaks	Papain-"commercial tenderizer"	13	1 hr/in.	Ambient	Sprinkle, fork
Mier *et al.* (1962) (B)	Beef, raw 3½ × 4 × 1 in.	Liquid tenderizer	1–10	1.5 hr	Ambient	Fork, pierce
Satterlee (1972) (C)	Beef, raw canner and cutter	Pancreas, extract-dried, "Viokase"	0.5	2 hr	5° and 13°C	Injection, postmortem
Moreau and Jankus (1963) (D)	Beef, raw	"Standard papain"	0.06	2 hr	70°C	Injection
Wang *et al.* (1958) (E)	Beef, freeze dried	Papain, "tropical plant"	0.03	30 min	Ambient	Rehydration
Goodwin and Waldroup (1970) (F)	Chicken	Papain, crystalline	0.01	—	Ice, slush	Injection, antemortem
Fry *et al.* (1966) (G)	Chicken	Papain, "Swift" 6 X, liquid"	0.2 2	— —	— —	Injection, antemortem, postmortem
Sosebee *et al.* (1964) (H)	Chicken, freeze dried	"Papain" Rhozyme-P-11	0.1	5 min	65°C	Rehydration
Smallings *et al.* (1971) (I)	Hog, live (to ham)	Papain "purified food grade"	0.25	—	—	Injection, antemortem

papain (Miyada and Tappel 1956; Huffman *et al.* 1961). This underlines the often stressed admonition that the amount of enzyme needed to produce optimum tenderization is much less than that required to detect *most* physical and chemical changes except those directly measuring texture parameters such as shear. However, even within this range, the amount of enzyme, the time and the temperature compatible with optimal tenderness are highly dependent upon the mode of application. The amount of enzyme per unit of protein decreased in the following order: dusting or soaking of fresh meat; forking, rehydration of freeze-dried meat with solution enzyme; postmortem injection; and finally the least amount of enzyme is used in antemortem injection.

Increased tenderness, as judged subjectively, is always accompanied by decrease in shear values. However, as has been pointed out, these measurements may not adequately represent all factors associated with tenderness such as chewability, ability to sink teeth into, etc. While consistent decreases in shear values accompanying increased tenderness is rather good *prima facie* evidence of the true effectiveness of meat tenderizing enzymes, adequate and optimal tenderness is not always accompanied by increased panel acceptability or preference (Smallings *et al.* 1971).

While Table 27.6 includes mostly results with papain-based tenderizers, other proteases have been used in quantitative testing for tenderness. In general it would appear that microbial-based enzyme tenderizers were in the past less effective than commercial papain-based ones (Wang *et al.* 1958; Sosebee *et al.* 1964). Of course this is highly dependent upon the proportion of inert ingredients added.

The effect of pancreatic enzymes has also been measured (Satterlee 1972). A pancreas enzyme preparation was most effective on USDA choice meats. That the collagenase activity of the pancreas preparation was involved was indicated by the requirement of double the amount of enzyme to tenderize Canner and Cutter beef from older animals. As previously mentioned, the collagen of older animals, by virtue of increased molecular crosslinking, is more resistant to collagenase action (Goll *et al.* 1964; Friedman 1977B).

Finally, it should be mentioned that amount and type of enzyme to be used may depend upon the method and conditions of cooking (Hay *et al.* 1953).

MEAT TENDERIZER TECHNOLOGY—STATE-OF-THE-ART

Sources, Production, Formulation

Historical evidence shows that application of meat tenderizing enzymes to meat antedates experimental systematic demonstration and modern usage (Wittmack 1878) by at least centuries if not by millennia. Chroniclers of the travels of the early exploration of the western hemisphere reported on the practice of cooking meat wrapped in papaya leaves. In modern practice, the main source of the meat tenderizing enzyme is the latex exuding from the cuts on the surface of the green unripe fruit of the papaya (Chapter 5). As previously indicated (Chapter 5) the latex of the fig (ficin), the pineapple fruit (bromelin) and dried hog pancreas (pancreatin, a mixture of proteolytic enzymes) and, of course, microbes, both molds and bacteria, and even yeast, have been proposed and may actually be employed (as judged from the patent and advertising literature) in meat tenderizing formulations. However, it is most likely that papaya latex far outstrips these other sources. Debecze (1970) estimated that one-third of the 300,000 kg of papain (dried papaya latex) imported per year into the United States was incorporated into tenderizer formulations intended for home use (see also Chapter 5). In 1958, approximately 75,000 liters of meat tenderizing solutions were sold (Ziemba 1958).

Production of crude papain and other meat tenderizing enzyme preparations has been dealt with in Chapter 5. The patent literature is rather sparse with regard to further processing or purification of the dried latex specifically for meat tenderizing purposes beyond formulating it for specific purposes. Hogan (1966) patented a series of alternative preliminary purification steps in preparation for antemortem injection. These include fractionation with salts or with organic solvents. In another patent Frederics (1965) described the preparation of crude proteolytic enzyme for meat tenderization from mesenteric lymph glands, said to be located around the casing of the lower intestine of edible animals. This is accomplished by simply covering the glands with several changes of acetone and drying (Chapter 5).

The state-of-the-art aspect of tenderization of meats and meat products as revealed by the patent literature is examined from the following viewpoints: source of enzymes in tenderizer formulations (Table 27.7), noninjection methods of applying tenderizer, and ante- and postmortem injection methods and problems.

Sources and Selective Action of Meat Tenderizing Enzymes.—Almost all of the patents mentioned papain and almost as frequently the plant proteinases, ficin and bromelin. In some instances the use of mixtures of papain and bromelin is recommended and illustrated. Thus, Sleeth and Campbell (1965) state that with such a mixture the myofibrillar proteins are attacked by papain and the connective tissue proteins by bromelin. A similar enzyme mixture was used for treatment of

TABLE 27.6

TENDERIZING ACTION OF PROTEINASES: RESULTS[1]

Ref.[2]	Sensory Data Scale[3]	Sensory Data Control	Sensory Data Treated	Shear Data Control	Shear Data Treated
A	0→10	7.1	8.4	26	18
B	6→1	4.0	1.8	23.9	15.8
C	1→5	2.2	3.0	14.3	12.2
E	1→10	5.5	7.0	—	—
F	1→9	4.60	6.05	—	—
G	1→8	2.5	3.7	23	15
H	−4→+4	−1.65	2.57	—	—
		−1.45	1.75		
I	1→9	6.6	7.8	6.24	−4.39

[1] Difference between "control" and "treated" significant at 5% level or lower.
[2] See Table 27.5; D not included here.
[3] Increasing tenderness in direction of arrow.

TABLE 27.7

ENZYME SOURCES OTHER THAN *C. PAPAYA* IN PATENT LITERATURE

Source	Patent No.[1]		
Plants:			
Pineapple (bromelin)	2,904,442		3,215,534
	2,961,324	3,188,123	3,533,803[2]
	3,156,566	3,188,803	3,549,987[2]
	163,540		
	3,166,423[2]	3,331,692	1,095,515 (Brit.)
Fig (ficin)	2,961,324	3,276,879	3,533,803
Milkweed (asclepain)	3,156,540	3,331,692	
Peanuts (arachain)	2,961,323		
Animals:			
Pancreas (pancreatin, trypsin)	3,183,097	3,533,803	978,885 (Brit.)
Mesenteric lymph	3,183,097		
Cathepsin	3,442,660	2,961,324	
Microorganisms:			
B. subtilis (subtilisin)	3,037,870		
Microbial collagenase	3,056,680		
Fungal—unspecified	2,825,654	2,904,442	
Fungal—*Trichoderma*			
Fungal—*Thamnidium, Aspergillus*	3,120,191		
Yeast	2,961,324		

[1] Unless specified, all are U.S. Patents.
[2] Mixture of plant proteases specified.

corned meat products (Bernholdt 1970). Mixtures of collagenase-elastase (of fungal origin) and myofibril-hydrolyzing plant protease comprise the basis of a tenderizer formulation by Underkofler (1959). Preferential action on scleroprotein is the basis for the use of a bacterial collagenase-elastase preparation (Williams 1962) and, it will be recalled, the basis for the action of pancreatic enzymes as tenderizers, according to Satterlee (1971). Application of the relatively heat-stable bacterial protease, subtilisin, as an alternative to papain is dictated by papain's undesirable stability to the temperatures used in the curing process for hams (Schleich and Arnold 1962; Frederics 1965). Spores of *Thamnidium* and *Aspergillus* molds were injected antemortem, presumably to provide proteases similar to those effective in long-aged beef for choice cuts of steak (Williams 1963).

Nonenzyme Ingredients of Meat Tenderizer Formulations.—Table 27.8 summarizes information on the ingredients used in meat tenderizing formula-

TABLE 27.8

NONENZYME INGREDIENTS IN TENDERIZER FORMULATIONS

Patent U.S.	NaCl	MSG	Poly.	Sugar	Other
2,805,163	–	+	–	–	–
2,825,654	+	+	–	+	Oil
2,961,324	–	+	–	–	PH
3,033,691	+			+	Pr. Gl.
3,037,870	–	–	+	–	–
3,076,712	–	–	–	–	Antibiot.
3,118,213	+	–	+	–	Pi
3,147,123	–	+	+	–	–
3,166,423	+	–	–	–	–
3,215,534	+	+	–	+	PH, fat Pr. Gl.
3,276,879	+	–	–	–	–
3,331,692	+	+	+	+	PH, Fat, Pi
Br. 900,994	+			+	–
1,036,094	+	–	–	–	Oleo
3,549,385	+				Glycerine

MSG = monosodium glutamate; Poly = polyphosphates; PH = protein hydrolysate; Pr. Gl. = Propylene glycol; Pi = orthophosphate.

tions other than the source of enzyme. These additives, especially sugars and salts in dry preparations, lend bulk and help standardize potency (see Chapter 5 for further purposes). Sugars are also said to stabilize activity of liquid preparations (Underkofler 1959). We have already alluded to the tenderization enhancing properties of both NaCl and the polyphosphates. Glutamate and protein hydrolysate as well as fat and spices may be incorporated for enhanced flavor and marbling.

Meat and Meat Products

Until quite recently meat tenderizers were applied mostly to steaks and thin slices of meat. In Table 27.9 are tabulated their use in other meat products, at least as revealed by the patent literature. With the advent of injection methods, tenderization of roasts became possible (see below).

TABLE 27.9

PRODUCTS FROM PROCESSES USING MEAT TENDERIZERS

Roasts, steaks, raw	Ground meat
Steaks, pieces, freeze-dried	"Enriched" meat
Cured meats, hams	Rendered products
Corned meats	Removal of meat from bones
Nitrite-free bacon	Sausage casing
Canned beef, pieces	Liver sausage
Poultry, lamb	

The special enzymatic requirements for tenderizing porcine muscle for ham production and for tenderized poultry have already been mentioned. Proteolytic enzymes may also be used to advantage in corned meats processing. By treatment of beef with a proteinase-containing pickling brine devoid of inhibitory additives, it was possible by dry roasting (*vs* the traditional simmering for many hours) to obtain a finished product within 2 hr. In addition, pickling time and final salt concentrations could be reduced (Bernholdt 1970). Freeze-dried meats can be tenderized by rehydration in a proteinase-containing liquid which comes with the package (White 1967) or by treatment of pieces of meat before freeze-drying (Cyclus 1962). Meat pieces destined for canning and freezing can also be treated with tenderizers. The attendant surface sloughing of the pieces during high-temperature short-time processing can be avoided by briefly submersing the meat in boiling water. This inactivates the enzyme (previously added *via* injection) at the surface. Remaining enzyme then tenderizes the meat in the interior of the pieces by exposing them to the optimum temperature of action of the tenderizing enzyme (Schack and Connick 1970). Pepsin is used in a process, also suitable for mention in Chapters 19 (color), 34 (proteins), or 36 (health), which describes the manufacture of a bright red nitrite-nitrate-free bacon product (Stead and Kennedy 1980).

Nontenderization Applications.—Pieces of meat were also incorporated in enzyme-treated ground meat to provide a special meat product of thermosetting character (Torr 1966). Comminuted mixtures of muscle and tendon treated with a fungal protease to yield an edible meat product are the basis of a Japanese patent. Recalling their use in preparation of "cod liver" oil in the 1920s, tenderizers may also be used in liver sausage products to liberate nutritionally unavailable vitamins and to increase oil emulsification, of significance in sausage manufacture (Dubois *et al.* 1972). Enzyme-hydrolyzed liver added to meat yields a unique product according to a patent by Laszlo Keller and Erdos (1970). Proteinases have been used to separate meat from bones and for recovering protein from meat rendering operations (Denton and Hogan 1966) and in the manufacture of collagen casings from limed hides (Fagan 1968).

Modes of Application

The following deals with alternative routes of administration of tenderizers to meat and meat products. We may divide these into two major categories: noninjection and injection procedures.

Noninjection.—Perhaps the simplest of the noninjection procedures is that of sprinkling dry tenderizing powders on the raw meat (Komarik 1964), a favorite procedure for home use. Alternative meat, both raw or freeze-dried, can be immersed in a solution of the tenderizer with or without accompanying multiple forking. A device and process for mechanical perforation of meat pieces which are submerged in a protease-containing liquid has been patented (Moreau 1971). In a rather ingenious arrangement, Eickerman (1964) interleaved thin slices of meat with enzyme impregnated paper, so that the meat juices extract the protease in the presence of air which wicks through the sheets. This patent contains a rather rare expression of enzyme dosage: "X lbs per ream of paper." Smythe *et al.* (1962) incorporated protease into an aerosol. The recent development of high pressure devices should lend itself to convenient application of meat tenderizers.

Antemortem Injection.—Tenderization of meat by antemortem injection involves intravenous injection of an appropriate solution of proteolytic enzyme before slaughter. In the first patent describing this mode of application, it is recommended that the enzyme be injected some 5 to 10 min prior to slaughter (Beuk *et al.* 1959). Thus it became possible to tenderize roasts as well as steaks and chops, and indeed apparently improved all cuts of low grade beef as well as poultry, lamb and mutton. Carcass available for such dry-heat cooking is more than doubled to over two-thirds of carcass weight. The amount of enzyme required to achieve optimal tenderness can be adjusted to the original toughness of the animal, from five parts ppm dried papaya latex (commercial papain) for prime heifers to 30 ppm for low grade steers. Presumably the action of the papainases supplements that of the endogenous proteases so that aging is accelerated.

This method of tenderizing had considerable commercial success during the 1960s. According to one source, about 5% of the total beef slaughtered in the United States in the early 1960s was tenderized and about one-third of the patentee's beef is now being tenderized (Mermelstein 1977).

However, certain disadvantages due mainly to over-tenderization arose. The necessity of keeping the meat cold to prevent further action of the enzyme was one difficulty. Even when kept cold, prolonged storage resulted in over-tenderization. Organs—the liver, kidney and heart—became mushy and imparted undesirable flavors. The muscles were not always tenderized uniformly due in some degree to the fact that some were more vascular than others and thus received a higher dosage of tenderizing enzyme. Adverse physiological reactions of the cattle were noted occasionally.

Since the original patent, two general approaches to the solution of the above-mentioned problems have developed: either modification and control of antemortem conditions or a shift to postmortem injection. The following specific measures to improve the antemortem procedure have been patented.

(1) The animal may be stunned and/or injected with a skeletal muscle relaxant to inhibit the voluntary reflex movement prior to slaughter (Reece 1970). U.S. Pat. 3,123,477.
(2) The enzyme is injected over a period equal to 60% of the time required for one blood circulation cycle and then promptly slaughtered. This prevents too much of the enzyme from reaching the organs and glands (Hogan and Bernholdt 1964). U.S. Pat. 3,163,540.
(3) Preinjection with an antibiotic (1.0 mg per kg of live weight). The rationale for this procedure in terms of effect on the control of tenderization is not clear (Williams 1963). U.S. Pat. 3,076,712.
(4) Expose the proteinase-treated meat to compressed carbon dioxide gas (200 to 1200 psi), or to oxygen for not more than 6 hr to apparently inactivate the added enzyme (Shank 1969). U.S. Pat. 3,422,660.
(5) Postmortem injection into the carcass of antemortem enzyme-injected animal of water (under pressure 30 to 100 psi) to the extent of 3% by weight (Williams 1964B). This is, of course, a variation of using water alone in the previously mentioned U.S. Pat. 3,122,446.
(6) Antemortem injection of spores of proteinase-producing microorganisms such as *Aspergillus* and *Thamnidium* (Williams 1964C) instead of active sources of proteinases. U.S. Pat. 3,128,191.
(7) Injection with reversibly inactivated papain. This may be accomplished by oxidizing the essential sulfhydryl group of the papain with catalase-liberated oxygen from H_2O_2 (Hogan 1960), via sulfhydryl-disulfide exchange (Kang *et al.* 1974) or simply by high pH, *ca* 11 (Beuk *et al.* 1977), via a reversible inactivation (Chapter 11). As mentioned, the enzyme does not act on the meat until it is reactivated by the sulfhydryl groups liberated during the early stages of cooking the meat (Hogan 1966).
(8) Injection with a papaya latex preparation treated to reduce but not eliminate the level of its chymopapain component, thus reducing adverse physiological reactions of the animal. Chymopapain can attack living tissues more readily than can the papain and papaya peptidase A components of the latex at the surface of the green fruit (Hogan 1966; Kang and Warner 1974).
(9) Treatments resulting in increased capillary permeability.
(10) Standardization and designation of the amount of enzyme to be injected, not on the basis of weight of tenderizer preparation, but upon the enzyme activity as measured by some convenient and realistic assay method (see below).

Postmortem Injection.—Alternative to improving antemortem injection one can inject the enzyme solution *after* slaughter. This can be done with techniques and injection devices readily accessible and frequently used in meat processing: jab or stitch pumping, multiple entry injection or multiple needle pumping. This should be done promptly after slaughter before rigor mortis sets in (Silberstein 1966). Williams (1964B), in a variation of patents previously alluded to, recommends injection under high pressure of a cold water solution of the enzyme. Sleeth and Campbell (1965) injected a mixture of papain and bromelin using a multiple entry technique.

Enzyme injection for corned beef production has already been mentioned. Williams (1962) avoided the use of papain by injecting a mixture of porcine elastase and classical *C. histolyticum* collagenase and exposing the beef to the endogenous, enzyme-potentiating temperature of 48°C before cooking. As a final note on the use of meat tenderizers, we should briefly like to bring up the problem of how to express the amount of enzyme actually used. In the foregoing tables we have expressed this as weight of enzyme per unit weight of protein assumed to be exposed to the enzyme in order to bring some uniformity to these tables so that qualitatively valid comparisons could be made.

Assessing Potency.—Until active site or immunoassay methods are applied to food enzymes (Chapters 37 and 39), a universal expression of potency on the basis of what the enzyme does rather than the weight of an impure preparation would be preferable. Indeed some of the patentees have expressed the amount of enzyme used on such a basis. Thus Hogan (1966) has defined a tyrosyl unit of plant proteinase activity—papain, bromelin and ficin—as that amount of enzyme which, under standard conditions, will catalyze the hydrolysis of the proteins in 1 hr in 1 g of meat at 40°C to the extent that the equivalent of 100 μg of tryosine will be transformed in a form soluble in trichloroacetic acid. This is similar to the assay of Moreau and Jankus (1963) for the detection and assay of papain in meat except that the latter defined activity in terms of mg of "standard" papain, presumably commercially dried papaya latex. Hogan found from four to 200 tyrosyl units per mg of plant protease. Methods of this type derive from the original assay of Anson (1938). Thus Hogan recommends the injection of 0.2 to 0.7 ml per kg of live animal of solution containing 1000 tyrosyl units per ml in 16 to 77 sec. Other acceptable methods for assessing enzyme activity—and hence enzyme potency—are visosimetric and milk-clotting methods. The latter is especially convenient and, according to Hogan, reflects the physiological effect of the enzyme on the animal since it is usually conducted at 40°C. Silberstein (1966) also expresses activity of enzyme to be used for an antemortem injection as milk-clotting units. Levels of enzyme injected in ante- and postmortem are shown in Table 27.10.

FUTURE POSSIBILITIES

Whitaker (1972B), in a discussion of the next 50 years in meat research, considered postmortem aging and enzyme addition to be areas of research likely to exert a major impact and suggested that it would be desirable to eliminate postmortem aging, partly because of the above-mentioned economic disadvantages. Perhaps this can be accomplished by an informed control of the meat enzymes, combined with addition of highly specific enzymes

TABLE 27.10

AMOUNTS OF MEAT TENDERIZING ENZYMES USED IN INJECTION PROCEDURES

Authors	Injection[1]	Enzyme	Amount
Beuk *et al.* (1959)	AM	Proteases	0.5 mg/lb LW
Williams (1964)	AM	Collagenase; Elastase	1 g/500 lb LW
Hogan (1966)	AM	Papain, plant[2] 1000 tyrosyl U/ml	0.1 to 0.35 tyrosyl[2] units/lb LW
Frederics (1965)	PM	Pancreas, dry; Mesenteric lymph, dry	0.06–0.75 oz/100 lb; 0.03–0.38 oz/100 lb
Sleeth and Campbell (1965)	PM	Papain, 0.004% sol.	*ca* 3% (v/w)
Silberstein (1966)	PM	Papain, other, 0.01–0.05 M.C.[3] units/ml	1–4% (v/w)

[1] AM, antemortem; PM, postmortem.
[2] Specific activity of enzymes = 40 to 200 tyrosyl unit per mg; see text for definition of units.
[3] Milk clotting units.

which will zero in on the offending polymers responsible for toughness. Among such enzymes, according to the Delphi study in the future of enzyme engineering (Hedén 1974), specific collagenases will be in widespread use by 1985. Perhaps such a collagenase will come from microorganisms isolated from cured hides (Lecroisey *et al.* 1975). This most active of all collagenases so far isolated is six times more active than that of the highly active *C. histolyticum.*

Since it is becoming quite clear that it is the quality rather than the quantity of the collagen as determined by degree and kinds of crosslinks that determines how tough a piece of meat will be, perhaps an enzyme that "de-ages" collagen in both senses of "aging" is needed. Such an enzyme has been sought for the purpose of preventing debilitating and aging effects in humans. Bjorksten (1977), in looking for such an anticrosslinking enzyme, isolated a Ca-activated "microprotease" from *B. cereus*, now available for research from enzyme supplyhouses. Its striking feature is its low molecular weight (it acts as a subunit of 5400 daltons) which apparently allows it to penetrate the collagen fibers. In this connection, it would be worthwhile to ascertain whether the active 10,000 dalton papain fragment remaining from the action of intestinal peptidase on papain which still retains its catalytic power after more than half of the original 102 amino acids are removed, would act on collagen. This fragment, first obtained 25 years ago by Hill and Smith (1956), might also be used to advantage in antemortem injection procedures. Other very active sources of proteolytic enzymes which might be considered are those present in metamorphizing insects whose muscle proteins are completely hydrolyzed within hours and the active enzymes in the viscera of fish, often waste products of fish processing.

Perhaps a more effective enzyme than a peptide bond splitter would be one that would hydrolyze or otherwise rupture the N-C linkage contained in most collagen crosslinks. The N comes from the epsilon-amino group of lysine and the C may come from a variety of amino acids (Stimler and Tanzer 1977).

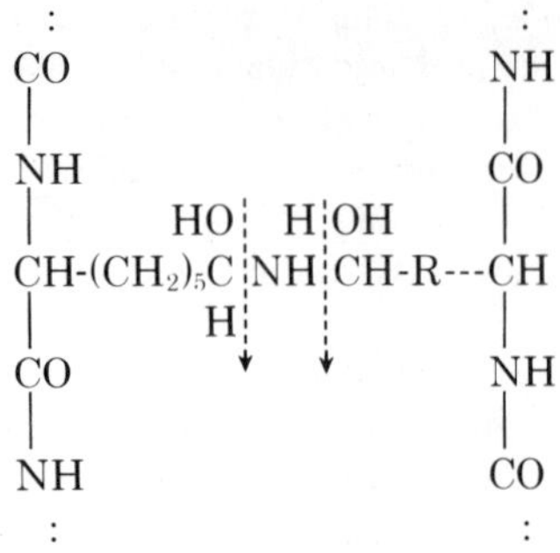

Enzymes which a search of the EC compendium reveals even remotely resemble such an "imine hydrolase" are prolidase and thiaminase (Chapter 35). Another enzyme which would split such a bond but so far has not been isolated is that which demethylates methylated purines such as caffeine (Schwimmer *et al.* 1971; Woolfolk 1975). In each case, the N is in a ring. Some crosslinks in both collagen and elastin are formed by lysine, not through its ε-amino group but through an oxidation product of lysine while it is still in the protein. This oxidation product formed via the action of a Cu-enzyme, lysyl oxidase (Stimler and Tanzler 1977) is α-amino adipic-δ-semialdehyde. Perhaps antemortem inhibition of this enzyme or its biosynthesis may prevent toughness in meat. Diets low in Cu might be helpful in this regard (Rucker *et al.* 1977). Undoubtedly increased understanding of how collagen is put together will be exploited for obtaining meat tenderizing enzymes. Finally information about "ground substances," how they are attached to collagen (Ristelli *et al.* 1976) and their breakdown should yield enzymes which might be helpful in conjunction with collagenases and anticrosslinking enzymes as a means of getting the collagen transforming enzymes to the site of their action. Furthermore, such knowledge may provide clues as to why such vanishingly small proteinase activity effectively tenderizes meat.

28

ROLES OF ENZYMES IN FISH TEXTURE AND TRANSFORMATION

In this chapter we will be discussing the action and consequences of both exogenous and endogenous enzymes upon the texture, functionality and modification of fish and fish products. Topics to be discussed include the texture of unfrozen and frozen fish and other marine animals, enzymatic aspects of the development of toughness in frozen fish, and the use and control of proteinases, both endogenous (autolysis) and exogenous in the form of added enzyme, as aids in the transformation of fish and other marine animals into new products.

ENZYME INVOLVEMENT IN FRESH FISH TEXTURE

As with meat there is a range in the texture of cooked fish. The texture corresponding to optimum consumer acceptability is more aptly described as "juicy" rather than tender. Fish texture instruments (Dassow *et al.* 1962) and correlates of fish toughness such as optical density of homogenates (Love 1962) and protein solubility (Awad *et al.* 1969) have been described. As to be expected, both similarities and differences in the muscle-to-food conversion process for meat and fish can be discerned. Some of the differences are due to the divergence in the intrinsic functions and corresponding biochemistry and morphology of the respective tissues. At least part of the contrasting enzymology is due to differences in response to the environment—fish are poikilothermic (cold-blooded) and essentially weightless. Perhaps more important differences reflect contrasting harvest (slaughter), holding, processing and storage practices. Thus, in harvesting fish via trawling, intense struggle with its attendant depletion of glycogen reserves is an almost inevitable precedent of death (Crawford 1972). Furthermore, fish are usually frozen either before or after rigor mortis; rigor mortis can occur within 1 hr after death (Nazir and Magar 1963). Since fish muscle meat becomes increasingly tough during frozen storage, most attention has been given to this aspect of fish texture although a few studies have dealt with the texture of fish stored in crushed ice. Thus, according to Moorjani *et al.* (1962) no toughness develops in fish stored aseptically under the latter conditions. No increases in nonprotein nitrogen action, indicative of proteinase, were detected. Nevertheless, scanning electron photomicrographs of trout muscle stored up to 4 days at 2°C exhibit dramatic changes in the transverse elements, including their collapse and shrinkage below the level of the rest of the myofibril (Schaller and Powrie 1971). On the other hand, excess softness ("sloppiness") is associated with starved fish and bacterial contamination in iced fish (presumably due to the action of proteases of the contaminating microbes) and can also be induced by dipping in polyphosphate solutions (Love 1968, 1980), which may raise the pH above neutrality (Love 1980).

Under certain conditions the action of proteolytic enzymes in fish is desirable, but only in order to create what is essentially a new food product. This aspect and the use of added proteases to fish will be discussed in the latter half of this chapter.

From a textural point of view, meat and fish muscle offer some rather interesting structural contrasts. The relatively short myofibrils present as layers (myotoma) which give rise to the "flakes" of cooked fish are separated by sheets of connective tissue (mycommata or septae) which are connected to and transmit most of the mechanical energy transduced via myosin ATPase to the spine. The mycommata are the site of all the stromal (sclero) protein which constitutes some 3 to 10% of the total protein of the muscle in contrast to that of mammalian skeletal muscle protein of which as much as 25% may be stromal. Although collagen crosslinking was invoked by Tran (1975) as a contributor to fish toughness and to explain the ameliorative effect of pyruvate (see below), most if not all of the scleroprotein fraction is converted to gelatin during cooking. The pH sensitivity of stored fish is due to properties of mycommata collagen. Another structural difference between beef and fish muscle is that the latter consists largely of white rather than red muscle.

The same classes of major myofibrillar proteins—actin, myosin troponin, actinin and tropomyosin—with presumably similar functions are present in about the same amounts as in meat muscles. In contrast to meat, the sarcoplasmic proteins of fish muscle are much more stable relative to those of the myofibrils. It appears that a great deal of this myofibrillar instability is due to the myosin moiety. In this connection, desirable textural quality (elasticity) of frozen fish paste for conversion into a gel-like food ("kamabaro") is dependent upon maintenance of the native myofibrillar protein structure (Fukumi *et al.* 1969). Bridgen (1972) concluded that the three cysteine residues of the trout actin molecule play no direct roles in polymerization, ATPase activity or Ca-binding, but may be involved in its stability and thus perhaps in the textural changes associated with toughness. Other related parameters which may play a role in fish in contrast with meat texture are the rate of aggregation and susceptibility to trypsin of the myofibrils, the instability of the ATPase activity and the very ready formation of actomyosin. Awad *et al.* (1969) observed a significant correlation between actomyosin extractability and tenderness of whitefish.

Parvalbumin.—Another interesting difference between meat and fish skeletal proteins is the subcellular distribution and level of what appears to be the equivalent of regulatory protein troponin C in mammals (Nockolds *et al.* 1972; Grandjean *et al.* 1977). In fish (and in amphibians as well) this Ca-binding protein is present as one of three closely related parvalbumins constituting some 20 to 30% of the sarcoplasmic myogen fraction, which, in mammals, it will be recalled, is a complex of several glycolytic and other sarcoplasmic enzymes. Parvalbumins share with these and other Ca-binding proteins, including the ubiquitous calmodulin, a close to identical amino acid sequence (homology). Although it is not yet implicated *directly* in fish texture (see below) we shall digress here briefly, adumbrating the properties of parvalbumins if for no other reason than that they are probably the most thoroughly understood of fish proteins. The complete amino acid sequence (molecular weight = 12,000) and structure as revealed by X-ray analysis have been elucidated. Some of its striking features are: solubility at very high (<90%) ammonium sulfate saturations; unusual amino acid composition, i.e., 10% phenylalanine and 20% alanine; the molecule is unusually spherical consisting largely of α-helices with no β-pleated sheets; the Ca-binding site consists of one glutamic and three aspartic acid residues in a tetrahedryl arrangement at the surface of the molecule; eight of the ten phenylalanine residues are buried in the core thus accounting for the core's extremely hydrophobic character. In keeping with its Ca-binding function there are no grooves or clefts characteristic of all enzymes so far investigated.

Rigor and Glycolysis

The sequence of events leading to rigor in fish seems to be similar to that of meat muscle but there are significant differences. For fresh fish consumption the texture associated with rigor (after cooking) is desirable whereas this texture is too stiff for filleting, which is therefore done either before or after maximum rigor. The glycogen content as well as ATP and creatine phosphate is usually lower and the lactic acid content higher due to the struggle of the fish at capture. Rigor mortis in white (but not in red) muscle does not appear to be so dependent on ATP depletion.

Lactic acid buildup continues beyond maximum rigor, the cut-off point for meat. In general, however, most fish exhibit a higher post-mortem pH than do warm-blooded animals. In general, the texture of stored fish with a pH greater than neutrality tends to be sloppy, and lactic acid formed due to glycolysis during storage can improve the texture by lowering the pH to the range of 6.6 to 6.9.

What pH decrease there is may be traced, at least in part, to the relative activity of the glycolytic enzymes. This causes changes in the water-holding capacity which can lead to a condition analogous to the PSE of pork (Chapter 26). This condition has been referred to as "chalky" and is characterized by a feeling of dryness as well as toughness and can be prevented by keeping the fish alive for 10 hr. The post-mortem fate and control of individual fish glycolytic enzymes have not been thoroughly investigated. However, a major difference between fish and meat glycolytic systems is the participation of hydrolases in degradation of glycogen as an alternate pathway to phosphorolysis. Glycogen is degraded to glucose which then enters the classical glycolytic pathway by being converted into glucose-6-phosphate:

$$\text{Glycogen} \xrightarrow{\text{Debranching enzyme}}$$
$$\alpha\text{-1,4-Glucan} \xrightarrow{\text{Amylases}} \text{Maltose,}$$
$$\text{Dextrins} \xrightarrow{\text{Oligosaccharidases}}$$
$$\text{Glucose} \xrightarrow[\text{(ATP)}]{\text{Hexokinase}}$$
$$\text{Glucose-6-phosphate} \longrightarrow \text{Glycolysis}$$

Glycolysis.—Of the ten remaining glycolytic enzymes in this pathway, lactic acid dehydrogenase (LDH) has been the one most thoroughly examined, especially with regard to its isozyme pattern. The number of isozymes of this enzyme has been reported to vary from one in brown trout (Melnick and Hultin 1970), to two in most other trout species and to 27 in a brook-brown trout hybrid; according to Tarr (1969) the number of isozymes varies with adaptation to surrounding temperatures. The brown trout enzyme is more readily solubilized than that from chicken. The coenzyme NADH has been detected in fish and has been implicated in fluorescent changes which take place in this food (Manohar 1969). Evidence for the existence and functioning of each individual glycolytic enzyme has been annotated (Tarr 1969).

The activity and control of the glycolytic enzymes post-mortem may be at least as important in determining flavor (Chapters 23 and 25) as texture of fish and also as an index of storage and freshness, and along with the TCA-cycle malic dehydrogenase, an indicator of previous freeze-thaw history (Chapter 37). This is also true of the catabolism of ATP which is ultimately converted to hypoxanthine via an enzyme system which apparently functions at temperatures as low as −30°C.

FROZEN FISH TEXTURE

Consonant with an ever-increasing proportion of the world's seafood catch conversion into frozen products (Finch 1970), studies on fish texture have more and more tended to focus on the one comprehensive textural problem which accompanies this freezing: The development of a decidedly undesirable texture upon prolonged storage. Descriptions of this toughness have included such adjectives as chewy, stringy and lack of succulence. When fish is comminuted and stored, the resulting product may turn spongy or cardboard-like and exude liquid to yield a "drip-dry" product. In addition, such stored fish and fish products lose otherwise desirable functional properties associated with their proteins, such as water holding capacity and fat emulsification ability (Chapter 34), and thus cannot, for instance, be processed into gelled products. They may also lose their nutritional value (Chapter 34).

A large body of evidence has arisen which strongly suggests that these changes are brought about principally by changes in the solubility via increased crystallinity, aggregation, denaturation and crosslinking of the myofibrillar fish protein molecules. These changes are reflected in the ever-increasing complicated supramolecular organization of the muscle proteins, finally sensed in the toughness of the product. Thus, Mao and Sterling (1970) list uncooked fish in the following order with respect to increasing degree of denaturation of their myosin: fresh, frozen, frozen-stored and dehydrated-stored. The circumstances leading to this toughness have been thoroughly reviewed by Sikorski *et al.* (1976) who divided these circumstances, parameters and causes of frozen fish toughness into the first five categories listed in Table 28.1. We shall concern ourselves with the enzymatic aspects of the primary events which eventually end up in altered texture and shall refer only

TABLE 28.1

FACTORS AND ENZYMES ASSOCIATED WITH FROZEN FISH TOUGHNESS

Adhesion of myofibrils
Decreased solubility of myofibrillar proteins
Rapid aggregation of myosin
Protein denaturation due to pH changes
Protein denaturation due to concentration of solutes (freezing, see Chapter 12)
Crosslinking of proteins by disulfide and lanthionine bridges
Preponderance of "free" water not bound to protein
Phospholipase action: resulting fatty acids interact with proteins
Phospholipase action: induction of protein-protein interaction
Lipoxygenase action: resulting lipohydroperoxide insolubilizes proteins
Trimethylamine-N-oxide demethylase: resulting formaldehyde crosslinks proteins

briefly to the secondary reactions far removed from these enzymatic events. The latter are dealt with in detail by Sikorski *et al.* They especially give attention to the large amounts of work done with model systems which ostensibly simulate these secondary reactions under controlled conditions. More precisely, we shall concern ourselves principally with the enzymes and the reactions they catalyze which produce the two classes of substances implicated as toughness reagents—free fatty acids and formaldehyde. The corresponding enzymes are phospholipase and the tentatively named TMAO demethylase.

Free Fatty Acids (FFA) and Phospholipases

This approach to the problem of toughness in frozen fish was initiated by the observations of Dyer and coworkers (Dyer 1951; Dyer and Dingle 1961) of rapid increases in fatty acids in fish such as plaice stored at −12°C, increases which seemed to be inevitably accompanied and paralleled by increases in toughness, protein denaturation and protein insolubility. Further work indicated that although there are active lipases in fish—a relatively recent example is a study of hydrolysis of triglycerides by anchovy lipases (Patton *et al.* 1975)—it would appear that the fatty acids which react with the protein are unsaturated FFAs arising from phospholipids rather than from triglycerides. Hence, the responsible enzymes are probably phospholipases A and B (Chapter 25). The substrate for their action has been assumed to be in membranes from which they are probably freed as a result of the freezing-induced disruption (Olley *et al.* 1962). Studies of the relevant enzymes are sparse. Preparation and properties of a phospholipase B from fish muscle were reported by Cohen *et al.* (1967). The enzyme was maximally active at pH 5 and was activated by freezing and sonication, thus suggesting a lysosomal origin and a role in texture of frozen fish. Properties of a phospholipase A recently isolated from pollock muscle are consonant with a role for this enzyme in freeze-toughening (Audley and Kinsella 1978). Braddock and Dugan (1972) measured the decrease in the level of the complex mixture of phospholipids in stored frozen fish. In particular, their results inferred a preferential enzymatic hydrolysis of certain phosphatidyl ethanolamines.

There are, of course, many factors which can influence the subsequent disposition of the enzymatically liberated FFA. As pointed out earlier by Olley *et al.* (1962) there is probably no simple cause-and-effect relationship. One important factor which could upset such a simple relationship is the presence of large amounts of fat in the fish. Thus, as a general rule the fatter the fish the less likely that it will develop toughness during frozen storage. It is not hard to imagine that the liberated FFA is readily dispersed in these fat depots which are located at the site of the myofibrillar proteins. Such an antitoughening effect becomes even more pronounced in frozen minces where the fat is more finely dispersed and evenly divided. Yet, as pointed out by Sikorski *et al.*, since some species of lean fish such as the spiny flathead resist toughness development (Bremner 1977), many unknown biological factors exist whose contributants to toughness may require alternative explanations.

Whether the FFAs are removed by solubilization into the fat phase or by reaction with the myofibrillar proteins, their presence—by virtue of their inhibition of phospholipase and also due to mass action in the water-limited environment in the frozen state (thus preventing resynthesis) (Chapter 12)—would tend to prevent phospholipase action. By the same token, FFA removal would tend to promote such action. Thus, the toughening action reinforces itself, tending to be self-perpetuating. Indeed it is conceivable, but by no means experimentally proven, that reinforcement of the toughness may come about as a result of the lipase-catalyzed esterification by the FFA of hydroxy groups of the constituent amino acids in the fish proteins. In this connection Mao and Sterling (1970) reported that denaturation of fish muscle myosin is accompanied by a rough doubling of the ester groups in the myosin. Undoubtedly the interactions between FFA and the myofibrillar proteins are complicated and multifaceted. Among the resulting crosslinks which may be formed are those in which Ca complexes with the FFA (Shenouda and Pigott 1977)

Actin . . . RCOO---Ca---OOCR . . . Actin
HI Crosslink HI

where HI stands for hydrophobic interaction (Chapters 4 and 34).

The above-mentioned parvalbumin could conceivably influence toughening, depending upon what happens to it at freezing temperatures and upon subsequent thawing and cooking, by regulating the amount of Ca available for such crosslinking. In this connection it will be recalled (Chapter 14) that Ca-linked interaction of phospholipids with proteins has been proposed as a general means for preventing enzyme action. Shenouda and Pigott (1977) did show that both polar and neutral lipids do interact with myosin and actin F (polymerized actin) without the aid of Ca but do so without denaturing or otherwise creating solubility changes. On the contrary, both Ca and Mg caused decreases in the degree of binding of polar lipids. Such binding, while not in itself resulting in insolubilization, could conceivably facilitate phospholipase action by delivering the substrate to the site of FFA-protein interaction. Sikorski *et al.* also called attention to a possible role of the non-FFA products of phospholipase action in toughness, viz., lysolecithin and phosphatides (Chapter 25). They can provide binding sites by virtue of hydrogen bonding through the newly exposed glycerol hydroxyl groups. Alternatively they might participate in fish toughening by inducing crosslinks of the ionic-hydrogen bond type.

Other Consequences of Lipase Action on Toughness.—Another secondary reaction which has been implicated in FFA-induced toughness and protein insolubility is the oxidation of polyunsaturated FFA to lipohydroperoxides (Chapter 24). Model experiments (Fig. 28.1) show that, for instance, linoleic acid hydroperoxide is ten-fold more effective than is linoleic acid as a muscle protein denaturant and precipitant (Jarenbäck and Liljemark 1975). This might explain why phospholipase action seems to be more effective than acylglycerol lipases in this regard. Also, further breakdown of the LOOH provides aldehydes (Chapter 20), especially malonaldehyde, in autoxidative reactions (Buttkus 1967) which are very effective crosslinking agents. Unlike other fat oxidative systems in which these secondary aldehyde products pile up and contribute to flavor (Chapter 25), both desirable (Chapter 21) and undesirable (Chapters 24 and 25), in fish they interact with the proteins and thus are removed. This explains, as suggested by Castell (1971), why lean fish (although containing unsaturated lipids at high levels) seldom become rancid. In spite of the opportunity for oxidation of sulfhydryl groups of the myofibrillar proteins, these appear to remain unchanged but may participate in initiation of oxidative changes. We deem it of more than passing, if incidental, interest to note that a C20:5 (Chapter 24) lipase-liberated FFA in mackerel has been pinpointed as the major factor in the reduction of cardiovascular disease incidence in Greenland Eskimos. The first event in the series of events leading to this reduction is the inhibition by this FFA of prostaglandin synthase acting on arachidonic acid, C20:4. This prevents the formation of the platelet-aggregating thromboxanes, according to Siess *et al.* (1980). Another enzymatic pathway which could conceivably influence toughness, in this case formaldehyde-induced, is that leading to the formation of choline through the action of phospholipase D (Chapter 26) on lecithin (see next section).

Formaldehyde (FA) and Trimethylamine-N-Oxide (TMAO) Demethylase

While amines—especially tri- and dimethylamines—coming from the breakdown of the uniquely piscine constituent trimethylamine-N-oxide (TMAO) have long been associated with fish quality deterioration due in part to their characteristic "fishy" odors, it is only in the last 15 or so years that the transformation of TMAO has been connected with unfavorable texture changes. Starting in 1963 (Amano *et al.* 1963) Japanese investigators have identified formaldehyde (HCHO) in fish and shown that it apparently arises from a special enzyme acting on TMAO and contributes to fish toughness in a limited number of fish and invertebrate sea species.

Sikorski *et al.* devoted much of their review on the contribution of HCHO to fish toughness showing that HCHO is present in such fish and that its presence correlates with toughness. They mod-

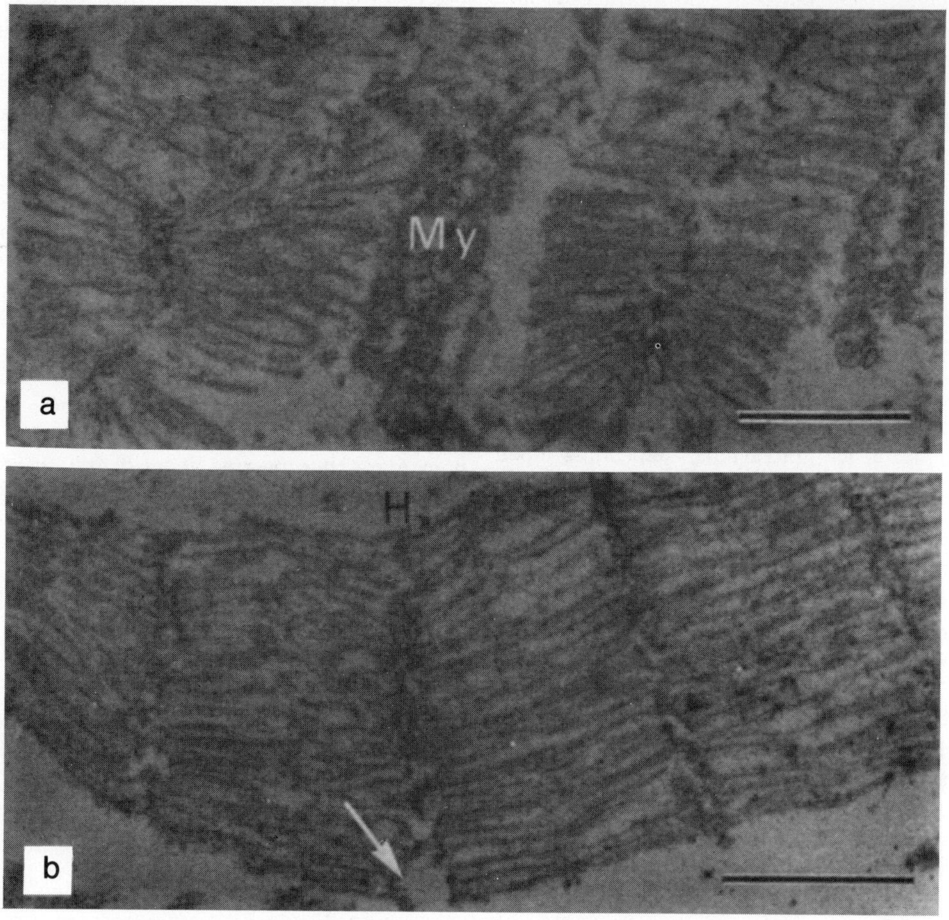

From Jarenbäck and Liljemark (1975)

FIG. 28.1. EFFECT OF UNSATURATED FATTY ACID ON CODFISH MUSCLE ULTRASTRUCTURE

Freeze-etch replica of residues from salt-extracted myofibrils incubated with (a) linoleic acid and (b) linoleic acid hydroperoxide. Note that insolubilization results in increased retention of residue in the original A-band; marker = 1 μm.

eled experiments showing that HCHO reacts with fish muscle proteins to yield modified proteins, whose properties would be reflected as toughness in cooked fish as well as minced freeze dried fish (Jensen 1979).

In essence, in addition to the statistical correlations between toughness and HCHO formation, the evidence is based on its well-known reactivity and on the ability of aldehydes in general to create crosslinks in proteins as in the now classical studies of Fraenkel-Conrat and Olcott (1948) on the specific reactions of HCHO with the individual constituent acids in proteins.

Due in part to its low molecular weight and hence relatively high diffusivity, HCHO is at least as effective as FFA as a toughening agent. Its formation and contribution seems to be predominant at relatively high temperatures and short times in the relatively few species where it is formed. These include gadoids, such as cod, Alaska pollock and hake, and invertebrates, such as clams—in unfilleted and in minced fish where the enzyme-bearing tissue has not been completely removed. HCHO is not formed in other common food fish such as mackerel, sole, flounder, halibut, plaice or squid (i.e., Cantoni *et al.* 1976). The nature of the interactions and linkage of FA with protein are still under investigation. For instance, it appears perhaps surprisingly that FA-induced toughness occurs without the formation of methylene crosslinks between protein chains (Connel 1975).

The Enzyme.—For some unknown reason, continuing progress in the enzymology of the requisite enzyme has appeared to come to a standstill since the pioneering but preliminary work of the Japanese investigators, who published their results in Japanese between 1964 and 1975 so that only abstracts have been available to the author. From these, some of which are cited by Sikorski *et al.*, and from the following additional citations (Tokunaga 1966; Amano and Yamada 1965; Harada 1970; Tomioka *et al.* 1974), it would appear that the enzyme is found in the pyloric ceca and kidney of gadoid fishes and some arthropods and echinoderms. Unlike other enzymes which form HCHO from *N*-methyl groups (sarcosine and dimethylglycine dehydrogenases) water does not appear in the overall equation (see below). It acts optimally at pH 5–6 in the presence of an endogenous, low molecular weight, heat stable "cofactor" or reducing agents (Fe^{2+}, SH-compounds, ascorbate), TMN (trimethylglycine) and betaine but is inhibited by Fe^{3+} and other *N*-methyl compounds such as choline and (reminiscent of the inhibition of L-cysteine sulfoxide lyases by S-substituted L-cysteine derivatives, Chapter 21), by trimethylamine.

In a short note without supporting data which has not to our knowledge been followed by a longer paper, Meyers and Zatman (1971) reported having isolated, after extensive purification, a similar HCHO-forming enzyme from *Bacillus* PM. The pH optimum and K_m (TMAO) for this 40,000-dalton single polypeptide chain (monomeric) enzyme were reported as 7.5 and 2.85 *mM*, respectively.

Mechanism of Formaldehyde Generation from TMAO.—The enzyme which catalyzes the transformation of TMAO to dimethylamine and HCHO may be considered to be a TMAO demethylase:

$$CH_3\cdot\overset{CH_3}{\underset{CH_3}{\dot{N}}}\rightarrow O \longrightarrow CH_3\cdot\overset{CH_3}{\dot{N}}\cdot H + HCHO$$

This intramolecular oxidation-reduction is reminiscent of that catalyzed by the ubiquitous glutathione-dependent glyoxalase, once considered to be an enzyme of glycolysis but more recently revived as a controversial factor in cell division and tumor genesis.

$$\underset{\text{Methyl glyoxal}}{CH_3COCO} + H_2O \longrightarrow \underset{\text{lactic acid}}{CH_3CHOHCOOH}$$

The "glyoxalase" reaction is actually catalyzed by two distinct enzymes, glyoxalase I, or lactoylglutathione (GSH) lyase:

$$CH_3COCO + GSH \longrightarrow CH_3CHOHCOSG$$

and glyoxylase II, or hydroxyacyl glutathione hydrolase:

$$CH_2CHOHCOSG + H_2O \longrightarrow CH_3CHOHCOOH + GSH$$

The glyoxalase I reaction apparently proceeds through an enediol proton transfer mechanism (Hall *et al.* 1976), which may be depicted as:

$$CH_3\text{-}\overset{O}{\overset{\|}{C}}\text{-}\overset{O}{\overset{\|}{C}}\text{-}H \longrightarrow CH_3\text{-}\overset{OH}{\overset{|}{C}}=\overset{O}{\overset{\|}{C}} \xrightarrow{+GSH} CH_3\text{-}\overset{OH}{\overset{|}{C}}=\overset{OH}{\overset{|}{C}}\text{-}SG \longrightarrow CH_3\overset{OH}{\overset{|}{CH}}\text{-}\overset{O}{\overset{\|}{C}}\text{-}SG \xrightarrow{H_2O} CH_3CHOHCOOH + GSH$$

The properties of the TMAO demethylase from fish (to which no EC number has yet been assigned), as revealed by the Japanese investigators and from a *Bacillus* but not those from a methylotrophic *Pseudomonas* (Large 1971), suggest to us the following related but not identical analogous mechanism:

$$(CH_3)_2\overset{O}{\overset{\|}{N}}\text{-}CH_3 \longrightarrow (CH_3)_2N(OH)=CH_2 \xrightarrow[E_1(?)]{+GSH} (CH_3)_2NH(OH)CH_2\text{-}SG \xrightarrow{-H_2O} (CH_3)_2NCH_2\text{-}SG \xrightarrow[E_2(?)]{-GSH,\ +H_2O} (CH_3)_2NCH_2OH \longrightarrow (CH_3)_2NH + HCHO$$

The overall rearrangement (which also occurs nonenzymatically at pH 11) thus goes through $(CH_3)_2NCH_2OH$ intermediate (Harada *et al.* 1975). Perhaps localized high pH in frozen fish could result in HCHO nonenzymatically (Spinelli and Koury 1979).

Trimethylamine.—An alternative pathway of TMAO catabolism is through catalysis by a microbial NAD-dependent trimethylamine-N-oxide reductase (Unmoto *et al.* 1965).

$$(CH_3)_3NO + NADH + H^+ \longrightarrow (CH_3)_3N + NAD^+ + H_2O$$

Many fish spoilage organisms are rich in this enzyme which as mentioned produces the "rotten fish" odor briefly alluded to in Chapter 25. This is the reason why measurement of trimethylamine levels in stored fish is a good indication, not only of poor flavor quality, but of overall wholesomeness. Yet, slight contamination would allow this reductase to decrease the level of TMAO so that the latter would not be available for degradation via TMAO demethylase.

There are, however, methylotrophic bacteria possessing a well characterized FA-liberating enzyme, trimethylamine dehydrogenase, which can utilize non-NAD hydrogen acceptors such as phenazine methosulfate (PMS):

$$(CH_3)_3N + H_2O + PMS \longrightarrow (CH_3)_2NH + HCHO + PMSH_2$$

This rather interesting Fe-S flavoprotein is an enzyme which may contain tightly-bound chromophoric nucleotide coenzyme, a pteridine (in folic acid) and ribitol (Colby and Zatman 1971; Beinert 1978; Steenkamp and Singer 1976). It is not likely that a similar coenzyme will be found in the TMAO demethylase since so far only the usual absorbance band at 280 nm has been observed.

A literature is developing on the microbial metabolism of the tertiary amines and, indeed, some consider TMAO to be a terminal electron acceptor in anaerobic bacterial metabolism. A TMA oxidase present in chicken liver prevents this malodorous volatile present in some feeds from getting into the eggs. TMAO itself is synthesized, at least in the red marine copepod *Calanus*, by an enzyme, now purified and well characterized, which is considered to be an NADPH-dependent tertiary-amine monooxygenase (Stroem 1980).

Putative Roles of Other Enzymes

There are probably other enzymes involved in frozen fish texture. Thus, the ATPase activity of the myofibrillar myosin and actomyosin are lost concomitantly with increased freezing-toughening. Enzymes which split or degrade the coenzyme nucleotide NAD are active in some frozen fish such as trout and not in others such as carp. This could conceivably affect texture indirectly through the action of dehydrogenase (Partmann 1973). Of course many of the factors affecting the action of enzymes in foods of low temperature which were subject to scrutiny in Chapter 12 are also at play in frozen fish. For instance, increased salt concentration could promote phospholipase action. Redistribution of enzymes and isozymes which occur upon freezing and during frozen storage afford an opportunity for quality control testing (Chapter 37).

A further role of enzymes in freeze-toughening could involve those whose action leads to reducing sugars which would react with protein amino groups and undergo Maillard nonenzymatic browning reaction leading to crosslinking of the myofibrillar proteins (Labuza 1975). Such sugars could accumulate from the action of many enzymes. The action of any one of these might not be significant. Breakdown of glycolysis could lead to such accumulation as a result of the action of the glycolytic glycogen hydrolases. Again, acquisition of knowledge and its application lie in the future.

Finally, crosslinking leading to toughness may arise from the action of deaminases which produce ammonia. The alkaline conditions produced by the presence of the ammonia leads to the formation of dehydroalanine which, as will be discussed in Chapter 34, not only forms the notorious lysinoalanine but other crosslinking in proteins such as that which gives rise to the lanthionines (Friedman 1977).

Prevention

Although the mechanisms by which toughness develops in frozen fish are, as we have seen, quite distinguishable from those that lead to some of the textural aberrations in mammalian meat (Chapter 26), some but not all of the approaches to prevention are quite similar. Insofar as they both involve enzyme action it is not surprising that precooking, which should inactivate the offending enzyme, has been advocated.

However, the drawback to precooking is that it may be necessary to apply more heat in order to inactivate the phospholipases than that required to cook the fish. These enzymes remain active after the triglyceride hydrolase lipases are destroyed (Olley and Lovern 1960; Bosund and Ganrot 1970). In this regard TMAO demethylase should ostensibly present no problem since in partially purified form it is readily inactivated below cooking temperatures at 40°C. Yet, Lall *et al.* (1975) observed that higher temperatures are required to inactivate this enzyme (as measured by dimethylamine appearance during subsequent storage) than those required to inactivate phospholipase.

Similar to their effect in PSE, many salts including NaCl and the polyphosphates are also effective in preventing freeze-toughening as observed by, among others, Lee and Toledo (1976) for comminuted fish muscle. The physical basis of these effects may lie in the water-binding capacities of these additives. Such binding would result in enhanced availability of interfaces required for interaction with the Interfacial Recognition Site of phospholipase, an interaction which is a prerequisite to the action of this enzyme as discussed in Chapter 25. Sikorski and his co-investigators attributed the stabilizing effects of pyrophosphate and glucose to the inhibition of trimethylamine oxide demethylase.

Departing from the parallelism with meats, toughening can also be prevented by amino acids, especially glutamate (Noguchi and Matsumoto 1970); a variety of organic and α-keto acids, especially pyruvate (Tran 1975); and by many, but again, not all, sugars (Noguchi *et al.* 1976). These substances may act as antidenaturation cryoprotectants and thus not be involved in affecting enzyme action in the cold. An exception might be in protecting ATPase activity of myofibrillar proteins. As is usually found in similar cases, combinations of these categories of substances exhibit synergistic effects (Bond 1975). In order to explain the protective effect action of the α-keto acids, Tran suggests that they form Schiff's bases. As with the pyrophosphates, such Schiff's bases would tend to provide less ligand for the Interfacial Recognition Site of the phospholipases.

Very little has been done to apply what little knowledge is available of the involved enzymes to prevention of freeze-toughening. As mentioned, the hydrolysis of the lecithin phospholipids by phospholipase D would release choline, a potent TMAO demethylase inhibitor. Among added enzymes which might be beneficial are the proteinases.

ROLES OF PROTEINASES IN FISH TEXTURE AND TRANSFORMATION

Although it has been suggested that the proteinases may continue to act in frozen fish, any significant action would probably take place only in nonfilleted fish which retains the most active proteolytic organs, or in comminuted minced fish where even small amounts of texture-transforming proteinases from vestiges of these organs left behind from filleting would be distributed throughout the mass of the product.

In general one would expect their action to be beneficial in terms of preventing toughness in that they would decrease the opportunity for crosslinking among the larger protein molecules. The fragments that did crosslink would be more soluble and the product less tough. Furthermore the resulting free amino acids and presumably smaller peptides would contribute to the just-mentioned cryoprotective effects. Such proteinase action would have to occur well in advance of the events which cause the proteins to become insoluble or aggregate, or to crystallize because the resulting supramolecular entities would be much more resistant to proteolysis. This suggests that potentiation of proteinase activity similar to that of the proteinases of meat (Chapter 26) might be effective. Furthermore such action would prevent the previously mentioned attachment of lipase substrates to the proteins.

Endogenous Proteinases.—While the role of proteolytic enzymes of fish muscles as determinants of the texture of frozen fish is problematical and speculative, a much more potent source of proteolytic activity, the viscera, is involved in fish spoilage (Finch 1970) and much more significantly, both without (autolysis) and with the aid of added commercial enzyme preparations, in the transformation of raw fish to new products in which the original fish has lost its identity and appearance as "fish." Proteinase action has thus become especially important in conjunction with alleviation of world hunger (Chapter 34).

Utilization of proteolytic action to transform the protein of fish and other marine creatures has been reviewed by Finch (1970) and Schwimmer (1975). Mackie (1974) provided a short and cogent review on the recovery of protein from waste fish by proteolytic enzymes. These enzymes, as mentioned, may be derived from microorganisms growing in fish fermentations, enzymes may be added, or the fishes' own enzymes can effect such transformations (autolysis). Undoubtedly autolytic processes act in concert with added proteolytic enzymes. Thus, Mackie (1974) stated that at least in one process raw fish were digested more rapidly than were cooked fish after the addition of an enzyme.

In some fish foods, especially in the fish sauces of Southeast Asia, all three processes may be operative (Van Veen 1965).

Autolytic Transformations

Fish are the principal food in which gross quantitative autolysis of the protein has played a major role in food transformation. This is because fish possess intense proteolytic activity and because the native fish proteins (in contrast to others whose utilization is still at the pilot plant stage) (Chapter 34) are highly susceptible to proteolysis.

Endogenous Gastrointestinal Proteinases of Seafood.—The principal focus of gastrointestinal proteinases in most marine creatures is the pyloric cecum corresponding to the "blind gut." It has been known for about 40 years that this gut contains a highly active trypsin-like enzyme (Johnston 1941). Johnston also found a pepsin-like proteinase active at pH 2 in the intestinal mucosa rather than in the stomach. Included among these proteases is a highly specific collagenase which acts only on the β-chain of this complex structural protein (Dugal and Raa 1978). As an example of enzyme potentiation taken to its extremes, so intense is this activity in the viscera that they liquify within 2 hr at 60°C and pH 4. The resulting products have been used as a feed. The work of Williams (1972) affords an interesting example of a more recent investigation of a marine proteinase, that of cecum of the starfish. The enzyme is membrane-bound rather than being secreted, has a very high pH optimum (above 9.00) and, contrary to most other proteinases, attacks native more readily than denatured hemoglobin.

Microbial Enzymes.—These act in concert with the fishes' enzymes in autolytic processes such as that in the manufacture of the Southeast Asian fish sauce, nuoc-mam. Most are salt tolerant halophiles. Illustrating biological adaptation at a molecular level, an interesting property of one of the proteases from one of these halophiles is its adsorption on hydroxylapatite in 3.4 *M* NaCl (Norberg and Van Hofsten 1970). It will be recalled (Chapter 6) that most enzymes are eluted from this adsorbent at far lower salt concentrations.

Cathepsins.—The action of these lysosomal enzymes in fish muscle is relatively weak compared with that of the viscera. Thus, they probably participate significantly in textural (if at all) and other quality changes only in filleted whole fish. These other quality attribute changes include flavor and wholesomeness. Thus, Reddi *et al.* (1972) found that NaCl apparently inactivated cathepsin activity in crude fractions prepared from fish. One reason NaCl is particularly effective in controlling fish spoilage microorganisms is that its inactivation of cathepsins may result in a dearth of amino acids used as nutrients by the spoilage organisms. That the cathepsins can supply such nutrients at adequate levels to support microbial growth is suggested by the observation that their overall activity, while relatively weak compared to that of the viscera, is about one order of magnitude greater than that of the beef cathepsins (Siebert 1958).

In 1964, Groninger accomplished a 2600-fold purification of an albacore muscle cathepsin with properties most closely resembling that of the pepsin-like cathepsin E of mammalian tissues (Table 26.5). Schmitt and Siebert (1967) purified a cod muscle cathepsin 3000-fold. This preparation had no specificity corresponding to the usual cathepsins. These same authors also found that there are at least five distinct dipeptidases associated with fish muscle. Ting *et al.* (1968) partially purified salmon muscle cathepsins. The major fraction acted optimally at pH 3.7 whereas minor components acting in the alkaline range were also present. Takahashi and Yamazawa (1969) described in detail two distinct types of cathepsins from carp muscle. One, most active at pH 8, was thermostable and insensitive to γ-irradiation, appeared to be a sulfhydryl enzyme and may be active in the presence of NH_3 generated by abundant deaminases of fish. This may also be the source of lanthionine (via elimination of a sulfur atom from the disulfide bond formed between two cysteine residues) implicated as alluded to previously in frozen fish toughness by virtue of myofibrillar crosslinking. The other, most active at pH 4, was thermolabile, very sensitive to γ-irradiation, not inhibited by sulfhydryl-binding mercurials and was inactivated by Mn^{2+}. Siebert and coworkers believe that these cathepsins play a role in the deteriorative changes in fish prior to processing and have attempted to inactivate them with ionizing radiation. Geist and Crawford (1974), from a study of the muscle cathepsins of Pacific sole, concluded that their results did support assumptions that these enzymes exert a substantial influence on the quality of marine food fish.

The squid is a rich source of lysosomes from which a 13,600 dalton proteinase has been isolated (Inaba *et al.* 1976). The investigators state that this enzyme might be the smallest of the proteinases so far isolated. It is, however, still larger than the microproteinase of *B. cereus* (Chapter 27).

The Autolytic Process.—Relatively few studies on autolytic processes in fish have been designed to obtain a fuller understanding of what actually occurs at the enzyme action level. The assumption that cecal enzymes are acting is attested to by the observation that "silage" (as the process is termed in the fish processing industry) of fillets alone liquefies poorly (Tatterson and Windsor 1974). As had been earlier observed with other autolytic processes to be discussed in Chapter 34, only recently has it been recognized that a portion of the protein in fish autolysates is always apparently inaccessible to further hydrolysis by the endogenous proteinases. For instance Raa and Gildberg (1976) isolated a proteinase-resistant fraction of cod viscera protein which is not attacked even after addition of further proteinase. It is present as an insoluble sediment of protein and lipid. Some lipase action probably also occurred and the absence of hydroxyproline would show that collagen is probably digested during autolysis. Some of the resistance would appear to be due to interactions between phospholipase-produced FFA and nonscleroprotein as discussed earlier in this chapter. As pointed out by these authors much remains to be learned about the enzymology of this process for more rational, less empirical, development of practical silage procedures.

Applications.—Table 28.2 outlines processes and studies of fish transformations in which autolytic proteinase action appears to play a dominant role. Some of the resulting food products, such as the nuoc-mam sauce of Vietnam, treasured more for their sensory attributes than as nutritional food, represent more than a mere change in texture and could have been discussed in chapters on enzyme-induced changes in appearance. Indeed we could also have presented a similar assemblage of processes which are principally dependent upon microbial fermentation, but have excluded it in interest of space and in keeping with our objective of a peripheral discussion only of *in vivo* microbial enzyme action.

The protein to be acted upon may be in cooked or in raw fish in a variety of starting materials, such as whole fish, ground-up or comminuted fish, fish waste, shrimp heads, trash fish, offal, the "sawdust" from filleting and the "stickwater" remaining from production of fish meal and oil. Almost 1 MMT (million metric tons) of menhaden are caught annually in the United States for this purpose (Burkholder *et al.* 1968). The enzymes may act in conjunction with a wide variety of processes from the time-honored production of sauces and partially dried fishes in pastes in Southeast Asia to modern methods involving solvent extractions for the production of FPC (fish protein concentrate). The end product may be, in addition to the above, a nongelling concentrate rich in protein (from fish stick-water) and/or amino acids or an odor-free impalatable powder. In some instances a limited proteolysis, only enough to solubilize denatured protein and to release lipid, is desired. The

TABLE 28.2

FISH PROTEIN PRODUCTS BY PROTEOLYTIC AUTOLYSIS

Reference	Product	Conditions, Comments
Meinke and Mattil (1973)	Protein isolates	Whole fish → soft creams; fillets → firm curds.
Koury *et al.* (1971)	FPC[1]	Autolysis rates studied—greatest at acid pH and 50°.
Lum (1969)	FPC	Ground fish self-digested at 50°–60°C; solvent to remove fat.
Linson (1968)	Edible concentrate	Ground fish + 1–2% citric acid; remove nonprotein by filtration, solvent, dry.
Tomiyama (1968)	FPC	Ground fish, pH 4.5, antibiotic, 50°C; used in milk substitute formulation.
VanVeen (1965)	Fish sauce, Southeast Asia, nuoc-mam	Whole fish kneaded, stored in salt for months; sauce tapped from bottom; halophile proteinase may participate in protein digestion.
McBride *et al.* (1961)	Liquid animal feed	Homogenates of whole herring liquified aseptically in 72 hr, 37°C, low pH; accelerated by added enzymes.
Tarr and Deas (1949)	Bacteriological peptones	Trypsin and pepsin of fish used to prepare peptones from fish proteins.

[1] Fish Protein Concentrate.

products may be used as is or may be incorporated into feeds and human foods or microbiological growth media. Even ionizing radiation has been combined with a limited autolysis for the isolation of fish protein concentrate FPC (Warrier *et al.* 1975).

In some processes autolysis can be a problem and lead to losses in protein yield and nutritional value. Thus, Koury *et al.* (1971) found an inverse correlation between yield of FPC prepared from hake and the degree of autolysis occurring prior to subsequent processing. They found that the autolysis was greater at acid pH's and above 50°C. In this case it appears that the catheptic enzymes and the pepsin of the intestinal mucosa rather than the caecal proteinases participated in the autolysis. Meinke and Mattil (1973) concluded that autolytic proteolysis may influence the yield of protein isolates from whole fish. Isolate, a somewhat purer form of protein than FPC, was obtained by isoelectric precipitation of extracts of whole fish. As expected, the catheptic enzymes of both the flesh and viscera were responsible for autolysis at low pH, whereas the extracellular visceral protease acted at or became activated during the alkaline extraction step of the isolation procedure.

Very little has been published about the effect of autolysis on the nutritive value of the protein of fish. According to Tomiyama (1968), that of FPC prepared from autolyzed protein is similar to that of isolated fish protein.

Autolytic proteinase action occurs during the marinading of herring (Sicho *et al.* 1972). They found that marinading too long leads to losses of valuable amino acids into the marinating pickle and presumably to overtenderizing. Ripening of the Dutch maatjes-cured herring occurs through the action of the proteases of the pyloric cecae of North Sea herring. Ruiter (1972) obtained satisfactory ripening by adding protease preparations to gutted herring.

Transformations by Added Proteinases

A large majority of investigations of the possibility of using commercial proteinases in connection with their action on fish proteins are directed to the preparation of the above-mentioned highly nutritious fish protein concentrates (FPC). Most nonenzymatic methods involve organic solvents and/or heat yielding denatured products with poor functional properties. Retention of the latter are important since FPC is intended for incorporation into formulated foods. Although of high nutritional quality the FPC prepared nonenzymatically is apt to have poor wetting and fat absorption characteristics (Chapter 34) in addition to being indispersible and insoluble. Table 28.3 briefly describes proposed processes in which commercial proteinase preparations are the principal agents for the conversion of fish proteins, lately to FPC.

In a statement by the Protein Advisory Committee (Anon. 1972) it was suggested that enzyme processing, although at that time less developed than alternative processing methods for the production of fish protein concentrate (FPC), could eventually result in the production of inexpensive milk analogs. This statement briefly describes three such processes. One process was proposed for the production of FPC in the United States. In the process, at that time at the pilot plant planning state, hake protein is first subjected to partial hydrolysis by enzymes followed by solvent extraction as an alternative to removal of the fat by centrifugation. The resulting product is said to be completely water soluble and has a slight meaty flavor.

A second process, then at the pilot plant stage, involved enzyme digestion of the whole fish, fish meal or press cake followed by oil separation to yield a product free of fish odor and had been used thus far as an animal feed. Re the third method, pilot plant studies of which were carried out in Chili, the PAG statement outlines a procedure in which eviscerated deboned fish are subjected to a "mild" enzyme treatment. This treatment is followed by addition of stabilized fat. The mixture is then pasteurized and spray dried to yield a light colored powder with no fish flavor which is readily dispersible in water.

In contrast to all of the above-described methods and those shown in Table 28.3 (the more recent of which apparently have not advanced beyond the pilot plant stage) according to Mackie (1974) the French are successfully engaged in a combined fishing-processing operation involving the addition of proteolytic enzymes. In this operation, fish waste and even whole fish are treated with proteinase and converted directly into hydrolysate aboard ship. After being held at 0°C on board until the ship returns to port, the hydrolysate is separated into oil and liquid. The latter is dried and used as FPC.

TABLE 28.3

FISH TRANSFORMATIONS BY ADDING PROTEINASE

Fish Substrate	Enzyme Sources	Conditions	Product/Study/Aim
Fish myofibrillar (a)[1]	Bromelin, pineapple succinylated	25°C, 10 min, pH 7.0	Modified functional properties
Various (b)	BASAP, others	pH 8.5, variable time	FPC
Fillet waste (F.W.), trash fish (c)	Bromelin; papain, 0.05%	60°C, 30 min	FPC
English Sole, F.W.	Pepsin	4 hr F.W.: H_2O—2.1	FPC
F. W. (d)	Pepsin	Hyd.: 37°C, 6 hr, pH 2 Synth: 45°C, 24 hr, pH4.5	Plastein, FPC
Red Hake FPC (e)	BASAP[2]	Variable	Process/kinetic study of solubilization
FPC (f)	Pronase, BASAP	Variable, ultrafiltration	Enzyme recycling
FPC (g)	Ficin, figs	40°, 2–7 hr	Solubilization
FPC (h)	Bromelin	57°, pH 6, 1 min	"High energy" FPC
FPC (i)	Trypsin	40°, pH 9	Soln. in membrane reactor
Marine life (j)	Pancreatin	Immerse in enzyme solution	Separation of edible tissue from edible flesh
Shredded, cooked (k)	Pancreatin; *A. oryzae*	45°C, 20 min, pH 6.5	Yeast-fermented FPC
Frozen (1)	"Proteinase," 0.1%	—	Sausage meat
Eviscerated (m)	Pineapple, papaya	Ambience, weeks	Cambodian fish paste

[1] Sources: (a) Miller and Groninger (1976); (b) Hale (1974); (c) Mackie (1974); (d) Onoue and Riddle (1973); (e) Archer *et al.* (1973); (f) Cheftel (1972); (g) Hevia *et al.* (1976); (h) Rutman (1971); (i) Bhurmiratana *et al.* (1977); (j) Fehmerlhing (1970); (k) Jeffreys (1970); (l) Bykova (1970); (m) Van Veen (1965).
[2] *Bacillus subtilis* alkaline protease.

Problems.—Among some of the problems which arise in conjunction with use of proteases are loss of protein owing to excessive enzyme action and microbial contamination. Archer *et al.* (1973) solved the latter problem by conducting the hydrolysis at 50°C and pH 8.8 using BASAP, *B. subtilis* alkaline protease (Monozyme). Another problem arises when pepsin is used in acid medium; quite a bit of salt can accumulate upon subsequent neutralization and concentration. Archer *et al.* (1973) used ammonia which is evaporated off during the drying step. Of course this would expose the protein to conditions in which there is a loss in nutritive value. Tarky *et al.* (1973) solved this problem by passing dilute hydrolysates through tubular ultrafiltration membranes. One of the texture-associated advantages of FPC preparations derived from processes in which proteinase action has occurred is elimination of gristly textural characteristics of FPC made by nonenzymatic products.

In this connection, fish protein hydrolysates as milk replacers are satisfactory, as measured by PER, only if not more than 40% of the casein is replaced when used for such purposes (Toulec 1973).

As with other hydrolysates, fish digests can be quite bitter due to the formation of bitter peptides (Mackie 1974). Fujimaki *et al.* (1973A), while testing various enzyme preparations, found that Pronase produced an acidic oligopeptide which had a desirable brothy flavor with a strong aftertaste. Free glutamic and aspartic acids were not present.

Analysis of bitter fractions from the action of bromelin, ficin or pronase (the latter yields the least bitter hydrolysate) on FPC suggested to Hevia and Olcott (1977) that the bitterness is due to the tripeptide leucyl (or glycyl) asparaginyl lysine. The plastein reaction which was at first designed to overcome such bitterness was applied to recovery of protein from fish (Onue and Riddle 1973). The formation of bitter peptides and of plastein as the results of proteinase action on other than fish proteins will be explored in Chapter 34.

With regard to the nutritional value, Hale (1974) commented on the loss of essential amino acids, especially isoleucine and phenylalanine from the hydrolysis of hake protein. Again BASAP came to the rescue. When hydrolyzed at a pH equal to or more than 8.5, the resulting FPC yielded the optimum balance of amino acids. The PER of a totally soluble FPC from alewives was found to be equal to that of casein. Tarky *et al.* (1973) found that FPC from waste fish had a low PER of 1.65 owing to a deficiency of tryptophan. After supplementation of the rations with an equal weight of casein, which has an excess of tryptophan, the PER was raised to 3.39 as compared to a value of 3.33 of casein alone.

According to Tomiyama (1968), the nutritional value of FPC from autolyzed protein is similar to that of fish protein isolate. As the result of a thorough safety evaluation of fish protein hydrolysate as prepared by Rutman (Table 28.3), involving a 10-month feeding and reproduction performance study, Ballester *et al.* (1977) concluded that this product was of high quality, safe and suitable for supplementing the protein intake in developing countries.

We complete our survey of added proteinases by mentioning rather esoteric uses of protease preparations for the express purpose of changing marine food texture: one a substitution of these enzymes for pounding to tenderize abalone (Cole and Stroh 1972); the other to reduce the undesirable gelatinous off-white curd in canned salmon arising from precipitation of heat-coagulated protein on the surface of the pink chunks (Yamamoto and Mackey 1981).

29

ENZYME ACTION AND PLANT FOOD TEXTURE

DETERMINANTS OF PLANT FOOD TEXTURE

While the texture of plant foods does not occupy the central area of consumer acceptance as does the texture of meat and fish, it is nonetheless a significant quality attribute of both fresh and cooked vegetables and fruits, and as such has received considerable attention as an object of scientific investigation. Plant foods exhibit a greater range of variability of both objective and texture-associated organoleptic responses than do most animal foods. This is due to their diversity and to the fact that they may be eaten both raw and cooked. Thus, in fresh fruits we think most often of the juiciness of oranges and peaches, the relative firmness of apples, whereas crispness, firmness, succulence and turgidity are most highly prized in fresh vegetables.

Objective measurements as standards of plant food texture are widely employed and accepted as subjective value judgments. As an example of such measurements on fruits, Ahmed *et al.* (1973) measured, with the same instrument, the resistances of citrus fruits to impact stresses as influenced by waxing and irradiation and as manifested by fruit puncture and rupture, peel oil sac rupture and peel shear forces. Shear force values decreased in the order of limes, lemons, Valencia oranges, Duncan grapefruit, tangerines, navel oranges and Temple oranges. As expected, subjection to ionizing radiation resulted in consistently lowered resistance to impact stress. Such objective measurements on vegetables are exemplified in a texture profile analysis (shown in Table 29.1) of fresh cucumber, a vegetable in which texture is of particular importance.

For both fruits and vegetables, texture maintenance has taken on an added significance because of the continuing expansion of mechanized harvesting. Frequently, harvesting of crops for which texture is an important consideration is being mechanized because of the shortage of labor, a shortage brought about in part perhaps because of the tediousness of handling needed to maintain this optimum texture. For instance, efforts in this direction have been directed towards cucumbers, sweet cherries and blueberries (Levin *et al.* 1975).

The above remarks pertain to raw foods. After cooking these fruits and vegetables, it is desirable

TABLE 29.1

A TYPICAL TEXTURE PROFILE: FRESH CUCUMBER[1]

Parameter	High Value	Low Value
Brittleness, kg	63.9	37.3
Hardness	72.5	40.3
Cohesiveness	0.034	0.024
Elasticity, mm	1.5	1.2
Gumminess, kg	2.46	0.95
Chewiness, kg min	3.75	1.15
Total work, cm^2	69.0	42.58

[1] Adapted from data of Breene *et al.* (1972) who used 24 cultivars exhibiting a broad genetic variability.

in many instances for the food to retain a recognizable structure (size, shape and form) with some but not excessive lowering of firmness and avoidance of sloughing. We are not concerned in this chapter with food products in which the original plant has been totally transformed in appearance or has taken on a new identity as a result of what frequently amounts to a change of state, such as with jellies, jams, gums, juices and sauces. These products will be discussed in Chapter 30.

The Cell and Cell Wall

On a workaday, macroscopic time scale, three chief factors associated with textural change of fruits and vegetables are maturity, other agronomical preharvest parameters and postharvest treatment. At the microscopic level the origin of textural properties resides in the characteristics of the plant cell. In particular, it is generally agreed that the cell wall predominates in determining texture, at least, or perhaps especially, in nonstarchy plants (Sterling 1963,1975; Reeve 1970).

The cells of most plant foods, especially vegetables, consist of parenchymatous tissue. Although displaying a wide range of morphology and phylogeny, all parenchyma cells have the same fundamental structure in that they are somewhat rigid, approximately 14-sided polyhedrons. Figure 29.1 is a simplified diagram of a mature parenchyma cell as revealed by the light microscope. Some of the structural, biochemical and chemical components of such a plant cell are listed in Table 29.2.

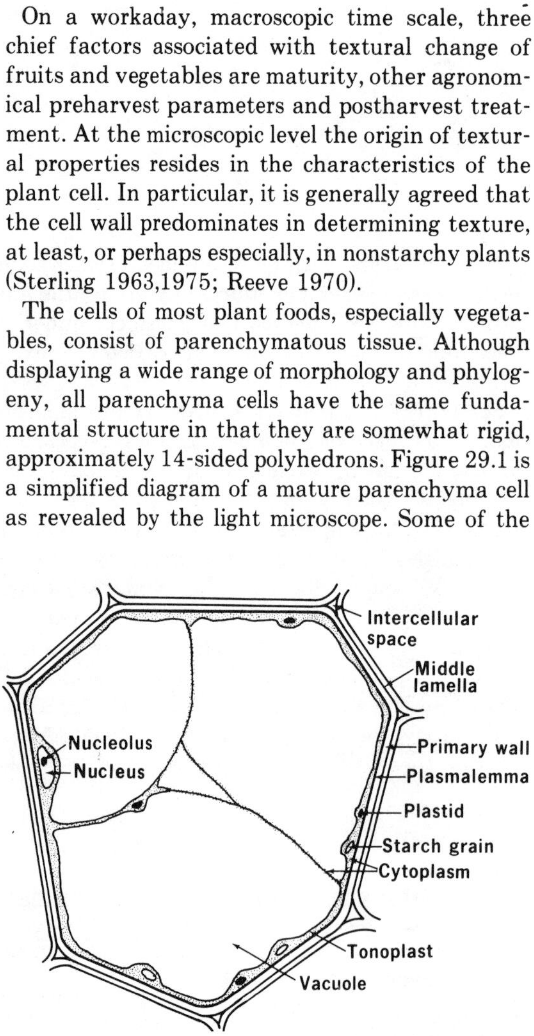

From Feinberg (1973)

FIG. 29.1. DIAGRAM OF A MATURE PARENCHYMATIC PLANT CELL COMMON TO MANY VEGETABLES (IDEALIZED)

In contrast to most animal cells, the living, actively metabolizing portion of the fully-grown parenchyma cell, the *protoplast*, constitutes but 5% or less of the total cell volume. The protoplast tissue is confined to a thin film (1 to 5 microns thick) between the *cell wall* and the *vacuole*, and to a few fine strands traversing the vacuole. Boundary membranes of the protoplast are the *plasmalemma* and the *tonoplast*, respectively bordering the cell wall and the vacuole.

Some plant physiologists consider the vacuole as a special liquid inclusion within the protoplast. Others have considered the vacuole to be the plant counterpart of the animal cell lysosome, site of the "acid" hydrolases. The vacuole plays a role in creating the textural attributes of crispness, firmness, succulence and turgidity which are so highly prized in raw salad vegetables. This turgor arises as the result of the interplay of osmotic pressure developed within the vacuole and the pressure exerted by the relatively rigid structure of the cell wall. The osmotic pressure of the vacuole can be traced to the selective permeability of the tonoplast, the membrane between the vacuole and the protoplast. When the cell is killed by heating, as in the thermal potentiation of enzymes, the entire protoplast loses its selective permeability, thus increasing the overall permeability to solute and water.

Other cellular determinants of texture include cell size and shape; intercellular space, cell wall turgor and pressure; cellular outgrowths and depositions; condition of the starch, if present within the cell (retrogradation, swelling, hydration); and cell and cell wall thickness and composition. It is generally agreed that as far as enzyme action and fresh plant foods are concerned, it is change in cell wall and intercellular composition which is the decisive factor in the texture achieved.

The Cell Wall and Intercellular Matrix.—The plant cell wall is a nonliving, semi-elastic structure enclosing the protoplast; it serves as a skeletal framework of the tissues. Parenchyma cells possess a primary wall in which cellulose is deposited as a network of more or less oriented microfibrils. Pectins, hemicelluloses and noncellulosic carbohydrates occur mainly as "encrusting substances" in the interstices between the cellulosic fibrils. The boundary between abutting walls of adjacent parenchyma cells, i.e., the middle lamellas, consists

TABLE 29.2

COMPONENTS OF THE CELL—STRUCTURAL AND BIOCHEMICAL

Structural	Biochemical
I. Vacuole	H_2O, inorganic salts, organic acids, oil droplets, sugars, water-soluble pigments, amino acids, vitamins
II. Protoplast	
A. Membrane	Protein, lipoprotein, phospholipids, phytic acid
Tonoplast (inner)	
Plasmalemma (outer)	
B. Nucleus	Nucleoprotein, nucleic acids, enzymes (protein)
C. Cytoplasm	
Active	
Chloroplasts	Chlorophyll
Other[1]	Enzymes, intermediary metabolites, nucleic acid
Mitochondria	Enzymes (protein), Fe, Cu, Mo, vitamins, coenzymes
Endoplasmic reticulum	Nucleoproteins, ribosomes, enzyme synthesis machinery; Golgi bodies (apparatus)
Golgi apparatus	Enzyme post-translational finishing, packaging, delivery
Inert	
Starch grains	Reserve carbohydrate (starch), phosphorus
Aleurone	Reserve protein
Chromoplast	Pigments (carotenoids)
Oil droplets	Triglycerides of fatty acids
Crystals	Calcium oxalate, etc.
III. Cell Wall	
Primary wall	Cellulose, hemicellulose, pectic substances and noncellulose polysaccharide
Middle lamella	Pectic substances and noncellulose polysaccharides, Mg, Ca
Plasmodesmata	Cytoplasmic strands interconnecting cytoplasm of cells through pores in the cell wall
Surface materials (cutin or cuticle)	Esters of long chain fatty acids and long chain alcohols

[1] Lysosomes (vacuoles?), peroxisomes, glyoxosomes.

mainly of pectic substances along with some proteinaceous material and other polysaccharides.

Specialized tissues, such as xylem and sclereids or fiber cells, possess a secondary wall in which the microfibrils are more highly oriented than in the primary layer. Such tissues can become lignified. These specialized cells normally make up an insignificant proportion of the tissue of most fruits and vegetables. Some vegetables, such as celery, also possess a collenchyma which serves mainly as mechanical support. Collenchyma cells, however, do not have true secondary walls but develop pronounced pectic and hemicellulosic thickenings mainly in the inner angles of their wall facets.

The composition and structure of plant cell walls, i.e., cellulosic microfibrils and "encrusting" substances, make possible a system that is both strong and plastic. These physical properties, combined with the natural turgor of living cells, are responsible for the various textural properties of raw vegetables. The capacity of cell wall carbohydrates to rehydrate is important to good texture.

However, major attention has been paid with good reason to the role of changes in the amount and transformation of the pectin of the intercellular matrix, usually catalyzed by enzymes, in relationship to textural changes during ripening, postharvest storage and subsequent processing.

Pectin

The term "pectin" usually designates the water-soluble fraction obtained from plants via extraction as a viscous heteropolysaccharide characterized by the presence of partially methoxylated D-galacturonic acid residues bonded by α-D-(1→4) linkages (Rexová-Benková and Marcovič 1976). Along with L-rhamnose residues, these constitute the backbone from which branch neutral sugars, usually D-galactose and L-arabinose (but also occasionally L-fucose and D-xylose). They are thus quite similar in overall structure but usually more abundant than their accompanying hemicelluloses, (Chapter 9). Partial removal of the methyl groups yields low methoxyl pectin (pectinic acid); the methyl-free product is pectic acid or pectate. At a supramolecular level these galacturonans exist as networks of fascicles and elementary fibrils, according to Dull and Leeper (1975). A fascicle of isolated pectic acid at pH 6 is shown in Fig. 29.2. In this figure each of the elementary fibrils in a fascicle has a mean width of 3 nm and a mean length of 10 nm.

Protopectin.—The fact that a significant portion of the anhydrogalacturonic acid residues in plants are water insoluble has given rise to the concept of "protopectin" (Joslyn 1962), part of the three-dimensional hemicellulosic network-like continuum in the intercellular space which merges with the cellulose-liginin cell wall proper. Protopectin may constitute side chains rich in galactose or the galactan may branch off from a run of anhydroga-

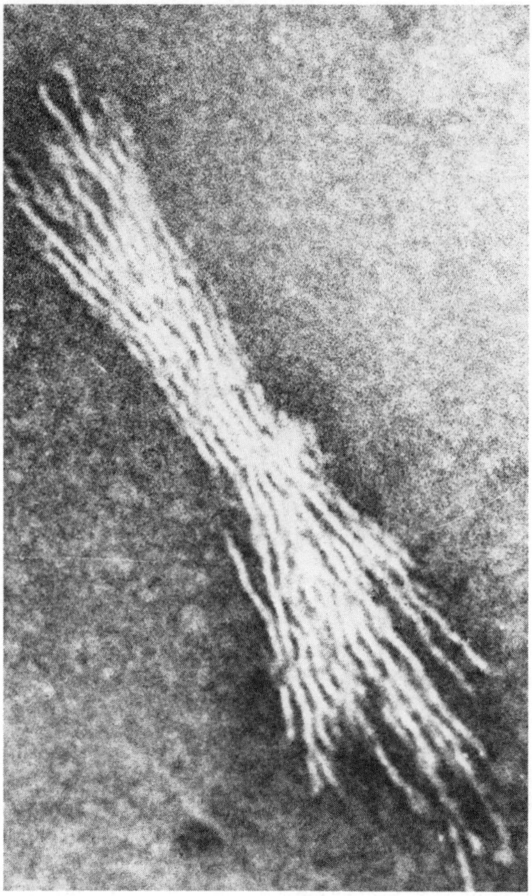

From Dull and Leeper (1975)

FIG. 29.2. FASCICLE OF ELEMENTARY FIBRILS OF PECTIC ACID

lacturonic acid residues constituting the backbone, as mentioned above. Even lignin may be part of the hemicellulose complex (Chapter 9). The stability of this network is maintained by an adequate supply of divalent cations, largely Ca^{2+} and Mg^{2+}.

Explanations of the insolubility of pectin polymers reviewed by Knee *et al.* (1975) include covalent association with a xyloglucan which is hydrogen bonded to cellulose microfibrils (Chapter 9) and linked through arabinogalactan attached to cell wall protein through the latter's serine. Frequently, investigators look upon protopectin as a "cellular adhesive" and as such it plays a crucial role in plant food texture. This is because the conversion of protopectin into a soluble form results in cell separation, reflected on a macroscopic scale in a loss of rigidity of the tissue and softening of the food. The enzymatic aspects of this solubilization will be examined shortly ("protopectinase").

Biosynthesis.—Formation of the glycosidic linkages in pectin is no different from that of most other polysaccharides via the uridinediphosphate sugar pathway, briefly alluded to in Chapter 17 and elsewhere (Hassid 1969). Some of the other known enzymes which may be involved in the synthesis of pectin are shown in Table 29.3. Included is UDPgalacturonosyl transferase, which could account for the presence of nongalacturonic (uronide) residues and even for the formation of "protopectin." It is of interest that both pectin and phytic acid, both involved in the texture of certain vegetables especially after cooking by virtue of their competing for Ca (see potatoes and dry legume seeds), appear to arise from the same precursor, *myo*-inositol. It seems fairly certain that uronide carboxyl is esterified after the glycosidic linkage is formed, probably from S-adenosyl methionine, although S-methyl cysteine has also been implicated. In addition to the references cited in Chapter 9, the biosynthesis of plant wall polysaccharides has been the topic of a symposium (Lamport 1972) and a review (Kauss 1974).

Faulty biosynthesis of pectin has been suggested as the etiology of textural disorder of tomato known as "blotchy ripening" (Hobson 1964).

PECTIN TRANSFORMING ENZYMES

Enzymes that act on pectins and substances derived therefrom are ubiquitously distributed among plants and microorganisms, but are not present in animals. After many years of controversy in attempting to classify the various types of action, it would appear from the excellent reviews of Rexová-Benková and Markovič (1976) and Rombouts and Pilnik (1972,1978) that there are three distinct types of reactions catalyzed by enzymes acting on the pectic substances. These are: the pectin esterases, the pectin or pectate depolymerases which comprise the hydrolases, and the lyases. The latter split pectins via a β-elimination mechanism involving the *removal* of water, in contrast to the hydrolases which *add* water (Fig. 29.3). These enzyme specificities can be further categorized, as with other glycan depolymerases, as to whether they attack from an end (exo-acting), proceed randomly via attack of internal linkages (endo-acting) or attack only short chain oligomers. Unlike the other glycanases, there is also specificity

TABLE 29.3

ENZYMES WHICH MAY BE INVOLVED IN PECTIN BIOSYNTHESIS

Enzyme	Substrates	Products
UDPglucose dehydrogenase	UDPglucose + 2 NAD^+ + H_2O	UDPgal + 2 NADH
myo-Inositol oxygenase	*myo*-Inositol + O_2	Gal + H_2O
UDPglucuronate 4-epimerase	UDPglucuronate	UDPgal
UDPgal transferase	Acceptor-β-gal + UDP	UDPgal + Acceptor
Gal kinase	Gal + ATP	Gal-1-P + ADP
UDPgal-polygal galtransferase	UDPgal + $(1,4\text{-}\alpha\text{-gal})_n$	$(1,4\text{-}\alpha\text{-gal})_{[n+1]}$ + ADP

[1] Gal = D-galacturonate or D-galacturonosyl.

with regard to whether the anhydrogalacturonide residues are esterified with methyl groups.

Pectin Esterase

The pectin esterases (PE), sometimes referred to as pectin methyl esterases (PME), catalyze the hydrolytic desterification of pectins to form partially demethoxylated pectins (pectinic acids) and methyl alcohol. Although universally distributed among the tissues of all plants and many microorganisms, the most frequently investigated richest source of PE is the tomato.

All esterases so far investigated prefer to start at the reducing end of the pectin chain and work their way down, leaving blocks of successive galacturonic acid residues with free carboxyl groups. The tomato enzyme(s) will also act at other loci, preferably where there is a lone methyl-esterified carboxyl group surrounded by stretches of free carboxyl groups. Indeed, in general, and except possibly for initiation of action down the chain, a requirement of PE action is the presence of an anhydrogalacturonide free carboxyl group adjacent to the ester linkage to be hydrolyzed. Some PEs adhere to the strictest specificity, starting only from the reducing end of the pectin chain and nowhere else. Much of this rather definitive information concerning the action pattern of the PEs comes from the laboratory of Macmillan and coworkers (Miller and Macmillan 1971).

The reaction usually stops abruptly before all the methoxyl groups are removed, in part due to rather intense product (pectate) inhibition and in part due to repulsion by the negatively charged carboxyls. On a monomer residue basis the competitive inhibition constant is of the same order of magnitude as the variously reported K_m's, about 1−18 m*M*. This repulsion can be overcome and inhibition allayed by high concentrations of neutral salts.

The pH optima for all of the food plant enzymes so far investigated tends to be in the slightly alkaline range, pH 7−9, whereas the range for enzymes prepared from microorganisms is broader and most optima are on the acid side. Where purified sufficiently, molecular weights of about 28,000 have been measured. They are all activated by NaCl or divalent cations, a property which may be important in the potentiation of their action, as discussed below. They are effectively inhibited by intermediate-length fatty acids—over 50% by myristic acid, C_{14} (Miller and McColloch 1959). As with most other enzymes, PE occurs in multiple molecular forms; at least four major isozymes have been identified in the tomato.

The enzyme *in vivo* appears to be localized at the site of its substrate at the cell wall. Except for the use of cell wall debris to adsorb the enzyme orange preferentially (Jansen *et al.* 1960), purification procedures have not to our knowledge yet included an affinity chromatography step. Such an approach might be helpful in resolving the question as to whether the PE activity from *Clostridium multifermans* resides in the same protein as that of pectate lyase (see below) which accompanies and cannot be separated from the PE during purification, using almost all of the known nonaffinity procedures (Macmillan and Vaughn 1964).

From the viewpoint of texture the important consequence of PE action is the facility with which the resulting polymeric end product at comparatively low levels forms a gel upon the addition of Ca^{2+}. Its action also has some health-related consequences in relation to the methanol which is released and also with regard to the effect on the "dietary fiber," as discussed in Chapter 36.

Polygalacturonases

The polygalacturonanases (PGs) or, alternatively, D-galacturonases, possess absolute specificity for the hydrolysis of the β-D-(1→4) linkages between galacturonic acid residues. Recently it was believed that although most PGs preferred desterified pectins as substrates, true polymethyl-galacturonases as hydrolases did exist. However, ac-

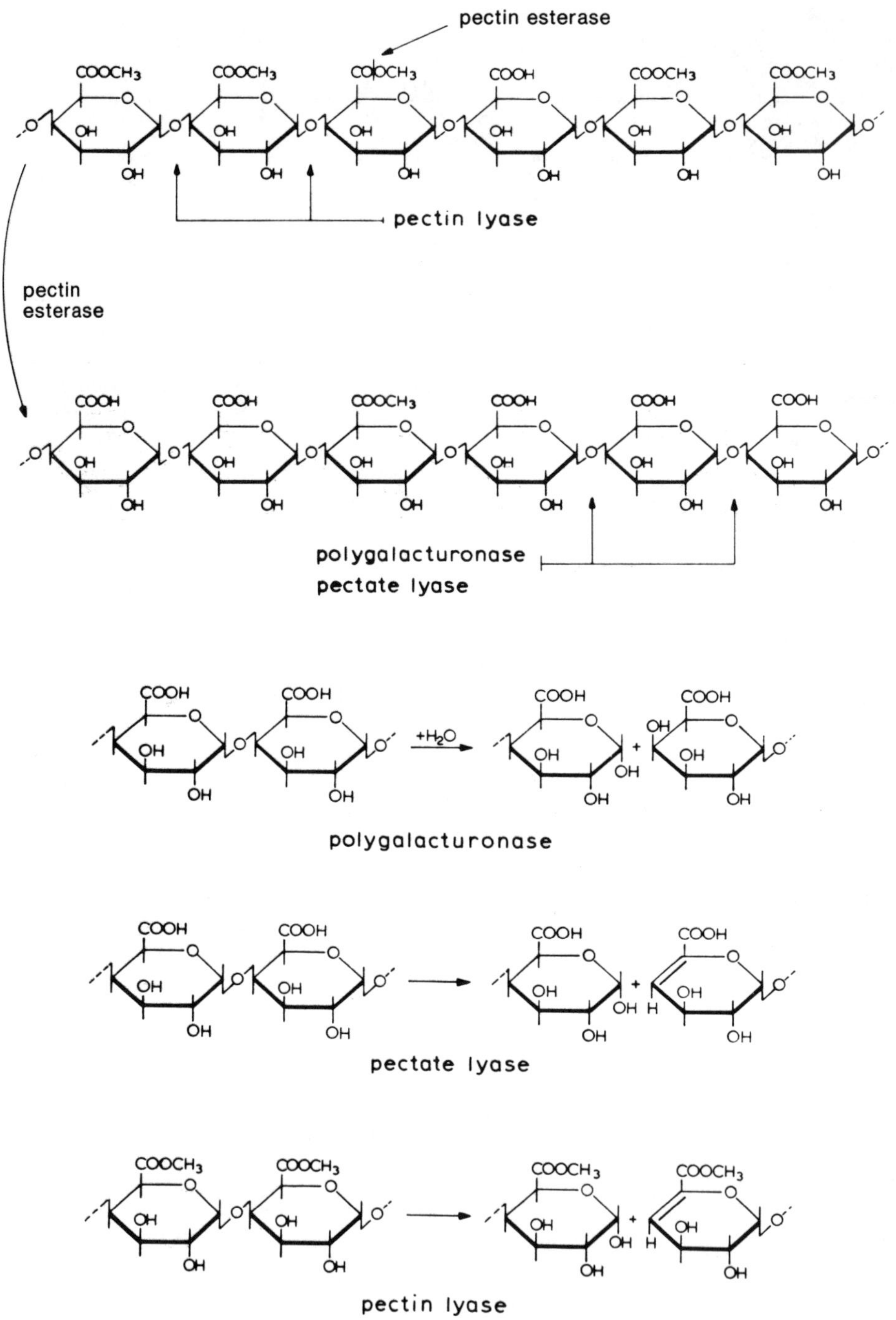

Courtesy of Pilnik and Rombouts (1978)

FIG. 29.3. ATTACK OF PECTIN BY POLYGALACTURONASE, PECTATE LYASE, PECTIN LYASE AND PECTIN ESTERASE cording to Rexová-Benková and Markovič, no such enzyme exists and the known PGs prefer pectic acid or very low-methoxyl pectins as substrates. Analogous to the specificity ranges of other glycanases such as the amylases and cellulases, PGs can be either endo- (random) or exo- (terminal) acting. Analogous to glucoamylase (Chapter 9), there are exo-enzymes which cleave off one galacturonic acid at a time from the reducing end. Others, in analogy to β-amylase, cleave the penultimate bond from the nonreducing end to form digalacturonides. In addition, there are enzymes which appear to prefer oligomers, splitting off one residue at a time.

Although all the endo PGs act on internal pectin linkages, resulting in a characteristic sharp decrease in viscosity (none will cleave the digalacturonic acid), individualistic specificity patterns emerge when comparing their action on the lower oligomers. For instance, the action of the predominant tomato PG on tetrasaccharide is characterized by an avid capacity for splitting both final and penultimate linkages with equal facility:

$$\text{(A)}\quad \text{Gal} \overset{1}{—} \text{Gal} \overset{2}{—} \text{Gal} \overset{3}{—} \text{Gal} \xrightarrow[-\text{Gal}]{+H_2O} \text{Gal} \overset{1}{—} \text{Gal} \overset{2}{—} \text{Gal} \xrightarrow[-\text{Gal}]{+H_2O} \text{Gal} \overset{1}{—} \text{Gal}$$

$$\text{(B)}\quad \text{Gal} \overset{1}{—} \text{Gal} \overset{2}{—} \text{Gal} \overset{3}{—} \text{Gal} \xrightarrow{H_2O} \text{Gal} \overset{1}{—} \text{Gal} + \text{Gal} \overset{3}{—} \text{Gal}$$

so that a mixture of all possible saccharides is formed. On the other hand, the endo PG of *Aspergillus niger*, present in most of the commercial pectinase preparations, will split the final bond of a tri- or tetrasaccharide (reaction A) only with difficulty.

A third type of specificity exemplified by the organism, *Erwinia carotovora*, found in rotting carrots can carry out reactions A and B as does the tomato enzyme, but it can also cleave the first as well as the second bond of the trisaccharide:

$$\text{(C)}\quad \text{Gal} \overset{1}{—} \text{Gal} \overset{2}{—} \text{Gal} \xrightarrow[-\text{Gal}]{+H_2O} \text{Gal} \overset{2}{—} \text{Gal}$$

Exo PGs.—With one exception, they attack the demethoxylated pectin molecule from the nonreducing end:

$$\text{Gal} \overset{1}{—} \text{Gal} \overset{2}{—} \text{Gal}\ldots \xrightarrow{H_2O} \text{Gal} + \text{Gal} \overset{2}{—} \text{Gal}\ldots$$

The only exception also constitutes the only exception to the rule that all pectin-acting enzymes are confined to plants and microorganisms, i.e., the exo PG in the gastrointestinal tract of some species of pathogenic insects.

Several endo PGs have been extensively purified and one has been crystallized (Uchino *et al.* 1966). Affinity chromatography has been used for their purification.

Pectin and Pectate Lyases

In the context of this chapter, the pectin and pectate lyases, formerly called pectin transeliminase, play only a pejorative role as determinants of solid plant food texture since they are largely confined to plant pathogenic organisms whose presence results in a deterioration in all aspects of food quality including, of course, texture. However, they do act in commercial preparations used for a variety of food-related purposes discussed in the next chapter.

Analogous to the esterases, lyase-like action occurs nonenzymatically at alkaline pH's. It would appear that this type of enzyme action occurs with acidic polysaccharides, the other well-established examples being lyases acting on hyaluronic acid, heparin, heparin sulfate and alginic acid.

The three specificities of the PGs, namely, exo-, endo- and oligo-, find their counterparts in the pectate lyases. However, unlike the PGs, no glucoamylase-like enzyme capable of splitting consecutive linkages in the long chain pectate molecule has been reported. Instead, there are definitely present, in some microorganisms, endo-enzymes which prefer the natural methyl-esterified D-galacturonan (pectin) to pectate as substrate. Endopectate lyases are produced only by bacteria and fusaria, usually have high pH optima and require Ca^{2+}. Exopectate lyases so far discovered are of the "β-amylase" type, splitting off di- "unsaturated" digalacturonides from the reducing end of the galacturonan. The oligopectate lyases split off terminal galacturonic residues of short chains to yield 4-deoxy-L-threo-hexose-5-ulosuronic acid (Fig. 29.3). Although several lyases have been isolated and shown to have molecular weights around 30,000, their isolation by affinity chromatography has not been reported to our knowledge as of this writing. An example of recent vintage (as of this writing) is that of Kamimiya *et al.* (1977), who isolated by nonaffinity chromatography procedures both pectate and pectin lyases from a favorite organism of investigation, *Erwinia aroideae.* With pectin as inducer (Chapters 4 and 5) they obtained a

36–38 kilodalton endo *pectate* lyase optimally active at pH 9 and in the presence of Ca^{2+}. When grown in the presence of an inhibitor of DNA replication, nalidixic acid, the organism produced a *ca* 30,000 dalton *pectin* lyase which was minimally stimulated, if at all, by divalent cations.

As mentioned, the pectin/pectate lyases may be major weapons in the armamentarium of plant predators in the microecological warfare of incursion, challenge and resistance to incursion played out at the molecular level of pectin enzymes. Relatively recently *in vivo* evidence has been forwarded of pectin-degrading enzyme action occurring during the pathogenesis of a rust disease (Welch and Martin 1975). Earlier, Albersheim and Anderson (1971) had isolated a protein, perhaps a lectin, from plant cell walls which quite effectively proscribed ($K_i = 1-2 \times 10^{-9}$) pectin-degrading enzyme action (but not that of other polysaccharide depolymerases) of plant pathogens.

Other Texture-modifying Enzymes

While the pectins and enzymes acting on them are of prime importance as determinants of plant food texture, undoubtedly other polymers and even nonpolymer food constituents and enzymes affecting plant foods contribute to their texture and to changes thereof. A partial list of such relevant enzymes would include: phenolase, active-oxygen generating oxygenases (Chapters 9, 15, 25 and elsewhere), the hemicellulases and/or "protopectinase," proteinase, cellulase, enzymes involved in lignin metabolism (Chapter 9), amylase, invertase, phytase, cutinase and callose metabolizing enzymes.

The first two or perhaps three classes of enzyme activity may affect texture by virtue of their relation to pectin itself. Thus the melanin formed as a result of phenolase action (Chapter 15) may block access of the pectin-degrading enzymes to pectin or to "protopectin" by intermeshing with the intracellular matrix (Joslyn and Deuel 1963), and perhaps also by inactivation of or inhibition of some of the cell wall associated glycanases (Bull 1970). Phytase can affect the textural properties of pectic substances by regulating the amount of divalent cations available for both pectic substance and pectin enzymes (see below).

Active Oxygen.—As discussed elsewhere on several occasions, foods have the opportunity to produce active oxgyen which can profoundly affect the food quality. These active forms of oxygen can give rise to a diversity of secondary reactions. Among these secondary reactions is the degradation of cellulose and pectin as shown in Fig. 9.6 (Kon and Schwimmer 1977; Schwimmer 1978).

In this particular case the depolymerization was shown to involve oxidative attack by ·OH, O_2^* and, to a lesser extent, ·OOH, all derived enzymatically from enzymatically produced O_2^-. Over 35 years ago, Kertesz (1943) suggested that "hydrogen peroxide" generated during the nonenzymatic oxidation of ascorbic acid might contribute to degradation of pectin in living plants. Of course, another way in which such destructive forms can give rise to textural changes is by destroying plant cell membranes. As pointed out by Sterling (1975), while pectin is responsible for much of the rigidity of plant foods, it is the membranes which control water relations within the living cell and, hence, that aspect of texture we recognize as turgor or succulence.

Protopectinases.—This term is invoked to explain or account for a transformation of galacturonosyl residues linked to the matrix in the middle lamella of plant cells into a soluble form, pectin. As previously mentioned, conversion of this "cellular adhesive" into a soluble form results in cell separation and a disruption of tissue rigidity that is reflected in softening.

The evidence that the bond split is actually enzyme-specific for the α-(1→4) anhydroglucuronosylglucuronide bonds has been inferential. Thus, investigators emphasize the frequently observed parallelism among development of the pectin-related enzymes we have just discussed, loss of "protopectin" and softening, inferring that these enzymes themselves are responsible for the conversion of the protopectin into its soluble form, pectin. However, if one looks upon protopectin as that part of the three-dimensional hemicellulose network rich in galacturonic acid residues, then "protopectinase" may be a complex of enzymes similar to the hemicellulases discussed in Chapter 9 (Joslyn 1962). Perhaps any enzyme action capable of loosening the structure of the middle lamella may be termed a "protopectinase." The term "macerase" (Chapter 30), occasionally used interchangeably with protopectinase, is really an operational term describing a macerating effect on the tissues. Any strategic glycosidic bond cleaved, be it by a galacturonanase, cellulase, α-arabinosidase or α-galactosidase (Pilnik and Voragen 1970), might result in pectin solubilization. Karr and Albersheim (1970) proposed that plants elaborate a "cell wall modifying" enzyme whose action is obligatory prior to the action of glycanases. On the other hand, Knee *et al.* (1975), noting that there was at that time no direct evidence concerning the mode of attachment of hemicelluloses to pectic polysaccharides, were able to liberate one-half of the uronic acid residues from isolated cortical cell walls of the apple fruit by a presumably hemicellulase-free polygalacturonase obtained from one of the most widely used sources of "macerase," the fungus *Sclerotinia fructigena.* A second *Sclerotinia* PG isozyme released more uronic acid covalently linked with a substantial proportion of hemicellulosic sugars. Significantly, preparations of hemicellulases galactanase or arabinosidase, although possessing small but detectable pectate lyase activity, did not liberate much uronic acid. Also, while protease (Pronase) treatment of walls liberated some polyuronide associated with glycopeptides, pretreatment of the walls with PG led to enhanced protease attack. Conversely PGs reportedly release cell wall-bound proteins and enzymes (Strand *et al.* 1976). While in the right direction, this approach should be pursued with enzymatically pure glycanases and peptidases obtained from the same plant tissue comprising the cells.

Other Enzymes.—While not as directly involved in texture as are the cell wall polysaccharides, cutin, the structural biopolymer of the cuticle of higher plants, does affect the appearance and resistance of the fruits to deterioration. Cutin, estolide, is a polyester of $C_{14}-C_{16}$ hydroxyfatty acids. Although it is not known if fruits contain cutin depolymerases, such enzymes have been purified from plant pathogenic fungi, especially *Fusarium* species. These esterases act specifically on the ester linkages between the carboxyl of one hydroxyfatty acid and the OH group of another (Soliday and Kolattudy 1976).

Of course, the amylases are important in the texture of starch-containing plant foods, especially the sweet potato after processing and perhaps in other starchy foods. Invertases impart aspects of texture as well as taste to dates. We shall briefly consider the role of lignin-associated enzymes in connection with pear and asparagus texture. Callose, an α-1,3-linked glucan implicated in food texture, will be considered in conjunction with tomato enzyme as will cellulose and cellulase. We now proceed to a discussion of how these endogenous enzymes influence texture of individual intact fruits and vegetables. Enzyme additions in transformed plant foods will be treated in Chapter 30.

ENZYME INVOLVEMENT IN PLANT FOOD TEXTURE: SPECIFIC FRUITS

The life cycles of fruits are characterized by the onset of the climacteric, which signals dramatic changes in their metabolism and is reflected in equally drastic changes in the appearance and texture of the fruit tissue. Ensuing senescence may be conceived as a programmed schedule of progressive cell disruption initiated and carried out by the cell wall polymer-degrading enzymes which are elaborated. Since the pioneering work of Appleman and coworkers some 50 years ago (Appleman and Conrad 1926) it has been repeatedly demonstrated, starting with apples and applied to most other fruits, that the softening, attendant ripening and senescence are invariably accompanied by a progressive transformation of the uronic acid or "pectic" substances.

Peaches.—To cite a more recent example, Schewfelt *et al.* (1971) uncovered striking parallelisms between ripening of peaches and *decreases* in: molecular weight of pectinic acid, the uronide fraction obtained by water extraction of the alcohol insoluble solids; degree of esterification of the soluble pectic constituents of the protopectin (insoluble uronide); and firmness as measured with a shear press. Pressey *et al.* (1971) further demonstrated that the softening was attended not only by a solubilization of the pectin but also by an increase in PG activity and a decrease in the size of the pectin solubilized in both tree and postharvest-ripened peaches. Interestingly, a feeble cellulase C_x activity (Chapter 9) also developed during ripening but reached a maximum *before* softening set in and did not increase greatly during postharvest ripening. A normalized version of these results is shown in Table 29.4. The Georgia workers suggest that cellulase may be involved in the initiation of processes leading to tissue softening, processes in which their results strongly suggest (but as they state do not prove) that PG action is functionally important. For instance the marked, well-known difference in texture between Freestone and Clingstone peaches is due to the presence in the Freestone and absence in ripening Cling fruits of endo PG activity (Pressey and Avants 1978). Both varieties exhibit ample exo PG activity. The action of the endo PG leads to high levels of water-soluble pectin in the Freestone which presumably results in the absence of the elastic-like texture associated with Clingstone.

Dates.—As with peaches, both cellulase and PG are absent in immature fruit and begin to develop

TABLE 29.4

PEACH RIPENING, TEXTURE AND ENZYME ACTIVITY

Parameter Measured	Tree Ripened Values[1] on July 8	July 24	July 28–31	Postharvest Ripening Days[2]
Firmness (shear)	100	33	6–8	2.5
Pectin, size	100	68	30–45	1.7
Pectin, % soluble	25	65	85–100	1.9
Polygalacturonase activity	0	46	78–100	2.2
Cellulase activity[3]	From 0 to 100 in 4 days			No change[4]

Source: Adapted from data of Pressey *et al.* (1971) on Freestone peaches.
[1] Percentage of maximum value observed between July 8 and 31.
[2] Time after harvest required for parameters to reach interpolated values equal to those obtained after 20 days of tree ripening; initial values for tree- and postharvest ripened peaches approximately the same.
[3] From data of Hinton and Pressey (1974) on Elberta peaches; cellulase activity measured from Aug. 3 to 7, after which the firmness was higher than the initial values measured for Freestone peaches.
[4] Firmness decreased to 18% of its freshly harvested value in 2 days.

during the early stage of ripening. Unlike peaches, activities of cellulase as well as of PG reach close to maximum values before softening sets in. Normal date ripening and upgrading poor quality (#2 dry) dates afford a striking illustration of the result of potentiation of the action of cell wall-associated enzymes as depicted in Fig. 29.4. The need to actually abolish cell walls to achieve desired chewy texture explains why, unlike most fruits, concomitant cellulase as well as pectin-degrading action is necessary. Indeed, of the three kinds of commercial food enzyme preparations added to poor quality tough dates—pectinase, invertase and cellulase—only the action of the enzymes in the cellulase preparations resulted in significant softening (Hasegawa and Smolensky 1971). Dates achieve an appropriate "eating" texture at a stage of senescence well beyond that of other edible fresh fruit.

In both immature and poor quality #2 ripe dates characterized by a tough, dry texture, the cell walls are quite distinct under the light microscope (Fig. 29.5). By allowing the good quality date to ripen normally or by exposing the poor quality date to thermal potentiating conditions—*ca* 50°C, in the presence of sufficient moisture—the cell walls are almost completely obliterated. Both fruits are converted from light-colored, tasteless, turgid fruit into dark, chewy, sweet, high quality dates.

Invertase.—Also unlike most fresh fruits, the texture of dates is determined in part by the action of invertase, which does not only sweeten the dates, but conversion of the sucrose into invert sugar imparts chewiness, water-holding properties and prevents grittiness due to the sucrose crystal formation at the low moisture levels which prevail. The presence of crystalline sucrose in certain defective types (sugar wall) results in this grittiness and also in a tendency to be harder than even No. 1 dry grade dates at equal moisture levels. In these types of dates, the latent invertase activity is not sufficient to permit an effective enzyme potentiation to proceed and prevent crystallization (Smolensky *et al.* 1975). Commercial food grade invertase preparations, when applied to such sugar wall dates as a spray, effectively correct this textural aberration at an invertase cost of about 4¢ per 100 kg of dates. Such treatment improves not only texture but appearance as well, converting the date from a dull-looking, crystalline fruit to the proper glistening translucent appearance and appropriate color, due in part to nonenzymatic browning initiated by the liberated reducing sugar (most of the desired color comes from the action of potentiated phenolase, as mentioned in Chapter 15). It is likely that a similar enzymology pertains to the attainment of appropriate qualities in other low moisture fresh fruits such as raisins and prunes. Perhaps in the future, cutinase preparations (see above) may be beneficially applied to regulate the drying and texture of such semidried fruits.

Pears.—One of the most complete assessments concerning the development of cell wall-associated enzymes in the developing pear is that of Yamaki and Matsuda (1977). As with other fruits, the activities of the enzymes tested for per unit fresh weight or per unit weight of DNA (measure of cell numbers) rose rapidly during ripening; endo-and exocellulases, PG, β-galactosidase, except for PE which remained constant.

However, pears have a unique aspect of textural quality which is due not to the polysaccharide portion of the cell wall but to the lignin. Lignin is

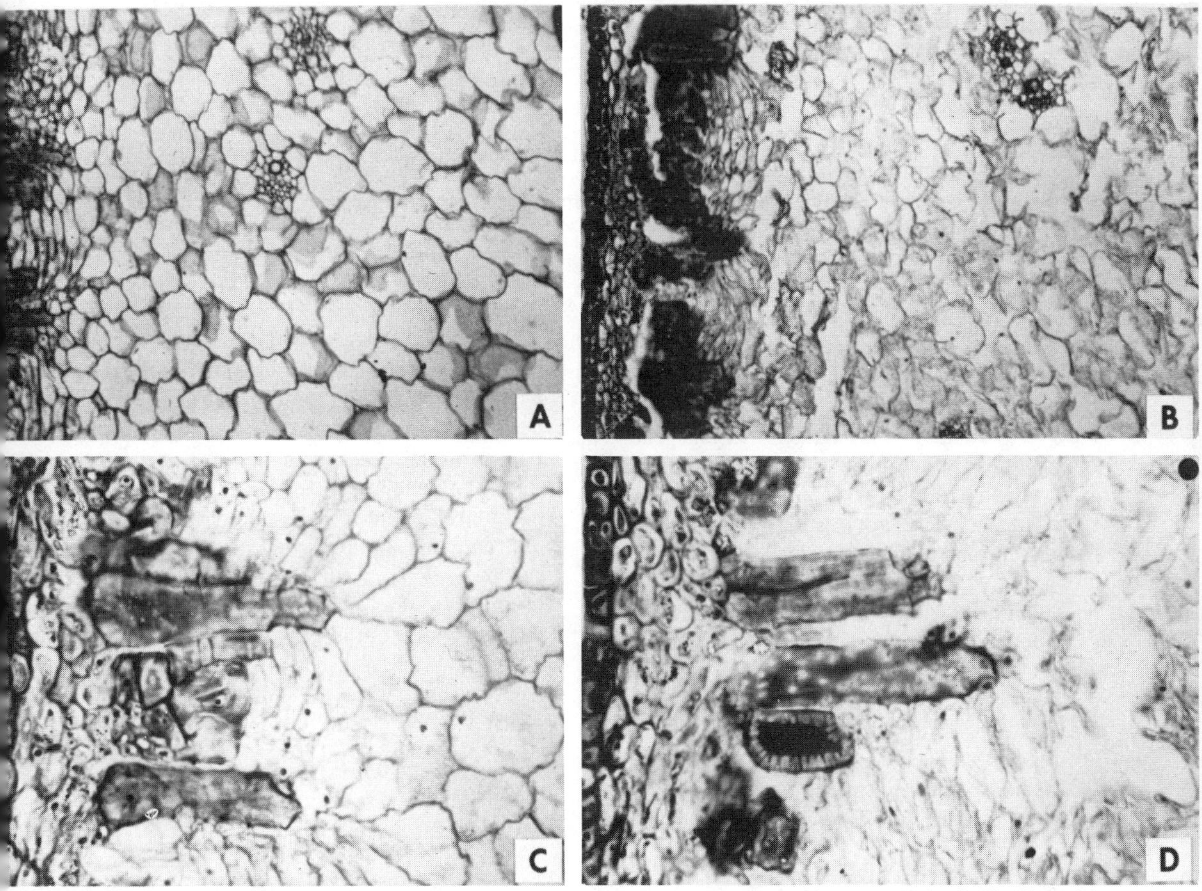

From Schwimmer (1978). These photographs supplied courtesy of Dr. C. W. Coggins, Univ. Calif., Riverside.

FIG. 29.4. CELL DISRUPTION IMPROVES DATE QUALITY

Photomicrographs (21×) of cross section of date tissue located midway from stem to stylar end of fruit: (A) Kimri stage of development, turgid and green; (B) Tree-ripe stage of development but not as ripe as a completely ripe date; (C) Number 2 dry grade of date; (D) Number 2 grade of date after filtration and incubation for 24 hr at 122°F.

the substance contributing to the hardness of the "stone" or "grit" cells, sclereids, whose presence imparts the rather pleasant textural sensation (if not present in excess) that is peculiarly associated with this fruit when eaten raw or cooked. The deposition of lignin has been considered to proceed via a peroxidase-catalyzed oxidation of the basic monomer coniferyl alcohol, 1-hydroxy, 2-methoxy, 3-hydroxy, *trans* propenyl benzene, or 3-methoxy, 4-hydroxy cinnamyl alcohol. Analogous to the pectin-pectic enzyme complex, peroxidase activity was found by Ranadive and Haard (1972) to be closely associated both temporarily and spatially with lignin deposition and content in these in the development of these grit cells in the pear. Also analogous to the action of the pectic-degrading enzymes, Ca^{2+} participates in this deposition and its regulation by virtue of its ability to release peroxidase from the cell wall, thus removing it from the site of lignin synthesis. With other plant foods, especially asparagus, the enzymes of lignin deposition present a problem in that the resulting lignin is laid down in the tissues during aging to impart an undesirable toughness (see below).

Citrus.—Not all fruits have been found to contain the full complement of pectin-degrading enzymes. It was believed for many years that citrus fruits, such as oranges (Macdonnel *et al.* 1945; Korner 1971), even the albedo, rich in pectin, and more recently grapefruit (Robertson 1976), were devoid of PG. Attempting to prepare a low methoxyl pectin by potentiation of the PE of orange peel—adsorbed in an affinity chromatographic enzyme-substrate complex to cell walls of the albedo (Jansen *et al.* 1960)—in the absence, ostensibly, of PG, we concluded that orange albedo must contain pectin depolymerases of some sort since after such thermal potentiation at 50°–60°C in NaCl solution, no pectin could be recovered (Schwimmer 1974). This was corroborated by Riov (1975), who showed that the C-1 aldehyde group in the galacturonic acid released by such depolymerization was ostensibly oxidized by a "uronic acid oxidase." The assay method for this oxidase still leaves open the possibility that the pectin in citrus may be degraded as the result of the action of an O_2^--generating enzyme as discussed earlier in this chapter. Indeed, as mentioned in the next chapter, such oxidative reactions have been implicated in cloud loss in citrus juices, an area in which pectolytic enzymes, both endogenous and added, play a more important role than they do in texture of whole citrus fruits. Yet the patent literature has recorded the formation of Ca pectate gels in citrus products. The secret to such success may be the adjustment of pH to slightly alkaline pH's which would allow potentiation of the PE without concomitant action of any PG which might be present (Buckley *et al.* 1976).

Apricots.—This is another fruit where the action of an endogenous PG as being responsible for softening and oversoftening associated with ripening has been questioned. As described by Luh *et al.* (1974,1978), such an oversoftening manifested only after canning and shaking was traced to the action of PG present in a contaminating fungus: *Byssochlamys fulva* in South Africa, *Rhizopus stolonifer* in Australia and *Rhizopus arrhizus* in the United States. Maintenance of sanitary conditions not conducive to mold growth and the use of Ca salts at 85°C (see below) have been recommended as means of retention and restoration of proper firmness of canned apricots (Luh *et al.* 1978).

Strawberry.—This is another fruit whose softening and senescence are not accompanied by dramatic increases in PG and whose oversoftening has been attributed to action of adventitious enzymes coming from invading microorganisms. However, Barnes and Patchett (1976) reported the absence of cellulase C_x activity (Chapter 9) in green strawberries, its presence in ripe and enhancement in overripe fruit. The enzyme relations in strawberry ripening and senescence are far from settled (Dennis 1977).

Bananas.—Probably because of the difficulty of demonstrating the PG activity in bananas, most of the attention of softening-associated enzymes has been directed to PE. Thus, Markovic *et al.* (1975) reported on the "detection" of endo and exo PGs in bananas and six multiple forms of PE. However, a previously reported apparent increase in activity of banana PE isozymes (Hultin and Levene 1965) was more recently shown to be really due to an increase in the extractability of this enzyme activity (Brady 1976) and the total activity apparently remains constant throughout ripening. Ease of extraction may have some relevance to the role of the enzyme in the softening process. "Chilling injury" of bananas, a quality defect, results in uneven softening and even at times a hardening of the pulp. This hardening has been attributed to inhibition of key mitochondrial oxidative enzymes (Murata and Ku 1966) and to temperature-induced anomalous kinetics of IAA oxidase, closely resembling that of allosteric enzymes (Haard 1973).

Cherries: Calcium; Preblanch, LT–LT and PE Potentiation.—Like apricots, cherries can turn overly soft during or after processing because of PG present in contaminating microorganisms (Lewis *et al.* 1963). Very little if any PG is present in the uncontaminated cherries but they do possess PE whose activity apparently increases with ripening (Al-DeLaimy *et al.* 1966). While this PE can create a textural problem, one of toughness in frozen and bruised cherries (Gee and McCready 1957; Van Buren 1974), its action has, in conjunction with Ca salts, been directed to improving brined cherry texture by a controlled firming. Evidence that improved chewiness of hot-fill red tart cherries by exposure to 55°–70°C in the presence of Ca salts equivalent to 0.4% Ca (LaBelle 1971) is due to activated (potentiated) PE is afforded by the investigation of van Buren (1974). The functions of Ca^{2+} in this "preblanch" or low temperature-long time (LT-LT) treatment, which, as we shall presently document, constitutes one of the most successful and widely adopted modes of enzyme potentiation in food processing, are probably to contribute to the activation of the PE once it has been liberated from the membrane destroyed by the enzyme-potentiating temperature range. One function of calcium is to combine with the PE-generated low methoxyl pectin to form a stable three-dimensional network of calcium pectate throughout the processed cherry (or other plant food). The mechano-elastic properties of this network are reflected as improved firmness and thus chewiness of cherries and prevention of sloughing in other processed fruits and vegetables. Of course, Ca^{2+} also can form gels with undemethoxylated pectin even after limited PG action so that in combination with the dehydrating effect of the high levels of sucrose present in canned cherries and other canned fruit, Ca^{2+} prevents the leakage of uronide-containing polymers out of the fruit into the syr (Chen and Joslyn 1967).

Ionizing Radiation.—Successful irradiation of fresh fruits to prolong storage is hampered by a rapid softening which, because it occurs so rapidly, has been assumed not to involve depolymerase action directly. Since membranes are the primary target of ionizing radiation (Chapter 10), the softening may be primarily due to loss of turgor which, as mentioned, is maintained by membrane viability. However, it is conceivable that the translocation of Ca^{2+} occasioned by membrane disruption may activate PG. Due to the latter's strategic position in the protopectin complex, the protopectin is exposed to "protopectinase" and other Ca-activatable depolymerase action. Certainly such dislocation of Ca^{2+} could account, along with potentiated PE action, for recovery of firmness frequently observed upon storage of irradiated fruit.

Apples.—The use of Ca salts to firm plant food was first applied to apples almost 35 years ago by Kertesz (1947). Since then it has become manifestly clear that Ca^{2+}, when used in conjunction with a preblanch under proper conditions, can be quite effective in maintaining firmness, not only in canning, but also in processing of frozen dehydrated and refrigerated apple slices.

When fruits are sulfited, as apples frequently are, Ca serves an additional textural function in that under certain conditions it counteracts the softening effect of sulfite (Ponting *et al.* 1971).

Ca-induced improved texture at pH 9 but not at 4–5 may be related to the circumstance that apple PG is activated by Ca^{2+} and has a pH optimum at 4.5–5 (Bartley 1978). On the other hand, the circumstance that PE potentiation works at all in apples may be due to the fact that apple PG is an exo-polygalacturonase; endo PG activity is absent. This would allow sufficiently large fragments of demethoxylated polyuronide to exist for sufficiently long periods to permit complexation with Ca^{2+}. Indeed, in general, it is the absence of high endo-PG activity or at least high PE: PG ratios at 60°–70°C which may be one of the prerequisites for successful Ca-preblanch firming effects, at least for fruits (see oranges).

A textural disorder of apples, internal breakdown, has been traced to enzymatic hydrolysis of pectin causing a decreased firmness followed by increased permeability of cell membranes causing loss of turgor (Pollard 1975). The observation that PE activity *increased* while a "tissue macerating factor" *decreased* during the development of this disorder would suggest that turgor loss may predominate.

VEGETABLE TEXTURE

What Is a Vegetable?

While in general the principles which apply to the role of enzyme action in fruit texture also hold for that of vegetables, there are significant differences. In order to show these differences we must first define what a vegetable is, since, after all, the vegetable is not recognized as a distinct biological entity in a strictly scientific sense.

In 1893, the Supreme Court, in rendering a decision that the tomato is a vegetable, defined a vegetable as a plant, or part thereof, which is "usually served at dinner in, with, or after the soup, fish, or meats which constitute the principal part of the repast. . . ." This definition, of course, is inconsistent with the botanical classification of a tomato as a fruit, i.e., the seed-bearing portion of the plant consisting of ripened ovary and its contents. All writers agree there is no scientific definition of "vegetable." Rather, usage, customs and mores (and legal opinions) have named a given food "fruit," "cereal," "nut" or "vegetable." However, perhaps we can eliminate those plant foods which are not considered vegetables. The naturally dried grass seeds (cereals), one-seeded dry fruits with a hard pericarp (nuts) and, finally, fruits may be disregarded. According to the food usage, fruits have most of the following characteristics: (1) all food fruits, in whole or in part, are also fruits botanically; (2) they can be eaten raw; (3) they are picked from either a woody plant such as a tree or bush, or a vine; (4) they are eaten mature; (5) they are sweet. Now taken separately any one of these criteria could be applicable to a given vegetable, but no vegetable passes all of them. The term "vegetable" therefore embraces a large and diverse category of plants and parts thereof. Inspection of Table 29.5, which shows a classification of vegetables based upon the gross morphology of the plant organs, reveals that one glaringly prominent texture-associated difference between fruits and nonfruit vegetables is that the latter comprise much more immature tissues, so that when eaten raw the turgor (membrane) component of texture predominates over the rigidity (pectin) component.

If there is one dominant theme which pervades the relation between enzyme action and textural modification in these diverse plants, tissues and organs, it is the recurrent implication of calcium, either naturally occurring or added as a step in processing, to modulate the physical state of the cell wall constituents both modified and unmodified. Unlike many fruits, the fate of calcium in this regard is, for many vegetables, in part determined by the presence of phytic acid and phytic acid transforming enzymes.

Tomatoes

Tomatoes are a rich source of PG which is implicated in softening during ripening and of PE needed for firming of processed products. In addition to pectin, cellulose and callose have also been implicated as modifiers of tomato texture.

Polygalacturonase.—Although pectate lyase and PG have been reported to be present in tomatoes, it is likely that only endo and exo PGs are present (Demain and Phaff 1957). More recent evidence of an endo PG activity is afforded by the investigation of Pressey and Avants (1971). They observed a progressive shift in its pH optimum from 4 to 2 with increasing activation by monovalent cations (Na^+, K^+) and decreasing initial substrate polymer substrate chain length when acted on by the major of two main isozymes. Both isozymes hydrolyze small substrates but only one is inhibited by pectate. One isozyme acting on long but not short chain galacturonans may be responsible for solubilizing some of the protopectin.

TABLE 29.5

CLASSIFICATION OF VEGETABLES

	Examples
Earth Vegetables	
Roots	Sweet potatoes, carrots
Modified stems	
Corms	Taro
Tubers	Potatoes
Modified buds	
Bulbs	Onions, garlic
Herbage Vegetables	
Leaves	Cabbage, spinach, lettuce
Petioles (leaf stalk)	Celery, rhubarb
Flower buds	Cauliflower, artichokes
Sprouts, shoots (young stems)	Asparagus, bamboo shoots
Fruit Vegetables	
Legumes	Peas, green beans
Cereals	Sweet corn
Vine fruits	Squash, cucumber
Berry fruits	Tomato, eggplant
Tree fruits	Avocado, breadfruit

Source: Feinberg *et al.* (1964).

Ripening.—As with other fruits (botanically speaking), ripening of green tomatoes is accompanied by increasingly rapid loss in firmness, pectin solubilization and, more frequently than not, increase in PG activity. Thus, Hobson (1968) observed a 200-fold increase in activity in going from green to orange. PG activity continued to increase for the next 2 weeks, during which the tomatoes turned deep red.

However, there is not always a one-to-one correspondence between softening and increase in PG activity. Besford and Hobson (1972) pointed out some softening occurs in green tomatoes in the apparent absence of conventional PG. This softening is not due to an alleged pectate lyase, which as mentioned appears to be confined to microorganisms. In two instances a reverse trend in the pectic enzyme activity has been observed, namely, in chill injury (Montes 1971) and after treatment with gibberellic acid (Lampe 1971). In the latter case the PG and PE activities were suppressed without influencing the rate of softening during otherwise normal ripening.

Apparently, repression of PG synthesis can also occur when picked tomatoes are exposed to a slightly elevated temperature, 33°C, and then brought back to ambient temperature, an effect which results in the extension of the storage life and maintenance of desirable firmness (Ogura *et al.* 1975). Another route to increased firmness prior to processing is via breeding. Buescher *et al.* (1976) describe tomato hybrids which, as shown in Table 29.6, exhibit lowered levels in all four texture parameters: softness, PE and PG activities, and water-soluble pectin.

Roles of Nonuronide Polysaccharides and Their Hydrolyases.—From time to time cellulase has been implicated as a causative factor in tomato texture change (Hobson 1968), whereas at other times this is denied. Thus, Sobotka and Watada (1971) could find no correlation between objective measurement of firmness and "cellulase" activity.

TABLE 29.6

GENETIC REDUCTION OF TEXTURE-AFFECTING ENZYMES IN TOMATOES[1]

Variable	Percentage of Normal, *rin* Mutant	C17; Hybrid with: *nor* Mutant
Softness	79	49
Pectin, water soluble	85	71
Pectinesterase	64	38
Polygalacturonase	45	70

Source: Buescher *et al.* (1976).
[1] Values 70 days after anthesis; unhybridized mutants do not ripen.

Simeonova (1976) observed, upon ripening, a sharp enhancement of C_x-cellulase activity in tomatoes picked unripe. Activity remained at a level which far exceeded the activity of this enzyme in fruits allowed to ripen on the vine. Whatever their role in texture, Sobotka and Stelzig (1974) clearly demonstrated that the ripening tomato fruit contains a rather active cellulase complex capable of completely degrading insoluble cellulase substrate to glucose and cellobiose, rivaling in this respect some of the cellulolytic microorganisms used for glucose production (Chapter 9).

Finally, Wallner and Walker (1975), after demonstrating the presence of β-glycosidases and susceptibility of cell walls to their action before maximum PG occurred during ripening, suggested that these glycosidases contribute to cell wall modifications which perhaps make pectin accessible to PG activity and lead to softening. They confirmed the central role of PG by showing that the ability of preparations from fruits at different stages of ripeness to attack isolated tomato fruit cell walls parallels increased PG activity. The significance of the presence of their β-1,3-glucanase in tomatoes will be discussed below in connection with callose formation during harvesting and processing.

Pectin Esterase.—As previously mentioned, tomatoes are probably the richest single source of PE among common food plants and the tomato is one of the few plant foods prepared as a source of commercial enzyme preparations (Weaver 1961). Its characteristic mode of action as outlined above, together with its high temperature stability (Pilnik and Voragen 1970) determines in part conditions for tomato canning. Like PG, the activity increases during ripening, especially as the tomato turns from green to red. However, the increase is not as spectacular as that for PG since PE (but not PG) is already present at appreciable levels in the green fruit. Indeed, according to Pressey and Avants (1972), the increase in activity, mainly due to a several-fold enhancement of one particular isozyme and in part to the appearance of a new iso-PE, is attenuated due to the disappearance of a third isozyme of PE.

One characteristic of these PEs, not in accordance with a role in firmness, is their apparent pH optima around neutrality, quite far removed from the acid pH in the tomato. Perhaps when immobilized to the cell wall *in situ*, activity-pH relations may be altered (Chapter 8) or they are still effectively active at these pH's. Like the citrus enzyme (see below), partially purified preparations of the tomato enzyme have been shown to cling increasingly tenaciously to isolated cell walls, in this case from tomatoes with increased ripeness (Nakagawa *et al.* 1971).

Pectin Esterase and Preblanch Firming of Canned Tomatoes.—It was in canned tomatoes that Ca pectate was probably first identified as the cause of firming in preblanch Ca treatment. The role of PE was surmised over 50 years ago (Appleman and Conrad 1928A,C) and confirmed by Kertesz *et al.* (1940). The fact that canned tomatoes firmed better than raw tomatoes was ascribed to the enhanced de-esterification occurring during blanching. Hsu *et al.* (1965) examined in some detail the role of PE, observed a maximum potentiation of PE activity and obtained the firmest tomatoes with a 30 sec exposure to 100°C and holding for 10 min prior to addition of $CaCl_2$ to a level of 0.05% at pH 4.4 followed by sealing and retorting. Firmness correlated highly ($r = 0.9$) with both Ca pectate level and PE activity. Further investigation by Deshpande *et al.* (1965) showed that more important for firming than total pectic (uronide) content were the molecular size of the soluble uronide and, of course, the presence of free carboxyl groups created in a *controlled* demethylation in order to achieve optimum textural quality. In 1970, the U. S. Food and Drug Administration amended identity standards for canned tomatoes to permit the increase of Ca salts to 0.1%, as Ca (Food and Drug Administration 1970), although canned tomatoes treated thus contain on the average 0.003% (Steagall 1970).

Callose and Callose Hydrolase.—A radically different view of determinants of postharvest and processing-associated changes in tomato texture considers callose, the amorphous 1,3-β-D-polyglucan formed upon bruising and heating, as a major contributor to texture in tomato fruit. This view, adopted by Dekazos (1972) is based upon nonquantitative fluorescence microscopy and requires independent confirmation and documentation. What little is known about callose enzymology suggests that a callose hydrolase activity (i.e., a β-D-1,3-glucanase) may be present in the green fruit and may also be present upon ripening (Wallner and Walker 1975).

Potatoes

Unlike most of the plant food we have thus far considered, potatoes are eaten only after exposure to high temperature. They also differ from fruits and vegetable fruit such as the tomato in experiencing a distinctly different preharvest development history since, being a modified stem, they do not ripen like a fruit. They also differ from most of the plant foods hitherto considered because of their high starch content. Upon cooking, potatoes may assume a variety of textural characteristics, terms for assessment of which are shown in Table 29.7. Implicit but not stated in these terms is the pervasive textural-associated cooked (especially canned) potato defect known as "sloughing" which is the result of tissue disintegration (classification 2) or lack of cohesiveness (classification 3). Sloughing and other aspects of potato texture have been frequently reviewed in the last 10 years, i.e., Linehan and Hughes (1969), Warren and Woodman (1974) and Reeve (1977). These reviews reflect the running "pseudo" controversy concerning the relative role of intracellular starch on the one hand and the cell wall and intercellular constituents on the other. The desired characteristic of mealiness ("flouryness" to the British) ascribed to by Warren and Woodman to the subjective perception of flow characteristics ("viscosity") of the cooled potato, is controlled primarily by solids content and therefore the starch. On the other hand, sloughing is viewed as the result of lack of cohesion (or adhesion) between cells brought about by excessive hydration of cell wall material and therefore involves mostly changes in pectin constituting at least one-half of the cell wall. In both cases, cations play a role as textural determinants. From the viewpoint of enzyme involvement, we wish to assess how their action may modify these polysaccharides and in turn how these modified polysaccharides and enzyme reactions, interacting with cations, ultimately cause changes in the texture.

TABLE 29.7

DESCRIPTIVE ASSESSMENT OF COOKED POTATO TEXTURE

Classification Scheme	Descriptive Terms		
Linehan and Hughes (1969)	Mealy	Dry	Structural
Scheme I	Disintegration	Consistency	Firmness
Scheme II	Floury	Close waxy	Soapy, mealy
Reeve (1972)	Firmness	Cohesiveness	Mealiness
	Sogginess	Stickiness	Gumminess

Pectin and PE.—If indeed pectin is involved in sloughing, it is not surprising that PE action and addition of Ca salts can help to contravene this textural problem. Bartolme and Hoff (1972), in investigating the mechanism of potato firming by preblanch heating at 60°–70°C, unearthed evidence which led them to the scheme depicted in Fig. 29.5 which, as we have seen, is echoed in the preblanch treatment of other foods. Seminal to the firming mechanism are the series of events starting with the potentiation of PE via cell disruption followed by desorption from cell wall and activation by divalent cations also released.

This basic study has its counterpart not only in canning (Talburt 1975) but also in other areas of potato processing. Thus, in a patent issued to Unilever (1966) for the preparation of a frozen potato product, new tubers are preheated between 40° and 65°C for 2 to 3 min. This preblanch is followed by a 20 to 40 sec exposure to 100°C, sufficient to inactivate surface enzymes, and the tubers are frozen. Interestingly, a *positive* test for PE activity (*vs* a negative test for blanching proper) is used, an index of adequate preblanching.

Starch and Phytate.—Enzymes involved in starch transformations in potatoes were examined in Chapter 17. Reeve (1976) adopts the eclectic viewpoint that both starch and pectin are involved in the preheating firming effect, the former in part by virtue of its retrogradation so that it does not participate in subsequent swelling leading to rounded, separated cells. Metal ions are considered to affect both starch and pectin. The role and effects of cations also have been investigated by Hughes *et al.* (1975) and by Keijbets *et al.* (1976). Interestingly, the latter found that, in general, ions catalyzed a nonenzymatic pectin lyase (β-elimination) reaction. Ca^{2+} hindered pectin (really uronide) solubilization, an effect which was enhanced by PE induced de-esterification and, in the context of the

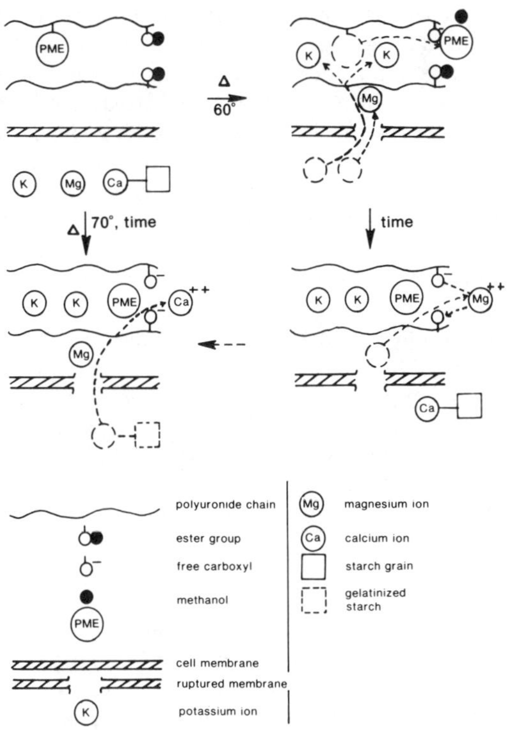

From Bartolme and Hoff (1972)

FIG. 29.5. CALCIUM INDUCED FIRMING OF POTATOES

Proposed mechanism to account for this treatment takes into account thermal potentiation of pectin esterase.

comparative roles in texture of starch and pectin, complexed more strongly with the resulting pectate. Another enzymatic factor which may enter into textural quality of cooked potatoes is the level of phytic acid which would release calcium after hydrolysis by phytase (Schwimmer 1956; Samotus and Schwimmer 1962C). According to Wager (1963), phytic acid leads to softening by acting as a calcium precipitant (see legume seeds). Phytase action is potentiated at preblanch temperature in absence of added Ca salt (Chang *et al.* 1977). Phytase is examined more closely in Chapter 36.

Green Beans.—Perhaps more than with any other vegetable, control and optimization of green (snap, string) bean texture has been a problem of quality control. One reason for this is the diversity of cellular structures. The problems encountered include: nonuniformity of raw material, sloughing of the skins in cooked and frozen products, stringiness or toughness due to overmaturity, loss of crispness due to overblanching and tendency of the beans to fragment. As with other vegetables, pectin and PE action are intimately involved as determinants of green bean texture, pectin being implicated from studies going back to the 1930s (Culpepper 1936). PE action was implicated somewhat later (Sistrunk and Cain 1960; Van Buren *et al.* 1960) as also was the use of $CaCl_2$ to correct textural quality defects (Van Buren *et al.* 1960; Kaczmarzyk *et al.* 1963). Since $CaCl_2$ appeared to exert a somewhat beneficial effect after inactivation of PE, Van Buren assigned a nonenzymatic role to the firming effect of this salt, although no evidence was tendered as to what this effect might be. In contrast to potatoes the role of phytase is probably negligible because of the comparatively low levels of phytate in pods as compared with that in seeds (Makower 1969). At first used extensively for canned beans, this LT-LT stepwise blanching approach involving PE potentiation appears to be effective in optimizing the texture of frozen green beans (Steinbuch 1976). One problem introduced by such treatment, excess leaching of useful constituents, can be circumvented by use of saturated blanching water and by the application of short-high steam treatments (Chapter 10). Another hypothetical potential problem posed in general by the use of Ca salts is, according to Alderton (1978), the resulting increased heat resistance of some microbial spores.

Dry Legume Seeds.—According to Morris and Seifert (1961), variations in cookability of dry beans were recognized, and explanations and solutions for these variations were proffered as early as the third century B.C. A PE firming effect is, in general, undesirable because of the resulting excess toughness and increasing cooking time. Apparently, phytase action takes on a more important role as a determinant of texture than in any other food so far considered. It acts in conjunction with PE in regulating and distributing available and added Ca (Mattson 1946; Morris and Seifert 1961). The activities of both PE and phytase can be thermally potentiated and the latter even when the beans are exposed to moisture-saturated atmosphere at 60°C (Chang *et al.* 1977).

Indeed, inadvertent storage of dry beans at high a_w (Chapter 13) above ambient temperature results in a pronounced toughening probably due to phytase potentiation. The softening of normal beans during cooking and the above-mentioned textural aberration can probably best be explained on the basis of the redistribution of Ca. In the uncooked bean most of the Ca is associated with pectin, whereas phytate is present in a water-soluble form (Chang *et al.* 1977). During cooking the Ca is driven from the pectin via interaction with phytic acid in the interior of the cell with which it forms an insoluble complex; in effect the Ca migrates from the cell wall into the cell. The cooled bean is thus free of Ca-pectin complexes which would tend to make the bean tough. In the absence of phytic acid, as would happen when the phytase action occurs, the resulting cooked beans are tough or at least require more cooking. This effect may be reinforced by PE action which would result in the formation of an even more rigid gel of Ca pectate throughout the bean.

While phytate or some other strong Ca-complexing agent is necessary to obtain easily cooked beans, its presence does not always ensure the desired texture because Ca can diffuse inward from the seed coat or from the cooking water and replace that taken up by the phytate (Crean and Haisman 1963). The interrelationships among Ca, enzymes, cell components and their localization may be depicted as follows:

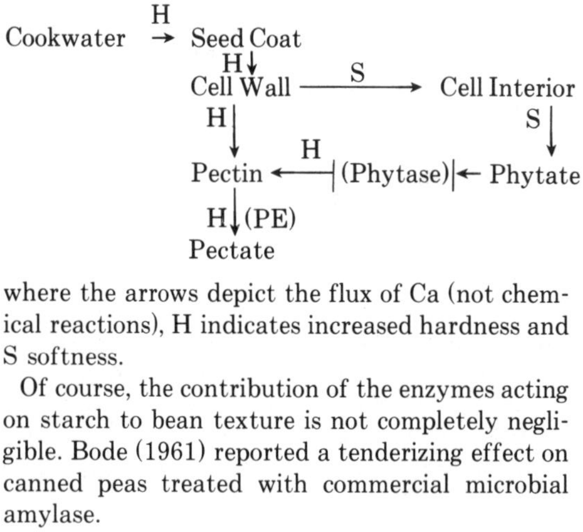

where the arrows depict the flux of Ca (not chemical reactions), H indicates increased hardness and S softness.

Of course, the contribution of the enzymes acting on starch to bean texture is not completely negligible. Bode (1961) reported a tenderizing effect on canned peas treated with commercial microbial amylase.

Cauliflower.—The PE of this flower vegetable can also be thermally potentiated to yield, in conjunction with Ca and LT-LT treatment, a superior canned product (Hoogzand and Doesburg 1961). Gradual enzymatically induced pectin transformation also appears to participate in texture changes attendant upon pickling of cauliflower. Saxton and Jewell (1969) pinpointed such action to the disruption of spheroromes, lysosome-like entities found within cauliflower cells. After fermentation the cauliflowers are normally "freshened" by rinsing in water and acidifying. The resulting increase in crispness, said to arise from such treatment, was shown by the investigators not to be a simple recovery in turgor pressure, but may have involved pectate and prior PE action since enhancement of elasticity and tissue cohesion occurred when sufficient Ca (2500 ppm) was added to the freshening solution.

Cucumbers.—In contrast to cauliflower, flower tissue in the cucumber plant can be a source of deleterious enzyme action with respect to retention of optimum texture of this pickled vegetable. Under most circumstances, the crispness of the fresh tissue is retained during the brine fermentation. Thus, as a rule the textural quality of such products can be reliably predicted by objective evaluation of the raw cucumber. Unlike the other vegetables we have considered, it would appear that endogenous PG activity is of minor importance in the softening of preserved cucumber, as concluded by Mustranta *et al.* (1976). Larger than average cucumbers, which yield pickles with impaired consistency, also exhibited impaired PE activity (high PG:PE ratios). As in carrots, peaches and tomatoes, the main pectic hydrolase in cucumber is an exo PG (tomatoes, of course, also contain an endo PG) which may explain why its action does not result in drastic changes in cucumber texture (Pressey and Avants 1975).

Oversoftening of cucumbers, which does occur sporadically in the pickling industry, was traced, as in the case of apricot and cherry softening, to enzymes containing fungi resident on the florets of the cucumber plant inadvertently introduced into the pickling brine (Etchells *et al.* 1958).

Although endo PG was singled out as the principal culprit in the softening, and probably rightfully so, other cell wall-degrading enzymes have also been implicated. Thus, as early as 1955, Etchells *et al.* and others detected floral cellulase activity. Later, Gremli and Neukom (1968) showed that a highly specific α-L-arabinosidase which acted on the arabinan linked to pectin but not on uronide linkages (and, hence, not a PG) could macerate cucumber tissue. Incidentally, this enzyme did not change the texture of potatoes, carrots or apples. In a similar vein, Bock *et al.* (1970), drawing conclusions from model experiments, implicated action of PE, cellulase, hemicellulase and even protease present in commercial microbial PG preparations in the enzymatic softening of pickled gherkins. Of course, these accompanying enzymes may simply facilitate the "macerating" action of the PG.

Recognition of the enzymatic origins of pickle softening led to a small literature on preventing the action of the offending enzymes, besides that of avoiding introduction of florets into brine. Most of these, introduced by Etchells, Bell and coworkers, involved treatment of the brine with forage tannins, nonspecific hydrolase inhibitors as mentioned in Chapter 14 (Etchells *et al.* 1958; Bell *et al.* 1965,1968). Cucumbers resistant to this type of textural deterioration appear to be so by virtue of similar endogenous tannins which inhibit PG (Mahadevan *et al.* 1965). However, Mustranta *et al.* did not consider PG inhibition to be important in texture. The practical development of such enzyme inhibitors may be hampered by the recent cautious approach concerning tannins as food additives. It is thus not surprising that more conventional means of preventing enzyme action such as thermal inactivation have been reexamined. Thus, Chavana and McFeeters (1977) conclude from a study of the thermal stability of the pectinases (most likely endo PGs) of eight fungi reported to commonly occur on cucumbers, fruits and flowers, that the currently recommended pasteurization treatment of fermentation brines is adequate for "pectinase" inactivation (78°C, 8 sec at pH 3.7).

Although, as mentioned, PE has not been considered to be important in cucumber softening, calcium salts have nevertheless been shown to exert a firming effect. The firming effect of sucrose and lactose (Jelen and Breene 1972) is probably due to their water-binding properties.

Miscellaneous.—Space and lack of information prevent an evaluation of the role of enzyme action in textural quality of the almost countless fruits and vegetables consumed throughout the world. Representative of studies on tropical fruits are the observations of Hussain and Shah (1975) that PG activity appeared at the initiation of ripening of the mango, increased rapidly during ripening and then decreased in the overripe fruit.

Avocado.—As an example of enzyme information or enzyme action in another type of "vegetable," the avocado, Zauberman and Nadel (1972) confirmed previous conclusions that PG was absent in the firm fruit and appeared only after about 1 week or more after harvest and reacted at a maximum at complete softening. The course of PE activity was exactly opposite that of PG; it decreased to a minimum, sometimes to an almost vanishing value coincidentally with fruit softening.

Olives.—Enzymes acting on *protopectin* are probably of prime importance in olive texture as available to the consumer since water-soluble pectin is leached out during the normal processing (curing and retorting) of this fruit vegetable (Chung *et al.* 1974). As a rule, oversoftness of olives harvested at the dark-black stage appears to be related to pectin enzyme action as reflected by total pectin content. They also traced a severe texture-affecting spoilage of California ripe olives during processing characterized by severe softening, skin rupture and flesh sloughing (as it was traced for brine cherries, cucumbers and apricots) to enzymes present in microorganisms. However, unlike the latter plant foods, the responsible enzymes appear not to be fungal pectin-degrading enzymes, but bacterial cellulases (Patel and Vaughn 1973). Interestingly, cellobiose, a potent cellulase inhibitor (Chapter 9), effectively suppresses microbial growth.

Asparagus and Enzymes of Lignin Biosynthesis.—Lignin, a textural determinant in pears, as mentioned, contributes to undesirable toughness in aging asparagus stems (but not heads) and in vegetables such as the swede (rutabaga). As pointed out by Joslyn and others, encrustation of hydrophobic lignin onto the surfaces of cellulose microfibrils and a similar physical association with other cell wall polysaccharides produces a network structure, the mechanical strength of which is dependent upon the degree of lignification.

The broad outline of how lignin is synthesized is fairly well understood. The ultimate precursors are sugars which, through the agency of enzymes of the aromatic (benzene) biosynthetic pathway (first worked out in detail for the amino acids tyrosine and phenylalanine) followed by the C_6:C_3 pathway (phenyl propane, cinnamic acid route to flavonoids) yields coniferyl alcohol and its congeners.

The only enzymes which appear so far to be associated with lignin biosynthesis from coniferyl alcohol via its oxidative polymerization are peroxidase and "laccase" (Chapter 15). According to Haard *et al.* (1974), the asparagus isoperoxidases pattern shifts coincident with increased lignin deposition although Freudenberg (1959) had shown earlier that laccase (Chapter 15) catalyzed the conversion of coniferyl alcohol into lignin-like polymers. Indeed, Sharma *et al.* (1975), citing Siegal (1962) and others, placed "peroxidase" synthesizing capability at the head of a list of the following enzymatically-related parameters involved in lignin biosynthesis: factors controlling rate of phenyl propane synthesis, polysaccharide framework of the cell wall and an oxidizing atmosphere. Rhodes *et al.* (1976) reported *no* change in peroxidase activity during the aging of swede roots, whereas the key enzyme at the entrance to the C_6:C_3 pathway, phenylalanine ammonia lyase (PAL), increased 30-fold. Similar increases for enzymes further down the pathway to coniferyl alcohol were observed. Undoubtedly, acquisition of such insight into lignin biosynthesis will be increasingly helpful in controlling unwanted lignin deposition. On the basis of a general mathematical model of competition for sugars between lignin biosynthesis and other pathways, Sharma *et al.* (1975) developed an equation designed to predict the degree of lignification during postharvest storage. How rapidly the biosynthetic enzymes can be mobilized to increase toughness is graphically illustrated in Fig. 29.6 where the upper half is a photograph of dry fibers from asparagus spears stored for 1 hr at 5°C and the bottom for 9 hr at 24.5°C. In the future, TSAI inhibitors (Chapter 14) of the enzymes unique to the lignin biosynthesis, which would have no effect on the consumers' enzymes, would be particularly effective.

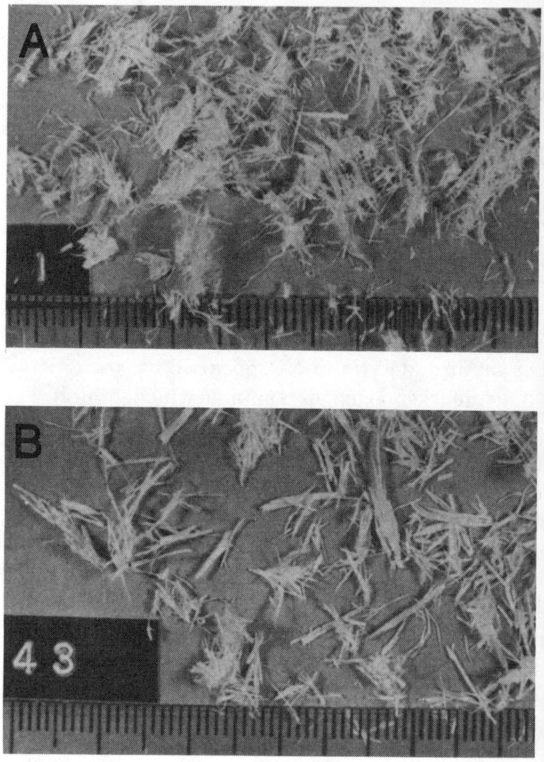

From Sharma et al. (1975)

FIG. 29.6. LIGNIN SYNTHESIS IN ASPARAGUS

Dry fiber from spears stored for (A) 1 hr at 45°C and (B) 9 hr at 24.5°C.

Sweet potatoes represent a case where thermal potentiation of an enzyme, α-amylase, in part of a processing procedure (Walter *et al.* 1976) is beneficial to texture whereas the potentiation of other enzyme action similar to PE prior to harvest or before processing leads to a textural disorder known as hardcore (Beuscher *et al.* 1976). This toughening of the inner part of the sweet potato, which can be induced by chilling at 2°C either before or after harvest and transferring after harvest to nonchilling temperatures before cooking, has been ascribed to either a potentiation of PE as described or a binding of newly synthesized phenolics to cell wall material, somewhat reminiscent of toughness due to lignin. In either case, increased permeability of the membrane is invoked.

It would be interesting to ascertain whether there is an enzymatic basis for the retention of the crisp texture after cooking of such delicacies of oriental cuisine as the water chestnut and bamboo shoots. In the case of the bamboo shoots the answer would appear to lie in the absence of hot water insoluble polysaccharides and, inferentially, protopectinase action. The bamboo shoot is not "tough" because of the absence of long laminated cellulose fibers. Abundance of cellulose microfibrils surrounded by arabinan and cemented together by xylan results in tissue low in tensile strength, and a tender and crisped cooked vegetable (Su 1969).

A discussion of the role of enzyme action in texture determination and modification should pay proper homage to the remarkable effects of controlled-atmosphere storage. Prominent among the fruit and vegetable qualities retained is that of texture. Typical shelf-life extensions are shown in Table 29.8. Apparently, the absence of normal oxygen tension and the increase in CO_2 levels delay the derepression of the genes coding for the synthesis of the polysaccharide-decomposing enzymes in most of the foods we have been discussing and for the enzymes of lignin biosynthesis in the case of asparagus (Sterling 1975). In the case of reduced atmosphere storage (subatmospheric, hypobaric)

TABLE 29.8

SHELF-LIFE OF FRESH PRODUCE

Produce	Air SL[3]	Controlled Atmosphere[1] CO_2	O_2	SL	Hypobaric[2,4] SL
Avocados	3–4 weeks	7%	1%	4–6 weeks	13–14 weeks
Beans, green	1–2	10	3	4–7	4–5
Lettuce	2–3	0	1–2	6–8	6–7
Apples	14–35	3–5	2–3	28–70	45
Bananas	1–2	5	1.5–2.5	3–8	13–21
Strawberries	<1	10	2	2–4	3–4

[1] Source: Brody (1970).
[2] Source: Burg (1975).
[3] Shelf-life in weeks.
[4] Practical development of the Dormavac Hypobaric System was achieved by Armour and Company and Grumman Corporation, for which they received the 1979 Industrial Achievement Award of the Institute of Food Technologists.

(Salunkhe and Wu 1975; Burg 1975) there is some controversy as to whether this absence of enzymes and improved texture is due principally to removal of ethylene. While such repression at the genetic level may be the principal enzyme-related effect, the possibility remains for effects of CO_2, if not on texture, then on other quality attributes such as color and flavor at the substrate level, both as enzyme affectors and as substrate (or product) itself. Pertinent enzymes in the latter regard which come to mind are those where mass action may play a role, such as the decarboxylases of the citrate cycle (Chapter 4) or the carboxylases involved in acid and sugar accumulation in fruits (Chapter 20). We suggest that CO_2 present as a replacement atmosphere in other packaged foods where enzymes are still active (i.e., meats) might affect these enzymes in such a way as to account for some of the beneficial effects of this gas as a packaging component.

Part VIII

ENZYME ACTION AND FOOD TRANSFORMATIONS: CHANGE-OF-STATE, APPEARANCE AND IDENTITY

30

ROLES OF ENZYME ACTION IN WINE AND FRUIT AND VEGETABLE JUICE MANUFACTURE AND APPEARANCE

So far our exploration of how enzymes influence texture has been limited to foods and processing/ postharvest/postmortem treatments as a result of which the general form, appearance and identity of the organism, tissue or other biological entity is retained in spite of enzyme action. In the next five chapters we shall explore the roles of enzyme action in initiating, guiding and facilitating the drastic changes in appearance and change-in-state. Thus, production of juices and alcoholic beverages (Chapter 31) entails a conversion of solid to liquid; that of cheese, liquid to solid (Chapter 33); and breadmaking (Chapter 32), complete transformation of identity of solid foods. Inevitably the subject matter spills over into other aspects of food quality and almost blends with or serves as a gateway to enzyme action in relation to health-associated attributes of foods, especially those related to protein quality and availability (Chapter 34).

This chapter will be devoted to gross transformations of plant foods to liquids in which nonendogenous enzymes play a major role. We shall deal here mainly with addition of enzymes modifying cell wall constituents in wine and juice prepared from fruits and vegetables, and also to a lesser extent with other aspects of other modified plant food, ranging from tea, coffee and cocoa to canned fruit, jams and intermediate moisture foods (see also Chapter 13).

For the most part, we shall discuss the following three roles of enzyme action: (1) removing cloud and clarification of clear fruit juices and wine musts, (2) the maintenance of the suspension in citrus juices (cloud retention) and (3) exploiting the macerating effect of cell wall modifying enzymes as aids in the manufacture of purees, vegetable juices and "unicellular" foods. In the first two employment modes, the intact plant is transformed to a liquid or semiliquid state by mechanical means and enzymes are involved as adjuncts in fining or maintenance operations; in the latter case, mostly in the "proposal" stage, enzyme action is conceived ultimately as the initiator of such drastic changes in state and appearance.

ENZYMES AS CLARIFICATION AND CLOUD RETENTION AIDS

Advantages/Disadvantages of Using "Pectinases"

Advantages.—One of the earliest triumphs of applied enzymology was the deliberate exploitation of PE and PG and, unbeknown to the earlier juice and wine producers, pectin/pectate lyases. Perhaps the principal circumstance that led to the ready adoption of enzymes in the respective industries is that resulting improvement in quality and stability of these food products goes hand in hand with substantial economic benefits including increased profits and lower consumer prices on the one hand, and ready adaptability to standard and advanced engineering plant procedures on the other. We deem it appropriate at this time to enumerate the more salient advantages before delving into the appropriate enzymological and manufacturing specifics. The use of enzymes for the manufacturing of wines and clear juices:

(1) facilitates the further disintegration of pulp prepared by mechanical means,
(2) increases yield of free run juice, wine musts and juice from pomace residues,
(3) increases rate of production of free run juice,
(4) decreases viscosity,
(5) facilitates and speeds up filtration of pulps and homogenates,
(6) decreases processing time at several steps in the manufacturing process,
(7) decreases energy requirements for heating, pressing, centrifuging, etc.,
(8) aids and is compatible with continuous processing,
(9) decreases labor requirements by replacement of hydraulic with continuous pressing,
(10) decreases processing temperatures, thus avoiding undesirable changes in flavor and color (i.e., caramelization) and in general stabilizes the juice,
(11) makes feasible manufacture of juice from raw materials (especially apples) otherwise unsuitable because of poor pressing characteristics, occasioned by variety or prolonged controlled atmosphere storage,
(12) obviates need for addition of gelatin (see below) which may give rise to haze development upon subsequent storage of an initially limpid product,
(13) improves color extraction in wine musts, and
(14) facilitates removal of tartrates by accelerating formation of argols, crystalline tartrate salts, in wine, brandy and grape juice manufacture.

In addition, enzymes may increase extractability and otherwise facilitate the manufacture of juices from dried fruits, although clarity is usually not the aim in manufacturing these juices.

This well-established enzyme technology, first suggested for apple juice and introduced in the United States by Kertesz (1930) and in Germany by Mehlitz (1930), has been the subject of frequent reviews. In addition to the book of Tressler and Joslyn (1971), reviews of two of the field's pioneers (Joslyn *et al.* 1952; Joslyn 1961B; Joslyn and Pilnik 1961; Amerine and Joslyn 1970) and Cruess (Cruess and Besone 1941; Cruess 1943; Cruess *et al.* 1955; Amerine *et al.* 1972), the following discussions, reviews and reports afford a comprehensive multifaceted overview with a special emphasis on wines: Amerine and Ough (1971); Blundstone *et al.* (1971); Fogarty and Ward (1972); Vos and Pilnik (1973). Reviews in non-English, of later vintage, include those of Montedoro (1976), Duerr and Schobringer (1976) and, as judged from the abstracts, a surge of Russian reviews and papers on the use of added enzymes in winemaking.

The degree of tissue and cell disruption attending the conversion of solid plant foods to the liquid state which proceeds without added liquid is a function of processing parameters, plant machinery and an engineering approach. By and large the gross morphology of the live tissue disappears. As we shall show in individual cases, cells, parts of cells and even cell walls may or may not survive. Only in the manufacture of juice from previously dry or semidry tissue (i.e., prune juice) is liquid added and the resulting juice obtained from dead tissue.

For the preparation of clear juices after comminution, large particles are first removed resulting in the cloud—a relatively stable, not easily removable colloidal suspension varying in appearance from translucent through opalescent, turbid, grainy, creamy to almost opaque.

Disadvantages.—The principal, potentially undesirable side effects of adding pectin-transforming enzyme preparations to fruit are the creation of certain color problems and accumulation of methyl alcohol (Chapter 35). Thus, one of the end products of PG action, the reducing sugar acid, galacturonic acid, may undergo nonenzymatic browning in white wines. An old problem of removal of desirable red color due to anthocyanase (Chapter 16) accompanying the pectin-transforming enzymes seems to have been circumvented. On the other hand, Montedoro and Bertuccioli (1976) observed decrease in color *tonality* (color nuances) which accompanied enzyme-induced increase in color intensity of red wines. PE action in these preparations gives rise to methanol (Chapter 35). Attempts to minimize this problem include alternative use of lyases (see below) and limiting PE action to the minimum required to allow PG action from the commercial enzymes to proceed efficiently (Dahodwala *et al.* 1974).

Nonenzymatic Cloud Removal.—The colloidal suspension can be removed by heat coagulation and freeze thawing—neither of which have been extensively used—and especially by the use of countercolloid or "fining" agents. In addition to gelatin, other fining agents include bentonite, polybases (polyethene amine), polyvinyl pyrrollidone (PVP) and, in general, polyelectrolytes which neutralize the electrostatic charge on the micron-sized particles comprising the cloud. In this connection PVP has been recommended as a supplement to enzyme treatment because it ties up fruit phenolics which, according to Vos and Pilnik (1973), would otherwise inhibit the added pectin-transforming enzymes. Alternatively one can potentiate the fruit's endogenous phenolase for the same purpose provided, of course, that the end products of phenolase action are less inhibitory than the phenolics. This is probably the rationale behind the interpolation of an "oxidizing" tank in the process.

Potentiation *vs* Addition.—As pointed out by Tressler and Joslyn (1971) spontaneous clarification of juices from apple, grape and especially lime (of the few clarified citrus products) can occur without added enzymes. The older literature suggests that this phenomenon may not be due to endogenous enzyme potentiation (activation) but to the enzymes of accompanying molds. Since this type of enzyme action is essentially unpredictable, the only advantage or use to which it might be put would be to harness its action in cooperation with that of added enzymes.

Involvement of Specific Enzymes in Fruit Juice Clarification

PG and PE.—The success attendant on the empirical use of commercial enzyme preparations containing pectolytic enzymes has not been conducive to a better understanding of just how and what enzymes are involved. In spite of fundamental investigations of apple juice clarification by Yamasaki *et al.* (1964) and Endo (1965) which strongly implicated PE and endo PG action, clarification procedures as pointed out by Dahodwala *et al.* (1974) are empirically optimized processes. It is thus unlikely to obtain a commercial preparation which is optimal for all fruits. In each case one has to change the enzyme or modify processing parameters. Neubeck (1975) stated that there was then-published literature on the effect of pectin enzymes on grapes but not on wine. Berk (1976) flatly asserted that the mechanism of clarification by pectin-degrading enzymes is not known and reiterated later that the mechanism remains to be elucidated. However, since these commercial enzyme preparations do possess, among others, intense pectin degrading activities, it would appear likely that the names given to these preparations (Table 30.1) do indeed reflect participation of pectin transforming enzymes. All measurements of related parameters such as decrease in viscosity and "pectin," and increase in galacturonic acid and methanol do not contradict such a supposition. Limited "model" experiments employing purified enzymes and carefully prepared "cloud" constituents have been carried out.

Thus, the lack of adequate PE to prepare the pectin for subsequent attack by PG can result in less than optimum conditions for clarification, a situation which can be overcome by adding tomato PE to the fungal PGs present in most commercial enzymes (Dahodwala *et al.* 1974). Endo (1965), who was able to effect clarification of apple juice using only purified fungal endo PG and PE, conceived of clarification as a three-step process: (1) solubilization of insoluble pectin, (2) decrease of viscosity of soluble pectin, and (3) flocculation of suspended particles. Only the first two steps are directly enzyme-dependent, the third involving a gelatin-like "refining" effect as in milk clotting (Chapter 33). The flocculation step occurs spontaneously, perhaps by a charge neutralization process involving protein, not too dissimilar from gelatin, which itself hastens the enzyme-induced clarification. However, unlike milk clotting, Endo found that Ca salts hinder rather than accelerate flocculation by inhibiting the first two stages. From a study of the behavior of model systems containing PE and PG and resuspended ultracentrifuged apple cloud precipitate, Yamasaki *et al.* (1964) proffered a mechanism in which the joint action of PE and PG strip the normally negatively charged cloud particle of its pectin coat (equivalent perhaps to Endo's insoluble pectin) exposing a core of pectin polymers intermingled with other glycans and protein. The molecules of the latter, being positively

TABLE 30.1

EXPERIMENTAL USES OF SOME COMMERCIAL PECTIN DEGRADING ENZYME PREPARATIONS[1]

Trade Names[2]	Producer/Supplier	Experimental Use/Study[3]	
Irgazymes	Ciba-Geigy, Basel	Citrus	—Increase fluidity, solids solids recovery in juice mfgr (A)
		Apples	—Maceration effect in nectar mfgr (B)
Klerzymes	Wallerstein, Illinois GB Fermentation, Iowa	Oranges	—Increase yield retention of cloud (C)
			—Also (B)
Pectavamorins	USSR	Grapes	—Immobilization for winemaking (D)
		Fruit	—In making fruit wines (E), wild rose juice (Q)
			—Gooseberry wine yields high MeOH (F)
Pectinexes	Novo, Copenhagen	Grapes (wine)	—Temperature, variety, skin contact time or yield, MeOH, flavor (G)
			—Also (A)
Pectinols	Rohm & Haas, Philadelphia	Grapes (wine)	—Early confirmation of benefits in winemaking noted; color loss due to "anthocyanase" (K)
		Prunes	—Juice production (L)
		Bananas	—Concentrated juice mfgr (M)
Pectin lyase	Kikkoman-Shoyu, Noda	Fruits	—Production of "Unicells" (H); mechanism of clarification (I)
		Apples	—Yields an unpressable applesauce (J)
Pectozymes	Fermco Biochemics, Ill.	Orange	—See (C)
Rapidases	Societe Rapidase, France; G.B.; Wallerstein	Bananas	—In juice extraction (N)
Rohapect, Rohament	Röhm, Darmstadt; Fermco	Apples	—See (B) vegetable hydrolysate (O); passion-fruit juice pulp (P)
Spark-L	Miles, Indiana	Bananas	—See (M)
Ultrazyme	Ciba-Geigy	Apples	—Poor quality fruit used for juice mfgr (J)
(100 Spezial)	Dr. Schubert AG Basel	Lemons	—To make clear juice

[1] This is what we believe to be a representive sampling and exclusion of other trade names does not imply judgment as to suitability for experimental or practical applications (i.e., Phylazim, a Hungarian (?) name).
[2] Most of these names are registered trademarks (®, ™) to contravene their use as generic names; in general the scientific publications mentioned here and elsewhere do not append these symbols to trade names.
[3] Sources: (A) Braddock *et al.* (1976); (B) Struebi *et al.* (1978); (C) Baker and Bruemmer (1972); (D) Pavlenko (1974); (E) Rokhlenko and Grebeshova (1977); (F) Rokhlenko (1972); (G) Ough and Crowell (1979); (H) Ishii and Takotsuka (1971); (I) Ishii and Yokotsuka (1973); (J) Vos and Pilnik (1973); (K) Cruess *et al.* (1955); (L) Bolin *et al.* (1975); (M) Tocchini and Lara (1977); (N) Munyanganizi and Coppens (1976); (O) Grampp (1969); (P) Lipitoa and Robertson (1977); (Q) Samsonova and Mel'yantseva (1978).

charged at the pH of the juice, are neutralized by the negatively charged soluble pectin and pectinate. Being colloids the oppositely charged entities will precipitate.

Immobilization.—Available literature on immobilizing pectic enzymes and on their use in juice making includes a thorough study of PE covalently coupled to porous glass particles (Weibel *et al.* 1975), a listing of "pectinase" as one of a wide variety of enzymes which can be immobilized on collagen (Bernath and Vieth 1974), immobilization of winemaking "enzymes" in a Russian commercial food enzyme preparation (Pavlenko *et al.* 1974); and linkage to a Sepharose-B matrix for the large-scale preparation of galacturonic acid oligomers (Van Houdenhoven *et al.* 1974). Young (1977) reported successful immobilization of PE, endo PG and pectin lyase present in a commercial "multipectic" enzyme system and the use of the immobilized system to clarify and reduce the viscosity of apple juice.

Pectin Lyase.—Commercial pectin-degrading enzymes, especially those manufactured from special strains of *Aspergillus*, have frequently been found to possess pectin lyase as well as hydrolase activity. In some model experiments in this field Iishi and Yokotsuka (1972,1973) showed that purified pectin lyase from *A. sojae* was effective in clarifying grape and apple juice. Comparison with effects of crude enzyme and with added endo PG led the investigators to suggest the presence of a special "clarification promoting factor" required for maximally efficient clarification. High pectin lyase-containing commercial enzyme preparations were recommended by Reymond and Bush (1972) for the clarification of highly acid fruit juices such as that from black currants. As expected, since pectin lyase does not require prior action of PE, very little methanol, as mentioned previously, is formed.

Unfortunately, the application of pectin lyase to apples for clarification purposes does not extend to improvement of pressability and filterability. On the contrary Vos and Pilnik (1973) reported that use of lyase yielded an unpressable applesauce-like product.

Cellulase and Hemicellulase.—From time to time participation of these enzymes in fruit juice clarification is suggested (Inoue *et al.* 1970; Montedero 1976). Earlier observation leading to such suggestions (Schubert 1952) may have been due to pectin lyase action. Hemicellulase action might also participate in the formation of soluble pectin involved as a component of the poised clarification systems in Endo's mechanisms and also in Iishi's promoting factor. As we have seen (Chapters 9 and 29), hemicellulases almost invariably accompany PG and other pectin transforming enzymes, especially in commercial preparations. An extensive Russian literature is also the source of a report of improvement of wine quality, but perhaps not clarification of grape juice by addition of "cytolytic" enzyme preparations rich in hemicellulases (Zinchenko *et al.* 1973).

Other Depolymerases and Postfiltration Haze.—Although proteinases have not been considered to be major participants in the events which lead to clarification of juices, scattered reports have so implicated it in, for instance, the preparation of green gage and cherry juices.

Proteolytic enzymes, however, have been recommended for the removal or prevention of afterfiltration haze due to formation of polyphenol-complexed to protein-containing particles in Sake wine (Tomoda 1964), cider (Bohuon and Drilleau 1976) and in an immobilized form in grape juice (Gaina *et al.* 1976). Ferns (1977) patented the use of "autodestroying protease" for similar purposes, analogous to "chill proofing" of beer. Similarly, addition of an amylase preparation has been suggested for prevention in apple juice of haze due to retrograded starch. Apparently this haze occurs frequently enough to prompt at least one commercial enzyme supplier, Miles, to provide an acid-stable α-amylase preparation derived from *A. niger* which is designed to withstand the low pH range encountered in the manufacture of juices from starchy fruits.

On the other hand α-amylase can also be used to reduce viscosity in fruit juices (in this case passion fruit), in which starch had gelatinized (Kwok *et al.* 1974). Gelatinized starch in cider may also be removed by α-amylase. Another special haze development, that due to slow deposition of hesperidin crystals in the syrup surrounding canned mandarin orange segments, is prevented in Japan by the use of the β-glycosidase hesperidinase (Okado and Ono 1969).

Haze formation is also induced in raspberry juice by pectin plus polyvalent cations and polyphenols (Latrasse *et al.* 1976). As mentioned (p. 537), potentiation of the fruits' own phenolase action, while hastening "enzyming" (Vos and Pilnik 1973), generates haze-producing polymerized "phenolics" according to Heatherbell (1976) who recommends both removal of phenolics and heat inactivation of phenolase (flash pasteurization) after clarification.

Soluble Instant Tea Haze and Tannase.—This diversion from fruit beverages is relevant here in that the haze in this beverage is also a problem and enzyme addition has been proposed for its solution. As mentioned, similar hazes appear in other beverages such as beer (Chapter 31). What such hazes all appear to have in common are oxidized polyphenolics or tannin components. As for beer, the haze, euphemistically referred to as "tea cream" in the industry, develops in instant tea infusions upon standing in the cold. It apparently originates from interaction of the heterogeneous oxidized gallotannins, theaflavins and thearubigins (Chapter 16) with caffeine, perhaps via hydrogen bonds to form coprecipitates (Rutter and Stainsby 1975). It thus differs from most other beverage hazes in that it apparently does not contain protein, whose function as a coprecipitant is taken on by caffeine. Tea haze can be avoided by addition of an enzyme, acyl tannin hydrolase, tannase. Apparently the phenolase-induced deesterification occurring during fermentation (Chapter 17) as usually carried out provides inadequate tannase activity for the prevention of tea cream formation.

Published information on the applications of microbial tannase preparations for the specific purpose of forestalling the development of this undesirable haze or cream seems to be confined to the patent literature (Tenco Brooke Bond Ltd. 1971; Takino 1976; Coggon and Sanderson 1973). In the latter patent "immobilized Tannase S"—for immobilized tannase see also Weetall and Detar (1974)—is added to a tea leaf homogenate in the absence of air, prior to the oxidative "fermentation" step. Tannase preparations for such purposes are probably produced from *Aspergillus oryzae* (other *Aspergillus* species are also rich in tannase activity) from which the enzyme has been isolated and well characterized (Iibuchi *et al.* 1972). Conditions for optimum tannase production do not coincide with those for maximum growth (Ganga *et al.* 1977). *Candida* yeasts also appear to be rich sources of tannase (Aoki *et al.* 1976) and should be suitable for food use. As a first approximation the action of tannase on instant creams may be written thus:

$$\text{Caffeine} + \text{Flavanyl Gallate}_{(ox)} \longrightarrow \text{Caffeine} \ldots\ldots \text{Tannin}_{ox} \text{ (Insoluble)}$$

$$\text{Tannin}_{ox} \xrightarrow{\text{Tannase}} \text{Gallic Acid}_{ox} \text{ (G)} + \text{Flavanol}_{ox} \text{ (F)}$$

$$\text{Caffeine} + \text{G (or F)} \longrightarrow \text{Caffeine} \ldots\ldots \text{G (or F) (Soluble)}$$

It should be stressed that solid information concerning the nature of tannase action in these heterogeneous complex mixtures is meager and the area of endeavor awaits further investigation.

Specific Applications

As with other enzymes used as food additives the number of organisms tested and found to be useful enzyme sources for juice clarification, mostly fungi, far exceed those which are actually used, partly due to legal considerations. In the United States only *Aspergillus niger* enzymes prepared under adequate sanitary safeguards are considered GRAS. Other countries allow various other *Aspergillus* species. Mikeladze *et al.* (1974) classified enzyme preparations used in the U.S.S.R. for wine and fruit juice production into three basic groups according to desired end results: (1) increase of juice yield and extractability during juice production, (2) increase in maceration and dispersibility without reduction of viscosity in manufacture of homogenized juices containing crushed pulp such as purees, some apple juices (see below), and (3) the present topic, increased yield and extractability and simultaneous clarification of the juice, such as wines and other clear fruit juices. Juices very low in pectin, or having adequate pectin enzyme activity of their own, can dispense with added enzyme according to Murch and Murch (1966). Apparently enzyme manufacturers do not generally produce special preparations for each of these types of applications for each commodity. However, they are usually packaged as liquids and frequently introduced in the processing line at two or more steps.

As illustrative of the actual use of such enzymes in winemaking is that adumbrated in *Food Processing* (Anon. 1971). One metric ton of destemmed grapes are heated at 60°C with 2.7 kg of liquid enzyme (specific activity not specified) pressed, filtered and additional enzyme (1.4 kg) added to the filtrate which is then concentrated, and refiltered and stored. Before further processing into wines, an additional 15 oz of liquid enzyme are added per 100 gal. juice which is filtered to remove tartrate after 6 weeks settling in the cold.

Experience with enzyme clarification of apples in Europe is discussed by Bielig *et al.* (1971) and in the previously mentioned paper by Vos and Pilnik (1973). The former obtained maximum yields at 45°C for 10 min when a centrifuge was used as a practical means of juice separation. The latter developed a pilot plant process for apples of poor pressing quality available in the Netherlands. Dutch apples release juice slowly at high pressures. This induces a blockage of capillaries through which juice from apples of good pressing quality escapes. The mashed apples are first mixed in an "oxidation" tank before adding one of a variety of commercial enzymes (4 g/100 kg) and feeding the oxidized enzyme-treated pulp into a continuous belt press.

Berries, due to a higher soluble pectin content than that of grapes and normal apples, yield (as do poor quality apples) pulps with little free run juice so that "enzyming" or other pectin-removing treatment is obligatory. Standard enzyme treatment of black currants consists of 2 hr at 55°C (Eid 1977). Even as mild a treatment as 0.5 hr at 20°C resulted in changed juice quality, including *increased* viscosity. As mentioned, pectin lyase has been recommended for black currant processing.

Enzyme preparations have also been employed to clarify juices prepared from tropical fruit such as guavas (Waldt and Mahoney 1967) and more recently from bananas in both South America (Tocchini and Lara 1977; Alroy *et al.* 1976) and in Africa (Munyangazi and Coppens 1976). The latter determined that best results were obtained at 50°C, pH 42 to 43 and 0.5% commercial enzymes. Banana juice can be drunk as is or fermented to an alcoholic beverage. Enzyme clarification works satisfactorily in the preparation of juice blends (Lopez and Johnson 1971).

The effectiveness of a "pectinase" preparation in increasing juice yields from Chenin Blanc and Muscat of Alexandria grapes is vividly illustrated in Fig. 30.1. It may be of interest to wine connoisseurs to mention that twice as many samples of wine from enzyme-treated Chenin Blanc grapes were judged (in paired sensory tests) to be of superior quality than were control wines, a result statistically significant at the 0.1% level.

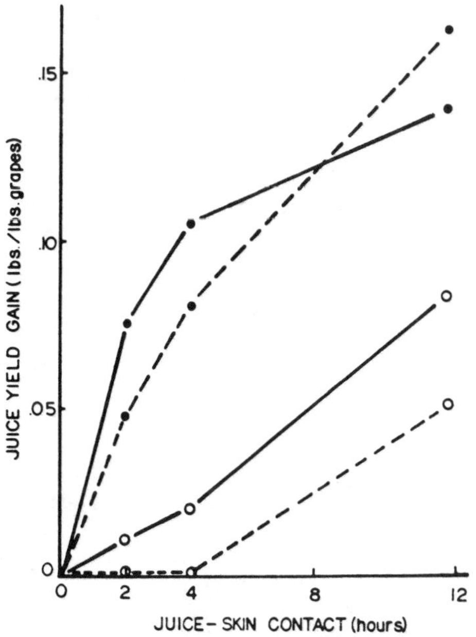

From Ough and Crowell (1979)

FIG. 30.1. WINEMAKING PECTINASE IMPROVES GRAPE JUICE YIELD

Open circles, controls, closed circles, enzyme-treated Chenin blanc, —, and Muscat of Alexandria,---, grape varieties.

Nonlimpid Juices

We now turn our attention to those beverages in which suspended matter in the form of haze, cloud, turbid suspension, nectars, purees, pastes and for jellies, gels and jams is the normal, desirable state of affairs and loss thereof via "collapse" or gelation is undesirable. Unfortunately in these poised colloidal systems even the relatively stable "clouds" are not indefinitely stable so that some effort has been directed towards stabilizing them by understanding the underlying factors contributing to this lability. The roles of enzyme action in this are as both causative and preventative agents.

Enzyme-associated Production of Prune Juice.—Addition of pectolytic enzymes to prunes is a key step in a process designed to streamline and modernize the manufacture of this time-honored beverage which may be looked upon as an intermediate between conventional purees and clear juices with regard to "appearance." Its manufacture is one of the few for fruit juices in which water is added as part of the process. The old diffusion process is eliminated by the use of commercial enzymes (0.045%) which, besides shortening process time, permits more efficient centrifugation and continuous processing. It also conserves energy and water and allows the re-introduction of vitamin A-rich fiber (Bolin *et al.* 1975). The process depicted in Fig. 30.2 and 30.3 yields a mild flavored beverage (Fig. 30.4) with diminished caramelized flavor.

Citrus Cloud Retention.—Most studies and comments on just what citrus cloud is chemically and what keeps the particles comprising the cloud suspended have been made in association with methods of stabilizing them. The now classic paper of Mizrahi and Berk (1970) stands out uniquely in this regard. The authors determined that Shamouti orange juice cloud consists of particles between 0.05 and 2 mm in diameter comprising negatively charged needle-like crystals of hesperidin (naringenin in grapefruit), chromoplastids (Chapter 25) and frequently stabilizing oil globules attached to amorphous rag particles. Contrary to previous ideas and other juices (see below), hydration of these particles rather than their electrical charge is responsible for stabilization. On the other hand, some investigators believe that, as in other fruit juices, most citrus juice clouds consist of protein and pectin (Baker and Bruemmer 1972; Lankveld 1973).

The participation of hesperidin crystals as a general constituent of orange cloud has been questioned (Baker and Breummer 1972), perhaps being peculiar to Shamouti oranges. Earlier Von Loesecke (1954) had reported that hesperidin crystals are present in the sludge settling out of canned orange juice from American varieties. Rothschild and Karsenty (1974) mention that freshly prepared orange juice is supersaturated with bioflavonoids. Indeed, as mentioned, the precipitation of such crystals is a problem in cans of orange segments. Baker and Bruemmer also minimize the importance of soluble pectin as a contributor to cloud stability. Indeed, in contrast to the role assigned to it in other fruit juices as previously

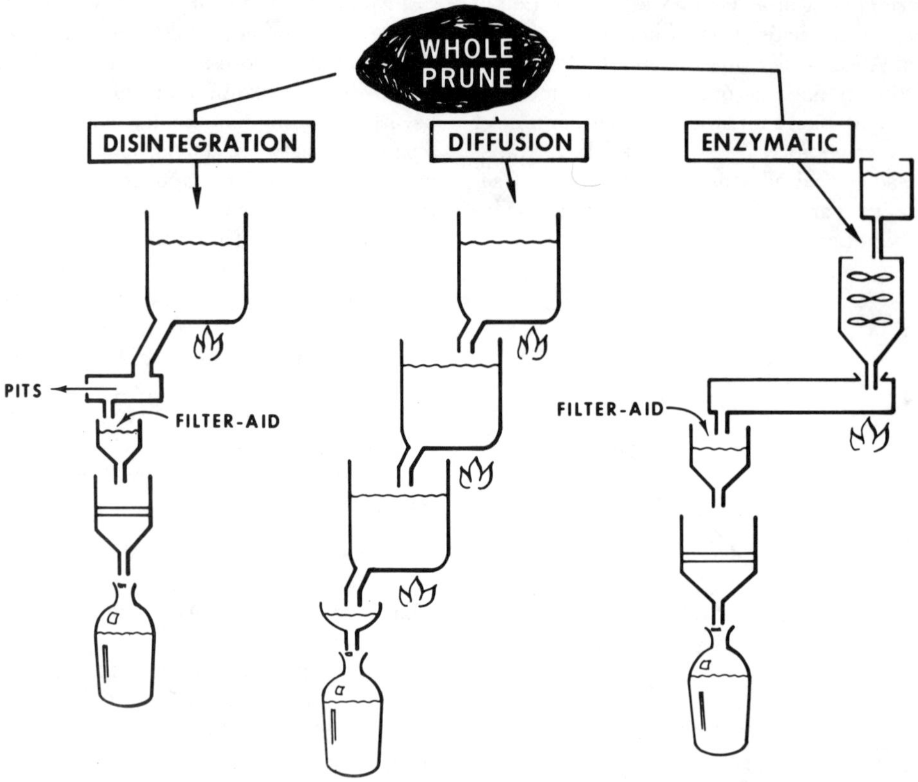

Courtesy of H. Bolin

FIG. 30.2. ALTERNATE ROUTES TO PRUNE JUICE PRODUCTION

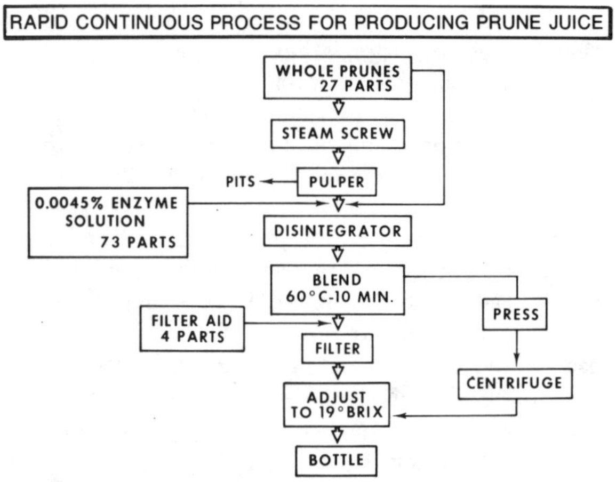

Courtesy of H. Bolin

FIG. 30.3. FLOWSHEET FOR PRUNE JUICE PRODUCTION USING ENZYMES

discussed, they present evidence which suggests that pectin may contribute to cloud collapse rather than to its stability by providing a handy PE substrate (see below). Indeed, adding pectin accelerates cloud loss, a phenomenon which has been used to evaluate cloud stability (Holland *et al.* 1976).

Cloud Collapse.—As mentioned, ideas about how the cloud particles remain suspended so long is largely arrived at by observing what makes them sink. For instance, Primo Yufere *et al.* (1962) in contrast to Baker and Bruemmer invoked a role, at least in orange juice concentrates, for *insoluble* pectin because they noted a parallelism between the action of PE on insoluble pectin and degree of clarification. Older theories (see below) stressed the loss of the protective effect of a hypothetical layer of pectin surrounding the suspended particle. Baker and Bruemmer stress the PE-induced formation of the gel of Ca and Mg pectate which drags

Courtesy of H. Bolin

FIG. 30.4. ENZYMATICALLY PRODUCED PRUNE JUICE

down occluded cloud particles. Strong evidence for this concept was provided by Kropp and Pilnik (1974). Still other suggestions invoke increased electrostatic attraction between protein and pectate as in prune juice (Lankveld 1973), and even the oxidative splitting of pectin via a hydroperoxide (O_2^-, O_2^* ?) formed during oxidation of ascorbic acid (Kiefer 1961).

Whatever the actual mechanism of subsequent events the *sine qua non* of flocculation is potentiation of PE action, known to be abundantly present in all citrus juices. Therefore, any explanation must account for the activation of a PE which as we have seen (Chapter 21) is normally quiescently attached to cell walls.

Prevention.—Since such cloudless citrus juices (except for that from lime) are usually unmarketable, substantial effort has been expended in preventing this unwanted clearing.

Some of these efforts in Europe especially for carbonated beverages (squash) have focused on the use of additives to keep the bioflavonoids in suspension (Zirlin 1974). In the United States the main attack on this problem was to heat the juice in a "pasteurization" step. However, it was known over 40 years ago that PE activity in orange juice is quite thermostable, surviving temperature up to 80°C and requiring more than 90°C for substantial periods for complete inactivation (Joslyn and Sedky 1940). Juices heated this much may lose much of their other desirable organoleptic properties. Even otherwise adequately pasteurized juices lose cloud attributed to PE which survives pasteurization upon prolonged storage (Rothschild and Karsenty 1974). In an attempt to maximize PE destruction with minimum quality loss, Eagerman and Rouse (1976) precisely determined PE inactivation parameters (Chapter 10) optimally compatible with pasteurization; for an $F_{(T)}$ of 1.0 min, z = 12.2 at 90°C for oranges and z = 9.3 at 85.5°C for grapefruit.

An alternative to heating was made available by Baker and Bruemmer (1972) who showed that judicious treatment with some but not all commercial pectolytic enzyme preparations were quite effective in preventing cloud loss (Fig. 30.5). This effect was attributed to the removal of pectate which could then no longer form gels. A temporary decrease in cloud intensity was attributed to increased release of PE from cell wall debris. However, this is ostensibly compensated for by the PG-catalyzed breakdown of the resulting pectate. Indeed they found that the most effective commercial enzyme preparations were those with high PG and "polymethylgalacturonase" activity (pectin lyase?). Effectiveness of PG (from yeast) for prevention of cloud collapse was confirmed by Krop and Pilnik (1974). They could not, however, obtain consistent improvement in cloud retention, as others had claimed to do, by treatment of citrus juices with preparations of proteolytic enzymes. Thus, this tended to eliminate theories of cloud loss

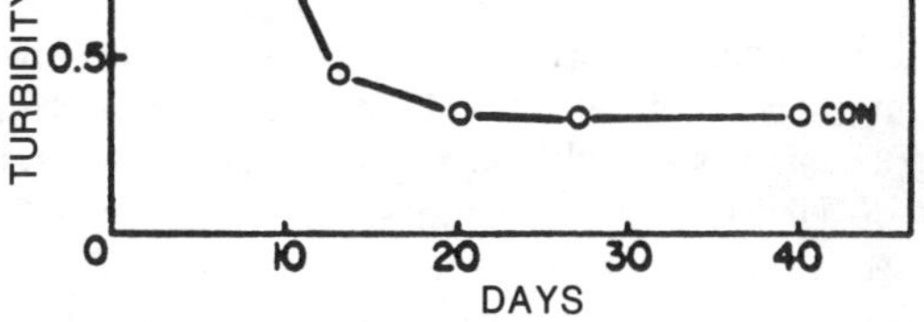

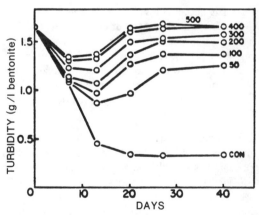

From Baker and Bruemmer (1972)

FIG. 30.5. ENZYMATICALLY INDUCED CLOUD RETENTION IN ORANGE JUICE

The effect of several concentrations of a commercial pectin degrading enzyme preparation on orange juice cloud stability.

invoking electrostatic interaction between protein and pectate to form a precipitate. The latter may still hold for other beverages such as apple juice as adumbrated above.

Another enzyme-assisted solution to the problem of cloud collapse is to replace the precipitated cloud with a stable colloidal suspension whose physical properties approximate those of the original cloud. One of the most widely used of such beverage bases and clouding agents is a by-product of citrus juice manufacture, finisher's pulp, whose yield and ease of recovery may be enhanced by inactivating PE and cautiously adding commercial pectolytic enzyme (Braddock and Kesterson 1976; Frusuma Europe 1975).

Gel Formation in Frozen Juice.—Phenomenologically related to cloud collapse is the undesirable gelling which may occur in unclarified concentrated citrus and other fruit juices. These concentrates form gels when reconstituted and the reconstituted juice will "break." Thus, orange juice concentrate remains indefinitely stable at −18°C, whereas at −15°C and −4°C the "break" occurs at 90 and 40 days. Again the cause of the "break" is probably enzymatic; viz, slow but unwanted action of PE even in the frozen state (Chapter 12).

To cause the formation of these gels, very little PE action is necessary because favorable pH, high sugar concentration and increased divalent cation concentration render even lightly demethylated pectin easily susceptible to gelation as in jams and jellies.

That PE action is involved is in accord with the observation that frequency of occurrence of gelling seems to be related to the amount of albedo fraction of peel, a rich source of PE (Chapter 29), which finds its way into the juice. It also occurs more frequently in seed-containing varieties where it is more difficult to clearly exclude the high PE-containing peel.

Other Fruit Juices.—Gel formation can be prevented by use of PG, not only in oranges but also in apples as was established 30 years ago by Walker *et al.* (1951). Loss of cloud stability of reconstituted lemon juice appears to originate from nonenzymatic causes, viz, an irreversible intermolecular bonding, perhaps through an entropy-increasing entanglement of pectin chains which occurs when the water is removed as a result of freezing (Chapter 12) below a critical pH-dependent a_w (Epstein and Mizrahi 1975). However, it should be pointed out that clarified lemon as well as lime juice is a useful food ingredient for some purposes and commercial pectinases are used in their manufacture.

Nectars.—Retaining the pulp in noncitrus fruits after mechanical disintegration and adding water, sugar and food acid results in fruit nectars. Such beverages, to be termed nectars, must contain at least 40% pulp. The manufacture of such nectars is described in detail in Tressler and Joslyn (1971) for the following fruits: apricots, cherries, guava, mango, papaya, passion fruit, peach, pear and plums. In contrast with citrus and other nonlimpid juices, rheological attributes related to flow are at least as important as cloud retention. For this reason some fruits such as the apple are unsuitable for nectar preparation unless enzyme-treated. Recently such treatment by some commercial pectolytic enzyme preparations, in a process developed by Struebi *et al.* (1978), appears to involve principally "protopectinase" action without concomitant extensive endo PG-induced depolymerization of the solubilized pectin or PE induced "saponification." Enzyme treatment transforms the pulp rheologically from a "mixed body type" fluid to a pseudoplastic fluid. This is in agreement with the concept that cell separation (but not cell wall disruption) occurred to impart the desired nectar-associated functional properties.

COMMINUTED TOMATO PRODUCTS

As for citrus juices and nectars, maintenance of cloud and prolonged suspension of solids with proper rheological properties and appearance (consistency) is a primary goal in the production of comminuted tomato products such as juices, soups, catsups, pastes, pulps and purees. Attainment and stable consistency are particularly imperative because in many of these products thickeners cannot be added. The principal factors contributing to consistency or its corollary, viscosity, of the serum of these products are: concentration, integrity or chain length of pectin molecules, the amount and nature of microcrystalline cellulose, and perhaps, as alluded to in Chapter 29, the development of callose, the β-1,3-glucan which develops upon bruising. As with other comminuted products, enzyme action and means of preventing such action profoundly influences manufacturing practices as well as final consistency. Earlier information and theories were reviewed by Joslyn (1961B) and some of the more practical aspects dealt with more recently by Gould (1974).

Impressive and supportive but not absolutely rigorous evidence has been amassed indicating that action of PGs prior to harvest (Chapter 29) is one of the factors contributing to final tomato juice consistency (Hobson 1963,1964; Hobson and Davies 1971). That this action resulting in a softer whole ripe tomato (Chapter 29) is also reflected in the consistency of the processed products is attested to by the observation that overripe, oversoft tomatoes yield lower consistency juices than do less ripe fruits (Luh *et al.* 1960). Although cellulase is present, according to Hobson and others, it does not appear to play a major role either in softening of the whole fruit or in juice consistency.

Another preprocessing parameter which may influence juice consistency appears to be the degree of bruising experienced by the tomato. Without referring to callose, York *et al.* (1967) observed that damaged tomatoes yielded juices with increased consistency and solids content. As alluded to in Chapter 29, both bruising and heat treatment of tomato fruits (50°C for 1 hr) stimulate the biosynthesis of the β-1,3-glucoside-linked callose which Dekazos considered might be the major polysaccharide contributing to the consistency of comminuted tomato products. However, this interesting suggestion has not been followed up with chemical analysis nor so far corroborated. If callose is indeed formed as evidenced by histological staining, it apparently is synthesized faster than it is hydrolyzed by the β-1,3-glucosidase present in tomatoes (Wallner and Walker 1975).

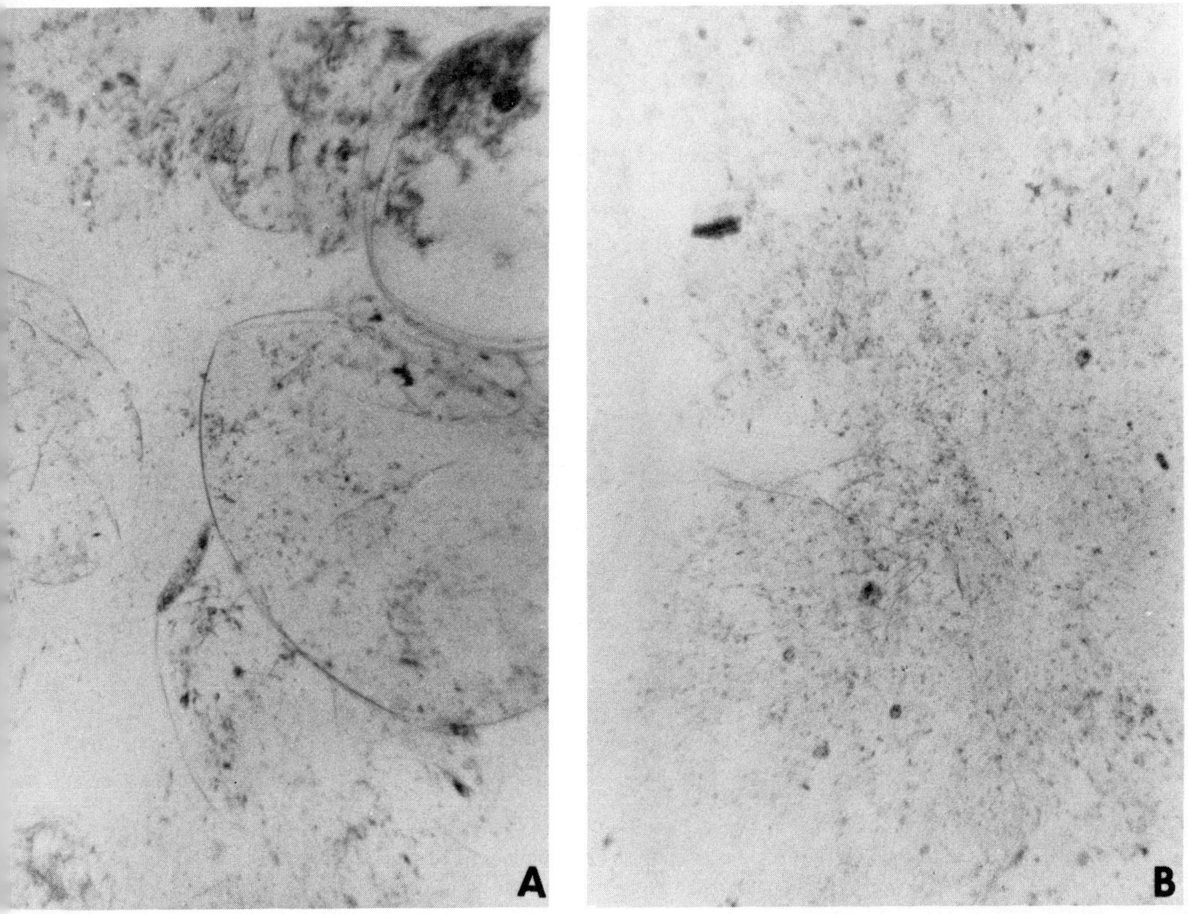

Courtesy of Becker et al. (1972)

FIG. 30.6. ACID TREATMENT OF TOMATOES DESTROYS CELLS

Photomicrographs of (A) intact cells present in commercial tomato juice and (B) cells from acid-treated tomatoes (both 81×).

While processing of tomatoes destroys the geometrical shape of the fruit, except canned whole tomatoes, the cells remain more or less intact as shown in Fig. 30.6, an electron micrograph of cells present in commercial tomato juice. The role of the cell and cell wall components such as the pectins and cellulose as determinants of juice consistency has been recognized for at least 25 years (Luh *et al.* 1954; Whittenberg and Nutting 1958; Joslyn 1961B; Twigg 1959; Becker *et al.* 1972).

Preventing Enzyme Action

There are three processing approaches which are the equivalents of blanching (Chapter 10) or proposed in tomato canning to prevent undesirable enzyme action: (1) the cold break, (2) the hot break and (3) the acid-hot break. The hot break is probably the most widely adapted approach, whereas the acid-hot break has been demonstrated as economically feasible but has not as yet been widely accepted.

The Cold Break.—In the Cold Break the tomatoes are first scalded to loosen the skins prior to their removal in order to prevent flesh from adhering to them. Parenthetically, enzyme potentiation via a fast-freezing of the surface of the tomato (Cagnoni 1957) has been suggested as means of facilitating skin removal. In this case the enzyme potentiated is PG since the lack of clean separation is probably due to adhesive properties of undegraded pectin. This method has been used extensively in Italy.

In the Cold Break the temperature at which the fruits are comminuted never exceeds 60°C. At this temperature the cell wall-degrading enzymes are not inactivated and their action allows the liberation into the serum of substantial amounts of cellulose fibrils adhering loosely to the still intact cell walls and as a result of the mechanical agitation attending the comminution process. When sloughed off from the cell wall these microscopic cellulose fibers become entangled with each other in a manner which contributes to an increase in consistency of the resulting tomato juice. Thus, in the Cold Break the contribution of cellulose overrides that of pectin, which tends to be broken down by the still active pectolytic enzymes. The main advantages of the Cold Break with regard to nonconsistency are improved retention of color and ascorbic acid. It is especially effective and useful for tomatoes which are unevenly colored and slightly underripe where the full complement of cell wall-degrading enzymes has not developed (Chapter 29). Another possible advantage is the expenditure of less energy than in the hot break process, although this may be negligible in comparison with the total energy extended in subsequent retorting. Another advantage might be avoidance or at least diminution of the problem of fouling of processing lines, especially heating tubes. At least this is the inference one might draw from the demonstration by Adams *et al.* (1955) that pectin-degrading enzymes can be added to prevent such a buildup of fouling deposits (Morgan 1967).

In any event it is important in the Cold Break process to pulp as rapidly as possible in order to prevent any further degradation of pectic substance which, while not the prime factor it is in the Hot Break, does contribute to consistency of Cold Break juice. The molecular weight of the pectin in the raw tomato should be high and the PE and PG activities low. Even with the best of conditions, Cold Break juices lack the dispersing effect of the cementing polymers (Luh and Daoud 1971).

The Hot Break.—In the Hot Break, it is imperative that the maximum temperature reached above 82°–83°C be achieved in minimum time (on a commercial scale, 15 min appears to be appropriate). This is because, since one of the advantages and purposes of the Hot Break is to inactivate the pectolytic enzymes, this effect would tend to be vitiated by the acceleration of the action of these enzymes during a prolonged warmup period. The linchpin to stoppage of enzyme action is the inactivation of PE since as explained in Chapter 29 both the endo and exo PGs in tomato require prior PE action to be effective. The necessary rapid raising of the temperature can be accomplished by direct steam injection with a rotary coil heat exchanger (Tressler and Joslyn 1971) but many ingenious designs have been devised to further minimize pectic enzyme action. Thus, Prosser (1974) first heated whole tomatoes at a minimum temperature of 88°C, then chopped them so as to allow the juice to flow into a specially designed vat mounted on a rotary heating element which conducted the tomatoes through a chute. A rotary cutter positioned immediately adjacent to the bottom open end of the chute disintegrated the rest of the pieces while they were being heated before cutting by the heated juice circulating within the vat. This points up one of the difficulties in attempting to understand the underlying enzymology of the Hot Break: the difficulty of scale-up or rather scale-down of heat exchanging and comminution apparatus to a laboratory scale where controlled experiments can be applied to large-scale experience. From the investigations of several experimenters, but especially from contributions of Luh and colleagues at the University of California at Davis, whose work we have previously cited, we may conclude that the role of pectin in tomato consistency in tomato products prepared from Hot Break juice is dictated, in the first instance, largely by the amount and integrity of the soluble pectin. These two variables are in turn dependent upon the efficiency of preventing the action of the endogenous PEs and PGs attendant on the mechanical disruption of the tissue.

The above conclusion, however, does not rule out the participation of nonsoluble pectin substances such as cellulose in achieving proper consistency especially of concentrated products such as tomato pastes. Thus, Smit and Norrtje (1958) showed that, as in the Cold Break, hemicellulase action takes place during a break-heat delay and confers increased consistency by permitting cellulose fibrils to slough off cell walls. As shown in Table 30.2, consistency of the paste, but neither the viscosity of the serum nor the pectate, increases with delay, with decreasing mesh size and with addition of hemicellulase enzyme preparation.

TABLE 30.2

TOMATO PASTE PARAMETERS INFLUENCED BY ENZYME-AFFECTING TREATMENTS

	% of No-delay Value[1]		
Measurement	½ Mesh[2]	No Enzyme	Enzyme[3]
Consistency of paste	148	127	230
Viscosity of serum	87	82	70
Volume of clear juice	140	220	340
Methoxyl content	108	97	—
Calcium pectate	86	51	—

Source: Adapted from data of Smit and Norrtje (1958).
[1] Three hour delay between break and preheat treatment of tomatoes.
[2] Comparison is with no-delay sample prepared with full mesh size screen.
[3] A hemicellulase-containing Pectinol added.

The Acid-Hot Break.—In an attempt to utilize the relatively neglected route of preventing enzyme action, acid inactivation, Wagner, Schultz and coworkers developed the Acid-Hot Break (Schultz *et al.* 1971). The pronounced increase in consistency in large scale field-processing trials is shown in Fig. 30.7. Although earlier laboratory-scale experiments led to the conclusion that increased consistency was due to more efficient enzyme inactivation, Becker *et al.* (1972) showed

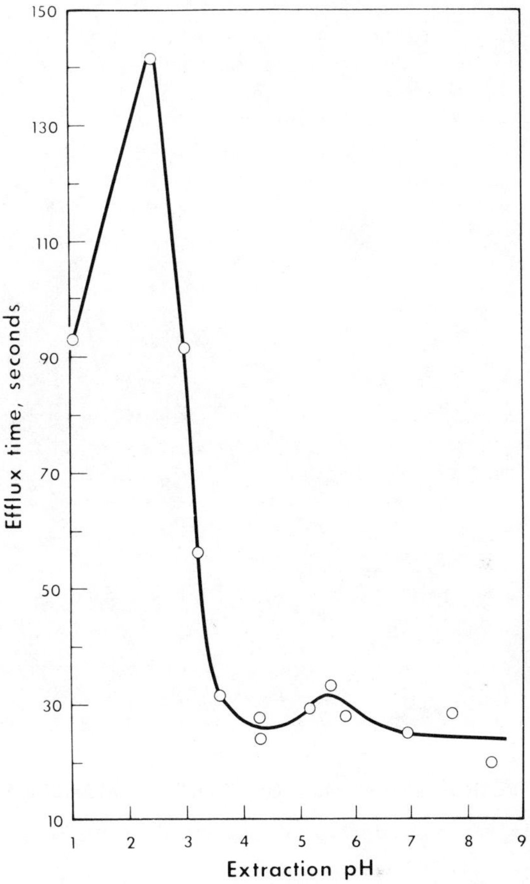

Courtesy of Wagner et al. (1968)

FIG. 30.7. ACID TREATMENT OF TOMATOES INCREASES JUICE CONSISTENCY

that treating Hot Break juices with acid solubilizes protopectin from the cell walls. The cell wall, rid of its intercellular cement, disintegrates upon being subjected to the extremely high shear forces of the distintegrators into a brush heap of cellulose fibrils (Fig. 30.8). These strands become entangled with strands from neighboring brush heaps and juice consistency increases severalfold.

The viscous serum helps keep the brush heaps distributed throughout the solution. The serum soluble viscosity component liberated by the acid treatment must tend to remain entrapped in the brush heaps until blending occurs. It then escapes to increase the serum viscosity.

Individually, neither the serum soluble polymers nor the cell wall fibers can cause gross consistency changes. In the presence of intact cells, substantial increases in the serum viscosity result in minor

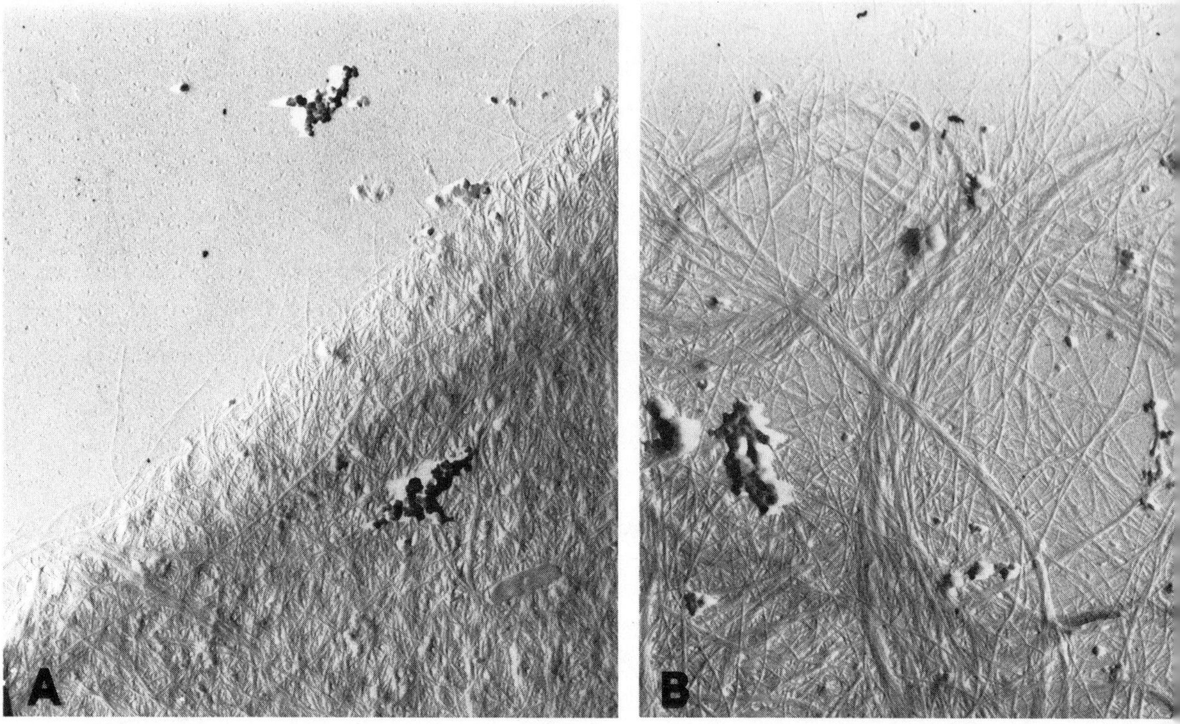

Courtesy of Becker et al. (1972)

FIG. 30.8. BRUSH-HEAP CELLULOSE MICROFIBRILS FROM TOMATOES

(A) Is photomicrograph of commercial tomato juice. (B) Subjected to extreme acid and mechanical treatment.

consistency increases. Increased amounts of dispersed cell wall fibers also result in only minor consistency increases unless a high viscosity serum is present. When a juice contains both large amounts of dispersed cell wall fibers and a high viscosity serum, the consistency is synergistically elevated.

Although pectolytic activity during the conversion of the tomatoes to juice and derived products is considered undesirable, both PG and PE action have also been exploited as analytical reagents and for quality control purposes such as in the preparation of the food for inspection for nonfood fragments (Chapter 37).

CELLULASE AND CELL-SEPARATING ENZYMES

We briefly touched upon the use of cell-separating enzymes in our discussion of hemicellulase and protopectinase in the previous chapter and also in our discussion of the effect of the use of commercial pectolytic enzymes in nectar production earlier in this chapter.

Vegetable Juices

Increased demand for "health" foods has directed the attention of several investigators outside of the United States to the use of such enzymes for the preparation of nonlimpid vegetable juice, especially raw carrot juice (Pilnik *et al.* 1975; Heinen and Van Twisk 1975; Dongowski and Bock 1977; Zetelaki-Horvath and Gatai 1977). Other juices produced by "total" enzymatic nonmechanical liquefaction or disintegration of plant tissue in these investigations include apples, apricots, cucumbers, oranges, peaches and pears. Several investigators consider these juices a new form of "unicellular" fruits and especially hitherto nonmechanically unliquefiable, hard vegetables.

While the idea of unicellular vegetables is perhaps the most novel approach, the practical approach may be to use tomato pulp as the source of cell separating enzymes. The general approach is to *not* disintegrate the tissue to the point of complete cell disruption but to the point where, ideally, separated cells are obtained. Since a large portion of the cells would be nevertheless disrupted at this juncture when the surviving cells are separated, the principle is to mechanically disrupt the tissue far short of this point and then to add enzymes to complete the cell separation task.

Other Food Applications of Cell Wall Degrading Enzymes

The varied texture-modifying effects have been given a variety of descriptive but not particularly enzymologically enlightening names: pectolytic enzymes, wall-separating enzymes, macerating enzymes, macerase, macerozyme, cytolytic enzymes, cytase, protopectinase, hemicellulase and, of course, cellulase (Chapters 9 and 29).

In the following we have sorted out applications not hitherto covered according to the predominant but probably, in most cases, not exclusive mode of enzyme action: pectolytic, cellulolytic and hemicellulolytic.

Pectolytic.—Both PE and the various PG are probably the principal enzymes whose actions are responsible for the following:

An aid in removal of coffee bean seed coat by exo PG secreted by *Erwinia* microorganisms during fermentation. This enzyme eliminates the mucilage layer between the berry and coat and speeds drying (Van Pee and Castelein 1972; Castelein and Pilnik 1976).

By utilizing the microbial pectolytic enzymes present in the "lights" from overfermented berries Butty (1973) reduced the time for mucilage degradation to 12 hr instead of the days normally required. Bock *et al.* (1973) advocated the addition of pectolytic systems of tomato and avocado to shorten the cooking time.

As far back as 1962 Blakemore observed that use of pectinase preparations for puree production softened fruit skins, that no cooking was required and that use of the enzyme prevented oil loss. On the other hand product recovery is enhanced when pectin-degrading enzymes are used in the manufacture of cold-pressed citrus oils (Platt and Poston 1962).

Other uses of pectolytic enzymes resulting in texture-associated changes include the preparation of high bulk density tea powders (Sanderson *et al.* 1974) and a nonfood use for a food waste product, fabrication of paper from pineapple fiber (Fujishige *et al.* 1977) and facilitation of fruit and vegetable peeling (Guigou 1974).

Two instances of unwanted pectolytic enzyme action in the present context are in the production of candied fruit due to adventitious microbial enzymes (Hildrum and Tjaberg 1972) and in the production of pectin from lemons due to endogenous enzyme action (Braddock *et al.* 1976). Starch, which can be a problem in pectin production from some fruits, can be readily removed by adding commercial amylase preparations (Lyutkanov *et al.* 1974).

Cellulolytic.—A review by Malmos (1978), available only as an abstract at the time of this writing, lists the manufacture of olive oil, soybean products, and cornstarch and vegetable and fruit products as instances where cellulase is added for texture-modifying action.

Toyama (1969) also mentions starch isolation as one of the promising places where cellulase can be used to advantage by decreasing mechanical damage to the starch granule. He also suggests that such enzyme action can be used in the preparation of instant tea, konbu jelly from seaweed, removal of soybean husk thus obviating the need for severe agitation and mechanical damage, and in dehydrated food which reconstitutes more readily and with improved morphological fidelity.

Patents by Kinki Yakurutu Company, Japan (1966) use cellulase preparations to facilitate extraction of the emulsion stabilizer alginic acid from algae, as a spray in water solution on mushrooms to prevent microbial contamination and to minimize sieving and grating vegetables which can be used without cooking.

Similar claims for the production of vegetable purees by the use of cellulase were made by Ghose and Pathak (1973) and Toyama (1969). Ghose and Pathak (1973) also used cellulase of *A. niger* Koji as an aid in removal of peel from oranges.

From the studies over the last decade by Italian investigators (Montedoro *et al.* 1975; Fantozzi *et al.* 1977; Leone *et al.* 1977) it is clear that cellulase action can expedite the production of olive oil and increase yield by improving drainage and freeing oil droplets from polyphenolic complexes.

Other suggestions for the use of cellulase probably, along with that of other cell-wall substances hydrolyzing enzymes, include: treatment of rice to facilitate grinding of starch granules in sake production (Komatsu 1965); production of mushroom seasoning (Oshikawa 1964); reducing the consistency of garlic purees (Li *et al.* 1967); and production of a sugar syrup from unpeeled pineapples.

Hemicellulolytic.—Uses of hemicellulases were reviewed in Japanese by Okasaki and Kubo (1977). Examples of their uses (frequently along with cellulase and pectin-transforming enzymes) from our files include: the speeding up of candied fruit production (Hori and Fugono 1969), utilization of spent coffee grounds using mannanase from *Rhizopus-Aspergillus* koji cultures (Hashimoto 1970), stabilizing liquid coffee against spoilage by presumably removing pentosan which would otherwise form a protective layer around the bacterium (Mohr 1972), production of edible aromatic extracts from rhubarb rhizomes (Tateo 1977), instant rice manufacture (Luh *et al.* 1980) and improvement of catsup and similar sauces by creating an enhanced creaminess (Fukumoto 1965).

Of course, one problem which the future use of such enzymes might create is the diminution of dietary fiber (Chapters 36 and 39).

Some applications, such as the removal of seed coat from coffee beans, may involve the combined action of enzymes from microbial flora, from the bean itself and from added enzymes.

When commercial cellulases are used, the actual cellulase action may be preceded by the action of other "macerating" enzymes. It would be of interest to ascertain in these applications how far the cellulose chain has to be degraded before its function as cell wall constituent is impaired and at what point this function completely disappears and gives rise to the observed effects.

It is generally agreed that a cell wall separating action involves at least a PG component. What is not clear, with the possible exception of cucumber tissue as our discussion of protopectinases indicated, is whether hemicellulase action is obligatory. Some studies indicate that rather highly specific hemicellulases, apparently completely free of enzymes capable of splitting galacturonide bonds in pectin, could nevertheless macerate cucumber tissue, but not that of potato, carrots or apples.

There are few careful studies available which are designed to answer some of these questions. It would appear that there are no hard and fast principles of cell separation. The type of cell and to what plant the cell wall belongs are important modifying parameters as to which enzymes are effective. Toyama (1969), who equated cell separating enzymes of *Rhizopus* with "protopectinase" (in his case largely endo PG), cited literature which he believed indicated a strong correlation between cells separating activity and endo PG. He suggested that random splitting by these endo PGs may release water soluble fragments. This presupposes the prior action of PE, unless a pectin lyase is present.

On the other hand, Byrde and Fielding (1962) reported that they separated a macerating factor from endo PG in a culture filtrate of *Sclerotinia* fungus. The powerful cell separating activity of pectin lyase was confirmed by Ishii and Yokotsuka (1971) who ensured the absence of PG, cellulase, protease and several hemicellulases in their lyase preparations.

The high degree of esterification of the pectin solubilized in apple nectar production, discussed below, may imply participation of pectin lyase present in the particular commercial preparation used. Figure 30.9 shows free-floating "unicells" completely lacking the characteristic described as being "cuddled up to each other with intercellular cementing materials." Pectin lyase was used to prepare such foods from cabbage, carrot, onion, potato, radish and spinach from 1 g pieces without prior mechanical treatment of the tissues. Complete maceration took place at 40°C in 10 vol of water in from 2 hr for onion to 8 hr for incomplete carrot tissue maceration.

For the future, the most significant application of these cell wall-associated enzymes may very well be production of "naked" protoplasts which can then potentially be used in a variety of ways: in genetic recombinant engineering, production of interspecific plant hybrids, intercellular transfer of subcellular organelles such as chloroplasts, isolation of useful mutants and uptake and genetic expression of useful viruses such as those which might carry the genes for the enzymes of nitrogen fixation (Carlson 1973).

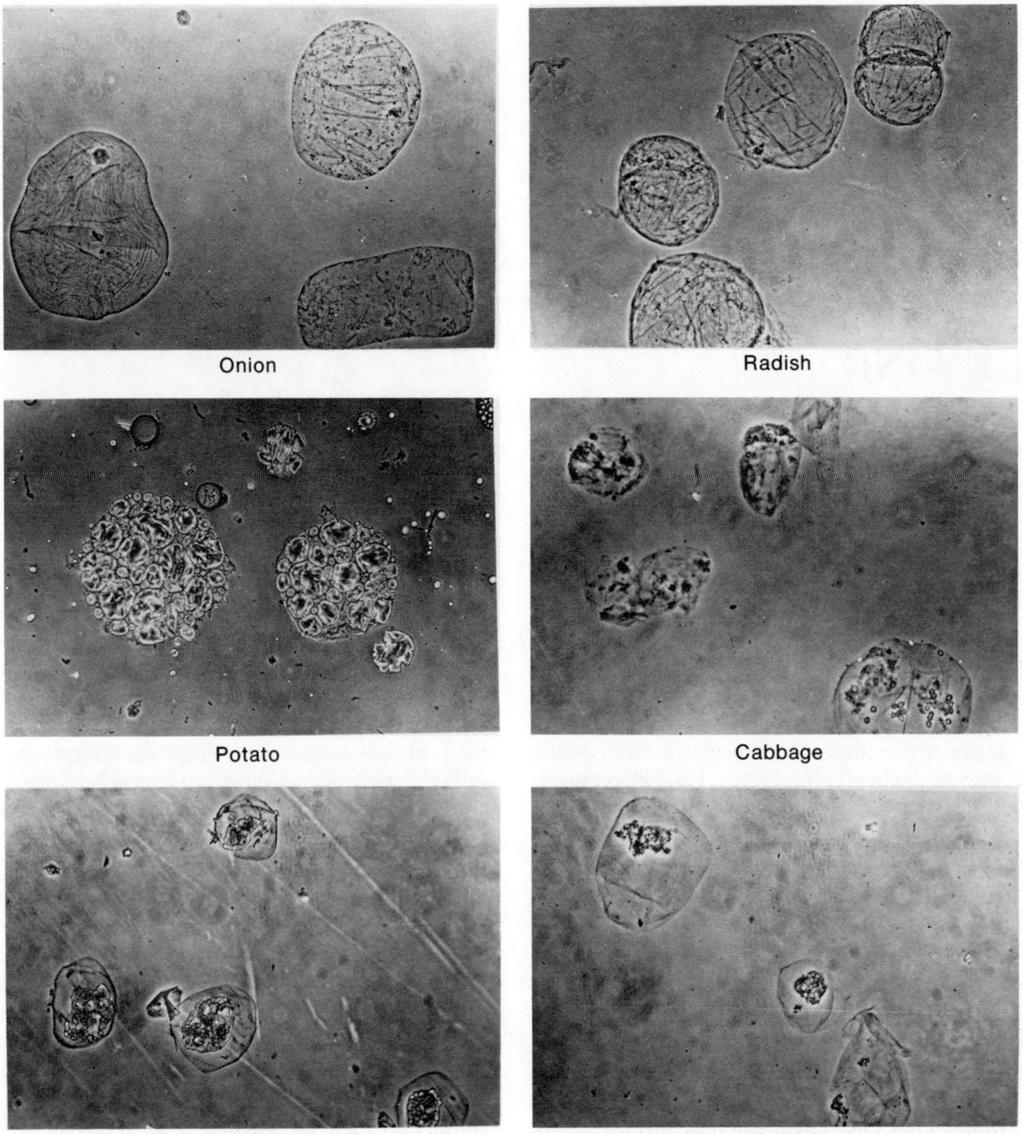

Onion

Radish

Potato

Cabbage

Spinach

Carrot

Courtesy of Ishii and Yokutsuku (1971)

FIG. 30.9. UNICELLS FROM FOOD PLANT TISSUE

Liberated by the action of purified pectate lyases from *Aspergillus sojae*.

31

ENZYME ACTION IN MALTING AND BREWING

We continue our exploration of enzymes as participants in and initiators and facilitators of modifications in food making where a change-in-state and appearance of the raw agricultural commodity occurs by turning attention to another class of beverages, those derived principally from grain, focusing in particular on beer. This beverage is usually considered as one of five categories of alcoholic beverages—the other four comprise nonfruit derived beverages, i.e., distilled spirits plus table and dessert wines and liqueurs (cordials).

Beer probably originated more recently than fruit-derived beverages. Wines probably arose in the pre-agricultural era as a consequence of berry gathering, whereas beer presumably had to await the rise of agriculture and cultivation of grain (Claudian 1970). Yet, beer can still be considered an "ancient" beverage since wine and beer were said to have been used as beverages and medicines in Egypt more than 6000 years ago (Leake and Silverman 1971; Gastineau *et al.* 1979).

From the data of Flynn (1975) we estimate that enough of this beverage is now being consumed to provide a short beer (*ca* 80 ml) every day for each inhabitant on this planet. In England about 100 years ago more than 30% of the per capita intake of calories was supplied by beer.

From the viewpoint of enzyme sources which participate in beer-making, history has seen a shift from what could be considered as a combination of the foods' own enzymes with those enzyme systems present in yeast which convert fermentable sugars to alcohol. This is different from sources of enzymes in starch-derived alcoholic beverages such as sake in whose manufacture both sugar- and alcohol-producing enzymes are supplied by living microorganisms, the former in fungi, the latter in yeast. In any case, except where it impinges on the general theme of nonliving enzyme action, we shall not discuss yeast fermentation or yeast-associated enzymes.

From the viewpoint of operations which lead to this change-of-state, beer differs from the beverages discussed in Chapter 30 in that water (rather than tissue disruption) added to a dry agricultural milled product is the initiator of endogenous enzyme action, if one considers malt to be the source of endogenous enzymes.

The progressively increasing replacement of 100% malt with more and more unsprouted grain (adjuncts), such as corn, sorghum and more recently, barley (see below), has given rise to the question as to whether malt should now be considered a source of exogenous enzyme acting on endogenously supplied starch in the adjunct. This dilemma has been compounded by the replacement and supplementation of the action of malt enzymes by a series of specific tailor-made microbial enzymes as exogenous food additives whose actions may be more in harmony with brewing parameters. Furthermore, they can be added and controlled more flexibly and can be used to diversify and make available a wide variety of modified products such as malt liquor and dietetic or "light" beers, as discussed below. As pointed out by Enari (1974) brewing has become a modern processing industry utilizing the achievements of biochemical and chemical technology. Indeed this conversion, lagging some 20 years behind the fruit juice industry, has been sparked by the adoption of controlled exogenous microbial enzyme action.

As late as 1975, Hysert, in his concise and informative overview of advances in enzymology of relevance to brewing, considered the impact of these advances to be rather limited at least in the United States, but having impressive potential at that time.

However, all this will probably not make malt obsolete and active research is proceeding on malt enzymes and on their roles in the brewing process. Table 31.1 presents a condensed summary of this brewing process, the types of enzymes involved and a brief synopsis of how the action of these enzymes may participate in and facilitate the process. The remainder of the chapter is devoted to a more in-depth survey of the roles of each enzyme (type).

It is, of course, no accident that the same types of enzyme activity are involved in both sweetener and beer production since the common main substrate is starch. However, the two principal differences from an enzymological point of view are: (1) the fermentable sugars and oligosaccharides in beer-making are further converted to alcohol by the enzymes of the yeast, and (2) the starting material for beer is not, as for sweeteners, almost pure starch, but the entire plant tissue or organ (in western culture, cereal grains) is the "substrate." Both differences lead to involvement of more than starch and carbohydrate transforming enzymes in the beer-making process, as indicated in Table 31.1. Another point of interest is that until quite recently, as pointed out by Pfisterer (1974), the major usage of enzymes was at first confined to the mashing process and relatively little attention had been paid to other steps. A survey of the research literature indicates enzyme addition is being rapidly expanded to facilitate their use in these other steps.

In keeping with our discussion of relevant endogenous enzymes, we shall examine in more detail barley enzymes which participate in brewing. We shall also dwell briefly on the use of enzymes in the manufacture of distilled alcoholic beverages and industrial alcohol.

α-AMYLASE

We had several occasions to refer to this starch-degrading enzyme. In particular we examined its specificity in connection with the enzymology of sugar sweetener production in Chapter 9. At that time, we referred to Table 9.3 which lists the known enzymes which catalyze the hydrolysis of the glucosidic linkages in the amylose and amylopectin moieties of starch. As further understanding of the delimitations of their specificites has ex-

TABLE 31.1

ENZYME ACTION IN BREWING

Step	Results In	Enzyme Action	Accomplishments
Cooking, mashing, malt + adjuncts + replacement enzymes[1]	The Mash	α-, β-Amylase	Provides sugar for yeast
		Cytases	Make starch available to amylases
		Hemicellulases	Facilitates mashing
		β-Glucanase	Degums mash
		Proteases	Provide N for yeast
Wet milling	Barley syrup	Proteases	Provide N for yeast
Lautering, hops added, clearing	Wort	Glucoamylase β-Amylase Pullulanase	Attenuate worts for producing malt liquor, ale, stout, and light beers
Fermentation of wort	Fermented wort	Glycolytic enzymes Glucanase	Ethanol, flavor, product
Filtration	Beer	β-Glucanase	Speed filtration
Pasteurization, -heat and/or millipore filter[2]			
Chillproofing	Beer	Papain	Maintain clarity at low temperatures
Carbonation, packaging	Beer	Papain	Continues to act

[1] For barley brewing.
[2] Originally used for canned draft beer.

panded, the number of types has contracted, as witnessed by the "demise" of enzymes corresponding to EC number 3.2.1.9 (amylopectin-1,6-glucosidase), changed to 3.2.1.41 (amylopectin 6-glucan hydrolase). This enzyme is now included or synonymous with the debranching enzyme pullulanase. Before we discuss these more recently discovered enzymes and their use in beer-making, we shall turn our attention to the classical enzymes, the α-and β-amylases.

Barley α–Amylase in Development and Germination (Malting)

In traditional brewing, the main enzyme responsible for hydrolysis of starch present in the malt and somewhat more recently in cereal adjuncts is α-amylase of germinated barley. α-Amylases are defined as endo-enzymes which catalyze the random hydrolysis of the α-D-(1→4) linkages in amylose, in shorter chain glucose oligomers, in amylopectin and in glycogen. The latter two substrates are branched molecules comprising relatively short α-D-(1→4) glucan chains branching off from each other via α-D-(1→6) linkages. Manners (1974) and Marshall (1975A), in reviews citing earlier work going back to 1970, distinguish three chain categories: A, B and C. A chains carry no substituent chains (singly) branched; B chains bear an average of three other substituent chains linked to primary glucosyl hydroxyl groups (multiple-branch); the C chain, one per amylopectin molecule, may be considered as the "final" or "first" chain. Its final (or first) glucose has a free reducing group. A portion of an amylopectin may be thus represented:

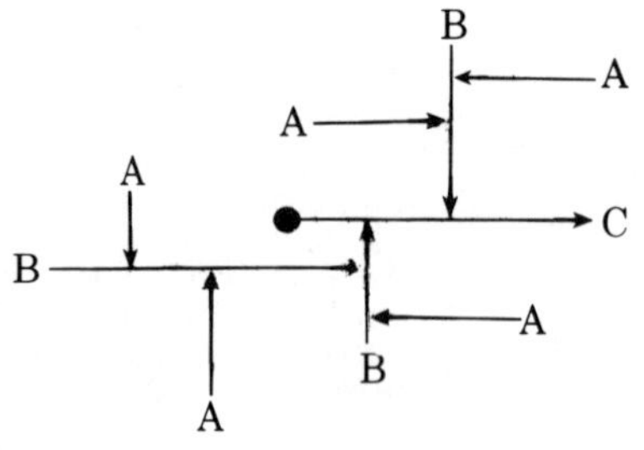

Illustrative of recent advances in our understanding of just how α-amylase may attack starch components are: (1) the computer-based conclusions of Banks and Greenwood (1977) that most plant α-amylases, including those of malt, hydrolyze amylose via a *preferred attack* in which only a single hydrolytic event occurs during each effective encounter between enzyme and substrate, thus eliminating the "zipper" or *multiple attack* theory; and (b) the work of Pfannemueller and Potratz (1977) on the production of "star" branched molecules which may be regarded as intermediate dextrins in the α-amylolysis of amylopectin.

The end-product of malt action and other α-amylases on amylose are the monosaccharide glucose, the disaccharide maltose and the trisaccharide maltotriose. Amylopectin yields, in addition, triaoses to hexaoses each containing one α-1,6 linkage, derived from an α-1,6-branch point.

α-Amylase is of considerable historical importance, being probably the first catalyst whose action was attributed to a nonvital agent. The term diastase or diastatic activity now limited to the combined action of the α- and β-amylases of barley was once used synonymously with "ferment" and later with "enzyme." The first known attempt to isolate an enzyme (Payen and Persoz 1833) involved fractionation of and precipitation of the protein in aqueous extracts of malt with ethanol to obtain enriched preparations of what even then was called "la diastase," whose importance to the "industrial art" was already well recognized. Although the existence of two distinct types of diastase in malt was recognized over 100 years ago (Sumner and Somers 1947), the prefixes "alpha" and "beta" did not come into general use until 1930. The prefixes refer to the early observations that the product produced by "takadiastase" (*Aspergillus oryzae* α-amylase) was considered to be α-maltose (really a mixture of α-dextrins which eventually yields, largely, a mixture of α- and β-maltoses due to mutarotation) whereas the product of what was then called "malt" amylase or diastase yielded mainly β-maltose, due to inversion of configuration at the newly-formed terminal C-1.

In Ungerminated Barley.—After establishment of the two distinct types of enzymes in malt, it was widely believed that α-amylase activity was completely absent in even the mature ungerminated barley grain (and in other grains such as wheat, as discussed in Chapter 32). Its reported detection in barley by Kneen (1944) and quantification of its activity during development of the wheat kernel (Schwimmer 1947B) was neglected until the late 1960s when its ontogenic and evolutionary significance with respect to the enzyme after germination and the ease of detection with immunochemical techniques led to a revival of interest. As indicated in Fig. 31.1, a small amount (about 1% of that present in the germinated seed) is present at anthesis. This leads to the circumstance that in the very immature kernel the specific activity in units per dry weight or of protein is quite high and easily detectable because of the low solids contents of the water-filled kernel. However, in the mature kernel large amounts of solids, proteins and β-amylase obscure its action. The observation of Kneen that the enzyme is not (as is the enzyme of malt) confined to the endosperm but is localized chiefly in the outermost layer of the kernel, (the endocarp-testa) was later confirmed (MacGregor *et al.* 1972; Banks *et al.* 1972; Allison *et al.* 1974).

The α-amylase may also complex, not with protein, but with its natural substrate, starch. Thus, Goering and Eslick (1976) and De Haas *et al.* (1978) described a self-liquefying waxy (high amylopectin:amylose ratio) barley starch which they suggest could replace corn syrup or even substantial amounts of malt in the brewing industry. Meredith and Jenkins (1973) said they detected heat stable and labile molecular forms of amylase which were neither α- nor β-amylase present not only in the pericarp but also in the flowers. Shifting isozyme patterns have been observed during development (Bilderback 1971) but MacGregor *et al.* (1974) reported on "*the* α-amylase" present in developing barley kernels which, after considerable purification, resembled in its physicochemical properties one of the minor iso-α-amylases of a particular variety of germinated barley malt. Obviously much remains to be learned about this interesting activity. Allison *et al.* (1974) suggested that the action of this α-amylase may influence the starch type and content of the mature barley grain and thus, indirectly, the beer-making process.

In Germinated Barley (Malt).—Upon germination, α-amylase content of cereal grain increases precipitously as indicated in Fig. 31.1. In what may have been the first instance of the use of affinity chromatography to purify an enzyme, Schwimmer and Balls (1949A) reported that isolation of a crystalline, homogeneous (by the standards of 30 years ago) preparation of malt α-amylase, the first of the amylases to be thus purified, thus bringing to fruition the efforts of Payen and Persoz 115 years earlier. This study also established the two noncatalytic hallmarks of α-amylases: the presence of Ca and heat stability relative to β-amylase, although plant α-amylases are more thermolabile

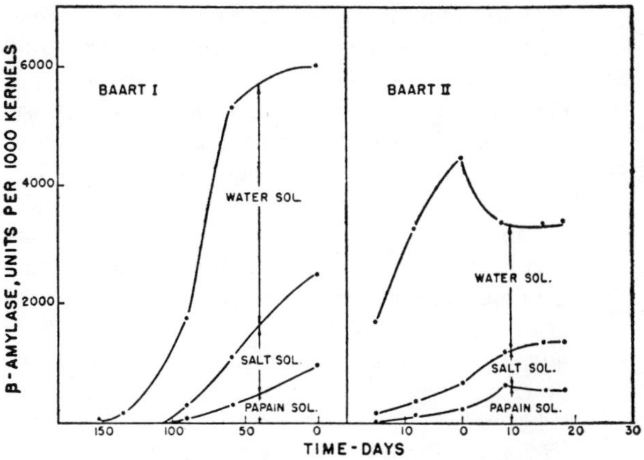

From Schwimmer (1947B)

FIG. 31.1. ACTIVITY AND SOLUBILITY OF BETA AMYLASE

During development and maturation of wheat kernels.

than fungal and especially bacterial α-amylases, as discussed in Chapter 9 and elsewhere.

MacGregor *et al.* (1971) separated this type of preparation into two isoenzymes; and Mitchell (1972), incorporating gel chromatography (Chapter 6) into a modification of the procedure of Schwimmer and Balls, isolated a homogeneous carbohydrate-containing α-amylase. Rodaway (1978) concluded that even such apparently homogeneous preparations are inhomogeneous with respect to glycosylation since the average enzyme molecule (41,400 daltons) had less than one sugar per molecule. The enzyme is singular with respect to its amino composition in that it is relatively rich in histidine and poor in glutamate and serine, thus resembling more closely microbial than eukaryotic proteins. Previous documentation of the presence and changes in the shifting iso-α-amylase spectra during malting and in malt can be found in the last three cited investigations.

In order to bring some order into the seemingly chaotic isoenzymology of barley α-amylase, MacGregor launched a program of actually isolating each isoenzyme and has isolated the most acidic one in addition to those mentioned earlier (MacGregor 1977). That knowledge is still rudimentary is attested to by the isolation of yet another barley amylase by Niku-Paavola (1977) who prefers to classify it as neither an α- nor a β-amylase nor as an endo- or exo-enzyme. Reminiscent of the "less random" cellulases (Chapter 9), its salient distinctive action appears to be the production of both more maltose and glucose and less intermediate dextrin from β-limit dextrin (that remaining after the removal of the outer chains by the action of β-amylase or amylopectin) than does α-amylase. It is thus a more efficient generator of substrates for limit dextrinases, enzymes hydrolyzing α-1,6-glucosidic linkages and thus of importance in malting and brewing operations. Illustrative of further applications of affinity chromatography using amylase substrates are the purification of α-amylases with crosslinked starch (Weber *et al.* 1976) and analysis of amylolytic enzymes using affinity chromatography with cycloamylose as ligand (Chapter 7).

After confirming Kneen's observation establishing that the enzyme is localized in the aleurone layer of the barley kernel, subsequent investigators have attempted to pinpoint the site of synthesis, storage and secretion within the aleurone cells. Illustrative of this approach is the conclusion of Firn (1975) that the gibberellic-induced enzyme (see below) may be secreted from the cells via membrane-bound vesicles. Using immunochemical localization techniques Jones and Chen (1976) concluded that the enzyme was present in the cytoplasm and in the perinuclear region, apparently in association with endoplasmic reticulum. The latter, it will be recalled, is the site of synthesis of most proteins, including enzymes. No enzyme was detected in or near the cell membrane. Immunochemical analysis was the experimental approach which led Daussant *et al.* (1974) to the conclusion that the bulk of the enzyme activity extracted from barley seeds at different stages of germination differed antigenically from amylases found in developing barley seeds.

Gibberellic Acid, α-Amylase and Malting

The adoption of gibberellic acid (Gibberellin A_3, hereafter referred to as GA) by the brewing industry constitutes an exception to the lag period between a basic discovery and its application for practical purposes. Indeed the initial hope in the basic sciences that GA would prove to be the Rosetta Stone that deciphers the mechanism of gene derepression in higher plants has not been fulfilled, whereas the applications to practical use have continued to expand. The gibberellins are diterpenoids (Fig. 31.2) and are thus related to the important class of flavorants discussed in Chapter 22.

Almost 40 years ago Hayashi (1940) showed that the active principle of "Bakanae" fungi infecting rice plants accelerated germination of barley and also enhanced α-amylase activity in the malt. The following is a thumbnail sketch of the salient features of GA, as gleaned from the reviews of Paleg (1965), Melcher and Varnar (1971) on its biochemistry and of Macleod (1967) and Palmer (1974) on its role in malting. GA is present in the embryo of all cereal seeds, including barley. Adding isolated GA to embryo-less seeds stimulates α-amylase formation and secretion. The major salt soluble protein formed is α-amylase (Varner *et al.* 1976).

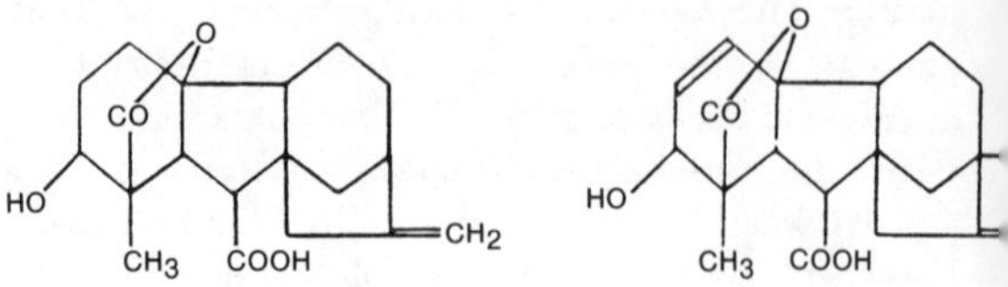

FIG. 31.2. GIBBERELLIN A_4 (LEFT) AND GIBBERELLIN A_3 (GIBBERELLIC ACID)

The site of synthesis of the α-amylase is the aleurone cells which make up the outermost layer of the endosperm of cereal grains. GA also induces α-amylase production in isolated aleurone cells. While it is generally agreed that α-amylase is synthesized *de novo* from amino acids, there has been some controversy as to whether GA actually induces synthesis of all of the α-amylase isozymes or triggers the release of the newly synthesized enzymes from the aleurone cells during barley steeping (Carlson 1972; MacGregor 1976; Rodaway 1978). Indeed many papers refer to GA-induced "release" rather than the "synthesis" of α-amylase. What may be synthesized is a specific protein which is somehow involved in this "release." On the other hand Brooks *et al.* (1976), in their comprehensive review of steeping of barley, a crucial step in the preparation of malt, cite evidence which indicates that no increase in amylase occurs during steeping. All agree that GA does stimulate release of all enzymes whose activity is increased. Another unsettled question is whether GA is involved in the biosynthesis of α-amylase during development and ripening of the seed as well as during germination (Duffus 1969; Bilderback 1971).

While the exact scenario by which GA derepresses the genes coding for *de novo* synthesis has not been written, it is generally agreed that it acts at later posttranscriptional stages of derepression, perhaps by activating latent messenger RNA. It appears to be involved in *turnover* (breakdown as well as biosynthesis) of α-amylase, other enzymes, protein and nucleic acids (Briggs 1968; Varner 1964; Taiz and Starks 1977; Schwimmer 1968B). Palmer (1974) cited evidence that GA is constantly required for the turnover of a "specific" type of soluble RNA. GA does not appear to induce formation of new protein-making machinery in the form of increased endoplasmic reticulum, but it does induce formation of vesicles, similar perhaps to those held responsible for α-amylase secretion from the aleurone cells (see below) during normal germination. Closely related to the question of protein-synthesizing stimulation *vs* secretory stimulating activities of GA is the localization of these enzymes within the aleurone cells. Evidence for cell walls, liposomes and their possible equivalents, the aforementioned vesicles and the significance of the localization to proposed mechanisms of GA action is presented, reviewed and discussed by Gibson and Paleg (1975,1976).

Another area which requires clarification is the question of how many GA-induced isozymes of α-amylase are formed and their relation to those in germinated and developing grain (Jacobsen *et al.* 1970; Tanaka and Akazawa 1970; Momotani and Kato 1974).

Practical Application of Gibberellic Acid.—Approval of GA as an additive (Federal Register 1960) was followed closely by its application in the brewing industry (Pollack 1962). By 1968 "gibberellin" malt assaying, with respect to diastatic activity (α + β amylase), double that of distillers malt (see below), was generally commercially available and presumably in widespread use.

Even in countries where GA addition was specifically prohibited as in Germany in 1966 (Weinfurtner *et al.* 1966), malting procedures were revised on the basis of obtaining the maximum natural physiological benefit of endogenous GA produced during germination. One of the techniques employed to this end has been "multiple steeping" which stimulates GA synthesis and secretion into the endosperm while suppressing continuation of further germination. Even where GA is allowed, it may be employed in conjunction with multiple steeping, i.e., as a spray after the second steep (MacWilliam and Reynolds 1966). Another stratagem is optimization of the short time-low temperature "curing," i.e., 1.5 hr at 70°C, preceding drying of the malt (Narziss and Rusitzka 1977).

The early observation that absence of embryo enhances GA-evoked α-amylase production in barley seeds, together with the realization that its physical separation would not be economically feasible, led to a variety of ingenious efforts to inactivate the embryo without removing it. These have included: acid, heat, freezing, organic solvents, crushing, abrasion of husk and pericarp at distal end of the grain, and turbulent agitation after steeping. Concerning abrasion, which is one of the more widely adopted of these methods, Lyall and Stowell (1977) showed that pericarp injury is not necessary and that the abrading process is effective simply by removal of husk. Of course abrasion without GA addition also accelerates water uptake and production of malt with enhanced α-amylase activity (Gothard 1974). Turbulent agitation after steeping, developed by Bloch and Morgan (1967), combined with elevated temperature resulted in a three-fold enhancement (Fig. 31.3). These investigators pointed out that one of the advantages of using GA is the very low levels needed to obtain effective results. They applied GA as a spray at the rate of 2 ppm of barley. Other procedures recommended levels as low as 0.2 ppm. In laboratory experiments concentrations as low as 10^{-7} *M* (0.04 ppm) will evoke α-amylase production.

As with other procedures in which enzyme action is exploited or, as in this case, enhanced (Chapter

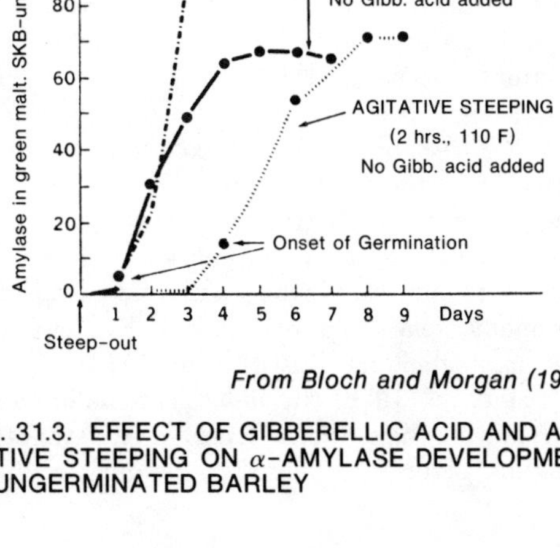

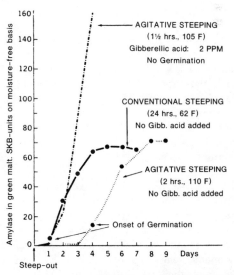

From Bloch and Morgan (1967)

FIG. 31.3. EFFECT OF GIBBERELLIC ACID AND AGITATIVE STEEPING ON α-AMYLASE DEVELOPMENT IN UNGERMINATED BARLEY

1), the decision to adopt GA commercially may have hinged upon entrainment to enhanced α-amylase activity of accrued benefits and advantages: reduction of steeping times by as much as one order of magnitude, use of less water, equipment capacity increase, increase in malt yield due to absence of rootlet formation and decreased respiration weight losses, increased wort fermentability, increased production of other enzymes needed or beneficial in brewing and reduction of acidity. When used for maltose syrup production, GA produces a product with improved filtration characteristics (Khryanin *et al.* 1976).

MacLeod (1967) listed the following disadvantages and problems attendant on the use of GA: increased respiration rate, but only if the grains are allowed to germinate; color formation; increased susceptibility to microbial contamination; and adverse foaming (head) potential. Perhaps another disadvantage, psychological, in the early years of its adoption was an unfamiliar pattern of analysis of the resulting malt, an analysis not traditionally indicative of a good malt.

Further efforts to improve GA-induced release of α-amylase in barley may be dependent upon more information concerning substances which enhance, inhibit or mimic its effects.

Enhancers, synergists or stimulators of GA-induced α-amylase production in plant seeds mentioned by Palmer (1974) in his comprehensive review include cyclic AMP (Chapter 4), ADP, glutamate, aspartate, mevalonic acid, helminthosporic acid, substances derived from the TCA cycle and certain amino acids. Stimulatory effects have also been reported with kinetin (Yakushkina and Dulin 1977), low pH (Palevitch and Thomas 1976), EDTA (Palevitch and Thomas 1976) and mixtures of trace metal salts—Mn, Zn, Co, Mo, Cu—(Mashev *et al.* 1977).

Suppression or inhibition of the α-amylase response to GA may be elicited by the plant growth regulator abscisic acid (Chrispeels and Varner 1966, some Cl-containing herbicides (Rao and Duke 1976), certain tannins (Jacobson and Corcoran 1977) and bromate used to modulate protease activity in some malts (Palmer 1974).

Cyclic AMP not only enhances but has been reported to imitate GA-like activity, as has the hormonally related prostaglandin $F_{2\alpha}$ but not other prostaglandins (Duffus *et al.* 1973).

Brooks *et al.* (1976) cite investigations in which the following have been reported to stimulate formation of or to mimic GA: 2,4-dichlorophenoxyacetic acid; 3-amino-1,2,4-triazole; *o*-phenyl phosphates; 2,3-benzoxazolenone; and chlorcholine chloride.

Other Effects of GA.—Although the major salt-soluble protein released from the aleurone cells is α-amylase, other enzymes of some importance to subsequent brewing are also released. For instance, one enzyme, an acid phosphatase, although not directly so involved, has been assayed as an indicator of the utilization of externally supplied GA in a barley endosperm bioassay (Brookes and Martin 1975).

The GA-induced release of the classical lysosomal enzyme, acid phosphatase—it is particularly suited as a sensitive GA indicator because it is released before α-amylase and other enzymes—bespeaks these subcellular organelles as one of the early loci of GA action. Some of the other enzymes (proteases, ribonucleases) appear to be synthesized *de novo* as a result of GA being present while others, especially β-1,3-glucanase, are apparently synthesized during water imbibition (steeping) without the aid of GA. The latter does trigger the release of the glucanase from aleurone cells whose walls are efficiently digested by the action of the released enzyme (Jones 1971; Palmer 1975). Other brewing-related enzymes, either synthesized or released, include limit dextrinase and α-glucosidase. Palmer (1974) also lists others secreted from the aleurone cells, some of which may be synthesized *de novo* within the cells in response to the presence of GA. A host of other enzymes whose actions are needed to prepare the seed for its new physiological functions are undoubtedly also elicited.

Other α-Amylases

Although ungerminated cereal adjuncts used in brewing contain detectable levels of α-amylase activity, these levels are so small that in contrast to their β-amylases they most probably do not contribute significantly to the α-amylase function in brewing. While Hysert (1975) lists pancreas as a source of α-amylase of interest in barley brewing, it is unlikely that it is now being used extensively for this purpose.

This leaves commercial microbial amylase preparations as the only other major amylases which function in some brewing operations, especially in Europe, where they are replacing some of the malt traditionally used in barley brewing. Both bacterial and, to a lesser extent, fungal amylases are now being used. It would appear from a perusal of several sources including advertisements of the enzyme suppliers that the bacterial α-amylase is obtained from *B. subtilis* and fungal α-amylase from the historical microorganism which really started the microbial enzyme revolution, Takamine's *Aspergillus oryzae*. Each is used for its own special purpose. Presumably they are manufactured in accordance with the principles outlined in Chapter 5. Frequently they are combined with other preparations rich in other enzymes used to perform other functions in brewing. Some of the uses of commercial α-amylase-containing preparations specifically designed or designated for brewing use, as gleaned from the manufacturer's own statements, are listed as follows.

Ciba-Geigy supplies Irgazyme B A10 containing α-amylase and β-glucanase for starch and gum liquefaction and degradation and for lowering viscosity. Enzymes allow control of process and product quality independent of raw material.

The Enzyme Development Corporation supplies Enzeco brewing Enzyme 35A2 as a specially formulated balanced enzyme system when malt is partially replaced by barley, and Adjuzyme as a standardized liquid saccharifying amylase for the production of completely fermented (low calorie) beer which also allows the brewer to use all the carbohydrate present thereby giving higher alcohol yields.

Grinsted Products offers Farinase, a fungal α-amylase which is free of protease activity and is thus useful in adjusting wort composition, as well as Alfamase which contains both bacterial α-amylase and β-glucanase to be used for adjunct cooking and as an amylase source in mashing.

Miles-Marschall has marketed Diazyme, which lets a brewmaster produce more fermentables for beer and malt liquor without increasing expensive malting facilities and without affecting the taste of the brew, and more recently Brew n Zyme, a combination of amylolytic and proteolytic enzymes to replace malt for a substantial cost saving while improving body, head and foaming properties and chill stability.

Novo recommends their fungal amylase, Fungamyl 800 L, for increasing attenuation, that is, fermentability of the wort, and their bacterial amylase preparations, Teramyl 60L and BAN 120, for the liquefaction of adjuncts such as maize, grits and rice.

Rohm and Haas manufactures a series of Rohzymes (S, 33 and H-39) for increased liquefaction of starch-containing foods.

Wallerstein produces bacterial and fungal amylase preparations whose names reflect their recommended use in the baking industry (Chapter 32).

Rohm, through Fermco, offers a series of Rohalases for starch processing, including the dextrinization of starch with malt when using unmalted grain for brewery use.

This list is by no means complete and exclusion of others does not in any way imply any value judgement concerning their suitability. Their function in specific steps in brewing and distilling, advantages and problems incurred with their use and how these problems have been overcome will be examined shortly.

As previously mentioned, microbial α-amylases are in general more heat stable and act optimally at higher temperatures in the order: malt $<$ fungal $<$ bacterial. These thermally related properties are in consonance with heating practices in brewing.

Function and Exploitation of α–Amylase Action in Brewing

Malt α–Amylase.—The principal functions of the α-amylases in beer-making are to thin the mash during the cooking process and to provide fermentable sugars, glucose, maltose and the next higher triaose, maltotriose, for the yeast. After grinding, a portion of the total malt to be used is mixed with adjuncts (as high as 40% by weight of total grain bill) such as rice, corn and more recently wheat (Hough *et al.* 1976). Water is added and the resulting mash is boiled to swell and gel and finally disrupt the starch granules. At the start of the heating, the α-amylase is acting on raw starch granules. Bathgate and Palmer (1973), in a thorough study of the degradation of both barley and malt starch granules under conditions simulating those of a conventional infusion mash, showed that the intact "raw" granules from malt are more susceptible to malt α-amylase than are granules from ungerminated barley. They attribute this difference as well as granule-size dependent differences in susceptibility to prior modification. The temperature of the mash infusion is then reduced to 65°C and the remainder of the malt is added. α-Amylase action in cooperation with that of β-amylase (see below) produces an optimum amount of fermentable sugars plus sizable amounts of dextrins, with the α-amylase controlling both rate and extent of starch degradation. For standard beers, one does not utilize the full potential of the combined action of these two amylases mainly because the alcohol level in the resulting beer would be too high. Dextrins, partially degraded amylose and amylopectin, which constitute about one-quarter of the total carbohydrate of a conventional brewers wort, traditionally impart "fullness of body" and contribute to foamability of the final product. Of the remaining carbohydrates in the wort 11% is glucose, 50% are disaccharides, mostly maltose, and 14% trisaccharides. About 5% of this fermentable carbohydrate becomes yeast and the remainder is converted into equal amounts of CO_2 and ethanol.

Malt Replacement with Microbial Amylase; Barley as Adjunct.—The existence of many suppliers of microbial enzymes to the brewery industry testifies to their successful application, if not in the United States, at least in most of the European countries. As late as 1965 it was believed that, in contrast to the alcohol distilling industry (see below), the incentive to switch to microbial α-amylase was minimal (Reed 1966). Ten years later Hysert (1975) stated that brewing with large proportions of barley and added enzymes had at that time not been widely adapted, presumably in the United States. Although one of the earliest patents on the use of fungal enzyme in the preparation of wort from barley was issued to Dennis and Quittenton (1962), Saletan (1968) stated that both bacterial and fungal α-amylase and even amyloglucosidase were employed in brewing at that time and less than 5 years later Krabbe (1972) reported that in Europe commercial microbial preparations were rapidly replacing malt as the sole source of α-amylase, or more frequently as a supplement to malt for conversion of ungerminated barley. Burgeoning of the literature led to frequent reviews, mostly in European-based journals. Thus, Stentebjerg-Olson (1971) cited numerous investigations covering laboratory, pilot plant and full-scale operations. More recent reviews in English or papers referring to the literature include those of Sorenson (1972), Button and Palmer (1974), Wieg (1975) Hysert (1975) Rao and Narasimham (1976) and Woodward (1978) as well as countless reviews in almost all of the European languages.

We wish to stress that while both bacterial *(B. subtilis)* and fungal α-amylases are used in beer-making in Europe they are not used interchangeably and not in the same step. Indeed the fungal enzyme really serves in ancillary or specialty functions (see below) in comparison with the bacterial enzyme whose use really constitutes a significant departure from traditional beer-making.

There are two distinct ways in which bacterial α-amylase is used in beer-making. One of them, for which the literature is rather sparse and presumably is not in widespread use, is in the preparation of "barley syrups" presumably at a site other than the brewery (Maule and Greenshields 1971; Wieg 1970,1975; Woodward and Bos 1973). These syrups are essentially condensed worts (80% solids) which are shipped to the brewery where they are diluted, yeast and hops are added, and the manufacture of the beer is completed. The commercial enzymes used in this manufacture contain, in addition to amylases, microbial proteinases to provide adequate N for the yeast and a very small amount of malt which provides the supplementary enzymes (β-amylase, limit dextrinase, see below) needed for obtaining the standard "attenuation" —the degree of starch hydrolysis.

The second and apparently the most widely adopted approach is to replace some, but usually not all, of the malt with barley and to use bacterial α-amylase for performing the major starch-degrading tasks. In terms of weight, however, the added enzyme constitutes a fraction of a percentage of the mash bill. Thus, Button and Palmer (1974) describe their experience with production scale brewing with up to 70% of malt replaced by barley. Included was 5−10% of preformed sugar added as a syrup. With the particular commercial enzyme preparation they used, the only relatively major adjustments in standard brewing practice were (1) increased expenditure of energy required to grind the barley, which is tough as compared to malt, and (2) extension of the time required to obtain a suitably attenuated wort with proper fermentability. Indeed Hysert (1975) considered this lack of fermentability (due to heat inactivation and lack of solubility of barley β-amylase and other supplementary starch-degrading enzymes present in normal malt) to be the most serious and frequently encountered difficulty in brewing with barley.

Many of these difficulties can be overcome by adding appropriate enzymes, both amylolytic and proteolytic (see below) to the mash along with the bacterial α-amylases.

As mentioned, fungal α-amylase may serve as a supplement to the bacterial enzyme at time of mashing but is more frequently added during the fermentation. This is because it is not as heat stable as the bacterial enzyme. It is probably used chiefly in the manufacture of dietetic beer and malt liquors (see below).

While Hysert pointed out some of these problems together with others such as lack of reliability at its then (1975) stage of development, the need for the enzymes used to meet trade and legislative (FDA) requirements, and recalcitrance on the part of consumers to accept other than so-called natural foods have led at least American brewers to look askance at brewing with added barley and microbial enzymes. Their use in most other countries suggest some advantages, the most important of which to the brewer is probably the reduction (7–17%) in cost without sacrificing of quality. Test after test has demonstrated that these beers are practically indistinguishable from malted beverages made in the traditional way. Another advantage is a more rational control of the level of α-amylase which is independent of agronomical considerations. In conjunction with endo-β-glucanase (see below) its use results in an easier run-off (see below). The enzyme process can be used in existing breweries. It lends itself to a greater degree of freedom in selection of mashing procedure and temperature programming, in part due to the higher thermal stability of the bacterial amylases (it also, however, leads to inactivation of barley β-amylase). In this connection the point of addition of the enzyme in the processing line can be varied at will. In conjunction with other glucanases, it allows the attainment of worts of any desired composition, reminiscent of the use of enzymes in the sugar sweetener industry (Chapter 9). Finally microbial α-amylase addition in beer-making is of advantage even when none of the malt is replaced by barley (Kieninger 1977).

Distilled Alcoholic Beverages.—In the preparation of distilled beverages, in contrast to traditional brewing practices, before the advent of fungal enzymes it was desirable to use a malt with the highest α-amylase activity possible, amounting to double that of brewing malt. This has largely been supplanted by gibberellin malt, when malt is used at all. A maximum conversion of starch to alcohol is the desired aim, using minimum malt, amounting to about 10–15% of the mash bill for an active standard distillers malt and about 5% for a gibberellin malt. The source of starch for these conversions includes not only most of the known edible cereal grains, but also potatoes, root vegetables such as yams, manioc and cassava, and probably, at least experimentally as mentioned in Chapter 30, starch fruits such as bananas (Munyanganizi and Coppons 1976).

In standard practice, unlike that of beer-making, the grain when used is mixed with water and heated under pressure to effect gelatinization and sterilization of the resulting mash. In contrast to brewing, only a minimal amount of malt (0.1%), if any, is used for liquification. Mash bills frequently contain rye which contributes yeast growth factors and flavor. Only after the mash has cooled to 60°C is the ground malt added to effect a rapid and fairly complete starch conversion to fermentable sugars. Also in contrast to conventional brewing, α-amylase action on the starch-derived dextrin intermediates continues after yeast is added to the cooled mash. Apparently there is a steady state of maltose production

$$\text{Dextrins} \xrightarrow{\alpha\text{-Amylase}} \text{Maltose} \xrightarrow{\text{Yeast}} \text{Ethanol}$$

until about 90–98% of the theoretical amount of alcohol is formed from the original starch.

During World War II industrial alcohol was produced from grain in the United States by using microbial α-amylase contained in "mold bran" (Underkofler 1966). Previous to this mold bran was extensively used outside of the United States, especially in Japan (Hansen *et al.* 1955; Reed 1966; Pieper 1970). Three principal variations of the application of mold enzymes in the distillery had been developed: the amylo process in which live culture of the fungus *Rhizopus delamar* grows and produces α-amylase and fermentable sugars simultaneously, the aforementioned mold bran process in which the moldy bran grown on flat trays is substituted for part of the malt, and a submerged culture of fungal amylase. Since then enzyme suppliers have been offering fungal enzymes for use in distilleries throughout the world. One of the problems encountered is that most of the α-amylase action occurs in the yeast fermentation vat so that there is the possibility of contamination of the yeast with the microorganisms still present in the enzyme preparation. There has been at least experimentation, if not implementation, with other enzymes in combination with fungal α-amylase. Thus, Jaeger (1976) considered the use of glucoamylase along with fungal amylase. Kreipe (1973) employed mixtures of Novo's heat-stable bacterial Teramyl 60L and Fungamyl at low pH's. He found that the latter was essential for satisfactory conversion in the manufacture of alcohol from potatoes.

One anticipated enzyme-related problem which did not arise was the possibility that the transglucosylase present in the quite crude commercial enzymes used would convert some of the amylase back to nonfermentable sugars, as in the case of glucose syrups (Chapter 9). Apparently the dynamics of the aforementioned steady state shift away from accumulation of nonfermentable sugar, perhaps *via* a reversal of the transglucosylase action. As with most such conversion processes that we have been examining in the last few chapters, the principal incentive to the use of microbial α-amylase and other enzymes in place of malt is economic.

OTHER BREWING-ASSOCIATED STARCH AND OTHER GLYCAN HYDROLASES

β-Amylase

Ungerminated Barley.—β-amylase, the exclusively maltose-producing α-1,4-glucanase, can be barely detected at the inception of anthesis of the cereal flowers (Fig. 31.1). Its activity in the developing seed increases rapidly to a more or less maximum constant value as a soluble enzyme between the tenth and twentieth day later (LaBerge *et al.* 1971; Duffus and Rosie 1973). As indicated in Fig. 31.1, while the total potential activity is more or less constant in the maturing kernel, the water soluble activity decreases due to its transformation into an insoluble inactive form.

Approximately half of the β-amylase of the ripe mature kernel is soluble in water or dilute buffer. Additional but not all of the enzyme can be solubilized with NaCl and other salts. To achieve full solubilization, as first noted by Ford and Guthrie (1908), it is necessary to use either reducing agents, usually sulfhydryl-containing compounds such as cysteine, mercaptoethanol or dithiothreitol (DTT), or to use proteolytic enzymes such as papain. Unlike other "immobilized" enzymes, this bound or "latent" β-amylase appears to be inactive and has been likened to a "zymogen-like" or "protected" storage form of the enzyme. It appears to be associated with the subcellular entities in the aleurone cells known as protein bodies (Tronier and Ory 1970).

Using sophisticated "tandem-crossed" immunoelectrophoretic techniques, Hejgaard and Carlsen (1977) were able to substantiate and correct some previous results and theories as to the cause of this insolubilization of β-amylase as well as to provide a thorough guide to the literature on the subject of bound β-amylase in the developing and maturing kernel. The bound and soluble forms of the enzyme are antigenically the same, as indicated previously by Tronier and Ory. Apparently after the maximum β-amylase has been elaborated, it associates with itself and with a special protein "Z" in the barley to form active aggregates which persist when the barley is extracted with water. These aggregates, which are apparently in part responsible for conflicting reports on the presence of variable isoenzymes, are formed through the interaction of free sulfhydryls present in both the β-amylase and in Z proteins. As new protein, especially glutelin, is laid down in the above-mentioned protein bodies, they too participate in SS-SH interchange with the β-amylase and Z-protein aggregates. Since this new protein is insoluble these aggregates also become insoluble. The associated β-amylase activity becomes both insoluble and inactive.

The four β-amylase isozymes of ungerminated barley grain have been isolated using nonaffinity methods as crystalline 57,200 dalton sulfhydryl protein complexes and thoroughly characterized by Visuri and Nummi (1972).

Malt.—Upon steeping and germination the total β-amylase per seed and per unit weight increases and more, but not all, becomes soluble and active, reaching a maximum at the end of the malting period. Together with the *de novo* synthesized α-amylase, β-amylase liberates the immobilized reserve energy needed for germination by hydrolyzing starch to metabolizable sugars (Meredith 1966 Enari 1974; Goswami *et al.* 1977). Although the isolation and crystallization of the enzyme from malt preceded that from ungerminated barley enzyme by over 20 years (Fischer *et al.* 1950), its isoenzymology has not been clarified. According to the more recent available information, again by Hejgaard (1978), two soluble forms of barley β-amylase, a "free" form and a form aggregated with the aforementioned Z protein, both present in ungerminated barley, survive and persist during malting. In addition, a third form of β-amylase is present, presumably a product of the barley enzyme modified by endogenous proteinase action. Unlike α-amylase, it does not appear to be synthesized *de novo* during germination. In the manufacture of malt, most of the small "diastase" losses occurring during kilning are due to loss of this relatively heat-labile β-amylase.

Action During Brewing.—In conventional malt brewing most of the action of β-amylase takes place during the heating-up period after the starch granules have gelled sufficiently to allow α-amylase action to proceed apace. The enzyme thus assists and accelerates the attainment of the proper attenuation by releasing fermentable maltose from nonreducing ends of dextrins newly formed by the action of α-amylase and other hydrolases. Indeed, rationalization and optimization of this conjoint action of α- and β-amylase was the aim of Toner and Potter (1977) in their computer simulation of the β-amylolysis of amylose and amylopectin. A possible application of such an approach would be maximization of maltose output for brewing dietetic beer (see below), perhaps in conjunction with adjustment of mashing conditions favoring β-amylase action. Such a dietetic beer prepared by a similar approach was used by Kozma (1976). He found that mashing at 60°–65°C and pH 4.1–5.0, conducive to maximizing β-amylase action, led to a dietetic "light" beer with less than 1% carbohydrate (see below). This benefit was traded off for increased mashing time. Sherman and Lewis (1975) suggested that a method for latent β-amylase which they developed might be useful in selecting barley for specific applications such as, perhaps, dietetic beer. Besides being inactivated by heat and acting under nonoptimal conditions during conventional brewing, diminution of the contribution of β-amylase as well as α-amylase to production of fermentable sugars is its inhibition by the tannins present in some cereal adjuncts, especially sorghum (Miller and Kneen 1947; Strumeyer and Malin 1969).

In the newer brewery processes employing bacterial amylases and barley adjunct, the opportunity for β-amylase action is further restricted by several circumstances. The larger dextrins formed by bacterial α-amylases provide fewer exposed nonreducing ends needed to serve as substrate for the β-amylase. The higher temperature at which mashing may be conducted to take advantage of the thermal stability of the added α-amylase results in even more rapid destruction of β-amylase. The full potential of the latter cannot be expressed because in the barley replacement it is partially present in a bound, inactive form, as just discussed. Therefore, the absence or removal of some of the malt proteinase prevents full release of this bound β-amylase.

Addition or presence of proteinase to commercial α-amylases intended for brewing may overcome this handicap. Interestingly, addition of commercial proteinase preparations to beer at a later stage (chillproof) is indirectly related to malt and barley amylase because some of the protein which is removed in this chillproofing is the aforementioned Z-protein which forms a dimer with barley β-amylase. In contrast to the latter, Z-protein is quite heat stable and some of it survives the brewing process to become a dominant beer protein (Hejgaard 1977).

Enzymes That Hydrolyze Interchain α-1,6-Linkages in Starch and Dextrins

In relation to brewing practice one may classify the series of "debranching" enzymes which preferably hydrolyze interchain α-1,6-linkages in glucans into those which (1) release α-1,4-glucans and oligosaccharides (dextrins) and are endo-acting, and (2) are exo-acting, glucoamylase-like enzymes which hydrolyze both consecutive α-1,4 and α-1,6-linkages in amylase and amylopectin.

α-Limit Dextrinase and Pullulanase.—This enzyme which catalyzes hydrolysis of α-1,6-D-glucosidic linkages in α-dextrins arising from the α-amylase catalyzed degradation of amylopectin is the particular α-1,6-hydrolyzing enzyme present in malt. The observation of Schwimmer (1951A) that enzymes in malt extract split four of seven bonds in an 8-glucose α-dextrin whereas α-amylase isolated from the same malt could split only three linkages suggested that malt contained a "limit" dextrinase. Some 25 years later Manners and Hardie (1977) isolated a purified α-limit dextrinase which rapidly hydrolyzed α-1,6-glucosidic linkages in both α- and β-limit dextrins derived from amylopectin as well as in pullulan. Somewhat reminiscent of the requirement for a soupçon of α-amylase to achieve maximal rate of starch hydrolysis by glucoamylase (Chapter 9), α-amylase is also needed to fully potentiate α-limit dextrinase action. Indeed, the authors conclude that there is no longer any justification for the view that germinated barley contains two distinct enzymes, one acting on fully branched undegraded amylopectin, the old "R" enzyme, and the other on branched α-dextrins. (See also debranching enzymes, Chapter 9.) This has been reflected in a revision of the EC classification of such enzymes. Manners and associates almost single-handedly are responsible for debranching enzymology in higher plants, mostly cereals (Manners 1975; Hardie *et al.* 1976). This limit dextrinase, like α-amylase, is present at low levels in ungerminated barley but increases by an order of magnitude upon germination or treatment with GA.

During normal beer-making, over 90% of the original α-1,6-glucosidic bonds present in the starch of the malt and adjuncts show up in the dextrins in the beer, about one-third in singly-branched dextrin (Marshall's A-chains) and two-thirds in multiple-branched (B-chain) dextrins (Enevoldsen and Schmidt 1974). This circumstance affords ample opportunity for the exploitation of debranching action by added enzymes. Enevoldsen (1975) was one of the first to determine, as previously mentioned, the degree of branching of these dextrins and to add a debranching enzyme, pullulanase, to produce a wort of high fermentability, or as a more suitable alternative to microbial α-amylase or glucoamylase under certain conditions. In the previously mentioned full-scale barley brewery trial of Button and Palmer (1974), it was found that pullulanase but not glucoamylase improved the relatively low attenuation they achieved with bacterial α-amylase alone replacing part of the malt. Undoubtedly commercial microbial pullulanase preparations will be used to speed up the brewing process, if this has not already been done. Patents have been issued on their use in the preparation of dietetic beers.

Glucoamylase and Specialty Beers.—In contrast to debranching enzymes or pullulanases, commercial brewing enzyme preparations containing glucoamylases are now available to the brewer after first being considered more than 10 years ago (Saletan 1968). Unlike other starch hydrolases hitherto considered in this chapter, barley malt apparently does not possess an enzyme with similar specificity. It is recommended for speeding up fermentation and especially for the manufacture of low calorie (dietetic, light) beer. Most of the published information for this use is confined to the patent literature.

Some of the patents involving enzymes are listed in Table 31.2. In order to obtain such a beer it is necessary to lower (and in the case of beer for diabetics, to remove) the α-dextrins which remain in conventional beer, without increasing the alcohol content. In general, there are three approaches to accomplish this removal: (1) use a yeast which ferments dextrins, (2) use malt and adjusted brewing conditions to achieve higher attenuation, and (3) add commercial microbial enzymes, usually during the fermentation. Now the sugars derive from hydrolysis of the dextrins are fermentable, s that the resulting beer would have a higher tha normal alcohol content. Indeed, specialty brew such as malt liquors achieve their higher alcoh contents by allowing supplementary enzyme ac tion, frequently that of added glucoamylase, t produce fermentable sugar from the α-dextrins Glucoamylase, it will be noted, is also most fre quently used in patents issued for manufacture o "light" beers. Its ready availability for beer un doubtedly derives from its successful adaptation i sweetener production and the distilling industr where its use was widespread by 1974 (Chapter 9 Unlike its end-use in the sweetener industry, th action results in a *reduction* in calorie intake. T achieve a normal alcohol level in dietetic (light beers, the wort may be partially fermented, th resulting alcohol boiled off, dextrin hydrolyzin enzyme added and the fermentation renewed. I general, commercial enzymes are allowed to ac during fermentation. Alternatively one can fer ment with a mixture of lactobaccilli which ca

TABLE 31.2

ENZYMATIC DEXTRIN REMOVAL FOR LOW CALORIE BEER AND MALT LIQUOR

A. MODULATION OF ENDOGENOUS MALT ENZYMES BY MODIFICATION OF STANDARD BREWING PRACTICE
- (1) Add "maltase"-rich malt infusion after cooking the mash to produce malt liquor (Glueck 1948).
- (2) Adjust initial mashing temperature to 64°C; maintain for 1 hr (Pauls and Quatz 1957).
- (3) Modify extraction procedure so that dextrin forming activity suppressed; add fermentable sugar (usually glucose) as source of alcohol (Sfa and Morton 1973).
- (4) Favor β-amylase activity to produce high-maltos wort by mashing at 60°–65°C and pH 4.1–5.4 (Kozma 1976).

B. ADDING GLUCOAMYLASE AND/OR AMYLASES
- (1) Enzyme added during mash or fermentation (Gablinger 1968).
- (2) Immobilized glucoamylase—acts in beer contain ing residual yeast or in clear beer followed b second fermentation (Woodward and Bennet 1976).
- (3) Enzyme-rich malt; mashing interrupted for 80 min; secondary fermentation supplemental wit yeast and diastase extracts.

C. ENZYMES ACTING ON α-1,6 LINKAGES (LIMI DEXTRINASE, PULLULANASE)
- (1) Diastase enriched with limit dextrinase, alcoho removed by boiling; multistage fermentation.

ransform both yeast-fermentable sugars and dexrins into nonalcohol end-products, thus killing two birds with one stone without having to add enymes.

In the future it should be possible to economically emove excess ethanol by means of immobilized alcohol dehydrogenase, once the problem of developing a process for inexpensive regeneration of NAD is solved. Nonpatent literature on this beer is are but there are a few reviews and discussions including that of Sfat and Bruce (1972) in English and others, i.e., Schur and Piendl (1977). Even here each covers only one particular process. Finally, the use of such enzymes may entrain other quality and processing changes as is so frequently he case. Thus, changes in the spectrum of carbohydrate nutrients available to the yeast could esult in altered nonethanol minor end-products of fermentation which affect beer flavor. The use of such enzymes may also contribute to prevention of certain types of hazes and to facilitation of filtration.

β-D-Glucanases

While some starch transforming enzymes may indeed prevent some haze and facilitate filtration in brewing, it is by the action of nonstarch polymer degrading hydrolases that these two crucial brewing-associated tasks are accomplished. The enzymes responsible for facilitation of filtration by inducing appropriate physical changes in the mash and wort are the endo-β-1,3-glucanases [β-glucansolubilase (Martin and Bamforth 1980)]. In addition to removing the supplementary service provided by β-amylase action, barley brewing also results in the undesirable intrusion into the wort of barley-derived gums and also in a loss of malt enymes. The presence of the gums, the β-glucans, reates a series of interrelated problems during brewing.

The β-Glucans.—Present in the cell walls of the endosperm of both barley and malt are polydisperse heterogeneous linear polymers of glucose inked through both β-1,4- (75%) and β,1-3 linkages (25%). They are water-soluble β-glucans with molecular weights ranging from 30,000 to over 00,000 (Parrish *et al.* 1960; Marshall 1975B; Bathgate and Dagliesh 1975; Narziss and Litzenerger 1977). Collectively they constitute 1.5 to .6% of barley and malt (Bourne and Pierce 1976). Slow-malting varieties of barley have more rigid cell walls and yield less extractable β-glucans (which are less susceptible to endogenous β-D-glucanase action) than more rapidly malting varieties. The question of extractibility during mashing, of considerable practical consequence in subsequent operations, is partially dependent upon endogenous barley enzymes in a manner somewhat analogous to the relation between protopectin and pectin in fruits and vegetables (Chapter 29). Once extracted, however, unlike the pectins they have no redeeming value, due mainly to their increasing the viscosities of mash and wort.

These gums may: interfere with malt grinding operations; reduce the dispersibility of the grind; interfere with availability of the starch to the amylases during mashing; interfere with attainment of full extract of proteins and nitrogen nutrients for the yeast; deposit as gelatinous precipitates when they originate from the final sparging or washing the mash with hot water in the preparation of the wort (the mash in which the starch-to-sugar conversion has taken place); and slow down the runoff at the lautering step through the slotted bottom of the vessel. They also slow down filtration at several other steps, including removal of yeast, and in membrane pasteurization filtration in the manufacture of bottled "draft beer." They can even clog filters. They affect foam formation adversely in the beer and can form gels in certain beers.

It is, thus, no wonder that considerable time and effort has gone into finding enzymes which can remove this troublesome gum.

Barley and Malt Endo-β-Glucanases.—Like most hydrolase activities, that of the endo-β-glucanases (activity measured by drop in viscosity, indicative of internal linkage splitting) rises sharply by at least one order of magnitude upon germination. The level of activity increases with increasing time of germination. However, unlike α-amylase, the response to GA is not clear-cut and, according to most investigators, seems to be due entirely to release of the responsible enzyme rather than to *de novo* synthesis.

The verdict reached from the accumulation of information in the last 10 years or so on the β-glucanases of the malt amassed by several investigators—a selection of whose papers, in addition to those cited, include those of a series by Luchsinger and coworkers (Luchsinger 1971), Marshall (1975B), Manners and Marshall (1969), Enari and Markkanen (1974) and Ballance and Meredith (1976)—is that they comprise a rather complex system of enzymes with what appears to be a rather bewildering range of specificities. The following three types of specificities have been fairly well-established: (1) an endo-1,4-β-glucanase which acts on barley β-glucans, (2) an endo-1,3-β-glucanase which appears to be a true "laminarinase" in that it can act on (hydrolyze) the β-1,3- linkage in laminarin in the *absence* of β-1,4- linkages, and (3) an endo-1,3-β-glucanase that will only act on these linkages in glucans which also contain, as do barley gums, β-1,4- linkages. Interestingly and parenthetically, in addition to higher plants and microorganisms (see below), an endo-1,3(4)-β-D-glucanase activity which acts on both β-1,3- and β-1,4- linkages is present in starfish and related molluscs. The barley enzyme complex is used to liberate protein from yeast (Chapter 32) to increase the yield of the extract in coffee substitute manufacture and to prevent the deposition of sticky droppings from chickens fed barley. In addition to these endo activities, exo-β-D-glucanases are present in malt which hydrolyze both -1,4- and -1,3- linkages. There does not seem to be a sharp delineation in specificities but rather an overlap between enzymes capable of hydrolyzing the β-1,4- linkages in barley gum and in cellulose.

Roles in Malting and Brewing.—Difficulties associated with the use of barley as an adjunct are, as mentioned, linked with the availability of β-glucan for extraction. The following sequence was proposed by Bathgate and Dagliesh (1975) to account for transformation of β-glucan during malting: (1) in analogy to protopectinase, release of insoluble hemicellulosic β-glucan by an unknown heat-stable enzyme system, (2) partial degradation of this glucan by endo-β-1,3-glucanase, and (3) degradation of all these smaller water soluble gums by endo-barley β-glucanase. Apparently this as yet undemonstrated enzyme system releases the soluble β-glucans from barley but there is not enough glucanase activity in the remaining malt to degrade them during the mashing. The inability of the existing barley enzymes to do so has been usually attributed to their inactivation during heating in relation to the amount of glucan which is extracted (Sparrow and Meredith 1969; Bourne and Pierce 1970; Chan and Baker 1976).

Some effort has been made to potentiate the action of these glucanases before they are heat-inactivated during the programmed heating that takes place during the mashing (see below). More recently Ballance and Meredith (1976) isolated an endo-β-1,3-glucanase from unkilned, germinated barley (green malt) which possessed considerable heat stability, at least more than had been previously observed for the activity referred to as "barley β-glucan endo-hydrolase." However, while stable at the early stage of mashing it was fairly unstable at 60°C in the pure state, rapidly inactivated at 65°C and almost instantaneously destroyed at the higher temperatures employed when bacterial amylase is used.

Use of Commercial Microbial Endo-β-Glucanases.—Enzyme supply houses now routinely supply brewers with preparations containing barley gum degrading enzymes. They are frequently supplied in combination with other beer-making enzymes. Enari and Markkanen (1975) determined activity and processing characteristics in eight commercial preparations, four microbial and four bacterial. These are displayed in Table 31.3. Figure 31.4 provides a graphic illustration of the effects of replacing malt with barley and also the effect of addition of one of the commercial glucanases on wort viscosity. These enzymes are useful not only where barley replaces malt but also where what the brewers term "poorly modified" malts are used.

As with other microbial glucanases, the β-glucanases can be readily induced by enrichment culture from safe, acceptable microorganisms, both fungi such as *Aspergillus niger* (Bernat 1977) and bacteria likely to be *Bacillus subtilis*. The activity in these preparations does not appear to be more thermostable than that of the purified enzyme of Ballance and Meredith (1976). Indeed it would appear that the β-glucanases frequently accompany α-amylase in commercial preparations and are offered as such by enzyme suppliers for use in cook-

TABLE 31.3

ENZYME REDUCTION OF RUNOFF TIMES IN BARLEY BREWING

Enzyme Origin	Fungal	Bacterial	All-Malt
Enzyme Activities Relative to Fungal Preparation			
Endo-β-glucanase	1	1	9
α-Amylase	1	900	*ca* 10^5
Protease	1	10	800
Percentage of Values Observed with 20% Barley in Mash			
Wort viscosity	84	87	89
Runoff time	51	50	33

Source: Adapted from Enari and Markkanen (1974).

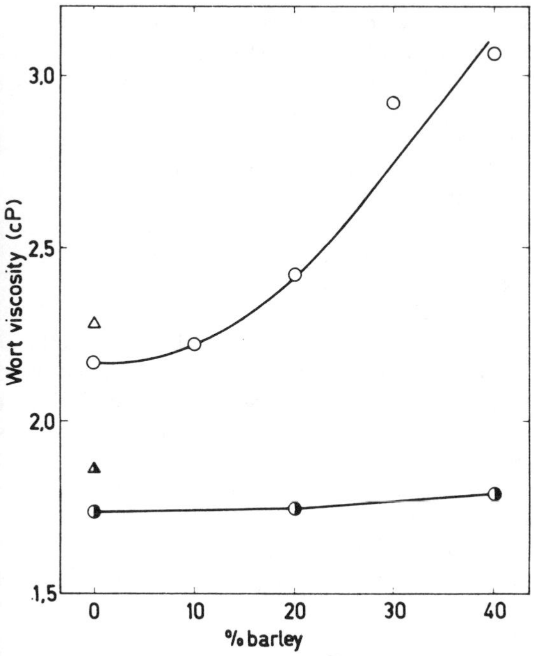

From Enari and Markkanen (1974)

FIG. 31.4. AVOIDANCE OF VISCOSITY PROBLEMS IN BARLEY BREWING BY ADDING β-GLUCANASE

The open and half open circles represent control and β-glucanase treatments, respectively, with varying proportions of unmalted barley in grain bill.

ing adjuncts. Amylase-free glucanase preparations which are also available are recommended for improvement of filtration only.

Addition of β-glucanase at fermentation improves the filtration rate of beer but not, of course, the lautering or runoff time. One possible objection to adding β-glucanase from some but not all sources during fermentation is the possible lysis of yeast walls. Such lysis may be of benefit in another context under different situations, viz, for isolation of protein from cell yeast (Chapter 34) (Monreal *et al.* 1967; Nagasaki *et al.* 1976). Indeed yeasts themselves may secrete β-glucanases (Farkas *et al.* 1973). Most of the literature subsequent to the commercial availability of these enzymes has been concerned with demonstrating the efficiency of the various preparations as applied to varying brewing practices in different countries (Wieg 1976). While undoubtedly very useful in ameliorating filtration and related difficulties, enzyme action alone cannot completely solve these problems; proper malting of the grains, design of the beer-house equipment and lautering technique are considered to be at least as important as lowering viscosity by enzyme action.

Other glucosidase activities in malt have been implicated in at least some aspects of conversion of grain to alcoholic beverages (Mitchell and Newman 1972). Invertase activity has been used as an index of optimal malt modification (Prentice 1973), and yeast invertase acts when sucrose is used to speed yeast growth.

While immobilization of these carbohydrate-transforming enzymes has been accomplished and as we have seen in Chapter 9 are used in sweetener manufacture, efforts towards immobilization in brewing has been concentrated on the next type of enzyme action we shall now examine, that of the proteinases.

PROTEINASE ACTION IN BREWING

The action of protein and peptide hydrolyzing enzymes contribute to the transformation of grains and other starchy foods to beer and related alcoholic beverages in at least three major and distinct aspects: (1) participation in conversion of barley to malt (modification), (2) providing free amino acids for yeast growth during fermentation, and (3) for elimination of chill haze in beer. In traditional brewing only endogenous grain enzymes are used as sources of proteases in (1) and (2) and added commercial enzymes in (3). With the advent of barley brewing with its concomitant reduction of malt, commercial proteolytic enzymes are now added to assist the remaining malt and adjunct proteinases.

Barley and Malt Proteases

Enzymes hydrolyzing peptide bonds perform several important functions in malt modification. As listed by Baxter *et al.* (1978), the most important are to provide soluble N for embryo growth and amino acids for the synthesis of carbohydrase, which ultimately provide the energy required for this growth. These amino acids are made available by the action of these enzymes on the principal reserve proteins of barley, the aleurone-localized hordein, about one-third of which disappears during malting.

Hordein and other malt proteins are now routinely identified by immunoelectrophoresis which has been adopted by the European Brewery Convention as the reference system for barley and malt protein (Enari 1971). Further refinement can be achieved by coupling this technique with affinity techniques (Chapter 7) using lectins with the carbohydrate moieties of individual proteins and enzymes as ligands (Boeg-Hansen *et al.* 1977).

Proteases contribute more directly to mobilization of carbohydrate reserves by attacking protein both in the membrane of endosperm cells (and thus render carbohydrate substrates in the cell interior accessible to carbohydrase) and in the interior of the cell, especially that surrounding the smaller starch granules. Another function, perhaps more important to the brewer than to the grain, is activation and solubilization of the latent β-amylase.

Furthermore, the free amino acids released by the proteases may participate in secondary nonenzymatic and nonmicrobially related transformations which may ultimately contribute to the flavor, color, foaming characteristics and body of the finished product—beer.

With such importance attached to malt modification it is perhaps not surprising that as early as 50 years ago it was already realized that the responsible enzymes constituted a rather complex proteolytic system consisting of soluble and unsoluble endo- and exo-polypeptide hydrolases acting on proteins as well as peptidases which do not act on proteins (Linderstrøm-Lang and Sato 1929; Mill and Linderstrøm-Lang 1929). Although many investigators have contributed to the further unraveling of this skein of activities, the papers from laboratories in Finland, as represented by Enari (1974), from Wisconsin as represented by Prentice *et al.* (1971) and more recently from England (Baxter *et al.* 1978) have been particularly pertinent. From these investigations it is clear that barley and malt contain (1) at least five endo peptidases comprising trypsin-like proteases, papain-like SH-enzymes and metallo-proteins; (2) at least three carboxypeptidases; and (3) at least seven peptide hydrolases with varying specificities and multiple forms.

In the view of Enari (1974) there are eight distinctly specific peptidases in barley malt, which may or may not act on proteins: three carboxypeptidase, three aminopeptidases, two alkaline non-trypsin-like alkaline peptidases and an alkaline BAPA-ase (BAPA = benzoyl arginyl *p*-nitroanilide) which may be a trypsin-like enzyme or an amidase. Baxter (1978) pointed out the importance of these proteinases and developed an assay using endogenous substrate, hordein.

Unlike the effect of germination on other enzymes we have been considering, malting results in only a modest rise in some of these activities. While some of these proteases, especially the carboxypeptidases, appear to be localized in the aleurone cells from which they are released or secreted by GA, some have been reported to be largely confined to the embryo.

Illustrative of the specificity-related complexity encountered is the report that the traditional substrate for trypsin-like enzymes, benzoyl arginine-*p*-nitroanilide (BAPA), is also hydrolyzed by one of the dipeptidases. Adding to the confusion is the presence of proteinase inhibitors some of which, unlike those from most other plants, act effectively on their own as well as on microbial and animal predator proteinases (Mikola and Enari 1970; Weiel and Hapner 1976; Bruhn and Djurtoff 1977).

The Wisconsin group is largely responsible for the available data on changes in protease activities which occur during development of the barley kernel. Some of these peptidases appear in the grain shortly after anthesis. The level of their activities per kernel, like that of α-amylase, remains roughly constant during development and ripening. Because of changes in inhibitors, disappearance and reappearance during maturation have been reported. With regard to their role in malt modification, because of their favorable pH optima, the action of the carboxypeptidase is probably more significant than that of other proteolytic and peptide hydrolyzing enzymes.

There is no serious loss of proteolytic activity as a consequence of the further treatment of the freshly germinated barley (green malt) during routine malting. Indeed kilning induces activation of at least the carboxypeptidases (Narziss and Linz 1975). Curing does, however, produce net loss in α-amino N, possibly through Maillard reactions with reducing sugars.

Proteases in Mashing

Traditional.—Much of the requisite action of the proteolytic enzyme in malt, adjunct or in commercial enzymes added to the malt (or all acting together), takes place during the mashing step.

Enari (1974) pointed out that because the free amino acids of wort amount to about 12% of that potentially available, stability of the malt proteinase is the limiting factor in the appearance of these amino acids. Narziss and colleagues in Germany (Narziss and Lintz 1975) and Enari and colleagues in Finland (Enari 1974) have been active in elucidating the contribution to the brewing process of malt proteinases during mashing. Thus, the German investigators have defined the limits of soluble N in wort compatible with a good beer. Too little N (<24 mg α-NH_2-N/100 ml) in the wort will not support yeast fermentation, creates flavor problems and exacerbates haze-related problems. An excessively unbalanced proportion of amino acids may lead to a beer which is too dark, does not have the proper "body" taste and foaming characteristics, and does not store properly.

Slight increases in peptidase activities occur during early stages and lower temperatures of mashing, followed by progressive and eventually complete loss of most of the activities. An exception is one of the carboxypeptidases which makes by far the preponderant contribution to amino acid liberation in mashing as well as in malting. They are not, however, the rate-limiting factor (Sopanen *et al.* 1980). Similar conclusions were arrived at by the Munich group, who also investigated such variables as cation addition and degree of malt modification.

Barley Brewing and Commercial Proteases.—To compensate for missing proteolytic enzymes in barley brewing, one may add commercial microbial enzymes or use malt from special high-proteinase barley germinated under special conditions or treated with GA. When adding microbial proteases to the mash to compensate for the missing malt enzyme—one of the first patents was issued just about 15 years ago to Bavisotto (1967)—it is necessary to be quite judicious in the selection of the correct amount and type to be added in order to maintain the proper amino acid level and balance in the wort. The proteinase action has to be somewhat more intense since the initial amino acid concentration in such mashes is lower due to substitution of barley for malt. Most enzyme suppliers offer special bacterial neutral proteinase preparations free of other brewing-related enzymes for this purpose (and the latter are usually offered free of proteinases). Alternatively they provide balanced mixtures of amylase and proteases which are particularly useful for preparation of the previously mentioned malt syrups. Such preparations, when added at appropriate levels, provide the proper balance of amino acids needed for the utilization of yeast (Enari 1974) which results in satisfactory beer with, for instance, a good head retention and excellent hop utilization. When too much protease is used, a less preferable full-bodied, caramel-flavored beer is produced. Papain as well as microbial enzymes may be used in the mash to conserve malt, according to Jones and Mercier (1974). This has become possible only with the development of commercially feasible, highly-refined papain as discussed in Chapter 5. Soluble nitrogen of a 100% malt mash was doubled when 0.05% of papain (as % of malt used) was added to the mash. Papain may also be added at the mash stage to chillproof the beer.

Chillproofing with Enzymes

Unless deliberately added, enzyme action from the grain and from commercial preparations has been effectively eliminated by the time the wort is filtered in the lautering step. In conventional brewing, proteinases are inactivated when the wort is heated immediately after the addition of the hops so that subsequently the only enzymes involved in the beer-making process are those present in the live yeast. After the yeast has been filtered, enzymes may again be added, usually before pasteurization, lagering (storage) and bottling or racking. This enzyme, papain, is frequently added to prevent the formation of chill haze which occurs when beer is chilled below 10°C. This haze is but one of the so-called nonbiological hazes which thwart achievement and preservation of the limpid, scintillating clarity which, as pointed out by Flynn (1975), is rapidly becoming the hallmark of a good beer not only in the United States but throughout the world. Other hazes and sediments which may develop during the brewing process or upon subsequent storage include those induced by metals, oxidation, oxalates and more rarely by insoluble polysaccharides.

What are termed "biological" hazes due to infection by wild yeast and bacteria are now seldom encountered since the institution in breweries of such antiseptic measures as pasteurization and sterile filtration.

The Chill-Haze Problem and Nonenzyme Solutions.—Chemically the haze consists largely of protein (15–65%) and polyphenol, usually tannins (10–35%); the remainder is carbohydrate.

Although a minority of investigators have contended that at least part of the protein comes from yeast, convincing immunochemical/immunoelectrophoretic evidence and that obtained by other modern techniques has clearly implicated malt as the major protein contributor and more precisely the hordein and β-globulin fractions of malt protein (Gorinstein 1978). It is clear that some of these proteins survive the heat treatment during brewing. One of the protein molecules which survives is the Z protein which forms a dimer with a β-amylase molecule as discussed earlier in this chapter.

The polyphenols/tannins come from both the grain and hops (Chapon *et al.* 1977; Knorr 1977). As with tea (Chapters 16 and 30) a series of complicated reactions are involved in the formation of these tannin-protein complexes. Oxidation of polyphenol usually has been assumed to be catalyzed by Fe^{3+}, but may also involve malt phenolase action (Jerumanis *et al.* 1976). While many polyphenols—catechin from barley "tannoid," lignin-associated methylated polyphenols, anthocyanogens (Chapter 16)—have been implicated, tracer and other new organic chemical separation and identification techniques have shown that the major polyphenolic components of haze are dimeric flavonoids in which two (+)-catechins are linked through their C atoms to form tannins, somewhat reminiscent of the dimeric oxidation products of tea flavonols (Chapters 16 and 22) (Eastmond and Gardner 1974; Gracey and Barker 1976).

Nonenzyme Removal.—Enzyme-induced removal or prevention of chill haze is but one of several options available to the brewer. In general these approaches involve removal of polyphenols/tannins and/or extraction or removal of proteins with or without the aid of commercial proteinases. Frequently both polyphenol and protein removers are used together (Kieninger 1977; Narziss and Bellmer 1976; McAdam *et al.* 1977) in what Flynn (1975) refers to as a "belt and braces [suspenders]" approach. The requisite operation is usually carried out during or after the filtration steps but may also take place at earlier stages. Kieninger (1977) pointed out that *complete* removal of either nitrogen (protein) or phenolic compounds is to be avoided because of resulting adverse affects on flavor and foaming characteristics.

The following substances have been used or recommended for polyphenol removal usually by precipitation in post-mash stages: tannin and tannic acid, polyvinyl pyrrolidone (PVP) largely replaced by PVPP (crosslinked PVP), diatomaceous earth and more recently nylon in various forms (Gracey and Barker 1976). Polyphenols may also be removed at the mash stage with formaldehyde or even earlier in the beer-making process by steeping of barley in alkaline solutions such as 0.1% KOH for a 2–4 hr period (Dadic *et al.* 1976).

Protein may be removed by nonenzymatic means in the post-mash stages, usually by adsorption to one or more of a large variety of adsorbents such as silicates, silica gel, hydrogel, kieselgur, stabifix, diatomite or bentonite. Alternatively they may be precipitated with finings such as "Magifloc" (Hough 1976; Gorinstein 1978). Some of these substances have been used in "immobilized" forms in association with filtration steps. This points up the statement of Flynn that the technologies associated with the use of these nonenzyme substitutes have penetrated existing and potential papain markets because they fit more easily into some existing brewing systems.

Eventually both enzymatic and nonenzymatic chillproofing may be rendered superfluous by the recent discovery of a barley mutant in which the polyphenolic haze moiety is absent (Von Wettstein *et al.* 1977).

Enzymatic Approaches.—Nevertheless use of papain has continued to expand ever since its introduction into the United States beer industry by Wallerstein (1911,1937). During the ensuing years patents continue to be issued on its use as a chillproofing agent, the more recent patents centering on stabilization of papain solutions with CO_2 reducing sugars such as $NaHSO_4$ and even sucrose (Stone and Saletan 1968; Ciba-Geigy 1977), and, of course, immobilization (see below). While the nonenzyme technologies offer severe competition, other nonproteolytic enzymes which have been proposed will probably not pose a serious threat, at least for awhile. These include the use of glucose oxidase (glucose provided by amylolysis) to exclude O_2 needed for haze formation—there are conflicting reports as to its effectiveness (Dadic and Van Gheluwe 1971)—and possibly tannase which is discussed in Chapter 30 effective in removal of a similar haze in tea.

On the basis of activity of the pure enzyme papain, which, as we discussed in Chapter 27, is but one of the proteinases present in papaya latex, it is added at a rate of about 0.2 ppm. One pound (0.45 kg) of 10 or 20% (w/v) crude papain, available to the brewer by enzyme suppliers as such, will chillproof 200 to 800 barrels of beer. The trade names of some of these preparations reflect the necessity for using them: Collopulin, Cristalase, Protesol and Scintillase. Indeed there is a continuing effort to devise assays which more accurately predict chillproofing efficacy. Knowing what enzymes are present through the future application of label immunoassay (Chapter 39) and active site titrant assay (Chapter 37) should clear up some of these problems.

The flurry of activity concerning both enzyme stability and assay has been stimulated by the adoption of proportioning equipment with which it is necessary to dilute solutions of papain beyond the previous standard 10%. With the advent of the use of millipore filters as an alternative to heat pasteurization for draft beer production, it is less likely that papain continues to act after bottling, suggesting that less enzyme may be needed to prevent chill haze in such beers. There has been only limited investigation on the action of papain on the chill-haze protein.

Immobilization.—An expanding effort has gone into the adaptation of papain as a chillproofing agent to an immobilized configuration, as has already been successfully done for nonenzymatic means of preventing chill-haze. Thus, Lieberman (1975) and Venkatasubramanian *et al.* (1975) employed collagen complexation as a means of immobilizing papain, the latter author using a spirally-wound reactor configuration for continuous operation; Basarova and Turkova (1977) immobilized papain onto hydroxyalkyl methacrylate (Spheron, Chapter 8); Monsan *et al.* (1978) employed treated glass beads, Spherosil; and Finley *et al.* (1979) chitin-glutaraldehyde (Fig. 8.4). All investigators report some degree of success, be it an observation on reduction of chill-induced haze in the beer, decrease in $(NH_4)_2SO_4$-precipitable N, prolonged enzyme stability or no deterioration in taste quality. Several potential advantages of immobilization are: flexibility with respect to stage in the brewing process (it should especially be useful for millipore filtered beer or after pasteurization of other beers), having some protein remain in the beer to have a satisfactory "head" of foam (the extent of protein breakdown may be more accurately controlled as mentioned); and as with all enzyme immobilization the enzyme should be able to be used repeatedly and continuously. Flynn (1975) reported a consensus of informed opinion including that of both enzyme suppliers and brewers that perfection and commercialization of immobilized papain for chillproofing could occur any time between 1980 and 1995.

Other Enzymes.—Because papain is fairly specific for haze proteins (Lieberman 1975) little effort has gone into finding other chillproofing proteases. The Kirin Brewing Company has developed what appears to be an effective chillproofing proteinase from a microorganism, a strain of *Serratia marcescens* (Takahashi *et al.* 1974). They observed that commercial proteases produced in Japan from other microorganisms such as *Trametes sanguinea* (Rapidase), Nagarse *(B. subtilis)* and Pronase *(Streptomyces)* could not serve as chillproof agents and that papain and the *Serratia* enzyme acted together synergistically, an effect which they attributed to the supplementation of proteolytic action by the *Serratia* enzyme of the "clotting" effect of the papain. This insight may explain why Stone (1973) found it necessary to add tannic acid along with "tannin-protein decomplexing" enzymes from *Aspergillus* spp. to achieve an effective chillproofing. However, a microbial rennet (Chapter 33) does seem to be effective (Nelson and Witt 1973).

A problem which appears to be a side effect of papain action, "winter gushing," may appear during the aging of bottled beer. It is suppressed by combining papain with microbial acid proteases (Amaha *et al.* 1978).

Use at the Mash Stages and Lipids.—While papain and microbial enzymes are added at mashing primarily to increase extract yield and to increase assimilable N for the yeast, it has been proposed to add papain as a chillproofing agent at the mashing stage, according to a series of private communications from several brewers to Flynn (1975). This illustrates the difficulty of assessing what is really going on from the literature alone in highly competitive food industries.

Since lipids affect the foaming properties of beer, a modicum of information is available on enzyme action on lipids in barley, malts and in beer (Morrison 1978).

32

ENZYME ACTION IN BREAD-MAKING AND OTHER TEXTURE-RELATED MODIFICATION OF CEREAL FOODS

We continue to pursue the theme of enzyme initiated and assisted transformations associated with textural and physical aspects of food identity and appearance by turning our attention to their participation in milling, baking and related processing of cereal grains, with special emphasis on bread-making. In so doing, we leave behind beverage-making and consider foods whose manufacture involves a series of textural transformations leading to food products which are usually solids. Their physical states range from moist-to-the-feel cakes to the hard crunch of crispy crackers. What they all have in common is a regime of operations which may be succinctly silhouetted as: cereal grain → milling → flour → water → kneading → dough → evolution of CO_2 → thermal treatment → food product. For bread-making and most, but not all, other cereal products the grain is wheat, rye or triticale; evolution of CO_2 arises via yeast fermentation and the thermal treatment is baking.

Bread and beer-making at first glance appear to be strikingly similar from the perspective of enzyme participation. In each case, the starch in the milled cereal is partially converted to sugar by amylase action; both amylase and proteinase may be supplemented or even replaced by external commercial microbial enzymes. Proteinase action improves the final product and also provides nitrogen for the yeast; fermentation by the yeast enzyme systems converts amylase-produced sugars into CO_2 and ethanol; endogenous starch and glycan gums which influence processing may be advantageously modified by added enzymes; these enzymes, while added primarily for texture and change-of-state associated purposes, also affect flavor, color and other aspects of appearance.

While participating in both bread and beer-making by acting on the same kinds of substrate and yielding similar products in the production of both bread and beer, these depolymerases are cast in decidedly different roles; their action in bread-making is muted, more a sculpturing with a fine chisel in comparison to the trip-hammer blasting of the structure of starch as effected by beer enzymes. Complete removal of starch and proteins is definitely not an end goal. Carbon dioxide rather than alcohol is the functional end-product of fermentation and unlike alcohol is used transitorily in bread-making and serves no function in the final processed food. Furthermore, the cast of enzymes is expanded in bread-making to include enzymes not normally considered as influential in beer-making *per se*: oxidoreductases, especially lipoxygenase, and perhaps lactase whose action on a milk constituent frequently used in baking constitutes a link to the subject matter of the next chapter.

These types of enzymes, as mentioned, can affect the textural quality of bread and other baked products. They can also influence color (Chapter 17), flavor (Chapter 24) and nutritional properties (Chapter 35).

STARCH-TRANSFORMING ENZYMES

Wheat α-Amylase

As with barley, further progress in rationalization in the baking industry hinges on ever-increasing insight into the nature of enzymes in the wheat kernel (Fig. 32.1). To this end, and for its own sake, considerable literature has evolved concerning the principal enzymatic participants; α-amylase has in particular been looked into thoroughly and active work is still going on.

Not surprisingly, the α-amylase picture in wheat closely parallels that of barley (Chapter 31). The enzyme present at anthesis in the newly formed wheat kernel, which is as active as a diastatic malt (Schwimmer 1947B), is localized within the pericarp (Banks *et al.* 1972; Kruger 1972A). Distribution within the mature wheat kernel of this and other enzymes of relevance to bread-making is shown in Table 31.1. Activity per kernel remains roughly constant during development of the kernel but tends to decrease during maturation.

From a consideration of changing patterns of immunoelectrophoretic analysis of proteinase-treated extracts, Olered and Jonsson (1970) suggested

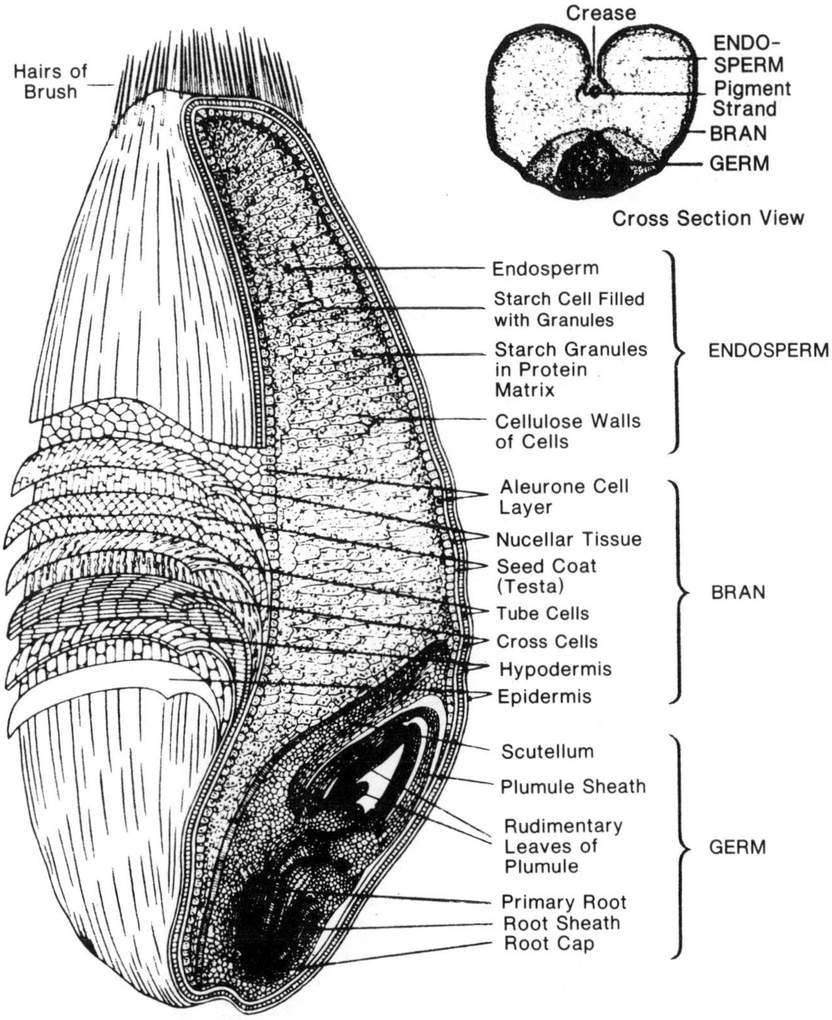

FIG. 32.1. THE WHEAT KERNEL

TABLE 32.1

DISTRIBUTION OF BREAD QUALITY-DETERMINING ENZYMES IN A GRAIN OF WHEAT

Activity Measured	Pericarp + Envelope	Aleurone	Embryo	Scutellum	Endosperm
	Percentage of Total Measured Activity in:				
α-Amylase	0	64	0	25	11
β-Amylase	0	11	0	23	67
Protease	0	15	31	48	6
Phytase	4	29	15	50	2
Lipase	0	61	35	—	4
Lipoxygenase	—	—	51	49	—

Source: Adapted from data of Drapron (1971).

that α-amylase is being continuously but reversibly inactivated during early stages of grain development and that the apparent loss in activity is highly dependent upon the moisture level as the kernel approaches maturity. Unlike β-amylase, it is not present in an inactive insoluble complex (Daussant and Corvazier 1970).

Frequently, it may be necessary to harvest wheat prematurely to avoid heavy rains, as in the northwest United States and in parts of Australia (Moos *et al.* 1972), to prevent germination. Treating such wheat with ethylene may hasten maturation (Hale *et al.* 1943). The effect on the α-amylase activity should be of interest since baking properties are considerably influenced by both maturity of the kernel and by ethylene treatment. Indeed, Olered (1967) showed that the α-amylase activity of high moisture kernels, as measured by the bread-making relevant "falling number" assay (Pyler 1973) was significantly greater than that of low moisture kernels. Wheat α-amylase proves no exception to the observation that enzyme level is frequently dependent upon variety (Kruger 1972A).

The latter Canadian investigator also established the existence of three isozymes of α-amylase in immature kernels. It is also owing to the work of Canadian investigators (Marchylo *et al.* 1976) that the α-amylases from immature wheat present at low levels in comparison with germinated wheat were first painstakingly purified and characterized. A summary of their properties in comparison with those from germinated and immature barley (Chapter 31) is shown in Table 32.2. Perhaps the most striking difference is the heat lability of the isozymes taken collectively. Even in the presence of stabilizing Ca^{2+}, its heat stability after purification is comparable to that of the heat-unstable β-amylase. Before purification its activity proved to be heat stable enough to permit the use of a heating step (Chapter 6) in its isolation. Another difference brought out in Table 32.2 is the large discrepancy in molecular weight between that measured by molecular sieve chromatography which is dependent upon molecular shape and that by detergent electrophoresis which is not. This difference is attributed to a highly compact structure which is destroyed in the detergent (sodium dodecyl sulfonate, SDS) in the latter measurement.

That there may be yet additional isoamylases in mature wheat is hinted at by the cogent immunoelectrophoretic observation of Daussant (1977) of a unique α-amylase antigen present in developing (immature) kernels not present in germinating seeds. It is apparently different from Kruger's purified isozyme; the latter may be the same as the most anodic antigen "(I)" of Daussant since both Daussant and Kruger found groups of isozymes in both germinated and immature wheat kernels. It is most likely from immunochemical and other considerations summarized by Marchylo *et al.* (1976) that these isozymes arise from slight variations in the genes coding for their syntheses. For instance,

TABLE 32.2

COMPARISON OF PROPERTIES OF α-AMYLASES FROM MATURE AND GERMINATED CEREAL GRAINS[1]

Properties	Mature Barley[2]	Mature Wheat[1]	Germinated Wheat[3]
Number of isozymes isolated	1	3	4
pH optimum range	5.5	3.6–5.8	5.5–5.7
Molecular weight range ($\times 10^{-3}$)	4.6	52–54	41.5–42.5
Activation energy (kcal/mol)	9.5	9.0–9.7	7.0–9.3
Isoelectric point (pI)	4.83	4.65–5.11	6.05–6.20
Heat stability (% remaining)	—	50% at 40°C	10% at 60°C

[1] Source: Adapted in part from Marchylo *et al.* (1976).
[2] Source: MacGregor *et al.* (1974); see Chapter 31.
[3] Source: Tkachuk and Kruger (1974).

three isozymes can be detected in all hexaploid wheats so far looked at whereas only two can be detected in tetraploid grains.

Germinated Wheat.—Upon germination the specific activity of α-amylase increases 100 to 1000-fold. Germination can occur at both preharvest and postharvest; in the latter case usually deliberately rather than adventitiously, in the form of wheat malt (Finney *et al.* 1971). The resolution and purification of consistent isozymes, of which there appear to be at least five and more probably eight, has stirred considerable interest and some controversy. Again much of this information comes from the laboratory of the above-mentioned Canadian investigator and from Daussant in France. As previously mentioned, both the Canadian and French investigators found some isozymes which are unique to the germinated grain and others which are present in ungerminated wheat as well (Kruger 1972B; Daussant and Renard 1972; Daussant 1977). An enzyme preparation which was resolved into similar isozyme components with similar if not identical properties as that from germinated hard wheat has been isolated from a soft wheat, durum, used in making pasta (Warchalewski and Tkachuk 1978). This finding has a genetic implication in that it suggests 42-chromosome hexaploid hard wheat (or at least the hard red spring cultivar, *Triticum aestivum*) is derived from durum, *Triticum durum*, a 28-chromosome tetraploid.

As with barley, gibberellic acid, GA (Chapter 31), induces *de novo* synthesis and/or release and secretion of α-amylase in and from the aleurone cells of wheat seeds. It is used in the baking industry not so much as a substitute for malting of wheat but rather as a stimulant, in conjunction with malting, of α-amylase biosynthesis/release. Thus, Jeffers and Rubenthaler (1974) report an approximate doubling of α-amylase activity when 5 ppm GA was added to the malting steep water. GA may also be useful for screening wheat cultivars for resistance to preharvest sprouting (see below); the greater the resistance, the less the stimulating effect of GA on α-amylase activity (Derera *et al.* 1977).

Among the relatively abundant wheat constituents which effectively inhibit α-amylase is phytic acid, apparently not due, as one might anticipate, to Ca-complexing (Sharma *et al.* 1978). Endosperm albumin protein fraction contains one group of wheat constituents which do not inhibit the α-amylase of wheat but do inhibit those of mammalian tissues (Pace *et al.* 1978) in analogy to the more well-known mammalian proteinase inhibitors from plants (Chapter 34). Significant anti-amylase activity may survive the baking process and interfere with normal digestion according to Silano (1977).

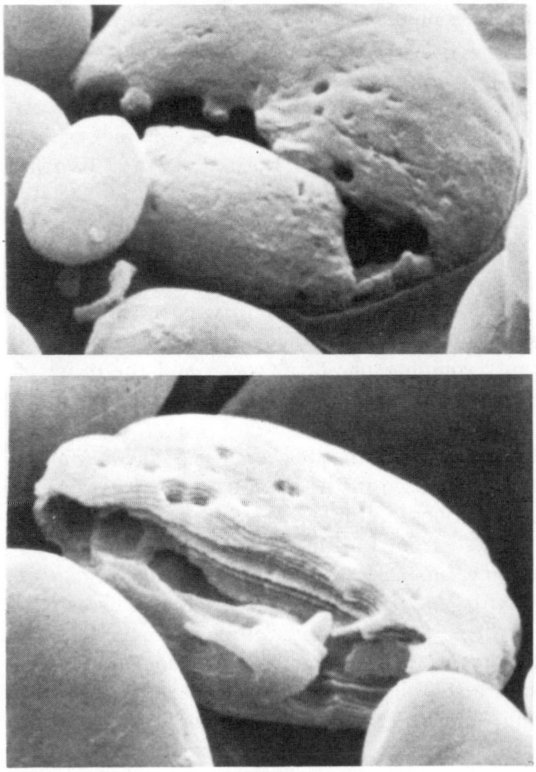

Courtesy of Bean et al. (1974)

FIG. 32.2. TWO VIEWS OF WHEAT STARCH GRANULES AFTER INCUBATION WITH MALTED WHEAT FLOUR
See text for explanation.

Finally, we should like to cite studies of α-amylase in other cereal seeds used in the baking industry: oats (Smith and Bennett 1974), rice (Tanaka *et al.* 1970), rye (Manners and Marshall (1971); Möttönen 1975) and of more than passing interest, the α-amylase of that manmade hybrid taxon derived from a cross between wheat and rye, triticale. The α-amylase in the developing kernel has been given special attention because of: (1) the high levels of activity in most lines of this hybrid as compared with rye or wheat (Table 32.3) which may necessitate inhibition of activity to improve resulting bread quality (see below), and (2) a more serious problem: that of shrivelling of the kernels of some lines at maturity. This shrivelling or shrinking is associated with and is probably caused by resurgence of α-amylase action as the result of a precocious germination. This phenomenon, resulting in an increased level of a complex of isozymes similar to those of the parent rye and wheat, is well documented in key papers by Lorenz (1972), Lorenz and Welsh (1976) and Silvanovich and Hill (1978), some of whose data are shown in Table 32.3. Silvanovich and Hill were able to dispense with nonaffinity steps in purifying the isoamylase system of triticale by the successive use of two affinity columns, one of glycogen and the other of a cycloamylose linked to a Sepharose derivative (Vretblad 1974).

As mentioned in Chapter 5, microbial α-amylases, also used in the baking industry as specific preparations suitable for food use, are usually obtained from the fungus *Aspergillus oryzae* for routine baking and from a bacterium, *Bacillus subtilis*, for the baking of specialty items as outlined below. A special precaution taken early in the history of their commercial development was to ensure the absence of proteolytic activity (Miller and Johnson 1954). They undoubtedly contain other starch transforming enzymes such as glucoamylase and debranching enzymes (Chapters 9 and 31).

β-Amylase

Not surprisingly, the enzymology of wheat β-amylase resembles closely that of the barley enzyme (Chapter 32). As with barley, at anthesis most of the starch hydrolyzing activity is due to α-amylase. Total β-amylase increases in the aleurone cells during development of the grain and reaches a maximum at the "milky" stage; it increases about five times faster than the kernel weight (Schwimmer 1947B). Concurrent with loss of water upon maturation, part of the amylase (as with barley) becomes "latent"—insoluble and inactive. Part of the activity can be extracted and restored with NaCl, the rest with papain. The salt effect may be related to the sensitivity of the enzyme molecule to changes in tertiary structure which can be observed with a fluorescent probe at low ($10^{-3}M$) concentrations of N-containing salts (Anisimov *et al.* 1977). It has been suggested that the enzyme's sulfhydryl groups are involved in the insolubilization (Roswell 1962).

β-Amylase deliberately adsorbed onto the glutenin fraction of wheat protein is not removed by salt or protease action (Rothfus and Kennel 1970). Apparently, the latent form can also exist in solution. Shinke and Nishira (1975) obtained such a form by extracting flour with urea to obtain what they termed active- and zymogen-β-amylases, the latter being at least a protomer (Chapter 4), activated by protease, reminiscent of the Z protein of barley (Chapter 31). Indeed, the same authors compared barley with wheat and concluded that the β-amylase distribution was the same in both cereals.

Wheat β-amylase was first purified to the point of crystallizability in the pre-isozyme era by Meyer

TABLE 32.3

α-AMYLASE IN DEVELOPING TRITICALE AND PARENTS

	4	3	2	1	0
Weeks to Maturity	Enzyme Activity Relative to Mature Wheat				
U/Dry Wt: Wheat	8.0	4.5	1.2	1.5	*1.0*
Rye	12.0	3.7	3.9	2.7	2.3
Triticale 6-TA-204	14.0	8.9	7.4	4.7	5.1
Triticale 6-TA-206	11.1	7.3	5.9	6.7	17.4
U/Protein[2]: Wheat	—	3.9	1.7	1.5	*1.0*
Rye	—	3.7	2.0	1.9	1.2
Trictical M-1019	—	3.8	2.2	2.2	1.7
Triticale-1	—	4.4	3.5	4.0	4.6
U/Kernel[2]: Wheat	—	1.6	1.5	1.4	*1.0*
Rye	—	1.9	1.5	1.5	0.9
Trictical M-1019	—	2.7	2.5	1.7	1.4
Triticale-1	—	2.0	3.5	3.5	3.6

[1] Source: Adapted from data of Lorenz and Welsh (1976); maturity (0 weeks) taken for wheat at harvest of spring cereal, Aug. 6; GTA-264 shriveled less than 6-TA20, *ca* 3 weeks before maturity; U = units of enzyme activity.
[2] Source: Adapted from data of Rao *et al.* (1976); maturity taken at harvest 31 days after anthesis; M-1019 shriveled less than M-1.

et al. (1953). Further purification studies were performed by Waldsmidt-Leitz and Dorfmann (1968) and by the Canadian investigators, Tkachuk and Tipples (1966). As with the α-amylases the latter detected isozymes which appeared and disappeared during development and germination. As with barley, upon germination of the mature wheat kernel, the latent amylase almost completely disappears but the *total* activity remains fairly constant (Kneen and Hads 1945; Schwimmer 1951B). Modifications which probably do occur (Daussant and Corvasier 1970; Kruger 1972B; Shinke and Nishira 1975) are not due to the synthesis of new enzyme molecules but come about as the result of the activation of (release) and proteolytic modification of a preexisting precursor. Thus, Daussant and Corvazier could detect no incorporation of radioactive leucine or glycine, the most abundant amino acids in the enzyme. Most investigators find fewer isozymes of β-amylase than of α-amylase, ranging from one to three or four at the most (Daussant and Abbot 1969; Tkachuk and Tipples 1966; Niku-Paavola *et al.* 1972). On the other hand, Romanian investigators first using electrophoresis and 5 years later immunoelectrophoresis (Alexandrescu *et al.* 1975) adhered to their first conclusion that β-amylase but not α-amylase occurs in wheat (as well as in rye and triticale) as a single molecular form. This enzyme has been subject to a benign neglect in the 1970s not only because it is always present at adequate levels for bread-making but also, perhaps, because of the burgeoning literature on microbial β-amylase, especially from *Bacillus cereus* touched upon in Chapters 9 and 31. The β-amylase of milled rice (Lorenz and Saunders 1978) has also been studied in relation to saké manufacture. Some information is available on triticale β-amylase including that of Lee and Unrau (1970) and in pertinent previously cited references.

Functions in Baking

The utility of amylase action in the baking of bread and other cereal products arises principally, but not entirely, from the requirement for a paced and measured evolution of CO_2 whose permeation of the dough gives bread, cakes and other leavened baked goods their characteristic spongy texture. The rate of CO_2 production has to be geared to the stretchability and gas-retaining properties of the gluten in the dough. This carbon dioxide can be added to the dough as carbonates in baking powder for baking of most sweet baked goods such as cakes, or can be evolved from the metabolism of sugars by a special microorganism, usually bakers' yeast, although bacteria are also involved in the fermentation of sugars in specialty items such as sourdough French breads and in bread baked by "natural" food adherents. The requisite fermentable sugar such as sucrose, glucose, fructose, maltose and lower maltodextrins may be added as such in "straight" dough processes in which all the components—flour, water, yeast, sugar, salt, fat, nonfat milk solid, soy flour (Chapter 34), and protease and other improvers (Pyler 1973)—are added simultaneously so that the sugar present added as sweetener can also be used as a source of the CO_2. Since skim milk solids are frequently used in these formulations, Pomeranz (1964) suggested that the unfermentable lactose in such doughs be converted by the enzyme lactase into fermentable glucose and galactose. It is in the "sponge" methods, in which the yeast produces CO_2 in the absence of any other ingredients other than water, flour and yeast food (mostly inorganic sources of mineral and nitrogen with starch as a filler amounting to less than 1% of total starch in the flour), that amylase action is essential for the production of fermentable sugars.

The source of substrate for this action is, primarily, a relatively minor fraction of the total starch, less than 4% of the broken starch granules for soft wheat flours (8–9% for hard wheat), damaged either deliberately during milling (Jones 1940; Anker and Geddes 1944; Farrand 1972) or in the field before and during harvesting, and during storage and handling. As previously noted, the starch in intact starch granules is less readily available to attack by amylases (Chapter 9). Figure 32.2 shows intact wheat starch granules after incubation in a malted wheat extract. The damaged portions of broken kernels have been digested away by the extract's amylases, but the intact kernels are not extensively attacked. It is not surprising, therefore, that a great deal of attention has been directed to predicting how much substrate is available as the result of action of the α- and β-amylases and the susceptibility of starch in the flour to attack. For a recent key to literature on the susceptibility of starch granules from various flours used in baking, see Fuwa *et al.* (1977). In addition to official methods approved by the AACC (1975) examples of activity and discussions in this area of investigation include papers by Williams and LaSeeleur (1970), Olered (1977), Pratt (1971) and Mathewson and Pomeranz (1978). For example, the latter investigators, in a study of hot paste viscosity and α-amylase susceptibility of hard red winter wheat flour, concluded that activity as measured by a standard colorimetric method did not reliably predict amylograph viscosity (an established prognosticator of bread-making quality) of the flours from various locations and environments.

As far as β-amylase action is concerned, it is generally agreed that while it undoubtedly participates along with α-amylase in providing fermentable sugars, there is an adequately high level of this enzyme in any given flour to allow considerable variability in activity without any perceptible effect on dough properties or bread quality. Thus, Ulig and Sprössler (1972) determined that β-amylase supplementation of flour does not increase fermentation. However, this is not true of α-amylase supplementation of flour made from most modern wheat.

An optimum level of α-amylase will produce along with the β-amylase present ("diastase") the equivalent of about 3 g of maltose per 100 g of flour corresponding to a CO_2 output (gassing power) producing a pressure of 500–600 mm Hg after 5 hr (Reed 1963). The joint action of the amylases is most effective during later stages of the fermentation but continues well into the baking phase. Audidier (1968) pointed out that the influence of baking time and the kinetics of temperature increase during baking are of primary importance in determining the role of amylolysis because the slower the rise not only is amylase action more prolonged but the susceptibility of the substrate—the starch in the swelling granules—to amylase action is also enhanced.

A combination of optimum starch granule damage with optimum α-amylase action contributes to the enhancement of several quality-related factors in bread-making. With respect to the dough, it supplies, as mentioned, sugar for fermentation and yields adequate gas; the rate of fermentation is accelerated. Faster dough mixing due to decreased viscosity, improved dough consistency (less "buckiness") due to release of water previously bound to amylase-susceptible starch also results in what is referred to as improved machinability of the dough.

With respect to the finished product, bread, optimum amylase action results in: optimum loaf volume, compressibility and fragility of crumb; increased smoothness and moistness in part due to α-amylase-produced α-dextrins and decreased staling by relocation of water and by preventing retrogradation of amylose. In addition, improved flavor and color due to nonenzymatic browning type reactions (Chapter 19) also occur.

α-Amylase Supplementation

For thousands of years, slow, inefficient growing and harvesting of wheat by hand and storage conditions encouraged incipient germination of the grains. This usually ensured an adequate if not uniformly predictable and optimum level of α-amylase. Starting about 100 years ago in the United States and later in other countries, a gradual deterioration in bread quality occurred. Mechanical harvesting and elevator storage of wheat resulted in mature undamaged and unsprouted kernels. By the early 1930s it was realized that this was due to the vanishing low level of α-amylase. As previously outlined, there is simply not enough α-amylase in such mature, sound wheat to produce enough α-dextrins for the abundant β-amylase to convert them to fermentable maltose.

Hence, for about the last 40 years, it has been routine practice to supplement wheat flour, usually at the mill, with commercial preparations of α-amylase from wheat or barley malt or from fungi, usually *Aspergillus oryzae*. By 1963, about 30 million kg of such malt was used in the United States for supplementation of wheat flour (Reed 1968). Especially useful is malt produced from triticale but not that from oats.

At the present time both malts and fungal enzyme preparations are used. A conspicuous difference between them is that fungal α-amylases are less stable so that they are essentially inoperative in the temperature range of 60°–70°C when starch in the dough swells and thus becomes more susceptible to the action of added malt and bacterial α-amylases which are more heat stable. Nevertheless, Miller *et al.* (1953) found that compressibility of the bread crumb was improved to the same extent by malt and fungal enzymes whereas bacterial α-amylase yielded overly soft crumbs. Whether malt or microbial enzyme preparations are used, Bracht (1973) found that conventional, strictly *"in vitro,"* oriented enzyme assays such as the traditional SKB method (Sanstedt *et al.* 1939) did not provide useful information on the effectiveness of these preparations whereas the baking designed tests—falling number and amylograms—accurately predicted their utility. Early attempts to supplement fungal with low levels of bacterial

α-amylase to retard staling apparently have not succeeded, probably because of the difficulty of adding just the right amount without over-softening occurring (Zobel 1973). The following random selection of commercial fungal α-amylases available to the miller and baker have been recommended for: improving the mixing and fermentation characteristics of dough (GB Fermentation Industries); improving gassing power during baking and retarding staling upon supplementation at mill or bakery (Takamyl Series); with reduced protease activity, Marschall (Miles); increasing the formation of fermentable sugars (Fungamyl 1600 S and 180, Novo); an enzyme concentrate, today's consistent replacement of yesterday's erratic malt (Doh-Tone, Pennwalt); baking, available with high and reduced protease activity (Asperzyme, Enzyme Development Corporation); modification of starch content of flour for improved machinability and fermentation of dough, better grain, texture, crumb structure and crust color of bread, and allows (via viscosity reduction) processing of precooked cereals at higher solids level with less water and time for evaporation (Rohzymes S and 33, Rohm and Haas); improving quality and uniformity of bread and other yeast-raised bakery products by stimulating fermentation, aiding in gluten development and improving machinability and makeup characteristics of the dough (Fermex); available with and without reduced protease activity (Wallerstein). This list of fungal enzyme preparations is probably not complete and in no way reflects on the suitability of their use as compared to others not listed. It does give an inkling of advantages in their almost universal use. As mentioned, precautions are taken to remove most of the protease which may be added later at activity levels in consonance with their requirement, as discussed below. Contributing prominently to adoption of the fungal enzymes were the pioneering investigations of Johnson and Miller, who in a review of their work (Johnson and Miller 1951) showed the superiority of fungal α-amylase for most bread baking purposes over that of cereal malts and the overly heat-stable bacterial α-amylase to be used in conjunction with glucoamylase. The latter probably accompanies most commercial fungal α-amylase preparations.

Nonsupplementation.—α-Amylase supplementation is not universally recommended. Under certain circumstances and for certain breads such as rye, reliable baking results may be obtained with α-amylase-poor flours by providing, for instance, sufficient quantity of what Drews (1972) termed swelling agents—gelatinized starch, guar flour, carob bean and some gums. Another alternative to supplementation of α-amylase-deficient flours *via* added enzyme may be to thermally potentiate endogenous enzymes by heating flour suspensions at 40°–50°C for 10–60 min (Popoditch *et al.* 1975).

Uses of Bacterial α-Amylase.—While not successful for use in bread-making because of their heat stability, commercial bacterial α-amylases are probably used in baked and otherwise cooked cereal products where extensive action on swollen starch in swollen granules is desirable to improve moistness and softness, such as in fruitcakes (Stone 1962), pie filling (Klis 1962) and baked confectionery mixtures (Baxter 1966). Tawfick and Atia (1972) hydrolyzed high amylopectin waxy maize starch with bacterial α-amylase for use as a formulation ingredient where nonretrograding starch is required. Whether such applications of bacterial α-amylase are in actual use is problematical. Available literature from enzyme suppliers specifically suggests: improving acceptability of a high-gluten-flour bean blend (Chapter 34) used as a baby food in developing countries (Bacterial Amylase, Novo); production of chemically leavened crackers of superior quality when blended with bacterial protease (Enzyme 4511-3, GB); and in delaying the onset of staling and moisture loss in coffee cakes, brownies and fruitcakes (Fresh-N, Wallerstein). Other enzyme producers also market bacterial α-amylase but stress its nonbaking applications.

Excessive α–Amylase Action

While an optimum degree of α-amylase action during the production of baked goods is needed to obtain maximum dough and baked goods' texture-related quality, the deterioration in functional quality which occurs when too much α-amylase action occurs is reflected in increases in several changes: the crumb is too fragile, gummy and sticky; loaf volume is reduced below optimal values; and excess gassing results in a soft dough and uneven texture or even cavitation of the loaf. Excess action may occur via over-supplementation with added commercial enzyme (more common in the United States) or via oversprouting, more likely to be the cause in other countries especially with baked products made from rye flour. However, Mecham (1975) pointed out that equating sprout damage solely with increase in α-amylase is a gross simplification. More recently, this problem has arisen again with United States-grown grains in connection with the introduction of the rye-wheat hybrid triticale, as discussed below, and also in connection with the expanding exportation of grain.

Problems caused in conjunction with the latter are illustrated by the export of western white wheat to Japan for the manufacture of noodles (Bean *et al.* 1974). While, because of high rainfall in the Pacific northwest, such wheat might be more likely to exhibit sprout damage and excess α-amylase activity, in this case apparently sound wheat caused problems in noodle-making which the California investigators traced to excess (for noodle-making) α-amylase action, manifested by sticky dough strings which stretch and break while being dried into noodles (Fig. 32.3).

The extreme sensitivity of noodles to excess α-amylase was illustrated by showing that noodles which yielded acceptable noodles were damaged by supplementation with wheat malts corresponding to levels below those required for adequate α-amylase for bread-making purposes. It was concluded, however, that the principal deleterious effect arose from components other than α-amylases.

Modulation.—As with other instances of unwanted enzyme action, means have been sought to suppress α-amylase. In this case, however, only a careful reduction of activity is required since continued α-amylase action is essential. Of course, an alternative is to remove all activity and then add measured quantities of commercial preparations at levels commensurate with maximum improvement of product. Proposals for modulating α-amylase activity include a comparatively rare instance of using pH extremes specifically for the purpose of inactivating an enzyme. Other instances constitute typical illustrations of the principles discussed in Chapters 1 and 14.

Older suggestions that rye flour be stored (aged) for several years and milling to low extraction value to exclude the aleurone (Table 32.1) are not considered practicable.

Bake small loaves; heat penetrates faster, causes faster α-amylase inactivation (Drapron 1971).

Adjust the pH of the dough to 4.2, add salt (Huber 1960) or the pH to 2.5 without salt (Meredith 1970); related is the use of HCl gas followed by NH_3 (Fuller *et al.* 1970).

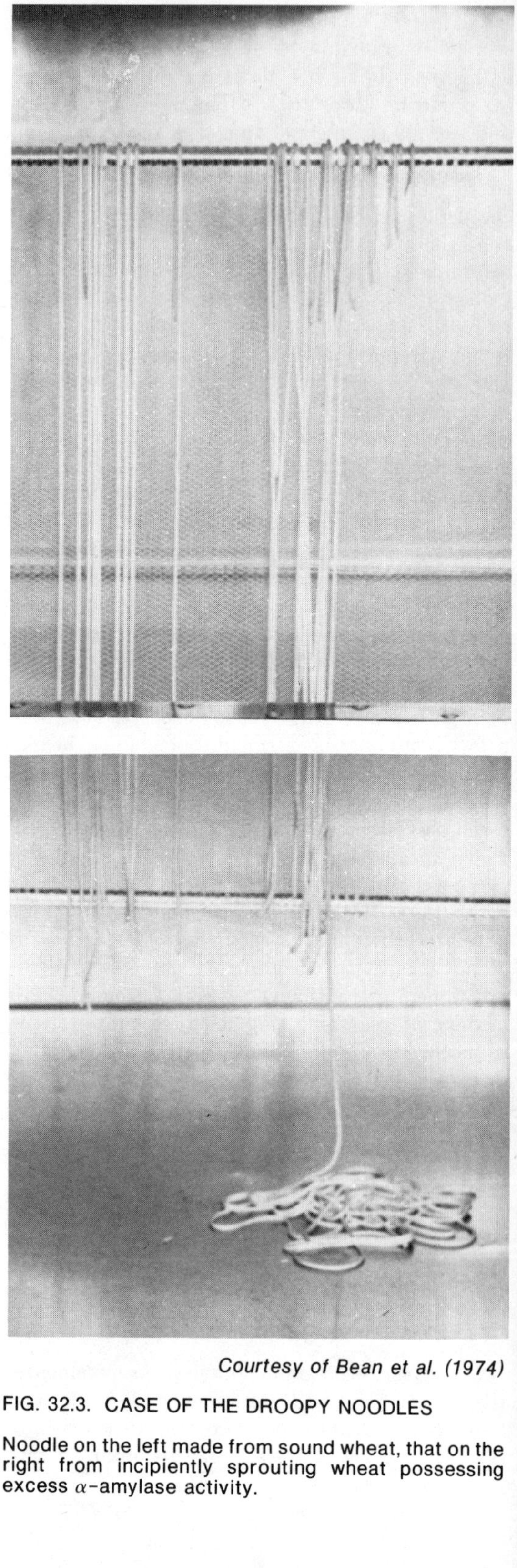

Courtesy of Bean et al. (1974)

FIG. 32.3. CASE OF THE DROOPY NOODLES

Noodle on the left made from sound wheat, that on the right from incipiently sprouting wheat possessing excess α-amylase activity.

Microwave heating (Eggebrecht 1972); this treatment also weakens the gluten. In HTST steam treatment of grain, heat penetrates only as far as the α-amylase-containing aleurone layers (Strom and Qvist 1963).

Add some general enzyme inhibitors such as that from Leoti sorghum (Wall and Blessin 1970; Lorenz 1972), phospholipids (Chapter 14) and phytic acid (Sharma *et al.* 1978).

Judiciously use sequestering agents for Ca, essential for α-amylase stability (Schwimmer 1951B). Westermarck-Rosendahl *et al.* (1979) selected the following as the most promising α-amylase inhibitors for baking purposes; Na di-, tri- and polyphosphates, Na dodecyl sulfate, Ca stearoyl lactate and citric acid. While most of these proposals were designed to inactivate flour α-amylase in conjunction with improvement of bread-making, also included are other suggestions based upon studies of amylase inhibition not so designed. However, any proposal, to be viable, is dependent on a rapid, accurate and convenient α-amylase assay such as the nephelometric procedure developed by Prasad *et al.* (1979). These and other α-amylase suppressants were evaluated with respect to their effect on dough and baking properties by Westermarck-Rosendahl *et al.* (1980).

The Pentosans, Pentosanase and the Glycanases

It is likely that significant levels of nonamylase glucanases are present in the enzyme supplements, from both seed malts and microbial sources, including glucoamylase, α-1,6-glucanase and cellulase and hemicellulase activity (Manners and Yellowlees 1973). It is conceivable that the action of these enzymes could influence bread-making by providing more fermentable sugars and altering the texture of the dough and final product. Indeed, addition of cellulase to improve bread quality has been patented (Tanabe Seiyaku 1968). However, the focus of attention on nonamylase enzyme action on carbohydrates in bread-making and other related cereal processing has been directed to the so-called pentosans, studies of which have been summarized and reviewed by Kulp (1968), D'Appolonia (1971) and more recently by Neukom (1976).

In analogy to barley gum (Chapter 31) the old term "flour gum" was applied to the soluble hemicellulose fraction of wheat endosperm. However, here the analogy ends. Rather than hexoses, the principal sugars of this gum, constituting some 3.6% of the wheat flour, are pentoses in a chain, with xylose constituting the backbone and arabinose making up stubby side chains (Chapter 9).

Unlike gums from barley or oats (Chapters 9 and 31) they are, even after isolation, covalently linked to proteins. Even more singularly, *ca* 10% of the ends of the pentosan chains are covalently linked to the monomethylated C_6:C_3-*o*-diphenol ferulic acid (Chapter 16). It has definitely been established that these chains are major cell wall constituents, and are probably insoluble in the intact wheat kernel and perhaps also in the wheat flour.

However, upon extraction with the mildest of solvents, water, part of this pentosan fraction either goes into solution or becomes soluble as the result of enzyme action during extraction. The insoluble part comprises "tailings" which settle on top of starch sediment in the separating of gluten or starch. It has been suggested that, being a major cell wall constituent of both the endosperm and the pericarp, some of the pentosan fraction may be rendered soluble, in analogy to the protopectin → pectin conversion (Chapters 29 and 30), via the action of hemicellulases and may even be transformed into polysaccharides with altered structure due to the action of endogenous transglycosylases and noncarbohydrases. Whatever happens enzymatically, the resulting soluble pentosans take up quite a bit of water, about 44 times their weight (Kim and D'Appolonia 1977). Most investigators are of the opinion that such a substance should affect dough and loaf properties although this subject apparently has not been settled. Thus, Neukom (1976) concluded that despite numerous studies on the effect of pentosans on dough and bread-making properties, the exact effect remains largely unresolved. One of the complications has to do with the interaction of this complex glucan-protein-polyphenol system with the oxidizing protein-modifying factors in the even more complex system represented by the dough. It is generally agreed that adding soluble pentosans exerts effects on the dough if not on the bread. For example, Kim and D'Appolonia reported that loaf volume was not affected by water-soluble pentosans. On the other hand, Patil *et al.* (1976) noted a pronounced improvement in bread volume when such fractions were added to dough lacking the water soluble fraction of flour and an adverse effect if bromate was also present (see discussion on oxidizing enzymes).

Carbohydrase relevance of the pentosans has been associated with the insoluble tailings, over one-third of which is protein. α-Amylase is used to clean up this fraction as a step in its isolation. Some investigators claimed that addition of these tailings to dough leads to coarser bread texture and dramatic decreases in loaf volume, one of the prime criteria for overall bread quality. Thus, Kim and D'Appolonia also reported a slight decrease in loaf volume upon addition of these tailings to flour, but point out that such a deleterious effect is not always observed. A redeeming feature of the presence of this pentosan fraction is its potent antistaling properties (due to its ability to slow down the rate of retrogradation of both amylose and amylopectin), much more pronounced than that of the soluble pentosan fraction. Part of the inconsistency may be in the action of the endogenous enzymes, as mentioned, which may change the structure and solubility of this fraction.

The other glycanase aspect of this problem has to do with results of experiments such as that of Kulp (1968) which suggested that removal via enzyme hydrolysis of the pentosans in this fraction could result in improved bread quality. Such use of "pentosanases" has been patented (Cooke and Johannson 1970). Indeed, even earlier in 1965, Rohm and Haas offered "gumase" among other purposes to utilize the pentosan-rich fraction of flour tailings and to increase cookie spread. After Kulp's investigation they suggested using such preparations (now called pentosanase-hexosanase) for modification of pentosan-rich tailings to improve baking characteristics as well as for improvement of cookie spread without alteration of shortening content. Until the effect is better understood, it is more likely that such enzyme preparations can be reliably used for increasing the yield of wheat starch in the production of this important food ingredient (Simpson 1955) and for more efficient degermination of corn, perhaps due to the action of the "hexosanases."

Other enzyme action which may affect the properties of this interesting proteoglycan fraction are the oxidases, as discussed below, and the proteolytic enzymes, whose action and effect on texture- and appearance-associated properties of manufactured cereal products will be explored later in this chapter.

PROTEASE ACTION AND BAKED GOODS QUALITY

Wheat Proteases

In contrast to the amylases of wheat, purification to the point of crystallization has not been achieved. Part of the reason for this is that in contrast to the amylases, which form two well characterized groups of isozymes, the α- and β-amylases whose specificities are sharply delineated and can be easily separated, knowledge of how many distinct groups of proteases exist has taken longer to unravel. For instance, an older controversy as to whether wheat contains a papain-like sulfhydryl enzyme (Balls and Hale 1938; Jørgensen 1939,1945), appears to be echoed in more recent publications. Although Skupin and Warchalewski (1971) concluded that they had isolated such isozymes from mature wheat, Madl and Tsen (1974) considered the evidence to be inconclusive. Another problem in unwrapping the enigma of the wheat protease complex is the choice of proper substrate. Gluten, the natural substrate, is not amenable to incorporation into assays because of its limited solubility. Hemoglobin has been used routinely. Some clarification might be achieved by follow-up of a rare attempt, published as a note, to use affinity chromatography (Chapter 7) to purify the enzyme using a hemoglobin-Sepharose column (Chua and Bushuk 1969).

Another part of the problem is the rather indiscriminate distribution of protease activity in the grain, also bespeaking several distinct groups. Quite a few studies have been made in this field. To cite an older investigation, Balls and Hale (1936) report proteolytic activity in descending order: germ > bran > whole wheat flour. In a summary of previous information, Drapron (1971) indicated that the proteases are present, with the exception of the outermost layers, throughout the kernel with a preponderance in the scutellum and embryo (see Table 32.1).

As with the β-amylases, the proteinases are not completely extractable with water, solubilization being dependent upon pH and the degree of association with gluten (McDonald and Chen 1964; Bernardin *et al.* 1965; Grant and Wang 1972; Madl and Tsen 1974).

In addition to papain-like enzymes, other endopeptidases detected in mature wheat grains have been categorized as acid, neutral and trypsin-like, and also as endopeptidases and exopeptidases (carboxypeptidases). The latter have been shown to be localized principally in the endosperm at maturity (Kruger and Preston 1977). That there are some heat acid-stable components is attested to by the observation of Wang and Grant (1969) that some activity survived prolonged exposure to 50°C and of Olcott (1950) that isolated gluten even after extended heat treatment undergoes slow degradation. As with other enzymes, their activity levels change during development, maturation and upon subsequent germination. Keys to the relevant literature include papers of Bushuk *et al.* (1971) and Kruger and Preston (1977). In general, the overall proteolytic activity of the whole wheat kernel, both hard and soft, declines steadily during development and maturation to a small but significant level of activity, estimated to be equivalent to a specific activity, about 0.01% of that of crystalline chymotrypsin (McDonald 1969). Its subcellular localization has been examined even less than the corresponding enzymes of barley (Chapter 31).

Upon germination the total protease activity increases considerably. Over 40 years ago Mounfield (1938) reported a ten-fold increase but wheat malt is about 50 times as active as patent flour. A three-fold increase in carboxypeptidase upon germination could not be accounted for by *de novo* synthesis stimulated by gibberellic acid (Preston and Kruger 1976, 1977). The same three isozyme components present in the mature kernel are present in the malted wheat and the increase is likely to be due to removal of inhibitor. Belitz and Lynen (1974) separated the proteolytic activity of wheat malt into two distinct groups: a relatively low molecular weight group acting optimally near pH 5, close to that of Kruger and Preston's carboxypeptidase, and a high molecular weight group of trypsin-like isoendoproteases acting at pH 7. Just where the papain-like protease fits in the wheat-malt protease picture is not clear. Malted wheat flour was used by Chua and Bushuk (1969) in their preliminary purification attempts using affinity chromatography.

Other Sources of Proteinases.—The overall proteolytic activity of flours from at least one variety of triticale was reported by Madl and Tsen to be elevated compared with that from its parent grains, wheat and rye; much of this activity seems to be concentrated in the outer endosperm (Singh and Katragadda 1980). Lorenz and Welsh (1976), however, reported that such activity of four triticale varieties was intermediate between those of the parental species. In any case, it is probably the excess α-amylase activity and not that of proteases, as previously discussed, which has posed a minor problem in triticale utilization.

As far as baking is concerned, it is the external sources of proteases that are of prime importance. Fungal preparations, almost exclusively from *A. oryzae*, suitably treated to remove most of the amylase (Miller and Johnson 1949) are now widely used, in fact, more widely than are the commercial α-amylases, according to several authorities. Bacterial proteinases, frequently admixed with α-amylase, are also offered by enzyme suppliers to the baking industry for special baked goods (see below). Bromelin and later papain (Chapter 5) have also been permitted but one should exercise particular caution in regulating their level (Feldberg and Baker 1971). Indeed, one of the principal problems in the use of proteinase, even more than with the α-amylases, is to establish a firm, reliable prognosticator assay (Chapter 37).

Occasionally some wheat may have a surplus of proteolytic activity due to a variety of causes to be discussed below.

Action in Cereal Processing.—Most authorities do not believe that there is enough protease activity in sound mature wheat to significantly alter the properties of the major wheat protein complex that gives to this seed those unique properties which permit it to be transformed into bread (Kasarda *et al.* 1971,1976). An exception is Hanford (1967), who suggested that endopeptidase action on gluten components would leave no detectable fragments but could still exert some influence on properties of dough and bread.

The end results of adding commercial protease preparations to appropriate doughs are quite similar to those of adding α-amylase preparation: improvement in texture compressibility of crumb, loaf volume and even bread color. However, it is probably widely adopted because of the economic advantages of improvement of the processibility of doughs, especially of bucky, tacky, hard-to-mix doughs. It may reduce mixing time, lower viscosity, increase extensibility and, according to Duquette (1972), reduce fermentation time. This "mellowing" of the dough appears to work especially well along with conditioning or aging agents such as bromates and other oxidizing agents, especially in sponge dough processes where the other ingredients, especially NaCl, a potent inhibitor of most proteases, are added after the enzyme has had a chance to act. Of course, this inhibition can be exploited positively to closely monitor and regulate proteolytic action.

One of the reasons their action has to be monitored more closely than that of amylase may be because this proteolytic action is not, as for α-amylase, limited by the amount of available substrate. On the other hand, Pratt (1971) stated that wheat gluten tolerates a fairly broad range of proteolytic supplementation.

Products.—Proteinases may be added in customary bread-making when the baker starts with a bucky dough which needs mellowing. Mellowing can also be obtained by mechanical means or by addition of reducing agents such as cysteine, and more recently, by specially processed dried torula yeast which, it is claimed, circumvents some of the shortcomings of proteases and other conditioners (Leinen 1978). Other products which require a slacker than normal dough and for which proteases have been recommended include hamburger buns, biscuits and English muffins. Bacterial proteinase has been recommended in the production of some cookies, bread sticks and crackers where a slackening of the dough promotes more uniform spreading out on the shallow metal baking sheets used. Thus, spreading allows the dough to continue to adhere to the baking pan after it is placed in the baking oven.

Mechanisms.—Since the nature of the forces and factors that contribute the unique viscoelastic properties of dough, beyond the consensus that they arise from the presence of gluten proteins, are only now in the process of being sorted out, it is perhaps not surprising that we really do not know why a limited proteolysis of flour proteins confers desirable properties on the dough. Some ideas which have been forwarded include the following.

Shorter polypeptide chains which are formed would permit realignment of the gluten protein matrix to facilitate interchange of S-S and -SH which until recently have been considered to intermolecularly link the protomers of glutenin, gliadin and other gluten proteins into a three-dimensional network.

More likely, according to Kasarda *et al.* (1976), is a specific aggregation of these protein subunits via well-defined noncovalent interactions to form microfibrillar structures. In any case these authors point out that disruption or formation of protein intramolecular S-S (covalent) bonds may stabilize conformations favorable to intermolecular noncovalent (secondary) interaction between subunits (Fig. 32.4), reminiscent of muscle protein interactions (Chapter 26). Thus, the scission of a few strategic peptide bonds could, without affecting disulfide bonds, change the properties of the dough and still permit concomitant action of "aging" agents.

Since the general improvements brought about by proteinases are so similar to those brought about by α-amylase, it has been suggested that proteolysis enhances amylolysis by facilitating migration of water, released from the gluten proteins as a result of proteolysis, to the starch granules. This in turn makes the latter more accessible to α-amylase attack by liberating bound β-amylase or even by providing amylases as contaminant. Some enzyme suppliers recommend a mixture of the two types of enzymes.

Insoluble gels formed by oxidative coupling of ferulic acid that is attached to pentosan-containing glycoproteins (see below) are readily dispersed by proteolytic enzymes (Neukom 1976). Such a complex can crosslink via oxidative coupling between gluten tyrosine and the ferulic acid. Such a complex would contribute to dough properties. Proteolysis could abolish this contribution. Proteolysis could also alter the structure of lipoprotein platelets created by a, to some extent, conjectural interaction of phospholipid and gluten protein, the so-called Grosskreutz model.

Disadvantages, Excesses and Correction Thereof.—According to Leinen (1978), proteases, even when used under optimal conditions, carry along with their action certain unavoidable inherent limitations. Loss of strength of the gluten is accompanied by the previously alluded-to lowering of water-retaining ability which may cause excessive slackening of the dough. This would be reflected in undesirable texture changes in the bread, including a more open grain. Using less water to compensate leads to other problems. Furthermore, according to Leinen, use of proteases does not reduce mixing time because enzyme action does not occur until the pH of the dough reaches 4.5 to 5.0 during fermentation, so that softening may be delayed as much as 2 hr.

These shortcomings are magnified when too much proteolytic action takes place to yield unacceptable end-products. This may inadvertently happen in several ways. Too much commercial enzyme may be introduced and certain varieties of wheat may contain high levels of protease. Bushuk *et al.* (1971) observed that high proteolytic levels in wheat yielded highly extended gluten from flour milled from the wheat. Germinated and immature wheat which may unavoidably have to be milled has, as we have seen, elevated levels of proteolytic activity. Long periods of storage may "activate" proteinase (via "incipient" sprouting) (Mounfield 1938; Fleming *et al.* 1960). The proteinases of infesting organisms, even insects, may find their way into the flour (Kretovich 1944; Pomeranz

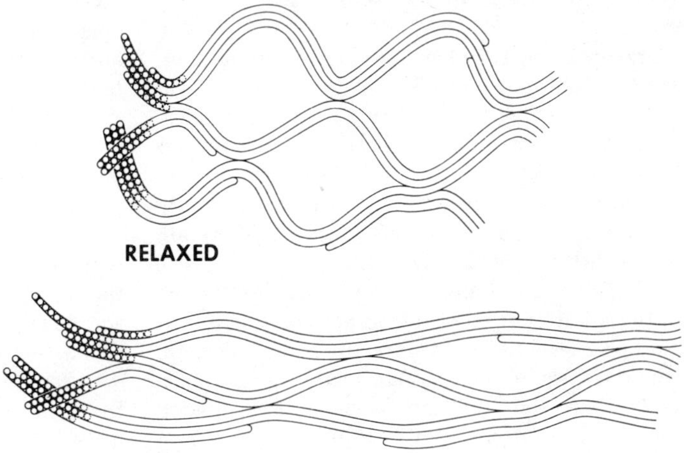

FIG. 32.4. PROPOSED ULTRASTRUCTURE OF GLUTEN

Depicted is the interaction of gluten fibrils to form a network resulting in gel; an alternative to SS-SH interactions to explain the mechanical properties of dough.

Courtesy of Kasarda et al. (1976)

1971). We have alluded to the possibility of high levels of activity in triticale.

There have been sporadic attempts to demonstrate improved bread quality as a result of modulation of proteinase action other than simply reduction of the duration of fermentation. These include inhibition by salt (Madl and Tsen 1974) and by "hydrothermal" treatment in which the wheat kernels are exposed to low pressure steam for less than 1 hr (Kandilis 1972).

LIPID-TRANSFORMING ENZYMES

Lipases and Related Hydrolases

Unlike lipoxygenase, discussed below, hydrolytic enzyme action on lipids has usually been considered to play a minor role, if any, in the conventional conversion of sound flour to baked goods, or the consideration, even if minor, has been one of concern (Pratt 1971) rather than of benefit. This concern is related to development of lipase-induced rancidity in stored grains and flours (Chapter 25), although Pyler (1969) states that the ultimate effect is a deterioration of the baking quality of the flour. Consideration of participation of lipids in the baking process and some results published in the patent literature suggest that a reexamination of possible positive effects of lipolysis, either endogenously or exogenously induced, may be merited. With regard to the former, purification attempts are usually made starting with a commercially supplied wheat germ lipase prepared according to the procedure of Singer (1948). Distinct acyl esterases which do not appear to be merely multiple forms of lipases but at least one of which is true lipase (acyl glycerol hydrolase) have been reported (Stauffer and Glass 1966; Fink and Hay 1969; Morrison 1978).

A notable property of the esterase(s) in wheat germ which hydrolyze(s) the convenient (for assay purposes) acyl esters of 4-methyl umbelliferone is their inhibition by aflatoxins (Pancholy and Lynd 1972). To our knowledge, little further, if any, purification has been attempted. Undoubtedly, the use of affinity chromatography will be helpful in the future. By contrast, considerable progress has been achieved in isolation, resolution into isozymes and characterization of rice lipases, notably that from the bran (Chapter 25).

Little if anything is known about acyl glycerol hydrolases (conventional lipases) of the developing wheat kernel. In the mature wheat kernel, lipase activity is concentrated in the aleurone layers and in the germ (Table 32.1), in agreement with distribution of activity in mill fractions. Although at a low level in the endosperm, it is detectable and contributes to more than half of the activity per kernel. The germ contributes less than one-fifth.

Although there is a considerable increase of lipase but not of other esterases upon germination (Drapron *et al.* 1969; Travener and Laidman 1972) which appears in the newly formed rootlets and coleoptile, Drapron *et al.* still found more *total* lipase activity in the endosperm of the residual kernel than in these other tissues. Maximum lipase activity corresponded to a plumule length of 7–9 cm.

Travener and Laidman (1972) could find no effect of GA on stimulation of synthesis, activation or secretion of lipase in the aleurone layer. Such stimulation, however, does appear to involve *de novo* protein synthesis and is dependent on non-GA factors emanating from the embryo, among which may be that ubiquitous "crossroads" metabolite which frequently accumulates in plant tissue (Chapter 4), glutamine. Another singular feature of the behavior of lipase during germination is that the lipase level increases markedly only in the germ when kernels are germinated in the dark, apparently in response to the need for FFA-related metabolism and growth which in the light is provided for by photosynthesis-associated events.

The Phospholipases and Related Enzymes.— The pattern of phospholipase B (Chapter 25) during seed development resembles that of soluble β-amylase in that it is not present at anthesis, rises to maximum during development and decreases to a constant minimum present in the mature kernel (Nolte *et al.* 1974).

In a rather drastic if not startling departure from the usually accepted route of enzyme formation, upon imbibition of water the aleurone cells in the mature wheat grain start to synthesize phospholipids, apparently by *de novo* formation of the requisite enzymes for phospholipid *synthesis*, and without any assist from GA or anything else emanating from the embryo, according to Varty and Laidman (1976). On the other hand, in contrast to the lipases, GA, along with other essential embryo-produced plant hormones, is needed for the formation and secretion of the enzymes of phospholipid *degradation*. GA, however, may play a role in vesiculation and release of the phospholipases and other embryo-originating factors responsible for *de novo* synthesis of these enzymes.

Among other esterases present in wheat whose action could conceivably play a role in breadmaking and texture are those catalyzing the hydrolysis of phosphate esters including the phosphodiesterase, phospholipase D, lipophosphodiesterase II (Nolte and Acker 1975) and the phosphomonoesterase phytase present in triticale as well as wheat (Singh and Sedah 1977). The connection between these enzymes lies in the activation of one of the phytases in wheat by lecithin and lysolecithin, its product of phospholipase action.

Role in Baking.—The principal influence of the lipases in the processing of cereal foods has, as mentioned, been traditionally one of concern due to formation of FFA responsible for hydrolytic rancidity during storage (Chapter 25). Its activity has been repeatedly correlated with unsound starting material. The undesirable overt manifestation of this activity—rancidity—may be exacerbated by prolonged storage, extremes of temperature and humidity, to infesting organisms, and incorporation of high lipase cereals such as rice and oats into the dough. However, we are not concerned with this aspect at this time.

As far back as 1936, Sullivan *et al.* concluded that while fresh wheat lipids from wheat germ had no effect on the functional properties of patent flour, the addition of free fatty acids resulted in diminished baking properties. Since the specific activity of bakers' yeast lipase is between two and three orders of magnitude greater than that of wheat flour (Pyler 1969) there would be, according to Miller and Kummerow (1948) at least ten times more lipase coming from the yeast than from the flour in a typical dough. Whether this yeast lipase is accessible to the wheat lipids or to those added is another question. Furthermore, in presence of oxidizing agents so frequently used as dough agers, much of the essential SH in lipase needed for activity (Singer 1948) might be oxidized. Nevertheless, undesirable excess of fatty acid production due to lipase action was reported by Miller and Kummerow and others.

However, it might still be possible that quite low levels of lipases could produce mono- and diglycerides from the added lipids to influence the antifirming and antistaling properties of the baked products. Such low levels of activity would be less likely to produce the FFAs, let alone hydrolytic rancidity. As is well known, monoglycerides are incorporated into breads in order to extend shelf-life (Ponte 1971). Caillat and Drapron (1974) found that such monoglycerides tend to be the end-products of wheat bran lipase action in model systems acting at water activities, a_w, in the range of 0.8–0.9 (Chapter 13).

Even more effective bread stabilizers are acyl esters of the sugars, also used as improved replacement for shortening and gluten in soy breads. These may be formed as the result of their synthesis by lipases present (Chung *et al.* 1976). It will be recalled (Chapter 25) that lipases can readily catalyze such syntheses. A sugar acyl hydrolyzing capability has been detected in wheat.

Surfactants would be expected to be formed from the action of phospholipases. Another function of the phospholipases might be to produce the unsaturated FFA needed for efficient action of the lipoxygenase action discussed below. Interaction between phospholipase and amylase is a possibility since Hanna and Lelievre (1975) showed that lysolecithin limits the attack by amylases on damaged starch granules because the starch is held within inclusion complexes of this phospholipase product. Phospholipase would also disturb the above-mentioned lipoprotein gluten platelets.

While this might be helpful where excess amylase action is a problem, lipase action has been forwarded as a means of deliberately forming the above-mentioned monoglycerides specifically to prevent bread staling. According to a patent issued to Johnson and Welch (1968), staling of baked goods is retarded by adding to the wheat dough an enzyme preparation possessing lipase (as well as protease and amylase) activity in an amount sufficient to increase the extractable monoglycerides content of bread by about at least 5 oz/100 lb of flour, in the range of monoglyceride concentration in dough formulations. In another related patent issued to Menzi (1970), the incorporation of lipase-containing preparations from pancreatin, wheat or corn germ into doughs is described. Pastas prepared from such treated doughs were said to possess superior elasticity and reduced stickiness as compared with control pastas. It would be interesting to ascertain if these ideas have ever been developed and utilized.

Apparently the best way to get rid of too much lipase is to "precook," especially when the lipase comes from other cereals incorporated with the wheat dough (Vaqueiro and Calderon 1975; Terebulina *et al.* 1977).

Lipoxygenase and Other Oxidoreductases

About 20 years after the use of soybean flour to bleach flour was first proposed (Chapter 17), Logan and Learmonth (1955) reported that soybean flour oxidized the gluten as well as carotene. A year later, Koch (1956) observed changes in dough behavior which accompanied this oxidation which he attributed to the oxidation of sulfhydryl groups in the gluten. Since then a substantial body of evidence has accumulated which by 1964 implicated lipoxygenase, from wheat or added as a pure enzyme or as bean flour (Chapter 24), in exerting a beneficial effect on doughs and baked goods processed therefrom (Kleinschmidt 1963; Bloksma 1964).

Reviews of the literature and general discussions to 1971 include those of Reed (1966), Guss *et al.* (1968), Mecham (1971), Reed and Thorn (1971) and Daniels *et al.* (1971). Not all investigators agreed that this enzyme participated significantly in bread-making. The evidence in favor of a substantial role of lipoxygenase as evidenced by the literature cited in these reviews may be summarized as follows:

Oxygen, one of the lipoxygenase cosubstrates, behaves in a manner similar to that of other conditioners. Furthermore, incorporation of oxygen into the dough is enzymatic.

Oxygen uptake is dependent in the presence of FFA and increases with increasing FFA levels.

Addition of fat increases the oxidation of the sulfhydryl (SH) groups of flour. Oxidation of some SH groups in flour is necessary to strengthen what was then considered the S-S crosslinked three-dimensional protein network comprising gluten in worked dough. Presumably the hydroperoxide formed via lipoxygenase played the same role in these constructs as do oxidizing agents used as dough conditioners. Interaction among O_2, LOOH and gluten S-S was depicted by Tsen and Hlynka (1962) as follows:

$$\text{-S-S-} \xleftarrow{\text{-SH}} O_2 \xrightarrow[\text{(Lipoxygenase)}]{\text{FFA}} \text{LOOH} \xrightarrow{\text{-SH}} \text{-S-S-}$$

Propyl gallate, considered to be a specific inhibitor of lipoxygenase, also caused a deterioration of the rheological properties of dough (Auerman *et al.* 1971).

When water is mixed with flour, the amount of readily nonpolar-solvent extractable lipid decreases and continues to decrease upon subsequent mixing. First established by Olcott and Mecham (1947), this binding appears to be due to a lipid's capacity to form bridges between gliadin and glutenin subunits to which they are bound through hydrophilic attraction and hydrophobic interaction, respectively (Chung and Tsen 1977). Dependence of *release* of lipid from gluten on the same factors influencing bleaching of carotene and other dough pigments, SH → S-S reactions and improvement in dough and bread characteristics, suggested that this release is instigated by lipoxygenase action. A proposed mechanism is depicted in Fig. 32.5. Improvement by lipoxygenase action could then come about through two distinct routes which both involve gluten: change in rheological properties of the gluten due to formation of disulfide bonds and increase in free lipid attendant upon an unspecified oxidation of gluten.

Since most patents employing lipoxygenase relate to the bleaching effect, few have been issued for the specific purpose of using this enzyme as a bread improver. Kleinschmidt and Viren (1970) incorporated soy flour into a friable dough additive comprising whey, sugar and an oil, such as a soybean containing a lipoxygenase substrate.

Arguing against the substantial role of lipoxygenase in bread-making are the following observations and inferences stemming from the literature cited in the above reviews: (1) A *decreased* tolerance to dough mixing upon the addition of FFA had been observed under certain conditions. (2) The oxygen uptake of the dough was considered to be much greater than can be required for this type of oxidation, and (3) by altering processing parameters (i.e., high mixing speed) improved texture-related dough properties could be obtained with defatted flour devoid of lipoxygenase substrate. Also, bound lipid could be irreversibly released in air only at high work levels; there were no changes in lipid binding at normal work levels. (4) A few investigators concluded that most of the FFA in flour dough may be oxidized via β-oxidation with lipoxygenases accounting for only a proportion of the oxygen consumed. (5) What were considered to be known lipoxygenase inhibitors, NDGA and tocopherol acetate, inhibit bleaching but do not produce concomitant loss of loaf volume under certain but not all conditions which might be pertinent to bread-making. (6) In some model systems, neither lipoxygenase nor hydroperoxides readily oxidized SH to S-S.

It was this sort of evidence, quite convincing at that time, that led some investigators cited by Reed and Thorn (1971) to conclude that the oxidative enzymes may play a minor role in bread-making.

While it is generally believed at the time of this writing that lipoxygenase plays more than a minor role, in a more recent review Morrison (1976), in Scotland, suggested that while chemical changes in fat are entirely due to lipoxygenase in the first instance, its practical benefit (other than bleaching) is that free lipid is released and some dough development is achieved without chemical additives. In another comprehensive review of this particular aspect of lipoxygenase action, Eskin *et al.* (1977) considered that the importance of lipoxygenase is in the prevention of lipid binding which ensures effectiveness of shortening fat in improving volume and softness of bread. Just why it is necessary to have free lipid (Daniels *et al.* 1975; Jacobsberg *et al.* 1976) is not completely understood, but the most plausible theory is that when lipids are bound they cannot serve their beneficial role in influencing the stability of the gas bubble structure during the expansion of the loaf (MacRitchie 1977). This aids in dough development by ensuring gas retention and thus promoting what is known as "spring back" during baking.

As indicated in Fig. 32.5, Daniels *et al.* (1971) postulated an "oxidized protein" presumably

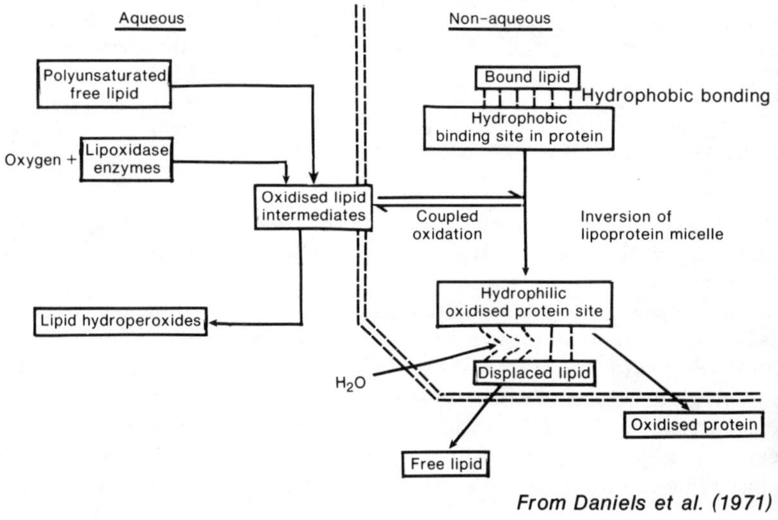

FIG. 32.5. ROLE OF LIPOXYGENASE ACTION IN DOUGH AND BREAD TEXTURE

By altering the degree of lipid binding to protein.

formed through the lipoxygenase catalyzed oxidation. A summary of their most recent work available (Frazier *et al.* 1977) is shown in Table 32.4. The data confirm many of the earlier conclusions outlined above favoring a lipoxygenase mediated role in baking as well as the early observation of Koch (1956) showing that reconstituting defatted flour with enzymatically peroxidized fat did not yield a dough with the same rheological properties as reconstituting with unoxidized but competent lipoxygenase substrates. While Koch suggested that the hydroperoxide was not involved directly but that a free radical may react with SH groups in the flour, Frazier *et al.* interpret the failure of LOOH to mimic the effect of lipoxygenase in terms of a coupled mechanism of oxidation of the SH groups which requires concomitant lipoxygenase action. They suggest a free radical lipid-containing intermediate (Chapter 25) which provides a link with protein oxidation sites before LOOH or the antioxidant NDGA can be oxidized. Previous work had suggested that NDGA is a true competitive inhibitor of lipoxygenase. They invoke S-S crosslinking as a consequence of incorporation of a free radical in gluten protein, as was found in model peroxidizing lysozyme systems (Schaich and Karel 1975). Similarly, Morrison (1976) speculated on the basis of tracer studies using ^{14}C-labeled lipoxygenase substrates that free radicals generated by lipoxygenase, not necessarily involving S-S, would

TABLE 32.4

SOY-TREATED WHEAT FLOUR DOUGH CONSIDERED AS A LIPOXYGENASE REACTION SYSTEM

Components of Soy-Dough LOX System[1]	Maximum Mechanical Dough Development: Mixing, min +LOX	Mixing, min −LOX	Relaxation, sec[3] +LOX	Relaxation, sec[3] −LOX
Complete System	13–14	10–11	38–40	13
plus inhibitor	10–11	7–8	38	22
less substrate	10	10	26	26
plus saturated FA	8	8	19–20	19–20
less cosubstrate[2]	20	20	12	8
" + product	—	20	—	7
" + product less substrate	—	20	—	14

Source: Adapted from data of Frazier *et al.* (1977).
[1] Complete enzyme system consists of full fat or reconstituted wheat flour (substrate) plus O_2 (cosubstrate); soybean flour as lipoxygenase (LOX); "less substrate," fat extracted flour; "less cosubstrate," N_2 for O_2; "inhibitor," antioxidant NDGA, 300 μg/g flour, "product," enzyme oxidized flour lipid, saturated FA, coconut oil.
[2] In absence of cosubstrate, O_2, no maximum observed in 20 min; relaxation time shown at 20 min mixing.
[3] Increase from 1 min of mixing to maximum relaxation time.

eventually cause the protein to be in a free radical form which could then crosslink and/or otherwise polymerize or dissaggregate.

A somewhat different point of view is that adopted by Graveland and coworkers (Graveland 1973; Graveland *et al.* 1978). Combining their earlier findings on the presence of hydroxyfatty acids (products of LOOH reduction) in wheat flour suspensions (Chapter 24) with an older suggestion of Mapson and Moustafa (1955) that low molecular weight thiols such as glutathione may be responsible for gluten oxidation, they suggested that the reduction of LOOH or related oxidized lipid is accomplished by these thiols:

$$LH + O_2 \xrightarrow{LOX} LOOH$$
$$LOOH + 2RSH \longrightarrow LOH + RSSR + H_2O$$

where LH, LOOH, LOH and LOX are linoleic acid, its peroxide, hydroxyacid and lipoxygenase, and RSH is a thiol such as glutathione. Such a scheme would be in accord with the conclusion of Gardner and Sessa (1977), whose comparison of LOOH degradation by cereals was alluded to in Chapter 24, that LOOH is reduced in water-flour mixtures by a mechanism *not* involving linoleic acid hydroperoxide isomerase.

The resulting disulfide, RSSR, may then oxidize the sulfhydryl in gluten proteins. This may result in the formation and interchange of crosslinked disulfide bonds which many investigators, i.e., Huebner *et al.* (1977), consider to be the key to protein functionality in general (Chapter 34) and quality and strength of wheat varieties in particular.

$$RSSR + P_1SH + P_2SH \longrightarrow P_1S\text{-}SP_2 + 2\,RSH$$

where P_1 and P_2 may be two different protein subunits in the same or different molecules or may be considered as belonging to the same subunit. When in the same subunit, the transfer of attachment of disulfide bonds from one protein chain to another may occur via disulfide-sulfhydryl interchange.

$$P_1S\text{-}SP_2 + P_3SH \longrightarrow P_1S\text{-}SP_3 + P_2SH$$

According to this view, interchange permits either a strengthening of protein network by chain elongation or relaxation of a too-rigid structure. It is a leavening thought that disulfide crosslinking and interchange have been invoked to explain aging of both flour and fauna (Chapter 27). This view, as mentioned in connection with protease action, is not universally accepted. Thus, according to the concept of Kasarda *et al.*, P_1SH and P_2SH would be part of the same molecule, and formation of intramolecular bonds would induce conformational changes (Chapters 3 and 4) which would be reflected in changes in noncovalent interactions among the subunits and manifested eventually as changes in rheological properties of the dough. In practice, reducing agents, which presumably loosen subunit binding, may accelerate the dough mixing process by slackening the dough to be followed by oxidizing agents which form disulfide bridges to restore the dough to its proper texture. Presumably, these desirable changes in gluten would work in cooperation with the benefits of freeing lipid, both lipoxygenase mediated. Certainly, the operations of both mechanisms are not mutually exclusive. However, in Scotland, Mann and Morrison (1975), since they could not demonstrate any increase in LOH when the thiol cysteine was added to dough, concluded that there was no connection between the level of accessible SH groups in dough and the reduction of LOOH to LOH.

Part of the problem in trying to assess the role of SH is that of its determination in flour. Thus, only recently, Graveland *et al.* (1978) reported that all previous SH values were low by a factor of six (correct values, 5–7 μmol SH per gram of flour). Using a revised analytical approach, they found no correlation between S-S or SH level and baking quality, but as shown in Table 32.5, dough mixing caused the SH content to decrease; the effect was strongest in the flour samples that had the best baking quality (loaf volume). Furthermore, in contrast to the above-mentioned results of the Scottish investigators, they report a strict stoichiometry between this decrease in SH and a corresponding increase in monohydroxyfatty acids. Accordingly, one molecule of LOOH will oxidize two SH groups. Thus, progress is being made towards an understanding of this serendipitous spin-off arising from the original application of lipoxygenase action in bean flour as a bleaching agent comprising mechanisms whose recalcitrance to elucidation and complexity rivals the viscoelastic properties of a bucky dough.

Other Oxidoreductases.—In discussions of possible mechanisms by which lipoxygenase improves dough and bread very little, if any, attention has been directed to the possibility that active oxygen species existing transiently but independently of the lipid substrate may be involved. Perhaps the oft-repeated observation that ascorbic acid at appropriate levels improves dough in a manner similar to bromate improvers (Prihoda *et al.* 1971), may be due to O_2^-, O_2^* and related free radicals (Chapters 16 and 24) rather than, as usually accepted, to dehydroascorbic acid or the half oxidized dehydroascorbate free radical coupling to gluten sulfhydryl oxidation (Jorgensen 1939).

Phenolases.—The oxidation of ascorbic acid is probably not initiated by ascorbate oxidase (Chapters 15 and 25) which, in spite of earlier reports, is probably not present in most cereals, but to its coupling to phenolase (or peroxidase) action:

$$o\text{-Diphenol} + \tfrac{1}{2}O_2 \xrightarrow{\text{Phenolase}} \text{Quinone} + H_2O$$

$$\text{Quinone} + \text{Ascorbate} \xrightarrow{\text{Nonenzymatic}} \text{Dehydroascorbate (DHA)}$$

TABLE 32.5

LIPOXYGENASE-RELATED LOSS OF SULFHYDRYLS AND FLOUR BAKING QUALITY

Measurement	Range, μmoles	Rank of Flour Samples	Deviation[1] from Rank
Baking quality	not applicable	1 2 3 4 5 6 7 8	0.0
Δ SH, per g protein, dough	6.7–24.1	2 1 3 4 5 6 8 7	0.5
SH, per g protein, dough	44.0–59.8	8 7 4 3 2 6 1 5	-1.5
SH, per g protein, flour	60.1–68.8	6 2 3 4 5 8 1 7	1.8
SH, per g flour	4.7–7.2	4 2 1 3 5 7 6 8	1.0
SS + SH, per g flour	15.4–20.2	6 1 2 3 4 7 5 8	1.5
SS, per g flour	10.7–14.0	7 3 2 1 4 5 8	2.0
SS, per g protein, flour	135–140	5 6 7 3 4 2 1 8	-2.1

Source: Calculated from the data of Graveland *et al.* (1978), showing that −ΔSS is equivalent to the number of monohydroxyacids derived from lipoxygenase via secondary reactions (see Chapter 24). ΔSH = change in going from flour to dough.

[1] Average deviation per sample from baking quality rank.

$$DHA + Glutathione\ (GSH) \longrightarrow GSSG + Ascorbate$$

$$GSSG + Gluten—(SH)_2 \rightleftharpoons Gluten—S\text{-}S + 2\ GSH$$

Normally, debranned wheat flour exhibits a low level of phenolase activity (Milner and Gould 1951). However, it is present at elevated levels in the bran and hence in whole wheat flour prepared from the dwarf "green revolution" wheat, and in wheat flour treated with mushrooms. Kuninori *et al.* (1976) showed that the observed increase in resistance to mixing and decrease in dough extensibility, similar to that of an oxidizing agent, was in part due to phenolase-mediated oxidation of SH groups. The investigators speculate on alternative effects of phenolase; oxidation of tyrosine residues in gluten protein (Chapter 34), formation of adducts of phenolase oxidation products with SH (Chapter 16) and oxidation of the proteopentosans through their constituent ferulic acid, as alluded to above and below.

Other Oxidases.—*Xanthine Oxidase in Milk.*—This, another O_2^- producing enzyme which is frequently adventitiously added to dough as a constituent of a nonflour ingredient, could conceivably influence dough and bread texture by virtue of its capability of producing O_2^- (Chapters 16 and 24). In this connection, the effect of adding SOD may be instructive. Also in this connection, starch-based pudding desserts which may be considered as either cereal or dairy foods are occasionally plagued by loss of firmness in the can. That this may be due to such oxygen species supplied by milk xanthine oxidase is suggested by unpublished observations of the author (Schwimmer 1974) that addition of purified xanthine oxidase to such a store-bought pudding completely destroyed its gel-like character and that SOD prevented this effect. Xanthine oxidase is not destroyed by pasteurization. Of course, other more practical means of preventing such action immediately suggest themselves, i.e., pasteurization to the point of xanthine oxidase inactivation, addition of commercially available food-grade catalase, etc.

An example of an oxidase other than lipoxygenase whose deliberate addition to dough to *improve* texture has been suggested is glucose oxidase. More specifically, Silverstein (1961) reported that addition of this enzyme enhanced and made more reproducible the improving effect of ascorbic acid via the H_2O_2 liberated.

Peroxidases.—The low level of phenolase in wheat flour suggests that peroxidases may also be involved. A key to the literature on the well-characterized wheat peroxidases is the review of Reed and Thorn (1971). Both peroxidase and catalase are concentrated in the germ which is removed in milling but can still be easily detected in white bread flour as well as in all mill fractions of wheat (Evans and Mecham 1971). Peroxidase as well as phenolase action (see above) has been invoked as a possible mechanism, as mentioned above, by which the ferulic acid attached to one proteopentosan molecule undergoes an oxidative coupling to the ferulic acid (Fig. 32.6) of another proteopentosan. This coupling is reminiscent of that induced by the action of tea phenolases acting on endogenous flavonols to form the dimeric bisflavanol, theaflavins (Chapter 16). This possibly oxidase-mediated process results in gels which may participate in dough and bread texture (Neukom 1976). In confirmation, Patil *et al.* (1976) conclude that rigidity of doughs reconstituted with these pentosans and the bread improver bromate may result in part from the oxidation of the pentosan-glycoprotein interaction product. However, they go on to indicate that this gelling does not appear to be related to the functional properties of the dough.

Catalase.—The biologically ubiquitous catalase has been reported as being present in wheat in studies extending from the 1930s (Sullivan 1946; Reed and Thorn 1971) but few such studies have been oriented toward a possible role in dough and bread texture. Thus, Hawthorne and Todd (1955) found that catalase, acting in its peroxidatic mode (Chapter 11), seemed to utilize slightly peroxidized linoleic acid as a substrate to bleach carotene. Primarily such a reaction could also affect lipid binding and dough texture as an "accessory lipoxygenase." It may also play a role as an O_2^- scavenger (Chapter 16). However, its predominance in the germ of the wheat kernel and its marked inactivation as a result of natural and artificial aging (Pyler 1969) suggest a very minor role in texture.

Finally, other oxidoreductases have been investigated as part of an overall program on their possible role in dough and bread texture but no direct experimental linkage was demonstrated. Thus, Honold and Stahmann (1968) suggested that NAD-linked dehydrogenases of wheat could, by accumulating NADH, alter the fermentation process in the dough and thus alter baking quality. Although we depicted the beneficial effect of as-

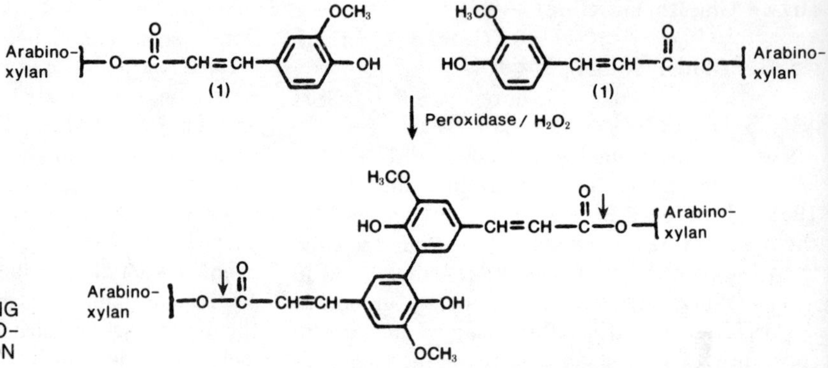

FIG. 32.6. CROSSLINKING IN WHEAT FLOUR PENTOSANS BY ENZYME ACTION

From Neukom (1976)

corbic acid addition on dough in terms of oxygen and/or phenolase-mediated reactions, we point out here that specific enzymes, one for oxidation (Tsen 1965; Grant and Sood 1980) and one for reduction (Redman 1974), have been proposed to account for such effects.

$$2\ \text{Ascorbate} + O_2 \xrightarrow{\text{oxidase}} \text{Dehydroascorbate (DHA)} + H_2O$$

$$2\ \text{DHA} + \text{Gluten}\ (SH)_2 \xrightarrow{\text{reductase}} \text{Gluten-S-S} + 2\ \text{Ascorbate}$$

While there is considerable progress in the empirical application and some understanding of how enzyme action can be exploited in cereal technology, many intriguing mechanisms and further rationalization of the use of enzymes lie in the future.

33

ENZYME ACTION IN CHEESEMAKING AND CHEESE TEXTURE

The domain of the role of enzymes in food texture was introduced with a detailed consideration of the manner in which subtle action of proteolytic enzymes participated in the conversion of mammalian muscle tissue to meat. In this chapter, we return to another mammalian-derived food in which subtle proteolytic enzyme action plays a major role in textural changes. We shall be discussing the role of proteases in the conversion of milk to cheese. Unlike the conversion of muscle to meat, the original tissue actually experiences a change-of-state as the result of minimal protease action involving, ideally, the hydrolysis of one peptide bond per molecule of one particular protein. Further textural alterations not involving phase changes take place mostly as the result of more widespread peptide bond splitting.

What bread, beer and cheesemaking have in common is the exploitation of the consequences of microbial growth and metabolism, on the one hand, and the concerted action of both discrete endogenous and exogenous enzymes on the other. Indeed Scott (1972), asking the more than academic question as to whether cheesemaking is applied enzymology or bacteriology, concluded that once the enzyme systems are present in the curd, whatever be their source—milk, commercial enzymes, starter or ripening microorganisms—living organisms are not necessary to complete that transformation into milk's "leap toward immortality" (Fadiman 1955). Like these ancient processed foods, the actual manufacture of cheese is a blend of almost pure art and applied empirical sciences with some attempts at rationalizing the process as scientists of several disciplines including enzymology strive to understand the molecular basis of the observed changes and to apply this insight so as to ever increasingly rationalize its production. Participation of proteases and other enzymes in the creation of other quality attributes of cheese, especially flavor, has been or will be dealt with in Chapters 20 and 23, and quality loss in Chapters 25 and 34.

Outline of Cheeses and Cheesemaking.—Short, general reviews emphasizing the roles of enzyme action in cheese texture include those of Sardinas (1976) and Scott (1972). More comprehensive reviews and books on cheeses and their manufacture include those of Sanders (1953), Brown (1955), Marquis and Haskell (1965) and a comprehensive overview of cheese chemistry in books edited by McKenzie (1971) and Webb *et al.* (1974).

The starting material for the manufacture of most cheeses in the United States is pasteurized whole milk. Other starting milks include nonfat skim milk for cottage and dietetic cheeses, whole milk to which cream has been added for Cream and Neufchatel. In addition to cows the following animals have supplied their milk for cheesemaking: asses (rare), buffalo, camels, goats, llamas, sheep, yaks and zebus; of these only goat cheese is in widespread use in the United States and western Europe.

After treatment with a "starter" culture of a lactic acid bacterium, usually *Streptococcus lactis* (primary phase of cheesemaking), to the milk is added a milk-clotting enzyme, rennet, to form a curd (secondary phase). This curd is cut up into slabs and held at about 39°C for over 1 hr ("the cooking" step). The resulting syneresed whey is drained off from the slabs of curd which are further subdivided (milling), treated with salt, pressed and ripened. Syneresis, it will be recalled, is the "weeping" or slow spontaneous movement and separation of liquid from a newly-formed colloidal semisolid mass. Neufchatel and cream cheeses are not ripened nor is cottage cheese renneted (see below). With other cheeses ripening (maturing, aging) proceeds under controlled conditions to encourage the growth of selected bacteria and fungi to give each cheese its final characteristic appearance, texture and flavor (tertiary phase). For instance, surface growth is encouraged for Camemberts, interior growth for blue cheeses. Muensters and Limburgers require moderately cool, moist atmospheres, whereas cheddar cheeses ripen at lower temperatures and water activities.

Since we are emphasizing the textural aspects of cheese quality in this chapter it may be well to point out at this juncture that a common classification of cheeses such as that in the U.S. Department of Agriculture Handbook No. *54* (Saunders 1953) is based upon their texture. Major categories include: *soft* (Camembert, cottage, Neufchatel); *semisoft* (blue, Muenster, Jack, Stilson); *hard* (Cheddar, Emmenthaler, Swiss); and *very hard* (Parmesan, Romano, Spolen).

With but few exceptions, such as the formation of the "eyes" of Swiss cheese and the use of lactase to prevent deposition of crystalline lactose, the saga of textural changes in cheese is that of the action of proteolytic enzymes. These proteases may be in the milk itself, but probably far more important are enzymes which are exogenous to the milk, added as rennets. Their action is all-decisive in converting liquid milk to a solid or at least semisolid state in the secondary phase of cheesemaking and, along with the enzymes from other sources, to textural changes during the tertiary phase of cheesemaking. We shall now examine these enzymes and assess their roles in cheesemaking.

MILK CLOTTING WITH CHYMOSIN (RENNIN)

The Enzyme

How important or prominent a given enzyme is depends upon one's point of view, as illustrated by the inclusion of chymosin in a symposium on "less familiar enzymes" (Anon. 1969) and by an offhand reference to it as another pepsin-like enzyme in a well-known reference (Fruton 1971). However, the specific properties of this enzyme which have made it the milk-clotting enzyme of choice for centuries are, of course, of considerable importance to food scientists in general and to dairy technologists in particular. Its action provides a striking illustration of how a minor chemical change, in this instance the splitting of a single peptide bond, in a large protein molecule can set into motion a cascade-like series of events leading to a drastic change in appearance and texture of a food. The "primary product" of enzyme action (Chapter 1) is also an intermediate whose "secondary" transformations confer changes in quality attributes. However, the "intermediate" is colloidally rather than chemically unstable and the secondary transformations are physiochemical events rather than true chemical changes brought about by breaking and making of covalent bonds.

The enzyme which accomplishes this remarkable transformation is known as rennin, an acid endopeptidase which is obtained from the fourth stomach (abomasum) of suckling or unweaned calves. Rennin—or as Foltmann (1966) suggested, chymosin, so as to not to confuse it with renin, the kidney protein involved in regulation of blood pressure—was first crystallized by Berridge (1945) and subsequently examined in detail at the Carlsberg Laboratory by Foltmann and colleagues (Foltmann 1966; Foltmann *et al.* 1973). That a research facility supported by a brewery foundation is responsible for many of most important early advances in our understanding of this enzyme of key importance to the dairy industry probably stems from Linderstrøm-Lang. His laboratory was a Mecca for young enzymologists in the years before and after World War II and he perspicaciously foreshadowed modern concepts by proposing about 50 years ago (Linderstrøm-Lang 1928) that the milk-clotting enzyme in calf rennet removes a colloid component of casein which stabilizes the milk suspension. Another Dane, Christian Hansen, whose name is still associated with commercial rennet, produced the first such enzyme for cheesemaking in 1874. As with other crystalline enzymes, the crystals of Berridge were found to contain several isozymes in addition to the major chymosin fraction. The preparation of commercial rennets containing chymosin was briefly alluded to in Chapter 5. Chymosin, like pepsin, secreted as an inactive 36,000 dalton proenzyme consisting of a single polypeptide chain, is readily converted to the active form at pH's below 5 by a limited proteolysis. As a result of this limited proteolysis, a 5000 dalton segment is cleaved from the N-terminus of the single polypeptide chain comprising chymosin, although dimers can be formed. Solutions of chymosin are stable at low pH's in the absence of a salt (Mickelsen and Ernstrom 1972). Its similarity to bovine pepsin with respect to specificity, stability and milk-clotting ability arises from homology in primary amino acid sequences of the two enzymes. Thus, a fragment of the chymosin chain has the sequence Asp-Thr-Gly-Ser-Ser as compared to a corresponding sequence in pepsin, Asp-Thr-Gly-Thr-Ser. Its action on synthetic substrates is about the same but more restricted. Optimum pH range of peptide splitting ability for most protein substrates is about 3.7–3.8 but for the substrate of interest, kappa casein, is in the neighborhood of 5 (Humme 1972; Miyoshi *et al.* 1976).

The Substrate

About 80% of the 4% of the protein present in milk is casein. Intact casein can readily be distinguished and separated from the remaining protein, mostly the lactose synthase regulatory protein, α-lactalbumin (Chapter 4), and from enzyme altered casein by its precipitability at pH 4.6. Casein is a complex of four phosphoproteins: α_{s1}-, about one-half of the total; β-, 25%; δ-, 5%, probably derived from β (see below); and κ, 15%, hereafter referred to as k-casein. Several other ill-defined minor protein constituents are also present. These proteins, along with an apatite-like mineral containing calcium and anions such as phosphate and citrate, are present in milk as supramolecular aggregates known as micelles. A rather comprehensive literature contributed by dairy and mammalian physiology-oriented investigators has developed. Reviews and keys to this literature can be found in papers by Swaisgood (1973), Slattery (1976,1978,1979) and Pepper and Farrel (1978).

Casein micelle formation represents another instance of self-assembly (Chapter 4) as indicated in Fig. 33.1, and probably performs the biological function of solubilizing and maintaining calcium and phosphorus in a disperse and readily assimilable form. Casein micelles are crenate-appearing spheres varying in radius from 25 nm to 150 nm containing about 450 subunits each comprising about 10,000 monomers of α_{s1}-, β- and k-caseins (Slattery 1976). The k-casein is the only one of these proteins that does not readily complex with the Ca^{2+}. Some but not all of its molecules contain considerable amounts of glycopeptide. The protein molecules of the micelle are in equilibrium with nonmicellar soluble casein molecules. A function of the k-casein is apparently to prevent precipitation of the other proteins.

The most recent available evidence indicates that these micelles are made up of submicelles which in turn comprise 25–30 casein monomers corresponding to a molecular weight of about 600,000 (Pepper and Farrel 1978). Although the k-caseins in the micelles are associated through intermolecular disulfide bonds, they are probably incorporated into the submicelles via noncovalent interactions and are subsequently fixed by disulfide bond formation. Bloomfield and Mead (1975) point out that the primary structure (the amino acid sequence) of the caseins which results in loose random coils with little or no helical content is related to their capacity to form micelles.

The concept of submicelles proposed by Slattery (1976) and coworkers is apparently being accepted as the most plausible in a series of previously proposed models of micellar structure. Previous models, detailed in Slattery's review and in the above-mentioned key citations, all center about the distribution of the k-casein molecules. Just about every conceivable position has been postulated: the exterior, the interior or evenly distributed. Anticipating the above-mentioned experimental conclusions of Pepper and Farrel, and eclectically combining aspects of these previous models, Slattery and Evard (1973) assumed the presence of soap-like submicelles consisting of the three major caseins. The unique feature of each micelle is that its surface is partly hydrophilic and partly hydrophobic, hence the term "amphiphilic" applied to these aggregates. The hydrophilic portion, represented by the dark areas on the surface of the submicelles (Fig. 33.1), is composed of the carbohydrate portions of the k-caseins. The light

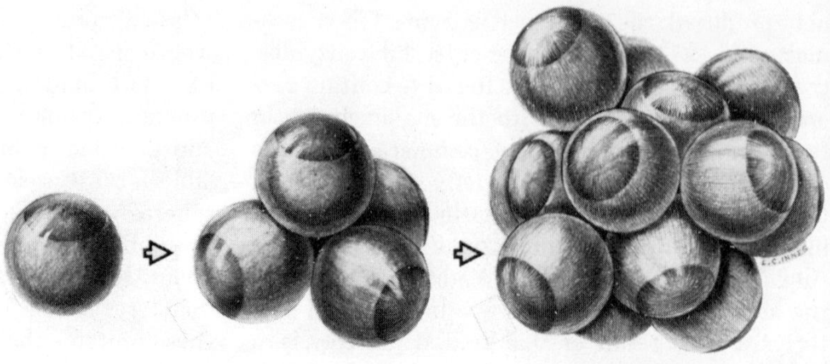

FIG. 33.1. SELF-ASSEMBLY OF CASEIN SUBMICELLES TO FORM MICELLES

Depicted are: a spherical submicelle of about 30 individual casein molecules (monomer) with hydrophobic patch on surface (left); an up to now undetected tetramer (center); an aggregate of 14 submicelles, the minimum micelle (right).

Courtesy of Slattery (1978)

portions represent nonpolar domains of α- and β-caseins which can form hydrophobic bonds after binding calcium. Thus, it is this asymmetric orientation and distribution of the k-casein molecules in one direction and the other caseins in the opposite direction within the submicelle that yields hydrophilic and hydrophobic areas on the subunit surface of the resulting amphophilic submicelles. Disulfide bonds appear to help create and maintain this orientation. As shown in Fig. 33.1, aggregation of submicelles through hydrophobic interactions results in growth until a minimum aggregate of 14 submicelles is reached to form a porous micelle. So far this model is in accord with and predicts a vast array of experimental data on micelle behavior including the prediction of an observed discontinuity in the size distribution of the micelles in milk. As an example of experimental data which such casein models must accommodate, Cheryan *et al.* (1975A), using immobilized carboxypeptidase, were able to demonstrate that the carboxy termini of at least some of the molecules of the three major caseins are present at the surface of the micelle. Application of peroxidase action as a topographical probe (Chapter 11) might be particularly helpful in this respect. In Chapter 10 we briefly alluded to the strengthening effect of ionizing radiation on casein micelles. Perhaps more to the point of the present discussion is that a good micelle model should adequately explain many of the chymosin-induced peculiarities of the milk clotting process.

The Primary Event: Hydrolysis of the Phe (105)–Met (106) Peptide Bond of Kappa-Casein

When rennin is added to milk at pH 5.2–5.4, after an elapsed time a curd is formed. Like all enzyme action the rate—in this case the reciprocal of the time required for the curd to become visible to the naked eye—is dependent upon the concentration of enzyme (Chapters 2, 3 and 37) but is frequently not directly proportional to this variable. In general there appears to be a time lag so that instead of enzyme concentration × time = a constant, the following relation is frequently observed, not only with rennin but also with other milk clotting enzymes.

$$\text{Rennet units} = [E] \times (T - t) = \text{a constant}$$

where T is the observed clotting time and t is an empirical correction factor. As with other enzyme reactions, the clotting time is quite sensitive to pH and salt concentration but is especially decreased by the addition of calcium salts.

Furthermore clotting does not occur at pH's higher than 7 when calcium already present is removed nor does it occur at lowered temperatures in the vicinity of 0°C. The longer the exposure to low temperature, the shorter the clotting time when the temperature of the milk is raised to ambience around 30°C. The time required for clotting at higher temperatures can be predicted from the Q_{10} for peptide bond splitting (see "cryptic accumulation," Fig. 12.3) according to the following equation

$$\text{Casein} \xrightarrow[\text{Low } Q_{10}]{\text{Enzyme hydrolysis}} \text{“Para-k-casein”} \xrightarrow[\text{High } Q_{10}]{\text{Aggregation}} \text{Clot}$$

$$t_x = -\frac{1}{Q_{Hyd}} \cdot t_0 + t_{30}$$

when t_x = time required for clot formation at 30°C after low temp; t_{30} = time for clotting without prior low temp; t_0 = time exposed to cold, and Q_{Hyd} = ratio, rate of peptide bond hydrolysis at 30°C to that at low temperature.

It is quite clear that the process consists of at least two distinct stages, one characterized by a limited proteolysis of one or more peptide bonds in one or more of the caseins, followed by nonenzymatic aggregation. As a result of the work of several investigators in the mid 1960s, especially that of previously cited Foltmann and colleagues, we know that the enzymatic trigger to milk clotting is the highly specific hydrolytic cleavage of the phenylalanyl methionine peptide bond between the 105th and 106th amino acids in the polypeptide chain comprising the 22,000 dalton k-casein to yield para-k-casein and an 8000 dalton glycopeptide (macropeptide) containing sialic acid, as shown in Fig. 33.2. Sialic [*N*-acetyl neuraminic] acid, a derivative of *N*-acetylglucosamine, is also present in the transport protein, Na^+/K^+-ATPase, also recently implicated in human weight control.

The importance of the carbohydrate portion in milk clotting is attested to by a variety of different observations. Among them are those of Wheelock and Penny (1972) that clotting time is inversely proportional to the rate and extent of liberation of the glycopeptide and that of Castle and Wheelock (1973) that the K_m for rennin action is dependent upon the carbohdyrate composition of the k-casein; it will be recalled that not all k-casein molecules are glycosylated. Clotting can be initiated by removing this carbohydrate portion with lysozyme, a hydrolase which liberates the sialic acid (Mullin and Wolfe 1974).

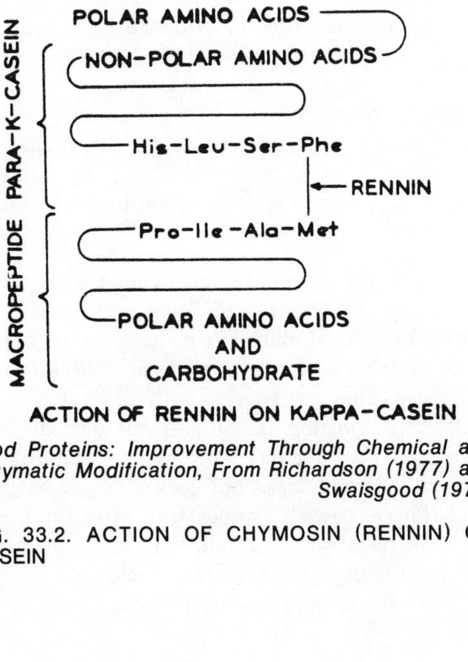

Food Proteins: Improvement Through Chemical and Enzymatic Modification, From Richardson (1977) and Swaisgood (1975)

FIG. 33.2. ACTION OF CHYMOSIN (RENNIN) ON CASEIN

Indeed k-casein has been studied as an isolated substrate not only to learn more about the nature of chymosin action in milk-clotting action under controlled conditions, but also as a more sophisticated and reliable measure of renneting power and to more fully understand casein micelle structure. Reciprocally, clues to the micelle structure should lead to a better understanding of the clotting mechanism. Thus, Humme (1972) measured the solubilized *N*-acetyl neuraminic acid present in the newly formed glycopeptide. Alternatively one conveniently measures the formation of para-k-casein by virtue of the circumstances that it is the only protein which, at the pH of milk, adsorbs to carboxymethyl cellulose and also contains all of the UV-absorbing, readily measurable amino acids of the parent molecule (Miyoshi *et al.* 1976). By measuring several changes induced in k-casein-chymosin reaction systems, Bingham (1975) was able to detect what were presumed to be intermediate stages in the formation of para-k-casein.

On the other hand Bingham also alluded to the difficulty of using k-casein as a substrate because of its tendency to form colloidal aggregates (which perhaps in analogy to the micelles of milk, are the true substrates) by uniting with other substrate molecules and also with the product. Some of these problems can be solved by using as substrates oligopeptides consisting of the amino acid sequences surrounding amino acids 105 and 106. A minimum of two amino acids on each side of this bond are required for such a peptide to serve as a substrate for chymosin (Raymond *et al.* 1973; Visser and Van Rooijen 1974). This peptide, Leu-Ser-Met-Ala-Ile, can be modified to be an even more sensitive substrate for assay of the milk clotting function of chymosin by substituting some but not all of the amino acids around the Phe-Met bond. The latter's hydrolysis is made easily detectable by insertion of a nitro group into the benzene ring of the Phe.

The tentative answer to the question of why the other caseins present in the micelle are apparently not attacked by chymosin during milk clotting is that, although they are also substrates—the peptide bond between Phe 23 and Phe 24 of isolated α_{s1}-casein is rapidly and specifically hydrolyzed (Hill *et al.* 1974; Mulvihill and Fox 1977A), and β-casein is also hydrolyzed, albeit much more slowly but nonetheless specifically (see below)—is that these other chymosin sensitive bonds are not accessible to the enzyme in the micelle during the initial stages of milk clotting. However, as we shall learn later in this chapter and in Chapter 34, subsequent degradation due to chymosin action in conjunction with that of milk and microbial proteases, may lead to significant changes in body, texture and flavor of cheese and other dairy products.

Post Enzyme Transformations

The enzymatic hydrolysis of the Phe(195)-Met(106) of k-casein is the trigger to the formation of the clot due to a series of events which have not been entirely clarified. It is generally agreed that enzyme action is not involved in this secondary stage of milk clotting, that this stage actually can be separated into two distinct phases: the primary phase or flocculation, and the secondary phase characterized by increase in firmness of the curd referred to as gelification. This firming, as mentioned, is usually accompanied by syneresis resulting in transfer of water from gel into whey. Also, the mechanism of formation of the clot probably does not, as formerly believed, involve crosslinking of the enzymatically formed para-k-casein through Ca^{2+}, calcium. Thus, Green *et al.* (1977) observed that other nonmetallic cations such as lysozyme and a salmine also reduce clotting time. From this and related observations they concluded that coagulation appears to occur by specific interactions between micelles modified by rennet.

That Ca is not directly involved in crosslinking to form the precipitate does not necessarily mean that Ca does not play an important role, either as part of the natural apatite present in the micelle, or when added, to accelerate the clotting.

The intimate mechanism involved in aggregation of micelles may involve electrostatic interactions between micelles due to altered charges on exposed parts of the newly formed para-k-casein. Thus, according to this view, flocculation is essentially an isoelectric precipitation because the isoelectric point of the para-k-casein is close to that of starter culture-treated milk (Kirchmeier 1972; Green 1972; Swaisgood 1975).

The mechanism suggested by Slattery's micelle model is that the newly exposed surfaces of the newly formed para-k-casein comprise extremely hydrophobic patches so that the micelles are attracted to each other via hydrophobic interaction. This interaction is manifested at first in the formation of fibrils and eventually in an increasingly firm clot. For the gelation phase, at least such a mechanism would be in accord with the increase in firmness of the clot during the secondary stage, an increase which is markedly affected by temperature, with variable Q_{10} values ranging to as high as 10.

Hydrophobic interaction is also in accord with the observation of Ruegg *et al.* (1974); changes in water-binding capacity play only a minor role in the secondary phase and syneresis is not accompanied by changes in the nature of hydration of the proteins of the coagulum.

The previously mentioned noncoagulability of renneted milk at temperatures near 0°C is also indicative of hydrophobic interaction in the secondary phase of milk clotting. From either point of view, the role of Ca is not that of crosslinking agent but rather to: (1) make the para-k-casein more positive (charge neutralization) as suggested by Green (1972), or (2) enhance hydrophobic interaction via a shielding effect (Slattery 1976). The net result in either case is a diminution of repulsion between the chymosin-altered micelles.

Again we encounter a situation where the enzymatic event leading to the formation of the primary product is fairly well elucidated but the more complex secondary reactions (in this case not involving the making or breaking of covalent bonds) leading to observable changes in food quality attributes still requires further investigation.

THE CURDLING STEP IN CHEESEMAKING

Standard Procedure

Even before the introduction of nonrennin rennets in cheesemaking, in principle any proteinase capable of splitting the Phe(195)-Met(106) peptide of k-casein will clot milk and serve as a rennet. However, from a practical viewpoint, only chymosin rennet was considered to be the rennet par excellence because it combines to an optimum degree the following four advantageous characteristics:

(1) Per unit weight it can clot milk much faster than any of the traditional proteinases.
(2) The ratio of milk clotting activity to peptide-splitting activities is at least double that of any other proteinase.
(3) After clotting the subsequent peptide bond splitting is much slower than that of the other proteinases, but does not cease entirely. This extremely low post-clot rate of proteolysis results in a stable firm curd needed in consequent steps leading to cheese.
(4) No bitter peptides are formed.

Rennet is usually added at about 25°–30°C in a liquid form, either batchwise or frequently metered, into the starter-treated milk at controlled rates. One runs across expressions such as "one cc of rennet per 1000 lb" or "0.2% by weight" of the milk. There is some but not too much latitude in the range of rennet to be added. For any given cheese too little rennet yields too soft a curd at normal cutting time whereas too much makes an overly soft cheese. Cheddar cheese made with a deficiency of rennet has an undesirable texture described as "curdiness."

Curd Tension.—Of prime importance is the correct assessment of the consistency of the curd as measured by "curd tension." Such measurements signal the end of the "cutting time," the interval between enzyme addition and the next step in cheesemaking, the cutting of the curd. Curd tension also serves as a reliable predictor of digestibility as a measure of the suitability of curdled milk as an ingredient in infant foods. Figure 33.3 shows the development of viscosimeter curd tension in whole milk after treatment with various rennets (see below) standardized to contain the same proteolytic activity.

Innovations and Variations

Space does not permit a complete survey of the many interesting, mostly novel, and adapted ideas and experiments designed to improve the efficiency of the curdling step in cheesemaking. By way of illustration we select three areas which seem to have had or potentially have particular enzymatic impact: cold renneting, manipulation of pH and immobilization.

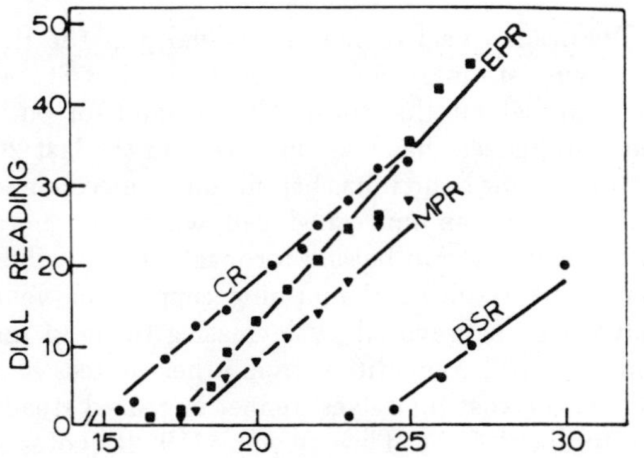

From Richardson et al. (1971)

FIG. 33.3. DEVELOPMENT OF CURD TENSION IN WHOLE MILK

Following addition of rennets prepared from calf (CR), *Mucor* (MPR), *Endothia* (EPR) and *Bacillus* (BSR) microorganisms.

Cold Renneting.—Exploitation of the previously discussed special behavior of the rennet-milk clotting system near freezing (Chapter 12) for continuous cheesemaking processes was first attempted by Berridge (1942). Subsequent developments have been detailed by Kosikowski (1975). Although there are several variations, they all have in common the initiation of clotting by warming up of starter-inoculated renneted milk in which no clot is visible. Departures from conventional cheesemaking are that part of the reaction occurred in cold milk during 4 to 24 hr storage, that final coagulation occurred instantaneously when the temperature was elevated to 20°C, and that the curds were drained and texturized at elevated temperature on moving belts. Although cold renneting is practiced in various countries such as the Netherlands for Edam cheese production and France for Camembert production, in 1980 cold renneting was not being used extensively in full-scale continuous cheesemaking.

Manipulation of Acidity.—Conventionally cottage cheese is made without renneting. Curdling occurs through development of low pH by allowing the starter organism to accumulate enough lactic acid to bring the pH down to 4.6, the isoelectric point of the caseins. This requires an extended period of time. In "direct acid" methods, detailed in patents assigned to Battelle and Swift, this period is claimed to be eliminated by preacidifying followed by renneting. A further advantage of methods combining acid and rennet coagulants is said to be a more precise control of the curd characteristic because the amount of enzyme added is not as critical as it is in conventional methods.

In the original Batelle method, recommended but not confined to the making of cottage cheese (Little 1968,1974), relatively low temperatures (less than 15°C) are used and in the Swift process (Foster and Cornwell 1965) relatively high temperatures up to 85°C are used in conjunction with renneting. In both cases the combination of lowered pH and temperature prevents coagulating until the rennet is added. A distinctive feature of the Battelle process is the use of acidification by slow release of acid from discrete particles coated with appropriate substances which retard dispersion and solubilization of the acid.

The Swift process is distinguished by the instantaneous curdling in a continuous flow of milk. In a related patent Roiner (1971) described a continuous process in which milk is heated to 26°−40°C, acidified to its isoelectric point and curdled by rennet addition while it is flowing. The striking differences between clotting in quiescent and moving milk has been the subject of considerable discussion. Thus, syneresis is accelerated in moving milk apparently due to reorientation of hydrophilic groups of the casein micelles (Motoc *et al.* 1970). In another modification of the Swift process, rennet is dispensed with and papain is used prior to acidification or addition of calcium salt, either of which coagulates the altered casein (Rice and Lantero 1974). Finally proteolytic enzymes may be dispensed with entirely according to some patents.

The principal barrier to the use of acid alone for curdling has been the texture of the resulting curd. This texture is quite appropriate for cottage cheese and Ricotta, in the production of which acid is used for curdling to form the typical soft granular texture characteristic of this type of cheese. In the process proposed by Rand and Hourigan (1975), commercial food enzymes are still used, not to clot milk directly, but to produce acid. Lactase is added to convert the endogenous lactose to galactose and glucose. Glucose oxidase is added to convert the glucose to δ-gluconolactone, which is frequently used in the dairy products as an acidulant. Rosenau *et al.* (1975) described a continuous process for the production of cheddar cheese in which a "coprecipitate" is formed by the addition of heat and acid to 2%-fat milk. Desired texture is achieved via extrusion and pressing. Direct acidification combined with continuous agitation reduces rennet requirements for Mozzarella cheese. Although not renneted, proteases might be added to such products to control one particular textural attribute, meltability (Lazaradis *et al.* 1981). Such alternatives will not obviate the need for renneting for many years to come and for most cheeses.

Immobilization.—In their review of coagulation of milk with immobilized enzymes Taylor *et al.* (1976) suggest the following advantages for them in cheesemaking. Immobilization conserves diminishing supplies of veal rennet (see below) since the enzyme does not become part of the cheese. Use of other proteases not heretofore considered as good rennets because of excess proteolysis might now become feasible. Absence of rennet action during the subsequent ripening would allow more controlled ripening times; the resulting greater storage stability would be especially advantageous with high-moisture cheeses. In the ripening process proteases could be selected for their beneficial effects on cheese ripening. Proteases used may constitute a compromise between curdling and cheese ripening propensities. Flexibility in reactor design—packed bed, stirred tank and fluidized bed reactors (Chapter 8), all applicable to continuous processing—has been used in conjunction with coagulation of milk by immobilized proteinases. Flow rate and contact times with immobilized rennets vary from 1 ml/min at 10°C to 250 ml/3 days at 4°C.

Another salient feature and potential advantage of immobilized rennet technology is its adaptation to continuous cold renneting in which both phase I and the enzyme are separated from the rest of the cheesemaking process. First suggested by Green and Crutchfield (1969), this technology as evidenced by the review of Taylor *et al.* (1976) is still in its early experimental stages. The reviewers conclude that the system developed in their laboratory, consisting of pepsin covalently attached to zirconium oxide-coated porous glass beads which are precoated with bovine serum albumin, was at that time the most advanced reported. While this approach is being perfected, such immobilized systems are being fruitfully used for elucidating the clotting mechanisms (Cheryan *et al.* 1975B).

RENNET ALTERNATIVES TO CHYMOSIN

Traditional veal rennet is a casualty of the rising living standard not only in the United States but throughout the world. The demand for both beef and cheese has risen markedly in the last 20 years. On one hand it has become quite uneconomic to slaughter an unweaned calf when the more mature animal can be sold for meat. On the other hand the resulting diminishing supplies of vells have been aggravated by increasing demand for cheese. Until competition from other sources had set in, the cost for calves' rennet increased steadily to about 1971. Thus, Wolnak (1972) quotes a three-fold price increase from 1963 to 1970−1971.

However, the advent of the microbial rennets caused the price to drop in the next year by some 25%. At that time about 20% of the cheese manufactured in the United States was made with rennets from sources other than vells. In the United States this trend may have been accelerated by the income tax structure which provided a shelter incentive for feeding calves to maturity. In other countries the shortage was occasioned by a general decrease in export of vells. Exploration of biological provenance for veal rennet substitutes for cheesemaking goes back several centuries and was looked into scientifically as far back as the 19th century. Not only higher mammals, plants and microorganisms but also such diverse sources as algae have been rennet candidates.

Animal Rennets

Other Animals.—One way to alleviate the rennet shortage would be to use the stomach of an animal other than the calf which is commonly slaughtered before weaning. This is exemplified by a Czechoslovakian evaluation of the possibility of preparing rennet from lambs' stomachs on an industrial scale. A 6–7 day extraction process was developed similar to that used for calves' vells in which 10% suspensions of lamb stomach were extracted with 5% NaCl and 2% boric acid in the presence of thymol. An acid soak is necessary to convert prorennin to rennin for both lambs and kids. More potent preparations can be obtained by incorporation of wood pulp which is said to adsorb inert mucus (Rosinec *et al.* 1977). Rennets from most other animals are largely derived from tissues high in pepsin rather than chymosin.

Pepsin Rennets as Chymosin Replacements and Supplements.—Most chymosin rennets already have pepsin in them. As much as 3% of the total milk clotting activity of the rennets from suckling calves may be accounted for as pepsin. An example of mammalian enzyme induction (Chapter 4), pepsin is present in the rennet from older calves or young calves which have been fed any food other than milk. Thus, the use of soybean protein as starter feed (Chapter 34) results in an inferior rennet. Common rennet may contain as much as 10–30% pepsin and pepsin extracts of the stomach of mature mammals contain appreciable amounts of chymosin. Both porcine and bovine pepsins (i.e., crude extracts of adult animal stomachs) are now widely accepted, usually as ingredients of blends with calf rennet. Bovine pepsin is considered to be somewhat higher in quality (Fox and Whalley 1971).

Although such blends may influence characteristics of the final cheese, the main consideration in their use appears to be price. Delforno and Gruev (1970) in a comparison of rennin (chymosin) and rennin-pepsin blends could find no differences in extent of proteolysis, yield, structure and organoleptic quality of the resulting cheeses. Fox and Whalley did detect a unique peptide in an acceptable cheese made with a pepsin rennet which was not present in a chymosin or chymosin-pepsin rennets.

Perhaps the major drawback in the use of such blends is mutual incompatability; each enzyme is unstable in the presence of the other presumably due their mutual action as proteases on each other.

As shown in Table 33.1, only at high pH's is there a tendency towards stabilization of pepsin in the presence of chymosin. O'Keefe *et al.* (1977) found that under normal cheesemaking conditions porcine pepsin and chymosin are equally stable. This is an area where some of the methods and principles of stabilization outlined in Chapter 5, especially glycosylation, may be profitably utilized.

Chicken Pepsin.—At first considered to yield an inferior cheese, chicken pepsin has experienced a renewed interest as a rennet, especially in Israel (Gutfeld and Rosenfeld 1975). A clue to the effectiveness of this rennet obtained from the chicken's equivalent of a stomach, the proventricula, is that it has the highest pH optimum (4) of all pepsins so far isolated. Glowing reports prompted the New Scientist to headline a brief report in 1971, "Chicken Lays Golden Eggs." Patents include those of Yeda (1971) and Horisberg *et al.* (1977). Chicken gizzard has also been proposed as a source of rennet (Horisberg *et al.* 1976). Immobilized pepsin has been used experimentally in continuous cheese manufacture (Ferrier *et al.* 1972; Cheryan *et al.* 1975C). It is likely that animal rennets, especially those containing pepsin, will continue to be used in cheesemaking in the future.

Plant Rennets.—The use of papain in conjunction with acid curdling was alluded to previously. The list of higher plant proteases shown in Table 5.1 includes many efficient milk clotters. Most of them were at one time or another considered as potential rennets for cheesemaking (Schwimmer 1954; Sardinas 1976). A few plant rennets have been used traditionally in cheesemaking. Thus, the extract of the cardoon flower petals *(Cunara cardunculus)* is still being used in Portugal for the preparation of Serra, a cheese from sheep milk. Potentially, plant rennets offer the advantage of the absence of toxins which are an ever-present consideration with fermentation products such as the microbial

TABLE 33.1

STABILITIES OF PEPSIN AND CHYMOSIN RENNET BLENDS[1]

pH	% Chymosin Remaining Alone	% Chymosin Remaining In Blend	% Pepsin Remaining Alone	% Pepsin Remaining In Blend
3.1	53	0	78	83
3.8	59	17	100	94
4.8	85	72	100	100
5.5	99	92	96	94
6.5	98	95	0	9

Source: Adapted from data of Mickelsen and Ernstrom (1972).
[1] Enzyme preparation stored for 48 hr at 30°C.

rennets. They can be used where taboos and mores proscribing such animal products are still in force.

However, like most other proteinases the ratio of milk clotting to proteolytic activity is, for most of these proteinases, considerably less than that for chymosin. Nevertheless, the search continues. Among the contenders which are of recent vintage or of renewed interest are prickly artichokes, insectivores, berries of *Withania coagulans* in the East Indies and the ash gourd, long used as a source of vegetable rennet. In most of these investigations NaCl and boric acid, used for extraction of chymosin rennet, are also used for extracting the plant enzymes.

Typical results with such experimental rennets are those of Gupta and Eskin (1977) on cheddar cheese made from ash gourd rennet. They reported that vegetable-renneted cheese is free from bitterness due to peptides and is highly acceptable, receiving an overall rating slightly lower than cheese produced with calf rennet, due perhaps to slight weakness of texture and coarseness. While plant sources may yet find their way into the catalogue of practical sources of rennets, in Israel the castor plant, once considered a promising source of such rennets (Zuckerman *et al.* 1963), has been replaced by chicken pepsin. Among the lower food plants, mushrooms of the *Tricholoma* genus are sources of caseinolytic proteases whose activities are enhanced by EDTA (Lamaison *et al.* 1980). An alga, sea lettuce, has been investigated as a source of milk-clotting enzymes (Abdel-Fattah and Edrees 1973). While this particular alga, which may be considered as both a plant and a collection of microorganisms, possesses only weak milk-clotting activity, many other microorganisms have been found which do indeed make excellent rennets.

Microbial Rennets

Production, Purification, Properties.—Although microbial rennets were recognized at the close of the 19th century (Gorini 1930), until about 15 years ago when the impending shortage of calf rennet was starting to be anticipated, only desultory attempts to develop commercially viable microbial rennet were recorded in the literature. It is largely due to the pioneering campaign mounted by Arima and coworkers at the University of Tokyo, that this cogent example of "tailor-made" microbial enzymes for a specific food use became commercially successful (Arima *et al.* 1967; Arima 1972). Of the over 800 microorganisms screened by this group, several dozen potentially acceptable milk clotters were unearthed. Fungal genera from which milk-clotting acid proteinases have been induced include *Absidia, Ascochyta, Aspergillus, Basidiobolus, Byssochlamys, Candida, Chaetomium, Dorthiorella, Endothia, Fomitopsis, Gliocladium, Humicola, Irpex, Lenzites, Mucor, Penicillium, Rhizopus, Sclerotium* and *Trametes*. Bacterial genera include *Alcaligenes, Bacillus, Corynebacterium, Escherichia, Lactobacillus, Pseudomonas, Serratia, Streptomyces, Thermoactinomyces* and *Vibrio*. References can be found in the concise review of rennet substitutes by Sardinas (1976). Of the promising organisms Arima's group selected one fungus, *Mucor pusillus*, for detailed study and utilization as a source of a practicable rennet (Arima 1972; Oka and Morihara 1974). The basic patent was issued to Arima and Iwasaki (1964). United States patents include that of Aunstrup (1976).

Rennets made from this microorganism and from two other fungi, *Mucor miehei* and *Endothia parasitica*, are now permitted for cheesemaking in most countries. They are all similar to chymosin in molecular weight, general clotting ability, pH optima and to some degree amino acid composition and sequence (homology).

Mucor pusillus.—From the studies at the University of Tokyo and from other subsequent studies with *M. pusillus* among which are those of Yu *et al.* (1971) and Oka and Morihara (1974), it is clear that the rennet of this microorganism belongs to a class of microbial acid proteinases with a pH optimum of 3−4. These acid proteases act on peptide bonds between aromatic, bulky and/or hydrophobic amino acids including, of course, phenyl alanine. Like chymosin, *M. pusillus* proteinase can utilize N-acylated peptides as substrates but much less efficiently than can pepsin. Since it is incapable of splitting Lys-X peptide bonds, it cannot activate trypsinogen as can some other microbial proteinases.

In this respect it differs from most other microbial acid proteinases. It also differs from chymosin in being more heat stable, more affected by Ca^{2+} and of course in not being as good a milk clotter per unit of proteolytic activity (Table 33.2).

Mucor miehei.—The rennets from this microorganism for cheesemaking were developed in the United States by Miles (Charles *et al.* 1970) and in Denmark by Novo (Aunstrup 1976). Its specificity is more restricted than that of *M. pusillus* rennet and of chymosin in that while chymosin prefers to split peptide bonds comprising aromatic amino acids, *only* those bonds containing aromatic amino acids are susceptible to the action of the *M. miehei* enzyme (Ottesen and Rickert 1970; Sternberg 1972; Oka and Morihara 1974). It has been demonstrated that like chymosin, it contains carbohydrate but is unique in having ornithine as a constituent amino acid.

Endothia parasitica.—This rennet was also developed in the United States at Pfizer. The fungus can be induced to produce high levels of enzyme activity by growing it on or in a medium containing milk and soybean meal. Studies on the enzyme

TABLE 33.2

MILK CLOTTING TO PEPTIDASE RATIOS OF SOME PROTEINASES[1]

Enzyme	Ratio	Enzyme	Ratio
Plant:		*Microbial:*	
Ficin	0.05	*M. pusillus*	0.6−0.7
Papain	0.05	*E. parasitica*	0.3−0.5
Chymopapain	0.1−0.2	*M. miehei*	0.3−0.6
Melon	0.01	*B. subtilis*	0.07
Animal:		*P. roqueforti*	0.07
Pepsin	0.2−1.0	*Rhizopus chinensis*	0.18
Chymotrypsin	0.01−0.05		
Trypsin	0.001		

Source: Values obtained from Arima (1972), Ilany and Metzer (1976), Matsubara and Feder (1971) and Puhan (1969).
[1] Relative to that of rennin = 1.0.

(Sardinas 1968; Whitaker 1970; Larson and Whitaker 1970) indicate that it has somewhat broader specificity than the *Mucor* enzymes and, unlike them but like chymosin, it can split a lysyl peptide bond and thus convert trypsinogen to trypsin. This property is used to advantage for amplification by as much as 1000-fold the sensitivity of its assay. It is some less sensitive to Ca^{2+} than are the other milk clotters.

Other Microbial Rennets.—While many other microorganisms (both fungi and bacteria) produce rennets and have been used to clot milk in cheesemaking, only the three fungal proteinases just discussed have been adopted to any large extent, if at all, in any country. Much of the experimentation with such enzymes during the 1960s and the first half of the 1970s has apparently tapered off. Examples of other fungi which apparently give fairly satisfactory rennets are *Penicillium citrinum* (Abdel-Fattah and El-Hawwary 1972), *Rhizopus* (Wang *et al.* 1969) and *Aspergillus ochraceus* (Foda *et al.* 1975).

Bacteria can also be coaxed into inducing milk clotting enzymes. *B. polymyxa* rennet production has been patented and a rennet made from this organism has been offered by the Wilson Laboratories under the trade name "Milcozyme." Attempts to use a variant mutant of the GRAS organism, *B. subtilis,* resulted in rennet which gave as soft but acceptable cheese (Puhan and Irvine 1973). The key to success was to increase the ratio of milk clotting capacity to proteolytic activity by lowering the pH. Earlier Murray and Kendall (1969) patented a rennet substitute from this bacterium in which the key claim was the partial isolation of both an acidic and a neutral peptidase.

Besides using safe organisms, successful adoption is predicated on the juxtaposition of several fortuities such as the absence or economical removal of nonclotting proteases. Thus, offending proteinases may be removed from *M. pusillus* rennets by sorption to silicate (Moelker and Mattijsen 1971). Casein has been used as an affinity adsorbent for the milk clotting enzyme of *B. subtilis* leaving unwanted protease in solution (Kuila *et al.* 1971). Removal of other esterases is discussed in Chapter 5.

In a syncretic synthesis of the very old and new, it may not be too far-fetched to assume that at least some enzyme manufacturers are indeed looking into the production of calf chymosin rennet by inserting the chymosin gene into an appropriate microorganism—via recombinant DNA technology. As mentioned in Chapter 5 this idea was one of the first proposals for the implementation of this technology. Since it is expected that one of the first applications of this technology will be the manufacture of insulin, and since established producers of insulin are hard at work towards this goal, it may be of interest to note that at least one of the principal suppliers of microbial rennets, Novo, also has been engaged in insulin production for many years in Europe where most national governments have not imposed strict rules for conducting such research.

Use in Cheesemaking.—In addition to the absolute requirements of safety and wholesomeness, in the final analysis successful adaptation of a microbial rennet will, as mentioned, be predicated on successful solutions of several problems.

That some of these problems have been overcome for the three microbial rennets shown in Table 33.3 is attested to by their widespread adoption throughout the world. As shown in Table 33.3, the trade names of these preparations are somewhat more palatable than the names of the organisms from which they were produced. Also included in Table 33.3 are cheeses which have been successfully made from these rennets. As judged from the abundant literature, in general, the cheeses are said to be indistinguishable from those made with animal rennets. Occasionally it will be stated that such cheese is actually superior, but more frequently that the experimental cheeses are slightly softer and ripen faster than do the chymosin rennet controls. However, slight reported differences may be due to variation of controls rather than to significant differences among rennets. Examples follow. It was reported that *Endothia*-derived Italian cheeses have a somewhat more sharp taste due to higher levels of polypeptide (Resmini *et al.* 1971). It has been stated that cheddars made with *Endothia* rennet should be consumed fresh and not be allowed to age too long. Slight increases in time or curd hardening, acidity development and whey drainage were observed when *M. miehei* rennet was used in making Camembert. Saint Paulin and Emmental cheeses were observed to be slightly softer, mellower and milder (Ramet and Alais 1973). Occasionally it is recommended that an unfavorable milk-clotting to proteolytic activity ratio can be compensated for by decreasing the rennet added and increasing by a small amount added $CaCl_2$ (Prins 1973).

Rennets from either *Endothia* or *Byssochlamys fulva* proved to be satisfactory for Edam-type continuous cheesemaking (Jedrychowski *et al.* 1975). More recently Wong *et al.* (1977), from a comparison of cheddar cheeses made with different rennets—*M. miehei, M. pusillus,* chymosin and chymosin-pepsin blends—concluded that no significant differences in the composition of the cheeses could be attributed to the type of clotting enzyme with one exception. One lot of *M. miehei* enzyme yielded a rancid cheese apparently due to the presence of the undesirable lipase which, as previously mentioned, can be economically removed. The authors made the cautionary admonition, which should be applicable to all food enzymes, that each new lot of enzyme should be carefully checked by both the manufacturer and the cheesemaker or other enzyme user for contaminating enzymes

TABLE 33.3

MICROBIAL RENNETS' TRADE NAMES, SUPPLIERS AND CHEESES

Trade Name and Supplier	Some Cheeses Made with Rennets
From *Mucor miehei*	
Fromase, Wallerstein, G. B. Ferm.	(1) Butter, (2) Brick, (3) Camembert,
Hannilase, Chr. Christian Lab.	(4) Cheddar, (5) Dolmati, (6) Edam
Marzyme, Miles/Marschall	(7) Gouda, (3) Italian-type, (9) Saint Pauli
Morcurd, Pfizer	(10) Trappist, (11) from buffalo milk
Rennilase, Novo	
From *Endothia parasitica*	
Sure Curd, Supraren,	(13) Brie, (3), (4), (14) Colby,
(international market),	(15) Emmenthaler, (8), (16) Muenster,
Pfizer	(17) Swiss (19), (18) Palputzei,
	(19) from lambs' milk
From *Mucor pusillus* Lindt	
Emporase, Dairyland	(1), (2), (3), (4), (20) Cottage, (6), (7), (8),
Food Laboratories	(20) Tilsit

(Chapters 5 and 36). The *M. miehei* enzyme has been immobilized and used experimentally in cheesemaking (Cheryan *et al.* 1975C).

Blends and Their Detection.—For a variety of reasons, it may advantageous to some cheesemakers to use blends of microbial and animal rennets. These advantages may include price, improved quality and quality control, and making cheese which could not be made with 100% microbial rennets. It would seem that pepsin is more frequently used than chymosin rennets for such blends (Richardson *et al.* 1967; de Koning 1972; Shovers *et al.* 1975; Greene and Stackpoole 1975). As with chymosin-pepsin blends, the microbial rennet-pepsin blends tend to lose activity due to self-digestion unless careful control and adjustment in conditions are instituted.

The use of such blends seems to have inspired a small literature on the detection or distinguishing of their individual proteinase components. Whereas such methods in the early 1970s were based upon electrophoretic physical separation and column techniques of the enzymes using traditional techniques such as column methods and gel electrophoresis (Shovers *et al.* 1972) and ion exchange (Garnot *et al.* 1972), more recent methods appear to rely on differential inactivation of the various enzymes. Thus, the z values (Chapter 10) for inactivation were found to be 4.52 for *M. miehei*, 4.62 for *M. pusillus* and 5.36 for rennin (Hyslop *et al.* 1975). Regnier (1977) determined that commercial *M. miehei* enzyme preparations were somewhat more heat resistant than those from *M. pusillus*. Pepsins can be easily distinguished from other rennets by their relative instability at high pH (Shovers *et al.* 1973) and at low pH in the presence of urea (Mulvihill and Fox 1977B). Carini and Todesco (1973) used electrophoresis to distinguish specific polypeptides released from β-casein by *M. miehei* but not by calf rennet. Some of the more biologically oriented techniques such as immunochemically-based separations used in the baking industry for the wheat amylases (Chapters 32 and 36) have been extensively explored for distinguishing the rennets in cheeses.

POSTCLOTTING CONTRIBUTION OF PROTEINASE ACTION TO CHEESE TEXTURE

The texture of the cheese-in-the-making curd after cutting, cooking, draining and matting of the rennet, for most cheeses, is more appropriate for chewing gum than for cheese. Conversion of rubbery resilient matted (or cheddared) curd to the final desired texture is determined during the pressing and subsequent curing operations. While the moisture content (*ca* 40%), fat (*ca* 25–30% for most full fat cheeses), pH and gas production all contribute to the textural attributes of the comestible final product, at least equal and, for some classes, paramount contribution is that of the physical and chemical state of the 15–30% protein present.

The disposition of the protein in the curing/aging/maturation phase of cheesemaking may be largely determined by the action of proteinases. This action when properly controlled is essential and obligatory not only for the purpose of attainment of proper texture and appearance but for its contribution to desirable flavor by providing (1) amino acid and oligopeptides in the proper proportion, (2) nutrients for the growth of flavor producing microorganisms which may also contribute to the protease pool, and (3) a means for accelerating autolysis of starter bacteria.

On the other hand too much proteolytic activity during the aging/curing of cheesemaking leads to undesirable off-flavor as well as textural attributes. The latter is manifested as an oversoftening and the former as a bitterness due to the presence of bitter peptides (Chapter 34). Thus, as in the aging of meat (Chapter 27), the attrite action has to be fine-tuned to the needs of the particular food characteristics to be acquired. While it is fairly certain that proteinases do indeed determine to a large extent the intricate sequence of events which lead to the desired, usually smooth, texture with varying degrees of firmness, their action is still but hazily perceived at the molecular level. A key to the early literature of these enzymes is the review of Ernstrom and Wong (1974).

While progress in this field has established that the consistency of cheese is related to undissolved N components (Noomen 1977), the relative contribution of the sources of proteases, peptidases and oligopeptidases responsible for dissolving these components is by no means settled. The three main sources usually considered significant are endogenase proteinase of the milk, that released by lysis of the starter bacteria and the surviving rennet enzymes carried over into the aging curd. In addition, some investigators consider microorganisms present in the milk before addition of starter as well as those responsible for ripening as minor sources whose action may affect cheese quality, including texture. A further source is commercial food enzyme added deliberately after clotting to modify and more fully utilize the final product. Undoubtedly for many cheeses all three principal proteinases are bought into play. Thus Harper *et al.* (1971) reported that in cheddar cheese slurries in which proteolytic events stretching over months of ripening are telescoped into a few days, proteinases from all three sources contributed to casein hydrolysis. The casein components were degraded at rates in the order, para-k-casein > β > α. Flavor development appeared to be directly related to disappearance of β-casein only. We deem it expedient to scrutinize each of these sources of protease separately.

Milk Protease System

Although first detected at the close of the 19th century, it is only relatively recently that substantial progress has been made in characterizing and assigning a role in cheesemaking to the milk enzymes. Selected key references testifying to this progress include those by Harper *et al.* (1971), Dulley (1972) and one of seven or more papers in a series by Reimerdes *et al.* (1976).

It is now well-established that the relatively feeble proteolytic activity of cows milk, probably originating in the mammary gland, is localized entirely within the casein micelles associated with α_{s1}- and β-caseins. The enzyme has been extensively purified and comprises at least two distinct proteases, a trypsin- or plasmin-like endopeptidase with a pH optimum of 7.5 to 8 with a high specificity for the lysyl and arginyl peptide linkages, and lower levels of a leucine exoaminopeptidase with a pH optimum at 7. Associated with milk whey proteins are proteinase inhibitors, 10–20% of whose activity is due to specific trypsin inhibitors.

Action in Milk and Cheese.—It has been repeatedly demonstrated that the preferred substrate of the major proteinase component in milk is β-casein. Thus, Reimerdes and colleagues confirmed that the purified enzyme hydrolyzed β-casein three times faster than it did total casein. Furthermore, β-casein was the only one of the isolated components of casein susceptible to its action.

Milk protease acts on β-casein in stored milk to yield what was at first thought to be a minor native component, γ-casein. It is now known that three specific bonds in β-casein are cleaved; Lys-(28)-Lys-(29) is especially susceptible to yield γ-casein fragment 29–209 and a 1–28 fragment which does not contain aromatic amino acids (thus escaping detection by the usual UV method); Lys-(105)-Lys-(106); and Lys-(107)-Glu-(108) to yield the TS-casein fragment from R-casein.

The level of activity of milk protease is so low that its contribution was considered to be of minor importance as compared to that of starter and rennet enzymes.

Its contribution to total proteolysis increases with increasing duration of ripening and its principal effect has been considered to be more on flavor than on texture. Thus, Visser and De Groot-Mostert (1977) conclude the contribution of milk protease to casein degradation is low in relation to that of rennet and starter. Its heat stability is such that it survives pasteurization and indeed its apparent activity is appreciably enhanced by heat treatment (Noomen 1977).

Creamer (1976) ascertained that the above-mentioned β-casein-derived peptides show up in Gouda cheese in 10 months and in cheddar cheese after 30 months. β-Casein is particular resistant to nonmilk proteases (Phelan *et al.* 1973). Noomen (1977), from model cheesemaking experiments, concluded that not only was weakening of cheese body very closely related to protein breakdown, but that the contribution of milk proteases to breakdown might be of more interest than frequently thought.

Rennet Proteinases

To ascertain the role, if any, of rennet proteinases in the post-clot stages of cheesemaking, one of the first questions to be asked is how much of the rennet is retained in the curd and how much of the retained enzyme survives during the long months of ripening. Some salient results of an investigation designed to answer these questions is shown in Table 33.4. Data are shown for survival of activity as function of pH and stage of cheesemaking. It will be noted that pepsin is the most unstable of the rennets tested and that the retention by the curd of chymosin but not of the microbial rennets was found to be pH-dependent. Holmes *et al.* (1977) concluded that the role of these enzymes in cheesemaking is nonexistent or minor as compared to that of proteinases released by cheese microorganisms. Similar conclusions were reached by Dulley (1974) based upon the lack of response of change in cheese quality to changes in rennet levels.

TABLE 33.4

PERCENTAGE SURVIVAL AND RETENTION OF RENNET ACTIVITIES IN CHEESEMAKING[1]

Stage of Cheesemaking	Cut		Dip		Press
Type of Rennet	Survived	Retained	Survived	Retained	Retained
Pepsin	21	*ca* 2	9	0	0
Chymosin	100	30	60	8	6
Pepsin-chymosin	76	23	22	5	4
Mucor pusillus	100	17	99	8	3
Mucor miehei	100	18	102	5	2

Source: Adapted from data of Holmes *et al.* (1977).
[1] Prior to aging.

Another question to be answered is what protein components, if any, are attacked by the rennet enzymes during the curing process. Chymosin appears to act readily on the isolated α_{s1} fraction of the casein micelle to yield a series of five major peptides, the Phe-(23)-Phe-(24) bond being one of the most susceptible (Hill *et al.* 1974). The specificity appears to be dependent upon pH and state of aggregation of the substrate (Mulvihill and Fox 1977A). Pronounced endopeptide hydrolysis of α_{s1} by chymosin to one of the major peptides (α_{s1}K) in aged cheese was strikingly demonstrated as part of the data of a control experiment, in a probing "cheese model" study of the contributions of ripening microorganisms to caseinolysis by Gripon *et al.* (1977) using aseptic curd (Fig. 33.4). In such curds slow hydrolysis of added δ-gluconolactone sustains acid production normally provided by starter microorganisms.

β-Casein, although hydrolyzable *in vitro*, appears to be particularly resistant to rennet proteinases in cheese, in part due to inhibition by NaCl and in part to inaccessibility of potentially hydrolyzable bonds occasioned by formation of aggregates of α- and β-caseins (Creamer 1976).

The lack of contribution of rennet action to protein breakdown in the post-clot phase of cheesemaking is not unanimously agreed upon. On the basis of previously cited results, Noomen (1977) suggested that protein breakdown in Noorhollandse Meshanger cheese could be attributed to the calf rennet used, in agreement with previous conclusions of Green and Foster (1974). Indeed proteolysis was greater in aseptic curds than in the presence of starter bacteria, and rennet enzyme acted with starter enzyme, perhaps synergistically.

Because they are more heat stable than are enzymes of calf and other animal rennets, the enzymes of microbial rennets are more likely to persist in milk and dairy products. Indeed their survival in wheys may create stability problems in sterile-based dietetic foods (Hyslop *et al.* 1979; Thunell *et al.* 1980).

Microbial Proteinases Arising from Cheese Microorganisms

Proteinases derived from microorganisms (other than the microbial rennets) could conceivably arise from the three sources. In decreasing order of putative contribution to quality changes during the ripening process they are: from added starter organisms such as *S. lactis* and *S. cremoris*, from organisms participating in the ripening process and from organisms adventitiously present in the milk.

Starter Protease Systems.—All starter organisms examined have been shown to possess considerable proteinase activity (Schmidt *et al.* 1977; Gripon *et al.* 1977). The protease systems of these organisms comprise (1) extracellular but surface bound proteinases, (2) intracellular casein-digesting enzymes whose formation is induced by casein, and (3) a ribosome bound dipeptidase whose action may be more important for flavor than texture. One of the proteinases has been extensively purified and characterized (Ohmiya and Sato 1975). Variations in specificity have been observed among enzymes from different strains of starters (Schmidt *et al.* 1977). Of particular interest is the spontaneous mutation of "fast" to "slow" acid producers. This phenomenon appears to be correlated with the organism's protease system (Westhoff and Cowman 1971). The broader specificity of the mutant slow strain results in an increase in free amino acids which may repress "fast" protease biosynthesis. These mutants may create both texture and health-related problems.

There is ample evidence that the starter proteinases are released from the microorganisms attendant on the latter's death and lysis. As mentioned, some of the other proteinases from other sources may cooperate in initiating this lysis. Autolysis and release of protease from starter bacteria was probably first convincingly demonstrated by Ohmiya and Sato (1972,1975) using both nutrient-rich culture media and aseptic rennet curd. From such experiments, from the presence of high concentration of cells in the curd and from the properties of the enzyme within the cell (they act on all casein fractions efficiently whereas milk and rennet proteinases do not) these and other investigators have concluded that starter microorganisms must be a major contributor to the proteolytic activity during ripening. The aseptic cheese model experiments of Gripon *et al.* (1977) provide ample evidence for such contribution by showing increase in casein-derived soluble nitrogen in general and specifically identified casein fragments in particular as a result of adding homogenates of starter microorganisms. Figure 33.4 vividly demonstrates that starter protease action is distinctly different from and complementary to that of added rennets.

Although, as pointed out by Gripon *et al.*, use of homogenates in these "model" cheesemaking experiments yielded somewhat slightly different products than those obtained with live starter, it is clear that a principal task of the starter endo/dipeptidase protease system is to produce high proportions of amino acids and oligopeptides. This suggests that starter contributes largely to flavor and that the rennet enzymes may be at least as important in texture and body-related functional properties of the finished cheese as proteinases from other sources. Indeed, Zevaco and Desmazeaud (1979) concluded that the action of a Ca-containing starter protease may complete the hydrolysis initiated by chymosins.

Proteinases of Ripening Organisms.—In their thorough study, Gripon *et al.* also showed that extracellular acid proteinase isolated from representative ripening *Penicillium* microorganisms, i.e., *roqueforti*, belonging to the same sub-subclass as the rennet enzymes, EC 3.4.23 (Modler *et al.* 1974), do not (as do the starter enzymes) liberate free amino acids and oligopeptides. They are, however, quite effective in converting β- to δ-casein (Fig. 33.4C). It could appear that ripener endoproteinases, along with the rennet exopeptidases, participate in taste development due in part to free amino acids and to their metabolic transforms. The involvement of enzymes in aroma genesis in cheese was discussed in Chapter 23.

From further results obtained by Gripon *et al.* with a neutral proteinase from *P. caseicolum* and from other investigations with proteases of other ripening organisms such as the micrococci (Moreno and Kosikowski 1973), aspergilli (Nakanishi and Ito 1973) and propionibacteria (Langsrud *et al.*

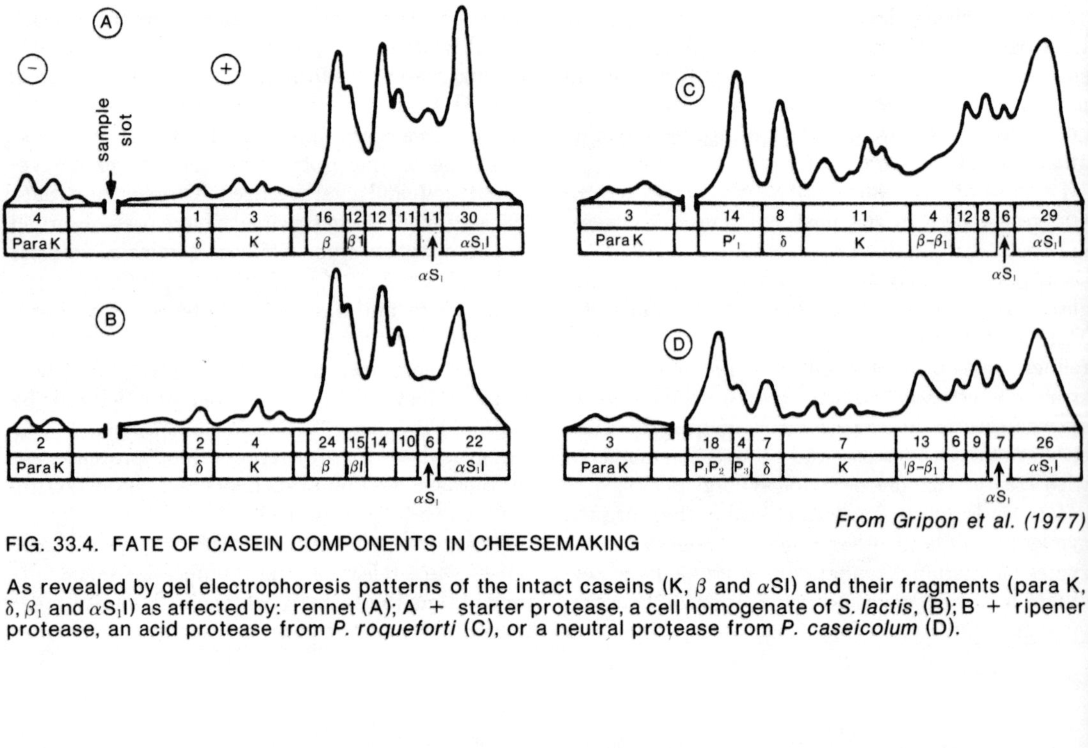

From Gripon et al. (1977)

FIG. 33.4. FATE OF CASEIN COMPONENTS IN CHEESEMAKING

As revealed by gel electrophoresis patterns of the intact caseins (K, β and αSI) and their fragments (para K, δ, β_1 and αS_1I) as affected by: rennet (A); A + starter protease, a cell homogenate of *S. lactis*, (B); B + ripener protease, an acid protease from *P. roqueforti* (C), or a neutral protease from *P. caseicolum* (D).

1977), it is clear that the degradation of β-casein by enzymes of ripening microorganisms is a major factor for texture and flavor. For instance the liberation of proline is responsible for the sweet taste of Swiss cheese.

Finally we should like to point out that not all microbial proteinase action is desirable. Especially damaging to firmness, according to Scott (1972), is bacterial protease activity in milk due apparently to proteolytic psychrophilic (cold-loving) microorganisms (Juffs 1973).

Phosphatase

Phosphatase activity in dairy products is important not only as an index of adequacy of pasteurization (Chapter 11) but also in cheese ripening. This participation of phosphatase is due to the circumstance that phosphate is covalently linked to the β-OH group of one or more serine residues of each of the major casein components: 8, 5 and 1 phosphates per molecule of α_{s1}-, β- and k-casein, respectively (Bingham and Farrel 1977). These investigators also provide a thorough review of phosphoproteins in milk, especially their formation as a posttranslation modification of casein, an event in the pathway leading to the self-assembly of the casein micelle. The caseins are phosphorylated by a phosphoprotein kinase localized in the Golgi bodies along with enzymes responsible for the glycosylation of the caseins. It will be recalled (Chapter 4) that cyclic AMP-dependent kinases are involved in translation of hormone into enzyme action via cascade regulatory systems.

It is also as a result of earlier work of Bingham and colleagues that it has been well-established that the caseins are dephosphorylated by a fairly specific phosphoprotein phosphatase present in milk (Bingham and Zittle 1963) which acts on phosphoserine residues in casein. Its pH optimum, 4.6, is distinctly different from that of the alkaline phosphatase (9−10) whose activity is used to monitor pasteurization. Its role in cheesemaking is suggested by the studies of Dulley and Kitchen (1973) which led them to conclude that the dephosphorylation of the caseins catalyzed by the phosphoprotein phosphatase is the rate-limiting step in the breakdown of casein during cheese ripening.

To summarize, the non-milk clotting action of enzymes in cheesemaking is an extremely complicated process in which rate and extent of protease and phosphatase action by enzymes from four and perhaps five biological sources—milk, rennet and starter, ripening and adventitious psychrophilic microorganisms (present in the milk)—act in mutually dependent reactions which eventually determine, to a large extent, quality attributes of cheese including texture and flavor. Rapid progress is currently being made towards clarification of the various contributions.

Addition of Enzymes

If indeed an acid phosphoprotein phosphatase is the pacemaking enzyme in the cheese ripening process, it should be possible to hasten this process by adding an external source of phosphatase, perhaps a special commercial food enzyme. As far as we are aware, no such enzyme has been experimented with nor is it available as a commercial food enzyme for such purposes. However, a limited literature indicates that some efforts have been made to partially replace the action of enzymes from ripening organisms by adding commercial food grade proteases with and without lipase (Kosikowski and Iwasaki 1975; Sood and Kosikowski 1979) or by the action of enzymes from special strains of *A. oryzae* (Nakanishi and Itoh 1974). Gripon *et al.* (1977), who suggested that penicillia proteases might also be added to improve quality and accelerate cheese ripening, point out that control of residual enzyme activity in ripened cheese during storage and distribution needs more development before such enzymes can be added.

Other proposals include patents for the use of proteinases as adjuncts in cheesemaking and in the manufacture of other dairy products for the purposes of modifying textural and functional properties. We alluded earlier in this chapter to adding proteinase prior to acid curdling. Representative proposals for the other uses of proteinases follow. Thus, rework cheese (trimmings, rejects, line changeovers, remainders) can be utilized. Significant functional properties of process cheese such as pumpability, flowability, sheetability (each related to cheesemaking steps), body and spreadability of the final product can be improved by adding appropriate commercial proteases (Kichline and Scharpf 1972). The enzyme preparation is added before pasteurization to the mixture of cheese, water and emulsification agents commonly employed in process cheese manufacture. Proteinases have also been used in the manufacture of margarine to prevent curdling of the latter when heated above its melting point (Wilding 1966). Hydrolysis of proteins in whey by enzymes derived from cheesemaking (Jost *et al.* 1976) and deliberately added to change its functional properties (Jost and Monti 1977) are additional uses of proteinase action which perhaps belong more properly in the ensuing chapter.

Lactase.—Since lactose, unlike sucrose, is not almost infinitely soluble, it can create textural problems in dairy products. In ice cream under certain circumstances it may cause grittiness due to the formation of lactose crystals. In frozen milk concentrate this crystallization somehow causes the flocculation of casein which does not occur when the lactose is hydrolyzed by lactase, as shown in Table 33.5. Perhaps it is not the presence of crystals per se that causes flocculation but rather destabilization of the casein micelles due to withdrawal of water during freezing (Chapter 12). This destabilization might be prevented by the presence of the products of lactase action on lactose, glucose and galactose. The use of lactase for this and related purposes was proposed by Pomeranz (1964), and Ahn and Kim (1977) proposed the use of a thermostable bacterial lactase for such purposes.

TABLE 33.5

LACTASE ACTION PREVENTS FLOCCULATION OF CASEIN IN FROZEN MILK CONCENTRATE

Lactose Hydrolyzed	Stability[1]
0%	45 days
5%	50
10%	85
20%	120
30%	210
50–85%	No flocculation

Source: Adapted from data of Rand and Hourigan (1975).
[1] Time required for 20% of the casein to flocculate.

The suggested use of lactase as a generator of substrate for glucose oxidase in cheese clotting was briefly alluded to earlier in this chapter. Further discussion of lactase utilization will be found in Chapter 34 and was alluded to in Chapter 9.

Part IX

ENZYME ACTION IN PROTEIN MODIFICATION AND HEALTH-ASSOCIATED ASPECTS OF FOOD QUALITY

34

PROTEIN FUNCTIONALITY AND NUTRITIOUSNESS AS MODIFIED BY ENZYME ACTION

So far our concern with the role of enzymes in and on foods has been confined primarily to the effect of their action on the hedonically-related organoleptic quality attributes of foods, as well as to some of the economic and environmental reasons for either promoting or suppressing their action. In this chapter we shall consider the effect of their action on attributes relating to the health and well-being of the consumer, more specifically, their role in the modification of one important nutrient—protein. In particular we shall assess the role of enzyme action on the availability and nutritional value of proteins present in what are considered to be potentially useful sources of inexpensive protein. In addition we shall examine those areas of enzymes' effects on proteins which have not thus far been explored. This chapter may be considered as a bridge between the discussion of texture-related and some of those health-oriented topics to be discussed in Chapters 35 and 36.

Related topics only briefly alluded to peripherally include: the enzymes of protein-consuming subjects, i.e., how the enzymes of the human digestive tract function in synthesis and turnover and their repression by the proteins in the diet; enzyme action arising as the result of deliberate microbial fermentation; enzyme action in relation to protein modification in production of conventional foods, especially those already covered in previous chapters; and the nutritional and other effects on the health of the food consumers' own enzymes such as toxins, antinutrients and trypsin inhibitors. We shall, however, have the opportunity of surveying methods in which enzymes are used to remove such substances in Chapter 36.

In this chapter the following areas will be explored:

(1) the effect on and use of commercial proteolytic enzyme preparations added to food in order to develop or influence properties or quality attributes which are (a) principally nonnutritive in thrust (functionality and related texture attributes and flavor), (b) nutritionally oriented—availability (extractability, solubility)—and related parameters associated with protein quality;
(2) autolytic processes in which endogenous enzymes, mostly proteolytic, modify the proteins and/or influence their availability;
(3) action of nonproteolytic enzymes (mostly glucanases) for improving protein availability via enhanced extractability;
(4) undesirable modification of proteins through the action of nonproteolytic enzymes and also how to prevent it; and
(5) enzyme action in the food which renders an otherwise potentially good nutritive source of

protein unavailable or unacceptable for human consumption, and how to prevent such action.

ADDITION OF PROTEOLYTIC ENZYMES

Some of the health-associated reasons for adding proteolytic enzymes to foods are to:

(1) increase extractability of proteins in foods;
(2) improve the digestibility of the food's protein;
(3) solubilize protein denatured in the course of processing of the food;
(4) maintain protein solubility in acid media so that the proteins (or the products of their partial degradation) can be incorporated into carbonated and other acid beverages;
(5) improve the manufacture of soy milk;
(6) prepare hydrolysates for parenteral nutrition (intravenous feeding)—as well as, perhaps, for the dubious and less documented but more profitable use of the so-called protein supplements for dieters;
(7) avoid alkali extraction of protein which renders some essential amino acids such as lysine (as lysinoalanine) and methionine nutritionally unavailable;
(8) partially hydrolyze proteins in foods and feedstuffs as a preliminary to acid hydrolysis in the quantitative determination of the amino acids in the protein (Chapter 39);
(9) assess the digestibility of food protein as a guide, indicator or index of its *in vivo* digestibility (Chapter 39);
(10) improve the nutritional quality of animal feeds by the addition of commercial preparations of proteolytic enzymes; and
(11) prepare protein hydrolysates for their synergistic effect in combination with phenolic antioxidants (Bishov and Henick 1975).

Other uses of proteases usually not associated with health and nutrition and not hitherto considered in detail include:

(12) impartation and improvement of desirable functional properties;
(13) preparation of soy sauce and related seasoning (Chapter 28); and
(14) deployment in the removal of unwanted substances such as fats and those possessing noxious odor and tastes and off-colors which are associated with the intact but not with partially hydrolyzed protein.

For Principally Nonnutritional Purposes

Most of the so-called nonnutritional purposes for adding proteolytic enzyme to foods have nutritional overtones and, indeed, it may be difficult at times to define the demarcation between these seemingly disparate reasons for adding proteases.

Improvement of functional and other properties makes the food more attractive and palatable and thus more available for nutritional purposes. This is especially true for many nonconventional sources of proteins; much research effort is being expended to make them acceptable items in the human diet.

Improvement of Functionality.—Modification of proteins to improve their functional properties has, as mentioned, nutritional reverberations. As pointed out by Adler-Nissen (1977A), enzymatic hydrolysis of food proteins offers a convenient means of improving functional properties of food proteins without deteriorating their nutritional value.

Functional properties or functionality of a protein or food are those physical properties usually associated with textural attributes in which appearance plays a strong role in the integrated perception, which impart to the protein a function(s) to perform. We have had occasion to observe examples of such changes in previous chapters of how processing changes the functional properties of endogenous proteins of many prepared foods. More frequently it is customary to associate special functional properties with specific food ingredients to be added to the food to impart these special properties. Such ingredients can also be carbohydrates (Chapter 9) and lipids (Chapter 32) as well as proteins. Occasionally the food consists mainly of the functionally modified proteins as, for instance, the textured plant proteins used in the production of imitation bacon and other simulated meat products (Gutcho 1973).

Included among functional properties are gelability, viscosity, moisture-retaining capacity, elasticity, dispersibility, whippability and the related foam stability. As food ingredients, these modified proteins are used as emulsifiers, emulsion stabilizers, cloudifiers, binders for cohesion and gelation agents. They are also added to foods to absorb water, to prevent moisture loss through syneresis, impart freeze/thaw stability to frozen foods and prevent crystallization of other food ingredients. Frequently a protein has to be modified so that its properties are compatible with the food into which it goes, i.e., improved bread with a high protein source rich in methionine (Betschart 1978A). Modified proteins can impart and improve such texture-associated functional properties as wettability, "mouth feel," "drop in" and "stir" (Hoer *et al.* 1972). They have become important ingredients of a wide variety of foods such as nougats and other confections, angel food mixes, soups, gravies, sauces, fish products, liver and other sausages, ham curing brines, cheeses, jams, marmalades and pie fillings (Wolf and Cowan 1971; Prendergast 1973). While it may be relatively easy to impart these properties to a protein or food containing functional protein, it is frequently more difficult to maintain the physical state represented by this particular functionality over the period corresponding to the shelf-life of the food or food ingredient. Hence, studies on functional properties usually include determinations of stability. Kinsella (1976) provides a comprehensive survey of the functional properties of proteins in foods.

Protein preparations in current use for their functionality include gelatin, unhydrolyzed protein of egg white (albumen) and soy albumen. The latter is a product obtained by limited proteolysis of soybean flour protein with an acid proteinase, usually pepsin, to a mixture of 3–4 kilodalton polypeptides. A proposed improvement in its preparation features ultrafiltration through selective membranes as part of a continuous process for controlled degradation to peptides of preselected chain length (Roozen and Pilnik 1973). Examples of the use of proteases to impart desired functionality of foods with high protein content other than soy include egg white (Grunden *et al.* 1974), milk whey (Kuehler and Stine 1974), fish (or FPC) (Chapter 28), rapeseed protein (Hermansson *et al.* 1974) and peanut protein (Beuchat 1977).

While pepsin is used routinely for the preparation of these modified proteins, other commercial food-use enzymes have been explored in detail. Some of these preparations are listed in Table 28.3. In connection with fish protein transformations, alkaline microbial protease preparations appear to have special promise but others are also useful.

Molecular Basis of Functionality.—Functional properties of proteins can be modified by physicochemical (restructuring) or by chemical means. Restructuring includes processes such as extrusion and spinning for producing texturized products and usually does not involve the making or breaking of covalent bonds. The latter can, for the present purposes, be further divided into purely chemical (nonenzymatic) and enzymatic transformations. Chemical and enzymatic modification have been combined (Table 28.3, first item). Thus, protease action is the major but not the only enzymatic avenue to protein modification.

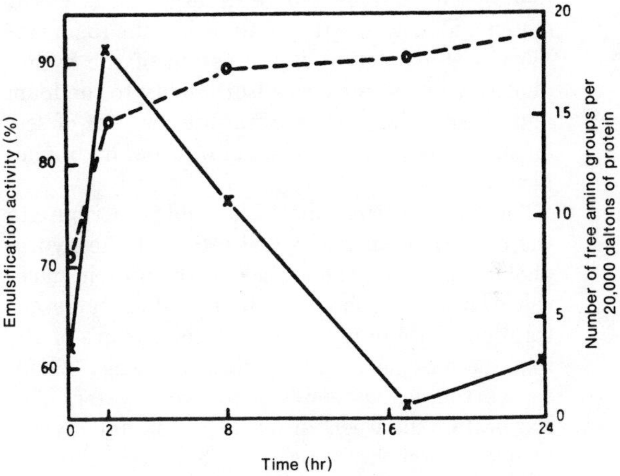

From Zakaria and McFeeters (1978)

FIG. 34.1. MAXIMUM FUNCTIONALITY OF A PROTEIN IS DEPENDENT UPON OPTIMAL DEGREE OF ENZYME ACTION

The functional property is emulsifiability, the protein is from soybean and the enzyme is pepsin catalyzing a limited hydrolysis at pH 3.

In general, a limited, optimum number of peptide bonds in a given protein molecule are split to yield the desired functionality. Further hydrolysis results in a disappearance or diminution of these functional properties as shown in Fig. 34.1 from a study by Zakaria and McFeeters (1978) on the improvement of the emulsification properties of soy protein by limited pepsin action. They suggest that either hydrolysis of a few key polypeptides results in relatively large changes in functional properties or that changes other than proteolysis occur which affect functionality. Of course these latter changes may occur as a *result* of proteolysis. Older papers hinted at a vague repeated reordering of the tertiary structure of the protein brought about by unfolding and refolding in a renaturation-like process akin, perhaps, to the regeneration of heat inactivated enzymes (Chapter 11).

Undoubtedly all of the forces that go into making up protein conformation and interaction—hydrogen bonding, Van der Waals forces, and ionic and hydrophobic interactions—all contribute. The question is which are the important influences for a given functionality. Thanks to more recent investigations such as that of Horiuchi *et al.* (1978) we are starting to obtain a clearer picture of these contributions. Upon comparison and correlation of the structure with the foam stability of five pepsin-hydrolyzed proteins—soybean, egg, casein, wheat gluten and gelatin—they found that the foam stability was a function of the extent of the hydrophobicity of the surface, adsorbability to the foam and of the total cystine (disulfide) but not of secondary protein structure nor of *internal* hydrophobicity.

The intramolecular disulfide bonds contribute to the conformation and stabilization of the hydrophobic regions on the surface of the protein molecule. This hydrophobicity in turn facilitates adsorption of these proteins onto the surface of the foam because, as in the oil droplet analogy (Chapter 3), there is a tendency to avoid water. This cooperation between disulfide bonds and hydrophobic interaction appears as a recurring theme in the creation of functional properties of important food proteins such as gluten in bread mixture (Chapter 32), casein in cheese texture (Chapter 33) and perhaps even troponin-T in meat texture (Chapter 26). In each case a limited number of peptide bonds are hydrolyzed to create stabilized hydrophobic surfaces whose interaction with other molecules and with the environment creates modified textural-functional properties in the food.

Nonproteolytic Enzymes.—The role of lipase in dough texture was alluded to in Chapter 32. Interaction of lipids with protein alters the latter, which in turn modifies the textural quality of a food (Karel 1973B). In at least one case, egg processing, such lipid-protein interaction can create problems. Minute quantities of yolk in separated egg white (Chapter 19) can dramatically reduce the foamability of the latter. This reduction can be prevented and stability recovered by addition of lipase (Murphy and Uhing 1959; Kobayashi *et al.* 1980). Catalase, used in conjunction with hydrogen peroxide as a suggested pasteurization treatment, is said to improve the functional properties of egg white (Vahedra and Nath 1973).

Soy Sauce and Related Seasoning.—Although not contributing a high proportion of the total protein intake of the diet, the conversion of high protein sources such as soy and fish condiments (Chapter 28) has been a long-established practice in much of Asia and the western Pacific area. Soy sauce is traditionally manufactured via the proteolytic action of the fungi *A. oryzae* or *A. sojae* grown on soybean meal or mixture of soy and wheat ("koji"). It is only within the last decade that an understanding of the protease system involved in this transformation has been developed. The individual enzymes comprising this system include a Zn-containing neutral proteinase and an alkaline proteinase (Nakadai *et al.* 1972; Sekine, 1972; Hayashi and Masaru 1972). In addition to these endopeptidases, amino exopeptidase action is responsible for the release of about 80% of the free glutamic acid of soy sauce and removal of most of the bitter peptides (see below). Glutamic acid is the main flavoring of this condiment (Chapter 20).

Attempts have been made to use commercial proteinase preparations to replace fermentation processes for soy sauce manufacture. However, acid hydrolysis is still the major alternative to the traditional koji microbial fermentation. For instance, in the United States acid-hydrolyzed protein is used in meat-like flavoring agents such as Vegamine (Griffiths) and Maggi Hydrolyzed Protein (Nestle). Yeast is also used as source of acid-hydrolyzed protein.

Proteases have also been used experimentally for the production of soy products such as miso and tofu which are more substantial sources of protein (Ebine 1972).

Bitter Peptides.—A side effect in applying the prescription of proteolysis to alleviate, hopefully, the world protein shortage, is the formation of bitterness in the protein hydrolysates of many of the more promising sources: soy and most legume seeds, peanuts and almonds but not those of egg white, gluten and filbert (Petritschek *et al.* 1972). In addition it may appear in more traditional foods such as cheese (Chapter 33), cocoa and saké, and in fish protein hydrolysates (Chapter 28). A medical problem, that of providing post-operative patients with a palatable high protein diet made with proteolysates of known foods, provided the spur which led to the first systematic demonstration by Murray and Baker (1952) that the bitterness is due to specific peptides derived from the action of the applied proteinases. Such predigested proteins are mandatory components of the diets of children with cystic fibrosis. Thus, the bitter peptides have posed problems for motley groups of consumers such as recuperating patients, affluent gourmets who can afford to appreciate the finer nuances of flavor in expensive cheeses and for the impoverished for whom an inexpensive source of available high-quality protein through the use of enzymes is still being sought and occupies most of our attention in this chapter.

With regard to cheese, bitter peptides in adequate but low levels probably contribute positively to overall flavor. Their flavor is probably partially masked by the presence of amino acids, astringent substances and glutamyl peptides. Occasionally they may reach levels high enough to be perceived as an objectionable taste, due to action of proteinases from adventitious or starter bacteria. Most of the peptides isolated from cheese have been identified as known sequences of amino acid of the milk proteins, especially the caseins (Chapter 33). By 1976, 61 bitter peptides had been isolated from proteolysates of foods or purified food proteins. Reference to these peptides and to further results and investigations mentioned in the ensuing discussion can be found in the comprehensive survey of Guigoz and Solms (1976).

Except for one 24-amino acid peptide, they vary in length from two to 15 amino acids and, of course, many of the shorter ones are derived from a larger one by hydrolysis. Of the 20 odd amino acids usually found in proteins, only methionine and cysteine are absent, and histidine was reported in only one peptide. The most frequently occurring of these amino acids are leucine (60 times), phenylalanine (55) and proline (50). In addition several cyclodipeptides (diketopiperazines) present in cocoa and peptides containing pyroglutamic acid (5-ketoproline)—both formed by elimination of water resulting from condensation of a COOH with an NH_2—are bitter. By 1971 it was apparent that the appearance of these bitter peptides was more dependent upon the nature of the substrate than on the specificity of the enzyme, and more on types of amino acids than on their sequence; although, as pointed out by Adler-Nissen (1977A), comparisons using different enzymes have not been carried out at a constant degree of hydrolysis.

Almost simultaneously Matoba and Hata (1972) and Ney (1971,1972) noted the high frequency of the occurrence of hydrophobic amino acids in bitter as compared with nonbitter peptides. Ney went further and proposed a set of rules which predicted that all peptides with "hydrophobicities" greater than 1400 would be bitter and those below a value of 1300 would not. The hydrophobicity of a peptide or a protein measured in calories per amino acid residue was defined as the sum of hydrophobicities of the side chains of the individual amino acids divided by the total number of amino acids per molecule. The hydrophobicity of the side chain of an amino acid is obtained by subtracting the transfer free energy of glycine from that of the amino acid. Transfer free energy is obtained from its solubility partition coefficient in a two solvent system or from its solubilities in two different solvents (in this case water, S_1, and ethanol, S_2) according to the well-known relation:

$$\text{Transfer Free Energy} = RT\ln(S_2/S_1)$$

About the same time bitter peptides were first isolated, Pardee (1951) used an almost identical relationship to predict the rate of movement of peptides on paper chromatograms. Thus, the more intensely bitter the peptide, the higher its R_F. The hydrophobicity of each of the three most frequently occurring amino acids in the protein-derived peptides listed by Guigoz and Solms—leucine, phenylalanine and proline—is greater than 2000 cal.

In general the bulkier and more hydrocarbon-like the side chain, the greater the hydrophobicity. This effect is counteracted by polar groups such as NH_2 as in asparagine and lysine, and OH in threonine and serine. The hydrophobicity values of all the side chains of amino acids found in proteins are positive, save that of glycine which has no side chain and is thus zero, and glutamine and asparagine (and perhaps serine) whose side chains possess negative hydrophobicities. Hydrophobicities of proteins reflect the hydrophobicities and bitterness of their proteolysates. Thus, casein (1605), zein (1480) and soybean protein (1540) yield bitter peptides upon enzymatic hydrolysis whereas meat protein (1300) and collagen (1280) and its derivative, gelatin, do not. Of the 61 food-derived and 145 synthetic bitter tasting peptides reviewed by Guigoz and Solms, only three, which contained alanine, had average hydrophobicities of less than 1300. As mentioned in Chapter 28 another peptide (or peptides) which may disobey the Ney rule is that obtained from FPC by Hevia and Olcott (1977), unless one does not count the asparagine. Astringency of some peptides in cheeses has been also ascribed to the presence of hydrophobic amino acids plus the phenolic OH of tyrosine (Harwalkar 1972). Hydrophobic peptide-protein interaction has also been evoked as a contributing factor to the sweetness of aspartyl dipeptide esters (Ariyoshi 1976) in this case to provide for binding to the sweet taste receptor site protein in the tongue (Chapter 20).

Preventing or Eliminating Bitterness.—Undoubtedly a low level of bitter peptides as mentioned, contributes to the overall flavor of many cheeses. Prevention of contamination by microorganisms possessing heat-stable protease systems such as that from a strain of *Pseudomonas fluorescens* (White and Marshall 1973) would close one avenue to bitter peptides. As discussed in Chapter 33 the protease system of starter microorganisms used in cheesemaking comprises both endo- and exopeptidases. The latter, if sufficiently active, could ensure removal or prevention of excess bitter peptide buildup. Indeed, one of the alleged flaws in model sterile cheesemaking experiments has been the development of bitter peptides when the starter organisms are replaced by δ-gluconolactone. Recent discussion on factors affecting bitterness and attempts to prevent bitter flavor stress the use of strains of starter bacteria with sufficient exopeptidase activity for this purpose (Stadhouders and Hup 1975; Visser 1977).

With regard to other high protein sources, it has been recommended that the hydrolysis be carried out to a lesser degree (Visser 1977). Another possibility would be to add substances such as the glutamic acid-rich peptide isolated from a digest of plastein (see below) which mask bitter peptide tastes (Noguchi *et al.* 1975). Another way would be to cleave the peptides with exopeptidases such as carboxypeptidase if they were commercially available. A rich source of exopeptidase, the kidney, was used by Clegg and McMillan (1974) to eliminate bitterness in endopeptidase-treated egg white and casein. Helbig *et al.* (1980) utilized the hydrophobicity of bitter peptides for debittering skim milk hydrolysate by adsorbing the peptides onto a hydrophobic chromatographic column (Chapter 6).

As discussed in Chapter 24 other off-flavors due to unwanted lipoxygenase and other enzyme action can be present in protein-rich food sources. Some of these flavors cling tenaciously to the undegraded protein even after the latter's denaturation and isolation. One way to remove these undesirable flavors would be to hydrolyze the proteins with appropriate enzymes. The release of these off-flavors was the primary objective of Fujimaki and coworkers at the University of Tokyo when they published the first (Fujimaki *et al.* 1968) of a series of papers on the sequel to this research, one of the relatively recent ones being that of Aso *et al.* (1977).

They succeeded in removing the objectionable odor associated with purified soybean protein by using both pure laboratory grade and commercial food grade endopeptidases, only to end up with bitter hydrolysates.

To resolve this new flavor problem they pursued two almost diametrically opposite approaches. One attack was to use an exopeptidase, carboxypeptidase. Especially effective was that present in a commercial enzyme, Molsin, which is rich in *Aspergillus* fungal acid carboxypeptidase activity.

Plastein.—The alternative approach was to abolish the C-terminal leucine which they suspected was the cause of the bitterness of these peptides by incorporating the peptides into ostensibly resynthesized protein-like molecules via the plastein reaction. Formation of plastein—first described almost 100 years ago (Danilewsi 1886)—in the presence of concentrated proteolysates by addition of fresh proteinase, was actively investigated as a controversial model for *in vivo* protein synthesis. Interest in the plasteins waned upon analysis of the free energy requirements for peptide bond synthesis and especially after publication of the landmark review by Lipmann (1941) on the role of ATP in providing this free energy. Nevertheless, the first clear indication of true peptide synthesis via condensation and elaboration of plastein was contained in a series of papers by Determann and coworkers in the 1960s (Determann and Koehler 1966). They showed that proteinases catalyzed the polymerization of pentapeptides of known amino acid sequence to form tri- to pentamers which could not be accounted for by transpeptidation.

Reviews attesting to the revived interest in plastein include those of the Tokyo group who initiated this revived interest (Fujimaki *et al.* 1971; Arai *et al.* 1975; Yamashita *et al.* 1976) and others (Schwimmer 1975; Eriksen and Fagerson 1976).

Using at first soybean proteolysates resulting from pepsin action, the Tokyo group showed that the two obligatory conditions for plastein synthesis were a substrate concentration of at least 20% oligopeptides, *ca* 0.2 *M* (optimum around 30%), and an optimum degree of hydrolysis amounting to the splitting of 70 to 90% of the available peptide bonds. Free amino acids do not readily participate, if at all, in plastein synthesis probably because it takes about 10 times as much energy to form a peptide bond from two amino acids than from two dipeptides. Of course, like all enzyme reactions, the Tokyo workers found temperature and pH optima characteristic for each of many proteinases. These, however, were not the same as those for the hydrolysis reactions, at least for most of the enzymes tried. The yield was highly dependent upon the enzyme concentration. The average molecular weights of original soybean protein, hydrolysate, and plastein were 20,000, 1500 and 5490, respectively, in apparent agreement with the findings of Determann. A schematic depiction of the process is shown in Fig. 34.2.

In spite of advances in experimental techniques and instrumentation, the controversy of the 1940s remains as to whether plastein is formed by net synthesis of peptide bonds via condensation:

$$E + (AA)_m \cdots COOH \longrightarrow E \cdot (AA)_m \cdots COOH,$$

$$E \cdot (AA)_m \cdots COOH + NH_2 \cdots (AA)_m \xrightarrow{-H_2O} E + (AA)_m \cdots CO\text{-}NH \cdots (AA)_n,$$

or by transpeptidation:

$$E + (AA)_m \cdots COOH \longrightarrow NH_2 \cdots AA_1 \cdots E + (AA)_{m-1} \cdots COOH,$$

$$(AA)_n \cdots COOH + NH_2 \cdots AA \cdot E \longrightarrow E + (AA)_n \cdots CO\text{-}NH \cdots (AA)_1, \text{etc.},$$

in which the number of peptide bonds is invariant. While the results of Determann and later of the Tokyo investigators undeniably demonstrate some direct synthesis of peptide bonds via condensation (and also that formation of intermolecular disulfide bonds are not involved), the more recent experiments of von Hoftsten and Lalasidis (1976) strongly suggest a constant reshuffling of amino acids resulting from transpeptidation and hydrolysis which are the principal reactions leading to plastein formation. Also in keeping with a major role of transpeptidation was the relatively early finding of the Tokyo workers that proteinase could catalyze the transesterification of amino acids from their alkyl esters to the peptides

$$R \cdot CHNH_2COR_1 + NH_2 \cdots (AA)_m \longrightarrow R \cdot CHNH_2 \cdot CO \cdot NH \cdots (AA)_m + R_1OH$$

$$R \cdot CHNH_2COR_1 + COOH \cdots (AA)_m \longrightarrow \begin{array}{l} R \cdot CHNH \cdot OC \cdots (AA)_m + R_1OH \\ \quad\; | \\ \quad COOH \end{array}$$

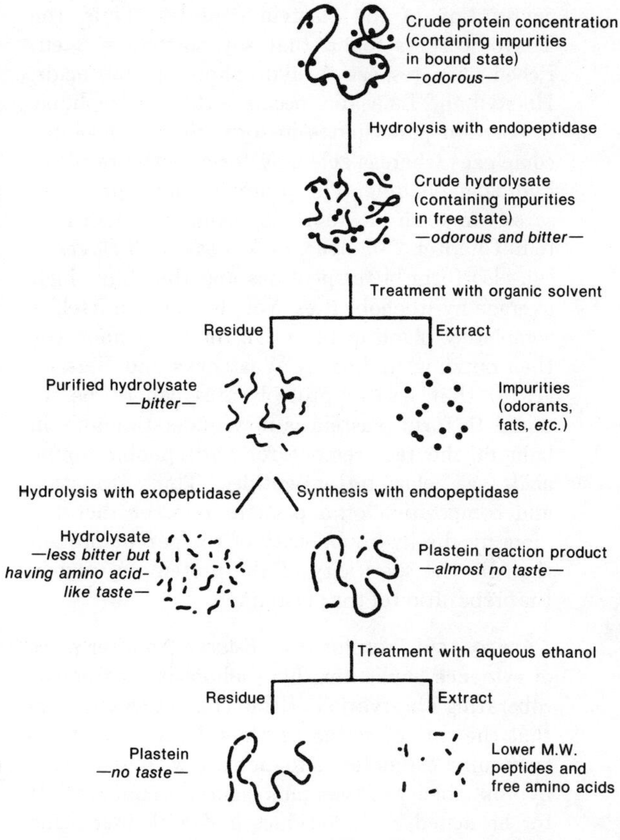

FIG. 34.2. PREPARATION AND APPLICATION OF PLASTEINS

As envisioned by the Tokyo team, the first step in chymotrypsin-catalyzed plastein formation is the interaction of the peptide with serine-195 in the active site of the enzyme, followed by a general base-catalyzed nucleophilic attack assisted by histidine 57 (also in the active site) of the resulting peptidyl chymotrypsin by another peptide to form a new peptide bond. Whether by condensation or transpeptidation plus hydrolysis the physical properties of some of the new peptides, not necessarily longer than the substrate peptides, formed by this constant reshuffling are such—precipitates and thixotropic gels at neutral but not at acid pH's, irreversible precipitability by the widely used protein precipitant trichloracetic acid—that they are incapable of further participation in the reshuffling and interact among themselves noncovalently.

Again, as with bitter peptides, hydrophobic interaction may play a major role in the formation and aggregation of the plastein peptides. Thus, the Tokyo workers found that soy plastein was enriched with respect to hydrophobic amino acids. Hofsten and Lalasidis speculate that hydrophobic interaction participates in formation of insoluble complexes whereas gels may form as the result of hydrogen bonding. The proteins whose proteolysates can form plastein—soybean, zein, fish protein (Chapter 28), algae, milk, yeast and clover—can also form bitter peptides and thus have high average hydrophobicities. Yet the plastein itself is completely bland in taste. On the other hand the then puzzling finding by Wasteneys and Borsook (1924) that gelatin proteolysates cannot be induced to form plastein is now understandable in light of the requirement for hydrophobic amino acids (see below, bitter peptides). Thus, formation and composition of a plastein may be dictated more by the hydrophobicity of the parent protein than by the specificity of the proteinase used in the preparation of the plastein.

Amino Acyl Transfer from Esters.—Another piece of evidence implicating hydrophobicity is the corroborating observation of the Tokyo investigators that the rate of amino acid uptake into plasteins from alkyl esters tends to increase with increasing hydrophobicity of the amino acid side chain except for branched chain residues and with increasing length of the alcohol moiety of the amino acid ester (Aso *et al.* 1977). Lynam and Satterlee (1980) evoke hydrophobic interaction to account for increased viscosity.

Increasing the chain length of the alcohol, which would also tend to increase the overall hydrophobicity of the amino acid ester, paradoxically permits the incorporation of large amounts of polar amino acids into plasteins with a resulting alteration of their functional properties. Thus, glutamic acid is effectively transferred by papain from its α, γ-diethyl ester to plastein to yield highly soluble products containing α-helices formed via repeated but limited pentapeptidyl transfer of amino acyl groups and some peptides. The Tokyo investigators suggest that by selection of amino acid esters it should be possible to impart a wide variety of physical properties to plastein. Thus, in addition to deodorizing, defatting and decoloring crude protein foods, it should also be possible to "tailor-make" plasteins with selected functional properties. A further step in this direction was taken as the result of the use of immobilized proteinases—papain or chymotrypsin immobilized on chitin—by Pallavicini *et al.* (1980) to produce plasteins from proteolysates of soybean, alfalfa leaf and wild grass proteins.

Plastein: Nutritional Aspects.—Whatever occurs in the formation of plastein, we know now that it leads to a product which is at least as nutritionally wholesome as the soybean protein from which it came. Thus, the chemical scores with respect to sulfur-containing amino acids and leucine were, respectively, 94.3 and 102.5% of the untreated protein. *In vitro* digestibilities were about the same (100%) and true digestibility of the plastein fed to rats was 90.3%. Biological value was 66.8, comparing favorably with the starting material. After supplementation with essential amino acids to give an amino acid pattern similar to that of casein, the ration yielded rates of growth equal to 96% of that of rats fed casein. For a key to methods of measuring protein quality, see Young *et al.* (1977).

The Tokyo group have applied the plastein reaction to enrich soybean protein with methionine supplement incorporated directly into the protein. This first enrichment was accomplished by either providing the proteolytic enzymes, as substrate for plastein synthesis, a mixture of soy protein hydrolysates and methionine-rich hydrolysates from methionine-rich protein such as ovalbumin or by supplementing the soybean protein hydrolysates with L-methionine esters. In the latter case they showed that the methionine adds on singly and not as oligomers, mostly to the carboxyl end of the peptides. By such means they raised the methionine level seven-fold. The advantage of incorporating methionine rather than adding it as a free amino acid is that this amino acid can be lost during processing *via* several routes. These include oxidation to the sulfoxide or to methional *via* a Strecker degradation (Chapter 19) and would not only destroy an essential amino acid, but would also give rise to undesirable flavors.

Too high a level of methionine, at least when fed in the free form, may also be deleterious nutritionally. At any rate, such high methionine-containing proteins can be diluted with proteins which are deficient in this amino acid. Furthermore, some evidence suggests that intestinal absorption of dietary nitrogen in the form of peptides or partially digested proteins may be preferable to absorption of free amino acids (Imondi and Stradley 1974; Chang *et al.* 1975).

On the other hand Moon *et al.* (1975) did not find significant differences between the nitrogen balances of young men fed egg albumin and those fed equivalent amounts of free amino acids. Also, the poultry industry has experienced no serious nutritional problems by supplementing chick feed with free methionine.

In addition to improving the methionine level of soybean protein, the Tokyo group was able, by the use of amino acid esters, to improve the nutritional quality of several other proteins by incorporating deficient essential amino acids into plasteins: gluten (lysine), zein (lysine, threonine and tryptophan) and, as shown in Fig. 34.3, deficient essential amino acids in proteins from photosynthetic organisms. The amounts recommended by FAO/WHO (Food and Agriculture Organization of the United Nations/World Health Organization) are shown as open columns in the background. Although these and other plasteins so far have been shown to possess superior nutritional qualities and have no toxicity, Eriksen and Fagerson point out that toxic peptides might be produced under certain circumstances, so that the safety of each plastein should be investigated in detail. This examination would be made only after the economic feasibility of producing plastein for human consumption has been demonstrated, while it has not and may not be for some time, if ever.

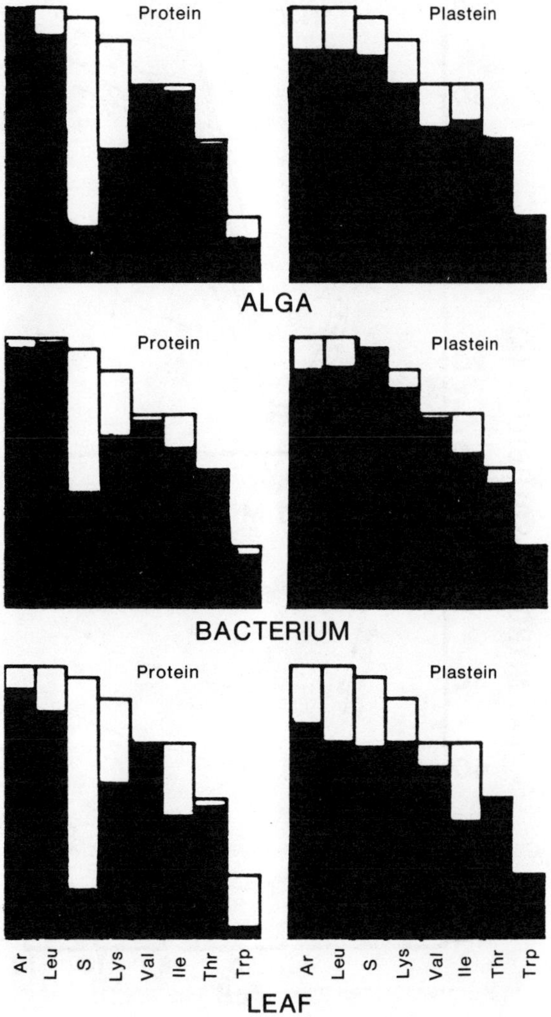

From Yamashita et al. (1976). Reprinted with permission from Agric. Food Chem. 24 (6) 1103. Copyright by Am. Chem. Soc.

FIG. 34.3. INCORPORATION OF ESSENTIAL AMINO ACIDS INTO PHOTOSYNTHETIC PROTEINS VIA THE PLASTEIN REACTION

Essential amino acids denoted by solid black columns; the background is the FAO/WHO pattern for these amino acids.

For Increasing Solubility and Nutritional Availability

Other approaches to upgrading the nutritional value of proteins in nonconventional potential sources have been rather limited but include expanding efforts to exploit the action of proteolytic enzymes. A good overview of the mechanics of proteolysis using commercial food enzymes from the viewpoint of the enzyme supplier is that of Adler-Nissen (1977A). In principle, solubilization and extraction may be regarded as yet another way of improving functionality. They are closely related to nutrition because if the protein is not extractable it is not nutritionally available. If extractable, the extraction procedure may be quite expensive or may lower the quality of the protein extracted due, in part, to destruction of essential amino acids. As in the case of FPC (Chapter 28) processing may render the protein insoluble.

Potentially, at least, the mild conditions of enzyme action circumvent some of these problems. However, control of solubility can be quite complicated. As illustrated in Fig. 34.4, solubilization of peanut protein is dependent upon the pH at which it was heated and also on the Ca^{2+} level.

One of the main general problems to be overcome is the constant danger of contamination during often long exposures of the nutrient-rich medium. One way to obviate this difficulty is to conduct the hydrolysis at pH extremes with enzymes of appropriate pH optima and stabilities—traditionally pepsin at low pH and more recently the proteinase originally designated for use at the high alkalinity of laundry detergent, i.e., Esperase of Novo which

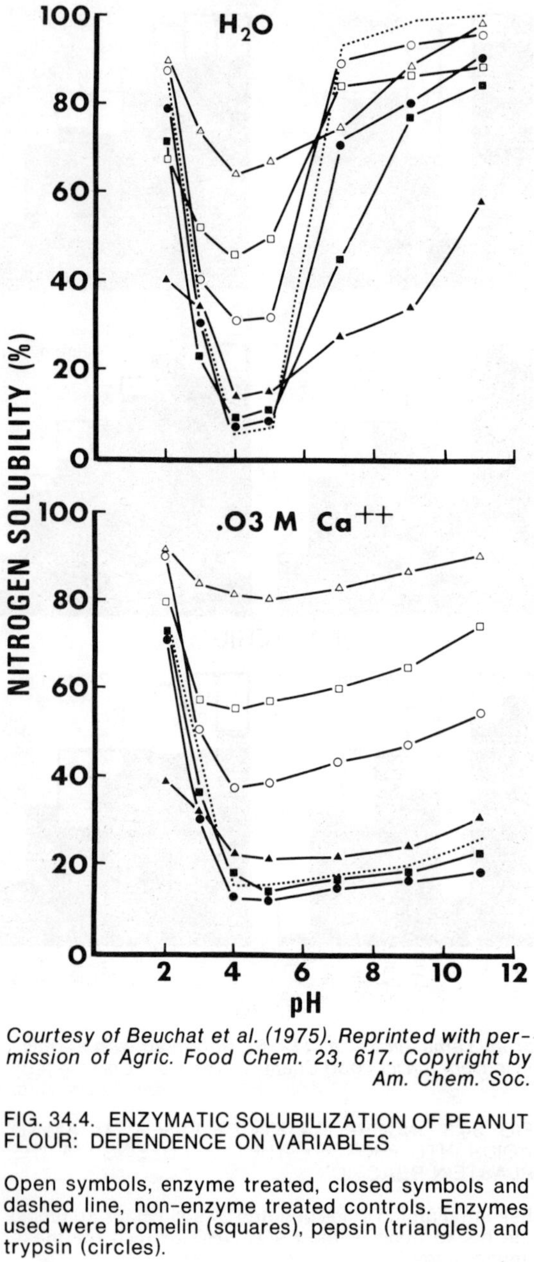

Courtesy of Beuchat et al. (1975). Reprinted with permission of Agric. Food Chem. 23, 617. Copyright by Am. Chem. Soc.

FIG. 34.4. ENZYMATIC SOLUBILIZATION OF PEANUT FLOUR: DEPENDENCE ON VARIABLES

Open symbols, enzyme treated, closed symbols and dashed line, non-enzyme treated controls. Enzymes used were bromelin (squares), pepsin (triangles) and trypsin (circles).

acts optimally at pH 10 and above (vonHofsten and Ladasidis 1976). Novo's Alcalase (pH 8) and BASAP (Chapter 28), but perhaps not Esperase, have been cleared for food use. Information concerning the solubility of soybean protein is available in reviews by Wolf and Cowan (1975) and Kinsella (1976).

Cottonseed Protein.—Rendering this protein suitable for human consumption constitutes an extreme example of the challenges to be met. Cottonseed is processed mainly for oil under conditions which result in the insolubilization and quality deterioration of its protein. Another hard nut to crack, common to other nonconventional high protein seeds, is the separation of protein from the hull of fibers. A special problem with cottonseed is that the gossypol it contains, in addition to being toxic, inhibits proteinase. Gossypol-less, glandless cottonseed has increased levels of suspected carcinogens which might find their way into protein preparations (Hendricks *et al.* 1980). Ideally one would wish to maximize oil recovery, protein yield and quality, to remove all of the gossypol, and to prevent the reserve storage proteins present in the subcellular organelles, the protein bodies, from being glued together by denatured cytoplasmic proteins. According to Martinez *et al.* (1970) these organelles should be free to separate from other cellular constituents.

Investigators at the U.S. Department of Agriculture have devised continuously improving solubilization procedures aimed at humanization of cottonseed protein, procedures in which enzyme action has played a pivotal role. A key to the literature on this sustained effort is a relatively recent report of this group by Childs and Forte (1976).

Using a heat-denatured protein fraction prepared from defatted cottonseed meal so as to have as much insoluble protein as possible, in their first effort they were able to solubilize 40–60% of the protein at optimal experimental conditions of 45°C in 5 hr. Out of ten commercial proteases tried, two from bacteria and bromelin were the most effective. As previously mentioned, microbial contamination of digests appears to be a problem where prolonged digestion at moderate pH's and temperatures prevail. In a revised procedure Childs extracted more than 50% of the protein from screw-expressed commercial cottonseed meal by the use of an enzyme-chemical method adapted from that used for coconut meal. Without preliminary enzymatic treatment (50°C, 1 hr, 25% suspension of cotton meal) only 15% of the protein was extracted by alkali (chemical) treatment. Interpolation of an ultrasonication step into the enzyme-chemical method increased the efficiency of total protein extraction some four-fold.

Other Seeds.—As with cottonseed, the heat generated in the extraction of oil from coconut not only denatures the protein of the resulting defatted meal, but has been reported to lower its nutritional value. In addition, the high fiber content of the meal interferes with the digestibility of the proteins and limits its application in dietary formulations. These problems could be solved by the above-mentioned combined enzyme-chemical treatment of coconut meal which was found to be superior to the separate use of either enzyme alone or alkali alone. Lachance and Molina (1974) were able to extract 80–90% of the protein on a pilot-plant scale. The nutritive value of this extracted protein was much higher than that of the original coconut meal.

Part of this superiority, in addition to removal of fiber, was due to relatively favorable distribution of essential amino acids, especially lysine. The PER values (corrected) of the original coconut meal, the enzyme-chemical treated extract, the extract supplemented with fiber, and of casein were, respectively, 2.32, 2.72, 2.55 and 2.50. The corresponding values for the net protein retention were 3.8, 4.5, 3.9 and 3.7. This paper also provides a summary of previous literature.

Enzyme processing of a variety of seed proteins—defatted sesame and peanut meals and four varieties of bean species, chickpeas, green gram, black gram and field bean—yielding increased solubility and nutritive values was examined by Sreekantiah *et al.* (1969). Some of their results, shown in Table 34.1, indicate that increased extractability and lysine content of sesame and chickpeas was obtained. The methionine appeared to be quite sensitive to their treatment consisting of incubating cooked meal suspensions with commercial fungal protease preparations of Japanese origin for 5 hr at 45°C. According to the authors, suitable blending of the hydrolysates should yield a protein-rich food with a reasonable balanced amino acid composition. The values they obtained are comparable to those for microbially fermented or predigested oriental foods such as tempeh and miso.

Sunflower seed is another source of potentially valuable protein. Saint-Rat (1971) points out that the biological value of the defatted seed cake is about the same as soybean meal. However, its digestibility is inferior due to the presence of hulls which are difficult to remove. As an alternative to the complex and expensive process of mechanical decortication, he incubated sunflower cake for short periods with a commercial enzyme preparation possessing proteinase, amylase, polygalacturonase and cellulase activities. The protein content of the resulting extract was 63.75% as compared with 42.5% for enzyme-treated defatted soymeal and 26% for similarly treated coconut meal. However, sunflower protein is very poor in lysine. For similar reasons, it might be worthwhile to apply enzyme processing to safflower seed, another potentially rich protein source with an intractable husk.

While an established staple food in the Soviet Union, oil from sunflower seeds is only now on the increase in the United States, a recognition of which is the official listing of sunflower seeds as a futures commodity to be traded on the Minneapolis Grain exchange. Undoubtedly the resultant increased production of a high protein by-product will spur increased research into the use of enzymes for its utilization.

TABLE 34.1

ENZYME PROCESSING OF SEEDS IMPROVES PROTEIN AVAILABILITY

	Seed Meal			
	Sesame		Chick Pea	
Measurement	Control	Processed	Control	Processed
Soluble protein (%)	4.63	31.31	7.38	19.00
Amino—N (%)	0.29	4.50	0.29	3.68
Lysine (mg/g)	14.9	18.5	26.0	17.1
Methionine (mg/g)	10.3	6.2	2.8	2.8
Protein solubilized (% of total)	—	64.4	—	86.12

Source: From the data of Sreekantiah *et al.* (1969).

Other Sources.—Proteases have been applied to leaf proteins (Edwards and Edwards 1974). Alfalfa or clover is first treated with alkali, then neutralized and digested with a mixture of pancreatin and bile. The presence of bile in conjunction with pancreatin suggests concomitant action of both lipase and proteinases.

Sekul and Ory (1977) describe a procedure for obtaining a pilot plant-produced papain-modified defatted peanut flour whose protein was 50% more soluble than that of the untreated flour.

The animal world has not been neglected as a source of unconventional protein for human use. We have already discussed the use of proteolytic enzymes in upgrading fish proteins in Chapter 28. Mammalian blood may also be a future source of large quantities of high grade protein. Thus, the plasma from the blood of slaughtered meat animals is used in Sweden and probably other countries for its excellent functional properties as an emulsifier and binder in sausage production. The red cells, mostly used now as an ingredient of animal feed can, according to Stachowicz *et al.* (1977), be upgraded to produce a light-colored edible protein product (except for a slight bitterness) and low in heme by the application of an alkaline proteinase (Alcalase) in conjunction with ultrafiltration in a hollow fiber membrane reactor (Chapter 8). Other proteins of animal origin whose hydrolysis by proteolytic enzyme has been recommended for increased nutritional availability include whey (Kuehler and Stine 1974) and keratin in various forms, especially feathers (Cherry *et al.* 1977; MacAlpine and Payne 1977).

Protein-containing Acidic Beverages.—There appears to be a highly competitive race to come up with a practical process for obtaining acid soluble protein (or its hydrolysis products) for use in acidic beverages by means of the action of added proteolytic enzymes. In addition to its yet latent benefit, as in providing a consumer-oriented product with high nutritional value, the economic potential of this activity is indicated by the fact that the preponderance of published information is available as patents as shown in Table 34.2. It will be noted that most of the raw material containing protein substrate is seed protein, but gelatin and milk whey are also included. Presumably the latter is designed to improve the nutritional value of the popular Russian beverage, kvass.

A survey of these and other procedures reveals that they almost all include, in addition to enzyme action, heat or alkali treatment, presumably to denature the protein in order to make it more susceptible to enzyme attack. In addition, all include a post-enzyme treatment for separation of acid insoluble residues, usually by centrifugation, sometimes combined with the use of an edible protein precipitant.

The Core.—Hydrophobicity, in addition to being involved in plastein synthesis and the bitterness of peptides, may also influence nutritional quality of amino acids in plant foods by contributing (along with the relatively low levels of some essential amino acids) to lower nutritional value of protein from plants as compared with that from animals.

Some 25 years ago, Birk and Bondi (1955) reported that after the *in vitro* action of proteolytic enzymes of the gastrointestinal tract on proteins present in a variety of plant feeds (soy, peanuts, wheat, maize, etc.), a TCA-precipitable "core" about 20 to 30 amino acids long persisted. Such a core was never found when animal proteins were subjected to the same enzyme action. Kirchgessner and Steenhart (1974) in a study of the distribution of amino acids in peptide cores remaining after partial hydrolysis of soybean protein with pepsin found that the essential amino acids threonine, valine, isovaline, leucine and phenylalanine are present. These amino acids might thus be less available than predicted from analysis. Furthermore, the relative dearth of such hydrophobic essential amino acids in plants is in accord with the more or less established general rule that hydrophobicity enhances proteolysis (Holtzer *et al.* 1980). This suggests that a fraction of the plant protein itself could conceivably contribute to "dietary fiber" but is not available for methionine incorporation via the plastein reaction.

Proteolytic Enzymes as Feed Additives.—The idea of adding enzymes to animal feeds, which now goes back half a century, is related to the fact that very young animals do not possess their full complement of proteinases. Furthermore, it was hoped that animals fattening on high caloric diets devoid of roughage could better utilize what protein they did receive in their ration by supplementing it with an external source of proteinase. There is no question that some commercial preparations of proteinases under certain conditions do stimulate growth of a variety of young animals including calves, pigs and chickens (Nelson and Catron 1960). On the other hand, no effects (Yang *et al.* 1962) or nega-

TABLE 34.2

ENZYME PROCESSING FOR PROTEIN BEVERAGE FORMULATIONS

Substrates	Enzymes	Conditions	Product/Remarks
Oleaginous seed	Various	Isoelectric	Drinks below pH 4.6[1]
Oilseed protein, heated	Various	pH 2.5–6	[2]
Plant protein	Acid protease	Below pH 4.6	Soluble when carbonated[3]
Gelatin and fruit juice	Papain, chymopapain	70°C, 10–20 min	Fruit type beverage[4]
Soybean protein, heat and alkali denatured	*B. subtilis* alkal. protease (BASAP)	38°–50°C, 48–100 hr	*ca* 75% protein solubilized[5]
Plant protein, acid- and heat-denatured	BASAP	pH 7.5–8.5, 6 hr + acid	75% Protein solubilized; bean-like flavor[6]
Oilseed protein, undefatted	From *Aspergillus* and *Bacillus* spp.	25°–75°C	[7]
Whey	"Terrazine"	37°C, 7–10 hr	Syrup for addn. to alcoholic ext., etc.[8]
Soybean protein isolate	Acid protease from *Trametes sanguinea*	pH 3 (lemon juice) 50°C, 8–10 hr	Lemon-flavored beverage containing protein[9]
Soybean protein	Alcalase S60, Novo, a BASAP	pH 8, 2 hr, 50°C; citric acid added to 4*M* (pH 3.5)	Yields, 68%; 3% with sucrose and lemon, makes beverage[10]
Plant and animal protein	Milezyme AFP, an acid fungal protease	pH opt. 2.5–3.5, 50°C	Enzyme readily sol; stable for 1 year[11]

[1] Source: Hawley (1974).
[2] Source: Hempenius *et al.* (1974).
[3] Source: Pour-El and Swenson (1973).
[4] Source: Appleman (1972).
[5] Source: Denault (1972).
[6] Source: Chiang and Sternberg (1972).
[7] Source: Sherba and Steiger (1972).
[8] Source: Romaskaya and Kalmysh (1971).
[9] Source: Sugimoto *et al.* (1971).
[10] Source: Adler-Nissen (1977B) and Olsen and Adler-Nissen (1979).
[11] Source: Anon. (1978).

tive effects have been reported (Lassiter *et al.* 1959). At least as indicated by the literature, there was a drop in interest which seemed to coincide with the widespread use of diethylstilbestrol (DES). One might expect a renewed interest as a result of the banning of DES as a possible carcinogen. Certainly there has been a renewed interest in eastern European countries as evidenced by reports, available only as abstracts, on its efficacy when added to feed for pigs (Wojcik *et al.* 1973), calves (Tomme *et al.* 1973), chicks (Trela 1977) and fish (Dabrowski and Glogowski 1977). That such treatment is still of interest in western Europe is evidenced by a West German patent (Saran and Hiller 1977) in which the enzyme preparation, a powder from *B. subtilis*, is suspended in soybean oil which is then sprayed onto the feed. It is not clear from available evidence that the nutritional benefit of adding commercial enzyme preparations to feed rations when advantageous (and not in all cases is it so) is due to the presence of the protease(s). Other factors such as the action of other enzymes ("amylase" is also added to feeds), the presence of unknown growth factors affecting the animal or its intestinal flora, or suppression of undesirable intestinal microbial growth by antibiotics could all contribute to beneficial effects obtained. At any rate, it appears that the addition of at least some of these enzyme preparations to young animal rations is empirically effective. Investigations with pure enzymes, although very expensive, should be undertaken to answer these questions.

PROTEIN MODIFICATION BY ENDOGENOUS PROTEASES—AUTOLYSIS

Autolysis is an alternative to adding enzyme to solubilize proteins or otherwise transform the food in which they are present. Its application is, of course, limited to those foods in which the proteolytic activity is sufficiently high to effect the desired changes. The food cannot be heated or otherwise treated to inactivate the proteinases prior to autolysis. During autolysis provision has to be made for protection against microbial contamination. One must anticipate that autolytic processes other than those due to endogenous proteinases may go on simultaneously. Also (Chapter

28) proteases may come from microorganisms and deliberately added sources such as pineapple (bromelin). The principal food source in which autolysis of the protein plays a major role in a gross transformation is fish, as examined in Chapter 28.

Autolysis of plants as means of releasing and making proteins available goes back to at least 1936 in a patent on the production of yeast protein autolysate for feed and food use (Weizmann 1936). In subsequent patents the yeast proteinase was supplemented by papain and seed protein (Weizmann 1939). Further implications of autolytic processes in yeast will be examined in connection with glucanase action later in this chapter.

Since soybeans possess proteolytic activity it should be possible to transform some soy protein *via* autolysis. Research into the nature and use of the proteinases of the soybean has been rather desultory. One reason is the experimental difficulty of separating the low level of activity from large amounts of background protein, which interferes with accurate assessment of the activity. Another problem is the presence of activators and inhibitors. This is probably the reason why Taguchi and Echigo (1968) found that conditions of soybean storage may determine whether the activity is detectable, stays the same or increases. Older literature was reviewed by Circle (1950) and more recent literature by Rackis (1972). The level of activity is of the same order of magnitude as that of wheat flour (Ofelt *et al.* 1955). Several distinct proteinases are present.

In the course of their investigations of soybean proteases Pinsky and Grossman (1969) observed a rather intriguing phenomenon which may have significance. Due to autolysis, the UV-absorbance of TCA solubles in a 10% suspension of defatted soybean which served as "no-enzyme-added" control increased, as expected, during the first 3 hr of incubation, but this increase was followed by a small but consistent transient decrease in TCA-soluble absorbance. This appears to be a general phenomenon since we have repeatedly observed it with defatted unhulled soybean meal (Schwimmer 1974) and a similar dip occurs during the autolysis of defatted cottonseed (Ory 1978) and peanut meal (Ory and Sekul 1977). In the latter case the initial increase in absorbance, usually indicative of proteolysis, was not accompanied by corresponding changes in the protein electrophoresis pattern, suggesting that the "autolysis" may not be due to protein hydrolysis. It is also of interest to note that TCA was actually used to separate the extract into six distinct proteinases, none of which had trypsin-like activity.

From the viewpoint of utilizing autolysis, it is of interest to note that the soy proteinases are maximally active below neutrality where some of the major lipoxygenase isoenzymes are practically inactive (Chapter 24).

Although autolysis as a distinct procedure or process has not been reported for the extraction and solubilization of soybean protein, it probably occurs in many processes. As in the case of some fish protein isolation procedures it may occur when it is not wanted. Such unwanted autolytic losses can also occur in recovery of proteins from plants. Thus, De Fremery *et al.* (1973) reported a 30% loss, apparently due to protease action, during the course of isolating protein from alfalfa. On the other hand, it appears to play an important role in soy milk preparation. Lo *et al.* (1968) report a ten-fold increase in nonprotein nitrogen during a 24 hr soak, a step in the preparation of soy milk. During the preparation of unheated soybean meal for extraction or for extrusion, temperatures of 50°–60°C in the range of maximum activity of the soybean proteases are used. Schwimmer (1974), using undehulled soybean meal, observed a limited proteolysis at low water:soybean ratios. When incubated at 60°C for at least 3 hr, a product with a wheat-like odor and altered functional properties was obtained.

The limited hydrolysis observed may reflect latency of the protease due to its association with stable intracellular particulates. Early observations that it required severe mechanical agitation to obtain maximum extraction of the proteolytic activity are also in line with the idea that these enzymes are located in rather stable subcellular organelles, the protein bodies of aleurone cells (Yatsu and Jacks 1968; Ory 1972). This suggests that cottonseed proteinase is not active as an autolytic agent in cottonseed because (aside from being heat-denatured) it is inhibited by the gossypol released when the cell is disrupted during oil extraction and in milling operations.

Common beans *(Phaseolus vulgaris)* have also been subjected to autolytic conditions. Although other interactions between natural substrate and enzyme were observed, Kon *et al.* (1973) did not detect proteolysis in 10% slurries of California Small White beans. Autolysis of this variety of bean has very little effect on PER when included in rat diets (Chang *et al.* 1979).

A promising high protein food source which does

have relatively high proteolytic activity is the peanut; Moseley and Ory (1973) isolated and characterized the principal proteolytic enzyme. They showed that autolysates prepared from 10% suspensions of peanut cotyledon after 12 hr at 37°C were hazy in appearance but remained essentially clear when refrigerated for a long time and should be suitable as a beverage supplement. As expected, there was a progressive increase in low molecular weight components.

IMPROVEMENT OF PROTEIN NUTRITION BY ACTION OF NONPROTEOLYTIC ENZYMES

Cellulases

For the most part, in this section we shall examine those processes and investigations in which there is substantial evidence that the availability and/or nutritional quality of proteins in foods is improved as the result of enzyme action other than that catalyzed by proteinases. The action may be autolytic, that is, due to the food's own enzyme, or may be deliberately evoked by adding an external source of enzyme. It is not always possible to make a clear distinction between effects due to proteolytic and nonproteolytic enzyme action, especially when the two types of activities contribute to the observed improvement.

There are some advantages to the use of nonproteolytic enzymes. One is that it is possible to obtain the protein in an undergraded form. Another is that it may not be necessary to denature the protein so that it becomes more susceptible to the proteinases. Therefore, the food does not have to be heated or cooked during processing. This aids in the clean separation of the protein bodies of seed cells (Ory 1972). Even when such enzyme processing is effective it is at times difficult to assign very precisely the contributions of various polysaccharidases in achieving this extraction. At any rate, the actions of these enzymes do frequently result in freeing the proteins from the polysaccharide matrix in which they may be entrapped in subcellular organelles, especially in the protein bodies of the aleurone cells of seeds.

Glucanase application is not always successful. Commercial cellulases were not effective in liberating protein in cottonseed meal (Molina and Lachance 1973) nor did a hemicellulase preparation help extract protein from coconut meal (Arzu *et al.* 1972). On the other hand, Saunders *et al.* (1972) were able to increase the *in vitro* digestibility of the high quality protein locked in the protein bodies of wheat bran by as much as 35% as the result of the action of the cellulases present in the commercial enzyme Pectinol 41P. Rats fed treated bran grew 25% faster than control rats. As shown in Fig. 34.5, cellulose of the bran cell walls was the substrate for these enzymes.

Partially purified fungal cellulases from *Trichoderma viridae (reesei)* as well as from *Aspergillus* and *Rhizopus* were effective in extracting 20 to 35% of the protein from sun-dried khesari *(Lathyrus sativum)* and from gram and soybean in a 3 hr treatment at 40°C (Ghose 1970). Several groups of Japanese investigators have been engaged in studies and applying the action of microbial "macerating" enzymes, probably a mixture of enzymes which hydrolyze cellulose, hemicelluloses and pectin. This action produces single cell suspensions and helps to liberate nutrients from the tissues and cells of vegetables and soybean (Kawai 1972) (see also Chapter 30). Instead of microbial enzymes, Bock *et al.* (1971) advocated the use of the polysaccharide-degrading enzyme complex of tomato and avocado as "macerase" to convert vegetables to stable single-cell suspensions. This presumably results in improved protein availability. According to Yamatsu *et al.* (1966) the combined action of the polymer-degrading enzymes from *Trametes sanguina* and soybean meal solubilizes almost 80% of the total nitrogen at pH 3. At this low pH no asepsis is necessary.

Many patents have been issued on the use of "macerase" enzymes in recovery of protein. Thus, Blanchon (1966) claimed to have separated the nutritive constituents contained in the cortical layer and envelope of cereal grains. Silberman (1971) converted by-products such as grass, leaves, roots, seeds and stems to end-products for use in soft drinks. With a commercial enzyme (Cellulase-36, Rohm and Haas) Hang *et al.* (1970) increased the nitrogen extracted from mung beans from 10 to 60% and that from peas from 30 to 60%.

In a more basic study, Mudgett *et al.* (1978) found that pretreatment of alfalfa leaves with cellulase plus pectinase buffered at pH 4.6 prior to subsequent mechanical expression of the cell contents enhanced protein recovery from the leaves by about 50% to a total of 70%. The investigators look upon this enzymatic pretreatment as a way of accelerating senescence by degrading structural polysaccharides of the cell wall and lamella, thus abolishing the rigidity and mechanical strength of

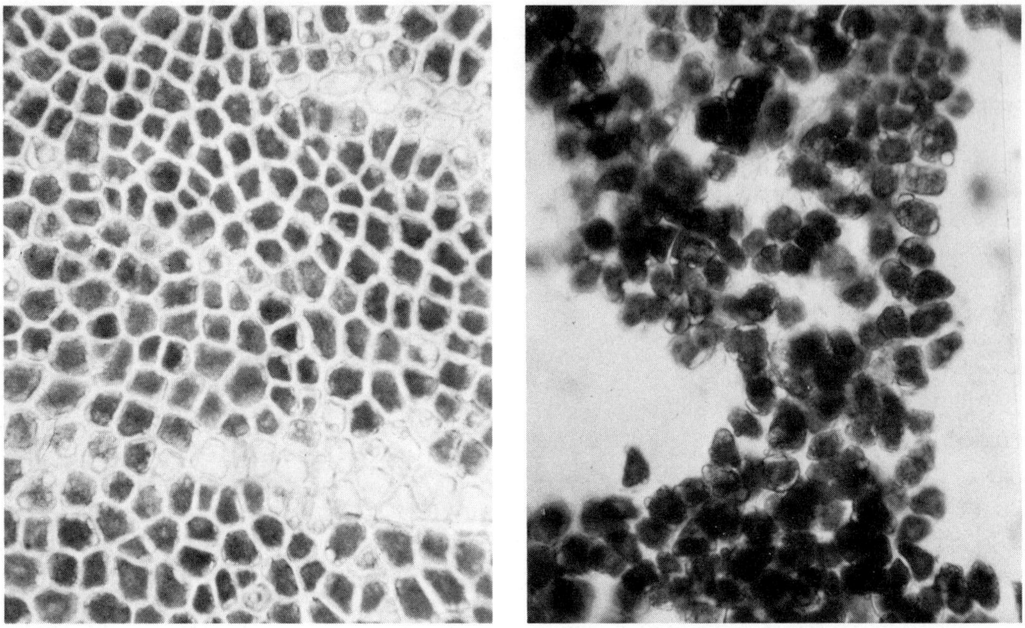

Courtesy of Saunders et al. (1972)

FIG. 34.5. ENZYMATIC PROCESSING OF BRAN ENHANCES NUTRIENT AVAILABILITY

The enzyme is cellulase and the nutrient is protein. Left, untreated aleurone layer; right, after treatment with 5% Pectinol. Note wall-less aleurone cells after treatment.

the plant tissue and allowing more facile mechanical leaf disruption leading to higher protein yield.

Noncellulolytic Enzymes

Lytic Glucanases of Yeast.—As mentioned in Chapter 9 a more grandiose role of cellulase in the production of single-cell protein (SCP) has been proposed. In these schemes, waste cellulose would be converted into glucose which would then serve as a food nutrient for the production of SCP.

Notwithstanding Weizman's early patents, one of the impediments to the practical utilization of SCP, especially that from yeast, is the barrier to protein availability posed by the cell envelope, mostly walls. Yeast walls are comprised of soluble components, mannan and glycogen, and noncellulosic water insoluble branched β-(1→3)- and β-(1→6)-glucans as well as, for some yeasts, galactomannans and an α-(1→3)-glucan (Ballou 1976; Manners and Meyer 1977). The key to removal of this envelope is the disintegration of an inner cell wall layer consisting primarily of alkali-insoluble β-(1→3)-glucan. Disruption of the cell wall by harsh chemical means may be injurious to the nutritional quality of the protein and physical means appears to be economically unfeasible (Tannenbaum 1977).

Fortunately, yeasts abound in glycan hydrolases, especially β-D-1,3-glucanases, some of which are responsible for the disintegration of the cell wall as part of the natural life cycle of the yeast cell, including both growth of the cell and the reproductive cycle (budding). In such times they serve as cell wall plasticizers. Much of our knowledge of these enzymes is due to the efforts of Phaff and colleagues. Key papers include those of Phaff (1977), Rombouts and Phaff (1976), Fleet and Phaff (1975) and La Chance *et al.* (1977). Japanese investigators appears to be responsible for much of the other activity, especially with nonyeast microbial glucanases. Key papers include those of Kitamura *et al.* (1972), Kobayashi *et al.* (1974) and Mori *et al.* (1977).

As with other glucanases (Chapters 9, 29–32) a multiplicity of endo- and exo-β-D-1,3-glucanases and debranching enzymes with varying specificities, at least as diverse as those of the starch-degrading enzymes (Chapters 9 and 23), have been described and isolated. They are localized in the periplasmic spaces between the cytoplasmic membrane and the cell wall. The pattern of specificities undergoes changes during the reproductive cycle of the yeast cell.

What is particularly singular and as yet inexplicable is the curious circumstance that some endoglucanases can cause lysis of the yeast cell and others with quite similar if not identical specificities cannot.

Lytic β-1,3-glucanases can be readily induced in a diversity of bacterial species and in a few fungi. Especially noteworthy are the lytic enzymes of *Bacillus circulans* and "Zymolase" from a strain of *Arthrobacter luteus*. This suggests that it should be economically feasible to produce lytic enzymes for the purpose of making SCP more nutritionally available, once the demand for SCP becomes great enough and other health-related problems such as the presence of excess purine (Chapter 35) and bitter thiamin (Chapter 25) are also practicably solved.

Typical of patents in this area is that of Dai Nippon (1977) in which both mechanical disruption and the action of lytic enzyme on *Candida* yeast present in a commercial enzyme, Cellulase-Onozuka, were utilized to prepare a high protein beverage. Alternatively, it should be possible to utilize the action of the yeast's own endogenous lytic glucanase to prepare SCP. The stratagem here is to persuade the β-glucanases to perform while keeping the proteinases in abeyance. Perhaps the so-called "alkali" separation of yeast proteins and impartation of functionality to them as embodied in several patents is predicated upon just such a differential potentiation of enzyme action. Thus, in a process patented by Komeji *et al.* (1978) pressure-homogenized yeast suspensions were extracted at the near potentiation temperature of 35°C for 5 hr and the pH was adjusted to 7.0 with alkali. Exploitation via autolysis awaits further insight into controls regulating the synthesis and activity of these enzymes during the yeast cell cycle. A step in this direction are observations that the biosynthesis of lytic glucanase may be induced by a specific polypeptide, α-factor, responsible for the formation of the copulatory outgrowth resembling Al Capp's Shmoo (Crandall *et al.* 1977; Nurse 1977).

Other Enzymes.—A powerful noncellulase glucanase from *Pestalotiopsis westerkijki* improved the extractability of protein in soybeans from 74 to 95% as a step in the preparation of soy milk which was used for infant feeding (Abdo and King 1967). Glucoamylase was used by Kazuo *et al.* (1979) to remove starch present in a corn-protein by-product of starch manufacturing as a step in soy sauce production.

In some instances the polysaccharidases are used to provide more calories per unit weight of the food by converting the insoluble polysaccharides to utilizable sugars. Thus, Uhlig and Grampp (1972) treated soy meal with pectinolytic enzyme alone or in combination with cellulase and/or hemicellulase in order to make available more soluble carbohydrate and thus improve the nutritional value of soybean. The small amount of starch in soybeans which appears to be associated with the protein and reduces its susceptibility to trypsin-catalyzed hydrolysis may be removed and protein digestibility improved by removing the starch with amylase (Boonvisut and Whitaker 1976). Digestibility of pea flour in milk replacers is improved by addition of amylase whose action provides the sugar required to replace milk lactose (Bell *et al.* 1974).

Finally, brief mention should be made of enzymes which do not degrade polymers, in relation to the effect of their action on the nutritive value of proteins. Naguib (1972) reports that treatment of milk with catalase (after hydrogen peroxide sterilization) does not affect the nutritional value of the milk protein. One can improve the protein content of fish simply by removing the fat. This can be accomplished by exposing the fish to lipase action (Burkholder *et al.* 1968).

PROTEIN IMPAIRMENT BY ENZYME ACTION AND ITS PREVENTION

Protein Linked Enzyme Action

In this particular section we shall explore the mechanisms and nutritional consequences of the action of endogenous food enzymes which result in deleterious modification of the protein. We previously pointed out that endogenous proteolytic action can, under some circumstances, lead to losses in yield and availability. This is an example of direct action of the enzyme on protein which serves as substrate. Another mode of modification is interaction of the protein with highly reactive products of enzyme action.

Polyphenol Oxidase (Phenolase).—The two principal enzyme types which have been implicated in such interactions with proteins are lipoxygenase and phenolase. In previous chapters the role of the latter enzyme in other aspects of food quality, especially color, was assessed. Here we confine ourselves to reactions in which there appears to be involvement of food protein.

In 1953, Sizer reviewed the experimental evidence for the oxidation of proteins by phenolase which is characterized by oxygen uptake, darkening and decrease in tyrosine. At least part of the oxidation proceeds beyond the DOPA stage to melanin (Chapter 15). Parenthetically, DOPA present in enzyme-treated rat feed interferes with psychophysiological studies of these rats (Hoeldtke *et al.* 1972). In the presence of extraneous protein, melanin is firmly attached to protein in the form of insoluble granules. We have examined the undesirable haze caused by the interaction of oxidized tannin with protein in beer (Chapter 31). From the viewpoint of nutrition this could add to the difficulty making the protein available in an inexpensive, acceptable form in other foods.

Insight into the nature of what happens is afforded by work of Pierpont (1969) and others on the interaction of phenolase-generated *o*-quinones with amino acids. Such quinones react, in descending preference, with sulfhydryl groups of cysteine, free ϵ-amino groups of lysine and N-terminal amino acids. About half of the lysine of bovine serum albumin reacted in 30 min at 30°C, pH 5.5. Thus, the nutritional quality may be impaired due to the unavailability of lysine and some of the sulfur amino acids, especially if the protein is methionine-deficient.

The most direct if meager evidence that impairment of the nutritional value of the protein does indeed occur is afforded by the experiments of Horigome and Kandatsu (1968). The nutritional implication of phenolase systems on protein quality is especially relevant to forage plants and presumably to future use of these plants as high-protein human foods. As shown in Table 34.3, casein recovered from solutions in which phenolase was oxidizing caffeic acid, isochlorogenic acid or the phenolic compounds of red clover leaves lost a significant amount of its nutritive quality.

In addition to major problems accompanying the Green Revolution caused by the introduction of dwarf wheat varieties is a rather minor one related to the present topic: The enzymatic browning and altered texture of dough and baked products because of the presence in these new wheat varieties of a highly active phenolase and increased levels of its natural phenolic substrates, alluded to in Chapter 32. Effects on the nutritive values of the wheat proteins have not as yet been reported.

In contrast to the above studies it appears that the interaction of some phenolic compounds with proteins without subsequent enzymatic oxidations is attended by an increase in the nutritive value of the protein (Van Sumere *et al.* 1975). Driedger and Hatfield (1972) found that daily weight gains, feed efficiency and nitrogen balance were all significantly greater in lambs fed soybean meal plus 10% tara tannins than lambs receiving soybean meal alone.

Such beneficial effects might be traced to the apparent "incipient" denaturation overtly manifested by an increase in susceptibility to proteinase attack and decrease in solubility as shown in Fig. 34.7. In keeping with this concept Neucere *et al.* (1978) noted that interaction of peanut protein with polyphenols is also accompanied by changes in protein conformation (α-helicity) and immunochemical specificity. This would appear as a mild manifestation of the more vigorous hydrophobic interaction of almost all tannins (also polyphenols) with proteins to precipitate them and for a few tannins, to fully denature them (Oh *et al.* 1980). Since they are present in foods such as persimmons and unripe fruit and to less extent in tea and coffee, this denaturation is perceived upon ingestion of such foods as an astringency. Proteins added to such food as milk to coffee and tea compete with

TABLE 34.3

NUTRITIONAL VALUE OF CASEIN EXPOSED TO PHENOLASE ACTION

Substrate (Endogenous)	Enzyme Source	Nutritional Parameters: Biol. Value	Available Lysine	Digestibility	Color
Caffeic acid	None	97.3	99.8	99.7	White
Caffeic acid	Orchard grass	89.4	87.3	96.3	Brown
i-Chlorogenic acid	Orchard grass	87.3	90.7	96.7	Green-brown
Red clover	Red clover	85.9	81.2	95.3	Brown

Source: Adapted from the data of Horigome and Kandatsu (1968).

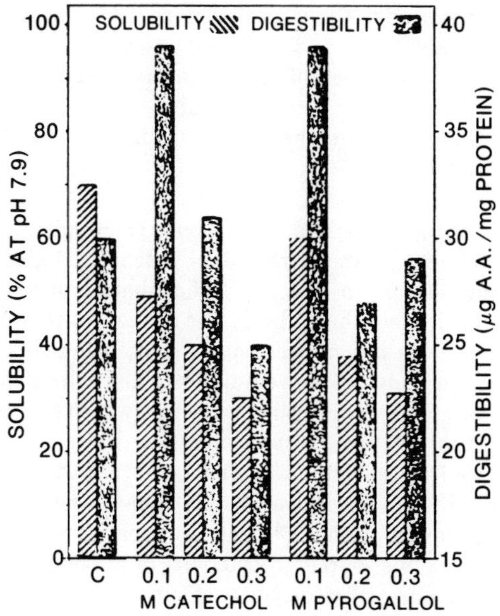

From Neucere et al. (1978)

FIG. 34.6. PROTEIN-POLYPHENOL INTERACTION MODIFIES SUSCEPTIBILITY TO PROTEOLYSIS

It also modifies the solubility of the protein.

the tongue proteins for the tannins, resulting in a decrease in astringency. In the case of tea, as we noted in Chapter 16, the products of phenolase action on the flavonols do retain their astringency, presumably because they still possess phenolic OHs.

A problem encountered in the utilization of sunflower and safflower seed protein (Betschart 1978B) is that they contain high levels of chlorogenic acid which reacts to impart an undesirable green color to the protein when the seed is processed in the conventional manner (Cater *et al.* 1972). This discoloration can be controlled by allowing the phenols to diffuse from dehulled kernels into aqueous solvents (Sosulski *et al.* 1972) or by alkaline extraction under nitrogen followed by ultrafiltration (O'Connor 1971). Processing alfalfa for protein recovery poses similar color-related problems (Free and Satterlee 1975).

Lipoxygenase and Other Oxidoreductases.— Another enzyme previously discussed whose action may alter or modify protein molecules is lipoxygenase. As mentioned (Chapter 21) the primary product of its action, existing at least momentarily, may be a highly reactive hydroperoxy free radical, which, in both enzymatic oxidation and nonenzymatic autoxidation, can self-propagate and interact with a variety of substances, including protein. For instance peroxidized or peroxidizing lipids interact with proteins to form free radicals in the latter (Karel *et al.* 1975) or induce the formation of insoluble crosslinked complexes involving protein-protein interactions (Roubal and Tappel 1966). Such hydroperoxides can destroy many essential amino acids in proteins, especially methionine and also cysteine (Gamage *et al.* 1973; Karel *et al.* 1975; Gardner *et al.* 1976) and form brown products (Chapter 19).

St. Angelo and Ory (1975) present evidence for a lipoxygenase-generated LOOH-protein interaction in stored peanuts. Indirect nutritional evidence of hydroperoxide damage to proteins is afforded by observations that antioxidant-treated herring meals protected the protein therein from deterioration of its nutritional properties (El-Lakany and March 1974).

There are undoubtedly many more enzyme reactions in foods which can give rise to highly reactive intermediates which in turn might damage the protein. For example, the enzymes responsible for the formation of the flavor of onion and crucifers form highly reactive primary products which appear to cause protein precipitation (Schwimmer 1960,1968A). Since the enzymes are no longer under the regulation of highly integrated metabolic pathways (Chapter 1), nonphysiological end-products may accumulate, or readily convertible substances such as hydrogen peroxide and active forms of oxygen (Chapter 24) may have the opportunity of reacting with the protein. Hydrogen peroxide in the presence of the ubiquitous peroxidase (Chapter 11) reacts with lysine and cysteine residues resulting in intermolecular crosslinking (Stahmann 1977). This would reduce the susceptibility of the protein to the digestive proteinase. Intermolecular crosslinking of proteins may also occur through the generation of bi-functional secondary products of lipoxygenase action, especially malonaldehyde. Both decrease in digestibility and loss of lysine would result in a food with lowered nutritional value. Prevention of the action of the above-discussed enzymes has been thoroughly explored in other chapters.

Prevention of Nonprotein Linked Undesirable Enzyme Action

In this section we include previously discussed enzymes whose actions make otherwise good sources of protein unfit to eat. Products of these enzyme reactions may be toxic, unnutritious or sensorily objectionable. Stratagems based upon the principles set forth in Chapters 10 through 14 have to be devised to prevent the action of these enzymes, or to remove these products.

Glucosinolase in Crucifer Seeds.—Glucosinolase was inspected in detail in Chapter 21 as an enzyme involved in imparting desirable flavors to cruciferous vegetables and seasonings. As briefly alluded to there and in Chapter 35, its action can also be undesirable and can create a barrier to the practical utilization of new sources of protein.

Three crucifer seeds have been investigated as sources of inexpensive quality protein: white mustard *(Sinapis alba)*, rapeseed *(Brassica compestris, B. napus)* and crambe seed *(Crambe abyssinia)*. The first, as previously noted, is in present use as a condiment and the latter two are grown for their oil, especially for the oil component erucic acid, a key starting material in nylon manufacture. Parenthetically other crucifers, usually from leaf tissue, are being looked into as high protein sources. Especially promising is the leafy green vegetable from the highlands of Ethiopia similar to mustard greens, *B. caranita* (Brown and Stein 1976). As noted, whatever the source, all the crucifers are bearers of glucosinolates or mustard oil glucosides as well as enzymes that attack them, the glucosinolases or thioglucosidases.

Prevention of the action of these glucosinolases is imperative for several reasons. First, it is generally agreed that the more or less pure protein isolates or concentrates obtained from these sources should be as free of flavor as possible. Such blandness would be compatible with their use as food ingredients. Second, as discussed in Chapter 35, some glucosinolase products possess toxic and disease-producing properties, goitrogenicity due to goitrin and baneful properties due to organic nitriles. Furthermore, as mentioned earlier in this chapter, the reactive products may interact with protein to make the latter more or less available for nutritional purposes.

One may distinguish among five general approaches, proposed singly and in combination, which have been used or at least experimentally tested to ensure the absence of such unwanted substances in the high protein products:

(1) allowing the enzyme to act and then remove the unwanted reaction products
(2) destroying the enzyme before it has a chance to act
(3) removing the substrate by extraction
(4) destroying the substrate, and
(5) breeding out the substrate and/or the enzyme.

Illustrative of approach (1), both Goering (1961) and Mustakas and Kirk (1965) potentiated glucosinolase action in mustard by incubation of seeds at 50°C for less than 1 hr. In the former process the enzyme was potentiated in the undefatted flaked seed at a moisture content of 30%. The goitrogenic end-products of glucosinolase action were removed by exposing the treated meal to direct steam, followed by solvent extraction to obtain the valuable oil. In a variation of this theme, the latter investigators added 0.3 part of unheated mustard flour (enzyme source) to 100 parts of heated, finely divided mustard seed (source of substrate).

An example of combining (2) and (3) is the approach discussed by Eapen *et al.* (1973). Whole rapeseeds were heated in water to destroy the enzyme, decorticated and the substrates leached out with water. A process was developed in Chile (Hill 1971) in which rapeseed meal is heated to inactivate the enzyme for 30 min at 80°–90°C and then extracted with water at 105°C for 5 hr. Kozlowska *et al.* (1972) tried various combinations of water, heat alcohol and alkali in extraction-diffusion processes for removing the substrate, principally progoitrin. As expected, each method had its advantages and disadvantages. Especially important were economic considerations. Josefsson (1975) provides further references which indicate that aqueous and organic solvents can be used to remove glucosinolates from rapeseed to yield high protein feeds which display high nutritive values in carefully controlled feeding tests.

Removal of the substrate by destroying it chemically (approach 4), i.e., with ferrous sulfate, has been successfully accomplished with crambe seed (Kirk *et al.* 1971). As expected, treatment of the cooked meal in various alkali solutions resulted in lowered nutritive value due to concomitant destruction of much of the lysine. Thus, rats fed a ferrous sulfate-treated crambe meal as 30% of a protein sufficient diet grew 70% as fast as those fed a basal diet. On the other hand, rats fed the same level of untreated crambe meal all died within 2 weeks, even though the meal had been heated.

With regard to approach 5, the Scandinavian countries have launched a breeding program whose aim is to lower the glucosinolates, especially the progoitrin component, of rapeseed. That they have succeeded in this aim is attested to by the finding of only 0.1 mg each of products of the glucosinolase reaction, oxazolidinethione and isothiocyanate, per gram of dry weight of the seed of *Bronowski* cultivar bred for the purpose of removing the glucosinolates. This is to be compared with values of 11.2 and 3.7 mg per g for a nonbred variety. The latter, as expected, was highly toxic to mice. They died within 8 days after the start of the diet. Unfortunately, these authors found that the low-glucosinolate variety, while not so toxic, did appear to contain high-molecular weight compounds which had exerted a detrimental effect on a nutritional value of the rapeseed meal. On the other hand a reproduction abnormality in rats fed protein concentrates from rapeseed and mustard was traced to zinc deficiency associated with a high level of phytic acid (McLaughlin *et al.* 1975). Addition of phytase should solve this problem.

Other Enzymes.—As discussed in Chapter 24, lipoxygenases may not only produce products which react with proteins but also unwanted substances whose flavors present troublesome snags in the more complete utilization of soybean protein from other legume seeds and in high protein cereal blends. Approaches for prevention of lipoxygenase action for this purpose were inspected in Chapter 24. Serendipitously, the use of acid for similar purposes also improved the digestibility of dry beans (Kon *et al.* 1974). The PER for acid-treated cooked pinto beans was 1.04 as compared to a value of 0.77 for whole beans cooked conventionally. It is likely that as a result of such conventional treatment in which the beans are soaked and cooked whole, the poor nutritional availability of both starch and protein is due to failure of cells to separate.

Another class of compounds which can undergo enzymatic transformation to both desirable and undesirable substances, in the latter instance HCN, are the cyanogenic glycosides discussed in Chapters 23 and 35. The intact glycosides are not considered to be toxic under most conditions. From the viewpoint of being a possible barrier to the utilization of an otherwise good source of protein, probably the only cyanogenic glycoside of concern is linamarin found in beans in general, but at a level which might cause concern only in lima beans in linseed if the latter is ever considered seriously as a protein source. Release of HCN in prunaceous fruits rarely, if ever, causes health-associated problems in the canning of such fruits.

Finally there are instances in food processing in which a step designed to stop undesirable enzyme action may result in a net decrease in the protein nutritional value of the food. As highlighted in Chapter 10, blanching results in the leaching out of proteins or its nutritional equivalent, amino acids, especially if hot water rather than steam is used.

To cite an example especially significant and vexing for protein nutrition in a developing country, shrimp account for a considerable proportion of the total seafood catch in Pakistan and in India. Blanching in combination with irradiation proposed as a step in the process for stabilizing shrimp resulted in a decrease in the dry weight protein content from 89 to 82%. Some loss of available lysine was reported, although in general the process was still superior to existing drying processes (Srinivas *et al.* 1974). Haq *et al.* (1969) found that the blanch water disposed of as the result of shrimp canning in Pakistan contained 16.6% extracted protein. This amounts to a loss of about 15,000 kg of protein in an 8 hr shift. On the other hand, in the processing of turnip greens, Meredith *et al.* (1974) found very little, if any, loss of lysine as a result of blanching per se, especially as compared with losses occurring in the draining steps. As annotated in Chapter 10 innovations in blanching technology, designed primarily to solve pollution and energy problems, also improve the retention of nutritional value of the food including the protein.

35

ENZYMES IN FOODS AS HEALTH AND SAFETY HAZARDS

In this chapter we continue our examination of how enzymes added to or present in foods either directly or indirectly affect the health of the food consumer. Since we surveyed their effect on the availability and quality of proteins in Chapter 34, this class of nutrients will not be discussed further. Adherence to the general policy of imposing an interdiction on discussions of the consumers' own enzymes and of enzymes of organized metabolic pathways of microorganisms associated with foods (Chapter 1) has to be relaxed in this chapter. The inner logic of the subject matter will necessitate some overstepping of the imposed boundaries from time to time, especially when these delineations tend to blur.

The following discussion is divided into two main categories: (1) baneful and allegedly antihealth effects of enzyme action in foods, and (2) the safety of commercial food grade enzyme preparations and of enzymes themselves. Effects of enzyme action deleterious to health may be further subdivided as follows:

(1) A putatively innocuous food constituent is converted directly or indirectly by enzymes into an undesirable substance which may adversely affect the health of the food consumer or exert some other pharmacological effect. It should be stressed that the levels of some of these putative toxicants may be so low as to cast considerable doubt as to their effect one way or the other on the consumer's health.

(2) A nonprotein nutrient is removed or reduced as a result of enzyme action. In each case we examine evidence bearing on the relative contributions of enzymatic and nonenzymatic processes to nutrient loss and survey proposals for retention of the nutrient.

GENERATION OF TOXIC AND UNWANTED PHARMACOLOGICALLY ACTIVE SUBSTANCES

Cyanide and Thiocyanate Ion from Cyanogenic Glucosides via Emulsin Action

In the last chapter we examined how the products of the action of the glucosinolases can indirectly prevent or delay the development of a technology leading to the utilization of a nutrient, in this case protein. The cyanogenic glucosides constitute another class of glucosides which can undergo endogenously enzyme-catalyzed transformations to both desirable (Chapter 22) and undesirable substances. The undesirable substances are, of course, the deadly hydrogen cyanide, HCN (cyanide, CN^-), and eventually thiocyanate ion, SCN^-, formed from CN^- by the consumers' own "detoxifying" enzymes. Unlike the glucosinolates in which the major (but not only) health hazard is due to transformation of one of many substrates (progoitrin), the CN^- is constant concomitant of the emulsin enzyme system (β-glucosidase-nitrilase) and thus arises independently of type of the substituent in the substrate. However, like the progoitrin transformation catalyzed by glucosinolase, the final end-product causes very similar if not identical diseases.

Emulsin substrates of particular significance as potential health hazards in foods are the acetone-derived linamarin usually accompanied by small amounts of lotaustralin (methyl linamarin) and the benzene or aromatic glycosides, amygdalin and dhurrin (Fig. 22.6).

Linamarin.—The two commonly used foods in which these are or may be present at levels high enough to have caused health problems are the lima bean and cassava. Potentially, linamarin may be a barrier to the use of another high protein plant tissue, flaxseed or linseed meal remaining from the production of linseed oil, as a source of human food. It may also create a health hazard because of its presence in white clover ingested by grazing cattle.

Lima Beans.—While most varieties of lima and other beans contain detectable but innocuously low levels of linamarin, equivalent to *ca* 15 mg/100 g of HCN (HCN-p) occasionally some varieties or a particular harvest of lima beans which caused toxic symptoms were found to contain 10 to 20 times this amount. However, it is not clear whether the toxicity was due to cyanide poisoning since lima beans are always cooked before eating, and cooking should destroy the β-glucosidase, in this case referred to as linamarase (Cooke *et al.* 1978). The above and other undocumented statements in the present discussion are cited in the comprehensive review of Montgomery (1969).

Cassava.—Although containing on the average less potential HCN than poisonous lima beans (*ca* 100 mg/100 g), far more baneful to a substantial segment of the earth's population is the catastrophe caused by presence of this cyanogenic glucoside in cassava. Also known as manioc and familiar to the western countries in the form of tapioca pudding, this tropical earth vegetable is the world's seventh most important food crop, following the major cereals, potatoes and yams (Dorozynski 1978). The presence of the linamarase system in cassava, along with a deficiency of iodine in the diet of users of this staple may be responsible for the widespread incidence in parts of Africa and in other tropical areas of goiter, cretinism and mental retardation.

From the notes of Dorozynski and Ononogbu (1980) reporting more recent evidence, from the previous literature cited by Montgomery and in papers reporting original investigations such as that of Osuntokun *et al.* (1970), it is clear that the culprit, as indicated before, is not CN^- but SCN^-, a potent inhibitor of iodine uptake by the thyroid gland. SCN^- presumably arises as a result of a normal detoxification mechanism in the human body by the action of rhodanese (Chapter 23) and/or 3-mercaptopyruvate cyanide sulfur transferase:

$$HS\text{-}CH_2\text{-}CO\text{-}COOH + CN^- \longrightarrow SCN^- + CH_3\text{-}CO\text{-}COOH$$

Logically we should have delayed the discussion of removal of toxicants by enzymatic means until Chapter 36 or perhaps not included it at all since it necessitates entering the human body to pinpoint enzymes which are part of the food consumer and not the food. We are, however, forced to step over these boundaries because of the puzzling circumstance, similar to that of lima beans, of the development of a food enzyme-derived toxicity despite traditional treatment of the food (except for tapioca) which should have prevented action of the enzymes. Except for tapioca, manioc is, at some step in the preparation of the many dishes in which it is used, heated (presumably inactivating the enzyme) usually after a grinding. Heat treatment apparently does not inactivate the enzyme system completely and enzyme-derived cyanide remains in these products but is present at low enough levels so as not to cause cyanide poisoning because it is efficiently "detoxified" to SCN^-. However, the constant presence of the latter in the bloodstream results in effective inhibition of iodine uptake by the thyroid. Alternatively, the undegraded substrate remaining after food is capable of producing CN^- after ingestion either nonenzymatically or perhaps with an assist from the β-glucosidases of the intestinal flora. These are areas which need clarification.

The Aromatic Cyanogenic Glycosides.—*Amygdalin.*—Unlike the acetone derivatives which may in themselves be toxic, amygdalin, the prototype of the cyanogenic glycosides, appears to be relatively harmless (but bitter) if ingested in absence of its enzyme-degrading system. However, ingestion in the presence of emulsin as in raw almond or apricot pits, as some children have done, and adults may be doing (see "laetrile" below) has been known to lead to severe cases of cyanide poisoning.

Another health-related problem generated by emulsin action on amygdalin, more closely allied to public health and pollution than to individual toxicity, is the presence of cyanide in the wash water discarded in traditional marzipan making (Chapter 22). Furthermore, as pointed out by Schab and Yannai (1973), the long soak period, lasting sometimes for days, allows the benzaldehyde, the coproduct of the nitrilase action, to be oxidized to the insoluble benzoic acid which might remain in the marzipan at more than safe levels.

Other potential sources of toxicity due to emulsin action are products made from stone fruits of the *Prunus* genus such as plums, apricots, cherry and peach. However, there is apparently very little threat from this source. Thus, Misselhorn and Adam (1976) conclude, from an analysis of free and bound cyanide in brandies made from the above-mentioned fruits, that the hazards of alcohol poisoning (including perhaps that of methanol) were greater than that from cyanide. Canned whole sweet cherries contain a maximum of 2.1 ppm HCN and canned depitted cherries no detectable amount but the cyanide content of apricot kernels ranged from 120 to 1770 ppm, presumably after allowing enzyme action to occur (Stoewsand *et al.* 1975). The use of cyanogenic glycosidase activity to diagnose a blanching failure was alluded to in Chapter 10. In canned goods any CN^- would tend to readily disappear by reaction with the can material and perhaps contribute to corrosion rather than to health problems.

Although not within the purview of food enzymology, discussion of the health aspects of amygdalin would not be complete without mentioning the continuing controversy generated by this cyanogenic glycoside in its guise as laetrile, a putative cure or alleviator of cancer. Indeed the rationale behind its alleged therapeutic effectiveness resides in the release (enzymatic ?) in human tumor cells of HCN.

However, it may be a health hazard if ingested with high β-glucosidase containing fresh foods such as lettuce, celery, peaches, beans, alfalfa sprouts and nuts according to Schmidt *et al.* (1978). The resulting nitrile apparently releases HCN spontaneously. As a result of feeding ten dogs amygdalin along with sweet almonds (it will be recalled that the sweet almond is quite low in substrate but has substantial emulsin activity) six died of cyanide poisoning and three developed neurological disorders. Reports on amygdalin poisoning appear with increasing frequency in medical journals.

Dhurrin.—Present in some sorghum varieties at levels as high as 250 mg/100 g of HCN-p, this tyrosine-derived cyanogenic glucoside could pose a health problem as a constituent of animal feeds (Gorz *et al.* 1977) and even as a constituent of a potential human source of lysine-rich grain protein (Wall and Jerrold 1978). Levels of HCN-p run about the same as in cassava, about 100 mg/100 g, about twice as high as that in grazing feed such as some sudan grasses. Its catabolism in injured *Sorghum bicolor* is depicted in Fig. 35.1. It will be recalled (Chapter 22) that dhurrin is the first cyanogenic glucoside and its catabolic enzyme system is the first emulsin whose morphologic and intracellular localization have been established.

Prevention and Correction.—While basic studies such as those of Conn and coworkers will undoubtedly provide the background to rationally control the action of emulsin systems responsible for the generation of HCN, present methods, some empirically arrived at, others suggested from the properties of the enzymes involved, have been devised to reduce HCN levels to innocuous levels or to obviate its baneful consequences. Montgomery in her thorough review recommends prolonged cooking to destroy substrate as well as enzymes and to drive off all possible HCN. Presumably the more acid the cooking medium, the more HCN will be released. Indeed in the preparation of the cassava dish known as gari, HCN is removed this way because of the low pH attained via the fermentation step in its preparation. Cyanide formation may also be reduced by removal of the substrate-rich outer portion by peeling, usually a step in preparation of most of the cassava dishes. Tapioca may not be as hazardous as other manioc-derived foods as anticipated on the basis of its HCN-p content because its preparation does not involve a heating step so that the enzyme may survive, be adsorbed and act; but the glucose, which is subsequently added, would tie up the cyanide, presumably via the reverse action of the still active nitrilase discussed in Chapter 22.

The most effective way of curing and preventing the endemic goiter caused by the eating of cassava has been to "vaccinate" the entire population with iodine, which counteracts the inhibition of iodine uptake into the thyroid by the accumulating SCN^- (Dorozynski 1978).

FIG. 35.1. HYDROGEN CYANIDE PRODUCTION FROM DHURRIN IN SORGHUM

The emulsin system consists of a β-glucosidase which hydrolyzes dhurrin (I) to *p*-hydroxy-(S)-mandelonitrile (II), which is converted nonenzymatically or more rapidly via the action of hydroxymandelonitrile lyase to *p*-hydroxybenzaldehyde (III) and HCN.

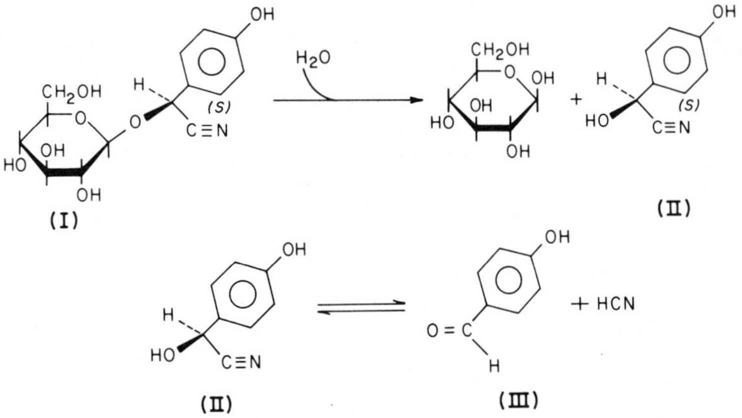

Courtesy of Kojima et al. (1979)

However, Dorozynski stresses the importance of finding better ways of detoxifying the tuber before consumption and of breeding for lines devoid of cyanogenic glycoside. To prevent the pollution of streams with HCN from apricot pits and the buildup of benzoic acid in marzipan preparation, Schab and Yannai (1973) recommended that emulsin action be optimally potentiated at pH 6.5 and 55°C so that the traditional 2 to 3 days soaking of the ground kernels can be completed in about 2 hr. They also recommend that the products of hydrolysis resulting from the complete breakdown of the bitter substrate amygdalin, HCN and benzaldehyde, be removed by steam distillation. In spite of the removal of the benzaldehyde, which presumably should contribute to the flavor of the marzipan, the investigators reported that the marzipan made with the debittered kernels was comparable organoleptically to that made with sweet almonds. In the future pollution caused by CN^- may be obviated by using immobilized enzymes as accomplished by Svenson and Anderson (1977). The enzymes immobilized were rhodanese, previously mentioned in this chapter and β-cyanoalanine synthase, present in legumes such as vetch and lupine (Hendrickson and Conn 1969).

$$HCN + \text{L-Cysteine} \longrightarrow \beta\text{-Cyanoalanine} + H_2S$$

We suggest that another effective enzyme to be added to foods or immobilized for environmental control is the aforementioned nitrilase acting in its condensing mode, along with an appropriate aldehyde or ketone cosubstrate. The latter could be part of the immobilization matrix. Further progress will be dependent not only on such basic work as that of Conn and coworkers at the University of California at Davis but also, as is a *sine qua non* in all of chemistry be it inorganic, biological or food, on more accurate analysis; in this case of assessing the cyanide potential, perhaps by the use of, say, laminarase as an analytical reagent (Chapter 37). This was the motivation of Cooke *et al.* (1978) to increase the purity of the cassava enzyme.

Methanol from Pectin via Pectin Esterase

Notwithstanding the jocular allusion in a drinking song of some years ago to sobering up on wood alcohol, ingestion or even inhalation of the vapors of this highly toxic substance, more formally designated as methanol (methyl alcohol), above sufficient levels causes at first blindness and then death. Since methanol is widely used as an industrial solvent, considerable attention has been given to its toxicity, especially when inhaled (Gleason *et al.* 1969; Gunter 1977). A level of 0.2% in the air is considered the maximum safe level when inhaled constantly during the working day. A lethal swallowed dose has been estimated at 2–8 oz. Many authorities believe that ingestion of relatively low levels of methanol does not pose any health problem.

Bertran (1974) suggested an upper level of safety of 4% by volume (31650 mg/liter or ppm). Lee *et al.* (1975) commented that the concern in various countries regarding its presence in wines appears to have little basis on toxicological grounds and that the danger of methanol poisoning from wine consumption would seem remote, since the highest levels reached are about two orders of magnitude less than the above-cited levels. Nevertheless Vos and Pilnik (1973) point out that little is known about the subacute toxicity of this normal constituent of many foods, especially fruits.

Methanol can be detected in the head space of many fruits, some crushed vegetables such as carrots and peanuts, but is particularly prevalent in many or most fruit juices, at levels from about 20 to 50 ppm in apple juice to as high as 200 to 3000 ppm in gooseberries. Methanol is also important in the field of food technology as a competitor to ethanol and petroleum hydrocarbons as nutrients for the microorganisms being considered as sources of single cell protein, SCP (Goldberg 1977; Minami *et al.* 1978).

Of special interest is its level in wines and other alcoholic beverages such as brandies where it can accumulate to levels approaching 0.1% (Dagher and Ruhayyim 1975). In both brandy, where it is a congener, and in wine it has, at appropriate levels, been considered to contribute to quality and distinctiveness. Castino (1975) used methanol as one of the parameters for the calculation of a discriminate function in evaluation of wine types. According to Amerine *et al.* (1972), the state of California has set maximum limit of 0.35 volume % (2765 ppm) and the USSR standard is set at 0.15 volume %. The processing of some fruit juices requires that methanol be maintained at a level of less than 50 ppm, according to Dahodwala *et al.* (1974).

Considerable effort has been expended and information garnered on the factors affecting the levels of methanol in juices and especially in fruit-derived alcoholic beverages such as wine and brandies as the results of investigation conducted outside of the United States in the past two or so decades. Selected references to some of this rather extensive literature can be found in papers by Lee *et al.* (1975) and by Gneckow and Ough (1976), both arising from United States investigations, rather rare in this field.

Role of Pectin Esterase (PE).—It is clear from this literature and from the independent experimental verification by the American investigators that the methanol present in these beverages arises from the action of PE on the pectin of the grapes during the maceration step in wine and brandy manufacture. Almost 30 years ago it was conclusively established that yeasts were not involved. While intuitively, the role of PE would seem to rest on firm foundation, the actual experimental evidence, by and large, is circumstantial, based upon three lines of information: red wines contain more methanol than do white wines; wines from heated contain more than those from unheated grapes and musts; and wine made from processes in which a commercial pectinase has been added contains more methanol than wine from processes in which such an enzyme is not employed.

From a survey of the literature on the methanol content of wines from all over the world, tabulated by Gneckow and Ough, the range of methanol in 239 samples of red wine varied from 0 to 635 ppm, averaging 124 ppm. The spread of values of 138 samples of white wine ranged from 0 to 120 ppm averaging out at 60 ppm. The average methanol content of 170 samples of experimental red wines made by Lee *et al.*, representing four distinct groups—Vinifera, French hybrids (Cabernet sauvignon, Pinot noir), Geneva hybrids and abrusca—was found to be 180 ppm (spread 40–465). The methanol contents of wines from white grape varieties of the same groups—Vinifera represented by Pinots, Rieslings and Gewirztraminers—averaged 64 ppm, ranging from 20 to 201 ppm. It will be recalled that white wines, regardless of the color of the grape, are made by separation of the skins before fermentation sets in, thus depriving the mash of a goodly portion of the PE substrate, pectin, much of which is associated with the skin and perhaps, to some extent, of PE also.

As an example, the effect of preheating on the grapes and hence presumably inactivating the enzyme (perhaps modulated by the increased availability of the pectin to surviving enzyme) is again afforded by the experiments of Lee *et al.* Thus, the methanol content of nine samples of wine prepared with a "hot press" averaged 30 ppm (range 18 to 57) whereas that from wine made with same lots of grapes in which fermentation occurred without heating prior to the separation of the skins (normal procedure for red winemaking) averaged 198 (129 to 465) ppm.

Confirmation of previous observations on the effect of using pectinases is shown in Fig. 35.2. Similar results were obtained by Gneckow and Ough. Vos and Pilnik (1973) reported a four- to five-fold increase in methanol content of apple juice due to the addition of commercial pectinase preparations. Where investigated the effect of adding more enzyme seems to be more a matter of increasing the rate rather than extent of methanol formation. This would be in keeping with the circumstance that the amount of available enzyme but not substrate is increased upon addition of

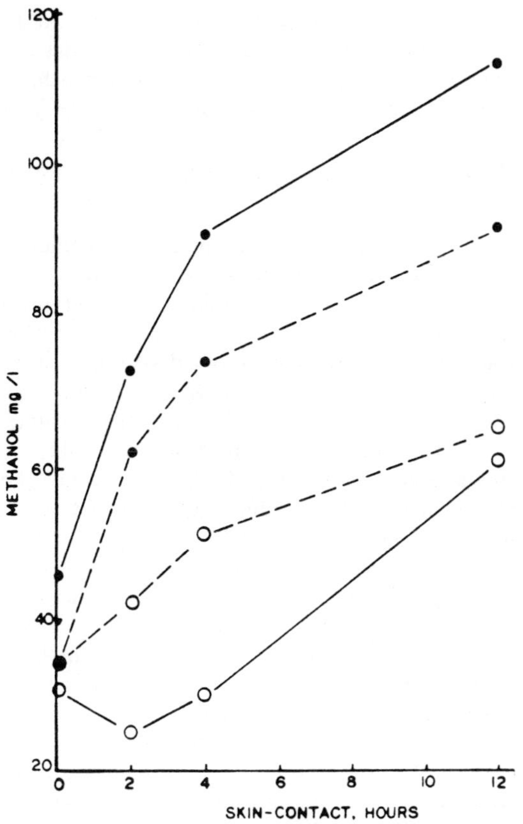

From Ough and Crowell (1979)

FIG. 35.2. ADDING PECTIC ENZYMES INCREASES METHANOL PRODUCTION DURING WINEMAKING

Control (open circles) *vs* pectic-enzyme-treated Chenin blanc (dashed lines) and Muscat of Alexandria (solid lines) grape varieties.

these preparations. In further support of participation of PE is the observation of Gneckow and Ough that during maceration methanol is readily formed during the early stages of maceration at the near potentiating temperatures of 38°C; practically none is formed at either 0°C or at 22°C. This suggests the potentiation of PE action is a necessary obligatory prelude to methanol formation.

Other than producing methanol, it is remotely possible that endogenous PE action might also create a health hazard indirectly. It will be recalled (Chapter 29) that Ca salts are added to some canned fruits and vegetables to improve their textural qualities after potentiation of the action of their endogenous PEs. Such Ca could increase the heat resistance of bacterial spores thus necessitating a rescheduling of retort programming.

Control and Suppression.—The experiments of Lee *et al.* show that inactivation of the grape PE by adequate heat treatment such as hot pressing can keep the level of methanol at vanishingly low levels. Branding high methanol contents of fruit brandies as primarily an export problem in Switzerland, Tanner (1972) recommends heating mashes from pears and stone fruit at 85°C for 30 min as a means of reducing the methanol content. For purposes of standardizing the methanol contents of various wines in Poland, Masior *et al.* (1976) removed that already formed from must (but not from wines) by a 2% vacuum distillation without affecting the quality of the wines made from these musts. Dahodwala *et al.* (1974) suggest a judicious adjustment of the PG:PE ratio using a fungal PG preparation as means of controlling methanol formation without interfering with PG action needed for efficiency of filtration. It will be recalled (Chapter 30) that most PGs require the prior removal of the methoxy group from the pectin to act efficaciously.

Another approach would be, if economically feasible, to remove the methanol enzymatically, perhaps with microbial methanol oxidase:

$$CH_3OH + O_2 \longrightarrow HCHO + H_2O_2$$

Presumably the formaldehyde would be readily transformed by other enzymes present and the hydrogen peroxide would be readily decomposed by catalase which accompanies most partially purified microbial enzyme preparations. Both primary products might be removed via peroxidase action with formaldehyde as the hydrogen donor. Catalase action was used by Barrati *et al.* (1978) in their immobilized microbial methanol oxidase designed to be deployed for antipollution purposes.

Biogenic Amines *via* Amino Acid Decarboxylases

A somewhat stronger case has been forwarded for the possible health hazard of biogenic amines in foods than for methyl alcohol from the viewpoint of food quality. Examples of their contribution to desirable flavor in milk and milk products and in flavor generation via Maillard nonenzymatic browning are given in Chapter 23. The secondary and tertiary amines are participants, along with nitrites, in the formation of carcinogenic nitrosamines. Since this formation is nonenzymatic and since biogenesis of these amines is so heterogeneous and experimental credentials relating the contribution of enzyme action to their formation is still far from established, we do not discuss them in any detail in this book.

Askar (1976) provides an overall view in German in his book on amines and nitrosamines in foods, the fourth in the series "Fortschritte in der Lebensmittelwissenschaft."

Definitions and Medical Problems.—Rice *et al.* (1976) define biogenic amines as aliphatic, alicyclic or heterocyclic organic bases of low molecular weight which arise as a consequence of metabolic processes in animals, plants and microorganisms. Historically and in practice they have been usually thought of as pharmacologically and physiologically active products of decarboxylation of the α-amino acids.

$$RCHNH_2COOH \longrightarrow RCH_2NH_2 + CO_2$$

The accepted nomenclature of the resulting amines in the biochemical literature is rather polyglot. The names of some of them derive from the parent amino acid (tyramine, histamine, tryptamine). Others are based upon their structure: phenylethylamine from phenylalanine, dopamine from dihydroxyphenylalanine (DOPA), isobutyl- and isoamylamines from valine and leucine, γ-aminobutyric acid (GABA, not strictly a base) from glutamic acid and still others more trivial if not macabre—cadaverine from lysine, putrescine from ornithine, agmatine from arginine and serotonin from 5-hydroxy tryptophan.

A particular biogenic amine may be vasoactive, psychoactive or both. Psychoactive amines such as GABA and serotonin participate in or otherwise affect the transmission of neural impulses in man. The most overt manifestation of the activity of the vasoactive amines is capillary vasodilation and accompanying lowering of blood pressure (hypotensive) by mainly histamine or, more widespread, raising of blood pressure by vasopressors such as tyramine, dopamine and serotonin.

A few years before the widespread attention was given to the "Chinese restaurant" syndrome occasioned in some diners due to an overdose of glutamate, a somewhat less distinct set of symptoms, which came to be known as the "cheese effect" was traced to the simultaneous ingestion of foods high in pressor amines, mainly tyramine, and a class of antidepressant drugs known as MAOI, the monoamine oxidase inhibitors. The presence of these drugs in the consumer's bloodstream prevented these amines, present in these foods at appreciable but at perfectly safe levels in the drug's absence, from being rapidly catabolized via the consumer's own monoamine oxidase which, like other flavoprotein oxidases, produces H_2O_2:

$$RCH_2NH_2 + H_2O + O_2 \longrightarrow RCHO + H_2O_2 + NH_3$$

The remarkable effect of the MAOIs is illustrated by the citation of Rice *et al.* (references to this and others in nonreferenced discussions in this section can be found in their food-oriented review) that marked elevation of blood pressure can be observed in individuals on MAOI upon oral ingestion of 6 mg of tyramine whereas it takes as much as 80 mg, injected subcutaneously, to obtain a comparable rise in the absence of MAOI. The pressor action of tyramine may be potentiated as much as 100-fold by MAOI.

Hypertensive crises occasionally leading to death were reported upon ingestion not only of cheese but also other foods such as liver, chocolate, pickled herring, all presumably due to tyramine; and broad bean, presumably due to dopamine derived from the naturally occurring DOPA.

Quite aside from the MAOI effect, apparently sufficiently excessive levels of tyramine and perhaps other aromatic monoamines which occasionally occur in foods can partially overcome the enzyme defenses of the consumer and cause some untoward hypertensive-related discomfort. Still under debate is the role of tyramine in triggering migraine headaches (Anet and Ingles 1976).

The other biogenic amine of major concern, histamine, is also present in cheese and especially in canned tuna fish previously exposed to microorganisms where, at sufficiently high levels, it appears to be the major cause of "scombroid" poisoning. The culprit organisms appear to be *P. morganii* and *E. aerogenes* (Taylor *et al.* 1978). This poisoning is characterized by a sudden lowering of the blood pressure, nausea, vomiting, facial flushing and headache, as well as by allergic reactions. Clinical symptoms are evoked by eating approximately 1 g of this amine in nonallergenic individuals. Histamine and other diamines may accumulate in individuals taking medications which are inhibitors of diamine oxidase (histaminase). Elevated levels of these amines in the blood, although not high enough to cause discomfort, may give rise to false positive tests for allergies and more serious diseases. Finally these diamines may be precursors to plant alkaloids and psychoactive poisons in some plant foods.

Plant Foods.—Among raw vegetables listed by Rice *et al.*, only spinach, tomatoes and avocado contained more than trace amounts of vasoactive biogenic amines. Fruits in which the amines are found include raspberry, pineapple (juice) and banana (mainly in the peel). Also, Stachelberger *et al.* (1977) report the following levels in μg/g (ppm) in bananas: serotonin, 77.5; the phenolase substrate (Chapter 15) dopamine, 650.5; and noradrenaline (105.5). In dates and figs only serotonin was detectable, at about 10 ppm. Among the 350 or so plants containing amine derivatives of tyrosine and phenylalanine listed by Smith (1977) are the following foods or potential foods: dasheen, spinach, black walnut, plantain (banana), granadilla, purslane, apple, white sapote, most of the citrus fruits, tomato, potato and tea. According to recent speculation, as reported in the newspapers, the presence of phenylalanine-derived phenylethyl amine accounts for the over-indulgence in chocolate—via an unconsciously motivated self-medication—by individuals undergoing the pangs of unrequited love.

Tryptamine and derivatives are found in similar foods but are more restricted. They have been detected in tomatoes but not in potatoes and occur frequently in Gramineae and Leguminosae. However, in very few of the plant foods eaten raw or cooked have they ever been seriously implicated as health hazards, i.e., plantain intake as a factor in widespread incidence of myocardial fibrosis. The only other foods derived from such plants in which amines, present at appreciable levels, have been implicated as potential health hazards are cocoa and sauerkraut and perhaps wine and nuts. Tyramine (Kenyhercz and Kissinger 1977) and phenethylamine present in cocoa-derived foods such as cocoa and chocolate at levels of 3–10 ppm, have been implicated in hypertensive crises and triggering of migraine. Sauerkraut can contain substantial quantities of tyramine, 100 μg/g, enough so that 6 g of sauerkraut, hardly enough for a frankfurter, could cause hypertensive crises in individuals taking MAOI drugs. Substantial quantities of histamine are also present. Beer may contain as much as 11 and wine 25 μg/ml of tyramine, enough to precipitate a hypertensive crises if imbibed immediately after taking MAOI.

Animal-derived Food.—Tyramine is present at substantial levels in most meats (100–300 μg/g) but may exert—along with histamine at maximally observed levels even in the absence of amine oxidase inhibitors—pharmacological effects at high levels of food intake in cheeses, pickled herrings and in some sausages. Maximum levels reported for these are *ca* 2000, 3000 and 1500 ppm, respectively. These are, of course, maximum values. The mean value of total biogenic amine determined for 26 samples of dry fermented sausage by Vandekerckhove (1977) was 540 ppm, with tyramine accounting for 61% of the total and histamine for only 7%. Sausages and probably other fermented foods to which MSG has been added may contain the neurotransmitter GABA, γ-amino-butyric acid. GABA, as mentioned, is the product of decarboxylation of glutamic acid.

Although histamine levels may rise to as high as about 0.2% in sausage and cheese, where its principal dangerous effect is probably to potentiate the action of the pressor amines, this diamine has, as mentioned occasionally, constituted a health hazard in tuna and related scombroid fishes which have undergone some microbial exposure where levels may rise to as high as 0.5%. Even at the higher ranges of minimum amounts required to cause clinical symptoms, eating of less than an ounce of this tuna could be toxic.

Formation: Microbial Amino Acid Decarboxylases.—It will be noted that just about all foods in which these amines are present at levels where they constitute a potential health hazard have gone through a fermentation. The reasonable inference is that they arise from the action of the enzymes of microorganisms, and that the particular enzymes involved are the L-amino acid decarboxylases. Indeed this class of enzymes, first described in microorganisms, are the first of the phosphopyridoxal enzymes found to contain vitamin B_6 as a prosthetic group (Gale and Epps 1944). Unlike the L-amino acid oxidases, which exhibit group specificity, each amino acid is decarboxylated by a specific decarboxylase. Because of this specificity, the amino acid decarboxylases have been used as analytical reagents (Chapter 37). The aromatic amino acid decarboxylases are constitutive but repressible by product amines and are widely distributed among such genera as *Achromobacter, Micrococcus, Staphylococcus* as well as the now traditional sources *Sarcina* (Nakazawa *et al.* 1977) and *Pseudomonas* (Denis *et al.* 1977). They are also present in higher plant food, i.e., alanine decarboxylase in tea plants (Takeo 1978) and glutamate decarboxylase in wheat (Brancher *et al.* 1974). Recent studies on decarboxylases are overwhelmingly concerned with the brain (Sourkes 1977) where an understanding of them holds promise of the alleviation of some mental diseases.

Many of the decarboxylases have been thoroughly purified and have been shown to consist of 800–900 kilodalton decamers which dissociate stepwise to dimers with loss of activity in the presence of Na^+ (Boeker 1978).

Information about microbial decarboxylases as causative factors in the buildup of biogenic amines in fermented foods is sparse and largely limited to cheese. Early work going back to the mid 1950s suggested that the decarboxylases may have been elaborated by coliforms and heat resistant streptococci present as contaminants in milk and not by the starter and ripening organisms. More recent work by Voigt and Eitenmiller (1978), however, showed that dairy-associated bacteria, both unwanted (i.e., *S. liquifaciens*) and wanted starters and ripeners, especially *S. lactis* and *Leuconostoc cremoris*, were highly potent sources of tyrosine decarboxylase.

They also went further and demonstrated variable but positive tyrosine decarboxylase activity, along with tyramine, in 143 out of 156 samples of different types of cheese. Histidine decarboxylase (as well as histamine) was detectable less frequently in the same samples. Since no correlation could be established between activities and levels of the corresponding amines, they suggest that a major factor limiting amine accumulation is the availability of the substrates. Since these are free amino acids, the amount formed, assuming that amines were not further degraded, would be regulated by action of exopeptidases and especially, perhaps, starter dipeptidases (Chapter 33). Proteolytic enzymes inducing autolysis of tuna tissue also control the accumulation of histamine formed via the action of histidine decarboxylase of contaminant microorganisms.

The investigations of Mayer *et al.* (1973) suggest that the decarboxylases of one particular microorganism *Pediococcus cerevisiae* may be responsible for the formation of amines in sauerkraut.

Removal and Prevention.—Rice *et al.*, after reviewing the limited information on microbial amine oxidases, suggested that the catabolism of the biogenic amines by bacteria may play an important role in the final concentration of these amines in foods. However, in a test of this hypothesis in the above-cited paper from the same laboratory by Voigt and Eitenmiller (1978) it was found that relatively few cheeses or dairy-associated bacteria possessed amine oxidase activity, both mono- and di-. They do, however, suggest that the possibility still exists of lowering amine contents by developing starter organisms with enriched amine oxidase levels. Perhaps in the future purified oxidase preparations appropriately stabilized (Chapter 5) could be added along with, for example, rennet. Preventing amine proliferation by coliforms by proper hygienic practice would also help. In the case of sauerkraut, suppressing growth of pediococci by early interruption of the cabbage fermentation at an adequately low pH appears to be an effective way of reducing amine levels. In dry fermented sausages, Vandekerckhove (1977) drew attention to variation in amine content in relation to methods of processing and to the flora. The use of highly specific monoamine oxidase inhibitors such as derivatives of pyrrolidone (Israili and Smissman 1977) and isophthalic acid (Endo *et al.* 1978) would be safe only if not allowed to enter the bloodstream, perhaps by anchoring them to unabsorbable bulky matrices.

Other Putative or Potential Health Hazards

Several other putatively hazardous substances which may be present in some foods may in principle, if not in experimental fact, be generated as the result of endogenous or added enzyme action. Some of these products could arise as a result of the "nonphysiological" pathway set up postmortem when the food cell disrupts (Chapter 1). Other "stress metabolites" may arise during preharvest or preslaughter as a consequence of imposing stresses on the living organism destined to be food (Haard and Cody 1978; Schwimmer 1978) and others, as we have seen, may arise from enzymes in microorganisms present either as normal components of fermented foods or as contaminants. Still others may originate, as in the case of methanol, from the action of commercial food enzymes added to foods for specific purposes. These include oxidized fats, 7-α-cholesterol oxide, hydrolysis products of milk proteins, pheophorbides and cyclodextrins.

That oxidized fats may be harmful or at least pharmacologically active is indicated, for instance, by the observation of Andia and Street (1975) that such oxidized lipids induce synthesis of microsomal xenobiotic or mixed function oxidase in the liver of subjects who ingest them. Yoshioka *et al.* (1974) concluded that the toxicity of autoxidized oils is dependent on the secondary degradation products rather than lipohydroperoxides (Chapter 24). In each instance just cited the lipids were oxidized nonenzymatically, but, as we have seen in previous chapters, enzymes (systems) such as the lipoxygenases in plant foods and NADPH or hemoglobin-dependent lipid oxidation systems in animal foods, discussed in Chapter 24 and elsewhere, catalyze the formation of LOOHs.

α-Cholesterol oxide, formed *in vivo* through faulty metabolism, has been implicated as a potent carcinogen according to evidence reviewed by Bischoff (1969). Although most of the investigations on this and similar derivatives published in cancer research journals were conducted on nonenzymatically autoxidized cholesterol, according to Smith and Kulig (1975), cholesterol is oxidized not by O_2 but by cholesterol 7-hydroperoxide. The latter, as shown by Smith and Teng (1974), can be formed by the action of the following enzyme-substrate systems: soybean lipoxygenase-ethyl linoleate, liver microsomes-NADPH and horseradish peroxidase-H_2O_2. Tsai and coworkers (1979) are investigating the possibility that such cholesterol derivatives accumulate in cholesterol-rich foods such as dried eggs. If indeed such substances can form in foods, it is somewhat comforting to know that the autoxidation of cholesterol is inhibited by cholesterol's constant companion in most foods, lecithins.

As mentioned in Chapter 17, pheophorbides resulting from chlorophyllase action on chlorophyll photosensitized albino rats. Since such products may be present in some leaf protein preparations, Holden (1974) suggested that some problems associated with chlorophyllase action could arise under special conditions. The previously recommended heating at 100°C, where rapid inactivation of chlorophyllase occurs rather than at 80°C, might prevent formation of such photosensitizers.

Another example of possible, but not probable, adverse effects on a minute fraction of the population occasioned by formation of a new substance from the action of added enzyme in a food, in this case cheese, is afforded by the observation of Spies *et al.* (1972) on the formation of new antigens generated by successive pepsin hydrolyses of bovine lactoglobulin. Another example of how the action of an extracellular enzyme from adventitious microorganisms may result in the formation of a toxic substance is the demonstration of cyclodextrin nephrosis in the rat (Frank *et al.* 1976). Cyclodextrins, it will be recalled (Chapters 9 and 32), are formed as a result of the action of a *Bacillus macerans* enzyme, the old Schardinger dextrinase, now cyclodextrin glucosyltransferase, on amylose. However, it should be emphasized that contamination of food with *B. macerans* is not likely.

Potentially, at least, the successful adoption of large-scale use of lactase for many of the purposes proposed for it (Chapters 9, 33, 36 and elsewhere) would entrain a new health-related problem, that of the development of cataracts of the eye due to excessive ingestion of galactose, one of the products of lactase action. Shukla (1975) suggests that fermentation of lactose would obviate this problem as well as those posed by the presence of lactose in milk and whey.

LOSS OF NUTRIENTS

Like entropy, with rare exception, deterioration of quality attributes in foods tends to increase with time, from harvesting to consumption unless precautions are taken to either convert the raw commodity to an entirely new product or, as in the case of beef, to allow it to age. Loss of health-associated quality attributes, especially loss of nutrients, is no exception. As has been frequently stressed, only vegetables consumed directly from the farm probably contain their full complement of vitamins. Kramer (1974), pointing out that nutrients are destroyed during food processing because of the sensitivity of these nutrients to pH, oxygen, light and/or heat, suggests that both trace elements and enzymes also catalyze nutrient loss. Extremes of these variables would tend to favor nonenzymatic reactions.

Other considerations of the effect of processing in general on nutrients include discussions by the American Medical Association (1974), Harris and Von Loesecke (1971), Wallace (1973) and the IFT (1974).

Not only can nutrients be lost due to the direct action of enzymes on the nutrients, but measures instituted to manage enzyme action. i.e, blanching, (Chapter 10), addition and potentiation could provide the environment conducive to nutrient loss not only during processing but also during subsequent storage. An example of how the addition of enzymes could adventitiously affect nutritional quality is provided by Jurics *et al.* (1971) who reported that a cellulase preparation added to various vegetables increased losses in vitamin B_2, nicotinic acid and ascorbate, as compared to conventionally cooked vegetables, whereas the carotene content was higher. However, one would expect that shorter cooking time resulting from adding the enzyme preparation would have tended to retain all of these nutrients. As we shall see, in some instances serious loss of nutrients due to direct enzyme action occurs only in special foods under special circumstances.

We shall now examine the evidence for their loss and measures instituted for their retention.

Vitamin A and Carotene

In the processing of most food, loss of these nutrients probably occurs nonenzymatically. Heat readily isomerizes all-*trans*-carotene to isomers with lower vitamin A activity (Sweeney and Marsh 1971). In processing of papaya some carotenoid is lost due to an acid-catalyzed isomerization (Chan *et al.* 1975), but this may have little effect on the nutritional value. An illustration of the light-induced carotene loss is afforded by the observation of Nogueira *et al.* (1973) that freeze-dried cherries and red papayas packed in clear glass containers lost β-carotene considerably faster than did fruit packed in brown containers.

We have already encountered lipoxygenase-catalyzed destruction of carotene in connection with color quality of wheat products (Chapter 17). Undoubtedly this enzyme participates in the loss of carotene in other foods, such as spinach, which are finely chopped or dried before cooking. Hauge *et al.* (1937) attributed the loss in carotene in dried alfalfa to enzymes, later shown to be principally lipoxygenase. Treatment of alfalfa with alkali in some of the protein-for-human food processes effectively destroys lipoxygenase (Chapters 21 and 34).

The effectiveness of blanching in stabilizing carotene in frozen vegetables presumably because lipoxygenase is inactivated was shown almost 40 years ago by Zimmerman *et al.* (1941). This investigation also illustrates the importance of ensuring enzyme inactivation prior to analysis of many food constituents in that they found that boiling acetone used for extraction was insufficiently hot to completely inactivate carotene oxidase (lipoxygenase). Enzyme action is more likely to be responsible for vitamin A loss in drying than in other types of processing (Labuza 1972). Enzyme action may be operational in the drying of apricots; Bolin and Stafford (1974) found that vacuum dehydration of apricots resulted in no loss whereas during sun drying 30% was lost.

Sood and Bhat (1974) attributed increased retention of carotene (and ascorbic acid) in pressure-cooked green leafy vegetables to diminished enzyme action occasioned not only by the higher cooking temperature and thus shorter cooking time but also to decreased enzyme oxidation (presumably lipoxygenase) during the decreased time lag between preparation and cooking and also due to the exclusion of oxygen. A discussion of vitamin A in relation to enzymes and food quality would not be complete without mentioning its possible self-destruction, due to its ability to destabilize lysosomal membranes and thus release degrading enzymes, as briefly alluded to in Chapter 26 (Sudhakaran and Kurup 1974).

Vitamin B_1 and Thiaminase I

As in the case of vitamin A, the causes of losses of vitamin B_1, thiamin, in foods are probably for most foods, nonenzymatic. One of the substances to which thiamin is quite sensitive is SO_2 in its various forms. The latter are effective aseptic agents partly because they are potent inhibitors of the thiamin pyrophosphate enzymes of the invading microorganisms. Other apparently nonenzymatic antivitamin factors present in foods investigated by Hilker (1976) include heat-stable factors, TMF, (mostly heme compounds) in marine food as the skipjack tuna and the crayfish (Rutledge and Ying 1972), some polyphenolics, especially 3,4-dihydroxycinnamic acid in blueberries, tannins and flavonoids in tea and presumably related polyphenols in betel nut, the latter creating a problem confined to Southeast Asia.

Also potentially capable of causing a vitamin deficiency, especially in Southeast Asia, has been the widespread consumption of fermented fish due to the presence of thiaminase I in the fish and possibly in some of the bacteria involved in the fermentation.

The action of this enzyme was considered to constitute a health problem for the Japanese who consume raw fish, sashimi, well-known in Japanese restaurants in the United States. Some thiamin loss may also occur in the occidental diet due to eating partially raw herring. Reviews on thiaminase I and other anti-vitamin B_1 factors include, besides the above-mentioned survey of her own investigations by Hilker (1976), earlier ones by Yudkin (1949) and Fujita (1954), in Japanese by Murata (1965) and more recently in Russian by Ostrovskii (1978). The paper by McCleary and Chick (1977) in English on the isolation and characterization of fern and mussel thiaminases also provides a good key to the literature.

Thiaminase I is a transferase which catalyzes the fission of thiamin, usually involving an amine as cosubstrate, via displacement of the methylene group of the pyrimidine moiety of this vitamin:

$$\underset{\text{Thiamin}}{Pyr \cdot CH_2 \cdot Thiaz^+} + \underset{\text{Amine}}{R \cdot NH_2} \longrightarrow \underset{\text{"Heteropyrithiamin"}}{Pyr \cdot CH_2 \cdot NHR} + \underset{= \text{4 Me-5-}\beta\text{-OH-Et-thiazole}}{Thiaz} + H^+$$

where Pyr and Thiaz are the pyrimidine and thiazole derivatives which, together with the methylene bridge, comprise thiamin.

The enzyme is present in some species of marine animals, in ferns such as the bracken fern, and the Australian nardoo. It is also present in many bacteria (Sera 1976). The fish and fern enzymes are strikingly similar in molecular weights (*ca* 100,000) pH optima (8.0–9.0) and in kinetic characteristics.

Overt symptoms of thiamin deficiency are foreshadowed and measured by a drop in the activity of one particular TPP-enzyme, blood transketolase.

Thiaminase I-caused vitamin B_1 deficiency has probably been more harmful to animals fed by humans than to humans directly (Evans 1975). Probably the earliest evidence for the harmful effect of thiaminase I was a development of "Chastek" paralysis in foxes fed raw fish heads (Yudkin 1949). In addition to fish fed to domestic animals, other sources of this enzyme in domestic animals are ferns ingested during grazing and perhaps the animal's intestinal flora i.e., *B. thiaminolyticus*. Among the domestic animals are chickens which have been fed anchovy or scallop viscera and pond-cultivated fish fed anchovy and herring (Lehmitz and Spannhoff 1977).

Grazing animals, including horses but especially sheep, may develop cerebral cortical necrosis as a result of thiaminase I action. This may be due in part to the intestinal flora since the enzyme activity appears in the feces even before symptoms appear. Indeed fecal thiaminase activity is now used to diagnose this disease.

Somewhat reminiscent of the effect of the monoamine oxidase inhibitor drugs and the biogenic amines, outbreaks of this disease may have been due to the administration of drugs to sheep, especially antihelminthics, amines which serves as extremely efficient thiaminase I cosubstrates. The sparse but developing literature in this area is exemplified by the paper of Linklater *et al.* (1977) which may be used as a key to others whose results have been mentioned.

It has been suggested that some of the loss of thiamin by intestinal microorganisms may not be due to thiaminase I but to the hydrolase, thiaminase II (Ikehata 1960; Evans 1975):

$$Pyr \cdot CH_2 \cdot Thiaz^+ + H_2O \longrightarrow Pyr \cdot CH_2OH + Thiaz + H^+$$

Prevention.—Fujita (1954) pointed out that thiamin can be altered chemically without seriously affecting its biological activity as vitamin B_1. Thus, exposure of thiamin to a freshly prepared garlic extract, as shown in Fig. 35.3, allows the alliinase-produced allicin (Chapter 21) to react with thiamin to form allithiamin, which, as indicated, has vitamin B_1 activity but is no longer a thiaminase substrate. Some multivitamin pills sold in Japan contain allithiamin instead of thiamin.

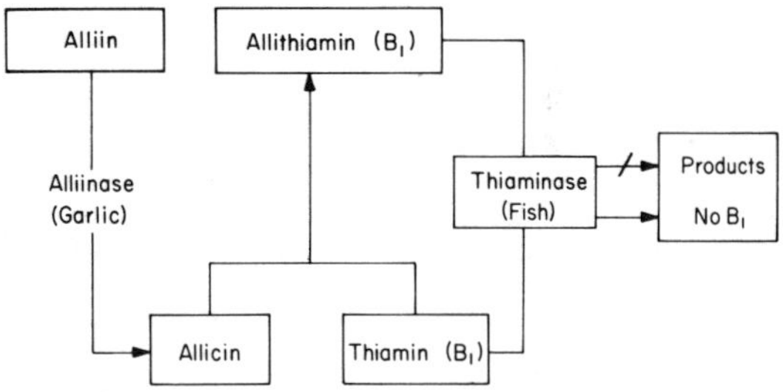

From Schwimmer (1969A)

FIG. 35.3. ALLIINASE ACTION PREVENTS THIAMINASE INDUCED LOSS OF VITAMIN B_1 ACTIVITY IN RAW FISH

Vitamin C.—One of the most unstable of the vitamins, ascorbic acid can be readily oxidized by a multiplicity of mechanisms, both enzymatic and nonenzymatic, operating in foods during storage and processing, as discussed in Chapter 16 and by Cook (1974). The specific enzyme catalyzing this oxidation, ascorbate oxidase, also examined in Chapter 16, is probably not involved in most instances of vitamin C activity loss in foods. Furthermore, oxidation of ascorbic acid does not in itself result in vitamin C loss since the dehydroascorbic acid formed is still biologically active. It is, it will be recalled, the hydrolysis of the lactone in dehydroascorbic acid to 2,3-diketogulonic acid which results in the irreversible abolition of biological activity as vitamin C:

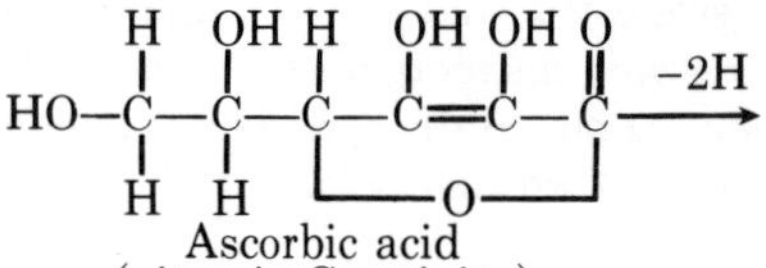

Ascorbic acid
(vitamin C activity)

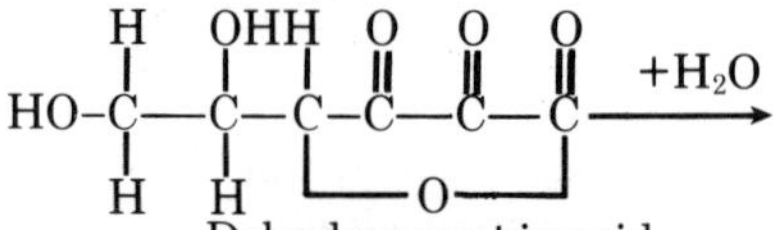

Dehydroascorbic acid
(vitamin C activity)

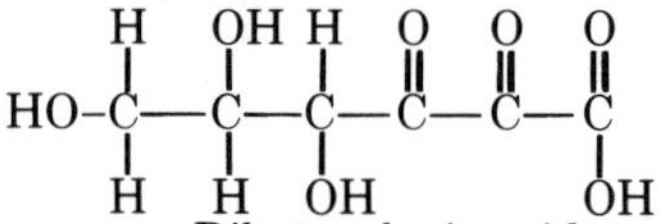

Diketogulonic acid
(No vitamin C activity)

This hydration occurs quite rapidly at pH's at or greater than neutrality, so that vitamin C activity can be retained despite oxidation by maintenance of highly acid conditions, not always feasible for many foods.

Loss of vitamin C activity through enzyme action may occur unwittingly or adventitiously as exemplified in the development of a mild anemia, a subclinical manifestation of vitamin C deficiency, in institutionalized Israeli infants fed a mixture of orange juice and mashed bananas. Since the presumably frequently used mixture was not always consumed immediately after preparation, as in the home, banana phenolase (indirectly via quinone rereduction, Chapter 16) oxidized the ascorbic acid in the orange juice. Presumably, the time lapse plus the buffering capacity of bananas which raised the pH sufficiently resulted in extensive conversion of the dehydroascorbic acid to diketogulonic acid.

Prevention of ascorbic acid loss was discussed in Chapter 16. At this point we only wish to point out that the stability of the biological activity of ascorbic acid may be affected by its presence in bound form, as in the peel of the potato and in cabbage as an ascorbigen generated from glucosinolase action (Chapter 12).

Other Nutrients.—Loss of protein by enzyme action was discussed in Chapter 34. Essential fatty acids and vitamin E, tocopherol, may be destroyed due to oxidation by lipoxygenase present in foods, especially in wheat flour during breadmaking (Drapron *et al.* 1974).

Vitamin B_6 is present in foods in three forms: pyridoxine, pyridoxamine and pyridoxal, the latter two usually phosphorylated. Pyridoxal phosphate is the coenzyme for many enzymes and is present in phosphorylase which serves as a reservoir for vitamin B_6 (Black *et al.* 1978). Presumably any enzyme which liberates the coenzyme from phosphorylase renders this B_6 congener susceptible to both enzymatic and nonenzymatic loss. Among several indications of how vitamin B_6 activity, once freed, may be lost by enzyme action during processing is a report on its oxidation by peroxidase (Hill 1970).

Phenolase action may also be involved since DOPA, a product of phenolase action on tyrosine, interacts with pyridoxal to make the latter nutritionally unavailable. Pyridoxal in the form of its phosphate can react with protein (Srncova and Davidek 1972). Indeed such reactions, speeded along by the liberation of this coenzyme from phosphorylase, rather than leaching could account for the 13–27% loss of vitamin B_6 in lima beans during blanching as reported by Raab *et al.* (1973), since the difference between steam and water blanching was not significant.

The loss of other nutrients via enzyme action which could conceivably occur but has not yet been experientially observed are those of folic acid by action of the glutamate-liberating hydrolases mentioned in Chapter 20, and of pantothenic acid by the amido-hydrolyzing pantothenase, especially if the food has been exposed to pseudomonads and retorted. Like peroxidase (Chapter 11) the thermally inactivated enzyme readily regenerates activity upon lowering of the temperature (Airas 1976). Finally, an adverse nutritional effect may come about from the circumstance that metals may not be as readily available as they would appear to be from analysis of a food because they may be too tightly bound to an enzyme. Thus, Wien *et al.* (1975) found that the bioavailability of iron in turnip greens is less than that of iron added to the diet in free form.

SAFETY OF ENZYMES AND ENZYME PREPARATIONS

Enzyme Preparations

In a sense the ensuing overview of the safety of enzyme preparations may be considered as a continuation of our exposition of the manufacture of enzymes in Chapter 5 in that before the enzyme preparation can be safely applied, it must pass rigorous safety tests.

Long established enzymes which are added to foods from plant and animal sources such as those from hog and cattle pancreas, liver catalase, stomach pepsin and chymosin, malt enzymes and bromelin and papain, and even older microbial commercial preparations such as the old "takadiastase" have been fairly well accepted as devoid of toxicity. Indeed, the proteinases prepared from these sources were long recommended and used for therapeutic uses as digestion aids (Chapter 36). The widespread application of commercial enzymes from new species and mutated strains of microorganisms has brought with it concern over the safety of these enzyme preparations, not so much with regard to the enzymes themselves, but to the accompanying nonenzyme material in these preparations—although under special circumstances the enzymes themselves may pose immunologically-related problems.

Commercial and industrial enzyme preparations from the bacteria *B. subtilis* and *B. licheniformis*, the fungi *Aspergillus oryzae* and *Aspergillus niger* and *Saccharomyces* yeasts are on the GRAS list and special dispensation has been given to glucose isomerase from *Streptomyces* (Chapter 9) and at least three microbial rennets. Enzyme preparations produced from these microorganisms are now part of the standard of identity of many foods.

Enzyme preparations are included in the specifications of the World Health Organization Additive Series No. 2 for identity and purity in 1972; *Rhizopus oryzae* for glucose forming enzymes; *Micrococcus lysodeikticus*, for bacterial catalase, mycelia of *A. niger* for amyloglucosidase and glucose oxidase, *Streptomyces* for glucose isomerase and *Trichoderma viride* for cellulase.

In general, studies on the safety of commercial preparations have, in addition to the routine short- and long-term feeding tests, been concerned with and checked by analyses for possible microbial toxins such as mycotoxins and β-nitroproprionate, microbiological contamination, heavy metals, antibiotics, presence of factors which stimulate both microbiological or animal growth and antibody and related immunological and allergenic effects.

Feeding Experiments.—Typical of results obtained are those of Van Logten *et al.* (1972) who could find no toxic effect of feeding the bacterial rennet from *Mucor pusillus* to rats at levels several orders of magnitude greater than the equivalent amount which a human consumer might consume. Similar results have been obtained by other workers in other countries with other enzyme preparations. Typical of results and discussions of enzyme safety are those of Fournier (1972) in France, Engst and Lewerenz (1972) in Germany and Skirko (1973) in the USSR. Van Logten *et al.* did find a stimulation of food intake and growth similar to that observed by Schwimmer and Kurtzman (1972) upon feeding to rats coffee decaffeinated by a fungal fermentation. More detailed results are shown in Table 35.1 which summarizes the findings on the toxicological evaluation of commercial food enzyme additives prepared by the Joint FAO/WHO Expert Committee on Food Additives. It is clear that the likelihood of hazard due to the ingestion of these and similar preparations is remote both from these and from other experiments in which the levels of enzyme fed may be several orders of magnitude greater than an equivalent amount which would ever be ingested by humans.

Sterility and Microbiological Contamination.— Typical of the results reported on the microbiological load in commercial enzyme preparations is that of Burzynska *et al.* (1977) who concluded that the microbiological quality of the 39 preparations obtained from Denmark and the Netherlands and made from *A. niger, B. subtilis, Mucor miehei* and papaya was well within the prescribed standards. Nevertheless, that assurance of sterility was still of concern is indicated by publication on the use of ionizing radiation (Delincee *et al.* 1975; Kawashima *et al.* 1976) and a patent on the use of low pH to sterilize enzyme preparations (Tower Pharmaceutical 1977).

Toxicants.—Although the possibility of mycotoxins such as aflatoxin in enzyme preparations was broached by Friedman (1964) and Fournier (1972), Van Logten *et al.* (1972) cited other investigators who could not detect mycotoxins in microbial rennets. The aspergilli and other microorganisms routinely used as food enzymes are, in general, not considered to be producers of aflatoxins. Ivanitskii (1973) did report some evidence of mycotoxins in strains of some fungi intended for enzyme production, especially in *Sclerotinia.*

A toxic fungal metabolite, β-nitropropionic acid, has been looked for in fungal enzyme preparations. It was reported present in one but not in any other enzyme preparation examined (Skirko 1973), and undetectable using an improved analytical method

TABLE 35.1

SAFETY OF SOME MICROBIAL ENZYME PREPARATIONS

Enzyme, Source	LD_{50}[2] mg/kg BW	Short-term Studies	Long-term Studies	Comments
Carbohydrases: *Aspergillus niger* Pectinase and amyloglucosidase preparations included	10,000	10% of diet	NDA	Available information is that *A. niger* not pathogenic; duckling and rat studies show no toxilogical effects at 10% level... meets WHO requirements
Glucose oxidase *A. niger*	NDA	10%	NDA	Same as for carbohydrases
Rennets: *Endothia parasitica*	2,000	40 mg/kg BW	100 mg	Studies extending over one year reveal no adverse effect; of a limited number of microbial metabolites tested for, none was detected; meets WHO requirements.
Mucor pusillus	5,000	0.1%	2%	No significant abnormalities; special aflatoxin production and reproduction studies revealed no abnormalities attributable to enzyme; meets WHO requirements.

Source: From toxicological evaluation based upon published results prepared by the Joint FAO/WHO Expert Committee on Food Additives (World Health Organ. 1975).

(Moskowitz and Cayle 1974). Burzynska *et al.* could find no accumulation of heavy metals. Teratogenic activity in a few preparations has also been looked for and not found. There is some evidence that some preparations contain feeble antibiotic activity (Kuzenchkin 1973; Burzynska *et al.* 1977).

Thus, in general it would appear that components of most of the commercial enzymes on the market today present a vanishingly small, if any, health hazard to most potential consumers when added to foods, especially when one considers that most tend to be removed by the time the consumer ingests the food. However, one precaution is suggested by the report of Priputina (1974) that rats fed a glucose oxidase:catalase preparation with milk showed no toxic symptoms but when apple juice was substituted for milk, changes were observed in blood lysozyme activity indicating to the author that the safety of these additives depends upon the product containing them.

The Enzymes Themselves as Putative Health Hazards

Immunological-associated Safety.—A few years ago there was a flurry of concern over allergenic and pulmonary disease putatively evoked by dried enzyme preparations used in laundry detergent formulations, especially with regard to possible harm to the home users of these detergents. Goldring *et al.* (1970) found, as is expected in all situations where individuals breathe in large amounts of foreign solid material, toxic effects can be produced. Thus, in the United Kingdom, antibody tests of *B. subtilis* workers involved in the production of laundry detergents showed that at least two-thirds of them showed positive results but no protease antibodies were found in users of the detergents (Pepys 1973).

The advocates of the application of immobilized enzymes stress that any safety problem, even if remote, brought about by adding enzymes to food could be further lowered by immobilization because in the first place the enzyme would not be left in the food and secondly the procedure of immobilization is likely to remove any toxicants which might be present in commercial enzyme preparations. While such considerations are undoubtedly of concern to workers in enzyme manufacturing, they may be of little concern to the consumers, because even if some enzyme is left in the foods, the conventional idea is that they are readily inactivated by the stomach and being protein, by their hydrolysis by the digestive proteases of the gastrointestinal tract.

Endogenous Food Enzymes.—While it has indeed been the reasonable, conventional and the experimentally upheld view that most, if not all, enzymes are inactivated by the time they get through to the intestinal tract, and even if they survive their molecules are too large to be absorbed into the bloodstream, evidence and speculation has been forwarded that some active enzymes may get through the tract. Matsumoto (1977), from a cytochemical study of the absorption of horseradish peroxidase through the human gastrointestinal tract, concluded that such passage of macromolecules may very well be a cause of food allergy and food poisoning.

However, much more attention has been drawn to the contention of Oster (Oster *et al.* 1974), backed up by impressive if indirect evidence that xanthine oxidase in homogenized milk is involved in the etiology of atherosclerosis. The involvement of this enzyme comes about, according to Oster, via its survival during travel through and absorption from the gastrointestinal tract, after which its target, via circulation, is ectopic sites in the walls of arteries and the myocardium upon which it is deposited. Once there, it oxidizes plasmalogens, fatty acid-like aldehydes alluded to briefly in Chapter 25 (it will be recalled that xanthine oxidase has also been termed aldehyde oxidase). The resulting active O_2^- and reactive oxygen species derived therefrom (Chapters 16, 18, 24 and elsewhere) presumably attack the delicate cell membrane of the arterial intima, resulting in the formation of scar tissue upon which cholesterol and atherosclerotic plaques can deposit. There has, of course, been considerable controversy concerning this intriguing hypothesis.

Ho and Clifford (1976), from model GI tract experiments, concluded that out of 100 *mg* of the enzyme present in several liters of milk only 20 *ng* were absorbed through the intestine. A considerable body of opinion maintains that the hypothesis has been amply refuted despite attempts to revive it (Anon. 1978). Recent available information on this controversy in a paper by Zikakis *et al.* (1977), is that considerable intact xanthine oxidase activity can indeed survive the digestive process but that additional research is needed to ascertain whether xanthine oxidase can get through the gastrointestinal tract intact. By the time this is being read, this question will probably have been answered. If this food enzyme is a health hazard, its elimination from the diet may be relatively easily accomplished by adequate heat treatment of the milk, by application of safe inhibitors, similar perhaps to allopurinol, the classical inhibitor of this enzyme, or perhaps by passing milk through an appropriate affinity column or more simply by dipping a bag of suitable ligand into the milk (Chapter 7).

36

ENZYMES AND THEIR ACTION IN FOODS AS HEALTH AND SAFETY BENEFITS

In this chapter we turn our attention to health-related beneficial effects of enzyme action as actually practiced successfully, as worked on experimentally, as proposed and as alleged. Inevitably some of the subject matter intrudes into the area of medical enzymology. Furthermore, some of the topics discussed in previous chapters, especially Chapters 34 and 35, could have been justifiably dealt with at this juncture, i.e., the use of enzymes to remove enzymatically-produced, potentially health-damaging substances from high protein foods. Nevertheless, the subject matter on how enzyme action can be beneficial tends to segregate into three distinct categories.

The first category is exploitation of enzyme action to remove or diminish the levels of substances which, above critical levels, may deleteriously affect the food consumers' health or affect physiological parameters in an undesirable way. If we define toxicants in this rather broad sense, then the action of these enzymes may be considered as one of detoxification.

The second category is involved with the relationships of enzymes and their action with enhancement and assurance of asepsis and wholesomeness of foods.

The third category includes health-related benefits incurred by ingestion of enzymes. It is in this area particularly that we skirt the edge of the field of enzymes as therapeutic agents. It will be recalled (Chapter 34) that proteinases, when added to animal feeds, appear to promote growth of the animals.

ENZYMES AS DETOXICANTS

What Are Toxicants?

A giant stride in the progress of preagricultural human culture ensued with adoption of cooking and leaching of previously inedible plants laboriously gathered in the course of the day's quest for food. According to Leopold and Ardrey (1972), innovations for removal of a variety of toxic substances made available to the food consumer an entirely new class of foodstuffs which had previously been unavailable as food because of naturally occurring toxicants. This profoundly influenced the direction of the evolution of social characteristics. While some endogenous enzyme action may have contributed to removal of these toxicants during leaching and heating, undoubtedly inactivation of biologically active proteins and washing out of low molecular weight toxicants were the major means of removal. It is only in the last few decades that serious rational campaigns to remove such substances by exploiting the properties of enzymes have been mounted.

As with other quality attribute enhancement, ridding foods of an unwanted health-affecting substance may be accomplished by the action of the food's own enzymes, autolysis (potentiation) or by adding enzymes from external sources. Examples to be discussed below are shown in Table 36.1.

Recognition that diet may play a role in the etiology of many diseases has recently been accepted by an increasingly large segment of the health research community and more importantly by those who fund such research, as exemplified by Senate hearings on proposals to put more money on diet-related aspects of cancer research. This recognition was crystallized through the publication of the book edited by Liener (1969) and by several discussions and reviews on naturally occurring toxicants in foods (National Academy of Sciences 1973; Coon 1974; Liener 1973, 1975; Strong 1974; IFT 1975; Schwimmer 1975).

The take-home lesson concerning putatively harmful substances in foods that came out of these discussions is that any one of the thousands "of chemicals" which comprise a food at a high enough level and under the right etiological circumstances could be toxic to a very limited number or to most food consumers. Thus, consumption of too much of any nutrient may have health-deteriorative effects quite aside from the added calories provided by some of them. Whatever their concentration, these levels can be regulated by harnessing enzymes.

Removal of Lactose from Milk by Lactase

The Health Problem.—By far the greatest research effort has been expended on the removal of lactose from milk and other dairy foods by the use of lactase. One reason for this concentration of effort is that several benefits other than health-related ones accrue, as we have had occasion to examine in Chapters 9, 33, 35 and elsewhere. The health-related benefit—a relatively recent suggested application of lactase—derives from the by now overwhelming evidence, first conclusively documented about 15 years ago (Bayless and Rosenzweig 1966), that large segments of the adult population of various ethnic groups are lactose-intolerant due to lactose malabsorption which in turn is due to a deficiency of lactase located in the brush cells of the intestinal (jejunum) epithelium. Indeed with respect to the world population as a whole, tolerance of adults to lactose appears to be the exception rather than the rule. However, it should be pointed out that malabsorption of lactose, measured by the appearance of elevated blood glucose after its ingestion, is not always accompanied by overt clinical lactose intolerance as evidenced by symptoms—bloat, cramps, flatulence, diarrhea—partly due to osmotic load and partly to decomposition of lactose by intestinal microflora which produces irritant acids and gases such as hydrogen which can be detected in the breath.

In some populations lactase deficiency may be caused by protein-calorie malnutrition, according to investigators at the University of Oklahoma led by Welsh (1978) who along with Paige and coworkers at Johns Hopkins University (Paige *et al.* 1977) have made major contributions in this area. Apparently these and other studies have not settled

TABLE 36.1

SUGGESTED ENZYMATIC REMOVAL FROM FOOD TOXICANTS AND ANTINUTRIENTS

Substance(s)	Food	Toxicity	Enzyme Action
Lactose	Milk	Intestinal upset	β-Galactosidase (Lactase)
Oligogalactosaccharides	Beans	Flatulence	α-Galactosidase
Nucleic acid	Single-cell protein	Gout	Ribonuclease
Lignan glycoside	Safflower seed	Catharsis	β-Glucosidase
Phytic acid	Beans, wheat	Mineral deficiency	Phytase (phosphate)
Trypsin inhibitor	Soybean	Protein not utilized	Urease (ammonia)
Ricin	Castor bean	Paralysis of respiratory and vasomotor systems	Proteinase
Cyanide	Fruit	Death	Rhodanese, cyanoalanine synthase, nitrilase (Chap. 35)
Tomatine	Green fruit	Alkaloid	Enzyme system of ripe fruit
Nitrite	Various	Carcinogen (?)	Nitrite reductase
Tannins	Various	Carcinogen (?)	Tannase
Caffeine	Coffee	Overstimulant	Purine demethylase systems in microorganisms
Cholesterol	Various	Atherosclerosis	Microbial enzymes
Saponin	Alfalfa	Cattle bloat	β-Glucosidase, new saponin combines with cholesterol
Cl-containing insecticides	Various	Carcinogens (?)	Glutathione S-transferase
Organophosphate	Various	Neurotoxin	Esterases

the question as to whether lactase deficiency in adults should be considered as: (1) one of the group of genetic diseases caused by failure of normal expression of a gene coding for (and resulting in failure to synthesize) a particular enzyme, (2) due to the inherently programmed normal repression of the lactase gene since most adults no longer need lactase, or (3) a consequence of lactase being an inducible enzyme (Chapter 4) so that the lactase gene turns off in response to the absence over long periods of lactose in the adult diet. Reviews include those of Kretchmer (1972) and Paige *et al.* (1975).

Prevention of Symptoms.—The uncomfortable consequences of lactase deficiency can be mitigated, not necessarily by completely eliminating lactose-containing food from the diet but, according to Welsh (1978), by taking them in small amounts or in the presence of liquid food with high osmolality which causes the food to empty from the stomach at a slower rate.

The other alternative, which will render obsolescent dietary management and therapy, is to remove the lactose with lactase. In addition to subjective relief of symptoms which could be partly psychological, illustrative of objective evidence of the effect of lactase action on the fate of lactose in milk are the findings shown in Table 36.2 (Payne-Bose 1977); a halving, on the average ($P < 0.002$), of hydrogen in the breath of malabsorbers even though two subjects put out *more* hydrogen.

As with the development of other microbial food enzymes, the field of dozens of possible candidates for lactase production has now narrowed down to a limited number, as of 1978 to two—the yeast *Saccharomyces lactis* and the fungus *Aspergillus niger.* By the time this appears a lactase from another yeast, *S. fragilis*, should be available for food use. One of the factors which eliminated other sources from further consideration was the intense product inhibition (Chapter 14) exhibited by their lactases. Review and comments on the development of their use in milk and whey treatment include those of Shukla (1975), Paige *et al.* (1977), McCormick (1976B) and Holsinger (1978).

Although highly active applicable commercial food grade preparations have been obtained from both the yeast and the fungal enzymes, the properties of the purified oligomeric (4 subunits) enzymes differ sharply. *A. niger* lactase acts optimally at 55°C and pH 4–5 and is stable between pH's 3 and 7. Corresponding values for both yeast enzymes are 35°C, *ca* pH 7, and pH's 6.0–8.5. Active exploration of other sources has continued (Wierzbicki and Kosikowski 1973B; Blankenship and Wells 1974; Griffiths and Muir 1980).

Investigations on the effectiveness and safety of adding preparations to milk have been underway for about 15 years. First available in the Netherlands from Gist-Brocade, the *S. lactis* enzyme is now sold to food technologists in the United States as a commercial preparation from the British Fermentation Industries as Maxilact which was used in the Netherlands to prepare powders from skim (Lactalac V) and from whole (Lactalac M) milk. Milk made by either reconstituting these powders or by adding *S. lactis* lactase preparations to whole milk without subsequent drying are somewhat sweeter than untreated milk, but at least as acceptable if not preferable as has been repeatedly demonstrated in formal sensory evaluations.

Based upon technology developed at the Eastern Regional Research Center of the U.S. Department of Agriculture, the *S. lactis* enzyme can now be added by the consumer to milk in the form of tablets of Lact-Aid, formulated and supplied by

TABLE 36.2

REDUCTION IN HUMAN BREATH HYDROGEN DUE TO TREATMENT OF MILK WITH LACTASE

Subject	Lactose Removed from Milk	Reduction in Breath Hydrogen	
1	95%	67% of control (3 hr)	6% (5 hr)
2	94	31	5
3	92	76	76
4	92	71	64
5	91	89	75
6	95	− 7	50
7	91	−12	−22

Source: Data from Payne-Bose *et al.* (1977); lactose ingestion of controls varied from 16.1 to 19.7 g, hydrogen output from 7 to 44 ppm.

the Sugarlo Company who have also, according to McCormick (1976B), developed a product under the trade name "Lactolo." Undoubtedly by the time this appears, lactase will have been made generally available by several suppliers. As of this writing, Novo plans to have a lactase preparation from *S. fragilis* available for food use in the not too distant future.

The cost, about 15 to 40¢ per quart, while probably quite acceptable to more affluent United States consumers, indicates that we have a long way to go before relief is at hand to would-be milk consumers in more impoverished lands.

The fungal lactase, available in the United States as lactase LP from Wallerstein, was preferred by Rand and Linklater (1973) because its superior heat stability permitted the enzyme to be added before and survive pasteurization, thus lowering costs. Its use on acid wheys was mentioned in Chapter 9. The conversion can be carried out overnight in the refrigerator with smaller amounts of enzyme or in 2 hr at about 30°–35°C with about twice as much as used at the low temperature.

Immobilization.—Lower costs undoubtedly will ensue when lactase is used on a large scale in the dairy before packaging and distribution. Hopefully this lowering in cost will be accelerated by immobilization. Various schemes have been tried and proposed for the incorporation of an immobilized enzyme system to remove lactose from dairy products and waste streams. In some of these studies lactase was used primarily as a test enzyme for demonstrating effectiveness and the specialized kinetic theory of immobilized enzymes in general (Hinberg *et al.* 1974; Hasselburger *et al.* 1974; Shukla 1975). In others, the investigators were primarily interested in the removal of lactose from whey (Wierzbicki *et al.* 1974; Coughlin 1977). The following schemes for immobilizing lactase have been reported: binding through covalent attachment to glass; to stainless steel, collagen particles of phenolformaldehyde resins and glutaraldehyde; to tannic acid, Sepharose and glutaraldehyde; *via* entrapment in hollow fibers, in acrylamide gel, porous cellulose sheets, cellulose triacetate; *via* acrylate and methacrylate polymerization, *via* ionizing radiation-induced polymerization and binding to magnetic supports.

Only limited studies are available on the actual application of immobilized lactase to remove lactose from milk. Woychick *et al.* (1974) felt that the problem of the rich medium for the growth of microorganisms provided by milk remains a major obstacle to commercial adaptation of immobilized lactase. On the other hand, Pastore *et al.* (1974) reported no external contamination of milk passing through a column of lactase entrapped in cellulose triacetate. They were able to hydrolyze 75% of the lactose in milk in 4 hr at 50°C. Illustrative of the problems one might encounter, Portelle and Thonart (1975) used immobilized *E. coli* lactase to prepare a lactose-free milk powder. The product was found to be highly hygroscopic and prone to nonenzymatic browning.

The pH of optimum *A. niger* lactase shifts down to 3.5 upon immobilization whereas that of the yeast enzyme is not altered (Kilara *et al.* 1977) but generally, the fungal enzyme appears to hold more promise, principally because of its greater lifetime during actual operation in immobilized systems and may well be used first for removal of lactose from whey. It is not likely that lactose-free milk from immobilized lactase action will be universally available before the mid-1980s.

Other Glycosidases.—*α-Galactosidase and Flatulence in Beans.*—In addition to lactose, there is in some foods a class of related compounds, the α-galactosyl oligosaccharides, which cause intestinal discomfort and flatulence when foods containing them are eaten. These foods include most legume seeds, soybean (other than its Oriental identities as tempeh and tofu) (Rackis 1975) and the various cultivars of the common beans and pulses. The compounds are the trisaccharide raffinose, 1-[6-α-D-galactosyl-α-D-glucosyl]-β-D-fructose, the tetrasaccharide stachyose, α-D-galactosyl raffinose and verbascose, galactosyl stachyose (Olson *et al.* 1975); and perhaps higher polysaccharides (Kurtzman and Halbrook 1970), whose presence may account for the synergistic effect on flatulence-related measurements produced by adding α-galactoside-free bean fractions to these oligosaccharides.

Two enzymatic approaches have been considered in their removal: *via* addition of an external source of enzyme or *via* potentiation of the action of the food's own enzyme (autolysis). α-Galactosidases are widely distributed in plants where they are present as lysosomal isozymes involved in the regulation of carbohydrate metabolism in general and galactolipids in leaves in particular (Thomas and Webb 1977). Becker *et al.* (1974) worked out optimum conditions for the *in situ* action of bean α-galactosidase in 10% bean slurries, subsequently partially purified by Kon and Wagner (1977). Loss of these α-galactosides was accompanied by decreased hydrogen evolution from rats fed these autolyzed beans as shown in Fig. 36.1. If indeed a synergy between α-galactoside and non-α-galactoside exists, then the dashed portion of Fig. 36.1 should be a curve with a decreasing slope. Schwimmer (1974) observed increase of the products of α-galactosidase action in soybean meal slurries. Kim *et al.* (1973) accomplished substantial removal of oligosaccharide from whole soybeans by a program of soaking, germination and resoaking at appropriate pH's. Undoubtedly the removal was facilitated by the action of endogenous α-galactosidase as well as by diffusion-extraction. The other approach, that of adding an extraneous source of α-galactosidase, is exemplified by the use of such a preparation from *Aspergillus saitoi* to remove almost quantitatively the oligosaccharides in soybean meal (Sugimoto and Van Buren 1970). Following the lead of the sugar refiners (Chapter 8) Thananunkul *et al.* (1976) removed some of the α-galactosides from soybean milk by the action of α-galactosidase present in polyacrylamide-entrapped mycelia of *Mortierella vinacea*, the same fungus used to remove raffinose from molasses. In another investigation, designed to remove raffinose from molasses (Chapter 8) but which may be applicable to bean products, Reynolds (1974) used α-galactosidase immobilized on nylon microfibrils in a continuous flow reactor. A blind alley stumbled into in the continuing search for an α-galactosidase for deflatulating legumes affords an illustration of the importance of not ignoring enzyme specificity. Galactose oxidase which, unlike glucose oxidase, catalyzes oxidation of its substrate both free and combined (Chapter 10), was used to monitor "removal" of galactosides, leading to the erroneous impression of the presence in certain microorganisms of high α-galactosidase activity.

We have already discussed the use of β-glucosidase to thoroughly hydrolyze cyanogenic glycosides as a health safety measure in Chapter 35. Other unwanted health-affecting substances in foods and feeds which may become foods which β-glucosidase action might effectively remove include bloat-causing saponins in alfalfa and the cathartic lignan glycoside in defatted safflower meal (Palter *et al.* 1972). An added serendipitous benefit of removing one of the two sugars linked to saponins is that the new glycoside may combine with cholesterol to prevent the latter from entering the bloodstream.

The Phosphoesterases

The actions of enzymes capable of catalyzing the hydrolysis of the C-O-P bonds in phosphate esters are, potentially at least, capable of improving the health-associated status of foods as diverse as single cell proteins (SCP), beans, bread and cottonseed meal. The two principal enzymes involved are the phosphodiesterase, ribonuclease, and the phosphomonoesterase, phytase.

Removal of Purines from SCP via Ribonuclease (RNase).—Among the many problems—chemical, economic and health-oriented—which have to be solved before single cell protein (SCP) can be considered as a potential source of protein in the

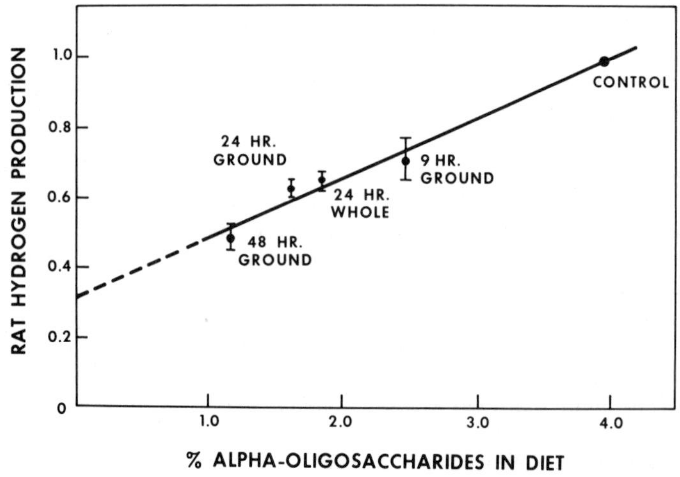

Courtesy of Becker et al. (1974)

FIG. 36.1. AUTOLYTIC REMOVAL OF α-GALACTOSIDES IN RAT BEAN DIET

Autolytic α-galactosidase action as measured by decrease in rat hydrogen production.

human diet is the reduction of the purine levels in such sources due to the high nucleic acid content of unicellular organisms (MacLaren 1975; Viikari and Linko 1977). This arises from the role of purines in the etiology of gout. Purine-containing nucleic acids can be removed by nonenzymatic methods such as that of Hedenskog and Mogren (1973) in which protein is separated from nucelic acid by precipitating the latter in hot strong alkali. One would expect a lowering of the nutritional value of the protein by such drastic treatment. Other chemical treatments are outlined by Viikari and Linko. A milder alternative, adopted by a group of investigators at MIT (Ohta *et al.* 1971), was to potentiate the nucleases of the unicellular organism. Thus, the yeast *Candida utilis* was subject to a heat shock at pH 4.0 at 68°C for 1 to 3 sec. This treatment probably damages the internal membranes of the yeast cells so that the ribonuclease has unlimited access to the RNA. Subsequent programmed incubation for somewhat longer periods at somewhat lower temperatures results in the degradation of the RNA to 3′-nucleotides. At first the latter accumulate in the cell but, probably because of the absence of selective membranes, the nucleotides diffuse out of the cell without a concomitant loss of protein. In this vein, Chao (1974) was granted a patent in which it is claimed that the nucleic acid content of SCP material is reduced to an acceptable level in food products by "physiological conditioning," followed by the action of endogenous ribonuclease.

The previously discussed plastein reaction has been applied to SCP (Fujimaki *et al.* 1973B). The reduced fluorescence and UV absorbance of the plastein as compared with the starting SCP suggest that the nucleic acid level was reduced. Proteinase action has also been used to solubilize SCP and leave behind insoluble RNA.

Since then various schemes have been forwarded for enzymatic RNA removal, most based upon the goal of more fully potentiating the action of the endogenous RNase. Thus, Trevelyan (1977) found that initiation of the autolytic degradation of RNA via membrane disorganization was promoted by the addition of ethanol (5% v/v). Alternatively, yeast which had first been air dried and rehydrated was incubated at 50°C in the presence of 0.5 *M* NaCl which, Trevelyan suggested, aided in liberating RNA from ribosomes. NaCl was also a key to unlocking RNA from disintegrated suspensions of yeast cells and protein in processes devised by Lindblom whose previous work is applied and cited in her paper on the use of yeast protein concentrates in breadmaking (Lindblom 1977). In addition to breaking up RNA-rich ribosomes, NaCl was also helpful by stabilizing the RNase during the exposure to 50°C and also in desorbing the enzyme from cell wall debris, thus allowing more contact with its substrate in solution. In another variation the NaCl-treated suspension was heated to 80°C for 14 sec, time enough to separate purine-containing compounds from precipitated protein. This suggests that some of the effectiveness of the MIT group's shock treatment may have been due to protein insolubilization as well as to RNase potentiation.

The investigations of Lindblom afford yet another illustration of the serendipitous entrainment of benefits accrued by adoption of a processing step. This involves enzyme action, in that she found that treating yeast to remove RNA also improved the breadbaking properties of the resulting yeast protein concentrates. This improvement came about as the result of decreased protein solubility.

Attempts, apparently so far not economically feasible, have been made to use exogenous sources of enzyme. Thus, the MIT group developed a process in which ribonuclease was added to briefly heat-shocked (80°C, 30 sec) cells to allow access of the added enzyme to its substrate (Castro *et al.* 1971). The action of barley malt phosphodiesterases was exploited in patents by Newell *et al.* (1975) and Haid *et al.* (1977), whereas Fazarkerley and Ebbon (1975) used the extracellular RNase secreted by *Candida* spp. as an enzyme additive to remove RNA from SCP prepared from these yeasts.

Viikari and Linko suggest the purine removed, whether as RNA by some chemical methods or as nucleotides via enzyme action, might be utilized to improve the overall economics of SCP production. Thus, instead of using pancreatic RNases which produce 3′-nucleotides, Kihlberg (1972) used microbial RNases which produce 5′-nucleotides (Fig. 2.1). The latter would find more use than 3′-nucleotides as biochemicals and perhaps as flavor potentiators (Chapter 20).

Phytate and Phytase

The Substrate.—The controversial status of phytic acid (phytate) in human nutrition is epitomized by the title of a review (in Polish and unfortunately unavailable to the author) translated as "phytic acid, a useful or harmful food factor"

(Kloczko and Rutowski 1977). Phytic acid, *myo*-inositol 1,2,3,4,5,6-hexakis dihydrogen phosphate, is ubiquitously distributed throughout the plant kingdom and is also present in the blood of amphibians, reptiles and birds. In plants it serves along with starch in potatoes and lysolecithin in cereal endosperms as a reservoir of P. It is a metabolism ballast ensuring dormancy. Thus, the P of potatoes, bound to starch in immature tubers, shifts to *myo*-inositol and appears thus as phytic acid P in developed and dormant tubers (Samotus and Schwimmer 1962A). In the blood of birds, etc., it plays the same role as corresponding 2,3-diphosphoglycerate in mammals as allosteric regulator via inhibition of the release of oxygen from hemoglobin (White 1976; Perutz 1978) and of AMP deaminase (Kruckeberg and Chilson 1976). This latter inhibition might be implicated in poultry flavor since the product of 5′-AMP deaminase is the flavor potentiator IMP (Chapter 20). Its apparent allosteric inhibition of microbial and plant α- but not β-amylase (Sharma *et al.* 1978) may have repercussions in its involvement as a determinant of processed vegetable texture (Chapter 29).

It is present at rather substantial levels in foods, mostly dry seeds, in which it has been implicated as a nutritional hazard, from about 1% for corn and beans to 5% in defatted seeds such as sesame. Its predominance as the major form of P in many of these foods led Lolas and Markakis (1975,1977) to suggest that total P is a reliable index of phytate in whole beans, but apparently not of grains such as triticale (Singh and Reddy 1977). In some grains it is largely localized in the bran aleurone cells, whereas in corn most of it accumulates in the germ.

Wherever it accumulates, by virtue of its highly ionized orthophosphate groups, it attracts and complexes with a variety of both cations and proteins that happen to be in the vicinity. Historically, phytic acid was isolated as the Ca_5Mg salt (phytin) but in most plant cells it is probably complexed largely to protein to form crenated golf ball-like structures about 1 to 2 μm in diameter (Ogawa *et al.* 1977). Quite aside from the action of enzymes on phytic acid, the latter is altered with respect to its localization and complexing partners during postharvest handling and processing.

The Health Question.—It is these latter changes, according to Rackis and Anderson (1977), resulting in the formation of altered protein-phytate-mineral complexes rather than phytic acid per se which is responsible for the reportedly reduced mineral availability of some protein isolates. Rackis and others (Reinhold *et al.* 1975) pointed out that dietary factors other than phytate such as dietary fiber, oxalates and phenolics may also contribute to reduced mineral availability. Nevertheless, phytate has been implicated as an antinutrient in beans, baked products prepared from whole wheat and potentially is of some concern in some of the newer sources of protein such as cottonseed and rapeseed. Historically the "phytate phobia" originated in pinpointing phytate as the putative culprit contributing to the high incidence of diseases related to calcium-deficient diets of children in World War II (McCance and Widdowson 1942). In an attempt to upgrade the nutritional quality of bread, phytate-rich whole wheat bread was substituted for branless and therefore phytate-poor white bread.

Since then more carefully controlled investigations using rats, pigs and especially chickens have demonstrated that phytate can, under certain conditions, limit the availability of phosphorus, calcium and perhaps of iron and especially of zinc, as summarized by Erdman and Forbes (1977). Exemplary of such findings and controversy are the findings of Nwokolo and Bragg (1977) that both fiber and phytate from palm kernel, soybean, cottonseed and rapeseed meals adversely affected the retention of all minerals tested—Ca, Cu, Mg, Mn P, Zn—when used as feed supplements for growing chickens. Yet, Erdman and Forbes (1977) showed that while zinc in whole fat soy flour was definitely not as well utilized by rats for both weight gain and uptake into the femur as was zinc carbonate added to a basal egg white protein diet, the presence of this soy flour did *not* affect the availability of zinc from other dietary sources. Wallis and Jaffee (1977) showed that the solubility of Fe in raw beans correlated inversely with the level of phytate and that there was also a weak but positive correlation between soluble Fe in soybeans, beans, rice, carrots and lettuce and absorption of ^{59}Fe by human subjects. Ferric phytate is unique in that it is the only common metal salt of phytate which is insoluble in acid.

Phytate may also be a problem in human nutrition where dietary intake in general is restricted and accompanied by a high proportion of high-phytate foods such as unleavened grain products consumed in the Middle East. The situation is complicated by the elusive action of intestinal phytase.

The Enzyme.—Among the factors which complicate a definitive assessment of an antinutrient role to phytate is the presence in the intestines of most monogastric mammals (the flora of ruminants readily remove phytate) including humans, of a feeble and variable phytate hydrolyzing activity inducible by phytate, perhaps in conjunction with vitamin A (Bitar and Reinhold 1972). This phytase is readily inhibited by such dietary components as Ca.

Somewhat more is known about plant and microbial phytases. In plants, phytic acid is almost invariably accompanied by but separated from enzymes capable of hydrolyzing phytic acid completely to free *myo*-inositol and orthophosphate ion.

The term "phytase" was at first officially designated by the Enzyme Commission as an enzyme which catalyzes the hydrolysis of phytate into free *myo*-inositol and six Pi's, EC 3.1.3.8. In 1972 this number was assigned to "1-phytase" and in 1976 to "3-phytase," whereas the term "phytase" was listed as "another" name for the recommended "6-phytase," EC 3.1.3.26, probably because of evidence corroborated recently by using tracer techniques that the first (but not the only) phosphate which comes off with all plant phytases so far looked at is linked to *myo*-inositol at its 6-position.

In addition to the older review of Sloane-Stanley (1961) key papers include those of Tomlinson and Ballou (1962) and Lim and Tate (1971) on wheat bran phytase and Maiti *et al.* (1974) on germinated legume phytase. All phytases are strongly inhibited by both excess substrate and by product Pi. Perhaps the most singular observation on the bran enzyme is that one of two isozymes separated during preparation is accompanied by an activator which can be removed and replaced by lecithin or lysolecithin.

Unlike most other plant phytases whose pH optima lie in the region of 5 to 6, those of wheat bran are 7.0–7.5. Although the 6-*myo*-inositol position is the first to be attacked, the subsequent pathways of degradation differ from plant to plant. Unlike wheat phytase, for the mung bean enzyme (*ca* 158,000 dalton oligomer) lecithin and phosphatide are slightly stimulatory below 30 μM. Above this concentration phosphatide, S*n*-3-glycerylphosphocholine (Chapter 25) is a potent inhibitor with a K_i of about $3 \times 10^{-6} M$. Phosphatide does not affect wheat phytases. In addition to wheat bran and seedlings, cells in suspension culture may prove to be a rich source of phytase (Olson 1972) as pointed out in Chapter 6.

In contrast with many purely enzymological investigations where the most active enzyme source is sought for purification purposes, in some food science investigations, as in the present case, it becomes necessary to demonstrate the presence and study the properties of enzymes; although these enzymes may be present in almost undetectable amounts, they may have profound effects on food quality. Thus, although Maiti *et al.* (1974) state that they demonstrated the absence of phytase in cotyledons of ungerminated mung bean seeds, Chang and Schwimmer (1977) showed that mung beans lost about half of their phytate when exposed to a water-saturated atmosphere. Crude bean extracts at first appeared to be devoid of phytase activity. It is only recently with the rise of the phytate phobia that attempts have been made to demonstrate, purify and characterize the phytases of dormant food legume seeds (Chang and Schwimmer 1977; Lolas and Markakis 1977).

Proving dormant bean phytase to be no exception, both groups of investigators found that the enzymes exhibited both substrate and product (Pi) inhibition, the latter shown in Fig. 36.2. These potent inhibitions, with the K_i of the product inhibition about the same as the K_m, have both metabolic and food-related consequences, the latter because of the inability to evoke the full potential of the endogenous phytase (see below). Undoubtedly in the future such hard-to-get-at enzymes will be purified by affinity chromatography (Chapter 7), not the least of whose virtues is the facile, gentle concentration of very dilute enzyme solutions. For phytase, a step in the right direction is the synthesis by Scheiner and Breitenbach (1976) of a substrate -spacer arm- matrix complex consisting of *myo*-inositol-2-phosphate coupled to aminohexyl-Sepharose. They suggest that phytic acid could also be used as a ligand.

Like most other hydrolases involved in transforming massive amounts of substrate, phytase can be readily induced in many microorganisms. Particularly rich sources are the fungi *Aspergillus ficuum* (Cosgrove 1977) and *A. terreus* (Yamamoto *et al.* 1973). Phytate can be transformed enzymatically by at least one other enzyme *via* a nonhydrolytic conversion to *myo*-inositol-pentaphosphate in a reverse kinase action (Biswas *et al.*

1978) perhaps through the intermediation of a transient phosphoprotein. The enzyme can also hydrolyze phytate in the absence of phytase:

$$AP_2(ADP) + IP_6 \longrightarrow AP_3(ATP) + IP_5$$

This irreversible reaction, catalyzed by inositol hexaphosphate adenosine diphosphate phosphotransferase present in germinating mung bean (Biswas *et al.* 1978), provides a ready source of useful free energy for the differentiating cells in the rapidly metabolizing tissue in the form of a nucleoside triphosphate. A similar enzyme utilizes GDP instead of ADP.

Removal of Phytate.—Nonenzymatic solutions to health and other problems that phytate may pose in foods include both nutritional and processing approaches. As an example of the nutritional approach, the supplementation of bean diets with zinc, Boloorforooshan and Markakis (1977) report what they consider to be a very modest increase in growth rate of 6% in PER. They also found that too much zinc induced excessive output of cholesterol. That processing can remove phytate is evidenced by its absence in such soy-derived foods as tempeh and tofu. It is the Ca salts used in the preparation of these foods that probably effects phytate removal by increasing its solubility in the region of the isoelectric point of the principal soy proteins (Ford *et al.* 1978). The latter investigators developed a lipid/protein process for the removal of phytate based upon this phenomenon. They report results derived from quantification of a more thorough characterization of the factors involved which can be applied to a large number of processes now being used for making edible soy proteins. In an alternative treatment, phytate was removed by alkaline extraction of soybean flour, removal of the insoluble phytate salts at pH 10.5, followed by isoelectric precipitation of the protein (Goodnight *et al.* 1978).

Suggestions for the enzymatic removal of phytate have included both the use of the food's own enzymes (autolysis, potentiation) and addition of exogenase phytase.

To avoid recurrence of calcium deficiency diseases which occurred in England during World War II, McCance and Widdowson (1942) showed that potentiation of phytase action in wheat bran (6 hr at 50°C and pH 4.5) resulted in a product which, when added to flour, yielded a bread from which Ca, Mg and P were much more readily available than from bread made with untreated bran. During normal breadmaking, in the absence of an overabundance of bran, much of the phytate disappears during the breadmaking process due to the action of phytase from both the wheat and the yeast (Ranhotra 1973; Reinhold 1975; Ferrel 1978). However, there has been some concern of the effect of product inhibition in some of the high-protein wheat products, especially those containing soy additives. Ranhotra and Lowe (1975) indicated that phytase from such products may continue to act in the GI tract after their consumption. Oats and corn are low in phytase so that little phytate is lost during processing of products containing these cereals.

Exploitation of endogenous phytase action has also been investigated as a means of removing phytate in legumes. To avoid leaching of nutrients and to minimize microbial contamination, Chang *et al.* (1977) succeeded in partially removing phytate from a navy bean cultivar by placing the dry beans in a water-saturated ("steaming") atmosphere in the usual enzyme-potentiating temperature range, as shown in Fig. 36.2. Complete hydrolysis was hampered by the above-mentioned inhibition by excess substrate and by product. Soaking whole beans in water at potentiating temperatures removes more than 90% of the phytate through both diffusion and enzyme action in small batches and about 50% on a large scale. In the latter case this treatment after cooking affected neither their protein efficiency ratio nor their apparent nitrogen digestibility to a great extent (Chang *et al.* 1979). This illustrates that, in principle at least, it is

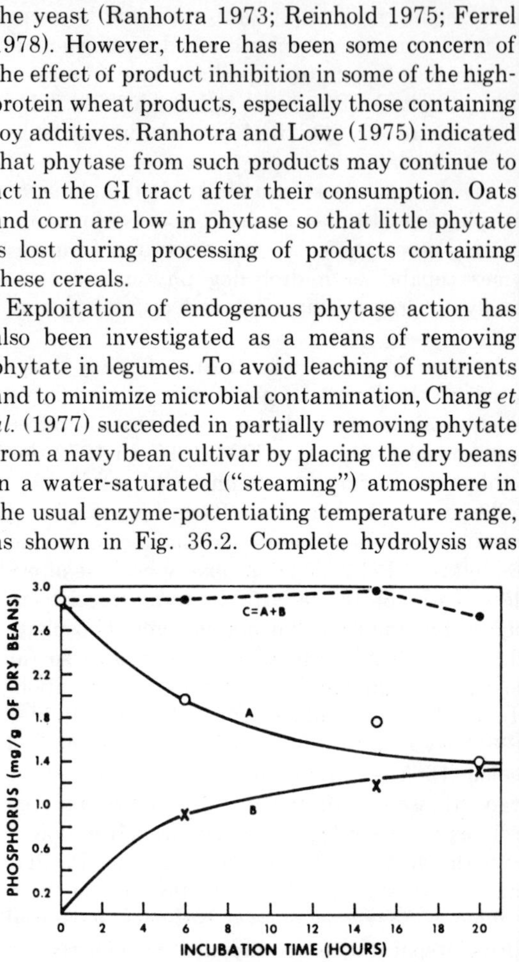

From Chang et al. (1977)

FIG. 36.2. TIME COURSE AND STOICHIOMETRY OF POTENTIATED PHYTASE ACTION IN BEANS

Navy beans were exposed to water-saturated air at 60°C. A, residual phytate phosphorus; B, inorganic phosphorus.

possible to potentiate enzyme action to remove a suspected antinutrient without affecting other important nutritional parameters.

Ferrel (1978) followed the appearance of mono-, di-, tri-, tetra- and penta-*myo*-inositol phosphates in suspensions of slurries (Table 36.3). The lower inositol phosphates from, for example, triphosphate on down are considered to be less severe antinutrients than the higher ones because they do not tie up metal ions so avidly.

TABLE 36.3

PHYTASE PRODUCTS IN AUTOLYZING BEANS

"X"-Inositol X	Distribution in Control[1]	Change Upon Slurrying[1]	"Steaming"[2]
Hexa- + Penta-	68%	−43%	−33%
Tetra- + Tri	23	+5	−1
Di- + Mono-	9	+28	
Orthophosphate	0	+10	+34

[1] Source: Calculated from data of Ferrel (1978); slurry incubated at 55°C for 24 hr.
[2] Source: Calculated from data of Fig. 36.2 combined with that of Ferrel.

Treatment of cottonseed flour provides an illustration of how processing may influence utilization of endogenous enzyme action in that phytase action occurs during heating of flour prepared from the glandless seed only after the latter has been air classified. This effect can be accounted for by the enrichment of both enzyme and substrate-bearing subcellular globoid bodies (Wozenski and Woodburn 1975).

An alternative way of removing phytate is to germinate the seed. Germination results in at least a ten-fold increase in activity and a simultaneous decrease in phytate, and, if done carefully, the seed can still be used for cooking in the customary manner (Reddy *et al.* 1978).

Very little has been done experimentally regarding the use of partially purified phytase preparation to remove phytate from foods; we are probably a long way off from practical application. Ferrel (1978) accelerated the removal of phytase from bean and wheat by the addition of a laboratory supply house wheat germ "phytase" and suggested that bean autolysis might be speeded up by the addition of a small amount of wheat.

Rojas and Scott (1969) and apparently Baugher and Campbell (1969) independently removed phytate from cottonseed meal by adding phytase in the form of cultures of *Aspergillus ficuum* and a strain of *Diplodia* fungus, respectively, and reported nutritional improvement. Serendipitously, they also nullified the toxic effect of gossypol by this treatment. This happened because of the now-vacant phytate attachment sites on the cottonseed proteins, sites which can also be occupied by the gossypol; bound gossypol is nontoxic.

Other Potential Toxicants and Antinutrients.— *Removal of Other Health-related Unwanted Substances.*—Proposals have been made and experiments conducted on enzyme-assisted removal from food of the following substances which are either toxicants or substances which at appropriate levels have been implicated in producing health-related hazards.

In addition to enhancing the palatability and protein nutritional quality of soybean, production of far eastern fermented soybean foods (Chapter 34) results in elimination of phytic acid. Wang *et al.* (1980) showed that this removal is most likely due to the presence of highly active intra- and extracellular phytases in most of the fungi involved, but not in that used to make sufu.

Trypsin Inhibitor in Soybeans.—The action of soybean's abundant urease, potentiated by spraying urea on full-fat unextracted soybeans followed by crushing, liberates enough ammonia according to Rambaud (1975) to alkali-inactivate the last traces of this antinutrient when followed by rolling, flaking and cooking in steam.

Ricin.—This extremely toxic protein, which in trace amounts may be responsible for the cathartic effects of castor oil, may be inactivated by enzymes of the castor seed via autolysis (Darzins 1960), thus presumably making castor bean protein potentially available as a feed or food.

Tomatine and Other Steroids.—The leaves and other green parts of the tomato plant constitute a potential source of high quality protein which cannot be used as feed or food because they contain the bitter and somewhat toxic steroidal alkaloid tomatine. Tomatine and tomatidine arising from glycosidase action, as shown in Fig. 36.3, are degraded to the innocuous "allopregnenolone" (a useful intermediate for the synthesis of steroid hormones) by an enzyme system in ripe tomatoes, (Heftmann and Schwimmer 1972) or can be degraded *in vivo* by enzymes of *Nocardia restrictus* (Belic and Socic 1972).

Glycosidase action to remove saponins, another class of steroidal alkaloids, was mentioned earlier

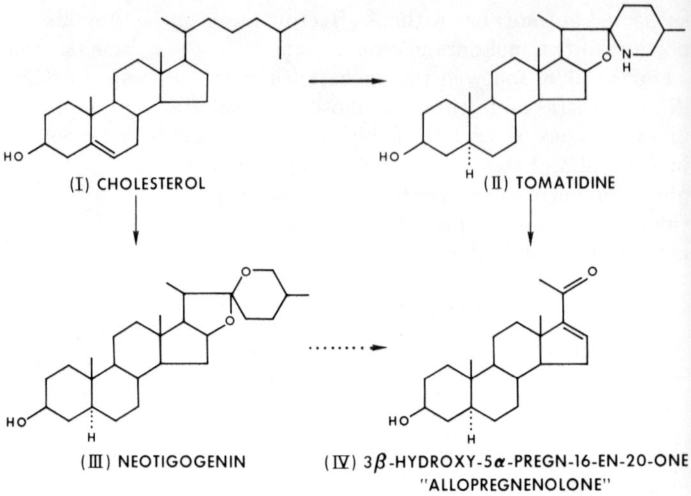

FIG. 36.3. BIODEGRADATION OF TOMATINE IN TOMATO

The enzyme system for biodegradation is in the ripe but not in the green fruit.

From Heftmann and Schwimmer (1972)

in this chapter. Cholesterol implicated in the etiology of atherosclerosis might be removed by microbial enzymes (Arima *et al.* 1970). Solanine (Fig. 36.4), the extremely toxic alkaloid which may occur in potatoes, may also someday be removed by enzyme action (Maga 1980).

L-DOPA.—An example of how the action of an enzyme in a food can produce a useful pharmaceutical is the formation by tyrosinase and accumulation in the velvet bean of the medically useful DOPA (Chapter 15). Just why DOPA accumulates in this food but not in other phenolase-containing foods is a subject for further investigation. Extracts of the bean do turn colored due apparently to a diphenol oxidase for which L-tyrosine is not a substrate (Zenin and Park 1978).

Adventitious Chemicals: Herbicides, Pesticides, Drugs.—Some plants develop resistance to herbicides by producing detoxifying enzyme systems which might be useful for removal of residual herbicides from plant foods; thus, glutathione S-transferase:

$$RX + G\text{-}SH \longrightarrow RS\text{-}G + HX$$

where RX is an organic halogen derivative, X a halogen, usually chlorine, GSH, glutathione, and R-SG, S-substituted glutathione. The enzyme was found in atrazine-resistant but not in susceptible plants (Lamoureux *et al.* 1970). Urea-derived herbicides such as Linuron are degraded by an herbicide-inducible acyl amidase (Engelhardt *et al.* 1971).

$$R\text{-}NHCO\text{-}R_1 + H_2O \longrightarrow R\text{-}NH_2 + HOOC\text{-}R_1$$

where R is a chlorinated or nitrated phenyl group and R_1 is an acyl, acylamino or alkamino acyl residue. Degradation of pesticides which also happen to be esters such as the phenyl carbamates (CIPC, IPC) and the esterase-inhibiting organophosphates (Chapters 2 and 37) by immobilized esterases has been looked into as part of environmental pollution control systems by Munnecke (1978,1980) who also pointed out that some *Hydromonas* spp. and DDT-resistant insects contain enzyme systems which degrade DDT to *p*-chlorophenylacetic acid. Other insecticides, drugs and poisons which find their way into foods may eventually be removed by using the previously mentioned powerful xenobiotic or mixed function-cytochrome P450 oxidase systems of the liver (Gillette *et al.* 1974; Guengerich 1977) perhaps in an immobilized configuration (Cohen *et al.* 1977).

Nitrite.—This reactant in the formation of carcinogenic nitrosamines may be removed from foods by the action of the Cu- and sometimes FAD-containing enzymes which reduce nitrite, especially the one with the recommended name "nitrite reductase"

$$HNO_2 + RH \longrightarrow H_2O + NO + R$$

where R is a hydrogen acceptor. Such an enzyme was used to lower the content of residual nitrite in sausage (Solov'ev and Prokosheva 1973) and in an immobilized enzyme study of nitrate and ni-

FIG. 36.4. STRUCTURE OF SOLANINE

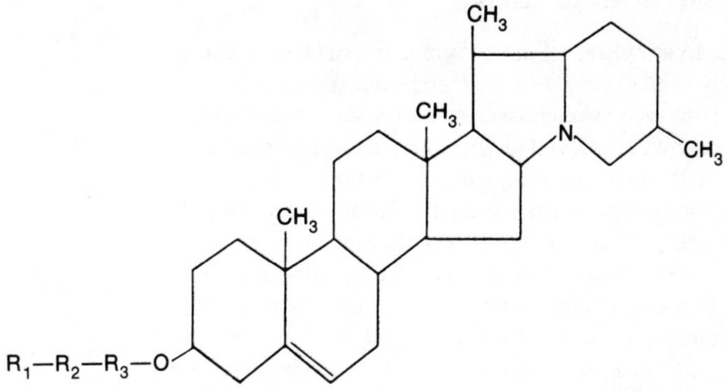

From Schwimmer and Burr (1967)

trite reduction using liquid membrane encapsulated whole cells of *Micrococcus denitrificans* (Mohan and Li 1975). Alternatively one might prevent nitrite formation from nitrate by use of specific inhibitors of nitrate reductase (Stulen *et al.* 1971).

Tannins.—Some condensed tannins such as that from the wattle-bark and carob pods which have been reported to exert growth-depressing effects and have been imputed to cause esophageal cancer when inhaled by leather workers may be altered in this respect by the action of tannase (Joslyn *et al.* 1968).

Caffeine.—This stimulant, a potent inhibitor of most phosphodiesterases, can be fermentatively removed from coffee by the action of an enzyme system in *Penicillium crustosum* whose first action is the removal of a methyl group at the 7 position of the xanthine ring (Schwimmer *et al.* 1971; Kurtzman and Schwimmer 1973). An even more active caffeine-degrading system, obtained by Woolfolk (1975) from a strain of *Pseudomonas putida*, apparently contains an enzyme which uniquely hydrolyzes caffeine to xanthine and methyl alcohol which, with uric acid, are considered to be toxicants. The use of such an isolated enzyme would yield a coffee containing 0.02% methanol (Chapter 35).

Undoubtedly enzyme approaches can be used for other toxicants or food substituents considered to be present in too large amounts such as cholesterol or fats. Indeed some research has been done on removing hydrogen from saturated fats with crude extracts of appropriate organisms. Cholesterol levels in the diet might be controlled by cholesterol-degrading enzyme such as that from *Arthrobacter simplex* (Arima *et al.* 1970), or as mentioned, by eating, along with cholesterol, enzymatically modified steroidal glycosides which combine with cholesterol, thus preventing its intestinal absorption.

Aflatoxins.—While the principal thrust of research on removal of aflatoxins has been channeled through chemical degradation, bisulfite and hypochlorite being leading candidates for this task, there exists a small body of evidence that lactoperoxidase and some microorganisms can degrade this carcinogen and that enzyme action is involved (Doyle and Marth 1978A).

Synthetic Dyes.—Another way that enzymes may be used to indirectly remove potentially undesirable substances from foods is in the search for cost effective, biologically-derived colorants to replace some synthetic dyes in foods. For instance, Schultz (1979) developed an inexpensive procedure for concentrating bixin, a well-known but fairly expensive yellow-orange colorant present in the aril of the seed of the tropical bush annato *(Bixa orellana).* A commercial PG is added to remove the aril pulp from the fresh seed surface and papain treatment breaks up the resulting emulsion. This allows the insoluble bixin to rapidly settle out and be readily removed.

PREVENTION OF MICROBIAL CONTAMINATION

Conceptually and to some extent in practice, enzymes may be involved in food asepsis in at least four ways: directly (1) as sterilizing agents and (2) as producers of products which possess antibiotic activity, and indirectly (3) by removing added sterilants and (4) as monitors of contamination (Chapter 37) or of adequacy of pasteurization (Chapter 11).

Enzymes as Sterilants

Lysozyme.—This enzyme, important in the historical development of enzymology as one of the first enzymes whose active-site structure and function were arrived at through a complete elucidation of its tertiary structure (Chapter 2), promises to also play an increasingly prominent role in keeping foods wholesome and free of contamination in the United States as it apparently already has in European countries. This prognosis is fortified by increasing emphasis on the use of "natural" food additives, especially preservatives, new leads on its inexpensive production and its potential clinical application as a bacteriolytic agent against infection, a therapeutic agent in wound treatment and as a potentiatior of several antibiotics.

The enzyme was first discovered by its ability to destroy bacteria (*Micrococcus lysodeikticus* being the classical and most sensitive organism and still used for enzyme assay) by decomposing their cell walls *via* the hydrolysis of β-1,4-links between derivatives of glucosamine comprising these cell walls. It also slowly attacks chitin, poly-*N*-acetylglucosamine. Indeed, it is the latter property which gives rise to the expectation that adequately active preparations can be supplied to the food industry at reasonable costs. Chitin, a poor substrate, and its deacetylated product, chitosan (not a substrate), constitute not only matrices for immobilizing enzymes but also inexpensive matrices/ligands for lysozyme concentration, purification and even isolation, if need be (Weaver *et al.* 1977; Muzzarelli *et al.* 1978).

Lysozymes are found in many microorganisms and in mammalian fluids such as milk (more in human than in cow), sweat and tears, and in egg white. In egg white it probably contributes to resistance to invading organisms (Eitenmiller *et al.* 1971; Yadav and Vadhera 1977). It may also contribute positively to food consumers' health by hastening the action of pepsin on milk protein (Kisza *et al.* 1974); see also Chapter 33.

A series of Japanese publications, mostly patents, in the late 1960s and early 1970s claim preserving action and/or increased digestibility by lysozymes as a preservative ingredient added to the following foods: plant-derived foods such as bean curd, flour, noodles and fresh vegetables; marine-derived foods such as oysters, shrimp, fish paste and lizard fish cakes; meats and sausages; dairy foods such as ice cream, cream and milk for infants (thus "humanizing" milk by raising the lysozyme level and also favoring growth of lactobacilli); sugar to prevent dental caries; liquid foods such as fruit juice and sake; and coated on food packaging film as an antiseptic barrier.

Caution concerning lysozyme application to foods may be warranted if a report by Kralovic (1973) on enhanced heat resistance of a strain of botulism-causing microorganisms through lysozyme treatment is substantiated.

It may be propitious at this time to point out that lysozyme may influence nonhealth-related food quality by influencing the stabilites of casein micelles (Chapter 33), egg white foam (Chapter 34) and egg white thickness. Its denaturation during storage may contribute to egg white thinning (Kato *et al.* 1978). It may control semihard cheese textures by suppressing organisms which produce excess butyric acid in "butyric acid blowout."

Lysozyme is used in Europe for "humanization" of human milk according to Scott (1975) who also states that apparently (in spite of the many patents and publications on its food-related uses) lysozyme is *not* permitted in foods in Japan. As a step toward the design of self-sterilizing immobilized enzyme columns, Mattiasson (1977) successfully delayed bacterial growth on columns of immobilized lactase from 10 to 80 hr by coimmobilizing lysozyme on the same column. The delay is much longer than that observed by using a lysozyme column in series with the lactase column.

Other Enzymes.—Some of the above-mentioned patents advocate the accompaniment of several oxidoreductases with lysozyme for increased effectiveness as a preservative. Oxidoreductases, especially glucose oxidase and peroxidase without lysozyme, have also been recommended as food preservatives. As discussed in Chapter 11, one of the natural functions of mammalian peroxidases may be as part of the body's defense against invading fungi, bacteria and viruses. A plant peroxidase was found by Urs and Dunleavy (1974) to destroy a soybean pathogen in the presence of KI, H_2O_2 and ascorbic acid. While this model study was directed toward understanding plant disease resistance, it suggests that the minimum amount of H_2O_2 needed for pasteurizing or sterilizing foods such as eggs and dairy products might be decreased when used in conjunction with such peroxidase-mediated antimicrobial systems. Bjorck (1978) exploited milk's own lactoperoxidase antibacterial properties by adding thiocyanate and H_2O_2 (Chapter 11) to effectively reduce the bacterial count and prevent multiplication of psychrotropic bacteria for up to 5 days. The treatment had no effect on measured milk properties and did not lead to accumulation of resistant bacteria.

Glucose oxidase, usually in combination with other enzymes, has also been used as a preservative in addition to many other applications suggested throughout this book. Thus, several Japanese and Russian publications combine, for purposes of food asepsis, the usual glucose oxidase-catalase-glucose oxygen absorbing system (Chapter 17) with lower than usual levels of chemical preservatives such as dehydroacetate, sorbate or sulfite. Bruchmann and Kolb (1973B) showed that commercial preparations of glucose oxidase, by virtue of its catalysis of H_2O_2 production, stabilized fermented fruit mashes. This preservative effect was enhanced ("synergized") by the addition of a cellulase preparation whose action provide a steady supply of glucose. Serendipitous bonuses accrued were higher alcohol and lower volatile acid yields of beverages distilled from such mashes. Another asepsis-related aspect of enzyme action in the food industry is the use of enzyme detergents to clean milking machines (Dunsmore *et al.* 1977). Finally, some enzymes elaborate products which have antibiotic properties, although little effort has been made to utilize these properties in terms of improvement of human health. One of the reasons a suspension of mustard seed flour can be stored at ambience is because of the antimicrobial action of the isothiocyanates and other products of glucosinolase action (Chapter 21). Interest in the products of alliinase and glucosinolase action (Chapter 21) as antibiotics (Virtanen 1958) has been revived, as exemplified by the report of Mantis *et al.* (1979), that garlic extract, containing allicin and other products of alliinase action (Chapter 21), prevented outgrowth of several food-poisoning bacterial spores.

In Chapter 4 we alluded to the cooperative action of the H_2O_2 supplied by oxidases of *Lactobacillus* and the use of this H_2O_2 by milk lactoperoxidase to destroy unwanted microorganisms in the intestinal flora to explain the alleged health-benefits of fermented milk products.

Enzymes for Removal of Sterilants

Catalase.—The principal enzyme used for this purpose is catalase, to remove H_2O_2 proposed as a pasteurizing agent or sterilant for foods such as eggs but now apparently used primarily in some dairy products. Although there is catalase in milk, its activity is too variable and subject to rapid substrate inactivation. Thus, Rosell (1961) recommended adding one part of 33% H_2O_2 to 1000 parts of milk at the farm, the same amount after the milk has been delivered to the dairy, followed by what would otherwise be a subpasteurizing heat treatment at 50°C for 30 min. Catalase is then added at 35°C to remove the residual H_2O_2.

A heat pasteurization along with the H_2O_2-catalase treatment is frequently used in cheese manufacture, especially that of Swiss cheese (Rheinhold 1972). This is because, according to some authorities, regulations in the United States at least in 1972 ruled that the H_2O_2 cannot be entirely substituted for heat pasteurization since assurance of total pathogen destruction was lacking. However, Chu *et al.* (1975) state that the upper limit of concentration of H_2O_2 which is permissible in the United States for the chemical sterilization of milk is 0.05%.

Catalase preparations for food use traditionally were prepared from beef livers, but microbial catalases are now available from Miles and probably other suppliers. *Aspergillus niger* catalase is much more resistant to extremes of pH and temperature and to substrate inactivation, especially in immobilized systems (Balcom *et al.* 1971; Altomare *et al.* 1974; Chu *et al.* 1975). Catalase, as noted in these references, has been subject to a large variety of immobilization modes.

Enzymatic Sulfite Removal.—Although no serious proposals have been made to remove SO_2 from wines and other foods by the use of enzymes, such enzymes have been obtained from mammalian liver and microorganisms. Like xanthine oxidase, the mammalian enzyme is a molybdenum-containing flavoprotein (Orme-Johnson *et al.* 1977). The Mo can be replaced by tungsten (Johnson *et al.* 1977) by feeding rats W. Since it is likely that commercial liver catalase for use in foods may also contain sulfite oxidase, we suggest that such preparations might be used to remove sulfite or SO_2 from foods when it is present in excess of that needed for the various functions it performs (Chapter 14) or to remove it altogether after it has served its purpose.

ENZYMES AS DIGESTIVE AND WEIGHT REDUCING AIDS

In taking up the following aspect of enzymes in food we intrude again into the medical, therapeutic and also quasinutritional fields which are peripheral to the principal themes in this book.

Yet, enzymes are added to foods or in conjunction with foods for alleged therapeutic and dietary purposes (other than those related to removal of toxicants and antinutrients) or generally for health improvement. We have discussed the application of enzymes in the making of low calorie beer (Chapter 31) and supplementation of animal feeds with enzymes for more efficient growth (Chapter 34). The preparation of low-methoxyl pectin to be used in low calorie foods would be another example of a health related use of an enzyme as would the weight-watching benefits brought about by enzyme action by the use of bacterial α-amylase in french fried potato production (Roan 1977). This action is said to reduce fat absorption during frying.

Another way of reducing would be to prevent starch from being hydrolyzed in the GI tract. The use of powerful α-amylase inhibitors was alluded to in Chapter 14 (Marshall *et al.* 1976; Schmidt *et al.* 1977; see Table 14.1). Kesler *et al.* (1970) recommend the use of "sugar" products, presumably modified amylodextrins prepared from hydroxypropylated starch which is said to be resistant to amylases of animal origin. A somewhat less drastic approach would be to modify the functional properties of the food so that calorie input per serving is lowered. Thus, Faber (1975) formulated a low calorie gelatin dessert with bromelin and papain which increased its whippability.

Other aspects of enzyme action hitherto not considered include release of nutrients such as vitamins from inactivating complexes or making them more stable by complexing as in the case of the formation of ascorbigen (Chapter 21). Illustrative of possibly similar effects is the observation that pronase digestion of foods increases vitamin B_{12} absorption through the intestine (Yamakago 1973). Pronase is thus a "digestive aid." The area of enzymes as "digestive aids" when taken orally is a no-man's-land between orthodox medical practice involved in supplementing diets with the digestive hydrolases of the GI tract in patients whose output of their enzymes may not be sufficient to handle the dietary load and the advocacy of "digestive aids," by means of advertisement, in which the word "natural" is prominent and which contain glowing testimonial praise of the spectacular benefits to be derived by the routine use of "enzymes" either in tablets or ingested in "Enzyme Foods."

In a sense, the discussion of lactase deficiency belongs in this section since it is, after all, deficiency of a digestive enzyme which can be "cured" by utilization of a "digestive" enzyme outside of the GI tract. Indeed even the various treatments for foods to improve protein availability (Chapter 34) may in a broad sense be considered as aiding the digestive processes of the food consumer. There is a scientifically respectable but controversial literature which indicates that oral administration of mixtures of proteinases and glucanases may be of benefit not only to domestic animals (Chapter 34) as feed supplements but also to humans who have overeaten or are suffering from a variety of GI ailments such as gastroenteritis and stomach resection (Abderhalden 1961; Goebell 1975). Exactly 50 years ago cellulase and hemicellulase of *A. oryzae* were advocated in a medical journal as therapeutic agents for rendering cell contents more available for the ordinary digestive enzymes (Grassmann and Rubenbauer 1931). The controversy as to whether individuals without specific disorders are helped is somewhat reminiscent of the need to take in more than an adequate amount (MDR) of vitamins by routine use of multivitamin pills or pills containing special vitamins such as C and E. "Enzymes" to be taken orally were universally popular in the early part of this century and are still in widespread use in Japan partly perhaps as a legacy of Takamine's work on "takadiastase," the enzymes of *Aspergillus*, and in other countries as well.

As recently as 1976 pharmaceutical studies on crude preparations of aminopeptidase from *A. japonicus* (the 118th in a series) suggested to Sugiura *et al.* (1976) that this preparation when used with other proteinases can function as a digestive aid. In recent years in the United States the sale of such preparations has been largely through health food outlets. These formulations frequently contain mixtures of papaya, papain, pepsin and takadiastase-like amylase-rich enzyme preparations. A pepsin chewing gum was sold in the United States not too long ago.

These preparations, the efficacy of which has some experimental basis, may be conceived to be half-way between that of oral use of enzyme for correcting real digestive enzyme deficiencies and claims such as that of C. Wade concerning "enzymes" in certain foods.

One can lose weight eating these foods according to Wade because these enzymes form a barrier against invasion of weight-causing fats, create a penetrating action to breakup, dissolve and melt away hard clumps of fat, burn up extra calories, attack waste materials, release excess fluids, stimulate the glands, dissolve cholesterol, etc. As pointed out by Whitaker (1977) in his review of Wade's book, the alleged misinformation proffered offers a few moments of chuckles or indignation with the liberties taken by the author with science—mitochondria lie within the cell nucleus; enzymes penetrate the nucleus to create a fat melting catalyst; tight belts can choke enzymes, HCl is an enzyme. There is also a more somber side because it shows how the public image of what an enzyme is and can do may be skewed; and along with detergent flap, may be falsely interpreted and used to public detriment. Also, as pointed out by Feinberg (1978), such nonscience-oriented spokesmen of the food and nutritional sciences may be influencing the public and their representatives to the point where they may be in policymaking positions in food science and nutrition research.

Furthermore, in spite of the misleading ideas proffered by the more flagrant violators of scientific evidence, as pointed out by Reed (1965), it is too easy to dismiss some of the more responsible advocacy as mere quackery because it seems to lack a basis in well-planned experimental work. There are nuggets of experimental evidence which, blown out of all proportion, can be interpreted to support some of such extravagant claims, i.e., the controversy concerning whether enzymes can get through the GI tract into the bloodstream. Matsumoto (1977), whose work was alluded to in Chapter 35, suggests that not only may absorption of macromolecules through the human GI tract constitute a potential health hazard, but that oral administration of some enzyme-containing medicines are effective because the ingested enzyme molecules are promptly absorbed through the intestinal metaplastic epithelium before digestion occurs.

That authentically proven use of enzymes as therapeutic drugs is possible is skillfully outlined in a popularized overview by Arehart-Treichel (1978). Urokinase and streptokinase, hydrolases which convert fibrinogen to fibrin (their names are exception to the "kinases" being ATP-transferring enzymes), are dramatically effective in dissolving clots in the blood circulating to and from the lungs (Fig. 36.5). Urushiol-cleaving enzymes may be of use for treatment of poison oak; several im-

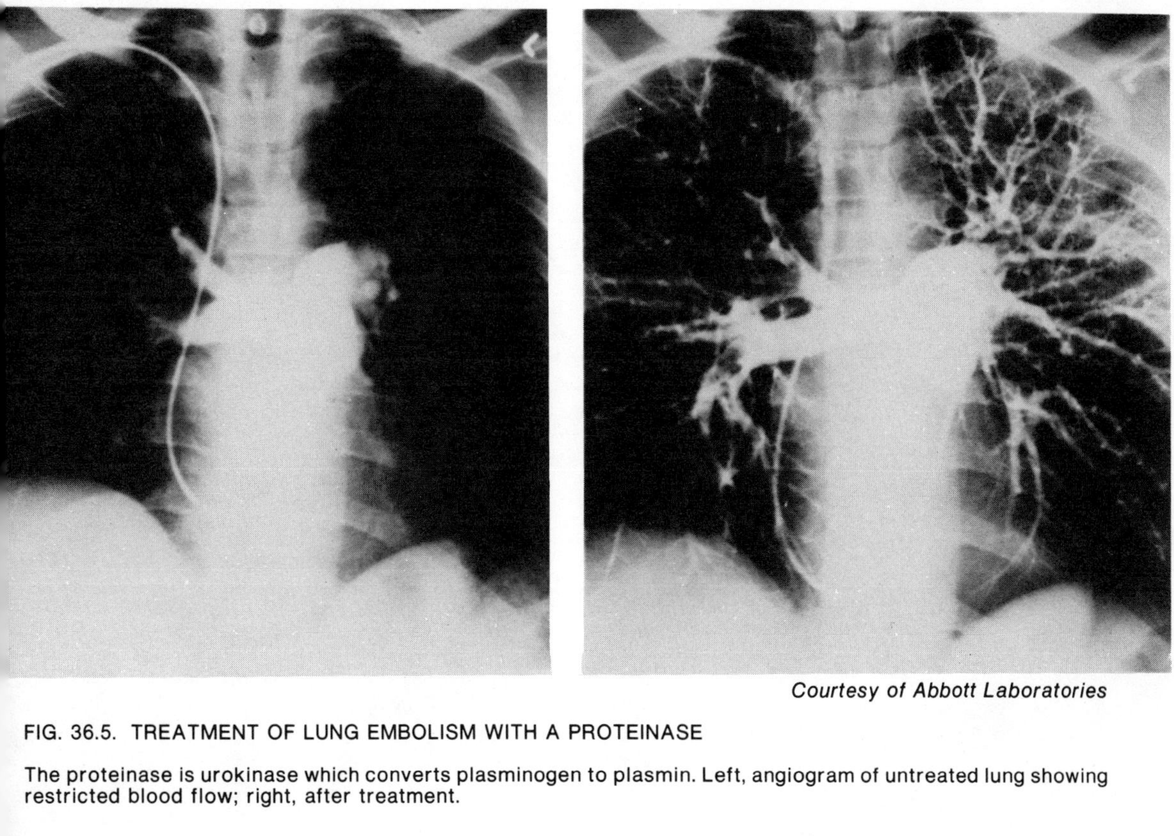

Courtesy of Abbott Laboratories

FIG. 36.5. TREATMENT OF LUNG EMBOLISM WITH A PROTEINASE

The proteinase is urokinase which converts plasminogen to plasmin. Left, angiogram of untreated lung showing restricted blood flow; right, after treatment.

mobilized enzymes may be used in the future for correcting enzyme deficiency diseases a bit more serious than that of lactase deficiency. Asparaginase was once, and may still be, considered a promising anti-tumor agent. A similar claim has been made for an enzyme with hitherto undetected specificifity, L-lysine α-oxidase, obtained from the now-classical cellulase-producing organism *Trichoderma viride (reesii)* by Kusakabe *et al.* (1980). But this has taken us far afield from the main theme of the book: enzyme action in and on foods. With regard to health, one of the outstanding benefits resulting from knowledge of enzyme action in both food science technology and in medical applications is the exploitation of enzymes as reagents in quantitative analysis and their action in related quality control testing, as subjected to scrutiny in the following chapters.

Part X

ENZYME ACTION IN THE QUALITY CONTROL LABORATORY—ASSAY, TESTING, ANALYSIS

37

ENZYME ASSAYS—PRINCIPLES AND APPLICATIONS IN FOOD QUALITY CONTROL

The keystone to understanding how enzyme processes are involved in and determine food quality, identity and improvement of foodstuffs rests, in the final analysis, largely on chemical analysis. The two intradoses comprising the arch briefly sketched in Chapter 1 (pp. 20–21) are determination of enzyme activity (assay) and estimation of food constituents, equivalent to "metabolite analysis" of the clinical laboratory, to be examined in Chapters 38 and 39. Indeed the present advanced state-of-the art of enzyme assay and analysis is largely due to the developments emerging from the clinical research laboratories. The food quality control laboratory thus has a massive reservoir to draw upon in applying enzyme analysis to the solution of food-oriented problems. However, the lack of familiarity and cost of such procedures combined with the necessity of adapting them from the relatively few tissues used in the clinical laboratory (i.e., blood and urine) to myriad types of samples the food scientist and analyst encounter, the necessity of running a large number of determinations or assays and the ever-increasing competition of sensitive selective nonenzyme methods (even for the determination of the presence of enzyme) have kept their adoption to a snail's pace.

USE OF ENZYMES IN OFFICIAL ANALYSES

Perhaps such acceptance can be gauged by examination of how enzymes are used in the latest available edition of the Official Methods of Analysis of the Association of Official Analytical Chemists (1975) using the categorization referred to in Chapter 1 as a guide. Analyses involving enzymes and enzyme assay occupy some 30 of the 1000 pages of this edition. More specialized manuals issued by individual food industries such as baking, dairy, brewing, etc., also describe procedures in which enzymes are used.

As indicators and correlates of food quality, history, etc., the uses of enzymes include: catalase and peroxidase as indices or indications of adequacy of blanching of processed vegetables; phosphatase as an index of adequacy of pasteurization not only of milk but also such dairy products as cream, butter, margarine, cheese, ice cream and other frozen desserts; α-amylase as an index of malt quality and strength in brewing; "diastatic," α- + β-amylase, for detection of adulteration or of heating of honey (see below); proteolytic activity of baking flour for general overall quality (Chapter 32); and "caseinolytic" activity, a qualitative test for the presence of chillproofing enzymes in beer.

In each of the above tests the food serves as a source of the enzyme and prepared substrate has to be added. Representing the category in which the enzyme and the substrate are present in the foods (see below) is the determination of the diastatic activity of flour suspensions as predictor of "gassing power" and other aspects of the bread texture (Chapter 32).

No test representing the use of isozymes is included. Standardization of a commercial enzyme preparation is represented by the assay of papain, unless one considers malt as a "commercial enzyme" additive in brewing. Of course an amylase assay in the breweries which use microbial α-amylases as the sole means of converting starch in cereal adjuncts would indeed belong in this category.

The list of food constituents in which enzymes are added to foods or prepared extracts of the foods for their estimation and detection is somewhat but not much more extensive than their use in enzyme assay. Essentially the method of Hudson and Harding (1915), developed 65 years ago, sucrose and raffinose are determined by the use of extracts of bottom and top yeasts as sources of invertase, and invertase and melibiase (α-galactosidase), respectively. Of somewhat more recent vintage is the use of the versatile glucose oxidase and peroxidase for the determination of glucose in sugar products and plant foods. Peroxidase in this analysis serves as an "auxiliary" enzyme (category B.5, Chapter 1). The determination of H_2O_2 with peroxidase goes back to the 19th century.

The use of enzymes in the determination of starch in food products is much more restricted than that of the common sugars. Purely enzymatic hydrolysis of starch is confined to grains and grain feeds. The enzyme used is a commercial diastase with substantial activities of both α-amylase and amyloglucosidase. For the determination of starch in less starchy foods, such as mustard, coffee, and spices in all of which starch may be added, enzymatic hydrolysis (by a freshly prepared malt extract) is combined with acid hydrolysis. Purely nonenzymatic analysis is used for the determination of starch in the following foods: desserts, dry milk, flour, meats, mustard (prepared), dressings, fruits, beer and peanut butter.

Two other analyses which fall into this category—the use of urease and uricase to detect contamination of flours, foods and fertilizer "quality"—might be just as frequently used in laboratories of governmental regulatory agencies as in food quality control laboratories in the food industry.

Procedures for the determination of food constituents by their effect on added, known enzymes is represented by the use of the inhibition of added choline esterase for the determination of pesticide residue and even this analysis is relegated to an alternative method.

The "short cut" category (B.3) contains two entries: The prediction of the protein digestibility quality of feeds by solubilization of nitrogen in these feeds due to the action of added pepsin and an estimate of brewing quality of grits as corn adjuncts by measuring the specific gravity of the extract resulting from adding amylase.

B.4, the use of enzymes as analytical aids, has the most and probably in our opinion more interesting entries than the other categories. Thus pectic enzymes are used to aid filtration—just as they do in food production on a large scale (Chapter 30)—in the preparation of tomato products for the determination of total solids. A commercial amylase preparation is used in the preparation of foods for vitamin B_1 analysis and biological assay. The amylase removes starch which interferes in the subsequent chemical determination, and pyrophosphatases present in the enzyme preparation remove the phosphate groups from thiamin pyrophosphate (cocarboxylase) to form thiamin. Probably the only really enzymatically pure enzyme used is carbonic anhydrase for speeding up the determination of CO_2 in wine by eliminating the need for distillation. Commercial proteinases are used as cleanup aids in the detection of "light filth" in pork and flour, and urease is used in a nonchemical analysis as an aid in screening microorganisms as part of the procedure for testing eggs for the presence of *Salmonella*. This brief survey indicates that very little of the potential of enzymes as aids in the quality control laboratory are used in official analysis, and where used they frequently involve cumbersome antiquated procedures. Thus, there are now readily available simple, reliable, well-established methods for catalase assay based upon strong absorbance of UV radiation by H_2O_2. Yet, while UV spectrometry is frequently used in non-enzyme AOAC analyses, the assay method described for this enzyme involves an outmoded cumbersome titration. The remainder of this chapter will be devoted to an examination of the uses and potential of enzyme assays not so much as research tools but for monitoring the problems the food technologist might encounter in the food quality control and development laboratories. We shall discuss enzyme assay principles, possible future application as well as proposed tests which have not yet found their way into official food laboratory manuals.

Most of the books published on enzyme assay and "metabolite" determination using enzymes are oriented to clinical and biochemical research laboratories. One of the first of such books is that of Stetter (1951) in German. Besides the continuing series of "Methods in Enzymology," since then the most comprehensive exposition and detailed instruction is the four volume 2600-page compendium edited by Bergmeyer (1974), a two-fold expansion of Bergmeyer's, the first edition in 1963. A later condensed version in German has been published. Other books include those of Guilbault (1976), primarily devoted to the use of enzymes as analytical reagents for the determination of individual chemical components of tissues, and the more restricted book of Lowry and Passonneau (1972) covering the use of the pyridine nucleotides for both enzyme assays and metabolite estimation and especially quantitative enzyme histochemistry. Reviews in *Analytical Chemistry* have routinely included sections on the use of enzymes in chemical analysis (Fishman *et al.* 1980). Reviews addressed primarily to food analysts include those of Roodyn (1969), Whitaker (1974B), Schormüller (1974), Beutler (1977) and Wiseman (1978). Reviews on food analysis in *Analytical Chemistry* usually include a section on enzyme assays and scattered examples of the use of enzymes in food constituent analysis (Foltz *et al.* 1977; Yeransian *et al.* 1979).

PRINCIPLES OF ENZYME ASSAY

Up to about 20 years ago, most enzymologists accepted as apodictic that estimation of enzyme activities could only be carried out by measuring the rate at which the substrate disappeared or the product appeared. Furthermore, it was assumed that one could not estimate the molar concentration of the enzyme (or rather that of its active site) without an estimation of the molar *activity* (katals per mole of enzyme) as described in Chapter 2 (Balls 1932; Bodansky 1937; Allison and Purich 1979; Whitaker 1974B). While it is still true that the only way we can obtain quantitative information concerning the catalytic efficiency of an enzyme is via the competency of its active site, it is no longer true that this has to be obtained using activity or rate assay. Molar concentrations can be estimated via analysis of the kinetics of inhibition by strong inhibitors or by the use of active site titrants, referred to briefly in Chapter 14 and elaborated upon later in this chapter. Absolute concentration of enzymes may also be gauged by immunochemical techniques which while identifying an enzyme as a unique protein do not necessarily tell us the state of the catalytic competency of this protein molecule. For now, we turn our attention to kinetic methods for the determination of enzyme activity. A tribute to the tremendous contribution of such assays to medicine is the overview discussion of Schmidt and Schmidt (1976).

To measure or determine enzyme activity, i.e., to perform an enzyme assay, a definite amount of reaction mixture containing, minimally, substrate, buffer, activators and stabilizers is mixed as rapidly as is feasible with an aliquot of the enzyme and the reaction is allowed to proceed (incubate) over a definite time interval at a constant temperature. The pH selected is usually a compromise between the pH optimum and the stability of the enzyme at the selected temperature. The pH selected is usually at or near the pH optimum, but the temperature is usually not optimal. One can stop the reaction and measure disappearance of substrate or appearance of product.

Ideally the reaction rate should follow zero-order kinetics, i.e., it should be constant over the period of measurement. This means, of course, that there should be a linear relation between concentration of substrate or of product and time. The rate should, for the accuracies required in the vast majority of assays, be independent of the substrate concentration when S is at least equal to 10 times K_m. At this concentration it can be readily shown from the Michaelis equation that the rate will be 91% of V_{max}.

Assuming that these ideal conditions are met, a minimum of two points on the t *vs* S or P curve is needed and may be chosen at predetermined or definite times for most assays of food enzymes in the range of minutes to hours (in the latter case it is necessary add as preservative an antimicrobial agent such as toluene); or after a definite increment of change of the reactants. In spectrophotometric assays, which most enzyme assays are, this increment should amount to a change in absorbance of about 0.1−0.3. The extent of reaction may be fixed by a sudden visible change such as the clotting of milk in the rennet assay or the sudden appearance of color in a peroxidase assay employing ascorbate and HI as competing hydrogen donors in the presence of starch (Schwimmer 1944A). Alternatively, continuous change such as absorbance may be recorded, as is routinely and more frequently done in well-equipped laboratories.

In noncontinuous assays, still of some utility, an aliquot of the enzyme-inactivating (stopping) reagent is added to the enzyme reaction mixture (or vice-versa) and the requisite aliquot is analyzed for product or substrate. Several alternatives are available for determination of the first point:

(1) A zero-point is determined by adding the stopping reagent aliquot of the action mixture immediately after addition of the last component (enzyme or substrate) of the reaction mixture.
(2) All ingredients, usually kept cold, less enzyme and/or substrate are mixed and treated with the stopping reagent and the missing ingredient is then added at "zero" time.
(3) As in (2) except that the reaction mixture minus the most stable but essential component is allowed to incubate for the same length of time and at the same temperature as the reaction mixture. This type of blank is especially useful for discontinuous assays where the substrate is unstable and undergoes parallel nonenzymatic transformations, or when impure extracts as enzyme sources are being used which undergo interfering changes.
(4) Use of the "flying start" in which the stopping reagent is added after a short (in comparison with total) incubation time. This is used when the reaction mixture may not be completely homogeneous or when there is a lag period before attainment of the steady state or constant reaction rate, i.e., lipoxygenase (Chapter 24).

In continuous assays no stopping reagent is required and the measurement of the changes in the system is made in the absence of either enzyme or substrate and then the ingredient is rapidly added and mixed. Figure 37.1 shows a semicontinuous measurement of the liberation of *p*-nitrophenol nonenzymatically followed by an acceleration due to the addition of carbonic anhydrase and subsidence of the reaction to the nonenzymatic rate upon addition of a stopping reagent. If enzyme or substrate contains impurities which cause interfering change, each can be eliminated by allowing the reaction to continue until no further change takes place, and then the substrate or enzyme is added.

Assay of Enzyme Activity in Absence of Linearity

The Problem.—In some situations it is difficult or impractical to achieve zero-order reaction rate and strict proportionality between enzyme concentration and rate for the following reasons:

(1) Failure to achieve saturating levels of substrate due to sparing solubility or cost of substrate, inhibition by excess substrate or to an unusually high K_m.
(2) Inhibition by reaction products.
(3) Instability of the enzyme due the choice of a temperature too close to the inactivating range. It will be recalled (Chapter 10) that the Q_{10} of enzyme inactivation due to thermal denaturation is about ten times the Q_{10} of the enzyme-catalyzed reactions.
(4) Instability due to pH, O_2, and other inactivators which may be present in the substrates or in other components of the reaction mixture.
(5) Instability due to reaction inactivation such as occurs with phenolase (Chapters 14 and 15) and lipoxygenase (Chapter 21).
(6) Deviation from Michaelis kinetics, especially when using allosteric enzymes which exhibit negative cooperativity (Chapter 4).
(7) Interference by other enzymes.
(8) Presence of endogenous reversible enzyme affectors, inhibitors or activators.

Solutions.—Over the years many approaches have been developed to correct for this lack of linearity of response. The following is a selection of some of the more widely used ones which we have found to be convenient and effective or may be of historical interest.

*First Order Constants.—*Just as the K_m can be neglected in the denominator of the Michaelis equation at high S so that the rate becomes independent of S, so S can be neglected in the denominator at low S so that the rate $v = (V_{max}S)/K_m = dP/dt = -dS/dt$. The rate is thus proportional to the substrate concentration and hence the reaction will follow first order kinetics:

$$K_f = 2.302/t \times \log S/(S-P)$$

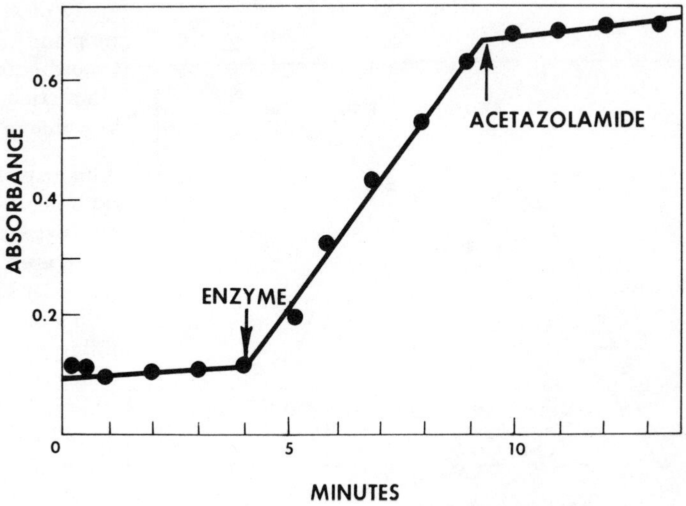

FIG. 37.1. STARTING AND STOPPING AN ENZYME ACTING ON AN UNSPECIFIC AND UNSTABLE SUBSTRATE

The enzyme is bovine carbonic anhydrase and the substrate is *p*-nitrophenyl acetate. The reaction is stopped with a highly specific inhibitor, acetazolamide.

From Schwimmer (1969B)

Historically, first-order reactions were the most frequently observed. The term "Katalase Fahigkeit" (K_f) refers specifically to the first-order rate constant and is still used to denote the activity of peroxidase and catalase in the 1975 edition of AOAC. It is probable that occasional observance of first-order kinetics at *high* S is an empirical circumstance of the combination of the above-listed factors which may contribute to lack of linearity.

Initial Rates.—It will be recalled that the relatively simple integrated Michaelis equation:

$$ket = K_m \ln[S/(S-P)] + P$$

is transformed to

$$ket = K_m(1 + S/Kp)\ln[S/(S-P)] + (1-K_m/K_p)P$$

when one of the products inhibits competitively (Schwimmer 1961; Darvey *et al.* 1975). For noncompetitively inhibiting products the integrated Michaelis equation is

$$ket = K_m(1 + S/K_p)\ln[S/(S-P)] + (1-K_m/K_p)P + P^2/2K_p$$

We display these equations at this time to show that the course of an enzyme reaction can become quite complicated so that it is frequently necessary to resort to more or less empirical measures to estimate the "initial rate" of the reaction. It has been our experience that, in general, the reaction should not go beyond 10% of completion. The initial rate has been estimated by measuring the maximum slope of the t *vs* S or t *vs* P plots as close to zero time or the origin as possible, extrapolating if necessary. Of course in this case more than two points are needed. The following are some of the procedures for estimation of initial rates.

Pseudolinearity.—The measurements are made over short increments of reaction course, each increment amounting to not more than 1% conversion of the substrate, so that the instrument used does detect the lack of linearity. For this approach to be effective, a low zero-time blank and a fairly sensitive measurement mode are prerequisites.

Mirror.—A favorite "eyeball" method is to place a straight-edged mirror perpendicular to the plane of the paper on which the curve is drawn with the line describing the straight edge running through the origin. The mirror is then pivoted on the origin until the curve and its image in the mirror form a smooth parabolic-like continuous curve. The initial rate is the slope of the straight line drawn orthogonally to the line describing the mirror's straight edge.

Extrapolation.—A somewhat more precise estimation may be obtained, as shown in Fig. 37.2, by plotting the ratio of increments of the ordinate (increase in product or decrease in substrate) per unit time against either P or t, and extrapolating the resulting curve to zero-time. The utility and

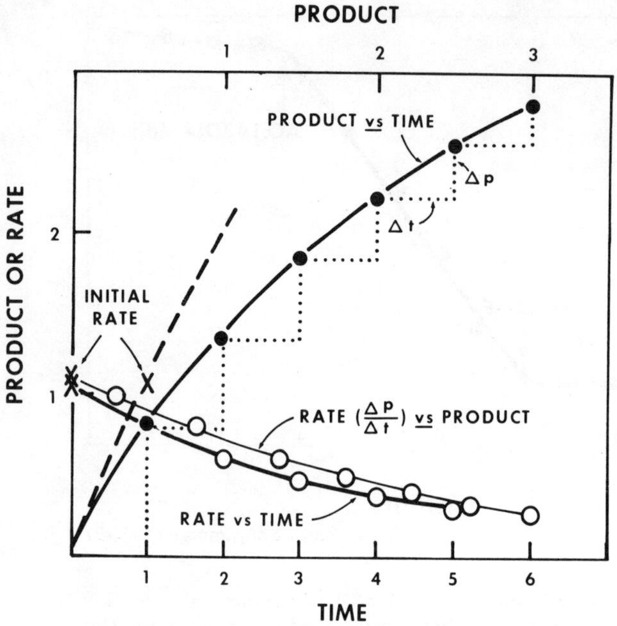

From Schwimmer (1976)

FIG. 37.2. ESTIMATION OF INITIAL RATES ON NON-LINEAR ENZYME REACTIONS BY THE INCREMENT METHOD

the basis for use of initial velocity curves as contrasted to progress curves are briefly dealt with by Atkins and Nimmo (1980).

Integrated Michaelis Equation.—Paradoxically, the integrated Michaelis equation was used by Johnson and Diven (1969) for estimation of initial rates. The initial rate, v, for an irreversible enzyme-catalyzed reaction obeying Michaelis kinetics is the absolute value of the slope of the straight line describing the following equation.

$$\ln(1-P/S)/S = -vt$$

The integrated Michaelis equation also provides the basis of the direct initial rate approximation by the graphic method developed by Cornish-Bowden (1975) based upon the same principles developed by the same group for the determination of enzyme kinetic constants *via* a direct linear plot as discussed in Chapter 2. Instead of values of S, one marks off points equal to values of (0.5P–S) on the abscissa. Corresponding values of P/t, the apparent rate, are marked off on the ordinate, which is drawn at an acute angle nonorthogonally to the abscissa in the fourth quadrant. Each x- and y-intercept is connected by a straight line to form a family of straight lines which should intersect at one point above the ordinate. The value of P/t on the ordinate on a line drawn from the point of intersection of the family of lines to the abscissa at the value of x = –S is the initial rate.

Linearity Between Rate and Enzyme Concentration (e).—It is possible to find a linear relation even with two-point rate determinations, if the curves are not too bent (Michal 1974). Ordinarily zero-order kinetics should assure linearity between e and v or initial rate. However, the presence of an inhibitor in the enzyme preparation will result in deviation from linearity because, in essence and inferred in Chapters 13 and 19, upon dilution of such an inhibitor-containing system present in some food extracts, the effect of the inhibitor on the rate will drop off faster than the intrinsic decrease in rate due to the catalytic action of the enzyme had there been no inhibitor present. It can easily be shown by simply dividing the Michaelis equation in absence of an inhibitor by that in its presence (v_o/v_i) and performing a few simple algebraic manipulations, that:

$$e/v_i = C_1e + C_2,$$

where C_1 and C_2 are constants.

By plotting e/v_i against e (setting e = 1 at the highest concentration of extract used) one then can obtain value of v_o, as shown in Fig. 37.3, where the glutamyl transpeptidase in an onion extract is inhibited to the extent of 83%. Similar plots for endogenous ampholytic inhibitor at various pH's is shown in Fig. 19.3. The maximum extrapolated initial rate for this endogenous potato invertase inhibitor model occurs at the predicted pH optimum of the inhibitor-free enzyme.

Such measurements may have applications as predictors of food quality and in food constituent analysis, but here they are presented as means of ensuring linearity between enzyme concentration and initial rate. Of course if one tries to correct for some of the causes of deviation listed above, such as using double distilled water to prevent progressive enzyme inactivation, one may dispense with some of these makeshift stratagems for ensuring linearity among rate, time and enzyme concentration.

Auxiliary Enzymes.—The measurement of enzyme activity frequently involves conversion of the product to a more easily measured substance or reconversion to substrate by the action of auxiliary

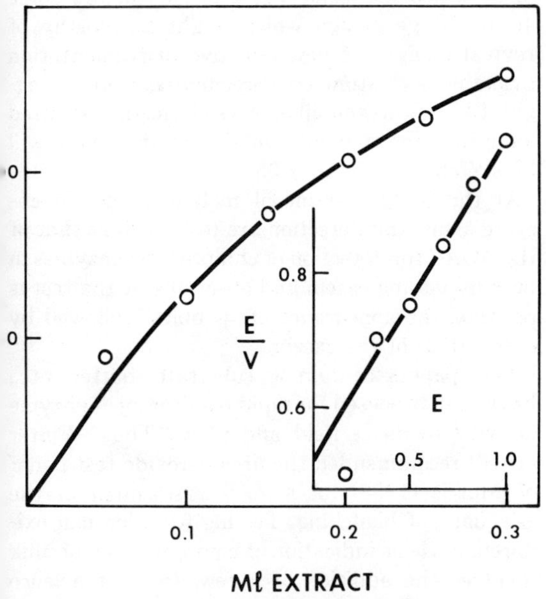

From Austin and Schwimmer (1971)

FIG. 37.3. ESTIMATION OF INITIAL RATE OF ENZYME REACTION IN PRESENCE OF ENDOGENOUS INHIBITOR

The enzyme reaction is the liberation of *p*-nitroaniline from γ-glutamyl nitroanilide catalyzed by enzyme(s) in crude onion extracts. See text for details.

enzyme or enzymes. The use of these auxiliary enzymes is particularly widespread and of prime importance in the determination of constituents (Chapter 38). At this juncture we only wish to point out that when such enzymes are used in determination of the activity of designated enzyme, (E_1) whose activity is being measured, the auxiliary enzyme E_2 should be present at levels which catalyze the conversion of its substrate at about 100 times the rate at which E_1 is acting. Bergmeyer (1963) showed that when the auxiliary enzyme acts as a "succeeding" indicator (Chapter 38) the error of measurement decreases drastically with increasing E_2:E_1 ratio. The actual excess needed depends upon the relative V_{max} and K_m value of the two enzymes. Thus, when V_2/K_2 is ten times V_1/K_1 (V and K refer to values of V_{max} and K_m), the inherent error of measurement in some setups is −23%. When the corresponding ratios vary by a factor of 100, the error is negligible for most enzyme assays, about −4%.

Enzyme Units

The idea of expressing the activity of enzymes in terms of units rather than in whatever measurement one happens to be using (cc, mg, OD, etc.) goes back to Northrop (Chapter 1) in his and his colleagues' work on crystalline enzymes. Basically a unit of activity is that amount of enzyme present in a definite volume of a defined reaction mixture incubating under rigidly controlled constant conditions which catalyzes a definite amount of change in a definite amount of time. Total activity of an extract is thus the number of units used in the assay times the ratio of the total volume of extract to the volume of extract used in the assay. Specific activity was defined by Northrop as the ratio of the activity of the enzyme divided by mg of protein N in the same volume of extract used in the assay. We are now somewhat more flexible, but mostly we frequently use units per mg of protein to express specific activity. Having defined a unit, one sets up a unitage chart or curve by plotting the amount of change against units. Ideally, this should but does not have to be a straight line, as discussed above. Any number of arbitrary unitage systems could be set up. For instance, Northrop used a dozen different activity units with an equal number of distinctly different assays in studying pepsin (Northrop *et al.* 1948).

This proliferation of arbitrary unitage systems caused quite a bit of confusion, especially when it came to comparing results of different investigators. Special calibrations and tables for comparison purposes were set up. The Enzyme Commission made an effort to put some order into this chaos in their first classification system in 1964. They defined a unit of enzyme as that amount which at 30°C under optimum conditions, preferably near saturating levels of the substrate (see above), will convert one micromole of substrate to product(s) per minute. The *concentration* of the enzyme was expressed as units per ml and the specific activity as units per mg of protein. Molecular activity, the old "turnover number," was defined as the number of molecules of substrate transformed/molecule of enzyme/min and the "catalytic center activity," that per unit active site. In the 1972 revision, the Enzyme Commission recommended the concept of "enzyme unit as a physically defined amount of enzyme" be abandoned. *Enzyme activity* is now the rate of transformation of the substrate that may be attributed to the catalysis by the enzyme. The new "unit" of enzymic activity is the *katal* which is that amount of enzyme action which transforms one mole of substrate per second; one katal equals 6×10^7 of the old units. Molecular activity has been replaced by "molar" activity. Happily, this was not changed in the 1978 revision.

Other Aspects of Enzyme Assay

While the techniques and instrumentation used to detect and assay enzymes are for the most part those used by analytical chemists in general, the most widely employed instrument has been the spectrophotometer. Indeed spectrophotometric instruments have been adapted to measurement of assay of 16 samples simultaneously. Performance of the enzyme- or fast-analyzers has been subjected to intense scrutiny. While many of these measurements have been made in the UV, more recently the more sensitive fluorescence spectroscopy has found increasing acceptance. In going from UV to fluorescence measurements, the sensitivity of the assay for NAD linked dehydrogenases may be increased as much as 100-fold, although we and others have found that its use is a bit fussy and requires some skill in getting reliable results (Leaback 1976). Assay of enzymes in clinical laboratories was given great impetus by the advent of automated analysis, which is only relatively recently finding its way into the food science laboratory. Thus, Hansen (1971) used an automated setup in assaying for β-N-acetylglucosaminidase in eggs as an index of adequacy of pasteurization (Chapter 11), and Vijayalakshmi *et al.* (1976) developed an automated procedure for the determination of pectin esterase activity in commercial enzyme preparations. Measurements in changes of pH and/or acid-base relations, once quite popular for enzyme assays, can still be used as a general method when the pK of the product differs from that of the substrate.

Radiometric isotopic analysis of enzymes was reviewed by Oldham (1973). Quite aside from their impact on the progress of biochemistry as tracers of metabolism, isotopes have proven to be valuable analytical adjuncts and their uses in conjunction with enzymes have been no exceptions. Radioactive substrates are employed in enzyme assay for several different reasons:

(1) when the enzyme to be assayed is elusive due to instability or is present in vanishingly low levels,
(2) when the substrate of the enzyme assay is expensive, and
(3) when the product of the enzyme reaction cannot be readily distinguished by other means.

Bioluminescence is used to assay for ATPase. Coming to the fore are ion selective, oxygen electrodes and, especially, microcalorimetry as perfectly general physical methods. Another physical measurement which had some vogue in enzymology some 40 to 50 years ago which might be worthy of revival in light of new sensitive instrumentation available is dilatometry (Sreenivasaya and Bhagvat 1937). Further silhouettes of enzyme-adapted analytical instrumentation and methodology will be sketched in Chapter 38.

At the other extreme of meticulousness in enzyme assay and detection are tests such as that of the AOAC for detection of chillproofing enzymes in beer by adding casein and observing, if the test is positive, the appearance of "pebbles" followed by coagulation of the casein.

Test papers containing substrate, buffer, etc., have been proposed for rapid monitoring of enzyme activity foods in field and plant. Thus, Winter (1976) recommended the urea peroxide test paper of Morris (1958) for a quick assessment of the adequacy of blanching. Testing for microbial oxidoreductase as indication of contamination of milk involves the addition of a few drops of a leuco methylene blue solution. It takes 3 hr for perfectly wholesome uncontaminated milk to turn this dye precursor blue. If it takes 1 hr, the milk is definitely contaminated.

As in other areas of enzyme analysis, the clinical laboratories are well ahead of the food quality control laboratories in packaging and tableting of enzymes and accessory reagents for the determination of enzyme activity. The use of reagent tablets is exemplified by their development for the assay among other enzymes of aspartate amino transferase (Guilbault 1976). The important consideration in such development is the stabilization of the reagents comprising the tablets with, for instance, mannitol, sucrose and a polyethylene glycol. Indicative of the importance of kits in clinical chemistry, the same enzyme was the object of a study of performance of kits of reagents for its assay (Barnett *et al.* 1976).

ENZYME ASSAYS AS INDICATORS, MARKERS, INDICES, MONITORS AND PREDICTORS

The rationale for use of enzyme assays as food quality indicators ranges from logical decisions based upon a one-to-one relationship between the assay and the observed quality attribute, to strictly empirical observations such as the finding of an enzyme whose heat stability characteristics mimic those of the death rate traits of target organisms as a test for adequacy of pasteurization (Chapter 11). In general we may discern the following categories of questions which we hope to answer by means of enzyme assay: Is the raw plant suitable for processing? This involves largely agronomical considerations. Can a suitable assay be found which will predict or serve as an index of processing parameters for maximum quality? Can the enzyme be used as a processing guide? Can something be learned of the past history of the processed food by the use of suitable tests?

As Indices of Suitability for Processing.—Ever since the recognition of the importance of enzymes in life processes, there has been a torrent of investigations striving to correlate the level of this or that enzyme with some biological function and/or activity. Thus, Crocker and Harrington (1918) concluded that catalase activity in seeds parallels physiological behavior much more than does "oxidase" activity but could find no correlation of either enzyme with seed viability. The Russian literature is still replete with extensive documentation of such attempts to correlate enzyme activity and the biological state of plants. Typical of conclusions reached from such studies are measurements of catalase and peroxidase activities of cucumbers as indices of inbreeding (Genchev and Mikhov 1975), mulching and frost resistance of young mandarin trees (Mikaberidze and Rosnadze 1974) and invertase as indicative of growing suitability of soil (Kuprevich and Shcherbakova 1971).

As Monitors and Predictors of Microbial Contamination.—While the non-Russian literature on such relations has abated somewhat, there is still a lively international literature on the relation of catalase and peroxidase activities of plants to their capacity to withstand infection, and also to predict such resistance in the vegetative phase of plant growth by determination of enzyme activities in the seed. This type of study is related to foods in that it aids in the selection of fruit and vegetable cultivars which are suitable for processing and it could be used to pinpoint the presence of incipient microbial contamination of individual lots of produce. Some typical examples of such studies are shown for plant and other foods in Table 37.1. Catalase as an indicator of bacterial contamination of milk was proposed about 75 years ago (König 1907). One may distinguish between detection of unique quality-changing enzyme activity, as in the lysolecithinase in milk, and some lesser empirical correlation between the level of activity of the host or invader enzyme and the presence of spoilage organisms. While many of these studies are directed to an understanding of the genesis of the infection, they should have some utilitarian value as indices and predictors of quality.

As Indicators of Micro and Macro Mineral Nutrients.—The leaves of plants deficient in trace metal nutrients usually have diminished activities of enzymes which contain these trace metals as integral parts of their active sites. These metals and corresponding enzymes with lower activities include: zinc, aldolase (O'Sullivan 1970) and carbonic anhydrase (Bar-Akiva and Lavon 1969); copper and ascorbate oxidase (Bar-Akiva *et al.* 1969); iron and catalase; and molybdenum and nitrate reductase (Bar-Akiva and Lavon 1968). From these and other studies it appears that the activity is low not only because an essential part of the active site is missing, but more importantly, the metal serves to derepress the gene coding for the biosynthesis of the protein moiety of the enzyme. Although the possibility that unstable rapidly turning over apoenzyme is formed in absence of trace metal has not been ruled out, apoenzyme elaboration would be poor cell economy. The status of macronutrients may also be gauged by enzyme assays. Thus, Besford (1975) found that assays of pyruvate kinase and phosphatase could be used as an indication of the suboptimal potassium levels in the leaves of both tomato and cucumber plants in the absence of the onset of visible deficiency symptoms. Both K^+ and Mg^{2+} are needed for full activation of pyruvate kinase, and phosphoenolpyruvate phosphatase is stimulated by Mg^{2+}. The ability to detect deficiencies before any visible damage is evident would allow the processor/grower to correct for the deficiency and hence render a particular harvest suitable for subsequent routine processing.

As Predictors of Protein and Crop Yield.—Another very important relation which has been investigated in detail but is still controversial is that between nitrate reductase activity of the leaves of young cereal plants and the eventual level of protein in the seeds used as food (Rao and Croy 1972; Dalling *et al.* 1975).

The rationale for the use of this particular enzyme as a predictor of protein yield is that the reaction catalyzed constitutes a bottleneck in the metabolic pathway leading to protein synthesis in plants. More recent results suggest that this enzyme is a good predictor of grain and even plant height as well as protein yield (Singh *et al.* 1976; Johnson *et al.* 1976). As the result of large-scale field tests the latter investigators reported correla-

TABLE 37.1

ENZYME ACTIVITIES AS DETECTORS OF MICROBIAL SPOILAGE

Food Plant	Spoilage Organism	Enzyme, Remarks
Citrus leaves	*Phoma tracleiphila*	Catalase, peroxidase positive in susceptible, negative in resistant variety[1]
Lemon	*Phytophthora citrophthora*	Catalase negative[2]
Milk	*Pseudomonas fluorescens*	Phospholipase positive[3]; heat stable protease[4]
Fruit	Several	Tannase[5]
Carnation stem	*Phialophora cinerescens*	Acid peroxidase isozyme, positive[6]
Sugar beet	*Cercospora* (leaf spot)	Phenolase, positive, then negative, intracellular redistribution[7]
Tomato	*Pyrenochaeta lysopersicae*	Pectin lyase; present in altered tomato tissue[8]
Shrimp	Unspecified	Decrease in activity of protease[9]
General	Fungi, fecal microorganisms	Alkaline phosphatase, trypsin-like protease[10]
Meat	Unspecified	Protease, as evidenced by presence of free tyrosine[11]

[1]Source: Tsiklauri (1972).
[2]Source: Cohen and Schiffmann-Nadel (1973).
[3]Source: Fox *et al.* (1975).
[4]Source: Malik and Swanson (1975).
[5]Source: Burkhardt (1977).
[6]Source: Grison *et al.* (1975).
[7]Source: Rautela and Payne (1970).
[8]Source: Davet (1975).
[9]Source: Soedigo *et al.* (1973).
[10]Source: Cook and Steers (1947).
[11]Source: Strange *et al.* (1977).

tion coefficients of +0.96 and +0.90 between the activity of this enzyme in the first expanded leaf of wheat seedlings 2 weeks after sowing of the seed, and protein and grain yield of mature wheat. Such correlations, if they hold up under a variety of situations, should be of great value in selecting lines of grain which make maximum use of limited resources. There would also be an environmental spin-off in that it would not be necessary to use high levels of fertilizers. This would reduce pollution of rivers and lakes due to runoff of excess nitrogen fertilizers.

In connection with the nutritional availability of protein, Camici *et al.* (1980) demonstrated that decrease in aldolase activity of alfalfa juice—due undoubtedly to its inactivation by proteases present—is a reliable index of protein extractability and hence of nutritional availability.

As Indicators and Markers of Development, Maturity and Senescence.—Earlier work in this area dealt with finding correlations of enzyme activity with gross fruit quality attributes such as texture. Thus, Nagel and Patterson (1967) found a correlation (r = 0.99) between total weight and decreasing levels of PE activity in developing pears. Predictors of senescence would be especially useful in fresh fruit storage and handling. Gorin and Heidema (1976) observed two peaks in time in the level of peroxidase in cold-stored Golden Delicious apples. The first peak, about twice the base level, occurred at the time of the climacteric and the second peak, about 100 days later, was less dramatic and conclusive (10–20% rise) at the onset of senescence. Thus, the time between the two events should give the warehouseman ample time to make a decision on when to sell the produce. Similar patterns were found for the GOT of peaches by Jen and Graham (1975). Panova and Bozova (1975) discerned a general decline in peroxidase and catalase in fruit and pears upon attaining what they call "consumer ripeness" at the end of the storage of these fruits.

A great deal of attention has been paid to finding enzymatic indices of maturity before harvesting. In general such maturity is signaled by the virtual disappearance of a particular enzyme; the early work of Pressey and coworkers shows that sucrose synthase activity in maturing potatoes (Chapter 19) followed such a pattern. Sowokinos (1973) studied this enzyme in detail with respect to its use as a predictor of potato tuber maturity. The enzyme activity dropped steadily. Thus the percentage of the total tuber weight attained by eight different varieties when they had reached the uniformly low level of 2000 units per average size tuber (maximum activities of immature tubers ranged from 28 to 42 thousand) ranged from 94.1 to 99.1%. There was a remarkable correspondence between accepted characteristics of a given variety as early or late maturing and time required to reach this level of enzyme activity. However, the enzyme assay could not distinguish between good and poor processing characteristics. Furthermore, illustrative of the ever-continuing competition between enzyme and nonenzyme methods mentioned above, Sowokinos later found an even better nonenzyme marker of maturity, the product of the sucrose synthase action, sucrose itself.

The decline in sucrose synthase activity appears to be due to total repression of the gene coding for its synthesis, not to an alteration of the catalytic and regulatory properties of the protein. In contrast, the PFK of tomatoes (Chalmers and Rowan 1971) and of bananas (Salminen and Young 1975) does appear to undergo subtle changes in regulatory properties. It will be recalled that this allosteric enzyme plays a pivotal role in the regulation of glycolysis (Chapters 4 and 26). The preclimacteric enzyme appears to be under stringent control and exhibits negative cooperativity, whereas the postclimacteric enzyme obeys simple Michaelis kinetics. Limited proteinase action, it will be recalled, also abolishes stringent control of PFK. Such considerations suggest that it may be worthwhile to exploit the regulatory properties of enzymes rather than activity levels as useful markers.

Of course not all enzyme activities decrease during ripening. As pointed out elsewhere ripening and maturity of fruits are signaled by the derepression of genes coding for enzymes which prepare the tissue for senescence. In addition to the glucanases, one such enzyme is invertase, especially the β-fructofuranosidase type which increases markedly during ripening. Although it has frequently been concluded from indirect evidence that such increases in activity are indeed due to *de novo* synthesis of the protein enzyme and not to removal of inhibitor, it is only recently with the advent of radioimmunoassay and its more recent off-shoot, enzyme immunoassay (Chapter 39), that such synthesis can readily be shown directly. Thus, Iwatsubo *et al.* (1976) found that activities of mature green, turning and red ripe tomatoes were 0.3, 1.0 and 4.7 units per g fresh weights, respectively, whereas the corresponding amounts of immunologically competent protein present were 2.3, 12.4 and 100 μg, thus showing that *new* protein which reacted immunologically like invertase was being synthesized.

Enzymes of Honey.—According to German and other countries' laws, honey is considered to be unacceptably spoiled if the "diastase" cannot be detected. It is thus no wonder that considerable European literature (FAO/WHO's *Codex Alimentaire* describes an official diastase procedure) has developed on the methodology of not only invertase but also amylase and phosphatase as indicators of honey quality (Dustmann 1972) and also of related parameters such as degree of ripeness, storage time and, as mentioned in connection with AOAC procedure, for adulteration. Variation in heat sensitivity of honey is in general due to variations in pH (Schade *et al.* 1958). Again illustrative of the competition from nonenzyme methods, a more sensitive and reliable indication of adulteration of honey with cane sugar or corn syrups is the determination of the ratio of ^{13}C to ^{12}C (Ziegler *et al.* 1977). Most honeys come from nectars of plants which use the C_3 type of photosynthetic CO_2 fixation, whereas most sweetener adulterants such as cane sugar and corn syrups come from plants which photosynthesize via the C_4 pathway. There is a slight but detectable discrimination of the stable ^{12}C isotope at the expense of ^{13}C. Even in this case, however, ultimately the reason why it works can be traced back to compartmentation of and isotopic discrimination by an enzyme discussed in Chapter 20, ribulose biphosphate carboxylase (Benedict 1978).

Enzyme Assays, Food Processing and Food Quality

In this section we take the opportunity of quickly scanning enzymes which have not been mentioned previously, with emphasis on their use as indices of specific food quality attributes and as guides to proper processing of the foods.

Cereal-derived Foods.—Peroxidase paucity was used early as an index of the purity of wheat starch (Falkenhausen 1935). More recently α-amylase has been extensively used as in an indicator of sprout damage, especially in rye flour. Frequently, correlation between enzyme activity and food quality can be improved by combining enzyme assay with a nonenzyme measurement. Thus, Drews (1973) found that loaf volume yield of rye bread correlated best with α-amylase activity and the amount of soluble pentosan. Phenolase activity has been used to detect bran in flour, to predict bread color (Altermann 1939), to test for the adequacy of heating rice for lipase inactivation (Fellers 1975), and to distinguish durum from other wheats (Lamkin *et al.* 1981). Previous storage history of wheat may be partially inferred from a glutamate decarboxylase assay since this enzyme is quite rapidly activated when the grains are wetted (Kott 1973). Although one of the main functions of malt in beermaking is to supply α-amylase for breaking down the starch, another enzyme, invertase, proves to be a better measure of overall malting quality of barley (Prentice 1975).

Vegetables and Other Plant-derived Foods.—The investigations of Hasling *et al.* (1973) on α-amylase assays as predictors of processing characteristics of sweet potatoes illustrate that some assays for the same enzyme are better indices than are other assays, and that a small change in assay procedure can greatly improve its utility. From the work of Buescher *et al.* (1975) it may be possible to ascertain if sweet potatoes have been treated with ethylene and to predict its adverse effect on the quality of the baked root (discoloration and poor flavor) by determination of phenol-oxidizing enzymes. High phenolase is also indicative of poor storage stability of sugar beet roots. Obversely, increased phenolase activity of coffee beans is indicative of improved quality of the beverage brewed from the roasted bean (De Amorim and Silva 1968; De Oliveira *et al.* 1977).

Lee (1969) found that a high PE activity in acetone powders prepared from apples was indicative of decreased softness of the fruit as measured by a shear press ($r = -0.85$, $n = 40$).

Dairy Foods.—Release of alkaline phosphatase and xanthine oxidase from the fat globule membranes has been used to assay the churning of the milk (Stannard 1975). pH-rate profiles may be used to differentiate human from cow bovine milk since the pH optima of their xanthine oxidases are distinctly different (Dougherty *et al.* 1977). We have already referred to the extensive use of enzyme assay for the detection of microbial contamination of dairy products.

Other Animal-derived Foods.—Succinic dehydrogenase of meat, as measured by the percentage of individual fibers which give a positive qualitative test for this enzyme, may be used as an objective predictor of some subjective judgments of frozen beef color and appearance, such as the marbling score (Schafer *et al.* 1973). The lysosomal hydrolases, aryl sulfatase, cathepsin and β-glucuronidase, were used as marker enzymes for ascertaining the inhibitory effect of nitrite on autolysis of lamb and fish tissue (Warrier *et al.* 1973). A substitute for nitrite should suppress autolysis in addition to inhibiting bacteria and improving color. Enzyme tests should be helpful in screening for such nitrite substitutes. The thermal history of a meat sample may be revealed by the susceptibility of its collagen to pronase attack (Snowden and Weidemann 1978). Decrease in activity of sulfhydryl proteases in shrimp is a good indication of spoilage (Soedigdo *et al.* 1973). The lysozyme content of egg white, (Chapters 34 and 36) predicts volume stability of foams made from this egg white (Sauter and Montoure 1972). Several independent investigations in both Germanys have established that elevated creatine kinase levels are highly predictive of PSE pork (Chapter 26) (Kallweit 1979). Instead of isozyme patterns (see below), extent of solubilization of cytochrome *c* oxidase was used to detect previous freezing and thawing of a variety of animal foods (Barbagli and Crenscenzi 1981). Not all suspected correlations turn out to be significant. Thus, Penner (1974) could not use a particular catalase assay to identify irradiated potatoes.

Interaction of Food Enzyme and Food Constituent.—Frequently results obtained by allowing the enzyme in food to act on its substrate which is also present in the food and determining a product(s) of the resulting enzyme action may also be helpful in quality control. The accumulation of free fatty acid in milk (Deeth and Fitzgerald 1976) and meat (Potthast and Hamm 1973) would constitute *prima facie* evidence that lipases are acting on the milk and meat triglycerides, giving rise eventually to hydrolytic rancidity (Chapter 25). Various nucleoside triphosphates, especially ATP, are used as indices of fish freshness (Chapters 24 and 28).

$$\text{ATP} \xrightarrow{\text{ATPase}} \text{ADP} \xrightarrow{\text{Adenylate kinase}} \text{AMP}$$

$$\xrightarrow{\text{AMP deaminase}} \text{IMP} \xrightarrow{\text{Nucleotidase}} \text{Inosine (+ Pi)}$$

$$\xrightarrow{\text{Nucleoside phosphorylase}} \text{Hypoxanthine}$$

$$\text{(+ ribose-5}'\text{-phosphate)}$$

In the above case one simply measures the end result of enzyme action during food storage; this is essentially a straightforward analysis of a food constituent. Enzyme involvement in the actual measurement may be a matter of preventing more enzyme action (i.e., by heat inactivation) during the preparation of the sample for analysis. In other cases one allows the enzyme, endogenous enzyme and substrate to get together under rigorously defined conditions such as temperature, pH, presence of stabilizers, etc., just as in routine enzyme assay except that substrate is not added. After a definite interval, the reaction, if rapid, is stopped and the product or the rate of its production is measured.

Frequently the results correlate with processing parameters and quality. The most prevalently used of such tests may be susceptibility of the starch in wheat to the amylases present, an AOAC method, and the enzymatically produced pyruvate in onion arising from alliinase action as a measure of onion odor strength, as discussed in Chapter 21 and shown in Fig. 37.4.

The determination of methanol resulting from the action of endogenous PE on pectin has been suggested as a test for the previous freezing history of strawberries (Leuprecht and Schaller 1968). The release and determination of the 30,000 dalton fragment resulting from the action of the calcium activated factor (CAF) during aging of beef as a predictor of meat texture (Chapter 26) belongs in this category (MacBride and Parrish 1977).

ISOZYME PATTERNS

While the analysis of isozyme patterns of human and plant tissues have proved to be valuable for clinical laboratories and for taxonomical and related botanical and plant physiology studies, they have not until recently found their way into the food quality control laboratory. One particular use where isozyme pattern analysis seems to have been adapted as a marker of food quality and past history is in determining if unfrozen meat and marine products have been previously frozen. The rationale for such application is based on release into the fluid obtained by standard treatment of the food of a particular isozyme resulting from cell disruption attendant upon freezing.

Thus, a normally intracellular and hence not easily centrifugible form of the malic enzyme (Chapters 4 and 20) from a frozen oyster finds its way into the supernatant fluid obtained by gently centrifuging a previously frozen shucked whole oyster (Gould 1973). The technique of choice for detection of such isoenzyme patterns is gel electrophoresis and the resulting patterns are usually referred to as zymo- or enzymograms. After electrophoretic separation such gels are treated with chromogenic substrates visualizing the positions of those protein bands which contain enzymes acting upon the added substrate.

Figure 37.5 shows the enzymograms of the malic enzyme present in the centrifuged tissue fluid of whole oyster meats. F is from a fresh unfrozen oyster, SC from a superchilled oyster where no cell disruption has occurred, X from a frozen oyster, and A and B from commercial oyster meats purchased in local supermarkets. Such techniques can also differentiate between spoilage due to autolysis and that due to freezing. Detection of prior freezing of beef, pork and lamb by examination of zymograms has also been looked into and appears to be quite promising.

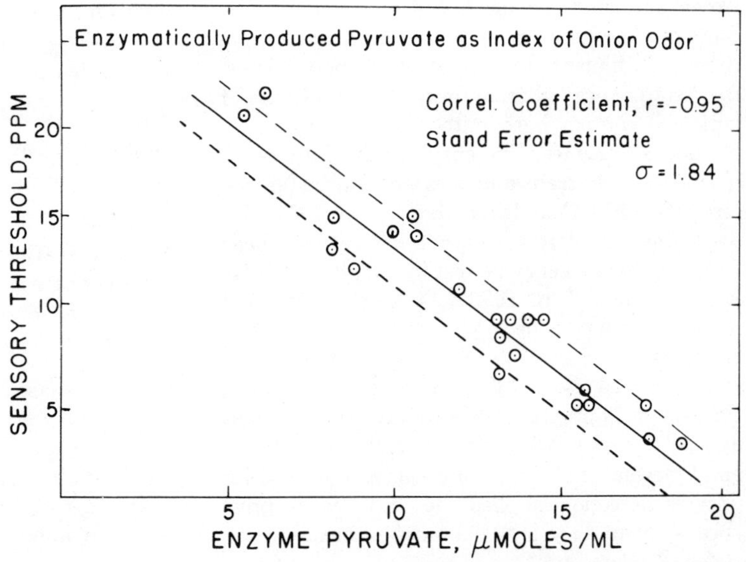

FIG. 37.4. ENZYMATICALLY PRODUCED PYRUVATE IN ONION HOMOGENATES AS INDEX OF ODOR STRENGTH

From Schwimmer and Guadagni (1962)

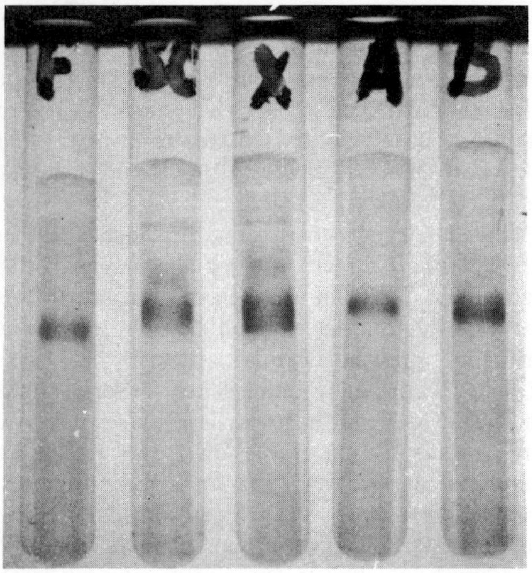

From Gould (1973)

FIG. 37.5. ENZYMOGRAMS OF THE MALIC ENZYME OF CENTRIFUGED OYSTERS AS A TEST FOR PRIOR FREEZING

See text for details.

The rationale behind such mammalian muscle food tests is that the press juice from unfrozen meats contains enzymes from the cell sarcoplasm only (Chapter 26), whereas juice from frozen meat contains enzymes from the mitochondria due to their freeze-induced disruption. The cells containing the mitochondria are, of course, disrupted during the preparation of the press juice, from both frozen and unfrozen meat. Since, in general, the proteins exhibiting the same enzyme activity from these subcellular sites differ electrophoretically, the resulting enzymogram from frozen beef but not pork will contain both mitochondrial and sarcoplasmic isoenzymes. The enzymes usually used as markers are the transaminases, aminotransferases, especially GOT (Vandekerckhove *et al.* 1972). Indeed, one can detect whether a meat has been frozen by such an enzyme test by simply using drip liquid (Guillot *et al.* 1973). At first the test did not work with pork because of mitochondrial damage in stress-susceptible pigs which, as discussed in Chapter 26, yield PSE meat, but refinement of the enzyme approach has allowed the method to be applied to pigs as well (Hamm *et al.* 1973). Beery and Lineweaver (1976), using a nongel method for isozyme detection, applied the GOT test to press juice of avian as well as mammalian meat.

There is a continuing effort directed toward simplifying and speeding up this technique, such as that of Burdette *et al.* (1976) in which the application of enzyme sensitive test paper permits reuse of the gel.

Other Application of Isozymology

The following examples will serve to illustrate the utility and versatility of isozymology not only directly in the food control laboratory but also in allied areas, especially on or associated with the farm. Some of these studies were not designed primarily for testing of any kind but we do believe that they might be adapted for such purposes.

Indicators of Fungal Infection.—Increase in peroxidase activity in fungal disease resistant lines of cereal and sweet potatoes was found to be associated with the appearance of new peroxidase isozymes only during challenge by the invading fungus (Seevers *et al.* 1971; Ru and Gau 1973; Haard and Marshall 1976). Fungal infection of peanuts induces change in the zymograms of several endogenous activities (Cherry 1977).

Specific Identification of Species.—Isoenzyme patterns of the esterases are used extensively taxonomically for identification and differentiation of species and varieties: potato (Desborough 1968), meat and fish (Thompson 1968). An elaborate dichotomous key to legumes encompassing 52 plants, 36 species and 13 genera was developed based upon isoesterase enzymograms (Chow 1975).

Isoenzyme patterns of peroxidase, phenolase, esterase and phosphatase were used as indicators of the vegetative (somatic) or generative origin of anther-derived plants (Corduan 1976).

Stage of Development.—Phosphatase enzymograms of tea leaves and tomato fruit were found to vary with the age of the plants (Baker and Takeo 1973).

Metal Insufficiency.—A new isozyme of lactate dehydrogenase develops in corn roots from plants suffering magnesium deficiency (Kudrev and Babalukova 1975).

Grains.—Classification of bread grains from a scan of α-amylase enzymograms as a substitute for the traditional "dropping number" was advocated by Olered and Jonsson (1967). Especially useful in the cereal field is the combination of electrophoresis with immunochemical detection techniques, referred to by Bøg-Hansen and Daussant (1974) as electro-immunoabsorption. Semolina contains an isozyme of phenolase not found in wheat. Feillet and Kobrehel (1974) developed a procedure which decimated the previously long time required for detection of adulteration of semolina with common wheat flour. Electrophoretic methods based primarily on protein but not necessarily enzyme patterns are also available for such purposes (Silano 1975).

Monitoring Food Processing.—Tirimanna (1972) found continuing changes in isoperoxidase patterns of tea leaves during fermentation and subsequent processing into the beverage (Chapter 16).

Wine Identification.—Invertase enzymograms of different wines could readily differentiate these beverages (Barna and Prillinger 1972).

STANDARDIZING AND MONITORING COMMERCIAL ENZYMES

Once a food grade enzyme preparation is added to a food it becomes part of the food, and until immobilized enzymes are used instead, it behooves the quality control technologist to know what and how much is being added. Upon receipt of an order of the enzyme preparation he or she should also know what is actually received and how much the company is paying per unit of enzyme received. To this end it is, or should be, customary to set up a rapid method for estimation of the "strength" of the sample by either a standard enzyme procedure or by using it in an analog of the process in which it will be employed. In some cases this may amount to a miniaturization of the large-scale operation.

The standard assay may come from special manuals available from the particular industry, from the enzyme manufacturer or from such broader manuals such as the AOAC, if one happens to have purchased papain or if one considers malt as a commercial α-amylase preparation. It is frequently advisable to modify the process analog procedure to more accurately reflect the enzyme's action in operation at a particular processing facility. On the other hand, one has to keep an eye open for constantly changing published procedures which may turn out to be more suitable than the one being used. For instance, release of dye from starch has greatly simplified α-amylase assay. Another example, taken from the dairy industry, is the need for a rapid test to evaluate the constantly evolving new rennets (Chapter 33), not only with regard to milk clotting, but also with regard to side effects such as the development of bitter peptides (Chapter 34). Richardson and Nelson (1968) were able to shorten the weeks it customarily would have taken to make such evaluations to 3 days by using levels of rennet highly in excess of that used in routine cheesemaking, thus permitting the early appearance and detection of bitter peptides.

Future Monitoring

Immunodiffusion and "Label" Immunoassay.—Although perhaps not practical in many quality control laboratories at the present time, it should be possible to detect adulteration of a given commercial enzyme preparation by other unwanted enzymes by the use of immunological related techniques combined with isozyme separation techniques such as the aforementioned electro-immunoabsorption. Thus, Daussant and Carfantan (1975) were able to detect adulteration of barley amylase with added bacterial α-amylase present at levels which contributed less than 1% to the total enzyme activity. Before such procedures become routine it will be necessary to have commercially available antibodies to the pertinent enzymes. Antirennets would be particularly useful for ascertaining the makeup of a given batch of purchased rennet.

An even more powerful approach, radioimmunoassay and its offshoot enzyme immunoassay, useful not only for the determination of the presence of enzymes in biological systems but also of almost any organic substance, will be examined in connection with enzymes as analytical aids in Chapter 39.

Operational Molarity *via* Active Site Titrations.—While some immunochemical methods are nonrate methods which can be used for the detection of enzymes, they do not answer the question of the catalytic competency of the enzyme which is present in the preparation. At least for the assay of enzymes available in substantial quantities, we believe we can look forward to the use of substrates which react specifically, irreversibly and covalently with the active site of the enzyme. Some of the various modes of such interaction were inspected in Chapter 14. Since the active site has to be in its catalytically competent configuration for the suicide substrate (hereafter referred to as active site titrant) to react, only those molecules which would catalyze the transformation of their normal substrates will participate. One simply adds enough titrant until all active enzyme molecules have reacted and then measures either one of the products not retained by the enzyme or else adds an excess of titrant and determines what is left.

Active site titrations are characterized by an almost instantaneous "burst" release of coproduct, usually colored. Principles of active site titration were lucidly exposited by Kédzy and Kaiser (1970). Historically the basis for active site titration of enzymes goes back to the classical work of Balls and Jansen briefly alluded to in Chapter 2. Upon its introduction into routine analytical chemistry the ion selective electrode has been quickly exploited for such titrations. Thus, Erlanger and Sack (1970), using a commercial fluoride electrode, were able to detect as little as 30 nanomoles of chymotrypsin, by liberation of F^-, as first detected by Balls and Jansen, from suitable organic fluoroorganophosphates. This amount still adds up to about 0.8 mg, enough for several hundred rate assays of chymotrypsin. By using 4-methylumbelliferyl-*p*-guanidinobenzoate as active site titrant, Brown *et al.* (1975) were able to quantitatively determine as little as 20 picomoles (0.05 mg) of acrosin, the trypsin-like proteinase of sperm, in the range of the most sensitive enzyme assays.

For the most part, operational molarity determinations have been confined to proteinases—trypsin, thrombin, papain, subtilisin and elastase—and related serine esterases such as cholinesterase. An entirely different kind of enzyme, carbonic anhydrase, was determined by "reverse burst" active site titration by adding an excess of the titrant *p*-nitrophenyl-*p*-sulfamyl benzoate (sulfonamides are potent reversible inhibitors of this enzyme) and remaining unreacted titrant estimated by release of nitrophenol with a nucleophile (Mendez and Kaiser 1975).

Such developments point to a radical departure in the assay of a variety of enzymes not only in partially purified commercial enzyme preparations but also the more active endogenous food enzymes. We shall have more to say about this approach in Chapter 39.

It may be apropos at this juncture to point out that it is possible, at least in principle, to use a rate assay to determine the molecular concentration of an enzyme. In the presence of a strong inhibitor, the I_{50} (Chapter 14) varies linearly with enzyme concentration (Dixon and Webb 1964).

$$I_{50} = K_i + (0.5/A) \cdot A \cdot e = K_i + (0.5/A) \cdot \text{kat}$$

where A is the molar activity, kat is the units of activity, expressed as katals in absence of inhibitor and e is the absolute concentration of inhibitor-binding sites. If there is one inhibitor binding site per molecule then e is of course the molar concentration of the enzyme. A can be calculated from the slope (0.5/A) of the plot of I_{50} *vs* kat and the value of e from A and the x- and y-intercepts.

In addition to topics related to enzyme testing which have been covered, there are several peripheral areas of investigation where enzymes are involved but may not be of direct concern to food quality control. Among these are use of enzymes in research biochemistry and organic chemistry to elucidate the structure of biologically-derived compounds including, of course, many food constituents: nucleic acids, proteins, carbohydrates, lipids, vitamins, coenzymes and fleeting intermediate metabolites. Whitaker (1974B) presents an interesting sampling which vividly illustrates what powerful tools enzymes are in structural studies. To cite a somewhat obscure but pertinent example with which the author has been involved, the specific cysteine lyases of the flowering shrub *Albizzia lophanta* contributed to an understanding of the nature of the covalent linkage of heme to protein in cytochrome *c* (Schwimmer and Kjaer 1960) and to elucidation of the structure of herbicide metabolites (Lamoureaux *et al.* 1973).

Another area of analytically-related enzymology is the application of histology and cytology to the localization of enzyme within cells. The classical approach was to immerse tissue slices in a solution of substrate to produce a colored insoluble product or to treat the tissue with solution of a substance which forms a visible precipitate with the product. Enzyme histochemistry has been considered as a link between biochemistry and morphology.

Such techniques can be quantitated so that it is possible to assay for activity within subcellular organelles. Among the instrumentation and techniques which have been useful are interference microscopy, radiochemical methods, cartesian divers, the quartz fiber fishpole balance and more recently the scanning electron microscope.

Perhaps it is a reflection of current trends that the 1963 edition of Bergmeyer has a 43 page chapter devoted to the histochemical detection of enzymes whereas histochemistry cannot be found even in the index of the 1974 edition. Nonetheless, enzyme histochemistry of enzymes is still a lively subject as witnessed by the book of Lowry and Passonneau (1972) and reviews of Meijer (1975), Schneider (1975), Zaki (1976) and Hardonk and Koudstaal (1976).

38

FOOD ANYLATES AS ENZYME SUBSTRATES

HISTORY AND PRINCIPLES

The use of enzymes as specific analytical reagents in quantitative analysis goes back some 135 years when Ossan (1846) determined hydrogen peroxide with a malt extract as source of peroxidase and a tincture of guaiacol as cosubstrate and indicator reagent, still used in one official inspection test for adequacy of blanching and still useful enough to be included in an automated setup (Chapter 11). Of substances which accumulate in living organisms the foodstuff sucrose was probably the first to be determined enzymatically, about 100 years ago. The name of the individual responsible is indelibly associated with food laboratories for his method for nitrogen analysis; he concluded that "cane sugar can be determined with great certainty with the aid of a few cc of the *invertine* (yeast extract) in the presence of virtually all the other carbohydrates (and other substances) with which it is ordinarily mixed" [author's translation] (Kjeldahl 1881). However, the true flowering of enzymatic analysis can be credited to its widespread adoption in the clinical laboratory. According to Schmidt and Schmidt (1976), clinical enzymology was born in 1908 when Wohlegmuth (1908), whose name is associated with the traditional amylase substrate, discovered that activity of this enzyme increases dramatically in serum and urine of individuals suffering from acute pancreatitis.

The use of deliberately added enzymes for metabolite analysis probably goes back to Marshall (1914) who used urease to determine urine urea. Clinical applications of enzyme analysis accelerated upon the dissemination of the UV optical test (Warburg 1948). However, the definitive breakthrough occurred when Sterkel *et al.* (1958), among others, showed that transaminase (aminotransferase) assays could be used as powerful diagnostic aids for hepatic and cardiac diseases. Somewhat earlier Stetter (1951) had listed 57 substances which could be analyzed with the aid of enzymes. Subsequent progress has been largely dependent upon the availability of enzymatically pure enzymes, substrates and cofactors, especially NAD and its congeners, at reasonable costs.

The main advantages in using enzymes as analytical reagents are accuracy, precision, speed, specificity and sensitivity. Their use usually combines accurate and precise results with speed of performance. Specificity leads to discrimination without, ideally, need for physical separation and without outside interference. This is especially true when one wants to exclude compounds which are closely related chirally to the unknown anylate. However, few enzymes are, as we have learned (Chapter 2), completely specific. It is customary in the present context to distinguish between *foreign* activity due to the presence of enzyme impurities and *side* activity due to the incomplete specificity of the pure enzyme, leading to unwanted enzyme-catalyzed change in constituents other than the unknown.

Specificity may be enhanced by the use of coupled reactions (see below). Many metabolites including important food constituents can be readily measured at levels as low as $10^{-16}M$ (Michal 1974). However, it should again be pointed out that in each of these categories, nonenzyme methods continually challenge, rival and frequently replace enzyme methods, with advanced separation techniques performing the equivalent of enzyme specificity. In general the enzymatic determination of constituent-as-substrate may be divided into three categories.

First, one allows the reaction to go to completion as far as possible and measures the total change (usually a color) representing or accompanying the disappearance of substrate or appearance of product. This approach is referred to as the *total change* or *end-point* method. Its principal advantage is that it is not necessary to rigorously control the host of variables affecting initial rate (Chapter 37). Inhibitors if not present at an excessive level do not have to be taken into account and it is usually preferable to use less than saturating amounts of unknown substrate.

Second, one can measure the rate at which the unknown is transformed by a precise amount of added enzyme—*kinetic analysis.*

Third, one measures the amount of product generated in a given time interval—*fixed time analysis.*

End-point Determinations

At the relatively low substrate concentration(s) employed in end-point determinations the reaction will approach completion asymptotically (creep reaction) so that excess enzyme is usually added to assure rapid completion. Creep can also be due to slow-acting side reactions. Furthermore in the "optical" test it is preferable to end up in the range of absorbance of 0.3–0.7. This means, for instance, an S on the order of $10^{-5}M$ (0.0004%) for the determination of sucrose with invertase and auxiliary enzymes. Since the K_m is $10^{-2}M$, the initial rate will be only 0.1% of that at the S concentration of $10 \times K_m$, traditionally recommended for determination of enzyme activity (Chapter 37) so that there is some sacrifice of time. Too much of even the "purest" of enzymatically pure enzymes could introduce foreign activity into the analysis. Some of these problems can be alleviated by switching to immobilized enzymes (see below).

The UV Optical Test with NAD(P)-linked Enzymes.—The main thrust of the search for suitable enzymes as analytical reagents has traditionally been to find an enzyme which converts the anylate-as-substrate into a product with altered optical properties. Even in the earliest days of enzyme analysis and assay, invertase action, for instance, was followed by changes in optical rotation. A more recent illustration, shown in Fig. 38.1, is the estimation of the flavoring potential of mustard and other cruciferous vegetables (Chapters 21 and 35). The molar extinction coefficient, ϵ, of the principal substrate in mustard, allyl glucosinolate, is 7800 at the spectral maximum, 227.5 nm. By contrast, the principal optically demonstrable product of glucosinolase action, allyl isothiocyanate, is practically transparent at 227.5 nm but absorbs maximally at 240 nm (Schwimmer 1961B).

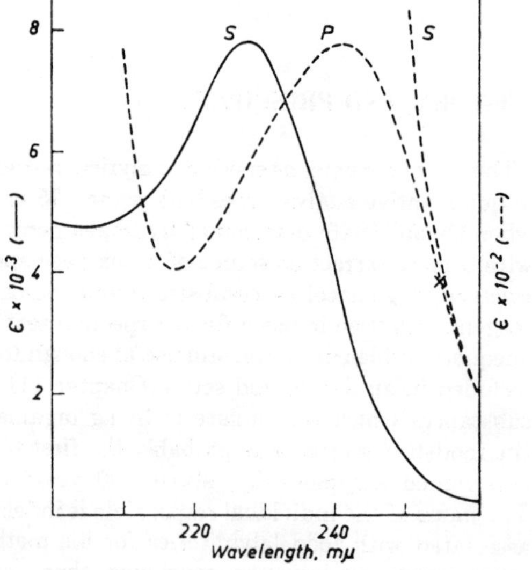

From Schwimmer (1961B)

FIG. 38.1. UV SPECTRA DIFFERENCE AS BASIS FOR ENZYMATIC ANALYSIS OF FOOD CONSTITUENTS

The substrate (S) is sinigrin or allyl glucosinolate (Chapter 21) and the product (P) is allyl isothiocyanate formed as a result of glucosinolase action.

Of course not all substances to be analyzed possess convenient visible or UV absorbance bands. To compensate for this one uses an indicator enzyme, especially an NAD(P)-dependent dehydrogenase. The NAD(P) becomes a cosubstrate (reaction partner) with the unknown in a two-substrate reaction. The spectra of NAD and NADH are shown in Fig. 38.2. The measurements are almost invariably run in the "optical test" at 365 or 340 nm. The dehydrogenases are thus *indicator* enzymes.

Auxiliary and Coupled Enzymes.—If one cannot find a suitable dehydrogenase, one may have to resort to a coupled system in which the unknown is first converted into a NAD(P)-dependent dehydrogenase substrate via (from the viewpoint of the optical test) an *auxiliary* enzyme. The first auxiliary enzyme to act on the substrate may be considered to be the "trigger" enzyme:

$$\textit{Anylate:}\ \xrightarrow[\text{enzyme}]{\text{trigger}} P_1 \xrightarrow{\text{auxiliary}} P_nH_2;$$

$$P_nH_2 + NAD^+ \xrightarrow[\text{dehydrogenase}]{\text{indicator}} P_n + H^+ + NADH \ \textit{or}$$

$$P_n + NADH + H^+ \xrightarrow[\text{reductase}]{\text{indicator}} P_nH_2 + NAD^+$$

For the most part these reactions are carried out at either 37°C for 10–20 min or, optionally, at 25°C for 30–60 min.

A supplier of purified enzymes, Boehringer-Mannheim (1976) has published a system of food analysis based almost completely on end-point optical tests. Of these, nine utilize only one trigger-indicator dehydrogenase, nine require separate triggers and indicators and ten use trigger-plus-auxiliary enzymes plus indicator dehydrogenases. In two analyses, for cholesterol and pyrophosphate, the indicator reactions are nonenzymatic.

In wines, but not in other foods, sorbitol determination requires trigger, auxiliary and indicator enzymes. Illustrative of the complications that enzyme specificity can lead to in the analysis of sorbitol in wine, an auxiliary enzyme, alcohol dehydrogenase, also serves to correct a *side* reaction of the indicator enzyme (G-6-P dehydrogenase), its catalysis of the reduction of the glycerol, also present in the wine. In general, a rather large excess of indicator as compared with auxiliary enzyme is used, the exact amount depending upon the enzymes' kinetic constants (Chapter 37).

Assurance of Reaction Completion.—One of the main problems in making the optical end-point test generally applicable is the question of assuring

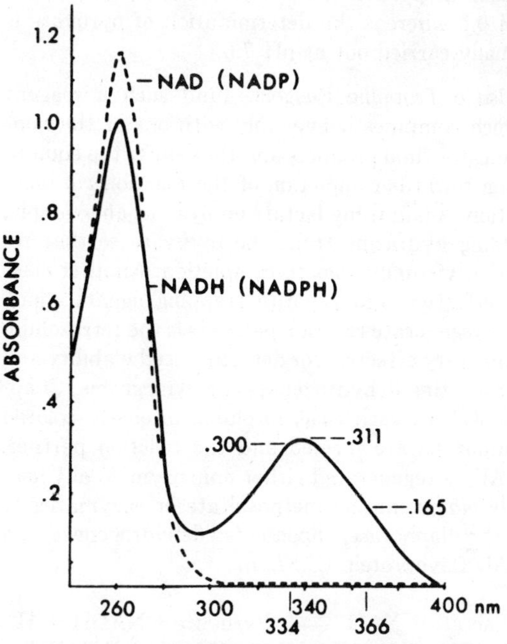

Courtesy of the Boehringer-Mannheim Corp.

FIG. 38.2. SPECTRA OF REDUCED AND OXIDIZED NICOTINAMIDE ADENINE DINUCLEOTIDES

These spectra form the basis for the optical or "Enzyme Test" ubiquitously used in enzymatic analysis of metabolites and food constituents. Shown are absorbances of $10^{-5}M$ solutions at neutrality of the oxidized (NAD, NADP) and reduced (NADH, NADPH) forms of these coenzymes.

completion of the reaction, avoiding the above-mentioned creep; or, in the absence of completing, applying the appropriate correction. The more common approaches are as follows.

Use Excess Reaction Partner.—Excess of added reaction partner, usually NAD(P) or NAD(P)H, is used in about five to ten-fold excess of the unknown. For the determination of lactate, a 50–100-fold excess is necessary to aid in driving the reaction to completion.

$$\text{Lactate} + NAD^+ \rightleftharpoons \text{pyruvate} + NADH + H^+$$

Too much reaction partner (using NADH) results in high readings in the spectrophotometer and the presence of trace amounts of enzyme inhibitors might become significant.

Change the pH.—Change the pH in order to change the equilibrium towards completion. Thus, the same enzyme used for the determination of pyruvate is also used for lactate. For the determination of pyruvate, the reaction is carried out at pH 9.5 whereas the determination of pyruvate is usually carried out at pH 7.6.

Use a Trapping Reagent.—Find such a reagent which combines irreversibly with one of the non-indicator final products and thus shifts the equilibrium towards completion of the reaction *via* mass action. Again using lactate analysis as an example, adding hydrazine traps the pyruvate so that reaction virtually goes to completion. Another class of effective and favorite trapping agents which also regenerate reaction partners is the tetrazolium salts, very effective for detecting seed viability and for locating dehydrogenases on zymograms (Chapter 37). In each case, insoluble intensely colored formazans are formed and the reaction partner, NAD, is regenerated either nonenzymatically usually *via* phenazine methosulfate or enzymatically with diaphorase, lipoamide dehydrogenase, an FAD flavoprotein (E.FAD):

$$\text{Lactate} + \text{NAD}^+ \rightleftharpoons \text{Pyruvate} + \text{NADH} + \text{H}^+{}_1$$
$$\text{NADH} + \text{H}^+ + \text{E.FAD} \rightleftharpoons \text{NAD}^+ + \text{E.FADH}_2$$
$$\text{E.FADH}_2 + \text{Tetrazolium} \rightarrow \text{E.FAD} + \text{Formazan}$$

The reactions can be quantitated by the use of appropriate formazan solubilizers (Coburn and Carroll 1973). The pyruvate may also be trapped by removing it with glutamate-pyruvate aminotransferase which has to be added to the analysis system along with L-glutamate:

$$\text{Pyruvate} + \text{L-glutamate} \rightleftharpoons \text{L-alanine} + \alpha\text{-ketoglutarate}$$

Change Reaction Partners.—Alternative cosubstrates can be used i.e., a coenzyme analog of NAD in which its niacin moiety is replaced by pyridine (Kaplan *et al.* 1956). With this reaction partner the equilibrium constant for dehydrogenase catalyzed reactions may be altered as much as 200-fold towards completion of the reaction.

Another special case of change of reaction partners is the substitution of arsenate for phosphate in a phosphorylase-catalyzed reaction:

$$\text{Starch} + \text{phosphate} \xrightleftharpoons{\text{phosphorylase}} \text{glucose-1-phosphate}$$

$$\text{Starch} + \text{arsenate} \rightarrow (\text{glucose-1-arsenate}) \xrightarrow{[\text{spontaneous}]} \text{glucose} + \text{arsenate}$$

Before glucoamylase became available such a reaction might have been applied to the determination of starch.

In theory and to a large extent in practice, one can use any number of auxiliary enzymes to transform the unknown to a substrate acted upon by an indicator enzyme. Thus, Michal (1974), in his excellent precis of enzymatic analysis, shows how with such *succeeding indicator* reactions it is possible to determine lactose, fructose, fructose-6-phosphate and the glucose phosphates in a single reaction mixture by adding successively the following five enzymes: glucose-6-phosphate dehydrogenase (indicator), phosphoglucomutase, hexokinase, glucose-6-phosphate isomerase and β-galactosidase. A succeeding indicator reaction scheme is also used for the determination of acetic acid (see below).

Kinetic (Rate) Determinations

Preceding indicator reactions are more frequently used in the enzymatic determination of substances by the alternative to the end-point approach, the rate or kinetic method. This approach is useful when used in conjunction with a regenerative auxiliary reaction because it becomes possible to maintain at a constant rate an anylate concentration which is a fraction of the K_m of the leading indicator enzyme:

$$\text{Anylate} + \text{Indicator Reaction Partner} \longrightarrow \text{Product}_1 + \text{Product}_2 \text{ (to be measured)}$$

$$\text{Product}_1 + \text{Auxiliary Reaction Partner} \longrightarrow \text{Product (Auxiliary)} + \text{Anylate (regenerated)}$$

These sets of reactions constitute a cycle. For each cycle the concentration of the measurable constituent increases by a multiple of the concentration of the unknown which remains constant. Just as important as keeping the reaction rate constant, this arrangement results in amplification of the original signal by factors ranging, according to Michal, from 100 to 100,000 so that as little as 10^{-11} moles of substance using the UV optical test and 10^{-14} moles using fluorescence measurements can be detected with NAD-dependent dehydrogenases as indicator enzyme. Enough indicator enzyme is added in this type of setup so that the rate of the reaction it catalyzes is about 1000 times that catalyzed by the auxiliary enzymes acting separately.

In spite of the attention to detail which kinetic methods require they have been in the ascendency compared with endpoint methods for several reasons besides increased sensitivity. These include: (1) increased rate of analysis (1 to 5 min); (2) decreased cost due to less enzyme being required; (3) foreign activity not as serious a complication; (4) reaction completion or correction for incompletion does not have to be a concern; (5) it lends itself more readily to automation; and (6) less technical assistance is required. Although inhibition by end-product has to be taken into account in most kinetic rate determinations, deliberate addition of competitive inhibitors which effectively increase the apparent K_m of the rate-limiting enzyme (Ependorf and Hine 1975) make it possible to measure these anylates at higher levels and thus more accurately and faster, as well as to swamp out the effects of endogenous inhibitors in the sample being analyzed.

There are, of course, alternatives to regeneration of substrate in using kinetic enzymatic analysis. Frequently it is not the anylate which is regenerated but a reaction product which couples the system in a manner similar to that of the formazan reaction. For example, Smith and Olson (1975) estimated ethanol by measuring the current generated at a carbon electrode as a result of the following series of reactions:

$$\text{Ethanol} + NAD^+ + H^+ \rightleftharpoons \text{NADH} + \text{acetaldehyde} \quad \text{(slow)}$$

$$\text{NADH} + \text{Redox Mediator (oxidized)} \xrightarrow[\text{(fast)}]{\text{Diaphorase}}$$

$$\text{Redox Mediator (reduced)} \xrightarrow{\text{Amperometric measurement}} \text{Redox Mediator (oxidized)}$$

The rate of reduction of the redox mediator, 2,6,-dichlorophenolindophenol (frequently used in ascorbic acid determinations and in peroxidase assay) was found to be proportional to the ethanol concentration. DiCesare and Atwood (1975) compared end-point and kinetic methods in procedures for optimizing trigger enzyme concentrations.

INSTRUMENTATION AND METHODOLOGY

With Nonimmobilized Enzymes

The last example shows that while UV optical methods have and undoubtedly will dominate enzyme analysis of substrates for some time, other powerful methods, instruments, techniques and approaches are rapidly being adopted. These instruments link enzyme reactions to the entire field of analytical chemistry wherever the latter can be applied at the moderate temperature and pH's demanded by enzyme reactions. Again the impetus is supplied by clinical research but some of these methods undoubtedly will find their way into the food quality control laboratory. We briefly adumbrate and highlight some of these evolving approaches as well as refinements of optical methods.

Optical.—As mentioned, spectrophotometry and its variants continue to be by far the most widely used and researched methods of enzyme analysis. About one-third of the papers reviewed by Fishman (1980) involve application of spectrophotometers of one sort or another. Especially useful, as mentioned in Chapter 37, are the centrifugal, fast or enzyme analyzers discussed below.

In addition to UV and visible absorbance measurements, both fluorescence and chemiluminescent measurements are finding acceptance mainly because of their higher sensitivity. Fluorimetric methods, used in conjunction with the above-mentioned signal amplification via recycling kinetics, allow detection and quantitative measurement of as little as 10^{-15} moles of NAD-linked substrates. The main problem posed by interference in fluorimetry is the presence of quenchers. ATP or any substance that can be linked to ATP can be measured by means of the luciferase-luciferin reaction at levels as low as 10^{-12} moles (Brolin *et al.* 1977). Luminescence may also be generated nonenzymatically as in the determination of glucose with glucose oxidase via the reaction of H_2O_2 with ferricyanide in the presence of luminol (Bostik and Hercules 1975) or with bis(2,4,6-trichlorophenyl) oxalate in the presence of perylene (Williams *et al.* 1976).

Manometry.—The measurement of pressure changes of gases (CO_2, O_2, N_2, NH_3) with time is a time-honored enzymological technique which served well in its day—mastery of the technique required to obtain useful information from the Warburg manometer was a minor triumph for a tyro enzymologist—but is being replaced by other methods. Its demise undoubtedly will be or has been hastened by the development of membrane gas sensing electrodes (see below).

Automated Analysis and Continuous Flow.—Automated analysis for handling large numbers of samples, first developed for clinical is now widely used in food laboratories. Special considerations which have to be taken into account when using enzymes in automated analysis have been discussed by Roodyn (1973). While it is being swiftly adapted by segments of the food industry it is not always suitable and frequently has to be modified, as expected, from standard clinical application, even sometimes at the expense of sensitivity. A case in point is the "step-backward" improvement in the automated enzymatic estimation of lactate and pyruvate in milk (Stahlhut-Klipp 1975), in which the sensitive fluorimetric estimation of NAD^+ was replaced by the UV "optical" test, presumably because of fluorescence quenchers in milk.

A sophistication of automated analysis resulting in enhanced efficiency which appears to be in the ascendency is the use of continuous flow techniques, first developed mostly for the determination of glucose (Lutz and Flückinger 1975), cholesterol (Papastathopoulos and Rechnitz 1975) and triglycerides (Coudon *et al.* 1974) in conjunction with mass testing for clinical purposes. Kinetic rather than end-point determination methods are particularly adaptable to flow-through automated analyses and especially with immobilized enzymes (see below).

Centrifugal Analyzer.—This highly advanced sophistication of instrumentation for automated analysis referred to by its developer as a "computer interfaced fast analyzer" (Anderson 1969A,B) frequently using lasers is now available commercially (Pesce 1974) and is being used increasingly in conjunction with enzymes (Bruce *et al.* 1978). This includes end-point, fixed time and kinetic analyses and also perhaps its first published use for the analysis of a food constituent, glutamate, in a variety of canned and dried soups, canned sauces, ready meals and seasonings (Arnoux *et al.* 1976). Pesce describes the device as a multicuvet spectrophotometer that simultaneously and rapidly measures a single substance in a large number of samples in the cuvets.

The cuvets, each accommodating less than 1 ml and arranged radially in a rotor, are moved rapidly past the light beam by means of centrifugation of the rotor, at about 1000 rpm which moves the samples and reagents, mixes them (with or without the aid of a vacuum), removes air bubbles, separates any liquid phases present and averages out errors.

With the first generation instrument developed by Anderson, who is also responsible for development of the zonal ultracentrifuge and other advanced biochemical implements, 15 biuret protein analyses were completed in as little as 30 sec after the start of centrifugation; end-point analysis of glucose of as many or more samples can be completed in 6 min. Principles of operation of one such analyzer is shown in Fig. 38.3. Pesce reported that centrifugal analyzers were available from three manufacturers in 1974 at as much as $55,000 per instrument.

Electrochemical.—All of the familiar electrochemical parameters have been used in conjunction with enzymes for constituent analysis: potentiom-

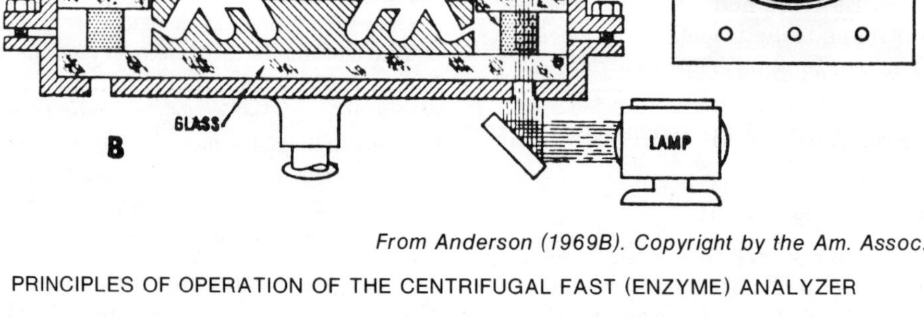

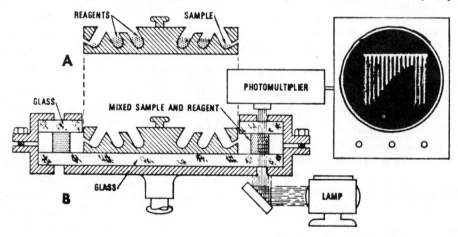

From Anderson (1969B). Copyright by the Am. Assoc. Adv. Sci.

FIG. 38.3. PRINCIPLES OF OPERATION OF THE CENTRIFUGAL FAST (ENZYME) ANALYZER etry and its variants, voltametry, amperometry, conductivity and polarography. The latter, a well established procedure in general analytical chemistry, was used, for example, for monitoring oxygen depletion due to the action of glucose oxidase in the determination of glucose (Nanjo and Guilbault 1974).

Microcalorimetry.—Because of increased sensitivity, sophisticated circuitry and the availability of commercial models of thermistors, calorimetry has become a powerful tool in biochemistry and recently in food science (Chapters 10 and 33). Calorimetry has the advantage of general applicability to all chemical reactions which either give off or take up heat. This ideal nonspecificity combined with the specificity of enzyme action yields, in principle, a flexible, broadly applicable analytical system for determination of food components. Among other potential advantages inherent in the thermochemical approach are the following: Empirical calibration is not used since one deals with enthalpy, a fundamental thermodynamic characteristic of the reaction system. Time consuming pretreatments are circumvented. It is usually not necessary to convert the product into a measurable secondary product, i.e., no auxiliary reactions are necessary. The primary trigger reaction is the indicator reaction. There is no interference from colored compounds or from too much protein. McGlothlin and Jordan (1975) described the use of Direct Injection Enthalpy (DIE) for enzymatic determination of glucose with hexokinase. Schmidt *et al.* (1976) devised an enzyme thermistor consisting of a flow-through device with an immobilized enzyme (see below) for the determination of glucose with glucose oxidase. We predict that microcalorimetry may yet give enzyme electrodes some degree of serious competition (Grime 1980).

Radioisotopes.—These are useful not only in enzyme assay as mentioned in Chapter 17, but also in enzyme analysis of constituents. Thus, oxalacetate, a fruit constituent as well as an important metabolic intermediate, was determined end-pointedly by measuring the radioactivity of citrate formed via citrate synthase as cited by Michal (1974):

$$\text{Acetyl CoA}^* + \text{Oxalacetate} \rightarrow \text{Citrate}^* + \text{CoA}$$

The key to success in this and in other radiometric analyses is that the product can be readily and completely separated from the substrate at the termination of the reaction, but need not be quantitatively recovered. In the above reaction the acetyl CoA is readily saponified, the reaction mixture acidified and the radioactive acetic acid driven off, leaving only a radioactive product. Although the procedure could not be used routinely for food analysis, an even more pervasive food constituent, lactose labeled in the 1-position [1-^{14}C] was estimated by the following series of reactions (Davies *et al.* 1975), after hydrolysis to labeled glucose:

$$[1\text{-}^{14}\text{C}]\,\text{glucose} \xrightarrow[\text{hexokinase}]{+\text{ATP}}$$

$$[1\text{-}^{14}\text{C}]\text{-glucose-6-phosphate} \xrightarrow[\text{GPDH}]{\text{NAD}^+}$$

$$[1\text{-}^{14}\text{C}]\text{-gluconate-6-phosphate} \xrightarrow[\text{EC 1.1.1.44}]{\text{NADP}^+}$$

$$\text{D-ribulose-5-phosphate} + [1\text{-}^{14}\text{C}]\,\text{CO}_2$$

The latter enzyme, phosphogluconate dehydrogenase (decarboxylating) is available from biochemical suppliers; GPDH is glucose-6-phosphate dehydrogenase. Radioactive CO_2 is easily separable and measurable. Availability of the nonradioactive isotope ^{13}C should before long make its debut in enzymatic as it has in nonenzymatic analysis of foods (Chapter 37).

Radioisotopes can also be used in kinetic determinations by a felicitous combination of isotope dilution methodology and Michaelis kinetics (Newsholme and Taylor 1968; Brooker and Appleman 1968). One makes two enzyme rate determinations, one in the presence of a known amount of labeled substrate alone and the other exactly the same except that the unlabeled unknown is included in the reaction mixture. This principle has been used in the determination of food constituents such as glycerol and glucose, using the respective kinases, and adenosine (Namm and Leader 1974), using the respective kinases.

Kits and Papers.—These have been long available not only for enzyme assay (Chapter 37) but also for constituent analysis for clinical use. Recently, Boehringer-Mannheim has provided kits for enzymatic food analysis emphasizing "simplicity, speed, and specificity" for the following: acetic, citric, glutamic acid, L-lactic and malic acids, glucose/fructose, sucrose/glucose, lactose/galactose, ethanol, glycerol and cholesterol. McCloskey and Replogle (1974) adapted a kit designed for serum alcohol analysis for analysis of alcohol in serum to its determination in wine. The abbreviated test consisted merely of adding the kit reagent (alcohol dehydrogenase + NAD + buffer) to an appropriate wine sample and measuring the usual absorbance increase at 340 nm. What the method lacked in accuracy was more than made up for by convenience and savings of time—a few minutes—as compared to the laborious distillation and oxidation by dichromate required in the official 1975 AOAC method. Clinical dipsticks, usually containing glucose oxidase, peroxidase and a peroxidase-cosubstrate leuco dye, available from Miles, Ames (Dextrostix), Eli Lilly and undoubtedly others have been used for some time for corn syrup, etc. Foreshadowing increased use of enzyme-impregnated strips specifically designed for food analysis, Jahns *et al.* (1976) have developed an enzyme paper for the semiquantitative determination of hypoxanthine whose presence, it will be recalled (Chapter 37), is an indication of lack of fish freshness. Its formation is itself an example of the result of action of endogenous enzymes on endogenous substrate. As applied here, hypoxanthine was estimated visually by Jahns *et al.* by dipping a test paper into a fish extract at ambience for 5 min. The test paper consisted of enzyme (xanthine oxidase, XO), stabilizer (gelatin), ammonium sulfate and visualizer dye, resazurin, which is oxidized by the XO-produced H_2O_2 to a purple substance. As stated by the authors, their strip could be used in a fish processing plant, at the dockside, or even aboard ship, in such a way that a simple visual change from blue to pink (*ca* 5 μM/g) could indicate to nonlaboratory personnel that the fish being tested is no longer fresh.

Immobilized Enzymes, Enzyme Electrodes and Reagentless Analysis

Immobilized enzymes lend themselves readily to automated analysis in part because of the same advantages their use engenders for other purposes discussed in Chapter 8. First proposed as part of enzyme electrodes (see below) about the time enzyme immobilization began to attract attention by Clark and Lyons (1962), their use for this purpose (especially in conjunction with high speed continuous flow systems) has given rise to a sizeable literature including some reviews (Guilbault and Sadar 1977; Horvath *et al.* 1975; Schifreen *et al.* 1977; Ngo 1980). It has also spawned a minor industry which supplies complete automated set-ups for the enzymatic determination of specific constituents combining enzyme immobilization with breakthroughs in electrochemical sensor techniques. These instruments have been described by Gray *et al.* (1977). At that time four commercial instruments were available for glucose analysis, using immobilized glucose oxidase. So keen is the competition from nonenzyme approaches that as this book goes to press, only two, Yellow Springs and Technicon, offer such glucose analyzers, and only the former is recommended by the designer, himself. Instruments (Fig. 38.4) designed especially for food analysis (Burns 1976) were also used for determination of lactose with lactase plus glucose oxidase and of sucrose with invertase plus mutarotase plus glucose oxidase. Yellow Springs Instruments can also determine galactose with galactose oxidase, cholesterol with its oxidase plus esterase; that of others could determine blood urea nitrogen with urease which produces ammonia which is sensed by the ammonia electrodes. In these and in similar "reagentless" analysis arrangements such as the laboratory setup shown in Fig. 38.5 for the determination of asparagine, and arginine and glutamine, the immobilized enzymes (asparaginase and arginase urease, or glutaminase) are used in series with an electrosensor (an ammonium ion selective electrode). In the future we may expect to see the exploitation of one particular advantage of immobilized enzymes, viz., their use for regenerating an expensive coenzyme, after the latter's level has been recorded as part of an enzyme indicator system, by inserting a second immobilized regenerating enzyme. In the case of NAD-dependent dehydrogenase this can be done by inserting into the automated system a second dehydrogenase which catalyzes the reduction by NADH (produced by the first enzyme) of an inexpensive substrate (Hornby *et al.* 1972; Guilbault 1976); or the NAD can be regenerated electrochemically (Coughlin and Alexander 1975).

Enzyme Electrodes.—As pointed out by Gray *et al.* most instrument manufacturers had, as of mid 1977, chosen to separate the enzyme and sensor functions to allow more flexibility in optimization of parameters affecting the performance of each. When they are closely coupled to each other the result is an enzyme electrode (Fig. 38.6). According to an officially recommended definition, an enzyme electrode is a detector in which an ion-sensitive electrode is covered with a coating that contains an enzyme which causes a reaction of a substance (substrate) to produce a species to which the elec-

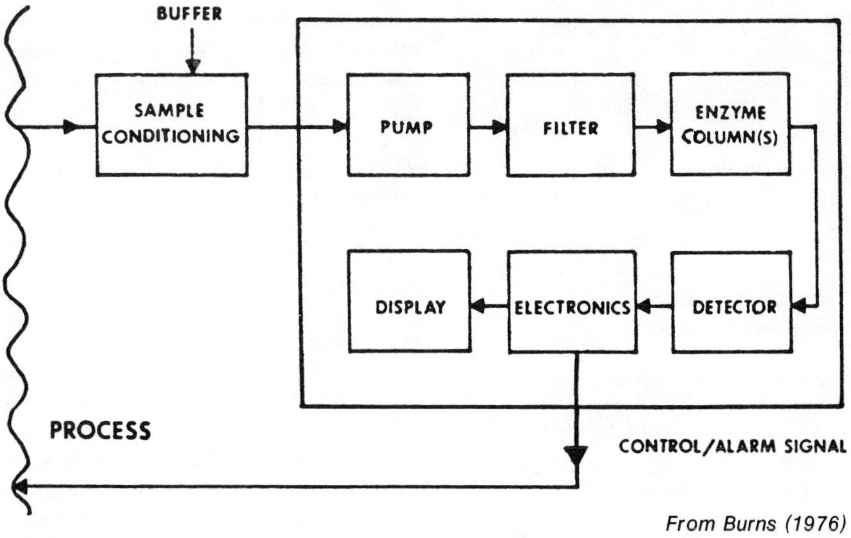

From Burns (1976)

FIG. 38.4. ELEMENTS OF AN ENZYME ANALYZER DESIGNED FOR USE IN FOODS

Shown are the elements of the Enzymax analyzer comprising liquid handling, enzyme reactor, amperometric detector and electronic display.

trode responds. The above-mentioned paper of Clark and Lyons on the first use of immobilized enzymes in analysis is also the first to suggest and describe the enzyme electrode, although the work of Updike and Hicks (1967) defined more clearly what we now recognize as an enzyme electrode. Reviews include, in addition to those on immobilization cited previously, those of Gough and Andrade (1973), Rechnitz (1975), Moody and Thomas (1975), Kessler *et al.* (1976) and Mosbach (1977).

These reviewers point out the usual advantages of enzymes and immobilization, especially selectivity, rapidity and sensitivity. While not at the present stage of development as sensitive as fluorescence measurements, determinations with enzyme electrodes can be carried out 10–100 times faster. On the other hand Guilbault and Tarp (1974) reported that the response time per sample of a urease electrode was 6 as compared to 3 min using an autoanalyzer. However, the cost per sample was slashed by over 90%.

Response time, a function of the geometry and design of the electrode, is more important in rate than in end-point determinations. Life time of the immobilized enzyme is also dependent upon the stability of the reference electrodes which, for instance, can leak mercury which would inactivate enzymes. Flowing systems prevent this sort of enzyme inactivation and prolong electrode life (Cordonnier *et al.* 1975).

The advantage of enzyme specificity is somewhat neutralized by lack of specificity of the electrode (Gray *et al.* 1977). There is also the limitation of range of substrates; since most nongas sensing elec-

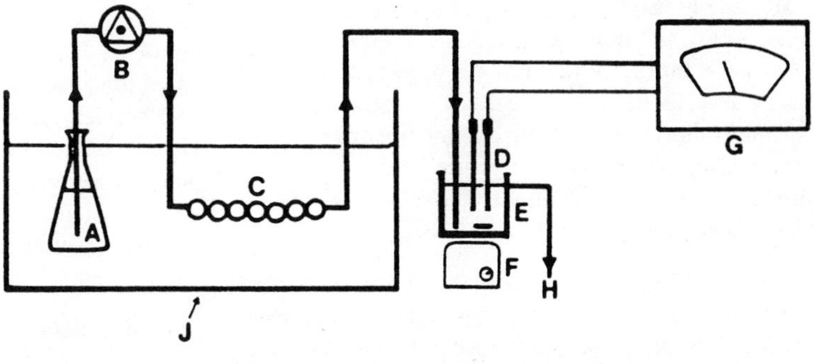

FIG. 38.5. REAGENTLESS ANALYSIS USING IMMOBILIZED ENZYME AND SENSOR IN SERIES

Used for the analysis of L-asparagine and L-arginine with immobilized asparaginase + urease and arginase. A, anylate solution; B, peristaltic pump; C, immobilized enzyme; D, pH or NH_4^+ electrode; E, thermostated vessel; F, magnetic stirrer; G, pH meter; H, drainage; J, water bath.

From Ngo (1976). Reproduced by permission of the Natl. Res. Counc. Can. from the Can. J. Biochem. 54, 63, Fig. 1.

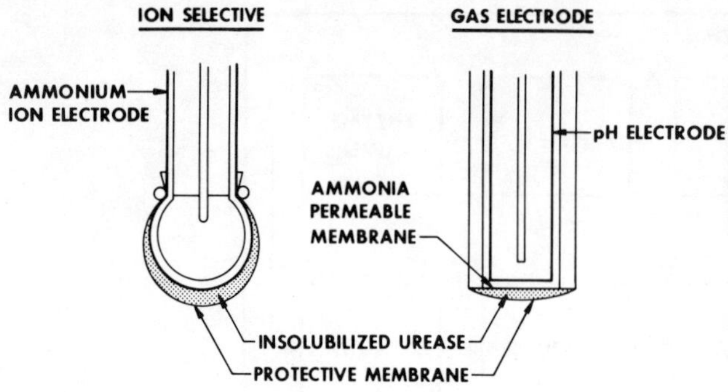

Courtesy of Gray et al. (1977). Reprinted with permission of Anal. Chem. 49 (12) 1068A, Fig. 2. Copyright by Am. Chem. Soc.

FIG. 38.6. REAGENTLESS ANALYSIS COMBINING IMMOBILIZED ENZYME AND SENSOR: THE ENZYME ELECTRODE
Shown are alternate arrangements for measuring urea.

trodes measure ions, it may be necessary to go through some rather convoluted transformations to end up with a measurable ion change. A rather special instance in which an ion was measured via a series of enzymatic transformations leading to a final NAD-dependent dehydrogenase action, but in which oxidation to NADH was not measured optically, is afforded by the application of phosphate electrodes to two systems developed by Guilbault and Cservalvi (1976) for the determination of inorganic phosphate. One of them starts with glycogen and utilizes the enzymes phosphorylase a, phosphoglucomutase and glucose 6-phosphate dehydrogenase:

$$\text{Glycogen} + \text{Pi} \rightarrow \text{G-1-P} \rightarrow \text{G-6-P} \; (+ \text{NADP}^+) \rightarrow \text{PGA} + \text{NADPH} + \text{H}^+$$

and the determination is effected by amperometric monitoring of NADPH.

Modes of enzyme immobilization encompass almost all of the now standard means discussed in Chapter 8 but emphasis has been placed on membrane associated systems: gel entrapment and confinement between and within membranes. In an interesting variation, magnetic enzyme membranes were fastened directly to CO_2 gas sensing electrodes (Cordonnier *et al.* 1975).

Selective Electrodes.—In general one may distinguish between ion selective and gas sensing electrodes. Ion selective electrodes are further categorized as: glass, which in addition to measuring the familiar H^+ can also be used for other cations; liquid membrane, in which the liquid is confined within an inert support of fritted glass or plastic, can be used (by varying the liquid composition) for the determination of Ca^{2+}, Cu^{2+}, Mg^{2+}, Mn^{2+}, Cl^-, NO_3^- and CO_3^{2-}; and crystalline membranes, which achieve selectivity by excluding ions from a crystalline lattice, are used for determining sulfide chloride, cyanide and cuprous ions.

Of these categories it would appear that the most promising for widespread adoption, especially for food analysis, are the enzyme electrodes utilizing gas sensors because among other reasons, as stated by Calvot *et al.* (1975), they are far superior to ion electrodes. Gas enzyme electrodes can measure any substrate which is transformed to ammonia, amines, SO_2, as well as O_2 and NH_3. Gray *et al.*, however, point to the severely limited utility of such electrodes, at least in 1977. Of course, a probe which one simply dips or inserts into the food to be analyzed would result in a drastic savings in time and space, giving rise not only to "reagentless" analysis but also to "sampleless" analysis. We believe that the pessimistic appraisal of Gray *et al.* that enzyme electrodes are of limited utility is a bit too gray and that at least some may be in at least limited use by the time this appears in print.

As a footnote to this discussion we wish to point out that similar instrumentation could be used for enzyme assay by immobilizing the anylate. More than 40 years ago the author semiquantitatively assayed for lipase (unpublished results) by dipping a pH meter-associated glass electrode coated with triglyceride into a lipase solution and noting the time required for the pH meter to recover functionality.

ANALYSIS OF FOOD CONSTITUENTS

This section consists of a roundup of what at the time of this writing is the current status of the use of enzymes as reagents for analysis of food constituents where those constituents are substrates of the added enzymes, excluding those investigations and applications used as illustrative examples previously in this chapter. At least two enzyme suppliers, Wallerstein and Boehringer-Mannheim, provide supplemental information concerning the use of their enzymes for food analysis.

Carbohydrates

Sugars.—Most of the above-mentioned illustrative material was taken from the field of sugar analysis and indeed, this is the first and continues to be the leading area of food analysis where enzymes are extensively used, some 100 years after Kjeldahl. Reviews include those of Trauberman (1975), Boehringer-Mannheim (1976) and Van Es (1977). Most of the current investigations on sucrose and glucose analysis are concerned with the application to particular groups of foods with the degree of quantitation and automation varying from application of available instruments mentioned above to sugar test sticks and kits for determination of glucose in beverages (Temperli *et al.* 1976).

In one case, that for the determination of sugars in sweetened cereals (Finley and Olson 1975), only invertase was used for sucrose conversion and a traditional chemical reduction colorimetric method was used to determine the resulting reducing sugars. Exemplificative of the ever-increasing competition of amazingly efficient nonenzymatic separation and detection techniques especially in mixtures is the conclusion of Zurcher and Hadorn (1976)—from a comparison of enzyme, GLC and chemical reduction methods for the determination of rather complex mixtures of sugars (glucose, fructose, sucrose, maltose, lactose and total trisaccharides) in cocoa-based cereal beverages and food concentrates—that the enzyme methods, while superior to chemical reduction methods and giving good agreement with GLC, possess less separation capability than does GLC. GLC gave the most accurate and fullest information over the whole spectrum of sugars. A sampling of other foods which have been analyzed for glucose, fructose and/or sucrose by automated enzyme methods include cereal products (Meuser 1972), wines (Schneyder and Plachy 1973), confectionery (Motz 1976) and granulated raw sugars (Sullivan *et al.* 1976).

Galactose and Derivatives.—Enzymes have been applied to analysis of lactose in the following foods: ice cream (Bitle 1977; Peeples and Hutcheson 1978), cheese (Frater *et al.* 1977) and milk-containing dried foods (Cheng and Christian 1977).

Since galactose oxidase, as previously mentioned, is not specific for the free monose only, galactose dehydrogenase is more frequently used for the determination of lactose via prior hydrolysis with β-galactosidase than is glucose dehydrogenase for glucose.

α-Galactosidase is used for analysis of raffinose in the sugar refining industry.

$$\text{Lactose} \xrightarrow{\beta\text{-galactosidase}} \text{Glucose} + \text{Galactose}$$

$$\text{Raffinose} \xrightarrow{\alpha\text{-galactosidase}} \text{Sucrose} + \text{Galactose}$$

$$\text{Galactose} + \text{NAD}^+ \longrightarrow \text{galactono-lactone} + \text{NADH} + \text{H}^+$$

Representative of the problems encountered in adapting an enzyme analysis to a particular food ingredient is the estimation of "absolute" raffinose in sugar refinery products. Galactose dehydrogenase cannot be used without further correctional information because these products are clarified with lead acetate (Hollaus *et al.* 1977). Of course the presence of two distinct glycosidic linkages in raffinose, galactosyl-1,6-glucose and glucosyl-1, 2-fructose, allows for flexibility in choice of enzymes both for hydrolysis and for detection of the resulting monose. Although free galactose in food is frequently estimated by GLC, the latter does not, as does galactose dehydrogenase, distinguish between D-galactose, the product of lactase action, and L-galactose, present ubiquitously in plants, including some foods (Roberts and Harrer 1973).

Fructose.—Among the many enzymatic alternatives to fructose analysis are those which permit the simultaneous determination of glucose:

$$\text{Fructose (or glucose)} + \text{ATP} \xrightarrow{\text{hexokinase}} \text{F-6-P (or G-6-P)}$$

$$\text{F-6-P} \xrightarrow{\text{phosphohexose isomerase}} \text{G-6-P}$$

$$\text{G-6-P} + \text{NADP}^+ \xrightarrow{\text{G-6-P-dehydrogenase}} \text{Phosphogluconate} + \text{NADPH} + \text{H}^+$$

Formation of NADPH is measured at 340 nm. An interesting one-enzyme alternative is to run the mannitol dehydrogenase reaction in its NADPH-oxidative or reductase mode (Yamanaka 1975):

$$\text{Fructose} + \text{NADH} + \text{H}^+ \longrightarrow \text{Mannitol} + \text{NAD}^+$$

Gluconate.—One of the products of the glucose oxidase action used for glucose analysis, gluconic acid, is an important food ingredient which itself may be determined enzymatically via gluconate kinase (I) and gluconate-6-phosphate dehydrogenase:

$$\text{D-Gluconate} \xrightarrow{+\text{ATP, I}} \text{gluconate-6-phosphate}$$

$$\xrightarrow{+\text{NADP}^+\text{, II}} \text{D-ribulose-5-phosphate} + \text{NADPH} + CO_2 + H^+$$

This method, used by Mergenthaler and Scherz (1978) for detection of illegal treatment of beer with glucose oxidase to remove oxygen, agreed moderately well with a newly developed nonenzymatic alternative GLC approach, again illustrating the nonstatic competition between "enzymatic" and "chemical" techniques in food analysis.

Starch.—The key enzyme is glucoamylase which should supplant, by the time this appears, α- and β-amylases designated for enzymatic starch analysis by the Association of Official Analytic Chemists (1975). The various published methods of which those of Boehringer-Mannheim and of Thiend *et al.* (1972) are representative, differ principally in means of estimating glucose (usually but not always enzymatically) and of solubilizing the starch in the sample in order to make it amenable to amyloglucosidase action, i.e., the use of dimethyl sulfoxide for this purpose in the enzymatic determination of starch in cereal foods (Libby 1970).

Apparently nonenzymatic hydrolysis is still the method of choice in analyzing for starch added to nonstarch foods such as meat (Lind 1977). The resulting glucose is determined enzymatically with glucose oxidase/peroxidase. This preference for acid hydrolysis probably stems from the abovementioned problem of preparing the starch as an accessible substrate for amyloglucosidase.

Other Glycans.—Solubilizing cellulose for its determination with cellulase is apparently so formidable (as discussed in Chapter 9) that solubilization has been abandoned in favor of a long incubation period—as long as 19 days, probably the record time-consuming enzymatic method for food analysis (Halliwell 1974). We suggest that some of the lessons learned from attempting to manufacture glucose from cellulose examined in Chapter 9 may be applied here, especially utilization of the "windshield wiper" effect achieved with cellobiose hydrolase and oxidoreductase. After this lengthy hydrolysis, cellulose may be estimated by decrease in insolubles, increase in soluble dry matter or by determination of glucose. The "enzyme" is the cellulase complex in the culture filtrates of *Trichoderma* fungi (Chapter 9).

Similarly, Halliwell used as the "enzyme" for the determination of hemicelluloses a culture filtrate from sheep rumen. An enzyme method has been proposed for the determination of dextran (Richards and Stokie 1974) whose formation is a problem in the sugar industry (Chapter 9).

Bacterial 1,3:1,4-β-glucan hydrolase (Chapter 34) was deployed by Anderson *et al.* (1978) for the enzymatic determination of such glucans in barley.

Triglycerides

Medical consensus that the serum triglyceride level is a factor in the etiology of heart diseases has accelerated the development of automated methods for its analysis and among these methods are those which rely on enzymes as reagents. In the most widely used enzymatic method the triglycerides are first hydrolyzed to FFA and glycerol and the latter is determined by the following sequence of reactions:

$$\text{Glycerol} + \text{ATP} \xrightarrow{\text{kinase}} \text{glycerol-1-phosphate} + \text{ADP}$$

$$\text{ADP} + \text{Phosphoenolpyruvate (PEP)} \xrightarrow{\text{kinase}} \text{ATP} + \text{Pyruvate}$$

$$\text{Pyruvate} + \text{NADH} + H^+ \xrightarrow{\text{LDH}} \text{lactate} + \text{NAD}^+$$

In actual performance of the analysis (Wahlefeld *et al.* 1973; Ziegenhorn 1975) all the ingredients of the reaction mixture are preincubated and then the glycerol kinase is added. The decrease of NADPH is measured by the "optical" test at 340 nm. This is a general method for analysis of any substance which can be phosphorylated by a kinase (Klose *et al.* 1975) and its salient feature applicable to other anylates is the indirect measurement of the ADP released in the first kinase reaction *via* PEP, PEP kinase (acting in the reverse direction) and LDH. The method has been recommended for application to food in the just cited references, almost all of which are patents. Lipases used include those from molds *(Rhizopus arrhizus)* and yeast (*Candida* spp.), pancreatin and, to include the lower triglycerides, less specific carboxylic esterases, particularly of advantage in food analysis.

Here again the rapid separation and detection of lipids by GLC, TLC and HPLC may restrict the use of enzymes in lipid analysis in foods to the use of lipases and phospholipases. Lipases continue to be useful for such tasks as ascertaining the amount and position of the fatty acid components in the triglyceride (Chapter 25) in connection with selection for processing and identity of vegetable oils (Mani and Wessels 1976) and animal food fats (Bracco and Winter 1976).

Alcohols, Aldehydes and Wine

The most analyzed of the alcohols is ethanol, and the most widely used enzyme is alcohol dehydrogenase. Its enzymatic analysis in wines yielded values closer than did a nonenzymatic method but still below theoretical calculated values (Holbach and Woller 1977). We referred to its use for ethanol analysis in wine in conjunction with kits. Alcohol dehydrogenase in series with aldehyde dehydrogenase has also been proposed and investigated for ethanol determination in foods of low alcohol contents such as fruit juices and preserves (Beutler and Michal 1977) and in kit form in nonalcoholic beverages (Buergin 1976).

Another enzyme proposed for ethanol determination which may have some limited use in "quickie" food analysis is the flavoprotein alcohol oxidase (Jansen and Ruelius 1968) which oxidizes not only ethanol but methanol and other alcohols to the corresponding aldehydes and H_2O_2 (Nanjo and Guilbault 1975; Axcell and Donninger 1977). The former devised an end-point enzyme electrode so that the only reagent required is buffer, whereas the latter used a kinetic approach; but both methods, which sense changes in oxygen, lack the specificity of the traditional ethanol dehydrogenase. The electrode might be useful in monitoring methanol in pectin enzyme treated fruit mashes (Chapter 30 and 35) and in PE assay.

Another important alcohol in wines, 2,3-butanediol and its immediate oxidation product acetoin (Chapter 25) can both be determined with a specific 2,3-butanediol dehydrogenase from the fungus of the *Sarcina* genus, the latter in conjunction with the use of the well-known redox dye, dichlorophenol indophenol (Muraki and Masuda 1976).

$$\underset{\text{Butane-2,3-diol}}{CH_3CHOH{\cdot}CHOHCH_3} + NADP^+ \rightleftharpoons \underset{\text{Acetoin}}{CH_3CHO{\cdot}CHOHCH_3} + NADPH + H^+$$

$$\text{Acetoin} + \text{Dye(ox)} \rightleftharpoons \text{Butane-2,3-diol} + \text{Dye (red)}$$

Acetoin presents less of a problem and the optical test can be applied directly.

Enzymatic analysis of sorbitol as well as xylitol in wines and dietetic foods (Chapter 9) requires a bit more manipulation in sample preparation before the sample is ready for subjection to enzyme action. The requisite dehydrogenase is not the indicator enzyme for sorbitol determination. Instead the fructose formed via sorbitol dehydrogenase

$$\text{D-Sorbitol} + NAD^+ \rightarrow \text{Fructose} + NADH + H^+$$

is linked to an optical test through the action of hexokinase, phosphoglucoisomerase, and G-6-P dehydrogenase (the hexokinase method). To drive the reaction to the right, NADH is removed with either alcohol (Boehringer-Mannheim 1976) or lactate (Beutler and Becker 1977) dehydrogenase (Chapter 9). To determine xylitol along with sorbitol in dietetic foods, the latter investigators used the same sorbitol dehydrogenase (SDH):

$$\text{Xylitol} + NAD^+ \xrightarrow{\text{SDH}} \text{Xylulose} + NADH + H^+$$

by measuring total NADH formed via the formazan reaction (Chapter 37).

Acids

Except for five-carbon compounds, methods have been developed which are specifically designed for enzyme analysis of the carboxylic acids present in foods all the way from C_1 (formic) to C_6 (citric and isocitric) acids. Without exception an NAD(P)-dependent dehydrogenase is the indicator enzyme but is not always also the triggering enzyme.

C_1.—Determination in molasses of formic acid, an important indicator of quality in this food product, with formate dehydrogenase (FDH) constitutes an example in which the enzyme is both trigger and indicator enzyme:

$$HCOOH + NAD^+ + H_2O \xrightarrow{FDH} H_2CO_3 + NADH + H^+$$

CO_2, the anhydride of the one-carbon carbonic acid, H_2CO_3, has also been determined enzymatically (Reinefeld and Bliesener 1977) with aid of PEP carboxylase (Chapter 20) but not yet in foods:

$$CO_2 + PEP \rightarrow Oxalacelate$$

The oxalacetate was determined colorimetrically in an automated assay.

C_2.—Acetic acid is not only the main component of vinegar but also contributes to and serves as an indicator of the quality of wines and sourdough breads. The principle of the particular enzymatic analysis of acetate used for these foods is the same as that used for analysis of glycerol; the ADP formed from the triggering kinase is eventually coupled to the UV optical test:

$$Acetate + ATP \xrightarrow{kinase} acetyl\ phosphate + ADP$$

McCloskey (1976) worked out optimal conditions with regard to its application to wines and fruit juices. Even without an automated setup, some 300 analyses per day per operator could be run and with greater accuracy, as compared to five per day using a standard procedure involving time-consuming and error-prone steam distillation.

Aficionados of sourdough french bread may be appreciative of the finding that enzymatic estimation of acetic (as well as D- and L-lactic) acids agreed with a less laborious but still time consuming paper chromatographic method used heretofore in Germany (Wutzel 1976). The lactic acid: acetic acid ratio in sourdough is used as a criterion of bread aroma.

Acetic acid may also be determined using only the kinase and estimating the other product, acetyl phosphate, by color developed after reaction with hydroxylamine and $FeCl_3$. Alternatively the Boehringer-Mannheim kit for acetic acid apparently utilizes completely different enzymes; citrate synthase (CS) in its reverse mode and malate dehydrogenase (MDH), and probably also acetate kinase (AK) and phosphotransacetylase (PT) in a rather elaborate preceding indicator reaction scheme (Bergmeyer 1974):

$$NAD^+ + malate \xrightarrow{MDH} NADH + H^+ + OAA$$

$$Acetate \xrightarrow[AK]{+ATP} acetyl\ phosphate \xrightarrow[PT]{+CoA} acetyl\ CoA \xrightarrow[CS]{+OAA, H_2O} citrate + CoA$$

What is actually measured is the appearance of NADH using the UV optical test, the MDH reaction being driven via the entrainment of oxalacetate by the citrate synthase action.

For the determination of acetic acid using single reagents, AK and PT appear to be replaced by the acetyl CoA synthetase, a class 6 enzyme:

$$Acetate + ATP + CoA \longrightarrow acetyl\ CoA + AMP + pyrophosphate$$

So far, oxalic acid, a flavor-contributing and calcium-sequestering constituent of some vegetables, has not to our knowledge been estimated enzymatically in such foods. Its level in urine has been estimated via the optical test with oxalate decarboxylase (Costello and Hatch 1976):

$$HOOC\text{-}COOH \xrightarrow[decarboxylase]{oxalate} HCOOH + CO_2$$

The formic acid is determined with formate dehydrogenase (see "C_1").

C_3.—Lactic acid is determined in fermented foods such as sourdough and beer (Drawert and Hagen 1970), wines (Postel *et al.* 1973) and cultured milk products (Blumenthal and Helbing 1971) as both L- and D-lactate because many microorganisms metabolize glucose through pathways other than glycolysis, resulting in D- instead of L-lactate. Specific NAD-dependent dehydrogenases are available for the analysis of each. Such information obtained *via* automated enzyme analyses is especially helpful in assessing the hygienic status of fresh (Suhren *et al.* 1977) and dried (Kohler 1974) milks. Enzymatic methods for analysis of lactate and other dairy product constituents was reviewed by Olling (1977).

C_4.—The determination of malic acid in apples with malate dehydrogenase (Gorin 1976) illustrates how an enzyme analysis may not always yield the precise information being sought, in this case "sourness," because not only malic acid but its salts are included in the analysis. In this case a titrimetric method was more reflective of the actual malic acid content.

Michal *et al.* (1976) applied the same principle used for glycerol and acetate, ADP coupling to UV, to determine succinic acid in foodstuffs with the same auxiliary and indicator enzyme but with a different triggering enzyme type. Instead of using a kinase (EC 4) the phosphorylation was catalyzed by a synthetase (EC 6) using GTP or ITP instead of ATP:

$$\text{Succinate} + \text{CoA} + \text{ITP} \xrightarrow[\text{synthetase}]{\text{succinyl CoA}} \text{succinyl CoA} + \text{GDP} + \text{Pi}$$

C_6.—The most important use of enzymatic analyses of C_6 carboxylic acid in foods is to detect adulteration of natural citrus juice with citric acid. Natural juices but not commercial citric acid preparations contain *iso*citric acid (*n*-citric: *iso*citric = 1:250, roughly) so that the addition of exogenous acid alters the ratio of the two isomeric acids. *Iso*citrate is determined via the NAD-dependent *iso*citrate dehydrogenase (Bergner-Lang 1977). Citrate in wines (Mayer and Pause 1969) and cheese (Olling 1977) was determined by first converting citrate to acetate and oxalacetate with citrate lyase:

$$\text{Citrate} \longrightarrow \text{oxalacetate} + \text{acetate}$$

Since oxalacetate is fairly unstable, spontaneously but slowly decarboxylating to pyruvate, both malate and lactate dehydrogenases are used:

$$\text{Oxalacetate} + \text{NADH} + \text{H}^+ \rightleftharpoons \text{malate} + \text{NAD}^+$$

$$\text{Pyruvate} + \text{NADH} + \text{H}^+ \rightleftharpoons \text{lactate} + \text{NAD}^+$$

It will be noted that almost all the enzyme methods for food acids have been developed in Europe, especially West Germany.

Flavor-related Substances

Flavor Potentiators.—Glutamate dehydrogenase (NAD-linked) has been incorporated in the development of automated analysis of glutamic acid, usually present as its monosodium salt, in soy sauce (Kikuchi *et al.* 1973), canned soups, sauces and seasonings by Armand *et al.* (1976) who also were involved in the use of a centrifugal analyzer for this purpose, as previously mentioned, and meat products by Inklaar (1977). (Parenthetically, another food additive in meat products, pyrophosphate, was also determined enzymatically via pyrophosphatase.) An alternative to glutamate dehydrogenase is glutamate decarboxylase which was incorporated into a glutamate electrode (Ahn *et al.* 1975). The other major flavor potentiators, the 5′-nucleotides, have been estimated by determining Pi release by the action of bull semen nucleotidase (Shimazono 1964). Boehringer-Mannheim describes the determination of GMP in soup via the kinase-ADP-PEP route mentioned first in connection with glycerol analysis.

A kit or "test combination" for glutamate, available from Boehringer-Mannheim, contains diaphorase (lipoamide oxidoreductase). Diaphorase, it will be recalled, is used for shifting the measurement of dehydrogenase reactions into the visible range.

The flavor potential of mustard flour has usually been laboriously determined by measuring enzymatic release of glucosinate-derived isothiocyanate (Chapters 31 and 35). Even the relatively simplified modification of Rosebrook and Burney (1968) of the original AOAC method still involves time-consuming distillation and back titration. More recently *total* glucosinolates in crucifers have been determined by measuring glucose arising from the action of glucosinolase (thioglucosidase) applied to the material retained on an ion exchange resin (Van Etten and Daxenbichler 1977). Individual glucosinolates were estimated by determining the isothiocyanates and other aglucones formed through thioglucosidase action with GLC. As mentioned earlier in this chapter the UV-absorbing characteristics of these aglucones (Schwimmer 1961B) might be profitably utilized for glucosinate analysis (Fig. 38.1).

Amygdalin, the source of almond flavor (Chapter 22) as well as a potential health hazard (Chapter 35) was determined with the aid of an enzyme electrode comprising β-glucosidase (+nitrilase?) coupled to a heterogeneous solid-state membrane cyanide electrode (Mascini and Liberti 1974) or more officially by distillation and titration of the HCN (Anon. 1976B). Enzymatically liberated HCN, measured colorimetrically after conversion to polymethin dye, was used as a measure of amygdalin in cherry products to which crushed pits are frequently added to accentuate the flavor (Daneschwar 1976).

Flavor producing amino acids and their α-glutamyl peptides may be determined by the total pyruvate released (pyruvate itself may be as we have seen enzymatically determined) by the combined action of endogenous alliinase and added γ-glutamyl transferase, as shown in Fig. 21.6.

Health, Nutrition and Safety

Amino Acids.—Both L- and D-amino acid oxidases as well as L-amino acid decarboxylase are available for the determination of amino acids in foods. Guilbault and Lubrano (1974) devised an amino acid enzyme electrode incorporating L-amino acid oxidase. Analyses with the flavoprotein oxidase, which unlike the decarboxylases is not specific for each amino acid, usually involve measurement of the H_2O_2 produced with peroxidase (Guilbault and Hieserman 1968).

The more specific amino acid decarboxylases should be particularly valuable for ascertaining the level and effect of processing of the nutritionally essential amino acids, especially L-lysine:

$$\text{L-Lysine} \xrightarrow{\text{L-Lysine decarboxylase}} \text{cadaverine} + CO_2$$

The reaction may be followed by measuring the diamine cadaverine with the aid of a diamine oxidase (Chapters 30 and 35) or by following CO_2 evolution (Roy 1979). Arginine may be measured optically with the aid of octopine dehydrogenase (Gaede and Grieshaber 1975):

$$\text{Pyruvate} + \text{L-arginine} + \text{NADH} + H^+ \rightleftharpoons \text{octopine} + \text{NAD}^+$$

As one might expect from its name, the enzyme first found in octopi was obtained from either of two delectable seafoods, scallops and clam, available in local fish markets. Octopine is α-*N*-D-propionyl-L-arginine.

Amino acids have also been determined in foods using the transaminases as trigger enzyme, i.e., aspartic acid in sugar beet (Burba and Kastning 1971): L-Aspartate + α-ketoglutarate $\rightarrow$ OAA + L-glutamate; L-glutamate + NADH + H^+ $\overset{\text{MDH}}{\rightleftharpoons}$ L-malate + NAD^+. A specific ammonia lyase enzyme electrode readily analyzes for histidine (Walters *et al.* 1980).

Vitamins

Vitamin C.—Ascorbate oxidase (Chapter 16) was used successfully, both in solution and immobilized, in conjunction with measurement of O_2 uptake for the determination of both ascorbic and dehydroascorbic acid in fresh canned spinach (Marchesini *et al.* 1974; Mattavelli *et al.* 1975; List and Knechtel 1980). Schwimmer (1961B) proposed that the highly specific activation of glucosinolase by ascorbate (Chapter 21) could be the basis for the estimation of ascorbic acid.

Essential Fatty Acids.—Defined for nutritional labeling purposes as the total *cis-cis* methylene-interrupted polyunsaturated fatty acids and thus essentially lipoxygenase substrates, these constitute one of the few classes of food components whose official analysis in various countries by GLC has received serious competition from or has been superseded by an enzymatic procedure, at least as an interim method (Waltking 1972; Prosser *et al.* 1977). The method, based on one adopted by the Canadian Food and Drug Directorate over 10 years ago, involves lipoxygenase-catalyzed disappearance of these fatty acids as measured by decrease in UV absorbance.

Cholesterol.—In addition to being a health-related food constituent, its detection in certain foods such as noodles and pasta is *prima facie* evidence of the presence of eggs in these products. The trigger enzyme is flavoprotein cholesterol oxidase; where it has so far been applied to such foods (Beutler and Michal 1976) and to milk fats (Grossmann *et al.* 1976), the H_2O_2 produced by this enzyme is measured via the following reactions:

$$H_2O_2 + \underset{\text{methanol}}{CH_3OH} \xrightarrow{\text{catalase}} \underset{\text{formaldehyde}}{HCHO} + H_2O$$

$$HCHO + \underset{\text{acetyl acetone}}{CH_3COCH_2COCH_3} + NH_3 \rightarrow \text{Lutidine dye (measured)}$$

Grossmann *et al.* concluded that the method is superior to all other nonenzyme methods where numerous analyses are required and is more reproducible than is GLC, but that the latter is best where other sterols accompany cholesterol. This as a general principle is probably true of enzyme *vs* nonenzyme methods.

Nitrates and Other Nitrogenous Food Constituents

The health-associated hazard attributed to nitrate and nitrite in foods has spurred development of fast analytical procedures, including those which utilize the NAD(P)-dependent dehydrogenases acting on nitrate and nitrite, usually in the reverse (reductase) direction (Kiang *et al.* 1975).

Enzymatic methods for determination of the biogenic amines whose presence in foods is usually associated with microbial enzyme action (Chapter 35) have been proposed utilizing the flavoprotein, H_2O_2-producing monoamine oxidase, for monoamine and a phosphopyridoxal enzyme, diamine oxidase, for diamine analyses (Guilbault *et al.* 1969). The latter enzyme was used as part of an enzyme electrode for the detection of incipient putrefaction in foods (Toul and Macholan 1975), and parenthetically figures prominently in establishing rape in forensic science (Suzuki *et al.* 1980).

Creatine and Creatinine.—These important meat and fish constituents (Chapters 26 and 28) have been estimated enzymatically for clinical analysis via an enzyme electrode containing an NH_3 sensor and a commercially available microbial creatine deiminase which yields N-methyl hydantoin and NH_3 (Thompson and Rechnitz 1973). Alternatively, Ettel and Tuor (1977) developed a method for "total creatinine" (creatine plus creatinine) in meat and meat-containing foods (soups, stocks, gravies, etc.) based on the same ADP/PEP/NADH principle cited previously for the determination of glycerol, acetate and succinate starting with the conversion of creatine to creatine phosphate by creatine kinase. Commercial creatininase, a hydrolase, is available for formation of creatine.

Seeds of *Lathyrus satilus* used illegally in India to adulterate widely consumed legume seeds, pulses and grams may be detected by virtue of their content of derivatives of L-α,β-diaminopropionic acid, neurotoxicants responsible for their proscription. After acid hydrolysis the liberated diamino acid may be estimated by the ammonia evolved upon addition of the highly specific diaminopropionate-ammonia lyase, a not yet officially recognized enzyme obtained from a pseudomonad (Rao *et al.* 1974).

Combining rate and end-point procedures, Yamada *et al.* (1979) developed an analysis for the ubiquitously distributed polyamines spermine and spermidine, using beef plasma amine oxidase as a specific reagent.

Mycotoxins.—These may be substrates of enzymes used in their analyses. Ochratoxins, peptide fungal secondary metabolites which have been detected in cereal and legumes, have been analyzed and distinguished on the basis of rate of change of their fluorescence excitation spectra at 450 nm upon cleavage with carboxypeptidase A (Hult *et al.* 1977). As little as 4 μg/kg barley could be estimated quantitatively. A somewhat more elaborate (in principle but simpler in practice) reaction sequence involving two enzymes, glutathione S-epoxytransferase (GSET) and glutathione (GSH) reductase (GSHR), were employed by Foster *et al.* (1974) for the estimation of 12,13-epoxytrichlorethecenes, a group of closely related toxic sesquiterpenoids (Chapter 22) secreted by some imperfect fungi into animal feeds. Excess GSH is used in the analysis:

$$\text{Epoxide} + \text{GSH} \xrightarrow{\text{GSET}} \text{GS-conjugate}$$

Excess GSH was nonenzymatically (n-e) estimated with the aid of Ellman's disulfide reagent (R_1SSR_1)

$$\text{GSH} + R_1SSR_1 \xrightarrow{\text{n-e}} \text{GSSR}_1 + R_1\text{SH (colored)}$$

$$\text{GSSR}_1 + \text{GSH} \xrightarrow{\text{n-e}} \text{GSSG} + R_1\text{SH (colored)}$$

$$\text{GSSG} + \text{NADPH} + H^+ \xrightarrow{\text{GSHR}} 2\text{GSH} + \text{NADP}^+$$

Thus, by regenerating GSH, both halves of Ellman's reagent are utilized for color development.

Daniel (1967) explored the use of liver microsomal reductase for the estimation of food coloring dyes.

Undoubtedly, many other proposals for the use of enzymes as analytical reagents in food control and research laboratories will have been proffered by the time this appears. Some of these will be adopted by at least segments of the food industry, others will fall by the wayside and yet others will enjoy a meteoric rise only to be superseded by a competing nonenzyme method.

The following may or may not be an indication as to the extent that enzymes will fit into food analysis in the future but the perhaps nonsignificant statistics are that of the 1000-odd papers cited in the reviews of analysis of food constituents in 1975 by Sloman *et al.*, 7% mentioned use of enzymes for one reason or another. In reviews of the same area 2 and 4 years later (Foltz *et al.* 1977; Yeranasian *et al.* 1979) only 2 and 4% of the 1000-odd papers in each review mention use of enzymes. In each review about half of these papers in which enzymes are mentioned deal with addition of purified enzymes as *reagents* for the quantitative estimation of specific anylates where the latter are *substrates* for the added enzyme, the subject matter of this chapter.

Enzyme analysis in general will eventually find a permanent, if small, niche as a useful weapon in the vast armamentarium of analytical procedure available to the food analyst and quality control technologist, as reagents in procedures more for repetitive analysis of a large number of samples of one or a limited number of anylates rather than a variety of components; and where the anylate is substrate. The same is probably true of anylates as enzyme affectors. The latter and other aspects of enzyme testing which hold more promise are explored in the ensuing and final chapter.

39

FURTHER ASPECTS OF THE UTILITY OF ENZYMES IN FOOD ANALYSIS

We wind up our exploration of the role of adding enzymes to foods or extracts thereof in food analysis by looking into: (1) their inhibition or activation by food constituents as a means of analyzing for these constituents; (2) how observation and measurement of this action of these added enzymes under special conditions can be used as "short cuts" for otherwise tedious, time-consuming, complicated assessments of food quality, usually biologically and nutritionally oriented; (3) their uses as analytical aids in cleanup procedures, preliminary treatments in preparation for analysis; and (4) as marker or label in enzyme immunoassay as an alternative to the Nobel Prize-winning related technique of radioimmunoassay.

ANYLATE AS ENZYME AFFECTOR

In principle, one can calculate the concentration of a food constituent if it inhibits (I) and if it activates (A) a known added enzyme at a given substrate concentration (S) from the well-known equations for enzyme inhibition and activation. For noncompetitive inhibition, one simply divides the Michaelis equation for no inhibition by that for noncompetitive inhibition and solves for I, in this case the food constituent to be analyzed (the anylate).

$$I = (v_o/v_i - 1)\, K_i$$

where v_o/v_i is the reciprocal of fractional activity remaining after adding the unknown. Here S does not have to be accurately added since it does not enter into this equation. This is not true for competitive inhibition

$$I = (v_o/v_i - 1) \cdot K_i \cdot (K_m + S)/K_m$$

For an obligatory activator, the rate or activity in its absence is zero so that

$$A = K_a\, (V_a/v_a - 1)^{-1} \cdot S$$

where V_a is the maximum rate reached with increasing amounts of activation at a constant level of S; K_a, the apparent $S_{0.5}$; and v_a, the observed rate.

Alternatively, one can read off the concentration of the anylate from a graph of the following straight line equations (Dixon plots):

Competitive:

$$\frac{1}{v_i} = \frac{K_m}{V \cdot S \cdot K_i} \cdot (I) + \left(\frac{K_m}{SV} + \frac{1}{V}\right)$$

Noncompetitive:

$$\frac{1}{v_i} = \frac{1}{V \cdot K_i}\left(1 + \frac{K_m}{S}\right) \cdot (I) + \left(\frac{K_m}{S \cdot V} + \frac{1}{V}\right)$$

Obligatory activation:

$$\frac{1}{v_a} = \frac{K_a(S + K_m)}{keS}\left(\frac{1}{A}\right) + \frac{S + K_m}{VS}$$

In all these equations, only A and v_i or v_a are the variables and the pertinent kinetic constants presumably can be obtained from the literature. For instance, a Dixon plot for the inhibition of carbonic anhydrase shown in Fig. 2.8 could be used to determine sulfide levels in foods.

These equations are unfortunately not always valid so that it is frequently necessary to resort to an empirical nonlinear calibration curve. In addition to sections and chapters in earlier cited books, Townshend (1973) provided a brief review of trace analysis by enzyme activation and inhibition.

Perhaps more than anylate-as-substrate, anylate-as-enzyme affector for trace analysis of food constituents appears to meet stiff competition from nonenzyme methods and there is a tendency to abandon them as other analytical instrumentation becomes more sensitive and inexpensive. Part of the reason for this hasty abandonment is that anylates-as-affectors compete more than do anylates-as-substrates with these methods and also that such enzyme analyses require constant recalibration. However, as we shall see, such methods will always find a niche in laboratories which cannot afford expensive instrumentation and for special cases such as for screening food samples.

Anylate as Enzyme Inhibitor

Pesticide Analysis.—Historically the most widespread application of anylate as inhibitor has been for the determination of pesticides by their ability to inhibit serine esterases, especially cholinesterase, not so much by laboratories in the food processing industry but more by those of food and environment regulatory agencies. They were the major means of pesticide analysis in the 1950s before the advent of GLC.

It should be pointed out that at least 90% of the papers devoted to pesticide analyses in the late 1970s have ignored enzymes. It is to this residue of enzyme-oriented studies dealing with pesticide residues in foods that we shall turn our attention. The principal classes of insecticide which have been estimated enzymatically are the organophosphates, organic chlorinated compounds, especially the chlorinated hydrocarbons, and the carbamates. Other residues from sprays applied to plants for their herbicidal activity have also been estimated by virtue of their abilities to inhibit enzymes.

The Organophosphorus Compounds.—The pioneering work of Balls and Jansen (1952) on the use of diisopropylfluorophosphate as a probe of an enzyme active site (Chapter 3) paved the way for the use of enzymes for the analysis of related organophosphorus compounds. Michel (1949) used such compounds to fully inhibit cholinesterase in order to apply a correction for nonenzymatic hydrolysis of acetylcholine in the assay of cholinesterase. Giang and Hall (1951) probably published the first thoroughly quantitative enzyme inhibition-based method for enzymatic analysis in food of commercial pesticide residues such as the parathions, alkyl esters of thiopyrophosphoric acid, and their oxygen analogs the paraoxons. These could be detected in the 1–100 μg range. However, the parathions are relatively weak esterase inhibitors and of course the method could not distinguish among the various insecticides if more than one was being used.

To correct for these deficiencies the following approaches were successfully adopted:

(1) Conversion of weak to strong inhibitors by chemical conversion, i.e., bromination to more potent inhibitors (Falscheer and Cooke 1956).
(2) Search for cholinesterases such as those extracted from the brain of the fly, boll weevil and especially the bee which were more sensitive to these compounds than was the human enzyme first used. The variable sensitivity of different insects to these compounds was used as a guide to the selection of their esterases.
(3) Use of noncholinesterase enzymes such as lipase and phosphatase.
(4) Separation of the pesticides by means of TLC followed by estimation of their esterase-inhibition capacity and hence their concentration while still on the chromatogram with appropriate spraying techniques (Miller *et al.* 1971; Mendoza 1973) including visualization of the products of the reaction, usually by diazotation or by the use of chromogenic substrates such as indoxyl acetate. As little as 50 picograms can be detected by this technique.

As an example of how this approach might be used in practice, at least in those laboratories without ready access to a good GLC instrument, Renvall (1975) used cholinesterase inhibition tests and TLC as well as GLC in the examination of 25 organophosphorus pesticides in 1600 samples of apples, cherries, citrus fruits, grapes, peaches, pears, plums, cabbage, carrots, cauliflower, cucumbers, dill, lettuce, potato, sweet pepper and tomato. It is of interest that virtually only imported produce exceeded the tolerance limit set by the Swedish government, and in most samples insecticide could not be detected.

Although cholinesterase inhibition continues to have a limited use in the screening of pesticide, for instance in milk samples (Konrad and Gabrio 1976), interest in its use has shifted from food to environmental protection. For instance, Goodson *et al.* (1973) devised an electrochemical system for continuously monitoring air and water supplies. In this setup the inhibition of immobilized cholinesterase is heralded by the ringing of a bell. Similarly, Baum and Ward (1971) employed ion selective enzyme procedures for organophosphate pesticide analysis.

Carbamates.—As of 1977, the analysis of carbamate pesticides via GLC had not advanced as far as analysis of the organophosphates so that enzymatic methods for carbamate analysis still enjoyed a fair degree of popularity judging from the literature and the very readable and informative review of Dorough and Thorstensen (1975). Carbamates, like the organophosphates, are inhibited by cholinesterase and much of the literature chronicled in the above review was directed to finding enzymes even more sensitive than from bee brain; horse liver, housefly head and even mammalian intestinal lipase have been looked at. Substrates such as indophenyl and 5-bromoindoxyl acetates lend themselves well to detection by product fluorescence in conjunction with TLC, in which the presence of inhibitor is signaled by the appearance of white on intensely colored backgrounds.

Chlorinated and Other Spray Residues.—Although GLC methods from the outset have been used for the analysis of these compounds there has been a small amount of attention devoted to their enzymatic analysis mainly because the criticism of the inaccuracies and repeatability of these methods (Miller *et al.* 1971) has led to sporadic investigation of the possibility of using enzyme methods of analysis via inhibition. More than almost 30 years ago Keller (1952) could detect as little as 2 μg of DDT in a 50 mg sample of wheat flour by measuring its capacity for inhibiting carbonic anhydrase.

It has been suggested that this particular inhibition is the reason that this chlorinated hydrocarbon prevents the deposition of calcium carbonate during the formation of bird eggshells, and thus the ultimate cause of the decline of bird populations in areas where DDT was used heavily as an agricultural spray.

Since then several other inhibitable enzymes have been proposed for the determination of organochlorinated pesticides. Thus, Guilbault and Sadar (1971) using the highly sensitive fluorescence-generating lipase substrate, 4-methyl umbelliferone, reported I_{50}'s in the range of $10^{-5}-10^{-4}M$ for commercial samples of aldrin, sevin, lindane, 2,4-D, heptachlor and DDT, thus permitting the detection of as little as 100 mg. As with the other pesticides, TLC methods have been developed to distinguish and separate them before detection and estimation of their inhibitory activity. Inhibition of phosphatase in such TLC setups has also been explored, especially for the detection of herbicides, with little or no successful application.

Perhaps utilization of what must be very powerful enzymes present in pesticide-resistant insects, enzymes which rapidly metabolize these pesticides as substrates, might at least push back the demise of enzymes as useful tools in their analysis as would the development of insecticide-as-substrate methods using the powerful cytochrome P-450-linked oxidases of liver microsomes (Chapter 36).

Other Enzyme Affectors

Naturally Occurring Antienzymes.—In contrast to the analysis of pesticide and herbicide residues in foods, analysis of the antienzymes is almost exclusively restricted to the use of the enzymes which they inhibit. The only alternative is via immunoassay (see below). For estimation of naturally occurring trypsin inhibitors widely distributed in many foods such as legumes, potatoes, wheat cereals and eggs, rate assays are routinely used in rather time-consuming, laborious procedures in which the activity of trypsin is compared with that in presence of an inhibitor. Thus, a "simplified" collaborative analysis of trypsin inhibitor in soybean (Kakade *et al.* 1974) involved the use of a minimum of 12 test tubes. Although not yet in use for routine analysis, active site titration (Chapters

3, 14 and 37) may yet prove to be a more convenient way of determining these antienzymes in foods (Zahnley and Davis 1970). Thus, as shown in Fig. 39.1, the liberation "burst" (O) of *p*-nitrophenol from the active site titrant, *p*-nitrophenyl-*p'*-guanidinobenzoate, resulting from interaction with trypsin, is subdued in the presence of chicken ovoinhibitor (OI), chicken ovomucoid (OM) and soybean trypsin inhibitor (STI). With the availability of fluorescent and other sensitive probes, such approach, especially when combined with chemical covalent transformation as proposed by Friedman *et al.* (1980), should become feasible for routine analysis of even traces of such antienzymes.

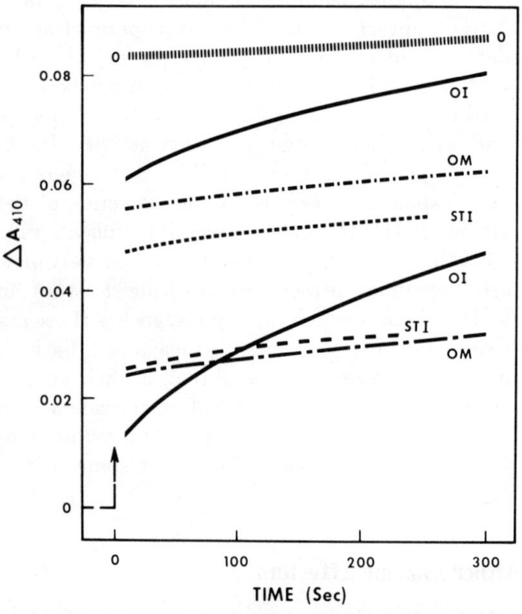

Courtesy of Zahnley and Davis (1970)

FIG. 39.1. BASIS FOR DEVELOPMENT OF NONRATE ACTIVE SITE TITRATION ANALYSIS FOR TRYPSIN INHIBITORS IN FOODS

Measured is the release of *p*-nitrophenol (A_{410}) from the active site titrant *p*-nitrophenyl-*p'*-guanidinobenzoate by trypsin in absence (O) and presence of chicken ovoinhibitor (OI), chicken ovomucoid (OM) and soybean trypsin inhibitor (STI). Arrow indicates time of addition of active site titrant.

We wish to remind the reader that enzyme inhibition can also be used as an index of food quality. As mentioned in Chapter 14, inhibition of amylase was once proposed as an index of tartness and astringency due to the presence of tannins and as a means of detecting the contamination of semolina devoid of or low in such inhibitors, with hard wheat.

The violent amatoxins present in poisonous mushrooms may be estimated by their potent inhibition of RNA polymerase (Faulstich and Cochet-Meilhac 1976). The relevant K_i's in the range of $10^{-9}M$ (<10 nanograms) ensure their ready detection in nonpoisonous mushrooms (1−16 nanograms per gram of fresh tissue) let alone in toxic *Amanita* species in which they can run as high as 100,000 × this range or in the area of about 0.2 mg/g. Ingestion of as little as 30 mg (*ca* 1/1000 of an ounce) of such mushrooms is usually fatal. This method is not as sensitive as radioimmunoassay which can pick up 0.05 nanograms.

Oxalic acid present in spinach, rhubarb and other vegetables might be estimated via enzyme inhibition because it is a less potent but still considerably effective inhibitor ($K_i = 5 \times 10^{-5}M$) of the malic enzyme (Sarawek and Davies 1976), as well as a transition state inhibitor ($K_i = 4 \times 10^{-6}$) of codfish oxalacetate decarboxylase (Chapter 3).

Inorganic Ions.—As with the advent of ever more convenient and sensitive methods for analysis of metals using instruments such as atomic absorption, X-ray fluorescence analyzers, and for the less affluent, ion selective electrodes discussed in Chapter 38, the use of enzyme inhibition and activation to analyze for these inorganic constituents has become practically moribund. Yet, such analyses are, we believe, a tribute to the ingenious uses to which enzymes can be put. Probably the first analytical application of the ability of substances to affect the activity of enzymes was that of Amberg and Loevenhart (1908) who used the inhibition of porcine liver esterase to detect ppm of fluoride in milk. Other examples of enzyme affector analysis for metals and anions are shown in Table 39.1.

Enzyme Activators.—As in the case of inhibitors, the use of enzyme activators for the detection and analysis of food substances has been largely superseded by nonenzymatic methods. We may distinguish among the four more or less distinct types of activation with increasing sensitivity:

(1) *Partial requirement*—the enzyme is active in the absence of activators,
(2) *Compulsory activation or absolute requirement*—the enzyme is completely inactive in absence of activator but as isolated does not contain it,
(3) *Reaction partner*—the anylate is a cosubstrate for the reaction being catalyzed, and

TABLE 39.1

ELEMENT ANALYSIS BY ENZYME INHIBITION[1]

Enzyme	Element (Sensitivity)	Enzyme	Element (Sensitivity)
Invertase	Hg (40); F (50)	Invertase	I (100); S (15)[2]
P-ase, alk.	Be (10); Ba (100)[2]	P-ase, acid	F (50); P (500)
Urease	Hg (50)	Hyaluronidase	Fe (500)
Deh., alcohol	Hg (10^{-3}); Ag (10^{-3})	Deh., i-citrate	Pb (1); In (1)
Catalase[3]	Cu (20)	Peroxidase	S (100)
Glucose oxidase	Hg (50)	Xanthine oxidase	Hg (50)

Source: Townsend (1973).
[1] Sensitivity or lower limit of analysis in nanograms; metals detected as their cations, I, S, F, as their anions and P as orthophosphate: P-ase, phosphatase; alk, alkaline; Deh., dehydrogenase.
[2] Estimation by their reversion of inhibition invertase by Hg and of alkaline phosphatase by Be.
[3] In presence of ascorbic acid.

(4) *Apoenzyme*—the tightly bound prosthetic group or coenzyme is removed from the isolated but usually unstable holoenzyme. Enzymes so used, mentioned and referenced by Townshend include isocitrate dehydrogenase (Mg, Zn, Mn), α-ketoglutarate dehydrogenase (Mg, Zn), DNase, creatine phosphokinase and luciferase (Mg).

Apoenzymes which are bereft of their tightly bound prosthetic groups but still activatable by them and used to determine anylates, which happen to be such prosthetic groups, include apoalkaline phosphatase (Zn, Co) and apophenolase (Cu).

Stabilization of apoenzyme during removal of prosthetic groups is improved by prior immobilization of the enzyme. Other metals which might be determined in this fashion include Mo (xanthine oxidase), Fe (lipoxygenase) and Se (glutathione peroxidase). Cl could be determined by the activation of mammalian α-amylase by Cl^-.

ADDING ENZYMES TO FOODS AS SHORTCUTS TO QUALITY ASSESSMENT

Digestibility and Availability of Food Constituents

Protein.—In Chapter 34 we discussed how added enzymes, especially proteinases, play significant roles in determining the nutritional values of many foods. The same or similar enzymes also are helpful in their roles as reagents for the assessment of the nutritional value of these same foods. Indeed, many of the conclusions reached concerning increased nutritional values in Chapter 34 were based on just such *in vitro* short cuts for longer *in vivo* testing. Now well enough established to be used as an official AOAC method for assessing the digestibility of animal feeds (Chapter 37), their use probably goes back to the investigations of Melnick *et al.* (1946) who used enzymes to assess the effect of heat processing of foods on the availability of amino acids. Since then it repeatedly has been shown that enzymatic release of amino acids by pepsin and/or pancreatin under appropriately controlled conditions correlates well with complicated, long drawn-out and more variable animal assays. Reviews on the subject include those of Mauron (1970) and Stahmann and Woldegiorgis (1975). The latter point to the following advantages over both biological and chemical scoring methods of using enzymes:

(1) It is less expensive, time-consuming and variable.
(2) It yields information concerning both overall protein quality as well as the pattern of essential amino acids.
(3) It is more suitable for monitoring possibly adverse effects of processing.
(4) Unlike nonenzyme chemical methods, it reflects the presence of proteinase inhibitors in the food.

The proteinases are also helpful in using the chemical methods by serving as analytical aids as discussed below.

One disadvantage of enzyme tests is that they do not reveal, as do most *in vivo* tests, the presence of toxicants not related to protein quality. A comparison of an enzyme method using pepsin and pancreatin with a chemical and a biological method is shown in Table 39.2. Research is continuing to find even shorter methods and better correlations. For instance, Hsu *et al.* (1977), using an automated multienzyme system consisting of trypsin,

TABLE 39.2

PROTEIN QUALITY OF PROCESSED FOODS AS MEASURED BY ENZYME DIGESTIBILITY AND NONENZYME METHODS[1]

	Enzyme Test Score	Enzyme Test PPD	Chemical Score	Animal Test PER
Egg, whole, standard	100	100	100	100
Beef	70	96	77	89
Cereal, breakfast	44	75	45	89
Cereal, protein fortified	72	93	60	63
Soy protein, texturized	53	76	66	56
Correlation with PER,[3] r	0.86	0.87	0.86	—

Source: Adapted from data of Stahmann and Woldegiorgis (1975).
[1] Values expressed relative to those of whole egg; actual value of PPD, pepsin pancreatin digest index value, 96; PER, protein efficiency ratio, 3.0; others, 100.
[2] Includes other foods not listed here.
[3] r is the correlation coefficient.

chymotrypsin and less pure porcine intestinal peptidase, simply measured the pH of a reaction mixture. After 10 min the pH dropped from slightly above neutrality for most of the 40 high protein foods tested to as low as 6.2. The correlation coefficient between the pH of these reaction mixtures and *in vivo* apparent digestibility was 0.90. This result includes variation from the standard pepsin or pepsin and hog pancreatin. Thus, Saunders and Kohler (1972) found improved correlation of enzyme tests with feeding tests (Fig. 39.2) for different wheat mill feeds as protein source, when they treated the latter with a fungal proteinase followed by chick pancreas acetone powder.

One enzyme which has not yielded correlatable data is papain. The enzyme most widely applied to measure protein digestibility of animal feed, largely supplanting pepsin used in the 1975 AOAC procedure, is "Pronase," one (or more) of the least specific of the proteinases, from *Streptomyces griseus*.

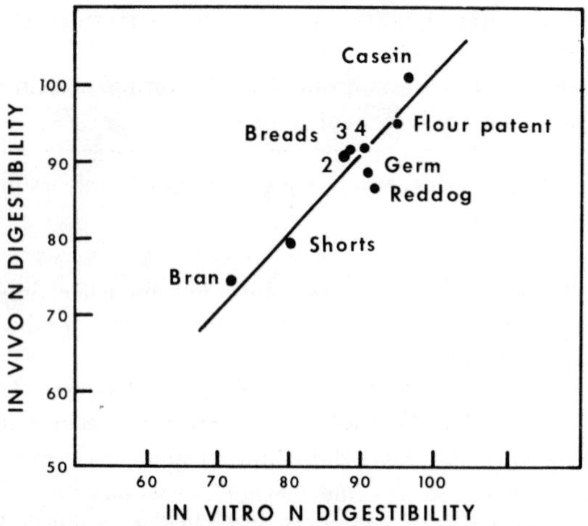

Courtesy of Saunders and Kohler (1972)

FIG. 39.2. ENZYME TEST AS A SHORTCUT FOR DIGESTIBILITY OF WHEAT MILL FRACTIONS

See text for enzyme used and other details.

As in the determination of food constituents as enzyme substrates and effectors, the use of enzymes in assessing protein quality continues to meet competition not so much from purely chemical methods but from improved biological assays. Thus, Splitter and Shipe (1976) compared purely chemical procedures and pepsin-pancreatin enzyme tests with the vole bioassay introduced less than 10 years ago for the nutritive value of high lysine opaque-2 maize and its normal counterpart. They found that unlike the chemical test, *available* lysine released in the enzyme test correlated more closely with the vole PER than did *total* lysine present in the maize.

Another up-and-coming competitor to enzyme tests is the revival and refinement of the microbiological assay proposed by Rockland and Dunn (1949) using the protozoan *Tetrahymena* whose essential amino acid requirement is similar to that of humans. Thus, Dryden *et al.* (1977) observed strong multiple correlation among the 2–4 week standard PER rat assay, the 4 day *Tetrahymena* bioassay and the 1 hr enzyme test of Hsu *et al.*—the vole bioassay takes about 6 days. Another advantage of these alternatives to rat assays is that they require much smaller quantities of sample. Dryden also used the above-mentioned test of Hsu *et al.* as an aid (see below) in preparation of their food protein samples for the *Tetrahymena* assay to overcome certain problems connected with this assay such as the partial dependence of growth of the organism on the presence of particulate matter as well as that of finding adequate cell counting procedures.

Carbohydrate Forages.—Cellulase may be added to forages to obtain more information about dry matter digestibility, of special importance in forages and feeds such as straw consumed by ruminants (Guggolz *et al.* 1971; Sundstoel *et al.* 1978; Hartley *et al.* 1974). Illustrative of the predictive value of such investigations, the latter authors found r = 0.978 for the correlation between *in vivo* digestibility of grasses and UV absorbance at 324 nm of filtrates obtained by incubating grass cell walls with commercial cellulase preparations for 16 hr. Instead of using a colorimetric method, the effect of the added cellulase was also measured by counting the percentage of cell walls digested which also gave a good correlation (r = 0.962) with *in vivo* digestibility of grasses but not of legumes.

Enzymes should also be useful as shortcuts for ascertaining the nutritional availability of other food constituents. Thus, the availability of lipids in fried foods where such food constituents may undergo polymerization and decomposition might be ascertained with the help of lipases.

Nonnutritionally-related Enzyme Tests

Damaged Starch.—As mentioned in Chapter 32, the degree of starch damage in wheat flour is a critical factor contributing to bread quality and in Chapter 37, the "gassing power" of flour is predictable by determining the extent of interaction between endogenous amylase and susceptible substrate. The actual starch available as substrate, i.e., the degree of starch damage, is measured by ensuring that there is an excess of amylase present, by *adding* α-amylase to which the damaged starch is more susceptible than is undamaged starch. Thus, Chiang *et al.* (1973) calculated damaged starch from differences in polarimeter readings between that obtained from a suitably prepared extract of a sample of flour suspension treated with α-amylase and that of an untreated control. Similarly, α-amylase was also used by Hansen and Jones (1977) to estimate the change in starch granules caused by heat-processing of wheat flour except that extent of change was observed through a microscope.

Other Tests.—"Diastase" is added to certain cereals to assess their suitability as starch adjuncts in beer-making. Also in the present category might be the detection and determination of skimmed milk powder in raw sausage by means of the enzymatic analysis of lactose via the UV optical test (Bahl 1972). Nordal and Rossebo (1972) described a simple and inexpensive enzymatic test for the specific detection of sodium caseinate in meat and other food products based on proteinase induced precipitation of casein in agar gel. Casein is also the target substrate in the use of added pepsin to milk to detect alteration in the protein due to freeze dry processing (Calgano and Maffeo 1971).

Analogous to the detection of skim milk in sausage via its enzymatically determined lactose, Promayon *et al.* (1976) proposed that chicory can be detected and estimated in soluble coffee via an enzymatic determination of fructose and glucose derived from inulin, a polymer of fructose also present in the Jerusalem artichoke, and from the bitter glycoside intybin, both present in chicory. As a final example of how adding enzymes can be used as a shortcut, Katou *et al.* (1976) estimated the mycelial content of shoyu koji via the action of β-1,3-glucanase and chitinase in the culture fluid of *B. macerans* strain.

ENZYMES AS ANALYTICAL AIDS

Admittedly the demarcation between the application of enzymes as analytical aids and other aspects of their analytically oriented exploitation is not sharp and indeed is a bit arbitrary. In general, however, the present category is meant to include instances of enzyme action on substances *other* than the anylate itself. With the exception of enzyme immunoassay the added enzyme action usually occurs before the final analytical procedure is employed.

Crude and Dietary Fiber

With the revival of interest in the putative health-related benefits of the presence of what was called roughage some 40 or 50 years ago, the need to distinguish "dietary" from "crude" fiber (Trowell 1976; Leveille 1976; Colmey 1978; James and Theander 1981) has led to an examination of enzymes as aids in this regard. Colmey cited a definition of *crude* fiber in a food as the residue remaining after treatment with hot sulfuric acid and alcohol, consisting mostly of cellulose, lignin and a trace of other glycans, whereas the *dietary* fiber is highly dependent on the "dieter," being that remnant of plant cells in the diet which is resistant to the action of the enzymes of a particular human's intestinal tract (indigestible residue). Depending on the diet and dieter it consists of variable proportions of complex carbohydrates such as lignin, cellulose, hemicelluloses, pectin, other glucans and, of late, even plant protein cores (Chapter 34). The standard measurement consists of weighing the residue after treatment with detergent.

In keeping with these definitions the enzymes so far used as aids in fiber determination *remove* "nonfiber" substance. Thus, Hellendoorn *et al.* (1975) in a "reverse" protein digestibility test using the same pepsin:pancreatin mixture considered the measured dry weight remaining to be dietary fiber. The values obtained coincided roughly with crude fiber for white cabbage, carrots and onions but diverged greatly for legume food seeds, cereal products and some vegetables such as potatoes. Similarly, Elchazly and Thomas (1976) investigated the treatment of plant foodstuffs with amylolytic and proteolytic enzymes as part of a procedure for the estimation of dietary fiber which they define as total water-insoluble unavailable carbohydrates. The insoluble residue remaining after this treatment removes the starch and protein is further analyzed nonenzymatically for cellulose, hemicellulose and crude lignin. The same investigators (Thomas and Elchazly 1976) also developed an enzyme method for the determination of crude fiber, using pancreatin and Rhozyme S which, it will be recalled (Chapter 30), is rich in polygalacturonase.

Amino Acids in Protein Analysis

The time-honored practice of hydrolyzing proteins with strong hot acid in preparation or analysis of the constituent amino acids has always posed problems of variable amino acid stability (tryptophan, asparagine and glutamine are particularly notorious in this respect) and the necessity of having to apply correction factors to these and other amino acids. Search for alternative mild methods, especially those involving proteinase action, at first enthusiastically pursued, became rather desultory because it has brought with it a new set of analytical problems. The pace of research into this area quickened only upon the realization that linkages between amino acids other than peptide bonds, although constituting a small fraction of the total bonds holding the protein together, possess high biochemical, textile, medical and nutritional significance. For food it is becoming increasingly clear that the presence of these nonpeptide bonds may constitute a record of the processing and storage history of the food especially with regard to heat and alkali treatment. It would appear that such treatment labilizes some of the serine, cysteine, cystine and threonine, resulting in the elimination of water therefrom and the formation of the highly active dehydroalanyl and methyldehydroalanyl residues still bound within the protein (Friedman 1977). For serine,

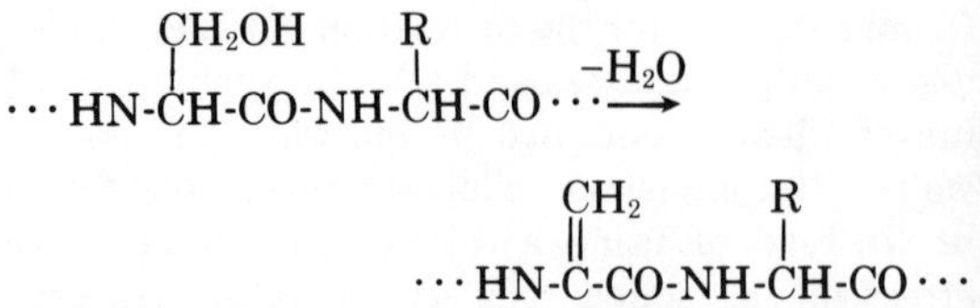

These active species can interact with a variety of side groups of intact amino acid residues but one interaction which has received the most attention has been with the epsilon group of lysine to form the notorious lysinoalanine (Friedman *et al.* 1981).

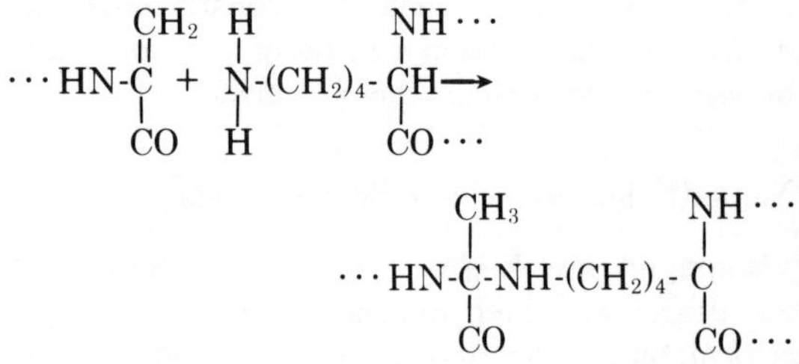

Such an isopeptide bond is not susceptible to the action of proteases, be they in the human intestine or in the reaction mixture of the food analyst who is trying to determine the amino acid content of a particular food. The failure of proteases to hydrolyze lysinoalanyl and other isopeptide bonds in food proteins renders the lysine and other chemically transformed essential amino acids nutritionally unavailable. The crosslinking which results from such interactions may also diminish protein digestibility; hence, the urgency of using enzymes to distinguish among the various types of linkages. Probably the first breakthrough was that of Hill and Smith (1962) who added to papain what was then called leucine aminopeptidase (now called simply aminopeptidase) and prolidase, which hydrolyzes the aminopeptide bond in aminoacyl-L-proline. It is a bond which is particularly recalcitrant to the action of proteinase, in contrast to the prolyl-L-amino acid peptide bond. The effect of crosslinking and other aspects of the complete hydrolysis of protein by enzymes are discussed by Milligan and Holt (1977).

Some of these auxiliary and relatively expensive enzymes can be at least partially dispensed with when the wide ranging pronase is used instead of papain. A somewhat different role as an analytical aid in amino acid determination was assigned to papain by Nair *et al.* (1976). They availed themselves of the circumstance that papain is stable and active in the general protein denaturant urea at concentrations as high as 9*M* (Lineweaver and Schwimmer 1941) to completely solubilize food protein in preparation for further action by appropriate proteases, preferably (they suggest but do not demonstrate) immobilized. A diligent search of the literature shows that as of this writing, the routine use of enzymes as alternative to the use of hot 6N HCl and the like for hydrolysis of food proteins in preparation for total amino acid analysis is virtually nonexistent. When they are so used they will probably be employed in a supplement to acid hydrolysis for individual analysis of specially labile amino acids such as cysteine and tryptophan, as was done by Purcell *et al.* (1978) for the determination of the amino acid composition of proteins of various sweet potato fractions.

Enzymes as Aids in Other Analyses

Proteinases.—The use of enzymes in clean-up procedures and the general utility of enzymes as analytical aids is probably not as appreciated as its contribution in this area merits. A perhaps surprising variety of substrates in foods have been analyzed or have been proposed to be analyzed with the aid of enzyme action. What usage of enzymes in these procedures have in common is that, as mentioned, they usually do not act on the substance to be analyzed or are not part of the final analytical procedure which yields the desired measurement. Frequently, they are used as cleanup aids. This is especially true of the proteinases as evidenced in the following sampling of such use.

—Papain was used as a cleanup aid in the determination of carboxymethylcellulose in food products. The anylate was hydrolyzed and measured colorimetrically (Graham 1971).

—Papain was also used to solubilize protein in the determination of available lysine in feedstuffs (Shorrock 1976); it will be recalled that presence of particulate matter stimulates *Tetrahymena* growth and leads to variable, spurious assay results.

—Crude pepsin (NF) was used as a general cleanup aid in the isolation of light filth in hamburgers (Alioto and Andreas 1976). Light filth is measured with the aid of a light microscope.

—Purified pepsin was used to release the anylate and as a cleanup aid in the determination of chlorogenic acid in leaf protein concentrates (Lahiry and Satterlee 1975). The anylate was determined by a standard colorimetric procedure. Large amounts of enzyme were required because the anylate inhibited the added pepsin.

—Takadiastase (a crude preparation of *Aspergillus* protease) was used for comparison purposes with a urea-acid system as medium of extraction for the determination of riboflavin (Roy *et al.* 1976). The anylate was determined by a simplified automated colorimetric analysis—both methods were more precise than the AOAC method.

—Pancreatin, presumably because of its proteolytic and perhaps lipolytic action, was used to remove nonimmunological reactive protein in the immunoprecipitation of immunologically active proteins in soft wheat, soy, beef, meats, eggs and dairy products (Flego 1977).

—Trypsin was used to release lipid from gluten in the determination of lipid in wheat flour. The anylate was determined by TLC (Terent'eva *et al.* 1973).

—α-Chymotrypsin was used in a nonfood analysis along with lipase to obviate the need for pretreatment of serum in the determination of serum lipid (Bucolo *et al.* 1975). The resulting glycerol was determined by means of a standard UV optical enzyme test (Chapter 38).

Other Enzymes.—The following is a selection of the use of enzymes other than proteinases:

—Carbonic anhydrase, as previously noted (Chapter 37), was used in the AOAC procedure to obviate the need for distillation in the determination in wines of CO_2.

—Amylase in a crude commercial food grade enzyme was used to remove starch which interfered with the AOAC determination, in foods, of vitamin B (thiamin) measured colorimetrically.

—A pyrophosphatase, as previously noted, also present in this amylase preparation, releases thiamin from cocarboxylase in the food so that the "total" thiamin can be estimated. In this case the enzyme acts on a substance closely related to anylate but its function, unlike that of the enzymes acting on anylate-as-substrate discussed in Chapter 38, is more that of an analytical aid to remove groups in the determination of "total" forms of the anylate.

—Phospholipase C was used in a similar manner for the determination of phospholipid in the lipid fraction of eggs so that the resulting acyl glycerols could be separated and measured by GLC-mass spectrometry (Gaskell and Brookes 1977).

—β-Glucuronidase, similarly, was used to release diethylstilbestrol for the latter's determination in cattle tissue in preparations for using GLC (Donoho *et al.* 1973).

—Amylase (Rapidase) was used to release the anylate from its complex with the amylose moiety of starch for the determination in spaghetti of monoglycerides measured and separated by GLC and TLC (Jodlbauer 1976).

—Glucoamylase was used to remove starch interfering with subsequent efficient action of added proteinases in the above-mentioned determination of the amino acids of the proteins of sweet potatoes (Purcell *et al.* 1978).

—Pectinase (Klerzyme) was used to speed filtration and aid in the isolation of fly eggs and maggots in tomato products as part of an inspection procedure for wholesomeness of tomato products (Vasquez and Cooper 1964).

—Glucose oxidase/catalase was used to remove the large excess of glucose for determination in honey of the relatively small amounts of sucrose measured enzymatically by the action of glucose oxidase and *peroxidase* (AOAC) on invertase-produced glucose (White 1977).

—D-Galactose dehydrogenase was used to remove D-galactose in some plant food tissues for the determination of the less well-known but widely distributed L-galactose by GLC (Roberts and Harrer 1973).

Finally, we should like to mention that enzymes can be analytical hazards as well as aids and it is almost always necessary to ensure their destruction prior to analysis for individual constituents by enzymatic or nonenzymatic means. How important this can be is attested to by the mentioning of "enzyme blocking" (by EDTA) in the title of a paper on the detection of polyphosphates in frozen chicken, fish fingers and hamburger using NMR spectroscopy (O'Neill and Richards 1978). Macerating the food sample with EDTA (thus tying up all the Mg^{2+} obligatory for polyphosphatases) while still frozen and allowing the sample to thaw in the NMR sample tube, permitted estimation with minimum sample preparation and with no loss of anylate due to enzyme action.

Enzyme Immunoassay (EIA)

Although virtually not used (with few exceptions) for food analysis at the start of this writing, we include this seminal, swiftly developing analytical technique, potentially of great importance in food analysis, among the category of "enzymes-as-analytical-aid" because the role enzymes play is not directly concerned with their catalytic specificity vis-à-vis the anylate to which, however, they are covalently bound. Their function is that of marker or label to signal the release of the anylate from its specific reactor. As labels, enzymes are interchangeable with nonenzyme markers such as bacteriophage, electron spin-label and, of course the progenitor to all and still most widely used of these, labeling with radioactive isotopes for RIA, radioimmunoassay, for which Yalow (1978) received the Nobel Prize in 1977, some 18 years after its introduction. This technique developed from the educated and correct surmise that the reason injected radioactive insulin disappeared less rapidly from the blood of diabetics and others who had previously received insulin than from those never hitherto injected with insulin was because the newly formed antibodies to insulin were binding the newly injected insulin molecules and preventing them from clearing the blood.

According to Scharpe *et al.* (1976) in their cogent and available review, EIA was introduced independently and apparently simultaneously less than ten years ago, by three groups of investigators. In addition to that of Scharpe *et al.* (1976) many other reviews have been, and will have been, written by the time this book appears. The abstracts indicate that discussions at the time of this writing far out-number issuance of papers reporting original results. They range from a two page overview by Barker (1977) to discussions of the future of enzyme immunoassay (as part of a series of reviews) by Jarvis (1979) and entire books (Thorell and Larson 1978; Vogt 1978) in just about every science associated language.

The anylates in themselves need not be intrinsically antigenic but can acquire antigenicity and thus elicit antibodies when they are conjugated to a protein (carrier) antigen and injected in an appropriate animal. Such haptens may be just about any organic molecule including those which are likely to be analyzed in foods. Perhaps one of the barriers to speedy acceptance of EIA for food analysis is the less than ready availability of the requisite hapten-elicited antibodies. The label (*) attached to the antibody through the anylate in the system described below is readily separable or possesses an altered signal from that of the free label, thus forming the basis for the quantitative determination of the unknown.

The basic principle of all label-immunoassay is competition between labeled and unlabeled anylate-as-antigen for its antibody:

Antigen, Unlabeled The unknown anylate (antigen)
\+ competes with
*Antigen, Labeled** known anylate, the reference sample,
\+ for
Antibody to Anylate anylate binding site on antibody
k_1 ⇅ k_2 to yield
[*Antibody-Anylate, Labeled** + Antibody-Anylate,] Unlabeled

where $k_1 >>> k_2$, i.e., $K_{assoc} = 10^{14} M^{-1}$.

For analytical purposes, the initial concentration of only the unlabeled unknown is variable so that the extent of this competition and hence change in the signal level is proportional to the amount of (unknown) anylate. The absolute level is determined with the aid of an appropriate calibration or standard curve using known amounts of unlabeled anylate.

This basic relation allows for great flexibility in analytical setups and the label can be placed on the antibody as well as the anylate.

Radioimmunoassays have to utilize the myriad available physical separation methods because the signal radioactivity is not affected by combination with antibody. While some of these separation techniques are not available for enzyme immunoassay because of the bulkiness of the enzyme, the signal does change (as it does in spin label immunoassay) so that one may, but does not have to, separate the two forms of the label. For EIA, this signal is turned off (i.e., the enzyme is inactive) when the label is associated with the antibody; but when on, the signal, (unlike that of RIA and spin label assay) is amplified. Other advantages claimed for EIA over RIA are mainly related to the instability of either the isotope itself or the antibody due to strong radioactivity, time-consuming counting of weak isotopes, laboratory and health hazards, and waste disposal problems. These considerations make it particularly more suitable than RIA for the food laboratory. On the other hand Yalow (1977) states that among the several disadvantages of EIA not possessed by RIA, the most important is the decreased sensitivity of the assay due to the sterically hindered antibody-antigen reaction occasioned by the presence of the bulky enzyme molecule attached to the known anylate. RIA, but not EIA, can pick up as little as 0.05 picomoles of the hormone gastrin. Scharpe *et al.* state that the EMIT method of EIA (see below) is more sensitive for the detection of psychoactive drugs than are TLC techniques but less sensitive than is RIA. The keen competition between the radio- and enzyme-immunoassays has been described by Ekins (1980) who states that a new ultrasensitive enzyme immunoassay (USERIA), which eclectically uses a radioactive compound as substrate for the signal enzyme, far outstrips RIA, allowing as little of 10^{-16} g (600 molecules) of cholera toxin to be detected.

EIA Without Separation (EMIT).—The workability of EIA in homogeneous systems depends, as mentioned, upon the circumstance that many enzymes to which an antigen-anylate is bound are active when free but become inactive when this complex binds to the antibody. It is only necessary to find a suitable chemical reaction which will bind the anylate to the signal enzyme without interfering with combination of anylate-as-antigen to the antibody nor with that of the enzyme-as-catalyst (and signal generator) to the substrate, i.e., the enzyme has to retain activity. This combination is usually accomplished by the same sort of reaction used for covalently immobilizing enzymes except, of course, in this enzyme-multiplied immunoassay system (EMIT), the enzyme has to be kept soluble. A common conjugation route is through derivitization of anylate (i.e., to oxime), followed by coupling to the enzyme through one of the well-known carbodiimide coupling reactions (Chapter 8).

Although peroxidase was at first frequently used in EIA, experience has shown, according to Scharpe *et al.*, that G-6-P and malate dehydrogenases and lysozyme are suitable. Lactase, alkaline phosphatase and glucoamylase are also used (Gibbons *et al.* 1980).

EIA with Separation.—Glucoamylase is also well suited as a label (Tateishi *et al.* 1976) for one of the many general separation techniques used, i.e., double antibody EIA in which the antibody-bound anylate is precipitated by the use of a second antibody which is usually elicited by injecting into animals the most abundant of the five major classes of immunoglobulins, IgG, thus eliciting an antibody to an antibody.

Other separation methods utilize the techniques of enzyme immobilization by insolubilizing the antibody which, in combination with double antibody EIA, is known as DASP (double antibody solid phase), or by immobilizing the anylate and labeling the antibody, known as ELISA (enzyme-linked immunosorbent assay). In one of its first uses in connection with a food, Converse (1978) adapted ELISA to detection of tomato ringspot virus in red raspberry plants. In one of the variations, illustrated in Fig. 39.3 (Nelboeck 1977), the sample is added to a test tube coated with antibody. After complete binding of the unknown anylate, a measured amount of peroxidase bound anylate-antigen is added. This displaces some of the unlabeled anylate, and only part of the label is immobilized. Increasing amounts of bound label, in this case peroxidase, indicate less and less unknown in the sample. This type of manipulation lends itself to adaptation to the centrifugal fast analyzer (Chapter 38). Perhaps one of the first food-oriented applications of EIA was the detection of the most common cause of food poisoning, staphylococcal enterotoxin A, in hot dogs, milk and mayonnaise by Saunders and Bartlett (1977). The ordinate of the corresponding DASP calibration curve used in this investigation shown in Fig. 39.4 is a measure of the product of the peroxidase action. In this case increasing activity of the remaining peroxidase (which was conjugated to the γ-globulin obtained from the enterotoxin antiserum) reflects increasing amounts of toxin.

Use of Enzyme Immunoassay in Enzyme Assays.—As previously suggested, the specificity of the antibody-antigen reaction affords an alternative to active site titration as direct nonrate estimation of the total number of immunologically competent and therefore presumably reasonably intact enzyme molecules. Yalow points out that assay of enzyme activity by RIA (and probably most EIA) is not influenced by enzyme affectors, that one can measure active enzyme, proenzyme and inactive enzyme in the same sample and that enzyme activity, in going from species to species, need not parallel immunological activity. Since even active-site titration may not always distin-

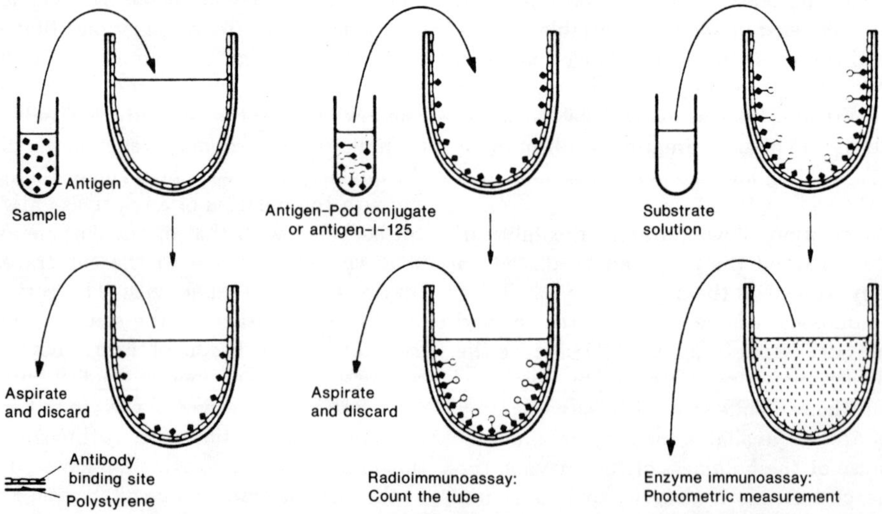

From Nelboeck (1977). Copyright by Academic Press, New York

FIG. 39.3. ENZYME- AND RADIOIMMUNOASSAY PRINCIPLES WITH ANTIBODY-COATED TUBES

Illustrated is the principle of the use of enzymes as markers as an alternative to radioactive markers.

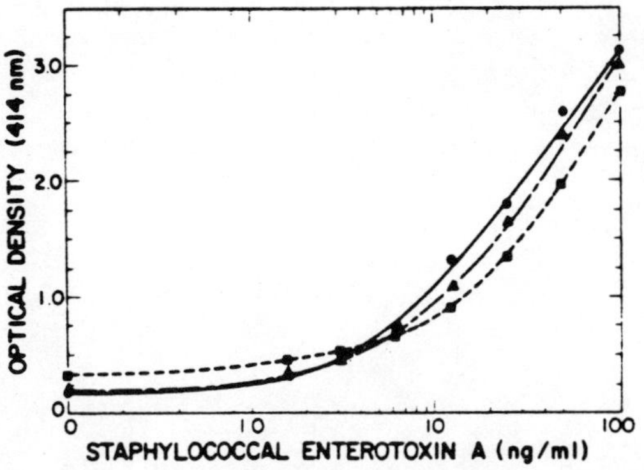

From Saunders and Bartlett (1977)

FIG. 39.4. ENZYME IMMUNOASSAY FOR FOOD POISONING

Shown are standard curves for detection of staphylococcal enterotoxin A in mayonnaise (circles), milk (squares) and hot dogs. Measured is the release of peroxidase in a double-antibody solid-phase enzyme immunoassay.

guish closely-related enzymes on the basis of their specificities it may be well to complement ASTs with EIA or RIA tests.

An example of the use of label immunoassay for determination of an enzyme concentration is afforded by the study of Iwatsubo *et al.* (1976), mentioned in Chapter 37, on the *de novo* synthesis of invertase in ripening tomatoes, as shown in Fig. 39.5. RIA standard curves for tomato invertase at mature green (solid triangles), turning (solid circles) and ripe (open triangles) states paralleled that isolated from the ripe fruit (open circles).

Of course, the label immunoassay principle is a special case of the broader competitive binding principle and hence is not necessarily dependent upon antibody for imparting specificity to the substance being competed for by the labeled competitors and unlabeled anylates. Yalow cites the use of specific protein receptors for the determination of vitamin B_{12} (intrinsic factor) and thyroxine (its binding globulin). Undoubtedly, vitamin B_{12} in foods could also (if it has not by the time this appears) be determined with an intrinsic factor. It should also be possible to use an enzyme or stable inhibitor of the enzyme as "competitee." Furthermore, EIA will undoubtedly meet increasing competition from other labels, especially spin label which is now being looked into for the determination of vitamin B_6 in foods (Brandon *et al.* 1980).

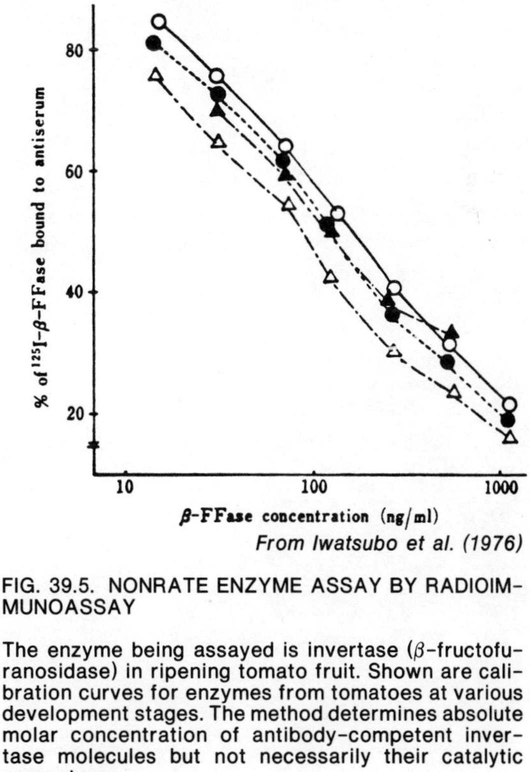

From Iwatsubo et al. (1976)

FIG. 39.5. NONRATE ENZYME ASSAY BY RADIOIMMUNOASSAY

The enzyme being assayed is invertase (β-fructofuranosidase) in ripening tomato fruit. Shown are calibration curves for enzymes from tomatoes at various development stages. The method determines absolute molar concentration of antibody-competent invertase molecules but not necessarily their catalytic competence.

BIBLIOGRAPHY AND SELECTED REFERENCES[1]

AACC. 1975. Cereal Laboratory Methods. Am. Assoc. Cereal Chem., St. Paul, Minn.

ABBOTT, J. A. 1972. Sensory assessment of food texture. Food Technol. *26*, 40–49.

ABDEL-FATTAH, A. F. and EDREES, M. 1973. Seasonal changes in the constituents of *Ulva lactuca*. Phytochemistry *12*, 481–485.

ABDEL-FATTAH, A. F. and EL-HAWWARY, N. M. 1972. Purification and proteolytic action of milk-clotting enzyme produced by *Penicillium citrinum*. J. Gen. Appl. Microbiol. *18*, 341–348.

ABDEL-FATTAH, A. F., MABROUK, S. S. and EL-HAWWARY, N. M. 1972. Production and some properties of rennin-like milk-clotting enzymes from *Penicillium citrinum*. J. Gen. Microbiol. *70*, 151–155.

ABDERHALDEN, R. 1961. Clinical Enzymology. Van Nostrand, Princeton, N.J.

ABDO, K. M. and KING, A. 1967. Enzymatic modification of the extractability of protein from soybeans, *Glycine max*. J. Agric. Food Chem. *15*, 83–87.

ABELES, R. H. and DOLPHIN, D. 1976. The vitamin B_{12} coenzyme. Acc. Chem. Res. *9*, 114–120.

ABELES, R. H. and MAYCOCK, A. L. 1976. Suicide enzyme inactivators. Acc. Chem. Res. *9*, 313–319.

ABERLE, E. D. and MERKEL, R. A. 1968. 5′-Adenylic acid deaminase in porcine muscle. J. Food Sci. *33*, 27–29.

ABRAMOVITZ, A. S. and MASSEY, V. 1976. Purification of intact Old Yellow enzyme using an affinity matrix for the sole chromatographic step. J. Biol. Chem. *251*, 5321–5326.

ABROL, Y. P., TIKOO, S., UPREY, D. C. and SACHAR, R. C. 1971. Studies on the browning of Indian bread. Cereal Sci. Today *16*, 304–306.

ACKER, L. 1962. Enzymic reactions in foods of low moisture contents. Adv. Food Res. *11*, 263–330.

ACKER, L. 1969. Water activity and enzyme activity. Food Technol. *23*, 1257–1270.

ACKER, L. and HUBER, L. 1969. Behavior of polyphenoloxidase in a water-deficient environment. Lebensm.-Wiss. Technol. *3*, 82–85. (German)

ACKER, L. and WIESE, R. 1972. The behavior of the lipase in systems of low water content. I. Influence of the physical state of the substrate on enzymatic lipolysis. Lebensm. Wiss. Technol. *5*, 181–184. (German)

ADAMS, D. M., BARACH, J. T. and SPECK, M. L. 1979. Inactivation of heat resistant proteases in ultra-high temperature treated milk. U.S. Pat. 4,175,141. Nov. 20.

ADAMS, H. W., NELSON, A. I. and LEGAULT, R. R. 1955. Film deposition of tomato juice on heat exchanger coils. Food Technol. *9*, 354–357.

ADAMS, H. W. and YAWGER, E. S. 1961. Enzyme inactivation and color of processed peas. Food Technol. *15*, 314–317.

ADAMS, J. L., BILLINGHAM, J. and SHAPIRA, J. 1974. A comparison of proposed methods for the *in vitro* synthesis of edible carbohydrate. *In* Immobilized Enzymes in Food and Microbial Processes. A.C. Olson and C.E. Cooney (Editors). Plenum Press, New York.

ADLER, S. P. and STADMAN, E. R. 1974. Cascade control of *Escherichia coli* glutamine synthetase. *In* Lipmann Symp. Energy, Regul. Biosynth. Mol. Biol. D. Richter (Editor). Walter de Gruyter, Berlin.

ADLER-NISSEN, J. 1977A. Enzymatic hydrolysis of food proteins. Process Biochem. *12* (7) 18–23, 32.

ADLER-NISSEN, J. 1977B. Preparation of polypeptides from soy protein. Fr. Demande 2,338,001. Aug. 12.

ADRIAN, J. 1973. Nutritional aspects of the Maillard reaction. IV. Aroma production. Ind. Alim. Agric. *90*, 559–564. (French)

AHLERS, J. 1974. Enzyme Kinetics. A Programmed Introduction to the Theory of Enzyme Kinetics and Its Practical Applications. Fischer, Stüttgart, Germany. (German)

AHMED, E.M., DENNISON, R.A., DOUGHERTY, R.H. and SHAW, P.E. 1978. Flavor and odor thresholds in water of selected orange juice components. J. Agric. Food Chem. *26*, 187–191.

AHMED, E. M., MARTIN, F. G. and FLUCK, R. C. 1973. Damaging stresses to fresh and irradiated citrus fruits. J. Food Sci. *38*, 230–233.

AHN, B. K., WOLFSON, S. K., JR. and YAO, S. J. 1975. An enzyme electrode for the determination of L-glutamic acid. Electrochem. Bioenerg. *2*, 142–153.

AHN, J. K. and KIM, H. U. 1977. Studies on the thermostable β-galactosidase EC-3.2.1.23 of *Bacillus coagulans* ATCC-8038 utilizable in dairy products. Korean J. Anim. Sci. *19*, 220–226.

AIBA, S., HUMPHREY, A. E. and MILLIS, N. F.

[1]Additional references may be found at the end of this section.

1973. Immobilized enzymes and alternative to whole cells of enzymes. *In* Biochemical Engineering, 2nd Edition. Academic Press, New York.

AIRAS, R. K. 1976. On the partial reactivation of inactivated pantothenase from *Pseudomonas fluorescens*. Biochim. Biophys. Acta *452*, 201–206.

AIZONO, Y., FUNATSU, M., FUJIKI, Y. and WATANABE, M. 1976. Purification and characterization of rice bran lipase. II. Agric. Biol. Chem. *40*, 317–324.

AKASHI, K., NISHIMURA, H. and MIZUTANI, J. 1975. Precursors and enzymatic development of caucas flavor components. Agric. Biol. Chem. *39*, 1507–1508.

AKATOV, V. A. and KHODAKOV, A. V. 1974. The effect of ultrasonics on the physicochemical properties of milk. FSTA *8*, 9 P1691. (Russian)

AKO, H., FOSTER, R. J. and RYAN, C. A. 1974. Mechanism of action of naturally occurring proteinase inhibitors. Studies with anhydrotrypsin purified by affinity chromatography. Biochemistry *13*, 132–139.

ALBERGHINA, F. A. M. 1964. Chlorogenic acid oxidase from potato tuber. Phytochemistry *3*, 65–72.

ALBERSHEIM, P. and ANDERSON, A. J. 1971. Protein from plant cell walls inhibit polygalacturonase secreted by plant pathogens. Proc. Natl. Acad. Sci. *68*, 1815–1819.

ALBERTSSON, P.-Å. 1962. Partition methods for fractionation of cell particles and macromolecules. Methods Biochem. Anal. *10*, 229–262.

AL-DELAIMEY, K. Å., BORGSTROM, G. and BEDFORD, C. L. 1966. Pectic substances and pectic enzymes of fresh and processed Montmorency cherries. Mich. State Univ. Agric. Exp. Stn. Q. Bull. *49*, 164–171.

ALDERTON, G. 1978. Personal communication. Albany, CA.

ALEXANDRESCU, V., MIHAILESCU, F. and PAUN, L. 1975. Amylases in the endosperms of wheat, rye, and triticale germinated seeds. II. Immunological and immunochemical investigations. Rev. Roum. Biochim. *12*, 61–69.

ALEXANDROV, V. Y. A. 1977. Cells, Molecules and Temperature. Conformation Flexibility of Macromolecules and Ecological Adaptation. Translated from Russian by V.A. Bernstam. Springer-Verlag, Berlin.

ALFORD, J. A. and SMITH, J. L. 1965. Production of microbial lipases for the study of triglyceride structure. J. Am. Oil Chem. Soc. *42*, 1038–1040.

ALI, M. and RICHARDS, J. F. 1975. Effect of gamma radiation on chicken liver catheptic activity and release of lysosomal cathepsin D. J. Food Sci. *40*, 47–49.

ALI, S. Y. and EVANS, L. 1969. Studies on the cathepsins in elastic cartilage. Biochem. J. *112*, 427–433.

ALIOTO, P. and ANDREAS, M. 1976. Collaborative study of an enzymatic digestion method for the isolation of light filth from ground beef and hamburger. J. Assoc. Off. Anal. Chem. *59*, 51–52.

ALLEN, J. C., ERIKSSON, C. and GALPIN, J. R. 1977. Affinity chromatography of lipoxygenases. Eur. J. Biochem. *73*, 171–177.

ALLEN, J. F. 1975. Two-step mechanism for the photosynthetic reduction of oxygen by ferredoxin. Biochem. Biophys. Res. Comm. *66*, 36–43.

ALLEN, R. C. 1975. The role of pH in the chemiluminescent response of the myeloperoxidase-halide-HOOH antimicrobial system. Biochem. Biophys. Res. Comm. *63*, 684–691.

ALLEN, T. H. and BODINE, J. H. 1941. Enzymes in ontogenesis (orthoptera). XVII. The importance of copper for protyrosinase. Science *94*, 443–444.

ALLEN, W. G. and DAWSON, H. C. 1975. Technology and uses of debranching enzymes. Food Technol. *29* (5) 70–72.

ALLEWELDT, G., DUERING, H. and WAITZ, G. 1975. Mechanism of sugar accumulation in growing grape berries. Angew. Bot. *49*, 65–73.

ALLISON, M. J., ELLIS, R. P. and SWANSTON, J. S. 1974. Tissue distribution of α-amylase and phosphorylase in developing barley grain. J. Inst. Brew. *80*, 488–491.

ALLISON, R. D. and PURICH, D. L. 1979. Practical considerations in the design of initial velocity enzyme rate assays. Methods Enzymol. *63*, 3–22.

ALROY, Y., LEVI, A., BRITO, I. and LIRA, M. 1976. Alcoholic fermentation of banana. Abstr. No. *56*, 36th Annu. Meeting, Inst. Food Technol. Anaheim, Calif., June 6–9.

AL-SHAIBANI, K. A., PRICE, R. J. and BROWN, W. D. 1977. Enzymic reduction of metmyoglobin in fish. J. Food Sci. *42*, 1156–1158.

ALTERMANN, W. 1939. The "flour indicator," a new apparatus for determining flour type. Z. Ges. Getreidew. *26*, 113–118. (German)

ALTERMANN, W. 1941. Testing the suitability of flour indicators. Z. Ges. Getreidew. *27*, 164–165. (German)

ALTMAN, P. L. and KATZ, D. T. 1976. Cell Biology. Biol. Handb. Ser. Fed. Am. Soc. Exp. Biol., Bethesda, MD.

ALTOMARE, R. E., GREENFIELD, P. F. and KITTRELL, J. R. 1974. Inactivation of immobilized fungal catalase by hydrogen peroxide. Biotechnol. Bioeng. *16*, 1675–1680.

ALWORTH, W. 1973. Stereochemistry and Its Applications in Biochemistry. Interscience Publishers, New York.

AMAHA, M., HORIUCHI, G. and YABUUCHI, S. 1978. Involvement of chill-proofing enzymes in the winter-type gushing of bottled beer. Tech. Q. Master Brew. Assoc. Am. *15*, 15–22.

AMANO, K., AMADA, K. and BITO, M. 1963. Detection of formaldehyde in gadoid fish. Nippon Suisan Gakkaishi *29*, 695–701. (Chem. Abstr. *63*, 3453d.) (Japanese)

AMANO, K. and YAMADA, K. 1965. The biological formation of formaldehyde in cod flesh. *In* The Technology of Fish Utilization. R. Krevzer (Editor). Fishing News, London.

AMBERG, S. and LOEVENHART, A. S. 1908. Further observations on the inhibiting effect of fluorides on the action of lipase, together with a method for detecting fluorides in food products. J. Biol. Chem. *4*, 149–164.

AMBERGER, A. and SCHALLER, K. 1974. Influence of variety and storage temperature on constituents involved in enzymic color change in potato tubers. Z. Lebensmittelunters. Forsch. *156*, 231–236. (German)

AMBIKE, S. H., BAXTER, R. M. and ZANID, N. D. 1970. The relationship of cytochrome P-450 levels and alkaloid synthesis in *Claviceps purpurea*. Phytochemistry *9*, 1953–1958.

AM. MED. ASSOC. 1974. Nutrients in Processed Foods, Vol. 2. Vitamins-Minerals. AMA Pub. Sci. Group, Acton, Mass.

AMERINE, M. A., BERG, H. W., KUNKEE, R. E., OUGH, C. S., SINGLETON, V. L. and WEBB, A. D. 1979. The Technology of Wine Making, 4th Edition. AVI Publishing Co., Westport, Conn.

AMERINE, M. A. and JOSLYN, M. A. 1970. Table Wines. The Technology of Their Production. University of California Press, Berkeley.

AMERINE, M. A. and OUGH, C. S. 1971. Recent advances in oenology. Crit. Rev. Food Technol. *2*, 407–515.

AMIR, J., WRIGHT, R. D. and CHERRY, J. H. 1971. Chemical control of sucrose conversion to polysaccharides in sweet corn after harvest. J. Agric. Food Chem. *19*, 954–957.

AMIR, V., KAHN, V. and UNTERMAN, M. 1977. Sugar accumulation in chemically debudded potato tubers during cold storage. Phytochemistry *16*, 1603–1604.

AMNEUS, H., DETLEF, G. and KASCHE, V. 1976. Resolution in affinity chromatography. The effect of the heterogeneity of immobilized soybean trypsin inhibitor on the separation of pancreatic proteases. J. Chromatogr. *120*, 391–397.

AMOORE, J. E. 1970. Molecular Basis of Odor. C.C. Thomas Co., Springfield, Ill.

AMORIM, H. V. and AMORIM, V. L. 1977. Coffee enzymes and coffee quality. *In* Enzymes in Food and Beverage Processing. R.O. Ory and A.J. St. Angelo (Editors). Am. Chem. Soc., Washington, D.C.

AMOTZ, S., NIELSEN, T. G. and THIESEN, N. O. 1976. Glucose isomerase product. Ger. Offen. 2,537,993. March 11.

ANDERSEN, K., SHANMUGAM, K. T., LIM, S. T., CSONKA, L. N., TAIT, R., HENNECKE, H., SCOTT, D. B., HOM, S. S. M., HAURY, J. F., VALENTINE, A. and VALENTINE R. C. 1980. Genetic engineering in agriculture with emphasis on nitrogen fixation. TIBS *5* (2) 35–39.

ANDERSON, E. O. 1939. Preventing development of oxidized flavor in milk through the addition of small amounts of pancreatic enzyme. Milk Dealer *29* (3) 32, 82.

ANDERSON, J. C. and JACKSON, D. S. 1972. The isolation of glycoproteins from Achilles tendon and their interaction with collagen. Biochem. J. *127*, 179–186.

ANDERSON, M., CHEESEMAN, G. E., KNIGHT, D. J. and SHIPE, W. F. 1972. The effect of aging cooled milk on the composition of the fat globule membrane. J. Dairy Res. *39*, 95–105.

ANDERSON, M. A., COOK, J. A. and STONE, B. A. 1978. Enzymatic determination of 1,3: 1,4-β-glucans in barley grains and other cereals. J. Inst. Brew. *84*, 233–239.

ANDERSON, N. G. 1969A. Analytical techniques for cell fractions. XII. A multiple-cuvet rotor for a new microanalytical system. Anal. Biochem. *28*, 545–562.

ANDERSON, N. G. 1969B. Computer interfaced fast analyzers. Science *166*, 317–324.

ANDIA, A. M. G. and STREET, J. C. 1975. Dietary induction of hepatic microsomal enzymes by thermally oxidized fats. J. Agric. Food Chem. *23*, 173–177.

ANDREN, R. K., MANDELS, M. and MODEIROS, J. E. 1976. Production of sugars from waste cellulose by enzymic hydrolysis: primary evaluation of substrates. Process Biochem. *11*, 2–11.

ANDRES, C. 1976. Sweetener outlook. Food Process. *37* (11) 46–48.

ANDRES, C. 1977. Enzyme modified cheeses. Food Process. (Chicago) *38* (1) 134.

ANDREWS, R. S. and PRIDHAM, J. B. 1967. Melanins from DOPA-containing plants. Phytochemistry *6*, 13–18.

ANESHANSLEY, D. J., EISNER, T., WIDOM, J. E. and WIDOM, B. 1969. Biochemistry at 100°C: Explosive secretory discharge of bombardier beetles (Brachinus). Science *165*, 61–63.

ANET, E. F. L. J. and INGLES, D. L. 1976. Do amines in foods trigger headache? CSIRO Food Res. Q. *36* (2) 28–31.

ANGELIDES, K. J. and FINK, A. L. 1978. Cryoenzymology of papain: Reaction mechanism with an ester substrate. Biochemistry *17*, 2659–2668.

ANISIMOV, A. A., ALEKSANDROVA, I. F., IUDINA, K. A., ABZEEV, SH. KH., KONDRATEVA, M. O. and BERESNEVA, G. G. 1977. Changes in the activity and tertiary structure of β-amylase from wheat seeds under the effect of salts. Biokhimiya (Moscow) *42*, 2217–2220. (Russian)

ANISIMOV, V. D., KASTL'EVA, T. B., LONGINOVA, L. N. and MOCHALKIN, A. I. 1975. Comparative study of some oxidoreductases of potato and *Phytophthora infestans*. Dokl. Akad. Nauk SSSR *222*, 480–482. (Russian)

ANKER, C. A. and GEDDES, W. F. 1944. Gelatinization studies upon wheat and other starches with the amylograph. Cereal Chem. *21*, 335–360.

ANON. 1969. Less familiar enzymes. Nature *223*, 1203–1204.

ANON. 1971. Liquid enzyme allows lower processing temperatures. Food Process. *32* (1) 30–31.

ANON. 1972. The potential of fish protein concentrate for developing countries. PAG Bull. *2*, 24–33.

ANON. 1973A. Prevention of deterioration of potatoes upon storage. Food Irradiat. Info. *2*, 43–50.

ANON. 1973B. The enzymatic recovery of candy scrap. Candy Industry Application and Data Sheet. Wallerstein Co., Travenol Lab., Deerfield, Ill.

ANON. 1974. Waste regulations focus attention on blanching. Food Prod. Manage. *96* (10) 14–15.

ANON. 1976A. Stabilization of industrial enzymes. Summaries or titles of papers presented at the symposium. Fermentation and Enzyme Technology Group, June 3. J. Appl. Chem. Biotechnol. *26*, 576–583.

ANON. 1976B. Pulses—determination of glycosidic hydrocyanic acid. Int. Organ. Stand. (ISO), 2164–2197.

ANON. 1976C. Enzymes, immobilized. Chem. Abstr. *85*, 688GS–689GS.

ANON. 1978. Products: Microbial enzymes. Food Technol. *23* (1) 50.

ANON. 1978. Xanthine oxidase. J. Food Prot. *41*, 548.

ANON. 1979. Enzyme Nomenclature. Recommendations 1978 of the Nomenclature Committee of the International Union of Biochemistry. Academic Press, New York.

ANON. 1980A. Enzyme for fuel cells. New Sci. *87*, 454.

ANON. 1980B. Symposium: Protein turnover in heart and skeletal muscle. H.E. Morgan and K. Wildenthal (Editors). Fed. Proc. *39*, 7–82.

ANSON, M. L. 1938. The estimation of pepsin, trypsin, papain and cathepsin with hemoglobin. J. Gen. Physiol. *22*, 79–89.

ANTONINI, E., FANELLI, M. R. R. and CHIANCONE, E. 1975. Properties of polymeric proteins bound to solid matrices: subunit exchange chromatography. *In* Protein-Ligand Interactions. H. Sund and G. Blauer (Editors). Walter de Gruyter, Berlin.

AOKI, K., SHINKE, R. and NISHIRA, H. 1976. Chemical composition and molecular weight of yeast tannase. Agric. Biol. Chem. *40*, 297–302.

AOSHIMA, H., KAJIWARA, T., HATANAKA, A., NAKATANI, H. and HIROMI, K. 1975. Biosynthesis of leaf alcohols. Kinetic study of lipoxygenase activation process caused by hydroperoxylinoleic acid. Agric. Biol. Chem. *39*, 2255–2257.

APPLEMAN, C. O. and CONRAD, C. M. 1926. Pectic constituents of peaches and their relation to softening of the fruit. Agric. Exp. Stn. Bull. *283*, 1–8.

APPLEMAN, C. O. and CONRAD, C. M. 1928. The pectic constituents of tomatoes and their relation to the canned product. Md. Agric. Exp. Stn. Bull. *291*, 1–17.

APPLEMAN, D. 1972. High protein fruit-type beverage and processing thereof. Can. Pat. 9,090,065. Jan. 21.

APPOLONIA, B. L. 1971. Role of pentosans in bread and dough. Bakers Dig. *45* (6) 20–23, 63.

AP REES, T., THOMAS, S. M., FULLER, W. A. and CHAPMAN, B. 1975. Location of gluconeogenesis from phosphoenolpyruvate in cotyledons of *Cucurbita pepo*. Biochim. Biophys. Acta *385,* 145–156.

ARAI, K. 1972. Mucor rennin. Proc. Intern. Symp. Convers. Manuf. Foodst. Microorganisms, 1971. Saikon Publishing Co., Tokyo.

ARAI, S., YAMASHITA, K. A. and FUJIMAKI, M. 1975. Plastein reaction and its application. Cereal Foods World *20,* 107–112.

ARAKAWA, N., FUJIKI, S., INAGAKI, C. and FUJIKI, M. 1976A. A catheptic protease active in ultimate pH of muscle. Agric. Biol. Chem. *40,* 1265–1267.

ARAKAWA, N., INAGAKI, C., KITAMURA, T., FUJIKI, S. and FUJIMAKI, M. 1976B. Some possible evidences for an alteration in the actin-myosin interaction in stored muscle. Agric. Biol. Chem. *40,* 1445–1447.

ARCHER, M. C., RAGNARSSON, J. O., WANG, F. C. and TANNENBAUM, S. R. 1973. Enzymatic solubilization of an insoluble substrate, fish protein concentrate: Process and kinetic considerations. Biotechnol. Bioeng. *15,* 181–196.

AREHART-TREICHEL, J. 1978. Enzymes: medicine's new gold mine. Sci. News *114* (4) 58–60.

ARIKI, M. and FUKUI, T. 1975. α-Glucan phosphorylase from sweet potato: Isolation and properties of the partially degraded enzyme. Biochim. Biophys. Acta *386,* 301–308.

ARIMA, K. 1972. Mucor rennin. *In* Proc. Intern. Symp. Manuf. Foodst. Microorganisms, 1971. Saikon Publishing Co., Tokyo.

ARIMA, K., BEPPU, T. and MATSUDA, Y. 1976. Microbial production of lipoxidase. Jpn. Kokai 76 29,290. Mar. 12.

ARIMA, K. and IWASAKI, S. 1964. Milk coagulation enzyme "microbial rennet" and method of preparation thereof. U.S. Pat. 3,151,039. Sept. 29.

ARIMA, K., IWASAKI, S. and TAMURA, G. 1967. Milk clotting enzyme from microorganisms. I. Screening test and identification of the potent fungus. Agric. Biol. Chem. *31,* 540–545.

ARIMA, K., NAGASAWA, M. B. and TAMURA, G. 1970. Microbial transformations of sterols. I. Decomposition of cholesterol by microorganisms. Rep. of the Noda Inst. No. *14,* 1–7. (Japanese)

ARIMA, S., DAIGUJI, H., SHIBUKAWA, M. and OHUCHI, S. 1977. Glutamic acid fermentation. Jpn. Pat. 77 6,359. Feb. 21.

ARIYOSHI, Y. 1976. The structure-taste relationships of aspartyl dipeptide esters. Agric. Biol. Chem. *40,* 983–992.

ARMAND, P., ABELLO, G. and GAYTE-SORBIER, A. 1976. Methods for the determination of sodium glutamate. II. Canned soups, sauces, ready meals, seasonings and aromas. Ann. Falsif. Expert. Chim. *59,* 545–550. (French)

ARMBRUSTER, F.C., HEADY, R. E. and CORY, R. P. 1974. Production of xylose (glucose) isomerase enzyme preparations. U.S. Pat. 3,813,318. May 28.

ARMSTRONG, D. L. and STANLEY, D. W. 1974. The utilization of microwave heating in the production of soymilk protein. Cereal Sci. Today *19,* 415.

ARNESON, R. M. 1970. Substrate induced chemiluminescence of xanthine oxidase and aldehyde oxidase. Arch. Biochem. Biophys. *136,* 352–360.

ARNESTARD, K. G. and BORDALEN, B. E. 1972. Caramel color and non-enzymatic browning. Tidsskr. Hermetikind. *58,* 177–180. (Norwegian)

ARNON, D. I. 1949. Copper enzyme in isolated chloroplasts. Polyphenoloxidase in *Beta vulgaris*. Plant Physiol. *24,* 1–15.

ARNOUX, A., ABELLO, G., GAYTE-SORBIER, A. and ARMAND, P. 1976. Methods for the determination of sodium glutamate. III. Utilization of the centrifugal fast analyzer. Ann. Falsif. Expert. Chim. *60,* 557–562. (French)

ARSENIS, C. and McCORMICK, D. B. 1964. Purification of liver flavokinase by column chromatography on flavin-cellulose compounds. J. Biol. Chem. *239,* 3093–3097.

ARZU, A., MAYORGA, H., GONZALEZ, J. and ROLZ, C. 1972. Enzymatic hydrolysis of cottonseed protein. J. Agric. Food Chem. *20,* 805–809.

ASADA, K. 76. Occurrence of superoxide dismutase in bovine milk. Agric. Biol. Chem. *40,* 1659–1660.

ASADA, K., URANO, M. and TAKAHASHI, M. 1973. Subcellular location of superoxide dismutase in spinach leaves and preparation and properties of crystalline spinach superoxide dismutase. Eur. J. Biochem. *36,* 257–266.

ASGHAR, K., REDDY, B. G. and KRISHNA, G. 1975. Histochemical localization of glutathione in tissues. J. Histochem. Cytochem. *23,* 744–779.

ASHMORE, C. R., CARROL, F., DOERR, L., TOMPKINS, G., STORES, H. and PARKER, W. 1973. Experimental prevention of dark-cutting meat. J. Anim. Sci. *36*, 33–36.

ASIMOV, I. 1959. Enzymes and metaphor. J. Chem. Educ. *36*, 535–538.

ASKAR, A. 1976. Amines and nitrosamines. Occurrence, significance, metabolism and determination. Advances in Food Science No. *4*. University Library of the Technical University, Berlin.

ASKAR, A. and BIELIG, H. J. 1976. Improving flavor of foods. II. Addition of enzymes. Alimenta *15*, 155–156, 158–159. (German)

ASLAM, S., JONES, D.P. and BROWN, T. R. 1976. Improved method for the removal of albumin from serum by affinity chromatography. Anal. Biochem. *75*, 329–335.

ASO, K., YAMASHITA, M., ARAI, S., SUZUKI, J. and FUJIMAKI, M. 1977. Specificity for incorporation of α-amino acid esters during the plastein reaction by papain. J. Agric. Food Chem. *25*, 1138–1141.

ASSOC. OFF. ANAL. CHEM. 1975. Official Methods of Analysis, 12th Edition. AOAC, Washington, D.C.

ATALLAH, M. T. and HULTIN, H. O. 1977. Preparation of soluble conjugates of glucose oxidase and catalase by cross-linking with glutaraldehyde. J. Food Sci. *42*, 7–11.

ATKINS, G. L. and NIMMO, I. A. 1975. Comparison of seven methods for fitting the Michaelis-Menten equation. Biochem. J. *149*, 775–777.

ATKINS, G. L. and NIMMO, I. A. 1980. Current trends in the estimation of Michaelis-Menten parameters. Anal. Biochem. *104*, 1–9.

ATKINSON, J. L. and FOLLETT, M. J. 1973. Biochemical studies on the discoloration of fresh meat. J. Food. Technol. *8*, 51–58.

ATTILA, A. 1971. Thin-layer-chromatographic changes in green peas during canning. Ind. Obst.-Gemeuseverwert. *56*, 335–337. (German)

ATTREP, K. A., MARIANI, J. M., JR. and ATTREP, M., JR. 1973. Search for prostaglandin A_1 in the onion. Lipids *8*, 484–486.

ATTRIDGE, T. H. 1974. Phytochrome-mediated synthesis of ascorbic acid oxidase in mustard cotyledons. Biochim. Biophys. Acta *362*, 258–265.

ATTRIDGE, T. H., STEWART, G. E. and SMITH, H. 1971. End-product inhibition of *Pisum* ammonia-lyase by the *Pisum* flavonoids. FEBS Lett. *17*, 84–86.

AUDIDIER, Y. 1968. Effects of thermal kinetics and weight loss kinetics on biochemical reactions in dough. Bakers Dig. *42* (5) 36–42.

AUDLEY, M. J. and KINSELLA, J. E. 1978. Isolation of phospholipase A from pollock muscle. J. Food Sci. *43*, 1771–1775.

AUERMAN, L., POPOV, M. P. and DUBTSOV, G. G. 1971. Effect of specific inhibitors of wheat flour lipoxygenase on physical properties of dough. Prikl. Biokhim. Mikrobiol. 7, 568–588. (Russian)

AUGUSTINSSON, K. B. 1948. Cholinesterase, a study in comparative enzymology. Acta Physiol. Scand. *15*, Supplement 52.

AUKRUST, L. E., NORUM, K. R. and SKALHEGG, B. A. 1976. Affinity chromatography of 3α-hydroxysteroid dehydrogenase from *Pseudomonas testosteroni*. Use of N,N-dimethylformamide to prevent hydrophobic interactions between the enzyme and the ligand. Biochim. Biophys. Acta *438*, 13–22.

AUNE, T. M., THOMAS, E. L. and MORRISON, M. 1977. Lactoperoxidase-catalyzed incorporation of thiocyanate ion into a protein substrate. Biochemistry *16*, 4611–4615.

AUNSTRUP, K. 1976. Preparation of a milk-coagulating enzyme. U.S. Pat. 3,988,207. Oct. 26.

AURAND, L. W., BOONE, N. H. and GIDDINGS, G. G. 1977. Superoxide and singlet oxygen in milk peroxidation. J. Dairy Sci. *60*, 363–367.

AURAND, L. and WOODS, A. E. 1959. Role of xanthine oxidase in the development of spontaneously oxidized flavor in milk. J. Dairy Sci. *42*, 1111–1117.

AUSTIN, S. J. and SCHWIMMER, S. 1971. L-γ-Glutamyl peptidase activity in sprouted onion. Enzymologia *40*, 273–285.

AVRAMEAS, S. 1969. Coupling of enzymes to proteins with glutaradehyde. Use of the conjugates for the detection of antigens and antibodies. Immunochemistry *6*, 43–52.

AVRON, 1977. Energy transduction in chloroplasts. Annu. Rev. Biochem. *46*, 143–155.

AWAD, A., POWRIE, W. D. and FENNEMA, O. 1969. Deterioration of fresh-water whitefish muscle during frozen storage at 10°C. J. Food Sci. *34*, 1–9.

AXCELL, B. C. and DONNINGER, C. 1977. Determination of ethanol in body fluids. Ger. Offen. 2,701,168. July 21.

AXELROD, B. 1947. Phosphatase activity as an index of pasteurization in citrus juices. Fruit Prod. J. *26*, 132–133.

AXELROD, B. 1974. Lipoxygenase. *In* Food Related Enzymes. J.W. Whitaker (Editor). Am. Chem. Soc., Washington, D.C.

AXELROD, B. and JAGENDORF, A. T. 1951.

The fate of phosphatase, invertase and peroxidase in autolyzing leaves. Plant Physiol. *26*, 406–410.

AYLWARD, F. and HAISMAN, D. R. 1969. Oxidation systems in fruits and vegetables—Their relation to the quality of preserved products. Adv. Food Res. *17*, 1–75.

AYRES, J. C., MUNDT, J. O. and SANDINE, W. E. 1980. Microbiology of Foods. W.H. Freeman and Co., San Francisco.

BABISH, J. G. and STOEWSAND, G. S. 1975. Hepatic microsomal enzyme induction in rats fed varietal cauliflower leaves. J. Nutr. *105*, 1592–1599.

BACHMAN, B. K. and LEE, C. -Y. 1976. Purification of human lactate dehydrogenase by general ligand affinity chromatography. Anal. Biochem. *72*, 153–160.

BACHMANN, M. 1973. Effect of climate on lipase activity of milk. Schweiz. Milchwirtsch. Forsch. *2* (June) 25–28. (German)

BADWEY, J. A. and KARNOVSKY, M. L. 1980. Active oxygen species and the functions of phagocytic leukocytes. Annu. Rev. Biochem. *49*, 695–726.

BAGSHAW, C. R. and TRENTHAM, D. R. 1974. The characterization of myosin-product complexes and of product release steps during the magnesium ion-dependent adenosine triphosphatase reaction. Biochem. J. *141*, 331–349.

BAHL, R. K. 1972. An enzymic method for the determination of skimmed milk powder in soup and sauce mixes. Analyst *1152*, 213–215.

BAILEY, M. J. and MARKANNEN, P. H. 1975. Use of mutagenic agents in the improvement of α-amylase production by *Bacillus subtilis*. J. Appl. Chem. Biotechnol. *25*, 73–79.

BAILEY, S. D., BAZINET, M. L., DRISCOLL, J. L. and McCARTHY, A. I. 1961. The volatile sulfur components of cabbage. J. Food Sci. *26*, 163–170.

BAKER, B. R. 1967. Design of Active-Site-Directed Irreversible Enzyme Inhibitors. John Wiley & Sons, New York.

BAKER, D. L. 1950. Deoxygenation process. U.S. Pat. 2,482,724. Sept. 20.

BAKER, J. E. 1976. Superoxide dismutase in ripening fruits. Plant Physiol. *68*, 644–647.

BAKER, J. E. and TAKEO, T. 1973. Acid phosphatase in plant issues. I. Changes in activity and multiple forms in tea leaves and tomato fruit during maturation and senescence. Plant Cell Physiol. *14*, 459–471.

BAKER, R. A. and BRUEMMER, J. H. 1972. Pectinase stabilization of orange juice clouds. J. Agric. Food Chem. *20*, 1169–1173.

BAKOWSKI, J. and MALESKI, W. 1968. Determination of peroxidase activity in blanched vegetables. Przem. Spozyiv. *22*, 407–408. (Polish)

BALASUBRANIUM, K., DEY, P. M. and PRIDHAM, J. B. 1976. α-Galactosidase from coconut kernel. Phytochemistry *15*, 1445–1446.

BALCOM, J., FOULKES, P., OLSON, N. F. and RICHARDSON, N. F. 1971. Immobilized catalase. Process Biochem. *6* (8) 42–44.

BALDWIN, R. L. 1975. Intermediates in protein folding reactions and the mechanism of protein folding. Ann. Rev. Biochem. *44*, 453–475.

BALDWIN, R. R., CAMPBELL, H. A., THIESSEN, R., JR. and LORANT, G. J. 1953. The use of glucose oxidase in the processing of foods with special emphasis on the desugaring of egg white. Food Technol. *7*, 275–282.

BALLANCE, G. M. and MEREDITH, W. O. S. 1976. Purification and partial characterization of the endo-β-1,3-glucanase from green malt. J. Inst. Brew. *82*, 64–67.

BALLESTER, D., YANEZ, E., BRUNSER, O., STEKEL, A., CHADUD, P., CASTANO, G. and MONCKEBERG, F. 1977. Safety evaluation of an enzymatic fish protein hydrolysate: 10-month feeding study and reproduction performance in rats. J. Food Sci. *42*, 407–409.

BALLOU, C. E. 1976. Structure and biosynthesis of the mannan component of the yeast cell envelope. Adv. Microb. Physiol. *14*, 93–158.

BALLS, A. K. 1932. The quantitative determination of enzyme action. J. Assoc. Off. Agric. Chem. *15*, 132–136.

BALLS, A. K. 1938. Enzyme action in food products at low temperatures. Proc. Gen. Conf. Refrig. Brit. Assoc. Refrig., July, London.

BALLS, A. K. 1939. Enzymes in foods and food preservation. *In* Yearbook of Agriculture. U.S. Dep. Agric., Washington, D.C.

BALLS, A. K. 1941. Protein-digesting enzymes of papaya and pineapple. Circ. *631*, U.S. Dep. Agric., Washington, D.C.

BALLS, A. K. 1942. The fate of enzymes in processed foods. Fruit Prod. J. *22*, 36–39.

BALLS, A. K. 1943. Desmo enzymes. Vortex *4* (10) 3–8.

BALLS, A. K. 1947. Enzyme actions and food quality. Food Technol. *1*, 245–251.

BALLS, A. K. 1948. What a foreman should know about enzyme action. Food Ind. *20*, 189–326.

BALLS, A. K. 1950. Enzymes and enzymology. *In* Encyclopedia of Chemical Technology, Vol. 5. R.E. Kirk and D.F. Othmer (Editors). The Interscience Encyclopedia, New York.

BALLS, A. K. 1960. Enzymes affecting proteins. *In* Food Enzymes: Symposium. H.W. Schultz (Editor). AVI Publishing Co., Westport, Conn.

BALLS, A. K. 1962. Catalysts and food technologists. Food Technol. *16*, 17–20.

BALLS, A. K. and ARANA, F. E. 1941. The curing of vanilla. Ind. Eng. Chem. *33*, 1073–1075.

BALLS, A. K. and HALE, W. S. 1932. The estimation of catalase in agricultural products. J. Assoc. Off. Agric. Chem. *15*, 483–490.

BALLS, A. K. and HALE, W. S. 1935. Process for inhibiting of the discoloration of fruits and vegetables. U.S. Pat. 2,011,465. Aug. 13.

BALLS, A. K. and HALE, W. S. 1936. Proteolytic enzymes of flour. Cereal Chem. *13*, 54–60.

BALLS, A. K. and HALE, W. S. 1938. The preparation and properties of wheat proteinase. Cereal Chem. *15*, 622–628.

BALLS, A. K. and JANSEN, E. F. 1952. Stoichiometric inhibition of chymotrypsin. Adv. Enzymol. *13*, 321–343.

BALLS, A. K. and LINEWEAVER, H. 1938. Action of enzymes at low temperatures. Food Res. *3*, 57–67.

BALLS, A. K., LINEWEAVER, H., and SCHWIMMER, S. 1940. Drying of papaya latex and stability of papain. Ind. Eng. Chem. *32*, 1277–1279.

BALLS, A. K., LINEWEAVER, H. and SCHWIMMER, S. 1941. Process for the preparation of papain. U.S. Pat. 2,257,218. Sept. 30.

BALLS, A. K. and SCHWIMMER, S. 1944. Digestion of raw starch. J. Biol. Chem. *156*, 203–210.

BALLS, A. K. and SWENSON, T. L. 1936. Dried egg white. Food Res. *1*, 319–325.

BALLS, A. K., WALDEN, M. K. and THOMPSSON, R. R. 1948. A crystalline β-amylase from sweet potatoes. J. Biol. Chem. *173*, 9–19.

BAMANN, E. and MYRBÄCK, K. 1941. Methods of Enzyme Research. (4 Vols). Georg Thiem, Verlag, Leipzig Photo Offset Reproduction, 1945, Academic Press, New York. (German)

BAMANN, E. and TRAPMANN, H. 1959. Metal-ion catalyzed reactions, especially in the area of the rare earth: A contribution to enzyme and enzyme model catalysis. Adv. Enzymol. *21*, 169–188. (German)

BANDYOPADHYAY, C. and TEWARI, G. M. 1973. Thin-layer chromatographic investigations of color developer in pinking of white onion purees. J. Agric. Food Chem. *21*, 252–254.

BANDYOPADHYAY, C. and TEWARI, G. M. 1976. Lachrymatory factor in sprouted onion *(Allium cepa)*. J. Sci. Food Agric. *27*, 733–735.

BANKS, W., EVERS, A. D. and MUIR, D. D. 1972. The location of alpha-amylase in developing cereal grains. Chem. Ind. 573–574.

BANKS, W. and GREENWOOD, C. T. 1977. Mathematical models for the action of alpha-amylase on amylase. Carbohydr. Res. *57*, 301–315.

BANQUI, S. M., AMTOO, A. K. and MODI, V. V. 1977. Glyoxylate metabolism and fatty acid oxidation in mango fruit during development and ripening. Phytochemistry *16*, 51–54.

BAR-AKIVA, A. and LAVON, R. 1969. Carbonic anhydrase activity as an indicator of zinc deficiency in citrus leaves. J. Hortic. Sci. *44*, 359–362.

BAR-AKIVA, A., LAVON, R. and SAGIV, J. 1969. Ascorbic acid oxidase activity as a measure of the copper nutrition requirement of citrus trees. Agrochemica *14*, 47–54.

BARANOWSKI, T. 1949. Crystalline glycerophosphate dehydrogenase from rabbit muscle. J. Biol. Chem. *180*, 535–541.

BARBOSA, M., VALLES, E., VASSAL, L. and MOCQUOT, G. 1976. Use of *Cynara cardunculus* L. extract as a coagulant in the manufacture of soft and cooked cheeses. Lait *56* (551/552) 1–17. (French)

BARFOED, H. C. 1976. Enzymes in starch processing. Cereal Food World *21*, 588–604.

BARKER, R. J. and LEHNER, Y. 1972. A look at honey bee gut functions. Am. Bee J. *112*, 336–338.

BARKER, S. A. 1975. High fructose syrups: New sweeteners in the food industry. Proc. Biochem. *10* (10) 39–40.

BARKER, S. A. 1976. Pure fructose syrups. Process Biochem. *11* (10) 20–25.

BARKER, S. A. 1977. Use of enzymes in analytical chemistry. Enzyme immunoassay. Proc. Anal. Dir. Chem. Soc. *14*, 103–104.

BARKER, S. A., HATT, B. W., SOMERS, P. J. and WOODBURY, R. R. 1973. The use of poly (4-vinylbenzeneboronic acid) resins in the fractionation and interconversion of carbohydrates. Carbohydr. Res. *26*, 55–64.

BARMAN, T. E. 1969. Enzyme Handbook, Vol. 1 and 2. Springer Verlag, New York.

BARMAN, T. E. 1974. Enzyme Handbook Supplement. Springer Verlag, New York.

BARMORE, C. R. and BIGGS, R. H. 1972. Acid-soluble nucleotides of juice vesicles of citrus fruit. J. Food Sci. *37*, 712–714.

BARNA, J. and PRILLINGER, F. 1972. Investigation of invertases in grape wines by polyacrylamide gel electrophoresis. Mitt. Rebe, Wein, Obstbau Früchtverw. *22*, 417–420. (German)

BARNELL, H. R. and BARNELL, F. 1945. Studies on tropical fruits. XVI. The distribution of tannins within the banana and the changes in their condition and amount during ripening. Ann. Bot. *9*, 77–99.

BARNES, M. F. and PATCHETT, B. J. 1976. Cell wall degrading enzymes and the softening of senescent strawberry fruit. J. Food Sci. *41*, 1392–1395.

BARNETT, R. N., EWING, N. S. and SKODON, S. B. 1976. Performance of "kits" used for clinical chemical analysis of GOT (aspartate aminotransferase). Clin. Biochem. *9*, 78–84.

BARRATI, J., COUDERC, R., COONEY, C. L. and WANG, D. I. C. 1978. Preparation and properties of immobilized methanol oxidase. Biotechnol. Bioeng. *20*, 333–348.

BARRETT, A. J. and DINGLE, J. T. 1972. The inhibition of tissue acid proteinases by pepstatin. Biochem. J. *127*, 439–441.

BARTA, E. J., LAZAR, M. E. and RASMUSSEN, C. L. 1973. Dehydration plant operations. *In* Food Dehydration, 2nd Edition, Vol. 1. W.B. Van Arsdel, M.J. Copley and A.I. Morgan, Jr. (Editors). AVI Publishing Co., Westport, Conn.

BARTLEY, I. M. 1978. Exo-polygalacturonase of apple. Phytochemistry *17*, 213–216.

BARTOLME, L. G. and HOFF, J. E. 1972. Firming of potatoes: Biochemical effects of preheating. J. Agric. Food Chem. *20*, 266–270.

BASAROVA, G. and TURKOVA, J. 1977. Properties, method of immobilization and application of papain bound to hydroxyalkylmethacrylate gel. Brauwissenschaft *30*, 204–209. (German)

BATA, J. and NESKOVIC, M. 1974. The effect of gibberellic acid and kinetin on chlorophyll retention in *Lemna trisulca* L. Z. Pflanzenphysiol. *73*, 86–88. (German)

BATHGATE, G. N. and DAGLIESH, C. E. 1975. Diversity of barley and malt β-glucans. Proc. Am. Soc. Brew. Chem. *33*, 32–36.

BATHGATE, G. N. and PALMER, G. H. 1973. The in vivo degradation of barley and malt starch granules. J. Inst. Brew. *79*, 402–406.

BAUGHER, W. L. and CAMPBELL, T. C. 1969. Gossypol detoxication by fungi. Science *164*, 1526–1527.

BAUM, G. and WARD, F. B. 1971. Ion-selective electrode procedure for organophosphate pesticide analysis. Anal. Chem. *43*, 947–948.

BAUM, G. and WROBEL, S. J. 1975. Affinity chromatography. Methods Enzymol. (N.Y.) *1*, 419–496.

BAUMRUCKER, C. R. 1980. Purification and identification of γ-glutamyl transpeptidase of milk membranes. J. Dairy Sci. *63*, 49–54.

BAUR, C., GROSCH, W., WIESER, H. and JUGEL, H. 1977. Enzymic oxidation of linoleic acid: formation of bitter-tasting fatty acids. Z. Lebensm. Unters. Forsch. *164*, 171–176.

BAVISOTTO, V. S. 1967. Process for producing brewers' wort with enzymes. U.S. Pat. 3,353,960. Nov 21.

BAXTER, E. D., BOOER, C. D. and WAINWRIGHT, T. 1978. Degradation of hordein by the proteolytic enzymes of barley and malt. J. Inst. Brew. *84*, 30–33.

BAXTER LABORATORIES. 1966. Confectionery additives. Br. Pat. 1,008,655. Aug. 7.

BAYLESS, T. M. and ROSENSWEIG, N. S. 1966. A racial difference in incidence of lactase deficiency: A survey of milk intolerance in healthy adult males. J. Am. Med. Assoc. *197*, 968–972.

BAYLISS, W. M. 1906. On some aspects of adsorption phenomenon. Biochem. J. *1*, 175–232.

BAYLISS, W. M. 1925. The Nature of Enzyme Action. Longmans, Green and Co., London.

BEAN, M. M., KEAGY, P. M., FULLINGTON, F. T. and MECHAM, D. K. 1974. Dried Japanese noodles. I. Properties of laboratory-prepared noodle doughs and damaged wheat flours. Cereal Chem. *51*, 416–427.

BEAUCAMP, K. and LILLY, H. 1977. Stabilizing agent for enzymes. Br. Pat. 1,488,988. Oct. 19.

BEAUCHAMP, C. O. and FRIDOVICH, I. 1973. Isozymes of superoxide dismutases. Biochim. Biophys. Acta *317*, 50–64.

BEAUX, Y. and DAPRON, R. 1975. Isolation and some physiochemical characteristics of lipoxygenase of the horsebean (*Vicia fava* L.) Ann. Technol. Agric. *23*, 309–321. (French)

BECK, C. L. and SCOTT, D. 1974. Enzymes in foods—for better or worse. *In* Food Related Enzymes. J.R. Whitaker (Editor). Am. Chem. Soc., Washington, D.C.

BECKER, R., MIERS, J.C., NUTTING, M. D., DIETRICH, W. R. and WAGNER, I. R. 1972. Consistency of tomato products. 7. Effects of acidification on cell walls and cell breakage. J. Food Sci. *37*, 118–125.

BECKER, R., OLSON, A. C., FREDERICK, D. A., KON, S., GUMBMANN, M. R. and WAGNER, J. R. 1974. Conditions for the autolysis of alpha galactosides and phytic acid in California

Small White beans. J. Food Sci. *39*, 766–769.

BECKER, S. 1958. The production of papain—an agricultural industry for tropical America. Econ. Bot. *12*, 62–79.

BECKER, W. and PFEIL, E. 1966. Continuous synthesis of optically active α-hydroxynitriles. J. Am. Chem. Soc. *88*, 4299–4300.

BECKHORN, E. J., LABBEE, M. D. and UNDERKOFLER, L. A. 1965. Production and use of microbial enzymes for food processing. J. Agric. Food Chem. *13*, 30–34.

BEDDOWS, C. G., ISMAIL, M. and STEINKRAUS, K. H. 1976. The use of bromelain in the hydrolysis of mackerel and the investigation of fermented fish aroma. J. Food Technol. *11*, 379–388.

BEERY, K. E. and LINEWEAVER, H. 1976. Objective methodology to differentiate between fresh and frozen-and-thawed meats. U.S. Army Natick Res. and Dev. Command, Tech. Rep. *TR75–102 FEL*, July.

BEHNKE, J. R., FENNEMA, O. and CASSENS, R. G. 1973. Rates of postmortem metabolism in frozen animal tissues. J. Agric. Food Chem. *21*, 5–16.

BEIDLER, L. M. 1966. Chemical excitation of taste and odor receptors. *In* Flavor Chemistry. Adv. Chem. Ser. *56*. Am. Chem. Soc., Washington, D.C.

BEINERT, H. 1978. Presence and function of metals in the mitochondrial electron transfer system. *In* Molecular Biology of Membranes, Symp. Proc. S. Fleischer, Y. Hatefi and D.H. MacLennan (Editors). Plenum Press, New York.

BELASCO, J. G. and KNOWLES, J. R. 1980. Direct observation of substrate distortion by triosephosphate isomerase using Fourier transform infrared spectroscopy. Biochemistry *19*, 472–477.

BELEHRADEK, J. 1954. Temperature and rate of enzyme action. Nature *173*, 70–71.

BELIC, I. and SOCIC, H. 1972. Microbiological degradation of tomatidine. Experientia *27*, 626.

BELITZ, H. D. and LYNEN, F. 1974. Proteolytic activity of wheat. Chem. Mikrobiol. Technol. Lebensm. *3*, 60–64. (German)

BELL, J. M., ROYAN, G. F. and YOUNGS, C. G. 1974. Digestibility of pea protein concentrate and enzyme-treated pea flour in milk replacers for calves. Can. J. Anim. Sci. *54*, 355–362.

BELL, T. A., ETCHELLS, J. L., SINGLETON, J. A. and SMART, W. G., JR. 1965. Inhibition of pectinolytic and cellulolytic enzymes in cucumber fermentations by *Sericea*. J. Food Sci. *30*, 233–239.

BELL, T. A., ETCHELLS, J. L. and SMART, W. W. G., JR. 1968. Enzyme inhibitor for preventing softening in brined foods. U.S. Pat. 3,374,099. Mar. 19.

BELLAMY, W. D. 1976. Applications of microbial enzyme technology to agricultural wastes. USA-ROC Symp. Enzyme Technol. Lehigh Univ., Aug. 9–14. Bethlehem, Pa.

BENDALL, J. R. 1954. The swelling effect of polyphosphates in lean meat. J. Sci. Food Agric. *5*, 468–475.

BENDALL, J. R. 1972. Consumption of oxygen by the muscles of beef animals and related species, and its effect on the colour of meat. J. Sci. Food Agric. *23*, 61–72.

BENDALL, J. R. 1973. Postmortem changes in muscle. *In* Structure and Function of Muscle, Vol. II, Part 2, 2nd Edition. G.H. Bourne (Editor). Academic Press, New York.

BENDALL, J. R. and TAYLOR, J. R. 1972. Consumption of oxygen by the muscles of beef animals and related species. II. Consumption of oxygen by post-rigor muscle. J. Sci. Food Agric. *23*, 707–719.

BENDER, M. L. 1971. Mechanisms of Homogeneous Catalysis from Protons to Proteins. John Wiley & Sons, New York.

BENDER, M. L. and BRUBACHER, L. J. 1973. Catalysis and Enzyme Action. McGraw-Hill, New York.

BENDER, M. L., CLEMENT, G. E., KEDZY, F. J. and HECK, H. D. 1964. The correlation of the pH (pD) dependence and stepwise mechanism of α-chymotrypsin-catalyzed reactions. J. Am. Chem. Soc. *86*, 3680–3690.

BENDER, M. L. and KOMIYAMA, M. 1978. Cyclodextrin Chemistry. Springer Verlag, Berlin.

BENDIX, G. H., HENRY, R. and STRODTZ, N. 1952. Preservation of green color in canned vegetables. U.S. Pat. 2,589,037. Mar. 11.

BENEDICKT, G. 1972. Influences on crust color in white bread. Dtsch. Lebensm.-Rundsch. *68*, 406. (German)

BENEDICT, C. R. 1978. The fractionation of stable carbon isotopes in photosynthesis. What's New in Plant Physiology *9* (4) 13–16.

BENEDICT, R. C., STRANGE, E. D. and SWIFT, C. E. 1975. Effect of lipid antioxidants on the stability of meat during storage. J. Agric. Food Chem. *23*, 167–173.

BENESCH, R., BENESCH, R. E. and YU, C. I. 1968. Reciprocal binding of human and diphosphoglycerate by human hemoglobin. Proc. Natl. Acad. Sci. *59*, 526–532.

BEN-ET, DOLEV, A. and TATARSKY, O. 1973. Compounds contributing to heat-induced off-flavor in avocado. J. Food Sci. *38*, 546–547.

BEN-GERA, I. and KRAMER, A. 1969. The utilization of food industries waste. Adv. Food Res. *17*, 77–152.

BENGTSSON, B. L. and BOSUND, I. 1966. Lipid hydrolysis in unblanched frozen peas *(Pisum sativum)*. J. Food Sci. *31*, 474–481.

BENGTSSON, B. L. and BOSUND, I. 1975. Processing vegetables. Ger. Pat. 1,949,648. Oct. 2.

BENGTSSON, B. L., BOSUND, I. and RASHUSSEN, I. 1967. Hexanal and ethanol formation in peas in relation to off-flavor development. Food Technol. *21*, 160A–164A.

BENGTSSON, G. and OLIVECRONA, T. 1977. Interaction of lipoprotein lipase with heparin—Sepharose. Evaluation of conditions for affinity binding. Biochem. J. *167*, 109–119.

BENJAMIN, N. D. and MONTGOMERY, M. W. 1973. Polyphenol oxidase of Royal-Ann cherries: Purification and characterization. J. Food Sci. *38*, 799–806.

BEN-NAIM, A. 1980. Hydrophobic Interactions. Plenum Press, New York.

BENNET, G. A., FREER, S. and SHOTWELL, O. D. 1976. Hydrolysis of corn oil by lipase from *Helminthosporium maydis*. J. Am. Soc. Oil Chem. Soc. *53*, 52–53.

BEN-SHALOM, N., KAHN, V., HAREL, E. and MAYER, A. M. 1977. Olive catechol oxidase—Changes during fruit development. J. Sci. Food Agric. *28*, 545–550.

BEREZIN, I. V. 1978. Enzymes as catalysts of electrochemical reactions. J. Mol. Catal. *14*, 719–723.

BERG, R. K., SLINDE, E. and FARSTAD, M. 1980. Discontinuities in Arrhenius plots due to the formation of mixed micelles and change in enzyme substrate availability. FEBS Lett. *109*, 194–196.

BERGE, R. K., SLINDE, E. and FARSTAD, M. 1980. Discontinuities in Arrhenius plots due to formation of mixed micelles and change in enzyme substrate availability. FEBS Lett. *109*, 194–196.

BERGERON, R. and CHANNING, M. A. 1976. The molecular disposition of *p*-nitrophenol and sodium *p*-nitrophenolate in the cyclohexaamylose cavity: a ^{13}C probe. Bioorg. Chem. *5*, 437–439.

BERGHEM, L. E. R., PETTERSSON, L. G. and AVIO-FREDRIKSSON, U. B. 1976. The mechanism of enzymatic cellulose degradation. Purification and some properties of two different 1,4-β-glucan glucan hydrolases. Eur. J. Biochem. *61*, 621–630.

BERGMEYER, H. -U. 1963. Methods of Enzymatic Analysis. Academic Press, New York.

BERGMEYER, H. -U. 1974. Methods of Enzymatic Analysis, Vol. 1–4, 2nd Edition. Academic Press, New York.

BERGNER-LANG, B. 1977. New observations on determination of isocitric acid in citrus fruit. Dtsch. Lebensm.-Rundsch. *73*, 211–216. (German)

BERGQUIST, D. H. 1973. Egg dehydration. *In* Egg Science and Technology. W.J. Stadelman and O.J. Cotterill (Editors). AVI Publishing Co., Westport, Conn.

BERK, Z. 1976. Braverman's Introduction to the Biochemistry of Foods. New Edition. Elsevier Scientific Publishing Co., Amsterdam.

BERKELEY, H. D. and GALLIARD, T. 1976. Substrate specificity of potato lipoxygenase. Phytochemistry *15*, 1481–1484.

BERLIN, E., KLIMAN, P. G. and PALLANSCH, M. J. 1970. Changes in state of water in proteinaceous systems. J. Colloid Interface Sci. *3*, 488–494.

BERMAN, H., RUBIN, B. H., CARRELL, H. L. and GLUSKER, J. P. 1974. Crystallographic studies of D-xylose isomerase. J. Biol. Chem. *249*, 3983–3984.

BERNARDI, G. 1971. Chromatography of proteins on hydroxyapatite. Methods Enzymol. *22*, 325–339.

BERNARDIN, J. E., MECHAM, D. K. and PENCE, J. W. 1965. Proteolytic action of wheat flour on nonfat dry milk proteins. Cereal Chem. *42*, 97–106.

BERNAT, J. A. 1977. Production of β-1,3-1, 4-glucanase by *Aspergillus niger*. Chem. Abstr. *88*, 20507r. (Polish)

BERNATH, F. R. and VIETH, W. R. 1974. Collagen as a carrier for enzymes: materials science and process engineering. *In* Immobilized Enzymes in Food and Microbial Processes. A.C Olson and C.L. Cooney (Editors). Plenum Press, New York.

BERNHARD, R. A. 1968. Comparative distribution of volatile aliphatic disulfides derived from fresh and dehydrated onions. J. Food Sci. *33*, 298–304.

BERNHARD, S. 1968. Enzymes: Structure and Function. W.A. Benjamin, New York.

BERNHOLDT, H. F. 1970. Method for producing corned meat products. U.S. Pat. 3,549,385. Dec. 22.

BERRIDGE, N. J. 1942. The second phase of rennet coagulation. Nature *149*, 194–195.

BERRIDGE, N. J. 1945. Purification and crystallization of rennin. Biochem. J. *39*, 423–428.

BERRY, J. A. 1971. The compartmentation of reactions in β-carboxylation photosynthesis. Carnegie Inst. Wash. Yearb. *69*, 648–671.

BERSET, C. and SANDRET, F. 1976. Sprout radio-inhibition of potato tubers. Comparison of β- or γ-radiation after effects on polyphenol content, phenoloxidase activity, and after cooking blackening. J. Food Sci. *9*, 85–90. (French)

BERTRAN CAPELLA, A. 1974. Methanol or methyl alcohol. Zacchia *10*, 233–247. (Spanish)

BESFORD, R. T. 1975. Pyruvate kinase as potential indicators of potassium and magnesium status of tomato and cucumber plants. J. Sci. Food Agric. *26*, 125–133.

BESFORD, R. T. and HOBSON, G. E. 1972. Pectic enzymes associated with the softening of tomato fruit. Phytochemistry *11*, 2201–2205.

BETSCHART, A. A. 1978A. Improving protein quality of bread-nutritional benefits and realities. *In* Nutritional Improvement of Food and Feed Proteins. M. Friedman (Editor). Plenum Press, New York.

BETSCHART, A. A. 1978B. Preparation of protein isolates from safflower seeds. U.S. Pat. 4,072,669. Feb 7.

BEUCHAT, L. R. 1977. Functional property modification of defatted peanut flour as a result of proteolysis. Lebensm. Wiss. Technol. *10*, 79–83.

BEUCHAT, L. R., CHERRY, J. P. and QUINN, M. R. 1975. Physiochemical properties of peanut flour as affected by proteolysis. J. Agric. Food Chem. *23*, 616–620.

BEUK, J. F., SAVICH, A. L. and GOESER, P. A. 1959. Method of tenderizing meat. U. S. Pat. 2,903,362. Sept. 8.

BEUK, J. F., WARNER, W. D. and KANG, C. K. 1977. Plant protease solutions. U.S. Pat. 4,024,285. May 17.

BEUTLER, H. O. 1977. Applications of enzymic analysis in food chemistry. Voedingmiddelen Technol. *10*, 18–19.

BEUTLER, H. O. and BECKER, J. 1977. Enzymic determination of D-sorbitol and xylitol in foods. Dtsch. Lebensm. Rundsch. *73*, 182–187. (German)

BEUTLER, H. O. and MICHAL, G. 1976. Determination of egg content in routine analysis: enzymic determination of cholesterol. Getreide Mehl Brot *30*, 116–118. (German)

BEUTLER, H. O. and MICHAL, G. 1977. New method for the enzymic determination of ethanol in foods. Fresenius' Z. Anal. Chem. *284*, 113–117. (German)

BHATIA, I. S. and ULLAH, M. R. 1968. Polyphenols of tea. IX. Qualitative and quantitative study of the polyphenols of different organs and some cultivated varieties of tea plant. J. Sci. Food Agric. *19*, 535–542.

BHUMIRATANA, S., HILL, C. G., JR. and AMUNDSON, C. H. 1977. Enzymatic solubilization of fish protein concentrate in membrane reactors. J. Food Sci. *42*, 1016–1021.

BIDLACK, W. R., OKITA, R. T. and HOCHSTEIN, P. 1973. The role of NADPH-cytochrome b_5 reductase. Biochem. Biophys. Res. Comm. *53*, 459–465.

BIEHL, B. 1972. Enzymological and cytological problems in cocoa preparation. Ann. Technol. Agric. *21*, 435–455. (German)

BIELIG, H. J., WOLFF, J. and BALCKE, K. J. 1971. Enzymic clarification of apple pulp for juice removal by rack and cloth presses or centrifuges. Flüssiges Obst. *38*, 408–414. (German)

BILD, G. S., RAMADOSS, C. S., PISTORIUS, E. K. and AXELROD, B. 1977. Variation in the response of lipoxygenase isoenzymes vis a vis the charge and polarity of the substrate. Fed. Proc. *35*, 833.

BILDERBACK, D. E. 1971. Amylases in barley seeds. Plant Physiol. *48*, 331–334.

BILLET, E. E. and SMITH, H. 1980. Control of phenylalanine ammonia-lyase and cinnamic acid 4-hydroxylase in gherkin tissues. Phytochemistry *19*, 1035–1041.

BINGHAM, E. W. 1975. Action of rennin on κ-casein. J. Dairy Sci. *58*, 13–18.

BINGHAM, E. W. and FARREL, H. M., JR. 1977. Phosphorylation of casein by the lactating mammary gland: A review. J. Dairy Sci. *60*, 1199–1207.

BINGHAM, E. W. and ZITTLE, C. A. 1963. Purification and properties of acid phosphatase in bovine milk. Arch. Biochem. Biophys. *101*, 471–480.

BIRAN, D. 1974. Effect of dry salting in the koshering process on the quality of pre-frozen beef packaged by different methods and materials. Master's Thesis. Dep. Food Sci., Rutgers Univ., New Brunswick, N.J.

BIRECKI, M., BIZIEN, H. and HENDERSON, H. M. 1971. Effect of culture, storage and variety on polyphenol oxidase and peroxidase activities. Am. Potato J. *48*, 255–261.

BIRK, Y. and BONDI, A. 1955. The action of proteolytic enzymes on protein feeds. Intermediary products precipitated by trichloracetic acid and phosphotungstic acid from peptic digests and pancreatic digests. J. Sci. Food Agric.

6, 549–555.

BISCHOFF, F. 1969. Carcinogenic effects of steroids. Adv. Lipid Res. *1*, 165–224.

BISHOV, S. J. and HENICK, A. S. 1975. Antioxidant effect of protein hydrolyzates in freeze-dried model systems. Synergistic action with a series of phenolic antioxidants. J. Food Sci. *40*, 345–348.

BISWAS, S., MAITY, I. B., CHAKRABARTI, S. and BISWAS, B. B. 1978. Purification and characterization of myo-inositol hexaphosphate-adenosine diphosphate phosphotransferase from *Phaseolus aureus*. Arch. Biochem. Biophys. *185*, 557–566.

BITAR, K. and REINHOLD, J. C. 1972. Phytase and alkaline phosphatase activities in intestinal mucosa of rats, chickens, calf and man. Biochim. Biophys. Acta *268*, 442–452.

BITLE, C. 1977. Enzymatic analyzer determines three sugars in ice cream mixes. Am. Dairy Rev. *39* (3) 28B,D,F.

BITO, M. 1976. Retention of the meat color of frozen tuna. Tokaiku Suisan Kenkyusho Kenkyu Hokoku *84*, 51–113. (Japanese)

BJORCK, L. 1978. Antibacterial effect of the lactoperoxidase system on psychrotrophic bacteria in milk. J. Dairy Res. *45*, 109–118.

BJÖRKMAN, R. and LÖNNERDAL, B. 1973. Studies on myrosinases. III. Enzymatic properties of myrosinases from *Sinapis alba* and *Brassica napus* seeds. Biochim. Biophys. Acta *327*, 121–131.

BJORKSTEN, J. 1977. Some therapeutic implications of the crosslinkage theory of aging. *In* Protein Crosslinking. M. Friedman (Editor). Plenum Press, N.Y.

BLACK, A. L., GUIRARD, B. M. and SNELL, E. E. 1978. The behavior of muscle phosphorylase as a reservoir for vitamin B_6 in the rat. J. Nutr. *108*, 670–677.

BLACKBURN, S. 1976. Enzymology, Vol. 3. Enzyme Structure and Function. Marcel Dekker, New York.

BLAIN, J. A. 1962. Moisture levels and enzyme activity. Recent Adv. Food Sci. *2*, 41–45.

BLAIN, J. A. 1970. Carotene-bleaching activity in plant extracts. J. Sci. Food Agric. *21*, 35–38.

BLAIN, J. A. 1975. Industrial enzyme production. *In* The Filamentous Fungi, Vol. 1. Industrial Mycology. J.E. Smith and D.R. Bergy (Editors). Halsted Press, Div. John Wiley & Sons, New York.

BLAIN, J. A. and SHEARER, G. 1965. Inhibition of soya lipoxidase. J. Sci. Food Agric. *16*, 373–378.

BLAIR, J. S. and AYERS, T. B. 1943. Protection of natural pigment in canning of peas. Ind. Eng. Chem. *35*, 85–98.

BLAKE, C. C. F. 1976. X-ray enzymology. FEBS Lett. *62*, Supplement, Feb. 4.

BLAKEMORE, S. M. 1962. Puree and method of making same. U.S. Pat. 3,031,307. Apr. 24.

BLANCHON, E. 1966. Process for separating the enzymes and nutritive constituents contained in the envelope and cortical layer of cereal grains. U.S. Pat. 3,255,015. June 7.

BLAND, J. 1976. Biochemical effects of excited state of molecular oxygen. J. Chem. Ed. *53*, 274–279.

BLANK, G. E. and SONDHEIMER, E. 1969. Influence of polyphenols on potato phosphorylase. Phytochemistry *8*, 823–826.

BLANKENSHIP, L. C. and WELLS, P. A. 1974. Microbial β-galactosidase. Survey for neutral pH optimum enzymes. J. Milk Food Technol. *37*, 199–202.

BLATT, W. F. 1971. Ultrafiltration for enzyme concentrations. Methods Enzymol. *22*, 39–49.

BLOCH, F. and MORGAN, A. I., JR. 1967. Germination inhibition in wheat and barley during steeping and alpha-amylase development in the presence of gibberellic acid. Cereal Chem. *44*, 61–69.

BLOCK, E., PENN, R. E., REVELLE, L. K. and BAZZI, A. A. 1979. The lachrymatory factor of the onion: Applications of microwave spectroscopy and flash vacuum pyrolysis techniques in organosulfur chemistry. Abstr., ACS/CSJ Chem. Congr. AGFD No. *408*.

BLOKSMA, A. H. 1964. The role of thiol groups and of flour lipids in oxidation-reduction reactions in dough. Bakers Dig. *38*, 53–60.

BLONDIN, G. A. and GREEN, D. E. 1975. A unifying model of bioenergetics. Chem. Eng. News *53* (45) 26–42.

BLOOMFIELD, V. A. and MEAD, R. J., JR. 1975. Structure and stability of casein micelles. J. Dairy Sci. *58*, 592–601.

BLUM, M. and THANASSI, J. W. 1977. Metal ion induced reaction specificity in vitamin B_6 model systems. The effect of Zn^{2+} and Cu^{2+} on the 5-deoxypyridoxal catalyzed reactions of α-phenyl-α-aminomalonic acid. Bioorg. Chem. *6*, 31–42.

BLUMENTHAL, A. and HEBLING, J. 1971. Contents of L- and D-lactic acid in various cultured milks. Mitt. Geb. Lebensmittelunters. Hyg. *62*, 159–166. (German)

BLUNDSTONE, H. A. W., WOODMAN, J. S. and ADAMS, J. B. 1971. Canned citrus products.

In The Biochemistry of Fruits and Their Products. Academic Press, London.

BOCK, W., DONGOWSKI, G. and KRAUSE, M. 1973. The use of tomato enzymes in producing potable macerates from vegetable and fruit. Nahrung *17*, 757–767. (German)

BOCK, W., KRAUSE, M. and DONGOWSKI, G. 1970. Characterization of processes affecting quality during the manufacture of pickled gherkins. Ernaehrungsforschung *15*, 403–415. (German)

BOCK, W., KRAUSE, M. and DONGOWSKI, G. 1971. Method for manufacture of vegetable macerates. E. Ger. (DDR) Pat. 84,317.

BODANSKY, O. 1937. The use of different measures of reaction velocity in the study of the kinetics of biochemical reactions. J. Biol. Chem. *120*, 555–574.

BODE, H. E. 1961. Enzymes act as tenderizers. *In* Practical New Canning and Freezing. Chilton Publications, Philadelphia.

BODNAR, D. A., HINMAN, C. W. and NELSON, W. J. 1972. Highly fermentable starch conversion syrups. U.S. Pat. 3,644,126. Feb. 22.

BODWELL, C. E. and PEARSON, A. M. 1964. The activity of partially purified bovine catheptic enzymes on various natural and synthetic substrates. J. Food Sci. *29*, 602–607.

BOEG-HANSEN, T. C., BJERRUM, O. J. and BROGREN, C. H. 1977. Identification and quantification of glycoproteins by affinity electrophoresis. Anal. Biochem. *81*, 78–87.

BOEKER, E. A. 1978. Arginine decarboxylase from *E. coli* B.: Mechanism and dissociation from the decamer to the dimer. Biochemistry *17*, 258–263.

BOEHRINGER-MANNHEIM CO. 1976. Enzymatic Food Analysis. Boehringer-Mannheim GmbH Biochemicals, Mannheim, West Germany.

BOGDANOV, V. P., MOROZKIN, A. D. and ABALIKHINA, T. A. 1974. Subunit structure of the glucose oxidase from Penicillium vitale. Mol. Biol. (U.S.S.R.) *10*, 14–16. (Russian)

BOGGS, M. M., DIETRICH, W. C., NUTTING, M. -D., OLSON, R. L., LINDQUIST, F. E., BOHART, G. S., NEUMAN, H. J. and MORRIS, H. J. 1960. Time-temperature tolerance of frozen foods. XXI. Frozen peas. Food Technol. *14*, 181–185.

BOGIN, E. and WALLACE, A. 1966. Organic acid synthesis in sweet and sour lemon fruits. Proc. Am. Soc. Hortic. Sci. *89*, 182–194.

BOGIN, E. and WALLACE, A. 1966B. The inhibition of citrate condensing enzyme by ATP. Biochim. Biophys. Acta *128*, 190–192.

BOHNENKAMP, C. G. 1979. Use of immobilized β-amylase/glucoamylase mixtures to produce high maltose syrups. Proc. Annu. Biochem. Eng. Symp. *9*, 81–86.

BOHUON, G. and DRILLEAU, J. -F. 1976. Non-biological cloudiness in ciders. Bios (Nancy) 7 (7/8) 13–19. (French)

BÖING, J. T. P. 1976. Progress in Biochemical Engineering. Naturwissenschaften *63*, 319–323.

BOKUCHARA, M. A. and SKOBELEVA, N. I. 1969. The chemistry and biochemistry of tea and tea manufacture. Adv. Food Res. *17*, 215–272.

BOLDINGH, J. 1975. Lipid metabolism in relation to human health. 11th Leverhume lecture. Chem. Ind. (London) (23) 984–993.

BOLIN, H. R., NURY, F. S. and FINKLE, B. J. 1964. An improved process for preservation of fresh peeled apples. Bakers Dig. *38* (3) 46–48.

BOLIN, H. R., PETRUCCI, V, and FULLER, G. 1975. Characteristics of mechanically harvested raisins produced by dehydration and by field drying. J. Food Sci. *40*, 1036–1038.

BOLIN, H. R. and STAFFORD, A. E. 1974. Effect of processing on provitamin A and vitamin C in apricots. J. Food Sci. *39*, 1034–1036.

BOLIN, H. R., STAFFORD, A. E and FULLER, G. 1975. Rapid enzymatic process yields nutritious, mild-flavored prune juice. Food Prod. Dev. *9*, (7) 92–94.

BOLOORFOROOSHAN, M. and MARKAKIS, P. 1977. Zinc supplementation of a bean diet for the rat. J. Food Sci. *42*, 1671, 1673.

BOMBEN, J. L., DIETRICH, W. C., FARKAS, D. F., HUDSON, J. S., DE MARCHENA, E. S. and SANSHUCK, D. W. 1973. Pilot plant evaluation of individual quick blanching (IQB) for vegetables. J. Food Sci. *38*, 590–594.

BOMBEN, J. L., DIETRICH, W. C., HUDSON, J. S., HAMILTON, H. K. and FARKAS, D. F. 1975. Yields and solids loss in steam blanching and cooling and freezing vegetables. J. Food Sci. *40*, 660–664.

BOMBEN, J. L., HUDSON, J. S., DIETRICH, W. C., DURKEE, E. L., FARKAS, D. F., RAND, R. and FARQUHAR, J. W. 1978. Vibratory spiral Blancher-Cooler. Off. Res. Dev., U.S. Environ. Prot. Agency, Cincinnati.

BOND, J. S. and BARRET, A. 1979. Degradation of aldolase by cathepsin B. Fed. Proc. *38*, 946.

BOND, R. M. 1975. Background paper on minced fish. Food Agric. Organ. U.N. Fish. Circ. *332*, 24.

BONE, D. 1970. Structuring intermediate moisture foods. *In* Innovations in Food Engineering.

Univ. Calif. Ext. Course, San Francisco.

BONE, D. 1973. Water activity in intermediate moisture foods. Developing shelf-stable formulations compatible with flavor, texture, and other aspects of food is a challenge to the food technologist. Food Technol. *24* (7) 71–76.

BONNER, J. 1975. Regulation of gene expression in higher organisms. How it all works. Ciba Found. Symp. (Structure and Function of Chromatin, 1974.) *28*, 315–335.

BOON, D. D. 1975. Discoloration in processed crabmeat. J. Food Sci. *40*, 756–761.

BOONVISUT, S. and WHITAKER, J. R. 1976. Effect of heat, amylase, and disulfide bond cleavage on the in vitro digestibility of soybean proteins. J. Agric. Food Chem. *24*, 1130–1135.

BORGSTROM, B. and ORY, R. L. 1970. Castor bean lipase: specificity of action. Biochim. Biophys. Acta *212*, 521–522.

BORGSTROM, G. 1961. Principles of Food Science, 2 Vols. Macmillan Company, New York.

BORISOVA, I. G., CHEPURENKO, N. V. and BUDNITSKAYA, E. V. 1977. Isoenzyme composition and some properties of pea lipoxygenase. Biokhimiya (Moscow) *42*, 2079–2095. (Russian)

BOROCHOV, E. H., CRAIG, T. W. and DUNNING, H. N. 1971. In situ conversion of starch. U.S. Pat. 3,617,300. Nov. 2.

BORODIN, J. 1882. Concerning chlorophyll crystals. Bot. Ztg. *40*, 608–610. (German)

BOROVSKY, D., SMITH, E. and WHELAN, W. J. 1976. On the mechanism of amylose branching by potato Q-enzyme. Eur. J. Biochem. *62*, 307–312.

BOSTICK, D. T. and HERCULES, D. M. 1975. Quantitative determination of blood glucose using enzyme induced chemiluminescence of luminol. Anal. Chem. *47*, 447–452.

BOSUND, J. and GANROT, B. 1970. Effect of pre-cooking Baltic herring on lipid hydrolysis during subsequent cold storage. Lebensm. Wiss. Technol. *2*, 59–61. (German)

BÖTTCHER, H. 1975. Enzyme activity and quality of frozen vegetables. I. Remaining residues of peroxidase. Nahrung *19*, 173–179. (German)

BOUNIAS, M. 1976. Kinetic parameters of the α-glucosidase activity of worker bees hemolymph. Apidologie *7*, 263–275. (French)

BOURNE, D. T., JONES, M. and PIERCE, J. S. 1976. Beta glucan and beta glucanases in malting and brewing. Tech. Q. Master Brew. Assoc. Am. *13*, 3–7.

BOURNE, D. T. and PIERCE, J. S. 1970. Beta-glucan and beta-glucanase in brewing. J. Inst. Brew. *76*, 328–335.

BOURNE, D. T. and PIERCE, J. S. 1972. β-Glucan and β-glucanase. Rev. Tech. Q. Master Brew. Assoc. Am. *9*, 151–157.

BOUTON, P. E., FORD, A. L., HARRIS, P. V., MACFARLANE, J. J. and O'SHEA, J.M. 1977. Pressure-heat treatment of postrigor muscle: Effects on tenderness. J. Food Sci. *42*, 132–135.

BOUTON, P. E., HARRIS, D. V., MACFARLANE, J. J. and O'SHEA, J. M. 1978. Pressure-heat treatment of postrigor muscle: effects on connective tissue. J. Food Sci. *43*, 301–303, 326.

BOUTON, P. E., HARRIS, D. V. and SHORTHOSE, W. R. 1976. Peak shear-force values obtained for veal muscle samples cooked at 50° and 60° C: Influence of aging. J. Food Sci. *41*, 197–198.

BOUVY, F. A. M. 1975. Applications for lactase-treated whey. Food Prod. Dev. *9* (2) 10–13.

BOWSKI, L., SAINI, R., RYU, D. Y. and VIETH, W. R. 1971. Kinetic modeling of the hydrolysis of sucrose by invertase. Biotechnol. Bioeng. *13*, 641–656.

BOYER, P.D. 1970–1976. The Enzymes, 3rd Edition, Vol. 1–13. Academic Press, New York.

BOYER, P. D., CHANCE, B., ERNSTER, L., MITCHELL, P., RACKER, E. and SLATER, E. C. 1977. Oxidative phosphorylation and photophosphorylation. Annu. Rev. Biochem. *46*, 955–1026.

BOYER, P. D., LARDY, H. A. and MYRBÄCK, K. 1959–1963. The Enzymes, 2nd Edition, Vol. 8. Academic Press, New York.

BOYER, P. D., STOKES, B. O., WOLCOTT, R. G. and DEGANI, C. 1975. Coupling of "high energy" phosphate bonds to energy transductions. Fed. Proc. *34*, 1711–1717.

BØG-HANSEN, T. C. and DAUSSANT, J. 1974. Immunochemical quantitation of isoenzymes. α-Amylase isoenzymes in barley malt. Anal. Biochem. *61*, 522–527.

BRACCO, U. and WINTER, H. 1976. Analytical identification of mixtures of animal fats. Rev. Fr. Corps Gras *23*, 87–93. (French)

BRACHT, T. -J. 1973. Diastatic activity of malt flours and enzyme preparations. Mühle und Mischfuttertecknik *110*, 346–348. (German)

BRADDOCK, R. J., CRANDALL, P. G. and KESTERSON, J. W. 1976. Pectin content of Meyer lemon. J. Food Sci. *41*, 1486.

BRADDOCK, R. J. and DUGAN, L. R. 1972. Phospholipid changes in muscle from frozen stored Lake Michigan Coho salmon. J. Food Sci. *36*, 426–429.

BRADDOCK, R. J. and KESTERSON, J. K. 1973. Enzymatic hydrolysis of fatty acids in orange juice phospholipids. J. Agric. Food Chem.

21, 318–319.

BRADDOCK, R. J. and KESTERSON, J. W. 1976. Enzyme use to reduce viscosity and increase recovery of soluble solids from citrus pulp-washing operations. J. Food Sci. *41,* 82–85.

BRADY, C. J. 1976. The pectinesterase of the pulp of the banana fruit. Aust. J. Plant Physiol. *3,* 163–172.

BRANCHER, E., STACHELBERGER, H. and WASHUETTL, J. 1974. Influence of γ-radiation on the glutamate decarboxylase activity in stored wheat. Mikrochim. Acta *3,* 381–384.

BRANDON, D. L., CORSE, J. and WINDLE, J. J. 1980. Homogeneous immunoassays for vitamin B_6. Abstr., 2nd Chem. Congr., N. Am. Cont., San Francisco. AGFD *110.*

BRATTAIN, M.G., MARKS, M. E. and PRETLOW II, T. G. 1976. The purification of horseradish peroxidase by chromatography on Sepharose-bound concanavalin A. Anal. Biochem. *72,* 346–352.

BRAVERMAN, J. B. S. 1963. Introduction to the Biochemistry of Foods. Elsevier Publishing Co., Amsterdam.

BREENE, W. M., DAVIS, D. W. and CHOU, H. -E. 1972. Texture profile analysis of cucumbers. J. Food Sci. *37,* 113–117.

BREMNER, H. A. 1977. Production and storage of mechanically separated fish flesh from Australian species. Food Technol. Aust. *29,* 89–93.

BREUIL, C. and KUSHNER, D. J. 1976. Cellulase induction and the use of cellulose as preferred growth substrate by *Cellvibrio gilvus.* Can. J. Microbiol. *22,* 1776–1781.

BREWER, G. J. 1970. An Introduction to Isozyme Techniques. Academic Press, New York.

BREWER, J. 1971. Enzyme purification by gel filtration. Process Biochem. *6* (9) 32–42.

BREWSTER, L. C., HASEGAWA, S. and MAIER, V. P. 1976. Bitterness prevention in citrus juices. Comparative activities and stabilities of the limonate dehydrogenase from *Pseudomonas* and *Arthrobacter.* J. Agric. Food Chem. *24,* 21–24.

BRIDGEN, J. 1972. The reactivity and function of thiol groups in actin. Biochem. J. *126,* 21–25.

BRIGGS, D. E. 1968. α-Amylase in germinating decorticated barley. Phytochemistry *7,* 513–519.

BRIGHT, H. J. 1974. Flavoprotein oxidases. *In* Food Related Enzymes. J.R. Whitaker (Editor). Am. Chem. Soc., Washington, D.C.

BRIGHTON, F. and HORNE, M. T. 1977. Influence of temperature on cyanogenic polymorphisms. Nature *265,* 437–438.

BRISKEY, E. J., CASSENS, R. G. and TRAUTMANN, J. C. 1966–1967. The Physiology and Biochemistry of Muscle as Food Volumes I, II. University of Wisconsin Press, Madison.

BRITTON, L. N. and MARKOVETZ, A. J. 1977. A novel ketone monoxygenase from *Pseudomonas cepacia.* J. Biol. Chem. *252,* 8561–8566.

BROCKERHOFF, H. and JENSEN, R. G. 1974. Lipolytic Enzymes. Academic Press, New York.

BROCKMANN, R. and ACKER, L. 1977. The behaviour of lipoxygenase systems of low water content. II. Investigation of reaction products of enzymic lipid oxidation. Lebensm. Wiss. Technol. *10,* 332–336. (German)

BRODNITZ, M. H. and PASCALE, J. V. 1971. Thiopropanal-S-oxide: A lachrymatory factor in onions. J. Agric. Food Chem. *19,* 261–272.

BRODY, A. L. 1970. Produce for all seasons—A prospectus. Presented at "Innovations in Engineering," Continuous Education in Engineering, March 23–26. University of California, San Francisco.

BROLIN, S. E., WETTENMART, G. and HAMMAR, H. 1977. Chemiluminescence microanalysis of substrates and enzymes. Strahlentherapie *153,* 124–131.

BROOKER, G. and APPLEMAN, M. M. 1968. The theoretical basis for the measurement of compounds. Biochemistry 7, 4182–4184.

BROOKES, P. A., LOVETT, D. A. and MacWILLIAMS, I. C. 1976. The steeping of barley. A review of the metabolic consequences of water uptake, and their practical implications. J. Inst. Brew. *82,* 14–26.

BROOKES, P. A. and MARTIN, P. A. 1975. Determination and utilization of gibberellic acid in germinating barley. J. Inst. Brew. (London) *81,* 357–363.

BROUN, G. B., MANECKE, G. and WINGARD, L. B., JR. 1978. Enzyme Engineering, Vol. 4. Plenum Press, New York.

BROUSSARD, L., HARMS, W. M. and SHIVE, W. 1976. Purification of the threonine-sensitive aspartokinase-homoserine dehydrogenase complex using chromatography on substituted Sepharose columns. Anal. Biochem. *72,* 16–23.

BROWN, A. G., EYTON, W. B., HOLMES, A. and OLLIS, W. D. 1969. Identification of the thearubigens as polymeric proanthocyanidins. Nature *221,* 742–744.

BROWN, A. J. 1902. Enzyme action. Trans. Chem. Soc. *81,* 373–379.

BROWN, C. R., ANDANI, Z. and HARTREE, E. F. 1975. Fluorimetric titration of operational molarity with 4-methylumbelliferyl *p*-guanidinobenzoate. Biochem. J. *149,* 147–154.

BROWN, D. E. 1976. Cellulase production by *Trichoderma viridae*. Biotechnol. Bioeng. Symp. *6*, 75–77.

BROWN, G. E., DIETRICH, W. C., HUDSON, J. S. and FARKAS, D. F. 1974. A reduced effluent blanch-cooling method using a vibratory conveyor. J. Food Sci. *39*, 696–700.

BROWN, H. E. and STEIN, E. R. 1976. There's protein in brassica. Agric. Res. *24* (10) 7.

BROWN, M. A. and CHURCH, J. A. 1972. Soya flour, the modern baking additive. Ind. Aliment. *11*, 109–110. (Italian)

BROWN, M. S. and MORALES, J. A. W. 1970. Determination of blanching condition for frozen pan-fried potatoes. Am. Potato J. *47*, 321–325.

BROWN, T. R. C. 1955. The Complete Book of Cheese. Gramercy Publishing Co., New York.

BRUCE, A. W., LEIENDECKER, C. M. and FREIER, E. F. 1978. Two-point determination of plasma ammonia with the centrifugal analyzer. Clin. Chem. *24*, 782–787.

BRUCHMANN, E. E. and KOLB, E. 1973A. Effect of inhibitors on the flavor forming system from raspberry calyx cone. Lebensm. Wiss. Technol. *6*, 107–110.

BRUCHMANN, E. E. and KOLB, E. 1973B. Using synergism between cellulase and glucose oxidase to stabilize fermented fruit must. Lebensm. Wiss. Technol. *6*, 158–164.

BRUEMMER, J. H. 1975. Aroma substances in citrus fruits and their biogenesis. *In* Odor and Taste Substances. F. Drawert (Editor). Verlag Hans Carl, Nurenberg. (German)

BRUEMMER, J. H., BAKER, R. A. and ROE, B. 1977. Enzymes affecting flavor and appearance of citrus. *In* Enzymes in Food and Beverage Processing. R.L. Ory and A.J. St. Angelo (Editors). Am. Chem. Soc., Washington, D.C.

BRUEMMER, J. H., and ROE, B. 1971. Substrate specificity of citrus alcohol NAD oxidoreductase. J. Agric. Food Chem. *19*, 266–268.

BRUEMMER, J. H. and ROE, B. 1976. Esterase activity and orange juice quality. Proc. Fla. State Hortic. Soc. *88*, 300–303.

BRUEMMER, J. H., ROE, B., BOWEN, E. R. and BUSLIG, B. 1976. Peroxidase reactions and orange juice quality. J. Food Sci. *41*, 186–189.

BRUENGER, F. W., STOVER, B. J. and ATHERTON, D. R. 1967. The incorporation of various metal ions into in vivo- and in vitro-produced melanin. Radiat. Res. *32*, 1–12.

BRUHN, L. C. and DJURTOFT, R. 1977. Protease inhibitors in barley. Trypsin inhibitors, chymotrypsin inhibitors and proteins analyzed by isoelectric focusing in polyacrylamide gel. Localization within the grain and effects of germination. Z. Lebensm. Unters. Forsch. *164*, 247–254.

BRUICE, T. C. 1976. Some aspects of mechanism as determined with small molecules. Ann. Rev. Biochem. *45*, 331–373.

BRUICE, T. C. and BENCOVIC, S. J. 1966. Bioorganic Mechanisms. W.A. Benjamin, New York.

BRUNAUER, S., EMMET, D. H. and TELLER, E. 1938. Adsorption of gases in multimolecular layers. J. Am. Chem. Soc. *60*, 309–319.

BUCHHOLZ, K. and REUSS, M. 1977. Coupling of material transport, reaction and deactivation by carrier-bound glucose oxidase and catalase. Chimia *31*, 27–30. (German)

BUCHNER, E. 1897. Alcoholic fermentation without yeast cells. Ber. Dtsch. Chem. Ges. *30*, 117. (German)

BUCK, P. A. and JOSLYN, M. A. 1953. Broccoli processing. Accumulation of alcohol in underscalded frozen broccoli. J. Agric. Food Chem. *4*, 309–312.

BUCKLE, K. A. and EDWARDS, R. A. 1969. Chlorophyll degradation products from processed pea puree. Phytochemistry *8*, 1901–1906.

BUCKLE, K. A. and EDWARDS, R. A. 1970. Chlorophyll degradation in frozen unblanched peas. J. Sci. Food Agric. *21*, 307–312.

BUCKLEY, K., MOWBRAY, M. and MITCHELL, J. R. 1976. Pectate gelled food product and method. U.S. Pat. 3,973,051. Aug. 3.

BUCOLO, G., YABUT, J. and CHANG, T. Y. 1975. Mechanized determination of triglycerides in serum. Clin. Chem. *21*, 420–424.

BUERGIN, D. 1976. Alcohol determination in "alcohol-free" beverages. Schweig. Brau. Rundsch. *87*, 174–175. (German)

BUESCHER, R. W., REITMEIER, C. and SISTRUNK, W. A. 1974. Association of phenylalanine ammonia lyase, peroxidase, and total phenolic content with brown-end discoloration of snap bean pods. HortScience *9*, 585.

BUESCHER, R. W. and SISTRUNK, W. A. 1976. Softening, pectolytic activity, and storage-life of *rin* and *nor* tomato hybrids. HortScience *11*, 603–604.

BUESCHER, R. W., SISTRUNK, W. A. and BRADY, P. L. 1975. Effects of ethylene on metabolic and quality attributes in sweet potato roots. J. Food Sci. *40*, 1018–1020.

BUESCHER, R. W., SISTRUNK, W. A. and KASAIAN, A. E. 1976. Induction of textural changes in sweet potato roots by chilling. J.

Am. Soc. Hortic. Sci. *101*, 516–519.

BULL, A. T. 1970. Kinetics of cellulase inactivation by melanin. Enzymologia *39*, 333–347.

BULL, H. B. and REESE, K. 1973. Thermal stability of proteins. Arch. Biochem. Biophys. *158*, 681–686.

BUNTON, C. A., KAMEGO, A. A., MINCH, M. J. and WRIGHT, J. L. 1975. Effect of changes in surfactant structure on micellar catalyzed spontaneous decarboxylations and phosphate ester hydrolysis. J. Org. Chem. *40*, 1321–1327.

BURBA, M. and KASTNING, M. 1971. Physiology of sugar beet metabolism during the growing period. I. Glutamine, glutamic acid, asparagine and aspartic acid. Zucker *24*, 386–396. (German)

BURDETT, P. E., KIPPS, A. E. and WHITEHEAD, 1976. A rapid techique for the detection of amylase isoenzymes using an enzyme sensitive "test paper." Anal. Biochem. *72*, 315–319.

BURG, S. P. 1975. Hypobaric storage and transportation of fresh fruits and vegetables. *In* Postharvest Biology and Handling of Fruits and Vegetables. N.F. Haard and D.K. Salunkhe (Editors). AVI Publishing Co. Westport, Conn.

BURI, R. and SOLMS, J. 1971. Ribonucleic acid—A flavor precursor in potatoes. Naturwissenschaften *58*, 56–57.

BURKHARDT, G. J., MERKEL, J. A. and SCOTT, L. E. 1973. Time and pressure regulated steam for peeling fruits and vegetables. HortScience *8*, 485–487.

BURKHARDT, R. 1973. Behavior of sucrase during the sweetening and bottling of wines. Mitteilungsbl. GDCh (Ges. Dtsch. Chem.) Fachgruppe Lebensmittelchemie Gerichtl. Chem. *27*, 1–5. (German)

BURKHARDT, R. 1977. Depside-splitting activity of fruit attacked by spoilage organisms. Effects on processing of spoiled produce. II. Dtsch. Lebensm.-Rundsch. *73*, 189–191. (German)

BURKHOLDER, L., BURKHOLDER, P. R., CHU, A., KOSTYK, N. and ROELS, O. A. 1968. Fish fermentation. Food Technol. *22*, 1278–1284.

BURNS, J. A. 1976. Food process instrumentation using immobilized enzymes. Cereal Foods World *21*, 594–598.

BURTIS, C. A., TIFFANY, T. O. and SCOTT, C. D. 1976. The use of a centrifugal fast analyzer for biochemical immunological analyses. Methods Biochem. Anal. *23*, 189–248.

BURTON, W. G. 1958. Experiments on the use of alcohol vapors to suppress the sprouting of stored potatoes. Eur. Potato. J. *1*, 42–51.

BURTON, W. G. 1975. Immediate effect of gamma irradiation upon the sugar content of potatoes previously stored at 2, 4, 5, 6, 10, and 15.5°. Potato Res. *18*, 109–115.

BURZYNSKA, H., URBANEK-KARLOWSKA, B., FONBERG-BROCZEK, M. and KWAST, M. 1977. Microbiological and chemical characterization of enzyme preparations. FSTA *10*, 6 T189. (Polish)

BUSCH, W. A., STROMER, M. H., GOLL, D. E. and SUZUKI, A. 1972. Ca^{2+}-specific removal of Z lines from rabbit skeletal muscle. J. Cell Biol. *52*, 367–381.

BUSHUK, W., HWANG, P. and WRIGLEY, C. W. 1971. Proteolytic activity of maturing wheat grain. Cereal Chem. *48*, 637–639.

BUTLER, J. A. V. 1945. Adsorption of enzymes on solid nucleic acids. Nature *156*, 781.

BUTTERY, R. G. 1977. The natural flavors of vegetables. Vortex *38* (10) 6–12.

BUTTERY, R. G. 1980. Vegetable and fruit flavors. *In* Flavor Research, 2nd Edition. R. Teranishi (Editor). Marcel Dekker, New York.

BUTTERY, R. G., LING, L. C. and BEAN, M. M. 1978. Coumarin off-odor in wheat flour. J. Agric. Food Chem. *26*, 179–180.

BUTTERY, R. G., SEIFERT, R. M. and LING, L. C. 1970. Characterization of some volatile potato components. J. Agric. Food Chem. *18*, 538–539.

BUTTKUS, H. 1967. The reaction of myosin with malonaldehyde. J. Food Sci. *32*, 432–434.

BUTTON, A. H. and PALMER, J. R. 1974. Production scale brewing using high proportions of barley. J. Inst. Brew. *80*, 206–213.

BUTTY, M. 1973. Rapid fermentation of coffee. Kenya Coffee *38*, 214–224.

BUTZOW, J. J. 1968. Probe of subunit structure in fungal *p*-diphenol oxidase by treatment with guanidine hydrochloride at high pH. Biochim. Biophys. Acta *168*, 490–506.

BUTZOW, J. J. and EICHHORN, G. L. 1975. Different susceptibility of DNA and RNA to cleavage by metal ions. Nature (London) *254*, 358–359.

BYKOVA, V. M. 1970. Means for improving the quality of sausage meat from frozen fish. Ryb. Khoz. *46*, 48–51. (Russian)

BYRDE, R. J. W. and FIELDING, A. H. 1962. Resolution of endopolygalacturonase and a macerating factor in a culture filtrate. Nature *196*, 1227–1228.

CAGNONI, D. 1957. Process for facilitating the peeling of tomatoes and apparatus for performing said process. U.S. Pat. 2,813,563. Nov. 19.

CAILLAT, J. -M. and DRAPRON, R. 1974. The lipase of wheat. Characteristics of its action in aqueous and low-moisture media. Ann. Technol. Agric. *23*, 273–286. (French)

CAIN, R. F. 1974. Closed loop blanching with hot water. *In* Waste Regulations Focus Attention on Blanching. Food Prod. Manage. *96* (10) 14–15.

CALCAGNO, L. and MAFFEO, G. 1971. Variation in cow milk protein in freeze drying processing. I. Casein digestion test by means of pepsin. Arch. Vet. Ital. *22*, 263–267. (Italian)

CALDWELL, K. A. 1970. Autolytic activity in aqueous extracts of chicken skeletal muscle. J. Agric. Food Chem. *18*, 276–279.

CALDWELL, K. A. and GROSJEAN, O. K. 1971. Lysosomal cathepsins of chicken skeletal muscle: distribution and properties. J. Agric. Food Chem. *19*, 108–111.

CALVOT, C., BERJONNEAU, A. -M., GELLF, G. and THOMAS, D. 1975. Magnetic enzyme membranes as active elements of electrochemical sensors. Specific amino acid enzyme electrodes. FEBS Lett. *59*, 258–262.

CAMICI, M., BALESRERI, E., TOZZI, M. G., BACCIOLA, D., SARCCHI, I., FELICIOLI, R. and IPATA, P. L. 1980. Relationships between enzyme levels and extractable proteins in alfalfa. J. Agric. Food Chem. *28*, 500–503.

CAMPBELL, A. 1937. Undesirable color change in frozen peas stored at insufficiently low temperature. Food Res. *2*, 55–57.

CAMPBELL, H. 1940. The scalding of cut corn for freezing. West. Canner Packer *32* (9) 51–55.

CAMPBELL SOUP CO. 1961. Proc. Flavor Chem. Symp., Campbell Soup Co., Camden, N.J.

CANONICO, P. G. and BIRD, J. W. 1970. Lysosomes in skeletal muscle tissue. Zonal centrifugation evidence for multiple cellular sources. J. Cell Biol. *45*, 321–334.

CANTAGALLI, P. and TASSI-MICCO, C. 1974. Effect of glucose oxidase on soluble proteins in semolina and alimentary paste. Riv. Soc. Ital. Sci. Aliment. *3*, 95–97. (Italian)

CANTONI, C., BIANCHI, M. A. and BERETTA, G. 1976. Formaldehyde contents in fishes, molluscs, and crustacea caught in the Mediterranean Sea. Arch. Vet. Ital. *27*, 145–148. (Italian)

CARDINI, G. and JURTSHUK, P. 1968. Cytochrome P-450 involvement in the oxidation of n-octane by cell-free extracts of *Corynebacterium*. J. Biol. Chem. *243*, 6070–6072.

CARERI, G., FASELLA, P. and GRATTON, E. 1979. Enzyme dynamics. The statistical physics approach. Annu. Rev. Biophys. Bioeng. *8*, 69–97.

CARINI, S. and TODESCO, R. 1973. Differentiation of microbial rennet from calf rennet. Latte *47*, 20–23. (Italian)

CARLIN, A. F. and AYRES, J. C. 1953. Effect of the removal of glucose by enzyme treatment on the shipping properties of dried albumen. Food Technol. *7*, 268–270.

CARLSON, P. S. 1972. Notes on the mechanism of action of gibberellic acid. Nature New Biol. *237*, 39–41.

CARLSON, P. S. 1973. The uses of protoplasts for genetic research. Proc. Natl. Acad. Sci. *70*, 598–602.

CARPENTER, J., SAFFLE, R. L. and KAMSTRA, L. D. 1961. Tenderization of beef by prerigor infusion of a chelating agent. Food Technol. *15*, 197–198.

CARR, P. and BOWERS, L. D. 1980. Immobilized Enzymes in Analytical Clinical Chemistry. John Wiley & Sons, New York.

CARRA, P. O. and COLLERAN, E. 1969. Coupled oxidation of myoglobin with ascorbate as a model of haem breakdown in-vivo. Biochem. J. *11*, 1–13.

CARRICK, D. B. 1929. Effect of freezing on the catalase activity of apples. Cornell Univ. Agric. Exp. Stn. Mem. No. *112*.

CARSON, J. F. 1967. Onion flavor. *In* Chemistry and Physiology of Flavors. H.W. Schultz, E.A. Day and L.M. Libbey (Editors). AVI Publishing Co., Westport, Conn.

CASEY, J. P. 1977. High fructose corn syrup. A case history of innovation. Staerke *29*, 196–204.

CASH, J. N., SISTRUNK, W. A. and STUTTE, C. A. 1976. Characteristics of Concord grape polyphenoloxidase involved in juice color loss. J. Food Sci. *41*, 1398–1402.

CASSENS, R. G., GREASER, M. L., ITO, T. and LEE, M. 1979. Reactions of nitrite in meat. Food Technol. *33* (7) 46–57.

CASTELEIN, J. M. and PILNIK, W. 1976. The properties of the pectate lyase produced by *Erwinia dissolvens*, a coffee fermenting organism. Lebensm.-Wiss. Technol. *9*, 277–283. (German)

CASTELFRANCO, P. A. and JONES, O. T. G. 1975. Protoheme turnover and chlorophyll synthesis in greening barley tissue. Plant. Physiol. *55*, 485–490.

CASTELL, C. H. 1971. Metal-catalyzed lipid oxidation and changes of proteins in fish. J. Am.

Oil Chem. Soc. *48*, 645–649.

CASTINO, M. 1975. Use of discriminant analysis in the evaluation of wine types. Atti Accad. Ital. Vite Vino, Siena *27*, 109–121. (Italian)

CASTLE, A. V. and WHEELOCK, J. V. 1973. Kinetics of rennin action on casein prepared by ultracentrifugation. J. Dairy Res. *40,* 77–84.

CASTRO, A. C., SINSKEY, A. J. and TANNENBAUM, S. R. 1971. Reduction of nucleic acid content in Candida yeast cells by bovine pancreatic ribonuclease A treatment. Appl. Microbiol. *22*, 422–427.

CATER, C. M., GHEYASUDDIN, S. and MATILL, K. F. 1972. The effect of chlorogenic, quinic, and caffeic acids on the solubility and color of protein isolates, especially from sunflower seeds. Cereal Chem. *49*, 508–514.

CATTELL, D. J. and NURSTEN, H. E. 1977. Fractionation and chemistry of ethylacetate-soluble thearubigins from black tea. Phytochemistry *15*, 1967–1970.

CAUGHEY, W. S., BARLOW, D. H., MAXWELL, J. C., VOLPE, J. A. and WALLACE, W. J. 1975. Reactions of oxygen with homoglobin, cytochrome oxidase and other heme proteins. Ann. N.Y. Acad. Sci. *244*, 1–9.

CAULINI, G., BINOTTI, I., ALESSANDRI, M., IPATA, P. L. and MAGNI, G. 1973. Preliminary observations on milks' nucleotidase. Bull. Soc. Ital. Biol. Sper. *48*, 890–893.

CERAMI, A. and KOENIG, R. J. 1978. Hemoglobin A_{IC} as a model for the development of the sequelae of diabetes mellitus. Trends Biochem. Sci. *3*, 73–75.

CHALMERS, D. J. and ROWAN, K. S. 1971. The climacteric in ripening tomato fruit. Plant Physiol. *48*, 235–240.

CHAMPAGNOL, F. 1976. Affinity chromatography of carbonic anhydrase. J. Chromatogr. *120*, 489–490.

CHAN, B. G., WAISS, A. C. and LIKEFAHR, M. J. 1978. Condensed tannin, an antibiotic chemical from *Gossypium hirsutum* L. J. Insect Physiol. *24*, 113–118.

CHAN, H. T., FLATH, R. A., FORREY, R. F., CAVALETTO, C. G., NAKAYAMA, T. O. M. and BREKKE, J. E. 1973. Development of off-odors and off-flavors in papaya puree. J. Agric. Food Chem. *21*, 566–570.

CHAN, H. T., KUO, M. T. -H., CAVALETTO, C. G., NAKAYAMA, T. O. M. and BREKKE, J. E. 1975. Papaya puree and concentrate changes in ascorbic acid, carotenoids and sensory quality during processing. J. Food Sci. *40*, 701–703.

CHAN, H. T. and YANG, H. Y. 1971. Identification and characterization of some oxidizing enzymes of the McFarlin cranberry. J. Food Sci. *36*, 169–173.

CHAN, H. Y. and BAKER, C. W. 1976. Influence of β-glucanase activity on malt modification. J. Am. Soc. Brew. Chem. *34*, 127–132.

CHAN, W. W. -C. and MAWER, H. M. 1972. Studies on protein subunits. II. Preparation and properties of active subunits of aldolase bound to a matrix. Arch. Biochem. Biophys. *149*, 136–145.

CHANCE, B. 1949. The composition of catalase-peroxide complexes. J. Biol. Chem. *179*, 1311–1330.

CHANCE, B. and MAEHLY, A. C. 1955. Assay of catalases and peroxidases. Methods Enzymol. *2*, 764–775.

CHANCE, B. and NISHIMURA, M. 1967. Sensitive measurements of changes of hydrogen ion concentration. Methods Enzymol. *10,* 641–650.

CHANDLER, B. V. 1971. A limonin-degrading enzyme in citrus albedo. J. Sci. Food Agric. *22,* 634–637.

CHANDLER, B. V. and NICOL, K. J. 1975. Debittering citrus products with enzymes. CSIRO Food Res. Q. *35*, 79–88.

CHANG, G. 1976. Personal communication. Berkeley.

CHANG, R., KENNEDY, B. M. and SCHWIMMER, S. 1979. Effects of autolysis on the nutritional properties of beans *(Phaseolus vulgaris).* J. Food Sci. *44*, 1141–1143.

CHANG, R. and SCHWIMMER, S. 1977. Characterization of phytase of beans *(Phaseolus vulgaris).* J. Food Biochem. *1*, 45–46.

CHANG, R., SCHWIMMER, S. and BURR, H. K. 1977. Phytate: removal from whole dry beans by enzymatic hydrolysis and diffusion. J. Food Sci. *42*, 1098–1101.

CHANG, Y. -H., RAYMUNDO, L. C., GLASS, R. W. and SIMPSON, K. L. 1977. Effect of high temperature on CPTA-induced carotenoid biosynthesis in ripening tomato fruits. J. Agric. Food Chem. *25*, 1249–1251.

CHANG, Y. -O., VARNELL, T. R. and MAK, I. T. 1975. Methionine induced toxicity in growing rats. Fed. Proc. *34*, 885.

CHAO, K. C. 1974. Enzymic degradation of nucleic acids. U.S. Pat. 3,809,776. May 7.

CHAPMAN, G. W., JR., ROBERTSON, J. A., PARKER, M. B. and BURDICK, D. 1976. Chemical composition and lipoxygenase activity in soybeans as affected by genotype and environment. J. Am. Oil Chem. Soc. *53*, 55–56.

CHAPON, L., LOUIS., C., CHAPON, S., MOLL, M. and KRETSCHMER, K. F. 1977. Hops and the equilibrium between proteins and tannins in beer. Monatsschr. Brau. *30*, 541–546. (German)

CHARALAMBOUS, G. and INGLETT, G. E. 1978. Flavor of Foods and Beverages. Academic Press, New York.

CHARI, C. N., NATARAJAN, C. P., MRAK, E. M. and PHAFF, H. J. 1948. The effect of blanching and dehydration on enzyme activity and storage quality of high moisture unprocessed prunes. Fruit Prod. J. *27*, 206–211.

CHARLES, M. 1976. Technology and economics of choosing supports for immobilized enzymes. Joint USA-ROC Symp. Enzyme Technol., Lehigh University, Bethlehem, Pa. Aug. 4–9.

CHARLES, R. L., GERTZMAN, D. P. and MELACHOURIS, N. 1970. Milk clotting enzyme product and process therefor. U.S. Pat. 3,549,390. Dec. 22.

CHARLWOOD, B. V. and BANTHORPE, D. V. 1978. The biosynthesis of monoterpenes. Prog. Phytochem. *5*, 65–126.

CHARM, S. E. and MATTEO, C. C. 1971. Scale-up of protein isolation. Methods Enzymol. *22*, 476–477.

CHASE, T., JR. 1974. Flavor enzymes. *In* Food Related Enzymes. J.R. Whitaker (Editor). Am. Chem. Soc., Washington, D.C.

CHAUVET, J. and ARCHER, R. 1976. Affinity chromatography of the serine proteases and their polypeptide inhibitors. *In* Anal. Control Immobilized Enzyme Syst., Proc. Intern. Symp., 1975. D. Daniel and J.-P. Kernevez (Editors). North-Holland Publishers, Amsterdam.

CHAVANA, S. and McFEETERS, R. F. 1977. Thermal inactivation of fungal pectinases in cucumber brines. Lebensm. Wiss. Technol. *10*, 290–294. (German)

CHEAH, K. S. 1976. Formation of nitrosomyoglobin in bacon involving lactate dehydrogenase. J. Food Technol. *11*, 181–186.

CHEAH, K. S. and CHEAH, A. M. 1971. Postmortem changes in structure and function of ox muscle mitochondria. I. Electron microscopic and polarigraphic investigations. J. Bioenergetics *2*, 85–93.

CHEAH, K. S. and CHEAH, A. M. 1976. The trigger for PSE condition in stress-susceptible pigs. J. Sci. Food Agric. *27*, 1137–1144.

CHEFTEL, C. 1972. Continuous enzymic solubilization of fish protein concentrate. Studies with recycling of enzymes. Ann. Technol. Agric. *21*, 423–433. (French)

CHEN, T. -S. and JOSLYN, M. A. 1967. The effect of sugars on the viscosity of pectin solutions. I. Comparison of corn syrup with sucrose solutions. J. Colloid Interface Sci. *23*, 399–406.

CHENG, C. -S. and PARRISH, F. C., JR. 1978. Molecular changes in the salt-soluble myofibrillar proteins of bovine muscle. J. Food Sci. *43*, 461–463.

CHENG, F. S. and CHRISTIAN, G. D. 1972. Rapid enzymatic determination of lactose in food products using amperometric measurements of the rate of depletion of oxygen. Analyst *102*, 124–131.

CHERNAYA, M. M., ADLI, K., LAVRENOVA, G. I., and STEPANOV, V. M. 1976. Biospecific chromatography of acid proteinases. Role of ion exchange and hydrophobic interactions. Biokhimiya *4*, 732–739. (Russian)

CHERRY, J. P. 1977. Oilseed enzymes as biological indicators for food uses and applications. *In* Enzymes in Food and Beverage Processing. R.L. Ory and A. St. Angelo (Editors). Am. Chem. Soc., Washington, D.C.

CHERRY, J. P., McWATTERS, K. H., MILLER, J. and SHEWFELT, A. L. 1977. Some chemical and nutritional properties of feather protein isolates containing various half-cystine levels. Adv. Exp. Med. Biol. *86B*, 503–530.

CHERYAN, M., RICHARDSON, T. and OLSON, N. F. 1975A. Surface structure of bovine micelles elucidated with insolubilized carboxypeptidase A. J. Dairy Sci. *58*, 651–657.

CHERYAN, M., VAN WYCK, P. J., OLSON, N. F. and RICHARDSON, T. 1975B. Secondary phase and mechanism of enzymic milk coagulation. J. Dairy Sci. *58*, 447–481.

CHERYAN, M., VAN WYCK, P. J., OLSON, N. F. and RICHARDSON, T. 1975C. Continuous coagulation of milk using immobilized enzymes in a fluidized bed reactor. Biotechnol. Bioeng. *17*, 585–598.

CHERYAN, M., VAN WYCK, P. J., RICHARDSON, T. and OLSON, N. F. 1976. Stability characteristics of pepsin immobilized on protein-coated glass used for continuous milk coagulation. Biotechnol. Bioeng. *18*, 273–279.

CHIANG, B. -Y., MILLER, G. D. and JOHNSON, J. A. 1973. Measuring damaged starch by the polarimetric method. Cereal Chem. *50*, 44–49.

CHIANG, P. C. and STERNBERG, M. Z. 1972. Soluble plant proteins. Ger. Offen. 2,220,299. Nov. 9.

CHICHESTER, C.O. 1972. The Chemistry of Plant Pigments. Academic Press, New York.

CHICHESTER, C. O., STADTMAN, F. H. and MACKINNEY, G. 1952. The products of the Maillard reaction. J. Am. Chem. Soc. *74*, 3418–3420.

CHICK, H., and MARTIN, C. J. 1912. Heat coagulation of proteins. J. Physiol. *45*, 261–295.

CHILDS, E. A. and FORTÉ, J. F. 1976. Enzymatic and ultrasonic techniques for solubilization of protein from heat-treated cottonseed products. J. Food Sci. *41*, 652–655.

CHILSON, O. P., COSTELLO, L. A. and KAPLAN, N. O. 1965. Effects of freezing on enzymes. Fed. Proc. *24* (2) Supplement, S55–S65.

CHIN, C. -K. and SACALIS, J. N. 1977. Metabolism of sucrose in cut roses. 3. Absorption of sugars by petal disks. J. Am. Hortic. Sci. *102*, 541–542.

CHISHOLM, M. D. 1973. Biosynthesis of 3-methoxycarbonylpropyl-glucosinolate in an *Erysimum* species. Phytochemistry *12*, 605–608.

CHISM, G. W., HAARD, N. F. and WEISS, P. 1975. Influence of temperature on kinetic properties of phosphorylase from two varieties of potato tuber. J. Food Sci. *40*, 94–96.

CHISM, G. W. and HULTIN, H. O. 1977. Solubilization of phosphofructokinase from the particulate fraction of chicken muscle. J. Food Biochem. *1*, 75–90.

CHOCK, P. B. and STADMAN, E. R. 1979. Intraconvertible enzyme cascades in metabolic regulations: A kinetic analysis. Fed. Proc. *38*, 295.

CHOW, H. 1975. A dichotomous key to the *Desmodium* species based upon esterase isozyme patterns. J. Singapore Natl. Acad. Sci. *4*, 121–124.

CHRISPEELS, M. J. and VARNER, J. E. 1966. Gibberellic acid-enhanced synthesis and release of α-amylase and ribonuclease by isolated barley aleurone layers. Plant Physiol. *42*, 398–406.

CHRISTENSEN, H. N. and PALMER, P. 1976. Enzyme Kinetics. A Learning Program for Students of the Biological and Medical Sciences, 2nd Edition. W.B. Saunders and Co., Philadelphia.

CHRISTENSEN, T. B., VEGARUD, G. and BIRKELAND, A. J. 1976. Stabilization of enzymes by glycosylation. Process Biochem. *11* (6) 25–26.

CHRISTIANSEN, J. A. 1957. An addendum to the Henri-Michaelis mechanism. Acta Chem. Scand. *11*, 660–662.

CHRISTISON, J. 1972. Preparation of immobilized glucoamylase. Chem. Ind. (London) *5*, 215–216.

CHU, H. D., LEEDER, J. G. and GILBERT, S. G. 1975. Immobilized catalase reactor for use in peroxide sterilization of dairy products. J. Food Sci. *40*, 641–643.

CHUA, G. K. and BUSHUK, W. 1969. Purification of wheat protease by affinity chromatography on hemoglobin-sepharose column. Biochem. Biophys. Res. Comm. *37*, 545–569.

CHUBEY, B. B. and DORELL, D. G. 1972. Enzymatic browning of stored parsnip roots. J. Am. Soc. Hortic. Sci. *97*, 107–109.

CHUNG, J., SCANU, A. M. and REMAN, F. 1973. Effect of phospholipids on lipoprotein lipase activation *in vitro*. Biochim. Biophys. Acta *296*, 116–123.

CHUNG, J. I., SAKAMURA, S. and LUH, B. S. 1974. Effect of harvest maturity on pectin and texture of canned black olives. Confructa *19*, 227–235.

CHUNG, O. K., POMERANZ, Y., GOFORTH, D. R., SHOGREN, M. D. and FINNEY, K. F. 1976. Improved sucrose esters in breadmaking. Cereal Chem. *53*, 615–626.

CHUNG, O. K. and TSEN, C. C. 1977. Functional properties of surfactants in breadmaking. II. Composition of lipids associated with doughs containing various levels of surfactants. Cereal Chem. *54*, 857–864.

CHURG, A. K. and MAKINEN, M. W. 1978. The electronic structure and coordination geometry of the oxyheme complex in myoglobin. J. Chem. Phys. *68*, 1913–1925.

CIBA-GEIGY A. -G. 1977. Additive protecting beers against chill haze. Fr. Pat. Appl. 2,336,478. July 22.

CILENTO, G. 1975. Dioxetanes in biological processes. J. Theor. Biol. *55*, 471–479.

CIRCLE, S. J. 1950. Proteins and other nitrogenous constituents. *In* Soybeans and Soybean Products, Vol. 1. K.S. Markley (Editor). John Wiley & Sons, New York.

CITRI, N. 1973. Conformational adaptability in enzymes. Adv. Enzymol. *37*, 397–648.

CLARK, L. C., JR. and LYONS, C. 1962. Electrode systems for continuous monitoring in cardiovascular surgery. Ann. N.Y. Acad. Sci. *102*, 29–45.

CLARK, L. C., JR. and LYONS, C. 1965. Studies of a glassy carbon electrode for brain polarography with observations on the effect of carbonic anhydrase inhibition. Ala. J. Med. Sci. *2*, 353–359.

CLARKE, F. M., MORTON, D. J. and WALSH, T. 1976–1977. Enzyme binding. *In* Report of Research, Division of Food Research, CSIRO, Sydney.

CLARKE, P. H 1976. Genes and enzymes. FEBS Lett. *62*, Supplement, 4. February.

CLAUDIAN, J. 1970. History of the usage of alcohol. *In* Alcohol and Derivatives. A. Tremolières (Editor). Pergamon Press, Oxford.

CLAYTON, R. K. and SMITH, C. 1960. *Rhodopseudomonas spheroides*: High catalase and blue-green double mutants. Biochem. Biophys. Res. Comm. *3*, 143–145.

CLEGG, M. K. and McMILLAN, A. D. 1974. Dietary enzymic hydrolysates of protein with reduced bitterness. J. Food Technol. *9*, 21–29.

CLELAND, W. W. 1967. The statistical analysis of enzyme kinetic data. Adv. Enzymol. *29*, 1–32.

CLELAND, W. W. 1970. Steady state kinetics. *In* The Enzymes, 3rd Edition, Vol. 2. P.D. Boyer (Editor). Academic Press, New York.

CLELAND, W. W. 1975. Partition analysis and concepts of net rate constants as tools in enzyme kinetics. Biochemistry *14*, 3220–3224.

CLEVELAND, L. and DAVIS, L. 1974. Superoxide dismutase activity of galactose oxidase. Biochim. Biophys. Acta *341*, 517–523.

CLINE, G. C. and RYEL, R. B. 1971. Zonal centrifugation. Methods Enzymol. *22*, 168–204.

CLYDESDALE, F. M. 1976. Instrumental techniques for color measurements of foods. Food Technol. *30* (10) 52–61.

CLYDESDALE, F. M. and FRANCIS, F. J. 1968. Chlorophyll changes in thermally processed spinach as influenced by enzyme conversion and pH adjustment. Food Technol. *22*, 792–796.

COBURN, H. J. and CARROLL, J. J. 1973. Improved manual and automated colorimetric determination of serum glucose, and use of hexokinase and glucose 6-phosphate dehydrogenase. Clin. Chem. *19*, 127–130.

COGGINS, C. W., JR., KNAPP, J. C. F. and RICKER, A. L. 1968. Post-harvest softening studies of Deglet Noor dates: Physical, chemical and histological changes. Date Grow. Inst. Rep. *45*, 3–6.

COGGON, P., MOSS, G. A., GRAHAM, H. N. and SANDERSON, G. W. 1973A. The biochemistry of tea fermentation: Oxidative degallation and epimerization of tea flavanol gallates. J. Agric. Food Chem. *21*, 727–733.

COGGON, P., MOSS, G. A. and SANDERSON, G. W. 1973B. Tea catechols oxidase: Isolation, purification and kinetic characterizations. Phytochemistry *12*, 1947–1955.

COGGON, P., ROMANCZYK, L. J. and SANDERSON, G. W. 1977. Extraction, purification and partial characterization of a tea metalloprotein and its role in the formation of black tea aroma constituents. J. Agric. Food Chem. *25*, 278–283.

COGGON, P. and SANDERSON, G. W. 1973. Manufacture of instant tea. Ger. Offen. 2,304,073. Aug. 16.

COHEN, C. 1975. The protein switch of muscle contraction. Sci. Am. *233* (5) 36–45.

COHEN, E. and SCHIFFMANN-NADEL, M. 1973. Changes in pectic substances of lemon albedo infected with *Phytophthora citrophthora*. HortScience *8*, 513.

COHEN, H., HAMOSH, M., ATIA, R. and SHAPIRO, B. 1967. Lysolecithinase in fish muscle. J. Food Sci. *32*, 179–181.

COHEN, S. R. 1968. "Best values" of Michaelis-Menten kinetic constants from experimental data. Anal. Biochem. *22*, 549–552.

COHEN, S. S. 1953. Studies on D-ribulose and its enzymatic conversion to D-arabinose. J. Biol. Chem. *201*, 71–84.

COHEN, W., BARICOS, W. H., KASTL, P. R. and CHAMBERS, R. P. 1977. Membrane immobilized liver microsome drug detoxifier. *In* Biomedical Applications of Immobilized Enzymes and Proteins. T.M.S. Chang (Editor). Plenum Press, New York.

COHN, E. J. *et al.* 1950. A system for the separation of the components of human blood: Quantitative procedures for the separation of the protein components of human plasma. J. Am. Chem. Soc. *72*, 465–474.

COLAS, A. and CHARLEGEGUE, A. 1974. Lipasic and lipoxygenasic activity of wheats and products of milling. Ann. Technol. Agric. *23*, 323–324. (French)

COLBY, J. and ZATMAN, L. J. 1971. The purification of a bacterial trimethylamine dehydrogenase. Biochem. J. *121*, 9P–10P.

COLE, I. and STROH, J. H. 1972. Treating abalone for making abalone steaks. Aust. Pat. 427,865. Sept. 28.

COLEMAN, J. E. 1974. Structure and mechanism of copper oxidase. *In* Food Related Enzymes. J.R. Whitaker (Editor). Am. Chem. Soc., Washington, D.C.

COLLINS, J. H. 1976. Homology of myosin DTNB light chain with alkali light chains, troponin C, and parvalbumin. Nature *259*, 699–700.

COLLINS, J. L. and McCARTY, I. E. 1969. Comparison of microwave energy with boiling water for blanching whole potatoes. Food Technol. *23*, 337–340.

COLMEY, J. C. 1978. High fiber foods in the

American diet. Food Technol. *32* (3) 42–47.

COLOWICK, S. P. and KAPLAN, N. O. 1955–1980. Methods in Enzymology, Vol. 1–69. Academic Press, New York.

COMTAT, J. and BARNOUD, F. 1976. Properties and mode of action of cellulases and xylanases. Physiol. Veg. *14*, 801–816.

CONN, E. E. 1969. Cyanogenic glucosides. J. Agric. Food Chem. *17*, 519–526.

CONN, E. E. 1973. Cyanogenetic glycosides. *In* Toxicants Occurring Naturally in Foods, 2nd Edition. Natl. Acad. Sci., Washington, D.C.

CONN, E. E. 1978. Cyanogenesis: The production of hydrogen cyanides by plants. *In* Effects of Poisonous Plants on Livestock. R.F. Keeler, K.R. Van Kampen and L.F. James (Editors). Academic Press, New York.

CONN, E. E. 1979. Biosynthesis of cyanogenic glycosides. Naturwissenschaften *66*, 28–34. (German)

CONN, E. E. and STUMPF, P. K. 1972. Outlines of Biochemistry, 3rd Edition. John Wiley & Sons, New York.

CONNEL, J. J. 1975. The role of formaldehyde as a protein crosslinking agent during the frozen storage of cod. J. Sci. Food Agric. *26*, 1925–1929.

CONSTANTIN, R. J., FONTENAT, J. E. and BARRIOS, E. P. 1974. Processing studies with eggplant. J. Am. Hortic. Sci. *99*, 505–507.

CONVERSE, R. 1978. Use of ELISA for detecting tomato ringspot virus in red raspberry plants. ELISA Workshop, Dec. 5–6. Corvallis, Oregon.

COOK, D. J. 1974. Nutritional losses in food processing—vitamin C. Process Biochem. *9* (5) 21–24.

COOK, J. W. and STEERS, A. W. 1947. Detection of fecal matter in food products. Investigation of the use of trypsin and alkaline phosphatase activity of feces as a measure of contamination. J. Assoc. Off. Agric. Chem. *30*, 168–181.

COOKE, A. and JOHANNSON, H. 1970. Baking additive and method for producing baked goods. U.S. Pat. 3,512,992. May 19.

COOKE, R. and FRANKS, K. 1980. All myosin heads form bonds with actin in rigor skeletal muscle. Biochemistry *19*, 2265–2269.

COOKE, R. D., BLAKE, G. G. and BATTERSHILL, J. M. 1978. Purification of cassava linamarase. Phytochemistry *17*, 381–383.

COOLER, F. W. 1976. Enzymes. Use and Control in Foods. Inst. Food Technol. IFT Short course. IFT, Chicago.

COOMBE, B. G. 1976. The development of fleshy fruits. Annu. Rev. Plant Physiol. *27*, 207–228.

COOMBS, J., MAW, S. L. and BALDRY, C. W. 1974. Metabolic regulation in C_4 photosynthesis: PEP carboxylase and energy charge. Planta *117*, 279–292.

COON, J. M. 1974. Natural toxicants. A perspective. Nutr. Abstr. Rev. *32*, 321–333.

COPSON, D. A. 1954. Microwave heating in freeze-drying, electronic. Microwave irradiation of orange juice concentrate for enzyme inactivation. Food Technol. *8*, 397–399.

CORDONNIER, M., LAWNY, F., CHAPOT, D. and THOMAS, D. 1975. Magnetic enzyme membranes as active elements of electrochemical sensors. Lactose, saccharose, maltose bienzyme electrodes. FEBS Lett. *59*, 263–267.

CORDUAN, G. 1976. Isoenzyme variation as an indicator for the generative or somatic origin of anther-derived plants of *Digitalis purpurea* L. Z. Pflanzenzücht. *76*, 47–55.

CORMIER, M. J., LEE, J. and WAMPLER, J. E. 1975. Bioluminescence. Annu. Rev. Biochem. *44*, 255–272.

CORNELL, R. and CHARM, S. E. 1976. Purification of carboxypeptidase G-1 by immunoadsorption. Biotechnol. Bioeng. *18*, 1171–1173.

CORNFORTH, J. W. 1976. Asymmetry and enzyme action. J. Mol. Catal. *1*, 145–157.

CORNISH-BOWDEN, A. 1975. The use of the direct linear plot for determining initial velocities. Biochem. J. *149*, 305–312.

CORNISH-BOWDEN, A. 1976. Principles of Enzyme Kinetics. Butterworths, London.

CORNISH-BOWDEN, A. 1976. The effect of natural selection on enzymic catalyses. J. Mol. Biol. *101*, 1–9.

CORNISH-BOWDEN, A. 1979. Fundamentals of Enzyme Kinetics. Butterworths Publishers, Woburn, Mass.

CORY, R. P. 1975. Enzymic isomerization product. U.S. Pat. 3,910,821. Feb. 19.

COSGROVE, D. J. 1977. Microbial transformations in the phosphorus cycle. Adv. Microbiol. Ecol. *1*, 95–134.

COSTELLO, J. and BOURKE, R. 1974. Modification of the "decarboxylase" method of glucose specific radioactivity determination. Anal. Biochem. *59*, 643–646.

COSTELLO, J. and HATCH, M. 1976. An enzymic method for the spectrophotometric determination of oxalic acid. J. Lab. Clin. Med. *87*, 903–908.

COTTER, W. P., LLOYD, N. E. and HINMAN,

N. E. 1971. Enzymically isomerizing glucose in syrups. U.S. Pat. 3,623,953. Nov. 30.

COUDON, B., BOUIGE, D. and GIRAUDET, P. 1974. Fully automated continuous flow system for determination of serum triglycerides. Clin. Chim. Acta *55*, 129–138. (French)

COUGHLIN, R. W. 1977. Immobilized lactase technology: Recent developments and applications. 37th Annu. Meet., Inst. Food Technol. Pap. *31*, Philadelphia.

COUGHLIN, R. W. and ALEXANDER, B. F. 1975. Simplified flow reactor for electrochemically driven enzymic reactions involving cofactors. Biotechnol. Bioeng. *17*, 1379–1382.

COWLING, E. B. 1975. Physical and chemical restraints in the hydrolysis of cellulose and lignocellulosic materials. Biotechnol. Bioeng. Symp. *5*, 163–181.

COXON, A. C., CURTIS, W. D., LAIDLER, D. A. and STODDART, J. F. 1979. Lock and key chemistry with crown compounds. J. Carbohydr. Nucleosides, Nucleotides *6*, 167–197.

COZZARELLI, N. R. 1980. DNA gyrase and the supercoiling of DNA. Science *207*, 953–960.

CPC INTERN. 1976. Immobilized enzyme systems. Neth. Appl. *74* 09,455. Jan. 14.

CRAMER, F. and HETTLER, H. 1967. Inclusion compounds of cyclodextrins. Naturwissenschaften *54*, 625–632.

CRANDALL, M. R. and MacKAY, V. L. 1977. Physiology of mating in three yeasts. Adv. Microb. Physiol. *15*, 307–398.

CRANE, F. L. 1977. Hydroquinone dehydrogenases. Ann. Rev. Biochem. *46*, 439–469.

CRAWFORD, L. 1972. Effort of premortem stress, holding temperatures and freezing on the biochemistry and quality of skipjack tuna. WOAA Tech. Rep. *NMFS SSRF-651.* U.S. Dep. Commer., Washington, D.C.

CREAMER, L. K. 1976. A further study of the action of rennin on β-casein. N. Z. J. Dairy Sci. *11*, 30–39.

CREAN, D. E. C. and HAISMAN, D. E. 1963. The interaction between phytic acid and divalent cations during the cooking of dried peas. J. Food Sci. Agric. *14*, 824–833.

CREASY, L. L. 1974. Sequence of development of autumn coloration in *Euonymus*. Phytochemistry *13*, 1391–1394.

CRIVELLI, G. and BONCORE, C. 1975. Nominal length of freezing and structural changes in vegetables. Freddo *29*, 3–8. (Italian)

CROCCO, S. C. 1975. Tomorrows promise: Enzymatic systems at work. Food Eng. *47* (7) 54–57.

CROCKER, W. and HARRINGTON, G. T. 1918. Catalase and oxidase content of seeds in relation to their dormancy, age, vitality and respiration. J. Agric. Res. *15*, 137–174.

CROSS, H. R. and BERRY, B. W. 1980. Beef hot-processing cuts costs. Agric. Res. *28* (7–8) 4–5.

CROTEAU, R. 1975. Biosynthesis of monoterpenes and sesquiterpenes. *In* Odor and Taste Substances. F. Drawert (Editor). Verlag Hans Carl, Nurenberg. (German)

CROTEAU, R. 1978. Biogenesis of flavor components: Volatile carbonyl compounds and monoterpenoids. *In* Postharvest Biology and Biotechnology. H.O. Hultin and M. Milner (Editors). Food & Nutrition Press, Westport, Conn.

CROTEAU, R. and KARP, F. 1977. Demonstration of a cyclic pyrophosphate intermediate in enzymatic conversion of neryl pyrophosphate to borneol. Arch. Biochem. Biophys. *184,* 77–86.

CRUESS, W. V. 1943. Role of enzymes in wine making. Adv. Enzymol. *3*, 349–386.

CRUESS, W. V. and BESONE, J. 1941. Observations on the use of pectic enzymes in wine making. Fruit Prod. J. *20*, 365–367.

CRUESS, W. V., QUACCHIA, R. and ERICSON, K. 1955. Pectic enzymes in wine making. Food Technol. *9*, 601–607.

CUATRECASAS, P. 1970. Protein purification by affinity chromatography. Derivatizations of agarose and polyacrylamide beads. J. Biol. Chem. *245*, 3059–3065.

CUATRECASAS, P. 1972. Affinity chromatography and purification of the insulin receptor of liver cell membranes. Proc. Natl. Acad. Sci. *69*, 1277–1281.

CUATRECASAS, P., WILCHEK, M. and ANFINSEN, C. B. 1968. Selective enzyme purification by affinity chromatography. Proc. Natl. Acad. Sci. *61*, 636–643.

CULLITAN, B. J. 1975. National Institutes of Health: The politics of taste and smell. Science *187*, 145–148.

CULPEPPER, C. W. 1936. Effect of stage of maturity of the snap bean on its composition and use as a food product. Food Res. *1*, 357–376.

CULVER, W. H. and CAIN, R. F. 1952. Nature, causes, and correction of discoloration of canned blackeyed and purple-hull peas (field peas). Texas Agric. Exp. Stn. Bull. *748*, 5–23.

CURL, A. L. and JANSEN, E. F. 1950. The effect of high pressures on pepsin and chymotrypsinogen. J. Biol. Chem. *185*, 713–723.

CURLING, J. 1978. Large scale ion exchange

chromatography of proteins. Proc. 12th FEBS Meet., July. Dresden.

CURRY, R. A. and TING, I. P. 1975. Heterogeneity of maize NAD malate dehydrogenase: Generation of multiple forms by incubation at pH 5.0. Arch. Biochem. Biophys. *167*, 774–776.

CURTIS, R., III. 1976. Genetic manipulation of *E. coli* for use as a safe industrial organism. Am. Soc. Microbiol. Conf. on Genetics and Molecular Biology of Industrial Microorganisms, Feb. Orlando.

CYCLUS, LTD. 1962. Production of dried meat. Br. Pat. 900,794. July 30.

DABROWSKI, K. and GLOGOWSKI, J. 1977. A study of the application of proteolytic enzymes to fish food. Aquaculture *12*, 349–360.

DADIC, M. and VAN GHELUWE, J. E. A. 1971. Potential antioxidants in brewing. Tech. Q. Master Brew. Assoc. Am. *8* (4) 182–190.

DADIC, M., VAN GHELUWE, J. E. A. and VALYI, Z. 1976. Alkaline steeping and the stability of beer. J. Inst. Brew. *82*, 273–276.

DAFFORN, A. and KOSHLAND, D. E., JR. 1973. Proximity, entropy and orbital steering. Biochem. Biophys. Res. Comm. *52*, 779–785.

DAFTARY, R. D. 1975. Wet process for making high-protein soybean products. U.S. Pat. 3,925, 569. Dec. 9.

DAGHER, S. M. and RUHAYYIM, I. G. 1975. Fusel oil and methanol content of Lebanese arak. J. Food Sci. *40*, 917–918.

DAGLETY FRANKLIN LTD. 1974. Pea processing. Br. Pat. 1,377,850. Feb. 27.

DAHLBERG, A. E. and DAHLBERG, J. E. 1975. Binding of ribosomal protein S of *E. coli* to the 3′-end of 16S rRNA. Proc. Natl. Acad. Sci. *72*, 2940–2944.

DAHODWALA, S., HUMPHREY, A. and WEIBEL, M. 1974. Pectic enzymes: Individual and concerted kinetic behavior of pectinesterase and pectinase. J. Food Sci. *39*, 920–926.

DAI-NIPPON SUGAR MANUFACTURING CO., LTD. 1977. Preparing protein-containing material. Br. Pat. 1,465,396. Mar. 23.

DALE, J. K. and LANGLOIS, D. P. 1940. Starch conversion sirup. U.S. Pat. 2,201,609. May 21.

DALLING, M. G., HALLORAN, G. M. and WILSON, J. H. 1975. The relation between nitrate reductase activity and grain nitrogen productivity in wheat. Aust. J. Agric. Res. *26*, 1–10.

DALZIEL, K. 1957. Initial steady state velocities in the evaluation of enzyme-coenzyme-substrate reaction mechanisms. Acta Chem. Scand. *11*, 1706–1723.

DANESCHWAR, M. 1976. Determination of amygdalin in different parts of the cherry fruit. Ind. Obst-Gemüseverwert. *61*, 374–376. (German)

DANIEL, J. W. 1967. Enzymic reduction of 230 food colorings. Food Cosmet. Toxicol. *5*, 533–556.

DANIELS, N. W., FRAZIER, P. J. and WOOD, P. S. 1971. Flour lipids and dough development. Bakers Dig. *45*, 20–25, 48.

DANILEWSKI, B. 1886. The Organoplastic Forces of the Organism. Charkoff, Moscow. (Russian)

D'APPOLONIA, B. L. 1971. Carbohydrates. *In* Wheat, Chemistry and Technology, 2nd Edition. Y. Pomeranz (Editor). Am. Assoc. Cereal Chem., St. Paul, Minn.

DARBYSHIRE, B. 1974A. Function of the carbohydrate units of three fungal enzymes in their resistance to dehydration. Plant Physiol. *54*, 717–721.

DARBYSHIRE, B. 1974B. Influence of dehydration on catalase stability. Comparison with freezing effects. Cryobiology *11*, 148–151.

DARBYSHIRE, B. 1975. The results of freezing and dehydration of horseradish peroxidase. Cryobiology *12*, 276–281.

DARBYSHIRE, B. and STEER, B. Y. 1973. Dehydration of macromolecules. I. Effect of dehydration-rehydration on indole-acetic acid oxidase, ribonuclease, ribulosediphosphate carboxylase and ketose-1-phosphate aldolase. Aust. J. Biol. Sci. *26*, 591–604.

DARVEY, I. G., SCHRAYER, R. and KOHN, L. D. 1975. Integrated steady state equations and the determination of individual rate constants. J. Biol. Chem. *250*, 4696–4701.

DARZINS, E. 1960. Edible castor cake product and method of producing the same. U.S. Pat. 2,920,963. Jan. 12.

DASSOW, J. A., McKEE, L. G. and NELSON, R. W. 1962. Development of an instrument for evaluating texture of fishery products. Food Technol. *16* (3) 108–110.

DASTOLI, F. R. 1974. Taste receptor proteins. Life Sci. *14*, 1417–1426.

DAUGHERTY, N. A. 1979. Isoenzymes. J. Chem. Educ. *56*, 442–445.

DAUSSANT, J. 1977. Immunochemical characterization of α-amylases in wheat seeds at different ontogenic steps. Cereal Sci. Today *22*, 468.

DAUSSANT, J. and ABBOTT, D. 1969. Immunochemical study of changes in the soluble proteins of wheat during germination. J. Sci.

Food Agric. *20*, 633–637.

DAUSSANT, J. and CARFANTAN, N. 1975. Electro-immunoabsorption in gel, application to enzyme studies (α- and β-amylases from barley). J. Immunol. Methods *8*, 373–382.

DAUSSANT, J. and CORVAZIER, P. 1970. Biosynthesis and modifications of α- and β-amylases in germinating wheat seed. FEBS Lett. *7*, 191–194.

DAUSSANT, J. and RENARD, M. 1972. Immunochemical comparisons of α-amylases in developing and germinating wheat seeds. FEBS Lett. *22*, 301–304.

DAUSSANT, J., SKAKOUN, A. and NIKUPAAVOLA, M. L. 1974. Immunochemical study on barley α-amylases. J. Inst. Brew. *80*, 55–58.

DAVET, P. 1975. Enzymes of *Pyrenochaeta lycopersici* contributing to parasitic degradations in some *Lysopersicum* species, and some host defense reactions. C.R. Hebd. Seances Acad. Sci., Ser. D *281*, 143–146. (French)

DAVEY, C. L. and GILBERT, K. V. 1968. Studies in meat tenderness. J. Food Sci. *33*, 343–348.

DAVEY, C. L. and GILBERT, K. V. 1974. The mechanism of cold-induced shortening in beef muscle. J. Food Technol. *9*, 51–58.

DAVEY, C. L. and GILBERT, K. V. 1975. The tenderness of cooked and raw meat from young and old beef animals. J. Sci. Food Agric. *26*, 953–960.

DAVIDEK, J., VELISEK, J. and JANICEK, G. 1972. Stability of inosinic acid, inosine and hypoxanthine in aqueous solutions. J. Food Sci. *37*, 789–790.

DAVIDEK, J., VELISEK, J. and ZELINKOUA, Z. 1975. Nucleotides and nucleosides in cultivated mushrooms. FSTA *8*, 4J 505. (Czech.)

DAVIE, E. W. and FUJIKAWA, K. 1975. Basic mechanisms in blood coagulation. Ann. Rev. Biochem. *44*, 799–829.

DAVIES, D. D. and PATIL, K. D. 1975. The control of NAD specific malic enzyme from cauliflower bud mitochondria by metabolites. Planta *126*, 197–211.

DAVIES, D. D., PATIL, K. D., UGOCHUKWU, E. N. and TOWERS, G. H. N. 1973. Aliphatic alcohol dehydrogenase from potato tubers. Phytochemistry *12*, 523–530.

DAVIES, D. R. 1974. Some aspects of sucrose metabolism. Ann. Proc. Phytochem. Soc. *10*, 61–81.

DAVIES, D. R., PRIDHAM, J. B. and RINTOUL, J. 1974. Control of starch synthesis in potato tuber (*Solanum tuberosum*). Biochem. Soc. Trans. *2*, 1112–1115.

DAVIES, D. R. and SEGAL, D. M. 1971. Protein crystallization: Microtechniques involving vapor diffusion. Methods Enzymol. *22*, 266–269.

DAVIES, E., BOURKE, E. and COSTELLO, J. 1975. An enzymic method for the determination of [1-^{14}C] lactose. Analyst (London) *100*, 758–760.

DAVIS, D. E. 1975. Increasing ham yields. Agric. Res. *23* (7) 12–13.

DAVIS, D. R., SISTRUNK, W. A. and MERRIL, S. C. 1975. Cyprex residues in spinach. Ark. Farm Res. *24* (4) 1–10.

DAVIS, W. R. 1942. Quantitative field test for estimation of peroxidase. Ind. Eng. Chem. Anal. Ed. *14*, 952–953.

DAWES, E. A. 1964. Enzyme kinetics. *In* Comprehensive Biochemistry, Vol. 12. M. Florkin and E.G. Stotz (Editors). Elsevier Publishing Co., New York.

DAWSON, C. R., STROTHKAMP, K. C. and KRUL, K. 1975. Ascorbate oxidase and related copper proteins. Ann. N.Y. Acad. Sci. *258*, 209–220.

DAWSON, C. R. and TARPLEY, W. B. 1951. Tyrosinase. *In* The Enzymes, 1st Edition, Vol. 2. G.B. Sumner and K. Myrbäck (Editors). Academic Press, New York.

DAXENBICHLER, M. E., VanETTEN, C. H. and SPENCER, G. F. 1977. Glucosinolates and derived products in cruciferous vegetables. Identification of organic nitriles from cabbage. J. Agric. Food Chem. *25*, 121–124.

DAY, E. A. 1967. Cheese flavor. *In* Symposium on Foods: The Chemistry and Physiology of Flavors. H.W. Schultz, E.A. Day and L.M. Libbey (Editors). AVI Publishing Co., Westport, Conn.

DAYTON, W. R., GOLL, D. E., ZEECE, M. G. and ROBSON, R. M. 1976. A Ca^{2+}-activated protease possibly involved in myofibrillar protein turnover. Purification from porcine muscle. Biochemistry *15*, 2150–2158.

De AMORIM, H. V. and SILVA, D. M. 1968. Relationship between the polyphenol oxidase activity of coffee beans and the quality of the beverage. Nature *219*, 381–382.

DEAN, P. D. G. and HARVEY, M. J. 1975. Applications of affinity chromatography. Process Biochem. *10* (9) 5–10.

DeBECZE, G. I. 1970. Food Enzymes. Crit. Rev. Food Technol. *1*, 479–518.

DECREAU, R. V. 1972. The process of microwave blanching. Food Prod. Manage. *95* (5) 12–15.

De DUVE, C. 1969. The function of peroxisomes. *In* Proc. Univ. of Miami Symp. on Membrane Function and Electron Transfer to Oxygen. Jan. Miami.

DEDYKHINA, V. P., LENTSOVA, L. V., NIKOLSKAYA, T. V. and NOVOZHILOUA, G. N. 1975. Changes in whale meat color during storage. FSTA *8*, 8 R501.

DEDYUKHINA, V. P., LENTSOVA, L. V., NIKO'SKAYA, T. V. and NOVOZHILOVA, G. N. 1975. Changes in whale meat color during storage. Izv. Vyss. Uchebn. Zaved. Pishch. Tekhnol. *5*, 145–146. (Russian)

DEETH, H. C. and FITZGERALD, C. H. 1976. Lipolysis in dairy products: A review. Aust. J. Dairy Technol. *31*, 53–64.

De FEKETE, M. A. R. 1968. The role of phosphorylase in metabolism of starch in plastids. Planta *79*, 208–221. (German)

De FREMERY, D. 1963. Relation between biochemical properties and tenderness of poultry. *In* Campbell Soup Co. Symp., Camden, N.J.

De FREMERY, D. 1966. Relationship between chemical properties and tenderness of poultry muscle. J. Agric. Food Chem. *14*, 214–217.

De FREMERY, D., MILLER, R. E., EDWARDS, R. H., KNUCKLES, B. E., BICKOFF, E. M. and KOHLER, G. O. 1973. Centrifugal separation of white and green protein fractions from alfalfa juice following controlled heating. J. Agric. Food Chem. *21*, 886–889.

De FREMERY, D. and STREETER, I. V. 1969. Tenderization of chicken muscle: The stability of alkali-insoluble connective tissue to postmortem aging. J. Food Sci. *34*, 176–180.

De GROOT, J. J. M. C., VELDINK, G. A., VLIEGENTHART, J. F. G., BOLDINGH, J., WEVER, R. and VAN GELDER, B. F. 1975. Demonstration by EPR spectroscopy of the functional role of iron in soybean lipoxygenase. Biochim. Biophys. Acta *377*, 77–79.

De HAAS, B. W., CHAPMAN, D. W. and GOERING, K. J. 1978. An investigation of the α-amylase from self-liquefying barley starch. Cereal Chem. *55*, 127–137.

DEINUM, J., LERCH, K. and REINHAMMAR, B. 1976. An EPR study of *Neurospora* tyrosinase. FEBS Lett. *69*, 161–164.

DEINUM, J., REINHAMMAR, B. and MARCHESINI, A. 1974. Stoichiometry of the three different types of copper in ascorbate oxidase from green zucchini squash. FEBS Lett. *42*, 241–245.

DEKAZOS, E. D. 1972. Callose formation by bruising and heating of tomatoes and its presence in processed products. J. Food Sci. *37*, 562–567.

De KETELAERE, A., DEMEYER, D., VANDEKERCKHOVE, P. and VERAEKE, I. 1974. Stoichiometry of carbohydrate fermentation during sausage ripening. J. Food Sci. *39*, 297–300.

DEKKER, R. F. H. and RICHARDS, G. N. 1976. Hemicellulases: Their occurrence, purification, properties, and mode of action. Adv. Carbohydr. Chem. Biochem. *32*, 277–352.

De La MAR, R. and FRANCIS, F. J. 1969. Carotenoid degradation in bleached papaya. J. Food Sci. *34*, 287–290.

De KONING, P. J. 1972 The replacement of calf rennet by microbial and other proteolytic enzymes during the clotting of milk. Ann. Technol. Agric. *21*, 357–366. (French)

DELFORNO, G. and GRUEV, P. 1970. Preparation of Bulgarian white cheese using rennet or an equal mixture of rennet and pepsin. Latte *44*, 631–635. (Italian)

DELINCEE, H. and PARTMANN, W. 1979. Eel muscle carnosinase: isoelectric point and molecular weight. J. Food Biochem. *3*, 43–52.

DELINCEE, H., MUNZNER, R. and RADOLA, B. J. 1975. Irradiation of industrial enzyme preparations for food processing. Lebensm. Wiss. Technol. *8*, 270–273. (German)

DELINCEE, H. and SCHAEFER, W. 1975. Influence of heat treatments of spinach at temperatures up to 100°C on important constituents. Heat inactivation of peroxidase isoenzymes in spinach. Lebensm. Wiss. Technol. *8*, 217–221.

DELOBBE, A., CHALMEAU, H. and GAY, P. 1975. Existence of two alternative pathways for fructose and sorbitol metabolism in *Bacillus subtilis* Marburg. Eur. J. Biochem. *51*, 503–510.

DEMAIN, A. L. 1973. The marriage of genetics and industrial microbiology—after a long engagement, a bright future. *In* Genetics of Industrial Microorganisms, Vol. 1. Bacteria. Z. Vansk, Z. Hostakelek and J. Cudlin (Editors). Elsevier Publishing Co., Amsterdam.

DEMAIN, A. L. and PHAFF, H. F. 1957. Softening of cucumber during curing. J. Sci. Food Agric. *5*, 60–64.

DEMBELE, S. and DUBOIS, P. 1973. Composition of the essences of shallots (*Allium cepa* L. var. aggregatums) Ann. Technol. Agric. *22*, 121–129. (French)

DENAULT, L. J. 1972. Soluble soybean protein. Ger. Offen. 2,219,712. Nov. 9.

DENG, J. C. and LILLARD, D. A. 1973. Effect of curing agents, pH, and temperature on the activity of porcine muscle cathepsins. J. Food Sci. *38*, 299–302.

DENIS, F., ALDELAIMI, K. and CHIRON, J. P. 1977. Study and demonstration of a lysine decarboxylase in *Pseudomonas aeruginosa*. C. R. Seances Soc. Biol. *171*, 484–487. (French)

DENNIS, C. 1977. The involvement of fungi in the breakdown of sulfited strawberries. Ann. Appl. Biol. *87*, 117.

DENNIS, G. E. and QUITTENTON, R. C. 1962. Enzymes in brewing. Can. Pat. 634,865. Apr. 31.

DENNISON, E. A. and AHMED, E. M. 1975. Irradiation of fruits and vegetables. *In* Postharvest Biology and Handling of Fruits and Vegetables. N.F. Haard and D.K. Salunkhe (Editors). AVI Publishing Co., Westport, Conn.

DENTON, A. E. and HOGAN, J. M. 1966. Enzymatic removal of meat from bones. U.S. Pat. 3,293,687. Dec. 27.

DE OLIVEIRA, J. C., SILVA, D. M., TEXEIRA, A. A. and AMORIM, H. V. 1977. Enzymic activity of polyphenoloxidase, peroxidase and catalase in beans of *Coffea arabica* L. related to the beverage quality. Turrialba *27*, 75–82. (Portuguese)

DERERA, N. F., BHATT, G. M. and McMASTER, G. J. 1977. On the problem of preharvest sprouting of wheat. Euphytica *26*, 299–308.

DESBOROUGH, S. 1968. Potato variety identification by use of electrophoretic patterns of tuber protein and enzymes. Am. Pot. J. *45*, 220–229.

DESHPANDE, S. N., KLINKER, W. J., DRAUDY, H. N. and DESROSIER, N. 1965. Role of pectic constituents and polyvalent ions in firmness of canned tomatoes. J. Food Sci. *30*, 594–609.

DESNUELLE, P. 1961. Pancreatic lipase. Adv. Enzymol. *23*, 129–161.

DESROSIER, N. W. 1976. Meat extender and process of making same. U.S. Pat. 3,952,111. Apr. 20.

DESROSIER, N. W. and DESROSIER, J. N. 1977. The Technology of Food Preservation, 4th Edition. AVI Publishing Co., Westport, Conn.

DESROSIER, N. W. and ROSENSTOCK, H. M. 1960. Radiation Technology in Food, Agriculture, and Biology. AVI Publishing Co., Westport, Conn.

DESROSIER, N. W. and TRESSLER, D. K. 1977. Fundamentals of Food Freezing. AVI Publishing Co., Westport, Conn.

DETERMANN, H. 1968. Gel Chromatography: Gel Filtration, Gel Permeation, Molecular Sieves. A Laboratory Handbook. Springer Verlag, New York.

DETERMANN, H. and KOEHLER, R. 1966. Plastein reaction. IX. Proline containing monomers and their enzymic conversion products. Ann. Chem. *690*, 197–202.

DE VILLIERS, O. T., MEYNHARDT, J. T. and DE BRUYN, J. A. 1974. The incorporation of carbon-14 dioxide into sorbitol and other compounds in Santa Rosa plum leaves. Agroplantae *6*, 55–58.

DEVORE, D. P. and SOLBERG, M. 1974. Oxygen uptake in postrigor bovine muscle. J. Food Sci. *39*, 22–27.

DEVORE, D. P. and SOLBERG, M. 1975. A study of the rate-limiting factors in the respiratory consumption of intact post-rigor bovine muscle. J. Food Sci. *40*, 651–652.

DHONUKSHE, B. and BHOWAL, J. G. 1974. Radiation-induced phenol-colour mutation in Dwarf wheat. Radiat. Res. *59*, 154–155.

DIAMALT-AKTIENGESELLSCHAFT. 1974. Use of L-cysteine and L-cystine in the preparation and improvement of foods. Lebensm.-Wiss. Technol. *7*, 64–71. (German)

DI CESARE, J. L. and ATWOOD, J. G. 1975. Kinetic substrate methods on an ultra microkinetic analyzer. Clin. Chem. *21*, 1031.

DICKEY, F. H. 1949. The preparation of specific adsorbents. Proc. Natl. Acad. Sci. *35*, 227–229.

DIEHL, H. C. 1932. A physiological view of freezing preservation. Ind. Eng. Chem. *24*, 661–665.

DIEHL, H. C., DINGLE, J. H. and BERRY, J. A. 1933. Enzymes can cause off-flavors when foods are frozen. Food Ind. *5*, 300–301.

DIERICKX, P., WAUTERS, C. and VENDRIG, J. 1975. On the requirement of a hydroxynitrile lyase in the conversion of orthonil to chloro-tolyl-acetic acid. Z. Pflanzenphysiol. *75*, 191–200.

DIETRICH, W. C. 1975. Recommendation of Special AFFI Working Group for Enzyme Research. Am. Frozen Food Inst., Burlingame, Cal.

DIETRICH, W. C., LINDQUIST, F. E., BOHART, G. S., MORRIS, H. J. and NUTTING, M.-D. 1955. Effect of degree of enzyme inactivation and storage temperature on quality retention in frozen peas. Food Res. *20*, 480–491.

DIETRICH, W. C., LINDQUIST, F. E. and BOGGS, M. M. 1957. Effect of maturity and storage temperature on quality of frozen peas. Food Technol. *11*, 485–487.

DIETRICH, W. C. and NEUMANN, H. J. 1965. Blanching brussel sprouts. Food Technol. *19*, 150–153.

DIETRICH, W. C., NUTTING, M. -D. F., BOGGS, M. M. and WEINSTEIN, N. E. 1962. Time-temperature tolerance of frozen foods. Food Technol. *16* (10) 123–128.

DIKEMAN, M. E., TUMA, H. J. and BEECHER, G. R. 1971. Bovine muscle tenderness as related to protein solubility. J. Food Sci. *36*, 190–193.

DILLARD, M. G., HENICK, A. S. and KOCH, R. B. 1961. Differences in reactivity of legume lipoxidase. J. Biol. Chem. *236*, 37–40.

DIMICK, K., PONTING, J. D. and MAKOWER, B. 1951. Heat inactivation of polyphenoloxidase in fruit juices. Food Technol. *5*, 237–241

DINELLI, D., MORISI, F., GIOVENCO, S. and PANSOLLI, P. 1973. Enzymic manufacture of fructose and syrups containing glucose and fructose. Ger. Offen. 2,303,872. Aug. 16.

DINGLE, J. I., JAQUES, P. J. and SHAW, I. H. 1979. Frontiers of Biology, Vol. 48: Lysosomes in Applied Biology and Their Therapeutics. North-Holland Publishing Co., Amsterdam.

DINGWALL, A. L. and CAMPBELL, H. 1975. New sweetener's potential. Food Process. Ind. *50* (11) 50–51.

DINIUS, D. A., OLTJEN, R. R. and SATTER, L. D. 1974. Influence of abomasally administered safflower oil on fat composition and organoleptic evaluation of bovine tissue. Anim. Sci. *38*, 887–892.

DIXON, B. 1975. Safeguards for microbial manipulators. New Sci. *68*, 618.

DIXON, M. and WEBB, E. C. 1964. Enzymes, 2nd Edition. Academic Press, New York.

DIXON, N. E., GAZZOLA, C., BLAKELY, R. L. and ZERNER, B. 1976. Metal ions in enzymes using ammonia or amides. Science *191*, 1144–1150.

DOIG, R. I., COLBORNE, A. J., MORRIS, G. and LAIDMAN, D. L. 1975. Induction of glyoxysomal enzyme activities in the aleurone cells of germinating wheat. J. Exp. Bot. *26*, 387–397.

DOLEY, S. G., HARVEY, M. J. and DEAN, P. D. G. 1976. The potential of Ultrogel, an agarose-polyacrylamide copolymer, as a matrix for affinity chromatography. FEBS Lett. 87–91.

Do NASCIMENTO, K. H., DAVIES, D. D. and PATIL, K. D. 1975. Unidirectional inhibition and activation of malic enzyme from potato by mesotartrate. Biochem. J. *149*, 349–356.

DONIGER, J. and GROSSMAN, L. 1976. Human correxonuclease. J. Biol. Chem. *251*, 4579–4587.

DONGOWSKI, G. and BOCK, W. 1977. Enzymatic liquefaction of carrots. Lebensm. Ind. *24*, 33–39. (German)

DONOHO, A. L., JOHNSON, W. S., SIECK, R. F. and SULLIVAN, W. L. 1973. Gas-chromatographic determination of diethylstilbesterol and its glucuronide in cattle tissues. J. Assoc. Anal. Chem. *56*, 785–792.

DONOVAN, J. 1977. A study of the baking process by differential scanning calorimetry. J. Sci. Food Agric. *28*, 571–578.

DONOVAN, J. W. 1979. Phase transitions of the starch-water system. Biopolymers *18*, 263–275.

DOROUGH, H. W. and THORSTENSON, J. H. 1975. Analysis for carbamate insecticides and metabolites. J. Chromatogr. Sci. *13*, 217–224.

DOROZYNSKI, A. 1978. Cassava may lead to mental retardation. Nature *272*, 120.

DORRELL, D. G. and CHUBEY, B. B. 1972. Acceleration of enzymatic browning in carrot and parsnip roots. J. Am. Soc. Hortic. Sci. *97*, 110–111.

DOUGAN, J. and HOWARD, G. E. 1975. Flavoring constituents of fermented fish sauces. J. Sci. Food Agric. *26*, 887–894.

DOUGHERTY, T. M., ZUKAKIS, J. P. and RZUCIDLO, S. J. 1977. A sensitive xanthine oxidase assay and its application to human milk and pig serum. Nutr. Rep. Int. *16*, 241–248.

DOUX-GAYAT, A. and AURIOL, P. 1976. Comparative study of the xylanases secreted by a pathogen placed under parasitic saprophytic conditions. Physiol. Veg. *14*, 663–675. (French)

DOUZOU, P. 1973. Enzymology at sub-zero temperatures. Mol. Cell. Biochem. *1*, 15–27.

DOUZOU, P. 1975. Recent progress in enzymology at sub-zero temperatures. Proc. Fed. Eur. Biochem. Soc. Meet. *40* (Enzymes/Electron Transp. Syst.), 99–112.

DOWNEY, W. K. 1975. Identity of the major lipolytic enzyme activity of bovine milk in induced lipolysis. Ann. Bull. Int. Dairy Fed. *86*, 80–89.

DOWNEY, W. K. 1976. Enzyme systems influencing processing and storage of milk and milk products. Proc. Int. Dairy Congr. *2*, 323–357.

DOWNING, D. T., AHERN, D. G. and BACHTA, M. 1970. Enzyme inhibition by acetylenic compounds. Biochem. Biophys. Res. Comm. *40*, 218–223.

DOWNTON, W. J. S. and HAWKER, J. S. 1975.

Evidence for the lipid-enzyme chilling sensitive plants. Phytochemistry *14*, 1259–1263.

DOYLE, M. P. and MARTH, E. H. 1978A. Degradation of aflatoxin by lactoperoxidase. Z. Lebensm. Unters. Forsch. *166*, 271–273.

DOYLE, M. P. and MARTH, E. H. 1978B. Aflatoxin is degraded by mycelia from toxigenic and nontoxigenic strains of *Aspergilli* grown on different substrates. Mycopathologia *63*, 145–153.

DRABIKOWSKI, W., KUZNICKI, J. and GRABAREK, Z. 1977. Similarity in calcium ion-induced changes between troponin-C and protein activator of 3′,5′-cyclic nucleotide phosphodiesterase and their tryptic fragments. Biochim. Biophys. Acta *485*, 124–133.

DRAPRON, B. 1979. Utilization of enzymes in the food industries: technological, hygienic and regulatory (legal) aspects. Cah. Nutr. Diet. *14*, 99–113. (French)

DRAPRON, R. 1971. Enzymes. Their role in the technology of wheat and its derivatives. Bull. Anciens Eleves Eleves Fr. Meun. No. *246*, 224–239. (French)

DRAPRON, R. 1972. Enzymatic reactions in systems of low moisture content. Ann. Technol. Agric. *21*, 487–499. (French)

DRAPRON, R., AHN, N. X., LAUNAY, B. and GUILBOT, A. 1969. Development and distribution of wheat lipase activity during the course of germination. Cereal Chem. *46*, 647–655.

DRAPRON, R., BEAUX, Y., CORMIER, M., GEFFROY, J. and ADRIAN, J. 1974. Effects of lipoxygenase in breadmaking. Destruction of carotenoids, tocopherol, and essential fatty acids and effects of on the flavor of the bread. Ann. Technol. Agric. *23*, 353–356. (French)

DRAPRON, R. and GUILBOT, A. 1962. Contribution to the study of enzymatic reactions in poorly hydrated biological environments: The degradation of starch as a function of water activity and of temperature. Ann. Technol. Agric. Part 2, *11*, 275–317. (French)

DRAWERT, F. 1975. Odour and Taste Substances. Intern. Symp. Bad Pyrmont, Germany, 1974. Verlag Hans Carl, Nurenberg.

DRAWERT, F. and HAGEN, W. 1970. Estimation of L(+)- and D(−)-lactic acid in beer. Brauwissenschaft *23*, 1–6. (German)

DRENTH, J., JANSONIUS, J. N., KOEKOEK, R. and WOLTHERS, B. G. 1971. The structure of papain. Adv. Protein Chem. *25*, 79–115.

DREWS, E. 1972. Quality characteristics of gluten as indices of the processing properties of rye for breadmaking. Getreide Mehl. Brot. *26*, 295–298. (German)

DREWS, E. 1973. Dependence of variations of meal quality (type 997) on rye quality. Chem. Abstr. *81*, 24403. (German)

DRIEDGER, A. and HATFIELD, E. E. 1972. Influence of tannins on the nutritive value of soybean meal for ruminants. J. Anim. Sci. *34*, 465–468.

DRIESSEN, F. M. and STADHOUDERS, J. 1975. The lipolytic enzymes and cofactors responsible for spontaneous rancidity in cow's milk. Ann. Bull. Int. Dairy Fed. *86*, 73–79.

DRUMM, H., BRUENING, K. and MOHR, H. 1972. Phytochrome-mediated induction of ascorbate oxidase in different organs of a dicotyledonous seedling (*Sinapis alba*). Planta *106*, 259–267.

DRYDEN, M. J., KENDRICK, J. G., SATTERLEE, L. D., SCHROEDER, L. J. and BLOCK, R. G. 1977. Predicting protein digestibility and quality using an enzyme *Tetrahymena pyriformis* W. bioassay. J. Food Biochem. *1*, 35–44.

DUBE, G., BRAMBLETT, V. D., JUDGE, M. D. and HARRINGTON, R. B. 1972. Physical properties and sulfhydryl content of bovine muscles. J. Food Sci. *37*, 23–36.

Du BOIS, M. W., ANGLEMIER, A. F., MONTGOMERY, M. W. and DAVIDSON, W. D. 1972. Effect of proteolysis on the emulsification characteristics of bovine skeletal muscle. J. Food Sci. *37*, 27–28.

DUCKWORTH, R. B. 1975. Water Relations of Foods. Academic Press, New York.

DUCLAUX, E. 1883. Traite de Microbiologie, Paris. Cited by Hoffmann-Ostenhoff (1953). Suggestions for a more rational classification and nomenclature of enzymes. Adv. Enzymol. *14*, 219–260.

DUDEN, R. 1971. Enzymic reactions in foods with very low water content. Lebensm. Wiss. Technol. *4*, 205–206. (German)

DUDEN, R., FRICKER, A., HEINTZE, K., PAULUS, K. and ZOHM, H. 1975. Influence of heat treatment of spinach at temperatures up to 100°C on important constituents. Lebensm.-Wiss. Technol. *8*, 147–150. (German)

DUDKIN, M. S., KAZANSKAYA, I. S. and KOZAREZ, E. I. 1976. Structure of wheat straw xylan and its enzymic hydrolysis. Khim. Drev., (6) 19–23. (Russian)

DUERR, P. and SCHOBINGER, U. 1976. Application of enzymes in the production of beverages from fruits and vegetables. Alimenta *15*, 143–149. (German)

DUFFUS, C. M. 1969. α-Amylase activity in the developing barley grain and its dependence on gibberellic acid. Phytochemistry *8*, 1205–1209.

DUFFUS, C. M., DUFFUS, J. H. and ROSIE, B. 1973. α-Amylase and its release by prostaglandin $F_{2\alpha}$ in barley endosperm slices. Experientia *29*, 962.

DUFFUS, C. M. and ROSIE, R. 1973. Starch hydrolyzing enzymes in the developing barley grain. Planta *109*, 153–160.

DUGAL, B. and RAA, J. 1978. Collagenase from fish: Contribution to autolytic degradation. IRCS Med. Sci.: Libr. Compend. *8*, 546.

DULL, G. G. and LEEPER, G. F. 1975. Ultrastructure of polysaccharides in relation to texture. *In* Postharvest Biology and Handling of Fruits and Vegetables. N.F. Haard and D.K. Salunkhe (Editors). AVI Publishing Co., Westport, Conn.

DULLEY, J. R. 1972. Bovine milk protease. J. Dairy Res. *39*, 1–9.

DULLEY, J. R. 1974. Contribution of rennet and starter enzymes to proteolysis in cheese. Aust. J. Dairy Technol. *29*, 65–69.

DULLEY, J. R. and KITCHEN, B. J. 1973. Acid phosphatases of cheddar cheese. Aust. J. Dairy Technol. *28*, 114–116.

DUNCAN, R. J. S. and TIPTON, K. F. 1969. Oxidation and reduction of glyoxylate by lactic dehydrogenase. Eur. J. Biochem. *11*, 58–61.

DUNN, B. M. and BRUICE, T. C. 1974. Physical organic models for the mechanism of lysozyme action. Adv. Enzymol. *37*, 1–60.

DUNN, G. 1974. A model for starch breakdown in higher plants. Phytochemistry *13*, 1342–1346.

DUNNILL, P. and LILLY, M. D. 1974. Recent developments in enzyme isolation processes. *In* Enzyme Engineering, Vol. 2. E.K. Pye and L.B. Wingard (Editors). Plenum Press, New York.

DUNSMORE, D. G., BALDOCK, T. G. and WHEELER, R. G. 1977. Simulator evaluations of interactions between components of milking machine cleaning systems. N.Z. J. Dairy Technol. *12*, 260–266.

DUPUY, P. and RIGAUD, J. 1970. Lipoxygenase in different varieties of peas. Abstr. 23, 3rd Intern. Congr. Food Sci. Technol., Washington, D.C.

DUQUETTE, N. C. 1972. Enzymatic baking composition. U.S. Pat. 3,650,764. Mar. 21.

DURBIN, R. D. and UCHYTIL, T. 1971. Purification and properties of alliin lyase from the fungus *Penicillium corymbiferum*. Biochim. Biophys. Acta *235*, 518–520.

DURMISHIDZE, S. V., PRIDZE, G. N. and KHACHIDZE, O. T. 1975. Isoenzyme composition and substrate specificity of the *o*-diphenol oxidase of grape vines. Chem. Abstr. *85*, 17075f. (Russian)

DUSTMANN, J. H. 1972. Influence of dialysis on the determination of invertase activity in honey. Lebensm. Wiss. Technol. *5*, 70–71. (German)

DUTTON, H. G., BAILEY, G. and KOHAKE, E. 1943. Dehydrated spinach. Changes in color and pigment during processing and storage. Ind. Eng. Chem. *35*, 1173–1177.

DWIVEDI, B. K. 1973. The role of enzymes in food flavors. I. Dairy products. Crit. Rev. Food Technol. *3*, 457–475.

DWIVEDI, B. K. 1975. Meat flavor. Crit. Rev. Food Technol. *5*, 427–535.

DYER, W. J. 1951. Protein denaturation in frozen and stored fish. Food Res. *16*, 522–527.

DYER, W. J. and DINGLE, J. R. 1961. Fish proteins with special reference to freezing. *In* Fish as Food, Vol. 1. G. Borgstrom (Editor). Academic Press, New York.

EAGERMAN, B. A. and ROUSE, A. H. 1976. Heat inactivation temperature-time relationships for pectinesterase inactivation in citrus juices. J. Food Sci. *41*, 1396–1397.

EAPEN, K. E., TAPE, N. W. and SIMS, R. P. A. 1973. Oilseed flour. U.S. Pat. 3,732,108. May 8.

EASTMOND, R. and GARDNER, R. J. 1974. Effect of various polyphenols on the rate of haze formation in beer. J. Inst. Brew. *80*, 192–200.

EBASHI, S. and KODOMA, A. 1966. Native tropomyosin-like action of troponin on trypsin-treated myosin B. J. Biochem. *60*, 733–734.

EBINE, H. 1972. Miso. *In* Conversion and Manufacture of Foodstuffs by Microorganisms. Proc. 6th Int. Symp. Int. Union Food Sci. Technol. Saikon Publishing Co., Tokyo.

EBNER, K. E. 1975. Enzymology, Vol. 2: Subunit Enzymes: Biochemistry and Functions. Marcel Dekker, New York.

ECHIGO, T. and TAKENAKA, T. 1973. Changes in erlose contents by honeybee invertase. J. Agric. Chem. Soc. Jpn. *47*, 177–183. (Japanese)

EDA, S., OHNISHI, A. and KATO, K. 1976. Xylan isolated from the stalk of *Nicotiania tabacum*. Agric. Biol. Chem. *40*, 359–364.

EDMONSON, D., MASSEY, V., PALMER, G., BEACHAM, L. M. III. and ELION, B. 1972. The resolution of active and inactive xanthine oxidase by affinity chromatography. J. Biol. Chem. *247*, 1597–1604.

EDSALL, J. T. 1976. James Sumner and the crystallization of urease. Trends in Biochem. Sci. *1*, 21.

EDWARDS, G. H. 1964. Effects of microwave radiation on wheat and flour. The viscosity of flour pastes. J. Sci. Food Agric. *15*, 108–114.

EDWARDS, G. W. and EDWARDS, A. W. 1974. Alfalfa extracts. U.S. Pat. 3,833,738. Sept. 10.

EDWARDS, V. H. 1972. Future directions in enzyme engineering. *In* Enzyme Engineering. L.B. Wingard (Editor). Interscience Publishers, New York.

EFFRONT, J. 1899. Enzymes and Their Applications. Dunot et Pinat, Paris. (French)

EFFRONT, J. and PRESCOTT, S E. 1917. Biochemical Catalysts in Life and Industry. Proteolytic Enzymes. John Wiley & Sons, New York.

EGAMI, F. 1975. Origin and early evolution of transition element enzymes. J. Biochem. *77*, 1165–1169.

EGELRUD, T. and OLIVECRONA, T. 1973. Purified bovine milk (lipoprotein) lipase: Activity against lipid substrates in the absence of exogenous serum factors. Biochim. Biophys. Acta *306*, 115–127.

EGGEBRECHT, E. 1972. Flour improvement. Ger. Pat. 1,275,499. June 17.

EICHELE, G., FORD, G. C. and JANSONIUS, J. N. 1979. Crystallization of pig mitochondrial aspartate aminotransferase by seeding with crystals of chicken mitochondrial isoenzyme. J. Mol. Biol. *135*, 513–516.

EICKERMAN, H. H. 1964. Tenderizing method. U.S. Pat. 3,150,982. Sept. 29.

EID, K. 1977. Noteworthy results from enzyme addition to black currants. Flüss. Obst. *44*, 83–86. (German)

EIGEL, W. N. and RANDOLPH, H. E. 1976. Comparison of calcium sensitivities of α_{s1}-B, β-A^2, and γ-A^2 caseins and their stabilization by K-casein A. J. Dairy Sci. *59*, 203–206.

EIGEN M. and DE MAEYER, L. 1958. Self-dissociation and protonic charge transport in water and ice. Proc. R. Soc. London Ser. A. *247*, 505–531.

EIGEN, M. and HAMMES, G. G. 1963. Elementary steps in enzyme reactions (as studied by relaxation spectrometry). Adv. Enzymol. *25*, 1–38.

EISENBERG, D. 1970. X-Ray crystallography and enzyme structure. *In* The Enzymes, 3rd Edition, Vol. 1. P.D. Boyer (Editor). Academic Press, New York.

EISENSTADT, M. E. 1978. Sweetening compositions. U.S. Pat. 4,085,232. Apr. 18.

EISENTHAL, R. and CORNISH-BOWDEN, A. 1974. The direct linear plot. A new graphical procedure for estimating enzyme kinetic parameters. Biochem. J. *193*, 715–720.

EITENMILLER, R. R., BARNHART, H. M. and SHAHANI, K. M. 1971. Effect of γ-irradiation on bovine and human milk lysozymes. J. Food Sci. *6*, 1127–1130.

EKINS, R. 1980. More sensitive immunoassays. Nature *284*, 14–15.

EL-BADAWI, A. A., ANGLEMEIER, A. F. and CAIN, R. F. 1964. Effects of soaking in water, thermal enzyme inactivation, and irradiation on the textural factors of beef. Food Technol. *18*, 1807–1810.

ELBEIN, A. D. 1974. Interactions of polynucleotides and other polyelectrolytes with enzymes and other proteins. Adv. Enzymol. *40*, 29–64.

ELCHAZLY, M. and THOMAS, B. 1976. A biochemical method for the determination of "dietary fiber" and its components in plant foods. Z. Lebensm. Unters. Forsch. *162*, 329–340. (German)

EL-DASH, A. A. 1971. The precursors of bread flavor: Effect of fermentation and proteolytic activity. Bakers Dig. *46* (6) 26–31.

ELDRIDGE, A. C., WARNER, K. and WOLF, W. J. 1977. Alcohol treatment of soybeans and soybean protein products. Cereal Chem. *54*, 1229–1237.

ELEY, D. D. 1955. Physical chemistry of enzymes. Nature *176*, 958–960.

EL-GARBAWI, M. and WHITAKER, J. R. 1963. Factors affecting enzymic solubilization. J. Food Sci. *28*, 168–172.

EL-LAKANY, S. and MARCH, B. E. 1974. Chemical and nutritive changes in herring meal during storage with and without antioxidant treatment. J. Sci. Food Agric. *25*, 899–906.

ELLIOT, M., FARNHAM, A. W., JANES, N. F., NEEDHAM, P. H. and PULMAN, D. A. 1974. Synthetic insecticide with a new order of activity. Nature *248*, 710–711.

ELLIS, G. P. 1959. The Maillard reaction. Adv. Carbohydr. Chem. *14*, 63–134.

EL-SHIBINY, S. and EL-SALAM, M. H. 1977. Action of milk-clotting enzymes on α-caseins from buffalo's and cow's milk. J. Dairy Sci. *60*, 1519–1521.

ELSTNER, E. F., STOFFER, C. and HEUPEL, A. 1975. Determination of superoxide free radical ion and hydrogen peroxide as products of photosynthetic oxygen reduction. Z. Naturforsch. *30C*, 53–57. (German)

EMBS, R. J. and MARKAKIS, P. 1965. The mechanism of sulfite inhibition of browning caused by polyphenol oxidase. J. Food Sci. *30*, 753–758.

EMERT, G. H., GUM, E. K., LANG, J. A., LIV,

T. H. and BROWN, R. D., JR. 1974. Cellulases. *In* Food Related Enzymes. J. Whitaker (Editor). Am. Chem. Soc., Washington, D.C.

ENARI, T. M. 1971. The second EBC reference serum for barley and malt proteins. J. Inst. Brew. *77*, 517–518.

ENARI, T. M. 1974. Malting, mashing, and the use of unmalted grain. *In* Industrial Aspects of Biochemistry. B. Spencer (Editor). Fed. Eur. Bio-Chem. Soc., Springer Verlag, New York.

ENARI, T. M. 1975. Amino acids, peptides, and proteins. Proc. Eur. Brew. Conv., Zeist, IEBC, Amsterdam, 73–89.

ENARI, T. M. and MARKKANEN, P. H. 1974. Microbial β-glucanases in brewing. Proc. Am. Soc. Brew. Chem. *33*, 13–17.

ENARI, T. M. and MARKKANEN, P. H. 1977. Production of cellulolytic enzymes by fungi. Adv. Biochem. Eng. *5*, 1–24.

ENDO, A. 1965. Studies on pectolytic enzymes of molds. 16. Mechanism of enzyme clarification of apple juice. Agric. Biol. Chem. *29*, 229–233.

ENDO, A., KITAHARA, N., OKA, H., MIENCHI-FUKAZAWA, Y. and TERAHARA, A. 1978. Isolation of 4,5-dehydroisophthalic acid, an inhibitor of brain glutamate decarboxylase by a *Streptomyces* species. Eur. J. Biochem. *82*, 257–259.

ENEVOLDSEN, B. S. 1975. Debranching enzymes in brewing. Proc. Eur. Brew. Conv., Nice. 683–697.

ENEVOLDSEN, B. S. and SCHMIDT. F. 1974. Dextrins in brewing. Singly-branched and multiply-branched dextrins in brewing. J. Inst. Brew. *80*, 520–533.

ENGAN, S. 1975. Oxidized flavor in beer. Mallasjuomat *5*, 123–139. (Danish)

ENGASSER, J. -M. and HORVATH, C. 1973. Effect of internal diffusion in heterogeneous enzyme systems: Evaluation of true kinetic parameters and substrate diffusivity. J. Theor. Biol. *42*, 137–155.

ENGASSER, J. -M. and HORVATH, C. 1974. Inhibition of bound enzymes. I–III. Biochemistry *13*, 3845–3858.

ENGASSER, J. -M. and HORVATH, C. 1976. Diffusion and kinetics with immobilized enzymes. Appl. Biochem. Bioeng. *1*, 127–220.

ENGEL, P. C. 1977. Enzyme Kinetics. Halstead Press, New York.

ENGELHARDT, G., WALLNÖFER, P. R. and PLAPP, R. 1971. Degradation of Linuron and some other herbicides and fungicides by a Linuron-inducible enzyme obtained from *Bacillus sphaericus*. Appl. Microbiol. *22*, 284–288.

ENGELMAN, D. M. and MOORE, P. B. 1976. Neutron-scattering studies of the ribosome. Sci. Am. *235* (4) 44–54.

ENGL, R. 1967. Influence of blanching on the quality and storage stability of dry vegetables. III. Storage browning with dried kohlrabis. Dtsch. Lebensm.-Rundsch. *63*, 35–40. (German)

ENGST, R. and LEWERENZ, H. J. 1972. Hygiene and toxicology of enzyme preparations. Ann. Technol. Agric. *21*, 619–627. (German)

EPEL, B. L., NEUMANN, J. and FISHMAN, R. 1973. Mechanism of the oxidation of ascorbate and manganese (2+) ion by chloroplasts. Role of the radical superoxide. Biochim. Biophys. Acta *325*, 520–529.

EPPENDORF, G. N. and HINE, N. 1975. Reagent method for the enzymic kinetic concentration determination of a substrate. Neth. Pat. Appl. 74 13,057. Apr. 8.

EPSTEIN, D. and MIZRAHI, S. 1975. Effect of concentration process on cloud stability of reconstituted lemon juice. J. Sci. Food Agric. *26*, 1603–1608.

EPSTEIN, H. F., SCHECHTER, A. N., CHEN, R. F. and ANFINSEN, C. B. 1971. Folding of staphylococcal nuclease: Kinetic studies in acid renaturation. J. Mol. Biol. *60*, 499–508.

ERDMAN, J. W., JR. and FORBES, R. M. 1977. Mineral availability from phytate-containing foods. Food Prod. Dev. *11* (10) 46–48.

ERICKSON, B. W. and MERRIFIELD, R. B. 1976. Solid phase peptide synthesis. *In* The Proteins. H. Neurath and R.L. Hill (Editors). Academic Press, New York.

ERIKSEN, S. and FAGERSON, I. S. 1976. The plastein reaction and its applications: A review. J. Food Sci. *41*, 490–493.

ERIKSSON, C. E. 1967. Alcohol: NAD oxidoreductase (E.C. 1.1.1.1) from peas. J. Food Sci. *33*, 525–532.

ERIKSSON, C. E. 1970. Nonenzymatic lipid oxidation by lactoperoxidase. Effect of heat treatment. J. Dairy Sci. *53*, 1649–1653.

ERIKSSON, C. E. 1975. Aroma compounds derived from oxidized lipids. Some biochemical and analytical aspects. J. Agric. Food. Chem. *23*, 126–128.

ERIKSSON, C. E. and SVENSSON, S. G. 1970. Lipoxygenase from peas, purification and properties of the enzyme. Biochim. Biophys. Acta *198*, 449–459.

ERIKSSON, C. E., QVIST, I. and VALLENTIN, K. 1977. Conversion of aldehydes to alcohols in liquid foods by alcohol dehydrogenase. *In* Enzyme in Food and Beverage Processing. R.L.

Ory and A.J. St. Angelo (Editors). Am. Chem. Soc., Washington, D.C.

ERIKSSON, C. E. and VALLENTIN, K. 1973. Thermal activation of peroxidase as a lipid oxidation catalyst. J. Am. Oil Chem. Soc. *50*, 264–268.

ERIKSSON, K. -E., PETTERSSON, B. and WESTENMARK, U. 1974. Oxidation: An important enzyme reaction in fungal degradation of cellulose. FEBS Lett. *49*, 282–285.

ERLANDSON, J. A. and WROLSTAD, R. E. 1972. Degradation of anthocyanins at limited water concentration. J. Food Sci. *37*, 592–595.

ERLANGER, B. F. and SACK, R. A. 1970. Operational normality of α-chymotrypsin solutions by a sensitive potentiometric technique using a fluoride electrode. Anal. Biochem. *33*, 318–322.

ERMAN, J. E. and YONETANI, T. 1975. Oxidation of cytochrome c peroxidase by hydrogen peroxide. Characterization of products. Biochim. Biophys. Acta *393*, 343–349.

ERNSTROM, C. A. and WONG, N. P. 1974. Milk-clotting enzymes and cheese chemistry. *In* Fundamentals of Dairy Chemistry. B.H. Webb, A.H. Johnson and J.A. Alford (Editors). AVI Publishing Co., Westport, Conn.

ESCHENBRUCH, R. and BONISH, P. 1976. Production of sulfite and sulfide by low and high-sulfite forming wine yeasts. Arch. Microbiol. *107*, 297–302.

ESKIN, N. A. M., GROSSMAN, S. and PINSKY, A. 1977. Biochemistry of lipoxygenase in relation to food quality. Crit. Rev. Food Sci. Nutr. *9*, 1–40.

ESKIN, N. A. M., HENDERSON, H. M. and TOWNSEND, R. J. 1972. Biochemistry of Foods. Academic Press, New York.

ESTERBAUER, H. and SCHAUENSTEIN, E. 1977. Formation of isomeric trihydroxyoctadecanoic acid by enzymic oxidation of linoleic acid by barley flour. Monatsh. Chem. *108*, 963–972. (German)

ETCHELLS, J. L., BELL, T. A. and WILLIAMS, C. F. 1958. Inhibition of pectinolytic and cellulolytic enzymes in cucumber fermentations by Scuppernong grape leaves. Food Technol. *12*, 204–208.

ETTEL, W. and TUOR, A. 1977. Enzymic determination of "total creatinine" in meat extract and meat extract-containing foods. Dtsch. Lebensm.-Rundsch. *73*, 357–361. (German)

ETTLINGER, M. G., DATEO, G. P., HARRISON, B. W., MABRY, T. J. and THOMPSON, C. P. 1961. Vitamin C as a coenzyme: The hydrolysis of mustard oil glucosides. Proc. Natl. Acad. Sci. *12*, 1875–1880.

ETTLINGER, M. G. and KJAER, A. 1968. Sulfur compounds in plants. *In* Recent Advances in Phytochemistry. T.J. Mabry (Editor). Appleton-Century-Crofts, New York.

EULER, H. V. and MYRBÄCK, K. 1927. Chemistry of the Enzymes. II. Special Chemistry of the Enzymes. J.F. Bergmann, Verlag, Munich. (German)

EVANS, J. J. 1969. Spectral similarities and kinetic differences of two tomato plant peroxidase isoenzymes. Plant Physiol. *45*, 66–99.

EVANS, J. J. and MECHAM, D. K. 1971. Occurrence and properties of peroxidase isoenzymes in wheat flour and milling fractions. Cereal Sci. Today *16*, 292.

EVANS, W. C. 1975. Thiaminases and their effects on animals. *In* Vitamins and Hormones. Advances in Research and Applications, Vol. 33. P. Munson *et al.* (Editors). Academic Press, New York.

EVANSON, J. M. 1971. Mammalian collagenases and their role in connective tissue breakdown. *In* Tissue Proteinases. A.J. Barrett and J.T. Dingle (Editors). North Holland Publishers, Amsterdam.

EVENTOFF, W., ROSSMAN, M. G., TAYLOR, S. S., TORFF, H. -J., MEYER, H., KEIL, W. and KILTZ, H. -H. 1977. Structural adaptations of lactate dehydrogenase isozymes. Proc. Natl. Acad. Sci. *74*, 2677–2681.

EVERSE, J., ZOLL, E. C., KAHAN, L. and KAPLAN, N. O. 1971. Addition products of diphosphopyridine nucleotides with substrates of pyridine nucleotide-linked dehydrogenases. Bioorg. Chem. *1*, 207–233.

EWING, E. E. and McADOO, M. H. 1971. An examination of methods used to assay potato tuber invertase and its naturally occurring inhibitor. Plant Physiol. *48*, 366–370.

EYTON, W. B. 1972. The chemistry of tea. Flavour Ind. *3*, 23–28, 36.

FABER, B. V. 1975. Process for preparing low calorie gelatin dessert with bromelain. U.S. Pat. 3,930,050. Dec. 30.

FADIMAN, C. 1955. Introduction. *In* The Complete Book of Cheese. R.C. Brown (Editor). Random House, New York.

FAGAN, P. V. 1968. Process for the manufacture of collagen casings from limed hides. U.S. Pat. 3,373,046. March 12.

FAIR, J. G., COLLINS, J. L., JOHNSTON, M. R. and COFFEY, D. L. 1973. Levels of DDT isomers in turnip greens after blanching and thermal processing. J. Food Sci. *38*, 189–191.

FAIRLEY, C. J. and SWAINE, D. 1973. Green tea fermentation by callus tissue enzyme preparations. Br. Pat. 1,318,035. May 23.

FALCONER, J. S. and TAYLOR, B. D. 1946. A specific property solubility test for protein purity and its application to the purification of pure liver esterase. Biochem. J. *40*, 835–843.

FALK, K. G. 1924. The Chemistry of Enzyme Actions, 2nd Edition. Am. Chem. Soc. Monogr. Ser., No. 1. Chemical Catalogue Co., New York.

FALKBRING, S. O., GÖTHE, P. O., NYMAN, P. O., SUNBERG, L. and PORATH, J. 1972. Affinity chromatography of carbonic anhydrase. FEBS Lett. *24*, 229–235.

FALKENHAUSEN, F. V. 1935. A new test for the estimation of purity of wet starch. Z. Spiritusind. *58*, 3–4. (German)

FALLSCHEER, H. O. and COOK, J. W. 1956. Report on enzymatic methods for insecticides. J. Off. Agric. Chem. *39*, 691–697.

FAN, M. L. 1975. Purification and properties of potato α-amylase. Taiwania *20*, 71–76.

FANG, T. T., FOOTRAKUL, P. and LUH, B. S. 1971. Effects of blanching, chemical treatment and freezing methods on quality of freeze-dried mushrooms. J. Food Sci. *36*, 1044–1048.

FANTOZZI, P., PETRUCCIOLI, G., and MONTEDORO, G. 1977. Enzyme treatment of olive pastes after single pressing extraction. Effect of cultivar, pressing time and storage. Riv. Ital. Sostanze Grasse *54*, 381–388. (Italian)

FARBER, L. 1957. The chemical evaluation of the pungency of onion and garlic by the content of volatile reducing substances. Food Technol. *11*, 621–624.

FARKAS, D. F. and GOLDBLITH, S. A. 1962. Kinetics of lipoxidase inactivation using thermal and ionizing energy. J. Food Sci. *27*, 262–276.

FARKAS, D. F., GOLDBLITH, S. A. and PROCTOR, B. E. 1956. Stopping storage off-flavors by curbing peroxidase. Food Eng. *28* (1) 52–53, 152.

FARKAS, V., BIELY, P. and BAUER, S. 1973. Extracellular β-glucanases of the yeast, *Saccharomyces cerevisiae.* Biochim. Biophys. Acta *321*, 246–255.

FARNHAM, M. G. 1950. Cheese modifying enzyme. U.S. Pat. 2,531,329. Nov. 21.

FARNHAM, M. G. 1957. Enzyme-containing powder and enzyme-modified product thereof. U.S. Pat. 2,794,743. June 4.

FARRAND, E. A. 1972. Controlled levels of starch damage in a commercial United Kingdom bread flour and effects on absorption, sedimentation value and loaf quality. Cereal Chem. *49*, 479–488.

FARVER, O., GOLDBERG, M. and PECHT, I. 1980. A circular dichroism study of the reactions of *Rhus* laccase with dioxygen. Eur. J. Biochem. *104*, 71–77.

FAULSTICH, H. and COCHET-MEILHAC, M. 1976. Amatoxins in edible mushrooms. FEBS Lett. *64*, 73–75.

FAURE, G. 1835. New observations on black mustard seeds. J. Pharm. *21*, 464–466. (French)

FAUST, O. and KARRER, P. 1929 The degradation of cellulose and cotton. Helv. Chim. Acta *12*, 414–417. (German)

FAVOROV, V. V. 1973. Purification of alginases by affinity chromatography on a Bio-Gel alginate column. Int. J. Biochem. *40*, 107–110.

FAZARKERLEY, S. and EBBON, G. P. 1975. Lowering the nucleic acid content of protein-containing materials. Ger. Offen. 2,444,990. Apr. 3.

FED. REGISTER. 1960. Food additives in the malting of barley. Fed. Regist. *25*, 2162.

FEDERICS, B. 1965. Process for tenderizing meat. U.S. Pat. 3,183,097. May 11.

FEENEY, R. E. 1976. Effects of temperature on enzymes. *In* Proc. Symp. Enzymes Food Processing Industry, Univ. Calif. Coop. Ext. Div., Jan. 14, Davis.

FEENEY, R. E., CLARY, J. J. and CLARK, J. R. 1964. A reaction between glucose and egg white proteins in incubated eggs. Nature *201*, 192–193.

FEENEY, R. E., VANDENHEEDE, J. and OSUGA, D. T. 1972. Macromolecules from cold-adapted antarctic fishes. Naturwissenschaften *59*, 22-29.

FEHMERLING, G. B. 1970. Separation of edible tissue flesh of marine creatures. U.S. Pat. 3,729,324. Apr. 24.

FEILLET, P. and KOBREHEL, K. 1974. Determination of common wheat content in pasta products. Cereal Chem. *51*, 203–209.

FEINBERG, B. 1973. Vegetables. *In* Food Dehydration, 2nd Edition, Vol. 2. W.B. Van Arsdel, M.J. Copley and A.L. Morgan (Editors). AVI Publishing Co., Westport, Conn.

FEINBERG, B. 1978. The mad world of nutrition. Presented at Aug. Meet., Inst. Food Technol., North Calif. Sec., Berkeley. (unpublished)

FEINBERG, B., SCHWIMMER, S., REEVE, R. and JUILLY, M. 1964. Vegetables. *In* Food Dehydration, Vol. 2. Products and Technology. W.B. Van Arsdel and M.J. Copley (Editors). AVI Publishing Co., Westport, Conn.

FELDBERG, C. and BAKER, W. B. 1971. The effect of papain in dough processing. Cereal Sci. Today *16*, 291.

FELL, H. B. and DINGLE, J. T. 1963. Mode of action of excess vitamin A. VI. Lysosomal protease and the degradation of cartilage matrix. Biochem. J. *87*, 403–408.

FELLER, U. K., SOONG, T. -S. and HAGEMAN, R. H. 1977. Leaf proteolytic activities and senescence during grain development of field-grown corn (*Zea mays* L.). Plant Physiol. *59*, 290–294.

FELLERS, D. A. 1975. Personal communication. Albany, California.

FELLERS, D. A. and BEAN, M. M. 1977. Storage stability of wheat-based foods. J. Food Sci. *42*, 1143–1147.

FENNEMA, O. R. 1973. Food Science, Vol. 3. Low Temperature Preservation of Foods and Living Matter. Marcel Dekker, New York.

FENNEMA, O. R. 1975A. Activity of enzymes in partially frozen aqueous systems. *In* Water Relations of Foods. R.B. Duckworth (Editor). Academic Press, New York.

FENNEMA, O. R. 1975B. Reaction kinetics in partially frozen aqueous systems. *In* Water Relations of Foods. R.B. Duckworth (Editor). Academic Press, New York.

FENNEMA, O. R. 1976. The U.S. frozen food industry: 1776–1976. J. Food Technol. *30* (6) 56–61, 68.

FERNS, R. S. 1977. Enzymatic clarification of liquids. U.S. Pat. 4,038,419. July 26.

FERRANTE, J. V. and NICHOLAS, D. J. D. 1976. Use of immuno-adsorbent affinity chromatography to purify component I of nitrogenase from extracts of *Azotobacter vinelandi.* FEBS Lett. *66*, 187–190.

FERREL, R. E. 1978. Distribution of bean and wheat inositol phosphate esters during autolysis and germination. J. Food Sci. *43*, 563–565.

FERRIER, L. K., RICHARDSON, T., OLSON, N. F. and HICKS, C. L. 1972. Characterization of insoluble pepsin used in a continuous milk clotting system. J. Dairy Sci. *55*, 726–734.

FERRY, J. G. and WOLFE, R. S. 1977. Nutritional and biochemical characterization of *Methanospirillum hungatii.* Appl. Environ. Microbiol. *34*, 371–376.

FEYS, M., NAESENS, W., TOBBACK, P. and MAES, E. 1980. Lipoxygenase activity in apples in relation to storage and physiological disorders. Phytochemistry *19*, 1009–1011.

FILLINGAME, R. H. 1980. The proton-translocating pump of oxidative phosphorylation. Annu. Rev. Biochem. *49*, 1079–1113.

FILNER, P. 1968. Regulatory mechanisms involving enzyme function synthesis and degradation. *In* Molecular Biology and Agriculture. Potential for Future Research. U.S. Dep. Agric., Agric. Res. Serv., Albany, Calif.

FILNER, P., WRAY, J. L. and VARNER, J. E. 1969. Enzyme induction in higher plants. Science *165*, 358–367.

FINCH, R. 1970. Fish protein for human food. Crit. Rev. Food Technol. *1*, 519–580.

FINCHAM, J. R. 1957. A modified glutamic acid dehydrogenase as a result of gene mutation in *Neurospora crassa*. Biochem. J. *65*, 721–728.

FINEAN, J. B., COLEMAN, R. and MICHEL, R. H. 1974. Membranes and Their Cellular Functions. Blackwell, Oxford.

FINK, A. L. 1976. Cryoenzymology: The use of sub-zero temperatures and fluid solutions in the study of enzyme mechanisms. J. Theor. Biol. *61*, 419–445.

FINK, A. L. and ANGELIDES, K. M. 1976. Papain-catalyzed reactions at subzero temperatures. Biochemistry *15*, 5287–5293.

FINK, A. L. and HAY, G. W. 1969. Isolation and purification of an esterase from wheat germ lipase. Can. J. Biochem. *47*, 135–142.

FINKLE, B. J. 1964. Treatment of plant tissue to prevent browning. U.S. Pat. 3,126,287. Mar. 24.

FINKLE, B. J. and NELSON, R. F. 1963. Enzyme reactions with phenolic compounds: Effect of *O*-methyltransferase on a natural substrate of fruit polyphenoloxidase. Nature *197*, 902–903.

FINLEY, J. W. and OLSON, A. C. 1975. Automated method for measuring added sucrose in sweetened cereal products with immobilized invertase. Cereal Chem. *52*, 500–505.

FINLEY, J. W., STANLEY, W. L. and WATTERS, G. G. 1979. Chill proofing beer with papain immobilized on chitin. Process Biochem. *14* (7) 12–13.

FINNEY, E. E., JR. 1972. Elementary concepts of rheology relevant to food texture studies. Food Technol. *26*, 68–77.

FINNEY, K. F., SHOGREN, M. and POMERANZ, I. 1971. Cereal malts in breadmaking. Cereal Sci. Today *16*, 303.

FINNOCCHIARO, T., RICHARDSON, T. and OLSON, N. F. 1980. Lactase immobilized on alumina. J. Dairy Sci. *63*, 215–222.

FIRN, R. D. 1975. Secretion of α-amylase by barley aleurone layers after incubation in gibberellic acid. Planta *125*, 227–233.

FISCHER, E. 1894. Influence of configuration

on the action of the enzymes. Ber. Dtsch. Grs. *27*, 2985–2993. (German)

FISCHER, E. H., MEYER, K. H., NOELTING, G. G. and PIGUET, A. 1950. Purification and crystallizations of malt β-amylase. Arch. Biochem. Biophys. *27*, 235–237.

FISHBEIN, W. N. and STOWELL, R. E. 1969. Studies on the mechanism of freezing damage to mouse liver using a mitochondrial enzyme assay. Cryobiology *6*, 227–234.

FISHMAN, M. M.and SCHIFF, H. F. 1976. Enzymes in analytical chemistry. Anal. Chem. *48*, 322R–332R.

FLAVIAN, S. and LEVI, A. 1970. A study of the natural disappearance of the limonin monolactone in the peel of Shamouti oranges. J. Food Technol. *5*, 193–195.

FLEET, G. H. and PHAFF, H. J. 1975. Glucanases in *Schizosaccharomyces.* Isolation and properties of an exo-β-glucanase from the cell extracts and culture fluid of *Schizosaccharomyces japonicus* var. versatilis. Biochim. Biophys. Acta *410*, 318–322.

FLEGO, L. 1977. Enzymic removal of non-reactive protein fractions from immunoprecipitation reactions. Boll. Chem. Lab. Provinc. *3*, 183–186. (Italian)

FLEMING., J. R., JOHNSON, J. A. and MILLER, B. S. 1960. Effect of malting procedure and wheat storage conditions on alpha-amylase and protease activities. Cereal Chem. *37*, 363–379.

FLETTERICK, R. J. and MADSEN, N. B. 1980. The structures and related functions of phosphorylase a. Annu. Rev. Biochem. *49*, 31–61.

FLICK, G. J., JR., AUNG, L. H., ORY, R. L. and ST. ANGELO, A. J. 1977. Nutrient composition and selected enzyme activities in *Sechium edule* Sw., the merliton. J. Food Sci. *42*, 11–13.

FLICK, G. J., JR., ORY, R. L. and ST. ANGELO, A. J. 1977. Comparison of nutrient composition of enzyme activity in purple, green and white eggplants. J. Food Sci. *25*, 117–120.

FLING, M., HOROWITZ, N. H. and HEINEMANN, S. F. 1963. The isolation and properties of crystalline tyrosinase from *Neurospora.* J. Biol. Chem. *238*, 2045–2053.

FLORKIN, M. and STOTZ, E. H. 1964. Comprehensive Biochemistry, Vol. 13, Revised Edition. Rpt. Comm. Enzymes of Intern. Union Biochem. Elsevier Publishing Co., Amsterdam.

FLORKIN, M. and STOTZ, E. H. 1973. Comprehensive Biochemistry, Vol. 13, 3rd Edition. Enzyme Nomenclature. Elsevier Scientific Publishing Co., New York.

FLURKEY, W. H. and JEN, J. J. 1978. Peroxidase and polyphenoloxidase activities in developing peaches. J. Food Sci. *43*, 1826–1831.

FLURKEY, W. H., YOUNG, L. W. and JEN, J. J. 1978. Separation of soybean lipoxygenase and peroxidase by hydrophobic chromatography. J. Agric. Food Chem. *26*, 1474–1476.

FLYNN, G. 1975. The market potential for papain. Rpt. *G99*. Trop. Prod. Inst., London.

FODA, M. S., ISMAIL, A. A. and KHORSHID, M. A. 1975. Production of a new rennin-like enzyme by *Aspergillus ochraceus.* Milchwissenschaft *30*, 598–601. (German)

FOGARTY, W. M. and GRIFFIN, P. J. 1974. Enzymes of *Bacillus* species. Process Biochem. *8* (6) 11–18, 24; (7) 27–25.

FOGARTY, W. M. and WARD, O. P. 1972. Pectic substances and pectolytic enzymes. Process Biochem. 7 (8) 13–15.

FOLTMANN, B. 1966. A review on prorennin and rennin. C.R. Trav. Lab. Carlsberg *35*, 143–231.

FOLTMANN, B., KAUFFMAN, D., PARL, M. and ANDERSEN, P. M. 1973. Comparison between the primary structures of chymosin (rennin), pepsin and their zymogens. Neth. Milk Dairy J. *27*, 288–297.

FOLTZ, A. K., YERANSIAN, J. A. and SLOMAN, K. G. 1977. Food. Anal. Chem. *49*, 194–220.

FONG, F. F. 1976. A successor to transition-state theory. Acc. Chem. Res. *9*, 433–438.

FOOD DRUG ADMIN. 1970. Canned tomatoes—standard of identity. Federal Food, Drug and Cosmetic Act. 53.40. Amended Feb. 27.

FORD, J. R., MUSTAKAS, G. C. and SCHMUTZ, R. D. 1978. Phytic acid removal from soybeans by a lipid protein concentrate process. J. Am. Oil Chem. Soc. *55*, 371–374.

FORD, J. S. and GUTHRIE, J. M. 1908. Contribution to the biochemistry of barley. J. Inst. Brew. *14*, 61–84.

FORNEY, F. W. and MARKOVETZ, A. J. 1971. The biology of methyl ketones. J. Lipid Res. *12*, 383–395.

FORREST, J. C., ABERLE, E. D., HEDRICK, H. B. and MERKEL, R. A. 1975. Principles of Meat Science. W.H. Freeman, San Francisco.

FORREY, R. R. and FLATH, R. A. 1974. Volatile components of *Prunus Salcina*, var. Santa Rosa. J. Agric. Food Chem. *22*, 496–498.

FORSTER, T. L., JENSEN, C. and PLATH, E. 1956. Preserving the original activity of milk lipase. J. Dairy Sci. *39*, 1120–1124.

FORSYTH, W. G. C. 1963. The mechanism of cacao curing. Adv. Enzymol. *25*, 457–492.

FORSYTH, W. G. C. and QUESNEL, V. C. 1957. Cacao glycosidase and color changes during fermentation. J. Sci. Food Agric. *8*, 505–509.

FOSTER, H. G., JR. and CORNWELL, E. H. 1965. Manufacture of cheese curd. U.S. Pat. 3,172,767.

FOSTER, J. M. 1979. Enzymes: Teaching biochemistry through laboratory experience. BioScience *29*, 539–544.

FOSTER, P. M. D., SLATER, T. F. and PATTERSON, D. S. P. 1974. A possible enzymic assay for trichothecene mycotoxins in animal feedstuffs. Biochem. Soc. Trans. *3*, 875–878.

FOSTER, R. 1979. The Nature of Enzymology. Croon Helm, London.

FOURCHE, J., JENSEN, H. and NEUZIL, E. 1977. α-Hydrazinophloretic acid, a competitive inhibitor of fungal tyrosinase. C.R. Hebd. Seances Acad. Sci., Ser. D *284*, 2133–2136.

FOURNIER, P. E. 1972. Purity of enzyme preparations. Mycotoxins. Ann. Technol. Agric. *21*, 607–617. (French)

FOX, C. W., CHRISOPE, G. L. and MARSHALL, R. T. 1975. Incidence and types of phospholipase C-producing bacteria in fresh and spoiled homogenized milk. J. Dairy Sci. *58*, 794.

FOX, C. W., CHRISOPE, G. L. and MARSHALL, R. T. 1976. Incidence and identification of phospholipase C-producing bacteria in fresh and spoiled homogenized milk. J. Dairy Sci. *59*, 1857–1864.

FOX, J. B., JR. 1966. The chemistry of meat pigments. J. Agric. Food Chem. *14*, 207–210.

FOX, J. B., JR., KEYSER, S. and NICHOLAS, R. A. 1975. Pyridine catalysis of ascorbate reduction of metmyoglobin. J. Food Sci. *40*, 435.

FOX, J. L. 1979. Science update: Progress in biochemistry. J. Chem. Educ. *57* (33) 22–23.

FOX, P. F. 1974. Enzymes in food processing. *In* Industrial Aspects of Biochemistry. B. Spencer (Editor). American Elsevier, North Holland, Amsterdam.

FOX, P. F. and WHALLEY, B. F. 1971. Bovine pepsin preliminary cheese making experiments. Ir. J. Agric. Res. *10*, 358–360.

FRAENKEL-CONRAT, H. and OLCOTT, H. S. 1948. The reaction of formaldehyde with proteins. V. Cross-linking between amino and primary amide or guanidinyl groups. J. Am. Chem. Soc. *70*, 2673–2684.

FRANCIS, F. J. and CLYDESDALE, F. M. 1975. Food Colorimetry: Theory and Applications. AVI Publishing Co., Westport, Conn.

FRANCIS, G. L. and BALLARD, F. J. 1980. Enzyme inactivation via disulphide-thiol exchange as catalyzed by a rat liver membrane protein. Biochemistry J. *186,* 581–590.

FRANCIS, M. and ALLCOCK, C. 1969. Geraniol β-Δ-glucoside; occurrence and synthesis in rose flowers. Phytochemistry *8*, 1339–1347.

FRANK, D. W., GRAY, J. E. and WEAVER, R. N. 1976. Cyclodextrin nephrosis in the rat. Am. J. Pathol. *83*, 367–382.

FRANKE, W. C. 1974. The correlation of metmyoglobin formation with the distribution of histochemically demonstrated enzyme activity in post-rigor muscle. Diss. Abstr. Int. B *34*, 5011–5012.

FRANZKE, C., KROLL, J. and PETZOLD, R. 1973. The glyceride structures of fats. VI. Specificity of the lipase of *Geotrichum candidum.* Nahrung *17*, 171–184. (German)

FRASER, T. H. and BRUICE, J. B. 1978. Chicken ovalbumin is synthesized and secreted by *Escherichia coli*. Proc. Natl. Acad. Sci. *75*, 5936–5940.

FRATER, S. G., DOUGLAS, G. and HUDDLE, B. 1977. An enzymatic method for the determination of lactose in cheese. Aust. J. Dairy Technol. *32*, 79–80.

FRAZIER, P. J., BRIMBLECOMBE, F. A., DANIELS, N. W. R. and EGGITT, P. W. R. 1977. The effect of lipoxygenase action on the mechanical development of doughs from fat-extracted and reconstituted wheat flours. J. Sci. Food Agric. *28*, 244–254.

FREAR, D. S., SWANSON, H. R. and TANARA, F. S. 1969. *N*-demethylation of substituted 3-(phenyl)-1-methyl ureas: isolation and characterization of a microsomal mixed function oxidase from cotton. Phytochemistry *8*, 2157–2169.

FREDERICS, B. 1965. Pancreatic and mesenteric lymph gland enzymes. U.S. Pat. 3,183,097. May 11.

FREE, B. L., and SATTERLEE, L. D. 1975. Biochemical properties of alfalfa protein concentrate. J. Food Sci. *40*, 85–89.

FREEMAN, G. G. and MOSSADEGHI, N. 1973. Studies on the relationship between water regime and flavour strength in watercress, cabbage and onion. J. Hort. Sci. *48*, 365–378.

FREEMAN, G. G. and WHENHAM, R. J. 1974. Changes in onion (*Allium cepa* L.). Flavour components resulting from some post-harvest processes. J. Sci. Food Agric. *25*, 499–515.

FREEMAN, G. G. and WHENHAM, R. J. 1975. A rapid spectrophotometric method of determination of thiopropanal S-oxide (lachrymator) in onion (*Allium cepa* L.) and its significance in flavour studies. J. Sci. Food Agric. *26*, 1529–1543.

FREEMAN, G. G. and WHENHAM, R. J. 1976A. Effect of overwinter storage at three temperatures on the flavor intensity of dry bulb onions. J. Sci. Food Agric. *27*, 37–42.

FREEMAN, G. G. and WHENHAM, R. J. 1976B. Nature and origin of volatile flavour components of onion and related species. Int. Flavours 7, 222–228.

FREEMAN, G. G., WHENHAM, R. J., SELF, R. and EAGLES, J. 1975. Volatile flavour components of parsley leaves (*Petroselinum crispum* (Mill.) Nyman). J. Sci. Food Agric. *26*, 465–470.

FRENKEL, C. 1971. Involvement of peroxidase and indole-3-acetic acid oxidase isozymes from pear, tomato, and blueberry fruit in ripening. Plant Physiol. *49*, 757–763.

FRENKEL, C. 1978. Role of hydroperoxides in the onset of senescence processes in plant tissues. *In* Postharvest Biology and Biotechnology. H.O. Hultin and M. Milner (Editors). Food and Nutrition Press, Westport, Conn.

FREUDENBERG, K. 1959. Biosynthesis and constitution of lignin. Nature *183*, 1152–1155.

FRICKER, A., DUDEN, R., HEINTZE, K., PAULUS, K. and ZOHM, H. 1975. Influence of heat treatment of spinach at temperatures up to 100°C on important constituents. VI. Total lipids and glycolipids. Lebensm. Wiss. Technol. *8*, 172–173.

FRIDKIN, M. and PATCHORNIK, A. 1974. Peptide synthesis. Annu. Rev. Biochem. *43*, 419–433.

FRIDOVICH, I. 1974. Superoxide dismutase. Adv. Enzymol. *41*, 35–97.

FRIED, M. and CHUN, P. W. 1971. Water-soluble nonionic polymers in protein purification. Methods Enzymol. *22*, 238–248.

FRIEDEN, E. and WALTER, C. 1963. Prevalence and significance of the product inhibition of enzymes. Nature *198*, 834–837.

FRIEDMAN, L. 1964. Adventitious toxic factors in protein concentrates. Food Technol. *18*, 1553–1563.

FRIEDMAN, M. 1977A. Crosslinking amino acids—stereochemistry and nomenclature. *In* Protein Crosslinking. Nutritional and Medical Consequences. M. Friedman (Editor). Plenum Press, New York.

FRIEDMAN, M. 1977B. Protein Crosslinking. Biochemical and Molecular Aspects. Plenum Press, New York.

FRIEDMAN, M., ZAHNLEY, J. C. and WAGNER, J. R. 1980. Estimation of the disulfide content of trypsin inhibitors as S-β-(2-pyridylethyl)-L-cysteine. Anal. Biochem. *106*, 27–37.

FRIEND, B. A., EITENMILLER, R. R. and SHAHANI, K. M. 1975. Role of lysine, tyrosine, and tryptophan residues in the activity of milk lysozymes. J. Food Sci. *40*, 833–836.

FRITZ, C. and BEEVERS, H. 1955. Oxidation of 2,3′,6-trichlorophenolindophenol by the lipoxidase system. Plant Physiol. *30*, 67–69.

FROMM, H. J. 1958. On the equilibrium and mechanism of adenylosuccinic acid synthesis. Biochim. Biophys. Acta *29*, 255–262.

FROMM, H. J. 1976. Initial Rate Enzyme Kinetics. Springer Verlag, Berlin.

FRUSUMA EUROPE. 1975. Process for obtaining a natural turbidifying ("clouding") agent. FSTA *8*, 9 H1523.

FRUTON, J. S. 1971. Pepsin. *In* The Enzymes, 3rd Edition, Vol. 3. P.D. Boyer (Editor). Academic Press, New York.

FRUTON, J. S. 1976. The emergence of biochemistry. Science *192*, 327–334.

FRUTON, J. S. and SIMMONDS, S. 1958. General Biochemistry, 2nd Edition. John Wiley & Sons, New York.

FRY, J. L., WALDROUP, P. W., AHMED, E. M. and LYDICK, H. 1966. Enzymic tenderization of poultry meat. Food Technol. *20*, 952–953.

FRYDMAN, R. B. and CARDINI, C. E. 1967. Studies on the biosynthesis of starch. II. J. Biol. Chem. *242*, 312–317.

FUJISHIGE, N., TSUBOI, H., KOBAYASHI, Y., MATSUO, R., and NISHIYAMA, M. 1977. Paper from pineapple fibers. Jpn. Kokai 77 118,004. Oct. 4.

FUJIMAKI, M., ARAI, S., YAMASHITA, M., KATO, H., and NOGUCHI, M. 1973A. Taste peptide fraction from a fish protein hydrolysate. Agric. Biol. Chem. *37*, 2891–2898.

FUJIMAKI, M., KATO, H., ARAI, S. and TAMAKI, E. 1968. Applying proteolytic enzymes on soybean. I. Effect on flavor. Food Technol. *22*, 889–893.

FUJIMAKI, M., KATO, H., ARAI, S. and YAMASHITA, M. 1971. Applications of microbial proteases to soybean and other material to prove acceptability, especially through the formation of plastein. J. Appl. Bacteriol. *39*, 119–131.

FUJIMAKI, M., UTAKA, K., YAMASHITA, M., and ARAI, S. 1973B. Production of higher-quality plastein from a crude single cell protein. Agric. Biol. Chem. *37*, 2303–2312.

FUJIMOTO, K., MIYASHIRO, M. and KANEDA, T. 1972. Enzymatic browning reaction of the Shiitake mushroom and its prevention. Mushroom Sci. *8*, 861–866.

FUJITA, K. 1954. Thiaminase. Adv. Enzymol. *15*, 389–422.

FUJIWARA, M., YOSHIMORA, M. and TSUNO, S. 1955. "Allithiamine", a newly found derivative of vitamin B12. III. On the allicin homologues in the plants of *Allium* species. J. Biochem. (Tokyo) *42*, 141–149.

FUKUMI, T., TAMOTO, K., NAKAMURA, M., KIDA, K., HIDESATO, T. and WATANABE, T. 1969. Freezing of fish meat paste and its application. X. Changes in the quality of fish meat during the manufacturing of frozen fish meat paste. Hokusuishi Geppo *26*, 687–698. (Japanese)

FUKUMOTO, J. 1965. Fruit and vegetable sauces. Jpn. Pat. 65 24,506. June 15.

FULLER, C. W., RUBIN, J. R. and BRIGHT, H. J. 1980. A simple procedure for covalent immobilization of NADH in a soluble and enzymatically active form. Eur. J. Biochem. *103*, 421–430.

FULLER, P., HUTCHINSON, J. B., McDERMOTT, E. E. and STEWART, B. A. 1970. Inactivation of α-amylase in wheat and flour with acid. J. Sci. Food Agric. *21*, 27–31.

FULLINGTON, J. G. 1967. Interaction of phospholipid-metal complexes with water-soluble wheat protein. J. Lipid Res. *8*, 609–614.

FULTON, S. P. and CARLSON, E. R. 1980. Dye-ligand affinity chromatography. Am. Lab. (Boston) *12* (10) 55–60.

FUNK, M. O., ISAAC, R. and PORTER, N. A. 1976. Preparation of lipid peroxides from arachidonic and γ-linolenic acids. Lipids *11*, 113–116.

FUWA, H., NAKAJIMA, M., HAMADA, A. and GLOVER, D. V. 1977. Comparative susceptibility to amylases of starches from different plant species and several single endosperm mutants and their double-mutant combinations with *opaque*-2 inbred Oh43 maize. Cereal Chem. *54*, 230–237.

GABLINGER, H. 1968. Preparation of a low dextrin beer by using glucosidase. U.S. Pat. 3,379,534. Apr. 23.

GADAMER, J. 1899. The ethereal oil of *Tropaeolum majus*. Arch. Pharm. *237*, 111–120. (German)

GADEN, E. L., MANDELS, M. H., REESE, E. T. and SPANO, L. A. 1976. Enzymatic Conversion of Cellulosic Materials. Biotechnol. Bioeng. Symp. 6. Interscience, New York.

GAEDE, G. and GRIESHABER, M. 1975. A rapid and specific method for the estimation of L-arginine. Anal. Biochem. *66*, 393–399.

GAINA, B. S., TYURINA, S. S., MINDADZE, R. K., PAVLENKO, N. M. and DATUNASHVILI, E. N. 1976. Changes in enzyme activity and protein content of grape juice during treatment with immobilized proteinases. FSTA *9*, H184. (Russian)

GAIND, K. N., GANDHI, K. S., JUNEJA, T. R., KJAER, A. and NIELSEN, B. J. 1975. 4,5,6,7-tetrahydroxydecyl isothiocyanate derived from a glucosinolate in *Capparia grandis*. Phytochemistry *14*, 1415–1418.

GAJZAGO, I., VAMOS-VIGYAZO, L. and NADUDVARI-MARKUS, V. 1977. Investigations into the enzymic browning of apricot cultivars. Acta Aliment. Acad. Sci. Hung. *6*, 95–114.

GALE, E. F. and EPPS, H. M. R. 1944. Studies on bacterial amino acid decarboxylases. 1. l-(+)-lysine decarboxylase. Biochem. J. *38*, 232–242.

GALEAZZI, M. A. M., CONSTANTINIDES, S. M. and SGARBIERI, V. C. 1976. Purification and characterization of banana polyphenoloxidase. Abstr. AGFD 134, 172nd Am. Chem. Soc. Meet. Port City Press, Baltimore.

GALLIARD, T. 1975. Degradation of plant lipids by hydrolytic and oxidative enzymes. *In* Recent Advances in the Chemistry and Biochemistry of Plant Lipids. T. Galliard and E.I. Mercer (Editors). Academic Press, London.

GALLIARD, T. and MATTHEW, J. A. 1975. Enzymic reactions of fatty acid hydroperoxides in extracts of potato tuber. I. Comparison 9-D- and 13-L-hydroperoxy-octadecadienoic acids as substrates for the formation of a divinyl ether derivative. Biochim. Biophys. Acta *398*, 1–9.

GALLIARD, T. and MATTHEW, J. A. 1976. The enzymic formation of long chain aldehydes and alcohols by α-oxidation of fatty acids in extracts of cucumber fruit *(Cucumis sativus)*. Biochim. Biophys. Acta *424*, 26–35.

GALLIARD, T. and MATTHEW, J. A. 1977. Lipoxygenase-mediated cleavage of fatty acids to carbonyl fragments in tomato fruits. Phytochemistry *16*, 339–343.

GALLIARD, T., MATTHEW, J. A., WRIGHT, A. J. and FISHWICK, M. J. 1977. The enzymic breakdown of lipids to volatile and nonvolatile carbonyl fragments in disrupted tomato fruits. J. Sci. Food Agric. *28*, 863–868.

GALLIARD, T., PHILLIPS, D. R. and REYNOLDS, J. 1976. The formation of *cis*-3-

nonenal, *trans*-2-nonenal and hexanal from linoleic acid hydroperoxide cleavage enzyme in cucumber *(Cucumis sativus)* fruits. Biochim. Biophys. Acta *441*, 181–192.

GALLOP, P. M., SEIFTER, S. and MEILMAN, E. 1957. Studies on collagen. 1. The partial purification, assay, and mode of activation of bacterial collagenase. J. Biol. Chem. *227*, 895–906.

GALPIN, J. R. and ALLEN, J. C. 1977. The influence of micelle formation on lipoxygenase kinetics. Biochim. Biophys. Acta *488*, 392–410.

GALPIN, J. R., TIELENS, G. M., VELDINK, G. A., VLIEGENHART, F. G. and BOLDINGH, J. 1976. On the interaction of some catechol derivatives with the iron atom of soybean lipoxygenase. FEBS Lett. *69*, 179–182.

GAMAGE, P. T. and MATSUSHITA, S. 1973. Interactions of autoxidized products of linoleic acids with enzyme proteins. Agric. Biol. Chem. *37*, 1–8.

GAMS, V. T. C. 1976. Chemical and technical hypotheses on the preparation of isomerized glucose syrups. Staerke *28*, 344–349. (German)

GANDOUR, R. D. and SCHOWEN, R. L. 1978. Transition States of Biochemical Processes. Plenum Publishing Corp., New York.

GANGA, P. S., NANDY, S. C. and SANTAPPA, M. 1977. Effect of environmental factors on the production of fungal tannase. Leather Sci. (Madras) *24*, 8–16.

GARDNER, H. W. 1975. Decomposition of linoleic hydroperoxides. Enzymic reactions compared with nonenzymic. J. Agric. Food Chem. *23*, 129–136.

GARDNER, H. W. and INGLETT, G. E. 1971. Food products from corn germ: Enzyme activity and oil stability. J. Food Sci. *36*, 645–648.

GARDNER, H. W., INGLETT, G. E. and ANDERSON, R. A. 1969. Inactivation of peroxidase as a function of corn processing. Cereal Chem. *46*, 626–634.

GARDNER, H. W. and SESSA, D. J. 1977. Degradation of fatty acid hydroperoxide by cereals and a legume: a comparison. Ann. Technol. Agric. *26*, 151–159.

GARDNER, H. W., WEISLEDER, D. and KLEIMAN, R. 1976. Addition of N-acetylcysteine to linoleic acid hydroperoxide. Lipids *11*, 127–134.

GARNOT, P., THAPON, J. L., MATHIEU, C. M., MAUBOIS, J. L. and DUMAS, B. R. 1972. Determination of rennin and bovine pepsins in commercial rennets and abomasal juices. J. Dairy Sci. *55*, 1641–1650.

GARRET, R. A. and WITTMAN, H. G. 1973. Structure and function of the ribosome. Endeavour *37*, 8–14.

GASCON, S. and OTTOLENGHI, P. 1967. Invertase isozymes and their localization in yeast. C.R. Trav. Lab. Carlsberg *36*, 85–93.

GASKELL, S. J. and BROOKES, C. J. W. 1977. Gas-liquid chromatography-mass spectrometry of phospholipid mixtures after enzymic hydrolysis. J. Chromatogr. *142*, 469–480.

GATFIELD, I. L. and STUTE, R. 1975. Enzymatic reactions in the presence of polymers. Influence of pH upon the interaction between horseradish peroxidase and ionic polymers. Lebensm.-Wiss. Technol. *8*, 121–122. (German)

GAWRON, O. and JONES, L. 1977. Structural basis for aconitase activity inactivation by butanedione and binding of substrates and inhibitors. Biochim. Biophys. Acta *484*, 453–464.

GEE, M. 1980. Some flavor and odor changes during low temperature dehydration of grapes. J. Food Sci. *45*, 146–147.

GEE, M., FARKAS, D. and RAHMAN, A. R. 1977. Some concepts for the development of intermediate moisture foods. Food Technol. *31* (4) 58–64.

GEE, M. and McCREADY, R. M. 1957. Texture changes in frozen Montmorency cherries. Food Res. *22*, 300–302.

GEIST, G. M. and CRAWFORD, D. L. 1974. Muscle cathepsins in three species of Pacific sole. J. Food Sci. *39*, 548–551.

GELFF, G. and BOUDRANT, J. 1974. Enzymes immobilized on a magnetic support. Preliminary study of a fluidized bed reactor. Biochim. Biophys. Acta *334*, 467–470.

GENCHEV, S. and MIKHOV, A. 1975. Effect of inbreeding on some physiological manifestations of cucumber (*Cucumis sativus*). Fiziol. Rast. (Sofia) *1*, 84–91. (Bulgarian)

GERBRANDY, S. J. 1974. Glycogen phosphorylase of potatoes. Purification and thermodynamic properties of the adsorption of glycogen. Biochim. Biophys. Acta *370*, 410–418.

GERBRANDY, S. J., SHANKAR, V., SHIVARAM, K. N. and STEGEMANN, H. 1975. Conversion of potato phosphorylase isozymes. Phytochemistry *14*, 2331–2333.

GERRITSEN, M., VELDINK, G. A., VLIEGENHART, A., JOHANNES, F. G. and BOLDINGH, J. 1976. Formation of α-and γ-ketols from oxygen-18-labeled hydroperoxide isomerase. FEBS Lett. *87*, 149–152.

GERWIN, B., BURSTEIN, S. R. and WESTLEY, J. 1974. Ascorbate oxidase. Inhibition, activation, and pH effects. J. Biol. Chem. *249*, 2005–2008.

GESTETNER, B. and CONN, E. E. 1974. The 2-hydroxylation of *trans*-cinnamic acids by chloroplasts from *Melilotus alba* Desr. Arch. Biochem. Biophys. *163*, 617–624.

GHELUWE, G. E. A. VAN and VALYI, Z. 1974. Beer flavor and implications of oxidation during brewing. Tech. Q. Master Brew. Assoc. Am. *11*, 184–192.

GHISLA, S., MASSEY. V., LHOSTE, J. -M. and MAYHEW, S. G. 1974. Fluorescence and optical characteristics of reduced flavines and flavoproteins. Biochemistry *13*, 589–597.

GHOSE, K. C. and HALDAR, D. P. 1970. Application of cellulase. III. Extraction of protein from *Khesari* and geam plants with fungal cellulases. J. Food Sci. Technol. 7, 160–161.

GHOSE, T. K. and KOSTICK, J. A. 1970. A model for continuous enzymatic saccharification of cellulose with simultaneous removal of glucose syrup. Biotechnol. Bioeng. *12*, 921–926.

GHOSE, T. K. and PATHAK, A. N. 1973. Cellulase. II. Applications. Process Biochem. *8* (5) 20–21, 24.

GIANG, P. A. and HALL, S. A. 1951. Enzymatic determination of organic phosphorus insecticides. Anal. Chem. *23*, 1830–1834.

GIBBONS, I., SKOLD, C., ROWLEY, G. L. and ULLMAN, E. F. 1980. Homogeneous enzyme immunoassay employing β-galactosidase. Anal. Biochem. *102,* 167–170.

GIBIAN, M. J. and GALAWAY, R. A. 1975. Kinetics of olefin oxygenation catalyzed by soybean lipoxygenase. Biochemistry *15*, 4209–4214.

GIBRIEL, A. Y., EL-SAHRIGI, F., KANDIL, S. H. and EL-MANSY, H. A. 1978. Effect of pH, sodium chloride, and sucrose on heat-inactivation and reactivation of peroxidases in certain foods. J. Sci. Food Agric. *29*, 261–266.

GIBSON, R. A. and PALEG, L. G. 1975. Further experiments on the α-amylase-containing lysosomes of wheat aleurone cells. Aust. J. Plant Physiol. *2*, 41–49.

GIBSON, R. A. and PALEG, L. G. 1976. Purification of gibberellic acid-induced lysosomes from wheat aleurone cells. J. Cell Sci. *22*, 413–425.

GIDDINGS, G. G. 1974. Reduction of ferrihemoglobin in meat. Crit. Rev. Food Technol. *5*, 143–173.

GIDDINGS, G. G. 1977A. The basis of color in muscle foods. Crit. Rev. Food Sci. Nutr. *9*, 81–114.

GIDDINGS, G. G. 1977B. Symposium: The basis of quality in muscle foods. The basis of color in muscle foods. J. Food Sci. *42*, 288–294.

GIERER, J. and OPARA, A. E. 1973. Studies on the degradation of lignin. The action of peroxidase and laccase on monomeric and dimeric model compounds. Acta Chem. Scand. *27*, 2909–2922.

GILBERT, J. and NURSTEN, N. E. 1972. Volatile constituents of horseradish roots. J. Sci. Food Agric. *23*, 527–539.

GILLET, C., EECKHOUT, Y. and VAES, S. 1977. Purification of procollagenase and collagenase by affinity chromatography on Sepharose-collagen. FEBS Lett. *74*, 126–128.

GILLETTE, J. R., MITCHELL, J. R. and BRODIE, B. B. 1974. Biochemical mechanisms of drug toxicity. Annu. Rev. Pharmacol. *14*, 271–288.

GINI, B. and KOCH, R. B. 1961. Study of a lipohydroperoxide breakdown factor in soy extracts. J. Food Sci. *26*, 359–364.

GINZBURG, A. S. 1973. Theoretical Principles and Techniques of Food Dehydration. Pishchevaya Promyshlennost, Moscow. (Russian)

GITENSHTEIN, B. M. 1974. Inactivation of enzymes derived from the pectin methylesterase of plums during processing into alcoholic beverages. FSTA 7 (4) H620. (Russian)

GLAZER, A. N., BAR-ELI, A. and KATCHALSKI, E. 1962. Preparation and characterization of polytyrosol trypsin. J. Biol. Chem. *237*, 1832–1838.

GLAZER, A. N. and SMITH, E. L. 1971. Papain and other sulfhydryl proteolytic enzymes. *In* The Enzymes, 3rd Edition, Vol. 3. P.D. Boyer (Editor). Academic Press, New York.

GLEASON, M. N., GASSELIN, R. E., HODGE, H. C. and SMITH, R. E. 1969. Clinical Toxicology of Commercial Products, 3rd Edition. Williams & Wilkins Co., Baltimore.

GLOSTER, J. and HARRIS, P. 1977. Fatty acid binding to cytoplasmic proteins of myocardium and red and white skeletal muscle in the rat. A possible new role for myoglobin. Biochem. Biophys. Res. Comm. *74*, 506–513.

GLUEK, A. C. 1948. Manufacture of a new malt liquor from cereal products. U.S. Pat. 2,442,806. June 8.

GMELIN, R. and VIRTANEN, A. L. 1959. A new type of enzymatic cleavage of mustard oil glucosides. Formation of allyl thiocyanate in *Thlaspi arvense* and benzylthiocyanate in *Lipidium*. Acta Chem. Scand. *13*, 1474–1475.

GNEKOW, B. and OUGH, C. S. 1976. Methanol in wines and musts. Am. J. Enol. Viticult. *27*, 1–6.

GOEBELL, H. 1975. Cut and punctureproof indication for therapy with enzyme preparations.

Chem. Abstr. *86*, 50384z. (German)

GOERING, K. J. 1961. Process of obtaining the proteinaceous feed material from mustard seed, rape seed and similar seeds. U.S. Pat. 2,987,399. June 6.

GOERING, K. J. and ESLICK, R. F. 1976. A self-liquefying waxy barley starch. Cereal Chem. *53*, 174–180.

GOLD, H. J. and WECKEL, K. G. 1959. Degradation of chlorophyll to pheophytin during sterilization of canned green peas by heat. Food Technol. *13*, 281–286.

GOLDBERG, B. and STERN, A. 1975. Generation of superoxide by the interaction of the hemolytic agent, phenylhydrazine, with human hemoglobin. J. Biol. Chem. *250*, 2401–2403.

GOLDBERG, B., STERN, A. and PEISACH, J. 1976. The mechanism of superoxide anion generation by the interaction of phenylhydrazine with hemoglobin. J. Biol. Chem. *251*, 3045–3051.

GOLDBERG, I. 1977. Production of SCP from methanol. Process Biochem. *12* (9) 12–18.

GOLDBERGER, A. and CAPLAN, S. R. 1976. Oscillatory enzymes. Annu. Rev. Biophys. Bioeng. *5*, 449–476.

GOLDBLITH, S. A. 1963. Radiation processing of foods and drugs. *In* Food Processing Operations, Vol. 1. M.A. Joslyn and J.H. Hyde (Editors). AVI Publishing Co., Westport, Conn.

GOLDENBERG, N. and MATHESON, H. R. 1975. Off-flavours in foods, a summary of experience: 1948–74. Chem. Ind. (London) *13*, 551–575.

GOLDRING, I. R., RATNER, I. M. and GREENBURG, L. 1970. Pulmonary hemorrhage in hamsters after exposure to proteolytic enzymes of *Bacillus subtilis*. Science *170*, 73–74.

GOLDSTEIN, J. L. and SWAIN, T. 1965. The inhibition of enzymes by tannins. Phytochemistry *4*, 185–192.

GOLDSTEIN, L. 1976. Kinetic behavior of immobilized enzyme systems. Methods Enzymol. *44*, 397–443.

GOLDSTEIN, L., LIFSHITZ, A. and SOKOLOVSKY, M. 1971. Water insoluble derivatives of naringinase. Int. J. Biochem. *2*, 448–456.

GOLL, D. E., ARAKAWA, N., STROMER, M. H., BUSCH, W. A. and ROBSON, R. M. 1970. Chemistry of muscle proteins as food. *In* The Physiology and Biochemistry of Muscle as a Food 2. E.J. Briskey, R.G. Cassens and B.B. Marsh (Editors). Univ. of Wisconsin Press, Madison.

GOLL, D. E., HOEKSTRA, W. G. and BRAY, R. W. 1964. Age-associated changes in bovine muscle connective tissue. II. Exposure to increasing temperature. J. Food Sci. *29*, 615–621.

GONG, C. -S., LADISCH, M. R. and TSAO, G. T. 1977. Cellobiase from *Trichoderma viride*. Biotechnol. Bioeng. *19*, 959–981.

GOODNIGHT, K. C., JR., HARTMAN, G. H., JR. and MARQUARDT, R. F. 1978. Low-phytate isoelectric-precipitated soybean protein isolate. U.S. Pat. 4,072,670. Feb. 7.

GOODSON, L. H. and JACOBS, W. B. 1973. An immobilized cholinesterase product for use in the rapid detection of enzyme inhibitors in air or water. Anal. Biochem. *51*, 362–367.

GOODWIN, T. L. and WALDROUP, P. W. 1970. Tenderizing meat from broiler breeder males by papain injection and aging in slush ice. Ark. Farm Res. *19* (3) 7.

GOODWIN, T. W. 1968. The Metabolic Roles of Citrate. Academic Press, London.

GOOSSENS, A. E. 1975. Protein flavour problems. Food Process. Ind. *50* (11) 29–30.

GORIN, H. 1976. Differences in L-malate determined enzymatically or titrimetrically in Golden Delicious apples. Z. Lebensm.-Unters. Forsch. *162*, 259–261. (German)

GORIN, N. and HEIDEMA, F. T. 1976. Peroxidase activity in golden delicious apples as a possible parameter of ripening and senescence. J. Agric. Food. Chem. *24*, 200–201.

GORINI, C. 1930. The chymase of *Bacterium prodigiosus*. Boll. Ital. Biol. Sper. *5*, 517–518. (Italian)

GORINSTEIN, S. 1978. Different forms of nitrogen and the stability of beer. J. Agric. Food Chem. *26*, 204–207.

GORKE, H. 1906. Concerning chemical reactions upon freezing of plants. Landw. Vers. Stat. *65*, 149–160. (German)

GORMLEY, T. R. and O'RIORDAN, F. 1976. Quality evaluation of fresh and processed mushrooms *(Pleurotus ostreatus)*. Lebensm.-Wiss. Technol. *9*, 75–78. (German)

GORNIAK, H. and KACZKOWSKI, J. 1974. Isolation and characterization of glucose oxidase from mycelium and nutrient of *Penicillium notatum* cultures. Bull. Acad. Pol. Sci. Ser. Sci. Biol. *22*, 351–355.

GORZ, H. J., HAAG, W. L., SPECHT, J. E. and NASKINS, F. A. 1977. Assay of *p*-hydroxybenzaldehyde as a measure of hydrocyanic potential in sorghums. Crop Sci. *17*, 578–582.

GOSWAMI, A. K., JAIN, M. K. and PAUL, B. 1977. α- And β-amylases in seed germination. Biol. Plant. *19*, 469–471.

GOTTSCHALK, A. 1950. Principles underlying

enzyme specificity in the domain of carbohydrates. Adv. Carbohydr. Chem. *5,* 49–78.

GOTHARD, P. G. 1974. Screening of barley varieties for alpha-amylase content. J. Inst. Brew. *80,* 387–390.

GOTOH, T. and SHIKIMA, K. 1976. Generation of the superoxide radical during autoxidation of oxymyoglobin. J. Biochem. (Tokyo) *80,* 397–400.

GOTTSCHALL, R. M. and KIES, M. W. 1942. Digestion of beef by papain. Food Res. *7,* 373–379.

GOUGH, D. A. and ANDRADE, J. D. 1973. Enzyme electrodes. Science *180,* 380–384.

GOULD, E. 1973. Collaborative study of a test to determine whether shucked oysters have been frozen or thawed. J. Assoc. Off. Anal. Chem. *56,* 541–543.

GOULD, R. F. 1966. Flavor Chemistry. Adv. Chem. Ser. *56.* Am. Chem. Soc., Washington, D.C.

GOULD, W. A. 1974. Tomato Production, Processing and Quality Evaluation. AVI Publishing Co., Westport, Conn.

GOVE, J. G. and HOYLE, M. C. 1975. The isozymic similarity of indoleacetic oxidase to peroxidase in birch and horseradish. Plant Physiol. *56,* 684–687.

GOVINDARAJAN, S. 1973. Fresh meat color. Crit. Rev. Food Technol. *4,* 117–140.

GOVINDARAJAN, S., HULTIN, H. O. and KOTULA, A. W. 1977. Myoglobin oxidation in ground beef: Mechanistic studies. J. Food Sci. *42,* 571–577.

GRACEY, D. E. F. and BARKER, R. L. 1976. Studies on beer haze formation. II. Dimeric flavonoids observed in profiles of beer: Nylon 66 adsorbants. J. Inst. Brew. *82,* 78–83.

GRAHAM, H. D. 1971. Determination of carboxymethylcellulose in food products. J. Food Sci. *36,* 1052–1055.

GRAMPP, E. 1969. Use of enzymes in fruit and vegetable processing. Ber. Wiss.-Tech. Komm., Int. Fruchtsaft-Union *9,* 73–108. (German)

GRANDJEAN, J., LASZLO, P. and GERDAY, C. 1977. Sodium complexation by the calcium binding site of parvalbumin. FEBS Lett. *81,* 376–380.

GRANROTH, B. 1970. Biosynthesis and decomposition of cysteine derivatives in onion and other *Allium* species. Ann. Acad. Sci. Fenn. Ser. A2 *154,* 1–171.

GRANROTH, B. 1974. Partial purification of cysteine synthase (*O*-acetyl sulfhydrase) from onion *(Allium cepa).* Acta Chem. Scand. *B28* 813–814.

GRANT, D. R. and SOOD, V. K. 1980. Studies of the role of ascorbic acid in chemical dough development. II. Partial purification and characterization of an enzyme oxidizing ascorbate in flour. Cereal Chem. *57,* 46–49.

GRANT, D. R. and WANG, C. C. 1972. Dialyzable components resulting from proteolytic activity in extracts of wheat flour. Cereal Chem. *49,* 201–207.

GRANT, N. H. and ALBURN, H. E. 1967. Reactions in frozen systems. VI. Ice as a possible model for biological structured-water systems. Arch. Biochem. Biophys. *118,* 292–296.

GRASSMANN, W. and RUBENBAUER, H. 1931. Cellulase and hemicellulase as therapeutic agents. Muench. Med. Wochenschr. *78,* 1817. (German)

GRAVELAND, A. 1970. Analysis of lipoxygenase nonvolatile reaction products of linoleic acid in aqueous cereal suspensions by urea extraction and gas chromatography. Lipids *47,* 352–361.

GRAVELAND, A. 1973. Enzymatic oxidation of linoleic acid in aqueous wheat flour suspensions. Lipids *8,* 606–611.

GRAVELAND, A., BOSVELD, P. and MARSEILLE, J. P. 1978. Determination of thiol groups and disulphide bonds in wheat flour and dough. J. Sci. Food Agric. *29,* 53–61.

GRAVES, D. J., SEALOCK, R. W. and WANG, J. H. 1965. Cold inactivation of glycogen phosphorylase. Biochemistry *4,* 290–295.

GRAY, D. N. and KEYES, M. H. 1977. Immobilized enzymes for chemical analysis. Chemtech. *7,* 642–648.

GRAY, D. N., KEYES, M. H. and WATSON, B. 1977. Immobilized enzymes in analytical chemistry. Anal. Chem. *49,* 1067A–1078A.

GRAY, J. I. and DUGAN, L. R. 1975. Inhibition of N-nitrosamine formation in model food systems. J. Food Sci. *40,* 981–984.

GREEN, M. L. 1972. On the mechanism of milk clotting by rennin. J. Dairy Sci. *39,* 55–63.

GREEN, M. L. and CRUTCHFIELD, G. 1969. Studies on the preparation of water-insoluble derivatives of rennin and chymotrypsin and their use in the hydrolysis of casein and the clotting of milk. Biochem. J. *115,* 183–190.

GREEN, M. L. and FOSTER, P. M. D. 1974. Comparison of the rates of proteolysis during ripening of Cheddar cheeses made with calf rennet and swine pepsin. J. Dairy Res. *41,* 269–282.

GREEN, M. L. and MARSHALL, R. J. 1977. The acceleration by cationic materials of the coagulation of casein micelles by rennet. J. Dairy Res. *44,* 521–531.

GREEN, M. L. and STACKPOOLE, A. 1975. Preparation and assessment of a suitable *Mucor pusillus* proteinase-swine pepsin mixture for cheddar cheesemaking. J. Dairy Res. *42*, 297–312.

GREENE, B. E. 1971. Oxidations involving the heme complex in raw meat. J. Am. Oil Chem. Soc. *48*, 637–639.

GREENE, B. E. and PRICE, L. G. 1975. Oxidation induced color and flavor changes in meat. J. Agric. Food Chem. *23*, 164–166.

GREENFIELD, P. F. and LAURENCE, R. L. 1975. Characterization of glucose oxidase and catalase on inorganic supports. J. Food Sci. *40*, 906–910.

GREENSHIELDS, R. N. and MACGILLIVRAY, A. W. 1972. Caramel. I. The browning reactions. Process Biochem. 7 (12) 11–13, 16.

GREENSTEIN, D. S. 1956. The calculation of reaction-velocity constants for reactions from the kinetics of the disappearance of enzyme-substrate compounds. Arch. Biochem. Biophys. *62*, 284–291.

GREMLI, H. and NEUKOM, H. 1968. Maceration of cucumber tissue by a purified α-1-arabinofuranosidase. Lebensm.-Wiss. Technol. *1*, 24–25. (German)

GREMLI, H. and WILD, J. 1974. Enzymatic flavor regeneration in processed foods. Proc. IV. Int. Congr. Food Sci. Technol. *1*, 158–161.

GRIPON, J. -C., DESAZEAUD, M. J., LE BARS, D. and BERGERE, J. -L. 1977. Role of proteolytic enzymes of *Streptococcus lactis, Penicillium roqueforti* and *Penicillium caseicolum* during cheese ripening. J. Dairy Sci. *60*, 1532–1538.

GRISON, R., GALLOIS, T., CHAPPET, A., DUBOUCHET, J. and PERESSE, M. 1975. Changes in the activity of an acid peroxidase along the carnation stem, after experimental inoculation with *Phialophora cinerescens*. C.R. Hebd. Seances Acad. Sci. Ser. D *281*, 131–134. (French)

GROBSTEIN, C. 1977. The recombinant-DNA debate. Sci. Am. *273* (7) 22–23.

GROMMECK, R. and MARKAKIS, P. 1964. The effect of peroxidase on anthocyanin pigments. J. Food Sci. *29*, 53–56.

GRONINGER, H. S., JR. 1964. Partial purification of a proteinase from albacore muscle. Arch. Biochem. *108,* 175–182.

GRONINGER, H. S. and SPINELLI, J. 1968. EDTA. Inhibition of inosine monophosphate dephosphorylation in refrigerated fishery products. J. Agric. Food Chem. *16*, 97–99.

GROSCH, W. 1969. Production of volatile alcohols in peas by action of lipoxygenase (EC 1.13.1.13) and alcohol dehydrogenase (EC 1.1.1.1.). Nahrung *13*, 393–401. (German)

GROSCH, W. 1972. The enzyme lipoxygenase—its properties and action in foods. Fette, Seifen, Anstrichm. *4*, 375–381. (German)

GROSCH, W. and LASKAWY, G. 1975. Differences in the amount and range of volatile compounds by lipoxygenase isoenzymes from soybeans. J. Agric. Food Chem. *23*, 791–794.

GROSCH, W., LASKAWY, G. and FISCHER, K. -H. 1974. Oxidation of linolenic acid in the presence of haemoglobin, lipoxygenase or by singlet oxygen. Identification of the volatile carbonyl compounds. Lebensm.-Wiss. Technol. *7*, 335–338.

GROSCH, W., LASKAWY, G. and KAISER, K. L. 1977. Co-oxidation of β-carotene and canthaxanthine by purified lipoxygenase from soybeans. Z. Lebensm.-Unters. Forsch. *165,* 77–81. (German)

GROSCH, W., LASKAWY, G. and WEBER, F. 1976. Formation of volatile carbonyl compounds and cooxidation of carotene by lipoxygenase from wheat, potato, flax, and beans. J. Agric. Food Chem. *24*, 456–459.

GROSJEAN, O. -K., COBB, B. F., III, MEBINE, B. and BROWN, W. D. 1969. Formation of a green pigment from tuna myoglobins. J. Food Sci. *34*, 404–407.

GROSS, A. F. and ELLIS, R. E. 1969. Lipase activity in spices and seasonings. Cereal Sci. Today *14*, 332–335.

GROSSMAN, S., BEN AZIZ, A., BUDOWSKI, P. P., ASCARELLI, I., GERTIER, A., BIRK, Y. and BONDI, A. 1969. Enzymic oxidation of carotene and linoleate by alfalfa: Extraction of the active fractions. Phytochemistry *8*, 2287–2293.

GROSSMAN, S., TROP, M., YARONI, S. and WILCHEK, M. 1972. Purification of soybean lipoxygenase by affinity chromatography. Biochim. Biophys. Acta *289*, 77–81.

GROSSMANN, A., TIMMEN, H. and KLOSTERMEYER, H. 1976. Enzymic estimation of cholesterol in milk fat. An alternative to the methods in current use. Milchwissenschaft *31*, 721–724. (German)

GRUBHOFER, N. and SCHLEITH, L. 1954. Coupling of proteins on diazotized polyaminostyrene. Z. Physiol. Chem. *297,* 108–122. (German)

GRUNDEN, L. P., VADEHRA, D. V. and BAKER, R. C. 1974. Effects of proteolytic enzymes on the functionality of chicken egg albumen. J. Food Sci. *39*, 841–843.

GUADAGNI, D. G., BOMBEN, J. L. and HUDSON, J. S. 1971. Factors influencing the development of aroma in apple peels. J. Sci. Food Agric. *22*, 110–115.

GUADAGNI, D. G. and NIMMO, C. C. 1958. Time-temperature of tolerance of frozen foods. XIII. Effect of regularly fluctuating temperatures in retail packages of frozen strawberries and raspberries. Food Technol. *12*, 306–310.

GUENGERICH, F. P. 1977. Separation and purification of multiple forms of microsomal cytochrome P-450. Activities of different forms of cytochrome P-450 towards several compounds of environmental interest. J. Biol. Chem. *252*, 3970–3979.

GUENOT, M. -C., PERRIOT, J. -J. and VINCENT, J. C. 1976. Development of the microflora and fatty acids of cocoa beans. Cafe-Cacao-The *20*, 53–58. (French)

GUENTHER, E. 1948–1952. The Essential Oils, Vol. 1–6. D. Van Nostrand Co., New York.

GUENTHER, F. and BURCKHART, O. 1967. Rapid method for determination of total alkaline phosphatase in egg yolks. Dtsch. Lebensm. Rundsch. *63*, 305–309. (German)

GUGGOLZ, J., SAUNDERS, R. M., KOHLER, T. J. and KLOPFENSTEIN, T. J. 1971. Enzymatic evaluation of processes for improving agricultural wastes for ruminant feeds. J. Anim. Sci. *33*, 167–170.

GUIDOTTI, G. 1976. The structure of membrane transport systems. Trends Biochem. Sci. *1*, 11–13.

GUIGOU, P. T. M. 1974. Process of paring and peeling fruits and vegetables. Fr. Pat. Appl. 2,207,657. Sept. 21.

GUIGOZ, Y. and SOLMS, J. 1976. Bitter peptides, occurrence and structure. Chem. Senses Flavor *2*, 71–84.

GUILBAULT, G. G. 1976. Handbook of Enzymatic Analyses. Marcel Dekker, New York.

GUILBAULT, G. G. and CSERFALVI, T. 1976. Ion selective electrodes for phosphate using enzyme systems. Anal. Lett. *9*, 277–289.

GUILBAULT, G. G. and HIESERMAN, J. E. 1968. Fluorimetric assay of amino acids. Anal. Biochem. *26*, 1–11.

GUILBAULT, G. G., KUAN, S. S. and BRIGNAC, P. 1969. Fluorimetric determination of oxidative enzymes. Analytical applications of the monoamine and diamine oxidase systems. Anal. Chim. Acta *47*, 503–509.

GUILBAULT, G. G. and LUBRANO, G. 1974. Amperometric enzyme electrodes. II. Amino acid oxidase. Anal. Chem. Acta *69*, 189–194.

GUILBAULT, G. G. and SADAR, M. H. 1971. Fluorimetric determination of pesticides. Anal. Chem. *41*, 366–368.

GUILBAULT, G. G. and SADAR, M. H. 1977. Analytical use of enzymes. Proc. Anal. Div. Chem. Soc. *14*, 302–306.

GUILBAULT, G. G. and TARP, M. 1974. Specific enzyme electrode for urea. Anal. Chim. Acta *73*, 355–365.

GUILLOT, G., LISCH, J. M. and ROSSET, R. 1973. Structural changes in frozen meat enabling the distinction between this and fresh or chilled meat. Characterization of GOT (glutamic-oxalacetic transaminase) present in muscles. Bull. Inst. Brt. Froid, Annexe *2*, 259–264. (French)

GUIRE, P. 1975. Photoreactive carrier derivatives for immobilization of enzymes and ligands. Fed. Proc. *34*, 690.

GUM, E. K., JR. and BROWN, R. D., JR. 1976. Structural characterization of a glycoprotein cellulase, 1,4-β-D-glucan cellobiohydrolase. Biochim. Biophys. Acta *446*, 371–386.

GUMBARIDZE, N. 1973. Transformation of quince polyphenols during fruit processing. Biol. Abstr. *58*, 007476. (Russian)

GUNSALUS, I., PEDERSON, T. C. and SLIGAR, S. G. 1975. Oxygenase-catalyzed biological hydroxylations. Annu. Rev. Biochem. *44*, 375–407.

GUNTER, B. J. 1977. Health hazard evaluation/toxicity determination report No. 76-23-319, Western Gear Corp. Govt. Rep. Ann. Index 77 (12) 101.

GUNTER, M. J., MANDER, L. N., McLAUGHLIN, G. M., MURRAY, K. S., BERRY, J. J., CLARK, P. E. and BUCKINGHAM, D. A. 1980. Towards synthetic models for cytochrome oxidase: A binuclear iron (III) porphyrin-copper (II) complex. J. Am. Chem. Soc. *102*, 1470–1473.

GUNTHER, R. C. 1979. Chemistry and characteristics of enzyme modified whipping proteins. J. Am. Oil Chem. Soc. *56*, 345–349.

GUPTA, C. B. and ESKIN, N. A. M. 1977. Potential use of vegetable rennet in the production of cheese. Food Technol. *31* (5) 62–66.

GUPTE, S. M. and FRANCIS, F. J. 1964. Effect of pH adjustment and high-temperature short-time processing on color and pigment retention in spinach. Food Technol. *18*, 1645–1648.

GUSS, P. L., RICHARDSON, T. and STAMANN,

M. A. 1968. Oxidation of various lipid substrates with unfractionated soybean and wheat lipoxidase. J. Am. Oil Chem. Soc. *45*, 272–276.

GUTCHO, M. 1973. Textured Foods and Allied Products. Noyes Data Corp., Park Ridge, N.J.

GUTCHO, M. 1974. Microbial Enzyme Production. Noyes Data Corp., Park Ridge, N.J.

GUTFELD, M. and ROSENFELD, P. P. 1975. The solution to Israel's rennet shortage. Dairy Ind. *40* (2) 52, 55.

GUTFREUND, H. 1965. An Introduction to the Study of Enzymes. Blackwell Scientific Publications, Oxford.

GUTFREUND, H. 1972. Enzymes. Physical Principles. John Wiley & Sons, New York.

GUTFREUND, H. 1976A. Enzymes: One Hundred Years. FEBS Lett. *62*, Suppl. 4, Feb.

GUTFREUND, H. 1976B. Kinetics: The grammar of enzymology. FEBS Lett. *62*, Suppl. 4, Feb. E13–E19.

GUTHRIE, J. P. and O'LEARY, S. 1975. General base catalysis by a steroidal enzyme model. J. Chem. *53*, 2150–2156.

GUTOWSKI, J. A. and LIENHARD, G. E. 1976. Transition state analogs for thiamin pyrophosphate-dependent enzymes. J. Biol. Chem. *251*, 2683–2686.

GUTTER, Y. 1973. Studies of the limonin-caused bitterness in early season citrus fruit. Research Summaries 1971–1973, Div. Fruit and Vegetable Storage, Volcani Insts. Res. Organ., Bet Dagan, Israel.

GUTTERMAN, B. M. 1956. Determination of acetaldehyde in frozen vegetables. J. Assoc. Off. Agric. Chem. *39*, 282–285.

GUTTERMAN, R. D., LOVEJOY, R. D. and BEACHMAN, L. M. 1951. Quality factors in processed vegetables. J. Assoc. Off. Agric. Chem. *34,* 231–232.

GUY, E. J. 1973. Ice cream manufacture with dairy products treated with lactase enzyme from *Saccharomyces lactis*. J. Dairy Sci. *56*, 627.

HAARD, N. F. 1971. Potential applications of hormones to plant and animal tissues as food. Crit. Rev. Food Technol. *2*, 305–353.

HAARD, N. F. 1972. Membrane-structure and cellular death in biological tissue. J. Food Sci. *37*, 504–512.

HAARD, N. F. 1973. Chilling injury of green banana fruit: Kinetic anomalies of IAA oxidase at chilling temperatures. J. Food Sci. *38*, 907–908.

HAARD, N. F. 1977. Physiological roles of peroxidase in postharvest fruits and vegetables. *In* Enzymes in Fruits and Beverages. R.L. Ory and A.J. St. Angelo (Editors). Am. Chem. Soc., Washington, D.C.

HAARD, N. F. and CODY, M. 1978. Stress metabolites in postharvest fruits and vegetables—role of ethylene. *In* Postharvest Biology and Biotechnology. H.O. Hultin and M. Milner (Editors). Food and Nutrition Press, Westport, Conn.

HAARD, N. F. and MARSHALL, M. 1976. Isoperoxidase changes in soluble and particulate fractions of sweet potato root resulting from cut injury, ethylene and black rot infection. Physiol. Plant Pathol. *8*, 195–205.

HAARD, N. F. and SALUNKHE, D. K. 1975. Postharvest Biology and Handling of Fruits and Vegetables. AVI Publishing Co., Westport, Conn.

HAARD, N. F., SHARMA, S. C., WOLFE, R. and FRENKEL, C. 1974. Ethylene induced isoperoxidase changes during fiber formation in postharvest asparagus. J. Food Sci. *39*, 452–456.

HAARD, N. F. and TOBIN, C. L. 1971. Patterns of soluble peroxidase in ripening banana fruit. J. Food Sci. *36*, 854–857.

HAAS, G. J., BENNET, D., HERMAN, E. B. and COLLETTE, D. 1975. Microbial stability of intermediate moisture foods. Food Prod. Dev. *10* (3) 86–94.

HAAS, L. W. and BOHN, R. M. 1934. Bleaching bread dough. U.S. Pat. 1,957,333. May 1.

HAAS, L. W. and RENNER, H. O. 1935. Oxidation products of fats and oils for use as shortening and bleaching agents in doughs, etc. U.S. Pat. 1,994,992. Mar. 19.

HABER, F. and WEISS, J. 1934. The catalytic decomposition of hydrogen peroxide by iron salts. Proc. Roy. Soc. London, Ser. A *147*, 332–351.

HADI, S. M., BAECHI, B., SHEPHERD, J. C. W., YUAN, R., INEICHEN, K. and BICKLE, T. A. 1979. DNA recognition and cleavage by the EcoP15 restriction endonuclease. J. Mol. Biol. *134,* 655–666.

HAGER, L. P., HOLLENBERG, P. F., RANDMEIR, T., CHIANG, R. and DOUBEK, D. 1975. Chemistry of peroxidase intermediates. Ann. N.Y. Acad. Sci. *244,* 80–93.

HAHN, P. F. 1943. Abolishment of alimentary lipemia following injection of heparin. Science *98*, 19–20.

HAI, D. Q., KOVACS, K., MATKOVICS, I. and MAYKOVICS, B. 1975. Peroxidase and superoxide dismutase contents of plant seeds. Biochem. Physiol. Pflanz. *167*, 357–359.

HAID, E., NELBOECK-HOCHSTETTER, M. and NAEHER, G. 1977. Proteins with low nucleic acid content from microorganisms. Ger. Pat. 2,622,982. Nov. 17.

HAISMAN, D. R. 1974. The effect of sulphur dioxide on oxidising enzyme systems in plant tissues. J. Sci. Food Agric. *25*, 803–810.

HAISMAN, D. R. and CLARKE, M. W. 1975. The interfacial factor in the heat-induced conversion of chlorophyll in green leaves. J. Sci. Food Agric. *26*, 1111–1126.

HAISMAN, D. R. and KNIGHT, D. J. β-Glucosidase activity in canned plums. J. Food Technol. *2*, 241–248.

HALDANE, J. B. S. 1930. Enzymes. Longmans, Green and Co., London.

HALDANE, J. B. S. and STERN, K. G. 1932. General Chemistry of the Enzymes. Verlag von Theodor Steinkopff, Dresden. (German)

HALE, M. B. 1974. Using enzymes to make fish protein concentrates. Mar. Fish. Rev. N.A.S.A. *36*, 15–18.

HALE, S. A., RICHARDSON, T., VON ELBE, J. H. and HAGEDORN, D. J. 1969. Isoenzymes of lipoxidase. Lipids *4*, 209–215.

HALE, W. S., SCHWIMMER, S. and BAYFIELD, E. G. 1943. Studies on treating wheat with ethylene. I. Effect on high moisture wheat. Cereal Chem. *20*, 224–233.

HALL, R. L. and EISS, M. 1976. Food irradiation: Is it the process of tomorrow—and will it always be? Abstr. *428*, 36th Meet. Inst. Food Technol. Anaheim, Ca.

HALL, S. S., DOWEYKO, A. M. and JORDAN, F. 1976. Glyoxylase I enzyme studies. 2. Nuclear magnetic resonance evidence for an enediol-proton transfer mechanism. J. Am. Chem. Soc. *98*, 7460–7461.

HALLAWAY, M., PHETHEAN, P. D. and TAGGART, J. 1970. A critical study of the intracellular distribution of ascorbate oxidase and a comparison of the kinetics of the soluble and cell-wall enzyme. Phytochemistry *9*, 935–944.

HALLIDAY, D. A. 1972. Sausage meat changes in cold storage. Process Biochem. *7* (5) 27–28.

HALLIWELL, B. 1974. Superoxide dismutase, catalase and glutathione peroxidase: Solutions to the problems of living with oxygen. New Phytol. *76*, 1075–1086.

HALLIWELL, B. and BUTT, V. S. 1974. Oxidative decarboxylation of glycolate and glyoxylate by leaf peroxisomes. Biochem. J. *138*, 217–224.

HALLIWELL, B. and FOYER, C. H. 1976. Ascorbic acid, metal ions, and the superoxide radical. Biochem. J. *155*, 697–700.

HALLIWELL, G. 1974. Cellulose. *In* Methods of Enzymatic Analysis, Vol. 3. H.U. Bergmeyer (Editor). Academic Press, New York.

HALLIWELL, G. and GRIFFIN, M. 1973. The nature and mode of action of the cellulolytic component C_1 of *Trichoderma koningii* on native cellulose. Biochem. J. *135*, 587–594.

HALVORSON, H. O. and DEMAIN, A. L. 1976. American Society of Microbiology Conference on Genetics and Molecular Biology of Industrial Microorganisms, Feb. Orlando, Fla.

HAMILTON, B. K., COLTON, C. K. and COONEY, C. C. 1974. Glucose isomerase: A case study of enzyme-catalyzed process technology. *In* Immobilized Enzymes in Food and Microbial Processes. A.C. Olson and C.C. Cooney (Editors). Plenum Press, New York.

HAMILTON, G. A. 1969. Mechanisms of two and four electron oxidations catalyzed by some metallo-enzymes. Adv. Enzymol. *32*, 55–96.

HAMILTON, G. A. 1974. Chemical models and mechanisms for oxygenases. *In* Molecular Mechanisms of Oxygen Activation. O. Hayashi (Editor). Academic Press, New York.

HAMM, R., TETZLAFF, L. and SCHEPER, J. 1973. Biochemical detection of frozen meat in watery pale pork. Z. Lebensm. Unters. Forsch. *152*, 1–7. (German)

HAMMES, G. G. 1968. Relaxation spectrometry of enzymatic reactions. Acc. Chem. Res. *1*, 321–329.

HAMMES, G. G. 1972. Regulation of enzyme catalysis. Proc. Fed. Eur. Biochem. Soc. *25*, 103–118.

HAMMES, G. G. and SCHIMMEL, P. R. 1970. Rapid reactions and transient states. *In* The Enzymes, 3rd Edition, Vol. 2. P.D. Boyer (Editor). Academic Press, New York.

HANAFUSA, N. 1972. Denaturation of enzyme protein by freeze-thawing and freeze-drying. I. Freeze-thawing and freeze-drying of myosin and some other muscle proteins. Contrib. Inst. Low Temp. Sci. Ser. B, Hokkaido Univ. *27*, 1–38.

HANAFUSA, N. 1974. Roles of additives protecting enzymes from inactivities during lyophilization of enzyme proteins. Teion Kagaku, Seibutsu-Hen *32*, 1–8. (Japanese)

HANFORD, J. 1967. The proteolytic enzymes of wheat and flour and their effect on bread quality in the United Kingdom. Cereal Chem. *44*, 499–511.

HANG, Y. D., WILKENS, W. F., HILL, A. S., STEINKRAUSE, K. H. and HACKLER, L. R.

1970. Enzymatic modification of nitrogenous constituents of pea beans. J. Agric. Food Chem. *18*, 1083–1085.

HANNA, T. G. and LELIEVRE, J. 1975. An effect of lipid on the enzymatic degradation of wheat starch. Cereal Chem. *52*, 670–697.

HANSEN, A. M., BAILEY, T. A., MALZAHN, R. C. and CORMAN, J. 1955. Plant scale evaluation of fungal amylase process for grain alcohol. J. Agric. Food Chem. *3*, 866–872.

HANSEN, L. U. 1971. Automation study of β-N-acetylglucosaminidase as an indicator for egg white pasteurization. J. Food Sci. *36*, 600–603.

HANSEN, L. U. and JONES, F. T. 1977. A microscopic view of thermal-processed wheat flour. J. Food Sci. *42*, 1236–1242.

HANSEN, R. J., MORIN, D. and HOPE, W. G. 1978. Amino acids inhibit myofibrillar proteinase from rat skeletal muscle. Fed. Proc. *37*, 540.

HANSEN, S. E., KJAER, A. and SCHWIMMER, S. 1959. A continuous chromogenic method for the assay of C-S lyases with S-(2,4-dinitrophenyl)-6-cysteine as substrate. C.R. Trav. Lab. Carlsberg *31*, 193–206.

HANSON, C. H. 1974. The Effect of FDA Regulations (GRAS) on Plant Breeding and Processing. Spec. Publ. No. *5*, Crop Sci. Soc. Am., Madison.

HAQ, Q. N., HANNAN, A. and RAHMAN, J. 1975. Studies on the bark of jute plant. Bangladesh J. Sci. Ind. Res. *10*, 191–196.

HAQ, S. K., SIDDIQUI, I. H. and KHAN, A. H. 1969. Blanched water, a waste product of the shrimp canning industry. Pak. J. Sci. Ind. Res. *12*, 49–51.

HARADA, K. 1970. Enzyme system catalyzing the formation of formaldehyde and dimethylamine in tissues of fish and shellfish. Suisan Dagakko Kenkyu Hokuku *25*, 163–241 (Japanese). Chem. Abstr. *83* 75670r.

HARADA, K., YOICHI, A. and HAYAISHI, K. 1975. Base-catalyzed intramolecular rearrangement of trimethylamine *N*-oxide. J. Fac. Agric., Kyushu Univ. *19*, 159–168. (Japanese)

HARDIE, D. G., MANNERS, D. J. and YELLOWLEES, D. 1976. The limit dextrinase from malted sorghum *(Sorghum vulgare)*. Carbohydr. Res. *50*, 75–85.

HARDONK, M. J. and KOUDSTAAL, J. 1976. Enzyme histochemistry as a link between biochemistry and morphology. Prog. Histochem. Cytochem. *8* (2) 1–68.

HAREL, E. and MAYER, A. M. 1968. Interconversions of sub-units of catechol oxidase from apple chloroplasts. Phytochemistry 7, 199–204.

HARISTOY, D. 1977. Study by γ-radiolysis of the reactivity of the superoxide ion in the oxyhemoglobin-methemoglobin system INIS. Atomindex *8* (11) Abstr. *309662*.

HARPER, W. J., CARMONA, A. and KRISTOFFERSON, T. 1971. Protein degradation in Cheddar cheese slurries. J. Food Sci. *36*, 503–506.

HARRIS, R. and VON LOESECKE, S. B. 1971. Nutritional Evaluation of Food Processing (Reprinted). AVI Publishing Co., Westport, Conn.

HARRIS, S. E., SCHWARTZ, R. J., TSAI, M. -J. and O'MALLEY, B. W. 1976. Effect of estrogen on gene expression in the chicken oviduct. J. Biol. Chem. *251*, 524–529.

HART, T. G. and SMITH, O. 1960. Potato quality. 27. The role of phosphorus in potato chip browning. Am. Potato J. *43*, 158–172.

HARTLEY, B. S. 1971. Pancreatic elastase. *In* The Enzymes, 3rd Edition, Vol. 3. P.D. Boyer (Editor). Academic Press, New York.

HARTLEY, R. D., JONES, E. C. and FENLON, J. S. 1974. Prediction of the digestibility of forages by treatment of their cell walls with cellulolytic enzymes. J. Sci. Food Agric. *25*, 947–954.

HARVEY, R. J. 1974. Method for modifying sour and bitter taste. U.S. Pat. 3,849,555. Nov. 19.

HARWALKAR, V. R. 1972. Characterization of an astringent flavor fraction from cheddar cheese. J. Dairy Sci. *55*, 735–741.

HASEGAWA, S. 1976. Metabolism of limonoids. Limonin D-ring lactone hydrolase activity in *Pseudomonas*. J. Agric. Food Chem, *24*, 24–26.

HASEGAWA, S. and MAIER, V. P. 1980. Polyphenol oxidase of dates. J. Agric. Food Chem. *28*, 891–893.

HASEGAWA, S. and SMOLENSKY, D. C. 1970. Date invertase properties and activity associated with maturity and quality. J. Agric. Food Chem. *18*, 902–904.

HASEGAWA, S. and SMOLENSKY, D. C. 1971. Cellulase in dates and its role in fruit softening. J. Food Sci. *36*, 966–967.

HASEGAWA, T., PEARSON, A. M., PRICE, J. F., RAMPTON, J. H. and LECHOWICH, R. V. 1970. Effect of microbial growth upon sarcoplasmic and urea-soluble proteins from muscle. J. Food Sci. *35*, 720–724.

HASHIMOTO, S. and FUNATSU, M. 1976. Fractionation of subunits in xylanases from *Trichoderma viride* with a new simple preparative polyacrylamide gel electrophoresis apparatus. Agric. Biol. Chem. *40*, 635–636.

HASHIMOTO, S., MURAMATSU, T. and FUNATSU, M. 1971. Studies on xylanase from

Trichoderma viride. I. Isolation and some properties of crystalline xylanase. Agric. Biol. Chem. *35*, 501–508.

HASHIMOTO, Y. 1970. Studies on the treatment of coffee beans. IV. Effect of salt concentration on the hydrolysis of mannan and spent coffee grounds by mannanase. J. Agric. Chem. Soc. *44*, 287–292. (Japanese)

HASLING, V. C., CATALANO, E. A. and DEOBALD, H. J. 1973. Modified method of analysis of sweet potato α-amylase. J. Food Sci. *38*, 338–339.

HASSELBURGER, F. X., ALLEN, B., PARUCHURI, M. C. and COUGHLIN, R. W. 1974. Immobilized enzymes: Lactase bonded to stainless steel and other dense carriers for use in fluidized bed reactors. Biochem. Biophys. Res. Comm. *57*, 1054–1062.

HASSID, W. Z. 1969. Biosynthesis of oligosaccharides and polysaccharides in plants. Science *165*, 137–144.

HATANAKA, A., KAJIWARA, K., SEKIYA, J. and KODA, T. 1978. Specificity of enzyme systems producing C_6-aldehyde in *Thea* and *Farugium* chloroplasts. Phytochemistry *17*, 548–549.

HATANAKA, A., KAJIWARA, T. and TAKAHIRO, H. 1975. Biosynthetic pathway of cucumber alcohol: *trans*-2,*cis*-6-nonadienol via *cis*-3,*cis*-6-nonadienal. Phytochemistry *14*, 2589–2592.

HATANAKA, C. and OMURA, H. 1973. Browning and antioxidant activity of a *p*-benzoquinone/glycine system. J. Jpn. Soc. Food Nutr. *26*, 457–462. (Japanese)

HATCH, M. D., MAU, S. -L. and KAGAWA, T. 1974. Properties of leaf NAD malic enzyme from plants with C_4 pathway photosynthesis. Arch. Biochem. Biophys. *165*, 188–200.

HATTON, M. W. C. and REGOECZI, E. 1976. The proteolytic nature of commercial samples of galactose oxidase. Purification of the enzyme by a simple affinity method. Biochim. Biophys. Acta *438*, 339–346.

HATTORI, S., YAMASUCHI, Y. and KASINAWA, T. 1974. Preliminary study on the microbiological formation of fruit flavors. Proc. IV. Int. Congr. Food Sci. Technol. *1a*, 58.

HATTULA, T. and GRANROTH, B. 1974. Formation of dimethyl sulfide from S-methylmethionine in onion seedlings. J. Sci. Food Agric. *25*, 1517–1521.

HAUGE, S. M., WILBUR, J. W. and HILTON, J. H. 1937. A further study of the factor in soybeans affecting the vitamin A value of butter. J. Dairy Sci. *20*, 87–91.

HAUPT, H., KUEDERLING, D., STARK, J., KÖNIG, J. and RUSCHIG, H. 1974. Purification and crystallization of glucose oxidase. Behring Inst. Mitt. *54*, 68–71. (German)

HAVIR, E. A., TAMIR, H., RATNER, S. and WARNER, R. C. 1965. Biosynthesis of urea. XI. Preparation and properties of crytalline arginosuccinase. J. Biol. Chem. *240*, 3079–3088.

HAWKER, J. S. 1969. Changes in the activities of malic enzyme, malate dehydrogenase, phosphopyruvate carboxylase and pyruvate decarboxylase during the development of a non-climacteric fruit (the grape). Phytochemistry *8*, 19–23.

HAWKER, J. S., WALKER, R. R. and RUFFNER, H. P. 1976. Invertase and sucrose synthase in flowers. Phytochemistry *15*, 1441–1443.

HAWLEY, R. L. 1974. Non-isoelectric protein. U.S. Pat. 3,830,942. Aug. 20.

HAWTHORN, J. and TODD, J. P. 1955. Catalase in relation to the unsaturated-fat oxidase activity of wheat flour. Chem. Ind. 446–447.

HAY, P. P., HARRISON, D. L. and VAIL, G. E. 1953. Effects of meat tenderizer on less tender cuts of beef cooked by four methods. Food Technol. *7*, 217–220.

HAYAISHI, O. 1962. Oxygenases. Academic Press, New York.

HAYAISHI, O., SHIMAZONO, H., KATAGIRI, M. and SAITO, Y. 1956. Enzymatic formation of oxalate and acetate from oxaloacetate. J. Am. Chem. Soc. *78*, 5126–5127.

HAYAKAWA, K. -I., TIMBERS, G. E. and STIER, E. F. 1977. Influence of heat treatment on the quality of vegetables: organoleptic quality. J. Food Sci. *42*, 1286–1289.

HAYASHI, K. and MASARU, T. 1972. Some characteristics of hydrolysis of synthetic substrates and proteins by the alkaline proteinase from *Aspergillus sojae*. Agric. Biol. Chem. *36*, 1755–1765.

HAYASHI, T. 1940. Biochemical studies on "Bakanae" fungus of rice. VII. Effect of gibberellin on the activity of amylase in germinated cereal grain. Bull. Agric. Chem. Soc. Jpn. *16*, 531–533.

HAYASHIDA, S., NOMURA, T., YOSHINO, E. and HONGO, M. 1976. The formation and properties of subtilisin-modified glucoamylase. Agric. Biol. Chem. *40*, 141–146.

HAYDAR, M., STULE, L. and HADZIYEV, D. 1975. Oxidation of pea lipids by pea lipoxygenase. J. Food Sci. *40*, 807–814.

HAYMAN, E. P., YOKOYAMA, H. and POLING,

S. M. 1977. Carotenoid induction in orange endocarp. J. Agric. Food Chem. *25,* 1251–1253.

HEADY, R. E. and JACAWAY, W. A. 1974A. Process for the production of levulose-bearing syrups. U.S. Pat. 3,847,740. Nov. 12.

HEADY, R. E. and JACAWAY, W. A. 1974B. Temperature-programmed process for the production of levulose-bearing syrups. U.S. Pat. 3,847,741. Nov. 12.

HEATHERBELL, D. A. 1976. Haze and sediment formation in clarified apple juice and apple wine. Alimenta *15,* 151–154.

HEATHERBELL, D. A. and WROLSTAD, R. E. 1971. The enzymatic regeneration of volatile flavor components in carrots. J. Agric. Food Chem. *19,* 281–284.

HEBER, U. and KRAUSE, G. H. 1980. What is the physiological role of photorespiration? TIBS *5* (2) 32–34.

HEDEN, C. -G. 1974. 1973 Henniker Delphi study. *In* Enzyme Engineering. 2. E.K. Pye and L.B. Wingard, Jr. (Editors). Plenum Press, New York.

HEDENSKOG, G. and MOGREN, H. 1973. Some methods for processing of single-cell protein. Biotechnol. Bioeng. *15,* 129–142.

HEDIN, S. G. 1907. A case of specific adsorption of enzymes. Biochem. J. *2,* 112–116.

HEDLUND, B. E., HAIRE, R. N. and ROSENBERG, A. 1975. Thermodynamic aspects of regulation of hemoglobin oxygenation by 2,3-DPG and other anions. Fed. Proc. *34,* 603.

HEFTMANN, E., KROCHTA, J. M., FARKAS, D. F. and SCHWIMMER, S. 1972. The chromatofuge, an apparatus for preparative rapid radial column chromatography. J. Chromatogr. *66,* 365–369.

HEFTMANN, E. and SCHWIMMER, S. 1972. Degradation of tomatine to 3β-hydroxy-5α-pregn-16-ene-20-one by ripe tomatoes. Phytochemistry *11,* 2783–2787.

HEIMANN, W. and KLAIBER, V. 1977. The lipoperoxidase-isomerase-system from oats. Z. Lebensm. Unters. Forsch. *165,* 131–136. (German)

HEIMANN, W., FRANZEN, K. H., RAPP, H. and ULLEMEYER, H. 1976. Studies on the development of volatile substances during the lipoxygenase-linoleic acid reaction. Z. Lebensm. Unters. Forsch. *162,* 109–114. (German)

HEINEN, E. A. and VAN TWISK, P. 1975. Using natural enzymes in the manufacture of carrot and beetroot juice. S. Afr. Food Rev. *2* (1) 43, 47.

HEINTZE, K., DUDEN, R., FRICKER, A., PAULUS, K. and ZOHM, H. 1975. Influence of heat treatment of spinach at temperatures up to 100° on important constituents. III. Changes of the N-fraction. Lebensm.-Wiss. Technol. *8,* 17–19.

HEJGAARD, J. 1977. Origin of a dominant beer protein. Immunochemical identity with a β-amylase-associated protein from barley. J. Inst. Brew. *83,* 94–96.

HEJGAARD, J. 1978. "Free" and "bound" β-amylases during malting of barley. Characterization by two-dimensional immunoelectrophoresis. J. Inst. Brew. *84,* 43–46.

HEJGAARD, J. and CARLSEN, S. 1977. Immunoelectrophoretic identification of a heterodimer β-amylase in extracts of barley grain. J. Sci. Food Agric. *28,* 900–904.

HELBIG, N. B., HO, L., CHRISTY, G. E. and NAKAI, S. 1980. Debittering of milk hydrolysates by adsorption for incorporation into acidic beverages. J. Food Sci. *45,* 331–335.

HELINSKI, D. R. 1976. Hybrid plasmids as molecular vehicles for the amplification of DNA and gene products. Am. Soc. Microbiol. Conf. Genetics Molec. Biol. Ind. Microorganisms, Orlando, Fla., Feb.

HELINSKI, D. R., HERSHFIELD, V., FIGURSKI, D. and MEYER, R. M. 1977. Construction and properties of plasmid cloning vehicles. *In* Recombinant Molecules: Impact on Science and Society. R.F. Beers, Jr. and E.G. Basset (Editors). Raven Press, New York.

HELLENDOORN, E. W., NOORDHOFF, M. G. and SLAGMAN, J. 1975. Enzymatic determination of the indigestible residue (dietary fibre) content of human food. J. Sci. Food Agric. *26,* 1461–1468.

HEMPENIUS, W. L., VALENTI, J. and MOSER, W. L. 1974. Process of making a base for protein beverages. U.S. Pat. 3,846,560. Nov. 12.

HENDERSON, H. M. 1968. Chlorogenic acid in relation to the metabolism of starch in potatoes. Am. Potato J. *45,* 41–45.

HENDERSON, H. M., HERGENROEDER, K. and STUCHLY, S. S. 1975. Effect of 2450-MHz microwave radiation on horseradish peroxidase. J. Microwave Power *10,* 27–35.

HENDERSON, H. M. and McEWEN, T. J. 1972. Effect of ascorbic acid on thioglucosidases from different crucifers. Phytochemistry *11,* 3127–3133.

HENDERSON, H. R. and CONN, E. E. 1969. Cyanide metabolism in higher plants. IV. Purification and properties of β-cyanoalanine synthase of blue lupine. J. Biol. Chem. *224,* 2632–2640.

HENDERSON, J. R., BUESCHER, R. W. and

MORELOCK, T. E. 1977. Influence of genotype and CO_2 content on discoloration, phenolic content, peroxidase, and phenolase activities in snap beans. HortScience *12,* 453–454.

HENDERSON, R. and WONG, J. H. 1972. Catalytic configurations. Ann. Rev. Biophys. Bioeng. *1,* 1–25.

HENKIN, R. I, and BRADLEY, D. F. 1969. Regulation of taste acuity by thiols and metal ions. Proc. Natl. Acad. Sci. *62,* 30–37.

HENRI, B. 1903. General Laws of the Action of Diastases. Hermann, Paris. (French)

HENRY, S., KOCZAN, J. and RICHARDSON, T. 1974. Bacteriocidal effectiveness of immobilized peroxidases. Biotechnol. Bioeng. *16,* 289–291.

HERBERT, D. and PINSENT, J. 1948. Crystalline bacterial catalase. Biochem. J. *43,* 193–202.

HERMANN, K. 1976. Discoloration of vegetables by phenolic constituents. Dtsch. Lebensm. Rundsch. *72,* 90–94. (German)

HERMANSSON, A. M., OLSSON, D. and HOLMBERG, B. 1974. Functional properties of proteins for foods—modification studies on rapeseed protein concentrate. Lebensm. Wiss. Technol. *7,* 176–181. (German)

HERRLINGER, F. and KIERMEIER, F. 1944. Inactivation and regeneration of heat treated peroxidase solutions. Biochem. Z. *317,* 1–12.

HERRLINGER, F. and KIERMEIER, F. 1948. Inactivation and regeneration of peroxidase in heat-treated plant tissues. Biochem. Z. *318,* 413–424. (German)

HERRMANN, K. 1968. Effect of storage, processing and preservation on the vitamin content of vegetables. Ernaehr.-Umsch. *15,* 81–83. (German)

HERS, H. G. 1976. The control of glycogen metabolism in the liver. Annu. Rev. Biochem. *46,* 167–189.

HERSHKO, A. and FRY, M. 1975. Post-translational cleavage of polypeptide chains: role in assembly. Annu. Rev. Biochem. *44,* 775–797.

HESSE, A. 1935. Use of enzymes in industry. I, II. Ergeb. Enzymforsch. *3,* 95–193; *4,* 147–172. (German)

HESSE, A. 1940. Technology of enzymes. *In* Handbook of Enzymology. F.F. Nord and R. Weidenhagen (Editors). Akademische Verlagesilloschaft Leipzig. (German)

HETRICK, J. H. and TRACY, P. H. 1948. Effect of high-temperature short-time heat treatments on some properties of milk. J. Dairy Sci. *31,* 867–887.

HEVIA, P. and OLCOTT, H. S. 1977. Flavor of enzyme-solubilized fish protein concentrate. J. Agric. Food Chem. *25,* 772–775.

HEVIA, P., WHITAKER, J. R. and OLCOTT, H. S. 1976. Solubilization of a fish protein concentrate with proteolytic enzymes. J. Agric. Food Chem. *24,* 383–385.

HEWITT, E. J. 1963. Enzymatic enhancement of flavor. J. Agric. Food Chem. *11,* 14–19.

HEWITT, E. J. and HASSELSTROM, T. 1960. Natural flavor of processed foods. U.S. Pat. 2,924,521. Feb. 9.

HEWITT, E. J., MACKAY, D. A. M., KONIGSBACHER, K. and HASSELSTROM, T. 1956. The role of enzymes in food flavors. Food Technol. *10,* 487–489.

HEYNEKER, H. L., SHINE, J., GOODMAN, H. M., BOYER, H. W., ROSENBERG, J., DICKERSON, R. E., NARANG, S. A., ITAKURA, K., LIN, S. and RIGGS, A. D. 1976. Synthetic *lac* operator DNA is functional *in vivo.* Nature *263,* 748–752.

HEYNS, K. and KLIER, M. 1968. Browning reaction and fragmentation of carbohydrates. Carbohydr. Res. *6,* 436–448. (German)

HEYNS, K., RÖPER, H. and KOCH, H. 1974. The problem of nitrosamine formation by the reaction of monosaccharides with amino acids (Maillard reaction) in the presence of sodium nitrite. Z. Lebensm. Wiss. Forsch. *154,* 193–200. (German)

HIATT, R. R. 1975. Hydroperoxide destroyers and how they work. Crit. Rev. Food. Sci. Nutr. *7,* 1–12.

HICKS, C. L., BUCY, J. and STOFER, W. 1979. Heat inactivation of superoxide dismutase in bovine milk. J. Dairy Sci. *62,* 529–532.

HICKS, E. W. 1944. Note on the estimation of diurnal temperature fluctuation of reaction rates in stored foodstuffs and other materials. J. Counc. Sci. Ind. Res. (Australia) *17,* 111–114.

HIGGINS, M. J. P., KORNBLATT, J. A. and RUDNEY, H. 1972. Acyl-CoA ligase. *In* The Enzymes, 3rd Edition, Vol. 8. P.D. Boyer (Editor). Academic Press, New York.

HIGUCHI, R. 1973. Prevention of deterioration by oxygen of cellulose packaging materials. Jpn. Kokai 73 59,089. Aug. 18.

HILDRUM, K. I. and TJABERG, T. B. 1972. Texture loss in fermented citron used for candying. J. Food Technol. *7,* 379–386.

HILKER, D. M. 1976. Thiamine-modifying properties of fish and meat products. J. Nutr. Sci. Vitaminol. *22,* Suppl., 3–6.

HILL, A. V. 1910. The combination of hemoglobin with oxygen and carbon monoxide. Biochem. J. *7*, 471–480.

HILL, C. J. 1971. Isothiocyanate and 5-vinyloxazolidine-2-thione content in rapeseed and rapeseed meal made in five Chilean oil factories. Oli, Grassi, Deriv. *7*, 2–7. (Spanish)

HILL, J. M. 1970. The oxidation of pyridoxal and related compounds by pea-seedling extracts or systems containing peroxidase. Phytochemistry *9*, 725–734.

HILL, R. D. 1975. Superoxide dismutase activity in bovine milk. Aust. J. Dairy Technol. *30*, 26–28.

HILL, R. D., LAHAV, E. and GIVOL, D. 1974. Rennin-sensitive bond in α_{s1} β-casein. J. Dairy Res. *41*, 147–153.

HILL, R. L. and BREW, K. 1975. Lactose synthetase. Structure and function. Adv. Enzymol. *43*, 411–485.

HILL, R. L. and SMITH, E. L. 1956. Crystalline papain. VI. Extensive stepwise hydrolysis without loss of proteolytic activity. Biochem. Biophys. Acta *19*, 376–377.

HILL, R. L. and SMITH, W. R. 1962. The complete enzymic hydrolysis of proteins. J. Biol. Chem. *237*, 389–396.

HILL, W. M. and SEBRING, M. 1973. Desugarization. *In* Egg Science and Technology. W.J. Stadelman and D.J. Cotterill (Editors). AVI Publishing Co., Westport, Conn.

HILTZ, D. F., FRASER, D., BISHOP, J. and DYER, W. J. 1974. Accelerated nucleotide degradation and glycolysis during warming and subsequent storage at −5°C of prerigor, quick-frozen adductor muscle of the sea scallop. J. Fish. Res. Board. Can. *31*, 1181–1187.

HIMMELHOCH, S. R. 1971. Chromatography of proteins on ion-exchange adsorbents. Methods Enzymol. *22*, 273–286.

HINBERG, I., KORUS, R. and O'DRISCOLL, K. F. 1974. Gel entrapped enzymes: Kinetic studies of immobilized β-galactosidase. Biotechnol. Bioeng. *16*, 943–963.

HINCHCLIFFE, J., VAISEY, M., McDANIEL, M. and ESKIN, N. A. M. 1974. Flavor characteristics of fava bean flours in relation to lipid composition. Cereal Sci. Today *19*, 414.

HINKLE, P. C. and McCARTY, R. E. 1978. How cells make ATP. Sci. Am. *228*, 104–121.

HINRICHS, J. R. and WHITAKER, J. R. 1962. Enzymatic degradation of collagen. J. Food Sci. *27*, 250–254.

HINTON, D. M. and PRESSEY, R. 1974. Cellulase activity in peaches during ripening. J. Food Sci. *39*, 783–785.

HIRANPRADIT, S. and LOPEZ, A. 1976. Activity of α- and β-amylase at temperatures from 4° to −23°C. J. Food Sci. *41*, 138–144.

HIRATA, F., SCHIFFMANN, E., VENKATASUBRAMANIA, K., SALOMON, D. and AXELROD, J. 1980. A phospholipase A_2 inhibitory protein in rabbit neutrophils induced by glucosteroids. Proc. Natl. Acad. Sci. *77*, 2533–2536.

HITCHCOCK, C. and NICHOLS, B. W. 1971. Plant Lipid Biochemistry. Academic Press, New York.

HO, C. Y. and CLIFFORD, A. J. 1976. Digestion and absorption of bovine milk xanthine oxidase and its role as an aldehyde oxidase. J. Nutr. *106*, 1600–1609.

HOAGLAND, R. E., McBRIDE, C. N. and POWICK, W. C. 1917. Changes in fresh beef during storage. U.S. Dep. Agric. Bull. *433*.

HOBSON, G. E. 1963. Influence of nitrogen and potassium fertilizer on pectic enzymes in tomato fruit. J. Sci. Food Agric. *14*, 550–554.

HOBSON, G. E. 1964. Polygalacturonase in normal and abnormal tomato fruit. Biochem. J. *92*, 324–332.

HOBSON, G. E. 1968. Cellulase activity during the maturation and ripening of tomato fruit. J. Food Sci. *33*, 588–592.

HOBSON, G. E. and DAVIES, J. H. 1971. The tomato. *In* The Biochemistry of Fruits and Their Products. A.C. Hulmes (Editor). Academic Press, London.

HOCHSTER, R. M. and WATSON, R. W. 1954. Enzymatic isomerization of D-xylose to D-xylulose. Arch. Biochem. Biophys. *48*, 120–129.

HODGE, J. E. 1953. Chemistry of browning reactions in model systems. J. Agric. Food Chem. *1*, 928–943.

HODGE, J. E. 1967. Nonenzymatic browning reactions. *In* Symposium on Foods: The Chemistry and Physiology of Flavors. H.W. Schultz, E. A. Day and L.M. Libbey (Editors). AVI Publishing Co., Westport, Conn.

HODGE, J. E., MILLS, F. D. and FISHER, B. E. 1972. Compounds of browned flavor derived from sugar amine reaction. Cereal Sci. Today *17*, 34–38.

HOEFERT, L. 1975. Tubule in dilated cisternae of endoplasmic reticulum of *Thlaspi arvensis* (cruciferae). Am. J. Bot. *62*, 756–768.

HOELDTKE, R., BALIGA, B., ISSENBURG, P. and WYRTMAN, R. J. 1972. DOPA in rat food containing wheat and oats. Science *175*, 761–762.

HOER, R. A., FREDERIKSEN, C. W. and HAW-

LEY, R. L. 1972. Enzyme modified protein process. U.S. Pat. 3,694,221. Sept. 26.

HOFFMANN-OSTENHOFF, O. 1953. Suggestions for a more rational classification and nomenclature of enzymes. Adv. Enzymol. *14*, 219–260.

HOGAN, J. M. 1966. Meat tenderizing compositions. U.S. Pat. 3,235,468. Feb. 15.

HÖGBERG, J., ORRENIUS, S. and O'BRIEN, P. J. 1975. Further studies on lipid peroxide formation in isolated hepatocytes. Eur. J. Biochem. *59*, 449–445.

HOGG, J. L., MORRIS, R., III and DURRANT, N. A. 1978. Proton inventories of serine protease charge-relay model in aprotic solvents. J. Am. Chem. Soc. *100*, 1590–1594.

HÖHN, E. and SOLMS, J. 1975. Taste compounds of Saccharomyces yeast. I. Isolation of taste compounds. Lebensm.-Wiss. Technol. *8*, 206–211. (German)

HOLBACH, B. and WOLLER, R. 1977. Determination of glycerol content of wine using Rebelin's method and an enzymic method. Wein-Wiss. *32*, 212–218. (German)

HOLBROOK, J., BUCHER, J. and PENNIAL, R. 1976. The binding by an NAD-affinity matrix of contaminating dehydrogenases in cytochrome oxidase. Z. Physiol. Chem. *357*, 623–627.

HOLCENBERG, J. S. and ROBERTS, J. 1977. Enzymes as drugs. Annu. Rev. Pharmacol. Toxicol. *17*, 97–116.

HOLDEN, M. 1961. The breakdown of chlorophyll by chlorophyllase. Biochem. J. *78*, 364–395.

HOLDEN, M. 1963. The purification and properties of chlorophyllase. Photochem. Photobiol. *2*, 175–180.

HOLDEN, M. 1965. Chlorophyll bleaching by legume seeds. J. Sci. Food Agric. *16*, 312–325.

HOLDEN, M. 1974. Chlorophyll degradation products in leaf protein preparations. J. Sci. Food Agric. *25*, 1427–1432.

HOLLAND, R. R., REEDER, S. K. and PRITCHETT, D. E. 1976. Cloud stability test for pasteurized citrus juices. J. Food Sci. *41*, 812–815.

HOLLANDER, P. M. and ERNSTER, L. 1975. Studies on the reaction mechanism of DT diaphorase. Action of dead-end inhibitors and the effects of phospholipids. Arch. Biochem. Biophys. *169*, 560–567.

HOLLAUS, F., WIENINGER, L. and BRAUNSTEINER, W. 1977. Experiences with enzymic determination of raffinose by means of galactose dehydrogenase in beets and sugar factory products. Zucker *30*, 653–658. (German)

HOLLUNGER, F. and NIKLASSON, B. 1975. The use of affinity gels for the study of the ligand binding properties of mammalian cholinesterase. Croat. Chem. Acta *47*, 361–369.

HOLMES, D. G., DUERECH, J. W. and ERNSTROM, C. A. 1977. Distribution of milk clotting enzymes between curd and whey and their survival during Cheddar cheese making. J. Dairy Sci. *60*, 862–869.

HOLSINGER, V. H. 1978. Lactose-modified milk and whey. Food Technol. *32* (3) 35–40.

HOLZER, H. and HEINRICH, D. C. 1980. Control of proteolysis. Annu. Rev. Biochem. *49*, 63–91.

HONOLD, G. R., FARKAS, G. L. and STAHMANN, M. A. 1967. The oxidation-reduction enzymes of wheat. II. A quantitative investigation of the dehydrogenases. Cereal Chem. *44*, 373–382.

HONOLD, G. R. and STAHMANN, M. A. 1968. The oxidation-reduction enzymes of wheat. IV. Qualitative and quantitative investigations of the oxidases. Cereal Chem. *45*, 99–108.

HOOD, D. E. 1975. Pre-slaughter injection of sodium ascorbate as a method of inhibiting metmyoglobin formation in fresh beef. J. Sci. Food Agric. *26*, 85–90.

HOOGZAND, C. and DOESBURG, J. J. 1961. Effect of blanching on texture and pectin of canned cauliflower. Food Technol. *15*, 160–163.

HOOVER, E. G. and KANDER, P. A. 1963. Influence of specific compositional factors of potatoes on chipping color. Am. Potato J. *40*, 17–24.

HORECKER, B. L., MELLONI, E. and PONTREMOLI, S. 1975. Fructose 1,6-biphosphatase: properties of the neutral enzyme and its modification by proteolytic enzymes. Adv. Enzymol. *42*, 193–226.

HORI, S. and FUGONO, T. 1969. Enzymes for candied fruits. U.S. Pat. 3,482,995. Dec. 9.

HORIGOME, T. and KANDATSU, M. 1968. Biological value of protein allowed to react with phenolic compounds in the presence of *o*-diphenol oxidase. Agric. Biol. Chem. *32*, 1093–1102.

HORIKAWA, H., OKASAYU, M., WADA, A. and KASKUSAMA, M. 1971. Green pigments formed from chlorogenic acid and amino acids. I. Isolation and purification of green pigments by Sephadex column chromatography. J. Food Sci. Technol. (Japan) *18*, 115–118. (Japanese)

HORISBERGER, M., SOZZI, T. and POUSAZ, R. 1976. Process for production of an enzyme for curdling milk. Swiss Pat. 582,196. May 29.

HORISBERGER, M., SOZZI, T. and POUSAZ, R. 1977. Milk-curdling enzyme. Br. Pat. 1,469,579. Oct. 17.

HORIUCHI, T., FUKUSHIMA, D., SUGIMOTO, H. and HATTORI, T. 1978. Studies on enzyme-modified proteins as foaming agents: Effect of structure on foam stability. Food Chem. *3*, 35–42.

HÖRL, W. H., JENNISSEN, H. P. and HEILMEYER, L. M. G., JR. 1978. Evidence for the participation of a Ca^{2+}-dependent protein kinase and a protein phosphatase in the regulation of the Ca^{2+} transport ATPase of the sarcoplasmic reticulum. Biochemistry *17*, 759–771.

HORMAN, I. and CAZENAUE, P. 1976. Increasing tea aroma. Ger. Offen. 2,462,303. Nov. 23.

HORNBY, W. E., INMAN, D. J. and McDONALD, A. 1972. The preparation of some immobilized dehydrogenases and their use in automated analysis. FEBS Lett. *23*, 114–116.

HORNSTEIN, I. and TERANISHI, R. 1967. The chemistry of flavor. Chem. Eng. News *45*, Apr. 3, 92–108.

HOROWITZ, R. 1964. Relations between the taste and structure of some phenolic glycosides. *In* Biochemistry of Phenolic Compounds. J.B. Harborne (Editor). Academic Press, New York.

HOROWITZ, R. and GENTILI, B. 1963. Dihydrochalcone derivatives and their use as sweeteners. U.S. Pat. 3,087,821. Apr. 30.

HORVATH, C. and ENGASSER, J. -M. 1974. External and internal diffusion in heterogeneous systems. Biotechnol. Bioeng. *16*, 909–923.

HORVATH, C., LEON, L., SANSUR, M. and SNYDER, L. 1975. The adaptation of immobilized enzymes for use in high speed continuous flow systems. Clin. Chem. *21*, 1017.

HORVATH, F. M., ZOLTAN, S., INCTEFI, I. and HORVATH, I. 1976. Isolation of enzymes with polymers. *In* 5th Intern. Ferment. Symp., Fed. Rep. Ger., Berlin.

HORWOOD, J. F. 1975. Cheese flavor: what is it? CSIRO Food Res. Q. *35*, 13–17.

HÖSEL, W. and NAHRSTEDT, A. 1975. Specific glucosidases for the cyanogenic glucoside triglochinin. Purification and characterization of β-glucoses from *Alocasia macrorrhiza* Schott. Z. Physiol. Chem. *356*, 1265–1275. (German)

HOSHINO, M., HIROSE, Y., SANO, K. and MITSUGI, K. 1975. Adsorption of microbial beta amylase EC-3.2.1.2 on starch. Agric. Biol. Chem. *39*, 2415–2416.

HOSTINOVA, E. and ZELINKA, J. 1975. Adsorption of α-amylase from *Streptomyces aureofaciens* on crosslinked starches. Staerke *27*, 343–346.

HOUGH, J. S. 1976. Silica hydrogels for chillproofing beer. Tech. Q. Master Brew. Assoc. Am. *13*, 34–39.

HOUGH, J. S., WADESON, A. and DANIELS, N. W. R. 1976. A new look at brewing flour. Brew. Guardian *105*, 38–39.

HOWLERDA, R. A., WHERLAND, S. and GRAY, H. B. 1976. Electron transfer reactions of copper proteins. Annu. Rev. Biophys. Bioeng. *5*, 363–396.

HRAZDINA, G. 1974. Reactions of anthocyanins in food products. Lebensm.-Wiss. Technol. *7*, 193–198. (German)

HSU, C. P., DESHPANDI, S. N. and DESROSIER, N. W. 1965. Role of pectin methyl esterase in firmness of canned tomatos. J. Food Sci. *30*, 583–588.

HSU, H. W., VAVAK, D. L., SATTERLEE, L. D. and MILLER, G. A. 1977. A multienzyme technique for estimating protein digestibility. J. Food Sci. *42*, 1269–1273.

HSU, L. H., FLORA, R. M. and BUNGAY, H. R. 1975. Affinity chromatography for removal of enzyme contaminants. Enzyme Technol. Dig. *4* (1) 4–12.

HSU, T. A., LADISCH, M. R. and TSAO, G.T. 1980. Alcohol from cellulose. CHEMTECH *10*, 315–319.

HSU, T. A. and TSAO, G. T. 1979. Convenient method for studying enzyme kinetic. Biotechnol. Bioeng. *21*, 2235–2246.

HUANG, H. I. 1955. Decolorization of anthocyanins by fungal enzymes. J. Agric. Food Chem. *3*, 141–146.

HUANG, H. T. and DOOLEY, J. G. 1976. Enhancement of cheese flavors with microbial esterases. Biotechnol. Bioeng. *18*, 909–919.

HUBER, H. 1960. Color changes by enzymes and acids. Brot Gebäck *14* (9) 165–172. (German)

HUBERT, P., DELLACHERIE, E., NEEL, J. and BAULIEU, E. -E. 1976. Affinity partitioning of steroid binding proteins. The use of polyethylene oxide-bound estradiol for purifying a 3-oxosteroid isomerase. FEBS Lett. *65*, 169–174.

HUDSON, C. S. and HARDING, T. S. 1915. Estimation of raffinose by enzymatic hydrolysis. J. Am. Chem. Soc. *37*, 2193–2198.

HUDSON, D. E. 1975. Shipping potatoes. Agric. Res. *23* (7) 15.

HUDSON, M. A., SHARPLES, V. J. and GREG-

ORY, M. E. 1974. Quality of home frozen vegetables. II. Effects of blanching and/or cooling in various solutions on conversion of chlorophyll. J. Sci. Food Agric. *9*, 105–114.

HUE, D., CORVOL, P., MENARD, J. and SICARD, P. J. 1976. Affinity chromatography in the purification of diagnostic enzymes. Process Biochem. *11* (6) 20–24.

HUEBNER, F. R., BIETZ, J. A. and WALL, J. S. 1977. Disulfide bonds: Key to wheat protein functionality. *In* Protein Crosslinking. Biochemical and Molecular Aspects. M. Friedman (Editor). Plenum Press, New York.

HUFFMAN, D. L., PALMER A. Z., CARPENTER, J. W. and SHIRLEY, R. L. 1961. The effect of antemortem injection of papain on tenderness of chickens. Poultry Sci. *40*, 1627–1630.

HUGHES, J. C., GRANT, A. and FAULKE, R. M. 1975. Texture of cooked potatoes. The effect of ion and pH on the compression strength of cooked potatoes. J. Sci. Food Agric. *26*, 731–738.

HULME, A. C. 1956. Carbon dioxide injury and presence of succinic acid in apples. Nature *178*, 218–219.

HULT, K., HÖKBY, E. and GATENBECK, S. 1977. Analysis of ochratoxin B alone and in the presence of ochratoxin A, using carboxypeptidase A. Appl. Environ. Microbiol. *33*, 1257–1277.

HULTIN, E. 1955. The influence of temperature on the rate of enzymic processes. Acta Chem. Scand. *9*, 1700–1710.

HULTIN, H. O. 1972. Enzymic activity and control as related to subcellular localization. J. Food Sci. *37*, 524–529.

HULTIN, H. O. 1974. Characteristics of immobilized multi-enzyme systems. J. Food Sci. *39*, 647–652.

HULTIN, H. O., ELMANN, J. D. and MELNICK, R. L. 1972. Modification of kinetic properties of lactic dehydrogenase by subcellular associations and possible role in the control of glycolysis. J. Food Sci. *37*, 269–273.

HULTIN, H. O. and HAARD, N. F. 1972. Biochemical control systems in food tissue. Symposium introduction. J. Food Sci. *37*, 503.

HULTIN, H. O. and LEVINE, A. S. 1965. Pectin methyl esterase in the ripening banana. J. Food Sci. *30*, 917–921.

HULTIN, H. O. and MILNER, M. 1978. Postharvest Biology and Biotechnology. Food and Nutrition Press, Westport, CT.

HULTIN, H. O. and PROCTOR, B. E. 1961. Banana aroma precursors. Food Technol. *16*, 111–113.

HUMME, H. E. 1972. The optimum pH for the limited specific hydrolysis of kappa-casein by rennin (primary phase of milk clotting). Neth. Milk Dairy J. *26*, 180–185.

HUNT, M. C., SMITH, R. A., KROPF, D. H. and TUMA, H. J. 1975. Factors affecting showcase color stability of frozen lamb in transparent film. J. Food Sci. *40*, 637–640.

HUNTER, I. R., WALDEN, M. K., SCHERER, J. R. and LUNDIN, R. E. 1969. Preparation and properties of 1,4,5,6-tetrahydroacetopyridine, a cracker-odor constituent of bread aroma. Cereal Chem. *46*, 189–195.

HUNTER, R. S. 1975. The Measurement of Appearance. John Wiley & Sons, New York.

HUNTING, W. M., GAGNON, M. and ESSELEN, W. B. 1959. New method for peroxidase determination. Anal. Chem. *31*, 143–144.

HUPKES, J. V. and VAN TILBURG, R. 1976. Production and properties of an immobilized glucose isomerase. Staerke *28*, 350–360.

HURST, D. T. 1972. Recent developments in the study of non-enzymic browning and its inhibition by sulphur dioxide. Sci. Tech. Surveys, Br. Food Manuf. Res. *75*.

HURST, R. L. 1977. Dry isomerase activation. U.S. Pat. 4,026,764. May 31.

HUSSAIN, A. and SHAH, A. H. 1975. Activity of pectic enzymes (pectinesterase and polygalacturonase) during the ripening of guava fruit. Pak. J. Agric. Sci. *12*, 191–194.

HUTCHINS, B. K., LIU, T. H. P. and WATTS, B. M. 1967. Effect of additives and refrigeration on reducing activity, metmyoglobin and malonaldehyde in raw ground beef. J. Food Sci. *32*, 214–217.

HUXLEY, H. E. 1965. The mechanism of muscular contraction. Sci. Am. *213* (6) 18–27.

HUXOLL, C. C., DIETRICH, W. C. and MORGAN, A. I., JR. 1970. Comparison of microwave with steam or water blanching of corn-on-the-cob. Food Technol. *47*, 84–87.

HWANG, D. H., CHABOT, J. F., HOOD, L. F. and KINSELLA, J. E. 1977. Ultrastructural changes in *Penicillium roqueforti* during germination and growth. J. Food Biochem. *1*, 3–14.

HYODO, H. and URITANI, I. 1967. Properties of polyphenol oxidases produced in sweet potato tissue after wounding. Arch. Biochem. Biophys. *122*, 299–309.

HYSERT, D. W. 1975. Recent advances in enzymology of relevance to brewing. Proc. Am. Soc. Brew. Chem. *33*, 114–118.

HYSLOP, D. B., SWANSON, A. M. and LUND, D. B. 1975. Heat inactivation of milk clotting enzymes. J. Dairy Sci. *58*, 795.

HYSLOP, D. B., SWANSON, A. M. and LUND, D. B. 1979. Heat inactivation of milk-clotting enzymes at different pH. J. Dairy Sci. *62,* 1227–1232.

ICHIKAWA, Y., SASA, H. and MICHI, K. 1973. Purification of ginger protease. Chem. Abstr. *80,* 92529g. (Japanese)

IDA, S., KOBAYAKAWA, K. and MORITA, Y. 1976. Ferredoxin-Sepharose affinity chromatography for the purification of assimilatory nitrite reductase. FEBS Lett. *65,* 305–307.

IDIEM POUTE, F. 1975. Lipase activity in germinating seedlings of *Cucumeropsis edulis.* J. Exp. Bot. *26,* 379–386.

IFT. 1974. The effects of processing on nutritional values. Sci. Status Summary, Inst. Food Technol. Expert Panel Food Safety Nutr. Food Technol. *28* (10) 77–80.

IFT. 1975. Naturally occurring toxicants in foods. Scientific Status Summary, Institute of Food Technol. Expert Panel Food Safety Nutr. Food Technol. *29* (3) 67–72.

IGLESIAS, H. A. and CHIRIFE, J. 1976. Prediction of the effect of temperature on water sorption isotherms of food material. J. Food Technol. *11,* 109–116.

IIBUCHI, S., MINODA, Y. and YAMADA, K. 1972. Hydrolyzing pathway, substrate specificity and inhibition of tannin acyl hydrolase of *Asp. oryzae* No. 7. Agric. Biol. Chem. *36,* 1553–1562.

IIZUKA, H., AYUKAWA, Y., SWEKANE, S. and KANNO, M. 1971. Production of extracellular isomerase by Streptomyces. U.S. Pat. 3,622, 463. Nov. 23.

IKEDIOBI, C. O. and SNYDER, H. E. 1977. Cooxidation of β-carotene by an isozyme of soybean lipoxygenase. J. Agric. Food Chem. *25,* 124–127.

IKEHATA, H. 1960. The purification of thiaminase II. J. Gen. Biol. *6,* 30–36.

ILANY, J. and METZER, A. 1976. Milk clotting activity of proteolytic enzymes. J. Dairy Sci. *52,* 43–46.

IMONDI, A. R. and STRADLEY, R. P. 1974. Utilization of enzymatically hydrolyzed soybean protein and crystalline amino acid diets by rats with exocrine pancreatic insufficiency. J. Nutr. *104,* 793–801.

INABA, T., SHINDO, N. and FUJI, M. 1976. Purification of cathepsin B from squid liver. Agric. Biol. Chem. *40,* 1159–1165.

INESI, G. and SCARPA, A. 1972. Fast kinetics of ATP-dependent Ca^{++}-uptake by fragmented sarcoplasmic reticulum. Biochemistry *11,* 356–359.

INGALLS, R. G., SQUIRES, R. G. and BUTLER, L. G. 1975. Reversal of enzymatic hydrolysis: Rate and extent of ester synthesis as catalyzed by chymotrypsin and subtilisin Carlsberg at low water concentrations. Biotechnol. Bioeng. *17,* 1627–1637.

INGLE, M. and HYDE, J. F. 1968. Effect of bruising on discoloration and concentration of phenolic compounds in apple tissue. Proc. Am. Soc. Hortic. Sci. *93,* 788–745.

INGLE, M. B. and BOYER, E. W. 1976. Production of industrial enzymes by Bacillus species. *In* Microbiology, 1976. D. Schlessinger (Editor). Am. Soc. Microbiol., Washington, D.C.

INGLETT, G. E. 1974. Symposium: Sweeteners. AVI Publishing Co., Westport, Conn.

INGLIN, M., FEINBERG, B. A. and LOWENBERG, J. R. 1980. Partial purification and characterization of a new intracellular β-glucosidase of *Trichoderma reesei.* Biochem. J. *185,* 515–519.

INGRAHAM, L. L. 1954. Reaction-inactivation of polyphenol oxidase. Temperature dependence. J. Am. Chem. Soc. *76,* 3377–3780.

INGRAHAM, L. L. 1962. Biochemical Mechanisms. John Wiley & Sons, New York.

INGRAHAM, L. L. and MAKOWER, B. 1954. Variation of the Michaelis constant with the concentrations of the reactants in an enzyme-catalyzed system. J. Phys. Chem. *58,* 266–270.

INKLAAR, P. A. 1977. Use of enzymes in analysis of additives in meat products. Voedingmiddelentechnol. *10* (6) 9–11. (Dutch)

INOUE, M., OKADA, S. and FUKOMOTO, J. 1970. Studies on juice-clarifying enzymes. III. Accelerating effect of hemicellulase III on the clarification of suspension of insoluble particles from orange juice. Nippon Nogei Kagaku Kaishi *44,* 8–14. (Japanese)

IOANNOU, J., CHISM, G. and HAARD, N. F. 1974. Molecular species of phosphorylase in postharvest potato tubers. J. Food Sci. *38,* 1022–1023.

IODICE, A. A. 1967. The carboxypeptidase nature of cathepsin A. Arch. Biochem. Biophys. *121,* 241–242.

IRVINE, G. N. 1955. Some effects of semolina lipoxidase activity on macaroni quality. J. Am. Oil Chem. Soc. *32,* 558–561.

IRVINE, G. N. and ANDERSON, J. A. 1953. Variation in principal quality factors of durum wheats with a quality prediction test for wheat or semolina. Cereal Chem. *30,* 334–342.

IRVINE, G. N. and WINKLER, C. A. 1950. Factors affecting the color of macaroni. Cereal Chem. *27,* 205–218.

ISHAY, J., FISCHL, J. and ALPERN, H. 1976.

Study of honeybee caste differentiation by glucose level differences during development. Insectes Soc. *23*, 23–28.

ISHERWOOD, F. A. 1973. Starch-sugar interconversion in *Solanum tuberosum.* Phytochemistry *12*, 2579–2591.

ISHERWOOD, F. A. 1976. Mechanism of starch-sugar interconversion in *Solanum tuberosum.* Phytochemistry *15*, 33–41.

ISHERWOOD, F. A. and BURTON, W. J. 1975. The effect of senescence, handling, sprouting, and chemical sprout suppression on respiratory quotient of potato tubers. Potato Res. *18*, 98–104.

ISHII, K., TAKAGI, S. and SADO, N. 1968. A study of the stability of ribonucleotides in dehydrated soups. Abstr. No. *90*, 28th Meet. Inst. Food Technol., Philadelphia. May 19–24.

ISHII, S. and YOKOTSUKA, T. 1971. Maceration of plant tissues by pectin *trans*-eliminase. Agric. Biol. Chem. *35*, 1157–1159.

ISHII, S. and YOKOTSUKA, T. 1972. Clarification of fruit juice by pectin transeliminase. J. Agric. Food Chem. *20*, 787–791.

ISHII, S. and YOKOTSUKA, T. 1973. Susceptibility of fruit juice to enzymatic clarification by pectin lyase and its relation to pectin in fruit juice. J. Agric. Food Chem. *21*, 269–272.

ISHIMARU, A. and YAMAZAKI, I. 1977. Hydroperoxide-dependent hydroxylation involving "H_2O_2-reducible hemoprotein" in microsomes of pea seeds. J. Biol. Osaka City Univ. *252*, 6118–6124.

ISHIMURA, Y., ULLRICH, V. and PETERSON, J. A. 1971. Oxygenated cytochrome P-450 and its possible role in enzymic hydroxylation. Biochem. Biophys. Res. Comm. *42*, 140–146.

ISMAIL, A. A., AHMED, N. S. and KHORSHID, M. A. 1975. Effect of gamma-ray irradiation on some enzyme activities of buffalo milk. Milchwissenschaft *30*, 423–424.

ISRAILI, K. H. and SMISSMAN, E. E. 1977. Synthesis of potential specific inhibitors of certain amino acid decarboxylases. J. Chem. Eng. Data *22*, 357–359.

IVANITSKII, A. M. 1973. Toxicological research of enzymatic preparations intended for use in the food industry. Vopr. Pitan. *32*, 39–46. (Russian)

IVERSEN, T. H. 1973. Myrosinase in cruciferous plants. Elec. Microsc. Enzymes. *1*, 131–149.

IWAMI, K., YASUMOTO, K. and MITTSUDA, H. 1975. Enzymatic cleavage of cysteine sulfoxide in *Lentinus edodes.* Agric. Biol. Chem. *39*, 1947–1955.

IWATA, S. and FUKUI, T. 1973. The subunit structure of α-glucan phosphorylase from potato. FEBS Lett. *36*, 222–226.

IWATSUBO, T., SEKIGUCHI, K., KURATA, K., TADA, T., IKI, K., NAKAGAWA, H., OGURA, N. and TAKEHANA, H. 1976. Increase of β-fructofuranosidase content in tomato fruit during the ripening process. Agric. Biol. Chem. *40*, 1243–1244.

IZUMI, K., ITO, T. and FUKAZAWA, T. 1977. Isometric tension development of glycerinated fibers prepared from normal and PSE porcine muscles and the effect of myosin irrigation on the tension development of "ghost" fibers. J. Food Sci. *42*, 113–116.

JAARMA, M. 1958. Influence of ionizing radiation on potato tubers. Ark. Kemi *13*, 97–105.

JACKS, J. T. and YATSU, L. Y. 1974. Phosphate-inhibition of lipase activity in peanuts. J. Am. Oil Chem. Soc. *51*, 112–113.

JACOBER, L. F. and RAND, A. G., JR. 1980. Characterization of the carbohydrate degrading enzymes in the surf clam crystalline style. J. Food Sci. *45*, 381–385.

JACOBSEN, J. J., YAMAGUCHI, Y. and MANN, L. K. 1968. An alkyl-cysteine sulfoxide lyase in *Tulbaghia violacea* and its relation to other alliinase-like enzymes. Phytochemistry *7*, 1099–1108.

JACOBSEN, J. J., SCANDALIOS, J. G. and VARNER, J. E. 1970. Multiple forms of amylase induced by gibberellic acid in isolated barley aleurone layers. Plant Physiol. *45*, 367–371.

JACOBSON, A. and CORCORAN, M. R. 1977. Tannins as gibberellin antagonists in the synthesis of α-amylase and acid phosphatase by barley seeds. Plant Physiol. *59*, 129–133.

JACOBY, M. 1900. On the aldehyde oxidizing ferment of the liver and the adrenal gland. Z. Physiol. Chem. *30*, 135–173. (German)

JADHAV, S. J. and SALUNKHE, D. K. 1974. Effects of certain chemicals on photoinduction chlorophyll and glycoalkaloid synthesis and on sprouting of potato tubers. Can. Inst. Food Sci. Technol. J. *7*, 178–182.

JAEGER, P. 1976. Saccharifying agents for distilleries—gram malt, cured malt, commercial enzymes. FSTA *8*, 8 H1399. (German)

JAEGGI, K. and KRASNOBAJE, W. 1977. Process for making flavorants from milk products and compositions containing same. U.S. Pat. 4,001,437. Jan. 4.

JAFFE, E. R. and NEUMANN, G. 1968. Hereditary methemoglobinemia, toxic methemoglobinemia, and the reduction of methemoglobin.

Ann. N.Y. Acad. Sci. *151*, 795–806.

JAHNS, F. D., HOWE, J. L., COUDURI, R. J., JR., and RAND, A. G., JR. 1976. A rapid visual enzyme test to assess fish freshness. Food Technol. *30* (7) 27–30.

JAKOB, M., HIPPLER, R. and LÜTHI, H. R. 1973. The influence of pectolytic enzyme preparations on the aroma of apple juice. Lebensm. Wiss. Technol. *6*, 138–141. (German)

JAKOBY, W. B. 1971A. Enzyme Purification and Related Techniques. Methods in Enzymology, Vol. 22. Academic Press, New York.

JAKOBY, W. B. 1971B. Crystallization as a purification technique. Methods Enzymol. *22*, 248–252.

JAKOBY, W. B. and WILCHEK, M. 1974. Affinity Techniques. Enzyme Purification: Part B. Methods in Enzymology, Vol. 34. Academic Press, New York.

JANKOV, S. I. and KIROV, M. B. 1972. Thermostability of polyphenoloxidases in juices from different grape varieties. Confructa *17*, 4–7. (German)

JANSEN, J. F. 1948. The isolation and identification of 2,2-dithiolisobutyric acid from asparagus. J. Biol. Chem. *176*, 657–664.

JANSEN, E. F. and BALLS, R. K. 1951. Enzymes in foods and feeds. *In* Yearbook of Agriculture, 1950, 1951. U.S. Dep. Agric., Washington, D.C.

JANSEN, E. F., JANG, R. and BONNER, J. 1960. Orange pectinesterase binding and activity. Food Res. *25*, 64–72.

JANSEN, E. F., NUTTING, M. F., JANG, R. and BALLS, A. K. 1950. Mode of inhibition of chymotrypsin by diisopropyl fluorophosphate. J. Biol. Chem. *185*, 209–225.

JANSEN, F. W. and RUELIUS, H. W. 1968. Alcohol oxidase, a flavoprotein from several *Basidiomycetes* species. Crystallization by fractional precipitation with polyethylene glycol. Biochim. Biophys. Acta *151*, 330–342.

JANSON, J. -C. 1977. Large scale chromatography of protein. Intern. Workshop Technol. Protein Separation. Reston, Va., Sept. 7–9. Pharmacia Fine Chemical AB, Uppsala.

JARENBÄCK, L. and LILJEMARK, A. 1975. Ultrastructural changes during frozen storage of cod. II. Effects of linoleic acid and hydroperoxides on myofibrillar proteins. Food. Technol. *10*, 437–452.

JARVIS, M., DALZIEL, J. and DUNCAN, H. J. 1974. Variation in free sugars and in different potato varieties during low temperature storage. J. Sci. Food Agric. *25*, 1405–1409.

JARVIS, R. F. 1979. The future outlook for enzymes in immunoassay. Antibiot. Chemother. (Basel) *26*, 105–117.

JAY, J. M. 1966. Influences of postmortem conditions on muscle microbiology. *In* Physiology and Biochemistry of Muscle as a Food. E.J. Briskey, R.G. Cassens and J.C. Trautman (Editors). Univ. of Wisconsin Press, Madison.

JEDRYCHOWSKI, L., POZNANSKI, S. and JAKBOWSKI, J. 1975. Use of enzyme preparations of microbial origin in continuous cheesemaking. Milchwissenschaft *30*, 688–673. (German)

JEFFERS, H. C. and RUBENTHALER, G. L. 1974. Effects of gibberellic acid on the amylase activity of malted wheat and comparison of methods for determining amylase activity. Cereal Chem. *51*, 772–779.

JEFFREYS, G. A. 1970. Fish protein concentrate. U.S. Pat. 3,547,652. Dec. 15.

JELEN, P. and BREENE, W. M. 1973. Texture improvement of fresh-pack dill pickles by addition of lactose and sucrose. J. Food Sci. *38*, 99–101.

JEN, J. J. and GRAHAM, J. 1975. Glutamic oxalacetic transaminase in peaches during maturation. J. Food Sci. *40*, 934–936.

JENCKS, W. P. 1975. Binding energy, specificity and enzymic catalyses: the Circe effect. Adv. Enzymol. *43*, 318–340.

JENEVEIN, E. P., JR. 1971. Antemortem method of tenderizing meat. U.S. Pat. 3,577,242. May 4.

JENSEN, M. H. 1979. Chemical and textural changes resulting from freeze drying of minced cod flesh. Lebensm.-Wiss. Technol. *12*, 342–345.

JENSEN, R. A. 1969. Metabolic interlock. Regulatory interactions exerted between biochemical pathways. J. Biol. Chem. *244*, 2816–2823.

JENSEN, R. G. and BAHR, J. T. 1977. Ribulose 1,5-bisphosphate carboxylase-oxygenase. Annu. Rev. Plant Physiol. *28*, 379–400.

JERUMANIS, J., VANHUYNH, N. and DEVREUX, A. 1976. Activity of polyphenoloxidase during malting of barley. Cerevisia *1*, 31–36.

JESSE, E. V. and ZEPP, G. A. 1977. Sugar policy options for the United States. U.S. Dep. Agric., Econ. Res. Serv. Agric. Econ. Rep. *351*.

JODLBAUER, H. D. 1976. Quantitative determination of monoglycerides in pasta products. Getreide Mehl Brot *30*, 181–187. (German)

JOFFE, F. M. and BALL, O. C. 1962. Kinetic and energetics of thermal inactivation and the regeneration of a peroxidase system. J. Food Sci. *27*, 587–592.

JOHNSON, B. and DANIELS, F. D., JR. 1974. Enzyme studies in experimental cryosurgery of the skin. Cryobiology *11*, 222–232.

JOHNSON, C. B., WHITTINGTON, W. J. and BLACKWOOD, G. C. 1976. Nitrate reductase as a possible predictive test of crop yield. Nature *262*, 133–134.

JOHNSON, D. A. and TRAVIS, J. 1976. Rapid purification of trypsin and chymotrypsin. Anal. Biochem. *72*, 573–576.

JOHNSON, D. C., NICHOLSON, M. D. and HAIGH, F. C. 1976. Dimethyl sulfoxide/paraformaldehyde: a nondegrading solvent for cellulose. Appl. Polymer Symp. *28*, 931–943.

JOHNSON, E. A., VILLA, T. G., LEWIS, M. J. and PHAFF, H. J. 1979. Lysis of the cell wall of the yeast *Phaffia rhodozyma* by a lytic enzyme complex from *Bacillus circulans* WL-12. J. Appl. Biochem. *1*, 273–282.

JOHNSON, F. H., EYRING, H. and POLISSAR, M. J. 1954. The Kinetics of Molecular Biology. John Wiley & Sons, New York.

JOHNSON, F. H., EYRING, H. and STOVER, B. J. 1974. The Theory of Rate Processes in Biology and Medicine. John Wiley & Sons, New York.

JOHNSON, J. A. and MILLER, B. S. 1951. Fungal Enzymes Help You Make Better Bread. McGraw Hill Publishing Co., New York.

JOHNSON, J. A. and MILLER, B. S. 1961. Browning in baked products. Bakers Dig. *35*, (5) 52–59.

JOHNSON, J. L., JONES, H. P. and RAJAGOPLAN, K. R. 1977. In vitro reconstitution of demolybdosulfite oxidase by a molybdenum cofactor from rat liver and other sources. J. Biol. Chem. *252*, 4494–5003.

JOHNSON, M. M. and DIVEN, W. F. 1969. An integrated rate equation for determining initial velocities. J. Theor. Biol. *25*, 331–338.

JOHNSON, R. H. and WELCH, E. A. 1968. Purified lipases for antistaling. U.S. Pat. 3,368,903. Feb. 13.

JOHNSTON, W. W. 1941. Tryptic enzymes from certain commercial fishes. J. Fish. Res. Board Can. *5*, 217–226.

JOLLEY, R. L., JR., EVANS, L. H., MAKINO, N. and MASON, H. S. 1974. Oxytyrosinase. J. Biol. Chem. *249*, 335–345.

JOLLEY, R. L., JR., ROBB, D. A. and MASON, H. S. 1969. The multiple forms of mushroom tyrosinase. J. Biol. Chem. *244*, 1593–1599.

JOLLY, R. C. and KOSIKOWSKI, F. V. 1975. Quantification of lactones in ripening pasteurized milk blue cheese containing added microbial lipases. J. Agric. Food Chem. *23*, 1175–1176.

JOLY, M. 1965. A Physical-chemical Approach to the Denaturation of Proteins. Molecular Biology, Vol. 6. Academic Press, New York.

JONES, C. R. 1940. The production of mechanically damaged starch in milling as a governing factor in the diastatic activity of flour. Cereal Chem. *17*, 133–169.

JONES, J. G. and MERCIER, P. L. 1974. Refined papain. Process Biochem. *9* (6) 21–24.

JONES, P. and MIDDLEMISS, D. N. 1974. Formation of catalase compound I by reaction with peroxyacetic acid: pH changes in unbuffered systems. Biochem. J. *143*, 473–474.

JONES, R. L. 1971. Gibberellic acid-enhanced release of β-1,3-glucanase from barley aleurone cells. Plant Physiol. *47*, 412–416.

JONES, R. L. and CHEN, R.-F. 1976. Immunohistochemical localization of α-amylase in barley aleurone cells. J. Cell Sci. *20*, 183–198.

JONES, S. B. 1977. Ultrastructure characteristics of beef muscle. Food Technol. *31*, 82–85.

JOSEFSSON, E. 1975. Effects of variation of heat treatment conditions on the nutritional value of low-glucosinolate rapeseed meal. J. Sci. Food Agric. *26*, 239–242.

JOSEPH, R. L. 1970. Production of tender beef. Process Biochem. *5* (11) 55–58.

JOSEPHSON, E. S. 1976. Overview of prospects for food irradiation. Abstr. 427, 36th Meet., Inst. Food Technol., June. Anaheim, CA.

JOSHI, P. R. and SHIRALKAR, N. D. 1977. Polyphenolases of a local variety of mango. J. Food Sci. Technol. *14*, 77–79.

JOSLYN, M. and PILNIK, W. 1961. Enzymes and enzyme activity. *In* The Orange: Its Biochemistry and Physiology. W.B. Sinclair (Editor). Univ. of Calif. Press, Berkeley.

JOSLYN, M. A. 1929. Some observations on the softening of dill pickle. Fruit Prod. J. *8* (8) 19–21; (9) 19–20.

JOSLYN, M. A. 1949. Enzyme activity in frozen vegetable tissue. Adv. Enzymol. *9*, 606–652.

JOSLYN, M. A. 1951. The action of enzymes in concentrated solution and in the dried state. J. Sci. Food Agric. *2*, 289–294.

JOSLYN, M. A. 1957A. Enzymes—in food products manufacture. West. Canner Packer *49* (11) 21–24, 29–32.

JOSLYN, M. A. 1957B. Role of amino acids in the browning of orange juice. Food Res. *22*, 1–14.

JOSLYN, M. A. 1961A. The freezing preservation of vegetables. Econ. Bot. *15,* 347–375.

JOSLYN, M. A. 1961B. Physiological and enzymological aspects of juice production. *In* Fruit and Vegetable Juice Processing Technology. D.K. Tressler and M.A. Joslyn (Editors). AVI Publishing Co., Westport, Conn.

JOSLYN, M. A. 1962. The chemistry of protopectin: A critical review of historical data and recent developments. Adv. Food Res. *11,* 1–107.

JOSLYN, M. A. 1963. Enzymes in food processing. *In* Food Processing Operations, Vol. 2. J.L. Heid and M.A. Joslyn (Editors). AVI Publishing Co., Westport, Conn.

JOSLYN, M. A. 1966. The freezing of fruits and vegetables. *In* Cryobiology. H.T. Meryman (Editor). Academic Press, New York.

JOSLYN, M. A. and BRAVERMAN, J. B. S. 1954. The chemistry and technology of the pretreatment and preservation of fruit and vegetable products with sulfur dioxide and sulfites. Adv. Food Res. *5,* 97–160.

JOSLYN, M. A. and CRUESS, W. V. 1929. Freezing of fruits and vegetables. Fruit Prod. J. *8* (7) 9–12.

JOSLYN, M. A. and DEUEL, H. 1963. The extraction of pectins from apple marc preparations. J. Food Sci. *28,* 65–83.

JOSLYN, M. A. and HEID, J. L. 1964. Food Processing Operations, Vol. 3. Their Management, Machines, Materials, and Methods. AVI Publishing Co., Westport, Conn.

JOSLYN, M. A. and MARSH, G. L. 1935. Browning of orange juice. Ind. Eng. Chem. *27,* 186–189.

JOSLYN, M. A., MIST, S. and LAMBERT, E. 1952. The clarification of apple juice by fungal pectic enzyme preparations. Food Technol. *6,* 133–139.

JOSLYN, M. A. and NEUMANN, H. J. 1963. Processed vegetable products. Peroxidase in frozen vegetables. J. Assoc. Off. Anal. Chem. *46,* 712–717.

JOSLYN, M. A., NISHIRA, H. and ITO, S. 1968. Leucoanthocyanins and related phenolic compounds of carob pods. J. Sci. Food Agric. *19,* 543–550.

JOSLYN, M. A. and PETERSON, R. G. 1958. Reddening of white onion bulb purees. J. Agric. Food Chem. *6,* 754–765.

JOSLYN, M. A. and PONTING, J. D. 1951. Enzyme-catalyzed oxidative browning of fruit products. Adv. Food Res. *3,* 1–44.

JOSLYN, M. A. and SANO, T. 1956. The formation and decomposition of green pigment in crushed garlic tissue. Food Res. *21,* 170–183.

JOSLYN, M. A. and SEDKY, A. 1940. Effect of heating on the clearing of citrus juices. Food Res. *5,* 223–232.

JOSLYN, M. A. and SHERRIL, M. 1933. Inversion of sucrose by invertase at low temperatures. Ind. Eng. Chem. *25,* 416–417.

JOST, R. and MONTI, J. C. 1977. Partial enzymatic hydrolysis of whey protein and trypsin. J. Dairy Sci. *60,* 1387–1393.

JOST, R., MONTI, J. C. and HILDAGO, J. 1976. Natural proteolysis in whey and susceptibility of whey proteins to acid proteases of rennet. J. Dairy Sci. *59,* 1568–1573.

JOVIN, J. M. 1976. Recognition mechanisms of DNA-specific enzymes. Annu. Rev. Biochem. *45,* 889–920.

JØRGENSEN, H. 1939. Further investigations into the nature of the action of bromates and ascorbic acid on the baking strength of wheat flour. Cereal Chem. *16,* 51–60.

JØRGENSEN, H. 1945. Studies on the Nature of the Bromate Effect. Einar Munksgaard, Copenhagen.

JUFFS, H. S. 1973. Proteolysis detection in milk. I. Interpretation of tyrosine value data for raw milk supplies in relation to natural variation, bacterial counts, and other factors. J. Dairy Res. *40,* 371–381.

JUNG, D. W. and LATIES, G. G. 1975. Trypsin induced ATPase activity in potato mitochondria. Plant Physiol. *36* (2) Supplement, 72.

JURD, L. 1972. Recent progress in the chemistry of flavylium salts. *In* Structure and Functional Aspects of Phytochemistry. V.R. Runeckles (Editor). Academic Press, New York.

JÜRGEN, H. 1975. Programmed Instruction Material for Food Technology Education. Fachbuchverlag, Leipzig, E. Germany.

JURICS, E. W., KOVATES, M. and DWORSCHAK, E. 1971. Effect of cellulase treatment on the nutrient contents of some vegetables. Elelmiszervizgalati Kozl. *17,* 199–208. (Hungarian)

KACZMARZYK, L. M., FENNEMA, O. and POWRIE, W. D. 1963. Changes produced in Wisconsin green snap beans by blanching. Food Technol. *17,* 943–946.

KADER, J. C. 1975. Proteins and the intracellular exchange of lipids. Biochim. Biophys. Acta *380,* 31–44.

KAESS, G. and WEIDMANN, J. F. 1967. Freezer burn as limiting factor in the storage of animal tissue. V. Experiments with beef muscle. Food Technol. *21,* 461–465.

KAGEN, L. J. 1973. Myoglobin—Biochemical,

Physiological and Clinical Aspects. Columbia Univ. Press, New York.

KAHL, G. and GAUL, E. 1975. In vivo and in vitro degradation of white potato phosphoglucomutase. Z. Pflanzenphysiol. *75,* 217–228. (German)

KAHLEM, G., CHAMPAULT, A., LOUIS, J. P., BAZIN, M., CHABIN, A., DELAIGUE, M., DAUPHIN, B., DURAND, R. and DURAND, B. 1976. Genetic determination and hormonal regulation of sexual differentiation in *Mercurialis annua*. Physiol. Veg. *13,* 763–769.

KAHN, V. 1976. Polyphenol oxidase isozymes in avocado. Phytochemistry *15,* 267–272.

KAHN, V. 1977. Some biochemical properties of polyphenoloxidase from two avocado varieties differing in their browning rates. J. Food Sci. *42,* 38–43.

KAKADE, M. L., RACKIS, J. J. and PASKI, G. 1974. Determination of trypsin inhibitor activity of soy products: A collaborative analysis of an improved procedure. Cereal Chem. *51,* 376–388.

KALBRENNER, J. E., WARNER, K. and ELDRIDGE, A. C. 1974. Flavors derived from linoleic and linoleic acid hydroperoxides. Cereal Chem. *51,* 406–415.

KALOYEREAS, S. A. 1947. The effect of various methods of blanching on ascorbic acid and soluble solids in cauliflower and spinach. Fruit Prod. J. *26,* 134–135.

KAMIMIYA, S., ITOH, Y., IZAKI, K. and TAKAHASHI, H. 1977. Purificaton and properties of a pectate lyase in *Erwinia aroideae*. Agric. Biol. Chem. *41,* 975–981.

KAMOGAWA, A., FUKUI, T. and NIKUNI, Z. 1968. Potato α-glucan phosphorylase: crystallization, amino acid composition and enzymatic reaction in the absence of added primer. J. Biochem. (Tokyo) *63,* 361–369.

KANDA, T., WAKABAYASHI, K. and NISIZAWA, K. 1976. Xylanase activity of an endocellulase of carboxymethyl-cellulase type from *Irpex lacteus*. J. Biochem. (Tokyo) *79,* 989–995.

KANDILIS, J. D. 1972. Quality improvement of Greek wheat by hydrothermal treatment. Hemika Hronika *37,* 241–251. (Greek)

KANG, C. K. and RICE, E. E. 1970. Degradation of various meat fractions by tenderizing enzymes. J. Food Sci. *35,* 563–565.

KANG, C. K. and WARNER, W. D. 1974. Tenderization of meat with papaya latex proteases. J. Food Sci. *39,* 812–818.

KANG, C. K., WARNER, W. D. and RICE, E. E. 1974. Tenderization of meat with proteolytic enzymes. U.S. Pat. 3,818,106. June 18.

KANISAWA, T. 1975. Use of microorganisms in the aroma industry. Hakko Kyokaishi *33,* 328–340. (Japanese). Chem. Abstr. *84,* 119964.

KANNER, J. and KAREL, M. 1976. Changes in lysozyme due to reactions with peroxidizing methyl linoleate in a dehydrated model system.

KANNER, J., MENDEL, H. and BUDOWSKI, P. 1976. Carotene-oxidizing factors in red pepper fruits. J. Food Sci. *41,* 183–185.

KAPLAN, N. O., CIOTTI, F. E. and STOLZENBACH, J. 1956. Reaction of pyridine nucleotide analogues with dehydrogenases. J. Biol. Chem. *221,* 833–844.

KAREL, M. 1973. Recent research and development in the field of low-moisture and intermediate-moisture foods. Crit. Rev. Food Technol. *3,* 329–373.

KAREL, M. 1973B. Protein-lipid interactions. J. Food Sci. *38,* 756–763.

KAREL, M. 1975. Free radicals in low moisture systems. *In* Water Relations of Foods. R.B. Duckworth (Editor). Academic Press, New York.

KAREL, M., SCHAICH, K. and RAY, R. B. 1975. Interaction of peroxidizing methyl linoleate with some proteins and amino acids. J. Agric. Food Chem. *23,* 159–163.

KARR, A. L., JR. and ALBERSHEIM, P. 1970. Polysaccharide-degrading enzymes are unable to attack plant cell-walls without prior action by a "wall modifying" enzyme. Plant Physiol. *46,* 69–80.

KARRER, P. and MANGELLI, O. 1929. Concerning the so-called Lilienfeld-Silk in relation to cellulase. Helv. Chim. Acta *12,* 989–990. (German)

KARRER, P., SCHUBERT, P. and WEHRLI, W. 1925. The enzymatic degradation of artificial silk and native cellulose. Helv. Chim. Acta *8,* 797–810. (German)

KARUBE, I., TANAKA, S., SHIRAI, T. and SUZUKI, S. 1977. Hydrolysis of cellulose in a cellulase-bead fluidized bed reactor. Biotechnol. Bioeng. *19,* 1183–1191.

KASARDA, D. D., BIRNARDIN, J. E. and NIMMO, C. C. 1976. Wheat proteins. Adv. Cereal Sci. *1,* 158–236.

KASARDA, D. D., NIMMO, C. C. and KOHLER, G. O. 1971. Proteins and the amino acid composition of wheat fractions. *In* Wheat Chemistry and Technology. Y. Pomeranz (Editor).

Am. Assoc. Cereal Chem., Minneapolis.

KASCHNITZ, R. M. and HATEFI, Y. 1975. Lipid oxidation in biological membranes. Electron transfer proteins as initiators of lipid autoxidation. Arch. Biochem. Biophys. *71*, 292–304.

KASTENSCHMIDT, L. L. 1970. The metabolism of muscle as a food. *In* Physiology and Biochemistry of Muscle as a Food. E.J. Briskey, R.G. Cassens and B.B. Marsh (Editors). Univ. of Wisconsin Press, Madison.

KATCHALSKI, E. 1970. Preparation and properties of enzymes immobilized in artificial membranes. *In* Symmetry Funct. Biol. Syst. Macromol. Level, Proc. 11th Nobel Symp. A. Engstrom (Editor). Almqvist, Wiksell, Stockholm.

KATO, A., WAKINGA, T., MATSUDOMI, N. and KOBAYASHI, K. 1978. Changes in lysozyme during egg white thinning. Agric. Biol. Chem. *42*, 175–176.

KATO, S. and MISAWA, T. 1974. Studies on the infection and multiplicity of plant viruses. VII. The breakdown of chlorophyll in tobacco leaves systemically infected with cucumber mosaic virus. Ann. Phytopath. Soc. Jpn. *40*, 14–21.

KATO, S., YANO, N., SUZUKI, I., ISHII, T., KURATA, T. and FUJIMAKI, M. 1974. Effect of L-cysteine on browning of egg albumen. Agric. Biol. Chem. *38*, 2425–2430.

KATOU, T., KOBAYASHI, K., IZUMI, Y. and HANAOKA, Y. 1976. An enzymatic method for the estimation of the mycelial content of shoyu koji. J. Agric. Chem. Soc. Jpn. *50*, 395–402. (Japanese)

KAUL, R. J. 1967. Curing of vanilla beans. U.S. Pat. 3,352,690. Nov. 14.

KAUZMANN, W. 1959. Some factors in the interpretation of protein denaturation. Adv. Prot. Chem. *14*, 1–63.

KAVANAU, J. L. 1950. Enzyme kinetics and the rate of biological processes. J. Gen. Physiol. *34*, 193–209.

KAWABATA, T. and SHAZUKI, H. 1972. Refuting the formation of N-nitrosamines by the Maillard reaction. J. Food Sci. Technol. *19*, 241–248. (Japanese)

KAWAI, M. 1972. Maceration of plant tissues by crude enzyme preparations. III. Fractionation of crude enzyme preparations. J. Ferment. Technol. *50*, 698–703.

KAWASHIMA, K., TANAKA, Y. and UMEDA, K. 1975. Irradiation of enzyme preparations. II. Radiosterilization of enzyme preparations by accelerated electron irradiation. FSTA *8* (1) A5. (Japanese)

KAY, H. D. and GRAHAM, W. R. 1935. The phosphatase test for pasteurized milk. J. Dairy Res. *5*, 191–203.

KAZENIAC, S. J. and HALL, R. M. 1970. Flavor chemistry of tomato volatiles. J. Food Sci. *35*, 519–530.

KAZUO, K. and MAOKI, M. 1979. Soy sauce. U.S. Pat. 4,180,590. Dec. 25.

KEARSLEY, M. W. 1974. Concentration of sugars by reverse osmosis. Food Trade Rev. *44*, 7–11.

KEAY, L., MOSELEY, M. A., ANDERSON, R. G., O'CONNOR, R. J. and WILDI, B. S. 1972. Production and isolation of microbial proteases. *In* Enzyme Engineering. L.B. Wingard, Jr. (Editor). John Wiley & Sons, New York.

KÉDZY, F. J. and KAISER, E. T. 1970. Principles of active site titration of proteolytic enzymes. Methods Enzymol. *19*, 3–20.

KEIJBETS, M. J. H., PILNIK, W. and VAAL, J. F. A. 1976. Model studies on behavior of pectic substances in the potato cell wall during boiling. Potato Res. *19*, 289–303.

KEILIN, D. and HARTREE, E. F. 1948. Properties of glucose oxidase (notatin). Biochem. J. *42*, 221–229.

KELLER, H. 1952. The determinaton of trace amounts of DDT by enzyme analysis. Naturwissenschaften *39*, 109. (German)

KELLEY, G. J., LATZKO, E. and GIBBS, M. 1976. Regulatory aspects of photosynthetic carbon metabolism. Annu. Rev. Plant Physiol. *27*, 181–205.

KELLEY, P. J. and CATLEY, B. J. 1976. A purification of trehalase from *Saccharomyces cerevisiae*. Anal. Biochem. *72*, 353–358.

KELLOGG, E. W., III and FRIDOVICH, I. 1975. Superoxide, hydrogen peroxide and singlet oxygen in lipid peroxidation by a xanthine oxidase system. J. Biol. Chem. *250*, 8812–8817.

KELLY, C. T. and FOGARTY, W. M. 1976. Microbial alkaline enzymes. Process Biochem. *11* (7) 3–9.

KELLY, S. H. and FINKLE, B. J. 1969. Action of a ring-cleaving oxygenase in preventing oxidative darkening of apple juice. J. Sci. Food Agric. *20*, 629–632.

KEMP, B. and SPINELLI, J. 1969. Comparative rates of IMP degradation in unfrozen and frozen-and-thawed (slacked) fish. J. Food Sci. *34*, 132–135.

KEMPNER, D. 1972. Temperature effects on enzyme reaction rates. Gilford Lab. Lett. *1* (11) 1.

KEMPNER, E. S. and MILLER, J. H. 1968. The molecular biology of *Euglena gracilis*. V.

Enzyme localization. Exp. Cell Res. *51*, 150–156.

KENDREW, J. C. 1963. Myoglobin and the structure of proteins. Science *39*, 1259–1266.

KENNEDY, M. G. H. and ISHERWOOD, F. A. 1975. Activity of phosphorylase in *Solanum tuberosum* during low temperature storage. Phytochemistry *14*, 666–670.

KENNEY, W. C. 1974. Molecular nature of isoenzymes. Horiz. Biochem. Biophys. *1*, 38–61.

KENTEN, R. H. 1957. Latent phenolase in extracts of broad bean (*Vicia faba* L.) leaves. I. Activation by acid and alkali. Biochem. J. *67*, 300–307.

KENYHERCZ, T. M. and KISSINGER, P. T. 1977. Tyramine from *Theobroma cacao*. Phytochemistry *16*, 1602–1603.

KEPLER, C. R. and TOVE, S. B. 1967. Biohydrogenation of unsaturated fatty acids. III. Purification and properties of a linoleate Δ^{12}-*trans*-isomerase from *Butyrovibrio fibrisolvens*. J. Biol. Chem. *242*, 5686–5692.

KERTESZ, D. 1957. State of copper in polyphenoloxidase (tyrosinase). Nature *180*, 506–507.

KERTESZ, Z. I. 1930. A new method for clarification of unfermented apple juice. Bull. *589*, N.Y. State Agric. Exp. Stn., Geneva.

KERTESZ, Z. I. 1943. A possible non-enzymatic mechanism of changes occurring in the pectic substances and other polysaccharides in living plants. Plant Physiol. *18*, 308–309.

KERTESZ, Z. I. 1947. Calcium improves pie apples. Food Packer *28*, 30–32.

KERTESZ, Z. I., TOLMAN, T. G., LOCONTI, J. D. and RUYLE, E. H. 1940. The use of calcium in the commercial canning of whole tomatoes. N.Y. State Agric. Exp. Stn. Tech. Bull. *252*.

KESLER, C. C., CAREY, P. L. and WILSON, O. G. 1970. Sugar products prepared from hydroxypropylated starch. U.S. Pat. 3,505,110. Apr. 7.

KESSLER, M., CLARK, L. C., JR., LUBBERS, D., SILVER, I. A. and SIMON, W. 1976. Ion and Enzyme Electrodes in Biology and Medicine. Univ. Park Press, Baltimore.

KEZDY, F. J. and KAISER, E. Y. 1970. Principles of active site titration of proteolytic enzymes. Methods Enzymol. *19*, 3–20.

KHALIDOVA, G. B. and KOSITIN, A. U. 1975. Effect of zinc deficiency on carbonic anhydrase activity of tomato chloroplasts. Bot. Zh. *60*, 522–558. (Russian)

KHAN, A. A. and KOLATTUKADY, P. E. 1974. Decarboxylation of chain fatty acids to alkanes by cell free extracts preparation of pea leaves *(Pisum sativum)*. Biochem. Biophys. Res. Comm. *61*, 1379–1386.

KHAN, A. W. 1971. Effect of temperature during post-mortem glycolysis and dephosphorylation of high energy phosphates on poultry meat tenderness. J. Food Sci. *36*, 120–126.

KHANTSIN, YA. G. 1976. Method for preparation of mustard. U.S.S.R. Pat. 501,747. Apr. 3.

KHOO, J. C., SPERRY, P. J., GILL, G. N. and STEINBERG, D. 1977. Activities of hormone-sensitive lipase and phosphorylase kinase by purified cyclic GMP-dependent protein kinase. Proc. Natl. Acad. Sci. *74*, 4843–4847.

KHRYANIN, V. N., GORSHKOVA, A. P. and ALEKHOVA, M. K. 1976. Use of gibberellin for the rapid growth of barley malt in malt kilns during maltose sirup production. Sakh. Promst. *11*, 65–68. (Russian)

KIANG, C. -H. KUAN, S. S. and GUILBAULT, G. G. 1975. A novel enzyme electrode method for the determination of nitrite based on nitrite reductase. Anal. Chim. Acta *80*, 209–214.

KICHLINE, T. P. and SCHARPF, L. G. 1972. Preparation of processed cheese. U.S. Pat. 3,635,733. Jan. 18.

KIEFER, F. 1961. A new oxidative mechanism in the deteriorative changes of orange juice. Food Technol. *15*, 302–305.

KIEFER, H. C., CONGDON, W. I., SCARPA, S. and KLOTZ, I. M. 1972. Catalytic acceleration of 10^{12}-fold by an enzyme-like synthetic polymer. Proc. Natl. Acad. Sci. *69*, 2155–2159.

KIENINGER, H. 1977. Present knowledge on the stabilization of beers. Brauwelt *37*, 1438–1448. (German)

KIERMEIER, F. 1949. Changes in enzyme activity due to freezing of plant tissues. Biochem. Z. *319*, 463–481. (German)

KIES, M. W. 1947. Complex nature of soybean lipoxidase. Fed. Proc. *6*, 267.

KIES, M. W., HAINING, J. L., PISTORIUS, E., SCHROEDER, D. H. and AXELROD, B. 1969. On the question of identity of soybean "lipoxidase" and carotene oxidase. Biochem. Biophys. Res. Comm. *36*, 312–315.

KIESVAARA, M. 1975. On the soluble nitrogen fraction of barrel-salted herring and semipreserves during ripening. Tech. Res. Cent. Finl. Mater. Process. Technol. Publ. *10*, 7–99.

KIHLBERG, R. 1972. The microbe as a source of food. Annu. Rev. Microbiol. *26*, 427–466.

KIKUCHI, M., OZAWA, Y. and SUZUKI, K. 1973. Automated measurement of L-glutamate in soy sauce using glutamate dehydrogenase.

Agric. Biol. Chem. *37*, 1673–1677.

KILARA, A. and SHAHANI, K. M. 1973. Removal of glucose from eggs: A review. J. Milk Food Technol. *36*, 509–514.

KILARA, A. and SHAHANI, K. M. 1979. The use of immobilized enzymes in the food industry: A review. Crit. Rev. Food Sci. Nutr. *12*, 161–198.

KILARA, A., SHAHANI, K. M. and WAGNER, F. W. 1977. Preparation and characterization of immobilized lactase. Lebensm.-Wiss. Technol. *10*, 84–88. (German)

KIM, S. K. and D'APPOLONIA, B. L. 1977. Bread staling studies. III. Effect of pentosans on dough, bread, and bread staling rate. Cereal Chem. *54*, 225–229.

KIM, W. J., SMIT, C. J. B. and NAKAYAME, T. O. M. 1973. The removal of oligosaccharides from soybeans. Lebensm.-Wiss. Technol. *6*, 201–204. (German)

KIMURA, H., MUKAI, M. and KITAMURA, T. 1974. Purification and properties of a protein activator of human pancreatic lipase. J. Biochem. (Tokyo) *76*, 1287–1292.

KING, A. D., JR., MICHENER, H. D., BAYNE, H. G. and MIHARA, K. L. 1976. Microbial studies on shelf life of cabbage and coleslaw. Appl. Environ. Microbiol. 404–407.

KING, C. J. 1974. Novel dehydration techniques. *In* Advances in Preconcentration and Dehydration of Foods. A. Spicer (Editor). John Wiley & Sons, New York.

KING, F. B., COLEMAN, D. A. and LECLERC, J. A. 1937. Report of the U.S. Department of Agriculture bread flavor committee. Cereal Chem. *14*, 49–58.

KING, M. M., LAI, E. K. and McCAY, P. M. 1975. Singlet oxygen production associated with enzyme-catalyzed lipid peroxidation in liver microsomes. J. Biol. Chem. *250*, 6496–6502.

KINJIRUSHIWASABI CO. 1974. Vegetable flavoring composition. Jpn. Pat. 74 39,825. Apr. 21.

KINKI YAKURUTU CO. 1966. Vegetable and fruit processing. Jpn. Pat. 66 10,221. June 1.

KINNUNEN, P. K. J., JACKSON, R. L., SMITH, L. C. and GOTTO, A. -M., JR. 1977. Activation of lipoprotein lipase by native and synthetic fragments of human plasma apolipoprotein C-II. Proc. Natl. Acad. Sci. *74*, 4848–4851.

KINSELLA, J. E. 1976. Functional properties of proteins in foods: a survey. Crit. Rev. Food Sci. Nutr. 7, 219–280.

KINSELLA, J. E. and HWANG, D. 1976. Biosynthesis of flavors by *Penicillium roqueforti*. Biotechnol. Bioeng. *18*, 927–938.

KINSELLA, J. E. and HWANG, D. 1976B. Enzymes of Penicillium roqueforti involved in the biosynthesis of cheese flavor. Crit. Rev. Food Sci. Nutr. *8*, 191–228.

KIRCHGESSNER, M. and STEINHART, H. 1974. Distribution of amino acids in blocks of molecular weights after pepsin in vitro digestion of soybean protein. Z. Tierphysiol. Tierernaehr. Futtermittelkd. *32*, 240–248. (German)

KIRCHMEIER, O. 1972. The secondary phase of milk coagulation by rennin. II. Changes in the electrochemical state of the micelle. Z. Lebensm.-Unters. Forsch. *149*, 211–217. (German)

KIRK, J. S. 1933. The concentration of soy bean urease. A new method for the purification of enzymes. J. Biol. Chem. *100*, 667–670.

KIRK, L. D., MUSTAKAS, G. C., GRIFFIN, E. L., JR. and BOOTH, A. N. 1971. Crambe seed processing: Decomposition of glucosinolates (thioglucosides) with chemical additives. J. Am. Oil Chem. Soc. *48*, 845–850.

KISHONTI, E. 1975. Influence of heat resistant lipases and proteases in psychrotrophic bacteria on product quality. Ann. Bull. Int. Dairy Fed. *86*, 121–124.

KISS, E. 1967. The use of glucose oxidase to increase the shelf life of egg powders. I. The enzymatic removal of glucose from eggs. Elelimiszertudomany *1*, 39–43. (Hungarian)

KISTIAKOWSKY, G. B. and LUMRY, R. 1949. Anomalous temperature effects in the hydrolysis of urea by urease. J. Am. Chem. Soc. *71*, 2006–2013.

KISZA, J., PANFIL, H. and ZIAJKA, S. 1974. Influence of added lysozyme on pepsin digestion of milk protein. Proc. 19th Int. Dairy Congr. *1E*, 565.

KISZA, J., SWITKA, J. and SURZYNSI, A. 1973. Utilization of manufacture of sweetened condensed milk. Lait *53*, 430–439.

KITAMURA, K., KANEKO, T. and YAMAMOTO, Y. 1972. Lysis of viable yeast cells by enzymes of *Arthrobacter luteus*. 1. Isolation of lytic strain and studies on its lytic activity. J. Gen. Appl. Microbiol. *18*, 57–71.

KIYOHARA, Y., TERAO, T., SHIORIRI-NAKANO, K. and OSAWA, T. 1976. Purification and characterization of β-N-acetylhexosaminidases and β-galactosidase from Streptococcus 6646K. J. Biochem. (Tokyo) *80*, 9–17.

KJAER, A. 1958. Secondary organic sulfur-compounds of plants (thiols, sulfides, sulfonium derivatives, sulfoxides, sulfones and isothiocyanates). *In* Encyclopedia of Plant Physiology. W. Ruhland (Editor). Springer-Verlag, Berlin.

KJAERGAARD, O. G. 1974. Spray drying of wash-active enzymes. *In* Advances in Preconcentration and Dehydration of Foods. A. Spicer (Editor). John Wiley & Sons, New York.

KJELDAHL, M. J. 1881. Research on carbohydrates of barley and of malt, especially from the viewpoint of the presence of cane sugar. C.R. Trav. Lab. Carlsberg *1*, 189–195. (French)

KLAUSTERMEYER, J. A. and MORRIS, L. L. 1975. The effects of ethylene and carbon monoxide on the induction of russet spotting of crisphead lettuce. Plant Physiol. *56* (Suppl.), 63.

KLEBANOFF, S. J. 1968. Myeloperoxidase-halide-hydrogen peroxide antibacterial system. J. Bacteriol. *95*, 2131–2138.

KLEBANOFF, S. J. 1974. Role of the superoxide anion in the myeloperoxidase-mediated antimicrobial system. J. Biol. Chem. *249*, 3724–3728.

KLEE, C. B., CROUCH, T. H. and RICHMAN, P. G. 1980. Calmodulin. Annu. Rev. Biochem. *49*, 489–515.

KLEIN, B. 1976. Isolation of lipoxygenase from split pea seeds, snap beans, and peas. J. Agric. Food Chem. *24*, 938–942.

KLEINSCHMIDT, A. W. 1963. Soya lipoxidase as a means of flavor improvement. Bakers Dig. *37* (5) 44–47.

KLEINSCHMIDT, A. W. and VIREN, S. T. 1970. Bread improving composition and method. U.S. Pat. 3,506,448. Apr. 14.

KLEINSCHMIDT, M. 1971. Fate of Di-Syston in potatoes during processing. J. Agric. Food Chem. *19*, 1196–1197.

KLEYN, D. H. and HO, C. -L. 1977. Rapid determination of alkaline phosphatase reactivation. J. Assoc. Off. Anal. Chem. *60*, 1389–1391.

KLINE, I., SONODA, T. T. and HANSON, H. L. 1954. Comparisons of the quality and stability of whole egg powders desugared by the yeast and enzyme methods. Food Technol. *8*, 343–349.

KLIS, J. B. 1962. Heat stable enzyme. Food Process. *23* (6) 70–71.

KLOCZKO, I. and RUTKOWSKI, A. 1977. Phytic acid—A useful or harmful food factor. Postepy Nauk Roln. *24*, 107–124. (Polish)

KLOPPING, H. L. 1971. Olfactory theories and odors of small molecules. J. Agric. Food Chem. *19*, 999–1004.

KLOSE, A. A., LUYET, D. J. and NENZ, L. J. 1970. Effect of contraction on tenderness of poultry muscle cooked in the prerigor state. J. Food Sci. *35*, 577–581.

KLOSE, S., WAHLEFELD, W. and HAGEN, A. 1975. Enzymatic analysis. Ger. Pat. 2,412,354. June 19.

KLOTZ, I. M., DARNALL, D. W. and LANGERMAN, N. 1975. Quaternary structure of proteins. *In* The Proteins, Vol. 1, 3rd Edition. H. Neurath and R.L. Hill (Editors). Academic Press, New York.

KLOTZ, I. M., ROYER, G. P. and SCARPA, I. S. 1971. Synthetic derivatives of polyethyleneimine with enzyme-like catalytic activity. Proc. Natl. Acad. Sci. *68*, 263–264.

KNAPP, F. W. 1965. Some characteristics of eggplant and avocado polyphenol oxidase. J. Food Sci. *30*, 930–936.

KNEE, M. 1975. Soluble and wall-bound glycoproteins of apple fruit tissue. Phytochemistry *14*, 2181–2188.

KNEE, M., FIELDING, A. H., ARCHER, S. A. and LABORDA, F. 1975. Enzymic analysis of cell wall structure in apple fruit cortical tissue. Phytochemistry *14*, 2213–2222.

KNEEN, E. 1944. A comparative study of the development of amylases in germinating cereals. Cereal Chem. *21*, 304–314.

KNEEN, E. and HADS, H. L. 1945. Effects of variety and environment on the amylases of germinated wheat and barley. Cereal Chem. *22*, 407–418.

KNORR, F. 1977. Polyphenols in the brewing process. Brauindustrie *62*, 1017–1028. (German)

KNOWLES, J. R. and ALBERY, J. W. 1977. Perfection in enzyme catalysis: the energetics of triosephosphate isomerase. Acc. Chem. Res. *10*, 105–111.

KNOWLES, J. R. and GUTFREUND, H. 1974. The functions of proteins as devices. *In* Chemistry of Macromolecules, Vol. 1. H. Gutfreund (Editor). University Park Press, Baltimore.

KNUCKLES, B. E., SPENCER, R. B., LAZAR, M. E., BICKOFF, E. M. and KOHLER, G. 1970. PRO-XAN process: Incorporation of sugar cane rollers in wet fractionation of alfalfa. J. Agric. Food Chem. *18*, 1086–1089.

KOBAYASHI, H. and SUSUKI, H. 1975. Kinetic studies of mold α-galactosidase on PNPG hydrolysis. Biotechnol. Bioeng. *17*, 1455–1465.

KOBAYASHI, Y., TANAKA, H. and OGASAWARA, N. 1974. Multiple β-1,3 glucanases in the lytic enzyme complex of *Bacillus circulans* WL 12. Agric. Biol. Chem. *38*, 959–965.

KOCH, G., EDLUND, K. and HOOGENDOORN, H. 1973. Lactoperoxidase in the prevention of plaque accumulation, gingivitis and dental caries. Odont. Rev. *24*, 367–372.

KOCH, R. B. 1956. Mechanisms of fat oxidation. Bakers Dig. *30* (2) 46–68.

KODAMA, S. 1913. On a procedure for separating inosinic acid. J. Tokyo Chem. Soc. *34,* 751–753. (Japanese)

KODENCHERY, U. K. and NAIR, M. P. 1972. Metabolic changes induced by sprout inhibiting doses of gamma irradiation in potatoes. J. Agric. Food Chem. *20,* 282–285.

KOENIGS, T. K. 1975. Hydrogen peroxide and iron: a microbial cellulolytic system? Biotechnol. Bioeng. Symp. *5,* 151–160.

KOHLER, J. 1974. Survey of pyruvate, L-(+)- and D-(–) lactate in dried milk powder. FSTA *8* (10) P1845. (German)

KOLKE, M. and HAMADA, M. 1971. Preparation of calcium phosphate gel deposited on cellulose. Methods Enzymol. *22,* 339–342.

KOJIMA, K. 1974. Safety evaluation of disodium 5′-inosinate, disodium 5′-guanylate and disodium 5′- ribonucleotide. Toxicology *2,* 185–206.

KOJIMA, M., POULTON, J. E., THAYER, S. S. and CONN, E. E. 1979. Tissue distribution of dhurrin and of enzymes involved in its metabolism in leaves of *Sorghum bicolor.* Plant Physiol. *63,* 1022–1028.

KOLATA, G. B. 1980. Genes in pieces. Science *207,* 392–393.

KOMARIK, S. L. 1964. Method of improving the tenderness of meat. U.S. Pat. 3,147,123. Sept. 1.

KOMATSU, S. 1965. Rice starch. Jpn. Pat. 65 11,900. June 1.

KOMEJI, T., YOSHIKAWA, N. and ASAHIRO, N. 1978. Separation of protein from yeast cell. Jpn. Kokai 78 03,592. Jan. 13.

KON, S. 1980. Effects of superoxide dismutase and other oxygen scavengers on enzymatic browning. J. Food Sci. *45,* 1066–1067.

KON, S. and DUNLAP, C. J. 1977. Snack foods from legumes. Food Prod. Dev. *11* (7) 77–78.

KON, S., FREDERICK, D. P., EGGLING, S. B. and WAGNER, J. R. 1973. Effect of treatment on phytate and soluble sugars in California small white beans. J. Food Sci. *38,* 215–217.

KON, S. and SCHWIMMER, S. 1977. Depolymerization of polysaccharides by active oxygen species derived from a xanthine oxidase system. J. Food Biochem. *1,* 141–152.

KON, S. and WAGNER, J. R. 1977. Partial separation and characterization of α-galactosidase from *Phaseolus vulgaris.* Lebensm.-Wiss. Technol. *10,* 106–108. (German)

KON, S., WAGNER, J. R. and BOOTH, A. N. 1974. Legume powders: preparation and some nutritional and physicochemical properties. J. Food Sci. *39,* 897–899.

KONIGSBACHER, K. S. 1974. Technology for shelf-stable foods. Food Prod. Dev. *8* (8) 28–30.

KONIGSBACHER, K. S. 1976. Enzymatic flavors propagation. *In* Enzymes: Use and Control in Foods. F.W. Cooler (Editor). Inst. Food Technol., Chicago.

KONIGSBACHER, K. S. and HEWITT, E. J. 1964. Enzymatic odor development. Ann. N.Y. Acad. Sci. *116,* 705–710.

KONRAD, H. and GABRIO, T. 1976. Rapid enzymic-colorimetric method for detecting residues of organophosphorus insecticides in milk. Nahrung *20,* 395–398.

KONZE, J. R. and ELSTNER, E. F. 1976. Pyridoxalphosphate-dependent ethylene production from methionine by isolated chloroplasts. FEBS Lett. *66,* 8–11.

KOPPENOL, W. H. 1976. Reactions involving singlet oxygen and the superoxide anion. Nature *262,* 420–421.

KORNBERG, A. 1976. For the love of enzymes. *In* Reflections on Biochemistry. In Honor of Severo Ochoa. A. Kornberg, B.L. Horecker, L. Cornudella and J. Oro (Editors). Pergamon Press, New York.

KORNBERG, R. D. 1977. Structure of chromatin. Annu. Rev. Biochem. *46,* 931–954.

KORNER, B. 1971. Enzymes of the orange and clarification of orange juice. Ph.D. Thesis, Technion-Israel Inst. Technol., Haifa.

KORNER, B. and BERK, Z. 1967. The mechanism of pink-red formation in leeks. Advan. Frontiers Plant Sci. *18,* 39–52.

KORTZ, J. 1973. Colour stability as influenced by sulfhydryl groups in fresh pork. FSTA *7* (5) 5662. (Polish)

KORUS, R. A. 1976. Personal communication. Albany, Calif.

KORUS, R. A. and O'DRISCOLL, K. F. 1975. The influence of diffusion on the apparent rate of denaturation of gel entrapped enzymes. Biotechnol. Bioeng. *17,* 441–444.

KORUS, R. A. and OLSON, A. C. 1977. Use of glucose isomerase in hollow fiber reactors. J. Food Sci. *42,* 258–260.

KOSHLAND, D. E., JR. 1960. The active site and enzyme action. Adv. Enzymol. *22,* 45–98.

KOSHLAND, D.E., JR. 1964. Conformation changes at the active site during enzyme action. Fed. Proc. *23,* 719–726.

KOSHLAND, D. E., JR. 1973. Protein shape and biological control. Sci. Am. *229* (4) 52–64.

KOSHLAND, D. E., JR. 1976. Role of flexibility in the specificity control and evolution of enzymes. FEBS Lett. *62*, Suppl., Feb. 4.

KOSIKOWSKI, F. V. 1975. Potential of enzymes in continuous cheesemaking. J. Dairy Sci. *58*, 994–1000.

KOSIKOWSKI, F. V. 1976. Flavor development by enzyme preparation in natural and processed cheddar cheese. U.S. Pat. 3,975,544. Aug. 17.

KOSIKOWSKI, F. V. and IWASAKI, T. 1975. Changes in cheddar cheese by enzyme preparations. J. Dairy Sci. *58*, 963–970.

KOTT, H. 1973. Effect of wetting temperature on glutamic acid decarboxylase activity in wheat embryos. Cereal Chem. *50*, 1–6.

KOURY, B., SPINELLI, J. and WIEG, D. 1971. Protein autolysis rates at various pH's and temperatures in hake, Merluccius productus, and Pacific herring, Clupea harengus pallasi, and their effect on yield in the preparation of fish protein concentrate. Natl. Oceanic Atmos. Admin. (U.S.), Fish Bull. *69*, 241–246.

KOVACS, K., HANUSZ, B. and MATKOVICS, B. 1975. Properties of enzymes. V. Catalase inhibition of glucose oxidase reaction. Enzyme *20*, 123–128.

KOZLOWSKA, H., SOSULSKI, F. W. and YOUNGS, C. G. 1972. Extraction of glucosinolates from rapeseed. Can. Inst. Food Sci. Technol. J. *5*, 149–154.

KOZMA, J. 1976. Dietetic beer; A new product. Soripar *23*, 3–7. (Hungarian)

KÖNIG, C. J. 1907. Biological and biochemical studies relating to milk. Milchwirtsch. Zentralbl. *3*, 233–261.

KRABBE, E. 1972. *Cited by* E.J. Beckhorn. Speculations on the commercial future of immobilized enzymes. *In* Enzyme Engineering. L.B. Wingard (Editor). Interscience Publishers, New York.

KRALOVIC, R. C. 1973. *Clostridium botulinum* type E spores increased heat resistance through lysozyme treatment. Abstr. Am. Soc. Microbiol. *73.*

KRAMER, A. 1972. Texture—its definition, measurement and relation to other attributes of food quality. Food Technol. *26*, 34–35.

KRAMER, A. 1974. Storage retention of nutrients. Food Technol. *28,* 50–60.

KRAMER, A. 1976. Use of color measurement in quality control of foods. Food Technol. *30* (10) 62–73.

KREIPE, H. 1973. Further tests with 2-day fermentation of potato and cereal mashes, using enzyme preparations as saccharification agents. Branntweinwirtschaft *113*, 121–123. (German)

KRETCHMER, N. 1972. Lactose and lactase. Sci. Am. *227* (4) 70–78.

KRETOVICH, V. L. 1944. Biochemistry of the damage to grain by the wheatbug. Cereal Chem. *21*, 1–16.

KREUZALER, F. and HAHLBROCK, K. 1975. Enzymic synthesis of an aromatic ring from acetate units. Partial purification and some properties of flavanone synthase from cell-suspension cultures of *Petroselinum hortense.* Eur. J. Biochem. *56*, 205–213.

KRIËBLE, V. K. 1921. Properties of oxynitrilase. J. Am. Chem. Soc. *43*, 164–175.

KRINSKY, N. I. 1974. Singlet excited oxygen as mediator of the antibacterial action of the leukocytes. Science *186,* 363–366.

KRISTOFFERSEN, T. 1973. Biogenesis of cheese flavor. J. Agric. Food Chem. *21,* 573–575.

KROCHTA, J. M. 1978. Air-cushioning harvesting of tomatoes. Food Technol. *32* (10) 78–82.

KROCHTA, J., CARLSON, R. A., OGAWA, J. M. and MANJI, B. T. 1977. Harvesting into foam reduces tomato losses. Food Technol. *31,* 42–46.

KROCHTA, J. M., TILLIN, S. J. and WHITEHAND, L. C. 1975. Ascorbic acid content of tomatoes damaged by mechanical harvesting. Food Technol. *29*, 28–30, 38.

KRONER, K. H. and KULA, M. R. 1978. Extraction of enzymes in aqueous two phase systems. Process Biochem. *13* (4) 7–9.

KROP, J. J. P. and PILNIK, W. 1974. Cloud loss studies in citrus juices. Cloud stabilization by a yeast polygalacturonase. Lebensm.-Wiss. Technol. 7, 121–124. (German)

KRUCKEBERG, W. and CHILSON, O. 1976. Erythrocyte AMP-deaminase: isolation, subunit size and regulation by phytic acid. Fed. Proc. *35*, 1470.

KRUGER, F. F. 1976. A note on the semiautomated determination of catalase in wheat. Cereal Chem. *53*, 796–801.

KRUGER, J. E. 1972A. Changes in the amylases of hard and spring wheat during growth and maturation. Cereal Chem. *49*, 379–390.

KRUGER, J. E. 1972B. Changes in the amylases of hard red spring wheat during germination. Cereal Chem. *49*, 391–398.

KRUGGEL, W. G. and FIELD, R. A. 1971. Soluble intramuscular collagen characteristics from stretched and aged muscle. J. Food Sci. *36*, 1114–1117.

KUBOWITZ, F. 1937. Chemical composition

of potato oxidase. Biochem. Z. *292*, 221–229. (German)

KUĆ, J., HENZE, R. E., ULLSTRUP, A. J. and QUACKENBUSH, F. W. 1956. Chlorogenic and caffeic acids as fungistatic agents produced by potatoes in response to inoculation with *Helminthosporium carbonium*. J. Am. Chem. Soc. *78*, 3123–3125.

KUCHINSKII, A. L. and YASSKAYA, A. L. 1976. Problem of determining the effectiveness of treatment for meat products. Chem. Abstr. *84*, 103910k. (Russian)

KUDREV, T. and BABALUKOVA, N. 1975. Influence of magnesium insufficiency on the fraction of easily soluble protein and on the isoenzyme composition of certain dehydrogenases. Agrochimica *19*, 336–347.

KUEHLER, C. A. and STINE, C. M. 1974. Effect of enzymatic hydrolysis on some functional properties of whey protein. J. Food Sci. *39*, 379–382.

KUHNE, W. V. 1876. Behavior of various organized and so-called unformed ferments. Trypsin (enzyme of the pancreas). Proc. Natl. Hist. and Med. Soc. of Heidelberg, 122–126. (German)

KUILA, R. K., DUTTA, S. M., BABBAR, I. J. and DUDANI, A. T. 1971. Procedure for separation of bacterial milk clotting activity from other proteolytic enzymes. Indian J. Exp. Biol. *9*, 510–511.

KULP, K. 1968. Pentosans of wheat endosperm. Cereal Sci. Today *13*, 414–417, 426.

KUMAZAKI, T., KASAI, K. and IISHI, S. 1976. Affinity chromatography of trypsin and related enzymes. II. An affinity adsorbent containing glycylglycyl-L-arginine. J. Biochem. (Tokyo) *79*, 749–755.

KUNDUG, W. 1976. The bacterial phosphoenolpyruvate phosphotransferase system. *In* The Enzymes of Biological Membranes, Vol. 3. Membrane Transport. A. Martonosi (Editor). Plenum Press, New York.

KUNINAKA, A., KIBI, M. and SAKAGUCHI, K. 1964. History and development of flavor nucleotides. Food Technol. *18*, 287–293.

KUNINORI, T., NISHIYAMA, J. and MATSUMOTO, H. 1976. Effect of mushroom extract on the physical properties of dough. Cereal Chem. *53*, 420–428.

KUNITAKE, T., SHINKI, S. and ASO, C. 1970. Imidazole catalyses in aqueous systems. III. Formation of the catalyst-substrate complex in the hydrolysis of a phenyl ester catalyzed by naphthylimidazole derivative. Bull. Chem. Soc. Jpn. *43*, 1109–1119.

KUNITZ, M. 1952. Crystalline inorganic pyrophosphatase isolated from baker's yeast. J. Gen. Physiol. *35*, 423–450.

KUPREVICH, V. F. and SHCHERBAKOVA, T. A. 1971. Soil Enzymes. Translated from Russian by the Indian National Scientific Documentation Centre, New Delhi.

KURACHI, K., SIEKIR, L. C. and JENSEN, L. H. 1976. Structures of triclinic mono- and di-N-acetylglucosamine:lysozyme complex—a crystallographic study. J. Mol. Biol. *101*, 11–24.

KURLAND, C. G. 1977. Structure and function of the bacterial ribosome. Annu. Rev. Biochem. *46*, 173–200.

KURODA, N. and HEGISHI, T. 1975. Contents of plasmalogen in foods. Chem. Abstr. *83*, 176978C. (Japanese)

KURONO, Y., STAMOUDIS, V. and BENDER, M. L. 1976. The deacylation of acyl-cycloamyloses. The catalytic effects of benzimidazole derivatives on the rate. Bioorg. Chem. *5*, 393–402.

KURTZMAN, R. H., JR. and HALBROOK, W. 1970. Polysaccharide from dry navy beans, *Phaseolus vulgaris*: its isolation and stimulation of *Clostridium perfringens*. Appl. Microbiol. *20*, 715–719.

KURTZMAN, R. H., JR. and SCHWIMMER, S. 1973. Decaffeination of beverages. U.S. Pat. 3,749,584. July 31.

KUSAKABE, H., KODAMA, K., KUNINAKA, A., YOSHINO, H., MISONO, H. and SODA, K. 1980. A new antitumor enzyme, L-lysine α-oxidase from *Trichoderma viride*. Purification and enzymological properties. J. Biol. Chem. *255*, 976–981.

KUZECHKIN, A. N. 1973. The effect of some enzyme preparations used in the food industry on the microflora of the albino rats intestine. Vopr. Pitan. *32*, 65–69. (Russian)

KÜHNE, W. V. 1876. Über das Verhalten verschiedener oraganisierter und sog. ungefarmter Fermente. Über das Trypsin (Enzym des Pankreas). Verhandlung. der Heidelberg. Nat. Hist. Med. Verein. pp. 122–126.

KWASNIEWSKI, R. 1975. Effect of depectinization of apple pulp on the yield and quality of juice. Ind. Aliment. Agric. *92*, 225–230. (French)

KWOK, S. C. M., CHAN, H. T., JR., NAKAYAMA, T. O. and BREKKE, J. E. 1974. Passion fruit starch and effect on juice viscosity. J. Food Sci. *39*, 431–433.

KWON, T. W., MENZEL, D. B. and OLCOTT, H. S. 1965. Reactivity of malonaldehyde with food constituents. J. Food Sci. *30*, 808–813.

LAAKONEN, E. 1973A. Factors affecting tenderness during heating of meat. Adv. Food Res. *20*, 257–324.

LAAKONEN, E. 1973B. Meat tenderization. Br. Pat. 1,307,420. Jan. 27.

LAAKONEN, E., SHERBON, J. and WELLINGTON, G. H. 1970. Low-temperature, long time heating of bovine muscle. J. Food Sci. *35*, 181–183.

LaBELLE, R. L. 1971. Heat and calcium treatments for firming red tart cherries in a hot-fill process. J. Food Sci. *36*, 323–326.

La BERGE, D. E., MacGREGOR, A. W. and MEREDITH, W. D. S. 1971. Changes in alpha- and beta-amylase activities during the maturation of different barley cultivars. Can. J. Plant Sci. *51*, 469–477.

LABUZA, T. P. 1972. Nutrient losses during drying and storage of dehydrated foods. Crit. Rev. Food Technol. *3*, 217–240.

LABUZA, T. P. 1975. Interpretation of sorption data in relation to the state of constituent water. *In* Water Relations of Foods. R.B. Duckworth (Editor). Academic Press, London.

LABUZA, T. P. 1979. A theoretical comparison of losses in food under fluctuating temperature sequences. J. Food Sci. *44*, 1162–1168.

LABUZA, T. P. 1980. Influence of water activity on chemical properties. Food Technol. *34* (4) 36–41.

LABUZA, T. P., CASSIL, S. and SINSKEY, A. J. 1972. Stability of intermediate moisture foods. 2. Microbiology. J. Food Sci. *37*, 160–162.

LABUZA, T. P., WARREN, R. M. and WARMBIER, H. C. 1977. The physical aspects with respect to water and non-enzymic browning. Adv. Exp. Med. Biol. *86B*, 379–418.

LACHANCE, P. A. and MOLINA, M. R. 1974. Nutritive value of a fiber-free coconut protein extract obtained by an enzymic-chemical method. J. Food Sci. *39*, 581–584.

LACHANCE, M.-A., VILLA, T. G. and PHAFF, H. J. 1977. Purification and partial characterization of an exo-β-glucanase from the yeast *Kluyveromyces aesturarii.* Can. J. Biochem. *55*, 1001–1006.

LADENSTEIN, R. and WENDEL, A. 1976. Crystallographic data of the selenoenzyme glutathione peroxidase. J. Mol. Biol. *104*, 877–882.

LADISCH, M. R., LADISCH, C. M. and TSAO, G. T. 1978. Cellulose to sugars: new path gives quantitative yield. Science *201,* 743–745.

LADISCH, M. R., EMERY, A. and RODWELL, V. W. 1977. Economic implications of purifications of glucose isomerase prior to immobilization. Ind. Eng. Chem. Process Res. Dev. *16*, 309–313.

LAHIRY, N. L. and SATTERLEE, L. D. 1975. Release and estimation of chlorogenic acid in leaf protein concentrate. J. Food Sci. *40*, 1326.

LAIDLER, K. J. 1958. The Chemical Kinetics of Enzyme Action. Oxford Univ. Press, London.

LaJOLLO, F., TANNENBAUM, S. R. and LABUZA, T. P. 1971. Reaction at limited water concentration. 2. Chlorophyll degradation. J. Food Sci. *36*, 850–853.

LAKSO, A. N. and KLIEWER, W. M. 1975. Physical properties of phosphoenolpyruvate carboxylase and malic enzyme in grape berries. Am. J. Enol. Vitic. *26*, 75–78.

LALEGERIE, P. 1974. β-Glucosidase of sweet almonds *(Amygdalis communis).* II. Catalytic properties. Biochimie *56*, 1297–1303. (French)

LALL, B. S., MANZER, A. R. and HILTZ, D. F. 1975. Preheat treatment and improvement of frozen fish storage stability at -10°C in fillets and minced flesh of silver hake *(Merluccius bilinearis).* J. Fish. Res. Board Can. *32*, 1450–1454.

LAMOUREUX, G. L., SHIMABUKURO, P. H., SWANSON, H. R. and FREAR, D. S. 1970. Metabolism of an atrazine in excized sorghum leaf section. J. Agric. Food Chem. *18*, 81–86.

LAMOUREUX, G. L., STAFFORD, L. E., SHIMABUKURO, R. H. and ZAYLSKIE, R. G. 1973. Atrazine metabolism in sorghum: Catabolism of glutathione conjugate of atrazine. J. Agric. Food Chem. *21*, 1023–1030.

LAMPE, C. H. 1971. Response of tomato fruits to certain growth regulators with emphasis on pectolytic enzymes, cellulase, and ethylene. Diss. Abstr. Int. B. *32*, 1308–1309.

LANDMANN, W. A. 1963. Enzymes and their influence on meat tenderness. Proc. Meat Tenderness Symp., Campbell Soup Co., Camden, N.J., 87–97.

LANGHURST, A. K., LONG, J. E. and HLAVACEK, R. G. 1976. High efficiency corn wet-milling plant expands to 140,000 bu/day. Food Process. *37* (6) 46–49.

LANGLYKKE, A. F., SMYTHE, C. V. and PERLMAN, D. 1952. Enzyme technology. *In* The Enzymes, 1st Edition, Vol. 2. G.B. Sumner and K. Myrbäck (Editors). Academic Press, New York.

LANGMUIR, I. and SCHAEFER, V. J. 1938. Activities of ureases and pepsin monolayers. J. Am. Chem. Soc. *60*, 1351–1360.

LANGSRUD, T., REINBOLD, G. W. and HAMMOND, E. G. 1977. Proline production by *Propionibacterium shermanii* P59. J. Dairy Sci. *60*, 16–23.

LANKVELD, J. M. G. 1973. Fruit juices as emulsion systems. *In* Proc. Symp. Emulsions Foams Food Technol. April, Ebeltoft, Denmark.

LAPANJE, S. 1978. Physicochemical Aspects of Protein Denaturation. John Wiley & Sons, New York.

LARDY, H. A. 1966. Regulation of energy-yielding processes in muscles. *In* Physiology and Biochemistry of Muscle as Food, Vol. 1. E.J. Briskey, R.G. Cassens and J.C. Trautmann (Editors). University of Wisconsin Press, Madison.

LARGE, P. J. 1971. Non-oxidative demethylation of trimethylamine *N*-oxide by *Pseudomonas aminovorans.* FEBS Lett. *18*, 297–300.

LARSON, M. K. and WHITAKER, J. R. 1970. *Endothia Parasitica* protease. Parameters affecting activity of the rennin-like enzyme. J. Dairy Sci. *53*, 253–261.

LASCELLES, J. 1965. The biosynthesis of chlorophyll. *In* Biosynthetic Pathways in Higher Plants. J.B. Pridham and T. Swain (Editors). Academic Press, New York.

LASSITER, C. A., FRIES, G. F., HUFFMAN, C. F. and DUNCAN, C. W. 1959. Effect of pepsin on growth and health of young dairy calves fed milk replacement rations. J. Dairy Sci. *42*, 666–670.

LASZO KELLER, A. and ERDOS, J. 1970. Meat products. Span. Pat. 358,936. May 16.

LATRASSE, A., SARRIS, J., AILLET, J. F. and FEUILLAT, M. 1976. Pectin and protein turbidity in depectinized raspberry juice. Ind. Aliment. Agric. *93*, 423–430. (French)

LAUER, F. and SHAW, R. 1970. A possible genetic source for chipping potatoes from 40°F storage. Am. Potato J. *47*, 275–278.

LAVINTMAN, N., TANDECARZ, J., CARCELLER, M., MENDIARA, S. and CARDINI, C. E. 1974. Role of uridine diphosphate glucose in the biosynthesis of starch. Eur. J. Biochem. *50*, 145–155.

LAWRENCE, R. C. 1965. Activation of spores of *Penicillium roqueforti.* Nature *208*, 801–803.

LAWRENCE, W. C. and COLE, E. R. 1968. Yeast sulfur metabolism and the formation of hydrogen sulfide in brewery fermentations. Wallerstein Lab. Commun. *31*, 95–115.

LAWRIE, R. A. 1966. Meat Science. Pergamon Press, Oxford.

LAYHEE, P. 1975. Engineered FF line yields 5 big production benefits. Food Eng. *47* (2) 61–62.

LAYTON, L. L., LUNDIN, R. E., CORSE, J. W. and BRANDON, D. L. 1979. Unpublished results. Albany, Calif.

LAZAR, M. E. 1972. Blanching and partial drying of foods with superheated steam. J. Food Sci. *37*, 163–166.

LAZAR, M. E., LUND, D. B. and DIETRICH, W. C. 1971. A new concept in blanching—IQB reduces pollution while improving nutritive value and texture of processed foods. Food Technol. *25*, 684–686.

LAZARUS, L., LEE, C.-Y. and WERMUTH, B. 1976. Applications of general ligand affinity chromatography for the mutual separation of deoxyribonuclease and ribonuclease free of protein contaminants. Anal. Biochem. *74*, 138–144.

LAZDUNSKI, M. 1974. Half "of the sites" reactivity and the role of subunit interactions in enzyme catalysis. Prog. Bioorg. Chem. *3*, 82–140.

LASZLO KELLER, A. and ERDOS, J. 1970. Meat products. Span. Pat. 385,936. May 16.

LEA, C. H., PARR, L. J. and CARPENTER, K. J. 1960. Chemical and nutritional changes in stored herring meal. 2. Br. J. Nutr. *14*, 91–113.

LEABACK, D. H. 1976. Applications of fluorimetric assays. FEBS Lett. *66*, 1–3.

LEAKE, C. D. and SILVERMAN, M. 1971. The chemistry of alcoholic beverages. *In* The Biology of Alcoholism. B. Kisslin and H. Bergleiter (Editors). Plenum Press, New York.

LECOQ, D., HERVAGAULT, J. F., BROUN, G., JOLY, G., KERNEVEZ, J. P. and THOMAS, D. 1975. The kinetic behavior of an artificial bienzyme membrane. J. Biol. Chem. *250*, 5496–5500.

LECROISEY, A., KEIL-DLOUHA, V., WOODS, D. R., PERRIN, D. and KEIL, B. 1975. Purification, stability and inhibition of the collagenase from *Achromobacter iophagus.* FEBS Lett. *59* (2) 167–171.

LEDWARD, D. A. 1972. Metmyoglobin reduction and formation during aerobic storage of beef at 1°C. J. Food Sci. *37*, 634–635.

LEE, C. K., MATTAI, S. E. and BIRCH, G. G. 1975. Structural functions of taste in the sugar series. J. Food Sci. *40*, 390–393.

LEE, C. M., CHICHESTER, C. O. and LEE, T. C. 1974. Physiological consequences of browned food products. Int. Congr. Food Sci. Technol. *7a*, 11–13.

LEE, C. M. and TOLEDO, R. T. 1976. Factors affecting textural characteristics of cooked comminuted fish muscle. J. Food Sci. *41*, 391–397.

LEE, C. Y. 1975. New blanching techniques. Korean J. Food Technol. 7, 100–106.

LEE, C. Y. and KAPLAN, N. O. 1976. General

ligand affinity chromatography in enzyme purification. J. Macromol. Sci. *A10* (1–2) 15–52.

LEE, C. Y., ROBINSON, W. B., VAN BUREN, J. P., AUREE, T. E. and STOEWSAND, G. G. 1975. Methanol in wines in relation to processing and variety. Am. J. Vitic. Enol. Viticult. *26*, 184–187.

LEE, D. D., LEE, Y. Y., REILLY, P. J., COLLINS, E. V., JR. and TSAO, G. T. 1976. Pilot plant production of glucose with glucoamylase immobilized to porous silica. Biotechnol. Bioeng. *18*, 253–267.

LEE, D. D., REILLY, P. J. and COLLINS, E. V. 1978. Pilot plant production of glucose with soluble α-amylase and immobilized glucoamylase. Enzyme Eng. *3*, 525–530.

LEE, F. A. 1958. The blanching process. Adv. Food Res. *8*, 63–109.

LEE, F. A. and MATTICK, L. R. 1961. Fatty acids of the lipides of vegetables. I. Peas. J. Food Sci. *26*, 273–275.

LEE, F. A. and WAGENKNECHT, A. C. 1958. Enzyme action and off-flavor in peas. II. The use of enzymes prepared from garden peas. Food Res. *23*, 584–589.

LEE, K. H. and LEE, H.-J. 1975. Studies on the production of fermented feeds from agricultural waste products. I. On the production and characteristic of xylanase by *Aspergillus niger*. Chem. Abstr. *85*, 3847x. (Korean)

LEE, M. L. and MUENCH, K. H. 1969. Prolyl transfer RNA synthetase. I. Purification and evidence for subunits. J. Biol. Chem. *244*, 223–230.

LEE, S. H. and LABUZA, T. P. 1975. Destruction of ascorbic acid as a function of water activity. J. Food Sci. *40*, 370–373.

LEE, W. Y. and UNRAU, A. M. 1970. Beta amylase of an alien genome combinant. Cereal Chem. *47*, 351–362.

LEE, Y. B., HARGUS, G. L., KIRKPATRICK, J. A., BERNER, D. L. and FORSYTHE, R. H. 1975. Mechanism of lipid oxidation in mechanically deboned chicken meat. J. Food Sci. *40*, 964–967.

LEE, Y. P. 1966. Potato phosphorylase. Methods Enzymol. *8*, 550–554.

LEE, Y. S. 1969. Measurements, characterization, and evaluation of pectinesterase activity in apple fruits. Ph.D. Dissertation, Univ. of Maryland, College Park.

LEE, Y. Y. and TSAO, G. T. 1974. Mass transfer characteristics of immobilized enzymes. J. Food Sci. *39*, 667–672.

LEGRAND, G. 1967. Ascorbic acid oxidation by potato tyrosinase. Bull. Soc. Fr. Physiol. Veg. *13*, 43–49. (French)

LEHMANN, H. 1965. Changes in enzymes at low temperatures; long-term preservation of blood. Fed. Proc. *24* (2) Suppl., S66–S69.

LEHMITZ, R. 1977. Transketolase activity and thiamin deficiency in the kidney of rainbow trout *(Salmo gairdneri)* fed crude herring. Arch. Tiernaer. *27*, 287–295. (German)

LEHMITZ, R. and SPANNHOFF, L. 1977. Transketolase activity and thiamin deficiency in the kidney of rainbow trout *(Salmo gairdneri)* fed crude herring. Arch. Tierernaehr. *27*, 287–295. (German)

LEHNINGER, A. L. 1975. Biochemistry, 2nd Edition. Worth Publishers, New York.

LEINEN, N. 1978. Natural dough conditioner reduces mix time 25–30% in breadsticks, crackers. Baking Ind. *145* (1772) 12–13.

LELOIR, L. F. and GOLDENBERG, S. H. 1962. Glycogen synthetase from rat liver. Methods Enzymol. *5*, 145–147.

LEMPERLE, E. 1977. Volatile aromas from grapes and their wines. Kali-Briefe *5* (3) 1–8. (German)

LEO, H. T. and TAYLOR, C. C. 1962. Low methoxyl pectins. U.S. Pat. 3,034,901. May 5.

LEONE, M., LAMPARELLI, F., LA NOTTE, E., LIUZZI, V. A. and PADULA, M. 1977. The use of enzymic pecto-cellulolytic system in olive oil making. Riv. Ital. Sostanze Grasse *54*, 514–530. (Italian)

LEOPOLD, A. C. and ARDREY, R. 1972. Toxic substances in plants and the food habits of early man. Science *176*, 512–513.

LERCH, K. 1976. *Neurospora* tyrosinase: molecular weight, copper content and spectral properties. FEBS Lett. *69*, 157–160.

LERMAN, L. S. 1953. A biochemically specific method for enzyme isolation. Proc. Natl. Acad. Sci. *39*, 232–236.

LERNER, A. B. 1953. Metabolism of phenylalanine and tyrosine. Adv. Enzymol. *14*, 49–77.

LERNER, H. R., MAYER, A. M. and HAREL, L. E. 1972. Evidence for conformational changes in grape catechol oxidase. Phytochemistry *11*, 2415–2421.

LEROUX, A., JUNIEN, C., KAPLAN, J. C. and BAMBERGER, J. 1975. Generalized deficiency of cytochrome b_5 reductase in congenital methemoglobinemia with mental retardation. Nature *258*, 619–620.

LEROUX, A. and KAPLAN, J. C. 1972. Presence of red-cell type of NADH-methemoglobin reductase (NADH-diaphorase) in human nonerythroid cells. Biochem. Biophys. Res. Comm. *47*, 945–950.

LESLEY, B. E. and SHUMATE, J. W. 1937. Process of preparing spinach or the like for canning. U.S. Pat. 2,097,198. Oct. 23.

LETTS, D. and CHASE, T. C., JR. 1973. Chemical modification of mushroom tyrosinase for stabilization to reaction inactivation. *In* Immobilized Biochemicals and Affinity Chromatography. R.B. Dunlop (Editor). Plenum Press, New York.

LEU, K. 1976. Formation of isomeric hydroperoxides from linoleic acid by lipoxygenase. Lebensm.-Wiss. Technol. *7*, 82–85. (German)

LEUPRECHT, H. and SCHALLER, A. 1968. Methodological studies of the measurement of pectin methyl esterase activity in thawed frozen strawberry puree. I. Fruchtsaft-Ind. *12*, 2–11. (German)

LEVEILLE, G. A. 1976. Dietary fiber. Cereal Food World *21*, 255–258.

LEVIN, J. H., TENNES, B. R. and MARSHALL, D. E. 1975. Mechanizing the harvest. Agric. Res. *23* (7) 7–10.

LEVITT, J. 1962. A sulfhydryl-disulfide hypothesis of frost injury and resistance in plants. J. Theor. Biol. *3*, 355–391.

LEVVY, G. A. and SNAITH, S. M. 1970. The inhibition of glycosidases by aldonolactones. Adv. Enzymol. *36*, 151–181.

LEWIN, R. 1975. Of beads, chromatin, and gene expression. New Sci. *66*, 308–310.

LEWIS, J. C., PIERSON, C. F. and POWERS, M. J. 1963. Fungi associated with softening of bisulfite-brined cherries. Appl. Microbiol. *11*, 93–99.

LEWIS, Y., NAMBUDIRI, E. S., KRISHNAMURTHY, N. and COIMBATORE, P. 1969. White pepper. Perfum. Essent. Oil Rec. *60*, 53–57.

LI, K. H., BUNDUS, R. H. and NOZNICK, P. P. 1967. Prevention of pink color in white onions. U.S. Pat. 3,352,691. Nov. 14.

LIBBY, R. A. 1970. Direct starch analysis using DMSO solubilization and glucoamylase. Cereal Chem. *47*, 273–281.

LIEBERMAN, E. 1975. Enzymes in the beer industry. Enzyme Technol. Dig. *4*, 69–75.

LIEBIG, J. VON. 1859. Familiar Letters on Chemistry, in Its Relations to Physiology, Dietetics, Agriculture, Commerce, and Political Economy, 4th Edition. J. Blyth (Editor). Walton and Maberly, London.

LIENER, I. E. 1969. Toxic Constituents of Plant Foodstuffs. Academic Press, New York.

LIENER, I. E. 1973. Naturally occurring toxicants of horticultural significance. HortScience *8*, 112–116.

LIENER, I. E. 1975. Effects of anti-nutritional and toxic factors on the quality and utilization of legume proteins. *In* Protein Quality of Foods and Feeds. Part 2. M. Friedman (Editor). Marcel Dekker, New York.

LIENHARD, G. E. 1973. Enzymatic catalysis and transistion-state theory. Science *180*, 149–154.

LIGHTBODY, H. D. and FEVOLD, H. L. 1948. Biochemical factors influencing the shelf life of dried whole eggs and means for their control. Adv. Food Res. *1*, 149–202.

LILLEY, D. M. J. and PARDON, J. F. 1979. Structure and function of chromatin. Annu. Rev. Genet. *13*, 197–234.

LIM, P. E. and TATE, M. E. 1971. The phytases. I. Lysolecithin-activated phytase from wheat bran. Biochim. Biophys. Acta *250*, 104–155.

LIN, L.-N. and BRANDTS, J. F. 1979. Evidence suggesting that some proteolytic enzymes may cleave only the *trans* form of the peptide bond. Biochemistry *18*, 43–47.

LIN, T.-S. and HULTIN, H. O. 1977. Oxidation of myoglobin in vitro mediated by lipid oxidation in microsomal fractions of muscle. J. Food Sci. *42*, 136–140.

LIND, L. 1977. Enzymatic determination of starch. Fleischwirtschaft *57*, 1496–1498. (German)

LINDBLOM, M. 1977. Bread baking properties of yeast protein concentrates. Lebensm.-Wiss. Technol. *10*, 341–345.

LINDEN, G., CHAPPELET-TORDO, D. and LAZDUNSKI, M. 1977. Milk alkaline phosphatase. Stimulation by Mg^{2+} and the properties of the Mg^{2+} site. Biochim. Biophys. Acta *483*, 100–106.

LINDERSTRØM-LANG, K. 1928. On the fractionation of casein. C.R. Trav. Carlsberg *17* (9) 1–114.

LINDERSTRØM-LANG, K. and SATO, M. 1929. The splitting of glycylglycine, alanylglycine and leucylglycine by intestinal and malt peptidases. Z. Physiol. Chem. *184*, 83–92. (German)

LINDLEY, M. G. and BIRCH, G. G. 1975. Structural functions of taste in the sugar series. J. Sci. Food Agric. *26*, 117–124.

LINDQUIST, R. N. 1975. The design of enzyme inhibitors: transition state analogs. *In* Drug Design, Vol. 5. E. J. Ariens (Editor). Academic Press, New York.

LINEHAN, D. J. and HUGHES, J. C. 1969. Texture of cooked potato. J. Sci. Food Agric. *20*, 110–123.

LINEWEAVER, H. 1939. Energy of activation of enzyme reactions and their velocity below 0°. J. Am. Chem. Soc. *61*, 403–408.

LINEWEAVER, H. and BURK, D. 1934. Determination of enzyme dissociation constants. J. Am. Chem. Soc. *56*, 658–666.

LINEWEAVER, H., JANG, R. and JANSEN, E. F. 1949. Specificity and purification of polygalacturonase. Arch. Biochem. Biophys. *20*, 137–152.

LINEWEAVER, H. and SCHWIMMER, S. 1941. Some properties of crystalline papain: stability toward heat, pH, and urea; pH optimum with casein as substrate. Enzymologia *10*, 81–86.

LINK, K. P. and WALKER, J. C. 1933. The isolation of catechol from pigmented onion scales and its significance in relation to disease resistance in onions. J. Biol. Chem. *100*, 379–383.

LINKLATER, K. A., DYSON, D. A., and MORGAN, K. T. 1977. Fecal thiaminase in clinically normal sheep associated with outbreaks of poliocephalomalacia. Res. Vet. Sci. *22*, 308–312.

LINSON, E. V. 1968. Edible marine protein concentrate. S. Afr. Pat. 6,705,900. June 18.

LIPITOA, S. and ROBERTSON, G. L. 1977. The enzymic extraction of juice from yellow passion fruit pulp. Trop. Sci. *19*, 105–112.

LIPMANN, F. 1941. Metabolic generation and utilization of phosphate bond energy. Adv. Enzymol. *1*, 99–162.

LIPMANN, F. 1973. Nonribosomal polypeptide synthesis on polyenzyme templates. Acc. Chem. Res. *6*, 361–367.

LIPSCOMB, W. N. 1972. Three-dimensional structures and chemical mechanisms of enzymes. Chem. Soc. Rev. *1*, 319–336.

LIPSCOMB, W. N. 1973. Enzymatic activities of carboxypeptidase A's in solution and in crystals. Proc. Natl. Acad. Sci. *70*, 3797–3801.

LITMAN, D. J. and CANTOR, C. R. 1974. Surface topography of the *Escherichia coli* ribosome. Enzymatic iodination of the 50S subunit. Biochemistry *13*, 512–518.

LITTLE, A. C. 1976. Physical measurements as predictors of visual appearance. Food Technol. *30* (10) 74–82.

LITTLE, A. C. and MACKINNEY, G. 1972. The color of foods. World Rev. Nutr. Dist. *14*, 59–84.

LITTLE, L. L. 1968. Process for making cheese by coagulating milk at a low temperature. U.S. Pat. 3,406,076. Oct. 15.

LITTLE, L. L. 1974. Manufacture of cheese curd. U.S. Pat. 3,792,171. Feb. 12.

LIU, H.-P. and WATTS, B. M. 1970. Catalysts of lipid peroxidation in meats. 3. Catalysts of oxidative rancidity. J. Food Sci. *35*, 570–596.

LLOYD, N. E., LEWIS, L. T., LOGAN, R. M. and PATEL, D. N. 1972. Process for isomerizing glucose to fructose. U.S. Pat. 3,694,314. Sept. 26.

LO, W. Y., STEINKRAUS, K. H., HAND, D. B., HACKLER, L. R. and WILKINS, W. 1968. Soaking soybeans before extraction as it affects chemical composition and yield of soymilk. Food Technol. *22*, 1188–1190.

LOBZOV, K. I. and VOLIK, V. G. 1972. Use of glucose oxidase and catalase in prolonging the storage time of egg powder. Chem. Abstr. *78*, 56563p. (Russian)

LOCKER, R. H. 1960. Degree of muscular contraction as a factor in tenderness of beef. Food Res. *25*, 304–307.

LOCKWOOD, R. M. 1970. Sonic jet drying. *In* Innovations in Food Engineering. Univ. of Calif. Ext. Course, San Francisco.

LODISH, H. F. 1976. Translational control of protein synthesis. Annu. Rev. Biochem. *45*, 39–72.

LOEF, H. W. and THUNG, S. B. 1965. Influence of chlorophyllase on the color of spinach during and after processing. Z. Lebensm. Unters. Forsch. *126*, 401–406. (German)

LOGAN, J. L. and LEARMONTH, E. M. 1955. Gluten oxidizing capacity of soya. Chem. Ind. (London), 1220.

LOHMANN, K. 1933. Synthesis of natural hexose-monophosphate from its components. Biochem. Z. *262*, 137–151. (German)

LOK, R. and COWARD, J. K. 1976. Steric constraints in intramolecular reactions at sp^3 carbon. Implication for methylase mechanisms. Bioorg. Chem. *5*, 169–175.

LOLAS, G. M. and MARKAKIS, P. 1975. Phytic acid and the other phosphorus compounds of beans (*Phaseolus vulgaris* L.) J. Agric. Food Chem. *23*, 13–15.

LOLAS, G. M. and MARKAKIS, P. 1977. The phytase of navy beans *(Phaseolus vulgaris).* J. Food Sci. *42*, 1094–1106.

LOOMIS, W. D. and BATTAILE, J. 1966. Plant phenolic compounds and the isolation of plant enzymes. Phytochemistry *5*, 423–438.

LOOMIS, W. D. and CROTEAU, R. 1973. Biochemistry and physiology of lower terpenoids. Recent Adv. Phytochem. *6*, 215–228.

LOPEZ, A. and BAGANIS, N. A. 1971. Effect of radiofrequency energy at 60 Mhz on food enzyme activity. J. Food Sci. *36*, 911–914.

LOPEZ, A., BLOCKET, M. F. and WOOD, C. B. 1959. Catalase and peroxidase in raw and blanched southern peas, *Vigna sinensis*. Food Res. *24*, 548–551.

LOPEZ, A. and JOHNSON, J. M. 1971. Apple-grapefruit juice products. Res. Div. Bull. Va. Polytech. Inst. State Univ. *75*.

LOPEZ, A. and QUESNEL, V. C. 1974. The contributions of sulfur compounds to chocolate aroma. Proc. 1st Int. Congr. Cocoa Chocolate Res., 92–104.

LOPEZ, A. S. and QUESNEL, V. C. 1976. Methyl-S-methionine sulfonium salt: precursor of dimethyl sulfide in cacao. J. Sci. Food Agric. *27*, 85–88.

LORENZ, K. 1972. Food uses of triticale. Food Technol. *26*, 66–74.

LORENZ, K. 1976. Microwave heating of foods—changes in nutrient and chemical composition. Crit. Rev. Food Sci. Nutr. *7*, 339–370.

LORENZ, K. and SAUNDERS, R. M. 1978. Enzyme activities in commercially milled rice. Cereal Chem. *55*, 77–86.

LORENZ, K. and WELSH, J. R. 1976. Alpha-amylase and protease activity of maturing triticale and its parental species. Lebensm.-Wiss. Technol. *9*, 7–10.

LORIMER, G. H., GEWITZ, H. S., VÖLKER, W., SOLOMONSON, L. P. and VENNESLAND, B. 1974. The presence of bound cyanide in the naturally inactivated form of nitrate reductase of *Chlorella vulgaris*. J. Biol. Chem. *249*, 6074–6079.

LOSTY, T., ROTH, J. S. and SHULTS, G. 1973. Effect of radiation and heating on proteolytic activity of meat samples. J. Agric. Food Chem. *21*, 275–277.

LOVE, R. M. 1962. The effect of freezing on fish muscle. *In* Recent Advances in Food Science, Vol. 2. J. Hawthorn and J.M. Leitch (Editors). Butterworths & Co., London.

LOVE, R. M. 1968. Histological observations on the texture of fresh muscle. *In* Soc. Chem. Ind. Monograph 27, Rheology and Texture of Foodstuffs. Soc. Chem. Ind., London.

LOVE, R. M. 1980. Biochemistry in the fishing industry. TIBS *5* (5) 3–6.

LOVERN, J. A. and OLLEY, J. 1962. Inhibition and promotion of post-mortem lipid hydrolysis in the flesh of fish. J. Food Sci. *27*, 551–559.

LOW, P. S. and SOMERO, G. N. 1975. Protein hydration changes during catalysis: a new mechanism of enzymic rate enhancement and ion activation/inhibition of catalysis. Proc. Natl. Acad. Sci. *72*, 3305–3309.

LOWE, C. R., HANS, M., SPIBEY, N. and DRABBLE, W. T. 1980. The purification of inosine 5′-monophosphate dehydrogenase from *Escherichia coli* by affinity chromatography on immobilized Procion dyes. Anal. Biochem. *104*, 23–28.

LOWEY, S., SLAYTER, H. S., WEEDS, A. G. and BAKER, H. 1969. Substructure of the myosin molecule. I. Subfragments of myosin by enzymic degradation. J. Mol. Biol. *42*, 1–29.

LOWRY, O. H. and PASSONNEAU, J. V. 1972. A Flexible System of Enzymatic Analysis. Academic Press, New York.

LU, A. T. and WHITAKER, J. R. 1974. Some factors affecting rates of heat inactivation and reactivation of horseradish peroxidase. J. Food Sci. *36*, 1173–1178.

LU, A. Y. H. and COON, M. J. 1968. Role of hemoprotein P-450 in fatty acid ω-hydroxylation in a soluble system from liver microsomes. J. Biol. Chem. *243*, 1331–1332.

LUCHSINGER, W. W., MAGREE, L. and VICTOR, J. F. 1971. The preparation of complex oligosaccharides by the action of a β-1,4-glucan hydrolase on barley-β-D-glucan. Cereal Sci. Today *16*, 292.

LUE, P. F. and KAPLAN, J. G. 1970. Metabolic compartmentation at the metabolic level: the function of a multienzyme aggregate in the pyrimidine pathway of yeast. Biochim. Biophys. Acta *222*, 365–372.

LUH, B. S. and DAOUD, H. N. 1971. Effect of break temperature and holding time on pectin and pectic enzymes in tomato pulp. J. Food Sci. *36*, 1039–1043.

LUH, B. S., DEMPSEY, W. H. and LEONARD, S. 1954. Consistency of paste and puree made from Pearson and Marzano tomatoes. Food Technol. *8*, 576–580.

LUH, B. S., OZBILGIN, S. and LIU, Y. K. 1978. Textural changes in canned apricots in the presence of mold polygalacturonase. J. Food Sci. *43*, 713–716.

LUH, B. S., PEUPIER, L. Y. and LIU, Y. K. 1974. Role of pectic enzymes on softening in canned apricots. Calif. Agric. *28* (7) 4–6.

LUH, B. S. and PHITHAKPOL, B. 1972. Characteristics of polyphenoloxidase related to browning in cling peaches. J. Food Sci. *37*, 264–268.

LUH, B. S., ROBERTS, R. L. and LI, C. F. 1980. Quick cooking rice. *In* Rice: Production and Utilization. B.S. Luh (Editor). AVI Publishing

Co., Westport, Conn.

LUH, B. S., VILLAREAL, F., LEONARD, S. J., and YAMAGUCHI, M. 1960. Effect of ripeness level on consistency of canned tomato juice. Food Technol. *14*, 635–639.

LUKES, T. M. 1959. Pinking of onions during dehydration. Food Technol. *13*, 391–393.

LUKES, T. M. 1971. Thin layer chromatography of cysteine derivations of onion flavor compounds and the lachrimatory factor. J. Food Sci. *36*, 662–664.

LULAI, E. C. and BAKER, C. W. 1976. Physiochemical characterization of barley lipoxygenase. Cereal Chem. *53*, 777–786.

LUM, K. C., JR. 1969. Fish concentrate. Br. Pat. 1,157,415. July 9.

LUMRY, R. 1959. Some aspects of thermodynamics and mechanism of enzyme catalysis. *In* The Enzymes, 2nd Edition, Vol. 1. P.D. Boyer, H.A. Lardy and K. Myrbäck (Editors). Academic Press, New York.

LUMRY, R. 1973. Some recent ideas about the nature of the interactions between proteins and liquid water. J. Food Sci. *38*, 744–754.

LUMRY, R. 1974. Search for mechanisms of enzyme catalysis. Enzymol. Pract. Lab. Med. Proc. Continuation Course 1972, 3–58. P. Blume and E.F. Freier (Editors). Academic Press, New York.

LUND, D. B. 1974. Wastewater abatement in canning vegetables by IOB blanching. Rep. EPA–660/2-74-006 for Off. Res. Dev., U.S. Environ. Protect. Agency, Washington, D. C.

LUND, D. B. 1977. Design of thermal processing for maximizing nutrient retention. Food Technol. *21*, 71–78.

LUND, D. B., FENNEMA, O. and POWRIE, W. D. 1969 Enzymic and acid hydrolysis of sucrose as influenced by freezing. J. Food Sci. *34*, 378–382.

LUNDBERG, K. 1972. Enzymes—Why Them? Fremad Vorleget, Copenhagen. (Danish)

LURIA, S. E. 1975. Colicins and the energetics of all membranes. Sci. Am. *233* (6) 30–37.

LUSE, R. A. and McLAREN, A. D. 1963. Mechanisms of enzyme inactivation by ultraviolet light and the photochemistry of amino acids. Photochem. Photobiol. *2*, 343–360.

LUTZ, R. A. and FLÜCKINGER, J. 1975. Kinetic determination of glucose with the GEMSAEC (ENI) centrifugal analyzer by the glucose dehydrogenase reaction and comparison with two commonly used procedures. Clin. Chem. *21*, 1372–1377.

LYALL, J. T. and STOWELL, K. C. 1977. The uptake of gibberellic acid by abraded barley. J. Inst. Brew. *83*, 35–36.

LYALL, N. 1976. Process for preparing a pollen-containing supplementary foodstuff. Br. Pat. 1,420,019. June 1.

LYAN, R. W. and TAYLOR, E. W. 1971. Mechanism of adenosine triphosphate hydrolysis by actinomyosin. Biochemistry *10*, 4617–4624.

LYNAM, E. K. and SATTERLEE, L. D. 1980. Dependency of plastein formation on hydrophobic interactions. Inst. Food Technol. 40th Annu. Meet. Abstr. *143.* June. New Orleans.

LYNCH, R. E., LEE, G. R. and CARTWRIGHT, G. E. 1976. Inhibition by superoxide dismutase of methemoglobin formation from oxyhemoglobin. J. Biol. Chem. *251*, 1015–1019.

LYNN, K. R. 1977. Cross linking in the radiolysis of some enzymes. *In* Protein Crosslinking. Biochemical Molecular Aspects. M. Friedman (Editor). Plenum Press, New York.

LYNN CO, Y. C. and SCHANDERL, S. 1967. Occurrence of 418 and 444 nm chlorophyll-type compounds in some green plant tissues. Phytochemistry *6*, 145–148.

LYUTSKANOV, N., PISHIISKII, I. and KRACHANOV, K. H. 1974. Enzymic purification of apple pectin. FSTA *8* (7) J1162. (Russian)

MAARSE, H. and TEN NOEVER DE BRAW, J. M. C. 1974. Another catty odour compound causing air pollution. Chem. Ind. (London) *1*, 36–37.

MACALLISTER, R. V. 1979. Nutritive sweeteners made from starch. Adv. Carbohydr. Chem. Biochem. *36,* 15–36.

MACALLISTER, R. V., WARDRIP, E. K. and SCHNYDER, B. J. 1975. Modified starches, corn syrups containing glucose and maltose, corn syrups containing glucose and fructose, and crystalline dextrose. *In* Enzymes in Food Processing, 2nd Edition. G. Reed (Editor). Academic Press, New York.

MACALPINE, R. and PAYNE, C. G. 1977. Hydrolyzed feather protein as a source of amino acids for broilers. Br. Poultry Sci. *18*, 265–273.

MACBRIDE, M. A. and PARRISH, F. C., JR. 1977. The 30,000 dalton component of tender bovine longissimus muscle. J. Food Sci. *42*, 1627–1629.

MACDONNELL, L. R., JANSEN, E. F. and LINEWEAVER, H. 1945. Properties of orange pectinesterase. Arch. Biochem. *6*, 389–401.

MACDOWALL, M. A. 1973. Action of proteinase A_2 of *Actinidia chinensis* on the B-chain of oxidized insulin. Biochim. Biophys. Acta *293*, 226–231.

MacGIBBON, D. B. and ALLISON, R. M. 1970. A method for the separation and detection of plant glucosinolases (myrosinases). Phytochemistry 9, 541–544.

MacGREGOR, A. W. 1976. A note on the formation of α-amylase in de-embryonated barley kernels. Cereal Chem. *53*, 792–796.

MacGREGOR, A. W. 1977. Isolation, purification and electrophoretic properties of an α-amylase from malted barley. J. Inst. Brew. *83*, 100–103.

MacGREGOR, A. W., GORDON, A. G., MEREDITH, W. O. S. and LACROIX, L. 1972. Site of α-amylase in developing kernels. J. Inst. Brew. *78*, 174–179.

MacGREGOR, A. W., La BERGE, D. E. and MEREDITH, W. O. S. 1971. Separation of alpha and beta amylase enzymes from barley malt by ion exchange. Cereal Chem. *48*, 490–498.

MacGREGOR, A. W., THOMPSON, R. G. and MEREDITH, W. O. S. 1974. Alpha-amylase from immature barley: Purification and properties. J. Inst. Brew. *80*, 181–187.

MACKAY, D. A. M. and HEWITT, E. J. 1959. Application of enzymes to processed foods. II. Comparison of the effect of flavor enzyme from mustard and cabbage upon dehydrated cabbage. J. Food Sci. *24*, 253–261.

MACKENZIE, G., SHAW, G. and THOMAS, S. 1976. Synthesis of analogs of 5-aminoimidazole ribonucleotides and their effects as inhibitors and substrates of enzymes involved in the biosynthesis of purine nucleotides. Chem. Soc. Chem. Commun. *12*, 453–455.

MACKIE, I. M. 1974. Proteolytic enzymes in the recovery of proteins from fish waste. Process Biochem. *9* (10) 12–14.

MACKINNEY, G. and JOSLYN, M. 1938. The rate of conversion of chlorophyll to pheophytin. J. Am. Chem. Soc. *60*, 1132–1136.

MACKINNEY, G. and JOSLYN, M. 1940. The conversion of chlorophyll to pheophytin. J. Am. Chem. Soc. *62*, 231–232.

MACKINNEY, G. and JOSLYN, M. 1941. Temperature coefficients of rate of pheophytin formation. J. Am. Chem. Soc. *63*, 2530–2531.

MACKINNEY, G. and LITTLE, A. C. 1962. Color of Foods. AVI Publishing Co., Westport, Conn.

MACKINNEY, G. and LITTLE, A. C. 1972. The coloring matters of food. World Rev. Nutr. Diet. *14*, 85–89.

MacLAREN, D. D. 1975. Single cell protein: new processes open wider food uses. Food Prod. Dev. *9* (6) 26–32.

MacLENNAN, D. H. and HOLLAND, D. C. 1976. The calcium transport ATPase of sarcoplasmic reticulum. *In* The Enzymes of Biological Membranes, Membrane Transport, Vol. 3. A. Martonosi (Editor). Plenum Press, New York.

MacLEOD, A. J. 1970. The chemistry of vegetable flavours. Flavour Ind. *1*, 665–672.

MacLEOD, A. J. 1976. Volatile flavor compounds of the Cruciferae. *In* Biol. Chem. Cruciferae Pap. Conf. J.G. Vaughn, A.J. Macleod, and B.M.G. Jones (Editors). Academic Press, London.

MacLEOD, A. M. 1967. Gibberellic acid and malting. Wallerstein Lab. Commun. *30*, 85–93.

MacLEOD, P. and MORGAN, M. E. 1955. Leucine metabolism of *Streptococcus lactis* var. *maltigenes.* I. Conversion of alpha-ketoisocaproic acid to leucine and 3-methylbutanol. J. Dairy Sci. *38*, 1208–1214.

MacLEOD, P. and MORGAN, M. E. 1956. Leucine metabolism of *S. lactis* var. *maltigenes.* II. Transaminase and decarboxylase activity of acetone powders. J. Dairy Sci. *39*, 1125–1133.

MacMILLAN, J. D. and VAUGHN, R. H. 1964. Purification and properties of a polygalacturonic acid-transeliminase produced by *Clostridium multifermentans.* Biochemistry *3*, 564–572.

MACRIS, B. J. and MARKAKIS, P. 1971. Postirradiation inactivation of horseradish peroxidase. J. Food Sci. *26*, 812–815.

MacRITCHIE, F. 1977. Flour lipids and their effects in baking. J. Sci. Food Agric. *28*, 53–58.

MacWILLIAM, I. C. and REYNOLDS, T. 1966. Some effects of the use of gibberellic acid in malting. J. Inst. Brew. *72*, 171–173.

MACY, R. L., JR., NAUMANN, H. D. and BAILEY, M. E. 1964. Water-soluble flavor and odor precursors of meat. II. Effects of heating on amino nitrogen constituents and carbohydrates in lyophilized diffusates from aqueous extracts of beef, pork and lamb. J. Food Sci. *29*, 142–148.

MADL, R. L. and TSEN, C. C. 1974. The proteolytic system of triticale. *In* Triticale: First Man Made Cereal. C.C. Tsan (Editor). Am. Assoc. Cereal Chem., St. Paul, Minn.

MADRIGAL, L. S., ORTIZ, A. N., COOKE, R. D. and FENANDEZ, R. H. 1980. The dependence of crude papain yields on different collection ("tapping") procedures for papaya latex. J. Sci. Food Agric. *31*, 279–285.

MADSEN, G. B. and NORMAN, B. E. 1973. New specialty syrups. *In* Molecular Structure and Function of Food Carbohydrate. Conference Proceedings. Applied Science Publishers, London.

MADSEN, R. F. 1974. Membrane concentration. *In* Advances in Preconcentration and Dehydration of Foods. A. Spicer (Editor). John Wiley & Sons, New York.

MAEHLY, A. C. and CHANCE, B. 1954. The assay of catalases and peroxidases. *In* Methods of Biochemical Analysis. D. Glick (Editor). Interscience Publishers, New York.

MAGA, J. A. 1974. Bread flavor. Crit. Rev. Food Technol. *5*, 55–142.

MAGA, J. A. 1976A. The role of sulfur compounds in food flavor. Crit. Rev. Food Sci. Nutr. *6*, 153–176, 241–270.

MAGA, J. A. 1976B. Lactones in foods. Crit. Rev. Food Sci. Nutr. *8*, 1–56.

MAGA, J. A. 1978. Cereal volatiles, a review. J. Agric. Food Chem. *26*, 175–178.

MAGAUDDA, G. 1973. The possibility of recognizing irradiated and non-irradiated potatoes by their weight loss. J. Food Sci. *38*, 1253–1254.

MAGEE, S. C. and EBNER, K. E. 1973. Inactivation of soluble bovine milk galactosyltransferase (lactose synthetase) by sulfhydryl reagents and trypsin. Protection by substrates and products. J. Biol. Chem. *249*, 6992–6998.

MAGGIO, E. T. 1980. Enzyme-immunoassay. CRC Press, Boca Raton, Fla.

MAHADEVAN, A., KUC, J. and WILLIAMS, E. B. 1965. Biochemistry of resistance in cucumber against *Cladosporium cucumerinum*. 1. Presence of a pectinase inhibitor in resistant plants. Phytopathology *55*, 1000–1003.

MAIER, H. G. 1970. Volatile flavoring substances in foods. Angew. Chem. Int. *9*, 917–926.

MAIER, V. P., BREWSTER, L. C. and HSU, A. C. 1973. Ethylene-accelerated limonoid metabolism in citrus fruits: a process for reducing bitterness. J. Agric. Food Chem. *21*, 490–494.

MAIER, V. P. and METZLER, D. M. 1965. Changes in individual date polyphenols and their relation to browning. J. Food Sci. *30*, 747–752.

MAIER, V. P. and TAPPEL, A. 1959. Rate studies of unsaturated fatty acid oxidation. J. Am. Oil Chem. Soc. *36*, 8–12.

MAIER, V. P., TAPPEL, A. L. and VOLMAN, D. H. 1955. Reversible inactivation of enzymes at low temperatures. Studies of temperature dependence of phosphatase and peroxidase-catalyzed reactions. J. Am. Chem. Soc. 77, 1278–1280.

MAILLARD, L. C. 1912. Action of amino acids on sugars. C.R. Acad. Sci. *154*, 66–68. (French)

MAITI, I. B., MAJUMDER, A. L. and BISWAS, B. B. 1974. Purification and mode of action of phytase from *Phaseolus Aureus*. Phytochemistry *13*, 1047–1051.

MAJOR, R. T. and THOMAS, M. 1972. Formation of 2-hexenal from linoleic acid by macerated *Ginkgo* leaves. Phytochemistry *11*, 611–617.

MAKINEN, K. K. 1979. Xylitol and oral health. Adv. Food Res. *25*, 413.

MAKINEN, K. K. 1976. Possible mechanisms for the cariostatic effect of xylitol. Int. J. Vitam. Res., Beih. *15*, 368–380.

MAKINO, N., McMAHILL, P. and MASON, H. S. 1974. The oxidation of copper in resting tyrosinase. J. Biol. Chem. *249*, 6062–6066.

MAKOWER, R. U. 1956. Influence of enzymes on the quality of processed fruits and vegetables. Econ. Bot. *10*, 38–41.

MAKOWER, R. U. 1964. Effect of nucleotides on enzymic browning in potato slices. Plant Physiol. *39*, 956–959.

MAKOWER, R. U. 1969. Changes in phytic acid and acid soluble phosphorus in metering pinto beans. J. Sci. Food Agric. *20*, 82–84.

MAKOWER, R. U. and BOGGS, M. M. 1960. Quality of cabbage dehydrated after chemical or steam inactivation of enzymes. Food Technol. *14*, 295–297.

MAKOWER, R. U. and SCHWIMMER, S. 1954. Inhibition of enzymic color formation by adenosine triphosphate. Biochim. Biophys. Acta *14*, 156–157.

MAKOWER, R. U. and SCHWIMMER, S. 1956. Method of inhibiting the browning of plant tissue. U.S. Pat. 2,738,280. Mar. 13.

MAKOWER, R. U. and SCHWIMMER, S. 1957. Enzymatic browning, reflectance measurements, and effect of adenosine triphosphate on color changes induced in plant slices by polyphenol oxidase. J. Agric. Food Chem. *5*, 768–773.

MALECKI, G. J. 1965. Blanching and canning green vegetables. U.S. Pat. 3,183,102. May 11.

MALIK, A. C. and SWANSON, A. M. 1975. Action of heat-stable *Pseudomonas fluorescens* protease on sterilized skim milk. J. Dairy Sci. *58*, 795.

MALKIN, R. and MALMSTROM, B. G. 1970. State and function of copper in biological systems. Adv. Enzymol. *33*, 177–244.

MALMOS, H. 1978. Enzyme applications in food, pharmaceuticals, and other industries: industrial applications of cellulase. AIChE Symp. Ser. *74* (172) 93–99.

MALMSTROM, B. G., ANDREASSON, L.-E. and REINHAMMAR, B. 1975. Copper-containing oxidases and superoxide dismutase. *In* The Enzymes, 3rd Edition, Vol. 12B. P.D. Boyer (Editor). Academic Press, New York.

MANDAL, S. K. and MUKHERJEE, S. K. 1974. Chemical changes of fish muscle during preservation with ammonia. J. Agric. Food Chem. *22*, 832–835.

MANDELS, M. 1975. Microbial sources of cellulase. Biotechnol. Bioeng. Symp. *5*, 81–105.

MANECKE, G. 1964. Serologically active protein resins and enzyme resins. Naturwissenschaften *51*, 25–34.

MANI, V. V. S. and WESSELS, H. 1976. Detection of esterified olive oil. Fette, Seifen Anstrichm. *78*, 351–359. (German)

MANN, D. L. and MORRISON, W. R. 1975. Effects of ingredients on the oxidation of linoleic acid by lipoxygenase in bread doughs. J. Sci. Food Agric. *26*, 493–505.

MANNERS, D. J. 1974. The structure and metabolism of starch. *In* Essays in Biochemistry. P.N. Campbell and F. Dickens (Editors). Academic Press, London.

MANNERS, D. J. 1975. Debranching enzymes in plant tissues. Biochem. Soc. Trans. *3*, 49–53.

MANNERS, D. J. and HARDIE, D. G. 1977. Studies on debranching enzymes. VI. The starch-debranching enzyme system of germinated barley. Tech. Q. Master Brew. Assoc. Am. *14*, 120–125.

MANNERS, D. J. and MARSHALL, J. J. 1969. Studies on carbohydrate-metabolizing enzyme. XXII. The β-glucanase system of malted barley. J. Inst. Brew. *75*, 550–561.

MANNERS, D. J. and MARSHALL, J. J. 1971. Studies on carbohydrate-metabolizing enzymes. XXIV. The action of malted-rye alpha-amylase on amylopectin. Carbohydr. Res. *18*, 203–209.

MANNERS, D. J. and MEYER, M. T. 1977. The molecular structures of some glucans from the cell walls of *Schizosaccharomyces pombe*. Carbohydr. Res. *57*, 189–203.

MANNERS, D. J., PALMER, G. H., WILSON, G. and YELLOWLEES, D. 1971. Effect of gibberellic acid on the development of some cereal carbohydrases. Biochem. J. *125*, 308–318.

MANNERS, D. J. and YELLOWLEES, D. 1973. Studies on debranching enzymes. I. The limit dextrinase activity of extracts of certain higher plants and commercial malts. J. Inst. Brew. *79*, 377–385.

MANNERVIK, B., JACOBSSON, K. and BOGGARAM, V. 1976. Purification of glutathione reductase from erythrocytes by the use of affinity chromatography on 2′,5′-ADP-Sepharose 4-B. FEBS Lett. *66*, 221–223.

MANOHAR, S. V. 1969. Some properties of fluorescence of fish muscle. J. Fish. Res. Board Can. *26*, 1368–1371.

MANSELL, R. L., BABBEL, G. R. and ZENK, M. H. 1976. Multiple forms and specificity of coniferyl alcohol dehydrogenase from cambial regions of higher plants. Phytochemistry *15*, 1849–1853.

MANTIS, A. J., KARIOANNOGLOU, P. G., SPANOS, G. P. and PANESTSOS, A. G. 1978. The effect of garlic extract on food poisoning in culture media. Lebensm.-Wiss. Technol. *11*, 26–28.

MANTIS, A. J., KOIDIS, P. A., KARAIOANNOGLOU, P. G. and PANESTOS, A. G. 1979. Effect of garlic extract on food poisoning bacteria. Lebensm.-Wiss. Technol. *12*, 330–332.

MAO, W. W. and STERLING, C. 1970. Parameters of texture changes in processed fish: cross-linkage of proteins. J. Texture Studies *1*, 484–490.

MAPSON, L. W. and MOUSTAFA, E. M. 1955. The oxidation of glutathione by a lipoxidase from pea seeds. Biochem. J. *60*, 71–80.

MAPSON, L. W., SWAIN, T. and TOMALIN, A. W. 1963. Influence of variety, cultural conditions and temperature of storage on enzymic browning of potato tubers. J. Sci. Food Agric. *14*, 673–684.

MAPSON, L. W. and TOMALIN, A. W. 1961. Preservation of peeled potatoes. III. Inactivation of phenolase by heat. J. Sci. Food Agric. *12*, 54–58.

MAPSON, L. W. and WARDALE, D. A. 1972. Role of indolyl-3-acetic acid in the formation of ethylene from 4-methylmercapto-2-oxo butyric acid by peroxidase. Phytochemistry *11*, 1371–1387.

MARCHESINI, A., MONTUORI, F., MUFFATO, D. and MAESTRI, D. 1974. Application and advantages of the enzymatic method for the assay of ascorbic and dehydroascorbic acids and reductones. J. Food Sci. *39*, 568–571.

MARCHYLO, B., KRUGER, J. E. and IRVINE, G. N. 1976. α-Amylase from immature hard red spring wheat. I. Purification and some chemical and physical properties. Cereal Chem. *53*, 157–173.

MARGALITH, P. and SCHWARTZ, Y. 1970. Flavor and microorganisms. Adv. Appl. Microbiol. *12*, 35–88.

MARIE, J., KAHN, A. and BOIVIN, P. 1976. L-type pyruvate kinase from human liver. Purification by double affinity elution, electrofocusing and immunological studies. Biochim. Biophys. Acta *438*, 393–406.

MARKAKIS, P. 1974. Anthocyanins and their stability in foods. Crit. Rev. Food Technol. *4*, 437–456.

MARKAKIS, P. 1975. Anthocyanin pigments in foods. *In* Postharvest Biology and Handling of Fruits and Vegetables. N.F. Haard and D.K. Salunkhe (Editors). AVI Publishing Co., Westport, Conn.

MARKAKIS, P. and EMBS, R. J. 1966. Effect of sulfite and ascorbic acid on mushroom phenol oxidase. J. Food Sci. *31*, 807–811.

MARKERT, C. L. 1968. Molecular basis for isozymes. Ann. N.Y. Acad. Sci. *151,* 14–40.

MARKERT, C. L. 1975. Biology of isoenzymes. *In* Isoenzymes, 3rd Int. Conf. 1974, Vol. 1. C.L. Markert (Editor). Academic Press, New York.

MARKERT, C. L. 1975. Isozymes, Vol. 1–4. Academic Press, New York.

MARKEY, P. E., GREENFIELD, P. F. and KITTRELL, J. R. 1975. Immobilization of catalase and glucose oxidase on inorganic supports. Biotechnol. Bioeng. *17*, 285–289.

MARKOVIC, O., HEINRICHOVA, K. and LENKEY, B. 1975. Pectolytic enzymes from banana. Collect. Czech. Chem. Commun. *40*, 769–774.

MARMSTAL, E. and MANNERVIK, B. 1979. Purification, characterization and kinetic studies of glyoxylase I from rat liver. Biochim. Biophys. Acta *556*, 362–370.

MARQUIS, V. and HASKELL, P. 1965. The Cheese Book. Simon and Schuster, New York.

MARRS, W. M. 1975. The properties and uses of natural and modified starches. Sci. Tech. Surv.—Br. Food Manuf. Ind. Res. Assoc. *85*, 1–21.

MARSH, B. B. 1966. Relaxing factor in muscle. *In* The Physiology and Biochemistry of Muscle as Food. E.J. Briskey, R.G. Cossens, and J.C. Trautman (Editors). Univ. of Wisconsin Press, Madison.

MARSH, B. B. 1977. The basis of tenderness in muscle foods. J. Food Sci. *42*, 295–297.

MARSHALL, E. K., JR. 1914. On soy bean urease: The effect of dilution, acids and alkalies and ethyl alcohol. J. Biol. Chem. *17*, 351–361.

MARSHALL, J. J. 1975A. Starch degrading enzymes, old and new. Staerke *27*, 377–383.

MARSHALL, J. J. 1975B. Degradation of barley glucan by a purified (1→4)-β-D-glucanase from the snail, *Helix pomatia.* Carbohydr. Res. *42*, 203–207.

MARSHALL, J. J., LAUDA, C. A. and WHELAN, W. J. 1976. Naturally occurring inhibitors of α-amylase. *In* Physiological Effects of Food Carbohydrates. A. Jeannes (Editor). Am. Chem. Soc., Washington, D. C.

MARSHALL, J. J. and RABINOWITZ, M. L. 1975. Enzyme stabilization by covalent attachment of carbohydrate. Arch. Biochem. Biophys. *167*, 777–779.

MARSHALL, R. O. and KOOI, E. R. 1957. Enzymatic conversion of D-glucose to D-fructose. Science *125*, 648–649.

MARSHALL, W. E. 1978. Enzymatic flavor enhancement in food products: A commercial view. Biotechnol. Bioeng. *18*, 921–925.

MARTELL, A. E. 1973. Artificial enzymes. *In* Metal Ions in Biological Systems, Vol. 2. Mixed-Ligand Complexes. H. Segel (Editor). Marcel Dekker, New York.

MARTENSSON, K. 1974. Preparation of an immobilized two enzymes system, β-amylase-pullulanase, on an acrylic co-polymer for the conversion of starch to glucose. II. Biotechnol. Bioeng. *16*, 579–591.

MARTIN, H. L. and BAMFORTH, C. W. 1980. The relationship between β-glucan solubilase, barley autolysis and malting potential. J. Inst. Brew. *86*, 216–221.

MARTIN, M. M. and MARTIN, T. S. 1979. The distribution and origins of the cellulolytic enzymes of the higher termite *Macroterme natalensis.* Physiol. Zool. *52,* 11–21.

MARTINEZ, W. H., BERARD, L. C. and GOLDBLATT, L. A. 1970. Cottonseed products—composition and functionality. J. Agric. Food Chem. *18*, 961–968.

MARTINS, C. B. and WHITAKER, J. R. 1968. Purification of cathepsin D and its action on actomyosin. J. Food Sci. *33*, 59–64.

MARX, J. 1980A. Calmodulin: A protein for all seasons. Science *208*, 274–276.

MARX, J. 1980B. Newly made proteins zip through the cell. Science *207*, 164–167.

MASCINI, M. and LIBERTI, A. 1974. Enzyme-coupled cyanide solid-state electrode. Anal. Chim. Acta *68*, 177–184.

MASHEV, N., RADNEV, R., DOKATANOVA, T. S. and GOSPODINOVA, M. 1977. Study of the mutual effect of trace elements and gibberellic acid on the biochemical and technological properties of Brewery barley. Chem. Abstr. *88*, 84449f. (Russian)

MASIOR, S., POGORZELSKI, E., CZYZYCKI, A. and MAREK, G. 1976. Removing methanol from musts and young wines made of hybrid grapes. Acta Aliment. Polonica *2*, 13–22.

MASON, H. S. 1955. Comparative biochemistry of the phenolase complex. Adv. Enzymol. *16*, 105–184.

MASON, H. S. 1957. Mechanisms of oxygen metabolism. Adv. Enzymol. *19,* 79–233.

MASSEY, V. 1953. Studies on fumarase. 4. The effects of inhibitors on fumarase activity. Biochem. J. *55,* 172–177.

MASSEY, V. and HEMMERICH, P. 1975. Flavin and pteridine monooxygenases. *In* The Enzymes, 3rd Edition, Vol. 12B. P.D. Boyer (Editor). Academic Press, New York.

MASURE, M. P. and CAMPBELL, H. 1944. Rapid estimation of peroxidase in vegetable extracts—an index of blanching adequacy for frozen vegetables. Fruit Prod. J. *23,* 369–374.

MATAVELLI, L., MARCHESINI, A. and MANITO, P. 1975. Ascorbic acid oxidase fixed on a solid support. Ann. Inst. Sper. Vallor. Technol. Prod. Agric. *6,* 47–50. (Italian)

MATHEIS, G. and BELITZ, H. D. 1975. Multiple forms of soluble monophenol, dihydroxyphenylalanine: oxygen-oxidoreductase (EC 1.14.18.1) from potato tubers. Z. Lebensm.-Unters. Forsch. *157,* 221–227. (German)

MATHESON, N. A. 1962. Enzymic activity at low moisture levels and its relation to deterioration in freeze-dried foods. J. Sci. Food Agric. *13,* 248–250.

MATHESON, N. K. and RICHARDSON, R. R. 1976. Starch phosphorylase enzymes in developing and germinating pea seeds. Phytochemistry *15,* 887–892.

MATHEW, A. G. and PARPIA, H. A. B. 1971. Food browning as a polyphenol reaction. Adv. Food Res. *19,* 75–145.

MATHEWSON, P. R. and POMERANZ, Y. 1978. Hot-paste viscosity and alpha-amylase susceptibility of hard red winter wheat flour. J. Food Sci. *43,* 60–63.

MATIKKALA, E. J. and VIRTANEN, A. I. 1967. On the quantitative determination of the amino acids and γ-glutamylpeptides of onion. Acta Chem. Scand. *21,* 2891–2893.

MATILE, P. and WINKENBACH, F. 1973. Function of lysosomes and lysosomal enzymes in the senescing corolla of the morning glory *(Ipomoea purpurea).* J. Exp. Bot. *22,* 759–771.

MATOBA, T. and HATA, T. 1972. Relation between bitterness of peptides and their chemical structures. Agric. Biol. Chem. *36,* 1423–1431.

MATOO, A. K. and MODI, V. V. 1970. Citrate cleavage enzyme in mango fruit. Biochem. Biophys Res. Comm. *39,* 895–904.

MATSUBARA, H. and FEDER, J. 1971. Other bacterial, mold and yeast proteinases. *In* The Enzymes, 3rd Edition, Vol. 3. P.D. Boyer (Editor). Academic Press, New York.

MATSUDA, Y., SATOH, T., BEPPU, T. and ARIMA, K. 1976. Purification and properties of Co^{2+}-requiring heme protein having lipoxygenase activity from *Fusarium oxysporum.* Agric. Biol. Chem. *40,* 963–976.

MATSUI, Y., YOKOI, T. and MOCHDA, K. 1976. Catalytic properties of a modified cyclodextrin. Chem. Lett. *10,* 1037–1040.

MATSUMOTO, T. 1977. A cytochemical study of the absorption of horseradish peroxidase through the human gastrointestinal tract. Okayama Igakki Zasshi *89,* 1358–1392. (Japanese)

MATSUO, R. R., BRADLEY, J. W. and IRVINE, G. N. 1968. Studies on pigment destruction during processing of spaghetti. Cereal Sci. Today *13,* 122.

MATSUSHITA, S., KOBAYASHI, M. and NITA, T. 1970. Inactivation of enzymes by linoleic acid hydroperoxides and linoleic acid. Agric. Biol. Chem. *34,* 817–824.

MATTAVELLI, L., MARCHESINI, A. and MANITTO, P. 1975. Ascorbic acid fixed on a solid support. Ann. Inst. Sper. Valor. Tecnol. *6,* 47–50.

MATTENHEIMER, H. 1966. The Theory of Enzyme Tests. Boehringer Mannheim Corp., New York.

MATTHEWS, B. W. and BERNHARD, S. A. 1973. Structure and symmetry of oligomeric enzymes. Annu. Rev. Biophys. Bioeng. *2,* 257–317.

MATTIASSON, B. 1977. The use of coimmobilized lysozyme in enzyme columns; a step toward the design of self-sterilizing enzyme columns. Biotechnol. Bioeng. *19,* 777–780.

MATTSON, S. 1946. The cookability of yellow peas. Acta Agric. Suec. *2,* 185–231.

MATYAS, J., PETROLZY, I., SOMFAI, E. and DAVID, A. 1974. Vegetable product coated with propionic acid impregnated particular material. U.S. Pat. 3,864,567. Nov. 5.

MATZ, S. A. 1962. Food Texture. AVI Publishing Co., Westport, Conn.

MAUK, A. G., MAUK, M. R. and TAKETA, F. 1973. Activation of methaemoglobin peroxidase by inositol hexaphosphate. Nature New Biol. *246,* 188–189.

MAULE, A. P. and GREENSHIELDS, R. N. 1971. Carbohydrate balance and its economies in brewing. Process Biochem. *6,* 28–31.

MAURON, J. 1970. Nutritional evaluation of protein by enzymatic methods. *In* Evaluation of Novel Protein Products. A.E. Bender *et al.* (Editors). Pergamon Press, Oxford.

MAXIMOW, N. A. 1904. The problem of respiration. Ber. Dtsch. Bot. Ges. *22,* 225–235, 488–489.

MAXWELL, J. C. and CAUGHEY, W. S. 1976. An infrared study of NO bonding to heme B and hemoglobin A. Evidence for inositol hexaphosphate induced cleavage of proximal histidine to iron bonds. Biochemistry *15,* 388–395.

MAY, S. W. and ZABORSKY, O. R. 1974. Ligand specific chromatography. Sep. Purif. Methods *3* (1) 1–86.

MAYER, K. and PAUSE, G. 1969. Malic acid, lactic acid and citric acid in Swiss wines. Vitis *8,* 38–49. (German)

MAYER, K., PAUSE, G. and VETCH, U. 1973. Formation of biogenic amines during sauerkraut fermentation. Ind. Obst. Gemeuseverwert. *58,* 307–309. (German)

MAZELIS, M. 1963. Demonstration and characterization of cysteine sulfoxide lyase in the cruciferae. Phytochemistry *2,* 15–22.

MAZELIS, M. 1975. The enzymatic cleavage of the C-S bond of substituted cysteines in higher plants. Phytochemistry *14,* 857–858.

MAZELIS, M. 1978. Personal communication. Albany, Calif.

MAZELIS, M. and CREWS, L. 1968. Purification of alliin lyase of garlic, *Allium sativum* L. Biochem. J. *108,* 725–730.

McADAM, R. L., BUTTERWORTH, E. R. and DUENSING, W. J. 1977. Adsorption stabilizing of beer. Tech. Q. Master Brew. Assoc. Am. *14,* 145–152.

McBRIDE, J. R., IDLER, D. R. and MACLEOD, A. A. 1961. The liquefaction of British Columbia herring by ensilage, proteolytic enzymes and acid hydrolysis. J. Fish. Res. Board Can. *18,* 93–112.

McCAIN, G. R., BLUMER, T. N., CRAIG, H. B. and STEEL, R. G. 1968. Free amino acids in ham muscles during successive aging periods and their relation to flavor. J. Food Sci. *33,* 142–146.

McCANCE, R. A. and WIDDOWSON, E. M. 1942. Mineral metabolism and dephytinized bread. J. Physiol. (London) *101,* 304–313.

McCAY, P. B., GIBSON, D. D., FONG, K. -L. and HORNBROOK, K. R. 1976. Effect of glutathione peroxidase activity on lipid peroxidation in biological membranes. Biochim. Biophys. Acta *431,* 459–468.

McCAY, P. B. and POYER, J. L. 1976. Enzyme generated free radicals as initiators of lipid peroxidation in biological membranes. *In* Enzymes of Biological Membranes, Vol. 4. Electron Transport Systems and Receptors. K.M. Martonosi (Editor). Plenum Press, New York.

McCLEARY, B. V. and CHICK, B. F. 1977. The purification and properties of a thiaminase I enzyme from nardoo *(Marsilea drummondii).* Phytochemistry *16,* 207–213.

McCLOSKEY, L. P. 1976. An enzyme assay for acetate in fruit juices and wines. J. Agric. Food Chem. *24,* 523–526.

McCLOSKEY, L. P. and REPLOGLE, L. L. 1974. Evaluation of an enzymatic method for estimating ethanol in wines using an enzyme kit. Am. J. Enol. Vitic. *25,* 194–197.

McCONNEL, J. E. W. 1956. Enzymes—heat resistance and regeneration. Review of literature on thermal inactivation of enzymes in foods. Natl. Canners Assoc. Publ. *D-252,* May 25.

McCORMICK, R. D. 1976A. Sublihydration: Improved color, texture, for reconstituted freeze-dried foods. Food Prod. Dev. *10* (3) 16, 18.

McCORMICK, R. D. 1976B. A nutritious alternative for the lactose-intolerant consumer. Food Prod. Dev. *10* (5) 17–18.

McDONALD, C. E. 1976. Lutein bleaching enzymes of durum wheat endosperm. Abstr. AGFC *133,* 172, Meet., Am. Chem. Soc., St. Paul, Minn.

McDONALD, E. C. 1969. Proteolytic enzymes of wheat and their relation to baking quality. Bakers Dig. *43,* 26–30, 72.

McDONALD, E. C. and CHEN, L. L. 1964. Properties of wheat flour proteinases. Cereal Chem. *41,* 443–455.

McFARLANE, I. J., LEES, E. M. and CONN, E. E. 1975. Biosynthesis of cyanogenic glycosides in higher plants. VI. In vitro biosynthesis of dhurrin, the cyanogenic glycoside of *Sorghum bicolor.* J. Biol. Chem. *250,* 4708–4713.

McFEETERS, R. F. 1975. Substrate specificity of chlorophyllase. Plant Physiol. *55,* 377–381.

McGLOTHLIN, C. D. and JORDAN, J. 1975. Enthalpimetric enzyme assay. Anal. Chem. *47,* 1479–1481.

McINTOSH, E. N. 1967. Effect of postmortem aging and enzyme tenderizers on mucoproteins of bovine skeletal muscle. J. Food Sci. *32,* 210–217.

McINTOSH, E. N. and CARLIN, A. F. 1963. The effect of papain preparations on beef skeletal muscle. J. Food Sci. *28,* 283–285.

McKENZIE, A. 1936. Asymmetric synthesis. Ergeb. Enzymforsch. *5,* 50–65.

McKENZIE, H. 1971. Milk Proteins, Chemistry and Molecular Biology. Academic Press, New York.

McLAREN, A. D. 1954. The adsorption and

reactions of enzymes and proteins on kaolinite. J. Phys. Chem. *58,* 129–137.

McLAREN, A. D. 1960. Enzyme action in structurally restricted systems. Enzymologia *21,* 356–364.

McLAREN, A. D. 1963. Enzymes in structurally restricted systems. II. The digestion of insoluble substrates by hydrolytic enzymes. Enzymologia *26,* 237–248.

McLAREN, A. D. 1974. Soil as a system of bound enzymes. Chem. Ind. *7,* 316–318.

McLAUGHLIN, J. M., JONES, J. D., SHAH, B. G. and BEARE-ROGERS, J. L. 1975. Reproduction in rats fed protein concentrate from mustard or rapeseed. Nutr. Rep. Inst. *15,* 327–335.

McPHERSON, A., JR. 1976. The growth and preliminary investigation of protein and nucleic acid crystals for x-ray diffraction analysis. Methods Biochem. Anal. *23,* 249–311.

McPHIE, P. 1971. Dialysis. Methods Enzymol. *22,* 23–32.

McWEENY, D. J. 1968. Reactions in food systems: negative temperature coefficients and other abnormal temperature effects. J. Food Technol. *3,* 15–30.

McWEENY, D. J., KNOWLES, M. E. and HEARNE, J. F. 1974. The chemistry of nonenzymic browning in foods and its control by sulphites. J. Sci. Food Agric. *25,* 735–746.

MEANS, A. R. and DEDMAN, J. R. 1980. Calmodulin—an intracellular calcium receptor. Nature *285,* 73–77.

MEANS, D. S., RYAN, R. and FEENEY, R. E. 1974. Protein inhibitors of proteolytic enzymes. Acc. Chem. Res. *7,* 315–320.

MECHAM, D. K. 1971. Lipids. *In* Wheat Chemistry and Technology, 2nd Edition. Y. Pomeranz (Editor). Am. Assoc. Cereal Chem., St. Paul, Minn.

MECHAM, D. 1975. Personal communication. Albany, Calif.

MEHLER, A. H. 1957. Introduction to Enzymology. Academic Press, New York.

MEHLITZ, A. 1930. On pectase activity. I. Enzymatic studies on favorable conditions of pectase coagulation. Biochem. Z. *221,* 217–231. (German)

MEIJER, A. E. F. H. 1975. Histochemistry of enzymes. Acta Histochem., Suppl. *14,* 33–46.

MEINKE, W. W. and MATTIL, K. F. 1973. Autolysis as a factor in the production of protein isolates from whole fish. J. Food Sci. *38,* 864–866.

MEISEL, P. and ULMANN, M. 1974. The Biosynthesis of Starch. Verlag Paul Parey Price, Hamburg. (German)

MEISTER, A. 1971–1979. Advances in Enzymology and Related Areas of Molecular Biology. John Wiley & Sons, New York.

MELCHER, U. and VARNER, J. E. 1971. Protein release by barley aleurone layers. J. Inst. Brew. *77,* 456–461.

MELLING, J. and PHILLIPS, B. W. 1975. Large scale extraction and purification. *In* Handbook of Enzyme Biotechnology. A. Wiseman (Editor). Ellis Horwood, Chichester, England.

MELNICK, D., OSER, B. L. and WEISS, S. 1946. Rate of enzymic digestion as a factor in nutrition. Science *103,* 326–329.

MELNICK, R. L. and HULTIN, R. L. 1970. Factors affecting distribution of lactate dehydrogenase between particulate and soluble phase of homogenized trout skeletal muscle. J. Food Sci. *35,* 67–72.

MELTZER, Y. L. 1973. Encyclopedia of Enzyme Technology. Future Stochastic Dynamics, Flushing, N.Y.

MENDENHALL, V. T. 1972. Oxidative rancidity in raw fish fillets harvested from the Gulf of Mexico. J. Food Sci. *37,* 547–550.

MENDEZ, W. M., JR. and KAISER, E. T. 1975. A "reverse burst" active site titration procedure for human carbonic anhydrase B. Biochem. Biophys. Res. Commun. *66,* 949–955.

MENDOZA, C. E. 1973. Thin-layer chromatography and enzyme inhibition techniques. J. Chromatogr. *78,* 29–40.

MENEFEE, E. 1976. Personal communication. Albany, Calif.

MENEZEZ, H. C. DE and VILLAS BOAS, H., JR. 1973. Electrophoretic separation of enzymic flavour-forming preparations from bananas and oranges. Rev. Bras. Tecnol. *4,* 221–223. (Portuguese)

MENON, I., KENDAL, R. Y., DEWAR, H. and NEWELL, D. J. 1968. Effects of onions on blood fibrinolytic activity. Br. Med. J. *3,* 351–352.

MENZI, R. 1970. Nonsticky and elastic pasta. U.S. Pat. 3,520,702. July 14.

MERCER, E. H. 1968. Cellular organization. *In* Molecular Biology and Agriculture. Potential for Future Research. U.S. Dep. Agric. *ARS-47,* Albany, Calif.

MEREDITH, F. I., GASKINS, M. H. and DULL, G. G. 1974. Amino acid losses in turnip greens (*Brassica rapa* L.) during handling and processing. J. Food Sci. *39,* 689–691.

MEREDITH, P. 1970. Inactivation of cereal alpha amylases by brief acidification: the pasting strength of the flour. Cereal Chem. *47,* 492–500.

MEREDITH, P. and JENKINS, L. D. 1973. Amylases of developing wheat, barley and oat grains. Cereal Chem. *50,* 243–254.

MEREDITH, W. O. S. 1966. Distribution of protein and free and bound amylase in cereal species. Proc. Am. Soc. Brew. Chem. 32–38.

MERGENTHALER, E. and SCHERZ, H. 1978. Gas chromatographic determination of gluconic acid in beer. Lebensm. Gericht. Chemie. *32,* 12. (German)

MERMELSTEIN, N. H. 1975. Immobilized enzymes produce high-fructose syrup. Food Technol. *29,* 20–26.

MERMELSTEIN, N. H. 1977. Enzyme process tenderizes beef before slaughter. Food Technol. *37,* 39–40.

MERMELSTEIN, N. H. 1979. Hypobaric transport and storage of fresh meats and produce earns 1979 IFT Food Technology Industrial Achievement Award. Food Technol. *33* (7) 32–40.

MERRIFIELD, R. B. 1963. Solid phase peptide synthesis. I. The synthesis of a tetrapeptide. J. Am. Chem. Soc. *85,* 2149–2154.

MERYMAN, A. 1966. Cryobiology. Academic Press, New York.

MESELSON, M. and YUAN, R. 1968. DNA restriction enzyme from *E. coli.* Nature *217,* 1110–1114.

MESSING, R. A. 1975. Immobilized Enzymes for Industrial Reactors. Academic Press, New York.

METZLER, D. E. 1977. Biochemistry: The Chemical Reactions of Living Cells. Academic Press, New York.

MEUSER, F. 1972. Accuracy of the enzymic determination of glucose. Getreide Mehl Brot *26,* 127. (German)

MEYER, D. D., VAIL, G. E., BRAMBETT, V. D., MARTIN, T. G. and HARRINGTON, R. B. 1967. Vitamin A supplement and hypoxanthine-uric acid and nucleotide content of selected beef muscles. J. Food Sci. *32,* 289–293.

MEYER, K. H., SPAHER, P. F. and FISCHER, E. H. 1953. Purification, crystallization and properties of beta-amylase of wheat. Helv. Chim. Acta *36,* 1924–1936. (French)

MEYER, R. I., TOKAY, L. and SUDEK, R. E. 1960. The effect of an oxygen scavenger packet, desiccant in packet system, on the stability of dry whole milk and dry ice cream mix. J. Dairy Sci. *43,* 144.

MEYER, W. L. and REED, J. P. 1975. An insoluble carboxypeptidase A-like activity of skeletal muscle. Fed. Proc. *34,* 511.

MEYERS, P. A. and ZATMAN, L. J. 1971. The metabolism of trimethylamine *N*-oxide by *Bacillus* PM 6. Biochem. J. *121,* 10P.

MEYNHARDT, J. T., De VILLIERS, O. T. and IRELAND, J. P. 1974. An investigation of the enzymes of the sugar accumulation cycle in Barlinka grape berries. Agroplantae *6,* 47–50.

MEYRATH, J. and VOLAVSEK, G. 1975. Production of microbial enzymes. *In* Enzymes in Food Processing, 2nd Edition. G. Reed (Editor). Academic Press, New York.

MICA, B. 1977. Change of sugar contents in selected potato varieties during storage. Staerke *11,* 368–372.

MICHAELIS, L. 1908. The adsorption of yeast-invertin. Biochem. Z. *7,* 488–492. (German)

MICHAELIS, L. 1922. Hydrogen Ion Concentration. Springer Verlag, Berlin. (German)

MICHAELIS, L. and DAVIDSOHN, H. 1911. The action of hydrogen ions on invertin. Biochem. Z. *35,* 386–412. (German)

MICHAELIS, L. and EHRENREICH, M. 1908. Adsorption analysis of enzymes. Biochem. Z. *10,* 283–299. (German)

MICHAELIS, L. and MENTEN, M. L. 1913. The kinetics of invertase action. Biochem. Z. *49,* 333–369. (German)

MICHAL, G. 1974. Enzymatic analysis. *In* Methodicum Chimicum (Chemical Methods), Vol. 1, Part B. F. Korte (Editor). Academic Press, New York.

MICHAL, G. 1979. Biochemical Pathways. Boehringer Mannheim Biochemicals, Indianapolis.

MICHAL, G., BENTLER, H. -O., LANG, G. and GÜNTHER, U. 1976. Enzymatic determination of succinic acid in foodstuffs. Z. Anal. Chem. *279,* 137–138. (German)

MICHEL, H. O. 1949. An electrometric method for the determination of red blood cell and plasma cholinesterase activity. J. Lab. Chin. Med. *34,* 1564–1568.

MICHELSON, A. A. 1976. Biological role of superoxide and superoxide dismutase in cellular metabolism. C.R. Soc. Biol. *170,* 1137–1146. (French)

MICHELSON, A. A. and MONOD, J. 1975. Superoxide dismutases and their applications as oxidation inhibitors. U.S. Pat. 3,920,521. Nov. 18.

MICKELSEN, R. and ERNSTROM, C. A. 1972. Effect of pH on the stability of rennin porcine pepsin blends. J. Dairy Sci. *55,* 294–297.

MIHALYI, V. and KOERMENDY, L. 1967. Changes in protein solubility and associated properties during ripening of Hungarian dry sausage. Food Technol. *21,* 1398–1401.

MIKABERIDZE, V. E. and ROSNADZE, G. R. 1974. Effect of soil mulching in young mandarin gardens on catalase and peroxidase activity and on frost resistance. Subtrop. Kul't. *4,* 53–57. (Russian)

MIKELADZE, G. G., BOL'FADORF, I. B. and KUTALTELADZE, L. 1974. Requirements of pectolytic enzymes preparations for use in production of juices, soft drinks and fruit wines. FSTA *8,* 7 H1272. (Russian)

MIKOLA, J. and ENARI, T. -M. 1970. Changes in the contents of barley proteolytic inhibitors during malting and mashing. J. Inst. Brew. *76,* 182–188.

MIKOLAJCZYK, M. and DRABOWICZ, J. 1978. Organosulfur compounds. 13. Optical resolution of chiral sulfinyl compounds via β-cyclodextrin inclusion complexes. J. Am. Chem. Soc. *100,* 2510–2515.

MILDVAN, A. S. 1970. Metals in enzyme catalysis. *In* The Enzymes, 3rd Edition, Vol. 2. P.D. Boyer (Editor). Academic Press, New York.

MILDVAN, A. S. 1974. Mechanism of enzyme action. Annu. Rev. Biochem. *43,* 357–399.

MILES LABORATORIES. 1976A. A bibliography on the use of alkylagaroses for Shaltiel hydrophobic chromatography. Miles Laboratories Inc., Elkhart, Indiana.

MILES LABORATORIES. 1976B. Research Products. Immobilized Biochemicals. Miles Laboratories, Elkhart, Indiana.

MILLER, A. L., FROST, R. G. and O'BRIEN, J. S. 1976. Purification of human liver β-D-galactosidase using affinity chromatography. Anal. Biochem. *74,* 537–545.

MILLER, B. S. and JOHNSON, J. A. 1949. Differential stability of alpha-amylase and proteinase. Cereal Chem. *26,* 359–371.

MILLER, B. S. and JOHNSON, J. A. 1954. Differential inactivation of enzymes. U.S. Pat. 2,683,682. July 13.

MILLER, B. S., JOHNSON, J. A. and PALMER, D. L. 1953. A comparison of cereal, fungal, and bacterial alpha-amylases as supplements for breadmaking. Food Technol. *7,* 38–42.

MILLER, B. S. and KNEEN, E. 1947. The amylase inhibitor of Leoti sorghum. Arch. Biochem. Biophys. *15,* 251–264.

MILLER, B. S. and KUMMEROW, F. 1948. The disposition of lipase and lipoxidase in baking and the effect of their reaction products on consumer acceptability. Cereal Chem. *25,* 391–398.

MILLER, G. J. and McCOLLOCH, R. J. 1959. Chainlength specificity for pectin-methylesterase inhibition by anionic detergents. Biochem. Biophys. Res. Comm. *1,* 91–93.

MILLER, J. C., SINK, J. D., SHERRITT, G. W. and ZIEGLER, J. H. 1971. Studies of a method for determining organochlorine pesticide residues in animal tissue. J. Food Sci. *36,* 880–887.

MILLER, L. and MACMILLAN, J. D. 1971. Purification and pattern of action of pectinesterase from *Fusarium oxysporum* f. sp. *vasinfectum.* Biochemistry *10,* 570–576.

MILLER, R. and GRONINGER, H. S., JR. 1976. Functional properties of enzyme-modified acylated fish protein derivatives. J. Food Sci. *41,* 268–272.

MILLER, R. E. 1976. Quantification of L-glutamine using *Escherichia coli* synthase, a sensitive fluorimetric assay. Anal. Chem. *75,* 91–99.

MILLER, W. H., MALLETTE, M. J., ROTH, L. J. and DAWSON, C. R. 1944. A new method for the measurement of tyrosinase-catecholase activity. II. Catecholase activity based on the initial reaction velocity. J. Am. Chem. Soc. *66,* 514–519.

MILLET, M. A., BAKER, A. J. and SATTER, L. D. 1976. Physical and chemical pretreatments for enhancing cellulose saccharification. Biotechnol. Bioeng. Symp. *6,* 125–153.

MILLIGAN, B. and HOLT, L. A. 1977. The complete enzymic hydrolysis of crosslinked proteins. *In* Protein Crosslinking. Nutritional and Medical Consequences. M. Friedman (Editor). Plenum Press, New York.

MILLIN, D. J. 1972. Improvements relating to fermentation of tea. U.S. Pat. 3,649,297. Mar. 14.

MILNER, M. and GOULD, M. R. 1971. The quantitative determination of phenol oxidase activity in dry plant tissue. Cereal Chem. *28,* 473–478.

MINAMI, K., YAMAMURA, M., SHIMIZU, S., OGAWA, K. and SEKINE, N. 1978. SCP production from methanol. I. A new methanol-assimilating, highly productive, thermophilic yeast. J. Ferment. Technol. *56,* 1–7.

MINTZ, M. S. 1970. Electrodialysis theory and applications. *In* Spec. Progr. on Innovations in Food Engineering. Univ. of Calif., Berkeley, Ext. Progr., March 23–26, San Francisco.

MIRANDA, A. F., SOMER, H. and DIMAURO, S. 1979. Isoenzymes as markers of differentiation. *In* Muscle Regeneration. A. Mauro (Editor). Raven Press, New York.

MISRA, H. P. and FRIDOVICH, I. 1972. The generation of superoxide radical during the autoxidation of hemoglobins. J. Biol. Chem. *247,* 6960–6969.

MISSELHORN, K. and ADAM, R. 1976. On the cyanide contents in stone fruit products. Branntweinwirtschaft *116,* 49–50. (German)

MITCHELL, E. D. 1972. Homogeneous alpha-amylase from malted barley. Phytochemistry *11,* 1673–1676.

MITCHELL, E. D. and NEWMANN, J. 1972. Glycosidases from malted barley. Phytochemistry *11,* 1341–1344.

MITCHELL, P. 1967. Proton-translocation phosphorylation in mitochondria, chloroplasts and bacteria: Natural fuel cells and solar cells. Fed. Proc. *26,* 1370–1379.

MITCHELL, P. 1979. Keilin's respiratory chain concept and its chemiosmotic consequences. Science *206,* 1148–1159.

MITCHELL, R. S. 1972. In-can blanching of green peas. J. Food Technol. *7,* 409–416.

MITZ, M. A. 1956. New insoluble active derivative of an enzyme as a model for study of cellular metabolism. Science *123,* 1076–1077.

MITADA, D. S. and TAPPEL, A. L. 1956. The hydrolysis of beef proteins by various proteolytic enzymes. Food Res. *21,* 217–225.

MIYAGA, K. and SUZUKI, K. 1964. Studies on taka-amylase A under high pressure. Arch. Biochem. Biophys. *105,* 297–302.

MIYOSHI, M., YOON, C. -H., IBUKI, F. and KANAMORI, M. 1976. Characterization of rennin action on κ-casein using CM-cellulose. Agric. Biol. Chem. *40,* 347–352.

MIZRAHI, S. and BERK, Z. 1970. Physicochemical characteristics of orange juice cloud. J. Sci. Food Agric. *21,* 250–253.

MOCK, W. L. 1976. Torsional-strain considerations in enzymology. Bioinorg. Chem. *5,* 403–414.

MODLER, H. W., BRUNNER, J. R. and STINE, C. H. 1974. Extracellular protease of *Penicillium roqueforti.* II. Characterization of a purified enzyme preparation. J. Dairy Sci. *57,* 528–534.

MOELKER, H. C. T. and MATTHIJSEN, R. 1971. Purification of microbial rennets. U.S. Pat. 3,591,388. July 6.

MOELLER, P. W., FIELDS, P. A., DUTSON, T. R., LANDMANN, W. A. and CARPENTER, Z. L. 1977. High temperature effects on lysosomal enzyme distribution and fragmentation of bovine muscle. J. Food Sci. *42,* 510–512.

MOHAMMED, A., FRAENKEL-CONRAT, H. and OLCOTT, H. S. 1949. The "browning" reaction of proteins with glucose. Arch. Biochem. *24,* 157–178.

MOHAN, R. R. and LI, N. N. 1975. Nitrate and nitrite reduction by liquid membrane encapsulated whole cells. Biotechnol. Bioeng. *17,* 1137–1156.

MOHR, E. 1972. Method for producing long-life liquid coffee. Ger. Offen. 2,063,489. Aug. 22.

MOLINA, M. R. and LACHANCE, P. A. 1973. Studies on the utilization of coconut meal. A new enzymic chemical method for fiber free protein extraction of defatted coconut flour. J. Food Sci. *38,* 607–610.

MOMOTANI, Y. and KATO, J. 1974. Effects of different gibberellins on the α-amylase induction and some properties of isoenzymes-3 and -8. *In* Plant Growth Subst., Proc. 8th Int. Conf., 1973. Hirowaka Publ. Co., Tokyo.

MONDY, N. I. and MUELLER, T. O. 1977. Potato discoloration in relation to anatomy and lipid composition. J. Food Sci. *42,* 14–18.

MONOD, J., WYMAN, J. and CHANGEUX, J. -P. 1965. On the nature of allosteric transitions: A plausible model. J. Mol. Biol. *12,* 88–118.

MONREAL, J., DEURUBURU, F. and VILLANUEVA, J. R. 1967. Lytic action of β(1-3)-glucanase on yeast cells. J. Bacteriol. *94,* 241–244.

MONSAN, P., DUTEURTRE, B., MOLL, M. and DURAND, G. 1978. Use of papain immobilized on Spherosil for beer chillproofing. J. Food Sci. *43,* 424–427.

MONTEDORO, G. 1976. Use of enzymatic preparations in red wine production. *In* Proc. 4th Int. Enol. Symp., Valencia, Spain. E. Lemperli and J. Frank (Editors). Int. Assoc. Mod. Winery Technol. Manage., Augustinberg, Germany.

MONTEDORO, G. and BERTUCCIOLI, M. 1976. Trials on the use of different enzyme preparations in red-wine making. Lebensm. Wiss. Technol. *9,* 225–231. (French)

MONTEDORO, G., BERTUCCIOLI, M. and PETRICCIOLI, G. 1975. Effect of treatments with enzymic additives and tannin-removing substances on the oil yield, rate of extraction, and analytical characteristics of the oils, vegetation waters, and waste waters obtained when extracting olive oils. Riv. Ital. Sostanze Grasse *52,* 255–265. (Italian)

MONTES, A. 1971. Physiological responses of tomato fruits, *Lycopersicum esculentum* Mill., subjected to chilling temperatures. Diss. Abstr. Int. B. *32,* 1400–1401.

MONTGOMERY, R. D. 1969. Cyanogens. *In* Toxic Constituents of Plant Foodstuffs. I.E. Liener (Editor). Academic Press, New York.

MOODY, G. J. and THOMAS, J. D. R. 1975. Analytical role of ion-selective and gas-sensing electrodes in enzymology. Analyst (London) *100,* 609–619.

MOON, W. -H., STUFF, J., BAILEY, L., MALZER, J. and CLARK, H. 1975. Nitrogen retention and plasma amino acids of men who consumed egg albumin or mixtures of amino acids. Fed. Proc. *33,* 712.

MOORE, K. 1980. Immobilized enzyme technology commercially hydrolyzes lactose. Food Prod. Dev. *14* (1) 50–51.

MOORE, K. K. 1977. Xylitol: Uncut gem among sweeteners. Food Prod. Dev. *11* (4) 66–70.

MOORJANI, M. N., BALIGA, B. R., VIJAYARANGA, B. and LAHIRY, N. L. 1962. Postrigor changes in nitrogen distribution and texture of fish during storage in crushed ice. Food Technol. *16,* 80–86.

MOOS, H. J., DERERA, N. F. and BALAAM, L. N. 1972. Effect of pre-harvest rain on germination in the ear and α-amylase activity of Australian wheat. Aust. J. Agric. Res. *23,* 769–777.

MOO-YOUNG, M., CHAHAL, D. S., SWAN, J. E. and ROBINSON, C. W. 1977. SCP production of *Chaetomium cellulolyticum,* a new thermotolerant cellulolytic fungus. Biotechnol. Bioeng. *19,* 527–538.

MORAWETZ, M. and HUGHES, W. L., JR. 1952. The interaction of proteins with synthetic polyelectrolytes. J. Phys. Chem. *56,* 64–69.

MOREAU, J. R. 1971. Meat tenderizing. Can. Pat. 874,377. June 29.

MOREAU, J. R. and JANKUS, E. E. 1963. An assay for measuring papain in meat tissue. Food Technol. *17,* 1047–1049.

MORENO, V. and KOSIKOWSKI, F. V. 1973. Degradation of β-casein by micrococcal cell-free preparations. J. Dairy Sci. *56,* 33–38.

MORGAN, A. I., JR. 1967. For minimum fouling: Evaporation concepts and designs. Food Technol. *21* (10) 63–68.

MORGAN, A. I., JR. and SCHWIMMER, S. 1965. Preparation of dehydrated food products. U.S. Pat. 3,170,803. Feb. 23.

MORGAN, M. E. 1976. The chemistry of some microbially induced flavor defects in milk and dairy foods. Biotechnol. Bioeng. *18,* 953–965.

MORI, H., YAMAMOTO, S. and NAGASAKI, S. 1977. Multiple forms of the lytic glucanase of *Flavobacterium dormitator* var. *glucanolyticae* and the properties of the main component enzyme. Agric. Biol. Chem. *41,* 611–613.

MORITA, Y. and YOSHIDA, C. 1970. Aromatic amino acid residues in Japanese-radish peroxidase a and apoenzyme. Agric. Biol. Chem. *34,* 590–598.

MORRIS, H. J. 1959. Detection of peroxidase activity in the processing of food materials. U.S. Pat. 2,905,594. Sept. 22.

MORRIS, H. M. 1958. Applications of peroxidase test paper in food processing. Food Technol. *12,* 265–267.

MORRIS, H. M. and SEIFERT, R. M. 1961. Constituents and treatments affecting cooking of dry beans. Proc. 5th Annu. Dry Bean Res. Conf., Denver. West. Reg. Res. Lab., U.S. Dep. Agric. Publ., Washington, D.C.

MORRIS, L. L., KLAUSTERMEYER, J. A. and KADER, A. A. 1975. Postharvest studies of lettuce. *In* Annu. Res. Proj. Rep., Iceberg Lettuce Res. Adv. Board, 52–59.

MORRISETT, J. D., JACKSON, R. L. and GOTTO, A. M., JR. 1975. Lipoproteins: Structure and function. Annu. Rev. Biochem. *44,* 183–207.

MORRISON, W. R. 1976. Lipids in flour, dough and bread. Bakers Dig. *50* (4) 29–36, 47–48.

MORRISON, W. R. 1978. Cereal Lipids. Adv. Cereal Sci. Technol. *2,* 205–348.

MORRISON, W. R. and PANPAPRAI, R. 1975. Oxidation of free and esterified linoleic and linolenic acids by wheat and soya lipoxygenases. J. Sci. Food Agric. *26,* 1225–1236.

MORROW, C. T. 1974. Psycho-physical analogues. Food Technol. *26* (2) 92–98.

MORROW, R. M., CARBONELL, R. G. and McCOY, B. J. 1975. Electrostatic and hydrophobic effects in affinity chromatography. Biotechnol. Bioeng. *17,* 895–914.

MORTON, R. K. 1950. Alkaline phosphatase of milk. 2. Purification of the enzyme. Biochem. J. *55,* 795–800.

MOSBACH, K. 1974. General introduction. *In* Insolubilized Enzymes. M. Salmona, C. Saronio and S. Garattini (Editors). Raven Press, New York.

MOSBACH, K. 1976. Immobilized enzymes. *In* Methods in Enzymology, Vol. 44. Academic Press, New York.

MOSBACH, K. 1977. "Togetherness" through immobilization. *In* Biotechnological Applications of Proteins and Enzymes. Pap. Conf. 1976. Z. Bohak and N. Sharon (Editors). Academic Press, New York.

MOSBACH, K. 1980. Future trends; immobilized enzymes. TIBS *5* (1) 4–8.

MOSBACH, K. and ANDERSON, L. 1977. Magnetic ferro fluids for preparation of magnetic polymers and their application in affinity chromatography. Nature *270,* 259–261.

MOSELEY, M. H. and ORY, R. L. 1973. Partial hydrolysis of proteins in peanut meals by endogenous proteolytic systems. J. Am. Peanut Res. Educ. Assoc. *5,* 201.

MOSKOWITZ, G. M. and CAYLE, T. 1974. A method for the detection of β-nitropropionic acid in crude biological extracts. Cereal Chem. *51,* 96–105.

MOSS, D. W. and BUTTERWORTH, R. J. 1974. Enzymology and Medicine. Pitman Medical, London.

MOSSEL, D. A. A. and SHENNAN, J. L. 976. Micro-organisms in dried foods: Their significance, limitation and enumeration. J. Food Technol. *11,* 205–220.

MOTOC, D., CONSTANTINESCU, S. and CONSTANTINESCU, A. 1970. Enzymatic coagulation of milk. Ind. Aliment. *21,* 69–71.

MOTZ, R. J. 1976. The enzymatic determination of sugars in the confectionery industry. CCB Rev. Chocolate, Confectionery and Bakery *1* (3) 40–41.

MOUNFIELD, J. D. 1938. The proteolytic enzymes of sprouted wheat. Biochem. J. *32,* 1675–1684.

MOURI, T., HASHIDA, W., SHIGA, I. and TERAMOTO, S. 1970. Nucleic acid decomposing enzymes of some mushrooms. FSTA *5,* 6J794. (Japanese)

MOUTOUNET, M. 1979. Co-oxydation of β-carotene in the presence of plum extract (*Prunus domestica* L. var. d'Ente). Lebensm. Wiss. Technol. *12,* 338–341. (French)

MOYER, J. C. and STOTZ, E. 1945. The electronic blanching of vegetables. Science *102,* 68–69.

MÖTTÖNEN, K. 1975. On the amylolytic proteins of rye. An electrofocusing study with liquid columns and gel slab. Staerke *27,* 346–352.

MUDGETT, R. E., RUFNER, R., BAJRACHARYA, R., KIM, K. and RAJAGOPALAN, K. 1978. Enzymatic effects on cell rupture in plant protein recovery. J. Food Biochem. *2,* 185–207.

MUELLER, T. J. and MORRISON, M. 1975. The transmembrane proteins in the plasma membrane of normal human erythrocytes. Evaluation employing lactoperoxidase and proteases. Biochemistry *14,* 5512–5516.

MUKERJI, S. K. 1974. Corn leaf phosphoenolpyruvate (PEP) carboxylase: activation by magnesium ions. Plant Sci. Lett. *2,* 243–248.

MULLENAX, D. C. and LOPEZ, A. 1975. Swine pancreatic lipase activity at low temperatures. J. Food Sci. *40,* 310–313.

MULLER, F. E. 1973. Sterilization process. Br. Pat. 1,337,593. Jan. 30.

MULLER, J. and PFLEIDERER, G. 1979. A new method of conjugating proteins for enzyme immunoassay. J. Appl. Biochem. *1,* 301–310.

MULLIN, W. J. and WOLFE, F. H. 1974. Disc gel electrophoresis of caseins treated with proteolytic and glycolytic enzymes. J. Dairy Sci. *57,* 9–14.

MULTON, J. L. and GUILBOT, A. 1975. Water activity in relation to the thermal inactivation of enzymic proteins. *In* Water Relations of Foods. R.B. Duckworth (Editor). Academic Press, New York.

MULVIHILL, D. M. and FOX, P. F. 1977A. Proteolysis of α_{S1}-casein by chymosin: Influence of pH and urea. J. Dairy Res. *44,* 533–540.

MULVIHILL, D. M. and FOX, P. F. 1977B. Selective denaturation of milk coagulants in 5M urea. J. Dairy Res. *44,* 319–324.

MUNETA, P. and WALRADT. J. 1966. Cysteine inhibition of enzymatic blackening with polyphenol oxidase from potatoes. J. Food Sci. *33,* 606–608.

MUNNECKE, D. M. 1978. Detoxification of pesticides using soluble or immobilized enzymes. Process Biochem. *13* (2) 14–17.

MUNNECKE, D. M. 1980. Enzymatic detoxification of waste organophosphate pesticides. J. Agric. Food Chem. *28,* 105–111.

MUNYANGANIZI, B. and COPPENS, R. 1976. Comparative study of two procedures for the extraction of banana juice applied to two different varieties. Ind. Aliment. Agric. *93,* 707–711.

MURAKI, M. and MASUDA, H. 1976. Enzymatic determination of butane,2,3-diol in wines. J. Sci. Food Agric. *27,* 345–350.

MURATA, K. 1965. Thiaminase. Rev. Japan. Lit. Beriberi Thiamine, 220–254. (Japanese)

MURATA, T. 1974. Enzymic mechanism of starch synthesis in plants. JARQ *8,* 127–132.

MURATA, T. 1977. Partial purification and some properties of ADP-glucose phosphorylase from potato tubers. Agric. Biol. Chem. *41,* 1995–2002.

MURATA, T. and KU, H. -S. 1966. Postharvest ripening and storage of bananas. V. Physiological studies on chilling injury of bananas. J. Food Sci. Technol. Tokyo *13,* 466–471. (Japanese)

MURAYAMA, W., KOIZUMI, S. and KOBAYASHI, N. 1976. Production of branched dextrin. Jpn. Kokai 76 88,645. Aug. 3.

MURCH, A. F. and MURCH, J. A. 1966. Production of polished juices. U.S. Pat. 3,236,655. Feb. 22.

MURPHY, J. F. and UHING, E. H. 1959. Method of processing egg albumin. U.S. Pat. 2,892,720. June 30.

MURR, D. P. and DENNIS, L. L. 1974. Effect of succinic acid-2-2-dimethylhydrazide on mushroom *o*-diphenol oxidase. J. Am. Soc. Hortic. Sci. *99,* 3–6.

MURR, D. P. and MORRIS, L. L. 1974. Influence of O_2 and CO_2 on *o*-diphenol oxidase in mushrooms. J. Am. Hortic. Sci. *99,* 155–158.

MURR, D. P. and MORRIS, L. L. 1975. Effect of storage temperature on postharvest changes in mushrooms. J. Am. Hortic. Sci. *100,* 16–19.

MURRAY, D. G. and LUFT, L. R. 1973. Low-D.E. corn starch hydrolysates. Food Technol. *27* (3) 32–40.

MURRAY, E. D. and KENDALL, M. S. 1969. Milk coagulating enzyme from bacteria. U.S. Pat. 3,482,997. Dec. 9.

MURRAY, J. M. and WEBER, A. 1974. The cooperative action of muscle proteins. Sci. Am. *236* (2) 58–70.

MURRAY, K. 1976. Biochemical manipulation of genes. Endeavour *35,* 129–133.

MURRAY, K. E., SHIPTON, J., WHITFIELD, F. B. and LAST, J. H. 1976. The volatiles of off-flavored unblanched green peas *(Pisum sativum).* J. Sci. Food Agric. *27,* 1093–1107.

MURRAY, R. E. and WHITFIELD, F. B. 1975. The occurrence of 3-alkyl-methoxy-pyrazines in raw vegetables. J. Sci. Food Agric. *26,* 973–986.

MURRAY, T. K. and BAKER, B. E. 1952. Studies on protein hydrolysis. 1. Preliminary observations on the taste of enzymic protein-hydrolysates. J. Sci. Food Agric. *3,* 470–475.

MURTHY, G. K., COX, S. and KAYLOR, L. 1976. Reactivation of alkaline phosphatase in ultra high-temperature, short-time processed liquid milk products. J. Dairy Sci. *59,* 1699–1710.

MUSTAKAS, G. C. and KIRK, L. D. 1963. Method of obtaining detoxified mustard seed products. U.S. Pat. 3,106,469. Oct. 8.

MUSTAKAS, G. C. and KIRK, L. D. 1965. Method of obtaining detoxified meal from seeds containing both isothiocyanate and thiooxazolidone. U.S. Pat. 3,173,792. Mar. 16.

MUSTRANTA, A., KIESVAARA, M. and KUUSI, T. 1976. Pectolytic changes in the composition of cucumbers during the harvest period. J. Sci. Agric. Soc. Finl. *48,* 407–414. (Finnish)

MUZZARELLI, R. A., BARTTINI, G. and ROCCHETTI, R. 1978. Isolation of lysozyme on chitosan. Biotechnol. Bioeng. *20,* 87–94.

MÜLLER, D. Studies on the new enzyme glucose oxidase. I. Biochem. Z. *199,* 136–170. (German)

NAGAI, Y. and HORI, H. 1972. Entrapment of collagen in a polyacrylamide matrix and its application in the purification of animal collagenases. Biochim. Biophys. Acta *263,* 564–573.

NAGASAKI, S., NISHIOKA, Y., MORI, H. and YAMAMOTO, S. 1976. Purification and properties of lytic β-1,3-glucanase from *Flavobacterium dormitator* var. *glucanolyticae.* Agric. Biol. Chem. *40,* 1059–1067.

NAGASHIMA, Z. and UCHIYAMA, M. 1959. The possibility that myrosinase is a single enzyme and mechanism of decomposition of mustard oil glucoside by myrosinase. Bull. Agric. Chem. Soc. Jpn. *23,* 555–556. (Japanese)

NAGEL, C. W. and PATTERSON, M. E. 1967. Enzymes and the development of the pear *(Pyrus communis).* J. Food Sci. *32,* 294–297.

NAGUIB, K. 1972. The effect of H_2O_2 treatment on the bacteriological quality and nutritive value of milk. Milchwiss. Ber. *12,* 758–762.

NAIR, B., OSTE, R., ASP, N. G. and DAHLQUIST, A. 1976. Enzymic hydrolysis of food protein for amino acid analysis. I. Solubilization of the protein. J. Agric. Food Chem. *24,* 386–389.

NAKADAI, T., SEIICHI, N. and NOBUYOSHI, I. 1972. The action of peptidases from *Aspergillus oryzae* in digestion of soybean proteins. Agric. Biol. Chem. *36,* 261–268.

NAKAGAWA, H., SEKIGUCHI, K., OZURA, N. and TAKEHANA, H. 1971. Binding of tomato pectinesterase and fructofuranosidase to tomato cell wall. Agric. Biol. Chem. *35,* 301–307.

NAKAJIMA, Y., ADACHI, T., ITO, J. and MIRUNO, H. 1974. Saccharification of starch in high maltose yields. Jpn. Kokai 74 55,857. May 30.

NAKAMURA, M. 1960. Some effects of ethanolic extracts of potatoes on phosphorylase activity. Bull. Agric. Soc. Jpn. *24,* 52–58.

NAKAMURA, S. and HAYASHI, S. 1974. Role of the carbohydrate moiety of glucose oxidase. Kinetic evidence for protection of the enzyme from thermal inactivation in the presence of sodium dodecyl sulfate. FEBS Lett. *41,* 327–330.

NAKANISHI, K. 1976. Studies on xylanase production. II. Inducers for xylanase production by

Streptomyces sp. Hakko Kogaku Zasshi *54,* 801–807. (Japanese)

NAKANISHI, T. and ITO, M. 1973. Enzymic studies on cheese ripening. VI. Effect of chemical modification on the turbidity in β-casein or a turbid substance solution obtained by the action of protease produced by *Aspergillus oryzae* var. B. Jpn. J. Dairy Sci. *22,* A65–A70.

NAKANISHI, T. and ITO, M. 1974. Enzymic studies on cheese ripening. X. Changes of volatile free fatty acids, carbonyl and sulfur compounds during ripening of cheese made with protease produced by *Aspergillus oryzae* strain B. Jpn. J. Dairy Sci. *6,* A181–186. (Japanese)

NAKANISHI, T. and SUYAMA, K. 1974. Long-chain 2,3-dialkyl acroleins in pork meat. Nippon Nogei Kagaku Kaishi *48,* 555–559 (Japanese). Chem. Abstr. *83,* 56873t.

NAKANO, H. and YASUI, T. 1976. Denaturation of myosin-ATPase as a function of water activity. Agric. Biol. Chem. *40,* 107–113.

NAKASHINI, K., YASUI, T. and KOBAYASHI, T. 1976. Studies on xylanase production. II. Inducers for xylanase production by *Streptomyces* spp. Hakko Kogaku Zasshi *54,* 801–807. (Japanese)

NAKAZAWA, H., SANO, K., KUMAGAI, H. and YAMADA, N. 1977. Distribution and formation of aromatic L-amino acid decarboxylase in bacteria. Agric. Biol. Chem. *41,* 2241–2247.

NAMM, D. H. and LEADER, J. P. 1974. A sensitive analytical method for the detection and quantitation of adenosine in biological samples. Anal. Biochem. *58,* 511–524.

NANJO, M. and GUILBAULT, G. G. 1974. Enzyme electrode for L-amino acids and glucose. Anal. Chim. Acta *73,* 363–373.

NANJO, M. and GUILBAULT, G. G. 1975. Amperometric determination of alcohols, aldehydes, and carboxylic acids with an immobilized alcohol oxidase enzyme electrode. Anal. Chim. Acta *75,* 169–180.

NARZISS, L. and BELLMER, H. -G. 1976. Effect of stabilizing beers with PVPP and bentonite on their polyphenol content and polymerization index. Brauwissenschaft *29,* 256–267. (German)

NARZISS, L. and LINTZ, B. 1975. Effect of mashing on enzyme activity and proteolysis. Brauwissenschaft *28,* 305–315. (German)

NARZISS, L. and LITZENBURGER, K. 1977. Experiments on the estimation of the molecular size of gums in wort and beer. Brauwissenschaft *30,* 330–332. (German)

NARZISS, L. and RUSITZKA, P. 1977. Effect of curing temperature on the behaviour of malt enzymes. Brauwissenschaft *30,* 101–111. (German)

NARZISS, L. and SEKIN, Y. 1974. Behavior of lipases during the malting and brewing processes. Brauwissenschaft *27,* 311–320. (German)

NASUNO, S. and NAKADAI, T. 1971. Formation of glutamic acid from defatted soybeans by *Aspergillus oryzae.* J. Ferment. Technol. *49,* 544–551.

NATL. ACAD. SCI. 1973. Toxicants Occurring Naturally in Foods, 2nd Edition. NAS, Washington, D.C.

NAU, H., LERCH, K. and WITTE, L. 1977. Amino acid sequence of the blocked N-terminal tryptic peptide of *Neurospora* tyrosinase by mass spectrometry. FEBS Lett. *79,* 203–206.

NAZIR, D. J. and MAGAR, N. G. 1963. Biochemical changes in fish muscle during rigor mortis. J. Food Sci. *28,* 1–7.

NEBESKY, E. A., ESSELEN, W. A., KAPLAN, A. M. and FELLERS, C. R. 1950. Thermal destruction and stability of peroxidase in acid foods. Food Res. *15,* 114–124.

NEILANDS, J. B. and STUMPF, P. K. 1955. Outlines of enzyme chemistry. John Wiley & Sons, New York.

NEILANDS, J. B. and STUMPF, P. N. 1958. Outlines of Enzyme Chemistry, Second Edition. John Wiley & Sons, New York.

NEIMS, A. H. and HELLERMAN, L. 1970. Flavoprotein catalysis. Annu. Rev. Biochem. *39,* 867–888.

NELBOECK, M. 1977. Some economic, enzymological and practical problems in the technological and analytical applications of immobilized enzymes. *In* Biotechnological Applications of Proteins and Enzymes. Z. Bohak and N. Sharon (Editors). Academic Press, New York.

NELSON, D. P., MILLER, W. D. and KIESOW, L. A. 1974. Calorimetric studies of hemoglobin function, the binding of 2,3-diphosphoglyceric acid and inositol hexaphosphate to human hemoglobin A. J. Biol. Chem. *249,* 4770–4775.

NELSON, H. 1972. Enzymatically produced flavors for fatty systems. J. Am. Oil Chem. Soc. *49,* 559–562.

NELSON, J. H., JENSEN, R. G. and PITAS, R. E. 1977. Pregastric esterase and other oral lipases: A review. J. Dairy Sci. *60,* 327–362.

NELSON, J. H. and WITT, P. R. 1973. Chillproofing beer with enzyme obtained from *Mucor pusillus* Lindt. U.S. Pat. 3,740,233. June 19.

NELSON, J. M. and DAWSON, C. R. 1944. Tyrosinase. Adv. Enzymol. *4,* 99–152.

NELSON, J. M. and GRIFFIN, E. G. 1916. Adsorption of invertase. J. Am. Chem. Soc. *38,* 1109–1115.

NELSON, J. M. and SCHUBERT, M. P. 1928. Water concentration and the rate of hydrolysis of sucrose by invertase. J. Am. Chem. Soc. *50,* 2188–2193.

NELSON, L. F. and CATRON, D. V. 1960. Comparison of different supplemental enzymes with and without diethylstilbestrol. J. Anim. Sci. *19,* 1279–1280.

NERAAL, R. and HAMM, R. 1977. The enzyme breakdown of tripolyphosphate and diphosphate in comminuted meat. XI. Influence of heating and freezing. Z. Lebensm. Unters. Forsch. *164,* 101–104. (German)

NESTEROV, N., TONCHEV, S., DIVOKA, G., GROSDANOV, A. and TSVETKOV, Z. 1977. Effect of freezing and freeze-drying on the activity of some lysosomal enzymes. Fleischwirtschaft *57,* 1335–1336. (German)

NEUBECK, C. 1975. Fruits, fruit products and wines. *In* Enzymes in Food Processing, 2nd Edition. G. Reed (Editor). Academic Press, New York.

NEUBECK, C. 1976. Practical aspects of enzymes in Food Processing. *In* Proc. Symp. on Enzyme in the Food Processing Industry. Univ. of Calif., Davis, Jan. 14.

NEUCERE, N. J., JACKS, T. J. and SUMRELL, G. 1978. Interactions of globular protein with simple polyphenols. J. Agric. Food Chem. *26,* 214–216.

NEUFELD, E. F., LIM, T. W. and SHAPIRO, L. J. 1975. Inherited disorders of lysosomal metabolism. Annu. Rev. Biochem. *44,* 357–401.

NEUKOM, H. 1976. Chemistry and properties of the non-starch polysaccharides (NSP) of wheat flour. Lebensm.-Wiss. Technol. *9,* 143–148.

NEUMANN, H., LAZAR, H. J., FARKAS, D. F. and RAHMAN, A. R. 1980. Dehydrated glycerol-treated celery. Unpublished results. Personal communication. Albany, CA.

NEURATH, H. 1976. Role of proteases in biological regulation. *In* Miami Winter Symposium *11.* Academic Press, New York.

NEURATH, H., GREENSTEIN, J. P., PUTNAM, F. W. and ERIKSON, J. O. 1944. The chemistry of protein denaturation. Chem. Rev. *34,* 157–265.

NEWBOLD, R. P. and SCOPES, R. K. 1971. Post-mortem glycolysis in a skeletal muscle: Effect of mincing and of dilution with or without addition of orthophosphate. J. Food Sci. *36,* 209–214.

NEWELL, J. A., SEELEY, E. A. and ROBBINS, E. A. 1975. Process of making yeast protein isolate having reduced nucleic acid content. U.S. Pat. 3,867,255. Feb. 18.

NEWSHOLME, E. A. and TAYLOR, K. 1968. A new principle for the assay of metabolites involving the combined effects of isotope dilution and enzymatic catalysis. Biochim. Biophys. Acta *158,* 11–24.

NEWTON, W. E. and OTSUKA, S. 1980. Molybdenum Chemistry and Biological Significance. Plenum Publishing Corp., New York.

NEY, K. H. 1971. Prediction of bitterness of peptides from their amino acid composition. Z. Lebensm. Unters. Forsch. *147,* 64–68. (German)

NEY, K. H. 1972. Amino acid composition of proteins and bitterness of their peptides. Z. Lebensm. Unters. Forsch. *149,* 321–323. (German)

NG, W. 1971. Method of improving color of cooked egg products. U.S. Pat. 3,598,612. Aug. 10.

NGO, T. T. 1976. Reagentless determination of L-asparagine and L-arginine via the combined use of immobilized enzymes and an ion-selective electrode. Can. J. Biochem. *54,* 62–65.

NICHOLS, R. and HAMMOND, J. B. W. 1975. The relationship between respiration, atmosphere and quality in intact perforated mushroom pre-packs. J. Food Technol. *10,* 427–435.

NICHOLSON, R. L. and McINTYRE, G. A. 1976. Production of xylanase by *Verticillium albo-atrum.* Proc. Indiana Acad. Sci. 1975 *85,* 324–333.

NICKERSON, J. T. and SINSKEY, A. J. 1972. Microbiology of Foods and Food Processing. American Elsevier, New York.

NICKLISCH, A., GESKE, W. and KOHL, J. G. 1976. Relevance of glutamate synthase and glutamate dehydrogenase to the nitrogen assimilation of primary leaves of wheat. Biochem. Physiol. Pflanz. *170,* 85–90. (German)

NIKU-PAAVOLA, M. L. 1977. Partial characterisation of a new barley amylase. J. Sci. Food Agric. *28,* 728–738.

NIKU-PAAVOLA, M. L. and HEIKKINEN, M. 1975. Comparison of a new amylase found in barley with the amylases of the fungi of barley husk. J. Sci. Food Agric. *26,* 239–242.

NIKU-PAAVOLA, M. L., NUMMI, M., KACHIN, A., DAUSSANT, J. and ENARI, T. -M. 1972. Isoelectric focussing electrophoresis of wheat β-amylases. Cereal Chem. *49,* 580–585.

NIMMO, I. A. and MABOOD, S. F. 1979. Nature of the random experimental error encountered when acetylcholine hydrolase and alcohol dehydrogenase are assayed. Anal. Biochem. *94,* 265–269.

NISHIKIMI, M. 1975. Oxidation of ascorbic acid with superoxide anion generated by the xanthine oxidase system. Biochem. Biophys. Res. Comm. *63,* 463–468.

NISHIMURA, H. and MIZUTANI, J. 1973. The C-S cleavage of *cis*-S-(1-propenyl)-L-cysteine sulfoxide by OH radicals from irradiated aqueous solutions. Agric. Biol. Chem. *37,* 213–217.

NISHIMURA, H. and MIZUTANI, J. 1975. Effect of γ-irradiation on development of lachrymator in onion. Agric. Biol. Chem. *39,* 2245–2246.

NISHIZAWA, K., OKAWA, T. and TOWAKI, S. 1971. Edible meat from fibrous meat. Jpn. Pat. 71 18,579. May 24.

NISHIZUKA, Y. 1971. S-alkyl-L-cysteine lyase *(Pseudomonas).* Methods Enzymol. *17,* 470–474.

NOCKOLDS, C. E., KRETSINGER, R. H., COFFEE, C. E. and BRADSHAW, R. A. 1972. Structure of a calcium-binding carp myogen. Proc. Natl. Acad. Sci. *69,* 581–584.

NODA, H., HORIGUCHI, Y. and ARAKAI, S. 1975. Studies on the flavor substances of "Nori," the dried laver Porphyra spp. II. Free amino acids and 5′-nucleotides. Bull. Jpn. Soc. Sci. Fish. *41,* 1299–1303.

NOGUCHI, M., YAMASHITA, M., ARAI, S. and FUJIMAKI, M. 1975. On the bitter-masking activity of a glutamic acid-rich oligopeptide fraction. J. Food Sci. *40,* 367–369.

NOGUCHI, S. and MATSUMOTO, J. J. 1970. Studies on the control of the denaturation of the fish muscle. I. Preventative effect of Na-glutamate. Bull. Jpn. Soc. Fish. *36,* 1078–1087.

NOGUCHI, S., GOSAWA, K. and MATSUMOTO, J. J. 1976. Studies on the control of denaturation of fish muscle proteins during frozen storage. VI. Preventative effects of carbohydrates. Nippon Suisan Gakkaishi *42,* 77–82. (Japanese)

NOGUCHI, T. and KANDATSU, M. 1971. Purification and properties of a new alkaline protease of rat skeletal muscle. Agric. Biol. Chem. *35,* 1092–1100.

NOGUEIRA, J. A., FONESECA, H. and LEME, J., JR. 1973. Effect of packaging on the preservation of ascorbic acid in lyophilized fruits. Solo *65,* 62–68. (Portuguese)

NOLTE, D. and ACKER, L. 1975. Phospholipase D—occurrence and properties. Z. Lebensm. Unters. Forsch. *159,* 225–233. (German)

NOLTE, D., REBMANN, H. and ACKER, L. 1974. Phosphatide-hydrolyzing enzymes in grain. Getreide Mehl Brot *28,* 189–191. (German)

NOMURA, K., YASUI, T., KIYOOKA, S. and KOBAYASHI, T. 1969. Xylanases of *Trichoderma viride.* II. Inhibition of enzymic xylan hydrolysis and a two-stage saccharification process. Hakko Kogaku Zasshi *47,* 313–317. (Japanese)

NONHEBEL, G. and MOSS, A. A. H. 1971. Drying of Solids in the Chemical Industry. CRC Press, Cleveland.

NOOMEN, A. 1977. A rapid method for the estimation of the dissolved nitrogen compounds in cheese. Neth. Milk Dairy J. *31,* 163–176.

NORBERG, P. and VON HOFSTEN, B. 1970. Chromatography of a halophilic enzyme on hydroxylapatite in 3.4M sodium chloride. Biochim. Biophys. Acta *220,* 132–133.

NORD, F. F. 1944–1970. Advances in Enzymology and Related Subjects of Biochemistry, Vol. 4–33. Interscience Publishers, New York.

NORD, F. F. and WEIDENHAGEN, R. 1932–1939. Ergebnisse der Enzymforschung (Results of Enzyme Research), Vol. 1–8. Akademische Verlagsgesellschaft, Leipzig. (German)

NORD, F. F. and WERKMAN, C. H. 1941–1943. Advances in Enzymology, Vol. 1–3. Interscience Publishers, New York.

NORDAL, J. and ROSSEBO, L. 1972. Enzymatic precipitation in agar gel as a method for the detection of sodium caseinate in raw and heated meat products. Z. Lebensm.-Wiss. Unters. Forsch. *148,* 65–69. (German)

NORDSTROM, K. 1964. Formation of esters from alcohols by brewers yeast. J. Inst. Brew. *70,* 328–336.

NORTHROP, J. H. 1919. The combination of enzyme and substrate. J. Gen. Physiol. *2,* 113–131.

NORTHROP, J. H., KUNITZ, M. and HERRIOTT, R. M. 1948. Crystalline Enzymes, 2nd Edition. Columbia Univ. Press, New York.

NOWAK, J. and SWIERZA, A. 1975. Proteolytic enzyme activity in stored potato tubers with different rest periods. Pol. Sci. Ser. Sci. Biol. *23,* 129–133.

NOWAK, T. and MILDVAN, A. S. 1972. Nuclear magnetic resonance studies of selectively hindered internal motion of substrate analogs at the active site of pyruvate kinase. Biochemistry *11,* 2813–2818.

NOYES, R. 1969. Dehydration Processes for Convenience Foods. Food Processing Review *2.*

Noyes Development Corp., Park Ridge, N.J.

NOZNICK, P. P. and BUNDUS, R. H. 1966. Preparation of garlic concentrate and powders. U.S. Pat. 3,258,343. June 28.

NURSE, P. 1977. Cell-cycle control in yeasts. Biochem. Soc. Trans. *5,* 1191–1193.

NURSTEN, H. E. 1970. Volatile compounds: The aroma of fruits. *In* The Biochemistry of Fruits and Their Products, Vol. 1. A.C. Holme (Editor). Academic Press, London.

NURSTEN, H. E. 1975. Chemistry of flavours—past, present and future. Int. Flavours Food Addit. *6,* 75–82.

NUTGEREN, D. H. 1975. Arachidonate lipoxygenase in blood platelets. Biochim. Biophys. Acta *380,* 299–307.

NWOKOLO, E. N. and BRAGG, D. B. 1977. Influence of phytic acid and crude fibre on the availability of minerals from four protein supplements in growing chicks. Can. J. Anim. Sci. *57,* 475–477.

NYE, W. and SPOEHR, H. A. 1943. The isolation of hexenal from leaves. Arch. Biochem. Biophys. *2,* 23–34.

NYIRI, L. K. 1974. Design considerations for animal cell cultures as sources of enzymes. *In* Enzyme Engineering, Vol. 2. E.K. Pye and L.B. Wingard, Jr. (Editors). Plenum Press, New York.

NYSTROM, J. M. and ANDREN, R. K. 1976. Pilot plant conversion of cellulose to glucose. Process Biochem. *11* (10) 26–34.

OAKENFULL, D. 1973. Effects of hydrophobic interactions on the kinetics of the reactions of long chain alkylamines with long chain carboxylic esters of 4-nitrophenol. J. Chem. Soc. Perkins Trans. *2,* 1005–1012.

OAKENFULL, D. and FENWICK, D. E. 1977. Hydrophobic interaction between the bile acids and long-chain alkyltrimethylammonium ions. Aust. J. Chem. *30,* 335–344.

OBANU, Z. A., LEDWARD, D. A. and LAWRIE, R. A. 1976. The proteins of intermediate moisture meat stored at tropical temperature. III. Differences between muscles. J. Food Technol. *11,* 187–196.

O'BRIEN, P. J. O. and RAHIMTULA, A. 1975. Involvement of cytochrome P-450 in the intracellular formation of lipid peroxides. J. Agric. Food Chem. *23,* 154–158.

O'CARRA, P., BARRY, S. and GRIFFIN, T. 1973. Spacer-arms in affinity chromatography: The need for a more rigorous approach. Biochem. Soc. Trans. *1,* 289–290.

O'CONNOR, D. E. 1971. Sunflower protein product. U.S. Pat. 3,622,556. Nov. 28.

ODLAND, D. and EHEART, M. S. 1975. Ascorbic acid mineral and quality retention in frozen broccoli in water, steam, and ammonia steam. J. Food Sci. *40,* 1004–1007.

OESTERGAARD, J. and KNUDSEN, S. L. 1976. Use of Sweetzyme in industrial continuous isomerization. Various process alternatives and corresponding product types. Staerke *28,* 350–356.

OGASAWARA, Y. 1967. Lysosomes. III. Influence of vitamin A on stability of lyosomes from rabbit liver. Nippon Nogei Kagaku Kaishi *41,* 492–497. (Japanese)

OGAWA, M., TANAKA, K. and KASAI, Z. 1977. Note on the phytin-containing particles isolated from rice scutellum. Cereal Chem. *54,* 1029–1034.

OGREN, W. L. and HUNT, L. D. 1978. Comparative biochemistry of ribulose bisphosphate carboxylase in higher plants. Basic Life Sci. *11,* 127–138.

OGURA, M. 1970. On the relation between food composition and serine dehydratase activity in rat liver. Agric. Biol. Chem. *34,* 585–589.

OGURA, N., NAKAGAWA, H. and TAKEHANA, H. 1975. Storage temperature of tomato fruits. II. Effect of the storage temperatures on changes in their polygalacturonase and pectinesterase activities accompanied with ripening. Nippon Nogei Kagaku Kaishi *49,* 271–274. (Japanese)

OH, H. I., HOFF, J. E., ARMSTRONG, G. S. and HAFF, L. A. 1980. Hydrophobic interaction in tannin-protein complexes. J. Agric. Food Chem. *28,* 394–398.

OHAD, I., FRIEDBERG, I., NE'EMAN, Z. and SCHRAMM, M. 1971. The fate of the amyloplast membrane during maturation and storage of potatoes. Plant Physiol. *47,* 465–477.

OHARA, M. and FUKUDA, J. 1978. Selective Reimer-Tiemann formylation of phenol via cyclodextrin inclusion complex. Pharmazie *33,* 467.

OHLHOFF, G. and THOMAS, A. F. 1971. Gustation and Olfaction. Int. Symp., Geneva, June, 1970. Academic Press, New York.

OHLMEYER, D. W. 1957. Use of glucose oxidase to stabilize beer. Food Technol. *11,* 503–509.

OHMIYA, K. and SATO, Y. 1972. Proteolytic action of dairy lactic acid bacteria. XII. Significant contribution of intracellular protease of lactic bacteria to casein hydrolysis in cheese ripening. Milchwissenschaft *27,* 417–422. (German)

OHMIYA, K. and SATO, Y. 1975. Preparation and properties of intracellular proteinase from *Streptococcus cremoris.* Appl. Microbiol. *30,* 738–745.

OHTA, S., MAUL, S., SINSKEY, A. J. and TANNENBAUM, S. R. 1971. Characterization of a heat shock process for reduction of the nucleic acid content of *Candida utilis.* Appl. Microbiol. *22,* 415–421.

OHTSUKI, K., KAWATABA, M. and TAGUCHI, K. 1976. Cellulase and xylanase activity in the culture medium of *Bacillus subtilis* var. natto. Chem. Abstr. *86,* 85839. (Japanese)

OHTSURU, M. and HATA, T. 1972. Molecular properties of multiple forms of plant tyrosinase. Agric. Biol. Chem. *36,* 2495–2503.

OHTSURU, M. and HATA, T. 1973. Studies on the activation mechanism of the myrosinase by L-ascorbic acid. Agric. Biol. Chem. *37,* 1971–1972.

OHTSURU, M. and HATA, T. 1975. Measurement of binding of ascorbic acid to myrosinase by rate of dialysis. Agric. Biol. Chem. *39,* 1505–1506.

OKA, T. and MORIHARA, K. 1974. Comparative specificity of microbial acid proteinases for synthetic peptides. Arch. Biochem. Biophys. *165,* 65–71.

OKADA, G. 1976. Enzymatic studies on a cellulase system of *Trichoderma viride.* IV. Purification and properties of a less-random type cellulase. J. Biochem. (Tokyo) *80,* 913–922.

OKADA, G. and NISIZAWA, K. 1975. Enzymic studies on a cellulase system of *Trichoderma viride.* III. Transglycosylation properties of two cellulase components of random type. J. Biochem. (Tokyo) *78,* 297–306.

OKADA, H. 1977. Review of studies on the fermentative production of glutamic acid and nucleic acid-related substances. Nippon Jozo Kyokai Zasshi *72,* 358–360. (Japanese)

OKADO, S. and ONO, S. 1969. Prevention of clouding of syrup in which canned orange segments are preserved. U.S. Pat. 3,428,255. Dec. 16.

OKASAKI, K. and KUBO, K. 1977. Uses of hemicellulase in the food industry. New Food Ind. *19* (12) 24–31. (Japanese)

OKAZAKI, T. and KORNBERG, A. 1964. Enzymatic synthesis of deoxyribonucleic acid. XV. Purification and properties of a polymerase from *Bacillus subtilis.* J. Biol. Chem. *239,* 259–268.

O'KEEFE, A. M., FOX, P. F. and DALY, C. 1977. Denaturation of porcine pepsin during cheddar cheese manufacture. J. Dairy Res. *44,* 335–344.

OKUBANJO, A. O. and STOUFFER, J. R. 1975. Postmortem glycolysis and isometric thaw tension development and decline in bovine skeletal muscle undergoing thaw rigor. J. Food Sci. *40,* 955–959.

OKUNTSOV, M. M. and PLOTNIKOVA, A. N. 1970. Correlation between the rhythmic movements of *Phaseolus multiflorus* leaves and the activity of some oxidative enzymes. Chem. Abstr. *78,* 40,615w. (Russian)

OLCOTT, H. S. 1950. Stabilization of solutions of wheat gluten in dilute acetic acid by brief heat treatment. Cereal Chem. *27,* 514–516.

OLCOTT, H. S. and MECHAM, D. K. 1947. Characterization of wheat gluten. I. Protein-lipid complex formation during doughing of flours. Lipoprotein nature of the glutenin fractions. Cereal Chem. *24,* 407–414.

OLDHAM, K. G. 1973. Radiometric methods of enzyme assay. Methods Biochem. Anal. *21,* 191–286.

OLERED, R. 1967. Development of alpha-amylase and falling number in wheat and rye during ripening. *In* Vaxtodling, Plant Husbandry, No. 23. Almquist and Widsells Boktryckeri AB, Uppsala, Sweden. (Swedish)

OLERED, R. 1977. Amylase and starch studies on bread grain. Sver. Utsaedesfoeren Tidskr. *87,* 91–101. (Swedish)

OLERED, R. and JONSSON, G. 1967. Electrophoretic studies on alpha amylase in wheat. Ber. Getreidechem.-Tag Detmold, 207–214. (German)

OLERED, R. and JONSSON, G. 1970. Electrophoretic studies of alpha-amylase in wheat. II. J. Sci. Food Agric. *21,* 385–392.

OLIVECRONA, T., EGELRUD, T., HERNELL, O., CASTBERG, H. and SOLBERG, P. 1975. Is there more than one lipase in bovine milk? Annu. Bull. Int. Dairy Fed. *86,* 61–72.

OLIVECRONA, T. and HERNELL, O. 1976. Human milk lipases, and their possible role in fat digestion. Paediatr. Paedol. *11,* 600–604.

OLIVERA, B. M. and LEHMAN, I. R. 1967. Linkage of polynucleotides through phosphodiester bonds by an enzyme from *Escherichia coli.* Proc. Natl. Acad. Sci. *57,* 1426–1433.

OLLEY, J., FARMER, J. and STEPHEN, E. 1969. The rate of phospholipid hydrolysis in frozen fish. J. Food Technol. *4,* 27–37.

OLLEY, J. and LOVERN, J. A. 1960. Phospholipide hydrolysis in cod flesh stored at various temperatures. J. Sci. Food Agric. *11,* 644–652.

OLLEY, J., PIRIE, R. and WATSON, H. 1962. Lipase and phospholipase activity in fish skeletal muscle and its relationship to protein denaturation. J. Sci. Food Agric. *13,* 501–516.

OLLING, C. C. J. 1977. Enzymic methods for analysis of dairy products. Voedingsmiddelentechnol. *10* (8) 10–15. (Dutch)

OLSEN, H. S. and ADLER-NISSEN, J. 1979. Industrial production and applications of a soluble enzymatic hydrolyzate of soy protein. Process Biochem. *14* (7) 6–11.

OLSON, A. C. 1972. Phytase secreted from tobacco cells grown in suspension culture. Abstr. *52,* 23rd Annu. Meet. Tissue Culture Assoc., In Vitro 7, 252.

OLSON, A. C., BECKER, R., MIERS, J. C., GUMBMANN, M. R. and WAGNER, J. R. 1975. Protein nutritional quality problems in the digestibility of dry beans. *In* Protein Nutritional Quality of Foods and Feeds, Part 2. M. Friedman (Editor). Marcel Dekker, New York.

OLSON, A. C. and COONEY, C. L. 1974. Immobilized Enzymes in Food and Microbial Processes. Plenum Press, New York.

OLSON, A. C., GRAY, G. M. and GUADAGNI, D. G. 1979. Naringin bitterness in grapefruit juice debittered with naringinase immobilized in a hollow fiber. J. Food Sci. *44,* 1361–1385.

OLSON, A. C. and KORUS, R. A. 1977. Immobilized enzymes. *In* Enzymes in Food and Beverage Processing. R.L. Ory and A.J. St. Angelo (Editors). Am. Chem. Soc., Washington, D.C.

OLSON, D. E. and PARRISH, F. C., JR. 1977. Relationship of myofibril fragmentation index to measures of beefsteak tenderness. J. Food Sci. *42,* 506–509.

OLSON, D. E., PARRISH, F. C., JR., DAYTON, W. R. and GOLL, W. R. 1977. Effect of postmortem storage and calcium activated factor on the myofibrillar proteins of bovine skeletal muscle. J. Food Sci. *42,* 117–124.

O'MALLEY, B. W. and SCHRADER, W. T. 1976. The receptors of steroid hormones. Sci. Am. *234* (2) 32–43.

OMURA, H., TOMITA, Y., MURAKAMI, H. and NAKAMURA, Y. 1974. Antitumor potentiality of the enzyme preparations, pumpkin ascorbate oxidase and shiitake mushroom polyphenol oxidase. J. Fac. Agric., Kyushu Univ. *18,* 191–200. (Japanese)

O'NEILL, I. K. and RICHARDS, C. P. 1978. Specific detection of polyphosphates in frozen chicken by combination of enzyme blocking and 31P F.T.N.M.R. spectroscopy. Chem. Ind. (London) (2) 65–67.

ONO, K. 1971. Lysosomal enzyme activation and proteolysis of bovine muscle. J. Food Sci. *36,* 838–839.

ONO, K., TOPEL, D. G. and ALTHEN, T. G. 1976. Cyclic AMP in longissimus muscle from control and stress susceptible pigs. J. Food Sci. *41,* 108–110.

ONONOGBU, I. C. 1980. The toxicity of cassava. TIBS *5* (9) 10–11.

ONUE, Y. and RIDDLE, V. M. 1973. Use of plastein reaction in recovering protein from fish. J. Fish. Res. Can. *30,* 1745–1747.

ONSLOW, H. 1915. A contribution to our knowledge of the chemistry of coat colour in animals and of dominant and recessive whiteness. Proc. Roy. Soc. *89B,* 36–52.

ONSLOW, M. W. 1921. Oxidizing enzymes. IV. The distribution of oxidizing enzymes among the higher plants. Biochem. J. *15,* 107–112.

ONSLOW, M. W. 1931. The Principles of Plant Biochemistry. Cambridge Univ. Press, Cambridge.

OPPENHEIMER, C. 1925–1936. The Ferments and Their Actions. Georg Thiem, Leipzig. (German)

ORME-JOHNSON, W. H., JACOB, G. S., HENZL, M. T. and AVERILL, B. A. 1977. Molybdenum in enzymes. *In* Bioinorganic Chemistry, 2nd Symp. Adv. Chem Ser. *162,* 389–401. Am. Chem. Soc., Washington, D.C.

ORSI, F. 1972. A derivatographic study of the thermal decomposition of glucose. Acta Aliment. Acad. Sci. Hung. *1,* 341–354.

ORY, R. L. 1972. Enzyme activities associated with protein bodies of seeds. *In* Symposium: Seed Proteins. G.E. Inglett (Editor). AVI Publishing Co., Westport, Conn.

ORY, R. L. 1978. Personal communication. New Orleans.

ORY, R. L., BARKER, R. H. and BOUDREAUX, C. J. 1964. Nature of the cofactor for the acid lipase of *Ricinus communis.* Biochemistry *3,* 2013–2016.

ORY, R. L. and ST. ANGELO, A. J. 1977. Enzymes in Food and Beverage Processing. Am. Chem. Soc., Washington, D.C.

ORY, R. L. and SEKUL, A. A. 1977. Spectrophotometric assay curves as anomolous indicators of proteolysis of oilseed proteins. J. Food Biochem. *1,* 67–74.

OSHIKAWA, Y. 1964. Production of seasonings from mushrooms with cellulase produced by microorganisms and the resulting product. U.S. Pat. 3,150,893. Sept. 29.

OSMUNDSEN, H. 1975. Computer program for the determination of kinetic constants of two enzymes acting simultaneously on the same substrate. Biochem. Biophys. Res. Commun. *67,* 324–330.

OSSAN, H. 1846. Detection of hydrogen peroxide in fermentation mixture. Poggendorfs Annalen *67,* 373–379. (German)

OSTER, K. A., OSTER, J. B. and ROSS, D. J. 1974. Immune response to bovine xanthine oxidase in atherosclerotic patients. Am. Lab. (Boston) *8* (6) 41–47.

OSTROVKII, Y. 1978. Antivitamins in food products. Vestn. Akad. Med. Nauk SSSR, (3) 72–78. (Russian)

O'SULLIVAN, M. 1970. Aldolase activity in plants as an indicator of zinc deficiency. J. Sci. Food Agric. *21,* 607–609.

OSUNTOKUN, B. O., LANGMAN, M. J. S., WILSON, J. and ALADETOYINBO, A. 1970. Controlled trial of hydroxycobalamin and riboflavone in Nigerian ataxic neuropathy. J. Neurol. Neurosurg. Psychiat. *33,* 663–666.

OTSUKA, Y., KATAKI, R. and FUJIMAKI, M. 1976. Purification and properties of an aminopeptidase from rabbit skeletal muscle. Agric. Biol. Chem. *40,* 2335–2342.

OTTESEN, M. and RICKERT, W. 1970. The acid protease of *Mucor mihei.* Methods Enzymol. *19,* 459–461.

OUGH, C. S. 1975. Further investigations with glucose oxidase-catalase systems for use with wine. Am. J. Enol. Vitic. *26,* 30–36.

OUGH, C. S. and CROWELL, E. A. 1979. Pectic-enzyme treatment of white grapes: Temperature, variety and skin-contact time factors. Am. J. Enol. Vitic. *30,* 22–27.

OUGH, C. S., NOBLE, A. C. and TEMPLE, D. 1975. Pectic enzymes effects on red grapes. Am. J. Enol. Vitic. *26,* 195–200.

OUTTRUP, H. 1976. The use of mutation in enzyme fermentations. Proc. ASM Conference on Genetics and Molecular Biology of Industrial Organisms. H.O. Halverson and A.L. Dennison (Cochairmen). Feb.

OVERHOLSER, E. L. and CRUESS, W. V. 1923. A study of the darkening of apple tissue. Calif. Agric. Exp. Stn. Tech. Pap. *7,* 1–40.

OWEN, J. E., HEWLETT, J. and LAWRIE, R. A. 1976. A note on the discoloration of frozen muscle, stored under fluorescent illumination, as influenced by an artificially induced high pH in the meat. J. Sci. Food Agric. *27,* 477–482.

OWENS, J. W. and STAHL, P. 1976. Purification and characterization of liver microsomal β-glucuronidase. Biochim. Biophys. Acta *438,* 474–486.

PACE, W., PARLAMENTI, R., RAB, A. U., SILANO, V. and VITTOZZI, L. 1978. Protein α-amylase inhibitors from wheat flour. Cereal Chem. *55,* 244–254.

PAEZ, L. E. and HULTIN, H. O. 1970. Respiration of potato mitochondria and whole tubers in relation to sugar accumulation. J. Food Sci. *35,* 46–51.

PAGINGTON, J. S. 1975. Flavouring and production of extruded soya proteins. Int. Flavour Food Add. *6,* 278–279.

PAIGE, D. M., BAYLESS, T. M., HUANG, S. -S. and WEXLER, R. 1975. Lactose hydrolyzed milk. Am. J. Clin. Nutr. *28,* 818–822.

PAIGE, D. M., BAYLESS, T. M., MELLITS, E. D. and DAVIS, L. 1977. Lactose malabsorption in preschool black children. Am. J. Clin Nutr. *30,* 1018–1022.

PALACIOS, R. 1976. *Neurospora crassa* glutamine synthetase. Purification by affinity chromatography and characterization of subunit structure. J. Biol. Chem. *251,* 4787–4791.

PALEG, L. G. 1965. Physiological effects of gibberellins. Annu. Rev. Plant Physiol. *16,* 222–291.

PALÉUS, S. and NEILANDS, J. B. 1950. Preparation of cytochrome c with the aid of ion exchange resin. Acta Chem. Scand. *4,* 1024–1030.

PALEVITCH, D. and THOMAS, T. H. 1976. Enhancement by low pH of gibberellin effects on dormant celery seeds and embryoless half-seeds of barley. Physiol. Plant. *37,* 247–252.

PALLA, J. -C. and VERRIER, J. 1974. Inhibition of broad bean *(Vicia faba)* lipoxygenase by hydroquinone derivatives. Ann. Technol. Agric. *23,* 367–373. (French)

PALLAVICINI, C., FINLEY, J. W. and STANLEY, W. L. 1980. Plastein synthesis with α-chymotrypsin immobilized on chitin. J. Sci. Food Agric. *31,* 273–278.

PALMER, G. H. 1973. Relationship between levels of gibberellic acid and the production and action of carbohydrases of barley. J. Inst. Brew. *79,* 513–518.

PALMER, G. H. 1974. The industrial use of gibberellic acid and its scientific basis: A review. J. Inst. Brew. *80,* 13–30.

PALMER, G. H. 1975. Influence of endosperm structure on extract development. Proc. Am. Soc. Brew. Chem. *33,* 174–180.

PALTER, R., LUNDIN, R. E. and HADDON, W. F. 1972. A cathartic glycoside isolated from *Catharthamus tinctorus.* Phytochemistry *11,* 2781–2784.

PANCHOLY, S. K. and LYND, J. Q. 1972. Characterization of wheat germ lipases. Phytochemistry *11,* 643–645.

PANOVA, R. and BOZOVA, L. 1975. Variations in the activity of some oxidative enzymes in growing and ripening apples and pears. Chem. Abstr. *84,* 28109a. (Bulgarian)

PANTHER, M. and WOLFE, F. H. 1972. Studies on the degradation of ascorbic acid by saskatoon-berry juice. Can. Inst. Food Sci. Technol. *5* (2) 93–96.

PAPAS, T. S. and MEHLER, A. H. 1968. Modification of the transfer function of proline transfer ribonucleic acid synthetase by temperature. J. Biol. Chem. *243,* 3767–3769.

PAPASTATHOPOULOS, D. S. and RECHNITZ, G. A. 1975. Enzymatic cholesterol determination using ion-selective membrane electrodes. Anal. Chem. *47,* 1792–1796.

PAPPENHEIMER, A. M., JR. 1977. Diphtheria toxin. Annu. Rev. Biochem. *46,* 69–94.

PARACELSUS. 1589. Volumen Medicinae Paramirum Theophrastsi de Medica Industria, Paranthesis Secunda Tractus de Ente Venini (Writings of Theophrastus on the medicine of the occult. Part 2. Treatise on internal poisons). J. Huserus (Editor). Conrad Waldkirch, Basel. (Latin)

PARDEE, A. B. 1951. Calculations on paper chromatography of peptides. J. Biol. Chem. *190,* 757–762.

PARDEE, A. B. 1973. Regulation of maximal enzyme synthesis. *In* Genetics of Industrial Microorganisms. Z. Vanek, Z. Hostalek and J. Cudin (Editors). Elsevier Publishing Co., Amsterdam.

PARISH, R. W. 1972. The intracellular location of phenol oxidases, peroxidase and phosphatases in the leaves of spinach beet. Eur. J. Biochem. *31,* 446–455.

PARK, H. K., MUGURUMA, M., FUKAZAWA, T. and ITO, T. 1975. Relationship between superprecipitating activity and constituents of myosin B prepared from normal and PSE porcine muscle. Agric. Biol. Chem. *39,* 1363–1370.

PARK, J. R., COLLINS, J. L., McCARTY, I. E. and JOHNSTON, M. R. 1971. Chemical changes and amylase activity of freshly shelled Southern peas, *Vigna sinensis.* J. Am. Soc. Hortic. Sci. *96,* 419–421.

PARK, Y. K., CHIKASI, C. and MORETTI, R. H. 1976. Characterization of cell bound isomerase of *Streptomyces bikiniensis.* J. Food Sci. *41,* 1383–1386.

PARKASH, O. and BHATIA, I. S. 1980. Graphical determination of p*K* values of the active-site groups of enzymes. Biochem. J. *185,* 609–610.

PARLIMENT, T. H., CLINTON, W. P., SCARPELLINO, R., SOUKUP, R. J. and EPSTEIN, M. F. 1976. Enhancement of coffee flavor. U.S. Pat. 3,962,321. Sept. 11.

PARRISH, F. C., GOLL, D. E., NEWCOMB, W. J., DE LUMEN, B. O., CHAUDRY, H. M. and KLINE, E. A. 1969. Molecular properties of post-mortem muscle, 7. J. Food Sci. *34,* 196–202.

PARRISH, F. W., PERLIN, A. S. and REESE, E. T. 1960. Selective enzymolysis of poly-β-D-glucans, and the structure of the polymers. Can. J. Chem. *38,* 2094–2104.

PARSEGIAN, V. 1978. Biophysical Discussions. Fast Biochemical Reactions in Solutions, Membranes and Cell. Rockefeller University Press, New York.

PARTMANN, W. 1973. Investigations on the breakdown of nicotinamide adenine dinucleotide (NAD) in fish muscle during frozen storage. Lebensm.-Wiss. Technol. *6,* 155–157. (German)

PARTMANN, W. 1975. Effects of freezing and thawing on food quality. *In* Water Relations of Foods. R.B. Duckworth (Editor). Academic Press, New York.

PARTRIDGE, S. M. 1966. Elastin. *In* Physiology and Biochemistry of Muscle as a Food, Vol. 1. E.J. Briskey, R.G. Cassens and J.C. Trautman (Editors). University of Wisconsin Press, Madison.

PASEK, A., HANUS, J. and SKACHOVA, H. 1977. Use of amylum derivatives for isolation of amylolytic enzymes. Nahrung *21,* 113–116. (German)

PASSON, P. G. and HULTQUIST, D. E. 1972. Soluble cytochrome b_5 reductase from human erythrocytes. Biochim. Biophys. Acta *275,* 62–73.

PASTAN, I. 1972. Cyclic AMP. Sci. Am. *227* (2) 97–105.

PASTORE, M., MORISI, F. and VIGLIA, A. 1974. Reduction of lactose of milk by entrapped galactosidase: II. Conditions for an industrial continuous process. J. Dairy Sci. *57,* 269–272.

PATEL, D. S. and PHAFF, H. J. 1960. Properties of tomato polygalacturonase. Food Res. *25,* 47–57.

PATEL, I. B. and VAUGHN, R. H. 1973. Cellulolytic bacteria associated with sloughing spoilage of California ripe olives. Appl. Microbiol. *25,* 62–69.

PATEL, N., LASKIN, A. I., DERELANKO, P. and FELIX A. 1979. Microbial production of methyl ketones. Purification and properties of a secondary alcohol dehydrogenase from yeast. Eur. J. Biochem. *101,* 401–406.

PATIL, B. C., SINGH, B. and SALUNKHE, D. K. 1971. Formation of chlorophyll and solanine in Irish potato *(Solanum tuberosum)* tubers and their control by gamma radiation. Lebensm.-

Wiss. Technol. *4,* 123–124.

PATIL, S. K., FINNEY, K. F., SHOGREN, M. D. and TSEN, C. C. 1976. Water-soluble pentosans of wheat flour. III. Effect of water-soluble pentosans on loaf volume of reconstituted gluten and starch doughs. Cereal Chem. *53,* 347–354.

PATTEE, H. E. and SINGLETON, J. A. 1977. Isolation of isomeric hydroperoxides from the peanut lipoxygenase-linoleic acid reactions. J. Am. Oil Chem. Soc. *54,* 183–185.

PATTERSON, B. D., HATFIELD, S. G. S. and KNEE, M. 1974. Residual effect of controlled atmosphere storage on the production of volatile compounds in two varieties of apples. J. Sci. Food Agric. *25,* 843–849.

PATTON, S. J., NEVENZEL, J. C. and BENSON, A. A. 1975. Specificity of digestive lipases in hydrolysis of wax esters and triglycerides studied in anchovy and other selected fish. Lipids *10,* 575–583.

PÄTZOLD, C. 1974. Effect of storage conditions on potato quality. Kartoffelbau. *25,* 62–63. (German)

PAUL, P., BRATZLER, L. J., FARWELL, E. D. and KNIGHT, K. 1952. Tenderness of beef. I. Rate of heat penetration. Food Res. *17,* 504–510.

PAUL, P. H., SYMONDS, A., VAROZZA, A. and STEWART, G. F. 1957. Effect of glucose removal on storage stability of egg yolk solids. Food Technol. *11,* 494–498.

PAULING, L. 1948. Nature of forces between large molecules of biological interest. Nature *161,* 707–709.

PAULING, L. 1975. Valence-bond theory of compounds of transition metals. Proc. Natl. Acad. Sci. *72,* 4200–4202.

PAULING, L. 1976. The Centennial celebration. C & E News *54* (17) 33–36.

PAULS, I. F. and QUATZ, R. R. 1957. Process for brewing a special beer. U.S. Pat. 2,783,147. Feb. 26.

PAULSON, A. T., VANDERDTOEP, J. and PORRIT, S. W. 1980. Enzymatic browning of peaches: Effect of gibberellic acid and ethepon on phenolic compounds and polyphenoloxidase activity. J. Food Sci. *45,* 341–345.

PAULUS, K., FRICKER, A., DUDEN, R., HEINTZE, K. and ZOHM, H. 1975. Influence of heat treatments of spinach at temperatures up to 100°C on important constituents. I. Introduction and description of the experiments employing treatments in water. Lebensm.-Wiss. Technol. *8,* 7–10. (German)

PAVLENKO, N. M. 1974. The application of immobilized enzymes in wine production. FSTA *8,* 7 H1262 (1976).

PAVLOVSKII, P. E. and SIMBIREVA, E. I. 1976. Changes in the activity of lysosomal collagenase during autolysis of intramuscular connective tissue. Prikl. Biokhim. Mikrobiol. *12,* 578–580. (Russian). FSTA *8,* 2 S242.

PAYEN, F. W. and PERSOZ, J. F. 1833. Treatise on diastase, the principal products of its reactions and their applications to industrial arts. Ann. Chim. Phys. *53,* 73–92. (French)

PAYNE-BOSE, D., WELSH, J. D., GEARHART, H. L. and MORRISON, R. O. 1977. Milk and lactose hydrolyzed milk. Am. J. Clin. Nutr. *30,* 695–697.

PEARLSTONE, J. R. and SMILLIE, L. B. 1977. The binding site of rabbit skeletal α-tropomyosin on troponin-T. Can. J. Biochem. *55,* 1032–1038.

PEDERSON, T. C. and AUST, S. D. 1973. The role of superoxide and singlet oxygen in lipid peroxidation by xanthine oxidase. Biochem. Biophys. Res. Comm. *52,* 1071–1078.

PEEPLES, M. L. and HUTCHESON, R. M. 1978. A method for determining lactose and sucrose in ice cream. J. Food Sci. *43,* 799–800.

PEILLET, P. and KOBREHEL, K. 1974. Determination of common wheat content in pasta products. Cereal Chem. *51,* 203–209.

PEISER, G. and YANG, S. F. 1975. Chlorophyll destruction in the presence of bisulfite. Plant Physiol. *56* (2) 13. (Suppl.)

PEKELHARING, C. A. 1902. Communication on pepsin. Z. Physiol. Chem. *35,* 8–30. (German)

PENNER, H. 1974. Identification of irradiated potatoes. V. Calorimetric determination of catalase. Z. Lebensm.-Unters. Forsch. *154,* 23–26. (German)

PENNY, I. F. 1974. The action of a muscle proteinase on the myofibrillar proteins of bovine muscle. J. Sci. Food Agric. *25,* 1273–1284.

PENNY, I. F., VOYLE, C. A. and DRANSFIELD, E. 1974. The tenderizing effect of a muscle proteinase on beef. J. Sci. Food Agric. *25,* 703–708.

PEPPER, L. and FARREL, H. M. 1978. Protein-protein interaction in the formation of casein micelles. Personal communication. Philadelphia.

PEPYS, J. 1973. Immunological and clinical findings in workers and consumers exposed to the enzymes of Bacillus subtilis. Proc. Roy. Soc. Med. *66,* 930–932.

PERHAM, R. N. 1975. Self-assembly of biological macromolecules. Philos. Trans. R. Soc. London, Ser. B. *272,* 123–126.

PERL, M. and DIAMANT, Y. 1963. Preparation and purification of soybean lipase. Isr. J. Chem. *1,* 192–193.

PERUTZ, M. F. 1978. Hemoglobin structure and respiratory transport. Sci. Am. *239* (6) 92–123.

PESCE, M. A. 1974. Topics in chemical instrumentation. 80. Centrifuge analyzers. New concept in automation for the clinical chemistry laboratory. J. Chem. Ed. *51,* A521–A534.

PETERKOFSKY, A. and GAZDAR, C. 1975. Interaction of Enzyme I of the phosphoenolpyruvate:sugar phosphotransferase system with adenyl cyclase of *Escherichia coli.* Proc. Natl. Acad. Sci. *72,* 2920–2924.

PETERSON, E. A. and SOBER, H. A. 1956. Chromatography of proteins. I. Cellulose ion-exchange adsorbents. J. Am. Chem. Soc. *78,* 751–755.

PETERSON, E. A. and SOBER, H. A. 1962. Column chromatography of protein: Substituted celluloses. Adv. Enzymol. *5,* 27–37.

PETRITSCHEK, A., LYNEN, F. and BELITZ, H. -D. 1972. Bitter peptides. II. Development of bitter taste in enzymatic hydrolysates of various proteins. Lebensm.-Wiss. Technol. *5,* 77–81. (German)

PEYNAUD, E. and RIBEREAU-GAYON, F. 1971. The grape. *In* The Biochemistry of Fruits and Their Products. A.C. Hulme (Editor). Academic Press, London.

PFANNEMÜLLER, B. and POTRATZ, C.H. 1977. Studies on modified branched polysaccharides. II. Star-shaped polymers with glycogen and amylopectin as structure models for starch. Staerke *29,* 73–80.

PFISTERER, E. 1974. Enzymes in fermentation. Tech. Q. Master Brew. Assoc. Am. *11,* 9–16.

PFLEIDERER, G. 1978. Lactate dehydrogenase. An example of the development of modern enzymology. Naturwissenschaften *65,* 397–406. (German)

PHAFF, H. J. 1977. Enzymatic cell wall degradation. *In* Proteins: Improvement Through Chemical and Enzymatic Modification. R.E. Feeney and J.R. Whitaker (Editors). Am. Chem. Soc., Washington, D.C.

PHELAN, J. A., GUINEY, J. and FOX, P. F. 1973. Proteolysis of β-casein in Cheddar cheese. J. Dairy Res. *40,* 105–112.

PHILIPPON, J. and ROUET-MAYER, M.A. 1973. Thawing of stone fruit. Prevention of enzymatic browning. Rev. Gen. Froid *64,* 487–493. (French)

PHILLIPS, D.R. and GALLIARD, T. 1978. Flavour biogenesis. Partial purification and properties of a fatty acid hydroperoxide cleaving enzyme from fruits of cucumber. Phytochemistry *17,* 355–358.

PICKETT, J. A. 1974. Estimation of nucleotides in beers and their effect on flavour. J. Inst. Brew. *80,* 42–47.

PIEPER, H. J. 1970. Microbial Amylases for the Production of Ethanol. Hans Ulmer, Stuttgart. (German)

PIERPONT, W. S. 1969. *o*-Quinones formed in plant extracts. Their reaction with bovine serum albumin. Biochem. J. *112,* 619–629.

PIETERSON, W. A., VIDAL, J. C., VOLWERK, J. J. and DE HAAS, G. H. 1974. Phospholipase A_2 and its zymogen from porcine pancreas. VII. Zymogen-catalyzed hydrolysis of monomeric substrates and the presence of recognition site for lipid-water interfaces in phospholipase A_2. Biochemistry *13,* 1455–1460.

PIFFERI, P. G., BALDASSARI, L. and CULTRERA, P. 1974B. Inhibition by carboxylic acids of an *o*-diphenol oxidase from *Prunus avium* fruit. J. Sci. Food Agric. *25,* 263–270.

PIFFERI, P. G. and CULTRERA, R. 1974. Enzymatic degradation of anthocyanins: the role of cherry polyphenol oxidase. J. Food Sci. *39,* 786–791.

PIFFERI, P. G., ZAMORINI, A. and MAZZOCCO, F. A. 1974A. Polyphenol oxidases in cherries. II. Forms active at neutral pH. Agrochimica *18,* 405–415. (Italian)

PIHASKI, K. and IVERSEN, T. -H. 1976. Myrosinase in Brassicaceae. I. Localization of myrosinase in cell fraction of roots of *Sinapis alba* L. J. Exp. Bot. *27,* 242–258.

PILNIK, W. and ROMBOUTS, F. M. 1978. Pectic enzymes. *In* Encyclopedia of Food Science. M.S. Peterson and A.H. Johnson (Editors). AVI Publishing Co., Westport, Conn.

PILNIK, W. and VORAGEN, A.G.J. 1970. Pectic substances and other uronides. *In* The Biochemistry of Fruits and Their Products, Vol. 1. A.C. Hulme (Editor). Academic Press, London.

PILNIK, W., VORAGEN, A. G. J. and VOS, L. DE. 1975. Enzymatic liquefaction of fruits and vegetables. Fluess. Obst *42,* 440–451. (German)

PINSKY, A. and GROSSMAN, S. 1969. Proteases of the soybean. II. Specificity. J. Sci. Food Agric. *20,* 74–75.

PINSKY, A., GROSSMAN, S. and TROP, M.

1971. Lipoxygenase content and antioxidant activity of some fruits and vegetables. J. Food Sci. *36,* 571–572.

PINTAURO, N. D. 1979. Food Processing Enzymes. Noyes Data Corp., Park Ridge, New Jersey.

PLACEK, C., BAVISOTTO, V. S. and JADD, E. C. 1960. Commercial enzymes by extraction (rennet). Ind. Eng. Chem. *52,* 2–8.

PLATT, W. C. and POSTON, A. L. 1962. Recovering citrus oil. U.S. Pat. 3,058,887. Oct. 16.

PLAYER, T. J. and HULTIN, H. O. 1977. Some characteristics of the NAD(P)H-dependent lipid peroxidation system in the microsomal fraction of chicken breast muscle. J. Food Biochem. *1,* 153–171.

PLOWMAN, K. M. 1976. Enzyme Kinetics. McGraw-Hill, New York.

POCKER, Y. and STONE, J. T. 1967. The catalytic versatility of carbonic anhydrase. III. Kinetic studies of the enzyme-catalyzed hydrolysis of *p*-nitrophenyl acetate. Biochemistry *6,* 668–678.

PODUSOLO, J. F. and BRAUN, P. E. 1975. Topographical arrangement of membrane proteins in the intact myelin sheath. Lactoperoxidase incorporation of iodine into myelin surface proteins. J. Biol. Chem. *250,* 1099–1105.

POKORNY, J., EL-ZEANY, B., KOLAKOWSKA, A. and JANICEK, G. 1974. Nonenzymic browning. IX. Correlation of autoxidation and browning reactions in lipid-protein mixtures. Z. Lebensm.-Unters. Forsch. *155,* 287–291. (German)

POKORNY, M., ZISSIS, E. and FLETCHER, H. G., JR. 1975. The inhibitory activities of 2-acetamido-2,3-dideoxy-D-hex-2-enonolactones on 2-acetamido-2-deoxy-β-D-glucosidase. Carbohydr. Res. *43,* 345–354.

POLING, S. M., HSU, W. -J. and YOKOYAMA, H. 1975. Structure-activity relationships of chemical inducers of carotenoid biosynthesis. Phytochemistry *14,* 1933–1938.

POLLARD, C. J. 1969. A survey of the sequence of some effects of gibberellic acid in the metabolism of cereal grains. Plant Physiol. *44,* 1227–1232.

POLLARD, J. E. 1975. Pectinolytic enzyme activity and changes in water potential components associated with internal breakdown in McIntosh apples. J. Am. Soc. Hortic. Sci. *100,* 647–649.

POLLOCK, C. J. and AP REESE, T. 1975A. Activities of enzymes of sugar metabolism in cold stored tubers of *Solanum tuberosum.* Phytochemistry *14,* 613–617.

POLLOCK, C. J. and AP REESE, T. 1975B. Cold-induced sweetening of tissue cultures of *Solanum tuberosum.* Planta *122,* 105–107.

POLLOCK, J. R. A. 1962. The nature of malting process. *In* Barley and Malt. A.H. Cook (Editor). Academic Press, London.

POLSON, A. 1953. Multimembrane electrodecantation and its application to isolation and purification of proteins and viruses. Biochim. Biophys. Acta *11,* 315–325.

POLYA, G. M. 1975. Purification and characterization of a cyclic nucleotide-regulated 5′-nucleotidase from potato. Biochim. Biophys. Acta *384,* 443–457.

POMAROLA, M. and SANDRET, F. 1973. Commercialization of potato irradiation in France. Food Irradiat. Inf. *12,* 35–42.

POMERANTZ, S. H. and WARNER, M. C. 1966. Identification of 3,4-dihydroxyphenylalanine as tyrosinase cofactor in melanoma. Biochem. Biophys. Res. Commun. *24,* 25–31.

POMERANZ, Y. 1964. Lactase (β-D-galactosidase). I. Occurrence and properties. II. Possibilities in the food industry. Food Technol. *18,* 682–687, 690–697.

POMERANZ, Y. 1971. Biochemical and functional changes in stored cereal grains. Crit. Rev. Food Technol. *2,* 45–80.

POMMIER, J., DEME, D., FIMIANI, E. and NUNEZ, J. 1975. In vitro thyroxine formation. Ann. Endocrin. *36,* 167–168.

PONTE, J. C., JR. 1971. Bread. *In* Wheat: Chemistry and Technology, 2nd Edition. Y. Pomeranz (Editor). Am. Assoc. Cereal Chem., St. Paul, Minn.

PONTING, J. D. 1944. Catechol test for frozen fruits. Quick Frozen Foods 7 (5) 31.

PONTING, J. D. 1954. Reversible inactivation of polyphenol oxidase. J. Am. Chem. Soc. *76,* 662–663.

PONTING, J. D. 1973. Osmotic dehydration of fruits—recent modifications and applications. Process Biochem. *8* (12) 18–20.

PONTING, J. D. and JACKSON, K. 1972. Prefreezing of Golden Delicious apple slices. J. Food Sci. *37,* 812–814.

PONTING, J. D., JACKSON, R., SANSCHUK, D. W. and HUXSOLL, C. C. 1975. Bulk storage and processing peaches. Agric. Res. *23* (7) 15.

PONTING, J. D., JACKSON, R. and WATTERS, G. 1971. Refrigerated apple slices: Effects of pH, sulfites and calcium on texture. J. Food Sci. *36,* 349–350.

PONTING, J. D. and JOSLYN, M. A. 1948. Ascorbic acid oxidation and the browning in apple-tissue extracts. Arch. Biochem. *19,* 47–63.

PONTREMOLI, S., DE FLORA, A., SALAMOND, E., MELLONI, E. and HORECKER, B. L. 1975. Hormonal effects on structure and catalytic properties of fructose 1,6-biphosphatase. Proc. Natl. Acad. Sci. *8,* 2969–2973.

POPADITCH, I. A., MIROVICH, A. I., MIRONOVA, N. I. and STOLBIKOVA, O. E. 1975. Effect of heat treatment on the activity of amylolytic enzymes. Khlebopek. Konditer. Prom. *5,* 5–11. (Russian)

PORATH, J. 1976. Bioaffinity chromatography-methodology and application. J. Macromol. Sci. Chem. *A10* (1–2), 1–14.

PORATH, J. and FLODIN, P. 1959. Gel filtration: A method for desalting and group separation. Nature *183,* 1657–1659.

PORATH, J. and KRISTIANSEN, T. 1975. Biospecific affinity chromatography. *In* The Proteins, 3rd Edition. H. Neurath, R. Hill and C.-L. Boeder (Editors). Academic Press, New York.

PORTELLE, D. and THONART, P. H. 1975. Immobilized β-galactosidase on insoluble carrier. II. Technological applications of immobilized β-galactosidase. Lebensm.-Wiss. Technol. *8,* 274–277. (French)

PORTER, M. C. 1972. Application of membranes to enzyme isolation and purification. *In* Enzyme Engineering. L.S. Wingard (Editor). Interscience Publishers, New York.

POSNER, I. and BERMUDEZ, D. 1977. Lipoprotein lipase stabilization by a factor of bovine milk. Acta Cient. Venez. *28,* 277–283.

POSSANI, L. D., BANERJEE, R., BALNY, C. and DOUZOU, P. 1970. Oxidation of haemoglobin by oxygen in light: Possible role of singlet oxygen. Nature *226,* 861–862.

POSTEL, W., DRAWERT, F. and HAGEN, W. 1973. Enzymatic investigations on the contents of L-(+)-and D(–)-lactic acid in wines. Z. Lebensm.-Unters. Forsch. *150,* 267–273. (German)

POSTON, J. M. 1977. Leucine 2,3-amino mutase: A cobalamine-dependent enzyme present in bean seedlings. Science *195,* 301–302.

POTTER, N. N. 1978. Food Science, 3rd Edition. AVI Publishing Co., Westport, Conn.

POTTHAST, K. and HAMM, R. 1973. Thin layer chromotographic method for the determination of lipolytic changes in meat. Z. Lebensm.-Unters. Forsch. *153,* 6–12. (German)

POTTHAST, K., HAMM, R. and ACKER, L. 1975. Enzymic reactions in low moisture foods. *In* Water Relations of Foods. R.B. Duckworth (Editor). Academic Press, New York.

POTTHAST, K., HAMM, R. and ACKER, L. 1977. Influence of water activity on glycolytic enzymes during storage. Z. Lebensm.-Unters. Forsch. *162,* 139–143. (German)

POTTY, V. H. 1969. Occurrence and properties of enzymes associated with mevalonic synthesis in the orange. J. Food Sci. *34,* 231–234.

POULSEN, P. B. and ZITTAN, L. E. 1976. Continuous glucose isomerization. Ger. Offen. 2,609,602. Sept. 23.

POUR-EL, A. 1979. Functionality and Protein Structure. Am. Chem. Soc., Washington, D.C.

POUR-EL, A. and SWENSON, T. C. 1973. Water-soluble protein. U.S. Pat. 3,741,771. June 26.

PRASAD, K., WATSON, C. A. and CARNEY, J. B., JR. 1979. Rapid nephelometric determination of alpha amylase activity in sprouted wheat kernels. Cereal Chem. *56,* 43–44.

PRATT, D. B., JR. 1971. Criteria of flour quality. *In* Wheat Chemistry and Technology. Y. Pomeranz (Editor). Am. Assoc. Cereal Chem., St. Paul, Minn.

PRATT, D. E. 1972. Water soluble antioxidant activity in soybeans. J. Food Sci. *37,* 322–323.

PRELOG, V. 1976. Chirality in chemistry. Science *163,* 17–24.

PRENDERGAST, K. 1973. Versatility of hydrolysed proteins. Food Manuf. *48* (4) 37, 39, 57.

PRENTICE, N. 1973. Invertase activities during the germination of barleys that differ in malting properties. Cereal Chem. *50,* 346–353.

PRENTICE, N. 1975. Invertase activity as a measure of malting quality. Cereal Chem. *52,* 650–655.

PRENTICE, N., MOELLER, M. and POMERANZ, Y. 1971. A note on changes in peptide hydrolase, esterase and amidase of maturing barley. Cereal Chem. *48,* 714–716.

PRESSEY, R. 1966. Separation and properties of potato invertase and invertase inhibitor. Arch. Biochem. Biophys. *113,* 667–674.

PRESSEY, R. 1970. Changes in sucrose synthetase activities during storage of potatoes. Am. Potato J. *47,* 245–251.

PRESSEY, R. 1972. Natural enzyme inhibitors in plant tissues. J. Food Sci. *37,* 521–523.

PRESSEY, R. and AVANTS, J. K. 1971. Effect of substrate size on the activity of tomato polygalacturonase. J. Food Sci. *36,* 486–489.

PRESSEY, R. and AVANTS, J. K. 1972. Multiple forms of pectinesterase in tomatoes. Phytochemistry *11,* 3139–3142.

PRESSEY, R. and AVANTS, J. K. 1975. Cucumber polygalacturonase. J. Food Sci. *40,* 937–939.

PRESSEY, R. and AVANTS, J. K. 1978. Difference in polygalacturonase composition of clingstone and freestone peaches. J. Food Sci. *43,* 1415–1417.

PRESSEY, R. and SHAW, R. 1966. Effect of temperature on invertase inhibitor and sugars in potato tubers. Plant Physiol. *41,* 1657–1661.

PRESSEY, R., HINTON, D. M. and AVANTS, J. K. 1971. Development of polygalacturonase activity and solubilization of pectin in peaches during ripening. J. Food Sci. *36,* 1070.

PRESTON, R. R. and KRUGER, J. E. 1977. Control of protein hydrolysis in germinating wheat. II. Effects of gibberellic acid. Cereal Foods World *22,* 468.

PRICE, J. F. and SCHWEIGERT, B. S. 1971. Science of Meat and Meat Products, 2nd Edition. W.H. Freeman Publishers, San Francisco.

PRICE, S. 1978. Anisole binding protein from dog olfactory epithelium. Chem. Senses Flavor *3,* 51–55.

PRICHAVUDHI, K. and YAMAMOTO, N. Y. 1965. Effect of drying temperatures on the chemical composition and quality of macadamia nuts. Food Technol. *19,* 1153–1156.

PRIDHAM, J. B. 1963. Enzyme Chemistry of Phenolic Compounds. Pergamon Press, New York.

PRIHODA, J., HAMPL, J. and HOLAS, J. 1971. Effect of ascorbic acid and potassium bromate on viscous properties of dough measured with a Hoeppler viscometer. Cereal Chem. *48,* 68–74.

PRIMO YUFERA, E., KOEN MOSSE, J. and ROY-IRANZO, J. 1962. Gelification in concentrated orange juices. VI. Relations between gelification, loss of cloud and evolution of pectins in pulp. Proc. 1st Int. Congr. Food Sci. Technol., London *2,* 337–343.

PRINS, J. 1973. Cheesemaking with the microbial rennet Rennilase. Rev. Esp. Lecheria *87,* 3–14. (Spanish)

PRIPUTINA, L. S. 1974. Glucose oxidase and catalase as food additives. Chem. Abstr. *81,* 103,345t. (Ukrainian)

PRIVALOV, P. L. 1974. Thermal investigations of biopolymers solutions and scanning microcalorimetry. FEBS Lett. *40* (Suppl.) 140–153.

PROHASKA, J. R. 1979. Mechanism for the glutathione peroxidase activity of rat liver glutathione-S-transferases. Fed. Proc. *38,* 390.

PROMAYON, J., BAREL, M., FOURNY, G. and VINCENT, J. -C. 1976. Determination of the chicory content in soluble mixtures of coffee and chicory. Cafe Cacao The *20,* 209–218. (French)

PROSSER, A. R., SHEPPARD, A. J. and HUBBARD, W. D. 1977. Modification of the Canadian Food and Drug Directorate method for polyunsaturated fatty acid determination. J. Off. Anal. Chem. *60,* 895–898.

PROSSER, W. L. 1974. Method and apparatus for obtaining tomato juice while minimizing enzymic action. U.S. Pat. 3,835,763. Sept. 17.

PRYME, I. F., JONER, P. E. and JENSEN, H. B. 1969. The appearance of phage associated lysozyme in *E. coli* B cells immediately after infection with phage T_2. Biochem. Biophys. Res. Commun. *36,* 676–681.

PUHAN, Z. 1968. N-Acetylneuraminic acid content of the nonprotein nitrogen fraction of milk formed during the primary reaction of curdling with rennet and rennet substitutes. Milchwissenschaft *23,* 331–333. (German)

PUHAN, Z. 1969. Protease composition of a rennet substitute from *Bacillus subtilis* and properties of its component proteases. J. Dairy Sci. *52,* 1372–1378.

PUHAN, Z. and IRVINE, D. M. 1973. Reduction of proteolytic activity of *Bacillus subtilis* by acidification of milk before Cheddar cheese manufacture. J. Dairy Sci. *56,* 323–327.

PULS, J., SINNER, M. and DIETRICHS, H. H. 1974. Carrier-bound xylanases. Holzforschung *28,* 106–113. (German)

PURCELL, A. E., WALTER, W. M. and GIESBRECHT, F. G. 1978. Proteins and amino acids of sweet potato (*Ipomea batatas* (L.) Lam.) fractions. J. Agric. Food Chem. *20,* 699–704.

PURICH, D. L. 1979. Enzyme Kinetics and Mechanism. Methods in Enzymology. Part A. Initial Rate and Inhibitor Methods. Academic Press, New York.

PURICH, D. L. 1980. Enzyme Kinetics and Mechanism. Methods in Enzymology. Part B. Isotopic Probes and Complex Enzyme Systems. Academic Press, New York.

PURR, A. 1950. Regeneration of plant peroxidases. Biochem. Z. *321,* 19–25. (German)

PURR, A. 1972. Enzymatic processes in cocoa beans during fermentation under vacuum filtration. Ann. Technol. Agric. *21,* 472–475. (German)

PYLER, E. J. 1969. Enzymes in baking. Theory and practice. Bakers Dig. *43* (4) 46–52.

PYLER, E. J. 1973. Baking Science and Technology, 2nd Edition. Siebel Publishing Co., Chicago.

QUESNEL, V. C. and JUMAHUNINGH, K. 1970. Browning reaction in drying cacao. J. Sci. Food Agric. *21,* 537–541.

RAA, J. and GILDBERG, A. 1976. Autolysis

and proteolytic activity of cod viscera. J. Food Technol. *11,* 619–628.

RAAB, C. A., LUH, B. S. and SCHWEIGERT, B. S. 1973. Effects of heat processing on the retention of vitamin B_6 in lima beans. J. Food Sci. *38,* 544–545.

RACKER, E. 1976. A New Look at Bioenergetics. Academic Press, New York.

RACKIS, J. J. 1972. Biologically active components. *In* Soybeans: Chemistry and Technology. A.K. Smith and S.J. Circle (Editors). AVI Publishing Co., Westport, Conn.

RACKIS, J. J. 1975. Oligosaccharides of food legumes. Alpha-galactosidase activity and the flatus problem. *In* Physiological Effect of Carbohydrates. ACS Symp. Series *15,* 207–222. Am. Chem. Soc., Washington, D.C.

RACKIS, J. J. 1977. Enzymes in soybean processing and quality control. *In* Enzymes in Food and Beverage Processing. R.L. Ory and A.J. St. Angelo (Editors). Am. Chem. Soc., Washington, D.C.

RACKIS, J. J. and ANDERSON, R. L. 1977. Mineral availability in soy protein products. Food Prod. Dev. *11* (10) 38–42.

RACKIS, J. J., McGHEE, J. E., HONIG, D. J., and BOOTH, J. N. 1975. Processing soybeans into foods: Selected aspects of nutrition and flavor. J. Am. Oil Chem. Soc. *52,* 249A–253A.

RADOLA, B.J. and DELINCÉE, H. 1972. The effect of ionizing radiation on enzymes. Ann. Technol. Agric. *21,* 472–486. (French)

RADOUCO-THOMAS, S. M. -A. 1962. Compositions and methods for improving meat. U.S. Pat. 3,042,529. July 3.

RAHN, O. 1932. Physiology of bacteria. P. Blakiston's & Son, Philadelphia.

RAISON, J. K., LYONS, J. M., MEHLHORN, R. J. and KEITH, A. D. 1971. Temperature induced phase change in mitochondrial membrane detected by spin labeling. J. Biol. Chem. *246,* 4036–4040.

RAJAGOPLAN, A. and GUNTER, A. 1976. The significance of isoenzymes in metabolic regulation and division processes. Biol. Zentralbl. *95,* 649–666. (German)

RALLS, J. W. 1974. Continuous in-plant hot-gas blanching of vegetables. Rep. *EPA-660/2 74-091,* Natl. Environ. Res. Cent., U.S. Environ. Prot. Agency, Corvallis, Oregon.

RALLS, J. W. and MERCER, W. A. 1973. Low water volume enzyme deactivation of vegetables before preservation. Rep. *EPA-R2-73-198,* for Off. Res. Monitor., U.S. Environ. Prot. Agency, Washington, D.C.

RAMADOSS, C. S., LUBY, L. J. and UYEDA, K. 1976. Affinity chromatography of phosphofructokinase. Arch. Biochem. Biophys. *175,* 487–494.

RAMADOSS, C. S., PISTORIUS, E. K. and AXELROD, B. 1976. Requirement for the presence of two isoenzymes of lipoxygenase in the coupled bleaching of carotene during linoleate oxygenation. Fed. Proc. *35,* 1652.

RAMASWAMY, N. K. and NAIR, M. 1974. Temperature and light dependency of chlorophyll synthesis in potatoes. Plant Sci. Lett. *2,* 249–256.

RAMASWAMY, S. and REGE, D. V. 1975A. Processing of brinjal. Indian Food Packer *29* (4) 15–21.

RAMASWAMY, S. and REGE, D. V. 1975B. Polyphenolic compounds in tissues of brinjals *(Solanum melangena).* Acta Aliment. Acad. Sci. Hung. *4,* 381–390.

RAMBAUD, M. 1975. Antitrypsin free products. U.S. Pat. 3,845,229. Oct. 29.

RAMET, J. P. and ALAIS, C. 1973. A coagulating protease produced by *Mucor mihei.* III. Utilization of Rennilase in the manufacture of semi hard and soft cheese. Lait *53,* 154–162.

RAMIREZ-MARTINEZ, J. R., LEVI, A., PADUA, H. and BAKAL, A. 1977. Astringency in an intermediate moisture banana product. J. Food Sci. *42,* 1201–1203.

RANADIVE, A. S. and HAARD, N. F. 1972. Peroxidase localization and lignin formation in developing pear fruit. J. Food Sci. *37,* 381–383.

RAND, A. D., JR. and HOURIGAN, J. A. 1975. Direct enzymic conversion of lactose in milk to acid. J. Dairy Sci. *58,* 1144–1150.

RAND, A. G. and LINKLATER, P. M. 1973. Use of enzymes for the reduction of lactose levels in milk products. Aust. J. Dairy Technol. *28,* 63–67.

RANDALL, C. F. and MACRAE, A. F. 1967. Hydrolytic enzymes in bovine skeletal muscles. 2. Proteolytic activity of water soluble proteins separated by starch gel electrophoresis. J. Food Sci. *32,* 182–184.

RANHOTRA, G. S. 1973. Factors affecting hydrolysis during breadmaking of phytic acid in wheat protein concentrate. Cereal Chem. *50,* 353–357.

RANHOTRA, G. S. and LOWE, R. J. 1975. Effect of phytase on dietary phytic acid. J. Food Sci. *40,* 940–942.

RANHOTRA, G. S., LOEWE, R. J. and PUYAT, L. V. 1974. Phytic acid in soy and its hydrolysis during breadmaking. J. Food Sci. *39,* 1023–1025.

RANJAN, S., PATNAIK, K. K. and LALORAYA, M. M. 1961. Enzymic conversion of meso-tartrate to dextro-tartrate in tamarind. Naturwissenschaften *48,* 406.

RANKINE, B. C. and POCOCK, K. F. 1969. Influence of yeast strain on binding of sulfur dioxide in wines and on its formation. J. Sci. Food Agric. *20,* 104–109.

RANKINE, B. C. and POCOCK, K. F. 1970. Dissolved oxygen in wine. Food Technol. Aust. *22,* 120–121, 123.

RAO, B. A. S. and NARASIMHAM, V. V. L. 1976. Brewing with enzymes. J. Food Sci. Technol. India *13,* 119–123.

RAO, D. R., HARIHARAN, K. and VIJAYALAKSHMI, K. R. 1974. A specific enzymatic procedure for the determination of neurotoxic components (derivatives of L-α,β-diaminopropionic acid) in *Lathyrus sativus.* J. Agric. Food Chem. *22,* 1146–1148.

RAO, N. N. and MODI, V. V. 1976. Fructose-1, 6-diphosphatase from *Mangifera indica.* Phytochemistry *15,* 1437–1439.

RAO, S. C. and CROY, L. I. 1972. Protease and nitrate reductase seasonal patterns and their relation to grain production of "high" vs "low" protein wheat varieties. J. Agric. Food Chem. *20,* 1138–1141.

RAO, V. R., MEHTA, S. L. and JOSHI, M. G. 1976. Peroxidase and amylase activity in developing grains of triticale, wheat and rye. Phytochemistry *15,* 893–895.

RAO, V. S. and DUKE, W. B. 1976. Effect of alachlor, propachlor, and prynachlor on GA3-induced production of protease and α-amylase. Weed Sci. *24,* 616–618.

RAPER, H. S. 1932. Tyrosinase. Ergeb. Enzymforsch. *1,* 270–280.

RATTAZZI, M. C., SCANDALIOS, J. G. and WHITT, G. S. 1976. Current Topics in Biological and Medical Research, Vol. 2. Alan R. Riss, New York.

RATTAZZI, M. C., SCANDALIOS, J. G. and WHITT, G. S. 1979. Isozymes. Current Topics in Medical Research. Alan R. Liss, New York.

RATTRIE, N. W. and REGENSTEIN, J. M. 1977. Action of crude papain on actin and myosin heavy chains isolated from chicken breast muscle. J. Food Sci. *42,* 1159–1163.

RAUTELA, G. S. and PAYNE, M. G. 1970. Relation of *o*-diphenol oxidase and peroxidase of sugar beets to *Cercospora* leaf spot. Phytopathology *60,* 238–245.

RAY, A. 1975. Steam blancher uses 50% less energy. Food Process. *37* (1) 64.

RAYMOND, M. N., BRICAS, E., SALESSE, R., GARNIER, J., GARNOT, P. and RIBIDEAU-DUMAS, B. 1973. Proteolytic unit for chymosin (rennin) activity based on a reference synthetic peptide. J. Dairy Sci. *56,* 419–422.

RECHNITZ, G. A. 1975. Membrane electrode probes for biological systems. Science *190,* 234–238.

REDDI, P. K., CONSTANTINIDES, S. M. and DYMSZA, H. A. 1972. Catheptic activity of fish muscle. J. Food Sci. *37,* 643–648.

REDDY, N. R., BALAKRISHNAN, C. V. and SALUNKHE, D. K. 1978. Phytate phosphorus and mineral changes during germination and cooking of black gram *(Phaseolus mungo)* seeds. J. Food Sci. *43,* 540–543.

REDMAN, D. G. 1974. Dehydroascorbic acid reductase in germinating grains. Chem. Ind. (London) *10,* 414–415.

REED, G. 1963. Malting of flour. Am. Soc. Bakery Eng. Bull. *171* (Aug.) 691–693.

REED, G. 1966. Enzymes in Food Processing. Academic Press, New York.

REED, G. 1968. Enzyme supplementation in baking. Bakers Dig. *41* (5) 84–87, 123.

REED, G. 1975. Enzymes in Food Processing, 2nd Edition. Academic Press, New York.

REED, G. 1976. The utility of enzymic processing. Cereal Foods World *21,* 578–580, 599.

REED, G. and THORN, J. A. 1971. Enzymes. *In* Wheat: Chemistry and Technology, 2nd Edition. Y. Pomeranz (Editor). Am. Assoc. Cereal Chem., St. Paul, Minn.

REED, L. J., PETITIT, F. H., ROCHE, T. E., PELLEY, J. W. and BUTTERWORTH, P. J. 1976. Structure and regulation of the mammalian pyruvate dehydrogenase complex. *In* Metabolic Interconversion of Enzymes, 4th Int. Symp., 1975. S. Shaltiel (Editor). Springer Verlag, Berlin.

REESE, E., CLAPP, R. C. and MANDELS, M. 1958. A new thioglucosidase in fungi. Biochem. Biophys. *75,* 228–242.

REESE, E. T. 1976. History of the cellulase program at the U.S. Army Natick Development Center. Biotechnol. Bioeng. Symp. *6,* 9–20.

REESE, E. T. and MANDELS, M. 1958. Enzyme action on partition chromatographic columns. J. Am. Chem. Soc. *80,* 4625–4627.

REESE, E. T., SIU, R. G. H. and LEVINSON, H. S. 1950. The biological degradation of soluble cellulose derivatives and its relationship to the mechanism of cellulase hydrolysis. J. Bacteriol. *59,* 485–497.

REESE, W. R. and DUNCAN, H. J. 1972. Studies on nucleotides and related compounds. II. An adenyl-cleaving enzyme in potato tubers. J. Sci. Food Agric. *23,* 345–351.

REEVE, R. M. 1970. Relationships of histological structure to texture of fresh fruits and vegetables. J. Texture Stud. *1,* 247–284.

REEVE, R. M. 1977. Pectin starch, and texture of potatoes: Some practical and theoretical implications. J. Texture Stud. *8,* 1–17.

REGNIER, J. M. 1977. Thermal resistance of clotting enzymes. Ann. Falsif. Expert Chim. *70,* 165–175. (French)

REICH, E., RIFKIN, D. B. and SHAW, E. 1975. Cold Spring Harbor Conference on Cell Proliferation, Vol. 2. Proteases and Biological Control. Cold Spring Harbor Laboratories, Cold Spring Harbor, N.Y.

REILAND, J. 1971. Gel filtration. Methods Enzymol. *22,* 287–321.

REILLY, P. J. 1977. Enzyme technology in the utilization of agricultural wastes. U.S. NTIS. *In* Govt. Rep. Announc., Index (U.S.) *77* (19) 11.

REIMERDES, E. H., KLOSTERMEYER, H. and SAYK, E. 1976. Milk proteinases. VII. Fractionation of components of the proteinase inhibitor systems of milk. Milchwissenschaft *31,* 329–334. (German)

REINECCIUS, G. A. 1979. Off-flavors in meat and fish: A review. J. Food Sci. *44,* 12–21.

REINER, J. M. 1969. Behavior of Enzyme Systems, 2nd Edition. Van Nostrand Reinhold Co., New York.

REINEFELD, E. and BLIESENER, K. -M. 1977. Enzymic determination of formic acid in its use in molasses. Zucker *30,* 650–652. (German)

REINHOLD, J. G. 1975. Phytate destruction by yeast fermentation in whole wheat meals. J. Am. Diet. Assoc. *66,* 38–41.

REINHOLD, J. G., ISMAIL-BEIGI, F. and FARADJI, B. 1975. Fiber *vs* phytate as determinant of the availability of calcium, zinc and iron of breadstuffs. Nutr. Rep. Int. *12,* 56–75.

REINHOLD, W. 1972. Swiss cheese varieties. Pfizer Cheese Monogr. *5,* 114–117.

RENDLEMAN, J. A. and HODGE, T. E. 1975. Complexes of carbohydrate with aluminate ion. Chromatography of carbohydrates on columns of anion-exchange resin (aluminate form). Carbohydr. Res. *44,* 155–167.

RENDLEMAN, J. A. and HODGE, T. E. 1977. Isomerization of aldoses to ketones on columns of aluminate anion exchange. Abstr. Am. Chem. Soc. Meet., Chicago, Aug.

RENVALL, S. 1975. Examination of phosphorous pesticides in fruits and vegetables. Var Foeda *27,* 310–312. (Swedish)

RESENDE, R., FRANCIS, F. J. and STUMBO, C. R. 1969. Thermal destruction and regeneration of enzymes in green bean and spinach puree. Food Technol. *23,* 63–66.

RESMINI, P., VOLONTERIO, G., SARACCHI, S. and ANNIBALDI, S. 1971. Experimental production of Parmigiano-Reggiano cheese with Pfizer milk-clotting enzyme (Supraren). II. Sci. Tecn. Latterio-Casearia *22,* 406–425. (Italian)

RESTREPO, F., SNYDER, H. E. and ZIMMERMAN, G. L. 1973. Calcium activation of soybean lipoxygenase. J. Food Sci. *38,* 779–782.

REXOVÁ-BENKOVÁ, L. and MARCOVIČ, O. 1976. Pectic enzymes. Adv. Carbohydr. Chem. Biochem. *33,* 323–385.

REXOVÁ-BENKOVÁ, L., MARKOVIČ, O. and FOGLIETTI, M. J. 1977. Separation of pectic enzymes from tomatoes by affinity chromatography on cross-linked pectic acid. Collect. Czech. Chem. Commun. *42,* 1736–1741.

REXOVÁ-BENKOVÁ, L. and TIBENSKY, V. 1972. Selective purification of *Aspergillus niger* endopolygalacturonase by affinity chromatography on cross-linked pectic acid. Biochim. Biophys. Acta *268,* 187–193.

REYMOND, D. and BUSH, D. A. 1972. Pectin transeliminases and their use in manufacture of fruit juices. Ann. Technol. Agric. *21,* 545–553. (French)

REYNOLDS, J. H. 1974. An immobilized alpha galactosidase continuous flow reactor. Biotechnol. Bioeng. *16,* 135–147.

RHODES, D. N. and MEEGUNGWAN, C. 1962. Treatments of meats with ionizing radiations. IX. Inactivation of liver autolytic enzymes. J. Sci. Food Agric. *13,* 279–282.

RHODES, M. J. C. 1973. Co-factor specificity of plant alcohol dehydrogenase. Phytochemistry *12,* 307–314.

RHODES, M. J. C., HILL, A. C. R. and WOOLTORTON, L. S. C. 1976. Activity of enzymes involved in lignin biosynthesis in swede root disks. Phytochemistry *15,* 707–710.

RICE, E. E. and LANTERO, O. J., JR. 1974. Cheese manufacture using proteolytic enzymes. Can. Pat. 943,810. Mar. 19.

RICE, S. L., EITENMILLER, R. R. and KOEHLER, P. E. 1976. Biologically active amines in foods: a review. J. Milk Food Technol. *30,* 353–358.

RICHARDS, G. N. and STOKIE, G. 1974. Analysis of dextran in sugar—an enzymatic method. Int. Sugar J. *76,* 103–107.

RICHARSON, C. C. and KORNBERG, A. 1964. A deoxyribonucleic acid phosphatase-exonuclease from *E. coli.* J. Biol. Chem. *239,* 242–250.

RICHARDSON, G. H., GANDHI, N. R., DIVATIA, M. A. and ERNSTROM, C. A. 1971. Continuous curd tension measurements during milk coagulation. J. Dairy Sci. *54,* 182–186.

RICHARDSON, G. H. and NELSON, J. H. 1968. Rapid evaluation of milk coagulating and flavor producing enzymes for cheese manufacture. J. Dairy Sci. *51,* 1502–1503.

RICHARDSON, G. H., NELSON, J. H., LUBNOW, R. E. and SCHWARZBERG, R. L. 1967. Rennin-like enzyme from *Mucor pusillus* for cheese manufacture. J. Dairy Sci. *50,* 1066–1072.

RICHARDSON, T. 1977. Functionality changes in proteins following action of enzymes. *In* Food Proteins: Improvement Through Chemical and Enzymatic Modification. R.E. Feeney and J.R. Whitaker (Editors). Am. Chem. Soc., Washington, D.C.

RICHTER, C., WENDEL, A. and WESER, U. 1975. Inhibition by superoxide dismutase of linoleic acid peroxidation induced by lipoxidase. FEBS Lett. *51,* 300–303.

RICHTER, E. and HANDKE, S. 1973. Influence of blanching and preservation by air drying at different temperatures, deepfreezing and freeze drying on oxalic acid of spinach. Z. Lebensm.-Unters. Forsch. *153,* 31–36. (German)

RICHTER, R. L. and RANDOLPH, H. E. 1971. Purification and properties of a bovine milk lipase. J. Dairy Sci. *54,* 1275–1281.

RIENITS, K. G., HARDT, H. and AVRON, M. 1974. ATP driven reverse electron transport in chloroplasts. FEBS Lett. *33,* 28–32.

RIKANS, L. E., GIBSON, D. D. and McCAY, P. B. 1979. Influence of dietary fat on micosomal monooxygenases. Fed. Proc. *3,* 865.

RINAUDO, M. T., PONZETTO, C., VIDANO, C., and MARLETTO, F. 1973. The origin of honey saccharase. Comp. Biochem. Physiol. B. *46,* 245–251. (German)

RIOV, J. 1975. Polygalacturonase activity in citrus fruit. J. Food Sci. *40,* 201–202.

RISTELLI, L., MYLLY, A. and KIVIRIKKO, K. I. 1976. Affinity chromatography of collagen glycosyltransferases on collagen linked to agarose. Eur. J. Biochem. *67,* 197–202.

ROAN, C. F. 1977. Enzyme-treated fried food. U.S. Pat. 4,058,631. Nov. 15.

ROBB, D. A., MAPSON, L. W. and SWAIN, T. 1964. Activation of the latent tyrosinase of broad bean. Nature *207,* 503–504.

ROBB, D. A., MAPSON, L. W. and SWAIN, T. 1965. On the heterogeneity of the tyrosinase of broad bean (*Vicia faba* L.). Phytochemistry *4,* 731–740.

ROBBINS, F. M. and COHEN, S. H. 1976. Effects of catheptic enzymes from spleen on the microstructure of bovine semimembranous muscle. J. Texture Stud. *7,* 137–142.

ROBERTS, E. A. H. 1942. The chemistry of tea fermentation. Adv. Enzymol. *2,* 113–134.

ROBERTS, E. A. H. 1952. The chemistry of tea fermentation. J. Sci. Food Agric. *3,* 193–198.

ROBERTS, E. H. 1962. Economic importance of flavonoid substances: Tea. *In* The Chemistry of Flavonoid Compounds. T.A. Geissmann (Editor). Pergamon Press, Oxford.

ROBERTS, G. R. 1972. Some observations on the nature of catechol oxidase in tea leaves. Tea Q. *43,* 164–167.

ROBERTS, R. J. 1976. Restriction endonucleases. Crit. Rev. Biochem. *4,* 123–164.

ROBERTS, R. M. and HARRER, E. 1973. Determination of L-galactose in polysaccharide material. Phytochemistry *12,* 2679–2682.

ROBERTS, T. C. and ECKHOFF, N. D. 1973. Possibility of determining protein in wheat by gamma ray capture. Feedstuffs *45* (18) 30–32.

ROBERTSON, C. R. 1976. Enzymatic processing in hollow fiber reactors. Joint USA ROC Symp. Enzyme Technol., Aug., Lehigh University, Bethlehem, Pa.

ROBERTSON, G. L. 1976. Pectinesterase in New Zealand grapefruit juice. J. Sci. Food Agric. *27,* 261–265.

ROBERTUS, J. D., KRAUT, J., ALDEN, R. A. and BIRKTOFT, J. 1972. Subtilisin: A stereochemical mechanism involving transition-state stabilization. Biochemistry *11,* 4293–4303.

ROBINSON, P. J., WHEATLEY, M. A., JANSON, J.-C., DUNNIL, P. and LILLY, M. D. 1974. Pilot scale affinity chromatography; Purification of β-galactosidase. Biotechnol. Bioeng. *16,* 1103–1112.

ROBSON, B. 1980. Designing biologically active polypeptides. TIBS *5,* 240–244.

ROCKLAND, L. B. 1957. A new treatment of hygroscopic equilibria: application to walnuts *(Juglans regia)* and other foods. Food Res. *22,* 1–25.

ROCKLAND, L. B. 1969. Water activity and storage stability. Food Technol. *23* (10) 1–8.

ROCKLAND, L. B. 1977. Personal communication. Berkeley, Calif.

ROCKLAND, L. B. and BEAVENS, E. A. 1957. De-bittering of citrus fruits. U.S. Pat. 2,816,835. Dec. 17.

ROCKLAND, L. B. and DUNN, M. S. 1949. Determination of the biological value of proteins with *Tetrahymena gelii.* Food Technol. *3,* 289–292.

ROCKLAND, L. B. and NISHI, S. K. 1980. Influence of water activity (A_w) on food product stability. Food Technol. *34* (4) 42–51.

ROCKLAND, L. B. and STEWART, G. F. 1981. Properties of Water in Relation to Food Quality and Stability. Academic Press, San Francisco.

RODAWAY, S. J. 1978. Composition of α-amylase secreted by aleurone layers of grains of Himalaya barley. Phytochemistry *17,* 385–389.

RODERS, M., GLENDE, E. A., JR. and RECKNAGEL, R. O. 1976. NADPH dependent lipid peroxidation of calcium bound microsomes. Res. Commun. Chem. Pathol. Pharmacol. *15,* 393–396.

RODNIOVA, N. A. and GORBACHEV, I. V. 1975. Isolation, purification and some physicochemical and catalytic properties of β-1,4-D-xylosidase. Chem. Abstr. *85,* 105784s. (Russian)

ROGOV, I. A., ZHUKOV, N. N. and SKRYABIN, V. P. 1975. Determination of the depth of penetration of infrared energy into food products. FSTA *8* (8) E225.

ROINER, F. X. J. 1971. Continuous production of cheese. Ger. Offen. 1,792,264. Apr. 23.

ROJAS, S. W. and SCOTT, M. L. 1969. Factors affecting the nutritive value of cottonseed meal as a protein source in chick diets. Poultry Sci. *48,* 819–835.

ROKHLENKO, S. G. 1972. The effect of enzyme treatment on the content of methyl alcohol in fruit wines. FSTA *5* (4) H612, H730. (Russian)

ROKHLENKO, S. G. and GREBESHOVA, R. N. 1977. Effect of enzyme treatment of fruit on quality of fruit wine. Prikl. Biokhim. Mikrobiol. *13,* 112–117. (Russian)

ROKUGAWA, K., FUJISHIMA, T., KUNINAKA, A. and YOSHINO, H. 1979. Immobilization of nuclease P_1 on cellulose. J. Ferment. Technol. *57,* 570–573.

ROMANI, R. J. 1972. Stress in the postharvest cell: the response of mitochondria and ribosomes. J. Food Sci. *37,* 513–517.

ROMANI, R. J. 1976. Enzymes in relation to maturation and postharvest handling of plant products. Proc. Symp. Enzymes in the Food Processing Industry. J.R. Whitaker (Chairman). University of California, Davis, Jan.

ROMASKAYA, N. N. and KALMYSH, V. C. 1971. Method of producing beverages from whey. USSR Pat. 322,173. FSTA *4* (8) H1217. (Russian)

ROMBOUTS, F. M. and PHAFF, H. J. 1976. Lysis of yeast cell walls. Lytic β-(1→3)-glucanases from *Bacillus circulans* WL-12. Eur. J. Biochem. *63,* 121–130.

ROMBOUTS, F. M. and PILNIK, W. 1972. Research on pectin depolymerases in the sixties: A literature review. Crit. Rev. Food Technol. *3,* 1–26.

ROMBOUTS, F. M. and PILNIK, W. 1978. Enzymes in fruit and vegetable juice technology. Process Biochem. *13* (8) 9–13.

ROODYN, D. B. 1968. Principles and practice of multiple enzyme analysis. *In* Automatic Analytical Chemistry Technicon Symposium (3rd, 1967) *2,* 233–237. Mediad, White Plains, N.Y.

ROODYN, D. B. 1969. Automated enzyme assays. Process Biochem. *4,* 57–61.

ROODYN, D. B. 1973. Use of enzymes in automatic analysis. Proc. Soc. Anal. Chem. *10,* 230–235.

ROOZEN, J. P. and PILNIK, W. 1970. Stability of adsorbed enzymes in water-poor systems. I. Stability of peroxidase at 25°C. Lebensm.-Wiss. Technol. *3,* 37–40. (German)

ROOZEN, J. P. and PILNIK, W. 1971. On the stability of adsorbed enzymes in water deficient systems. V. The effect of storage and irradiation by electrons on the stability of alkaline phosphatase. Lebensm.-Wiss. Technol. *4,* 196–200.

ROOZEN, J. P. and PILNIK, W. 1973. Ultrafiltration controlled enzymatic degradation of soy protein. Process Biochem. *8* (7) 24–34.

ROREM, E. S. and SCHWIMMER, S. 1963. Double pH optima of potato invertase. Experientia *19,* 150–153.

ROSE, I. A. 1972. Enzyme reaction stereospecificity: A critical review. Crit. Rev. Biochem. *1,* 95–148.

ROSEBROOK, D. P. and BURNEY, J. E. 1968. Investigation of the determination of volatile isothiocyanates in mustard seeds and flour. J. Assoc. Off. Agric. Chem. *51,* 633–636.

ROSELL, J. M. 1961. Hydrogen peroxide-catalase method for treatment of milk. Can. Dairy Ice Cream J. *40* (8) 50–52.

ROSENAU, J. R., ANDERSON, J. C. and MORRIS, H. A. 1975. Production of cheddar-like cheeses via nonfermentative pH manipulation. J. Food Sci. *40,* 890–891.

ROSINEC, J., HERIAN, K., KRCAL, Z. and RIZMAN, M. 1977. Liquid rennet with increased enzyme activity. Czech. Pat. 170,672. Dec. 15.

ROSS, K. D. 1979. Definition of bound water by water activity depression. Lebensm.-Wiss. Technol. *12,* 172–176.

ROSWELL, E. V. and GOAD, L. J. 1962. Latent beta-amylase of wheat: Its mode of attachment to glutelin and its release. Biochem. J. *84*, 73P–74P.

ROTH, G. R., STANFORD, N. and MAJORS, P. W. 1975. Acetylation of prostaglandin synthesis by aspirin. Proc. Natl. Acad. Sci. *72,* 3073–3076.

ROTHE, G. M. 1975. Intracellular localization and some properties of two aldehyde oxidases isoenzymes in potato tubers. Biochem. Physiol. Planz. *167,* 411–418. (German)

ROTHFUS, J. A. and KENNEL, J. 1970. Properties of wheat beta-amylase adsorbed on glutenin. Cereal Chem. *47,* 140–146.

ROTHSCHILD, G. and KARSENTY, A. 1974. Influence of holding time before pasteurization, pasteurization and concentration on the turbidity of citrus juices. J. Food Sci. *39,* 1042–1044.

ROUBAL, W. T. and TAPPEL, A. L. 1966. Polymerization of proteins induced by free radical peroxidation. Arch. Biochem. Biophys. *113,* 150–155.

ROY, A. K. and BANASCHAK, H. 1971. Catalytic acceleration of methemoglobin reduction in human erythrocytes by polyphenols and quinones. Acta Biol. Med. *26,* 289–298. (German)

ROY, R. B. 1979. An improved semiautomated enzymic assay of lysine in foods. J. Food Sci. *44,* 480–482, 487.

ROY, R. B., SALTPETER, J. and DUNMIRE, D. L. 1976. Evaluation of urea-acid system as a medium of extraction for B group vitamins. J. Food Sci. *41,* 996–1000.

ROZIER, J. 1971. The role of meat catalase in the manufacture of dry sausage. Fleischwirtschaft *51,* 1063–1066. (German)

RU, K. and GAU, C. -P. 1973. Studies on the relations between rice blast disease resistance and enzymes. Chem. Abstr. *84,* 40,857a. (Russian)

RUBIN, B. A., KHANDOBINA, L. M. and GERASKINA, G. V. 1976. Distribution and functions of the phenol oxidases. Chem. Abstr. *85,* 15,946b. (Russian)

RUCHTI, J. and McLAREN, A. D. 1964. Enzyme reactions in structurally restricted systems. V. Further observations on the kinetics of yeast β-fructofuranosidase (invertase) activity in viscous media. Enzymologia *27,* 185–198.

RUCKER, R. B., MURRAY, J. and RIGGINS, R. S. 1977. Nutritional copper deficiency and penicillamine administration: some effects on bone collagen and elastin crosslinking. *In* Protein Crosslinking. Nutritional and Medical Consequences. M. Friedman (Editor). Plenum Press, New York.

RUEGG, M., LUSCHER, M. and BLANC, B. 1974. Hydration of native and rennin coagulated caseins as determined by differential scanning calorimetry and gravimetric sorption measurements. J. Dairy Sci. *57,* 387–393.

RUFFNER, H. P., HAWKER, J. S. and HALE, C. R. 1977. Temperature and enzymic control of malate metabolism in berries of *Vitis vinifera.* Phytochemistry *15,* 1877–1880.

RUFFNER, H. P. and KLIEWER, W. M. 1975. Phosphoenolpyruvate (PEP) carboxykinase in grape berries. Plant Physiol. *56,* 67–71.

RUITER, A. 1972. Substitution of proteases in the enzymic ripening of herring. Ann. Technol. Agric. *21,* 597–605. (French)

RUPP, H., and WESER, U. 1976. Copper I and copper II in complexes of biochemical significance studied by x-ray photoelectron spectroscopy. Biochim. Biophys. Acta *446,* 151–165.

RUSSELL, D. W. 1971. The metabolism of aromatic compounds in higher plants. X. Properties of the cinnamic acid 4-hydroxylase of pea seedlings and some aspects of its metabolic and developmental control. J. Biol. Chem. *246,* 3070–3078.

RUTLEDGE, J. E. and YING, L. C. 1972. Reduction of antithiamine activity in crayfish by heat treatment. J. Food Sci. *37,* 497–498.

RUTMAN, M. 1971. Process for preparing high-energy fish protein concentrate. U.S. Pat. 3,561,973. Feb. 9.

RUTTER, P. and STAINSBY, G. 1975. The solubility of tea cream. J. Sci. Food Agric. *26,* 455–463.

RYALL, A. L. and LIPTON, C. M. 1979. Handling, Transportation and Storage of Fruits and Vegetables, Vol. 1. AVI Publishing Co., Westport, Conn.

SAENGER, W. 1976. α-Cyclodextrin inclusion complexes: Mechanism of adduct formation and intermolecular reactions. Jerusalem Symp. Quantum Chem. Biochem. 1975. *8,* 265–305.

SAENGER, W., NOLTMEYER, M., MANOR, P. C., HINGERTY, B. and KLAR, B. 1976. Topography of cyclodextrin inclusion compounds. IX. "Induced-fit" complex formations of the model enzyme. Bioorg. Chem. *5,* 187–195.

SAGE, B. A. and O'CONNOR, J. D. 1976. An affinity column for ecdysone binding proteins. Anal. Biochem. *73,* 240–246.

SAGHIR, A. R., COWAN, J. N. and SALJI, J. P. 1966. Goitrogenic activity of onion volatiles. Nature *211,* 87.

SAID, W. I., KHALIL, H. and SALAMA, B. 1973. Sprout inhibition of potatoes for local consumption by treatments with sprout inhibitors. Agric. Res. Rev. *51,* 91–99.

ST. ANGELO, A. J. and ORY, R. L. 1975. Effects of lipoperoxides on proteins in raw and processed peanuts. J. Agric. Food Chem. *23,* 141–145.

SAINT-RAT, L. 1971. Enzymatic extraction of proteins from various oilseed cakes. C.R. Seances Acad. Agric. Fr. *57,* 826–830. (French)

SAITO, Z. 1974. The milk lipases. IX. Effects of gamma-irradiation on lipase activities. Jpn. J. Dairy Sci. *23,* 153–158. (Japanese)

SAKAI-IMAMURA, M. 1975. Fatty acid oxidation and chlorophyll bleaching. Nat. Sci. Rep. Ochanomizu Univ. *26,* 109–125.

SAKAMURA, S., SHIBUSA, S. and OBATA, Y. 1966. Separation of polyphenol oxidase for anthocyanin degradation in egg plant. J. Food Sci. *31,* 317–319.

SALE, A. J. H. 1976. A review of microwaves for food processing. J. Food Technol. *11,* 319–329.

SALEH, B. and WATTS, B. M. 1968. Substrates and intermediates in the enzymatic reduction of metmyoglobin in ground beef. J. Food Sci. *33,* 353–358.

SALEH, N. A. M., FRITSCH, H., KREUZALER, F. and GRIESBACH, H. 1978. Flavanone synthase from cell suspension cultures of *Haplopappus gracilis* and comparison with synthase from parsley. Phytochemistry *17,* 183–186.

SALETAN, L. T. 1968. Carbohydrases of interest in brewing, with particular reference to amyloglucosidase. Wallerstein Lab. Commun. *31* (104) 33–42.

SALMINEN, S. O. and YOUNG, R. E. 1975. The control properties of phosphofructokinase in relation to the respiratory climacteric in banana fruit. Plant Physiol. *55,* 45–50.

SALUNKHE, D. K. and DO, J. Y. 1976. Biogenesis of aroma constituents of fruits and vegetables. Crit. Rev. Food Sci. Nutr. *8,* 161–190.

SALUNKHE, D. K. and WU, M. T. 1974. Subatmospheric storage of fruits and vegetables. Lebensm.-Wiss. Technol. *7,* 261–275. (German)

SALUNKHE, D. K. and WU, M. T. 1975. Subatmospheric storage of fruits and vegetables. *In* Postharvest Biology and Handling of Fruits and Vegetables. AVI Publishing Co., Westport, Conn.

SAMEJIMA, H. 1974. Recent trends in enzyme engineering in Japan. *In* Enzyme Engineering, Vol. 2. E.K. Pye and L.B. Wingard (Editors). Plenum Press, New York.

SAMEJIMA, H., NAGANO, Y., OTA, S., KANZAKI, Y., MATSUO, H. and KURODA, K. 1972. Process for preparing coloring agents for foods and beverages. U.S. Pat. 3,658,557. Apr. 25.

SAMOTUS, B. 1971. Storage of potato tubers under water. Preliminary investigations. Potato Res. *14,* 145–149.

SAMOTUS, B. and SCHWIMMER, S. 1962A. Predominance of fructose accumulation in cold-stored immature potato tubers. J. Food Sci. *27,* 1–4.

SAMOTUS, B. and SCHWIMMER, S. 1962B. Effect of maturity and storage on distribution of phosphorus among starch and other components of potato tuber. Plant Physiol. *37,* 519–522.

SAMOTUS, B. and SCHWIMMER, S. 1962C. Phytic acid as a phosphorus reservoir in the developing potato tuber. Nature *194,* 578–579.

SAMOTUS, B. and SCHWIMMER, S. 1963. Changes in carbohydrate and phosphorus content of potato tubers during storage in nitrogen. J. Food Sci. *28,* 1–5.

SAMSONOVA, A. N. and MEL'YANTSEVA, E. G. 1978. Effect of enzyme preparations on the yield of wild rose juice. Chem. Abstr. *89,* 4816f. (Russian)

SAMUELSSON, B., GOLDYNE, M., GRANSTROM, E., HAMBERG, M., HAMMARSTROM, S. and MALMSTEN, C. 1978. Prostaglandins and thromboxanes. Annu. Rev. Biochem. *47,* 997–1029.

SANDERS, G. P. 1953. Cheese varieties and descriptions. U.S. Dep. Agric. Handb. *54,* 1–151.

SANDERSON, G. W. 1975. Black tea aroma and its formation. *In* International Symposium, Odor and Taste Substances. F. Drawert (Editor). Verlag Hans Cord, Nurenberg, Germany. (German)

SANDERSON, G. W., BERKOWITZ, J. E., CO., H. and GRAHAM, H. N. 1972. Biochemistry of tea fermentation: Products of the oxidation of tea flavonols in a model tea system. J. Food Sci. *37,* 399–404.

SANDERSON, G. W., and COGGON, P. 1977. Use of enzymes in the manufacture of black tea and instant tea. *In* Enzymes in Food Beverage Processing. R.L. Ory and A.J. St. Angelo (Editors). Am. Chem. Soc., Washington, D.C.

SANDERSON, G. W. and GRAHAM, H. N. 1973. On the formation of black tea aroma. J. Agric. Food Chem. *21,* 576–585.

SANDERSON, G. W., SIMPSON, W. S. and

SHAW, W. 1974. Pectinase enzyme treating process for preparing high bulk density tea powders. U.S. Pat. 3,787,582. Jan. 22.

SANDRET, F., MICHIELS, L. and BERSET, C. 1974. Irradiated potato tubers. Identification by sprouting and tissue culture. EURATOM [Rep] EUR 5126 d/e/f/i/n, 217–219. (French)

SANDSTEDT, R. M., KNEEN, E. and BLISH, M. J. 1939. A standardized Wohlegemuth procedure for alpha-amylase activity. Cereal Chem. *16,* 712–723.

SANDSTROM, W. M. 1946. The general chemistry of enzymes. *In* Enzymes and Their Role in Wheat Technology. J.A. Anderson (Editor). Interscience Publishers, New York.

SANNER, T., KOVACS-PROSZT, G. and WITOWSKI, S. 1974. Aspects of the effect of ionizing radiation on enzymes. *In* Improvement of Food Quality by Irradiation. Proc. Conf. Int. Atomic Energy Agency, Vienna.

SAPERS, G. M., ABBOT, T., MASSIE, D., WATADA, A. and FINNEY, E. E., JR. 1977. Volatile composition of McIntosh apple juice as a function of maturity and ripeness indices. J. Food Sci. *42,* 44–47.

SAPERS, G. M. and NICKERSON, J. T. R. 1962. Stability of spinach catalase. I–III. J. Food Sci. *27,* 272–290.

SARAN, R. and HILLER, G. 1977. Enzyme-containing animal fodder composition. Ger. Offen. 2,602,260. Aug. 4.

SARAWEK, S. and DAVIES, D. D. 1976. The identification and properties of a naturally occurring inhibitor of malic enzyme. Phytochemistry *15,* 479–481.

SARDINAS, J. L. 1968. Rennin enzyme of *Endothia parasitica*. Appl. Microbiol. *16,* 248–255.

SARDINAS, J. L. 1976. Calf rennet substitutes. Process Biochem. *11* (4) 10–17.

SASAKI, R., IKURA, K., KATSURA, S. and CHIBA, H. 1976. Regulation of human AMP deaminase by ATP and 2,3-biphosphoglycerate. Agric. Biol. Chem. *40,* 1797–1803.

SASAKI, T., TADOKORO, K. and SUZKI, S. 1973. Phosphofructokinase of *Solanum tuberosum* tuber. Phytochemistry *12,* 2834–2849.

SATO, K., HEGARTY, G. R. and HERRING, H. K. 1973. The inhibition of warmed-over cooked meats. J. Food Sci. *38,* 398–403.

SATO, M. 1967. Metabolism of phenolic compounds by the chloroplasts. III. Phenolase as an enzyme concerning the formation of esculetin. Phytochemistry *6,* 1363–1373.

SATO, M. 1976. Association by 2,3-dihydroxybenzaldehyde of monomeric phenolase in spinach chloroplasts. Phytochemistry *15,* 1665–1667.

SATO, M. and HASEGAWA, M. 1976. The latency of spinach chloroplast phenolase. Phytochemistry *15,* 61–65.

SATO, T. 1973. Sprout inhibition by irradiation. J. Food Sci. Technol. *20,* 26–36. (Japanese)

SATO, Y., WATANABE, K. and ASAI, A. 1967. Tenderization of meat by *Aspergillus* protease. Nippon Shokuhin Kogy Gakkaishi *14,* 235–240. (Japanese)

SATOH, T., MATSUDA, Y., TAKASHIO, M., SATOH, K., BEPPU, T. and ARIMA, K. 1976. Isolation of lipoxygenase-like enzyme from *Fusarium oxysporum.* Agric. Biol. Chem. *40,* 953–961.

SATOUCHI, K., MORI, T. and MATSUSHITA, S. 1974. Characterization of inhibitor protein for lipase in soybean seeds. Agric. Biol. Chem. *38,* 97–101.

SATTAR, A. and DEMAN, J. M. 1975. Photoxidation of milk and milk products. A review. Crit. Rev. Food Sci. Nutr. *1,* 73–138.

SATTERLEE, L. D. 1971. Effect of a porcine pancreatic collagenase on muscle connective tissue. J. Food Sci. *36,* 130–132.

SATTERLEE, L. D. 1972. Tenderization of beef using an extract of porcine pancreas. Lebensm. Wiss. Technol. *4,* 163–166.

SATTERLEE, L. D., WILHELM, M. S. and BARNHART, M. 1971. Low dose gamma irradiation of bovine metmyoglobin. J. Food Sci. *36,* 549–551.

SAUNDERS, G. C. and BARTLETT, M. L. 1977. Double-antibody solid-phase enzyme immunoassay for the detection of staphylococcal enterotoxin. J. Appl. Environ. Microbiol. *34,* 518–522.

SAUNDERS, J. A. and CONN, E. E. 1978. Presence of the cyanogenic glucoside dhurrin in isolated vacuoles from *Sorghum.* Plant Physiol. *61,* 154–157.

SAUNDERS, R. M., CONNOR, M. A., EDWARDS, R. H. and KOHLER, G. O. 1972. Enzymatic processing of wheat bran. Cereal Chem. *49,* 436–443.

SAUNDERS, R. M. and KOHLER, G. O. 1972. In vitro determination of protein digestibility in wheat millfeeds for monogastric animals. Cereal Chem. *49,* 39–103.

SAUTER, E. and MONTOURE, J. E. 1972. The relationships of lysozyme content of egg white to volume stability of foam. J. Food Sci. *37,* 918–920.

SAWADA, Y., OHYAMA, T. and YAMAZAKI, I. 1973. Green pea superoxide dismutase. *In*

Oxidases and Related Redox Systems. T.E. King, H.M. Mason and M. Morrison (Editors). University Park Press, Baltimore.

SAWYER, R. L. 1975. Sprout inhibition. *In* Potato Processing, 3rd Edition. W.F. Talburt and D. Smith (Editors). AVI Publishing Co., Westport, Conn.

SAXTON, C. A. and JEWELL, G. G. 1969. The morphological changes produced in cauliflower stems during pickling, and their relationship to texture parameters. J. Food Technol. *4,* 363–375.

SAYRE, R. N. 1968. Post-mortem changes in extractibility of myofibrillar protein from chicken pectoralis. J. Food Sci. *33,* 609–612.

SAYRE, R. N. 1970. Chicken myofibril fragmentation in relation to factors influencing tenderness. J. Food Sci. *35,* 7–10.

SAYRE, R. N., BRISKEY, E. J. and HOEKSTRA, W. G. 1963. Effect of excitement, fasting, and sucrose feeding on porcine muscle phosphorylase and postmortem glycolysis. J. Food Sci. *26,* 472–477.

SCANDALIOS, J. G. 1974. Isozymes in development and differentiation. Annu. Rev. Plant Physiol. *25,* 225–258.

SCHAB, R. and YANNAI, S. 1973. Improved method for debittering apricot kernels. J. Food Sci. Technol. *10,* 57–59.

SCHABORT, J. C. 1978. Cucurbacitin 19-hydroxylase in *Cucurbita maxima.* Phytochemistry 17, 1062–1064.

SCHACK, W. R. and CONNICK, F. G. 1970. Preparation of meat pieces and products. U.S. Pat. 3,533,803. Oct. 13.

SCHADE, J. E., MARSH, G. L. and ECKERT, J. E. 1958. Diastase activity and hydroxymethyl-furfural in honey and their usefulness in detecting heat alteration. Food Res. *23,* 446–463.

SCHAFER, D. E. and KROPF, M. E. 1973. Objective predictors of frozen beef color. J. Anim. Sci. *37,* 271.

SCHAICH, K. M. and KAREL, M. 1975. Free radicals in lysozyme reacted with peroxidizing methyl linoleate. J. Food Sci. *40,* 456–459.

SCHALLER, D. R. and POWRIE, W. D. 1971. Scanning electron microscopy of skeletal muscle from rainbow trout, turkey and beef. J. Food Sci. *36,* 552–559.

SCHALLER, K. 1974. Model studies of black spots on potatoes. Z. Lebensm.-Unters. Forsch. *155,* 339–341. (German)

SCHALLER, K. and AMBERGER, A. 1974. Relations between enzymic browning of potatoes and several constituents of the tuber. Qual. Plant.-Plant Foods Hum. Nutr. *24,* 183–190. (German)

SCHANBACHER, F. L. and SMITH, K. L. 1975. Formation of unusual whey proteins and enzymes: Relation to mammary function. J. Dairy Sci. *58,* 1048–1062.

SCHANDERL, S. H., CHICHESTER, C. O. and MARSH, G. L. 1962. Degradation of chlorophyll and several derivatives in acid solution. J. Org. Chem. *27,* 3865–3868.

SCHANDERL, S. H. and LYNN CO, Y. C. 1966. Changes in chlorophylls and spectrally related pigments during ripening of *Capsicum frutescens.* J. Food Sci. *31,* 141–145.

SCHANDERL, S. H., MARSH, A. L. and CHICHESTER, C. O. 1965. Color reversion in processed vegetables. J. Food Sci. *30,* 312–324.

SCHARPE, S. L., COOREMAN, W. M., BLOMME, W. J. and LAEKEMAN, G. M. 1976. Quantitative enzyme immunoassay: Current status. Clin. Chem. *22,* 733–738.

SCHATZKI, T. F., WITT, S. C., WILKINS, D. E. and LENKER, D. H. 1980. Characterization of growing lettuce from density contours. Pattern Recognition *12.* (in press)

SCHEINER, O. and BREITENBACH, M. 1976. Synthesis of Sepharose derivatives for the affinity chromatography of enzymes of the metabolism of *myo*-inositol phosphates. Monatsh. Chem. *107,* 581–586. (German)

SCHERRER, R. and EATON, R. F. 1981. Elastic network model of the cell wall isolated from *Bacillus megaterium.* J. Bacteriol. *144.* (in press)

SCHIEMANN, D. A. 1976. Proposed standards for alkaline phosphatase in the pasteurized milk determined by automated procedure. J. Milk Food Technol. *39,* 263–268.

SCHIEMANN, D. A., and BRODSKY, M. H. 1976. Studies of Scharer's original method for alkaline phosphatase in milk with a modification utilizing an organic buffer. J. Milk Food Technol. *39,* 191–195.

SCHIFREEN, R. S., HANNA, D. A., BOWERS, L. D. and CARR, P. W. 1977. Analytical aspects of immobilized enzyme columns. Anal. Chem. *49,* 1929–1939.

SCHIMKE, R. T. 1971. Control of enzyme levels in mammalian tissues. Adv. Enzymol. *37,* 135–187.

SCHINDLER, M. and SHARON, N. 1976. Reversible inactivation of lactose synthase by the modification of HIS 32 in human α-lactalbumin. Biochem. Biophys. Res. Commun. *69,* 167–173.

SCHINELLER, D. J., DOUGHERTY, R. H. and BIGGS, R. H. 1972. Influence of 5′-nucleotides on flavor threshold of octanal. J. Food Sci. *37,* 935–937.

SCHLEGEL, S. 1975. Anaerobic treatment of highly concentrated effluents from the food industry. Wasser Luft Betr. *19,* 447–450. (German)

SCHLEICH, H. and ARNOLD, R. S. 1962. Composition and methods for processing meat products. U.S. Pat. 3,037,870. June 5.

SCHLÜTER, M. and GMELIN, R. 1972. Abnormal enzymatic splitting of 4-methylthioglucosinolate in fresh plants of *Eruca sativa.* Phytochemistry *11,* 3427–3441. (German)

SCHMID, P. 1971. Idaein-splitting enzymes in apple peel. Z. Lebensm.-Unters. Forsch. *146,* 198–202. (German)

SCHMIDT, D. D., FROMMER, W., JUNGE, B., MÜLLER, W., WINGENDER, W., TRUSCHEIT, E. and SCHÄFER, D. 1977. α-Glucosidase inhibitors. New complex oligosaccharides of microbial origin. Naturwissenschaften *64,* 535–536.

SCHMIDT, E. and SCHMIDT, F. W. 1976. Clinical enzymology. FEBS Lett. *62* (Suppl. 4) E62–E79.

SCHMIDT, E. S., NEWTON, G. W., SANDERS, S. M., LEWIS, J. P. and CONN, E. E. 1978. Laetrile toxicity studies in dogs. J. Am. Med. Assoc. *239,* 943–947.

SCHMIDT, H. -L., KRISAM, G. and GRENNER, G. 1976. Microcalorimetric methods for substrates determination in flow systems with immobilized enzymes. Biochim. Biophys. Acta *429,* 283–290.

SCHMIDT, R. H., MORRIS, H. A. and McKAY, L. L. 1977. Species differences and effect of incubation time on lactic streptococcal intracellular proteolytic enzyme activity. J. Dairy Sci. *60,* 1677–1682.

SCHMITT, A. and SIEBERT, G. 1967. Distinguishing aliphatic dipeptidases from cod muscle. Z. Physiol. Chem. *348,* 1009–1016. (German)

SCHNEIDER, F. 1975. Cytochemistry of enzymes. Acta Histochem. Suppl. *14,* 33–46.

SCHNEYDER, J. and PLACHY, L. 1973. Enzymic determination of sugars in wines. Mitt. Rebe, Wein, Obstbau Früchtverwertung *23,* 177–178. (German)

SCHOENBEIN, C. F. 1861. Some separations produced by the capillary attraction of paper. Poggendorf's Ann. *114,* 275–280.

SCHOENBEIN, C. F. 1863. On the catalytic activity of organic materials and its distribution in the plant and animal world. J. Prakt. Chemie *89,* 323–339. (German)

SCHOOT-UITERKAMP, A. J. M., EVANS, L. H., JOLLEY, R. L. and MASON, H. S. 1976. Absorption and circular dichroism spectra of different forms of mushroom tyrosinase. Biochim. Biophys. Acta *453,* 200–204.

SCHORMÜLLER, J. 1974. The importance of enzyme activity in food chemistry. *In* Methods of Enzymatic Analyses. H. Bergmeyer (Editor). Academic Press, New York.

SCHRAMM, M. and LOYTER, A. 1966. Purification of α-amylase by precipitation of amylase-glycogen complexes. Methods Enzymol. *8,* 533–537.

SCHREIER, P. 1975. Sulphur-containing flavour substances. *In* International Symposium Odour and Taste Substances. F. Drawert (Editor). Verlag Hans Carl, Nürnberg, Germany.

SCHREIER, P., DRAWERT, F. and JUNKER, A. 1976. Identification of volatile constituents from grapes. J. Agric. Food Chem. *24,* 331–336.

SCHREIER, P. and HEIMANN, W. 1971. Lipoxygenase—"lipoperoxidase"—system in cereals. II. Characterization of the hydroperoxide decomposing enzymes. Helv. Chim. Acta *54,* 2803–2809.

SCHREINER, H. C. 1969. Method for acceleration of the ageing of slaughtered meat. U.S. Pat. 3,451,825. June 24.

SCHREYEN, L., DIRINCK, P., VAN WASSENHOVE, F. and SCHAMP, H. 1976. Analysis of leek volatiles by headspace condensation. J. Agric. Food Chem. *24,* 1147–1152.

SCHUBERT, E. 1952. Differentiation of polygalacturonase of *Aspergillus niger.* Nature *168,* 931–932.

SCHULTZ, A. W. 1960. Food Enzymes. AVI Publishing Co., Westport, Conn.

SCHULTZ, H. W., DAY, E. A. and LIBBEY, L. M. 1967. Symposium on Foods. The Chemistry and Physiology of Flavors. AVI Publishing Co., Westport, Conn.

SCHULTZ, W. G. 1979. An enzyme extraction process for the tropical annato farmer. Abstr. *123,* 39th Annu. Meet. Inst. Food Technol., St. Louis, June.

SCHULTZ, W. G., GRAHAM, R. P., ROCKWELL, W. C., BOMBEN, J. L., MIERS, J. C. and WAGNER, J. R. 1971. Field processing of tomatoes. 1. Process and design. J. Food Sci. *36,* 397–399.

SCHUR, F. and PIENDL, A. 1977. Dextrins in beer. Brauwissenschaft *30* (2) 46–50. (German)

SCHUSTER, L. 1971. Preparative acrylamide

gel electrophoresis: Continuous and disc techniques. Methods Enzymol. *22,* 412–433.

SCHUTTE, L. 1974. Precursors of the sulfur-containing flavor compounds. Crit. Rev. Food Technol. *4,* 457–505.

SCHWARTZ, J. and MARGALITH, P. 1972. Production of flavor enhancing materials by *Streptomyces* strain improvement and fermentation. J. Appl. Bacteriol. *35,* 271–278.

SCHWIMMER, S. 1943. The determination and thermal properties of plant peroxidase in relation to food processing. Ph.D. Dissertation. Georgetown University, Washington, D.C.

SCHWIMMER, S. 1944A. Regeneration of heat-inactivated peroxidase. J. Biol. Chem. *154,* 487–495.

SCHWIMMER, S. 1944B. Comparison of crude and purified preparations of a leucylpeptidase associated with beef muscle. J. Biol. Chem. *154,* 361–366.

SCHWIMMER, S. 1945. Role of maltase in the digestion of raw starch. J. Biol. Chem. *161,* 219–234.

SCHWIMMER, S. 1947A. Purification of malt alpha-amylase. Cereal Chem. *24,* 315–325.

SCHWIMMER, S. 1947B. Development and solubility of amylase in wheat kernels throughout growth and ripening. Cereal Chem. *24* (3) 167–179.

SCHWIMMER, S. 1950. Kinetics of malt α-amylase action. J. Biol. Chem. *186,* 181–193.

SCHWIMMER, S. 1951A. On the dextrinase activity of crystalline amylase preparations. Cereal Chem. *28,* 77–78.

SCHWIMMER, S. 1951B. The malt amylases Brew. Dig. *26,* 29T–48T.

SCHWIMMER, S. 1953A. Column procedures for the salt fractionation of enzymes. Nature *171,* 443.

SCHWIMMER, S. 1953B. Enzymes, enzyme systems of the white potato. J. Agric. Food Chem. *1,* 1063–1069.

SCHWIMMER, S. 1953C. Enzymatic activities in a microbial preparation of glucose oxidase. Fed. Proc. *12,* 266–267.

SCHWIMMER, S. 1953D. Evidence for the purity of Schardinger dextrinogenase. Arch. Biochem. Biophys. *43,* 108–117.

SCHWIMMER, S. 1954. Industrial production and utilization of enzymes from flowering plants. Econ. Bot. *8,* 99–113.

SCHWIMMER, S. 1956. Phytic and other acids of the potato tuber. Fed. Proc. *15,* 351.

SCHWIMMER, S. 1957. Nonoxidative, nonproteolytic enzymes. Annu. Rev. Biochem. *26,* 63–96.

SCHWIMMER, S. 1958. Influence of polyphenols and potato components on potato phosphorylase. J. Biol. Chem. *232,* 715–721.

SCHWIMMER, S. 1960. Myrosin-catalyzed formation of turbidity and hydrogen sulfide from sinigrin. Acta Chem. Scand. *14,* 1439–1441.

SCHWIMMER, S. 1961A. Chronometric integrals of product-inhibited enzyme reactions and the hydrolysis of S-ethyl-L-cysteine. Biochim. Biophys. Acta *48,* 132–138.

SCHWIMMER, S. 1961B. Spectral changes during the action of myrosinase on sinigrin. Acta Chem. Scand. *15,* 535–544.

SCHWIMMER, S. 1962. Theory of double pH optima of enzymes. J. Theor. Biol. *3,* 102–110.

SCHWIMMER, S. 1963. Alteration of the flavor of processed vegetables by enzyme preparations. J. Food Sci. *28,* 460–466.

SCHWIMMER, S. 1964A. Controlling enzyme action leads to better foods for more people. Canner Packer *133,* 48A–50A.

SCHWIMMER, S. 1964B. L-Cysteine sulfoxide lyases. Competition between enzyme and substrate for added pyridoxal phosphate. Biochim. Biophys. Acta *81,* 377–385.

SCHWIMMER, S. 1965. Support of DNA synthesis by reconstituted nucleohistones. Life Sci. *4,* 124–125.

SCHWIMMER, S. 1966. DNA polymerase activity of mung bean seedlings. Phytochemistry *5,* 791–794.

SCHWIMMER, S. 1967. Susceptibility to nucleases of DNA synthesized with nucleohistone primer. Biochim. Biophys. Acta *134,* 59–68.

SCHWIMMER, S. 1968A. Enzymatic conversion of *trans*(+)-S-1-propenyl-L-cysteine sulfoxide to bitter and odor bearing components of onion. Phytochemistry *7,* 401–404.

SCHWIMMER, S. 1968B. Templates in a test tube. Enzymatic replication of DNA. *In* Molecular Biology and Agriculture. Potential for Future Research. West. Exp. Stn. Collab. Conf., Albany, Calif, March.

SCHWIMMER, S. 1968C. Differential effects of putrescine, cadaverine, and glyoxal-bis (guanylhydrazone) on DNA- and nucleohistone-supported DNA synthesis. Biochim. Biophys. Acta *166,* 251–254.

SCHWIMMER, S. 1968D. A kinetic model for nucleohistone-dependent DNA synthesis. Experientia *24,* 887–888.

SCHWIMMER, S. 1968E. Inhibition of in vitro DNA synthesis by auxins. Plant Physiol. *43,* 1008–1010.

SCHWIMMER, S. 1969A. Trends and perspectives in the enzymology of foods. Lebensm.-

Wiss. Technol. *2,* 97–103.

SCHWIMMER, S. 1969B. Inhibition of carbonic anhydrase by mercaptans. Enzymologia *37,* 163–173.

SCHWIMMER, S. 1969C. *In situ* acrylamide polymerization. Effect on appearance and rehydration of dehydrated vegetables. Food Technol. *23* (7) 115–116.

SCHWIMMER, S. 1969D. Characterization of S-propenyl-L-cysteine sulfoxide as the principal endogenous substrate of L-cysteine sulfoxide of onion. Arch. Biochem. Biophys. *130,* 312–320.

SCHWIMMER, S. 1971A. S-alkyl-L-cysteine sulfoxide lyase [*Allium cepa* (onion)]. Methods Enzymol. *17,* 475–478.

SCHWIMMER, S. 1971B. Enzymatic conversion of γ-glutamyl cysteine peptides, a coupled reaction for enhancement of onion flavor. J. Agric. Food Chem. *19,* 980–983.

SCHWIMMER, S. 1972. Biochemical control systems: Cell disruption and its consequences in food processing. J. Food Sci. *37,* 530–535.

SCHWIMMER, S. 1973. Flavor enhancement of Allium products. U.S. Pat. 3,725,085. Apr. 3.

SCHWIMMER, S. 1974. Transformations of vegetable proteins. Rep. to U.N. Ind. Dev. Organ. DP/ISR/64/511/11–12, April.

SCHWIMMER, S. 1975. Effects of enzymes on the nutritional quality and availability of proteins. *In* Protein Nutritional Quality of Foods and Feeds. Part 2. Quality Factors—Plant Breeding, Composition, Processing and Antinutrients. M. Friedman (Editor). Marcel Dekker, New York.

SCHWIMMER, S. 1976. Enzymatic analysis for quality control in the food industry. Proc. Joint ROC-USA Symp. Enzyme Technol., Lehigh Univ., Bethlehem, PA, August.

SCHWIMMER, S. 1978. Enzyme action and modification of cellular integrity in fruits and vegetables: Consequences for food quality during ripening, senescence and processing. *In* Postharvest Biology and Biotechnology. H.O. Hultin and M. Milner (Editors). Food and Nutrition Press, Westport, Conn.

SCHWIMMER, S. 1980A. Influence of water activity on enzyme reactivity and stability. Food Technol. *34* (5) 64–74, 82.

SCHWIMMER, S. 1980B. Water activity and nutrient stability: Enzyme action. Abstr. 2nd Chem. Congr. N. Am. Cont., San Francisco. AGFD *37.*

SCHWIMMER, S. and ALDERTON, G. 1960. Unpublished results. Albany, Calif.

SCHWIMMER, S. and ARONSON, A. K. 1966. Apparent synthesis of single-stranded DNA by DNA polymerase at high template concentrations. Life Sci. *5,* 1415–1422.

SCHWIMMER, S. and ARONSON, A. 1967. Susceptibility to nucleases of DNA synthesized with nucleohistone primer. Biochim. Biophys. Acta *134,* 59–68.

SCHWIMMER, S. and AUSTIN, S. J. 1971A. Gamma glutamyl transpeptidase of sprouted onion. J. Food Sci. *36,* 807–811.

SCHWIMMER, S. and AUSTIN, S. J. 1971B. Enhancement of pyruvic acid release and flavor in dehydrated *Allium* powders by gamma glutamyl transpeptidase. J. Food Sci. *36,* 1081–1085.

SCHWIMMER, S. and BALLS, A. K. 1949B. Isolation and properties of crystalline α-amylase from germinated barley. J. Biol. Chem. *179,* 1063–1074.

SCHWIMMER, S. and BALLS, A. K. 1949B. Starches and their derivatives as adsorbents for malt α-amylase. J. Biol. Chem. *180,* 883–894.

SCHWIMMER, S. and BEVENUE, A. 1956. Reagent for the differentiation of 1,4 and 1,6-linked glucosaccharides. Science *123,* 533–544.

SCHWIMMER, S., BEVENUE, A. and WESTON, W. J. 1956. Separation of hexose phosphates by paper electrophoresis in borate buffers. Arch. Biochem. Biophys. *60,* 270–283.

SCHWIMMER, S., BEVENUE, A., WESTON, W. J. and POTTER, A. L. 1954. Survey of major and minor sugar and starch components of the white potato. J. Agric. Food Chem. *2,* 1284–1290.

SCHWIMMER, S. and BONNER, J. 1965. Nucleohistone as template for the replication of DNA. Biochim. Biophys. Acta *108,* 67–72.

SCHWIMMER, S. and BURR, H. C. 1975. Structure and chemical composition of the potato tuber. *In* Potato Processing, 3rd Edition. W.F. Talburt and O. Smith (Editors). AVI Publishing Co., Westport, Conn.

SCHWIMMER, S., BURR, H. C., HARRINGTON, W. O. and WESTON, W. J. 1957A. Gamma irradiation of potatoes: effects on sugar content, chip color, germination, greening, and susceptibility to mold. Am. Potato J. *34,* 31–41.

SCHWIMMER, S., CARSON, J. F., MAKOWER, R. U., MAZELIS, M. and WONG, F. F. 1960. Demonstration of alliinase in a protein preparation from onion. Experientia 16, 449–452.

SCHWIMMER, S. and CURL, A. L. 1951. The nature and preparation of enzymes. *In* Yearbook of Agriculture, 1950–1951. U.S. Dep. Agric., Washington, D.C.

SCHWIMMER, S. and FRIEDMAN, M. 1972. Genesis of volatile sulphur-containing food flavors. Flavour Ind. *3,* 137–145.

SCHWIMMER, S. and GARIBALDI, J. 1952. Further studies on the production, purification, and properties of the Schardinger dextrinogenase of *B. macerans.* Cereal Chem. *29,* 108–122.

SCHWIMMER, S. and GRANROTH, B. 1975. Unpublished observations. Albany, Calif.

SCHWIMMER, S. and GUADAGNI, D. G. 1962. Relation between olfactory threshold concentration and pyruvic acid content of onion juice. J. Food Sci. *27,* 94–97.

SCHWIMMER, S. and GUADAGNI, D. G. 1967. Kinetics of the enzymic development of pyruvic acid and odor in frozen onions treated with cysteine C-S-lyase. J. Food Sci. *33,* 193–196.

SCHWIMMER, S., HENDEL, C. E., HARRINGTON, W. O. and OLSON, R. L. 1957B. Interrelation among measurements of browning of processed potatoes. Am. Potato J. *34,* 119–132.

SCHWIMMER, S. and INGRAHAM, L. L. 1980. Effective temperature is independent of reaction order in a thermally fluctuating system. J. Food Sci. *45,* vi–vii, 1462. (Letter to Ed.)

SCHWIMMER, S., INGRAHAM, L. L. and HUGHES, H. M. 1955. Temperature tolerance in frozen food processing. Effective temperatures in thermally fluctuating systems. Ind. Eng. Chem. *47,* 1149–1151.

SCHWIMMER, S., KABAT, S. and FILNER, P. 1965. Retention by membrane filters of phosphatase-produced deoxyadenosine in DNA polymerase assay. Biochim. Biophys. Acta *108,* 150–151.

SCHWIMMER, S. and KJAER, A. 1960. Purification and specificity of the C-S-lyase of *Albizzia lophanta.* Biochim. Biophys. Acta *42,* 316–324.

SCHWIMMER, S. and KURTZMAN, R. H., JR. 1972. Fungal decaffeination of roast coffee infusions. J. Food Sci. *37,* 921–924.

SCHWIMMER, S., KURTZMAN, R. H., JR. and HEFTMANN, E. 1971. Caffeine metabolism by *Penicillium roqueforti.* Arch. Biochem. Biophys. *147,* 109–113.

SCHWIMMER, S., MAKOWER, R. U. and ROREM, E. S. 1961. Invertase and invertase inhibitor in potato. Plant Physiol. *36,* 313–316.

SCHWIMMER, S. and MAZELIS, M. 1963. Characterization of alliinase of *Allium cepa* (onion). Arch. Biochem. Biophys. *100,* 66–73.

SCHWIMMER, S., MAZELIS, M. and CARSON, J. F. 1961. Enzymological basis of flavor development in onions. Abstr. *106,* 21st Annu. Meet., IFT, New York, June.

SCHWIMMER, S. and OLCOTT, H. S. 1953. Reaction between glycine and hexose phosphates. J. Am. Chem. Soc. *75,* 4855–4856.

SCHWIMMER, S. and OLIVERA, B. M. 1966A. Electrophoretic properties of enzymatically synthesized DNA. J. Mol. Biol. *20,* 585–587.

SCHWIMMER, S. and OLIVERA, B. M. 1966B. Electrophoresis of products of DNA—and nucleohistone-supported DNA synthesis. Biopolymers *4,* 953–955.

SCHWIMMER, S. and PARDEE, A. B. 1953. Principles and procedures in the isolation of enzymes. Adv. Enzymol. *14,* 375–409.

SCHWIMMER, S. and ROREM, E. S. 1960. Biosynthesis of sucrose by preparations from cold and room temperature stored potatoes. Nature *187,* 1113–1114.

SCHWIMMER, S., RYAN, C. A. and WONG, F. F. 1964A. Specificity of L-cysteine sulfoxide lyase and partially competitive inhibition by S-alkyl-L-cysteines. J. Biol. Chem. *239,* 777–782.

SCHWIMMER, S., VENSTROM, D. W. and GUADAGNI, D. G. 1964B. Relation between pyruvate content and odor strength of reconstituted onion powders. Food Technol. *18,* 121–124.

SCHWIMMER, S. and WESTON, W. J. 1956. Effect of phosphate and other factors in potato extracts on amylose formation by phosphorylase. J. Biol. Chem. *220,* 143–155.

SCHWIMMER, S. and WESTON, W. J. 1958. Chlorophyll formation in potato tubers as influenced by gamma irradiation and by chemicals. Am. Potato J. *35,* 542–543.

SCHWIMMER, S. and WESTON, W. J. 1961. Enzymatic development of pyruvic acid in onion as a measure of pungency. J. Agric. Food Chem. *9,* 301–304.

SCHWIMMER, S., WESTON, W. J. and MAKOWER, R. U. 1957. A survey of biochemical changes in potatoes initiated by Co^{60} irradiation in the decakilorep range. Fed. Proc. *16,* 244.

SCHWIMMER, S., WESTON, W. J. and MAKOWER, R. U. 1958. Biochemical effects of gamma radiation on potato tubers. Arch. Biochem. Biophys. *75,* 425–434.

SCOTT, D. 1953. Glucose conversion in the preparation of albumen solids by glucose oxidase-catalase system. J. Agric. Food Chem. *1,* 727–730.

SCOTT, D. 1964. Removal of glucose by enzymatic action. U.S. Pat. 3,162,537. Dec. 22.

SCOTT, D. 1975. Glucose oxidase. *In* Enzymes in Food Processing, 2nd Edition. G. Reed (Editor). Academic Press, New York.

SCOTT, D. and HAMMER, F. E. 1962. Deoxygenating method and products. U.S. Pat. 3,016,336. Jan. 9.

SCOTT, D. and KLIS, J. B. 1962. Produce Salmonella-free yolk and eggs. Food Process. *23* (9) 76–77.

SCOTT, R. 1972. Cheesemaking—enzymology or bacteriology? Process Biochem. 7(11) 33–36.

SCRUTTON, M. C. and UTTER, M. F. 1965. Pyruvate carboxylase. II. Some physical and chemical properties of the highly purified enzyme. J. Biol. Chem. *240,* 1–9.

SEEVERS, P. M., DALY, J. M. and CATEDRAL, F. F. 1971. The role of peroxidase isozymes in resistance to wheat stem rust disease. Plant Physiol. *48,* 353–360.

SEGAL, H. L. 1959. The development of enzyme kinetics. *In* The Enzymes, 2nd Edition, Vol. 1. P.D. Boyer, H.A. Lardy and K. Myrbäck (Editors). Academic Press, New York.

SEGEL, I. H. 1975. Enzyme Kinetics: Behavior and Analysis of Rapid Equilibrium and Steady State Enzyme Systems. John Wiley & Sons, Chichester.

SEIFERT, R. M., BUTTERY, R. G., LUNDIN, R. W., HADDON, W. F. and BENSON, M. 1978. Identification of a thiamin odor compound from photolysis of thiamin. J. Agric. Food Chem. *26,* 1173–1176.

SEIFTER, S. and HARPER, E. 1971. The collagenases. *In* The Enzymes, 3rd Edition, Vol. 3. P.D. Boyer (Editor). Academic Press, New York.

SEIGEKER, D. S. 1975. Isolation and characterization of naturally occurring cyanogenic compounds. Phytochemistry *14,* 9–29.

SEITZ, E. W. 1974. Industrial application of microbial lipases. A review. J. Am. Oil Chem. Soc. *51,* 12–16.

SEKINE, H. 1972. Some properties of neutral proteinase I and proteinase II of Aspergillus sojae as zinc containing metallo enzyme. Agric. Biol. Chem. *36,* 2143–2150.

SEKIYA, M., AOSHIMA, H., KAJIWARA, T., TOGO, T. and HATANAKA, A. 1977. Purification and some properties of potato tuber lipoxygenase and detection of linoleic acid radical in the enzyme reaction. Agric. Biol. Chem. *44,* 827–832.

SEKIYA, J., NUMA, S., KAJIWARA, T. and HATANAKA, A. 1976. Biosynthesis of leaf alcohol. Agric. Biol. Chem. *40,* 185–190.

SEKUL, A. A. and ORY, R. L. 1977. Rapid enzymatic method for partial hydrolysis of oilseed proteins for food use. J. Am. Oil Chem. Soc. *54,* 32–35.

SEKURA, R. and MEISTER, A. 1977. Gamma glutamyl synthetase: Further purification, half-of-the-sites reactivity, subunits and specificity. J. Biol. Chem. *252,* 2599–2605.

SELTZER, E., HARRIMAN, A. J. and HENDERSON, R. W. 1961. Process for converting green tea extract. U.S. Pat. 2,975,057. Mar. 14.

SELVENDRAN, R. R., REYNOLDS, J. and GALLIARD, T. 1978. Production of volatiles by degradation of lipids during manufacture of black tea. Phytochemistry *17,* 233–236.

SERA, Y. 1976. A study on the method for identification of bacterial thiaminase. Chem. Abstr. *86,* 51988m. (Japanese)

SESSA, D. J., WARNER, K. and HONIG, D.H. 1974. Soybean phosphatidylcholine develops bitter taste on autoxidation. J. Food Sci. *39,* 69–72.

SETSER, C. S., HARRISON, D. L., KROPF, D. H. and DAYTON, A. D. 1973. Radiant energy-induced changes in bovine muscle pigment. J. Food Sci. *38,* 412–417.

SFAT, M. R. and MORTON, B. J. 1972. Novel raw material and mashing system to produce lower calorie beer. Tech. Q. Master Brew. Assoc. Am. *9,* 89–94.

SFAT, M.R. and MORTON, B.J. 1973. Low carbohydrate beer. U.S. Pat. 3,717,471. Feb. 20.

SHAHANI, K. M., ARNOLD, R. G., KILARA, A. and DWIVEDI, B. K. 1976. Role of microbial enzymes in flavor development in foods. Biotechnol. Bioeng. *18,* 891–917.

SHAHANI, K. M., HARPER, W. J., JENSEN, R. G., PARRY, R. M., JR. and ZITTLE, C. A. 1974. Enzymes in bovine milk: A review. J. Dairy Sci. *56,* 531–543.

SHAIN, Y. and MAYER, A. M. 1968. Activation of enzymes during germination. Trypsin-like enzyme in lettuce. Phytochemistry *7,* 1491–1498.

SHALLENBERGER, R. S. 1971. Molecular structure and taste. *In* Gustation and Olfaction. Int. Symp., Geneva, June 1970. G. Olhoff and A.F. Thomas (Editors). Academic Press, New York.

SHALTIEL, S. 1975. Hydrophobic chromatography. Use in the resolution, purification and probing of proteins. Fed. Eur. Biochem. Soc. Meet. (Proc.) *40,* 117–127.

SHANKARANARAYANA, M. L., RAGHAVAN, B. and ABRAHAM, K. 1974. Volatile sulfur compounds in food flavors. Crit. Rev. Food Technol. *4,* 395–435.

SHANKARANARAYANA, M. L., RAGHAVAN, B., ABRAHAM, K. O. and NATARJAN, C. P. 1973. Volatile sulfur compounds in food flavors.

Crit. Rev. Food Technol. *4,* 395–435.

SHANNON, B. M. 1966. Native and added peroxidase activity, ascorbic acid, and palatibility of frozen vegetables. Master's Thesis. Dept. Home Econ., University of Illinois, Bloomington.

SHANNON, S., YAMAGUCHI, M. and HOWARD, F. D. 1967. Precursors involved in the formation of pink pigments in onion purees. J. Agric. Food Chem. *15,* 423–426.

SHARMA, C. B., GOEL, M. and IRSHAD, M. 1978. Myoinositol hexaphosphate as a potential inhibitor of α-amylases. Phytochemistry *17,* 201–204.

SHARMA, S. C., WOLFE, R. R. and WANG, S. S. 1975. Kinetic analysis of post-harvest texture changes in asparagus. J. Food Sci. *40,* 1147–1151.

SHARMAN, I. M. 1976. Fructose and xylitol. The Turku dental studies. Nutr. Food Sci. *43,* 20–23.

SHARON, M. and MAYER, A. M. 1967. The effect of sodium chloride on catechol oxidase from apples. Isr. J. Chem. *5,* 275–280.

SHARON, N. 1977. Lectins. Sci. Am. *236,* 108–119.

SHARPE, P. J. H. and GOESCHL, J. D. 1975. Thermodynamics of plant metabolism; model of temperature effects. Plant Physiol. *56* (2) Suppl., 79.

SHASTRY, B. S. and RAO, R. R. 1975. Lipoxgygenase from rice bran. Cereal Chem. *52,* 597–603.

SHENOUDA, S. Y. K. and PIGOTT, G. M. 1977. Fish myofibrillar protein and lipid interaction in aqueous media crosslinking. *In* Protein Crosslinking: Biochemical and Molecular Aspects. M. Friedman (Editor). Plenum Press, New York.

SHERBA, S. E. and STEIGER, R. B. 1972. Soybean fractionation employing a protease. U.S. Pat. 3,640,725. Feb. 8.

SHERMAN, P. 1972. Structure and textural properties of foods. Food Technol. *26,* 69–79.

SHERMAN, R. A. and LEWIS, M. J. 1975. Determination of the latent β-amylase of barley and malt. Proc. Am. Soc. Brew. Chem. *33,* 26–28.

SHEWFELT, A. L., PAYNTER, V. A. and JEN, J. J. 1971. Textural changes and molecular characteristics of pectic constituents in ripening peaches. J. Food Sci. *36,* 573–575.

SHIBAI, H., ENEI, H. and HIROSE, Y. 1978. Purine nucleoside fermentations. Process Biochem. *13* (11) 6–8, 32.

SHIMAZONO, H. 1964. Distribution of 5′-ribonucleotides in foods and their application to foods. Food Technol. *18,* 294–303.

SHIMIZU, C. and MATSUURA, F. 1968. Purification and properties of "methemoglobin reductase." Agric. Biol. Chem. *5,* 587–592.

SHIMIZU, C. and MATSUURA, F. 1971. Occurrence of a new enzyme reducing metmyoglobin in dolphin muscle. Agric. Biol. Chem. *35,* 468–475.

SHIMIZU, J. and KAGA, T. 1972. Apparatus for continuous hydrolysis of raffinose. U.S. Pat. 3,664,927. May 23.

SHIMOYAMA, M., SAKAMOTO, M., NASU, S., SHIGEHISA, S. and UEDA, I. 1972. Identification of the 3′-5′-cyclic phosphodiesterase inhibitor. Biochem. Biophys. Res. Comm. *48,* 235–241.

SHINE, W. E. and STUMPF, P. K. 1974. Fat metabolism in higher plants. LVIII. Plant α-oxidation systems. Arch. Biochem. Biophys. *162,* 147–157.

SHINKE, R. and NISHIRA, H. 1975. Barley and malt amylases. XX. Comparison of wheat zymogen α-amylase with barley zymogen. Chem. Abstr. *83,* 24074s. (Japanese)

SHIO, I. 1979. Regulation of phosphoenolpyruvate carboxylase by synergistic action of aspartate and 2-oxoglutarate. Agric. Biol. Chem. *43,* 2479–2485.

SHIPE, W. F., LEE, E. C. and SENYK, G. F. 1975. Enzymic modification of milk flavor. J. Dairy Sci. *58,* 1123–1126.

SHIPMAN, J. W., RAHMAN, A. R., SEGARS, R. A., KAPSALIS, J. G. and WESTCOTT, D. E. 1972. Improvement of the texture of dehydrated celery by glycerol treatment. J. Food Sci. *37,* 568–571.

SHIVARAM, K. N. 1976. Purification and properties of potato phosphorylase isozymes. Z. Naturforsch. *31,* 424–432.

SHORENSTEIN, R. G., PRATT, C. S., HSU, C. and WAGNER, E. 1968. A model system for the study of equilibrium hydrophobic bond formation. J. Am. Chem. Soc. *90,* 6199–6207.

SHORROCK, C. 1976. An improved procedure for the assay of available lysine in feedstuffs using *Tetrahymena pyriformis.* Br. J. Nutr. *35,* 333–341.

SHOVERS, J., DEDERICH, C. B. and STEINKE, P. K. 1975. Milk-coagulating composition for use in cheesemaking. Ger. Offen. 2,422,005. May 7.

SHOVERS, J., FOSSUM, G. and NEAL, A. 1972. Procedure for the electrophoretic separation and visualization of milk-clotting enzymes in milk coagulants. J. Dairy Sci. *55,* 1532–1534.

SHOVERS, J., KORNOKOWSKI, R. and FOSSUM, G. 1973. Differential assay for milk coagulants in mixtures with porcine pepsin. J. Dairy Sci. *56,* 994–997.

SHRIKHANDE, A. J. and FRANCIS, F. J. 1974. Effect of flavonols on ascorbic acid and anthocyanin stability in model systems. J. Food Sci. *39,* 904–906.

SHUKLA, T. P. 1975. Beta-galactosidase technology: A solution to the lactose problem. Crit. Rev. Food Technol. *5,* 325–356.

SICHO, V., KAS, J., CEPICKA, J. and CELIKOVSKY, J. 1972. Study of proteolytic processes in fish meat in the course of marinading. FSTA *6* (1) R34. (Russian)

SIEBERT, G. 1958. Protein-splitting enzyme activity of fish flesh. Experientia *14,* 65–66.

SIEGEL, D. G., THENO, D. M., SCHMIDT, G. R. and NORTON, H. W. 1978. Meat massaging: The effects of salt, phosphate and massaging on cooking loss, binding strength and exudate composition in sectioned and formed ham. J. Food Sci. *43,* 331–333.

SIEGEL, M. I., McCONNELL, R. T., PORTER, N. A. and CUATRECASAS, P. 1980. Arachidonate metabolism via lipoxygenase and 12L-hydroperoxy-5,8,10,14-icosatetraenoic acid peroxidase sensitive to anti-inflammatory drugs. Proc. Natl. Acad. Sci. *77,* 308–312.

SIEGELMAN, H. W. 1953. Brown discoloration and shrivel of cherries. Proc. Am. Soc. Hortic. Sci. *61,* 265–269.

SIEGELMAN, H. W. 1955. Detection and identification of polyphenol oxidase substrates in apple and pea skins. Arch. Biochem. Biophys. *56,* 97–102.

SIESO, V., NICOLAS, M., SECK, S. and CROUZET, J. 1977. Volatile constituents of tomato: Evidence for and formation by enzymatic route of *trans*-hexen-2-ol. Agric. Biol. Chem. *40,* 2349–2353. (French)

SIGMAN, D. S. and MOOSER, G. 1975. Chemical studies of enzyme active sites. Annu. Rev. Biochem. *44,* 889–931.

SIGMUND, V. 1968. Removing sugar from eggs. Czech. Pat. 125,969. Jan. 15.

SIKES, R. W., CHRONISTER, R. B. and WHITE, L. E., JR. 1977. Origin of the direct hippocampus anterior Golgi thalamic bundle in the rat. A combined horseradish peroxidase analysis. Exp. Neurol. *57,* 379–395.

SIKORSKI, Z., OLLEY, J. and KOSTUCH, S. 1976. Protein changes in frozen fish. Crit. Rev. Food Sci. Nutr. *8,* 97–129.

SILANO, V. 1975. Determination of soft wheat in semolina and pasta foods. Boll. Chim. Unione Ital. Lab. Prov. *1,* 338–354. (Italian)

SILANO, V. 1978. Biochemical and nutritional significance of wheat albumin inhibitors of α-amylase. Cereal Chem. *55,* 722–731.

SILANO, V. and ZAHNLEY, J. C. 1978. Association of *Tenebrio molitor* L. α-amylase with two protein inhibitors from wheat flour. Differential scanning calorimetric comparison of heat stabilities. Biochim. Biophys. Acta *533,* 181–185.

SILBERMAN, H. C. 1971. Enzymatic by-product conversion. U.S. Pat. 3,615,721. Oct. 26.

SILBERSTEIN, O. O. 1966. Method of tenderizing meat. U.S. Pat. 3,276,879. Oct. 4.

SILBERT, D. F. 1975. Genetic modification of membrane lipid. Annu. Rev. Biochem. *44,* 315–339.

SILVANOVICH, M. P., and HILL, R. D. 1977. α-Amylases from triticale 6A190: Purification and characterization. Cereal Chem. *54,* 1270–1281.

SILVERSTEIN, O. 1961. Heat stable α-amylase in baking. Bakers Dig. *35* (6) 44–48.

SIMEONOVA, I. 1976. Activity of enzymes which break down tomato cell walls. III. Content of pectin and cellulose substances and pectolytic and cellulase enzyme activities following the short-term storage of tomatoes. Chem. Abstr. *86,* 42006n. (Russian)

SIMPSON, F. J. 1955. Separation of starch and gluten. VII. The application of bacterial pentosanases to the recovery of starch from wheat flour. Can. J. Technol. *33,* 33–40.

SINDEN, S. L. 1971. Control of potato greening with household detergents. Am. Potato J. *48,* 53–56.

SINGER, T. P. 1948. On the mechanism of enzyme inhibition by sulfhydryl reagents. J. Biol. Chem. *174,* 11–21.

SINGER, T. P. and KENNEY, W. C. 1974. Biochemistry of covalently bound flavines. Vitam. Horm. N.Y. *32,* 1–45.

SINGH, A., SRINIVASAN, R. A. and DODANI, A. J. 1976. Role of lipolytic bacteria in degradation of fat of experimental cheddar cheese during ripening. Indian J. Dairy Sci. *29,* 22–26.

SINGH, B., HABIB, G. and SADEH, G. 1977. The characteristics of phytase and acid phosphatase and its relationship to certain minerals in triticale. Cereal Sci. Today *22,* 468.

SINGH, B. and REDDY, N. R. 1977. Phytic acid and mineral compositions of triticales. J. Food Sci. *42,* 1077–1083.

SINGH, B. and SEDEH, H. G. 1977. The characteristics of phytase and its relationship to certain minerals in triticale. Cereal Food World *22,* 468.

SINGH, R. and SHEORAN, I. S. 1972. Enzymic browning of whole wheat meal flour. J. Sci. Food Agric. *23,* 121–125.

SINGH, R. P. and WANG, C. Y. 1977. Quality of frozen foods. J. Food Process Eng. *1,* 97–127.

SINGH, S. and SANWAL, G. G. 1976. Multiple forms of α-glucan phosphorylase in banana fruits: Properties and kinetics. Phytochemistry *15,* 1447–1451.

SINGLETON, J. A., PATTEE, H. E. and SANDERS, T. H. 1975. Some parameters affecting volatile production in peanut homogenates. J. Food Sci. *40,* 386–389.

SINGLETON, J. A., PATTEE, H. E. and SANDERS, T. H. 1976. Production of flavor in enzyme and substrate enriched peanut homogenates. J. Food Sci. *41,* 148–151.

SINNER, M. 1975. β-1,4-Xylan-4-xylanohydrolases (E.E. 3.2.1.8) and Theory of Action on Deciduous Tree Xylans. Kommissionverlag Buchhandlung Max Wiedbusch, Hamburg, Germany. (German)

SINSHEIMER, R. L. 1977. Recombinant DNA. Annu. Rev. Biochem. *46,* 415–438.

SISTRUNK, W. A. and BAILEY, F. L. 1965. Relation of processing procedure to discoloration of canned blackeye peas. Food Technol. *19,* 189–191.

SISTRUNK, W. A. and BRADLEY, G. A. 1975. Quality and nutritional value of canned turnip greens as influenced by processing technique. Ark. Farm Res. *24* (2) 5.

SISTRUNK, W. A. and CAIN, R. F. 1960. Chemical and physical changes in green beans during preparation and processing. Food Technol. *14,* 357–362.

SISTRUNK, W. A. and CASH, J. N. 1974. Processing factors affecting quality and storage stability of Concord grape juice. J. Food Sci. *39,* 1120–1123.

SISTRUNK, W. A. and CASH, J. N. 1975. Spinach quality attributes and nitrate-nitrite levels as related to processing and storage. J. Am. Soc. Hortic. Sci. *100,* 307–309.

SIU, R. G., BORZELLECA, J. F., DAY, H. G., IRVING, G. W., JR., LA DU, B. N., McCOY, J. R., MILLER, S. A., PLAA, G. L., SHIMKIN, M. B. and WOOD, J. L. 1977. Evaluation of health aspects of GRAS food ingredients: lessons learned and questions unanswered. Fed. Proc. *37,* 2525–2562.

SIZER, I. W. 1952. Oxidation of proteins by tyrosinase and peroxidase. Adv. Enzymol. *14,* 129–162.

SIZER, I. W. and JOSEPHSON, E. S. 1942. Kinetics as a function of temperature of lipase, trypsin, and invertase activity from −70 to 50°C (−95 to 122°F). Food Res. *7,* 201–209.

SKELTON, G. S. 1969. Development of protolytic enzymes in growing papaya fruit. Phytochemistry *8,* 57–60.

SKINNER, F. T. 1975. Enzymes technology. Chem. Eng. News *53,* Aug. 18, 22–41.

SHIRKO, B. K. 1973. Comparative public health evaluation of complex enzyme preparations from Aspergillus oryzae. Vopr. Pitan. 5, 69–72. (Russian)

SKUJINS, J. J. and McLAREN, A. D. 1967. Enzyme reaction rates at limited water activities. Science *158,* 1569–1570.

SKUPIN, J. and WARCHALEWSKI, J. 1971. Isolation and properties of protease A from wheat grain. J. Sci. Food Agric. *22,* 11–15.

SLATTERY, C. W. 1976. Review: Casein micelle structure; examination of models. J. Dairy Sci. *59,* 1547–1556.

SLATTERY, C. W. 1978. Variation in the glycosylation pattern of bovine κ-casein with micelle size fits relationship to a micelle model. Biochemistry *17,* 1100–1104.

SLATTERY, C. W. 1979. A phosphate-induced submicelle equilibrium in reconstituted casein micelle systems. J. Dairy Res. *46,* 253–258.

SLATTERY, C. W. and EVARD, R. 1973. A model for the formation and structure of casein micelles from the subunits of variable composition. Biochim. Biophys. Acta *317,* 529–538.

SLABNICK, E. and FRYDMAN, R. B. 1970. A phosphorylase involved in starch biosynthesis. Biochim. Biophys. Acta *317,* 529–538.

SLABNICK, E. and FRYDMAN, R. B. 1970. A phosphorylase involved in starch biosynthesis. Biochem. Biophys. Res. Commun. *38,* 709–714.

SLAYTER, R. O. and TAYLOR, S. A. 1960. Terminology in plant- and soil-water relations. Nature *187,* 922–924.

SLEETH, R. B. and CAMPBELL, J. F. 1965. Tenderization of meat. U.S. Pat. 3,166,423. Jan. 19.

SLOAN, A. E., SCHLEUTER, D. and LABUZA, T. P. 1977. Effect of sequence and method of addition of humectants and water on a_w lowering ability on an IMF system. J. Food Sci. *42,* 94–96.

SLOANE-STANLEY, G. H. 1961. Phytase. *In* Biochemists Handbook. C. Long (Editor). D. Van Nostrand Co., Princeton, N.J.

SLOMAN, K. G., FOLTZ, A. K. and YERANSIAN, A. 1975. Food. Anal. Chem. *47,* 56R–85R.

SMALLINGS, J. B., KEMP, J. D., FOX, J. D. and MOODY, W. G. 1971. Effect of antemortem injection of papain on the tenderness and quality of dry-cured hams. J. Anim. Sci. *32,* 1107–1112.

SMIT, C. B. J. and NORRTJE, B. K. 1958. Observations on the consistency of tomato paste. Food Technol. *12,* 356–358.

SMITH, E. L., LIGHT, A. and KIMMEL, J. R. 1962. Protein structure in relation to enzymic activity, with special reference to papain. Biochem. Soc. Symp. *21,* 88–103.

SMITH, G. C., ARANGO, T. C. and CARPENTER, Z. L. 1971. Effects of physical and mechanical treatments on the tenderness of the beef longissimus. J. Food Sci. *36,* 445–449.

SMITH, J. B. and BENNETT, M. D. 1974. Amylase isozymes of oats (*Avena sativa* L.). J. Sci. Food Agric. *25,* 67–71.

SMITH, L. L. and KULIG, M. J. 1975. On the derivation of carcinogenic steroids as from cholesterol. Cancer Biochem. Biophys. *1,* 179–184.

SMITH, L. L. and TENG, J. I. 1974. Sterol metabolism. XXIX. On the mechanism of lipid peroxidation in the rat liver. J. Am. Chem. Soc. *96,* 2640–2641.

SMITH, M. D. and OLSON, C. L. 1975. Differential amperometric determination of alcohol in blood or urine using alcohol dehydrogenase. Anal. Chem. *47,* 1075–1077.

SMITH, M. T. and BRIGGS, D. E. 1980. Externally applied gibberellic acid and α-amylase formation in grains of barley. Phytochemistry *19,* 1025–1033.

SMITH, T. 1977. Review. Tryptamine and related compounds in plants. Phytochemistry *16,* 171–175.

SMITH, T. J., 1975. Dry blanching apparatus and product. U.S. Pat. 3,910,175. Oct. 7.

SMITH, W. L. and LANDS, W. E. M. 1972. Oxygenation of unsaturated fatty acids by soybean lipoxygenase. J. Biol. Chem. *247,* 1038–1074.

SMITH, W. L., and ROBE, K. 1973. Saves 300–4000 gpm water, improves vegetable quality. Food Process. *34* (3) 36–37.

SMOLARSKY, M. 1980. Mechanism of papain catalysis: Studies of active-site acylation and deacylation by the stopped-flow technique. Biochemistry *19,* 484–487.

SMOLENSKY, D. C., RAYMOND, W. R., HASEGAWA, S. and MAIER, V. P. 1975. Enzymatic improvement of date quality. Use of invertase to improve texture and appearance of "sugar wall" dates. J. Sci. Food Agric. *26,* 1523–1528.

SMYTHE, C. V. 1955. Enzymes. *In* Handbook of Food and Agriculture. F.C. Blanck (Editor). Reinhold, New York.

SMYTHE, C. V., NEUBECK, C. E. and ROBBINS, E. A. 1962. Meat tenderizing compositions. U.S. Pat. 3,033,691. May 8.

SNAUERT, F. and MARKAKIS, P. 1976. Effect of germination and gamma irradiation on the oligosaccharides of navy beans (*Phaseolus vulgaris* L.). Lebensm.-Wiss. Technol. *9,* 93–95.

SOBER, H. A. and PETERSON, E. A. 1954. Chromatography of proteins on cellulose ion-exchangers. J. Am. Chem. Soc. *76,* 1711–1712.

SOBOTKA, F. E. and STELZIG, D. A. 1974. An apparent cellulase complex in tomato (*Lycopersicon esculentum* L.) fruit. Plant Physiol. *53,* 759–763.

SOBOTKA, F. E. and WATADA, A. E. 1971. Cellulase in high pigment and crimson tomato fruit. J. Am. Soc. Hortic. Sci. *96,* 705–707.

SOC. CHEM. IND., LONDON. 1961. Monograph No. 11. Production and application of Enzyme Preparations in Food Manufacture. Soc. Chem. Ind., London, and Macmillan Co., New York.

SOEDIGDO, P., IJAD, R. and TAN, H. L. 1973. Shrimp spoilage evaluation based on enzyme activity determination. Chem. Abstr. *83,* 106930. (Indonesian)

SOLBERG, M. 1968. Factors affecting fresh meat color. Proc. Meat Inst. Found. Symp. *1,* 22–38.

SOLBERG, M. 1970. The chemistry of color stability in meat—a review. Can. Inst. Food Sci. Technol. *33,* 55–62.

SOLIDAY, C. L. and KOLATTUKUDY, P. E. 1976. Isolation and characterization of a cutinase from *Fusarium roseum culmorum* and its immunological comparison with cutinases from *F. solani pisi.* Arch. Biochem. Biophys. *176,* 334–343.

SOLOV'EV, V. I. and PROKOSHEVA, G. A. 1973. Use of bacterial nitrite reductase in sausage manufacture. Prikl. Biokh. Mikrobiol. *9,* 512–515.

SOMERVILLE, C. R. and OGREN, W. L. 1980. Photorespiration mutants of *Arabidopsis thaliana* deficient in serine-glyoxylate aminotransferase activity. Proc. Natl. Acad. Sci. *77,* 2648–2687.

SOMKUTI, G. A. 1974. Control of lipase content in *Mucor* rennet preparations. J. Dairy Sci. *57,* 898–899.

SONDHEIMER, E. 1952. Synthesis of violet leaf perfume, 2 *(trans),* 6 *(cis)*-nonadienal. J. Am. Chem. Soc. *74,* 4040–4043.

SONE, T. 1972. Consistency of Foodstuffs. (S. Matsumoto, Translator from Japanese). R. Reidel Publishing Co., Dordrecht, Holland.

SOOD, R. and BHAT, C. M. 1974. Changes in ascorbic acid and carotene content of vegetables of cooking. J. Food Sci. Technol. *11,* 131–133.

SOOD, V. K. and KOSIKOWSKI, F. V. 1979. Accelerated Cheddar cheese ripening by added microbial enzymes. J. Dairy Sci. *62,* 1865–1872.

SOPANEN, T., TAKKINEN, P., MIKOLA, J. and ENARI, T. -M. 1980. Rate-limiting enzymes in the liberation of amino acids in mashing. J. Inst. Brew. *86*, 211–215.

SORENSEN, S. A. 1972. Microbial enzymes and their use in brewing. Ind. Aliment. Agric. *89,* 1267–1276.

SORENSEN, S. A. and FULLBROOK, P. 1972. Application of microbial enzymes in brewing industry. Tech. Q. Master Brew. Assoc. Am. *9,* 166–172.

SOSEBEE, M. E., MAY, K. N. and POWERS, J. J. 1964. The effects of enzyme addition on the quality of freeze dehydrated chicken meat. Food Technol. *18,* 551–554.

SOSULSKI, F. W., McLEARY, C. W. and SOOLIMAN, F. S. 1972. Diffusion extraction of chlorogenic acids from kernels. J. Food Sci. *37, 253,* 253–256.

SOURKES, T. L. 1977. Enzymology of aromatic amino acid decarboxylases. Mod. Pharmacol. Toxicol. *10,* 477–496.

SOUTHARD, J. H. and HULTIN, H. O. 1969. Glycolytic activity of chicken breast muscle mitochondria. J. Food Sci. *34,* 622–623.

SOWOKINOS, J. R. 1973. Maturation of *Solanum tuberosum.* I. Comparative sucrose and sucrose synthetase levels between several good and poor processing varieties. Am. Potato J. *50,* 234–247.

SOWOKINOS, J. R. 1976. Pyrophosphorylases in *Solanum tuberosum.* Plant Physiol. *57,* 63–68.

SPALLANZANI. 1784. Experiment on digestion. Cited by Effront and Prescott (1917). (French)

SPARROW, A. H. and CHRISTENSEN, E. 1954. Improved storage quality of potato tubers following exposure to gamma radiation from Cobalt 60. Nucleonics *12* (3) 16–22.

SPARROW, D. H. B. and MEREDITH, W. O. S. 1969. Malt cytolytic activity of barley and its relation to maltability. J. Inst. Brew. *75,* 237–242.

SPENCER, B. 1976. Industrial Aspects of Enzymes. Symp. Proc. Fed. Eur. Biochem. Soc., Springer Verlag, New York.

SPENCER, M. D., PANGBORN, R. M. and JENNINGS, W. G. 1978. Gas chromatographic and sensory analysis of volatiles from cling peaches. J. Agric. Food Chem. *26,* 725–732.

SPERLING, R., FURIE, B. C., BLUMENSTEIN, M., KEYT, B. and FURIE, B. 1978. Metal binding properties of γ-carboxyglutamic acid: Implications for the vitamin-K-dependent blood coagulation proteins. J. Biol. Chem. *253,* 3898–3906.

SPICER, A. 1974. Advances in Preconcentration and Dehydration of Foods. John Wiley & Sons, New York.

SPIES, J. R., STEVAN, M. A. and STEIN, W. J. 1972. The chemistry of allergens. 21. Eight new antigens generated by successive pepsin hydrolyses of bovine β-lactoglobulin. J. Allergy Clin. Immunol. *50,* 82–91.

SPILLER, G. A. and AMEN, R. J. 1976. Fiber in Human Nutrition. Plenum Press, New York.

SPIRO, T. G. and BURKE, J. M. 1976. Protein control of porphyrin conformation. Comparison of heme proteins with mesoporphyrin IX analogues. J. Am. Chem. Soc. *98,* 5482–5491.

SPLITTER, J. L. and SHIPE, W. F. 1976. Enzymatic hydrolysis and vole bioassay for estimation of the nutritive quality of maize. J. Food Sci. *41,* 1387–1391.

SREEKANTIAH, K. R., EBINE, H., OHTA, T. and NAKANO, M. 1969. Enzyme processing of vegetable protein foods. Food Technol. *23,* 1055–1061.

SREENIVASAYA, M. and BHAGVAT, K. 1937. Dilatometry and its applications in the study of enzymes. Ergeb. Enzymforsch. *6,* 235–243.

SREERANGACHAR, H. B. 1943. Studies on the fermentation of Ceylon tea. Biochem. J. *37,* 653–674.

SRERE, P. A. 1974. Controls of citrate synthase activity. Life Sci. *15,* 1695–1710.

SRINIVAS, U. K., VAKIL, U. K. and SRINIVASAN, A. 1974. Nutritional and compositional changes in dehydro-irradiated shrimp. J. Food Sci. *39,* 807–811.

SRINIVASAN, V. R. 1976. Enzymatic saccharification of cellulose. USA-ROC Symp. on Enzyme Technology. Aug 9–14. Lehigh University, Bethlehem, Pa.

SRINIVASAN, V. R. and BUMM, M. W. 1974. Isolation and immobilization of β-D-glucosidase from *Alcaligenes faecalis.* Biotechnol. Bioeng. *16,* 1413–1418.

SRNCOVA, V. and DAVIDEK, J. 1972. Reaction of pyridoxal with proteins. J. Food Sci. *37,* 310–312.

STACEY, B. E. 1974. Plant polyols. Annu. Rev. Proc. Phytochem. Soc. *10,* 47–59.

STACHELBERGER, H., BANCHER, E., WASHUTTL, J., RIEDERER, P. and GOLD, A. 1977. Quantitative determination of some biogenic amines in bananas, dates and figs. Qual. Plant.-Plant Foods Hum. Nutr. *27,* 287–291.

STACHOWICZ, K. J., ERIKSSON, C. E. and TJELLE, S. 1977. Enzymatic hydrolysis of ox-blood hemoglobin. *In* Enzymes in Food and Beverage Processing, R.L. Ory and A.J. St. Angelo (Editors). Am. Chem. Soc., Washington, D.C.

STADHOUSERS, J. and HUP, G. 1975. Factors affecting bitter flavour in Gouda cheese. Neth. Milk Dairy J. *29,* 335–353.

STADTMAN, E. R. 1948. Nonenzymatic browning in fruit products. Adv. Food Res. *1,* 325–372.

STADTMAN, E. R. and GINSBURG, A. 1974. The glutamine synthetase of *E. coli.* Structure and control. *In* The Enzymes, 3rd Edition, Vol. 10. P.D. Boyer (Editor). Academic Press, New York.

STAGNI, N. and DE BERNARD, B. D. 1968. Lysosomal activity in rat and beef skeletal muscle. Biochim. Biophys. Acta *170,* 129–139.

STAHLHUT-KLIPP, H. 1975. Enzymic determination of pyruvate and lactate in milk by automatic continuous analyses. Dtsch. Molk.-Ztg. *96,* 622–626. (German)

STAHMANN, M. A. 1977. Cross-linking of protein by peroxidase. *In* Protein Crosslinking: Biochemical and Medical Aspects. M. Friedman (Editor). Plenum Press, New York.

STAHMANN, M. A. and WOLDEGIORGIS, S. 1975. Enzymatic methods for protein quality determination. *In* Protein Nutritional Quality of Foods. M. Friedman (Editor). Marcel Dekker, New York.

STANLEY, W. L. and OLSON, A. C. 1974. The chemistry of immobilizing enzymes. J. Food Sci. *39,* 660–666.

STANLEY, W. L., WATTERS, G. G., KELLY, S. H. and OLSON, A. C. 1978. Glucoamylase immobilized on chitin with glutaraldehyde. Biotechnol. Bioeng. *20,* 135–140.

STANNARD, D. J. 1975. The use of marker enzymes to assay the churning of milk. J. Dairy Res. *42,* 241–246.

STARK, G. R. and DAWSON, C. R. 1963. Ascorbic acid oxidase. *In* The Enzymes, 2nd Edition, Vol. 8. P.D. Boyer, H.A. Lardy and K. Myrbäck (Editors). Academic Press, New York.

STARKENSTEIN, E. 1910. Properties and mode of action of the diastatic ferments of warm-blooded animal. Biochem. Z. *24,* 191–209. (German)

STARKLE, M. 1924. Methyl ketones in oxidative decomposition of triglycerides (also fatty acids) by molds with regard to the rancidity of cocoa fat. Biochem. Z. *51,* 371–382.

STAUFFER, C. E. and GLASS, R. L. 1966. The glycerol esterhydrolases of wheat germ. Cereal Chem. *43,* 644–657.

STEAGALL, E. F. 1970. EDTA titration of calcium and magnesium: determination of calcium in canned tomatoes. J. Assoc. Off. Anal. Chem. *53,* 720.

STEENKAMP, D. J. and SINGER, T. P. 1976. On the presence of a novel covalently bound oxidation-reduction cofactor, iron and labile sulfur in trimethylamine dehydrogenase. Biochem. Biophys. Res. Comm. *71,* 1289–1295.

STEERS, E., CUATRACASAS, E. and POLLARD, H. 1971. The purification of β-galactosidase from *E. coli* by affinity chromatography. J. Biol. Chem. *246,* 196–200.

STEINBUCH, E. 1976. Technical note: Improvement of texture of frozen vegetables by stepwise blanching treatments. J. Food Technol. *11,* 313–316.

STENFLO, J. 1978. Vitamin K, prothrombin, and γ-carboxyglutamic acid. Adv. Enzymol. *46,* 1–131.

STENLID, G. and SAMORODOVA-BIANKI, G. B. 1966. Effect of some flavonoids on ascorbic acid oxidation in plants. Chem. Abstr. *71* (10) 304r. (Russian)

STENTEBJERG-OLSEN, B. 1971. Microbial enzymes in brewing. Process Biochem. *6* (4) 29–32.

STEPHENS, N. L. 1977. The Biochemistry of Smooth Muscle. University Park Press, Baltimore.

STERKEL, R. L., SPENCER, J. A., WOLFSON, S. K., JR. and WILLIAMS-ASHMAN, H. G. 1958. Serum isocitrate dehydrogenase activity with particular reference to liver disease. J. Lab. Clin. Med. *52,* 176–184.

STERLING, C. 1963. Texture and cell-wall polysaccharides. Rec. Adv. Food Sci. *3,* 259–281.

STERLING, C. 1975. Anatomy of toughness in plant tissues. *In* Postharvest Biology and Handling of Fruits and Vegetables. N.F. Haard and D.K. Salunkhe (Editors). AVI Publishing Co., Westport, Conn.

STERN, D. J., LEE., W. H., McFADDEN, W. H. and STEVENS, K. L. 1967. Volatiles from grapes. Identification of volatiles from Concord essence. J. Agric. Food Chem. *15,* 1100–1102.

STERNBERG, D. 1976A. A method for increasing cellulase production by *Trichoderma viride.* Biotechnol. Bioeng. *18,* 1751–1760.

STERNBERG, D. 1976B. β-Glucosidase of *Trichoderma:* Its biosynthesis and role in saccharification of cellulose. Appl. Environ. Microbiol. *31,* 648–654.

STERNBERG, M. 1972. Bond specificity, active site, and milk clotting mechanism of *Mucor mihei* protease. Biochim. Biophys. Acta *285,* 383–392.

STERNBERG, M. 1976. Purification of industrial enzymes with polyacrylic acids. Process Biochem. *11* (7) 11–12.

STETTER, H. 1951. Enzymatic Analysis. Verlag Chemie, Weinheim. (German)

STEVENS, K. L., BREKKE, J. E. and STERN, D. J. 1970. Volatile constituents in guava. J. Agric. Food Chem. *18,* 598–599.

STEVENS, K. L., GUADAGNI, D. G. and STERN, D. J. 1970. Odour character and threshold values of nootkatone and related compounds. J. Sci. Food Agric. *21,* 590–592.

STEVENSON, A. and SWARTZ, K. 1942. Art of preserving green foodstuffs and maintaining the color thereof. U.S. Pat. 2,305,643. Dec. 22.

STIMLER, N. P. and TANZER, M. L. 1977. Location of the intermolecular crosslinking sites in collagen. *In* Protein Crosslinking. M. Friedman (Editor). Plenum Press, New York.

STOCKMAN, D. R., HALL, T. C. and RYAN, D. S. 1976. Affinity chromatography of the major seed protein of the bean (*Phaseolus vulgaris* L.). Plant Physiol. *58,* 292–294.

STOEWSAND, G. S., ANDERSON, J. L. and LAMB, R. C. 1975. Cyanide content of apricot kernels. J. Food Sci. *40,* 1107.

STOLL, A. and SEEBECK, E. 1951. Chemical investigations on alliin, the specific principle of garlic. Adv. Enzymol. *11,* 377–400.

STOLLER, D. and LEVINE, L. 1963. Two-dimensional immunodiffusion. Methods Enzymol. *6,* 848–854.

STONE, E. I., HALL, R. M., FORSYTHE, R. H. and KAZENIAC, S. J. 1976. Formation of volatile compounds from U-^{14}C-labeled linoleic acids in green beans. Abstr. AGFD140, 172nd Am. Chem. Soc. Meet., San Francisco.

STONE, E. J., HALL, R. M. and KAZENIAC, S. J. 1975. Formation of aldehydes and alcohols in tomato fruit from U-^{14}C-labeled linolenic acid and linolenic acids. J. Food Sci. *40,* 1138–1141.

STONE, I. M. 1962. Process of producing baked confections and the products resulting therefrom by alpha-amylase. U.S. Pat. 3,026,205. Mar. 20.

STONE, I. M. and SALETAN, L. T. 1968. Stability of diluted chillproofing enzyme solutions. Wallerstein Lab. Commun. *31* (104) 45–49.

STONE, I. W. 1973. Chillproofing fermented malt beverages. U.S. Pat. 3,749,582. July 31.

STONE, N. and MEISTER, A. 1962. Function of ascorbic acid in the conversion of proline to hydroxyproline. Nature *194,* 555–557.

STRAIN, H. H. 1941. Unsaturated fat oxidase: Specificity, occurrence, and induced oxidation. J. Am. Chem. Soc. *63,* 3542.

STRAND, L. L., RECHTORIS, C. and MUSSELL, H. 1976. Polygalacturonases release cell-wall-bound proteins. Plant Physiol. *58,* 722–725.

STRANDBERG, G. W. and SMILEY, K. L. 1972. Glucose isomerase covalently bound to porous glass beads. Biotechnol. Bioeng. *14,* 509–513.

STRANGE, E. D., BENEDICT, R. C., GUGGER, R. E., METZGER, V. G. and SWIFT, C. E. 1974. Simplified methodology for measuring meat color. J. Food Sci. *39,* 988–992.

STRANGE, E. D., BENEDICT, R. C., SMITH, J. L. and SWIFT, C. E. 1977. Evaluation of rapid tests for monitoring alterations in meat quality during storage. J. Food Protect. *40,* 843–847.

STRAUB, T. S. and BENDER, M. L. 1972. Cycloamyloses as enzyme models. The decarboxylation of benzoylacetic acids. J. Am. Chem. Soc. *94,* 8881–8888.

STREITEL, R. H., OCKERMAN, H. W. and CAHILL, V. R. 1977. Maintenance of beef tenderness by inhibition of rigor mortis. J. Food Sci. *42,* 583–585.

STRIZHEVSKAYA, A. Y. 1975. Dynamics of xylanase activity in different species of fungus of the genus *Fusarium*. Chem. Abstr. *85,* 188967c. (Russian)

STROLLE, E. O. 1977. Moisture sorption by whey-soy powders with sugars in the amorphous state. Abstr. *57,* 37th Annu. Meet., IFT, Dallas.

STROM, G. K. and QVIST, O. 1963. Treatment of rye to inactivate α-amylase. Getreide Mehl *13* (1) 7–12.

STROMER, M. H., GOLL, D. E. and ROTH, L. E. 1967. Morphology of rigor-shortened bovine muscle and the effect of trypsin on pre- and post-rigor myofibrils. J. Cell Biol. *34,* 431–445.

STRONG, F. M. 1974. Toxicants occurring naturally in foods. Nutr. Rev. *32,* 225–231.

STROTHER, G. K. and ACKERMAN, E. F. 1961. Physical factors influencing catalase rate constants. Biochim. Biophys. Acta *47,* 317–326.

STROTHKAMP, K. G. and DAWSON, C. R. 1974. Quaternary structure of ascorbate oxidase. Biochemistry *13,* 430–434.

STROTHKAMP, K. G., JOLLEY, R. L. and MA-

SON, H. S. 1976. Quaternary structure of mushroom tyrosinase. Biochem. Biophys. Res. Commun. *70,* 519–524.

STRUEBI, P., ESCHER, F. E. and NEUKOM, H. 1978. Use of a macerating pectic enzyme in apple nectar processing. J. Food Sci. *43,* 260–263.

STRUMEYER, D. H. and MALIN, M. J. 1969. Identification of the amylase inhibitor from seeds of *Leoti sorghum.* Biochim. Biophys. Acta *184,* 643–645.

STRYER, L. 1975. Biochemistry. W.H. Freeman and Co., San Francisco.

STULEN, I., KOCH-BOSMA, T. and KOSTER, A. 1971. An endogenous inhibitor of nitrate reductase in radish cotyledons. Acta Bot. Neerl. *20,* 386–389.

STUMPF, P. K. and CONN, E. E. 1980. The Biochemistry of Plants, 8 Vols. Academic Press, New York.

SU, J. -C. 1969. Carbohydrate metabolism in the shoot of bamboo *Leleba oldham.* VIII. Fine structure of the fibers. Chung Kuo Nung Yeh Hua Hsueh Hui Chih (Spec. Issue), 25–34.

SUDHAKARAN, P. R. and KURUP, P. A. 1974. Vitamin A and lysosomal stability in rat liver. J. Nutr. *104,* 1466–1475.

SUDYINA, O. G. 1963. Chlorophyll reaction in the last stage of biosynthesis of chlorophyll. Photochem. Photobiol. *2,* 181–190.

SUGIHARA, J. and CRUESS, W. V. 1945. Observations on the oxidase of garlic. Fruit Prod. J. *24* (10) 297–298.

SUGIMOTO, H. and VAN BUREN, J. P. 1970. Removal of oligosaccharides from soybean milk by an enzyme from Aspergillus saitoi. J. Food Sci. *35,* 655–660.

SUGIMOTO, H., VAN BUREN, J. P. and ROBINSON, W. B. 1971. Enzymic process for a protein containing beverage based on soybean protein. J. Food Sci. *36,* 729–731.

SUGISAWA, H. and EDO, H. 1964. Thermal polymerization of glucose. Chem. Ind. (London), 892.

SUGIURA, M., SUZUKI, M., ISHIKAWA, M. and SASAKI, M. 1976. Studies on enzymes. 118. Pharmaceutical studies on aminopeptidase from *Aspergillus japonica* I. Chem. Pharm. Bull. *24,* 2286–2293. (Japanese)

SUHREN, G., HEESCHEN, W. and TOLLE, A. 1975. Free fatty acids in milk and bacterial activity. Ann. Bull. Int. Dairy Fed. *86,* 51–57.

SUHREN, G., HEESCHEN, W. and TOLLE, A. 1977. Automated and manual determination of lactate in milk. Milchwissenschaft *32,* 709–712. (German)

SULLIVAN, B. 1946. Oxidizing enzymes of wheat and flour. *In* Enzymes and Role in Wheat Technology. J.A. Anderson (Editor). Interscience, New York.

SULLIVAN, B., NEAR, C. and FOLEY, G. H. 1936. The role of lipids in relation to flour quality. Cereal Chem. *13,* 318–324.

SULLIVAN, R. A. and INFANTINO, D. 1975. Method for rapid curing of cheese. U.S. Pat. 3,859,446. Jan. 7.

SULLIVAN, J. P., PATEL, K. and WASILEWSKI, J. C. 1976. Enzymatic determination of glucose: applications of the Enzymax Glucose Analyzer. Proc. Annu. Meet. Sugar Ind. Technol. *34,* 33–39.

SUMNER, J. B. 1926. Note. The recrystallization of urease. J. Biol. Chem. *70,* 97–98.

SUMNER, J. B. 1933. Parallel adsorption of crystalline pepsin and peptic activity upon casein and ovalbumin. Proc. Soc. Exp. Biol. Med. *31,* 204–206.

SUMNER, J. B. and MYRBÄCK, K. 1950–1952. The Enzymes, 1st Edition, 4 Vols. Academic Press, New York.

SUMNER, J. B. and O'KANE, D. J. 1948. The chemical nature of yeast saccharase. Enzymologia, *12,* 251–253.

SUMNER, J. B. and SOMERS, G. F. 1947. Chemistry and Methods of Enzymes, 2nd Edition. Academic Press, New York.

SUMNER, J. B. and SUMNER, R. J. 1940. The coupled oxidation of carotene and fat by carotene oxidase. J. Biol. Chem. *134,* 531–533.

SUNDSTOEL, F., KOSSILA, V., THEANDER, O. and THOMSEN, K. V. 1978. Evaluation of the feeding value of straw. A comparison of laboratory methods in the Nordic countries. Acta Agric. Scand. *28,* 10–16.

SUNG, S. K., ITO, T. and FUKAZAWA, T. 1976. Relationship between contractility and some biochemical properties of myofibrils prepared from normal and PSE porcine muscle. J. Food Sci. *41,* 102–107.

SUOMALAINEN, H. and RONAKAINER, P. 1963. Keto acids in baker's yeast and in fermentation solution. J. Inst. Brew. *69,* 478–483.

SURENDRANATHAN, K. K. and NAIR, P. M. 1976. Stimulation of the glyoxylate shunt in gamma-irradiated banana. Phytochemistry *15,* 371–373.

SUSTEK, E., DILL, C. W. and HERLICK, S. A. 1975. Lipolytic activity with the membrane fraction of bovine skim milk. J. Dairy Sci. *58,* 1519–1520.

SUTTER, H. 1936. Polyphenol oxidase. Ergeb. Enzymforsch. *5,* 273–284.

SUTTIE, J. W. 1977. Role of Vitamin K in the synthesis of clotting factors. Adv. Nutr. Res. *1,* 1–22.

SUTTIE, J. W. 1980. Mechanism of action of vitamin K: Synthesis of γ-carboxyglutamic acid. Crit. Rev. Biochem. *8,* 175–223.

SUZUKI, O., OYA, M., KATSUMATA, Y., MATSUMOTO, T. and YADA, S. 1980. A new enzymic method for the demonstration of spermine in human seminal stains. J. Forensic Sci. *25,* 99–102.

SVENSON, A. and ANDERSSON, B. 1977. The application of cyanide-metabolizing enzymes to environmental control. 1. Preparation and characterization of rhodanese and β-cyanoalanine synthase, immobilized on solid supports. Anal. Biochem. *83,* 739–745.

SVENSSON, S. G. and ERIKKSON, C. E. 1974. Thermal inactivation of lipoxygenase from peas *(Pisum sativum).* 3. Activation energy obtained from single heat treatment experiments. Lebensm.-Wiss. Technol. *1,* 142–144. (German)

SWAIN, T. 1962. Economic importance of flavonoid compounds: foodstuffs. *In* The Chemistry of Flavonoid Compounds. E.A. Geissman (Editor). Pergamon Press, Oxford.

SWAISGOOD, H. E. 1973. The caseins. Crit. Rev. Food Technol. *3,* 375–414.

SWAISGOOD, H. E. 1975. Primary sequence of kappa-casein. J. Dairy Sci. *58,* 583–592.

SWAISGOOD, H. E. 1977. Kinetics of immobilized enzymes. The Catalyst (Philadelphia) (A newsletter for teachers of food science), Spring issue.

SWAISGOOD, H. E. and HORTON, H. R. 1975. Characteristics of soluble and immobilized sulfhydryl oxidase. Presented at RANN Conference, Washington, D.C.

SWAISGOOD, H. E. and PATEE, H. E. 1968. Peanut alcohol dehydrogenase. 2 Physical-chemical and kinetic properties. J. Food Sci. *33,* 400–405.

SWARDT, G. H., DE and DUVENAGE, A. J. 1971. Malate dehydrogenase in the ripening tomato. Agroplantae *3,* 69–72.

SWEENEY, A. C. and MARSH, A. C. 1971. Effect of processing on provitamin A in vegetables. J. Am. Diet. Assoc. *59,* 238–243.

SWEENEY, J. P. and MARTIN, M. 1958. Determination of chlorophyll and pheophytin in broccoli by various procedures. Food Res. *23,* 635–647.

SWEIGART, R. D. 1979. Industrial applications of fiber-entrapped enzymes. Appl. Biochem. Bioeng. *2,* 209–218.

SWOBODA, B. E. P. and MASSEY, V. 1965. Purification and properties of the glucose oxidase from *Aspergillus niger.* J. Biol. Chem. *240,* 2209–2215.

SWOBODA, P. A. T. and PEERS, K. E. 1977. Metallic odour caused by vinyl ketones formed in the oxidation of butterfat. The identification of octa-1,*cis*-5-dien-3-one. J. Sci. Food Agric. *28,* 1019–1024.

SWORDS, G., BOBBIO, P. A. and HUNTER, G. K. L. 1978. Volatile constituents of jack fruit *(Arthocarpus heterophyllus).* J. Food Sci. *43,* 639–641.

SZARKOWSKI, J. W. 1957. Trypsin-induced cresolase activity of tyrosinase. Bull. Acad. Pol. Sci. *5,* 1–3. (French)

SZCZESNIAK, A. S. 1963. Classification of textural characteristics. J. Food Sci. *28,* 385–389.

SZCZESNIAK, A. S. 1972. Instrumental methods of texture measurement. Food Technol. *26,* 50–62.

SZCZESNIAK, A. S., LOEW, B. J. and SKINNER, E. Z. 1975. Consumer texture profile. J. Food Sci. *40,* 1253–1256.

SZEGO, C. M. 1974. Lysosomes as mediators of hormone action. Recent Prog. Horm. Res. *30,* 171–233.

TABUSHI, I., KURODA, Y. and MOCHIZUKI, A. 1980. The first successful carbonic anhydrase model prepared through a new route to regiospecifically bifunctionalized cyclodextrin. J. Am. Chem. Soc. *102,* 1152–1153.

TAGUCHI, H. and ECHIGO, T. 1968. Changes in soybean proteins during storage. II. Sulfhydryl groups and proteinase activity. Chem. Abstr. *70,* 66929m. (Japanese)

TAINTER, M. L. 1951. Papain. Ann. N.Y. Acad. Sci. *54,* 143–296.

TAIZ, L. and STARKS, J. E. 1977. Gibberellic acid enhancement of DNA turnover in barley aleurone cells. Plant Physiol. *60,* 182–189.

TAKAHASHI, N., SASAKI, R. and CHIBA, H. 1979. Enzymatic improvement of food flavor. IV. Oxidation of aldehydes in soybean extracts by aldehyde oxidase. Agric. Biol. Chem. *43,* 2557–2561.

TAKAHASHI, R., ODA, K., OKADA, K. and ARIMA, K. 1974. Production and purification of a new chillproofing enzyme and its properties. Agric. Biol. Chem. *38,* 1685–1692.

TAKAHASHI, T. and YAMAZAWA, M. 1969. Carp muscle cathepsin. Chem. Abstr. *74,* 83351s. (Japanese)

TAKASAKI, Y. 1972. Enzymatic method for manufacture of fructose. U.S. Pat. 3,689,362. Sept. 5.

TAKASAKI, Y. 1974. Method for separating fructose. U.S. Pat. 3,806,363. Apr. 23.

TAKASAKI, Y. 1976. Productions and utilizations of β-amylase and pullulanase from *Bacillus cereus* var. *mycoides*. Agric. Biol. Chem. *40,* 1515–1522.

TAKASAKI, Y., KOSUGI, Y. and KANBAYASHI, A. 1969. Streptomyces glucose isomerase. *In* Fermentation Advances. Paper presented at Inst. 3rd Fermentation Symp., 1968. D. Perlman (Editor). Academic Press, New York.

TAKASAKI, Y. and TANABE, O. 1971. Enzymatic method for converting glucose in glucose syrups to fructose. U.S. Pat. 3,616,221. Oct. 26.

TAKEDA PHARM. CO. 1966. Seasoning composition. Jpn. Pat. 66 28,066. July 22.

TAKENISHI, S. and TSUJISAKA, Y. 1975. On the modes of action of the three xylanases produced by a strain *Aspergillus niger* van Tieghem. Agric. Biol. Chem. *39,* 2315–2323.

TAKEO, T. 1978. L-Alanine decarboxylase of *Camellia sinensis*. Phytochemistry *17,* 313–314.

TAKINO, Y. 1976. Enzymatic solubilization of tea cream. U.S. Pat. 3,959,497. May 25.

TALBURT, W. F. 1975. Canned white potatoes. *In* Potato Processing, 3rd Edition. W.F. Talburt and O. Smith (Editors). AVI Publishing Co., Westport, Conn.

TALSKY, G. 1971. The anomalous temperature dependence of enzyme-catalyzed reactions. Angew. Chem. Int. Ed. Engl. *10,* 548–554.

TAMIR, M. and ALUMOT, E. 1969. Inhibition of digestive enzymes by condensed tannins from green and ripe carobs. J. Sci. Food Agric. *20,* 199–202.

TAMMAN, G. 1895. On the activity of unformed ferments. Z. Physik. Chem. *18,* 426–442. (German)

TAMURA, M., USHIRO, S. and HASAGAWA, S. 1976. Process for the continuous isomerization of dextrose. U.S. Pat. 3,960,663. June 1.

TAMURA, Y. and MORITA, Y. 1975. Thermal denaturation and regeneration of Japanese-radish peroxidase. J. Biochem. (Tokyo) *78,* 561–571.

TAN, C. T. and FRANCIS, F. J. 1962. Effect of processing temperature on pigments and color of spinach. J. Food Sci. *27,* 232–241.

TANABE SEIKAKU, LTD. 1968. Bread improvement with cellulase and hemicellulase. Jpn. Pat. 68 5,701. May 5.

TANAKA, K. and O'BRIEN, P. J. 1975. Mechanisms of H_2O_2 formation by leukocytes. Properties of the NAD(P)H oxidase of intact leukocytes. Arch. Biochem. Biophys. *169,* 436–442.

TANAKA, M., AMAYA, J., LEE, T. -C. and CHICHESTER, C. O. 1974. Effects of the browning reaction on the quality of protein. Int. Congr. Food Sci. Technol. *7a,* 31–32.

TANAKA, V. and AKAZAWA, T. 1970. Alpha-amylase enzymes in gibberellic acid-treated barley half seeds. Plant Physiol. *46,* 586–591.

TANAKA, V., ITO, K. and AKAZAWA, T. 1970. Enzymic mechanism of starch breakdown in germinating rice seeds. III. Alpha-amylase isozymes. Plant Physiol. *46,* 650–654.

TANCHEV, S. S. 1974. Kinetics of the thermal degradation of anthocyanins during the sterilization and storage of blueberry and blackberry juices. Nahrung *18,* 303–308. (German)

TANDECARZ, J., LAVINTMAN, N. and CARDINI, C. E. 1977. A primer independent activity of muscle phosphorylase b. Mol. Cell. Biochem. *16,* 141–148.

TANEJA, S. R. and SACHAR, R. C. 1976. Enzyme synthesis by conserved messengers in germinating wheat embryos. Phytochemistry *15,* 1589–1594.

TANFORD, C. 1980. The Hydrophobic Effect. Formation of Micelles and Biological Membranes, 2nd Edition. John Wiley & Sons, New York.

TANG, C. -S. 1973. Localization of benzyl glucosinolate and thioglucosidase in *Carica papaya* fruit. Phytochemistry *12,* 769–773.

TANI, N., OHTSURU, M. and HATA, T. 1974. Purification and general characteristics of bacterial myrosinase produced by *Enterobacter cloacae*. Agric. Biol. Chem. *38,* 1623–1630.

TANNENBAUM, S. R. 1977. Single-cell protein. *In* Food Proteins. J.R. Whitaker and S.R. Tannenbaum (Editors). AVI Publishing Co., Westport, Conn.

TANNER, H. 1972. Manufacture of fruit brandies with low methanol contents. Alkohol-Ind. *85* (2) 27–28. (German)

TAO, K. -L. and KHAN, A. A. 1975. Occurrence of some enzymes in starchy endosperm and hormonal regulation of isoperoxidase in aleurone of wheat. Plant Physiol. *59,* 797–800.

TAPPEL, A. L. 1966A. Effects of low temperatures and freezing on enzymes and enzyme systems. *In* Cryobiology. H.T. Meryman (Editor). Academic Press, New York.

TAPPEL, A. L. 1966B. Lysosomes: Enzymes and catabolic reactions. *In* The Physiology and Biochemistry of Muscle as a Food. E.J. Briskey *et al.* (Editors). University of Wisconsin Press, Madison.

TAPPEL, A. L. 1970. Lysosomal enzymes and other components. *In* Lysosomes in Biology and Pathology, Vol. 2. J.T. Dingle and H.B. Fell (Editors). North Holland Publishers, Amsterdam.

TAPPEL, A. L. 1973. Lipid peroxidation damage to cell components. Fed. Proc. *32,* 1870–1874.

TAPPEL, A. L., LUNBERG, W. O. and BOYER, P. D. 1953. Effect of temperature and antioxidants upon the lipoxidase-catalyzed oxidation of sodium linoleate. Arch. Biochem. Biophys. *42,* 293–304.

TAPPEL, A. L., MIYADA, D. J., STERLING, C. and MAIER, V. P. 1956. Meat tenderization. II. Factors affecting the tenderization of beef by papain. Food Res. *21,* 375–383.

TARANTOWICZ-MAREK, E., BRALCYK, J. and KLECZOWSKI, K. 1975. RNA synthesis in isolated maize seedlings nuclei. Effect of GA_3 and cAMP. Bull. Acad. Pol. Sci. Ser. Sci. *23,* 227–232.

TARKY, W., AGARWALA, O. P. and PIGOTT, G. M. 1973. Protein hydrolysate from fish waste. J. Food Sci. *38,* 917–918.

TARLADGIS, B. G. 1962. Interpretation of the spectra of meat pigments. II. Cured meats. The mechanism of colour fading. J. Sci. Food Agric. *13,* 485–491.

TARR, H. L. A. 1966. Post-mortem changes in glycogen, nucleotides, sugar phosphates and sugars in fish muscles. A review. J. Food Sci. *31,* 846–854.

TARR, H. L. A. 1969. Contrasts between fish and warm blooded vertebrates in enzyme systems of intermediary metabolism. Fish. Res. Symp. 1968. O.W. Neuhaus (Editor). Academic Press, New York.

TARR, H. L. A. and DEAS, C. P. 1949. Bacteriological peptones from fish flesh. J. Fish. Res. Board Can. *7,* 552–560.

TASSINARI, T. and MACY, C. 1977. Differential speed two roll mill pretreatment of cellulosic materials for enzymatic hydrolysis. Biotechnol. Bioeng. *19,* 1321–1330.

TATE & LYLE LTD. and DE WHALLEY, C. S. 1944. Inversion of sugar liquors. Br. Pat. 564,270. Sept. 20.

TATE, W. P. and CASKEY, C. T. 1974. Polypeptide chain termination. *In* The Enzymes, 3rd Edition, Vol. 10. P.D. Boyer (Editor). Academic Press, New York.

TATEISHI, K., YAMAMOTO, H., OGIWARA, T., HAYAISHI, C. and KITAGAWA, M. 1976. Enzyme immunoassay of testosterone using the testosterone-glucoamylase complex. J. Biochem. (Tokyo) *80,* 191–194.

TATEO, F. 1977. Preparation of vegetable extracts by mechanical and enzyme processes. Riv. Ital. Sostanze Grasse *54,* 114–115.

TATSUMI, Y., CHACHIN, K. and OGATA, K. 1972. Browning of potato tubers by gamma irradiation. II. Relationship between browning and changes in *o*-diphenol, and ascorbic acid contents and in polyphenol oxidase activities in irradiated potato tubers. J. Food Sci. Technol. *19,* 508–513. (Japanese)

TATTERSON, I. N. and WINDSOR, M. L. 1974. Fish silage. J. Sci. Food Agric. *25,* 369–379.

TAUBER, H. 1940. Unsaturated fat oxidase. J. Am. Chem. Soc. *62,* 2251.

TAUBER, H. 1949. The Chemistry and Technology of Enzymes. John Wiley & Sons, New York.

TAWFIK, M. E. and ATIA, R. M. 1972. Action of bacterial α-amylase on gelatinization characteristics of waxy rice flour. Cereal Chem. *49,* 343–345.

TAYLOR, M. J., RICHARDSON, T. and OLSON, N. F. 1976. Coagulation of milk with immobilized proteases: A review. J. Milk Food Technol. *39,* 864–871.

TAYLOR, N. F., HILL, L. and EISENTHAL, R. 1975. Specificity of oxidase and kinase preparations from *Pseudomonas fluorescens* towards deoxyfluoromonosaccharides. Can. J. Biochem. *53,* 57–64.

TAYLOR, S. L., GUTHERTZ, L. S., LEATHERWOOD, M., TILLMAN, F. and LIEBER, E. R. 1978. Histamine production by food-borne bacterial species. J. Food Safety *1,* 173–178.

TEISSON, C. 1972. Investigations on the internal browning of the pineapple. Fruits *27,* 603–612. (French)

TEMPERLI, A., SCHAERER, H. and KUENSCH, U. 1976. Enzymic method for the determination of glucose in beverages. Fluess. Obst *43,* 257–258. (German)

TENCO BROOKE BOND, LTD. 1971. Enzymic solubilization of tea cream. Br. Pat. 1,249,932. Oct. 13.

TERANISHI, R., HORNSTEIN, I., ISSENBERG, P. and WICK, E. L. 1971. Flavor Research Principles and Techniques. Marcel Dekker, New York.

TEREBULINA, N. A., BAIKOV, V. G., NECHAEV, A. P., SOSEDOV, N. I. and GAVRICHENKOV, Y. D. 1977. The effect of drying temperature on the content and composition of lipids in wheat grain. Vopr. Pitan. *2,* 70–73. (Russian). Chem. Abstr. *86,* 169,499f.

TERENT'EVA, G. N., VAKAR, A. B. and NECHAEV, A. P. 1973. Enzymic method for the isolation and determination of strongly bound lipids of gluten. Prikl. Biokhim. Mikrobiol. *9,* 502–505. (Russian)

TERMOTE, F., ROMBOUTS, F. M. and PILNIK, W. 1977. Stabilization of cloud in pectinesterase active orange juice by pectic acid hydrolysates. J. Food Biochem. *1,* 15–34.

TEWARI, G. M. and BANDYOPADHYAY, C. 1975. Quantitative evaluation of lachrymatory factor in onion by thin layer chromatography. J. Agric. Food Chem. *23,* 645–647.

TEWARI, G. M. and BANDYOPADHYAY, C. 1977. Pungency and lachrymatory factor as a measure of flavour strength of onions. Lebensm.-Wiss. Technol. *10,* 94–96.

THALER, H. and GEIST, G. 1939. The chemistry of ketone rancidity. Decomposition of fatty acids by *Penicillium glaucum.* Biochem. Z. *30,* 121–136. (German)

THANANUNKUL, D., TANAKA, M., CHICHESTER, C. O. and LEE, T. -C. 1976. Degradation of raffinose and stachyose in soybean milk by α-galactosidase from *Mortiella vinacea.* Entrapment of α-galactosidase within polyacrylamide gel. J. Food Sci. *41,* 173–175.

THAYER, P. S. 1953. The amylases of *Pseudomonas saccharophila.* J. Bacteriol. *66,* 656–663.

THEORELL, H., HOLMAN, R. T. and AKESON, A. 1947. Crystalline lipoxidase. Acta Chem. Scand. *1,* 571–576.

THIJSSEN, H. A. C. 1974. Fundamentals of concentration processes. *In* Advances in Preconcentration and Dehydration of Foods. A. Spicer (Editor). John Wiley & Sons, New York.

THIVEND, P., MERCIER, C. and GUILBOT, A. 1972. Determination of starch with glucoamylase. Methods Carbohydr. Chem. *6,* 100–105.

THOMA, J. A. and KOSHLAND, D. E. 1960. Competitive inhibition by substrate during enzyme action. Evidence for the induced-fit theory. J. Am. Chem. Soc. *82,* 3329–3333.

THOMAS, B. and ELCHAZLY, M. 1976. The problem of crude fiber determination. Getreide Mehl Brot *30,* 252–255. (German)

THOMAS, B. and WEBB, J. A. 1977. Multiple forms of α-galactosidase in mature leaves of *Cucurbita pepo.* Phytochemistry *16,* 203–206.

THOMAS, P. and JANAVE, M. T. 1973. Polyphenol oxidase activity and browning of mango fruits induced by gamma irradiation. J. Food Sci. *38,* 1149–1152.

THOMAS, W. E. 1928. A process for bleaching vegetables. U.S. Pat. 1,685,511. Sept. 11.

THOMOPOULOS, C. 1975. Effect of certain sodium salts on the blanching of green beans. Ind. Aliment. Agric. *92,* 531–534. (French)

THOMPSON, E. H., WOLF, I. D. and ALLEN, C. E. 1973. Ginger rhizome. New source of proteolytic enzyme. J. Food Sci. *38,* 652–655.

THOMPSON, H. and RECHNITZ, G. A. 1973. Ion electrode based enzymatic analysis of creatine. Anal. Chem. *44,* 246–249.

THOMPSON, K. N., JOHNSON, R. A. and LLOYD, N. E. 1974. Process for isomerizing glucose to fructose. U.S. Pat. 3,788,945. Jan. 29.

THOMPSON, M. H. and FARRAGUT, R. N. 1969. Problem of the "green" frozen raw breaded shrimp. Fish Ind. Res. *5* (1) 1–10.

THOMPSON, R. R. 1968. An enzyme (esterase) method for identification of animal and fish species. J. Assoc. Off. Anal. Chem. *514,* 746–748.

THORELL, J. I. and LARSON, S. M. 1978. Radioimmunoassay and Related Techniques. Mosby Publ., St. Louis, Mo.

THUNG, S. B. 1963. Browning of heated spinach. Annu. Rep. Inst. Res. Storage and Proc. Hort. Produce. 1BUT-Wageningen, Netherlands, 43–49.

THUNELL, R. K., ERNSTROM, C. A. and HARTMANN, G. H., JR. 1980. Effects of small concentrations of *Mucor mihei* protease on stability of sterile milk-based dietetic foods. J. Dairy Sci. *63,* 32–36.

THURSTON, C. F. 1972. Disappearing enzymes. Process Biochem. 7 (8) 18–22.

TIMELL, T. E. 1965. Wood hemicelluloses. Adv. Carbohydr. Chem. *20,* 410–483.

TING, C. -Y., MONTGOMERY, M. and ANGLEMIER, A. F. 1968. Partial purification of salmon muscle cathepsins. J. Food Sci. *33,* 617–621.

TING, S. V. and ATTAWAY, J. A. 1971. Citrus fruits. *In* The Biochemistry of Fruits and Their Products. A.C. Hulme (Editor). Academic Press, London.

TIRIMANA, A. S. L. 1972. Starch gel electrophoresis of the peroxidase isozymes of the tea leaf. J. Chromatogr. *65,* 587–588.

TISHEL, M. and MAZELIS, M. 1966. The accumulation of sugars in potato tubers at low temperature and some associated enzymatic activities. Phytochemistry *5,* 895–902.

TKACHUK, R. and KRUGER, J. E. 1974. Wheat α-amylase. II. Physical characterization. Cereal Chem. *51,* 508–529.

TKACHUK, R. and TIPPLES, R. H. 1966.

Wheat beta-amylases. II. Characterization. Cereal Chem. *43,* 62–69.

TOCCHINI, R. P. and LARA, J. C. C. 1977. Manufacture of natural and concentrated banana juice. FSTA *10* (3) H377. (Portuguese)

TOKUNAGA, T. 1966. The development of dimethylamine and formaldehyde in Alaskan pollock muscle during frozen storage. III. Effect of various kinds of additives. Chem. Abstr. *69,* 105157j. (Japanese)

TOLBERT, N. 1973. Activation of polyphenoloxidase of chloroplasts. Plant Physiol. *51,* 234–244.

TOLMASOFF, J. M., ONO, T. and CUTLER, R. G. 1980. Superoxide dismutase: Correlation with life-span and specific metabolic rate in primate species. Proc. Natl. Acad. Sci. *77,* 2777–2781.

TOMB, W. H. and WEETALL, H. H. 1977. Enzyme carriers. U.S. Pat. 4,025,667. May 24.

TOME, D., NICOLAS, J., and DRAPRON, R. 1978. Influence of water activity on the reaction catalyzed by polyphenoloxidase from mushrooms in organic liquid media. Lebensm.-Wiss. Technol. *11,* 38–41. (German)

TOMIOKA, K., OGUSHI, J. and ENDO, K. 1974. Studies on dimethylamine in foods—II. Enzymatic formation of dimethylamine from trimethylamine oxide. Bull. Jpn. Soc. Fish. *40,* 1021–1026. (English)

TOMIYAMA, T. 1968. Preparation of fish protein concentrate by autolysis. Zesz. Probl. Pastepov. Nauk Roln. *80,* 385–398. (English)

TOMLINSON, R. V. and BALLOU, C. E. 1962. Myoinositol polyphosphate intermediates in the dephosphorylation of phytic acid by phytase. Biochemistry *1,* 166–171.

TOMME, M. F., DEVYATKIN, A. I. and YADRINTSEV, A. YA. 1973. Use of the enzyme preparation pektawamorin Pkh when fattening calves on beet pulp. Biol. Abstr. *57* (11) 790. (Russian)

TOMODA, K. 1964. Acid protease produced by *Trametes sanguinea,* a wood destroying fungus. II. Physical and enzymological properties of the enzyme. Agric. Biol. Chem. *28,* 774–778.

TONER, M. C. and POTTER, O. E. 1977. Computer simulation of the β-amylolysis of starch components. J. Inst. Brew. *83,* 78–81.

TONG, M. -M. and PINCOCK, R. E. 1969. Denaturation and reactivity of invertase in frozen solutions. Biochemistry *8,* 908–913.

TOOKEY, H. L. and GENTRY, H. S. 1969. Proteinase of *Jarilla chocola,* a relative of papaya. Phytochemistry *8,* 989–991.

TOOKEY, H. L., VAN ETTEN, C. H. and DAXENBICHLER, M. E. 1977. Glucosinolates. *In* Toxic Constituents of Plant Foodstuffs, 2nd Edition. I.F. Liener (Editor). Academic Press, New York.

TORR, D. 1966. Method of preparing meat products. U.S. Pat. 3,276,880. Oct. 4.

TORREY, M. 1974. Dehydration of Fruits and Vegetables. Food Technology Review. Noyes Data Corp., Park Ridge, N.J.

TOUL, Z. and MACHOLAN, L. 1975. Enzyme electrode for rapid determination of biogenic polyamines. Collect. Czech. Chem. Commun. *40,* 2208–2217.

TOULEC, R. 1973. *Cited by* Mackie (1974). Proteolytic enzymes in the recovery of proteins from fish waste. Process Biochem. *9* (10) 12–14.

TOWER PHARMACEUTICAL CO. 1977. Process for the purification and sterilization of acidophilic biologicals. Br. Pat. 1,471,336. Apr. 21.

TOWNSHEND, A. 1973. Trace analysis by enzyme inhibition and activation. Process Biochem. *8* (3) 22, 24.

TOYAMA, N. 1969. Applications of cellulases in Japan. Adv. Chem. Ser. *95,* 359–390.

TOYAMA, N. 1976. Feasibility of sugar production from agricultural and urban cellulosic wastes with *Trichoderma viride.* Biotechnol. Bioeng. Symp. *6,* 207–219.

TOYOMIZU, M., NAKAMURA, T. and SHONO, T. 1976. Fatty acid composition of lipid from horse mackerel muscle. Bull. Jpn. Soc. Sci. Fish. *42,* 101–108. (Japanese)

TRAN, V. D. 1975. Effect of sodium pyruvate on texture of frozen stored cod fillets. J. Food Sci. *40,* 888–889.

TRAUBERMANN, L. 1975. Immobilized enzymes put to work in analysis and control. Food Eng. *47,* 58–60.

TRAVENER, R. J. A. and LAIDMAN, D. L. 1972. The induction of lipase activity in the germinating wheat grain. Phytochemistry *11,* 989–997.

TRAVERSO-RUEDA, S. and SINGLETON, V. L. 1973. Catecholase activity in grape juice and its implications in wine making. Am. J. Enol. Vitic. *24,* 103–109.

TRELA, S. 1977. Study of the improved use of protein in grain components of concentrate feed mixtures. Chem. Abstr. *87,* 1327776k. (Polish)

TRESSL, P. and KOSMAN, D. J. 1980. o,o-Dityrosine in native and horseradish-activated galactose oxidase. Biochem. Biophys. Res. Comm. *92,* 781–786.

TRESSL, R. 1975. The formation of volatile flavour substances by Maillard-reactions. *In* International Symposium, Odor and Taste Substances. F. Drawert (Editor). Verlag Hans Carl, Nürnberg.

TRESSL, R. and DRAWERT, F. 1973. Biogenesis of banana volatiles. J. Agric. Food Chem. *21,* 560–565.

TRESSL, R., HOLZER, M. and KOSSA, T. 1977. Formation of flavor components in asparagus. J. Agric. Food Chem. *25,* 459–463.

TRESSL, R. and RENNER, R. 1976. Formation of aroma substances in foods. Dtsch. Lebensm. Rundsch. *72* (2) 37–44. (German)

TRESSLER, D. K. and JOSLYN, M. A. 1971. Fruit and Vegetable Juice Processing. AVI Publishing Co., Westport, Conn.

TRESSLER, D. K., VAN ARSDEL, W. B. and COPLEY, M. J. 1968. The Freezing Preservation of Foods, 4th Edition, Vol. 1–4. AVI Publishing Co., Westport, Conn.

TREVELYAN, W. E. 1977. Induction of autolytic breakdown of RNA in yeast by addition of ethanol and by drying/rehydration. J. Sci. Food Agric. *28,* 579–588.

TROLLER, J. A. 1980. Influence of water activity on microorganisms found in foods. Food Technol. *34* (5) 76–82.

TROLLER, J. A. and CHRISTIAN, J. H. B. 1978. Water Activity and Food. Academic Press, New York.

TROMMER, W. E., BLUME, H. and KAPMEYER, H. 1976. Synthesis of a "transition state" analog of the lactate dehydrogenase catalyzed redox reaction. Justus Liebigs Ann. Chem., 856–878.

TRONIER, B. and ORY, R. L. 1970. Association of bound beta amylase with protein bodies in barley. Cereal Chem. *47,* 464–471.

TROWBRIDGE, C. G., KREHBIEL, A. and LASKOWSKI, M., JR. 1963. Substrate activation of trypsin. Biochemistry *2,* 843–850.

TROWELL, H. 1976. Definition of dietary fiber and hypotheses that it is a protective factor in certain diseases. Am. J. Clin. Nutr. *29,* 417–427.

TSAI, L. S., IJICHI, K., HUDSON, C. A. and MEEHAN, J. J. 1980. Cholesterol oxides in commercial dried egg products. J. Food Sci. *45.* (in preparation)

TSCHESCHE, H. 1974. Biochemistry of natural proteinase inhibitors. Angew. Chem. Int. Ed. *13,* 10–28.

TSEN, C. C. 1965. The improving mechanism of ascorbic acid. Cereal Chem. *42,* 86–97.

TSEN, C. C. and HLYNKA, I. 1962. The role of lipids in oxidation of doughs. Cereal Chem. *39,* 209–214.

TSIKLAURI, M. S. 1972. Changes of some biochemical indexes in leaves of various varieties infected with *Phoma tracheiphila.* Chem. Abstr. *83,* 175385k. (Russian)

TSONG, T. Y. 1973. Detection of three kinetic phases in the thermal unfolding of ferricytochrome c. Biochemistry *12,* 2209–2214.

TSUKUDA, N. 1970. Discoloration of red fishes. VI. Partial purification and specificity of the lipoxidase like enzyme responsible for carotenoid discoloration in fish skin. Nippon Suisan Gakkaishi *36,* 725–733. (Japanese)

TSUKUDA, N. and AMANO, K. 1972. Effects of sodium bisulfite on prevention of blackening of prawn and amounts of sulfur dioxide remaining in prawn. Chem. Abstr. *80,* 359385. (Japanese)

TSUKUDA, N. and KITAHARA, T. 1974. Esterified fatty acid composition of astaxanthin diester in the skin of seven red fish. Chem. Abstr. *81,* 166612y. (Japanese)

TSUMURA, N. and SATO, T. 1965. Enzymatic conversion of D-glucose to D-fructose. V. Partial purification and properties of the enzyme from *Aerobacter cloacae.* Agric. Biol. Chem. *29,* 1123–1128.

TSURUO, I., and HATA, T. 1968. Studies on myrosinase in the mustard seed. IV. Sugars and glycosides as competitive inhibitors. Agric. Biol. Chem. *21,* 1420–1424.

TURKOVA, J. 1979. Affinity chromatography. *In* Handbook of Chromatography and Allied Methods. O. Mikes (Editor). Horwood Publishers, Chichester, England.

TURNER, J. F. and TURNER, D. H. 1975. The regulation of carbohydrate metabolism. Annu. Rev. Plant Physiol. *26,* 159–186.

TUTT, D. E. and SCHWARTZ, M. A. 1971. Model catalysts which simulate penicillinase. V. The cycloheptaamylose-catalyzed hydrolysis of penicillins. J. Am. Chem. Soc. *93,* 767–772.

TWIGG, B. A. 1959. Consistency and serum separation of catsup. University Microfilms, Ann Arbor, Mich.

TYSON, C. A. 1975. 4-Nitrocatechol as a colorimetric probe for non-heme dioxygenases. J. Biol. Chem. *250,* 1765–1770.

UCHINO, F., KURONO, R. and DOI, S. 1966. Crystallization of an endopolygalacturonase from *Acrocylindrium* sp. Agric. Biol. Chem. *30,* 1066–1068.

UEDA, S. and OHBA, R. 1976. Pullulanase responsible for digesting raw starch. Staerke *28,* 20–22.

UHLIG, H. and GRAMPP, E. 1972. Soya meal treatment. U.S. Pat. 3,640,723. Feb. 8.

UHLIG, H. and SPRÖSSLER, B. 1972. The natural enzymes of cereals and their supplementation with microbiological preparations. Muehle *109,* 221–223. (German)

ULBRICH, N., LIN, A., TODOKORO, K. and WOOL, I. G. 1980. Identification by affinity chromatography of the rat liver ribosomal proteins that bind to *Escherichia coli* 5 S ribosomal ribonucleic acid. J. Biol. Chem. *255,* 797–801.

ULLRICH, V. and DUPPEL, W. 1975. Iron- and copper-containing monooxygenases. *In* The Enzymes, 3rd Edition, Vol. 12B. P.D. Boyer (Editor). Academic Press, New York.

ULRICH, R. 1970. Organic acids. *In* The Biochemistry of Fruits and Their Products. A.C. Hulme (Editor). Academic Press, London.

UNDERKOFLER, L. A. 1959. Meat tenderizer. U.S. Pat. 2,904,442. Sept. 15.

UNDERKOFLER, L. A. 1966. A contribution. *In* Enzymes in Food Processing. G. Reed. Academic Press, New York.

UNDERKOFLER, L. A. 1972. Enzymes. *In* Handbook of Food Additives, 2nd Edition. T.O. Furia (Editor). CRC Press, Cleveland.

UNDERKOFLER, L. A. 1976. Mass production and purification of enzymes. USA-ROC Symp. on Enzyme Technology. Aug. 9–14. Lehigh Univ., Bethlehem, Pa.

UNEMOTO, T., HAYASHI, M., MIYAKI, K. and HAYASHI, M. 1965. Intracellular localization and properties of trimethylamine-*N*-oxide reductase in *Vibrio parahaemolyticus.* Biochim. Biophys. Acta *110,* 319–328.

UNILEVER, N. V. 1966. Treatment of new potatoes. Neth. Pat. 6,605,033. Oct. 17.

UPDIKE, S. J. and HICKS, G. P. 1967. The enzyme electrode. Nature *214,* 986–988.

URITANI, I. 1978. The biochemistry of host response to infection. Prog. Phytochem. *5,* 29–64.

URS, N. V. and DUNLEAVY, J. V. 1975. Bactericidal activity of horseradish peroxidase on *Xanthomonas phaseoli* var *sojensis.* Phytopathology *64,* 542–545.

URQUIDI, R. L. 1975. Enzymes—food catalysts are key to new foods, new processes. Food Process. *36* (10) 38–45.

U.S. DEP. AGRIC., AGRIC. MKTG. SERV. 1975. Enzyme inactivation tests (frozen vegetables). Technical Inspection procedures for the use of USDA inspectors. USDA Agric. Mktg. Serv., Washington, D.C.

UTSUMI, I. 1974. Preservation of packed materials. Jpn. Kokai *74* 86,285. Aug. 19.

UY, R. and WOLD, F. 1977. Posttranslational covalent modification of proteins. Science *198,* 890–896.

UYEDA, M. and PEISACH, J. 1975. Optical difference spectroscopy of tyrosyl groups of kangaroo, horse and sperm whale myoglobins. Fed. Proc. *34,* 598.

UYEDA, M., YOSHIDA, S., SUZUKI, K. and SHIBATA, M. 1976. Isolation of an alkaline protease inhibitor, AP-I, produced by *Streptomyces pseudogriseolus* strain No. KTO–332. Agric. Biol. Chem. *40,* 1237–1238.

VAES, G. 1972. Release of collagenase as an inactive proenzyme by bone explants in culture. Biochem. J. *126,* 275–289.

VAHEDRA, D. V. and NATH, K. R. 1973. Eggs as a source of protein. Crit. Rev. Food Technol. *4,* 193–309.

VAKS, B. and LIFSHITZ, A. 1975. An enzyme that attacks limonin at acid pH. Lebensm. Wiss. Technol. *8,* 236. (German)

VALLE-VEGA, P., YOUNG, C. T. and SWAISGOOD, H. E. 1980. Arginine decarboxylase plug flow reactor for determination of arginine and peanut maturity. J. Food Sci. *45,* 1003–1007.

VAN ARSDEL, W. B. 1957. The time-temperature tolerance model of frozen foods. I. The problem and the attack. Food Technol. *11* (1) 28–33.

VAN ARSDEL, W. B., COPLEY, M. J. and MORGAN, A. I., JR. 1973. Food Dehydration, Vol. 1 and 2. AVI Publishing Co., Westport, Conn.

VAN BUREN, J., DE VOS, L. and PILNIK, W. 1976. Polyphenols in golden delicious apple juice in relation to method of preparation. J. Agric. Food Chem. *24,* 448–451.

VAN BUREN, J. P. 1974. Heat treatments and the texture and pectins of red tart cherries. J. Food Sci. *39,* 1203–1205.

VAN BUREN, J. P., MOYER, J. C., WILSON, W. B. and HAND, D. B. 1960. Influence of blanching conditions on sloughing, splitting and firmness of canned snap beans. Food Technol. *14,* 233–236.

VANDEKERCKHOVE, P. 1977. Amines in dry fermented sausages. J. Food Sci. *42,* 283–285.

VANDEKERCKHOVE, P., DEMEYER, D. and HENDERICKX, H. 1972. Evaluation of a method to differentiate between nonfrozen and frozen-and-thawed meat. J. Food Sci. *37,* 636–637.

VAN DER HEIJDEN, R. S., BRUSSEL, L. B. P. and PEER, H. G. 1978. Chemoreception of sweet tasting dipeptide esters, a third binding site. Food Chem. *3,* 207–212.

VAN DONGEN, D. B. and COONEY, D. O. 1977. Hydrolysis of salicin by β-glucosidase in a hollow fiber reactor. Biotechnol. Bioeng. *19,* 1253–1258.

VAN ES, A. 1977. Enzymic determination of carbohydrates. Voedingsmiddelen-technol. *10* (11) 24–27. (Dutch)

VAN ETTEN, C. H. and DAXENBICHLER, M. E. 1977. Glucosinolates and derived products in cruciferous vegetables: Total glucosinolate by retention on anion exchange resin and enzymatic hydrolysis to measure released glucose. J. Assoc. Off. Anal. Chem. *60,* 946–949.

VAN ETTEN, C. H., DAXENBICHLER, M. E., WILLIAMS, P. H. and KWOLEK, W. F. 1976. Glucosinolates and derived products in cruciferous vegetables: Analysis of the edible part from 22 varieties of cabbage. J. Agric. Food Chem. *24,* 452–455.

VAN ETTEN, C. H., DAXENBICHLER, M. E. and WOLFF, I. A. 1969. Natural glucosinolates (thioglucosides) in foods and feeds. J. Agric. Food Chem. *17,* 483–491.

VAN ETTEN, R. L., SEBASTIAN, J. F., CLOWES, G. A. and BENDER, M. L. 1967. Acceleration of phenyl ester cleavage by cycloamyloses. A model for enzyme specificity. J. Am. Chem. Soc. *89,* 3242–3253.

VAN HOUDENHOVEN, F. E. A., DE WIT, P. J. G. M. and VISSER, J. 1974. Large-scale preparation of galacturonic acid oligomers by matrix-bound polygalacturonase. Carbohydr. Res. *34,* 233–239.

VAN LEEMPUTTEN, E. and HORISBERGER, M. M. 1974. Immobilization of enzymes on magnetic particles. Biotechnol. Bioeng. *16,* 385–396.

VAN LOGTEN, M. J., DEN TONKELAAR, E. M., KROES, R. and VAN ESCH, G. J. 1972. Toxicity studies on a microbial rennet. Food Cosmet. Toxicol. *10,* 649–654.

VAN PEE, W. and CASTELEIN, J. M. 1972. Study of the pectinolytic microflora, particularly the enterobacteriaceae, from fermenting coffee in the Congo. J. Food Sci. *37,* 171–174.

VAN SUMERE, C. F., ALBRECHT, J., DEDONDER, A., DE POOTER, H. and PÉ, I. 1975. Plant proteins and phenolics. *In* The Chemistry and Biochemistry of Plant Proteins. J.B. Harborne and C.F. Van Sumere (Editors). Academic Press, London.

VAN TWISK, P., MELTLER, B. W. and CORMACK, R. H. 1976. Production of glucose from maize grits on commercial scale. Staerke *28,* 23–25.

VAN VEEN, A. G. 1965. Fermented and dried seafood products in Southeast Asia. *In* Fish and Food, Vol 3. G. Borgstrom (Editor). Academic Press, New York.

VAQUEIRO, G. C. and CALDERON, P. R. 1975. High-protein flours made from rice and wheat. Tecnol. Aliment. *10,* 159–169.

VARNER, J. E. 1964. Gibberellic acid controlled synthesis of α-amylase in barley endosperm. Plant Physiol. *39,* 413–415.

VARNER, J. E., FLINT, D. and MITRA, R. 1976. Characterization of protein metabolism in cereal grain. *In* Genetic Improvement of Seed Proteins, Proc. Workshop 1974. Natl. Acad. Sci., Washington, D.C.

VAROQUAUX, P., CLOCHARD, A., SARRIS, J., AVISSE, C. and MORFEAUX, J. N. 1975. Automatic measurement of heat destruction and regeneration of peroxidase. Lebensm.-Wiss. Technol. *8,* 60–63. (French)

VAROQUAUX, P. and SARRIS, J. 1979. The effect of ascorbic acid on the activity of polyphenoloxidase. Lebensm.-Wiss. Technol. *12,* 318–320. (French)

VARTY, K. and LAIDMAN, D. L. 1976. The pattern and control of phospholipid metabolism in wheat aleurone tissue. J. Exp. Bot. *27,* 748–758.

VASQUEZ, A. W. and COOPER, J. C. 1964. Enzymatic analysis of tomato products for the determination of fly eggs and maggots. J. Assoc. Off. Anal. Chem. *47,* 531–533.

VENKATASUBRAMANIAN, K., SAINI, R. and VIETH, W. R. 1975. Immobilization of papain EC-3.4.4.10 on collagen and the use of collagen papain membranes in beer chill proofing. J. Food Sci. *40,* 109–113.

VENTER, J. C., VENTER, B. R., DIXON, J. E. and KAPLAN, N. O. 1975. Possible role of glass bead immobilized enzymes as therapeutic agents. Immobilized uricase as enzyme therapy for hyperuricemia. Biochem. Med. *12,* 79–91.

VERGER, R. and DE HAAS, G. H. 1976. Interfacial enzyme kinetics of lipolysis. Annu. Rev. Biophys. Bioeng. *5,* 77–117.

VERSTEEG, C., MARTENS, L. J. H., ROMBOUTS, F. M., VORAGEN, A. G. J. and PILNIK, W. 1977. Enzymatic hydrolysis of naringin in grapefruit juice. Lebensm.-Wiss. Technol. *10,* 268–272.

VESTERBERG, O. 1971. Isoelectric focussing of

proteins. Methods Enzymol. *22,* 289–411.

VETTER, J. L., NELSON, A. I. and STEINBERG, M. P. 1959. Heat inactivation of peroxidase in HTST processed whole kernel corn. Food Technol. *13,* 410–413.

VICK, B. and ZIMMERMAN, D. 1976. The identification of linoleate hydroperoxide lyase in germinating watermelon seedlings. Plant Physiol. *36* (2, Suppl.) 84.

VIDIGAL, C. C. C. and CILENTO, G. 1975. Evidence for the generation of excited methylglyoxal in the myoglobin catalyzed oxidation of acetoacetate. Biochem. Biophys. Res. Comm. *62,* 184–190.

VIIKARI, L. and LINKO, M. 1977. Reduction of nucleic acid content of SCP. Process Biochem. *12* (5) 17–19, 35.

VIJAYALAKSHMI, M. A., SARRIS, J. and VAROQUAUX, P. 1976. Automatic determination of the pectin esterase activity in commercial enzyme preparations. Lebensm.-Wiss. Technol. *9,* 21–23. (French)

VINES, H. M. 1968. Citrus enzymes II. Mitochondrial and cytoplasmic malic dehydrogenase from grapefruit vesicles. Proc. Am. Soc. Hortic. Sci. *92,* 179–184.

VIRTANEN, A. I. 1958. Antimicrobiological substances in our cultivated plants and their significance for the plants and for human and animal nutrition. Schweiz. Z. Path. Bakt. *21,* 970–993. (German)

VISSER, F. M. W. 1977. Contribution of enzymes from rennet, starter bacteria and milk to proteolysis and flavor development in Gouda cheese. 5. Some observations on bitter extracts from aseptically made cheeses. Neth. Milk Dairy J. *31,* 265–276.

VISSER, F. M. W. and DE GROOT-MOSTERT, A. E. A. 1977. Contribution of enzymes from rennet, starter bacteria and milk to proteolysis and flavor development in Gouda cheese. 4. Protein breakdown: A gel electrophoresis study. Neth. Milk Dairy J. *31,* 247–264.

VISSER, S. and VAN ROOIJEN, P. J. 1974. Synthetic substrates for chymosin. Z. Physiol. Chem. *355,* 1264. (German)

VISURI, K. and NUMMI, M. 1972. Purification and characterization of crystalline β-amylase from barley. Eur. J. Biochem. *28,* 555–565.

VITZTHUM, O. G., WERKHOFF, P. and HUBERT, P. 1975. Volatile components of roasted cocoa: Basic fraction. J. Food Sci. *40,* 911–916.

VOGT, W. 1978. Enzyme Immunoassay: Principles and Practical Use. Verlag Thieme, Stuttgart, Germany. (German)

VOIGT, M. N. and EITENMILLER, R. R. 1978. Role of histidine and tyrosine decarboxylases and mono- and diamine oxidases. J. Food Prot. *41,* 182–186.

VOIROL, F. 1972. The blanching of vegetables and fruits. A study of the literature and some practical applications. Food Process. Ind. *41* (490) 27–33.

VOLGER, H. G. and HEBER, U. 1975. Cryoprotective leaf proteins. Biochim. Biophys. Acta *412,* 335–349.

VOLKENSTEIN, M. W. and GOLDSTEIN, B. N. 1966. A new method for solving the problems of stationary kinetics of enzymological reactions. Biochim. Biophys. Acta *115,* 471–477.

VON HOFSTEN, B. and LALASIDIS, G. 1976. Protease-catalyzed formation of plastein products and some of their properties. J. Agric. Food Chem. *24,* 460–465.

VON LOESECKE, H. W. 1954. Orange juice. *In* Chemistry and Technology of Fruit and Vegetable Juice Products. D.K. Tressler and M.A. Joslyn (Editors). AVI Publishing Co., Westport, Conn.

VON LOESECKE, H. W. 1955. Drying and Dehydrated Foods, 2nd Edition. Reinhold Publishing Co., New York.

VON STOCKAR, U., DER YANG, R. and WILKE, C. R. 1977. Computation of the fraction of induced cells in enzyme induction systems. Biotechnol. Bioeng. *19,* 445–458.

VON WETTSTEIN, D., JENDE-STRID, B., AHRENST-LARSEN, B. and SOERENSEN, J. A. 1977. Biochemical mutant in barley renders chemical stabilization of beer superfluous. Carlsberg Res. Commun. *42,* 341–351.

VON WILLERT, D. J. 1975. The role of inorganic phosphate in the regulation of phosphoenolpyruvate carboxylase of *Mesembryanthemum crystallinum* L. Planta *122,* 273–280. (German)

VOS, L. DE and PILNIK, W. 1973. Pectolytic enzymes in apple juice extraction. Process Biochem. *8* (8) 18–19.

VOSE, J. R. 1972. The fractionation of two glucosinolases from *Sinapis alba* seed by isoelectric focussing. Phytochemistry *11,* 1649–1653.

VREMAN, H. and CORSE, J. 1976. Personal communication. Albany, California.

VRETBLAD, P. 1974. Immobilization of ligands for biospecific affinity chromatography via their hydroxyl groups: the cyclohexa-amylose-α-amylase system. FEBS Lett. *47,* 86–89.

WAGENKNECHT, A. C. and LEE, F. A. 1958. Enzyme action and off-flavor in frozen peas. Food Res. Sci. *23,* 25–31.

WAGENKNECHT, A. C., LEE, F. A. and BOYLE, F. P. 1952. The loss of chlorophyll in green peas during frozen storage and analysis. Food Res. *17,* 343–350.

WAGER, H. G. 1963. Role of phytin in the texture of cooked potatoes. J. Sci. Food Agric. *14,* 583–586.

WAGGONER, P. E. and DIMOND, P. E. 1957. Altering the pigments produced by tyrosinase and *o*-hydroxyphenols with a *m*-hydroxyphenol *4*-chlororesorcinol. Plant Physiol. *32,* 240–242.

WAGNER, G., YANG, J. C. and LOEWUS, F. A. 1975. Stereoisomeric characterization of tartaric acid produced during L-ascorbic acid metabolism in plants. Plant Physiol. *55,* 1071–1073.

WAGNER, J. R., MIERS, J. C., SANSHUK, D. W. and BECKER, R. 1968. Consistency of tomato products. 4. Improvement of the acidified hot break process. Food Technol. *22,* 1484–1489.

WAHLEFELD, A. W., MÖLLERING, H., GRUBER, W. and BERNT, E. 1973. Method and reagents for determining triglycerides. Ger. Pat. 2,229,849. Dec. 20.

WAKABAYASHI, Y., IWASHIMA, A. and NOSE, Y. 1976. Affinity chromatography of thiamin pyrophosphokinase of rat brain. Biochim. Biophys. Acta *429,* 1085–1087.

WAKU, K. and NAKAZAWA, Y. 1972. Hydrolyses of 1-O-alkyl-, 1-O alkenyl-, and 1-acyl-2 [1-^{14}C]-linoeoyl-glycero-3-phosphorylcholine by various phospholipases. J. Biochem. (Tokyo) *72,* 149–155.

WALDSCHMIDT-LEITZ, E. and DORFMANN, T. 1968. Grain proteins XVI. Beta amylase from wheat. Z. Physiol. Chem. *349,* 153–156. (German)

WALDT, L. M. and MAHONEY, R. D. 1967. Depectinizing guava juice with fungal pectinase. Food Technol. *21,* 87–89.

WALEY, S. G. 1980. Kinetics of suicide substrates. Biochem. J. *185,* 771–773.

WALKER, G. C. 1964. Color deterioration in French beans. Food Sci. *29,* 360–383.

WALKER, J. R. 1976. The control of enzymatic browning in fruit juices by cinnamic acids. J. Food Technol. *11,* 341–345.

WALKER, L. H., NIMMO, C. C. and PATTERSON, D. C. 1951. Frozen apple juice concentrate. Food Technol. *5,* 148–151.

WALL, J. A. and BLESSIN, C. H. 1970. Composition of sorghum plant and grain. *In* Sorghum Production and Utilization. J.A. Wall and W.M. Ross (Editors). AVI Publishing Co., Westport, Conn.

WALL, J. S. and JERROLD, W. 1978. Corn and sorghum grain proteins. Adv. Cereal Sci. Technol. *2,* 135–219.

WALLACE, G. M. 1973. Nutrient losses during food processing. Food Technol. N.Z. *8* (3) 11–19.

WALLACE, J. M. and WHEELER, E. L. 1972. Lipoxygenase inactivations in wheat protein concentrate by heat-moisture treatments. Cereal Chem. *49,* 92–98.

WALLACE, J. M. and WHEELER, E. L. 1979. Two lipoxygenase isoenzymes and an activator in wheat germ. Phytochemistry *18,* 389–393.

WALLERSTEIN, L. 1911. Preventing the clouding of beer on chilling by adding papain. U.S. Pat. 995,825. June 20.

WALLERSTEIN, L. 1950. Manufacture of crystalline dextrose. U.S. Pat. 2,531,999. Nov. 28.

WALLIS, V. and JAFFE, W. G. 1977. Soluble iron in vegetables. Arch. Latinoam. Nutr. *27,* 195–204.

WALLNER, S. J. and WALKER, J. E. 1975. Glycosidases in cell wall-degrading extracts of ripening tomato fruits. Plant Physiol. *55,* 94–98.

WALSH, D. E., YOUNGS, V. L. and GILLES, K. A. 1970. Inhibition of durum wheat lipoxidase with L-ascorbic-acid. Cereal Chem. *47,* 119–125.

WALTER, W. M., JR., PURCELL, A. E. and HOOVER, M. W. 1976. Changes in amyloid carbohydrates during preparation of sweet potato flakes. J. Food Sci. *41,* 1374–1377.

WALTERS, C. L., BURGER, I. H., JEWELL, G. G., LEWIS, D. F. and PARKE, D. V. 1975. Mitochondrial enzyme pathways and their possible role during curing. Z. Lebensm.-Wiss. Forsch. *158,* 193–203.

WALTKING, A. E. 1972. A reliability of the enzymic method for measuring essential fatty acid. Nutr. Rep. Int. *5,* 17–26.

WANG, C. C. and GRANT, D. R. 1969. Proteolytic enzymes in wheat flour. Cereal Chem. *46,* 537–544.

WANG, H. and MAYNARD, N. 1955. Studies on enzymatic tenderization of meat. I. Basic techniques and histological observation of enzymatic action. Food Res. *20,* 587–597.

WANG, H., WEIR, E., BIRKNER, M. and GINGER, B. 1958. Studies on enzymatic tenderization of meat. III. Histological and panel analysis of enzyme preparations from three distinct sources. Food Res. *23,* 423–438.

WANG, H. L., RUTTLE, D. I. and HESSELTINE, C. W. 1969. Milk-clotting activity of proteinases produced by *Rhizopus.* Can. J. Microbiol. *15,* 99–104.

WANG, H. L., SWAIN, E. W. and HESSELTINE, C. W. 1980. Phytase of molds used in oriental food fermentation. J. Food Sci. *45,* 1262–1266.

WANG, L. H. and KUO, Y. C. 1976. Enzymatic saccharification of bagasse pith. USA-ROC Symp. on Enzyme Technol. Aug. 9–14. Lehigh University, Bethlehem, Pa.

WANG, S. S. and DIMARCO, G. R. 1972. Isolation and characterization of the native, thermally inactivated and regenerated horseradish peroxidase isozymes. J. Food Sci. *37,* 574–578.

WANG, S. S., GALLILI, G. E., GILBERT, S. G. and LEEDER, J. G. 1974. Inactivation and regeneration of immobilized catalase. J. Food Sci. *39,* 338–341.

WANG, S. S., HAARD, N. F. and DIMARCO, G. R. 1971. Chlorophyll degradation during controlled atmosphere storage. J. Food Sci. *36,* 657–661.

WARBURG, O. 1948. Hydrogen-transporting Enzymes. Verlag Dr. W. Saenger, Berlin. (German)

WARCHALEWSKI, J. R. and TKACHUK, R. 1978. Durum wheat α-amylases: Isolation and purification. Cereal Chem. *55,* 146–156.

WARDALE, D. A. and GALLIARD, T. 1975. Subcellular localization of lipoxygenase and lipolytic acyl hydrolase enzymes in plants. Phytochemistry *14,* 2323–2329.

WARDLE, N. E. 1978. Effect of aspirin and phenols on lipo-oxygenase. IRCS Med. Sci.: Libr. Compend. *8,* 543.

WARREN, O. A. and WOODMAN, J. S. 1974. The texture of cooked potatoes. A review. J. Sci. Food Agric. *25,* 129–138.

WARRIER, S. K. B., DOKE, S. N., GORE, M. A., HARIKUMAR, P. and KUMTA, U. S. 1975. A new method for isolation of fish protein concentrate from Bombay duck *(Harpodon nehreus)* by gamma radiation and heat treatment. Lebensm.-Wiss. Technol. *8,* 87–88.

WARRIER, S. K. B., NINJOOR, V., SAWANT, P. L., HIRLEKAR, G. M. and KUMTA, U. S. 1973. Effect of nitrite and mild heat treatment on the inhibition of autolytic enzymes in lamb and in muscle tissue of the fresh water fish *Tilapia mossambica.* Fleischwirtschaft *53,* 980–982. (German)

WARWICK, M. J. and SHEARER, G. 1980. The identification and quantitation of some nonvolatile oxidation products of fatty acids developed during prolonged storage of wheat flour. J. Sci. Food Agric. *31,* 316–318.

WASTENEYS, H. and BOORSOOK, H. 1924. The enzymic synthesis of protein. I. The synthesizing action of pepsin. J. Biol. Chem. *62,* 15–25.

WATADA, A. E. 1975. Quality evaluation of fruits and vegetables by spectrophotometric technique. *In* Postharvest Biology and Handling of Fruits and Vegetables. N.F. Haard and D.K. Salunkhe (Editors). AVI Publishing Co., Westport, Conn.

WATARI, H. and ISOGAI, Y. 1976. A new plot for allosteric phenomena. Biochem. Biophys. Res. Commun. *69,* 15–18.

WATSON, J. D. 1976. Molecular Biology of the Gene, 3rd Edition. W.A. Benjamin, Menlo Park, Calif.

WATTERSON, J. G., SCHAUNBAUM, C., LOCHER, R., DI PIERRI, S. and KUTZER, M. 1975. Temperature-induced transitions in the conformation of intermediates in the hydrolytic cycle of myosin. Eur. J. Biochem. *56,* 79–90.

WATTS, B. M., KENDRICK, J., ZIPSER, M. W., HUTCHINS, B. and SALEH, B. 1966. Enzymic reducing pathways in meat. J. Food Sci. *31,* 855–862.

WEAST, C. A. and MACKINNEY, G. 1940. Chlorophyllase. J. Biol. Chem. *133,* 551–558.

WEAVER, C. and CHARLEY, H. 1974. Enzymic browning of ripening bananas. J. Food Sci. *39,* 1200–1202.

WEAVER, E. A. 1961. Preservation of tomatopectase. U.S. Pat. 2,982,697. May 2.

WEAVER, G. L., KROGER, M. and KATZ, F. 1977. Deaminated chitin affinity chromatography: A method for the isolation, purification and concentration of lysozyme. J. Food Sci. *42,* 1084–1087.

WEAVER, M. L. and HAUTALA, E. 1970. Study of hydrogen peroxide, potato enzymes and blackspot. Am. Potato J. *47,* 457–468.

WEAVER, M. L., HAUTALA, E. and REEVE, R. M. 1971. Distribution of oxidase enzymes in potato tubers relative to blackspot susceptibility. Am. Potato J. *48,* 16–20.

WEAVER, M. L., REEVE, R. M. and KNENEMAN, R. W. 1975. Frozen French fries and other frozen potato products. *In* Potato Processing, 3rd Edition. W.F. Talburt and O. Smith (Editors). AVI Publishing Co., Westport, Conn.

WEAVER, M. L., ROBERTS, J., GEORGE, J. E., JENSEN, M. and SANDAR, N. 1965. Harvesting potatoes into water in a water-tight truck bed. Am. Potato J. *42,* 147–162.

WEAVER, M. O., FANTA, G. F., DOANE, W. M. and BAGLEY, E. B. 1976. An honor for Super Slurper. Agric. Res. *24* (7) 12–13.

WEBB, A. D. and NOBLE, A. C. 1976. Aroma

of sherry wines. Biotechnol. Bioeng. *18,* 939–952.

WEBB, B. H., JOHNSON, A. H. and ALFORD, J. A. 1974. Fundamentals of Dairy Chemistry. AVI Publishing Co., Westport, Conn.

WEBB, J. L. 1963–1966. Metabolic Inhibitors, Vol. 1–3. Academic Press, New York.

WEBER, F., ARENS, D. and GROSCH, W. 1973A. Identification of lipoxygenase isoenzymes as carotene oxidases. Z. Lebensm.-Unters. Forsch. *152,* 152–154. (German)

WEBER, F., LASKAWY, G. and GROSCH, W. 1973B. Enzymatic carotene destruction in peas, soybeans, wheat and flaxseed. Z. Lebensm.-Unters. Forsch. *52,* 324–331. (German)

WEBER, G. 1963–1975. Advances in Enzyme Regulations, Vol. 1–13. Pergamon Press, Elmsford, N.Y.

WEBER, K. and OSBORN, M. 1969. The reliability of molecular weight determination by dodecyl sulfate-polyacrylamide gel electrophoresis. J. Biol. Chem. *244,* 4406–4412.

WEBER, M., FOGLIETTI, M. J. and PERCHERON, F. 1976. Purification of α-amylases by affinity chromatography on crosslinked starch. Biochimie *58,* 1299–1302. (French)

WEBER, M. M. 1971. Activation of enzymes in meat. U.S. Pat. 3,561,976. Feb. 9.

WEETAL, H. H. 1972–1975. Immobilized Enzymes: A Compendium of References from the Recent Literature Prepared for Corning Glass Works by the New England Application Center, Univ. of Conn., Storrs.

WEETALL, H. H. 1980. Enzyme Engineering, Vol. 5. Plenum Press, New York.

WEETAL, H. H. and DETAR, C. C. 1974. Immobilized tannase. Biotechnol. Bioeng. *16,* 1095–1102.

WEIBEL, M. K., BARRIUS, R., DELLOTO, R. and HUMPHREY, A. E. 1975. Immobilized enzymes: pectin esterase covalently coupled to porous glass particles. Biotechnol. Bioeng. *17,* 85–98.

WEIEL, J. and HAPNER, K. D. 1976. Barley proteinase inhibitors: a possible role in grasshopper control? Phytochemistry *15,* 1885–1887.

WEIL, L. and SEIBLES, J. S. 1955. Photooxidation of crystalline ribonuclease in the presence of methylene blue. Arch. Biochem. Biophys. *54,* 368–377.

WEINFURTNER, F., WULLINGER, F. and PIENDL, A. 1966. Barley variety—cultivation conditions—wort composition—fermentation pattern. 5. Mechanism of enzyme formation in germinating barley. Brauwissenschaft *19,* 390–395. (German)

WEISER, H. H., MOUNTNEY, G. J. and GOULD, W. A. 1971. Practical Food Microbiology and Technology, 2nd Edition. AVI Publishing Co., Westport, Conn.

WEIZMANN, C. 1936. Yeast or protein preparations. Br. Pat. 450,529. July 20.

WEIZMANN, C. 1939. Protein products. Br. Pat. 509,495. July 17.

WELCH, B. I. and MARTIN, N. F. 1974. Evidence of pectinase activity between Cronartium ribicola and Pinus monticola. Phytopathology *64,* 1287–1289.

WELLNER, D. 1967. Flavoproteins. Annu. Rev. Biochem. *36,* 669–690.

WELLS, W. D. and JAMES, G. P. 1976. Rapid dextrans formation in stale cane and its processing consequences. Proc. Queensl. Soc. Sugar Cane Technol. *43,* 287–301.

WELSH, J. D. 1978. Diet therapy in adult lactose malabsorption: present practice. Am. J. Clin. Nutr. *31,* 592–596.

WENGENMAYER, H., EBEL, J. and GRISEBACH, H. 1976. Enzymatic synthesis of lignin precursors. Purification and properties of a cinnamoyl-CoA-NADPH reductase from cell suspensions of soybeans *(Glycine max).* Eur. J. Biochem. *65,* 529–536.

WEST, S. B., HUANG, M. Y. and LU, H. 1977. Liver microsomal DT diaphorase of benzo[A]-pyrene. Fed. Proc. *36,* 959.

WESTERMARK, U. and ERIKSSON, K. -E. 1975. Purification and properties of cellobiose: quinone oxidoreductase from *Sporotrichum pulverulentum.* Acta Chem. Scand. *B29,* 419–424.

WESTERMARK-ROSENDAHL, C., JUNNILA, L. and KOIVISTOINEN, P. 1979. Efforts to improve baking properties of sprout-damaged wheat by reagents reducing α-amylase activity. I. Screening tests by the Falling Number method. Lebensm.-Wiss. Technol. *12,* 321–324.

WESTHEIMER, F. A. 1959. Enzyme models. *In* The Enzymes, 2nd Edition, Vol. 1. P.D. Boyer, H.A. Lardy and K. Myrbäck (Editors). Academic Press, New York.

WESTHOFF, D. C. and COWMAN, R. A. 1971. Substrate specificity of the intracellular proteinase from a slow acid producing mutant of *Streptococcus lactis.* J. Dairy Sci. *54,* 1265–1296.

WESTICK, W. J., ALLSOP, J. and WATTS, R. W. E. 1974. The effect of gold salts on the purine:pyrophosphatephosphoribosyltransferase enzymes of human blood cells. Biochem. Pharmacol. *23,* 163–165.

WESTLEY, J. 1969. Enzyme Catalysis. Harper & Row, New York.

WETLAUFER, D. B. and RISTOW, S. 1973.

Acquisition of the three dimensional structure of proteins. Annu. Rev. Biochem. *42,* 135–158.

WEURMAN, C. 1961. Gas-liquid chromatographic studies on the formation of volatile compounds in raspberries. Food. Technol. *15,* 531–536.

WEURMAN, C. and SWAIN, T. 1953. Chlorogenic acid and enzymic browning of apples and pears. Nature (London) *172,* 678.

WHEELER, E. L. and WALLACE, J. M. 1978. Kinetics of wheat germ lipoxygenase adsorbed to hydrophobic surfaces. Phytochemistry *17,* 41–44.

WHEELOCK, J. V. and PENNY, J. P. 1972. Role of the primary phase of rennin action in the clotting of milk. J. Dairy Res. *39,* 23–36.

WHITAKER, J. R. 1970. Protease of *Endothia parasitica.* Methods Enzymol. *19,* 436–445.

WHITAKER, J. R. 1972A. Principles of enzymology for the Food Sciences. Marcel Dekker, New York.

WHITAKER, J. R. 1972B. The next 50 years in meat science research. J. Anim. Sci. *34,* 957–959.

WHITAKER, J. R. 1973. Some recent developments in enzymology. Food Technol. *27* (4) 16–26.

WHITAKER, J. R. 1974A. Food Related Enzymes. Am. Chem. Soc., Washington, D.C.

WHITAKER, J. R. 1974B. Analytical applications of enzymes. *In* Food Related Enzymes. J.R. Whitaker (Editor). Am. Chem. Soc., Washington, D.C.

WHITAKER, J. R. 1975. Formal course in food enzymology. Catalyst *1* (2) 1–9.

WHITAKER, J. R. 1976. Development of flavor, odor, and pungency in onion and garlic. Adv. Food Res. *22,* 73–133.

WHITAKER, J. R. 1977. Review of: "The New Enzyme-Catalyst Diet: Amazing Way to Quick Permanent Weight Loss" by C. Wade. Food Technol. *31* (11) 118.

WHITE, B. B. and OUGH, C. S. 1973. Oxygen uptake studies on grape juice. Am. J. Enol. Vitic. *24,* 148–152.

WHITE, C. H. and MARSHALL, R. T. 1973. Heat stable protease from *Pseudomonas-fluorescens* P-26 degrades ultra high temperature pasteurized milk. J. Dairy Sci. *56,* 624.

WHITE, J. 1977. Specific determination of sucrose in honey. J. Assoc. Off. Anal. Chem. *60,* 669–672.

WHITE, J. L. 1967. Packaged meat product. U.S. Pat. 3,331,692. June 18.

WHITE, S. L. 1976. Titration of the carboxyhemoglobin tetramer-dimer equilibrium by inositol hexaphosphate. J. Biol. Chem. *251,* 4763–4769.

WHITESIDES, G. M., LAMOTTE, A., ADALSTEINSSON, O. and COLTON, C. K. 1976. Covalent immobilization of adenylate kinase and acetate kinase in a polyacrylamide gel: enzymes for ATP regeneration. Methods Enzymol. *44,* 887–897.

WHITHEY, R. J. 1969. Lanthanum ion catalysis of nucleophilic displacement reactions of monoesters of methylphosphonic acid. Can. J. Chem. *47,* 4383–4387.

WHITTENBERG, B. A., WHITTENGER, J. B. and CALDWELL, P. R. B. 1975. Role of myoglobin in the oxygen supply to red skeletal muscle. J. Biol. Chem. *250,* 9038–9043.

WHITTENBERGER, R. T. and NUTTING, C. G. 1958. High viscosity of cell wall suspensions prepared from tomato juice. Food Technol. *12,* 421–424.

WICKLOW, D. T., DETROY, R. W. and ADAMS, F. 1980. Differential modification of the lignin and cellulose composition in wheat straw from fungal colonists of ruminant digestion: Ecological implications. Mycologia *72,* 1065–1076.

WICKREMASINGHE, R. L. 1974. The mechanism of operation of climatic factors in the biogenesis of tea flavor. Phytochemistry *13,* 2057–2063.

WIEG, A. J. 1970. Technology of barley brewing. Process Biochem. *5* (8) 46–50.

WIEG, A. J. 1975. Enzymic treatment of raw grains. Brew-N-zymes for enzymic treatment of raw grains. Fermentation *71,* 175–184. (French)

WIEG, A. J. 1976. Improvement of beer filtration by the addition of enzymes. Process Biochem. *11* (6) 27–29.

WIEKER, H. J., JOHANNES, K. J. and JESS, B. 1970. Computer programs for the determination of kinetic parameters from sigmoidal steady-state kinetics. FEBS Lett. *8,* 178–185.

WIELAND, H. 1972. Food Processing Review No. *23.* Enzymes in Food Processing and Products. Noyes Data Corp., Park Ridge, N.J.

WIEN, E. M., VAN CAMPEN, D. R. and RIVERS, J. M. 1975. Factors affecting the concentration and availability of iron in turnip greens to rats. J. Nutr. *105,* 459–466.

WIENER, C. 1974. The flavourist as biochemist. Flavour Ind. *5,* 237–238.

WIERZBICKI, L. E., EDWARDS, V. H. and KO-

SIKOWSKI, F. V. 1974. Hydrolysis of lactose in acid whey by lactase bound to porous glass particles in tubular reactors. J. Food Sci. *39,* 374–378.

WIERZBICKI, L. E. and KOSIKOWSKI, F. V. 1973A. Kinetics of lactose hydrolysis in acid whey by β-galactosidase from *Aspergillus niger.* J. Dairy Sci. *56,* 1396–1399.

WIERZBICKI, L. E. and KOSIKOWSKI, F. V. 1973B. Lactase potential of various microorganisms grown in whey. J. Dairy Sci. *56,* 26–32.

WIESEMANN, M. L. and BINKLEY, F. 1977. Investigation of γ-glutamyl transferase in human milk. Fed. Proc. *36,* 792.

WILCHEK, M. and HEXTER, S. 1976. The purification of biologically active compounds by affinity chromatography. Methods Biochem. Anal. *23,* 347–385.

WILD-ALTAMIRANO, C. 1969. Enzymic activity during growth of vanilla fruit. 1. Proteinase, glucosidase, peroxidase, and polyphenoloxidase. J. Food Sci. *34,* 325–329.

WILDENRADT, H. L. and SINGLETON, V. L. 1974. Production of aldehydes as a result of oxidation of polyphenolic compounds and its relation to wine aging. J. Enol. Viticult. *25,* 19–26.

WILDER, C. J. 1962. Factors affecting heat inactivation and partial reactivation of peroxidase purified by ion-exchange chromatography. J. Food Sci. *27,* 567–573.

WILDING, M. D. 1960. Non-curdling margarine. U.S. Pat. 3,250,628. May 10.

WILEY, R. C. 1977. Uses of endogenous enzymes in fruit and vegetable processing. *In* Enzymes in Food and Beverage Processing. R.I. Ory and A.J. St. Angelo (Editors). Am. Chem. Soc., Washington, D.C.

WILKE, C. R. 1975. Cellulose as a Chemical and Energy Resource. Biotechnol. Bioeng. Symp. 5. John Wiley & Sons, New York.

WILKE, C. R. and MITRA, G. 1976. Conversion of cellulosic materials to sugar. U.S. Pat. 3,972,775. Aug. 3.

WILKE, C. R. and YANG, R. D. 1975. Process for enzymatic hydrolysis of newsprint. Presented at 8th Cellulose Conf. on Wood Chemical, May. Syracuse, N.Y.

WILKIE, K. C. B. 1979. The hemicelluloses of grasses and cereals. Adv. Carbohydr. Chem. Biochem. *36,* 215–264.

WILKIE, N. M. 1976. Restriction endonuclease technology. Heredity *36,* 287.

WILKINS, W. F. 1961. The isolation and identification of the lachrymogenic compounds of onion. Cornell Univ. Ph.D. Thesis. University Microfilms, Ann Arbor, Mich.

WILKINSON, B. R. and DORRINGTON, R. E. 1975. Lysozyme (muramidase) from waste egg white. Process Biochem. *10* (2) 24–25.

WILKINSON, K. D. and ROSE, I. W. 1980. Glucose exchange and catalysis by two crystalline hexokinase·glucose complexes. J. Biol. Chem. *255,* 7569–7574.

WILKINSON, T. J., MARIANO, P. S. and GLOVER, G. I. 1976. Affinity chromatographic separations of chemically modified α-chymotrypsins from α-chymotrypsin. Sep. Sci. *11,* 383–385.

WILLIAMS, A. A. and TUCKNOTT, O. G. 1978. The volatile aroma of fermented ciders: minor neutral components from the fermentation of sweet coppin apple juice. J. Sci. Food Agric. *29,* 381–397.

WILLIAMS, B. E. 1962. Processes for the tenderizing of meat. U.S. Pat. 3,056,680. Oct. 2.

WILLIAMS, B. E. 1963. Tenderizing and preserving meat. U.S. Pat. 3,076,712. Feb. 5.

WILLIAMS, B. E. 1964A. Processes for improving the tenderness of meat using a cold water buffered enzyme. U.S. Pat. 3,156,566. Nov. 10.

WILLIAMS, B. E. 1964B. Processes for tenderizing meat employing water and gas under pressure. U.S. Pat. 3,119,696. Jan. 28.

WILLIAMS, B. E. 1964C. Process for improving the texture of meat by aqueous injection containing cellulose gum. U.S. Pat. 3,147,122. Sept. 1.

WILLIAMS, D. C. 1972. Starfish proteolytic enzymes. Abstr., Pacific Slope Biochemical Conf. June. Davis, Calif.

WILLIAMS, D. C., III, HUFF, G. F. and SEITZ, W. R. 1976. Evaluation of peroxyoxalate chemiluminescence for determination of enzyme generated peroxide. Anal. Chem. *48,* 1003–1006.

WILLIAMS, P. C. and LA SEELEUR, G. C. 1970. Determination of damaged flour: Cooperative study of present day procedures. Cereal Sci. Today *15,* 4–9.

WILLIAMS, P. J. and STRAUSS, C. R. 1978. The influence of film yeast activity on the aroma volatiles of flor sherries—a study of volatiles isolated by headspace sampling. J. Inst. Brew. *84,* 148–152.

WILLSTÄTTER, R. 1922. On enzyme isolation. Ber. Dtsch. Chem. Ges. *55,* 3601–3623. (German)

WILLSTÄTTER, R. and ROHDEWALD, M. 1932. On desmo- and lyo-trypsin of colorless blood corpuscles. Z. Physiol. Chem. *204,* 181–192. (German)

WILLSTÄTTER, R. and WALDSCHMIDT-LEITZ, E. 1923. Concerning pancreatic lipase. Z. Physiol. Chem. *125,* 132–198. (German)

WILLSTÄTTER, R., WALDSCHMIDT-LEITZ, E. and HESSE, A. 1923. Adsorption behavior of pancreas amylase. Z. Physiol. Chem. *125,* 93–104. (German)

WILSON, A. K., SALUNKHE, D. K. and WILCOX, E. B. 1973. Effects of water composition on quality of canned apple slices, tart cherries, and green beans. Confructa *18,* 21–31.

WILSON, C. M. 1975. Plant nucleases. Annu. Rev. Plant Physiol. *26,* 187–208.

WILSON, G. D. 1960. Factors influencing quality of fresh meats. *In* The Science of Meat and Meat Products. Am. Meat Inst. Found. (Editor). W.H. Freeman and Co., San Francisco.

WILSON, I. B. and CABIB, E. 1956. Acetylcholinesterase: enthalpies and entropies of activation. J. Am. Chem. Soc. *78,* 202–207.

WINGARD, L. B., KATCHALSKI-KATZIR, E. and GOLDSTEIN, L. 1976. Immobilized Enzyme Principles. Academic Press, New York.

WINGARD, L. M., JR., BEREZIN, I. V. and KLYOSOV, A. A. 1980. Enzyme Engineering. Future Directions. Plenum Press, New York.

WINGER, R. J. and FENNEMA, O. 1976. Tenderness and water holding properties of beef muscle as influenced by freezing and subsequent storage at -3 or 15°C. J. Food Sci. *41,* 1433–1438.

WINTER, E. 1969. Behavior of peroxidase during blanching of vegetables. Z. Lebensm.-Unters. Forsch. *141,* 201–208. (German)

WINTER, F. H. 1976. Critical points in food processing. *In* Proc. Symp. on Enzymes in the Food Processing Industry, Jan. Univ. Calif. Extension, Davis.

WISEMAN, A. 1971. Microbial intracellular enzymes. Process Biochem. *6,* 27–28.

WISEMAN, A. 1973. Industrial enzyme stabilization. Process Biochem. *8* (8) 14–15.

WISEMAN, A. 1975. Industrial practice with enzymes. *In* Handbook Enzyme Biotechnology. A. Wiseman (Editor). Ellis Horwood, Chichester, England.

WISEMAN, A. 1978. Enzymic methods in food analysis. Dev. Food Anal. Tech. *1,* 179–195.

WISEMAN, A. and GOULD, B. 1968. New enzymes for industry. New Sci. *38* (66) 66–68.

WISEMAN, A. and WOODWARD, J. 1975. Industrial yeast invertase stabilization. Process Biochem. *10* (6) 24–30.

WITTENBERG, B. A., WITTENBERG, J. B. and CALDWELL, R. B. 1975. Myoglobin in muscle oxygen consumption. Fed. Proc. *34,* 653.

WITTMACK, C. 1878. The fermentative action of the juice of the fruit of *Carica papaya.* Pharm. J. *9,* 449–451.

WOHLGEMUTH, J. 1908. A new method for the quantitative determination of amylolytic ferments. Biochem. Z. *9,* 1–9. (German)

WOJCIK, S., JANAS, J., POLONIS, A., SABA, L. and BIALKOWSKI, Z. 1973. Hematological and biochemical indices in the blood of young fattening cattle receiving enzymatic preparations. Pol. Arch. Weter. *16,* 245–254. (Polish)

WOLF, W. J. 1975. Lipoxygenase and flavor of soybean products. J. Agric. Food Chem. *32,* 136–141.

WOLF, W. J. and COWAN, J. C. 1975. Soybeans as a Food Source, Revised Edition. CRC Press, Boca Raton, FL.

WOLFENDEN, R. 1972. Analog approaches to the structure of the transition state in enzyme reactions. Acc. Chem. Res. *5,* 10–18.

WOLFENDEN, R. 1976. Transition state analog inhibitors and enzyme catalysis. Annu. Rev. Biophys. Bioeng. *5,* 271–306.

WOLFENDEN, R. 1978. Transition state affinity as a basis for the design of enzyme inhibitors. *In* Transition State Biochemical Processes. R.D. Gandour and R.L. Schowen (Editors). Plenum Press, New York.

WOLNAK, B. 1972. Present and future technological status of enzymes. U.S.N.T.I.S.P.B. Rep. *219636/8,* Springfield, Va.

WOLNAK, B. 1974. Survey of the enzyme industry. *In* Enzyme Engineering, Vol. 2. E.K. Pye and L.B. Wingard (Editors). Plenum Press, New York.

WOLPERT, J. S. and ERNST-FONBERG, M. L. 1975. Dissociation and characterization of enzymes from a multienzyme complex involved in CO_2 fixation. Biochemistry *14,* 1103–1107.

WONG, J. T. 1975. Kinetics of Enzyme Mechanisms. Academic Press, New York.

WONG, J. T. and HANES, C. S. 1962. Kinetic formulations for enzymic reactions involving two substrates. Can. J. Biochem. Physiol. *40,* 763–804.

WONG, N. P., LACROIX, D. E., VESTAL, J. H. and ALFORD, J. A. 1977. Composition of Cheddar cheese made with different milk clotting enzymes. J. Dairy Sci. *60,* 1522–1527.

WONG, T. 1975. Color changes and their control. Presented at Symposium on Enzyme Problems in the Fruit and Vegetable Industry. Univ. Calif., Davis, Jan. 14.

WOOD, B. J. B. and INGRAHAM, L. L. 1965. Labelled tyrosinase from labelled substrate. Nature *203,* 291–292.

WOOD, T. 1961. The browning of ox-muscle extracts. J. Sci. Food Agric. *12,* 61–69.

WOOD, T. M. 1975. Properties and mode of action of cellulases. Biotechnol. Bioeng. Symp. *5,* 111–137.

WOOD, W. I. 1976. Tables for the preparation of ammonium sulfate solutions. Anal. Biochem. *73,* 250–257.

WOODROOF, J. G. and LUH, B. S. 1975. Commercial Fruit Processing. AVI Publishing Co., Westport, Conn.

WOODS, A. E. 1902. Observations on plant peroxidase. U.S. Dep. Agric. Bull. *18.*

WOODWARD, J. D. 1978. Enzymes in practical brewing. Brew. Dig. *53* (5) 38–44.

WOODWARD, J. D. and BENNETT, A. B. 1976. Improvements in or relating to beer production. Br. Pat. 1,421,955. Jan. 21.

WOODWARD, J. D. and BOS, C. 1973. Preparation of barley syrup by means of bacterial enzymes. Schweiz. Brau.-Rundsch. *84,* 26–29. (German)

WOOLFOLK, C. A. 1975. Metabolism of *N*-methylpurines by a *Pseudomonas putida* strain isolated by enrichment on caffeine as the sole source of carbon and nitrogen. J. Bacteriol. *123,* 1088–1106.

WORLD HEALTH ORGAN. 1972. WHO Food Additives Series No. 2. Specifications for the identity and purity of some enzymes and certain other substances. World Health Organ., Geneva.

WORLD HEALTH ORGAN. 1975. WHO Food Additive Series, No. 6. Toxicological evaluation of some food colors, enzymes, flavour enhancers, thickening agents, and certain food additives. World Health Organ., Geneva.

WORTHINGTON BIOCHEMICAL CORP. 1972, 1976. Worthington Enzyme Manual. Freehold, N.J.

WOYCHIK, J. H., WONOLOWSKI, M. V. and DAHL, K. J. 1974. Preparation and application of immobilized beta-galactosidase. *In* Immobilized Enzymes in Food and Microbial Processes. A.C. Olson and C.L. Cooney (Editors). Plenum Press, New York.

WOZENSKI, J. and WOODBURN, M. 1975. Phytic acid and phytase activity in four cottonseed protein products. Cereal Chem. *2,* 665–669.

WU, A. C. M., EITENMILLER, R. R. and POWERS, J. J. 1975. Yields from chymotrypsin and lysozyme under fluctuating temperature treatments. J. Food Sci. *40,* 840–843.

WU, C. -S. C. and SAYRE, R. N. 1971. Myosin stability in intact chicken muscle and a protein component released after aging. J. Food Sci. *36,* 133–137.

WU, C. -S. and YANG, J. T. 1975. Conformation of Ca^{++} BIN binding muscle proteins: Troponin C and DTNB light chains of myosin. Fed. Proc. *34,* 538.

WU, M. T. and SALUNKHE, D. K. 1972A. Control of chlorophyll and solanine syntheses and sprouting of potato tubers by hot paraffin wax. J. Food Sci. *37,* 629–630.

WU, M. T. and SALUNKHE, D. K. 1972B. Control of chlorophyll and solanine formation in potato tubers by oil and diluted oil treatments. HortScience *7,* 466–467.

WUTZEL, H. 1976. Enzymic analysis for the control of sour dough. Getreide Mehl Brot *30,* 70–72. (German)

WYKES, J. R., DUNNILL, P. and LILLY, M. D. 1971. Conversion of tyrosine to L-dihydroxyphenylalanine using immobilized tyrosinase. Nature New Biol. *230,* 187.

WYNN, C. H. 1973. The Structure and Function of Enzymes. Edward Arnold, London.

YABUMOTO, K., JENNINGS, W. G. and YAMAGUCHI, M. 1977. Volatile constituents of canteloupe, *Cucumis melo,* and their biogenesis. J. Food Sci. *42,* 32–37.

YADAV, N. K. and VADHERA, D. V. 1977. Mechanism of egg white resistance to bacterial growth. J. Food Sci. *42,* 97–99.

YAGI, K. and YAMANO, T. 1980. Flavins and Flavoproteins, Vol. 6. University Park Press, Baltimore.

YAHIRO, A. T., LEE, S. M. and KIMBLE, D. O. 1964. Bioelectrochemistry. I. Enzyme utilizing bio-fuel cell studies. Biochim. Biophys. Acta *88,* 375–383.

YAKUSHKINAN, I. and DULIN, A. F. 1977. Photosynthetic activity of barley seedlings treated with kinetin and gibberellin. Chem. Abstr. *87,* 1109a. (Russian)

YALOW, R. S. 1978. Radioimmunoassay: a probe for the fine structure of biologic systems. Science *200,* 1236–1254.

YAMADE, H., ISOBE, K., TANI, Y. and HIROMI, K. 1979. A differential determination for spermine and spermidine with beef plasma oxidase. Agric. Biol. Chem. *43,* 2487–2491.

YAMAGATA, M., HORIMOTO, K. and NAGAOKA, C. 1971. Accuracy of predicting occurrence of greening in tuna based on content of trimethylamine oxide. J. Food Sci. *36,* 55–57.

YAMAGUCHI, M., HUGHES, D. L. and HOWARD, F. D. 1960. Effect of color and intensity of fluorescent lights and applications of chemicals and waxes on chlorophyll development of White Rose potatoes. Am. Potato J. *37,* 229–236.

YAMAGUCHI, M., PERDUE, J. W. and MAC GILLVRAY, J. H. 1960. Nutrient composition of White Rose potatoes during growth and after storage. Am. Potato J. *37,* 73–76.

YAMAGUCHI, M. H., SHANNON, S., HOWARD, F. D. and JOSLYN, M. A. 1965. Factors affecting the formation of a pink pigment in purees of onions. Proc. Am. Soc. Hortic. Sci. *86,* 475–483.

YAMAGUCHI, M. H., TIMM, H. and CLEGG, H. D. 1966. Effect of maturity and postharvest conditions on sugar conversion and drip quality of potato tubers. Proc. Am. Soc. Hortic. Sci. *89,* 456–463.

YAMAGUCHI, N. and FUJIMAKI, M. 1973. Browning reaction from reducing sugars and amino acids. XII. Changes in antioxidative activities of browning reaction products of D-xylose and L-amino acids during storage. J. Food Sci. Technol. (Tokyo) *20,* 489–491. (Japanese)

YAMAKAGO, T. 1973 . Effect of pronase on vitamin B-12 absorption. J. Kurume Med. Assoc. *36,* 1159. (Japanese)

YAMAKI, S. and MATSUDA, K. 1977. Changes in the activities of some cell wall-degrading enzymes during development and ripening of Japanese pear fruit. Plant Cell Physiol. *18,* 81–93.

YAMAMOTO, H. Y., STEINBERG, M. P. and NELSON, A. T. 1962. Kinetic studies of the heat inactivation of peroxidase. J. Food Sci. *27,* 113–119.

YAMAMOTO, K., SAMEJIMA, K. and YASUI, T. 1977. A comparative study of the changes in hen pectoral muscle during storage at 4° and at -20°. J. Food Sci. *42,* 1642–1645.

YAMAMOTO, O. 1977. Ionizing radiation-induced crosslinking in proteins. *In* Protein Crosslinking, Part A. M. Friedman (Editor). Plenum Press, New York.

YAMAMOTO, S., MINODA, Y. and YAMADA, K. 1973. Formation and some properties of the acid phosphatase in *Aspergillus terreus.* Agric. Biol. Chem. *37,* 2719–2726.

YAMAMOTO, Y., KADOWAKI, Y., ENDO, K. and KISHIDA, K. 1967. Studies on the taste substances in foods. IV. Purification and properties of an alkaline phosphatase in carp muscle. Chem. Abstr. *68,* 2061y. (Japanese)

YAMANAKA, H., TAKAMIZAWA, M. and AMANO, K. 1973. Relation between the color of tuna meat and the activity of metmyoglobin reductase I. and II. Bull. Jpn. Soc. Fish. *39,* 667, 673. (Japanese)

YAMANAKA, K. 1975. Specific micromethod for the determination of fructose with D-mannitol dehydrogenase. Chem. Abstr. *83,* 175066s. (Japanese)

YAMANAKA, Y. 1975. Effect of β-glucosidase on the enzymatic hydrolysis of cellulose. Lawrence Berkeley Lab. Publ. *LBL-4413.* Ph.D. Dissertation. Dep. Chem. Eng., Univ. California, Berkeley.

YAMANAKA, Y., CARROAD, P. A., RIAZ, M. and WILKE, C. R. 1977. Decomposition of lignin and cellobiose in relation to the enzymatic hydrolysis of cellulose. Lawrence Berkeley Lab. Publ. *LBL-5960.* University of California, Berkeley.

YAMANISHI, T., KAWATSU, M., YOKOYAMA, T. and NAKATANI, Y. 1973. Methyl jasmonate and lactones including jasmine lactone in Ceylon tea. Agric. Biol. Chem. *37,* 1075–1078.

YAMASAKI, I. 1974. Peroxidase. *In* Molecular Mechanisms of Oxygen Activation. O. Mayashi (Editor). Academic Press, New York.

YAMASAKI, M., YASUI, T. and ARIMA, K. 1964. Pectic enzymes in the clarification of apple juice. 1. Study on the clarification in a simplified model. Agric. Biol. Chem. *28,* 779–787.

YAMASAKI, Y., SUZUKI, Y. and OZAWA, J. 1977. Purification and properties of two forms of glucoamylase from *Penicillium oxalicum.* Agric. Biol. Chem. *41,* 755–762.

YAMASHITA, I., IINO, K., NEMOTO, Y. and YOSHIKAWA, S. 1977. Studies on flavor development in strawberries. 4. Biosynthesis of volatile alcohol and esters from aldehyde during ripening. J. Agric. Food Chem. *25,* 1165–1168.

YAMASHITA, I., NEMOTO, Y. and YOSHIKAWA, S. 1975. Formation of volatile esters in strawberries. Agric. Biol. Chem. *39,* 2303–2307.

YAMASHITA, I., NEMOTO, Y. and YOSHIKAWA, S. 1976. NAD-dependent alcohol dehydrogenase and NADP-dependent alcohol dehydrogenase from strawberry seeds. Agric. Biol. Chem. *40,* 2331–2335.

YAMASHITA, M., ARAI, S. and FUJIMAKI, M. 1976. Plastein reaction for food protein improvement. J. Agric. Food Chem. *24,* 1100–1104.

YAMATSU, O., TOBARI, M. and SHIMAZONO,

H. 1966. Studies on enzymes produced by *Trametes sanguina.* J. Ferment. Technol. *44,* 847–853.

YANG, F. F. 1980. Regulation of ethylene biosynthesis. HortScience *15,* 238–243.

YANG, J. C. and LOEWUS, F. A. 1975. Metabolic conversion of L-ascorbic acid to oxalic acid in oxalate-accumulating plants. Plant Physiol. *56,* 283–285.

YANG, M. G., BUSH, L. J. and ODELL, G. U. 1962. Enzyme supplementation of rations of dairy calves. J. Agric. Food Chem. *10,* 332–334.

YANG, S. 1976. Biochemistry of ethylene formation. Presented at USDA West. Reg. Res. Cent., Sept.

YATSU, L. Y. and JACKS, T. J. 1968. Association of lysosomal activity with aleurone grains in the plant seeds. Arch. Biochem. Biophys. *124,* 466–471.

YEATMAN, J. N. 1972. Physiological aspects of texture perception, including mastication. Food Technol. *26,* 141–147.

YEDA RES. & DEV. CO. 1971. Chicken pepsin. Isr. Pat. 30,520. July 3.

YEUNG, K. K. and CARRICO, R. J. 1976. Purification of malic enzymes by affinity chromatography on immobilized n^6-(6-aminohexyl)-adenosine 2′,5′-bisphosphate. Anal. Biochem. *74,* 369–375.

YOKOE, Y. 1963. Animal cellulases. Seibatsu Kagaku *15,* 151–156. (Japanese)

YOKOTSUKA, T., IWAASA, T. and FRIJU, M. 1974. Process for producing a protein hydrolysate. U.S. Pat. 3,852,479. Dec. 3.

YOKOYAMA, H. 1975. Better citrus with bioregulators. Agric. Res. *23* (7) 6.

YORK, G. K., O'BRIEN, M., TROMBROPOULOS, D., WINTER, F. H. and LEONARD, S. J. 1967. Relation of fruit damage to quality and consistency of tomato concentrates. Food Technol. *21,* 69–71.

YOSHIDA, S. 1969. Biosynthesis and conversion of aromatic amino acids in plants. Annu. Rev. Plant Physiol. *20,* 41–62.

YOSHIKAWA, S., NISHIMARU, S., TASHIRO, T. and YOSHIDA, M. 1970. Collection and classification of words of the description of food texture. J. Texture Stud. *1,* 437–463.

YOSHIOKA, M., TACHINABA, K., and KANEDA, T. 1974. Toxicity of autoxidized oils. IV. Impairments of metabolic functions induced by autoxidized methyl linoleate. Yukagaku *23,* 327–331. (Japanese)

YOSHITAKE, J., SHIMAMURA, M., ISHIZAKI, I. and IRIE, Y. 1976. Xylitol production by *Enterobacter liquefaciens.* Agric. Biol. Chem. *40,* 1493–1503.

YOUNG, L. S. 1976. Preparation, characterization, and performance of an immobilized multipectic enzyme system. Diss. Abstr. Int. B *37,* 1166–1167.

YOUNG, V. R., RAND, W. M. and SCRIMSHAW, N. S. 1977. Measuring protein quality in humans. A review and proposed method. Cereal Chem. *54,* 929–948.

YOUNT, R. G., FRYE, J. S. and O'KEEFE, K. R. 1973. Inhibition of heavy meromyosin by purine disulfide analogs of adenosine triphosphate. Cold Spring Harbor Symp. Quant. Biol. *37,* 113–119.

YU, J., TAMURA, G. and ARIMA, K. 1971. Milk-clotting enzymes form microorganisms. IX. Agric. Biol. Chem. *35,* 1398–1401.

YU, M. H., WU, M. T., WANG, D. J. and SALUNKHE, D. K. 1974. Nonenzymatic browning in synthetic systems containing ascorbic acid, amino acids, organic acids, and inorganic salts. Can. Inst. Food Sci. Technol. *7,* 279–282.

YUAN, R., BICKLE, T. A., EBBERS, W. and BRACK, C. 1976. Multiple steps in DNA recognition by restriction endonucleases from *E. coli* K. Nature *256,* 556–560.

YUDKIN, W. H. 1949. Thiaminase, the Chastek-paralysis factor. Physiol. Rev. *29,* 389–402.

ZABORSKY, O. R. 1973. Immobilized Enzymes. CRC Press, Cleveland.

ZABORSKY, O. R. 1974. Immobilization of enzymes with imidoester-containing polymers. *In* Food and Microbial Processes. A.C. Olson and C.L. Cooney (Editors). Plenum Press, New York.

ZABORSKY, O. R. and OGLETREE, J. 1974. Immobilization of glucose oxidase via activation of its carbohydrate residues. Biochem. Biophys. Res. Commun. *61,* 210–216.

ZAHNLEY, J. E. and DAVIS, J. G. 1970. Determination of trypsin-inhibitor complex dissociation by use of the active site titrant, *p*-nitrophenyl *p*′-guanidinobenzoate. J. Am. Chem. Soc. *9,* 1428–1433.

ZAKARIA, F. and McFEETERS, R. F. 1978. Improvement of the emulsification properties of soy protein by limited pepsin hydrolysis. Lebensm.-Wiss. Technol. *11,* 42–44.

ZAKI, F. G. 1976. Subcellular localization of enzymes. *In* Manual of Procedures: Seminar on Clinical Enzymology. F.W. Sunderman (Editor). Inst. Clin. Sci., Philadelphia.

ZARSKA, B. and LEWAK, S. 1976. The role of lipases in the removal of dormancy in apple seeds. Planta *132,* 177–181.

ZAUBERMAN, G. and SCHIFFMANN-NADEL, M. 1972. Pectin methylesterase and polygalacturonase in avocado fruit at various stages of development. Plant Physiol. *49,* 864–865.

ZECHMEISTER, L. 1950. Progress in Chromatography, 1938–1947. Chapman and Hall, London.

ZECHMEISTER, L. 1951. Some aspects of chromatography. Fortschr. Chem. Org. Naturst. *8,* 341–365.

ZEFFREN, E. and HALL, P. H. 1973. The Study of Enzyme Mechanisms. Wiley-Interscience, New York.

ZELITCH, I. 1975. Pathways of carbon fixation in green plants. Annu. Rev. Biochem. *44,* 123–145.

ZENDER, R., LATASTE-DOROLLE, C., COLLET, R. A., ROWINSKI, P. and MOUTON, R. F. 1958. Aseptic autolysis of muscle: Biochemical and microscopic modifications occurring in rabbit and lamb muscle during aseptic and anaerobic storage. Food Res. *23,* 305–326.

ZENIN, C. T. and PARK, Y. K. 1978. Isoenzymes of polyphenol oxidase from L-DOPA containing velvet bean. J. Food Sci. *43,* 646–647.

ZEPPEZAUER, M. 1971. Formation of large crystals. Methods Enzymol. *22,* 253–266.

ZETELAKI-HORVATH, K. and GATAI, K. 1977. Disintegration of vegetable tissues by endopolygalacturonases. Acta Aliment. Acad. Sci. Hung. *6,* 225–237. (Hungarian)

ZEVACO, C. and DESMAZEAUD, M. J. 1979. Hydrolysis of β-casein and peptides by intracellular neutral protease of *Streptococcus diacetylactis.* J. Dairy Sci. *63,* 15–24.

ZGLICZYNSKI, J. M. and STELMASZYNSKA, T. 1975. Chlorinating ability of human phagocytosing leucocytes. Eur. J. Biochem. *157,* 157–162.

ZIEGENHORN, J. 1975. Improved method for enzymic determination of serum triglycerides. Clin. Chem. *21,* 1627–1629.

ZIEGLER, H., STICHLER, W., MAURIZIO, A. and VORWOHL, G. 1977. Use of isotopes for the characterization of honeys, their origin and adulteration. Apidologie *8,* 337–347. (German)

ZIEMBA, J. V. 1958. Quick tenderizing poses challenge to packers. Food Eng. *30* (4) 120–121, 123.

ZIENTY, M. F. 1971. Stabilization of glucose isomerase in Streptomyces olivaceus cells. Ger. Offen. 2,223,340. May 13.

ZIGMAN, S. 1977. Human lens modification. Fed. Proc. *36,* 293.

ZIKAKIS, J. P., RZUCIDLO, S. J. and BIASOTTO, N. O. 1977. Persistence of bovine milk xanthine oxidase activity after gastric digestion in vivo and in vitro. J. Dairy Sci. *60,* 533–541.

ZIMMERMAN, D. C. and VICK, B. A. 1970. Hydroperoxide isomerase, a new enzyme of lipid metabolism. Plant Physiol. *46,* 445–453.

ZIMMERMAN, G. L. and SNYDER, H. E. 1969. Meat pigment changes in intact beef samples. J. Food Sci. *34,* 258–260.

ZIMMERMAN, W. I., TRESSLER, D. K. and MAYNARD, L. A. 1941. Determination of carotene in fresh and frozen vegetables by an improved method. II. Carotene content of asparagus and green lima beans. Food Res. *6,* 57–78.

ZIMMERMANN, U. and STEUDLE, E. 1978. Physical aspects of water relations in plant cells. Adv. Bot. Res. *6,* 46–117.

ZINCHENKO, V. I., SALMANOVA, L. S. and MINCHUK, F.L . 1973. Enzyme treatment of grape pulp for wine production. U.S. Pat. 3,737,324. June 5.

ZIRLIN, A. 1974. Prevention of crystallization of sparingly soluble flavonoids in food systems. U.S. Pat. 3,832,475. Aug. 27.

ZITTLE, C. A. 1953. Adsorption studies of enzymes and other proteins. Adv. Enzymol. *14,* 319–374.

ZOBEL, H. F. 1973. Review of bread staling. Bakers Dig. *47* (5) 52–61.

ZOBEL, M. 1975. Handbook of Starches in Monograph Form, Vol. 5, Part 5. The Use of Starches in the Food Processing Industry. Verlag Paul Parey, Berlin. (German)

ZOUEIL, M. E. and ESSELEN, W. B. 1959. Thermal destruction rates and regeneration of peroxidase in green beans and turnips. Food Res. *24,* 119–133.

ZUCKERMAN-STARK, S. and LEIBOWITZ, J. 1962. Researches on milk-clotting enzymes from Palestinian sources. Enzymologia *25,* 252–260.

ZUCKERMAN-STARK, S. and LEIBOWITZ, J. 1963. Milk-clotting enzymes from Palestinian sources. IV. Enzymologia *26,* 294–296.

ZUBER, H. 1976. Enzymes and Proteins from Thermophilic Microorganisms. Structure and Function. Proc. Int. Symp, Zurich, July, 1975. Birkhauser Verlag, Basel, Switzerland.

ZÜRCHER, K. and HADORN, H. 1976. Comparative sugar determination in food concentrates and breakfast beverages. Mitt. Geb. Lebensmittelunters. Hyg. *67,* 379–388. (German)

ADDITIONAL REFERENCES

ADELSTEIN, R. S. and EISENBERG, E. 1980. Regulation and kinetics of the actin-myosin interaction. Annu. Rev. Biochem. *49*, 921–956.

AL-BAKIR, A. and WHITAKER, J. R. 1978. Purification and characterization of invertase from dates. J. Food Biochem. *2*, 133–160.

ANDERSSON, L. and WOLFENDEN, R. 1980. Transition state affinity jump chromatography. J. Biol. Chem. *255*, 1106–1107.

ANDRES, C. 1980. Instant lipase-modified butterfat imparts rich butter flavor. Food Process. (Chicago) *41* (11) 104–105.

ARKCOLL, D. B. and HOLDEN, M. 1973. Changes in chloroplast pigments during the preparation of leaf protein. J. Sci. Food Agric. *24*, 1217–1227.

BARBAGLI, C. and CRESCENZI, G. S. 1981. Influence of freezing and thawing on the release of cytochrome oxidase from chicken's liver and from beef and trout muscle. J. Food Sci. *46*, 491–493.

BOHAK, Z. and SHARON, N. 1977. Biotechnological Applications of Proteins and Enzymes. Academic Press, New York.

CHOCK, P. B., RHEE, S. G. and STADTMAN, E. R. 1980. Interconvertible enzyme cascades in cellular regulation. Annu. Rev. Biochem. *49*, 813–843.

CLOUGHLEY, J. B. 1980. The effect of temperature on enzyme activity during the fermentation phase of black tea manufacture. J. Sci. Food Agric. *31*, 920–923.

DAHENY, J. P. and WOLNAK, B. 1980. Enzymes: The Interface Between Technology and Economics. Marcel Dekker, New York.

FINLEY, J. W., KROCHTA, J. M. and HEFTMANN, E. 1978. Rapid preparative separation of amino acids with the chromatofuge. J. Chromatogr. *157*, 435–439.

FIRSHT, A. R. 1977. Enzyme Structure and Mechanism. W.H. Freeman & Co., Reading, England.

FISHMAN, M. M. 1980. Enzymes in analytical chemistry. Anal. Chem. *52*, 185R–199R.

FRIEDMAN, M., ZAHNLEY, J. C. and MASTERS, P. M. 1981. Relationship between in vitro digestibility of casein and its content of lysinoalanine and D-amino acids. J. Food Sci. *46*, 127–131.

GALLIARD, T. 1980. Degradation of acyl lipids: hydrolytic and oxidative enzymes. *In* Lipids: Structure and Function, Vol. 4. The Biochemistry of Plants. P.K. Stumpf (Editor). Academic Press, New York.

GASTINEAU, C. F., DARBY, W. J. and TURNER, T. T. 1979. Fermented Food Beverages in Nutrition. Academic Press, New York.

GRAY, G. R. 1980. Affinity chromatography. Anal. Chem. *52*, 9R–15R.

GRIFFITHS, M. W. and MUIR, D. D. 1980. Hydrolysis of lactose by a thermostable β-galactosidase immobilized on DEAE-cellulose. J. Sci. Food Agric. *31*, 397–404.

GRIME, J. K. 1980. Biochemistry and clinical analysis by enthalpimetric measurements—A realistic alternative approach? Anal. Chem. Acta *118*, 191–225.

HALL, P. L. 1980. Enzymic transformations of lignin. 2. Enzyme Microbiol. Technol. *2*, 170–176.

HALLIWELL, G. and GRIFFIN, M. 1978. Affinity chromatography of the cellulase system of *Trichoderma koningii*. Biochem. J. *169*, 713–715.

HELMREICH, E. J. M. and KLEIN, H. W. 1980. The role of pyridoxal phosphate in the catalysis of glycogen phosphorylases. Angew. Chem. Int. Ed. Engl. *19*, 441–455.

HENDRICKS, J. D., SINNHUBER, R. O., LOVELAND, P. M., PAWLOWSKI, N. E. and NIXON, J. E. 1980. Hepatocarcinogenicity of glandless cottonseeds and cottonseed oil to rainbow trout *(Salmo gairdnerii)*. Science *208*, 309–311.

HICKS, C. L. 1980. Occurrence and consequence of superoxide dismutase in milk products: A review. J. Dairy Sci. *63*, 1199–1204.

HOLTZER, H. and HEINRICH, D. C. 1980. Control of proteolysis. Annu. Rev. Biochem. *49*, 63–91.

JAKSCHIK, B. A. and LEE, L. H. 1980. Enzymatic assembly of slow reacting substance. Nature *287*, 51–52.

JAMES, W. P. T. and THEANDER, O. 1981. The Analysis of Dietary Fiber in Food. Marcel Dekker, New York.

KAHN, V., GOLDSCHMIDT, S., AMIR, J. and GRANIT, R. 1981. Soluble and bound potato tuber peroxidase. J. Food Sci. *46*. *(in press)*

KALLWEIT, E. 1979. Some blood parameters as predictors of meat quality. Acta Agric. Scand. Suppl. *21*, 396–400.

KAUSS, H. 1974. Biosynthesis of pectin and hemicelluloses. Annu. Proc. Phytochem. Soc. *10*, 191–205.

KEYES, M. H. 1980. Enzymes, immobilized. *In* Kirk-Othmer Encyclopedia of Chemical Technology, 3rd Edition, Vol. 9. John Wiley & Sons, New York.

KOBAYASHI, T., KATO, I., OHMIYA, K. and

SHIMIZU, S. 1980. Recovery of foam stability of yolk-contaminated egg white by immobilized lipase. Agric. Biol. Chem. *44*, 413–418.

LAMAISON, J. -L., POURRAT, H. and POURRAT, A. 1980. Purification and properties of a neutral protease from *Tricholoma columbetta*. Phytochemistry *19*, 1021–1023. (French)

LAMKIN, W. M., MILLER, B. S., NELSON, S. W., TRAYLOR, D. D. and LEE, M. S. 1981. Polyphenol oxidase activities of hard red winter, soft red winter, hard red spring, white common, club, and durum wheat cultivars. Cereal Chem. *58*, 27–31.

LAMPORT, D. T. A. 1972. Symposium on the biogenesis of plant cell wall polysaccharides. Abstr. 164th Meet. Am. Chem. Soc., New York. *AGFD 1–6*.

LAZARADIS, H. N., ROSENAU, J. R. and MAHONEY, R. R. 1981. Enzymatic control of meltability in a direct acidified cheese product. J. Food Sci. *46*, 332–335.

LILJENBERG, C. 1977. Chlorophyll formation: the phytylation step. *In* Lipids and Lipid Polymers in Higher Plants. M. Tevini and H.K. Lichtenthaler (Editors). Springer-Verlag, New York.

LIST, D. and KNECHTEL, W. 1980. Determination of ascorbic acid by immobilized ascorbate oxidase. Fluess. Obst *47*, 57–61. (German)

LOWE, J. and INGRAHAM, L. L. 1974. An Introduction to Biochemical Reaction Mechanisms. Prentice-Hall, Englewood Cliffs, N.J.

LYONS, J. M., GRAHAM, D. and RAISON, J. K. 1979. Low Temperature Stress in Crop Plants. The Role of the Membrane. Academic Press, New York.

MAEDA, H. and SUZUKI, H. 1977. Preparation of immobilized enzymes by radiation. Process Biochem. *12* (6) 9–12, 32.

MAGA, J. A. 1980. Potato alkaloids. CRC Crit. Rev. Food Sci. Nutr. *12*, 371–405.

MILL, C. K. and LINDERSTRØM-LANG, K. 1929. Note on the proteolytic enzymes in green malt. C.R. Trav. Lab. Carlsberg *17*, 1–14.

MIYAKE, T. 1980. Process for producing a sweetener. U.S. Pat. 4,219,571. Aug. 26.

NGO, T. T. 1980. Bioanalytical applications of immobilized enzymes. Int. J. Biochem. *11*, 459–465.

OLIVECRONA, T. 1980. Biochemical aspects of lipolysis in bovine milk. Bull. Int. Dairy Fed. *118*, 19–25.

PITCHER, W. H., JR. 1980. Immobilized Enzymes for Food Processing. CRC Press, Boca Raton, Fla.

PREISS, J. and LEVI, C. 1980. Starch biosynthesis and degradation. *In* The Biochemistry of Plants, Vol. 3. Carbohydrates: Structure and Function. J. Preiss (Editor). Academic Press, New York.

ROBERTS, J. L., JR. and SAWYER, D. T. 1981. Facile degradation by superoxide ion of carbon tetrachloride, chloroform, methylene chloride, and *p,p'*-DDT in aprotic media. J. Am. Chem. Soc. *103*, 712–714.

SATO, M. 1966. Metabolism of phenolic substances by chloroplasts. II. Conversion by the isolated chloroplasts of *p*-coumaric to caffeic acid. Phytochemistry *5*, 385–389.

SCHWIMMER, S. 1947C. Sources of beta-amylase as supplements to barley malts in saccharification and fermentation. Cereal Chem. *24*, 70–78.

SCOTT, J. 1981. Natural selection in the primordial soup. New Sci. *89*, 153–155.

SIESS, W., ROTH, P., SCHERER, B., KURZMANN, I., BOEHLIG, B. and WEBER, P. C. 1980. Platelet membrane fatty acids—platelet aggregation and thromboxane formation during a mackerel diet. Lancet *1* (8166) 441–444.

SINGH, B. and KATRAGADDA, R. 1980. Proteolytic activity and its relationship to other biochemical characteristics and bread quality of triticale. Lebensm. Wiss. Technol. *13*, 237–242.

SNOWDEN, J. M. and WEIDEMANN, J. F. 1978. A morphological and biochemical examination of the hydrothermal denaturation of collagen. Meat Sci. *2*, 1–18.

SPINELLI, J. and KOURY, B. 1979. Nonenzymic formation of dimethylamine in fish products. J. Agric. Food Chem. *27*, 1104–1108.

STEAD, E. T. and KENNEDY, R. A. 1980. Bacon processed product. U.S. Pat. 4,218,492. Aug. 19.

STROEM, A. R. 1980. Biosynthesis of trimethylamine oxide in *Calanus finmarchicus*. Properties of the soluble trimethylamine monooxygenase. Comp. Biochem. Physiol. *65B*, 243–249.

STURGEON, C. M. and KENNEDY, J. F. 1981. A quick-reference summary of recent literature on molecular immobilization and bioaffinity phenomena. Enzyme Microbiol. Technol. *3*, 76–80.

THIES, W. 1979. Quantitative analysis of glucosinolates after their desulfation on ion exchange columns. Proc. 5th Int. Rapeseed Conf. *1*, 136–139.

TOBKIN, H. E., JR. and MAZELIS, M. 1980. Alliin lyase: Preparation and characterization for onion bulbs. Arch. Biochem. Biophys., 157–193.

TOM, R. A. and CARROAD, P. A. 1981. Effect of reaction conditions on hydrolysis of chitin by *Serratia marcescens* QBM1466 chitinase. J. Food Sci. *46*, 646–647.

WALSH, C. 1979. Enzymatic Reaction Mechanisms. W.H. Freeman & Co., San Francisco.

WALTERS, R. R., JOHNSON, P. A. and BUCK, R. P. 1980. Histidine ammonia lyase enzyme electrode for the determination of L-histidine. Anal. Chem. *52*, 1684–1690.

WARDALE, D. A. and LAMBERT, E. A. 1980. Lipoxygenase from cucumber fruit: Localization and properties. Phytochemistry *19*, 1013–1016.

WEIDEMANN, J. F., KAESS, G. and CARRUTHERS, L. D. 1967. The histology of pre-rigor and post-rigor bovine muscle before and after cooking and its relation to tenderness. J. Food Sci. *32*, 7–13.

WESTERMARCK-ROSENDAHL, C., JUNILLA, L. and KOIVISTOINEN, P. 1980. Efforts to improve the baking properties of sprout-damaged wheat by reagents reducing α-amylase activity. III. Effects on technological properties of flour. Lebensm. Wiss. Technol. *13*, 193–197.

YAMAMOTO, M. and MACKEY, J. 1981. An enzymic method for reducing curd formation in canned salmon. J. Food Sci. *46*, 656–657.

YERANSIAN, J. A., SLOMAN, K. G. and FOTZ, A. K. 1979. Food. Anal. Chem. *51*, 105R–134R.

General Subject Index (*See also* Enzyme Index)

Abalone, proteinase(p) tenderizing of, 510
Aboard-ship fish processing with proteinase(p), 508
Abomasum, chymosin rennet from, 91, 595
Abortion of β-oxidation, 371
Abortive complexes in affinity elution, 132
Abrasives in enzyme purification, 110
Abscisins, as a gibberellins supplement, 558
  LOX and synthesis of, 299
  sugar accumulation, invertase and, 358
Absidia spp., milk clotting proteinases in, 602
Absolute Reaction Rate Theory, 51–52, 65
Absolute requirement, in enzyme activation analysis, 707
Absorption, -desorption at low a_w's, 239
  intestinal, of enzymes, 665
  of light, and food color, 268
Acacia farnesiana, cysteine C-S lyase of, 380
Accuracy, enzyme analysis and, 685
Acerbity, an off-flavor, 349
Acetal phosphatides. *See* Plasmalogens
Acetaldehyde, as index of maturity, 394
  as product inhibitor, 255
  in cheese volatiles, 407
  off-flavor of frozen foods and, 431
Acetic acid (acetate), cysteine synthase product, 376
  enzymatic analysis for, 691, 698
  in enzyme purification, 117
  in food aromas, 410
Acetoin, diacetyl formation and, 408
α-Acetolactate, cheese aroma enzymes and, 408, 409
Acetone, in enzyme production, purification, 116, 117
  powders, 109, 116, 416
Acetyl CoA, in citric acid accumulation, 354
  in the glyoxalate cycle, 351
  mevalonate from, 403
Acetyl receptor protein, homology with ATPases, 84
O-Acetyl serine, cysteine synthase substrate, 376
N-Acetyl salicylate. *See* Aspirin
N-Acetyl tyrosine, 238
Acetylacetone, for HCHO estimation, 700
Acetylation of enzyme active site by aspirin, 437
N-Acetylglucosamine, 44
  in chitin, 662
  in glucose oxidase, 662
N-Acetylglucosamine-4-lactone, as TSAI, 254
Acetyl-1-phenylalanylaminoacetaldehyde, as TSAI, 254
Acetylphenylalanylglycinal, as TSAI, 254
N-Acetyl-L-tryptophanamide, 49
Achilles tendon collagen, collagenase substrate, 487
Achromobacter spp., amino acid decarboxylases in, 641
Acid(s), as alternative to enzymes, 19
  carotenoid isomerization by, 644
  excess α-amylase in flour, 581
  for inactivating, LOX in pinto beans, 633
  for removing emulsin generated HCN, 636
  for sterilizing enzyme(p), 647
  GRAS, 247
  in food processing and products, 263
  in gibberellic acid malting, 557
  organic, fish toughness prevention with, 505
  organic, in fruits, 389
  pheophytinization with, 305
  sourness of, a non-volatile flavor, 349
  with rennets, in direct acid cheesemaking, 599–600
Acid accumulation, plant enzymes of, 349–356
Acid cooked, an off-odor of unblanched foods, 431
Acid degree value (ADV), and dairy flavor, 446–447
Acid enzyme process, in corn syrup production, 154, 155, 158
Acid foods, peroxidase heat resistance of, 206
Acid Hot Break, in tomato processing, 547–548
Acid hydrolysis for glutamate production, 363
Acid hydrolyzed proteins, 616
Acid inactivation of enzymes, 213, 263–264

Acid juices, pectin lyase(p) for clarification, 537–538
Acid oxidized, an off-odor description in unblanched frozen foods, 431
Acidity, control in fruits, 254
Acid-to-sugar ratio, 349
Acinetobacter spp., limonin dehydrogenase from, 454
Aconitate, and prochirality, 33
Acridness, an off-odor of unblanched frozen vegetables, 430
Acrylamide gel as immobilized enzyme support, 142, 653
Acrylamide-ultragel, matrix in affinity chromatography, 126
Acrylonitrile as immobilized enzyme support, 145
Actin, a muscle thin filament component, 464, 466
  fish, 498
  meat texture, texture defects, 468, 469, 470, 476
Actinidia chinensis. *See* Gooseberry, Chinese
Actinins, meat, Z-disc proteinase, fish and, 464, 466, 498
Actinomyces spp., α-amylase(p) for maltose production, 157
Actiplanes missouriensis, glucose isomerase source, 161, 163
Activation, enzyme, 18, 274, 289, 442–444, 515
  by freezing, 500
  by ionizing radiation, 337
  by sonication, 500
  in blanching, 191, 194
  in enzyme potentiation, 16
  of matrix, in affinity chromatography, 274, 275
  See also Activator(s), Potentiation, and specific enzymes
Activators, enzyme, *See also* specific activators
  as anylates in enzymatic analyses, 704, 705–706
  from NEB, 328
Active center. *See* Active site
Active oxygen. *See* Oxygen, active
Active site, enzymes, affinity chromatography and, 125, 128–129, 134
    Arrhenius-plot discontinuity and, 220
    catalytic efficiency and, 50–59
    domain of, 255
    immobilization of enzymes and, 138
    inhibition and, 38–42, 249–254, 255–256
    ionizing radiation, sensitivity to, 198
    isozymes and, 83
    modification, 40–41, 56, 129
    molecular architecture and key amino acids of, 40, 41–42, 49, 56, 485
    organophosphate as probes for, 704
    purification of enzymes and, 119
    reagents, off-flavor prevention with, 437
    representations of, 42, 57, 58, 83, 485
    specificity and, 38–42
    tertiary structure and, 40
    titration, 20, 39, 40
      analysis for antienzymes by, 705–706
      EIA and, 714–715
      operational molarity of enzyme(p) by, 683–684
  water and, 54, 220
Actomyosin, cold shortening, overlap in, 469
  fish texture and, 498
  PSE and, 40
Acyl carrier protein, 69
Acyl-papain, 226
Adenine, structure, 363–364
Adenosine, 364
Adenosine diphosphate glucose (ADPG), 332, 333, 356
Adenosine monophosphate. *See* AMP
Adenosine monophosphate cyclic, second messenger. *See* Cyclic AMP
Adenosine-5′-phosphate. *See* AMP
Adenosine triphosphate. *See* ATP
Adenylic acid (adenylate). *See* AMP
Adenylylation, in cascade regulation of glutamine synthetase, 72
S-Adenosyl mercaptan, 286
S-Adenosyl methionine, as enzyme cofactor, 46
  DNA methylation, restriction DNases and, 97
  in ethylene biosynthesis, 286
  in pectin biosynthesis, 514
  phenolase action prevention by, 284
Adhesion, fish myofibrils, 500
Adhesion, potato textural attribute, 520
Adhesions, proteinase(p) for treatment, 4
Adhesives, manufacture of with amylase(p), 5
Adiabatic expansion, IQB and, 192
Adjectives describing subjective textural quality, 462–463
Adjuncts. *See* Brewing, adjuncts
ADP, *See also Enzyme Index*
  Arrhenius law temperature discontinuity, 220
  gibberellic acid synergist, 558
  hexokinase produced, use of in enzymatic analysis, 696, 697, 698, 699
  ligand in affinity elution, 132
  Sepharose derivatives for affinity chromatography, 128
Adrenalin, meat color and, 321
  meat tenderization *via*, 482
Adsorbents. *See* Adsorption and adsorbents
Adsorption and adsorbents, affinity chromatography, 124, 126–128
  for chillproofing beer, 570
  in enzyme purification, 118–119, 124, 126–128
  isotherm and affinity chromatography, 121
  of enzymes, 138
  of pectin esterase to cell debris, 552
  on gas (foaming) for enzyme purification, 113
Adsorptive forces in taste perception, 346
Adulteration detection involving enzymes, of, beer, with illegally used GOX, 696
  citrus juice, with citric acid, 699
  coffee, with chicory, 709
  durum, with hard wheat flour, 683, 706
  grams and pulses, with toxic *Lathyrus* seeds, 701
  honey, with corn syrups, *vs* ^{13}C, 679
  malt amylase(p), with microbial amylase(p), 683
Aerobic organisms, SOD and, 286
Aerosol, meat tenderizer, 493
Aerosol OT, 264
Affigel, affinity chromatography matrix, 126
Affinity chromatography, Chapter 7 (123–136), 107, 248, 368, 385
  adaptations, non-chromatographic, non-purification, non-enzyme, 133–134
  adsorption, 125–131
    classification of kinds of, 125–126
    ligands, active site directed, 128–129

ligands, non-active site directed, 129–131
matrices and adsorbents, 126–128, 449, *See also* Matrices
advantages, 132–133
desorption-elution, 131–133
enzymes mentioned in other than Chapter 8, lipoprotein lipase, 425
lipoxygenase, 425
lysozyme, 662
olfactory receptor protein, 346
phytase, 657
polygalacturonase, 517
rennets, to remove non-curdling proteinases, 603
triticale α-amylase, 576
troponin subunits, tropomyosin binding, 446
xylanase, 178
historical, 123–125
macromolecular substrates as ligands-matrices, 123–124, 129, 178, 517, 552, 576
scale-up and food applications, 134–137
Affinity electrophoresis for enzyme purification, 125
Affinity elution, 132–133
Affinity labeling, 40
Affinity packets for removal of enzyme contaminants, 134–135, 449
Affinity partitioning, 133–134
Aflatoxin, enzymatic degradation, of, 661
Aflatoxins, toxicological evaluation of enzyme(p), 647, 648
After-cooking darkening, Irad effects on, 18, 200, 285
Agarose, in affinity chromatography, 124, 126, 127, 135
in enzyme immobilization, 144, 145
in non-affinity chromatography, 118, 119, 120
Age, human, and cathepsin, 480
meat texture and, 463
SOD and life-span, 286
Aggregation, platelets, cardiovascular disease, LOX, prostaglandin synthase and, 501
Aggregation, fish myofibril protein, texture and, 499, 500
Aging, beef, post-rigor, acceleration of, 483–484
affinity chromatography and, 136
aroma genesis and, 415
cathepsin in, the case against, 475–478, 479
cathepsin in, the case for, 18, 473–475, 478–480
in muscle-to-meat conversion, 473, 475–480
vs postharvest curing, 36
Aging, cheese, 594
Aging, milk, and induced rancidity, 448
Aging, rye flour, to prevent excess α-amylase, 580
Aging, wheat flour, inactivation of enzymes by, 586, 591
Agmatine, a decarboxylase-generated biogenic amine, 640
Agricultural waste, xylan in, 176
AICAR, adenylosuccinate lyase inhibitor in 5′-nucleotide accumulation, 366
"Ain" as enzyme suffix, 28
Ainunegi, 378
Air, blanching, cooling and expulsion, 189, 194
classification, accelerated enzyme action by, 659
classification, enzyme purification by, 109
color, food, *via* bubbles and expulsion of, 268, 269
cushioning, control of enzymatic browning by, 281
pesticides in, analysis, 707
Alanine, immobilized enzyme production of L-, 150
Alanine, Swiss cheese flavor precursor, 409
Alaska pollock, TMAO demethylase of, 503
Albacore, muscle cathepsin, 506
Albedo, citrus, accumulation of bitter limonoid in, 454
debittering enzyme from, 454–455
frozen juice gelling and, 544
PG and PE in, 521–522
Albizzia lophanta, cysteine lyase of, 32, 376, 380
Albumens, 339, 615
Albumin, blood, dimer formation by Irad, 198
Albumin, egg desugaring and NEB of, 339
Alcaligenes spp., milk clotting enzymes in, 602
Alchemy and alchemists, 3, 395
Alcohol. *See* Ethanol
Alcholic beverages, distilled, 369, 552, 561–562, 638
Alcohols, :aldehyde ratio aroma, and enzymatic adjustment of, 393–394, 423
aliphatic, as sprout inhibitors, 337
aromatic, terpene as potato ADH substrates, 393
as aldehyde precursors, 371
as aroma constituents, 371, 389, 393–394, 413
enzymatic analysis for, 697
intermediate, genesis of, 393–394
long chain, from hydrocarbons *via* xenobiotic oxidase, 412
quinones, reaction with, 401
unsaturated, as pea ADH substrate, 393
Aldehyde, active, in diacetyl evolution, 408
Aldehydes, aliphatic, 369–373
aroma, compared with alcohols, 393
aroma of wine, 401
as LOX secondary products, 427, 428
as TSAIs, 254
branched, off-odor of, 407
carbonic anhydrase hydration of, 31
crosslinking, LOOH, and fish toughness, 501, 502
enzymatic analysis for, 697
enzymatic reduction *vs* formation in frozen foods, 432
enzymatic removal of objectionable, 423
in aroma of beverages, 8, 400–401, 402–403
in aroma of bread, 413
in aroma of cheese, 407
in aroma of fruits, 388, 395
in aroma of vegetables, 370–373
from alcohols *via* phenolase action, 371, 400
from Strecker degradation, 401, 402
off-odors, 407, 424, 446
Aldol condensation of plasmalogens, 446
Aldonolactones as enzyme inhibitors, 254
Ale, enzymes and, 553
Aleurone layer of cereal kernels, 626, 628
enzymes in, 556, 568, 574, 576, 581
HTST penetration of, to manage α-amylase, 581
morphology, 656
phytate in, 562, 567
proteins and protein bodies in, 626–627, 628, 652, 657
separation of enzyme and substrate, 13
Alfalfa, an off-odor of unblanched frozen foods as, 430
chicken color and, 301
enzymes of, β-glucosidase in sprouts, a potential health hazard, 636
hydroperoxide isomerase, 430

lipoxygenase, 430
nutritional loss by alkali treatment of, 644
protein, autolytic loss of, 626
enzyme(p) processing of, 624, 627
polyphenol discoloration of, 631
Algae, 353
cellulase(p) treatment of, 549
rennet from, 602
Alginic acid, as immobilized enzyme support, 145
as matrix-ligand in affinity packet, 135
in affinity chromatography, 124
Alkali(es), *See also* pH extremes
autolysis, enzyme purification by, 109
barley treatment for chillproof beer, 570
for chlorophyll retention, 307–308
for phytate removal, 58
for purine removal from SCP, 655
formaldehyde in fish *via*, 503
glucose isomerization by, 153–154
GRAS, 247
in enzyme purification, 110
in food processing and products, 263
in tortilla and olive processing, 263
in hominy grits making, 263
peptide bond labilizing by, 710
to inactivate, alfalfa LOX, 644
antitrypsin *via* urease(p), 651, 659
nucleotidases, 367
Alkalinity, slight, control of enzymatic browning by, 290–291
Alkaloids, xenobiotic monooxygenase and, 412
2-Alkanones, in aroma volatiles, 371
amino acid stimulated production of, 409
enzymatic ester formation from, 392
Alkenals, aroma volatiles, 370, 371
enzymatic removal of, 373
Alkenols, aroma volatiles, 371
as orange ADH substrates, 393, 394
Alkenones, aroma volatiles, 370, 371, 402
S-Alk(en)yl cysteine(s), 32, 250, 377
S-Alk(en)yl-l-cysteine sulfoxides, onion aroma precursors, alliinase substrates, 250, 374–381
Alkyl fluorophosphates, and enzyme regeneration, 213
Alkylamines, as enzyme models, 61
Alkyldiamines, as spacer attachments in affinity chromatography, 127
S-Alkyl-L-cysteines, as alliinase inhibitors, 377
Alkylsulfenate, 375
Allergenicity of enzyme(p), 647, 648
Allergies, from food enzyme passage to blood, 649
Alliaceous foods, 260
Allicin, alliinase substrate, 380
as antibiotic, 663
for prevention of thiaminase I caused vitamin B_1 deficiency, 645
in *Allium* spp., 378
Alliin, garlic aroma precursor, 374, 376
Allithiamin, not a thiaminase I cosubstrate, 645
Allium cepa, 312, *See also* Onion
A. porrum, 312, *See also* Leeks
A. sativum, 312, 374, 380, *See also* Garlic
A. victorialis, caucus, 378
D-Allose, product of alkaline glucose isomerization, 154
Allosteric enzymes and allostery, 67–73
in affinity chromatography and elution, 129–130, 131–132
biological development stage, evaluation of, 679
glutamine synthetase and cascade control, 72–73
in glycolysis and gluconeogenesis enzymes, 70–72
kinetics, 67–68, 70, 333, 352
in muscle contraction, 466, 467
S-Allyl cysteine sulfoxide. *See* Alliin
Allyl disulfide, 378
Allyl esters, 374
Allyl glucosinolate, 686
Allyl isothiocyanate, 383, 686
Allyl sulfenic acid, 375
Almonds, amygdalin, emulsin and essential oil from, 396, 398, 399, 636
Altitude, lipase-caused off-flavors and, 447
Alum, plus blanching, discoloration prevention, 195
Alumina, as immobilized enzyme support, 143, 144
C_γ and other hydrates, adsorbents in enzyme purification, 110, 118
Aluminate as model glucose isomerase, 60
Amadori rearrangement in NEB and diabetes, 334
Amatoxins, RNA polymerase inhibitor, 706
Ambrettolide, musk odor lactone, 411
Amine(s), aromatic, in affinity chromatography, 128
as thiaminase I cosubstrates, 645
gas sensing electrode, 694
in cheese aroma, 408, 409
in NEB, 327
Amino acid pool, and glutamate accumulation, 362, 363, 378
Amino acid side chains, 40
branched, as aroma precursors, 390, 409
cyanogenic glucosides, side groups from, 396
enzyme-pH rate profiles and, 262
glucosinolate, side groups from, 382–383
in photosensitization of enzymes, 200
nucleophilic, in enzyme immobilization, 144
sulfur-containing, as codfish off-flavor precursors, 457
D-Amino acids, enzyme inhibitors, 31
L-Amino acid(s), as aldehyde precursors, 394
as aroma precursors, 369, 386, 400, 409, 414
as dairy off-odor precursors, 452
as ester precursors, 390
as flavorants, non-volatile, 349
as gibberellic acid stimulants, 558
as methoxypyrazine precursors, 452
as yeast food, proteinase-generated, 567, 572
enzymatic analysis for, 700
essential, in fish proteolysates, 509
essential, in plant proteolysate core, 624
fatty acid relations in cheese aroma, 409
for preventing oxidative rancidity, 451
hydrophobicity, and peptide bitterness, 617–618
hydrophobicity, proteinase action enhanced by, 624
in enzyme biosynthesis (translation), 78
in Irad effects on proteins, 196
in onion pinking, 312–313
in vacuoles of plant cells, 513
NEB and, 327
non-nutritive, 263
of proteins, proteinase(p) as aids in analysis for, 710–711
preventing frozen fish toughness, 505

production from DL with immobilized enzymes, 150, 151
reaction with LOX-generated LOOH, 631
reaction with phenolase-generated *o*-quinones, 401
Strecker degradation of, 393
transfer from esters to plasteins, 620
vs D, enzyme preference for, 31
vs peptides, nutritiousness of, 620
ε-Amino group, lysine, enzyme reaction inactivation and, 273
pK_a, 262
reaction with phenolase-generated *o*-quinones, 6
Amino transfer, *via* phosphopyridoxal enzymes, 45
2-Aminoacrylate, 375
Aminoacyl sites in translation step of enzyme biosynthesis, 79
α-Aminoadipic-δ-semialdehyde, lysyl oxidase product, 496
4-Aminoantipyrine, H-donor in peroxidase assay, 208
γ-Aminobutyric acid (GABA), psychoactive amine, 640
6-Aminocaproic acid, cathepsin B inhibitor, 474
1-Aminoglucose as a TSAI, 254
Aminohexyl derivatives, for affinity chromatography, 425
1-Aminopropanecarboxylic acid, in ethylene biogenesis, 286
β-Aminopropionitrile, for tenderizing meat, 482
Aminostyrene polymers as immobilized enzyme supports, 145, 338
3-Amino-1,2,4-triazole, for antigreening, 311
gibberellic acid mimic, 558
Ammonia, alliinase product, 32, 375
cellulolysis pretreatment, 176
for oxalate removal, 189
gas sensing enzyme electrode, 694
in blanching, 195
in cheese aroma, 458
in feed and food processing, 263
LOX inactivation, 301
manometry of, in enzymatic analysis, 689
Ammonium phosphate, enzyme production clarification aid, 100
Ammonium sulfate, use in enzyme purification, 100, 110, 115, 116, 117, 118
Amniocentesis, enzyme assay for birth defect, prediction, 4
AMP, *See also Enzyme Index*
as allosteric affector, 71, 356, 471–472
as group specific ligand, 125, 128–129
as ligand in affinity elution, 132
as phosphorylase b activator, 332
conversion to IMP gustator, 150, 151, 364
cAMP. *See* Cyclic AMP
Amperometry in enzymatic analysis, 689, 691
Amphilicity of casein micelles, 595
Ampholytes, enzyme inhibitors as, and double pH optima, 329, 331
Amplification, signal, by proenzyme activation, 603
in EIA *vs* RIA, 713
recycling in kinetic enzymatic analysis, 688, 689
Amplitude, effect on reaction rates at fluctuating temperature, 229
Amygdalin, 396, 398, 635
Amylo process, α-amylase for ethanol production, 561
Amylograph, *vs* α-amylase assay as bread quality predictor, 578
Amylopectin, β-amylase substrate, 154
branching, structure of, 554
glucoamylase substrate, 154, 155, 159
pullulanase substrate, 154, 159
synthesis by Q-enzyme, 332, 333
Amyloplast membranes, potato sweetening and disintegration of, 334–335
Amylose, as enzyme substrate, 250
double helix of, 332, 333
retrogradation of, in bread, 578
retrogradation of, in starch syrup production, 157
synthesis, phosphate and chain length during, 330
Anaerobes, blanching effluent treatment, 195
cheese aroma, hydrogen and, 409
protection by SOD, 287
Anaerobic packaging of meats and MRA, 322
Analog inhibitors, 249–254. *See also* TSAI
Analysis, affinity chromatography in, 132
Analysis, chemical, immobilized enzymes in, 140
Analysis, chemical, enzyme(p) as aids in, 709–715
Analysis, enzymatic. *See* Enzymatic analysis
Anana sativa. *See* Pineapple
Anaplerotic enzymes, plant acid accumulation enzymes as, 351, 352
Anchimeric assistance and orbital steering, 55
Anchoring, catalytic efficiency of enzymes and, 55
Anchovies, lipase of, 500
lipid of, effect of SOD, 286
thiaminase I of and vitamin B_1 deficiency, 644
Anemia, mild, a subclinical symptom of vitamin C deficiency, 646
Anhydride hydrolysis, spontaneous, 225
Anhydrogalacturonic residues in pectin, 513
Animals, as food enzyme(p) sources, 91
Animals, cheese from milk of, 594
Anionic detergents, 264
Anisole, olfactory receptor protein of, 346
Anisotropic diffusive ultrafilters, 114
Anomaly, low temperature, of enzymes, 219–220
Anosmia, 345
Antarafacial insertion, of oxygen in LOX reaction mechanism, 425
Antemortem injection of ascorbate for meat color retention, 323
Antemortem injection of meat tenderizer proteinase(p), 481–484, 494
Anthesis, enzymes in plant development following, 352, 555, 576
Anthocyanidins, relation to catechins, 293
Anthocyanins, accumulation in stressed plants, 269
as enzyme feedback inhibitors, 256
as food pigments, 269
autumn leaf color and, 268
decoloration, 18, 237, 270, 277
leakage of prevented with alum, 195
Anthranilic acid as ligand in affinity elution, 132
Anti-anxiety receptors, affinity chromatography of, 132
Anti-autolysis, 482
Antibiotics, antemortem papain(p) injection and, 494
as analog inhibitors, 249
as transition state analog inhibitors, 254
cerulenin, for preventing citrus bitterness, 456

flavor enzyme products as, 663
glucose oxidase as, 338
in enzyme(p) toxicological evaluation, 647, 648
in meat tenderizer formulations, 492
in 5′-nucleotide production, 367
in onion and garlic, 376
in proteinase(p) used as feed additive, 625
isothiocyanates as, 382
peroxidases as, 204
Antibodies, as enzyme inhibitors, 258
in affinity chromatography, 125, 130, 131, 134
in enzyme and other immunochemical procedures, 683, 712–714
in enzyme(p) toxicological evaluation, 647
peroxidase as topological probe of, 205
Anti-boll weevil factor, 259
Anti-coagulants, inhibitors of DT diaphorase, 317
Anti-caries effect of xylitol, lactoperoxidase and, 177
Anticodon, in enzyme biosynthesis, 76
Anticrosslinking, by enzymes, search for, 496
Antidepressants, pressor amines in cheese and, 640
Antienergistic interaction in immobilized enzyme action, 251
Antienzymes, as naturally-occurring polymers, as transition state analog inhibitors, 254, 257
enzymatic analysis for, 705–706
of α-amylase, for weight control, 254, 257
of α-amylase, in wheat, 575
of invertase, in potatoes, 329, 331
of proteinases, 131, 254, 257, 331, 436, 568, 575, 705–706
removal by blanching, 189
Antifoaming agents as enzyme stabilizers, 103
Antigen(s), as anylate in enzyme immunoassay, 713, 714, 715
from pepsin action on lactalbumin, 643
grain amylases as, 556, 574, 577
Antihelminthics, as thiaminase I cosubstrates, 645
Antimetabolites, enzyme inhibition by, 249
Antioxidants, as enzyme stabilizers, 103
as lipoxygenase inhibitors, 437
from legumes, 437
GRAS, 247
meat color retention with, 322
NEB-generated, 327, 437
off-flavor of stored cereal and, 429
stabilization with SOD, 285–286
Antipain, a TSAI, 254
Antirennet, 683
Anylates, food as enzyme affectors, 703–707
Anylates, food as enzyme substrates, 694–702. *See also* Food anylates
AOAC, use of enzymes in Official Methods of Analysis, 669–671
Apatite, in casein micelles, 595
Apigenin, naringin oxidase product, 456
Apoenzymes, 45
biosynthesis of metalloenzymes and, 677
in enzymatic activation analysis, 607
in enzyme regeneration, 213, 214
Appearance, enzymes and food quality and identity, 267, 268, 533–535
Apple nectar, pectinase(p) for making, 538
Apples, aroma and character impact esters, 389
canned, preblanch, PE, Ca and firming of, 523
enzymatic browning, pacers and control of, 279, 285, 286
enzyme assay as for predicting senescence of, 678
enzymes of, ascorbate oxidase reported, 289
catalase, loss in frozen, 223
β-galactosidase, 296
laccase reported, 271
lipase of seed, 444
phenolase, reaction inactivation of, 273
succinic dehydrogenase, 356
superoxide dismutase, 287
maturity of, acetaldehyde as index, 394
nonmaceration with hemicellulase(p), 550
peeling, aroma release by, 393
pesticides in, GLC-enzymatic analysis of, 706
scald, an off-color disorder, 269, 319
sorbitol sweetness of, 360
succinic acid in controlled atmospheres, 356
texture (elasticity) of, 462
Apple juice, enzymatic flavor improvement of, 423, 439
enzyme(p) for making, 540
gel formation in and prevention by PG, 544
nonlimpid, enzyme(p) in making, 548
pectinase(p) safety and blood lysozyme, 648
PE:PG ratio adjustment, 537
spontaneous clarification of, 537
starch-caused haze in, 539
Applesauce, pectin lyase(p) for making, 539
use of enzyme derived syrups in, 152
Applied enzymology, non-food, 3, 4–6
Apricot(s), dried, enzymatic off-flavors, in, 237
emulsin-amygdalin systems in pits of, as health hazard, 636
enzymatic browning of, 269, 275, 279
enzyme-related nutrient loss/retention, 644
oversoftening of canned by resident pectinase, 6, 189, 522
solar energy and dried, 233
Apricot juice, enzyme(p) in making, 548
Apricot nectars, enzymes in making, 544
Aquacides in enzyme concentration, purification, 109
Arabinans, in bamboo shoot texture, 530
Arabinogalactan, protopectin component, 514
L-Arabinose, in pectin side chains, 513
in wheat flour pentosan side chains, 581
Arachidonic acid, as prostaglandin synthase substrate, 427, 437, 501
LOOHs from LOX action on, 299, 427
Arachis hypogea. See Peanut
Archeus, as enzyme, 3
Arginine, 69
antioxidant from heating with xylose, 437
as agmatine precursor, *via* decarboxylase, 640
enzymatic analysis for, 700
esters as trypsin assay substrates, 240
guanidinium of, pK_a, 262
Argols, in wine making, 536
Arils, enzymatic removal of from seed surface, 661
Aroma, cellulase (for) extract of, 550
lipoxygenase and, 299
LOX-generated, 299
potentiation of, by GMP, 364
preformed, biosynthesized *in vivo*, 360, 368, 370

Aroma genesis, enzymatic aspects of, bread and other baked goods, 413–414
classifications, 368–369
essential oils, and flavorants, 388, 395–405
fruits, 389, 395
in cheeses and other dairy products, 407–413
in vegetables, 368–381, 444
meats and other animal foods, 414–416
wines, 389–395
Aromagrams, demonstrating flavorese effect, 418–419
Aromatic amines, 260
Aromatic carboxylic acids, control of enzymatic browning with, 282
Aromatic nitro groups as nitrate reductase substrates, 260
Arrhenius discontinuity, 219–230, 334
Arrhenius temperature rate-relation, energy of activation, Q_{10} and, 51, 184
fluctuating temperatures and, 227–228
frozen/liquid state and, 221
in chill-sensitive plants, 334
low temperature deviation from, by enzymes, 219–230, 334
Arrowhead myosin, from PSE pigs, 471
Arsenate, in enzymatic analysis, 163, 688
Arterial intima, 649
Arthritis, lysosomes, cathepsins, and, 480
Arthrobacter spp., glucose isomerase(p) source, 16
A. luteus, lytic D-glucanases in, 629
A. simplex, cholesterol degrading enzyme system in, 661
Arthropods, TMAO, demethylase in, 503
Artichoke, as a flower bud vegetable, 524
Artichokes, off-odors of unblanched, frozen, 430, 431
Artillery, of bombardier beetle, peroxidase, catalase and, 205
Aryl-alkyl amines, as gene regulators, 269
Aryl-methyl glucosinolates, 38
Asclepias mexicana. See Milkweed
Ascochyta, milk clotting proteinases in, 602
L-Ascorbate, *See also* L-Ascorbic acid, Vitamin C
and blanching, 190, 194, 195
and polyphenol biosynthesis, 276
as enzyme inactivator, 263
as LOX inhibitor, 300, 437
as metmyoglobin reductant *in vivo*, 317
as onion pinking inhibitor, 313
as quinone reductant, 260
as TMAO demethylase activator, 503
bound forms of, 646
browning, a type of NEB, 327
carotene oxidation by, 302
dough conditioning by, phenolase and, 590–591
enzymatic browning, control of, 12, 279, 282, 283, 284–285
for chlorophyll retention, 307
for stabilizing enzyme(p), 102
free radical of, 284
glucosinolase activation by, 385, 386
in breadmaking, 300
in fluctuating temperature—stored food, 228
in peroxidase assays, 207, 208
in phenolase reaction inactivation, 273–274
meat color and, 322, 323, 325
off-flavor lipid peroxidation, and, 435
retention in food processing, 194, 237, 546
superoxide dismutase, O_2^-, and, 285
L-Ascorbic acid, Vitamin C, *See also* Ascorbate
as contributor to fruit sourness, 351
biosynthesis of, 351
enzymes of, accumulation of, 354, 355
precursor to tartaric and oxalic acids, 354
Ascorbigen, glucosinolase and, 384, 646
Ascorbyl stearate, and meat color, 322
"Ase" as enzyme suffix, 28
Asepsis, partial and blanching, 189
Aseptic curds, model cheesemaking and, 607
Ash gourd, rennet from, 94, 602
Asparagine, immobilized enzymatic reagentless analysis for, 693
in sugar-enzyme linkages, 338
peptide bitterness and, 617
Asparagus, as a young stem vegetable, 524
blanching of, 189
calcium and chlorophyll retention in, 310
dithiolanes from, 374
etiolation of, 311
freezing and frozen, 189, 430, 431
GMP gustator and nuclease potentiation in, 364, 366
off-color, 269
toughening, peroxidase, and lignin biosynthesis, 204, 529–530
L-Aspartate, as a gibberellin synergist, 558
aspartic acid, 33
β-carboxyl, pK_a of, 262
in sweet dipeptide esters, 618
Aspergillus, A. ficuum, rich phytase(p) source, 657, 659
A. flavus, meat tenderizing proteinase from, 486
A. japonicus, aminopeptidase(p) as digestive aid, 664
A. niger, enzyme(p) from, GRAS, 9, 10, 95, 647
catalase, 663
cellulase, 549
endo-β-glucanase(p), for brewing, 566
glucose oxidase(p), in the United States, 338
lactase, 180, 652, 653
pectinase, 540
enzymes of, anthocyanase, 296
glucose oxidase, purification of, 110
isopullulanase, 154
PG, specificity, 517
thioglucosidases, 385
xylanase, 179
mycelia, enzyme(p) from, sanctioned by WHO, 647, 648
A. ochraceus, rennet from, 603
A. oryzae, enzyme(p) from GRAS, α-amylase, 9, 95, 155, 559, 576, 583
proteinase, 10, 486, 509, 569, 609, 625
takadiastase, 89, 155, 554
tannase, 440
in soy sauce manufacture, 616
maltase of, 158
proteinases of, 486
A. saitoi, α-galactosidase(p) from, 9, 654
A. sojae, pectin lyase(p) of, for juice clarification, 538
proteinases and peptidases in soy sauce production, 616

A. sydowi, thioglucosidases of, 385
A. terreus, a rich phytase source, 657
Aspergillus species, 9, 10, 89, 95
enzyme(p) from, uses, acid carboxypeptidase, 618
cellulose, 627
proteinase, 571, 608
milk clotting proteinases of, 602
spores for tenderizing meat, 492
Aspirin, as enzyme inhibitor and suicide substrate, 253, 254, 437
Asporogenesis, and secrecy in enzyme(p) production, 96
Assays, enzymes, *See also* specific enzymes
amplification by proenzyme activation, 603
as diagnostic aids in medicine, 4, 671, 685
as indicators, markers, indices, monitors and predictors, 558, 567, 578, 581, 676–681
by active site titration, 684–685, 705–706
by immunochemical techniques, 683, 714–715
by isozyme patterns, 681–683
for adequacy of heat processing, 191, 205–212
for standardizing and monitoring enzyme(p), 495, 682, 683–684
heat stability and, 110
in official analyses, 669–671
principles of, 670–676
Asses, cheese from milk of, 594
Association-dissociation of enzyme subunits in ice, 223
Astaxanthins, enzymatic liberation, in fish color, 296, 302, 325
Astringency, 347, 349
of blanched bananas, 190
of cheese, 457
of eggplant, suppression, 195
of tyrosine peptides, 617
phenolase action and, 457
polyphenol-protein interaction and, 630–631
Asymmetric syntheses, enzymatic, 396
Atavism, peroxidase as, 204
Atherosclerosis, xanthine oxidase absorption and, 649, 660
Atomic absorption, elemental analysis by, 706
ATP, *See also Enzyme Index*
affinity chromatography and elution, 129, 132
as allosteric affector, 71, 72, 73
as ligase obligatory substrate, 30
biosynthesis of, 85
chicken texture and temperature, 469
electron transport reversal by, 285
enzymatic browning, control of, 285
ester biosynthesis, 390
fish quality and catabolism of, 489, 499, 680
free energy common currency, 10
hydrolysis at low a_w, 241
hydrolysis in muscle contraction, 3, 285, 316, 466–467
immobilized enzyme regeneration, 285
in bioluminescence measurements, 689
in citric acid accumulation, 354
in meat texture, 467, 468, 469, 470
in MRA and meat color, 322
in muscle, replenishment of, 467–468
in starch biosyntheses and sugar interconversions, 328, 332
in sucrose translocation, 357
muscle, plasticizing effect on, 466
potato respiration, 336–337
Atrazine, structural analyses and resistance to, 380, 660
Attar aroma, glycosidase flavorese, and, 399
Attenuation of beer worts by enzyme(p), 553, 559, 563, 564
Aubergines. *See* Eggplants
Augustinsson-Edie plot, 36–37
Autoanalyzers. *See* Automated analysis, 693
Autolysate, fish, PER of, 510
Autolysis, enzymes, surviving, 203
for 5′-nucleotide seasoning, 364
for solubilizing glucose isomerase, 164
in enzyme purification, 109
in FPC production, protein loss, 508
in invertase production, 160
in Irad effects on enzymes, 200
limited, 506–507
meat texture and, 473
nitrite-induced, enzyme assay as monitor, 680
of fish, 506–508
of high protein sources, 625–627
starter proteinase release by, 607–608
vs freeze-thawing, 681
Autolytic enzymes, problem in fructose syrup production, 164
Automated analysis, for *in vitro* protein digestibility, 707–708
in blanching adequacy peroxidase assay, 207
in enzymatic analysis, 685, 689–700
compared with enzyme electrode, 693
for glutamate in foods, 699
for sugars in foods, 695
immobilized enzymes used in, 692
kinetic mode of, suitability, 689
in pasteurization adequacy enzyme assay, 212
Autophagosome, 82
Autoxidation of lipids, lysozyme free radical *via*, 285, 589
prevention by SOD, 286
red meat color and, 320
water activity and, 237
Autoxidation of phenols, and aroma, 401
Autumn foliage color, 269
Auxiliary enzymes in enzymatic analysis, 686
Auxiliary enzymes in enzyme assay, 670, 674–675
Availability of proteins, enzymes for *in vitro* assessment of, 707
Avicel, as cellulase substrate, 169, 170
Avidin-biocytin complex, in affinity chromatography, 131
Avocado(s), as a tree fruit vegetable, 524
biogenic amines, trace in, 641
blanching bitterness of, 421
enzymatic browning, pacers of, 13, 279
PE, PG, and texture enzymology of, 529
phenolase isozymes of, 274, 275
SOD of, 287
with tomato, macerase(p) from, 627
Azaserine as analog inhibitor, 249–250
Azaserine as suicide substrate, 253
Azo dyes, spacer arm activation in affinity chromatography, 124
Azophenols, in affinity chromatography, 124

Baby foods, enzyme(p), enzyme-derived ingredients in, 9, 152, 579
Bacillus cereus, β-amylase, in, 154, 577
as rennet source, 10
microproteinase in, 507
B. circulans, lytic β-D-glucanase of, 302, 629
B. coagulans, glucose isomerase source, 161, 166
heat-stable α-amylase, 154
B. licheniformis, enzyme(p), GRAS, from, 154, 163, 646
B. macerans, cyclodextrin glucotransferase from, 63, 95
B. PM, TMAO demethylase from, 503
B. polymyxa, β-amylase of, 154
heat-stable α-amylase, 154
rennet from, 603
B. stearothermophilus, source of heat-stable α-amylase(p), 154
B. subtilis, affinity separation of α-amylase and proteinase by affinity methods, 135
α-amylase(p), supplement, 559
enzyme(p), GRAS, 95, 625, 647
mutation of, for tailor-made enzyme(p), 96, 97
proteinase(p), usually alkaline, uses, 96, 509, 571, 621, 622, 625
rennet from, 603
xylanase of, 179
B. thiaminolyticus, thiaminase I in grazing and, 645
Bacillus species, milk clotting enzymes in, 602
Bacon, imitation, enzyme-imparted protein functionality, 614
LDH action in curing of, 324
nitrite-free, tenderization of, 493
preservation by Irad, 13
Bacteria, and enzymatic browning, 287
enzyme(p) production from, 94–100
essential amino acids, 621
growth of at low a_w's, 235
in specialty breadmaking, 577
meat discoloration cause, 318–319
peroxidase-caused death of, 204
rennet proteinases in genera, 602
ripening, *See* Cheese
thiaminases in, 645
Bacteriophage, separation of, immunoassay for, 117, 712
Bacteriostatic agents, in meat tenderization, 483
Bagasse, enzymatic biomass conversion, 167, 176, 177
Bakanae fungi, gibberellins from, 556
Baked goods, *See also* Bread, Breadmaking, individual items
enzymatic off-flavor removal, 423
enzymatic off-flavors in, 237
enzyme-derived sweeteners in, 152
identity of, 267
texture enzymology of, 9, 572–592
Baking, and egg desugaring, 339
Baking, enzyme survival, 445
Baking enzymology, 17, 55, 234, 572–592. *See also* Bread, Breadmaking
Baking powder, 577
Ball milling, for stimulating cellulolysis, 177
Bamboo shoots, young stem vegetable, and texture, 524, 530
Banana(s), and orange, vitamin C loss in, 646
aroma, constituents of, 389
astringency of blanched, 189
enzymatic browning, pacers of, 299
enzymes of, flavorese, 418
LOX, other aroma-generating systems, 287, 299, 450
PE, PG, other texture-affecting enzymes, 522
phenolase, and enzymatic browning, 275, 276, 299, 646
phosphorylase, 331
green, and chlorophyll, 302
Irad effects on enzymes in, 199
juice, pectinase(p) in making, 537, 540
peels, biogenic amines in, 641
sulfiting, 189
texture of (elasticity), 462
BAPA, trypsin substrate, 568
Barbiturates, in 5′-nucleotide production, 367
Barium, enzymatic inhibition analysis for, 707
Barium salts, in enzyme purification, 112, 116
Barley, alkali treatment for chillproofing beer, 570
as browning adjunct, 552, 553
brewing, microbial enzyme(p) in, 560–561
enzymes of, α-amylase, 554–555
β-amylase, 562
invertase, assay quality index, 680
laminarase, 556
lipase, and off-flavor, 445
LOX, and off-flavor, 299, 429
nucleases, 366
xylanase, 188
germinated, *See* Malt
β-D-glucans in, enzymatic analysis for, 696
lipids, enzymes, beer and, 571
malt, *See* Malt
proline in, as malt quality index, 414
sprouted, *See* Malt
syrup, enzyme(p)-produced, malt replacement, 560
Base-catalyzed hydrolysis, 225
Base-pairing in enzyme biosynthesis, 75, 78, 79
Basket centrifuge in enzyme(p) production, 120
Batch adsorption, in enzyme purification, 118
Batch reactors, 141
Beaniness, a LOX-caused off-odor, 424
Beans, acid treatment of, 264, 436
autolytic modification of, 626
blanching, inadequacy of catalase test, 211
enzymatic removal of α-galactosides from, 653
enzymes of, alcohol dehydrogenase, 393
flavorese, 417
α-galactosidase, and flatulence, 18, 653
β-glucosidase, potential health hazard, 636
LOX, and off-odor, 428
phenolase, 275
phytase, texture, nutrition, 18, 653, 657, 658–659
horse, *See* Horse beans
Irad of, effect on enzymes of, 199
lentil, *See* Lentil beans
lima, *See* Lima beans
odor, constituents of, 371, 386
off-colors, 269
PER of protein from acid-treated, 633
phytate as antinutrient in, 656
protein isolation by affinity chromatography, 130
scarlet runner, *See* Scarlet runner beans
seedlings, enzymes in, 45

snap, *See* Snap beans, 279
sprouts, etiolation of, 311
texture, enzymology of, 527–528
Bee food, honeybee invertase for preparation, 358
Beef, *See also* Meat
as bovine skeletal muscle, 463
color, 315–325
color of frozen, enzyme assay as predictor of, 680
dried, 243
plasma, amine oxidase(a) from, 701
protein digestibility of, 708
water sorption isotherm of, 232
zymograms, to detect freeze-thawing, 681, 682
Beer, as an alcoholic beverage, 552
biogenic amines as potential toxins in, 640
body, effect of enzyme(p) on, 9, 569
brilliance and clarity retention with GOX, 342
chillproofing with enzymes(p), 9, 19, 569–571, 669
dextran contamination of, 360
dietetic (light, low-calorie), 552, 553, 664
enzymatic removal of off-flavors from, 421
enzymatically generated off-flavors of, 368, 421, 423, 445
enzymology of, 552–571, *See also* Brewing, Malt
flavor of, 364
gluconate in, from illegal GOX use, 696
haze, types and composition, 569–570, 630
lipids and foaming, 571
short, 553
sulfur off-odors genesis in, 457
Z-protein in, 563
Beet sugar refining, use of immobilized enzymes in, 150–151
Beets, geosmin and aroma of, 387
Belt-and-braces approach to beer chillproofing, 570
Benincasa cerifera. *See* Ash gourd
Bentonite, adsorbent in enzyme purification, 118
as a fining agent in juice-making, 537
control of enzymatic browning with, 281
immobilized enzyme support, 144
Benzaldehyde, a product of emulsin action, 396
in *Sorbus* and *Prunus* aromas, 389, 399
oxidized during marzipan making, 636, 637
Benzene boronic acid as a TSAI, 254
Benzoic acid, folic acid component, 249
in marzipan making, 636, 637
Benzotropolones, and tea color, 293
2,3-Benzoxazolenone, as gibberellin mimic, 558
Benzoxycarbonyl peptides, and induced fit, 44
Benzoyl arginineamide, cathepsin substrate, 474
Benzyl benzoate, in cranberries, 389, 390
Benzyl isothiocyanate, in papaya, 382, 395
Berkeley process, glucose production from cellulose, 174–175
Berry fruits as vegetables, 524
Berry juices, enzyme(p) requirement in making of, 540
Beryllium, enzymatic inhibition analysis for, 707
Betacyanins, beet pigmentation, 313
Betaine, as TMAO demethylase cofactor, 503
Betel nut, antithiamins in, 644
Beverages, 370, *See also* specific beverages
acidic, *See also* High protein acidic beverages
acidic enzyme(p)-solubilized protein in, 625
alcoholic, *See* Alcholic beverages, Distilling, specific beverages
identity of, appearance, color, 261, 267
oral lipases used in, 412
enzyme(p)-derived sweeteners used in, 152
fermented, appearance and color, 267
oxygen removal with glucose oxidase-catalase, 9
turbidity, 267. *See also* Haze
BHA, antioxidant and LOX inhibitor, 301, 437
BHT, antioxidant, 437
Bi-bi kinetics, 272
Bifunctional reagents, crosslinkers in enzyme immobilization, 143
Bifunctional reagents, matrix activators in affinity chromatography, 127
Bilberry, character impact esters of, 389
Bile salts, for enzyme extraction, 110
lipase activation, 442, 448
Binders, food, enzymatic(p) production of, 152
functional proteins as, 614, 624
Binding, enzyme immobilization by, 142–144
Biocytin, coenzyme of fatty acid metabolism, 45
ligand in affinity chromatography, 131
Bioenergetics, 84–86
Bioflavonoids, cloud retention, suspensions of, 451, 453
Biogel P, in enzyme purification, 120. *See also* Agarose
Biogenic amines, 639–641
amino acid decarboxylase, 640, 641–642
removal and prevention, 642
toxicity of in foods, 639–641
Biological functions, of enzymes *vis-à-vis* other proteins, 84–86
of LOX, 299
of myoglobin, 316–318
of peroxidase, 203–205
of vitamins, 45–46, 276, 289–290, 345, 385, 644, 657
Bioluminescence, by luciferase and myeloperoxidase, 203
dioxetanes and, 373
in enzymatic analysis, 689
in enzyme assays, 676
Biomass conversion—enzymatic, cellobiase in, 172–173
cellobiose:quinone oxidoreductase in, 172, 276
cellulase in, 167–176
cellulose as substrate, 167
glucose production by, 174–175
xylanase in, 176–180
Biopolymer degradation, hydrogen bonds in, 172
Biospecific complementarity, 137
Biospecific eluants in affinity chromatography, 131–132
Biosynthesis, of chlorophyll, regulation by light, 309–310. *See also* specific substance, biosynthesis of
Biosynthesis of enzymes and other proteins, and mutation, 95–96
apparatus, as Irad target, 199–200
feedback gene repression, regulation of, 76–77
in microbial enzyme(p) production, 94–100
of α-amylase, and gibberellins, 556–557, 575
of invertase, in tomato ripening, 679, 715
of lytic β-glucanase, yeast, α-factor, 629
of phenolase, masked RNA, regulation, 274
posttranslational, cellular aspects, 81–82
transcription, regulation in, 74–77
translation, regulation, 77–80
Biosynthetic flavors, 348, 368

Biotin, as coenzyme part, 45
as ligand in affinity chromatography, 128
2,3-Biphosphoglycerate. *See* 2,3-Diphosphoglycerate
Birch, xylanase from, 177
Birds, phytic acid in blood of, 318, 364
Birth defects, enzyme assay as aid in detecting, 4
Biscuits, fungal proteinase(p) for, 484
Bitter almond oil, 396
Bitter end, milking to, and hydrolytic rancidity, 449
Bitter peptides, evaluating microbial rennets for, 683
hydrophobicity and bitterness of, 55, 345, 616–618
in cheese, 451, 605
in fish proteolysates, 509
in soybean products, 424
removal, prevention, 423, 618
Bitterness, alliinase-induced in onion, 375
an enzyme-induced off-flavor, 423
citrus, 452–456
coumarin derivatives, 457
of peptides, *See* Bitter peptides
of thiamin, 457
phenolase-induced, in eggplant, 457
in enzymatic browning, 270, 275
prevention in onion processing, 262
Bixin, colorant, enzymatic(p) extraction of from natto, 661
LOX-induced bleaching of, 299
Black currant, ascorbate oxidase in, 289
Black gram protein, enzyme(p) solubilizing of, 623
Black Heart, 278
Black Spot, enzyme involvement, 278, 288
Black tea, 291–294
Black walnut, biogenic amines in, 641
Blackening of prawns, 325
Black-eyed peas, blanching with alum, 195
Blanching, advantages, 187–189
combined with other treatments, 191, 195–196, 226, 258, 336
cooling, in, 194
disadvantages, problems, 189–191
effluent, 195
environment, energy impact of, 190, 192, 193, 195
enzyme inactivation by and assays for adequacy of, 190–191, 206–211, 280, 526, 609, 639, 669
enzyme regeneration after, 215–217
food color, quality and, 268, 305
in dehydration processes, 237, 245
IQB and other steam-based innovations, 192–193
LT-LT (Preblanch), *See also* Hot-, Cold-break, Precooking, HTST
nutrient loss, prevention, 190, 195, 633, 644, 646
process optimization, trade-offs, 191–192, 207
specific foods or food classes, cucumber pickles, to prevent oversoftening, 529
flour, to inactivate excess enzymes, 581, 587
fruits, 191, 281, 393
high protein seeds, 436, 632
meat extenders, 245
shrimp, nutritional loss, 633
tomato processing, 546–548
vegetables, 18
types of heat treatment, 187–188
vs pasteurization, z-values for, 211
water innovations in, 192
waterless, 193–194
Bleaching, agents, GRAS, 247
of β-carotene and other pigments, by LOX isozymes, 425, 435
of wheat flour with LOX(p) in bean flours, 298, 427, 428
Blenders, in enzyme purification, 109
Blind gut, pyloric cecum as, 506
Blistering, bread, prevention with proteinase(p), 10
Bloat, a lactase deficiency symptom, 651
Blood, avian, phytic acid in, 318, 364
enzyme assays and enzymatic analysis of, 4, 669, 685
enzyme inactivation in frozen, 223
fatty acids, coffees and cascades, 442
milk lipoprotein lipases and, 448
protein, enzyme(p) processing of, 624
proteins, separation of, 116, 117
Blood clotting, anti-, factor in onion, 376
enzymatic(p) clot dissolution of, in emphysema, 665
enzymes, and red meat color, 321
enzymes, proteins of, in cascade, 46, 65, 74
phosphatides of, 256
supercontraction by, and meat texture, 483
vitamin K and γ-carboxyglutamate, 45–46
Bloom, of red meat and MRA, 322
Blotchy ripening, a tomato texture defect, 53, 514
Blowout, enzymatic(p) prevention of, 662
Blue cheese, aroma genesis in, 410, 411
classified as semisoft, 594
lipase-treated for aroma enhancement, 412
Blue dextrans as ligands in affinity chromatography, 129
Blueberries, antithiamins in, 644
character impact ester of, 389
mechanized harvest, texture and, 511
Blueing of crab, 325
Body, beer, enzymatic(p)-improvement of, 19, 559, 569
cheese, and milk proteinase, 606
processed cheese, enzyme(p)-improved, 609
Boiled clothes, a fish off-flavor, 436
Boll weevil, cholinesterase(a) from, 704
Bombardier beetle, catalase and peroxidase roles in, 205
Bonds, disulfide (S-S), *See* Disulfide bonds
hydrogen. *See* Hydrogen bonds
Bone explants, collagenase in, 487
Bones, enzymatic(p) removal of meat from, 10, 493
Bonita. *See* Tuna
Boric acid for rennet preparation, 91, 202
Borneol, structure, 404
Boron compounds as model glucose isomerase, 60
Botulism and Ca and lysozyme(p) treatment, 527, 662
Bourbon process, vanilla bean curing by, 399–400
Bracken fern, thiaminase I in, and grazing, 645
Brain, as food, 463
color perception by, 268
flavor perception by, 345
phosphatides in, 256
Bran, wheat, enzymatic browning due to, 275
in white flour, enzyme assay for detecting, 679
kernel morphology, 573
phenolase in, 275, 591
phytate in, nutrition and, 656
Branch points in amylopectin, enzyme action on, 154, 155
Branched chain amino acids and aromas, 369, 373, 391
Brandies, enzymatically-generated HCN in, 636

enzymatically(p)-generated methanol in, 638
enzymatic(p) mash treatment improves, 663
Brassica, C-S lyases of, 379, 380
B. campestris, 632
B. caranita, protein from, 632
B. napobrassica, C-S lyase of, 380
B. napus, 632. *See also* Rapeseed
Bread, *See also* Baked goods, Breadmaking
aroma genesis, 413–414
as a fermented food, 406
crumbling, and α-amylase, 326
curling prevented by proteinase(p), 10
identity of, 267
loaf volume improvement with enzymes(p), 10, 578, 583
malt lipase off-flavors in, 445
SCP-fortified, enzymatic purine removal from, 655
slicing, α-amylase action and, 326
staling, *See* Staling
sticks, bacterial proteinase(p) for, 584
Breadfruit as tree fruit vegetable, 524
Breadmaking enzymology, 9, 237, 572–592
α-amylase, 572, 573–576, 577–581
β-amylase, 572, 576–578
compared to brewing, 572
cost of added enzymes, 104
lipase and phospholipase, 585–587
LOX and other oxidoreductases, 572, 587–592
pentosanase, 582–585
proteinase, 572, 582–585
See also Baking, Bread, Cereal foods, and specific products
Breakfast cereals, oral lipase(p) used in, 412
Breath hydrogen, as measure of unassimilable galactose, 652, 654
Breeding, for controlling enzyme, 12, 437, 524, 525, 632, 633
Brevibacterium, for gustator production, 363, 367
Brewing, enzymology of, 552–571
α-amylase of barley and malt, 553–559
α-amylase in the brewing process, 559–561
compared to syrup and breadmaking, 9, 553, 572
non-α-amylase glucanases, 9, 562–567
nonenzymatic alternatives and other enzymes, 553, 567, 569–570, 571
papain(p) in chillproofing, 569–571
proteinases, in the brewing process, 567–569
See also Beer, other brewing products
Brewing adjuncts, enzymatic assessments of suitability of, 670, 709
Brick, as immobilized support, 144
Brick and brie cheeses, made with microbial rennets, 604
Brightness, color perception component, 268
Brilliance, imparted by enzyme-derived sweeteners, 153
Brine, pickling, enzymatic aspects of, 282, 525
Brinjals, 189. *See also* Eggplants
Brittleness, a vegetable texture, 511
Broad bean LOX and flour bleaching, 437
Broccoli, ammonia-steam blanch of, 195
flavorese and, 417
peroxidase of, 210
Broiling, meat texture and, 483
Broken cream, a phospholipase C-caused flavor defect, 449
Bromate, in flour texture, pentosans and, 581, 591
Bromelia pinguin. *See* Maya
Brothiness, a fish proteolysate flavor, 509
Brown end, snap bean, enzyme-caused discoloration, 279–280
Brown rot fungi, cellulose degradation by, 172
Brownie points, for bacterial α-amylase(p), 579
Browning, *See also* Discoloration, Off-color
apple scald, 269
chlorophyll containing foods, 303–305
of green vegetables, 305–309
of meats, 315–321
phenolase-related. *See* Enzymatic browning
Bruising, enzymatic impaired quality by, 277, 437, 545
Brunauer-Emmet-Teller layer (BET), and a_w, 241, 323
Brush heap cellulose, acid-hot break, consistency and, 547–548
as cellulase substrate, 169
Brussels sprouts, blanching, solids loss in, 190
catalase assay as blanching index, 211
microwave blanching, 194
off-odor of unblanched, frozen, 430
pink center, 277
residual peroxidase in blanched, 207
BTA, antioxidant, meat color retention with, 322
Buckiness, proteinase(p)-remedied dough texture, 483, 590
Budding of yeast, role of lytic glucanases in, 628
Buds, flower, as vegetables, 524
Buffalo, cheese from milk of, 594, 604
Buffers and buffering capacity, GRAS, 247
in enzyme assays, 671
of bananas, vitamin C deficiency and, 646
salts, for enzyme(p) standardization, 103
water removal, enzymes and, 223, 244
Bulbs, as vegetables, 525
Bulk density, enzyme(p) treatment for increasing, 549
Bull semen, 5′-nucleotidase(a) from, 699
Bundle sheath, site of C_3 pathway in C_4 plants, 352, 353
Burns, treatment with collagenase, 4
Burnt flavor in milk, 382
Burst release of product in active site titration, 684
2,3-Butanediol, enzymatic analysis for in wines, 697
Butanol in enzyme purification, 112
Butt reddening, an enzyme-caused lettuce off-color, 277
Butter, lipase and lipase(p) in flavor of, 412, 447
Butter cheese, made with microbial rennets, 604
Butterfat, enzyme caused flavor changes in, 413, 423
Butterfly wing color, 268
Buttermilk, sourness of, 349
Butyric acid, an off-odorant, 445, 446
blowout in cheese, suppression by lysozyme(p), 662
Butyric lactone, in dairy product aromas, 411
By-products, cellulose- and xylan-containing, 167
Byssochlamys fulva, pectinase and apricot oversoftening, 522
Byssochlamys spp., milk clotting enzymes in, 602

C_3 photosynthesis, by bee plants, and honey adulteration, 679
enzymes in acid accumulation, 352
plant sugar accumulation *via*, 356

C_4 photosynthesis, of corn, HFCS, and honey adulteration, 679
pathway in plant acid, sugar accumulation, 351, 352
PEP carboxylase in, 352
sugar accumulation *via* PEP carboxykinase, 356
C_4-C_9 aliphatic volatiles from foods, 371
C_6:C_3 pathway, in lignin biosynthesis, 529, 530
^{13}C in enzymatic analysis, 691
^{13}C:^{14}C ratio, honey adulteration, enzyme assay and, 679
Cabbage, as a leaf vegetable, 524
blanching with ascorbate, 195
enzyme-caused off-flavors in, 6, 422, 423, 575
enzymes of, ADH, pH and flavor, 394
alliinase, 380
ascorbate oxidase, 289
flavorese, 416, 417
glucosinolase, 382, 417
lipoxygenase, 430
fermentation (sauerkraut), 406, 642
isothiocyanate and nitriles in, 283, 382
pectin lyase(p)-produced unicellular food from, 550, 551
Cabernet Sauvignon wine, methanol in, 638
Cacao. *See* Cocoa
Cadaverine, a biogenic amine, 640
enzymatic analysis for, 700
Cadmium salts as cellulolysis pretreatment, 176
Cadoxin, cellulolysis stimulation with, 173, 177
Caffeic acid(s), as phenolase substrates, 276, 282, 630
biosynthesis of *cis- via* phenolase, 276
structure, 350
Caffeine, as cAMP phosphodiesterase inhibitor, 73, 442
enzymatic demethylation of, 496
microbial enzyme degradation of, 657, 661
-tannin complexes as tea cream, 359, 540
Caffeoyl shikimic acid, endogenous phenolase substrates, 276
Cake mixes, spray-dried, functionality, off-flavor in, 235, 237, 615
Cakes, improvement by enzymes(p), 10, 579
Calanus, TMAO synthase in, 504
Calcium carbonate, immobilized enzyme support, 144
Calcium ion, as enzyme co-factor affector, 46, 425, 442, 443, 466, 478, 575, 603
as nutrient, phytate and, 656
ATPase release, in the cold, 227
binding, by chelators and meat tenderizing, 482
by pectate, LT-LT, PE, and firmness, 522–523, 525, 526, 527–528
color perception and, 268
electrode, analysis for, 694
fish toughness, crosslinking and, 501
in casein micelles and renneting role, 595, 598
in muscle function and texture, 466, 467, 469, 470, 478
lignin deposition and, 521
lipid for oxidation, off-flavor and, 435
5′-nucleotide accumulation and, 364
protopectin and, 514
red meat color, 321
redistribution in cooked vegetables, 526, 527–528
spore heat resistance, PE and, 639
Calcium oxalate, 513
Calcium oxide as cellulolysis pretreatment, 176
Calcium phosphate, immobilized enzyme support, 144
gel, enzyme adsorbent, 118
Calcium phosphatide, enzyme complex as inhibitor, 257
Calcium salts, as enzyme clarification aids, 100
firming effect on pickles, 529, *See also* Calcium ions
for green color retention in vegetables, 307
in blanching, 190
in cheesemaking, 600
in juice clarification, 537
influence on solubility of enzyme-processed proteins, 621, 622
Calcium stearoyl lactate to inactivate excess flour α-amylase, 581
Calcium-binding proteins, homology of, 73, 466, 467, 498
Caldariomyces, chloroperoxidase of, 204
Callose, canned tomato texture and, 518, 525–526, 544
Calmodulin, 73, 467, 499
Calorimetry, 54. *See also* Microcalorimetry
Calves, proteinase(p)-containing feed for, 624
suckling, chymosin rennin from abomasum of, 10, 91, 494, 595
Calvin pathway, photosynthetic. *See* C_3 photosynthesis, 352
Calyx, berry, flavorese from, 418, 419
Calyx cone as flavorese source, 418, 419
Cambia, ADH isozymes, 394
Cambodian fish paste, 509
Camellia sinensis, 292. *See also* Tea
Camembert cheese, 594, 595, 599, 604
cold renneting of, 599
Camels, cheese from milk of, 594
(+)-Camphor, structure, 404
Cancer, enzyme(p) for treatment of, 5, 661, 666
Candida spp., enzymatic removal of purines from, 655
enzyme(p)-processed protein from, 629
lipase(a) from, 697
milk clotting enzymes in, 602
tannase(p) from, 541
Candied fruit, hemicellulase(p) facilitated making of, 551
Candle-like, an enzyme-caused off-flavor description, 434
Candy. *See* Confections, also creams, fondants, etc.
Cane sugar. *See* Sugar cane
Canned foods and canning, blanching, 189, *See also* Blanching
enzyme regeneration in, 216
fruits, enzyme-related firmness of, 522–523
green vegetables, chlorophyll retention in, 305–309
hams, phosphatase as a pasteurization index of, 212
orange juice, prevention of NEB by enzyme(p), 291, 341
post-blanch cooking in, 194
radappertization and sterilization, 196
steriflamme, 194
vegetable, enzyme-associated texture of, 525–529
Canner and cutter meat, collagenase(p) tenderization of, 490
Cantaloupe, 13, 94, 279, 389, 391
CAP. *See* Cyclic AMP receptor site
Caper, flavor and proteinase(p) from, 94, 382
Capillarity, role in enzyme action at low a_w, 242
Cappiness, an oxidative rancidity description, 450
Capric acid, an off-odorant, 446
Caprylic acid, an off-odorant, 446
Capsicum annuum, 302. *See* Peppers, Paprika
Caramel and caramelization, a beer flavor enhanced by

proteinase(p), 569
a type of NEB, 326, 327
as food pigments, 269
prevention by using enzyme(p), 536
Carbamates, enzymatic analysis for, 705
Carbanion, formed in GOX action, 339
Carbodiimides, for coupling in EIA, 713
in affinity chromatography immobilization, 127, 128, 144
Carbohydrates, 12, 257, *See also* individual carbohydrates
as aroma precursors, 408, 415
as enzyme constituents, 46, 47–48, 125, 203, 338
digestibility by *in vitro* enzyme(p) tests, 708
enzymatic analysis for, 694–696
functionality of, 614
plant sugar accumulation *via* degradation, 356
Carbon dioxide (CO_2), carbonic anhydrase as aid in analysis for, 711
control of enzymatic browning with, 283
enzymatic analysis for manometry of, 698
in breadmaking, 571, 572
in controlled atmosphere storage, 531
in packaging, 311, 531
in waterless blanching, 194
Carbon monoxide, control of enzymatic browning with, 281
mitochondrial reducing power and, 324
Carbonate determination, enzyme electrode for, 694
Carbonation, a brewing step, papain(p) effects on, 553
Carbonium ion in transition state, 254
Carbonyls, in fruits, 389
in onion pinking, 312, 313
NEB and, 327
Carborundum, in enzyme purification, 110
S-(2-Carboxyethyl)-L-cysteine, in OFP biosynthesis, 133, 376
γ-Carboxyglutamic acid, blood clotting, 45
Carboxyglutamyl tyrosine, a proteinase substrate, 474
Carboxylic acids, aromatic, control of enzymatic browning with, 282
Carboxyls, glutamate, pK_a's of, 262
Carboxymethyl cellulose (CMC), as cellulase substrate, 169, 172
for tenderizing meat, 484
in chymosin assay, 597
in dry food model systems, 241
papain(p) used in analysis for, 711
prophenolase activation by, 274
Carboxymethyl hydrazide, for enzyme immobilization, 158
Carboxymethyl xylans, as xylanase substrates, 179
Carcass positioning, and meat texture, 463, 468
Carcinogens, 622, 639, 661
Cardboard, a barley off-odor, 299, 429, 445
a fish texture defect description, 499
Cardiovascular disease, onion eating and, 376
prostaglandin synthase inhibition and, 501
Cardoon flower petals, rennet from, 601
Carica papaya. *See* Papaya
Caries, prevention by xylitol, 177
Carlsberg Laboratory, 594
Carob bean, as alternative to α-amylase(p) in breadmaking, 579
tannin from, 661
Carotene(s), blanching loss and retention, 189, 190, 263
bleaching by LOX, 298, 299, 425, 427, 428, 431, 435, 578
color of, 268
non-LOX bleaching of, 300
reaction with quinones, 401
Carotenoids, baking aroma genesis and degradation of, 414
biosynthesis, regulators of, 269
conversion to abscisins, 299
enzyme-related loss of, 644
in fish discoloration, 325
mevalonate and, 403
reactions with quinones, 400, 401
See also Carotene
Carrageenan as enzyme inhibitor, 258
Carrots, aroma, constituents of, 371, 386, 387, 638
bitter substance in, 457
blanching, losses and enzyme assays for, 190, 211
color, control of, 269, 275
enzymes of, exoPG, 528
flavorese, 416, 417
nucleases, 366
peroxidase, 264
phenolase, and enzymatic browning, 366
juice, enzymatic(p) making of, 548, 550, 551
phytate as antinutrient in, 655
rotting PG in, 517
suberization of, 278
texture (elasticity) of, 462
Carvacrol, structure, 404
Cascade regulation of enzymes, blood clotting, 64, 65, 74
cyclic AMP, prostaglandins, calmodulin in, 73
glutamine synthetase, 72–73
glycolytic enzymes, 71, 74
hormones, second and third messengers, 73
lipase, caffeine and, 442
Casein, as a micellar complex of 4 phosphoproteins, 595
as affinity adsorbent for rennets, 598
as phosphoproteins and phosphatase action on, 609
flocculation in frozen milk, 610
hydrophobicity and, 596, 598, 647
in milk clotting, 55, 225, 595–598
insusceptibility of non-k-caseins to rennet(p), 597–598
involvement in cheese texture and flavor, 409, 606, 609, 616, 662
nutritiousness impaired by phenolase, 630
α_{s1}-Casein, localization of milk proteinase in, 606
β-Casein, attack by ripener proteinase, cheese texture and, 608–609
contribution to cheese texture, 606, 608
peptides from, and rennet blends, 604
preferential hydrolysis by milk proteinases, 606
κ-Casein, substrate for milk clotting proteinases, 595, 596
Casein micelles, and Irad effects on enzymes, 198
hydrolytic rancidity and, 448
in milk clotting, 598
localization of milk proteinases in, 606
structure, composition and self-assembly of, 595–596
Caseinate in meat, enzymatic detection of, 709
Caseinogen, in affinity chromatography, 123–124
Casing, sausage, enzyme tenderization, 493
Cassava, enzyme generated HCN and CNS^- as toxin

from, 634–635
Castor beans, glyoxylate cycle in, 351
lipase activator in, 443
Castor plant, rennet from, 602
Catabolism of amino acids in feedback repression, 76
Catabolite repression, 76, 95, 169
Catalysis and catalysts, 3, 28. *See also* Catalytic efficiency
Catalytic competency and operational molarity, 671, 683–684
Catalytic efficiency of enzymes, active site, tertiary structure and, 54–59
and induced fit, 44, 53
and quaternary structure, 66–67
comparison with nonenzyme catalysts, 50
in definitions of "Enzyme," 27, 28
metals, role in, 57–59
model compounds for, 59–62
organic physical chemical theory, 55–57
Cataracts, from excess dietary galactose, 643
Catechins, as phenolase substrates, 276, 292
condensed, in beer chill haze, 570
Catechol, in onion skin, 276
Catechols, interaction with protein, 631
melanin from DOPA as, 277
phenolase substrate, 273
Cation, metal, as enzyme cofactor, 45, 46
Cationic detergents, 264
Catsup, consistency of, 550, 554
Cattle, enzyme-aided analysis for diethylstilbestrol in, 712
Cattle pancreas, enzyme(p) from, considered safe, 647
Caucus, flavor precursor of, 378
Cauliflower, as a flower bud vegetable, 524
blanching, losses, ascorbate, and enzymes, 190, 195, 207
enzyme induction in, 77
enzymes of, 399, 430
frozen, effects of fluctuating temperature, 228
purpling, a cold stress response, 227, 269
texture enzymology of, 528
Celery, as a petiole vegetable, 524
collenchyma of and texture of, 513
enzymes of, flavorese, 416
β-glucosidase, a potential health hazard, 636
peroxidase, 264
phenolase and browning of, 275
etiolation of, 311
Cell(s), adhesive, propectin as, 518
and enzymes, definition, 27
as Irad targets, 196, 199
blood, *See* Erythrocytes, Leucocytes, Platelets
fixed, as supports for immobilized enzymes, 138, 145, 164
function, traced by inhibitors, 247
in enzyme biosynthesis, 81–82
sclereids and pear texture, 521
structure, organelles of, 15, 27, 81–82, 306, 353, 465, 502, 512, 583
texture and, 512–513
See also Cell disruption, specific cell organelles
Cell disruption, and alkali treatment, 263
and date quality, 520, 521
and digestibility, 436
and enzyme action in dry foods, 237
and enzyme action in frozen foods, 224, 226–227, 336
and flavorese effects, 420
by ferredoxin-mediated lipid oxidation, 434
in blanching, 189–192, 237, 416
in enzymatic color changes, 227, 278, 292
in enzymatic flavor changes, 6, 347, 368–369, 395, 400, 415, 420
in enzyme isolation, 109–110
in freeze-thaw detection, 682
in juice and wine making, 536
lipid peroxidation, flavor and, 372, 435
NEB and, 327
Cell envelopes, 15, 141, 196, 336
Cell membrane(s), *See also* Cell, Cell disruption, Cell membrane destabilization
and penicillin, 249
and plant food texture, 518, 524
and tea flavor, 400
as Irad target, 199
chill-sensitivity, 334
fragility of, as enzymatic browning pacer, 279
in enzyme biosynthesis, 77, 82
in enzyme regulation, 65
in fish texture, 500
maintenance and modification, modes of enzyme management, 15, 16, 27, 246, 335, *See also* Cell membrane destabilization
milk fat, enzymes of, hydrolytic rancidity, 448, 449, 451
permeability and color perception, 268
in muscle function, 467
PE and hardcore, 530
peroxidase as topological probe of, 205
phospholipids, 15, 25, 279, 433
starfish cecum, proteinase of, 506
yeast, 629
Cell membrane destabilization, and flavoreses, 418
and PE potentiation, 725
by freezing, consequences, 227
by Irad, and fruit texture, 523
and meat texture, 473
by lipid oxidation, 434
by vitamin A, 446, 482
in cocoa making, and aroma, 402
in SCP purine removal, 655
of starch granules, in cold sweetening, 334
Cell separation, by enzyme(p), 548–551
in cooked potatoes, 526
in nectar production, 544
in tomato products, 546
protopectin in, 515
See also Cell, Cell wall
Cell wall(s), Ca redistribution, texture and, 528
cell separation and, 526
debris, as adsorbent affinity chromatography matrix, 191, 515, 522
decomposing enzyme(p) for, protein processing, 627, 629
protoplast isolation, 398
texture related food products, processing, 535, 547–548
enzymes of and sugar translocation, 289, 357
in hard core disorder of sweet potatoes, 530
in tomato processes, 546, 547

of the wheat kernel, 573
of yeast, 628
structure, composition and function, 512–513
Cellobiose, as cellobiase (β-glucosidase) substrate, 170
as cellulase inducer, 169
as cellulase inhibitor, 171, 173, 529
for prevention of olive oversoftening, 529
Cellodextrins, as cellulase, cellobiase substrates, 170
Cellohexaose, 170, 171
Cellophane, colored, for antigreening, 311
Cellotetraose, synthesis by cellulases, 170
Cellulolysis, cellobiase in, 172–173
cellulase systems for, 169–172
hydrogen bonds and resistance to, 172
oxidative, 172, 173, 518
pretreatments of cellulose for, 176
reactors for glucose production, 174–176
Cellulomonas spp., 176
Cellulose, and derivatives, in enzyme immobilization, 144, 145, 653
in enzyme purification, 119, 124, 127, 132
as cellulase inducer, 169
as dietary fiber, 710
brush-heap, shear, acid-hot break and, 547–548
cell wall and protopectin, 512, 513, 514
derivatives, as cellulase substrate, 169–170, 176
enzymatic analysis for, 696
for capillarity in model low a_w systems, 240
hydrolytic degradation, *See* Cellulolysis
microcrystalline, consistency and, 544, 547
microfibrils, *See* Microfibrils, Cellulose
pretreatment, for cellulase action on, 176
SCP from, 628
with xylan and lignin, in by-products, 167
Cementing polymers, and consistency, 546
Centralized operations and enzyme action, 319
Centrifugation, and gel filtration, *See* Chromatofuge
fast analyzer in enzymatic analysis and EIA, 690, 699, 714
in enzyme production and purification, 107, 114–115
time-saving with enzyme(p), 541
Cephalin, dried, NEB, eggs and, 339–340
Cephalosporin, a TSAI, 254
Cercospora spp., enzyme assay for detection of, 678
Cereal products, dried, 237
processing enzymology, 10, 572, 592
protein digestibility of, 708
sugars and starch in, enzymatic analysis for, 695, 696
See also Bread, specific products, Cereals
Cereals, *See also* Grains, specific cereals
as vegetables, 524
enzymes in, ascorbate oxidase, 289
lipases, 209
LOOH isomerase, 303
prophenolase, 274
in whiskey making (distilling), 561
milled, off-flavor in, 237
the earth's most used foods, 635
zymograms, applications of, 682–683
Cerebral cortical necrosis, a thiaminase I-caused disease, 645
Cerebrosides, off-flavor and, 446
Cerulenin, inhibitor of flavonone synthase, 456
Ceylon mite, and tea flavor, 15
Chaetomium cellulolyticum, as poor cellulase source, 169
Chaetomium spp., milk clotting enzyme in, 602
Chain length, cell-separating enzymes and, 550
Chain length, of pectin, consistency and, 544
Chain types in amylopectin, 554
Chalkiness, a fish texture defect, 499
Change-of-state, and foods, 533, 535, 552, 595
Chaotropic salts in affinity chromatography, 131
Chapaties, enzymatic browning of, 278, 281
Character impact, fruit aroma, esters of, 389
Charcoal, in affinity chromatography, 123
invertase adsorption for invert sugar, 160
Charge neutralization, and juice clarification, 537
in milk clotting mechanism, 598
Charge transfer chromatography and affinity chromatography, 125
Charge-activated relay in metalloenzyme action, 56–57
Charybdis. *See* Scylla, 328
Chastek paralysis, and thiaminase I, 645
Cheddar cheese, cheddaring, a step in cheesemaking, 605
classified as hard, 594
curdiness of, a texture defect, 599
enzyme of aroma genesis in, 409
microbial rennets for, 604
oral lipase(p) used in, 412
Cheese, as fermented foods, 406
biogenic amines as potential toxins in, 641
bitter peptides in, 617
blow-out, prevention with lysozyme(p), 662
flavor, 362, 369, *See also* Cheese aroma, Cheesemaking
functional proteins in, 615, 616
lactose, enzymatic analysis for, 695
microbial rennets used in making, 604
processed, enzyme(p) in making, 609–610
spray drying of, 235
texture-based classification of, 594
Cheese aromas, components of, 392, 407
enzymes of carbohydrate metabolism and, 408–409
enzymes of lipid metabolism, 409–412
enzymes of protein metabolism, 409
lactones, 411–412
lipase, 410, 412–413
sources and costs of enzymes(p), 104, 407
Cheese manufacture. *See* Cheesemaking
Cheese texture. *See* Cheesemaking
Cheesecake, oral lipase(p) use in, 412
Cheesecloth, in enzyme purification and immobilization, 111, 143
tearing by rennet cellulase contaminant, 105
Cheesemaking, and cheese texture enzymology, 6, 10, 19, 234, 593–610, 616
chymosin, rennet (veal) and rennet alternatives, 594–595, 600–605
continuous cold renneting in, 599
post curdling proteinase contribution to texture, 605
the curdling step, 598–600
the primary event: cleavage of Phe(105)-Met(106) of k-casein, 596–598
the substrate, kappa-casein in micelles, 595–596
water removal in, 234
Chelators, as enzyme stabilizers, 102
control of enzymatic browning, 281

for preventing milk oxidative rancidity, 451
for tenderizing meat, 482–483
in subunit dissociation, 274
in taste acuity, 346
Chemical analysis, enzyme(p) as aids in, 709–715
Chemical oxygen demand (COD) in blanching, 194
Chemical score *vs* enzyme test for protein quality, 708
Chemicals, adventitious, removal by blanching, 189
combined with blanching, 194–195
control of enzyme action by, 11, 247–264, *See also* Inhibitors, Inhibition
manufacture, use of enzymes in, 6
Chemiosmosis in ATP biosynthesis, 85, 290
Chemotactic peptides, phospholipase A_2 and leucocytes, 443
Chemotaxis, and methylation, 86
Chenin blanc grapes, increased wine yield by enzyming, 540
Cherries, amygdalin in product, enzymatic analysis for, 699–700
ascorbate oxidase reported in, 289
HCN generating enzyme system in pits of, 636
light-induced loss of nutrients in, 644
maraschino, use of enzyme derived maltose in, 152
mechanized harvest, and texture, 511
microwave blanching of, 281
nectar, enzymes in manufacture, 544
pesticides, in enzymatic-TLC analysis for, 705
phenolase isozymes and enzymatic browning, 274, 275
stem discoloration, 277
texture-affecting enzymes and processed, 522–523
Chewiness, a fish texture defect, 499
of cucumbers, 511
Chewing. *See* Mastication
Chewing gum, pepsin(p) in, 4, 664
Chicken(s), cathepsins, autolysis and texture, 473
enzyme blocking for NMR analysis of polyphosphate in, 712
liver, enzyme and Irad of, 199
ovoinhibitor, active site titration of, 706
pepsin(p), as rennet in Israel, 10, 261, 601
proteinase(p)-containing feeds for, 624
skin color, 301, 315
vitamin B_1 deficient, thiaminase I and, 645
water in frozen fluctuating storage, 228
zymograms of to detect previous freeze-thawing, 682
Chick-peas (garbanzos), enzyme(p) for solubilizing protein, 623
Chicory, enzymatic detection in coffee, 709
Chill haze, nature of and enzymatic(p) removal, 567, 570–571
Chill sensitivity and membranes, 334
Chilling, off-flavor prevention by, 437
Chilling injury, a fruit texture defect, 522, 524
Chillproofing beer, 553, 569–571
Chinese chives, flavor precursor of, 378
Chinese restaurant syndrome, 640
Chirality, analog inhibition and, 249
and enzyme specificity, 31–33
in LOX action mechanism, 425
of cyanogenic glycosides, 396
of glucosinolates, 396
of model enzymes, 59
of onion flavor precursors, 374
of pentoses, hexoses and HFCS, 154, 162
Chitin, as immobilized enzyme support, 143, 145, 158, 571, 662
hydrogen bonds and enzyme degradation of, 172
Chitosan, as lysozyme affinity chromatography matrix, 124, 662
in adsorption affinity chromatography, 124
Chives, dehydrated, rehydration of, 246
flavor of, 378, 382
Chloramphenicol, for yield of microbial enzymes, 97
Chloride ion, as enzyme affector, 46, 317, 474, 484
as peroxidase substrate, 203, 204
determination, selective electrode for, 694
suggested enzyme activation analysis for, 707
Chlorinated hydrocarbons, enzymatic analysis for, 704, 705
Chlorinated pesticides, enzymatic analysis for, 704, 705
Chlorine, 300
Chlorine dioxide, peroxidase donor, acceptor, 31
Chlorocholine chloride, a gibberellin mimic, 558
S-β-Chloroethyl-L-cysteine, 33
Chloroform autolysis, in enzyme purification, 106, 109
Chlorogenic acid, as a plant acid, 350
as endogenous phenolase substrate, 276, 283, 293, 295
decrease of, as enzymatic browning pacer, 279
pepsin(p), used in analysis for, 711
quinone, phosphorylase inhibitor, 330
structure, and melanins from, 277
iso-Chlorogenic acid, 630
Chlorophyll, and blanching, 189
as food pigments, 268, 269
biosynthesis, and solanine biosynthesis, 311
chlorophyllase in, 307, 310
Irad inhibition of, 337
prevention of, 310–311
scheme and regulation of, 307–310
color loss in foods, in canned, 305–306
in dehydrated, 309
in fresh, postharvest, 309–311
in frozen, 228, 303–305
prevention of, 305, 306–308, 310–311
destruction by bleaching enzymes, 269, 299, 309
destruction by ethylene, 309
in fluctuating temperature storage, 228
pheophytinization of, 302–305, 306, 311
retention, 305, 308, 310–311
Chlorophyllide, 302, 305, 310
Chlorophyllin, 269
Chlorophylls a and b, 302
Chloroplasts, and photosynthesis, 352–354
as plant cell component, 513
ATP-driven reverse electron transport of, 285, 322
chlorophyll changes in, 306, 309–310
ethylene generation and, 286
food color and, 268
glutamate synthase of, 362
isolation *via* cell-wall degrading enzymes, 550
lipid peroxidation, off-flavor and, 435
of mesophyll cells, site of C_4 photosynthesis, 352, 353
phenolase in, 274, 275
plant acid accumulation and, 352
SOD of, 287
tea flavor and, 15, 400
ubiquinone in, 46

Chocolate, oral lipase(p) used in, 412, *See also* Confections
quality, enzymes and, 401–403
syrup, α-amylase for viscosity control, 9
toxic effects of tyramine in, 640
Cholemyoglobin, green meat pigment, 325
Cholera toxin, 600 molecules detectable by USERIA, 713
Cholesterol, biosynthesis of, 405
blood, onions and cysteines, 376
degradation by microbial enzyme system, 660
dietary feedback repression and, 76
enzymatic analysis for, 691, 700
LOX-induced oxidation of, 299
prevention of intestinal absorption, 651, 654, 661
7-α-Cholesterol oxide, as an enzyme-generated toxin in foods, 642, 643
Choline, as TMAO inhibitor, 503
Chorleywood no-time process, in breadmaking, 305
Chromatin in adsorption affinity chromatography, 124
Chromatofuge, in enzyme purification, 122
Chromatography, gas-liquid, 349, 364, 369, 704, 705
hydrophobic, 119
molecular sieve, 574
non-affinity in enzyme purification, 118–122
Chromatography, affinity. *See* Affinity chromatography
Chromatoplast, a plant cell organelle, 513
Chromogenic assays, 207, 336, 673, 684
Chronometric integrals, 250, 251, 253, 255
Churning, lipase and, 447
Cider, aroma constituents, 389
as fermented beverage, 407
peroxidase, heat resistance of, 206
phenolase in color and flavor of, 270, 278
starch haze removal by α-amylase(p), 537
1,8-Cineole, structure, 404
Cinnamic acid(s), as TSAIs, 255
control of enzymatic browning with, 282
metabolism by xenobiotic oxidase, 412
phenolase in biosynthesis of, 275
route. *See* C_6:C_3 pathway
Cinnamon, a bread off-flavor, 421
Cinnamyl alcohol, ADH substrate, 394
"Circe" effect, 49
Circular dichroism (CD), 44, 271, 273
Circulatory dialysis in enzyme purification, 113
Cisternae, endoplasmic reticulum, enzymes in, 387, 467
Citramalic acid, structure, 350
Citrate, 33, *See also* Citric acid
as allosteric affector, 71
as enzyme inhibitor, 256, 289
for standardizing enzyme(p), 103
in casein micelles, 595–596
prochirality of, 33
tenderizing meat with, 482
Citrate cycle. *See* TCA cycle
Citric acid, accumulation in fruits, 354
as a blanching adjunct, 195
as a fruit flavor, and structure of, 349, 350
as enzyme inactivator, 263, 281, 581
as onion pinking inhibitor, 313
as sequestrant, astringency suppressant, 195
control of enzymatic browning with, 281, 285
enzymatic analysis kit for, 691
meat color retention with, 323
iso-Citric acid, structure, 350
Citronellol, structure, 404
Citrus fruit(s), biogenic amines in, 641
enzyme assay for monitoring spoilage, 678
enzymes of, *See also* Citrus juice
ascorbate oxidase in rind, 289
PE and PG in albedo, 521–522
phenolase not active in, 275
lycopene synthesis in, 269
monoterpenoid synthesis in, 403
pesticides on, enzymatic-TLC analysis for, 705
sourness of, 349
uronic acid oxidase, 522
See also Orange(s), other citrus fruits, Citrus juice
Citrus juice, aroma and flavor of, 349, 354, 385, 390, 393–394, 395, 403
bitterness, enzyme involvement in, 452–456
cloud, retention and collapse, 522, 538, 541–543
enzymatic(p) removal of off-flavors from, 423
PE and PG in thermal stability, 209
See also Orange juice, other specific juices
Citrus leaves, water potential isotherm, 233
Citrus oils, cold-pressed, pectinase(p), 549
Cladosporium resinae, source of exo-pullulanase, 154
Clam(s), enzyme-caused toughening in, 502
flavor of, 416
luciferase, 203
octopine dehydrogenase(a) from, 700
style, potential glucanase(p) from, 91
tenderization of with proteinase(p), 10
Clarification and clarity, aided by Steriflamme, 194
of beer, by enzyme(p), 553, 569–571
of fruit juices, by enzyme(p), 535–539, 540
of instant tea, 539–540
phenolase a problem in enzyme(p) for, 275
promotion factor, pectin lyase as, 357
spontaneous, 536
See also Juices, specific beverages
Classification, nomenclature, of enzymes and problems in, 28–31, 270–271, 375, 440–441, 657. *See also Enzyme Index*
Classification of cheese, based on texture, 594
Claude Bernard's dictum, 63
Clearing, a brewing step, enzyme(p) used in aiding, 553, 555
Clearing factor. *See* Heparin
Climacteric, enzyme and isozyme changes during, 288, 352
Climate, tea flavor and, 13, 15
Clingstone peaches, endo PG and texture of, 519
Clinical enzymology, 685
Closed loop steam blanching, retrieval of nutrient solids by, 194
Clostridium histolyticum, collagenase of, 487, 495, 496
C. multifermentans, PE, pectate lyases from, 515
Cloth as enzyme immobilization support, 141
Clots, blood. *See* Blood clots
Clotting. *See* Blood, Cheesemaking, Milk Clotting
Cloud(s) and clouding, agents, 545, 614
collapse, retention of, 522, 535, 538, 541–543, 544
composition and physiochemical properties of, 536–537

enzymatic(p) removal of, 535, 538, 539, 540–541
nonenzymatic removal of, 537
See also Haze, specific beverages
Clover, linamarin in, grazing cattle and, 635
phenolase-impaired nutritiousness of, 630
white, CN-glycosides, and ecology, 422
Cloves, aroma constituents of, 392
C-N bond, nonpeptide, enzyme for, 496
CoA. *See* Coenzyme A
Cobalamine, as coenzyme and affinity chromatography ligand, 45, 128
Cobalt, as cofactor, activator, Zn replacement, 45, 46, 163, 301, 385
enzyme activation analysis for, 707
Cocarboxylase. *See* Thiamin pyrophosphate
Cockroaches, wood-eating, cellulase of, 168
Cocoa, anthocyanase in making of, 296
aroma genesis, enzymes in, 270, 369, 401, 402, 406, 407
as a fermented food, 406
biogenic amines as potential toxins in, 641
bitter peptides in, 616, 617
enzymes, resident, in production of, 14, 294–296
lipase(p) added for flavor, 413
phenolase in color, flavor, 8, 14, 270, 275, 276, 278
sugar in products, enzymatic analysis for, 695
Coconut(s), an oxidative rancidity description, 450
aroma constituents, 392
oil, alkanones in rancid, 411
protein, enzyme(p) processing of, 622, 627
Codex Alimentaire of FAO/WHO, 679
Codfish, enzyme-caused toughness in, 502
flavor of, 416
off-odor genesis in, 457
oil, proteinase(p) for extracting, 493
oxalacetate decarboxylase, TSAI inhibitor of, 52, 706
water sorption isotherm of, 232
Codons, 74, 75
Coenzyme A, 45. *See also Enzyme Index* for CoA-related enzymes
Coenzyme F420, in methane producing enzyme systems, 46
Coenzyme M, 46. *See also* 2-Mercaptoethane sulfonic acid
Coenzyme Q, 46. *See also* Ubiquinone
Coenzymes, as enzyme cryoprotectant, 224
as ligands in affinity chromatography, 125, 128–129
definition and listing, 45, 46, *See also* specific enzymes
freeze-thaw enzyme inactivation and, 222
in acid enzyme inactivation, 263
Cofactors of enzymes, definition, listing, 45–47, *See also* specific cofactors
in affinity chromatography, 131
in enzyme purification, 113, 114
Coffee, berry, enzymes in removal of seed coat, 6, 549, 550
cellulase(p) for preservation, 551
enzymatic decaffeination of, 647
enzymatic detection of chicory in, 709
flavor and flavorants, 370, 413
grounds, hemicellulase(p), utilization of, 550
instant, 234
phenolase and quality of, 278, 403, 680
spray drying of, 235
water sorption isotherm of, 232
Coffee cakes, use of bacterial α-amylase(p) in, 579
Coffee whiteners, oral lipases used in, 412
Cohesion, imparted to foods by functional proteins, 614
potato texture and, 526
Cohesiveness, imparted to food with enzyme-derived syrups, 153
potato texture and, 511, 526
Colby cheese, made with microbial rennets, 604
Cold, enzyme action in. *See* Low temperature
Cold break, in tomato processing, 546
Cold hardiness, and cell membranes, 334
Cold renneting, 599
Cold shock, in enzyme purification, 110
Cold shortening, a meat textural aberration, 227, 469–470
Cold sterilization by peroxidase, 205
Cold sweetening, enzymes in and theories of, 333–334
peas, 359–360
potatoes, 328, 331–337, 402
prevention of, 336–337
See also Starch-sugar interconversion
Cole slaw, enzyme-caused off-flavor of, 6, 7, 422, 423
Coliforms, in milk, cheese amines, amino acid decarboxylases and, 642
Collagen, as immobilized enzyme support, 141, 571, 653
crosslinks and meat texture, 476
diminution agents for tenderizing meat, 482
hydrogen bonds and enzyme degradation of, 172
hydrophobicity of and nonbitterness of proteolysates, 617
in affinity chromatography, 124, 129
in enzyme action at low a_w's, 243
meat history *via* pronase susceptibility to, 680
meat tenderization *via* enzymatic attack, 281, 282, 463, 476
post-translational cleavage of, 65
Collapse, of clouds, 541, 542
Collodion, as immobilized enzyme support, 143
Colloids, juice making and, 536, 537
Colneleic acid. *See* Vinyl ethers
Color, food, *See also* Off-color, Discoloration, Browning, etc.
and appearance, an aspect of, 267, 268
as a food quality attribute and guide, 267–269
definitions of, 267–268
developer, in onion pinking, 312
food colorants and pigments, 268–269
losses, 189, 298, 299, *See also* specific cases, Off-colors, etc.
measurement of, 268
nonphysiological, 269, 277
of bread, 573
of fruits and vegetables, *See* specific foods
of meats, 315–325
retention, fruits, *See* Enzymatic browning, prevention of
green color of plant foods, *See* Chlorophyll
imparted by enzyme-derived sweeteners, 153
meats, 321–325
paprika, 301
tomatoes, in Cold Break, 547
vegetables, *See* Chlorophyll retention, Enzymatic browning, prevention of

wine, by use of GOX(p), 341
reversion, green color retention *via*, 300
Colorants, biological, bioregulators and, 269, 661
enzymatic analysis for, 701
enzyme(p)-aided extraction from natto, 536
food, endogenous, 268–269, *See also* Pigments
GRAS, 247
LOX-caused loss of, 298
Colostrum, lipase, 448
peroxidase asepsis and, 205
Come-up period in canning, 305
Commercial food enzyme preparations, safety and toxicology of, 634, 646–648, *See also* Enzyme preparations
Comminuted fish, for autolytic products, 507
Comminuted products, processing enzymology of, 535–551
tomato products, enzymology of, 545–548
Compaction of muscle in rigor, 468
Compartmentalization. *See* Compartmentation
Compartmentation, cellular, of enzyme and substrates, 15, 16, 17, 19, 357, 387, 405, 410. *See also* Localization, Decompartmentation
Competition, between enzyme producers and producer-users, 157
Competitive inhibition. *See* Inhibition, competitive
Complement, and cascade mechanisms, 74
Complementary oxidation, of fats and hemes in meats, color/flavor losses, 320, 322–323
of myoglobin and acetoacetic acid, 320
Composted, unblanched frozen foods off-odor description, 430
Compound I, catalase, 220
Compressibility, an objective component of food texture, 462
Computer, cascade regulation as, 73
Computer interfaced fast analyzer, 690
Computer programs for enzyme kinetics, 36
Concentrates, food, 234
protein. *See* Protein concentrates
Concentration, by water removal, 234–235
of solute, effects on enzyme behavior, 238
Concentration *vs* dehydration, 234
Conconavalin A, as ligand in affinity chromatography, 130
Condensed milk, use of lactase(p) in making, 180
Conditioning of cold-stored potatoes, 328
Conductivity, in enzyme analysis, 691
Cones, eye, color perception, 268
Confectionery (candy making), α-amylase(p) for candy scrap recovery, 9
bacterial α-amylase(p) in making, 579
candied fruit, multiple enzyme action in making, 106
enzyme-modified starches for texture of, 152
enzyme-produced sweeteners in, 152
functional proteins in, 615
hemicellulases in making, 550
invertase(p) in making, 9, 19, 159
lipase(p) in making, 9, 412
PG action, undesirable in, 549
sugars, enzymatic analysis for, 695
xylitol in, to prevent caries, 177
Confinement, enzyme immobilization by, 141–142
Conformational changes in proteins and enzymes, and enzyme inhibition, 259
cold-induced effect on enzymes, 220
color perception and, 268
in ATP biosynthesis, 85, 285
in enzyme immobilization, 145
in enzyme regeneration, 213, 214
in gluten, SS-SH relationships, and dough texture, 584, 585, 590
in MRA and meat color, 322
in muscle function, 463–467
in protein denaturation, 186
induced fit and nature of enzyme action, 42–44
laccase action and, 273
of biologically active receptors, 85–86
of isozymes, in regulation, 48, 49, 83
of lipase interfacial recognition site, 444
proenzyme activation and, 275
Coniferyl alcohol, lignin from, ADH substrate, 394, 521, 529, 530
Connective tissue, muscle, meat, texture and tenderizers, 463, 468, 491
Connectors, in affinity chromatography, 126, 127
Conserved messengers. *See* RNA
Consistency, a potato texture description, 526
cellulase(p) for reducing, 549
of comminuted tomato products, 544
PG assay for monitoring retention of, 29
Constitutive *vs* inducible enzymes and enzyme biosynthesis, 75
Consumer, end of the food processing chain, 11
Continuous procedures, flow, in automated kinetic enzyme analysis, 689–690
HFCS production with immobilized glucose isomerase, 166
in food processing, use of enzymes(p) in, 536
mix processes, bleaching in breadmaking, 300
with immobilized rennet in cheesemaking, 600
Contractile apparatus, meat tenderization *via*, 481
Control of enzyme action, 11, Chapters 10, 12–14. *See also* control of specific quality attributes, processes, and foods
Controlled atmosphere storage, 278, 281, 310–311, 530–531, 536
Convection drying, 235
Converter enzymes in cascade regulation, 73
Cooked off-flavor of milk, treatment with sulfhydryl oxidase(p), 452
Cookies, lipase(p) for improved flavor of, 412
pentosanase(p) for increased spread of, 582
proteinase(p) for improved, 10, 584
Cooking, a brewing step, enzymes in, 553
human culture development and, 650
meat texture and method of, 463, 483
of lima beans, destroys HCN-enzyme system, 636
time shortened by enzyme(p), 549
Cooling, after blanching, losses in, 190, 194
freeze-thaw, enzyme-caused meat tenderization by, 483
milk, spontaneous rancidity and, 447
Cooperativity, in enzyme action, inactivation, 44, 68, 186, 206, 213
negative, 66
Coordination sphere, role in metalloenzyme catalysis, 58
C-O-P and C-O-P-O-C bonds, 366
Copepods, TMAO synthesis in, 504

Copolymerization and enzyme immobilization, 141
Copper, ammonium cellulose, cellulase substrate, 168, 169
as nutrient, phytate, fiber, 656
chelates for preventing oxidative rancidity, 451
compounds in enzyme purification, 112, 113
cupric pheophytin, chlorophyllide, in color reversion, 307
electrodes, selective for determination of, 694
enzymatic inhibition/activation analysis for, 707
enzymes, in evolution and photosynthesis, 59
in enzymes, oxidation state, mechanisms, 46
and photosynthesis, 59
ascorbate oxidase, 46, 288
cytochrome oxidase, 46
galactose oxidase, 339
hexose oxidase, liver, 46
laccase, 273
lysyl oxidase, 496
nitrite reductase, 46
phenolase, 271, 272, 282
superoxide dismutase, 287
in hemocyanins, oxidation state of, 271
pyridoxal phosphate, as model enzyme, 59
Coprecipitation, in tea cream, 539
Copulatory outgrowth, lytic glucanase and, 629
Core, remaining after proteolysis of plant protein, 624
Corms as vegetables, 524
Corn, *See also* Sweet corn
as brewing adjunct, 552
C_4 photosynthesis in, 352
enzymatic off-flavor in products, 237
enzymes of, lipase and rancidity of, 445–446
lipoxygenase of, 429–430
peroxidase thermal stability of, 207
xylanase of, 178
grits, quality of and peroxidase inactivation, 209
hemicellulase(p)-assisted degermination of, 9, 582
high lysine, 209
hydrolysates, spray drying of, 235
on-the-cob, blanching of, 193
Refiners Association, syrup composition table, 153
scalding of, 207
starch, *See* Starch
syrups, *See* Syrups, starch-derived
zymograms as metal deficiency indicator in, 682
Corn leaf blight, and lipase, 445
Corncob xylan, 176
Corrosivity factor, 191
Corrugation, to prevent enzymatic browning, 281, 294
Cortisone, and meat texture, 482
Corynebacterium spp., for glutamate production, 363
milk clotting enzymes in, 602
Cost, unit of enzyme(p) for food use, 104, 635
Cosubstrate as anylate in enzyme activation analysis, 706–707
Cosubstrate as eluant in affinity procedures, 131, 132
Cottage cheese, 19, 594, 599–600
Cotton (fibers) as cellulase substrate, 169, 173, 176
Cottonseed, hulls, in biomass conversion, 167
phytase action in, 659
protein, autolytic modification of, 626, 627
protein, enzymatic(p)-processing of, 622, 627
p-Coumaric acid, phenolase product, 275, 282
Coumarins, bitter, phenolase in biosynthesis of, 276, 457
p-Coumaryl CoA, a flavonone synthase substrate, 455
Countercurrent dialysis, in enzyme purification, 113
Countercurrent two phase systems, in enzyme(p) production, 117
Coupled enzymes, in enzymatic analysis, 686
Coupling, oxidative, of flour proteoglycans, 591, 592
Covalent affinity chromatography, 128
Covalent bond breaking, and enzymes, 27
Covalent bonding, enzyme immobilization by, 142–144
Covalent chromatography, 132
Cowiness, a lipase-caused off-odor, 447
Cows, and effect of grazing, 422
and effect of lactation on spontaneous hydrolytic rancidity, 447
and effect of udder ultrasonication on enzymes, 201
and goiter, 382
cheese from milk of, 593, 598–600
milk, humanizing, 9, 662
pepsin(p), rennet replacement, supplement, 10, 601, 602
Crab, blueing of, 325
Crackers, bacterial α-amylase(p) in making, 579
bacterial proteinase(p) in making, 10, 584
lipase(p) for flavor improvement, 412
texture of, 572
Crambe abyssinia. See Crambe seed
Crambe seed, protein and glucosinolase from, 632
Cranberry, character impact ester of, 389
phenolase and enzymatic browning of, 275
Crassulaceae, 359
Crayfish, antithiamins in, 644
Crayfish, prophenolase in, 274
Cream, lipids, and lipase specificity, 413
lysozyme(p)-preserved, 662
separation, and avoidance of induced lipolysis, 449
xanthine oxidase in, 451
Cream cheese, 593
Creaminess, a cloud attribute, 536
a texture, enhancement with cellulase(p), 550
Crease, of the wheat kernel, 573
Creatine, enzymatic analysis for, 701
Creatine, in muscle ATO replenishment, 467
Creep, in end-point enzymatic analysis, 686, 687
Cretinism, due to enzyme-generated toxins in cassava, 635
Crispiness, a cracker texture attribute, 572
Crocodiles, melanin pigmentation of, 277
Crop yield, prediction of by enzyme assay, 677–678
Crosslinking and crosslinks, enzymes, inhibition by penicillin, 249
enzyme activation and, 204
for enzyme(p) stabilization, 103
in enzyme immobilization by, 142–143
of collagen, aging (chronological), meat texture and, 496
and elastin, lysyl oxidase in, 482
cleavage of, 476
proteinase action and, 463, 496
of dextrans (Sephadex), for enzyme purification, 120
of proteins, at low a_w's, 243
by Irad-induced dimerization of, 198
in milk clotting, 598
nutritiousness decreased by LOX-induced, 631
of alkali treated, 711

of fish, enzyme-induced toughness due to, 500, 501–504, 505
of wheat dough proteins, and texture, 584, 585, 589
starch, in adsorption affinity chromatography, 124, 556
Crossover plot, in muscle-to-meat, 472
Crotonaldehyde, an oxidase substrate, 373
Crucifer(s), alliinase in, 379
enzymology of glucosinolates in, 370, 381–386, 387
seeds, protein source, glucosinolase and, 632–633
total glucosinolates in, enzymatic analysis for, 699
Crude fiber, enzyme(p) as aid in determining, 710
Crumb firmness, enzyme(p) improvement of, 326, 578, 583
Crustacea, phenolase in, 275
Cryoadaptation, 229–230
Cryobiology, 218
Cryoenzymology, 218–230
Cryolysis, 213
Cryophiles, 230
Cryoprotectants, frozen fish anti-tougheners as, 505
and persistence of enzyme action, 224
Cryoprotection, 229–230
Cryosolvents, 226
Cryptic intermediates, accumulation in the cold, 224, 225, 229
Crystalline cellulose, as cellulase substrate, 169, 172
Crystalline glucose hydrate, 157–158
Crystallinity of fish myofibril protein, texture and, 498, 499
Crystallization, in enzyme purification, X-ray analysis, 69, 117–118
of specific enzymes, 117, 118, 299, 517, 555, 576–577, 595
prevention, of food ingredients, by functional proteins, 614–615
of ice cream lactose, by lactase(p), 610
of sucrose by invertase(p), 9
Crystals, inclusion within plant cells, 513
Cuaguayote, proteinase source, 94
Cucumber(s), an off-odor description in potatoes, 431
aroma constituents of fresh, 369, 370, 371
as vine fruit vegetable, 524
ascorbate oxidase of, 13, 287
fatty acid α-oxidation by enzymes of, 435
glyoxylate cycle in, 351
juice, enzymatic(p) manufacture of, 548
LOX and aroma of freshly crushed, 299, 370, 371, 372
pickled, acid sourness in, *via* fermentation, 263, 349, 406
color of, chlorophyll and, 302
cooperative enzyme sources in making, 6
dill, 259
flavor of, microbial lipase and, 413
peroxidase, heat resistance, 206
texture enzymology of, 6, 528–529
texture profile of, 511
Cucumber melon aroma, 372
Cucumis melo. *See* Melon
Cucumis spp., ascorbate oxidase of, 13, 289. *See also* specific species
Cucurbita pepo. *See* Pumpkin
Cucurbits. *See Cucumis* spp. and specific species
Cucuropsis edulis, seeds, lipase-caused off-flavor of, 445
Cumulative feedback, regulation, 72
repression, of enzyme biosynthesis, 71
Cunara cadunculus, rennet from, 601
Curare, and meat texture, 463
Curd tension, rennet assay by measurement of, 599
Curdiness, a cheese texture defect, 599
Curdling, of butter prevented by proteinase(p), 609
of milk in cheesemaking, 597, 598, 600
Curing—enzymatic aspects of, ham, 323–325, *See also* Ham
of cacao seed in cocoa/chocolate making, 294–295
of cheese (aroma), 407–411, (texture), 605–610
of malt (kilning), 557, 568
of meats, 315, 493, *See also* Meats, cured
of sweet potatoes, 359
of vanilla beans, 399–400
postharvest, of food plants, 368
Curling of bread, prevention by proteinase(p), 10
Currants, black, enzyme(p) for making juice from, 537–538, 540
Custard, milk, rennet for home preparation, 8
Cuticle of plant cell, 513
Cutins, biosynthesis and food texture, 390, 392, 513, 519
Cutting time, curd tension and, in cheesemaking, 598
Cyanide. *See* HCN and cyanide
Cyanidin-3-α-L-arabinoside, cocoa anthocyanase substrate, 296
Cyanocobalamin as coenzyme, 45, 46. *See also* Vitamin B_{12}
Cyanogen, antibiotic effect from peroxidase action, 205
Cyanogen bromide (CNBr), in affinity chromatography, 127, 128
Cyanogenic glycosides, 396–399, 634–637
biosynthesis of, 396–397
chemical structure and occurrence, 396, 398
ecology, polymorphism, grazing and, 422
emulsin and, enzymology of, 397–398
-generated HCN as toxin from, 416, 634–637
removal of bitter glucosides with, 423
flavor- and aroma-related aspects, 388, 399, 423
health related aspects of, 634–637
in high protein potential foods, 633, 635
intracellular compartmentation of, and emulsin, 398
Cyathus stercoreus, lignin degradation, cellulose liberation by, 169
Cyclic AMP, and 5′-nucleotide gustators, 366
and prostaglandin and calmodulin, 73
as a gibberellin synergist, 558
catabolite repression and, 76
glycolysis, regulation *via* cascade and, 71
hormones (the second messenger), 73
in affinity chromatography, 128
in cheesemaking, 408–409
in meat texture and PSE, 470
lipase, coffee, and blood FA, 442
phosphorylase activation and, 332
receptor site (CAP), transcription, and RNA polymerase, 76
Cyclic GMP, yin-yang, in lipase regulation, 442
Cyclic nucleotides, 73
2′,3′-Cyclic nucleotides, 366
Cycloalliin, 374, 376
Cycloamyloses, as enzyme-generated health hazard, 643
as model enzymes, 60, 61
in affinity chromatography, 576
Cyclodextrins. *See* Cycloamyloses
Cyclodioxygen fatty acid, prostaglandin synthase

product, 435
Cyclodipeptides, bitter, 617
Cycloserine, as analog inhibitor, 249
p-Cymene, 404
Cynera cardunculus. See Prickly artichoke
Cysteine, absent in bitter peptides, 617
and enzyme stereospecificity, 32, 33
antagonists, for meat tenderizing, 482
as antioxidant, 437
as cheese aroma precursor, 409
as enzyme(p) stabilizer, 103, 110
as green cheese off-odorant precursor, 457
enzymatic browning control with, 283
for dough mellowing, 584
in proteins, alkali lability of, 710
lanthionine from, 263
mercapturic acid, detoxification, 77
prevention of NEB in dried eggs, 340
reaction with H_2O_2, 631
reaction with phenolase-produced *o*-quinones, 630
sulfhydryl, pK_a of, 262
sulfhydryl at action site, 56
D-Cysteine, and enzyme stereospecificity, 32, 33
L-Cysteine sulfoxides, as alliinase substrates, 374–381
L-Cysteine-globin, green pigment of tuna, 316, 325
L-Cysteines, as cysteine lyase substrates, 32, 33, 379–380, 684
Cysternae of endoplasmic reticula, enzyme localization in, 386, 467
Cystic fibrosis, predigested proteins, bitter peptides and, 616
Cystine, L-cystine, as cysteine lyase substrate, 32, 380
in protein, alkali lability, 710
protein functionality and content of, 616
Cytochrome(s), 28. *See also* specific cytochromes
Cytochrome b_5, in myoglobin reduction and red meat color, 317
Cytochrome c, heat-induced aggregation of, 213, 214
heme protein linkage and cysteine lyase, 380
in carotene bleaching, 330
milk oxidative rancidity and, 451
purification by ion exchange, 119
reduction with GOX, 339
triphasic kinetics in denaturation of, 206
Cytochrome P450, *See also* Xenobiotic oxidase
and LOOH formation in meats, 320
and monophenol hydroxylation, 427, 428
in mammalian detoxification, 77
lipid oxidation, off-flavor and, 435
proposed removal of environmental toxins with, 660
Cytology, photomicrographs of food organism cells, 306, 353, 465, 502, 521
Cytophaga spp., cellulose digestion by, 169
isoamylase from, 155
Cytoplasm, invertase isozymes of, sugar accumulation, 357, 513
Cytoplasmic membranes. *See* Cell membranes

D_E, in Irad of foods, definition of, 197
D_{37} values, in Irad target theory, 196, 197
DABCO, singlet oxygen quencher, 173
Dairy products, *See also* Cheese
aroma constituents, origin, 371
as fermented foods, 406
change-of-state of milk and, 267, 335, 593
enzyme assays for pasteurization adequacy, 186, 187, 211–212, 669
enzyme(p) for off-flavor removal, 423
lactose in, enzymatic analysis for, 695
oxygen removal for off-flavor prevention, 438
tallowing in dried, 237
Damkohler number, in immobilized enzyme kinetics, 148
Dark-cutting meat, 316, 321
Dasheen, biogenic amines in, 641
DASP, a variation of EIA, 714
Dates, invertase of and invertase(p) for sugar wall, 9, 358
phenolase and enzymatic browning of, 18, 275, 276
rehydration for quality upgrade, 246
texture-modifying enzymes of, 14, 221, 519–520
DDT, enzymatic analysis for, 705
DDT-resistance and DDT-degrading enzymes, 660
De novo biosynthesis of enzymes, *See also* Biosynthesis of enzymes
distinguished from activation by EIA, 715
following non-sterilizing Irad, 199–200, 275
of phospholipid biosynthesis enzymes by water inhibition, 586
or release, of α-amylase by gibberellins, 556, 575
Deacetylation, Arrhenius discontinuity, trypsin, and, 220
Dead-end complexes, in affinity elution, 132
DEAE cellulose, in enzyme purification, 119
DEAE Sephadex, as support in enzyme immobilization, 150
De-aging collagen, 496
Death, programmed, cathepsin and, 480
Debittering, enzymatic, citrus juices, 444, 454, 455–456
cyanogenic glycerides, 423
high protein sources, 423, 618–619, 654
olives, 263
shark meat, 423
yeast, 457
2,4-Decadienal, an off-flavorant, 424
Decadienoates, Bartlett pear, 389
δ-Decalactone, a tea flavorant, 400
Decantation, in enzyme(p) production, 120
Decarboxylation, enzymes catalyzing, *See Enzyme Index*
in melanin formation, 276
of amino acids in cheese and other foods, 409, 642–643
of fatty acids to hydrocarbons, 435
pyridoxal phosphate and, 45
Decimal reduction time. *See* D-value
Decolorization, of anthocyanins, 295–297
of proteins *via* plastein, 619, 620
Decompartmentation, enzyme, substrate, flavor genesis in meat, aging and, 415
in aroma genesis in salad vegetables, 16, 372, 373
in enzyme potentiation, 14–17
in induced milk rancidity, 448
5′-nucleotide accumulation and, 366
See also Compartmentation, Enzyme potentiation
Deep fat frying, meat texture and, 483
Defatting of proteins *via* plastein, 619, 620
Degallation in tea production, 294
Degenerate pi orbitals, 287

Degermination of corn with hemicellulase(p), 582
Degumming, enzymatic, 5, 553, 556–557
Dehydrated foods, 231–246
 enzyme behavior and action in, 237–245
 enzyme management of, 245–246
 enzyme stability in, 243–245
 enzyme(p)-facilitated rehydration of, 549
 flavorese(p) treatment of, 416–417
 microbiology of, 235–236
 nonenzyme chemical changes in, 236–237, 431
 physics and engineering of, 231–235
 See also Dehydration, IMF, Freeze-dried foods
Dehydration, *See also* Dehydrated foods, Dried food, Drying
 engineering, methods and processes, 234–235
 home, 15
 in frozen foods, enzyme action and, 222
 of vegetables, *See* Vegetables, dehydrated
 onion, 237, 378
 retention of desirable enzymes, 246
 surface, in microwave blanching, 193
Dehydroacetate, partial replacement with enzyme preservatives, 663
Dehydroalanine, residues of, in proteins at high pH, 129, 263, 710
Dehydroascorbate, in enzymatic browning prevention, 283
 free radical, in dough conditioning, 590
 lactone hydrolysis and vitamin C loss, 645, 646
Dehydrolimonoic acid A-ring lactone, from limonoate dehydrogenase, 453, 454
Delay between harvesting and processing, and food quality, 432
Delayed bitterness, citrus, 452
Delphinidin, 293
Demethylation, of caffeine, 496, 661
 of TMAO, and fish texture, 501–504
Denaturants, protein, 186
Denaturation, protein, agents of (denaturants), 186, 369
 and heat inactivation of enzymes, 186–187
 and meat texture, 464, 471
 and polyphenol astringency, 630–631
 and unfolding of tertiary structure, 187
 as evidence of GOX(p) use, 340
 by high pressures, 201
 by microwaves, 194
 for immobilized enzyme support, 145
 in enzyme purification, 106, 112
 incipient, by polyphenols, 630
 of fish myofibrils and texture, 499, 500
 prophenolase activation by agents of, 274
 reversal of, 212, 214, *See also* Regeneration, enzyme
 surface, for enzyme inactivation, 200, 201
Density gradients in enzyme purification, 114–115
Dental plaque, dextrans in, and enzyme(p) for reduction of, 342, 360. *See also* Caries
Deodorization of proteins *via* plastein, 618, 620
Deoxyfluoroglucose, as GOX substrate, 338
4-Deoxy-L-threo-hexose-5-ulosuronic acid, 517
Deoxyribonucleic acid. *See* DNA
Depolarization, electrical, in muscle function, 467
Derepression. *See* Gene derepression
Desglutamyl lentinic acid, cysteine lyase substrate, 380
Desiccation, irreversible enzyme inactivation in frozen systems by, 223
Design, of food-related enzyme inhibitors, 248
 of processing plants to minimize adverse enzyme action, 282
Desmoenzymes, and enzyme immobilization, 138
Desorption, 232
Desserts, enzyme modified starches in, 152
Desugaring, of cold sweetened potatoes, 336, 337
 of eggs, 339–341
Detection, enzymatic of, bran residues in white flour, 679
  caseinate in meat, 709
  Cercospora spp., 678
  of chicory in coffee, 709
  of eggs in "pasta" products, 700
  of freeze-drying of milk, 709
  of freeze-thawed animal foods, 709
  of skim milk in sausages, 709
  See also Adulteration
Detergents, *See also* Surfactants
 anionic, as blanching alternative, supplement, 195, 264
 chromatography, 121, 125
 dietary fiber assessment using, 710
 enzyme subunit dissociation by, 274
 enzyme(p) as, 5, 621–622, 648, 663
 greening prevention with, 311
 in chlorophyllase assay, 307
 in enzyme purification, 112
 in pre-blanch cold wash, 195
 to inactivate excess flour α-amylase, 581
Detoxification, 77, 636–637. *See also* Toxicants, Xenobiotic oxidase
Development, of cereal grains, β-amylase solubility and, 562, 576
 enzyme assay as marker of, 678
 zymograms indicative of, 682
Dextran, enzymatic analysis for, 696
 problem in sugar refining, brewing and teeth, 630
Dextrans, derivative in enzyme production, purification, 117, 120
Dextrins, absent in low a_w amylolysates, 242
 as α-amylase products, 155
 from non-α,β-amylases in malt, 556
 in beer, removal for dietetic, 564
 in bread, function of, 578
 structures of, in beer, syrups, 154, 554, 560, 564
Dextrose equivalents, of corn syrups, 153
Dhurrin, structure and enzyme-generated HCN, 398, 636, 637
Dhuti, 382
Diabetes, GOX treatment of, 342
 NEB in sequelae of, 334
 xylitol, 177
Diacetyl as a beer off-flavor and removal of, 421, 423
Diacetyl in cheese aroma, enzymatic genesis of, 408–409
Dialdehyde starch as crosslinker support in enzyme immobilization, 143
Dialkylacrolein, an enzyme-caused pork off-odorant, from plasmalogen, 446
Diallyl disulfide, garlic and, 374
Dialysis, in enzyme purification, 113
Diamines, enzymatic analysis for, 700, *See also* Biogenic

amines
precursors to plant alkaloid, 640
L-α,β-Diaminopropionic acid, neurotoxin, 701
o-Dianisinidine, H-donor in peroxidase assay, 208
Diatomaceous earth, for chillproofing beer, 570
for immobilized enzyme support, 144
Diazobenzidines, as crosslinkers in enzyme immobilization, 143
2,6-Dibromoquinone-4-chloridate, 211
4,4-Dibutyl-γ-butyric lactone, in dairy product aromas, 411
α-Dicarbonyls, Strecker degradation and, 401
Dicarboxylic acids, direct oxidation by peroxidase, 31
2,6-Dichlorophenol indophenol, as H-donor in peroxidase assay, 208
as redox mediator in enzymatic analysis, 689, 697
2,4-Dichloroxyacetic acid, as a gibberellin mimic, 558
Dicoumarol, inhibitor of DT diaphorase, 317
Dielectric constant, 225
Dietary fiber, enzyme(p) as aids in determining, 709–710
PE action and, 515
plant proteolysate core as, 624
possible removal by enzyme(p) in foods, 550
Dietetic, beer, enzymatic production of, 564
foods, sorbitol and xylitol in, enzymatic analysis for, 697
foods, surviving rennet in, 607
Diethylpyrocarbonate, a nucleotidase inhibitor, 367
Diethylstilbestrol, enzyme(p) feed as alternatives, enzymatic analysis, 482, 625, 712
Diethylthiocarbamate, phenolase inhibitor, 281
Diets and dieting, and meat texture, 463
and naringin taste, 455
enzyme induction and gene repression by, 75, 76
enzyme(p), protein functionality and, 664
veal enzyme inhibitors, 254, 257
Diffusion, effect on immobilized enzymes, 147–148
Diffusional inhibition, 251
Digalactosyl glycerides, and blanching, 191
Digestibility, protein, after phenolase action on, 630
enzyme(p), for *in vitro* assessment of, 21, 707–708
modified by polyphenol complexes, 630, 631
of acid-treated beans, 436, 633
of phytase-potentiated beans, 658
Digestion, role in enzymology history, 1, 2
Digestive aid, enzyme(p) as, 663, 664
Digitonin as cellulase synthesis stimulant, 169
Dihydrochalcone sweetener, 361
Dihydroxyacetone phosphate, 39
2,3-Dihydroxybenzaldehyde, phenolase dimerization by, 274
3,4-Dihydroxycinnamic acid, as an antithiamin, 344
Dihydroxyfatty acids, 242
Dihydroxyfumaric acid, peroxidase substrate, 31
Dihydroxyheptacenes, in blanched avocado, 421
5,6-Dihydroxyindole, from phenolase action, 270, 276
3,4-Dihydroxylphenylamine. *See* Dopamine
Dihydroxyphenylalanine. *See* DOPA
Diisoamyl disulfide, in chocolate aroma, 402
Diisocyanates, as crosslinkers in enzyme immobilization, 143
Diisopropylfluorophosphate, active site probe, 40, 704
2,3-Diketogulonate, and NEB, 327
2,3-Diketogulonic acid, and loss of vitamin C, 645–646
Diketopiperazines, bitter, 617
Dilatometry, for enzyme assay, 676
Dill pickles, softening of, prevention, 259
Diluents used in standardizing enzyme(p), 103
Dimerization, of enzymes by Irad, 198
of phenolase by substrate, 274
N,N-Dimethyl dodecyl glycine, enzyme model, 61
N,N-Dimethyl formamide, hydrophobic interaction, in affinity chromatography, 127
Dimethyl sulfide, as aroma and off-odorant, 387, 416, 457
Dimethyl sulfoxide, a cryosolvent, 226
for starch solubilization, in enzymatic analysis, 696
in cellulolysis pretreatment, 176
Dimethyl propiothetin, a salmon off-odorant precursor, 457
Dimethylpyrazine, in aromas, 351
S-(2,4-Dinitrophenyl)-L-cysteine, 33
Dioxane, in enzyme purification, 116
Dioxetones, in LOX mechanism, 373
o-Diphenols, as phenolase substrates, 30, 272
titanite complexes of borate, 283
p-Diphenols, as laccase substrates, 271, 273, 274, 276, 529
2,3-Diphosphoglycerate, allosteric affector of Hb, 318, 364, 655
meat color retention and, 322
Diploidia spp., strain of as phytase(p) source, 659
Dipsticks, in enzymatic analysis, 692
Direct acid methods, for cottage cheese making, 599
Direct enzymatic flavors, 348
Direct injection enthalpy (DIE) in enzymatic analysis, 691
Direct linear plot, enzyme initial rates by, 674
Discoloration, black-eyed peas, 195, *See also* Enzymatic browning, Off-color, Pinking, Purpling
of red meat, 315, 318–321. *See also* Meat, color
Discontinuities, in enzyme Arrhenius plots, 219–220, 221
in relation to chill-sensitivity, 334
in water sorption isotherms, 232
Discriminate function, methanol in wine, contribution to, 638
Discrimination, enzyme, 43–44. *See also* Specificity
Dispersibility, a proteinase(p)-aided protein functionality, 10, 508, 614
Disruption, cell. *See* Cell disruption
Distilled alcoholic beverage manufacture, 552, 560, 561–562
Distiller's malt, 561
Distiller's wash, flavorese-generated aromas in, 418
Distinctness-of-image, and color and appearance, 268
Distribution, of food in the food processing chain, 10
Disulfide, and derivatives, in *Allium* food odors, 375
bonds, bridges, in fish toughness, 560
in pasta proteins from GOX(p) action, 340
in phenolase function, Cu and, 340
in plastein formation, 618
in protein functionality, 616
in protein structure, 41
in regeneration of inactivated enzymes, 213
in subunit dissociation, 274
reagent, Ellman's, 701

-thiol interchange, affinity chromatography, 125
and meat texture, 464
enzyme catalyzing, 260
in cheese aroma, 409
in cocoa aroma, 402
in frost injury, enzyme behavior, 223
in gluten, enzyme(p) facilitation, 484
SOD effect on, 286
Diterpenoids, gibberellins as, 556
Dithiocarbamates, antioxidants, 437
Dithiolanes, in aroma, 370
Dithiolisobutyric acid, 387
Dithiothreitol, an enzyme stabilizer, 103
O,O-Dityrosine, in galactose oxidase, 240
Diversity of food constituents, enzymatic basis of, 19, 228
Divinyl sulfone activation, in affinity chromatography, 127
Djenkolic acid, endogenous C-S cysteine lyase substrate, 32, 33, 380
DNA, biosynthesis and replication, 57, 75, 377, 518
H-bonds and polymer enzymolysis, 172
Irad and, 198, 377
isozymes, evolution and, 48
nucleus packing of, histones in, 75, 76
recombinant technology, 6, 12, 85, 86, *See also* Genetic engineering
topological rearrangements, enzymes of, 75, 76
transcription, and enzyme biosynthesis, 74–76
Zn^{2+}-catalyzed degradation of, in evolution, 59
Dodecalactones, in aroma, 411
Dog foods, pancreatic lipase(p) used in, 412
Dolmati cheese made with microbial rennets, 604
Domain, of active site, 250
DOPA (Dopa), in Parkinson's disease treatment, 273–274, 660
in rat feed for psychophysiology research, 630
phenolase substrate, pharmaceutical production of, 276, 278
via immobilized phenolase, 273–274
vitamin B_6 loss by complexing with, 646
Dopa quinone, phenolase, melanin and, 270, 275–276
Dopachrome, phenolase, melanin and, 270, 276
Dopamine, endogenous phenolase substrate, bananas, 276, 277, 279
Dormancy, lipase and, 444
Dorthiorella spp., milk clotting enzyme in, 602
Double affinity chromatography, 130, 131
Double antibody EIA, 714
Double bond reduction by ADH, 394
Double helix, of amylose, Q-enzyme and, 332, 333
of DNA, 74
Double pH optima, of one enzyme, 261, 299, 329, 331
Double reciprocal plots, 36
Dough, 10, 55, 572, 580, 587. *See also* Breadmaking, Flour, Gluten
Doughnuts, GOX(p)-treated dried eggs in, 340
Downward flow fixed bed reactors, 165
Dressings, as IMF, 234
Dried foods, *See also* Drying, specific foods
eggs, production and quality of, 236, 339–340
enzymology of, *See* Water activity, low
freeze-dried in, for flavor improvement, 246
fruits, enzyme(p)-prepared juices from, 536, *See also* Fruit juices
meat, 237
microbial growth in, a_w, 235–236
milk, 237
rehydration of, *See* Rehydration of foods
relation to frozen foods, 231
soups, 233
Drip-dry, a fish texture defect, 449
Droopy noodles, and excess α-amylase, 580
Drop in, a textured attribute imparted by functional proteins, 615
Dropping number, zymograms as replacement for, 682
Drosera rotundifolia. See Sundew
Dry blanching, 194
Dry cleaning, with enzymes(p), 5
Dry heat-extrusion cooking, enzyme inactivation by, 436
Dry ice, in enzyme purification, 409
Dry legume seeds, texture enzymology of, 527. *See also* Bean
Dry mixes, lipase(p) used in, 412
Dryers, types of, 235
Drying, control of undesirable enzyme action by, 11, 210, 245–246, 267, *See also* Water activity, low
food, engineering and methods of, 234–235
rates, effect on enzymes, 246
Dryness, a fish flavor defect, 499
a potato texture, 526
Dry-to-the-touch foods, 234
DSC (differential scanning calorimetry), 186
Durum wheat, pasta from, enzymology of, 301, 575, 679–680
D-value, definition, 187
Dwarf wheat, phenolase and browning of, 278, 591, 630
Dyes, food, enzymatic analysis for, 701
LOX-induced bleaching of, 299, 301
food enzyme(p), as alternative to, 661
Dynomill, in large-scale enzyme purification, 110

E-. *See trans*
Earth vegetables, 524
EC. *See* Enzyme Commission
Echinoderms, TMAO demethylase in, 503
Ecology, CN^- glycosides and, 422. *See also* Environment, Pollution
Economic aspects, of enzyme(p) use, 8, 89–90, 92, 104, 160, 420
ECTEOLA-cellulose, in enzyme purification, 119
Edam cheese, renneting of, 599, 604
Eddy transfer, in IMF preparation, 234
EDTA, as enzyme affector, 256, 367, 445, 714
blanching with, 195
Eel, flavor enzymology of, 416
Effective kinetic parameters, of immobilized enzymes, 145
Effectiveness factor (η), in immobilized enzyme kinetics, 148–149
Effluent, blanching, treatment of, 194, 195
Egg(s), antitrypsins in, 706
enzyme-aided analysis for phospholipids in, 712
enzyme-caused off-flavors in, 237, 423, 457
glucose removal to prevent NEB in dried, 6, 9, 326,

339–341
hard-boiled, off-color of, 341
protein, nutritiousness of, 708
water sorption isotherms of solids, 232
Egg albumin, improved by enzyme(p) whippability, 615
Egg white, a functional protein, 615, *See also* Albumen
antitrypsin of, 254
color of, 268
desugaring of, 339–341
glucose removal *via* fermentation, 340
lysozyme in and functionality of, 92, 662
proteinase(p) treatment of for functionality, 615
Egg yolk, color, 301
desugaring *via* fermentation, 340
in egg white, removal with enzyme(p), 9, 412
Eggplant(s), as a berry fruit vegetable, 524
blanching and astringency of, 189, 195
LOX of, 430
phenolase and enzymatic browning of, 271, 275, 276, 296
phenolase-caused bitterness in, 457
processing, PG assay as guide, 209
Eggshells, and DDT, 705
EI. *Enzyme Index*
EIA. *See* Enzyme immunoassay
Elasticity, an objective texture quality attribute, 232, 462, 463, 468, 473, 551
Elastin, proteinase(p) action and meat texture, 463, 487–488
Electric potential, color perception and, 268
in muscle function, 467
Electrical stimulation, and meat texture, 469, 484
Electrochemical approaches, in enzymatic analysis, 689, 690–691
Electrodecantation, in enzyme purification, 115
Electrodes, enzyme. *See* Enzyme electrodes
Electrodialysis, in enzyme purification, 115, 235
Electro-immunoabsorption, in breadmaking, 682–683
Electron acceptor, TMAO as terminal, 504
Electron orbital(s), and flavor, 345
of singlet oxygen, 287
Electron transfer, in model enzyme reactions, 59
Electron transport, poisons and reverse, 285, 335
Electronegativity, and sweetness perception, 346
Electrons, as ionizing radiation, 196, 197, 200
Electrophiles, immobilized enzyme supports as, 144
Electrophoresis, and enzyme purification, 115
gel, *See* Gel electrophoresis
of peptides to distinguish rennets in blends, 605
Electrostatic interaction and immobilized enzyme action, 146
Elemental analysis, by enzyme inhibition and activation, 706–707
β-Elimination. *See* Pectin lyase-like action
ELISA, EIA with separation, 714
Ellman's reagent, for disulfides, 701
Elongation of peptide chains in protein biosynthesis, 79
Elution, in affinity procedures, 131, 132
in non-affinity enzyme purification, 118, 120, 122
Embryo, wheat, enzymes, gibberellins, in malting, 556, 557, 574
EMIT, EIA without separation, 713, 714
Emmenthaler cheese, 594, 604
Emphysema, cathepsin in, and enzyme(p) for treatment, 480, 665
Emulsion(s) and emulsification, and stability of, imparted by functional proteins, 614, 615, 624
enzymatic(p) dispersal of, 661
forming capacity, frozen fish and, 499
in cheesemaking, 609
lipase activation by, 442
Endocarp-testa, α-amylase in, 555
Endogenous enzyme(s), *See also* Autolysis, Potentiation
for toxicant removal, 651, 653–654
in model low a_w systems, 240
safety of food, 648
Endogenous inhibition, increased at low a_w, 239
of γ-glutamyl transferase, 380, 675
of invertase and double pH optima, 331
of nitrate reductase, 259
rate *vs* enzyme nonlinearity, due to, 674
Endogenous substrates, of phenolase, 276, 277. *See also* other enzymes
Endoplasmic reticulum, cysternae and other extensions, 82, 385
glucosinolase in, 385, 387
in biosynthesis of enzymes, 81
in plant and animal cells, 82, 373, 385, 386, 387, 467, 513
malt α-amylase biosynthesis and, 556
Endorphins and enkephalins, 134
Endosperm, wheat, enzymes in, 535, 573, 574, 585–586
Endothermic reaction of xylitol, cooling effect of, 167
Endothia parasitica, rennets from, 95, 602, 603, 604, 647, 648
Endothia spp., milk clotting enzymes in, 602, 605
End-point enzymatic analysis, principles of, 686–688, 690
Energy, fuel, enzyme-associated expenditure/conservation, 5–6, 193–194, 195–196, 205, 436, 536, 546
enzyme-associated production of, 5–6, 167, *See also* Biomass conversion
Energy coupling, and enzyme catalyses, 86
Energy of activation, of enzyme inactivation, 18, 206, 436–437
of enzyme reactions, catalytic efficiency and the transition state, 50–51
deviation from the Arrhenius temperature, 225
in Absolute Reaction Rate theory, 51
of protein denaturation, 186, 205, 213
of specific enzymes, α-amylases, 335, 574
β-amylase, immobilized, 148–149
catalase, 210
peroxidase, 205–206
trypsin, 225
Energy transduction, enzymes, other proteins as devices for, 86
in MRA, 322
in muscle function, 466–467
English muffins, fungal proteinase(p) for, 583
Enriched meat, enzyme(p) tenderization of, 493
Enrichment and yield, in enzyme purification, 111–112
Enterobacter aerogenes, decarboxylase, scombroid poisoning due to, 640
E. cloacae, thioglucosidase in, 385
E. liquefaciens, xylan reductase of, 177
Entrainment of benefits, in decision to use enzymes, 340, 447, 557–558. *See also* Serendipity
Entrapment, a mode of enzyme immobilization,

141–142, 164
Entropy, in enzyme specificity and catalysis, 44, 49
of activation of enzyme inactivation, 206
of protein denaturation, 186
pectin chain entanglement, cloud and, 544
Environment, enzymes for monitoring and controlling, 286, 660. *See also* Air, Pollution, Water
Environmental impact of blanching, 190–192
Enzymatic activation analysis, anylates as activators, 704, 706–707
Enzymatic activity, determination of, *See* Assay of enzymes
expression of, 675
Enzymatic analysis, anylates as substrates, Chapter 38
chemical analysis in which the anylate is substrate, activator, inhibitor, or product of enzyme(s) = [enzyme(a)] action, 685–707
Enzymatic browning, phenolase related, NEB as, and as NEB, 326–327
of plant foods, desirable, 291–295
undesirable, ascorbate involvement, 284–285
endogenous substrates of, 276
non-phenolase enzymes, involvement in, 285–286, 287–288
pacers of, 278–280
physiological disorders leading to, 18, 200, 227, 278, 285
prevention, control of, 254, 256, 273, 280–283, 284–285, 290–291
pigments and melanins, 270, 276–277, 631
Enzymatic inhibition analysis, anylate as inhibitor, 703, 704–706
Enzyme(s), activation of, *See* Activation of enzymes
activity, determination of, *See* Assay, enzyme
expression of, 674–675
analyzers, 676, 680
as charge separating catalysts and transducers, 86
as Irad targets, 196–198
as proteins, chemical and biological aspects, 27, 85–86
assays, *See* Assay(s) of enzyme
biological control, regulation of, Chapter 4, *See also* Regulation of enzymes
biosynthesis of, *See* Biosynthesis of enzymes and other proteins, *De novo*, Gene derepression, Translation
catalytic efficiency, *See* Catalytic efficiency
classification schemes, *See* Classification, nomenclature
definitions of, 27–28
electrodes, enzymatic analysis, 140, 692–694, 697, 699, 700, 701
immobilization, *See* Immobilization of enzymes and immobilized enzymes
immunoassay (EIA), 712–715
for nonrate enzyme assays, 714–715
with separation (ELISA), 714
without separation (EMIT), 715
inactivation, *See* Inactivation of enzymes
inactive, separation from active by affinity chromatography, 133
kinetics, *See* Kinetics, enzyme
management, 8, *See also* Management of enzymes
models, *See* Models of enzymes
modified, purification by affinity chromatography, 133
multiple forms of, 46–48, *See also* Isozymes
non-amino acid components of, 45–47
"perfection," and evolution, 50
potentiation of endogenous, *See* Potentiation of enzyme action
purification and isolation, *See* Purification of enzymes
regeneration, *See* Regeneration of enzyme activity
synthesis of, organic, 80, 133
theory of life, 4
units, molar activity and enzyme assay, 675
See also Coenzymes, Cofactors, Prosthetic groups, specific components
Enzyme action, control, non-biological, and prevention of, *See* Chapters 10–14, *See also* specific enzymes, foods and processes
endogenous, 14–18, *See also* Potentiation, specific enzymes, foods
NEB precursors *via*, 326–342
potentiation of, *See* Potentiation of enzyme action
slow, in foods during storage, 6, 218, 237–239, 430–431
vs nonenzyme action, activation energy and catalytic efficiency, 50–51
as error source in chemical analysis, 712
at low temperatures, 225
at low water activities, 235–236
chlorophyll destruction, 303–305
enzyme models, 60–62
glucose isomerization, 60, 153–154
in cheesemaking, 599–600
in cole slaw deterioration, 6, 7
in hydrolyzed protein manufacture, 616
in nonenzymatic browning, 326–327
interrelation in, distinguishing between, in foods, 18–19
invert sugar production, 159–160
Enzyme Commission (EC). *See* Classification and *Enzyme Index*
Enzyme histochemistry, 684
Enzyme preparations, amino acid liberation in protein analysis, 710–711
as analytical aids, 709–715
dietary and crude fiber analysis, 709–710
enzyme immunoassay, 712–715
various others not cited elsewhere, 711–712
Enzyme preparations, as chemotherapeutic agents, pharmaceuticals, 4, 134, 663–666
Enzyme preparations [enzyme(a)] as specific reagents in chemical analysis, Chapter 38, 703–707. *See* Enzymatic analysis
Enzyme preparations [enzyme(p)] commercial/ industrial for food use, costs of using, 103–104
health related benefits of using, 508–510, 621–625, 627–629, 637
list of, and uses, 9–10, *See* also specific enzyme(p) and uses
safety and toxicology of, 646–649
sources, production and purification, 89–100, 105–108
stabilization and handling, 100–103
standardization and monitoring of, 683–684
trade names of, *See* Trade names of enzyme(p)
why use them? 90–91

Enzyme(p) producers and suppliers for, baked goods and cereal processing industries, 579, 581
brewing industry, 559
dairy industry, 412, 594, 603, 604, 652–653
distilling industry, 561
juice and winemaking industries, 538, 539, 540
protein processing, 622, 624
sweetener production, 157, 161, 162, 163, 164, 165, 166
Enzyme-modified starch, for food rehydration, 235
Enzyme-protein interaction, inhibition by, 257–258
Enzyme-substrate complex, and active site, 38–44
direct evidence for, 38–39, 53–54, 202–203, 209, 220
in enzyme kinetics, 33–38, *See also* Kinetics, enzyme
in specificity, lock-and-key, template-mold metaphors for, 39, 42
unproductive, Absolute Reaction Rate theory, enzyme regulation, 64–65
Enzymic. *See* Enzymatic
Enzyming, fruit juice preparation by, 540
Enzymograms, 281, 282. *See also* Isozyme patterns, Zymograms
Enzymology, access to literature and information, 22–24, *See also* specific enzymes and topics
applied, non-food related, 4–6, 134, 663–666, 669, 685, 696, 701
history, association with foods, 3–4, *See also* specific enzymes and topics
X-ray, 41–42
Epicatechins as hot beverage phenolase substrates, 293, 295
Epichlorhydrin, crosslinker, 120
Epidermal cells, substrate localization, 81, 398, 399
Epimerization, in alkaline conversion of glucose, 154
Epinephrine, and cascade regulation, 73
and complementary oxidation of hemes, 320
and PSE, 470, 471
Epithelium, localization of human lactase in jejunal, 651
Epithio specifier protein, epithiobutanes, from glucosinolase action, 384
Epoxidation, of LOOH, 427, 428, 429
Epoxy activation, in affinity chromatography, 127
Equilibrium, chemical, not affected by enzyme, 27
ER. *See* Endoplasmic reticulum, 81
Erlose, in honey, 358
Erucic acid, crucifer seed oil, nylon from, 632
Erwinia aroideae, pectin and pectate lyases of, 517–518
E. carotovora, specificity of PG of, 517
Erwinia spp., exo PG of, in coffee beans dehulling, 549
Erythrocytes, as edible protein source *via* enzyme(p), 624
as enzyme source, 128, 317, 450
as immobilized enzyme supports, 145
enzymatic lysis of, 442
lysis by phospholipase B, 442
Erythrose-4-phosphate, in aromatic amino biosynthesis, 84
Escherichia coli, enzymes of, 63–68, 100
Escherichia spp., milk clotting enzymes in, 602
Esculetin, 276
Espresso machine, lipase action in, 447
Essential amino acids. *See* Amino acids, essential
Essential fatty acids, enzymatic analysis for, 700
enzyme-related loss and retention of, 246, 646
Essential oils, enzyme involvement in aroma, production, 9, 369, 370, 374, 392, 395–396, 403, 419–420
Esterification of hydroxyamino acids, in fish toughness, 500
Esters, as aldehyde precursors, 371
in aromas, biosynthesis of, 370, 388–391, 393
in cheese aroma, 410
in green tea, 402
proteinase catalyzed synthesis of, 238–239
sweet, of aspartyl dipeptide, 618
Estolides. *See* Cutin
Ethanal. *See* Acetaldehyde
Ethanol, as an antioxidant, 437
beer, genesis from carbohydrates, 553, 560
distillery production of, 561
enzymatic analysis for, 21, 691, 697
enzyme sterilant-enhanced yield of, in brandies, 663
for off-flavor prevention and removal, 436, 439
from breadmaking, 572
in enzyme(p) production and purification, 116, 117, 398
product inhibition by, 255
purine removal from SCP facilitated by, 655
Raoult's law deviation by, 241
Ethyl acetate, in enzyme purification, 100, 109
Ethyl butyrate in orange aroma, 389
Ethyl isovalerate, a fruity odorant, 452
S-Ethyl-L-cysteine sulfone, and sulfoxides, C-S lyase substrates, 33
Ethylene, and chlorophyll destruction, 309
as mRNA unmasker in fruit ripening, 278
biosynthesis, 204, 286, 299
citrus debittering with, 455
hypobaric storage and, 531
in vanillin production, 400
treatment of wheat, to prevent germination, 574
Ethylene chlorohydrin decomposition, 225
Ethylene diamine tetraacetic acid. *See* EDTA
Ethylene glycol, derivatives as matrices in affinity chromatography, 127
manufacture of by immobilized DNA recombinant enzymes, 6, 7
Ethylene imine (EI), mutant for enzyme production, 97
Ethylene oxide, enzymatic production from ethylene, 6
Ethylene polymers as immobilized enzyme supports, 145
Ethyl-2-methyl butanoate, in cantaloupe aroma, 389, 391
Ethyl-3-methyl butyrate, in blueberry aroma, 389
Etiolation, of vegetables, 311
Eugenol, synthetic vanillin from, 400
Eukaryotes, enzyme biosynthesis by, 76
Euphorbia cerifera. *See* Caper
Eutectic changes, and enzyme action in frozen food, 224
Evaporation, and ultrasonication, 200
enzyme(p) improvement of, 10
Evolution, α-amylases, in cereal, 554–555
and feedback gene repression, 76
and nonenzymatic browning, 326
and peroxidase function, 204
and product inhibition, 250
and selective metal ion catalysis, 59
enzyme perfection, induced fit, transition state in, 50
isozymes and genetic drift in, 48

of lysozyme to α-lactalbumin, 361
Zn-catalyzed DNA hydrolysis and, 55
Excited oxygen. *See* Oxygen, active
Excreta, as proteinase(p) in leather making, 5
Exercise, meat texture and, 463
Exolysis, in enzyme purification, 110
Expanded bed reactors, 165
Explosive decompression in enzyme purification, 110
Extensibility, a texture related breadmaking parameter, 583
of muscle during rigor, 468
Extracellular, enzymes, 138
polyanions, 258
Extracorporeal shunts, enzymes in, 342
Extract release volume, 489
Extraction, selective for enzyme crystallization, 116
Extraction agents in enzyme purification and production, 110–111
Extrapolation for enzyme initial rates, 673–674
Extrusion, for protein functionalization, 615
Eyeball methods, for enzyme initial rate estimation, 673
Eyes, energy transduction by proteins of, 268
lens, γ-glutamyl transferase in, 380
melanin pigmentation of, 277
of potatoes, Irad killing of, 200
of Swiss cheese, 594

Fabric cleaning with enzymes, 5
Fabricated foods, enzyme modified starches for texturization of, 153
α-Factor, lytic β-glucanase biosynthesis inducted by, 629
FAD. *See* Flavin adenine dinucleotide
Falcons, digestion of meat by, historical, 17
Falling number, *vs* α-amylase assay as bread quality predictor, 578
FARCE, freezing at reactive center of enzyme, 55
Farnesene, in apple scald, 314
Farnesol, in monoterpene biosynthesis, 405
Fascicles, of pectic acid, 513, 514
Fast analyzers, in enzyme assay, 676, 690
Fat(s), a breadmaking ingredient, 577
analysis for by ultrasonication, 200
as meat aroma precursors, 415
contribution to cheese texture, 605
globules, of milk and induced rancidity, 448
neutral, as lipase substrates, 440
of fish, antitoughening effect of, 500
Fatty acid ketols, from LOOHs, 427, 428, 429
Fatty acids, *See also* Lipids
amino acids, reciprocal relations in cheese, 409
as flavorant repositories, 410
as PE inhibitors, 515
as precursors to aldehydes in vegetables, 371
as precursors to esters in fruit, 391
as precursors of sour acids, 351–352
blood, lipase, cAMP and, 73–74
combination with blood and muscle proteins, 118, 319
in pheophytinization, 303
in plants, non-LOX associated transformations of, 410–412, 433–436
lipase-generated hydrolytic rancidity, 436, 444–449, 586, 680
LOOH decomposition catalyzed by, 450
migration and lipase specificity, 441
reactions with quinones, 401
Fatty acids, lower, concentration and desirability of, 446–447
in cheese aroma, 407
Fatty acids, metabolism, convergence with amino acids, 382
Fatty acids, polyunsaturated, *See also* Fatty acids, Lipoxygenase
accumulation, relative to saturated, 226
as essential, enzymatic analysis for, 700
as LOX substrates, 372–373, 424–425
in fish and meat off-flavors, 433, 435–436
in milk oxidative rancidity, 450
inhibition of prostaglandin synthase by C20:5, 501
LOX-induced improvement of dough rheology by, 587
LOX-induced loss of, and food quality, 299, 431
meat color, 319
non-LOX transformation of, 433–436
phospholipase-generated, in cell disruption, 16
FAVA bean, LOX in, 428
phenolase-isozymes of, 274
FDP, fructose-1,6-diphosphate. *See* F-6-P + FDP
Feathers, as immobilized enzyme support, 145
melanin pigmentation of, 276
Fecal organisms, enzyme assay for the detection of, 678
Feedback inhibition, 70–74
Feedback repression, in eukaryotes, 76
Feeds, as food off-flavor source, 422
carbohydrate digestibility of, 709
enzyme(p) supplementation of, 624–625
enzyme-aided analysis for lysine, 711
liquid, *via* fish autolysis, 507
poultry, carotenoids in, 302
radappertization of, 196
soybean, urease, urea and, 13
starter, pepsin-chymosin relations and, 601
Felinine, 33
β-Fenchol, structure, 404
Fenchone, structure, 404
Fenton reagent as cellulolysis pretreatment, 176
Fenton's reagent, 172
Fermentation, an aroma genesis, 406, 413
for egg desugaring, 340
gustator production by, 363, 367
in breadmaking, 572, 591
in brewing, 553, 560, 563
in soy sauce production, 616
in tea making, 291, 292
of cacao seed, 294
of coffee berries to remove seed coat, 6, 549, 550
of coffee to remove caffeine, 647, 661
of vegetables, 527, 528, 642
of whey to remove lactose, 643
Fermented foods, 349, 407. *See also* Fermentation
Ferments, enzymes as, in history of enzymology, 4, 28, 554
Ferns, thiaminase I in, grazing and, 645
Ferredoxin, an enzyme cofactor, 46
hydrogen sulfide from, 456
in affinity chromatography, 129

in ethylene biogenesis, 286
in lipid oxidation *via*, and off-flavor, 434
in meat color and flavor, 320
in nitrogen fixation, 46
Ferric chloride decomposition, 225
Ferric hydroxide, enzyme adsorbent, 118
Ferricyanide, as peroxidase H-acceptor, 203
Ferriprotoporphyrin III, 203
Ferrocytochromes, peroxidase substrates, 31
Ferrous ion, LOOH reduction by, 429
Ferrous sulfate for glucosinolate destruction, 633
Ferryl hemoglobin ligand, 316
Fertilizer needs lessened by enzyme assay, 678
Ferulic acid, a proteopentosan flour component, 591, 592
phenolase inhibition and color control by, 282
FFA, free fatty acids. *See* Fatty acids
f-Function, Michaelis, in enzyme-pH theory, 262
Fiber, dietary, and mineral availability, 656
enzyme-aided determination, 709–710
enzyme(p)-assisted conversion to paper, 649
Fibrils, cellulose, 512, 513
Fibrin, formation of by cascade proteinases, 665
Fibrinogen, cascade conversion to fibrin by urokinase, 665
Fibrinolytic activity, of onions, 376
Ficus carica. See Figs
Field bean protein, enzyme(p) solubilization of, 623
Figs, ficin(p) from, 13, 93, 94, 486, 492
Figs, phenolase-produced desirable color of, 275
Filleting, in glycolysis, rigor and texture, 498–499, 502
proteinase removal by, 505
sawdust, proteolysis of, 507
Filter, paper, as cellulase substrate, 168, 169
press, in enzyme purification, 110
pressure leaf, in fructose syrup production, 165
Filter-cel, in enzyme purification, 113
Filth, pepsin-aided detection of, 711
Filtration, in brewing, enzymatic aspects of, 553, 565–567
in enzyme production and purification, 101, 105, 107
in juice making, enzymatic aspects of, 536
Fining agents in beverage making, 281, 535, 537, 570
Firing in tea production, 292
Firmness, a fruit texture attribute, 14, 511
First order kinetics, a solution to nonlinear enzyme assay, 672–673
in catalase and peroxidase assays, 207, 211
in fluctuating temperature systems, 228
in reaction inactivation, 273
inapplicable to enzyme regeneration, 216
Fish, autolytic transformation of, 505–508
cold-adapted enzymes of Antarctic, 229, 230
color, loss and discoloration, 299, 301, 302, 325
flavor genesis, enzymes of, 415–416
freshness, enzymatic aspects, 364, 366, 499, 680
lipase(p) for increasing protein content of, 629
off-flavor, enzymatic aspects, 423, 435–436, 446
texture, description and defects of, 497–498, 499
endogenous proteinase in, 505–506
glycolytic and myofibrillar enzymes in, 497–499
of frozen fish, enzymes, FFA in, 55, 227, 327, 500–501, 502
TMAO lyase-produced HCHO in, 501–504
prevention of, 504–505
Fish meal, 507
Fish pastes, 498, 507
Fish preservation, 263
Fish processing wastes as food proteinase source, 91
Fish products, blocking enzymes in, for NMR analysis of, 712
functional proteins as, 615
from proteinase(p) treatments, 506–510
Fish protein concentrates (FPC), enzymatic aspects, 8, 10, 235, 423, 505, 507, 508–510, 620
Fish proteinase(p), for transforming, 508–510
Fish sauces, 262, 406, 416
Fish smoked, lipase(p) treatment of, 412
Fish thiaminase I, and vitamin B_1 deficiency, 644–645
Fish water, enzyme(p) facilitated, evaporation of, 10
Fish zymograms, for species identification, 662
Fish nets, preservation, enzymatic, 259
Fishiness, an off-odor of unblanched frozen vegetables, 423, 431
Fixation, intracellular, as enzyme immobilization mode, 141, 154
Fixed point *vs* fixed time, enzymatic analysis, 686, 690
Flatulence, symptom of α- and β-galactosidases deficiencies, 651, 653
Flavan-3,4-diols, off-color, 269
Flavanols as phenolase substrates, 276
Flavin adenine dinucleotide, as coenzyme or prosthetic group of, diaphorase, 688
role in catalytic efficiency of enzymes, 58
glucose oxidase, 338–339
hydroxynitrilase, 398
ketone monooxygenase, 392
monoamine oxidase, 701
sulfite oxidase, 663
xenobiotic oxidase (flavoprotein-linked monooxygenase), 77, 427, 660
xanthine oxidase, 285, 451, 649
role in 5′-nucleotide accumulation, 336
role in superoxide anion generation, 286, 452
Flavin mononucleotide, coenzyme, active oxygen generator, 45, 125, 451
Flavins as food pigments, 269
Flavolan, 296
Flavolan tannins, glycosidase inactivation by, 296
Flavonoids, as ascorbate oxidase affectors, 289
as feedback inhibitors, 256
as food pigments and off-colors, 269
astringency of, 609
of tea, 292–294, 609
pacers of enzymatic browning, 279
phenolase in biosynthesis of, 275
See also Anthocyanins
Flavon-3-ols, and tea color, 293
Flavonone(s), biosynthesis and oxidation, 455, 456
from tea phenolase action, 400
glycosides in citrus bitterness, 452, 453, 455–456
Flavoproteins, 69. *See also* FAD, FMN
Flavor defects. *See* Off-flavor
Flavorants, as food ingredients, 247, 349
nonvolatile, accumulation of, 349–367
Flavors, desirable, enzymatic aspects—acid, accumulation in plants, 349–356
animal foods, 414–416, 480

as guide to food wholesomeness, 34
beverages, hot, 400–403
blanching, role in, 190
bread beer, and related foods, 362, 412, 413, 572–573
cheese and dairy products, 9, 362, 407–412, 605, 608–609
flavorese for restoration of, 416–420, *See also EI*
gustator (flavor potentiators), 346, 361–367, 699
of vegetables, 369–387, 698, 700
sugar and other sweeteners, 152–179, 356–361, 694–695
via nonenzymatic browning, 327
See also Aroma, Taste, Odors
preformed, *See* Preformed flavors
undesirable, 421–457, 616–618. *See also* Off-flavor
Flaxseed, protein from and glucosinolase of, 635
Flexibility of enzymes, 44
Flies, cholinesterase(a) from, 704
Flip-flop, a regulation of enzymes, 66
Flocculation, in cloud removal, 537
in the primary phase of milk clotting, 598
of frozen milk casein, prevention by lactase(p), 610
Floppy body, in enzyme catalysis, transition state as, 4
Florets, source of pickle oversoftening enzymes, 528–529
Flotation, in enzyme purification, 109
Flounder, texture of frozen, 502
Flour, semolina, 301, 414, *See also* Semolina
soybean, as enzyme(p), 299, 300, 301, 401, 424, 428, 589
wheat, enzyme assays for gassing power, 669
enzyme testing for quality of, 681
for standardization of enzyme(p), 103
high gluten blend, use of bacterial α-amylase(p) in, 579
in kernel morphology, 573
LOOH epoxidation, 427, 428
LOX(p)-treated, effects of, 301, 401, 424, 428, 589
lysozyme(p)-preserved, 662
thiols, 429
Flour milling, a breadmaking step, 573
ultrasonication in, 200
Flouriness, and flaw in potato texture, 526
Flowability of processed cheese, improved by proteinase(p), 609
Flower buds, as vegetable, 524
Flower fragrances, flavorese enhancement of, 419–420
Fluctuating temperatures, reaction rates, and food quality, 227–229
Fluidized bed, drying, 235
reactor, 176
for immobilized enzyme action, 141, 176, 235
for milk clotting with immobilized enzymes, 600
Fluorescence, in enzyme assay and analysis, 676, 688, 706
Fluorescent probes, and tertiary enzyme structure, 576
Fluoride, as NCI and enzymatic analysis for, 255, 707
1-Fluoro-2-nitro-4-azidobenzene, photosensitization for enzyme immobilization, 144
Flush, in tea production, 292
Flying start, in enzyme assays, 672
FMN. *See* Flavin mononucleotide
Foam, for control of enzymatic browning, 281
Foam mat drying, 235, 236
Foam stability, with enzyme(p), 153, 614, 616
Foaming, a protein functionality, 615, 616
and lysozyme, 615, 662
for enzyme purification, 113
in GOX egg treatment, 341
of beer, enzymes in, 558, 559, 571, *See also* Gushing
of milk, lipase-induced, 447
Folding, of proteins, and Arrhenius discontinuity, 220
proteins in enzyme regeneration, 214
Folic acid and folate(s), as enzyme cofactor, 46
in cheese amino acid biosynthesis, 409
inhibition of biosynthesis of, 249
loss by enzyme action, 646
Fomitopsis spp., milk clotting enzyme in, 602
Food, additives, as enzyme inhibitors, 247–248
essential oils as, 395
anylates, 694–702
alcohols, aldehydes, and acids, 697–699
carbohydrates, 694–695, 709
elemental analysis, 706–707
flavor related, 699–700
health, nutrition and safety related, 696–697, 700, 704–706, 714–715
nitrogen-containing, 701–702
colorants, *See* Colorants, food
enzymes, *See* Enzyme(p)
gums, enzyme(p) in manufacture of, 9, 11
identity, enzymes, change-of-state and, 21, 267, 535
ingredients, immobilized enzymes in production of, 150–151, *See also* specific enzyme, ingredients
microbiology, 6
pigments (list), 268–269, *See also* specific pigments
processing, chain, enzyme action in, 7, 11, 12, 18
enzyme assays and zymograms as guides in, 679–681, 683
significance of enzymes for, 12, 18, *See also* specific cases
wastes, for enzymatic biomass conversion, 167
quality, impact of enzymes, 14–18, *See also* specific cases
yield, enzyme assays for predicting, 12, 677–678
Forages, digestibility *via* enzyme test, 709. *See also* Feeds
Force, measurement and food texture, 462
Forced air (centrifugal stirred) drying, 235
Foreign activity, *vs* side, in enzymatic analysis, 685, 686
Forensic science, enzyme testing in, 207–208, 701
Formaldehyde, for chillproofing beer, 570
TMAO demethylase, fish texture and, 501–504
Formate, enzymatic analysis in molasses, 697
N-Formyl-L-tryptophan, 44
Fortified wheat blends, enzyme management of, 245
Fouling, tubes, prevention by enzyme(p) and Cold Break, 546
Fountain syrups, use of enzyme-derived sweeteners in, 152
Foxes, fishheads and thiaminase I-caused paralysis, 645
F-6-P + FDP, as allosteric affectors, 71, 471
as biospecific eluants, 132
enzymatic analysis for, 688
in cold sweetening and starch biosynthesis, 333, 334
in glycolysis and gluconeogenesis, 33, 72, 472
in muscle texture, 471, 472
in nonenzymatic browning, 334

in plant sugar accumulation, 357
FPC. *See* Fish protein concentrate
Fragmentation index. *See* Myofibrillar fragility
Fragrances, 374, 378, 419–420. *See also* Aroma(s)
Free energy, and homologous inhibition, 250
in ATP biosynthesis, 85
in ester synthesis at low a_w, 239
in peptide bond synthesis, 618
of hydrogen bonds, 186
specificity, entropy, ES and, 44
Free fatty acids (FFA). *See* Fatty acids
Free radical(s), ascorbate, 284, 285
fatty acid, direct aldehyde precursor, 427
from Irad, 13, 198
from lipid oxidation during breadmaking, 598
hydroxy, *See* Hydroxy free radical
in cellulolysis, 176
in oxidase reaction mechanism, 339, 425, 426
in proteins from reaction with LOOH, 631
of LOOHs, 427
O_2^-, *See* Superoxide anion
propagation chain, breaking of, 283, 437
stabilization at low a_w's, 236
Freestone peaches, endo PG, texture and, 519
Freeze drying, β-carotene in cherries during, 644
enzyme survival during, 244, 245, 273
in enzyme purification, 109
meat tenderizer action on, 490
of beef, 480
Freeze toughening, 504
Freeze-dried foods, as a low moisture food, 234
for flavor improvement, 246, 420
in model low a_w systems, 240
Irad effects on enzymes of, 199
Freeze-thawing, and milk off-flavor, 447
detection *via* enzyme assay, zymograms, 680, 681–682
enzyme behavior in, 220–222, 224
in enzyme purification, 108
juice cloud removal, 531
stability, imparted by functional proteins, 614
Freezer burn and ATPase, 321
Freezing, advantages of, enzyme relation to, 9, 218
and cellular disruption, 226–227
and meat texture, 463, 480
as a special case of drying, 223–224
enzyme effects during, 221–224
enzyme potentiation by, 366
fish toughness, 504, *See also* Fish, frozen
in gibberellin malting, 557
of onions, alliinase and, 279–280
post-blanch cooling and, 194
French hybrid wines, methanol in, 638
Frequency, effect on reaction rates in fluctuating temperatures, 228
Freshening, of cauliflower, and turgor, 528
Freundlich's adsorption isotherm, and affinity chromatography, 124
Fried foods, lipase(p) for fat digestibility, 708
Frogs, melanin pigmentation of, 277
Frost injury, and sulfhydryls, 223
Frozen fish texture, 499–505. *See also* Fish, texture
Frozen foods, effective temperatures during storage of, 227–229
enzyme behavior and fats in, 221, 223, 225–227, 229
enzyme regeneration in blanched, 212, 215–216
enzyme related nutrient changes, 644
enzyme(p) treatment of sugar cap, 291
flavorese treatment of, 416–417
literature on, 218
nonenzymatic reactions in, 225
off-flavor enzymology of, 221, 223, 225, 288, 299, 430–433, 444
Frozen milk, enzymatic(p) prevention of casein flocculation in, 610
Frozen potato product, PE and texture of, 526
α-Fructose as a competitive product inhibitor, 255
Fructose, enzymatic analysis for, 688, 695
fructose syrups, enzymology of production, 160–167
Fructose-6-phosphate. *See* F-6-P + FDP
Fruit juices, clarification enzymology of, 9, 104, 535–539, 540–541
decolorization of anthocyanins, 296
enzymatic analysis for alcohol in, 697
lysozyme(p)-preserved, 662
non-limpid, cloud retention of, 540–544
P.E.-generated methanol in, 638
spray drying of, 235
syrup, adulteration of, 162
See also Clarification, Cloud, Nectars, specific juices
Fruit vegetables, 524
Fruitcakes, use of bacterial α-amylase(p) in, 10, 579
Fruitiness, as off-odors, 431, 452
Fruits, acid-to-sugar ratio, 349
botanical, as vegetables, 524
color enzymology, 13, 209, 235, 238, 241, 251, 269, 275, 278, 279, 280–282, 284–285, 290–291, 296–297
definition of, as foods, 523
enzyme assay for monitoring spoilage of, 678
flavor enzymology, 349–358, 360, 388–395, 420, 630
flavor restoration with flavorese, 416–417
Irad effects on, 523
pesticides in, enzymatic-TLC analysis for, 705
texture enzymology, 227, 519–523
See also specific fruits, attributes, processing, enzymes, assays, and analysis
Frying, and dieting, enzyme(p) for, 664
and meat texture, 483
as a blanching process, 294
as a food dehydration process, 234
L-Fucose, in pectin side chains, 513
and stereospecificity of glucose isomerase, 162
Full-fat soybean products, off-flavor in, 424
Fullness of body. *See* Body
Functional groups, odor perception and, 345
Functional properties of protein. *See* Functionalities
Functionalities of proteins, and dieting, 664
impartation of, *via* enzyme action, 620, 624, 629
improvement of, *via* proteinase(p), 614–616
molecular basis of, 615–616
NEB effects, on egg, 339–340
of meat, affected by tenderizers, 489, 493
varieties of, 614–615
Fungal infection, indicated by enzyme tests, 677, 678, 682
Fungi, alpha-amylase(p) for baking, 578–579, *See also* specific uses
and blanching, 189
cultivated, as source of termite cellulase, 168

enzyme(p) production from, 95–96
milk clotting acid proteinases in genera of, 602
onion, alliinase from, 380
PG of, as cause of oversoftening, 522, 528–529
proteinase, as meat tenderizers, 486
specificity of acid proteinases of, 486
wood-rotting, xylanase of, 178
xerophilic, 235
Furanones, in fruit aromas, 389
Furanoterpenoids, peroxidase and, 204
Furs, enzyme(p) treatment of, 5
Fusarium spp., cutin depolymerase in, 519
pectate and pectin lyase in, 517
xylanase induction in, 178
Futures commodities, sunflower seed as, 623
F-value, for food enzyme inactivation, 182, 187, 206, 209, 543

Gadoids, enzyme-caused toughening of, 502, 503
α-Galactans, in beans, 653
Galactolipids, metabolism of and α-galactosidase, 653
Galactomannan, yeast cell wall component, 628
Galactonolactone, as biospecific eluant, 132
D-Galactose, enzymatic analysis for, 695
as an enzyme-generated toxin, 643
fermentation in breadmaking, 577
in pectin and protopectin, 513
lactase inhibition by, 255
L-Galactose in plants, 695
Galactosyl-β-1,4-*N*-acetylglucosamine, by α-lactalbumin, 361
Galactosyl diglycerides, 442
Galacturonans, pectin as, 513
D-Galacturonic acid, and pectin structure, 513
from PG action, a NEB problem, 536
in wine making, 537
oxidation of by, citrus enzyme, 522
Gallocatechins, tea, 294
Galloyl esters, tea phenolase substrates, 292
Gamma rays. *See* Ionizing radiation, 196
Gangliosides, enzyme caused off-flavor and, 446
Garbanzos, LOX of, 430. *See also* Chick-peas
Gari, removal of HCN by acidity of, 636
Garlic, aroma enzymology, 373–374
as a bulb vegetable, 524
dehydration, microbial survival, 237
purees, cellulase(p) treatment of, 549
green pigmentation of, 269, 313–314
rehydration of dehydrated, 246
Garlic mustard, thiocyanates from, 384
Gas, bubbles in baking, lipid, gluten and LOX, 588
injection drying, 235
liquid chromatography, *See* GLC
production and cheese texture, 605
Gassing power of flours, enzymatic aspects of, 579, 709
Gastric juice, as enzyme(p) source, 108
Gastrin, RIA of, 713
Gastrointestinal tract, alleged enzyme passage through, 648–649
enzyme(p) production from, 91–92
enzyme(p) treatment of diseases of, 664
of marine animals, proteinases of, 91, 506, 507–508
of termites, cellulase in, 168
Gel chromatography, scale-up, 120–122
Gel density electrophoresis in enzyme purification, 115
Gel electrophoresis, for isozyme patterns, 681, *See also* Zymograms
with immunochemical techniques, 622, 683
Gel filtration, in enzyme purification, 120–121
with centrifugation and. *See* Chromatofuge
Gelability, a functional property, 614
Gelatin, a fining agent, juice clarification, 536, 537
as a functional protein, 615
enzyme(p) treated for beverages, 625
from fish myocommata by cooking, 498
hydrophobicity, and plastein, 620
nonbitterness of hydrolysates, 617
in enzyme immobilization, 165
lipase(p) in manufacture of, 9
proteinase(p) for removal, in photography, serigraphy, 5
texture (elasticity) of, 462
Gelatinization of starch, 200
Gelation agents, functional proteins as, 614
Gels, formation and prevention in frozen juices, 194, 544
formation from clouds, 541
from oxidation of flour proteoglycan-phenol, 584, 590, 591
in enzyme purification, 118
loss of in starch-based puddings, 591
low-methoxyl pectin, canned food texture and, 523, 525–526
quick-setting, use of enzyme modified starches in, 153
thixotropic, of plastein, 620
Gene derepression of biosynthesis of enzymes, autumn foliage color, senescence and, 269
by ethylene, 309
by glutamate, nitrate reductase and, 263
controlled atmosphere storage and, 530–531
during seed germination, 359
gibberellins and, 556–557
in chlorophyll degradation, 309
in hormone regulation, 77
of acid accumulation, 355
of carotenoid biosynthesis by alkylaryl amines, 269
of cheese ripeners, 408, 409
of cold sweetening system in potatoes, 329–330
Gene expression. *See* Gene derepression
Gene repression of biosynthesis of enzymes, 12
by amines, products of amino acid decarboxylase, 641
by amino acids in fast cheese starters (proteinases), 607
by light (lipase), 586
by substrate, lactose, in the *lac* operon, 74–75
determining texture (PG), 524
feedback, 76
in hypobaric storage, 350
in lactose malabsorption (lactase), 652
management of enzymes by, 12
of mineral deficient plants, 677

of valine biosynthesis, 409
upon maturity (potato sucrose synthase), 678
Generative origin, zymograms as indicator of, 682
Genes, in biosynthesis of enzymes, 74–75
mutation of for tailor-made enzyme(p), 95–97
of the *lac* operon and enzyme regulation, 75–76
structural, discontinuous in eukaryotes, 76
See also DNA, Evolution, Genetic engineering, P-gene, etc.
Genetic code, in enzyme biosynthesis, 40, 74
Genetic drift, and variation as source of isozymes, 47, 48
Genetic engineering, DNA recombinant, for chymosin(p) production, 603–604
DNA topoisomerases in, 76
for enzyme(p) in ethylene glycol manufacture, 7
for N-fixation, enzyme(p), protoplasts and, 550
methylation and restriction DNases, 85–86, 97–98
principles of, for enzyme(p) production, 97–99
Genetic regulation of enzymes, 74–81
Genetic stabilization of enzymes, 101
Genetically independent proteins, as isozymes, 47
Genetics. *See* Breeding, Mutations
Geneva hybrid wines, PE-generated methanol in, 638
Genome, 309, 408. *See also* Genes, Gene derepression
Gentiobiose, 396
Gentiobiosyl mandelonitrile, 396. *See also* Amygdalin
Geometric chirality *(cis-trans)* in analog inhibition, 249
Geometric isomerism, of LOX-produced LOOHs, 425, 426
Geosmin, an intense fragrance, 369, 387
Geotrichum candidum, lipase of, 441
Geranial, flavorant, 403, 404
Geraniol, structure and biosynthesis, 403, 404, 405
Geraniol β-D-glucoside, rose fragrance precursor, 420
Geranium, as a dairy product off-flavor, 452
Geranyl acetone, tea flavorant, 400
Geranyl pyrophosphate, a biosynthetic branch point, 403, 405
Germ, wheat. *See* Wheat germ
Germinated barley. *See* Malt
Germinated wheat. *See* Wheat malt
Germination, *See also* Sprouting
for removal of phytate, 659
in enzyme purification, 108
lipase and, 444
management of enzyme action by, 12
precocious, in triticale kernel shrivelling, 576
Gherkins, pickled, enzymatic oversoftening of, 529
Gibberellic acid and gibberellins, α-amylase biosynthesis/release in cereals, 556–557, 585
biosynthesis of, 405
control of enzymatic browning with, 280
glyoxosome formation, 355
in biosynthesis/release of other enzymes, 558, 589
malting with, 557–558
RNA synthesis enhanced by, 77
tomato PG and PE repression with, 524
Gibberellin malts, for distillery use, 561
Ginger, proteinase(p) from, 94
Ginkgo leaf, aroma, 373
Gizzards, chicken, rennet from, 601
Glandless cottonseed, 622
Glandular tissue, 435
Glass beads, as immobilized enzyme supports, 144, 653
in enzyme purification, 110
siliconized, lipase activation by, 442
GLC *vs* enzymatic analysis, 695, 697, 700, 704
and preformed S volatiles in vegetables, 374
in enzymatic analysis for glucosinolates, 699
in flavorese investigations, 419
utility limitation, in flavor research, 349, 369
Gliadin, as part of the gluten network, 484, 587
Gliocladium spp., milk clotting enzymes in, 602
γ-Globulin as antibody in EIA, 714
Globulins, in enzyme purification, 113, 116
Gloss, in food color and appearance, 26
α-1,3-Glucans, juice consistency and, 544, 545
α-1,4-Glucans, phosphorolysis of, 331. *See also* Amylose, Starch
β-D-Glucans, as yeast cell wall components, 628
in barley and malt brewing, 565, 566
Glucobrassicin, 384
Gluconeogenesis, and lysosomes, 82
enzymes and allosteric regulation of, 71, 353
Irad effects on, in bananas, 199
regulation of by proteinase, 65
sugar accumulation *via*, 356–357
Gluconic acid, as food quality determinant, 341
enzymatic analysis for, 695–696
GOX action end-product, 338
in tartaric acid accumulation, 354
γ-Gluconolactone, as GOX primary product, 338
in model cheesemaking systems, 607, 618
Gluconolactones, as copper chelators, 418
Glucoproteic acceptor, in starch synthesis, 332
Glucosamine, acylation of, 390
D-Glucose, as α-amylase end product, 155, 242, 554, 556
as glucoamylase product, 154
as glucose oxidase substrate, 338
as nutrient for SCP organisms, 628
catabolite repression by, 76, 169
enzymatic analysis of, 205, 208, 342, 670, 691
enzymatic(p) removal from eggs, 340–341
enzyme(p)-produced syrups, crystals, from starch, 157–159
enzyme(p)-produced, from cellulose, 167, 174–176
glucose isomerase substrate, for HFCS, 9, 162–167
-hexokinase, isolatable ES complex, 39
in brewers' wort, 560
in fish texture, 499
in nonenzymatic browning, 20, 330
nonenzymatic isomerization of, 60, 153–154
nonenzymatic production of, 153
α-D-Glucose, as a NCI, 255
β-D-Glucose, as a competitive product inhibitor, 255
preferred GOX substrate, 337
Glucose-1,6-diphosphate, intermediary in phosphoglucomutase action, 40
Glucose-1-phosphate, in fish texture, 499
in glycolysis and gluconeogenesis, 71
in starch metabolism, 380–382
in sugar accumulation in plants, 283
Glucose-6-phosphate, in glycolysis, in G-6-P, 71
in NEB, 344
Glucose phosphates, enzymatic analysis for, 688
Glucosides, α- and β-, enzyme exclusivity for either, 31
as source of aroma volatiles, 396. *See also* specific glucosides

Glucosinolates (thioglucosides, mustard oil glucosides), and glucosinolase, 384–386, 387
biosynthesis structure, 382–383, 396, 411, 455
enzymatic analysis for, 699
flavor related aspects, 381–382, 386, 417
health related aspects, 384, 632–633
syn- and *anti*-isomers of, 383
Glucosyl thichydroxamate, 383
Glucuronic acid, as ligand in affinity chromatography, 129
as xylan component, 177, 180
glucuronides in detoxification, 77
Glucuronides, in detoxification, 77
L-Glutamate and L-glutamic acid, and folic acid loss in foods, 646
as flavor potentiator (gustator), 361–363
as GABA precursor and decarboxylase substrate, 640, 641
as gibberellin synergist, 558
γ-carboxyl, pK_a of, 262
carboxylation of, in proteins, 45–46
enzymatic analysis for, 690, 691, 699
enzymes of accumulation of, 361–363
frozen fish toughness prevention by, 505
in aminotransferase reactions, 33–34, 351
in meat tenderizer formulations, 492
meats, color retention by, 322
production of, enzyme(p) *vs* fermentation, 363
vegetables, chlorophyll retention by, 301
Glutamine, as NH_2 donor in biosynthetic pathways, 363
azaserine as antimetabolite of, 249
hydrolysis of, 381
pivotal metabolite in enzyme regulation, 72–73
stimulation of wheat lipase biosynthesis by, 586
γ-Glutamyl cycle, 380
Glutamyl peptides, from proteolysis, flavor-bearing, 362, 618
in gustator accumulation, 361, 362
γ-Glutamyl peptides, as precursors to flavor precursors, 15, 237, 378, 379
Glutaraldehyde, as matrix activator in affinity chromatography, 127
for immobilizing enzymes, 143, 158, 164, 653
α-Glutaric acid, in the TCA cycle, 351
Glutathione (GSH), as enzyme cofactor, 46
in affinity elution, 132
in dough development, 591
in glutamate gustator, accumulation, 362
in the glyoxylase reaction, 503
meat color retention with, 323, 324
meat tenderization with, 482
melanin modification and color control by, 283
metmyoglobin reduction *in vivo* by, 317
Glutelins. *See* Glutenin
Gluten and bread texture, as a functional protein, 610
in proteinase(p) flour treatment, 583–584
oxidoreductases-mediated oxidation/reduction of, 587–592
ultrastructure and function, models for, 584, 585, 587
Glutenin, as part of the gluten network, 484
β-amylase solubility and, 562
LOOH-transforming enzymes in, 429
Glycerol, as enzyme reaction transport medium, 241
enzymatic analysis of, 691
in enzyme purification, 110
in meat tenderizer formulations, 492
in Raoult's law deviations, 240, 241
Glycine, and calculation of hydrophobicity, 617
and detoxification mechanisms, 77
in serine biosynthesis, 409
Glycogen, branching structure of, 554
changes in freeze-dried meat, 241, 242
changes in frozen meat, 227
in affinity chromatography, 124, 576
in fish texture and glycolysis, 499
removal in enzyme purification, 113
substrate for isoamylase, not pullulanase, 155
yeast cell wall component, 628
Glycolic acid, in oxalic acid accumulation, 350, 355
Glycolysis and glycolytic enzymes, as a multienzyme complex, 69, 117–118
as protomers (oligomericity), 71
excess, in cheese starters, and flavor, 408
gluconeogenesis, and enzyme regulation, 70–72, 356
in brewing, 553
in fish and quality of, 498–499
in frozen food off-flavor, 431
in meat, enhanced tenderness *via*, 469, 471–473, 480, 482
in meat textural aberrations, 469–471
in muscle-to-meat, 464, 467–469
in plant acid accumulation, 351
in potato cold sweetening, 333–334
LDH isozymes and regulation of, 84
NEB and, 334
purification by affinity elution, 132
Glycopeptide, product of chymosin action on k-casein, 597
Glycopeptides, protopectinase and, 519
Glycoproteins, and cryoprotection, 229–230
biosynthesis, sugars function, 81
k-casein as, 595
enzymes as, 46
alliinase as, 376
α-amylase as, 555
endocellulase, 170
glucose oxidase, 38, 46
glucosinolase as, 385
in affinity chromatography, 130
invertase as, 46, 103, 358
peroxidases, 203
phenolase as, 274
tannase, 46
stability of, 103
Glycosides, cyanogenic. *See* Cyanogenic glycosides
Glycosylation, for stabilizing enzymes, 103
Glycyl phenylalanineamide, cathepsin substrate, 474
Glycyl-L-tyrosine, carboxypeptidase substrate, 44
Glyoxal, methoxypyrazines from, 386
Glyoxosome, and the glyoxylate cycle, 351, 355, 357
Glyoxylate, an inhibitor of fatty acid oxidation, 352
Glyoxylate cycle, in acid and sugar accumulation and, 199, 351, 356
Glyoxylic acid, conversion to oxalic acid, 350, 355
GMP, guanosine monophosphate, gustator, 364
Goats, cheese from milk of, 594
Goiter, due to enzyme-generated toxin, 376, 382, 384, 632, 635
Goitrin, 382, 384
Goitrogens, 376, 632

Gold salts, 224
Golden Delicious apples, 281, 310
Golgi apparatus or body, post-translation, localization and lysosomes, 81, 82, 373, 376, 609
Gooseberries, Chinese, proteinase source, 94
  PE-generated methanol in, 638
  sourness of, 252
Gossypol, apparent detoxification by phytase, 659
  inhibitor of cottonseed proteinase, 622, 626
Gouda cheese, 604, 606
Gourds, 279
Gout, prevention by removal of purines, 655
Gradient, pH, and ATP biosynthesis, 85, 290
Gradients, in enzyme purification, 115, 116
Grain(s), *See also* Cereals and specific grains
  AOAC enzymatic analysis of starch in, 670
  beer from, 552
  enzyme assay for storage history, 680
  laccase of, 271
  ripening and solar energy, 233
Graininess, a cloud attribute, 536
Gramineae, biogenic amines in members of, 641
Grams, adulteration, detection of, 701
Granadilla, biogenic amines in, 641
Granules, starch. *See* Starch granules
Grape leaves, ascorbate oxidase of, 289
Grape vines, xylan in, 177
Grapefruit, bitterness and naringinase debittering of, 453, 455–456
  malate dehydrogenase of, 352
  pitting of, a cold stress response, 227
  texture of, 511
Grapes, *See also* Raisins, Winemaking
  acid accumulation enzymology, 351, 352, 354
  aroma of, 371, 388
  blanching of, 281–282
  color change enzymology, 274, 275, 296, 297
  juice-making enzymology, 104, 393, 537, 539, 540–541
  pesticides in, enzymatic-GLC analysis for, 705
  skin, PE in, methanol from, 638
  sugar accumulation enzymology, 356, 357
  varieties in winemaking, 281, 282, 541, 638, 639
GRAS, chemicals, as enzyme inhibitors, 247–248, 256
  enzyme(p) as, 486, 647
Grass, off-odor of unblanched frozen foods, 430
  phenolase-impaired nutritiousness of, 630
Grasshopper eggs, prophenolase of, 274
Grassiness, a LOX-caused off-odor, 424
Gravies, creatine in, enzymatic analysis for, 701
  functional protein in, 615
Grazing, enzyme-generated unwanted substances ingested, 422, 635, 636
Greasiness, diminution with enzyme-derived sweeteners, 153
Green beans, aroma of fresh, 371
  as a legume vegetable, 524
  ascorbate oxidase of, 289
  blanching of, 190, 195, 207
  chlorophyll in, 302, 304, 307
  off-odors of unblanched, 430
  PE-Ca induced firming of canned, 527
  peroxidase, heat stability of, 205, 207
Green color of vegetables. *See* Chlorophyll
Green discolorations, 313–314, 325
Green leaf aldehyde. *See* Leaf aldehyde
Green revolution, dwarf wheat, phenolase and, 275, 591, 638
Green tea, 292, 401–402
Green vegetables, ascorbate oxidase in, 289
Greengage juice, proteinase(p) in making, 539
Greens. *See* Turnip greens
Grinding, in enzyme purification, 109, 110
Gristliness, an enzyme-solved FPC texture problem, 509
Grit, removal by blanching, 189
Grit cells. *See* Sclereids
Grittiness, in ice cream, prevented by enzyme(p), 9, 153, 610
Ground meat, color, 315
Group specific ligands, in affinity chromatography, 130
Growth factors, in enzyme(p), 647
Growth regulators, plant hydroxynitrile lyase in synthesis, 398
Growth stimulation, by feeding enzyme preparations, 647
Guaiacol, in peroxidase assay, 207, 208, 219, 224, 685
Guanidine, a reversible enzyme inactivator, 213
  as H-bond breaking inhibitor, 259
  prophenolase activation by, 274
  protein denaturants, 186
Guanidinium, arginine, pK_a of, 262
  salts as eluant in affinity chromatography, 131
Guanine, structure, 363–364
Guanosine 5′-phosphate, gustator, 364. *See also* GMP
Guanosine triphosphate (GTP) in translation, 78, 79, 80
Guanylic acid (guanylate). *See* GMP
Guar, controlled hydrolysis by hemicellulases, 9
Guar flour, as alternative to α-amylase(p) in breadmaking, 579
Guava juice, nectar, enzyme(p) in making, 540, 544
Gumminess, a potato texture, 526
  an excess α-amylase-caused baked goods texture defect, 579
Gums, as alternatives to α-amylase(p) in breadmaking, 579
  of flour, pentosans as, 581
  β-D-glucans, in brewing, 565–566
Gunpowder tea, 292
Gushing, enzyme(p) prevention in beer, 571
Gustators, enzymatic analysis for, 699–700, *See also* specific gustators
  immobilized enzyme production of, 150, 151
  in flavor genesis, 361–367, 369

Haemagglutinins. *See* Lectins
Hake, enzyme caused toughening, 502
  enzyme produced FPC, 509
Half-of-the-sites reactivity, 66, 212
Half-reactions, in GOX action mechanism, 339
Halibut, texture of frozen, 502
Hallochrome, 276
Halogen anions, peroxidase donors, 31, 204, 205
Halophiles, fish sauce proteinase, 506
Ham(s), curing brines functional proteins, 615

curing enzymology, 323–325
enzyme assay for pasteurization adequacy, 212
enzyme(p) tenderization, 493
lipase, smoked flavor, 413
Hamburger, and onions, 376
blocking enzymes, for NMR, 71
buns, proteinase(p) in making, 584
enzyme(p) tenderization, 493
filth in, enzyme-aided detection, 711
turkey, color, 315
Hammermill pretreatment of cellulose, 176
Handling, of foods, 13
Hanging drop, enzyme crystallization, 118
Haptens, and EIA, 713
Hard spring wheat. *See* Wheat
Hardcore, a sweet potato texture defect, 530
Hardness, a vegetable texture, 511
Harvesting, and processed food quality, 11, 13, 100, 346
Haze(s), and food color, appearance, 267, 268
beer, enzyme(p) for removal, 569–571
fruit juice, enzyme(p) for removal, 537–541
enzyme(p) for retention, 544
in teas, 292
phenolase, composition, 270, 630
postfiltration, 539
starch, enzyme(p) for removal, 9, 539
See also Cloud(s), individual beverages
Hb. *See* Hemoglobin
HCN and cyanide, and mitochondrial reducing power, 324
enzymatic analysis for, 699–700
enzymatic removal, 399, 416, 637
enzyme-inhibitor, 248, 256, 282
enzyme-generated toxicant, 633, 634–637
thiol proteinase reactivation, 213
HCN-p, definition, values, 635, 636
Health foods, enzyme(p) for making, 548–549
Health-related anylates, enzymatic analysis, 700
Heart, as food, LDH of, 83, 463
Head, beer. *See* Foaming, Gushing
Head-to-tail fusion, terpenoid biosynthesis, 403
Heat inactivation, for detecting rennet blends, 605
Heat processing, enzymatic assessment of, 709. *See also* Blanching
Heat resistance of enzymes in foods, 206, 209
Heat shock, for SCP purine removal, 655
Heat stability, of enzymes, 185–186, 206, 209, 210, 555, *See also* individual enzymes
proteins, proteinase susceptibility, 187
Heat treatments, and peroxidase assay, 206–209
categories, 187–188
Heating, selective, in enzyme purification, 112
Heating rates, and enzyme regeneration, 215
Heat-up, in blanching, enzyme activation, 215
Heavy metal ions, as enzyme inhibitors, 249
enzymatic affector analysis, 707
α-Helix (helices), protein, in plasteins, 620
in protein-protein interaction, 630
in secondary protein structure, 40
nucleation, in enzyme regeneration, 263
Helminthosporic acid, gibberellin synergist, 558
Helminthosporum maydis, lipase, and corn blight, 445
Heme compounds, as antithiamins, 644
Heme proteins, as enzymes, bleachers, 46, 203, 213, 435, 645
Hemicelluloses. *See* Hemicellulase, Protopectin, Pentosan, Xylan
Hemin, as model peroxidase, 219
in blueing of crab, 325
Hemoglobin, as non-enzyme protein, 28
carotene bleaching by, 300
enzyme(p) treated, for human food, 624
in affinity chromatography, 124, 582
in diabetes, and NEB, 317, 334
in erythrocytes, 81, 317
in meat color, 315–316, 318, 320
oxygenation regulation by phosphates, 322, 364, 656
Hemolymph, bee, invertase, gluco-, 358
Hemolysis, and meat off-flavors, 433–434
Hemoproteins. *See* Heme proteins
Heparin, inhibitor, activator, 258, 443, 447–448
Hepatogastric organ, snail, cellulase, xylanase in, 178
Heptanol, an off-odorant, 424
4-*cis*-Heptenal, a fish off-odorant, 436
in cheese aroma, 408, 410
Herbage vegetables, 254
Herbicides, as gibberellin suppressants, 558
attempted enzymatic analysis for, 705
removal by blanching, 189
removal by enzyme(p), proposed, 660
metabolism by xenobiotic oxidase, 412
Herring, antioxidant protection, 631
as fermented food, 369
flavor, enzymatic, 416
marinading, proteinase action during, 508
pickled, biogenic amines in, 640, 641
thiaminase I in, vitamin deficiency, 644
Hesperedin, structure, enzymology, 453, 539, 541
Heterocyclics, in food aromas, 374, 387, 416
Heterogeneous catalysis, immobilized enzyme kinetics, 147–149
Heteropolymers, as isozymes, 47
Heteropyrithiamin, thiaminase I co-product, 645
Hexametaphosphate, blanching with, 95
tenderizing meat with, 482, 483
Hexanal LOX generated aroma, off-odor, 401, 424, 431, 432
Hexane, for off-flavor removal, 439
Hexanoic acid, papaya puree off-odor, 445
Hexanol, an off-odorant, 424
Hexenals, double bond reduction by ADH, 394
LOX generated aroma, off-odors, 370, 371, 372, 424
trans-2-Hexenol, aroma and level, 424
cis-3-Hexen-1-yl pyruvate, in celery aroma, 387
Hide, potent microbial collagenase from, 496
High capacity operation, enzyme purification, 112
High fructose corn syrups, HFCS. *See* Fructose syrups
High glucose syrups, 157–158
High lysine corn, peroxidase and oil of, 209
High performance liquid chromatography (HPLC), 427, 697
High pressure, enzyme inactivation, meat tenderization, 201, 484, 493
High protein foods, *See also* Proteins
acidic beverages, enzyme(p) for, 624–625
enzyme-related problems, solutions, 423, 445, 507–510, 613, 633
enzyme(p)-derived flavorants in, 415
spray drying of, 235
High temperature conditioning of beef, 480

High temperature-short time processing. *See* HTST
Hippuric acid, detoxification product, 77
Histamine, enzyme generated toxicant, 640, 642
Histidine, as inhibitor, 367, 393
allosteric affector, 82
enzymatic analysis for, 700
histamine precursor, 640, 642
imidazole, and photosensitization of enzymes, 200
pK_a, 262
in enzyme active site, 56
Histochemistry of enzymes, 684
Histological changes, by meat tenderizers, 488–489
Histones, and DNA, 75, 86, 162
Hogs, pancreatin from, 647
Holozyme, and enzyme regeneration, 214
Hollow fiber membrane reactor, enzyme immobilization, configuration variations, applications, 143, 153, 164
for debittering juices with naringinase, 456
for lactose removal with lactase(p), 653
for protein processing, with enzyme(p), 615
Home dehydrators, 236
Hominy grits, 263
Homofermentation, 408
Homogenization of milk, and induced lipolysis, 448
Homogenizers, in enzyme purification, 110
Homoiothermy, 63
Homology, of protein amino acid sequence, calcium-binding proteins, enzymes, 476
Na/K ATPase, non-enzyme proteins, 85
proteinases in rennets, 602
organic chemical, and inhibition, 250
Homolytic reactions. *See* Free radicals
Honey, adulteration HFCS, 162
enzyme assays as indicator of quality, heating, 669, 679
enzyme-aided analysis for sucrose in, 712
enzymes of sugar accumulation, 358
Hops, in beer flavor, aroma, haze, 405, 553, 570
Hordein, fate in brewing, 567–568, 569
Hormones, affinity chromatography, 134
as non-enzyme proteins, 28
in enzyme biosynthesis, 77
in enzyme regulation, 63, 73, 442, *See also* Cyclic AMP
in meat texture, 463, 471, 482
membranes, site of action of, 15–16
receptor sites, 15, 74, 134
Horse beans, off-colors, -odors, 227, 430
Horse nettles, proteinase, 94
Horseradish, peroxidase, 203, 205, 431
pungency, 383
Horses, thiaminase I in forage, 645
Horticultural practice, to prevent nonenzymatic browning, 336
Hot break, in tomato processing, enzyme inactivation, 546–547
Hot dogs, EIA of enterotoxin in, 714, 715
Hot fill, and PE action, 522
Hot gas blanching, 194–195
Hot oil cooking, as blanching, 194
Hot pressing, of grapes and PE inactivation, 683
Hot processing, tenderizing meat by, 484
HTST, for removing excess α-amylase, 581
in blanching, 192, 195, 306
in enzyme regeneration, 215
pasteurization by, 192
Hue, a color perception component, 268
Humane slaughter, and PSE, 471
Humanization, of cows' milk with lysozyme(p), 9, 662
of high protein feed *via* enzymes, 507–510, 622–624, 626
Humectancy, control with enzyme-derived sweeteners, 153
Humectants, in IMF, 234
Humicola spp., milk clotting enzymes in, 602
Hura crepitans. *See* Jabillo
HVP, hydrolyzed vegetable proteins, 19, 616
Hybrid(s), from enzyme(p)-produced protoplasts, 550
of tomato mutants, and texture, 524, 525
triticale as, 575–576
Hydration, and cloud stabilization, 541
of substrate, in substrate inhibition, 238
Hydrocarbon length, in hydrophobic chromatography, 119
Hydrocarbons, *via* fatty acids, and off-flavor, 435
Hydrocortisone, and phospholipase A_2, 443
Hydrocyanic acid. *See* HCN and cyanide
Hydrogel, for chillproofing beer, 570
Hydrogen bonds and bonding, breaking, in enzyme inactivation, 186–187, 244, 249, 263
in biopolymer enzyme susceptibility, 172, 187
in enzyme catalytic efficiency, 54, 56–57
in fish texture, *via* crosslinks, 501
in plastein gelation, 620
in protein functionality, 616
in protein secondary structure, 41
in substrate inhibition, 238
in sweetness perception, 346
in tea cream, 539, 540
super-, in Arrhenius discontinuity, 219
Hydrogen cyanide. *See* HCN and cyanide
Hydrogen donors, in peroxidase assays, 207, 208
Hydrogen, D_R-, in LOOH formation by LOX, 372
Hydrogen, L_S-, LOOH and LOX, 372
Hydrogen ion, *See also* pH, Proton
as enzyme cofactor, affector, 46, 249, 260–262
catalysis of hydrolysis, 50
in pheophytinization, 305, 306
mobility in ice, 222
Hydrogen peroxide, accumulation in bombardier beetle, 205
as enzyme inactivator, 213
catalase(p) for removal of, 663
enzymatic analysis for, 670, 685, 699, 700
enzymatic decomposition, *See* Catalase, Peroxidase
for off-flavor removal, 281, 452
from Irad, 198
in doughs, from added GOX(p), 340, 591
in food discolorations, 300, 318, 325, 341
non-food industrial applications, 5
production, *via* O_2^- intermediation, *See* Superoxide anion
without O_2^- intermediation, 338
with peroxidase(p), as sterilant, 62
scavengers, in meat color retention, 323, 324
Hydrogen pro-L_S, abstraction in LOX action, 425
Hydrogen sulfide, in aromas, off-odors, 413, 423, 456, 457
Hydrolysis, acid-, base-catalyzed, 225, 255
enzyme catalyzed. *See Enzyme Index*

Hydrolytic rancidity, *See also* Lipase
enzyme management of, 245
in fish, 436
in situ ES estimation of, 680
in meats, 446, 680
in milk, prevention, control, 449
products, 446–449
the enzymes, and spontaneous, 446–448
the problem, 446
the substrates, and induced, 447, 448–449
in plant foods, 444–446
in rice bran, oil, 445
microbial, 447, 449
Hydrolyzed vegetable protein (HVP), 19, 616
Hydromonas spp., DDT-degrading system of, 660
Hydronium ion, H_3O^+, 306
Hydroperoxy free radical, 198, 631
Hydrophilic agents, for stabilizing enzyme(p), 102
Hydrophilicity, of casein micelles, 595
Hydrophobic chromatography, 55, 119, 618
Hydrophobic interaction, *See also* Hydrophobicity
and plastein formation, properties, 620
in acid inactivation, 263
in active site and enzyme catalysis, 55, 56, 61
in enzyme inhibition, 250
in enzyme purification, 119, 127, 618
in enzyme regeneration, 263
in enzyme subunit interaction, 66, 68
in fish toughness, 501
in lignin deposition, 529
in milk clotting, 595, 596, 598
interface with food science, 55–56
lipid-gluten, and LOX, 588
polyphenol-tannin interaction, 630
Hydrophobicity, and plastein formation, 620
and protein nutritiousness, 624
in enzyme immobilization, 146
in protein denaturation, 187
in protein functionalization, 616
of bitter peptides, 617–618
of calmodulin core, 498
of casein micelle surface, 595, 596
of lignin, and toughness, 529
of proteins, proteolysis susceptibility, 624
Hydroquinone, laccase substrate, 273
Hydroquinones, off-flavor prevention, 437
Hydrostatic sealed steam blanching, 193
Hydrothermal treatment, for enzyme inactivation, 585
Hydroxy free radical (·OH), antioxidants as scavengers, 437
from Irad, 198
from O_2^-, *See* Superoxide anion
in cellulose degradation, 172
in chlorophyll degradation, 305
in ethylene biogenesis, 286
β-Hydroxyalcohols, in cheese aroma, 411
2-Hydroxyalkenyl glucosinolate, and goiter, 384
Hydroxyalkyl isothiocyanates, 384
Hydroxyalkylmethacrylate, in chillproofing beer, 571
Hydroxyamino acids, in fish texture, 500
Hydroxyapatite, as immobilized enzyme support, 144
in enzyme purification, 118–119, 506
o-Hydroxybenzoic acid methyl ester, a flavorant, 399
p-Hydroxybenzyl isothiocyanate, 382
2-Hydroxybutenyl isothiocyanate, goitrin precursor, 382
Hydroxyethyl cellulose, cellulase substrate, 170
Hydroxyfatty acids, cutin polymer of, 519
from LOOH reductase, 438
from LOX, 299, 395
SH oxidation in doughs, 242, 589–590
ω-, *via* xenobiotic oxidase, 412
Hydroxylamine-ester exchange *via* trypsin, 224
2-Hydroxymethyl methacrylate, immobilized enzyme support, 142
β-Hydroxynitriles, in cyanogenic glucoside synthesis, 396
m-Hydroxyphenols, in melanin lightening, 283
Hydroxypropyl starch, in dieting, and amylase, 644
8-Hydroxyquinoline, phenolase inhibitor, 281
5-Hydroxytryptophan, serotonin precursor, 640, 641
Hygroscopicity, control with enzyme-derived sweeteners, 153
of lactase-treated milk, of xylitol, 177, 653
Hy-I. *See* Hydrophobic interaction
Hyperbolic relation, enzyme kinetics, odor perception, 32–35, 346
Hyperbranching, amylopectin and Q-enzyme, 332
Hyperoside, apples, 296
Hypertensive crises, *via* biogenic amines, 640
Hypobaric storage, 310, 531
Hypochlorite, for aflatoxin destruction, 66
Hypoxanthine, enzyme test for, fish freshness, 366, 499, 680

Ice, proton mobility in, 22. *See also* Freezing
Ice cream, enzymatic analysis for lactose in, 695
enzyme(p), for de-sanding, 9, 153, 610
lysozyme(p)-preserved, 662
Ice crystals, barrier to enzyme action, 222
Ice surfaces, as catalysts, 225
Icing of fish, and texture, 497–498
Idaein, apple anthocyanin, 296
I-gene, of the *lac* operon, 75, 97
IgG, antibody as antigen in EIA, 714
Imbibition. *See* Steeping
Imidazole(s), active site nucleophile, 56
derivatives as enzyme models, 61–62
of histidine, pK_a of, 262
UV sensitivity, in proteins, 200
Immature wheat, and ethylene, 309
Immobilization, of anylate in enzymatic analysis, 694
Immobilization and immobilized enzymes, advantages problems, 139, 140
applications, analytical, in enzyme electrodes, reagentless analysis, 140, 596, 692–694
to foods, for amino acids in proteins, 711
carbohydrates, 654, 692
N-compounds, ascorbate, 701
other, on-going, 5, 150–151, 164–167
proposed and experimental, for cold sterilization, 205
for dietetic (light) beer, 564, 565
for Dopa manufacture, 273, 274

for ethylene glycol, 6
for inversion of sucrose, 160
for plastein production, 620
for removal of, cyanide, environmental, 637
glucose, oxygen [with GOX(p)], 338, 342
haze from beer (chillproofing), 570, 571
lactose from milk, whey, 180, 653
naringin bitterness from citrus juices, 455–456
nitrate, nitrite, 660–661
off-flavor from milk, 451–452
off-flavors from other foods, 439, 443
pectin, 538
pollutants, 5, 286, 660, 705
tea cream from soluble tea, 539
in cheesemaking, 600, 605
in flavorese, 420
for standardizing enzyme(p), 103
history, 138–139
in regeneration of ATP, $NAD(P)^+$, 285, 692
in vivo, a biological regulation mode, 65
kinetics, 146–149, 251, 336
literature, 137
properties, non-kinetic, 144–146
stability and safety, 140, 149–151, 662
See also individual enzyme, *Enzyme Index*
Immune systems radappertized food for patients lacking, 196
Immunoassays, 679, 712–715. *See also* EIA, RIA
Immunochemical procedures, *See also* EIA, RIA
as enzyme assay, 683
for amylases in cereals, 555, 556, 562, 573
for malt, beer proteins, 567–568, 569
for rennets in blends, 605
in affinity chromatography, 120, 130, 134, 368
of polyphenol-protein complexes, 630
Immunoenzymology, 20
Immunoglobulins, 65, 130
Immunology, and enzyme(p) safety, 74, 647, 648
IMP (inosine-5′-phosphate), as gustator, 364
degradation in fish, 367
immobilized enzyme for producing, 150, 151
in purine biosynthesis, 363–366
In vitro synthesis of enzymes, 80
In vivo testing, *in vitro* short cuts to, *via* enzymes, 707–709
Inactivation of enzymes, and microbial death, 191
and temperature-activity profiles, 183–184
by acid, 213, 263–264, *See also* Acid(s), pH extremes
by energy input, 181–201
heat, thermal, 181–195, 200, 206, 436
Irad, 196–200
non-thermal, non-Irad, 200–201
by enzyme reaction products, 260
by enzymes, 260
by organic solvents, 436
by pH extremes, 263, 436
by substrate (suicide), 663
during frozen food storage, 221–222
in meat cooking, and texture, 483
of α-amylase during wheat development, 574
Q_{10} of, 184–185, *See also* Energy of activation
reaction, syncatalytic, *See* Reaction inactivation
reversible, *See also* Regeneration
See also individual enzymes, foods, attributes
In-can blanching, 192, 455
Incipient denaturation, 630
Incompatibility, of rennet blend enzymes, 601
Indeculator dialysis, 133
Indicator enzymes, in enzymatic analysis, 685, 686–687, 691
Indium, enzymatic inhibition analysis for, 707
Individual quick blanching (IQB), 192–193
Indole, 39
Indole, Trp, in nuclease active site, 263
Indole acetic acid, and peroxidase, 31, 204
Indole-5,6-quinone, melanin monomer, 270, 276
Indolylmethyl alcohol, 384
3-Indolylmethyl glucosinolate, 384
Indophenyl acetate, chromogenic substrate, 705
Indoxyl acetate, chromogenic substrate, 241, 704
Induced fit, in ATP biosynthesis, 84–85
in enzyme catalytic efficiency, 53
in enzyme regulation, 83–84
in enzyme specificity, NCI, 42–44, 256
in evolution, 44, 50
in nonenzyme proteins, 85
in solute inhibition, 259
Induced rancidity. *See* Hydrolytic rancidity
Inducers. *See* Induction, inducible enzymes
Induction, inducible, enzymes; calf pepsin, by diet, 601
in eukaryotes, 76, 77
in lactose malabsorption, 652
of β-galactosidase in prokaryotes, 75–76
of microbial cellulases, 169
of microbial xylanases, 178
of serine dehydratase, and protein quality, 76, 77
of yeast lytic β-D-glucanases, 629
Infant digestion, and milk lipase isozymes, 448
Information sources on enzymes, 22–24. *See also* specific areas
Infrared radiation (micronization), for enzyme inactivation, 199
Infrared spectroscopy, 53
Inhibition, inhibitors of enzymes, *See also* Inhibitors, food related
active site directed (analog, competitive), 249–254
active site reagents as, 249, 437
allosteric, 67–68, 69–72, 256, 364, *See also* Allosteric enzymes
and affinity procedures, 131, 132, 134, 248
and chirality, *See also* individual enzymes, inhibitors
and low temperature enzyme behavior, 219, 224
applications of, nonenzyme, 247–248
as anylates in enzymatic analysis, 703–706
as blanching alternative, 263–264
as criterion of enzyme presence, 282
as food additives, 247–248
by transition state analogs, TSAIs, 52, 253–254, 437
classification, modes, 248–249, 250
competitive, as anylates in enzymatic analysis, equation, 703
equations for, derivation, 39
in kinetic substrate enzymatic analysis, 689
endogenous, *See* Endogenous inhibition
feedback, 70, 72, 74, 256, *See also* Allosteric enzymes
for operational molarity estimation, 684
hydrogen ions as, 260–264
in flavor investigations, 369, 370, 418

medical applications, 249–250
non-active site directed, 255–260
noncompetitive (NCI), discussion, examples, 254, 255–256, 282, 329
equations for, derivation, 42–43
in induced fit, 43–44
of immobilized enzymes, 145, 147–149, 251
product, 250–253, 255–256, 515
shotgun, 248
substrate, 64–65, 238, 239, 657, 762
via polymer-polymer interaction, 257–258
via protein-water perturbation, 259–260
Inhibitors, food related, medical; and flavor, 367, 393, 437
as antioxidants, 437
as natural products, 248
as NEB products, 328
as weight reducing aids, 254, 257, 664
design of, 248
enzymatic browning control with, 281, 282
-enzyme dissociation constant, *See* K_i
for meat color retention, 332
from grapes, a winemaking problem, 537
from malt, a brewing problem, 563
in mackerel, and heart disease, 501
in meat texture, 469, 470
pharmaceuticals as, 248, 249–250, 253
pickle oversoftening prevention, 259
polypeptide, *See* Antienzymes
surfactants as, 259–260
to prevent milk rancidity, 448
Initial rates, in enzyme assay, 673
Initiation, of cellulose hydrolysis (C_1-), 171–172
Initiation factors, in enzyme, protein, biosynthesis, 28
Injection of enzymes, and meat texture, 463, 464
Injury, plant, and enzymatic browning, 275, 276
Inorganic immobilized enzyme supports, 144
Inorganic ions, enzymatic analysis for, 706–707
Inosine, 241
Inosine-5′-phosphate, inosinate, inosinic acid. *See* IMP
myo-Inositol, pectin, phytate precursor, 514, 656
Inositol phosphates, 659. *See also* Phytic acid
In-package desiccation, 235
Insect(s), and tea flavor, 15, 400
damage and enzymatic browning, 13, 277
flour proteinase excess from, 584–585
hormones, affinity chromatography, 127
larvae, Irad effect on, 14, 197
PG in, 517
prophenolase in, 275
termites, cellulase, 169
xylanases in, and cellulases in, 178
Insecticides, 13, 253, 277. *See also* Pesticides
Insectivores, rennets from, 602
Insolubility, a substrate requirement for lipase, 443–444
Insoluble glycan hydrolysis, stimulation, 157
Insoluble supports, in enzyme immobilization, 144–145
Instant tea, enzyme(p) for making, improving, 294, 549
haze. *See* Tea cream
Instrumentation, in color assessment, quality, 268
in enzymatic analysis, immobilized, 692–697
unimmobilized, 689–692
in flour quality assessment, 578
in texture quality assessment, 462
Insulin, and glycolysis, 71
and radioimmunoassay, 712
receptor, affinity chromatography, 134
Integrated kinetic equations, 250–252
Integuments, and melanin, 277
Intercellular cement, and the Acid-Hot Break, 547
Intercellular substance, and food color, 268
Interconversion, of plant acids, 350–356
of plant sugars, 329, 330, 356
Intercrosslinking, and Irad, 198
Interfacial recognition site, and C_1-cellulase, 172
of lipase, 443–444
Intermediary metabolism, 7, 369. *See also* specific instances
Intermediate ES complexes, and relaxation spectroscopy, 53–54
Intermediate moisture foods (IMF), 189, 231, 234, 237, 462
Internal breakdown, a PG-induced apple texture defect, 523
Internal diffusion, and immobilized enzymes, 149
Intestinal flora, and thiaminases, 645
Intracellular enzymes and immobilization, 138
Intracellular lipid, and meat color, 319
Intracellular localization of enzymes, α-amylase in malt, 556
glucosinolase in crucifers, 385, 387
lipase in plants, 44
lipoxygenase, in plants, 430
Intracellular membranes. *See* Cell membranes
Intrinsic kinetic parameters, immobilized enzymes, 145
Intybin, bitter chicory glucoside, 709
Invert sugar, in honey, 358
production, 19, 151, 152, 153–154, 159–160. *See also* Fructose
Invertebrates, IMP accumulation in, 364
Iodide, and ascorbate in peroxidase assay, 207
as peroxidase donor, 203, 204, 207
enzyme activation by, 317
Iodination, and reaction inactivation, 273
Iodine, and emulsin-produced goiter, 636
enzymatic inhibition analysis for, 707
Ion exchange columns, for GI isomerization, 164–165
Ion exchange resins, as immobilized enzyme supports, 141, 142
for establishing ascorbate oxidase, 289
in enzyme purification, 106, 118, 119, 120
Ion transport, and enzyme inhibition, 248
Ionic-hydrogen bond crosslinks, fish toughness, 501
Ionizing radiation, irradiation, *See also* Irad in subentries
and fruit texture, 511, 523
and meat color, 316, 320
and meat texture, 463, 473, 480
and milk lipase, 199
and onion flavor, 378
and potato color storage, 278, 285
and superoxide anions, 320
as cellulolysis pretreatment, 176
enzyme assay to ascertain, 680
enzymes surviving, prevention, 198–199
food applications, 11, 13–14, 197, 337
for enzyme immobilization, 653
for enzyme(p) sterilization, 198, 647
in enzymes purification, 110

on fish cathepsins, 506
postirradiation effect, enzymes, 198, 199–200, 275
target theory, 196–199
to prevent greening, NEB, 311, 337
with blanching, 633
Ions, as enzyme cofactors affectors, 46
IQB. *See* Individual quick blanching
Irad. *See* Ionizing radiation
Irish moss, 258
Iron, as enzyme cofactor, activator (non-heme), 46, 57–58, 287, 317, 425, 426
as nutrient, 656, 677
-copper derivatives as enzyme models, 61
-cysteine catalysis of LOOH decomposition, 434
enzymatic inhibition analysis for, 707
oxidation states, in meat pigments, peroxidase, 202, 316
particles, in shrimp greening, 325
β-Irone, phenolase and tea aroma, 401
Irpex lacteus, xylanase of, 179
Irpex spp., milk clotting enzymes from, 602
Irradiation. *See* Ionizing radiation
Isoamylamine, a decarboxylase generated toxin, 640
(−)-Isoborneol, a terpenoid, 404
Isobutanol, in enzyme purification, 112
Isobutyl aldehyde, in cocoa aroma, 401
Isobutyl djenkolic acid, 33
Isobutylamine, a decarboxylase-generated toxin, 640
Isoelectric focusing, in enzyme purification, 113, 117
Isoelectric points, of caseins, in milk clotting, 598
of meat proteins, in texture, 468
of milk, in direct acid cheesemaking, 600
of wheat, malt β-amylases, 574
phytate removal *via*, 658
Isoenzymes. *See* Isozymes
Isolates, soybean, 263
Isolation. *See* Purification
Isoleucine, loss from FPC *via* enzymes, 509
Isomaltose, 106
Isomerization, of galactose, for sweetener, 179
of glucose, enzymatic, 160–166
nonenzymatic, 60, 153–154
D-Isomers, as enzyme activators, 249
Isopentanol, an off-odorant, 424
3-Isopentoate, in terpenoid biosynthesis, 405
Isopentyl acetate, in banana aroma, 389
Isopeptide bonds, 710
Isophthalate, monoamine oxidase inhibitor, 642
Isoprene unit, in biosyntheses, 403
Isopropanol, in enzyme purification, 116, 117
off-flavor removal with, 439
Isopropyl-*N*-chlorophenyl carbamate (CIPC), sprout inhibitor, 337
Isotherm, adsorption, affinity chromatography, 124
water sorption, 232
Isothiocyanates, analysis for, for breeding out glucosinolates, 633
and flavorese, 417
and glucosinolase, 382, 383–384
enzyme-produced, as sterilants, 633
in affinity chromatography, 127
Isothujone, terpenoids, 404
Isotope analysis, in enzyme assays, 676
Isotope dilution, enzyme kinetics, in enzymatic analysis, 691
Isotope discrimination, in honey adulteration detection, 679
Isotopes, in enzyme mechanism elucidation, 272
Isozymes, and blanching, 185, 190, 207
and enzyme regeneration, 213
and polyploidy, 575
basics, history, specificity, 47–48
fingerprints, 20, *See also* Isozymes, patterns
in enzyme regulation, biology, 28, 82–84
in starch sugar interconversion, 331
interconversion protein, 360
of specific enzymes, aldolase, 43, 83
aminotransferase (GOT), 682
α-amylase, 555–556, 574, 575, 682
β-amylase, 562, 577
aspartate kinase, 83
carboxyl esterase, 682
cellobiase, cellulase, 171
cysteine lyase, 379
α-galactosidase, 360
β-glucosidase, 398
patterns, 288, 681–683
separation, by affinity procedures, 130, 132, 133
by electrophoresis (zymograms), 115
-subunit interchange, 47, 83–84, 274
invertase, 84, 357–358, 682
lactate dehydrogenase (LDH), 47, 83, 84, 682
lipase, 44, 440–448, 586
lipoxygenase, 48, 299, 305, 428, 430, 433
malate dehydrogenase, 352
pectin esterase, 515, 524
peroxidase, 215, 288, 678, 682, 683
phosphatase, 357, 382
phosphorylase, 331, 358
phytase, 656
polygalacturonase, 519, 524
proteinases, 568, 583, 626
xylanase, 180
Italian cheeses, aroma, rennets for making, 604

Jabillo, proteinase source, 94
Jack bean, and urease crystallization, 13
Jack cheese, 594
Jack fruits, aroma of, 389
Jams, as IMF, 234
functional proteins in, 615
Jarilla chocola, proteinase source, 94
Jasmine, flavorese-enhanced yield, 420
Jejunum, intestinal lactase in, 651
Jellies, enzyme-derived sweeteners in, 152
Jelly making, pectinase(p) for, 9
Jerusalem artichoke, inulin in, 709
Jet airplane engine, for food drying, 201
Juice manufacture, enzymology, 9, 535–551. *See also* Fruit juice, and others
Juices, spray drying, 235
Juiciness, a fruit texture attribute, 511
an optimal fish texture attribute, 497
Jute bark, xylan, 177

K_i—enzyme-inhibitor dissociation constant, derivation, estimation, 39, 42–43
for RNA polymerase-amatoxin, 706
in enzymatic inhibition analysis, 703
in modes of inhibition, 250, 256
of TSAIs *vs* non-TSAIs, 52, 254
See also Inhibition, Kinetics, individual enzymes
K_m, estimation, 35–38. *See also* individual enzymes
in assays, and in kinetic enzyme analysis, 670, 688
Kale, ascorbate oxidase in, 289
Kambaro, fish paste, 498
Kaolin, immobilized enzyme support, 144
Kaolinite, kinetics of enzymes adsorbed on, 138
Katal, molar activity, and enzyme units, 675
Katalase Fahigkeit (K_f), a first order assay, 673
Keratin, enzyme solubilization of, 624
Kernels, fruit, blanching, and β-glucosidase, 191
α-Keto acids, for preventing fish toughness, 505
β-Ketoacyl CoA, in cheese aroma genesis, 141
α-Ketoglutarate, an allosteric affector, 72, 73, 362
α-Ketoglutaric acid, structure, 350, 351
Ketols, LOOH transforms, 429, 430, 434
Ketones, and aromas, 414, *See also* 2-Methyl ketones
from LOOH, 427, 428
Ketose-aldose reactions, 163
Khesari, 627
Kidney, as a food, 463
enzyme(p) for bitter peptide removal, 618
for onion flavor enhancement, 381
fish, TMAO demethylase in, 503
Kids, oral lipase(p) from, 9, 412
Kieselguhr, for chillproofing beer, 570
Kilning, and malt enzymes, 562, 568
Kimri, date ripeness stage, 521
Kinaesthetic aspects of food texture, 462
Kinetic enzymatic analysis, 688–689, 690
Kinetics, allosteric (non-Michaelis), 67–68, 333, 352
and absolute reaction rate theory, 51–52
bi-bi, *See* Kinetics, two-substrate
computer programs for, 36
inhibition, *See* Inhibition
integrated, chronometric integrals, 250, 251–252, 672–673, 674
literature on, 38
Michaelis, hyperbolic, steady state, 34–35, 65
of adsorbed enzymes, 138
of immobilized enzymes, *See* Immobilization
of inactivation, *See* Inactivation
of invertase, 238, *See also* other individual enzymes
of onion odor development, 379
of regeneration, 213, 214, 215
ping-pong, *See* Kinetics, two-substrate
transforms, equations. *See* K_m, K_i
transient state, relaxation spectroscopy, 50–52
two-substrate, 37–38, 212, 317, 339
with isotope dilution, in enzymatic analysis, 691
Kinetins, and chlorophyll retention, 311
as gibberellin stimulants, 558
Kits, for enzyme assay, enzymatic analysis, 676, 691, 697, 699
Kiwi, 13, 275, 279
Kneading, in breadmaking, 572
Kohlrabi, 224, 430
Koji process, for enzyme(p) production, 99, 100
for ethanol production, 561
for glucose production *via* cellulase, 176
for soy sauce production, 616
Rhizopus-Aspergillus culture in, 555
Konbu jelly, enzyme(p) for making, 549
Koshering, and meat texture, 484, 643
Kvass, 624

Label immunoassay, 683
Labrusca wines, PE-generated methanol in, 638
Lachrymator, onion, 275, 300, 375–376, 395
Lacquer tree, laccase source, 271
α-Lactalbumin, and evolution, 361
antigenicity of hydrolyzed, 643
in affinity chromatography, 130
specifier protein in lactose synthesis, 76
D-Lactate, in fermented foods, enzymatic analysis for, 698
L-Lactate, as product inhibitor, 255
enzymatic analysis for, 688, 690, 691, 698
Lactation, and hydrolytic rancidity, 447
Lactic acid, :acetic acid ratio, in sourdough aroma, 698
as food flavorant, 349
enzymatic analysis for, *See* L-Lactate
from starter bacteria, dairy products, 408
in PSE, 470
structure, 350
Lactobacilli, as starter organisms in cheesemaking, 408, 607–608, 694
for dietetic beer, 564–565
lysozyme-stimulated growth of, 662
milk clotting enzymes in, 602
peroxidases of, as sterilants, 204–205, 663
Lactones, cheese aroma, enzyme of, 408, 411, 412
hydrolysis, negative Q_{10} of, 225
in citrus bitterness, 452, 453
in fruit aroma, 389, 395
in hot beverage aroma, 400
in orbital steering, 55
Lactose, as gene derepressant, 75
biosynthesis, 69–70
cheese aroma precursors, 408
enzymatic analysis for, 688, 691
enzyme(p), for eliminating sandiness due to, 179, 610
for removal from milk, 104, 651–653
for standardizing enzyme(p), 103
in casein flocculation, 610
intolerance, 651–652
pickle-firming effect of, 529
removal from whey, 9, 180, 643, 653
Lactulose, 179
Laetrile, 636. *See also* Amygdalin
Lag period, in enzyme assays, 208, 220, 425, 672
in peroxidase regeneration, 213
Lambs, as meat, color of, 315
texture of, 463, 493
zymograms of freeze-thawed, 687
cheese from milk drunken by, 604
lipase(p) (oral), from, 9, 412
protein nutrition of, 630
rennet from, 601
Lamella, and enzyme(p) treatment of alfalfa, 627

Laminarin, 566. *See also* β-D-Glucans, Lichenans
Lanthanum salts, as model enzymes, 59
Lanthionine, in fish toughness, 500, 504, 506
in high pH-treated foods, 263
Larvae, no invertase in bee, 358
pupation and phenolase, 275
Lasers, in enzymatic analysis, 690
Latency, of lysosomal enzymes, 226
of RNA, *See* RNA
of soybean proteinase, and protein modification, 626
of wheat, barley β-amylases, 562, 576
Lathyrus satilus, adulterant seed, 701
L. sativum. *See* Khesari
Laundering, enzymes in, 5, 621, 648
Lauric acid, lipase-produced off-odorant, 445
Lautering, and β-D-glucanase, in brewing, 553, 567
LD_{50}, of enzyme(p), 648
Leaching, and blanching, 189, 190, 336
in human cultural development, 650
to prevent NEB, 336
Lead, enzymatic inhibition analysis for, 707
Lead salts, in enzyme purification, 112
Leaf aldehyde, 370, 371, 372
Leaf protein, 624. *See also* Alfalfa
Leaf stalks, as vegetables, 524
Leafy vegetables, blanching, testing for adequacy, 211
Leaky effect, in blanching, 191
Leather making, proteinase(p) used in, 5
Leather workers, and tannins, 661
Leaven. *See* yeast
Leaves, as vegetables, 524
Lecithin, *See also* Phospholipids
activation of phytase by, 657
dimerization, soy off-flavor, 424
inhibitor of cholesterol autoxidation, 643
loss, in noodles, 237
NEB, and dried eggs, 236
phospholipase substrate, 441
Lectins, affinity chromatography, 125, 126, 130, 131, 568
inhibition of PG, 518
removal by blanching, 189
Leeks, flavor precursor, 378
pigmentation of, 313
rehydration of dehydrated, 246
Legal action, in enzyme patent disputes, 157
Legumes, antioxidants, LOX inhibitors in, 437
as vegetables, 524
bitter peptides in proteolysates, 616
α-galactosides, flatulence, 651, 653–654
off-flavor and LOX, 8, 299, 424–429
phenolase, 275
texture enzymology of, 527–528
Leguminosae, biogenic amines in, 641
Lemon juice, pectinase(p) for, 537, 544
Lemon peel, for pectin production, PG, 549
Lemons, citrate synthase, citric acid, in, 354
Lenape potato, solanine in, 38
Lentil beans, LOX of, 430
Lenthionine, from shiitake mushroom, 379
Lentinus edodis. *See* Shiitake mushrooms
Lenzites spp., milk clotting enzymes, 602
Leoti sorghum, α-amylase inhibitor in, uses, 581
Less-random specificity, of glucanases, 170–171, 555
Lettuce, as a leaf vegetable, 524
chloroplasts, 285
β-glucosidase, potential health hazard, 636
methoxypyrazine, and aroma, 386
phenolase and off-colors, 275, 277, 278
phytate in, 656
Leucine, biogenic amine precursor, 640
flavor precursor, 369, 391, 402, 409
off-flavor precursor, 452, 617
Leucoanthocyanins, phenolase substrates, 276
Leucocyanidins, cocoa phenolase substrate, 295
Leucocytes, as bacteriocide, and luminescence, 203
and peroxidase, 17, 204
chemotactic enzymes, 86
phospholipase A_2, 443
Leuconostoc cremoris, cheese amines *via* enzymes of, 642
Leuconostoc spp., and cheese aroma, 407
Leupeptines, cathepsin B TSAIs, 254, 474
Lewis acids, and metalloenzyme catalysis, 57
Lichenans, historical cellulase substrates, 167, 168
Ligand-induced conformation. *See* Induced fit, conformational changes
Ligands, in affinity chromatography, 127, 128–129
Light beer, enzyme(p) in production of, 564
Light, and chlorophyll biosynthesis, 310
and color perception, 268
and Irad effects, 311
and milk oxidative rancidity, 450
and red meat color, 320
and sun-struck beer flavor, 440
and wheat germ lipase, 586
β-carotene loss, 644
with FMN, active oxygen generator, 287, 450
Light filth, enzymatic cleanup, for detection, 670
Lightness, a color perception component, 268
Lights, from overfermented coffee berries, 549
Lignin and lignification, and pear texture, 513, 518, 520–521
and protopectin, 513
as dietary fiber component, 710
as immobilized enzyme support, 144
biosynthesis in controlled atmosphere storage, 531
enzyme in, 204, 394
-cellulose interrelations, 174
degradation, enzymes of, 172, 276
in asparagus toughness, 521, 529–530
in enzyme purification, 106
Liliaceae, cysteine lyases of, 379, 380
Lima beans, aroma components, 371
color retention of green, 302
cyanogenic glycosides in, 633, 635
off-odor of frozen unblanched, 430, 431
vitamin loss from, 645
Lime juice, enzyme(p) for clarification, 544
Limes, texture of, 511
β-Limit dextrin, from β-amylase, and pullulanase, 154, 155, 556
Limonene, in bread aroma, 413
in citrus, structure, 401, 403, 404, 405
Limonin, bitter citrus limonoid, 403, 452–455
Limonoate ion, 453
Limonoic acid A-ring lactone, 453
Limonoid bitterness, prevention removal, 423, 454–455
Limpid juices, enzyme(p) used in making, 535–539
Linalool, an aroma component, 401, 403–404, 405
Linamarin, a cyanogenic glycoside, 633, 635

Line changeover, enzymes, and process cheese, 609
Lineweaver-Burk equation, 35–36, 148, 149
Linoleic acid, as aldehyde precursor, 372
as LOX substrate, 298, 299, 434
enzymatic isomerization of, 438
hydroperoxides. *See* Lipohydroperoxides
Linolenic acid(s), as LOX substrate, 292, 298, 424, 427
as prostaglandin precursor, 435–436
hydroperoxides, *See* Lipohydroperoxides
singlet oxygen action on, 373
Linseed meal, protein, glucosinolase, 635
Lipemia, and lipoprotein lipase, 447–448
Lipid autoxidation. *See* Autoxidation
Lipids, *See also* specific lipids
and carotene color, 268
as aroma precursors, 369, 371, 411
enzyme-aided analysis for, 711
functionality of, 614
-heme interaction, and meat color, 315, 319–320, 323
in Arrhenius discontinuity, 220
in beer, and foaming, 571
in dough, and LOX, 587–588
in off-odor genesis, 422–436, 440–449
in NEB, 327–328
protein process, and phytate, 658
Lipohydroperoxides (LOOHs), and proteins, interaction, 501, 502, 631
and transforms, as health hazards, 642–643
as aroma precursors, 371, 372
as product of LOX action, 242, 371–373, 423, 425–426
decomposition of, 372, 428, 434, 437
in ascorbate oxidation, 271, 372
in carotene bleaching, 300, 301
in cell disruption, 16
in cereal processing, 300, 301, 414, 587–588
in fish quality, 436, 501, 502
in meat color, 315, 319–320, 323
in milk, dairy products quality, 450–452
in NEB, 237
in off-odor genesis, 423, 424, 452
isomerism of, 371, 372, 424, 425, 427
nondegradative transformations, 411, 427, 428, 429, 435, 436, 438, 587, 591
non-LOX mediated formation of, 320, 433–436
polymerization, 242, 428, 429, 438
Lipoic acid, 46, 69, 128, 387
Lipolysis. *See* Lipase, Hydrolytic rancidity
Lipoprotein, and gluten, 584, 587
as lipoprotein lipase substrate, milk quality, 441, 443, 447–448
plant, 513
Liposomes, and immobilized enzymes, 142, 145
Liquefaction, by enzyme(p), in brewing, 559, 560, 561
in juice making, 548
Liquid membranes, and immobilized enzymes, 661
Liver, enzyme inactivation in frozen, 223
enzymes and Irad of, 199
a meat food, 463
sausage, enzyme(p) tenderization, 493
tyramine, toxic effects, 640
enzymes of, catalase(p), 9, 91, 663
glutathione peroxidase, 320
xenobiotic oxidase, and color, 320
Lizard fish cakes, lysozyme(p) preserved, 662
Llamas, cheese from milk, 594
Loaf volume, a texture-related bread quality parameter, 576, 579
enzyme assay for predicting, 679
optimization with enzyme(p), 583
Lobsters, pigments, off-colors, 269
xylanase in, 178
Lock-and-key metaphor, and induced fit, 42–44
enzyme models of, 39, 61
odor perception, 345
Locust bean gum, controlled enzyme-hydrolysis, 9
LOH. *See* Hydroxyfatty acid(s)
Longevity, and SOD, 286
LOOH. *See* Lipohydroperoxides
Lossen rearrangement, and glucosinolase, 384, 385
Louisiana process, for cellulolysis, 176–177
Love, unrequited, and psychoactive amines in chocolate, 641
Low acid foods, peroxidase heat resistance in, 206
Low moisture, enzyme behavior. *See* Water activity
Low moisture foods, classification, 233–234
Low temperature anomaly, 219–220
Low temperature drying, 236, 237
Low temperature enzyme behavior, Arrhenius law deviation, 219–230
cellular disruption, 226–227
cryoprotection, cryoadaptation, 229–230
cryptic intermediates, 225–229
fluctuating temperatures, 227–229
freezing and thawing, 220–222
in cold-renneting, 599–600
inactivation, 220–223
minimizing factors, 219–223
persistence factors, 223–230
solute concentration, 224–225
specificity changes, 444–445
Low temperatures, in enzyme purification, 116
Low temperature storage, enzymatic browning, 278, 279
sugar accumulation. *See* Cold sweetening
Low temperature-long time processing (LT-LT), in vegetable canning, 522
Low-calorie beer. *See* Dietetic beer
Low-methoxyl pectin, in dieting, enzyme(p) for, 664
PE, and firming, 93, 513, 522, 523, 525–526
Luciferin, in bioluminescence, 689
Luminol, for chemiluminescence in enzymatic analysis, 698
Lung, as food, 463
Lupines, β-cyanoalanine synthase in, 637
LOX in, 430
Luster, in food color, appearance, 268
Lutidine dye, HCHO measured with, 700
Lyase-like action of high pH on polysaccharides, 517
Lycopene biosynthesis, 269
lymph glands, meat tenderizer from, 491
Lye peeling, and enzymatic browning, 277
Lysine, ε-amino, in active site, enzymatic analysis for, 700
in reaction inactivation, 273, 274
pK_a of, 262
and peptide bitterness, 617
as peroxidase H donor, 631
cadaverine from, *via* decarboxylase, 640
enzyme-aided analysis for, 711

lysinoalanine from, 263
nutritional availability, 630, 708
reaction with phenolase derived quinones, 630
Lysinoalanine, and fish toughness, 504
from alkali inactivation, 263
in proteins, isopeptides, 711
Lysis, in lysozyme assay, 663
of cheese starter organisms, 607–608
of yeast, in brewing, 566
Lysolecithin, activation of phytase by, 657
and hydrogen bonding, fish texture, 501
and phospholipases, 442
in starch, and chill sensitivity, 334
-starch complex, and enzymes in breadmaking, 587
Lysosomes, disruption, enzyme activation by, 16, 227–228, 473, 483
enzymes in, cathepsins, 466, 473–475
α-galactosidase, 683
β-galactosidase, 474, 475
lipase, 444
phosphatase, acid, 558
in biology, medicine, enzyme regulation, 65, 82
in high temperature beef conditioning, 480
in muscle-to-meat conversion, 464, 475
latency of enzymes in, 226
pH optima, 261, 477, 478
plant spherosomes as, 528
vacuole as, 512, 513

Macadamia nuts, brown centers, 246
Macaroni, color and LOX, 301
enzymatic off-flavor, 237
Maceration, by enzyme(p), *See* Macerase in *Enzyme Index*
in pickle oversoftening, 529
in tea production, 292
Machinability of flour, enzyme(p) improved, 579, 583
Mackerel, and cardiovascular disease, 501
flavor of, 416
texture of, 502
Maclura prunifera. *See* Osage orange
Macroanions, inhibition by, 245, 258–259
prophenolase activation by, 274
Macroligands, in affinity chromatography, 123–124, 127
Macromolecules, and rehydrability, 189
Macrophages, cathepsins, and meat texture, 474
Maggots, enzyme-aided detection of, 712
Magnesium (salts), and regenerated phosphatase, 216–217
and protopectin, 514
as allosteric affectors, 73
as enzyme activator, 466, 712
as enzyme cofactor, 31, 46, 211, 216, 224
-ATP, as ATPase substrate, 482
electrode, 694
enzyme activation analysis for, 707
fatty acid crosslinks, and fish texture, 501
for tenderizing meat, 483, 484
in chlorophyll, 302–303
in phytin, 656
Magnetic polymers, in affinity immobilization procedures, 127, 653
Maillard reaction, in NEB, 326, 327
Malate, PEP carboxylase inhibitor, 253
Maleic anhydride copolymer, enzyme immobilization, purification, 106, 146
Maleic hydrazide, sprout inhibitor, 337
Malic acid, and glyoxylate cycle in fruits, 350, 351
for control of nonenzymatic browning, 281
Malonaldehyde, from LOX, quality impairment measure, 237, 501, 631
Malonyl CoA, flavonone synthase cosubstrate, 455
Malt, as enzyme(p) source, 154, 647
beer haze source, 569–570
enzyme(p) replacement, *See* Barley brewing
enzymes of, α-amylases, 555–556, 557–560
β-amylases, 562–563
debranching (limit dextrinase, pullulanase), 563–564
β-glucanases, 558, 565–566
β-glucosidase, 558
phenolase, 558
phosphatase, acid, 558
phosphodiesterase, 655
proteinase, peptidases, 567–568
for whiskey manufacture, 561
historical, as diastase, 3, 9
in brewing, 552–553, 555, 559–560, 563, 566, 568–569
poorly modified, β-D-glucanase(p) for, 566
quality, invertase assay as indicator of, 567
Malt, barley. *See* Malt
wheat. *See* Wheat, germinated
Malt liquor, enzyme(p) for brewing, 553, 559, 564
enzyme(p)-derived sweeteners in, 152
Maltiness, a dairy flavor defect, 452
Malting, 556–558
Maltol, in bread aroma, 413
Maltose, accumulation in food plants, 359
at low a_w's, temperatures, 221, 242
β-amylase optimization, for dietetic beer, 563
as product of, α- and β-amylases, 155, 554, 556
in brewing, 559, 563
in whiskey manufacture, 561
syrups, β-amylase key to, 152, 154
α-Maltose, α-amylase product, 154, 554
β-Maltose, β-amylase product, 154
Maltotriose, fermentable, in brewing, 559
production with debranching enzymes, 157
Maltotrioses, end-products of α-amylase, 155, 242
Mammary gland, and milk enzymes, 606
Management of enzymes, 181–264
Manganese, as enzyme activator, 317
as enzyme allosteric affector, 73
as enzyme cofactor, 46, 287, 415
electrode, 694
enzyme activation analysis for, 707
in enzyme purification, 113
Mangoes, citrate synthase of, 354
enzymatic browning, 275, 278
fatty acid oxidation by mitochondria, 352
glyoxylate cycle enzyme, 351
Irad of, 278

nectar, enzyme(p) for making, 544
phenolase isozymes of, 271, 274, 275
texture enzymes, 529
Manioc. *See* Cassava
Mannan(s), invertase moiety, 159
yeast cell wall component, 628
Mannitol, in salting-in, -out, 240, 241
·OH scavenger, 172, 173, 324
stabilizer, in enzyme assay, 676
Mannose, and glucose isomerization, 154
Manometry, in enzymatic analysis, 689
Manton-Gaulin device, in enzyme purification, 110
Manure, as source of biomass-converting enzymes, 169
MAOI drugs, and food biogenic amines, 641
inhibitors of monoamine oxidase, 640, 641
Maple syrup, imitation, enzyme(p) for, 180
Marble ball dialysis, 113
Marinading, proteolytic autolysis during, 508
Marine food, color of, 628, *See also* Fish
lysozyme(p)-preserved, 662
Marmalades, functional proteins in, 615
Marzipan, emulsin, and HCN, 399, 636, 637
Mashing, bill, for distilleries, 561
enzyme function, in brewing, 553, 560, 568–569
enzyme(p) added to, 569, 571
β-D-glucans, and viscosity, 565
in dietetic beer production, 564
Masked messengers. *See* RNA
Masking, of pigmentation by enzymatic browning, 277
Mass transfer, in immobilization, 142, 145–146, 147–148
Massaging, and meat texture, 463, 484
Mastication, and flavor genesis, 368, 369, 375
Matrices, bulky, and food enzyme inhibitors, 642
in affinity chromatography, 127, 131, 657
Matting, a cheesemaking step, and texture, 605
Maturation of wheat, and α-amylase, 573
Maturing, of cheeses, 594
Maturing agents, 274
Maturity, acid:sugar ratio criterion, 349–350
color as guide to, 269
enzyme assays as indicators of, 20, 333, 678–679
Mayonnaise, as oily emulsion, 6
EIA of enterotoxins in, 714
McLaren kinetics, and starch granule digestion, 336
MDR, of vitamins, and enzyme(p) as digestive aids, 664
Mealiness, a potato texture, 526
Mealworm, prophenolase in, 274
Meat, aging, *See* Aging
biogenic amines in, 641
color, discoloration, brown, 227, 298, 318–321
non-brown, 325
generation of, 315, 323–325
of frozen, 227
pigments, associated enzymes, 315–318
retention, 321–323
cured, 315–325
dark-cutting, 316
dehydrated, 237, 241
enzyme assay to detect spoilage in, 678
flavor genesis, 407, 414–416
freeze-dried, 242
from muscle, *See* Muscle-to-meat conversion
frozen, 227, 480, 680, 681–682
Irad treatment, 197, 198, 480
koshering of, 463, 484
mechanically deboned, 433
off-flavors, associated enzymes, 237, 433, 466, 680
packaging, and enzymes, 322, 531
roasted, antioxidants generated in, 437
sweet-and-sour, 263
textural aberrations of, 180, 469–471, 494
texture, enzyme assay for predicting, 680
factors affecting, 463, *See also* Meat tenderization, Muscle
optimization of. *See* Meat tenderization
Meat extender, blanching of, 245
Meat processing and products, blanching, 189, 209
enzymatic analysis for constituents, 699, 701–709
enzyme regeneration in, 210
hydrolysates, bitterness, hydrophobicity, 617
proteinase(p) for tenderizing other uses, 493
Meat tenderization by endogenous enzymes, antemortem treatment, 481–482
contractile apparatus modification, 482–483
future possibilities, 495–496
proteinase potentiation, 483–484
Meat tenderization with proteinase(p), action *in situ*, 488–489
action *in vitro*, 486–488
action on meat proteins, 484–490
enzymology of meat tenderizers, 484–486
future possibilities, 495–496
papain, 484–485
technology, details, 491–495
Mechanization of harvest, and texture, 511
Medical enzymology, 5, 448, 501, 585, 650–666, 667, 685, 696, 712, 715
Medicine, beer and wine as, 552. *See also* Pharmaceuticals
Melanin(s), as natural pigment, food color, 269, 273, 276–277
formation *via* phenolase, 270, 276
lightening of, 283
phenolase-caused bitterness in, 457
-protein granules, 630
Mellowing of dough, by enzyme(p), 583
Melons, ADHs of, 394
aroma constituents, 370, 371, 391
milk clotting proteinase in, 94, 603
Meltability, of nonrennet cheese with enzyme(p), 600
Melt-down, in frozen desserts, control of, 153
Melting points, food related enzymes, 186–187
Membrane reactors, hollow fiber, 264
Membrane separations, in enzyme purification, 113
Membranes, cell, *See* Cell Membranes
magnetic, for enzyme electrodes, 694
Menhaden, stickwater, 507
Mental retardation, enzyme-caused, 317, 635
Mentha piperita, peppermint, 403
Mercaptans, as off-odorants, 32, 457. *See also* Thiol(s)
2-Mercaptoethane sulfonic acid, coenzyme M, 46
Mercapturic acids, in detoxification, 77
Mercerized cellulose, as cellulase substrate, 169
Mercury, enzymatic inhibition analysis for, 707
Mercury salts, as enzyme inhibitors, 248, 356
in enzyme purification, 112
Merliton, 13

Meromyosin, 466
Mesityl oxide and off-flavor, 421
Mesophyll cells, emulsin localization, 398, 399
site of C_4 photosynthesis, 352–353
Messenger RNA (mRNA). *See* RNA
Metabolic pathways, 7, 10, 69–74. *See also* specific pathways, cycles
Metabolic regulation, 12, 69–74
Metabolism, vectorial, in ATP biosynthesis, 85
Metabolism, intermediary. *See* Intermediary metabolism
Metabolite analysis, enzymes as reagents, 669
Metal cations, as enzyme models, 59
Metal(s) (ions), as gibberellin stimulants, 558
ascorbate oxidation catalyzed by, 259
cell separation and texture, 526
chelates, in milk oxidative rancidity, 450
chelators, as NCIs, 258
enzyme assay as indicator of insufficiency, 677
enzyme-bound, as unavailable nutrient, 646
heavy, tested for in enzyme(p), 647
in acid inactivation, 263
in beer haze, 569
in taste acuity, 346
Metallic odors, enzyme-caused, 423
Metalloglass, immobilized enzyme support, 144
Metalloprotein catalysts, nonenzyme, 401, 402, 433–434
Metaplastic epithelium, and intestinal absorption, 665
Methacrylate derivatives, as immobilized enzyme supports, 145
as matrices in affinity chromatography, 127
Methanol, as nutrient, in production of, 536
as PE product, 515
as putative beverage toxin, 637–639
caffeine demethylase product, 661
enzymatic analysis for, 697
in enzyme purification, concentration, 101, 116
in wines and winemaking, 536, 537
PE-generated, freezing history, 681
via Irad, 198
Methemoglobin, and meat color, 316
Methemoglobinemia, 316
Methional, aroma constituent, 370, 387
loss of dietary methionine as, 620
Methionine, absent in bitter peptides, 617
and enzyme specificity, 33
and ethylene biogenesis, 286, 299
as methional precursor, 370, 387, 620
decarboxylation, by peroxidase, 31
immobilized enzyme production of L-, 150
into proteins *via* plastein, 620
Methodology, in enzymatic analysis, with immobilized enzymes, 692–694
without immobilized enzymes, 689–692
Methoxy groups, and pectin structure, 513
Methoxypyrazines, aroma constituents, 369, 370, 380, 406
as dairy off-odorants, 452
Methyl alcohol. *See* Methanol
Methyl allyl disulfide, 378
Methyl anthranilate, in Concord grapes, 389
2-Methyl butanal, a malty odorant, 452
3-Methyl butanal, Cheddar cheese aroma, 373, 409
2-Methyl butyrate, in apple aroma, 389
Methyl cellulose, xanthine oxidase degradation of, 172–173
α-Methyl-DL-cysteine, 33
Methyl disulfide, 378
Methyl djenkolic acid, 33
Methyl esters, as off-odorants, 392–393
S-Methyl glutathione, 33
Methyl glyoxal, and myoglobin oxidation, 320
glyoxalase I substrate, 503
Methyl group transfer, by folate enzymes, 46
2-Methyl heptane, in cheese aroma, 410
N-Methyl hydantoin, 701
2-Methyl ketones, aroma constituents, 371, 392, 410, 414
biogenetic enzyme of, 410–411, 414
conversion to esters, 392
β-Methyl lanthionine, 3, 380
Methyl linoleate, and enzymatic browning, 283
peroxidizing, 245
Methyl methionine sulfonium, odorant precursor, 387, 457
Methylation, de-, in sensing, adapting to environment, 85–86
of polyphenols, 284, 290
3-Methyl-1-butyl acetate, in bilberry, 389
S-Methyl-L-cysteine, and cysteine lyases, 33, 251, 376
biosynthesis, 376
in pectin biosynthesis, 514
sulfoxide, onion and cysteine lyases, 33, 378, 380
Methylene blue, enzyme photosensitization, 200
Methyl-β-methiopropionate, in pineapple aroma, 389
N-Methyl-*N′*-nitrosoguanidine, for enzyme mutant production, 97
Methylotrophs, trimethyl dehydrogenase in, 504
Methylthioalk(en)yl glucosinolates, 362
S-Methylthio-α-ketobutyric acid, bread aroma, 413
Methyl-β-xyloside, xylanase inducer, 178
Metmyoglobin, in frozen meat, 227
in meat browning, 318–321
ligand, color of, 316
reducing activity, MRA, color retention, generation, 321–322
reduction of, *in vivo*, 317–318
-S-S-R, a green tuna discoloration, 269, 325
Mevalonic acid, aroma precursor, 369, 387, 403
as gibberellin synergist, 558
in limonoid biosynthesis, 454
Mexican process, for vanilla production, 399, 400
Micelles, casein, in lipase action, 443
in low temperature enzyme behavior, 220
in LOX assay, 425
Michaelis constant. *See* K_m
Michaelis enzymes, sensitivity to substrate, 65
Michaelis f-function, and pH-rate profiles, 262
Michaelis kinetics, deviation as development stage indicator, 679
deviation of by LOX, *See also* Kinetics
modification by immobilization, 146–149
Michaelis-Menten equation, and endogenous inhibition, 239
Microbial contamination, absent in off-flavor cole slaw, 6, 457
adventitious enzyme action, 236, 449, 452

and enzyme management, 13
as problem in using immobilized enzymes, 653
bitter peptides from, 618
during protein processing with enzymes, 509, 621, 622, 625
enzyme assays, zymograms, as predictors, monitors, 676, 682
enzyme-related prevention of, 423, 661–663
from α-amylase(p) in whiskey manufacture, 561
in gibberellin malting, 55
in toxicological evaluation of enzyme(p), 647
of fish, cathepsin, and salt, 506
prevention with enzyme-derived syrups, 153
source of excess flour proteinase, 584
sulfur odors as indication of, 374
Microbial enzymes, as flavoreses, 420
considered safe as enzyme(p), 647, 648
effects of, *See* Microbial contamination, Microbial growth, Fermentation
prehistoric use, 113
production of, 94–104
Microbial fermentation, for egg desugaring, 340
Microbial flora, in coffee seed coat removal, 6, 549, 550
Microbial growth, 12, 218, 235–236
energy of activation of, 225
Microbial metabolites, in safety evaluation of enzyme(p), 647, 648
Microbial rancidity. *See* Hydrolytic rancidity
Microbial rennets, 95. *See also* Rennets, *Enzyme Index*
Microbial spoilage. *See* Microbial contamination
Microcalorimetry, in enzymatic analysis, assay, 676, 691
Microcapsules, in immobilization mode, 141, 164
and flavorese, 417, 420
Micrococcus denitrificans, nitrite reductase of, 660–661
M. lysodeikticus, as enzyme(p) source, 95, 647
lysozyme in, assay for lysozyme with, 662
Micrococcus spp., amino decarboxylases in, 641
cheese ripeners, proteinase, texture, 608
Microcrystalline cellulose, 169, 544, 547
Microdomains, liquid water in frozen foods, 223, 224
Microdroplets, 223, 224
Microencapsulation, 141, 164, 417, 420
Microenvironment, in frozen foods, 223, 224
in immobilization, 145
Microfibrils, cellulose, and protopectin, 514
in bamboo shoot texture, 530
in cell walls, 512, 513
in the Cold Break, 546
in the Hot-Acid Break, 547–548
gluten as, 484
Micronization. *See* Infra-red
Micronutrients, assay for nutritional status of, 677–678
Microorganisms, and pasteurization, 211. *See also* Microbial
Microporous ultrafilters, 114
Microsomes, muscle, and meat color retention, 324
xenobiotic oxidase of liver, 705
Microstructure of foods, and texture, 462
Microwave (blanching), advantages, disadvantages, 193–194
and meat texture, 483
for removal of excess α-amylase, 581
no effect *per se* on enzymes, 193
of soybeans, 436
water removal by, 235
Middle lamella, in plant cell structure, 512, 513
protopectinase loosening, 518, 519
Milk, and flavorese, 416
animals providing cheeses, 594
as active oxygen generator, 451
as bread component, 573
casein micelles, 595–596
clotting, *See* Cheesemaking
cold sterilization and catalase(p), 9
enzymatic analysis for cholesterol in fat, 700
enzyme assay for detection spoilage of, 678
enzymes of, catalase, 663
lactoperoxidase, antibiosis of, 204
lipoprotein lipase, 447–448
lysozyme, 662
proteinases, cheese texture, 605
superoxide dismutase, 287, 450, 451
xanthine oxidase, atherosclerosis, off-flavor, 204, 451, 649
homogenized, 448, 649
humanizing, with lysozyme(p), 9, 662
lactose, *See* Lactose
moving, *vs* quiescent, 600
off-flavors, 237, 382, 422, 423, 446–452, 680
pasteurization, 186, 211–212, 449
pesticide, in screening for, 705
skim, 451
spray drying, 235
udder ultrasonication and enzymes of, 201
Milk, soybean. *See* Soybean milk
Milk clotting, *See also* Cheesemaking
acid *vs* rennet, 19, 594, 599–600
and cloud flocculation, compared, 537
and curd tension, 599–600
as a limited proteolysis, 65
assay, a variable time-fixed reaction assay, 670
Irad effects on, 198
low temperature nonoccurrence, 225–226, 599
:peptidase ratio of proteinases, 603
substrate of, kappa-casein, 595–596
unit cost of rennet for, 104
Milk products, *See also* Cheese, other products
assay for regenerated phosphatase in, 216
concentrated, lactase(p) for making, 9
condensed, lactase(p) for making, 180
custard, rennet for home use, 8
dried, 10, 237
evaporated, enzyme(p) for, 9, 10
frozen, enzyme(p) for preventing flocculation, 610
lactase(p) for sweetness, digestibility, 9, 10
Milk substitutes, from enzyme(p)-produced, FPC, 509
Milkweed, proteinase of, asclepain, 94, 492
Milky stage of wheat kernel development, β-amylase, 576
Milling, a cellulolysis pretreatment, 176
a step in cheesemaking, 594
of flour, *See* Flour milling
wet, in brewing, 553
Millipore filter, and chillproofing beer, 571
Mimosa, flavorese enhanced fragrance yield, 420
Minced fish, enzyme-caused toughening of, 502
Mineral availability, and dietary phytate, 656

Miniaturization, and enzyme(p) evaluation, 683
Miracle wheat, and enzymatic browning, 275, 278, 280
Miraculin, and taste acuity, 346
Mirror method, for initial rates, 673
Miso, enzyme(p), nutritional value, 616, 623
Mitochondria, and enzyme localization, 385
  and flavorese, 417, 418
  as microbial symbionts, 282
  chilling injury, oxidoreductases of, 522
  fatty acid oxidation by, 352
  in ATP biosynthesis, oxidative phosphorylation, 85, 285
  in cold shortening of beef, 469
  in cold sweetening of potatoes, 344
  in ester biosynthesis, 391
  in freeze-thawing, 682
  in frozen foods, 431
  in meat-to-muscle conversion, 468
  in myoglobin function, 316–317
  in pig PSE, 471, 682
  malate dehydrogenase, 352
  preparation of, 110
  SOD in, 287
  with ATP, for enzymatic browning control, 285
Mixing speed, of doughs, and LOX, 588
M-M. *See* Michaelis-Menten
Mobility, and transport of substrate at low a_w's, 233, 241
  of protons, 225
Models, food systems, in enzyme behavior studies, 226, 228, 237–242
Models of enzymes, enzyme action, anions, inorganic, as, 60
    cations, metal, as, 59
    cycloamyloses as, 60, 61
    imidazole(s) as, 60, 61, 62
    NEB as, 326
    OH^- as, 153–154
    various other, 55, 56, 58, 59
Moistness, bacterial α-amylase(p), baked goods, 579
Moisture, in temperature-fluctuated chicken storage, 228
Moisture-retention, a protein functional property, 614. *See also* Water
Molar activities, of enzymes, 36, 209, 210, 670
Molasses, invertase(p) in making, 159
Mold bran. *See* Koji
Molds, and enzymatic browning, 277
  as enzyme(p) sources, 95
Molecular architecture of the active site, 49
  peroxidase as probe of, 205
Molecular biology of mutation, 97
Molecular exclusion chromatography. *See* Gel chromatography
Molecular Relaxation, alternative to Transition-State, 52
Molecular sieve, chromatography, for molecular weights, 594
  in enzyme purification, 110
Molecular structure, enzyme as probes of, 864
Molluscs, endo-1,3(4)-β-glucanase in, 566
Molybdenum, as enzyme cofactor, 46
  enzyme activation analysis for, 707
  nutritional status by enzyme assay, 677
Monogalactosyl glycerides, 191
Monoglycerides, lipase specificity, bread texture, 413, 441, 586
2′-Monoglycerides, lipase inhibitors, 252
Monohydroxyfatty acids, 242
Monolayers, and lipase action, 443
  of enzymes, 138
  of water in foods, 232–233
Monomechanism, enzyme reaction as, 56
Monomer polymerization, during food rehydration, 235
Monophenols, as phenolase substrates, 282
  hydroxylation by LOOHs, 428, 429
Monosodium glutamate. *See* Glutamic acid, Glutamate
Monoterpene alcohols, ADH substrates, 394
Monoterpenes, in aromas, 370, 401, 403
Monoterpenoids, structure, biosynthesis, 403–405
Morphology, and affinity chromatography, 134
Mortar-and-pestle, in enzyme purification, 110
Mortierella vinacea, α-galactosidase(p) from, 151, 654
Mouse-ear cress, photorespiration mutants of, 361
Mouth feel, an enzyme(p)-produced texture, 153, 234, 360, 370, 615
Mozzarella cheese, made by direct acidification process, 600
MRA. *See* Metmyoglobin reducing activity
mRNA. *See* RNA
MSG. *See* L-Glutamate
Mucilage, enzymatic removal from coffee berry, 549
Mucin, removal in enzyme purification, 113
Mucopolysaccharides, 82, 482
Mucoproteins, and collagen, 496
  proteinase action on, 488
Mucor miehei, as rennet source, 10, 105, 602–603, 604, 605, 647
  M. pusillus, lipase of, and cocoa aroma, 403
Mucor spp., milk clotting enzymes in, 95, 602–605
Muenster cheese, 594, 604
Multienzyme complexes, 69–70, 353
Multipectic enzyme system. *See* Pectinase
Multiphase system in enzyme purification, 117
Multiple attack, *vs* single, and α-amylase mechanism, 554
Multiple forking, a meat tenderizer application mode, 493
Multiple needle pumping, for antemortem treatment, 495
Multivitamin pills, containing allithiamin, 645
Mung beans, enzymes in, 366, 657
Muscat of Alexandria, grapes, increased wine yield by enzyming, 541
Muscle—vertebrate, ATP, ATPase, *See* Muscle, proteins of, function
  function (contraction), 285, 322, 466–468, 449
  glycolysis, ATP replenishment, 467–477, 499
  hexokinase, 108, *See also* other specific muscle enzymes
  IMP accumulation in, 364, *See also* IMP
  LDH isozymes, enzyme regulation, 83, 499
  proteins of, contractile apparatus, 316–317, 463–464, 484, 497, 498
  structure, 464–466, 497, 498
Muscle relaxants, tenderizing meat with, 482, 483
Muscle-to-meat conversion, aging, 472, 473, 496
  calcium activated factor (CAF) in, 478, 479

glycolysis, rigor in, 468–469, 471–473
postmortem progression of events in, 464
proteinases, other nonglycolytic enzymes, 473–480
See also Aging, Meat tenderization
Mushroom(s), a dairy off-odor, 452
aroma, and genesis, 379, 380, 392
ascorbate-aided blanching of, 195
enzyme(p) for preservation of, 549
nucleases and gustators in, 18, 366
phenolase, and enzymatic browning, 275, 276, 279, 282, 285
in texture of bread, dough, 591
isozymes of, 274
purification, properties, 108, 270, 271, 272, 274–275
polyvinyl wrapped, 279
rennets from, 602
toxins, detection by enzymatic analysis, 706
Mustard, and flavorese, 417
flour, as seasoning, 381, 386, 699
greens, 632
oil glucoside, *See* Glucosinolates
oils, 396
seed, enzyme-generated off-flavor, of, 445
food protein and glucosinolase from, 632
Mustiness, an enzyme-caused off-odor, 423
Musts, wine, removal of methanol from, 639
Mutarotation, and amylase, 554
Mutations, enzymatic browning control by, 280
molecular biology of, 97
nonsense, and affinity chromatography, 133
of barley, for beer chillproofing, 570
of cheese starters, and proteinase, 607
of microorganisms, for enzyme(p), 95–97, 647
of tomato, PG, PE, texture, 525
Mutton, as sheep skeletal muscle, 463
Mycelia, fungal, for enzyme(p) production, 163, 388, 647
shoyu koji, enzymatic shortcut to counting, 709
Mycotoxins, enzymatic analysis for, 701
in enzyme(p) safety evaluation, 647, 648
Myelin sheath, peroxidase for topological probe of, 205
Myocardial fibrosis, biogenic amines in foods implicated in, 641
Myocardium, xanthine oxidase and atherosclerosis, 649
Myocommata, fish, 498
Myofibrils, beef, cooking and texture, 483
protein, fragility, texture, 468, 476, 477, 478, 479
function, structure, 464–465
tenderizer(p) on, 486–488, 491
fish, 498, 499–500, 502
Myogen, 486, 498
Myogen A, 117
Myoglobin, biological function, structure, 316–318
heat induced aggregation, 213, 214
in meat colors, 318–325
in meat off-flavors (rancidity), 433–434
reduction, *in vitro* and color retention, *See* MRA *in vivo*, 317–318
Myosin (thick filament), arrowhead, from PSE pigs, 471
fish, and texture, 498, 499, 500, 501
function, structure, 464, 465–466
low temperature behavior, 220, 223
proteinase(p) action on, 486, 487
sulfhydryl of, and meat texture, 473
Myrcene, structure, 404
Myristic acid, a PE inhibitor, 515
Myronate, 382

NAD, NAD^+, *See also* NAD(P), $NAD(P)^+$
and 5′-nucleotide accumulation, 366
as enzyme cosubstrate, coenzyme, 30, *See also* specific enzymes
as product inhibitor, 255
in muscle-to-meat conversion, 472
UV spectrum and optical test, 686–687
NAD^+, $NADP^+$, reduced. *See* NADH, NADPH
NADH, *See also* NAD(P)H
as dehydrogenase product, reductase cosubstrate, 30, 255
in bread quality, 591
in enzymatic browning prevention, 285
in fresh meat retention, 322
in ham curing, 324
$NAD(P)^+$, depletion in red meat discoloration, 321
in affinity chromatography, 128, 132
in enzyme analysis, availability, 685, 687–688
regeneration, immobilization, analysis, 140, 439, 688, 692
to NAD(P)H ratios and fruit aroma, 394
NADPH, and LOOH formation, 634
NAD(P)H, and oxidative phosphorylation reversal, 285
and singlet oxygen, 392
NAD^+-acetaldehyde adduct, as TSAI, 254
NAD^+-pyruvate adduct, as TSAI, 254
NAD^+-sulfite complex, 257
Nadir, temperature, effect on enzymes, 221, 228
Nalidixic acid, and pectin lyase, 518
Naphthalene acetic acid methyl ester (MENA) as sprout inhibitor, 337
Nardoo, Australian, thiaminase I in, and grazing, 645
Naringenin, 453, 541
Naringin, structure, removal from citrus, 423, 453, 455–456
Naritutin, a citrus flavonone glycoside, 453
Natick process, glucose from cellulose, 174, 175
Natto seed, enzyme(p) for extraction of bixin from, 661
Navel oranges, limonin and bitterness in, 452, 454
Navy beans, enzymes and Irad of, 199
NCI. *See* Noncompetitive inhibitors
NDGA, antioxidant LOX inhibitor, 301, 347, 587
NEB. *See* Nonenzymatic browning
Nectars, enzyme(p) for making, 544
Negative temperature coefficients, 225
Neochlorogenic acid, cocoa phenolase substrate, 295
Neohesperidose, citrus flavonone glycoside, 453
Neotigogenin, 660
Nephelometry, in assay of α-amylase, 581
Nerol, structure, 404
Nerves, wild game meat off-flavor and, 446
Nessler test for NH_3, in enzyme purification, 116
Neufchatel cheese, 549, 594, 595
Neurospora crassa, glutamate phosphorylase, 220
Neurospora spp., phenolase in, 271, 273, 274
Neurotransmission, enzyme inhibition and, 248

Neurotransmitters, affinity chromatography of, 134
Newsprint, for enzymatic(p) biomass conversion, 167, 173, 177
Niacin, blanching loss, 190, 643
  for red meat color retention, 322
  NAD(P)$^+$ moiety, 46
Nibs, cacao seed, and enzymes, 402
Nickel, as enzyme cofactor, 46
Nickel oxide as immobilized enzyme support, 144
Nicotinamide adenine dinucleotide. *See* NAD, NAD$^+$
Nicotinamide adenine dinucleotide phosphate. *See* NADP, NADP$^+$
Nitrate, blanching loss of, 189, 190
  enzymatic analysis for, 694, 701
Nitriles, enzyme-generated off-flavorants, health hazards, 384, 423, 632
Nitrite, and NEB, 328
  as preservative, 315
  enzymatic(p) removal of, 660–661
  in ham curing, 315, 324, 473
  in nitrosamine formation, 20, 225, 639
  salmon roe treatment with, 325
Nitrochatechol, a LOX inhibitor, 425
Nitrogen, for yeast, enzyme-generated, 553, 573
  yield prediction by enzyme assay, 677–678
Nitrogen balance, 630
Nitrogen fixation enzymes, affinity chromatography of, 129
Nitrogen gas, for chlorophyll retention, 307
  for enzymatic browning control, 283
  for frozen food, 303
  for NEB prevention, 336–337
  in dry blanching, 194
  manometry in enzymatic analysis, 689
Nitrogen oxide, 300
Nitrogen trichloride, 300
Nitrogen-gamma capture, Irad and N analysis, 200
p-Nitrophenol, burst, in active site titration, 706
  in chromogenic enzyme assays, 672, 684
Nitrophenyl esters, a hydrolysis by carbonic anhydrase, 31
p-Nitrophenyl acetate, a chromogenic substrate, 58, 61
p-Nitrophenyl glucoside, as glucosinolase substrate, 385
p-Nitrophenyl methyl phosphonate, 59
p-Nitrophenyl phosphate, alkaline phosphatase substrate, 211
p-Nitrophenyl-*p*-sulfamyl benzoate, active site titrant, 684
β-Nitropropionate, in toxicological evaluation of enzyme(p), 647, 648
Nitrosamines, 315, 328, 639, 660
Nitrosyl hemochrome, cooked cured meat pigment, 324, 613
Nitrosyl hemoglobin in cured meats, 315, 316, 318
Nitrosyl metmyoglobin, reduction in ham curing, 324
Nitrosyl myochrome, 316
Nitrosyl myoglobin in cured meats, 315, 316, 320, 324
Nitrotetrazolium blue, active oxygen acceptor, 451
NMR analysis, enzyme action as error source in, 57, 712
Nojirimycin, a TSAI, 254
Nomenclature of enzymes. *See* Classification, and nomenclature, of enzymes
Nonactive site-directed ligands, in affinity chromatography, 129
Non-affinity chromatography, in enzyme purification, 118–122
δ-Nonanoic lactone, in dairy products aromas, 411
Nonanol, as sprout inhibitor, 337
Nonbiological changes in dehydrated foods, 236–237
Nonbrown discolorations of meat and fish, 325
Nonenzymatic browning, α-amylase and bread color, aroma, 578
  and aldolase, 243
  and enzyme evolution, 326
  and melanins, 327
  and purine degrading enzymes, 415
  antioxidants generated by, 437
  aromas generated by, 387, 414, 578, 639
  categories of, 327–328
  enzymatic browning as, 327
  in diabetes, 334
  in dried eggs, other foods, prevention, 235, 236, 237, 338–342
  in fish texture, 327, 504
  in fructose syrups, 162
  in lactase(p)-treated products, 653
  in potato products, 328–337
  of orange juice, prevention, 291
  PG action in wine and juice, 536
Nonenzymatic reactions, in blanching, 189
  in frozen foods, 225
Nonequilibrium processes, for water removal, 234
Nonfermented beverages, appearance and color of, 267
Non-hyperbolic kinetics, 66–68
Nonionic polymers, 117, 258
Nonionizing, nonthermal energy, for enzyme control, 200–201
Nonlimpid juices, enzymology of, 540–551
Nonlinearity in enzyme assay, 672–675
Non-Michaelis kinetics, and nonlinearity in enzyme assay, 672
  of adsorbed enzymes, 138. *See also* Kinetics
Nonpeptide bonds, in proteins, 710–711
Nonsense mutations, isolation of inactive enzyme of, 133
Noodles, enzymatic quality loss, 237, 580
  lysozyme(p) preserved, 662
Noorhollandse Meshanger cheese, 607
Nootkatone, biosynthesis of, 405
Noradrenaline, in banana peel, 641
Nougats, functional proteins in, 615
Nucellar tissue, in the wheat kernel, 573
Nucleation, in enzyme regeneration, 263
Nucleic acids, affinity chromatography, 134, 138
  as enzyme inhibitor, 258
  as Irad target, 199
  in enzyme biosynthesis, 74, 75, 76
  in enzyme purification, 112, 113, 117
  in genetic engineering, enzyme(p) production, 97, 98
  in SCP, removal of, 654–655
Nucleohistone, ion exchange isolation of, 119
Nucleophile(s) and nucleophilic attack, by metal ion in enzymes, 57–58, 59
  for enzyme regeneration, 213
  in active site and catalytic efficiency, 56, 57, 59
  in enzyme classification schemes, 30
  in enzyme immobilization, 144
  in ice, 225
  in LOOH reduction, 426, 427, 428

in plastein formation, 619
Nucleoproteins, 199, 513
Nucleoside diphosphate glucose, in cold sweetening, 333
Nucleoside 5′-phosphates, 364. *See also* 5′-Nucleotides
Nucleoside triphosphates, in fish freshness, 680. *See also* specific ones
in 5′-nucleotide accumulation, 366
Nucleosides, 364
Nucleotides, 18. *See also* specific ones
3′-Nucleotides, 31
5′-Nucleotides, and enzyme specificity, 31
as gustators, 31, 241, 349, 363
endogenous enzymes producing, 363–367
enzymatic analysis for, 699
enzymes degrading and prevention, 366–367
from yeast, 665
in fish freshness, 364, 499, 680
production of, 150, 151, 367
Nucleus, cell, as Irad target, 196, 199
in plants, 513
Nuoc-mam sauce, an enzyme-produced fish product, 6, 507
Nutrient(s), as toxins, 651
deficiencies, enzyme-caused, 299, 629–631, 644–646, 656
enzymatic analysis for, 700
losses, in blanching, enzyme management practices, 190, 626, 643–644
retention, countermeasures, 189, 191, 631–632, 645
status, assays, zymograms for assessing, 677–678, 682
supplements, GRAS, 247
See also Nutritional availability, Nutritiousness
Nutrition and flavor, 345
Nutritional quality. *See* Nutritiousness
Nutritional value. *See* Nutritiousness
Nutritiousness—protein, *See also* PER
and protein functionality, 614
and serine dehydrase induction (derepression), 76, 77
enzyme assays as indices of, 678
enzyme-assisted *in vitro* assessment of, 20, 670, 707–709
impairment by endogenous enzyme and prevention, 299, 509, 629–633
improvement, by non-proteinase enzymes, 9, 627–629
by proteinase(p), 9, 509–510, 620–625
via solubilization by endogenous enzymes, 507–508, 625–627
of bread, 573
plastein, 620–621, 624
Nuts, biogenic amines in, 641
phenolase in, enzymatic browning, 275
ripening, and solar energy, 233
Nylon, control of enzymatic browning with, 282
for chillproofing beer, 570
from erucic acid, 632
in affinity chromatography, 127
microfibrils, as immobilized enzyme support, 145, 654

Oats, α-amylase of, 575
enzymatically generated bitterness in, 237
LOX of, 430
Occam's Razor, 43, 476
Occlusion of enzymes, 141
Octanoic acid, off-odorant in papaya purees, 445
Octanol, as sprout inhibitor, 337
odor threshold, 364
1-Octenol, an off-flavorant, 424
Oct-en-3-one, in mushroom aroma, 392
Octopine, 700
Octopus, octopine and dehydrogenase from, 700
Odor receptor, affinity chromatography of, 134
Offal, fish, autolytic transformation of, 507
Off-colors, as nonphysiological pigments, 269. *See also* specific food, Off-color, Color, Discoloration, Enzymatic browning, NEB, etc.
Off-flavors, and enzyme regeneration, 216
bitter peptides, 616–618
dairy-associated, 447–452
enzymatic *vs* nonenzymatic, 421–422
in citrus fruits, 452–456
in dried eggs and other products, 237, 339
in fish and meat, 433–436
in frozen foods, 430–433
in plant (non-citrus) foods, 237, 369, 402, 424–432
lipase, involvement in, 440–449
lipid, involvement in, 402, 422–436, 440–449
LOX involvement in, 299, 424–430
non-lipase, non-LOX involvement, 421, 433–436, 450, 451–452
prevention and removal, by enzymes, 9, 421, 423, 437–439, 449–451, 454, 455–456, 457, 618
nonenzymatic, 189, 436–437, 449, 618
z-value of, 206–207
OFP, onion flavor precursor, 14, 15, 376–379
O-gene, mutation, enzyme production, 75, 97
Oil(s), animal, enzyme(p) for extracting, 5
citrus, increased yield with enzyme(p), 549
corn, high lysine, assay as index, 209
droplets, and hydrophobic interaction, 616
enzymatic prevention of off-flavors in, 423
essential, *See* Essential oils
in meat tenderizer formulations, 492
off-flavor due to lipase in spices, 445
vegetable, enzyme(p) in production of, 549
Oil of bitter almonds, 396, 430
Oilseeds, enzyme caused off-flavors in, 382, 384–385, 428, 429–430, 618
Oilseeds, protein *via* enzyme processing of, 622, 623, 626, 632–633
Oily cardboard, an off-flavor, 450
Oldness, a lipase-caused off-odor, 447
Olfactory cells, and flavor, 345
Oligomers, oligomericity. *See* Subunits
Oligopeptides, fish, flavor of, 509
Oligosaccharides, *See also* Dextrins
as enzyme components, 46–47, 338
as glucanase substrates, products, 154, 155, 170, 171, 178, 517
as α-glucosidase TSAI, and weight reduction, 254
as hexose oxidase substrates, 339
as transglycosylation products, 106, 157, 238
Olive(s), alkali in processing of, 263
debittering, 263
enzymatic browning pacers, 279
oil, and lipase action in the cold, 444–445

enzyme(p)-assisted production, 549
oversoftening by resident cellulase, 529
Onions, alliinase, flavor-producing enzyme of, 260, 261, 376–377, 631
amino acids and peptides, flavor precursors in, 371
and flavorese, 416, 417
as a bulb vegetable, 524
bitterness in, 423
dehydrated, rehydration of, 246
dehydration, 237, *See also* Onion processing
and microbial survival, 236
E,S interaction, for flavor intensity estimation of, 681
flavor, control, retention, 378
precursor, 376–377
precursor precursor, 380–381
δ-glutamyl transpeptidase in, 237, 380–381
lachrymator, 275, 300, 313, 375–376, 381
pectin lyase(p)-produced unicellular food from, 550, 551
pharmacologically active substances in, 376
pinking of, 312–313
processing, and flavor loss, 378, 379–380
skin, catechol in, 276
sprouted, 237
Ontogeny, of barley and malt amylases, 554
Oospora suaveolens, strawberry flavorese source, 420
Opacity, a cloud attribute, 536
Opalescence, a color and cloud attribute, 269, 536
Opaque whiteness, 268
Opaque-2-maize, enzyme test for lysine availability, 708
Operational molarity of enzymes, 683–684, 704–705
Operon, *lac*, of *E. coli*, 75
Optical instrumentation, methodology in enzymatic analysis, 689
Optical rotation, in enzymatic analysis, assay, 686
Optical rotatory dispersion (ORD), induced fit, denaturation, 44, 186
Orange juice, cell-separating enzyme(p) for making, 549
cloud retention with pectinase(p), 539
enzymatic deoxygenation of, 290
enzyme-caused off-flavors of, 393, 394, 423, 445
frozen concentrate, gel in, and prevention, 194, 544
GMP (gustator) in, 364, 366
pasteurization of, 212, 543
processing, enzymology of, 541–544
sacs, limonin in, 455
vitamin C loss, with banana, 646
water sorption isotherm of, 132
Orange Pekoe tea, 292
Orange segments, enzyme(p) for separation of, 9
Oranges, *See also* Orange juice
and flavorese, 416
aroma and genesis, 371, 381
color, 269, 302
enzyme(p) for peeling, 549
enzymes of, acetyl CoA synthetase, in vesicles, 391
alcohol dehydrogenase, and aroma, 394
geraniol dehydrogenase, 403
limonin dehydrogenase, 454–455
lipase, 455
nucleases, 366
PE, PG, 521–522
peroxidase, 423
phosphatase, 212
uronic acid oxidase, 522
texture of, 511, 521–522
Orbital steering, in enzyme catalysis, 55, 59
ORD. *See* Optical rotatory dispersion
Organelles—intracellular, *See also* Lysosomes, other specific ones
as Irad targets, 196
in enzyme potentiation, 15, 16
in gibberellin induced α-amylase release, 558
in post-translation enzyme biosynthesis, 81–82
intercellular transfer of, 550
Organic acids, 513
Organic peroxides, destruction by sulfite, 257
Organic phosphates, hemoglobin binding, and meat color and, 321
Organic polymers, effect on D_{37} values of enzymes, 197
Organic solvents, enzyme inhibition by, 259
in enzyme production, purification, 116–117, 160
in gibberellin malting, 557
in low a_w model food systems, 240, 241
in seed protein processing, 436, 632
Organic synthesis of enzymes, 80
Organomercurials in affinity chromatography, 125
Organophosphates, as active site probes, 40
as ligands in affinity chromatography, 128
enzymatic inhibition analysis for, 704–705
irreversible inhibition by, 253
Orientation of substrate, in enzyme catalysis, 55
Ornithine, 640
in *Mucor miehei* rennet proteinase, 603
Orthophosphate, inorganic (Pi) salts, ion; and amylose chain length, 330
as allosteric regulator, 71, 73, 471, 472
as enzyme inhibitor, 332, 445, 492, 656
enzyme electrode analysis for, 694
enzyme(p) standardization with, 103
in acid sugar accumulation, 353, 357
in NEB, 328
in starch synthesis, and ATP, 332
Osage orange, proteinase source, 94
Oscillation, an enzyme regulation mode, 74
Osmolality, and lactose malabsorption, 652
of enzyme derived sweeteners, 153, 162
Osmosis, direct, reverse, 235
Osmotic drying, 234, 235
Osmotic shock in enzyme purification, 110
Osmovac process, 282
Out-of-hand food, 231
Ovalbumin, in affinity chromatography, 124
synthesis by genetic engineering, 98
Oven roasting, and meat texture, 483
Overblanching, and peroxidase test, 207
and pheophytinization, 305
Overmilking, and lipase-induced rancidity, 449
Overripening, of fruit on flavor texture, 420, 529
of tomatoes on product consistency, 545
Overshoot, enzyme action in fluctuating temperatures and, 228
Oversoftening, cheese, by excess proteinase, 605
cucumber, by resident fungal pectinase, 528–529
dough, by excess α-amylase, 579
fruits, by Irad, and protopectin, 523
olives, by resident fungal pectinases, 259, 522, 528
tomatoes, green, ripe, 524, 545
Oversprouting, cause of excess α-amylase, 579
Ovomucoid, analysis by active site titration, 706

Oxalacetate, enzymatic analysis for, 691
  in enzyme chiral specificity, 33
Oxalacetic acid, in citric acid accumulation, 350, 354
Oxalic acid (oxalates, OAA), acid accumulation regulation by, 355
  as a TSAI, 52, 254
  biosynthesis from ascorbic acid, 351
  enzymatic analysis for, 698, 706
  in meat color retention, 323
  removal from foods, 189
  structure, 350
Oxalis. *See* Wood sorrel
Oxazines, 384
Oxazolidinethione, analysis, breeding out glucosinolase, 633
Oxidation, as cause of beer haze, 569
  mutual, of myoglobin and lipid in meats, 319, 433–434
α-Oxidation of fatty acids, 204, 390, 395
β-Oxidation (spiral) of fatty acids, abortive shunting, in ester biosynthesis, 390, 391
    in methyl ketone biosynthesis, 392
  and off-flavor, 435
  in aldehyde genesis, 371
  in banana aroma genesis, 395
  in cheese aroma genesis, 410–411
  in dough, *vs* LOX action, 588
Oxidation-reduction, internal, 30
Oxidative flavors, 348
Oxidative off-flavors, enzymatic prevention of, 423
Oxidative phosphorylation, and ATP biosynthesis, 85
  reversal and food color, 285, 322
Oxidative polymerization, in lignin biosynthesis, 529–530
Oxidative rancidity, and antioxidants, 437
  enzymatic, prevention, 423
  enzymatic estimation of, 680
  enzymatic *vs* nonenzymatic, 624
  in corn, 429–430
  in dried foods, at low a_w, 237
  in fish, 435
  in meats, 433–434
  of milk, 450–451
  via LOX action, 245, 299, 423
Oxidized linseed oil, an off-odor of unblanched frozen vegetables, 431
Oxidizing tank, in juice and wine making, 537
2-Oxoglutaric acid, structure, 33, 350
Oxonium ion, enzyme transition state as, 254
Oxygen (O_2), *See also* specific oxidoreductases
  a dearth of, and tissue death, 6
  activation in LOX mechanism, 425
  and dehydrated food off-flavor, 431
  and Irad sensitivity, 196, 198
  and red meat color, 318, 320–321
  and sulfhydryls, in frozen foods, 223
  as dough conditioner and LOX, 587
  as enzyme inhibitor, 260
  as RDP carboxylase substrate, 30
  by non-catalase peroxidases, 203
  cellulolysis stimulation by, 174
  enzyme assay and, 672
  for meat tenderization, 483
  in enzymatic analysis, 689, 694
  in enzymatic browning, 277
  myoglobin transport of, 317
  singlet, triplet, 20, 287. *See also* Singlet oxygen
Oxygen—active, *See also* Singlet oxygen, Superoxide anion, Hydroxy free radical
  and aroma, 395, 401
  and atherosclerosis, 649
  and off-flavors, 286, 378, 438, 452
  and sugar accumulation, 358
  cellulolysis by, 172
  deactivation (quenching), 173, 285–286, 430, 450
  forms and generators, 16, 286–287, 451
  in carotene blanching, 300
  in citrus juice cloud collapse, 543
  in dough improvement by ascorbate, 590
  in ethylene biogenesis, 286
  in plant food texture, 173, 518
  in red meat discoloration, 320
  on antioxidants, 437
Oxygen—removal (exclusion, scavenging), and red meat color retention, 323
  from foods with GOX, 12, 291, 341–342, 438, 450
  from milk to prevent oxidative rancidity, 450
  from orange juice by phenolase, 291
  from potatoes, to prevent enzymatic browning, 291
    to prevent sugar accumulation, NEB, 336
  off-flavor prevention, 423, 438
Oxyhemoglobin, 315, 316
Oxymyoglobin, 315, 316, 317, 321, 322
Oxytocin, cathepsin C inhibitor, 474
Oysters, blanching of, 190
  flavor of, 416
  lysozyme(p) preserved, 662
  shucking by micronization, 200
  zymograms of, to detect freeze-thawing, 681, 682

^{32}P, in potatoes, and NEB, 328
Packaging, and delivery, of newly synthesized enzymes, 81
  as off-flavor source, 421
  enzymatic browning control by proper, 278, 281
  in carbon dioxide, 311, 531
  lysozyme(p)-treated, 662
  of beer, and papain(p) action, 553
  of red meat, MRA and color, 322
Packed bed plugged flow reactors, 165, 166, 600
Packets, affinity, enzyme removal with, 135
Pain, as a component of flavor, 347
Paint grinder, in large-scale enzyme purification, 110
Paintbrush test, for enzymes in columns, 138
Pale Soft Exudates (PSE), a pork textural defect, 470–471
Palm kernels, aroma constituents, 392
Pancake mix, lipase(p) used in preparing, 412
Pancakes, improvement with proteinase(p), 10
Pancreas, as food, 463
  enzyme(p) from considered safe, 647
  enzymes of, α-amylase(p) from, in barley brewing, 558
    as desmoenzymes, 138
    lipases, specificity, with milk and flavor, 314, 440, 447

proteinases, 486, 487–488, *See also* Carboxypeptidase
meat tenderizer(p) from, 492
pancreatin from, 91
Pantothenic acid, 46, 646
Papaya(s), in digestive aid formulations, 664
isothiocyanate from, 382, 395
leaves, for tenderizing meat, 491
nectar, enzyme(p) in manufacture, 545
nutrient loss, in red light, 644
off-flavors of purees, 393, 445
papainases (thiol proteinases) of, 484–486
papain(p) from, 92–94
polyphenols, no phenolase in, 13, 275
Paper, enzyme impregnated, meat tenderization, 493
enzyme(p) in manufacture of, 549
GOX(p) as anti-yellowing agent in, 5
immobilized amylase in effluent from making, 5
test, in enzymatic analysis, 691
Paprika, color loss, 302
color retention, 301
Paraformaldehyde as cellulolysis pretreatment, 176
Para-k-casein, primary product of chymosin action on k-casein, 20, 596, 597
Paraoxons, enzymatic analysis for, 704
Parathions, enzymatic analysis for, 704
Parenchyma, of plant cells, 512, 513
Parmesan cheese, 594
Parsley, aroma of, 370
ascorbate oxidase in, 289
flavonone synthase in, 455
naringin oxidase in, 456
Parsnips, phenolase browning, suberization, 275, 278
Partial requirement, in enzyme activation analysis, 706
Partially competitive inhibition, 250, 251
Partitioning, affinity, 133–134
in enzyme immobilization, 145, 146–149
Parvalbumin, a fish Ca-binding protein, 467, 498, 501
Passion fruit nectar, enzyme(p) in manufacture, 544
Pasta, *See also* Durum
carotenoid retention, 301
enzymatic off-flavors, 237
GOX(p) treatment of, 340
hard wheat adulteration of, 679
lipase(p) improved, 587
Pastes, tomato, consistency of, 544, 547
Pasteur effect, in muscle-to-meat conversion, 468
Pasteurization, 187
by Irad (radurization), 14, 196
eggs, ham, 212
fruit juices, HTST, flash, 192, 212, 539
in brewing, 557
of cucumbers, to prevent oversoftening, 529
of milk, and catalase(p) pretreatment, 633
and HTST, 192
and rancidities, 449, 450, 451
by ultrasonics, 200
enzymes surviving, 211–212, 449, 450, 591, 606
phosphatase assay for adequacy of, 186, 211–212, 216
vs blanching, z-values, 211
of wine, 200
vs blanching, 187
Pathogen, plant, phenolase as defense against, 276
Peaches, aroma constituents, 389
blanching, 189, 281
enzymatic browning, control, pacers, 13, 269, 275, 279, 281
enzyme assay for predicting senescence, 678
enzymes of, β-glucosidase, potential health hazard, 636
GOT, 678
laccase, reported, 271
peroxidase, heat resistance, 206
PGs, cellulase, and texture of, 519, 520
phenolase, 275
juice, enzyme(p) in making, 548
nectars, enzyme(p) in making, 544
pinking of, 189
texture enzymology, 462, 519, 520
Pea pods, methoxypyrazines in, 386
Peas, and flavorese, 417
aroma, aroma constituents, 371
blanching, 190, 192, 194, 195, 199, 207, 237
C_{18} fatty acid derivatives in, 428, 429
chilling, 437
chlorophyll, color of, 302
enzymes of, ADH, 393
α-amylase, and sugar accumulation, sweetness, 359–360
catalase, 303
LOOH isomerase, 304
LOX, and off-flavor, -color, 223, 303, 433
peroxidase, 207, 303
frozen, and acetaldehyde level, 451
chlorophyll, color, 228, 302
enzymes, fate, stability, 223, 226, 341–342
off-odors of unblanched, 430, 431, 432–433
GMP as gustator in, 364
Irad, and blanching, 199
off-odor, from bruising, 436
texture enzymology, 195, 528
Peanuts, aroma, aroma constituents, 371, 392, 419, 430, 638
and flavorese, 417, 419
enzymes of, ADH, and off-flavor, 393
lipase, 455
LOX, and aroma of fresh, 430
PE, and methanol, 638
peroxidase, 682
proteinase (arachain), 94, 492
glyoxylate cycle in, 351
protein, autolytic modification of, 626, 627
bitter peptides in proteolysates, 616
enzyme(p) solubilization of, 624
-LOOH interaction in stored, 631
proteinase(p) functionalization of, 615
zymogram, as infection indicator, 682
Pear, vegetable, 13
Pears, aroma constituents of, 389
blanching, 189
enzyme assay for maturity, 678
enzymes of, cellulase, β-galactosidase, PG, 520
PE, 520, 678
peroxidase, sclereids, and texture, 204, 206, 521
phenolase, 275, 276
methanol in brandy from, and prevention, 639
nectars, enzyme(p) in making purees, 544, 548
off-colors, 189, 269, 276
sorbitol accumulation, sweetness of, 360

texture, lignin, stone cells (sclereids), 204, 462, 520–521
Pebbles, removal by blanching, 189
Pectate as PG substrate, PE inhibitor, 515, 516
texture, Ca^{2+}, canned food firming, 522–523, 526–529
Pectic acid, fascicles, 513, 514
Pectin, and protopectinase, 518, 519
as dietary fiber, 710
as immobilized enzyme support, 145
as intercellular component, 512, 513
as macroion enzyme inhibitor, 258
biosynthesis, 514
crosslinked, for affinity chromatography, 124
degradation by active oxygen, 173, 518
in cell-separating enzyme applications, 544
in citrus debittering enzyme(p), 454
in fruit texture, 519, 520–523
in juice making, 537–539, 540–544
in orange juice cloud, 541
in vegetable texture, 13, 524, 526, 527, 528–529
lemon, production of, 549
low methoxyl, 93, 135, 517, 523, 664, *See also* Pectinic acid
maintenance in tomato processing, comminution, 545–548
structure, properties, 513, 514
Pectin lyase-like action, of cations, 526
Pectinic acid, 513, 515, 519
Pediococcus cerevisiae, and sauerkraut, biogenic amines, 642
Peel, alkali for removal, 263, *See also* Skin
citrus, texture measurements of, 511
enzymatic browning control by removal, 281
ester biosynthesis in removal, 392
lemon, for pectin production, 549
Penetrometers, for objective texture assessment, 462
Penicillamines, 33, 346
Penicillin, as enzyme inhibitor, 249, 254
in glutamate gustator production, 363
Penicillium amagaskinense, source of GOX(p) in Japan, 338
P. citrinum, RNase(p) and rennets from, 151, 603
P. corymbiferum, cysteine lyase of, 380
P. crustosum, caffeine removal by, 661
P. roqueforti, and cheese aroma, texture, 410, 608
P. vitale, source of GOX(p) in U.S.S.R, 338
Penicillium spp., milk clotting enzymes in, 95
proteinase(p) from, 59, 609
Pennycress, thiocyanates, glucosinolase of, 384–385, 386
Pentadiene hydroperoxides, 299
cis-cis-Pentadienes, and LOX specificity, 298, 425
Pentane, a LOX secondary product, 427, 428, 430
Pentathiapanes, mushroom aroma, 379
Pentosan(s), in flour, enzyme(p) removal of, bread texture, 581–582
with α-amylase, co-predictor of loaf volume, 679
Pentose(s), in flour pentosans, *See also* specific pentoses
and stereospecificity of glucose isomerase, 162
Pentose phosphate shunt, in acid, sugar accumulation, 333–334, 352, 357
2-Pentylfuran, as soy off-flavorant, 424
PEP. *See* Phosphoenolpyruvate
Peppermint, monoterpenes and flavor, 403
Peppers, aroma, methoxypyrazine, in, 386, 387
as a dairy off-flavor description, 452
color of, and LOX action, 302
piquancy of, 347
retting, enzymatic, 405
water potential isotherm, 233
Pepstatin, cathepsin D inhibitor, 257, 474
Peptides, as chelate ligand, in oxidative rancidity, 451
as nonvolatile flavorants, 349, 618
astringent, tyrosine in, 617
biosynthesis, 77, 78, 80
bitter, *See* Bitter peptides
chemotactic, phospholipase A_2 activators, 443
enkephalins, 134
in cocoa aroma genesis, 402
in plastein formation, 618–620
lipoprotein lipase activators, 448
synthesis, *in vitro, de novo*, 56, 80
Peptidoglycan synthesis, inhibition by penicillin, 249
Peptidoglycans, as antiamylases, 257
Peptones, *via* fish proteolytic autolysis, 507
PER (Protein Efficiency Ratio), enzyme shortcuts for, 708
of enzyme-processed beans, 626, 685
of enzyme-solubilized seed proteins, 623
of proteinase-produced FPC, 509–510
Perfume manufacture, enzyme(p) used in, 5, 420
Perhydroxyl, from singlet oxygen, 287
in phenolase, laccase action mechanisms, 272, 273
Pericarp, cereal grains, enzymes in, 13, 55, 574
Periderm, formation *via* wounding, 337
Perinucleus, α-amylase in, 556
Periodate, affinity chromatography matrix activator, 127
and oxidation of glycoprotein enzymes, 338
Periplasma, site of yeast lytic glucanases, 629
Permeation chromatography. *See* Gel chromatography
Peroxidizing methyl linoleate, enzyme activator, 260
Peroxisomes, photorespiration, acid accumulation, 210, 355
Perry, enzyme-derived sweeteners in, 152
Persimmons, astringency of, 347, 630
Pervaporation, 109, 113, 235
Perylene, for luminescent enzyme assays, 689
Pestalotiopsis westerkijki, enzyme(p), for soy milk, 629
Pesticides, blanching removal of, 189
enzymatic analysis for, 20, 670, 704–705
enzymatic(p) removal of (proposed), 660
Irad as (radicide), 196
synthesis, with hydroxynitrilase, 398
Pet food, as IMF, 234
Petals, as flavorese source, 420
Petioles, as vegetables, 524
P-gene (promoter gene), 75, 76, 97
pH, pH changes, enzyme associated; -activity profiles, theory examples, 261–262, 331, 475
and color of green vegetables, 303–305, 307
and color of red meat, 318, 321, 323
and enzyme potentiation, 360
digestibility of proteins measured by, 708
enzyme activation by, 274
gradients, in ATP biosynthesis, purification, color, 85, 117, 290, 332

high, *See* Alkali(es)
in enzymatic analysis, 687–688
in enzyme assay, 671, 672
in enzyme inactivation, 107, 436, *See also* Acids, Alkalies
in enzyme processing of protein, 621, 622
in enzyme purification, 113, 114, 116, 117
in food flavor, dehydrogenase equilibria, flavorese, 394, 418
lactone formation, citrus bitterness, 452–453
off-odor perception, 431
sourness, and acid accumulation enzymes, 349–356
in gibberellin effects, 558
in honey quality, 679
in immobilized enzyme systems, 146, 147, 148
in texture of foods, bread, 584
cheese, 601
fish, 499
low, *See* Acid(s)
vs blanching, 195, 263–264
pH optima, and heat stability, 191
basic theory, 261–262
double, 299, 329, 331
effects of, immobilization, 146–147
low temperature, 220
water activity, 238
food-related significance, 262
in enzyme assays, 670
lactase, 652, 653
LOX isozymes, 299, 433
microbial rennet proteinases, 603
neutral collagenase, 475
nucleases, 366
papain, 485
peroxidase, 203
PEs, 515
phosphorylase isozymes, potato, 334
proteinases, further, 506, 606, 619, 626
TMAO lyase, 503
of specific enzymes—enzyme types, alkaline phosphatase, milk, 609
alkaline proteinases, detergents, 622
casein phosphoprotein phosphatase, 609
cathepsins, 474, 478, 506
chymopapain, 486
chymosin, pepsins, 601, 602
endopectate lyase, 518
of wheat, malt α-amylase isozymes, 574
pH stability, of chymosin, pepsin, in blends, 601, 602, 605
of enzymes, 261, 338, 663
Phage, contamination with, 363
Phagocytes, lysosomes, and aging of meats, 474
Pharmaceuticals, enzymes as, 4, 663–666
enzymes in manufacture of, 5
water removal from, 235
Pharmacology, enzyme inhibition in, 248
Phaseolus coccina, scarlet runner bean, 428–429
P. vulgaris, 626. *See also* Beans
Phaseolus spp., 436
Phe(105)-Met(106), hydrolysis, primary milk clotting event, 596, 597
Phenacetaldehyde, oxidation by peroxidase, 31
Phenacetyl chymotrypsin, 133
Phenazine methosulfate, for NAD^+ regeneration, 688
2-Phenethyl isothiocyanate, 382
Phenol-formaldehyde resins, immobilized enzymes supports, 143, 653
Phenolic glycosides, taste and structure, 346
Phenolics, phenols. *See* Polyphenols
Phenolphthalein phosphate, alkaline phosphatase substrate, 211
Phenyl borate, affinity chromatography ligand, 132
Phenyl hydrazine, and meat rancidity, 433–434
Phenyl phosphate, alkaline phosphatase substrate, 211
o-Phenyl phosphates, gibberellin mimics, 558
Phenyl propane pathway. *See* C_6:C_3 pathway
Phenyl pyruvate, direct oxidation by peroxidase, 31
Phenylalanine (Phe), biosynthetic enzymes, affinity chromatography, 130
immobilized enzyme production of L, 150
in bitter peptides, 617
in lignin biosynthesis, 529
loss, from FPC, 509
Phenylethyl amine, an enzyme-generated toxicant, 640, 641
Pheophorbides, from chlorophyllides, as photosensitizers, 305, 643
Pheophytin, pheophytinization, 302, 303–304, 305, 306, 310–311
Phialophora cinerescens, enzyme assay for detection of, 678
Phoma tracleiphila, a spoilage organism, 678
Phosphagens, in muscle ATP replenishment, 467
Phosphate. *See* Orthophosphate, inorganic
Phosphate esters, and red meat color, 318
as antioxidants, 437
Phosphatides, acetal, *See* Plasmalogens
as general enzyme inhibitors, 256
as phospholipase substrates, 440, 442
as phytase affectors, 657
hydrogen bonding, fish texture, and, 501
3-*sn*-Phosphatidyl aminoethanol. *See* Cephalin
Phosphatidyl choline, an enzyme-generated off-flavorant, 124
ethanolamines, in fish, 500. *See also* Cephalin
3′-Phosphoadenosine-5′-phosphosulfate (PAPS), cofactor, 46, 383
Phosphocellulose, affinity elution matrix, 132
Phosphocreatine, in muscle ATP replenishment, 467–468. *See also* Phosphagen
Phosphoenol pyruvate (PEP), as biospecific eluant, 132
enzyme inhibition by analogs of, 255
in biosynthesis of aromatic amino acids, 84
in plant acid accumulation, 352–354
in plant sugar accumulation, 358–359
structure, 350
3-Phospho-D-glycerate, 30
3-Phosphoglyceric acid, structure, 350
Phosphoglycerides. *See* Phosphatides
2-Phosphoglycollate, 30, 355
Phospho-2-keto-deoxyheptanoate (PKDA), 84
Phospholipids, and meat discoloration, 319
as general enzyme inhibitors, 501, 581
as lipoprotein lipase activators, 443
as phytase affectors, 657
disoriented, in membranes, 15, 16
enzymes as aids in analysis for, 712

in frozen fish texture, 500–501
level, as enzymatic browning pacer, 279
melting points of, and cold hardiness, 334
mitochondria-microsome, cold sweetening, 334
Phospholipoproteins, as lipase activators, 448
Phosphoproteins, casein as, 595
in phosphotransferase action, 658
Phosphopyridoxal. *See* Pyridoxal phosphate
Phosphoric acid, as food flavorant, 349
Phosphorus, and dietary phytate, 656
enzymatic inhibition analysis for, 707
Photobacterium leignathi, SOD source, 286
Photochemical activation, catalysis, of myoglobin destruction, 320
for enzyme immobilization, 144
Photorespiration, and chlorophyll destruction, 305
and plant acid accumulation, 355
functions, intervention in, 361
Photosensitization, for enzyme inactivation, 200
Photosynthesis, and light-induced lipase repression, 586
ferredoxin in, 46
lipid peroxidation, and off-flavor, 435
Photosynthetic pathways, and emulsin localization, 398
enzymes, in acid accumulation, 351
Phthalides, aroma volatiles, 370, 387
Physiological conditioning, 655
function, shift, persistence of enzyme cold action, 224
Phytic acid (phytate), α-amylase inhibition by, 575, 581
as antinutrient in plant foods, removal, 565–569
as HbO_2 allosteric regulator, meat color, 318, 320
as vegetable texture determinant, 513, 514, 524, 527
in glucosinolase-free rapeseed, 633
in 5′-nucleotide accumulation, 364
phobia, in World War II, and bran, 656
to remove excess flour α-amylase, 581
Phytin, isolation of phytase as, 656
Phytochrome, and ascorbate oxidase, 290
Phytofluoene, LOX-induced bleaching of, 299
Phytol, and mevalonate, 304, 403
Phytophthora citrophthora, a citrus spoilage fungus, detection, 678
Phytyl group, and chlorophyllase, 304
Pi. *See* Orthophosphate, inorganic
Pickled vegetables, as fermented foods, 302, 406, 413
Pickles. *See* Cucumbers
Pie fillings, α-amylase(p) for making, 579
as IMF, 234
functional proteins in, 615
Pigments, pigmentation, *See also* Color, Discoloration, Pinking
biological, non-physiological, 269
in cured ham, 324
in red meats, 315–318
loss, enzymatic, 270, 296–297, 298–310
of fish skins, 325
of foods, 268–269
of onion, other alliaceous foods, 312–314
Pigs, pepsin(p) rennets from, 601
proteinases(p) in feed of, 624
Pike, flavor of, 416
Pileus mexicanus. *See* Cuaguayote
Pineapple, and flavorese, 417
and internal browning, 278
aroma constituents, 381, 389
biogenic amines in, 641
bromelin from, 13, 93, 486, 492
fiber from, *via* enzyme(p), 549
Pinenes, structures, 404
Ping-pong, bi-bi kinetics, 317, 339
Pink center, of Brussels sprouts, 277
rib, of lettuce, 278
Pinking, of onions, and alliinase, 312–313
of peaches, pears, *via* blanching, 189
Pinocytosis, 82
Pinot noir, PE generated methanol in, 638
Pipelines, dairy, and lipase-caused off-flavors, 447
Pitcher plant, proteinase source, 94
pK_a's, of acids, and sourness of foods, 349
of amino acid side groups, 262
PKDA, 84
Plaice, texture of frozen, 502
Plant(s), cold hardy, membranes of, 15
flowering, as enzyme(p) sources, 91
growth inhibitor, synthesis *via* LOX, 299
pathogens, and pectinases, 518
peroxidase-directed destruction of, 662
rennets from, 94, 601
Plant space, reduction *via* immobilized enzyme use, 140
Plantain, biogenic amines in, 641
Plasmalemma of plant cells, 512, 513
Plasmalogens, phospholipase-caused off-flavors from, 446
xanthine oxidase oxidation, and atherosclerosis, 649
Plasmodesmata, of plant cell walls, 513
Plastein, 140, 618–621, 655
Plastids, and starch biosynthesis, 332
membrane disintegration and cold sweetening, 334–335
Plateaus, in yield from enzyme action in ice, 229
Platelets, blood, aggregation and prostaglandin synthase, 501
gluten-phospholipid, and phospholipase, 584, 586, 587
β-Pleated sheets, in enzyme regeneration, secondary structure, 40, 214
PLP. *See* Pyridoxal phosphate
Plucking, in tea production, 292
of poultry, and texture, 463
Plugflow reactors, in immobilized enzyme action, 141
Plums, aroma constituents of, 389
β-galactosidase, 191
HCN-generating system in pits of, 636
nectars, enzyme(p) for making, 544
phenolase, enzymatic browning, sorbitol sweetness of, 257, 360
Plumule leaves, of the wheat kernel, 573
Poikilothermy, and fish texture, 497
Poison oak, enzyme(p) treatment of, 665
Poisons, and enzyme inhibitors, 248
Polarimetry (ORD, CD), and induced fit evidence, 44
and invertase assay, 159
Polarography, in enzymatic analysis, 691
Pole beans, 302. *See also* Green beans
Pollen, and honey sugar enzymes, 358
Pollock, phospholipase, texture of frozen, 500
Pollution, enzyme-related control of, by cyanide, 637
by excess nitrate in fertilizer, 678
by pesticides, 286, 660
by septic tanks, 5

by sewage, 196
by whey, 653
by white water in papermaking, 5
in blanching, 189, 190–191, 195
in water, air, enzymatic analytical monitoring, 705
Polyacrylamide, affinity chromatography matrix, 126
Polyacrylates, as coating, in meat tenderization, 483
as enzyme inhibitors, 258
Polyamides, as affinity chromatography matrices, 127
Polyanions, as enzyme inhibitors, 258
Polybases, as fining agents, 537
Polycations, as enzyme inhibitors, 258
polycations, for stabilizing enzyme(p), 102
Polyelectrolytes, as enzyme inhibitors, 258–259
Polyenoic acids, as dairy off-odor precursors, 452
Polyesters, as affinity chromatography matrices, 127
Polyethylene glycol, and humectancy of IMF, 242
in enzyme production, 109, 113, 117
stabilizer in analysis kits, 676
Polyethylene oxide, affinity partitioning ligand, 134
1,3,-D-Polyglucans. *See* Callose, Laminarin
Polylysine, as enzyme inhibitor, 258
in affinity chromatography, 127
Polymer elasticity, Raoult's law deviation, 232
Polymerization, for entrapment, immobilization, 142
Irad-induced, of casein micelles, 198
oxidative, in lignin biosynthesis, 529, 530
Polymer-polymer interaction, inhibition by, 257–258
Polymers, as IMF humectants, 234
biological, and texture, 462
in enzyme purification, 117
in low a_w model systems, 240
Polymethene dyes, for HCN estimation, 699
Polymorphism, 422
Polypeptide inhibitors, 267, *See also* Antienzymes, 257
Polyphenols, as antithiamins, 644
as chill haze constituents, 569
as enzyme(p) preservative, 100
astringency of, 630–631
cell wall-bound, PE and hard core, 530
from quinones, 284, 289, 297, 371, 401
in flour proteopentosan, 581
Irad, and synthesis of, 200
-oil complex, enzyme(p) for release, 549
-pectin complex, a postfiltration haze, 539
PG inhibition by, 537
phenolase, and enzymatic browning, 13, 276, 277–279, 284, 289, 290, 292–293, 294, 295
-protein complex, and postfiltration haze, 539
improved nutrition, 630
PVP for removal of, from juices, 537
Polyphosphate, and fish sloppiness, 479
enzymes as error source in analysis for, 712
for frozen fish toughness prevention, 505
in meat tenderizer formulations, 492
tenderizing meat with, 483–484
to inactivate excess flour α-amylase, 581
Polyploidy, of wheat, and α-amylase isozymes, 575
Polyporus spp., laccase source, 271
Polystyrene, immobilized enzyme support, 145
Polythene amines, as fining agents, 537
Polythiolanes, 374
Polytyrosyl trypsin, 138
Polyunsaturation. *See* Fatty acids, polyunsaturated
Polyvinyl(s), affinity chromatography matrices, 127
packaging, 278
pyrollidone (PVP), fining agent, phenolics remover, 241, 537
in low a_w model systems, 241
Pome fruits, sourness enzymology, 302, 349, 352
P-O-P bonds, 366
Pore size, of support, and immobilized enzyme action, 149
Pork, as skeletal muscle, 463
color of, 315, 323, *See also* Ham
enzyme-aided filth detection in, 670
mitochondria, 324
off-flavors, 446, 470–471
soft exudate (PSE), and fish texture, 499, 505
enzyme assay for predicting, 680
hormone treatment for preventing, 482
tenderizing with enzyme(p), 489
zymograms, for detecting freeze-thawing, 681
Porphyrin c, 33
in chlorophylls, and vegetable color, 303, 304, 310
in peroxidases, 204
myoglobin oxidation, meat discoloration, 325
Positional chirality *(syn-anti)*, and analog inhibition, 249
Positional isomerism, of LOX-produced LOOHs, 425–426
Postharvest, interval to processing, 435
storage, and lignification, 530. *See also* specific instances
Post-irradiation, effects on enzymes, 198
Postmortem, and meat flavor genesis, 415
chemical application for tenderizing meat, 484
injection for meat tenderization, 495
Post-operative patients, and bitter peptides, 616
Postrigor, stage of muscle-to-meat conversion, 464
Posttranslational events, in biosynthesis of enzymes, and gibberellin action, 557
casein phosphorylated, glycosylated, 597
cellular aspects, packaging, delivery, 81–82
modifications, *via* additions, 81
via limited proteolysis, 65
signal peptidase in, 81
Potassium (salts), as activator, 46, 332, 446, 677
enzyme assay for nutrient status, 677
ferricyanide, negative Q_{10} of reduction of, 225
in soil, potato phenolase enhanced by, 278
role in enzyme catalysis, 58
Potatoes, antiproteinases in, 331
aroma of baked, 387
as dairy off-odor description, 452
ascorbate, bound, lost, 190, 646
blanching, leaching, 190, 193, 195, 330
cold sweetening of, *See* Cold sweetening
enzymatic browning, *See* Enzymatic browning
enzymes of, ADHs, 393
aldehyde oxidase, 373
α-amylase, 335–336
ascorbate oxidase, absent, 289
geraniol dehydrogenase, 403
invertase, anti-invertase, 328–329, 330, 331
LOX, LOOH-transforming enzymes, 427, 430
nucleosidases, 367
of glycolysis, 333–334

of the pentose phosphate shunt, 333–334
of the starch-sugar cycle, 329–334
peroxidases, 288
phenolases, 274, 331
RNase, 18
flavor of, 364, 431
green color, prevention, 311
Irad treatment, 4, 200, 337
lye-peeling, 200, 263
maturity, estimated by enzyme assay, 678–679
NEB of products, *See* Nonenzymatic browning
spray drying of, 235
sprouting, 335
sweet, *See* Sweet potatoes
texture of cooked, canned, 462, 526–527
Potentiation, of drugs, antibiotics, 662
Potentiation, of enzymes; *See also* Activation, Autolysis
and product inhibition, 252
approaches, general discussion, 12, 14, 17–18
by underblanching, 191
for gustator enhancement, 365
for toxicant removal, 651, 653–654
in barley, to prevent droppings deposition, 566
in cauliflower, for freshening, 528
in cereals, beans, 653–654, 658–659
in citrus debittering, 454
in dates, 520, 521
in fish, 506–507
in flours, 579
in fruits, 392, 522–523
in marzipan making, 399
in meat, for texture optimization, *See* Aging, Meat
in olive debittering, 263
in protein processing, 505–508, 625–627, 655
in sweet potato, *via* starch susceptibility, 359–360
in tea production, 292, 400
in vanilla curing, 399
in vegetables for color, texture, 305, 525, 526, 527, 529
in winemaking, 537, 639
thermal, 18, *See also* specific cases
vs enzyme(p), 18, 537, 579
Potentiometry, in enzymatic analysis, 691
Poultry, as avian food, 463
enzyme(p) tenderization of, 493
feed, free methionine in, 621
xanthophyll in, 302
flavor, phytate, IMP deaminase, 364, 656
muscle-to-meat conversion, 464
Power requirement, in large-scale enzyme purification, 110
Power stroke, in muscle contraction, 466–467
PPD (pepsin-pancreatin digest), for protein digestibility, 708
Prawns, blackening of, 325
IMP, GMP gustator accumulation, 364. *See also* Shrimp
Preblanching. *See* LT-LT
Preceding indicator, in enzymatic analysis, 688, 698
Precipitation, steps in enzyme purification, 116
Precision, in enzymatic analysis, 685
Precooked cereals, enzyme(p) used in making, 579
Precooking, to inactivate enzymes in dough, 587
to prevent fish toughness, 504–505
Predators, and enzymatic browning, 211
Predigested proteins, bitter peptides in, 616
Preferred attack, *vs* multiple, in α-amylase mechanism, 554
Preformed flavors, as off-flavors, 422
at extreme end of flavor continuum, 406
essential oils as, 395
in fruits, 388
in parsley, and flavorese, 417
in tea, 40
isothiocyanate, in capers, 382
nonvolatile, accumulation enzymology, 349–467
terpenoids as, biosynthesis, 402–405
vs enzyme-potentiated, 348, 349
Prefreezing, and enzymatic browning, 281
Preharvest, and enzymatic browning, 280
sprouting, resistance, 575
Prenyl mercaptan, beer off-flavor, 368, 421
Prepeeled potatoes, and browning, 281
Preprocessing, control of enzymes by, 11, 12–14
Prerigor cooking, for tenderization, 483
in muscle-to-meat conversion, 464
Preservatives, enzyme(p) as, 662, 663
GRAS, 247
in enzyme assays, 670
in enzyme(p), 100
Preserves, enzymatic analysis for alcohol in, 697
Presses, in enzyme purification, 109
Pressor amines, 640, 641
Pressure, and enzymatic browning, 277
and meat texture, 463, 469
cooking, nutrient retention by, 644
enzyme inactivation by high, 201
leaf filter, in HFCS production, 165
Prickly artichoke, rennet from, 94, 602
Primary structure, of papain, 485. *See also* Protein(s)
Priming, of phosphorylase, 331, 332
Prism effects, and food color, 268
Proanthocyanidins, and color, 269, 294
Process analog, and enzyme(p) assessment, 683
Process cheese, proteinase(p) in making, 609
Process lines, in enzyme production, 107
Process optimization, 191–192, 283
Processed foods, restoration of flavor, 416–420
Processing suitability, enzyme assay as guide, 677
Processing time, reduction with enzyme(p), 536
Prochirality, and enzyme specificity, 33
Product form, of enzyme(p), 101
Product inhibition, active site directed, 250–253
and nonlinearity in enzyme assays, 672
and plant sugar accumulation, 358
chronometric integrals of, 252
in frozen foods, 222
in kinetic enzyme analysis, 689
non-active site-directed, 255–256
Product modification, and enzyme action control, 11, 12, 282
Proenzymes. *See Enzyme Index*
Prognostications, on food enzymology, 23–24, 495–496
Progoitrin, 384, 635
Progress curves, *vs* initial rates, in enzyme assays, 674
Prokaryotes, enzyme biosynthesis regulation in, 75–76
Pro-L_S hydrogen, in LOX action mechanism, 425
Proline, as bread aroma precursor, 414, 415
as malting, brewing quality index, 414
associated peptidases, 415

in bitter peptides, 617
Promoter gene. *See* P-gene
Propanol, in enzyme purification, 110, 116
Propanolol, and dark cutting meat, 321
S-trans-(+)-(1-Propenyl)-L-cysteine-*S*-oxide. *See* OFP
S-[(+)-2-Propenyl]-L-cysteine sulfoxide. *See* Alliin
Propenyl sulfenic acid, 375
Propenylpropyl disulfide, 378, 379
Propinquity, in enzyme catalytic efficiency, 54
of heme and fats, in meat browning, 319
Propionibacteria, in cheese texture, 608
Propionic acid, in Swiss cheese flavor, 409
vapors, for inactivating enzymes, 264
Propiothetin, seafood flavor precursor, 416
S-Propyl-L-cysteine, 33
S-Propyl-L-cysteine sulfoxide, 378
Propyl disulfide, 375
Propyl gallate, antioxidant, 437
LOX inhibitor, dough rheology affector, 587
meat color retention with, 322
Propylene glycol, in meat tenderizer formulations, 492
Prostaglandins, biosynthesis, 435, 437, 501
the third messenger, in onions, 73, 376
Prosthetic groups, definition, 45
in acid inactivation, 263
in enzyme regeneration, 213, 214
Protamines, as enzyme inhibitors, 258
in enzyme purification, 113
Protective peptides, 216
Protein(s), amino acid analysis, enzyme-aided, 710
and NEB, 327
as enzyme stabilizers, 102, 110
as flavor, off-flavor precursors, 408, 409, 457
as haze, cloud components, 537, 541, 569
as reserve metabolite, 12, 362, 622, 626
biosynthesis, *See* Biosynthesis of enzymes
blanching losses, 633
blood, 116, *See also* Blood clotting
bodies, cottonseed, 622, 626, *See also* Aleurone
cryoprotective, 229
enzyme-, protein-processing, 9, 10, 18, 436, 614–629
enzyme assay for predicting yield of, 12, 677–678
functional, *See* Functionalities of proteins
α-helix of, *See* α-Helix
homologies of amino acid sequence, 85, 467, 602
hydrolysis, hydrolysates of, 9, 19, 492, 616, *See also* Proteins, enzyme, protein processing, *Enzyme Index*
in definitions of enzymes, 27
in fish texture, *See* Fish
in meat texture, *See* Meat, Muscle-to-meat, Meat tenderization
modification, *See* specific applications, cases
nonenzyme, functions of, 25, 27, 28, 85, 86, *See also* specific proteins
nutritional aspects, *See* Nutritiousness, protein, PER
plant, as food source, *See* Protein, enzyme-, protein-processing
core of, as dietary fiber, 710
solubility, equation for, 115
structure, 40–42, 55
turnover, 200, 362, 480
Protein complexes—food quality import with, fatty acids, formaldehyde, and fish texture, 500–503
in muscle-to-meat conversion, 563
melanin, and color, 630
pectin, and cloud collapse, 543
pentosan-polyphenol, and bread texture, 581–582
phospholipid, and texture, 501, 564, 581
quinone, and nutritiousness, 277, 630
tannins, and astringency, hazes, health, 258–259, 569, 630
Protein concentrates, and isolates, 424, 436, 507, 508, *See also* specific cases
enzyme-aided analysis for chlorogenic acid in, 712
Protein efficiency ratio. *See* PER
Protein-protein interaction, and hydrophobic interaction, 55
at low temperatures, 220, 223, 229
at low water activities, 242, 244
in breadmaking, 584
in casein micelles, milk clotting, 595, 596–598
in enzyme purification, 115–116
Proteolysates, as foods, in medicine, 616
plastein from, 619. *See also* Protein, hydrolysis
Proteolysis. *See* Proteins, hydrolysis
Proteopentosans, 590, 591
Proteus morganii, histidine decarboxylase in, 640
Protochlorophyllide, in chlorophyll biosynthesis, 310
Protohemins, 203, 310
Protomers. *See* Subunits
Proton, *See also* Hydrogen ion
concerted transfer, in enzyme catalytic efficiency, 56
gradient, in ATP biosynthesis *via* oxidative phosphorylation, 85, 284
in ice, mobility, 225
in sweetness perception, 346
Protopectin, plant food texture, 513–514, 518, 524, 581
Protoplasts, enzyme(p) for preparing, 398, 550
Protoporphyrins, 204, 303, 310
Prototropy, and pH optima, 262
Protozoa, termites, cockroaches, cellulase in, 168. *See also Tetrahymena*
Provitamin A, 297. *See also* Carotene
Provolone cheese, aroma *via* oral lipase(p), 412
Proximity. *See* Propinquity
Prunasin, from amygdalin *via* β-glucosidase, 396
Prunes, color, phenolase, 275
juice, enzyme(p) in producing, 536, 537, 541
off-flavor, phenolase, 237, 423, 457
texture enzymology, 520
Prunin, a bitter flavonone glycoside, 452
Prunus amygdalus var. *amara*, bitter almond, essential oil from, 399
var. *dulcis*, sweet almond, 399
Prunus spp., aroma of fruits of, 389, 399
canned fruits, HCN disappearance in, 636
PSE (pale, soft exudate). *See* Pork
Pseudolinearity, in enzyme assay, 673
Pseudomonas fluorescens, bitter peptides—prevention, 618
GOX(p) source, 338
in milk, enzyme assay for detection, 678
P. fragi, in dairy off-flavors, 452
P. putida, caffeine demethylase in, 661
flavoprotein linked monoxygenase (xenobiotic oxidase) in, 412
P. stutzerei, exo-maltotetraohydrolase in, 154
Pseudomonas spp., cysteine lyase, 380
diaminopropionate lyase, 701

enzymes in, amino acid decarboxylases, 641
glucose isomerase, 162
in aroma genesis, 369
in dairy off-flavors, 452
isoamylase, 154
limonoid catabolyzing, 454
milk clotting (rennet) proteinases, 602
of ester biosynthesis *via* ketones, 392
Pseudoplasticity, of nectars, 544
Pseudosubstrates, 56
Psychoactive amines, 640, 641
Psychology, and color perception, 268
Psychrophiles, in proteinase oversoftening due to, 609
Psychrotrophs, pasteurization survival of, 449
Pteridine, in trimethylamine dehydrogenase, 504
Pteroyl glutamic acid, carboxypeptidase G substrate, 362
Puddings, starch-based, and xanthine oxidase, 591
Puff-drying, 235
Pullulan, pullulanase substrate, 154, 563
Pulp, disintegration, filtration with enzymes, 536
finisher's, a citrus clouding agent, 544
in nectars, 544
tomato, and product consistency, 544
Pulse, α-galactosides in, 653
Lathyrus seed adulteration, detection, 701
Pulse radiolysis, 53, 198
Pumpability, process cheese, and proteinase(p), 609
Pumping, milk, and hydrolytic rancidity, 447
Pumpkin, ascorbate oxidase of, 289
proteinase of, 94
Puncture, and texture, 511
Pupation, and phenolase, 275
Purees, fruit, off-odors of, 420, 445
garlic, enzyme(p) for thinning, 549
pea, for enzyme, storage studies, 436–437
tomato, consistency of, 544
vegetable, enzyme(p) for producing, 536, 540, 549
Purging, with N_2, to prevent browning, 283
Purification of enzymes, and immobilization, 140
by affinity chromatography, 123–136
by chromatographic, non-affinity procedures, 118–122
by crystallization, 117–118
general considerations, enrichment *vs* yield, 108, 111–112
in food science, special problems, considerations, 107
selective, high capacity operations, 112–115
solubility methods, 115–117
Purine(s), as 5′-nucleotide gustators, 363–364, 366
as toxicants in, removal from, SCP sources, 18, 629, 655
biosynthesis, degradation, 249–250, 364–366
dimerization, 225
enzymes acting on and fish quality, 366, 499, 680
in nucleic acids, and synthesis of enzymes, 74, 75, 78, 79
purity, criteria of, in enzyme isolation, 115
Purple peas, blanching with alum, 195
Purpling, of cauliflower, 227
Purpurogallin Zahl (P.Z.), in peroxidase assay, 207
Purslane, biogenic amines in, 641
Push-pull, in enzyme catalytic efficiency, 56
Putrefaction, incipient, enzymatic analysis for, 701
Putrescine, a decarboxylase-produced biogenic amine, 640
PVP, in enzyme purification, beer chillproofing, 259, 289, 570
PVPP, in beer chillproofing, 570
Pycnometer, in enzyme purification, 116
Pyloric ceca, TMAO demethylase in, 503
Pyrazines, in food aromas, 369, 401, 402, 414
Pyrenochaeta lysopersicae, enzyme assay for detection, 678
Pyridine derivatives, in food aromas, 402, 414
Pyridoxal phosphate, as alliinase coenzyme, 376–377
as amino acid decarboxylases coenzyme, 641
as dehydratases coenzyme, 393
as ethylene synthase coenzyme, 286
as threonine aldolase coenzyme, 409
as vitamin B_6 reservoir, 646
in affinity chromatography, 128
in phosphorylase, 220, 331
in reaction inactivation, 379
versatility as coenzyme, 45
Pyridoxamine, pyridoxine, vitamin B_6 congeners, 646
Pyrimidine(s), as moiety of thiamin, 645
as nucleic acid base, 74, 75, 78
biosynthesis, 69
nucleoside diphosphate sugars, enzyme cofactors, 46
Pyrogallol, and proteins, 635
for meat color retention, 323
peroxidase substrate, 207
Pyroglutamic acid, in bitter peptide, 617
Pyrolysis, and flavor genesis, 348, 407
Pyrophosphate, as product inhibitor, 253, 256, 358
enzymatic analysis for, 699, 701, 709
for sweetening sweet corn, 358–359
in plant sugar accumulation, 356
Pyrroles, biosynthesis of, 309
ring scission and food color, 309
Pyrrolidones, as monoamine oxidase inhibitors, 642
Pyruvate (pyruvic acid), as alliinase product, 32, 374
as biospecific eluant, 132
as decarboxylases product, 52
as product inhibitor, 255
enzymatic analysis for, 690
fish toughness prevention, 498, 505
in diacetyl accumulation in cheese, 408
in glycolysis, and LDH isozymes, 84
in macerated onions, a measure of pungency, 681
in muscle-to-meat conversion, 468
in N_2-stored potatoes, 337
structure, 350

Q_{10}, rate temperature coefficient, 184–186, 197, 318, 596
Quality assessment, enzymes as aids in, 707–709
Quality control, enzyme involvement in laboratories of, Chapters 37–39
Quality maximum, and residual peroxidase, 207
Quanta, light, and food color, 268
Quantum tunneling, alternate to transition state theory, 52

Quasiequilibrium, in evaporative processes, 234
Quasifrozen state, 221, 242
Quasinutrition, and enzymes, 663
Quaternary structure of enzymes, proteins, *See also* Subunits
and acid inactivation, 263
and catalytic efficiency, 66–67
and heat stability, 186, 209, 210
and inhibition, 249, 259
and isozyme frequency (polymorphisms), 47–48
as oligomers, 53
in prophenolase activation, 274, 275
Quenchers, free radical, 103, 172, 286–287, 591
Quenching, in fluorimetry, 689
Quercitin, ascorbate oxidase inhibitor, 289
Quince, phenolase, browning, 275, 276, 289
Quinic acid, 350
Quinol, laccase substrate, 273
o-Quinones, as enzyme inactivator, 260
as phenolase products, 273, 293, 590
browning, a type of NEB, 327
carotene, terpenoid degradation by, 401
in beverage aroma genesis, 400–403
in breadmaking, 590
in Strecker degradation, 401, 402
interaction with amino acids, 630
modification, and enzymatic browning, 283
-protein interactions, 277, 630
re-reduction, 283, 284–285, 289, 295, 297, 401, 646
secondary transformations of, 20
tea phenolase and, 293
toxic, from peroxidase action, 205

Rabbit, glyceraldehyde-3-phosphate dehydrogenase, 230
Rad, definition, 196
Radappertization, prevention of enzyme action after, 198–199
sterilization Irad equivalent, 196
Radicidation, Irad equivalent of insecticide, 198
Radioimmunoassay (RIA), as enzyme assay replacement, 679, 714, 715
de novo biosynthesis of enzymes ascertained by, 679
for enzyme(p) standardization, 683
of amatoxins, *vs* enzymatic analysis, 706
vs enzyme immunoassay, 712
Radioisotopes, as label in immunoassays, 712, 713, 714
in aroma research, 369, 372
in enzymatic analysis, 691
Radiolysis, 198
Radiosensitivity, of enzymes in foods, 200
Radish(es), flavor-producing enzymes in, 260, 369
pectin lyase(p), unicellular food from, 550, 551
Radurization, and meat texture, 13, 463
Irad equivalent of pasteurization, 196
Raffinose, enzymatic analysis for, 695
enzymatic removal from beans, 653–654
removal in sugar refining with enzymes, 150, 151
Rag, as orange juice cloud component, 541
citrus, removal and bitterness, 545
Raisins, and solar energy, 233
color enzymology of, 275, 283
texture enzymology of, 520
Raman spectroscopy, and enzymes, 4
Rami, cellulose of, 176
Rancidity. *See* Hydrolytic rancidity, Oxidative rancidity
Randomness, of glucanase action, estimate, 171
RANN, and immobilized enzyme support, 138
Raoult's law, and enzyme behavior at low a_w's, 232, 240, 241
Rape, and enzymatic analysis in forensic science, 701
Rapeseed, enzyme(p) functionalization of, 615
enzyme-caused off-flavor of, 455
protein, and glucosinolase, 632
Rapid radial centrifugal chromatography, 122
Raspberries, and flavorese, 418–419
aroma volatiles, 371, 389, 418
biogenic amines in, 641
color, ascorbate, in fluctuating storage, 228
EIA of virus, 714
Raw meat color, stabilization, 315
Rayons, historical cellulase substrates, 168
α-Rays, an ionizing radiation, 198
β-Rays, treatment of potatoes with, 285
Reaction, cessation in the frozen state, 322
completion assurance, in end-point enzymatic analysis, 687–688
Reaction inactivation, and nonlinearity in enzyme assay, 672
by suicide products, 260
LOX-induced off-flavor prevention by, 273, 438
of LOX, in frozen foods, 303
of onion alliinase, 379
of phenolase, 273, 274
Reaction order, in enzyme assays, 671
in thermally fluctuating systems, 227
Reaction partner, as anylate in enzyme activation analysis, 706–707
Reaction pathway, enzyme, altered at low a_w, 242
Reaction rates, and fluctuating temperatures, 227–229
ratios, and activation energy differences, 225
Reaction reversal, at low a_w's, temperatures, 22, 238–239
by immobilized enzymes, 140
Reaction time-space, and off-flavor development, 432
Reactivation. *See* Regeneration
Reactors, in immobilized enzyme use, 141, 165, 176, 654
Reagentless analysis, 692, 693
Receptors—biological, affinity chromatography and, 134, 346
as labels in immunoassays, 715
for cyclic AMP, 71, 73, 76
hormones, 63
of flavor perception, 345, 346
of food quality attributes, as proteins, 85
visual, and food color quality, 345
Recliner, in induced fit analogy, 53
Recombinant DNA. *See* Genetic engineering
Reconstitutibility. *See* Rehydration
Red meat, discoloration, 318–321
Reddening, of leeks, 313
of potatoes, 270
Redolence, 389
Redox mediators, in enzymatic analysis, 689

Reduced atmosphere storage. *See* Hypobaric storage
Reducing agents, as enzyme stabilizers, 103
Reducing milieu, *via* ATP, 285
Reducing sugars, in cold-stored potatoes, and NEB, 238, 239, 327
Reduction, of glucose for egg desugaring, 340
Reflection, refraction and food color, 268
Regeneration, of $NAD(P)^+$, 140
  of substrate, product, in enzymatic analysis, 688, 689
Regeneration of enzymes—heat inactivated, and protein functionalization, 615
  at low, fluctuating temperatures, 213, 229, 649
  in HTST, 192
    in processed foods, 118, 191, 215–216
    milk alkaline phosphatase, testing for, 216–217
    peroxidase, catalase, 213–215
  non-temperature-related, other enzymes, 112, 213, 263
Regiospecificity, 61
Regulation of enzymes *in vivo, See also* specific enzyme, food
  and hormones, 70, 72, 73–74
  and induced fit, 84–85
  and isozymes, 82–83
  as integral, built-in facet of enzymes, 28, 63
  by allostery, 67–68
  by limited proteolysis, proenzymes, 65, 82, 273
  genetic, including biosynthesis, 74–78
  hierarchy of, 63
  immobilization as a means of, 65
  in definitions of enzymes, 27
  in feedback control of metabolic pathways, 70–74
  in glycolysis, gluconeogenesis, 69, 70–72
  lipase, phospholipases, 442, 443, *See also* other enzymes
  of Michaelis enzymes, affector sensitivity, 64–65, 67
  via cascades, 72–74
  via multienzyme complex, 69–70
  via oscillation of activity, 74
Regulatory proteins, 69–70. *See also* Specifier proteins
Regulatory sites, 68, 69, 125
  in affinity chromatography, 125
Rehydration of foods, and blanching, 189
  and enzyme(p)-modified starch, 152
  enzyme(p)-facilitated, 159
  nonenzyme-related facilitation of, 235
  other enzymatic aspects of, 246
Rejects, cheese, enzyme(p) treatment of, 609
Relative humidity, and water activity, 231–232
Relaxation, of doughs, and LOX, 589
  of muscle, 466, 467
  spectroscopy, temperature jump, and ES intermediates, 53–54
Renaturation, 212, 214, 615. *See also* Regeneration
Rendered meat, tenderization of, 493
Renin, confusion with rennin, 594
Renneting, cold. *See also* Milk clotting, Cheesemaking
Replication, of DNA, 75
Repolymerization, during glucose production, 158
  enzymatic, of xylooligosaccharides, 179
Reporter groups, and affinity labeling, 56
Repressors. *See* Gene repression
Reproductive cycle, yeast, and lytic β-glucanase, 629
Re-reduction. *See* Quinones
Resazurin, visualizer dye, in enzyme test paper, 692
Reserve carbohydrate, cellulose, polysaccharide, 167, 357, 513
Reshuffling, of peptides in plastein, 619
Resonance destabilization, in enzyme catalysis, 4
Respiration rate, and enzymatic browning, 278
Restraint, and meat texture, 463
Restricted diffusional chromatography. *See* Gel filtration
Restructuring, of proteins for functionalization, 615
Retina, and color perception, 268
Retorting, and blanching, 189
Retrogradation, of amylose, in bread, enzyme(p) prevention, 578
    in maltose production, 157
    in postfiltration haze, 539
  of xylans, 179
Retting, enzyme involvement in, 405
Reversal, of enzyme action. *See* Reaction reversal
Reverse burst, active site titration, 684
Reverse osmosis, for egg degusaring, 340
Reverse protein digestibility, for dietary fiber, 710
Reversible inactivation. *See* Regeneration
Rework cheese, into processed, *via* enzyme(p), 609
Rhamnose, in citrus flavonone glycosides, 453
  in pectin structure, 513
Rheology, of nectars, 545
Rhizomes, rhubarb, aroma of, 550
Rhizopus arrhizus, apricot softening by PG of, 642
    lipase(a) from, for enzymatic analysis, 697
  R. delemar, enzyme(p) from, for whiskey production, 561
  R. niveus, glucoamylase(p) source, 158
  R. oryzae, enzyme(p) from, sanctioned by WHO, 647
  R. stolonifer, apricot oversoftening by PG of, 522
Rhizopus spp., enzyme(p) sources, 95, 550, 602, 603, 627
Rhodopseudomonas spheroides, catalase, 97
Rhodopsins, 85, 268, 345
Rhubarb, acerbity of, 349
  as a petiole vegetable, 524
  enzyme(p)-aided aroma extraction from, 551
  oxalic acid accumulation enzymes, 354
Rhus, Chinese lacquer tree, laccase source, 271, 274
RIA. *See* Radioimmunoassay
Ribitol, in trimethylamine dehydrogenase, 504
Riboflavin, enzyme-aided analysis for, 711
  in coenzymes, 45
  retention during blanching, 189, 190
Ribonucleic acid. *See* RNA
Ribonucleosides, 364
Ribose, as flavor precursor, 367, 415
Ribosomal RNA. *See* RNA
Ribosomes, as Irad targets, 196, 199
  destabilization in purine removal, 655
  in biosynthesis of enzymes, 77–80
  in rough endoplasmic reticulum, 386, 387
  ion exchange chromatography of, 119
  peroxidase as topological probe of, 205
D-Ribulose-1,5-diphosphate (RBP, RDP), 30
Rice, β-amylase, and saké, 576
  enzyme assay as heating adequacy index, 679
  improvement during storage, 237
  instant, enzyme(p) used in making, 550
  lipase of bran, and rancidity, 445

lipoxygenase, 429
phytate as antinutrient in, 656
stickiness, 237, 246
Ricin, castor bean, inactivation of, 651, 659
Ricinoleic acid, polymer, *via* LOX action at low a_w, 242
tetramer, a lipase activator, 443
Ricotta cheese, made without renneting, 600
Rigidity, and vegetable texture, 524
Rigor mortis, and meat and fish texture, 464, 468, 497, 498–499
Ripening, *See also* Aging, Curing, Maturation
of cheese, flavor enzymology, 362, 407, 409
texture enzymology, 594, 600, 608–609
of fruits, flavor enzymology, 338, 349, 351, 355, 356
texture enzymology, 519–522, 529
of fruits, vegetables, enzyme assay as markers of, 678–679
of herring, by endogenous enzymes, 508
of vegetables, texture enzymology, 524–525, 529
of wheat, barley, and the amylases, 573–574, 576
RNA, and Irad of potatoes, 337
degradation, by RNases, and specificity, 31, 32
for purine removal, from SCP sources, 629, 655
in 5′-nucleotide accumulation, 346
in transcription, 74–75, 76
messenger (mRNA)—masked (conserved, latent) and unmasking, by light, 309
by plant growth regulators, 278, 309, 358, 557
in senescence, 269
in translation, 77–80
of invertase, 358
of phenolase, 275, 278
ribosomal (rRNA), 77, 78
transfer (tRNA), in translation, 77–80
Roasts, color, flavor texture, 315, 437, 493
Rocking, rock-and-roll, dialysis, 113
Rods, eye, in color perception, 268
Roe, color of, 301, 325
Roll-of-the-dice stratagems, for enzyme(p) production, 97
Romano cheese, 412, 594
Roots and rootlets, as vegetables, 524
in malting, and gibberellins, 558
vegetable, in whiskey production, 561
suberization of, 278
wheat kernel, 573
Rosaceae, sorbitol sweetness in fruits of, 360
Roses, petals, attar *vs* enzyme(p), 399
flavorese enhanced attar yield from, 420
Rosetta Stone, of gene derepression, 556
Rotary dialysis, 113
Rough endoplasmic reticulum, 385, 387
Rubber, biosynthesis, 403
production, enzyme(p) used in, 5
Rubberiness, and enzyme-induced off-odor, 424
Run-off, in brewing, enzyme(p)-assisted, 561, 566–567
Russet spotting, of lettuce, 278
Rust disease, and PG action, 517
Rutabagas, glucosinolase, goitrin, goiter, 382
Rutinose, a citrus disaccharide, 453
Rye, β-amylase of, 571
bread, α-amylase and quality of, 579
aroma genesis in, 413
flour, enzyme assay for quality of, 679
in rye whiskey production, 561
nucleases, 336

Saccharomyces cerevisiae, enzyme(p) source, 95
S. fragilis, lactase(p) source, 95, 652
S. lactis, lactase(p) source, 652, 653
var. *maltigenes*, and dairy off-odors, 452
Saccharomyces spp., enzyme(p) from, as GRAS, 647
Sacs, oranges. *See* Orange juice
enzymatic(p) detoxification of, 654
Safety, of food enzymes, enzyme(p), 646–649
Safety valve, diacetyl formation as, 408
Safflower seed, discoloration, 631
Sage, monoterpene aroma, 403
Saint Paulin cheese, 604
Sake, bitter peptides, 616
enzyme(p)-assisted making of, 549
lysozyme(p)-preserved, 662
production of, 552, 577
Salad dressings, enzyme-derived sweeteners in, 152
oxygen removal with GOX(p), 9
Salads, flavor of, 349
Salicin, β-glucosidase substrate, 185
Salicylate, as enzyme inhibitor, 253, 254
as protein denaturant, 186
Salivary gland, cellulase in termites', 168
Salmine, as Ca replacement in milk clotting, 598
in enzyme purification, 113
Salmon, canned, curd removal with enzyme(p), 510
off-odor genesis, 457
roe, color, pigments in, 325
Salmonella, GOX(p)-desugared eggs free of, 340
growth medium, SOD(p)-preserved, 286
thermal destruction curve, 212
Salmonids, astaxanthin in, 302
Salt(s), as inhibitors, 259
for standardizing enzyme(p), 103
in breadmaking, 577
in enzyme purification, 115–116
in meat tenderizers, 492
prophenolase activation by, 274
tolerance, 353
Salting, osmotic drying, 234
Salting-in and -out, and IMF, 233, 240, 241
Salvia officinalis, sage, 403
Sambungrin, a cyanogenic glycoside, 396
Sand, in enzyme purification, 110
removal by blanching, 189
Sandiness, enzymatic(p) prevention of, 179, 610
prevention by enzyme-derived sweeteners, 153
Sanitation, and blanching, 193, 522
Saponins, enzymatic removal of, 651, 654, 659–660
Sarcina, amino acid decarboxylases in, 641
2,3-butanediol dehydrogenase from, 697
Sarcomeres, in muscle-to-meat conversion, 468
Sarcoplasm, isozymes in freeze-thaw detection, 682
Sarcoplasmic reticulum, and meat texture, 464, 467, 477
Sarcosomal proteins, 463
Sarcotestae, papaya, enzymes in, 395
Sarracenia purpurea, pitcher plant, 94

Sarret unit, for GOX assay, 340
Sashimi, thiaminase I, and vitamin B_1 deficiency, 645
Saskatoon berry, ascorbate oxidase in, 289
Saturation, color perception component, 268
Sauces, cellulase(p)-enhanced creaminess of, 550
fish, as fermented food, 406
functional proteins in, 615
Sauerkraut, acid in, *via* fermentation, 263
biogenic amines in, as potential toxicants, 641, 642
lipase and flavor of, 413
production, role of blanching, 195
taste of, 349
Sausage(s), biogenic amines in, 641
enzyme(p) in production of, 10, 508
enzyme-processed proteins in, 624
fish, *via* proteinase (p), 509
flavor enzymology, 414, 415–416
off-color of, and H_2O_2, 325
slow drying of, 246
Sawdust, from fish filleting, autolysis, 507
Scald, apple, and LOX, 298, 314
Scalding, 188, 207, 237. *See also* Blanching
Scale, reduction by enzyme(p), 10. *See also* Fouling
Scale-down, for enzyme(p) suitability, 546, 547
Scale-up, in enzyme(p) production, 99–100
Scallops, AMP degradation in, 366
octopine dehydrogenase(a) from, 700
viscera, thiaminase I in, 645
Scanning electron micrographs, of muscle structure, 465, 469, 497
Scarlet runner bean, LOX isozymes of, 428–429
Scattering, of light, and food color perception, 268
Scavengers, free radical. *See* Free radicals
Schiff's bases, and α-keto acids as anti-tougheners, 506
in NEB, 334
in PLP pyridoxal phosphate enzyme action, 376, 377
Schutz's law, 149, 168
Sclereids, lignin, and pear texture, 513, 520–521
Scleroproteins, beef, tenderness, *via* carcass orientation, 460
via meat tenderizing proteinase(p), 481, 487–488, 492
function in muscle, 463, 498
of fish (myocommata), 498
Sclerotinia spp., enzyme(p) from, toxicological evaluation, 647
macerating factor from, 519, 550
milk clotting enzymes in, 602
Sclerotinization, and phenolase, 275
Scombroid poisoning, by histamine in foods, 640
Scopoletin, 277, 457
Scorched pork odor, of organic thiocyanates, 385
SCP. *See* Single cell protein, Yeast
Scrambled eggs, from GOX(p)-treated, 340
Scutellum, of wheat germ, 573, 754
Scylla, and Charybdis, 328
Sea bacteria, SOD(p) source, 286
Sea lettuce, proteinase of, 94
Sea squirt, melanin pigmentation of, 277
Seafood, AMP degradation in, 366
flavor genesis in, 415–416
Seasonings, cellulase(p)-aided preparation of, 549
enzymatic analysis for glutamate in, 690, 699
gustators in, and AMP deaminase, 364
retention *via* enzyme inhibition, 367
rehydration of dry, 246
Seaweed, cellulase(p)-aided processing of, 549
5′-nucleotide gustator accumulation in, 364
Second order kinetics, 213
Secondary metabolites, 422
Secondary products of enzyme action, 16, 19–21, 224, 260
Secondary structure, protein, in acid inactivation, 263
Secretion, *vs* biosynthesis, of malt α-amylase, 556–557
Sedanolides, aroma volatiles, 370
Seeds, *See also* specific food seed
coat, Ca^{2+}, and bean texture, 528
wheat, 573
enzyme-, protein-processing of, 9, 615, 617, 621–627
germinated, enzymes in, 65, 204, *See also* Germination, Sprouting, Malting
in enzyme purification, 108
Selective extraction, in enzyme purification, 116
Selenium, in enzymes, 46
Self-assembly, molecular, supramolecular, 66, 69, 213, 215, 595, 596
Self-digestion, 139–140, 507. *See also* Autolysis
Self-hydrolysis, and enzyme immobilization, 149
Self-medication, and psychoactive amines in foods, 641
SEM. *See* Scanning electron micrographs
Semolina flour, 301, 683
Senescence in plants, acceleration *via* enzyme(p), 627–628
and color, 269
and off-flavors, 276, 347, 423
sugar accumulation (sweetening) and, 335, 337, 359
endopeptidase participation in, 16
enzymes assays as markers of, 678–679
gene derepression during, 12, 269
glutamate accumulation in, 362
inhibitors, 311
Sensing, biological, methyl transfer enzymes in, 85
Sepals, flavorese from, 418, 419
Sephadex, Sepharose, derivatives, in enzyme purification, immobilization, 119, 120, 126, 127, 128, 150, 653
Septae, fish, 498
Sequestrants, as noncompetitive inhibitors, 255
enzymatic browning control by, 285
GRAS, 247
in blanching, 195
of Ca^{2+}, to remove excess α-amylase, 58
Serendipity, 12, 654. *See also* Entrainment
Serine, acetylated, *N*-terminal, of proteins, 273
alkali lability of, in proteins, 710
and bitter peptides, 617
biosynthesis of, 409
in active site, 39, 56, 129, *See also* Serine proteinases
in plant cell walls, 249, 514
phosphorylated, in casein, 609
Serotonin, a decarboxylase-generated psychoactive amine, 640, 641
and meat texture, 482
Serra, ewe's milk cheese, made with plant rennet, 601
Serratia marcescens, proteinase(p) from, for chillproofing, 571
Serratia spp., enzymes of, in cheese making, 410, 602
Serum, of comminuted tomato products, 547

Serum albumin, dimer of *via* Irad, 198
Sesame seed protein, enzyme(p) solubilizing of, 623
Sesquiterpenoids, from mevalonate, 403
quinone-aided degradation of, 401
Sewage, radappertization for feed use, 196
Sex differentiation, and invertase, 358
Shallots, flavor of, 246, 378
Shamouti, an orange variety, 541
Shark meat, urea removal from *via* urease(p), 423
Sharpness, an unblanched frozen food off-odor, 430
Shear, and enzymatic browning, 277
and tomato products consistency, 547
force, an objective texture assessment, 462, 479, 511
in enzyme purification, 110
Sheep, cheese from milk of, 594
rumen, culture filtrate, as enzyme(a), 696
thiaminase I and diseases of, 645
Sheetability, of processed cheese, enzyme(p)-improved, 609
Sherry, aroma volatiles of, 389
Shiitake mushrooms, flavor enzymology of, 364, 366, 379, 380
Shikimic acid, in gooseberry sourness, 350, 352
Shivering, and meat texture, 463
Short beers, 551
Shortcuts, to food quality assessments *via* enzymes, 707–709
Shortening, of muscle in rigor, 468
Shotgun approach, to enzyme inhibition, 248
to obtaining mutants for enzyme(p) production, 96
Shrimp, AMP degradation in, 366, *See also* Prawns
blanching, protein losses in, 633
enzyme assay to detect spoilage of, 678, 680
GOX(p) treatment of, 323
greening of, 325
lysozyme(p)-preserved, 662
Shrinkage, prevention by blanching, 189
Shriveling, an amylase-associated triticale defect, 575–576
Sialic acid, in k-casein, 597
Side activity, specificity, in enzymatic analysis, 685–686, 687
Side products, accumulation in frozen foods, 224
Sieves, molecular, in enzyme concentration, purification, 109
Sieving, minimized by cellulase(p), 549
Signal amplification, in enzymatic analysis, EIA, 688, 689, 713
Silage, an unblanched frozen food off-odor, 430
and fish autolysis, 507
Silastic resin, as immobilized enzyme support, 144
Silent coppers, in phenolase, 271
Silica, as enzyme adsorbent, 118
as immobilized enzyme support, 144, 165
gel, for chillproofing beer, 570
Irad effect countermeasure, 198
Silicates, for chillproofing beer, 570
for removing nonrennet enzymes from rennets, 603
Silk, as immobilized enzyme support, 145
enzyme(p)-aided degumming of, 5
Silver, enzymatic inhibition analysis for, 707
"-sin", as enzyme name suffix, 28
Sinapis alba, white mustard, 632
Single cell protein, and lytic glucanase of yeast, 628
bitterness, 423
methanol as nutrient for, 638
purine removal, 18, 629, 655
via chitinase(p), 180
Single cell suspensions, of vegetables, 627
Singlet oxygen (O_2^*), and red meat color, 316, 320
as a form of active oxygen, 17, 20, 287
from O_2^-, and SOD for quenching, 17, 20, 173, 286, 287, 591
in aroma genesis, ester biosynthesis, 287, 373
in LOOH and off-flavor formation, 435
in LOX reaction mechanism, 425
in milk off-flavors, 450, 451, 452
Sizing, and amylase(p) in papermaking, textiles, 5
Skim milk, and Irad effects on enzymes, 199
solids, in breadmaking, 577
Skin(s), *See also* Peel
fruit, softening with enzyme(p), 549
grape, PE in, methanol from, 638
human, enzymes of, 275, 279
Slackening, slackness, of dough, and proteinase, 487
Slivovitz, Irad effects on, 198
Sloppiness, a fish texture defect, 497
Sloughing, a potato texture defect, 526
and consistency of tomatoes, 547
of canned olives, 259
Sludge, radappertization for feed use, 196
reduction with enzyme(p), 100
Slush drying, 235
Snails, cellulase and xylanase of, 177
Snake venom, L-amino acid oxidase source, 108
phospholipases source, 442
Snap beans, enzymatic browning of, 279–280
Soapiness, soapy odors, lipase-caused, 445, 447
Sodium, as enzyme cofactor, 46, 85, 446, 515
as enzyme inhibitor, 259, 282, 585, 642
Sodium aurothiomalate, inhibition abolished by, 224
Sodium bicarbonate, and blanching, 195
meat color retention with, 323
Sodium chloride, and halophile proteinases, 506
fish cathepsin inactivation by, 506
for chlorophyll retention, 306
for control of enzymatic browning, 282
for control of flour proteinase, 583, 585
for purine removal from SCP organisms, 655
in enzyme purification, 116
in salting-in, -out, 240, 241
solubilization of grain β-amylase with, 562
stabilizing enzyme(p) with, 102
tenderizing and koshering meat with, 484
Sodium dodecyl sulfate, as enzyme inhibitor, 259
prophenolase activation by, 274
Soft beverages, 263, 625. *See also* Beverages, Juices
Softening, imparted to sweet baked goods by bacterial α-amylase(p), 237, 576, 579
softness, by sulfite. *See also* Oversoftening
Sogginess, a potato texture defect, 526
Soil, as an immobilized enzyme system, 13, 138
enzyme assay as indicator of quality of, 677
Soja hispidus. *See* Soybeans
Solanine, potato alkaloid, 311, 382, 660, 661
Solanum eleagnifolium, *See* Horse nettles
S. melongena. *See* Eggplant(s)
Solar energy, and low moisture foods, 233

heat *via*, 234
utilization *via* enzymatic biomass conversion, 167
Sole, cathepsins of, 506
texture of frozen, 502
Solid food systems, models of, 239–242
Solid phase, in lipase action, 443–444
peptide synthesis, 80
Solka floc, a cellulase substrate, 173
Solubility, methods, in enzyme purification, 115–117
of proteins, equation for, 115
Solutes, as inhibitors, 259
concentration, in persistence of enzyme action, 224–225, 238, 244
Solvents, extraction, drying with, 235
in enzyme purification, 109
organic, inactivation, purification with, 103, 108, 436
removal of off-flavors with, 439
Sonic vibration, for drying foods, 201
Sophorose, as cellulase inducer, 169
Sorbate, for preserving enzyme(p), 102
Sorbitol, accumulation in fruits, enzymology of, 360
enzymatic analysis for, in wines, 697
Sorbus spp., aroma volatiles from fruits of, 399
Sorghum *(Sorghum bicolor)*, amylase inhibitory tannins in, 563, 581
as brewing adjunct, 552
dhurrin in, as toxicant precursor, 636
protoplasts, and enzyme, substrate compartmentation, 399
Sorrel, acerbity of, 349
Soups, creatine in, enzymatic analysis for, 701
dried, 233
functional proteins in, 615
glutamate in enzymatic analysis for, 690, 699
lipase(p) used in making, 412
tomato, consistency of, 544
Sour cream, as an emulsion, 6
Sourdough French bread, flavor constituents, 349, 577, 698
Sourness, as a desirable flavor, *See* Acid(s), pH
as an enzyme-caused off-flavor, 423, *See also* Acid(s), pH
objective determinants of, 349
Sour-stale, an unblanched frozen food off-odor, 431
Soy sauce, glutamate in, 362
proteinase(p)-aided production of, 616
Soybean(s), antitrypsin (trypsin inhibitor), enzymatic analysis for, 705, 706
curd, lysozyme(p)-preserved, 662
enzymes of, α-galactosidase, 654
lipase, 445
LOX, 209, 424–428
proteinase, 626
urease, 13, 46, 423
flatulence of, 653, 654
flour, in breadmaking, 9, 300, 577
milk, enzyme-aided production of, 412, 626, 654
off-flavor prevention, 263, 423, 436, 437, 438–439
off-flavors of, 424
oil, enzyme(p)-aided production of, 549
phytate, as antinutrient, 656
protein, albumen function protein from, 614, 615
autolytic modification of, 626
bitter peptides from, 616, 617, 618
isolates, enzyme(p)-treated for acidic beverages, 624
plastein from, 619, 620
starter feed, for calves, 601
textured, 462, 708
proteinase in, 94
starch, interference with proteolysis, 629
Spacer arm, in affinity, hydrophobic chromatography, 119, 127
Spaghetti, enzyme-aided analysis for lipid in, 712
Spears, asparagus, texture of, 530
Species, identification *via* zymograms, 682
Specific displacement elution. *See* Affinity elution
Specific gravity, measured in enzyme purification, 116
Specificity, and the active site, 38–42
apparent change at low temperatures, a_w's, 226, 242–243
entropy and, 44
enzyme classification, nomenclature and, 28–33
in enzymatic analysis, 685–686
isozymes and, 47–48
non-amino acid enzyme components and, 45–47
Specifier proteins, 70, 83, 384
Spectrophotometers, in enzyme assay, analysis, 676, 689
Spermine, spermidine, enzymatic analysis for, 701
Spherosil, for immobilizing enzymes, 571
Spherosomes, in cauliflower. *See also* Lysosomes
Sphingomyelin, and enzyme-caused off-flavors, 466
Spices, cellulase(p) treatment of, 9
lipase-caused rancidity in, 246
rehydration of dry, 246
Spin label(ing), and affinity labeling, 56
immunoassay, 712, 713
Spinach, as leaf vegetable, 524
ascorbate, enzymatic analysis for, 700
biogenic amines in, 641
chlorophyll and green color of, 302, 306, 307
enzymes of, ascorbate oxidase(p), 289
catalase, 210, 223
chlorophyllase, 307–308
glycolate oxidase, and oxalic acid in, 355
nuclease, 366
peroxidase, 190, 206
prophenolase, 274
SOD, 287
flavorese and, 416
galactosyl glycerides in, 191
nutrient losses, 190, 644
off-odor of unblanched, 430
oxalic acid, and sourness of, 349
enzymatic inhibition analysis for, 706
pectin lyase(p)-produced unicellular food from, 550
Spinal injury, chymotrypsin treatment of, 4–5
Spinning, for protein functionality, 615
Spin-paired Cu, in phenolase, 271
Spiny flathead, 500
Spiral catabolic pathway. *See* Oxidation
Spirits, distilled, 553. *See also* Whiskey
Spleen, as food, 463
Spolen cheese, 594
Sponge methods, in breadmaking, 577
Sponginess, a fish texture defect, 499
Spontaneous clarification, fermentation for egg desugaring, 340
rancidity, *See* Hydrolytic rancidity, milk
via endogenous PG, 537
Spores, injection of, to tenderize meat, 494

microbial, heat resistance of, 187, 639
outgrowth, prevention of, 663
purification by two-phase systems, 206
thermal destruction curve, 206
Spray drying, 101, 235
Spray residues, enzymatic analysis for, 253, 705
Spreading, of dough, bacterial proteinase(p) for, 584
Spring-back, of dough, and LOX, 588
Sprouted barley. *See* Malt
Sprouts and sprouting, as vegetables, 524
damage, and excess flour α-amylase, 580
incipient, in triticale, 575–576, 584
inhibition of growth by Irad, 14, 337
inhibitor to prevent NEB, 336
of onions, 380
Squash, ascorbate oxidase of, 289
beverage, cloud retention, 453
gluconeogenesis in, 357
off-flavor of unblanched frozen, 430
vine fruit vegetable, 524
Squid, cathepsin of, 506
melanin pigmentation of, 277
SS-SH interchange. *See* Disulfide-thiol interchange
Stabifix, for chillproofing enzymes, 570
Stability and stabilization of enzymes, α-amylases, 559
enzyme(p), 101–103
GOX(p), 338
heat, 46–47
immobilized enzymes, 149–150, 342
in enzyme assays, 672
in enzyme purification, 107, 112, 113, 120
lipoprotein lipases, by heparin, 448
low water activities and, 242, 243–245
NEB and, 328
substrates as stabilizers, 101, 191
Stabilizers, GRAS, 247
Stachyose, and flatulence, 653
Stainless steel, as immobilized enzyme support, 144, 653
Staleness, an unblanched frozen food off-odor, 430, 447
Staling, bread, delay with bacterial α-amylase(p), 326, 579
Standardization, of enzyme(p), 101–102, 103
Staphylococcal enterotoxin A, EIA of, in foods, 714
Staphylococcus aureus, lipase, 441
Staphylococcus spp., amino acid decarboxylases, and biogenic amines, 641
Star dextrins, 554
Starch, affinity chromatographic use of, 124, 556
biosynthesis, *See* Starch-sugar interconversion cycle
composition, structure, 154, 156, 554, 555, 579
damage, enzymatic assessment of, 708
derivatives, as immobilized enzyme supports, 145
-derived sugar syrups. *See* Syrups, Sugars, specific sugars
enzymatic analysis for, 670, 696
enzyme(p)-facilitated grinding, isolation, 549, 582
for enzyme(p) standardization, 103
gel, water sorption, isotherm of, 232
gelatinized, as α-amylase(p) alternative, in breadmaking, 579
via micronization, 200
granules, raw, attack by enzymes, 157, 158, 355, 559, 577, 587
haze, 9, 539
hydrogen bonds, and amylolysis, 172
hydroxypropyl, amylase resistance of, 664
in breadmaking, cereal processing, 573, 575, 577–581
in brewing, distilling, 553, 554–555, 559–562, 563, 564
in low a_w model systems, 241
liquefaction, saccharification of, *See* Amylases *(Enzyme Index)*
micronization of, 200
modified *via* debranching enzyme(p), 155, 156
modifiers, GRAS, 247
non-food uses of, 5
nutritional availability of, 633
retrogradation, *See* Retrogradation
soybean, proteinase inhibition by, 629
-sugar interconversion cycle, 328–336, 337, 359
sulfate, as enzyme inhibitor, 258
vegetable structure and, 512, 513, 519, 526
waxy, barley, maize, 156, 555, 579
Starch-based puddings, liquefaction *via* xanthine oxidase, 591
Starfish, endo-1,3(4)-D-glucanase in, 566
proteinase of, 506
Starter bacteria, culture, in cheesemaking, 408, 594, 607
enzymes of, amine oxidases, 642
amino acid decarboxylases, 409, 642
aminotransferases, 409
diacetyl-producing enzymes, 408–409
α-ketoacid decarboxylase, 409
lactase, 408
peptidases, 618
proteinase, 406, 409, 605, 607–608
growth inhibition by milk lipase, 409, 447
Starvation, and fish flavor, texture, 436, 497
Steady state. *See* Kinetics
Steaks, color of, 315
enzyme(p) tenderization of, 490, 493
Steam blanching, and enzyme regeneration, 215
vs water blanching, 192–193
Steeping, gibberellins and, in malting, 556–558
β-1,3-glucanase biosynthesis during(p), 558
Stems, as vegetables, 524
Step-backward improvements, in automated enzymatic analysis, 690
Stereospecificity, 31–33. *See also* Chirality
Steric hindrance, in affinity chromatography, 127
in muscle contraction, 466
Steriflamme, hot gas blanching, 194
Sterilants, removal by, replacement with, enzyme(p), 9, 13, 205, 661–663
Sterility, maintenance during enzyme(p) treatment, 90, 140, 341
Sterilization, and radappertization, 196
vis-à-vis blanching, 187, 191
Steroids, imidazole complexes as model enzymes, 61
mevalonate in biosynthesis of, 403
toxic effects of, putative, in foods, 311, 382, 659–660, 661
xenobiotic oxidase in metabolism of, 77
Stevioside, α-glucosyl transferase action on, 361
Stickiness, in confection control with enzyme-derived sweeteners, 153
of doughs, rectified with lipase(p), 587
potato texture and, 526
Stickwater, and fish protein autolysis, 507

Stilson cheese, 594
Stir, a functional protein-imparted texture, 615
Stirred tank reactor, for immobilized enzyme action, 139, 140, 600
Stitch pumping, of meat tenderizers, 495
Stochastic approach, for mutants for enzyme(p) production, 96
Stone cells. *See* Sclereids
Stone fruit, blanching of for brandy making, 639
Storage, controlled atmosphere, 530–531
  hypobaric, *See* Hypobaric storage. *See also* specific foods, processes
Stout, enzyme(p) for producing, 553
Strawberries, alcohol dehydrogenases of, 394, 403
  aroma constituents and their biosynthesis, 389
  as a dairy off-odor, 453
  frozen, ascertaining past history of, 68
    color and ascorbate during storage, 228
  leaves and autumn coloration, 269
Strecker degradation, alcohols accumulation *via*, 393
  in bread aroma, 414
  loss of dietary methionine *via*, 620
  quinone-catalyzed, 402
  tea, cocoa aromas *via*, 401, 402
Streptococcus lactis, and proteinase of cheese texture, 594, 608
  S. liquefaciens, cheese amine *via* enzymes of, 642
Streptococcus spp., anti-amylase in, 257
  as starters, 408, 594
  glucose isomerase(p), considered safe, 647
  production, use, 161, 163, 164, 166
  milk contaminants, cheese amines *via*, 642
Streptomyces griseus, pronase from, 509, 571, 708
Streptomyces spp., milk clotting enzymes in, 602
  vegetable aroma components *via* enzymes in, 369
  xylanase induction in, 178
Streptomycin, in enzyme purification, 113
Stress, and struggling, animal, fish, and texture, 463, 470
String beans. *See* Green Beans
Stringent control, metabolic regulation *via* allosteric enzymes, 12, 679
Stringiness, a fish texture defect, 499
Stylosanthes humilis, legume, xylanase in, 178
Stress in plant, metabolite(s), and enzyme potentiation, 14
    limonin as, 455
    off-flavors as, 422
    onion flavor precursor as, 378
    peroxidase in formation of, 204
    tea flavor aroma volatiles as, 40
    toxicants as, 642
  response to, controlled atmosphere storage, 311
    Irad, 14, 199, 337
    phenolase, PAL, increases as, 275
    sugar accumulation as, 356
Subatmospheric storage. *See* Hypobaric storage
Suberin, and suberization, 278
Subfreezing temperatures, effect on enzymes, 223
Sublihydration, 235
Submerged culture, for enzyme(p) production, 99
  for ethanol production, 561
Submicelles, casein, and self-assembly, 55, 595
Substrate(s), absence of as enzyme control, 13
  access restriction, for enzyme control, 11, 279
  affinity chromatography and, 125, 131–132
  analogs, as enzyme inhibitors, 249–254
  chromogenic, for assays, zymograms, 207, 669, 673, 684
  delocalization, and Irad effects, 200
  distortion by enzyme, 53
  endogenous, phenolase, 276, 277, 293, *See also* other specific sources
  enzyme inhibitors, *See* Substrate inhibition
  enzyme products as, *See* Product inhibition
  enzyme stabilizers, 101, 191
  enzyme xero- and cryo-protectants, 224, 244
  inhibition, enzyme regulation by, 64–65
    in dry foods, at low a_w's, 238
    in frozen foods, low temperatures, 22
    modification, manipulation, to manage enzymes, 11, 12, 260, 281–283, 645
    non-active site-directed, 256
  -product adducts, as TSAIs, 254
  removal, to control enzymes, 282, 632
  solubility, size, and enzyme action, 241
  suicide, 253
Subunits, affinity chromatography and, 125, 130
  dissociation, in acid inactivation, 186
    in chelator-, detergent-inactivation, 260, 274
    in low temperature enzyme behavior, 220, 222, 230
    in thermal inactivation, 186
  enzyme quaternary structure and, 66
  hydrophobic interaction of, 55, 67
  interchange, 174, 274
  isozymes and, 48, 83–84
  of gluten, and LOX-induced dough change, 590
  of troponin, and meat texture, 464, 466, 467, 478
Succeeding indicators, in enzymatic analysis, 688
Succinic acid and succinate, accumulation in controlled atmosphere storage, 355
  enzymatic analysis for, 699
  in cheese aroma genesis, 411
  monosubstituted esters as model enzymes, 59
  structure, 350
Succinic acid-2,2-dimethyl hydrazide, browning control with, 283
Succinyl group, for ligand activation in affinity chromatography, 128
Succulence, a fish texture, 499
Succulence, a vegetable texture, 511
Sucrose, accumulation in non-potato plants, 357–358
  accumulation in potatoes, 328, 329, 332, 337
  acid hydrolysis of, 19, 159, 221
  as enzyme cryo- and xero-protectant, 224, 244
  enzymatic analysis for, 670, 676, 691, 692, 712
  enzymatic browning control, 282
  esters of, synthesized by lipase, and bread texture, 586
  floatation, density gradient medium, in enzyme purification, 109
  fructose production from, 160
  hydration of, and substrate inhibition, 238, 239
  in breadmaking, 577
  pickle firming effect of, 529
  potatoes, maturity indicator of, 679
  translocation of, 333
  See also Sugar, Sugar refining
Sufu, nonphytase involvement in making, 659
Sugar(s), accumulation enzymology, in cold-stored

potatoes, 328–336
in honey, 358
in other food plants, 358–361
prevention, 336–337
aroma genesis involvement, 369
color of, 268
enzymatic analysis for, 21, 694–695
for osmotic drying, 235
for stabilizing enzyme(p), 102
in meat tenderizer formulations, 492
invert, *See* Invert sugar
lysozyme(p)-treated for caries prevention, 662
-to-shikimic acid pathway, and plant acids, 352
Sugar beets, enzyme assay to detect spoilage, 678
phenolase and enzymatic browning of, 275
Sugar burn, mitigation and lipases, 413
Sugar cane, C_4 photosynthesis in, 352
PEP carboxykinase in, 357
Sugar cap, GOX(p) incorporation in frozen, 291
Sugar refining, enzymatic analysis for dextran, 696
enzymatic analysis for raffinose, 695
enzyme(p) for removing dextrans, 10, 360
enzyme(p) for speeding filtration, 10
immobilized enzyme for raffinose removal, 150–151
prevention of tassel invertase action, 13, 360
ultrasonication in, 200
Sugar syrups, starch derived, 131, 152–167
Sugar wall, a date texture defect, 520
Suicide substrates, affinity labelling and, 56, 260
Sulfated glycans, as enzyme inhibitors, 258
Sulfenic acid(s), 32, 379
Sulfhydryl. *See* Thiol(s)
Sulfite and sulfur dioxide, aflatoxin degradation with, 661
as an antioxidant, 437
as enzyme inhibitor, 257, 281, 283, 289
asepsis, and TPP-enzyme inhibition, 644
chlorophyll bleaching by, 304
enzymatic browning control by, 246, 281, 283
for dried fruits, 246
for fish discoloration, 325
gas sensing electrode, 694
in fructose syrup production, 163
vitamin B_1 loss by, 644
Sulfolipids, 442
Sulfonamides, as enzyme inhibitors, 284, 684
Sulfur, enzymatic analysis for, 707
volatiles, odor and aroma, 369, 370, 373–381, 409
Sulfur dioxide. *See* Sulfite
Sulfuric acid, as enzyme extraction agent, 110
as sprout inhibitor, 337
Sun drying, enzyme-related nutrient loss by, 644
Sunbeam peach, 279
Sundew, proteinase source, 94
Sunflower seed, protein, 623, 631
Sunstruck beer, 368, 421
Supercoiling of DNA, 75
Supercontraction, in meat texture defects, 470, 483
Superheated steam drying, 235
Superhydrogen bonding, 219, 229
Superoxide anion (O_2^-), as enzyme inhibitor, inactivator, 260
as SOD substrate, 20, 59, 286, 287, 451
ascorbate oxidation by, 289
biopolymer degradation by, 172–173, 518, 522, 591
blanching, pasteurization and, 17, 450
enzymatic browning and, 285
for waste disposal, 286
generation of, 198, 285–286, 287, 320, 339
immobilized enzyme stability and, 150
in alcohol accumulation, 393
in biopigment bleaching, 300, 305
in breadmaking, 590, 591
in ethylene biogenesis, 286
in lipid peroxidation, off-flavor, 20, 435, 450
in milk off-flavors, 237, 450, 451
nucleotide accumulation and, 367
plant acid accumulation and, 355
red meat color and, 316, 320
xanthine oxidase, atherosclerosis and, 649
Superslurper, as immobilized enzyme support, 144
Supersubstrate binding site, 444. *See also* Interfacial recognition site
Supports for immobilized enzymes, 139, 141–144, 145, 146, 149, 165. *See also* individual enzymes
Supramolecular edifices, peroxidase as topological probe of, 205
Supreme Court, and vegetables, 523
Surface culture in food enzyme production, 99
Surface denaturation, 107
Surfactants, *See also* Detergents
as blanching alternative, 264
as inhibitors, 259–260
GRAS, 247
in glutamate production and, 363
lipase activation by, 442
prophenolase activation by, 274
wheat phospholipase, and bread texture, 587
Sweat, lysozyme in, 662
Sweating, in cocoa production, 402
in vanillin production, 400
Swede (rutabaga), toughness, texture and lignin in, 529
Sweet almonds, emulsin in, 636
Sweet birch bark, essential oil from, 399
Sweet corn, as a vegetable, 524
nuclease, potentiation of, 366
peroxidase, energy of activation, 205
sweetening of, 253, 256
Sweet potatoes, α-amylase potentiation in making dehydrated, 18, 246
as enzyme source, 154
as root vegetable, 524
color control of, 269
enzyme-aided amino acid analysis of, 711
ethylene treatment of, 278
PAL, phenolase of, 275, 276
peroxidase, 682
sugar accumulation, enzymology, 359–360
texture enzymology, 530
zymograms and resistance to infection, 682
Sweeteners, *See also* Sugars, individual sweeteners
accumulation enzymology, 356–361
GRAS, 247
production, compared with brewing, 553
production enzymology, 152–180, 618
Sweetening of potatoes, a cold stress response, 227. *See also* Cold sweetening
Sweetness, an off-odor of unblanched frozen foods, 430
Sweetness perception, theories of, 346
Swiss cheese, 409, 594, 604

Swivel, impairment in PSE, 471
  of myosin and energy transduction, 467
Symbiosis, among organelles, and the cell, 282
  in aroma genesis, 366, 369, 386, 388, 406
  in cellulolysis by termites, 168
Syn-anti isomerism, of glucosinates, 385
  of onion lachrymator, 375
Syncatalytic inactivation. *See* Reaction inactivation
Syneresis, in cheesemaking, 594, 600
  prevention with functional proteins, 614
Synergy, among cellulase system components, 171–172
  among LOX isozymes, 299, 304
  between enzyme(p), for chillproofing, 177
    for preservation, 663
  between α-galactosides, other flatulence producers, 653–654
  between gibberellins, other α-amylase derepressants, 558
  between gustators, 364
  in low temperature enzyme behavior, 229
  of combined enzyme inactivation treatments, 199
Synzymes, 50, 59. *See also* Model enzymes
Syrup, chocolate. *See* Chocolate, syrup
  pineapple, cellulase(p) in making, 549
Syrups, corn, 151, 152–167
  dextrin, 156–157
  dextrose equivalents, 153
  enzymatic production of, 154–167
  fructose, *via* immobilized glucose isomerase, 161–167
  glucose, *via* glucoamylase(p), 157–158
  high fructose corn, 161–167
  maltose, 157, 558
  modified starch, and nonsweet dextrins, 155
  nonenzymatic production of, 153–154
  nonsweet properties imparted to foods, 153
  starch hydrolyzing enzyme(p) in, 154
  unit cost of enzyme(p) used in making, 104

T. *See* Temperature
Tailings, wheat flour pentosan in, 581
Tailor-made enzymes, by mutations, 96
Tallowing, a dried milk off-flavor, 237, 450
Tamarind fruit, (+)- and meso-tartaric acids in, 355
Tandem-crossed immunoelectrophoresis, 362
Tannic acid, immobilized enzyme support, 145, 653
  in chillproofing beer, 571
  in enzyme purification, 113, 145, 653
Tannin(s), as antithiamins, 644
  as enzyme inhibitors, 258–259, 296, 529, 563, 706
  as gibberellin antagonists, 558
  as haze components, 539–540, 570
  as health benefit (tara), 630
  as health hazard (condensed), 661
  as plant insecticides, 259
  astringency, tartness of, 190, 347, 630, 706
  enzyme assay, interference by, 259
  in banana blanching, 190
  in brewing, 563, 570
  in tea, cocoa, 293, 294, 295, 296, 539–540, 644
  phenolase in biosynthesis, transformation of, 276, 292
  protein complexes, 630–631
  tannase(p) for removal of, 539–540, 570, 661
Tap water, in enzyme purification, 114
Tapioca, nontoxicity, hydroxynitrilase and cassavas, 635, 636
Target enzyme, in cascade regulation, 73
  in enzyme inactivation, 260
Target theory, of ionizing radiation effects, 196–199
Taro, as a corm vegetable, 524
L-Tartaric acid, as food flavor, structure and biosynthesis, 350, 351
Tartness, amylase inhibition as index of, 706
Tartrate, as malic enzyme affector, 352
Tartrates, enzymatic removal in winemaking, 536
Taste, acuity, receptors of, 345, 346, 347
Taxon of rye and wheat triticale, 575
TBA values, 304
TCA cycle, *See* Tricarboxylic acid cycle, 351
Tea (leaves), antithiamin polyphenols in, 644
  aroma genesis enzymology, 15, 369, 400–401, 406, 407, 419
  astringency of, 631
  biogenic amines in, 641
  color genesis enzymology, 292–294
  cream, nature, enzymatic removal, 539–540
  enzymes of, ADHs, 394
    esterase, 401
    LOX, 372
    phenolase, 270, 292, 293–294, 400–401
  instant powder, soluble, *See* Instant tea
  production of, 291–292
  zymograms, guide to plucking fermentation progress, 682, 683
TEAE-cellulose, in enzyme purification, 119
Tempeh, flavor, 413
  α-galactosides in, 653
  nutritional value of, 623, 658
Temperature, activity profiles of enzymes, 183–184
  and meat texture, 463
  and red meat color, 328
  as enzymatic browning pacer, 279
  coefficients, negative, 225, 444–445
    positive, 197, 318, 596
  control in HFCS production, 165–166
  fluctuation, effect on enzymes, foods, 19, 216, 224, 227–229, 305, 447
  in enzyme assay, 671
  in immobilized enzyme systems, 148–149
  jump, in relaxation spectroscopy, 54
  low, *See* Low temperature enzyme behavior
  optimum and heat stability, 183–184
Template-mold paradigm, 42. *See also* Lock-and-key metaphor
Tenderay, meat tenderization, 483
Tenderization, meat, *See* Meat tenderization
  of abalone with proteinase(p), 510
Tenderness, a meat textural attribute, 463, 476, 477. *See also* Meat texture, Meat tenderization
Tendons, and meat tenderizers, 493
Tensile strength, and bamboo shoot texture, 530
Tension, and meat texture, 463
Teratogens, tested for in safety evaluation of enzyme(p), 648
Termination, in translation, enzyme biosynthesis, 80

Termites, cellulases of, 168
Ternary complexes, 132, 134
Terpenes and terpenoids, aroma, in essential oils, 369, 370
in fruits, 388, 389
biogenesis, 403–405
from carotenoids, 414
reaction with quinones, 401
α-Terpineal, 403
Terpinenes, structures, 404
α-Terpineol, structure, 404
Terpinin-4-ol, structure, 404
Terpinolene, structure, 404
Tertiary structure of proteins, and enzyme active site, 40–42, 54–59
in acid inactivation, 263
in proenzyme activation, 275
phenolase, 272
reordering of in protein functionalization, 615
Test papers, for enzyme analysis, 692
for enzyme assay, 676
for zymograms, 682
Test stick, 695
Testa, of the wheat kernel, 573
Testing, enzymes used in, 20, Chapters 37–39
Tetrachloronitrobenzene, as sprout inhibitor, 337
Tetrahedral intermediates, in enzyme catalysis, 56
Tetrahedral pyramid, enzyme transition state as, 254
Tetrahydroxybenzyl, reactions with quinones, 401
Tetrahydroxydecyl isothiocyanate, caper aroma, 382
Tetrahymena vs enzyme tests for protein digestibility, 708, 711
Tetrapyrroles, in food pigments, 269
Tetrazolium salts, trapping agents in enzymatic analysis, 688
Textile, desizing with amylase(p), 5
Textile manufacture, catalase(p), for H_2O_2 removal, 5
Texture, and protein functionality, 614–616
blanching-caused loss of, 189
definitions, perception of, 461–463
enzyme tests for predicting, 681
enzymology of individual food classes, bread, other cereal products, 572–592
cheese, other dairy products, 409, 593–610
fish, 497–505
fruits, vegetables, 227, 376, 511–531
meats, poultry, 463–495
instruments for measuring, 462
protein S-S, hydrophobicity and, 616
Thamnidium spp., and spores, for tenderizing meat, 484, 492
Thaw rigor, a meat textural defect, 470
Thawing, enzymatic browning during, 281
enzyme behavior during, 219, 220–222. *See also* Freeze-thawing
Theaflavins, tea, 293, 591
Thearubigins, tea, 293
Theaspirone, in tea from phenolase action, 401
Therapeutic agents, as enzyme inhibitors, 248
Thermal destruction curves, spores *vs* peroxidase, 206
Thermal energy, control of enzymes by, 183–196
Thermoactinomyces spp., milk clotting enzymes in, 602
Thermocycle blanching, 194
Thermodynamics, and enzymes, 27
Thermosetting meat products, proteinase(p) in, 493
Thermostability. *See* Stability, Stabilization
Thiaesters, in fruit aromas, 388
Thiamin, as biospecific eluant, 132, *See also* Vitamin B_1
as coenzyme moiety, 45
bitter taste of, 421, 457
thiaminase I, fission of, 645
Thiamin modifying factors. *See* TMF
Thiamin monophosphate in affinity chromatography, 129
Thiamin pyrophosphate, affinity chromatography and, 128, 131
and vitamin B_1 analysis, 670
as coenzyme, 45, 254, 408
sulfite cleavage of, 257
Thiazole, moiety of thiamin, 645
Thiazoline pyrophosphate, as a TSAI, 254
Thiazol(in)es, in aroma, 370, 374, 387
Thick filament, myosin, muscle structure and, 464
Thickeners, use of enzyme-derived syrups in, 153
Thickness and density, Irad used for, 200
Thiele modulus, in immobilized enzyme kinetics, 148
Thin filaments, muscle structure and, 464, 465, 466
Thin layer chromatography. *See* TLC
Thinning of egg white, and lysozymes, 662
Thiocyanate ion, as enzyme-generated toxin, 384, 634–637
as peroxidase substrate, sterilant, 205, 662–663
from papain activation, 213
Thiocyanates, *via* glucosinolase action, 384–385
Thiogalactoside, in affinity elution, 132
Thioglucosides. *See* Glucosinolates
Thiohydroxamic *O*-sulfonate, 383
Thiol(s), *See also* Disulfides
affinity chromatography and, 126
β-amylase solubilization, 562, 576
bread texture and, 587–590, 591
cathepsins, and emphysema, 480
enzyme activators, 213, 474, 484–485, 494, 503
enzymes, sulfite inhibition of, 257
low temperature enzyme behavior and, 220, 223, 224, 230
meat texture and, 472–473, 474, 476, 482
pK_a, in cysteine, 262
reaction with LOOH, 429
reaction with quinones, 402
taste acuity and, 346
Thiolactose, as ligand in chromatography, 130
Thiophenes, from onion lachrymator, 374, 375, 395
syn-Thiopropanal-*S*-oxide, onion lachrymator, pinking precursor, 269, 313, 375–376
Thiosulfinates, onion, 375
Thiourea, and mustard flavor, 417
Third order kinetics, in enzyme regeneration, 213, 214
Thixotropic gels, of plastein, 619
Thompson Seedless grapes, blanching, 282
Threonine, in cheese aroma genesis, 409
in peptide bitterness, 617
Threose-2,4-diphosphate, enzyme inhibitor, 255
Thromboxanes, prostaglandin synthase products, 501
(−)-Thujol, (−)-thujone, thymol, structures, 404
Thyroid, peroxidase of, 203, 204
Thyroxin, synthesis by thyroperoxidase, 205
Tilsit cheese, made with microbial rennets, 604

Time-scale, of secondary reactions, 369
Tissue culture, plant, 336, 657
Tissue comminution, and aroma, 369
Tissue differentiation, and glutamate accumulation, 362
Tissue regeneration, cathepsin in, 480
Tissue, whole, as Irad target, 199
Titanate, enzymatic browning control with, 282
Titania-alumina, immobilized enzyme support, 164
Titanium, as immobilized enzyme support, 144, 165
TLC, in enzymatic analysis of pesticide residues, 704, 705
  in lipid analysis, 697
TMAO. *See* Trimethylamine *N*-oxide
TMF, non-thiaminase antithiamins, 644
Tobacco, a phenolase-caused off-odor in prunes, 237, 423, 457
  laccase in, 271
α-Tocopherol, ineffectiveness in meat color retention, 322
α-Tocopherol acetate, and LOX, in breadmaking, 588
Tofu making, enzymology of, 6, 616, 653, 658
Toluene, for autolysis in enzyme purification, 109
  in fruit aroma, 388
Tomatidine, 659, 660
Tomatine, degradation by ripe tomato enzyme system, 651, 659–660
Tomato(es), aroma constituents, genesis, 370, 371, 373, 374
  as berry fruit vegetable, 524
  biogenic amines in, 641
  comminuted products, enzymology of, 189, 194, 264, 545–548
  enzyme assays for spoilage in processing, 209, 678
  enzyme(p) from, 549, 627
  enzyme-aided analysis, inspection, of products, 670, 712
  enzyme-induced firming of canned, 525–526
  enzymes of, alcohol dehydrogenases, 394
    ascorbate oxidase, 289
    cellulase, 524–525, 545
    β-1,3-glucanase, 526, 545
    LOX, 430
    PE, 525
    peroxidase, heat resistance, in juice, 206
    phenolase, 275
    phosphofructokinase, allostericity as maturity indicator, 679
    polygalacturonases, 209, 524, 545, 546, 548, 549
    SOD, 287, 289
  flavorese treatment of, 417–418
  GMP, gustator in, 364
  green, chlorophyll, tomatine in, 310, 320, 659
  juice, production, enzymology of, 544–548
  lycopene, and color of, 269
  pickled, as a fermented food, 406
  ripening, delayed by Irad, 14
  taste, 349
  texture, ripening enzymology of, 524
  tomatine degradation, by ripe, 651, 659–660
  water potential *vs* content, 233
Tonality, a wine color attribute and pectinase(p), 537
Tongue proteins, polyphenols, and astringency, 630, 631
  taste receptor, 346
Tonoplast, 512, 513
Top agar layer method, in enzyme(p) production, 96
Topology, peroxidase as probe of biological, 205
Tortillas, alkali in preparation of, 263
Torula yeast, for dough mellowing, 583
Total-change enzymatic analysis. *See* End-point
Toughening, of dry beans, 527–528
  of frozen cherries, 522
  of frozen fish, 499–505
Toxicants and toxins, enzymatic inhibition analysis for, 704
  enzymatic removal, 636, 637, 639, 650–661
  enzyme generated, in foods, 634–643
  enzyme immunoassay of, 715
  enzyme inhibitors, 256
  removal by blanching, 189
  removal by immobilized enzymes, 139
  removal with O_2^-, 286
Toxicology, and enzyme inhibition, 247
  of enzyme(p), 647–648
TPP. *See* Thiamin pyrophosphate
Trade names of enzyme(p) for use, as analytical aids, 710, 712
  in breadmaking and other cereal processing, 300, 579
  in brewing, 559
  in cheesemaking (rennets), 603, 604
  in chillproofing beer, 570
  in distilleries, whiskey making, 561
  in protein processing, 509, 618, 621, 622, 625, 627
  in removing lactose from milk, 652, 653
  in sugar refining, 361
  in wine and juice making, cell-separating operations, 538
Trade-offs, in enzyme-related decisions, 12
Trametes sanguinea, acid proteinase from, 625
Trametes spp., enzyme(p) from for candying fruit, 406
  milk clotting enzymes in, 602
Tranquilizers, and meat texture, 463
Transcription. *See* RNA
Transduction, 316
  enzymes, other proteins, as devices for, 86
  in ATP biosynthesis, 85, 285, 316, 322
  in food color perception, 268
  in muscle function, 4, 466–467
Transesterification, of amino acids to plasteins, 619, 620
Transfer free energy, hydrophobicity, and peptide bitterness, 617
Transfer restriction, in enzyme immobilization, 145
Transfer, RNA. *See* RNA
Transglycosylation, a problem in corn sweetener production, 106, 157
  amylopectin biosynthesis *via*, 332–333
Transition metals, and taste acuity, 346
Transition state analog inhibitors (TSAIs), 52, 253–254, 282, 437
Transition state in enzyme catalysis, and entropy, 44, 51, 206
  as floppy body, 4
  energy of activation, lowering, 50–52
  of enzyme models, 61
  role of metals in lowering, 57
Translation. *See* Biosynthesis of enzymes
Translocation in plants, in sucrose accumulation, and invertase, 357–358

of acids, 350
of citric acid, 354
of fructose and sorbitol, 360
Translucency, and food color, 269
Transpeptidation, in plastein formation, 619, 620
Transport, active, 16, 360
Transportation, in the food processing chain, 11, 13
Trapping of intermediates, and active site, 56
Trapping reagents, in end-point enzymatic analysis, 688
Trappist cheese, made with microbial rennets, 604
Tree fruits, as vegetables, 524
Triacetin, as lipase substrates, 440
Triaoses, 559
Tricarboxylic acid cycle and enzymes of, and LDH isozymes, 84
derivatives, as gibberellin stimulants, 558
diacetyl and cheese aroma, 408
in controlled atmosphere storage, 531
in freeze-thawing, 682
in muscle texture, 460, 469
in N_2-stored potatoes, 337
in plant acid accumulation, 352, 354
link with glycolysis, 69, 468
prochiral specificity and, 33
Trichlorethylene, for flotation in enzyme purification, 109
Trichloroacetic acid, 369, 624
Trichoderma koningii, cellulase of, 173
T. reesei, cellulase(p) from, 95, 174–175, 627, 647
cellulolytic enzymes of, 169–174
in large-scale cellulolysis to glucose, 174–175
lysine α-oxidase from, 666
xylanase of, 178, 179
Trichoderma spp., proteinase(p), meat tenderizer, 492
Tricholoma spp., mushroom, rennet from, 602
Trigger enzyme, in enzymatic analysis, 687, 689, 691
Triggering step, of muscle contraction, 466
Triggers, of sugar accumulation in potatoes, 328–336
Triglochinin, a cyanogenic glycoside, 398
Triglycerides, as cheese aroma precursors, 408
as enzyme substrates, *See* Lipase
in adsorption affinity chromatography, 124
Trigonal pyramids, in enzyme catalysis, 57, 254
Trilaurin, as enzyme substrate, 241
Trimethyl glycine, 503
Trimethylamine, a fish off-flavorant, 436
enzymology of, 503, 504
Trimethylamine-*N*-oxide (TMAO), frozen fish texture and, 501–504
off-color, off-flavor precursor, 325, 436
Trimethylamine oxide. *See* Trimethylamine *N*-oxide
Trimmings, cheese, treatment with proteinase(p), 609
Triolein, 241
Triphasic kinetics, in cytochrome *c* denaturation, 206
Triplet state, of oxygen, 287
Trisodium phosphate, to inactivate excess flour α-amylase, 581
Trisulfides, in onion aroma, 375
Triterpenoid dilactone. *See* Limonin
Triterpenoids, 403, 405
Trithiolane, 387
Triticale, amylases of, 575–577
Triticum aestivum, polyploidy and α-amylase isozymes of, 575
T. durum, polyploidy and α-amylase isozymes of, 575
tRNA. *See* Ribonucleic acid
Tropomyosin, fish, 498
in muscle function, meat texture, 464, 466, 478
Troponins, in muscle function, 464, 466, 498
Troponin-T, in beef aging, texture, 478–479, 681
Trout, flavor of, 416
Trp. *See* Tryptophan
Trypsin inhibitor. *See* Antienzymes
Trypsinogen. *See* Proenzymes, *Enzyme Index*
Tryptamine, an enzyme-generated food, 640, 641
Tryptophan, affinity chromatography, 129
derivatives, in meat texture, 482
immobilized enzyme production of L-, 150
in enzyme active site, 263
loss from FOC, 509
metabolism, peroxidase in, 205
tryptamine from, *via* decarboxylase, 640, 641
TSAI. *See* Transition state analog inhibitors
T-shirts, silk screening, use of enzyme(p) in, 5
Tuberculosis, and milk pasteurization, 211
Tubers, as vegetables, 524
Tubulin, affinity chromatography of, 134
Tulbaghia violacea, lyase of, 379, 380
Tuna, histamine as potential toxin in, 641
IMP-histidine complex, 367
5′-nucleotide accumulation in, 364
off-colors, 269, 275
pigments in, 325
Tunaxanthin, esters of, 325
Tungsten, as Mo replacement in enzymes, 663
Tunneling, quantum alternative to transition state theory, 52
Turbidity, a cloud attribute, 536
and appearance of foods, 267
Turbulence, 107
Turbulent film drying, 235
Turgidity, a vegetable texture, 511
Turgor, cell walls, and vegetable texture, 512, 513, 524
pressure and plant food texture, 524
Turkey, color of, 315
Turnip greens, bioavailability of iron in, 646
blanching, lysine loss in, 189, 194, 633
Turnips, enzymes of, 213, 430
Turnover number, definition, 36
Tweens, lipase activation by, 442
Twigs, essential oils *via* enzymes from, 399
Twitch switch, in muscle contraction, 467
Two-phase systems for enzyme purification, 117, 133–134
Tyramine, an enzyme generated food toxicant, 640, 641
Tyrosine, and lignin biosynthesis, 259
as tyramine precursor, and decarboxylase substrate, 640, 642
derivatives and meat texture, 482
in affinity chromatography, 129
in astringent peptides, 617
in protein, phenolase-caused loss of, 630
of gluten, phenolase, and bread texture, 591
phenolase substrate, melanin precursor, 270, 272, 273, 274, 276, 277, 279

Ubiquinone, in chloroplast electron transfer system, 46
Udders, and hydrolytic rancidity in milk, 201
UDP, UDPG, in sucrose biosynthesis, accumulation, 333, 334, 357, 358
UDP-sugar pathway, for pectin biosynthesis, 514
Ultrafiltration, and enzyme immobilization, 141
as nonevaporative drying process, 235
in enzyme protein processing, 509, 615, 624
in enzyme purification, 101, 114
Ultrasonics and ultrasonication, and resonance destabilization by enzymes, 4
food-related uses, other, 200–201
for enzyme inactivation, 200–201
for intracellular enzyme solubilization, 110, 163–164
for pasteurization, 200
in enzyme protein processing, 622, 623
Ultraviolet radiation. *See* UV
Ulva lactuca. See Sea lettuce
Umbelliferone derivatives, for chromogenic assay and active site titration, 684
Uncleanliness, a lipase-caused off-odor, 447
Underblanching, 191, 211, 225, 237, 305
Undershoot, and enzymes in fluctuating temperatures, 228
Unfolding, enzyme protein, and inhibition, inactivation, 187, 206, 214, 260, 263
refolding, in protein functionalization, 615
Unformed ferment, 28
Ungerminated barley. *See* Barley, ungerminated
Unicellular food, *via* pectin lyase(p), 110, 535, 548, 550, 551
Unidirectional flow, effected by enzyme immobilization, 140
Unit costs, of using enzyme(p), 104
Units of enzyme activity, expression of, 674–675
Universal catalyst, catalase, 210
Unleavened bread, 278
Unmasking of RNA, in chlorophyll biosynthesis, 309
Unwholesomeness, and off-odor, 345
Uranyl phosphate, enzyme adsorbent, 118
Urea, *See also* Urease
as protein denaturant, 186
enzymatic analysis for, 692
enzymatic removal from shark meat, 423
H-bond breaking inhibitor, 213, 259
in protein amino acid analysis, 711
peroxide, in peroxidase test paper, 207, 676
phenolase activation by, 274
salting-in and -out, 240, 241
soybean antitrypsin inactivation by, 657, 659
Wöhler's synthesis of, 396
Uric acid, product of caffeine demethylase, 661
Uridine diphosphate. *See* UDP
Uridine diphosphate glucose. *See* UDPG
Uridine triphosphate. *See* UTP
Urine, enzymatic anylates of, 4, 685
postprandial odor of, 387
Urushiol, enzyme(p) degradation of, and poison oak (ivy), 665
USDA choice beef, meat tenderizing enzyme and, 490
USERIA, ultrasensitive EIA using radioactive substrates, 713
UTP, as allosteric regulator, 73
UV, absorbance, H_2O_2 for catalase assay, 211
catalyzed reactions, 225
enzyme inactivation by, 200
optical methods, in enzymatic analysis, 685, 686–688, 689

Vacuoles, as subcellular components, 82, 512, 513
enzyme localization in, 398, 399
isoinvertases, and sugar translocation, accumulation, 357
Vacuum-dried foods, 234, 644
Valeryl aldehyde, in cocoa aroma, 402
Valine, and flavorese, 418
as aroma precursor, 369, 376, 402, 409
as off-flavor precursor, 452
biogenic amine from, *via* decarboxylase, 640
excess glycolysis reservoir, 408
immobilized enzyme production of L-, 150
in glucosinolate biogenesis, 382–383
van der Waal's forces, in enzyme catalysis, 55
in enzyme purification, 119
in protein functionality, 616
Vanilla beans, curing, enzymology of, 399–400
Vanillin, glycosidase-produced flavorant, 400–401
Varieties, zymogram, for identification of, 682
Vascular bundles, phenolase localization, 275
Vasoactive, vasodilator amines, 640, 641
Vectorial metabolism, and ATP biosynthesis, 85
Vegetable(s), aroma genesis enzymology, 370–387, 638
blanching, processing, 193–197, 205–211
casein, 382, 396, 398
color enzymology, 275–280, 302–314
definition, classification, 523–524
flavor restoration with flavorese, 417
juices, purees, enzymology of, 235, 535, 544–548
off-flavor enzymology, 6, 7, 18, 428–432, 434, 435, 438, 439
oils, *See* Oils, vegetable
pear, *See* Merliton
pesticides in, enzymatic-TLC analysis for, 705
texture, enzymology of, 9, 523–531, 662
Vegetative origin, zymograms, as indication of, 682
Vells, dried calf abomasum, rennet from, 600
Venturi effect, in thermocycle blanching, 194
Verbascose, 653
Verticillium albo-atrum, xylanase in, 179
Very low density lipoproteins, 443
Vesicles, fruit, site of acid accumulation enzymes, 351, 354
malt α-amylase secretion from, 556
Vesicular tissue, 435
Vesiculation, gibberellin induced, 586
Vetch, β-cyanoalanine synthase in, 637
Vibrating IQB, 193
Vibrio spp., milk clotting enzymes in, 602
Vicinal hydroxyls, in affinity chromatography, 127
Vine fruits, as vegetables, 524
Vinegar, from enzyme-derived glucose syrups, 158
in mustard manufacture, 386
Vinifera wines, 638
Vinyl ethers, from potato LOOH, 427, 428, 430
4-Vinyl guaiacol, a soybean off-flavorant, 424
Vinyl ketones, as geranium- and mushroom-like aromas, 452

as off-flavorants, 423, 424
Vinyl off-flavor, in peas, 191
5-Vinyl oxazolidine-2-thione, goitrin, 384
Violaxanthin, LOX induced bleaching of, 299, 325
Violet leaf perfume, 372
Virus, in EIA of, 714
isolation by multiphase techniques, 117
Viruses, in enzymatic browning, 277
Viscera, of scallops, enzymes in, 505, 506, 645
Viscoelasticity, a texture related dough rheological parameter, 590
Viscosity, an enzyme(p)-induced protein functionality, 614, 615
and enzyme behavior in frozen food, 222
chocolate syrups, control with enzyme(p), 9
dough, amylograph *vs* α-amylase assay, 578
fruit juices, reduction with enzyme(p), 536, 537, 539
of xylitol solutions, 177
plastein, due to hydrophobic interaction, 620
potato texture and, 526
substrate inhibition due to, 238–239
worts, reduction in brewing with enzyme(p), 565–567
Visible radiation, for enzyme control, 200. *See also* Photosensitization
Vital force, in enzyme history, 4
Vitamin A, and mevalonate, 403
and odor perception, 345
enzyme-related loss of, 644
intestinal phytase and, 657
LOX destruction of, 229, 299, 644
lysosome fragility, and meat texture, 482, 644
Vitamin B_1, blanching caused loss of, 190, *See also* Thiamin
enzyme aided analysis for, 670
Vitamin B_2, enzyme-related loss of, 643. *See also* Riboflavin
Vitamin B_6, enzyme-related loss of, 646, *See also* Pyridoxine
spin label immunoassay of, 715
Vitamin B_{12}, enzyme-caused increased uptake of, 664
receptor as label in RIA, 715
Vitamin C, deficiency, and meat tenderizing, 482. *See also* Ascorbic acid
enzyme-related loss and retention of, 643, 645–646
Vitamin E, lysosomes, and meat texture, 482
Vitamin H. *See* Biotin
Vitamin K, as coenzyme, 45, 46
Vitamins, as coenzymes, 45–46
enzymatic analysis for, 700
enzymatic release, and binding, 664
tenderizing meat, 482
VK_m-space, in enzyme kinetics, 37
Volatile flavors. *See* Aromas
Vole bioassay *vs* enzyme test for protein digestibility, 708
Voltametry, in enzymatic analysis, 691

Waffles, improvement by papain(p), 10
improvement by proteinase(p), 10
Walseth cellulose, as cellulase substrate, 169
Warburg manometer, 689
Warfare, microecological, and pectinases, 518
Waste, disposal, reduction; cellulose, biomass, *via* cellulase(p) systems, 167, 173–176
in blanching, 189, 190–191, 195
whey, *via* lactase(p), 286
xylan, agricultural, *via* xylanase(p), 176, 177, 179–180
See also Pollution, Environment
Water, as enzyme extractant, 110
as lipase inhibitor, 446
evaporation processes for, 234–235
imbibition by seeds, and phospholipid biosynthesis, 586
in blanching, 192–195
in enzyme catalytic efficiency, 54
in enzyme inactivation, protein denaturation, 187, 263
microdroplets, in frozen food systems, 240
pesticides in, monitored by enzymatic analysis, 705
potato browning controlled with, 281, 337
removal in frozen foods, 222. *See also* Freezing, Frozen Food, Drying
restricted systems, *See* Water activity, low, Drying
state of, and enzyme behavior, 220, 223, 238, 243, 259
texture and, 462, 484, 520, 521, 605
transport medium at low a_w's, 241
Water activity—low, altered reaction pathways and specificity(p), 242–243, 659
and blanching, 191, 245
and enzyme stability, 243–245, 436, 437
and lipase action, 445, 446, 586
definition, significance, 231–233
enzyme behavior, consequences, in foods, 191, 236, 239–242, 367, 544, 586
in model systems, 237–242, 390
Water potential, 233
Water sorption isotherms, 232, 240, 241
Water-binding, absorption, (-holding capacity); a protein functional property, 614
and fish toughness, prevention, 499, 505
by sucrose, and substrate inhibition, 298, 299
in rennet milk clotting, 598
of muscle, enzymes, and meat texture, 468, 489
of sugars, and pickle firming, 529
Watercress, glucosinolate as stress metabolite in, 14
Waterless blanching, 193–195
Watermelon, ascorbate oxidase in, 289
glyoxylate cycle in, 351
Wattle bark, tannin from, 259, 661
Wax, plant, synthesis of, 390, 392
Waxing, of fruit, 511
of potatoes to prevent greening, 311
Waxy barley starch, self-liquefying, 156, 555
Weeping, in cheesemaking, 594
Weight control, reduction. *See* Diet(s)
Weight loss, in blanching, control of, 194
Wet cardboard, a fish off-flavor, 436
Whale meat color, bacterial, 319
Wheat, as phytase(p), 659
bran, enzyme(p) processing of, 627, 628
in koji, 616
bread aroma precursors, 407
distribution of enzymes in grain, 574
durum, *See* Durum wheat

dwarf, *See* Dwarf wheat
enzyme assay to ascertain history, quality, 680, 681
enzymes of, α-amylase, 573–576
β-amylase, 574, 576–578
glutamate decarboxylase, 679–689
lipase, 245, 445–446, 574
lipoxygenase, 301, 429, 587
nitrate reductase, 675–676
nucleases, 245, 366
peroxidase, 591
phenolase, 275, 414, 591, 679
proteinases, 17, 583
superoxide dismutase (SOD), 287
flour, *See* Flour, wheat
germ, 573, 585
germinated, as α-amylase(p), 575–576, 579
hard red spring, 575
heating of kernels, effects on enzymes, 244
miracle, *See* Dwarf wheat
morphology of kernel, 573
off-flavor and lipases of, 445–446
storage stability of intact grains, 237
See also Bread, Breadmaking, Cereals, Flour
Wheat malt. *See* Wheat, germinated
Whey and enzyme(p) for utilization, antiproteinases in, 606
functionalization of proteins of, 610
in cheesemaking, 594
incorporation into kvass, 625
lactose removal, 9, 643
resolubilization of protein of, 624
sweetener production, 180
Whippability, a protein functional property, 614
increase of, as dieting aid, 664
Whiskey production, 552, 561–562
White Muscat grapes, blanching of, 282
White mustard seed, protein, enzyme, from, 632
White water, clarification with immobilized enzyme, 5
Wholesomeness, food color as guide to, 21, 268
Wild game meat, off-flavor of, 446
Wild onion, flavor precursor of, 378
Wild rose juice, pectinase(p) in making, 537
Wilting, a blanching benefit, 188
and enzyme potentiation, 400
Windburn, enzymatic browning *via*, 287
Windshield wiper dialysis, in enzyme purification, 113
Windshield wiper effect, of supplementary enzymes in glucoanalysis, 158–159, 172–173, 696
Winemaking, enzymology of, 342, 360, 535–539, 540. *See also* Fruit juices, Grapes
Wines, biogenic amines, 641
bouquet, aroma genesis, enzymology of, 388, 389, 400, 413
classification, 552
color, loss, retention, enzymology of, 267, 275, 283, 296, 341, 696
methanol, enzyme produced, 637–639
off-flavor, aroma loss, enzymology of, 393, 423, 457
pasteurization of, by ultrasonication, 200
Winter gushing, of beer, 571
Wintergreen, glycosidase, and aroma of, 399
Withania coagulens, berry, rennet from, 94, 602
Withering, in tea production, aroma, 292, 400
Wood eating insects, xylanase, in, 178
Wood sorrel, oxalic acid sourness of, 354
Wool, as immobilized enzyme support, 145
Work, and dough texture, 511
Worker bee, invertase in hemolymph of, 358
Wort, in brewing, enzyme(p) treatment of, 553, 560, 561, 566–567
Wound debridement, with proteinase, 4
Wound periderm, 337
Wrinkled pea, high amylose and LOX in, 437

Xanthophyll, LOX-bleached, and retention, 263, 299, 301–302
Xenobiotics, 77. *See also* Xenobiotic oxidase *(Enzyme Index)*
X-ray crystallography, and cellulase action, 168
X-ray enzymology, and enzyme tertiary structure, 41–42
enzyme crystals for, 118
induced fit, evidence by, 44
myoglobin and, 317
vs active site modification, 56
X-ray fluorescence analysis, 706
Xylan, as affinity chromatography matrix, 178
cellulose accompaniment in agricultural wastes, 167
enzymatic(p) xylose production from, 179–180
enzymolysis of, 177–180
in bamboo shoot texture, 530
Xylem, of plant cells, 513
Xylene, in fruit aroma, 388
Xylitol, anticaries, and peroxidase, 177, 205
Xylose, glucose accompaniment, 176
glucose isomerase inducer, 163
in pectin side chains, 513
production, *via* xylanase(p), 179–180
with arginine, antioxidant formation, 437

Yaks, cheese from the milk of, 594
Yams, as the sixth most used food, 635
in distilling, 56
Yeast(s), acid-hydrolyzed protein from, 616
as enzyme(a) for sucrose, raffinose, 670
astaxanthin from *via* enzymes, 302
dextrin-fermenting, for dietetic beers, 564
fermentation, for egg desugaring, 340
growth and a_w, 235
in baking, 572, 577
in brewing, 552
in meaning of "enzyme," 4
lytic glucanases of, 329, 567, 628–629
off-flavor of, caused by, 421, 456
protein from, *via* autolysis, 626, 629
proteinase(p) for tenderizing meat, 492
tannase(p) from, 540
torula, for dough mellowing, 583
Yield, crop. *See* Crop yield
of malt, increased by gibberellins, 555

of oil, enzyme-caused increase of, 549
Yin-yang hypothesis, lipase activation and, 442
Yogurt, as a fermented food, 407
in meat flavoring manufacture, 415
lactoperoxidase and antibiosis, 204
Yolks, egg, color of, 301
egg, desugaring by fermentation, 340
Young modulus, 462

Z=. *See cis-*
Z-disc, 478, 479
Zebu, cheese from the milk of, 594
Zein, hydrolysates, bitterness and hydrophobicity of, 617
plastein from hydrolysates of, 620
Zenith, temperature, 228
Zero- or First-point, in enzyme assays, 672
Zero order kinetics, in enzyme assays, 208, 211, 670
in fluctuating temperatures, 227
in immobilized enzyme systems, 139
Z-gene, of β-galactosidase operon *(lac)*, 75
Zinc (salts), as enzyme cofactor, 45, 46, 57, 255, 287, 393, 616
as nutrient, 633, 656
enzyme activation analysis for, 707
enzyme assay for plant nutritional status, 677
in cellulolysis pretreatment, 176
in enzyme purification, 116
in model enzymes, 59
pheophytin complexes, 307
Zingiber officale. See Ginger
Zip code, signal peptidase as, 81
Zipper theory, of α-amylase action mechanism, 554
Zirconium, as immobilized enzyme support, 144, 165
Z-line, disc, muscle, and meat texture, 464, 465, 470, 476, 478
Zonal centrifugation, 114–115, 690
Zone electrophoresis, 260
Zone melting, in enzyme concentration, 109
Zone refining and affinity chromatography, 134
Z-protein, in chill haze, and β-amylase, 562, 563, 576
z-Values, definition, 187
z-Values of enzymes in foods, 206, 209, 211, 543
Zymogens. *See* Proenzymes *(Enzyme Index)*
Zymograms. *See* Isozyme patterns

Enzyme Index

The four-number code at the extreme right of the same line as each main (enzyme name) entry is the "EC number" assigned in Anon. (1979). An incomplete code denotes either (1) an entire class, subclass or sub-subclass, or (2) that the indexed enzyme has not been assigned an EC serial number, i.e., is not listed in Anon. (1979), but nevertheless belongs to the designated sub-subclass, in the author's judgment.

N.A. denotes either "not applicable" or "numbers not assigned" by either Anon. (1979) or by the author. The context of the entry should make clear which of the meanings applies.

More than one serial (fourth) number indicates that the same term has been assigned to more than one EC code; parentheses denote a degree of equivocacy or ambiguity.

The terms enzyme(a), = "(a)" and enzyme(p), = "(p)" are defined in both the *General Subject* and *Enzyme Indexes*. Terms which already denote enzyme(p), such as Pancreatin, Rennet, and Pronase, are not suffixed.

Acetate kinase (a), for acetic acid, acetate, 698 — 2.7.2.1
Acetoin dehydrogenase, in aromas, off-odors, 408 — 1.1.1.5
Acetoin racemase, and cheese aroma, 408 — 5.1.2.4
Acetolactate decarboxylase, and cheese aroma, 408 — 4.1.1.5
Acetolactate mutase, and cheese aroma, 408 — 5.4.99.3
Acetolactate synthase, and cheese aroma, 408 — 4.1.3.18
Acetylcholine esterase, low temperature behavior, 219 — 3.1.1.7
Acetyl-CoA cocarboxylase, biocytin as coenzymes, 45 — 6.4.1.2
Acetyl-CoA synthetase, as (a), for acetate, 698 — 6.2.1.1
  in ester biosynthesis, 390
β-*N*-Acetyl-D-glucoseaminidase, pasteurization index — 3.2.1.30
Acid α-amylase, in *Aspergillus* spp., 154 — 3.2.1.1
Acid hydrolyses, in lysosomes, 82, 474, 475 — 3.-.-.-
Acid microbial proteinases(p), uses, 602, 625 — 3.4.23.6
Acid phosphatase, assay, gibberellins, 558 — 3.1.3.2
  inhibition of, 254, 258
  milk, Irad effects, 199
  sugar, in translocation, 357
Acid proteinases, pepstatin inhibition, 257. *See also* Proteinases, Rennets — 3.4.23.-
Aconitase, in plant acid accumulation, 350 — 4.2.1.3
  prochiral specificity, 33
Acrosin, active site titration, 684 — 3.4.21.10
Actinidin, Chinese gooseberry proteinase, 94 — 3.4.22.14
Acyl amidase, herbicide degradation, 660 — 3.5.1.4

Acyl glycerol hydrolases. *See* Lipase
Acyl hydrolases, cycloamylose as model, 61 — 3.1.-.-
Acyl-CoA dehydrogenase, in β-oxidation, 391 — 1.3.99.3
Acyl-CoA long chain transferase, cutin synthesis, 390 — 2.3.1.-
Acyl-CoA reductase, in ester synthesis, 390, 391 — 1.2.1.42
Acyl-CoA thiohydrolases, as transferases, 391 — 3.1.2.-
  in ester synthesis, 382, 390, 392
Adenine deaminase, seafood, hypoxanthine, and, 366 — 3.5.4.2
Adenine phosphoribosyl transferase, cold, 224 — 2.4.2.7
Adenosine triphosphatases. *See* ATPases — 3.6.1.3,5,8
S-Adenosyl methionine regeneration enzyme, 286 — 4.2.1.(22)
Adenyl cyclase, in cascades, 73 — 4.6.1.1
Adenylate deaminase. *See* AMP deaminase — 3.5.4.6
Adenylate kinase, and fish quality, 680 — 2.7.4.3
  stability, purification, 112, 185, 263
Adenylosuccinate lyase, and 5′-nucleotides, 365–366 — 4.3.2.2
Adenylosuccinate synthetase, 5′-nucleotides, 365 — 6.3.4.4
ADH. *See* Alcohol dehydrogenase — 1.1.1.1
ADPglucose pyrophosphorylase, sugar-starch syntheses, 253, 334, 358 — 2.7.7.27
ADPDstarch glucosyl transferase. *See* Starch synthase — 2.4.1.21
Alanine decarboxylase, in tea leaves, 641 — 4.1.1.-
Alcohol dehydrogenase, as (a), 687, 691, 697, 707 — 1.1.1.1
  biochemistry, properties, 39, 112, 254, 255, 393, 427
  immobilized, for light beer, 565
  in flavor, 393–394, 418, 419, 423, 427, 439
  in plants used for food, beverages, 393–394
Alcohol dehydrogenase (NAD(P)$^+$), in melons, 394 — 1.1.1.71
Alcohol dehydrogenase (NADP$^+$), in fruits, tea, 46, 394 — 1.1.1.2
Alcohol oxidase(a), for alcohols, 697 — 1.1.3.13
Aldehyde dehydrogenase, as (a), for alcohols, 697 — 1.2.1.3
Aldehyde oxidase, liver, potato, 373, 424 — 1.2.3.1
Aldehyde reductase, 390, 393–394 — 1.1.1.1
Aldolase, affinity chromatography, 130 — 4.1.2.13
  as index, indicator, 670, 677
  low temperature behavior, 223, 230
  meat texture, 471, 480
Alginate lyase, 517 — 4.2.2.3
Alkaline α-amylase, 54, 261 — 3.2.1.1
Alkaline phosphatase, apo- as (a), for Zn, Cu, 707 — 3.1.3.1
  assay for contamination detection, 678
  biochemistry, properties, 216, 224, 258, 609
  fish muscle, 376, 506
  milk, dairy product technology, 211–212, 216–217, 680
S-Alkylcysteine lyase, 32, 36–37, 308, 376, 684 — 4.4.1.6
Alliin lyase. *See* Alliinase — 4.4.1.4
Alliinase, aroma genesis, 370, 374, 375, 380, 381 — 4.4.1.4
  biochemistry properties, 31, 32, 186, 374, 376–377, 380
  enzyme(a), for flavor potential, 671, 700
  enzyme(p), for vitamin B_1 adequacy, 645
  in garlic, onion, 374–376
  in off-color, 312–314
  in off-flavor, 423, 456
  in processing, 236, 261, 379
Amidophosphoribosyl transferase, and glutamate, 363 — 2.4.2.14
Amine oxidase, as (a), for spermine, spermidine, 701 — 1.4.3.4
  as enzyme(p), for amine removal, 642
Amino acid *N*-acyl hydrolase. *See* Aminoacylase
Amino acid ammonia lyases, Mn as cofactor, 46 — 4.3.1.-
Amino acid decarboxylases, biogenic amines, 640, 641–642 — 4.1.1.-
  in brain research, 641–642
  in cheese, sausage aroma, 409, 415
  in ester biosynthesis, 390, 391, 394
L-Amino acid dehydrogenase, in aroma, 373 — 1.4.1.5
D-Amino acid oxidase, 39, 700 — 1.4.3.3
L-Amino acid oxidase, as (a), for L-amino acids, 700 — 1.4.3.2
Amino acid transferases. *See* Aminotransferases
Amino acid-tRNA synthetases, affinity elution, 132, 220 — 6.1.1.-
Aminoacylase, in L-amino acid production, 150, 151 — 3.5.1.14
Aminoexopeptidases, in soy sauce making, 616 — 3.4.11.-
5-Aminolevulinate synthase, chlorophyll synthesis, 310 — 2.1.1.37
Aminopeptidase, in protein amino acid analysis, 710 — 3.4.11.11
Aminotransferases, *See also* individual enzyme — 2.6.1.-
  aroma genesis, 371, 373, 394, 401, 409, 413
  assays, diagnostic aids, 685
  flavoreses, 418
  glucosinolate biosynthesis, 382
  glutamate accumulation, 362
  off-odor production, 452
  plant acid accumulation, 351
  stereospecificity, 31, 32
  zymograms, for freeze-thaw detection, 682

AMP deaminase, allosteric inhibition by phytate, 656 — 3.5.4.6
enzyme(p), for seasoning, 364
muscle-to-meat conversion, 472, 475
5′-nucleotide accumulation, manufacture, 150, 151, 346
Amylase(s), *See also* Diastase, α-Amylase, β-Amylase — 3.2.1.-
as desmoenzymes, 138
enzyme(p) non-food uses, 5, 104
fish texture and, 449, 504
joint action, in cereal processing, 563, 578
NEB, color aroma and, 362, 415
specificities, 155, 556
stabilities, 113, 185, 264, 555, 556
texture-modifying enzymes, plant foods, 518, 519
α-Amylase, action, in dried foods, 237, 242 — 3.2.1.1
in frozen foods, systems, 221
in glucoamylase, pullulanase action, 158, 159, 563
on starch granules, 158, 159, 335
analytically related uses, aid in assessing dietary fiber, 710
flour quality, starch susceptibility, 681
for Cl⁻, as apoenzyme(a), 707
inhibition, as tartness measure, 706
assay, as predictor, 577–578, 581, 669, 679, 680
by immunochemical methods, 683
in clinical diagnosis, 695
of enzyme(p), and projected use, 100
bacterial, and NEB, 326
heat stability, species, 154, 155, 156, 185
uses of, in baking, 237, 576, 579
in brewing, distilling, 558–559, 561
other uses, 9, 155, 156, 335, 579, 664
bee, honey, 358, 679
biochemistry, isozymology, 554–555, 574, 575
kinetics, action mechanism, 34, 250, 257, 554, 575
properties, calcium in, 46
specificity, other, 46, 154, 155, 556
breadmaking, cereal processing, 194, 264, 577–581
brewing, 559–561
cereals, barley, malts, 9, 112, 554–556
rice, triticale, others, 273, 575
wheat, wheat malt, 573–576, 579
distilling, (whiskey production), 561–562
enzyme(p), sources mixed, unspecified, uses of, 539, 578–579, 582
fungal, acid, from *Aspergillus* spp., 154
acid inactivation differential, 263
acid stable, from *A. niger*, 539
enzyme(p), for alcoholic beverages, 5, 560, 561
for breadmaking, 576, 578–579
for other uses, 9, 576
of *Rhizopus delamar*, in Amylo process, 561
takadiastase, 155
immobilized, 5, 138, 157
in malting, 556–558
mammalian, pancreatic, 46, 257, 548, 575, 707
microbial, other aspects, enzyme(p) from; addition to, detection in, malt, 683
as feed additive, 625
for maltose production, 157
for protein processing, 623, 629
for tenderizing canned peas, 528
in non-amylase enzyme(p), 106, 337, 568
in pectin production, 549
Irad of, for safety, 198
milk, Irad effects, 199
potatoes, and cold sweetening, 335–336
sweet potato, others, potentiation, 359–360
weight control *via* inhibition of, 254, 257
zymograms, in food quality control, 682
β-Amylase, affinity chromatography, other purification, 112, 124 — 3.2.1.2
as enzyme(a), for starch, 696
as enzyme(p), 157, 560, 564, 577
bacterial, 124, 154, 155, 577
barley, malt, and in brewing, 560, 562–563, 564
immobilized, energy of activation, 148–149
in frozen foods, 221
in maltose syrup production, 157
latent, in developing grains, 562, 577
β-limit dextrin from, 154
molar activity, 50
other sources, 359, 577
stability, 112, 185
wheat, and in baking, breadmaking, etc., 576–578
Amyloglucosidase. *See* Glucoamylase — 3.2.1.3
Amylopectin-6-glucan hydrolase. *See* Pullulanase — 3.2.1.41
Amylopectin-1,6-glucosidase. *See* Pullulanase
Amylose phosphorylase. *See* Phosphorylase
Anomer isomerases, 338 — 5.3.1.-
Anserinase, in fish flavor, 416 — 3.4.13.5
Anthocyanases, in cocoa production, 294 — 3.2.1.-
in pectinase(p), and wine color, 536
secondary transformations of, 20
Anthranilate synthase, affinity chromatography, 129–130 — 4.1.3.27
Apyrase. *See* Adenosine triphosphatase — 3.6.1.5
Arabinosidase(s), as protopectinase component, 518 — 3.2.1.55
in cocoa manufacture, 296

microbial, in pickle oversoftening, 528
Arachain, peanut proteinase, 94, 492 3.4.99
Arginase, affinity chromatography, 124 3.5.3.1
induction, dietary overload, 76
Mn as cofactor of, 46
Arginosuccinase, cold-induced activation, 220 4.3.2.1
Aryl sulfatase, 257, 680 3.1.6.1
Asclepain, milkweed proteinase, 94, 492 3.4.22.7
Ascorbate oxidase, and nonenzymatic browning, 237 1.10.3.3
as enzyme(a), for ascorbate, dehydroascorbate, 700
assay, for plant nutritional status, 677
biochemistry, properties, 209, 251, 288–290
confusion with other oxidases, 289
Cu in, 46, 288
in enzymatic browning, 285
source reported in, 289–290
Ascorbic acid oxidizing enzyme, in breadmaking, 592 1.-.-.-
Asparaginase, purification for antitumor agent, 5, 134, 665 3.5.1.1
Aspartate aminotransferase (GOT), assay, kits for, as markers of development, 678, 686 2.6.1.1
stable ES complex of, 54
zymogram to detect freeze-thawing, 682
Aspartate kinase, allosteric regulation of, 83 2.7.2.4
ATP hydrolase. *See* Adenosine triphosphatase
ATPases (Adenosine triphosphatase, Apyrase), ATP synthesis (oxidative phosphorylation), 85, 285, 322 3.6.1.3,5,8
biochemistry, other, 6, 38, 46, 85, 366, 467, 597
in fish flavor, texture, 498, 505, 680
in meat color, flavor, 322, 415
in muscle-to meat conversion, texture, 284–285, 334
in potato color problems, 284–285, 334

BASAP (*Bacillus subtilis* alkaline proteinase), 509, 625 3.4.21.14
1,2-Benzenediol oxidase. *See* Phenolase
Benzoyl decarboxylase, cycloamylose as model of, 61 4.1.1.-
Bromelain, Bromelin, pineapple proteinase; enzyme(p), in breadmaking, 587 3.4.22.4
long considered safe, 647
meat tenderizer, 485, 492
protein processing, dieting, 509, 622, 664
inactivation, reactivation, 213
2,3-Butanediol dehydrogenase, as enzyme(a), 697 1.1.1.4
L-(+)-Butanediol dehydrogenase, and cheese aroma, 408 1.1.1.76

CAF. *See* Calcium activated factor
Caffeine demethylase, microbial, 661 N.A.
Calcium activated factor, for predicting meat texture, 680 N.A.
in aging of beef, and properties of, 478
in meat discoloration, 321
in meat protein functionality, 489
Callose hydrolase, in canned tomato firming, 525–526 3.2.1.58
Carbamoyl phosphate synthetase, allosteric regulation of, 69 6.3.4.16
Carbohydrases. *See* Glycosylases
Carbon monoxide dehydrogenase, Ni in, 46 N.A.
Carbonic anhydrase, active site, Zn and inhibition, 31, 46, 50, 58, 256 4.2.1.1
affinity chromatography of, 197
as aid in wine CO_2 analysis, 670, 711
assay by active site titration, 684
assay of, for plant Zn status, 677
molar activity, high, 50, 209–210
Carboxydismutase. *See* Ribulose bisphosphate carboxylase
Carboxylases, Carboxy-lyases, 28, 30, 257 4.1.1.-
Carboxylesterase, (carboxylic-, carboxy-), *See also* Carboxylic ester hydrolases 3.1.1.1
immobilized for pesticide removal, 660, 705
in orange juice off-odors, 393
lipase and, 440
zymograms of, in food quality control, 682
Carboxylic ester hydrolases, and flavorese, 418 3.1.1.-
as enzyme(a), for triglycerides, 697
as enzyme(p), 9, *See also* Lipase
as pectinase(p) contaminant, 393
cutin depolymerase as, 519
in aroma genesis, 371, 390, 393, 400, 410
lipase as, 440
low a_w behavior, 237, 241, 242, 244, 245
suicide substrates for, 253
Carboxypeptidase(s), as enzyme(p), and uses, 423, 618, 701 3.4.17.-
biochemistry, catalytic efficiency, 31, 41, 42, 46, 47, 52, 57
immobilized, for probing casein micelle, 596
in brewing, 569
in meat texture, 476

sources of, 569, 582, 618
Carboxypeptidase A, in beef muscle, 480 — 3.4.17.1
Carboxypeptidase G and G-1, affinity chromatography of, 134 — 3.4.22.12
in glutamate accumulation, 362
Carnosinase, and eel flavor, 416 — 3.4.13.3
Catalase, as unorganized ferment (historical), 4 — 1.11.1.6
assays, alternative methods, 670
as blanching adequacy index, 210–211, 669
as food plant quality indices, 677, 678
to identify irradiated potatoes, 337, 680
biochemistry—properties, as enzyme stabilizer, 103
catalysis, specificity, Fe in, 31, 46, 50, 203
enzyme-substrate complex of, 203, 220
inhibition, inactivation, regeneration, 213, 257, 258, 663
thermal stability, inactivation, 50, 51, 209, 216, 225
breadmaking, 591
citric acid accumulation in citrus, 354
color and, 288, 303, 323, 324
enzyme(a), for elemental analysis, H_2O_2, 700, 707
enzyme(p), for H_2O_2 removal from cheese, other, 5, 629, 639, 663
in glucose oxidase(a,p), 338, 341, 342
low a_w, dried food, behavior, 237
low temperature, frozen food, behavior, 220, 222, 223, 224
other uses, suggested, 323, 324, 591, 663
production, purification, safety, 91, 92, 95, 97, 110, 112, 117, 647
sources, mammalian, 9, 91, 92, 644, 663
microbial, 95, 97, 110, 210, 223, 591, 647
Catecholase, catechol oxidase, 270, 271 — 1.10.3.1
Cathepsin(s), affinity chromatography of, 136 — 3.4.-.-
as enzyme(p), for tenderizing meat, 492
assay, to monitor autolysis, 680
biology, classification, properties, 130, 473–475, 476, 480
fish, and FPC, 506–507, 508
in frozen meat, behavior, release, 221, 483
in meat aging, texture, 466, 474, 476, 478–480, 482
in red meat discoloration, 321
Irad effects on, in liver, 199
Cathepsin B, and meat texture, 473, 474, 482 — 3.4.22.1
Cathepsin C, 473, 474, 480 — 3.4.14.1
Cathepsin D, 473, 474, 480 — 3.4.23.5
Cell separating enzymes, as enzyme(p), 548–549, 551 — N.A.
Cell wall degrading enzymes, as enzyme(p), 548–551 — N.A.
Cell wall modifying enzymes, protopectinase as, 518–519
Cellobiase, properties, role in cellulolysis, 171, 172, 173, 174 — 3.2.1.21
as enzyme(a) for cellulose, 696
Cellobiohydrolase (cellobiosylhydrolase). *See* Exo-cellobiohydrolase
Cellobiose:quinone oxidoreductase, 172, 276 — N.A.
Cellulose, affinity chromatography, in purification, 114, 124 — 3.2.1.4
biochemistry, properties, 167–172
C_1-, cellulolysis initiator, 171–172 — 3.2.1.91
C_x-, endocellulase, 171–172, 258, 519, 522, 524
enzyme(a), for cellulose, 696
enzyme(p), and preservation, 259, 549, 663
cell-separating uses of, 398, 548–550
for cellulolysis (biomass conversion) to glucose, 9, 174–176
for date improvement, 520
for nutritional assessment, improvement, 627–628, 629, 643, 664, 709
in breadmaking, baking, 9, 581
in confection making, 9, 106
miscellaneous uses of, 9
production, safety of, 95, 168–169, 549, 647
in texture, 518, 519, 522, 524, 528–529, 545
microbial rennet contaminant, 105
sources, fruits, 519, 520, 522
fungi, 95, 168–169, 259, 528, 545, 627
invertebrates, 168, 178
vegetables, 524, 528, 529, 545
Cellulolysin, enzyme(p) for protoplast preparation, 398 — N.A.
Chalcone isomerase, and classification problems, 30 — 5.5.1.6
Chitinase, 180, 709 — 3.2.1.14
Chloroperoxidase, chloride peroxidase, 203 — 1.11.1.10
Chlorophyll *a* dehydrogenase, 311 — N.A.
Chlorophyll *b* reductase, 309 — N.A.
Chlorophyllase, and color retention in green vegetables, 18, 305, 307, 308 — 3.1.1.14
as potential health hazard, 643
as biochemistry, properties, 307–308, 310
thermal potentiation of, 186, 308
Cholesterol esterase, 442 — 3.1.1.13
Cholesterol oxidase, as enzyme(a), for cholesterol, 700 — 1.1.3.6
Choline aliesterase. *See* Choline esterase
Choline esterase, homologous inhibition of, 250 — 3.1.17.8
in active site quest, 40, 134
in affinity chromatography, 128, 134

in enzymatic inhibition analysis, for pesticides, 704–705
Chorismate mutase, affinity chromatography of, 130 5.4.99.5
Chymopapain as enzyme(p), as a rennet, 603 3.4.22.6
for making acidic protein beverages, 625
for treating spinal injuries, 4–5
in tenderizing meat, 484, 485–486
Chymosin, action on casein and biochemistry, 594–595 3.4.23.4
as enzyme(p), *See* Rennet(s)
gene, for rennet production *via* genetic engineering, 603–604
Chymotrypsin, active site quest, catalytic efficiency, 39, 40, 52, 60, 254 3.4.21.1
as clean-up aid in analysis, 711
assay *via* active site titration, 684
immobilization, pH-activity profiles, 146, 147
in affinity chromatography, 128, 129, 131, 133, 135
in plastein formation, 619
inhibition, inactivation, 40, 187, 201, 274, 704
isozymes from one chain, 48
protein digestion assessment with, 708
specificity, substrates for, 49, 486, 603
Chymotrypsinogen. *See* Proenzymes
Cis-trans isomerases, in terpene biosynthesis, 404–405 5.2.-.-
Citrate cleavage enzyme, in plant acid accumulation, 350 4.1.3.8
Citrate lyase, as enzyme(a), for acetate, citrate, 698, 699 4.1.3.6
Citrate synthase, as enzyme(a), for oxalacetate, 691 4.1.3.7
in plant citric acid accumulation, 350, 354
Clearing factor lipase. *See* Lipoprotein lipase 3.1.1.34
Coenzyme hydrolases, in aroma genesis, 415, 418 3.-.-.-
Collagen glycosyl transferase, affinity chromatography of, 129 2.4.-.-
Collagenase, affinity chromatography of, 124, 126 3.4.24.3 3.4.24.7
as enzyme(p), for tenderizing meat, 487, 492, 495
for treating burns, 4
fish, 506
from cured hide microorganisms, 496
in meat aging, 475
thermal potentiation, for tenderization, 18, 186, 463, 483, 484
Correxonuclease, purification of, 116 3.1.11.6
Creatinase(a), for creatinine, creatine, 701 3.5.3.3
Creatine kinase, apo-, as enzyme(a) for Mg, 707 2.7.3.2
assay, to predict PSE, 680
for muscle ATP replenishment, 467–468
Creatinine deiminase, enzyme electrode, for creatinine 3.5.4.21
Cresolase, *vis-à-vis*, catecholase, phenolase, 270, 274 1.14.18.1
C-S lyases. *See* *S*-Alkylcysteine lyases, Alliinase
Cucurbitacin 19-hydroxylase, for debittering squash, 456 1.13.-.-
Cutinase, in fruit drying, texture, 518, 519, 520 3.1.1.-
β-Cyanoalanine synthase, for HCN removal, 637 4.4.1.9
Cyclic AMP phosphodiesterase, Ca-transport regulation by, 467 3.1.4.17
inhibitors of, 73, 332, 442
protein activator of in cascades, 467
Cyclodextrin glucanotransferase, 95, 643 2.4.1.19
Cyclooxygenases, 40. *See also* Prostaglandin synthase 1.14.99.-
Cysteine lyase, yeast, and H_2S in beer, 423 4.4.1.10
Cysteine lyases. *See* Alliinases, *S*-Alkylcysteine lyases
Cysteine synthase, 376 4.2.99.8
Cytase, 49, 168. *See also* specific glycanase 3.2.1.-
Cytochrome b_5 reductase, in disease, in off-flavor, 317, 435 1.6.2.2
Cytochrome *c* oxidase, affinity chromatography of, 133 1.9.3.1
assay to detect freeze-thawing, 680
Cu in, metalloporphyrin models of, 46, 61
in ham curing, 324
Cytochrome *c* peroxidase, 203, 204 1.11.1.5
Cytochrome *c* reductase, 113, 223, 324 1.6.99.3
Cytochrome P450-linked oxidase. *See* Xenobiotic oxidase
Cytolytic enzymes. *See* Cell-separating enzymes

Deacylase. *See* Aminoacylase
Deaminases, and fish toughness, 504, 506 4.3.1.-
Debranching enzymes. *See* Isoamylase, Pullulanase
Decarboxylases, controlled atmosphere storage, 531 4.1.1.-
enzyme models for, 61
Dehydratases, hydro-lyases, 393 4.2.1.-
Dehydroascorbate reductase, wheat, in breadmaking, 592 N.A.
Dehydrogenases, *See also* specific enzymes 1.-.1.- (mostly)
and red meat color, 320–321

as enzyme(a), and the Optical Test, 686–687
freeze-thawing, activity during, 222
immobilized, 140, 439
in affinity chromatography, 128, 129, 133
in off-flavor genesis, removal, 423, 452
NAD^+, $NADP^+$, as coenzymes, cosubstrates, 45, 46
Dehydropeptidase, purification. *See also* Aminoacylase 3.5.1.14
Depolymerase, products as substrates, inhibitors, 250, 255 N.A.
Dextranase(p), in sugar refining, 360 3.2.1.11
Diacetyl reductase. *See* Acetoin dehydrogenase 1.1.1.5
Diacylglycerol lipase, 441. *See also* Lipoprotein lipase 3.1.1.34
Diamine oxidase, as enzyme(a), for diamines, lysine, 700, 701 1.4.3.6
as enzyme(p) for diamine removal, 642
inhibitors, medication, 640
L-α,β-Diaminopropionate lyase(a), for *Lathyrus* toxin, 701 4.3.1.-
Diaphorase(a), for NAD^+ regeneration, UV-to-visible, 688, 699 1.6.4.3
Diastase [α- + β- (+ gluco-) amylases], as enzyme(a), for brewing adjunct assessment, 709 3.2.1.1–3
for starch, 670
assay, for flour gassing power prediction, 669
for honey quality, history, 669, 679
historical connotation of term, 3, 4, 5
Dimethylallyl transferase, in terpenoid biosynthesis, 403 2.5.1.1
Dimethylglycine dehydrogenase, a HCHO producer, 503 1.5.99.2
Dioxygenases, nitrocatechol inhibition a mark of, 425 1.13.11.-
Dipeptidases, and cheese biogenic amines, 642 3.4.13.11
in cheese starter bacteria, 607, 608
o-Diphenol oxidase, *o*-Diphenolase, 270. *See also* Phenolase 1.10.3.1
p-Diphenolase, 270, 271. *See also* Laccase 1.10.3.2
Diphosphatase. *See* Inorganic pyrophosphatase 3.6.1.1
Disulfide-thiol interchange enzyme, and blanching, 195 5.3.4.1
DNA gyrase. *See* DNA topoisomerases 5.4.99.-
DNA joinase (ligase) in genetic engineering, 9, 28, 97, 98, 99 6.5.1.1
DNA repair exonuclease. *See* Correxonuclease
DNA replicase, 28 2.7.7.7
DNA topoisomerases, synaptase, 28, 76 5.4.99.-
DNases, affinity chromatography, 128 3.1.21.- to 3.1.25.-
apo-, as enzyme(a), for Mg, Zn, 707
in 5′-nucleotide accumulation, 365
site-specific, restriction, in genetic engineering, 97, 98
DT-diaphorase, in red meat color retention, 323 1.6.99.2

Elastase, assay, *via* active site titration, 684 3.4.21.11
low-temperature behavior, intermediates, 226
meat texture, 487–488
specificity of, 486
Emulsin (β-Glucosidase + Hydroxynitrile lyase, as enzyme(a) electrode, for amygdalin, 699
as enzyme(p), for debittering *Prunus* products, 423
biochemistry, properties, preparation, 185, 396, 397–399
HCN from, as food toxicant, 634–637
in history of enzymology, 4, 28, 396
in marzipan making, 399
Endoamylases. *See* α-Amylase
Endo-1,3-β-D-glucanase, lytic, 302, 628, 629 3.2.1.39
Endo-1,3(4)-β-D-glucanase, in barley in brewing, 565–567, 696 3.2.1.6
Endocellulase, 170, 172, 520. *See also* Cellulase 3.2.1.4
Endonucleases, in 5′-nucleotide accumulation, 366 3.1.26.-
Endopectate lyase, 517 4.2.2.2
Endopeptidases. *See also* Proteinases 3.4.99.-
barley, 568
in soy sauce production, 616
in wheat, 582, 583
milk, 606
Endopolygalacturonase, EndoPG, *See also* Polygalacturonase 3.2.1.15
as cell-separating enzyme(p), 550
immobilization of, 539
in Freestone and Clingstone peach textures, 519
in juice clarification, 537–539
in nectar making, ineffective, 544
tomato, in Hot Break, 546
tomato, in ripening, texture of, 524
Endoxylanase. *See also* Xylanase, 178 3.2.1.32
Enolase, purification, models of, 61, 112 4.2.1.11
Enoyl-CoA hydratase, in hydroxyacid biosynthesis, 411 4.2.1.17
Enzyme(a), enzyme preparations for enzymatic analysis. *See* Enzymatic analysis, *General Subject Index*
Enzyme(p), enzyme preparation(s) added or proposed to be added to foods, as medicines. *See* Enzyme preparations in *General Subject Index*

Esterase, oral, pregastric. *See* Lipase, oral
Esterases. *See* Carboxylesterase, Carboxylic ester hydrolases
Ethanolamine kinase, Na as cofactor, 46 — 2.7.1.82
Ethylene synthase, 286 — N.A.
Euphorbain, caper proteinase, 94 — 3.4.99.7
Exo-amylases. *See* β-Amylase, Glucoamylase
Exo-cellobiohydrolase, Exo-cellulase, 170, 171–172, 173 — 3.2.1.91
Exo-α-1,4-glucan hydrolases, 106, 154 — 3.2.1.1,3
Exo-1,3-β-D-glucosidase, 525–526 — 3.2.1.58
Exo-1,4-α-D-glucosidase, *See* Glucoamylase — 3.2.1.3
Exo-maltohexaohydrolase, 154 — 3.2.1.98
Exo-maltotetraohydrolase, 154 — 3.2.16.-
Exonucleases, 366 — 3.1.4.1
Exopectate lyase, Exopolygalacturonate lyase, 515, 516, 517–518 — 4.2.2.9
Exo-peptidases, *See also* Carboxypeptidases — 3.4.17.-, 3.4.23.6
  and biogenic amines in cheese, 642
  bitter peptide removal with, 618
  glutamate in soy sauce *via*, 616
  in meat texture, 475
  syn distortion of *trans* substrates by, 52
Exopolygalacturonase, biochemistry, properties, 516, 517 — 3.2.1.67
  in fruit texture, 523, 528
  in vegetable texture, consistency, 528, 546
  microbial enzyme(p), for seed coat separation, 549
Exo-pullulanase, 154, 158, 159. *See also* Pullulanase — 3.2.1.41
Exo-xylanase, 179. *See also* Xylanase — 3.2.1.32

Fatty-acid peroxidase, in aroma genesis, 204, 395, 411, 435 — 1.11.1.3
Fattyacyl NADPH-linked reductase (desaturase), 395 — N.A.
Ferriglobin reductase, 317 — 1.6.2.4
Ficin, fig proteinase, 93, 94 — 3.4.22.3
  action on meat, 488
  as enzyme(p), for fish solubilization, 509
    for meat tenderizing, 485, 492
  milk clotting:peptidase ratio, 603
Flavokinase, affinity chromatography of, 124
Flavonone oxidase, 456 — N.A.
Flavonone synthase, 455, 456 — N.A.
Flavoprotein oxidases, as enzyme destabilizers, 103 — 1.-3.-.-, 1.14.-.- (mostly)
  as H_2O_2 generators, 203
  as superoxide anion generators, 286, 287
  ascorbate oxidation mediated *via*, 289
  in biogenic amine catabolism, 640
  in lipid oxidation, off-flavor, 345
Flavoprotein-linked monooxygenase. *See* Xenobiotic oxidase — 1.14.14.1
Flavorese, and dairy off-odors, 452 — N.A.
  and enzyme regeneration after blanching, 196, 216
  concept, progress, prognosis, 416–420
Formate dehydrogenase, Se in, 46 — 1.2.1.2
  as enzyme(a), for formic, oxalic acids, 697, 698
Fructofuranosidase, β-D, not in honey, 358, *See also* Invertase — 3.2.1.26
  increase during ripening, 679
Fructose-bisphosphatase (-diphosphatase, in plant sugar accumulation, 356 — 3.1.3.11
  regulation in gluconeogenesis, 71, 82
Fumarase, in enzyme stereospecificity, 33 — 4.2.1.2

GAD. *See* Glyceraldehyde phosphate dehydrogenase
Galactanases, in protopectinase complex, 519 — 3.2.1.89,90
Galactose dehydrogenase, (a), for galactose, 695 — 1.1.1.48
  (p), for D-galactose removal in determining L-galactose, 712
Galactose oxidase, as enzyme(a), not for galactose, and specificity, 695 — 1.1.3.9
  as enzyme(p), for galactose, galactoside removal, 654
  immobilized, 692
  in affinity chromatography, 124, 135
  inhibition, activation, 204, 339
α-Galactosidase and α-D-galactosidase, as enzyme(a), for raffinose, 670, 695 — 3.2.1.22
  immobilized in fungi, in sugar refining, 150, 151, 360
  in beans, and flatulence, 199, 651, 653–654
  in honey, and α-glucosidase, 358
  in plant lysosomes, protopectinase, 518, 653
β-Galactosidase, *See also* Lactase
  analytically related uses of, 474, 475, 688, 692, 695, 714 — 3.2.1.23
  emulsin and, 398
  enzyme(p), uses, *See* Lactase
  non-enzyme(p) in cocoa-, cheese-making, 296, 408
  properties, provenance, 130, 226, 255, 296, 408, 520, 651, 695

regulation, genetic, lactose synthesis, 70, 75	
Galactosyl transferase. *See* Lactose synthase	2.4.1.22
Galacturonokinase, in pectin biosynthesis, 515	2.7.1.44
Geraniol dehydrogenase, 403	1.1.1.-
GI. *See* Glucose isomerase	5.3.1.18
Globin reductase, in red meat color retention, 323	1.6.2.4
Glucan hydrolases, listing, uses of enzyme(p), 9, 154	3.2.1.-
β-Glucan solubilase, 565	3.2.1.6
Glucanase(s), in clam style, as enzyme(p), 91	3.2.1.-
in cocoa manufacture, 294, 403	
in *Trametes*, for protein solubilization, 627	
β-Glucanase(s). *See* individual enzymes	
β-1,3-Glucanase, for mycelial content of shoyu koji, 709	3.2.1.58
Glucoamylase, as enzyme(a), for starch, 696	3.2.1.3
as enzyme(p), as analytical aid, 712, 714	
in baking, cereal processing, 576, 581	
in brewing, distilling, light beer, 553, 561, 564	
in glucose, maltose production, 157–158	
miscellaneous uses, 9, 342, 629	
biochemistry, properties, 154, 158, 159, 294	
production, purification of, 95, 106, 135	
safety of, 647, 648	
Glucoinvertase, in honey, 358	3.2.1.20
Gluconate kinase(a), for gluconate in beer, 695–696	2.7.1.12
Gluconate-6-phosphate dehydrogenase(a), for gluconate, 696	1.1.1.69
Gluconolactonase, in GOX(p), 338	3.1.1.17
Glucosamine phosphate isomerase, in glutamate accumulation, 363	5.3.1.10
Glucose dehydrogenase, immobilized, for NAD^+ regeneration, 140	1.1.1.47
Glucose isomerase, as enzyme(p), in fructose syrup production, 166–167	5.3.1.5,18
with invertase(p), for pure fructose, 160	
biochemistry, properties, 30, 46, 58, 162–263, 185	
immobilization and immobilized, 150, 151, 164–165	
production, purification, 95, 162, 163–164	
safety of, 647	
Glucose oxidase, as enzyme stabilizer, 103	1.1.3.4
as enzyme(a), 4, 34, 670, 692, 696, 707	
as enzyme(p), usually with catalase(p), as chemical analysis aid, 712	
for egg white desugaring, 340–341	
for food color control, 291, 323, 326, 337, 342	
for food flavor control, 423, 450	
in beer, wine, 291, 570, 696	
in breadmaking, 591	
in cheesemaking, 600	
non-food uses, 5, 342	
other food uses, 9, 341–342, 662	
production, purification, 110, 135	
safety of, 647, 648	
biochemistry, properties, 46, 244, 338–339, 398	
immobilization, immobilized, 338, 342, 692, 693	
Glucose-6-phosphatase, in GOX(p), 337	3.1.3.9
Glucose-6-phosphate dehydrogenase, as enzyme(a), 687, 688, 691, 694	1.1.1.49
in sucrose accumulation in plants, 375	
isozymes of *via* genetic variation, 47	
label in enzyme immunoassay, 713	
Glucose-6-phosphate isomerase, 159	5.3.1.9
α-D-Glucosidase, honey invertase as, 358	3.2.1.20
in potato cold sweetening, 336	
in raw starch granule digestion, 159	
oligosaccharide TSAI inhibitor of, general, 254	
See also Maltase	
β-D-Glucosidase, *See also* Cellobiase	3.2.1.21
as emulsin component, 398, 399	
an enzyme(p) for detoxifying safflower meal, 651, 654	
as flavorese, 420	
in food, as potential health hazard, 635–636	
inactivation, survival, inhibition, 185, 191, 254	
low temperature behavior, 226	
naringinase component, 455	
vanillin production, 399, 400	
Glucosinolase (Thioglucosidase), as enzyme(a), for ascorbate, glucosinolate, 686, 699, 700	3.2.3.1
as enzyme(p), flavorese, 416, 417, 418	
as health benefit, 633	
as health hazard, 384, 436, 445, 632–633	
ascorbate activation of, ascorbinogen, 45, 384, 385, 386, 646, 700	
biochemistry, transformations, 380, 383–384, 385, 467	
in flavor genesis, 382, 383, 386, 417, 456, 686	
α-Glucosyl transferase, in off-flavor genesis, 243, 430, 456	2.4.1.-
prevention of action, 436, 632, 633	
stevioside glucoside from, 361	
β-D-Glucuronidase, affinity chromatography, other purification, 108, 130	3.2.1.31
as aid in analysis for diethylstilbestrol, 712	
assay, to monitor autolysis, 680	

double pH optima of, 261
Glutamate decarboxylase, as enzyme(a), for glutamate, 699 — 4.1.1.15
cold-induced inactivation of, 220
wheat, and assay for storage history, 641, 680
Glutamate dehydrogenase, as (a), for glutamate, 699 — 1.4.1.2
in gustator accumulation, 362
isozymes of, *via* subunit polymerization, 48
Glutamate synthase, in gustator accumulation, 362, 363 — 1.4.7.1
Glutamate-pyruvate aminotransferase(a), 688 — 2.6.1.2
Glutamic-oxalacetic transaminase, GOT. *See* Aspartate aminotransferase
Glutaminase, in gustator accumulation, 363, 381 — 3.5.1.2
in meat texture, 362
Glutamine synthetase, affinity elution of, 132 — 6.3.1.2
regulation feedback, at genetic, substrate levels, 70, 76
regulation *via* cascade control, 73–74
Glutamine utilizing enzymes, 362, 363 — N.A.
γ-Glutamyl carboxylase, vitamin K coenzyme for, 45–46 — 4.1.1.-
γ-Glutamyl hydrolase, 362, 381 — 3.4.22.11
γ-Glutamyl transferase, transpeptidase, as (a), for *Allium* flavor potential, 700 — 2.3.2.2
as (p), and flavorese, 381, 417
in milk flavor, 451
in onion, other *Allium* spp., mushroom, flavor, 237, 380–381
in plant stress response, 30
specificity, 15
Glutathione *S*-epoxytransferase(a), for mycotoxins, 701 — 2.5.1.18
Glutathione peroxidase, and flavor, 438
LOOH as co-substrate of, 427
metmyoglobin reduction and meat color, 317, 320, 324
Se in, and biological functions, 46, 203, 204, 707
Glutathione reductase, affinity chromatography of, 128 — 1.6.4.2
as (a), for mycotoxins, 701
Glutathione *S*-transferase, herbicide degradation by, 660 — 2.5.1.18
Glycanase, inhibition by melanins, 518 — 3.2.1.-
Glyceraldehyde phosphate dehydrogenase, as glycolytic oligomer, 71 — 1.2.1.2
cold adaptation of, in cold sweetening, 230, 333
in muscle-to-meat, 472
in myogen A, a multienzyme complex, 69, 117
NCIs of, product and analog, 255
Glycerol ester hydrolase. *See* Lipase
Glycerol kinase(a), for triglycerides, 696–697 — 2.7.1.30
Glycine reductase, Se in, 46 — 1.4.2.1
Glycogen synthase, a gluconeogenetic enzyme, 70 — 2.4.1.11
Glycolate oxidase, in acid accumulation in plants, 350, 355 — 1.1.1.31
in chlorophyll degradation, 305
in lipid oxidation, off-flavor, 435
in photorespiration, 305, 355, 361
Glycosidase(s), Glycosylase(s), Glycoside hydrolases, *See also* individual enzymes, classes — 3.1.1.-
anthocyanases as, in cocoa, 294, 295, 296
as flavoreses, in off-flavor removal, 418, 427, 539
behavior at low a_w's, 238, 240
enzyme(p), miscellaneous uses of, 20
in essential oil, flavorant production, 399
in NEB, 327
in tomato texture, 525
inhibition by chiral analogs, 249
with both glucosidase, galactosidase activities, 358
Glyoxalase I, affinity elution, 132 — 4.4.1.5
resemblance to TMAO demethylase, 503
Glyoxalase II, 503 — 3.1.2.6
GMP synthetase, 363 — 6.3.4.1
GOT. *See* Aspartate aminotransferase
GOX. *See* Glucose oxidase
Gumase. *See* Pentosanase

Hemicellulases, *See also* Pentosanase, Protopectinases, Xylanase — 3.2.1.-
as enzyme(a), sheep rumen filtrate, 696
as enzyme(p), as digestive aids, 664
cell-separating applications, 548, 550
commercial readily available in Japan, 104
for starch isolation, 582
in protein processing, 627, 629
various other uses, 9
as protopectinase component, 518, 519
in brewing, 553
in juice clarification, 539
in oversoftening of gherkin pickles, 528
in vegetable texture modification, 518, 519, 528, 547
in wheat and pentosan solubilization, 581
Heparin lyase, Heparin sulfate lyase, 517 — 4.2.2.7,8
Hesperidinase(p) (β-Glucosidase + β-Rhamnosidase), for hesperidin crystals removal, 9 — 3.2.1.21, 3.2.1.43

Irad of, 198
value of, 104
Hexenal isomerase, 372 5.3.3.-
Hexokinase, as enzyme(a), for sugars, hexitols, 688, 691, 696, 697 2.7.1.1
-ADP PEP route of analysis, 696, 697, 699, 701
as glycolytic oligomer, 70
in sucrose accumulation, translocation, 357
in texture of meat, fish, 471, 499
purification of muscle, 108
stable ES complex, 39
Hexose diphosphatase, a glycolytic oligomer, 70 3.1.3.11
Hexose oxidase, liver, Cu in, 46 1.1.3.5
Hexose phosphate isomerase, in sucrose accumulation, 357 5.3.1.19
Histaminase, and toxicity of histamine, 640 1.4.3.6
Histidine ammonia lyase(a), enzyme electrode, 700 4.3.1.3
Histidine decarboxylase and biogenic amine in cheese, 642 4.1.1.22
Hurain, jabillo proteinase, 94 3.4.99.9
Hyaluronate lyase, 517 4.2.2.1
Hyaluronidase(a), for elemental analysis, 703
Hydratases, in accumulation of alcohols, 393 4.2.1.-
Hydrogen bondase, in polymer degradation, 172 N.A.
Hydrolases, lysosomal, 82, *See also* Lysosomes, enzymes 1.-.-.-
as transferases, 30
Hydroperoxide isomerase, 300, 301, 302, 303, 304, 427, 428, 429–430, 438 5.3.99.1
Hydroxyacyl glutathione hydrolase. *See* Glyoxalase II 3.1.2.6
Hydroxymandelonitrile lyase. *See* Hydroxynitrile lyase 4.1.2.11
Hydroxymethyl glutaryl CoA reductase, and feedback, 76 1.1.1.88
in terpenoid biosynthesis, 403
Hydroxymethyl glutaryl synthase, and terpenoids, 403 4.1.3.5
Hydroxynitrile lyase, emulsin component, 398–399, 636 3.5.5.1

D-Iditol dehydrogenase, 360 1.1.1.5
Imine hydrolase, 496 3.5.99.-
Iminodipeptidase, 415 3.4.13.8
Incorporealase. *See* DNA topoisomerases
Indole acetic acid oxidase, a peroxidase, 31, 204, 522 1.11.1.7
Inorganic pyrophosphatase, as enzyme(a), for pyrophosphate, 699 3.6.1.1
in amylase(p), as aid in vitamin B_1 analysis, 670, 711
in monoterpene biosynthesis, 403
in 5′-nucleotide accumulation, 366
myo-Inositol kinase, 30 2.7.1.64
myo-Inositol oxygenase, in pectin biosynthesis, 515 1.13.99.1
Invertase [Invertin(e)], aroma genesis, 402, 405 3.2.1.26
as enzyme (a), for elemental analysis, 707
for sucrose, 292, 670, 685, 686, 692
as enzyme (p), for date upgrading, 358, 520
for invert sugar confection, other uses, 5, 9, 19, 159–160
production purification, 95, 118, 130, 160
as glycoprotein, 46, 103
as texture-modifying enzyme, in dates, 358, 518, 519
assay of as correlates, markers, indices of, barley malting quality, 680
black spots of potatoes, 288
honey quality, history, 679
ripening of plant foods, 288 (by EIA), 715
cut flower longevity, 358
de novo synthesis during ripening, 679
gluco-, an α-glucosidase in honey, 358 3.2.1.20
glucosyl transfer by, 238
immobilized, 138, 160, 183, 292
in sucrose translocation, accumulation, 357–358
in sugar cane tassels, 360
in yeast, 95, 160, 360–361
inhibition, inhibitors, 255, 328–329, 331
isozymes, 84, 357
kinetics-related, 50, 138, 238
low a_w behavior, 238–239, 240
potato, and NEB of products, 327, 329, 331
temperature-activity profile, 183
Isoamylase, in cold sweetening, 336 3.2.1.68
in fish texture, 449
specificity, source of enzyme(p), 154, 155
Isocitrate dehydrogenase(a), apo-, for Mg, Zn, Mn 1.1.1.41
for isocitrate, juice adulteration, 699
Isocitrate lyase, Irad effects on, bananas, 199 4.1.1.31
Isomerases, specificities, metal role, 29, 30, 58 5.-.-.-
Isopullulanase, 154 3.2.1.57

Ketoacyl CoA thiolase, in β-oxidation, 391 — 2.3.1.16
β-Ketoacyl CoA hydrolase, in 2-methyl ketone synthesis, 411 — 3.1.2.-
β-Ketoacyl CoA transferase, in cheese aroma, 411 — 2.8.3.5
α-Ketodecarboxylase, and aldehyde genesis, 371, 373 — 4.1.1.1
  and cheese aroma, off-odor, 409, 452
  and hot beverage aroma, 401
α-Ketodecarboxylases, and 2-methyl ketones, 411 — 4.1.1.-
α-Ketoglutarate dehydrogenase, 69, 707 — 1.2.4.2
Ketone monooxygenase, in ester biosynthesis, 392, 711 — 1.14.13.-
Kinases, Mg in, in affinity chromatography, 58, 128 — 2.7.-.-

Laccase, in lignin biosynthesis, 276, 529 — 1.10.3.1
  properties, provenance, 30, 46, 271, 273, 274
  relation to ascorbate oxidase, 288
Lactase(p), *See also* β-Galactosidase — 3.3.1.23
  breadmaking uses, 577
  dairy products uses, 450, 594, 600, 610, 643, 652–653
  health-related uses, problems, 643, 652–653, 664
  immobilized, 180, 251, 292, 653, 692
  production, provenance, 98, 252, 610, 645, 652
  sweetener production with, 180
Lactate dehydrogenase as (a), 599, 688, 696, 697 — 1.1.1.27
  fish glycolysis, texture and, 499
  Irad of potatoes and, 199
  isozymes, regulation, testing, 48, 70, 83, 84, 499, 682
  meat color, texture, 324, 471
  plants, oxalic acid *via*, 355
  properties, purification, 113, 131, 255, 299
  zymograms, and Mg deficiency, 682
Lactic dehydrogenase. *See* Lactate dehydrogenase
Lactoperoxidase, aflatoxin degradation by, 661 — 1.11.1.(7,10)
  lipid-oxidizing capacity of inactivated, 450
  longevity and, 663
  sterilizing use of enzyme(p), 662–663
  xylitol, anticaries, and salivary, 177
Lactoyl glutathione lyase, 132, 503 — 4.4.1.5
Laminarinase, 566, 629 — 3.2.1.(6,39)
LDH. *See* Lactate dehydrogenase
Lecithinases. *See* Phospholipases
Less random cellulases, 170 — 3.2.1.4,91
Leucine 2,3-aminomutase, a cobalamin enzyme, 45 — 5.4.3.-
Leucine aminopeptidase, 46, 475, 606
Ligases, 30, 31 — 6.-.-.-
Limit dextrinases, 113, 560. *See also* Pullulanase — 3.2.1.(10,41)
Limonoate dehydrogenase, and citrus bitterness, 454 — 1.1.1.-
Limonoate D-ring lactone hydrolase, in citrus, 454 — 3.1.1.36
Linamarinase, an emulsin, 635, 637 — N.A.
Linoleate isomerase, and off-flavor, 438 — 5.2.1.5
Lipase, aroma, 17, 19, 124, 290, 369, 372, 400, 402, 409, 410, 412–413, 415 — 3.1.1.3
  as enzyme(a), assays, and related, 239, 679, 680, 694, 696–697, 704, 705, 708
  as enzyme(p), mostly oral, 5, 9, 405, 409, 412–413
  biochemistry, properties, 15, 31, 73, 110, 186, 242, 415, 440–446, 586
  biology, function, 15, 42, 73, 415, 444, 448, 586
  cereals, cereal processing, 209, 239, 243, 245, 429, 445–446, 568, 574, 585–587, 679
  color, 302, 304–305, 325
  control, factors affecting, 186, 198, 221, 228, 238, 239, 242, 243, 255, 257, 445, 541
  dairy related, 122, 199, 407, 409, 410, 412–413, 446–449, 452, 609, 680
  fish, 325, 446, 499–501, 629
  interfacial recognition site, 443–444
  meat, 9, 415, 446
  microbial, 95, 122, 145, 252, 410, 413, 415, 441, 445, 449
  nutritional, 629, 708
  off-flavor, 17, 209, 243, 245, 407, 423, 445, 446–449, 451, 452, 587, 680
  oral (pregastric), *See* Lipase, as enzyme(p)
  pancreatic, 257, 441, 447
  plant, plant foods, 242, 302, 304–305, 390 (non-cereal), 402, 413, 431, 444–445
  texture, 17, 239, 413, 499–501, 586–587, 609, 679
Lipoamide dehydrogenase, for NAD^+ regeneration, 688 — 1.6.4.3
Lipohydroperoxidase isomerase. *See* Hydroperoxide isomerase
Lipohydroperoxide lyase, 372–373, 427, 467
Lipohydroperoxide peroxidase (LOOH acceptor), 204 — 1.11.1(9)
  carotene bleaching, by, 300
  inhibition, 253, 437
  monophenol hydroxylation by, 428, 429
Lipohydroperoxide reductase, Lipohydroperoxidase. *See* Lipohydroperoxide peroxidase
Lipophosphodiesterase II, 586 — 3.1.1.4

Lipoprotein lipase, 258, 441, 443, 447 3.1.13.4
Lipoxygenase, analytical-, assay-related, 425, 642, 643, 672, 700, 707 1.13.11.12
animal, -derived foods, 325, 436, 437, 500, 643
aroma, 8, 18, 267, 371–372, 392, 393, 395, 401, 413, 414, 415, 419, 430
as enzyme(p), in baking, 300, 428, 587–589
biochemistry, biology, function, 286, 299, 437
inhibition inactivation, 245, 263, 264, 273, 300, 304, 437, 587
control, factors affecting, 198, 199, 205, 209, 223, 226, 240, 242, 253, 257, 260, 301–302, 436–438, 558, 587
properties, 46, 243, 289, 299, 301, 373, 392, 424–430, 433, 442
purification, 118, 119, 135, 425
cereal, -processing, 243, 245, 300–301, 428–430, 574, 587–590
color, 8, 9, 298, 299–303, 304, 314, 320, 325, 327, 425, 428
food quality (general), 20, 205, 298–299, 431
health-related, 299, 437, 630, 642, 643, 644, 646
legumes, -derived foods, 209, 245, 299, 419, 424–429, 433, 437, 626, 631
microbial, 301
off-flavor, 299, 423, 430–432, 370, 438
plants, -derived foods, other, 263, 298, 299, 303, 305, 359, 370–373, 401, 430
texture, 299, 500, 501, 502, 587–589
transformations of primary products, 427–429
Long chain acyl CoA hydrolase, 220 3.1.2.-
LOOH isomerase. *See* Hydroperoxide isomerase
LOOH reductase. *See* Lipohydroperoxide peroxidase
LOX. *See* Lipoxygenase
Luciferases, other, as (a), 203, 204, 226, 689, 707 1.13.12.5–7
Lyases, 29, 30 1.-.-.-
carbon-sulfur, 32, 45 4.4.-.-
oligopectate, 517
Lysine decarboxylase, 642, 700 4.1.1.8
Lysine α-oxidase(p), for cancer treatment, 666 1.13.12.2
Lysolecithin acyl hydrolase, Lysolecithinase, Lysophospholipase. *See* Phospholipase B 3.1.1.5
Lysozyme, as (p), sterilant, other, 9, 92, 93–94, 103, 104, 164, 210, 423, 602 3.2.1.17
assay, as predictor, etc., 648, 662, 680, 714
biochemistry, 40, 41–42, 44, 45, 56, 70, 187, 254, 361
inhibition, inactivation, 198, 245, 254, 589
purification of, in purification, 110, 124
egg, microbial, 41–42, 210, 680
milk, dairy, 9, 70, 361, 597, 598, 662
Lysyl oxidase, collagen crosslinks, meat texture, 482, 496 1.13.12.(2)
Lytic enzymes, 164 N.A.
Lytic glucanase, Lytic β-1,3-glucanase. *See* Endo-1,3-β-D-Glucanase

Macerase, macerating enzymes, cell-separation enzyme(p), 398, 519, 549, 550, 627 3.2.1.-
Malate dehydrogenase, as (a), 698–699 1.1.1.37
label in immunoassay, 713
plant acid accumulation *via*, 350, 352
Malate synthase, acid accumulation and, 199, 351, 355 4.1.3.2
Malic enzyme, affinity purification, 132, 133
as enzyme(a), for oxalic acid, 706
in fruit acid accumulation, 350, 352
zymograms, to detect freeze-thawing, 681, 682
Maltase, in honey, malt liquor production, 358, 564 3.2.1.20
Mannanase(p), coffee grounds utilization, 550 3.2.1.25
Mannitol dehydrogenase(a), for fructose, 695 1.1.1.67
Melibiase. *See* α-Galactosidase
3-Mercaptopyruvate sulfur transferase, 635 2.8.1.2
Methanogenase, Ni in, 46 N.A.
Methanol oxidase, 639 1.1.3.(13)
Methionine *S*-methyl transferase, browning control, 290 2.1.1.12
Methyl transferases, and bitter polyphenols, 457 2.1.1.-
Metmyoglobin reductase, and meat color retention, 317–318 1.6.2.(4)
Mexicanain, cuaguayote thiol proteinase, 94 3.4.99.14
Microbial rennets, in cheesemaking, 602–605, 607 3.4.23.6
Microprotease, *Bacillus cereus*, 496 3.4.99.-
Milk clotting enzyme. *See* Rennets, individual proteinases
Mixed function oxidase, phenolase as, 272
See also Xenobiotic oxidase
Monoacyl glycerol lipase, 441 3.1.1.23
Monoamine oxidase, biogenic amines, 640, 641, 642, 701 1.4.3.4
Monophenol monooxygenase, 270. *See also* Phenolase 1.14.18.1
Monophenolase, 274. *See also* Phenolase, Tyrosinase 1.14.18.1

Muramidase. *See* Lysozyme
Mutarotase, in GOX(p), 338, 692 5.1.3.3
Myeloperoxidase, leucocytes, 203, 204 1.11.1.(8,10)
Myogen A (GAD + aldolase) crystals, 117 N.A.
Myokinase. *See* Adenylate kinase 2.7.4.3
Myosin ATPase. *See* ATPase
Myrosin, 28, 383. *See also* Glucosinolase 3.2.3.1
Myrosinase, Myrosulfatase. *See* Glucosinolase

NAD^+-degrading enzymes, fish toughness, 504
$NAD(P)^+$ nucleosidase, and dehydrogenase dysfunction, 322 3.2.2.(6)
$NAD(P)^+$ peroxidase, 203 1.11.1.2
$NAD(P)^+$ transhydrogenase, in ham curing, 324
NAD(P)H dehydrogenase (quinone), 317 1.6.99.2
NADPH-cytochrome P450 reductase. *See* Xenobiotic oxidase
$NAD(P)^+$-linked (-dependent) dehydrogenases, and aroma, 317 1.-.1.-
  assay, in enzymatic analysis, 676, *See also* specific cases
  breadmaking application (suggested), 591
  inhibition, 257, 259
Na^+/K^+ ATPase, homology with enzymes, proteins, 85 3.6.1.3
  in human weight control, 597
Naringin oxidase, 456 N.A.
Naringinase, (β-L-Rhamnosidase + β-Glucosidase), 455–456 N.A.
Nitrate reductases, biochemistry, 46, 77, 129, 363 1.6.6.1
  inhibition, 160, 257, 259
  secondary transformation of product, 20
Nitric oxide reductase, Mo in, 46 1.7.99.2
Nitrilase. *See* Hydroxynitrile lyase
Nitrite reductase, 46, 661 1.7.99.3
Nitrogenase, Mo, Fe in, 46 1.18.2.1
Notatin. *See* Glucose oxidase
Nuclease P_1. *See* RNase
Nucleases, acid inactivation mechanism, 263 3.1.11.- to 3.1.31.-
  5′-gustator accumulation *via*, 365–366
  in enzyme purification, affinity, methods, 113, 124–125, 138
Nucleosidase, in meat color, food flavor, 322, 366–367 3.2.2.1
Nucleoside kinase, and food flavor, 364 2.7.4.4
Nucleoside phosphorylases, fish quality, 680 2.4.2.1,2
  flavor, purification, 113, 367
Nucleotidase, fish quality, 680 3.1.3.5
  flavor, as enzyme(a), 367, 415, 699
Nucleotide kinase(s), as (a), in meat texture, 468, 699 2.7.4.6
Nucleotidyl transferases, Zn in, 57 2.7.7.-

Octopine dehydrogenase(a), for arginine, 700
Oleinase, and oxidative rancidity, milk, 450 N.A.
Oligopectate lyase, 517 4.2.2.6
Oligopeptidase, and cheese texture, 605, 608 3.4.-.-
Oxalacetase, oxalic acid from, 355
Oxalacetate decarboxylase, as (a), TSAI of, 52, 706 4.1.1.3
Oxidase, flavoprotein, and O_2^-, 20, 393 N.A.
Oxidoreductases, as oxygen scavengers, 12 1.-.-.-
  enzyme(p) as (list), 9
  for electric current generation, 5–6
  in shrimp greening, 325
Oxygenases, cold-induced acceleration of, 224 1.13.-.-
  as texture-modifying enzymes, 518
Oxynitrilase. *See* Hydroxynitrile lyase

PAL. *See* Phenylalanine ammonia lyase 4.3.1.5
Palmitoyl (long chain acyl) CoA hydrolase, 220 3.1.2.2
Pancreatin, pancreatic enzyme(p), hydrolases 1.-.-.-
  analytical-, assay-related uses, 697, 707, 711
  in cereal processing, 586
  nonfood uses, 5
  production, 91
  protein modification applications, 492, 509, 624
Pantothenase, and nutritional losses, 696 3.5.1.22
Papain, as enzyme(p), as meat tenderizer, 92, 486–489, 491, 492, 494, 496 3.4.22.2
  beer brewing, 9, 92, 569, 570–571
  cereal processing, 10, 583
  color retention, recovery, 9, 323, 661
  dairy-related, 266, 600, 601, 603
  extraction uses (non-protein), 9, 10, 160, 576, 661
  health-related, 493, 625, 647, 664
  production, purification, 92–93, 104, 113
  protein processing, 509, 620, 625, 627, 661

biochemistry, 28, 40, 41, 56, 116, 484–485, 486, 603
control, factors affecting, 110, 226, 254, 395
immobilization, 147, 570, 571, 670
Papainases. *See* Thiol proteinases 3.4.22.-
Papaya peptidase A, 485 3.4.22.-
PE. *See* Pectin esterase
Pectate lyase. *See* Exopectate lyase 4.2.2.9
Pectic enzymes, Pectinolytic enzymes, Pectic hydrolases, Pectin-degrading enzymes. *See* Pectinase
Pectin esterase (demethoxylase), assay as marker, detector, 526, 678, 680 3.1.1.11
enzyme(p), uses, in pectinase(p), 522, 536, 550
firming of intact processed plant foods, 522–523, 524, 525, 526, 528
flavor, 352, 402
fruit, vegetable ripening, storage, texture, 227, 402, 520, 523, 528, 529, 680
immobilization, fungal, 575, 639
juice production, stability, 537–538, 542–543, 544
methanol, alcoholic beverages, 198, 533, 637–639
purification, properties, 46, 153, 194, 197, 198, 209, 515
Pectin lyase, 515, 516, 517–518 4.2.2.10
as cell-separating enzyme, 550–551
as (p), in juice- and wine-making, 536, 537, 539
Pectin methylesterase. *See* Pectin esterase
Pectinase (PE + PG ± pectin lyase), endogenous, resident microbial, analytical, 670, 710, 712 N.A.
beverages, 296, 455, 535, 546
candy and pectin making, 549
color and, 275, 296
immobilized, 538
in seedcoat removal, 6, 549, 550
oversoftening apricot, vegetable, 6, 18, 189, 259, 522, 527–528, 529
protopectinase as, in part, 518, 519
removal from nonpectinase(p), 455
enzyme(p), color and aroma, 296, 536, 537, 636, 637, 661
in wine-, juice-making, 537–539, 540, 543–544, 638
nonpectin-related enzymes in, 359, 360, 536, 537
sources, safety, 95, 98, 549, 648
suppliers, trade names, 538, 539, 540
various other uses of, 9, 454, 520, 546, 549, 629, 661
Penicillinase, 61, 95 3.5.2.6
Pentosanase, in breadmaking, 582 3.2.1.-
PEPcarboxykinase. *See* Phosphoenolpyruvate carboxykinase 4.1.1.32
PEPcarboxylase. *See* Phosphoenolpyruvate carboxylase 4.1.1.31
PEP-HPr phosphotransferase, sugar accumulation, 358, 360 2.7.3.9
PEPkinase. *See* Phosphoenol kinase 2.7.1.40
Pepsin, as enzyme(p), as a rennet, 10, 600, 601, 602, 605 3.4.23.1
chicken, 10, 261, 601
other uses, 5, 614, 615, 664
production-, health-related, 4, 92, 601, 614, 615, 643, 647
assay, analysis, testing, 605, 670, 675, 707–708, 709, 710, 711
fish, and FPC losses, 506, 508
in milk clotting, 226, 595, 600, 601, 603
purification, properties, 4, 115, 124, 138, 257, 261
Pepsinogen. *See* Proenzymes
Peptidase(s) and α-Peptide hydrolase(s), cheese texture and, 605 3.4.9.-
for *in vitro* digestibility tests, 708
in glutamate accumulation, 262
in malt, brewing, 568
signal, in posttranslation, 81
soy sauce production and, 616. *See also* Exopeptidase
Permease, and active transport, 28, 75 N.A.
Peroxidase, as (a), in EIA, 205, 670, 685, 700, 707, 713, 715 1.11.1.7
as (p), 294, 622, 639, 663
assay for adequacy, 207–208, 209, 676
beverages, 199, 206, 237, 293, 294
biochemistry, 31, 74, 84, 130, 202–206, 286, 529–530
blanching, assay, (non-blanching), 669, 670–671, 676, 677, 678, 679
inactivation of, 185, 190, 194, 205–207, 212–213, 258, 264, 437
color and, 285, 287, 288, 293, 300, 428
flavor, LOOH, and, 204, 237, 400, 423, 429, 431, 434, 477
fruits, other foods, 206, 288, 400, 591
health and, 205, 631, 639, 643, 646, 649, 662, 663
Irad and, 191, 197, 198, 199
isozymes, zymograms, 84, 190, 213, 288, 682, 683
low temperature, frozen foods, 219, 221, 222, 226, 244, 431–432
low a_w's, dried foods, 237, 240, 241, 244
regeneration of inactivated, 212, 213–216
vegetables, 190, 199, 203, 206, 207, 209, 219, 258, 264, 288, 431, 521, 529–530

Peroxygenase. *See* Lipohydroperoxide peroxidase
PFK. *See* Phosphofructokinase 2.7.1.11
PG. *See* Polygalacturonase 3.2.1.15
Phenol formylase, cycloamylose as model of, 61 N.A.
Phenol hydroxylase, 271 1.14.13.7
Phenolase, affinity, other purification, 108, 112, 119, 124 1.10.3.2, 1.14.13.8
as (a), assay of as index, etc., 209, 276, 680, 707
as (p), 291, 294
biochemistry, 30, 46, 217, 238, 270–276, 293
control, factors affecting, 128, 199, 209, 217, 238, 254, 255, 263, 281–282
cereal, cereal processing, 275, 278, 414, 590–591, 630
enzymatic browning, 275, 276–278, 292–295, 327, 630
control of, 280–283, 284–285, 290–291, 341
NEB and, 327, 341
fruit, 8, 18, 237, 274, 275, 278, 423, 457, 646
health-related aspects, 290, 629–631, 646
immobilization, 273–274, 278
low a_w behavior, dried foods, 237, 238, 240, 241
low temperature, frozen foods, 294, 341
non-plant organisms, 273, 275
off- and de-colorations, non-brown, 275, 277, 296–297, 313, 325, 536, 630
off-flavors, 237, 276, 423, 457
other enzymes and, 270–271, 275, 289, 296, 331, 341
reaction inactivation, regeneration, 217, 260, 273
tannins, tannase and, 276, 293, 295, 539
tea, cocoa, coffee, 8, 18, 290–295, 400–403
vanilla curing, aroma, 271, 400, 402, 403
vegetables, 8, 238, 263, 271, 275, 277–278, 278–279, 280, 283, 297, 313, 423
wine, juice, 296–297, 537, 539
zymograms, isozymes, 274, 682, 683
Phenylalanine ammonia lyase, color, vegetables, 269, 279–289, 288 4.3.1.5
texture, stress, lignin, 256, 275, 530
Phenylesterase, imidazoles as models of, 61 3.1.1.-
Phosphatases. *See* Phosphomonoesterase, specific enzymes 3.1.3.-(+)
Phosphatide acyl hydrolase. *See* Phospholipase A_2 3.1.1.4
Phosphodiesterase, and flavor, 365, 366, 449 3.1.4.1
and health related aspects, 654, 655
caffeine inhibition and cAMP, 73, 442
lanthanum salts as models of, 39
milk, 449
See also RNases
Phosphoenolpyruvate carboxykinase, in acid accumulation in food plants, 350, 352, 353–354 4.1.1.(32,38)
in sugar accumulation, 356, 357, 358–359
Phosphoenolpyruvate carboxylase, as (a), for CO_2 4.1.1.31
in food plant acid accumulation, 350, 351, 352, 353
Mg as cofactor, 46
Phosphoenolpyruvate kinase, as (a), for lipids, 696 2.7.1.40
for 5′-nucleotides, 699
Phosphoesterases, as detoxicants, 651, 654–659 3.1.-.-
Phosphofructokinase, allostericity, oscillation, 71, 74, 471, 679 2.7.1.11
food plant sugar accumulation, 333, 356
meat texture, 471, 482
purification *via* affinity methods, 129, 132, 133
Phosphoglucoisomerase(a), for sorbitol, 697 5.3.1.9
Phosphoglucomutase, active site, in glycolysis, 40, 70 2.7.5.1
as (a), for Pi, sugars, 688, 694
in cold sweetening, 334
Phosphogluconate dehydrogenase(a), for lactose, 691 1.1.1.44
Phosphoglyceraldehyde dehydrogenase. *See* GAD
3-Phospho-D-glycerate carboxylase, 30 4.1.1.39
Phosphoglycerokinase, 70, 113 2.7.2.3
Phosphoglyceromutase, a glycolytic enzyme, 70
Phosphohexose isomerase, 70, 695 5.3.1.8,9
Phospho-2-keto-deoxyheptanoate aldolases, 83 4.1.2.15
Phospholipase(s), *See also* specific enzymes 3.1.1.-
cereal, cereal processing, 237, 567, 586–587
classification, biochemistry, 440–441, 442
color, flavor, 327, 444, 446
dried, frozen foods, 227, 229, 236, 241, 501
fish, texture, 221, 227, 499–500, 505, 507, 544
membrane, enzyme potentiation, 16
Phospholipase A_1, 441 3.1.1.32
Phospholipase A_2, biochemistry, 440, 441, 443 3.1.1.4
chemotactic peptides, leucocytes and, 443
fish texture and, 500
Phospholipase B, 440, 441, 500 3.1.1.5
Phospholipase C, as analytical aid, 712 3.1.4.3

as (p), to prevent oxidative rancidity, 423, 449, 450, 451
Phospholipase D, and fish toughness, 501, 505 — 3.1.4.4
as (p), off-flavor prevention with, 423
Phosphomonoesterases, assay, as enzyme(a), 677, 705 — 3.1.3.-
cheesemaking, as enzyme(p), 609
control, factors affecting, 46, 219, 269, 355, 366
meat, color aroma, 321, 322, 415
5′-nucleotide accumulation, 255, 360
zymograms, food quality control, 682
Phosphoprotein phosphatase, cheese texture pacer, 609 — 3.1.3.16
Phosphoribomutase, purification, 113 — 2.7.5.6
Phosphoribosylaminoamidazole carboxylase, 249 — 4.1.1.21
Phosphorylase, as (a), 688, 694 — 2.4.1.1
banana, 331, 359
biochemistry, 48, 73, 119, 163, 330–332
pyridoxal phosphate in, 45, 228, 331, 646
low temperature, frozen foods, 220, 227, 229
muscle-to-meat, texture, 227, 470, 471, 472
potato, cold sweetening, 329–333
Phosphotransacetylase(a), for acetate, 698 — 2.3.1.8
Phytase(s), beans, health, removal, 658–659 — 3.1.3.(2,8,26)
other vegetables, texture, 518, 527–528
cereal, cereal processing, 574, 586, 587
potentiation, properties, 10, 186, 251, 657–658
6-Phytase, 675 — 3.1.3.26
Phytase ADP phosphotransferase, 657–658 — 2.7.1.-
Pinguinain, maya proteinase, 94 — 3.4.99.18
Polygalacturonase, affinity methods, assay, 124, 125, 209 — 3.2.1.15
beverages, 402, 543, 545, 546
biochemistry, 515–517, 524
control, factors affecting, 195, 215, 261, 518
color, flavor, 252, 402, 536
fruit, texture, ripening, 209, 519–520, 522, 529
fungal, enzyme(p), macerase, 517, 519, 537, 639, *See also* Pectinase
health, 639
immobilization, protopectinase component, 518, 519, 538
vegetable, texture, 209, 524, 525, 527–528
Polymethylgalacturonase, pectate lyase?, 515–516 — N.A.
Polyphenol oxidase, 270. *See also* Phenolase — 1.10.3.2
Polyphosphatase, 482, 712 — 3.6.1.10
Polysaccharidases. *See* Glycanases
Pomiferin, osage orange proteinase, 94 — 3.4.99.-
Proenzymes, 81, 98, 201, 274, 443, 603 — N.A.
Prolidase, and aroma genesis, texture, 415, 496 — 3.4.13.9
in amino acid analysis of proteins, 710
Proline aminopeptidase, 415 — 3.4.11.5
Proline carboxypeptidase, and aroma, 415 — 3.4.16.2
Proline dipeptidase, 415 — 3.4.13.9
Prolyl dipeptidase, 415 — 3.4.13.8
Pronase [*Streptomyces griseus* neutral proteinase(p)], analytically-related uses, 680, 708, 710 — 3.4.24.4
food processing, other uses, 509, 519, 571, 664
Prostaglandin synthase, inhibition, and health, 253, 501 — -.14.99.1
meat color, flavor, 298, 320, 415, 435, 437
Proteases, Protein hydrolases. *See* Proteinases
Protein kinase, cAMP, cascade, lipase, 71, 73, 442 — 2.7.1.37
casein, muscle contraction, 467, 609
Proteinase(s), assay, analytically-related, 21, 669, 670, 678, 683, 684, 687, 707–708, 710–711 — 3.4.16.-to-3.4.99.-
biochemistry, properties, 4, 29, 30, 40, 41, 42, 46, 56, 140, 187, 216, 219, 260, 274, 474–475, 484–486
affinity, other purification, 110, 116, 124, 128, 133, 135, 588
biological functions, regulation, 16, 17, 81, 82, 334, 463, 480, 568, 679
control, factors affecting, 7, 19, 212, 219, 229, 254, 257, 331, 707
limited proteolysis by, 65, 81, 274, 331, 334, 443, 478, 479, 487, 496, 558, 596–597, 615, 626, 679
food quality attributes affected, color, 237, 319, 327
flavor, 164, 334, 362, 403, 409, 410, 415, 416, 616, 618
health, 480, 496, 613–614, 621–627, 642, 647, 648, 659, 664, 665, 707–708, 710
texture, 473–480, 483, 496, 505, 528
foods, processing, baking, cereals, 17, 124, 135, 414, 484, 574, 582–587
blanching, other energy input, 195, 198, 199, 200, 201
brewing, malt, beer, 553, 558, 559, 567–571, 669
cheese, 409, 410, 451, 594–605, 605–610
liquid, other, 400, 403, 505, 539, 616, 624, 625

meat, 319, 414, 463, 473–480, 483–496
plant-derived, other, 16, 254, 334, 492, 518, 519, 568, 625–627, 659, 679
protein processing, 506–510, 613–626
immobilized, 164, 238–239, 600
microbial, 5, 95, 506, 554, 602–604, 605–609, 616 — 3.4.21.14, 3.4.23.6, 3.4.24.4, 3.4.99.-
See also Proteinase(p), individual enzymes
Proteinase(p), color control, 341, 510
feed additive use, 525–526
food and food ingredient production uses, beer, 9, 405, 570–571
cheese, other dairy products, 10, 409, 450–451, 577, 598–602, 604, 605
fish, other marine food, 10, 508–510
flavorants, related, 10, 362, 363, 405, 409, 616
liquid foods, other, 535, 543, 616, 624, 625
meat, 10, 492, 491–495
protein processing, 10, 508–510, 614–615, 621–625
sugar refining, 10
health related use, *See* Proteinase, health
non-food, non-health uses, 4–5, 519
off-flavor prevention, removal, 423, 439, 450–451, 618
production safety, 10, 91, 92–94, 95, 198, 491–492
in, interaction with other enzyme(p), 9, 10, 102, 105, 106, 129, 133, 135, 160, 164, 559, 566, 670
See also specific proteinase, proteinases(p), Pancreatin, Rennets, etc.
Protopectinase, solubilizing *in situ* pectin, 514, 518–519, 529, 530, 544, 549, 550 — N.A.
Pullulanase, as enzyme(p), 157, 553, 564 — 3.2.1.41
biochemistry, 152, 154, 157, 554
in brewing, breadmaking, 563–564, 576
Purine deaminase. *See* Adenine deaminase
Purine demethylase, 661 — N.A.
Pyrophosphatase. *See* Inorganic pyrophosphatase
Pyruvate carboxylase, behavior in the cold, 46, 220, 224 — 6.4.1.1
in plant acid accumulation, 350
Pyruvate dehydrogenase biochemistry, 46, 69, 254 — 1.2.4.1
in aroma genesis, 394, 408
Pyruvate kinase, K, Mg, in and use, 46, 58, 70, 677 — 2.7.1.40
potato sweetening, meat texture, 333, 471

Q-enzyme, amylopectin synthase, 332–333 — 2.4.1.18

Raffinase. *See* α-Galactosidase
RBP-, RDP-carboxylase-oxygenase. *See* Ribulosebisphosphate carboxylase-oxygenase
Relaxing factor. *See* ATPase
Rennets—milk-clotting enzyme(p), assay, 596, 597, 599, 683 — 3.4.23.-
cheesemaking roles, curdling, 8, 19, 594, 598–599, 603–605
post-curdling, texture, 605–606, 606–607
redolence, taste, contribution, 10, 362, 407, 409
control, factors affecting, 184, 198, 605
immobilization, 600, 605
low temperature behavior, 20, 225, 226, 596–597
other alternatives, uses, enzymes, 450, 571, 599
production—sources, animal, 9, 10, 92, 412, 494, 595, 601
microbial, 10, 95, 122, 198, 410, 602–603 — 3.4.23.6
plant, 6, 94, 402
safety, economics, 9, 104, 140, 600–601, 647
Rennin. *See* Chymosin — 3.4.23.4
R-enzyme. *See* Pullulanase
Restriction DNases, in genetic engineering, 85–86 — 3.1.23–25.-
β-L-Rhamnosidase, a naringinase moiety, 455 — 3.2.1.43
Rhodanese, and flavorese, 416 — 2.8.1.1
immobilized, for HCN removal, 637
toxicity of HCN, NCS^- and, 635
Ribonuclease. *See* RNase
Ribulose bisphosphate carboxylase-oxygenase, carbon isotope discrimination, adulteration, 679 — 4.1.1.39
photorespiration, photosynthesis, 30, 361
plant acid accumulation, 355

stability, 244
RNA polymerase (-nucleotidyl transferase), as (a) (inhibition), for amatoxins, 706 — 2.7.7.6
in enzyme(p) production, 97
in transcription (biosynthesis of enzymes), 40, 74, 75, 76
t-RNA synthetases. *See* Amino acid-tRNA synthetases — 6.1.1.-
RNase(s), biochemistry, 31, 32, 40, 42, 54, 55 — 3.1.13.26,27
affinity, other purification, 112, 124–125, 128
control, factors influencing, 112, 187, 198, 206, 213, 244, 258, 286
enzyme(p) uses, 150–151, 655
flavor, 18, 150–151, 364, 415, 423
immobilization, 150, 151, 367, 423
meat, animal, 263, 415
plant-derived food, 18, 366, 588
SCP purine removal *via*, 654–655

Sarcosine dehydrogenase, formaldehyde *via*, 503 — 1.5.99.1
Schardinger dextrinogenase. *See* Cyclodextrin glucanotransferase
Sealase, 28 — 6.5.1.1
Secondary alcohol dehydrogenase, and bread aroma, 414 — 1.1.1.1
and cheese aroma, 411
Serine dehydratase, and protein quality, 76, 77 — 4.2.1.13
Serine proteinases, as (a), for pesticides, 704–705 — 3.4.21.-
mechanism of action, 52, 56–57
Sinigrinase. *See* Glucosinolase — 3.2.3.1
SOD. *See* Superoxide dismutase
Solanain, horse nettle proteinase, 94 — 3.4.99.21
Sorbitol dehydrogenase, as (a), for wine sorbitol, 797 — 1.1.1.14
in plant sweetener accumulation, 360
Sorbitol phosphatase, 360 — 3.1.3.-
Soyin, soybean proteinase, 94 — 3.4.22.-
Starch synthase, in sweetening, 30, 220, 332, 334 — 2.4.1.21
Starch-hydrolyzing, -debranching enzymes, 154–155 — 3.2.1.-
Steroid isomerase, affinity partitioning of, 134 — 5.3.3.1
Streptodornase, a medical proteinase(p), 95 — 3.1.21.1
Streptokinase, a medical enzyme(p), 95, 665 — 3.4.22.10
Subtilisin, and charge relay, 57 — 3.4.2.14
immobilized, pH optima of, 147
meat tenderizer(p), 492
Succinate (succinic) dehydrogenase, biochemistry, 46, 249 — 1.3.99.1
assay, beef quality predictor, 680
biochemistry, 46, 249
controlled atmosphere storage and, 355–356
red meat discoloration and, 321
Succinyl CoA synthetase, as (a), for succinate, 699 — 6.2.1.4,5
Sucrose hydrolase. *See* Invertase
Sucrose phosphatase, in sucrose accumulation, 357 — 3.1.3.24
Sucrose phosphate synthase, in potatoes, 333 — 2.4.1.14
in sucrose accumulation, 357
Sucrose synthase, in cold sweetening, 333, 334 — 2.4.1.13
in plant sucrose accumulation, 357
in potatoes, assay as maturity marker, 678–679
Sugar phosphatase, in sucrose translocation, 357 — 3.1.3.23
Sulfatase, imidazoles as models, 61 — 3.1.6.11
Sulfhydryl (thiol) oxidase, milk off-flavor, 423, 450–451 — 1.8.3.2
Sulfite oxidase, as (p), for SO_2 removal, 257, 663 — 1.8.99.-
Superoxide dismutase (SOD), biochemistry, 46, 257, 286–287, 361, 451 — 1.15.1.1
blanching, enzyme regeneration, 216
color and, 285, 290, 317, 323, 324
enzyme(p), as antioxidant (suggested uses), 103, 285–286, 323, 423, 438, 450
flavor, 216, 280, 286, 407, 423, 438, 450
meat, 317, 323, 324
milk, cheese, 407, 450, 451
plant foods, 216, 285, 286
Synthases, reverse lyases, 29, 30 — 6.-.-.-

Takadiastase, *Aspergillus oryzae*, α-amylase(p), uses, 89, 554, 647, 677, 711 — 3.4.23.6
Tannase (tannin hydrolase, -esterase), a glycoprotein, 46
as enzyme(p), immobilized, 539–540, 661
assay to detect spoilage, 678
phenolase as, in tea making, 294
Tannin-protein decomplexing enzyme, for chillproofing, 571 — N.A.
Tertiary amine monooxygenase (TMAO synthase), 504 — 1.13.11.-
Thiamin pyrophosphate carboxylases, 644 — 4.1.-.-

Thiamin pyrophosphokinase, affinity methods, 128, 132 — 2.7.6.2
Thiaminase I, an imine hydrolase, 496 — 2.5.1.2
  for removing thiamin bitterness, 457
  vitamin B_1 deficiency, sashimi and, 644–645
Thiaminase II, vitamin deficiency and, 645 — 3.5.99.2
Thiocyanate isomerase, 384 — 5.99.1.1
Thioglucosidase, microbial, 383, 385, 420. *See also* Glucosinolase — 3.2.3.1
Thiol proteinases, 484–485, 570, 582, 587. *See also* Papain, other specific enzymes — 3.4.22.-
Thiosulfate cyanide sulfur transferase. *See* Rhodanese — 2.8.9.1
Threonine aldolase, and cheese aroma, 409 — 4.1.2.5
Thrombin, assay *via* active site titration, 684 — 3.4.21.5
Thyroperoxidase, 203, 204 — 1.11.1.19
TMAO demethylase. *See* Trimethylamine oxide aldolase — 4.1.2.32
Transaminases. *See* Aminotransferases
Transferases (synthases), 29 — 2.-.-.-
Transglucosidase (Transglucosylase), removal from glucoamylase(p), 106, 135, 157, 562 — 2.4.1.25
Transmethylases, enzyme models for, 61 — 2.1.1.-
Trehalase, affinity chromatography of, 130 — 3.2.1.28
Triacyl lipase. *See* Lipase — 3.1.1.3
Trimethylamine dehydrogenase, 504 — 1.5.99.7
Trimethylamine oxidase (TMAO synthase), 504 — 1.13.99.-
Trimethylamine oxide aldolase, and fish texture, 503–505 — 4.1.2.32
Trimethylamine oxide reductase, rotten fish odor, 504 — 1.6.6.9
Triosephosphate isomerase, 50, 70 — 5.3.1.1
Tripolyphosphatase, and meat texture, 483
Trypsin, affinity, other purification, 110, 113, 123, 129, 131, 135 — 3.4.21.4
  analytical-, assay-related, 684, 711
  as enzyme(p), 5, 450–451, 509, 629, *See also* Pancreatin
  biochemistry, properties, 28, 48, 138, 198, 204, 225, 249, 274, 451, 486
  immobilization, 147, 400, 452
  in milk, 606
  low temperature behavior, 219, 220, 224
  protein processing, 498, 509, 629
  texture, flavor influenced by, 450–451, 497
Trypsinogen. *See* Proenzymes
Tryptophan pyrollase, a peroxidase, 204 — 1.13.11.11
Tryptophan synthase, immobilization, 140 — 4.2.1.20
  stable ES complex, 39
Twistase. *See* DNA topoisomerases
Tyrosinase, 270, 271, 272, 274. *See also* Phenolase — 1.10.3.1, 1.14.18.1
Tyrosine decarboxylase, in cheese starters, 642 — 4.1.1.25

UDP glucuronate 4-epimerase, in pectin biosynthesis, 515 — 5.1.3.6
UDPG dehydrogenase, in pectin biosynthesis, 515 — 1.1.1.22
UDPG transferase, in glucosinolate biosynthesis, 383 — 2.4.1.-
UDPgalacturonate polygalacturonate transferase, 515 — 2.4.1.43
UDPgalacturonosyl transferase, 515 — 2.4.1.75
Urease, affinity, other purification, 4, 13, 130 — 3.5.1.5
  analytical-, assay-related, 655, 670, 685
  as (p), potentiation, in protein foods, 423, 651, 659
  biochemistry, history, 4, 13, 31, 46, 50, 58, 138, 257
  low-temperature, -a_w behavior, 117, 219, 240, 241
Uricase, as (a), in fertilizers, contaminant detection, 670 — 1.7.3.3
Urokinase(p), as therapeutic agent, 665 — 3.4.21.31
Uronic acid oxidase, in citrus, 522 — N.A.

Verdoperoxidase, 203, 204 — 1.11.1.(8,10)

Wall separating enzymes, 549. *See also* Cell separating enzymes — 3.2.1.-

Xanthine oxidase, analytical-, assay-related, 680, 692, 707 — 1.2.3.1
  biochemistry, 46, 133, 451
    as O_2^- producer, 280, 285, 287, *See also* Superoxide anion
  breadmaking, 591

flavor, off-flavor, 367, 435, 450, 451, 591, 680
immobilization, 286
intestinal absorption, atherosclerosis, 649
milk, 435, 450, 451, 649, 680
texture, 173, 591
Xenobiotic oxidase (flavoprotein-linked monooxygenase), aroma, off-flavor, 399, 412, 434 1.14.14.1
as (a), for pesticides (suggested), 705
biochemistry, function, 6, 77, 412, 642
immobilization anti-pollution, 660
lipid oxidation, health, 642
off-flavor, 412, 434
plants, 412
Xylanases, biomass conversion, xylitol, 124, 176, 178–180 3.2.18.32+
Xylitol dehydrogenase (xylulose reductase), 177
Xylose isomerase, and glucose isomerase, 162 5.3.1.5
β-Xylosidase (xylobiase), 178, 179

Zingibain, ginger proteinase, 94 3.4.22.-
Zymase (glycolytic pathway enzyme), 4 N.A.
Zymogens. *See* Proenyzmes
Zymolase, 629 3.2.1.6

Other AVI Books

DRYING AND STORAGE OF AGRICULTURAL CROPS
Hall
ELEMENTARY FOOD SCIENCE 2nd Edition
Nickerson and Ronsivalli
ELEMENTS OF FOOD TECHNOLOGY
Desrosier
ENCYCLOPEDIA OF FOOD ENGINEERING Vol. 1
Hall, Farrall, Rippen
ENCYCLOPEDIA OF FOOD TECHNOLOGY Vol. 2
Johnson and Peterson
ENCYCLOPEDIA OF FOOD SCIENCE Vol. 3
Peterson and Johnson
FOODBORNE AND WATERBORNE DISEASES: THEIR EPIDEMIOLOGIC CHARACTERISTICS
Tartakow and Vorperian
FOOD LAW HANDBOOK
Schultz
FOOD PRODUCTS FORMULARY: CEREALS, BAKED GOODS, DAIRY AND EGG PRODUCTS
Tressler and Sultan
FOOD PRODUCTS FORMULARY: FRUIT, VEGETABLE AND NUT PRODUCTS
Tressler and Woodroof
FOOD PRODUCTS FORMULARY: MEATS, POULTRY, FISH AND SHELLFISH
Komarik, Tressler, Long
FOOD SCIENCE 3rd Edition
Potter
HANDBOOK FOR FOOD PROCESSORS AND ENGINEERS
Farrall
HANDBOOK OF REFRIGERATING ENGINEERING Vol. 1 and 2 4th Edition
Woolrich
PRESCOTT & DUNN'S INDUSTRIAL MICROBIOLOGY 4th Edition
Reed
SOURCE BOOK FOR FOOD SCIENTISTS
Ockerman
THE TECHNOLOGY OF FOOD PRESERVATION 4th Edition
Desrosier and Desrosier
TROPICAL & SUBTROPICAL FRUIT
Nagy and Shaw